D0023557

ELECTRONIC DEVICES AND CIRCUITS

FOURTH EDITION

Theodore F. Bogart, Jr.
University of Southern Mississippi

Prentice Hall
Upper Saddle River, New Jersey *Columbus, Ohio*

Library of Congress Cataloging-in-Publication Data
Bogart, Theodore F.
Electronic devices and circuits / Theodore F. Bogart, Jr.—4th ed.
p. cm.
Includes index.
ISBN 0-13-393760-7
1. Electronic circuits. 2. Electronic apparatus and appliances.
I. Title.
TK7867.B57 1997
621.3815—dc20 96-28816
CIP

Cover photo: FPG International
Editor: Linda Ludewig
Production Editor: Stephen C. Robb
Design Coordinator: Julia Zonneveld Van Hook
Cover Designer: Proof Positive/Farrowlyne & Associates
Production Manager: Deidra M. Schwartz
Illustrations: Steve Botts; Precision Graphics
Marketing Manager: Debbie Yarnell

This book was set in Times Roman and Helvetica by Bi-Comp, Inc. and was printed and bound by R. R. Donnelley & Sons Company. The cover was printed by Phoenix Color Corp.

Simon & Schuster/A Viacom Company
Upper Saddle River, New Jersey 07458

Printed in the United States of America

10 9 8 7 6 5 4 3 2 1

ISBN: 0-13-393760-7

Prentice-Hall International (UK) Limited, *London*
Prentice-Hall of Australia Pty. Limited, *Sydney*
Prentice-Hall Canada, Inc., *Toronto*
Prentice-Hall Hispanoamericana, S. A., *Mexico*
Prentice-Hall of India Private Limited, *New Delhi*
Prentice-Hall of Japan, Inc., *Tokyo*
Simon & Schuster Asia Pte. Ltd., *Singapore*
Editoria Prentice-Hall do Brasil, Ltda., *Rio de Janeiro*

PREFACE

Electronic Devices and Circuits is a modern, thorough treatment of the topics traditionally covered during a two- or three-semester course in electronic device theory. The minimum preparation for students beginning their study of this material is a course in dc circuit analysis. While the first eight chapters do not require an extensive knowledge of ac circuit theory, many subsequent chapters assume that the student understands impedance concepts and can perform phasor computations. Therefore, students who begin their study without having completed a course in ac circuit analysis should be taking that course as a corequisite. Calculus is neither used nor required for the development of theoretical principles, but the discussion of electronic differentiators and integrators uses some of the standard notation from calculus as a symbolic aid. The practical applications of differentiators and integrators are emphasized, as in filtering and waveshaping.

A principal consideration in selecting topics for the book was the significance of each in modern industrial applications and the impact they are likely to have in emerging technologies. Consequently, integrated-circuit theory is covered extensively, as are field-effect devices and their applications in large-scale integration, the theory of operational amplifiers, including many important applications of those versatile devices, optoelectronics, switching regulators, and class D amplifiers.

Each new concept in *Electronic Devices and Circuits* is introduced from a systems, or block-diagram, approach. For example, the effect of input and output resistance on the voltage gain of an amplifier is developed by regarding the amplifier as a functional block, rather than as a particular circuit. Once the fundamental principle has been thoroughly discussed, it is applied to each and every amplifier circuit studied thereafter. A similar approach is used to develop the theory of feedback, frequency response, multistage amplification, amplifier bias, distortion, linearity, oscillation, filters, voltage regulation, and modulation.

Most chapters are accompanied by SPICE examples and exercises. Appendix A contains SPICE instructional material that parallels the sequence in which electronic device theory is introduced in the book, so new simulation skills can be developed at the same pace that new theory is taught.

FEATURES OF THE FOURTH EDITION

1. Chapter 14 (Applications of Operational Amplifiers) has been reorganized so that its scope is now more accurately reflected by its title. A circuit for a current-controlled voltage source whose input has a return path to ground has been added to Chapter 14. Some topics that were originally covered in this chapter (voltage comparators and clipping, clamping, and rectifying circuits) have been moved to a new chapter (15).
2. Chapter 15 (Special-Purpose Circuits) is new. In addition to the topics covered originally in Chapter 14, this chapter contains topics not covered in earlier editions, including logarithmic and antilogarithmic amplifiers, transconductance amplifiers, phase-locked loops, and integrated-circuit function generators.
3. Explicit equations for calculating the exact cutoff frequency of two low-pass or two high-pass filters in cascade have been added to Chapter 10. Derivations are provided in a new Appendix C.
4. In response to suggestions from reviewers, modifications and clarifications in terminology, derivations, and illustrations have been made throughout the book to improve readability.

ACKNOWLEDGMENTS

The author gratefully acknowledges the following reviewers for their insightful suggestions: Alan H. Czarapata, Dave Krispinsky, David A. O'Brien, Mike Wilson, Edward Peterson, and Francis M. Turner.

CONTENTS

1 INTRODUCTION 1

1–1 The Study of Electronics 1
1–2 The Use of Computers 2
SPICE and PSpice 2
1–3 Circuit Analysis and Circuit Design 3

2 SEMICONDUCTOR THEORY 5

2–1 Introduction 5
2–2 Atomic Structure 5
2–3 Semiconductor Materials 8
2–4 Current in Semiconductors 9
Holes and Hole Current 11 ▪ Drift Current 12 ▪ Diffusion Current 15
2–5 P- and N-Type Semiconductors 16
2–6 The PN Junction 18
2–7 Forward- and Reverse-Biased Junctions 21
The Diode Equation 23 ▪ Breakdown 25 ▪ Temperature Effects 27
2–8 Schottky Diodes 30
Ohmic Contacts 31
2–9 Summary and Overview 31
Semiconductor Materials 31 ▪ Current in Semiconductors 32 ▪ P and N Materials 33 ▪ The PN Junction 33
2–10 A Semiconductor Glossary 36
Exercises 40

3 THE DIODE AS A CIRCUIT ELEMENT 45

3–1 Introduction 45
3–2 The Diode as a Nonlinear Device 46
3–3 AC and DC Resistance 48
3–4 Analysis of DC Circuits Containing Diodes 52
3–5 Analysis of Small-Signal Diode Circuits 54
The Loan Line 56
3–6 Analysis of Large-Signal Diode Circuits 61
Rectifiers 62 ■ Elementary DC Power Supplies 64
3–7 Diode Switching Circuits 67
3–8 Diode Types, Ratings, and Specifications 74
Exercises 80

4 BIPOLAR JUNCTION TRANSISTORS 87

4–1 Introduction 87
4–2 Theory of BJT Operation 88
I_{CBO} Reverse Current 91
4–3 Common-Base Characteristics 93
Common-Base Input Characteristics 94 ■ Common-Base Output Characteristics 96 ■ Breakdown 99
4–4 Common-Emitter Characteristics 100
I_{CEO} and Beta 101 ■ Common-Emitter Input Characteristics 104 ■ Common-Emitter Output Characteristics 104
4–5 Common-Collector Characteristics 109
4–6 Bias Circuits 111
Common-Base Bias Circuit 112 ■ Common-Emitter Bias Circuit 116 ■ Common-Collector Bias Circuit 121
4–7 Design Considerations 123
CB Bias Design 123 ■ CE Bias Design 124 ■ CC Bias Design 125
4–8 The BJT Inverter (Transistor Switch) 126
Inverter Design 128 ■ The Transistor as a Switch 129
4–9 Transistor Types, Ratings, and Specification 130
4–10 Transistor Curve Tracers 133
Exercises 136

5 SMALL-SIGNAL BJT AMPLIFIERS 143

5–1 AC Amplifier Fundamentals 143
Amplifier Gain 143 ■ Input and Output Resistance 144 ■

Source Resistance 146 ■ Load Resistance 148 ■ The Purpose of Bias 149 ■ Coupling Capacitors 150

5–2 Graphical Analysis of the Small-Signal CE Amplifier 151
The Effect of Q-Point Location on AC Operation 155 ■ Linearity and Distortion 156 ■ The Effect of Load Resistance on AC Operation 158

5–3 Amplifier Analysis Using Small-Signal Models 163
Small-Signal Parameters 163 ■ Small-Signal CB Amplifier Model 164 ■ Small-Signal CE Amplifier Model 170 ■ The Effects of AC Load Resistance 174 ■ Small-Signal CC Amplifier Model 179

5–4 The Dependence of Small-Signal Parameters on DC Bias Conditions 187
Transconductance 187 ■ Output Resistance 187 ■ The Transconductance Model 188

5–5 Small-Signal Parameter Equivalents: *h* Parameters 190

Exercises 193

6 BIAS DESIGN IN DISCRETE AND INTEGRATED CIRCUITS 199

6–1 The Effects of Parameter Variability 199
Thermal Runaway 199 ■ Bias Stabilization in Discrete Circuits 200

6–2 Stability Factors 201

6–3 Calculating the Bias Point of the Stabilized Circuit 204

6–4 Small-Signal Performance of the Bias-Stabilized Circuit 207

6–5 Stabilization by Collector Feedback 214

6–6 Discrete Circuit Bias Design 215

6–7 Integrated Circuits 218
Applications and Classifications of Integrated Circuits 218 ■ Crystal Growth 219 ■ PN Device Fabrication—The Photolithographic Process 220 ■ Epitaxial and Buried Layers 224 ■ Resistors 224 ■ Capacitors 228 ■ Component Interconnections 229 ■ Packaging 230

6–8 Integrated-Circuit Bias Techniques 231
Current Sources and Sinks 231 ■ Current Mirrors 234

Exercises 238

7 FIELD-EFFECT TRANSISTORS 245

7–1 Introduction 245

7–2 Junction Field-Effect Transistors 245
Transfer Characteristics 251

7–3 JFET Biasing 254
Fixed Bias 254 ■ General Algebraic Solution—Self-Bias 259 ■

Voltage-Divider Bias 261 ▪ General Algebraic Solution—Voltage-Divider Bias 264

7–4 JFET Bias Design 267

7–5 Manufacturer's Data Sheets 271

7–6 Metal-Oxide-Semiconductor FETs 273
Depletion-Type MOSFETs 273 ▪ Enhancement-Type MOSFETs 278 ▪ Enhancement MOSFET Transfer Characteristic 282 ▪ Enhancement MOSFET Bias Circuits 282 ▪ General Algebraic Solution 284 ▪ Feedback Bias 287 ▪ General Algebraic Solution—Feedback Bias 288

7–7 Integrated-Circuit MOSFETs 289

7–8 VMOS and DMOS Transistors 290
VMOS Transistors 290 ▪ DMOS Transistors 291

Exercises 292

8 FET CIRCUITS AND APPLICATIONS 299

8–1 The Common-Source JFET Amplifier 299
Small-Signal JFET Parameters 299 ▪ The Common-Source JFET Configuration 302 ▪ Bias-Stabilized JFET Amplifiers 305 ▪ Unbypassed Source Resistor 312

8–2 The Common-Drain and Common-Gate JFET Amplifiers 313
Common-Drain Amplifier 313 ▪ Common-Gate Amplifier 316

8–3 The JFET Current Source 317

8–4 The JFET as an Analog Switch 319
The JFET Chopper 322

8–5 Small-Signal MOSFET Amplifiers 323

8–6 VMOS and DMOS Amplifiers 326
VMOS Amplifiers 326 ▪ DMOS Amplifiers 326

8–7 MOSFET Inverters 327
The MOSFET as a Resistor 327 ▪ The MOSFET Inverter with MOSFET Load 328

8–8 Capacitive Loading of Switching Circuits 332
Capacitive Loading of the MOSFET Inverter 332

8–9 CMOS Circuits 336
CMOS Inverters 336 ▪ The CMOS Analog Switch 338

Exercises 339

9 *h* AND *y* PARAMETERS 347

9–1 Two-Port Networks 347
Two-Port Network Parameters 348

9–2 Hybrid- (h-) Parameter Definitions 348

9–3 Hybrid Equivalent Circuits 353
Gain and Impedance Computations Using h Parameters 355

9–4 Transistor h Parameters 359

9–5 BJT Amplifier Analysis Using h Parameters 364
Common-Emitter Amplifiers 364 ▪ CE h-Parameter Approximations 368 ▪ Common-Collector and Common-Base Amplifiers 371 ▪ CC and CB Approximations 375

9–6 y Parameters 377

9–7 y-Parameter Equivalent Circuits 380

9–8 FET y Parameters 382

Exercises 387

10 FREQUENCY RESPONSE 393

10–1 Definitions and Basic Concepts 393
Amplitude and Phase Distortion 395

10–2 Decibels and Logarithmic Plots 396
Decibels 396 ▪ Semilog and Log-Log Plots 400 ▪ One-nth Decade and Octave Intervals 403

10–3 Series Capacitance and Low-Frequency Response 404

10–4 Shunt Capacitance and High-Frequency Response 414
Thevenin Equivalent Circuits at Input and Output 419 ▪ Miller-Effect Capacitance 421

10–5 Transient Response 424

10–6 Frequency Response of BJT Amplifiers 426
Low-Frequency Response of BJT Amplifiers 426 ▪ Design Considerations 433 ▪ High-Frequency Response of BJT Amplifiers 436

10–7 Frequency Response of FET Amplifiers 439
Low-Frequency Response of FET Amplifiers 439 ▪ High-Frequency Response of FET Amplifiers 441

Exercises 446

11 MULTISTAGE AMPLIFIERS 455

11–1 Gain Relations in Multistage Amplifiers 455
Frequency Response of Cascaded Stages 461

11–2 Methods of Coupling 462

11–3 RC-Coupled BJT Amplifiers 464

11–4 Direct-Coupled BJT Amplifiers 474
The Darlington Pair 479 ▪ Cascode Amplifier 483

11–5 Transistor Arrays 487

11–6 Multistage FET Amplifiers 489
Direct-Coupled FET Amplifiers 491 ▪ A Bifet Amplifier 493

11–7 Transformer Coupling 494

11–8 Limitations of Practical Transformers 500
Copper Losses 500 ▪ Eddy Currents 500 ▪ Hysteresis Losses 501 ▪ Transformer Efficiency 501 ▪ Leakage Flux and Coupling Coefficients 502 ▪ Loading Effects 504 ▪ Frequency Response 507 ▪ Mutual Inductance 509

Exercises 512

12 INTEGRATED DIFFERENTIAL AND OPERATIONAL AMPLIFIER CIRCUITS 525

12–1 Introduction 525
Difference Voltages 525

12–2 The Ideal Differential Amplifier 526
The FET Differential Amplifier 534

12–3 Common-Mode Parameters 526

12–4 Practical Differential Amplifiers 537
Bias Methods in Integrated Circuits 539 ▪ Active Loads in Integrated Circuits 545

12–5 Introduction to Operational Amplifiers 547

12–6 Circuit Analysis of an Operational Amplifier 548

Exercises 552

13 OPERATIONAL AMPLIFIER THEORY 557

13–1 The Ideal Operational Amplifier 557
The Inverting Amplifier 558 ▪ The Noninverting Amplifier 561

13–2 Feedback Theory 564
Feedback in the Noninverting Amplifier 564 ▪ Feedback in the Inverting Amplifier 568

13–3 Frequency Response 573
Stability 573 ▪ The Gain-Bandwidth Product 573 ▪ User-Compensated Amplifiers 576

13–4 Slew Rate 578

13–5 Offset Currents and Voltages 586
Input Offset Current 586 ▪ Input Offset Voltage 591 ▪ The Total Output Offset Voltage 592

13–6 Operational Amplifier Specifications 593

Exercises 604

14 APPLICATIONS OF OPERATIONAL AMPLIFIERS 613

14–1 Voltage Summation, Subtraction, and Scaling 613
Voltage Summation 613 ▪ Voltage Subtraction 616

14–2 Controlled Voltage and Current Sources 622
Voltage-Controlled Voltage Sources 622 ▪ Voltage-Controlled Current Sources 623 ▪ Current-Controlled Voltage Sources 627 ▪ Current-Controlled Current Sources 628

14–3 Integration, Differentiation, and Waveshaping 630
Electronic Integration 630 ▪ Practical Integrators 634 ▪ Electronic Differentiation 637 ▪ Practical Differentiators 639 ▪ Waveshaping 645

14–4 Instrumentation Amplifiers 649

14–5 Oscillators 654
The Barkhausen Criterion 655 ▪ The RC Phase-shift Oscillator 657 ▪ The Wien-bridge Oscillator 660 ▪ The Colpitts Oscillator 663 ▪ The Hartley Oscillator 665

14–6 Active Filters 666
Basic Filter Concepts 666 ▪ Active Filter Design 670

Exercises 680

15 SPECIAL-PURPOSE CIRCUITS 687

15–1 Voltage Comparators 687
Hysteresis and Schmitt Triggers 690 ▪ An Astable Multivibrator 693

15–2 Clipping and Rectifying Circuits 696
Clipping Circuits 696 ▪ Precision Rectifying Circuits 701

15–3 Clamping Circuits 705

15–4 Logarithmic and Antilogarithmic Amplifiers 707

15–5 Transconductance Amplifiers 711
The 3080 Programmable Transconductance Amplifier 713

15–6 Phase-Locked Loops 715
FM Demodulation 716 ▪ Frequency Synthesizers 717 ▪ The 565 Integrated-Circuit PLL 718

15–7 The 8038 Integrated-Circuit Function Generator 718

Exercises 720

16 POWER AMPLIFIERS 727

16–1 Definitions, Applications, and Types of Power Amplifiers 727
Large-Signal Operation 728

16–2 Transistor Power Dissipation 728

16–3 Heat Transfer in Semiconductor Devices 731
Conduction, Radiation, and Convection 731 ▪ Thermal Resistance 732 ▪ Derating 735 ▪ Power Dissipation in Integrated Circuits 736

16–4 Amplifier Classes and Efficiency 736
Class-A Amplifiers 736 ▪ Efficiency 736 ▪ Transformer-Coupled Class-A Amplifiers 742 ▪ Class-B Amplifiers 746

16–5 Push-Pull–Amplifier Principles 747
Push-Pull Amplifiers with Output Transformers 747 ▪ Class-B Efficiency 749

16–6 Push-Pull Drivers 752

16–7 Harmonic Distortion and Feedback 754
Harmonic Distortion 754 ▪ Using Negative Feedback to Reduce Distortion 755

16–8 Distortion in Push-Pull Amplifiers 759
Cancellation of Even Harmonics 759 ▪ Crossover Distortion 760 ▪ Class-AB Operation 761

16–9 Transformerless Push-Pull Amplifiers 762
Complementary Push-Pull Amplifiers 762 ▪ Quasi-Complementary Push-Pull Amplifiers 770 ▪ Integrated-Circuit Power Amplifiers 771

16–10 Class-C Amplifiers 772
Amplitude Modulation 775

16–11 MOSFET and Class-D Power Amplifiers 779
MOSFET Amplifiers 779 ▪ Class-D Amplifiers 780

Exercises 784

17 POWER SUPPLIES AND VOLTAGE REGULATORS 789

17–1 Introduction 789

17–2 Rectifiers 790
Half-Wave Rectifiers 790 ▪ Full-Wave Rectifiers 790 ▪ Full-Wave Bridge Rectifiers 793

17–3 Capacitor Filters 797
Percent Ripple 800 ▪ Repetitive Surge Currents 804

17–4 RC and LC Filters 806
RC Filters 806 ▪ LC π Filters 810

17–5 Voltage Multipliers 811
Half-Wave Voltage Doubler 811 ▪ Full-Wave Voltage Doubler 812 ▪ Voltage Tripler and Quadrupler 812

17–6 Voltage Regulation 814
Line Regulation 816

17–7 Series and Shunt Voltage Regulators 816
The Zener Diode as a Voltage Reference 817 ▪ Series Regulators 817 ▪ Current Limiting 822 ▪ Foldback Limiting 824 ▪ Shunt Regulators 825

17–8 Switching Regulators 827

17–9 Three-Terminal Integrated-Circuit Regulators 830

17–10 Adjustable Integrated-Circuit Regulators 833

Exercises 839

18 SPECIAL ELECTRONIC DEVICES 845

18–1 Zener Diodes 845
The Zener-Diode Voltage Regulator 846 ▪ Temperature Effects 848 ▪ Zener-Diode Impedance 850

18–2 Four-Layer Devices 851
Silicon Controlled Rectifiers (SCRs) 851 ▪ Shockley Diodes 855 ▪ SCR Triggering 855 ▪ Half-Wave Power Control Using SCRs 858 ▪ Silicon Controlled Switches (SCSs) 861 ▪ DIACs and TRIACs 863

18–3 Optoelectronic Devices 867
Photoconductive Cells 868 ▪ Photodiodes 870 ▪ Phototransistors 873 ▪ Solar Cells 876 ▪ Light-Activated SCR (LASCR) 879 ▪ Light-Emitting Diodes (LEDs) 879 ▪ Optocouplers 883 ▪ Liquid-Crystal Displays (LCDs) 885

18–4 Unijunction Transistors 892
Programmable UJTs (PUTs) 899

18–5 Tunnel Diodes 901

18–6 Voltage-Variable Capacitors (Varactor Diodes) 903

Exercises 906

19 DIGITAL-TO-ANALOG AND ANALOG-TO-DIGITAL CONVERTERS 899

19–1 Overview 911
Analog and Digital Voltages 911 ▪ Converting Binary Numbers to Decimal Equivalents 912 ▪ Some Digital Terminology 913 ▪ Resolution 914

19–2 The *R*-2*R* Ladder DAC 914

19–3 A Weighted-Resistor DAC 918

19–4 The Switched Current-Source DAC 920

19–5 Switched-Capacitor DACs 922

19–6 DAC Performance Specifications 925
An Integrated-Circuit DAC 926

19–7 The Counter-Type ADC 927
Tracking A/D Converter 928

19–8 Flash A/D Converters 929

19–9 The Dual-Slope (Integrating) ADC 931

19–10 The Successive-Approximation ADC 933

19–11 ADC Performance Specifications 935
Integrated-Circuit A/D Converters 936

Exercises 937

A APPENDIX: SPICE AND PSPICE 939

A–1 Introduction 939

A–2 Describing a Circuit for a SPICE Input File 940
The Title 940 ▪ Nodes and Component Descriptions 940 ▪ Specifying Numerical Values 941 ▪ DC Voltage Sources 941 ▪ DC Current Sources 943

A–3 The .DC and .PRINT Control Statements 943
The .DC Control Statement 943 ▪ Identifying Output Voltages and Currents 944 ▪ The .PRINT Control Statement 945 ▪ The .END Statement 945 ▪ Circuit Restrictions 947

A–4 The .TRAN and .PLOT Control Statements 948
The .TRAN Control Statement 948 ▪ The .PLOT Control Statement 948 ▪ .DC Plots 949 ▪ LIMPTS and the .OPTIONS Control Statement 949

A–5 The SIN and PULSE Sources 949
The SIN Source 950 ▪ The PULSE Source 950

A–6 The Initial Transient Solution 952

A–7 Diode Models 954

A–8 BJT Models 956

A–9 The .TEMP Statement 958

A–10 AC Sources and the .AC Control Statement 958
The .AC Control Statement 959 ▪ AC Outputs 959 ▪ AC Plots 960 ▪ Small-Signal Analysis and Distortion 960

A–11 Junction Field-Effect Transistors (JFETs) 964

A–12 MOS Field-Effect Transistors (MOSFETs) 965

A–13 Controlled (Dependent) Sources 966

A–14 Transformers 968

A–15 Subcircuits 969

A–16 Probe and Control Shell 971
Probe 971 ■ Control Shell 973

A–17 The PSpice Library 974

B **APPENDIX: STANDARD VALUES OF 5% AND 10% RESISTORS** 977

C **APPENDIX: DERIVATION OF FREQUENCY-RESPONSE EQUATIONS** 979

C–1 Cutoff Frequency of Two Isolated High-Pass RC Filters in Cascade 979

C–2 Cutoff Frequency of Two Isolated Low-Pass RC Filters in Cascade 980

ANSWERS TO ODD-NUMBERED EXERCISES 981

INDEX 993

1 INTRODUCTION

1–1 THE STUDY OF ELECTRONICS

The field of study we call *electronics* encompasses a broad range of specialty areas, including audio systems, digital computers, communications systems, instrumentation, and automatic controls. Each of these areas in turn has its own specialty areas. However, all electronic specialties are the same in one respect: they all utilize electronic *devices*—transistors, diodes, integrated circuits, and various special components. The electrical characteristics of these devices make it possible to construct circuits that perform useful functions in many different kinds of applications. For example, a transistor may be used to construct an audio amplifier in a high-fidelity system, a high-frequency amplifier in a television receiver, an electronic switch used in a computer, or a current controller in a dc power supply. Regardless of one's specialization, a knowledge of device theory is a vital prerequisite to understanding and applying developments in that area.

The study of electronic devices is now almost wholly synonymous with the study of *semiconductor* devices. Semiconductor material is widely abundant, yet unique in terms of its electrical properties because it is neither a conductor nor an insulator. *Silicon,* an element found in ordinary sand, is now the most widely used semiconductor material. As we shall discover in a subsequent chapter, it must be subjected to many closely controlled manufacturing processes before it acquires the properties that make it useful in the fabrication of electronic devices.

If any one trend characterizes modern electronics, it is the quest for miniaturization. Each year we learn of new technological advances that result in ever more sophisticated circuitry being packaged in ever smaller integrated circuits. The primary reasons for this trend are improved reliability, reduced costs, and, in the case of high-frequency and digital computer circuitry, increased speed due to the reduction of interconnecting paths.

The effort to miniaturize has been so successful, and the complexity of the circuitry contained in a single package is now so great, that some students and educators despair, or question the validity, of studying the fundamental entities of which these devices are composed. There is a temptation to abandon the study of semiconductor structure and the properties of individual transistors, diodes, and similar building blocks of integrated circuits. However, what is now the "microscopic" nature of electronic circuits should be studied for numerous compelling reasons, not the least of which is the many career opportunities in the design and fabrication of the devices themselves. Furthermore, the intelligent application of integrated circuits depends on a knowledge of certain underlying limitations and special characteristics, as well as insights that can be gained only through a study of their structure. Finally, there are many applications in which ultraminiature electronic devices cannot and never will be used—applications involving heavy power dissipation. In these applications, only fundamental, building-block–type components, such as *discrete* transistors, can be used, and an understanding of their behavior is paramount.

1–2 THE USE OF COMPUTERS

The study and understanding of electronics demands certain mathematical skills, because circuit behavior can be described in practical terms only by equations. Quantitative (numerical) results obtained from solving equations are the principal means we have for comparing, predicting, designing, and evaluating electronic circuits. In this book, the reader is given the opportunity to use computers as an aid in obtaining those results.

There are basically two ways that computers are used in the study of electronics. First, we can choose to write our own programs in one of the standard computer languages, such as BASIC or FORTRAN. Each program we write will be designed to solve a specific circuit problem. For example, we might wish to determine the magnitude of the output voltage in a certain single-transistor amplifier. Our program will be designed to produce that result, and only that result, for that circuit, and only that circuit. (However, it is usually very easy to change the *values* of the components in the circuit and to reuse the program with those changes.) When a computer is used this way, programmers must have a good knowledge of electronic circuit theory, since they must be able to select and rearrange the theoretical equations that apply to the circuit. At the same time, they must be skilled programmers, having a good knowledge of the computer language and the ability to use it to solve various kinds of equations.

On the other hand, in the second way of using computers, we rely on someone else's programming skills, through acquisition of a program specially designed to solve electronic circuit problems. A popular example of such a program is SPICE (Simulation Program with Integrated Circuit Emphasis), developed at the University of California, Berkeley. To use this kind of program, we need only specify the components in the circuit we wish to study, describe how they are interconnected, and "tell" the program what kind of results we want (output voltage, etc.). Very little programming skill is required because we simply supply *data* to a program that is already in computer memory. Also, very little knowledge of electronics theory is required: we need just enough to be able to describe the components in the circuit and how they are connected.

SPICE and PSpice

In this book, discussion of electronic circuit theory is accompanied by numerous examples and exercises using SPICE to solve circuit problems. For reference and/or review purposes, Appendix A contains a summary of SPICE programming techniques. It also contains material on the use of PSpice, a popular microcomputer version of the original (Berkeley) SPICE. With a few minor exceptions, any circuit simulation written for SPICE can be executed successfully by PSpice, although the reverse is not true. All SPICE programs in examples and exercises in this book have been checked and found to run successfully in PSpice and in version 2G.6 of SPICE on a mainframe computer (Honeywell DPS 90).

1–3 CIRCUIT ANALYSIS AND CIRCUIT DESIGN

In the context of electronic circuits, *analysis* generally means finding voltages, currents, and/or powers given device characteristics and the component values in a circuit. Circuit *design,* or *synthesis,* turns that process around by finding component values and selecting devices so that certain voltages, currents, and/or powers are developed at specific points in a circuit. As a simple example, we can analyze a voltage divider to determine the voltage it develops, given the voltage across it and the values of its resistors. A typical design problem is to select resistor values so that a specified voltage is developed.

In circuit analysis, we usually derive equations for voltage, current, or power in terms of component values. Thus, circuit design is often performed by solving such equations for component values in terms of voltage, current, or power. However, there are no hard-and-fast rules for teaching design. It is generally agreed that a thorough understanding of analysis techniques is a vital prerequisite to developing design skills. Accordingly, the principal thrust of this book is electronic circuit analysis. Nevertheless, because of the practical importance of design, numerous examples are given to show how analysis techniques can be turned around and used to create circuits having specific properties. In some cases, circuit design can be accomplished only by trial-and-error repetitions of the analysis procedure. Computers are very useful in that type of design activity, as we shall see in forthcoming examples.

2 SEMICONDUCTOR THEORY

2–1 INTRODUCTION

Semiconductor devices are the fundamental building blocks from which all types of useful electronic products are constructed—amplifiers, high-frequency communications equipment, power supplies, computers, control systems, to name only a few. It is possible to learn how these devices—diodes, transistors, integrated circuits—can be connected together to create such useful products with little or no knowledge of the semiconductor theory that explains how the devices themselves operate. However, the person who understands that theory has a greater knowledge of the capabilities and limitations of devices and is therefore able to use them in more innovative and efficient ways than the person who does not. Furthermore, it is often the case that the success or failure of a complex electronic system can be traced to a certain peculiarity or operating characteristic of a single device, and an intimate knowledge of how and why that device behaves the way it does is the key to reliable designs or to practical remedies for substandard performance.

In Sections 2–2 through 2–7, we will study the fundamental theory underlying the flow of charged particles through semiconductor material, the material from which all modern devices are constructed. We will also learn the theory of operation of the most fundamental semiconductor device: the diode. The theory presented here includes numerous equations that allow us to compute and assign quantitative values to important atomic-level properties of semiconductors. **Readers who do not need or wish to study semiconductor theory in such depth may prefer to go directly to Section 2–9, "Summary and Overview," which is a condensed version of the sections that follow.** Although Section 2–9 does not contain the technical detail of Sections 2–2 through 2–7, it does provide adequate background for all subsequent discussions of device theory.

2–2 ATOMIC STRUCTURE

A study of modern electronic devices must begin with a study of the materials from which those devices are constructed. Knowledge of the principles of material composition, at the level of the fundamental structure of matter, is an important prerequisite in the field of study we call *electronics,* because the ultimate concern of that field is predicting and controlling the flow of atomic charge. Towards developing an appreciation of how atomic structure influences the electrical properties of materials, let us begin with a review of the structure of that most fundamental of all building blocks: the atom itself.

Every chemical element is composed of atoms, and all of the atoms within a *single* element have the same structure. Each element is unique because that common structure of its atoms is unique. Each atom is itself composed of a central *nucleus* containing one or more positively charged particles called *protons.* When an atom is complete, its nucleus is surrounded by negatively charged particles, called *electrons,* equal in number to the quantity of protons in the nucleus. Since the positive charge on each proton is equal in magnitude to the negative charge on each electron, the complete atom is electrically neutral. Depending on the element that a particular atom represents, the nucleus may also contain particles called *neutrons* that carry no electrical charge. An atom from the simplest of all elements, hydrogen, contains exactly one electron and has a nucleus composed of one proton and no neutrons. All other elements have at least one neutron.

Figure 2–1(a) is a diagram of the structure of an isolated atom of the element *silicon,* the material used most often in the construction of modern electronic devices. Rather than showing a cluster of individual protons and neutrons in the nucleus, the figure simply shows that the nucleus contains 14 (positively charged) protons and 14 neutrons. Notice that the atom is neutral, since it contains a total of 14 (negatively charged) electrons. These electrons are shown distributed among three distinct *orbits* around the nucleus. The electrons within a given orbit are said to occupy an electron *shell,* and it is well known that each shell in an atom can contain no more than a certain maximum number of electrons. If the first four shells are numbered in sequence beginning from the innermost shell (shell number 1 being that closest to the nucleus), then the maximum number (N_e) of electrons that shell number n can contain is

$$N_e = 2n^2 \tag{2–1}$$

Figure 2–1

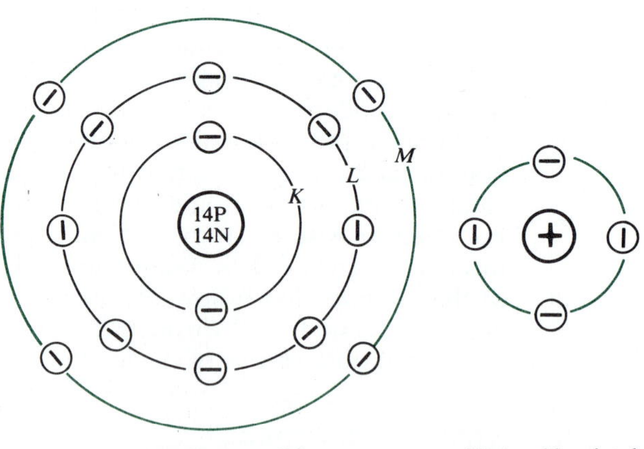

(a) A diagram of the silicon atom, showing its nucleus and electron shells. P = proton; N = neutron.

(b) An abbreviated diagram of the silicon atom, showing valence electrons only.

Notice in Figure 2–1(a) that shell number 1 (labeled K) is filled because it contains $2 \times 1^2 = 2$ electrons and shell number 2 (labeled L) is filled because it contains $2 \times 2^2 = 8$ electrons. However, shell number 3 (M) is not filled because it has a capacity of $2 \times 3^2 = 18$ electrons but contains only 4.

Each shell is divided into *subshells,* the nth shell having n subshells. The first subshell in an electron shell can contain 2 electrons. If the shell has a second subshell, it can contain 4 additional, or 6 total, electrons; and if there is a third subshell, it can contain four more, or 10, electrons. The first shell of silicon has only one subshell ($n = 1$) and it is filled (2 electrons); the second shell has two ($n = 2$) filled subshells (containing 2 and 6 electrons); and the third shell has one filled subshell containing 2 electrons. The second subshell of the third shell can contain 6 electrons but has only 2. In practice, shells are given letter designations rather than numerical ones, so shells 1, 2, 3, 4, . . . , are referred to as shells $K, L, M, N,$ . . . , as shown in Figure 2–1(a). Subshells are designated s, p, d, f.

Example 2–1

The nucleus of a germanium atom has 32 protons. Assuming that each shell must be filled before a succeeding shell can contain any electrons, determine the number of electrons in each of its shells and in the subshells of each shell.

Solution. There are 32 electrons, and they fill the shells as shown in the *contents* column of the following table:

Shell	Capacity ($2n^2$)	Contents
K ($n = 1$)	2	2
L ($n = 2$)	8	8
M ($n = 3$)	18	18
N ($n = 4$)	32	4
	Total:	32

The contents of the subshells in each shell are shown in the following table:

Shell	Subshells	Capacity	Contents
K	s	2	2
L	s	2	2
	p	6	6
M	s	2	2
	p	6	6
	d	10	10
N	s	2	2
	p	6	2
	d	10	0
	f	14	0
		Total:	32

Not every electron in every atom is constrained forever to occupy a certain shell or subshell of an atomic nucleus. Although electrons tend to remain in their shells because of their force of attraction to the positively charged nucleus, some of them acquire enough energy (as, for example, from heating) to break away from their "parent" atoms and wander randomly through the material. Electrons that have escaped their shells are called *free* electrons. Conductors have a great many free electrons, while insulators have relatively few.

The outermost shell in an atom is called the *valence* shell, and the number of electrons in the valence shell has a significant influence on the electrical properties of an element. The reason that the number of valence electrons is important is that electrons in a nearly empty valence shell or subshell are more easily dislodged (freed) than electrons in a filled or nearly filled shell. Moreover, valence electrons, being farther from the nucleus than electrons in the inner shells, are the ones that experience the least force of attraction to the nucleus. Conductors are materials whose atoms have very few electrons in their valence shells (copper atoms have only one), and, in these materials, the heat energy available at room temperature (25°C) is enough to free large numbers of those loosely bound electrons. When an electrical potential is applied across the ends of a conductor, free electrons readily move from one end to the other, creating a transfer of charge through the conductor, i.e., an electrical current. Valence electrons in insulators, on the other hand, are tightly bound to their parent atoms.

Since we will be concerned primarily with the behavior of electrons in the valence shell, we will abbreviate all future diagrams of atoms so that only the nucleus and the valence electrons are shown. The silicon atom is shown in this abbreviated manner in Figure 2–1(b).

2–3 SEMICONDUCTOR MATERIALS

Virtually all modern electronic devices are constructed from *semiconductor* material. As the name implies, a semiconductor is neither an electrical insulator (like rubber or plastic) nor a good conductor of electric current (like copper or aluminum). Furthermore, the mechanism by which charge flows through a semiconductor cannot be entirely explained by the process known to cause the flow of charge through other materials. In other words, a semiconductor is something more than just a conductor that does not conduct very well, or an insulator that allows some charge to pass through it.

The electrical characteristics of a semiconductor stem from the way its atoms interlock with each other to form the structure of the material. Recall that conductors have nearly empty valence shells and tend to produce free electrons, while insulators tend to retain valence electrons. The valence shell of a semiconductor atom is such that it can just fill an incomplete subshell by acquiring four more electrons. For example, we have already noted that the p subshell of the M shell in silicon contains 2 electrons. Given four more, this subshell would be filled. A semiconductor atom seeks this state of stability and achieves it by *sharing* the valence electrons of four of its neighboring atoms. It in turn shares each of its own four electrons with its four neighbors and thus contributes to the filling of their subshells. Every atom duplicates this process, so every atom uses four of its own electrons and one each from four of its neighbors to fill its p subshell. The result is a stable, tightly bound, lattice structure called a *crystal.*

The interlocking of semiconductor atoms through electron sharing is called *covalent bonding.* It is important to be able to visualize this structure, and Figure 2–2 shows a two-dimensional representation of it. Of course, a true crystal is a three-dimensional object—the covalent bonding occurs in all directions—but the

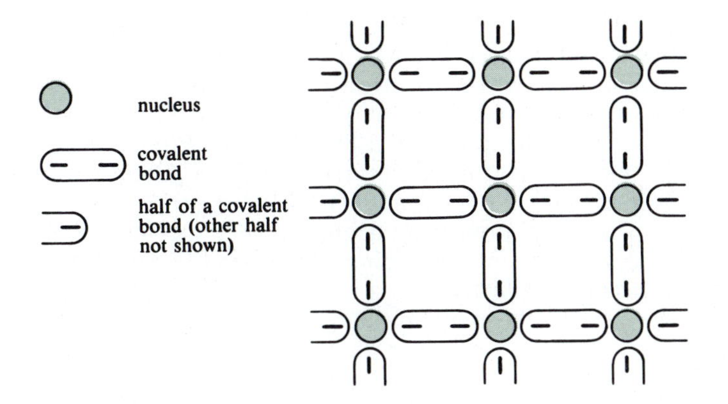

Figure 2–2
Covalent bonding in a semiconductor crystal

figure should help clarify the concept of covalent bonding. Remember that we show only the valence electrons of each atom.

In Figure 2–2, each pair of shared electrons forms a *covalent bond,* which is shown as two electrons enclosed by an oval. Note that only the center atom is shown with a complete set of 4 covalent bonds, but all of the other atoms would be similarly interlocked with their neighbors. As previously described, the center atom uses its own four electrons and one from each of four neighbors, so it effectively has 8 electrons in its *M* shell. The atom directly above the center one similarly has four of its own electrons and one from each of four neighbors (although only three are shown), so it too has 8.

Germanium (symbol Ge) is another element whose valence shell enables it to establish covalent bonds with its neighbors and form a crystalline structure. In Example 2–1, we saw that its *N* shell contains 4 electrons, 2 of which are in the *p* subshell. Therefore, it too can complete a subshell by the acquisition of 4 electrons. Like silicon (symbol Si), germanium atoms interlock and form a semiconductor material that is used to construct electronic devices. (However, because of temperature-related properties that we will discuss later, germanium is not now so widely used as silicon.) Note that the diagram in Figure 2–2 applies equally to germanium and silicon, since only valence shell electrons are shown. Carbon also has a valence shell that enables it to establish covalent bonds and form a crystal, but it assumes that form only after being subjected to extreme heating and pressure. A carbon crystal is in fact a *diamond,* and it is not used in the construction of semiconductor devices.

2–4 CURRENT IN SEMICONDUCTORS

We have mentioned that the source of electrical charge available to establish current in a conductor is the large number of free electrons in the material. Recall that an electron is freed by acquiring energy, typically heat energy, that liberates it from a parent atom. Electrons are freed in semiconductor materials in the same way, but a greater amount of energy is required, on the average, because the electrons are held more tightly in covalent bonds. When enough energy is imparted to an electron to allow it to escape a bond, we say that a covalent bond has been *ruptured.*

It is instructive to view the electron liberation process from the standpoint of the quantity of energy possessed by the electrons. The unit of energy that is conventionally used for this purpose is the *electronvolt* (eV), which is the energy acquired by one electron if it is accelerated through a potential difference of one volt. One eV equals 1.602×10^{-19} joules (J). According to modern quantum theory, an electron in an isolated atom must acquire a very specific amount of energy in order to be freed, the amount depending on the kind of atom to which it belongs and the shell it occupies. Electrons in a valence shell already possess considerable energy, because a relatively small amount of additional energy will liberate them. Electrons in inner shells possess little energy, since they are strongly attracted to the nucleus and would therefore need a great deal of additional energy to be freed. Electrons can also move from one shell to a more remote shell, provided they acquire the distinct amount of energy necessary to elevate them to the energy level represented by the new shell. Furthermore, electrons can *lose* energy, which is released in the form of heat or light, and thereby fall into lower energy shells. Free electrons, too, can lose a specific amount of energy and fall back into a valence shell.

When atoms are in close proximity, as they are when they are interlocked to form a solid material, the interactions between adjoining atoms make the energy levels less distinct. In this case, it is possible to visualize a nearly continuous *energy band,* and we refer to electrons as occupying one band or another, depending on their roles in the structure. Energy band diagrams are shown in Figure 2–3. Free electrons are said to be in the *conduction band* because they are available as charge carriers for the conduction of current. *Valence-band* electrons have less energy and are shown lower in the energy diagram. The region between the valence and conduction bands is called a *forbidden band,* because quantum theory does not permit electrons to possess energies at those particular levels. The width of the forbidden band is the energy *gap* that electrons must surmount to make the transition from valence band to conduction band. Note in Figure 2–3(a) that a typical insulator has a large forbidden band, meaning that valence electrons must acquire a great deal of energy to become available for conduction. For example, the energy gap for carbon is 5.4 eV. Energy gaps for semiconductors depend on temperature. As shown in Figure 2–3, the room-temperature values for silicon and germanium are about 1.1 eV and 0.67 eV, respectively. The energy gap for a conductor is quite small ($\leq$0.01 eV) or nonexistent, and the conduction and valence bands are generally considered to overlap.

Figure 2–3
Energy band diagrams for several different materials

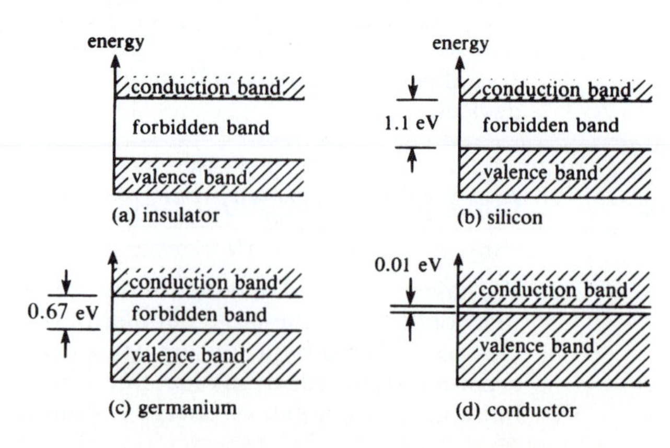

As might be surmised from the foregoing discussion, the number of free electrons in a material, and consequently its electrical *conductivity,* is heavily dependent on temperature. Higher temperatures mean more heat and therefore greater electron energies. At absolute zero (−273°C, or 0 K), all electrons have zero energy. But as the temperature is raised, more and more electrons acquire sufficient energy to cross the gap into the conduction band. For a semiconductor, the result is that conductivity increases with temperature (resistance decreases), which means that a semiconductor has a *negative* temperature coefficient of resistance. Although the number of conduction-band electrons in a conductor also increases with temperature, there are so many more of these than in a semiconductor that another effect predominates: their number becomes so vast that they collide frequently and interfere with each other's progress when under the influence of an applied electric potential. The increased heat energy imparted to them at higher temperatures also makes their motion more erratic and compounds the problem. Consequently, it becomes more difficult to establish a uniform flow of charge at higher temperatures, resulting in a positive temperature coefficient of resistance for conductors.

Holes and Hole Current

What really distinguishes electrical current in a conductor from that in a semiconductor is the existence in the latter of another kind of charge flow. Whenever a covalent bond in a semiconductor is ruptured, a *hole* is left in the crystal structure by virtue of the loss of an electron. Since the atom that lost the electron now has a net positive charge (the atom becomes what is called a positive *ion*), we can regard that hole as representing a unit of positive charge. The increase in positive charge is, of course, equal to the decrease in negative charge, i.e., the charge of one electron: $q_e = 1.6 \times 10^{-19}$ coulombs (C). *If a nearby valence-band electron should now enter the hole, leaving behind a new hole, then the net effect is that a unit of positive charge has moved from the first atom to the second.* This transfer of a hole from one atom to another constitutes a flow of (positive) charge and therefore represents a component of electric current, just as electron flow contributes to current by the transfer of negative charge. We can therefore speak of *hole current* in a semiconductor as well as electron current. Figure 2–4 illustrates the concept.

Figure 2–4 illustrates the *single* repositioning of a hole, but it is easy to visualize still another valence electron entering the new hole (at B in the figure), causing the hole to move again, and so forth, resulting in a hole path through the crystal. Note that holes moving from left to right cause charge transfer in the same direction as electrons moving from right to left. It would be possible to analyze semiconductor current as two components of electron transfer, but it is conventional to distinguish between conduction-band electron flow and valence-band hole flow. Here is an

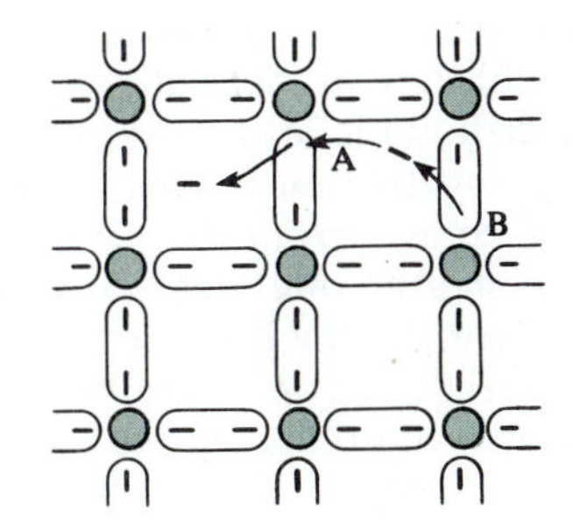

Figure 2–4
Hole current. When the electron in position A is freed, a hole is left in its place. If the electron in position B moves into the hole at A, the hole, in effect, moves from A to B.

important point that is worth repeating, because it is a source of confusion to students and is not emphasized enough in most textbooks: Hole current occurs at the *valence*-band level, because valence-band electrons do not become free electrons when they simply move from atom to atom. Electron current always occurs in the *conduction* band, and involves only the flow of free electrons. If a conduction-band electron falls into a hole (which it may), this does *not* constitute current flow; indeed, such an occurrence is a cancellation of charge, and we say that a hole–electron pair has been *annihilated,* or that a *recombination* has occurred. Since charge transfer can take place by the motion of either negatively charged electrons or positively charged holes, we refer to electrons and holes collectively as charge *carriers.* Note that hole current does not occur in a conductor.

Since holes in a (pure) semiconductor are created by electrons that have been freed from their covalent bonds, the number of free electrons must equal the number of holes. This equality applies to the semiconductor materials we have studied so far, because we have assumed them to be pure (in the sense that they are composed exclusively of atoms from one kind of element). Later we will study semiconductor materials that have been made impure purposely in order to change the balance between holes and electrons. Pure semiconductor material is said to be *intrinsic.* It follows that the electron *density,* in electrons/m^3, equals the hole density, holes/m^3, in an intrinsic semiconductor. The subscript i is used to denote an intrinsic property; n_i refers to intrinsic electron density, and p_i is intrinsic hole density. Thus,

$$n_i = p_i \tag{2–2}$$

At room temperature, the charge carrier densities for germanium and silicon are approximately $n_i = p_i = 2.4 \times 10^{19}$ carriers/m^3 for germanium and $n_i = p_i = 1.5 \times 10^{16}$ carriers/m^3 for silicon. These figures seem to imply vast numbers of carriers per cubic centimeter, but consider the fact that a cubic centimeter of silicon contains more than 10^{22} atoms. Thus, there are approximately 10^{12} times as many atoms as there are carriers in silicon. Consider also that the conductor copper contains approximately 8.4×10^{28} carriers (free electrons) per cubic meter, a carrier density that is immensely greater than that of either germanium or silicon. Note that the carrier density of germanium is greater than that of silicon because the energy gap, as shown in Figure 2–3, is smaller for germanium than for silicon. At a given temperature, the number of germanium electrons able to escape their bonds and enter the conduction band is greater than the number of silicon electrons that can do likewise.

Drift Current

When an electric potential is applied across a semiconductor, the electric field established in the material causes free electrons to drift in one direction and holes to drift in the other. Because the positive holes move in the opposite direction from the negative electrons, these two components of current *add* rather than cancel. The total current due to the electric field is called the *drift* current. Drift current depends, among other factors, on the ability of the charge carriers to move through the semiconductor, which in turn depends on the type of carrier and the kind of material. The measure of this ability to move is called *drift mobility* and has the symbol μ. Following are typical values for hole mobility (μ_p) and electron mobility

(μ_n) in germanium and silicon:

Silicon	Germanium
$\mu_n = 0.14\ \text{m}^2/(\text{V} \cdot \text{s})$	$\mu_n = 0.38\ \text{m}^2/(\text{V} \cdot \text{s})$
$\mu_p = 0.05\ \text{m}^2/(\text{V} \cdot \text{s})$	$\mu_p = 0.18\ \text{m}^2/(\text{V} \cdot \text{s})$

Note that the units of μ are square meters per volt-second. Recall that the units of electric field intensity are V/m, so μ measures carrier velocity (m/s) per unit field intensity: $(\text{m/s})/(\text{V/m}) = \text{m}^2/(\text{V} \cdot \text{s})$. It follows that

$$v_n = \overline{E}\mu_n \quad \text{and} \quad v_p = \overline{E}\mu_p \tag{2–3}$$

where $\overline{E}$ is the electric field intensity in V/m and v_n and v_p are the electron and hole velocities in m/s. Although the value of μ depends on temperature and the actual value of the electric field intensity, the values listed above are representative of actual values at low to moderate field intensities and at room temperature.

We can use carrier mobility to compute the total *current density J* in a semiconductor when the electric field intensity is known. Current density is current per unit cross-sectional area.

$$\begin{aligned} J = J_n + J_p &= \overbrace{nq_n\mu_n\overline{E}}^{J_n} + \overbrace{pq_p\mu_p\overline{E}}^{J_p} \\ &= nq_nv_n + pq_pv_p \end{aligned} \tag{2–4}$$

where
J = current density, A/m^2
n, p = electron and hole densities, carriers/m^3
$q_n = q_p$ = unit electron charge $= 1.6 \times 10^{-19}$ C
μ_n, μ_p = electron and hole mobilities, $\text{m}^2/(\text{V} \cdot \text{s})$
$\overline{E}$ = electric field intensity, V/m
v_n, v_p = electron and hole velocities, m/s

Equation 2–4 expresses the fact that the total current density is the sum of the electron and hole components of current density, J_n and J_p. For intrinsic material, equation 2–4 can be simplified as follows:

$$\begin{aligned} J &= n_iq_n\overline{E}(\mu_n + \mu_p) = p_iq_p\overline{E}(\mu_n + \mu_p) \\ &= n_iq_n(v_n + v_p) = p_iq_p(v_n + v_p) \end{aligned}$$

An analysis of the units of the first of equation 2–4 shows that

$$\overbrace{\left(\frac{\text{carriers}}{\text{m}^3}\right)}^{n}\overbrace{\left(\frac{\text{coulombs}}{\text{carrier}}\right)}^{q}\overbrace{\left(\frac{\text{V}}{\text{m}}\right)}^{\overline{E}}\overbrace{\left(\frac{\text{m}^2}{\text{V} \cdot \text{s}}\right)}^{\mu} = \frac{\text{C}}{\text{s} \cdot \text{m}^2} = \frac{\text{A}}{\text{m}^2} = \text{current density}$$

Example 2–2

A potential difference of 12 V is applied across the ends of the intrinsic silicon bar shown in Figure 2–5. Assuming that $n_i = 1.5 \times 10^{16}$ electrons/m^3, $\mu_n = 0.14\ \text{m}^2/(\text{V} \cdot \text{s})$, and $\mu_p = 0.05\ \text{m}^2/(\text{V} \cdot \text{s})$, find

1. the electron and hole velocities;
2. the electron and hole components of the current density;

Figure 2–5
(Example 2–2)

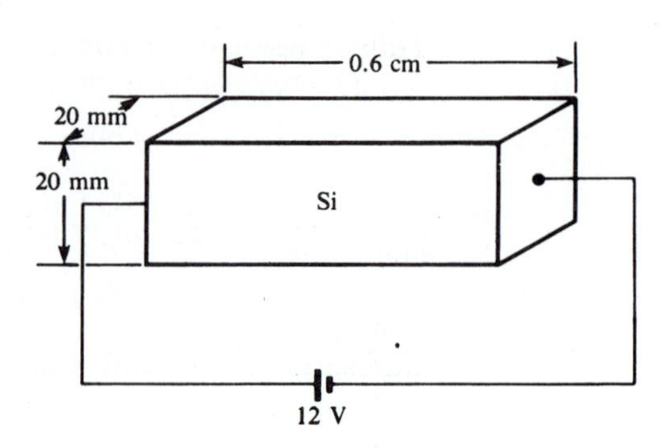

3. the total current density; and
4. the total current in the bar.

Solution. We will assume that the electric field is established uniformly throughout the bar and that all current flow is along the horizontal axis of the bar (in the direction of the electric field).

1. $\overline{E} = (12\text{ V})/(0.6 \times 10^{-2}\text{ m}) = 2 \times 10^3$ V/m. From equation 2–3,

$$v_n = \overline{E}\mu_n = (2 \times 10^3\text{ V/m})\,[0.14\text{ m}^2/(\text{V}\cdot\text{s})] = 2.8 \times 10^2\text{ m/s}$$
$$v_p = \overline{E}\mu_p = (2 \times 10^3\text{ V/m})\,[0.05\text{ m}^2/(\text{V}\cdot\text{s})] = 10^2\text{ m/s}$$

2. Since the material is intrinsic,

$$p_i = n_i = 1.5 \times 10^{16}\text{ carriers/m}^3$$

and

$$J_n = n_i q_n v_n = (1.5 \times 10^{16})(1.6 \times 10^{-19})(2.8 \times 10^2) = 0.672\text{ A/m}^2$$
$$J_p = p_i q_p v_p = (1.5 \times 10^{16})(1.6 \times 10^{-19})(10^2) = 0.24\text{ A/m}^2$$

3. $J = J_n + J_p = 0.672 + 0.24 = 0.912$ A/m²
4. The cross-sectional area A of the bar is $(20 \times 10^{-3}\text{ m}) \times (20 \times 10^{-3}\text{ m}) = 4 \times 10^{-4}\text{ m}^2$. Therefore, $I = JA = (0.912\text{ A/m}^2)(4 \times 10^{-4}\text{ m}^2) = 0.365$ mA.

Recall that the resistance of any body can be calculated using

$$R = \frac{\rho l}{A} \tag{2–5}$$

where
R = resistance, ohms (Ω)
ρ = *resistivity* of the material, Ω · m
l = length, m
A = cross-sectional area, m²

Conductance, which has the units of siemens (S), is defined to be the reciprocal of resistance, and *conductivity* is the reciprocal of resistivity:

$$\sigma = \frac{1}{\rho} \tag{2–6}$$

Thus, the units of conductivity are $1/(\Omega \cdot m)$, or siemens/meter (S/m).
The conductivity of a semiconductor can be computed using

$$\sigma = n\mu_n q_n + p\mu_p q_p \tag{2-7}$$

Note that it is again possible to identify a component of conductivity due to electrons and a component due to holes.

Example 2–3

1. Compute the conductivity and resistivity of the bar of intrinsic silicon in Example 2–2 (Figure 2–5).
2. Use the results of (1) to find the current in the bar when the 12-V potential is applied to it.

Solution

1. $n = p = n_i = p_i = 1.5 \times 10^{16}/\text{m}^3$. From equation 2–7,

$$\begin{aligned}\alpha &= (1.5 \times 10^{16})(0.14)(1.6 \times 10^{-19}) \\ &\quad + (1.5 \times 10^{16})(0.05)(1.6 \times 10^{-19}) \\ &= 4.56 \times 10^{-4}\ \text{S/m}\end{aligned}$$

Then $\rho = 1/\sigma = 1/(4.56 \times 10^{-4}) = 2192.98\ \Omega \cdot \text{m}$.

2. $R = \rho l/A = (2192.98)(0.6 \times 10^{-2})/(4 \times 10^{-4}) = 32.89\ \text{k}\Omega$

$I = E/R = 12/(32.89 \times 10^3) = 0.365\ \text{mA}$

The current computed this way is the same as that computed in Example 2–2.

Diffusion Current

Under certain circumstances, another kind of current besides drift current can exist in a semiconductor. Whenever there is a concentration of carriers (electrons or holes) in one region of a semiconductor and a scarcity in another, the carriers in the high density region will migrate toward the low density region, until their distribution becomes more or less uniform. In other words, there is a natural tendency for energetic carriers to disperse themselves to achieve a uniform concentration. Visualize, in an analogous situation, a small room crowded with people who are constantly squirming, crowding, and elbowing each other; if the room were suddenly to expand to twice its size, the occupants would tend to nudge and push each other outward as required to fill the new space. Another example of this natural expansion of energetic bodies is the phenomenon observed when a fixed quantity of gas is injected into an empty vessel: The molecules disperse to fill the confines of the container.

During the time that carriers are migrating from the region of high concentration to the one of low concentration, there is a transfer of charge taking place, and therefore an electric current. This current is called *diffusion* current, and the carriers are said to diffuse from one region to another. Diffusion is a transient (short-lived) process unless the region containing the higher concentration of charge is continually replenished. In many practical applications that we will study later, carrier replenishment does occur and diffusion current is thereby sustained.

2–5 P- AND N-TYPE SEMICONDUCTORS

Recall that intrinsic semiconductor material has the same electron density as hole density: $n_i = p_i$. In the fabrication of semiconductor materials used in practical applications, this balance between carrier densities is intentionally altered to produce materials in which the number of electrons is greater than the number of holes, or in which the number of holes is greater than the number of electrons. Such materials are called *extrinsic* (or *impure*) semiconductors. They are called *impure* because, as we will discuss presently, the desired imbalance is achieved by introducing certain impurity atoms into the crystal structure. Materials in which electrons predominate are called *N-type* materials, and those in which holes predominate are called *P-type* materials.

Let us first consider how N-type material is produced. Suppose that we are able, by some means, to insert into the crystal structure of a semiconductor an atom that has 5 instead of 4 electrons in its valence shell. Then 4 of those 5 electrons can (and do) participate in the same kind of covalent bonding that holds all the other atoms together. In this way, the impurity atom becomes an integral part of the structure, but it differs from the other atoms in that it has one "excess" electron. Its fifth valence electron is not needed for any covalent bond. Figure 2–6 illustrates how the structure of a silicon crystal is modified by the presence of one such impurity atom. An impurity atom that produces an excess electron in this way is called a *donor* atom, because it donates an electron to the material. The nucleus of the donor atom is labeled *D* in Figure 2–6. When a large number of donor atoms are introduced into the material, a correspondingly large number of excess electrons are created. Materials used as donor impurities in silicon include *antimony, arsenic,* and *phosphorus.*

Extrinsic semiconductor material is said to have been *doped* with impurity atoms, and the process is called *doping.* The impurity material is called a *dopant.* In Chapter 6 we will discuss various ways in which doping is accomplished. Since the silicon material illustrated in Figure 2–6 has been doped with donor atoms and therefore contains an excess of electrons, it now constitutes N-type material. Note that the electrons are in "excess" only in the sense that there are now more electrons than holes; the material is still electrically neutral, because the number of protons in each donor nucleus still equals the total number of electrons that the donor atom brought to the material. Each excess electron is, however, very loosely bound to a donor atom and is, for all practical purposes, in the conduction band. The donor

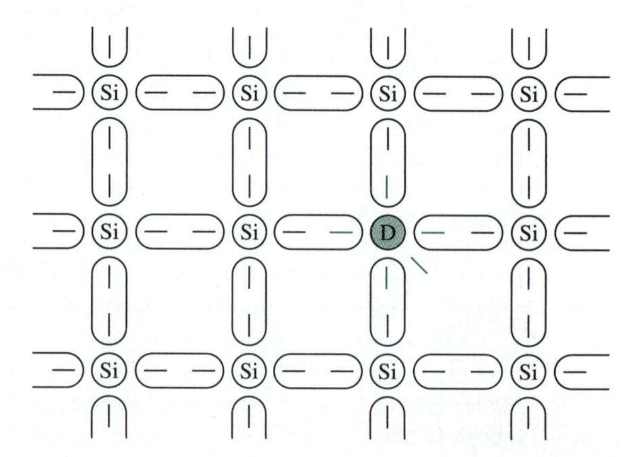

Figure 2–6
Structure of a silicon crystal containing a donor atom. The donor's nucleus is labeled D and the nuclei of the silicon atoms are labeled Si. The donor electrons are shown by blue dashes. Note the excess electron.

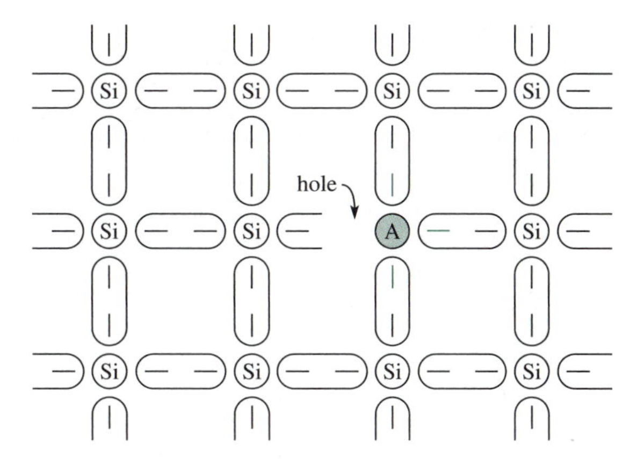

Figure 2–7
Structure of a silicon crystal containing an acceptor atom. The acceptor's nucleus is labeled A and the nuclei of the silicon atoms are labeled Si. The acceptor electrons are shown by blue dashes. Note the incomplete bond and resulting hole caused by the acceptor's presence.

atom itself therefore becomes a positive ion. In all subsequent computations, we will assume that all impurity atoms are thus ionized.

P-type material is produced by doping a semiconductor with impurity atoms that have only three electrons in their outermost shells. When this kind of impurity atom is introduced into the crystal structure, an electron deficiency results, because the impurity atom contributes only three of the required four electrons necessary for covalent bonding. In other words, a hole is created everywhere the impurity atom appears in the crystal. Such impurity atoms are called *acceptors,* because the holes they produce can readily accept electrons. Figure 2–7 shows a single acceptor atom in a silicon crystal. Materials used for doping silicon to create P-type material include *aluminum, boron, gallium,* and *indium.* Note once again that P-type material, like N-type, is electrically neutral because the electron deficiency exists only in the sense that there are insufficient electrons to complete all covalent bonds.

Although electrons are more numerous than holes in N-type material, there are still a certain number of holes present. The extent to which electrons dominate depends on the level of the doping: The more heavily the material is doped with donor atoms, the greater the degree to which the number of electrons exceeds the number of holes. In N-type material, electrons are said to be the *majority* carriers and holes the *minority* carriers. Similarly, the degree of acceptor doping controls the number of holes in P-type material. In this case, holes are the majority carriers and electrons the minority carriers.

An important relationship between the electron and hole densities in most practical semiconductor materials is given by

$$np = n_i^2 \tag{2–8}$$

where n = electron density
p = hole density
n_i = intrinsic electron density

Equation 2–8 states that the product of electron and hole densities equals the square of the intrinsic electron density. Since $n_i = p_i$, equation 2–8 is, of course, equivalent to $np = p_i^2 = n_i p_i$. All of the theory we have discussed so far in connection with mobility, conductivity, and current density is applicable to extrinsic as well as intrinsic semiconductors. The carrier densities used in the computations are often found using equation 2–8.

Example 2–4

A bar of silicon with intrinsic electron density 1.4×10^{16} electrons/m^3 is doped with impurity atoms until the hole density is 8.5×10^{21} holes/m^3. The mobilities of the electrons and holes are $\mu_n = 0.14$ m^2/(V · s) and $\mu_p = 0.05$ m^2/(V · s).

1. Find the electron density of the extrinsic material.
2. Is the extrinsic material N-type or P-type?
3. Find the extrinsic conductivity.

Solution

1. From equation 2–8,

$$n = \frac{n_i^2}{p} = \frac{(1.4 \times 10^{16})^2}{8.5 \times 10^{21}} = 2.3 \times 10^{10} \text{ electrons/m}^3$$

2. Since $p > n$, the material is P-type.
3. From equation 2–7,

$$\begin{aligned}\sigma &= n\mu_n q_n + p\mu_p q_p \\ &= (2.3 \times 10^{10})(0.14)(1.6 \times 10^{-19}) + (8.5 \times 10^{21})(0.05)(1.6 \times 10^{-19}) \\ &= 5.152 \times 10^{-10} + 68 \approx 68 \text{ S/m}\end{aligned}$$

Note in the preceding example that the conductivity, 68 S/m, is for all practical purposes determined exclusively by the component of the conductivity due to holes, which are the majority carriers in this case. In practice, this is almost always the case: The conductivity essentially depends only on the majority carrier density. This result is due to a phenomenon called *minority carrier suppression.* To illustrate, suppose that the majority carriers are electrons, and that there are substantially more electrons than holes. Under these conditions, there is an increased probability that an electron–hole recombination (annihilation) will occur, thus eliminating both a free electron and a hole. Since there are very many more electrons than holes, the resultant percent decrease in electrons is much less than the percent decrease in holes. This effective suppression of minority carriers is reflected in equation 2–8 and leads to the following approximations, valid in most practical cases, for computing conductivity:

$$\begin{aligned}\sigma &\approx n\mu_n q_n \quad \text{(N-type material)} \\ \sigma &\approx p\mu_p q_p \quad \text{(P-type material)}\end{aligned} \tag{2–9}$$

When one carrier type has a substantial majority, it is apparent from these equations that the conductivity of a semiconductor increases in direct proportion to the degree of doping with impurity atoms that produce the majority carriers.

2–6 THE PN JUNCTION

When a block of P-type material is joined to a block of N-type material, a very useful structure results. The region where the two materials are joined is called a *PN junction* and is a fundamental component of many electronic devices, including transistors. The junction is not formed by simply placing the two materials adjacent to each other, but rather through a manufacturing process that creates a transition from P to N within a single crystal. Nevertheless, it is instructive to view the formation of the junction in terms of the charge redistribution that would occur if two dissimilar materials were, in fact, suddenly brought into very close contact with each other.

Let us suppose that a block of P-type material on the left is suddenly joined to a block of N-type material on the right, as illustrated in Figure 2–8(a). In the figure, the acceptor atoms and their associated "excess" holes are shown in the P region. Remember that the P region is initially neutral because each acceptor atom has the same number of electrons as protons. Similarly, the donor atoms are shown with their associated "excess" electrons in the N region, which is likewise electrically neutral. Remember also that *diffusion* current flows whenever there is a surplus of carriers in one region and a corresponding lack of carriers of the same kind in another region. Consequently, at the instant the P and N blocks are joined, electrons from the N region diffuse into the P region, and holes from the P region diffuse into the N region. (Recall that this hole current is actually the repositioning of holes due to the motion of *valence*-band electrons.)

For each electron that leaves the N region to cross the junction into the P region, a donor atom that now has a net positive charge is left behind. Similarly, for each hole that leaves the P region (that is, for each acceptor atom that captures an electron), an acceptor atom acquires a net negative charge. The upshot of this process is that negatively charged acceptor atoms begin to line the region of the junction just inside the P block, and positively charged donor atoms accumulate just inside the N region. This charge distribution is illustrated in Figure 2–8(b) and is often called *space charge.*

It is well known that accumulations of electric charge of opposite polarities in two separated regions cause an electric field to be established between those regions. In the case of the PN junction, the positive ions in the N material and the negative ions in the P material constitute such accumulations of charge, and an electric field is therefore established. The direction of the field (which by convention is the direction of the force on a positive charge placed in the field) is from the positive N region to the negative P region. Figure 2–9 illustrates the field $\overline{E}$ developed across a PN junction.

Note that the direction of the field is such that it *opposes* the flow of electrons from the N region into the P region, and the flow of holes from the P region into the N region. In other words, the positive and negative charges whose locations

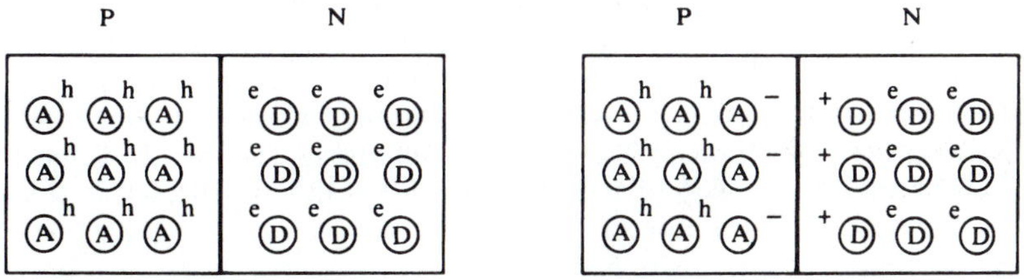

(a) Blocks of P and N materials at the instant they are joined; both blocks are initially neutral.

(b) The PN junction showing charged ions after hole and electron diffusion.

Figure 2–8
Formation of a PN junction. A = acceptor atom; h = associated hole; D = donor atom; e = associated electron; + = positively charged ion; − = negatively charged ion.

Figure 2–9
The electric field $\overline{E}$ across a PN junction inhibits diffusion current from the N to the P side. There are no mobile charge carriers in the depletion region (whose width is proportionally much smaller than that shown).

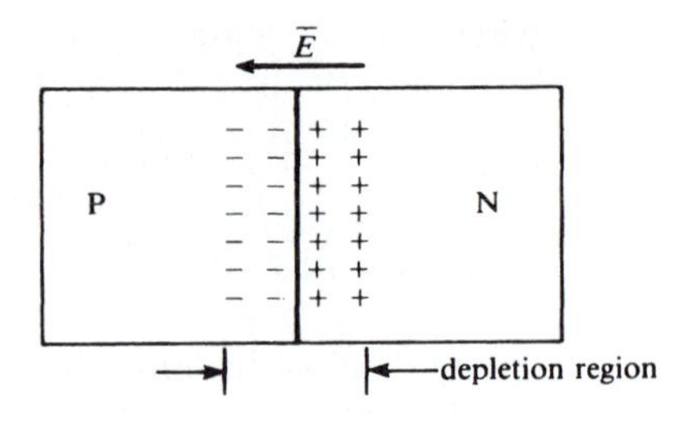

were *caused* by the original diffusion current across the junction are now inhibiting the further flow of current across the junction. An equivalent interpretation is that the accumulation of negative charge in the P region prevents additional negative charge from entering that region (like charges repel), and, similarly, the positively charged N region repels additional positive charge. Therefore, after the initial surge of charge across the junction, the diffusion current dwindles to a negligible amount.

The direction of the electric field across the PN junction enables the flow of *drift* current from the P to the N region, that is, the flow of electrons from left to right and of holes from right to left, in Figure 2–9. There is therefore a small drift of *minority* carriers (electrons in the P material and holes in the N material) in the opposite direction from the diffusion current. This drift current is called *reverse* current, and when *equilibrium* conditions have been established, the small reverse drift current exactly cancels the diffusion current from N to P. The net current across the junction is therefore 0.

In the region of the junction where the charged atoms are located, there are no mobile carriers (except those that get swept immediately to the opposite side). Remember that the P-region holes have been annihilated by electrons, and the N-region electrons have migrated to the P side. Because all charge carriers have been depleted (removed) from this region, it is called the *depletion* region. See Figure 2–9. It is also called the *barrier* region because the electric field therein acts as a barrier to further diffusion current, as we have already described. The width of the depletion region depends on how heavily the P and N materials have been doped. If both sides have been doped to have the same impurity densities (not the usual case), then the depletion region will extend an equal distance into both the P and N sides. If the doping levels are not equal, the depletion region will extend farther into the side having the smaller impurity concentration. The width of a typical depletion region is on the order of 10^{-16} m. In the practical PN junction, there is not necessarily the abrupt transition from P- to N-type material shown in Figure 2–8. The junction may actually be formed, for example, by a gradual increase in the donor doping level of one block of P-type material, so that it gradually changes its nature from P-type to N-type with increasing distance through the block.

The electric field shown in Figure 2–9 is the result of the *potential difference* that exists across the junction due to the oppositely charged sides of the junction. This potential is called the *barrier* potential because it acts as a barrier to diffusion current. (It is also called a *junction* potential, or *diffusion* potential.) The value of the barrier potential, V_0, depends on the doping levels in the P and N regions, the type of material (Si or Ge), and the temperature. Equation 2–10 shows how these variables affect V_0:

$$V_0 = \frac{kT}{q} \ln \left(\frac{N_A N_D}{n_i^2} \right) \tag{2–10}$$

where V_0 = barrier potential, volts
k = Boltzmann's constant = 1.38×10^{-23} J/K
T = temperature of the material in kelvin (K = 273 + °C; note that the correct SI unit of temperature is kelvin, *not* °K or degrees Kelvin.)
q = electron charge = 1.6×10^{-19} C
N_A = acceptor doping density in the P material
N_D = donor doping density in the N material
n_i = intrinsic electron density

Note that the barrier potential is directly proportional to *temperature.* As we shall see throughout the remainder of our study of semiconductor devices, temperature plays a very important role in determining device characteristics and therefore has an important bearing on circuit design techniques. The quantity kT/q in equation 2–10 has the units of volts and is called the *thermal voltage,* V_T:

$$V_T = \frac{kT}{q} \text{ volts} \qquad \textbf{(2–11)}$$

Substituting equation 2–11 into equation 2–10,

$$V_0 = V_T \ln\left(\frac{N_A N_D}{n_i^2}\right) \qquad \textbf{(2–12)}$$

Example 2–5

A silicon PN junction is formed from P material doped with 10^{22} acceptors/m^3 and N material doped with 1.2×10^{21} donors/m^3. Find the thermal voltage and barrier voltage at 25°C.

Solution. $T = 273 + 25 = 298$ K. From equation 2–11,

$$V_T = \frac{kT}{q} = \frac{(1.38 \times 10^{-23})(298)}{1.6 \times 10^{-19}} = 25.7 \text{ mV}$$
$$n_i^2 = (1.5 \times 10^{16})^2 = 2.25 \times 10^{32}$$

From equation 2–12,

$$V_0 = V_T \ln\left(\frac{N_A N_D}{n_i^2}\right)$$
$$= 0.0257 \ln\left(\frac{10^{22} \times 1.2 \times 10^{21}}{2.25 \times 10^{32}}\right)$$
$$= (0.0257)(24.6998) = 0.635 \text{ V}$$

2–7 FORWARD- AND REVERSE-BIASED JUNCTIONS

In the context of electronic circuit theory, the word *bias* refers to a dc voltage (or current) that is maintained in a device by some externally connected source. We will discuss the concept of bias and its practical applications in considerable detail in Chapter 3. For now, suffice it to say that a PN junction can be biased by connecting a dc voltage source across its P and N sides.

Recall that the internal electric field established by the space charge across a junction acts as a *barrier* to the flow of diffusion current. When an external dc source is connected across a PN junction, the polarity of the connection can be such that it either opposes or reinforces the barrier. Suppose a voltage source V is connected as shown in Figure 2–10, with its positive terminal attached to the P side of a PN junction and its negative terminal attached to the N side. With the polarity of the connections shown in the figure, the external source creates an electric field component across the junction whose direction *opposes* the internal field established by the space charge. In other words, the barrier is reduced, so diffusion current is enhanced. Therefore, current flows with relative ease through the junction, its direction being that of conventional current, from P to N, as shown in Figure 2–10. With the polarity of the connections shown in the figure, the junction is said to be *forward biased.* (It is easy to remember that a junction is forward biased when the *P*ositive terminal of the external source is connected to the *P* side, and the *N*egative terminal to the *N* side.)

When the PN junction is forward biased, electrons are forced into the N region by the external source, and holes are forced into the P region. As free electrons move toward the junction through the N material, a corresponding number of holes progresses through the P material. Thus, current in each region is the result of majority carrier flow. Electrons diffuse through the depletion region and recombine with holes in the P material. For each hole that recombines with an electron, an electron from a covalent bond leaves the P region and enters the positive terminal of the external source, thus maintaining the equality of current entering and leaving the source.

Since there is a reduction in the electric field barrier at the forward-biased junction, there is a corresponding reduction in the quantity of ionized acceptor and donor atoms required to maintain the field. As a result, the depletion region *narrows* under forward bias. It might be supposed that the forward-biasing voltage V could be increased to the point that the barrier field would be completely overcome, and in fact reversed in direction. This is not, however, the case. As the forward-biasing voltage is increased, the corresponding increase in current causes a larger voltage drop across the P and N material outside the depletion region, and the barrier field can never shrink to 0.

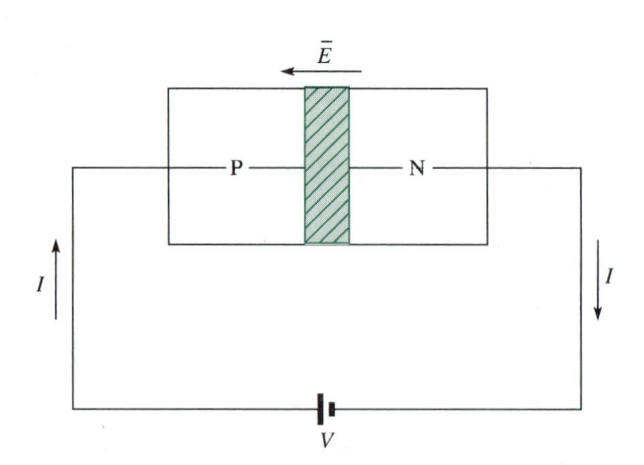

Figure 2–10
A voltage source V connected to forward bias a PN junction. The depletion region (shown shaded) is narrowed.

The Diode Equation

As we will soon learn, a common application of a PN junction is in the construction of a *diode.* The relationship between the voltage V across a PN junction and the current I through it is given by the so-called diode equation:

$$I = I_s(e^{V/\eta V_T} - 1) \tag{2–13}$$

where I = current, A
V = voltage, V (positive, for forward bias)
I_s = *saturation* current, A
η = *emission coefficient* (a function of V whose value depends also on the material; $1 \leq \eta \leq 2$)
V_T = thermal voltage (see equation 2–11)

Equation 2–13 reveals some important facts about the nature of a forward-biased junction. The value of V_T at room temperature is about 0.026 V (see Example 2–6). The value of η for silicon is usually assumed to be 1 for $V \geq 0.5$ V and to approach 2 as V approaches 0. So when V is greater than $2V_T$, or about 0.05 V, the term $e^{V/\eta V_T}$ begins to increase quite rapidly with increasing V. For $V > 0.2$, the exponential is much greater than 1. Consequently, equation 2–13 shows that the current I in a silicon PN junction increases dramatically once the forward-biasing voltage exceeds 200 mV or so. The *saturation current* I_s in equation 2–13 is typically a very small quantity (we will have more to say about it when we discuss *reverse* biasing), but the fact that I_s is multiplied by the exponential $e^{V/\eta V_T}$ means that I itself can become very large very quickly. Figure 2–11 shows a plot of I versus V for a typical forward-biased silicon junction. Note the rapid increase in the current I that is revealed by the plot and the accompanying table of values. For this figure we assumed that $I_s = 0.1$ pA ($= 10^{-13}$ A).

Suppose now that the connections between the PN junction and the external voltage source are reversed, so that the positive terminal of the source is connected to the N side of the junction and the negative terminal is connected to the P side. This connection *reverse biases* the junction and is illustrated in Figure 2–12.

The polarity of the bias voltage in this case *reinforces,* or strengthens, the internal barrier field at the junction. Consequently, diffusion current is inhibited to an even greater extent than it was with no bias applied. The increased field intensity must be supported by an increase in the number of ionized donor and acceptor atoms, so the depletion region widens under reverse bias.

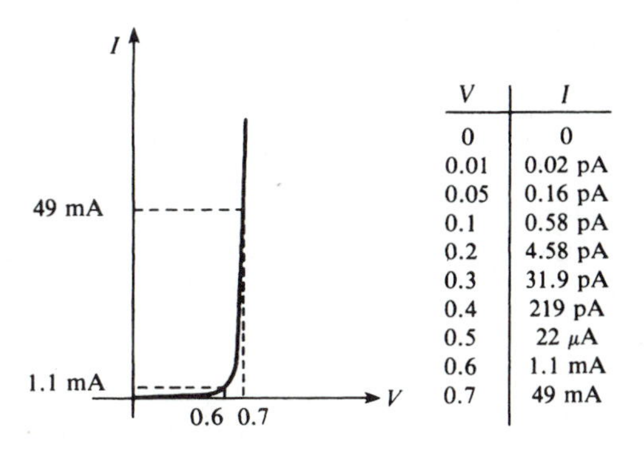

V	I
0	0
0.01	0.02 pA
0.05	0.16 pA
0.1	0.58 pA
0.2	4.58 pA
0.3	31.9 pA
0.4	219 pA
0.5	22 μA
0.6	1.1 mA
0.7	49 mA

Figure 2–11
Current versus voltage in a typical forward-biased silicon junction. $I_s = 0.1$ pA.

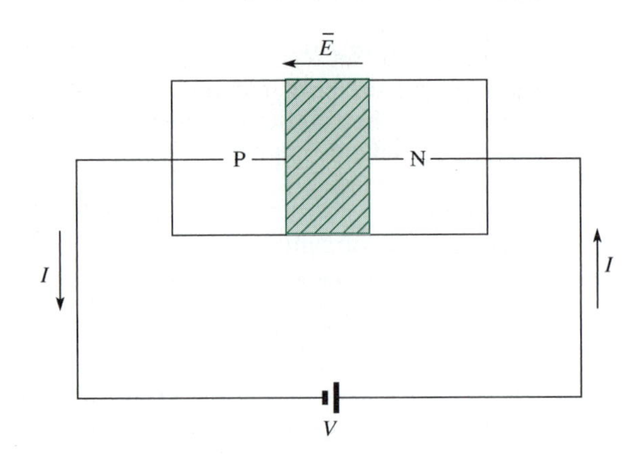

Figure 2–12
A voltage source V connected to reverse bias a PN junction. The depletion region (shown shaded) is widened. (Compare with Figure 2–10.)

Recall that the unbiased PN junction has a component of drift current consisting of minority carriers that cross the junction from the P to the N side. We discussed the fact that this reverse current is the direct result of the electric field across the depletion region. Since a reverse-biasing voltage increases the magnitude of that field, we can expect the reverse current to increase correspondingly. This is indeed the case. However, since the current is due to the flow of minority carriers only, its magnitude is very much smaller than the current that flows under forward bias (the forward current).

It is this distinction between the ways a PN junction reacts to bias voltage—very little current flow when it is reverse biased and substantial current flow when it is forward biased—that makes it a very useful device in many circuit applications. In one of the most common applications, a single PN junction is fitted with a suitable enclosure, through which conducting terminals are brought out so that electrical connections can be made to the P and N sides. This device is called a (discrete) *diode.* The P side of the diode is called its *anode* and the N side is called its *cathode.* Figure 2–13(a) shows the standard symbol for a PN junction diode. Figure 2–13(b) shows the diode connected to an external source for forward biasing, and 2–13(c) shows reverse biasing. Diode circuits will be studied in detail in Chapter 3.

Returning to our discussion of the reverse-biased junction, we should mention that it is conventional to regard reverse voltage and reverse current as *negative* quantities. When this convention is observed, equation 2–13, repeated here, can be used to compute reverse current due to a reverse-biasing voltage:

$$I = I_s(e^{V/\eta V_T} - 1) \qquad \textbf{(2–14)}$$

To illustrate, suppose $\eta V_T = 0.05$, $I_s = 0.01$ pA, and the reverse-biasing voltage is $V = -0.1$ V. Then

$$I = 10^{-14}(e^{-0.1/0.05} - 1) = 10^{-14}(-0.8647) = -0.8647 \times 10^{-14} \text{ A}$$

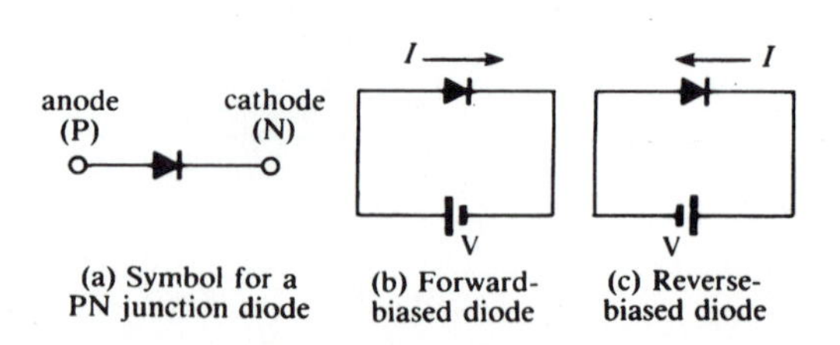

Figure 2–13
Diode symbols and bias circuits

From the standpoint of plotting the I-versus-V relationship in a PN junction, the sign convention makes further good sense. If forward current is treated as positive (upward), then reverse current should appear below the horizontal axis, i.e., downward, or negative. Similarly, forward voltage is plotted to the right of 0 and reverse voltage is plotted to the left of 0, i.e., in a negative direction. Figure 2–14 shows a plot of I versus V in which this convention is observed. Note that the current scale is exaggerated in the negative direction, since the magnitude of the reverse current is so very much smaller than that of the forward current.

When V is a few tenths of a volt negative in equation 2–14, the magnitude of the term $e^{V/\eta V_T}$ is negligible compared to 1. For example, if $V = -0.5$, then $e^{V/\eta V_T} \approx 4.5 \times 10^{-5}$. Of course, as V is made even more negative, the value of $e^{V/\eta V_T}$ becomes even smaller. As a consequence, when the junction is reverse biased beyond a few tenths of a volt,

$$I \approx I_s(0 - 1) = -I_s \tag{2–15}$$

Equation 2–15 shows that the reverse current in the junction under these conditions is essentially equal to I_s, the saturation current. This result accounts for the name *saturation* current: The reverse current predicted by the equation never exceeds the magnitude of I_s.

Equation 2–14 is called the *ideal* diode equation. In real diodes, the reverse current can, in fact, exceed the magnitude of I_s. One reason for this deviation from theory is the existence of *leakage current,* current that flows along the surface of the diode and that obeys an Ohm's law relationship, not accounted for in equation 2–14. In a typical silicon diode having $I_s = 10^{-14}$ A, the leakage current may be as great as 10^{-9} A, or 100,000 times the theoretical saturation value.

Breakdown

The reverse current also deviates from that predicted by the ideal diode equation if the reverse-biasing voltage is allowed to approach a certain value called the reverse *breakdown* voltage, V_{BR}. When the reverse voltage approaches this value, a substantial reverse current flows. Furthermore, a very small increase in the reverse-bias voltage in the vicinity of V_{BR} results in a very large increase in reverse current. In other words, the diode no longer exhibits its normal characteristic of maintaining a very small, essentially constant reverse current with increasing reverse voltage. Figure 2–15 shows how the current–voltage plot is modified to reflect breakdown. Note that the reverse current follows an essentially vertical line as the reverse

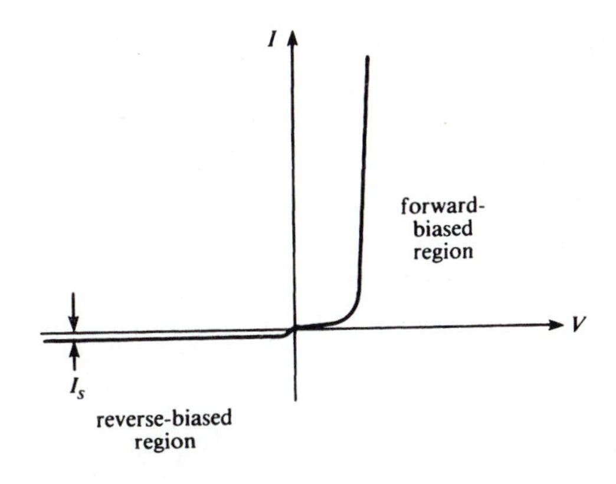

Figure 2–14
Current–voltage relations in a PN junction under forward and reverse bias. The negative current scale in the reverse-biased region is exaggerated.

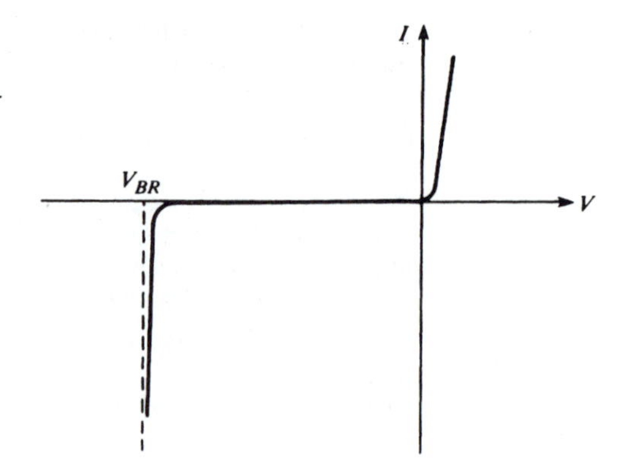

Figure 2–15
A plot of the I–V relation for a diode, showing the sudden increase in reverse current near the reverse breakdown voltage

voltage approaches V_{BR}. This part of the plot conveys the fact that large increases in reverse current result from very small increases in reverse voltage in the vicinity of V_{BR}.

In ordinary diodes, the breakdown phenomenon occurs because the high electric field in the depletion region imparts high kinetic energy (large velocities) to the carriers crossing the region, and when these carriers collide with other atoms they rupture covalent bonds. The large number of carriers that are freed in this way accounts for the increase in reverse current through the junction. The process is called *avalanching.* The magnitude of the reverse current that flows when V approaches V_{BR} can be predicted from the following experimentally determined relation:

$$I = \frac{I_s}{1 - \left(\frac{V}{V_{BR}}\right)^n} \tag{2–16}$$

where n is a constant determined by experiment and has a value between 2 and 6.

Certain special kinds of diodes, called *zener* diodes, are designed for use in the breakdown region. The essentially vertical characteristic in the breakdown region means that the voltage across the diode remains constant in that region, independent of the (reverse) current that flows through it. This property is useful in many applications where the zener diode serves as a *voltage reference,* similar to an ideal voltage source. Zener diodes are more heavily doped than ordinary diodes, and they have narrower depletion regions and smaller breakdown voltages. The breakdown mechanism in zener diodes having breakdown voltages less than about 5 V differs from the avalanching process described earlier. In these cases, the very high electric field intensity across the narrow depletion region directly forces carriers out of their bonds, i.e., strips them loose. Breakdown occurs by avalanching in zener diodes having breakdown voltages greater than about 8 V, and it occurs by a combination of the two mechanisms when the breakdown voltage is between 5 V and 8 V. The characteristics and special properties of zener diodes are discussed in detail in Chapter 17.

Despite the name *breakdown,* nothing about the phenomenon is *inherently* damaging to a diode. On the other hand, a diode, like any other electronic device, is susceptible to damage caused by overheating. Unless there is sufficient current-limiting resistance connected in series with a diode, the large reverse current that would result if the reverse voltage were allowed to approach breakdown could cause excessive heating. Remember that the power dissipation of any device is

$$P = VI \text{ watts} \tag{2–17}$$

where V is the voltage across the device and I is the current through it. At the onset of breakdown, both V (a value near V_{BR}) and I (the reverse current) are liable to be large, so the power computed by equation 2–17 may well exceed the device's ability to dissipate heat. The value of the breakdown voltage depends on doping and other physical characteristics that are controlled in manufacturing. Depending on these factors, ordinary diodes may have breakdown voltages ranging from 10 or 20 V to hundreds of volts.

Temperature Effects

The ideal diode equation shows that both forward- and reverse-current magnitudes depend on temperature, through the thermal voltage term V_T (see equations 2–11 and 2–13). It is also true that the saturation current, I_s in equation 2–13, depends on temperature. In fact, the value of I_s is more sensitive to temperature variations than is V_T, so it can have a pronounced effect on the temperature dependence of diode current. A commonly used rule of thumb is that I_s doubles for every 10°C rise in temperature. The following example illustrates the effect of a rather wide temperature variation on the current in a typical diode.

Example 2–6

A silicon diode has a saturation current of 0.1 pA at 20°C. Find its current when it is forward biased by 0.55 V. Find the current in the same diode when the temperature rises to 100°C.

Solution. From equation 2–11, at $T = 20°\text{C}$,

$$V_T = \frac{kT}{q} = \frac{1.38 \times 10^{-23}(273 + 20)}{1.6 \times 10^{-19}} = 0.02527 \text{ V}$$

From equation 2–13, assuming that $\eta = 1$,

$$I = I_s(e^{V/\eta V_T} - 1) = 10^{-13}(e^{0.55/0.02527} - 1) = 0.283 \text{ mA}$$

At $T = 100°\text{C}$,

$$V_T = \frac{1.38 \times 10^{-23}(273 + 100)}{1.6 \times 10^{-19}} = 0.03217 \text{ V}$$

In going from 20°C to 100°C, the temperature increases in 8 increments of 10°C each: $(100 - 20) = 80$; $80/10 = 8$. Therefore, I_s doubles 8 times, i.e., increases by a factor of $2^8 = 256$. So, at 100°C, $I_s = 256 \times 10^{-13}$ A.

$$I = 256 \times 10^{-13}(e^{0.55/0.03217} - 1) = 0.681 \text{ mA}$$

In this example, we see that the forward current increases by about 240% over the temperature range from 20°C to 100°C.

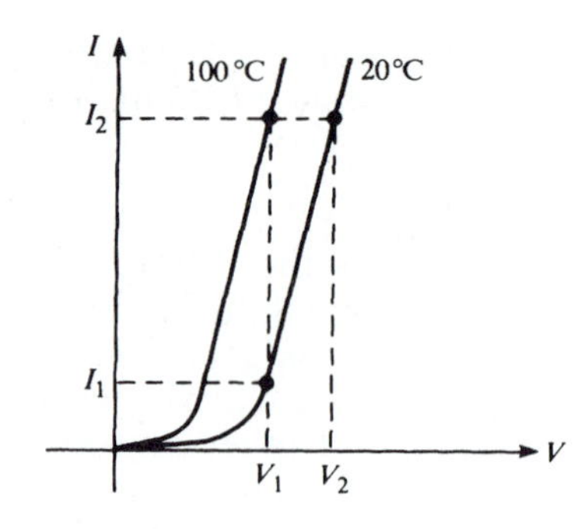

Figure 2–16
Increasing temperature causes the forward I–V characteristic to shift left

Example 2–6 illustrates that forward current in a diode increases with temperature when the forward voltage is held constant. This result is evident when the *I–V* characteristic of a diode is plotted at two different temperatures, as shown in Figure 2–16. At voltage V_1 in the figure, the current can be seen to increase from I_1 to I_2 as the temperature changes from 20°C to 100°C. (Follow the vertical line drawn upward from V_1—the line of constant voltage V_1.) Note that the effect of increasing temperature is to shift the *I–V* plot toward the left. Note also that when the *current* is held constant, the voltage *decreases* with increasing temperature. At the constant current I_2 in the figure, the voltage can be seen to decrease from V_2 to V_1 as temperature increases from 20°C to 100°C. (Follow the horizontal line drawn through I_2—the line of constant current I_2.) As a rule of thumb, the forward voltage decreases 2.5 mV for each 1°C rise in temperature when the current is held constant.

Of course, temperature also affects the value of reverse current in a diode, since the ideal diode equation (and its temperature-sensitive factors) applies to the reverse- as well as forward-biased condition. In many practical applications, the increase in reverse current due to increasing temperature is a more severe limitation on the usefulness of a diode than is the increase in forward current. This is particularly the case for germanium diodes, which have values of I_s that are typically much larger than those of silicon. In a germanium diode, the value of I_s may be as great as or greater than the reverse leakage current across the surface. Since I_s doubles for every 10°C rise in temperature, the total reverse current through a germanium junction can become quite large with a relatively small increase in temperature. For this reason, germanium devices are not so widely used as their silicon counterparts. Also, germanium devices can withstand temperatures up to only about 100°C, while silicon devices can be used up to 200°C.

Example 2–7

SPICE

Use SPICE to obtain a plot of diode current versus diode voltage for a forward-biasing voltage that ranges from 0.6 V through 0.7 V in 5-mV steps. The diode has saturation current 0.01 pA and emission coefficient 1.0.

Solution. Figure 2–17(a) shows a diode circuit that can be used by SPICE to perform a .DC analysis and generate the required plot. Note that VDUM is a dummy voltage source used as an ammeter to determine diode current. The polarity of VDUM is such that positive current values, I(VDUM), will be plotted. (If we plotted I(V1), current values would be negative.) Although the .MODEL statement

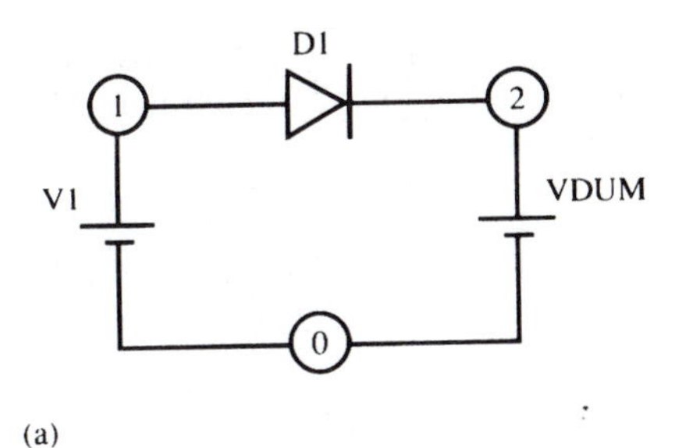


```
DIODE CHAR
V1 1 0
VDUM 2 0
D1 1 2 DIODE
.MODEL DIODE D IS=1E-14 N=1
.DC V1 0.6 0.7 5MV
.PLOT DC I(VDUM)
.END
```

(a)

```
DIODE CHAR
****        DC TRANSFER CURVES                         TEMPERATURE =    27.000 DEG C
******************************************************************************
   V1            I(VDUM)
                      0.000D+00        2.000D-03        4.000D-03        6.000D-03   8.000D-03
                     - - - - - - - - - - - - - - - - - - - - - - - - - - - - - - - - - - - -
6.000D-01   1.188D-04 .*                .                .                .                .
6.050D-01   1.441D-04 .*                .                .                .                .
6.100D-01   1.749D-04 .*                .                .                .                .
6.150D-01   2.122D-04 .*                .                .                .                .
6.200D-01   2.574D-04 . *               .                .                .                .
6.250D-01   3.123D-04 . *               .                .                .                .
6.300D-01   3.789D-04 .  *              .                .                .                .
6.350D-01   4.597D-04 .  *              .                .                .                .
6.400D-01   5.578D-04 .   *             .                .                .                .
6.450D-01   6.767D-04 .    *            .                .                .                .
6.500D-01   8.211D-04 .     *           .                .                .                .
6.550D-01   9.962D-04 .      *          .                .                .                .
6.600D-01   1.209D-03 .       *         .                .                .                .
6.650D-01   1.466D-03 .         *       .                .                .                .
6.700D-01   1.779D-03 .           *     .                .                .                .
6.750D-01   2.159D-03 .                 .*               .                .                .
6.800D-01   2.619D-03 .                 .    *           .                .                .
6.850D-01   3.177D-03 .                 .        *       .                .                .
6.900D-01   3.855D-03 .                 .               *.                .                .
6.950D-01   4.677D-03 .                 .                .     *          .                .
7.000D-01   5.675D-03 .                 .                .           *    .                .
                     - - - - - - - - - - - - - - - - - - - - - - - - - - - - - - - - - - - -
```

(b)

Figure 2–17
(Example 2–7)

specifies the saturation current, IS, and emission coefficient, N, the values used are the same as the default values so these could have been omitted from the statement. The .DC statement causes V1 to be stepped from 0.6 V through 0.7 V in 5-mV increments.

Figure 2–17(b) shows the resulting plot. (Portions of the complete printout produced by SPICE have been omitted to conserve space.) The plot has values of diode current, I(VDUM), scaled along the horizontal axis, and diode voltage, V1, along the vertical axis. By rotating the plot 90° counterclockwise, we see the conventional portrayal of current versus voltage, similar to that in Figure 2–11. Note that the forward current ranges from 0.1188 mA to 5.676 mA as the forward voltage ranges from 0.6 V to 0.7 V.

PSPICE

In PSpice, the dummy voltage source, VDUM, in Example 2–6 is not required, because the current in the diode can be specified directly as I(D1) (or i(d1)). Thus, the circuit can be described with just two nodes:

```
diode char
v1 1 0
d1 1 0 diode
.model diode d is=1e-14 n=1
.dc v1 0.6 0.7 5mv
.plot dc i(d1)
.end
```

This input file is written in lowercase letters for illustrative purposes, but PSpice, unlike the original Berkeley SPICE, does not distinguish between lowercase and uppercase letters, so either (or both) could have been used.

2–8 SCHOTTKY DIODES

It is possible to create a junction having properties similar to those of a PN junction by bonding aluminum (Al) to suitably doped N-type silicon. The junction that results is called a *metal–semiconductor* (MS) junction. Like a PN junction, the MS junction presents a low resistance to current flow when it is forward biased (metal positive with respect to the N-type silicon) and a high resistance when reverse biased. A depletion region and a barrier potential are established in a metal–semiconductor junction by a mechanism similar to that described for PN junctions, except in this case carrier diffusion consists only of electrons diffusing from the semiconductor to the metal. The electrons accumulate at the metal surface and the depletion region exists only in the semiconductor side of the junction.

In Chapter 3 we will discuss applications of diodes in *switching* circuits, where the bias on a diode is changed very rapidly from reverse to forward, or vice versa. In these applications it is necessary for a diode to *respond* very rapidly, that is, to change its nature very quickly from that of a high resistance to that of a low resistance, or vice versa. Metal–semiconductor junctions are able to respond more rapidly in these situations than are their PN counterparts because only majority carriers (electrons in the N-type silicon) are involved in the process. Diodes formed and used this way are called *Schottky barrier* diodes, or simply Schottky diodes. Figure 2–18 shows the special symbol used to represent a Schottky diode.

A metal–semiconductor junction with Schottky-diode properties can also be formed by bonding gold (Au) to P-type germanium. This device is called a *gold-bonded diode* and responds very rapidly in switching applications.

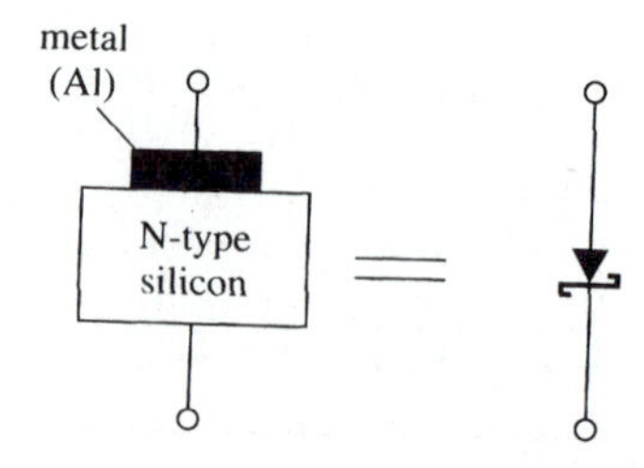

Figure 2–18
Construction and schematic symbol for a Schottky barrier diode

Figure 2–19
Ohmic contacts formed by using aluminum and N^+ or P^+ material

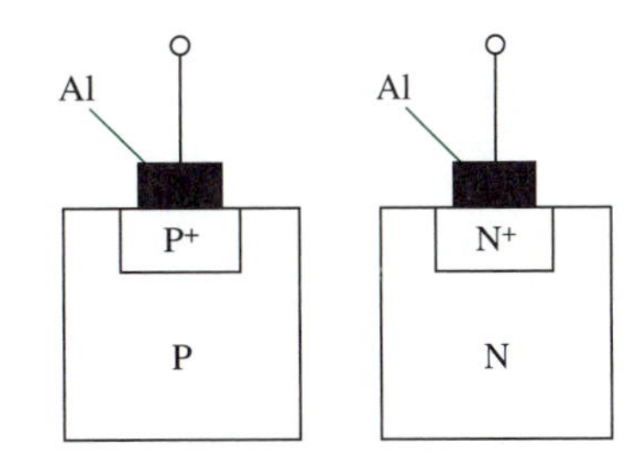

Ohmic Contacts

As we shall study in Chapter 6, aluminum is widely used in the fabrication of electronic devices to provide electrical contacts at semiconductor surfaces, where terminals can be attached or where interconnections can be made to other devices. When aluminum is bonded to P-type silicon, no diode junction is formed. The bond is simply called an *ohmic contact,* because it exhibits a resistance that is independent of the voltage polarity across it.

To form a low-resistance ohmic contact, it is often the practice to join aluminum to a region of very heavily doped (highly conductive) semiconductor material. Usually the heavily doped P or N material is embedded in a region of more lightly doped material of the same type. Heavily doped N material is designated by N^+, and heavily doped P material by P^+. A bond between aluminum and N^+ silicon forms an ohmic contact rather than a Schottky diode. Ohmic contacts are illustrated in Figure 2–19.

2–9 SUMMARY AND OVERVIEW

This section is a condensed version of the material covered in Sections 2–1 through 2–8. It does not contain the technical detail of those sections and is provided here for readers who do not need or wish to study semiconductor theory in great depth. It does, however, provide sufficient background for all subsequent discussions of electronic device theory.

Semiconductor Materials

Modern electronic devices—diodes, transistors, and integrated circuits—are constructed from a special class of materials called *semiconductors.* As the name implies, a semiconductor is neither a good conductor of electrical current nor an insulator. However, it is not the ability or lack of ability to conduct current that makes semiconductors useful. Rather, it is their ability to form *crystals* having special electrical properties that makes them so valuable. The element *silicon* (Si) is the most widely used semiconductor material, followed by *germanium* (Ge).

Since neither silicon nor germanium occurs naturally in a state suitable for use as semiconductor material, each must be subjected to a complex manufacturing process in which crystals are "grown" from a batch of melted, highly purified material. Figure 2–2 is a diagram of the atomic structure of a semiconductor crystal. The only electrons shown in this example are those furthermost from the center (the nucleus) of each atom. These electrons are called *valence* electrons and are the ones most responsible for the electrical properties of the material. Notice that each atom *shares* valence electrons with four of its neighbors, as indicated by the ovals enclosing electrons around each nucleus. Shared electrons tend to stay bound to their "parent" atoms in *covalent bonds*, which are symbolized in the figure by the ovals. Although the interlocking structure of the crystal is highly stable, it is possible for a valence electron to acquire enough energy (usually heat energy) to overcome the covalent bond and thus escape its parent atom. We say that a covalent bond has been *ruptured*. The escaped electron becomes a *free* electron that can wander about in the material. There is enough heat energy at room temperature to free a large number of electrons, which are then available for the conduction of electrical current through the crystal, just as the flow of electrons through a conductor constitutes current. (However, there are vastly more free electrons in a conductor than in a semiconductor.)

When an electron escapes a covalent bond, it leaves behind a *hole* in the crystal structure. Since electrically neutral atoms have as many positively charged *protons* in their nucleus as they do negatively charged electrons outside the nucleus, an atom that has a hole from having lost an electron has a net *positive* charge. The atom is called a positive *ion*. It is also possible for a wandering electron to fall into a hole, thus returning the atom to a neutral state. In those cases, we say that a *recombination*, or *annihilation*, has occurred.

Current in Semiconductors

We have already noted that free electrons are available in a semiconductor to establish current flow through it. These electrons are called charge carriers, because they carry negative charge from one location to another when they move. Holes are also charge carriers—in this case, positive charge carriers. Recall that an atom having a hole is positively charged. If a valence electron leaves one atom to occupy a hole in an adjacent atom, the atom it left becomes positively charged and the atom it joined becomes neutral. In effect, positive charge has moved from one atom to another. See Figure 2–4. The movement of holes through a crystal in this way is called *hole current*. Thus, there are two types of current in a semiconductor: electron current and hole current. Note that the movement of positive charge in one direction is equivalent to movement of negative charge in the opposite direction, so the two components *add* to equal the total current in the material. Also note that a recombination (free electron falling into a hole) does not leave a hole behind, so there is no hole flow in that case.

There are no holes in a conductor, so electron current is the only type that can exist. This current results from the presence of an *electric field* through the material, created, for example, by an externally connected voltage source. The electric field drives electrons from the negative terminal of the voltage source through the material to the positive terminal. Current that exists due to the force of an electric field is called *drift* current. In a semiconductor, both electron current and hole current are created by electric field forces, so both electron drift and hole drift occur when a voltage source is connected across the ends of a semiconductor. A given electric field will cause electrons to flow in one direction and holes to flow in the opposite direction. If the same number of electrons and holes are subjected to the same electric field, the electrons will move faster than the holes (electrons are said to have greater *mobility*), so the component of drift current due to electrons will be greater than that due to holes.

Another type of current that can exist in a semiconductor is called *diffusion* current. Diffusion occurs when there is an imbalance in the number of carriers of a given type between two different regions of a semiconductor. Carriers tend to migrate from a region where there are many of their own type to a region where there are fewer, thus correcting the imbalance. If, for example, there are many more electrons at one end of a bar of semiconductor material than there are at the other end, electrons diffuse from the high-density end to the low-density end until their distribution is more or less uniform throughout. Unless the region containing the excess electrons is replenished with new electrons (as it often is in practical devices), diffusion current ceases when the imbalance has been corrected. Hole current can also be of the diffusion type.

P and N Materials

Pure semiconductor material is said to be *intrinsic.* In practice, certain impurities are introduced into intrinsic material during the manufacturing process to give it new properties. The process of introducing impurities is called *doping,* and material that has been doped is called *extrinsic.* The purpose of doping is to create a semiconductor that has more electrons than holes (N material) or more holes than electrons (P material). To create N material, the semiconductor is doped with impurity atoms that have five instead of four valence electrons. When each such atom joins the crystal structure, four of its valence electrons create the usual covalent bonds with other atoms, and the fifth electron is free. This type of impurity is called a *donor,* because every such atom donates one free electron to the material. Figure 2–6 illustrates a donor atom in a crystal structure. To create P material, the semiconductor is doped with atoms that have three instead of four valence electrons. Each such atom forms covalent bonds with three neighbors only and thus creates one hole in the structure. These atoms are called *acceptors,* because they can each accept one electron. Figure 2–7 illustrates this case.

It is important to note that both N and P materials are electrically *neutral.* Although every donor atom donates a free electron, the donor atom's nucleus brought with it just the right number of positively charged protons to neutralize the charge carried by all its electrons. Similarly, each acceptor atom has the same number of protons as electrons and is also electrically neutral.

Although N material has more electrons than holes, it does have some holes, and P material does have some electrons. In N material, electrons are called the *majority* carriers and holes, the *minority* carriers. In P material, holes are majority carriers and electrons are minority carriers. N material that has been very heavily doped and that is, therefore, very conductive is said to be N^+, and heavily doped P material is P^+.

The PN Junction

When a block of N material is constructed adjacent to a block of P material, the boundary between the two is called a PN junction. At the junction, holes diffuse from the P material into the N material and electrons diffuse from the N material into the P material. (Remember that diffusion occurs when there is an imbalance of carriers of a given type.) Every electron crossing the junction leaves behind a donor atom with a net positive charge, and every hole leaves behind an acceptor atom with a net negative charge. See Figure 2–8. Consequently, after the diffusion there is a thin layer of positive ions on the N side of the junction and a thin layer of negative ions on the P side. There are no mobile charge carriers in this region, and it is called the *depletion* region because it is depleted of such charge. See Figure 2–9. The layers of opposite charge established an electric field (like a voltage source) directed from the N material toward the P material. The direction of the field *opposes* the flow of further electron current from N to P and hole current from P to N. The voltage difference between the charged regions is therefore called a *barrier* voltage. The value of the barrier voltage depends on temperature and doping levels, but it is typically about 0.7 V for silicon and 0.3 V for germanium.

Suppose now that an external voltage source is connected across the P and N material, as shown in Figure 2–10. Notice that the positive terminal of the source is connected to the P material and the negative terminal is connected to the N material. The polarity of the external source thus *opposes* the barrier voltage at the PN junction and enhances the flow of current through the material and across the junction. The PN junction is said to be *forward biased.* If the polarity of the external source is reversed (positive to N and negative to P), as shown in Figure 2–12, the barrier voltage is reinforced and very little current flows. In this case, the junction is said to be *reverse biased.*

A *diode* is a PN junction and therefore has the property just described: It permits a generous flow of current in one direction (when forward biased) and permits virtually no current to flow in the opposite direction (when reverse biased). This property is responsible for many useful applications of diodes in electronic circuits. The P side of a diode is called the *anode* and the N side is called the *cathode.* Figure 2–13 illustrates forward and reverse biasing of a diode and shows the standard symbol for the device.

The *diode equation* permits us to compute the current through a diode as a function of the voltage V connected across it:

$$I = I_s(e^{V/\eta V_T} - 1) \qquad \textbf{(2–18)}$$

where I = current through the diode, A

V = anode-to-cathode voltage across the diode, V (positive for forward bias, negative for reverse bias)

I_s = *saturation current,* A

η = *emission coefficient* (a function of V whose value also depends on the material: $1 \le \eta \le 2$).

V_T = *thermal voltage,* V

$= \dfrac{kT}{q}$

k = Boltzmann's constant = 1.38×10^{-23} J/K

T = temperature in kelvin (K = 273 + °C)

q = electronic charge = 1.6×10^{-19} C

The current through a reverse-biased diode, called *reverse current,* is very small but not totally zero. The saturation current, I_s, in equation 2–18 is the current that flows through the reverse-biased diode (from cathode to anode) when voltage V is a few tenths of a volt negative. As we shall see in Example 2–8, equation 2–18 produces a negative (reverse) current when voltage V is negative (reverse bias). When voltage V is a few tenths of a volt positive (forward bias), equation 2–18 will show that the positive (forward) current becomes quite large. The forward current becomes large when the forward voltage approaches about 0.7 V in silicon and about 0.3 V in germanium. Figure 2–11 illustrates this fact for a forward-biased silicon diode. Figure 2–14 shows diode current versus voltage under both forward- and reverse-bias conditions.

Example 2–8

A silicon diode has saturation current 1 pA. Using the values of η given here and assuming the temperature is 25°C ("room temperature"), find the current in the diode when

1. it is reverse biased by 0.1 V ($\eta = 2$);
2. it is reverse biased by 1.0 V ($\eta = 2$);
3. the anode is shorted to the cathode ($\eta = 2$);
4. it is forward biased by 0.5 V ($\eta = 1$);
5. it is forward biased by 0.7 V ($\eta = 1$).

Solution. We first calculate the thermal voltage at $T = 273 + 25°\text{C} = 298$ K:

$$V_T = \frac{kT}{q} = \frac{(1.38 \times 10^{-23})(298)}{1.6 \times 10^{-19}} = 0.0257 \text{ V}$$

1. Since the diode is reverse biased, we substitute $V = -0.1$ V in equation 2–18:

$$\begin{aligned} I &= I_s(e^{V/\eta V_T} - 1) = (1 \text{ pA})(e^{(-0.1 \text{ V})/2(0.0257 \text{ V})} - 1) \\ &= (1 \text{ pA})(0.143 - 1) = -0.857 \text{ pA} \end{aligned}$$

   The negative result tells us that the current is reverse current, as expected. Note that the value of the exponential term (0.143) is relatively small but not zero.

2. Substituting $V = -1$ V in (2–18), we find

$$I = (1 \text{ pA})(e^{(-1 \text{ V})/2(0.0257 \text{ V})} - 1) = (1 \text{ pA})(3.55 \times 10^{-9} - 1) = -1 \text{ pA}$$

   Notice that the value of the exponential term is now, for all practical purposes, equal to 0 and $I = -I_s$. For all larger reverse voltages, I will remain at -1 pA.

3. Since the anode is shorted to the cathode, $V = 0$ and

$$\begin{aligned} I &= (1 \text{ pA})(e^{(0 \text{ V})/2(0.0257 \text{ V})} - 1) \\ &= (1 \text{ pA})(1 - 1) = 0 \text{ A} \end{aligned}$$

   As expected, the diode equation shows that the current through the diode is 0 when the voltage across it is 0.

4. Since the diode is forward biased, we substitute $V = 0.5$ V in (2–18) to obtain

$$\begin{aligned} I &= (1 \text{ pA})(e^{(0.5 \text{ V})/1(0.0257 \text{ V})} - 1) \\ &= (1 \text{ pA})(2.814 \times 10^{8} - 1) \approx (10^{-12} \text{ A})(2.814 \times 10^{8}) = 0.2814 \text{ mA} \end{aligned}$$

   Notice that the value of the exponential term (2.814×10^8) in this case is so much larger than 1 that, for all practical purposes, $I = I_s e^{V/\eta V_T}$. For all forward-biasing voltages greater than a few tenths of a volt, this will be the case.

5. Substituting $V = 0.7$ V in (2–18) gives

$$\begin{aligned} I &= (1 \text{ pA})(e^{(0.7 \text{ V})/1(0.0257 \text{ V})} - 1) \\ &= (1 \text{ pA})(6.75 \times 10^{11}) = 675 \text{ mA} \end{aligned}$$

   Notice how fast the current now increases for a very small increase in forward voltage—from 0.1035 mA to 675 mA when the forward voltage increases by just 0.1 V.

We can tell from the presence of temperature T in the diode equation that diode current depends on temperature. Furthermore, the value of I_s itself depends heavily on temperature. As a rule, I_s approximately doubles in value for every 10°C rise in temperature. Much of the art of electronic circuit design using semiconductors is concerned with compensating for the effects of temperature changes on circuit performance.

If the reverse-biasing voltage across a diode is increased to a value called the *breakdown voltage,* the reverse current through the diode will no longer be limited to the small saturation value, I_s. When breakdown occurs, the diode conducts heavily in the reverse direction, limited only by whatever resistors or other components are in series with it. Breakdown does not necessarily result in permanent damage to a diode. If the reverse current is limited so that the power dissipation rating of the diode is not exceeded ($P = VI$), then no irreversible damage occurs. Figure 2–15 shows the increase in reverse current when the reverse voltage is near the breakdown value, V_{BR}. The value of V_{BR} depends on doping and other physical characteristics of a diode and may range in value from about 10 V to several hundred volts.

2–10 A SEMICONDUCTOR GLOSSARY

Acceptor An impurity atom used in the doping process to create a hole in a semiconductor crystal; contains 3 electrons in its outermost valence shell.

Annihilation See *recombination.*

Anode The P side of a PN junction diode.

Avalanching Large current flow through a reverse-biased diode when it breaks down; caused by a high electric field imparting high velocities to electrons that then rupture covalent bonds.

Band, energy See *energy band.*

Barrier diode See *Schottky diode.*

Barrier potential Potential (voltage) established by the presence of layers of charge lying on opposite sides of a PN junction.

Barrier voltage See *barrier potential.*

Bias Connection of an external voltage source across a PN junction. See also *forward bias* and *reverse bias.*

Boltzmann's constant (k) Constant used in the diode equation; $k = 1.38 \times 10^{-23}$ J/K.

Breakdown voltage The reverse-biasing voltage across a diode that causes it to conduct heavily in the reverse direction (cathode to anode).

Carrier See *charge carrier.*

Carrier density The number of carriers (holes or electrons) per cubic meter of semiconductor material. See also *electron density* and *hole density.*

Cathode The N side of a PN junction diode.

Charge carrier An electron, which carries one unit of negative charge, q, or a hole, which carries one unit of positive charge.

Charge of an electron (q) $q = 1.6 \times 10^{-19}$ coulombs (C).

Conduction band The energy band of free electrons.

Conductivity (σ) A measure of the ability of a particular material to conduct current; units: S/m; the reciprocal of resistivity.

Covalent bond Shared electrons held to a parent atom in a crystal.

Current density (J) Current per unit cross-sectional area, A/m^2.

Depletion region Region in a PN junction where there are no mobile charge carriers.

Diffusion current Migration of carriers from a region where there are a large number of their own type to a region where there are fewer.

Diffusion potential See *barrier potential.*

Diode A PN junction. A discrete diode is a PN junction fitted with an enclosure and terminals for connection to its P and N sides (anode and cathode).

Diode equation Equation for computing the current through a diode as a function of anode-to-cathode voltage across it. See equation 2–13.

Donor An impurity atom used in the doping process to create a free electron in a semiconductor crystal; contains 5 electrons in its outermost valence shell.

Doping The process of introducing impurity atoms into a crystal to create P- or N-type material.

Drift current Current created by the motion of charge carriers under the influence of an electric field.

Electron, charge on (q) See *charge of an electron.*

Electron current The flow of (free) electrons.

Electron density (n) The number of free electrons per cubic meter of material.

Electron, free See *free electron.*

Electron-volt (eV) Unit of energy; the energy acquired by an electron when accelerated through a potential difference of 1 V: 1 eV = 1.602×10^{-19} J.

Emission coefficient (η) Coefficient used in the diode equation; its value depends on voltage and type of material; $1 \leq \eta \leq 2$.

Energy band A range of energies possessed by electrons. See also *conduction band, forbidden band,* and *valence band.*

Energy gap The width of the forbidden band; the difference in energy levels between the conduction and valence bands.

eV See *electron-volt.*

Extrinsic Impure; refers to semiconductor material that has been doped with impurity atoms.

Forbidden band The range of energy levels that electrons cannot possess in a particular type of material.

Forward bias Connection of the positive terminal of a voltage source to the P side of a PN junction and the negative terminal to the N side (in a diode, positive terminal to anode and negative terminal to cathode).

Forward current Current that flows from P to N in a forward-biased PN junction (anode to cathode in a diode).

Free electron An electron available to serve as a charge carrier in the creation of electrical current; a free electron is in the conduction energy band and is not bound to a parent atom.

Gap, energy See *energy gap.*

Ge Symbol for the element germanium.

Hole A vacant position in the structure of interlocking valence electrons in a crystal; the absence of an electron in a covalent bond; considered to have 1 unit of positive charge equal in magnitude to q.

Hole current The flow of holes, which is actually the repositioning of holes due to electrons moving from hole to hole.

Hole density (p) The number of holes per cubic meter of semiconductor material.

Impurity atom Atom introduced into a semiconductor crystal to create either N or P material; see also *donor* and *acceptor.*

Intrinsic Pure; refers to undoped semiconductor material.

I_s See *saturation current.*

J See *current density.*

Junction See *PN junction* or *MS junction.*

k See *Boltzmann's constant.*

Kelvin (K) Unit of temperature; K = 273 + °C.

Leakage current Cathode-to-anode (reverse) current that flows over the surface of a diode; not accounted for by the diode equation.

Majority carrier The carrier type whose number or density exceeds that of the other type in a particular material; electrons in N material and holes in P material.

Minority carrier The carrier type whose number or density is less than that of the other type; holes in N material and electrons in P material.

Mobility (μ) A measure of the ability of a carrier to move within a semiconductor; units: $m^2/V \cdot s$.

MS junction A metal–semiconductor junction that has diode properties; see also *Schottky diode.*

n See *electron density.*

N_A Acceptor doping density; the number of acceptor atoms per cubic meter in P-type material.

N_D Donor doping density; the number of donor atoms per cubic meter in N-type material.

n_i Electron density of intrinsic semiconductor material.

N^+ Heavily doped N material.

Neutron Atomic particle in the nucleus of an atom; electrically neutral.

N material Semiconductor material that has been doped (with donor atoms) so that it contains more free electrons than holes.

Nucleus The center of an atom; consists of protons and neutrons.

Ohmic contact A junction of dissimilar materials that does not have diode properties; examples include aluminum and N^+ material and aluminum and P material.

Orbit Path occupied by electrons around the nucleus of an atom.

p See *hole density.*

p_i Hole density of intrinsic semiconductor material.

P$^+$ Heavily doped P material.

P material Semiconductor material that has been doped (with acceptor atoms) so that it contains more holes than free electrons.

PN junction Boundary between adjoining regions of P and N material; forms a diode.

Proton Atomic particles in the nucleus of an atom; has one unit of positive charge equal in magnitude to the charge of one electron (q).

q See *charge of an electron.*

q_n Same as q.

q_p Positive charge carried by one proton (or hole); has the same magnitude as q.

Recombination The occurrence of a free electron falling into a hole.

Resistivity (ρ) A measure of the ability of a particular material to create resistance; the resistance per unit length and cross-sectional area of the material; $\rho = 1/\sigma$.

Reverse bias Connection of the positive terminal of a voltage source to the N side of a PN junction and the negative terminal to the P side (in a diode, positive terminal to cathode and negative terminal to anode).

Rupture, of a covalent bond The release of an electron from the bond; creates a free electron and a hole.

Saturation current (I_s) The reverse current through a PN junction that flows when the junction is reverse biased by a few tenths of a volt.

Schottky diode A diode formed by an MS junction consisting of aluminum and N-type silicon.

Semiconductor A crystalline material that is neither a good conductor nor an insulator; current flow in a semiconductor is the motion of both electrons and holes.

Shell An orbit around the nucleus of an atom; can contain a specific maximum number of electrons.

Si Symbol for the element silicon.

Space charge A layer of charged particles lying in the P and N sides of a PN junction.

Subshell A subdivision of a shell; can contain a specific maximum number of electrons.

Thermal voltage (V_T) A component of the barrier potential at a PN junction; value depends on temperature.

Valence band Band of energies possessed by valence electrons in an atom.

Valence electron An electron in the outermost shell of an atom.

Valence shell Outermost electron shell of an atom.

V_{BR} See *breakdown voltage.*

V_T See *thermal voltage.*

Zener diode A diode designed to break down at a specific reverse voltage and to be used in the breakdown region.

EXERCISES

Section 2–2
Atomic Structure

2–1. An electron subshell in a certain atom is filled to its capacity of 10 electrons. The next subshell contains 4 electrons. In what shell are these subshells contained? What is the total electron capacity of that shell? (Assume that each shell must be filled before a succeeding one can contain any electrons.)

2–2. The nucleus of a gallium atom has 31 protons. Assuming that each shell must be filled before a succeeding one can contain any electrons, determine the number of electrons in each of its shells and in the subshells of each shell.

2–3. Repeat Exercise 2–2 for an atom of phosphorus, whose nucleus contains 15 protons.

Section 2–4
Current in Semiconductors

2–4. How much energy is acquired by 24 μC of charge when it is accelerated through a potential difference of 12 mV? Express your answer in both joules and eV. (One coulomb carries a charge equivalent to 6.242×10^{18} electrons.)

2–5. What is the width of the energy gap in silicon, expressed in joules? Repeat for germanium.

2–6. What is the electron velocity in a bar of silicon at room temperature when an electric field intensity of 1800 V/m is established in it? What is the hole velocity?

2–7. A bar of intrinsic germanium 6 cm long has a potential difference of 12 V applied across its ends. If the electron velocity in the bar is 73 m/s, what is the electron mobility?

2–8. A bar of intrinsic silicon 4.8 cm long has a potential difference of 60 V applied across its ends. If the hole velocity in the bar is 77.5 m/s, what is the hole mobility?

2–9. If $p_i = 1.5 \times 10^{10}$ holes/cm^3, $\mu_n = 0.15$ m^2/(V $\cdot$ s), and $\mu_p = 0.04$ m^2/(V $\cdot$ s) in a bar of intrinsic silicon in which the electric field intensity is 500 V/m, find

a. the component of current density in the bar due to hole flow;
b. the component of current density in the bar due to electron flow; and
c. the total current density in the bar.

2–10. If $n_i = 1.6 \times 10^{16}$ electrons/m^3 in a bar of intrinsic silicon in which the electron and hole velocities are 140 m/s and 50 m/s, respectively, find

a. the component of current density in the bar due to hole flow;
b. the component of current density in the bar due to electron flow; and
c. the total current density in the bar.

2–11. A bar of intrinsic silicon having a cross-sectional area of 3×10^{-4} m^2 has an electron density of 1.5×10^{16} electrons/m^3. If $\mu_n = 0.14$ m^2/(V $\cdot$ s) and $\mu_p = 0.05$ m^2/(V $\cdot$ s), how long should the bar be in order that the current in it be 1.2 mA when 9 V is applied across its ends?

2–12. A bar of intrinsic germanium 4 cm long has $p_i = 2.4 \times 10^{19}$ holes/m^3. The electron and hole mobilities are 0.38 m^2/(V $\cdot$ s) and 0.18 m^2/(V $\cdot$ s), respectively. What should be the dimensions of the cross-section if it is to be square in shape and if 3.2 mA are to flow in the bar when a 60-V potential difference is applied across its ends?

2–13. By doing a *dimensional analysis* (similar to that done in the text to verify that the units of equation 2–4 are A/m^2), show that the units of equation 2–7 are S/m.

2–14. Find the conductivity and resistivity of intrinsic silicon if $n_i = 1.5 \times 10^{16}$ electrons/m^3 and the electron and hole mobilities are 0.14 m^2/(V $\cdot$ s) and 0.05 m^2/(V $\cdot$ s), respectively.

2–15. Find the total resistance between ends A and B of the intrinsic germanium bar shown in Figure 2–20. Assume the nominal, room-temperature values given in the text for germanium parameters.

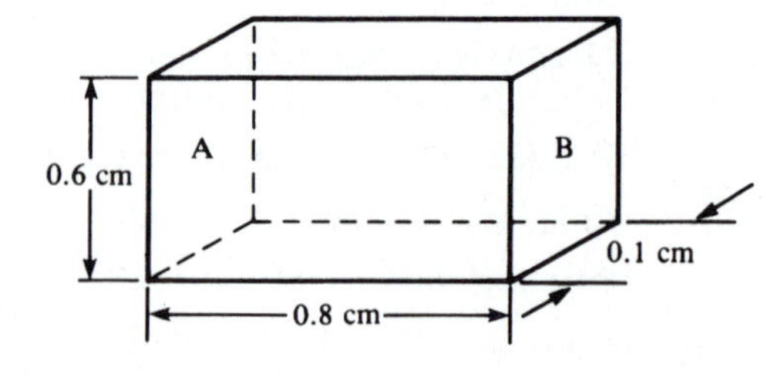

Figure 2–20
(Exercise 2–15)

Section 2–5
P- and N-Type Semiconductors

2–16. How many free electrons are in 1 in^3 of N-type silicon if the intrinsic electron density is 1.5×10^{16} electrons/m^3 and the extrinsic hole density is 0.82×10^{11} holes/m^3? (One inch = 2.54 cm.) Assume that all donor atoms are ionized.

2–17. How many free electrons are in a bar of extrinsic germanium measuring (4 mm) × (50 mm) × (1.5 mm) if the intrinsic hole density is 2.4×10^{19} holes/m^3 and the extrinsic hole density is 7.85×10^{14} holes/m^3? Assume that all donor atoms are ionized.

2–18. One cubic centimeter of silicon has been doped with 1.8×10^{14} atoms of arsenic. If the intrinsic carrier density is 1.6×10^{16} carriers/m^3, what are the electron and hole densities in the doped material? Assume that all impurity atoms are ionized.

2–19. What is the current density in extrinsic silicon whose hole density is 4.5×10^{18} holes/m^3, when an 8-kV/m electric field intensity is established in it? Assume nominal, room-temperature values for all silicon parameters.

2–20. Find the total current in the extrinsic silicon bar shown in Figure 2–21. The electron density in the bar is 2.6×10^{20} electrons/m^3. Assume nominal, room-temperature values for all silicon parameters.

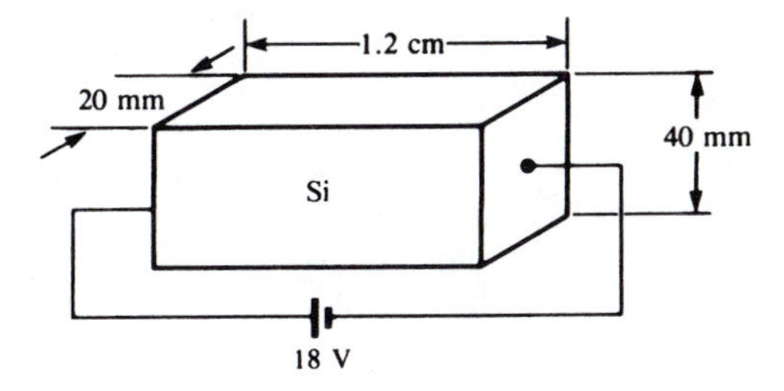

Figure 2–21
(Exercise 2–20)

2–21. A bar of P-type silicon has a majority carrier density of 7.62×10^{22} carriers/m^3. Its cross-sectional area is 2.4×10^{-6} m^2. How long should the bar be in order that it have a resistance between its ends of 8.2 Ω? Assume nominal, room-temperature values for all silicon parameters.

2–22. A bar of silicon 0.1 cm long has a cross-sectional area of 8×10^{-8} m^2 and is heavily doped with phosphorus. What should be the majority carrier density resulting from the doping if the bar is to have a resistance of 1.5 kΩ? Assume nominal, room-temperature values for all silicon parameters.

Section 2–6
The PN Junction

2–23. A silicon PN junction is formed from N material doped with 2.5×10^{21} donors/m^3 and P material doped to have the same impurity density. Find the thermal voltage and barrier voltage at 40°C.

2–24. To what temperature would the junction in Exercise 2–23 have to be raised (in °C) in order that the thermal voltage be 30 mV? What would be the barrier voltage at that temperature?

Section 2–7
Forward- and Reverse-Biased Junctions

2–25. A silicon PN junction has a saturation current of 1.8×10^{-14} A. Assuming that $\eta = 1$, find the current in the junction when the forward-biasing voltage is 0.6 V and the temperature is 27°C.

2–26. Repeat Exercise 2–25 when the forward-biasing voltage is 0.65 V.

2–27. The forward current in a PN junction is 1.5 mA at 27°C. If $I_s = 2.4 \times 10^{-14}$ A and $\eta = 1$, what is the forward-biasing voltage across the junction?

2–28. The forward current in a PN junction is 22 mA when the forward-biasing voltage is 0.64 V. If the thermal voltage is 26 mV and $\eta = 1$, what is the saturation current?

2–29. A bar of silicon is doped so that one side (side A) has 1.85×10^{22} electrons/m^3 and the other side (side B) has 2.66×10^{10} electrons/m^3. If the bar is to be used as a junction diode, which side should be the anode and which side should be the cathode?

2–30. A junction diode has an external voltage source of 0.15 V connected across it, with the positive terminal of the source connected to the cathode of the diode. The saturation current is 0.02 pA, the thermal voltage is 26 mV, and $\eta = 2$. Find the theoretical (ideal) diode current. Repeat for a source voltage of 0.3 V, and again for a source voltage of 0.5 V.

2–31. A junction diode is connected across an external voltage source so that the negative terminal of the source is connected to the anode of the diode. If the external voltage source is 5 V and the saturation current is 0.06 pA, what is the theoretical (ideal) diode current?

2–32. The reverse breakdown voltage of a certain diode is 150 V and its saturation current is 0.1 pA. Assuming that the constant n in equation 2–16 is 2, what is the current in the diode when the reverse-biasing voltage is 149.95 V?

2–33. In an experiment designed to investigate the breakdown characteristics of a certain diode, a reverse current of 9.3 nA was measured when the reverse voltage across the diode was 349.99 V. If the breakdown voltage of the diode was 350 V, and its saturation current was known to be 1.0 pA, what value of the constant n in equation 2–16 is appropriate for this diode?

2–34. The manufacturer of a certain diode rates its maximum power dissipation as 0.1 W and its reverse breakdown voltage as 200 V. What maximum reverse current could it sustain at breakdown without damage?

2–35. A certain diode has a reverse breakdown voltage of 100 V and a saturation current of 0.05 pA. How much power does it dissipate when the reverse voltage is 99.99 V? Assume that n in equation 2–16 is 2.5.

2–36. A diode has a saturation current of 45 pA at a temperature of 373 K. What is the approximate value of I_s at $T = 273$ K?

2–37. When the voltage across a forward-biased diode at $T = 10$°C is 0.621 V, the current is 4.3 mA. If the current is held constant, what is the voltage when $T = 40$°C? Repeat for $T = -30$°C.

Section 2–8

Schottky Diodes

2–38. Name two specific materials that can be joined to form an MS junction. In which material does a depletion region form in such a junction?

2–39. What is the principal application for Schottky diodes? What aspect of the device's operating principle is responsible for the characteristic that makes it suitable for that application?

2–40. Give two examples of metal–semiconductor junctions that do not have diode properties.

Section 2–9

Summary and Overview

2–41. What two elements are the most widely used semiconductors? What property makes them useful for constructing electronic devices?

2–42. What is a covalent bond? What is the result of a covalent bond being ruptured?

2–43. How is an atom having a hole in a crystal structure charged (positively or negatively)? What is a recombination?

2–44. What is the difference between drift current and diffusion current?

2–45. What is the process used to create P and N materials called? What are the impurity atoms used to create each type of material called?

2–46. What are the majority carriers in N material? What are the minority carriers in P material? What does the notation N^+ mean?

2–47. What is the voltage across a PN junction established by the diffusion of carriers across it called? What is the direction of the electric field due to that voltage?

2–48. Which terminals of a voltage source should be connected to which sides of a PN junction to reverse bias it?

2–49. A silicon diode has saturation current 2.4 pA and its temperature is 30°C.

a. Find the current through the diode when it is reverse biased by 0.09 V. Assume $\eta = 2$.

b. Find the current when the diode is reverse biased by 0.9 V. Assume $\eta = 2$.

c. Find the current when the diode is forward biased by 0.55 V. Assume $\eta = 1$.

d. Find the current when the diode is forward biased by 0.67 V. Assume $\eta = 1$.

2–50. A silicon diode has saturation current 1 pA and its temperature is 25°C. What value of forward-biasing voltage will cause a current of 10 mA to flow through it? Assume $\eta = 1$.

2–51. The saturation current in a certain diode at 25°C is 1.5 pA. What is its approximate value at 55°C?

2–52. A certain diode breaks down when its reverse voltage reaches 100 V. If it conducts 50 mA at breakdown, how much power does it consume?

SPICE EXERCISES

Note: In the exercises that follow, assume that all parameters have their default values unless otherwise specified.

2–53. Use SPICE to obtain a plot of diode current versus diode voltage for a forward-biasing voltage that ranges from 0.5 V through 0.75 V in 5-mV steps. The diode has saturation current 0.1 pA and emission coefficient 1.5.

2–54. Use SPICE to obtain a plot of diode current versus diode voltage for a reverse-biasing voltage that ranges from 10 V to 100 V in 5-V steps. The diode has a saturation current of 0.01 pA, emission coefficient 1.0, and reverse breakdown voltage 200 V.

2–55. Use SPICE to obtain a plot of diode current versus diode voltage for a reverse-biasing voltage that ranges from 199 V to 200 V in 50-mV steps. The diode has a saturation current of 0.01 pA, emission coefficient 1.0, reverse breakdown voltage 199.9 V, and the reverse current at breakdown is 1 mA.

2–56. Use SPICE to determine the total *change* in the current of a diode that is forward biased by 0.65 V when the temperature is changed from 20°C to 70°C. The diode has saturation current 0.5 pA and emission coefficient 1.0.

2–57. By repeated runs of SPICE programs, determine the maximum temperature at which a diode can be operated if the forward current through it is not allowed to exceed 1 A when it is forward biased by 0.65 V. The diode has saturation current 0.2 pA and emission coefficient 1.0.

3 THE DIODE AS A CIRCUIT ELEMENT

3–1 INTRODUCTION

In Chapter 2 we studied the construction and properties of a PN junction and mentioned that a semiconductor diode is an example of an electronic device that contains such a junction. A *discrete* diode is a single PN junction that has been fitted with an appropriate case (enclosure) and to which externally accessible leads have been attached. The leads allow us to make electrical connections to the anode (P) and cathode (N) sections of the diode. Semiconductor diodes are also formed inside *integrated circuits* as part(s) of larger, more complex networks and may or may not have leads that are accessible outside the package. (Integrated circuits have a large number of semiconductor components embedded and interconnected in a single crystal. We will study these devices in detail in Chapter 6.)

In this chapter, we will investigate current and voltage relationships in circuits that contain diodes. We will learn how to analyze a circuit containing a diode by replacing the diode with a simpler *equivalent* circuit element (such as a resistance). We will see that our choice of the circuit element(s) used to replace the diode depends on the voltage and current levels in the circuit we are analyzing and on the accuracy that we desire from our analysis. This approach will serve as our introduction to this standard and widely used method of electronic circuit analysis: Replace actual devices by simpler equivalent circuits in order to obtain solutions that are sufficiently accurate for the application in which they are used. Understand that this is an *analysis* method; that is, we "replace" the diode by its equivalent circuit on paper only, in order to simplify calculations.

We will study the diode and its behavior as a circuit element in considerable detail. A thorough analysis of this, the most fundamental building block in semiconductor electronics, is motivated by the following important facts:

1. The diode is an extremely useful device in its own right. It finds wide application in practical electronic circuits.
2. Many electronic devices that we will study later, such as transistors, contain PN junctions that behave like diodes. A solid understanding of diodes will help us understand and analyze these more complex devices.
3. A number of standard techniques used in the analysis of electronic circuits of

all kinds will be introduced in the context of diode circuit analysis. Important concepts such as linearity, small- and large-signal operation, quiescent points, bias, load lines, and equivalent circuits are best learned by applying them to the analysis of the relatively simple diode circuit.

In a later chapter, we will apply our knowledge of diode circuit analysis to the study of several practical circuits in which these versatile devices are used.

3–2 THE DIODE AS A NONLINEAR DEVICE

Linearity is an exceptionally important concept in electronics. For our purposes now, we can best understand the practical implications of this rather broad concept by restricting ourselves to the following definition of a linear electronic device: *A device is linear if the graph relating the voltage across it to the current through it is a straight line.*

If we have experimental data that shows measured values of voltage and the corresponding values of current, then it is a simple matter to plot these and determine whether the linearity criterion is satisfied. Often we have an equation that relates the voltage across a device to the current through it (or that relates the current to the voltage). If the equation is in one of the general *forms*

$$V = K_1 I + K_2 \tag{3–1}$$

or

$$I = K'_1 V + K'_2 \tag{3–2}$$

where K_1, K_2, K'_1, and K'_2 are any constants, positive, negative, or 0, then the graph of V versus I is a straight line and the device is linear. In equation 3–1, K_1 is the *slope* of the line and has the units of ohms. In equation 3–2, K'_1 is the slope and has the units siemens (formerly mhos).

In fundamental circuit analysis courses, we learn that resistors, capacitors, and inductors are all linear electronic devices because the voltage–current equation for each is in the form of equation 3–1 with $K_2 = 0$:

1. For resistance R, $V = IR$; $K_1 = R$ in equation 3–1.
2. For capacitance C, $V = IX_C$, where X_C is the (constant) capacitive reactance, so $K_1 = X_C$ in equation 3–1.
3. For inductance L, $V = IX_L$, where X_L is the (constant) inductive reactance, so $K_1 = X_L$ in equation 3–1.

In these voltage–current equations, we of course assume that any other circuit characteristics such as frequency are held constant. In other words, only the magnitudes of voltage and current are regarded as variables. In each case, all else being equal, increasing or decreasing the voltage causes a proportional increase or decrease in the current. Figure 3–1 is a plot of the voltage V across a 200-Ω resistor versus the current I through it. The linearity property of the resistor is clearly evident, and follows from the Ohm's law relation $V = 200I$. Note that the slope of the line equals the resistance, $r = \Delta V/\Delta I = 200$, and that the linear relation applies to *negative* voltages and currents as well as positive. Reversing the directions (polarities) of the voltage across and current through a linear device does not alter its linearity property. Note also that the slope of the line is everywhere the same: No matter where along the line the computation $\Delta V/\Delta I$ is performed, the result equals 200 Ω.

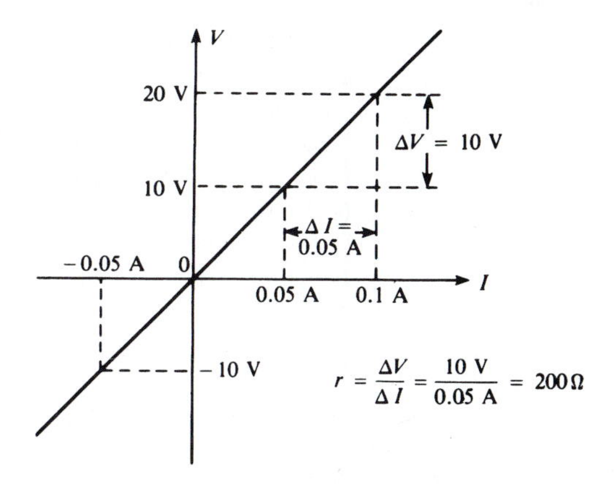

Figure 3–1
The graph of V versus I for a resistor is a straight line. A resistor is a linear device, and the value of ΔV/ΔI is the same no matter where it is computed.

When displaying the voltage–current relationship of an electronic device on a graph, it is conventional to plot current along the vertical axis and voltage along the horizontal axis—the reverse of that shown in Figure 3–1. Of course, the graph of a linear device is still a straight line; reversal of axes is equivalent to expressing the V–I relation in the form of equation 3–2, with slope having the units of conductance, $G = \Delta I/\Delta V = 1/R$ (siemens) instead of resistance.

In Chapter 2 we stated that the current–voltage relation for a PN junction (and therefore for a diode) is

$$I = I_s(e^{V/\eta V_T} - 1) \text{ amperes} \qquad (3\text{–}3)$$

where I_s = saturation current

V_T = thermal voltage (equation 2–11)

η = a function of V, whose value ranges between 1 and 2

Equation 3–3 is clearly not in the form of either equation 3–1 or 3–2, so the diode's voltage–current relation does not meet the criterion for a linear electronic device. We conclude that a diode is a *nonlinear* device. Figure 3–2 is a graph of the I–V characteristic of a typical silicon diode in its forward-biased region. The graph is most certainly not a straight line.

Figure 3–2 shows how values of ΔV and ΔI are found at two different points on the I–V curve. Using these values, we can calculate the resistance of the diode at the two points from $r = \Delta V/\Delta I$. At the point where $V = 0.65$ V and $I = 30$ mA, we find

$$r = \frac{\Delta V}{\Delta I} = \frac{0.015 \text{ V}}{20 \times 10^{-3} \text{ A}} = 0.75\ \Omega$$

At the point where $V = 0.58$ V and $I = 2.2$ mA,

$$r = \frac{\Delta V}{\Delta I} = \frac{0.04 \text{ V}}{4 \times 10^{-3} \text{ A}} = 10\ \Omega$$

We see that there is greater than a 10-to-1 change in the resistance of the diode when the voltage across it is changed from 0.65 V to 0.58 V! Unlike a linear device, the resistance of a nonlinear device depends on the voltage across it (or current

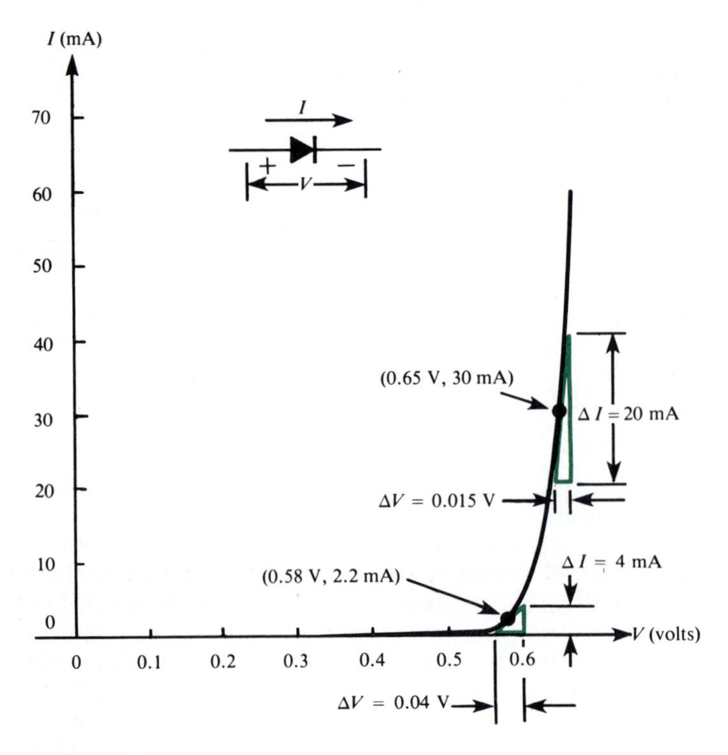

Figure 3–2
A forward-biased diode characteristic. The value of $\Delta V/\Delta I$ depends upon the location where it is computed.

through it)—i.e., the resistance depends on the point where the values of ΔV and ΔI are calculated. In the case of a diode, we further note that the I–V characteristic becomes very nearly horizontal at low values of current and in the reverse-biased region (see Figure 2–14). Therefore, in these regions, large changes in voltage, ΔV, create very small changes in current, ΔI, so the value of $r = \Delta V/\Delta I$ is very large.

The region on the I–V curve where the transition from high resistance to low resistance takes place is called the *knee* of the curve. In Figure 3–2, the knee is in the vicinity of $I \approx 1$ mA to $I \approx 5$ mA. When the diode current is significantly greater or less than that in the vicinity of the knee, we will say that it is biased *above* or *below* the knee, respectively.

3–3 AC AND DC RESISTANCE

The resistance we calculated in Section 3–2 using the relation $\Delta V/\Delta I$ is called the *ac* (or *dynamic*) *resistance* of the diode. It is called *ac* resistance because we consider the small *change* in voltage, ΔV, such as might be generated by an ac generator, causing a *change* in current, ΔI. In using this graphical method to calculate the ac resistance, the changes ΔV and ΔI must be kept small enough to avoid covering sections of the I–V curve over which there is an appreciable change in slope. For example, in Figure 3–2 we would not want to calculate ac resistance between $V =$

0.55 V and $V = 0.65$ V because the slope of the curve changes appreciably between those values.

Henceforth we will refer to the ac resistance of the diode as r_D, where the lowercase r is in keeping with the convention of using lowercase letters for ac quantities. Thus,

$$r_D = \frac{\Delta V}{\Delta I} \text{ ohms} \quad \textbf{(3–4)}$$

When a dc voltage is applied across a diode, a certain dc current will flow through it. The *dc resistance* of a diode is found by dividing the dc voltage across it by the dc current through it. Thus the dc resistance, also called the *static* resistance, is found by direct application of Ohm's law. We will designate dc diode resistance by R_D:

$$R_D = \frac{V}{I} \text{ ohms} \quad \textbf{(3–5)}$$

Like ac resistance, the dc resistance of a diode depends on the point on the I–V curve at which it is calculated. For example, in Figure 3–2 we find that the dc resistance at the point located near the knee is $R_D = (0.58 \text{ V})/(2.2 \text{ mA}) = 263.6\ \Omega$, while the dc resistance at the point well above the knee is $R_D = (0.65 \text{ V})/(30 \text{ mA}) = 21.6\ \Omega$. For a diode like the one whose curve is shown in Figure 3–2, the reverse-biased current is approximately $-1\ \mu\text{A}$ when $V = -1$ V, so the dc resistance in this case is $R_D = (-1 \text{ V})/(-10^{-6} \text{ A}) = 1\ \text{M}\Omega$. We see that the diode is nonlinear in both the dc and the ac sense; that is, both its dc and ac resistances change over a wide range.

When analyzing or designing diode circuits, it is often the case that the I–V characteristic curve is not available. In most practical work, the ac resistance of a diode is not calculated graphically but is found using a widely accepted approximation. In virtually every application where it is necessary to compute the ac resistance of a diode, the dc current in the diode has a value somewhere above the knee. It can be shown that when the dc current is above the knee, the ac resistance is closely approximated by $r_D \cong V_T/I$, where V_T is the thermal voltage (equation 2.11), and I is the dc current in amperes. For $T = 300$ K, V_T is about 0.026 V, so at room temperature

$$r_D \cong \frac{0.026}{I} \text{ ohms} \quad \textbf{(3–6)}$$

This approximation is valid for both silicon and germanium diodes.

To illustrate the use of equation 3–6, consider the point shown above the knee on the I–V curve in Figure 3–2. At this point, the dc current is 30 mA, so according to equation 3–6, $r_D = 0.026/(30 \times 10^{-3}) = 0.86\ \Omega$. This value agrees reasonably well with that we calculated previously using the graphical technique: 0.75 Ω.

There is one additional component of diode resistance that should be mentioned. The resistance of the semiconductor material and the contact resistance where the external leads are attached to the PN junction can be lumped together and called the *bulk resistance* r_B of the diode. Usually less than 1 Ω, the bulk resistance also changes with the dc current in the diode, becoming quite small at high current levels. The total ac resistance of the diode is $r_D + r_B$ but at low current levels r_D is so much greater than r_B that r_B can usually be neglected. At high current levels, r_B is typically on the order of 0.1 Ω.

When a diode is connected in a circuit in a way that results in the diode being forward biased, there should always be resistance in series with the diode to limit the current that flows through it. The following example illustrates a practical circuit that could be used to determine I–V characteristics.

Example 3–1

The circuit shown in Figure 3–3 was connected to investigate the relation between the voltage and current in a certain diode. The adjustable resistor R was set to several different values in order to control the diode current, and the diode voltage was recorded at each setting. The results are tabulated in the table in Figure 3–3.

1. Find the dc resistance of the diode when the voltage across it is 0.56, 0.62, and 0.67 V.
2. Find the ac resistances presented by the diode to an ac signal generator that causes the voltage across the diode to vary between 0.55 V and 0.57 V, between 0.61 V and 0.63 V, and between 0.66 V and 0.68 V.
3. Find the approximate ac resistances when the diode voltages are 0.56 V, 0.62 V, and 0.67 V. Assume bulk resistances of 0.8 Ω, 0.5 Ω, and 0.1 Ω, respectively.

Solution

1. It will be necessary to calculate the current in the diode for each of the resistance settings shown in the data table. We know that the voltage drop across the resistor R must equal 5.0 V minus the drop across the diode: $V_R = 5 - V_D$, where V_R = voltage drop across the resistor and V_D = voltage drop across the diode. The current in the resistor, which is the same as the current in the diode, is therefore $I = (5 - V_D)/R$. Using this equation, the diode currents at each setting are calculated as follows:

$$I_1 = \frac{(5 - 0.55)\text{ V}}{6312\ \Omega} = 0.705\text{ mA} \qquad I_2 = \frac{(5 - 0.56)\text{ V}}{4269\ \Omega} = 1.04\text{ mA}$$

Figure 3–3
(Example 3–1)

Measurement Number	*R (ohms)*	*V (volts)*
1	6312	0.55
2	4269	0.56
3	2877	0.57
4	599	0.61
5	405	0.62
6	274	0.63
7	85.0	0.66
8	57.5	0.67
9	39.0	0.68

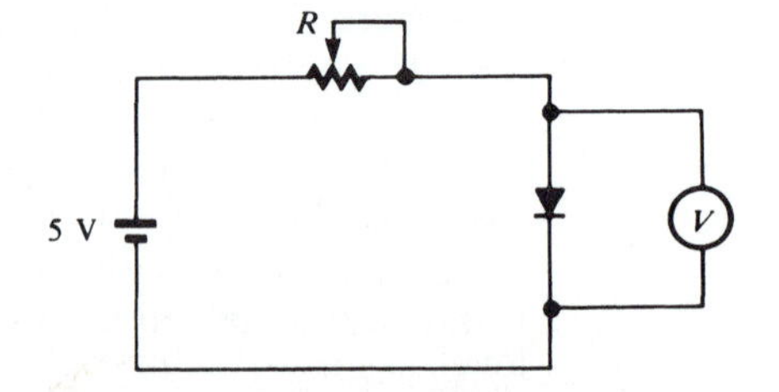

$$I_3 = \frac{(5 - 0.57)\text{ V}}{2877\ \Omega} = 1.54\text{ mA} \qquad I_4 = \frac{(5 - 0.61)\text{ V}}{599\ \Omega} = 0.733\text{ mA}$$

$$I_5 = \frac{(5 - 0.62)\text{ V}}{405\ \Omega} = 10.8\text{ mA} \qquad I_6 = \frac{(5 - 0.63)\text{ V}}{274\ \Omega} = 15.9\text{ mA}$$

$$I_7 = \frac{(5 - 0.66)\text{ V}}{85\ \Omega} = 51.1\text{ mA} \qquad I_8 = \frac{(5 - 0.67)\text{ V}}{57.5\ \Omega} = 75.3\text{ mA}$$

$$I_9 = \frac{(5 - 0.68)\text{ V}}{39.0\ \Omega} = 110.8\text{ mA}$$

The dc diode resistances at the voltages specified are found from equation 3–5, $R_D = V/I$. At $V = 0.56$ V,

$$R_D = \frac{0.56\text{ V}}{1.04 \times 10^{-3}\text{ A}} = 538.5\ \Omega$$

At $V = 0.62$ V,

$$R_D = \frac{0.62\text{ V}}{10.8 \times 10^{-3}\text{ A}} = 57.4\ \Omega$$

At $V = 0.67$ V,

$$R_D = \frac{0.67\text{ V}}{75.3 \times 10^{-3}\text{ A}} = 8.90\ \Omega$$

2. The ac diode resistances are found from equation 3–4, $r_D = \Delta V/\Delta I$, as follows:

$$r_D = \frac{(0.57 - 0.55)\text{ V}}{(1.54 - 0.705) \times 10^{-3}\text{ A}} = \frac{0.02\text{ V}}{0.835 \times 10^{-3}\text{ A}} = 23.95\ \Omega$$

$$r_D = \frac{0.02\text{ V}}{8.57 \times 10^{-3}\text{ A}} = 2.33\ \Omega$$

$$r_D = \frac{0.02\text{ V}}{59.7 \times 10^{-3}\text{ A}} = 0.34\ \Omega$$

3. The approximate ac resistances are found using the relation (3.6) $r_D \cong 0.026/I$ and adding the bulk resistance r_B. At $V = 0.56$ V,

$$r_D = \frac{0.026}{I_2} + r_B = \frac{0.026}{1.04 \times 10^{-3}\text{ A}} + 0.8\ \Omega = 25.8\ \Omega$$

At $V = 0.62$ V,

$$r_D = \frac{0.026}{I_5} + r_B = \frac{0.026}{10.8 \times 10^{-3}\text{ A}} + 0.5\ \Omega = 2.91\ \Omega$$

At $V = 0.67$ V,

$$r_D = \frac{0.026}{I_8} + r_B = \frac{0.026}{75.3 \times 10^{-3}\text{ A}} + 0.1\ \Omega = 0.445\ \Omega$$

Note that each ac resistance calculated in part 3 is at a diode voltage in the middle of a range (ΔV) over which an ac resistance is calculated in part 2. We can therefore expect the approximations for r_D to agree reasonably well with the values calculated using $r_D = \Delta V/\Delta I$. The results bear out this fact.

3–4 ANALYSIS OF DC CIRCUITS CONTAINING DIODES

In virtually every practical dc circuit containing a diode, there is one simplifying assumption we can make when the diode current is beyond the knee. We have seen (Figure 3–2, for example) that the I–V curve is essentially a vertical line above the knee. *The implication of a vertical line on an I–V characteristic is that the voltage across the device remains constant, regardless of the current that flows through it.* Thus the voltage drop across a diode remains substantially constant for all current values above the knee. This fact is responsible for several interesting applications of diodes. For present purposes, it suggests that the diode is equivalent to another familiar device that has this same property of maintaining a constant voltage, independent of current: a voltage source! Indeed, our first simplified equivalent circuit of a diode is a voltage source having a potential equal to the (essentially) constant drop across it when the current is above the knee.

For a silicon diode, depending on small manufacturing variations and on the actual current flowing in it, the voltage drop above the knee is around 0.6 to 0.7 V. In practice, it is usually assumed to be 0.7 V. For germanium diodes, the drop is assumed to be 0.3 V. Therefore, for analysis purposes, we can replace the diode in a circuit by either a 0.7-V or a 0.3-V voltage source whenever the diode has a forward-biased current above the knee. Of course, the diode does not store energy and cannot produce current like a true voltage source, but the voltages and currents in the rest of a circuit containing the forward-biased diode are exactly the same as they would be if the diode were replaced by a voltage source. (The substitution theorem in network theory justifies this result.) Figure 3–4 illustrates these ideas.

In Figure 3–4(a), we assume that a forward-biased silicon diode has sufficient current to bias it above its knee and that it therefore has a voltage drop of 0.7 V. Applying Kirchhoff's voltage law around the loop, we have

$$E = IR + 0.7 \tag{3–7}$$

from which we find $I = (E - 0.7)/R$ amperes. Figure 3–4(b) shows the equivalent circuit with the diode replaced by a 0.7-V source. Note that the polarity of the equivalent source opposes that of the voltage source E. Therefore the net voltage across R is $E - (0.7\text{ V})$. Consequently, $I = (E - 0.7)/R$ amperes as before.

Our assumption that the voltage drop across the diode is constant above the knee is usually accompanied by the additional assumption that the current through the diode is zero for all lesser voltages. For analyzing the dc voltage and current in a circuit containing a diode, we therefore, in effect, replace the I–V characteristic by one of those shown in Figure 3–5.

The idealized characteristic curves in Figure 3–5 imply that the diode is an *open* circuit (infinite resistance, zero current) for all voltages less than 0.3 V or 0.7 V and becomes a short circuit (zero resistance) when one of those voltage values is reached. These approximations are quite valid in most real situations. Note that

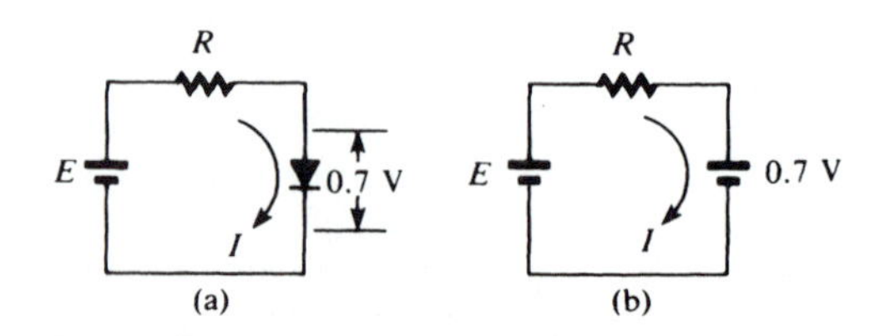

Figure 3–4
For analysis purposes, the forward-biased diode in (a) can be replaced by a voltage source, as in (b).

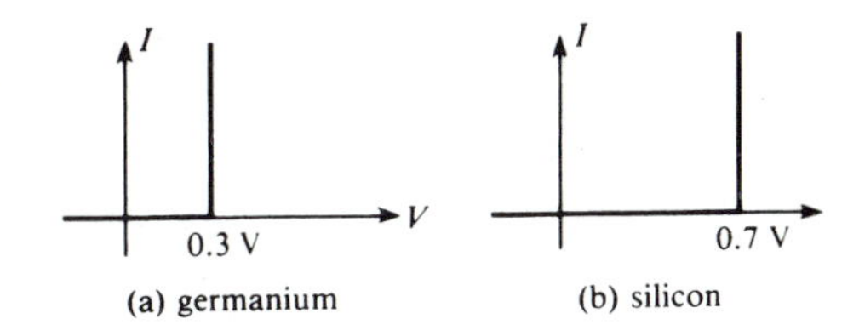

Figure 3–5
Idealized characteristic curves. The diodes are assumed to be open circuits until the forward-biasing voltages are reached.

it is not possible to have, say, 5 V across a forward-biased diode. If a diode were connected directly across a +5-V source, it would act like a short circuit and damage either the source, the diode, or both. When troubleshooting a circuit that contains a diode that is supposed to be forward biased, a diode voltage measurement greater than 0.3 V or 0.7 V means that the diode has failed and is in fact open.

Example 3–2

DESIGN

Assume that the silicon diode in Figure 3–6 requires a minimum current of 1 mA to be above the knee of its I–V characteristic.

1. What should be the value of R to establish 5 mA in the circuit?
2. With the value of R calculated in (1), what is the minimum value to which the voltage E could be reduced and still maintain diode current above the knee?

Solution

1. If I is to equal 5 mA, we know that the voltage across the diode will be 0.7 V. Therefore, solving equation 3–7 for R,

$$R = \frac{E - 0.7}{I} = \frac{(5 - 0.7)\text{ V}}{5 \times 10^{-3}\text{ A}} = 860\ \Omega$$

2. In order to maintain the diode current above the knee, I must be at least 1 mA. Thus,

$$I = \frac{E - 0.7}{R} \geq 10^{-3}\text{ A}$$

Therefore, since $R = 860\ \Omega$,

$$\frac{E - 0.7}{860} \geq 10^{-3}\text{ A}$$

or

$$E \geq (860 \times 10^{-3}) + 0.7$$

$$E \geq 1.56\text{ V}$$

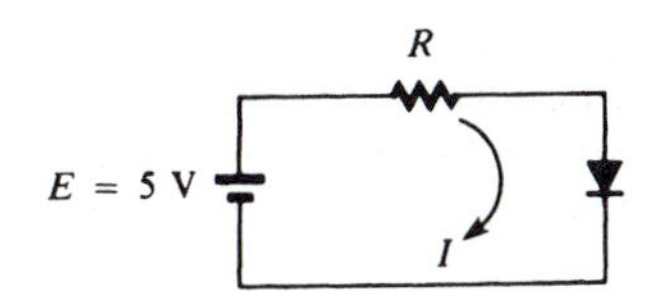

Figure 3–6
(Example 3–2)

In some dc circuits, the voltage drop across a forward-biased diode may be so small in comparison to other dc voltages in the circuit that it can be neglected entirely. For example, suppose a circuit consists of a 25-V source in series with a 1-kΩ resistor in series with a germanium diode. Then $I = (25 - 0.3)/(1\ \text{k}\Omega) = 24.7$ mA. Neglecting the drop across the diode, we would calculate $I = 25/(1\ \text{k}\Omega) = 25$ mA, a result that in most practical situations would be considered close enough to 24.7 mA to be accurate.

3–5 ANALYSIS OF SMALL-SIGNAL DIODE CIRCUITS

Generally speaking, electronic devices can be considered to "operate" (be used) in one of two distinct kinds of applications: *small signal* and *large signal.* In small-signal applications, the current and voltage changes in the device occur over a very limited range of its I–V characteristic curve. In other words, the quantities ΔV and ΔI to which we have referred in earlier paragraphs are very small in comparison to the overall voltage and current ranges through which the device is capable of operating. In practice, small-signal operation can be considered to be that which occurs when the voltage and current changes are over a range of the I–V curve that is essentially linear. Note in Figure 3–2 that the diode's characteristic curve is essentially linear for currents greater than about 5 mA, that is, in the region above the knee.

In contrast, large-signal operation occurs when the voltage and current changes in a device occur over substantially its entire I–V characteristic. Large-signal applications usually require voltage and current changes over a region of the I–V curve in which there is a significant change in its slope. For example, a circuit in which the voltage across a silicon diode varies from -5 V to $+0.7$ V would be considered a large-signal application. In this case the diode would change in nature from an extremely high resistance when reverse biased (negative voltage region) to a moderately high resistance when slightly forward biased, to a very small resistance when biased above the knee. In this section we will study equivalent circuits for a diode operated in small-signal applications, and in the next section we will do the same for large-signal applications.

Consider the circuit shown in Figure 3–7(b). Note that the circuit contains a dc voltage source in series with an ac voltage source. The dc voltage source produces the constant voltage E volts, and the ac voltage source produces the sinusoidal wave $e = A \sin \omega t$, where the peak value A is assumed to be small in comparison to E, and ω is the frequency in radians per second. Therefore, the total voltage $v(t)$ applied to the series combination of the resistor R and the diode is the sum of the dc and ac voltages: $v(t) = E + A \sin \omega t$. The total voltage $v(t)$ is called an ac voltage with a *dc level* of E volts. It is sketched in Figure 3–7(a). Note that $v(t)$ has maximum value $E + A$ and minimum value $E - A$. We wish to determine the voltage across

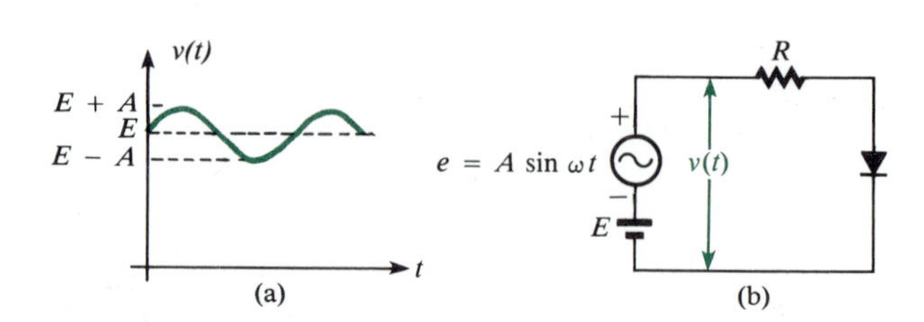

Figure 3–7
The voltage $v(t)$ in (b) is the sum of a dc and an ac component: $v(t) = E + A \sin \omega t$. As shown in (a), $v(t)$ has a minimum value $E - A$ and maximum $E + A$.

Figure 3–8
The ac equivalent circuit of Figure 3–7(b). Note that the diode is replaced by its ac equivalent resistance, r_D.

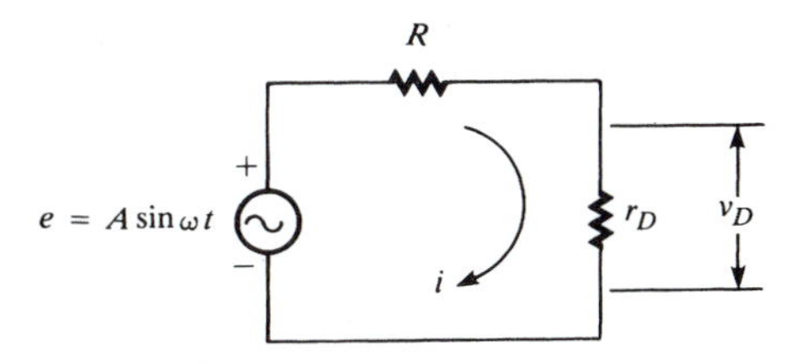

and current through the diode, under the assumption that the voltage and current *changes* are small enough to be considered small-signal operation.

Recall that a circuit containing two voltage sources can be analyzed using the principle of *superposition.* That is, we can determine the current in the circuit due to each source acting alone and then add these results to find the actual current when both sources are present. The superposition principle can be applied only when all the circuit components are linear, and, since we are assuming small-signal operation, we are justified in applying it to the present problem.

We should first determine the dc diode current in Figure 3–7(b) because this current is needed to calculate the ac resistance r_D (equation 3–6). Eliminating (shorting) the ac voltage source in Figure 3–7(b), we obtain a circuit like that shown in Figure 3–4, and therefore, assuming a silicon diode, $I = (E - 0.7)/R$. We can then find the ac resistance of the diode using equation 3–6: $r_D = 0.026/I$. Eliminating the dc voltage source, the ac equivalent circuit is then as shown in Figure 3–8. Note that the ac equivalent circuit represents the diode by a single equivalent resistance equal to r_D ohms. We see that the ac current in Figure 3–8 can be calculated from a direct application of Ohm's law:

$$i = \frac{e}{R + r_D} = \frac{A}{R + r_D} \sin \omega t$$

The ac diode voltage v_D can be calculated from $v_D = ir_D$ or by application of the voltage-divider rule:

$$v_D = \frac{r_D}{R + r_D} e = \frac{r_D A \sin \omega t}{R + r_D}$$

Combining the results of our dc and ac analyses, we find the total current and voltage in the diode to be

$$i(t) = I + i = \frac{E - 0.7}{R} + \frac{A}{R + r_D} \sin \omega t \tag{3–8}$$

and

$$v_D(t) = 0.7 + \frac{r_D \text{A} \sin \omega t}{R + r_D} \tag{3–9}$$

Example 3–3

Assuming that the silicon diode in Figure 3–9 is biased above its knee and has a bulk resistance of 0.1 Ω, find the total current in and total voltage across the diode. Sketch the current versus time.

Solution. Shorting the ac source, we find the dc current to be

$$I = \frac{(6 - 0.7)\text{ V}}{270\ \Omega} = 19.63\text{ mA}$$

Figure 3–9
(Example 3–3)

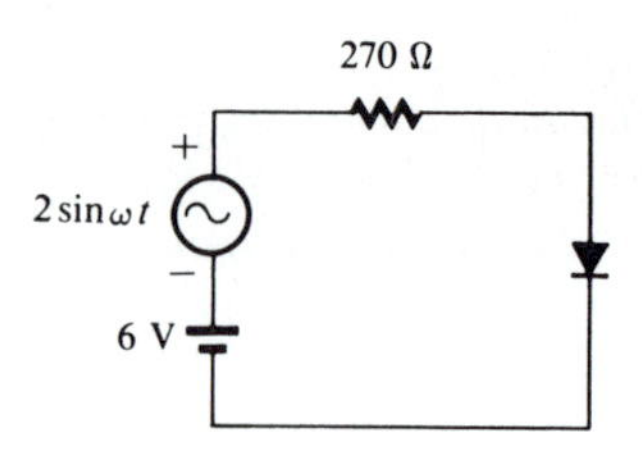

The ac resistance is therefore

$$r_D = \frac{0.026}{I} + r_B = \frac{0.026}{19.63 \times 10^{-3}\text{ A}} + 0.1\ \Omega = 1.42\ \Omega$$

The ac current is

$$i = \frac{e}{R + r_D} = \frac{2 \sin \omega t}{271.42} = 7.37 \sin \omega t \text{ mA}$$

and the ac voltage is

$$v_D = \frac{r_D}{R + r_D} e - \left(\frac{1.42}{271.42}\right) 2 \sin \omega t = 0.01 \sin \omega t \text{ V}$$

Finally, the total current and voltage are

$$i(t) = 19.63 + 7.37 \sin \omega t \text{ mA}$$

and

$$v_D(t) = 0.7 + 0.01 \sin \omega t \text{ V}$$

Figure 3–10 is a sketch of the total current. Note that the maximum current is 19.63 + 7.37 = 27 mA and the minimum current is 19.63 − 7.37 = 12.26 mA. It would be difficult to sketch $v_D(t)$ accurately because its dc level (0.7 V) is so much greater than its ac component. The ac variation is only ±10 mV because the diode's ac resistance is so small.

The Load Line

Small-signal diode analysis can be performed graphically using the diode's *I–V* characteristic curve. Although this method is not frequently used in practice, it deserves to be studied because it provides insights to the combined static and dynamic (dc and ac) behavior of the diode circuit. Consider the circuit shown in Figure 3–11. Here we show the dc equivalent circuit that results when the ac source

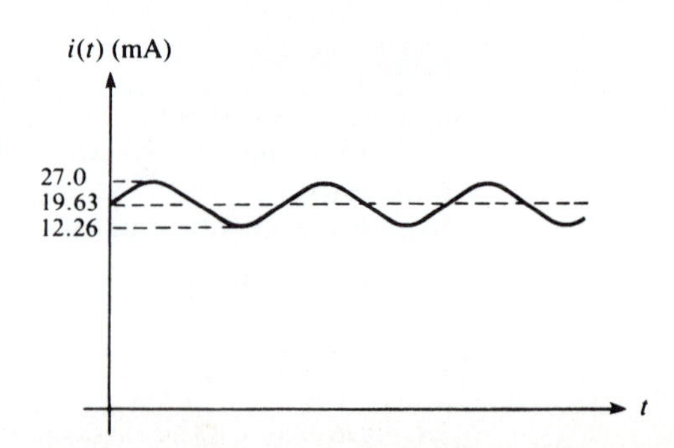

Figure 3–10
Current in the circuit of Figure 3–9 (Example 3–3). The ac component varies ±7.37 mA above and below the dc component of 19.63 mA.

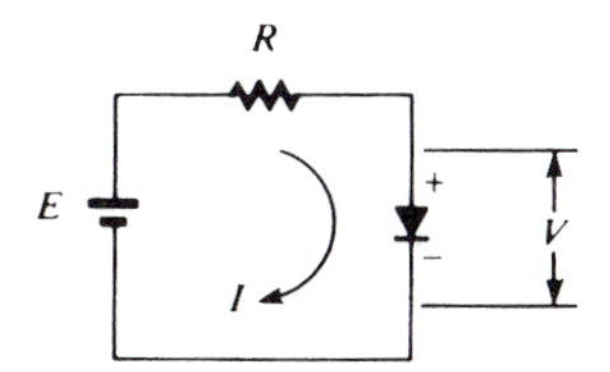

Figure 3–11
The diode current I and diode voltage V are regarded as variables.

in Figure 3–7 is shorted, except instead of treating the dc diode voltage as a constant, we consider it to be the variable quantity V.

By Kirchhoff's voltage law, $E = IR + V$. Solving for I, we obtain

$$I = \frac{-V}{R} + \frac{E}{R} \qquad \textbf{(3–10)}$$

In equation 3–10, we regard I and V as *variables,* whereas E and R are constants. For example, if equation 3–10 were applied to Figure 3–9, we would have

$$I = \frac{-V}{270} + \frac{6}{270}$$

or

$$I = -(3.7 \times 10^{-3})V + 0.0222 \qquad \textbf{(3–11)}$$

Compare equations 3–10 and 3–11 with the general form (equation 3–2) of a linear relation between I and V. Recall also that the general equation for the graph of a straight line plotted in an xy coordinate system is

$$y = mx + b \qquad \textbf{(3–12)}$$

where m is the slope of the line and b is its y-intercept.

$$I = \left(-\frac{1}{R}\right)V + \frac{E}{R}$$
$$y = \quad mx \quad + b$$

We see that the variable I in (3–10) corresponds to the variable y in (3–12), while the variable V in (3–10) corresponds to the variable x in (3–12). Further comparison of equations 3–10 and 3–12 reveals that the slope m in (3–12) corresponds to $-1/R$ in (3–10), while the y-intercept b in (3–12) corresponds to the I-intercept E/R in (3–10). We conclude that the graph of equation 3–10 is a straight line on I–V axes and has slope $-1/R$ and I-intercept E/R. This line is called the *dc load line.* In the example (equation 3–11) we note that the load line has slope -3.7×10^{-3} and I-intercept 0.0222 A = 22.2 mA.

Figure 3–12 shows the graph of the load line corresponding to equation 3–11. Note that a straight line having a negative slope is always a line that goes downward for increasing values of the horizontal variable (V in this case). Note also that the V-intercept (V_0) of the graph can be found by setting I equal to 0 in equation 3–10 and solving for V:

$$0 = \frac{-V_0}{R} + \frac{E}{R}$$

$$\frac{V_0}{R} = \frac{E}{R}$$

$$V_0 = E$$

In Figure 3–12, the V-intercept is $V_0 = E = 6$ V.

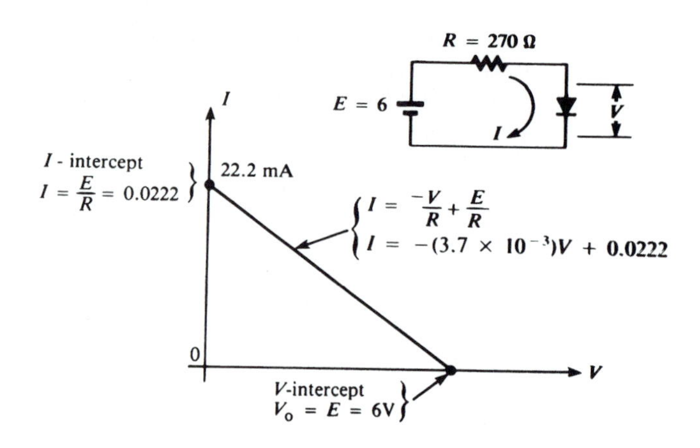

Figure 3–12
A plot of the load line $I = -(3.7 \times 10^{-3})V + 0.0222$. The load line shows all possible combinations of diode current I and diode voltage V, for a fixed E and R. The actual values of I and V depend on the diode inserted in the circuit.

The significance of the dc load line is that every *possible* combination of the current I and voltage V in the circuit of Figure 3–11 is a point that lies somewhere on the line. Given a particular diode, whose characteristic I–V curve we happen to have, our objective is to find the current–voltage combination that results when that diode is inserted in the circuit. We can find that point by plotting the dc load line on the same I–V axes that contain the diode's characteristic curve. The intersection of the dc load line and the characteristic curve gives us the actual diode current and voltage that result when the diode is used in the circuit. What we have accomplished, in effect, is a graphical solution to the two simultaneous equations

$$I = -(1/R)V + E/R \quad \text{(the load line)}$$

and

$$I = I_s(e^{V/\eta V_T} - 1) \quad \text{(the characteristic curve)}$$

Figure 3–13 shows the solution obtained by plotting the load line for the circuit of Figure 3–9 (i.e., equation 3–11) on the same set of axes as a hypothetical I–V characteristic. In the figure, we see that the diode voltage is 0.66 V and that the corresponding current is 19.8 mA. This point of intersection, labeled Q, is called the *quiescent point* (Q-point), or the *operating point* of the diode. It represents the dc current and voltage in the circuit when only the dc source voltage $E = 6$ V is present in the circuit, that is, when the ac source voltage in Figure 3–9 is 0. Recall that all of our analysis so far has been under the assumption that the ac source is shorted out. The quiescent point is sometimes called the *bias point* because it represents the voltage and current in the diode when it is forward biased by the dc source.

Figure 3–13
The intersection of the load line with the diode characteristic (at the point labeled Q) determines the diode voltage (0.66 V) and diode current (19.8 mA).

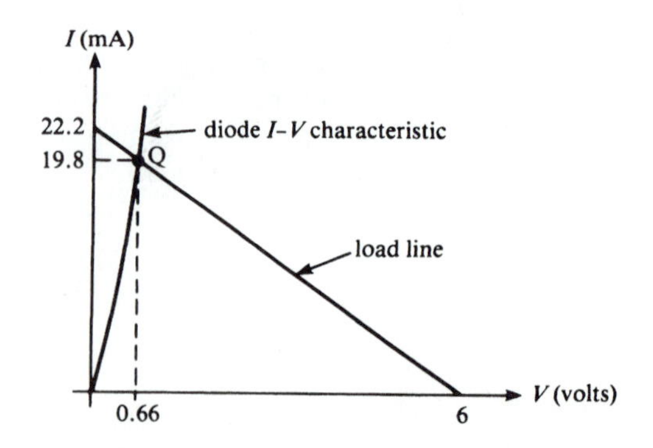

Let us now restore the ac source to the circuit, shown in series with the dc source in Figures 3–7 and 3–9. As we saw earlier, the total applied voltage is then $v(t) = E + e = E + A \sin \omega t$. We can think of this time-varying voltage as generating a whole series of new load lines: one for each of all the (infinitely many) instantaneous values that $v(t)$ has between its minimum, $E - A$ volts, and its maximum, $E + A$ volts. If, for example, we were to freeze time at the point where $v(t) = E - A$ volts, then it would be the same as if the dc circuit shown in Figure 3–11 had a voltage source of $E - A$ volts instead of E volts. This would create a load line that intersects the I-axis at $(E - A)/R$ amperes and the V-axis at $E - A$ volts, in other words, a line parallel to and lying below the original load line. Similarly, when $v(t) = E + A$ volts, there is a corresponding load line lying above and parallel to the original load line. All other instantaneous values create load lines lying between these two extremes. Figure 3–14 illustrates these ideas for the example circuit shown in Figure 3–9.

As the total voltage $v(t)$ varies between 4 and 8 V, the intersections of all the corresponding load lines with the I–V characteristic curve generate all the current–voltage combinations that occur in the circuit. Consequently, we can visualize circuit operation as a point moving continuously along the I–V curve between the two points A and B in Figure 3–14. As this point moves along the curve, the current in the circuit varies sinusoidally between 12.5 mA and 27 mA, as shown in the figure. Also, the diode voltage $v_D(t)$ is seen to vary between 0.65 V and 0.67 V. We have therefore solved the problem posed in Example 3–4 using a graphical rather than a computational method.

We should note that we can also find the ac resistance of the diode when the circuit operates between points A and B in Figure 3–14:

$$r_D = \frac{\Delta V}{\Delta I} = \frac{(0.67 - 0.65)\text{ V}}{(27 - 12.5) \times 10^{-3}\text{ A}} = 1.38\ \Omega$$

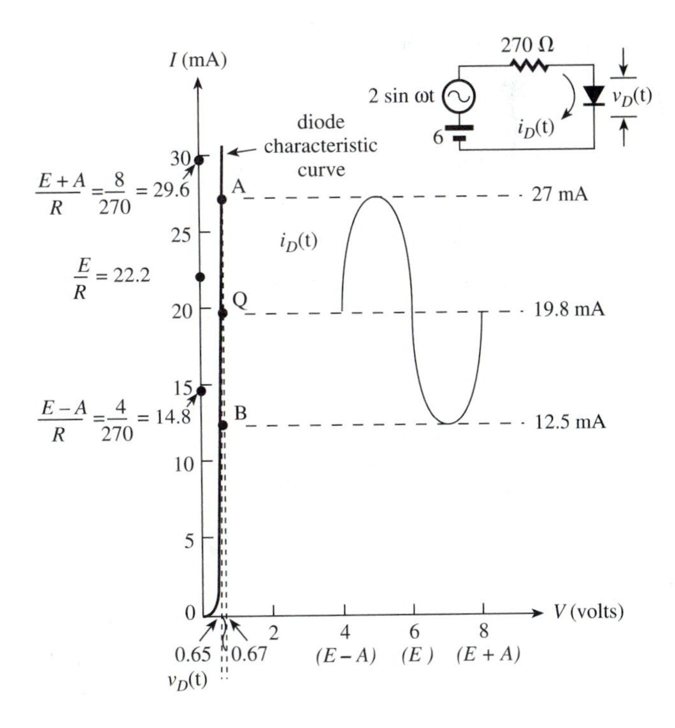

Figure 3–14
The effect of an ac source in the diode circuit can be analyzed by thinking of it as creating a series of parallel load lines. In this way, the minimum and maximum voltage and current can be determined.

All our graphical solutions, including the ac resistance, compare favorably with those computed in Example 3–4.

Example 3–4

SPICE

Use SPICE to obtain a plot of the diode current in Figure 3–15(a) versus time. Assume all diode parameters have their default values.

Figure 3–15
(Example 3–4)

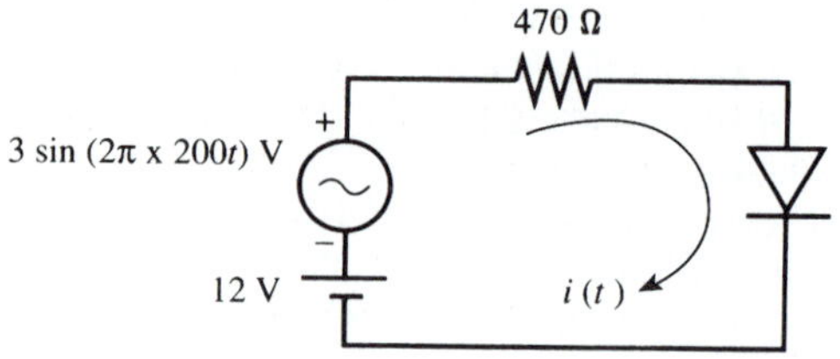

(a)

```
DIODE CURRENT
V1 1 0 12V
V2 2 1 SIN(0 3 200)
R1 2 3 470
VDUM 3 4
D1 4 0 DIODE
.MODEL DIODE D
.TRAN 0.25MS 5MS
.PLOT TRAN I(VDUM)
.END
```

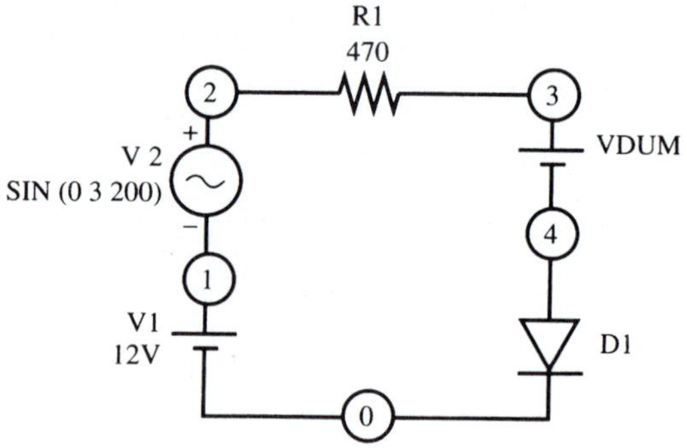

(b)

```
DIODE CURRENT
****       TRANSIENT ANALYSIS                  TEMPERATURE =    27.000 DEG C
*****************************************************************************
   TIME         I(VDUM)
                   1.500D-02        2.000D-02        2.500D-02        3.000D-02   3.500D-02
                    - - - - - - - - - - - - - - - - - - - - - - - - - - - - - - - - - - -
0.000D+00   2.396D-02 .                .          *  .               .                .
2.500D-04   2.593D-02 .                .             .     *         .                .
5.000D-04   2.770D-02 .                .             .          *    .                .
7.500D-04   2.911D-02 .                .             .              *.                .
1.000D-03   3.001D-02 .                .             .               *                .
1.250D-03   3.032D-02 .                .             .               .*               .
1.500D-03   3.001D-02 .                .             .               *                .
1.750D-03   2.911D-02 .                .             .              *.                .
2.000D-03   2.770D-02 .                .             .          *    .                .
2.250D-03   2.593D-02 .                .             .     *         .                .
2.500D-03   2.396D-02 .                .          *  .               .                .
2.750D-03   2.200D-02 .                .     *       .               .                .
3.000D-03   2.023D-02 .                .*            .               .                .
3.250D-03   1.882D-02 .            *   .             .               .                .
3.500D-03   1.792D-02 .        *       .             .               .                .
3.750D-03   1.761D-02 .       *        .             .               .                .
4.000D-03   1.792D-02 .        *       .             .               .                .
4.250D-03   1.882D-02 .            *   .             .               .                .
4.500D-03   2.023D-02 .                .*            .               .                .
4.750D-03   2.200D-02 .                .     *       .               .                .
5.000D-03   2.396D-02 .                .          *  .               .                .
                    - - - - - - - - - - - - - - - - - - - - - - - - - - - - - - - - - - -
```

Solution. Figure 3–15(b) shows the circuit with a dummy voltage source inserted to obtain positive values of diode current. Notice that we require a *transient* (.TRAN) analysis to obtain a plot versus time and that the sinusoidal source is modeled by a SIN source. (Specifying an AC source and performing an .AC analysis would not produce a plot versus time.) Since the frequency of the source is 200 Hz, one period occupies 1/200 = 5 ms. Thus the control statement .TRAN 0.25MS 5MS will cause 5 ms/0.25 ms = 20 values of current to be plotted over one full period.

Shown next is the plot produced by a program run. (Portions of the complete printout produced by SPICE have been omitted to conserve space.) The peak value of the current is seen to be 30.32 mA, occurring at t = 1.25 ms. We can compare this result with that calculated using equation 3–8: Since the dc current is (12 − 0.7 V)/470 Ω = 24 mA, the approximate value of r_D is 0.026/(24 mA) = 1.08 Ω. Thus, by equation 3–8, the peak value of the current is

$$\frac{E-0.7}{R}+\frac{A}{R+r_D}=\frac{11.3\text{ V}}{470\ \Omega}+\frac{3\text{ V}}{471\ \Omega}=30.37\text{ mA}$$

We see that there is very favorable agreement between the SPICE plot (30.32 mA) and equation 3–8 (30.37 mA).

3–6 ANALYSIS OF LARGE-SIGNAL DIODE CIRCUITS

As we indicated in Section 3–5, a diode is said to operate under large-signal conditions when the current and voltage changes it undergoes extend over a substantial portion of its characteristic curve, including portions where there is a significant change in slope. In every practical large-signal application, the diode is operated both in the region where it is well forward biased (above the knee) and into the region where it is either reverse biased or biased near zero volts. We have seen that such large excursions will change the resistance of the diode from very small to very large values.

When the resistance of a diode changes from a very small to a very large value, it behaves very much like a *switch.* An ideal (perfect) switch has zero resistance when closed and infinite resistance when open. Similarly, an *ideal* diode for large-signal applications is one whose resistance changes between these same two extremes. When analyzing such circuits, it is often helpful to think of the diode as a *voltage-controlled switch:* a forward-biasing voltage closes it, and a zero or reverse-biasing voltage opens it. Depending on the magnitudes of other voltages in the circuit, the 0.3- or 0.7-V drop across the diode when it is forward biased may or may not be negligible. Figure 3–16 shows the idealized characteristic curve for a

Figure 3–16
Idealized silicon diode characteristics, used for large-signal analysis

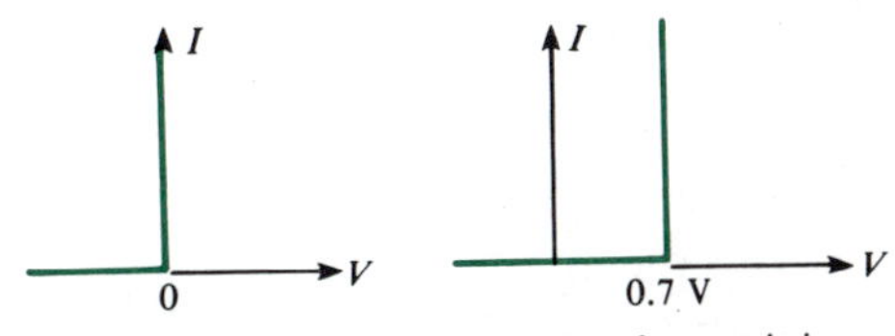

(a) I–V characteristic when the diode is treated as a perfect switch having 0 voltage drop.

(b) I–V characteristic when the diode is treated as a perfect switch that closes when V = 0.7 volts.

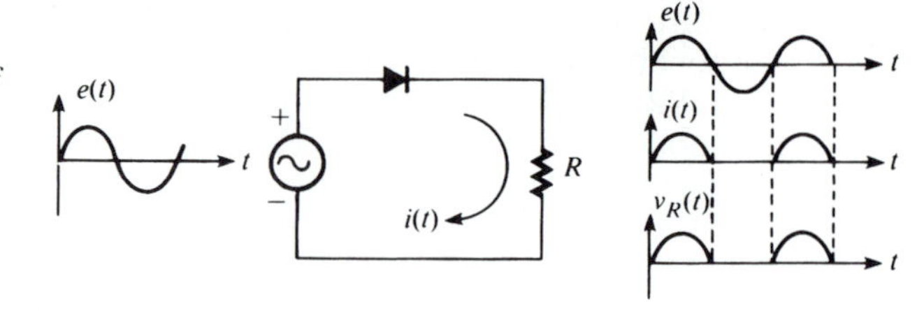

Figure 3–17
The diode used as a rectifier. Current flows only during the positive half-cycle of the input.

silicon diode (a) when the 0.7-V drop is neglected and (b) when it is not. In case (a), the characteristic curve is the same as that of a perfect switch.

Rectifiers

One of the most common uses of a diode in large-signal operation is in a *rectifier* circuit. A rectifier is a device that permits current to flow through it in one direction only. It is easy to see how a diode performs this function when we think of it as a voltage-controlled switch. When the anode voltage is positive with respect to the cathode, i.e., when the diode is forward biased, the "switch is closed" and current flows through it from anode to cathode. If the anode becomes negative with respect to the cathode, the "switch is open" and no current flows. Of course, a *real* diode is not perfect, so there is in fact some very small reverse current that flows when it is reverse biased. Also, as we know, there is a nonzero voltage drop across the diode when it is forward biased (0.3 or 0.7 V), a drop that would not exist if it were a perfect switch.

Consider the rectifier circuit shown in Figure 3–17. We see in the figure that an ac voltage source is connected across a diode and a resistor R, the latter designed to limit current flow when the diode is forward biased. Notice that no dc source is present in the circuit. Therefore, during each positive half-cycle of the ac source voltage $e(t)$, the diode is forward biased and current flows through it in the direction shown. During each negative half-cycle of $e(t)$ the diode is reverse biased and no current flows. The waveforms of $e(t)$ and $i(t)$ are sketched in the figure. We see that $i(t)$ is a series of positive current pulses separated by intervals of zero current. Also sketched is the waveform of the voltage $v_R(t)$ that is developed across R as a result of the current pulses that flow through it. Note that the net effect of this circuit is the conversion of an ac voltage into a (pulsating) dc voltage, a fundamental step in the construction of a dc *power supply.*

If the diode in the circuit of Figure 3–17 is turned around, so that the anode is connected to the resistor and the cathode to the generator, then the diode will be forward biased during the negative half-cycles of the sine wave. The current would then consist of a sequence of pulses representing current flow in a counter-clockwise, or negative, direction around the circuit.

Example 3–5

Assume that the silicon diode in the circuit of Figure 3–18 has a characteristic like that shown in Figure 3–16(b). Find the peak values of the current $i(t)$ and the voltage $v_R(t)$ across the resistor when

1. $e(t) = 20 \sin \omega t$, and
2. $e(t) = 1.5 \sin \omega t$. In each case, sketch the waveforms for $e(t)$, $i(t)$, and $v_R(t)$.

Figure 3–18
(Example 3–5)

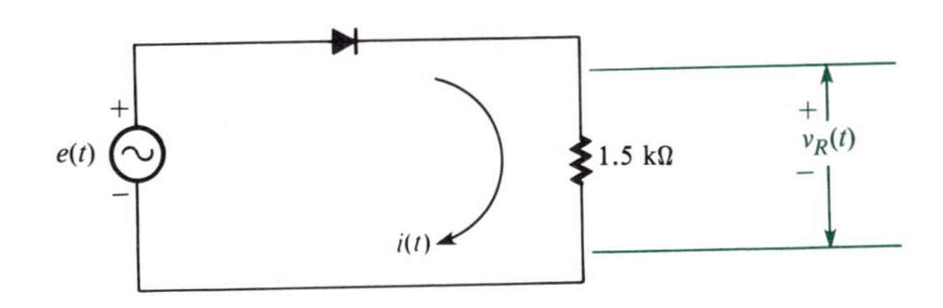

Solution

1. When $e(t) = 20 \sin \omega t$, the peak positive voltage generated is 20 V. At the instant $e(t) = 20$ V, the voltage across the resistor is $20 - 0.7 = 19.3$ V, and the current is $i = 19.3/(1.5\ \text{k}\Omega) = 12.87$ mA. Figure 3–19 shows the resulting waveforms. Note that because of the characteristic assumed in Figure 3–16(b), the diode does not begin conducting until $e(t)$ reaches +0.7 V and ceases conducting when

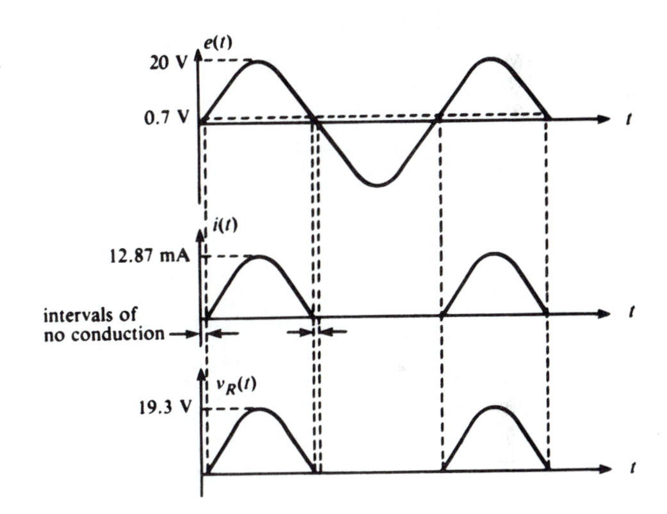

Figure 3–19
Diode current and voltage in the circuit of Figure 3–18. Note that the diode does not conduct until e(t) reaches 0.7 V, so short intervals of nonconduction occur during each positive half-cycle.

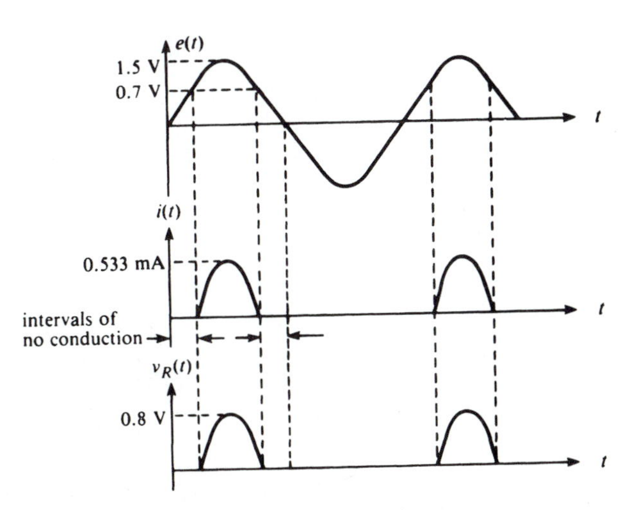

Figure 3–20
Diode current and voltage in the circuit of Figure 3–18 when the sine-wave peak is reduced to 1.5 V. Note that the intervals of nonconduction are much longer than those in Figure 3–19.

e(t) drops below 0.7 V. The time interval between the point where e(t) = 0 V and e(t) = 0.7 V is very short in comparison to the half-cycle of conduction time. From a practical standpoint, we could have assumed the characteristic in Figure 3–16(a), i.e., neglected the 0.7-V drop, and the resulting waveforms would have differed little from those shown.

2. When e(t) = 1.5 sin ωt, the peak positive voltage generated is 1.5 V. At that instant, $v_R(t) = 1.5 - 0.7 = 0.8$ V and $i(t) = (0.8 \text{ V})/(1.5 \text{ k}\Omega) = 0.533$ mA. The waveforms are shown in Figure 3–20. Note once again that the diode does not conduct until e(t) = 0.7 V. However, in this case, the time interval between e(t) = 0 V and e(t) = 0.7 V is a significant portion of the conducting cycle. Consequently, current flows in the circuit for significantly less time than one-half cycle of the ac waveform. In this case, it clearly would *not* be appropriate to use Figure 3–16(a) as an approximation for the characteristic curve of the diode.

Elementary DC Power Supplies

As already mentioned, an important application of diodes is in the construction of dc power supplies. It is instructive at this time to consider how diode rectification and waveform filtering, the first two operations performed by every power supply, are used to create an elementary dc power source. (If desired, this entire discussion can be deferred to a more detailed theoretical analysis in Chapter 16.)

The single diode in Figure 3–17 is called a *half-wave* rectifier, because the waveforms it produces ($i(t)$ and $v_R(t)$) each represent half a sine wave. These half–sine waves are a form of pulsating dc and by themselves are of little practical use. (They can, however, be used for charging batteries, an application in which a steady dc current is not required.) Most practical electronic circuits require a dc voltage source that produces and maintains a *constant* voltage. For that reason, the pulsating half–sine waves must be converted to a steady dc level. This conversion is accomplished by *filtering* the waveforms. Filtering is a process in which selected frequency components of a complex waveform are *rejected* (filtered out) so that they do not appear in the output of the device (the filter) performing the filtering operation. The pulsating half–sine waves (like all periodic waveforms) can be regarded as waveforms that have both a dc component and ac components. Our purpose in filtering these waveforms for a dc power supply is to reject *all* the ac components.

The simplest kind of filter that will perform the filtering task we have just described is a capacitor. Recall that a capacitor has reactance inversely proportional to frequency: $X_C = 1/2\pi fC$. Thus, if we connect a capacitor directly across the output of a half-wave rectifier, the ac components will "see" a low-impedance path to ground and will not, therefore, appear in the output. Figure 3–21 shows a filter

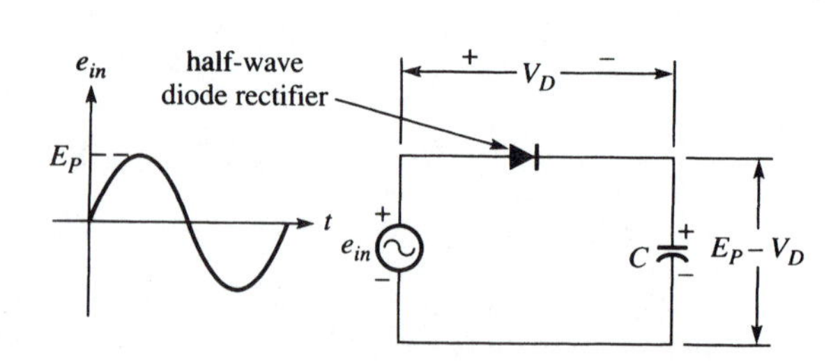

Figure 3–21
Filter capacitor C effectively removes the ac components from the half-wave rectified waveform.

capacitor, C, connected in this way. In this circuit the capacitor charges to the peak value of the rectified waveform, V_{PR}, so the output is the dc voltage V_{PR}. Note that $V_{PR} = E_P - V_D$, where E_P is the peak value of the sinusoidal input and V_D is the dc voltage drop across the diode (0.7 V for silicon).

In practice, a power supply must provide dc current to whatever load it is designed to serve, and this load current causes the capacitor to discharge and its voltage to drop. The capacitor discharges during the intervals of time between input pulses. Each time a new input pulse occurs, the capacitor recharges. Consequently, the capacitor voltage rises and falls in synchronism with the occurrence of the input pulses. These ideas are illustrated in Figure 3–22. The output waveform is said to have a *ripple voltage* superimposed on its dc level.

When the peak-to-peak value of the output ripple voltage, V_{PP}, is small compared to V_{dc} (a condition called *light loading*), we derive (in Chapter 16) the following equations for V_{dc} and V_{PP}:

$$V_{dc} = \frac{V_{PR}}{1 + \dfrac{1}{2f_r R_L C}} \tag{3–13}$$

$$V_{PP} = \frac{V_{dc}}{f_r R_L C} \tag{3–14}$$

where V_{PR} = peak value of the rectified waveform $(E_P - V_D)$
f_r = frequency of the rectified waveform
R_L = load resistance
C = filter capacitance

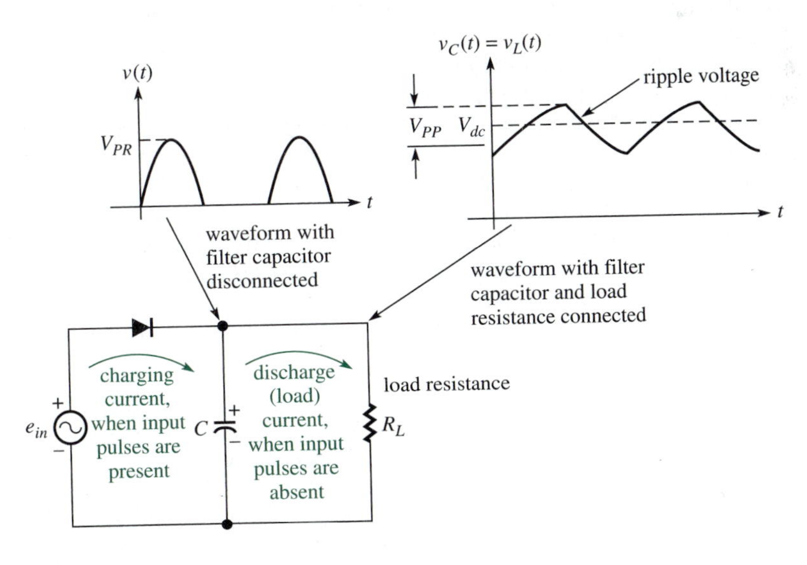

Figure 3–22
When load resistance R_L is connected across the filter capacitor, the capacitor charges and discharges, creating a load voltage that has a ripple voltage superimposed on a dc level.

Example 3–6

The sinusoidal input, e_{in}, in Figure 3–22 is 120 V rms and has frequency 60 Hz. The load resistance is 2 kΩ and the filter capacitance is 100 μF. Assuming light loading and neglecting the voltage drop across the diode,

1. find the dc value of the load voltage;
2. find the peak-to-peak value of the ripple voltage.

Solution

1. The peak value of the sinusoidal input voltage is $E_P = \sqrt{2}$ (120 V rms) = 169.7 V. Since the voltage drop across the diode can be neglected, $V_{PR} = E_P =$ 169.7 V. From equation 3–13,

$$V_{dc} = \frac{169.7\text{ V}}{1 + \dfrac{1}{2(60\text{ Hz})(2\text{ k}\Omega)(100\ \mu\text{F})}} = 162.9\text{ V}$$

2. From equation 3–14,

$$V_{PP} = \frac{162.9\text{ V}}{(60\text{ Hz})(2\text{ k}\Omega)(100\ \mu\text{F})} = 13.57\text{ V}$$

A *full-wave* rectifier effectively inverts the negative half-pulses of a sine wave to produce an output that is a sequence of positive half-pulses with no intervals between them. Figure 3–23 shows a widely used full-wave rectifier constructed from four diodes and called a full-wave diode *bridge.* Also shown is the full-wave rectified output. (See Figures 16–6 and 16–7 for a detailed explanation of how the bridge operates.) Note that on each half-cycle of input, current flows through *two* diodes, so the peak value of the rectified output is $V_{PR} = E_P - 2V_D$ or $E_P - 1.4$ V for silicon.

As in the half-wave rectifier, the full-wave rectified waveform can be filtered by connecting a capacitor in parallel with load R_L. The advantage of the full-wave rectifier is that the capacitor does not discharge so far between input pulses, because a new charging pulse occurs every half-cycle instead of every full cycle. Consequently, the magnitude of the output ripple voltage is smaller. This fact is illustrated in Figure 3–24.

Equations 3–13 and 3–14 for V_{dc} and V_{PP} are valid for both half-wave and full-wave rectifiers. Note that f_r in those equations is the frequency of the *rectified*

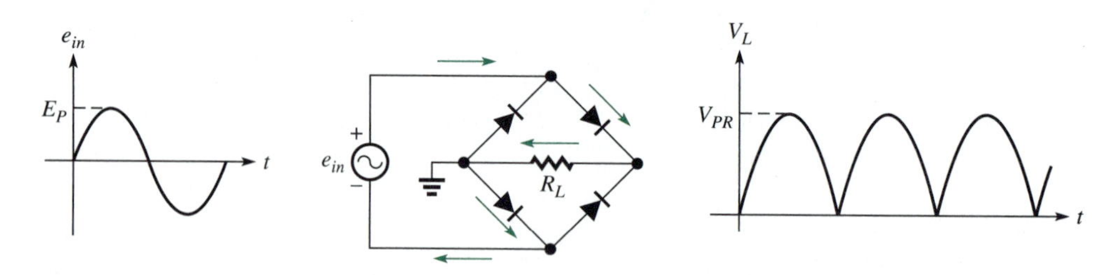

Figure 3–23
The full-wave bridge rectifier and output waveform. The arrows show the direction of current flow when e_{in} *is positive.*

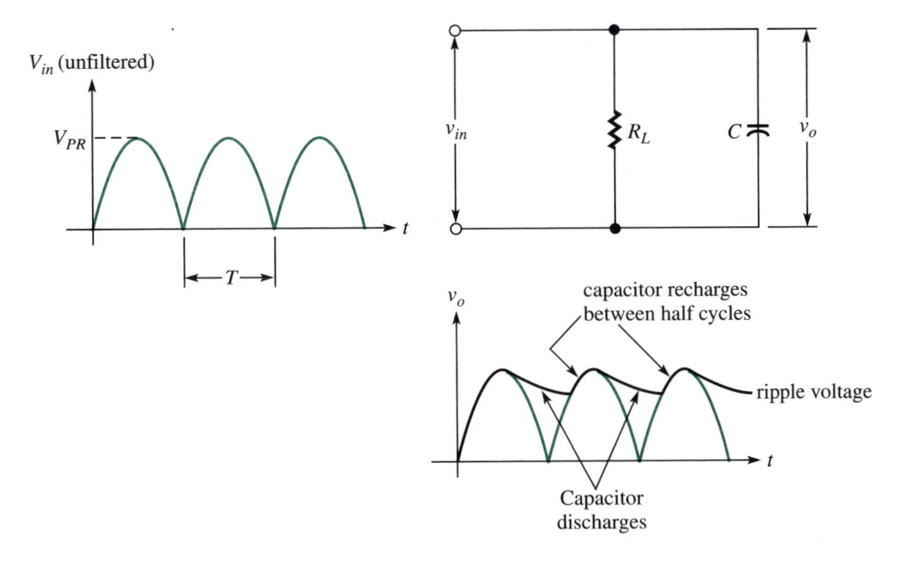

Figure 3–24
The ripple voltage in the filtered output of a full-wave rectifier is smaller than in the half-wave case because the capacitor recharges at shorter intervals. T *equals the period of the full-wave rectified waveform (one-half the period of the unrectified sine wave).*

waveform, which, in a full-wave rectifier, is *twice* the frequency of the unrectified sine wave (see Figure 3–24). If the same input and component values used in Example 3–6 are used to compute V_{dc} and V_{PP} for a full-wave rectifier (f_r = 120 Hz), we find

$$V_{dc} = \frac{169.7\text{ V}}{1 + \dfrac{1}{2(120\text{ Hz})(2\text{ k}\Omega)(100\ \mu\text{F})}} = 166.2\text{ V}$$

and

$$V_{PP} = \frac{166.2\text{ V}}{(120\text{ Hz})(2\text{ k}\Omega)(100\ \mu\text{F})} = 6.92\text{ V}$$

Note that the peak-to-peak value of the ripple voltage is one-half that found for the half-wave rectifier.

Although the elementary power supplies we have described can be used in applications where the presence of some ripple voltage is acceptable, where the exact value of the output voltage is not critical, and where the load does not change appreciably, more sophisticated power supplies have more elaborate filters and special circuitry (voltage regulators) that maintain a constant output voltage under a variety of operating conditions. These refinements are discussed in detail in Chapter 16.

3–7 DIODE SWITCHING CIRCUITS

In another very important large-signal application of diodes, the devices are switched rapidly back and forth between their high-resistance and low-resistance states. *Digital logic* circuits, the building blocks of digital computers, are a typical example.

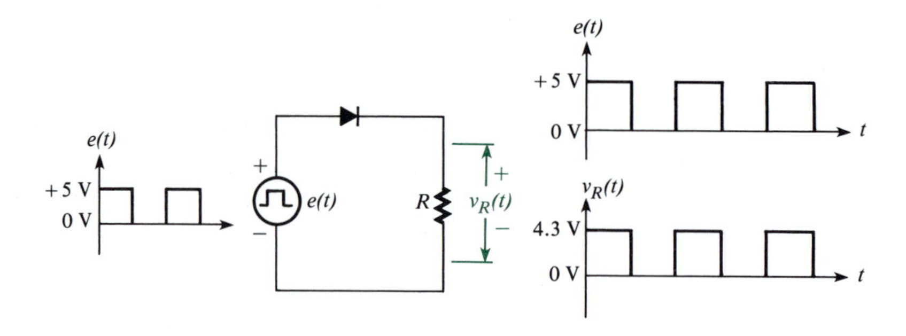

Figure 3–25
The diode is forward biased when the square-wave voltage is +5 V. Note that the (silicon) diode voltage is 0.7 V when it conducts.

In these applications, the circuit voltages are pulse-type waveforms, or square waves, that alternate between a "low" voltage, often 0 V, and a "high" voltage, such as +5 V. These essentially instantaneous changes in voltage between low and high cause the diode to switch between its "off" and "on" states. Figure 3–25 shows the voltage waveform that is developed across a resistor in series with a silicon diode when a square wave that alternates between 0 V and +5 V is applied to the combination. When $e(t) = +5$ V, the diode is forward biased, or "ON," so current flows through the resistor and a voltage equal to $5 - 0.7 = 4.3$ V is developed across it. When $e(t) = 0$ V, the diode is in its high-resistance state, or "OFF," and, since no current flows, the resistor voltage is zero. This operation is very much like rectifier action. However, we study digital logic circuits in just the extreme cases where the voltage is either low or high. In other words, we assume that every voltage in the circuit is at one of those two levels. Because the diode in effect performs the function of switching a high level into or out of a circuit, these applications are often called *switching circuits.*

Diode switching circuits typically contain two or more diodes, each of which is connected to an independent voltage source. Understanding the operation of a diode switching circuit depends, first, on determining which diodes, if any, are forward biased and which, if any, are reverse biased. The key to this determination is remembering that a diode is forward biased only if its anode is positive *with respect to* its cathode. The important words here (the ones that usually give students the most trouble) are "with respect to." Stated another way, the anode voltage (with respect to ground) must be more positive than the cathode voltage (with respect to ground) in order for a diode to be forward biased. This is of course the same as saying that the cathode voltage must be more negative than the anode voltage. Conversely, in order for a diode to be reverse biased, the anode must be negative with respect to the cathode, or, equivalently, the cathode positive with respect to the anode. The following example should help clarify these ideas.

Example 3–7

Determine which diodes are forward biased and which are reverse biased in each of the configurations shown in Figure 3–26. The schematic diagrams in each part of Figure 3–26 are drawn using the standard convention of omitting the connection

Figure 3–26
(Example 3–7)

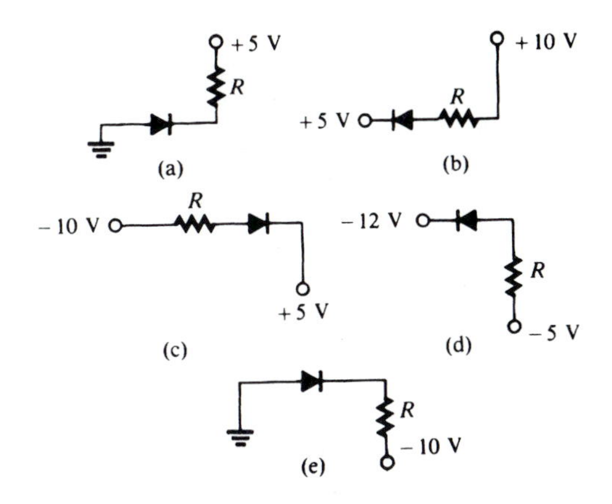

line between one side of a voltage source and ground. In this convention, it is understood that the *opposite* side of each voltage source shown in the figure is connected to ground. If the reader is not comfortable with this convention, then he or she should begin the process now of becoming accustomed to it, for it is widely used in electronics. As an aid in understanding the explanations given below, redraw each circuit with all ground connections included. For example, Figure 3–27 shows the circuit that is equivalent to Figure 3–26(c).

Solution

1. In (a) the anode is grounded and is therefore at 0 V. The cathode side is positive by virtue of the +5-V source connected to it through resistor R. The cathode is therefore positive with respect to the anode; i.e., the anode is more negative than the cathode, so the diode is reverse biased.
2. In (b) the anode side is more positive than the cathode side ($+10\text{ V} > +5\text{ V}$), so the diode is forward biased. Current flows from the 10-V source, through the diode, and into the 5-V source.
3. In (c) the anode side is more negative than the cathode side, so the diode is reverse biased. Note that (essentially) no current flows in the circuit, so there is no drop across resistor R. Therefore, the total reverse-biasing voltage across the diode is 15 V. (See also Figure 3–27, and note that the sources are series-aiding.)
4. In (d) the cathode side is more negative than the anode side ($-12\text{ V} < -5\text{ V}$), so the diode is forward biased. Current flows from the −5-V source, through the diode, and into the −12-V source.
5. In (e) the anode is grounded and is therefore at 0 V. The cathode side is more negative than the anode side ($-10\text{ V} < 0\text{ V}$), so the diode is forward biased. Current flows from ground, through the diode, and into the −10-V source.

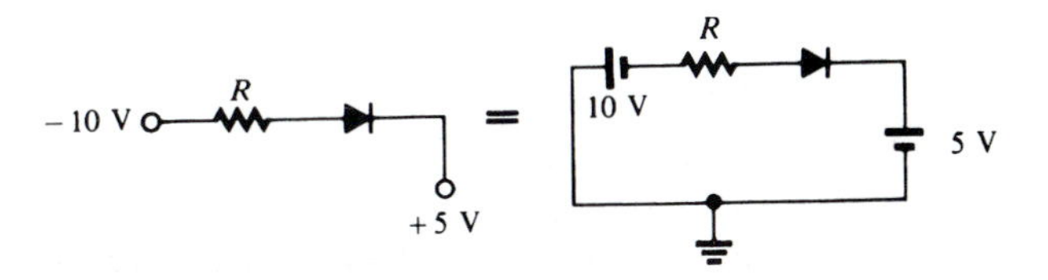

Figure 3–27
The circuit of Figure 3–26(c) is redrawn in an equivalent form that shows all ground connections.

Figure 3–28
A typical diode switching circuit like those used in digital logic applications. The equivalent circuit, showing all ground paths, is shown in (b).

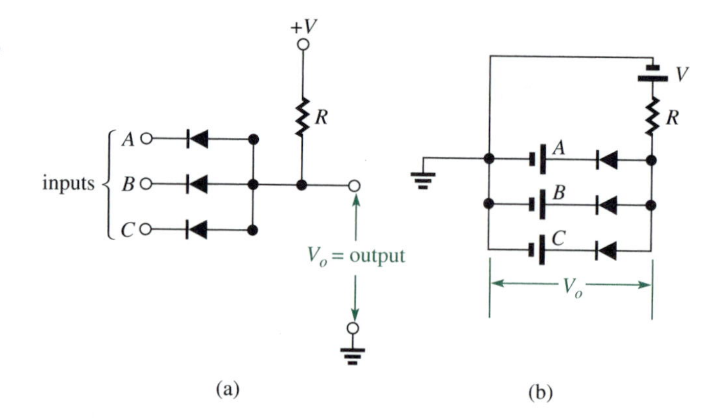

Figure 3–28 shows a diode switching circuit typical of those used in digital logic applications. It consists of three diodes whose anodes are connected together and whose cathodes may be connected to independent voltage sources. The voltage levels connected to the cathodes are called *inputs* to the circuit, labeled A, B, and C, and the voltage developed at the point where the anodes are joined is called the *output* of the circuit. All voltages are referenced to the circuit's common ground. The voltage source V is a fixed positive voltage called the *supply* voltage. The figure shows the conventional way of drawing this kind of circuit (a), and the complete equivalent circuit (b).

Let us assume that the inputs A, B, and C in Figure 3–28 can be either +5 V (high) or 0 V (low). Suppose further that the supply voltage is $V = +10$ V. If A, B, and C are all +5 V, then all three diodes are forward biased (+10 > +5) and are therefore conducting. Current flows from the 10-V source, through the resistor, and then divides through the three diodes. Suppose the dc "ON" resistance of each diode is 300 Ω and the resistor R is 1 kΩ. Figure 3–29 shows the resulting equivalent circuits. The parallel combination of the three 5-V sources, each having a 300-Ω series resistor, is equivalent to a single 5-V source in series with a 300/3 = 100-Ω

Figure 3–29
The diode circuit of Figure 3–28 when all inputs are +5 V

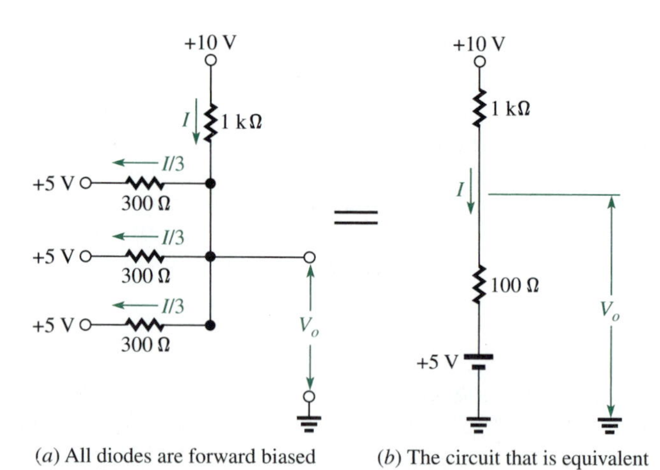

(*a*) All diodes are forward biased when all inputs are +5V.

(*b*) The circuit that is equivalent to (*a*), showing all inputs replaced by a single equivalent resistor and voltage source.

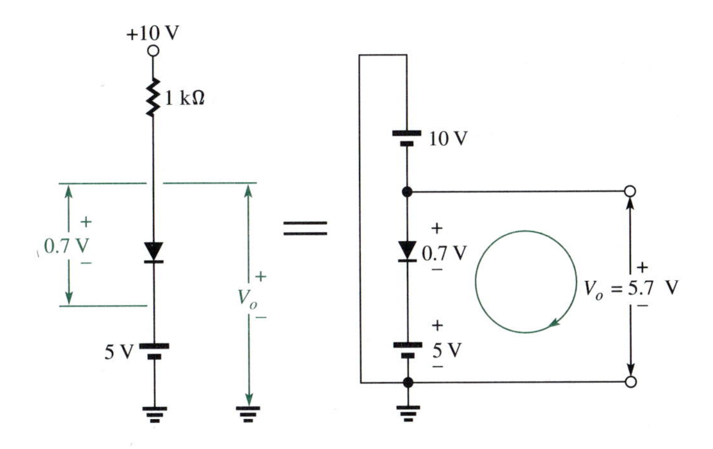

Figure 3–30
The circuit that is equivalent to Figure 3–29(a) when the diodes are assumed to have a fixed 0.7-V drop

resistor (Millman's theorem). The circuit is therefore equivalent to that shown in Figure 3–29(b). The current I is found by dividing the total series resistance into the net voltage across it:

$$I = \frac{(10 - 5)\text{ V}}{(1000 + 100)\ \Omega} = \frac{5\text{ V}}{1100\ \Omega} = 4.5\text{ mA}$$

The output voltage is

$$V_o = 5\text{ V} + I(100\ \Omega) = 5 + (4.5 \times 10^{-3}\text{ A})\ (100\ \Omega) = 5.45\text{ V}$$

When the dc resistance of the diode is not known, an approximate solution can be obtained by assuming a 0.7-V drop across each (silicon) diode. Under this assumption, the equivalent circuit appears as shown in Figure 3–30. Using this equivalent circuit, we can write Kirchhoff's voltage law around the clockwise loop to obtain $V_o = 5\text{ V} + 0.7\text{ V} = 5.7\text{ V}$.

Suppose now that input $A = 0$ V and $B = C = +5$ V, as shown in Figure 3–31(a). It is clear that the diode connected to input A is forward biased. If we

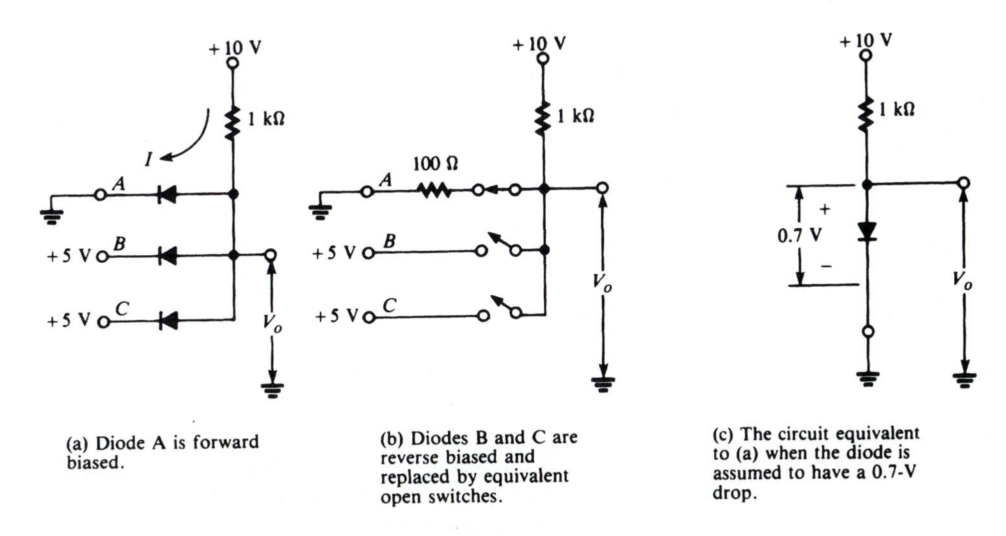

Figure 3–31
The circuit of Figure 3–28 when input A *is 0 V and inputs* B *and* C *are +5 V*

temporarily regard that "ON" diode as a perfect closed switch, then we see that the anode side of all diodes will be connected through this closed switch to 0 V. Therefore, the other two diodes have +5 V on their cathodes and 0 V on their anodes, causing them to be reverse biased. In reality, the "ON" diode is not a perfect switch, so it has some small voltage drop across it (across the 300 Ω) and the anodes are *near* 0 V rather than exactly 0 V. The net effect is the same: one diode is forward biased and the other two are reverse biased.

Figure 3–31(b) shows the equivalent circuit that results if we treat the reverse-biased diodes as open switches. Notice that we now show the dc "ON" resistance as 100 Ω rather than 300 Ω. This is an assumption, based on the fact that the diode now carries substantially more current than before. We have seen that the diode resistance decreases with increasing current. The output voltage can now be calculated using the voltage-divider rule:

$$V_o = \left(\frac{100}{1000 + 100}\right)(10\ \text{V}) = 0.9\ \text{V}$$

This result is improbably large but is, after all, only as good as our knowledge or assumption of the "ON" resistance. Figure 3–31(c) shows the equivalent circuit that results when we assume that the diode voltage drop is 0.7 V. In this case we see that $V_o = 0.7$ V.

If input B is at 0 V while $A = C = +5$ V, then it should be obvious that the output voltage V_o is exactly the same as in the previous case. Any combination of inputs that causes one diode to be forward biased and the other two to be reverse biased has the same equivalent circuits as shown in Figure 3–31.

When any two of the inputs are at 0 V and the third is at +5 V, then two diodes are forward biased and the third is reverse biased. It is left as an exercise to determine the output voltage in this case.

Finally, if all three inputs are at 0 V, then all three diodes are forward biased. The equivalent circuits are shown in Figure 3–32. The three parallel resistors are equivalent to 100 Ω, so

$$V_o = \left(\frac{100}{1000 + 100}\right)(10\ \text{V}) = 0.9\ \text{V}$$

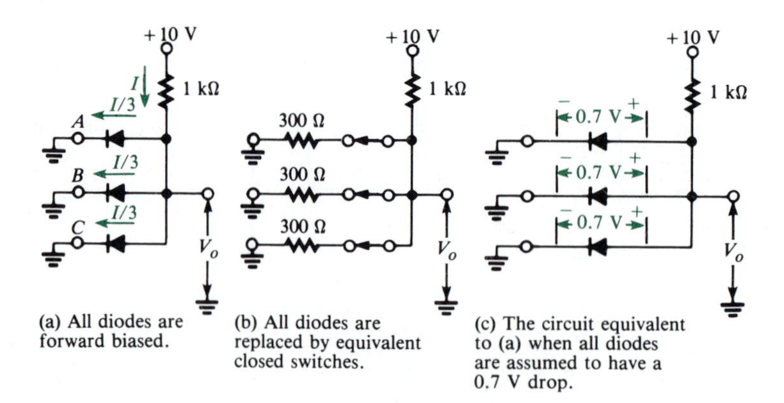
+10 V
1 kΩ
I
I/3
A
B
C
V_o
300 Ω
0.7 V
(a) All diodes are forward biased.
(b) All diodes are replaced by equivalent closed switches.
(c) The circuit equivalent to (a) when all diodes are assumed to have a 0.7 V drop.

Figure 3–32
The circuit of Figure 3–28 when all inputs are 0 V

If we regard the drop across each diode as 0.7 V, then $V_o = 0.7$ V, as shown in Figure 3–32(c). The diode circuit we have just analyzed is called a diode AND gate because the output is high if and only if inputs *A and B and C* are all high.

We have seen that the first step in this kind of analysis is to determine which diodes are forward biased and which are reverse biased. This determination is best accomplished by temporarily regarding each diode as a perfect, voltage-controlled switch. At this point, one might legitimately question how we determined that the forward-biased diode shown in Figure 3–31 is the only one that is forward biased. After all, the other two diodes appear to have their anode sides more positive (10 V) than their cathode sides (5 V), and seem therefore to meet the criterion for forward bias. However, if this were the case, then these "ON" diodes would, as closed switches, connect +5 V to the anodes, which in turn would forward bias the third diode (since its cathode is at 0 V). The third diode would then act as a closed switch connecting 0 V to the anodes. Here, then, is the dilemma: how can the anodes be simultaneously at 0 V and at 5 V? Obviously they cannot, and this contradiction proves that only the one diode shown in Figure 3–31 can be forward biased. The fact that it *is* forward biased, and therefore connects 0 V to the anodes, ensures that the other two diodes are reverse biased.

A rule that is useful for determining which diode is truly forward biased is to determine which one has the greatest forward-biasing potential measured from the supply voltage to its input voltage. For example, in Figure 3–31, the net voltage between the supply and input *A* is 10 V − 0 V = 10 V, while the net voltage between the supply and inputs *B* and *C* is +10 V − 5 V = 5 V. Therefore, the first diode is forward biased and the other two diodes are reverse biased.

Example 3–8

Determine which diodes are forward biased and which are reverse biased in the circuits shown in Figure 3–33. Assuming a 0.7-V drop across each forward-biased diode, determine the output voltage.

Solution

1. In (a) diodes D_1 and D_3 have a net forward-biasing voltage between supply and input of 5 V − 5 V = 0 V. Diodes D_2 and D_4 have a net forward-biasing voltage

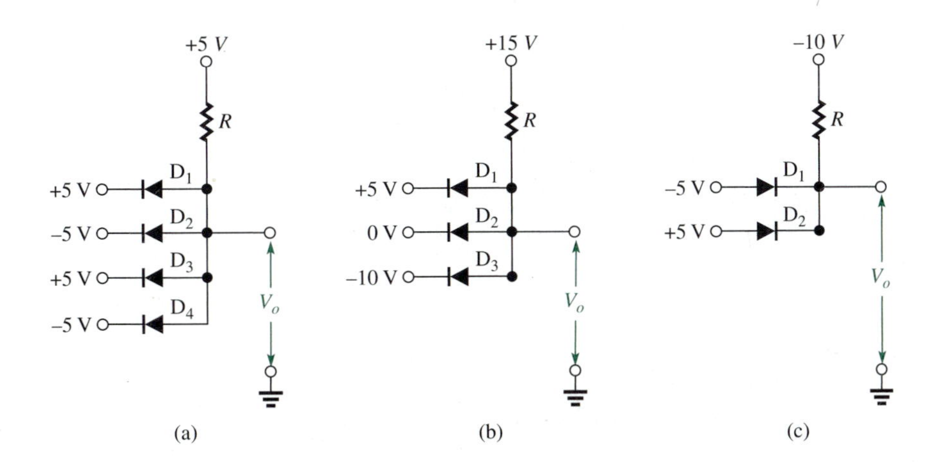

Figure 3–33
(Example 3–8)

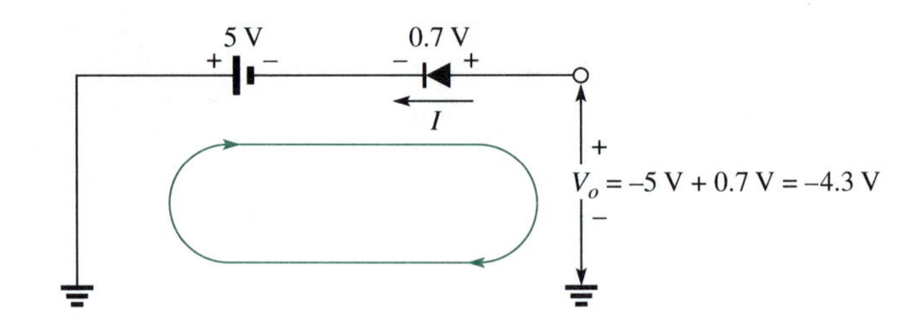

Figure 3–34
The circuit equivalent to Figure 3–33(a)

of 5 V − (−5 V) = 10 V. Therefore D_2 and D_4 are forward biased and D_1 and D_3 are reverse biased. Figure 3–34 shows the equivalent circuit path between input and output. Writing Kirchhoff's voltage law around the loop, we determine $V_o = -5\text{ V} + 0.7\text{ V} = -4.3\text{ V}$.

2. In (b) the net forward-biasing voltage between supply and input for each diode is

$$D_1:\quad +15\text{ V} - (+5\text{ V}) = +10\text{ V}$$
$$D_2:\quad +15\text{ V} - 0\text{ V} = +15\text{ V}$$
$$D_3:\quad +15\text{ V} - (-10\text{ V}) = +25\text{ V}$$

Therefore, D_3 is forward biased and D_1 and D_2 are reverse biased.

$$V_o = -10\text{ V} + 0.7\text{ V} = -9.3\text{ V}$$

3. In (c) the net forward-biasing voltage between supply and input for each diode is

$$D_1:\quad -10\text{ V} - (-5\text{ V}) = -5\text{ V}$$
$$D_2:\quad -10\text{ V} - (+5\text{ V}) = -15\text{ V}$$

Notice that the diode positions are reversed with respect to those in (a) and (b), in the sense that the cathodes are joined together and connected through resistor R to a negative supply. Thus, the diode for which there is the greatest negative voltage between supply and input is the forward-biased diode. In this case, that diode is D_2. D_1 is reverse biased, by virtue of the fact that its cathode is near +5 V and its anode is at −5 V. Figure 3–35 shows the equivalent circuit path between input and output. Writing Kirchhoff's voltage law around the loop, we see that $V_o = 5\text{ V} - 0.7\text{ V} = 4.3\text{ V}$.

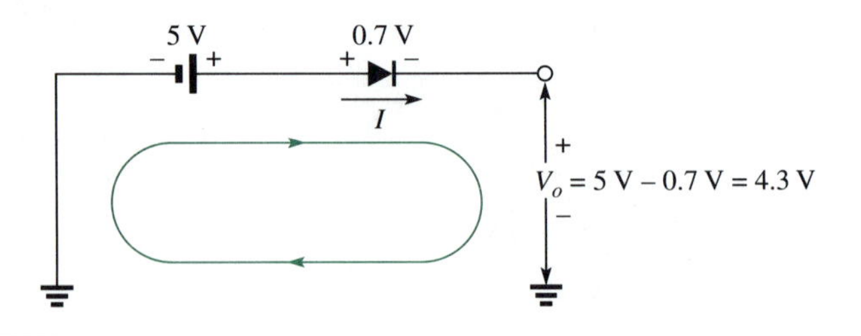

Figure 3–35
The circuit equivalent to Figure 3–33(c)

3–8 DIODE TYPES, RATINGS, AND SPECIFICATIONS

Discrete diodes—those packaged in individual cases with externally accessible anode and cathode connections—are commercially available in a wide range of

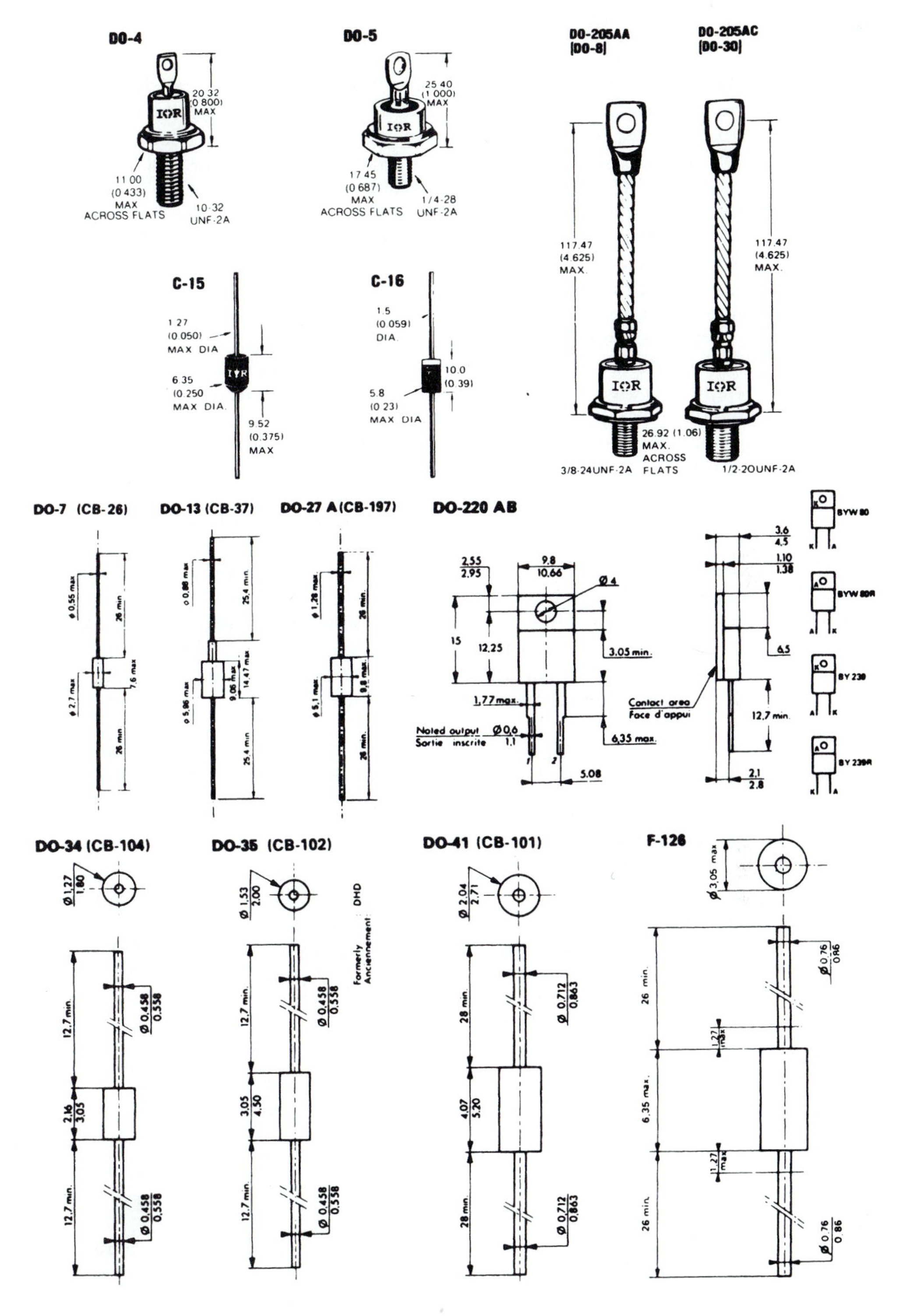

Figure 3–36
Discrete diode case styles (Courtesy of International Rectifier Corp. and Thomson-CSF Components Corp.)

types designed for different kinds of service and for a variety of applications. We find, for example, *switching* diodes designed specifically for use in logic circuit applications, like those discussed in the last section. These diodes typically have low power dissipation ratings, are small in size, and are designed to respond rapidly to pulse-type inputs, that is, to switch between their "ON" and "OFF" states with minimum delay. *Rectifier,* or *power,* diodes are designed to carry larger currents and to dissipate more power than switching diodes. They are used in power supply applications, where heavier currents and higher voltages are encountered. *Small-signal* diodes are general-purpose diodes used in applications like those discussed in Section 3–5.

Figure 3–36 illustrates the variety of sizes and shapes that commercially available diodes may have. Each of those shown has a designation that identifies the standard case size it has (DO-4, DO-7, etc.). Materials used for case construction include glass, plastic, and metal. Metal cases are used for large, rectifier-type diodes to enhance the conduction of heat and improve their power dissipation capabilities.

There are two particularly important diode ratings that a designer using commercial, discrete diodes should know when selecting a diode for any application: the *maximum reverse voltage* (V_{RM}) and the *maximum forward current.* The maximum reverse voltage, also called the *peak inverse voltage* (PIV), is the maximum reverse-biasing voltage that the diode can withstand without breakdown. If the PIV is exceeded, the diode "breaks down" only in the sense that it readily conducts current in the reverse direction. As discussed in Chapter 2, breakdown *may* result in permanent failure if the power dissipation rating of the device is exceeded. The maximum forward current is the maximum current that the diode can sustain when it is forward biased. Exceeding this rating will cause excessive heat to be generated in the diode and will lead to permanent failure. Manufacturers' ratings for the maximum forward current will specify whether the rating is for continuous, peak, average, or rms current, and they may provide different values for each. The symbols I_o and I_F are used to represent forward current.

Example 3–9

DESIGN

In the circuit of Figure 3–37, a rectifier diode is used to supply positive current pulses to the 100-Ω resistor load. The diode is available in the combinations of ratings listed in the table portion of the figure. Which is the least expensive diode that can be used for the application?

Figure 3–37
(Example 3–9)

V_{RM}	*Max I_o (average)*	*Unit Cost*
100 V	1.0 A	$0.50
150 V	2.0 A	1.50
200 V	1.0 A	2.00
200 V	2.0 A	3.00
500 V	2.0 A	3.50
500 V	5.0 A	5.00

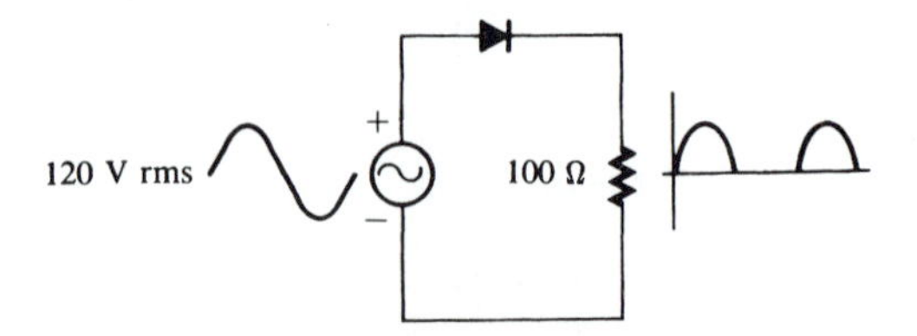

Solution. The applied voltage is 120 V rms. Therefore, when the diode is reverse biased by the *peak* negative value of the sine wave, it will be subjected to a maximum reverse-biasing voltage of (1.414)(120) = 169.7 V. The V_{RM} rating must be greater than 169.7 V.

The average value of the current is one-half the average value of a single sinusoidal pulse: $I_{AVG} = (½)(0.637I_p)$ A, where I_p is the peak value of the pulse. (Note that the factor ½ must be used because the pulse is present for only one-half of each full cycle.) The peak forward current in the example (neglecting the drop across the diode) is $I_p = (169.7\text{ V})/(100\ \Omega) = 1.697$ A. Therefore, the average forward current through the diode is $I_{AVG} = (½)(0.637)(1.697) = 0.540$ A.

The least expensive diode having ratings adequate for the peak inverse voltage and average forward current values we calculated is the one costing $2.00.

Figure 3–38 shows a typical manufacturer's specification sheet for a line of silicon small-signal diodes. Like many other manufactured electronic components, diodes are often identified by a standard *type* number in accordance with JEDEC (Joint Electron Devices Engineering Council) specifications. Diode type numbers have the prefix 1N, like those shown in the leftmost column of Figure 3–38. (Not all manufacturers provide JEDEC numbers; many use their own commercial part numbers.) The second column in the specification sheet shows the maximum reverse voltage, V_{RM}, for each of the diode types. Note that V_{RM} ranges from 20 V to 200 V for the diodes listed. The third column shows the rated average forward current I_o of each diode in mA, and these range from 0.1 mA to 200 mA. The next two pairs of columns list values of reverse current I_R for different values of reverse voltage V_R and ambient temperature T_{amb}. The next column gives capacitance values in pF, an important specification in high-frequency and switching applications. The column headed t_{rr} lists the reverse recovery time of each diode, in nanoseconds. This specification relates to the time required for a diode to switch from its ON to its OFF state and is another important parameter in switching circuit design. Finally, the maximum rated power dissipation is given in mW. The product of diode voltage and diode current should never exceed this rating in any application (unless there is some auxiliary means for removing heat, such as a cooling fan).

Figure 3–39 shows a typical specification sheet for a line of silicon rectifier diodes. Note that the forward current ratings for these diodes are generally larger than those of the small-signal diodes. The current ratings are given as $I_{F(AV)}$ (average), and I_{FSM}, each in units of amperes. I_{FSM} is the maximum non-repetitive forward current that the diode can sustain, that is, the maximum value of momentary or *surge* current it can conduct. Note that the I_{FSM} values are much larger than the $I_{F(AV)}$ values. The voltage ratings are specified by V_{RRM}, the maximum repetitive reverse voltage that each diode can sustain. Also note the large physical sizes and the metal cases of the stud-mounted rectifiers that are capable of conducting currents from 12 to 40 A.

silicon signal diodes

Type	V_R-V_{RM} max (V)	I_O $V_F = 1V$ min (mA)	I_R / (µA)	V_R (V)	I_R (µA)	/ T_{amb} (°C)	C max (pF)	t_{rr} (ns)	P_{tot} (mW)	Case
GENERAL PURPOSE AND HIGH SPEED SWITCHING										T_{amb} = 25°C
1N 456	30	40	0,025	25	5	150			250	
1N 456A	30	100	0,025	25	5	150			250	
1N 457	70	20	0,025	60	5	150			250	
1N 457A	70	100	0,025	60	5	150			250	
1N 458	150	7	0,025	125	5	150			250	
1N 458A	150	100	0,025	125	5	150			250	
1N 461	30	15	0,5	25	30	150			250	
1N 461A	30	100	0,5	25	30	150			250	
1N 462	70	5	0,5	60	30	150			250	
1N 462A	70	100	0,5	60	30	150			250	
1N 463	200	1	0,5	175	30	150			250	
1N 464	150	3	0,5	125	30	150			250	
1N 464A	150	100	0,5	125	30	150			250	
1N 482	40	100 •	0,25	30	30	150			250	
1N 482A	40	100	0,025	30	15	150			250	
1N 482B	40	100	0,025	30	5	150			250	
1N 483	70	100 •	0,25	60	30	150			250	
1N 484	130	100 •	0,25	125	30	150			250	
1N 484A	130	100	0,025	125	15	150			250	
1N 484B	130	100	0,025	125	5	150			250	
1N 914	100	10	0,025	20	50	150	4	4	250	
1N 914A	100	20	0,025	20	50	150	4	4	250	
1N 914B	100	100	0,025	20	50	150	4	4	250	
1N 916	100	10	0,025	20	50	150	2	4	250	
1N 916A	100	20	0,025	20	50	150	2	4	250	
1N 916B	100	30	0,025	20	50	150	2	4	250	
1N 3062	75	20	0,1	50	100	150	1	2	250	
1N 3063	75	10	0,1	50	100	150	2	4	250	
1N 3064	75	10	0,1	50	100	150	2	4	250	DO 35 glass
1N 3066	75	50	0,1	50	100	150	6	50	250	
1N 3070	200	100	0,1	150	100	150	5	50	250	
1N 3071	200	200	0,1	125	500	125	8	3000	500	
1N 3595	150	200	0,1	50	100	150	2,5	4	250	
1N 3600	50	10	0,1	50	100	150	1	2	250	
1N 3604		50	0,05	50	50	150	2	2	250	
1N 3605		0,1••	0,05	30	50	150	2	2	250	
1N 3606		0,1••	0,05	50	50	150	2	2	250	
1N 4148	100	10	0,025	20	50	150	4	4	500	
1N 4149	100	10	0,025	20	50	150	2	4	500	
1N 4150	50	200	0,1	50	100	150	2,5	4	500	
1N 4151	75	50	0,05	50	50	150	2	2	500	
1N 4152	40	0,1••	0,05	30	50	150	2	2	500	
1N 4153	75	0,1••	0,05	50	50	150	2	2	500	
1N 4154	25	30	0,1	25	100	150	4	2	500	
1N 4244	20	20	0,1	10	100	150	0,8	0,75	250	
1N 4305	75	10	0,1	50	100	150	2	4	500	
1N 4446	100	20	0,025	20	50	150	4	4	500	
1N 4447	100	20	0,025	20	50	150	2	4	500	
1N 4448	100	100	0,025	20	50	150	4	4	500	
1N 4449	100	30	0,025	20	50	150	2	4	500	
1N 4450	40	200	0,05	30	50	150	4	4	500	
1N 4454	75	10	1	50	100	150	2	4	500	

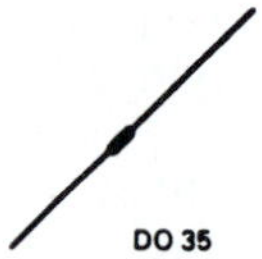

•$V_F = 1,1V$ ••$V_F = 0,55V$

Figure 3–38
A typical diode data sheet (Courtesy of Thomson-CSF)

General Purpose Stud Mounted Silicon Rectifiers

SILICON RECTIFIERS — GENERAL PURPOSE

Silicon rectifiers are available from less than 1 Amp to 3000 Amps, and voltages to 3000 volts. In addition to the standard industry packages, there are a number of other packages available as required. Two of these are shown below. These are ruggedly built devices with an excellent reputation for reliability and performance.

AXIAL LEAD SILICON RECTIFIERS — 750 mA TO 6 AMPS

$I_{F(AV)}$ (A) @ Max T_C (°C)	750 mA @ 25	1 @ 75	1.5 @ 40	1.5 @ 40	2 @ 100	3 @ 125	4 @ 120	6 @ 95
I_{FSM} (A)	22	30	50	50	50	150	200	400
Notes	(1)(2)	(1)(2)	(1)(2)	(1)(2)	(2)	(2)	(2)	(2)
Case Style	DO-41	DO-41	DO-41	DO-41	DO-41	C-12	C-16	C-15
PART NUMBERS								
V_{RRM} 50 Volts	--	1N4001	10D05	1N4816	20D05	–	–	60S05
100 Volts	–	1N4002	10D1	1N4817	20D1	30S1	40D1	60S1
200 Volts	1N2069 A	1N4003	10D2	1N4818	20D2	30S2	40D2	60S2
300 Volts	–	–	10D3	1N4819	–	30S3	–	–
400 Volts	1N2070 A	1N4004	10D4	1N4820	20D4	30S4	40D4	60S4
500 Volts	–	–	10D5	1N4821	–	30S5	–	60S5
600 Volts	1N2071 A	1N4005	10D6	1N4822	20D6	30S6	40D6	60S6
700 Volts	--	–	–	1N5052	–	–	–	–
800 Volts	–	1N4006	10D8	1N5053	20D8	30S8	40D8	60S8
1000 Volts	--	1N4007	10D10	1N5054	20D10	30S10	–	60S10

(1) Temperature given is ambient temperature. (2) Also available on tape reel.

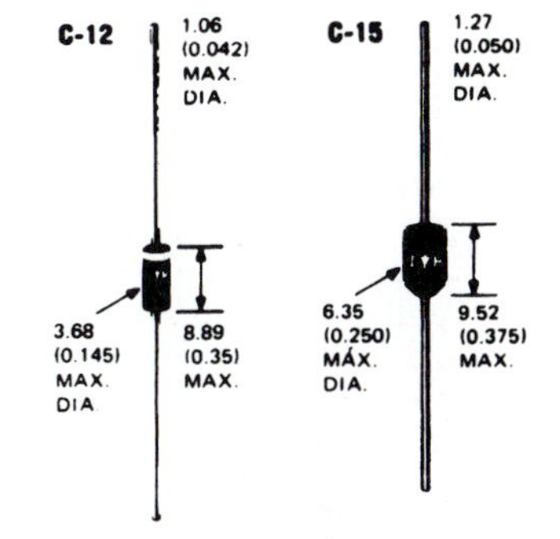

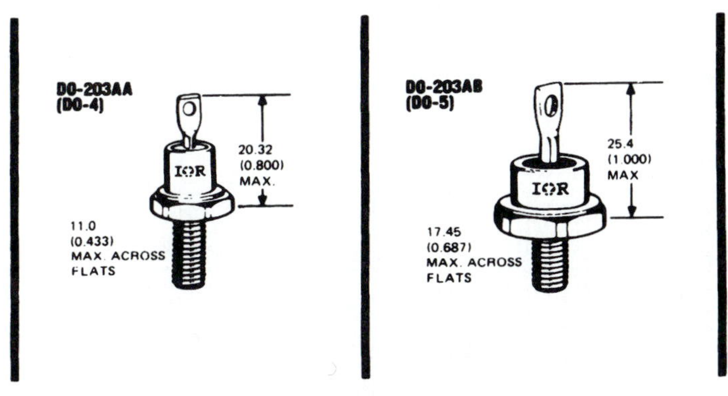

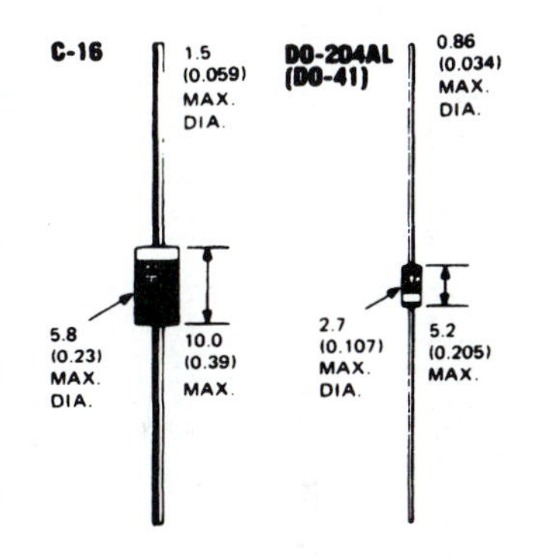

STUD MOUNTED[4] SILICON RECTIFIERS — 12 TO 40 AMPS

$I_{F(AV)}$ (A) @ Max T_C (°C)	12 @ 150	12 @ 150	12 @ 150	15 @ 150	16 @ 150	35 @ 140	40 @ 140	40 @ 150
I_{FSM} (A)	200	240	250	250	300	500	500	800
Notes	(5)	(5)	(5)	(5)	(5)	(5)	(5)	(5)
Case Style	DO-4	DO-4	DO-4	DO-5	DO-4	DO-5	DO-5	DO-5
PART NUMBERS								
V_{RRM} 50 Volts	12F5	1N1199A	12F5B	1N3208	16F5	1N1183	40HF5	1N1183A
100 Volts	12F10	1N1200A	12F10B	1N3209	16F10	1N1184	40HF10	1N1184A
150 Volts	–	1N1201A	–	–	–	1N1185	–	1N1185A
200 Volts	12F20	1N1202A†	12F20B	1N3210	16F20	1N1186	40HF20	1N1186A
300 Volts	–	1N1203	–	1N3211	–	1N1187	40HF30	1N1187A
400 Volts	12F40	1N1204A†	12F40B	1N3212	16F40	1N1188	40HF40	1N1188A
500 Volts	–	1N1205A		1N3213		1N1189	40HF50	1N1189A
600 Volts	12F60	1N1206A†	12F60B	1N3214	16F60	1N1190	40HF60	1N1190A
700 Volts	–	1N3670A	–	–	–	1N3765	40HF80	
800 Volts	12F80	1N3671A	12F80B		16F80	1N3766	40HF100	–
900 Volts	–	1N3672A	–	–	–	1N3767	40HF120	–
1000 Volts	12F100	1N3673A	12F100B		16F100	1N3768		

(4) Metric threads available on some stud packages.

(5) Cathode-to-stud. For anode-to-stud, add "R" to base number (example 12FLR, 40HFLR, 1N3889R).

† JAN and/or JAN-TX types available.

OTHER CONNECTIONS

In addition to the flex leads shown in the case style drawings, IR also has threaded stud, threaded hole and flag terminal top connections available for some case styles. Contact your local IR Distributor, IR Field Office or IR El Segundo offices for more information.

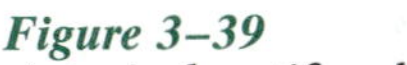

Figure 3–39
A typical rectifier data sheet (Courtesy of International Rectifier)

EXERCISES

Section 3–2
The Diode as a Nonlinear Device

3–1. Make a sketch of the *I–V* characteristic curve for a 10-kΩ resistor when current *I* is plotted along the vertical axis and voltage *V* along the horizontal axis. What is the slope of the characteristic? Be certain to include units in your answer.

3–2. Make a sketch of the *I–V* characteristic curve for a 0.1-μF capacitor when the rms value of the current *I* is plotted along the vertical axis and the rms value of the voltage *V* is along the horizontal axis. Assume that the frequency is fixed at 1000 Hz. What is the slope of the characteristic? Be certain to include units in your answer. What would be the slope if the frequency were changed to 10 kHz?

Section 3–3
AC and DC Resistance

3–3. Using the diode *I–V* characteristic shown in Figure 3–40, find (graphically) the approximate ac resistance when the current in the diode is 0.1 mA. Repeat when the voltage across it is 0.64 V. Is the diode silicon or germanium?

3–4. Using the diode *I–V* characteristic shown in Figure 3–40, find (graphically) the approximate value of the dynamic resistance when the current in the diode is 0.2 mA. Repeat when the voltage across the diode is 0.62 V. What is the approximate maximum knee current?

3–5. Find the dc resistance of the diode at each point specified in Exercise 3–3.

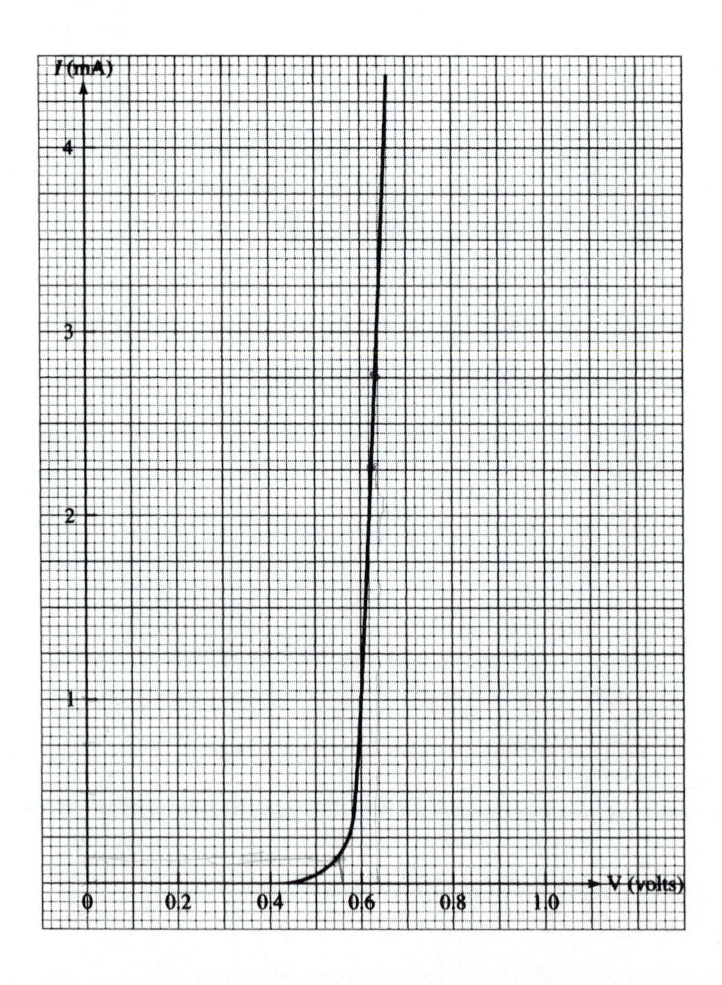

Figure 3–40
(Exercises 3–3 and 3–4)

3–6. Find the static resistance of the diode at each point specified in Exercise 3–4.

3–7. Neglecting bulk resistance, use equation 3–6 to find the approximate ac resistance of the diode at each point specified in Exercise 3–3.

3–8. Assume that the bulk resistance of the diode whose *I–V* characteristic is shown in Figure 3–40 is 0.1 Ω when the current is greater than 1.5 mA and 0.5 Ω when the current is less than 1.5 mA. Use equation 3–6 to find the approximate dynamic resistance of the diode at each point specified in Exercise 3–4.

3–9. A certain diode conducts a current of 440 nA from cathode to anode when the reverse-biasing voltage across it is 8 V. What is the diode's dc resistance under these conditions?

3–10. When the reverse-biasing voltage in Exercise 3–9 is increased to 24 V, the reverse current increases to 1.20 μA. What is its dc resistance in this case?

3–11. In the test circuit shown in Figure 3–3, a voltage of 0.68 V was measured across the diode when the resistance *R* was adjusted to 230 Ω. A voltage of 0.69 V was measured across the diode when the resistance *R* was adjusted to 150 Ω. In each case, the fixed dc voltage source was set to 10 V.

- **a.** What is the dc resistance of the diode at each measurement?
- **b.** What is the ac resistance of the diode when the voltage across it changes from 0.68 V to 0.69 V?

3–12. In the circuit shown in Figure 3–41, the current *I* is 34.28 mA. What is the voltage drop across the diode? What is its dc resistance?

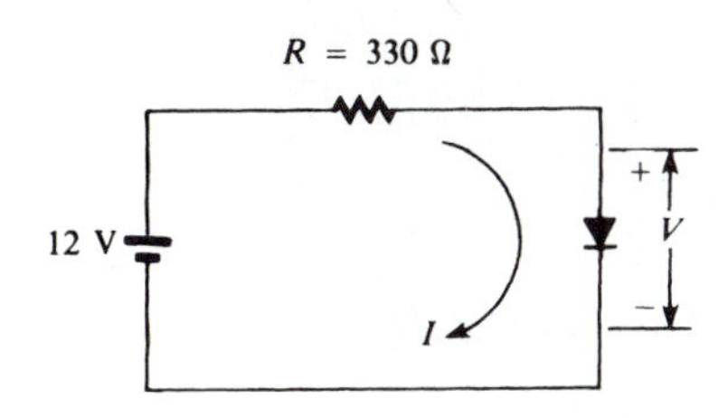

Figure 3–41
(Exercise 3–12)

3–13. Repeat Exercise 3–12 if the resistor *R* is 220 Ω and the current *I* is 51.63 mA.

Section 3–4
Analysis of DC Circuits Containing Diodes

3–14. Assume that the voltage drop across a forward-biased silicon diode is 0.7 V and that across a forward-biased germanium diode is 0.3 V.

- **a.** If D_1 and D_2 are both silicon diodes in Figure 3–42, find the current *I* in the circuit.
- **b.** Repeat if D_1 is silicon and D_2 is germanium.

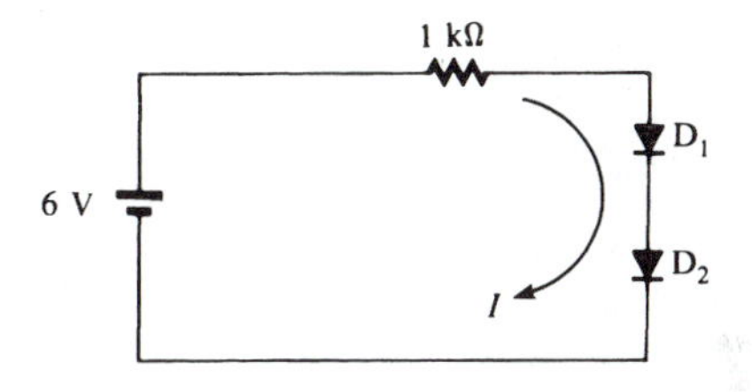

Figure 3–42
(Exercise 3–14)

3–15. Repeat Exercise 3–14 when the constant source voltage is changed to 9 V.

3–16. In the circuit shown in Figure 3–43, the diode is germanium. Find the percent error caused by neglecting the voltage drop across the diode when calculating the current *I* in the circuit. (Assume that a forward-biased germanium diode has a constant voltage drop of 0.3 V.)

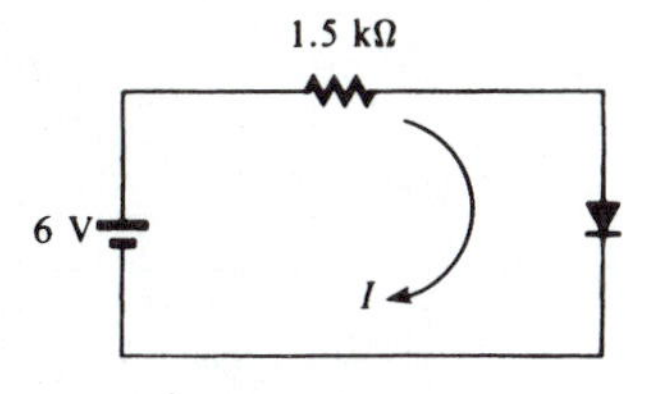

Figure 3–43
(Exercise 3–16)

3–17. Repeat Exercise 3–16 when the source voltage is changed to 3 V and the resistor is changed to 470 Ω.

Section 3–5
Analysis of Small-Signal Diode Circuits

3–18. Assume that the diode shown in Figure 3–44 has a 0.65-V drop across it.

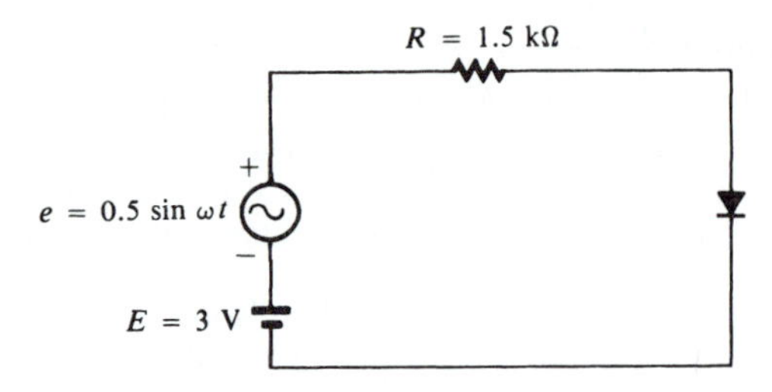

Figure 3–44
(Exercise 3–18)

a. Find the dc current in the diode.
b. Find the ac resistance of the diode. Assume that the diode is at room temperature.
c. Write mathematical expressions (functions of time) for the total current through and voltage across the diode.
d. What are the minimum and maximum values of the current through the diode?

3–19. Repeat Exercise 3–18 when $E = 2.0$ V, $e = 0.25 \sin \omega t$, and $R = 1.25$ kΩ.

3–20. Figure 3–45 shows the I–V characteristic for the diode in the circuit of Figure 3–44.
a. Write the equation for the load line and draw it on the figure.
b. Graphically determine the diode voltage and current at the quiescent point.
c. Determine the dc resistance at the quiescent point.
d. Graphically determine the minimum and maximum values of the current through the diode.
e. Determine the ac resistance of the diode.

3–21. Repeat Exercise 3–20 for the circuit values given in Exercise 3–19.

Section 3–6
Analysis of Large-Signal Diode Circuits

3–22. In the circuit shown in Figure 3–18, the 1.5-kΩ resistor is replaced by a 2.2-kΩ resistor. Assume that the silicon diode has a characteristic curve like that shown in Figure 3–16(b). If $e(t) = 2 \sin \omega t$, find the peak value of the current $i(t)$ and the voltage $v_R(t)$ across the resistor. Sketch the waveforms for $e(t)$, $i(t)$, and $v_R(t)$.

3–23. The silicon diode in Figure 3–46 has a characteristic curve like that shown in Figure 3–16(b). Find the peak values of the current

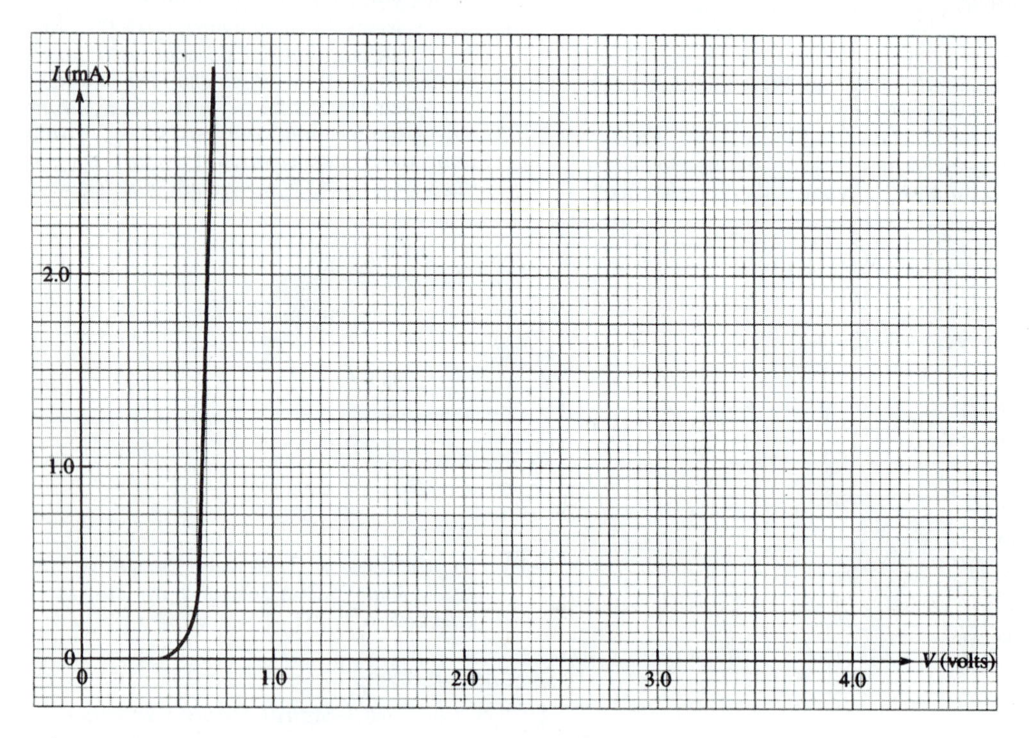

Figure 3–45
(Exercise 3–20)

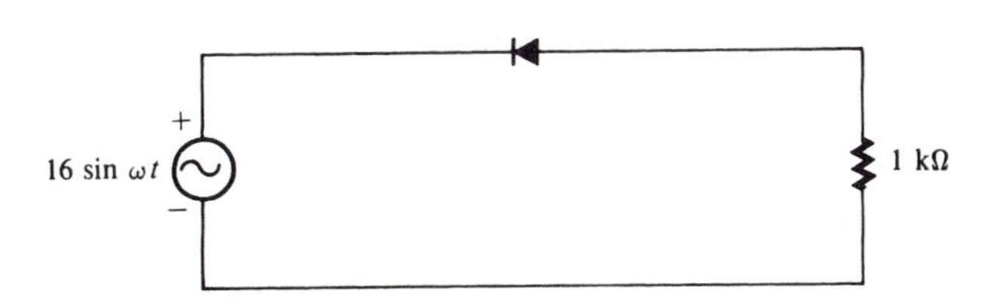

Figure 3–46
(Exercise 3–23)

$i(t)$ and the voltage $v_R(t)$ across the resistor. Sketch the waveforms for $e(t)$, $i(t)$, and $v_R(t)$.

3–24. The half-wave rectifier in Figure 3–22 has a 250-μF filter capacitor and a 1.5-kΩ load. The ac source is 120 V rms with frequency 60 Hz. The voltage drop across the silicon diode is 0.7 V. Assuming light loading, find
a. the dc value of the load voltage;
b. the peak-to-peak value of the ripple voltage.

3–25. The half-wave rectifier in Exercise 3–24 is replaced by a silicon full-wave rectifier, and a second 250-μF capacitor is connected in parallel with the filter capacitor. Assume light loading and do not neglect the voltage drop across the diodes. Find
a. the dc value of the load voltage;
b. the peak-to-peak value of the ripple voltage.
c. If the load resistance is decreased by a factor of 2, determine (without recalculating) the approximate factor by which the ripple voltage is changed.

Section 3–7
Diode Switching Circuits

3–26. In the circuit of Figure 3–25, the voltage source is replaced by a square-wave generator whose output alternates between +2.5 V and −2.5 V. If the diode is germanium and $R = 330\ \Omega$, find the peak voltage across and current through the resistor. Sketch $v_R(t)$ and $i(t)$.

3–27. Determine which of the diodes in Figure 3–47 are forward biased and which are reverse biased.

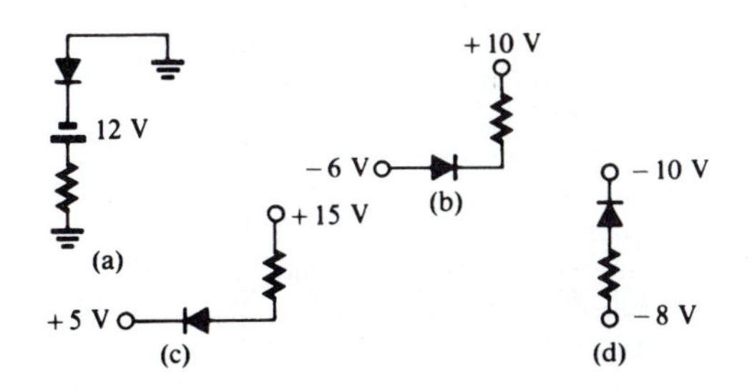

Figure 3–47
(Exercise 3–27)

3–28. Determine which of the diodes shown in Figure 3–48 are forward biased and which are reverse biased.

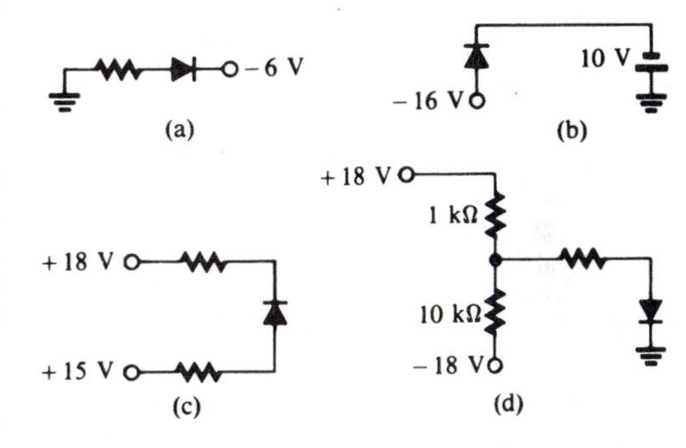

Figure 3–48
(Exercise 3–28)

3–29. The inputs A and B in Figure 3–49 can be either 0 V or +10 V. Each diode is silicon and has resistance 400 Ω when it is forward biased. Find V_o for each of the following cases:
a. $A = 0$ V, $B = 0$ V
b. $A = 0$ V, $B = +10$ V
c. $A = +10$ V, $B = 0$ V
d. $A = +10$ V, $B = +10$ V

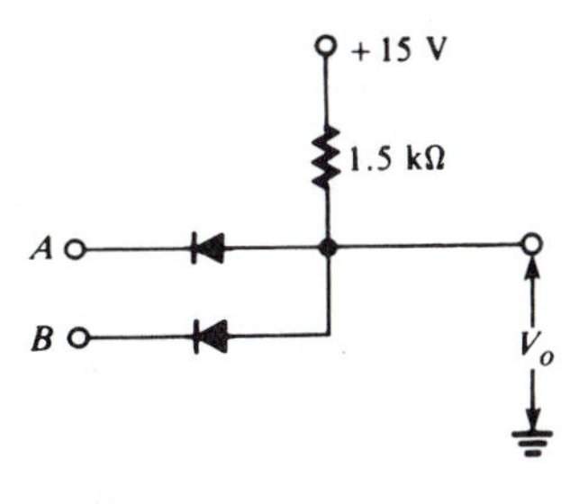

Figure 3–49
(Exercise 3–29)

3–30. The inputs A, B, and C in Figure 3–50 can be either +10 V or −5 V. Each silicon diode

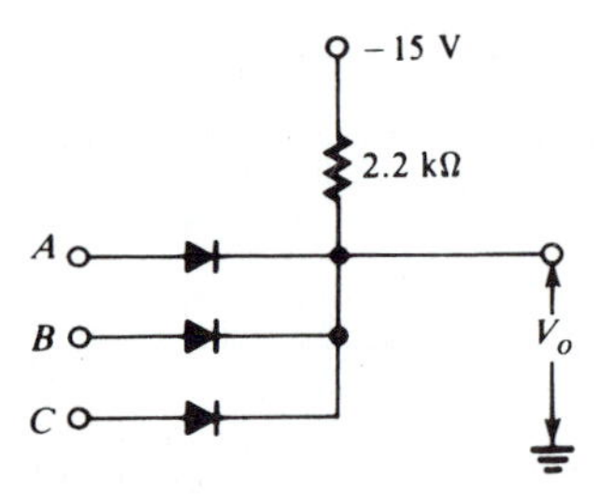

Figure 3–50
(Exercise 3–30)

has resistance 1200 Ω when it is forward biased. Find V_o when

a. $A = B = C = -5$ V
b. $A = C = -5$ V, $B = +10$ V
c. $A = B = +10$ V, $C = -5$ V
d. $A = B = C = +10$ V

3–31. In the circuit of Exercise 3–30, A, B, and C can be either 0 V or −5 V. Assuming that the forward $V_D = 0.7$ V, find V_o when

a. $A = B = C = 0$ V
b. $A = B = 0$ V, $C = -5$ V
c. $A = C = -5$ V, $B = 0$ V
d. $A = B = C = -5$ V

3–32. In the circuit of Exercise 3–29, A and B can be either 0 V or −5 V. Assuming that the forward $V_D = 0.7$ V, find V_o when

a. $A = B = -5$ V
b. $A = -5$ V, $B = 0$ V
c. $A = 0$ V, $B = -5$ V
d. $A = B = 0$ V

Section 3–8

Diode Types, Ratings, and Specifications

3–33. In the circuit shown in Example 3–9 (Figure 3–37), suppose the load resistor R is changed to 47 Ω. What then is the least expensive of the diodes listed in the example that can be used in this application?

3–34. In the circuit shown in Example 3–9 (Figure 3–37), suppose the ac voltage is 100 V rms and the load resistor is changed to 68 Ω. What then is the least expensive of the diodes listed in the example that can be used in this application?

3–35. A small-signal diode is to be used in an application where it will be subjected to a reverse voltage of 35 V. It must conduct a forward current of 0.01 A when the forward-biasing voltage is 1.0 V. The reverse current must not exceed 30 mA when the reverse voltage is 30 V. Select a diode type number from Figure 3–38 that meets these requirements.

3–36. A silicon diode is to be used in an application where it will be subjected to a reverse-biasing voltage of 85 V. The forward current will not exceed 100 mA, but it must have a 0.5-W power dissipation rating. Select a diode type number from Figure 3–38 that meets these requirements.

3–37. A rectifier diode is to be used in a power supply design where it must repeatedly withstand sine wave reverse voltages of 250 V rms and must conduct 0.6 A (average) of forward current. The forward surge current through the diode when the supply is first turned on will be 25 A. It is estimated that the diode case temperature (T_c) will be 30°C. Select a diode type number from Figure 3–39 that meets these requirements.

3–38. A rectifier diode is to be used in a large power supply where it must be capable of withstanding repeated reverse voltages of 450 peak volts. The forward current in the diode will average 13.5 A. Select a diode type number from Figure 3–39 that meets these requirements.

DESIGN EXERCISES

3–39. The silicon diode in Figure 3–6 requires a minimum current of 5 mA to be above the knee of its I–V characteristic. If the dc voltage source is changed to 10 V, what is the largest value of resistance R that could be used in the circuit if operation must be at or above the knee?

3–40. In the circuit shown in Figure 3–7(b), $E = 9$ V, $R = 1$ kΩ, and $e = A \sin 1000t$ V. The circuit must be operated so that the peak value of the ac voltage drop across the diode is 10 mV. If the dc drop across the diode is fixed at 0.7 V, what should be the peak value (A) of the ac signal e?

3–41. In the rectifier circuit shown in Figure 3–17, $e(t) = A \sin 2\pi \times 100t$ V. Assume the voltage drop across the diode when it is conducting is 0.7 V. If conduction must begin during each positive half-cycle at an angle no greater than 5°, what is the minimum peak value A that the ac source must produce?

3–42. A full-wave rectifier with a capacitor filter supplies a dc value of 60 V to a 2-kΩ load. The ac voltage source driving the rectifier has frequency 60 Hz. What minimum value of filter capacitance must be used if the peak-to-peak ripple voltage cannot exceed 1 V?

3–43. In the diode AND gate shown in Figure 3–49 (Exercise 3–29), each diode has a dc resistance of 100 Ω when conducting. The inputs can be either 0 V or +5 V. Under all conditions in which the output voltage is supposed to be low, its value can be no greater than 1 V. What is the smallest value of resistance that can be used to replace the 1.5-kΩ resistance in the circuit?

SPICE EXERCISES

Note: In the exercises that follow, assume that all device parameters have their default values, unless otherwise specified.

3–44. Use SPICE to obtain a plot of the diode current in Figure 3–15(a) versus time when the circuit is modified as follows: The dc voltage source is changed to 10 V, the ac voltage source is changed to $2 \sin(2\pi \times 1000t)$ V, and the resistor is changed to 620 Ω. The plot should cover at least one full cycle of the ac signal. Compare the peak value of the current computed by SPICE with that predicted by equation 3–8.

3–45. Use SPICE to obtain plots of the voltages, versus time, across the diode and across the resistor in the modified circuit described in Exercise 3–44.

3–46. Use SPICE to obtain a plot of voltage versus time across R in Figure 3–17 when $R = 1$ kΩ and

a. $e(t) = 2 \sin(2\pi \times 100t)$ V
b. $e(t) = 40 \sin(2\pi \times 100t)$ V.

Each plot should cover at least one full cycle of $e(t)$. Use the results to determine the peak value of the output in each case and comment on the validity of neglecting the drop across the diode in each case.

3–47. Use SPICE to obtain a plot of the diode current versus time in Figure 3–17 when $R = 500$ Ω and $e(t) = 1.5 \sin 6283t$ V. Use the results to determine the first *angle* (after 0°, at $t = 0$) at which significant conduction begins, assuming that any current less than 90 μA is essentially nonconduction.

3–48. Use SPICE to determine the dc output of the diode AND gate in Figure 3–28(a) when $R = 1$ kΩ, $V = +10$ V, and

a. $A = 0$ V, $B = 0$ V, $C = 0$ V
b. $A = 0$ V, $B = 0$ V, $C = +5$ V
c. $A = 0$ V, $B = +5$ V, $C = +5$ V
d. $A = +5$ V, $B = +5$ V, $C = +5$ V

Set the "ohmic resistance" parameter of the diode to 100 Ω. Do the results confirm that the circuit behaves like an AND gate, that is, that the output is high only when all inputs are high?

4 BIPOLAR JUNCTION TRANSISTORS

4–1 INTRODUCTION

The workhorse of modern electronic circuits, both discrete and integrated, is the *transistor.* The importance of this versatile device stems from its ability to produce *amplification,* or *gain,* in a circuit. We say that amplification has been achieved when a small variation in voltage or current is used to create a large variation in one of those same quantities, and this is the fundamental goal of most electronic circuits. As a means for creating gain, the transistor is in many ways analogous to a small valve in a large water system: By expending a small amount of energy (turning the valve), we are able to control—increase or decrease—a large amount of energy (in the flow of a large quantity of water). When a device such as a transistor is used to create gain, we supply a small signal to it and refer to that as the *input;* the large current or voltage variations that then occur at another point in the device are referred to as the *output.*

The two most important kinds of transistors are the *bipolar junction* type and the *field-effect* type. The bipolar junction transistor (BJT) is so named (*bi polar*—two polarities) because of its dependence on both holes and electrons as charge carriers. We will study the theory and applications of "bipolars" in this and the next chapter, and devote equal time to field-effect transistors (FETs), which operate under completely different principles, in later chapters.

We should note that bipolars are studied first for historical reasons, and because the theory of these devices follows naturally from a study of PN junctions. Their precedence in our study should in no way infer that they are of greater importance than field-effect transistors. Bipolars were the first kind of transistors to be widely used in electronics, and they are still an important segment of the semiconductor industry. Most people in the industry still use the word *transistor* (as we shall) with the understanding that a bipolar transistor is meant. However, field-effect technology has now evolved to the point where FETs are at least as important as bipolars, and are, in fact, used in greater numbers than BJTs in integrated circuits for digital applications.

4–2 THEORY OF BJT OPERATION

A bipolar junction transistor is a specially constructed, *three-terminal* semiconductor device containing *two* PN junctions. It can be formed from a bar of material that has been doped in such a way that it changes from N to P and back to N, or from P to N and back to P. In either case, a junction is created at each of the two boundaries where the material changes from one type to the other. Figure 4–1 shows the two ways that it is possible to alternate material types and thereby obtain two junctions.

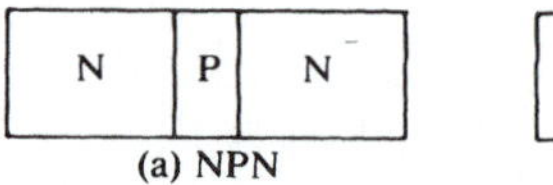

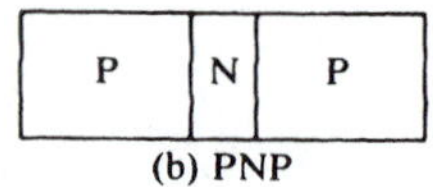

Figure 4–1
NPN and PNP transistor construction

When a transistor is formed by sandwiching a single P region between two N regions, as shown in Figure 4–1(a), it is called an *NPN* type. Figure 4–1(b) shows the *PNP* type, containing a single N region between two P regions.

The middle region of each transistor type is called the *base* of the transistor. Of the remaining two regions, one is called the *emitter* and the other is called the *collector* of the transistor. Let us suppose that terminals are attached to each region so that external electrical connections can be made between them. (In integrated circuits, no such accessible terminals may be provided, but it is still possible to identify the base, emitter, and collector and to form conducting paths between those regions and other internal components.) Figure 4–2 shows terminals attached to the regions of each transistor type. The terminals are labeled according to the region to which they connect.

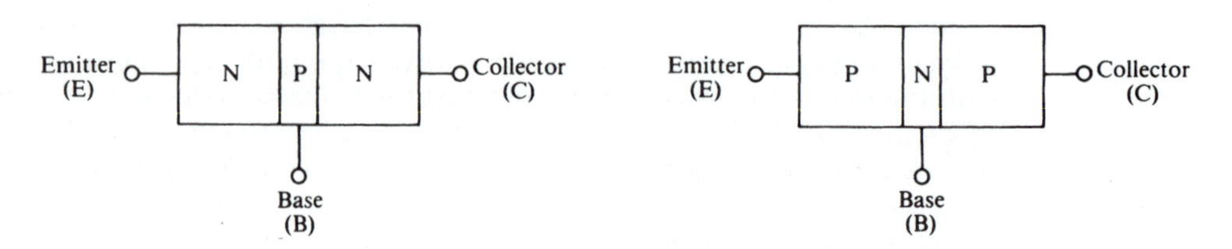

Figure 4–2
Base, emitter, and collector terminals of NPN and PNP transistors

The physical appearance of an actual transistor bears little resemblance to the figures we have shown so far. However, these kinds of diagrams are very helpful in understanding transistor theory, and Figures 4–1 and 4–2 are representative of actual transistors in at least one respect: the base region is purposely shown *thinner* than either the emitter or collector region. For reasons that will become evident soon, the base region in an actual transistor is made even thinner in proportion to the other regions than is depicted by the figures. Also, the base region is much more lightly doped than the other regions.* Both these characteristics of the base

* In integrated circuits, the emitter is usually heavily doped (N^+) material. Most integrated-circuit BJTs are of the NPN type.

are important for the transistor to be a useful device and to perform what is called normal "transistor action."

For the sake of clarity, we will continue our discussion of transistor theory in terms of the NPN type only. The underlying theory is equally applicable to the PNP type and can be "translated" for PNPs simply by changing each carrier type mentioned in connection with NPNs and reversing each voltage polarity. To obtain normal transistor action, it is necessary to bias both PN junctions by connecting dc voltage sources across them. Figure 4–3 illustrates the correct bias for each junction in the NPN transistor. In the figure, we show the result of biasing each junction separately, while in practice both junctions will be biased simultaneously by one external circuit, as will be described presently.

As shown in Figure 4–3(a), the emitter-base junction is forward biased by the dc source labeled V_{EE}. Note that the *N*egative terminal of V_{EE} is connected to the *N* side of the NP junction, as required for forward bias. Consequently, there is a substantial flow of diffusion current across the junction due to the flow of the majority carriers (electrons) from the N-type emitter. This action is exactly that which we discussed in connection with a forward-biased PN junction in Chapter 2. The depletion region at this junction is made narrow by the forward bias, as also described in Chapter 2. The width of the base region is exaggerated in the figure for purposes of clarity. When the majority electrons diffuse into the base, they become minority carriers in that P-type region. We say that minority carriers have been *injected* into the base.

Figure 4–3(b) shows that the collector–base junction is reverse biased by the dc source labeled V_{CC}. The *P*ositive terminal of V_{CC} is connected to the *N*-type collector. As a result, the depletion region at this junction is widened, and the only current that flows from base to collector is due to the minority electrons crossing the junction from the P-type base. Recall from Chapter 2 that *minority* carriers readily cross a reverse-biased junction under the influence of the electric field, and they constitute the flow of reverse current in the junction.

Figure 4–4 shows the NPN transistor the way it is biased for normal operation, with both dc sources V_{EE} and V_{CC} connected simultaneously. Note in the figure that the negative terminal of V_{CC} is connected to the positive terminal of V_{EE} and that both of these are joined to the base. The base is then the "ground," or common point of the circuit and can therefore be regarded as being at 0 V. The emitter is

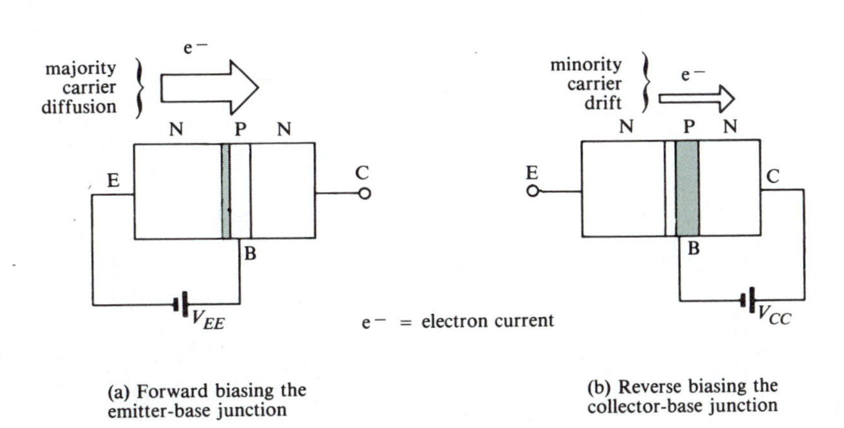

Figure 4–3
Biasing the two PN junctions in an NPN transistor

Figure 4–4
The NPN transistor with both bias sources connected

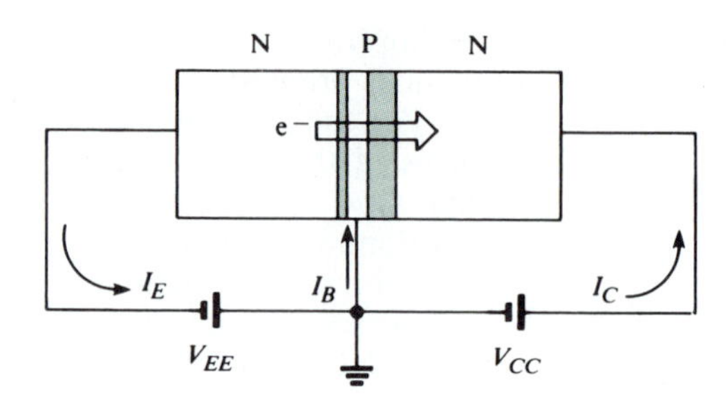

negative *with respect to the base,* and the collector is positive *with respect to the base.* These are the conditions we require in order to forward bias the emitter–base junction and to reverse bias the collector–base junction.

Since the base region is very thin and is lightly doped (so there are relatively few holes in it), very few of the electrons injected into the base from the emitter recombine with holes. Instead, they diffuse to the reverse-biased base–collector junction and are swept across that junction under the influence of the electric field established by V_{CC}. Remember, again, that the electrons injected into the base are the minority carriers there, and that minority carriers readily cross the reverse-biased junction. We conclude that electron flow constitutes the dominant current type in an NPN transistor. For a PNP transistor, in which everything is "opposite," hole current is the dominant type.

Despite the fact that most of the electrons injected into the base cross into the collector, a few of them do combine with holes in the base. For each electron that combines with a hole, an electron leaves the base region via the base terminal. This action creates a very small base current, about 2% or less of the electron current from emitter to collector. As we shall see, the smaller this percentage, the more useful the transistor is in practical applications.

Note in Figure 4–4 that arrows are drawn to indicate the direction of *conventional* current in the NPN transistor. Of course, each arrow points in the opposite direction from the electron flows that we have described. Conventional current flowing from V_{CC} into the collector is called *collector current* and designated I_C. Similarly, current into the base is I_B, the *base current,* and current from V_{EE} into the emitter is *emitter current,* I_E. Figure 4–5(a) shows the standard electronic symbol for an NPN transistor, with these currents labeled alongside. Figure 4–5(b) shows the same block form of the NPN that we have shown earlier and is included as an aid for relating the physical device to the symbol. Figure 4–6 shows the standard symbol for a PNP transistor and its equivalent block form. Comparing Figures 4–5 and 4–6, we note first that the emitter of an NPN transistor is represented by an

Figure 4–5
Equivalent NPN transistor diagrams

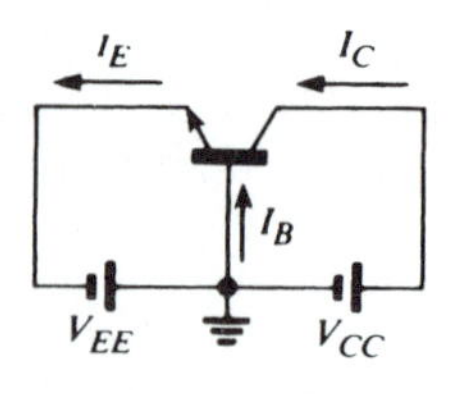

(a) NPN transistor symbol and bias currents

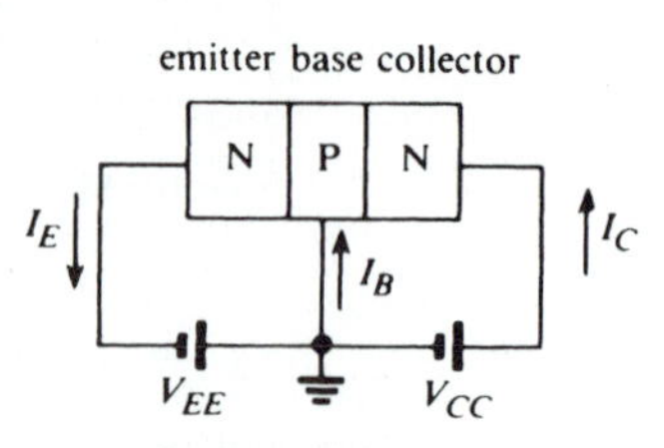

(b) The block form of the NPN transistor corresponding to (a)

Figure 4–6
Equivalent PNP transistor diagrams

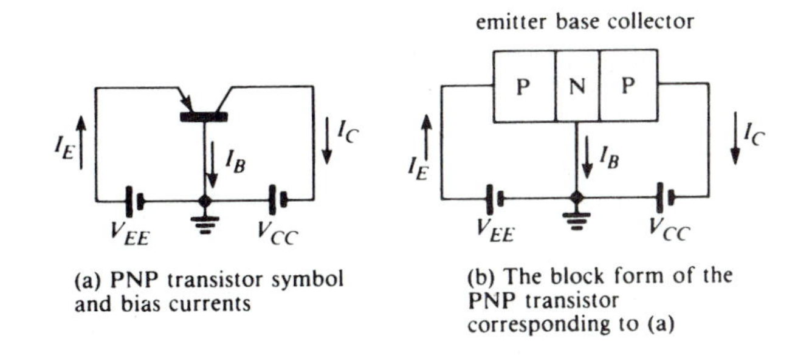

(a) PNP transistor symbol and bias currents

(b) The block form of the PNP transistor corresponding to (a)

arrow pointing out from the base, whereas the emitter of a PNP transistor is shown as an arrow pointing into the base. It is easy to remember this distinction by thinking of the arrow as pointing in the direction of conventional current flow, out of or into the emitter of each type of transistor. We further note that the polarities of the bias sources for the PNP transistor, V_{EE} and V_{CC}, are the opposite of those for the NPN transistor. In other words, the positive and negative terminals of each source in Figure 4–6 are the reverse of those in Figure 4–5. These polarities are, of course, necessary in each case to maintain the forward and reverse biasing of the junctions, as we have described. Note, for example, that the negative terminal of V_{CC} is connected to the P-type collector of the PNP transistor. Recapitulating, here is the all-important universal rule for biasing transistors for normal operation (memorize it!): *The emitter–base junction must be forward biased, and the collector–base junction must be reverse biased.*

To emphasize and clarify an important point concerning transistor currents, Figure 4–7 replaces each type of transistor by a single block and shows the directions of currents entering and leaving each. Applying Kirchhoff's current law to each of Figures 4–7(a) and (b), we immediately obtain this important relationship, applicable to both NPN and PNP transistors:

$$I_E = I_C + I_B \tag{4–1}$$

I_{CBO} Reverse Current

Recall from Chapter 2 that a small reverse current flows across a PN junction due to *thermally* generated minority carriers that are propelled by the barrier potential. When the junction is reverse biased, this reverse current increases slightly. For moderate reverse-bias voltages, the reverse current reaches its saturation value I_s. Since the collector–base junction of a transistor is reverse biased, there is likewise a reverse current due to thermally generated carriers. Of course this "reverse" current, in the context of a transistor, is in the same direction as the main (collector) current flowing through the device due to the injection of minority carriers into the base. The total collector current is, therefore, the *sum* of these two components: the injected minority carriers and the thermally generated minority carriers.

Suppose that the external connections between the base and emitter are left open and that the collector–base junction has its normal reverse bias, as shown in

Figure 4–7
Each transistor type is replaced by a single block to highlight current flows in and out of the devices.

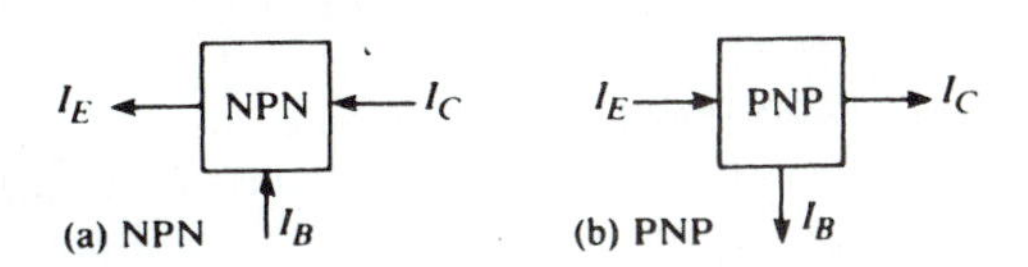

(a) NPN

(b) PNP

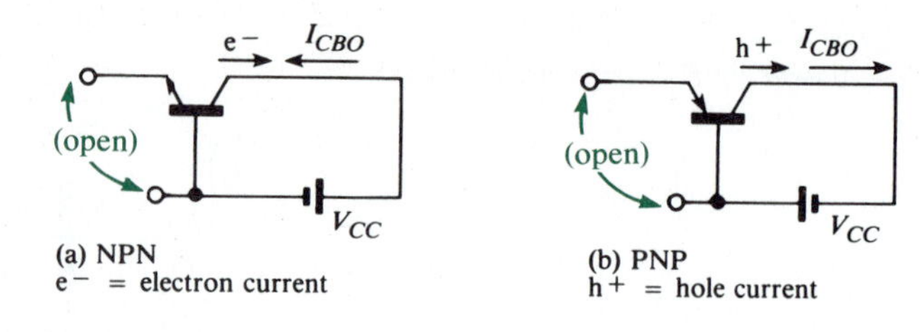

Figure 4–8
I_{CBO} is the collector current that flows when the emitter is open.

Figure 4–8. (See also Figure 4–3(b).) Since the emitter is open, there can be no carriers injected into the base. Consequently, the only current that flows must be that "reverse" component due to thermally generated carriers. This current is designated I_{CBO}, the *C*ollector-to-*B*ase current, with the emitter *O*pen. Therefore, in normal operation, with the emitter circuit connected, the total collector current can be expressed as

$$I_C = I_{C(INJ)} + I_{CBO} \tag{4–2}$$

where $I_{C(INJ)}$ is the component of collector current due to carriers injected into the base.

An important transistor parameter, called *alpha,* is defined as the ratio of the collector current resulting from carrier injection to the total emitter current:

$$\alpha = \frac{I_{C(INJ)}}{I_E} \tag{4–3}$$

Thus, α measures the portion of the emitter current that "survives," after passage through the base, to become collector current. Clearly, α will always be less than 1, since some of the emitter current is drained off in the base through recombinations. Generally speaking, the greater the value of α (the closer it is to 1), the better the transistor, from the standpoint of many practical applications that we will explore later. In other words, we want a transistor to be constructed so that its base current is as small as possible, since that makes $I_{C(INJ)}$ close to I_E and α close to 1. Typical transistors have values of α that range from 0.95 to 0.995.

From equation 4–3, we have $I_{C(INJ)} = \alpha I_E$. Substituting in equation 4–2, we obtain

$$I_C = \alpha I_E + I_{CBO} \tag{4–4}$$

This equation states that the total collector current is that portion of the emitter current that makes it through the base (αI_E) plus the thermally generated collector current (I_{CBO}).

In modern transistors, particularly silicon, I_{CBO} is so small that it can be neglected for most practical applications. Remember that I_{CBO} is exactly the same as the reverse diode current we discussed in Chapter 2. We saw that its theoretical value, as a function of temperature and voltage, is given by equation 2–14, and that it is quite sensitive to temperature variations. Since the collector–base junction in a transistor is normally reverse biased by at least a volt or so, the theoretical value of I_{CBO} is for all practical purposes equal to its saturation value (I_s, in Chapter 2). Remember that I_s approximately doubles for every 10°C rise in temperature; so we can say the same about I_{CBO} in a transistor. This sensitivity to temperature can become troublesome in some circuits, if high temperatures and large power dissipations are likely. We will explore those situations in more detail later.

We should also remember that the theoretical value of I_{CBO}, like reverse diode current, is usually much smaller than the reverse *leakage* current that flows across the surface. In silicon transistors, this surface leakage may so completely dominate the reverse current that temperature-related increases in I_s remain negligible. In fact, it is conventional in most texts and product literature to refer to I_{CBO} as the (collector-to-base) *leakage current.* Finally, we should note that in some literature the (unfortunate) notation I_{CO} is used to mean I_{CBO}.

Since I_{CBO} is negligibly small in most practical situations, we can set it equal to 0 in equation 4–4 and obtain the good approximation $I_C \approx \alpha I_E$, or

$$\alpha \approx \frac{I_C}{I_E} \tag{4–5}$$

Example 4–1

The emitter current in a certain NPN transistor is 8.4 mA. If 0.8% of the minority carriers injected into the base recombine with holes and the leakage current is 0.1 μA, find (1) the base current, (2) the collector current, (3) the exact value of α, and (4) the approximate value of α, neglecting I_{CBO}.

Solution

1. I_B = (0.8% of I_E) = (0.008)(8.4 mA) = 67.2 μA.
2. From equation 4–1, $I_C = I_E - I_B$ = 8.4 mA − 0.0672 mA = 8.3328 mA.
3. From equation 4–2, $I_{C(INJ)} = I_C - I_{CBO} = 8.3328 \times 10^{-3}\text{A} - 10^{-7}\text{A}$ = 8.3327 mA. By equation 4–3, $\alpha = I_{C(INJ)}/I_E$ = (8.3327 mA)/(8.4 mA) = 0.9919881.
4. By approximation 4–5, $\alpha \approx I_C/I_E$ = (8.3328 mA)/(8.4 mA) = 0.992.

For the conditions of this example, we see that the exact and approximate values of α are so close that the difference between them can be entirely neglected.

4–3 COMMON-BASE CHARACTERISTICS

In our introduction to the theory of transistor operation, we showed a bias circuit (Figure 4–4) in which the base was treated as the ground, or "common" point of the circuit. In other words, all voltages (collector-to-base and emitter-to-base) were *referenced* to the base. This bias arrangement results in what is called the *common-base* (CB) configuration for the transistor. It represents only one of three possible ways to arrange the external circuit to achieve a forward-biased base-to-emitter junction and a reverse-biased collector-to-base junction, since any one of the three terminals can be made the common point. We will study the other two configurations in later discussions.

The significance of having a common point in a transistor circuit is that it gives us a single reference for both the *input* voltage to the transistor and the *output* voltage. In the CB configuration, the emitter–base voltage is regarded as the input voltage and the collector–base voltage is regarded as the output voltage. See Figure 4–9. For an NPN transistor, V_{BE} is positive, and for a PNP transistor, V_{EB} is positive. Similarly, V_{CB} is a positive output voltage for an NPN, and V_{BC} is positive for a PNP. Emitter current is input current and collector current is output current.

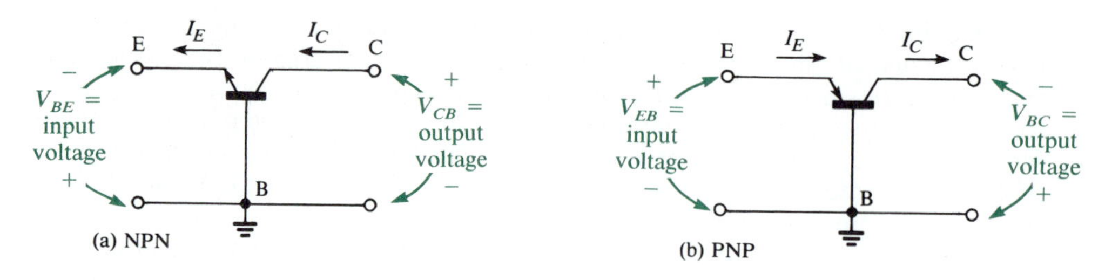

Figure 4–9
Input and output voltages in NPN and PNP common-base transistors

In our analysis of the CB configuration, the "input" voltage will of course be the same as the emitter–base *bias* voltage (V_{EE}), and the "output" voltage will be the same as the collector–base bias voltage (V_{CC}). In Chapter 5, we will adopt a more realistic viewpoint in which we will regard small (ac) variations in the emitter–base and collector–base voltages as the input and output, respectively. For the time being, we will concern ourselves only with the effects of changes in V_{EE} and V_{CC} on the behavior of the transistor. Do not be confused by the fact that the "input" current in the NPN circuit (Figure 4–9(a)) flows *out* of the emitter. Again, it will be small changes in the magnitude of I_E that we will ultimately regard as the "input."

Our objective now is to learn how the input and output voltages and the input and output currents are related to each other in a CB configuration. Toward that end, we will develop sets of characteristic curves called *input* characteristics and *output* characteristics. The input characteristics show the relation between input current and input voltage *for different values of output voltage,* and the output characteristics show the relation between output current and output voltage *for different values of input current.* As these statements suggest, there is in a transistor a certain *feedback* (the output voltage) that affects the input, and a certain "feed-forward" (the input current) that affects the output.

Common-Base Input Characteristics

Let us begin with a study of the CB input characteristics of a typical NPN transistor. Since the input is across the forward-biased base-to-emitter junction, we would expect a graph of input current (I_E) versus input voltage (V_{BE}) to resemble that of a forward-biased diode. That is indeed the case. However, the exact shape of this I_E–V_{BE} curve will, as we have already hinted, depend on the reverse-biasing output voltage, V_{CB}. The reason for this dependency is that the greater the value of V_{CB}, the more readily minority carriers in the base are swept through the base-to-collector junction. (Remember that the reverse-biasing voltage enhances such current.) The increase in emitter-to-collector current resulting from an increase in V_{CB} means that the (input) emitter current will be greater for a given value of (input) base-to-emitter voltage. Figure 4–10 shows a typical set of input characteristics in which this feedback effect can be discerned. Figure 4–10 is our first example of a *family* of transistor curves, a very useful way to display transistor behavior graphically and one that can provide rewarding insights if studied carefully. Although characteristic curves are seldom used in actual design or analysis problems, they convey a wealth of information and we will see many more of them in the future. Each set should be scrutinized and dwelled upon at length. Try to visualize how currents and/or voltages change when one quantity is held constant and the others are varied. Note

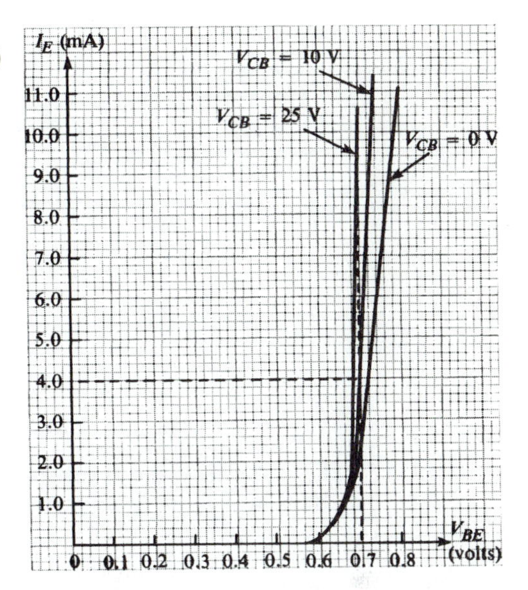

Figure 4–10
Common-base input characteristics (NPN)

that a family of curves can show the relationships among *three* variables: two represented by the axes and the third represented by each curve. In Figure 4–10, each curve corresponds to a different value of V_{CB} and therefore each shows how emitter current varies with base-to-emitter voltage for a fixed value of V_{CB}. A good way to view this family is to think of an experiment in which the reverse-biasing voltage V_{CB} is fixed and a set of measurements of I_E is made for different settings of V_{BE}. Plot these results, and then set V_{CB} to a new value and repeat the measurements. Each time V_{CB} is set to a new value, a new curve is obtained.

Note in Figure 4–10 that each curve resembles a forward-biased diode characteristic, as expected. For a given value of V_{BE}, it can be seen that I_E increases with increasing V_{CB}. To illustrate, the vertical dashed line shown in the figure is a line of constant V_{BE} ($V_{BE} = 0.7$ V) that intersects the curve corresponding to $V_{CB} = 10$ V at $I_E = 4$ mA. Following the line upward, it is clear that I_E has a larger value at $V_{CB} = 25$ V. This variation has already been explained in terms of the way V_{CB} promotes minority carrier flow. We see in the figure that there is actually little difference in the shapes of the curves as V_{CB} is changed over a fairly wide range. For that reason, the effect of V_{CB} on the input is often neglected in practical problems. An "average" forward-biased diode characteristic is assumed.

The CB input characteristics for a PNP transistor will of course have the same general appearance as those shown for an NPN in Figure 4–10. However, in a PNP transistor, a forward-biasing input voltage is positive when measured *from* emitter *to* base. Therefore, the positive horizontal scale would be labeled V_{EB} rather than V_{BE} (see Figure 4–9). Some data sheets show negative values for PNP voltages and/or currents because these quantities have directions that are the opposite of the corresponding NPN quantities. For example, if the horizontal axis in Figure 4–10 were labeled V_{BE} for a PNP transistor, then all scale values would be negative. These sign conventions (rather, this *lack* of consistency) can be confusing but can always be resolved by remembering the fundamental rule for transistor bias: base–emitter forward, and base–collector reverse. Finally, we should mention that some authors and some data sheets refer to the CB input characteristics as the *emitter characteristics* of a transistor.

Example 4–2

The transistor shown in Figure 4–11 has the characteristic curves shown in Figure 4–10. When V_{CC} is set to 25 V, it is found that $I_C = 8.94$ mA.

Figure 4–11
(Example 4–2)

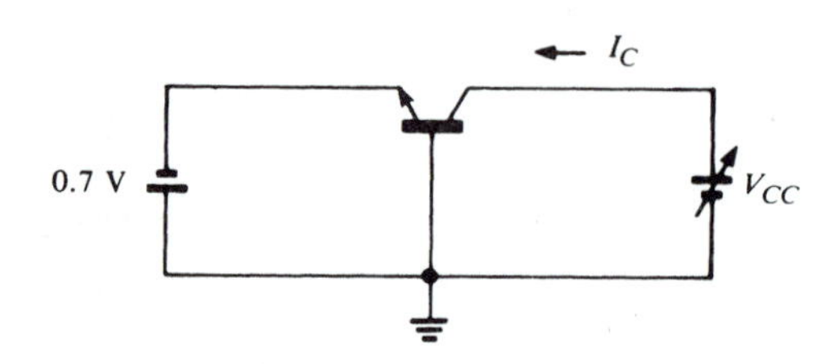

1. Find the α of the transistor (neglecting I_{CBO}).
2. Repeat if $I_C = 1.987$ mA when V_{CC} is replaced by a short circuit to ground.

Solution

1. In Figure 4–11, we see that $V_{BE} = 0.7$ V. From Figure 4–10, the vertical line corresponding to $V_{BE} = 0.7$ V intersects the $V_{CB} = 25$ V curve at $I_E = 9.0$ mA. Therefore, $\alpha \approx I_C/I_E = (8.94 \text{ mA})/(9.0 \text{ mA}) = 0.9933$.
2. When V_{CC} is replaced by a short circuit, we have $V_{CB} = 0$. From Figure 4–10, $I_E = 2$ mA at $V_{CB} = 0$ V and $V_{BE} = 0.7$ V. Therefore, $\alpha \approx I_C/I_E = (1.987 \text{ mA})/(2.0 \text{ mA}) = 0.9935$.

Common-Base Output Characteristics

Consider now an experiment in which the collector (output) current is measured as V_{CB} (the output voltage) is adjusted for fixed settings of the emitter (input) current. Figure 4–12 shows a schematic diagram and a procedure that could be used to conduct such an experiment on an NPN transistor. Understand that Figure 4–12 does not represent a practical circuit that could be used for any purpose other than investigating transistor characteristics. Practical transistor circuits contain resistors and have input and output voltages that are different from the dc bias voltages. However, at this point in our study of transistor theory we are interested in the *transistor* itself. We are using characteristic curves to gain insights to how the voltages and currents relate to each other in the *device,* rather than in the external circuit. Once we have gleaned all the device information we can from studying characteristic curves, we will have a solid understanding of what a transistor really is and can proceed to study practical circuits. When I_C is plotted versus V_{CB} for different values of I_E, we obtain the family of curves shown in Figure 4–13: the *output* characteristics for the CB configuration. A close examination of these curves will reveal some new facts about transistor behavior.

We note first in Figure 4–13 that each curve starts at $I_C = 0$ and rises rapidly for a small positive increase in V_{CB}. In other words, I_C increases rapidly just as V_{CB} begins to increase slightly beyond its initial negative value. Since each curve represents a fixed value of I_E, this means that while I_C is increasing, the ratio I_C/I_E must also be increasing. But I_C/I_E equals α, so the implication is that the value of α for a transistor is not constant. Alpha starts at 0 and increases as V_{CB} increases. The reason for this fact is that a very small portion of the emitter current is able

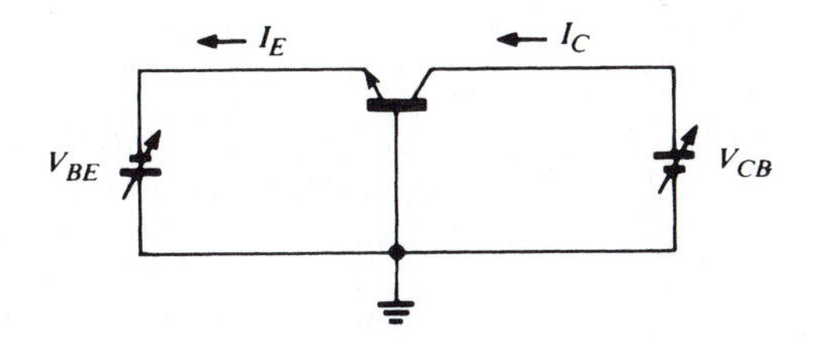

I_E	V_{CB}	I_C
1 mA	−1	
1 mA	0	
1 mA	5	
1 mA	10	
1 mA	15	
1 mA	20	

I_E	V_{CB}	I_C
2 mA	−1	
2 mA	0	
2 mA	5	
2 mA	10	
2 mA	15	
2 mA	20	

— etc. —

I_E	V_{CB}	I_C
9 mA	−1	
9 mA	0	
9 mA	5	
9 mA	10	
9 mA	15	
9 mA	20	

1. With V_{CB} set to −1 V, adjust V_{BE} to obtain $I_E = 1$ mA. Measure and record I_C.
2. Increase V_{CB} positively in small steps, each time measuring I_C. Adjust V_{BE} as necessary to maintain the original value of I_E. Continue until V_{CB} has reached 20 V. Plot I_C versus V_{CB}.
3. Repeat step 1, with V_{BE} adjusted to produce a new, slightly larger value of I_E. Then repeat step 2.

Repeat step 3 until the fixed value of I_E has reached 9 mA.

Figure 4–12
An experiment that could be used to produce the output characteristics shown in Figure 4–13

to enter the collector region until the reverse-biasing voltage V_{CB} is allowed to reach a value large enough to propel all carriers across the junction. When V_{CB} is negative, the junction is actually forward biased, and minority carrier flow is inhibited. The proportion of the carriers that are swept across the junction (α) depends directly on the value of V_{CB} until V_{CB} no longer forward biases the junction. The portion of the plot where V_{CB} is negative is called the *saturation region* of the transistor. By definition (no matter what the transistor configuration), a transistor is *saturated* when *both* its collector-to-base junction *and* emitter-to-base junction are forward biased.

Once V_{CB} reaches a value large enough to ensure that a large portion of carriers enter the collector (close to 0 in the figure), we see that the curves more or less level off. In other words, for a fixed emitter current, the collector current remains essentially constant for further increases in V_{CB}. Note that this essentially constant value of I_C is, for each curve, very nearly equal to the value of I_E represented by the curve. In short, the ratio I_C/I_E, or α, is very close to 1 and is essentially constant. These observations correspond to what we had previously assumed about the nature of α, and the region of the plot where this is the case is called the *active region.* In its active region, a transistor exhibits those "normal" properties (transistor action) that we have associated with a forward-biased emitter–base junction and a reverse-biased collector–base junction. Apart from some special digital-circuit applications,

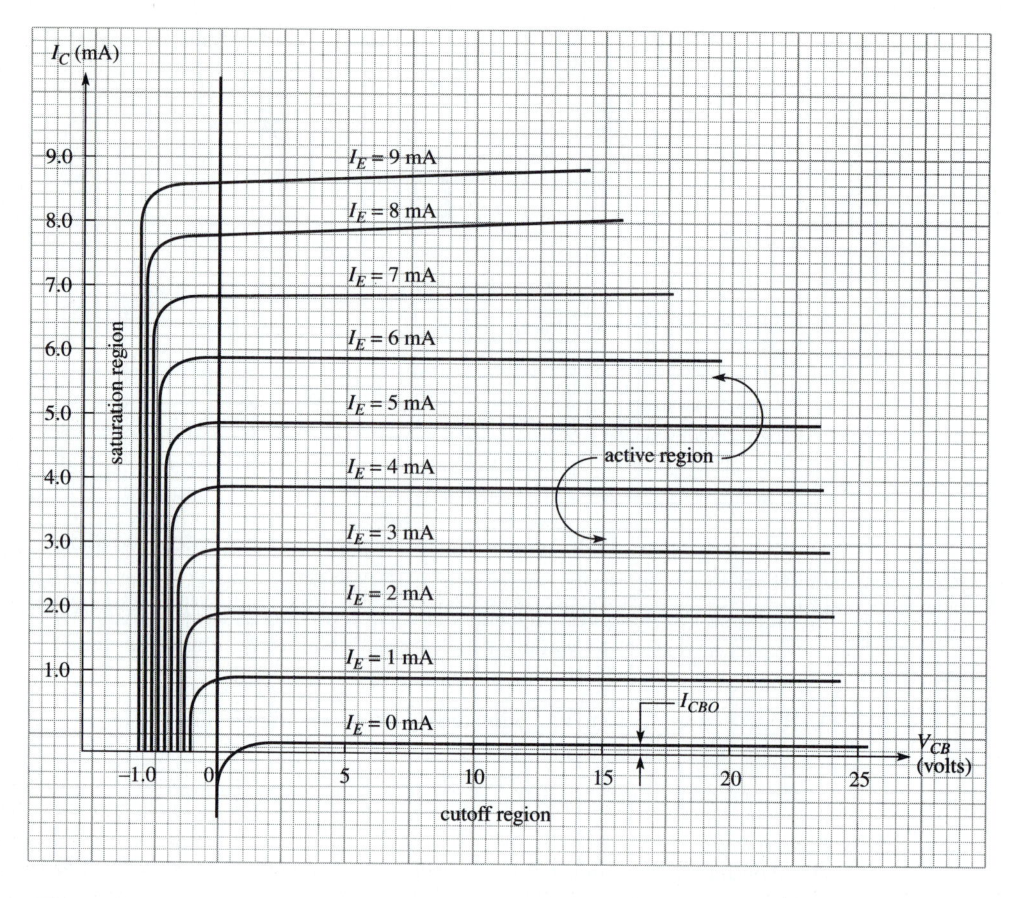

Figure 4–13
Common-base output characteristics (NPN). Note that the negative V_{CB} scale is expanded.

a transistor is normally operated (used) in its active region. Note that we can detect a slight rise in the curves as they proceed to the right through the active region. Each curve of constant I_E approaches a horizontal line that intersects the I_C-axis at a value equal to I_E, implying that I_E approaches I_C, and that α approaches 1, for increasing V_{CB}. This we attribute to the increased number of minority carriers swept into the collector, which increases the collector current, as the reverse-biasing value of V_{CB} is increased.

There is one other region of the output characteristics that deserves comment. Note that the curve corresponding to $I_E = 0$ is very close to the horizontal axis, i.e., to the $I_C = 0$ line. When the emitter current is made 0 (by opening the external emitter circuit), no minority carriers are injected into the base. Under those conditions, the only collector current that flows is the very small leakage current, I_{CBO}, as we have previously described (see Figure 4–8). With the scale used to plot the output characteristics in Figure 4–13, a horizontal line corresponding to $I_C = I_{CBO}$ coincides with the $I_C = 0$ line, for all practical purposes. The region of the output characteristics lying below the $I_E = 0$ line is called the *cutoff* region, because the collector current is essentially 0 (cut off) there. A transistor is defined to be cut off when *both* the collector–base and emitter–base junctions are reverse biased. Except for special digital circuits, a transistor is not normally operated in its cutoff region.

Example 4–3

A certain NPN transistor has the CB input characteristics shown in Figure 4–10 and the CB output characteristics shown in Figure 4–13.

1. Find its collector current when $V_{CB} = 10$ V and $V_{BE} = 0.7$ V.
2. Repeat when $V_{CB} = 5$ V and $I_E = 5.5$ mA.

Solution

1. From Figure 4–10, we find $I_E = 4$ mA at $V_{BE} = 0.7$ V and $V_{CB} = 10$ V. In Figure 4–13, the vertical line $V_{CB} = 10$ V intersects the $I_E = 4$ mA curve at approximately $I_C = 3.85$ mA. The collector current under these conditions is therefore 3.85 mA.
2. The conditions given require that we *interpolate* the output characteristics along the vertical line $V_{CB} = 5$ V, between $I_E = 5$ mA and $I_E = 6$ mA. The value of I_C that is halfway between the $I_E = 5$ mA and $I_E = 6$ mA curves is approximately 5.4 mA. Note that high accuracy is not possible when using characteristic curves in this way. In most practical situations, we could simply assume that $I_C = I_E =$ 5.5 mA without seriously affecting the accuracy of other computations.

Breakdown

As is the case in a reverse-biased diode, the current through the collector–base junction of a transistor may increase suddenly if the reverse-biasing voltage across it is made sufficiently large. This increase in current is typically caused by the avalanching mechanism already described in connection with diode breakdown. However, in a transistor it can also be the result of a phenomenon called *punch through.* Punch through occurs when the reverse bias widens the collector–base depletion region to the extent that it meets the base–emitter depletion region. This joining of the two regions effectively shorts the collector to the emitter and causes a substantial current flow. Remember that the depletion region extends farther into the lightly doped side of a junction, and that the base is more lightly doped than the collector. Furthermore, the base is made very thin, so the two junctions are already relatively close to each other. Punch through can be a limiting design factor in determining the doping level and base width of a transistor. Figure 4–14 shows

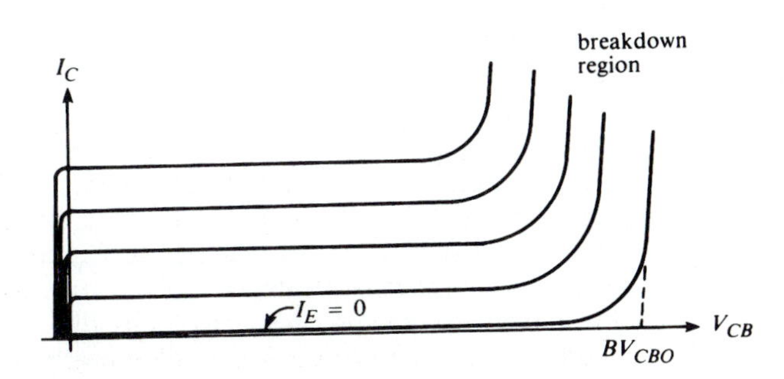

Figure 4–14
Common-base output characteristics showing the breakdown region

how the CB output characteristics appear when the effects of breakdown are included. Note the sudden upward swing of each curve at a large value of V_{CB}. The collector-to-base breakdown voltage when $I_E = 0$ (emitter open) is designated BV_{CBO}. As can be seen in Figure 4–14, breakdown occurs at progressively lower voltages for increasing values of I_E.

Although the base–emitter junction is not normally reverse biased, there are practical applications in which it is periodically subjected to reverse bias. Of course, it too can break down, and its reverse breakdown voltage is usually much less than that of the collector–base junction. Base–emitter breakdown is often destructive, so designers must be aware of the manufacturer's specified maximum reverse base–emitter voltage.

As a final note on transistor operation, we should mention that some transistors can be (and occasionally are) operated in what is called an *inverted* mode. In this mode, the emitter is used as the collector and vice versa. Normally, the emitter is the most heavily doped of the three regions, so unless a transistor is specifically designed for inverted operation, it will not perform well in that mode. The α in the inverted mode, designated α_I, is generally smaller than the α that can be realized in conventional operation.

4–4 COMMON-EMITTER CHARACTERISTICS

The next transistor bias arrangement we will study is called the *common-emitter* (CE) configuration. It is illustrated in Figure 4–15. Note that the external voltage source V_{BB} is used to forward bias the base–emitter junction and the external source V_{CC} is used to reverse bias the collector–base junction. The magnitude of V_{CC} must be greater than V_{BB} to ensure that the collector–base junction remains reverse biased, since, as can be seen in the figure, $V_{CB} = V_{CC} - V_{BB}$. (Write Kirchhoff's voltage law around the loop from the collector, through V_{CC}, through V_{BB}, and back to the collector.) The emitter terminal is, of course, the ground, or common, terminal in this configuration.

Figure 4–16 shows that the input voltage in the CE configuration is the base–emitter voltage (V_{BE} for an NPN and V_{EB} for a PNP), and the output voltage is the collector–emitter voltage (V_{CE} for an NPN and V_{EC} for a PNP). The input current is I_B and the output current is I_C. The common-emitter configuration is the most useful and most widely used transistor configuration and we will study it in considerable detail. In the process we will learn some new facts about transistor behavior.

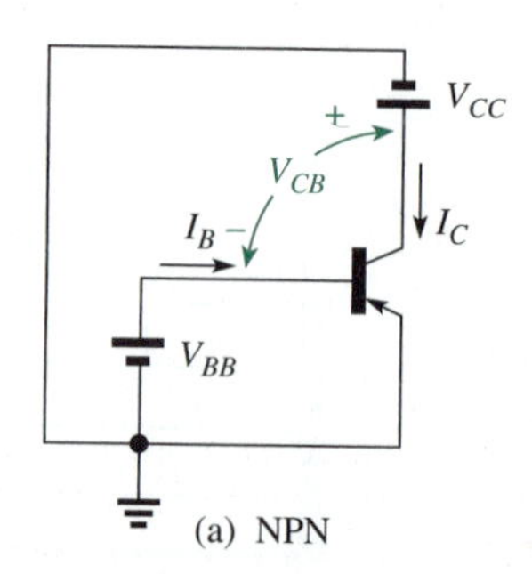

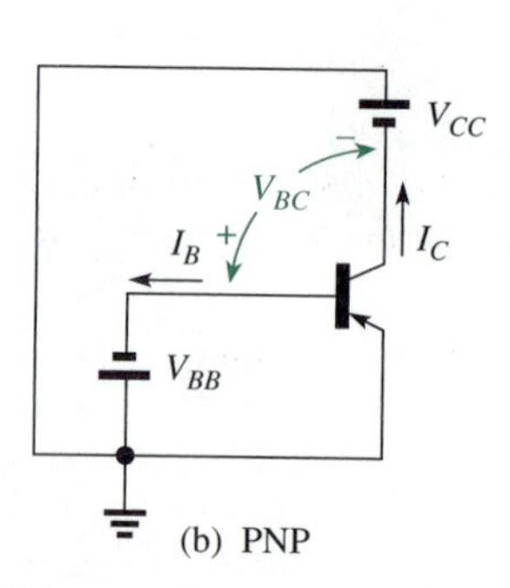

Figure 4–15
Common-emitter (CE) bias arrangements

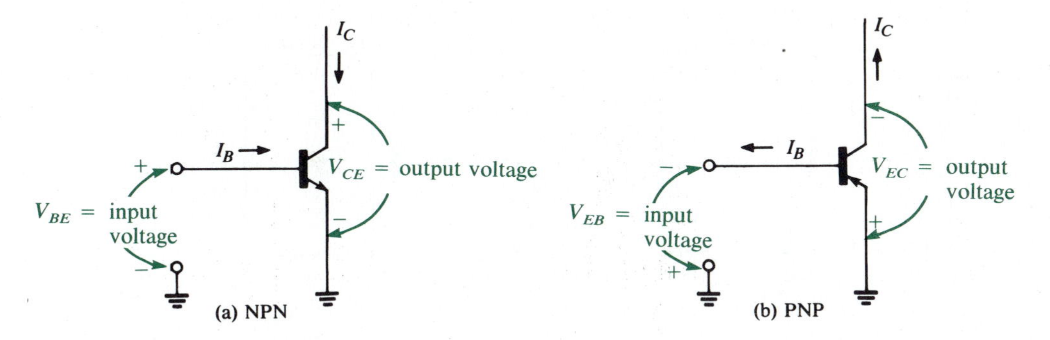

Figure 4–16
Input and output voltages and currents for NPN and PNP transistors in the CE configuration

I_{CEO} and Beta

Before investigating the input and output characteristics of the CE configuration, we will derive a new relationship between I_C and I_{CBO}. Although this derivation does not depend in any way on the bias arrangement used, it will provide us with some new parameters that are useful for predicting leakage in the CE configuration and for relating CE input and output currents. Equation 4–4 states that

$$I_C = \alpha I_E + I_{CBO}$$

or

$$I_C - I_{CBO} = \alpha I_E$$

Dividing through by α,

$$\frac{I_C}{\alpha} - \frac{I_{CBO}}{\alpha} = I_E$$

Substituting $I_B + I_C$ for I_E on the right-hand side,

$$\frac{I_C}{\alpha} - \frac{I_{CBO}}{\alpha} = I_B + I_C$$

Collecting the terms involving I_C leads to

$$I_C\left(\frac{1}{\alpha} - 1\right) = I_B + \frac{I_{CBO}}{\alpha}$$

But $1/\alpha - 1 = (1 - \alpha)/\alpha$, so

$$I_C = \frac{\alpha I_B}{1 - \alpha} + \frac{I_{CBO}}{1 - \alpha} \quad \textbf{(4–6)}$$

Using equation 4–6, we can obtain an expression for reverse "leakage" current in the CE configuration. Figure 4–17 shows NPN and PNP transistors in which the base–emitter circuits are left open while the reverse-biasing voltage sources remain connected. As can be seen in Figure 4–17, the only current that can flow when the base is left open is reverse current across the collector–base junction. This current flows from the collector, through the base region, and into the emitter. It is designated I_{CEO}—Collector-to-*E*mitter current with the base *O*pen. (Note, once again,

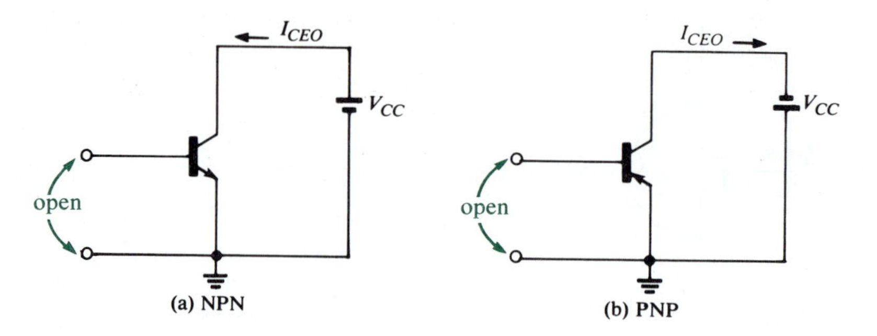

Figure 4–17
Collector–emitter leakage current I_{CEO}

that this "reverse" current is in the same direction as normal collector current through the transistor.) Since I_B must equal 0 when the base is open, we can substitute $I_B = 0$ in equation 4–6 to obtain

$$I_{CEO} = \frac{I_{CBO}}{1 - \alpha} = \left(\frac{1}{1 - \alpha}\right) I_{CBO} \tag{4–7}$$

Since the α of a transistor is close to 1, $1 - \alpha$ is close to 0, and so $1/(1 - \alpha)$ can be quite large. Therefore, equation 4–7 tells us that CE leakage current is much larger than CB leakage current. For example, if $I_{CBO} = 0.1\ \mu\text{A}$ and $\alpha = 0.995$, then $I_{CEO} = (0.1\ \mu\text{A})/0.005 = 20\ \mu\text{A}$. In effect, collector–base leakage current is amplified in the CE configuration, a result that can cause problems in high-temperature circuits, particularly those containing germanium transistors.

Returning to equation 4–6, let us focus on the factor $\alpha/(1 - \alpha)$ that multiplies I_B. This quantity is another important transistor parameter, called *beta:*

$$\beta = \frac{\alpha}{1 - \alpha} \tag{4–8}$$

Beta is always greater than 1 and for typical transistors ranges from around 20 to several hundred. When α is close to 1, a small increase in α causes a large increase in the value of β. For example, if $\alpha = 0.99$, then $\beta = 0.99/(1 - 0.99) = 99$. If α is increased by 0.005 to 0.995, then $\beta = 0.995/(1 - 0.995) = 199$. Because a small change in α causes a large change in β, small manufacturing variations in transistors that are supposed to be of the same type cause them to have a wide range of β values. It is not unusual for transistors of the same type to have betas that vary from 50 to 200.

In terms of β, equation 4–6 becomes

$$I_C = \beta I_B + \frac{I_{CBO}}{1 - \alpha} \tag{4–9}$$

or

$$I_C = \beta I_B + I_{CEO} \tag{4–10}$$

Although I_{CEO} is much greater than I_{CBO}, it is generally quite small in comparison to βI_B, especially in silicon transistors, and it can be neglected in many practical circuits. Neglecting I_{CEO} in equation 4–10, we obtain the approximation $I_C \approx \beta I_B$. This approximation is widely used in transistor circuit analysis, and we will often write it as an equality in future discussions, with the understanding that I_{CEO} can be neglected:

$$I_C = \beta I_B \qquad (I_{CEO} = 0) \tag{4–11}$$

Example 4–4

A transistor has $I_{CBO} = 48$ nA and $\alpha = 0.992$.

1. Find β and I_{CEO}.
2. Find its (exact) collector current when $I_B = 30\ \mu A$.
3. Find the approximate collector current, neglecting leakage current.

Solution

1. $$\beta = \frac{\alpha}{1 - \alpha} = \frac{0.992}{0.008} = 124$$

$$I_{CEO} = \frac{I_{CBO}}{1 - \alpha} = \frac{48 \times 10^{-9}}{0.008} = 6\ \mu A$$

2. $I_C = \beta I_B + I_{CEO} = (124)(30\ \mu A) + 6\ \mu A = 3726\ \mu A = 3.726$ mA
3. $I_C \approx \beta I_B = 124(30\ \mu A) = 3.72$ mA

Equation 4–8 tells us how to find the β of a transistor, given its α. It is left as an exercise at the end of this chapter to show that we can find α, given β, by using the following relation:

$$\alpha = \frac{\beta}{\beta + 1} \tag{4–12}$$

Example 4–5

As we will learn in a later discussion, the β of a transistor typically increases dramatically with temperature. If a certain transistor has $\beta = 100$ and an increase in temperature causes β to increase by 100%, what is the percent change in α?

Solution. A 100% increase in β means that β increases from 100 to 200. Let α_1 = value of α when $\beta = 100$ and α_2 = value of α when $\beta = 200$. From equation 4–12,

$$\alpha_1 = \frac{100}{100 + 1} = 0.990099$$

$$\alpha_2 = \frac{200}{200 + 1} = 0.995025$$

Thus, the percent change (increase) in α is

$$\frac{\alpha_2 - \alpha_1}{\alpha_1} \times 100\% = \frac{0.995025 - 0.990099}{0.990099} \times 100\% = 0.488\%$$

We see that a large change in β (100%) corresponds to a small change in α (0.488%).

Common-Emitter Input Characteristics

Since the input to a transistor in the CE configuration is across the base-to-emitter junction (see Figure 4–16), the CE input characteristics resemble a family of forward-biased diode curves. A typical set of CE input characteristics for an NPN transistor is shown in Figure 4–18. Note that I_B increases as V_{CE} decreases, for a fixed value of V_{BE}. A large value of V_{CE} results in a large reverse bias of the collector–base junction, which widens the depletion region and makes the base smaller. When the base is smaller, there are fewer recombinations of injected minority carriers and there is a corresponding reduction in base current. In contrast to the CB input characteristics, note in Figure 4–18 that the input current is plotted in units of *micro*amperes. I_B is, of course, much smaller than either I_E or I_C ($I_B \approx I_C/\beta$). The CE input characteristics are often called the *base characteristics.*

Common-Emitter Output Characteristics

Common-emitter output characteristics show collector current I_C versus collector voltage V_{CE}, for different *fixed* values of I_B. These characteristics are often called *collector characteristics.* Figure 4–19 shows a typical set of output characteristics for an NPN transistor in the CE configuration.

The approximate value of the β of the transistor can be determined at any point on the characteristics in Figure 4–19 simply by dividing the values I_C/I_B at the point. As illustrated in the figure, at V_{CE} = 5 V and I_B = 50 μA, the value of I_C is 5 mA, and the value of the β at that point is therefore I_C/I_B = (5 mA)/(50 μA) = 100. Clearly β, like α, is not constant, but depends on the region of the characteristics where the transistor is operated. The region where the curves are approximately horizontal is the *active* region of the CE configuration. In this region, β is essentially constant, but increases somewhat with V_{CE}, as can be deduced from the rise in each curve as V_{CE} increases to the right.

As in the CB configuration, the collector current in the CE configuration will

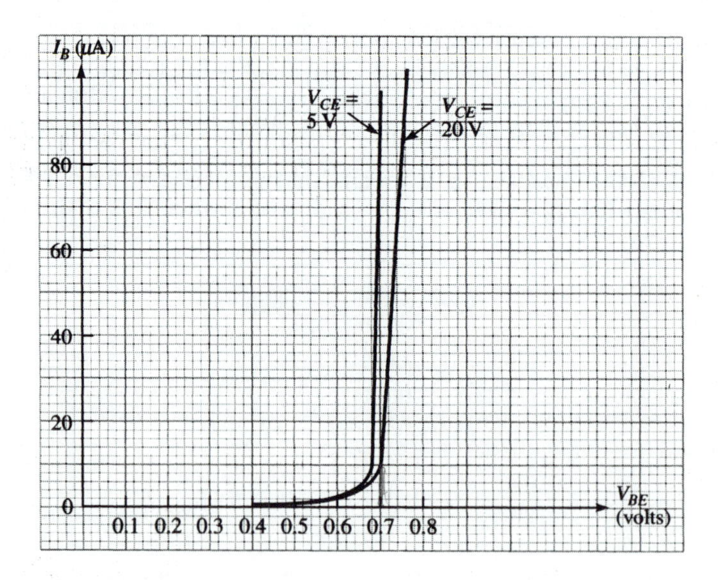

Figure 4–18
Common-emitter input characteristics

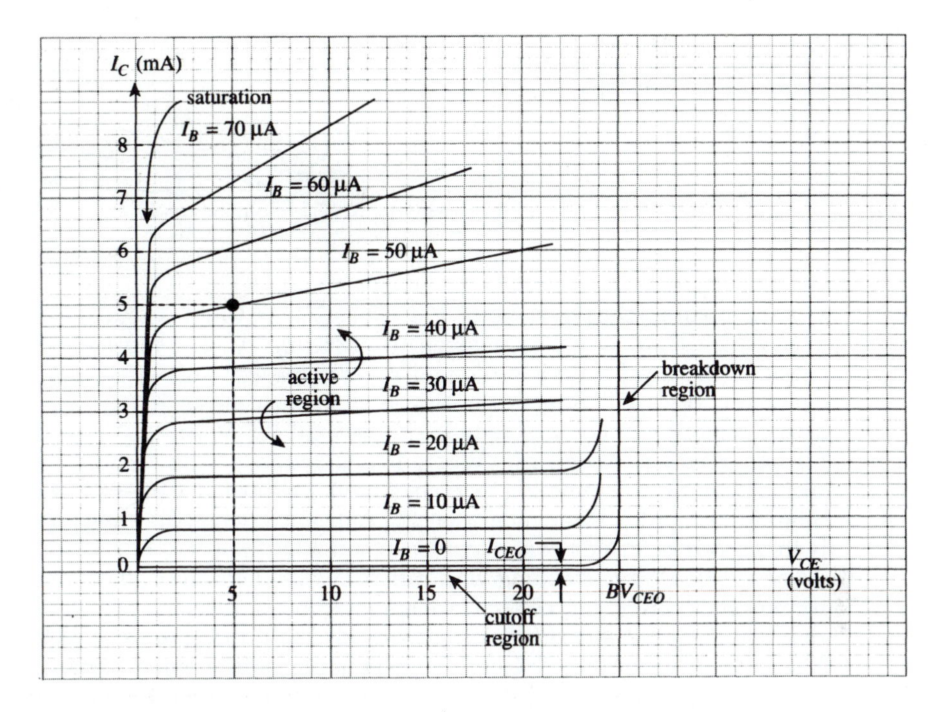

Figure 4–19
Common-emitter output characteristics

increase rapidly if V_{CB} is permitted to become large. When $I_B = 0$ (base open), the collector-to-emitter current at which breakdown occurs is designated BV_{CEO}. The value of BV_{CEO} is always less than that of BV_{CBO} for a given transistor. BV_{CEO} is sometimes called the "sustaining voltage" and denoted LV_{CEO}.

When interpreting the characteristics of Figure 4–19, it is important to realize that each curve is drawn for a small, essentially constant value of V_{BE} (about 0.7 V for silicon). Figure 4–20 illustrates this point.

Note in Figure 4–20 that the total collector-to-emitter voltage V_{CE} (which is the same as V_{CC} in our case) is the sum of the small, forward-biasing value of V_{BE}

Figure 4–20
$V_{CE} \approx V_{CB} + 0.7$ V for silicon. When V_{CE} is reduced to about 0.7 V, then $V_{CB} \approx 0$ and the collector–base junction is no longer reverse biased.

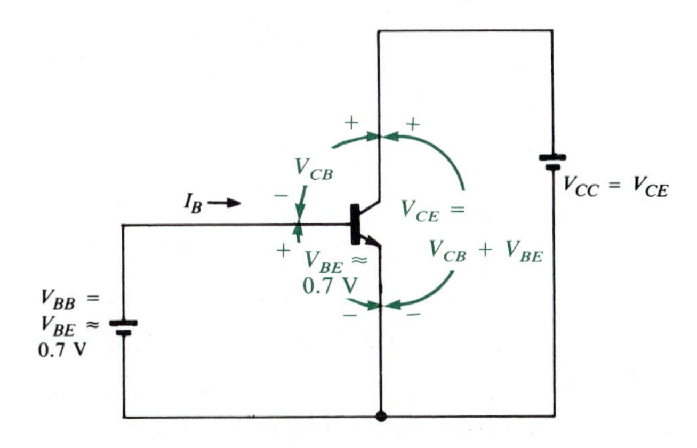

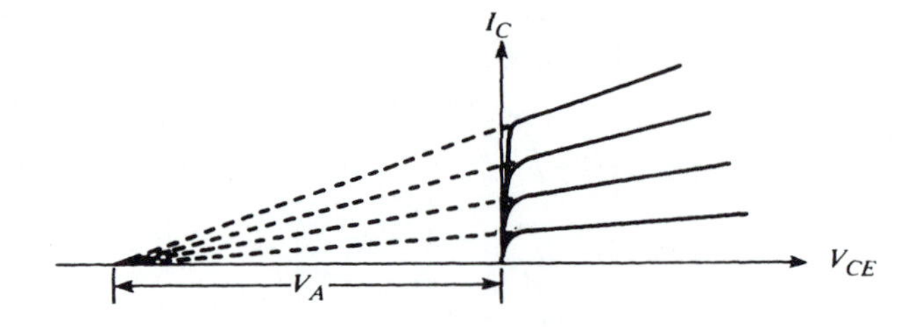

Figure 4–21
The Early voltage, V_A, is the intersection of the characteristic curves with the horizontal axis.

and the reverse-biasing value of V_{CB}. Thus $V_{CE} = V_{CB} + V_{BE}$, or, for silicon, $V_{CE} \approx V_{CB} + 0.7$. Thus, if V_{CE} is reduced to about 0.7 V, V_{CB} must become 0, and the collector–base junction is no longer reverse biased. This effect can be seen in the characteristics of Figure 4–19. Notice that each curve is reasonably flat (in the active region) until V_{CE} is reduced to around 0.5 V to 0.7 V. As V_{CE} is reduced further, I_C starts to fall off. When V_{CE} is reduced below about 0.2 V or 0.3 V, the collector–base junction becomes well *forward* biased and collector current diminishes rapidly. Remember that V_{CB} is negative when the collector–base junction is forward biased. For example, if $V_{CE} = 0.2$ V, then $V_{CB} = V_{CE} - V_{BE} \approx 0.2 \text{ V} - 0.7 \text{ V} = -0.5$ V. In keeping with our previous definition, the transistor is said to be *saturated* when the collector–base junction is forward biased. The saturation region is shown on the characteristic curves. The saturation value of V_{CE}, designated $V_{CE(sat)}$, ranges from 0.1 V to 0.3 V, depending on the value of base current.

Notice in Figure 4–19 that I_C does not equal 0 along the line corresponding to $I_B = 0$. When $I_B = 0$, the collector current is the same as that which flows when the base circuit is open (see Figure 4–17), that is, I_{CEO}. The region below $I_B = 0$ is the *cutoff* region.

Comparing the CB output characteristics in Figure 4–13 with the CE output characteristics in Figure 4–19, we note that the curves rise more steeply to the right in the CE case. This rise simply reflects the fact we have already discussed in connection with the CE input characteristics: The greater the value of V_{CE}, the smaller the base region, and, consequently, the smaller the base current. But, since base current is constant along each curve in Figure 4–19, the effect appears as an increase in I_C. In other words, the fact that there are fewer recombinations occurring in the base means that a greater proportion of carriers cross the junction to become collector current.

It is apparent in Figure 4–19 that the characteristic curves corresponding to large values of I_B rise more rapidly to the right than those corresponding to small values of I_B. If these lines are projected to the left, as shown in Figure 4–21, it is found that they all intersect the horizontal axis at approximately the same point. The point of intersection, designated V_A in Figure 4–21, is called the *Early voltage,* after J. M. Early, who first investigated these relations. Of course, a transistor is never operated with V_{CE} equal to the Early voltage. V_A is simply another useful parameter characterizing a transistor's behavior. It is especially useful in computer simulation programs such as SPICE that analyze transistor circuits and require complete descriptions of transistor characteristics.

Example 4–6

A certain transistor has the output characteristics shown in Figure 4–19.

1. Find the percent change in β as V_{CE} is changed from 2.5 V to 10 V while I_B is fixed at 40 μA.

2. Find the percent change in β as I_B is changed from 10 μA to 50 μA while V_{CE} is fixed at 7.5 V.

Neglect leakage current in each case.

Solution

1. At the intersection of the vertical line $V_{CE} = 2.5$ V with the curve $I_B = 40$ μA, we find $I_C \approx 3.8$ mA. Therefore, the β at that point is approximately (3.8 mA)/(40 μA) = 95.

   Traveling along the curve of constant $I_B = 40$ μA to its intersection with the vertical line $V_{CE} = 10$ V, we find $I_C \approx 4.2$ mA. Therefore, $\beta \approx$ (4.2 mA)/(40 μA) = 105. The percent change in β is

$$\frac{105 - 95}{95} \times 100\% = 10.53\%$$

2. At the intersection of the vertical line $V_{CE} = 7.5$ V with the curve $I_B = 10$ μA, we find $I_C \approx 0.8$ mA. Therefore, $\beta \approx$ (0.8 mA)/(10 μA) = 80.

   Traveling up the vertical line of constant $V_{CE} = 7.5$ V to its intersection with the curve $I_B = 50$ μA, we find $I_C \approx 5.2$ mA. Therefore, $\beta \approx$ (5.2 mA)/(50 μA) = 104. The percent change in β is

$$\frac{104 - 80}{80} \times 100\% = 30\%$$

Example 4–7

SPICE

(Computer Generation of Characteristic Curves) Use SPICE to obtain a set of output characteristics for an NPN transistor in a CE configuration. The ideal maximum forward β is 100 and the Early voltage is 200 V. The characteristics should be plotted for V_{CE} ranging from 0 V to 50 V in 5-V steps and for I_B ranging from 0 to 40 μA in 10-μA steps. Use the results to determine the actual value of β at $V_{CE} = 25$ V and $I_B = 30$ μA.

Solution. Figure 4–22(a) shows the circuit and the SPICE input file. Note that a constant-current source is used to supply base current. The transistor .MODEL statement specifies that $\beta = 100$ and that the (forward) Early voltage VAF = 200 V.

Figure 4–22
(Example 4–7)

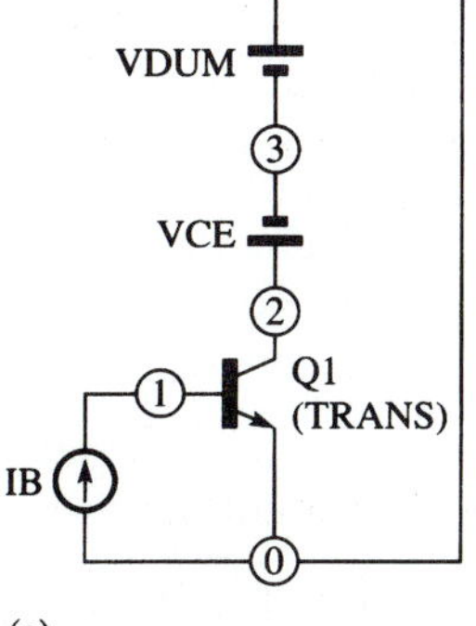


```
EXAMPLE 4.7
IB 0 1
VDUM 0 3
VCE 2 3
Q1 2 1 0 TRANS
.MODEL TRANS NPN BF=100 VAF=200
.DC VCE 0 50 5 IB 0 40U 10U
.PLOT DC I(VDUM)
.END
```

(a)

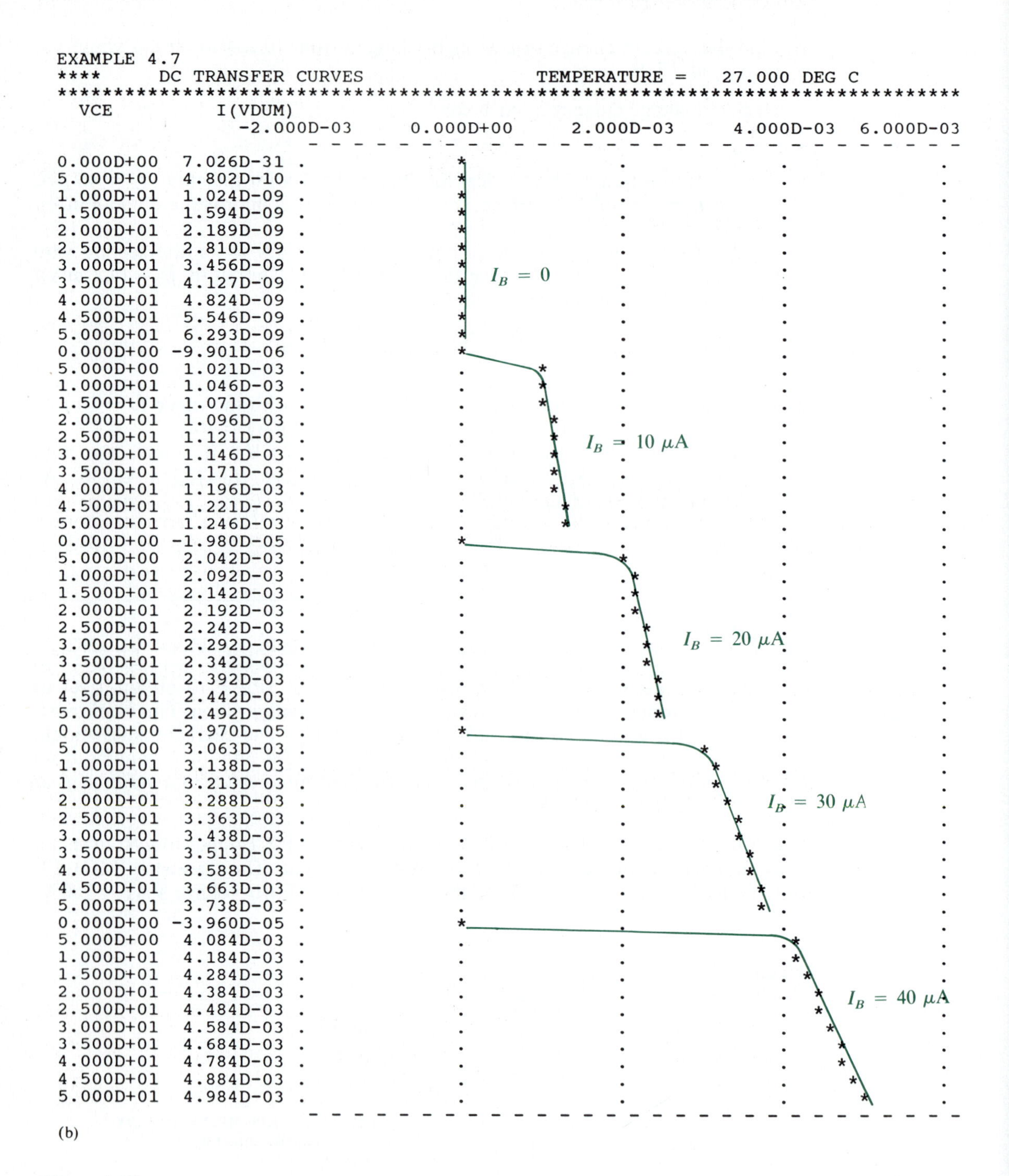

```
EXAMPLE 4.7
****     DC TRANSFER CURVES                  TEMPERATURE =   27.000 DEG C
******************************************************************************
  VCE           I(VDUM)
                  -2.000D-03      0.000D+00      2.000D-03      4.000D-03   6.000D-03
```

VCE	I(VDUM)
0.000D+00	7.026D-31
5.000D+00	4.802D-10
1.000D+01	1.024D-09
1.500D+01	1.594D-09
2.000D+01	2.189D-09
2.500D+01	2.810D-09
3.000D+01	3.456D-09
3.500D+01	4.127D-09
4.000D+01	4.824D-09
4.500D+01	5.546D-09
5.000D+01	6.293D-09
0.000D+00	-9.901D-06
5.000D+00	1.021D-03
1.000D+01	1.046D-03
1.500D+01	1.071D-03
2.000D+01	1.096D-03
2.500D+01	1.121D-03
3.000D+01	1.146D-03
3.500D+01	1.171D-03
4.000D+01	1.196D-03
4.500D+01	1.221D-03
5.000D+01	1.246D-03
0.000D+00	-1.980D-05
5.000D+00	2.042D-03
1.000D+01	2.092D-03
1.500D+01	2.142D-03
2.000D+01	2.192D-03
2.500D+01	2.242D-03
3.000D+01	2.292D-03
3.500D+01	2.342D-03
4.000D+01	2.392D-03
4.500D+01	2.442D-03
5.000D+01	2.492D-03
0.000D+00	-2.970D-05
5.000D+00	3.063D-03
1.000D+01	3.138D-03
1.500D+01	3.213D-03
2.000D+01	3.288D-03
2.500D+01	3.363D-03
3.000D+01	3.438D-03
3.500D+01	3.513D-03
4.000D+01	3.588D-03
4.500D+01	3.663D-03
5.000D+01	3.738D-03
0.000D+00	-3.960D-05
5.000D+00	4.084D-03
1.000D+01	4.184D-03
1.500D+01	4.284D-03
2.000D+01	4.384D-03
2.500D+01	4.484D-03
3.000D+01	4.584D-03
3.500D+01	4.684D-03
4.000D+01	4.784D-03
4.500D+01	4.884D-03
5.000D+01	4.984D-03

(b)

Figure 4–22
(Continued)

The .DC statement causes VCE to be stepped in 5-V increments from 0 through 50 V and I_B to be stepped in 10-μA increments from 0 through 40 μA.

The plot generated by SPICE is shown in Figure 4–22(b). This presentation is somewhat different than that shown in Figure 4–19, since the individual curves corresponding to different values of I_B must be displaced vertically by the printer. The curves have been filled in between the plotted points (asterisks) and labeled with the values of I_B to clarify the presentation. At $V_{CE} = 25$ V and $I_B = 30$ μA, we see that I(VDUM), which equals I_C, is 3.636 mA. Thus,

$$\beta = \frac{I_C}{I_B} = \frac{3.636 \text{ mA}}{30 \ \mu\text{A}} = 121.2$$

Note how the curves rise for increasing values of V_{CE} due to the effect of the Early voltage, meaning that β increases with increasing V_{CE} (and with increasing I_B).

4–5 COMMON-COLLECTOR CHARACTERISTICS

In the third and final way to arrange the biasing of a transistor, the collector is made the common point. The result is called the *common-collector* (CC) configuration and is illustrated in Figure 4–23. It is apparent in Figure 4–23(a) that

$$V_{CE} = V_{CB} + V_{BE} \tag{4–13}$$

or $V_{CB} = V_{CE} - V_{BE}$. In our case, $V_{CE} = V_{CC}$ and $V_{CB} = V_{BB}$, so $V_{BB} = V_{CC} - V_{BE}$. Now V_{BE} is the small, essentially constant voltage across the forward-biased base-to-emitter junction (about 0.7 V for silicon). Thus,

$$V_{BB} = V_{CB} \approx V_{CC} - 0.7 \tag{4–14}$$

Therefore, in order to keep the collector–base junction reverse biased ($V_{CB} > 0$), it is necessary that V_{BB} be larger than $V_{CC} - 0.7$.

Figure 4–24 shows that the base–collector voltage is the input voltage and the base current is the input current. The emitter–collector voltage is the output voltage and the emitter current is the output current.

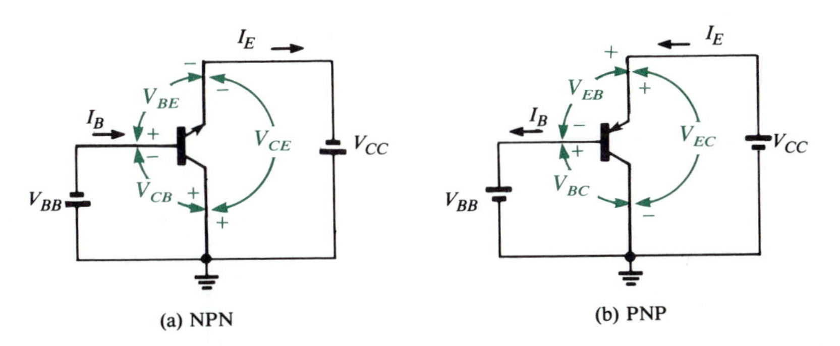

Figure 4–23
The common-collector (CC) bias configuration

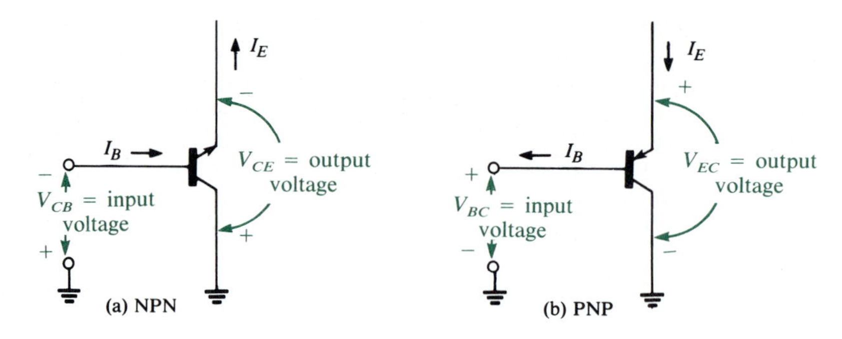

Figure 4–24
Input and output voltages and currents in the CC configuration

Figure 4–25 shows a typical set of input characteristics for an NPN transistor in the CC configuration. It is clear that these are not the characteristics of a forward-biased diode, as they were in the CB and CE configurations. We can see that each curve is drawn for a different fixed value of V_{CE} and that each shows the base current going to 0 very quickly as V_{CB} increases slightly. This behavior can be explained by remembering that V_{BE} must remain in the neighborhood of 0.5 V to 0.7 V in order for any appreciable base current to flow. But, from equation 4–13,

$$V_{BE} = V_{CE} - V_{CB} \tag{4–15}$$

Therefore, if the value of V_{CB} is allowed to increase to a point where it is near the value of V_{CE}, the value of V_{BE} approaches 0, and no base current will flow. In Figure 4–25 refer to the curve that corresponds to $V_{CE} = 5$ V. When $V_{CB} = 4.3$ V, we have, from equation 4–15, $V_{BE} = 5 - 4.3 = 0.7$ V, and we therefore expect a substantial base current. In the figure, we see that the point $V_{CB} = 4.3$ V and $V_{CE} = 5$ V yields a base current of 80 μA. If V_{CB} is now allowed to increase to 5 V, then $V_{BE} = 5 - 5 = 0$ V, and the base–emitter junction is no longer forward biased. Note in the figure that $I_B = 0$ when $V_{CE} = V_{CB} = 5$ V.

Figure 4–26 shows a typical set of CC output characteristics for an NPN transistor. These show emitter current, I_E, versus collector-to-emitter voltage, V_{CE}, for different fixed values of I_B. Note that these curves closely resemble the CE output characteristics shown in Figure 4–19. This resemblance is expected, since the only distinction is that I_E in Figure 4–26 is along the vertical axis instead of I_C, and $I_E \approx I_C$.

Figure 4–25
Common-collector input characteristics

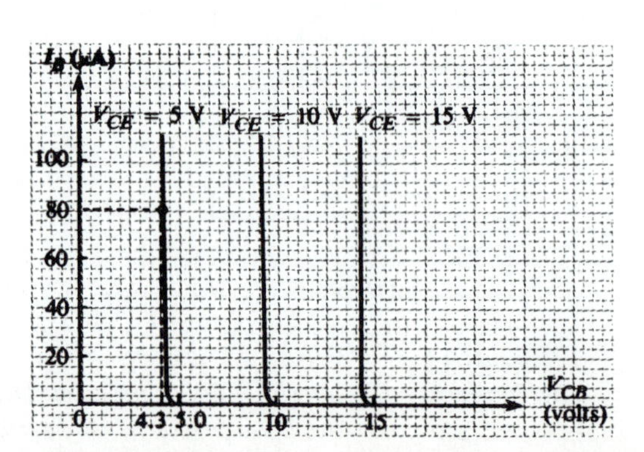

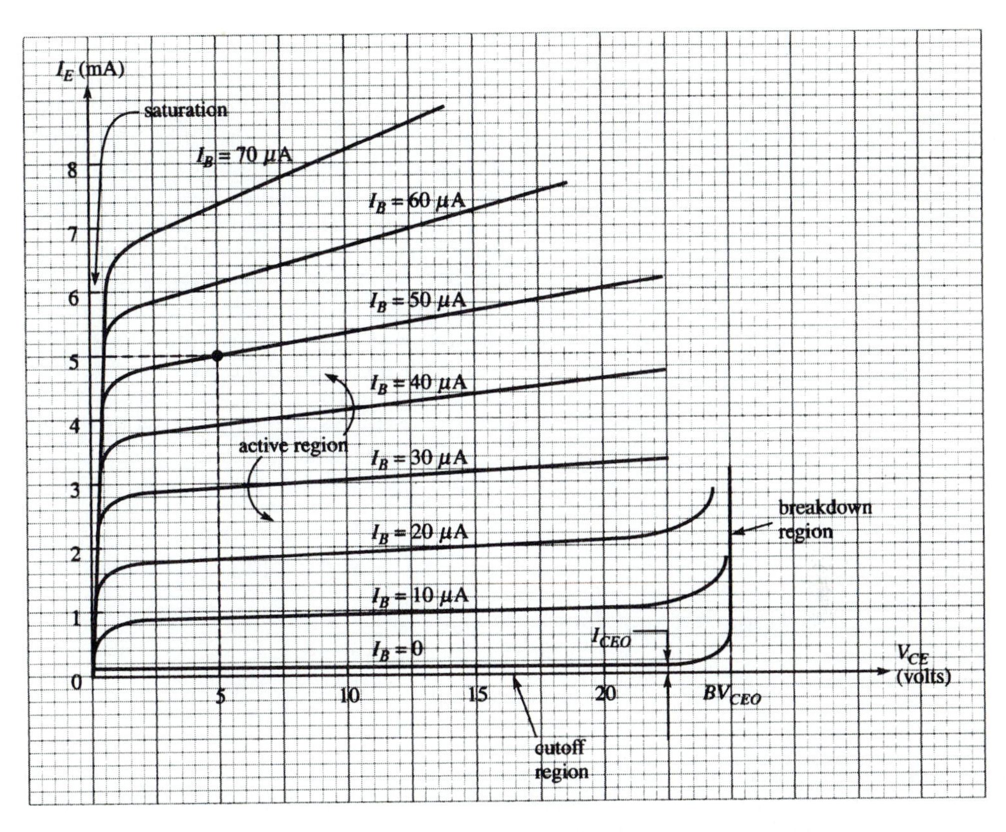

Figure 4–26
Common-collector output characteristics

When leakage current is neglected, recall that $I_C = \beta I_B$. But $I_E = I_C + I_B = \beta I_B + I_B$. Therefore,

$$I_E = (\beta + 1)I_B \tag{4–16}$$

Equation 4–16 relates the input and output currents in the CC configuration.

4–6 BIAS CIRCUITS

In our discussion of BJT theory up to this point, we have been using the word *bias* simply to specify the polarity of the voltage applied across each of the PN junctions in a transistor. In that context, we have emphasized that the base–emitter junction must be forward biased and the collector–base junction must be reverse biased, to achieve normal transistor action. We wish now to adopt a more restrictive interpretation of "bias." Henceforth we will be concerned with adjusting the *value* of the bias, as needed, to obtain *specific* values of input and output currents and voltages. In other words, we accept the fact that both junctions must be biased in the proper *direction* and concentrate on a practical means for changing the *degree* of bias, so that the output voltage, for example, is exactly the value we want it to be. When we have achieved a specific output voltage and output current, we say that we have set the bias *point* to those values.

Common-Base Bias Circuit

In practical circuits, we control the bias by connecting external resistors in series with the external voltage sources V_{CC}, V_{EE}, etc. We can then change resistor values instead of voltage source values to control the dc input and output voltages and currents. The circuit used to set the bias point this way is called a *bias circuit.* Figure 4–27 shows common-base bias circuits in which a resistor R_E is connected in series with the emitter and a resistor R_C is in series with the collector. Notice that we still regard emitter current as input current and base–emitter voltage as input voltage, as in past discussions of the CB configuration. See Figure 4–9. Likewise, collector current and collector–base voltage are still outputs. The only difference is that the input voltage is no longer the same as V_{EE}, because there is a voltage drop across R_E, and the output voltage is no longer the same as V_{CC}, due to the drop across R_C. The external voltage sources V_{EE} and V_{CC} are called *supply* voltages. Of course, the characteristic curves are still perfectly valid for showing the relationships between input and output voltages and currents.

Writing Kirchhoff's voltage law around the collector–base loop in Figure 4–27(a), we have

$$V_{CC} = I_C R_C + V_{CB} \tag{4–17}$$

Rearranging equation 4–17 leads to

$$I_C = \frac{-1}{R_C} V_{CB} + \frac{V_{CC}}{R_C} \tag{4–18}$$

When we regard I_C and V_{CB} as variables, and V_{CC} and R_C as constants, we see that equation 4–18 is the equation of a straight line. When plotted on a set of I_C–V_{CB} axes, the line has slope $-1/R_C$, and it intercepts the I_C-axis at V_{CC}/R_C. Equation 4–18 is the equation for the (NPN) CB *load line.* This load line has exactly the same interpretation as the diode load line we studied in Chapter 3: It is the line through *all possible* combinations of voltage (V_{CB}) and current (I_C). The actual bias point must be a point lying somewhere on the line. The precise location of the point is determined by the inputs V_{BE} and I_E.

We can find the point where the load line intercepts the V_{CB}-axis by setting $I_C = 0$ in equation 4–18 and solving for V_{CB}. Do this as an exercise, and verify that the V_{CB}-intercept is $V_{CB} = V_{CC}$. Thus, the load line can be drawn simply by drawing a line through the two points $V_{CB} = 0$, $I_C = V_{CC}/R_C$ and $I_C = 0$, $V_{CB} = V_{CC}$.

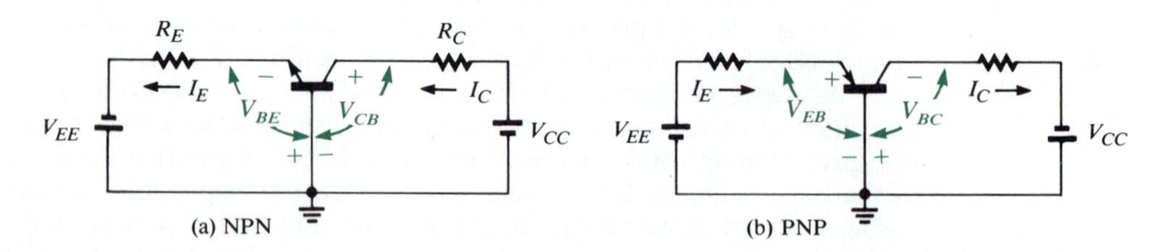

Figure 4–27
Practical CB bias circuits

Example 4–8

Determine the equation of the load line for the circuit shown in Figure 4–28. Sketch the line.

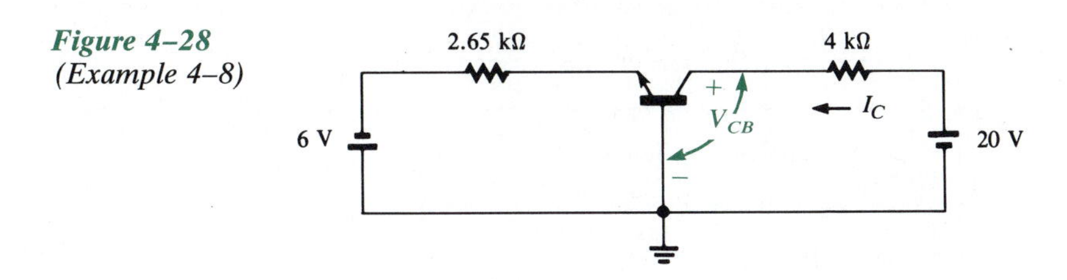

Figure 4–28
(Example 4–8)

Solution

$$I_C = \frac{-1}{R_C} V_{CB} + \frac{V_{CC}}{R_C}$$
$$= \frac{-1}{4 \times 10^3} V_{CB} + \frac{20}{4 \times 10^3}$$
$$= -2.5 \times 10^{-4} V_{CB} + 5 \times 10^{-3}\ \text{A}$$

The load line has slope -2.5×10^{-4} S, intercepts the I_C-axis at 5 mA, and intercepts the V_{CB}-axis at 20 V. It is sketched in Figure 4–29.

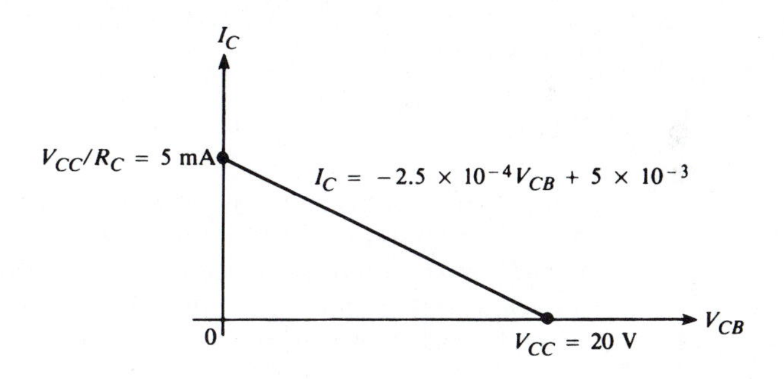

Figure 4–29
(Example 4–8) Load line for the bias circuit shown in Figure 4–28

We can determine the bias point by plotting the load line on the output characteristics of the transistor used in the circuit. To illustrate, the load line determined in Example 4–8 is shown drawn on a set of CB output characteristics in Figure 4–30.

To locate the bias point on the load line shown in Figure 4–30, we must determine the emitter current I_E in the circuit of Figure 4–28. One way to find I_E would be to draw an *input* load line on an input characteristic and determine the value of I_E where the line intersects the characteristic. This is the same technique used in Chapter 3 to find the dc current and voltage across a forward-biased diode in series with a resistor, which is precisely what the input side of the transistor circuit is. However, this approach is not practical for several reasons, not the least of which is the fact that input characteristics are seldom available.

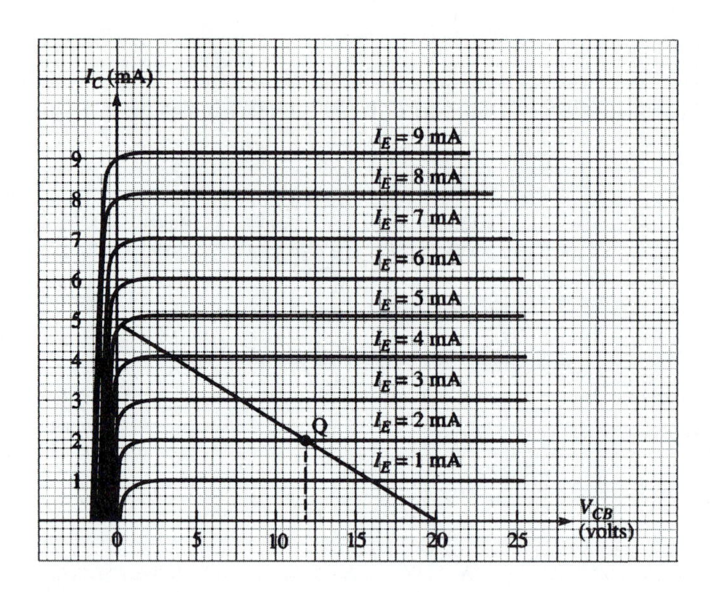

Figure 4–30
Load line plotted on CB output characteristics. The bias point (or quiescent point, labeled Q) is the intersection of the load line with the I_E = 2 mA curve.

The most practical way to determine I_E is to regard the base–emitter junction as a forward-biased diode having a fixed drop of 0.7 V (silicon) and solve for the diode current, the way we did in Chapter 3. Refer to Figure 4–31. In Figure 4–31, it is evident that

$$I_E = \frac{V_{EE} - 0.7}{R_E} \tag{4–19}$$

Note that we are neglecting the "feedback" effect of V_{CB} on the emitter current. Also note that the positive side of V_{EE} is connected to the circuit common, or ground, so it would normally be referred to as a *negative* voltage. However, in our equations, we treat V_{EE} as the absolute value of that voltage. Returning to our example circuit of Figure 4–28, we apply equation 4–19 to find

$$I_E = \frac{(6 - 0.7)\ \text{V}}{2.65\ \text{k}\Omega} = 2\ \text{mA}$$

Figure 4–31
The input side of the transistor in a CB configuration can be regarded as a forward-biased diode, for purposes of calculating I_E.

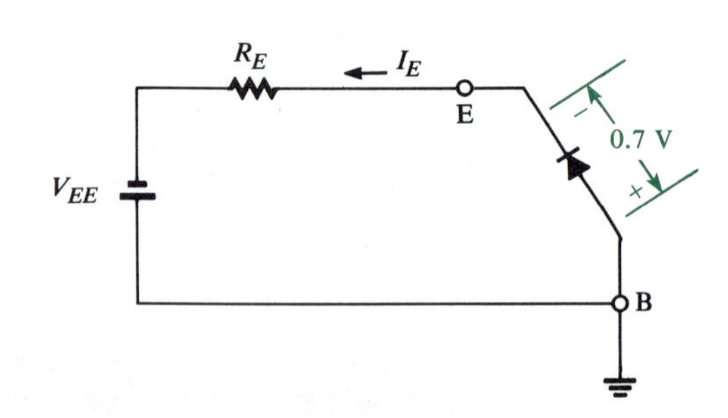

In Figure 4–30, the bias point, labeled Q, is seen to be the intersection of the load line with the curve $I_E = 2$ mA. At that point, $I_C \approx 2$ mA and $V_{CB} = 12$ V.

The bias point is often called the *quiescent* point, *Q*-point, or *operating* point. It specifies the dc output voltage and current when no ac voltage is superimposed on the input. As we shall discover in Chapter 5, the circuit is used as an *ac amplifier* by connecting an ac voltage source in series with the emitter. As the ac voltage alternately increases and decreases, the emitter current does the same. As a result, the output voltage and current change *along the load line* over a range determined by the change in I_E values.

Transistor input and output characteristics are useful for gaining insights to transistor behavior and, when used with load lines, they help us visualize output current and voltage variations. However, they are seldom used to design or analyze transistor circuits. One reason they are not is that transistors of the same type typically have a wide variation in their characteristics. For that reason, manufacturers do not (cannot) publish a set of curves that could be used for every transistor of a certain type. Furthermore, the accuracy that can be obtained using approximations and purely algebraic methods of analysis (as opposed to graphical methods) is almost always adequate for practical applications. We have already seen an example of this kind of algebraic approximation, when we regarded the input to the CB transistor as a forward-biased diode having a fixed voltage drop. We will now show how the entire bias circuit can be analyzed without the use of characteristic curves.

Since $\alpha \approx 1$ and $I_C = \alpha I_E$, it is true that $I_C \approx I_E$. Therefore, once we have determined I_E using equation 4–19, $I_E = (V_{EE} - 0.7)/R_E$ (silicon), we have a good approximation for I_C. We can then use equation 4–17 to solve for V_{CB}:

$$V_{CB} = V_{CC} - I_C R_C \qquad \textbf{(4–20)}$$

Example 4–9

Determine the bias point of the circuit shown in Figure 4–28 without using characteristic curves.

Solution. We have already shown (equation 4–19) that

$$I_E = \frac{V_{EE} - 0.7}{R_E} - \frac{(6 - 0.7)\text{ V}}{2.65\text{ k}\Omega} = 2\text{ mA}$$

Then, using $I_C \approx I_E$, we have from equation 4–20

$$\begin{aligned} V_{CB} &= V_{CC} - I_C R_C \\ &= 20\text{ V} - (2 \times 10^{-3}\text{ A})(4 \times 10^{3}\ \Omega) = 20\text{ V} - 8\text{ V} = 12\text{ V} \end{aligned}$$

Note that the bias point computed this way, $I_C = 2$ mA and $V_{CB} = 12$ V, is the same as that found graphically in Figure 4–30.

Summarizing, here are the four equations that can be used to solve for all input and output currents and voltages in the NPN, CB bias circuit of Figure 4–27(a):

$$\begin{aligned} V_{BE} &= 0.7\text{ V (Si)},\quad 0.3\text{ V (Ge)} \\ I_E &= \frac{V_{EE} - V_{BE}}{R_E} \\ I_C &= I_E - I_B \approx I_E \\ V_{CB} &= V_{CC} - I_C R_C \end{aligned} \qquad \textbf{(4–21)}$$

Equations 4–21 can be used for PNP transistors (Figure 4–27(b)) by substituting V_{EB} for V_{BE} and V_{BC} for V_{CB}. Substitute the *absolute value* of V_{CC} in the equation. For example, if we have $I_C = 1$ mA, $R_C = 1$ kΩ, and $V_{CC} = -15$ V in the PNP circuit of Figure 4–27(b), then $V_{BC} = 15 - (1\text{ mA})(1\text{ k}\Omega) = +5$ V. Since the base terminal is common in this circuit, the output voltage is often expressed with the base as reference, i.e., as V_{CB}. Of course, $V_{CB} = -V_{BC}$, so $V_{CB} = -5$ V in this example.

Common-Emitter Bias Circuit

Figure 4–32 shows practical bias circuits for NPN and PNP transistors in the common-emitter configuration. Notice that these bias circuits use only a single supply voltage (V_{CC}), which is a distinct practical advantage. The values of R_B and R_C must be chosen so that the voltage drop across R_B is greater than the voltage drop across R_C, in order to keep the collector–base junction reverse biased. The selection of values for R_B and R_C is part of the procedure used to design a CE bias circuit, which we will cover in Chapter 6.

As discussed in Chapter 3, it is important to be able to visualize the operation of electronic circuits when their diagrams are drawn without the ground paths shown, since that is the usual practice. The schematic diagrams in Figure 4–32 are useful for identifying closed loops, around which Kirchhoff's voltage law can be written, but the diagrams are rarely drawn as shown. Figure 4–33 shows the conventional way for drawing schematic diagrams of the CE bias circuits. As we have done in Figure 4–33, we will hereafter omit the ground paths in transistor schematics.

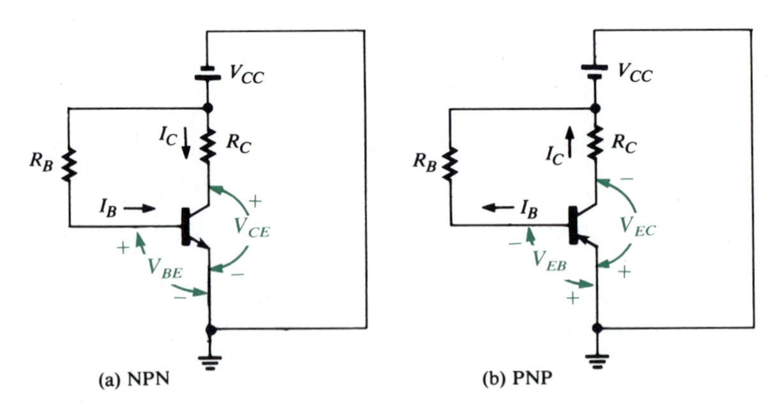

Figure 4–32
Practical CE bias circuits

Figure 4–33
Schematic diagrams of the CE bias circuits shown in Figure 4–32, with ground paths omitted

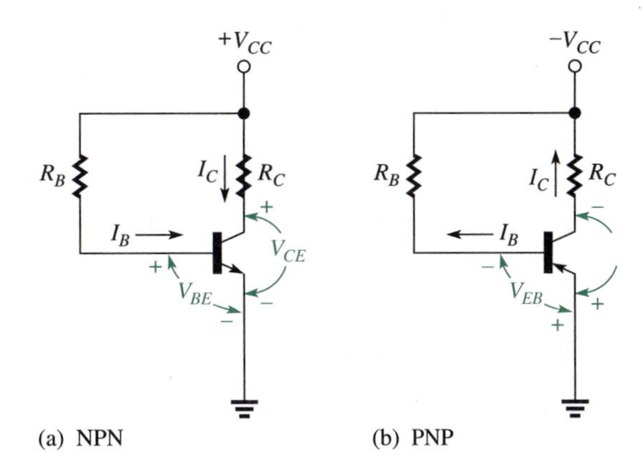

When studying an example or working an exercise, feel free to draw in any omitted ground paths whose presence would aid in understanding the circuit.

By writing Kirchhoff's voltage law around the output loop in Figure 4–32(a) or 4–33(a), we can obtain the equation for the load line of an NPN transistor in a CE configuration: $V_{CC} = I_C R_C + V_{CE}$, or

$$I_C = \frac{-1}{R_C} V_{CE} + \frac{V_{CC}}{R_C} \tag{4–22}$$

Note the similarity of equation 4–22 to the equation for the CB load line (4–18). The CE load line has slope $-1/R_C$, intercepts the I_C-axis at V_{CC}/R_C, and intercepts the V_{CE}-axis at V_{CC}. Figure 4–34 shows a graph of the load line plotted on I_C–V_{CE}-axes.

When the CE load line is plotted on a set of CE output characteristics, the bias point can be determined graphically, provided the value of I_B is known. To determine the value of I_B, we can again regard the input side of the transistor as a forward-biased diode having a fixed voltage drop, as shown in Figure 4–35. From Figure 4–35, we see that

$$I_B = \frac{V_{CC} - V_{BE}}{R_B} \tag{4–23}$$

where $V_{BE} = 0.7$ V for silicon and 0.3 V for germanium. Once again, we neglect the feedback effect of V_{CE} on I_B (the effect shown in Figure 4–18). As in the case

Figure 4–34
The CE load line for the bias circuit in Figure 4–33(a)

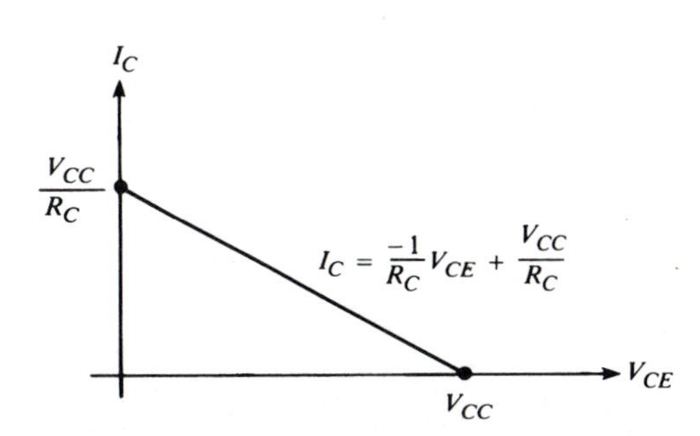

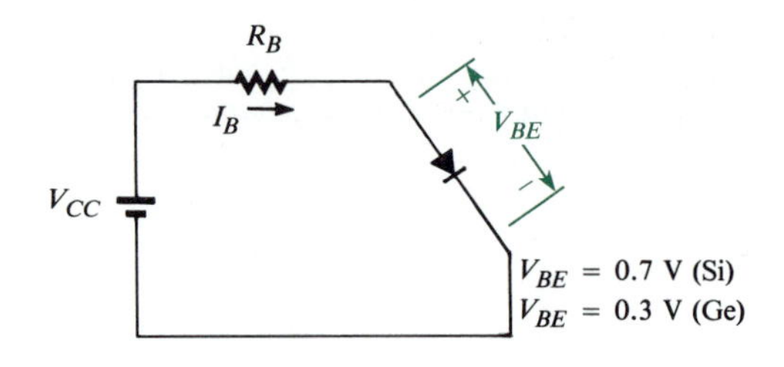

Figure 4–35
The input to an NPN transistor in the CE configuration can be regarded as a forward-biased diode having a fixed voltage drop.

of the CB bias circuit, the CE bias point can be determined algebraically. The equations for an NPN circuit are summarized as follows:

$$V_{BE} = 0.7\text{ V (Si)},\quad 0.3\text{ V (Ge)}$$
$$I_B = \frac{V_{CC} - V_{BE}}{R_B}$$
$$I_C = \beta I_B$$
$$V_{CE} = V_{CC} - I_C R_C \tag{4–24}$$

Equations 4–24 can be applied to a PNP bias circuit (Figure 4–32(b)) by substituting V_{EC} for V_{CE} and V_{EB} for V_{BE}. Substitute the absolute value of V_{CC}. Remember that V_{EC} is positive in a PNP circuit and V_{CE} is negative ($V_{CE} = -V_{EC}$).

Example 4–10

The silicon transistor in the CE bias circuit shown in Figure 4–36 has a β of 100.

1. Assuming that the transistor has the output characteristics shown in Figure 4–37, determine the bias point graphically.
2. Find the bias point algebraically.
3. Repeat (1) and (2) when R_B is changed to 161.43 kΩ.

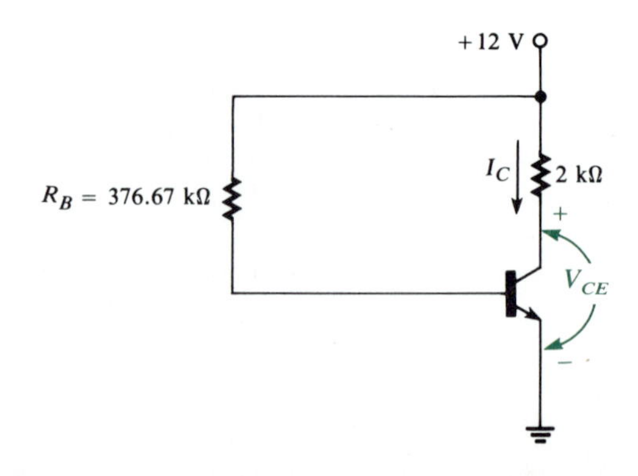

Figure 4–36
(Example 4–10)

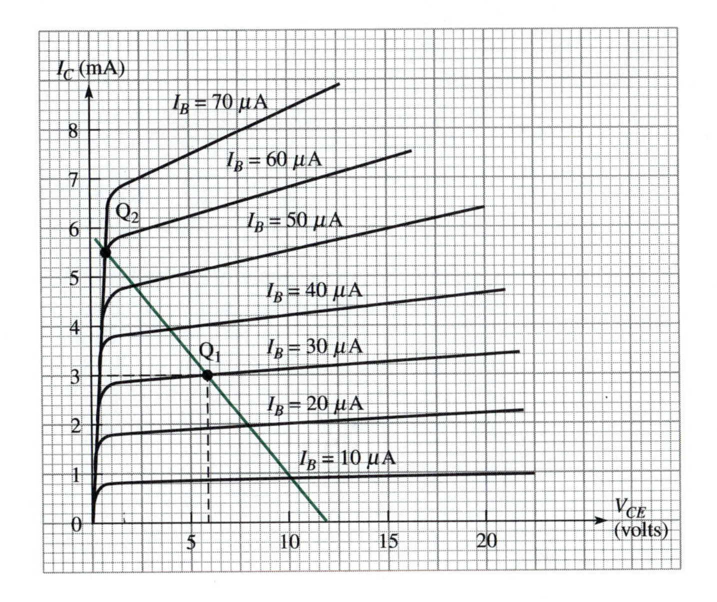

Figure 4–37
Load line plotted on CE output characteristics (Example 4–10). The bias point is shifted (into saturation) by the change in R_B.

Solution

1. The equation of the load line is

$$I_C = -\frac{1}{2 \times 10^3}V_{CE} + \frac{12}{2 \times 10^3}$$
$$= -0.5 \times 10^{-3}V_{CE} + 6 \times 10^{-3}$$

The load line intersects the I_C-axis at 6 mA and the V_{CE}-axis at 12 V. It is shown plotted on the output characteristics in Figure 4–37. To locate the bias point, we find I_B:

$$I_B = \frac{(12 - 0.7)\text{ V}}{376.67\text{ k}\Omega} = 30\ \mu\text{A}$$

At the intersection of the $I_B = 30\ \mu$A curve with the load line, labeled Q_1 in Figure 4–37, we see that the bias point is $I_C \approx 2.95$ mA and $V_{CE} = 6$ V.

2. From equations 4–24, we find

$$V_{BE} = 0.7\text{ V}$$
$$I_B = \frac{(12 - 0.7)\text{ V}}{376.67\text{ k}\Omega} = 30\ \mu\text{A}$$
$$I_C = (100)(30\ \mu\text{A}) = 3\text{ mA}$$
$$V_{CE} = 12\text{ V} - (3\text{ mA})(2\text{ k}\Omega) = 6\text{ V}$$

These results are in good agreement with the bias values found graphically.

3. Changing R_B to 161.43 kΩ has no effect on the load line. Note that the load line equation (4–22) does not involve R_B. However, the value of I_B is changed to

$$I_B = \frac{(12 - 0.7)\text{ V}}{1161.43\text{ k}\Omega} = 70\ \mu\text{A}$$

Thus, the bias point is shifted along the load line to the point labeled Q_2 in Figure 4–37. We see that Q_2 is now in the saturation region of the transistor. At Q_2, $I_C \approx 5.7$ mA and $V_{CE} \approx 0.5$ V. This result illustrates that the bias point can be changed by changing the value of external resistor(s) in the bias circuit.

Using equations 4–24 to find the new bias point, we have

$$I_C = \beta I_B = (100)(70\ \mu\text{A}) = 7\ \text{mA} \quad (!)$$
$$V_{CE} = 12\ \text{V} - (7\ \text{mA})(2\ \text{k}\Omega) = -2\ \text{V} \quad (!)$$

These are clearly erroneous results, since the maximum value that I_C can have is 6 mA, and the minimum value that V_{CE} can have is 0 V. The reason equations 4–24 are not valid in this case is that the bias point is not in the active region. Remember that β decreases in the saturation region, and, in this example, can no longer be assumed to equal 100. (As an exercise, use Figure 4–37 to calculate the value of β at Q_2.)

Example 4–11

SPICE

Use SPICE to find the bias point of the CE circuit in Figure 4–36.

Solution. The circuit is redrawn for analysis by SPICE as shown in Figure 4–38(a). A .DC analysis is performed and the values of V(2), which is V_{CE}, and I(VDUM),

(a)

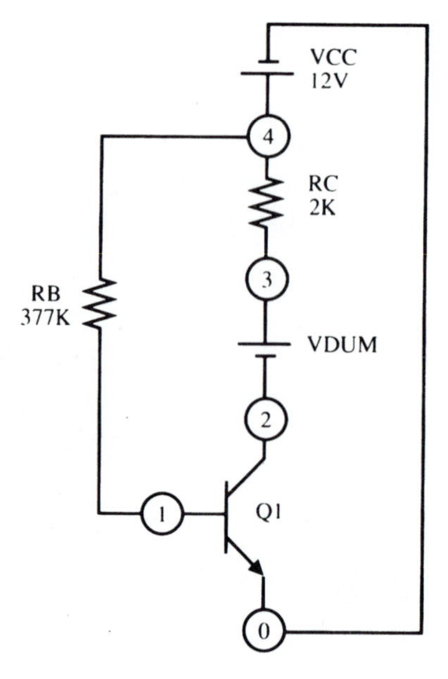


```
EXAMPLE 4.11
Q1 2 1 0 TRANS
RB 4 1 377K
RC 3 4 2K
VDUM 3 2
VCC 4 0
.MODEL TRANS NPN BF=100
.DC VCC 12 12 1
.PRINT DC V(2) I(VDUM)
.END
```

(b)

```
EXAMPLE 4.11
****     DC TRANSFER CURVES                         TEMPERATURE =    27.000 DEG C
*******************************************************************************
  VCC           V(2)          I(VDUM)
 1.200E+01      6.060E+00     2.970E-03
```

Figure 4–38
(Example 4–11)

which is I_C, are printed. The results shown in Figure 4–38(b) reveal that $V_{CE} = 6.06$ V and $I_C = 2.97$ mA, in close agreement with the values calculated in Example 4–10.

Common-Collector Bias Circuit

Figure 4–39 shows common-collector bias circuits for NPN and PNP transistors. Once again, the load line for Figure 4–39(a) can be derived by writing Kirchhoff's voltage law around the output loop:

$$V_{CC} = I_E R_E + V_{CE}$$

$$I_E = \frac{-1}{R_E} V_{CE} + \frac{V_{CC}}{R_E} \qquad \textbf{(4–25)}$$

Recall that the output characteristics for the CC configuration are, for all practical purposes, the same as those for the CE configuration. Therefore, we will not present another example showing the load line plotted on output characteristics.

As in the previous configurations, we must find I_B in order to determine the bias point. Figure 4–40 shows a circuit that is equivalent to the loop in Figure 4–39(a) that starts at V_{CC}, passes through R_B, through the base–emitter junction,

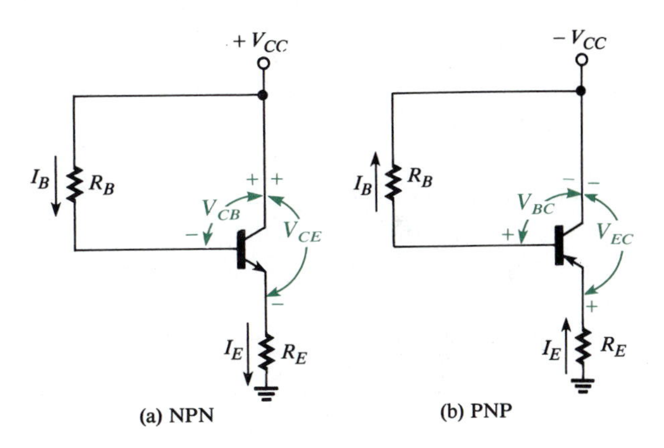

Figure 4–39
Common-collector bias circuits

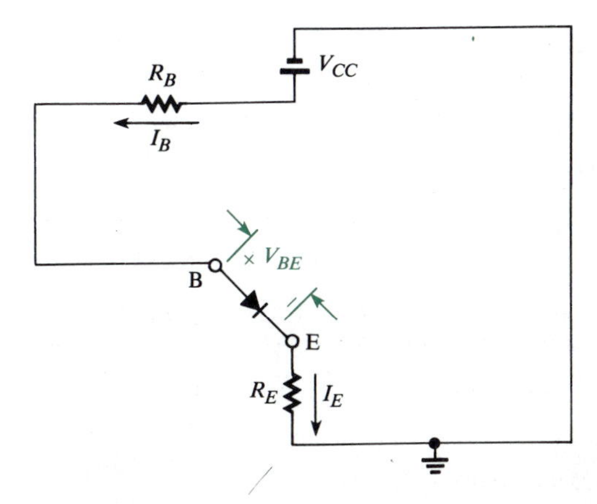

Figure 4–40
A circuit equivalent to the input side of Figure 4–39(a)

through R_E, and back to V_{CC}. Writing Kirchhoff's voltage law around the loop in Figure 4–40, we have

$$V_{CC} = I_B R_B + V_{BE} + I_E R_E \tag{4–26}$$

By equation 4–16, $I_E = (\beta + 1)I_B$. Substituting for I_E in equation 4–26, we obtain

$$V_{CC} = I_B R_B + V_{BE} + (\beta + 1)I_B R_E$$

or

$$V_{CC} - V_{BE} = I_B[R_B + (\beta + 1)R_E]$$

Solving for I_B,

$$I_B = \frac{V_{CC} - V_{BE}}{R_B + (\beta + 1)R_E} \tag{4–27}$$

Summarizing, the equations for determining the bias point in an NPN CC configuration are

$$\boxed{\begin{aligned} V_{BE} &= 0.7\text{ V (Si)}, \quad 0.3\text{ V (Ge)} \\ I_B &= \frac{V_{CC} - V_{BE}}{R_B + (\beta + 1)R_E} \\ I_E &= (\beta + 1)I_B \\ V_{CE} &= V_{CC} - I_E R_E \end{aligned}} \tag{4–28}$$

For a PNP CC bias circuit, substitute V_{EC} for V_{CE} and V_{EB} for V_{BE}, and use the absolute value of V_{CC}.

Example 4–12

Find the bias point of the germanium transistor in the circuit of Figure 4–41. Assume that $\beta = 120$.

Figure 4–41
(Example 4–12)

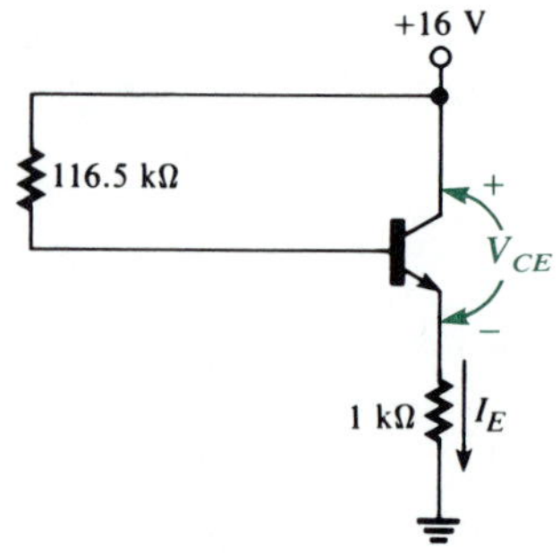

+16 V
116.5 kΩ
V_{CE}
1 kΩ
I_E

Solution. From equations 4–28,

$$V_{BE} = 0.3\text{ V}$$

$$I_B = \frac{16\text{ V} - 0.3\text{ V}}{116.5 \times 10^3\ \Omega + 121 \times 10^3\ \Omega} = 66.105\ \mu\text{A}$$

$$I_E = (121)(66.105\ \mu\text{A}) = 8\text{ mA}$$

$$V_{CE} = 16\text{ V} - (8 \times 10^{-3}\text{ A})(10^3\ \Omega) = 8\text{ V}$$

4–7 DESIGN CONSIDERATIONS

CB Bias Design

Although a bias circuit for the common-base configuration can be designed using a single dc power supply, we will consider only the two-source design (Figure 4–27) at this point in our discussion. In the usual design scenario, the supply voltages V_{EE} and V_{CC} are fixed, and we must choose values for R_E and R_C to obtain a specified bias current I_E and bias voltage V_{CB}. Letting $I_C = I_E$, equations 4–21 are easily solved for R_E and R_C in terms of I_E and V_{CB}:

$$R_E = \frac{V_{EE} - V_{BE}}{I_E}$$
$$R_C = \frac{V_{CC} - V_{CB}}{I_E} \qquad \textbf{(4–29)}$$

In practical discrete-circuit designs, it is often necessary to use standard-valued resistors. The standard values closest to the values calculated from equations 4–29 are used, and the circuit is analyzed to determine the resulting values of I_E and V_{CB}. If variation from the desired bias values is a critical consideration in a particular application, it may be necessary to use precision resistors or to calculate the total possible variation that could arise from using resistors that have a specified tolerance. The next example demonstrates these ideas.

Example 4–13

DESIGN

A common-base bias circuit is to be designed for an NPN silicon transistor to be used in a system having dc power supplies +15 V and −5 V. The bias point is to be $I_E = 1.5$ mA and $V_{CB} = 7.5$ V.

1. Design the circuit, using standard-valued resistors with 5% tolerance.
2. What are the actual bias values if the resistors selected have their nominal values?
3. What are the possible ranges of I_E and V_{CB}, taking the resistor tolerances into consideration?

Solution

1. From equations 4–29,

$$R_E = \frac{(5 - 0.7)\ \text{V}}{1.5 \times 10^{-3}\ \text{A}} = 2867\ \Omega$$
$$R_C = \frac{(15 - 7.5)\ \text{V}}{1.5 \times 10^{-3}\ \text{A}} = 5000\ \Omega$$

Appendix B contains a table of the standard values of resistors having 5% and 10% tolerances. The standard 5% resistors with values closest to those calculated for R_E and R_C are $R_E = 3$ kΩ and $R_C = 5.1$ kΩ.

2. From equations 4–21,

$$I_E = \frac{(5 - 0.7)\ \text{V}}{3\ \text{k}\Omega} = 1.43\ \text{mA}$$
$$V_{CB} = 15\ \text{V} - (1.43\ \text{mA})(5.1\ \text{k}\Omega) = 7.69\ \text{V}$$

3. The ranges of possible resistance values for R_E and R_C are

$$R_E = 3\text{ k}\Omega\ 5 \pm 0.05(3\text{ k}\Omega) = 2850\text{–}3150\ \Omega$$
$$R_C = 5.1\text{ k}\Omega \pm 0.05(5.1\text{ k}\Omega) = 4845\text{–}5355\ \Omega$$
$$I_{E(min)} = \frac{V_{EE} - V_{BE}}{R_{E(max)}} = \frac{(5 - 0.7)\text{ V}}{3150\ \Omega} = 1.365\text{ mA}$$
$$I_{E(max)} = \frac{V_{EE} - V_{BE}}{R_{E(min)}} = \frac{(5 - 0.7)\text{ V}}{2850\ \Omega} = 1.509\text{ mA}$$
$$V_{CB(min)} = V_{CC} - I_{E(max)}R_{C(max)}$$
$$= 15\text{ V} - (1.509\text{ mA})(5355\ \Omega) = 6.92\text{ V}$$
$$V_{CB(max)} = V_{CC} - I_{E(min)}R_{C(min)}$$
$$= 15\text{ V} - (1.365\text{ mA})(4845\ \Omega) = 8.39\text{ V}$$

We see that considerable variation from the desired bias point is possible when using standard-valued resistors.

CE Bias Design

In Chapter 5 we discuss the design of a voltage-divider bias circuit that has certain properties superior to the design shown in Figure 4–32. However, when simplicity and minimization of the number of components are the primary considerations, the circuit of Figure 4–32 is used. Assuming the supply voltage V_{CC} is fixed, we solve equations 4–24 for R_B and R_C in terms of the desired bias values for I_C and V_{CE}:

$$R_B = \frac{V_{CC} - V_{BE}}{I_C/\beta}$$
$$R_C = \frac{V_{CC} - V_{CE}}{I_C} \qquad \textbf{(4–30)}$$

The practical difficulty with this design is that the bias point depends heavily on the value of β, which varies considerably with temperature. Also, there is typically a wide variation in the value of β among transistors of the same type. Consequently, this design is not recommended for applications where wide temperature variations may occur or for volume production (where different transistors are used). The next example demonstrates this point.

Example 4–14

DESIGN

An NPN silicon transistor having a nominal β of 100 is to be used in a CE configuration with $V_{CC} = 12$ V. The bias point is to be $I_C = 2$ mA and $V_{CE} = 6$ V.

1. Design the circuit, using standard-valued 5% resistors.
2. Find the range of possible bias values if the β of the transistor can change to any value between 50 and 150 (a typical range). Assume that the 5% resistors have their nominal values.

Solution

1. From equations 4–30,

$$R_B = \frac{(12 - 0.7)\ \text{V}}{2 \times 10^{-3}\ \text{A}/100} = 565\ \text{k}\Omega$$

$$R_C = \frac{(12 - 6)\ \text{V}}{2 \times 10^{-3}\ \text{A}} = 3\ \text{k}\Omega$$

From Appendix C, the standard 5% resistors having values closest to those calculated are $R_B = 560\ \text{k}\Omega$ and $R_C = 3\ \text{k}\Omega$.

2. From equations 4–24,

$$I_B = \frac{(12 - 0.7)\ \text{V}}{560\ \text{k}\Omega} = 20.18\ \mu\text{A}$$

$$I_{C(min)} = \beta_{(min)} I_B = 50(20.18\ \mu\text{A}) = 1.01\ \text{mA}$$

$$I_{C(max)} = \beta_{(max)} I_B = 150(20.18\ \mu\text{A}) = 3.03\ \text{mA}$$

$$V_{CE(min)} = V_{CC} - I_{C(max)} R_C = 12\ \text{V} - (3.03\ \text{mA})(3\ \text{k}\Omega) = 2.92\ \text{V}$$

$$V_{CE(max)} = V_{CC} - I_{C(min)} R_C = 12\ \text{V} - (1.01\ \text{mA})(3\ \text{k}\Omega) = 8.97\ \text{V}$$

In most practical applications, the possible variation of V_{CE} from 2.92 V to 8.97 V would be intolerable.

CC Bias Design

To obtain resistor values for the common-collector bias circuit (Figure 4–37), we solve equations 4.28 for R_E and R_B in terms of the desired bias values I_E and V_{CE}:

$$R_E = \frac{V_{CC} - V_{CE}}{I_E}$$

$$R_B = \frac{(\beta + 1)}{I_E}(V_{CC} - V_{BE} - I_E R_E) \tag{4–31}$$

Example 4–15

DESIGN

An NPN silicon transistor having $\beta = 100$ is to be used in a CC configuration with $V_{CC} = 24$ V. The desired bias point is $V_{CE} = 16$ V and $I_E = 4$ mA.

1. Design the bias circuit using standard-valued 5% resistors.
2. Find the actual bias point when the standard resistors are used, assuming they have their nominal values.

Solution

1. From equations 4–31,

$$R_E = \frac{(24 - 16)\ \text{V}}{4\ \text{mA}} = 2\ \text{k}\Omega$$

$$R_B = \frac{101}{4 \times 10^{-3}\ \text{A}}[24\ \text{V} - 0.7\ \text{V} - (4\ \text{mA})(2\ \text{k}\Omega)] = 386{,}325\ \Omega$$

From Appendix B, the standard 5% resistors having values closest to those calculated are $R_E = 2\ \text{k}\Omega$ and $R_B = 390\ \text{k}\Omega$.

2. From equations 4–28,

$$I_B = \frac{(24 - 0.7)\ \text{V}}{390\ \text{k}\Omega + 101(2\ \text{k}\Omega)} = 39.358\ \mu\text{A}$$

$$I_E = 101(39.358\ \mu\text{A}) = 3.98\ \text{mA}$$

$$V_{CE} = 24\ \text{V} - (3.98\ \text{mA})(2\ \text{k}\Omega) = 16.04\ \text{V}$$

4–8 THE BJT INVERTER (TRANSISTOR SWITCH)

Transistors are widely used in digital logic circuits and switching applications similar to those described in Chapter 3. Recall that the waveforms encountered in those applications periodically alternate between a "low" and a "high" voltage, such as 0 V and +5 V. The fundamental transistor circuit used in switching applications is called an *inverter,* the NPN version of which is shown in Figure 4–42. Note in the figure that the transistor is in a common-emitter configuration, but there is no bias voltage connected to the base through a resistor, as in the CE bias circuits studied earlier. Instead, a resistor R_B is connected in series with the base and then directly to a square or pulse-type waveform that serves as the inverter's input. In the circuit shown, V_{CC} and the "high" level of the input are both +5 V. The output is the voltage between collector and emitter (V_{CE}), as usual.

When the input to the inverter is high (+5 V), the base–emitter junction is forward biased and current flows through R_B into the base. The values of R_B and R_C are chosen (designed) so that the amount of base current flowing is enough to *saturate* the transistor, that is, to drive it into the saturation region of its output characteristics. Figure 4–43 shows a load line plotted on a set of CE output characteristics and identifies the point on the load line where saturation occurs. Note that the value of V_{CE} corresponding to this point, called $V_{CE(sat)}$, is very nearly 0 (typically about 0.1 V). The current at the saturation point is called $I_{C(sat)}$ and is very nearly equal to the intercept of the load line of the I_C-axis, namely, V_{CC}/R_C. When the transistor is saturated, it is said to be *ON.* This analysis has shown that a high input to the inverter (+5 V) results in a low output ($\approx$ 0 V).

When the input to the transistor is low, i.e., 0 V, the base–emitter junction has no forward bias applied to it, so no base current, and hence no collector current,

Figure 4–42
An NPN transistor inverter, or switch

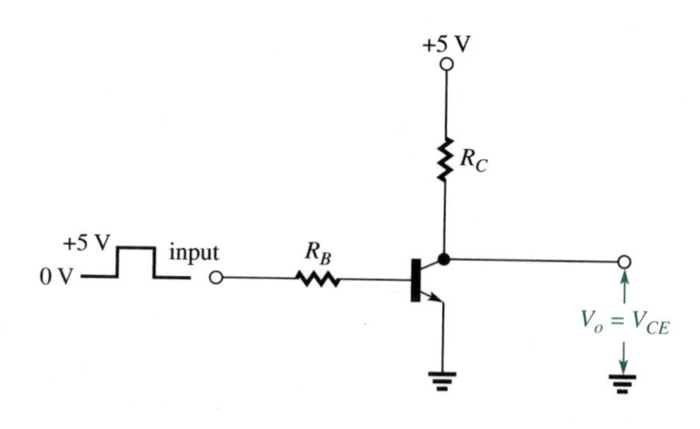

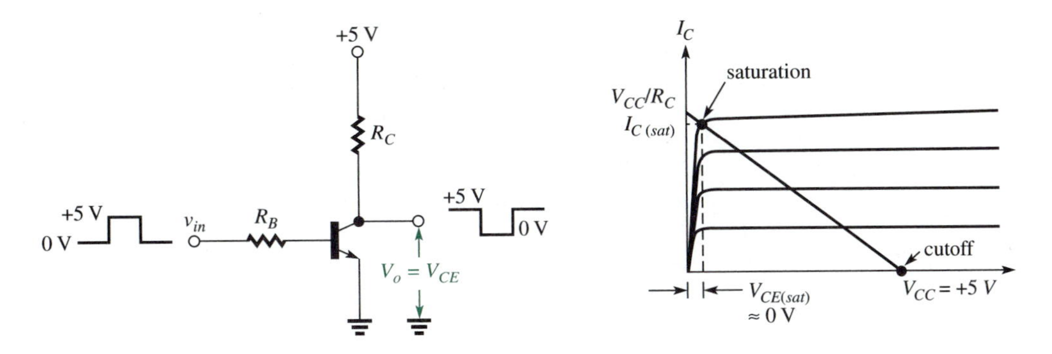

Figure 4–43
When the input to the inverter is high (+5 V), the transistor is saturated and its output is low (≈ 0 V). When the input to the inverter is low, the transistor is cut off and its output is high.

flows. There is, therefore, no voltage drop across R_C, and it follows that V_{CE} must be the same as V_{CC}: +5 V. This fact is made evident by substituting $I_C = 0$ in the equation for V_{CE} (equation 4–24): $V_{CE} = V_{CC} - I_C R_C = V_{CC} - (0)(R_C) = V_{CC}$. In this situation, the transistor is in the cutoff region of its output characteristics, as shown in Figure 4–43, and is said to be *OFF.* A low input to the inverter results in a high output, and it is now obvious why this circuit is called an inverter.

In designing and analyzing transistor inverters, it is usually assumed that $I_{C(sat)} = V_{CC}/R_C$ and that $V_{CE(sat)} = 0$ V. These are very good approximations and lead to results that are valid for most practical applications. Under these assumptions, we can easily derive the voltage–current relations in a transistor inverter. Since the transistor is cut off when the input is low, regardless of the values of R_B and R_C, the equations we will study are those that apply when the input is high. Actually, these equations are precisely those we have already derived for a CE transistor, for the special case $I_C = I_{C(sat)}$. Thus,

$$I_C = I_{C(sat)} = \frac{V_{CC}}{R_C} \tag{4–32}$$

$$I_B = \frac{I_{C(sat)}}{\beta} = \frac{V_{CC}}{\beta R_C} \tag{4–33}$$

$$I_B = \frac{V_{HI} - V_{BE}}{R_B} \tag{4–34}$$

where V_{HI} is the high level of the input voltage, usually the same as V_{CC}.

Example 4–16

Verify that the circuit in Figure 4–44 behaves like an inverter when the input switches between 0 V and +5 V. Assume that the transistor is silicon and that $\beta = 100$.

Solution. It is only necessary to verify that the transistor is saturated when $V_{in} = +5$ V. From equation 4–34,

$$I_B = \frac{(5 - 0.7)\text{ V}}{430\text{ k}\Omega} = 10\ \mu\text{A}$$

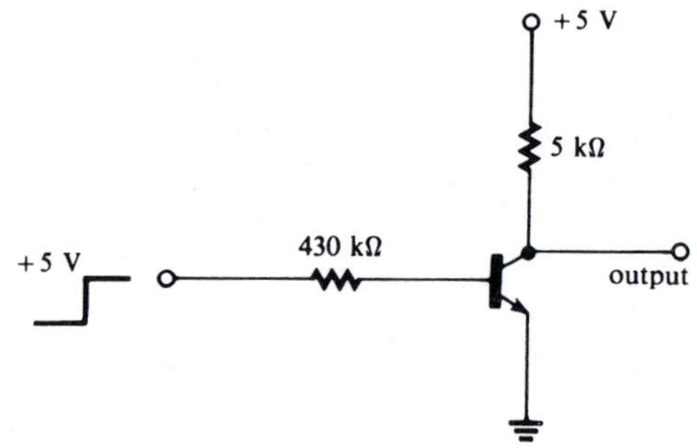

Figure 4–44
(Example 4–16)

Then $$I_C = \beta I_B = 100(10\ \mu\text{A}) = 1\text{ mA}$$
and $$V_{CE} = 5 - (1\text{ mA})(5\text{ k}\Omega) = 5 - 5 = 0\text{ V} = V_{CE(sat)}$$

Inverter Design

To design a transistor inverter we must have criteria for specifying the values of R_B and R_C. Typically, one of the two is known (or chosen arbitrarily), and the value of the other is derived from the first. Using equations 4–33 and 4–34, we can obtain the following relationships between R_B and R_C:

$$R_B = \frac{V_{HI} - V_{BE}}{I_B} = \frac{(V_{HI} - V_{BE})\beta R_C}{V_{CC}} \tag{4–35}$$

$$R_C = \frac{V_{CC}R_B}{\beta(V_{HI} - V_{BE})} \tag{4–36}$$

Equation 4–35 can be used to find R_B when R_C is known, and equation 4–36 to find R_C when R_B is known. However, since these equations are valid only for a specific value of β, they are not entirely practical. We have already discussed the fact that the β of a transistor of a given type is liable to vary over a wide range. If the actual value of β is smaller than the one used in the design equations, the transistor will not saturate. For this reason, the β used in the design equations should always be the *smallest* possible value that might occur in a given application. In other words, equations 4–35 and 4–36 are more practical when expressed in the form of inequalities, as follows:

$$R_B \leq \frac{(V_{HI} - V_{BE})\beta R_C}{V_{CC}} \tag{4–37}$$

$$R_C \geq \frac{V_{CC}R_B}{\beta(V_{HI} - V_{BE})} \tag{4–38}$$

These inequalities should hold for the entire range of β-values that transistors used in the inverter may have. This will be the case if the minimum possible β-value is used.

We should note that when a transistor has a higher value of β than the one for which the inverter circuit is designed, a high input simply drives it deeper into saturation. This *overdriving* of the transistor creates certain new problems, including the fact that it slows the speed at which the device can switch from ON to OFF, but the output is definitely low in the ON state.

Example 4–17

DESIGN

An inverter having $R_C = 1.5\ \text{k}\Omega$ is to be designed so that it will operate satisfactorily with silicon transistors whose β-values range from 80 to 200. What value of R_B should be used? Assume that $V_{CC} = V_{HI} = +5$ V.

Solution. Using equation 4–35 with $\beta = \beta_{(min)} = 80$, we find

$$R_B = \frac{(V_{HI} - V_{BE})\beta_{(min)}R_C}{V_{CC}} = \frac{(4.3)(80)(1.5\ \text{k}\Omega)}{5} = 103.2\ \text{k}\Omega$$

The Transistor as a Switch

A transistor inverter is often called a transistor *switch.* This terminology is appropriate because the ON and OFF states of the transistor correspond closely to the closing and opening of a switch connected between the collector and the emitter. When the transistor is ON, or saturated, the voltage between collector and emitter is nearly 0, as it would be across a closed switch, and the current is the maximum possible, V_{CC}/R_C. When the transistor is OFF, zero current flows from collector to emitter and the voltage is maximum, as it would be across an open switch. The switch is opened or closed by the input voltage: a high input closes it and a low input opens it. Figure 4–45 illustrates these ideas.

In many switching applications, the emitter may be connected to another circuit, or to another voltage source, instead of to ground. When analyzing such complex digital circuits, it is quite helpful to think of the transistor as simply a switch, for then it is easy to understand circuit operation in terms of the collector circuit being

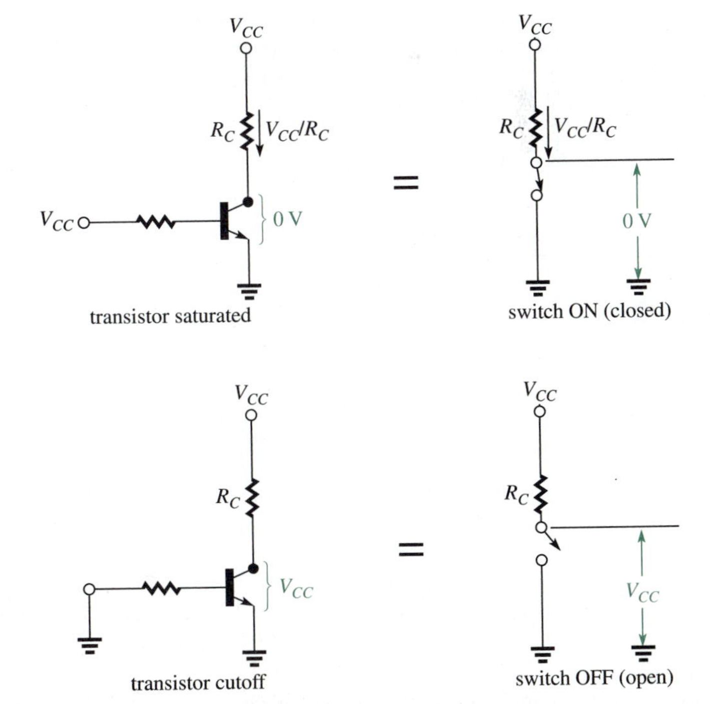

Figure 4–45
The transistor as a voltage-controlled switch. A high input closes the switch and a low input opens it.

connected to or disconnected from the emitter circuit. For example, if the emitter in the basic inverter circuit were connected to −5 V instead of ground, then the output would clearly switch between +5 V and −5 V instead of between +5 V and 0 V.

4–9 TRANSISTOR TYPES, RATINGS, AND SPECIFICATIONS

In modern electronic circuits, discrete transistors are used primarily for applications in which only one or a small number of devices are required, and in applications where substantial power is dissipated. Although older designs, composed entirely of discrete devices, can still be found in large numbers, most new circuits containing a large number of transistors are constructed in integrated-circuit form. In many applications, both discrete and integrated components are used. In these applications, the integrated circuit typically performs complex, low-level *signal conditioning,* and a discrete transistor then drives a power-consuming load such as an indicator lamp or an audio speaker. This use of the transistor is an example of *interfacing;* it provides a link between a device having limited power capabilities and a load that requires large voltages or currents.

Discrete transistors are packaged in a wide variety of metal and plastic enclosures (cases). Figure 4–46 shows some of the standard case types, which are identified

TO-204 (TO-3)
PACKAGE SUFFIX A

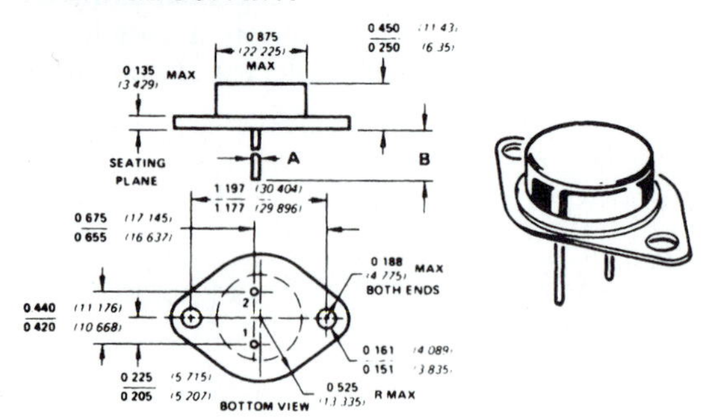

All Dimensions in Inches
(All Dimensions in Millimeters)

TO-220
PACKAGE SUFFIX D

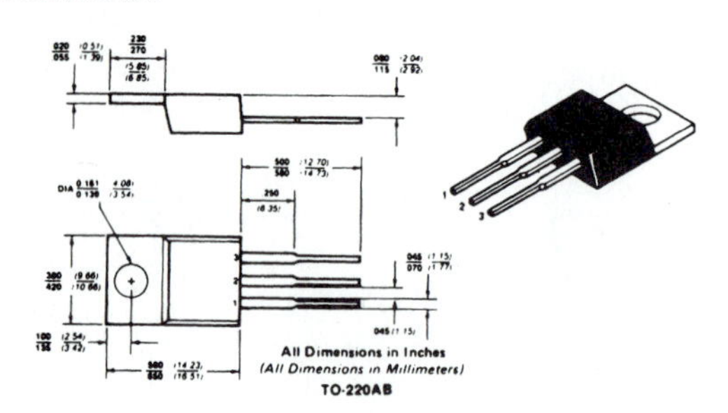

TO-205 (TO-39)
PACKAGE SUFFIX B

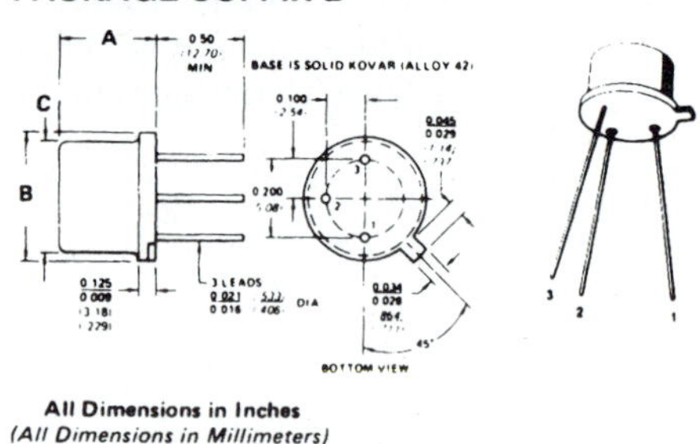

All Dimensions in Inches
(All Dimensions in Millimeters)

TO-92-18 (WITH TO-18 LEAD FORM)
PACKAGE SUFFIX L

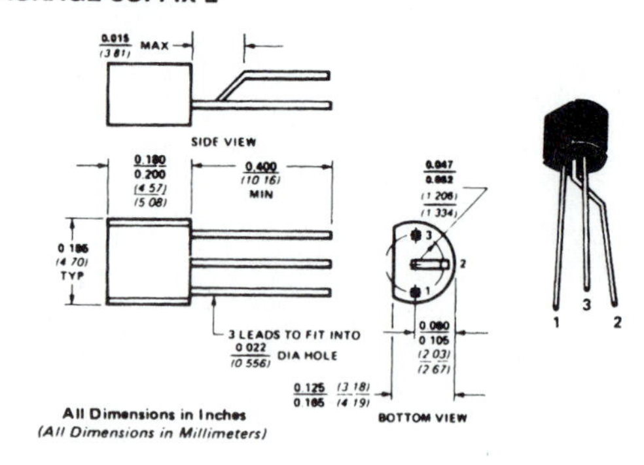

Figure 4–46
A few of the standard transistor case (enclosure) types, with TO-designations (old JEDEC numbers in parentheses) (Courtesy of Siliconix Inc.)

by standard TO-numbers. Three leads are brought out through each enclosure to permit external connection to the transistor's emitter, base, and collector. In some power transistors, rated for high power dissipation, the collector is attached and electrically common to the metal case. (The majority of the power dissipated in a transistor occurs at the collector–base junction, since the collector voltage is usually the largest voltage in the device.) A transistor manufacturer uses a consistent scheme that can be followed to identify the base, emitter, and collector terminals for a given case type. For example, in the TO-39 case, the three leads are attached in a semicircular cluster and a metal tab on the case is adjacent to the emitter. The base is the center lead in the cluster and the collector is the remaining lead.

A discrete transistor of a specific type, having registered JAN (military) specifications, is identified by a number with the prefix 2N. While all transistors having the same number may not be identical, they are all designed to meet the same performance specifications related to voltage and current limits, power dissipation, operating temperature range, and parameter variations. More than one manufacturer may produce a transistor with a given 2N number. Many manufacturers also produce "commercial"-grade devices that do not have 2N designations.

Figure 4–47(a) and 4–47(b), see page 132, shows parts of a typical set of

2N2218,A/2N2219,A
2N2221,A/2N2222,A
2N5581/82

JAN, JTX, JTXV AVAILABLE

2N2218,A
2N2219,A
CASE 79-02
TO-39 (TO-205AD)

2N2221,A
2N2222,A
CASE 22-03
TO-18 (TO-206AA)

2N5581
2N5582
CASE 26-03
TO-46 (TO-206AB)

GENERAL PURPOSE TRANSISTOR

NPN SILICON

MAXIMUM RATINGS

Rating	Symbol	2N2218 2N2219 2N2221 2N2222	2N2218A 2N2219A 2N2221A 2N2222A	2N5581 2N5582	Unit
Collector-Emitter Voltage	V_{CEO}	30	40	40	Vdc
Collector-Base Voltage	V_{CBO}	60	75	75	Vdc
Emitter-Base Voltage	V_{EBO}	5.0	6.0	6.0	Vdc
Collector Current — Continuous	I_C	800	800	800	mAdc
		2N2218,A 2N2219,A	2N2221,A 2N2222,A	2N5581 2N5582	
Total Device Dissipation @ T_A = 25°C Derate above 25°C	P_D	0.8 4.57	0.4 2.28	0.6 3.33	Watt mW/°C
Total Device Dissipation @ T_C = 25°C Derate above 25°C	P_D	3.0 17.1	1.2 6.85	2.0 11.43	Watts mW/°C
Operating and Storage Junction Temperature Range	T_J, T_{stg}	−65 to +200			°C

FIGURE 1 – NORMALIZED DC CURRENT GAIN

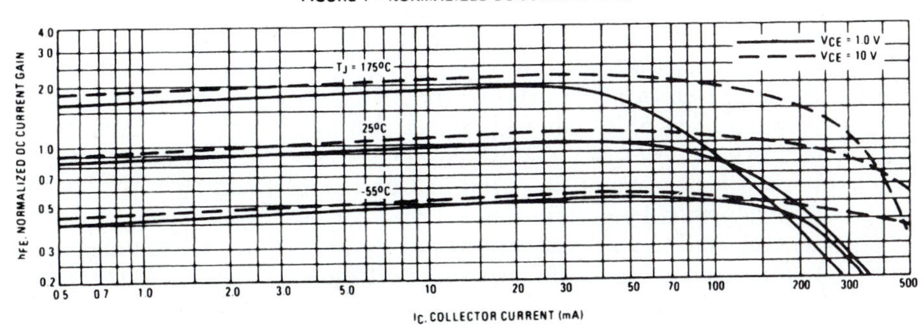

Figure 4–47(a)
Typical transistor specifications, showing maximum ratings and beta (h_{FE}) variation (Courtesy of Motorola Corporation)

ELECTRICAL CHARACTERISTICS (T_A = 25°C unless otherwise noted.)

Characteristic		Symbol	Min	Max	Unit
OFF CHARACTERISTICS					
Collector-Emitter Breakdown Voltage		$V_{(BR)CEO}$			Vdc
(I_C = 10 mAdc, I_B = 0)	Non-A Suffix		30	—	
	A-Suffix, 2N5581, 2N5582		40	—	
Collector-Base Breakdown Voltage		$V_{(BR)CBO}$			Vdc
(I_C = 10 μAdc, I_E = 0)	Non-A Suffix		60	—	
	A-Suffix, 2N5581, 2N5582		75	—	
Emitter-Base Breakdown Voltage		$V_{(BR)EBO}$			Vdc
(I_E = 10 μAdc, I_C = 0)	Non-A Suffix		5.0	—	
	A-Suffix, 2N5581, 2N5582		6.0	—	
Collector Cutoff Current (V_{CE} = 60 Vdc, $V_{EB(off)}$ = 3.0 Vdc)	A-Suffix, 2N5581, 2N5582	I_{CEX}	—	10	nAdc
Collector Cutoff Current		I_{CBO}			μAdc
(V_{CB} = 50 Vdc, I_E = 0)	Non-A Suffix		—	0.01	
(V_{CB} = 60 Vdc, I_E = 0)	A-Suffix, 2N5581, 2N5582		—	0.01	
(V_{CB} = 50 Vdc, I_E = 0, T_A = 150°C)	Non-A Suffix		—	10	
(V_{CB} = 60 Vdc, I_E = 0, T_A = 150°C)	A-Suffix, 2N5581, 2N5582		—	10	
Emitter Cutoff Current (V_{EB} = 3.0 Vdc, I_C = 0)	A-Suffix, 2N5581, 2N5582	I_{EBO}	—	10	nAdc
Base Cutoff Current (V_{CE} = 60 Vdc, $V_{EB(off)}$ = 3.0 Vdc)	A-Suffix	I_{BL}	—	20	nAdc
ON CHARACTERISTICS					
DC Current Gain		h_{FE}			—
(I_C = 0.1 mAdc, V_{CE} = 10 Vdc)	2N2218,A, 2N2221,A, 2N5581(1)		20	—	
	2N2219,A, 2N2222,A, 2N5582(1)		35	—	
(I_C = 1.0 mAdc, V_{CE} = 10 Vdc)	2N2218,A, 2N2221,A, 2N5581		25	—	
	2N2219,A, 2N2222,A, 2N5582		50	—	
(I_C = 10 mAdc, V_{CE} = 10 Vdc)	2N2218,A, 2N2221,A, 2N5581(1)		35	—	
	2N2219,A, 2N2222,A, 2N5582(1)		75	—	
(I_C = 10 mAdc, V_{CE} = 10 Vdc, T_A = −55°C)	2N2218A, 2N2221A, 2N5581		15	—	
	2N2219A, 2N2222A, 2N5582		35	—	
(I_C = 150 mAdc, V_{CE} = 10 Vdc)(1)	2N2218,A, 2N2221,A, 2N5581		40	120	
	2N2219,A, 2N2222,A, 2N5582		100	300	

Figure 4–47(b)
Typical transistor specifications for electrical (dc) characteristics (Courtesy of Motorola Corporation)

manufacturer's transistor specifications. Included are the "maximum ratings" and "electrical characteristics" of the 2N2218,A, 2N2219,A, 2N2221,A, 2N2222,A, and 2N5581/82 series of silicon transistors. Notice the various case styles in which these transistors are available.

The maximum ratings in Figure 4–47(a) show the maximum voltages that each device can sustain between different sets of transistor terminals and the maximum power dissipation, P_D, at 25°C. Notice that the power dissipation is *derated* by a "derating factor" for operation at higher temperatures. For example, with the derating factor of 2.28 mW/°C, the 2N2221A would have a maximum rated power dissipation at 55°C equal to $P_D = (0.8\text{ W}) - (55°\text{C} - 25°\text{C})[(2.28\text{ mW})/°\text{C}] = (0.8\text{ W}) - (0.0684\text{ W}) = 0.7316\text{ W}$. The effect of temperature on transistor specifications will be covered in detail in Chapter 15 (Section 15–3). A transistor circuit designer must be certain that a transistor used in a particular application will not be subjected to voltages or power dissipations that exceed the specified maximums; failure to do so may result in severe performance degradation or permanent damage.

The graph labeled "Normalized DC Current Gain" shows how β varies with V_{CE}, junction temperature (T_J), and collector current. (In the CE configuration, current *gain* is the ratio of output current to input current, or I_C/I_B, which is approximately β. As we shall learn later when h parameters are discussed, β is also designated by h_{FE}.) Notice that the curves are *normalized,* a typical practice in transistor data sheets. For example, at $I_C = 3$ mA, $V_{CE} = 10$ V, and $T_J = 25°C$, the normalized β (h_{FE}) equals 1, so the normalized β at another combination of I_C, V_{CE}, and T_J is interpreted as a multiple of the β value at 3 mA, 10 V, and 25°C. To illustrate, the curves show that the normalized β at $I_C = 3$ mA, $V_{CE} = 10$ V, and $T_J = 175°C$ is approximately 2. Therefore, if a particular transistor had a β of 60 at 3 mA, 10 V, and 25°C, we could expect it to have a β of 120 at 3 mA, 10 V, and 175°C. The chart illustrates a typical transistor characteristic: For a fixed collector current, *the value of β increases with increasing temperature.*

The *electrical characteristics* listed in Figure 4–47(b) show important transistor parameters associated with the dc operation of each device. Included are breakdown voltages and reverse leakage currents (called "cutoff" currents in the specifications). Notice that the breakdown voltages previously referred to as BV_{CBO} and BV_{CEO} are listed as $V_{(BR)CBO}$ and $V_{BR(CEO)}$, respectively.

Figure 4–46 does not show some other specifications that are usually furnished by a manufacturer, including small-signal characteristics. Small-signal characteristics are associated with the ac operation of a transistor, which we will cover in Chapter 5. Other specifications often furnished by the manufacturer include graphs showing additional parameter variations with temperature, voltage, and current.

4–10 TRANSISTOR CURVE TRACERS

We have mentioned that characteristic curves are seldom included in transistor specifications. These vary widely among transistors of a given type and are rarely used for circuit design purposes. However, in areas such as component testing, preliminary circuit development, and research, it is often useful to be able to study the characteristic curves of a single device and to obtain important parameter values from the curves. Recall that parameters such as α, β, BV_{CBO}, BV_{CEO}, leakage currents, saturation voltages, and the Early voltage can be discerned from appropriate sets of characteristic curves.

The most widely used method for obtaining a set of characteristic curves is by use of an instrument called a *transistor curve tracer.* A curve tracer is basically an oscilloscope equipped with circuitry that automatically steps the currents (or voltages) in a semiconductor device through a range of values and displays the family of characteristic curves that result. Selector switches allow the user to set the maximum value and the increment (step) value of each current or voltage applied to the device. For example, to obtain a family of transistor collector characteristics, the user might set the base current increment to be 10 μA, the maximum collector voltage to 25 V, and the number of steps to be 10. The characteristics would then be displayed as a family of curves showing I_C versus V_{CE} for $I_B = 0$, 10 μA, 20 μA, ... Figure 4–48 shows a typical curve tracer.

Figure 4–49(a) is a photograph of a curve tracer display showing a typical set of NPN collector characteristics. The horizontal sensitivity of the display was set for 2 V/division, so the horizontal axis (V_{CE}) extends from 0 to about 13 V. The vertical sensitivity was set for 1 mA/division, so the vertical axis (I_C) extends from 0 to about 6.5 mA. The base current increment is 10 μA. Using this display, we can determine, for example, that the β of the transistor at $V_{CE} = 4$ V and $I_B = 40$ μA is approximately

$$\beta = \frac{I_C}{I_B} = \frac{3\text{ mA}}{40\ \mu\text{A}} = 75$$

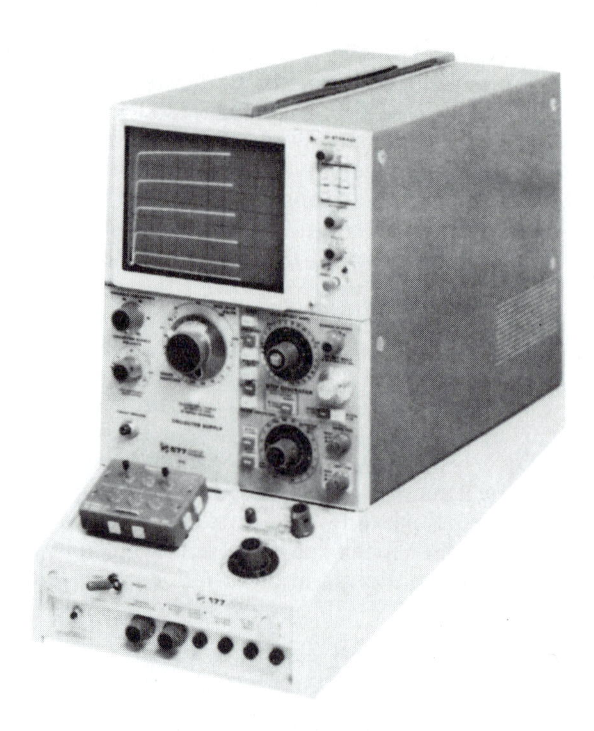

Figure 4–48
A typical transistor curve tracer (Tektronix Model 577) (Courtesy of Tektronix, Inc.)

The curve tracer from which this display was obtained permits the user to select a value of series collector resistance (R_C), which, for the display shown, was set to 2 kΩ. Notice that the base current curves become shorter with increasing current. *An imaginary line connecting the right-hand tips of each curve represents the load line for the circuit.* This load line is seen to intersect the V_{CE}-axis at 13 V and the

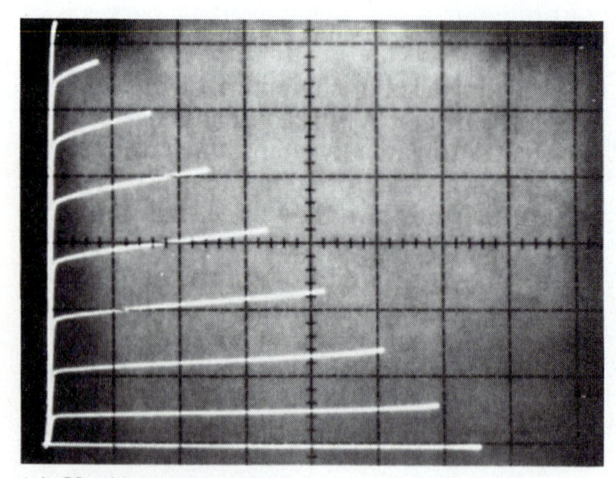

(a) Horizontal: 2 V/div
Vertical: 1 mA/div
ΔI_B = 10 μA

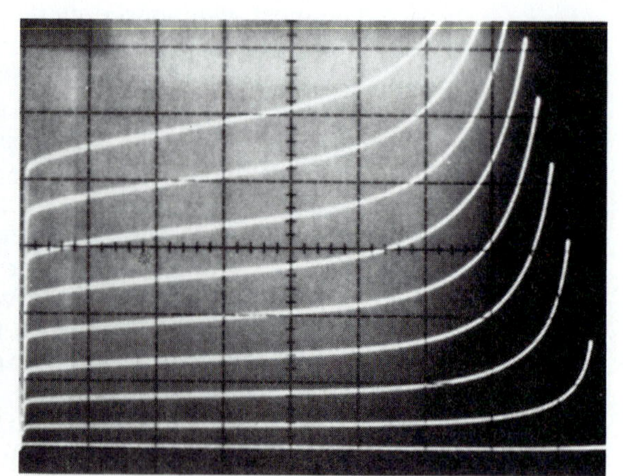

(b) Horizontal: 2 V/div
Vertical: 0.2 mA/div
ΔI_B = 2 μA

Figure 4–49
Photographs of curve tracer displays

I_C-axis at 6.5 mA. Thus, the value of V_{CC} used in this circuit is 13 V, and the load line intersects the I_C-axis at the value expected:

$$I_C = \frac{V_{CC}}{R_C} = \frac{13\text{ V}}{2\text{ k}\Omega} = 6.5\text{ mA}$$

One convenient feature of a curve tracer is that it permits a user to expand or contract the display in different regions of the characteristic curves by adjusting the sensitivity and range controls.

Figure 4–49(b) shows collector characteristics of the same transistor when the curve tracer settings are adjusted to generate larger values of V_{CE}. In this example, the horizontal sensitivity is 2 V/division, the vertical sensitivity is 0.2 mA/division, and the base current increment is 2 μA. With these settings, the breakdown characteristics are clearly evident. For example, at $I_B = 12\ \mu$A and $V_{CE} = 12$ V, it can be seen that the transistor is in its breakdown region and that the collector current is approximately 0.84 mA.

Most curve tracers can be used to obtain characteristic curves for devices other than transistors. Some even have special adapters that allow the testing of integrated circuits. Figure 4–50 shows photographs of diode characteristics that were obtained from a curve tracer display. The forward and reverse characteristics are shown in Figure 4–50(a), with a horizontal sensitivity of 10 V/division. It can be seen that the diode enters breakdown at a reverse voltage of about 25 V. With this scale the forward characteristic essentially coincides with the vertical axis. However, when the sensitivity is set to 0.2 V/division, the forward characteristic appears as shown in Figure 4–50(b) and can be examined in detail. (The origin of the axes is at the center of the display.) We see that the knee of the characteristic occurs at about 0.62 V, and there is sufficient detail to compute dc and ac resistances in the forward region.

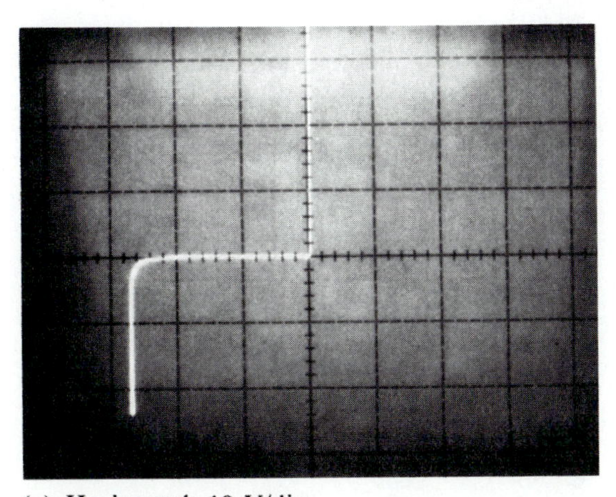

(a) Horizontal: 10 V/div

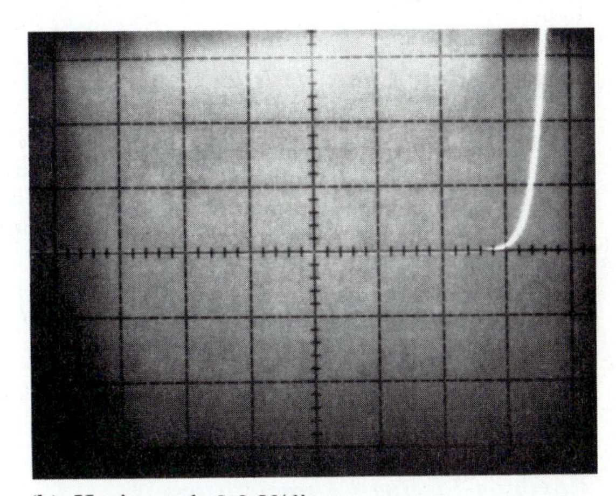

(b) Horizontal: 0.2 V/div

Figure 4–50
Curve tracer displays of diode characteristics

EXERCISES

Section 4–2
Theory of BJT Operation

4–1. In a certain transistor, the emitter current is 1.01 times as large as the collector current. If the emitter current is 12.12 mA, find the base current.

4–2. A (conventional) current of 26 μA flows out of the base of a certain transistor. The emitter current is 0.94 mA. What is the collector current and what kind of transistor is it (NPN or PNP)? Draw a transistor symbol and label all current flows, showing directions and magnitudes.

4–3. In a certain transistor, 99.5% of the carriers injected into the base cross the collector–base junction. If the leakage current is 5.0 μA and the collector current is 22 mA, find
a. the exact α;
b. the emitter current; and
c. the approximate α when I_{CBO} is neglected.

4–4. A germanium transistor has a surface leakage current of 1.4 μA and a reverse current due to thermally generated minority carriers of 1.2 nA at 10°C. If $\alpha = 0.992$ and $I_E = 0.8$ mA, find I_C at 10°C and at 90°C. (Assume that surface leakage is independent of temperature.)

4–5. Using equation 4–2 and neglecting I_{CBO}, derive the following approximation: $I_B \approx (1 - \alpha)I_E$.

Section 4–3
Common-Base Characteristics

4–6. A certain transistor has the CB input characteristics shown in Figure 4–10. It is desired to hold I_E constant at 9.0 mA while V_{CB} is changed from 0 V to 25 V. What change in V_{BE} must accompany the change in V_{CB}?

4–7. A transistor has the CB input characteristics shown in Figure 4–10. If $\alpha = 0.95$, find I_C when $V_{BE} = 0.72$ V and $V_{CB} = 10$ V.

Section 4–4
Common-Emitter Characteristics

4–8. A transistor has an α of 0.98 and a collector-to-base leakage current of 0.02 μA.
a. Find its collector-to-emitter leakage current.
b. Find the β of the transistor.
c. Find I_C when $I_B = 0.04$ mA.
d. Find the approximate I_C, neglecting leakage current.

4–9. A transistor has $I_{CBO} = 0.1$ μA and $I_{CEO} = 16$ μA. Find its α.

4–10. Derive the relation $\alpha = \beta/(\beta + 1)$. (*Hint:* Solve equation 4–8 for α.)

4–11. Under what condition is the approximation $I_{CEO} \approx \beta I_{CBO}$ valid?

4–12. A transistor has the CE output characteristics shown in Figure 4–19.
a. Find the emitter current at $V_{CE} = 12.5$ V and $I_B = 60$ μA.
b. Find the α at that point (neglecting leakage current).

4–13. An NPN transistor has the CE input characteristics shown in Figure 4–18 and the CE output characteristics shown in Figure 4–19.
a. Find I_C when $V_{BE} = 0.7$ V and $V_{CE} = 20$ V.
b. Find the β of the transistor at that point (neglecting leakage current).

4–14. Using graphical methods, determine the approximate value of the Early voltage for the transistor whose CE output characteristics are shown in Figure 4–51.

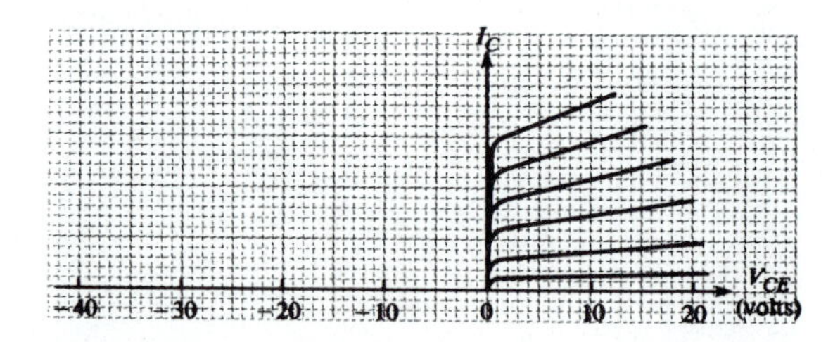

Figure 4–51
(Exercise 4–14)

4–15. In a certain experiment, the collector current of a transistor was measured at different values of collector–emitter voltage, while the base current was held constant. The results of the experiment are summarized in the following table:

$I_B = 100\ \mu A$

V_{CE}	I_C
5 V	15 mA
10 V	16 mA
15 V	17 mA
20 V	18 mA

$I_B = 200\ \mu A$

V_{CE}	I_C
5 V	30 mA
10 V	32 mA
15 V	34 mA
20 V	36 mA

$I_B = 300\ \mu A$

V_{CE}	I_C
5 V	45 mA
10 V	48 mA
15 V	51 mA
20 V	54 mA

Plot the experimental data and graphically determine approximate values for the following:

a. β at $V_{CE} = 8$ V and $I_B = 100\ \mu A$
b. β at $V_{CE} = 14$ V and $I_B = 250\ \mu A$
c. the Early voltage

Section 4–5
Common-Collector Characteristics

4–16. A certain transistor has the common-collector output characteristics shown in Figure 4–26. Neglecting leakage current, find approximate values for
a. β at $V_{CE} = 12.5$ V and $I_B = 20\ \mu A$;
b. I_E at $V_{CE} = 12.5$ V and $I_B = 45\ \mu A$; and
c. α at $V_{CE} = 2.5$ V and $I_B = 70\ \mu A$.

4–17. Prove that equation 4–16 is equivalent to $I_E = I_B/(1 - \alpha)$.

Section 4–6
Bias Circuits

4–18. Determine the equation for the load line of the circuit shown in Figure 4–52. Sketch the line and label the values of its intercepts.

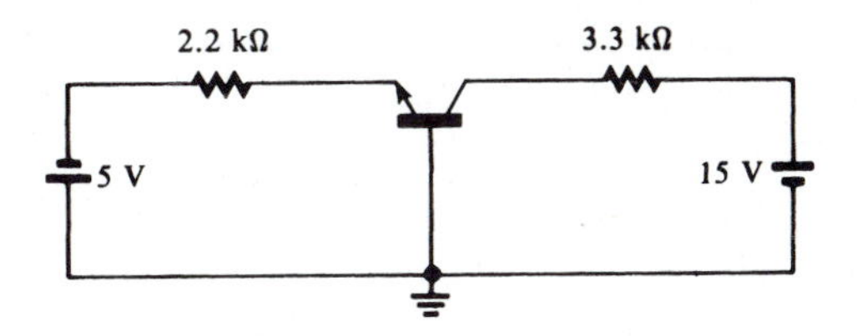

Figure 4–52
(Exercise 4–18)

4–19. In the circuit of Figure 4–52, find
a. I_C when $V_{CB} = 10$ V; and
b. V_{CB} when $I_C = 1$ mA.

4–20. In the circuit shown in Figure 4–53, find
a. I_C when $V_{BC} = 20$ V; and
b. V_{BC} when $I_C = 4.2$ mA.

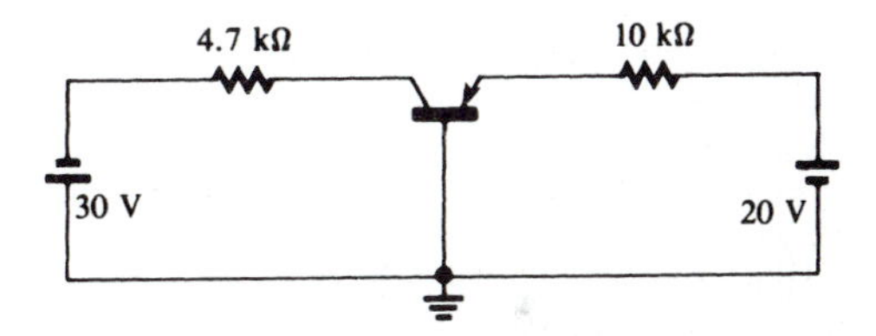

Figure 4–53
(Exercise 4–20)

4–21. The silicon transistor shown in Figure 4–54 has the CB output characteristics shown in Figure 4–55.
a. Draw the load line on the characteristics and graphically determine V_{CB} and I_C at the bias point.
b. Determine the bias point without using the characteristic curves.

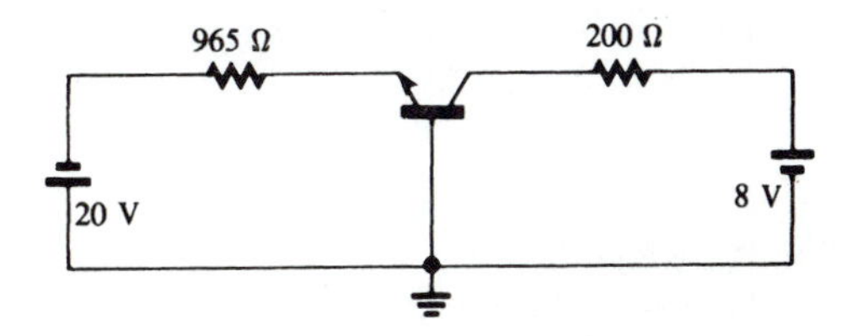

Figure 4–54
(Exercise 4–21)

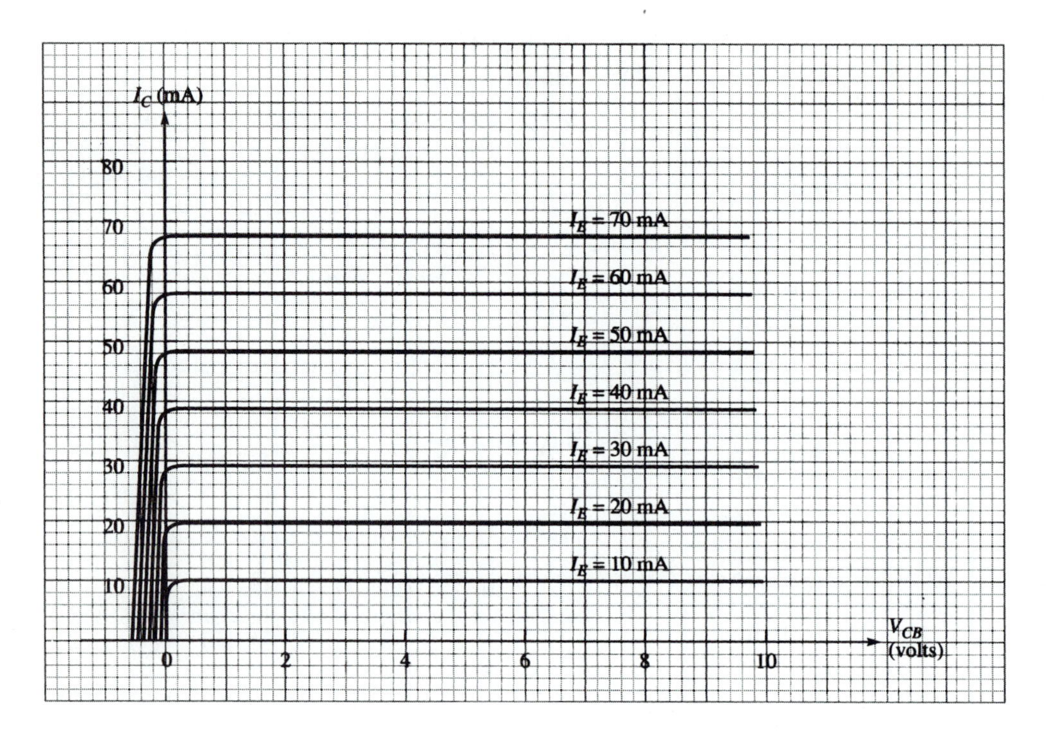

Figure 4–55
(Exercise 4–21)

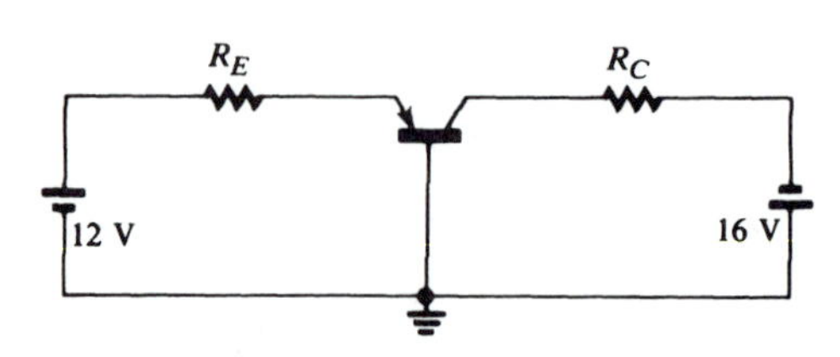

Figure 4–56
(Exercise 4–22)

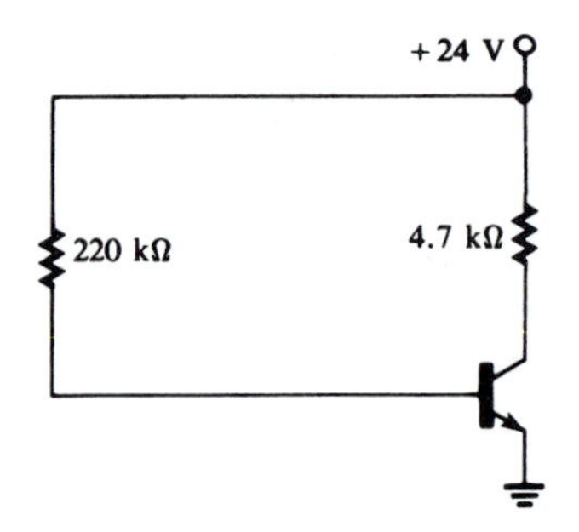

Figure 4–57
(Exercise 4–23)

4–22. The transistor shown in Figure 4–56 is germanium.
  a. If $R_C = 1\ \text{k}\Omega$, what value of R_E will cause V_{BC} to equal 0 V?
  b. If $R_E = 1.5\ \text{k}\Omega$, what value of R_C will cause V_{BC} to equal 0 V?

4–23. In the circuit shown in Figure 4–57, find
  a. V_{CE} when $I_C = 1.5$ mA,
  b. I_C when $V_{CE} = 12$ V, and
  c. V_{CE} when $I_C = 0$.

4–24. In the circuit shown in Figure 4–58, find
  a. V_{EC} when $I_C = 12$ mA;
  b. I_C when $V_{EC} = 2.5$ V; and
  c. V_{EC} when $I_C = 0$.

4–25. The silicon transistor shown in Figure 4–59

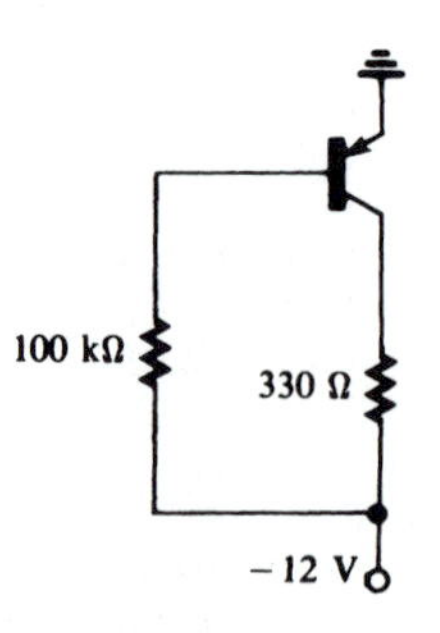

Figure 4–58
(Exercise 4–24)

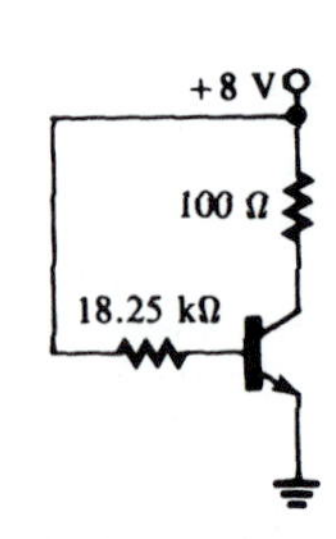

Figure 4–59
(Exercise 4–25)

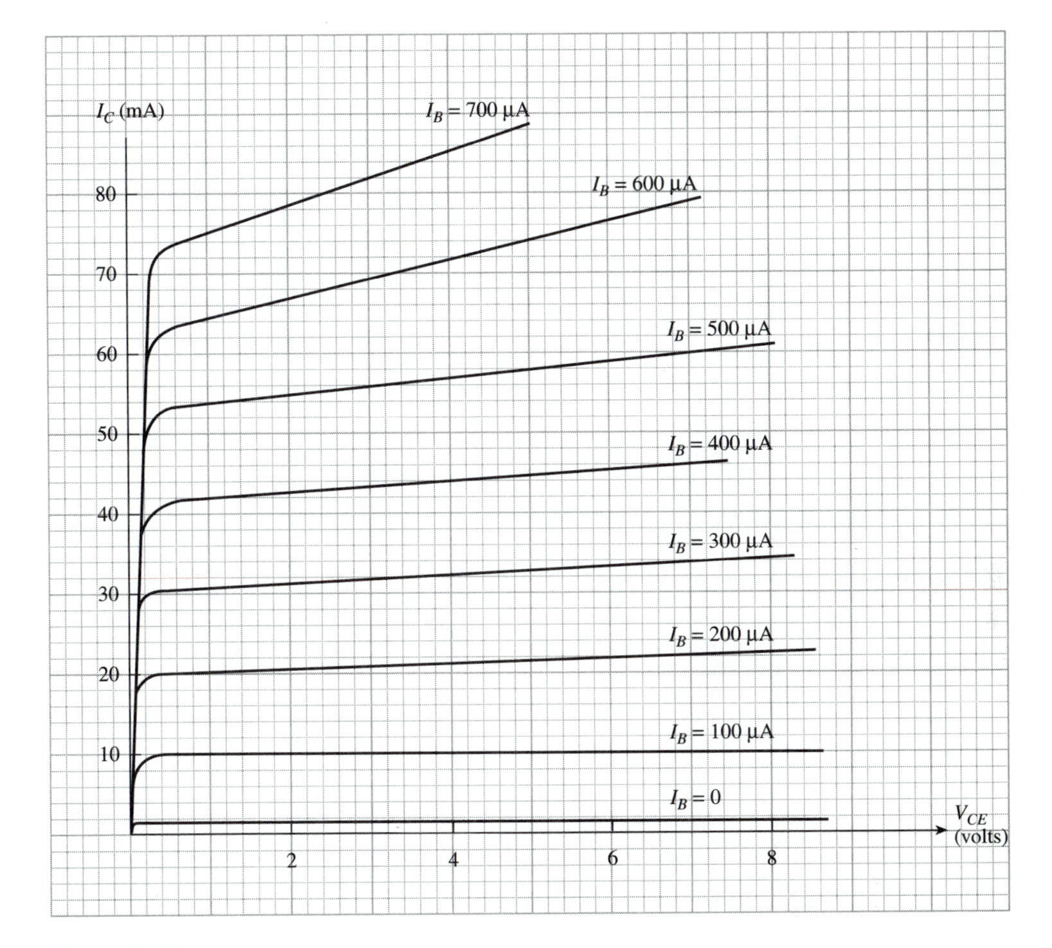

Figure 4–60
(Exercise 4–25)

has the CE output characteristics shown in Figure 4–60. Assume that $\beta = 105$.

a. Draw the load line on the characteristics and graphically determine V_{CE} and I_C at the bias point.

b. What is the approximate value of I_{CEO} for this transistor?

c. Calculate V_{CE} and I_C at the bias point without using the characteristic curves.

4–26. Assuming that $\beta = 150$, find the bias point of the germanium transistor shown in Figure 4–61.

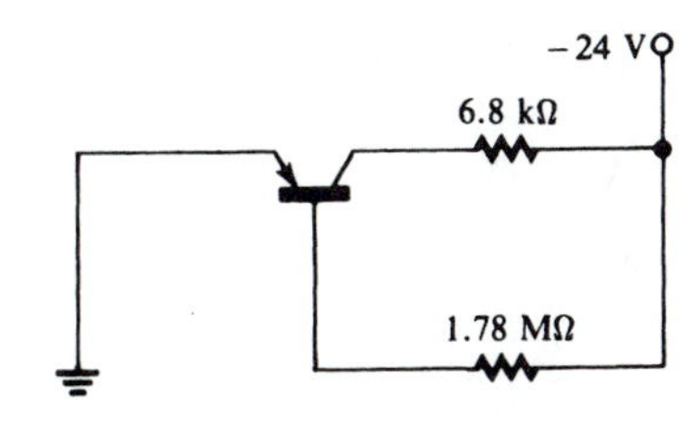

Figure 4–61
(Exercise 4–26)

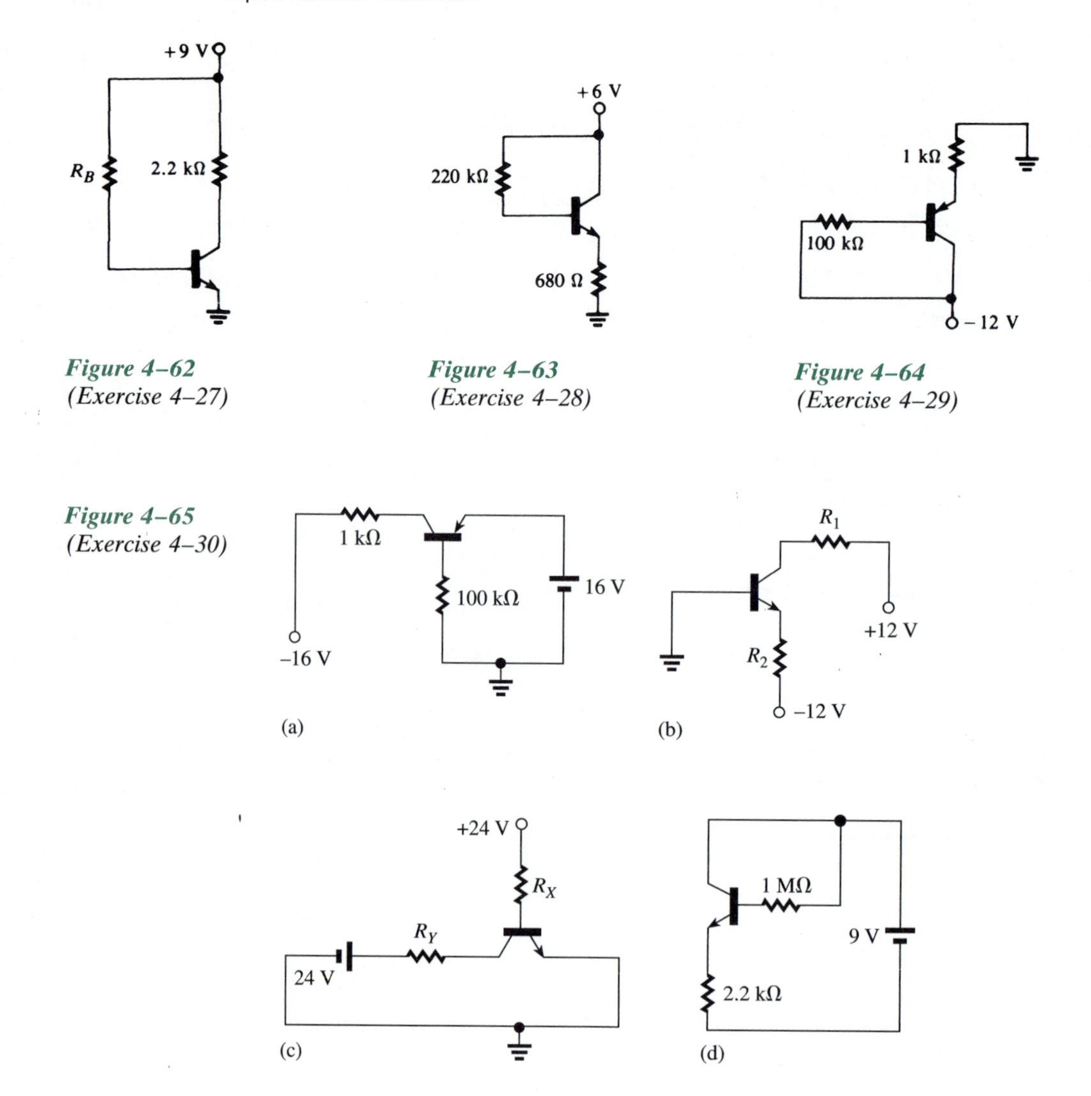

Figure 4–62
(Exercise 4–27)

Figure 4–63
(Exercise 4–28)

Figure 4–64
(Exercise 4–29)

Figure 4–65
(Exercise 4–30)

4–27. What value of R_B in the circuit shown in Figure 4–62 will just cause the silicon transistor to be saturated, assuming that $\beta = 100$ and $V_{CE(sat)} = 0.3$ V?

4–28. Calculate the bias point of the silicon transistor shown in Figure 4–63. Assume that $\beta = 80$.

4–29. Calculate the bias point of the silicon transistor shown in Figure 4–64. Assume that $\beta = 100$.

4–30. Determine the bias configuration (CB, CE, or CC) of the transistor in each part of Figure 4–65.

DESIGN EXERCISES

4–31. **a.** Design a bias circuit for an NPN silicon transistor in a common-base configuration. The bias point should be $I_E = 2$ mA and $V_{CB} = 9$ V. Supply voltages are +20 V and −10 V. Use standard-valued resistors with 5% tolerance and draw a schematic diagram of your design.

b. Calculate the range of possible values that the bias point could have, taking the resistor tolerances into consideration.

4–32. **a.** Design a bias circuit for a PNP silicon transistor in a common-emitter configuration. The nominal β of the transistor is 80 and the supply voltage is -24 V. The bias point is to be $I_C = 5$ mA, and $V_{CE} = -10$ V. Use standard-valued resistors with 10% tolerance and draw a schematic diagram of your design.
b. Calculate the actual bias point assuming the 10% resistors have their nominal values.
c. Calculate the range of possible values that the bias point could have if the value of β changed over the range from 50 to 100. Assume the resistors have their nominal values.

4–33. **a.** Design a bias circuit for an NPN silicon transistor in a common-collector configuration. The nominal β for the transistor is 100, and the supply voltage is 30 V. The bias point is to be $I_C = 10$ mA, and $V_{CE} = 12$ V. Use standard-valued resistors having 5% tolerance and draw a schematic diagram of your design.
b. Calculate the *minimum* value that V_{CE} could have if *both* the resistor tolerances and a variation in β from 60 to 120 are taken into account. (*Hint:* Use equations 4–28 to derive the expression

$$V_{CE} = V_{CC} - \frac{(V_{CC} - V_{BE})}{\dfrac{R_B}{(\beta + 1)R_E} + 1}$$

4–34. **a.** Design a bias circuit for an NPN silicon transistor having a nominal β of 100, to be used in a common-emitter configuration. The bias point is to be $I_C = 1$ mA, and $V_{CE} = 5$ V. The supply voltage is 15 V. Use standard-valued 5% resistors and draw a schematic diagram of your design.
b. Calculate the possible range of values of the bias point taking into consideration *both* the resistor tolerances and a possible variation in β from 30 to 150. Interpret and comment on your results.

Section 4–8
The BJT Inverter (Transistor Switch)

4–35. The input to the circuit shown in Figure 4–66 alternates between 0 V and 10 V. If the silicon transistor has a β of 120, verify that the circuit operates as an inverter.

4–36. What would be the output voltages from the inverter in Figure 4–66 if

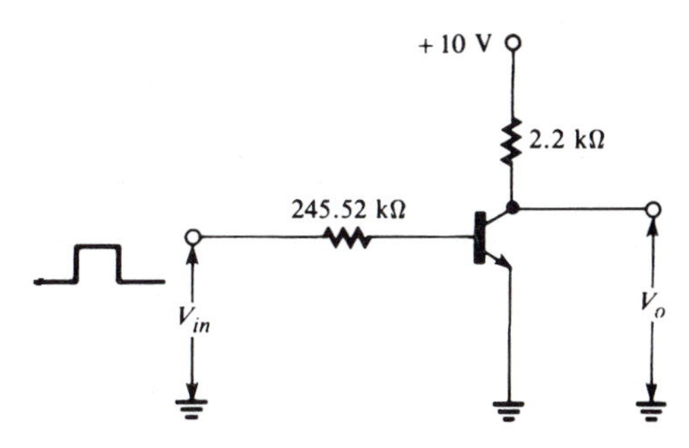

Figure 4–66
(Exercise 4–35)

a. the input voltage levels were changed to -5 V and $+10$ V?
b. the input voltage levels were changed to 0 V and $+15$ V?
c. the β of the transistor were changed to 150?

4–37. A transistor inverter is to be designed using a silicon transistor whose β may vary from 60 to 120. If the series base resistance is to be 100 kΩ, what should be the value of R_C? Assume that $V_{CC} = V_{HI} = 4.5$ V.

4–38. What is the minimum value of β for which the silicon transistor in Figure 4–67 will operate satisfactorily as an inverter?

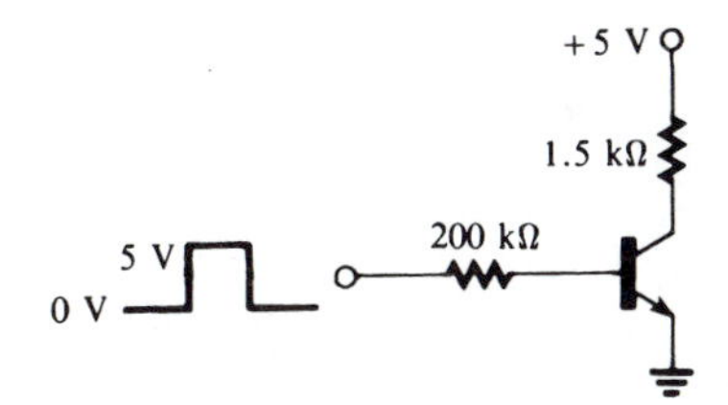

Figure 4–67
(Exercise 4–38)

Section 4–9
Transistor Types, Ratings, and Specifications

Note: In Exercises 4–39 through 4–40, refer to the manufacturer's specification sheets given in Section 4–9.

4–39. The input to the 2N2222A transistor in Figure 4–68 is a square wave that alternates between $\pm V$ volts. Assuming that no base current flows when the input is at $-V$ volts (so there is no drop across R_B), what is the maximum safe value for V? Assume that a safe value for V is one that does not exceed 80% of the manufacturer's rated maximum.

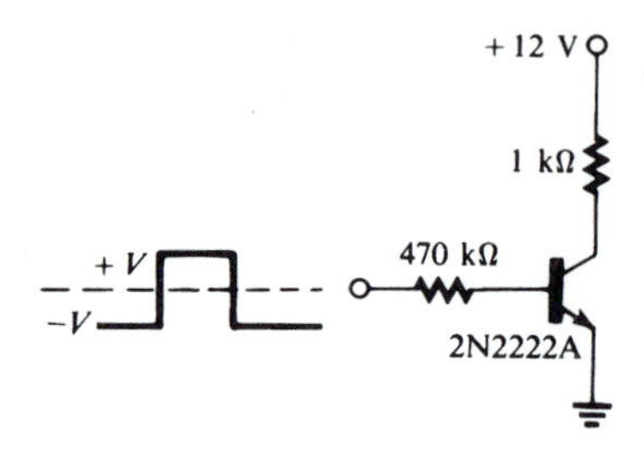

Figure 4–68
(Exercise 4–39)

4–40. A 2N2221A transistor is to be operated at 75°C. What is its maximum rated power dissipation at that temperature?

4–41. The β of a 2N2218A transistor is 40 at T = 25°C, V_{CE} = 10 V, and I_C = 3 mA. What typical value for β could be expected at
a. T = −55°C, I_C = 5 mA, and V_{CE} = 10 V?
b. T = 175°C, I_C = 80 mA, and V_{CE} = 10 V?

4–42. The β of a 2N2219A transistor is 140 at T = 175°C, V_{CE} = 10 V, and I_C = 80 mA. What typical value for β could be expected at
a. T = 25°C, I_C = 3 mA, and V_{CE} = 10 V?
b. T = −55°C, I_C = 5 mA, and V_{CE} = 10 V?

4–43. What is the manufacturer's rated maximum value for the current I in Figure 4–69? Assume that T = 25°C.

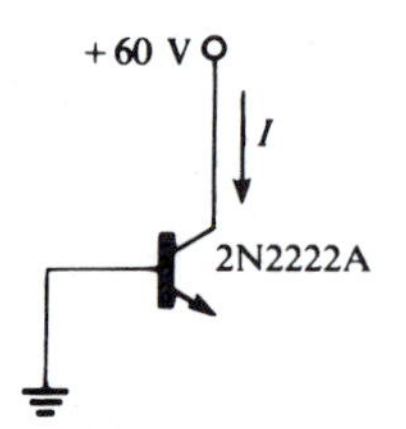

Figure 4–69
(Exercise 4–43)

4–44. A certain circuit is designed so that it will operate satisfactorily at T = 25°C if the transistor has a β of at least 50 when I_C = 10 mA and V_{CE} = 10 V. Which transistor(s) in the 2N2218A–2N2222A series can be used for the application?

Section 4–10
Transistor Curve Tracers

4–45. Using the curve tracer display of the collector characteristics shown in Figure 4–49(b), find the approximate β of the transistor when
a. V_{CE} = 7 V and I_B = 8 μA, and
b. V_{CE} = 8 V and I_B = 12 μA.

4–46. Using the curve tracer display of the diode characteristic shown in Figure 4–50, find the approximate dc resistance of the diode when it is forward biased by 0.64 V.

SPICE EXERCISES

Note: In the exercises that follow, assume all device parameters have their default values unless otherwise specified.

4–47. By using SPICE to generate common-emitter output characteristics of an NPN transistor at T = 0°C and at T = 90°C, determine the change in the values of I_{CEO} and β over that temperature range. V_{CE} should be stepped from 0 V to 20 V in 2-V increments, and I_B should be stepped from 0 to 500 μA in 50-μA increments. Determine the values of β at V_{CE} = 10 V and I_B = 100 μA. Determine the values of I_{CEO} at V_{CE} = 16 V and I_B = 0. Set the "beta temperature exponent" (XTB) to 1.5, and the forward Early voltage (VAF) to 131. (These are typical parameter values for the 2N2222 transistor.) Set the "ideal maximum forward beta" (BF) to 100.

4–48. By using SPICE to generate common-base output characteristics of an NPN transistor at T = 0°C and at T = 90°C, determine the change in the values of I_{CBO} and α over that temperature range. V_{CB} should be stepped from 0 V to 50 V in 5-V increments and I_E should be stepped from 0 to 10 mA in 2-mA increments. Determine the values of α at V_{CB} = 25 V and I_E = 8 mA. Determine the values of I_{CBO} at V_{CB} = 40 V and I_E = 0. Use the transistor parameter values given in Exercise 4–47.

4–49. Use SPICE to determine the bias point of the circuit shown in Figure 4–41 (Example 4–12). Set the "ideal maximum forward beta" (BF) to 120 and then repeat the analysis with BF = 240. Comment on the effect that the change in beta has on the bias point.

4–50. To investigate the effect of temperature changes on the bias point of an NPN transistor, use SPICE to determine I_C and V_{CE} in the circuit shown in Figure 4–32(a) at T = 0°C, T = 27°C, and T = 90°C. Use V_{CC} = 24 V, R_B = 470 kΩ, and R_C = 2.2 kΩ. Set the transistor parameter values given in Exercise 4–47. Using your results, calculate the percent changes in V_{CE} and I_C over the temperature range from 0°C to 90°C.

5 SMALL-SIGNAL BJT AMPLIFIERS

5–1 AC AMPLIFIER FUNDAMENTALS

In Chapter 4, we studied the dc, or *static,* characteristics of transistors and used these to determine the dc output voltages and currents that resulted from the application of dc inputs. We wish now to investigate the extent to which small *changes* in the input voltage and current cause changes in the corresponding output quantities. We know, for example, that increasing the input voltage (V_{BE}) to an NPN transistor in a CE configuration will cause the input current I_B to increase, which will then cause the output current I_C to increase, since $I_C = \beta I_B$. Likewise, decreasing the input voltage causes a decrease in output current. When the input variations are small enough to confine the variation in output voltage and current to the active portion of a transistor's output characteristics, we say that the device is operating under *small-signal conditions.* More precisely, we will define small-signal operation to occur when the output variations are so small that there is negligible change in the values of the device *parameters* (α, β, etc., about which we will have more to say later). Small-signal operation is studied from the standpoint of the transistor's behavior as an *ac amplifier.* Before discussing the ac characteristics of transistors, we will introduce some important general concepts that apply to all ac amplifiers.

Amplifier Gain

When the total change in the output voltage from a device is greater than the total change in the input voltage that caused it, the device is said to be an *ac voltage amplifier.* The ac *voltage gain,* designated A_v, is defined to be the ratio of the change in output voltage to the change in input voltage:

$$A_v = \frac{\Delta V_o}{\Delta V_{in}} \tag{5–1}$$

Thus, an ac voltage amplifier has $A_v > 1$. Figure 5–1 illustrates the concept. Note carefully in Figure 5–1 that only the ac components of the input and output voltages are used to compute the ac voltage gain. Both the input and output signals are shown superimposed on dc levels, but these dc values have no bearing on the value

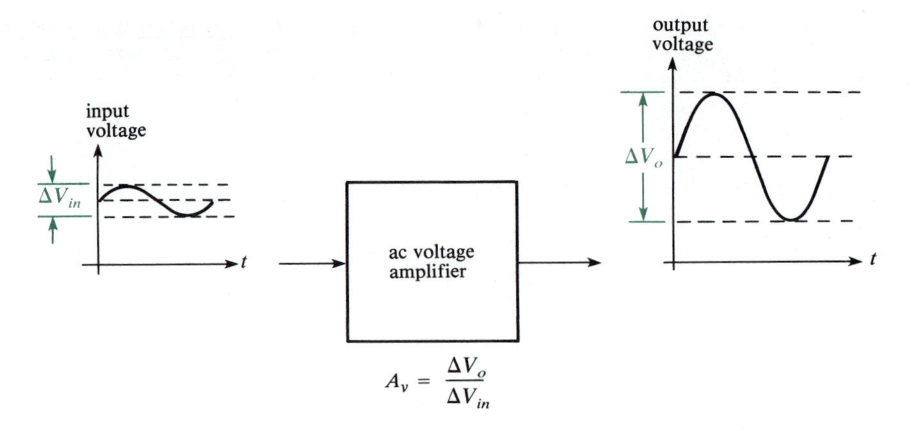

Figure 5–1
In an ac voltage amplifier, the total variation in output voltage, ΔV_o, is greater than the total variation in input voltage, ΔV_{in}.

of the ac voltage gain. The rms values of the ac input and output components can also be used to compute A_v:

$$A_v = \frac{v_o\,(\text{rms})}{v_{in}\,(\text{rms})} \tag{5–2}$$

Unless it is necessary to emphasize that we are referring to rms values, we will hereafter drop the rms notation, with the understanding that v_o and v_{in} mean rms values.

AC current gain, A_i, is defined to be the ratio of the total output current variation to the total input current variation:

$$A_i = \frac{\Delta I_o}{\Delta I_{in}} = \frac{i_o\,(\text{rms})}{i_{in}\,(\text{rms})} \tag{5–3}$$

A device having $A_i > 1$ is an *ac current* amplifier. In general, an ac amplifier may have $A_v > 1$, or $A_i > 1$, or both; in other words, it may amplify either voltage or current, or both. The *power* gain, A_p, is defined to be the ratio of output power to input power, and may be computed as the product of voltage gain and current gain:

$$A_p = P_o/P_{in} = A_v A_i \tag{5–4}$$

Although the word *gain* implies that there is an increase in signal level, it is possible to have a value of gain less than 1. For example, if an amplifier has a voltage gain equal to 0.5, it simply means that the ac output voltage variation is one-half that of the input. In this case, we say that the amplifier *attenuates* (reduces) the signal voltage applied to it.

Input and Output Resistance

The *input resistance* to an amplifier is the total equivalent resistance at its input terminals. The dc input resistance, R_{in}, is the resistance that a dc source would "see" when connected to the input terminals, and the ac resistance, r_{in}, is the resistance that an ac input source would see at the terminals. In either case, the input resistance can be computed as the ratio of input voltage to input current:

$$R_{in} = \frac{V_{in}}{I_{in}} \text{ (dc)} \qquad r_{in} = \frac{v_{in}}{i_{in}} \text{ (ac)} \tag{5–5}$$

The ac input power can be computed using any of the familiar power relations:

$$P_{in} = [v_{in}\text{ (rms)}][i_{in}\text{ (rms)}] = \frac{v_{in}^2\text{ (rms)}}{r_{in}} = [i_{in}^2\text{ (rms)}]r_{in} \tag{5–6}$$

The output resistance of an amplifier is the total equivalent resistance at its output terminals. Output resistance is the same as the Thevenin equivalent resistance that would appear in series with the output if the amplifier were replaced by its Thevenin equivalent circuit. Like input resistance, output resistance can be defined as a dc resistance R_o, or as an ac resistance, r_o. Output power can be computed using equation 5–6, by substituting the subscript *o* (*out*) for *in* in each term.

Example 5–1

Figure 5–2 shows the conventional symbol for an amplifier: a triangular block with output at the vertex. As shown in the figure, the input voltage to the amplifier is $v_{in}(t) = 0.7 + 0.008 \sin 10^3 t$ V. The amplifier has an ac current gain of 80. If the input current is $i_{in}(t) = 2.8 \times 10^{-5} + 4 \times 10^{-6} \sin 10^3 t$ A, and the ac component of the output voltage is 0.4 V rms, find (1) A_v, (2) R_{in}, (3) r_{in}, (4) i_o (rms), (5) r_o, and (6) A_p.

Solution

1. v_{in} (rms) $= 0.707(0.008 \text{ V-pk}) = 5.66 \times 10^{-3}$ V rms

$$A_v = \frac{v_o\text{ (rms)}}{v_{in}\text{ (rms)}} = \frac{0.4 \text{ V}}{5.66 \times 10^{-3}\text{ V}} = 70.7$$

2. The dc input resistance is the ratio of the dc component of the input voltage to the dc component of the input current:

$$R_{in} = \frac{V_{in}}{I_{in}} = \frac{0.7 \text{ V}}{2.8 \times 10^{-5}\text{ A}} = 25 \text{ k}\Omega$$

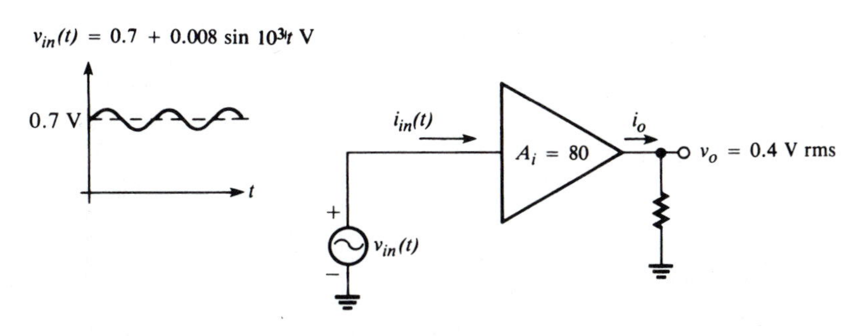

Figure 5–2
(Example 5–1)

3. The ac input resistance is the ratio of the ac components of the input voltage and current:

$$r_{in} = \frac{v_{in}}{i_{in}} = \frac{0.008 \text{ V-pk}}{4 \times 10^{-6} \text{ A-pk}} = 2 \text{ k}\Omega$$

4. $i_o \text{ (rms)} = A_i i_{in} \text{ (rms)} = 80(0.707)(4 \times 10^{-6} \text{ A-pk}) = 0.226 \text{ mA rms}$

5. $r_o = \dfrac{v_o \text{ (rms)}}{i_o \text{ (rms)}} = \dfrac{0.4 \text{ V}}{0.226 \times 10^{-3} \text{ A}} = 1770 \ \Omega$

6. $P_{in} = \dfrac{v_{in}^2 \text{ (rms)}}{r_{in}} = \dfrac{(5.66 \times 10^{-3} \text{ V})^2}{2 \times 10^3 \ \Omega} = 1.6 \times 10^{-8} \text{ W}$

$$P_o = \frac{v_o^2 \text{ (rms)}}{r_o} = \frac{(0.4 \text{ V})^2}{1770 \ \Omega} = 9.04 \times 10^{-5} \text{ W}$$

$$A_p = \frac{p_o}{P_{in}} = \frac{9.04 \times 10^{-5} \text{ W}}{1.6 \times 10^{-8} \text{ W}} = 5650$$

Note that the power gain can also be computed in this example as the product of the voltage and current gains: $A_p = A_v A_i = (70.7)80 = 5656$. The small difference between the two results is due to roundoff error.

Source Resistance

Every signal source has internal resistance (its Thevenin equivalent resistance), which we will refer to as *source resistance,* r_S. When a signal source is connected to the input of an amplifier, the source resistance is in series with the input resistance, r_{in}, of the amplifier. Notice in Figure 5–3 that r_S and r_{in} form a voltage divider across the input to the amplifier. The input voltage at the amplifier is

$$v_{in} = v_S \left(\frac{r_{in}}{r_S + r_{in}} \right) \qquad \textbf{(5–7)}$$

Now,

$$v_o = A_v v_{in} = A_v v_S \left(\frac{r_{in}}{r_S + r_{in}} \right)$$

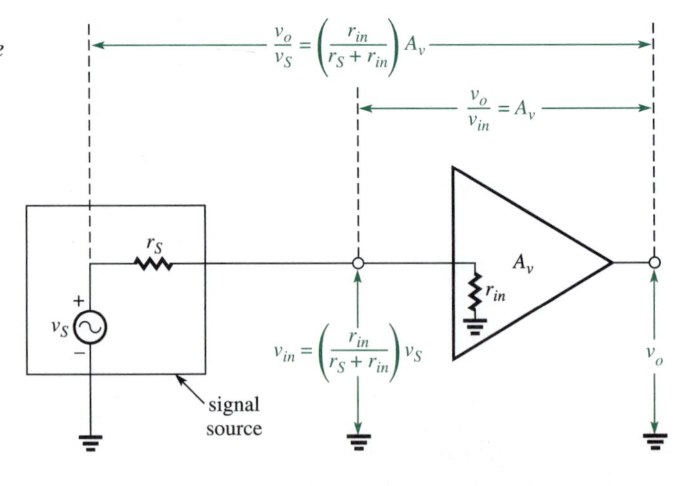

Figure 5–3
r_S and r_{in} form a voltage divider across the amplifier input. The voltage gain from source to output is reduced by the factor $r_{in}/(r_S + r_{in})$.

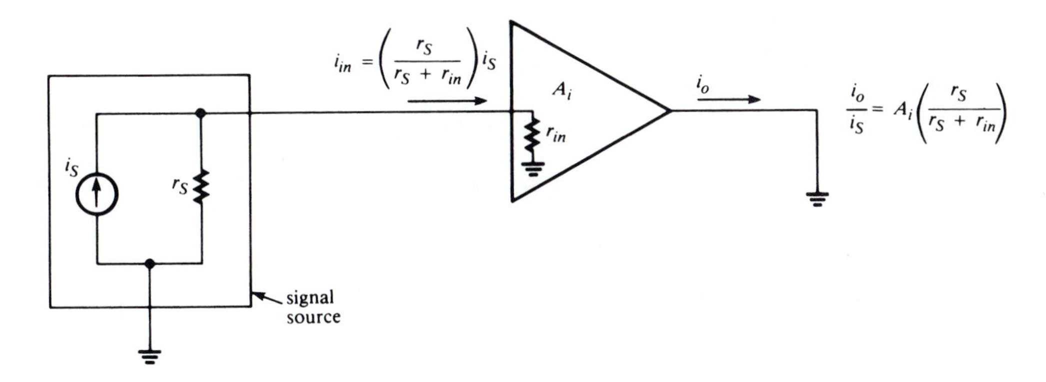

Figure 5–4
A current amplifier should have a small input resistance in order to make the quantity $r_S/(r_S + r_{in})$ close to 1.

So

$$\frac{v_o}{v_S} = A_v\left(\frac{r_{in}}{r_S + r_{in}}\right) \quad (5\text{–}8)$$

Equation 5–8 shows that the overall voltage gain *between source voltage and amplifier output*, v_o/v_S, equals the amplifier voltage gain *reduced* by the factor $r_{in}/(r_S + r_{in})$.

If r_{in} is much larger than r_S, then $r_{in}/(r_S + r_{in}) \approx 1$, so there is little reduction in the overall voltage gain caused by the voltage-divider effect. It is therefore desirable, in general, for a voltage amplifier to have as large an input resistance as possible.

On the other hand, if current amplification is desired, the amplifier should have as *small* an input resistance as possible. When r_{in} is small, the majority of the current generated at the signal source will be delivered to the amplifier input. This fact is illustrated in Figure 5–4, where the source is shown as a (Norton) equivalent current source. (For purposes of computing gain, the output is shown grounded; this ensures that all amplifier current is delivered to the output.)

As shown in Figure 5–4, the current delivered to the amplifier input is the source current i_S reduced by the factor $r_S/(r_S + r_{in})$. Therefore, r_{in} should be much less than r_S to make the quantity $r_S/(r_S + r_{in})$ close to 1. The overall current gain from source to output is

$$\frac{i_o}{i_S} = A_i\left(\frac{r_S}{r_S + r_{in}}\right) \quad (5\text{–}9)$$

Example 5–2

The amplifier shown in Figure 5–5 has $A_v = 10$ and $A_i = 10$. It is driven by a signal source that has source resistance 1000 Ω. Find the overall voltage gain and current gain, from source to output, when (1) $r_{in} = 10$ kΩ, and (2) $r_{in} = 100$ Ω. (Assume that the output is open for computing voltage gain, and grounded for computing current gain.)

Figure 5–5
(Example 5–2)

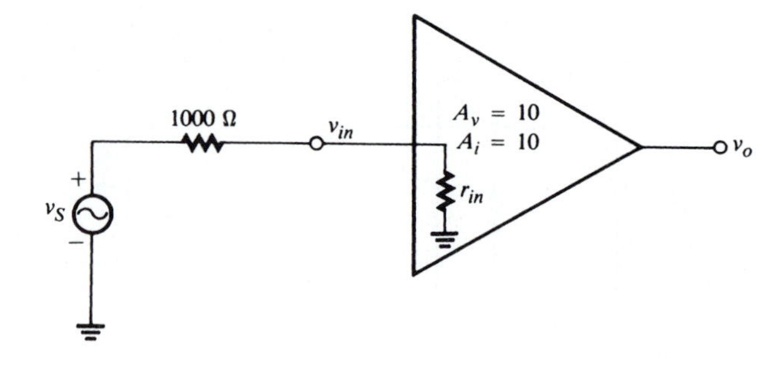

Solution

1. $r_{in} = 10\ \text{k}\Omega \quad (r_{in} = 10r_S)$

$$\frac{v_o}{v_S} = A_v\left(\frac{r_{in}}{r_S + r_{in}}\right) = 10\left(\frac{10^4}{10^3 + 10^4}\right) = 9.09$$

$$\frac{i_o}{i_S} = A_i\left(\frac{r_S}{r_S + r_{in}}\right) = 10\left(\frac{10^3}{10^3 + 10^4}\right) = 0.909$$

2. $r_{in} = 100\ \Omega \quad (r_{in} = 0.1r_S)$

$$\frac{v_o}{v_S} = 10\left(\frac{100}{10^3 + 100}\right) = 0.909$$

$$\frac{i_o}{i_S} = 10\left(\frac{10^3}{10^3 + 100}\right) = 9.09$$

This example shows that when $r_{in} = 10r_S$, the voltage gain is reduced by about 10% and the current gain is reduced by about 90%; when $r_{in} = 0.1r_S$, the voltage gain is reduced by about 90% and the current gain by about 10%.

Load Resistance

An ac amplifier is always used to supply voltage, current, and/or power to some kind of *load* connected to its output. The load may be a speaker, an antenna, a siren, an indicating instrument, an electric motor, or any one of a large number of other useful devices. Often the load is the input to another ac amplifier. Amplifier performance is analyzed by representing its load as an equivalent load resistance (or impedance). When a load resistance R_L is connected to the output of an amplifier, there is again a voltage division between the output resistance of the amplifier and the load. Figure 5–6 shows a Thevenin equivalent circuit of the output of an ac

Figure 5–6
The output voltage from an ac amplifier divides between r_o and the load resistance R_L.

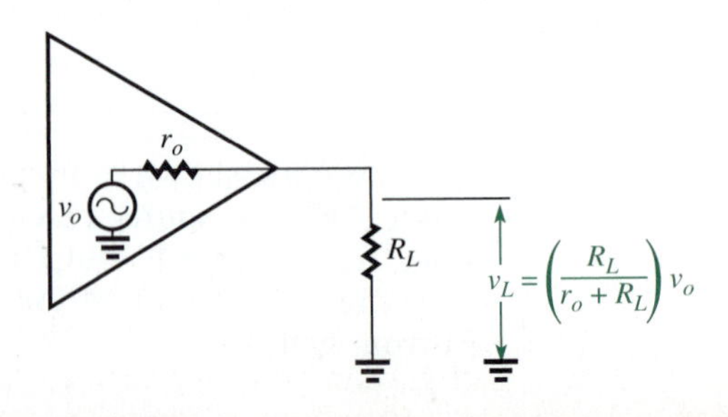

amplifier in which the output voltage v_o is produced by a voltage source in series with r_o. As can be seen in the figure, the load voltage v_L is determined by

$$v_L = \left(\frac{R_L}{r_o + R_L}\right) v_o \qquad \textbf{(5–10)}$$

For a voltage amplifier, r_o should be much smaller than R_L in order to maximize the portion of v_o that appears across the load. By converting the amplifier output to a Norton equivalent circuit, we can show that

$$i_L = \left(\frac{r_o}{r_o + R_L}\right) i_o \qquad \textbf{(5–11)}$$

When the effects of both r_S and R_L are taken into account, the overall voltage gain from source to load becomes

$$\frac{v_L}{v_S} = A_v \left(\frac{r_{in}}{r_S + r_{in}}\right)\left(\frac{R_L}{r_o + R_L}\right) \qquad \textbf{(5–12)}$$

Similarly, the overall current gain is

$$\frac{i_L}{i_S} = A_i \left(\frac{r_S}{r_S + r_{in}}\right)\left(\frac{r_o}{r_o + R_L}\right) \qquad \textbf{(5–13)}$$

where i_S is the (Norton) equivalent source current, v_S/r_S.

If a signal source has fixed resistance r_S, then from the *maximum power transfer theorem,* maximum power is transferred from the source to an amplifier when $r_{in} = r_S$. Similarly, if the amplifier has fixed output resistance r_o, then maximum power is transferred from the amplifier to a load when $R_L = r_o$. The amplifier is said to be *matched* to its source when $r_{in} = r_S$ and matched to its load when $R_L = r_o$. (We should also note that if the values of r_S and r_o can be controlled, maximum power transfer occurs when $r_S = 0$ and $r_o = 0$, irrespective of the values of r_{in} and R_L.)

The Purpose of Bias

In most single-transistor amplifiers, the output voltage must always be positive or must always be negative. When that is the case, it is not possible for the output to be a pure ac waveform, since, by definition, an ac waveform is alternately positive and negative. The purpose of bias in a transistor amplifier is to set a dc output level somewhere in the middle of the total range of possible output voltages so that an ac waveform can be superimposed on it. Figure 5–7 illustrates the point. The ac input causes the output voltage to vary above and below the bias voltage, but the instantaneous values of the output are always positive (in this example). In other words, the output is of the form

$$v_o(t) = V_B + A \sin \omega t \qquad \textbf{(5–14)}$$

where V_B is the bias voltage, or dc component, of the output, and A is the peak value of the sinusoidal, ac component. We have studied this kind of waveform in Chapter 3 and know that its values range from $V_B - A$ to $V_B + A$.

It is apparent that the values of V_B and A must be such that $V_B + A$ is not greater than the maximum possible output voltage and $V_B - A$ is not less than the minimum possible output voltage. If these conditions are not satisfied, then the

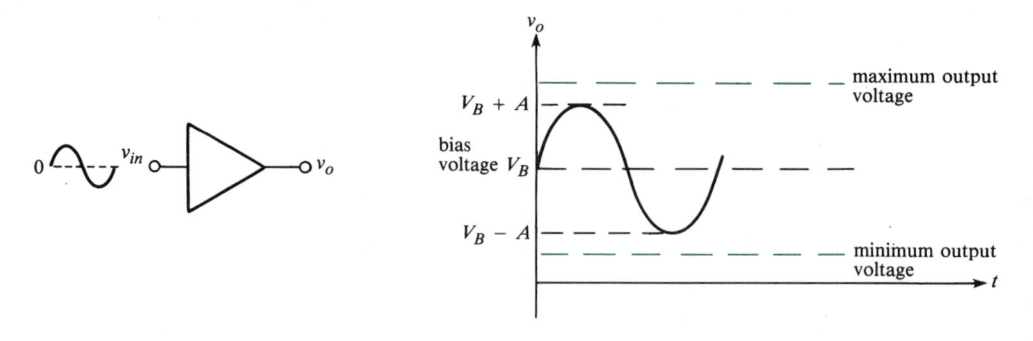

Figure 5–7
The purpose of bias is to provide a dc level about which ac variations can occur.

output voltage will reach its minimum or maximum extremes before the total ac variation can take place. The result is a flattening of the output waveform called *clipping.* Figure 5–8 illustrates positive and negative clipping caused by values of V_B that are too large or too small and by values of A that are too large. When clipping is caused by the amplitude A being too large, as shown in Figure 5–8(c), the amplifier is said to be *overdriven.*

The purpose of an ac amplifier is to produce an output waveform that is an amplified version of the input waveform. Since clipping defeats this purpose, it is said to *distort* the signal, and clipping is an example of amplitude *distortion.* In a transistor amplifier, the minimum and maximum output voltages are typically the voltages at saturation and cutoff, respectively. Thus, the minimum output may be a saturation voltage of a few tenths of a volt, and the maximum output may be a cutoff voltage equal to the supply voltage.

Coupling Capacitors

In many amplifier applications, the source or the load, or both, cannot be subjected to a dc voltage or be permitted to conduct dc current. For example, an electromagnetic speaker is designed to respond to ac fluctuations only and may not operate properly

Figure 5–8
Clipping caused by improper setting of the bias voltage and by overdriving

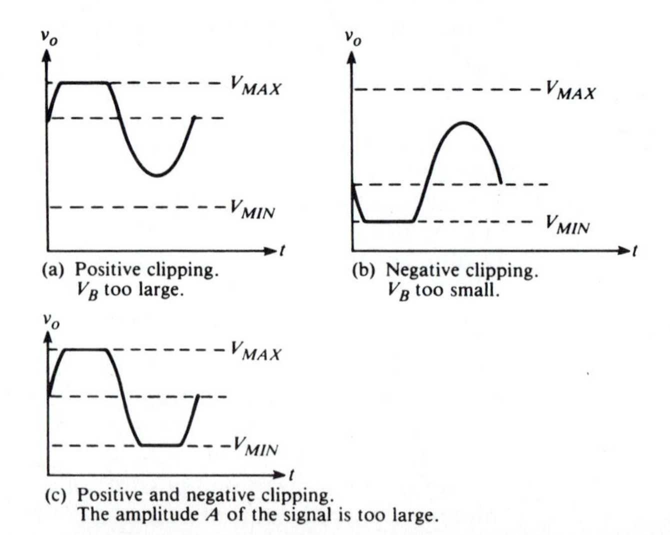

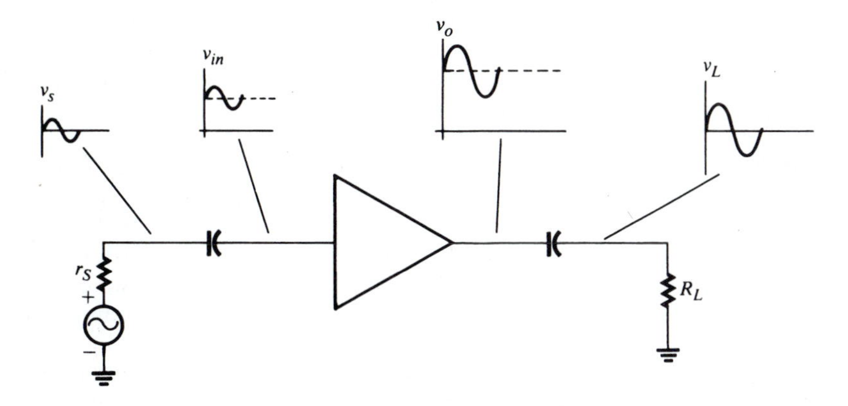

Figure 5–9
Coupling capacitors are used to block the flow of dc current between the amplifier and the signal source and between the amplifier and load.

if it conducts a dc current. To prevent the dc component of an amplifier's output voltage from producing dc current in the load, a capacitor is connected in series with the load. Similarly, to prevent the flow of dc current from the amplifier into the signal source (or vice versa), a capacitor is connected in series with the source. These connections are shown in Figure 5–9. The capacitors are called *coupling* capacitors, or *blocking* capacitors, because they block the flow of dc current. The capacitors must be large enough to present negligible impedance to the ac signals.

5–2 GRAPHICAL ANALYSIS OF THE SMALL-SIGNAL CE AMPLIFIER

Our study of transistor amplifiers begins with an analysis of the common-emitter circuit, because that is the most widely used configuration. Figure 5–10 shows the CE bias circuit we studied in Chapter 4 modified by the inclusion of an ac signal source in series with the base. For the sake of completeness we include a coupling capacitor, but we will assume for the time being that it is large enough to have negligible effect on the ac signal. We will also temporarily postpone discussion of source and load resistance.

Notice in Figure 5–10 that we now designate input and output voltages and

Figure 5–10
A common-emitter amplifier

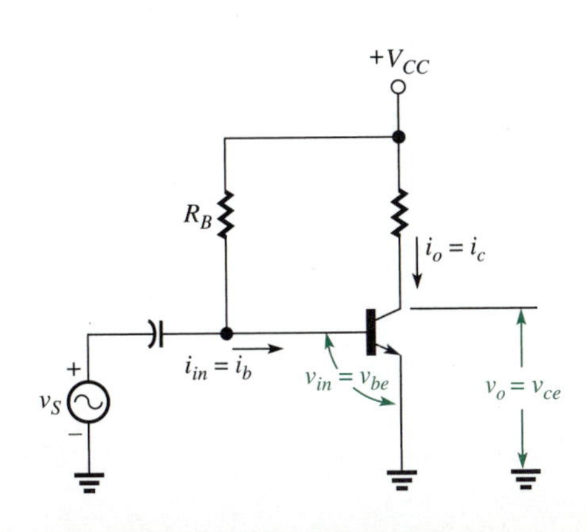

currents with lowercase letters to represent ac quantities. The signal source causes small variations in the transistor input voltage, which in turn cause small variations in the base current. As the base current decreases and increases, the collector current does the same. Since the collector current is approximately β times as large as the base current, we achieve current gain between input and output. The transistor's ability to produce gain can be attributed to the difference in impedance levels at input and output: A small current into a low-resistance, forward-biased junction controls current flow across a high-resistance, reverse-biased junction.

It is instructive to view ac operation in terms of the variation in output voltage and output current that occurs along a load line plotted on a set of transistor output characteristics. As a specific example, we will perform a graphical analysis of the circuit shown in Figure 5–11.

To determine the total variation in base current caused by the signal source in Figure 5–11, we will use the input characteristic shown in Figure 5–12. Here, we neglect the feedback effect of V_{CE} on the input characteristics and show only a single, average I_B–V_{BE} characteristic for the transistor. Under the assumption that the base–emitter junction is forward biased by 0.65 V, the $\pm$0.03-V variation of v_S causes V_{BE} to vary between 0.62 V and 0.68 V. We see in Figure 5–12 that the input voltage variation causes I_B to vary sinusoidally between a minimum value of 20 μA and a maximum value of 40 μA.

The load line for the circuit in Figure 5–11 intersects the V_{CE}-axis at V_{CC} = 18 V and the I_C-axis at (18 V)/(3 kΩ) = 6 mA. It is shown plotted on a set of output characteristics in Figure 5–13. The Q-point is located at the intersection of the load line with the base current curve corresponding to *the base current that flows when* $v_S = 0$. By definition, the Q-point locates the *dc* bias values of V_{CE} and I_C, which are the output values when no ac signal is present. Since Figure 5–12 shows that I_B = 30 μA when V_{BE} = 0.65 V (v_S = 0), the Q-point is the intersection of the load line with the I_B = 30 μA curve. We see in Figure 5–13 that the transistor is biased at V_{CE} = 9 V and I_C = 3 mA.

Recall that the load line is a plot of all *possible* combinations of I_C and V_{CE}. Therefore, as the base current alternates between its extremes of 20 μA and 40 μA, the values of I_C and V_{CE} change along the load line between its intersections with the two curves I_B = 20 μA and I_B = 40 μA. We see in Figure 5–13 that the collector current changes between I_C = 2 mA and I_C = 4 mA as the base current changes between 20 μA and 40 μA. Since the base current is changing sinusoidally, the collector current is doing likewise. The sinusoidal i_c waveform is sketched in the figure.

Figure 5–11
An ac amplifier whose performance is analyzed graphically

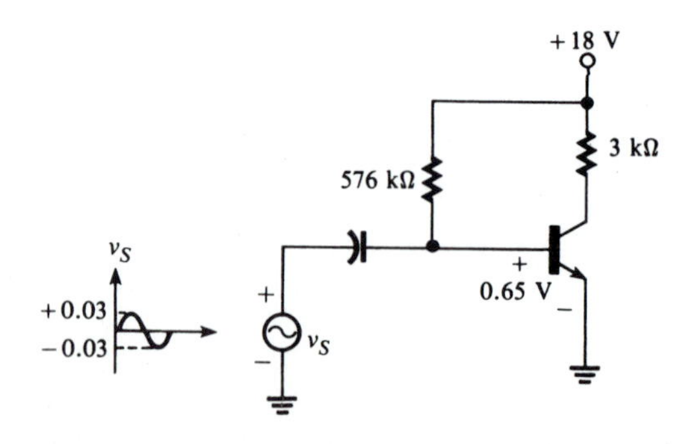

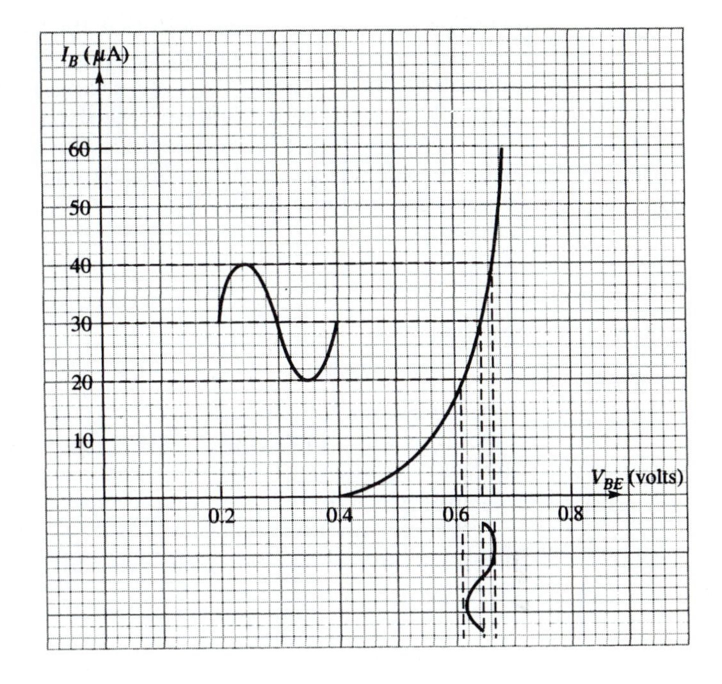

Figure 5–12
The base–emitter voltage varies from 0.62 V to 0.68 V as v_S varies ± 0.03 V about the bias voltage of $V_{BE} = 0.65$ V. This input voltage variation causes I_B to vary between 20 μA and 40 μA.

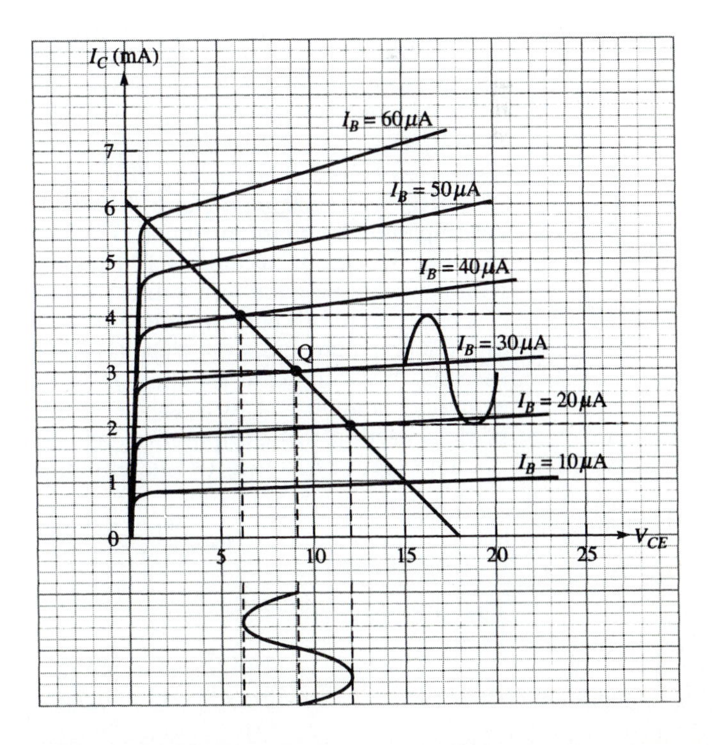

Figure 5–13
As the base current varies between 20 and 40 μA, the collector–emitter voltage varies between 6 V and 12 V and I_C varies between 2 mA and 4 mA.

We also see in Figure 5–13 that the values of V_{CE} change between 6 V and 12 V as the base current changes between 20 μA and 40 μA. Note carefully that V_{CE} *decreases* when I_B and I_C are increasing, and vice versa. Therefore, the sinusoidal voltage v_{ce} is 180° out of phase with each of the sinusoidal currents i_b and i_c. Since i_b is in phase with v_{be}, we conclude that v_{ce} is also 180° out of phase with v_{be}. In other words, *the ac output voltage from a common-emitter amplifier is 180° out of phase with the ac input voltage.* This fact can also be deduced from the load line equation:

$$I_C = \frac{-1}{R_C}V_{CE} + \frac{V_{CC}}{R_C}$$

or

$$V_{CE} = V_{CC} - I_C R_C \qquad \textbf{(5–15)}$$

Since the term $I_C R_C$ in equation 5–15 subtracts from the constant V_{CC}, we observe that an increase in I_C causes a decrease in V_{CE}. Thus, as input voltage increases, I_B increases, I_C increases, and V_{CE} decreases. The common-emitter voltage amplifier is said to cause *phase inversion,* or to *invert* voltage.

We can use the values we have obtained graphically to compute important characteristics of the amplifier. The current gain A_i is

$$A_i = \frac{i_o}{i_{in}} = \frac{\Delta I_C}{\Delta I_B} = \frac{(4 \text{ mA}) - (2 \text{ mA})}{(40\ \mu\text{A}) - (20\ \mu\text{A})} = \frac{2 \times 10^{-3}}{20 \times 10^{-6}} = 100$$

(This computation neglects the fact that a very small portion of the input current produced by v_S in Figure 5–11 is shunted through the 576-kΩ base resistor and does not therefore reach the transistor base.) The voltage gain A_v is

$$A_v = \frac{v_o}{v_{in}} = \frac{\Delta V_{CE}}{\Delta V_{BE}} = \frac{(6 - 12) \text{ V}}{(0.68 - 0.62) \text{ V}} = \frac{-6 \text{ V}}{0.06 \text{ V}} = -100$$

In this computation of A_v, note that the change in V_{CE} is written as $\Delta V_{CE} = (6 - 12)$ V $= -6$ V rather than $\Delta V_{CE} = (12 - 6)$ V $= +6$ V. $V_{CE} = 6$ V when $V_{BE} = 0.68$ V, and $V_{CE} = 12$ V when $V_{BE} = 0.62$ V. This computation, of course, leads to the *negative* value -100 for A_v. The negative sign simply denotes the phase-inverting property of the amplifier, and it is conventionally included with any gain that is accompanied by phase inversion. The *magnitude* of the voltage gain is 100, meaning that the ac output voltage is 100 times as great as the ac input voltage. Do not confuse negative gain with values of gain that are less than 1.

Note that both current gain and voltage gain are greater than 1 in a common-emitter amplifier, meaning that it amplifies both current and voltage. As we shall see, this is not the case for the CB and CC configurations. The power gain of the example amplifier is $A_p = A_v A_i = (100)(100) = 10{,}000$. Note that only the magnitude of A_v is used in this computation, as power gain is always positive.

The input resistance of the amplifier is

$$r_{in} = \frac{v_{in}}{i_{in}} = \frac{\Delta V_{BE}}{\Delta I_B} = \frac{0.06 \text{ V}}{20\ \mu\text{A}} = 3000\ \Omega$$

The output resistance is

$$r_o = \frac{v_o}{i_o} = \frac{\Delta V_{CE}}{\Delta I_C} = \frac{6 \text{ V}}{2 \text{ mA}} = 3000\ \Omega$$

The Effect of Q-Point Location on AC Operation

Let us consider now how the location of the Q-point affects the ac operation of the amplifier. Suppose the value of the base resistor R_B in Figure 5–11 is changed from 576 kΩ to 3.47 MΩ. The quiescent value of the base current will then become

$$I_B = \frac{(18 - 0.65)\ \text{V}}{3.47 \times 10^6\ \text{V}} = 5\ \mu\text{A}$$

Figure 5–14 shows that the Q-point in this case is shifted down the load line to a point where it intersects the $I_B = 5\ \mu$A curve. At the new Q-point, $I_C = 0.5$ mA and $V_{CE} = 16.5$ V. When the base current increases 10 μA beyond the Q-point to 15 μA, it can be seen in Figure 5–14 that the collector current increases to 1.5 mA and V_{CE} decreases to 13.5 V. However, when the base current decreases 10 μA below the Q-point (to (5 μA) − (10 μA) = −5 μA), the transistor enters its cutoff region. Clearly I_C cannot be less than 0 and V_{CE} cannot be greater than $V_{CC} = 18$ V.

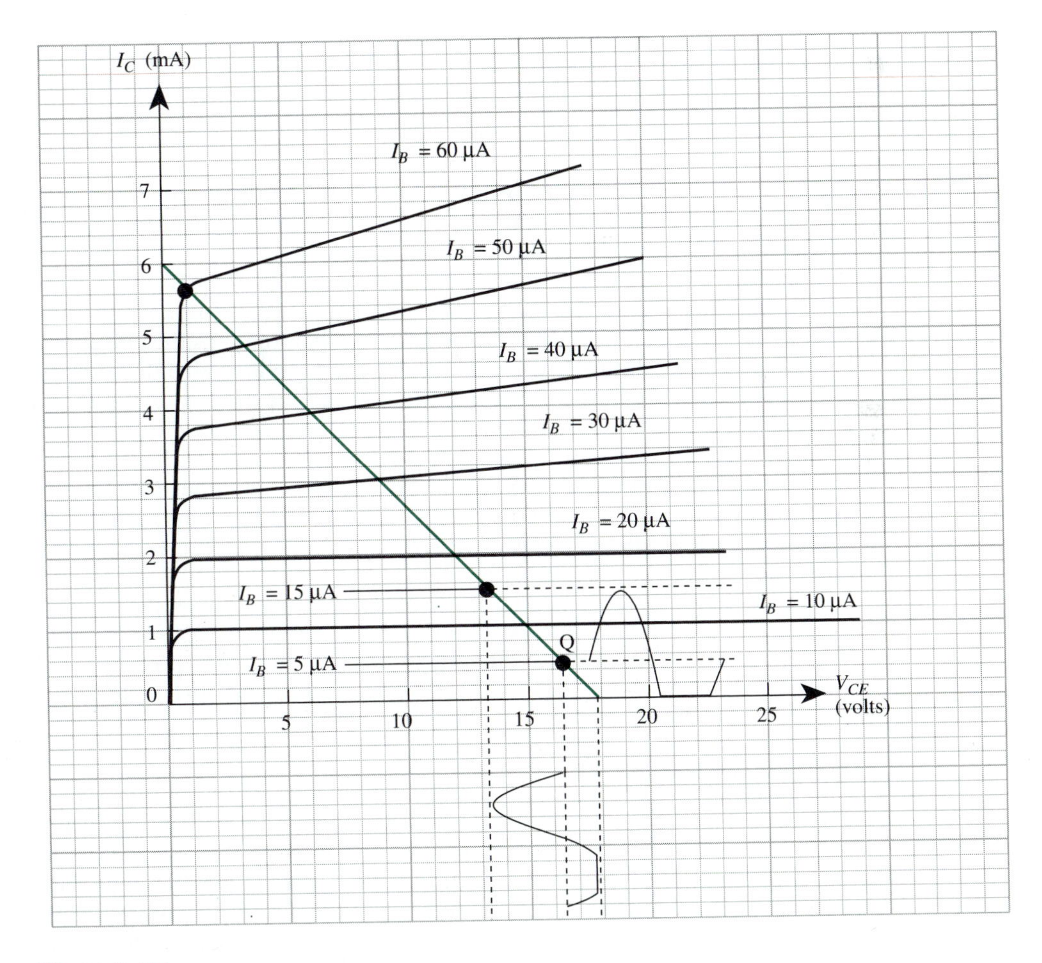

Figure 5–14
When the base resistance is increased, the Q-point moves down the load line and the signal causes the amplifier to cut off, resulting in clipping.

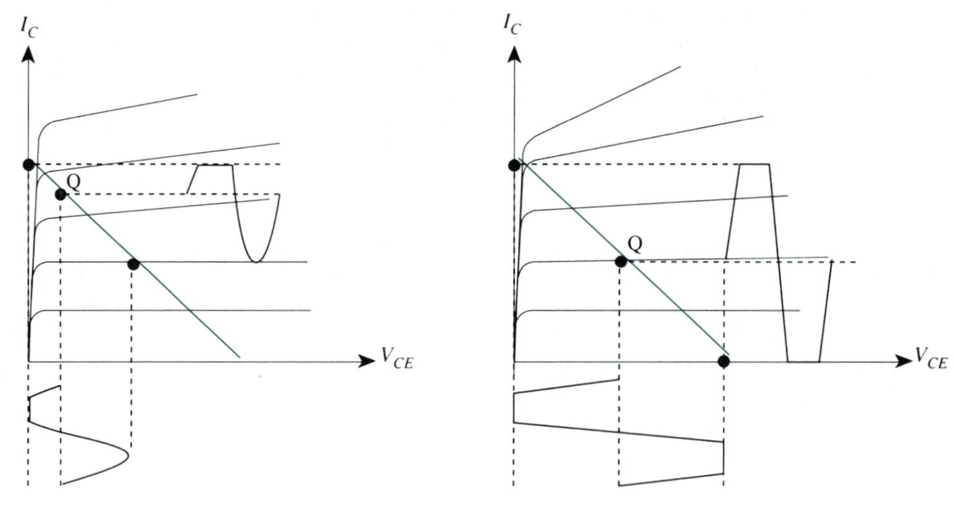

(a) The Q-point is located too close to saturation and the output voltage shows negative clipping.

(b) Too large an input signal causes both positive and negative clipping.

Figure 5–15

As shown in the figure, the output current prematurely becomes 0 in the sine-wave cycle, and clipping results. At the same time, V_{CE} reaches its limit of 18 V and the output waveform shows positive clipping. With the Q-point in its new location, the output voltage change cannot exceed 18 V − 16.5 V = 1.5 V without positive clipping occurring. Thus, the maximum peak-to-peak output voltage is 2 × 1.5 = 3 V, and we say that the amplifier has a maximum output *swing* of 3 V. This reduced swing limits the usefulness of the amplifier and illustrates the importance of locating the bias point somewhere near the center of the load line.

If the Q-point is located too far *up* the load line, the output swing will be limited by the onset of saturation. This fact is illustrated in Figure 5–15(a). In this case, a substantial increase in base current beyond its quiescent value causes the transistor to saturate. The collector current cannot exceed its saturation value and V_{CE} cannot be less than 0. Consequently, the output voltage is a negatively clipped waveform, as shown in the figure.

Even if the Q-point is located at the center of the load line, positive and negative clipping can occur if the input signal is too large. Figure 5–15(b) shows what happens when the total change in base current is so great that the transistor is driven into saturation at one end and cut off at the other. We see that both positive and negative clipping occur due to the amplifier being *overdriven.*

Linearity and Distortion

To be useful, an amplifier's output waveform must be a faithful replica of the input waveform (or a phase-inverted replica of the input). That clearly is not the case when clipping occurs. Apart from clipping, the degree to which the output waveform has the same shape as the input depends upon the amplifier's *linearity.* To be linear, any change in output voltage must be *directly proportional* to the change in input voltage that created it. For example, if $\Delta V_o = 1$ V when $\Delta V_{in} = 0.01$ V, then ΔV_o must equal 2 V when $\Delta V_{in} = 0.02$ V, and ΔV_o must equal 0.5 V when $\Delta V_{in} = 0.005$ V. The linearity of a transistor can be determined by examining the extent to which equal increments of base current correspond to equally spaced curves on the CE output characteristics. If we assume that input current is directly proportional to input voltage (i.e., that the base–emitter junction is linear, in the sense discussed in Chapter 3), then changes in input voltage should cause the output voltage to vary to a proportional extent along the load line. This will be the case only if the curves of constant base current are equally spaced. To demonstrate this fact, Figure 5–16 shows a set of CE output characteristics that have been intentionally distorted to exaggerate nonuniform spacing. Notice that the distance between the curves increases for larger values of base current. In the figure, the base current is assumed to vary sinusoidally 20 μA below and 20 μA above its quiescent value of 50 μA. The variation from 50 μA to 70 μA causes a change in I_C from 4 mA to 8 mA, but the variation from 50 μA to 30 μA only causes I_C to change from 4 mA to 2 mA. Similarly, the change in V_{CE} is (12.5 V) − (2.5 V) = 10 V when I_B increases

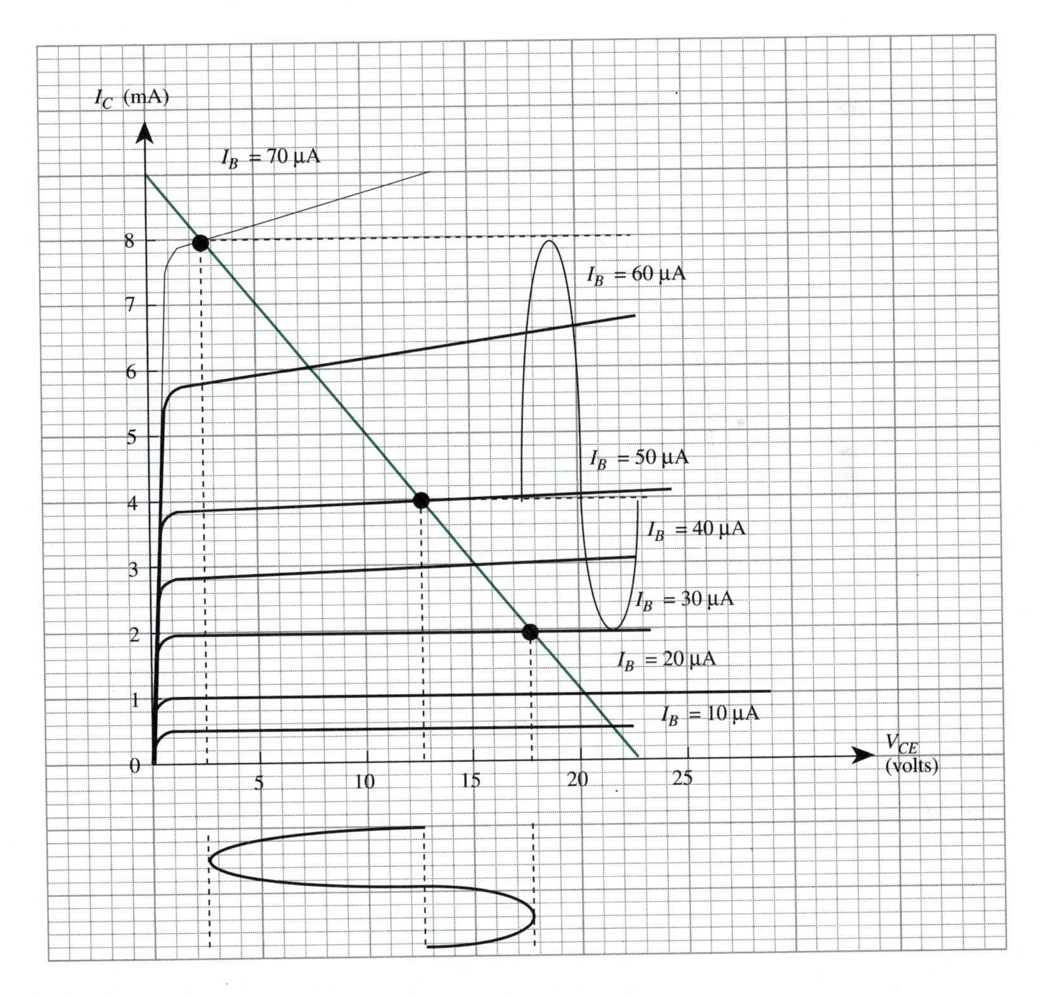

Figure 5–16
Unequal spacing between equal intervals of base current represents a nonlinear characteristic that causes a distorted output.

to 70 μA but is only (17.5 V) − (12.5 V) = 5 V when I_B decreases to 30 μA. We see that the output is a distorted, nonsymmetrical waveform.

The active region of a transistor's output characteristics is the region where the base current curves are generally found to have equal or nearly equal spacing. For this reason, the active region is often called the *linear* region. Of course, the characteristics shown in Figure 5–16 are decidedly nonlinear. In this example, the nonlinearity is due to the fact that device parameters (such as β) change significantly over the region of operation. Small-signal analysis of this device would thus be restricted to a very small range of operation along the load line.

The Effect of Load Resistance on AC Operation

Let us now consider the effect of connecting a load resistor R_L across the output of the CE amplifier. Figure 5–17 shows the circuit we studied earlier, modified to include a capacitor-coupled load resistor of 6 kΩ. It is important to realize that *as far as ac performance is concerned, R_L is in parallel with R_C. A dc source is a short circuit to ac signals,* so the 3-kΩ resistor in Figure 5–17 is effectively grounded through the 18-V source, and the ac voltage at the collector "sees" 3 kΩ in parallel with 6 kΩ. Another way of viewing this is from the standpoint of analysis by superposition: If we are interested in the output voltage due only to the ac source, v_S, we replace all other voltage sources by short circuits.

The ac load resistance, designated r_L, is the parallel combination of R_C and R_L:

$$r_L = R_C \| R_L = \frac{R_C R_L}{R_C + R_L} \quad \textbf{(5–16)}$$

In our example, r_L = (3 kΩ) ‖ (6 kΩ) = 2 kΩ. The dc load resistance is, of course, still equal to 3 kΩ, because the coupling capacitor blocks the flow of dc current into the 6-kΩ resistor.

The existence of an ac load that differs from the dc load means that the output voltage is no longer determined by variations along a load line based on R_C = 3 kΩ. Instead, the output is determined by variations along an *ac load line,* based on r_L = 2 kΩ. The load line based only on the value of R_C will hereafter be called the *dc load line.* Since the ac load line represents all possible combinations of collector voltage and current, it must include the point where the ac input goes through 0. That point is, of course, the Q-point on the dc load line, so we conclude that the

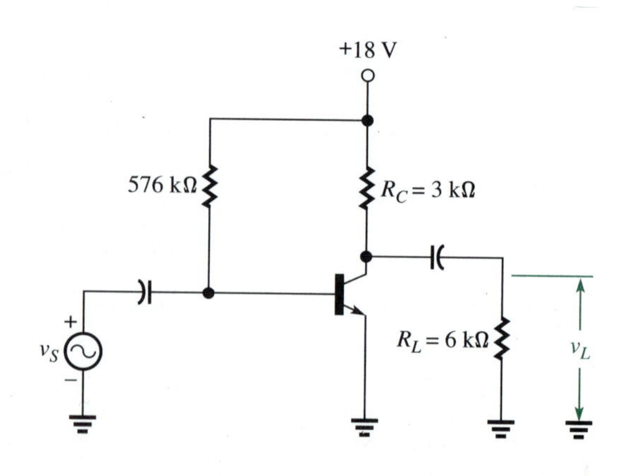

Figure 5–17
The CE amplifier circuit of Figure 5–11, modified to include a capacitor-coupled load resistor

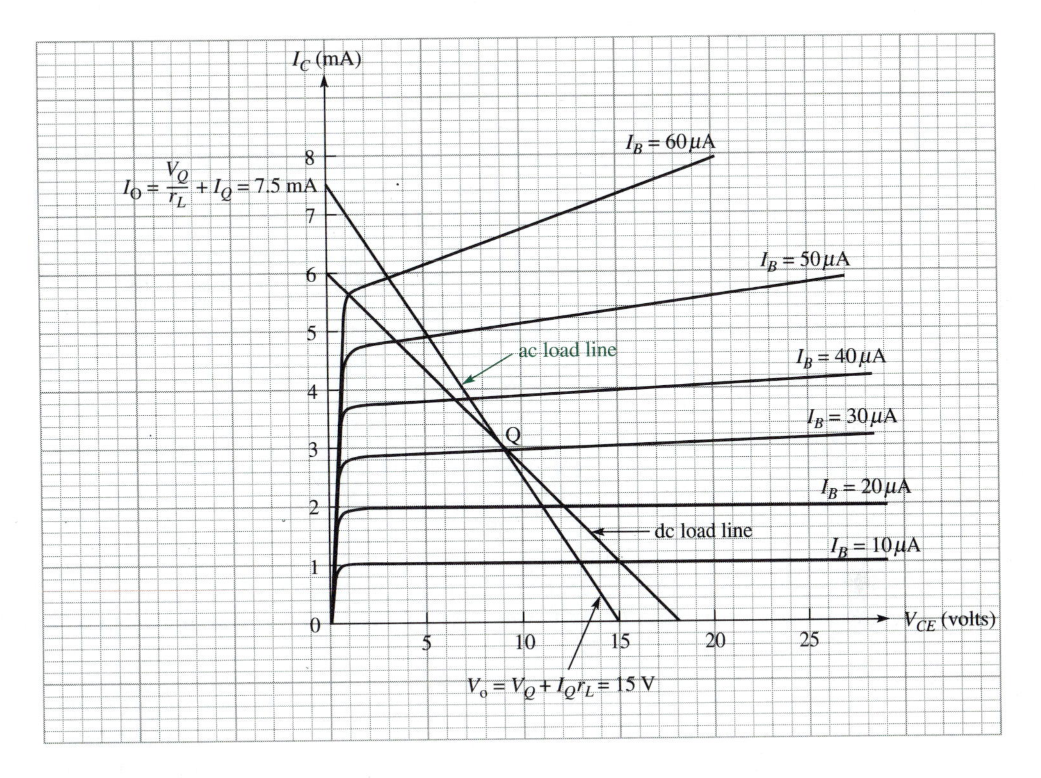

Figure 5–18
DC and ac load lines

dc and ac load lines intersect at the Q-point. Figure 5–18 shows both the dc and ac load lines of this example plotted on the output characteristics.

Note that the ac load line is steeper than the dc load line. Remember that the slope of the dc load line is $-1/R_L$, whereas that of the ac load line is $-1/r_L$. Since $r_L < R_L$, the latter slope is greater than the former. The intercepts of the ac load line are

$$I_0 = \frac{V_Q}{r_L} + I_Q \qquad \textbf{(5–17)}$$

and

$$V_0 = V_Q + I_Q r_L \qquad \textbf{(5–18)}$$

where
I_0 = intercept of the ac load line on the I_C-axis
V_0 = intercept of the ac load line on the V_{CE}-axis
V_Q = quiescent value of V_{CE}
I_Q = quiescent value of I_C

For our example, we have

$$I_0 = \frac{9\text{ V}}{2 \times 10^3\ \Omega} + 3 \times 10^{-3}\text{ A} = 7.5\text{ mA}$$

$$V_0 = 9\text{ V} + (3 \times 10^{-3}\text{ A})(2 \times 10^3\ \Omega) = 15\text{ V}$$

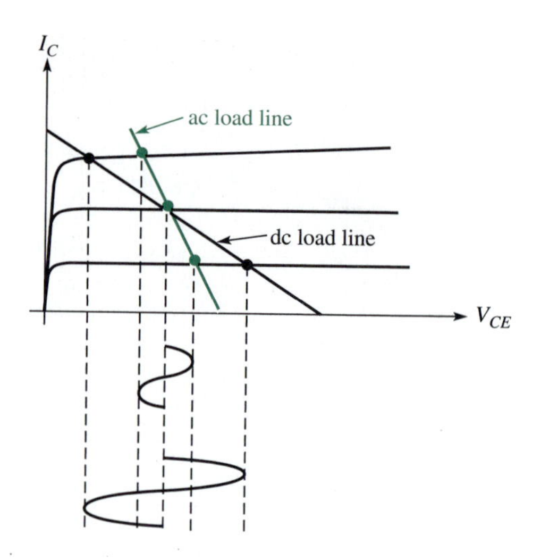

Figure 5–19
The output voltage determined by the ac load line is smaller than it would be if it were determined by the dc load line.

It is an exercise at the end of this chapter to show that equations 5–17 and 5–18 also apply to the dc load line when R_C is substituted for r_L (i.e., for the case $R_L = 1$).

It must be emphasized again that the ac load line represents all possible combinations of collector–emitter voltage and collector current, and that the dc load line no longer applies. It is a common mistake to believe that the dc load line governs the voltage across R_C while the ac load line governs the voltage across R_L. Remember that the current through and voltage across R_L are pure ac waveforms that go both positive and negative, since the capacitor blocks the dc component of the collector waveform. The only difference between v_L and the collector voltage is the dc component in the latter.

The practical implication of the ac load line is that it makes the magnitude of the ac output voltage smaller than it would be if the output variations were determined by the dc load line. This fact is illustrated in Figure 5–19, where the output voltages determined by both dc and ac load lines are plotted. The same base current variation is assumed for both, and it can be seen that the steeper slope of the ac load line results in a smaller output. The connection of a load across the output of an amplifier always reduces the amplitude of its ac output.

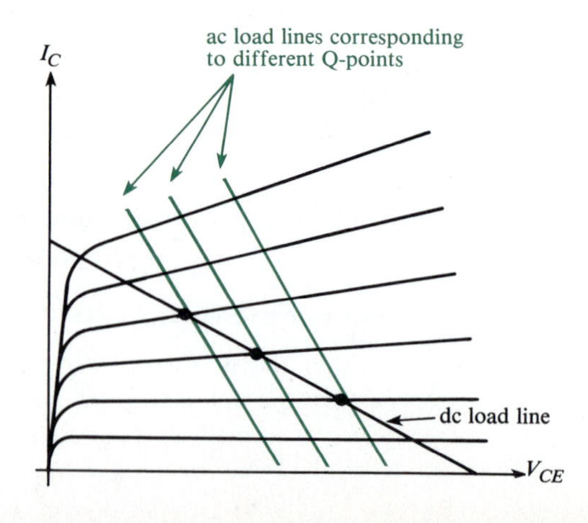

Figure 5–20
If the Q-point shifts along the dc load line, the ac load line shifts to a parallel location.

If the base resistance R_B is changed, the Q-point will shift to a new location on the *dc* load line. Since the ac load line passes through the Q-point, it too will shift. As illustrated in Figure 5–20, ac load lines corresponding to different Q-points are parallel to each other, since all have the same slope, $-1/r_L$.

Example 5–3

The silicon transistor shown in Figure 5–21 has the output characteristics shown in Figure 5–22. Assuming that $\beta = 75$ and the input resistance to the transistor is 2600 Ω, find the voltage gain when (1) $R_L = 3$ kΩ, and (2) $R_L = 12$ kΩ.

Figure 5–21
(Example 5–3)

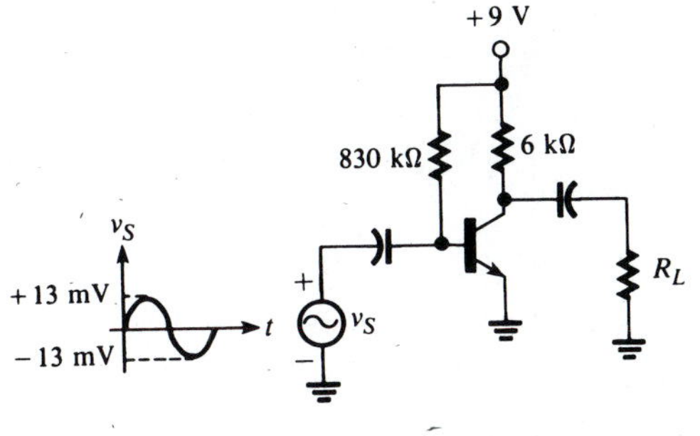

Solution. The dc load line intersects the I_C-axis at $V_{CC}/R_C = (9\text{ V})/(6\text{ k}\Omega) = 1.5$ mA. It intersects the V_{CE}-axis at $V_{CC} = 9$ V. The dc load line is shown plotted on the output characteristics in Figure 5–22.

The quiescent point is found using equations 4–24:

$$I_B = \frac{9 - 0.7\text{ V}}{830 \times 10^3\text{ V}} = 10\ \mu\text{A}$$

$$I_C = 75\,(10\ \mu\text{A}) = 0.75\text{ mA}$$

$$V_{CE} = 9 - (0.75\text{ mA})(6\text{ k}\Omega) = 4.5\text{ V}$$

The quiescent point is labeled Q in Figure 5–22.

1. When $R_L = 3$ kΩ, we have $r_L = (3\text{ k}\Omega) \parallel (6\text{ k}\Omega) = 2$ kΩ. From equation 5–18, the ac load line intersects the V_{CE}-axis at $V_0 = V_Q + I_Q r_L = 4.5\text{ V} + (0.75\text{ mA})(2\text{ k}\Omega) = 6$ V. The ac load line also passes through the Q-point and is plotted on the characteristics in Figure 5–22.

The maximum base current is the quiescent base current added to the maximum input voltage divided by the input resistance:

$$i_b(\text{max}) = (10\ \mu\text{A}) + \frac{v_{in}(\text{max})}{r_{in}} = (10\ \mu\text{A}) + \frac{13\text{ mV}}{2.6\text{ k}\Omega} = 15\ \mu\text{A}$$

Similarly, the minimum base current is

$$i_b(\text{min}) = (10\ \mu\text{A}) - \frac{13\text{ mV}}{2.6\text{ kV}} = 5\ \mu\text{A}$$

As can be seen in Figure 5–22, when i_b varies between 5 μA and 15 μA, V_{CE} varies between 3.8 V and 5.3 V. Thus,

$$A_v = \frac{v_o}{v_{in}} = \frac{(3.8 - 5.3)\text{ V}}{26\text{ mV}} = -57.69$$

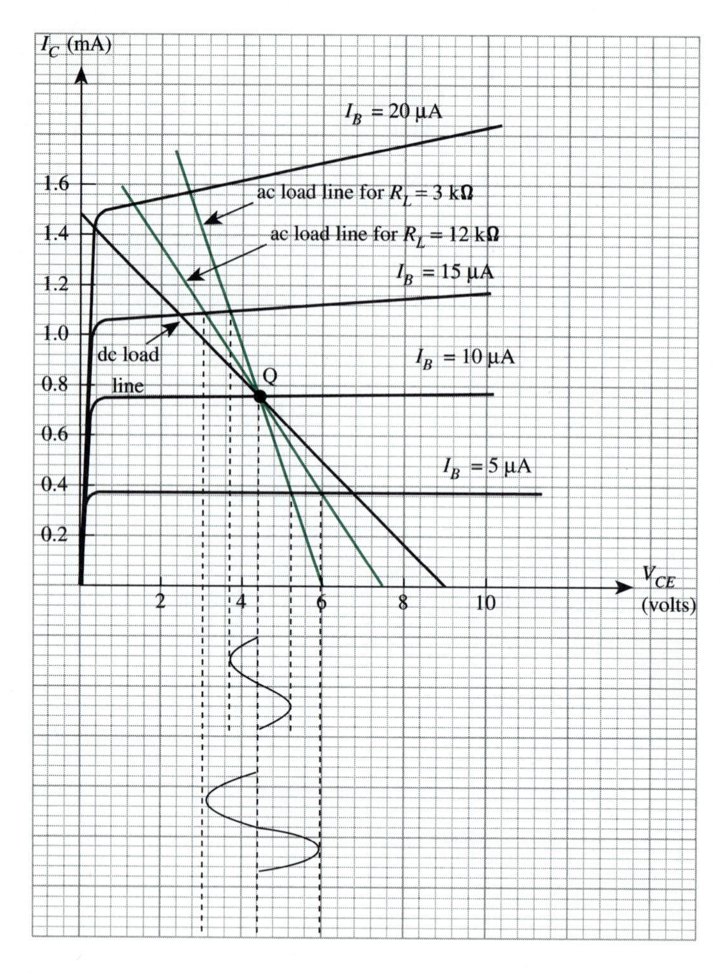

Figure 5–22
(Example 5–3)

2. When $R_L = 12\ \text{k}\Omega$, $r_L = (12\ \text{k}\Omega) \parallel (6\ \text{k}\Omega) = 4\ \text{k}\Omega$. Thus, $V_0 = V_Q + I_Q r_L = 4.5 + (0.75\ \text{mA})(4\ \text{k}\Omega) = 7.5\ \text{V}$. The ac load line corresponding to $r_L = 4\ \text{k}\Omega$ is plotted in Figure 5–22. The total base current variation is the same as in (1): 5 μA to 15 μA. As can be seen in Figure 5–22, the variation in V_{CE} in this case is from 3.2 V to 6 V. Therefore,

$$A_v = \frac{v_o}{v_{in}} = \frac{(3.2 - 6)\ \text{V}}{26\ \text{mV}} = -107.7$$

Note the increase in voltage gain over that found in (1). The larger value of R_L in (2) results in a larger value of v_L and an ac load line that is less steep.

5-3 AMPLIFIER ANALYSIS USING SMALL-SIGNAL MODELS

Small-Signal Parameters

Since transistor circuits are usually analyzed using algebraic rather than graphical methods, it is convenient to have an equivalent circuit that can be substituted for the transistor wherever it appears. Many different kinds of equivalent circuits have been developed for transistors, each of which has special features that make it more useful or more accurate than others for a particular kind of analysis. The form that an equivalent circuit takes depends on the transistor *parameters* that are chosen as the basis for the circuit. A transistor parameter is simply a transistor characteristic or property that can be given a numerical value. For example, α and β are transistor parameters. The latter are examples of *derived* parameters: they are computed from a numerical relationship between two quantities (the ratio of two currents, in this case). Transistor parameters can also specify inherent physical characteristics, such as the resistance of the base region, or the width of the collector–base depletion region.

Small-signal parameters are parameters whose values are determined under small-signal (ac) operating conditions. For example, the small-signal value of β is defined to be

$$\beta = \frac{i_c}{i_b}\bigg|_{V_{CE} = \text{constant}} \tag{5–19}$$

Equation 5–19 states that small-signal β is the ratio of ac collector current to ac base current at a specified (fixed) value of V_{CE}. Small-signal β can be determined from a set of collector characteristics by constructing a vertical line (a line of constant V_{CE}) and finding $\Delta I_C/\Delta I_B$ along that line. (As an exercise, use Figure 5–13 to find the small-signal β at $V_{CE} = 10$ V when I_B varies from 20 μA to 40 μA.) Up to now, we have computed the (approximate) value of β by taking the ratio of two *dc* currents: $\beta \approx I_C/I_B$. To distinguish this value from the small-signal value, many authors use the notation β_{DC} ("dc beta") when referring to the ratio of dc currents. In most practical applications, the small-signal and dc values of β are close enough to be assumed equal, and we will hereafter use the notation β_{DC} only when it is necessary to emphasize that we are referring strictly to the dc value. Like small-signal β, small-signal α is defined in terms of ac currents:

$$\alpha = \frac{i_c}{i_e}\bigg|_{V_{CB} = \text{constant}} \tag{5–20}$$

An important physical parameter of a transistor is its small-signal resistance from emitter to base, called the *emitter resistance* and *designated* r_e. This resistance is the same as the small-signal input resistance of the transistor in its common-base configuration. It is defined as

$$r_e = \frac{v_{be}}{i_e}\bigg|_{V_{CE} = \text{constant}} \tag{5–21}$$

Since the emitter–base junction can be regarded as a forward-biased diode, an approximate value for r_e can be found in the same way in which we found the dynamic resistance of a diode (Chapter 3). Recall that $r_D \approx V_T/I \approx 0.026/I$ at room temperature, where I is the dc current in the diode. Similarly,

$$r_e \approx \frac{0.026}{I_E} \text{ ohms} \tag{5–22}$$

where I_E is the dc emitter current.

The small-signal collector resistance r_c is the ac resistance from collector to base. It is the same as the output resistance of a transistor in its common-base configuration and typically has a value of several megohms, because it is across a reverse-biased junction.

$$r_c = \left.\frac{v_{cb}}{i_c}\right|_{I_E = \text{constant}} \text{ ohms} \tag{5–23}$$

Small-Signal CB Amplifier Model

An equivalent circuit for an electronic device is called a *model.* Using just the parameters we have discussed so far, we can construct a simple but reasonably accurate small-signal model for a transistor. Figure 5–23 shows a transistor in the common-base configuration and its approximate small-signal model. Remember that all voltages and currents are ac quantities, so all polarities periodically alternate. The polarities shown in the figure should be interpreted as reference directions for instantaneous values. For example, Figure 5–23(a) shows that an increase of current into the emitter terminal is accompanied by an increase of current out of the collector terminal.

Notice that the model in Figure 5–23 includes a *controlled* ac current source that produces a current equal to αi_e. Thus, the collector current i_c is equal to αi_e, so the model accurately reflects the relationship between i_e and i_c. The model does *not* show the feedback effect we discussed in Chapter 4; that is, it does not reflect the fact that the value of i_e depends on the value of V_{CB}. In a later chapter, we will discuss the more accurate and sophisticated *hybrid model* and will include the feedback effect then. For most practical design and analysis problems, the feedback relationship can be ignored.

It is clear that the transistor input resistance equals r_e in our approximate CB model. This is a relatively small value, usually less than 100 Ω.

$$r_{in} = r_e \qquad \text{(common base)} \tag{5–24}$$

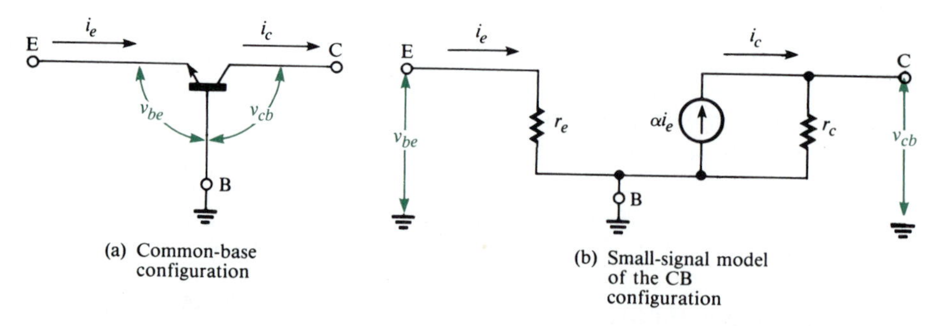

Figure 5–23
A CB transistor and its small-signal model

The transistor output resistance, on the other hand, can be seen to equal the large value r_c:

$$r_o = r_c \qquad \text{(common base)} \tag{5–25}$$

To illustrate how the small-signal model can be used to analyze a practical amplifier, we will incorporate it into the CB amplifier circuit shown in Figure 5–24(a). Figure 5–24(b) shows the ac equivalent of just that part of (a) that is external to the transistor. Note that all dc sources are treated as ac short circuits to ground, in accordance with our previous discussion. Also, the coupling capacitor is assumed to have negligible impedance and is replaced by a short in the ac equivalent circuit. Finally, Figure 5–24(c) shows the complete ac equivalent when the transistor is replaced by its small-signal model.

In Figure 5–24(c), we see that R_E is in parallel with r_e and that r_c is in parallel with R_C. In practical circuits, R_E is usually much greater than r_e ($R_E \gg r_e$), so the parallel combination $R_E \parallel r_e$ essentially equals r_e. Also, $r_c \gg R_C$ in practical circuits, so $r_c \parallel R_C$ essentially equals R_C. Thus, Figure 5–24(c) can be replaced by the practical equivalent shown in Figure 5–25.

In Figure 5–25, it is clear that

$$v_S = v_{in} = i_e r_e \quad \text{and} \quad v_o = i_c R_C = \alpha i_e R_C \tag{5–26}$$

or, since $\alpha \approx 1$, $v_o \approx i_e R_C$. Therefore, the voltage gain is

$$A_v = \frac{v_o}{v_{in}} \approx \frac{i_e R_C}{i_e r_e} = \frac{R_C}{r_e} \tag{5–27}$$

Figure 5–24
Evolution of an ac equivalent circuit for a common-base amplifier

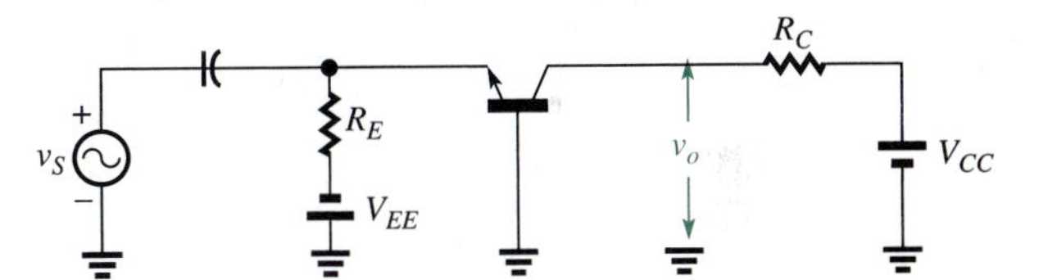

(a) A CB amplifier circuit driven by a small-signal voltage source, v_S

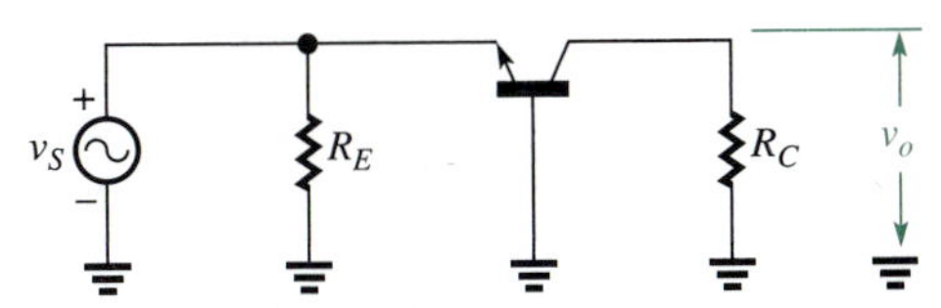

(b) The CB amplifier of (a) when the circuit external to the transistor is replaced by its ac equivalent

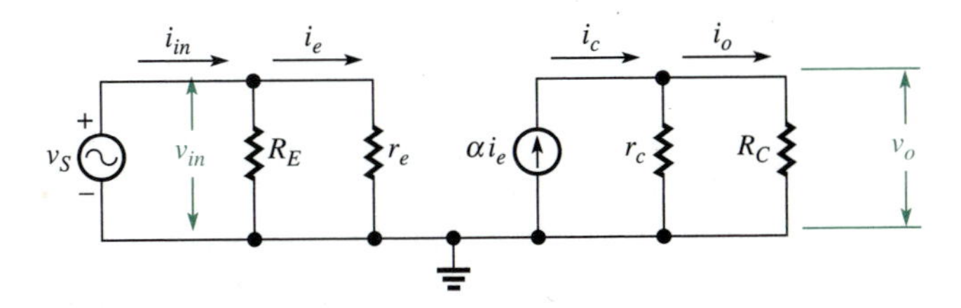

(c) The complete, small-signal, CB equivalent circuit, when the transistor in (b) is replaced by its model

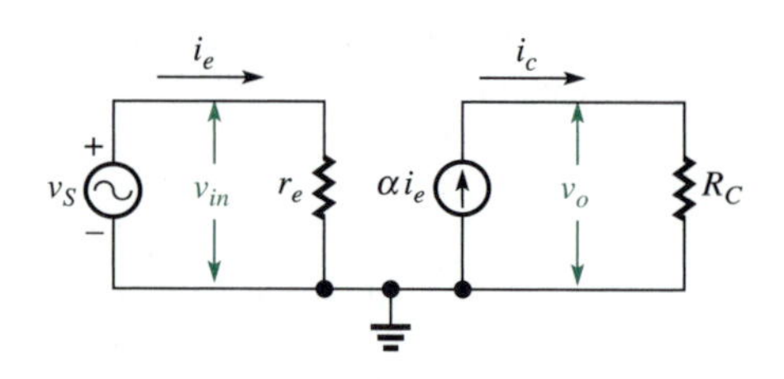

Figure 5–25
A practical CB equivalent circuit that incorporates the (usual) condition that $r_e \parallel R_E \approx r_e$ and $R_C \parallel r_c \approx R_C$

The current gain in Figure 5–25 is

$$A_i = \frac{i_o}{i_{in}} = \frac{i_c}{i_e} = \alpha \tag{5–28}$$

Thus, the current gain of a CB amplifier is always less than 1.

Figure 5–26(a) shows a CB amplifier that is driven by a source having internal resistance r_S. The amplifier load is the resistor R_L. Figure 5–26(b) shows the ac equivalent circuit that results when the assumptions $R_E \parallel r_e \approx r_e$ and $r_c \parallel R_C \approx R_C$ are once again imposed. Figure 5–26(c) shows the amplifier circuit when the transistor is replaced by a single block having the parameters derived in equations 5–26 through 5–28. Note that the voltage source in Figure 5–26(c) is the (Thevenin) equivalent of the current source in Figure 5–25.

Figure 5–26
The CB amplifier with source and load resistances included

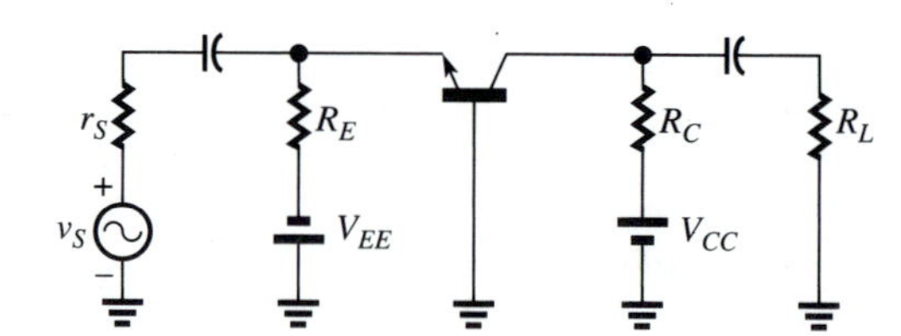

(a) The CB amplifier having load R_L and driven by a source with resistance r_S

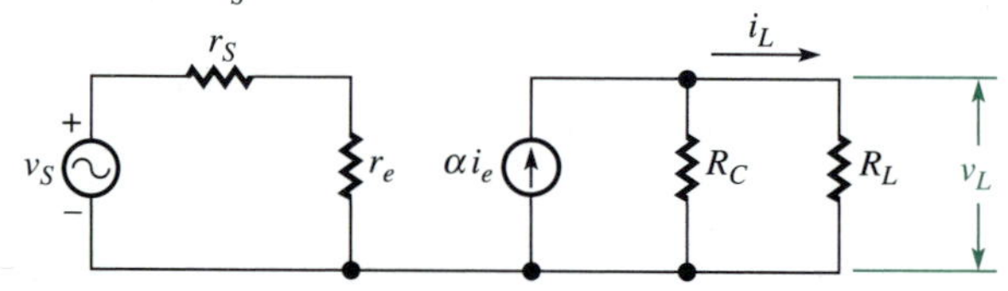

(b) The ac equivalent circuit of (a)

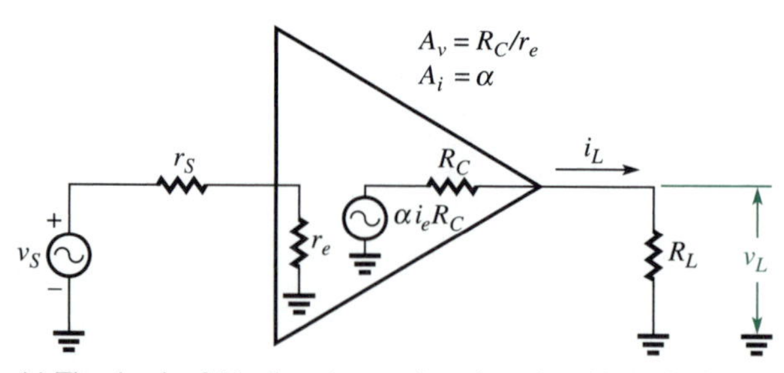

(c) The circuit of (b) when the transistor is replaced by a single amplifier block

Equations 5–12 and 5–13 in Section 5–1 give the overall voltage and current gains of an ac amplifier, from source to load. Applying these equations to Figure 5–26(c), we have

$$\frac{v_L}{v_S} = \frac{R_C}{r_e}\left(\frac{r_e}{r_S + r_e}\right)\left(\frac{R_L}{R_C + R_L}\right) \tag{5–29}$$

and

$$\frac{i_L}{i_S} = \alpha\left(\frac{r_S}{r_S + r_e}\right)\left(\frac{R_C}{R_C + R_L}\right) \tag{5–30}$$

where i_S is the (Norton) equivalent source current v_S/r_S.

The ac output voltage from a CB amplifier is *in phase* with the ac input voltage. We can deduce this fact by rewriting equation 4–17 to obtain the following equivalent form:

$$V_{CB} = V_{CC} - I_C R_C \tag{5–31}$$

An increase in emitter-to-base (input) voltage reduces the forward bias on the emitter–base junction and thus reduces the emitter current. But a decrease in emitter current causes a decrease in collector current, since $I_C = \alpha I_E$. Decreasing I_C causes the quantity $I_C R_C$ in equation 5–31 to decrease and therefore causes V_{CB} to increase. Recapitulating, an increase in emitter (input) voltage causes an increase in collector (output) voltage, and we conclude that input and output are in phase.

When a transistor is connected in a circuit to produce gain, the transistor and all the associated external components it needs to operate properly (such as bias resistors) are referred to collectively as an amplifier *stage.* It is important to distinguish between the input and output resistances of the transistor alone and those of the stage of which it is a part. Hereafter, we will use the notation (*stage*) after the symbol r_{in} or r_o when we wish to emphasize that it refers to a stage characteristic. Figure 5–27 illustrates these distinctions in a CB amplifier stage.

Figure 5–27
r_{in}(stage) and r_o(stage) refer to the input and output resistances of the overall amplifier stage.

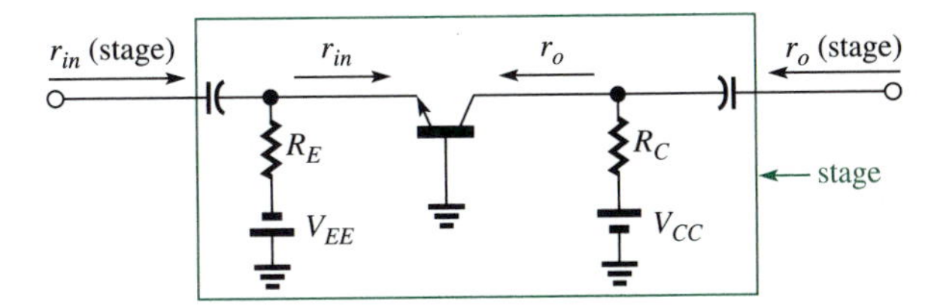

(a) A CB amplifier stage

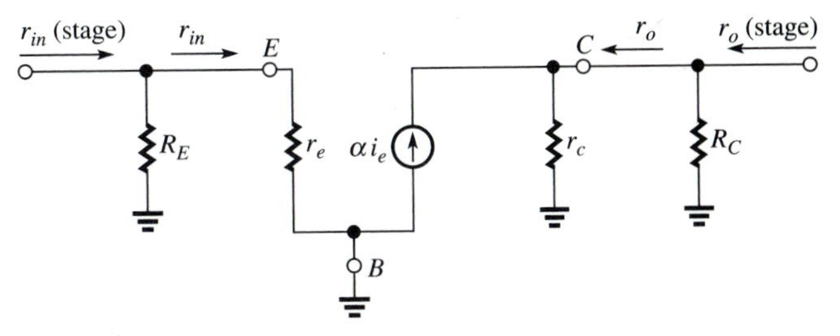

(b) The ac equivalent circuit of (a)

We note that

$$r_{in} = r_e$$
$$r_{in}(\text{stage}) = r_e \| R_E \approx r_e \tag{5–32}$$
$$r_o = r_c$$
$$r_o(\text{stage}) = r_c \| R_C \approx R_C \tag{5–33}$$

Equations 5–34 summarize the small-signal analysis equations for a common-base amplifier stage, based on the approximate CB transistor model.

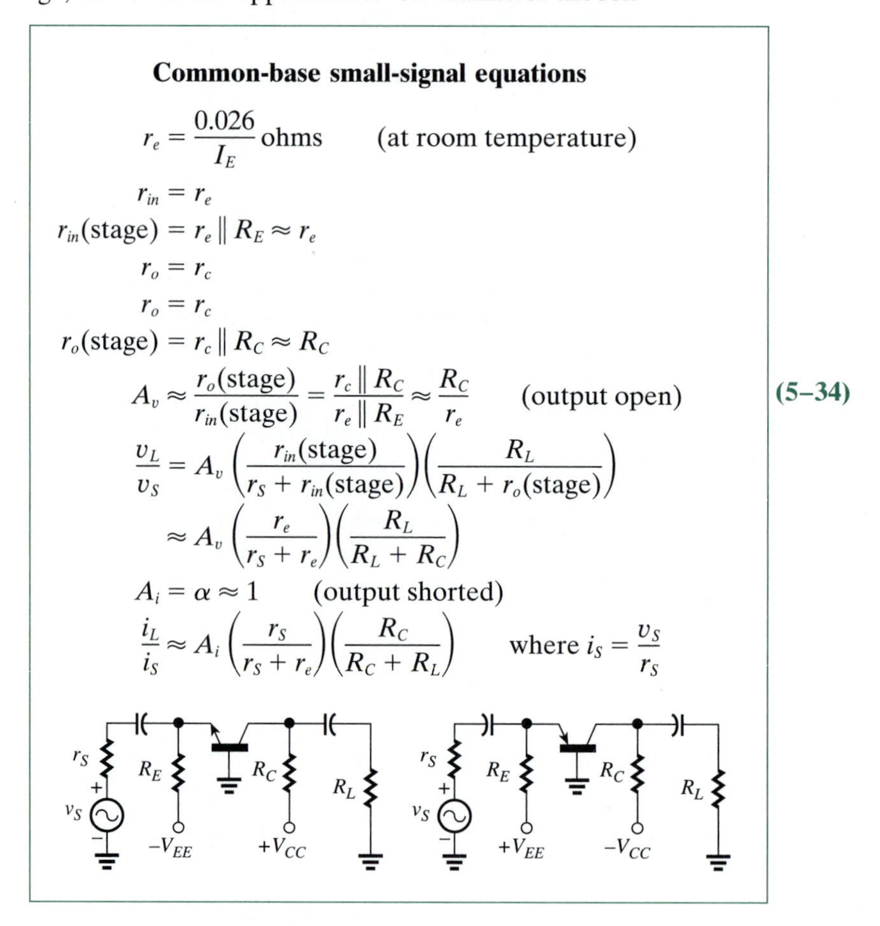

Common-base small-signal equations

$$r_e = \frac{0.026}{I_E} \text{ ohms} \qquad \text{(at room temperature)}$$
$$r_{in} = r_e$$
$$r_{in}(\text{stage}) = r_e \| R_E \approx r_e$$
$$r_o = r_c$$
$$r_o = r_c$$
$$r_o(\text{stage}) = r_c \| R_C \approx R_C$$
$$A_v \approx \frac{r_o(\text{stage})}{r_{in}(\text{stage})} = \frac{r_c \| R_C}{r_e \| R_E} \approx \frac{R_C}{r_e} \qquad \text{(output open)} \tag{5–34}$$
$$\frac{v_L}{v_S} = A_v\left(\frac{r_{in}(\text{stage})}{r_S + r_{in}(\text{stage})}\right)\left(\frac{R_L}{R_L + r_o(\text{stage})}\right)$$
$$\approx A_v\left(\frac{r_e}{r_S + r_e}\right)\left(\frac{R_L}{R_L + R_C}\right)$$
$$A_i = \alpha \approx 1 \qquad \text{(output shorted)}$$
$$\frac{i_L}{i_S} \approx A_i\left(\frac{r_S}{r_S + r_e}\right)\left(\frac{R_C}{R_C + R_L}\right) \qquad \text{where } i_S = \frac{v_S}{r_S}$$

Example 5–4

For the circuit in Figure 5–28, find (1) r_{in}, (2) r_{in}(stage), (3) A_v, (4) v_L, (5) i_L, (6) i_L/i_S (assume that $\alpha = 1$), and (7) i_L, using the result of (6).

Solution

1. From equations 4–21,

$$I_E = \frac{(6 - 0.7)\text{ V}}{2 \times 10^3 \text{ V}} = 2.65 \text{ mA}$$

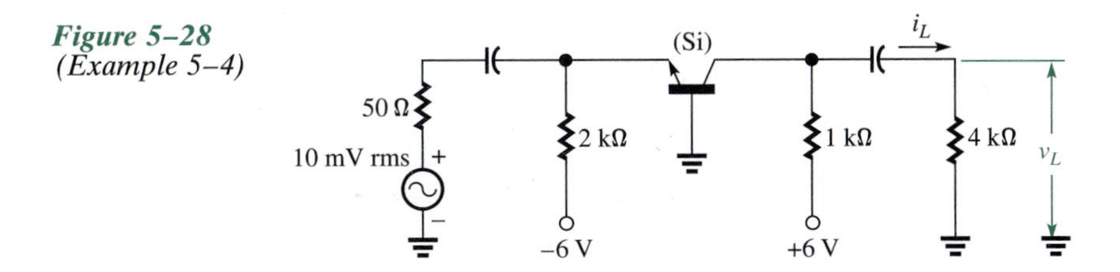

Figure 5–28
(Example 5–4)

Therefore,

$$r_{in} = r_e = \frac{0.026 \text{ V}}{2.65 \times 10^{-3} \text{ A}} = 9.81 \ \Omega$$

2. $r_{in}(\text{stage}) = r_e \| R_E = \dfrac{(9.81)(2000)}{2009.81} = 9.76 \ \Omega$

This result confirms equation 5–32, since, for all practical purposes,

$$r_{in}(\text{stage}) = r_e = 9.81 \ \Omega$$

3. $A_v = R_C/r_e = 1000/9.81 = 101.9$

4. $\dfrac{v_L}{v_S} = 101.9\left(\dfrac{r_e}{r_S + r_e}\right)\left(\dfrac{R_L}{R_L + R_C}\right)$

$$= 101.9\left(\frac{9.81}{50 + 9.81}\right)\left(\frac{4 \times 10^3}{1 \times 10^3 + 4 \times 10^3}\right) = 13.37$$

Therefore,

$$v_L = \frac{v_L}{v_S} v_S = 13.37(10 \text{ mV rms}) = 133.7 \text{ mV rms}$$

5. $i_L = \dfrac{v_L}{R_L} = \dfrac{133.7 \text{ mV rms}}{4000 \ \Omega} = 33.4 \ \mu\text{A rms}$

6. $\dfrac{i_L}{i_S} = \alpha\left(\dfrac{r_S}{r_S + r_e}\right)\left(\dfrac{R_C}{R_C + R_L}\right)$

$$= 1\left(\frac{50}{50 + 9.81}\right)\left(\frac{1 \times 10^3}{1 \times 10^3 + 4 \times 10^3}\right) = 0.167$$

7. To compute i_L using the value 0.167 for i_L/i_S, we must first find the source current i_S. Remember that i_S is the Norton equivalent source current (the current source that results from converting the input voltage source to an equivalent current source). Figure 5–29 shows the input source conversion.

As shown in the figure, $i_S = 0.2$ mA rms. Therefore, $i_L = 0.167 i_S = 0.167$ (0.2 mA rms) $= 33.4 \ \mu$A rms. This is the same value obtained in (5) for i_L.

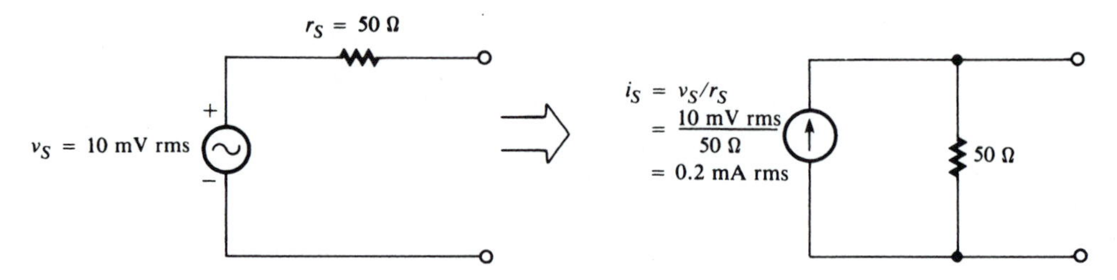

Figure 5–29
Conversion of the input voltage source to an equivalent current source (Example 5–4)

Small-Signal CE Amplifier Model

To develop a model for the transistor in its common-emitter configuration, we will first investigate the input resistance in that configuration. Figure 5–30 shows the CE input circuit with r_e drawn inside the emitter terminal, to emphasize that it is an internal transistor parameter. The ac input resistance is $r_{in} = v_{be}/i_b$. Recalling that $i_e = (\beta + 1)i_b$, we have

$$r_{in} = \frac{v_{be}}{i_e/(\beta + 1)} = (\beta + 1)\frac{v_{be}}{i_e}$$

But $v_{be}/i_e = r_e$, so

$$r_{in} = (\beta + 1)r_e \approx \beta r_e \qquad \text{(common emitter)} \qquad \textbf{(5–35)}$$

Equation 5–35 shows that the input resistance in the CE configuration is approximately β times greater than that in the CB configuration. The quantity $(\beta + 1)r_e$ is often designated as r_π in texts and data sheets. It can also be shown that the output resistance of a transistor in its CE configuration is approximately β times *smaller* than it is in the CB configuration: $r_o \approx r_c/\beta$. Because the CE input resistance is β times greater and the output resistance is β times smaller than the corresponding values in the CB configuration, the common-emitter amplifier is inherently better suited for voltage amplification than its CB counterpart.

Figure 5–31 shows an approximate model for the common-emitter transistor. Note that the controlled current source has value βi_b, reflecting the fact that $i_c = \beta i_b$. Once again, we neglect the feedback effect, whereby the value of i_b depends somewhat on V_{CE}.

Figure 5–30
The input circuit for a transistor in the CE configuration

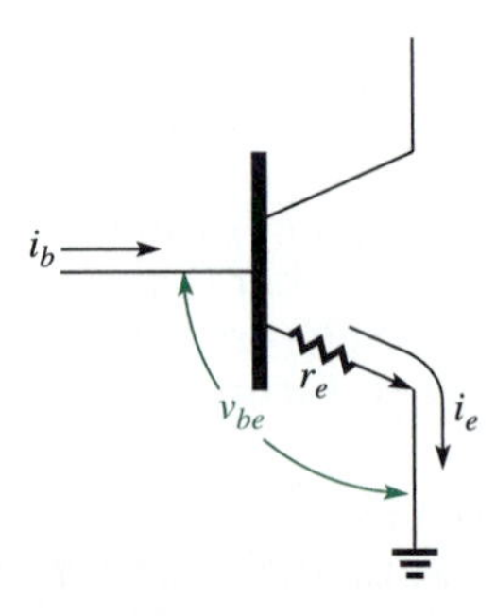

Figure 5–31
Approximate small-signal model for a transistor in the common-emitter configuration

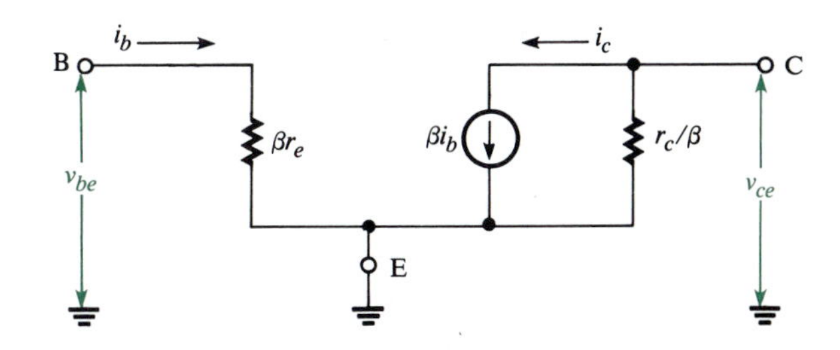

Figure 5–32 shows a common-emitter amplifier stage and its ac equivalent circuit. Note that the dc supply voltages are treated as ac short circuits, as before. In Figure 5–32(b), it can be seen that r_c/β is in parallel with R_C. In most practical circuits, $r_c/\beta \gg R_C$ and so $(r_c/\beta) \parallel R_C \approx R_C$. For example, typical values are $r_c = 10$ MΩ, $\beta = 100$, and $R_C = 1$ kΩ, for which $(r_c/\beta) \parallel R_C = (100 \text{ k}\Omega) \parallel (1 \text{ k}\Omega) = 990\ \Omega$. For the approximate analysis methods we are developing now, we will assume that the parallel combination equals R_C. Under that assumption, the output resistance of the stage is

$$r_o(\text{stage}) = R_C \tag{5–36}$$

and the output voltage is

$$v_o = i_c r_o(\text{stage}) = \beta i_b R_C \tag{5–37}$$

It is clear from Figure 5–32(b) that

$$v_{in} = (\beta r_e) i_b \tag{5–38}$$

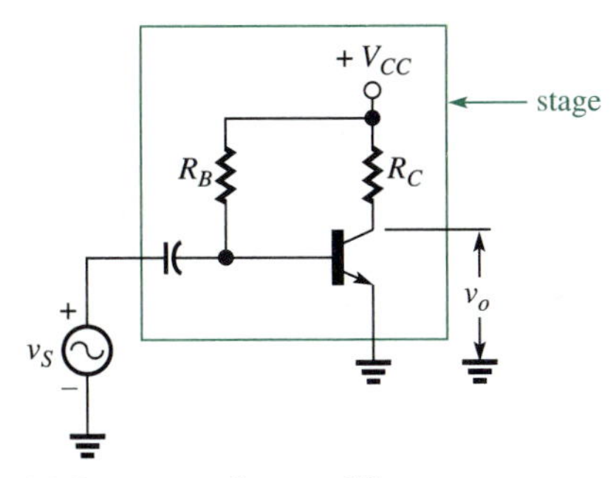

(a) Common-emitter amplifier

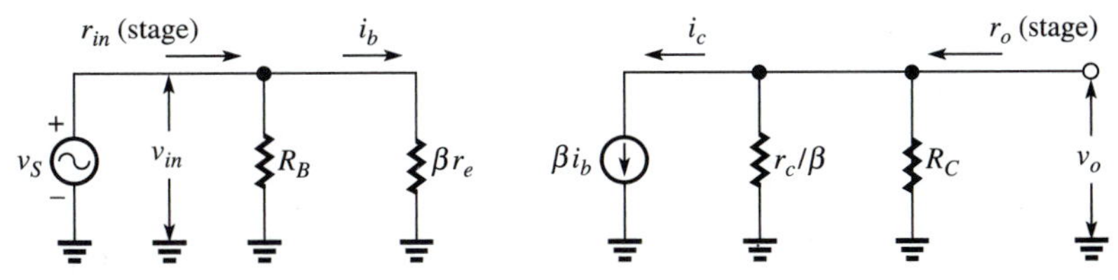

(b) The ac equivalent circuit of (a), with the model of Figure 5.31 incorporated

Figure 5–32
A common-emitter amplifier and its ac equivalent circuit

Therefore,

$$A_v = \frac{v_o}{v_{in}} = \frac{-\beta i_b R_C}{\beta r_e i_b} = \frac{-R_C}{r_e} \quad \textbf{(5–39)}$$

where the minus sign is inserted to denote phase inversion between input and output, as discussed in Section 5–2. The current gain of the transistor is the ratio of its output current i_c to its input current i_b:

$$A_i = \frac{i_C}{i_b} = \beta \quad \textbf{(5–40)}$$

Equations 5–39 and 5–40 show that the CE amplifier can provide voltage and current gains that are *both* greater than 1.

The input resistance of the amplifier stage in Figure 5–32(b) is

$$r_{in}(\text{stage}) = R_B \parallel (\beta r_e) \quad \textbf{(5–41)}$$

In many practical circuits of the design shown in Figure 5–32(a), R_B is much larger than βr_e and equation 5–41 reduces to $r_{in}(\text{stage}) \approx \beta r_e$. However, in some improved bias circuits we will study later, that is not always the case.

Figure 5–33 shows a CE amplifier stage that includes source and load resistances. Applying equations 5–12 and 5–13 to find the overall voltage and current gains, we find

$$\begin{aligned} \frac{vL}{v_S} &= A_v \left[\frac{r_{in}(\text{stage})}{r_S + r_{in}(\text{stage})}\right]\left[\frac{R_L}{R_L + R_C}\right] \\ &= \frac{-R_C}{r_e}\left[\frac{R_B \parallel (\beta r_e)}{r_S + R_B \parallel (\beta r_e)}\right]\left[\frac{R_L}{R_L + R_C}\right] \end{aligned} \quad \textbf{(5–42)}$$

$$\begin{aligned} \frac{i_L}{i_S} &= A_i \left[\frac{r_S \parallel R_B}{r_S \parallel R_B + r_{in}}\right]\left[\frac{R_C}{R_L + R_C}\right] \\ &= A_i \left[\frac{r_S \parallel R_B}{r_S \parallel R_B + \beta r_e}\right]\left[\frac{R_C}{R_L + R_C}\right] \end{aligned} \quad \textbf{(5–43)}$$

where $i_S = v_S/r_S$.

Equations 5–44 summarize the small-signal analysis equations for a common-emitter amplifier stage, based on the approximate CE transistor model.

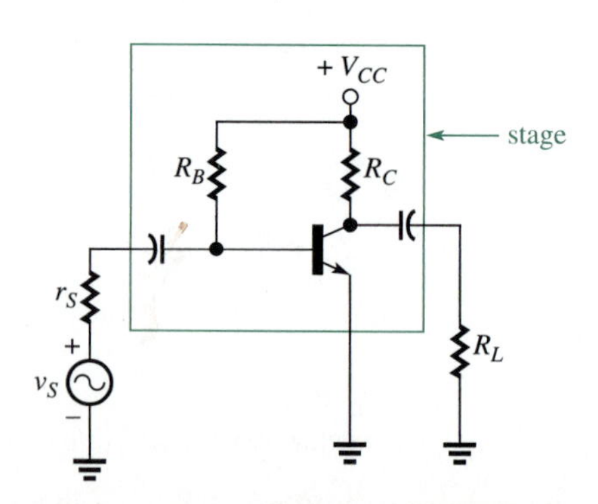

Figure 5–33
A common-emitter amplifier stage having source and load resistances

Common-emitter small-signal equations

$$
\begin{aligned}
r_e &= \frac{0.026}{I_E} \text{ ohms} \qquad \text{(at room temperature)}\\
r_{in} &= (\beta + 1)r_e = r_\pi \approx \beta r_e\\
r_{in}(\text{stage}) &= R_B \parallel r_{in} \approx R_B \parallel (\beta r_e)\\
r_o &= r_c/\beta\\
r_o(\text{stage}) &= R_c \parallel (r_c/\beta) \approx R_C\\
A_v &= \frac{-r_o(\text{stage})}{r_e} \approx \frac{-R_C}{r_e} \qquad \text{(output open)}\\
\frac{v_L}{v_S} &= A_v \left[\frac{R_B \parallel (\beta r_e)}{r_S + R_B \parallel (\beta r_e)}\right]\left(\frac{R_L}{R_L + R_C}\right)\\
A_i &= \beta\\
\frac{i_L}{i_S} &= A_i \left(\frac{r_S \parallel R_B}{r_S \parallel R_B + \beta r_e}\right)\left(\frac{R_C}{R_C + R_L}\right) \qquad \text{where } i_S = \frac{v_S}{r_S}
\end{aligned}
\tag{5–44}
$$

Example 5–5

Find the output voltage of the amplifier shown in Figure 5–34. Assume that $r_S = 0$.

Solution. To find r_e, we must first find the quiescent emitter current:

$$I_B = \frac{V_{CC} - 0.7}{R_B} = \frac{14.3 \text{ V}}{200 \times 10^3 \ \Omega} = 71.5 \ \mu\text{A}$$

$$I_E \approx I_C = \beta I_B = 100(71.5 \ \mu\text{A}) = 7.15 \text{ mA}$$

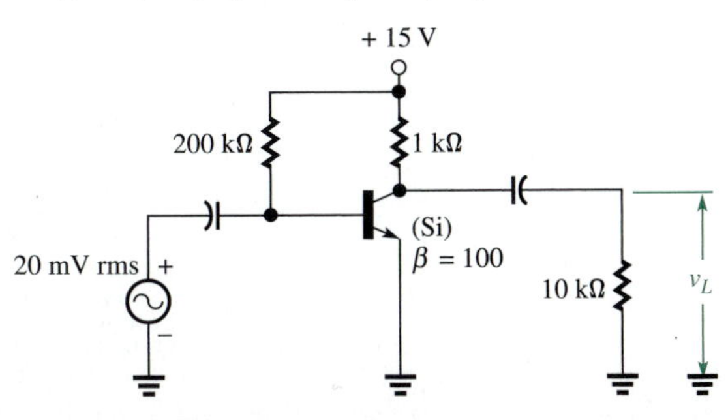

Figure 5–34
(Example 5–5)

Therefore, $r_e = (0.026\text{ V})/(7.15 \times 10^{-3}\text{ A}) = 3.64\ \Omega$. From equation 5–39, $A_v = -R_C/r_e = -10^3/3.64 = -274.7$. Since $r_S = 0$, equation 5–42 becomes

$$\frac{v_L}{v_S} = A_v\left(\frac{R_L}{R_L + R_C}\right) = -274.7\left[\frac{10\text{ k}\Omega}{(10\text{ k}\Omega) + (1\text{ k}\Omega)}\right] = -249.7$$

Therefore, the magnitude of v_L is $v_L = (249.7)(20\text{ mV rms}) = 4.99\text{ V rms}$.

The Effects of AC Load Resistance

Another way to view the voltage gain equation is to consider how gain varies as a function of *ac load resistance* r_L. In order to focus on that aspect alone, let us assume that a CE amplifier is driven by a source having $r_S = 0$, so equation 5–42 becomes

$$\frac{v_L}{v_S} = \frac{-R_C}{r_e}\left(\frac{R_L}{R_L + R_C}\right) \tag{5–45}$$

Equation 5–45 may be written

$$\frac{v_L}{v_S} = \frac{-1}{r_e}\left(\frac{R_L R_C}{R_L + R_C}\right) = \frac{-R_L \parallel R_C}{r_e} = \frac{-r_L}{r_e} \tag{5–46}$$

Applying equation 5–46 to Example 5–5, we find

$$\frac{v_L}{v_S} = \frac{-(10 \times 10^3\ \Omega) \parallel (1 \times 10^3\ \Omega)}{3.64\ \Omega} = -249.7$$

Then, the magnitude of v_L is $v_L = (249.7)v_S = 249.7(20\text{ mV rms}) = 4.99\text{ V rms}$.

Equation 5–46 shows that the amplifier voltage gain is directly proportional to the ac load resistance $r_L = R_L \parallel R_C$. Since the value of $R_L \parallel R_C$ is always less than the value of R_C alone, the effect of connecting a load R_L to the amplifier is always to reduce the voltage gain. This is the same effect we saw in graphical analysis of the CE amplifier (see Figure 5–19 and Example 5–3).

Another practical consequence of a capacitor-coupled load R_L is that it reduces the maximum permissible voltage swing at the output of the amplifier. Recall from

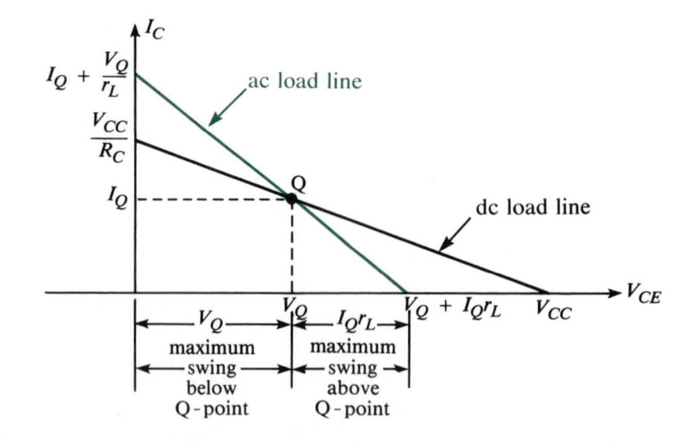

Figure 5–35
The ac load line limits the output voltage swing to the minimum of V_Q and $I_Q r_L$.

equations 5–17 and 5–18 that the ac load line intersects the V_{CE}-axis at $V_0 = V_Q + I_Q r_L$ and intersects the I_C-axis at $I_0 = I_Q + V_Q/r_L$, where I_Q and V_Q are the Q-point coordinates. It is therefore clear that the positive-going output voltage cannot exceed the bias voltage by more than $I_Q r_L$ volts. This fact is illustrated in Figure 5–35. The output voltage can swing $I_Q r_L$ volts above the Q-point and V_Q volts below the Q-point. Thus the maximum output voltage variation is the *minimum* of the two values V_Q and $I_Q r_L$. If the output attempts to exceed V_Q by more than $I_Q r_L$ volts, its positive peak will be clipped; if it attempts to swing more than V_Q volts below V_Q, its negative peak will be clipped.

Example 5–6

Find the maximum peak-to-peak voltage that the source in Figure 5–36 can produce without causing clipping at the amplifier output.

Figure 5–36
(Example 5–6)

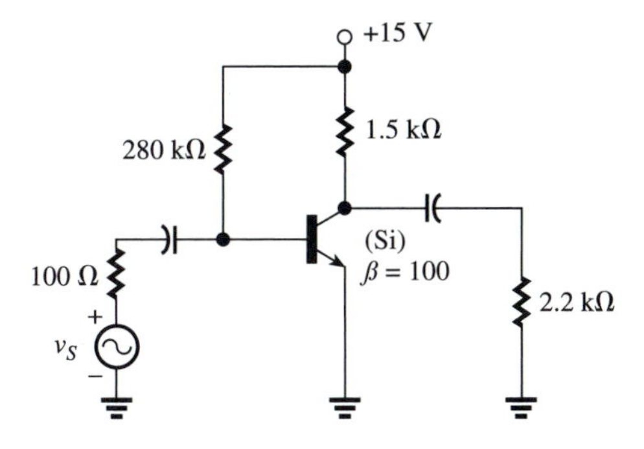

Solution. Solving first for the quiescent values, we have

$$I_B = \frac{(15 - 0.7)\text{ V}}{280 \times 10^3\ \Omega} = 51.1\ \mu\text{A}$$

$$I_C = I_Q = \beta I_B = 100(51.1\ \mu\text{A}) = 5.11\text{ mA}$$

$$V_Q = V_{CC} - I_Q R_C = 15\text{ V} - (5.11\text{ mA})(1.5\text{ k}\Omega) = 7.34\text{ V}$$

$$r_L = R_C \| R_L = (1.5\text{ k}\Omega) \| (2.2\text{ k}\Omega) = 892\ \Omega$$

Therefore, the maximum output swing *above* $V_Q = 7.34$ V is $I_Q r_L = (5.11\text{ mA})(892\ \Omega) = 4.56$ V. Since $V_Q = 7.34$ V, the maximum output variation from the Q-point is the minimum of 4.56 V and 7.34 V, or 4.56 V. Thus, the maximum peak-to-peak ac output is (2)(4.56) = 9.12 V p–p.

To determine the source voltage that produces a 9.12-V p–p output, we must find the voltage gain of the stage.

$$I_E \approx I_C = 5.11\text{ mA}$$

$$r_e = \frac{0.026\text{ V}}{I_E} = \frac{0.026}{5.11 \times 10^{-3}\text{ A}} = 5.09\ \Omega$$

To apply the voltage gain equation (5–42), we must compute $r_{in}(\text{stage}) = R_B \| (\beta r_e)$:

$$r_{in}(\text{stage}) = (280 \times 10^3\ \Omega) \| (100)(5.09)\ \Omega = (280 \times 10^3\ \Omega) \| 509\ \Omega \approx 509\ \Omega$$

Applying equation 5–42 now, we have

$$\frac{v_L}{v_S} = \frac{-1.5 \times 10^3}{5.09}\left(\frac{509}{100 + 509}\right)\left(\frac{2.2 \times 10^3}{1.5 \times 10^3 + 2.2 \times 10^3}\right) = -146.45$$

Therefore, the maximum peak-to-peak source voltage is

$$v_S(\text{max, p–p}) = \frac{v_L(\text{max, p–p})}{146.45} = \frac{9.12}{146.45} = 62.27 \text{ mV p–p}$$

Example 5–7

SPICE

Use SPICE to determine the quiescent point, the voltage gain v_L/v_S, and the current gain i_L/i_S of the amplifier shown in Figure 5–36. Assume each of the coupling capacitors is 10 μF and that the signal frequency is 10 kHz. (These values ensure that the reactance of the capacitors is negligible and will have no effect on amplifier gain.) Assume the magnitude of the signal-source is 50 mV.

Solution. Figure 5–37(a) shows the circuit redrawn for analysis by SPICE and the input data file. Note that we have inserted a dummy voltage source to measure the current in R_L, so that the current gain i_L/i_S can be determined. A .DC analysis is not required, since SPICE automatically provides "operating point information" when performing an .AC analysis of a circuit containing a transistor.

The results of a program run are shown in Figure 5–37(b). We see that the operating (quiescent) point is $V_{CE} = -7.402$ V and $I_C = 5.07$ mA. The results of the analysis show that the phase angle of the output, VP(6), is very nearly 180°,

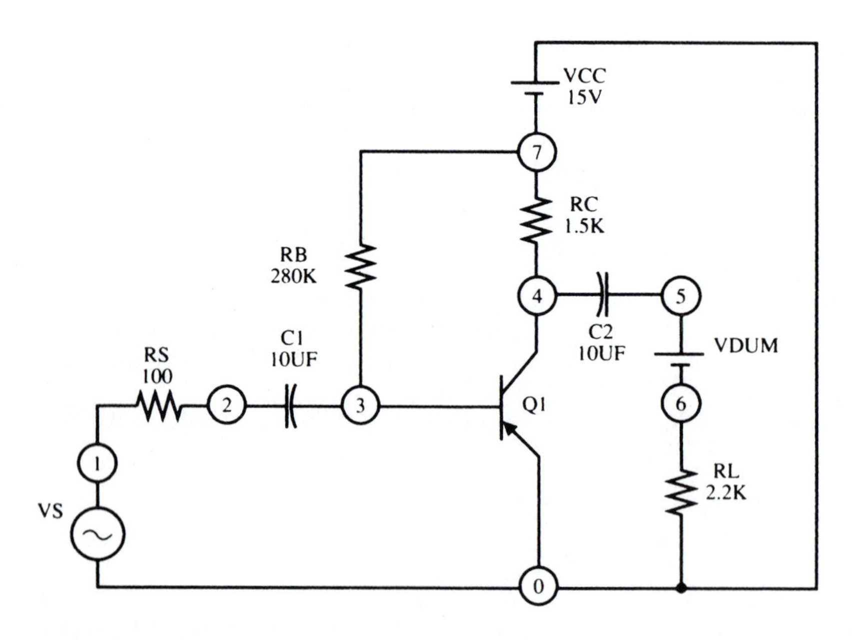


```
EXAMPLE 5.7
VS 1 0 AC 50MV
VCC 0 7 15V
RS 1 2 100
C1 2 3 10UF
RB 7 3 280K
RC 7 4 1.5K
C2 4 5 10UF
RL 6 0 2.2K
VDUM 5 6
Q1 4 3 0 TRANS
.MODEL TRANS PNP BF=100
.AC LIN 1 10K 10K
.PRINT AC I(VS) I(VDUM)
V(6) VP(6)
.END
```

Figure 5–37(a)
(Example 5–7)

```
EXAMPLE 5.7
****      OPERATING POINT INFORMATION       TEMPERATURE =    27.000 DEG C
******************************************************************************
**** BIPOLAR JUNCTION TRANSISTORS
              Q1
MODEL       TRANS
IB         -5.07E-05
IC         -5.07E-03
VBE            -.816
VBC            6.585
VCE           -7.402
BETADC       100.000
GM          1.96E-01
RPI         5.11E+02
RX          0.00E+00
RO          1.00E+12
CPI         0.00E+00
CMU         0.00E+00
CBX         0.00E+00
CCS         0.00E+00
BETAAC       100.000
FT          3.12E+18
*******02/21/89 ********  SPICE 2G.6    3/15/83 ********13:32:50*****
EXAMPLE 5.7
****      AC ANALYSIS                       TEMPERATURE =    27.000 DEG C
******************************************************************************
    FREQ        I(VS)       I(VDUM)      V(6)        VP(6)
 1.000E+04    8.202E-05   3.319E-03   7.302E+00  -1.798E+02
```

Figure 5–37(b)
(Example 5–7)

confirming that the amplifier performs phase inversion. The voltage and current gains are

$$\frac{v_L}{v_S} = \frac{-V(6)}{50\text{ mV}} = \frac{-7.302\text{ V}}{50\text{ mV}} = -146.04$$

$$\frac{i_L}{i_S} = \frac{\text{I(VDUM)}}{\text{I(VS)}} = \frac{3.319\text{ mA}}{82.02\text{ A}} = 40.46$$

The values found by SPICE for the quiescent point and the voltage gain agree very closely with those found in Example 5–6.

Figure 5–20 in our discussion of graphical amplifier analysis shows that the ac load line shifts with changes in the Q-point. It is obvious that shifting the ac load line causes it to have different intercepts on the horizontal and vertical axes. It should therefore be possible to locate the Q-point in such a way that the maximum positive swing equals the maximum negative swing, and thereby attain the maximum possible peak-to-peak output swing. As we can see in Figure 5–35, this condition occurs when

$$V_Q = I_Q r_L \tag{5–47}$$

When the dc load line equation is rewritten and evaluated at $V_{CE} = V_Q$ and $I_C = I_Q$, we obtain

$$V_Q = V_{CC} - I_Q R_C \tag{5–48}$$

Solving equations 5–47 and 5–48 simultaneously (see Exercise 5–19) leads to the following equations for the optimum Q-point coordinates:

$$\text{(optimum) } I_Q = \frac{V_{CC}}{R_C + r_L} \tag{5–49}$$

$$\text{(optimum) } V_Q = V_{CC} - \left(\frac{V_{CC}}{R_C + r_L}\right) R_C \tag{5–50}$$

Example 5–8

PSPICE

Use PSpice to obtain a plot of v_{ce} versus time in the circuit shown in Figure 5–38(a). The plot should cover two full cycles of output. Also find the quiescent values of V_{CE} and I_C.

Solution. To obtain a plot of v_{ce} versus time, we must use a SIN source and perform a .TRAN analysis (see Appendix Sections A–4 and A–5). In order for the plot to cover two full cycles, we set *TSTOP* in the .TRAN statement to two times the period of the sine wave: 2(1/10 kHz) = 0.2 ms. The circuit is redrawn in

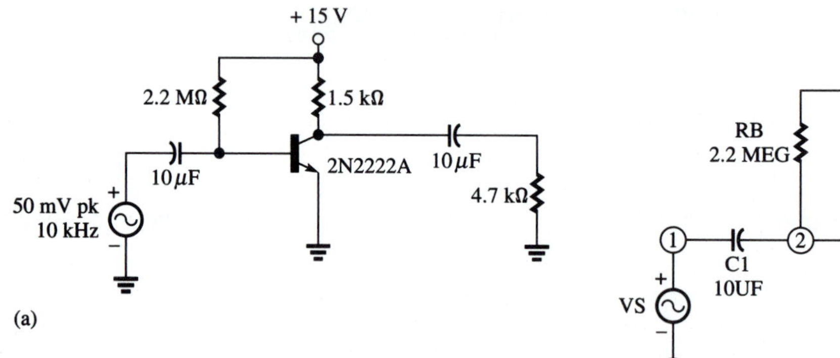

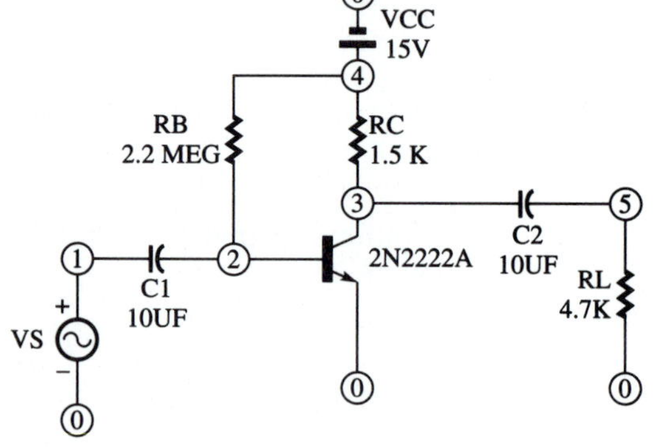


```
EXAMPLE 5-8
VS 1 0 SIN(0 50MV 10KHZ 0 0)
VCC 4 0 15V
C1 1 2 10UF
C2 3 5 10UF
RB 2 4 2.2MEG
RC 3 4 1.5K
RL 5 0 4.7K
Q1 3 2 0 Q2N2222A
.LIB EVAL.LIB
.TRAN 500NS 0.2MS
.DC VCC 15V 15V 1
.PRINT DC IC(Q1) VCE(Q1)
.PLOT TRAN V(3)
.END
```

(EVAL.LIB for Evaluation version of PSpice only)

(b)

Figure 5–38
(Example 5–8)

Figure 5–38 *(Continued)*

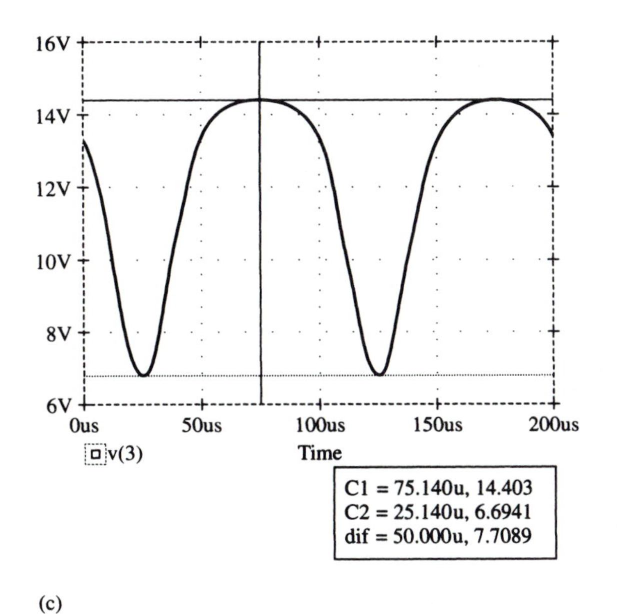

(c)

a PSpice format in Figure 5–38(b). Also shown is the PSpice input circuit file. Note that we use the PSpice library to access the model statement for the 2N2222A transistor (see Appendix Section A–17). Also note that the Q-point can be obtained in a PSpice .PRINT statement by requesting outputs IC(Q1) (the dc collector current of Q1) and VCE(Q1) (the dc collector-to-emitter voltage of Q1).

Execution of the program reveals that IC(Q1) = 1.125 mA and VCE(Q1) = 13.31 V. Since V_{CC} = 15 V, this quiescent point is close to the cutoff region of the transistor. We might therefore expect positive clipping to occur if the input to the amplifier is sufficiently large (see Figures 5–8(a) and 5–14). Figure 5–38(c) shows a plot produced by the PSpice Probe option (Section A–16), and we see that clipping does indeed occur. The Probe cursors are set at the minimum and maximum values of v_{ce}, C1 = 14.403 V and C2 = 6.6941 V. Thus, positive clipping occurs at v_{ce} = 14.403 V.

Small-Signal CC Amplifier Model

Figure 5–39(a) shows a common-collector amplifier and Figure 5–39(b) shows the corresponding small-signal equivalent circuit. Note that the collector in Figure 5–39(b) is shown grounded, since V_{CC} in Figure 5–39(a) is connected directly to the collector and dc sources are once again treated as ac short circuits. Some authors refer to the CC transistor as a "grounded collector" configuration, which is correct in the ac sense. A load resistance R_L is not included at this time.

As in the CE model, the resistance between base and emitter is seen to be βr_e, which is again an acceptable approximation for $r_\pi = (\beta + 1)r_e$. The total input resistance at the base of the transistor (between base and ground) is derived as follows:

$$r_{in} = \frac{v_{in}}{i_{in}} = \frac{v_{in}}{i_b} \tag{5–51}$$

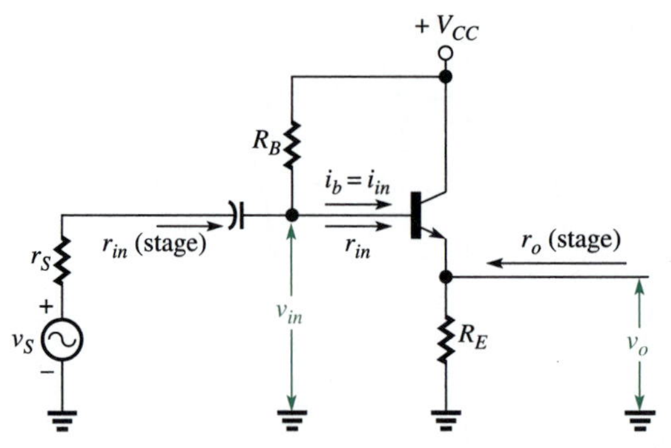

(a) A common-collector amplifier

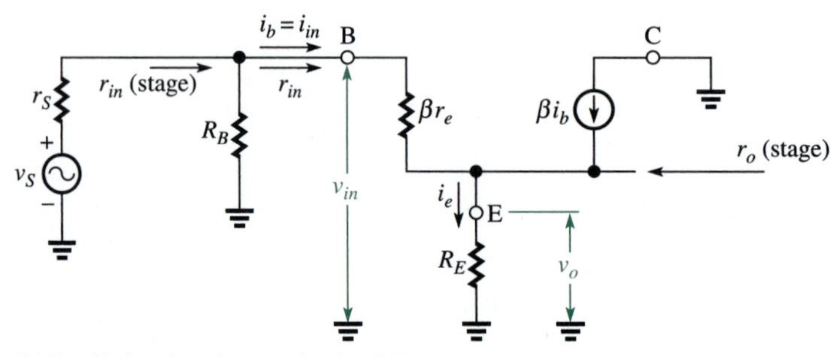

(b) Small-signal equivalent circuit of (a)

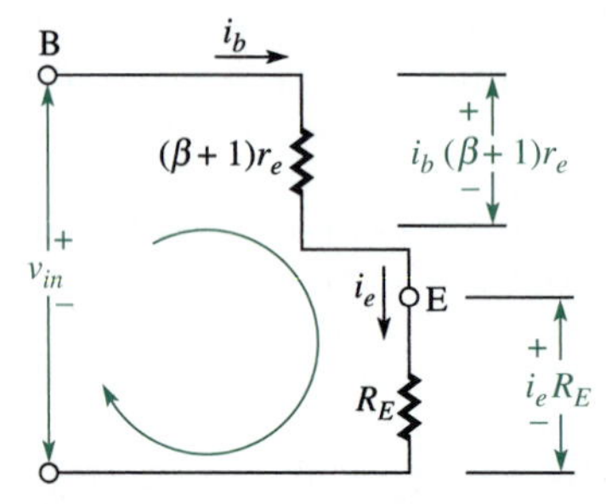

(c) Replacing βr_e with the more accurate value $(\beta + 1)r_e$ and writing Kirchhoff's voltage law around the loop gives $v_{in} = i_b\,(\beta + 1)r_e + i_e R_E$.

Figure 5–39
A common-collector amplifier and its small-signal equivalent circuit

Remember that in a CC circuit v_{in} *is the ac base-to-collector voltage, which the figure clearly shows is the same as the ac base-to-ground voltage.* The input current i_{in} is the same as the base current i_b.

For purposes of this derivation, let us temporarily replace βr_e in Figure 5–39(b) by the more accurate value $(\beta + 1)r_e$. See Figure 5–39(c). Then, writing Kirchhoff's voltage law from B to ground in Figure 5–39(c), we obtain

$$
\begin{aligned}
v_{in} &= i_b(\beta + 1)r_e + i_e R_E \\
&= i_b(\beta + 1)r_e + i_b(\beta + 1)R_E \\
&= i_b(\beta + 1)(r_e + R_E) \qquad \textbf{(5–52)}
\end{aligned}
$$

Substituting (5–52) into (5–51), we find

$$r_{in} = \frac{i_b(\beta + 1)(r_e + R_E)}{i_b} = (\beta + 1)(r_e + R_E) \tag{5–53}$$

Again invoking the approximation $\beta + 1 \approx \beta$, we have

$$r_{in} = (\beta + 1)(r_e + R_E) \approx \beta(r_e + R_E) \tag{5–54}$$

In many practical circuits, $R_E \gg r_e$, so r_{in} can usually be computed simply as

$$r_{in} \approx \beta R_E \tag{5–55}$$

It is apparent from Figure 5–39(b) that

$$r_{in}(\text{stage}) = R_B \parallel r_{in} \approx R_B \parallel \beta R_E \tag{5–56}$$

Equations 5–54 and 5–55 reveal the single most important feature of the CC amplifier in practical applications: its input resistance can be made very large in comparison to the other configurations. For example, using the typical values $R_E = 1\ \text{k}\Omega$ and $\beta = 100$, we have $r_{in} \approx 100\ \text{k}\Omega$. Later in the chapter we will discuss an application in which this feature is very valuable.

Remember from Chapter 4 that the output voltage in the CC configuration is the collector–emitter voltage. Since the collector is grounded, the ac output voltage is the same as the ac emitter-to-ground voltage (see Figure 5–39(b)). We can therefore derive the voltage gain of the CC transistor as follows:

$$A_v = \frac{v_o}{v_{in}} = \frac{i_e R_E}{v_{in}} \tag{5–57}$$

Substituting from equation 5–52 for v_{in} in equation 5–57, we obtain

$$\begin{aligned} A_v &= \frac{i_e R_E}{i_b(\beta + 1)(r_e + R_E)} \\ &= \frac{i_b(\beta + 1)R_E}{i_b(\beta + 1)(r_e + R_E)} = \frac{R_E}{r_e + R_E} \end{aligned} \tag{5–58}$$

Since $r_e + R_E > R_E$, equation 5–58 shows that *the CC transistor always has a voltage gain less than 1.* As mentioned before, it is usually the case that $R_E \gg r_e$, so the following approximation is often used:

$$A_v = \frac{R_E}{r_e + R_E} \approx \frac{R_E}{R_E} = 1 \tag{5–59}$$

Equation 5–59 states that v_o/v_{in} is (approximately) 1, from which it follows that $v_o \approx v_{in}$. It is not difficult to visualize why the ac output voltage is approximately the same as the ac input voltage in a CC configuration: the input and output are "separated" only by the small ac resistance of the forward-biased base–emitter junction. In other words, the input (base-to-ground) voltage is the same, except for the small drop across the junction, as the output (emitter-to-ground) voltage. Note that there is no phase inversion between input and output. As the base-to-ground voltage increases, so does the emitter-to-ground voltage. Since the output voltage is essentially the same as the input voltage, in magnitude and phase, the emitter is said to *follow* the base. In that context, a transistor in the CC configuration is often called an *emitter follower* (more often, in fact, than it is called a common-collector circuit).

Example 5–9

The common-collector amplifier in Figure 5–39(a) has $V_{CC} = 15$ V, $R_B = 75$ kΩ, and $R_E = 910$ Ω. The β of the silicon transistor is 100. Find

1. r_{in}(stage);
2. A_v.

Solution

1. From equations 4–28,

$$I_B = \frac{V_{CC} - V_{BE}}{R_B + (\beta + 1)R_E} = \frac{15\text{ V} - 0.7\text{ V}}{75\text{ k}\Omega + 101(910\ \Omega)} = 85.7\ \mu\text{A}$$
$$I_E = (\beta + 1)I_B = (101)(85.7\ \mu\text{A}) = 8.57\text{ mA}$$

Then,

$$r_e = \frac{0.026\text{ V}}{I_E} = \frac{0.026}{8.57\text{ mA}} = 3.03\ \Omega$$

From equation 5–54,

$$r_{in} = (\beta + 1)(r_e + R_E) = 101(3.03\ \Omega + 910\ \Omega) = 92.2\text{ k}\Omega$$

Note that $r_{in} \approx \beta R_e = 100(910\ \Omega) = 91$ kΩ. From equation 5–56,

$$r_{in}(\text{stage}) = R_B \| r_{in} = \frac{(75\text{ k}\Omega)(92.2\text{ k}\Omega)}{75\text{ k}\Omega + 92.2\text{ k}\Omega} = 41.36\text{ k}\Omega$$

2. From equation 5–58,

$$A_v = \frac{R_E}{r_e + R_E} = \frac{910\ \Omega}{3.03\ \Omega + 910\ \Omega} = 0.997$$

Note that $A_v \approx 1$.

Since the common-collector input and output currents are i_b and i_e, respectively, we find the current gain for the transistor in this configuration to be

$$A_i = \frac{i_e}{i_b} = \frac{(\beta + 1)i_b}{i_b} = \beta + 1 \approx \beta \quad \textbf{(5–60)}$$

Thus, while the CC voltage gain is somewhat less than 1, the current gain is substantially greater than 1, and so, therefore, is the power gain:

$$A_p = A_v A_i \approx A_i \quad \textbf{(5–61)}$$

Rather than presenting a lengthy derivation for the output resistance of the CC transistor, let us make an intuitive observation, from which we can formulate a general rule for determining output resistance of CC amplifiers. We have already seen how the relationship $i_e = (\beta + 1)i_b$ leads to the result that the resistance looking into the base is $(\beta + 1)$ times greater than the actual resistance connected through the base to ground (equation 5–54). Conversely, that same relationship implies that resistance looking into the emitter is $(\beta + 1)$ times *smaller* than the resistance from the emitter back to the signal source. Now, the output resistance of the CC stage in Figure 5–39, r_o(stage), is the resistance looking into the emitter in parallel with R_E. Therefore, applying the foregoing rule, we find

$$r_o(\text{stage}) = R_E \left\| \left[\frac{(\beta+1)r_e + R'_B}{\beta+1} \right] \right. \tag{5–62}$$

where R'_B is the resistance looking from the base toward the signal source. Figure 5–40 shows how R'_B is defined in the ac equivalent of the base circuit. R'_B is computed by shorting the signal source to ground, so

$$R'_B = R_B \parallel r_S \tag{5–63}$$

Substituting (5–63) into (5–62), we obtain

$$\begin{aligned} r_o(\text{stage}) &= R_E \left\| \left[\frac{(\beta+1)r_e + R_B \parallel r_S}{\beta+1} \right] \right. \\ &= R_E \left\| \left(r_e + \frac{R_B \parallel r_S}{\beta+1} \right) \approx R_E \right\| \left(r_e + \frac{R_B \parallel r_S}{\beta} \right) \end{aligned} \tag{5–64}$$

Equation 5–64 shows that the output resistance of an emitter follower can be quite small. For example, for the typical values $R_E = 1\ \text{k}\Omega$, $r_e = 25\ \Omega$, $R_B = 100\ \text{k}\Omega$, $r_S = 50\ \Omega$, and $\beta = 100$, we find

$$\begin{aligned} r_o(\text{stage}) &\approx (1 \times 10^3) \parallel \{25 + [(100 \times 10^3) \parallel 50]/100\} \\ &= (1 \times 10^3) \parallel (25 + 0.5) = (1 \times 10^3) \parallel 25.5 \approx 25\ \Omega \end{aligned}$$

When $r_S = 0$, notice that $R_B \parallel r_S = 0$ and equation 5–64 becomes

$$r_o(\text{stage}) \approx R_E \parallel r_e \approx r_e \tag{5–65}$$

Figure 5–41 shows an emitter follower having a load resistance R_L that is capacitor-coupled to the emitter. Since the total ac resistance connected between emitter and ground is now $r_L = R_E \parallel R_L$, the input resistances specified by equations 5–54 and 5–56 become

$$r_{in} = (\beta+1)(r_e + r_L) \approx \beta(r_e + r_L) \tag{5–66}$$

$$r_{in}(\text{stage}) = R_B \parallel (\beta+1)(r_e + r_L) \approx R_B \parallel \beta(r_e + r_L) \tag{5–67}$$

The approximations in these equations are valid under the usual circumstance:

$$\beta + 1 \approx \beta \tag{5–68}$$

Figure 5–40
R'_B is the (Thevenin) equivalent resistance looking from the base toward the source. $R'_B = R_B \parallel r_S$.

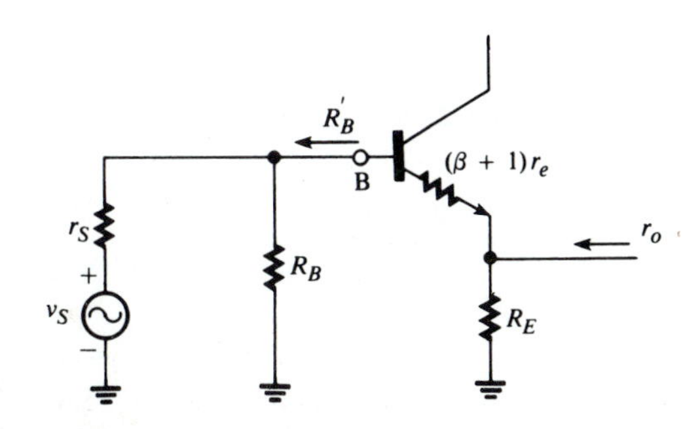

Figure 5–41
An emitter follower having load resistance R_L

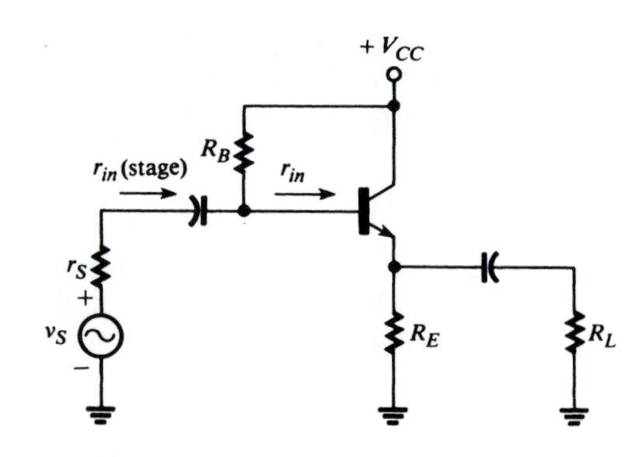

The overall voltage gain of the emitter-follower stage, taking load and source resistances into account, can be determined as follows:

$$\frac{v_L}{v_S} = \frac{r_L}{r_e + r_L}\left[\frac{r_{in}(\text{stage})}{r_S + r_{in}(\text{stage})}\right] \tag{5–69}$$

where $r_L = R_E \parallel R_L$.

For small r_S, equation 5–69 is approximately equivalent to

$$\frac{v_L}{v_S} \approx \frac{R_E}{r_e + R_E}\left[\frac{R_L}{R_L + r_o(\text{stage})}\right] \tag{5–70}$$

(Equation 5–70 is exact when $r_S = 0$.)

Although the emitter follower by itself has a voltage gain less than 1, it can be used to improve the voltage gain of a larger amplifier system. Because of its large input resistance, it does not "load" the output of another amplifier. In other words, the load presented by an emitter follower to another amplifier does not appreciably reduce the voltage gain of that amplifier. Also, because it has a small output resistance, the emitter follower can drive a "heavy" load (small resistance) whose presence would otherwise reduce voltage gain. For these reasons, an emitter follower is valuable as an intermediate stage between an amplifier and a load. When an emitter follower is used this way, it is called a *buffer* amplifier, or an *isolation* amplifier, because it effectively isolates another amplifier from the loading effect of R_L. This use is illustrated in the next example.

Example 5–10

An amplifier having an output resistance of 1 kΩ is to drive a 50-Ω load, as shown in Figure 5–42(a). Assuming that the amplifier has a voltage gain of $A_v = 140$ (with no load connected), find

1. the voltage gain with the load connected, and
2. the voltage gain when an emitter follower is inserted between the amplifier and the load, as shown in Figure 5–42(b).

Solution

1. The voltage gain with the 50-Ω load connected is

$$\frac{v_L}{v_S} = A_v\left(\frac{R_L}{r_o + R_L}\right) = 140\left(\frac{50}{1000 + 50}\right) = 6.67$$

The 50-Ω load severely reduces the voltage gain.

Figure 5–42
(Example 5–10)

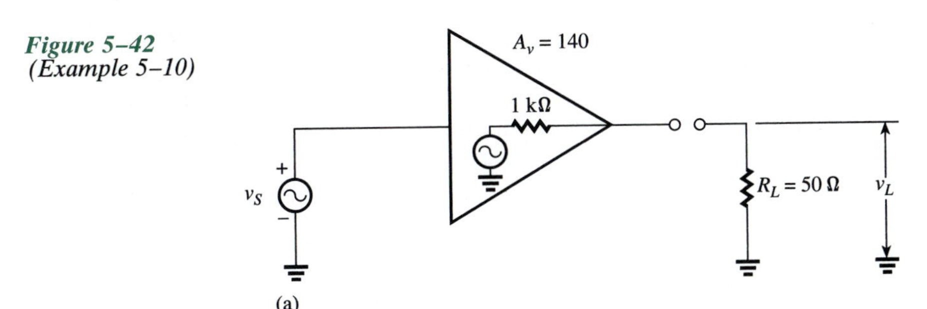

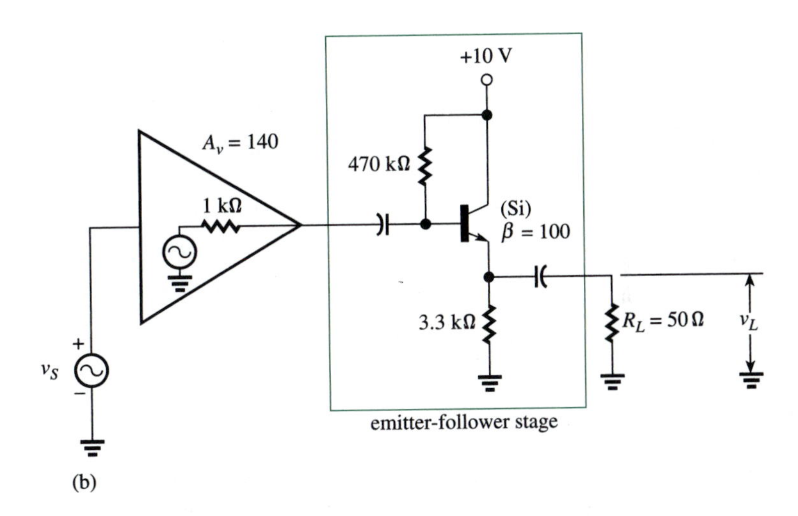

2. To find the input resistance of the emitter follower, we must find r_e, which means that we must first find the dc bias current I_E:

$$I_B = \frac{V_{CC} - 0.7}{R_B + (\beta + 1)R_E} = \frac{9.3\ \text{V}}{470 \times 10^3\ \Omega + 101(3.3 \times 10^3\ \Omega)} = 11.58\ \mu\text{A}$$

$$I_E = (\beta + 1)I_B = 101(11.58\ \mu\text{A}) = 1.16\ \text{mA}$$

Therefore,

$$r_e = \frac{0.026\ \text{V}}{I_E} = \frac{0.026\ \text{V}}{1.16 \times 10^{-3}\ \text{A}} = 22.4\ \Omega$$

The ac load resistance for the emitter follower is

$$r_L = R_E \parallel R_L = (3.3\ \text{k}\Omega) \parallel (50\ \Omega) \approx 50\ \Omega$$

From equation 5–67,

$$r_{in}(\text{stage}) = (470 \times 10^3) \parallel 100(22.4 + 50) = (470 \times 10^3\ \Omega) \parallel (7.24 \times 10^3\ \Omega)$$
$$= 7.13\ \text{k}\Omega$$

As far as the emitter follower is concerned, the source resistance is the output resistance (1 kΩ) of the amplifier driving it. Therefore, from equation 5–69, the overall gain of the emitter follower is

$$\frac{v_L}{v_S} = \frac{50}{22.4 + 50}\left(\frac{7.13 \times 10^3}{1 \times 10^3 + 7.13 \times 10^3}\right) = 0.605$$

The system gain from amplifier to load is then (0.605)(140) = 84.7. We see that insertion of the emitter follower improved the voltage gain from 6.67 to 84.7, a 1170% increase. It is an exercise at the end of this chapter to verify that very nearly the same result is obtained using approximation 5–70 instead of equation 5–69, even though r_S is relatively large (1 kΩ).

Equations 5–71 summarize the small-signal analysis equations for a common-collector amplifier stage, based on the approximate CC model.

Common-collector small-signal equations

$$r_e \approx \frac{0.026}{I_E} \text{ ohms} \qquad \text{(at room temperature)}$$

$$r_{in} = (\beta + 1)(r_e + R_E) \approx \beta(r_e + R_E) \qquad \text{(output open)}$$

$$r_{in}(\text{stage}) = (\beta + 1)(r_e + r_L) \parallel R_B \approx \beta(r_e + r_L) \parallel R_B$$

$$\approx \beta r_L \parallel R_B \quad (r_L \gg r_e) \qquad \text{where } r_L = R_E \parallel R_L$$

$$r_o(\text{stage}) = R_E \left\| \left(r_e + \frac{R_B \parallel r_S}{\beta + 1} \right) \right.$$

$$= R_E \parallel r_e \quad (r_S = 0)$$

$$\approx r_e \quad (R_E \gg r_e)$$

$$A_i = \beta$$

$$A_v = \frac{R_E}{r_e + R_E} \qquad \text{(output open)} \tag{5–71}$$

$$\approx 1 \quad (R_E \gg r_e)$$

$$\frac{v_L}{v_S} = \frac{r_L}{r_e + r_L} \left[\frac{r_{in}(\text{stage})}{r_S + r_{in}(\text{stage})} \right]$$

$$= \frac{r_L}{r_e + r_L} \qquad (r_S = 0)$$

$$\approx 1 \quad (r_L \gg r_e)$$

$$\frac{v_L}{v_S} \approx \frac{R_E}{r_e + R_E} \left[\frac{R_L}{R_L + r_o(\text{stage})} \right] \qquad (r_S \approx 0)$$

+V_{CC} −V_{CC} R_B r_S v_S R_E R_L

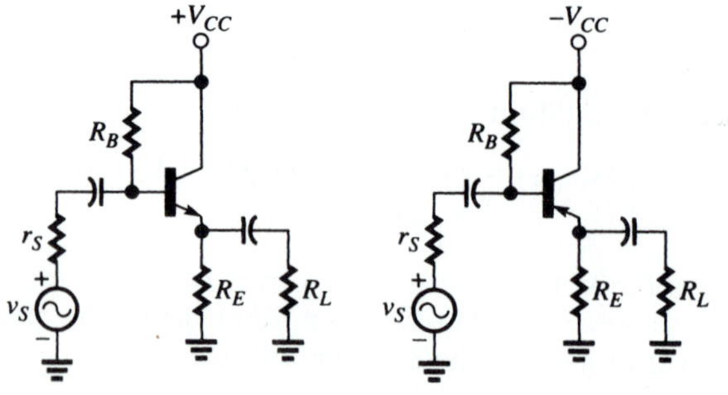

5–4 THE DEPENDENCE OF SMALL-SIGNAL PARA

Transconductance

We have seen that one of the important pa
of the last section, the emitter resistance r_e
of the relation

$$r_e = \frac{V_T}{I_E} \approx \frac{0.026}{I_E}\,\Omega \qquad ($$

In fact, all small-signal parameters depe
greater extent than others. Our goal n
reflects that dependency better than the ...
the advantages of the model is that it permits us to perfo...
signal analysis based entirely on a knowledge of the dc characteristics of a ...
We begin with a discussion of a new small-signal parameter, *transconductance,* whose approximate value can be found using only dc quantities.

Transconductance is another *derived* parameter that is widely used in the analysis of electronic devices of all kinds. It is designated g_m and is defined as the ratio of a small-signal *output* current to a small-signal *input* voltage, with dc output voltage held constant:

$$g_m = \left.\frac{i_o}{v_{in}}\right|_{V_o = \text{constant}} \qquad \textbf{(5–72)}$$

Since g_m is the ratio of a current to a voltage, its units are those of conductance, i.e., siemens. It is called *trans*conductance because it relates input and output quantities (*across* the device).

For a BJT, transconductance is defined in terms of the common-emitter configuration; that is, input voltage is v_{be} and output current is i_c:

$$g_m = \left.\frac{i_c}{v_{be}}\right|_{V_{CE} = \text{constant}} \qquad \textbf{(5–73)}$$

For v_{be} less than about 10 mV, it can be shown that the value of g_m is closely approximated by the following:

$$g_m \approx \frac{I_C}{V_T} \approx \frac{I_C}{0.026} \quad \text{(at room temperature)} \qquad \textbf{(5–74)}$$

where I_C is the dc collector current and V_T is the thermal voltage kT/q. Since $I_E \approx I_C$, we see that $g_m \approx I_E/0.026 = 1/r_e$ (from equation 5–22).

Output Resistance

The output resistance of a CE transistor can also be determined using dc bias values. Recall that

$$r_o = \left.\frac{v_{ce}}{i_c}\right|_{I_B = \text{constant}} \qquad \textbf{(5–75)}$$

Equation 5–75 states that the output resistance is the reciprocal of the slope of a line of constant base current on a set of CE output characteristics. See Figure 5–43.

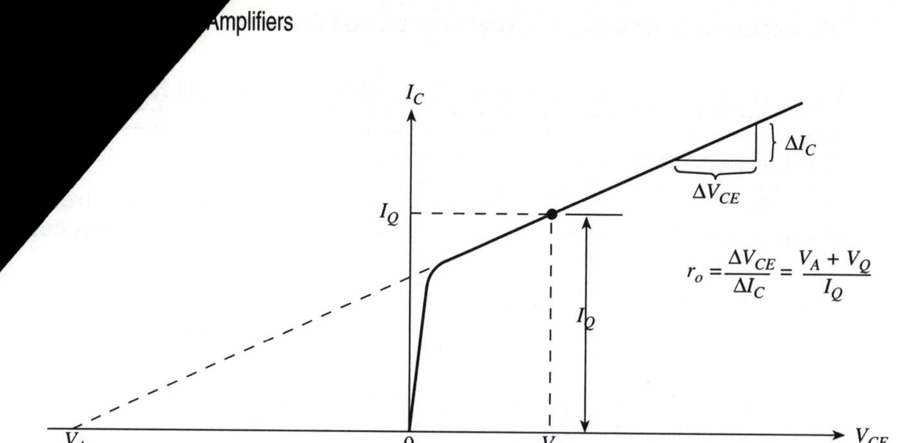

Figure 5–43
Common-emitter output resistance is the reciprocal of the slope of a line of constant base current on the CE output characteristics.

From Figure 5–43, it is apparent that

$$r_o = \frac{V_A + V_Q}{I_Q} \tag{5–76}$$

where
V_A = Early voltage
V_Q = quiescent voltage
I_Q = quiescent current

Instead of output resistance, the output *conductance* g_o is used in many transistor models:

$$g_o = \frac{1}{r_o} = \frac{I_Q}{V_A + V_Q} \tag{5–77}$$

The Transconductance Model

Figure 5–44 shows a CE transistor model based on the parameters we have discussed in this section. Note that the current source representing collector current is now labeled $g_m v_{be}$. From equation 5–73, we have $i_c = g_m v_{be}$, so the current source is correctly labeled. Because the model utilizes g_m, it is often called a *transconductance model.*

Figure 5–44
Transconductance model of a CE transistor

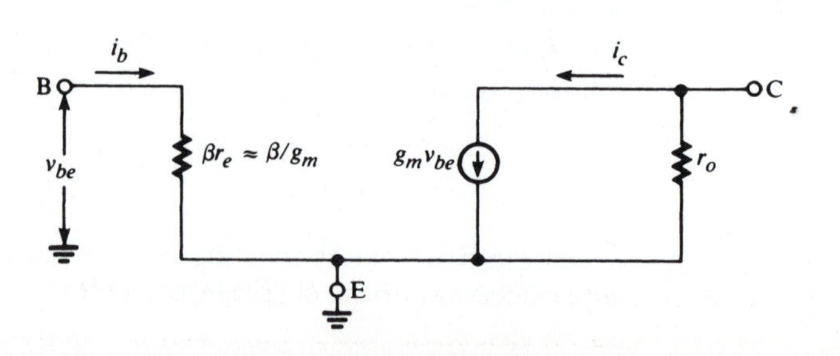

Example 5–11

Use the transconductance model of Figure 5–44 to determine the load voltage in the circuit of Figure 5–45. Assume that β = 80 and V_A = 140 V.

Figure 5–45
(Example 5–11)

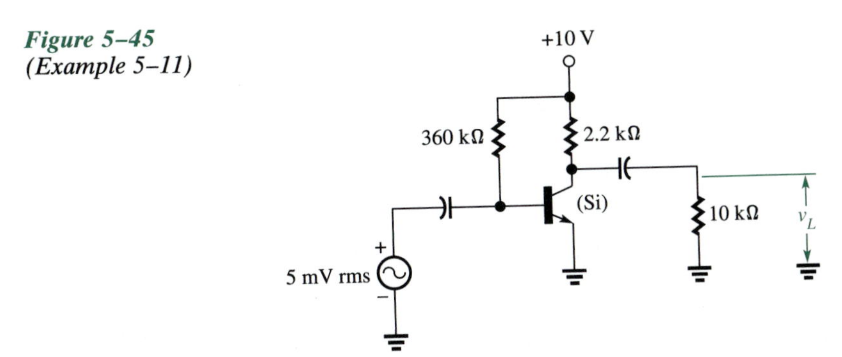

Solution

$$I_B = \frac{V_{CC} - 0.7}{R_B} = \frac{10 - 0.7\text{ V}}{360 \times 10^3\ \Omega} = 25.83\ \mu\text{A}$$

$$I_Q = I_C = \beta I_B = 2.07\ \mu\text{A} \approx I_E$$

$$r_e = \frac{0.026\text{ V}}{I_E} = \frac{0.026}{2.07 \times 10^{-3}\text{ A}} = 12.56\ \Omega$$

$$r_\pi \approx \beta r_e = (80)(12.56\text{ V}) \approx 1\text{ k}\Omega$$

From equation 5–74,

$$g_m \approx \frac{I_C}{0.026} = \frac{2.07 \times 10^{-3}\text{ A}}{0.026} = 79.6\text{ mS}$$

$$V_Q = V_{CC} - I_Q R_C = 10\text{ V} - (2.07\text{ mA})(2.2\text{ k}\Omega) = 5.45\text{ V}$$

From equation 5–76,

$$r_o = \frac{V_A + V_Q}{I_Q} = \frac{(140 + 5.45)\text{ V}}{2.07 \times 10^{-3}\text{ A}} = 70.26\text{ k}\Omega$$

Figure 5–46 shows the ac equivalent circuit of the amplifier with the transconductance model incorporated. The total equivalent resistance across the output is (70.26 kΩ) ∥ (2.2 kΩ) ∥ (10 kΩ) ≈ (1.8 kΩ). Therefore, the load voltage is v_L = $(79.6 \times 10^{-3} v_{be})(1.8\text{ k}\Omega) = (79.6 \times 10^{-3})(5\text{ mV rms})(1.8 \times 10^3) = 0.716$ V rms.

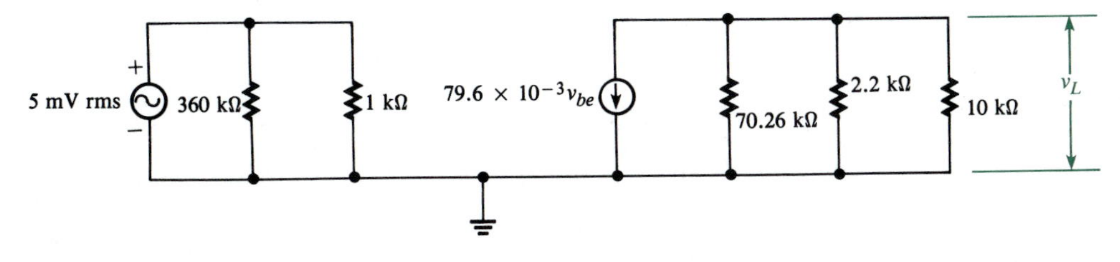

Figure 5–46
The ac equivalent circuit of Figure 5–44 (Example 5–11)

The results of the preceding example show that the voltage gain of the amplifier is

$$\frac{v_L}{v_S} = \frac{-0.716 \text{ V}}{5 \times 10^{-3} \text{ V}} = -143.2$$

where the minus sign is inserted, as usual, to denote phase inversion. Using equation 5–46, which gives the voltage gain based on the model we developed in the previous section, we find

$$\frac{v_L}{v_S} \approx \frac{-r_L}{r_e} = \frac{-1.8 \times 10^3 \ \Omega}{12.56 \ \Omega} = -143.3$$

We see that there is good agreement between the calculations.

5–5 SMALL-SIGNAL PARAMETER EQUIVALENTS: *h* PARAMETERS

In Chapter 9 we will study the theory and application of a set of derived parameters called *h parameters.* These are used as the basis for constructing *hybrid models* of electronic devices of many different kinds. Transistor data sheets often provide values for small-signal *h* parameters instead of or in addition to the physical parameters we have discussed so far. For that reason, and because some electronic devices courses do not have time to cover *h* parameters in depth, Table 5–1 shows a list of approximate parameter equivalents and conversions. In *h*-parameter notation, the second subscript attached to the letter *h* identifies the parameter as a CB, CC, or CE parameter: b = CB; c = CC; and e = CE. For example, h_{ie}, h_{re}, h_{fe}, and h_{oe} are common-emitter parameters, while h_{ib}, h_{rb}, h_{fb}, and h_{ob} are common-base parameters.

Figure 5–47 shows the manufacturer's small-signal specifications for the 2N2218A–2N2222A series transistors whose dc specifications were given in Chapter 4. The graphs showing how the *h* parameters vary with collector current clearly demonstrate the dependence of parameter values on dc conditions. Note also that a significant variation is possible among devices of the same type. Each graph shows the *h*-parameter variation for a typical "high-gain" unit (labeled 1) and for a typical "low-gain" unit (labeled 2).

Table 5–1

Model Parameter	Transconductance Model Parameter (CE)	Approximate *h*-Parameter Equivalent	Typical Value
r_e	α/g_m	h_{re}/h_{oe} ($\approx h_{ib}$)	25 Ω
α	α	h_{fb}	0.99
r_c	$r_o/(\beta + 1)$	$1/h_{ob}$	10 MΩ
$r_p = (\beta + 1)r_e$	β/g_m	h_{ie}	2.5 kΩ
β	β	h_{fe}	100
$r_c/(\beta + 1)$	r_o	$1/h_{oe}$	100 kΩ
$(\beta + 1)/r_c$	g_o	h_{oe}	50 μS
$1/r_c$	$1/(\beta + 1)r_o$	h_{ob}	0.1 μS
α/r_e	g_m	h_{fe}/h_{ie}	40 mS

SMALL-SIGNAL CHARACTERISTICS

Characteristic		Symbol	Min	Max	Unit
Current-Gain — Bandwidth Product(2) (I_C = 20 mAdc, V_{CE} = 20 Vdc, f = 100 MHz)	All Types, Except 2N2219A, 2N2222A, 2N5582	f_T	250 300	— —	MHz
Output Capacitance(3) (V_{CB} = 10 Vdc, I_E = 0, f = 100 kHz)		C_{obo}	—	8.0	pF
Input Capacitance(3) (V_{EB} = 0.5 Vdc, I_C = 0, f = 100 kHz)	Non-A Suffix A-Suffix, 2N5581, 2N5582	C_{ibo}	— —	30 25	pF
Input Impedance (I_C = 1.0 mAdc, V_{CE} = 10 Vdc, f = 1.0 kHz)	2N2218A, 2N2221A 2N2219A, 2N2222A	h_{ie}	1.0 2.0	3.5 8.0	kohms
(I_C = 10 mAdc, V_{CE} = 10 Vdc, f = 1.0 kHz)	2N2218A, 2N2221A 2N2219A, 2N2222A		0.2 0.25	1.0 1.25	
Voltage Feedback Ratio (I_C = 1.0 mAdc, V_{CE} = 10 Vdc, f = 1.0 kHz)	2N2218A, 2N2221A 2N2219A, 2N2222A	h_{re}	— —	5.0 8.0	X 10^{-4}
(I_C = 10 mAdc, V_{CE} = 10 Vdc, f = 1.0 kHz)	2N2218A, 2N2221A 2N2219A, 2N2222A		— —	2.5 4.0	
Small-Signal Current Gain (I_C = 1.0 mAdc, V_{CE} = 10 Vdc, f = 1.0 kHz)	2N2218A, 2N2221A 2N2219A, 2N2222A	h_{fe}	30 50	150 300	—
(I_C = 10 mAdc, V_{CE} = 10 Vdc, f = 1.0 kHz)	2N2218A, 2N2221A 2N2219A, 2N2222A		50 75	300 375	
Output Admittance (I_C = 1.0 mAdc, V_{CE} = 10 Vdc, f = 1.0 kHz)	2N2218A, 2N2221A 2N2219A, 2N2222A	h_{oe}	3.0 5.0	15 35	μmhos
(I_C = 10 mAdc, V_{CE} = 10 Vdc, f = 1.0 kHz)	2N2218A, 2N2221A 2N2219A, 2N2222A		10 25	100 200	

h PARAMETERS

V_{CE} = 10 Vdc, f = 1.0 kHz, T_A = 25°C

This group of graphs illustrates the relationship between h_{fe} and other "h" parameters for this series of transistors. To obtain these curves, a high-gain and a low-gain unit were selected and the same units were used to develop the correspondingly numbered curves on each graph.

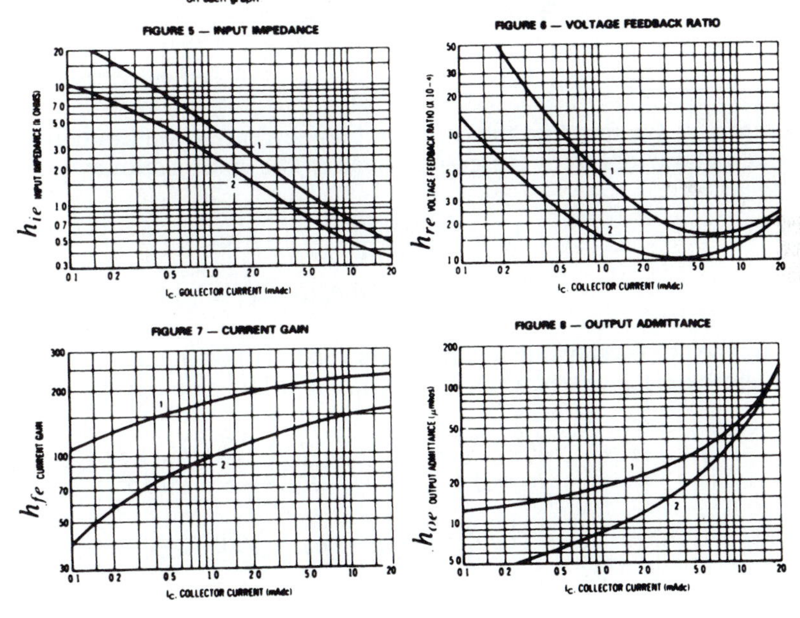

Figure 5–47
Typical small-signal specifications (courtesy of Motorola Corporation)

Example 5–12

Use the h-parameter specifications given in Figure 5–47 to find the quantities listed, in connection with the circuit shown in Figure 5–48. (Assume a "low gain" device having $\beta_{DC} = 100$.)

1. the small-signal β of the transistor
2. the transconductance of the transistor
3. the emitter resistance r_e
4. the output resistance of the amplifier stage
5. the voltage gain v_L/v_S

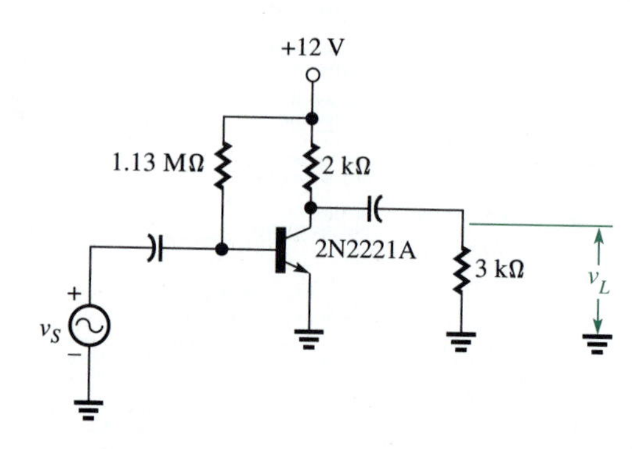

Figure 5–48
(Example 5–12)

Solution. We first find the collector bias current, so we can determine the small-signal parameters from the graphs given in the specifications:

$$I_B = \frac{12 - 0.7\text{ V}}{1.13 \times 10^6\ \Omega} = 10\ \mu\text{A}$$

$$I_C = \beta_{DC} I_B = 100(10\ \mu\text{A}) = 1\text{ mA}$$

Using the graphs showing h parameters versus collector current, we find the following approximate values when $I_C = 1$ mA in a low-gain device:

$$h_{fe} \approx 100 \qquad h_{ie} \approx 2.5\text{ k}\Omega$$

$$h_{re} \approx 1.5 \times 10^{-4} \qquad h_{oe} \approx 8.5 \times 10^{-6}\text{ S}$$

1. From Table 5–1, the small-signal $\beta = h_{fe} = 100$.
2. From Table 5–1,

$$g_m = \frac{h_{fe}}{h_{ie}} = \frac{100}{2.5\text{ k}\Omega} = 0.04\text{ S}$$

3. From Table 5–1,

$$r_e = \frac{h_{re}}{h_{oe}} = \frac{1.5 \times 10^{-4}}{8.5 \times 10^{-6}\text{ S}} = 17.6\ \Omega$$

4. From Table 5–1,

$$r_o = \frac{1}{h_{oe}} = \frac{1}{8.5 \times 10^{-6}\,\text{S}} = 117.6\ \Omega$$

Therefore, $r_o(\text{stage}) = (117.6\ \text{k}\Omega) \parallel (2\ \text{k}\Omega) \approx 2\ \text{k}\Omega$.

5.

$$\frac{v_L}{v_S} = \frac{-R_C}{r_e}\left(\frac{R_L}{R_C + R_L}\right) = \frac{-r_L}{r_e} = \frac{-(2\ \text{k}\Omega) \parallel (3\ \text{k}\Omega)}{17.6\ \Omega} = -68.2$$

Table 5–2 compares, in a very general way, the important small-signal characteristics of the three configurations.

Table 5–2

	Common Base	Common Emitter	Common Collector
Voltage Gain	Large (noninverting)	Large (inverting)	≈ 1 (noninverting)
Current Gain	≈ 1	Large	Large
Power Gain	Moderate	Large	Small
Input Resistance	Low	Moderate	High
Output Resistance	High	Moderate	Low

EXERCISES

Section 5–1
AC Amplifier Fundamentals

5–1. An ac amplifier has a voltage gain of 55 and a power gain of 456.5. The ac output current is 24.9 mA rms and the ac input resistance is 200 Ω. Find
a. the current gain;
b. the rms value of the ac input current;
c. the rms value of the ac input voltage;
d. the rms value of the ac output voltage;
e. the ac output resistance; and
f. the output power.

5–2. An ac amplifier has a current gain of 0.95 and a voltage gain of 100. The ac input voltage is 120 mV rms and the ac input resistance is 25 Ω. Find the output power.

5–3. The signal source connected to the input of an ac amplifier has an internal resistance of 1.2 kΩ. The voltage gain of the amplifier from its input to its output is 140. What minimum value of input resistance should the amplifier have if the voltage gain from signal source to amplifier output is to be at least 100?

5–4. The amplifier in Figure 5–49 has current gain $A_i = 80$ from amplifier input to amplifier output. Find the load current i_L.

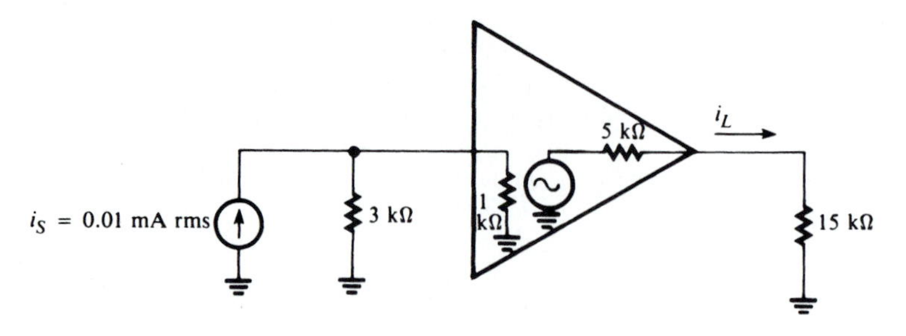

Figure 5–49
(Exercise 5–4)

5–5. An ac amplifier is driven by a 20-mV-rms signal source having internal resistance 1 kΩ. The output resistance of the amplifier is 50 Ω. The voltage gain of the amplifier from its input to its (open-circuit) output (A_v) is 150. What power is delivered to the load if the amplifier is matched to its source and matched to its source and load?

Section 5–2
Graphical Analysis of the Small-Signal CE Amplifier

5–6. Suppose the source voltage in Figure 5–11 causes the transistor base current to vary sinusoidally between 10 μA and 50 μA. Assuming that the transistor has the output characteristics shown in Figure 5–13, what are the total (peak-to-peak) variation in V_{CE} and the total variation in I_C?

5–7. Assuming that the transistor shown in Figure 5–50 has the input characteristics shown in Figure 5–12 and the output characteristics shown in Figure 5–13, find

a. V_{CE} and I_C at the Q-point; and
b. the voltage gain.

5–8. If R_B in Figure 5–50 is changed to 1.935 MΩ, what is the maximum peak-to-peak output voltage without clipping?

5–9. At V_{CE} = 10 V, the CE output characteristics of a certain transistor show that I_C has the values listed in the following table for different values of I_B:

I_B	I_C
10 μA	0.5 mA
15 μA	0.75 mA
20 μA	1.00 mA
25 μA	1.25 mA
30 μA	2.10 mA
35 μA	3.15 mA
40 μA	4.00 mA

At the quiescent voltage V_{CE} = 10 V, what maximum range of I_B values results in the output being a linear reproduction of the input?

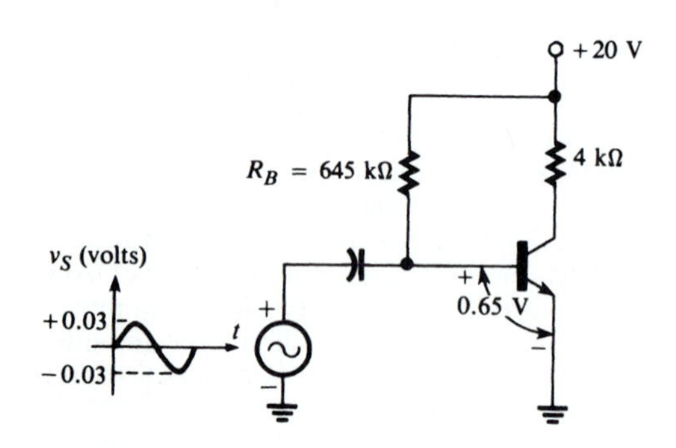

Figure 5–50
(Exercises 5–7 and 5–8)

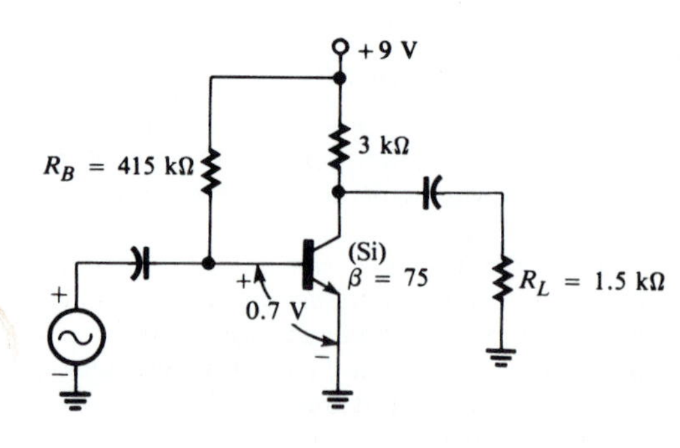

Figure 5–51
(Exercise 5–11)

5–10. Show that equations 5–17 and 5–18 correctly give the intercepts of a dc load line when R_C is substituted for r_L. (This shows that the dc load line is a special case of the ac load line, the case where $R_L = \infty$.)

5–11. The transistor shown in Figure 5–51 has the CE output characteristics shown in Figure 5–52.

a. Plot the dc and ac load lines on the output characteristics.

b. Determine the voltage gain of the circuit if a 24-mV p–p input voltage causes a 20-μA p–p variation in base current.

5–12. Repeat Exercise 5–11 when R_L is changed to 10.5 kΩ.

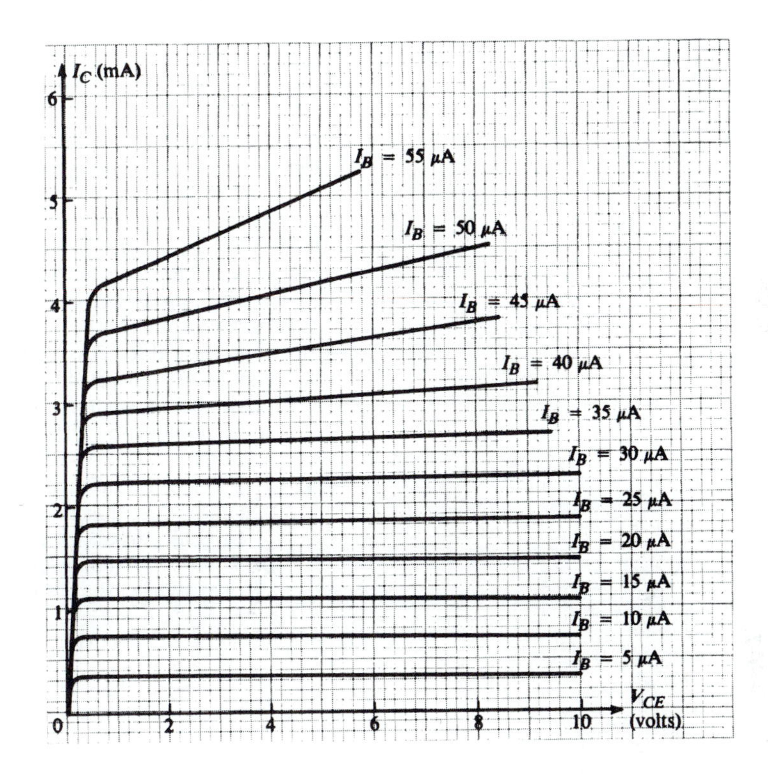

Figure 5–52
(Exercise 5–11)

Section 5–3
Amplifier Analysis Using Small-Signal Models

5–13. The silicon transistor in the amplifier stage shown in Figure 5–53 has an α of 0.99 and collector resistance $r_c = 2.5$ MΩ. Find

a. the input resistance of the amplifier stage;

b. the output resistance of the amplifier stage;

c. the voltage gain of the amplifier; and

d. the current gain of the amplifier.

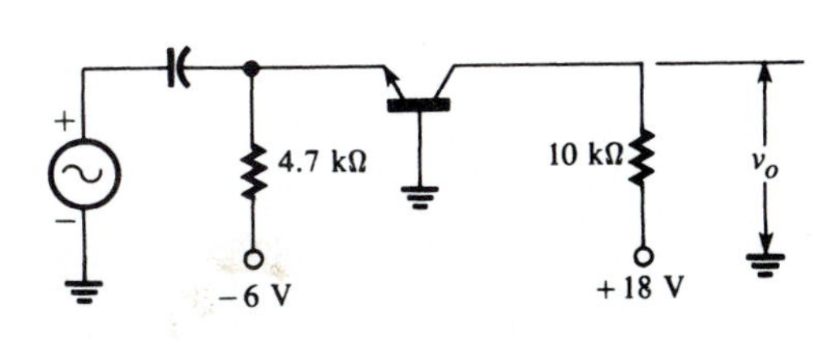

Figure 5–53
(Exercise 5–13)

5–14. The silicon transistor in the amplifier stage shown in Figure 5–54 has a collector resistance $r_c = 1.5\ \text{M}\Omega$. Find
a. the input resistance of the amplifier stage;
b. the output resistance of the amplifier stage; and
c. the rms load voltage v_L.

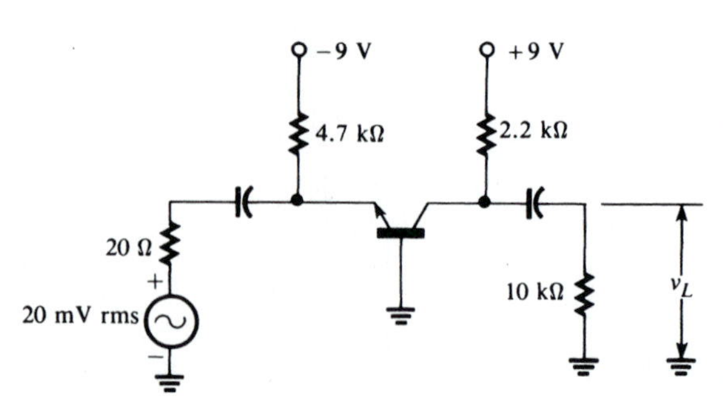

Figure 5–54
(Exercise 5–14)

5–15. Find the voltage gain of the amplifier shown in Figure 5–55. The transistor is germanium.

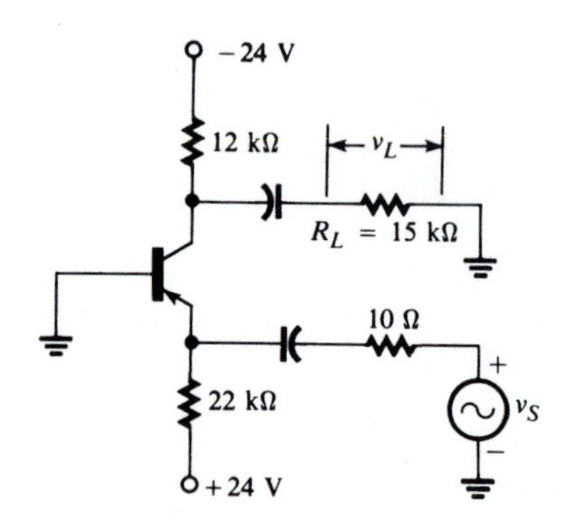

Figure 5–55
(Exercise 5–15)

5–16. What value of R_L in the amplifier of Exercise 5–15 would result in a voltage gain of 100?

5–17. Find the rms load voltage v_L in the amplifier of Figure 5–56 when R_L equals (a) 1 kΩ, (b) 10 kΩ, and (c) 100 kΩ. Assume that $\beta = 100$.

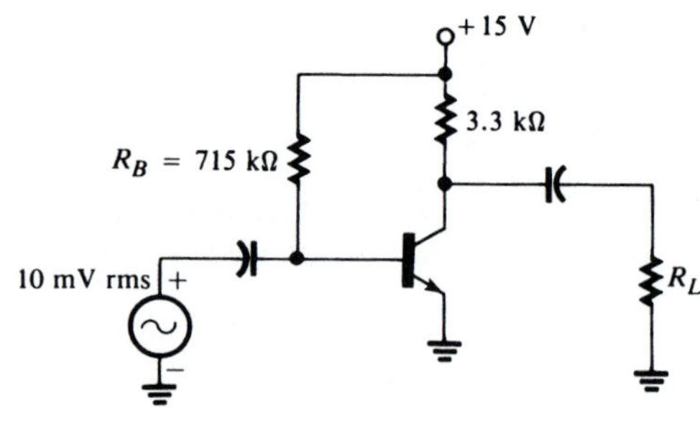

Figure 5–56
(Exercise 5–17)

5–18. Find the maximum peak-to-peak output voltage swing for each value of R_L in Exercise 5–17.

5–19. Solve equations 5–47 and 5–48 simultaneously to derive the expressions for I_Q and V_Q at the bias point where maximum peak-to-peak output can be achieved.

5–20. Find the value of R_B in the amplifier of Exercise 5–17 that results in maximum peak-to-peak output swing when $R_L = 1\ \text{k}\Omega$. What is the maximum swing when that value of R_B is used?

5–21. **a.** For the amplifier in Figure 5–57, find the value of R_B that results in maximum peak-to-peak output swing.
b. What is the maximum peak-to-peak value of v_S when R_B is set to the value found in (a)?

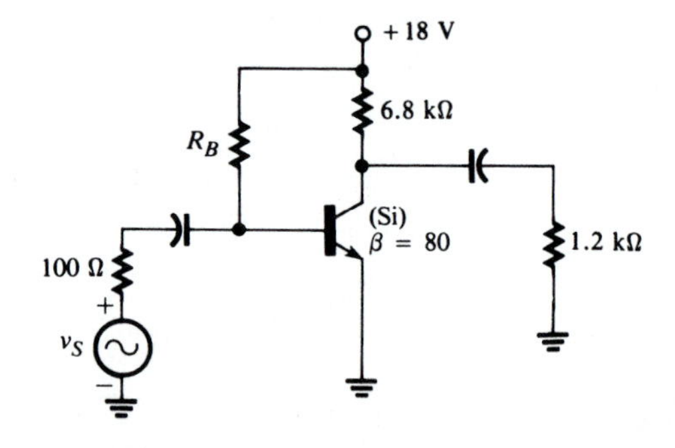

Figure 5–57
(Exercise 5–21)

5–22. For the amplifier circuit shown in Figure 5–58, find
a. r_{in}(stage);
b. r_o(stage);
c. A_v; and
d. v_L/v_S.

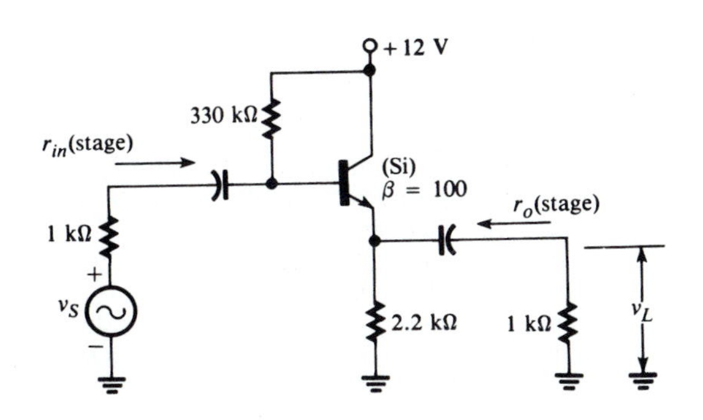

Figure 5–58
(Exercise 5–22)

5–23. Find the overall voltage gain of the amplifier system in Example 5–9 using approximation 5–70 instead of equation 5–69.

5–24. Find the output voltage v_L in Figure 5–59.

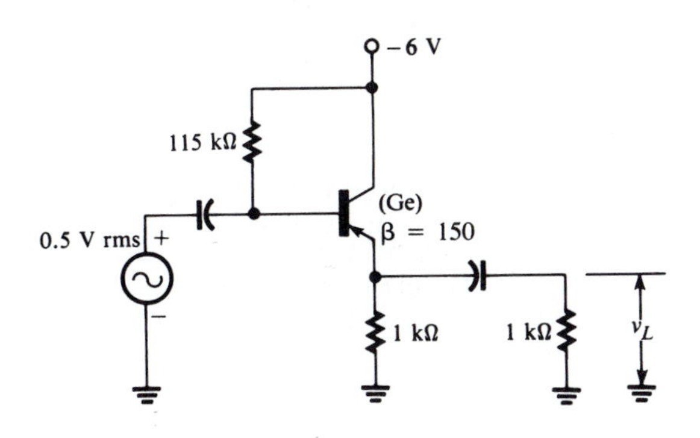

Figure 5–59
(Exercise 5–24)

Section 5–4
The Dependence of Small-Signal Parameters on DC Bias Conditions

5–25. Find the (approximate) transconductance of the transistor in the circuit shown in Figure 5–60, at room temperature, when
a. $R_B = 330$ kΩ and $\beta = 50$;
b. $R_B = 330$ kΩ and $\beta = 150$; and
c. $R_B = 220$ kΩ and $\beta = 50$.

5–26. Assuming that the Early voltage of the transistor in Exercise 5–25 is 100 V, find the output resistance of (a) the transistor, and (b) the stage, for each of the combinations of parameters given in Exercise 5–25.

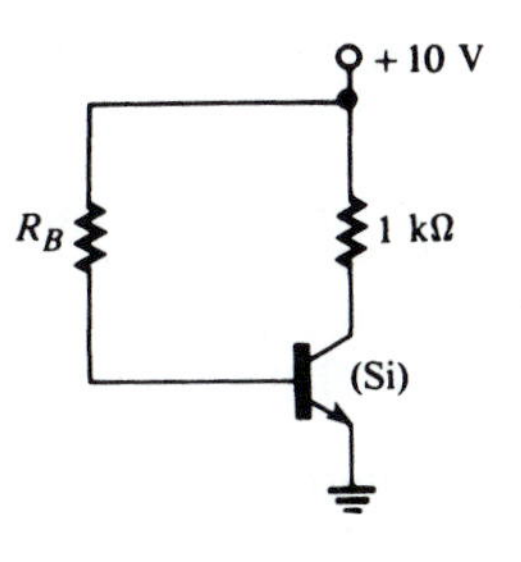

Figure 5–60
(Exercise 5–25)

5–27. Draw the ac equivalent circuit of the amplifier shown in Figure 5–61, using the transconductance model for the transistor. Label all component values in the circuit and the model. Assume that $\beta = 100$ and $V_A = 100$ V.

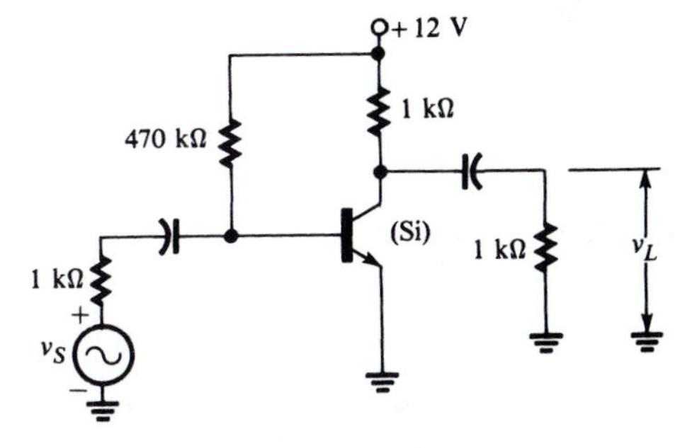

Figure 5–61
(Exercise 5–27)

5–28. Use the equivalent circuit drawn for Exercise 5–27 to find v_L/v_S.

Section 5–5
Small-Signal Parameter Equivalents: **h** *Parameters*

5–29. Using the manufacturer's small-signal specifications furnished in Section 5–5, find the percent change in h_{fe}, h_{re}, h_{ie}, and h_{oe} for the 2N2222A transistor, when the dc collector current is changed from 0.5 mA to 2.0 mA. Assume a "high-gain" unit.

5–30. Find the percent change in each of the following parameters of the transistor in Exercise 5–29, when I_C is changed from 0.5 mA to 2.0 mA: β, g_m, r_e, r_π, g_o.

5–31. A circuit has been designed using the 2N2221A transistor to amplify 1-kHz signals. After each unit is manufactured, the bias point of the transistor is adjusted so that $I_C = 1$ mA and $V_{CE} = 10$ V. What are the ranges of values of β and g_m that could be expected for the transistors that are used in the circuit?

5–32. The transistor shown in Figure 5–62 has the following h-parameters: $h_{fe} = 120$, $h_{ie} = 2.2$ kΩ, $h_{oe} = 40\ \mu$S, and $h_{re} = 6 \times 10^{-4}$. Draw the complete ac equivalent circuit and find the value of the load voltage v_L.

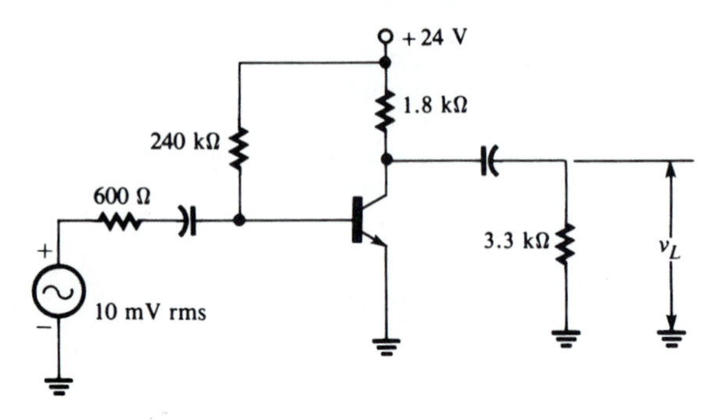

Figure 5–62
(Exercise 5–32)

SPICE EXERCISES

Note: In the exercises that follow, assume that all devices have their default values, unless otherwise specified. The capacitor values and frequencies specified in these exercises are such that capacitive reactances are negligible and have no effect on amplifier performance.

5–33. Use SPICE to determine the quiescent point, voltage gain v_L/v_S, and current gain i_L/i_S of the common-emitter amplifier shown in Figure 5–33. The transistor has a beta of 200. Component values are: $r_S = 1$ kΩ, $R_B = 470$ kΩ, $R_C = 1$ kΩ, and $R_L = 10$ kΩ. Both coupling capacitors are 10 μF, and $V_{CC} = 18$ V. The analysis should be performed at 10 kHz.

5–34. Use SPICE to determine the quiescent point, voltage gain v_L/v_S, and current gain i_L/i_S of the common-base amplifier shown in Figure 5–26(a). The transistor has a beta of 80. Component values are: $r_S = 100$ Ω, $R_E = 2.2$ kΩ, $R_C = 6.2$ kΩ, and $R_L = 10$ kΩ. The input coupling capacitor is 100 μF, and the output coupling capacitor is 10 μF. $V_{CC} = 36$ V and $V_{EE} = 6$ V. The analysis should be performed at 10 kHz.

5–35. Use SPICE to determine the quiescent point, voltage gain v_L/v_S, and current gain i_L/i_S of the common-collector amplifier shown in Figure 5–41. The transistor has a beta of 150. Component values are: $r_S = 1$ kΩ, $R_B = 180$ kΩ, $R_E = 1$ kΩ, and $R_L = 300$ Ω. The input coupling capacitor is 10 μF, and the output coupling capacitor is 100 μF. $V_{CC} = 9$ V.

5–36. Use SPICE to determine r_{in}(stage) and r_{in} (see Figure 5–39(a)) of the common-collector amplifier in Exercise 5–35. (*Hint:* Find ac voltages and currents at appropriate points in the circuit.)

6 BIAS DESIGN IN DISCRETE AND INTEGRATED CIRCUITS

6–1 THE EFFECTS OF PARAMETER VARIABILITY

We have seen that the values of transistor parameters cannot be treated as absolute constants, since they change with both temperature and bias conditions. Conversely, changes in parameter values can change the bias point of a transistor. The parameter β is a prime example: it increases with temperature as well as with collector current (see Figure 4–47(a)), and an increase in β in turn causes a further increase in collector current. This last assertion is especially apparent in a bias circuit designed to maintain a constant dc base current I_B, for in that case the equation $I_C = \beta I_B$ shows that I_C increases directly with β.

Since we have also seen that the ac performance of a transistor amplifier depends heavily on the location of the Q-point, *it is desirable to make the dc bias as independent as possible of the parameters that affect it the most.* The common-emitter bias circuit we have been using up to now (Figure 5–32) is very poor in that respect, since it holds base current constant at the value $I_B = (V_{CC} - V_{BE})/R_B$ and thus makes I_C directly dependent on the highly variable value of β: $I_C = \beta I_B$. A bias circuit that maintains constant base current is sometimes called a *fixed-bias* design. (This terminology is misleading because it implies that the Q-point is fixed, which is most certainly *not* the case.)

Thermal Runaway

Another parameter whose variability affects the bias point is the collector-to-base leakage current, I_{CBO}. Recall that I_{CBO} has a temperature-sensitive component that approximately doubles for every 10°C rise in temperature. Since the collector current in a CE stage is given by $I_C = \beta I_B + I_{CEO} \approx \beta I_B + \beta I_{CBO}$ (equations 4–9 and 4–10), we see that an increase in I_{CBO} results in an increase in I_C. In modern silicon transistors, which have very small reverse currents, this increase in collector bias current due to an increase in temperature is not usually a serious problem. However, in some applications involving power transistors, large leakage currents, high temperatures, and heavy power dissipation, there is a possibility that a large change in I_{CBO} can lead to the destructive phenomenon called *thermal runaway.* Thermal runaway occurs when an increase in I_{CBO} causes an increase in I_C that in turn increases the power dissipation, and hence the temperature, of the collector–base junction. The increase in temperature further increases the value of I_{CBO}, which further increases I_C, which further increases the temperature, and so forth, until the dissipation exceeds the rated maximum for the transistor. This headlong plunge toward destruction is compounded by the fact that increased temperatures also cause β to increase, which causes I_C to increase even more rapidly. If the transistor is biased so that an increase in I_C causes an increase in power dissipation (not always the case), then thermal runaway is possible. *Thermal runaway will not occur in a CE amplifier stage if the bias value of V_{CE} is less than $(1/2)V_{CC}$.* The reason for this is that the dc power dissipation, $P = I_C V_{CE}$, is maximum when $V_{CE} = V_{CC}/2$. An increase in I_C will result in less power dissipation if V_{CE} is initially less than $V_{CC}/2$, but greater dissipation if V_{CE} is initially greater than $V_{CC}/2$.

Another temperature-sensitive parameter that affects the bias point to some extent is the dc voltage drop across the base–emitter junction, V_{BE}. *V_{BE} decreases approximately 2.5 mV for each °C rise in temperature.* In the common-emitter "fixed-bias" circuit, $I_B = (V_{CC} - V_{BE})/R_B$, so a decrease in V_{BE} causes an increase in I_B. The increase in I_B in turn increases I_C: $I_C = \beta I_B$. In summary, an increase in temperature always increases the dc collector current because β and I_{CBO} increase with temperature and because V_{BE} decreases with temperature.

Bias Stabilization in Discrete Circuits

Our objective now is to develop a common-emitter bias circuit that is less sensitive to transistor parameter changes, particularly changes in β. A circuit designed so that the Q-point is not heavily dependent on the value of β has the added advantage that transistors having a wide range of β-values can be used in it. Furthermore, bias circuits that are relatively unaffected by changes in β also tend to be independent of changes in other temperature-sensitive parameters, such as I_{CBO}. The process of designing a bias circuit to make it insensitive to parameter changes is called bias *stabilization.* The stabilized circuit that is most widely used in *discrete* BJT designs is shown in Figure 6–1. Bias designs used in integrated circuits are substantially different and will be discussed in a later section.

Note in Figure 6–1 that the base-to-ground voltage V_B is determined by the voltage-divider network composed of resistors R_1 and R_2. (This design is often called the *voltage-divider bias* method.) We will discuss the proper selection of these component values shortly, but for the moment let us assume that they establish an essentially constant base-to-ground voltage. The most important stabilization component in Figure 6–1 is the emitter resistor R_E. We can visualize how R_E tends

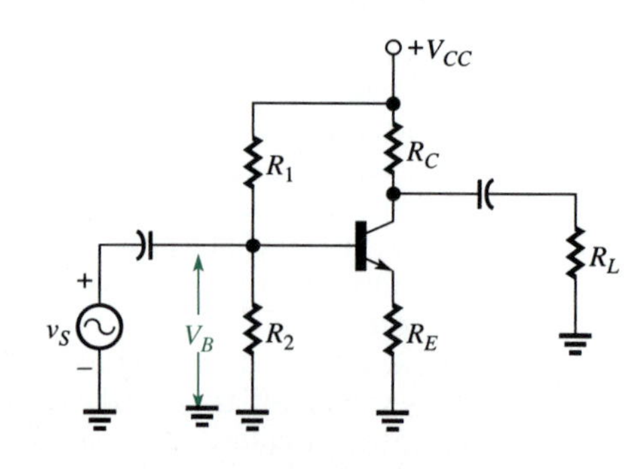

Figure 6–1
A stabilized CE bias circuit

Figure 6–2
Any change in I_C affects V_{IN} in a way that has the opposite effect on I_C, thus tending to keep I_C constant.

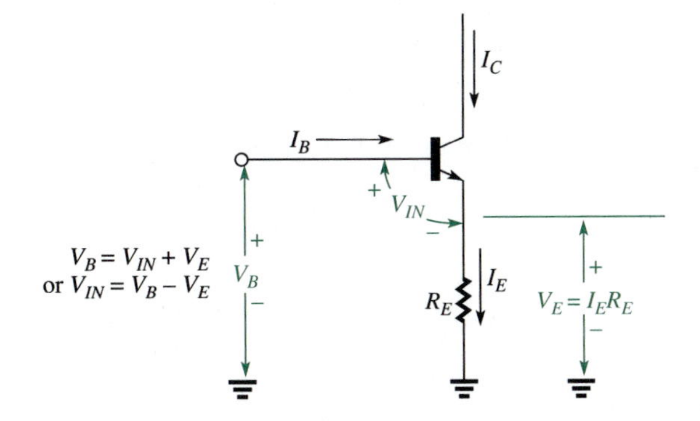

to make the bias point independent of parameter changes by considering its effect on the CE input voltage V_{IN}. Referring to Figure 6–2, note that

$$V_{IN} = V_B - V_E \tag{6–1}$$

where V_E is the emitter-to-ground voltage, and

$$V_E = I_E R_E \tag{6–2}$$

Now, any parameter change that causes I_C to increase will cause I_E to increase by about the same amount. By equation 6–2, an increase in I_E causes an increase in V_E. But if V_B is essentially constant, then equation 6–1 shows that V_{IN} decreases when V_E increases. The reduction in V_{IN} reduces I_B, which then reduces I_C, thus compensating for the parameter change that tried to increase I_C. Of course, any parameter change that tends to reduce I_C is similarly compensated for by an increase in I_B. In a nutshell, changes in the bias value of I_C automatically change the input voltage in a way that has the opposite effect on I_C, thus tending to restore I_C to its original value. The use of an emitter resistor to stabilize the bias point is called *emitter stabilization* or *current feedback.*

6–2 STABILITY FACTORS

The ratio of a change in collector current to the change in the parameter value that caused it is called a *stability factor.* A stability factor is thus a measure of how sensitive collector bias current is to changes in a parameter value. Stability factors can be defined for each of the three parameters we have discussed:

$$S(I_{CBO}) = \frac{\Delta I_C}{\Delta I_{CBO}} \tag{6–3}$$

$$S(V_{BE}) = \frac{\Delta I_C}{\Delta V_{BE}} \tag{6–4}$$

$$S(\beta) = \frac{\Delta I_C}{\Delta \beta} \tag{6–5}$$

To illustrate, if the value of $S(I_{CBO})$ for a certain bias circuit is 20, then a change in I_{CBO} from 0.01 μA to 0.02 μA ($\Delta I_{CBO} = 0.01\ \mu$A) will cause a change in I_C equal to $\Delta I_C = S(I_{CBO})\Delta I_{CBO} = 20(0.01\ \mu\text{A}) = 0.2\ \mu$A. An ideal stability factor would have value *zero,* implying *no* change in I_C, for *any* change in parameter value. The actual value depends on the components used in the bias circuit and can never be 0. We therefore seek bias designs that make the stability factors as small as practical.

Using calculus, it can be shown that approximate values of the stability factors for the circuit in Figure 6–1 are found from

$$S(I_{CBO}) = \frac{(\beta + 1)(1 + R_B/R_E)}{(\beta + 1) + R_B/R_E} \approx \frac{R_E + R_B}{R_E + R_B/\beta} \quad \textbf{(6–6)}$$

$$S(V_{BE}) = \frac{-\beta}{R_B + R_E(\beta + 1)} \quad \textbf{(6–7)}$$

$$S(\beta) \approx \frac{I_{C1}S_2(I_{CBO})}{\beta_1\beta_2} \quad \textbf{(6–8)}$$

where

$$R_B = R_1 \parallel R_2$$
$$I_{C1} = \text{initial value of } I_C$$
$$\beta_1 = \text{initial value of } \beta$$
$$\beta_2 = \text{larger value of } \beta$$
$$S_2(I_{CBO}) = \text{value of } S(I_{CBO}) \text{ for } \beta = \beta_2$$

Note that $S(V_{BE})$ is negative, so when an increase in temperature causes V_{BE} to decrease, ΔV_{BE} is itself a negative quantity and $\Delta I_C = S(V_{BE})(\Delta V_{BE})$, being the product of two negative quantities, is a positive quantity.

Equations 6–6 through 6–8 show that each stability factor is reduced (improved) when R_E is made large. This confirms our intuitive analysis of the beneficial contribution of emitter resistance in providing current feedback. $S(I_{CBO})$ and $S(\beta)$ are improved by making R_B small. As we shall see presently, there are other considerations that prevent us from making R_E arbitrarily large and/or R_B arbitrarily small. As a rule of thumb, satisfactory stability can be realized by making the ratio R_B/R_E less than 10.

The stability equations can be applied to the "fixed-bias" circuit we studied earlier by letting R_E equal 0 and R_B equal the value of the series base resistor. Each of the stability equations gives the stability factor related to variation in one parameter only, so the total change in collector current over a certain temperature range can be approximated by

$$\Delta I_C \approx \Delta I_{CBO}S(I_{CBO}) + \Delta V_{BE}S(V_{BE}) + \Delta\beta S(\beta) \quad \textbf{(6–9)}$$

where ΔI_{CBO}, ΔV_{BE}, and $\Delta\beta$ are the total changes in the respective parameter values over the temperature range. The expression is an approximation because all three parameters are changing simultaneously with temperature. To compute an individual stability factor whose value depends on another parameter, use a midrange or average value of the parameter. The next example illustrates the computation.

Example 6–1

The following table shows the parameter values of a typical transistor at temperatures 10°C and 100°C:

	10°C	**100°C**
I_{CBO}	0.01 μA	1.2 μA
V_{BE}	0.74 V	0.54 V
β	60	140

Figure 6–3
(Example 6–1)

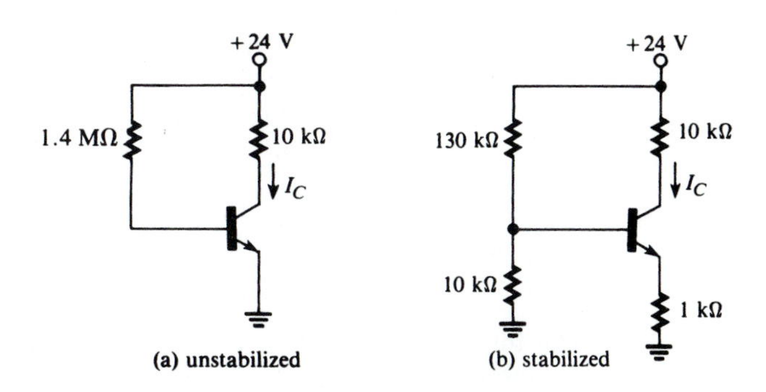

Find the total change in the dc collector current over the temperature range from 10°C to 100°C for each of the circuits shown in Figure 6–3. Each transistor is biased at $I_C = 1$ mA at 10°C.

Solution. Applying equations 6–6 through 6–8 to the unstabilized circuit of Figure 6–3(a), with $R_E = 0$ and $R_B = 1.4$ MΩ, we find

$$S(I_{CBO}) \approx \frac{R_E + R_B}{R_E + R_B/\beta} = \frac{R_B}{R_B/\beta} = \beta$$

The average β over the given temperature range is $(60 + 140)/2 = 100$, so we use $S(I_{CBO}) \approx 100$.

$$S(V_{BE}) \approx \frac{-\beta}{R_B + R_E(\beta + 1)} = \frac{-\beta}{R_B} = \frac{-100}{1.4\ \text{M}\Omega} = -7.14 \times 10^{-5}$$

$$S(\beta) \approx \frac{I_{C1} S_2(I_{CBO})}{\beta_1 \beta_2} = \frac{(1 \times 10^{-3})(140)}{(60)(140)} = 1.67 \times 10^{-5}$$

Now, $\Delta I_{CBO} = (1.2\ \mu\text{A}) - (0.01\ \mu\text{A}) = 1.19\ \mu\text{A}$, $\Delta V_{BE} = (0.54\ \text{V}) - (0.74\ \text{V}) = -0.2$ V, and $\Delta\beta = 140 - 60 = 80$. So, from equation 6–9,

$$\Delta I_C \approx (1.19\ \mu\text{A})(100) + (-0.2)(-7.14 \times 10^{-5}) + (80)(1.67 \times 10^{-5})$$
$$= (0.119\ \text{mA}) + (0.014\ \text{mA}) + (1.336\ \text{mA}) = 2.47\ \text{mA}$$

We see that the collector current increases by about 1.47 mA in the unstabilized circuit (to 2.47 mA, more than double its original value). The greatest part of the change in I_C is due to the change in β.

For the emitter-stabilized circuit of Figure 6–3(b), we have $R_B = (130\ \text{k}\Omega) \parallel (10\ \text{k}\Omega) = 9.29\ \text{k}\Omega$. Applying the stability equations, we find

$$S(I_{CBO}) \approx \frac{(1\ \text{k}\Omega) + (9.29\ \text{k}\Omega)}{(1\ \text{k}\Omega) + \dfrac{9.29\ \text{k}\Omega}{100}} = 9.42$$

$$S(V_{BE}) \approx \frac{-100}{(9.29\ \text{k}\Omega) + (101\ \text{k}\Omega)} = -9.07 \times 10^{-4}$$

$$S(\beta) \approx \frac{(1 \times 10^{-3}) S_2(I_{CBO})}{(60)(140)}$$

where

$$S_2(I_{CBO}) = \frac{(1\ \text{k}\Omega) + (9.29\ \text{k}\Omega)}{(1\ \text{k}\Omega) + \dfrac{9.29\ \text{k}\Omega}{140}} = 9.65$$

so

$$S(\beta) \approx \frac{(1 \times 10^{-3})(9.65)}{(60)(140)} = 1.15 \times 10^{-6}$$

Applying equation 6–9,

$$\begin{aligned}\Delta I_C &\approx (1.19\ \mu\text{A})(9.42) + (-0.2)(-9.07 \times 10^{-4}) + (80)(1.15 \times 10^{-6}) \\ &= (0.0112\ \text{mA}) + (0.181\ \text{mA}) + (0.092\ \text{mA}) = 0.284\ \text{mA}\end{aligned}$$

The increase in collector current in the stabilized circuit is about 28%, much less than that of the unstabilized circuit over the same temperature range. Note particularly how stabilization has reduced the change in I_C caused by the change in β —1.336 mA in the unstabilized circuit versus 0.092 mA in the stabilized circuit. We concluded that the stabilized circuit could be used without difficulty if transistors having a range of β from 60 to 140 were substituted into it.

6–3 CALCULATING THE BIAS POINT OF THE STABILIZED CIRCUIT

If the stabilized bias design in Figure 6–1 truly makes the bias point independent of parameter values, then it should be possible to determine the quiescent current and voltage without using any parameter values in the computations. The analysis procedure that follows will demonstrate that this is indeed the case. Refer to Figure 6–4.

Note in Figure 6–4 that the base-to-ground voltage V_B is determined by the voltage-divider network composed of resistors R_1 and R_2 connected between V_{CC} and ground. If we neglect the loading effect of R_{in} looking into the base, then

$$V_B \approx \left(\frac{R_2}{R_1 + R_2}\right) V_{CC} \tag{6–10}$$

Assuming a silicon transistor, it is apparent from the figure that $V_E = V_B - 0.7$. Then the emitter current is the emitter-to-ground voltage V_E divided by R_E:

$$I_E = \frac{V_E}{R_E} = \frac{V_B - 0.7}{R_E} \tag{6–11}$$

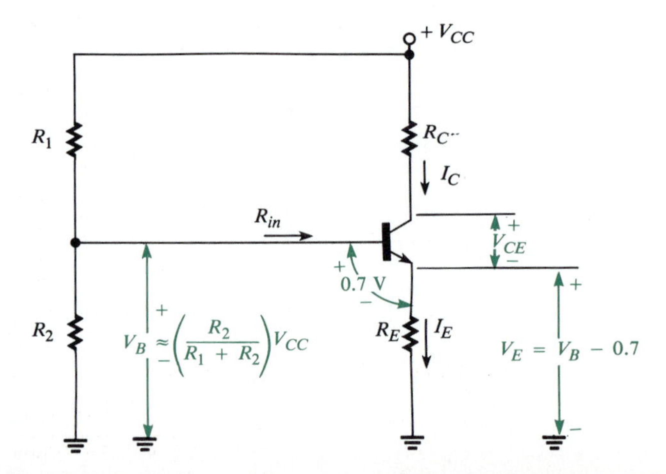

Figure 6–4
Computing dc voltages and currents in the stabilized bias circuit

This computation gives us the quiescent emitter current, which is very nearly the same as the quiescent collector current:

$$I_C = \alpha I_E \approx I_E \tag{6–12}$$

By constructing a Thevenin equivalent circuit of the voltage-divider network and writing Kirchhoff's voltage law through it and the base-emitter junction, it can be shown that the exact value of the quiescent collector current is

$$I_C = \beta I_B = \beta \left(\frac{\left(\dfrac{R_2}{R_1 + R_2}\right) V_{CC} - V_{BE}}{R_1 \| R_2 + (\beta + 1) R_E} \right) \tag{6–12a}$$

To find the quiescent value of V_{CE}, we write Kirchhoff's voltage law from V_{CC} through R_C and R_E to obtain

$$V_{CC} = I_C R_C + V_{CE} + I_E R_E \tag{6–13}$$

Thus,

$$V_{CE} = V_{CC} - I_C R_C - I_E R_E \tag{6–14}$$

or, since $I_C \approx I_E$,

$$V_{CE} \approx V_{CC} - I_C(R_C + R_E) \tag{6–15}$$

Note that the equation of the dc load line is

$$I_C = -\left(\frac{1}{R_C + R_E}\right) V_{CE} + \frac{V_{CC}}{R_C + R_E} \tag{6–16}$$

Example 6–2

Assuming that the bias circuit shown in Figure 6–5 is well stabilized against variations in β, find I_C and V_{CE} at the Q-point.

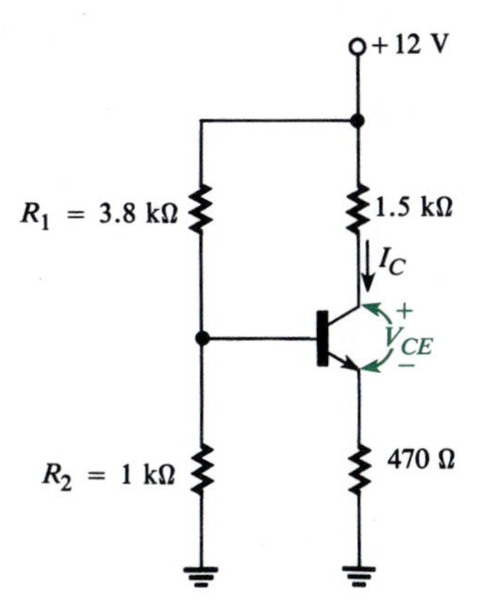

Figure 6–5
(Example 6–2)

Solution. From equation 6–10,

$$V_B \approx \left(\frac{1 \times 10^3}{1 \times 10^3 + 3.8 \times 10^3}\right) 12\text{ V} = 2.5\text{ V}$$

Then $V_E = V_B - 0.7\text{ V} = 1.8\text{ V}$ and $I_E = (1.8\text{ V})/(470\ \Omega) = 3.83\text{ mA} \approx I_C$. Thus, the quiescent collector current is about 3.83 mA. From equation 6–15, $V_{CE} \approx 12 - (3.83 \times 10^{-3}\text{ A})(1.5\text{ k}\Omega + 470\ \Omega) = 4.45\text{ V}$. Notice that no transistor parameter values were used in the preceding computations.

Recall that we neglected the loading effect of R_{in} (Figure 6–4) on the computation of V_B (equation 6–10). Following a procedure similar to that used to derive equation 5–53, it is easy to show that

$$R_{in} = (\beta + 1)R_E + V_{BE}/I_B \approx \beta R_E \tag{6–17}$$

If R_{in} is not neglected, the value of V_B is (approximately)

$$V_B \approx \left(\frac{R_2 \parallel R_{in}}{R_1 + R_2 \parallel R_{in}}\right) V_{CC} \tag{6–18}$$

Since R_{in} depends on β, equation 6–18 shows that V_B and hence the Q-point do depend to some extent on β. However, in a well-stabilized circuit, R_E is large and $R_B = R_1 \parallel R_2$ is small, so $R_2 \parallel R_{in} \approx R_2$. The whole point here is that R_1 and R_2 should be small enough in comparison to βR_E that the voltage V_B is unaffected by changes in β. In effect, the voltage divider should look as much like an ideal voltage source as possible (i.e., hold V_B constant), regardless of how the load βR_E changes. V_B stays essentially constant if R_2 is small and R_E is large. (Draw a Thevenin equivalent circuit looking out from the base and convince yourself of this fact.) This perspective is simply another way of viewing the already-established result that R_E should be large and R_B small to achieve good β-stability. The next example shows how the stability deteriorates when the voltage-divider resistors have the same ratio as in the previous example, but are each increased tenfold in value.

Example 6–3

Suppose R_1 and R_2 in Figure 6–5 are increased to 38 kΩ and 10 kΩ, respectively. Find I_C when (1) $\beta = 50$ and (2) $\beta = 150$.

Solution

1. When $\beta = 50$, $R_{in} \approx \beta(R_E) = (50)(470\ \Omega) = 23.5\text{ k}\Omega$. From equation 6–18,

$$V_B \approx \left[\frac{(10 \times 10^3) \parallel (23.5 \times 10^3)}{38 \times 10^3 + (10 \times 10^3) \parallel (23.5 \times 10^3)}\right] 12\text{ V}$$

$$= \left(\frac{7 \times 10^3}{45 \times 10^3}\right) 12\text{ V} = 1.87\text{ V}$$

$$V_E = (1.87 - 0.7)\text{V} = 1.17\text{ V}$$

$$I_E \approx I_C = \frac{V_E}{R_E} = \frac{1.17\text{ V}}{470\ \Omega} = 2.49\text{ mA}$$

2. When $\beta = 150$, $R_{in} \approx \beta R_E = (150)(470\ \Omega) = 70.5\text{ k}\Omega$. Then

$$V_B \approx \left[\frac{(10 \times 10^3) \parallel (70.5 \times 10^3)}{38 \times 10^3 + (10 \times 10^3) \parallel (70.5 \times 10^3)}\right] 12\text{ V}$$

$$= \left(\frac{8.76 \times 10^3}{46.76 \times 10^3}\right) 12 = 2.25\text{ V}$$

$$V_E = (2.25 - 0.7)\text{V} = 1.55 \text{ V}$$

$$I_E \approx I_C = \frac{V_E}{R_E} = \frac{1.55 \text{ V}}{470 \,\Omega} = 3.30 \text{ mA}$$

With the larger values of R_1 and R_2, the increase in β causes I_C to increase by about 33%. Note that the ratio R_B/R_E in this example is $(R_1 \parallel R_2)/R_E = 16.84$, which is somewhat large for good stability.

6–4 SMALL-SIGNAL PERFORMANCE OF THE BIAS-STABILIZED CIRCUIT

Figure 6–6(a) shows the bias-stabilized circuit incorporated into an ac amplifier stage. Both source resistance and load resistance are included. Figure 6–6(b) shows the ac equivalent circuit of the amplifier. Note that R_1 and R_2 are both connected to *ac ground* and are therefore in parallel as far as ac signals are concerned. These are combined into $R_B = R_1 \parallel R_2$ in the ac equivalent circuit. The total resistance from collector to ground is the parallel combination of the transistor output resistance r_o, the collector resistor R_C, and the load resistor R_L: $r_L = r_o \parallel R_C \parallel R_L \approx R_C \parallel R_L$.

In Figure 6–6(b), + and − symbols are used to show instantaneous polarities and to emphasize the fact that the ac emitter voltage v_e is in phase with v_{be} and

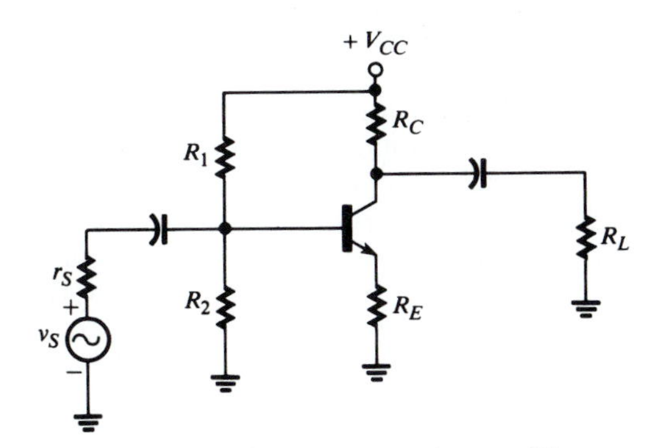

(a) The bias-stabilized common-emitter amplifier

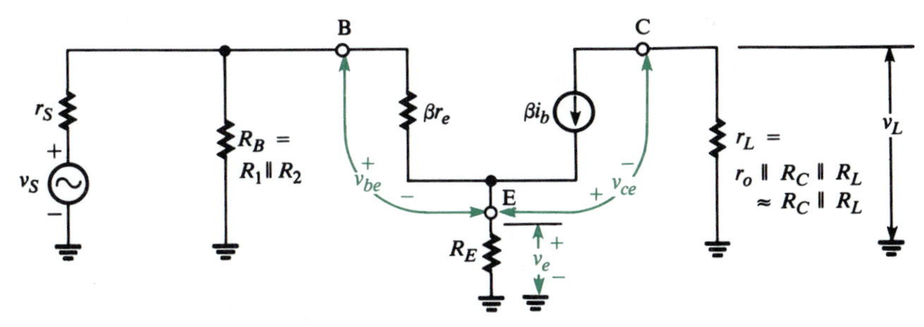

(b) The ac equivalent circuit of (a)

Figure 6–6
The bias-stabilized CE transistor incorporated into an ac amplifier

that v_{ce} is 180° out of phase with v_{be}. It can be seen in the figure that v_L (which is also out of phase with v_{be}) is

$$v_L = v_{ce} - v_e \tag{6–19}$$

The significance of equation 6–19 is that the ac load voltage is *reduced* by the amount v_e from what it would otherwise be if R_E were not present. This reduction in load voltage due to R_E is called *degeneration.* Using the same derivation that was used to show that the voltage gain A_v equals r_L/r_e when the only resistance in the emitter circuit was r_e, we can show that the voltage gain when the emitter circuit resistance is $r_e + R_E$ is given by

$$A_v = \frac{-r_L}{r_e + R_E} \tag{6–20}$$

Since the denominator in equation 6–20 now includes R_E, it is clear that the voltage gain is considerably smaller than it would be if R_E were 0. Thus, equation 6–20 shows the extent to which gain is reduced by degeneration. In most practical circuits, $R_E \gg r_e$, so a good approximation for the voltage gain is

$$A_v \approx \frac{-r_L}{R_E} \tag{6–21}$$

While it is desirable for R_E to be as large as possible to achieve good bias stability, we see that large values of R_E reduce the amplifier voltage gain. The fact that voltage gain must be sacrificed for bias stability, or vice versa, is a good illustration of the *trade-off principle* that underlies all electronic design problems: It is inevitably necessary to trade one desirable feature for another. In other words, the improvement of one aspect of a circuit's performance is achieved only at the expense of another aspect. The *art* of electronic circuit design is the ability to make reasonable compromises that satisfy all the requirements of a given application.

One desirable implication of equation 6–21 is that it makes the *ac voltage gain essentially independent of the transistor parameter* r_e, and thus independent of the transistor used in the circuit as well as its bias point. Note that A_v now depends only on *external* resistor values: $A_v \approx -r_L/R_E = -(R_C \parallel R_L)/R_E$. Thus, by sacrificing the *magnitude* of the voltage gain, we achieve *predictability.* In some applications it may be far more important to have an amplifier whose voltage gain will only vary from, say, 9.5 to 10.5 when different transistors are used than it would be to have a large gain that could vary from 90 to 200.

One way to retain the desirable effect of R_E on bias stability and still achieve a large voltage gain is to connect a capacitor in parallel with R_E, as shown in Figure 6–7. The capacitor should be large enough to have an impedance that is negligible in comparison to R_E at all frequencies of the ac signal. When this is the case, *the emitter is at ac ground* because the ac resistance in the emitter circuit is again simply r_e. The capacitor effectively "shorts out" R_E for ac signals, and it is called an emitter *bypass* capacitor, because it bypasses ac signals around R_E to ground. Note that the dc input resistance at the base is still $R_{in} \approx \beta R_E$, because the capacitor is an open circuit to dc, and bias stabilization is therefore maintained.

Let us now consider the effect of the voltage-divider resistors R_1 and R_2 in Figure 6–6(a) on the ac performance of the amplifier. As shown in Figure 6–6(b), the parallel combination $R_1 \parallel R_2 = R_B$ appears between base and ground as far as ac signals are concerned. It therefore reduces the overall input resistance to the amplifier stage, as shown by equations 6–22 and 6–23:

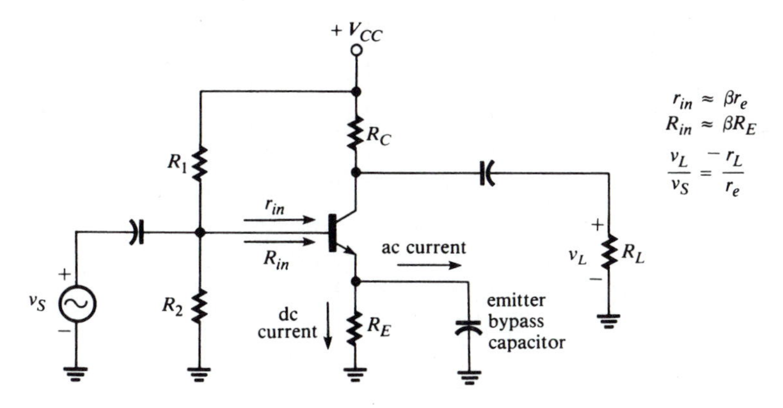

Figure 6–7
The emitter bypass capacitor effectively connects the emitter to ac ground.

(a) emitter unbypassed

$$r_{in}(\text{stage}) = R_B \| \beta(r_e + R_E) \quad \textbf{(6–22)}$$

(b) emitter bypassed

$$r_{in}(\text{stage}) = R_B \| \beta r_e \quad \textbf{(6–23)}$$

As usual, r_S and r_{in}(stage) form a voltage divider across the input of the amplifier. Therefore, the reduction in r_{in}(stage) caused by the presence of R_B reduces the overall voltage gain of the amplifier. Note that R_B also reduces the overall current gain, because it provides a path to ground for some of the ac input current that would otherwise enter the base. Clearly, R_1 and R_2 should both be as large as possible to prevent a serious deterioration in gain. Since it is desirable to have R_B small from the standpoint of bias stability, we see that there is again a trade-off between stability and gain.

Equations 6–24 summarize the equations used to determine bias conditions and ac performance of the bias-stabilized CE amplifier.

Bias and small-signal equations for the emitter-stabilized CE amplifier

$$R_{in} = (\beta + 1)R_E \approx \beta R_E$$

$$V_B \approx \left(\frac{R_2 \| R_{in}}{R_1 + R_2 \| R_{in}}\right) V_{CC}$$

$$\approx \left(\frac{R_2}{R_1 + R_2}\right) V_{CC} \qquad (R_{in} >> R_2)$$

$$V_E = V_B - 0.7 \qquad \text{(Si)}$$

$$I_C \approx I_E = \frac{V_E}{R_E}$$

$$I_C = \beta \left(\frac{\left(\frac{R_2}{R_1 + R_2}\right) V_{CC} - V_{BE}}{R_1 \| R_2 + (\beta + 1)R_E} \right)$$

$$V_{CE} = V_{CC} - I_C R_C - I_E R_E$$

$$\approx V_{CC} - I_C(R_C + R_E) \quad \textbf{(6–24)}$$

$$A_v = -\frac{R_C}{R_E + r_e} \approx -\frac{R_C}{R_E} \qquad (R_E >> r_e,\ R_E \text{ unbypassed})$$

$$= -\frac{R_C}{r_e} \qquad (R_E \text{ bypassed})$$

$$r_{in} = (\beta + 1)(R_E + r_e)$$

$$\approx \beta R_E \qquad (R_E >> r_e,\ R_E \text{ unbypassed})$$

$$\approx \beta r_e \qquad (R_E \text{ bypassed})$$

$$r_{in}(\text{stage}) = R_B \parallel r_{in} \qquad \text{where } R_B = R_1 \parallel R_2$$

$$\frac{v_L}{v_S} = A_v \left[\frac{r_{in}(\text{stage})}{r_S + r_{in}(\text{stage})}\right]\left(\frac{R_L}{R_C + R_L}\right)$$

$$= \left(\frac{r_L}{R_E + r_e}\right)\left[\frac{r_{in}(\text{stage})}{r_S + r_{in}(\text{stage})}\right] \qquad (R_E \text{ unbypassed})$$

$$= \left(\frac{r_L}{r_e}\right)\left[\frac{r_{in}(\text{stage})}{r_S + r_{in}(\text{stage})}\right] \qquad (R_E \text{ bypassed})$$

Example 6–4

Find r_{in}(stage) and the voltage gain v_L/v_S for each of the amplifiers shown in Figure 6–8. Note that both amplifiers are identical with the exception of the emitter bypass capacitor in 6–8(b). Assume that $\beta = 100$ for each transistor.

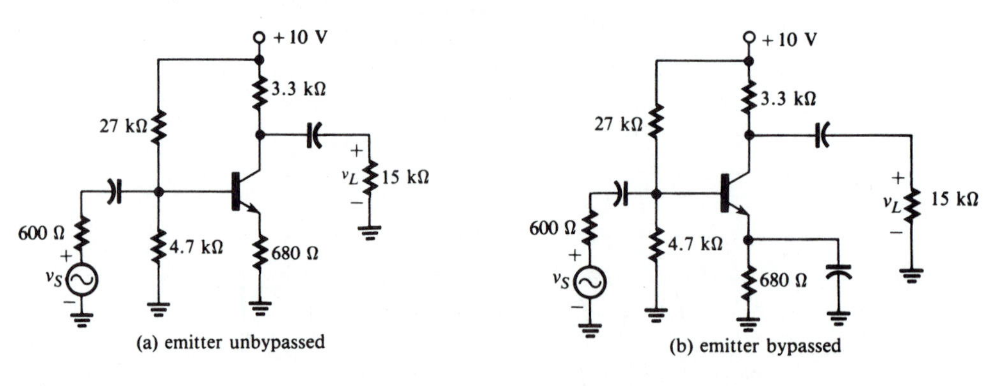

Figure 6–8
(Example 6–4)

Solution. Since the dc circuits are identical, we calculate the dc emitter current for each of Figure 6–8(a) and (b) as follows:

$$R_{in} \approx \beta R_E = 100(680\ \Omega) = 68\ \text{k}\Omega$$
$$R_{in} \| R_2 = (68\ \text{k}\Omega) \| (4.7\ \text{k}\Omega) = 4.39\ \text{k}\Omega$$
$$V_B = \left(\frac{R_{in} \| R_2}{R_1 + R_{in} \| R_2}\right) 10\ \text{V} = \left(\frac{4.39 \times 10^3}{27 \times 10^3 + 4.39 \times 10^3}\right) 10\ \text{V} = 1.39\ \text{V}$$
$$V_E = V_B - 0.7 = (1.39 - 0.7)\text{V} = 0.69\ \text{V}$$
$$I_E = \frac{V_E}{R_E} = \frac{0.69\ \text{V}}{680\ \Omega} \approx 1\ \text{mA}$$

Therefore, $r_e \approx 0.026/(1\ \text{mA}) = 26\ \Omega$.

For Figure 6–8(a), we have

$$r_{in}(\text{stage}) = R_1 \| R_2 \| \beta(R_E + r_e)$$
$$= (27 \times 10^3) \| (4.7 \times 10^3) \| 100(680 + 26) = 3.79\ \text{k}\Omega$$

The overall voltage gain is

$$\frac{v_L}{v_S} = \left(-\frac{R_C}{R_E + r_e}\right)\left[\frac{r_{in}(\text{stage})}{r_S + r_{in}(\text{stage})}\right]\left(\frac{R_L}{R_C + R_L}\right)$$
$$= \left(-\frac{3.3 \times 10^3}{706}\right)\left(\frac{3.79 \times 10^3}{600 + 3.79 \times 10^3}\right)\left(\frac{15 \times 10^3}{3.3 \times 10^3 + 15 \times 10^3}\right)$$
$$= -(4.67)(0.862)(0.82) = -3.3$$

For Figure 6–8(b)

$$r_{in}(\text{stage}) = R_1 \| R_2 \| \beta r_e = (27 \times 10^3) \| (4.7 \times 10^3) \| 2600 = 1.58\ \text{kV}$$

and

$$\frac{v_L}{v_S} = \left(-\frac{3.3 \times 10^3}{26}\right)\left(\frac{1.58 \times 10^3}{600 + 1.58 \times 10^3}\right)\left(\frac{15 \times 10^3}{3.3 \times 10^3 + 15 \times 10^3}\right)$$
$$= -(126.9)(0.72)(0.82) = -74.9$$

Note that the voltage gains in both cases are considerably lower than those of previous examples. However, both circuits are well stabilized: $R_B/R_E \approx 5.9$. Note also that bypassing R_E with the capacitor in Figure 6–8(b) did not restore *all* the voltage gain that the transistor itself is capable of producing [(3.3 kΩ)/(26 Ω) = 126.9]. Bypassing R_E reduced r_{in}(stage) from 3.79 kΩ to 1.58 kΩ, which, as the computations show, caused the voltage division at the input to change from 0.862 to 0.72. The relatively large value of r_S (600 Ω) is responsible for a greater loss when r_{in}(stage) is reduced.

Example 6–5

SPICE

By trial-and-error runs of SPICE programs, determine the maximum signal level v_S that can be used to drive the amplifier in Figure 6–8(b) without creating serious visible distortion in the load voltage, v_L. Assume that the coupling capacitors are each 10 μF, the emitter bypass capacitor is 100 μF, and the signal frequency is 10 kHz. (These values ensure that the capacitive reactances are negligible and have no effect on amplifier gain.)

Solution. Figure 6–9(a) shows the circuit redrawn for analysis by SPICE and the input data file. Since we must observe the output waveform to detect distortion, it is necessary to perform a .TRAN analysis and specify a SIN source rather than an .AC analysis with a source designated AC. Figure 6–9(b) shows the results of a program run when the peak value of the SIN source is 0.1 V. It is clear that the load voltage is severely distorted due to clipping. Repeating trial-and-error runs using progressively smaller amplitudes for v_S leads to the results shown in Figure 6–9(c). A peak value of about 0.01 V for v_S is found to be the largest that can be

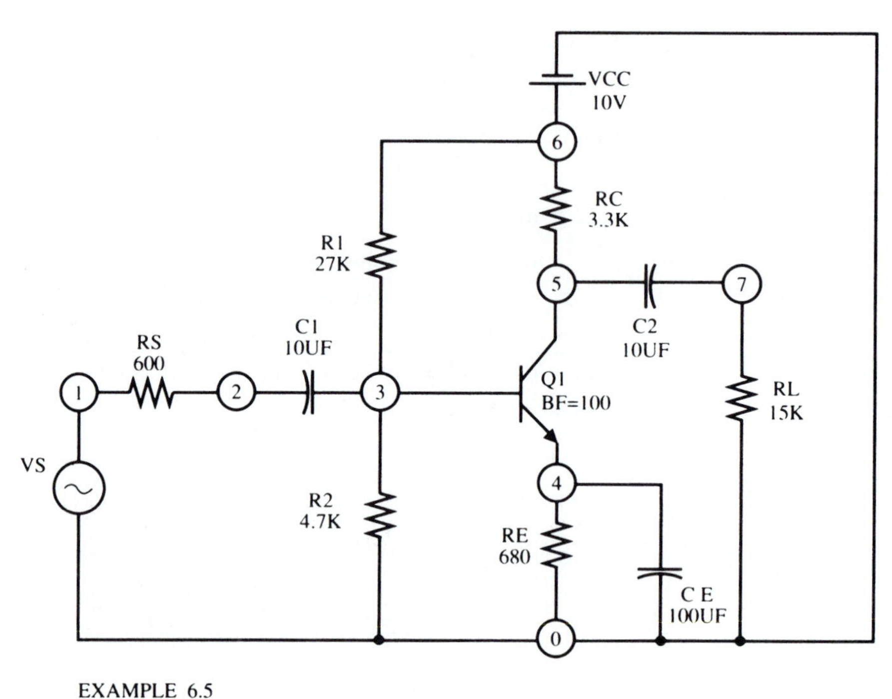


```
EXAMPLE 6.5
VS  1 0 SIN(0 0.1 10K)
VCC 6 0 10V
RS  1 2 600
C1  2 3 10UF
R2  3 0 4.7K
R1  6 3 27K
RC  6 5 3.3K
RE  4 0 680
CE  4 0 100UF
C2  5 7 10UF
RL  7 0 15K
Q1  5 3 4 TRANS
.MODEL TRANS NPN BF=100
.TRAN 5US 0.1MS
.PLOT TRAN V(7)
.END
```

(a)

Figure 6–9
(Example 6–5)

```
EXAMPLE 6.5
****     TRANSIENT ANALYSIS                        TEMPERATURE =    27.000 DEG C
****************************************************************************
   TIME       V(7)
                 -1.000D+01      -5.000D+00        0.000D+00       5.000D+00  1.000D+01
                  - - - - - - - - - - - - - - - - - - - - - - - - - - - - - - - - -
0.000D+00  0.000D+00 .             .                *               .             .
5.000D-06 -3.239D+00 .             .       *        .               .             .
1.000D-05 -5.969D+00 .          *  .                .               .             .
1.500D-05 -6.002D+00 .          *  .                .               .             .
2.000D-05 -6.011D+00 .          *  .                .               .             .
2.500D-05 -6.013D+00 .          *  .                .               .             .
3.000D-05 -6.010D+00 .          *  .                .               .             .
3.500D-05 -6.000D+00 .          *  .                .               .             .
4.000D-05 -5.966D+00 .          *  .                .               .             .
4.500D-05 -3.064D+00 .             .       *        .               .             .
5.000D-05  7.459D-02 .             .                *               .             .
5.500D-05  1.621D+00 .             .                .    *          .             .
6.000D-05  2.221D+00 .             .                .     *         .             .
6.500D-05  2.439D+00 .             .                .      *        .             .
7.000D-05  2.514D+00 .             .                .      *        .             .
7.500D-05  2.533D+00 .             .                .      *        .             .
8.000D-05  2.513D+00 .             .                .      *        .             .
8.500D-05  2.436D+00 .             .                .      *        .             .
9.000D-05  2.216D+00 .             .                .     *         .             .
9.500D-05  1.603D+00 .             .                .   *           .             .
1.000D-04  6.162D-02 .             .                *               .             .
                  - - - - - - - - - - - - - - - - - - - - - - - - - - - - - - - - -
```

(b)

```
EXAMPLE 6.5
****     TRANSIENT ANALYSIS                        TEMPERATURE =    27.000 DEG C
****************************************************************************
   TIME       V(7)
                 -1.000D+00      -5.000D-01        0.000D+00       5.000D-01  1.000D+00
                  - - - - - - - - - - - - - - - - - - - - - - - - - - - - - - - - -
0.000D+00  0.000D+00 .             .                *               .             .
5.000D-06 -2.371D-01 .             .         *      .               .             .
1.000D-05 -4.651D-01 .             .*               .               .             .
1.500D-05 -6.556D-01 .         *   .                .               .             .
2.000D-05 -7.819D-01 .     *       .                .               .             .
2.500D-05 -8.252D-01 .    *        .                .               .             .
3.000D-05 -7.782D-01 .     *       .                .               .             .
3.500D-05 -6.490D-01 .         *   .                .               .             .
4.000D-05 -4.563D-01 .             .*               .               .             .
4.500D-05 -2.276D-01 .             .          *     .               .             .
5.000D-05  9.118D-03 .             .                *               .             .
5.500D-05  2.289D-01 .             .                .     *         .             .
6.000D-05  4.132D-01 .             .                .           *   .             .
6.500D-05  5.506D-01 .             .                .               .*            .
7.000D-05  6.343D-01 .             .                .               .    *        .
7.500D-05  6.620D-01 .             .                .               .     *       .
8.000D-05  6.324D-01 .             .                .               .    *        .
8.500D-05  5.468D-01 .             .                .               .*            .
9.000D-05  4.076D-01 .             .                .          *    .             .
9.500D-05  2.219D-01 .             .                .     *         .             .
1.000D-04  1.332D-03 .             .                *               .             .
                  - - - - - - - - - - - - - - - - - - - - - - - - - - - - - - - - -
```

(c)

Figure 6–9
(Continued)

used without creating visible distortion in v_L. Note that the magnitude of the voltage gain in that case is

$$\left|\frac{v_L}{v_S}\right| = \frac{0.6620 - (-0.8252)\text{ V p–p}}{0.02\text{ V p–p}} = 74.36$$

which is in excellent agreement with the result calculated in Example 6–4.

6–5 STABILIZATION BY COLLECTOR FEEDBACK

Figure 6–10 shows another bias design that is sometimes used to stabilize a CE amplifier against parameter changes. Note that the resistor R_B is connected between collector and base and thus provides a feedback path between output and input. For this reason, the design is called *collector feedback or voltage feedback.* The figure also shows an emitter resistor R_E that provides current feedback, as previously discussed, but it is omitted in some designs.

Any parameter change that causes I_C to increase in Figure 6–10 also causes the collector voltage to decrease, because of the increased drop across R_C. But this decrease in collector voltage is reflected through R_B as a decrease in base voltage. The decrease in base voltage reduces I_C and thus compensates for the change that originally caused I_C to increase.

Like emitter stabilization, collector feedback also causes a loss in voltage gain. The ac collector voltage "opposes" the ac input voltage at the base. Remember that the collector and base voltages are 180° out of phase. This is an example of *negative feedback,* which has many offsetting benefits that we will study in more detail in a later chapter.

We will not take time to derive the bias and small-signal equations for the collector-stabilized circuit but summarize them here for reference. Note that the dc equations can be applied to the case where R_E in Figure 6–10 is omitted, simply by setting $R_E = 0$. R_E can also be set to 0 in the small-signal equations if it is omitted in the design or if it is bypassed with a capacitor.

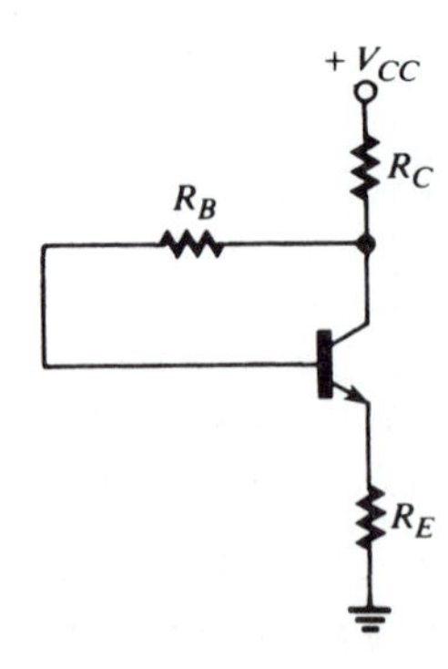

Figure 6–10
A bias design that is stabilized by collector (voltage) feedback as well as emitter (current) feedback

Bias and small-signal equations for the collector-stabilized CE amplifier

$$I_C \approx \frac{V_{CC} - V_{BE}}{R_C + R_E + R_B/\beta}$$

$$V_{CE} = V_{CC} - I_C R_C - I_E R_E$$

$$\approx V_{CC} - I_C(R_C + R_E)$$

$$r_{in}(\text{stage}) = \left(\frac{R_B}{|A_v| + 1}\right) \| (\beta + 1)(r_e + R_E)$$

$$\approx \frac{R_B}{|A_v|} \| \beta R_E \qquad (|A_v| \gg 1, R_E \gg r_e) \tag{6–25}$$

$$A_v = \frac{-\left(\dfrac{R_B}{r_e + R_E} - 1\right)}{(R_B/r_L) + 1}$$

$$\approx \frac{-r_L}{r_e} \qquad (R_E = 0, \text{ or } R_E \text{ bypassed})$$

where $r_L = R_C \| R_L$

$$\frac{v_L}{v_S} \approx A_v\left[\frac{r_{in}(\text{stage})}{r_S + r_{in}(\text{stage})}\right]$$

6–6 DISCRETE CIRCUIT BIAS DESIGN

Because the voltage-divider bias circuit (Figure 6–11) has been so widely used, certain practical, time-tested guidelines have evolved for its design. It has been found that reasonable compromises between stability, gain, and output swing can be achieved by designing the bias circuit in accordance with two basic criteria. First, the emitter-to-ground voltage should be approximately one-tenth of the V_{CC} supply voltage. Second, the dc current in R_2 should be approximately 10 times greater than the dc base current I_B. The latter criterion is met if the input resistance looking into the base is 10 times greater than R_2. As discussed earlier, this resistance ratio must be large to make the base voltage insensitive to β variations, and to allow us

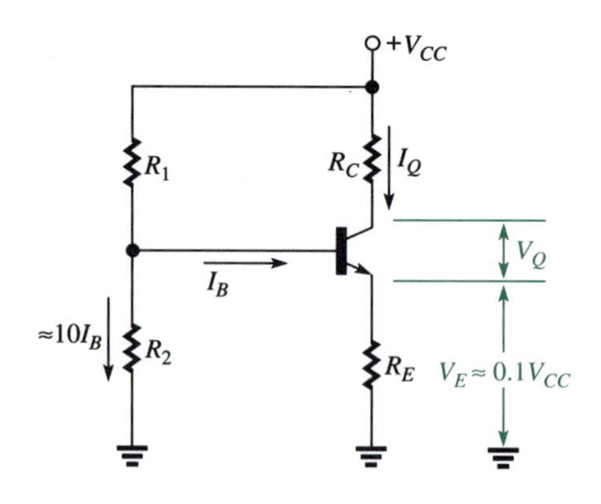

Figure 6–11
Design criteria for the voltage-divider bias

to neglect the loading effect of the base on the voltage divider. The two criteria are met if the following approximations are satisfied by the design:

$$V_E \approx 0.1V_{CC} \tag{6–26}$$

$$R_2 \approx \beta(\text{min})R_E/10 \tag{6–27}$$

where β(min) is the minimum value of β that transistors used in the circuit may have.

In the absence of other constraints imposed by a particular application, the conditions expressed by equations 6–26 and 6–27 will lead to a design that has good dc and ac performance characteristics. Our development of a design procedure based on these conditions begins with the assumption that the quiescent values of V_{CE} and I_C (V_Q and I_Q in Figure 6–11) are specified, i.e., known constants. The objective is to find values for R_C, R_E, R_1, and R_2 that set the required Q-point and simultaneously satisfy the two design criteria. In most practical design problems, the value of the supply voltage, V_{CC}, is predetermined by other system considerations, and we will therefore treat it as a known constant. The first step is to find V_E using the first of the two conditions: $V_E = 0.1V_{CC}$. Then, since $I_E \approx I_Q$,

$$R_E = \frac{V_E}{I_Q} \tag{6–28}$$

Solving equation 6–15 for R_C leads to

$$R_C = \frac{V_{CC} - V_Q - V_E}{I_Q} \tag{6–29}$$

Now,

$$V_B = V_E + V_{BE} \tag{6–30}$$

and, neglecting base loading,

$$V_B = \left(\frac{R_2}{R_1 + R_2}\right)V_{CC} \tag{6–31}$$

Equating (6–30) to (6–31),

$$V_E + V_{BE} = \left(\frac{R_2}{R_1 + R_2}\right)V_{CC} \tag{6–32}$$

Condition 6–27 requires that

$$R_2 = \beta(\min)R_E/10 \tag{6–33}$$

Solving (6–32) for R_1 in terms of R_2 leads to

$$R_1 = R_2\left(\frac{V_{CC}}{V_E + V_{BE}} - 1\right) \tag{6–34}$$

Example 6–6

Design a voltage-divider bias circuit in which $V_{CC} = 24$ V, $V_Q = 12$ V, and $I_Q = 1$ mA. The circuit should perform satisfactorily using silicon transistors whose values of β range from 50 to 200.

Solution. From equation 6–26, $V_E = 0.1(24) = 2.4$ V. From (6–28), $R_E = (2.4\text{ V})/(1\text{ mA}) = 2.4\text{ k}\Omega$. From (6–29),

$$R_C = \frac{(24 - 12 - 2.4)\text{ V}}{1\text{ mA}} = 9.6\text{ k}\Omega$$

From (6–33),

$$R_2 = \frac{50(2.4\text{ k}\Omega)}{10} = 12\text{ k}\Omega$$

From (6–34),

$$R_1 = (12 \times 10^3)\left(\frac{24}{2.4 + 0.7} - 1\right) = 80.9\text{ k}\Omega$$

The completed design, using the closest standard 5% resistors, is shown in Figure 6–12.

Once a bias circuit has been designed, it should be analyzed to verify that the design specifications are satisfied, particularly if the resistors used have standard (nonprecision) values that differ from the design calculations. In the case of this example, we find

$$V_B \approx \left(\frac{12 \times 10^3}{12 \times 10^3 + 82 \times 10^3}\right) 24\text{ V} = 3.06\text{ V}$$

$$V_E \approx (3.06 - 0.7)\text{ V} = 2.36\text{ V}$$

Figure 6–12
(Example 6–6)

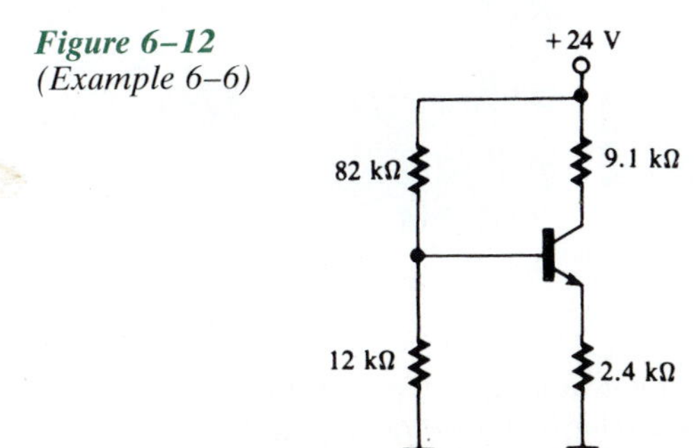

$$I_E \approx I_Q = \frac{2.36 \text{ V}}{2.4 \text{ k}\Omega} = 0.983 \text{ mA}$$

$$V_Q \approx 24 \text{ V} - (0.983 \text{ mA})[(9.1 \text{ k}\Omega) + (2.4 \text{ k}\Omega)] = 12.69 \text{ V}$$

We see that the actual quiescent values agree well with those required of the design.

6–7 INTEGRATED CIRCUITS

The methods used to bias amplifiers in integrated circuits are substantially different from those used in discrete circuits. To understand why this is the case and to appreciate the constraints imposed by integrated circuit technology on circuit design in general, it is necessary to learn some fundamental facts about how integrated circuits are constructed.

Applications and Classifications of Integrated Circuits

We have mentioned in several previous discussions that diodes and transistors are found in both *discrete-* and *integrated*-circuit form. Discrete components are packaged in individual enclosures and have leads that allow electrical connections to be made between component terminals, such as emitter, base, and collector. Discrete circuits are composed of discrete components, including resistors, capacitors, and/or inductors, interconnected by wires or through conducting paths on a printed circuit board. Figure 6–13(a) shows a typical discrete transistor circuit. In contrast, the components of an integrated circuit are all constructed on a single, tiny piece of semiconductor crystal, called a *chip,* which may contain hundreds of diodes, transistors, resistors, and/or capacitors. (Inductors are not constructed in integrated-circuit form.) The conducting paths that interconnect the components

(a)

(b)

Figure 6–13
(a) Discrete and (b) integrated circuits. The ant holds in its mouth an integrated-circuit chip that could contain all the circuitry in the discrete circuit ((b): courtesy of Philips Science and Industry Division).

of an integrated circuit are contained entirely within the device, and the only leads that are brought out are those necessary for power supply connections, grounds, and circuit inputs and outputs. Figure 6–13(b) shows a typical integrated-circuit chip of the size that might well contain all of the discrete circuitry shown in Figure 6–13(a). An integrated circuit (IC) fabricated entirely on a single silicon chip is called a *monolithic* IC. A *hybrid* integrated circuit contains one or more monolithic circuits interconnected with external resistors and capacitors using *thin-film* or *thick-film* techniques.

One obvious advantage of integrated circuits over their discrete counterparts is the fact that very complex circuits can be constructed in very small packages, with attendant savings in wiring, assembly, cooling requirements, and material costs. Besides the economic benefits derived from miniaturization, integrated circuits are inherently more reliable than discrete circuits. Because the IC components are contained in a single rigid structure, an IC is not so susceptible as a discrete circuit to the kinds of mechanical failures that afflict the latter: ruptures and shorts in interconnecting paths, caused by shock and vibration; connector misalignments; solder-joint failures; and so forth. Furthermore, a good deal of back-up or redundant circuitry can be included in an IC at very little additional cost, thus ensuring satisfactory performance in the event that some components do fail. Back-up circuitry is feasible because circuit complexity is not so significant a cost factor as it is in discrete circuits, due to the nature of the manufacturing process. As we shall see in later discussions, the number of components that can be fabricated on a single chip is limited more by the need for maintaining isolation between components than by their cost. Although numerous steps are required to manufacture an IC, a large number of identical devices can be fabricated simultaneously, contributing to cost reduction.

Finally, integrated circuits are advantageous in high-frequency applications and in high-speed computer circuits because of the small distances that electrical signals travel between individual components. Long signal paths create delays and phase shifts that limit the frequency at which such circuits can operate.

Crystal Growth

The starting point in the manufacturing procedure for both discrete components and integrated circuits is the production of a single symmetric crystal, most often a silicon crystal. Silicon is obtained from certain chemical compounds, but is not initially in a form suitable for semiconductor devices. In its natural form, silicon is said to be *polycrystalline,* because it is composed of a large number of crystals having different orientations. To obtain a single crystal of uniform orientation, it is necessary to melt the polycrystalline silicon and then allow it to cool and solidify under certain closely controlled conditions. The location in the "melt" where cooling takes place and the rate at which cooling occurs are particularly critical. In one manufacturing process, a crucible of molten silicon is pulled slowly through a furnace and the crystal forms where the cooling melt emerges.

The development of a single crystal by controlled cooling of molten material is called *crystal growth.* A *seed* crystal is often used as a "starter" upon which the crystal is grown. In the process most often used for IC production, called the *Czochralski technique,* the seed crystal is brought into contact with the molten material and then slowly extracted while being rotated. The crystal formed using the Czochralski technique is in the shape of a cylinder up to 5 in. in diameter and several feet long. This cylindrical *ingot* is then sliced, using a diamond cutter, into thin *wafers,* about 0.5 mm thick. Figure 6–14(a) shows a typical cylindrical ingot

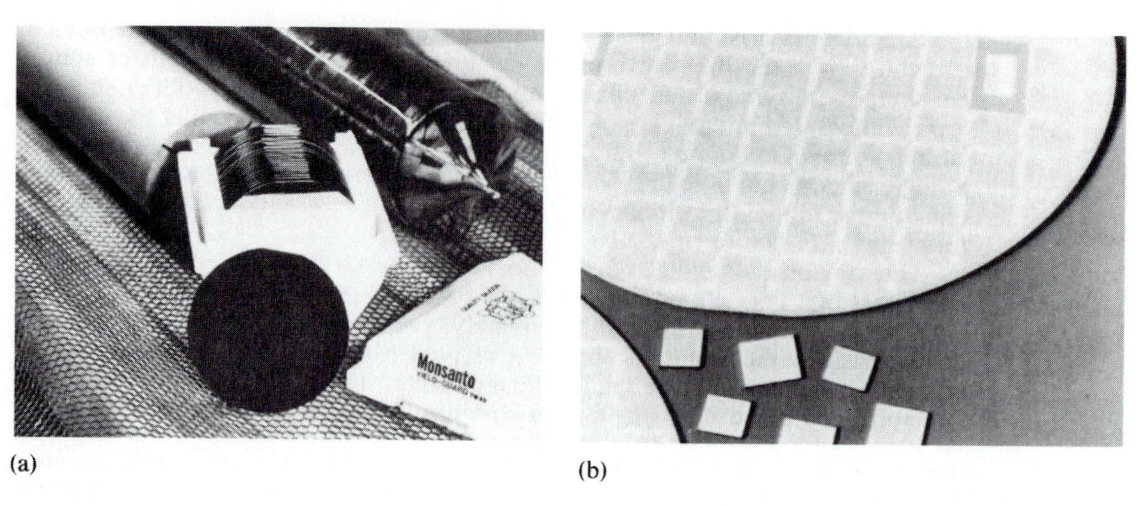

(a) (b)

Figure 6–14
(a) An ingot of silicon crystal and wafers (courtesy of Monsanto Electronic Materials Company). (b) A wafer showing the individual chips (dies) cut from it (courtesy of Gould AMI Semiconductors, Santa Clara, CA, 1985).

and a wafer obtained from it. The wafers are polished and each is used to produce a hundred or more identical IC chips. The finished chips, called *dies* or *dice,* are cut from the wafer (Figure 6–14(b)) and mounted in individual enclosures after the attachment of leads.

PN Device Fabrication—The Photolithographic Process

The method most widely used for integrated-circuit fabrication is now also used to produce discrete devices. The process to which we refer, called *batch production* using *photolithographic* methods, is a truly remarkable blend of ingenuity and precision, and is responsible for many dramatic achievements in the field of microminiature electronics. The manufacturing procedure is called *batch processing* because it permits the simultaneous production of many identical chips from a single wafer. This procedure has been made possible by technological advances in a number of diverse fields, including chemical process control, photochemistry, and computer-aided design and manufacturing (CAD/CAM).

The essence of the photolithographic process is the use of photographic methods to alter the characteristics of a special coating applied to the surface of a wafer. The coating is altered in an intricate pattern so that tiny regions of the crystal will become P- or N-type material when the wafer is later subjected to certain doping treatments. One such doping treatment is called *impurity diffusion,* whereby donor or acceptor impurities are allowed to diffuse through the surface pattern and enter the crystals at the desired locations. We will discuss some details of this important process and then return to the details of photolithography.

Recall that *carrier* diffusion is the migration of holes or electrons from a region where there is a surplus of one or the other to a region where there is a corresponding scarcity. The diffusion process used in batch production is a similar phenomenon, except that it is impurity *atoms* such as boron or phosphorus that are allowed to diffuse into semiconductor material to make it either P- or N-type. To enable atoms to diffuse into silicon, the material must be raised to a high temperature, on the order of 1000°C. At that temperature, silicon atoms leave the crystal structure, thus making it possible for migrating impurity atoms to occupy the vacancies. Recall that boron atoms entering the crystal structure create holes and therefore produce P-type material, while phosphorus atoms supply excess electrons to create N material.

The first step in the photolithographic processing is to create a thin layer of silicon dioxide (SiO_2) on the surface of a silicon wafer. This is accomplished by placing the crystalline wafer in a furnace containing oxygen gas (O_2), which then reacts chemically with the Si to produce SiO_2. Impurity atoms cannot diffuse through the SiO_2 layer and into the Si crystal, so the SiO_2 is used to prevent the creation of P- and N-type regions where they are not desired. To create P and N regions where they are desired, "windows" must be made in the SiO_2 layer; that is, the SiO_2 must be removed at selected locations to permit impurity atoms to diffuse into the crystal. We say that windows are "opened" where the SiO_2 is removed.

The next step in the process is to deposit a coating of *photoresist* (PR) material on top of the SiO_2 layer. The photoresist is a photosensitive organic material that changes its composition (becomes polymerized) wherever it is exposed to ultraviolet light. The PR coating is exposed to ultraviolet light through a glass *mask* that *prevents* exposure (is opaque) at any location where a window is desired. Thus the mask creates a pattern of unexposed PR corresponding to regions where impurity diffusion will be allowed to occur. The unexposed PR material is removed and hydrofluoric acid (HF) is applied to the surface. The acid does not affect the regions of exposed PR but etches away the SiO_2 where it is not protected by the PR, thus creating the windows. Once the windows in the SiO_2 have been opened by the HF acid treatment, the exposed PR material is removed and the structure is ready for the impurity diffusion step. The wafer is placed in an oven and impurity atoms diffuse through the windows into the silicon crystal. The depth to which the diffused P or N layer penetrates the crystal, and its concentration (doping density), are determined by closely controlling the temperature of the process and the length of time it is allowed to occur. Because diffusion occurs only at those locations where windows have been opened, the process is often called *selective diffusion.* Figure 6–15 illustrates how the steps we have described so far could be used to create two PN junctions in an N-type crystal.

The final steps in the photolithographic processing are those required to deposit metal contact surfaces where terminal leads can be attached and, in the case of integrated circuits, any metallic paths needed to interconnect devices. This phase of the procedure is called *metallization.* The metal most often used is aluminum (Al) because it adheres well to Si and SiO_2. Gold is sometimes used, but additional steps are then necessary to ensure adhesion. Vaporized aluminum is first deposited over the entire surface of a wafer that has undergone the diffusion processing described earlier. A layer of photoresist is then applied on top of the aluminum and exposed through a mask that defines the metallization pattern. As before, the unexposed PR material is removed and the aluminum is etched away from all regions where it is not covered by exposed PR.

The process we have described could be used to obtain hundreds of discrete PN diodes from a single wafer. The SiO_2 layer is generally left intact and the wafer is scribed, or cut, with a laser to separate the individual devices. The steps required for the fabrication of a more complex device, such as a transistor, are like those for a single PN junction, but involve repeated oxidations and diffusions to create alternating layers of P and N regions. The crystal itself may be used as one of those

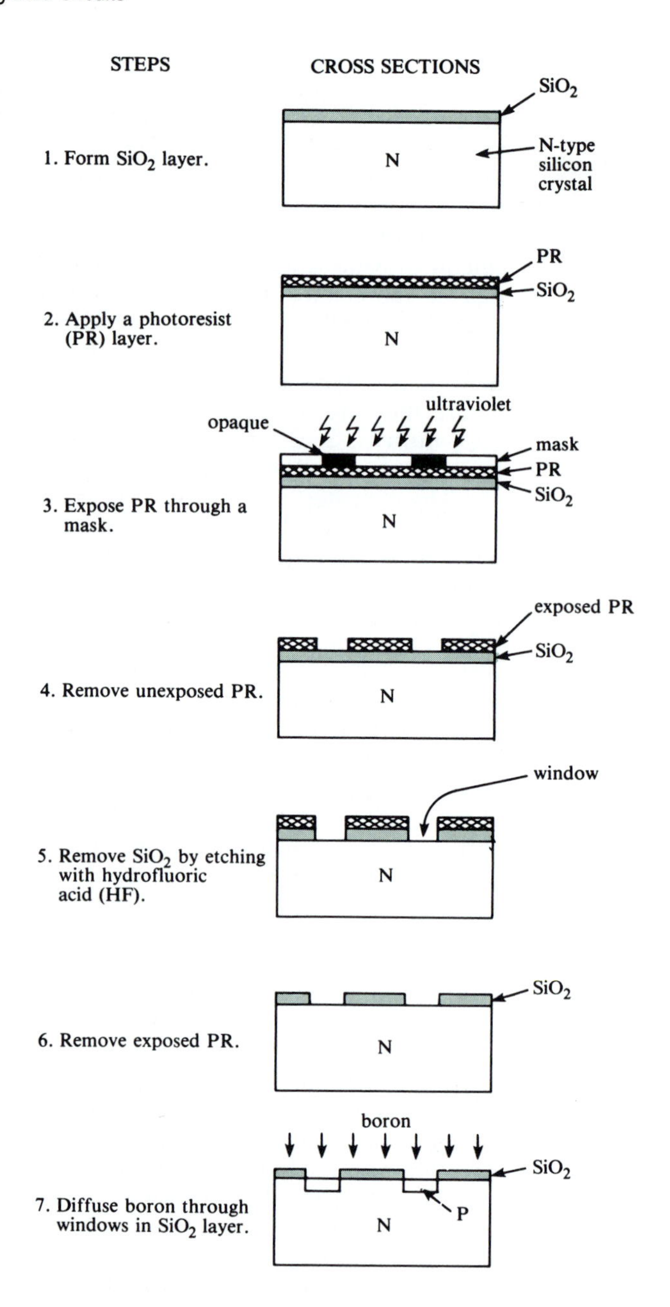

Figure 6–15
The creation of two PN junctions in an N-type crystal using photolithography (The thickness of all layers is exaggerated for clarity.)

layers (for example, as the N-type collector in an NPN transistor). A transistor whose base, emitter, and collector regions lie in parallel planes, one on top of the other, is said to be a *planar* transistor. Most integrated-circuit transistors are of this type. Figure 6–16 summarizes the steps involved in the fabrication of a planar NPN transistor. Note that three oxidation steps, two diffusion steps, and one metallization step are required. Four masks are required.

All the steps we have described for fabricating PN devices are applicable to batch production of integrated-circuit chips, with a few variations and some addi-

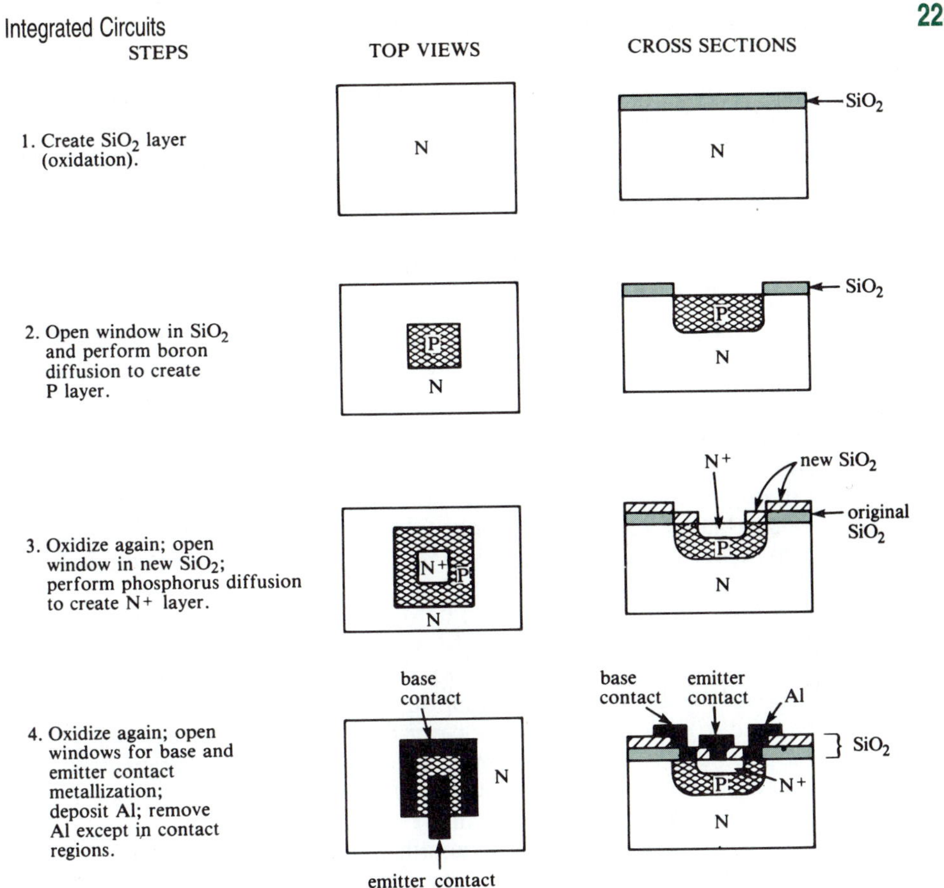
STEPS
TOP VIEWS
CROSS SECTIONS
1. Create SiO2 layer (oxidation).
N
N
SiO2
2. Open window in SiO2 and perform boron diffusion to create P layer.
P
N
SiO2
P
N
3. Oxidize again; open window in new SiO2; perform phosphorus diffusion to create N+ layer.
N+
P
N
N+
new SiO2
original SiO2
P
N
4. Oxidize again; open windows for base and emitter contact metallization; deposit Al; remove Al except in contact regions.
base contact
N
emitter contact
base contact
emitter contact
Al
SiO2
P
N+
N

Figure 6–16
Summary of steps involved in the fabrication of a planar NPN transistor

tional steps we will describe presently. Of course, the diffusion and metallization masks necessary to produce a large number of complex devices must be very intricate and very precise. In fact, the preparation of these masks is one of the most expensive and time-consuming steps in the process. In many cases, the required precision can be obtained only through the use of computer-aided design techniques. Newer techniques employ a computer-controlled electron beam that "writes" a pattern directly on the wafer surface, thus eliminating the need for masking altogether. Much finer line widths can be achieved this way. Similar computer techniques are now also being employed to create very fine masks used in conventional photolithographic procedures.

As an alternative to impurity diffusion, a modern method called *ion implantation* is also being used to create P and N regions in a crystal. In this method, impurity ions are given high energies by an acceleration tube and literally driven through the surface of the wafer and embedded in the crystal. Since ion implantation is blocked by photoresist (or metal), the photolithographic methods we have already described can be used to control the pattern where P and N regions are formed. One advantage of ion implantation is that it can be performed at room temperature. The average depth to which the ions penetrate the crystal, called the *projected range,* can be closely controlled by adjusting the energy imparted to the ions in the accelerator tube. This degree of control is a second advantage of ion implantation: the doping level and depth of doping can be set quite accurately. Very thin regions of P or N material can be created very precisely, as is necessary in some IC designs. Ion implantation is used when these factors are critical to the operation of a circuit.

One disadvantage of ion implantation is that the high-energy ions rupture the crystal structure to a certain extent. The damage can be repaired by heating, a process called *annealing.*

Epitaxial and Buried Layers

In the fabrication of most monolithic integrated circuits, the processes we have described for producing tiny P and N regions in a silicon crystal are not applied directly to the wafer but rather to a thin crystal layer grown on the surface of the wafer. The original (wafer) crystal is called the *substrate* for the circuit, and the thin layer is called an *epitaxial* layer. Typically, the substrate is P-type silicon and the epitaxial layer is N-type silicon.

The substrate (wafer crystal) serves as the seed crystal for the growth of the epitaxial layer. The procedure for growing this new layer, called *epitaxy,* is a special case of *chemical vapor deposition* (CVD), in which gaseous chemicals are used to deposit a layer of solid material on a crystal. Besides an epitaxial layer, CVD can be used to deposit SiO_2 and polycrystalline silicon. Epitaxy is a special case because the solid deposited is a uniformly oriented crystal that aligns itself with the orientation of the crystal in the substrate. It occurs when the gas used is *silane* (SiH_4) and the temperature is greater than 1000°C.

In integrated circuits, the metallization pattern required to interconnect devices is deposited on the *surface* of the chip. For an NPN transistor, which is the most common type fabricated in ICs, the N-type epitaxial layer serves as the collector. To reduce the collector resistance in the path between the epitaxial collector region and the surface, a region of heavily doped (highly conductive) N^+ material is created *beneath* the epitaxial layer, and another from the epitaxial layer to the surface, where contact is made to the collector. Because of its location, the N^+ region lying beneath the epitaxial layer is called a *buried layer.* It is diffused into the substrate wafer before the epitaxial layer is created. In a later step, an N^+ region is diffused into the epitaxial layer from the wafer surface to provide the low-resistance collector-contacting path. Figure 6–17 summarizes the steps involved in the fabrication of a single NPN transistor in an epitaxial layer with a buried N^+ layer. The several oxidations and the masking steps required to perform all the diffusions are not shown explicitly in the figure. Diodes are fabricated in integrated circuits in exactly the same way as transistors. In fact, a diode is often formed by shorting the collector of a transistor to its base. Figure 6–18 illustrates this use.

Resistors

In a bipolar integrated circuit, each resistor is formed by diffusing a certain quantity of P- or N-type material into the epitaxial layer on the surface of an electrically isolated "island." The total resistance of each diffused resistor is determined by the geometry of the region in which it is formed (length, width, and depth) and the conductivity of the diffused material, i.e., the degree of doping. Once the

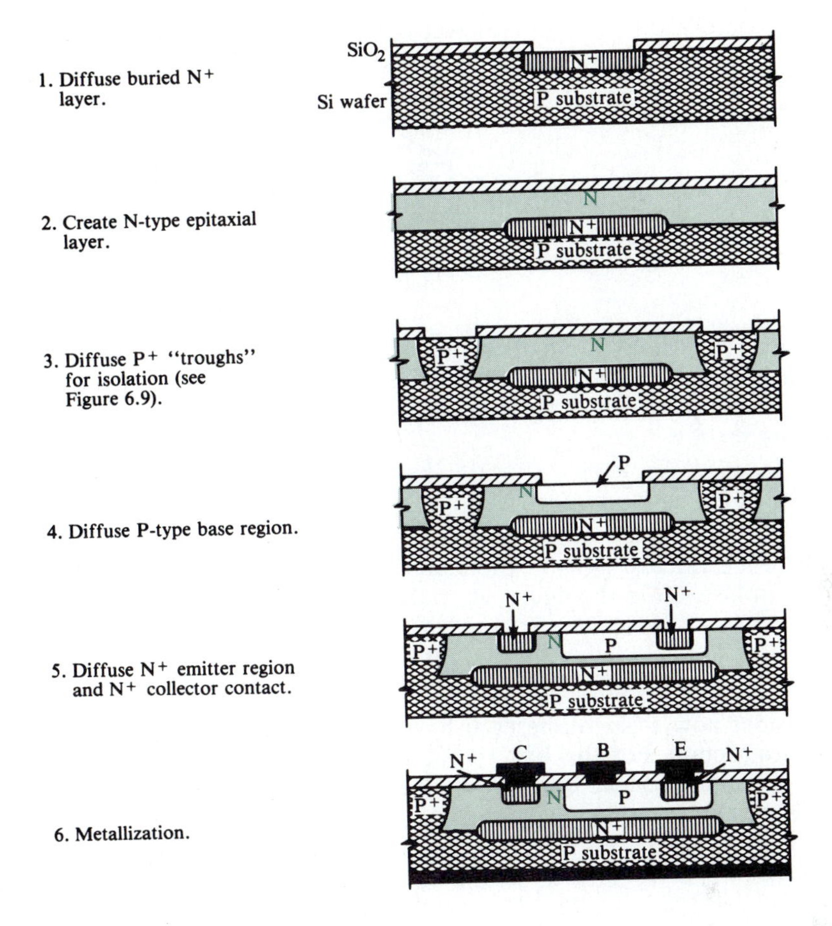

Figure 6–17
A summary of the steps required to fabricate a planar NPN transistor in an epitaxial layer. Note that the buried N^+ layer is created before the epitaxial layer.

resistors have been formed, they are interconnected with other components, as required, by metallization.

It is desirable to have all resistor diffusions occur at the same time that the diffusions for transistor base or emitter regions occur, thus avoiding the need for additional masking and diffusion steps. Therefore, the design of a resistor amounts to the specification of its *geometry* rather than the doping level of the diffused material, as the latter will be the same for all devices undergoing a diffusion step.

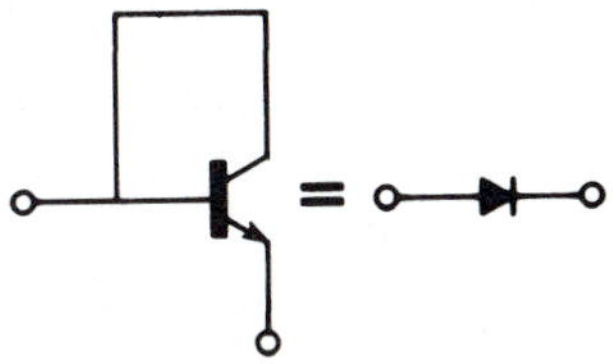

Figure 6–18
In integrated circuits, a diode is often fabricated by shorting the collector of a transistor to its base.

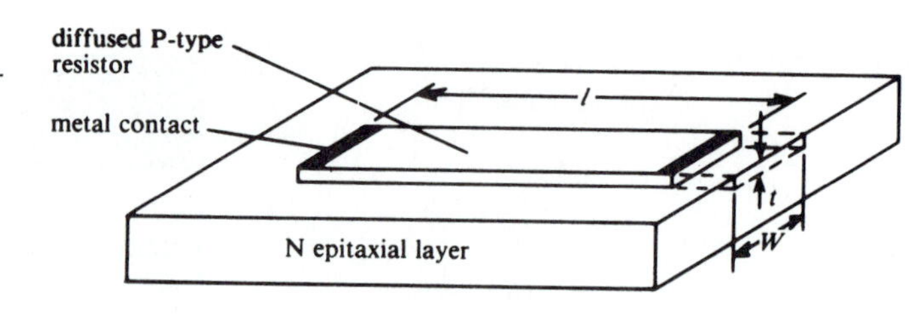

Figure 6–19
A resistor formed by diffusing P-type material into an N-type epitaxial layer. The dimensions shown are used in computing the resistance.

Figure 6–19 shows a bar of diffused material forming a P-type resistor in an N-type epitaxial layer and identifies the dimensions that determine its total resistance.

Recall (Chapter 2) that the resistance R between the ends of the bar shown in Figure 6–19 is

$$R = \frac{\rho l}{A} \text{ ohms} \tag{6–35}$$

where ρ is the resistivity of the material in $\Omega \cdot$ m and A is the cross-sectional area. As can be seen in the figure, $A = tW$, so

$$R = \frac{\rho l}{tW} \text{ ohms} \tag{6–36}$$

Consider now a bar of material having fixed thickness t and having $l = W$, i.e., a square section. For this bar, the quantities l and W in equation 6–36 cancel, and we obtain the special quantity R_s, called the *sheet resistance* (also called the *sheet resistivity*):

$$R_s = \frac{\rho}{t} \text{ ohms/square} \tag{6–37}$$

Note that R_s has the same value for *any* square section of a diffused layer having constant thickness and uniform resistivity. Note also that the units of ρ and t must be consistent in order for R_s to have the proper units. For example, if ρ is in ohm · meters, then t must be in meters. The thickness of an integrated-circuit layer is often expressed in *microns,* where 1 micron = 1 μm = 10^{-6} m. It is the sheet resistance of a diffused layer that is of primary interest to an integrated-circuit designer. Its value, which can be measured using special techniques, is typically in the range from 2 to 10 Ω/square for emitter diffusions and 100 to 200 Ω/square for base diffusions.

Example 6–7

A diffused P layer has a sheet resistance of 125 Ω/square. If the P material has a conductivity of 2000 S/m, find the thickness of the layer in microns.

Solution. From equation 2–6, $\rho = 1/\sigma = 1/2000 = 5 \times 10^{-4}\ \Omega \cdot \text{m}$. From equation 6–37,

$$t = \frac{\rho}{R_s} = \frac{5 \times 10^{-4}}{125} = 4 \times 10^{-6} \text{ m} = 4\ \mu\text{m}$$

The ratio of the length of a diffused resistor to its width, l/W (see Figure 6–19), is called its *aspect ratio, a.* From equation 6–36, we have

$$R = \left(\frac{\rho}{t}\right)\left(\frac{l}{W}\right) = R_s a \qquad \textbf{(6–38)}$$

Since the sheet resistance R_s is generally fixed by the requirements for base and emitter diffusion, it is the aspect ratio that designers control to obtain a resistance of desired value.

Example 6–8

It is desired to fabricate a 1.5-kΩ resistor using a diffused P layer having sheet resistance 200 Ω/square.

1. What aspect ratio should the resistor have?
2. What should be the total length of the diffused region if its width is 30 μm?

Solution

1. From equation 6–38, $a = R/R_s = 1500/200 = 7.5$.
2. From the definition of aspect ratio,

$$a = \frac{l}{W} = 7.5$$

$$l = 7.5W = 7.5(30\ \mu\text{m}) = 225\ \mu\text{m} = 0.225\ \text{mm}$$

For a diffused resistor to have a large resistance, it is clear that it must have a large aspect ratio, l/W. This can be accomplished by using a zigzag or meandering pattern to increase its length, such as shown in Figure 6–20. In general, very large resistance values are difficult to obtain in integrated circuits and are avoided where possible. In some cases, large resistance values are obtained using ion implantation, which makes possible a very thin impurity layer (small value of t), and a corresponding large value for R_s. One advantage of this method is the saving of chip space that would otherwise be needed to obtain a large aspect ratio.

A resistor can be isolated from other IC components by connecting a suitable reverse-biasing potential around it. For example, the N material in Figure 6–20 could be connected to the highest positive voltage in the circuit. An alternative structure for an integrated-circuit resistor is the *double-diffused* resistor illustrated in Figure 6–21. To produce this structure, a layer of P material is diffused into the

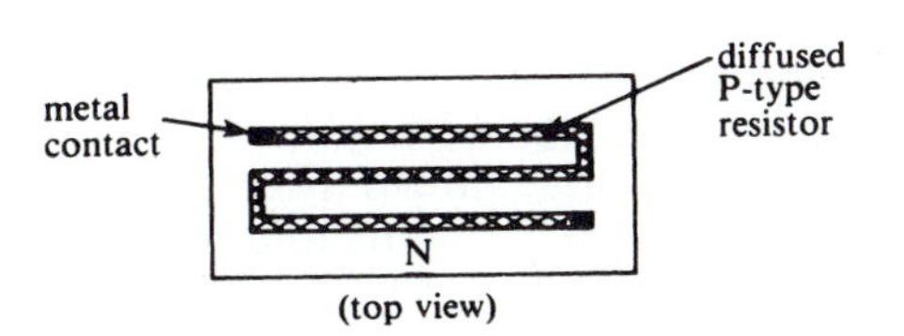

Figure 6–20
The resistance of a diffused resistor can be increased by using a zigzag pattern that increases its aspect ratio.

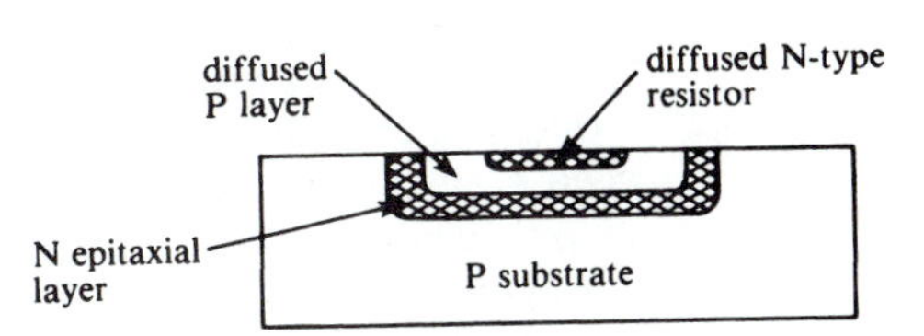

Figure 6–21
A double-diffused integrated-circuit resistor

epitaxial layer at the same time that the base diffusion occurs, and the resistor is then formed as an N-type layer diffused into the P material when the emitter diffusion occurs.

It is difficult to obtain precise resistance values in integrated circuits. Tolerances are typically ±20%. On the other hand, two or more resistors of nearly equal value can be obtained with less difficulty. The production of components having equal or very similar characteristics is called *matching,* and IC resistors can be matched to within ±5% or better. Good matching is possible because resistors are created at the same time using the same diffusion material. In general, integrated-circuit designers avoid design procedures that require precise resistor values and take advantage of techniques that depend on matched values.

Capacitors

We have mentioned that a reverse-biased diode has capacitance. Recall that capacitance exists when two conducting regions are separated by an insulator, or dielectric. In the case of a reverse-biased PN junction, there is capacitance between the P and N sides because each side is conductive and the depletion region between them acts as a dielectric. Therefore, when a capacitor is required in an integrated-circuit design, it can be obtained by fabrication of a reverse-biased diode. (A forward-biased diode also has capacitance, but clearly does not block the flow of dc current like a capacitor.) Either the base–emitter or the base–collector junction of a transistor can be used, though the latter is preferred because of its higher reverse breakdown voltage. Capacitance values obtained this way are quite small, usually less than 20 pF. Tolerances are generally large, ±20%, but, as in the case of resistors, reasonably good matching can be achieved.

Recall the fundamental capacitance equation:

$$C = \frac{\varepsilon A}{d} \tag{6–39}$$

where ε is the permittivity of the dielectric, A is the area of the parallel conducting surfaces, and d is the distance the surfaces are separated. The capacitance of a reverse-biased junction therefore depends directly on the area of the junction and inversely on the width of the depletion region. Since the width of the depletion region changes with the value of the reverse-biasing voltages (increases with increasing voltage), the capacitance also changes with bias voltage. This characteristic is undesirable for integrated-circuit applications but is exploited in some special applications where the diode serves as a voltage-controlled capacitor, called a *varicap diode,* or *varactor* (discussed in Chapter 17).

Integrated-circuit capacitors can also be constructed using an area of metallization as one conducting surface, silicon dioxide as the dielectric, and an N^+ diffused layer created during emitter diffusion as the other conducting surface. These capacitors have excellent characteristics, including low leakage and high breakdown voltage. Capacitance values are again small, comparable to those of junction capacitors. Large capacitors are impractical in integrated circuits and are avoided in circuit design.

Figure 6–22 shows a cross-sectional view of an NPN transistor, a resistor, and a capacitor all constructed on a P substrate and interconnected to form a common-emitter amplifier. The capacitor is the SiO_2 dielectric type.

It is important to note that it is generally easier and more economical to

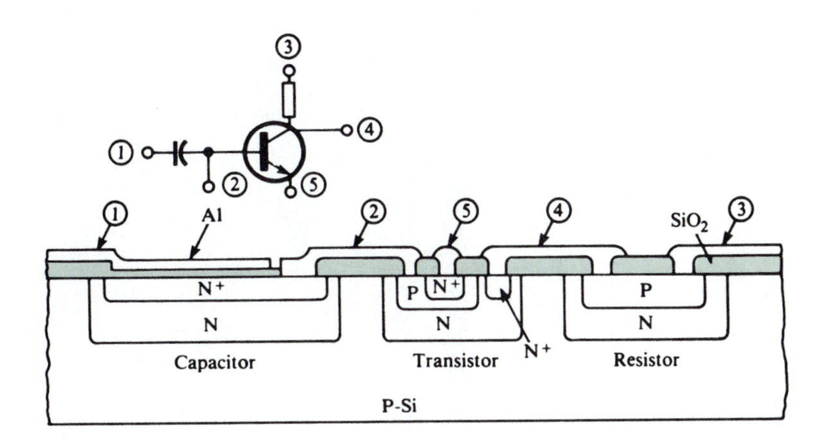

Figure 6–22
An integrated circuit containing interconnected transistor, resistor, and capacitor

construct transistors and diodes in integrated circuits than it is to construct resistors and capacitors. For this reason, designers use transistors and/or diodes in innovative ways to perform circuit functions that could be performed more easily in discrete circuits using resistor and capacitor networks. For example, in integrated circuits, transistors are used as *current sources* to provide bias for other transistors, thus avoiding the resistor–capacitor networks discussed in Chapter 5 for bias designs. This technique will be discussed in detail in a later section. In Chapter 7 we will discuss field-effect transistors and learn that diffused resistors are almost never used in those kinds of integrated circuits, because a transistor itself can serve as a resistor.

Component Interconnections

We have already discussed how photolithographic methods are used to deposit an aluminum metallization pattern on the surface of a chip to interconnect components. As can be seen in Figure 6–22, the metallization makes contact with component terminals and crosses the SiO_2 layer where required to electrically join other component terminals. The metallization pattern is also used to create large metal areas called *pads* along the edges of the chip where external leads can be attached. These pads are electrically connected by the metallization to IC terminals that must be accessible from outside the package, such as power supply terminals, ground, input(s), and output(s). Figure 6–23 shows an enlarged top view of an IC chip in which the metallization pattern and the pads are clearly visible.

Interconnecting paths can also be formed using highly conductive polycrystalline silicon. This method is especially useful for joining components in complex structures involving several layers of circuitry. *Crossovers,* points where one conducting path crosses another without electrically contacting it, are avoided whenever possible by careful selection of the location of each component with respect to others (i.e., component layout, also called circuit *topology*). If a crossover is unavoidable, it is made at a resistor location, because the resistors are covered by SiO_2 and the conducting path can be placed on top of that protective coating. Figure 6–24 is a diagram illustrating this concept.

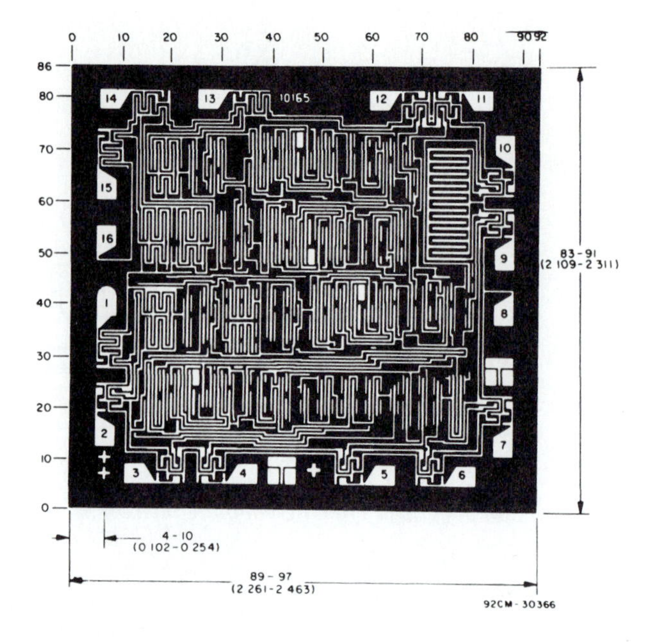

Figure 6–23
Enlarged top view of an integrated-circuit chip. The white areas are metallization. Note the pads along each side (courtesy RCA Corporation).

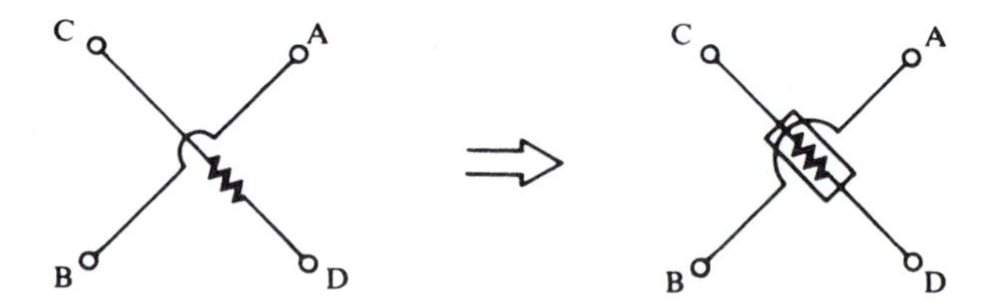

Figure 6–24
If conducting path AB must cross conducting path CD without contacting it, as shown in (a), it can be accomplished as shown in (b), by routing path AB over a resistor covered with SiO_2.

Packaging

Although some integrated circuits are available in cylindrical metal cases, such as the TO-5 enclosures used to package discrete transistors, the majority of modern ICs are packaged in the case type called a *dual in-line package* (DIP). These rectangular cases have rigid metal tabs, called *pins,* that protrude from along each side of the package (see Figure 6–25). Each pin is electrically connected to one of the IC pads and therefore serves as a terminal to which external circuit connections can be made. The package can be conveniently inserted into a plugboard for experimentation or into holes drilled in a printed circuit board. DIP packages are available with several standard numbers of pins ranging from 4 to 64. Integrated circuits are also packaged in "flatpak" form, where the pins protrude straight out from the case instead of at right angles to it.

Figure 6–25 shows the standard DIP pin numbering system used by manufacturers to identify specific terminals and in schematic diagrams. Note that a groove or dot on one end of the case identifies the top and that numbering proceeds down the left side and up the right side.

Like discrete transistors and diodes, integrated circuits are given numbers for identification purposes. However, the numbers do not have standard prefixes such as 1N or 2N. A particular integrated circuit may be produced by several different manufacturers and the identification number, which is usually printed on the case,

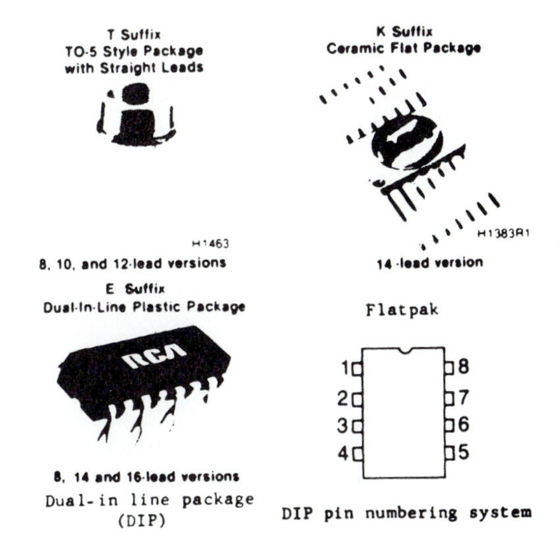

Figure 6–25
Integrated-circuit packages (courtesy RCA Corporation)

will have certain additional letters or codes to identify the manufacturer. For example, the 741 monolithic IC operational amplifier is manufactured by RCA as a CA741 and by National Semiconductor as an LM741.

6–8 INTEGRATED-CIRCUIT BIAS TECHNIQUES

As already mentioned, it is not practical to bias integrated-circuit BJT amplifiers using resistor–capacitor networks, as is the practice in discrete circuits. In the first place, the coupling and bypass capacitors in those designs are far too large to be constructed in a monolithic IC. For that reason, most IC amplifiers are *dc amplifiers,* meaning both *d*irect-*c*oupled and *d*irect-*c*urrent, i.e., there are no coupling capacitors to restrict the lowest frequency at which they can be operated. Second, a good discrete bias circuit requires 3 or 4 resistors per transistor. Since integrated-circuit resistors are more expensive, more difficult to construct, and more space-consuming than IC transistors, the discrete designs are again inappropriate.

Current Sources and Sinks

Integrated-circuit designers use transistor–diode combinations instead of resistors to provide bias for transistor amplifiers. They also take advantage of the fact that transistors and diodes fabricated on the same chip are well *matched,* due to having been created by the same diffusion processes. For example, the voltage drops across all forward-biased PN junctions in one chip are nearly equal and tend to *track* each other with temperature (i.e., increase or decrease by the same amount as temperature changes). The fundamental bias circuit in an IC is a dc *constant-current source,* a transistor circuit that provides nearly constant current to a load. Often only one such circuit is constructed in a monolithic IC and its characteristics are reflected to numerous loads using *current "mirroring,"* which we will discuss presently.

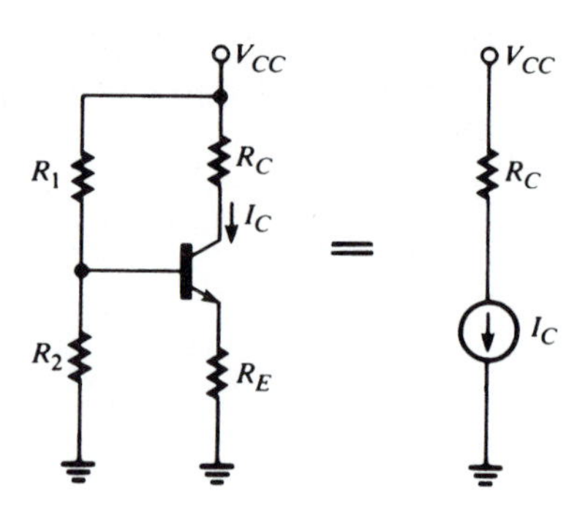

Figure 6–26
Voltage-divider bias method. The transistor acts like a constant-current source for the load represented by R_C.

The voltage-divider method for biasing a CE transistor is redrawn in Figure 6–26. Recall that the base-to-ground voltage V_B can be found from

$$V_B = \left(\frac{R_2 \| R_{in}}{R_1 + R_2 \| R_{in}}\right) V_{CC} \tag{6–40}$$

where R_{in} is the dc resistance looking into the base:

$$R_{in} = (\beta + 1)R_E \tag{6–41}$$

The emitter-to-ground voltage V_E is

$$V_E = V_B - V_{BE} \tag{6–42}$$

The dc emitter current is

$$I_E = V_E/R_E \tag{6–43}$$

Also,

$$I_C = \alpha I_E \tag{6–44}$$

By combining equations 6–40 through 6–44, we can obtain an explicit expression for I_C:

$$I_C = \alpha \left\{ \frac{\left[\dfrac{R_2 \| (\beta + 1)R_E}{R_1 + R_2 \| (\beta + 1)R_E}\right] V_{CC} - V_{BE}}{R_E} \right\} \tag{6–45}$$

The important point to notice in equation 6–45 is that *the resistance R_C does not appear.* In other words, the current I_C that flows through R_C does not depend on the value of R_C. If we regard R_C as a dc load resistance, then we have a device that behaves like a constant-current source: the current flowing through the connected load does not change when the load is changed.

It is important to remember that equation 6–45 is valid only so long as the transistor is operated in its active region. Therefore, the range of values of load resistance R_C for which the transistor can serve as a constant-current source is limited by the requirement that the collector-to-ground voltage V_C satisfy $V_C \geq V_B$. To be a good current source, the value of the constant current (I_C) should depend as little as possible on the values of β and V_{BE}, and hence not be affected by changes in β and V_{BE} caused by temperature variations. Since I_C is the quiescent current in the circuit, the same criteria that we discussed for good Q-point stability apply to the constant-current source. Recall that good stability is achieved by making R_E as large as possible. The trade-off is that large values of R_E reduce the value of the constant current that can be maintained.

Example 6–9

1. Determine the value of the current in the constant-current source shown in Figure 6–27.
2. What is the largest value of R_C that could be used in this circuit without affecting its operation as a current source?

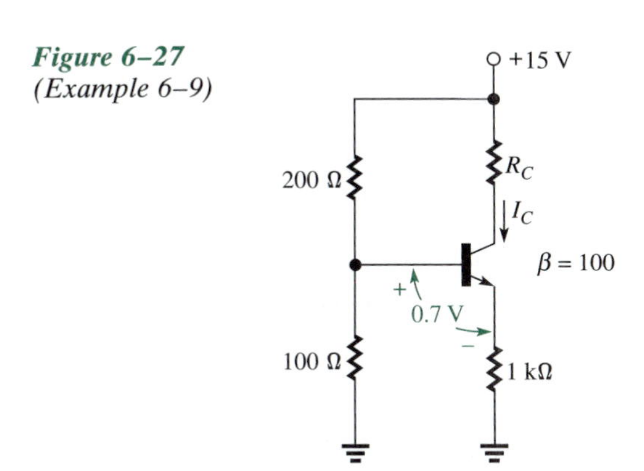

Figure 6–27
(Example 6–9)

Solution

1. From equation 6–40,

$$V_B = \left[\frac{100 \parallel (101 \times 10^3)}{200 + 100 \parallel (101 \times 10^3)}\right] 15\text{ V} = 4.997\text{ V}$$

From equation 4–12, $\alpha = \beta/(\beta + 1) = 100/101 = 0.99$. From equation 6–45,

$$I_C = \frac{0.99(4.997 - 0.7)}{1 \times 10^3} = 4.254\text{ mA}$$

2. For the circuit to operate as a current source, we require that $V_C \geq V_B$. Therefore, we must have

$$V_{CC} - I_C R_C \geq V_B$$

$$15\text{ V} - 4.254 \times 10^{-3} R_C \geq 4.997\text{ V}$$

$$-4.254 \times 10^{-3} R_C \geq -10.003$$

$$R_C \leq \frac{10.003}{4.254 \times 10^{-3}} = 2.35\text{ kV}$$

Thus, R_C must be less than 2350 Ω.

Some authors refer to the circuit in Figure 6–26 as a current *sink,* because it draws, or "drains," current from the supply voltage through the load resistor R_C. In this context, a current *source* is regarded as a device that delivers current *to* a load, i.e., "pushes" the current from the supply voltage out to the load. The PNP

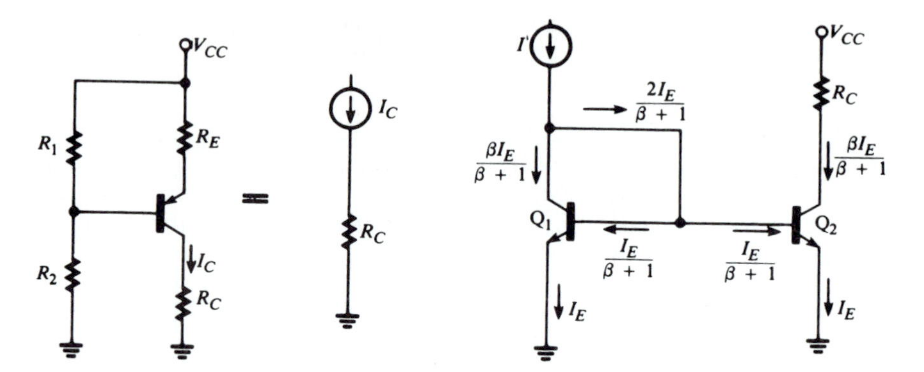

Figure 6–28
A PNP current source

Figure 6–29
An integrated-circuit current mirror

transistor circuit shown in Figure 6–28 is an example. However, the most common practice is to refer to both configurations as current sources, as we will do hereafter.

Current Mirrors

The objective of using a current source in biasing is to establish a constant quiescent current I_C, and hence a constant V_{CE}, in another transistor used as an amplifier. This objective is accomplished using the *current mirror* shown in Figure 6–29. In the figure, a current source producing constant current I is shown connected to transistor Q_1, whose collector is shorted to its base. Q_1 is a *diode-connected transistor.* We will show that the current I is essentially reproduced (mirrored, or reflected) in the collector of Q_2, meaning that the bias of Q_2 is set and maintained by the constant-current source.

We assume that Q_1 and Q_2 are fabricated on the same chip and are therefore closely matched. Since the base–emitter junctions of Q_1 and Q_2 are in parallel, the value of V_{BE} is the same for each transistor. Since the transistors are matched, the equal values of V_{BE} produce equal values of I_E. Again because of matching, Q_1 and Q_2 have the same β and therefore the same base current. Recall that

$$I_B = \frac{I_E}{\beta + 1} \tag{6–46}$$

The collector current in each of Q_1 and Q_2 is β times the base current in each:

$$I_C = \frac{\beta I_E}{\beta + 1} \tag{6–47}$$

As can be seen in Figure 6–29, the total current I delivered by the current source is the sum of the collector current in Q_1 and the two base currents:

$$I = \frac{\beta I_E}{\beta + 1} + \frac{2I_E}{\beta + 1} = \frac{I_E(\beta + 2)}{\beta + 1} \tag{6–48}$$

The ratio of the collector current in Q_2 to the source current I is, therefore,

$$\frac{I_{C2}}{I} = \frac{\beta I_E/(\beta + 1)}{I_E(\beta + 2)/(\beta + 1)} = \frac{\beta}{\beta + 2} = \frac{1}{1 + 2/\beta} \tag{6–49}$$

Assuming that β is large, the quantity $1/(1 + 2/\beta)$ is approximately equal to 1, meaning that

$$I_{C2} \approx I \tag{6–50}$$

This analysis shows that the source current I is closely mirrored in the collector of Q_2. Note that the emitters of Q_1 and Q_2 could both be connected to a voltage source instead of ground without changing this result.

Example 6–10

Assuming that $\beta = 150$, find the (1) approximate and (2) exact values of I_C and V_{CE} in the circuit shown in Figure 6–30.

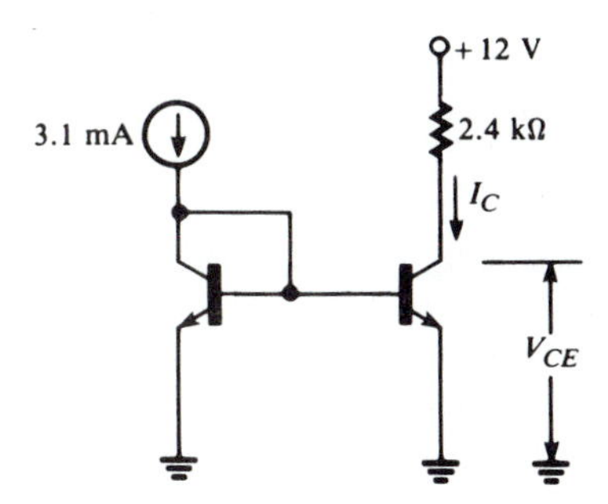

Figure 6–30
(Example 6–10)

Solution

1. The current mirror makes I_C approximately equal to the value of the constant-current source: $I_C \approx 3.1$ mA. Therefore, $V_{CE} \approx 12 - (3.1\text{ mA})(2.4\text{ k}\Omega) = 4.56$ V.
2. The exact value of I_C (assuming exactly matched transistors) is found from equation 6–49:

$$I_C = I\left(\frac{1}{1 + 2/\beta}\right) = (3.1\text{ mA})\left(\frac{1}{1 + 2/150}\right) = 3.059\text{ mA}$$

Therefore, $V_{CE} = 12 - (3.059\text{ mA})(2.4\text{ k}\Omega) = 4.66$ V.

A single current source can be used with a current mirror to set the bias for numerous amplifiers in the same integrated circuit. The next example illustrates this point.

Example 6–11

Find the approximate value of V_{CE} in each of Q_1, Q_2, and Q_3 in Figure 6–31.

Solution. The current source is mirrored in each of Q_1, Q_2, and Q_3. Therefore,

$$I_{C1} = 1.5\text{ mA} \qquad V_{CE1} = 10 - (1.5\text{ mA})(2.5\text{ k}\Omega) = 6.25\text{ V}$$
$$I_{C2} = 1.5\text{ mA} \qquad V_{CE2} = 10 - (1.5\text{ mA})(3\text{ k}\Omega) = 5.5\text{ V}$$
$$I_{C3} = 1.5\text{ mA} \qquad V_{CE3} = 10 - (1.5\text{ mA})(4.2\text{ k}\Omega) = 3.7\text{ V}$$

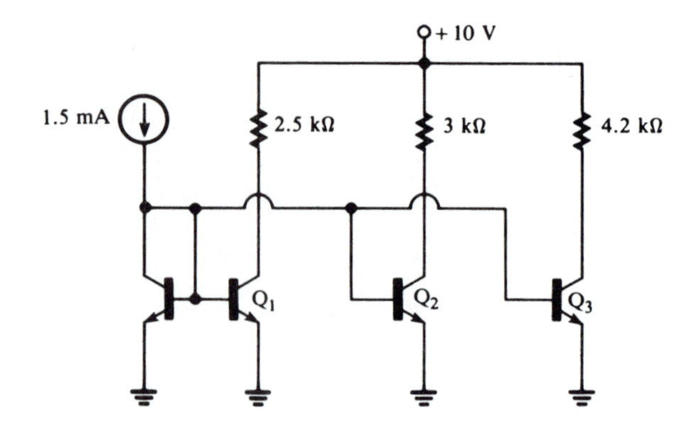

Figure 6–31
(Example 6–11)

When current mirroring is used to bias several transistors from a single current source, each collector current is actually somewhat smaller than it would be if the mirror were applied to just one transistor. The greater the number of transistors, the smaller the collector current in each. It is easy to show that the ratio of any collector current, I_C, to the source current, I, is given by

$$\frac{I_C}{I} = \frac{1}{1 + \frac{n+1}{\beta}} \tag{6–51}$$

where n is the total number of transistors biased from a single source. If β is large and n is not too large, then the ratio given by (6–51) is still close to 1. Exercise 6–41 requires the derivation of equation 6–51 for the case $n = 3$.

Example 6–12

PSPICE

Use PSpice to find V_{CE} in Q_1, Q_2, and Q_3 of Figure 6–31. Assume the β of all transistors is 200 and let the other parameters default.

Solution. The PSpice circuit and input circuit file are shown in Figure 6–32. Note that the collector-to-emitter voltage (V_{CE}) of each transistor is the same as the collector-to-ground voltage (V_C) of each, so the PSpice .PRINT statement can request VC(Q1), VC(Q2), and VC(Q3) to obtain those values. Execution of the program reveals that VC(Q1) = 6.324 V, VC(Q2) = 5.588 V, and VC(Q3) = 3.824 V. These differ somewhat from the approximate values calculated in Example 6–11 but agree very closely with theoretically exact values calculated using equation 6–51:

$$I_C = \frac{1.5 \text{ mA}}{1 + \frac{4}{200}} = 1.47 \text{ mA}$$

$$V_{CE}(Q1) = 10 \text{ V} - (1.47 \text{ mA})(2.5 \text{ k}\Omega) = 6.325 \text{ V}$$

$$V_{CE}(Q2) = 10 \text{ V} - (1.47 \text{ mA})(3 \text{ k}\Omega) = 5.59 \text{ V}$$

$$V_{CE}(Q3) = 10 \text{ V} - (1.47 \text{ mA})(4.2 \text{ k}\Omega) = 3.826 \text{ V}$$

Figure 6–32
(Example 6–12)

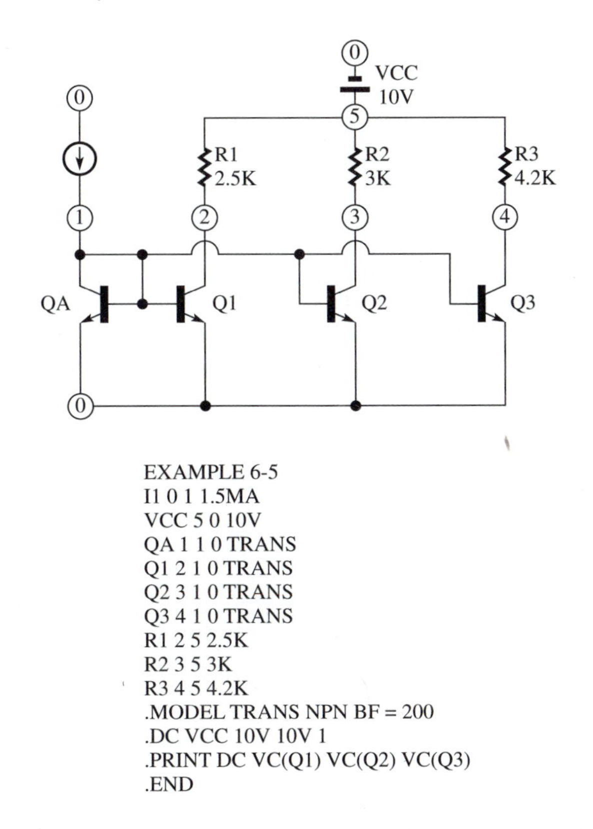


```
EXAMPLE 6-5
I1 0 1 1.5MA
VCC 5 0 10V
QA 1 1 0 TRANS
Q1 2 1 0 TRANS
Q2 3 1 0 TRANS
Q3 4 1 0 TRANS
R1 2 5 2.5K
R2 3 5 3K
R3 4 5 4.2K
.MODEL TRANS NPN BF = 200
.DC VCC 10V 10V 1
.PRINT DC VC(Q1) VC(Q2) VC(Q3)
.END
```

Because of the extensive use of current mirrors in integrated circuits, a different convention is sometimes used in schematic diagrams to show transistors having their base terminals connected together, such as those in Figure 6–31. The horizontal line representing the base terminal is extended through the vertical bar of the symbol so that base connections can be conveniently shown on both the right and left sides of the transistor. Figure 6–33 shows how the circuit of Example 6–11 (Figure 6–31) could be redrawn using this convention. The diode-connected transistor is also frequently shown on schematics using just a diode symbol.

Of course, the current-mirroring techniques we have discussed can also be used in discrete circuits. However, good current mirroring requires the use of well-matched transistors, which makes it a less practical and more expensive alternative in discrete circuits.

Figure 6–33
The circuit of Figure 6–31 redrawn to show a convention used in some schematic diagrams

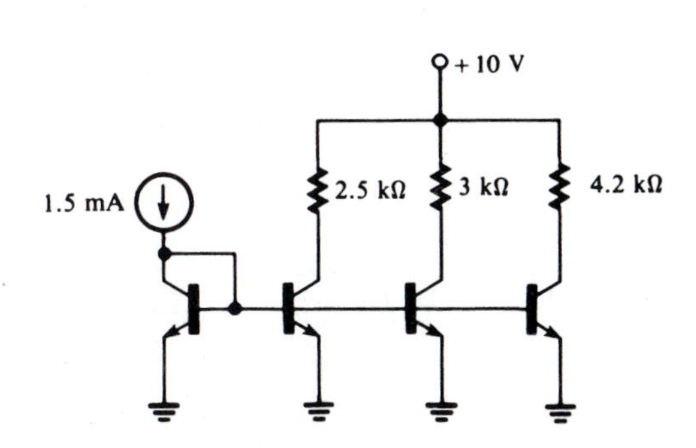

EXERCISES

Section 6–2
Stability Factors

6–1. When the temperature of the transistor in a certain circuit increases from 10°C to 70°C, the change in dc collector current caused by the change in V_{BE} alone is 0.14 mA. What is the value of $S(V_{BE})$ for the circuit?

6–2. When the temperature of a certain transistor increases, V_{BE} changes from 0.685 V to 0.660 V, I_{CBO} changes from 0.12 μA to 0.26 μA, and β changes from 85 to 120. If the collector current is 3.820 mA at the lower temperature, find its approximate value at the higher temperature, given that $S(V_{BE}) = -12.1 \times 10^{-4}$, $S(I_{CBO}) = 65$, and $S(\beta) = 2.2 \times 10^{-5}$.

6–3. Find $S(V_{BE})$, $S(I_{CBO})$, and $S(\beta)$ for the circuit shown in Figure 6–34 when a temperature increase causes β to change from 50 to 100. Assume that the dc collector current at the lower temperature is 1.65 mA.

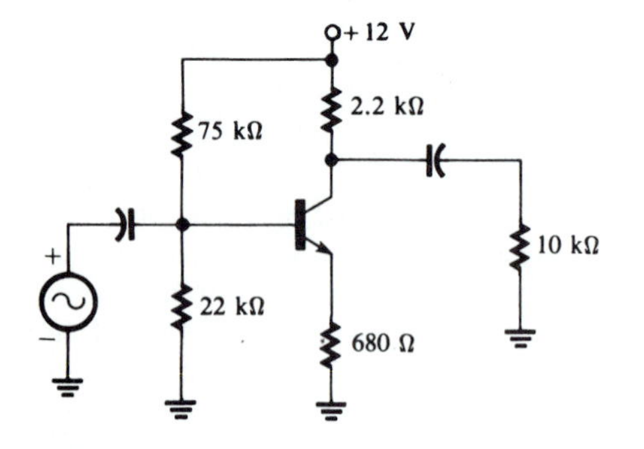

Figure 6–34
(Exercise 6–3)

6–4. If the temperature change that causes β to increase from 50 to 100 in the transistor of Exercise 6–3 also causes V_{BE} to decrease 45 mV and I_{CBO} to increase 0.2 μA, find the approximate collector current at the higher temperature.

Section 6–3
Calculating the Bias Point of the Stabilized Circuit

6–5. Find the quiescent values of I_C and V_{CE} in the circuit shown in Figure 6–35,
a. neglecting the loading effect of the dc resistance looking into the base; and
b. taking the loading effect into account.
What percent error in I_C is caused by neglecting the loading effect?

6–6. Repeat Exercise 6–5 when R_1 is changed to 68 kΩ and R_2 is changed to 15 kΩ.

6–7. What will be the value of I_C if the β of the transistor in Figure 6–35 is changed from 120 to 300? What is the corresponding percent change in I_C? (Note that the loading effect of the resistance looking into the base should not be neglected in this computation.)

6–8. Suppose R_1 in Figure 6–35 is changed to 68 kΩ and R_2 is changed to 15 kΩ. If the β changes from 120 to 300, what will be the percent change in I_C? (Note that the loading effect of the resistance looking into the base should not be neglected in this computation.)

Figure 6–35
(Exercise 6–5)

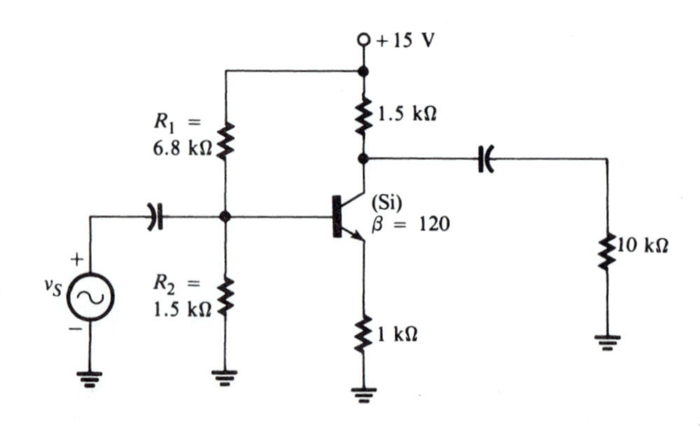

Section 6–4
Small-Signal Performance of the Bias-Stabilized Circuit

6–9. **a.** Find the approximate value of the overall voltage gain (v_L/v_S) of the amplifier stage shown in Figure 6–36.
b. By what percent would the voltage gain change if the quiescent value of the current increased 10%?

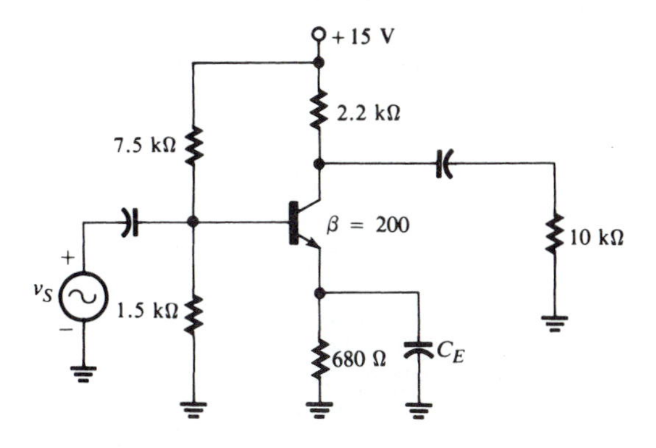

Figure 6–36
(Exercise 6–9)

6–10. Repeat Exercise 6–9 if the emitter bypass capacitor C_E is removed.

6–11. **a.** Find the approximate rms load voltage v_L in the circuit of Figure 6–37.
b. Repeat, if the emitter bypass capacitor is removed.

6–12. Repeat Exercise 6–11 if the values of R_1 and R_2 are changed to 68 kΩ and 15 kΩ, respectively.

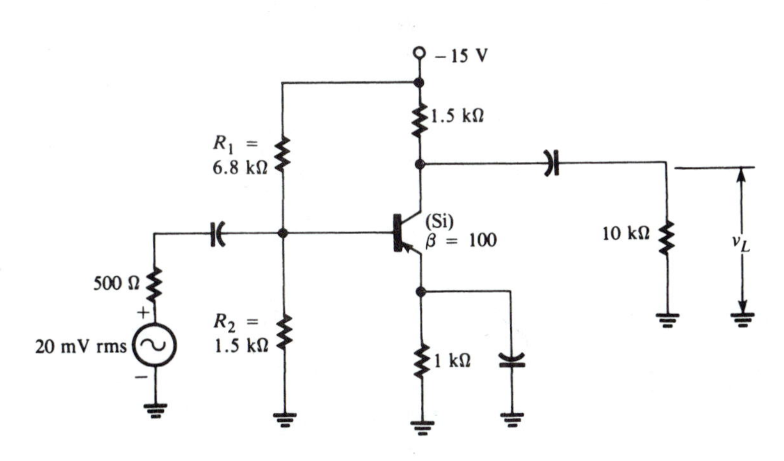

Figure 6–37
(Exercise 6–11)

6–13. A common-emitter amplifier is to have a voltage gain of 5 when it drives a 10-kΩ load. To ensure good bias stability, the ratio R_B/R_E should equal 4, where $R_B = R_1 \parallel R_2$. The quiescent value of V_{CE} is to be 10 V ± 2 V. Design the amplifier, assuming that $\beta = 200$, $V_{CC} = 20$ V, and the signal source driving it has zero source resistance.

Section 6–5
Stabilization by Collector Feedback

6–14. For the circuit shown in Figure 6–38, find
a. the quiescent values of I_C and V_{CE}
b. A_v;
c. r_{in}(stage); and
d. v_L/v_S.

6–15. Repeat Exercise 6–14 when $R_E = 0$.

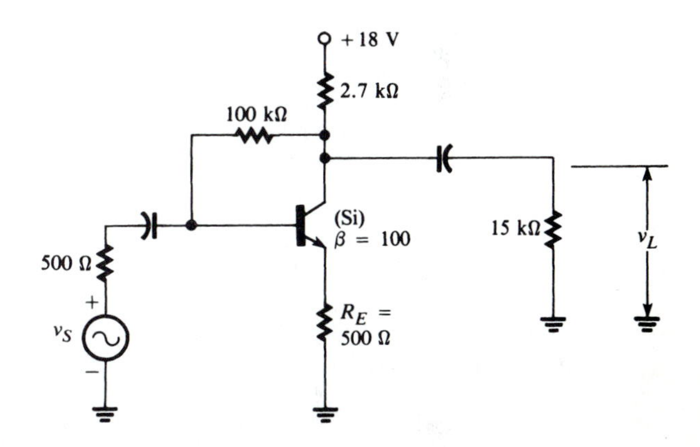

Figure 6–38
(Exercise 6–14)

DESIGN EXERCISES

Section 6–6
Discrete Circuit Bias Design

6–16. Design a voltage-divider bias circuit that sets the quiescent values of V_{CE} and I_C to 7 V and 1.2 mA. The supply voltage is 18 V. Assume that the silicon transistors have values of β that range from 80 to 200. Use standard 5% resistor values closest to your design calculations and analyze the final circuit to find the actual Q-point.

6–17. Design a voltage-divider bias circuit that sets the quiescent values of V_{CE} and I_C to 10 V and 2 mA. The supply voltage is 30 V. Assume that the smallest value of β that the silicon transistor used in the circuit might have is 100. Use standard 10% resistor values closest to your design calculations and analyze the final circuit to find the actual Q-point. Assuming that $r_S = 0$ and $R_L = \infty$, calculate the ac voltage gain of the circuit.

6–18. Show that the ac voltage gain of an amplifier whose bias circuit is designed using the procedure given in Section 6–6 is $A_v \approx 9 - 10V_Q/V_{CC}$, assuming that $r_S = 0$, $R_L = \infty$, and R_E is unbypassed.

Section 6–7
Integrated Circuits

6–19. What are two major advantages of integrated circuits in comparison to discrete circuits? Can you think of an application in which a discrete circuit would be more appropriate than an integrated circuit? Explain.

6–20. What is the difference between polycrystalline silicon and the silicon used in integrated-circuit fabrication? Briefly describe one process used to convert one type to the other.

6–21. Integrated-circuit manufacturing involves the production of *ingots, wafers,* and *chips.* How are these related to each other? What is another name for a chip?

6–22. How is impurity diffusion different from carrier diffusion? What environmental requirement is necessary to achieve impurity diffusion?

6–23. What material is used to shield a silicon crystal from impurity diffusion? How is this material applied to the surface of a wafer?

6–24. Briefly define and/or describe the nature of each of the following terms and discuss its relation to batch processing using photolithographic methods:
- **a.** window
- **b.** photoresist
- **c.** mask
- **d.** hydrofluoric acid
- **e.** metallization

6–25. Name a principal material used in each of the following processes:
- **a.** oxidation
- **b.** etching
- **c.** P-layer diffusion
- **d.** N-layer diffusion
- **e.** metallization

6–26. Name and describe a way of creating impurity layers in a crystal without using impurity diffusion. What are the principal advantages of this method in comparison to impurity diffusion? What is one disadvantage?

6–27. What is an epitaxial layer? Describe one way in which it can be created.

6–28. What is a buried layer and what is its purpose? When and how is it formed?

6–29. The conductivity of the P material used to construct the diffused resistor shown in Figure 6–39 is 1250 S/m.
- **a.** Find the sheet resistance of the diffused layer.
- **b.** Find the resistance of the resistor.

6–30. The diffused layer used to construct the resistor shown in Figure 6–40 has sheet resistance 50 Ω/square. What is the resistance of the resistor?

6–31. A diffused resistor is to be fabricated using P material having resistivity $4 \times 10^{-4}\ \Omega \cdot \text{m}$. The thickness of the layer is 5 μm. If the resistor has width 0.025 mm and is to have

Figure 6–39
(Exercise 6–29)

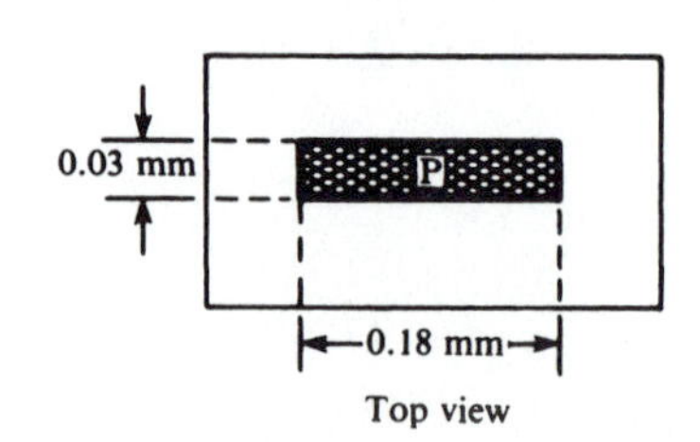

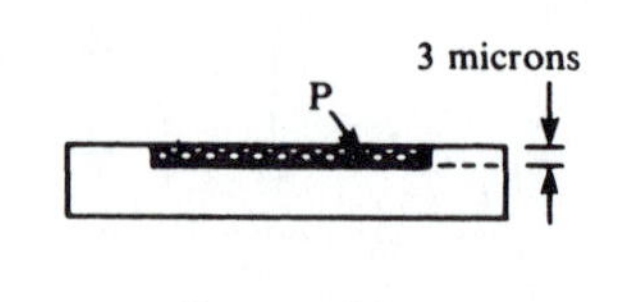

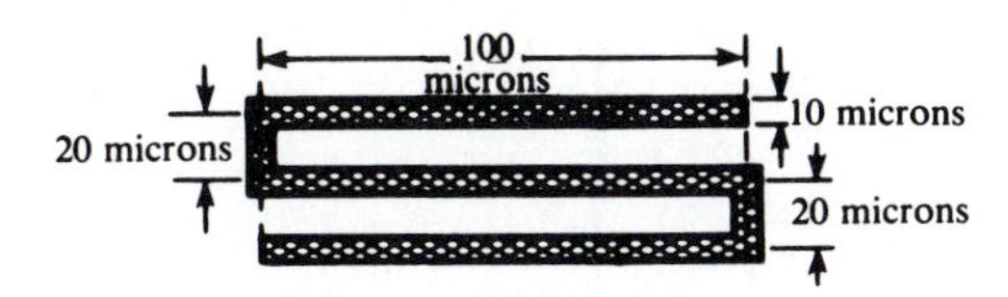

Figure 6–40
(Exercise 6–30)

total resistance 800 Ω, what should its total length be? What is its aspect ratio?

6–32. What technique is used to create an impurity layer having a very large sheet resistance? What capability of this technique makes high sheet resistance possible?

6–33. Describe two methods used to fabricate capacitors in monolithic integrated circuits.

6–34. Find the capacitance of a capacitor consisting of two conducting surfaces separated by a layer of silicon dioxide 1 μm thick. The surfaces have dimensions 80 μm × 80 μm and the permittivity (dielectric constant) of silicon dioxide is 6.6×10^{-11} F/m.

6–35. Draw a schematic diagram of the circuit corresponding to that shown in the cross-sectional view in Figure 6–41. (No buried layers are shown.)

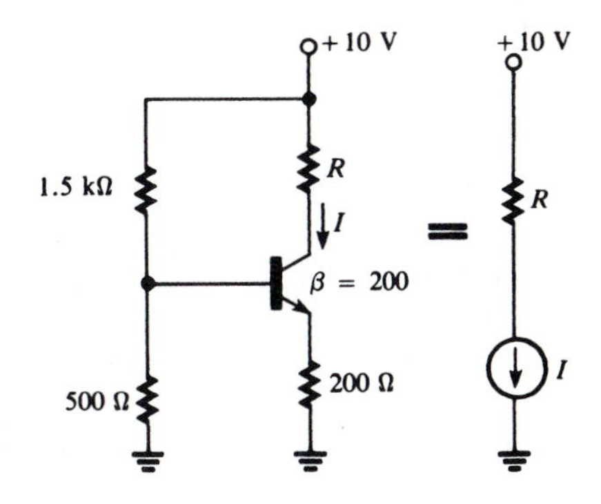

Figure 6–42
(Exercises 6–37 and 6–38)

6–39. Find the value of the current I produced by the current source shown in Figure 6–43. Assume that the transistor is operated in its active region and that $V_{BE} = 0.7$ V.

6–40. Assuming that the transistors are perfectly matched and that $I_E = 1.2$ mA, find the exact current in the base, collector, and emitter of each transistor shown in Figure 6–44. Also find the value of the constant current I in the current source. Assume that $\beta = 100$.

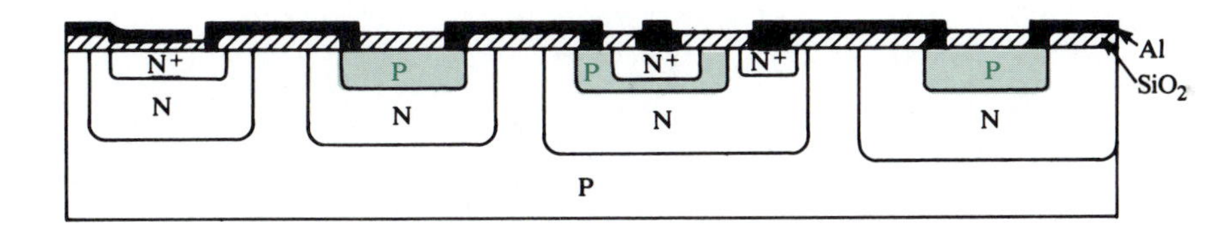

Figure 6–41
(Exercise 6–35)

Section 6–8
Integrated-Circuit Bias Techniques

6–36. Give two reasons why the bias designs used in discrete circuits are not appropriate for integrated circuits.

6–37. Find the value of the current I produced by the constant-current source shown in Figure 6–42. Assume that the transistor is operated in its active region and that $V_{BE} = 0.7$ V.

6–38. What is the largest value of R that can be used in Figure 6–42 without affecting the operation of the circuit as a constant-current source?

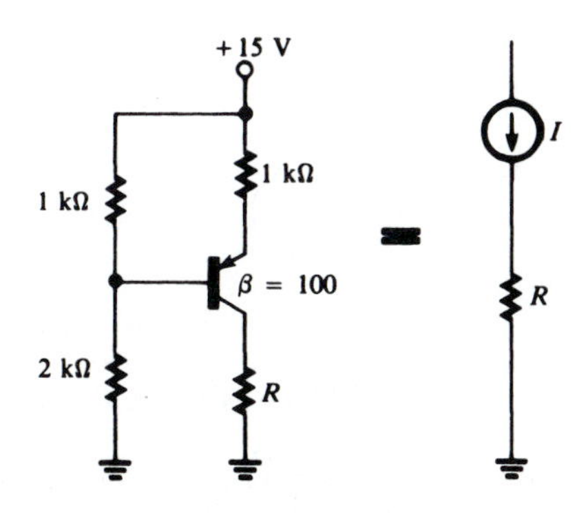

Figure 6–43
(Exercise 6–39)

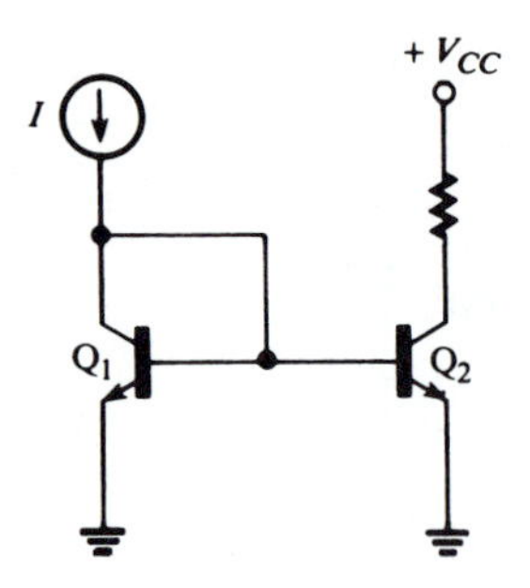

Figure 6–44
(Exercise 6–40)

6–41. Assuming that the transistors are perfectly matched and that $I_E = 2.4$ mA, find the exact current in the base, collector, and emitter of each transistor shown in Figure 6–45. Also find the value of the constant current I in the current source. Assume that $\beta = 200$.

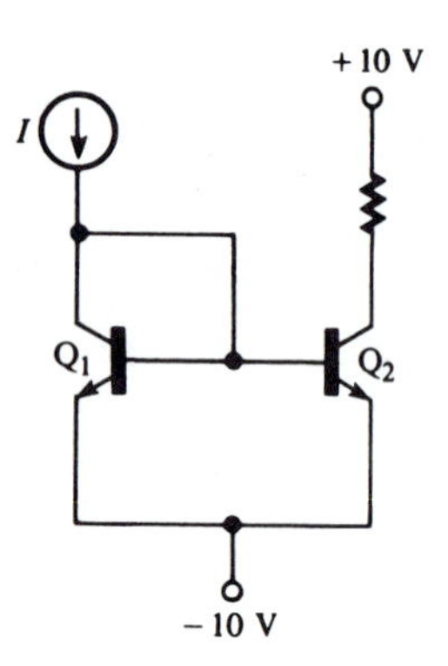

Figure 6–45
(Exercise 6–41)

6–42. Assuming that transistors Q_1 and Q_2 in Figure 6–46 are exactly matched, find (a) the approximate and (b) the exact values of I_C and V_{CE}. Assume that $\beta = 180$.

6–43. Assuming that transistors Q_1 and Q_2 in Figure 6–47 are exactly matched, find (a) the approximate and (b) the exact values of I_C and V_{CE}. Assume that $\beta = 120$.

6–44. Assuming perfectly matched transistors, find the approximate values of I_C and V_{CE} in each of transistors Q_1, Q_2, and Q_3 in Figure 6–48.

6–45. Assuming perfectly matched transistors, find the approximate values of I_C and V_{CE} in each of transistors Q_1, Q_2, and Q_3 in Figure 6–49.

6–46. When three transistors are biased from a single current source, such as those shown in Figures 6–48 and 6–49, show that the ratio

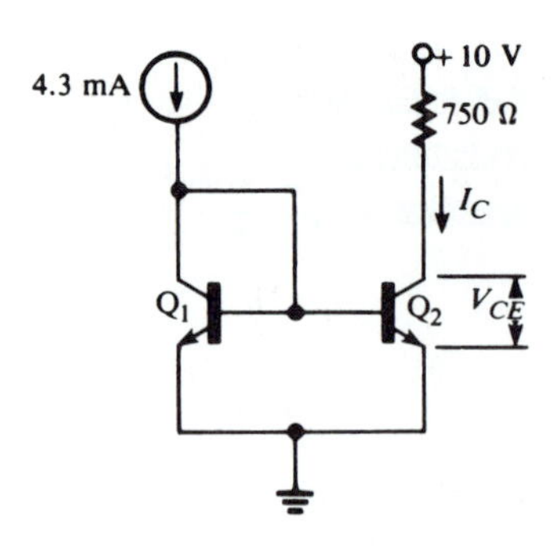

Figure 6–46
(Exercise 6–42)

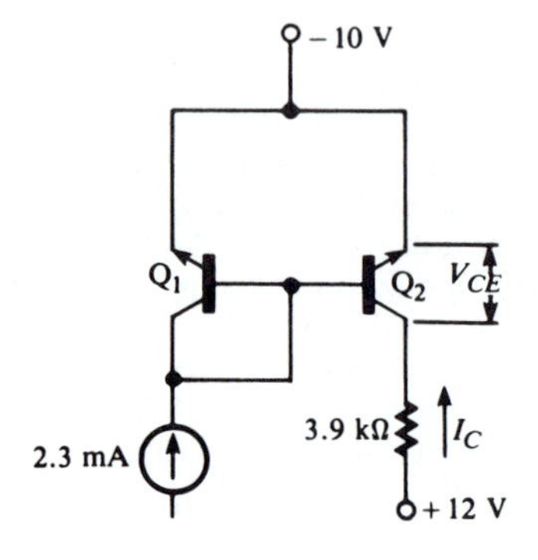

Figure 6–47
(Exercise 6–43)

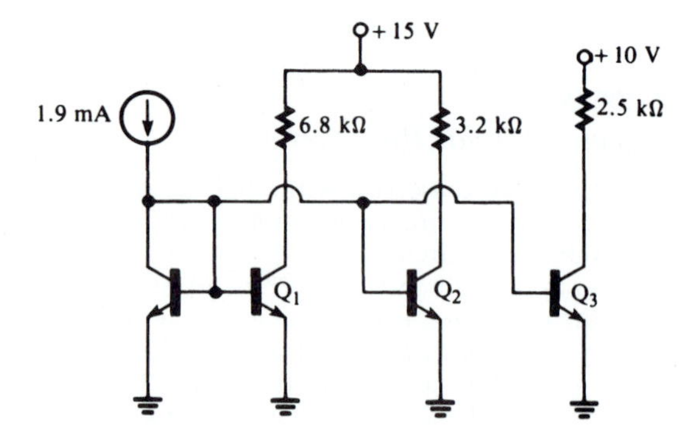

Figure 6–48
(Exercise 6–44)

of the collector current I_C in any one of them, to the source current I, is given by $I_C/I = 1/(1 + 4/\beta)$. Assume that all transistors are perfectly matched. (*Hint:* Follow the same procedure used to derive equation 6–49.)

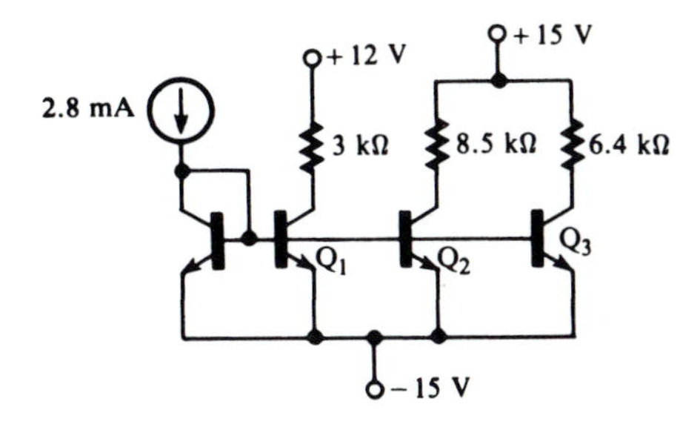

Figure 6–49
(Exercise 6–45)

SPICE EXERCISES

Note: In the exercises that follow, assume that all devices have their default parameters, unless otherwise specified. The capacitance values and frequencies given in the exercises are such that capacitive reactances are negligible and have no effect on amplifier performance.

6–47. Use SPICE to determine (approximately) the maximum temperature that the transistor in Figure 6–6(a) can be operated without creating visible distortion (clipping) in the load voltage across R_L when the input is a sine wave having peak value 1.8 V and frequency 10 kHz. Component values in the circuit are: r_s = 100 Ω, R_1 = 24 kΩ, R_2 = 4.7 kΩ, R_C = 1 kΩ, R_E = 220 Ω, and R_L = 10 kΩ. Both coupling capacitors are 10 μF. V_{CC} = 24 V. The "ideal maximum forward beta" of the transistor is 100, the "forward Early voltage" is 130 V, and the "forward and reverse beta temperature exponent" is 1.5.

6–48. Use SPICE to determine the quiescent point and voltage gain, v_L/v_S, of the collector-stabilized CE amplifier shown with equations 6–25. Component values are: r_S = 0, R_B = 100 kΩ, R_C = 2.7 kΩ, R_E = 470 Ω, and R_L = 10 kΩ. Both coupling capacitors are 10 μF. Note that R_E is not bypassed by a capacitor. V_{CC} = 15 V. The frequency of the input is 10 kHz. Compare the results obtained from SPICE with the results obtained by solving the theoretical equations (6–25).

6–49. Repeat Exercise 6–48 for the case where the emitter resistor is bypassed by a 100-μF capacitor. Also use SPICE to determine r_{in}(stage) for that case. Compare the results obtained from SPICE with the results obtained by solving the theoretical equations (6–25).

6–50. To investigate the effect of β on the current mirror shown in Figure 6–31 (Example 6–11), use SPICE to determine V_{CE} in Q_1, Q_2, and Q_3 for the case where the β of every transistor is 100 and again for the case where every β is 200. In each case, compare the results obtained from SPICE with the results shown in the example for the ideal current mirror.

7 FIELD-EFFECT TRANSISTORS

7–1 INTRODUCTION

The field-effect transistor (FET), like the bipolar junction transistor, is a three-terminal semiconductor device. However, the FET operates under principles completely different from those of the BJT. A field-effect transistor is called a *unipolar* device because the current through it results from the flow of only one of the two kinds of charge carriers: holes or electrons. The name *field effect* is derived from the fact that the current flow is controlled by an electric field set up in the device by an externally applied voltage.

There are two main types of FETs: the junction field-effect transistor (JFET) and the metal-oxide-semiconductor FET (MOSFET). We will study the theory and some practical applications of each. Both types are fabricated as discrete components and as components of integrated circuits. The MOSFET is the most important component in modern digital integrated circuits, such as microprocessors and computer memories.

7–2 JUNCTION FIELD-EFFECT TRANSISTORS

Figure 7–1 shows a diagram of the structure of a JFET and identifies the three terminals to which external electrical connections are made. As shown in the figure, a bar of N-type material has regions of P material embedded in each side. The two P regions are joined electrically and the common connection between them is called the *gate* (G) terminal. A terminal at one end of the N-type bar is called the *drain* (D), and a terminal at the other end is called the *source* (S). The region of N material between the two opposing P regions is called the *channel.* The transistor shown in the figure is therefore called an *N-channel* JFET, the type that we will study initially, while a device constructed from a P-type bar with embedded N regions is called a *P-channel* JFET. As we develop the theory of the JFET, it may be helpful at first to think of the drain as corresponding to the collector of a BJT, the source as corresponding to the emitter, and the gate as corresponding to the base. As we shall see, the voltage applied to the gate controls the flow of current between drain and source, just as the signal applied to the base of a BJT controls the flow of current between collector and emitter.

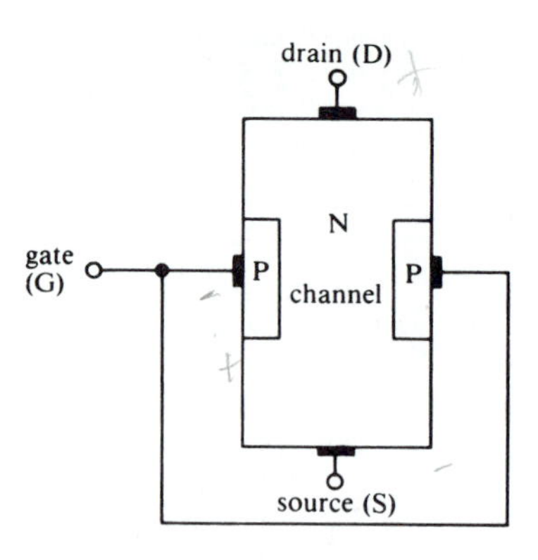

Figure 7–1
Structure of an N-channel JFET

When an external voltage is connected between the drain and the source of an N-channel JFET, so that the drain is positive with respect to the source, current is established by the flow of electrons through the N material from the source to the drain. (The source is so named because it is regarded as the origin of the electrons.) Thus, *conventional* current flows from drain to source and is limited by the resistance of the N material. In normal operation, an external voltage is applied between the gate and the source so that the PN junctions on each side of the channel are reverse biased. Thus, the gate is made negative with respect to the source, as illustrated in Figure 7–2. Note in the figure that the reverse bias causes a pair of depletion regions to form in the channel. The channel is more lightly doped than the gate, so the depletion regions penetrate more deeply into the N-type channel than into the P material of the gate.

The width of the depletion regions in Figure 7–2 depends on the magnitude of the reverse-biasing voltage V_{GS}. The figure illustrates the case where V_{GS} is only a few tenths of a volt, so the depletion regions are relatively narrow. (V_{DS} is also assumed to be relatively small; we will investigate the effect of a large V_{DS} presently.) As V_{GS} is made more negative, the depletion regions expand and the width of the channel decreases. The reduction in channel width increases the resistance of the channel and thus decreases the flow of current I_D from drain to source.

To investigate the effect of increasing V_{DS} on the drain current I_D, let us suppose for the moment that the gate is shorted to the source ($V_{GS} = 0$). As V_{DS} is increased slightly above 0, we find that the current I_D increases in direct proportion to it, as

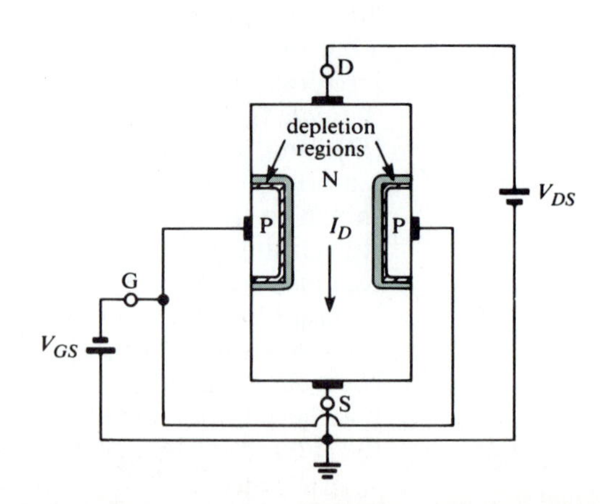

Figure 7–2
Reverse biasing the gate-to-source junctions causes the formation of depletion regions. V_{GS} is a small reverse-biasing voltage for the case illustrated.

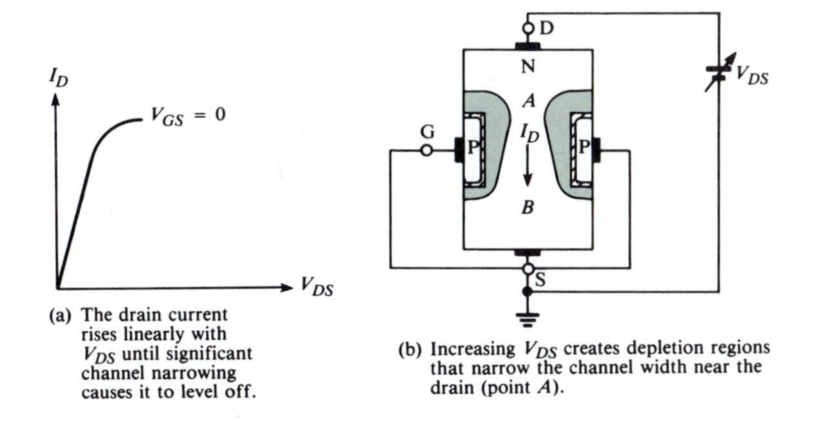

(a) The drain current rises linearly with V_{DS} until significant channel narrowing causes it to level off.

(b) Increasing V_{DS} creates depletion regions that narrow the channel width near the drain (point *A*).

Figure 7–3
Effects of increasing V_{DS} while the gate is shorted to the source ($V_{GS} = 0$)

shown in Figure 7–3(a). This is as we would expect, since increasing the voltage across the fixed-resistance channel simply causes an Ohm's law increase in the current through it. As we continue to increase V_{DS}, we find that noticeable depletion regions begin to form in the channel, as illustrated in Figure 7–3(b). Note that the depletion regions are broader near the drain end of the channel (in the vicinity of point A) than they are near the source end (point B). This is explained by the fact that current flowing through the channel creates a voltage drop along the length of the channel. Near the top of the channel, the channel voltage is very nearly equal to V_{DS}, so there is a large reverse-biasing voltage between the N channel and the P gate. As we proceed down the channel, less voltage is available because of the drop that accumulates through the resistive N material. Consequently, the reverse-biasing potential between channel and gate becomes smaller and the depletion regions become narrower as we approach the source. When V_{DS} is increased further, the depletion regions expand and the channel becomes very narrow in the vicinity of point A, causing the total resistance of the channel to increase. As a consequence, the rise in current is no longer directly proportional to V_{DS}. Instead, the current begins to level off, as shown by the curved portion of the plot in Figure 7–3(a).

Figure 7–4(a) shows what happens when V_{DS} is increased to a value large enough to cause the depletion regions to meet at a point in the channel near the drain end. This condition is called *pinch-off.* At the point where pinch-off occurs, the gate-to-channel junction is reverse biased by the value of V_{DS}, so (the negative of) this value is called the *pinch-off voltage, V_p*. The pinch-off voltage is an important JFET parameter, whose value depends on the doping and geometry of the device. V_p is always a negative quantity for an N-channel JFET and a positive quantity for a P-channel JFET. Figure 7–4(b) shows that the current reaches a maximum value at pinch-off and that it remains at that value as V_{DS} is increased beyond $|V_p|$. This current is called the *saturation* current and is designated I_{DSS}—the *D*rain-to-*S*ource current with the gate *S*horted.

Despite the implication of the name pinch-off, note again that current continues to flow through the device when V_{DS} exceeds $|V_p|$. The value of the current remains constant at I_{DSS} because of a kind of self-regulating or equilibrium process that controls the current when V_{DS} exceeds $|V_p|$: Suppose that an increase in V_{DS} *did* cause I_D to increase; then there would be in the channel an increased voltage drop

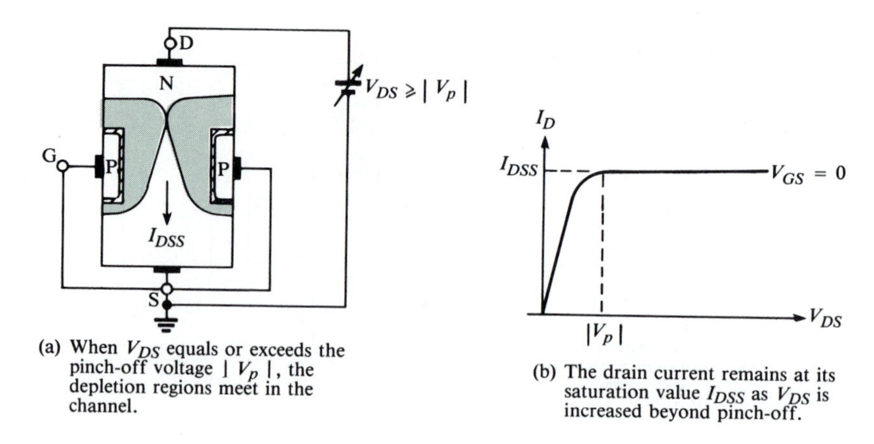

(a) When V_{DS} equals or exceeds the pinch-off voltage $|V_p|$, the depletion regions meet in the channel.

(b) The drain current remains at its saturation value I_{DSS} as V_{DS} is increased beyond pinch-off.

Figure 7–4
The N-channel JFET at pinch-off

that would expand the depletion regions further and reduce the current to its original value. Conversely, if current ceased to flow at pinch-off, the depletion region would shrink and current flow would resume. Of course, this change in current never actually occurs; I_D simply remains constant at I_{DSS}.

A typical set of values for V_p and I_{DSS} are -4 V and 12 mA, respectively. Suppose we connect a JFET having those parameter values in the circuit shown in Figure 7–5(a). Note that the gate is no longer shorted to the source, but a voltage $V_{GS} = -1$ V is connected to reverse bias the gate-to-source junctions. The reverse bias causes the depletion regions to penetrate the channel farther along the entire length of the channel than they did when V_{GS} was 0. If we now begin to increase V_{DS} above 0, we find that the current I_D once more begins to increase linearly, as shown in Figure 7–5(b). Note that the slope of this line is not as steep as that of the $V_{GS} = 0$ line, because the total resistance of the narrower channel is greater than before. As we continue to increase V_{DS}, we find that the depletion regions again approach each other in the vicinity of the drain. This further narrowing of

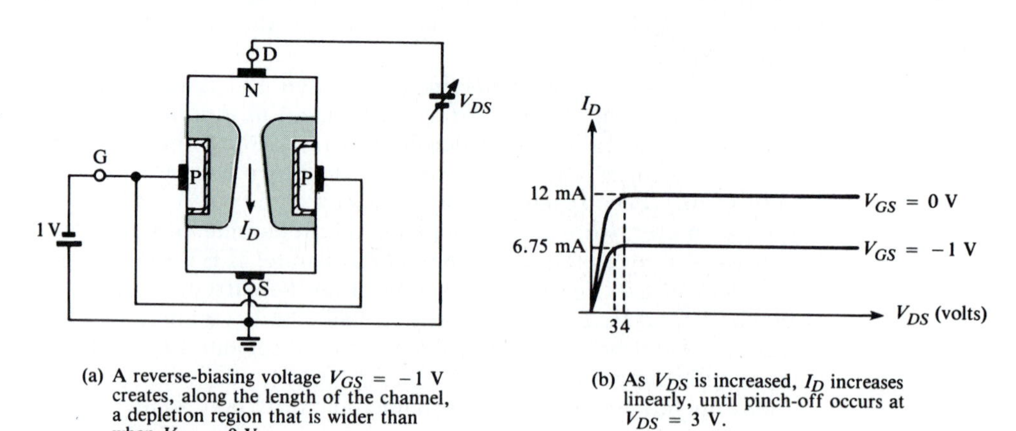

(a) A reverse-biasing voltage $V_{GS} = -1$ V creates, along the length of the channel, a depletion region that is wider than when $V_{GS} = 0$ V.

(b) As V_{DS} is increased, I_D increases linearly, until pinch-off occurs at $V_{DS} = 3$ V.

Figure 7–5
Effects of increasing V_{DS} when $V_{GS} = -1$ V

the channel increases its resistance and the current again begins to level off. Since there is already a 1-V reverse bias between the gate and the channel, the pinch-off condition, where the depletion regions meet, is now reached at $V_{DS} = 3$ V instead of 4 V ($V_{DS} = V_{GS} - V_p$). As shown in Figure 7–5(b), the current saturates at the lower value of 6.75 mA as V_{DS} is increased beyond 3 V.

If the procedure we have just described is repeated with V_{GS} set to -2 V instead of -1 V, we find that pinch-off is reached at $V_{DS} = 2$ V and that the current saturates at $I_D = 3$ mA. It is clear that increasing the reverse-biasing value of V_{GS} (making V_{GS} more negative) causes the pinch-off condition to occur at smaller values of V_{DS} and that smaller saturation currents result. Figure 7–6 shows the family of characteristic curves, the *drain characteristics,* obtained when the procedure is performed for $V_{GS} = 0, -1, -2, -3$, and -4 V. The dashed line, which is parabolic, joins the points on each curve where pinch-off occurs. A value of V_{DS} on the parabola is called a *saturation voltage* $V_{DS(sat)}$. At any value of V_{GS}, the corresponding value of $V_{DS(sat)}$ is the difference between V_{GS} and V_p: $V_{DS(sat)} = V_{GS} - V_p$, as we have already described. The equation of the parabola is

$$I_D = I_{DSS}\left(\frac{V_{DS(sat)}}{V_p}\right)^2 \tag{7–1}$$

To illustrate, we have, in our example, $V_p = -4$ V and $I_{DSS} = 12$ mA; so at $V_{DS(sat)} = 3$ V we find

$$I_D = (12\text{ mA})\left(\frac{3}{-4}\right)^2 = 6.75\text{ mA}$$

which is the saturation current at the $V_{GS} = -1$ V line (see Figure 7–5(b)). Note in Figure 7–6 that the region to the right of the parabola is called the *pinch-off region.* This is the region in which the JFET is normally operated when used for small-signal amplification. It is also called the *active* region, or the *saturation* region. The region to the left of the parabola is called the *voltage-controlled–resistance* region, the *ohmic* region, or the *triode* region. In this region, the resistance between drain and source is controlled by V_{GS}, as we have previously discussed, and we can

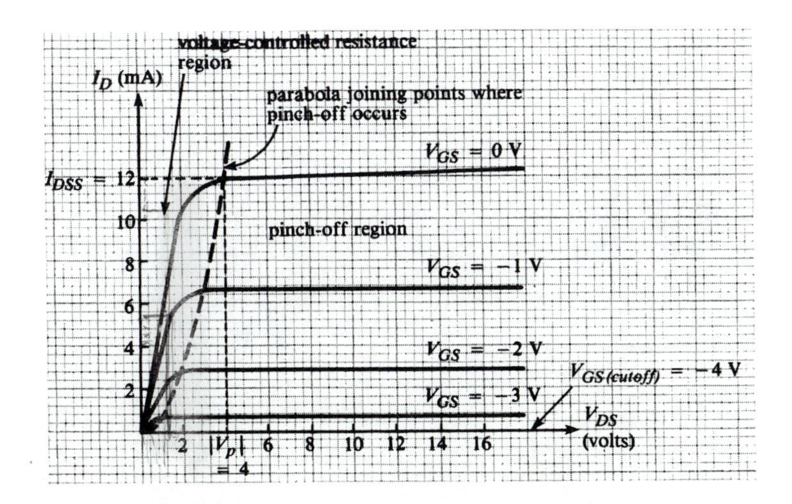

Figure 7–6
Drain characteristics of an N-channel JFET

see that the lines become less steep (implying larger resistance) as V_{GS} becomes more negative. The device acts like a voltage-controlled resistor in this region, and there are some practical applications that exploit this characteristic.

The line drawn along the horizontal axis in Figure 7–6 shows that $I_D = 0$ when $V_{GS} = -4$ V, regardless of the value of V_{DS}. When V_{GS} reverse biases the gate-to-source junction by an amount equal to V_p, depletion regions meet along the entire length of the channel and the drain current is cut off. Since the value of V_{GS} at which the drain current is cut off is the same as V_p, the pinch-off voltage is also called the *gate-to-source cutoff voltage.* Thus, there are two ways to determine the value of V_p from a set of drain characteristics: It is the value of V_{DS} where I_D saturates when $V_{GS} = 0$, and it is the value of V_{GS} that causes all drain current to cease, i.e., $V_p = V_{GS(cutoff)}$.

One property of a field-effect transistor that makes it especially valuable as a voltage amplifier is the very high input resistance at its gate. Since the path from gate to source is a reverse-biased PN junction, the only current that flows into the gate is the very small leakage current associated with a reverse-biased junction. Therefore, very little current is drawn from a signal source driving the gate, and the FET input looks like a very large resistance. A dc input resistance of several hundred megohms is not unusual. Although the gate of an N-channel JFET can be driven slightly positive, this action causes the input junction to be forward biased and radically decreases the gate-to-source resistance. In most practical applications, the sudden and dramatic decrease in resistance when the gate is made positive would not be tolerable to a signal source driving a FET.

Figure 7–7 shows the structure and drain characteristics of a typical P-channel JFET. Since the channel is P material, current is due to hole flow, rather than electron flow, between drain and source. The gate material is, of course, N type. Note that all voltage polarities are opposite those in the N-channel JFET. Figure 7–7(b) shows that positive values of V_{GS} control the amount of saturation current in the pinch-off region.

Figure 7–8 shows the schematic symbols used to represent N-channel and P-channel JFETs. Note that the arrowhead on the gate points into an N-channel JFET and outward for a P-channel device. The symbols showing the gate terminal off-center are used as a means of identifying the source: the source is the terminal

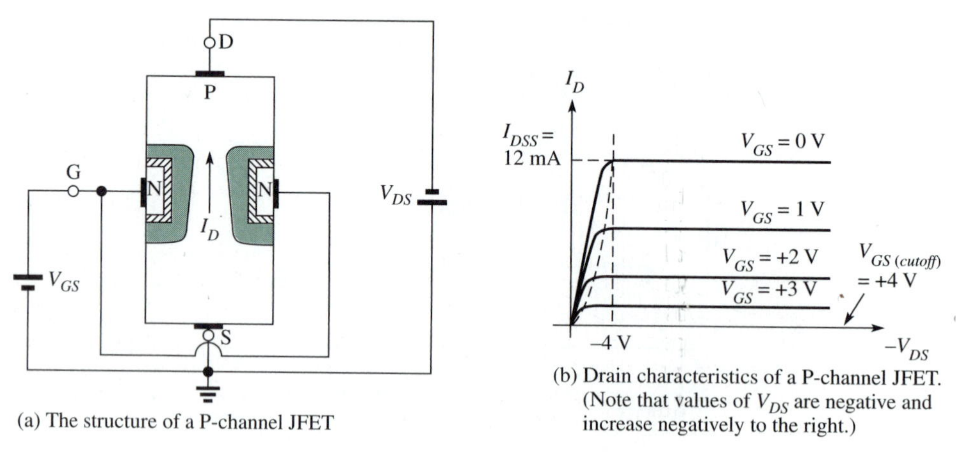

(a) The structure of a P-channel JFET

(b) Drain characteristics of a P-channel JFET. (Note that values of V_{DS} are negative and increase negatively to the right.)

Figure 7–7
Structure and characteristics of a P-channel JFET

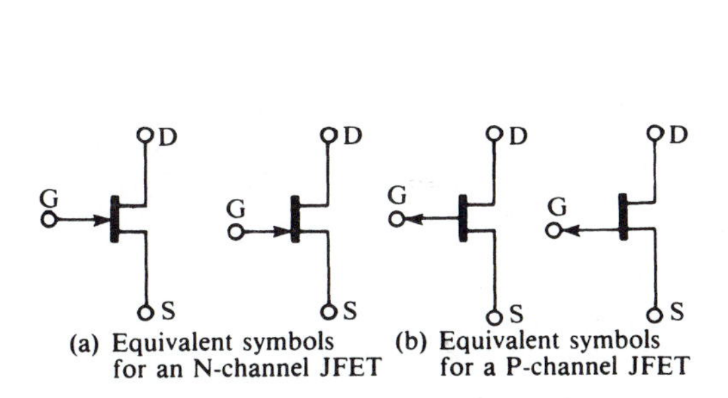

Figure 7–8
Schematic symbols for JFETs

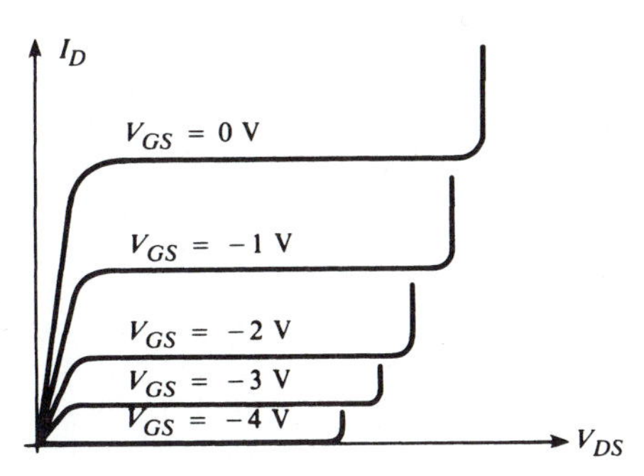

Figure 7–9
Breakdown characteristics of an N-channel JFET

drawn closest to the gate arrow. Some JFETs are manufactured so that the drain and source are interchangeable, and the symbols for these devices have the gate arrow drawn in the center.

Figure 7–9 shows the breakdown characteristics of an N-channel JFET. Breakdown occurs at large values of V_{DS} and is caused by the avalanche mechanism described in connection with BJTs. Note that the larger the magnitude of V_{GS}, the smaller the value of V_{DS} at which breakdown occurs.

Transfer Characteristics

The *transfer characteristic* of a JFET is a plot of output current versus input voltage, for a fixed value of output voltage. When the input to a JFET is the gate-to-source voltage and the output current is drain current (common-source configuration), the transfer characteristic can be derived from the drain characteristics. It is only necessary to construct a vertical line on the drain characteristics (a line of constant V_{DS}) and to note the value of I_D at each intersection of the line with a line of constant V_{GS}. The values of I_D can then be plotted against the values of V_{GS} to construct the transfer characteristic. Figure 7–10 illustrates the process.

In Figure 7–10, the transfer characteristic is shown for $V_{DS} = 8$ V. As can be seen in the figure, this choice of V_{DS} means that all points are in the pinch-off region. For example, the point of intersection of the $V_{DS} = 8$ V line and the $V_{GS} = 0$ V line occurs at $I_D = I_{DSS} = 12$ mA. At $V_{DS} = 8$ V and $V_{GS} = -1$ V, we find $I_D = 6.75$ mA. Plotting these combinations of I_D and V_{GS} produces the parabolic transfer characteristic shown. The nonlinear shape of the transfer characteristic can be anticipated by observing that equal increments in the values of V_{GS} on the drain characteristics ($\triangle V_{GS} = 1$ V) do not produce equally spaced lines. (Recall from our discussion of BJT output characteristics that this situation creates output signal distortion when the device is used as an ac amplifier; practical JFET circuits incorporate a means for reducing distortion, at the expense of gain, that we will discuss in a later chapter.) Note that the intercepts of the transfer characteristic are I_{DSS} on the I_D-axis and V_p on the V_{GS}-axis.

The equation for the transfer characteristic *in the pinch-off region* is, to a close approximation,

$$I_D = I_{DSS}\left(1 - \frac{V_{GS}}{V_p}\right)^2 \tag{7–2}$$

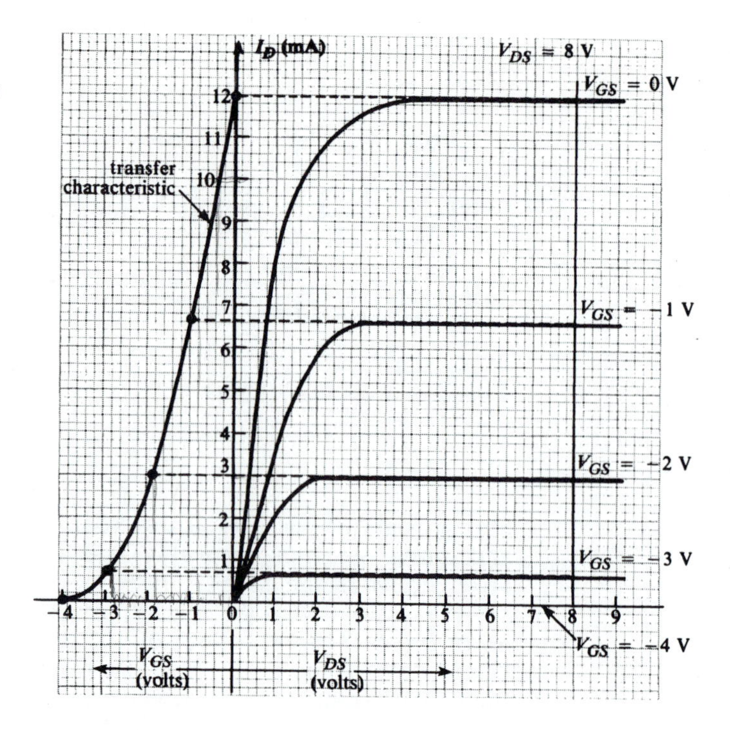

Figure 7–10
Construction of an N-channel transfer characteristic from the drain characteristics

Note that equation 7–2 correctly predicts that $I_D = I_{DSS}$ when $V_{GS} = 0$ and that $I_D = 0$ when $V_{GS} = V_p$. The transfer characteristic is often called the *square-law* characteristic of a JFET and is used in some interesting applications to produce outputs that are nonlinear functions of inputs.

Example 7–1

An N-channel JFET has a pinch-off voltage of −4.5 V and $I_{DSS} = 9$ mA.

1. At what value of V_{GS} in the pinch-off region will I_D equal 3 mA?
2. What is the value of $V_{DS(sat)}$ when $I_D = 3$ mA?

Solution

1. We must solve equation 7–2 for V_{GS}:

$$(1 - V_{GS}/V_p)^2 = I_D/I_{DSS}$$
$$1 - V_{GS}/V_p = \sqrt{I_D/I_{DSS}}$$
$$V_{GS} = V_p(1 - \sqrt{I_D/I_{DSS}})$$
$$V_{GS} = -4.5\left[1 - \sqrt{(3\text{ mA})/(9\text{ mA})}\right] = -1.9\text{ V}$$

2. Equation 7–1 relates I_D and $V_{DS(sat)}$. Solving for $V_{DS(sat)}$, we find

$$V_{DS(sat)} = \sqrt{(V_p)^2 I_D/I_{DSS}} = \sqrt{(4.5)^2(3\text{ mA})/(9\text{ mA})} = 2.6\text{ V}$$

Note that we use the positive square root, since V_{DS} is positive for an N-channel JFET. For a P-channel JFET, we would use the negative root. The value of $V_{DS(sat)}$ could also have been determined from the fact that $V_{DS(sat)} = V_{GS} - V_p = -1.9 - (-4.5) = 2.6$ V.

Example 7–2

SPICE

Use SPICE to obtain a plot of the transfer characteristic of an N-channel JFET having I_{DSS} = 10 mA and V_p = −2 V. The characteristic should be plotted for V_{DS} = 10 V.

Solution. Figure 7–11(a) shows a SPICE circuit that can be used to obtain the desired characteristic. The value of BETA in the .MODEL statement is found

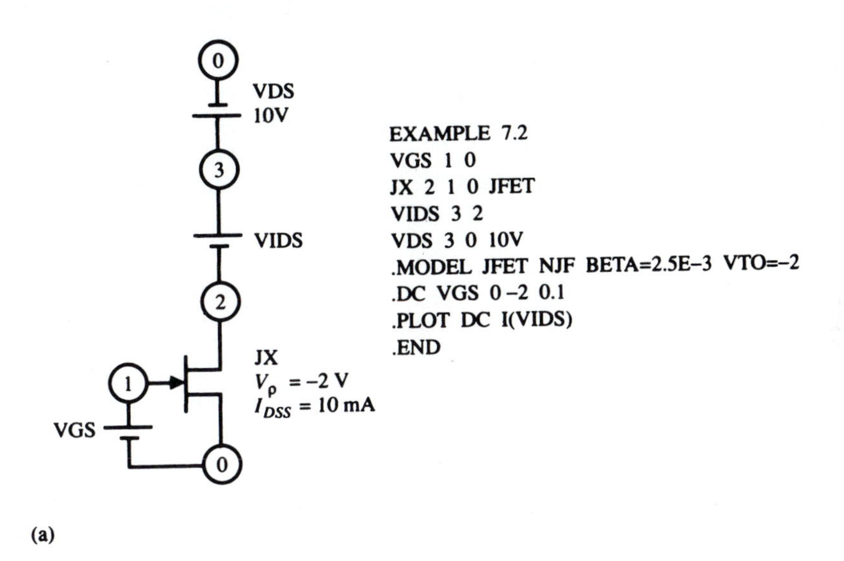
0
VDS
10V
3
VIDS
2
JX
V_p = −2 V
I_{DSS} = 10 mA
1
VGS
0
EXAMPLE 7.2
VGS 1 0
JX 2 1 0 JFET
VIDS 3 2
VDS 3 0 10V
.MODEL JFET NJF BETA=2.5E-3 VTO=-2
.DC VGS 0 -2 0.1
.PLOT DC I(VIDS)
.END
(a)

```
EXAMPLE 7.2
****      DC TRANSFER CURVES                       TEMPERATURE =    27.000 DEG C
*******************************************************************************
  VGS          I(VIDS)
                  0.000D+00       5.000D-03      1.000D-02      1.500D-02  2.000D-02
                - - - - - - - - - - - - - - - - - - - - - - - - - - - - - - - - - -
0.000D+00  1.000D-02 .              .              *              .              .
1.000D-01  9.025D-03 .              .           *  .              .              .
2.000D-01  8.100D-03 .              .         *    .              .              .
3.000D-01  7.225D-03 .              .      *       .              .              .
4.000D-01  6.400D-03 .              .    *         .              .              .
5.000D-01  5.625D-03 .              .  *           .              .              .
6.000D-01  4.900D-03 .              *              .              .              .
7.000D-01  4.225D-03 .            * .              .              .              .
8.000D-01  3.600D-03 .          *   .              .              .              .
9.000D-01  3.025D-03 .        *     .              .              .              .
1.000D+00  2.500D-03 .       *      .              .              .              .
1.100D+00  2.025D-03 .      *       .              .              .              .
1.200D+00  1.600D-03 .   *          .              .              .              .
1.300D+00  1.225D-03 .  *           .              .              .              .
1.400D+00  9.000D-04 .  *           .              .              .              .
1.500D+00  6.250D-04 . *            .              .              .              .
1.600D+00  4.000D-04 .*             .              .              .              .
1.700D+00  2.250D-04 .*             .              .              .              .
1.800D+00  1.000D-04 *              .              .              .              .
1.900D+00  2.500D-05 *              .              .              .              .
2.000D+00  1.201D-11 *              .              .              .              .
                - - - - - - - - - - - - - - - - - - - - - - - - - - - - - - - - - -
```

(b)

Figure 7–11
(Example 7–2)

from

$$\beta = \frac{I_{DSS}}{V_p^2} = \frac{10 \times 10^{-3}\ \text{A}}{(-2\ \text{V})^2} = 2.5 \times 10^{-3}\ \text{A/V}^2$$

Note that we step VGS from 0 to −2 V in 0.1-V increments. The voltage source labeled VGS has its positive terminal connected to the gate of the FET, but the stepped voltages should all be negative. (Alternatively, we could reverse the polarity of VGS and step it through positive voltages.) The voltage source labeled VIDS is a dummy source used to obtain positive values of drain current. The .PLOT statement produces a plot of I(VIDS) (i.e., I_{DS}) versus V_{GS}, which constitutes a transfer characteristic.

Figure 7–11(b) shows the plot produced by SPICE. (Rotate it 180° to obtain the orientation shown in Figure 7–10.) Note that $I_D = 10$ mA $= I_{DSS}$ when $V_{GS} = 0$ and that $I_D \approx 0$ when $V_{GS} = V_p$. Although the values of V_{GS} printed down the left margin should all be negative, SPICE prints only the absolute values of the independent variable associated with a .PLOT statement, which is V_{GS} in this example.

7–3 JFET BIASING

Fixed Bias

Like a bipolar transistor, a JFET used as an ac amplifier must be biased in order to create a dc output voltage around which ac variations can occur. When a JFET is connected in the *common-source* configuration, the input voltage is V_{GS} and the output voltage is V_{DS}. Therefore, the bias circuit must set dc (quiescent) values for the drain-to-source voltage V_{DS} and drain current I_D. Figure 7–12 shows one method that can be used to bias N-channel and P-channel JFETs.

Notice in Figure 7–12 that a dc supply voltage V_{DD} is connected to supply drain current to the JFET through resistor R_D, and that another dc voltage is used to set the gate-to-source voltage V_{GS}. This biasing method is called *fixed bias* because the

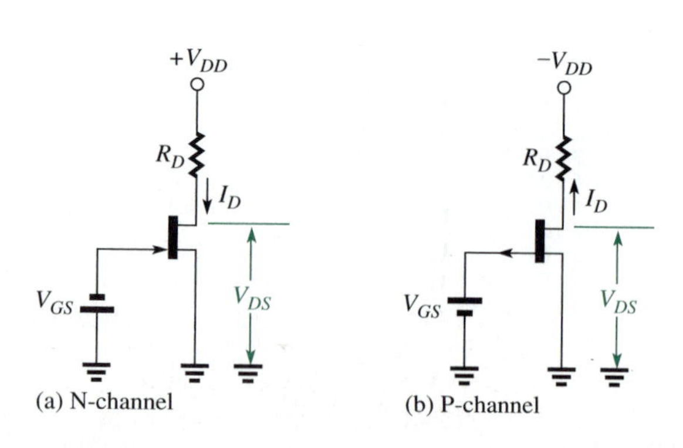

Figure 7–12
Fixed-bias circuits for N- and P-channel JFETs

gate-to-source voltage is fixed by the constant voltage applied across those terminals. Writing Kirchhoff's voltage law around the output loops in Figure 7–12, we find

$$V_{DS} = V_{DD} - I_D D_D \qquad \text{(N-channel)}$$
$$V_{DS} = -V_{DD} + I_D R_D \qquad \text{(P-channel)} \qquad \textbf{(7–3)}$$

When using these equations, always substitute a positive value for V_{DD} to ensure that the correct sign is obtained for V_{DS}. V_{DS} should always turn out to be a positive quantity in an N-channel JFET and a negative quantity in a P-channel JFET. For example, in an N-channel device where V_{DD} is +15 V from drain to ground, if $I_D = 10$ mA, and $R_D = 1$ kΩ, we have $V_{DS} = 15 - (10\text{ mA})(1\text{ k}\Omega) = +5$ V. For a P-channel device where V_{DD} is −15 V from drain to ground, $V_{DS} = -15 + (10\text{ mA}) \times (1\text{ k}\Omega) = -5$ V. Equations 7–3 can be rewritten in the form

$$I_D = -(1/R_D)V_{DS} + V_{DD}/R_D \qquad \text{(N-channel)}$$
$$I_D = (1/R_D)V_{DS} + V_{DD}/R_D \qquad \text{(P-channel)} \qquad \textbf{(7–4)}$$

Equations 7–4 are the equations of the dc load lines for N- and P-channel JFETs and each can be plotted on a set of drain characteristics to determine a Q-point. This technique is the same as the one we used to determine the Q-point in a BJT bias circuit. The load line intersects the V_{DS}-axis at V_{DD} and the I_D-axis at V_{DD}/R_D.

Example 7–3

The JFET in the circuit of Figure 7–13 has the drain characteristics shown in Figure 7–14. Find the quiescent values of I_D and V_{DS} when (1) $V_{GS} = -1.5$ V, and (2) $V_{GS} = -0.5$ V.

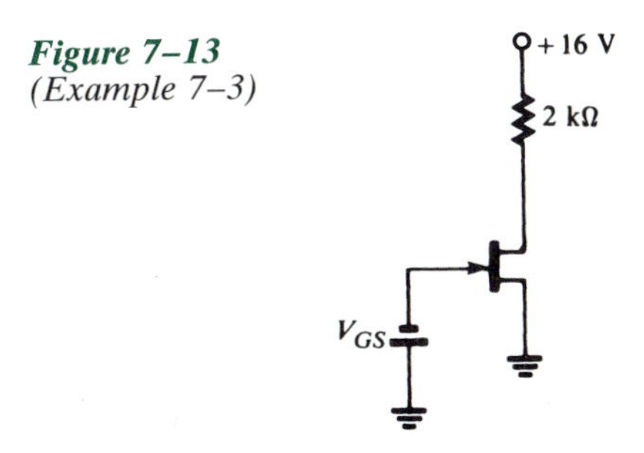

Figure 7–13
(Example 7–3)

Solution

1. The load line intersects the V_{DS}-axis at $V_{DD} = +16$ V and the I_D-axis at $I_D = (16\text{ V})/(2\text{ k}\Omega) = 8$ mA. It is plotted on Figure 7–14.
   At the intersection of the load line with $V_{GS} = -1.5$ V (labeled Q_1), we find the quiescent values $I_D \approx 4$ mA and $V_{DS} \approx 8$ V.
2. The load line is, of course, the same as in part (1). Changing V_{GS} to −0.5 V moves the Q-point to the point labeled Q_2 in Figure 7–14. Here we see that $I_D \approx 6.8$ mA and $V_{DS} \approx 2.4$ V.

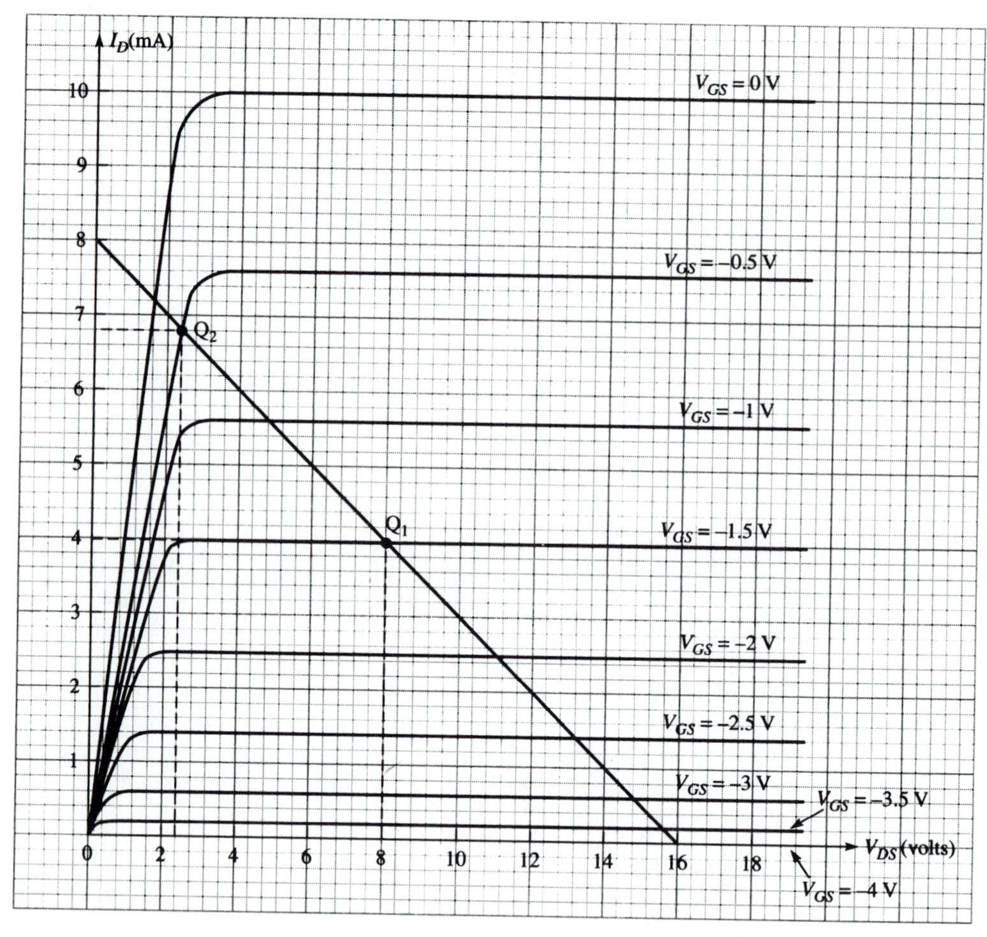

Figure 7–14
(Example 7–3)

Part 2 of the preceding example illustrates an important result. Note that changing V_{GS} to -0.5 V in the bias circuit of Figure 7–13 caused the Q-point to move out of the pinch-off region and into the voltage-controlled–resistance region. As we have already mentioned, the Q-point must be located in the pinch-off region for normal amplifier operation. *To ensure that the Q-point is in the pinch-off region, the quiescent value of* $|V_{DS}|$ *must be greater than* $|V_p| - |V_{GS}|$. The pinch-off voltage for the device whose characteristics are given in Figure 7–14 can be seen to be approximately -4 V. Since $|V_{GS}| = 0.5$ V and the quiescent value of V_{DS} at Q_2 is only 2.4 V, we do not satisfy the requirement $|V_{DS}| > |V_p| - |V_{GS}|$. Q_2 is therefore in the variable-resistance region.

Of course, the quiescent value of I_D can also be determined using the transfer characteristic of a JFET. Since the transfer characteristic is a plot of I_D versus V_{GS}, it is only necessary to locate the V_{GS} coördinate and read the corresponding value of I_D directly. The value of V_{DS} can then be determined using equation 7–3. While graphical techniques for locating the bias point are instructive and provide insights to the way in which the circuit variables affect each other, the quiescent values of I_D and V_{DS} can be calculated using a straightforward computation, if the values of V_p and I_{DSS} are known. The next example illustrates that the square-law characteristic is used in this computation.

Example 7–4

Given that the JFET in Figure 7–13 has $I_{DSS} = 10$ mA and $V_p = -4$ V, compute the quiescent values of I_D and V_{DS} when $V_{GS} = -1.5$ V. Assume that it is biased in the pinch-off region.

Solution. From equation 7–2,

$$I_D = I_{DSS}(1 - V_{GS}/V_p)^2 = (10\,\text{mA})\left(1 - \frac{-1.5}{-4}\right)^2 = 3.9\,\text{mA}$$

From equation 7–3, $V_{DS} = V_{DD} - I_D R_D = 16 - (3.9\text{ mA})(2\text{ k}\Omega) = 8.2$ V. These results are in close agreement with those obtained graphically in Example 7–3. Note that it was necessary to assume that the JFET is biased in the pinch-off region, to justify the use of equation 7–2. If the computation had produced a value of V_{DS} less than $|V_p| - |V_{GS}| = 2.5$ V, we would have had to conclude that the device is not biased in pinch-off and would then have had to use another means to find the Q-point.

The values of I_{DSS} and V_p are likely to vary widely among JFETs of a given type. A variation of 50% is not unusual. When the fixed-bias circuit is used to set a Q-point, a change in the parameter values of the JFET for which the circuit was designed (caused, for example, by substitution of another JFET) can result in an intolerable shift in quiescent values. Suppose, for example, that a JFET having parameters $I_{DSS} = 13$ mA and $V_p = -4.3$ V is substituted into the bias circuit of Example 7–3 (Figure 7–13), with V_{GS} once again set to -1.5 V. Then

$$I_D = (13\,\text{mA})\left(1 - \frac{-1.5}{-4.3}\right)^2 = 5.51\,\text{mA}$$

$$V_{DS} = 16 - (5.51\,\text{mA})(2\,\text{k}\Omega) = 4.98\,\text{V}$$

These results show that I_D increases 41.3% over the value obtained in Example 7–3 and that V_{DS} decreases 68.7%. Note also that the value of V_{DS} (4.98 V) is now perilously near the pinch-off voltage (4.3 V). We conclude that the fixed-bias circuit does not provide good Q-point stability against changes in JFET parameters.

Figure 7–15 shows a bias circuit that provides improved stability and requires only a single supply voltage. This bias method is called *self-bias,* because the voltage drop across R_S due to the flow of quiescent current determines the quiescent value of V_{GS}. We can understand this fact by realizing that the current I_D in resistor R_S creates the voltage $V_S = I_D R_S$ at the source terminal, with respect to ground. For the N-channel JFET, this means that the source is positive with respect to the gate, since the gate is grounded. In other words, the gate is negative (by $I_D R_S$ volts) with respect to the source, as required for biasing an N-channel JFET: $V_{GS} = -I_D R_S$. For the P-channel device, the gate is positive by $I_D R_S$ volts, with respect to the source: $V_{GS} = I_D R_S$.

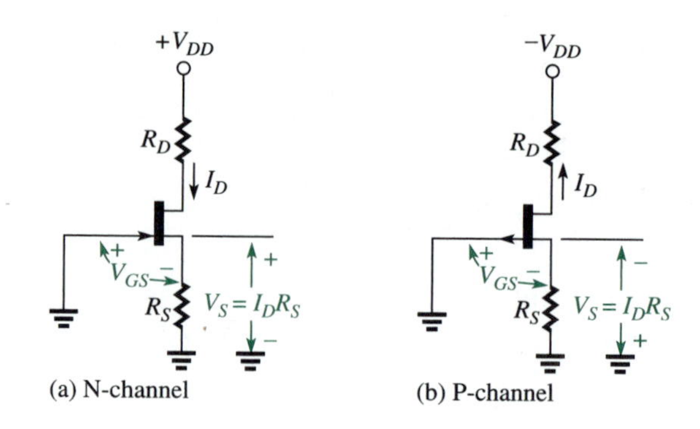

Figure 7–15
Self-bias circuits

The equations

$$V_{GS} = -I_D R_S \qquad \text{(N-channel)} \tag{7–5}$$

$$V_{GS} = I_D R_S \qquad \text{(P-channel)} \tag{7–6}$$

describe straight lines when plotted on V_{GS}–I_D-axes. (Verify these equations by writing Kirchhoff's voltage law around each gate-to-source loop in Figure 7–15.) Each line is called the *bias line* for its respective type. The quiescent value of I_D in the self-bias circuit can be determined graphically by plotting the bias line on the same set of axes with the transfer characteristic. The intersection of the two locates the Q-point. In effect, we solve the bias-line equation and the square-law equation simultaneously by finding the point where their graphs intersect. The quiescent value of V_{DS} can be found by summing voltages (writing Kirchhoff's voltage law) around the output loops in Figure 7–15:

$$\begin{aligned} V_{DS} &= V_{DD} - I_D(R_D + R_S) \qquad \text{(N-channel)} \\ V_{DS} &= -V_{DD} + I_D(R_D + R_S) \qquad \text{(P-channel)} \end{aligned} \tag{7–7}$$

The next example illustrates the graphical procedure.

Example 7–5

The transfer characteristic of the JFET in Figure 7–16 is given in Figure 7–17. Determine the quiescent values of I_D and V_{DS} graphically.

Solution. Since $R_S = 600\ \Omega$, the equation of the bias line is $V_{GS} = -600 I_D$. It is clear that the bias line always passes through the origin ($I_D = 0$ when $V_{GS} = 0$), so

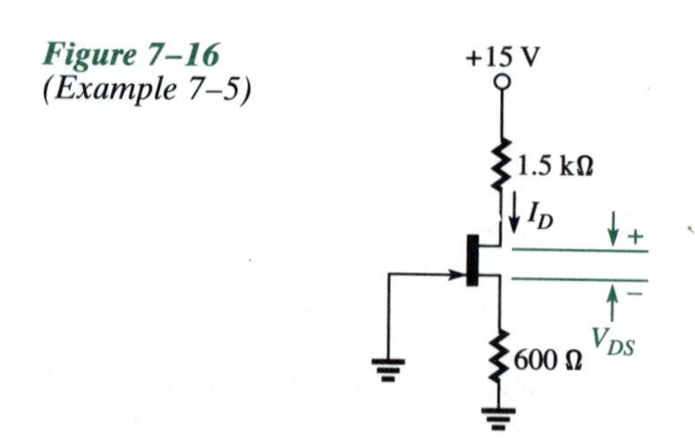

Figure 7–16
(Example 7–5)

Figure 7–17
(Example 7–5)

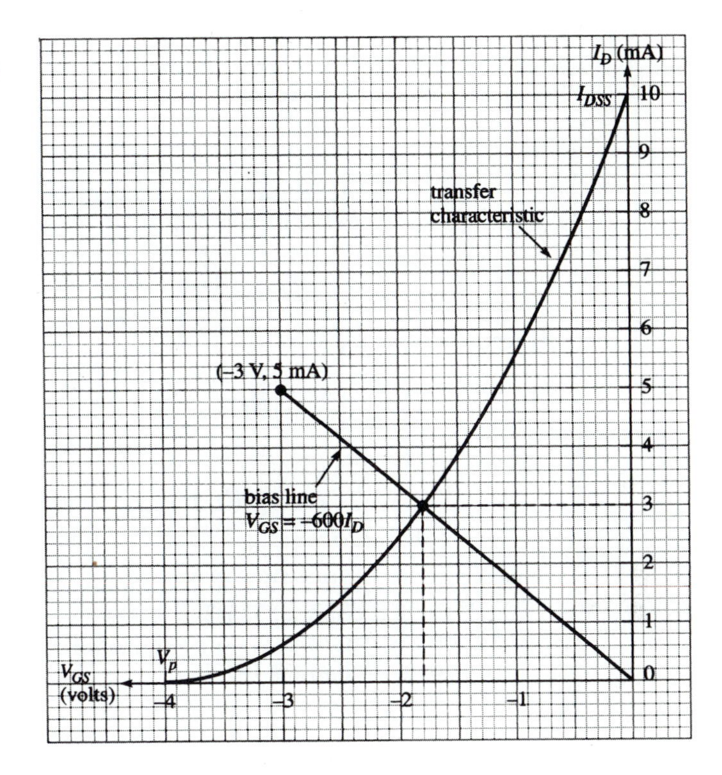

(0,0) is one point on the line. To determine another point on the line, choose a convenient value of V_{GS} and solve for I_D. In this example, if we let $V_{GS} = -3$, then

$$I_D = \frac{-V_{GS}}{600\ \Omega} = \frac{-(-3\ \text{V})}{600\ \Omega} = 5\ \text{mA}$$

Thus, (−3 V, 5 mA) is another point on the bias line. We can then draw a straight line between the two points (0,0) and (−3 V, 5 mA) and note where that line intersects the transfer characteristic. The line is plotted on the transfer characteristic shown in Figure 7–17. We note that it intersects the characteristic at $I_D \approx 3$ mA, which is the quiescent drain current. The corresponding value of V_{GS} is seen to be approximately −1.8 V. The quiescent value of V_{DS} is found from equation 7–7:

$$V_{DS} = 15\ \text{V} - (3\ \text{mA})[(1.5\ \text{k}\Omega) + (0.6\ \text{k}\Omega)] = 8.7\ \text{V}$$

General Algebraic Solution—Self-Bias

The quiescent values of I_D and V_{GS} in the self-bias circuit can also be computed algebraically by solving the bias-line equation and the square-law equation simultaneously. To perform the computation, we must know the values of I_{DSS} and V_p. As in the fixed-bias case, the results are valid only if the Q-point is in the pinch-off region, i.e., if $|V_{DS}| > |V_p| - |V_{GS}|$. We must therefore assume that to be the case, but discard the results if the computation reveals the quiescent value of $|V_{DS}|$ to be less than $|V_p| - |V_{GS}|$. Equations 7–8 give the general form of the algebraic solution for the quiescent values of I_D, V_{DS}, and V_{GS} in the self-bias circuit. Since absolute values are used in the computations, the equations are valid for both P-channel and N-channel devices.

General algebraic solution for the bias point of self-biased JFET circuits

$$I_D = \frac{-B - \sqrt{B^2 - 4AC}}{2A}$$

where
$$A = R_S^2$$
$$B = -\left(2|V_p|R_S + \frac{V_p^2}{I_{DSS}}\right)$$
$$C = V_p^2$$
$$|V_{DS}| = |V_{DD}| - I_D(R_D + R_S) \quad \text{See note 1.}$$
$$|V_{GS}| = I_D R_S \quad \text{See note 2.} \qquad \textbf{(7–8)}$$

Note 1. V_{DS} is positive for an N-channel JFET and negative for a P-channel JFET.
Note 2. V_{GS} is negative for an N-channel JFET and positive for a P-channel JFET.

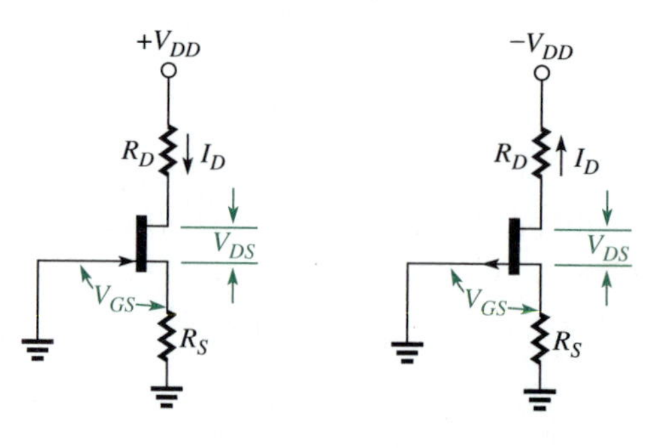

Example 7–6

Use equations 7–8 to find the bias point that was determined graphically in Example 7–5.

Solution. As shown in Figure 7–16, $R_S = 600\ \Omega$ and $R_D = 1.5\ \text{k}\Omega$. Also, the transfer characteristic in Figure 7–17 shows that $I_{DSS} = 10$ mA and $V_p = -4$ V. Thus, with reference to equations 7–8, we find

$$A = R_S^2 = (600)^2 = 3.6 \times 10^5$$
$$B = -\left(2|V_p|R_S + \frac{V_p^2}{I_{DSS}}\right) = -\left[2(4)(600) + \frac{(-4)^2}{10 \times 10^{-3}}\right] = -6.4 \times 10^3$$
$$C = V_p^2 = (-4)^2 = 16$$

$$I_D = \frac{-B - \sqrt{B^2 - 4AC}}{2A}$$
$$= \frac{6.4 \times 10^3 - \sqrt{40.96 \times 10^6 - 4(3.6 \times 10^5)(16)}}{2(3.6 \times 10^5)} = 3.0 \text{ mA}$$
$$|V_{DS}| = |V_{DD}| - I_D(R_D + R_S) = 15 \text{ V} - 3 \text{ mA}(1.5 \text{ k}\Omega + 600 \ \Omega) = 8.7 \text{ V}$$
$$|V_{GS}| = I_D R_S = (3 \text{ mA})(600 \ \Omega) = 1.8 \text{ V}$$

Since the JFET is N-channel, $V_{GS} = -1.8$ V. These results agree well with those found in Example 7–5. Since $|V_{DS}| = 8.7 \text{ V} > |V_p| - |V_{GS}| = 4 \text{ V} - 1.8 \text{ V} = 2.2$ V, we know the bias point is in the pinch-off region and the results are valid.

To demonstrate that the self-bias method provides better stability than the fixed-bias method, let us compare the shift in the quiescent value of I_D that occurs using each method, when the JFET parameters of the previous example are changed to $I_{DSS} = 12$ mA and $V_p = -4.5$ V. In each case, we will assume that the initial bias point (using a JFET with $I_{DSS} = 10$ mA and $V_p = -4$ V) is set so that $I_D = 3$ mA and that a JFET having the new parameters is then substituted in the circuit. We have already seen that $I_D = 3$ mA when $V_{GS} = -1.8$ V, so let us suppose that a fixed-bias circuit has V_{GS} set to -1.8 V. When I_{DSS} changes to 12 mA and V_p to -4.5 V, with V_{GS} fixed at -1.8 V, we find that the new value of I_D in the fixed-bias circuit is

$$I_D = I_{DSS}\left(1 - \frac{V_{GS}}{V_p}\right)^2 = (12 \times 10^{-3} \text{ A})\left(1 - \frac{1.8}{4.5}\right)^2 = 4.32 \text{ mA}$$

This change in I_D from 3 mA to 4.32 mA represents a 44% increase.

Suppose now that the JFET parameters in the self-bias circuit change by the same amount: $I_{DSS} = 12$ mA and $V_p = -4.5$ V. Using equations 7–8, we find $I_D = 3.46$ mA. In this case, the increase in I_D is 15.3%, less than half that of the fixed-bias design.

Figure 7–18 shows the transfer characteristic of the JFET having $I_{DSS} = 10$ mA and $V_p = -4$ V, and the transfer characteristic corresponding to the JFET with $I_{DSS} = 12$ mA and $V_p = -4.5$ V. The self-bias line $V_{GS} = -600I_D$ is shown intersecting both characteristics. Note that it intersects these characteristics at the values of I_D previously calculated for the self-bias circuit: 3 mA and 3.46 mA. Also shown is the vertical line corresponding to $V_{GS} = -1.8$ V—i.e., the line corresponding to the fixed-bias condition. This line intersects the characteristics at the two values previously calculated for the fixed-bias circuit: 3 mA and 4.32 mA. It is apparent in this figure why the self-bias method produces a smaller change in I_D than the fixed-bias method when a change in JFET parameters results in a different transfer characteristic: The smaller the slope of the bias line, the smaller the change in I_D.

Voltage-Divider Bias

Our interpretation of Figure 7–18 reveals that good bias stability against changes in JFET parameters can be achieved by making the slope of the bias line as small as possible. The slope of this line becomes smaller as R_S is made larger, but large values of R_S can result in unacceptably small values of I_D. One way to obtain a bias

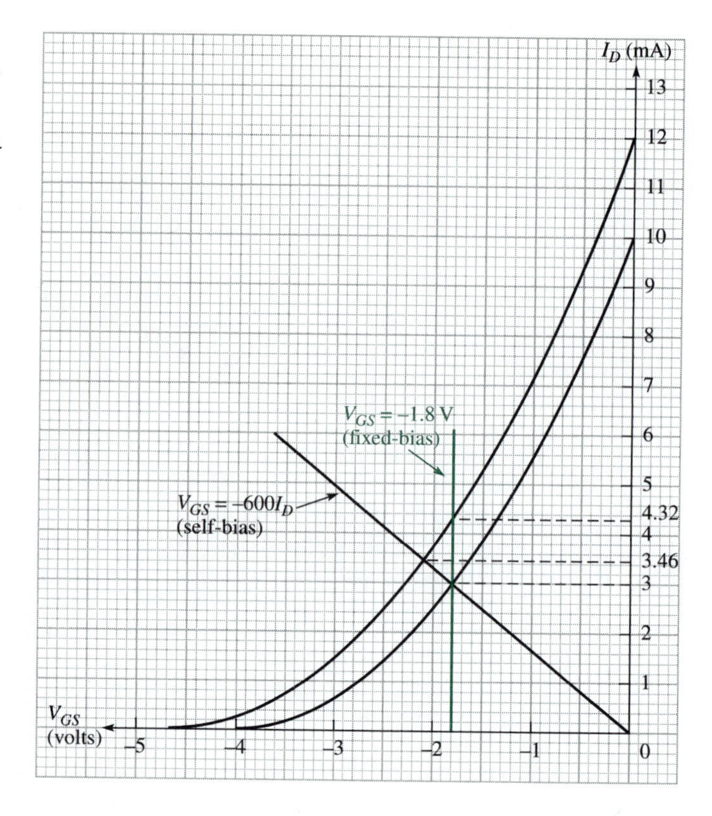

Figure 7–18
The quiescent value of I_D is initially 3 mA for both the fixed- and self-bias circuits. When the transfer characteristic changes, there is a smaller change in I_D for the self-bias circuit than for the fixed-bias circuit.

line having a small slope (large value of R_S) and still maintain a respectable amount of drain current is to connect a *positive* voltage V_{GG} to the gate (of an N-channel JFET) in the self-bias circuit.

This arrangement is shown in Figure 7–19(a). The effect of V_{GG} is to shift the intercept of the bias line on the horizontal axis to V_{GG}, as shown in Figure 7–19(b). The equation of this bias line is

$$V_{GS} = V_{GG} - I_D R_S \tag{7–9}$$

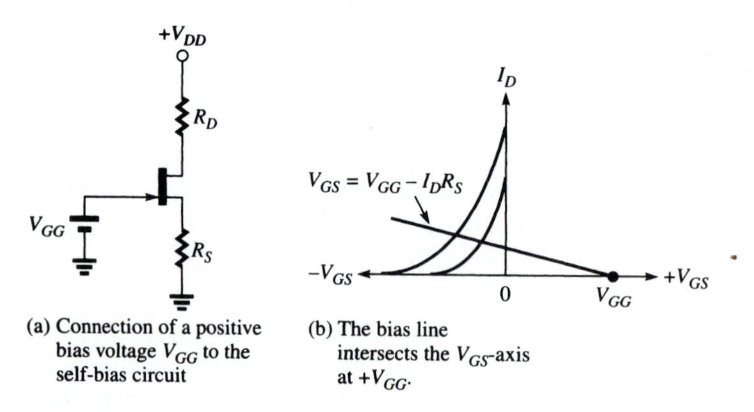

(a) Connection of a positive bias voltage V_{GG} to the self-bias circuit

(b) The bias line intersects the V_{GS}-axis at $+V_{GG}$.

Figure 7–19
A positive gate voltage V_{GG} reduces the slope of the bias line and improves bias stability.

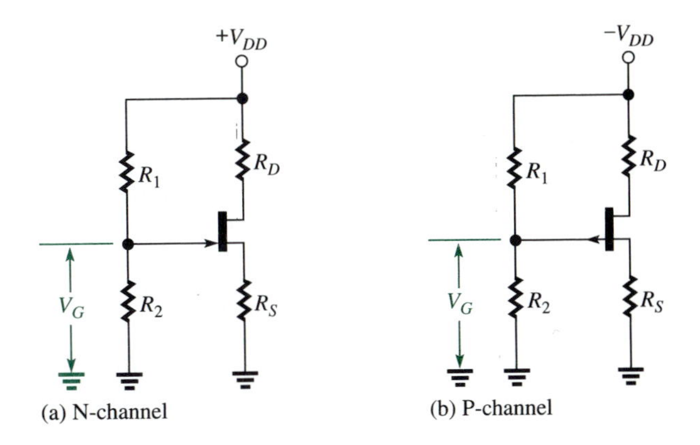

Figure 7–20
Biasing the gate using a voltage divider rather than a separate V_{GG}

In practice, the positive gate voltage is obtained from a voltage divider connected across the gate from the positive supply voltage V_{DD}. For a P-channel JFET, the gate voltage is made negative and is obtained from the negative supply $-V_{DD}$. These connections are shown in Figure 7–20. Since the input resistance at the gate is very large, it is not necessary to consider its loading effect on the voltage divider. Thus, the gate-to-ground voltage V_G for the N-channel JFET is determined from

$$V_G = \left(\frac{R_2}{R_1 + R_2}\right) V_{DD} \qquad \textbf{(7–10)}$$

(For a P-channel device, $V_G = -R_2V_{DD}/(R_1 + R_2)$.) The bias-line equations for N- and P-channel devices are then

$$V_{GS} = V_G - I_DR_S \qquad \text{(N-channel)} \qquad \textbf{(7–11)}$$

$$V_{GS} = V_G + I_DR_S \qquad \text{(P-channel)} \qquad \textbf{(7–12)}$$

Note that V_G is a positive number in equation 7–11 and a negative number in equation 7–12.

Example 7–7

The circuit in Figure 7–20(a) has V_{DD} = 10 V, R_1 = 3.3 MΩ, R_2 = 680 kΩ, R_S = 500 Ω, and R_D = 1 kΩ. If V_{DS} = 3 V, find I_D and V_{GS}.

Solution. Since $V_{DS} = V_{DD} - I_D(R_D + R_S)$, we have

$$3\text{ V} = 10\text{ V} - I_D(1\text{ k}\Omega + 500\ \Omega)$$

$$I_D = \frac{7\text{ V}}{1.5\text{ k}\Omega} = 4.67\text{ mA}$$

From equation 7–10,

$$V_G = \left(\frac{R_2}{R_1 + R_2}\right) V_{DD} = \left(\frac{680\text{ k}\Omega}{3.3\text{ M}\Omega + 680\text{ k}\Omega}\right) 10\text{ V} = 1.71\text{ V}$$

From equation 7–11,

$$V_{GS} = V_G - I_DR_S = 1.71\text{ V} - (4.67\text{ mA})(500\ \Omega) = -0.625\text{ V}$$

General Algebraic Solution—Voltage-Divider Bias

The general form of the algebraic solution for the bias point in a voltage-divider bias circuit can be found by solving the square-law equation (7–2) simultaneously with the bias-line equation (7–11 or 7–12). The results, shown in equations 7–13, are valid for both P-channel and N-channel devices, since absolute values are used in the computations. The computed values must be checked to verify that the solution is in the pinch-off region: $|V_{DS}| > |V_p| - |V_{GS}|$. Note that equations 7–13 reduce to equations 7–8 for self-bias when 0 is substituted for V_G.

General algebraic solution for the bias point of JFET circuits using voltage-divider bias

$$I_D = \frac{-B - \sqrt{B^2 - 4AC}}{2A}$$

where

$$A = R_S^2$$

$$B = -\left[2(|V_p| + |V_G|)R_S + \frac{V_p^2}{I_{DSS}}\right]$$

$$C = |V_p| + |V_G|)^2$$

$$|V_G| = \frac{R_2}{R_1 + R_2}|V_{DD}|$$

$$|V_{DS}| = |V_{DD}| - I_D(R_D + R_S) \quad \text{See note 1.}$$

$$|V_{GS}| = |V_G| - I_D R_S \quad \text{See note 2.} \qquad (7\text{–}13)$$

Note 1. V_{DS} is positive for an N-channel JFET and negative for a P-channel JFET.

Note 2. V_{GS} is negative for an N-channel JFET and positive for a P-channel JFET.

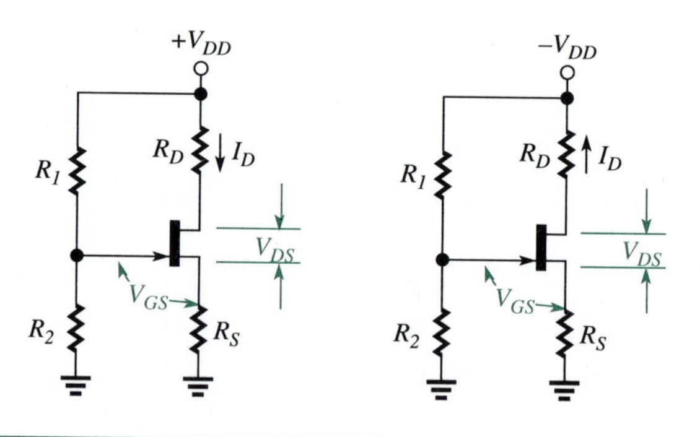

Example 7–8

The P-channel JFET in Figure 7–21 has the transfer characteristic shown in Figure 7–22. Determine the quiescent value of I_D (1) graphically and (2) algebraically.

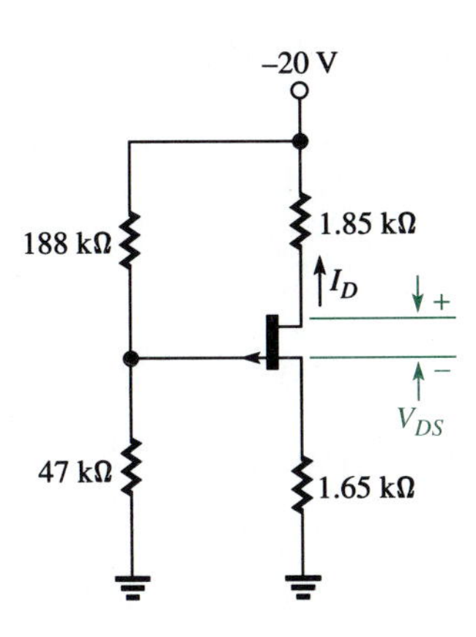

Figure 7–21
(Example 7–8)

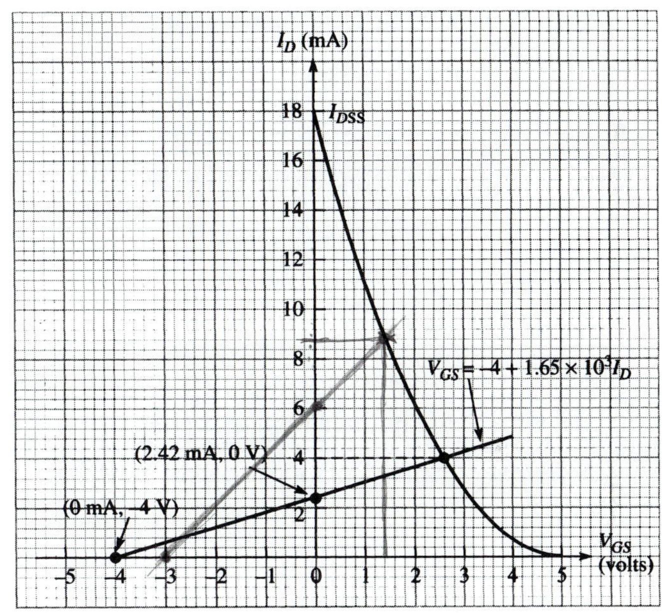

Figure 7–22
(Example 7–8)

Solution

1. To find the equation of the bias line, we must first find the gate-to-ground voltage V_G:

$$V_G = \left(\frac{47 \times 10^3}{188 \times 10^3 + 47 \times 10^3}\right)(-20 \text{ V}) = -4 \text{ V}$$

Then, from equation 7–12, the bias line is

$$V_{GS} = -4 + 1.65 \times 10^3 I_D \tag{7–14}$$

This line intersects the V_{GS}-axis at -4 V. A second point that is easy to find on the line is its intercept on the I_D-axis. Let $V_{GS} = 0$, and solve for I_D:

$$I_D = \frac{4}{1.65 \times 10^3} = 2.42 \text{ mA}$$

The bias line joining the points (0 mA, −4 V) and (2.42 mA, 0 V) is shown drawn on the transfer characteristic in Figure 7–22. It can be seen that the bias line intersects the characteristic at the quiescent value $I_D \approx 4$ mA.

2. From Figure 7–21, $R_D = 1.85$ kΩ, $R_S = 1.65$ kΩ, and $V_{DD} = 20$ V. From the transfer characteristic in Figure 7–22, we see that $V_p = 5$ V and $I_{DSS} = 18$ mA. We have already found (in part 1) that $V_G = -4$ V. With reference to equations 7–13,

$$A = R_S^2 = (1.65 \times 10^3)^2 = 2.7225 \times 10^6$$

$$B = -\left[2(|V_p| + |V_G|)R_S + \frac{V_p^2}{I_{DSS}}\right]$$

$$= -[2(5 + 4)(1.65 \times 10^3) + 5^2/18 \times 10^{-3}] = -31.09 \times 10^3$$

$$C = (|V_p| + |V_G|)^2 = (5 + 4)^2 = 81$$

Substituting these values into the equation for I_D, we find $I_D = 4.02$ mA. Then

$$|V_{DS}| = |V_{DD}| - I_D(R_D + R_S)$$

$$= 20\text{ V} - 4.02\text{ mA}\,(1.85\text{ k}\Omega + 1.65\text{ k}\Omega) = 5.93\text{ V}$$

Since the JFET is P-channel, we know $V_{DS} = -5.93$ V.

$$|V_{GS}| = |V_G| - I_D R_S = 4\text{ V} - (4.02\text{ mA})(1.65\text{ k}\Omega) = 2.63\text{ V}$$

Since 5.93 V > 5 V − 2.63 V = 2.37 V, we know the solution is valid.

Example 7–9

SPICE

Use SPICE to determine the bias point of the JFET in Example 7–8.

Solution. Figure 7–23(a) shows the SPICE circuit and input data file. Note that

$$\text{BETA} = \beta = \frac{I_{DSS}}{V_p^2} = \frac{18 \times 10^{-3}\text{ A}}{(5\text{ V})^2} = 7.2 \times 10^{-4}\text{ A/V}^2$$

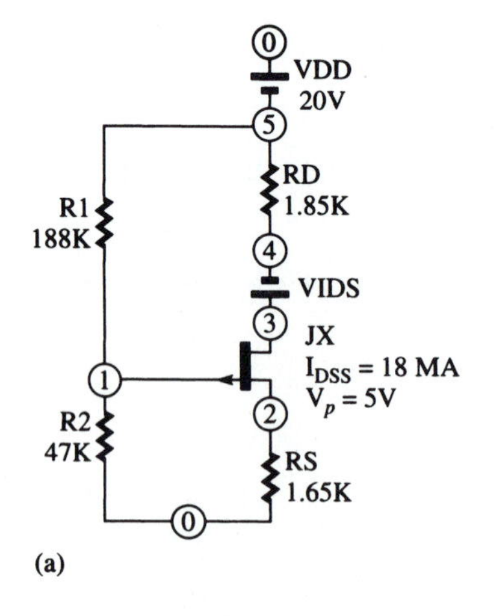


```
EXAMPLE 7.9
JX 3 1 2 JFET
R1 5 1 188K
R2 1 0 47K
RS 2 0 1.65K
VIDS 3 4
RD 5 4 1.85K
VDD 0 5 20V
.MODEL JFET PJF BETA=7.2E-4 VTO=-5
.DC VDD 20 20 1
.PRINT DC I(VIDS) V(3,2)
.END
```

Figure 7–23
(Example 7–9)

```
EXAMPLE 7.9
****        DC TRANSFER CURVES              TEMPERATURE = 27.000 DEG C
******************************************************************************
   VDD          I(VIDS)        V(3, 2)        V(1, 2)
2.000E+01     4.022E-03    -5.923E+00     2.636E+00
```
(b)

Figure 7–23 (*Continued*)

Also note that we must specify the value of VTO at −5 rather than +5, despite the fact that the JFET is a P-channel device.

Figure 7–23(b) shows the results of a program run. We see that I(VIDS), the drain current, is 4.022 mA; V(3, 2), the value of V_{DS}, is −5.923 V; and V(1, 2), the value of V_{GS}, is 2.636 V. These values are in close agreement with those found in Example 7–8.

7–4 JFET BIAS DESIGN

The design of a JFET bias circuit requires that we find values for R_D, R_S, and/or R_1 and R_2 that produce specified values of I_D and V_{DS}, given a supply voltage V_{DD}. Equations 7–15 for the self-bias circuit are found by solving equation 7–7 for R_D and by solving equation 7–5 or 7–6 simultaneously with the square-law equation for R_S. The results can be used for either N-channel or P-channel designs, since absolute values of V_p and V_{DD} are used in the computations.

Design of JFET self-bias circuits

$$R_S = \frac{-B - \sqrt{B^2 - 4AC}}{2A}$$

where

$$A = I_D^2$$

$$B = -2|V_p|I_D$$

$$C = V_p^2\left(1 - \frac{I_D}{I_{DSS}}\right)$$

$$R_D = \frac{|V_{DD}| - |V_{DS}| - I_D R_S}{I_D} \qquad (7\text{–}15)$$

Equations 7–16 can be used to find values for R_D, R_S, R_1, and R_2 in the voltage-divider bias circuit for N- and P-channel JFETs. Note that a value for V_G can be assumed or can be determined if the *range* through which the bias point can be allowed to change is specified. Figure 7–24 illustrates the computation. The characteristic curves represent the range over which the JFET characteristic may change. A line drawn through the bias points that are desired when the characteristic changes, (V_{GS1}, I_{D1}) and (V_{GS2}, I_{D2}), will intersect the horizontal axis at the required value of V_G. This value can be computed from the slope of the bias line, as shown in Figure 7–24.

Given a value for V_G, the voltage-divider resistor R_1 can be found as shown in equations 7–16, by assuming a value for R_2. R_2 will usually be smaller than R_1, so R_2 sets an upper limit on the input resistance of the circuit.

Design of voltage-divider bias circuits for N- and P-channel FETs

$$|V_G| = \frac{I_{D1}(|V_{GS2}| - |V_{GS1}|)}{I_{D2} - I_{D1}} - |V_{GS1}|$$

where (V_{GS1}, I_{D1}) and (V_{GS2}, I_{D2}) are the allowed range of the bias point (see Figure 7–24).

$$R_S = \frac{-B - \sqrt{B^2 - 4AC}}{2A}$$

where $A = I_D^2$

$$B = -2(|V_p| + |V_G|)I_D$$

$$C = (|V_p| + |V_G|)^2 - V_p^2\frac{I_D}{I_{DSS}}$$

$$R_D = \frac{|V_{DD}| - |V_{DS}| - I_D R_S}{I_D} \qquad (7\text{–}16)$$

Choose R_2. Then

$$R_1 = \frac{R_2(|V_{DD}| - |V_G|)}{|V_G|}$$

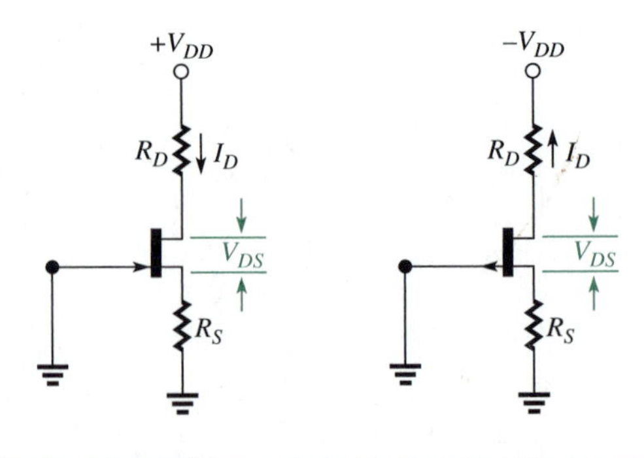

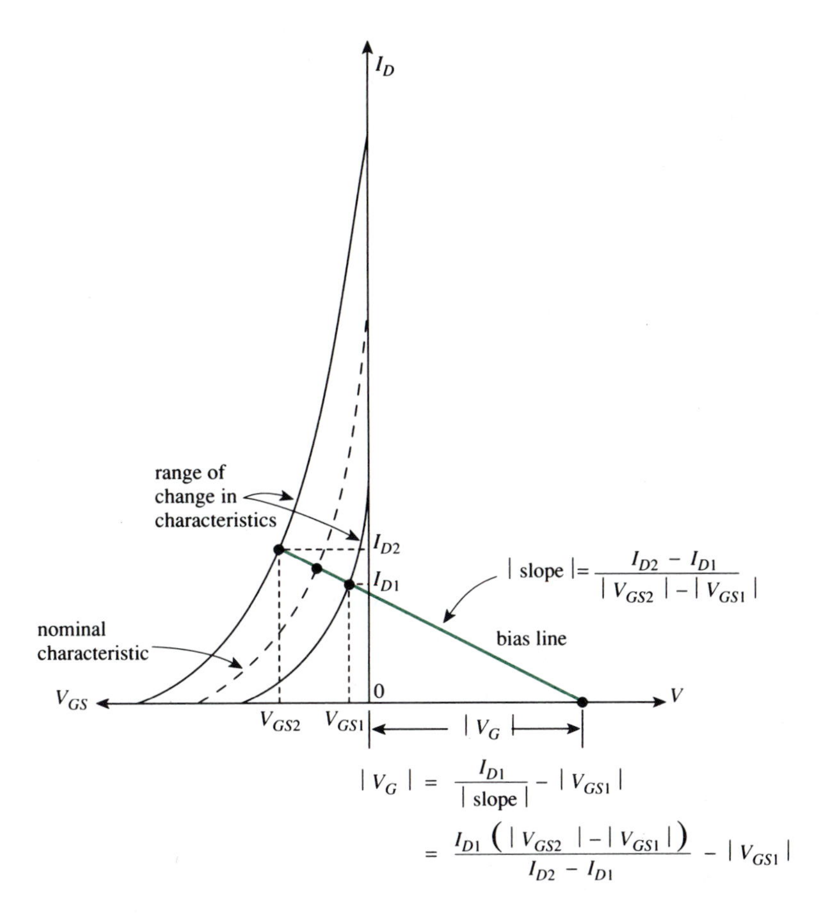

Figure 7–24
Computing the value of $|V_G|$ for a voltage-divider bias circuit when a specified range of bias points, (V_{GS2}, I_{D2}) to (V_{GS1}, I_{D1}), is allowed

Example 7–10

DESIGN

An N-channel JFET is to be biased at $V_{DS} = 6$ V using a 15-V supply. The nominal FET characteristic has $V_p = -3.5$ V and $I_{DSS} = 13.5$ mA. The quiescent drain current should not vary more than ± 0.5 mA from a nominal value of 6 mA when the JFET characteristic changes from $V_p = -3$ V to $V_p = -4$ V, with a corresponding change in I_{DSS} from 12 mA to 15 mA. Find values for R_D, R_S, R_1, and R_2 in a voltage-divider bias design.

Find the actual range of I_D and V_{DS} over the range of the JFET characteristic when the standard-valued 5% resistors closest to the calculated values are used in the design. Assume the resistors have their nominal values.

Solution. We must first find a value for V_G using the procedure illustrated in Figure 7–24. Thus, we must find the values of V_{GS2} and V_{GS1} corresponding to $I_{D2} = 6.5$ mA and $I_{D1} = 5.5$ mA when the JFET characteristic changes through the specified range. Solving the square-law equation for V_{GS} (see Example 7–1) we have

$$V_{GS} = V_p\left(1 - \sqrt{\frac{I_D}{I_{DSS}}}\right)$$

Thus,

$$V_{GS1} = (-3\text{ V})\left(1 - \sqrt{\frac{5.5\text{ mA}}{12\text{ mA}}}\right) = -0.969\text{ V}$$

and

$$V_{GS2} = (-4\text{ V})\left(1 - \sqrt{\frac{6.5\text{ mA}}{15\text{ mA}}}\right) = -1.367\text{ V}$$

Therefore, from Figure 7–24,

$$V_G = \frac{I_{D1}(|V_{GS2}| - |V_{GS1}|)}{I_{D2} - I_{D1}} - |V_{GS1}|$$
$$= \frac{5.5\text{ mA}(1.367\text{ V} - 0.969\text{ V})}{6.5\text{ mA} - 5.5\text{ mA}} - 0.969\text{ V} = 1.22\text{ V}$$

Using equations 7–16, we find

$$A = I_D^2 = (6 \times 10^{-3})^2 = 36 \times 10^{-6}$$
$$B = -2(|V_p| + |V_G|)I_D = -2(3.5 + 1.22)(6 \times 10^{-3}) = -56.64 \times 10^{-3}$$
$$C = (|V_p| + |V_G|)^2 - \frac{V_p^2 I_D}{I_{DSS}} = (3.5 + 1.22)^2 - (3.5)^2\left(\frac{6\text{ mA}}{13.5\text{ mA}}\right) = 16.83$$

Then

$$R_S = \frac{-B - \sqrt{B^2 - 4AC}}{2A}$$
$$= \frac{56.64 \times 10^{-3} - \sqrt{(56.64 \times 10^{3})^2 - 4(36 \times 10^{-6})(16.83)}}{2(36 \times 10^{-6})}$$
$$= 398\ \Omega$$

and

$$R_D = \frac{|V_{DD}| - |V_{DS}| - I_D R_S}{I_D} = \frac{15\text{ V} - 6\text{ V} - (6\text{ mA})(398\ \Omega)}{6\text{ mA}} = 1.1\text{ k}\Omega$$

Letting $R_2 = 330\text{ k}\Omega$, we have

$$R_1 = \frac{R_2(|V_{DD}| - |V_G|)}{|V_G|} = \frac{(330\text{ k}\Omega)(15\text{ V} - 1.22\text{ V})}{1.22\text{ V}} = 3.7\text{ M}\Omega$$

The standard-valued 5% resistors closest to the calculated values are $R_S = 390\ \Omega$, $R_D = 1.1\text{ k}\Omega$, $R_1 = 3.6\text{ M}\Omega$, and $R_2 = 330\text{ k}\Omega$. Using these values in equations 7–18, we find that I_D ranges from 5.65 mA to 6.65 mA and V_{DS} ranges from 5.09 V to 6.58 V, over the range of the JFET characteristic.

7–5 MANUFACTURER'S DATA SHEETS

Figures 7–25 and 7–26 show typical data sheets for a series of N-channel JFETs: the 2N4220, 2N4221, and 2N4222. Note in particular the specification for the range of values of I_{DSS} for each device, as shown in Figure 7–25. We see, for example, that I_{DSS} for the 2N4222 can range from 5 mA to 15 mA. The pinch-off voltage (designated $V_{GS(off)}$) for the 2N4222 is seen to have a maximum value of −8 V. Devices having small values of I_{DSS} will have smaller values of $V_{GS(off)}$. Another important static characteristic shown in Figure 7–25 is I_{GSS}, the *gate reverse current,*

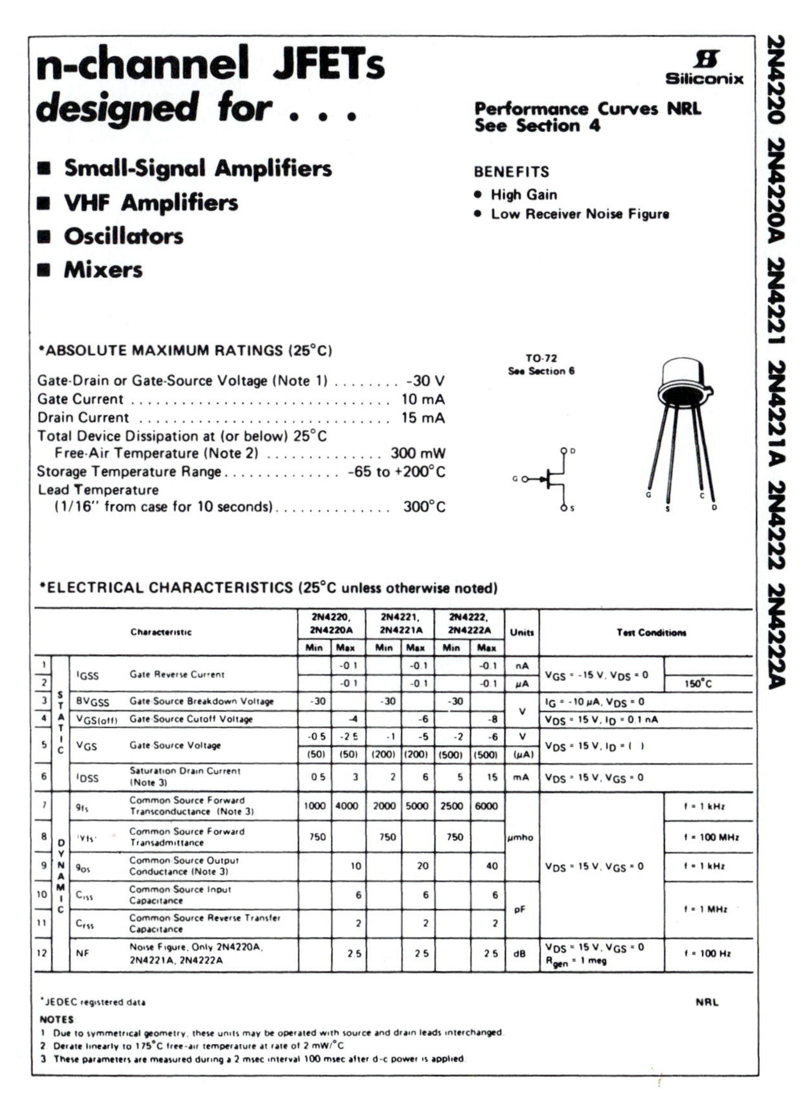

n-channel JFETs designed for . . .

Siliconix

2N4220 2N4220A 2N4221 2N4221A 2N4222 2N4222A

- Small-Signal Amplifiers
- VHF Amplifiers
- Oscillators
- Mixers

Performance Curves NRL See Section 4

BENEFITS
- High Gain
- Low Receiver Noise Figure

*ABSOLUTE MAXIMUM RATINGS (25°C)

Gate-Drain or Gate-Source Voltage (Note 1) .... -30 V
Gate Current .... 10 mA
Drain Current .... 15 mA
Total Device Dissipation at (or below) 25°C
Free-Air Temperature (Note 2) .... 300 mW
Storage Temperature Range.............. -65 to +200°C
Lead Temperature
(1/16" from case for 10 seconds).............. 300°C

TO-72
See Section 6

G D S

*ELECTRICAL CHARACTERISTICS (25°C unless otherwise noted)

			Characteristic	2N4220, 2N4220A Min	2N4220, 2N4220A Max	2N4221, 2N4221A Min	2N4221, 2N4221A Max	2N4222, 2N4222A Min	2N4222, 2N4222A Max	Units	Test Conditions	
1	STATIC	I_{GSS}	Gate Reverse Current		-0.1		-0.1		-0.1	nA	V_{GS} = -15 V, V_{DS} = 0	
2					-0.1		-0.1		-0.1	μA		150°C
3		BV_{GSS}	Gate-Source Breakdown Voltage	-30		-30		-30		V	I_G = -10 μA, V_{DS} = 0	
4		$V_{GS(off)}$	Gate-Source Cutoff Voltage		-4		-6		-8		V_{DS} = 15 V, I_D = 0.1 nA	
5		V_{GS}	Gate-Source Voltage	-0.5	-2.5	-1	-5	-2	-6	V	V_{DS} = 15 V, I_D = ( )	
				(50)	(50)	(200)	(200)	(500)	(500)	(μA)		
6		I_{DSS}	Saturation Drain Current (Note 3)	0.5	3	2	6	5	15	mA	V_{DS} = 15 V, V_{GS} = 0	
7	DYNAMIC	g_{fs}	Common Source Forward Transconductance (Note 3)	1000	4000	2000	5000	2500	6000	μmho	V_{DS} = 15 V, V_{GS} = 0	f = 1 kHz
8		$\|y_{fs}\|$	Common Source Forward Transadmittance	750		750		750				f = 100 MHz
9		g_{os}	Common Source Output Conductance (Note 3)		10		20		40			f = 1 kHz
10		C_{iss}	Common Source Input Capacitance		6		6		6	pF		f = 1 MHz
11		C_{rss}	Common Source Reverse Transfer Capacitance		2		2		2			
12		NF	Noise Figure, Only 2N4220A, 2N4221A, 2N4222A		2.5		2.5		2.5	dB	V_{DS} = 15 V, V_{GS} = 0, R_{gen} = 1 meg	f = 100 Hz

*JEDEC registered data

NRL

NOTES:
1. Due to symmetrical geometry, these units may be operated with source and drain leads interchanged.
2. Derate linearly to 175°C free-air temperature at rate of 2 mW/°C
3. These parameters are measured during a 2 msec interval 100 msec after d-c power is applied.

Figure 7–25

Manufacturer's specifications for a series of N-channel JFETs (Courtesy of Siliconix, Inc.)

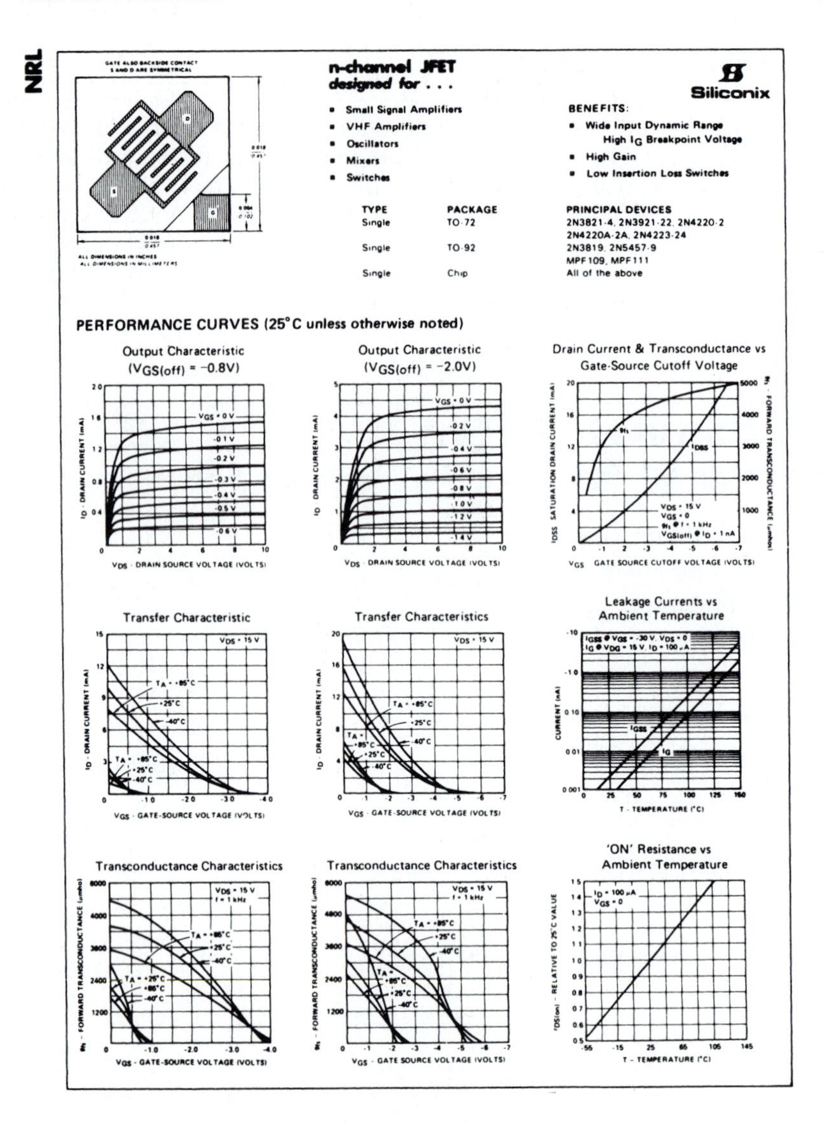

Figure 7–26
Typical manufacturer's performance curves for the N-channel JFETs of Figure 7–23 (Courtesy of Siliconix, Inc.)

which is the gate current when the gate-source junction is reverse biased, the normal mode of operation. This current provides a measure of the dc input resistance of the device, from gate to source. We see that the maximum specified value for the magnitude of I_{GSS} is 0.1 nA when $V_{GS} = -15$ V and $V_{DS} = 0$. Thus, the minimum gate-to-source resistance under those conditions is $R = (15\text{ V})/(0.1 \times 10^{-9}\text{ A}) = 150 \times 10^{9}\ \Omega$.

Figure 7–26 shows typical "performance curves" for the 2N4220 series of JFETs. The curves are also applicable to a number of similar JFETs produced by the manufacturer and show how wide a range of characteristics is possible. Note that two sets of output characteristics (drain characteristics) are shown, one representing a device having a small I_{DSS} ($\approx$ 1.4 mA) and another having a larger I_{DSS} ($\approx$ 4.2 mA).

Two sets of transfer characteristics are also shown in the figure. These show the range that can be expected in the characteristic among devices of the same type, as well as variations due to temperature. The transfer characteristic in the center of the figure is applicable to the 2N4222 JFET. At 25°C, we see that any given 2N4222 could have a value V_p between about -2 V and -5 V. An interesting temperature phenomenon of JFETs is revealed by these characteristics. Note that the (maximum) value of I_{DSS} at 25°C is less than the value at -40°C, but greater than the value at 85°C. Although it is difficult to see in the figure, this result is accounted for by the fact that the three temperature curves intersect and cross through each other near the V_p end of the characteristics. For the 2N4222, this point can be seen to occur at about $V_{GS} = -4.5$ V. Thus, *the 2N4222 characteristics have zero temperature coefficient when the bias point is set at (about) -4.5 V.* For every JFET, there is a value of V_{GS} near V_p that results in a zero temperature coefficient.

7–6 METAL-OXIDE-SEMICONDUCTOR FETs

The metal-oxide-semiconductor FET (MOSFET) is similar in many respects to its JFET counterpart, in that both have drain, gate, and source terminals, and both are devices whose channel conductivity is controlled by a gate-to-source voltage. The principal feature that distinguishes a MOSFET from a JFET is the fact that the gate terminal in a MOSFET is *insulated* from its channel region. For this reason, a MOSFET is often called an *insulated-gate* FET, or IGFET. There are two kinds of MOSFETs: the *depletion* type and the *enhancement* type, also referred to as depletion-*mode* and enhancement-*mode* MOSFETs. These names are derived from the two different ways that the conductivity of the channel can be altered by variations in V_{GS}, as we shall see.

Depletion-Type MOSFETs

Figure 7–27 shows the structure of an N-channel, depletion-type MOSFET. A block of high-resistance, P-type silicon forms a *substrate,* in which are embedded two heavily doped N-type wells, or pockets, labeled N^+. A thin layer of silicon dioxide (SiO_2), which is an insulating material, is deposited along the surface. Metal contacts penetrate the silicon dioxide layer at the two N^+ wells and become the drain and source terminals. Between the two N^+ wells is a more lightly doped region of N

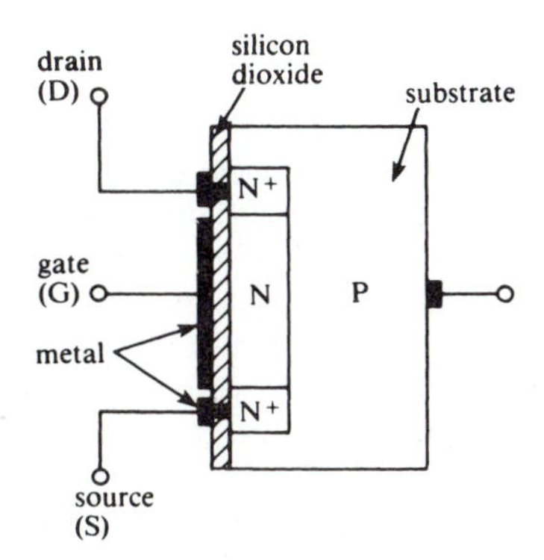

Figure 7–27
Structure of an N-channel depletion-type MOSFET

material that forms the channel. Metal (aluminum) is deposited on the silicon dioxide opposite the channel and becomes the gate terminal. Note that the silicon dioxide insulates the gate from the channel. Going from gate to channel, we encounter *m*etal, *o*xide, and *s*emiconductor, in that sequence, which accounts for the name *MOSFET*. Notice that there is no PN junction formed between gate and channel, as there is in a JFET.

Figure 7–28 shows the normal mode of operation of a depletion-type, N-channel MOSFET. A voltage V_{DS} is connected between drain and source to make the drain positive with respect to the source. The substrate is usually connected to the source, as shown in the figure. When the gate is made negative with respect to the source by V_{GS}, the electric field it produces in the channel drives electrons away from a portion of the channel near the SiO_2 layer. This portion is thus *depleted* of carriers and the channel width is effectively narrowed. The narrower the channel, the greater its resistance and the smaller the current flow from drain to source. Thus, the device behaves very much like an N-channel JFET, the principal difference being that the channel width is controlled by the action of the electric field rather than by the size of the depletion region of a PN junction. Since there is no PN junction, the voltage V_{GS} can be made *positive* without any concern for the consequences of forward biasing a junction. In fact, making V_{GS} positive attracts more electrons *into* the channel and increases, or *enhances,* its conductivity. Thus, the gate voltage in a depletion-type MOSFET can be varied through both positive and negative voltages and the device can operate in both depletion and enhancement modes. For this reason, the depletion-type MOSFET is also called a *depletion-enhancement* type.

Although there is a PN junction between the N material and the P substrate, this junction is always reverse biased and very little substrate current flows. The substrate has little bearing on the operation of the device and we will hereafter ignore its presence, other than to show its usual connection to the source. The resistance looking into the gate is extremely large, on the order of thousands of megohms, because there is no PN junction and no path for current to flow through the insulating layer separating the gate and the channel.

Because of the similarity of a depletion-type MOSFET to a JFET, we would expect it to have similar parameters and operating characteristics. This is indeed the case, as shown by the drain characteristics in Figure 7–29. Note that current increases linearly with increasing V_{DS} until a pinch-off condition is reached. Beyond pinch-off, the drain current remains constant at a saturation value depending on V_{GS}. More negative values of V_{GS} cause pinch-off to be reached sooner and result in smaller values of saturation current. If $V_{GS} = 0$, the drain current saturates at I_{DSS} when $V_{DS} = -V_p$ volts. If V_{GS} is made sufficiently negative to deplete the entire

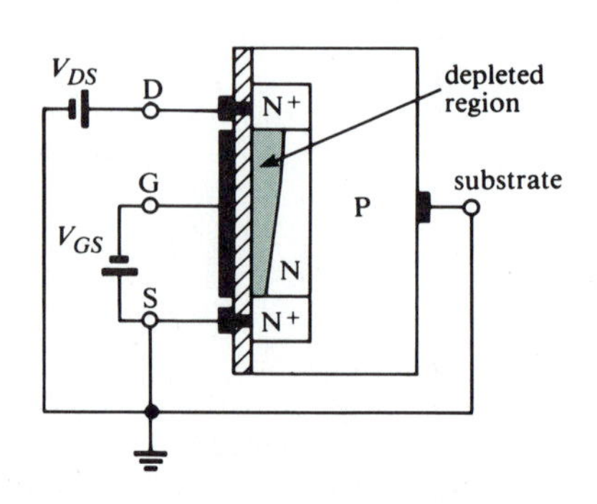

Figure 7–28
Operation of an N-channel, depletion-type MOSFET. The electric field produced by V_{GS} creates in the channel a region that is depleted of carriers.

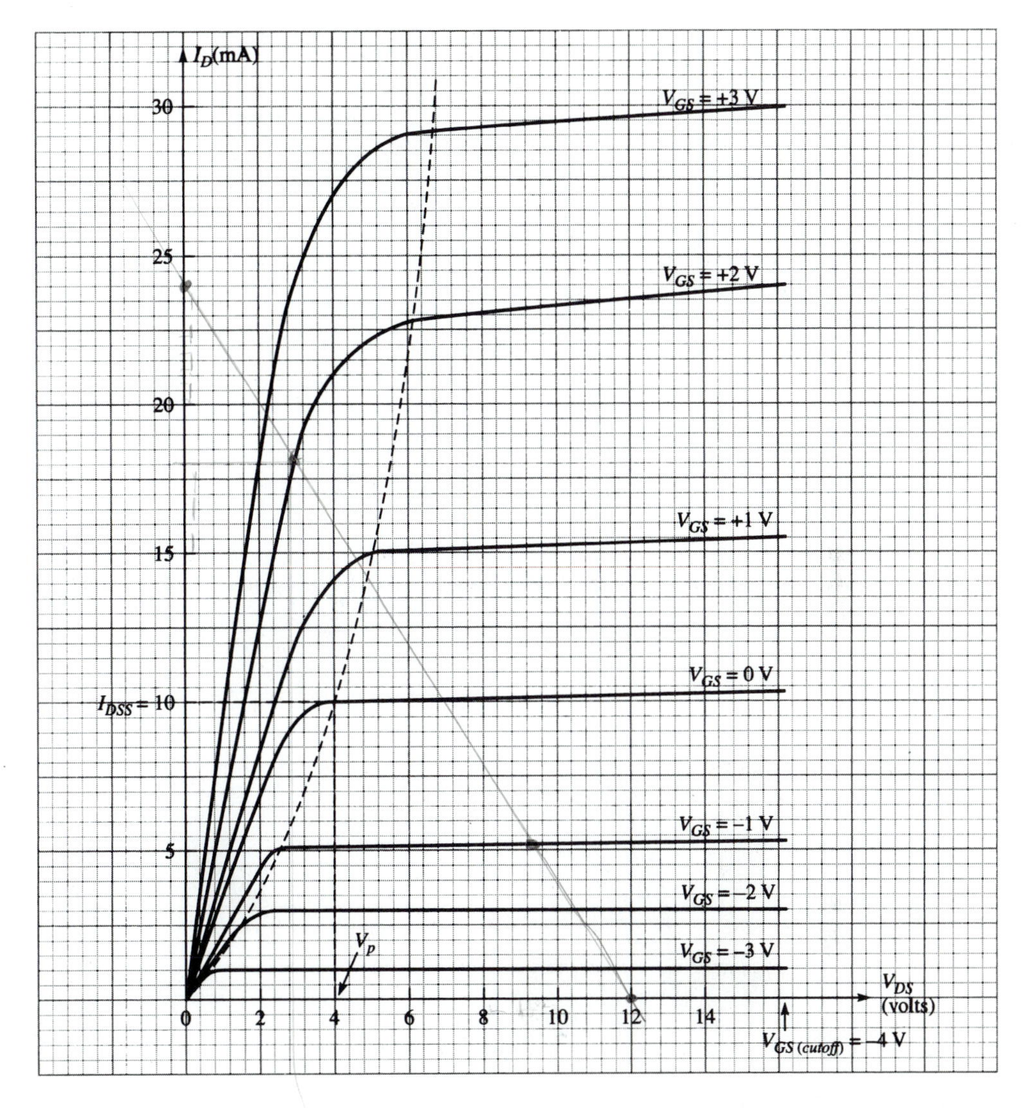

Figure 7–29
Drain characteristics of an N-channel depletion-type MOSFET, showing operation in the depletion and enhancement modes

channel, the drain current is completely cut off. The value of V_{GS} at which this occurs is the gate-to-source cutoff voltage, $V_{GS(cutoff)} = V_p$. Note that the characteristics in Figure 7–29 also show operation in the enhancement mode, where V_{GS} is positive. Figure 7–30 shows the drain characteristics of a P-channel, depletion-type MOSFET. Notice that depletion occurs in this device when V_{GS} is positive and enhancement when V_{GS} is negative.

Figure 7–31 shows the schematic symbols for N- and P-channel depletion-type MOSFETs. The arrow on each symbol is drawn on the substrate terminal and its direction indicates whether the device is N-channel or P-channel. It points into the device for an N-channel and outward for a P-channel. Notice that the gate terminal

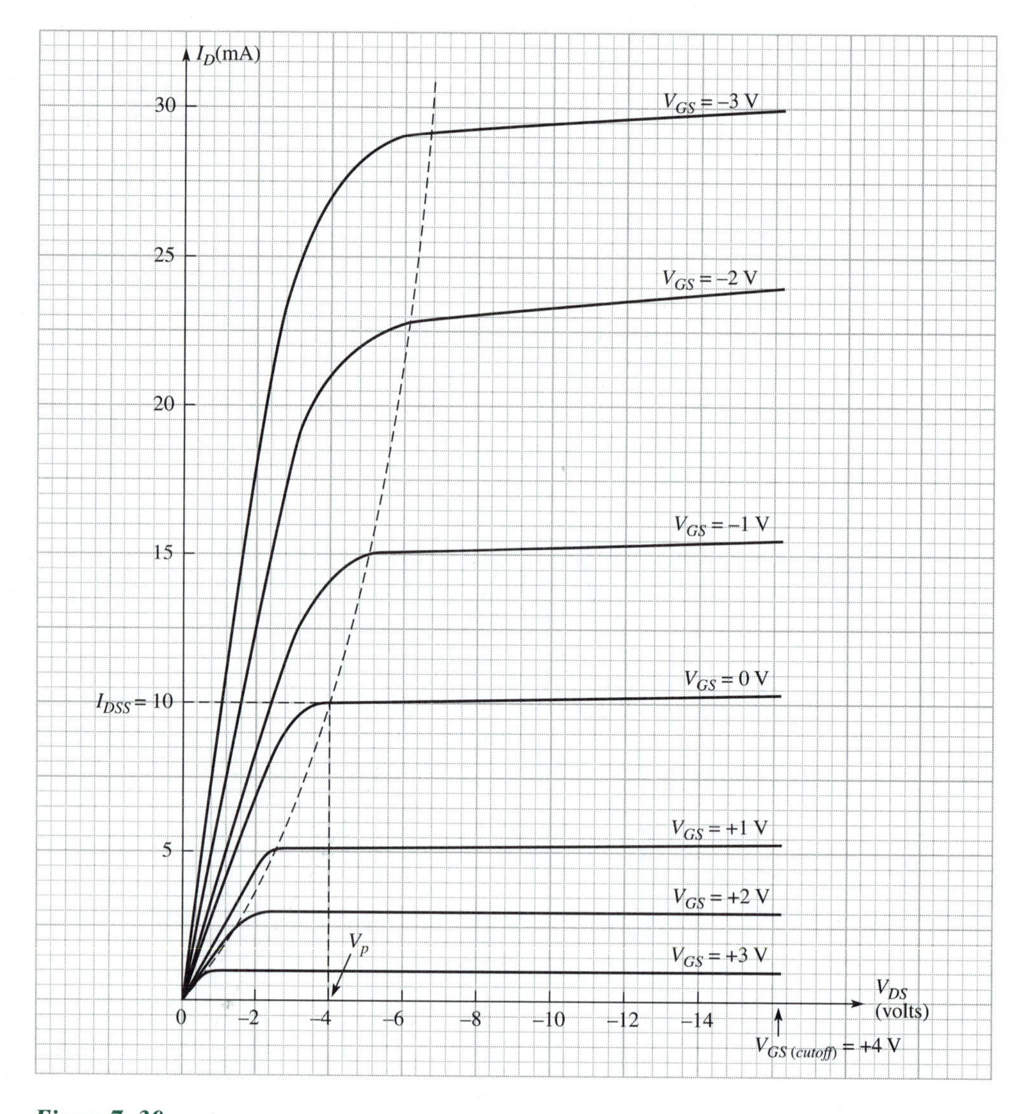

Figure 7–30
Drain characteristics of a P-channel depletion-type MOSFET, showing operation in the depletion and enhancement modes

Figure 7–31
Schematic symbols for N- and P-channel depletion-type MOSFETs

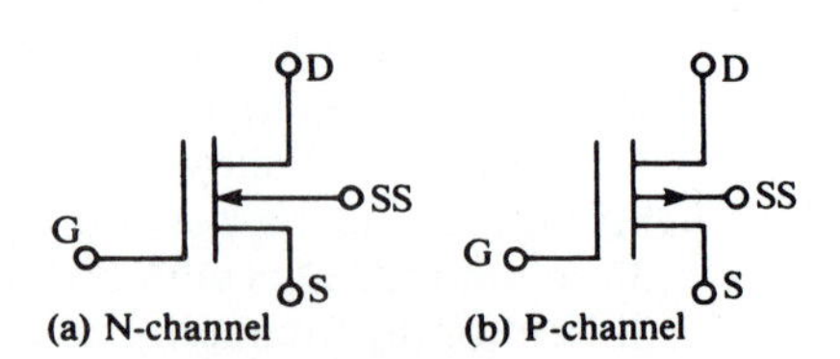

is attached to a line that is separated from the rest of the symbol, to emphasize that the gate is insulated from the channel.

The square-law equation for the transfer characteristic of a depletion-type MOSFET is identical to that for a JFET:

$$I_D = I_{DSS}\left(1 - \frac{V_{GS}}{V_p}\right)^2 \tag{7–17}$$

This equation correctly predicts I_D when the depletion-type MOSFET is operated in the enhancement mode. Note, for example, that when V_{GS} is positive in an N-channel device, $(1 - V_{GS}/V_p) > 1$ (since V_p is negative), and therefore I_D is greater than I_{DSS}. Figure 7–32 shows transfer characteristics for N- and P-channel devices. Note in each case that I_D exceeds I_{DSS} in the enhancement mode.

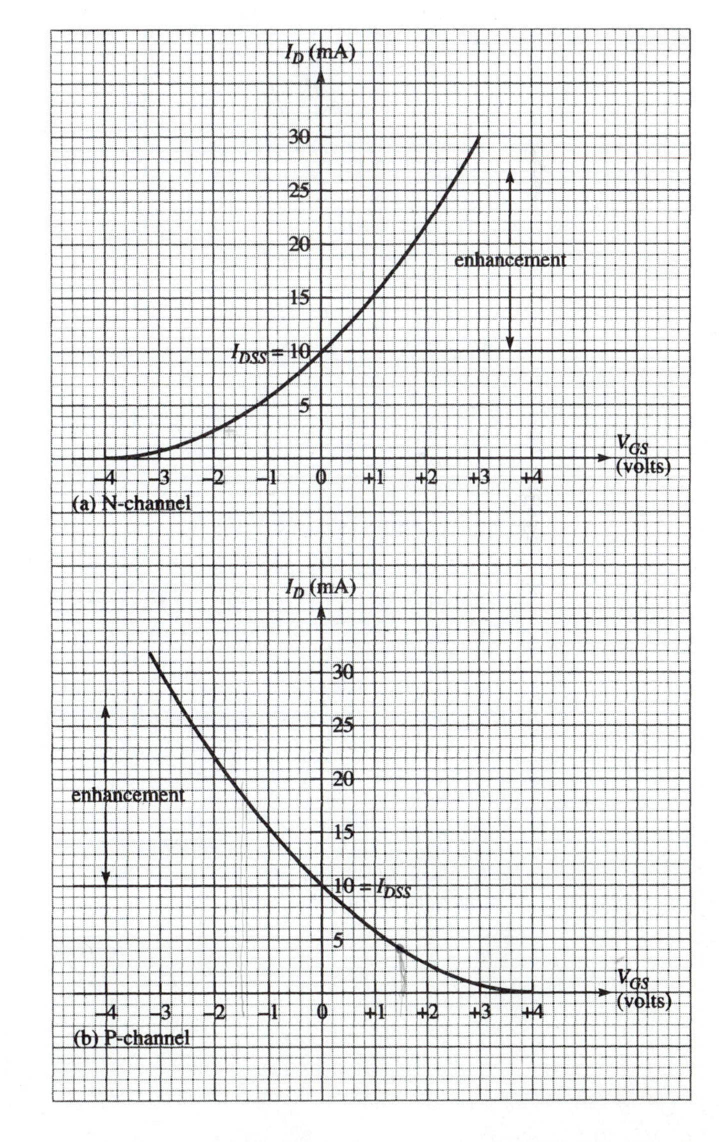

Figure 7–32
Transfer characteristics of N-channel and P-channel depletion-type MOSFETs

The bias techniques we have already discussed for junction FETs are wholly applicable to depletion-type MOSFETs, since the characteristics of the two devices are so similar. We will not repeat our discussion of those techniques.

Example 7–11

1. An N-channel depletion-type MOSFET has I_{DSS} = 18 mA and V_p = −5 V. Assuming that it is operated in the pinch-off region, find I_D when V_{GS} = −3 V, and again when V_{GS} = +2.5 V.
2. Repeat (1) if the MOSFET is P-channel and V_p = +5 V.

Solution

1. From equation 7–17, for V_{GS} = −3 V,

$$I_D = (18\text{ mA})\left(1 - \frac{-3}{-5}\right)^2 = (18\text{ mA})(0.4)^2 = 2.88\text{ mA}$$

For V_{GS} = +2.5 V,

$$I_D = (18\text{ mA})\left(1 - \frac{2.5}{-5}\right)^2 = (18\text{ mA})(1.5)^2 = 40.5\text{ mA}$$

2. From equation 7–17, for V_{GS} = −3 V,

$$I_D = (18\text{ mA})\left(1 - \frac{-3}{5}\right)^2 = (18\text{ mA})(1.6)^2 = 46.08\text{ mA}$$

For V_{GS} = +2.5 V,

$$I_D = (18\text{ mA})\left(1 - \frac{2.5}{5}\right)^2 = 4.5\text{ mA}$$

Enhancement-Type MOSFETs

Recall that the channel of an N-channel depletion MOSFET is the region of N material between the drain and the source (see Figure 7–27). In the enhancement MOSFET, there is no N-type material between the drain and the source; instead, the P-type substrate extends all the way to the SiO_2 layer adjacent to the gate. This structure is shown in Figure 7–33. Apart from the absence of the N-type channel, the construction is the same as that of the depletion MOSFET.

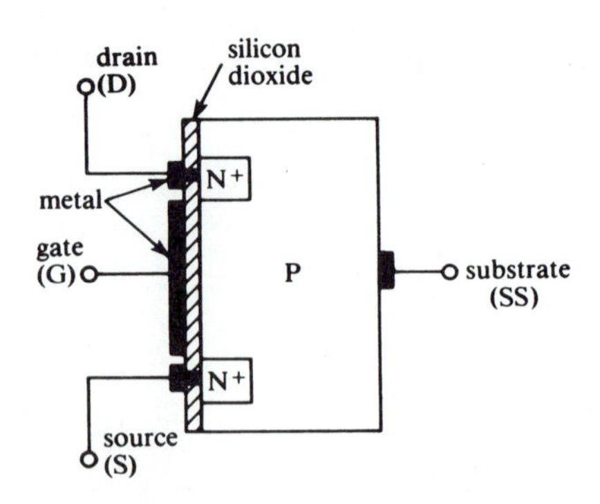

Figure 7–33
Enhancement-type MOSFET. The structure is similar to that of a depletion MOSFET, but note the absence of N-type material between drain and source.

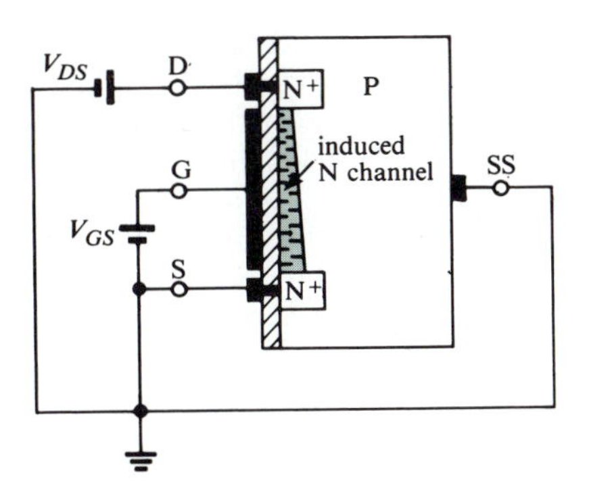

Figure 7–34
The positive V_{GS} induces an N-type channel in the substrate of an enhancement MOSFET.

Figure 7–34 shows the normal electrical connections between drain, gate, and source. As in the depletion MOSFET, the substrate is usually connected to the source. Notice that V_{GS} is connected so that *the gate is positive with respect to the source.* The positive gate voltage attracts electrons from the substrate to the region along the insulating layer opposite the gate. If the gate is made sufficiently positive, enough electrons will be drawn into that region to convert it to N-type material. Thus, an N-type channel will be formed between drain and source. The P material is said to have been *inverted* to form an N-type channel. If the gate is made still more positive, more electrons will be drawn into the region and the channel will widen, making it more conductive. In other words, making V_{GS} more positive *enhances* the conductivity of the channel and increases the flow of current from drain to source. Since electrons are induced into the channel to convert it to N-type material, the MOSFET shown in Figures 7–33 and 7–34 is often called an *induced N-channel* enhancement-type MOSFET. When this device is referred to simply as an N-channel enhancement MOSFET, it is understood that the N channel exists only when it is induced from the P substrate by a positive V_{GS}.

The induced N channel in Figure 7–34 does not become sufficiently conductive to allow drain current to flow until V_{GS} reaches a certain *threshold* voltage, V_T. In modern silicon MOSFETs, the value of V_T is typically in the range from 1 to 3 V. Suppose that $V_T = 2$ V and that V_{GS} is set to some value greater than V_T, say, 10 V. We will consider what happens when the drain-to-source voltage is gradually increased above 0 V. As V_{DS} increases, the drain current increases because of normal Ohm's law action. The current rises linearly with V_{DS}, as shown in Figure 7–35. As V_{DS} continues to increase, we find that the channel becomes narrower at the drain end, as illustrated in Figure 7–34. This narrowing occurs because the *gate-to-drain* voltage becomes smaller when V_{DS} becomes larger, thus reducing the positive field at the drain end. For example, if $V_{GS} = 10$ V and $V_{DS} = 3$ V, then $V_{GD} = 10 - 3 = 7$ V. When V_{DS} is increased to 4 V, $V_{GD} = 10 - 4 = 6$ V. The positive gate-to-drain

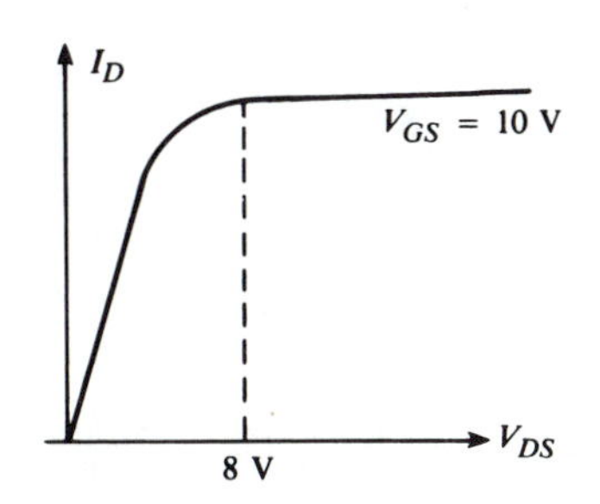

Figure 7–35
The drain current in an N-channel enhancement MOSFET increases with V_{DS} until $V_{DS} = V_{GS} - V_T$ (= 10 − 2 = 8 V in this example).

voltage decreases by the same amount that V_{DS} increases, so the electric field at the drain end is reduced and the channel is narrowed. As a consequence, the resistance of the channel begins to increase, and the drain current begins to level off. This leveling off can be seen in the curve of Figure 7–35. When V_{DS} reaches 8 V, then $V_{GD} = 10 - 8 = 2\text{ V} = V_T$. That is, the positive voltage at the drain end reaches the threshold voltage and the channel width at that end shrinks to zero. Further increases in V_{DS} do not change the shape of the channel and the drain

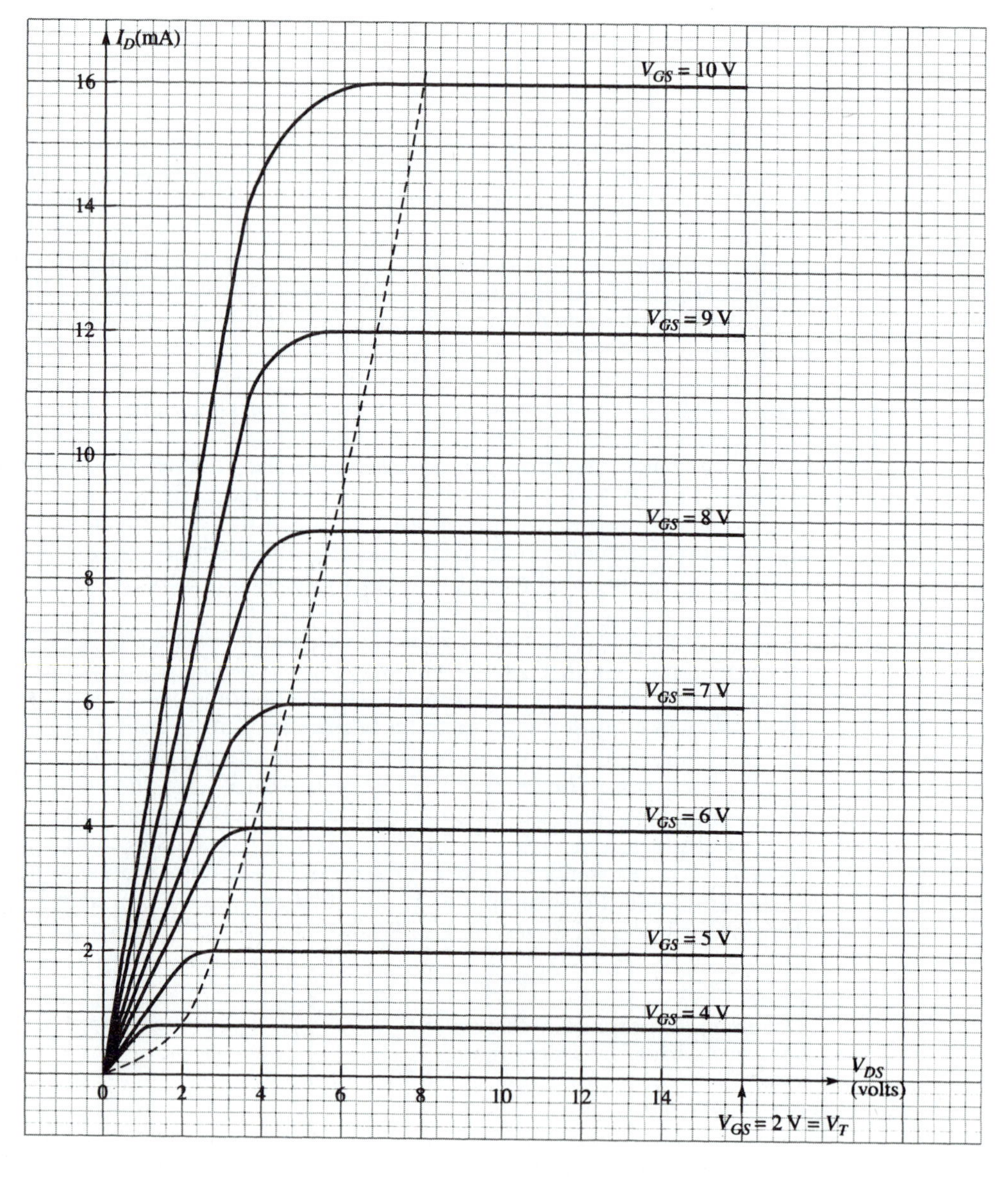

Figure 7–36
Drain characteristics of an induced N-channel enhancement MOSFET. Note that all values of V_{GS} are positive.

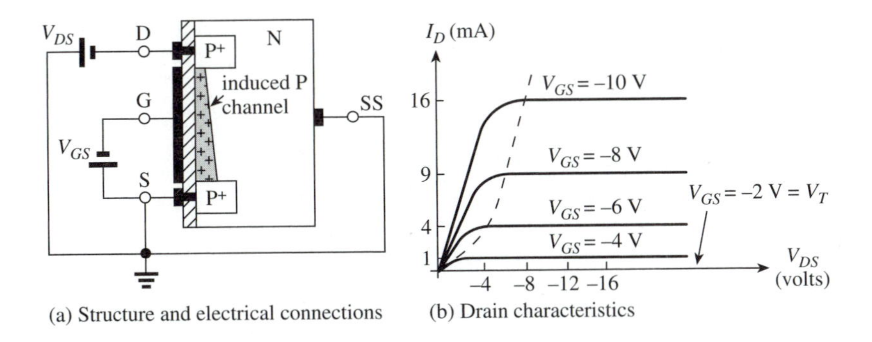

(a) Structure and electrical connections (b) Drain characteristics

Figure 7–37
The induced P-channel enhancement MOSFET

current does not increase any further; i.e., I_D saturates. This action is quite similar to the saturation that occurs at pinch-off in a junction FET.

When the process we have just described is repeated with V_{GS} fixed at 12 V, we find that saturation occurs at $V_{DS} = 12 - 2 = 10$ V. Letting $V_{DS(sat)}$ represent the voltage at which saturation occurs, we have, in the general case,

$$V_{DS(sat)} = V_{GS} - V_T \tag{7–18}$$

Figure 7–36 shows a set of drain characteristics resulting from repetitions of the process we have described, with V_{GS} set to different values of positive voltage. When V_{GS} is reduced to the threshold voltage $V_T = 2$ V, notice that I_D is reduced to 0 for all values of V_{DS}. The drain characteristics are similar to those of an N-channel JFET, except that all values of V_{GS} are positive in the case of the enhancement MOSFET. The enhancement MOSFET can be operated only in an enhancement mode, unlike the depletion MOSFET, which can be operated in both depletion and enhancement modes. The dashed, parabolic line shown on the characteristics in Figure 7–36 joins the saturation voltages, i.e., those satisfying equation 7–18. As in JFET characteristics, the region to the left of the parabola is called the *voltage-controlled–resistance* regions where the drain-to-source resistance changes with V_{GS}. We will refer to the region to the right of the parabola as the *active* region. The device is normally operated in the active region for small-signal amplification.

Figure 7–37(a) shows the structure of a P-channel enhancement MOSFET and its electrical connections. Note that the substrate is N-type material and that a P-type channel is induced by a negative V_{GS}. The field produced by V_{GS} drives electrons away from the region near the insulating layer and inverts it to P material. Figure 7–37(b) shows a typical set of drain characteristics for the P-channel enhancement MOSFET. Note that all values of V_{GS} are negative and that the threshold voltage V_T is negative. N-channel and P-channel MOSFETs are often called *NMOS* and *PMOS* devices for short.

Figure 7–38 shows the schematic symbols used to represent N-channel and P-channel enhancement MOSFETs. As in previous FET symbols, the arrow is drawn

Figure 7–38
Symbols for enhancement-type MOSFETs

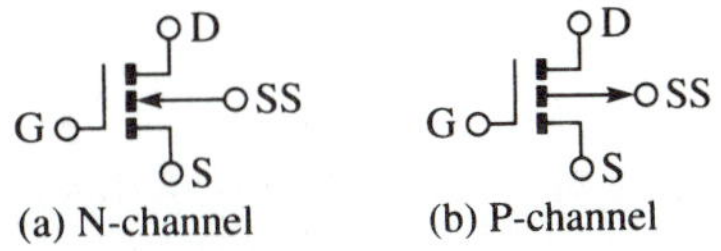

D G SS S D G SS S

(a) N-channel (b) P-channel

pointing into the device for an N-channel FET and outward for a P-channel FET. The broken line symbolizes the fact that the channel is induced rather than being an inherent part of the structure.

Enhancement MOSFET Transfer Characteristic

In the active region, the drain current and gate-to-source voltage are related by

$$I_D = 0.5\beta(V_{GS} - V_T)^2 \qquad V_{GS} \geq V_T \tag{7–19}$$

where β is a constant whose value depends on the geometry of the device, among other factors. A typical value of β is 0.5×10^{-3} A/V^2. Figure 7–39 shows a plot of the transfer characteristic of an N-channel enhancement MOSFET for which $\beta = 0.5 \times 10^{-3}$ A/V^2 and $V_T = 2$ V.

Enhancement MOSFET Bias Circuits

Although enhancement MOSFETs are most widely used in digital integrated circuits (and require no bias circuitry in those applications), they can, and occasionally do, find applications in small-signal amplifiers. Figure 7–40 shows one way to bias an enhancement NMOS for such an application. This circuit appears to be identical to the bias circuit we used for an N-channel JFET (Figure 7–20), but it is quite

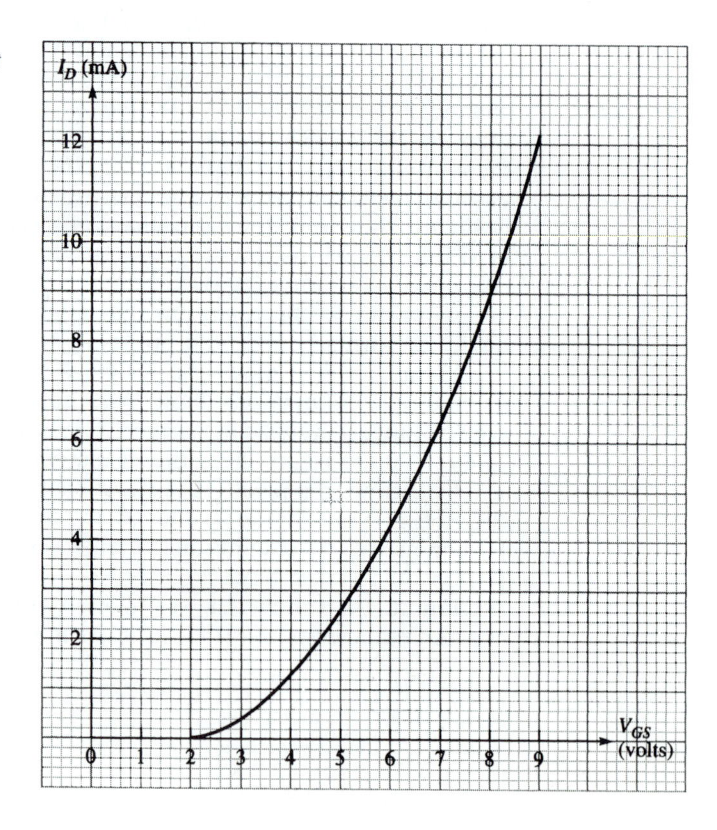

Figure 7–39
Transfer characteristic for an enhancement NMOS FET. $\beta = 0.5 \times 10^{-3}$; $V_T = 2$ V.

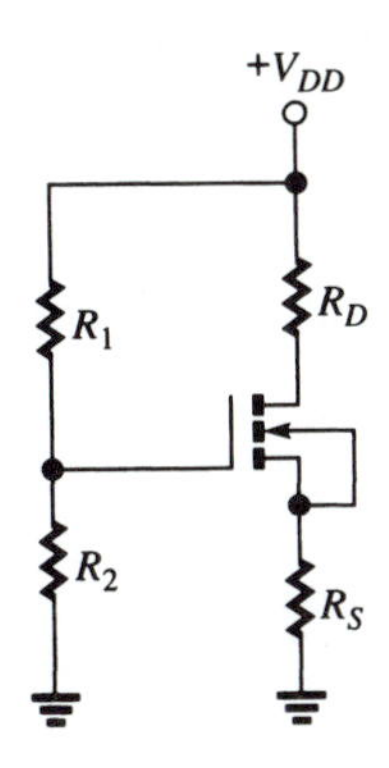

Figure 7–40
A bias circuit for an enhancement MOSFET

different in principle. The resistor R_S does not provide self-bias as it does in the JFET circuit. Self-bias is not possible with enhancement devices; it can occur only in depletion devices. In Figure 7–40, the resistor R_S is used to provide feedback for bias stabilization, in the same way that the emitter resistor does in a BJT bias circuit. The larger the value of R_S, the less sensitive the bias point is to changes in MOSFET parameters caused by temperature changes or by device replacement. Recall that R_S in a JFET self-bias circuit also provides this beneficial effect: We saw (Figure 7–18) that the greater the value of R_S, the less steep the bias line.

Figure 7–41 shows the voltage drops in the enhancement MOSFET bias circuit. R_1 and R_2 form a voltage divider that determines the gate-to-ground voltage V_G:

$$V_G = \left(\frac{R_2}{R_1 + R_2}\right) V_{DD} \tag{7–20}$$

The voltage divider is not loaded by the very large input resistance of the MOSFET, so the values of R_1 and R_2 are usually made very large to keep the ac input resistance of the stage large. Writing Kirchhoff's voltage law around the gate-to-source loop, we find

$$V_{GS} = V_G - I_D R_S \qquad \text{(NMOS)} \tag{7–21}$$

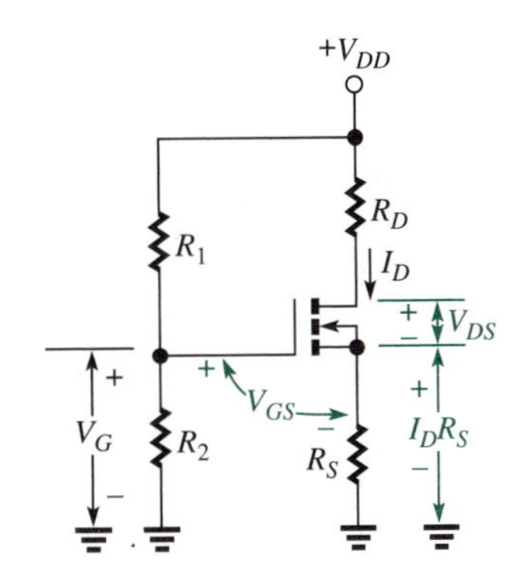

Figure 7–41
Voltage drops in the enhancement NMOS bias circuit

For a PMOS device, V_G and V_{GS} are negative, so equation 7–21 would be written

$$V_{GS} = V_G + I_D R_S \qquad \text{(PMOS)} \tag{7–22}$$

(Note that I_D is considered positive in both equations.) Writing Kirchhoff's voltage law around the drain-to-source loop, we find

$$V_{DS} = V_{DD} - I_D(R_D + R_S) \qquad \text{(NMOS)} \tag{7–23}$$

Again regarding I_D as positive in both the NMOS and PMOS devices, the counterpart of equation 7–23 for a PMOS device is

$$V_{DS} = -|V_{DD}| + I_D(R_D + R_S) \qquad \text{(PMOS)} \tag{7–24}$$

V_{DS} is negative in a PMOS circuit; note that the absolute value of V_{DD} must be used in equation 7–24 to obtain the correct sign for V_{DS}.

Equation 7–21 can be rewritten in the form

$$I_D = -(1/R_S)V_{GS} + V_G/R_S \tag{7–25}$$

Equation 7–25 is seen to be the equation of a straight line on I_D–V_{GS}-axes. It intercepts the I_D-axis at V_G/R_S and the V_{GS}-axis at V_G. The line can be plotted on the same set of axes as the transfer characteristics of the device, and the point of intersection locates the bias values of I_D and V_{GS}.

General Algebraic Solution

We can obtain general algebraic expressions for the bias points in PMOS and NMOS circuits by solving equation 7–19 simultaneously with equation 7–21 or 7–22 for I_D. The results are shown as equations 7–26 and are valid for both NMOS and PMOS devices.

General algebraic solution for the bias point of NMOS and PMOS circuits

$$|V_G| = \frac{R_2}{R_1 + R_2}|V_{DD}|$$

$$I_D = \frac{-B - \sqrt{B^2 - 4AC}}{2A}$$

where

$$A = R_S^2$$

$$B = -2\left[(|V_G| - |V_T|)R_S + \frac{1}{\beta}\right]$$

$$C = (|V_G| - |V_T|)^2$$

$$|V_{DS}| = |V_{DD}| - I_D(R_D + R_S) \qquad \text{See note 1.}$$

$$|V_{GS}| = |V_G| - I_D R_S \qquad \text{See note 2.}$$

(7–26)

1. V_{DS} is positive for an NMOS FET and negative for a PMOS FET.

2. V_{GS} is positive for an NMOS FET and negative for a PMOS FET.

Example 7–12

The transfer characteristic of the NMOS FET in Figure 7–42 is given in Figure 7–43 ($\beta = 0.5 \times 10^{-3}$ and $V_T = 2$ V). Determine values of V_{GS}, I_D, and V_{DS} at the bias point (1) graphically and (2) algebraically.

Solution

1. From equation 7–20,

$$V_G = \left(\frac{22 \times 10^6}{47 \times 10^6 + 22 \times 10^6}\right) 18 \text{ V} = 5.74 \text{ V}$$

Substituting in equation 7–25, we have

$$I_D = -2 \times 10^{-3} V_{GS} + 11.48 \times 10^{-3}$$

This equation intersects the I_D-axis at 11.48 mA and the V_{GS}-axis at V_G = 5.74 V. It is shown plotted with the transfer characteristic in Figure 7–43. The two plots intersect at the quiescent point, where the values of I_D and V_{GS} are approximately $I_D = 2.0$ mA and $V_{GS} = 4.6$ V. The corresponding quiescent value of V_{DS} is found from equation 7–23:

$$V_{DS} = 18 - (2.0 \text{ mA})[(2.2 \text{ k}\Omega) + (0.5 \text{ k}\Omega)] = 12.60 \text{ V}$$

Figure 7–42
(Example 7–12)

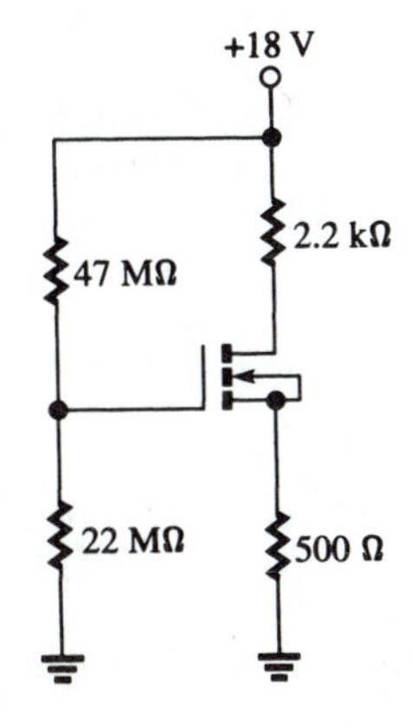

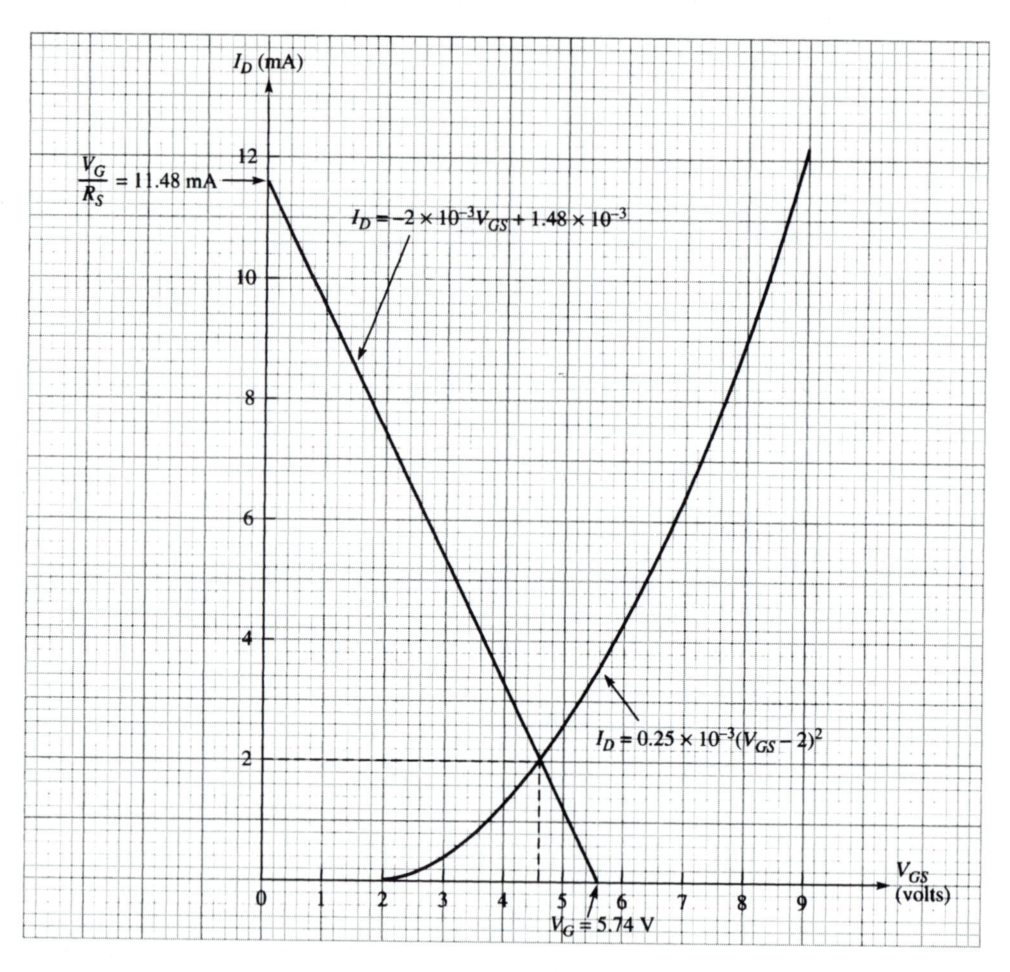

Figure 7–43
(Example 7–12)

In order for this analysis to be valid, the Q-point must be in the active region; that is, we must have $V_{DS} > V_{GS} - V_T$. In our example, we have $V_{DS} = 12.60$ V and $V_{GS} - V_T = 2.6$ V, so we know the results are valid. The validity criterion can be expressed for both NMOS and PMOS FETs as $|V_{DS}| > |V_{GS} - V_T|$.

2. We have already found $V_G = 5.74$ V. Using $R_S = 500\ \Omega$, $R_D = 2.2$ kΩ, $V_{DD} = 18$ V, $V_T = 2$ V, and $\beta = 0.5 \times 10^{-3}$, we have, with reference to equations 7–26,

$$A = (500)^2 = 2.5 \times 10^5$$
$$B = -2[(5.74 - 2)500 + 1/(0.5 \times 10^{-3})] = -7.74 \times 10^3$$
$$C = (5.74 - 2)^2 = 13.9876$$

Substituting these values into the equation for I_D, we find $I_D = 1.927$ mA. Then, $V_{DS} = 18\text{ V} - (1.927\text{ mA})(2.2\text{ k}\Omega + 500\ \Omega) = 12.8$ V and $V_{GS} = 5.74\text{ V} - (1.927\text{ mA})(500\ \Omega) = 4.78$ V. These results agree well with those obtained graphically in part 1.

Example 7–13

SPICE

Use SPICE to find I_D, V_{DS}, and V_{GS} in Example 7–11.

Solution. The SPICE circuit and input data file are shown in Figure 7–44(a). Note that the parameter V_T is entered in the .MODEL statement as VTO = 2 and that β is entered as KP = 0.5E-3 (see Section A–12). All other parameter values are allowed to default. The results of the analysis are shown in Figure 7–44(b). We see that I_D = I(VIDS) = 1.926 mA, V_{DS} = V(3, 2) = 12.8 V and V_{GS} = V(1, 2) = 4.776 V, in good agreement with the previous example.

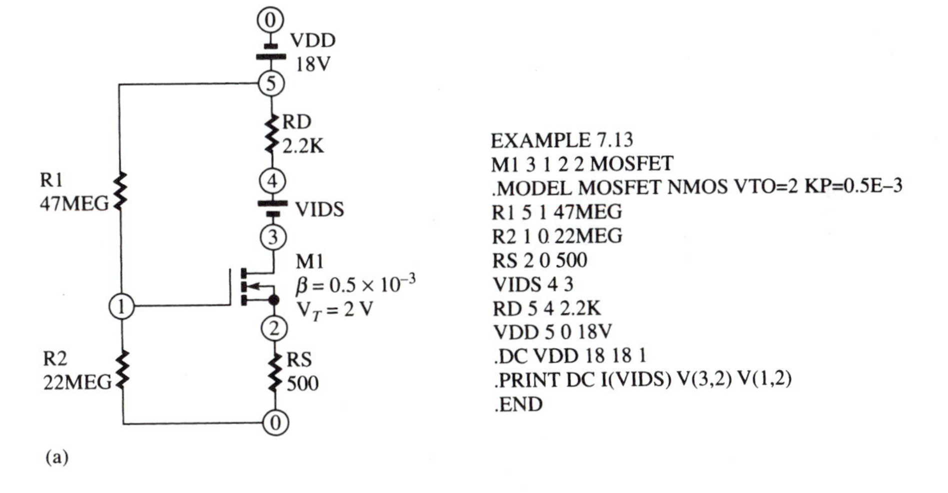


```
EXAMPLE 7.13
****        DC TRANSFER CURVES                    TEMPERATURE = 27.000 DEG C
*****************************************************************************
    VDD           I(VIDS)         V(3,2)          V(1,2)
1.800E+01       1.926E-03       1.280E+01       4.776E+00
```
(b)

Figure 7–44
(Example 7–13)

Feedback Bias

Figure 7–45 shows another way to bias an NMOS FET. The resistor R_G, which is usually very large, is connected between drain and gate and carries no current because of the very large (essentially infinite) resistance at the gate. Since there is no voltage drop across R_G, $V_{GS} = V_{DS}$. We can therefore be sure that $V_{DS} > V_{GS} - V_T$, which ensures that the device is biased in the active region. R_G provides negative feedback to stabilize the bias point. For example, if I_D increases for any reason, then there is a greater drop across R_D and V_{DS} decreases. But, since $V_{GS} = V_{DS}$, this means that V_{GS} also decreases. The decrease in V_{GS} causes a decrease in I_D, thus counteracting the original increase in I_D.

From Figure 7–45, it is clear that

$$V_{DS} = V_{DD} - I_D R_D \tag{7–27}$$

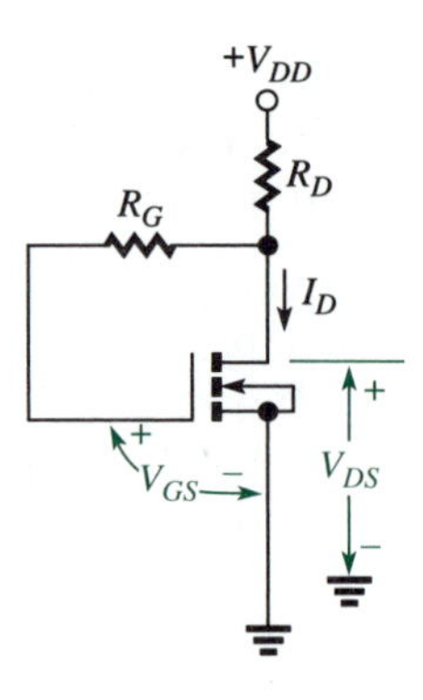

Figure 7–45
Use of a feedback resistor R_G to bias an enhancement MOSFET

Since $V_{GS} = V_{DS}$, the equation of the transfer characteristic (equation 7–23) can be written as

$$I_D = 0.5\beta(V_{DS} - V_T)^2 \qquad \textbf{(7–28)}$$

General Algebraic Solution—Feedback Bias

By solving equations 7–27 and 7–28 simultaneously for I_D, we can obtain the general algebraic solution for the bias point in a MOSFET circuit employing feedback bias. (The derivation is Exercise 7–40.) Equations 7–29 show the results, valid for both NMOS and PMOS circuits.

General algebraic solution for the bias point of PMOS and NMOS circuits with feedback bias

$$I_D = \frac{-B - \sqrt{B^2 - 4AC}}{2A}$$

where $A = R_D^2$

$$B = -2\left[(|V_{DD}| - |V_T|)R_D + \frac{1}{\beta}\right]$$

$$C = (|V_{DD}| - |V_T|)^2$$

$$|V_{DS}| = |V_{GS}| = |V_{DD}| - I_D R_D$$ (positive for NMOS, negative for PMOS) **(7–29)**

7–7 INTEGRATED-CIRCUIT MOSFETs

By far the greatest number of MOSFETs manufactured today are in integrated circuits. The enhancement-type MOSFET has a very simple structure (Figure 7–33) that makes its fabrication in a crystal substrate a straightforward and economical procedure. Furthermore, a very great number of devices can be fabricated in a single chip. Enhancement MOSFETs account for the vast majority of very large scale integrated (VLSI) circuits manufactured, and they are the primary ingredients of digital ICs such as microprocessors and computer memories.

The fabrication of integrated-circuit MOSFETs is accomplished using the same photolithographic techniques and batch production methods we discussed in Chapter 6. Because tens of thousands of components may be fabricated in a single VLSI chip, the techniques we described for producing very fine masks and for direct writing of patterns using electron beams are particularly appropriate to VLSI technology. Ion implantation, which allows close control of impurity concentration and layer depth, is widely used to control the values of threshold voltages and other MOSFET characteristics.

Figure 7–46 shows cross-sectional views of PMOS and NMOS FETs embedded in crystal substrates. Note that a layer of polycrystalline silicon is deposited over the gate of an NMOS device to form the gate terminal. This layer improves device performance but adds to the complexity of the manufacturing procedure. PMOS devices are less expensive to produce but do not perform as well as NMOS circuits, primarily because the mobility of the majority carriers (holes) in P material is smaller than that of the majority carriers (electrons) in N material. NMOS circuits are generally preferred and can be produced with the greatest number of components per chip for a given performance capability.

Another type of digital integrated circuit using enhancement MOSFETs has both PMOS and NMOS devices embedded in the same substrate. These circuits are called *complementary* MOS, or CMOS, circuits. They are more difficult to

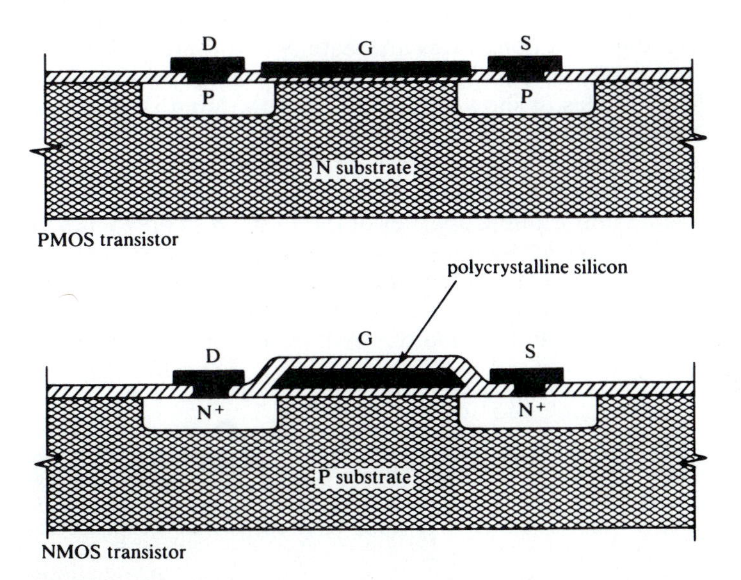

Figure 7–46
Cross-sectional views of integrated-circuit MOSFETs

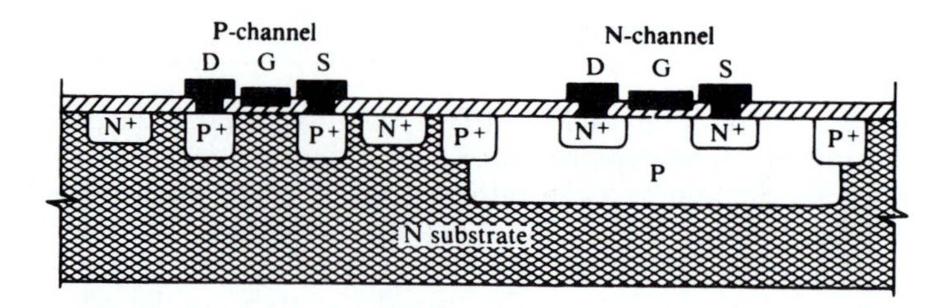

Figure 7–47
Complementary MOS (CMOS) integrated circuit containing both NMOS and PMOS devices

construct than either PMOS or NMOS circuits, but they have the best performance characteristics, especially in terms of switching speed. We will discuss applications of CMOS circuitry in Chapter 8. Figure 7–47 shows a cross-sectional view of a CMOS circuit containing one PMOS and one NMOS transistor. Note that it is necessary to embed a P-type layer in the N substrate for the NMOS transistor. This P layer, frequently called a "tub," is necessary for the formation of the *induced* N channel of the NMOS transistor. Also note the N^+ and P^+ regions used to isolate the transistors. The CMOS structure can be made with either polycrystalline silicon or aluminum gate electrodes. The more complex structure of a CMOS IC is evident in the figure.

7–8 VMOS AND DMOS TRANSISTORS

VMOS Transistors

Still another variation in MOS structure is called VMOS, which is used to produce both N-channel and P-channel enhancement MOSFETs. The name is derived from the appearance of the cross-sectional view (Figure 7–48), in which it can be seen that a V-shaped groove penetrates alternate N and P layers. (In reality, the device is formed in a crater shaped like an inverted pyramid.) As can be seen in the N-channel VMOS transistor shown in Figure 7–48, the length of the induced N channel is determined by the *thickness* of the P layer. The layers are formed using epitaxial growth and diffusion methods, which provide good control over thickness and therefore good control of channel length. Since the aspect ratio of the channel determines some important properties of the FET, this control is a valuable feature of the method. Also, the technique conserves space on the chip surface because the channel can be made longer simply by making the P region thicker. As a result, a greater number of devices can be created in one chip using conventional photolithographic methods. Finally, VMOS transistors have greater current-

Figure 7–48
VMOS structure. Note that the length of the channel depends on the thickness of the diffused P layer.

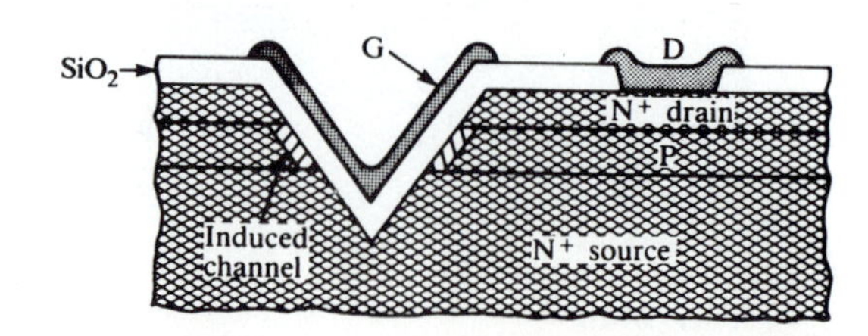

handling capabilities than their planar counterparts and are finding use in power-amplifier applications.

DMOS Transistors

DMOS is an FET structure created specifically for high-power applications. It is a planar transistor whose name is derived from the double-diffusion process used to construct it. Figure 7–49(a) shows a cross-sectional view. Note that drain-to-source current flows through a P-type channel region (that is inverted by a positive gate-to-source voltage), an N^- (lightly doped) epitaxial region, and an N^+ substrate. The channel length can be closely controlled in the diffusion process and is typically very short (a few micrometers). For this reason, it is called a *short-channel* MOS transistor. DMOS transistors are widely used for *switching* heavy currents and high voltages, as in class D amplifiers (Section 15–11) and switching regulators (Section 16–8).

The DMOS transistor has a *parasitic* diode between its drain and source. A parasitic component of a semiconductor device is one that exists as a result of the structure of the device rather than by design. In other words, the parasitic diode in a DMOS transistor is inherent in its structure of P and N layers: Its existence is inevitable. In Figure 7–49(a), the N^+ source and the P channel (called the *body*) form an *electrical bond* rather than a PN junction. The parasitic diode is the PN junction between the body and drain. Since the body is electrically connected to the source, the parasitic diode is effectively connected across the drain and source terminals, with anode connected to source and cathode connected to drain. The schematic symbol for the DMOS transistor often includes the parasitic diode, as shown in Figure 7–49(b).

Figure 7–49
The DMOS transistor

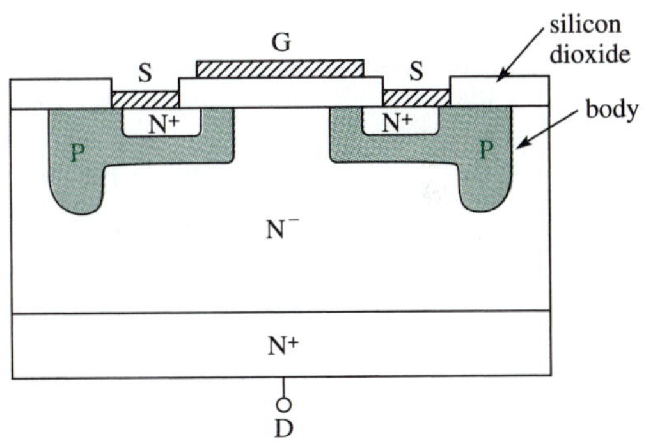

(a) Cross-sectional view.

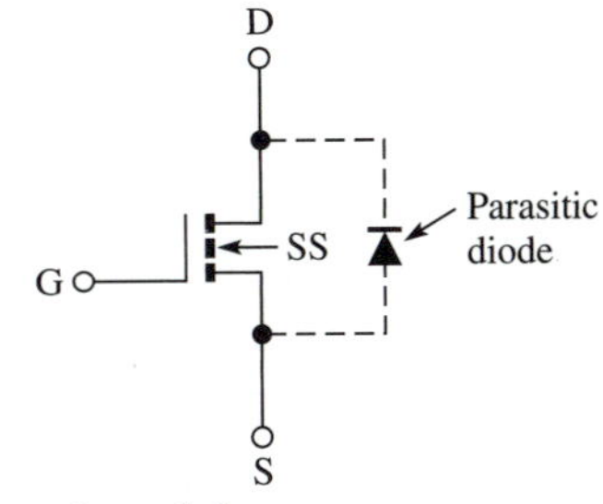

(b) Schematic symbol.

One property shared by short-channel power FETs is that they have a *linear* transfer characteristic. That is, when the gate-to-source voltage is greater than the threshold voltage, the drain current is a linear function of the gate-to-source voltage rather than the nonlinear function illustrated in Figure 7–39. The linear transfer characteristic of short-channel devices is the result of a phenomenon called *velocity saturation,* whereby the velocity of charge carriers reaches but does not exceed a certain value as drain-to-source current increases. The linear transfer characteristic is illustrated and discussed in more detail in Section 8–5.

EXERCISES

Section 7–2
Junction Field-Effect Transistors

7–1. An N-channel JFET has the drain characteristics shown in Figure 7–6. Find the approximate dc resistance between drain and source when $V_{DS} = 1$ V and (a) $V_{GS} = 0$ V, (b) $V_{GS} = -1$ V, and (c) $V_{GS} = -2$ V. Explain why these results confirm that these points lie in the voltage-controlled–resistance region of the characteristics.

7–2. An N-channel JFET has $I_{DSS} = 16$ mA and $V_p = -6$ V.

a. What is the value of $V_{DS(sat)}$ when $V_{GS} = -4$ V?

b. What is the saturation current at $V_{GS} = -4$ V?

7–3. A P-channel JFET has a pinch-off voltage of 8 V. At what value of V_{GS} does $V_{DS(sat)} = -3$ V?

7–4. Using the transfer characteristic shown in Figure 7–10, find approximate values for (a) I_D, when $V_{DS} = 8$ V and $V_{GS} = -1.6$ V; and (b) V_{GS}, when $V_{DS} = 8$ V and $I_D = 10$ mA.

7–5. Using the transfer characteristic shown in Figure 7–10, find the total change in I_D (ΔI_D) when (a) V_{GS} changes from -2.8 V to -1.8 V, and (b) V_{GS} changes from -1.8 V to -0.8 V. What do these results tell you about the linearity of the device?

7–6. An N-channel JFET has a pinch-off voltage of -5.8 V and $I_{DSS} = 15$ mA. Assuming that it is operated in its pinch-off region, find the value of I_D when (a) $V_{GS} = 0$ V, (b) $V_{GS} = -2$ V, and (c) $V_{GS} = -6.5$ V.

7–7. An N-channel JFET having a pinch-off voltage of -3.5 V has a saturation current of 2.3 mA when $V_{GS} = -1$ V. What is its saturation current when (a) $V_{GS} = 0$ V, and (b) $V_{GS} = -2$ V?

7–8. A P-channel JFET has a pinch-off voltage of 6 V and $I_{DSS} = 18$ mA. At what value of V_{GS} in the pinch-off region will I_D equal 6 mA? What is the value of V_{DS} at the boundary of the pinch-off and voltage-controlled–resistance regions when $I_D = 6$ mA?

Section 7–3
JFET Biasing

7–9. The JFET in the circuit of Figure 7–50 has the drain characteristics shown in Figure 7–14. Find the quiescent values of I_D and V_{DS} when (a) $V_{GS} = -2$ V, and (b) $V_{GS} = 0$ V. Which, if either, of the Q-points is in the pinch-off region?

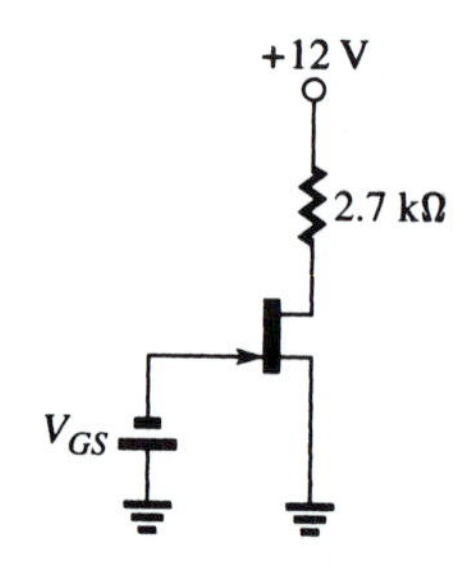

Figure 7–50
(Exercise 7–9)

7–10. Using Figure 7–14, determine the value of V_{GS} in Exercise 7–9 that would be required to obtain $V_{DS} = 8$ V.

7–11. The JFET shown in Figure 7–51 has $I_{DSS} = 14$ mA and $V_p = -5$ V. Algebraically determine the quiescent values of I_D and V_{DS} for (a) $V_{GS} = -3.6$ V, (b) $V_{GS} = -3$ V, and (c) $V_{GS} = -1.7$ V. In each case, check the validity of your results by verifying that the quiescent point is in the pinch-off region. Identify any cases that do not meet that criterion and for which the results are therefore not valid.

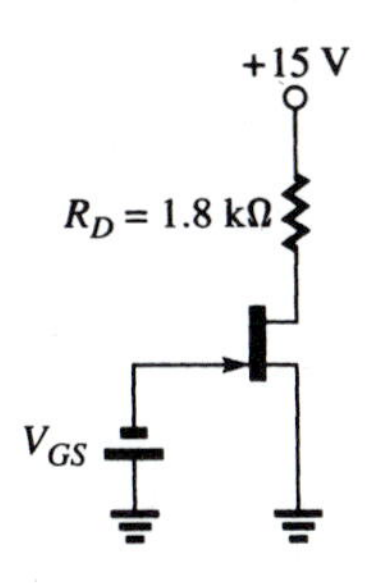

Figure 7–51
(Exercise 7–11)

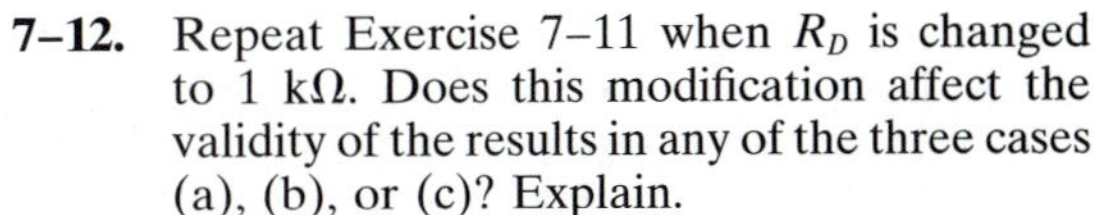

7–12. Repeat Exercise 7–11 when R_D is changed to 1 kΩ. Does this modification affect the validity of the results in any of the three cases (a), (b), or (c)? Explain.

7–13. Figure 7–53 shows the transfer characteristic of the JFET in the circuit of Figure 7–52. Graphically determine the quiescent values of I_D and V_{GS}. Compute the quiescent value of V_{DS} based on your results, and verify their validity.

7–14. Algebraically determine the quiescent values of I_D, V_{GS}, and V_{DS} in the circuit of Exercise 7–13. (Refer to Figure 7–53 to obtain values for I_{DSS} and V_p.)

7–15. Figure 7–55 shows the transfer characteristic for the JFET in the circuit of Figure 7–54.

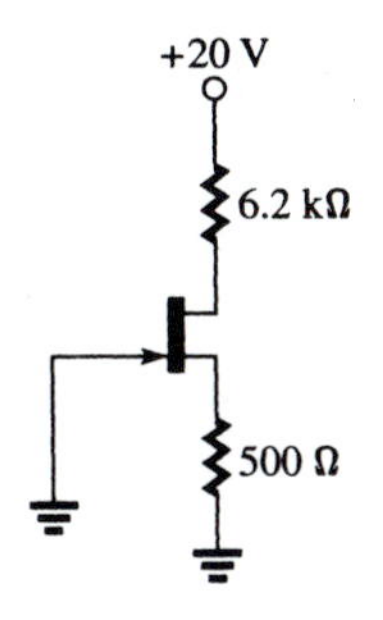

Figure 7–52
(Exercise 7–13)

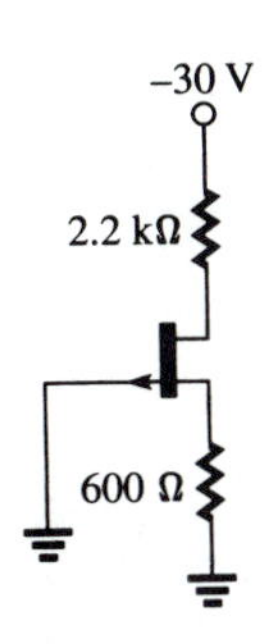

Figure 7–54
(Exercise 7–15)

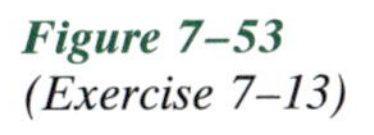

Figure 7–53
(Exercise 7–13)

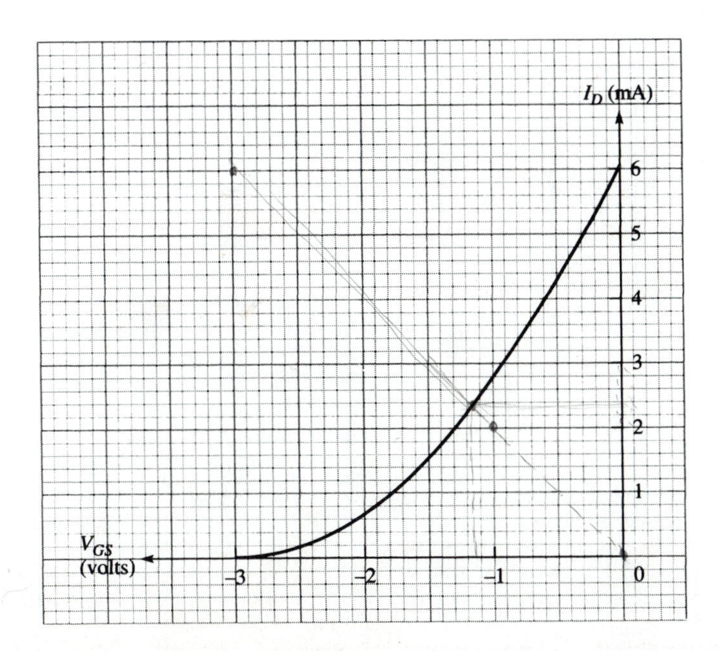

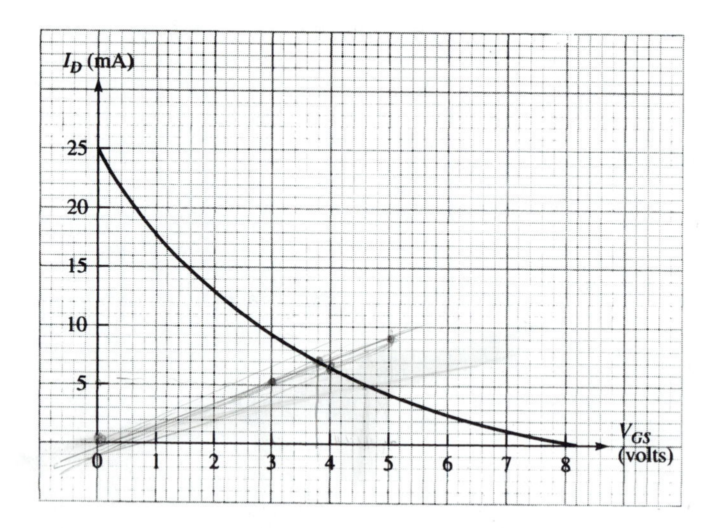

Figure 7–55
(Exercise 7–15)

a. Graphically determine the quiescent values of I_D and V_{GS}. Compute the quiescent value of V_{DS} based on your results, and verify their validity.

b. Determine the quiescent values of I_D, V_{GS}, and V_{DS} algebraically.

7–16. Figure 7–18 shows the possible range of the transfer characteristics for each of the JFETs in Figure 7–56. Find the change in I_D that could be expected in each bias circuit if the JFET characteristics changed over their possible range.

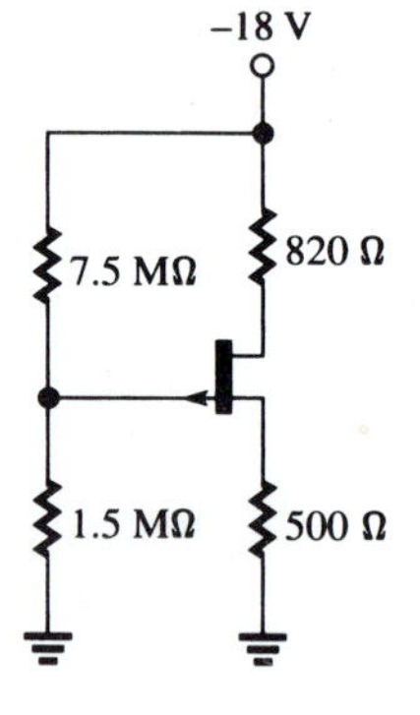

Figure 7–57
(Exericse 7–17)

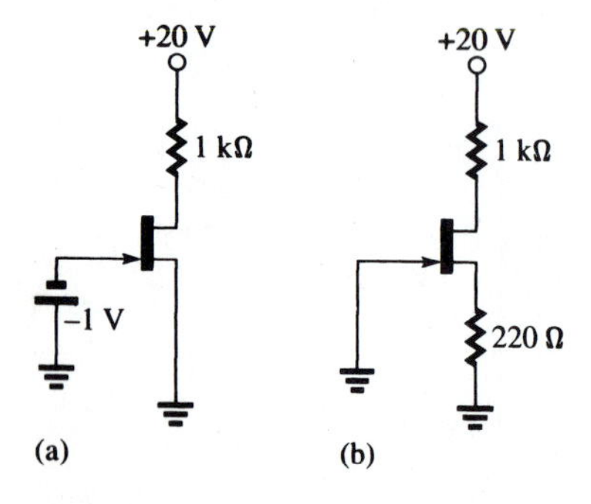

Figure 7–56
(Exercise 7–16)

7–17. The JFET in the circuit of Figure 7–57 has the transfer characteristic shown in Figure 7–22. Graphically determine the quiescent value of I_D and use it to determine the quiescent value of V_{DS}. Verify the validity of your results.

7–18. Algebraically determine the quiescent values of I_D and V_{DS} in Exercise 7–17. (Refer to Figure 7–22 to determine I_{DSS} and V_p.)

7–19. Figure 7–59 shows the transfer characteristic of the JFET in the circuit of Figure 7–58.

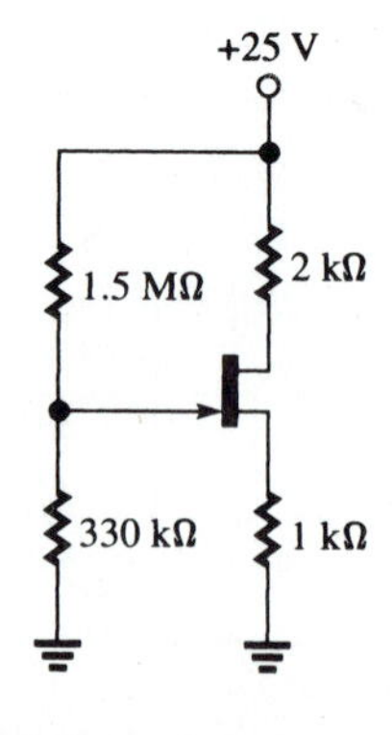

Figure 7–58
(Exercise 7–19)

Figure 7–59
(Exercise 7–19)

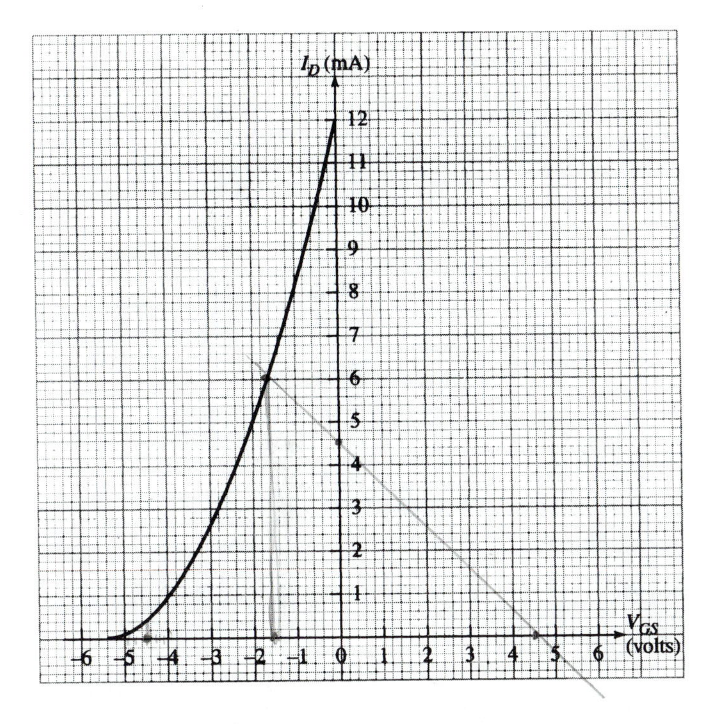

a. Graphically determine the quiescent value of I_D. Verify the validity of your result.

b. Algebraically determine the quiescent values of I_D and V_{DS}. Verify the validity of your results.

7–20. It is desired to modify the circuit of Exercise 7–19 (Figure 7–58) so that the bias line intersects the transfer characteristic at $V_{GS} = -2.8$ V and the V_{GS}-axis at +3 V. Assuming that the JFET has the same transfer characteristic (Figure 7–59), answer the following:

a. If R_1 is to retain its original value (1.5 MΩ), what should be the new value of R_2?

b. What should be the new value of R_S?

c. What is the new value of V_{DS}?

d. Is the JFET biased in its pinch-off region?

Section 7–4
JFET Bias Design

DESIGN EXERCISES

7–21. For an N-channel JFET, $I_{DSS} = 12$ mA and $V_p = -3$ V. Using a 15-V supply voltage, design a self-bias circuit that gives a Q-point of $I_D = 6$ mA and $V_{DS} = 9$ V. Verify that the Q-point in your design is in the pinch-off region of the JFET.

7–22. In a P-channel JFET, $I_{DSS} = 15$ mA and $V_p = 4$ V. Using a 20-V supply voltage, design a self-bias circuit that gives a Q-point of $V_{DS} = -8$ V and $I_D = 4$ mA. Use 5% resistor values that are closest to your calculated values. Then find the actual bias point, assuming the resistors have their nominal values. Verify that this bias point is in the pinch-off region of the JFET.

7–23. An N-channel JFET with $V_p = -2$ V and $I_{DSS} = 10$ mA is to be biased using a voltage divider that sets V_G to 2 V. The required bias point is $I_D = 4$ mA and $V_{DS} = 4$ V. The supply voltage is 15 V. Find values for R_D, R_S, R_1, and R_2.

7–24. A P-channel JFET is to be biased at the nominal values $I_D = 3$ mA and $V_{DS} = -5$ V. The quiescent value of I_D should not vary outside the range from 2.8 mA to 3.5 mA when the JFET characteristic changes over a range in which $I_{DSS} = 10$ mA and $V_p = 1.5$ V to $I_{DSS} = 15$ mA and $V_p = 3$ V. (The nominal characteristic has $I_{DSS} = 12$ mA and $V_p = 2$ V.)

a. Design the circuit, using an 18-V supply.

b. Find the actual range of I_D and V_{DS} over the range of the JFET characteristic when standard valued 5% resistors closest to the calculated values are used. Assume the resistors have their nominal values.

c. Determine if the bias point remains in the pinch-off region over the range of variation when the resistors in (b) are used.

Section 7–5
Manufacturer's Data Sheets

7–25. By referring to the manufacturer's data sheets for the 2N4220–2N4222 series of JFETs, and assuming that $T = 25°C$, determine the following:

a. What is the maximum permissible drain current in each device?

b. What is the maximum pinch-off voltage for 2N4221 transistors?

c. What is the maximum permissible value of V_{GS} for each device?

d. For which device is the variation in possible values of I_{DSS} the greatest (in terms of the ratio of maximum to minimum values)?

7–26. Refer to the manufacturer's data sheet in Figure 7–26 to determine the following:

a. Locate the transfer characteristic that shows the maximum value of I_{DSS} to be 12 mA at 40°C. What is the approximate value of V_{GS} at the point of zero temperature coefficient? What is the approximate maximum value of I_{DSS} at 25°C?

b. Locate the drain characteristics for which I_{DSS} is approximately 1.4 mA. What is the approximate dc resistance between drain and source when $V_{DS} = 0.5$ V and $V_{GS} = 0$ V? At what value of V_{DS} will $I_D = 1.2$ mA when $V_{GS} = -0.1$ V?

Section 7–6
Metal-Oxide-Semiconductor FETs

7–27. The MOSFET shown in Figure 7–60 has the drain characteristics shown in Figure 7–29.

a. Graphically determine the quiescent values of I_D and V_{DS} when $V_{GS} = -1$ V.

b. Repeat, when $V_{GS} = +2$ V.

c. Which, if either, of these Q-points is in the pinch-off region?

d. At which, if either, of these Q-points is the device operating in an enhancement mode?

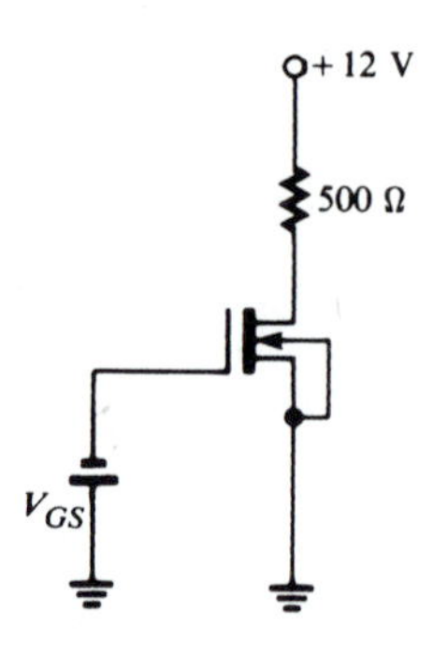

Figure 7–60
(Exercise 7–27)

7–28. A depletion-type MOSFET has the transfer characteristic shown in Figure 7–32(b).

a. Graphically determine the value of I_D when $V_{GS} = +1.5$ V and when $V_{GS} = -1.5$ V.

b. Algebraically determine the value of I_D when $V_{GS} = +1.5$ V and when $V_{GS} = -1.5$ V.

7–29. An induced N-channel enhancement MOSFET has the drain characteristics shown in Figure 7–36. At what value of V_{DS} would a curve corresponding to $V_{GS} = 7.35$ V intersect the parabola?

7–30. The induced N-channel enhancement MOSFET whose drain characteristics are shown in Figure 7–36 is to be operated in its active region with a drain current of 10.4 mA. What value should V_{DS} exceed?

7–31. An enhancement NMOS FET has $\beta = 0.5 \times 10^{-3}$ and $V_T = 2.5$ V. Find the value of I_D when (a) $V_{GS} = 6.14$ V, and (b) $V_{GS} = 0$ V.

7–32. An enhancement PMOS FET has $\beta = 0.5 \times 10^{-3}$ and $V_T = -2$ V. What is the value of V_{GS} when $I_D = 10.32$ mA?

7–33. An enhancement NMOS FET has the transfer characteristic shown in Figure 7–39.

a. Graphically determine V_{GS} when $I_D = 6.4$ mA.

b. Algebraically determine V_{GS} when $I_D = 6.4$ mA.

7–34. In the bias circuit of Figure 7–40, $R_1 = 2.2$ MΩ, $R_2 = 1$ MΩ, $V_{DD} = 28$ V, $R_D = 2.7$ kΩ, and $R_S = 600$ Ω. If $V_{GS} = 5.5$ V, find (a) I_D and (b) V_{DS}.

7–35. In the bias circuit of Figure 7–40, $R_1 = 470$ kΩ, $V_{DD} = 20$ V, $R_D = 1.5$ kΩ, $R_S = 220$ Ω, $I_D = 6$ mA, and $V_{GS} = 6$ V. Find (a) V_{DS} and (b) R_2.

7–36. The MOSFET shown in Figure 7–61 has the transfer characteristic shown in Figure 7–39. Graphically determine the quiescent values

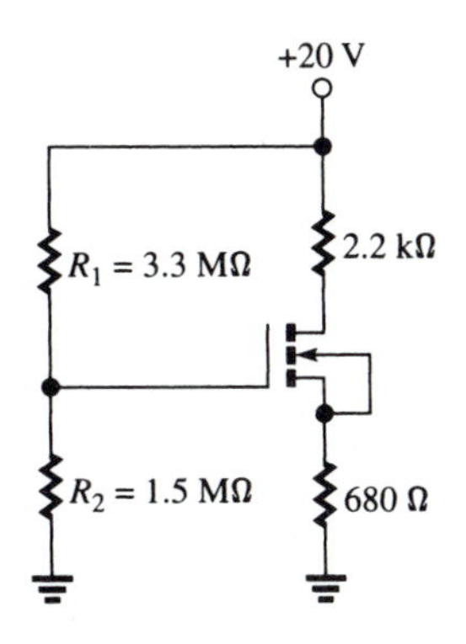

Figure 7–61
(Exercise 7–36)

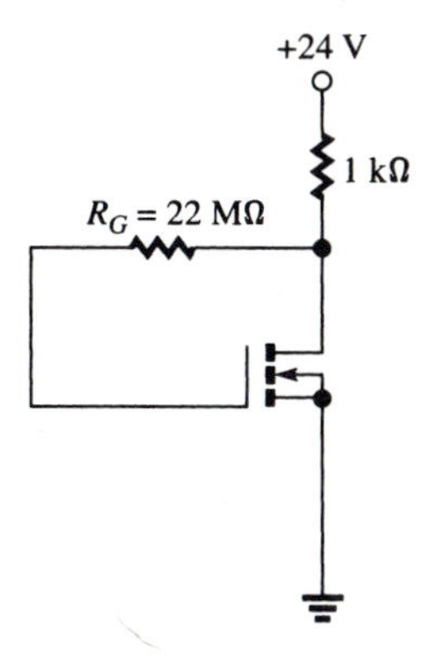

Figure 7–63
(Exercise 7–41)

of I_D and V_{GS}. Find the quiescent value of V_{DS}.

7–37. Algebraically determine I_D, V_{GS}, and V_{DS} in Exercise 7–36. Verify that the MOSFET is operating in its active region.

7–38. It is desired to modify the circuit of Exercise 7–36 so that the load line intersects the transfer characteristic (Figure 7–39) at $V_{GS} = 5$ V and the V_{GS}-axis at 7 V.

a. If R_2 is to retain its original value (1.5 MΩ), what should be the new value of R_1?
b. What should be the new value of R_S?
c. What is the new value of V_{DS}?
d. Is the new bias point in the active region?

7–39. The MOSFET in Figure 7–62 has $\beta = 0.62 \times 10^{-3}$ and $V_T = -2.4$ V. Algebraically determine the quiescent values of I_D, V_{GS}, and V_{DS}. Verify the validity of your results.

7–40. Derive equations 7–29.

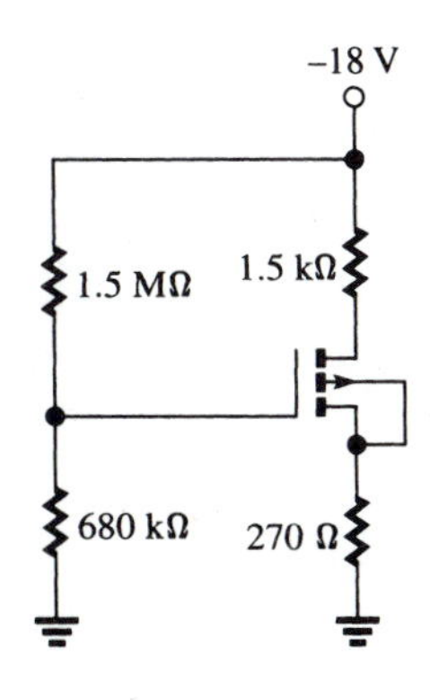

Figure 7–62
(Exercise 7–39)

7–41. The MOSFET in Figure 7–63 has $\beta = 0.5 \times 10^{-3}$ and $V_T = 2$ V. Find the quiescent values of V_{DS}, V_{GS}, and I_D. Is the device in the active region?

7–42. How would the solution to Exercise 7–41 be affected if R_G were reduced from 22 MΩ to 10 MΩ?

Section 7–7
Integrated-Circuit MOSFETs

7–43. What does CMOS stand for? Why is it so named?

7–44. List the three types of MOS integrated circuits, NMOS, PMOS, and CMOS, in increasing order of manufacturing complexity and again in increasing order of performance capabilities.

Section 7–8
VMOS and DMOS Transistors

7–45. List and briefly describe three advantages of VMOS technology.

7–46. What process is used to construct DMOS transistors? What is their primary application? What characteristics make them different from conventional MOSFETs?

SPICE EXERCISES

Use SPICE to solve:
7–47. Exercise 7–14
7–48. Exercise 7–18
7–49. Exercise 7–37
7–50. Exercise 7–39
7–51. Exercise 7–41

8 FET CIRCUITS AND APPLICATIONS

8–1 THE COMMON-SOURCE JFET AMPLIFIER

Like bipolar transistors, field-effect transistors can be connected as ac amplifiers to achieve voltage, power, and/or current gain. For example, in the most useful JFET configuration, the *common-source* amplifier, a small variation in the gate-to-source (input) voltage creates a large variation in the drain-to-source (output) voltage, when the device is operated in its pinch-off region. We will develop equations that allow us to compute the magnitude of the gain in this and other FET configurations when the input variation is small compared to its total possible range, i.e., under small-signal conditions. We begin with a discussion of the FET small-signal parameters that directly affect gain magnitude computations.

Small-Signal JFET Parameters

Recall that transconductance is defined to be the ratio of a small-signal output current (i_o) to the small-signal input voltage (v_{in}) that produced it. The transconductance of a JFET is defined in the context of a *common-source* configuration, for which the output current is i_d and the input voltage is v_{gs}:

$$g_m = \frac{i_d}{v_{gs}} = \frac{\Delta I_D}{\Delta V_{GS}}\bigg|_{V_{DS} = \text{constant}} \text{ siemens} \qquad \textbf{(8–1)}$$

where ΔI_D is the *change* in drain current caused by ΔV_{GS}, the *change* in gate-to-source voltage. Note that we resume the use of lower case letters in keeping with the usual convention for representing small-signal quantities. The value of g_m can be found graphically using a JFET transfer characteristic, since the latter is a plot of drain current versus gate-to-source voltage. Transconductance is thus the *slope* of the transfer characteristic at the operating point, as illustrated in Figure 8–1.

As shown in the figure, g_m is calculated graphically by drawing a line tangent to the characteristic at the quiescent point and then measuring its slope. Note that transconductance increases with increasing values of I_D, that is, at Q-points lying farther up the curve toward I_{DSS}, because the curve becomes steeper in that direction. Since the current in a JFET amplifier varies above and below the Q-point, and

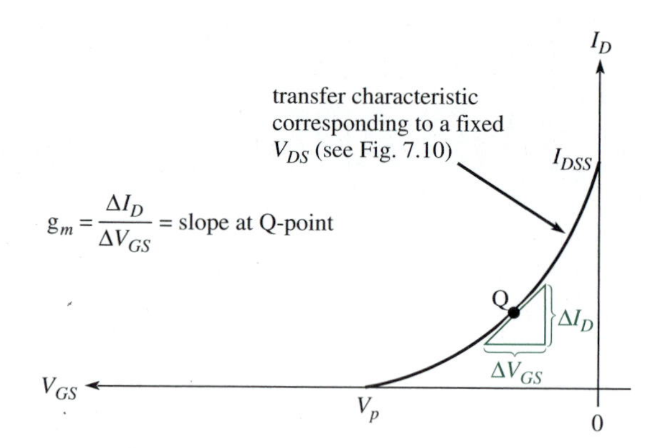

Figure 8–1
Transconductance is the slope of the transfer characteristic. Note that the slope, and hence the value of g_m, increases with increasing I_D.

since the transconductance changes at every point along the characteristic, small-signal analysis is valid only if ΔI_D is small enough to make the change in g_m negligible.

It can be shown that under small-signal conditions the value of g_m is approximately given by

$$g_m = \frac{2I_{DSS}}{|V_p|}\left(1 - \left|\frac{V_{GS}}{V_p}\right|\right) \text{ siemens} \tag{8–2}$$

where V_{GS} is the quiescent value of the gate-to-source voltage. The use of absolute values in equation 8–2 makes it valid for both N-channel and P-channel devices. Equation 8–2 can be used in conjunction with the equation for the transfer characteristic (Exercise 8–4) to express g_m in terms of I_D, the quiescent value of the drain current:

$$g_m = \frac{2I_{DSS}}{|V_p|}\sqrt{\frac{I_D}{I_{DSS}}} \text{ siemens} \tag{8–3}$$

Equation 8–3 clearly shows that the value of g_m increases with increasing I_D. It is a maximum when $I_D = I_{DSS}$, for which it is given the special symbol g_{mO}. From equation 8–3, with $I_D = I_{DSS}$, we have

$$g_{mO} = \frac{2I_{DSS}}{|V_p|} \text{ siemens} \tag{8–4}$$

Although a large value of g_m is desirable, and results in a large voltage gain (as we shall presently see), we would never bias a JFET at $I_D = I_{DSS}$ for small-signal operation. Obviously, no signal variation above that Q-point could occur if that were the case.

Example 8–1

An N-channel JFET has the transfer characteristic shown in Figure 8–2. Determine, both (1) graphically and (2) algebraically, the value of its transconductance when $I_D = 4$ mA. Also determine the value of g_{mO}.

Solution

1. As shown in Figure 8–2, a line is drawn tangent to the transfer characteristic at the point where $I_D = 4$ mA. Using values obtained from the figure, the slope of

Figure 8–2
(Example 8–1)

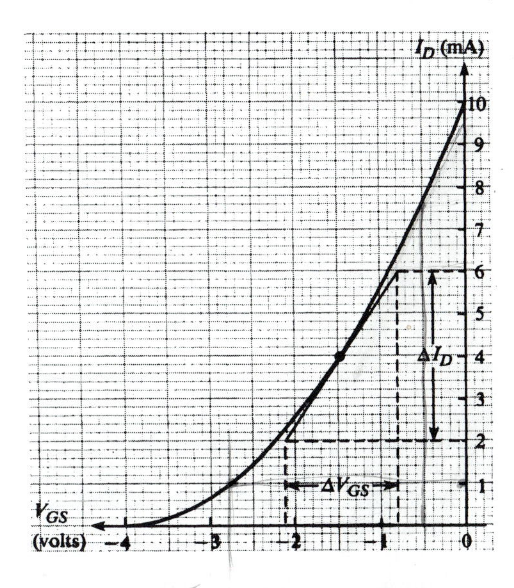

the tangent line is determined to be

$$g_m = \frac{\Delta I_D}{\Delta V_{GS}} = \frac{(6\ \text{mA}) - (2\ \text{mA})}{(2.1\ \text{V}) - (0.8\ \text{V})} = \frac{4 \times 10^{-3}\ \text{A}}{1.3\ \text{V}} = 3.077 \times 10^{-3}\ \text{S}$$

2. From Figure 8–2, it is evident that $V_p = -4$ V and $I_{DSS} = 10$ mA. Using equation 8–3 with $I_D = 4$ mA, we find

$$g_m = \frac{2(10 \times 10^{-3}\ \text{A})}{4\ \text{V}} \sqrt{\frac{4 \times 10^{-3}}{10 \times 10^{-3}}} = 3.16 \times 10^{-3}\ \text{S}$$

We see that there is good agreement between the graphical and algebraic solutions.

The value of g_{mO} is computed using equation 8–4:

$$g_{mO} = \frac{2(10 \times 10^{-3}\ \text{A})}{4\ \text{V}} = 5 \times 10^{-3}\ \text{S}$$

The small-signal output resistance of a common-source JFET is defined by

$$r_o = \frac{v_{ds}}{i_d} = \frac{\Delta V_{DS}}{\Delta I_D}\bigg|_{V_{GS} = \text{constant}} \text{ohms} \tag{8–5}$$

This parameter is also called the *drain resistance* r_d, or r_{ds}. It can be determined graphically from a set of drain characteristics (Figure 7–6), but the lines are so nearly horizontal in the pinch-off region that it is difficult to obtain accurate values. In any case, the value is generally so large that it has little effect on the computation of voltage gain in practical circuits. Values of r_d range from about 50 kΩ to several hundred kΩ in the pinch-off region.

Figure 8–3 shows the small-signal equivalent circuit of the common-source JFET, incorporating the parameters we have discussed. The current source having value $g_m v_{gs}$ is a voltage-controlled current source, since the current it produces

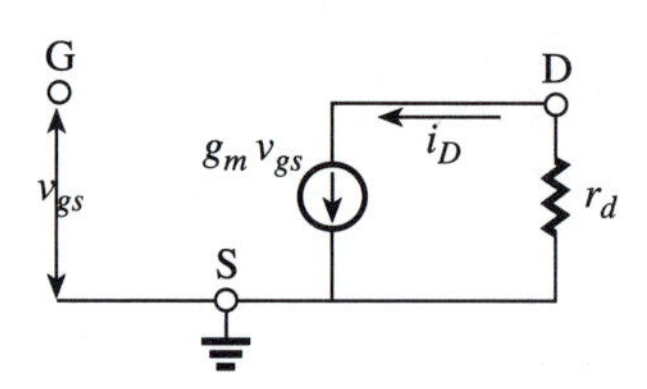

Figure 8–3
Small-signal equivalent circuit of a common-source JFET

depends on the (input) voltage v_{gs}. It is apparent in the figure that $i_d = g_m v_{gs}$, in agreement with the definition of g_m (equation 8–1):

$$g_m = \frac{i_d}{v_{gs}} \Rightarrow i_d = g_m v_{gs}$$

Notice that there is an open circuit shown between the gate and source terminals. Since the gate-to-source junction is reverse biased in normal operation, the extremely large resistance between those terminals can be assumed to be infinite in most practical situations.

The Common-Source JFET Configuration

Figure 8–4 shows a common-source JFET amplifier with fixed bias V_{GG}. R_G is a large resistance connected in series with V_{GG} to prevent the dc source from shorting the ac signal to ground. (Recall that a dc source is an ac short circuit.) The input resistance of the JFET is so large that there is negligible dc voltage division at the gate; that is, the major part of V_{GG} appears from gate-to-source instead of across R_G, so $V_{GS} \approx V_{GG}$. From another viewpoint, the dc gate current is so small that there is negligible drop across R_G. The input coupling capacitor serves the same purpose it does in a BJT amplifier, namely, to provide dc isolation between the signal source and the FET. For the moment, we ignore any signal-source resistance (r_S) and assume that the output is open ($R_L = \infty$).

The total gate-to-source voltage is the sum of the small-signal source voltage, v_S, and the bias voltage V_{GG}. For example, if $V_{GG} = -2$ V and v_S is a sine wave having peak value 0.1 V, then $v_{gs} = -2 + 0.1 \sin \omega t$ volts. Thus, v_{gs} varies between the extreme values $-2 - 0.1 = -2.1$ V and $-2 + 0.1 = -1.9$ V. When v_{gs} goes more positive (toward -1.9 V), the drain current increases. This increase in i_d causes the output voltage v_{ds} to *decrease,* since

$$V_{DS} = V_{DD} - I_D R_D \qquad \textbf{(8–6)}$$

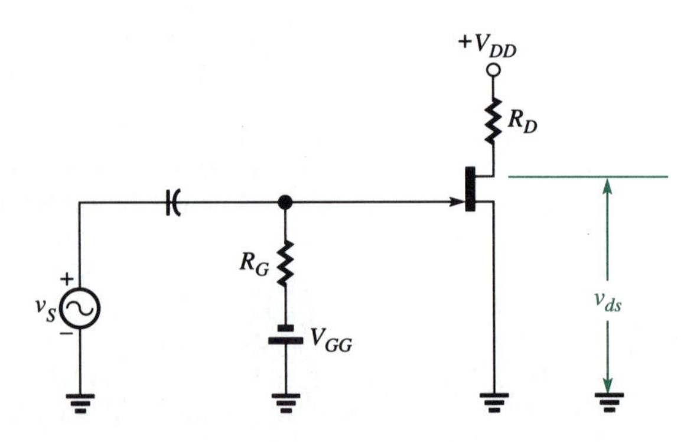

Figure 8–4
A common-source amplifier with fixed bias

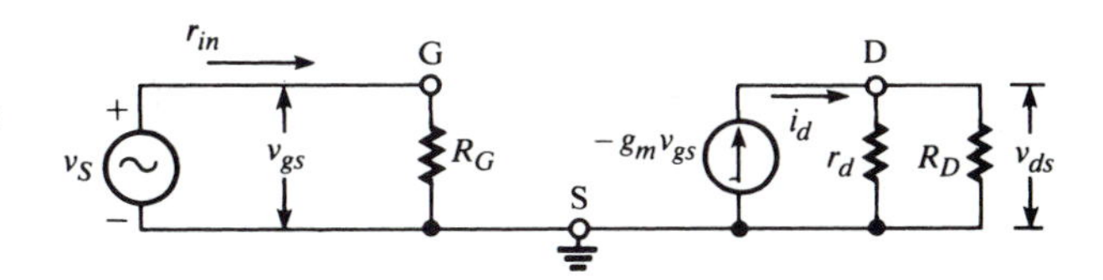

Figure 8–5
The small-signal equivalent circuit of the common-source amplifier shown in Figure 8–4

We conclude that an increase in the input signal voltage causes a decrease in the output voltage and that the output is therefore 180° out of phase with the input.

Figure 8–5 shows the equivalent circuit of the common-source amplifier in Figure 8–4, with the JFET replaced by its small-signal equivalent. The coupling capacitor is assumed to have negligible impedance at the signal frequency we are considering (for now), so it is replaced by a short circuit. As usual, all dc sources are treated as ac short circuits to ground. Notice that the arrow in the controlled current source can be shown reversed (from Figure 8–3) with a minus sign attached. Note: If arrow is reversed, minus sign *must* be attached (not "can be"). The minus sign denotes the phase inversion between input and output, as we have described.

It is clear from Figure 8–5 that

$$v_{ds} = i_d(r_d \parallel R_D) \tag{8–7}$$

Since $i_d = -g_m v_{gs}$, we have

$$v_{ds} = -g_m v_{gs}(r_d \parallel R_D) \tag{8–8}$$

There is no signal-source resistance, so $v_S = v_{gs}$, and the voltage gain is therefore

$$A_v = \frac{v_{ds}}{v_S} = \frac{v_{ds}}{v_{gs}} = -g_m(r_d \parallel R_D) \tag{8–9}$$

In most practical amplifier circuits, the value of r_d is much greater than that of R_D, so $r_d \parallel R_D \approx R_D$ and a good approximation for the voltage gain is

$$\frac{v_{ds}}{v_{gs}} \approx -g_m R_D \tag{8–10}$$

It is important to remember that a JFET amplifier is operated in its pinch-off region. Therefore, the bias point and the voltage variations around the bias point must always satisfy $|V_{DS}| \geq |V_p| - |V_{GS}|$.

So long as the signal source does not drive the gate positive with respect to the source, the reverse-biased gate-to-source resistance is very large and the input resistance to the amplifier is essentially R_G. The equivalent circuit in Figure 8–5 is based on this assumption, and it can be seen that $r_{in} = R_G$.

Example 8–2

The JFET in the amplifier circuit shown in Figure 8–6 has $I_{DSS} = 12$ mA, $V_p = -4$ V, and $r_d = 100$ kΩ.

1. Find the quiescent values of I_D and V_{DS}.
2. Find g_m.
3. Draw the ac equivalent circuit.
4. Find the voltage gain.

Figure 8–6
(Example 8–2)

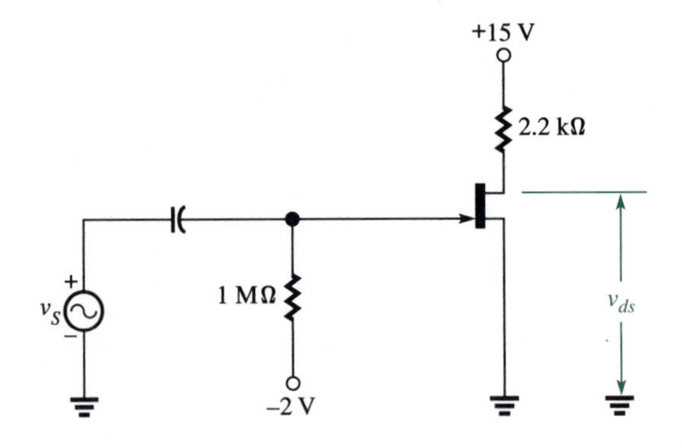

Figure 8–7
(Example 8–2)

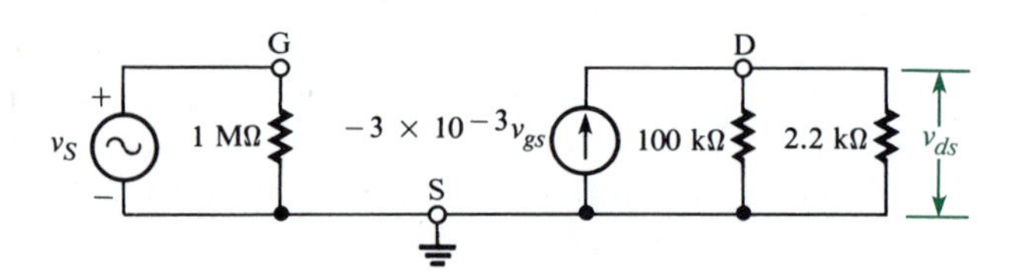

Solution

1. I_D is found using the equation of the transfer characteristic:

$$I_D = I_{DSS}\left(1 - \frac{V_{GS}}{V_p}\right)^2 = (12\text{ mA})\left(1 - \frac{-2\text{ V}}{-4\text{ V}}\right)^2 = 3\text{ mA}$$

From equation 8–6, $V_{DS} = 15 - (3\text{ mA})(2.2\text{ k}\Omega) = 8.4\text{ V}$. Since $V_{DS} = 8.4\text{ V}$ and $|V_p| - |V_{GS}| = 4 - 2 = 2\text{ V}$, we have $|V_{DS}| > |V_p| - |V_{GS}|$, confirming that the JFET is biased in its pinch-off region.

2. From equation 8–3,

$$g_m = \frac{2(12\text{ mA})}{4\text{ V}}\sqrt{\frac{3\text{ mA}}{12\text{ mA}}} = 3 \times 10^{-3}\text{ S}$$

3. The ac equivalent circuit is shown in Figure 8–7.

4. Since $r_d \parallel R_D = (100\text{ k}\Omega) \parallel (2.2\text{ k}\Omega) \approx 2.2\text{ k}\Omega$, we have, from equation 8–10,

$$\frac{v_{ds}}{v_s} = \frac{v_{ds}}{v_{gs}} \approx (-3 \times 10^{-3}\text{ S})(2.2 \times 10^3\ \Omega) = -6.6$$

As the preceding example illustrates, the voltage gain obtainable from a JFET amplifier is generally smaller than that which can be obtained from its BJT counterpart. The principal advantage of the JFET amplifier is its very large input resistance. In a later chapter, we will study *multistage* amplifiers and learn that JFETs and BJTs are combined (BIFET circuits) to achieve large input resistance and high gain.

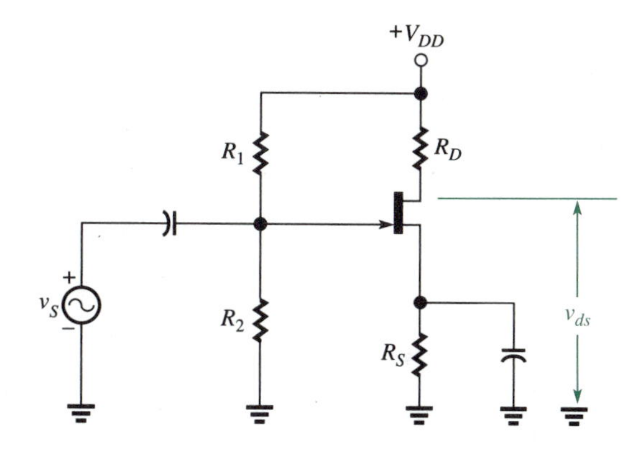

Figure 8–8
A common-source amplifier with improved bias stabilization

Bias-Stabilized JFET Amplifiers

The JFET amplifier we have been studying employs the fixed-bias method for setting the Q-point. Recall from Chapter 7 that this method makes the Q-point sensitive to parameter changes and is therefore undesirable in any application where JFET parameters may vary. Figure 8–8 shows a common-source JFET amplifier using the improved bias method that incorporates self-bias and a voltage divider across the gate. Note the *source bypass capacitor* connected in parallel with R_S. This capacitor serves the same purpose as the emitter bypass capacitor in a BJT common-emitter amplifier, namely, to eliminate the ac degeneration that would otherwise occur due to part of the output signal being dropped across the resistor. As far as the ac signal (v_{ds}) is concerned, the source is grounded, and there is no loss across R_S. Of course, the *dc* voltage V_{DS} is unaffected by the capacitor, so R_S continues to serve its role in providing self-bias for the JFET.

Figure 8–9 shows the ac equivalent circuit of the bias-stabilized amplifier in Figure 8–8. Notice that R_1 and R_2 appear in parallel to ac signals, so $r_{in} = \mathrm{R}_1 \parallel R_2$. These resistors are made quite large to maintain a large input resistance to the amplifier.

The voltage gain of the bias-stabilized amplifier is derived in exactly the same way and has exactly the same value as the fixed-bias amplifier, assuming no signal-source resistance: $A_v = -g_m(r_d \parallel R_D) \approx -g_m R_D$.

When the JFET amplifier is biased using only the self-biasing resistor (no voltage divider), the voltage gain is again computed using $A_v = -g_m(r_d \parallel R_D)$. A resistor R_G is connected between gate and ground to provide dc continuity for the gate circuit. This resistor can be made very large to maintain a large input resistance to the amplifier. The reverse current through the reverse-biased gate–source junction is very small, so the voltage drop across R_G is negligible. However, if an extremely large resistance is used with a JFET having excessive leakage current, the drop should be taken into account. For example, if $R_G = 10\ \mathrm{M}\Omega$ and $I_{DS(reverse)} = 0.1\ \mu\mathrm{A}$, the gate voltage will be increased by $(10\ \mathrm{M}\Omega)(0.1\ \mu\mathrm{A}) = 1$ V

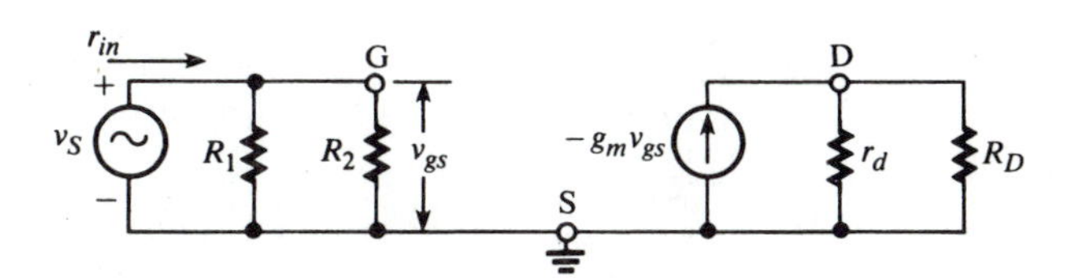

Figure 8–9
The ac equivalent circuit of the bias-stabilized common-source amplifier

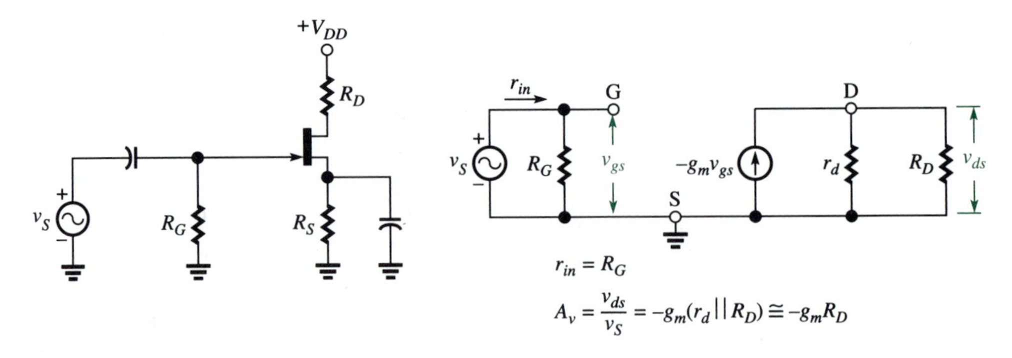

Figure 8–10
The self-biased JFET amplifier and its ac equivalent circuit

with respect to ground. Figure 8–10 shows the self-biased JFET amplifier and its ac equivalent circuit.

To avoid confusion, we will hereafter refer to resistance associated with the signal source as *signal-source resistance,* which we will designate by r_S, and we will continue to refer to resistance in series with the source terminal of the JFET as simply *source resistance,* R_S. When signal-source resistance is present, there is the usual voltage division between r_S and the input resistance of the amplifier, r_{in}. When a load resistor R_L is capacitor-coupled to the output, there is also a voltage division between the amplifier output resistance $r_d \| R_D$ and R_L. Thus,

$$\frac{v_L}{v_s} = \left(\frac{r_{in}}{r_S + r_{in}}\right) A_v \left(\frac{R_L}{R_L + r_d \| R_D}\right) \tag{8–11}$$

Figure 8–11 shows the common-source amplifier in each of the three bias arrangements with load and signal-source resistances included. Also shown are the ac equivalent circuits of each. Note in each case that r_S is in series with the amplifier input and R_L is in parallel with the amplifier output.

For the fixed-bias amplifier (Figure 8–11(a)), equation 8–11 becomes

$$\frac{v_L}{v_S} = \left(\frac{R_G}{r_S + R_G}\right)[-g_m(r_d \| R_D)]\left(\frac{R_L}{R_L + r_d \| R_D}\right) \tag{8–12}$$

$$\approx \left(\frac{R_G}{r_S + R_G}\right)(-g_m R_D)\left(\frac{R_L}{R_L + R_D}\right)$$

$$= \left(\frac{R_G}{r_S + R_G}\right)(-g_m)(R_D \| R_L) \tag{8–13}$$

For the voltage-divider/self-biased amplifier (Figure 8–11(b)), equation 8–11 becomes

$$\frac{v_L}{v_S} = \left(\frac{R_1 \| R_2}{r_S + R_1 \| R_2}\right)[-g_m(r_d \| R_D)]\left(\frac{R_L}{R_L + r_d \| R_D}\right) \tag{8–14}$$

$$\approx \left(\frac{R_1 \| R_2}{r_S + R_1 \| R_2}\right)(-g_m R_D)\left(\frac{R_L}{R_L + R_D}\right)$$

$$= \left(\frac{R_1 \| R_2}{r_S + R_1 \| R_2}\right)(-g_m)(R_D \| R_L) \tag{8–15}$$

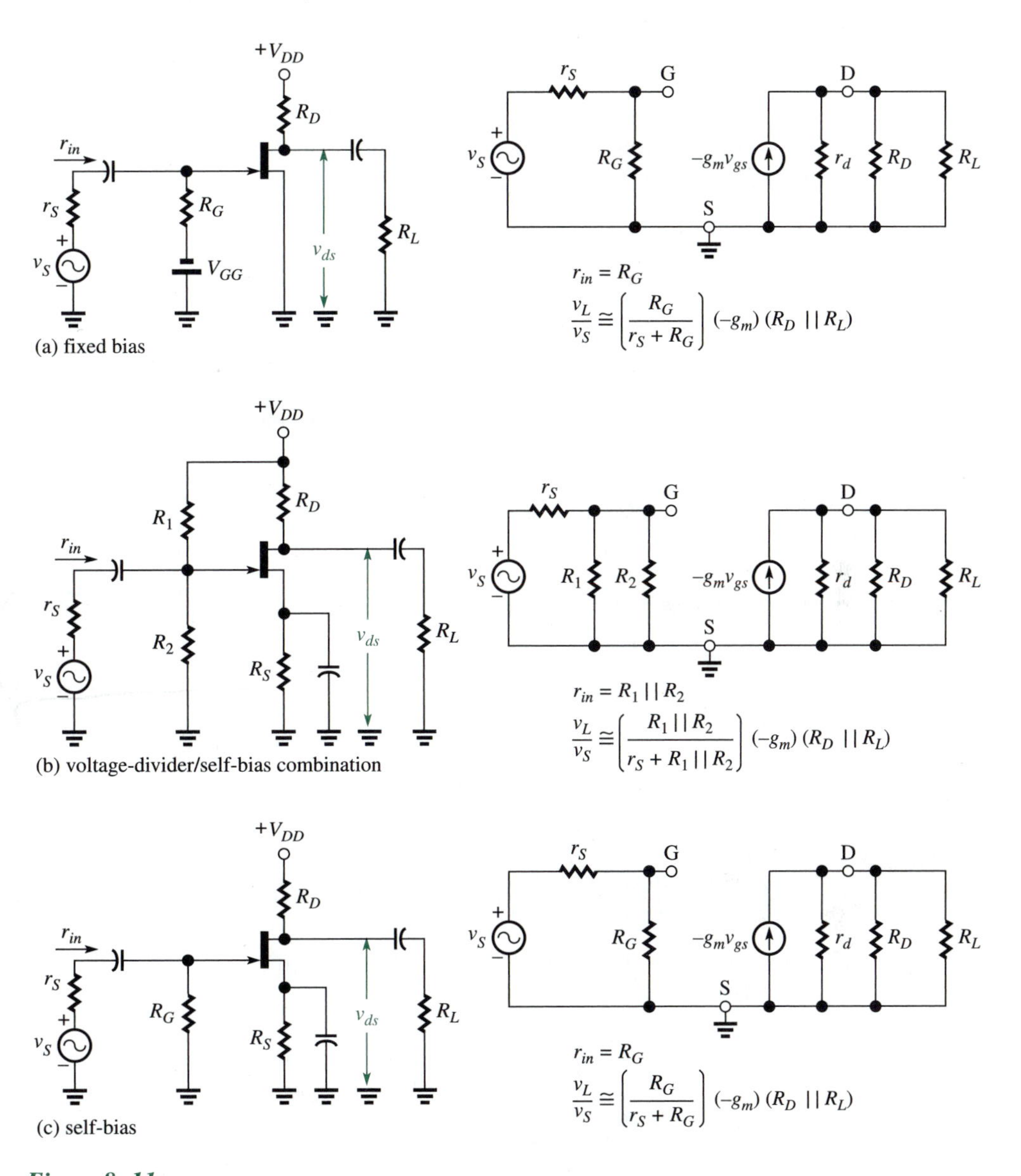

Figure 8–11
Common-source amplifiers with load and signal-source resistances included and their ac equivalent circuits

For the self-biased amplifier (Figure 8–11(c)), equation 8–11 becomes

$$\frac{v_L}{v_S} = \left(\frac{R_G}{r_S + R_G}\right)(-g_m)(r_d \| R_D)\left(\frac{R_L}{R_L + r_d \| R_D}\right) \qquad \textbf{(8–16)}$$

$$\approx \left(\frac{R_G}{r_S + R_G}\right)(-g_m)(R_D \| R_L) \qquad \textbf{(8–17)}$$

Of course, all of the gain relations derived above are exactly the same for P-channel JFET amplifiers.

Example 8–3

The JFET shown in Figure 8–12 has transconductance 4000 μS at its bias point. Its drain resistance is 100 kΩ. Assuming small-signal conditions, find the overall voltage gain, v_L/v_S.

Figure 8–12
(Example 8–3)

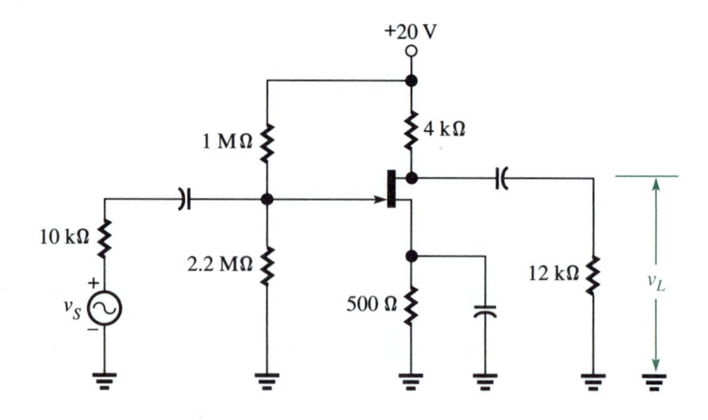

Solution

$$R_1 \| R_2 = (1\ \text{M}\Omega) \| (2.2\ \text{M}\Omega) = 687.5\ \text{k}\Omega$$
$$r_d \| R_D = (100\ \text{k}\Omega) \| (4\ \text{k}\Omega) = 3.846\ \text{k}\Omega$$

Using equation 8–14, we find

$$\frac{v_L}{v_S} = \left(\frac{687.5\ \text{k}\Omega}{10\ \text{k}\Omega + 687.5\ \text{k}\Omega}\right)[(-4\ \text{mS})(3.846\ \text{k}\Omega)]\left(\frac{12\ \text{k}\Omega}{12\ \text{k}\Omega + 3.846\ \text{k}\Omega}\right) = -11.48$$

Assuming that the approximations that lead to equation 8–15 are valid, we obtain from that equation

$$\frac{v_L}{v_S} = \left(\frac{687.5\ \text{k}\Omega}{10\ \text{k}\Omega + 687.5\ \text{k}\Omega}\right)(-4\ \text{mS})(4\ \text{k}\Omega \| 12\ \text{k}\Omega) = -11.83$$

The results found using the exact and approximate equations are quite close, certainly close enough for all practical work.

Example 8–4

SPICE

1. Determine the voltage gain v_L/v_S of the amplifier shown in Figure 8–13(a) by obtaining a plot of the load voltage when v_S is a 0.1-V peak sine wave with frequency 10 kHz. The JFET has $V_p = -5$ V and $I_{DSS} = 18$ mA. Repeat when the peak voltage of the input is changed to 3 V.
2. Find the input resistance, r_{in}(stage), of the amplifier.

Solution

1. The redrawn SPICE circuit and the input data file are shown in Figure 8–13(b). Note that BETA $= I_{DSS}/V_p^2 = 18$ mA/25 $= 7.25 \times 10^{-4}$ A/V^2. The output is

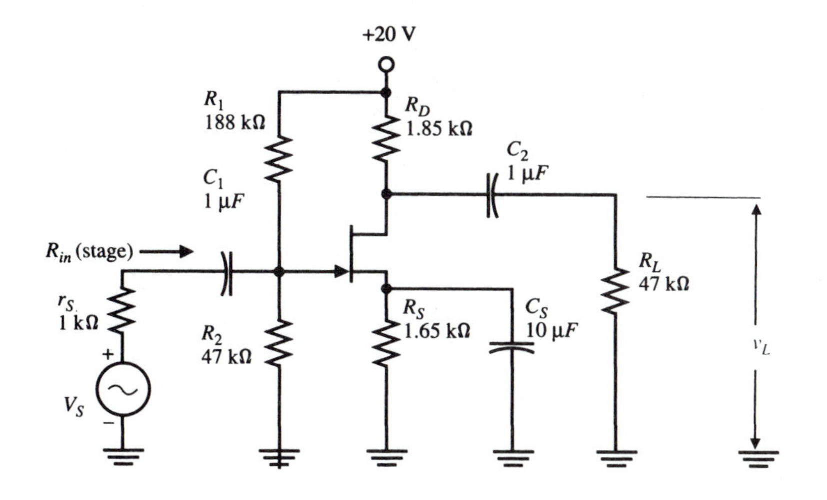
+20 V
R1
188 kΩ
RD
1.85 kΩ
C2
1 μF
C1
1 μF
Rin (stage)
rS
1 kΩ
R2
47 kΩ
RS
1.65 kΩ
CS
10 μF
RL
47 kΩ
vL
VS

(a)

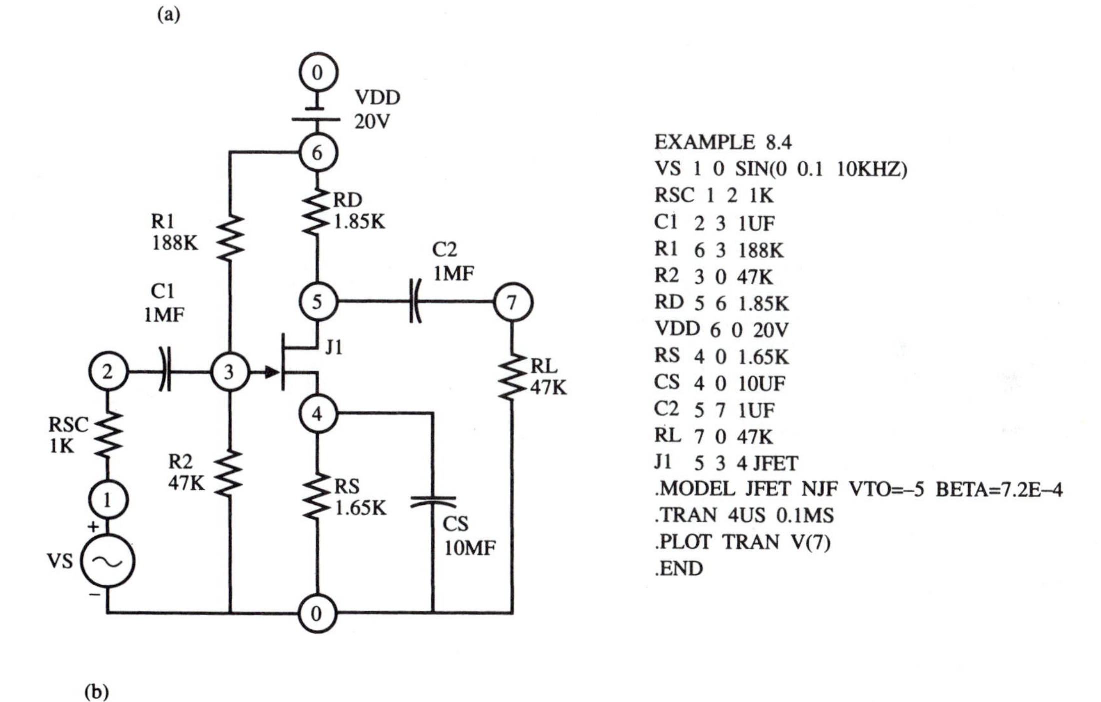
VDD
20V
RD
1.85K
R1
188K
C2
1MF
C1
1MF
J1
RL
47K
RSC
1K
R2
47K
RS
1.65K
CS
10MF
VS
EXAMPLE 8.4
VS 1 0 SIN(0 0.1 10KHZ)
RSC 1 2 1K
C1 2 3 1UF
R1 6 3 188K
R2 3 0 47K
RD 5 6 1.85K
VDD 6 0 20V
RS 4 0 1.65K
CS 4 0 10UF
C2 5 7 1UF
RL 7 0 47K
J1 5 3 4 JFET
.MODEL JFET NJF VTO=-5 BETA=7.2E-4
.TRAN 4US 0.1MS
.PLOT TRAN V(7)
.END

(b)

Figure 8–13
(Example 8–4)

plotted in 26 points over one full cycle ($1/10^4$ Hz = 0.1 ms). The plot, appearing in Figure 8–14(a), shows that the peak-to-peak output is 0.5972 V − (−0.5965 V) = 1.1937 V. Thus, the voltage gain is −(1.193 V p–p)/(0.2 V p–p) = −5.97, where the minus sign reflects the phase inversion evident in the plot.

Figure 8–14(b) shows the plot that results when the peak value of the input is changed to 3 V. Clearly, the amplifier is overdriven by that input, and severe

```
EXAMPLE 8.4
****     TRANSIENT ANALYSIS               TEMPERATURE =    27.000 DEG C
********************************************************************************
   TIME      V(7)
                -1.000D+00    -5.000D-01     0.000D+00      5.000D-01  1.000D+00
                      - - - - - - - - - - - - - - - - - - - - - - - - - - - - -
0.000D+00  0.000D+00 .              .             *             .             .
4.000D-06 -1.472D-01 .              .         *   .             .             .
8.000D-06 -2.862D-01 .              .     *       .             .             .
1.200D-05 -4.080D-01 .              .  *          .             .             .
1.600D-05 -5.044D-01 .              *             .             .             .
2.000D-05 -5.687D-01 .            * .             .             .             .
2.400D-05 -5.965D-01 .           *  .             .             .             .
2.800D-05 -5.859D-01 .            * .             .             .             .
3.200D-05 -5.376D-01 .             *.             .             .             .
3.600D-05 -4.550D-01 .              .*            .             .             .
4.000D-05 -3.437D-01 .              .   *         .             .             .
4.400D-05 -2.114D-01 .              .       *     .             .             .
4.800D-05 -6.676D-02 .              .           * .             .             .
5.200D-05  8.084D-02 .              .             . *           .             .
5.600D-05  2.221D-01 .              .             .     *       .             .
6.000D-05  3.485D-01 .              .             .         *   .             .
6.400D-05  4.525D-01 .              .             .            *.             .
6.800D-05  5.282D-01 .              .             .             .*            .
7.200D-05  5.712D-01 .              .             .             . *           .
7.600D-05  5.792D-01 .              .             .             . *           .
8.000D-05  5.518D-01 .              .             .             .*            .
8.400D-05  4.905D-01 .              .             .             *             .
8.800D-05  3.986D-01 .              .             .          *  .             .
9.200D-05  2.815D-01 .              .             .       *     .             .
9.600D-05  1.459D-01 .              .             .   *         .             .
1.000D-04  7.405D-05 .              .             *             .             .
                      - - - - - - - - - - - - - - - - - - - - - - - - - - - - -
```

(a)

```
EXAMPLE 8.4
****     TRANSIENT ANALYSIS               TEMPERATURE =    27.000 DEG C
********************************************************************************
   TIME      V(7)
                -5.000D+00     0.000D+00     5.000D+00      1.000D+01  1.500D+01
                      - - - - - - - - - - - - - - - - - - - - - - - - - - - - -
0.000D+00  0.000D+00 .              *             .             .             .
4.000D-06 -3.721D+00 .    *         .             .             .             .
8.000D-06 -4.418D+00 . *            .             .             .             .
1.200D-05 -4.683D+00 .*             .             .             .             .
1.600D-05 -4.824D+00 *              .             .             .             .
2.000D-05 -4.899D+00 *              .             .             .             .
2.400D-05 -4.927D+00 *              .             .             .             .
2.800D-05 -4.915D+00 *              .             .             .             .
3.200D-05 -4.862D+00 *              .             .             .             .
3.600D-05 -4.754D+00 .*             .             .             .             .
4.000D-05 -4.558D+00 .*             .             .             .             .
4.400D-05 -4.167D+00 . *            .             .             .             .
4.800D-05 -2.153D+00 .        *     .             .             .             .
5.200D-05  2.087D+00 .              .     *       .             .             .
5.600D-05  5.046D+00 .              .             *             .             .
6.000D-05  6.620D+00 .              .             .    *        .             .
6.400D-05  7.133D+00 .              .             .     *       .             .
6.800D-05  7.161D+00 .              .             .     *       .             .
7.200D-05  7.161D+00 .              .             .     *       .             .
7.600D-05  7.160D+00 .              .             .     *       .             .
8.000D-05  7.160D+00 .              .             .     *       .             .
8.400D-05  7.159D+00 .              .             .     *       .             .
8.800D-05  6.968D+00 .              .             .     *       .             .
9.200D-05  5.949D+00 .              .             .  *          .             .
9.600D-05  3.670D+00 .              .         *   .             .             .
1.000D-04 -2.936D-02 .              *             .             .             .
                      - - - - - - - - - - - - - - - - - - - - - - - - - - - - -
```

(b)

Figure 8–14
(Example 8–4)

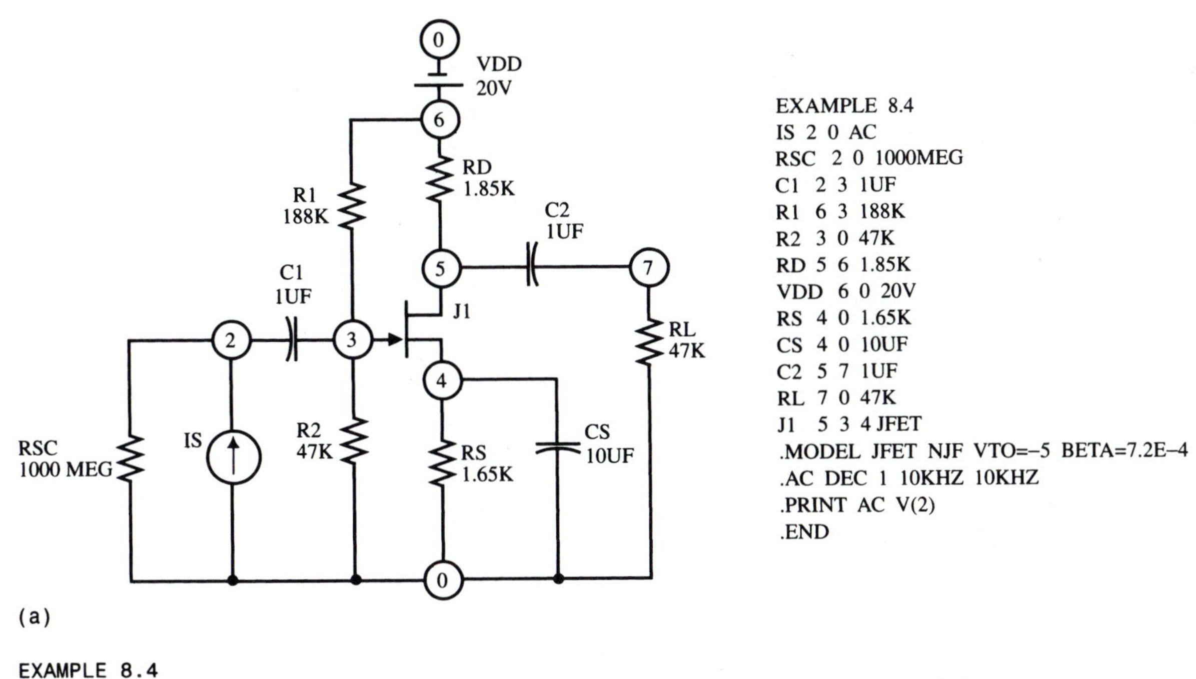
VDD
20V
RD
1.85K
R1
188K
C2
1UF
C1
1UF
J1
RL
47K
RSC
1000 MEG
IS
R2
47K
RS
1.65K
CS
10UF
EXAMPLE 8.4
IS 2 0 AC
RSC 2 0 1000MEG
C1 2 3 1UF
R1 6 3 188K
R2 3 0 47K
RD 5 6 1.85K
VDD 6 0 20V
RS 4 0 1.65K
CS 4 0 10UF
C2 5 7 1UF
RL 7 0 47K
J1 5 3 4 JFET
.MODEL JFET NJF VTO=-5 BETA=7.2E-4
.AC DEC 1 10KHZ 10KHZ
.PRINT AC V(2)
.END
(a)

```
EXAMPLE 8.4
****     AC ANALYSIS                              TEMPERATURE =   27.000 DEG C
*************************************************************************************
  FREQ          V(2)
1.000E+04     3.760E+04
(b)
```

Figure 8–15
(Example 8–4)

clipping at approximately −4.8 V and +7.2 V is apparent in the output. Of course, the voltage gain is unchanged, but the clipped waveform is not useful for determining its value. The waveform does reveal that the maximum permissible peak-to-peak output is approximately 9.6 V (twice the value of the smallest clipping level).

2. Figure 8–15(a) shows the circuit and input data file used to determine r_{in}(stage). Note that we have replaced the SIN generator and its internal resistance of 1 kΩ by an ac current source, I_S, whose value is allowed to default to 1 A. Since

$$r_{in} = \frac{v_{in}}{i_{in}} = \frac{v_{in}}{1\text{ A}} = v_{in}$$

the input resistance is numerically equal to v_{in} at node 2.

Note this important point: The current source must have a resistance in parallel with it because SPICE open-circuits all current sources when it computes the "initial transient solution," and there must be a dc path to ground from every node. In this example, we connect the very large resistance $R_{SC} = 1000$ MΩ in parallel with the source to provide that path. R_{SC} is so much larger than r_{in}(stage) that its presence has negligible effect on the computation.

The results of the ac analysis in Figure 8–15(b) show that $V(2) = v_{in} = r_{in}(\text{stage}) = 3.76 \times 10^4\ \Omega$, which equals the parallel combination of R_1 and R_2. As discussed in Example A–3 of Appendix A (note 5), practical voltage limitations and distortion are not factors in an .AC analysis performed by SPICE, so the unrealistically large input voltage does not affect the validity of the result.

Unbypassed Source Resistor

Figure 8–16 shows a common-source amplifier with an unbypassed source resistor and the small-signal equivalent circuit of the amplifier. For the moment, we do not consider the effects of signal-source resistance and load resistance. Assuming $r_d \gg R_D$,

$$v_o = i_d R_D = -g_m v_{gs} R_D \quad \textbf{(8–18)}$$

From the figure, using Kirchhoff's voltage law, we see that

$$v_{in} = v_{gs} + i_d R_S = v_{gs} + g_m v_{gs} R_S \quad \textbf{(8–19)}$$

Dividing (8–18) by (8–19), we find the voltage gain:

$$\frac{v_o}{v_{in}} = \frac{-g_m v_{gs} R_D}{v_{gs} + g_m v_{gs} R_S} = \frac{-g_m v_{gs} R_D}{v_{gs}(1 + g_m R_S)} = \frac{-g_m R_D}{1 + g_m R_S} \quad \textbf{(8–20)}$$

Figure 8–16
Common-source amplifier with unbypassed source resistor and the small-signal equivalent circuit

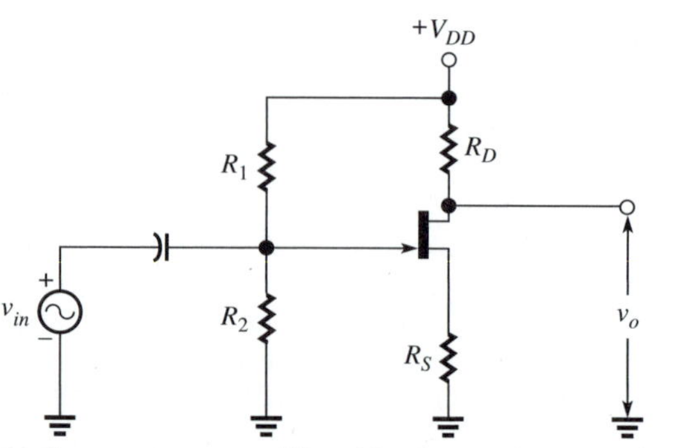

(a) Common-source amplifier with unbypassed source resistor R_S.

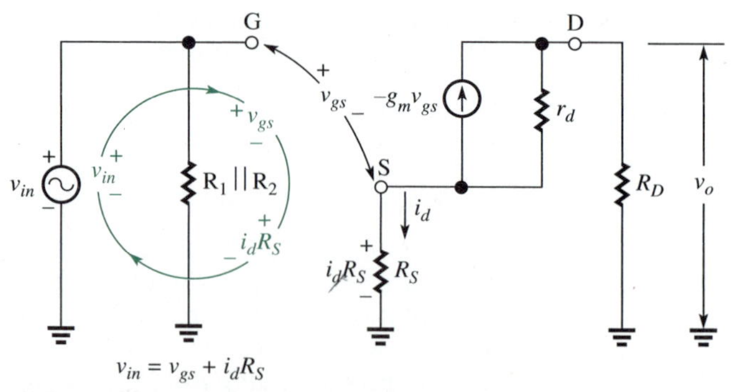

(b) The small-signal equivalent circuit of (a).

Since the voltage gain of the amplifier *with* bypassed source resistor is $-g_m R_D$, we see that the effect of omitting the bypass capacitor is to reduce gain by the factor $1/(1 + g_m R_S)$. Using the typical values $g_m = 4000\ \mu S$, $R_D = 3\ k\Omega$, and $R_S = 500\ \Omega$, the gain with R_S bypassed is $-g_m R_D = -(4000\ \mu S)(3\ k\Omega) = -12$. This value is reduced in the unbypassed case by the factor $1 + g_m R_S = 1 + (4000\ \mu S)(500\ \Omega) = 3$ to the value $-\frac{12}{3} = -4$.

When signal-source resistance r_S and load resistance R_L are considered, the usual voltage division takes place at input and output, and we have

$$\frac{v_L}{v_s} = \left(\frac{r_{in}}{r_S + r_{in}}\right)\left(\frac{-g_m R_D}{1 + g_m R_S}\right)\left(\frac{R_L}{R_L + R_D}\right)$$
$$= \left(\frac{r_{in}}{r_S + r_{in}}\right)\left(\frac{-g_m}{1 + g_m R_S}\right)(R_D \| R_L) \qquad \textbf{(8–21)}$$

where $r_{in} = R_1 \| R_2$ (or R_G if fixed- or self-biased).

8–2 THE COMMON-DRAIN AND COMMON-GATE JFET AMPLIFIERS

Common-Drain Amplifier

Figure 8–17 shows a JFET connected as a common-drain amplifier. Note that the drain terminal is connected directly to the supply voltage V_{DD}, so the drain is at ac ground. Since the input and output signals are taken with respect to ground, the drain is common to both, which accounts for the name of the configuration. Since the ac output signal is between the source terminal and ground, it is the same signal measured from source to drain as measured across R_S. Note that the JFET is biased using the combined self-bias/voltage-divider method, though either of the other two bias methods we studied for the common-source configuration could also be used. As can be seen from the figure,

$$v_{in} = v_{gs} + v_L \qquad \textbf{(8–22)}$$

Therefore, the gate-to-source voltage is

$$v_{gs} = v_{in} - v_L \qquad \textbf{(8–23)}$$

Figure 8–17
A common-drain JFET amplifier

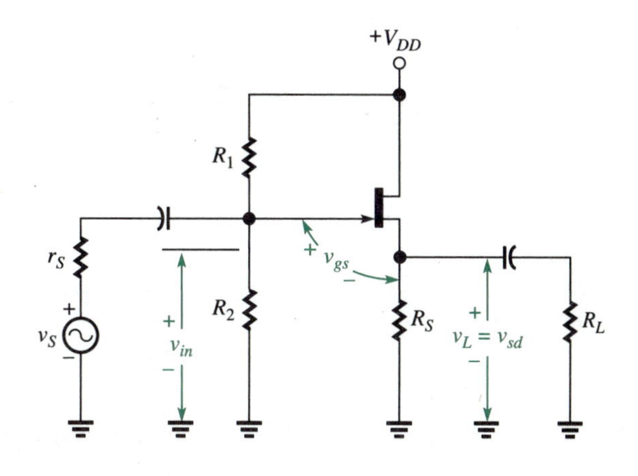

Figure 8–18 shows the ac equivalent circuit of the common-drain amplifier. Although this circuit appears at first glance to be the same as the common-source configuration (Figure 8–11(b)), note that v_{in} is *not* v_{gs}. Rather, v_{in} is related to v_{gs} by equations 8–22 and 8–23. Neglecting the signal-source resistance r_S for the moment, we derive the voltage gain v_L/v_{in} as follows:

$$v_L = g_m v_{gs}(r_d \parallel R_S \parallel R_L) \tag{8–24}$$

Substituting $v_{gs} = v_{in} - v_L$ (equation 8–23) into (8–24), we obtain

$$\begin{aligned} v_L &= g_m(v_{in} - v_L)(r_d \parallel R_S \parallel R_L) \\ &= g_m v_{in}(r_d \parallel R_S \parallel R_L) - g_m v_L(r_d \parallel R_S \parallel R_L) \\ v_L + g_m v_L(r_d \parallel R_S \parallel R_L) &= g_m v_{in}(r_d \parallel R_S \parallel R_L) \end{aligned}$$

Solving for v_L/v_{in}, we obtain

$$\frac{v_L}{v_{in}} = \frac{g_m(r_d \parallel R_S \parallel R_L)}{1 + g_m(r_d \parallel R_S \parallel R_L)} \tag{8–25}$$

It is clear from equation 8–25 that the voltage gain of the common-drain configuration is always less than 1. Note that there is no phase inversion between input and output. When $g_m(r_d \parallel R_S \parallel R_L) \gg 1$ (often the case), equation 8–25 shows that

$$\frac{v_L}{v_{in}} \approx 1 \tag{8–26}$$

In other words, the load voltage is approximately the same as the input voltage, in magnitude and phase, and we say that the output *follows* the input. For this reason, the common-drain amplifier is called a *source follower.*

The input resistance of the source follower can be seen from Figure 8–17 to be the same as it is for a common-source amplifier:

$$r_{in} = R_1 \parallel R_2 \tag{8–27}$$

Similarly, r_{in} is the same as it is for a common-source amplifier when the other bias arrangements are used. When the voltage division between signal-source resistance and r_{in} is taken into account, we find the overall voltage gain to be

$$\frac{v_L}{v_S} = \left(\frac{r_{in}}{r_S + r_{in}}\right)\left[\frac{g_m(r_d \parallel R_S \parallel R_L)}{1 + g_m(r_d \parallel R_S \parallel R_L)}\right] \tag{8–28}$$

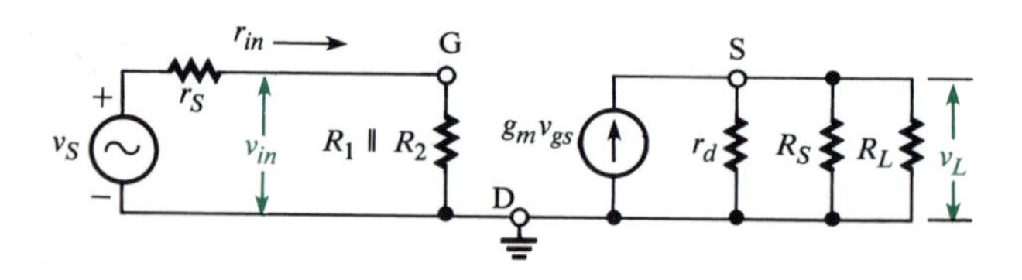

Figure 8–18
The small-signal equivalent circuit for the common-drain amplifier

The output resistance of the source follower (r_o(stage), looking from R_L toward the source terminal) is R_S in parallel with the resistance "looking into" the JFET at the source terminal. The resistance looking into the JFET at the source terminal can be found using equation 8–25. Since we are looking to the left of R_S and R_L, the parallel combination of those resistances is not relevant in the equation. Substituting r_d for $r_d \| R_S \| R_L$, equation 8–25 becomes

$$A_v = \frac{v_o}{v_{in}} = \frac{g_m r_d}{1 + g_m r_d}$$

Now,

$$r_o = \frac{v_o}{i_o} = \frac{A_v v_{in}}{i_o} = \frac{A_v}{g_m} = \frac{r_d}{1 + g_m r_d}$$

Thus,

$$r_o(\text{stage}) = R_S \left\| \left(\frac{r_d}{1 + g_m r_d} \right) \right. \qquad \textbf{(8–29)}$$

It is almost always true that $g_m r_d \gg 1$, so (8–29) reduces to

$$r_o(\text{stage}) \approx R_S \left\| \left(\frac{1}{g_m} \right) \right. \qquad \textbf{(8–30)}$$

Like the BJT emitter follower, the source follower is used primarily as a buffer amplifier because of its large input resistance and small output resistance.

Example 8–5

The JFET in Figure 8–19 has $g_m = 5 \times 10^{-3}$ S and $r_d = 100$ kΩ. Find

1. the input resistance;
2. the voltage gain; and
3. the output resistance of the amplifier stage.

Solution

1. $r_{in} = R_1 \| R_2 = (1.8\ \text{M}\Omega) \| (470\ \text{k}\Omega) = 372.7\ \text{k}\Omega$.
2. Since $r_d \| R_S \| R_L = (100\ \text{k}\Omega) \| (1.5\ \text{k}\Omega) \| (3\ \text{k}\Omega) = 990\ \Omega$, we have, from equation 8.28,

$$\frac{v_L}{v_S} = \left(\frac{372.7\ \text{k}\Omega}{10\ \text{k}\Omega + 372.7\ \text{k}\Omega} \right) \left[\frac{(5 \times 10^{-3}\ \text{S})(990\ \Omega)}{1 + (5 \times 10^{-3}\ \text{S})(990\ \Omega)} \right) = 0.81$$

3. From equation 8–29,

$$r_o(\text{stage}) = (1.5\ \text{k}\Omega) \left\| \left(\frac{100\ \text{k}\Omega}{1 + (5 \times 10^{-3}\ \text{S})(100\ \text{k}\Omega)} \right) \right.$$
$$= (1.5\ \text{k}\Omega) \| (199.6\ \Omega) = 176.2\ \Omega$$

Figure 8–19
(Example 8–5)

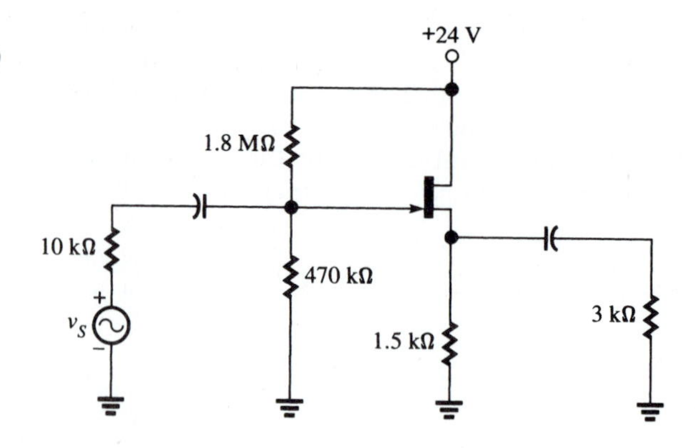

Note that the approximation for r_o (equation 8–30) gives a nearly equal result:

$$r_o(\text{stage}) \approx R_S \parallel (1/g_m) = (1.5\ \text{k}\Omega) \parallel 200 = 176.5\ \Omega$$

Common-Gate Amplifier

Figure 8–20 shows a common-gate amplifier and its small-signal equivalent circuit. We see that the load voltage is developed across the parallel combination of r_d, R_D, and R_L:

$$v_L = -i_d r_L = g_m v_{gs} r_L \tag{8–31}$$

Figure 8–20
The common-gate amplifier

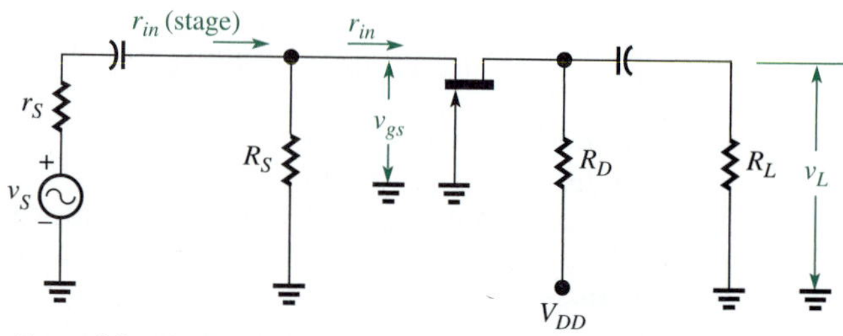

(a) amplifier circuit

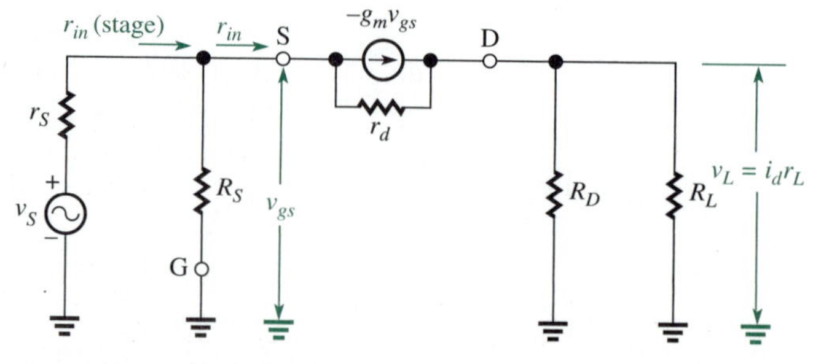

(b) small-signal equivalent circuit

where $r_L = R_D \| R_L$. The voltage gain v_L/v_{gs} is, therefore,

$$\frac{v_L}{v_{gs}} = \frac{g_m v_{gs} r_L}{v_{gs}} = g_m r_L \tag{8–32}$$

We see that the magnitude of this gain is calculated in the same way as it is for the common-source amplifier. Note, however, that the gain is a positive quantity, so there is no phase inversion between input and output.

When signal-source resistance r_S is taken into account, voltage division occurs at the input, and the gain v_L/v_S becomes

$$\frac{v_L}{v_S} = \left(\frac{r_{in}(\text{stage})}{r_S + r_{in}(\text{stage})}\right) g_m r_L \tag{8–33}$$

The input resistance, r_{in}(stage), is the parallel combination of R_S and the resistance r_{in} looking into the source terminal. As can be seen in Figure 8–20(b),

$$r_{in} = \frac{v_{gs}}{i_s} = \frac{v_{gs}}{i_d} = \frac{v_{gs}}{g_m v_{gs}} = \frac{1}{g_m} \tag{8–34}$$

Therefore,

$$r_{in}(\text{stage}) = r_{in} \| R_S = \frac{1}{g_m} \bigg\| R_S \tag{8–35}$$

The input resistance of the common-gate amplifier is much smaller than its common-source and common-drain counterparts. Using the typical values $g_m = 4000\ \mu\text{S}$ and $R_S = 500\ \Omega$, we find $r_{in}(\text{stage}) = (1/4000\ \mu\text{S}) \| (500\ \Omega) = (250\ \Omega) \| (500\ \Omega) = 166.7\ \Omega$. To avoid the large reduction in voltage gain predicted by equation 8–33 when such a small value of r_{in}(stage) is used, it is clear that signal-source resistance r_S must be very small.

8–3 THE JFET CURRENT SOURCE

BJT constant-current sources were discussed extensively in Chapter 6 (Section 6–8). A JFET can be used to supply constant current to a variable load by connecting its gate directly to its source, as illustrated in Figure 8–21. Here, the resistor R_D is regarded as the (variable) load resistance. To be able to supply a current that is independent of R_D, the JFET must remain in its pinch-off region. Recall that the condition for pinch-off is $|V_{DS}| > |V_p| - |V_{GS}|$. Since $V_{GS} = 0$ in this case, the condition reduces to

$$|V_{DS}| > |V_p| \tag{8–36}$$

The constant current produced by the JFET is then $I_D = I_{DSS}$, because that is the drain current in the pinch-off region when $V_{GS} = 0$.

So long as the JFET is in its pinch-off region, the line corresponding to $V_{GS} = 0$ is essentially horizontal, meaning that the same current flows regardless of V_{DS}. See Figure 8–22. In reality, the line rises slightly to the right, so the current source is not perfect. Of course, no current source is perfect. The JFET current source would be perfect if r_d were infinite, which would be the case if the line were horizontal: $r_d = \Delta V_{DS}/\Delta I_D$ with $\Delta I_D = 0$. The JFET can also be used to supply a constant current equal to some value less than I_{DSS} by biasing it appropriately.

Figure 8–21
JFET constant-current sources

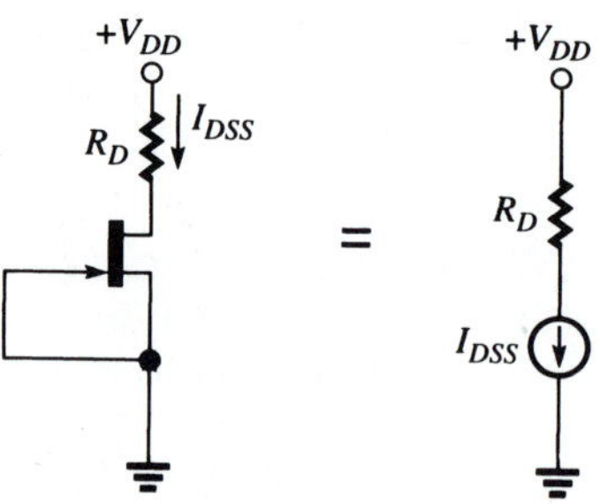

(a) N-channel JFET current source

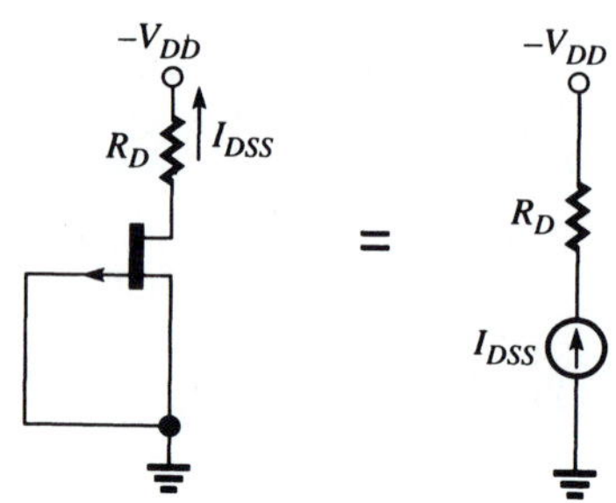

(b) P-channel JFET current source

Figure 8–22
A JFET current source maintains an essentially constant current equal to I_{DSS} in the pinch-off region. If the characteristic were perfectly flat, then ΔI_D would equal zero and r_d would be infinite.

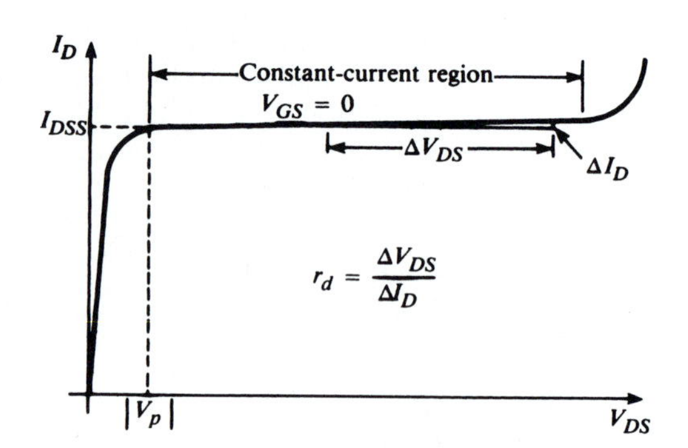

Example 8–6

The JFET shown in Figure 8–23 has $V_p = -4$ V and $I_{DSS} = 14$ mA. What is the maximum value of R_L for which the circuit can be used as a constant-current source?

Solution. To keep the JFET operating in the pinch-off region, we require

$$|V_{DS}| > |V_p|$$

$$18 - (14\text{ mA})R_L > 4$$

$$-14 \times 10^{-3} R_L > -14$$

$$R_L < \frac{14}{14 \times 10^{-3}} = 1\text{ k}\Omega$$

Thus, R_L must be less than 1 kΩ.

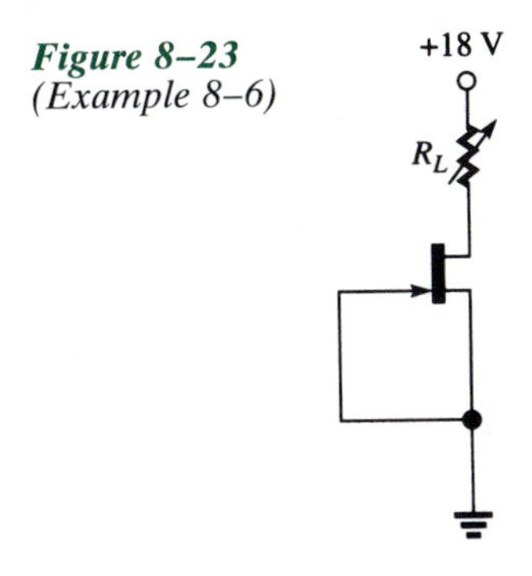

Figure 8–23
(Example 8–6)

8–4 THE JFET AS AN ANALOG SWITCH

An *analog switch* is an electronically controlled device that will either pass or shut off a continuously varying analog-type signal. Figure 8–24 illustrates the concept. By way of contrast, a *digital switch* is one whose output switches between only two possible levels (low or high), such as the BJT inverter we discussed in Chapter 4. As illustrated in Figure 8–24, the analog switch is "opened" or "closed" by a digital-type input. Depending on the nature of the device, a high input may close the switch and a low input may open it, or vice versa. An analog switch is also called a *digital analog switch* (DAS) because a digital input controls the switching of an analog signal.

A JFET can be used as an analog switch by connecting it as shown in Figure 8–25. Note that the analog signal (v_d) is connected to R_D, where a fixed supply voltage (V_{DD}) would normally be connected. The digital signal that opens and closes the switch is the gate-to-source voltage V_{GS}. V_{GS} is either 0 V, which causes the JFET to conduct, or V_p, a negative voltage (for an N-channel JFET) that cuts the JFET off. The output voltage of the switch, v_o, is the drain-to-source voltage, which will be either v_d (when the JFET is cut off) or close to 0 (when the JFET is conducting). Note that the switching arrangement is somewhat different from that

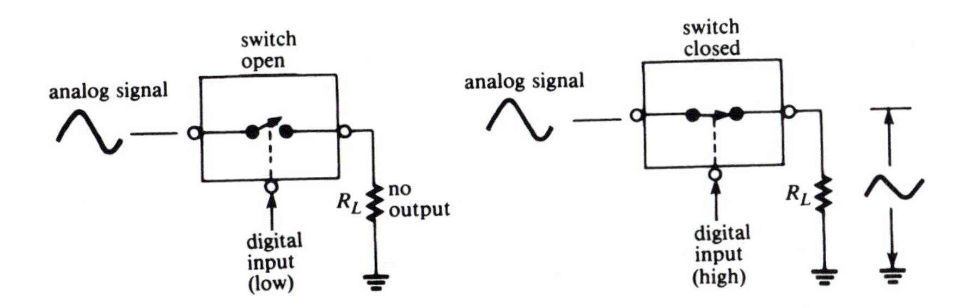

Figure 8–24
An analog switch connects or disconnects a variable (analog) signal, depending on the level of a digital input. In the arrangement shown, the switch is in series with the load R_L.

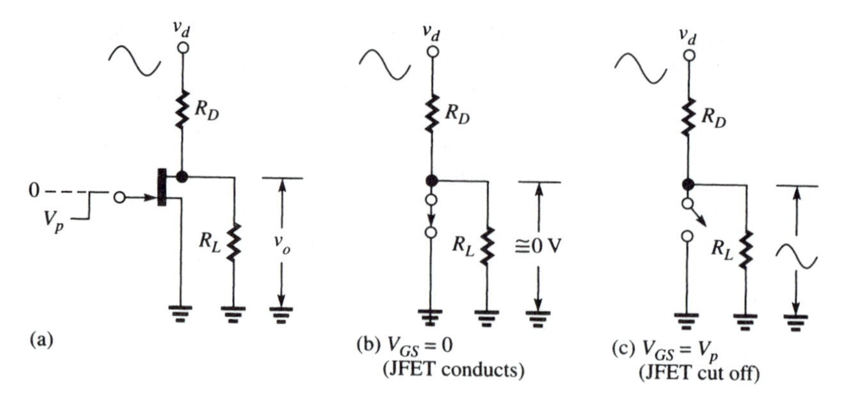

Figure 8–25
The JFET as an analog switch. Note that the switch is in parallel with R_L.

shown in Figure 8–24, because the switch is now in *parallel* with the load resistance R_L. When the switch is closed (JFET on), it effectively shorts out R_L; when the switch is open (JFET cut off), the short is removed.

When used as an analog switch, the JFET is operated in its voltage-controlled–resistance region rather than in pinch-off. As an aid in understanding how the JFET operates as a switch, refer to Figure 8–26, which shows a portion of the drain characteristics for $V_{GS} = 0$ and for $V_{GS} = V_p$. Only the rising portion of the $V_{GS} = 0$ curve in the voltage-controlled–resistance region is shown. The line corresponding to $V_{GS} = V_p$ coincides with the horizontal axis, since $I_D = 0$ in this case. Imagine that the variation in v_d creates a series of parallel load lines, each intersecting the V_{DS}-axis at an instantaneous value of v_d, just as a load line would intersect at V_{DD} if a fixed drain supply voltage were present. Thus, when $V_{GS} = V_p$, the values of V_{DS} are the same as the variations in v_d. This condition corresponds to that shown in Figure 8–25(c). When $V_{GS} = 0$, the operating point moves to the $V_{GS} = 0$ curve and the output voltage (V_{DS}) is very small. This corresponds to Figure 8–25(b). As long as V_{GS} remains at 0, the variations in I_D and V_{DS} are traced by a point that moves up and down the $V_{GS} = 0$ curve. For small variations, the curve is nearly linear and can be seen to be quite steep. The resistance ($\Delta V/\Delta I$) in this region is therefore very small and approximates the short-circuit condition we discussed for the case when the FET is "on" (conducting).

Note in Figure 8–26 that the $V_{GS} = 0$ curve extends into the third quadrant: the region where V_{DS} is negative and I_D is negative. This is the region of operation when the analog signal v_d goes negative and the current through the channel reverses direction. The reversal of polarity causes the gate-to-source junction to be forward

Figure 8–26
Operation of the JFET as an analog switch can be viewed as a variation in $V_{DS} = v_d$ along the horizontal axis (when $V_{GS} = V_p$), or as a variation along the $V_{GS} = 0$ curve when the JFET is conducting.

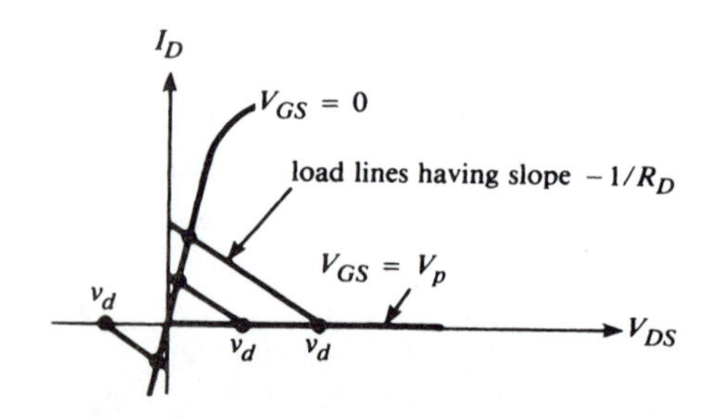

biased but does not affect the *channel* resistance, so the slope of the $V_{GS} = 0$ curve is unchanged. The total variation in v_d must be small so that operation takes place over a small, nearly linear portion of the $V_{GS} = 0$ curve on either side of the origin. Also, R_D must be large enough to ensure that the variation occurs along the lower portion of the $V_{GS} = 0$ curve; that is, the load line should not be steep.

When the JFET is conducting, the small resistance $V_{DS}/I_D \approx v_{ds}/i_d$ in the region around the origin is called the *ON resistance,* $R_{D(ON)}$. Typical values range from 20 to 100 ohms. The smaller the value of $R_{D(ON)}$, the more nearly ideal the switch. While a BJT switch has a lower ON resistance, the JFET switch has the advantage that $i_d = 0$ when $v_d = 0$.

Example 8–7

The JFET in Figure 8–27 has $R_{D(ON)} = 50\ \Omega$. If $v_d = 100$ mV, what is the load voltage v_L (1) when $V_{GS} = V_p$, and (2) when $V_{GS} = 0$ V?

Solution

1. When $V_{GS} = V_p$, the JFET is cut off, and the circuit is equivalent to that shown in Figure 8–28. By the voltage-divider rule,

$$v_L = \left[\frac{100\text{ k}\Omega}{(100\text{ k}\Omega) + (10\text{ k}\Omega)}\right](100\text{ mV}) = 90.9\text{ mV}$$

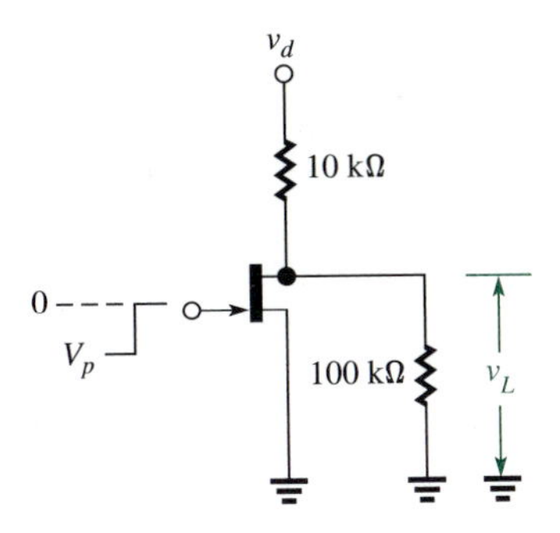

Figure 8–27
(Example 8–7)

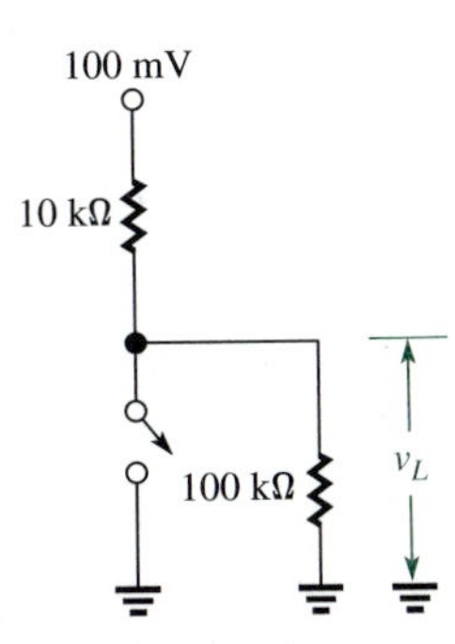

Figure 8–28
(Example 8–7) The circuit equivalent to Figure 8–27 when $V_{GS} = V_p$ and the JFET is cut off

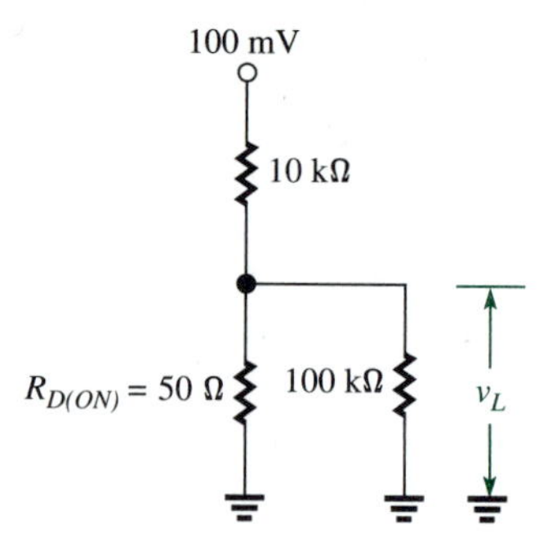

Figure 8–29
(Example 8–7) The circuit equivalent to Figure 8–27 when $V_{GS} = 0$ and the JFET is conducting

2. When $V_{GS} = 0$, the circuit is equivalent to that shown in Figure 8–29.

$$R_L \parallel R_{D(ON)} = (100\ \text{k}\Omega) \parallel (50\ \Omega) \approx 50\ \Omega$$

Therefore,

$$v_L = \left(\frac{50\ \Omega}{50\ \Omega + 10\ \text{k}\Omega}\right)(100\ \text{mV}) = 0.497\ \text{mV}$$

The JFET Chopper

A *chopper* is an analog switch that is turned on and off at a rapid rate by a periodic sequence of pulses, such as a square wave. It is used to convert a slowly varying signal into a series of pulses whose amplitudes vary slowly in the same way as the signal. Figure 8–30 illustrates the concept. A chopper is an example of a *modulator,* in this case, a pulse-amplitude modulator.

Figure 8–31 shows a JFET connected as a chopper. In this variation, the analog switch is in series with the load resistor, R_L, across which the chopped waveform is developed. When the switch is closed (JFET on), current flows from the analog signal source and into R_L. When the switch is open (JFET cut off), no current flows and the output voltage is 0.

Applying the voltage-divider rule to the circuit of Figure 8–31 when the JFET is on, we find

$$v_L = \left(\frac{R_L}{R_L + R_{D(ON)} + r_s}\right) v_d \qquad \textbf{(8–37)}$$

If R_L is much greater than $R_{D(ON)} + r_S$, then v_L is approximately the same as v_d. Thus, the amplitude of the output (pulse) follows the analog input during each interval when the JFET is conducting.

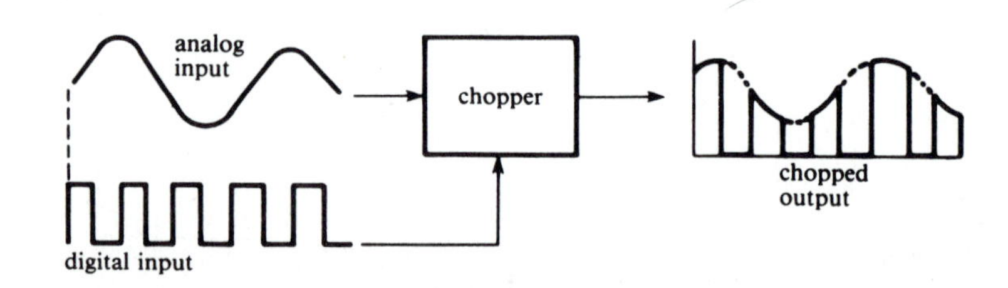

Figure 8–30
A chopper produces a series of pulses whose amplitudes follow the variations of an analog input signal.

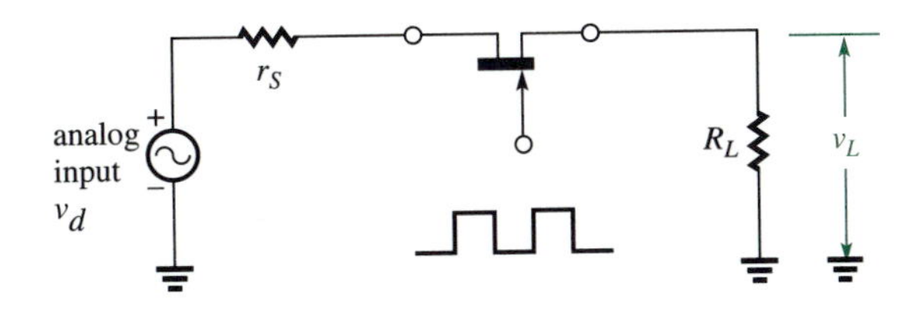

Figure 8–31
The JFET connected as a chopper. Note that the switch is in series with the load.

8–5 SMALL-SIGNAL MOSFET AMPLIFIERS

The most convenient small-signal model for a MOSFET, like that of a JFET, incorporates the transconductance of the device. Recall that the characteristics of a depletion-type MOSFET are quite similar to those of a JFET, the only difference being that the depletion MOSFET can be operated in both the depletion and enhancement modes. As we might therefore expect, the small-signal model for the depletion MOSFET is identical to the JFET model, and it is used in the same way to analyze a depletion-type MOSFET amplifier. The value of the transconductance can be found graphically and algebraically in the same way it is found for a JFET.

The transconductance of the *enhancement*-type MOSFET can be found graphically using the transfer characteristic and the definition:

$$g_m = \left.\frac{\Delta I_D}{\Delta V_{GS}}\right|_{V_{DS} = \text{constant}} \text{siemens} \tag{8–38}$$

Figure 8–32 shows how g_m is computed as the slope of a line drawn tangent to the characteristic at the operating or Q-point. It is clear from the figure that the slope of the characteristic, and hence the value of g_m, changes as the Q-point is changed. Therefore small-signal analysis requires that the signal variation around the Q-point be confined to a limited range over which there is negligible change in g_m, i.e., to an essentially linear segment of the characteristic. Also, to ensure that the device is operated within its active region, the variation must be such that the following inequality is always satisfied:

$$|V_{DS}| > |V_{GS} - V_T| \tag{8–39}$$

It can be shown that under small-signal conditions the transconductance of an enhancement MOSFET can be determined from

$$g_m = \beta(V_{GS} - V_T) \text{ siemens} \tag{8–40}$$

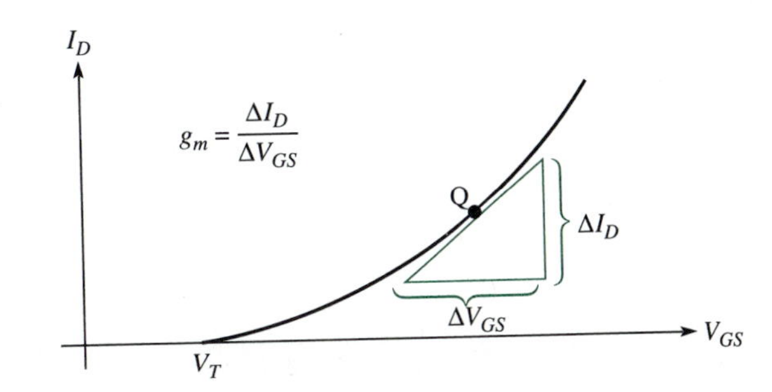

Figure 8–32
The transconductance of an enhancement MOSFET is the slope of a line drawn tangent to the transfer characteristic at the Q-point.

Figure 8–33
A common-source NMOS amplifier and its equivalent circuit

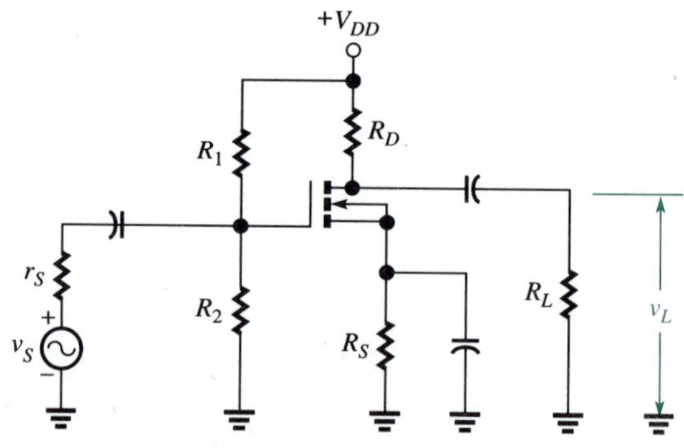

(a) An N-channel enhancement MOSFET (NMOS) amplifier with voltage-divider bias

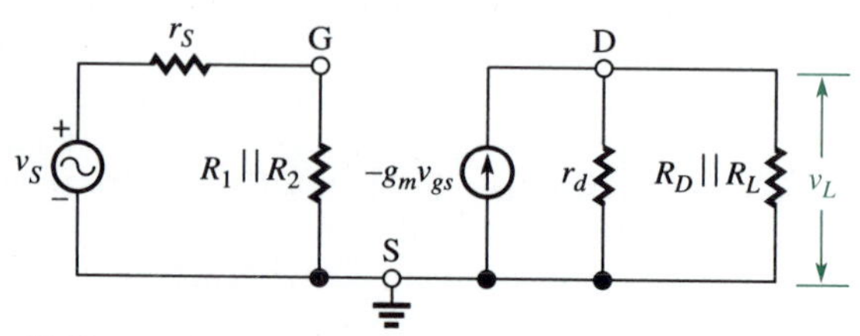

(b) The small-signal equivalent circuit of (a)

Figure 8–33 shows a common-source enhancement MOSFET amplifier and its small-signal equivalent circuit. The MOSFET is biased using the voltage-divider method discussed in Chapter 7 (Figure 7–40).

Notice that the small-signal equivalent circuit of the NMOS amplifier is identical to that of the JFET amplifier (Figure 8–11(b)). Consequently, all gain and impedance relations are the same as those derived for the JFET amplifier:

$$r_{in} = R_1 \| R_2 \tag{8–41}$$

$$\frac{v_L}{v_S} = \left(\frac{R_1 \| R_2}{r_S + R_1 \| R_2}\right)(-g_m)(r_d \| R_D \| R_L) \tag{8–42}$$

Example 8–8

The MOSFET shown in Figure 8–34 has the following parameters: $V_T = 2$ V, $\beta = 0.5 \times 10^{-3}$, $r_d = 75$ kΩ. It is biased at $I_D = 1.93$ mA.

1. Verify that the MOSFET is biased in its active region.
2. Find the input resistance.
3. Draw the small-signal equivalent circuit and find the voltage gain v_L/v_S.

Solution

1. $V_{DS} = V_{DD} - I_D(R_D + R_S) = 18\text{ V} - (1.93\text{ mA})(2.2\text{ k}\Omega + 500\ \Omega) = 12.78\text{ V}$

$$V_G = \left(\frac{22\text{ M}\Omega}{47\text{ M}\Omega + 22\text{ M}\Omega}\right)18\text{ V} = 5.74\text{ V}$$

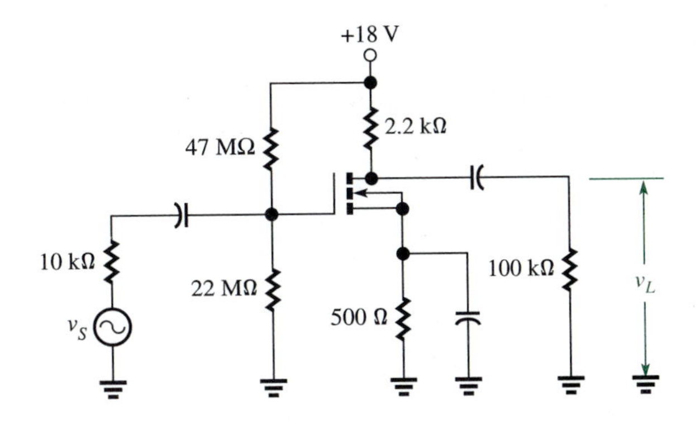

Figure 8–34
(Example 8–8)

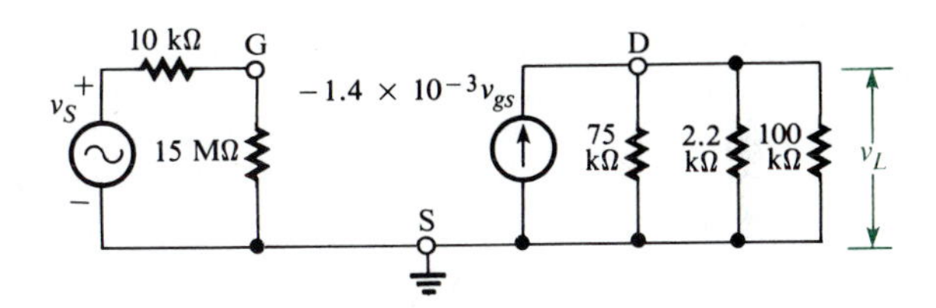

Figure 8–35
(Example 8–8) The small-signal equivalent circuit of Figure 8–34

Using equation 7–21 to find V_{GS}, we have

$$V_{GS} = 5.74 \text{ V} - (1.93 \text{ mA})(500\ \Omega) = 4.78 \text{ V}$$
$$|V_{GS} - V_T| = |4.78 - 2| = 2.78 \text{ V}$$

Therefore, condition 8–39 is satisfied:

$$12.78 = |V_{DS}| > |V_{GS} - V_T| = 2.78$$

and we conclude that the MOSFET is biased in its active region.

2. $r_{in} = R_1 \parallel R_2 = (47 \text{ M}\Omega) \parallel (22 \text{ M}\Omega) = 15 \text{ M}\Omega$.
3. From equation 8–40,

$$g_m = 0.5 \times 10^{-3}(4.78 \text{ V} - 2 \text{ V}) = 1.4 \times 10^{-3} \text{ S}$$

The small-signal equivalent circuit is shown in Figure 8–35. From equation 8–42,

$$\frac{v_L}{v_s} = \left(\frac{15 \times 10^6}{10 \times 10^3 + 15 \times 10^6}\right) \times (-1.4 \times 10^{-3})[(75 \times 10^3) \parallel (2.2 \times 10^3) \parallel (100 \times 10^3)]$$
$$= (0.999)(-1.4 \times 10^{-3})(2.09 \times 10^3) = -2.92$$

Like JFETs, MOSFETs can be operated in common-drain and common-gate configurations. Since the JFET and MOSFET small-signal models are identical, the gain and impedance equations are also identical.

8–6 VMOS AND DMOS AMPLIFIERS

VMOS Amplifiers

The results of Example 8–8 show that the voltage gain of a MOSFET amplifier, like that of a JFET amplifier, is generally rather small. Voltage gains of FET amplifiers are small because transconductance values are small, typically 3 to 4 mS. On the other hand, VMOS FETs (Figure 7–48) have comparatively large transconductance values, on the order of 100 mS. Thus, amplifiers constructed with these devices have voltage gains 20 to 30 times greater than those based on other types of FETs.

DMOS Amplifiers

As discussed in Section 7–8, short-channel power FETs have a linear transfer characteristic. Figure 8–36 shows an example, the transfer characteristic of the MTM15N35 discrete power FET at 25°C. Note that values of drain current are significantly higher than those shown on transfer characteristics for the small-signal MOSFETs discussed earlier. The maximum drain-to-source voltage for this device is specified to be 350 V. We should note that, technically speaking, the MTM15N35 is not a MOSFET, since it has a silicon gate rather than a metal gate. Silicon gates are commonly found in FETs designed for switching-type applications because of certain advantages they enjoy.

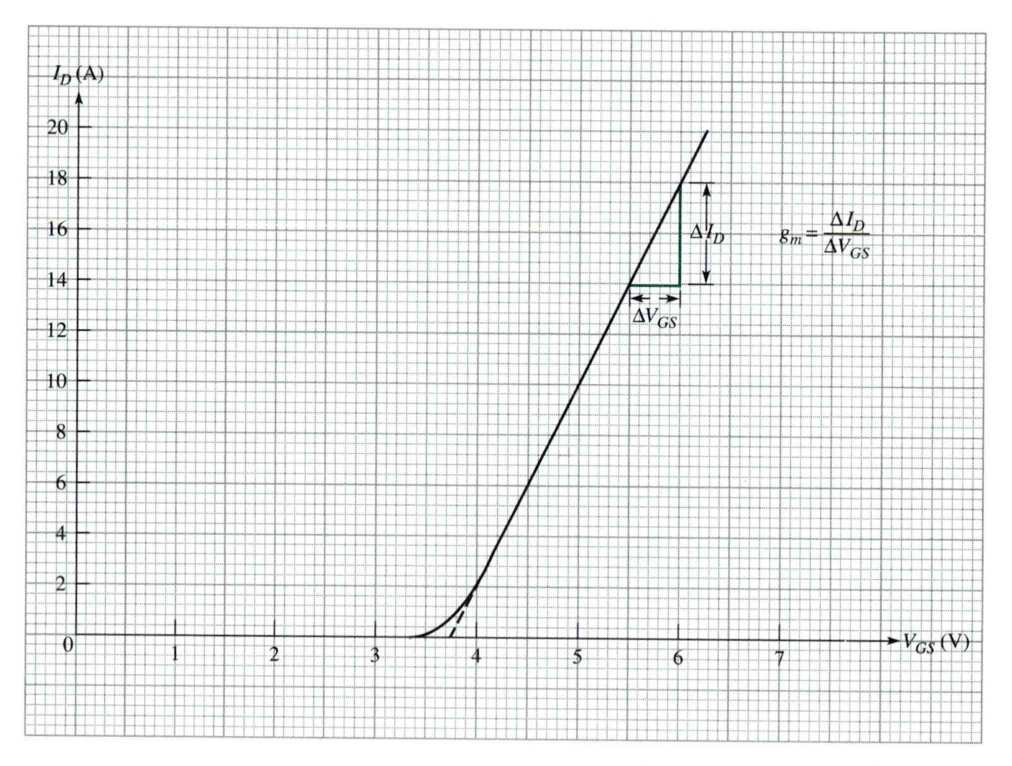

Figure 8–36
Linear transfer characteristic of the MTM15N35 DMOS transistor

The equation for the transfer characteristic of a short-channel power FET is

$$I_D = g_m(V_{GS} - V_T) \qquad V_{GS} \geq V_T \tag{8–43}$$

where V_T is the threshold voltage and g_m is the transconductance. In practice, the threshold voltage is generally assumed to equal the value where an extension of the linear characteristic intersects the V_{GS}-axis, as shown by the dashed line in Figure 8–36.

Example 8–9

1. Find the equation of the transfer characteristic of the MTM15N35 power FET.
2. Use the equation to determine the drain current when the gate-to-source voltage is 5 V.

Solution

1. Examination of the transfer characteristic in Figure 8–36 reveals that the dashed line intersects the V_{GS}-axis at $V_T \approx 3.75$ V. The slope of the linear characteristic is

$$g_m = \frac{\Delta I_D}{\Delta V_{GS}} = \frac{18\text{ A} - 14\text{ A}}{6\text{ V} - 5.5\text{ V}} = \frac{4\text{ A}}{0.5\text{ V}} = 8\text{ S}$$

Thus, from equation 8–43,

$$I_D = 8(V_{GS} - 3.75)\text{ A}$$

2. Substituting $V_{GS} = 5$ V into the equation for the transfer characteristic, we find

$$I_D = 8(5\text{ V} - 3.75\text{ V}) = 10\text{ A}$$

8–7 MOSFET INVERTERS

The MOSFET as a Resistor

In Chapter 6 we discussed the fact that it is easier and more efficient to construct a transistor than it is to construct a resistor in an integrated circuit. This is especially true in enhancement MOSFET circuits because of the simple structure of those devices. For this reason, MOSFETs are connected so that they serve the roles of resistors in integrated circuits. One way an enhancement MOSFET can be connected as a resistor is shown in Figure 8–37.

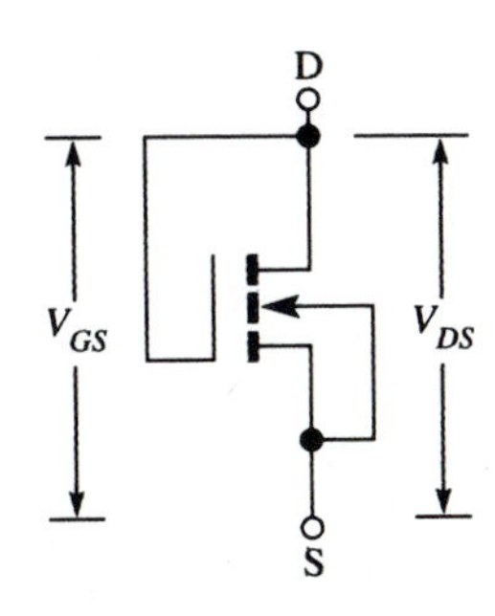

Figure 8–37
An NMOS FET connected as a resistor. Note that $V_{DS} = V_{GS}$.

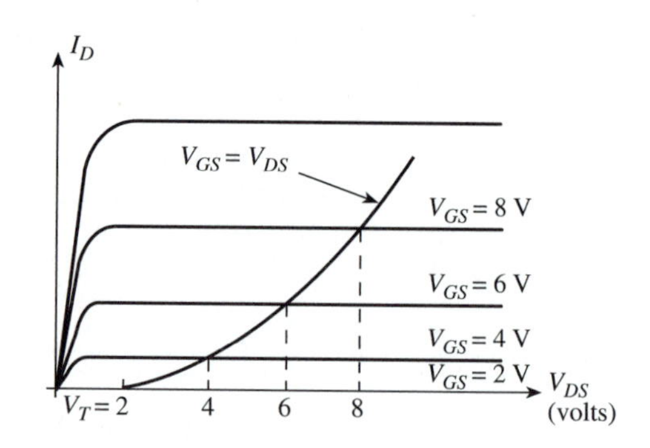

Figure 8–38
A plot of $V_{GS} = V_{DS}$ on a set of NMOS drain characteristics, showing the nonlinear nature of the enhancement MOSFET connected as a resistor (Figure 8–37)

Note in Figure 8–37 that the drain is connected directly to the gate. It is apparent in this case that $V_{DS} = V_{GS}$. It is therefore true that $|V_{DS}| > |V_{GS} - V_T|$, so the device is always in its active region. Figure 8–38 shows a plot of the curve $V_{DS} = V_{GS}$ on a typical set of NMOS drain characteristics and it can be seen that the curve lies entirely in the active region. Notice that the curve is nonlinear, meaning that the MOSFET acts like a nonlinear resistor. Since $V_{GS} = V_{DS} =$ the voltage V across this nonlinear resistor, its I–V characteristic is clearly

$$\begin{aligned} I &= 0.5\beta(V - V_T)^2 \qquad V > V_T \\ &= 0 \qquad\qquad\qquad\quad V \le V_T \end{aligned} \tag{8–44}$$

From (8–44), it can be shown using calculus that the small-signal resistance at any value of I is

$$r = \frac{1}{\sqrt{2\beta I}} \qquad (V > V_T) \tag{8–45}$$

Solving (8–44) for V/I gives the dc resistance:

$$\frac{V}{I} = R = \sqrt{\frac{2}{\beta I}} + \frac{V_T}{I} \qquad (V > V_T) \tag{8–46}$$

Equations 8–45 and 8–46 show that the resistance depends on the value of β, which in turn depends on the aspect ratio (length/width) of the channel. Large values of resistance (small values of β) are achieved by making the channel long and narrow.

The MOSFET Inverter with MOSFET Load

Figure 8–39 shows an NMOS inverter circuit having an NMOS FET used as a load resistor. The theory of operation of this inverter is similar to that of the BJT inverter we studied in Chapter 4, in that the MOSFET (Q_1) acts like a voltage-controlled switch. The principal difference is that the BJT inverter we studied before had a fixed, linear load resistor, while the NMOS inverter has another MOSFET acting as a nonlinear resistor whose value depends on whether Q_1 is ON or OFF. Note that the substrate of Q_2 is shown grounded, but it is not connected to the source of Q_2. If it were connected to the source, then the drain of Q_1 would be grounded, since the substrate is common to all devices in an integrated circuit. The circuit

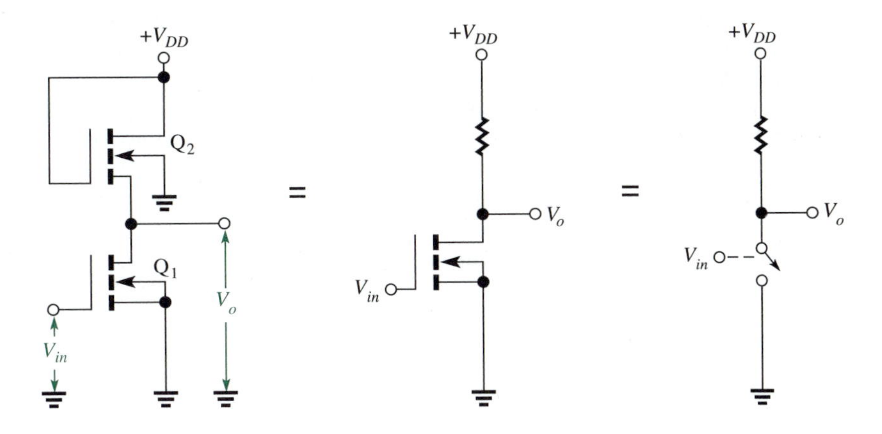

Figure 8–39
An NMOS inverter circuit using an NMOS FET (Q_2) as a load resistor. Q_1 acts as a voltage-controlled switch that is opened or closed by V_{in}.

arrangement has the effect of making the threshold voltages for Q_1 and Q_2 slightly different, but we will neglect this difference in our discussion.

Figure 8–40(b) shows a typical set of NMOS drain characteristics for transistor Q_1 in Figure 8–40(a). Superimposed on these characteristics is the nonlinear load line that results when transistor Q_2 is used as the load resistor. In the discussion that follows, we will use the subscripts 1 and 2 to designate voltages and currents in Q_1 and Q_2, respectively. Thus, for example, V_{DS1} is the drain-to-source voltage of Q_1, which is the output voltage of the inverter. When Q_1 is off, $I_{D1} \approx 0$ and V_{DS1} is a large voltage. These are the conditions at point B in Figure 8–40(b). When Q_1 is conducting, V_{DS1} (identified in the figure as V_{ON}) is small, and Q_1 is in its voltage-controlled–resistance region. These conditions correspond to point A in Figure 8–40(b).

As shown in Figure 8–40, the output voltage switches between V_{ON} (low) and $V_{DD} - V_T$ (high). The *input* to the inverter is assumed to alternate between these same two levels, since the output of another, similar circuit is normally the input

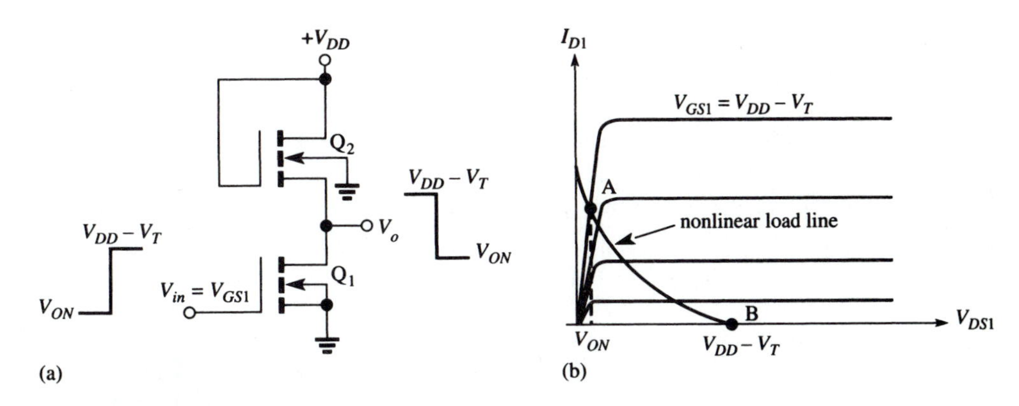

Figure 8–40
The NMOS inverter switches between point A (low) and point B (high) on the nonlinear load line representing Q_2.

to an inverter. When the input is at V_{ON}, which is a small voltage less than V_T, Q_1 does not conduct and the output is high ($V_{DD} - V_T$, at point B). When the input is high, $V_{GS1} = V_{DD} - V_T$, so Q_1 conducts and the output is low (V_{ON}, at point A).

The aspect ratios of Q_1 and Q_2 in the inverter circuit are designed so that the resistance of Q_2 is greater, typically ten times greater, than the resistance of Q_1. This means that β_2 is typically one-tenth of β_1.

To determine the value of V_{ON}, we must analyze the circuit for the case where $V_{GS1} = V_{DD} - V_T$. See Figure 8–41, and notice that $V_{DS2} = V_{GS2}$. By writing Kirchhoff's voltage law from V_{DD} to ground, we find $V_{DD} = V_{GS2} + V_{ON}$, or

$$V_{GS2} = V_{DD} - V_{ON} \quad \textbf{(8–47)}$$

Thus,

$$\begin{aligned} I_{D2} &= \frac{1}{2}\beta_2(V_{GS2} - V_T)^2 \\ &= \frac{1}{2}\beta_2[(V_{DD} - V_{ON}) - V_T]^2 \end{aligned} \quad \textbf{(8–48)}$$

It can be shown that the current in an enhancement MOSFET *in its voltage-controlled–resistance region* is given by

$$I_D = \beta[(V_{GS} - V_T)V_{DS} - 0.5V_{DS}^2] \quad \textbf{(8–49)}$$

Applying equation 8–49 to Q_1, with $V_{GS1} = V_{DD} - V_T$, we have

$$I_{D1} = \beta_1[(V_{DD} - V_T - V_T)V_{ON} - 0.5V_{ON}^2] \quad \textbf{(8–50)}$$

But $I_{D1} = I_{D2}$, so the right-hand sides of equations 8–48 and 8–50 can be equated to each other, giving

$$\frac{1}{2}\beta_2[(V_{DD} - V_{ON}) - V_T]^2 = \beta_1[(V_{DD} - 2V_T)V_{ON} - 0.5V_{ON}^2] \quad \textbf{(8–51)}$$

Expanding, rearranging, and collecting terms in equation 8–51 leads to

$$\left(1 + \frac{\beta_1}{\beta_2}\right)V_{ON}^2 - 2\left[(V_{DD} - V_T) + \frac{\beta_1}{\beta_2}(V_{DD} - 2V_T)\right]V_{ON} + (V_{DD} - V_T)^2 = 0 \quad \textbf{(8–52)}$$

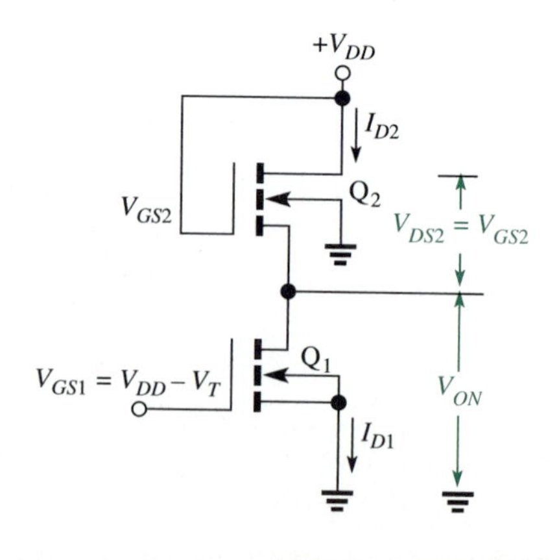

Figure 8–41
Voltage relationships when the inverter input is high ($V_{DD} - V_T$) and its output is low (V_{ON})

This equation can be solved for V_{ON} using the quadratic formula:

$$V_{ON} = \frac{-B - \sqrt{B^2 - 4AC}}{2A} \tag{8–53}$$

where

$$A = 1 + \frac{\beta_1}{\beta_2}$$

$$B = -2\left[(V_{DD} - V_T) + \frac{\beta_1}{\beta_2}(V_{DD} - 2V_T)\right]$$

$$C = (V_{DD} - V_T)^2$$

Example 8–10

The NMOS inverter in Figure 8–40(a) has the following parameters: $V_{T1} = V_{T2} = 2$ V, $\beta_1 = 0.25 \times 10^{-3}$, $\beta_2 = 0.025 \times 10^{-3}$, and $V_{DD} = 10$ V. Find the high and low output voltages.

Solution. The high output voltage is

$$V_{HI} = V_{DD} - V_T = 10\text{ V} - 2\text{ V} = 8\text{ V}$$

To find the low voltage, V_{ON}, we use equation 8–53 to calculate

$$A = 1 + \frac{0.25 \times 10^{-3}}{0.025 \times 10^{-3}} = 11$$

$$B = -2\left[(10\text{ V} - 2\text{ V}) + \frac{0.25 \times 10^{-3}}{0.025 \times 10^{-3}}(10\text{ V} - 4\text{ V})\right] = -136$$

$$C = (10\text{ V} - 2\text{ V})^2 = 64$$

and

$$V_{ON} = \frac{136 - \sqrt{(136)^2 - 4(11)(64)}}{2(11)} = 0.49\text{ V}$$

Since $V_{ON} = 0.49$ V is less than the threshold voltage, $V_T = 2$ V, we can be sure that the MOSFET will remain off when its input is the voltage V_{ON} produced by another, identical inverter.

We have used NMOS devices in all the figures and examples we have presented so far for our discussion of MOSFET inverters. The theory is wholly applicable to PMOS devices. Note, however, that a "high" level in a PMOS inverter is a voltage near 0, while a "low" level is a large negative voltage, near $-V_{DD}$. In modern practice, NMOS devices are more widely used than PMOS devices in integrated circuits because NMOS switching speeds are faster. Neither NMOS nor PMOS devices are used to fabricate general-purpose logic circuitry (logic gates that can be accessed externally and connected in various ways to suit various applications), because the switching speeds of both are slow compared to CMOS circuitry, which we will discuss later. NMOS devices are used in *dedicated* VLSI circuits: those that perform specific functions in a larger system. Examples include computer memories and microprocessors.

8–8 CAPACITIVE LOADING OF SWITCHING CIRCUITS

The output of a switching circuit, such as a MOSFET inverter, always has a certain amount of capacitance in parallel with it. This capacitance may be inherent in the input of the device that serves as the load for the switching circuit, and it may be *stray* capacitance associated with conducting paths and terminal connections on the output side of the circuit. Capacitance in parallel with the output is called *capacitive loading* and is responsible for slowing the speed at which the circuit can switch from high to low and/or from low to high. This fact is illustrated in Figure 8–42. To minimize the detrimental effects of capacitive loading, the output resistance of the switching circuit should be as small as possible.

As can be seen in Figure 8–42, the time required for the output of the switching circuit to change from low to high depends on the time constant R_oC_L that governs the charging of capacitance C_L through resistance R_o. Recall that approximately 5 time constants must elapse for the capacitor to be essentially fully charged. Therefore, the smaller the value of R_o, the smaller the time constant, and the more quickly the output voltage can reach its high output level. Similarly, when the output goes low, the capacitance must discharge through R_o and the time required for the output to reach its low level again depends on the time constant R_oC_L.

Capacitive Loading of the MOSFET Inverter

In many switching circuits, the output resistance when the output is changing from low to high is different from the output resistance when the output is changing from high to low. Thus, the device may be able to turn ON faster than it can turn OFF, or vice versa. This is the case in the MOSFET inverter, as illustrated in Figure 8–43, because the resistance of Q_2 is usually much greater than that of Q_1. When Q_1 is switched OFF by a low input, the output of the inverter rises as the load capacitance is charged through the large resistance of Q_2. On the other hand, when Q_1 is switched ON by a high input, the load capacitance discharges rapidly through the small ON resistance of Q_1.

In characterizing the time required for a waveform to change from low to high and from high to low, it is common practice to specify its *rise time*, t_r, and *fall time*, t_f. As illustrated in Figure 8–44, t_r is the total time required for the waveform to rise from 10% of its final (high) value to 90% of its final value, and t_f is the total time to fall from 90% of its high value to 10% of its high value.

It can be shown that the following expressions are good approximations for

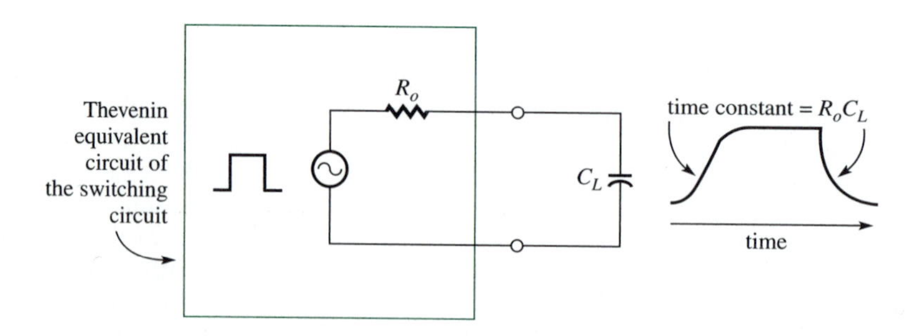

Figure 8–42
The time constant determined by the output resistance and load capacitance limits the speed at which the output can change levels.

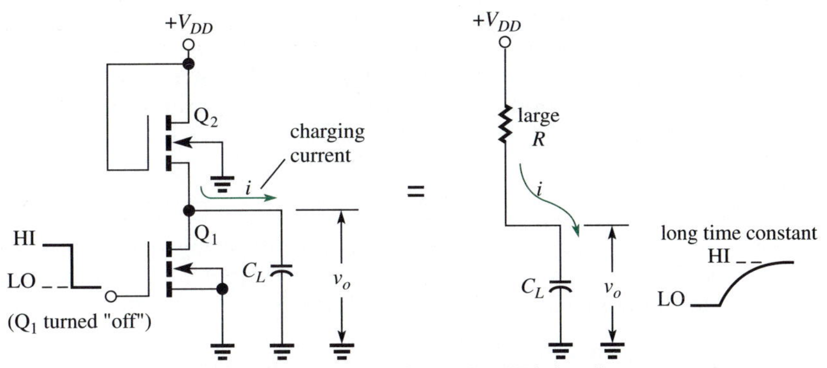

(a) The time required for the output to change from LO to HI is long because the time constant is large when the capacitance charges through Q_2.

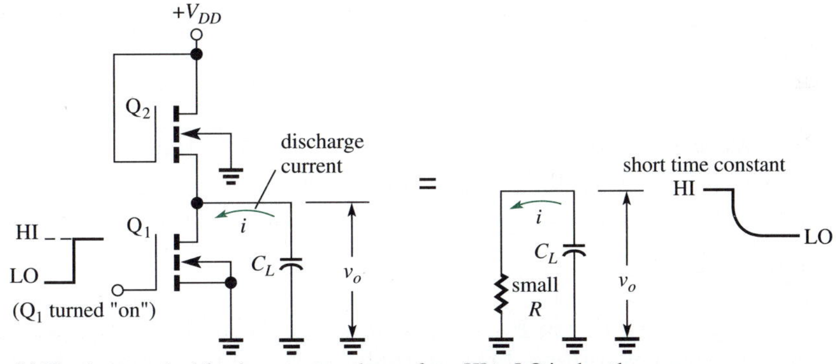

(b) The time required for the output to change from HI to LO is short because the time constant is small when the capacitance discharges through Q_1.

Figure 8–43
A comparison of switching times (LO to HI versus HI to LO) in an NMOS inverter

the rise and fall times at the output of an NMOS inverter with an NMOS load:

$$t_r \approx \frac{17.8C_L}{\beta_2(V_{DD} - V_T)} \tag{8–54}$$

$$t_f \approx \frac{1.6C_L}{\beta_1(V_{DD} - 2V_T)} \tag{8–55}$$

Figure 8–44
The rise time, t_r, and fall time, t_f, of a waveform

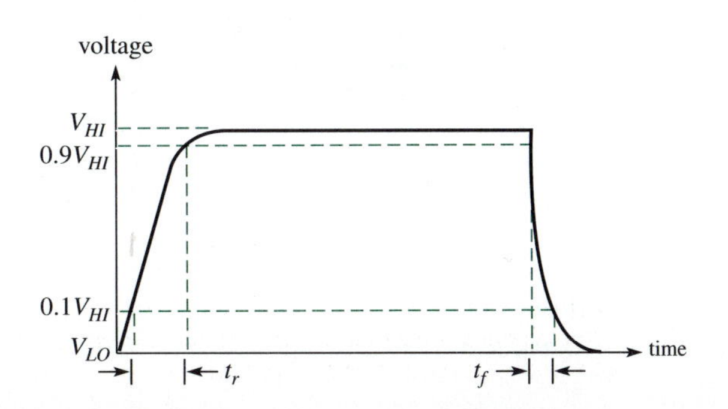

These equations show that t_r and t_f are directly proportional to C_L and inversely proportional to β. The dependency on β follows from the fact that channel resistance is inversely related to β. Since β_1 is typically 10 times greater than β_2, the channel resistance of Q_1 is about $\frac{1}{10}$ that of Q_2. The next example illustrates the disparity between the rise and fall times of an NMOS inverter.

Example 8–11

An NMOS inverter (Figure 8–43) has the following parameters: $\beta_1 = 40 \times 10^{-6}$, $\beta_2 = 4 \times 10^{-6}$, $V_T = 1$ V, and $V_{DD} = 10$ V. If it has a capacitive load of 1 pF, calculate the approximate rise and fall times at its output.

Solution. From equation 8–54,

$$t_r \approx \frac{17.8 \times 1 \times 10^{-12}}{(4 \times 10^{-6})(10 - 1)} = 494 \text{ ns}$$

From equation 8–55,

$$t_f \approx \frac{1.6 \times 1 \times 10^{-12}}{(40 \times 10^{-6})(10 - 2)} = 5 \text{ ns}$$

We see that the rise time is almost one hundred times longer than the fall time, a typical ratio in NMOS switching circuits.

Example 8–12

SPICE

The NMOS inverter in Example 8–10 has a 10-pF capacitive load and is driven by a rectangular waveform that alternates between 0 V and +8 V with a frequency of 400 kHz. The pulsewidth is 0.5 μs. Obtain a plot of the output over one full cycle of input.

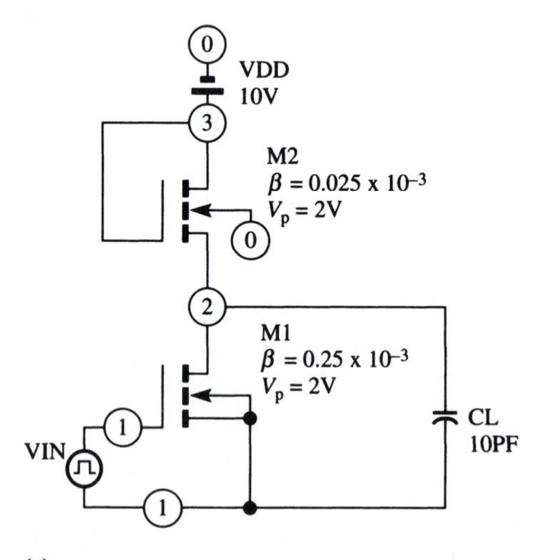


```
EXAMPLE 8.12
M1 2 1 0 0 MOSFET1
M2 3 3 2 0 MOSFET2
.MODEL MOSFET1 NMOS KP=0.25E-3 VTO=2
.MODEL MOSFET2 NMOS KP=0.025E-3 VTO=2
VDD 3 0 10
CL 2 0 10PF
VIN 1 0 PULSE(0 8 0 1PS 1PS 0.5US 2.5US)
.TRAN 0.1US 2.5US
.PRINT TRAN V(2)
.END
```

Figure 8–45
(Example 8–12)

```
EXAMPLE 8.12
****      TRANSIENT ANALYSIS                           TEMPERATURE =    27.000 DEG C
********************************************************************************
   TIME      V(2)
                 0.000D+00     2.000D+00     4.000D+00     6.000D+00  8.000D+00
                     - - - - - - - - - - - - - - - - - - - - - - - - - - - - - -
0.000D+00  7.999D+00 .             .             .             .             *
1.000D-07  4.900D-01 .  *          .             .             .             .
2.000D-07  4.900D-01 .  *          .             .             .             .
3.000D-07  4.900D-01 .  *          .             .             .             .
4.000D-07  4.900D-01 .  *          .             .             .             .
5.000D-07  4.900D-01 .  *          .             .             .             .
6.000D-07  4.128D+00 .             .             .*            .             .
7.000D-07  5.416D+00 .             .             .         *   .             .
8.000D-07  6.053D+00 .             .             .             *             .
9.000D-07  6.437D+00 .             .             .             .  *          .
1.000D-06  6.693D+00 .             .             .             .    *        .
1.100D-06  6.877D+00 .             .             .             .     *       .
1.200D-06  7.016D+00 .             .             .             .      *      .
1.300D-06  7.124D+00 .             .             .             .       *     .
1.400D-06  7.210D+00 .             .             .             .       *     .
1.500D-06  7.281D+00 .             .             .             .        *    .
1.600D-06  7.341D+00 .             .             .             .        *    .
1.700D-06  7.391D+00 .             .             .             .         *   .
1.800D-06  7.434D+00 .             .             .             .         *   .
1.900D-06  7.471D+00 .             .             .             .         *   .
2.000D-06  7.504D+00 .             .             .             .          *  .
2.100D-06  7.533D+00 .             .             .             .          *  .
2.200D-06  7.559D+00 .             .             .             .          *  .
2.300D-06  7.582D+00 .             .             .             .          *  .
2.400D-06  7.603D+00 .             .             .             .          *  .
2.500D-06  7.622D+00 .             .             .             .          *  .
                     - - - - - - - - - - - - - - - - - - - - - - - - - - - - - -
```

(b)

Figure 8–45
(Continued)

Solution. The SPICE circuit and input data file are shown in Figure 8–45(a). The period of the input is

$$T = \frac{1}{400 \times 10^3 \text{ Hz}} = 2.5\ \mu\text{s}$$

Thus, the .TRAN statement produces one full cycle of output at 0.1-μs intervals. The PULSE specification for V1 shows that the rise and fall times of the input pulse are each set to 1 ps, which is very small compared to the pulse width. These specifications make the input pulse very nearly ideal. If the rise and fall times of the input were set to 0 or allowed to default, SPICE would set them equal to the much longer time $TSTEP = 0.1\ \mu$s. Note that two .MODEL statements are required in the data file, since the MOSFETs have different values of β.

The plot of the output is shown in Figure 8–45(b). Note that the rise time of the output is much longer than the fall time. Also note that the low output level is $V_{ON} = 0.49$ V, in agreement with the calculations in Example 8–10.

8–9 CMOS CIRCUITS

CMOS Inverters

Recall that *complementary* MOS (CMOS) integrated circuits contain both NMOS and PMOS devices embedded in the same chip (see Figure 7–47). The CMOS design is now widely used in general-purpose digital circuits because it provides faster switching times than either PMOS or NMOS devices alone. It is also used in special-purpose MSI and LSI circuits. Figure 8–46 shows a CMOS inverter, the fundamental building block of the CMOS logic family. Notice that there is but a single, positive supply voltage, V_{DD}, and that the *source* of the PMOS transistor is connected to V_{DD}. This connection is necessary for the proper operation of a PMOS device, whose drain must be negative *with respect to* its source. (In Chapter 7 we showed the isolated PMOS transistor with a negative supply voltage connected to its drain; the same bias is achieved by making the source positive with respect to the drain.) Notice in Figure 8–46 that the substrate terminal of Q_2 is shown connected to $+V_{DD}$. This P-channel device has an N-type substrate, which must be connected to the most positive voltage in the circuit to ensure that the junction between the substrate and the induced P channel is reverse biased. The "substrate" of the N-channel device (Q_1) is actually the P-type tub in which the device is embedded, as shown in Figure 7–44. Therefore, the substrate terminal of Q_1 is shown connected to the most negative voltage (ground) to ensure reverse bias.

As an aid in understanding the operation of a CMOS inverter, let us review and summarize the gate-to-source conditions that cause enhancement NMOS and PMOS transistors to either conduct (turn ON) or turn OFF. These conditions are given in the following table:

Gate-to-Source Polarity

	To Cause Conduction (Turn ON)	**To Turn OFF**
NMOS	Gate positive with respect to source (more positive than V_T)	Gate zero or negative with respect to source (less positive than V_T)
PMOS	Gate negative with respect to source (more negative than V_T)	Gate zero or positive with respect to source (more positive than V_T)

To illustrate how the conditions summarized in the table apply to CMOS circuits, let us first suppose that an NMOS transistor has $V_T = +2$ V and that its source is

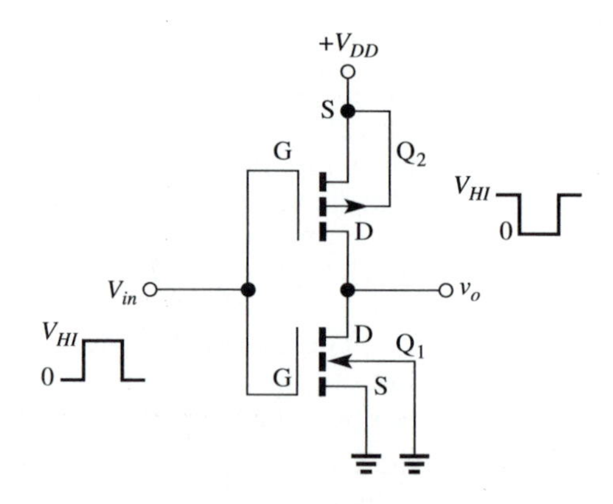

Figure 8–46
A CMOS inverter

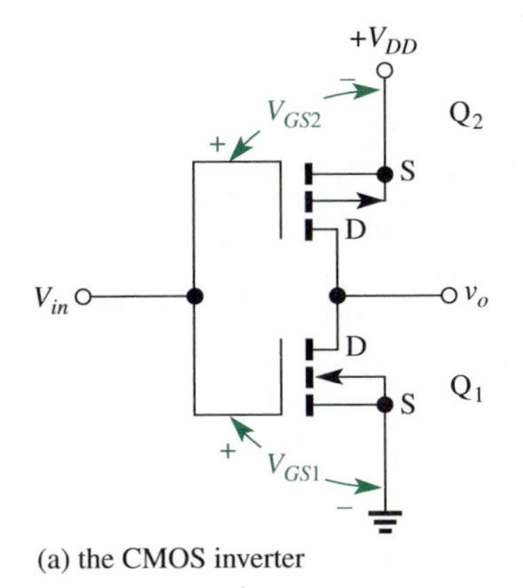

(a) the CMOS inverter

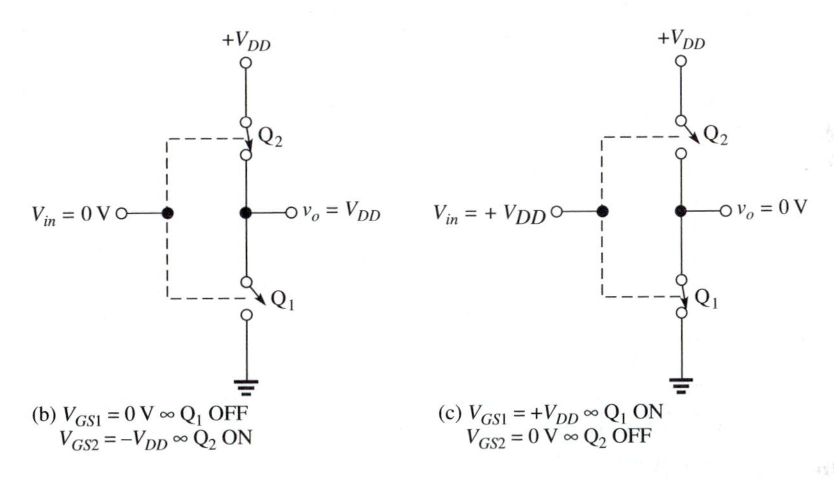

(b) $V_{GS1} = 0$ V ∞ Q_1 OFF
$V_{GS2} = -V_{DD}$ ∞ Q_2 ON

(c) $V_{GS1} = +V_{DD}$ ∞ Q_1 ON
$V_{GS2} = 0$ V ∞ Q_2 OFF

Figure 8–47
CMOS inverter operation illustrated in terms of voltage-controlled switches

grounded, such as Q_1 in Figure 8–46. Then 0 V applied to its gate will turn it OFF because $V_{GS} = 0 < V_T$. When +10 V is applied to its gate, it will turn ON (conduct) because $V_{GS} = +10 > V_T$. Consider now a PMOS transistor that has $V_T = -2$ V and that *has +10 V connected to its source,* such as Q_2 in Figure 8–46. Then 0 V applied to its gate will turn it *ON,* because $V_{GS} = -10$ V, which is more negative than −2 V. When +10 V is applied to its gate, the result is $V_{GS} = 0$ V, and it is turned OFF because 0 V is more positive than −2 V.

Notice in Figure 8–46 that the gates of Q_1 and Q_2 are connected together at the input to the inverter. Therefore, when V_{in} is high (say, +10 V), Q_1 is ON and Q_2 is OFF. On the other hand, when V_{in} is low (0 V), Q_1 is OFF and Q_2 is ON. To see how these facts relate to inverter operation, think of each device as a switch that is closed when it is ON and open when it is OFF. See Figure 8–47. As shown in the figure, a low input closes Q_2 and opens Q_1, so the output is connected through Q_2 to $+V_{DD}$. A high input closes Q_1 and opens Q_2, so the output is connected through Q_1 to ground (0 V). Thus, one device is always ON, and the output level is always the opposite of the input level.

The input and output levels of CMOS inverters alternate between a low near 0 V and a high near $+V_{DD}$. One advantage of CMOS circuitry is that the high level can be anywhere from about 3 V to about 18 V, a very wide range in comparison to other digital logic families. The higher the level of V_{DD}, the faster the switching time and the more immune the circuit is to electrical *noise* (random fluctuations in signal levels), but the greater the power dissipation. In general, CMOS circuitry is considered to have very low power dissipation.

The reason that CMOS circuits have faster switching times than their NMOS and PMOS counterparts is that *the output resistance is small both when the output is changing from low to high and when it is changing from high to low.* When the output is changing from low to high, the load capacitance is charged through the small ON resistance of Q_2 (Figure 8–47(b)). When the output is changing from high to low, the load capacitance discharges through the small ON resistance of Q_1 (Figure 8–47(c)). Thus, the time constant is small in both cases, and the rise and fall times are about equal.

The CMOS Analog Switch

A CMOS analog switch, also called a *bilateral transmission gate,* is formed by connecting an NMOS device in parallel with a PMOS device. As we shall presently explain, this combination acts like two parallel switches, both of which are simultaneously closed or simultaneously open. Actually, when the switches are closed, both conduct small positive and negative signals, but only one conducts large positive signals, and the other conducts large negative signals.

Figure 8–48 shows how the devices are connected and the schematic symbol for the CMOS analog switch. Note that the PMOS substrate terminal is connected to the most positive voltage and the NMOS substrate to the most negative voltage, labeled $-V_{SS}$.

Notice in Figure 8–48 that there are two inputs, labeled C and $\overline{C}$. $\overline{C}$ (read: "C-bar") is the standard representation for the *complement,* or inverse, of C. Thus, if C is high, then $\overline{C}$ is low, and vice versa. These two inputs are the *control* inputs for the switch and determine whether the analog signal is passed from input to output or is shut off. Think of C and $\overline{C}$ as being obtained from the input and output

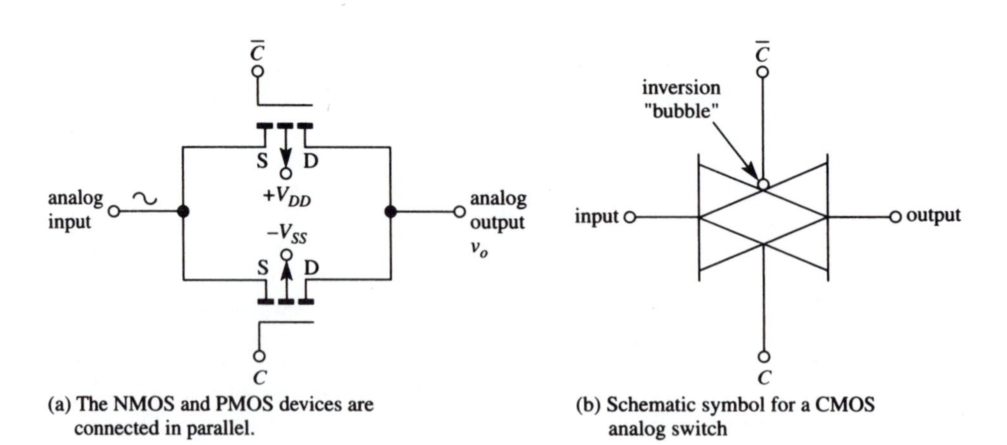

(a) The NMOS and PMOS devices are connected in parallel.

(b) Schematic symbol for a CMOS analog switch

Figure 8–48
The CMOS analog switch, or bilateral transmission gate

Figure 8–49
The range of analog input voltages over which a MOSFET conducts when the CMOS analog switch is ON ($C = +5$ V). Note that both devices conduct in the range from -3 V to $+3$ V. (NMOS $V_T = +2$ V; PMOS $V_T = -2$ V.)

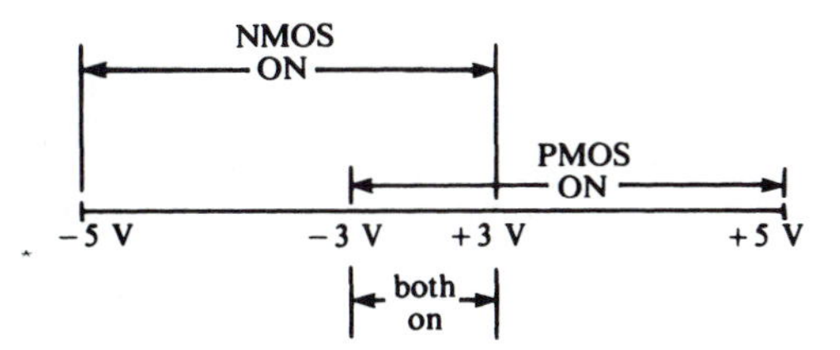

of an inverter, not shown, and understand that we need only change the level of C to control the switch; the level of $\overline{C}$ will automatically be the opposite of that of C. In the schematic symbol, the circular "bubble" at the $\overline{C}$ input is used to show that the input there is inverted. (For the sake of convention, we will follow this unfortunate and misleading notation; however, to be accurate, C instead of $\overline{C}$ should be shown connected to the bubble, since the bubble itself is usually regarded as performing the inversion operation.)

It is helpful to analyze the operation of the CMOS analog switch in terms of specific parameter values, so let us suppose that V_T is +2 V for the NMOS FET, −2 V for the PMOS FET, $V_{DD} = +5$ V, and $V_{SS} = -5$ V. Assume that the analog input can vary continuously between +5 V and −5 V, and that C and $\overline{C}$ are either +5 V or −5 V. When $C = -5$ V, the gate of the NMOS FET is at −5 V and it is therefore OFF for all input voltages in the range −5 V to +5 V. (To be ON, the input would have to be more negative than −7 V.) Since $\overline{C} = +5$ V, the gate of the PMOS FET is at +5 V and it is also OFF for all input voltages in the range −5 V to +5 V. (To be ON, the input would have to be more positive than +7 V.) We conclude that $C = -5$ V causes the analog switch to be open. Now suppose that $C = +5$ V. Then the gate of the NMOS FET is at +5 V and it is ON for input voltages in the range −5 V to +3 V. Since $\overline{C} = -5$ V, the PMOS FET is ON for all input voltages in the range −3 V to +5 V. Between the two devices, there is a conducting path from input to output for the entire range of input voltages from −5 V to +5 V, and we conclude that $C = +5$ V closes the analog switch. Figure 8–49 illustrates the overlapping of the ranges of input voltages for which each device conducts. It can be seen that both are conducting in the range of input between −3 V and +3 V.

EXERCISES

Section 8–1
The Common-Source JFET Amplifier

8–1. **a.** What is the value of the transconductance of a JFET when its gate-to-source voltage equals its pinch-off voltage?
b. At what value of I_D does $g_m = 0$?

8–2. Find the transconductance of the JFET whose transfer characteristic is shown in Figure 8–2, when $I_D = 1$ mA, both (a) graphically and (b) algebraically.

8–3. Find the transconductance of the JFET whose transfer characteristic is shown in Figure 8–2, when $V_{GS} = -0.5$ V, both (a) graphically and (b) algebraically.

8–4. Use equation 8–2 and the equation of the transfer characteristic of a JFET to derive equation 8–3.

8–5. The maximum transconductance of a certain N-channel JFET is 9.8×10^{-3} S. If $I_{DSS} = 18$ mA, what is its pinch-off voltage?

8–6. The following measurements were taken from a curve tracer display of the output characteristics of a P-channel JFET along the line V_{GS} = +2.5 V: I_D = 4.2 mA at V_{DS} = −4.5 V; I_D = 4.3 mA at V_{DS} = −12 V. What is the drain resistance at V_{GS} = 2.5 V?

8–7. The JFET in the amplifier shown in Figure 8–50 has I_{DSS} = 16 mA, V_p = −4.5 V, and r_d = 80 kΩ.
a. Find the quiescent values of I_D and V_{DS}.
b. Find g_m.
c. Draw the ac equivalent circuit.
d. What is the peak value of the ac component of v_{ds}?

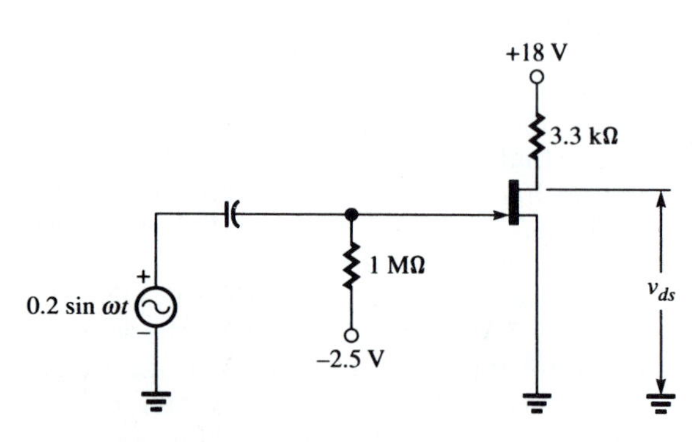

Figure 8–50
(Exercise 8–7)

8–8. The JFET in the amplifier shown in Figure 8–51 has I_{DSS} = 14 mA, V_p = 3 V, and r_d = 120 kΩ.
a. Find the quiescent values of I_D and V_{DS}.
b. Find g_m.
c. Draw the ac equivalent circuit.
d. Find the voltage gain.

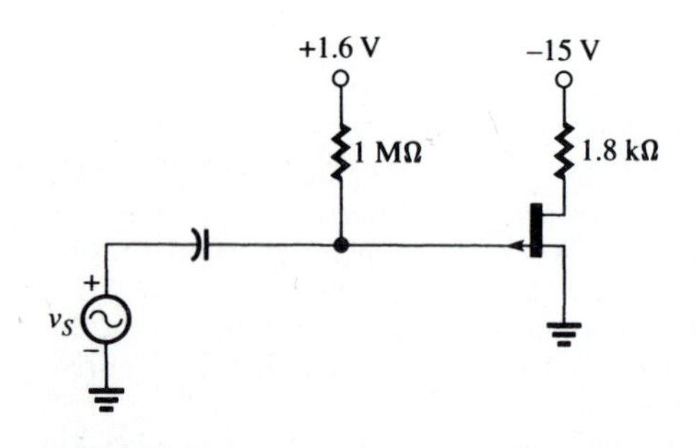

Figure 8–51
(Exercise 8–8)

8–9. The JFET in the amplifier shown in Figure 8–52 has $g_m = 4.2 \times 10^{-3}$ S and r_d = 98 kΩ.
a. Find the input resistance of the amplifier.
b. Find the voltage gain.

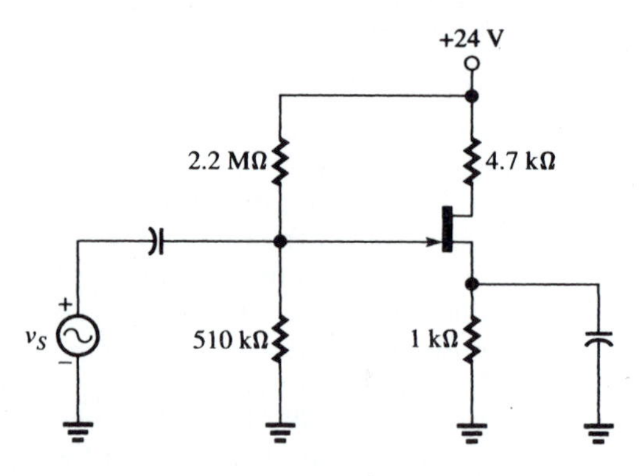

Figure 8–52
(Exercise 8–9)

8–10. The JFET in the amplifier shown in Figure 8–53 has r_d = 50 kΩ, V_p = 5 V, and I_{DSS} = 15 mA. The dc source-to-ground voltage is −5.15 V.
a. Find the voltage gain.
b. Find the input resistance of the amplifier.

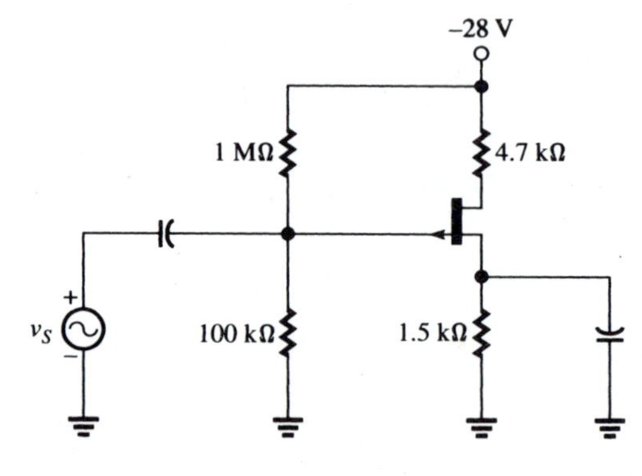

Figure 8–53
(Exercise 8–10)

8–11. The JFET in the amplifier shown in Figure 8–54 has r_d = 75 kΩ, V_p = −3.6 V, and I_{DSS} = 9 mA. The dc voltage drop across the 3.3-kΩ resistor is 6.38 V. Find the voltage gain of the amplifier.

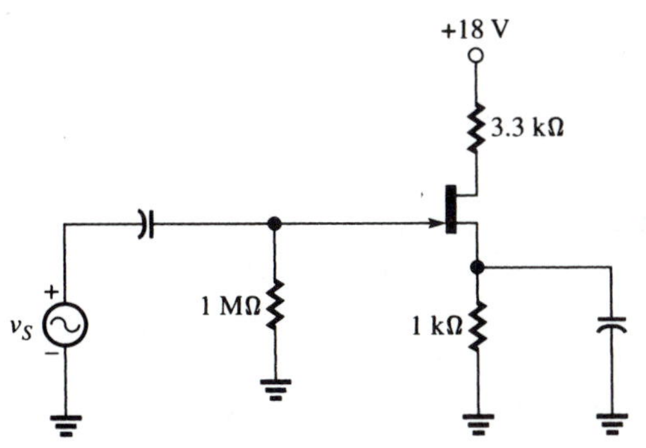
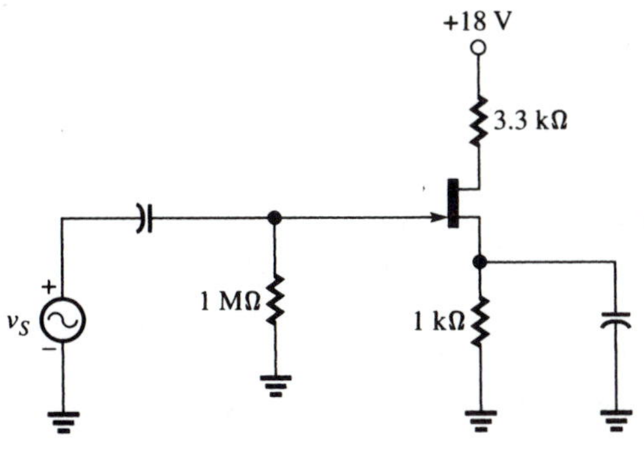

Figure 8–54
(Exercise 8–11)

8–12. The JFET in the amplifier shown in Figure 8–55 has r_d = 100 kΩ and g_m = 2871 μS. Find the voltage gain v_L/v_S.

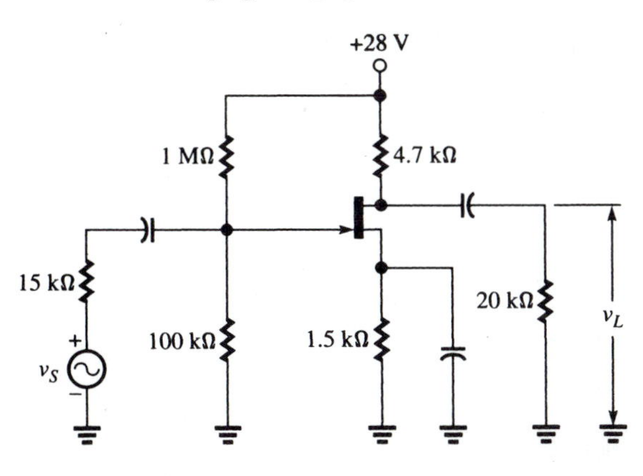

Figure 8–55
(Exercise 8–12)

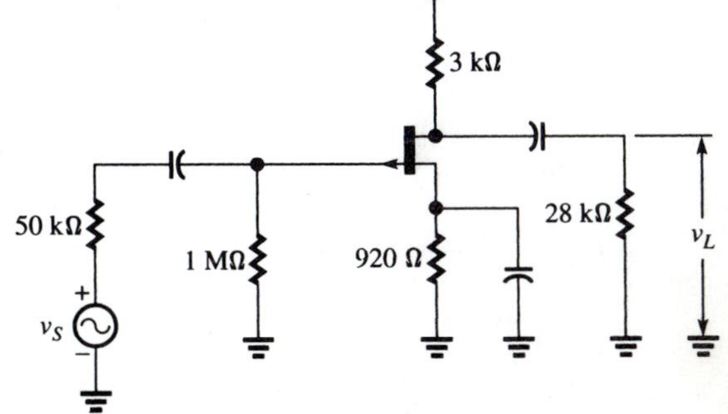

Figure 8–56
(Exercise 8–13)

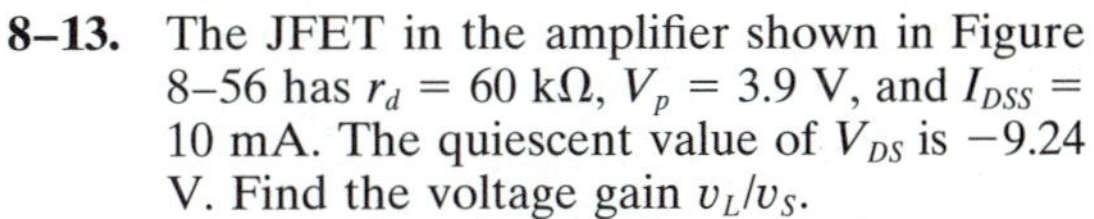

8–13. The JFET in the amplifier shown in Figure 8–56 has r_d = 60 kΩ, V_p = 3.9 V, and I_{DSS} = 10 mA. The quiescent value of V_{DS} is −9.24 V. Find the voltage gain v_L/v_S.

8–14. Repeat Exercise 8–11 when the source-bypass capacitor is removed.

8–15. Repeat Exercise 8–12 when the source-bypass capacitor is removed.

Section 8–2

The Common-Drain and Common-Gate JFET Amplifiers

8–16. The JFET in the amplifier shown in Figure 8–57 has g_m = 0.004 S and r_d = 90 kΩ.

a. Find the voltage gain v_L/v_S.
b. Find the input resistance.

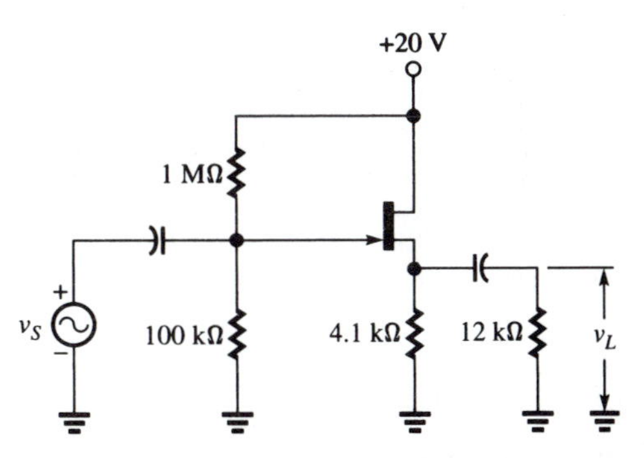

Figure 8–57
(Exercise 8–16)

8–17. The JFET in the amplifier shown in Figure 8–58 has g_m = 5200 μS and r_d = 80 kΩ.

a. Find the voltage gain v_L/v_S.
b. Find the output resistance.

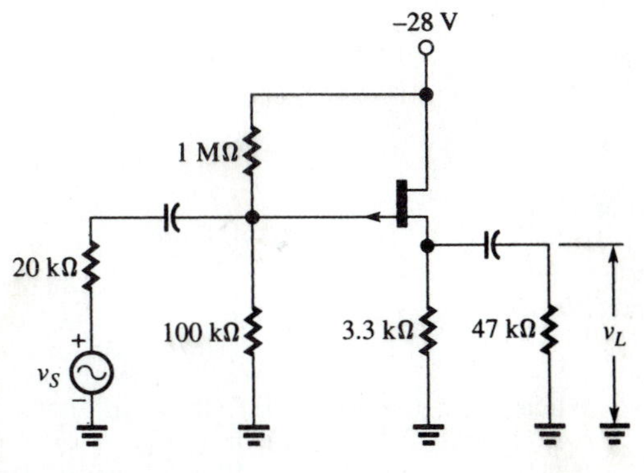

Figure 8–58
(Exercise 8–17)

8–18. The JFET in the amplifier in Figure 8–59 has g_m = 3.58 mS and r_d = 100 kΩ. Find the voltage gain v_L/v_S.

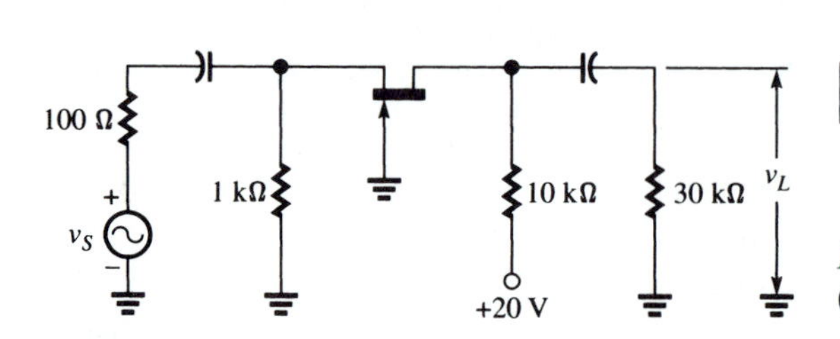

Figure 8–59
(Exercise 8–18)

8–19. The amplifier in Figure 8–20(a) has r_S = 50 Ω, R_S = 680 Ω, R_D = 2 kΩ, R_L = 4 kΩ, and V_{DD} = 15 V. The pinch-off voltage of the JFET is −3.9 V and the saturation current is 15 mA. The dc voltage drop across R_D is 6.24 V. Find the voltage gain v_L/v_S. Drain resistance r_d can be neglected.

Section 8–3
The JFET Current Source

8–20. The JFET shown in Figure 8–60 has V_p = −5 V and I_{DSS} = 12 mA. What is the maximum value of R_L for which the circuit can be used as a constant-current source?

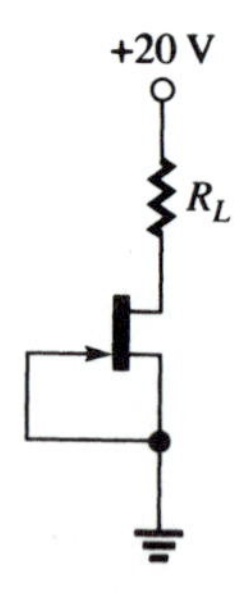

Figure 8–60
(Exercise 8–20)

8–21. When V_{GS} = 0 V, the JFET shown in Figure 8–61 begins to break down at V_{DS} = 15 V. If V_p = −3.5 V and I_{DSS} = 10 mA, over what range of R_L can the circuit be used as a constant-current source?

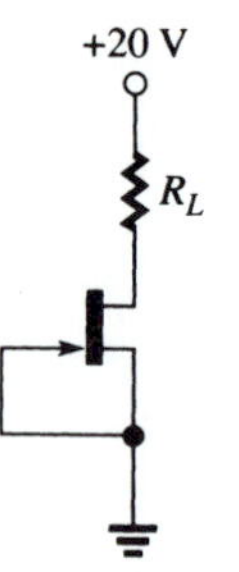

Figure 8–61
(Exercise 8–21)

8–22. When V_{GS} = 0 V, the JFET shown in Figure 8–62 begins to break down at V_{DS} = 18 V. If V_p = 4 V and I_{DSS} = 12 mA, over what range of R_L can the circuit be used as a constant-current source?

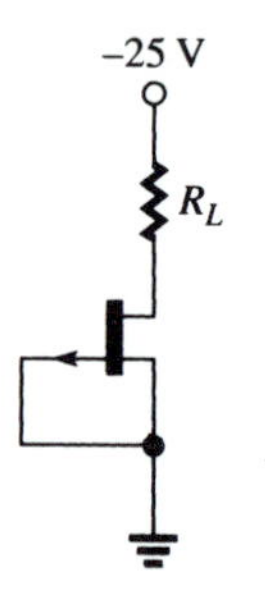

Figure 8–62
(Exercise 8–22)

Section 8–4
The JFET as an Analog Switch

8–23. The drain-to-source resistance when the JFET in Figure 8–63 is conducting is 80 Ω. If v_d = 0.16 V, find the load voltage v_L (a) when V_{GS} = V_p and (b) when V_{GS} = 0 V.

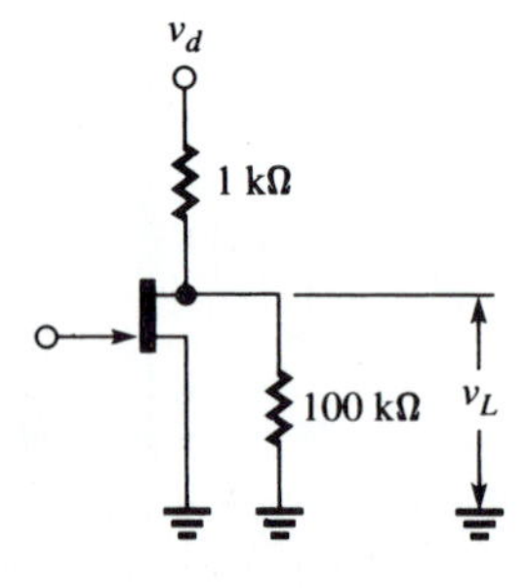

Figure 8–63
(Exercise 8–23)

8–24. The JFET in Figure 8–64 has $R_{D(ON)} = 100\ \Omega$ and $V_p = 4$ V. If $v_{in} = 0.1 \sin 1000t$, write the expression for v_L (a) when $V_{GS} = 4$ V and (b) when $V_{GS} = 0$ V.

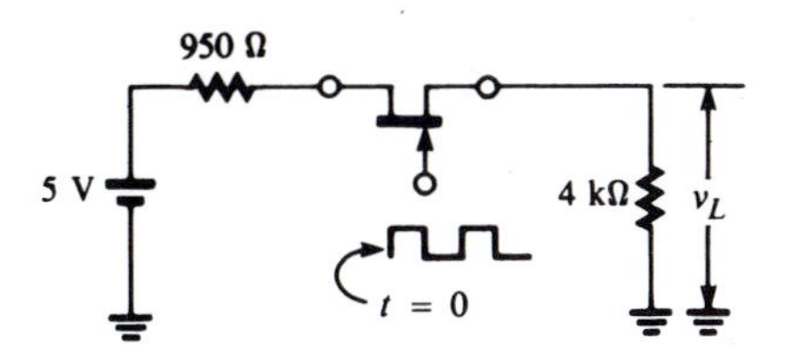

Figure 8–65
(Exercise 8–25)

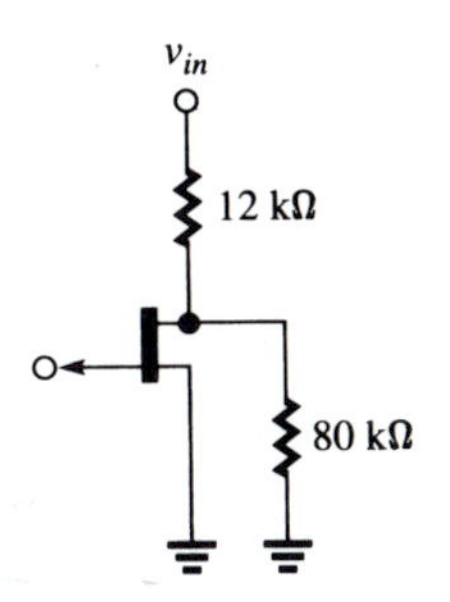

Figure 8–64
(Exercise 8–24)

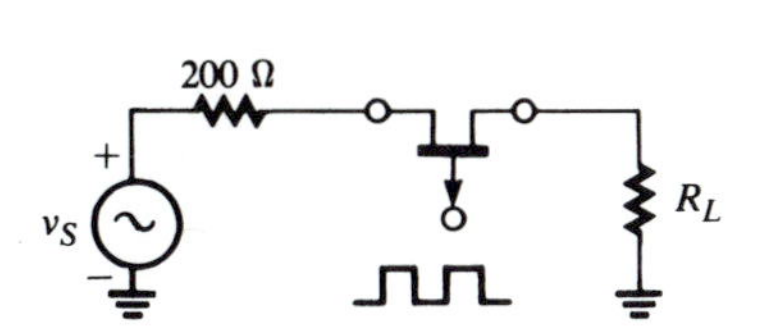

Figure 8–66
(Exercise 8–26)

8–25. The square wave shown in Figure 8–65 turns the chopper ON when high and OFF when low. If it has frequency 1 kHz, sketch v_L over a time period of 3 ms. Assume that $R_{D(ON)}$ for the JFET is 50 Ω.

8–26. The JFET shown in Figure 8–66 has $R_{D(ON)} = 40\ \Omega$. If the output voltage is to be no less than 90% of the signal voltage when the JFET is ON, what is the minimum permissible value of R_L?

Section 8–5
Small-Signal MOSFET Amplifiers

8–27. Graphically determine the transconductance of the MOSFET whose transfer characteristic is shown in Figure 8–67 when $V_{GS} = 7$ V.

8–28. Algebraically determine the transconductance of the MOSFET in Exercise 8–27, given that its β is 0.5×10^{-3}.

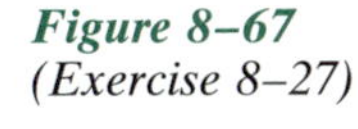

Figure 8–67
(Exercise 8–27)

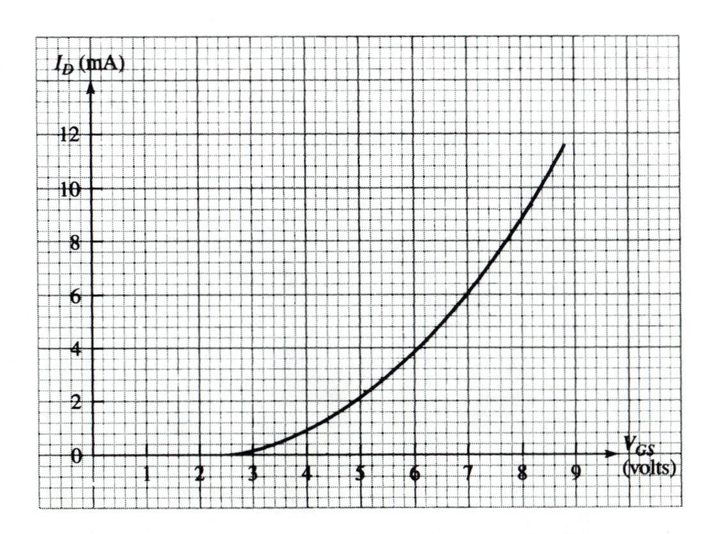

8–29. The MOSFET shown in Figure 8–68 has $r_d = 100$ kΩ and a transconductance of 3×10^{-3} S.
a. Draw the ac equivalent circuit.
b. Find the voltage gain v_L/v_S.

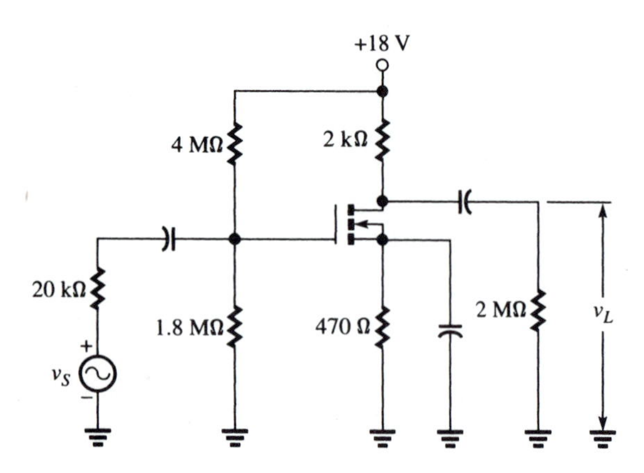

Figure 8–68
(Exercise 8–29)

8–30. The MOSFET shown in Figure 8–69 has $V_T = 2.4$ V, $\beta = 0.62 \times 10^{-3}$, and $r_d = 120$ kΩ. It is biased at $I_D = 5$ mA.
a. Find the input resistance.
b. Find the voltage gain v_L/v_S.

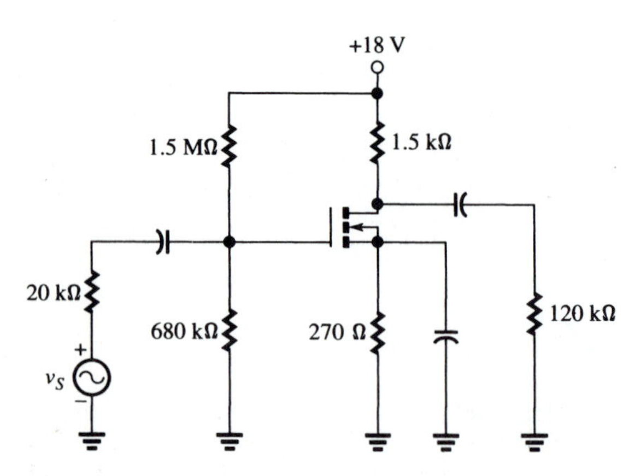

Figure 8–69
(Exercise 8–30)

Section 8–6
VMOS and DMOS Amplifiers

8–31. A DMOS transistor has threshold voltage 3.5 V. When $V_{GS} = 5$ V, $I_D = 3$ A. What is I_D when $V_{GS} = 10$ V?

8–32. When a DMOS transistor is operated above its threshold voltage, it is found that $I_D = 10$ A when $V_{GS} = 6$ V and $I_D = 20$ A when $V_{GS} = 8$ V. Find the transconductance and threshold voltage of the transistor.

Section 8–7
MOSFET Inverters

8–33. An enhancement MOSFET used as a resistor in an integrated circuit has its gate connected directly to its drain. The MOSFET has $\beta = 0.1 \times 10^{-3}$ and $V_T = 2$ V.
a. What is the current in the MOSFET when the voltage across it is 8 V?
b. What is the ac resistance of the MOSFET when the voltage across it is 10 V?
c. What is the dc resistance of the MOSFET when the voltage across it is 10 V?

8–34. An NMOS FET used as a resistor in an integrated circuit has its gate connected directly to its drain. The MOSFET has $\beta = 0.1 \times 10^{-3}$ and $V_T = 2$ V. What is the voltage across the MOSFET when the current through it is 1 mA? Assume that $V > V_T$.

8–35. Using equation 8–44 for the case $V > V_T$, derive equation 8–46.

8–36. An enhancement MOSFET inverter (having another MOSFET as a load resistor) has $\beta = 0.4 \times 10^{-3}$ and $V_T = 2$ V. The supply voltage for the inverter circuit is 12 V, and the output low voltage of the inverter is 0.5 V. What is the current in the inverter when it is ON?

8–37. The gate-to-source voltage of Q_2 in Figure 8–70 is 9.7 V. The threshold voltages of Q_1

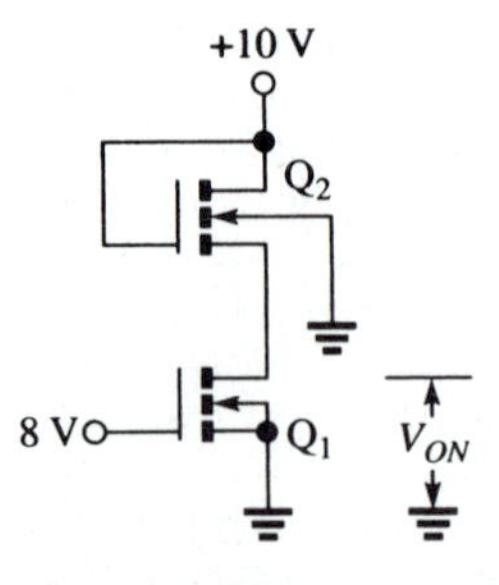

Figure 8–70
(Exercise 8–37)

and Q_2 are 2 V. If $\beta_1 = 0.2 \times 10^{-3}$, find (a) V_{ON} and (b) I_D.

8–38. The parameters of Q_1 and Q_2 in Figure 8–71 are $\beta_1 = 0.2 \times 10^{-3}$, $\beta_2 = 0.02 \times 10^{-3}$, and $V_{T1} = V_{T2} = 2.2$ V. Assuming that the inverter input levels are the same as its output levels and that $V_{DD} = 12$ V, find (a) the output high voltage and (b) the output low voltage.

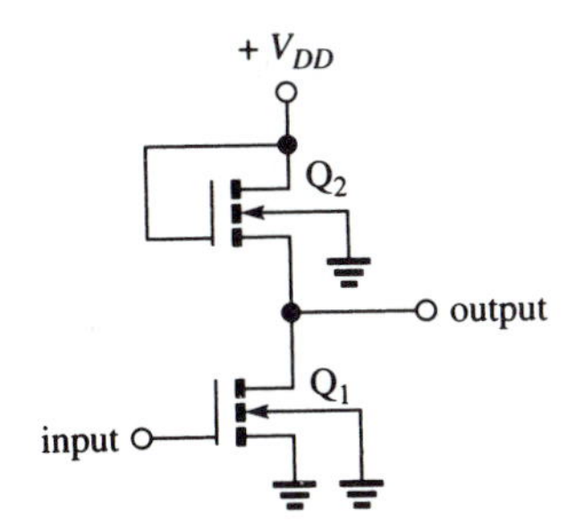

Figure 8–71
(Exercise 8–38)

8–39. Repeat Exercise 8–38 if V_{DD} is changed to 10 V.

Section 8–8
Capacitive Loading of Switching Circuits

8–40. The output resistance of a switching circuit is 20 kΩ. If the output has a capacitive load of 1.5 pF, how long does it take for the output level to reach (essentially) its high output level of 5 V when it switches from low to high?

8–41. If the switching circuit in Exercise 8–40 has an output resistance of 1200 Ω when it is switching from high to low, how long does it take the output to reach (essentially) 0 V when it switches?

8–42. Find the rise time (t_r) of the output of the switching circuit in Exercise 8–40. (*Hint:* Write the equation for the voltage across a charging capacitor as a function of time and solve for the times at which it reaches 10% and 90% of its final value.)

8–43. The MOSFETs in the inverter shown in Figure 8–72 have $V_T = 1.8$ V, $\beta_1 = 86 \times 10^{-6}$, and $\beta_2 = 9 \times 10^{-6}$. Find the approximate rise and fall times at the output.

8–44. What is the maximum load capacitance that the inverter of Exercise 8–43 can drive if the rise time can be no greater than 400 ns and the fall time can be no greater than 50 ns?

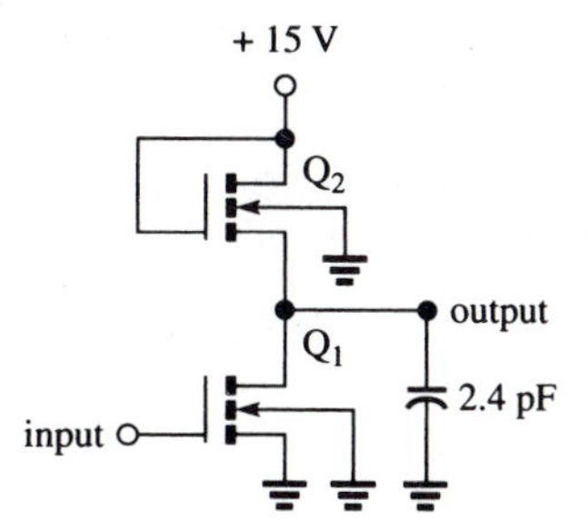

Figure 8–72
(Exercise 8–43)

Section 8–9
CMOS Circuits

8–45. The input and output voltages of the inverter shown in Figure 8–73 are either 0 V or 10 V. Find the input voltage that will cause (a) Q_1 to turn OFF, (b) Q_1 to turn ON, (c) Q_2 to turn OFF, and (d) Q_2 to turn ON.

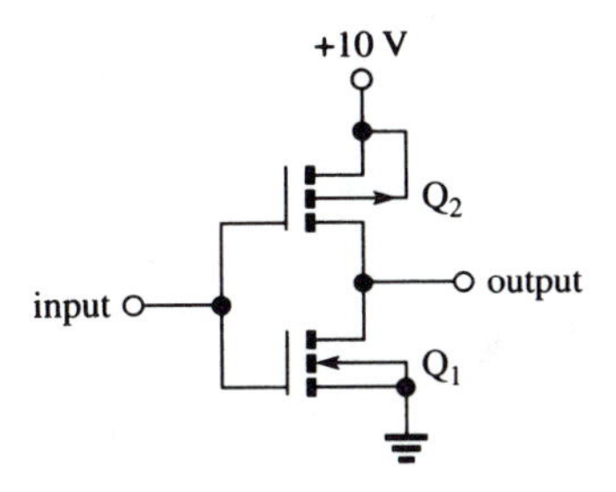

Figure 8–73
(Exercise 8–45)

8–46. Suppose the input to the circuit in Figure 8–73 alternated between −5 V and +15 V. Find V_{GS1} and V_{GS2} (a) when the input is −5 V and (b) when the input is +15 V. Which transistor is ON and which is OFF in each of (a) and (b)? What would be the output voltage in each case?

SPICE EXERCISES

8–47. Use SPICE to find the voltage gain, v_L/v_S, and input resistance, r_{in}(stage), of the amplifier shown in Figure 8–56. The JFET has $V_p = 3.9$ V and $I_{DSS} = 10$ mA. The input and output coupling capacitors are each 1 μF and the source bypass capacitor is 10 μF. The input signal v_S has frequency 10 kHz.

8–48. The JFET in the amplifier shown in Figure 8–58 has $V_p = 2.5$ V and $I_{DSS} = 12$ mA. Use SPICE to find the voltage gain v_L/v_S and the input resistance, r_{in}(stage). The input coupling capacitor is 1 μF and the output coupling capacitor is 10 μF. The input signal v_S has frequency 10 kHz.

8–49. **a.** Use SPICE to verify the voltage gain obtained in Example 8–8 for the MOSFET amplifier. Obtain a plot of the load voltage over one full cycle when the input, v_S, is a 20-kHz sinewave voltage with peak value 0.5 V. The input and output coupling capacitors are 1 μF and the source bypass capacitor is 10 μF.

b. Obtain the plot described in (a) when the source bypass capacitor is removed, and comment on the results.

8–50. The CMOS inverter in Figure 8–46 is driven by a 1-MHz square wave that alternates between 0 V and +10 V. Use SPICE to obtain a plot of the output over one full cycle of input. Allow all MOSFET parameters to have their default values. Then obtain the same plot when the output drives a 10-pF capacitive load. Comment on the results.

9 *h* AND *y* PARAMETERS

9–1 TWO-PORT NETWORKS

A *port* is a pair of terminals in an electrical network. It is convenient to think of the input terminals of an electronic circuit as one port and the output terminals as another, so, for analysis purposes, the circuit is regarded as a *two-port network.* Figure 9–1 shows the input and output ports of a general network: one that might be composed of a single semiconductor device or of any number of interconnected components, such as an amplifier stage.

Note the polarities assigned to the input and output voltages in Figure 9–1. Also note that current flowing *into* each + terminal is assumed to be positive. These conventions are standard in two-port network theory, but they do not mean that actual voltages and currents in a network being analyzed must have those polarities. As usual, a negative value simply implies that a polarity or direction is opposite to that assumed. Do not be confused by the fact that currents seem to be flowing from open circuits into the ports of the network; as we shall soon see, the terminals at a port are often shorted for analysis purposes, thus providing a path for current flow, or else they are left open as shown (in which case the currents are zero). When the currents and voltages are ac quantities, the polarities are used as instantaneous references.

In most electronic circuits analyzed as two-port networks, one terminal of each port is common to both input and output. Figure 9–2 illustrates this fact in the circuit of a common-source JFET, using a circuit diagram we have already studied in detail. Note that the source terminal is common to both the input port and the output port. The common terminal of each port is the circuit ground.

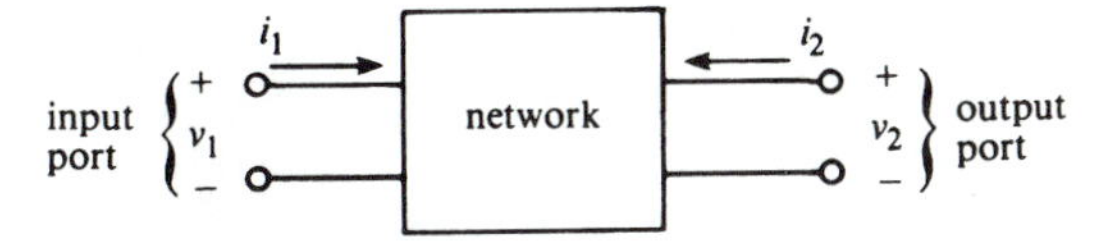

Figure 9–1
A general two-port network. Note the (assumed positive) polarities of input and output voltages and currents.

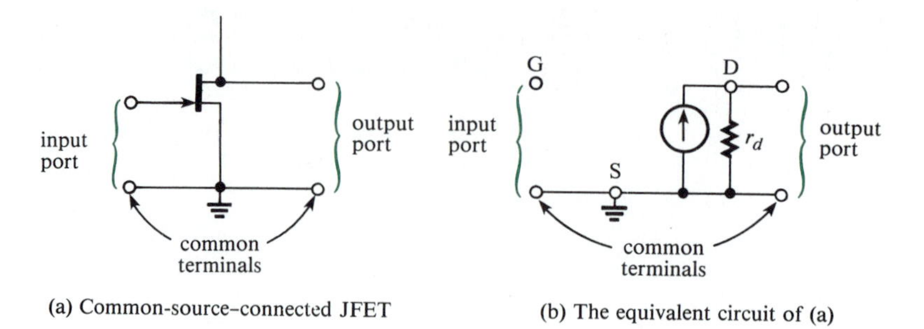

(a) Common-source-connected JFET

(b) The equivalent circuit of (a)

Figure 9–2
The source terminal is common to both the input port and the output port. This common terminal is the circuit ground, or reference.

Two-Port Network Parameters

In an earlier chapter, we distinguished between inherent physical parameters of a device and *derived* parameters that are based on algebraic ratios, such as α and β in a bipolar transistor. Two-port parameters are derived parameters defined by certain ratios of input and output voltages and currents. The specific ratios that define each parameter depend on the parameter *set* from which they are derived. We shall study two important sets of parameters: *hybrid,* or *h,* parameters and *admittance,* or *y,* parameters. Both of these sets are widely used to describe characteristics of electronic devices.

The important feature of two-port parameters is that they completely characterize a network for which their values are given. In other words, if all the two-port parameter values associated with a network are known, then every important characteristic of that network can be determined, including its voltage and current gains and its input and output resistances. Furthermore, any circuit having equal parameter values is equivalent to and can be substituted for the actual network. The most important application of this fact is that it is possible to construct an accurate circuit *model* of an electronic device based on its parameter values. The performance of the device in a larger system can then be evaluated using the equivalent circuit model to represent the device.

9–2 HYBRID- (*h*-) PARAMETER DEFINITIONS

The h parameters of a two-port network are designated h_{11}, h_{12}, h_{21}, and h_{22}. Their values for any network are such that the following two network equations are always satisfied:

$$v_1 = h_{11}i_1 + h_{12}v_2 \tag{9–1}$$

$$i_2 = h_{21}i_1 + h_{22}v_2 \tag{9–2}$$

In other words, for *any* possible combination of input and output voltage and current that might ever exist in a given network, equations 9–1 and 9–2 will be satisfied if the h-parameter values of that network are used in the equations. Think and dwell for a moment on the implications of the previous statement: It should seem rather startling that such a powerful set of parameters can exist. However, as we shall presently see, the two equations themselves provide us with the means for defining the parameters in a way that forces them to satisfy the equations.

Note that the left side of equation 9–1 is a voltage. Therefore, each term on the right-hand side must also be a voltage. Similarly, each term in equation 9–2 must be a current. It follows that the units of all the h parameters are different. For example, h_{11} must have the units of ohms, since it multiplies the current i_1 to produce a voltage term, while the units of h_{22} must be siemens, since it multiplies the voltage v_2 to produce a current term. This mixture of units accounts for the name *hybrid* parameters.

If we set v_2 equal to 0 in equation 9–1, we obtain

$$v_1 = h_{11} i_1 \tag{9–3}$$

Solving (9–3) for h_{11} gives us the *definition* of that parameter:

$$h_{11} = \frac{v_1}{i_1}\bigg|_{v_2 = 0} \text{ ohms} \tag{9–4}$$

In words, h_{11} is the ratio of input voltage to input current when the output voltage (v_2) is 0, i.e., *when the output is short-circuited.* See Figure 9–3. It is clear from Ohm's law that h_{11} must have the units of ohms, and it is called the *short-circuit input resistance.* For the time being, we will consider all voltages and currents to be ac quantities, so we will refer to h_{11} more properly as the short-circuit input *impedance.*

If we set $i_1 = 0$ in equation 9–1, we obtain

$$v_1 = h_{12} v_2 \tag{9–5}$$

Solving for h_{12} gives the definition of h_{12}:

$$h_{12} = \frac{v_1}{v_2}\bigg|_{i_1 = 0} \tag{9–6}$$

In words, h_{12} is the ratio of input voltage to output voltage when the input current is 0, i.e., *when the input is open-circuited.* Since h_{12} is the ratio of two voltages, it is a dimensionless quantity (it is sometimes said to have the units of volts per volt, V/V). Since we are accustomed to thinking of voltage gain as the ratio of output to input voltage, and h_{12} is just the reverse, it is called the *open-circuit reverse voltage ratio.* (It is sometimes referred to as the reverse voltage *transfer* ratio.)

Setting $v_2 = 0$ in equation 9–2, we obtain

$$i_2 = h_{21} i_1 \tag{9–7}$$

or

$$h_{21} = \frac{i_1}{i_1}\bigg|_{v_2 = 0} \tag{9–8}$$

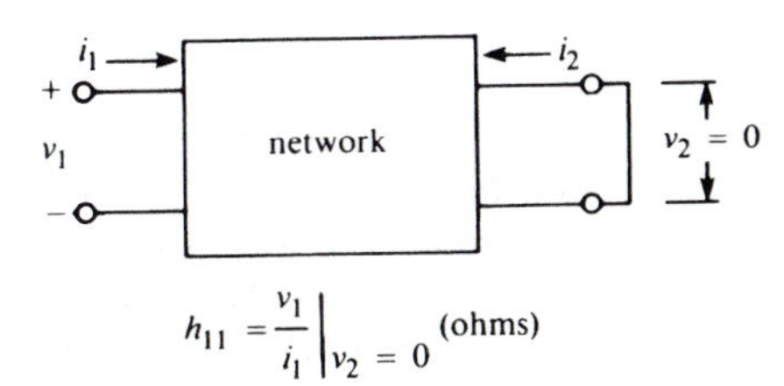

Figure 9–3
h_{11} is the input impedance when the output is short-circuited, i.e., when $v_2 = 0$.

Table 9–1

Parameter	Definition	Name
h_{11}	$\left.\frac{v_1}{i_1}\right\vert_{v_2 = 0}$	Short-circuit input impedance
h_{12}	$\left.\frac{v_1}{v_2}\right\vert_{i_1 = 0}$	Open-circuit reverse voltage ratio
h_{21}	$\left.\frac{i_2}{i_1}\right\vert_{v_2 = 0}$	Short-circuit forward current ratio
h_{22}	$\left.\frac{i_2}{v_2}\right\vert_{i_1 = 0}$	Open-circuit output admittance

Since h_{21} is the ratio of the output to input current with the output short-circuited, it is a dimensionless quantity called the *short-circuit forward current ratio* (sometimes called the forward current *transfer* ratio).

Setting $i_1 = 0$ in equation 9–2, we obtain

$$i_2 = h_{22}v_2 \tag{9–9}$$

or

$$h_{22} = \left.\frac{i_2}{v_2}\right\vert_{i_1 = 0} \text{siemens} \tag{9–10}$$

Since h_{22} is the ratio of output current to output voltage with the input open-circuited, it has the units of siemens and is called the *open-circuit output admittance* (or conductance, in the case of resistive circuits).

Table 9–1 summarizes the definitions and names of the h parameters. Note that *short-circuit* always refers to the condition at the output port, and *open-circuit* always refers to the condition at the input port.

The next two examples demonstrate procedures that can be used to find the values of the h parameters of some simple networks.

Example 9–1

Find the h parameters of the network shown in Figure 9–4.

Solution. Figure 9–5 shows the network with the output short-circuited ($v_2 = 0$), which is the condition required in the definitions of h_{11} and h_{21}.

Since h_{11} is the resistance looking into the input and since the 40-Ω resistor is

Figure 9–4
(Example 9–1)

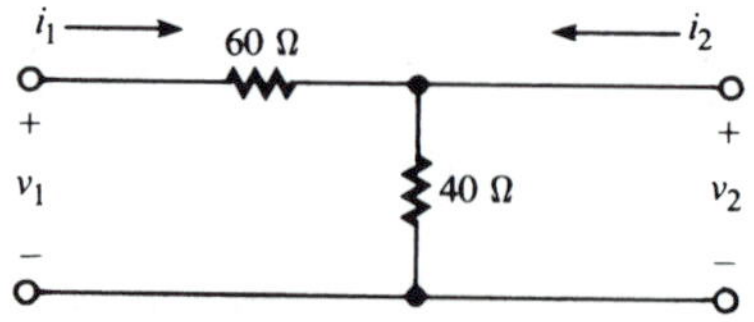

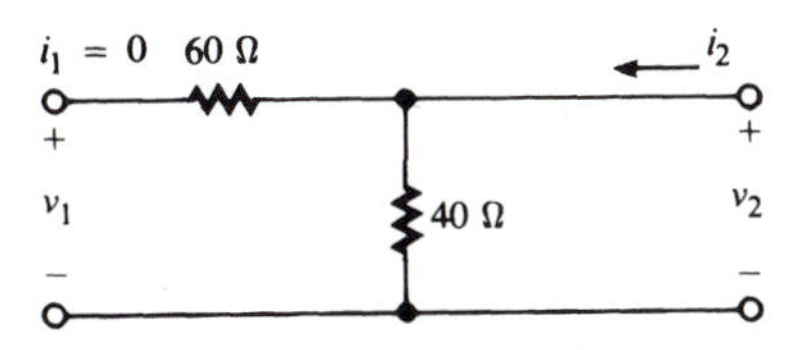

Figure 9–5
(Example 9–1) The output is shorted when calculating h_{11} and h_{21}.

shorted, it is clear that

$$h_{11} = \left.\frac{v_1}{i_1}\right|_{v_2 = 0} = 60\ \Omega$$

Values of the h parameters can always be found without assigning explicit values to v_1, i_1, v_2, or i_2. However, it is sometimes easier to perform the computations by *assuming* a value for one of those quantities and then solving for another. For example, h_{11} in Figure 9–5 could have been found by assuming that v_1 equals (say) 60 V and then solving for i_1. It is obvious that $i_1 = 1$ A when $v_1 = 60$ V, so

$$h_{11} = \frac{60\text{ V}}{1\text{ A}} = 60\ \Omega$$

Since the 40-Ω resistor is shorted, all the current i_1 will flow in the short. Thus, the magnitudes of i_1 and i_2 are equal. Note, however, that current flows *out* of port 2, that is, opposite to the assumed positive direction of i_2. Therefore, $i_2 = -i_1$, and

$$h_{21} = \left.\frac{i_2}{i_1}\right|_{v_2 = 0} = \frac{-i_1}{i_1} = -1$$

Figure 9–6 shows the network with the input open-circuited ($i_1 = 0$), the condition necessary for finding h_{12} and h_{22}.

To find h_{12}, we must find the ratio of v_1 to v_2 in Figure 9–6. Since $i_1 = 0$, there is no voltage drop across the 60-Ω resistor. Therefore, the voltage v_1 must be the same as the voltage v_2 developed across the 40-Ω resistor, i.e., $v_1 = v_2$. It follows that

$$h_{12} = \left.\frac{v_1}{v_2}\right|_{i_1 = 0} = 1$$

The value of h_{12} could have been found by imagining that a voltage source, say, 20 V, was applied across the output terminals. With the input open-circuited, it would again be apparent that $v_1 = v_2 = 20$ V, leading to the same value for h_{12}: $v_1/v_2 = 1$. Note, however, that it would be *wrong* to apply a voltage v_1 across the input, calculate v_2, and then find v_1/v_2. If a voltage source v_1 were applied across the input, then current would flow into the circuit, violating the condition that $i_1 = 0$.

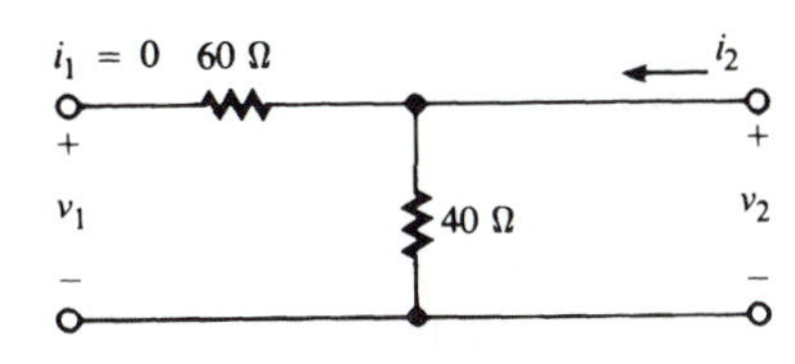

Figure 9–6
(Example 9–1) The input is open-circuited when calculating h_{12} and h_{22}.

The value of h_{22} can be found in Figure 9–6 by recognizing that it equals the output conductance of the network, that is, the reciprocal of the resistance looking into the output terminals. Since one side of the 60-Ω resistor is open-circuited, the output resistance is clearly 40 Ω, and

$$h_{22} = \frac{i_2}{v_2}\bigg|_{i_1 = 0} = \frac{1}{40} = 0.025 \text{ S}$$

Summarizing, we have found

$$h_{11} = 60\ \Omega \qquad h_{12} = 1$$
$$h_{21} = -1 \qquad h_{22} = 0.025 \text{ S}$$

Example 9–2

Find the h parameters of the network shown in Figure 9–7. Note that the capacitor has impedance $0 - j500\ \Omega$, or $500\angle{-90^\circ}\ \Omega$, at the frequency for which the analysis is to be performed. The values found for the h parameters will therefore be valid only at that one frequency of operation.

Solution. When the output of Figure 9–7 is shorted, the capacitive reactance is in parallel with the 1-kΩ resistance. See Figure 9–8. Therefore,

$$h_{11} = \frac{v_1}{i_1}\bigg|_{v_2 = 0} = 10^3 + 10^3 \parallel (-j500)$$
$$= 10^3 + \frac{-j0.5 \times 10^6}{10^3 - j500} = 10^3 + \frac{0.5 \times 10^6\angle{-90^\circ}}{1118\angle{-26.56^\circ}}$$
$$= 10^3 + 447.23\angle{-63.43^\circ} = 1000 + 200 - j400$$
$$= 1200 - j400\ \Omega = 1.265 \times 10^3\angle{-18.44^\circ}\ \Omega$$

To find h_{21}, we use the current-divider rule to find the current i (Figure 9–8) that flows in the right-hand branch:

$$i = \left(\frac{-j500}{1000 - j500}\right) i_1 = \left(\frac{500\angle{-90^\circ}}{1118\angle{-26.56^\circ}}\right) i_1 = 0.447\angle{-63.44^\circ} i_1$$

Notice that i in Figure 9–8 is in the opposite direction of the assumed direction for i_2, so $i_2 = -i$. Therefore,

$$i_2 = -i = -0.447\angle{-63.44^\circ} i_1 = 0.447\angle{116.56^\circ} i_1$$

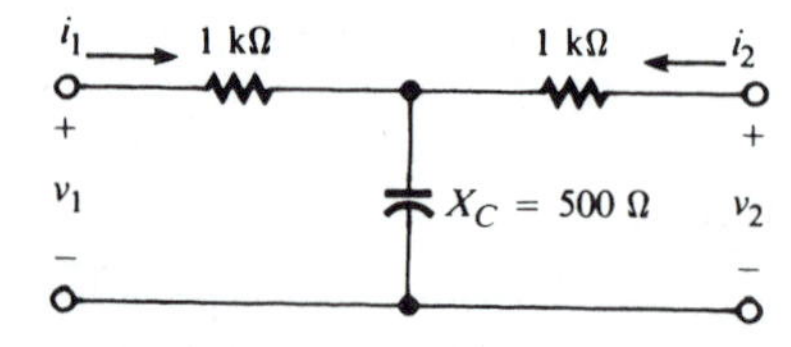

Figure 9–7
(Example 9–2)

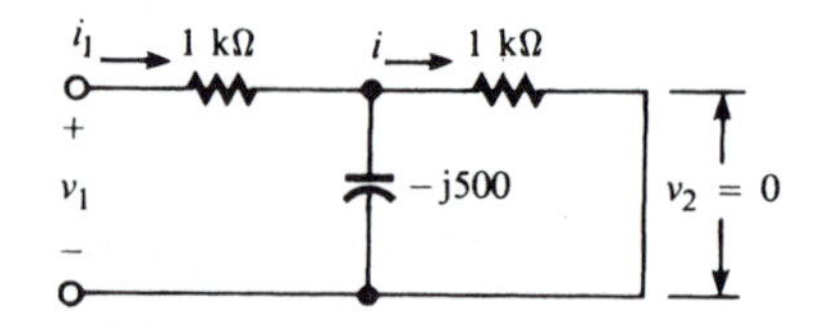

Figure 9–8
(Example 9–2) The network of Figure 9–7 with $v_2 = 0$, as required to find h_{11} and h_{21}

Thus, dividing through by i_1, we find

$$h_{21} = \left.\frac{i_2}{i_1}\right|_{v_2 = 0} = 0.447\underline{/116.56^\circ}$$

The value of h_{12} can be found by assuming that a voltage v_2 is applied across the output terminals of Figure 9–7. Since the input terminals are open-circuited when computing h_{12}, the voltage v_1 will be the voltage developed across the capacitor, which can be found using the voltage-divider rule:

$$v_1 = \left(\frac{-j500}{1000 - j500}\right) v_2 = \left(\frac{500\underline{/-90^\circ}}{1118\underline{/-26.56^\circ}}\right) v_2 = 0.447\underline{/-63.44^\circ} v_2$$

Thus,

$$h_{12} = \left.\frac{v_1}{v_2}\right|_{i_1 = 0} = 0.447\underline{/-63.44^\circ}$$

Finally, h_{22} is the output admittance of the network with the input open-circuited. The output impedance under that condition is clearly the sum of 1000 ohms of resistance and 500 ohms of capacitive reactance, so

$$h_{11} = \frac{1}{1000 - j500} = \frac{1}{1118\underline{/-26.56^\circ}} = 8.94 \times 10^{-3}\underline{/-26.56^\circ}\ \mathrm{S}$$

We emphasize again that the h-parameter values computed in this example are only valid at the frequency for which $X_C = 500\ \Omega$. The h-parameter values for any network having reactive components will always depend on frequency.

9–3 HYBRID EQUIVALENT CIRCUITS

If the h parameters of a circuit are known, then it is possible to construct another circuit that

1. has a current source and a voltage source (whose values are determined by the known h-parameter values); and
2. is entirely equivalent to the original circuit. Figure 9–9 shows the *hybrid equivalent circuit* that can be substituted for any circuit whose h parameters are known. Note that it contains a controlled voltage source whose output is $h_{12}v_2$ volts and a controlled current source whose output is $h_{21}i_1$ amperes. An impedance equal to h_{11} ohms is in series with the voltage source, and an impedance equal to $1/h_{22}$ ohms is in parallel with the current source.

It is easy to verify that the hybrid equivalent circuit shown in Figure 9–9 has the same h parameters as the circuit to which it is equivalent. If we set $v_2 = 0$ in

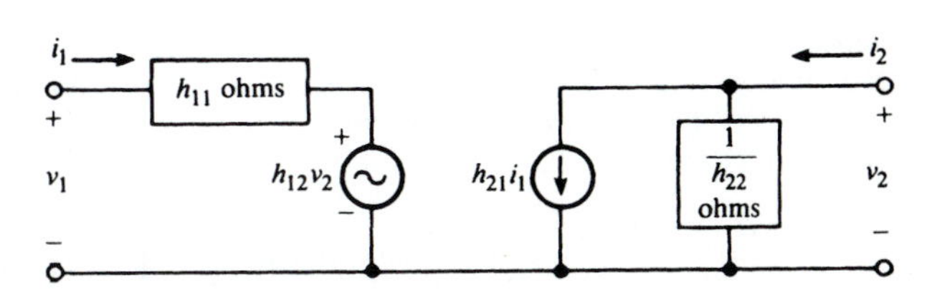

Figure 9–9
The hybrid equivalent circuit

the equivalent circuit, by shorting the output port, then the voltage source on the input side produces 0 V (i.e., it is a short circuit), and the input resistance is then clearly h_{11} ohms. Also, with the output shorted, it can be seen that $i_2 = h_{21}i_1$, so $h_{21} = i_2/i_1$, as required. It is left as an exercise at the end of this chapter to verify that the reverse voltage ratio and the output admittance in Figure 9–9 are h_{12} and h_{22}, respectively.

It is important to realize that even though the hybrid circuit contains active sources, it is a valid equivalent for a network that contains only passive components. Of course, for very simple networks there is no reason to use the hybrid equivalent circuit as an analysis tool. For complex devices such as transistors, the hybrid circuit is a very useful means for obtaining a relatively simple but accurate model, provided the values of the device's h parameters are known. The next example demonstrates the validity of using the hybrid circuit for a simple, resistive network.

Example 9–3

Using the h-parameter values found in Example 9–1 for the network shown in Figure 9–10, verify that the hybrid equivalent circuit can be used to determine the voltage v_2 across the 24-Ω resistor when the input is 15 V.

Solution. Let us first use simple network analysis to find v_2. The parallel combination of the 40-Ω and 24-Ω resistors is equivalent to 15 Ω, so by the voltage-divider rule,

$$v_2 = \left(\frac{15\,\Omega}{60\,\Omega + 15\,\Omega}\right)(15\text{ V}) = 3\text{ V}$$

Figure 9–11 shows the hybrid equivalent circuit of the network, with $v_1 =$ 15 V and a 24-Ω resistor connected across the output. (The controlled voltage source is shown as a dc source, since we are modeling a dc circuit.) Note that $h_{12}v_2 = 1v_2 = v_2$ and $h_{21}i_1 = -i_1$.

Figure 9–10
(Example 9–3)

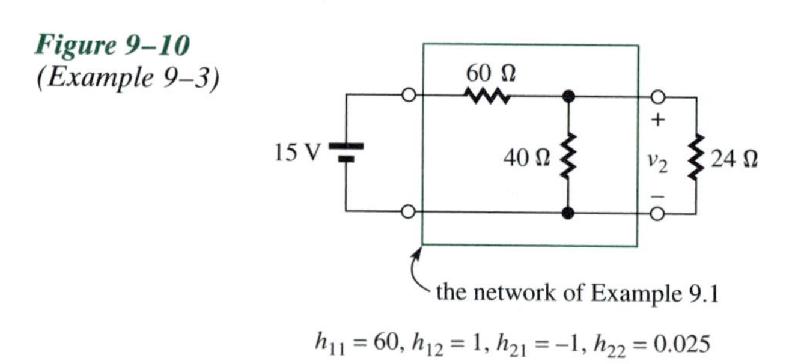

Figure 9–11
(Example 9–3) The resistive network of Example 9–1 is replaced by its hybrid equivalent circuit.

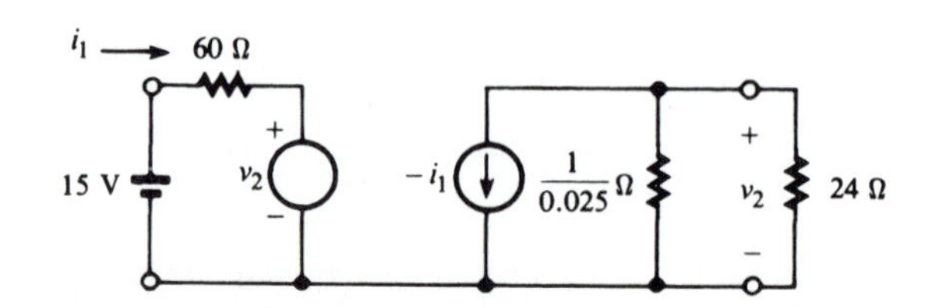

At the input side of Figure 9–11, we see that

$$i_1 = \frac{15 - v_2}{60} \qquad \textbf{(9–11)}$$

The current source is unchanged if we reverse its direction and assign it the value $-(-i_1) = i_1$. The output voltage v_2 is, therefore,

$$v_2 = i_1\left(\frac{1}{0.025}\,\Big\|\,24\right) = 15i_1$$

Substituting i_1 from equation 9–11, we obtain

$$v_2 = 15\left(\frac{15 - v_2}{60}\right) = \frac{15 - v_2}{4}$$

$$4v_2 = 15 - v_2$$

$$v_2 = 3 \text{ V}$$

Although the computations are laborious, we have demonstrated that analysis using the hybrid equivalent circuit produces the same result ($v_2 = 3$ V) as the direct analysis.

As a memory aid, h parameters are given special subscripts in place of the double-number subscripts we have been using. Following is a list of those special designations:

$h_{11} = h_i$ (for *i*nput impedance)
$h_{12} = h_r$ (for *r*everse voltage ratio)
$h_{21} = h_f$ (for *f*orward current ratio)
$h_{22} = h_o$ (for *o*utput admittance)

Figure 9–12 shows the hybrid equivalent circuit using these new designations. We will use this notation in all future discussions.

Gain and Impedance Computations Using *h* Parameters

We will now derive h-parameter expressions for the voltage gain, current gain, input impedance, and output impedance of a circuit that is driven by a signal source having signal-source impedance and that has a load impedance connected across its output. Figure 9–13 shows the general hybrid equivalent circuit with signal source and load impedance connected. We should note that the quantities we seek are *not* equal to the h parameters themselves, since the values of those parameters are derived only under the very special conditions $v_2 = 0$ or $i_1 = 0$.

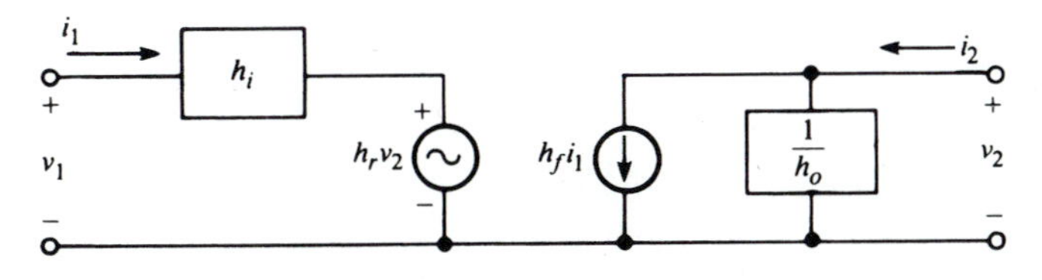

Figure 9–12
The hybrid equivalent circuit using the h_i, h_r, h_f, h_o notation

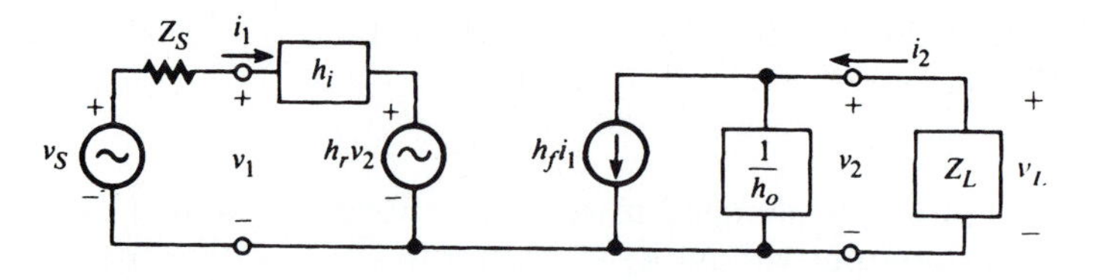

Figure 9–13
The hybrid equivalent circuit with signal source and load connected

We begin our analysis by finding the voltage gain $A_v = v_2/v_1$, i.e., the gain between the input of the hybrid circuit and the load, omitting v_S and its impedance for the time being. The input current is the voltage difference across h_i divided by h_i:

$$i_1 = \frac{v_1 - h_r v_2}{h_i} \tag{9–12}$$

Applying Kirchhoff's current law at the output, we obtain

$$i_2 = h_f i_1 + h_o v_2 \tag{9–13}$$

Substituting $i_2 = -v_2/Z_L$, and i_1 from (9–12), into (9–13) gives

$$\frac{-v_2}{Z_L} = h_f\left(\frac{v_1 - h_r v_2}{h_i}\right) + h_o v_2 \tag{9–14}$$

Equation 9–14 can be solved for v_2/v_1 with the result that

$$A_v = \frac{v_2}{v_l} = \frac{-h_f Z_L}{h_i(1 + h_o Z_L) - h_r h_f Z_L} \tag{9–15}$$

If the output is open ($Z_L = \infty$), equation 9–15 becomes

$$A_v = \frac{-h_f}{h_i h_o - h_r h_f} \tag{9–16}$$

To find the current gain $A_i = i_2/i_1$, we apply the current-divider rule to determine the portion of the current $h_f i_1$ in the current source that flows in the load:

$$i_2 = \left[\frac{(1/h_o)}{(1/h_o) + Z_L}\right] h_f i_1 \tag{9–17}$$

Solving (9–17) for i_2/i_1 and simplifying gives

$$A_i = \frac{i_2}{i_1} = \frac{h_f}{1 + h_o Z_L} \tag{9–18}$$

If $Z_L = \infty$, then $A_i = 0$, since there is no load current in that case. If $Z_L = 0$ (output short-circuited), then A_i reduces to h_f, as would be expected from the definition of h_f.

The input impedance Z_i looking into the hybrid circuit is found by dividing v_1 by i_1. From equation 9–12, we have $v_1 = h_i i_1 + h_r v_2$. Dividing by i_1,

$$\frac{v_1}{i_1} = h_i + \frac{h_r v_2}{i_1} \tag{9–19}$$

Substituting $v_2 = -i_2 Z_L$ into (9–19) gives

$$\frac{v_1}{i_1} = h_i - \frac{h_r i_2 Z_L}{i_1} \tag{9–20}$$

The factor i_2/i_1 on the right side of (9–20) is the current gain, which we have already derived (equation 9–18). Substituting (9–18) into (9–20) gives

$$Z_i = \frac{v_1}{i_1} = h_i - \frac{h_r h_f Z_L}{1 + h_o Z_L} \text{ ohms} \tag{9–21}$$

If $Z_L = \infty$, then (9–21) becomes

$$Z_i = h_i - \frac{h_r h_f}{h_o} \tag{9–22}$$

The overall voltage gain v_L/v_S can now be determined by taking into account the voltage division between Z_S and Z_i at the input:

$$\frac{v_L}{v_S} = \frac{Z_i}{Z_S + Z_i} A_v \tag{9–23}$$

It is left as an exercise at the end of this chapter to show that (9–23) can be written

$$\frac{v_L}{v_S} = \frac{-h_f Z_L}{(Z_S + h_i)(1 + h_o Z_L) - h_r h_f Z_L} \tag{9–24}$$

If $Z_L = \infty$, then (9–24) becomes

$$\frac{v_L}{v_S} = \frac{-h_f}{(Z_S + h_i)h_o - h_r h_f} \tag{9–25}$$

Just as the load impedance Z_L on the output side affects the input impedance, the source impedance Z_S on the input side affects the output impedance. To include Z_S in the analysis, we must set v_S to 0. In that case, the input current i_1 is the voltage $-h_r v_2$ divided by the total impedance in the input loop:

$$i_1 = \frac{-h_r v_2}{h_i + Z_S} \tag{9–26}$$

Substituting (9–26) in (9–13) gives

$$i_2 = \frac{-h_f h_r v_2}{h_i + Z_S} + h_o v_2 \tag{9–27}$$

Dividing through by v_2 gives the output admittance Y_o:

$$Y_o = \frac{i_2}{v_2} = \frac{-h_f h_r}{h_i + Z_S} + h_o \tag{9–28}$$

Therefore,

$$Z_o = \frac{1}{Y_o} = \frac{1}{h_o - \dfrac{h_f h_r}{h_i + Z_S}} \tag{9–29}$$

The results of our derivations are summarized in equations 9–29a.

$$
\begin{aligned}
A_v = \frac{v_2}{v_1} &= \frac{-h_f Z_L}{h_i(1 + h_o Z_L) - h_r h_f Z_L} \\
&= \frac{-h_f}{h_i h_o - h_r h_f} \qquad (Z_L = \infty) \\
\frac{v_L}{v_S} &= \frac{-h_f Z_L}{(Z_S + h_i)(1 + h_o Z_L) - h_r h_f Z_L} \\
&= \frac{-h_f}{(Z_S + h_i)h_o - h_r h_f} \qquad (Z_L = \infty) \\
A_i = \frac{i_2}{i_1} &= \frac{h_f}{1 + h_o Z_L} \\
&= 0 \qquad (Z_L = \infty) \\
&= h_f \qquad (Z_L = 0) \\
Z_i = \frac{v_1}{i_1} &= h_i - \frac{h_r h_f Z_L}{1 + h_o Z_L} \\
&= h_i - \frac{h_r h_f}{h_o} \qquad (Z_L = \infty) \\
&= h_i \qquad (Z_L = 0) \\
Z_o = \frac{v_2}{i_2} &= \frac{1}{h_o - \dfrac{h_f h_r}{h_i + Z_S}} \\
&= \frac{1}{h_o - \dfrac{h_f h_r}{h_i}} \qquad (Z_S = 0)
\end{aligned}
\qquad \textbf{(9–29a)}
$$

Example 9–4

Find the input impedance Z_i and the rms output voltage v_L of the circuit shown in Figure 9–14.

Solution. Comparing Figure 9–14 with Figure 9–13, we see that $Z_S = 500\ \Omega$, $h_i = 10^3$, $h_r = 2 \times 10^{-4}$, $h_f = 100$, $h_o = 1/(50\ \text{k}\Omega) = 2 \times 10^{-5}$, and $Z_L = 2 \times 10^3\ \Omega$. From equation 9–21,

$$Z_i = 10^3 - \frac{(2 \times 10^{-4})(100)(2 \times 10^3)}{1 + (2 \times 10^{-5})(2 \times 10^3)} = 961.54\ \Omega$$

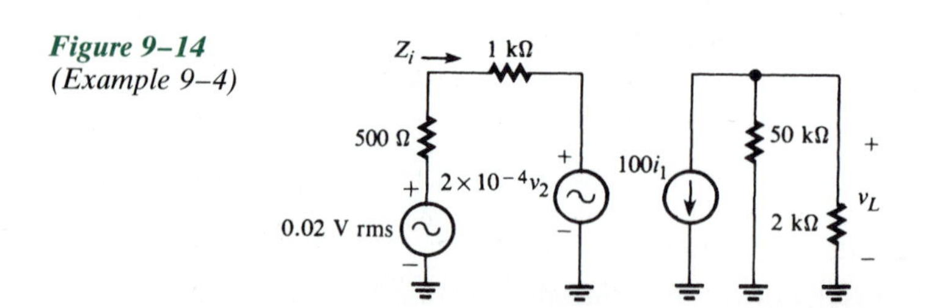

Figure 9–14
(Example 9–4)

From equation 9–15,

$$A_v = \frac{v_2}{v_1} = \frac{-100\,(2 \times 10^3)}{10^3[1 + (2 \times 10^{-5})(2 \times 10^3)] - (2 \times 10^{-4})(100)(2 \times 10^3)}$$
$$= -200$$

The minus sign means that there is a phase inversion between input and output.
The overall voltage gain is found from equation 9–23:

$$\frac{v_L}{v_S} = \left(\frac{961.54}{500 + 961.54}\right)(-200) = -131.58$$

Thus, the rms output voltage is $v_L = (0.02)(131.58) = 2.63$ V rms. Verify that this same result can be obtained using equation 9–24.

Example 9–5

Find the input current delivered by the source in Figure 9–14 and the current in the load. Find the current gain using the computed values of input and load current, and verify that equation 9–18 gives the same result.

Solution. We see from Figure 9–13 that

$$i_1 = \frac{v_s}{Z_S + Z_i}$$

From Example 9–4, $Z_i = 961.54\ \Omega$. Therefore,

$$i_1 = \frac{0.02 \text{ V rms}}{500\ \Omega + 961.54\ \Omega} = 13.68\ \mu\text{A rms}$$

Also from Example 9–4, $v_L = 2.63$ V rms, so

$$i_L = \frac{v_L}{Z_L} = \frac{2.63 \text{ V rms}}{2\text{ k}\Omega} = 1.315 \text{ mA rms}$$

Therefore,

$$A_i = \frac{i_L}{i_1} = \frac{1.315 \text{ mA}}{13.68\ \mu\text{A}} = 96.1$$

Using equation 9–18, we obtain the same result:

$$A_i = \frac{h_f}{1 + h_o Z_L} = \frac{100}{1 + (2 \times 10^{-5})(2 \times 10^3)} = 96.1$$

9–4 TRANSISTOR *h* PARAMETERS

Bipolar transistor specifications often include h-parameter values (see Figure 5–47), and these values are frequently used in the design and analysis of BJT amplifier circuits. Field-effect–transistor specifications more often include y-parameter values, which we will discuss later in the chapter.

Since there are three possible BJT configurations (CE, CC, and CB), there are three different ways that the input and output can be defined, and therefore three corresponding sets of h parameters. A second subscript is added to each h-parameter symbol to show the configuration for which it applies: e for CE, c for CC, and b for CB. Thus, the 12 h parameters for a bipolar transistor are designated

h_{ie}, h_{fe}, h_{re}, h_{oe}: common-emitter h parameters
h_{ic}, h_{fc}, h_{rc}, h_{oc}: common-collector h parameters
h_{ib}, h_{fb}, h_{rb}, h_{ob}: common-base h parameters

If all of the h-parameter values in one configuration are known, then the values corresponding to any other configuration can be determined. The common-emitter values are the ones most often given.

The values specified for transistor h parameters are almost always small-signal quantities. Occasionally, a dc value will be given, in which case it is common practice to use capital-letter subscripts, as, for example, h_{FE}. It is important to remember that all small-signal BJT parameter values are affected by dc (quiescent) current, as discussed in Chapter 5. Each small-signal parameter is the ratio of certain ac voltages and currents, and each is valid only in a small region over which there is negligible change in device characteristics.

Figure 9–15 shows how common-emitter h-parameter values can be determined graphically using characteristic curves. In the common-emitter configuration, note that

$$v_1 = v_{be} \qquad i_1 = i_b$$

and

$$v_2 = v_{ce} \qquad i_2 = i_c$$

Therefore, from the definitions of the h parameters, the computation of h_{ie} and h_{fe} must be performed with $v_{ce} = 0$. Note that requiring the *ac* quantity v_{ce} to be 0 is the same as requiring that the *dc* quantity V_{CE} be held constant. Similarly, the computation of h_{re} and h_{oe} requires that $i_b = 0$, which is satisfied by requiring that I_B remain constant. A typical set of h-parameter values for a silicon NPN transistor is

$$h_{ie} = 1200\ \Omega \qquad h_{re} = 2 \times 10^{-4}$$
$$h_{fe} = 100 \qquad h_{oe} = 20 \times 10^{-6}\ \text{S}$$

Notice that

$$h_{fe} = \frac{i_c}{i_b}\bigg|_{V_{CE} = \text{constant}}$$

is the small-signal BJT parameter β. Also,

$$h_{fb} = \frac{i_c}{i_e}\bigg|_{V_{CB} = \text{constant}}$$

is the small-signal BJT parameter α.

Figure 9–16 shows the h-parameter equivalent circuit of a transistor in its common-emitter configuration. Recall that we discussed in Chapter 4 how the output voltage of a transistor affects its input characteristics, and we saw evidence of that fact in the family of curves of I_B versus V_{BE} generated by different values

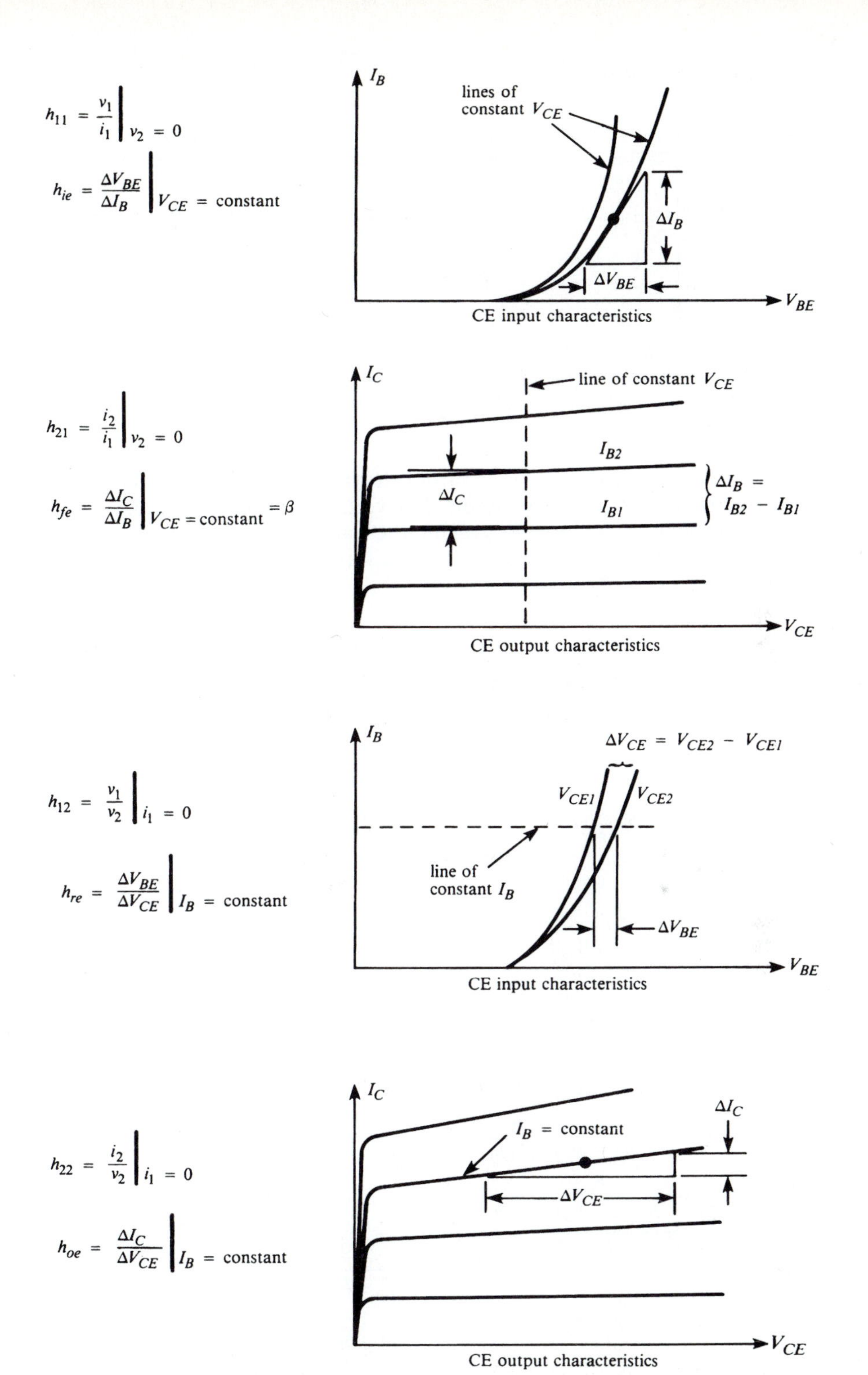

Figure 9–15

Graphical determination of common-emitter h parameters. Notice that the ac quantities v_{ce} and i_b are 0 when the dc quantities V_{CE} and I_B are held constant.

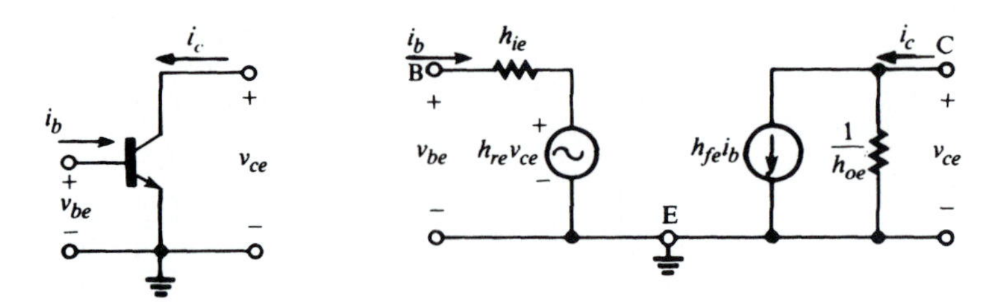

Figure 9–16
The h-parameter equivalent circuit of a common-emitter transistor. Notice that the output voltage (v_{ce}) affects the input current (i_b), and vice versa.

of V_{CE}. It should now be apparent that this feedback effect is accounted for in the h-parameter equivalent circuit: The voltage source $h_{re}v_{ce}$ in the input side of Figure 9–16 opposes v_{be}, so the greater the value of v_{ce}, the smaller the current i_b. See Figure 4–18. The value of h_{re} is a measure of how significant the feedback influence

Table 9–2
h-Parameter Conversion Equations

		CE	CB	CC
CE	h_{ie}	1400 Ω	$\dfrac{h_{ib}}{(1+h_{fb})(1-h_{rb})+h_{ib}h_{ob}}$	h_{ic}
	h_{re}	2×10^{-4}	$\dfrac{h_{ib}h_{ob}-h_{rb}(1+h_{fb})}{(1+h_{fb})(1-h_{rb})+h_{ib}h_{ob}}$	1 mi h_{rc}
	h_{fe}	100	$\dfrac{-h_{fb}(1-h_{rb})-h_{ob}h_{ib}}{(1+h_{fb})(1-h_{rb})+h_{ib}h_{ob}}$	$-(1+h_{fc})$
	h_{oe}	2×10^{-5} S	$\dfrac{h_{ob}}{(1+h_{fb})(1-h_{rb})+h_{ib}h_{ob}}$	h_{oc}
CB	h_{ib}	$\dfrac{h_{ie}}{(1+h_{fe})(1-h_{re})+h_{ie}h_{oe}}$	14 Ω	$\dfrac{h_{ic}}{h_{ic}h_{oc}-h_{fc}h_{rc}}$
	h_{rb}	$\dfrac{h_{ie}h_{oe}-h_{re}(1+h_{fe})}{(1+h_{fe})(1-h_{re})+h_{ie}h_{oe}}$	4×10^{-5}	$\dfrac{h_{fc}(1-h_{rc})+h_{ic}h_{oc}}{h_{ic}h_{oc}-h_{fc}h_{rc}}$
	h_{fb}	$\dfrac{-h_{fe}(1-h_{re})-h_{ie}h_{oe}}{(1+h_{fe})(1-h_{re})+h_{ie}h_{oe}}$	−0.99	$\dfrac{h_{rc}(1+h_{fc})-h_{ic}h_{oc}}{h_{ic}h_{oc}-h_{fc}h_{rc}}$
	h_{ob}	$\dfrac{h_{oe}}{(1+h_{fe})(1-h_{re})+h_{ie}h_{oe}}$	2×10^{-7} S	$\dfrac{h_{oc}}{h_{ic}h_{oc}-h_{fc}h_{rc}}$
CC	h_{ic}	h_{ie}	$\dfrac{h_{ib}}{(1+h_{fb})(1-h_{rb})+h_{ib}h_{ob}}$	1400 Ω
	h_{rc}	$1-h_{re}$	$\dfrac{1+h_{fb}}{(1+h_{fb})(1-h_{rb})+h_{ib}h_{ob}}$	1
	h_{fc}	$-(1+h_{fe})$	$\dfrac{h_{rb}-1}{(1+h_{fb})(1-h_{rb})+h_{ib}h_{ob}}$	−101
	h_{oc}	h_{oe}	$\dfrac{h_{ob}}{(1+h_{fb})(1-h_{rb})+h_{ib}h_{ob}}$	2×10^{-5} S

is, and since h_{re} is usually very small, we have been able to neglect the effect in earlier discussions. The h-parameter circuit provides us with a more accurate analysis model. Of course, the important fact that input current (i_b) affects output voltage is also reflected in the h-parameter model, by virtue of the controlled current source $h_{fe}i_b$. This source is the same one we used in our original model for the transistor: the current source we labeled βi_b. See Figure 5–31.

Table 9–2 gives conversion equations that can be used to obtain h-parameter values for any configuration, given the values in another configuration. The table also lists typical values of each parameter in each configuration.

Example 9–6

Using Table 9–2 and the common-emitter output characteristics shown in Figure 9–17, find approximate values for h_{fe}, h_{oe}, h_{fb}, h_{ob}, h_{fc}, and h_{oc} when $I_C = 4.2$ mA and $V_{CE} = 15$ V. The transistor whose characteristics are shown has $h_{ie} = 1500\ \Omega$ and $h_{re} = 2 \times 10^{-4}$ at the operating point.

Solution. The operating point where $I_C = 4.2$ mA and $V_{CE} = 15$ V is labeled Q in Figure 9–17. To find h_{fe}, we determine ΔI_C along the vertical line through Q (a line of constant V_{CE}) and the corresponding value of ΔI_B. As shown in the figure, $\Delta I_C \approx (5.6\text{ mA}) - (3.1\text{ mA}) = 2.5$ mA and $\Delta I_B = (50\ \mu\text{A}) - (30\ \mu\text{A}) = 20\ \mu$A. Therefore,

$$h_{fe} = \left.\frac{\Delta I_C}{\Delta I_B}\right|_{V_{CE} = 15\text{ V}} \approx \frac{2.5\text{ mA}}{20\ \mu\text{A}} = 125$$

To find h_{oe}, we determine the slope of the line corresponding to $I_B = 40\ \mu$A (a line of constant I_B). As shown in the figure, $\Delta I_C \approx (4.4\text{ mA}) - (4\text{ mA}) = 0.4$ mA and $\Delta V_{CE} \approx (28\text{ V}) - (2\text{ V}) = 26$ V.

Therefore, $$h_{oe} = \left.\frac{\Delta I_C}{\Delta V_{CE}}\right|_{I_B = 40\ \mu\text{A}} \approx \frac{0.4 \times 10^{-3}}{26} = 15.38\ \mu\text{S}$$

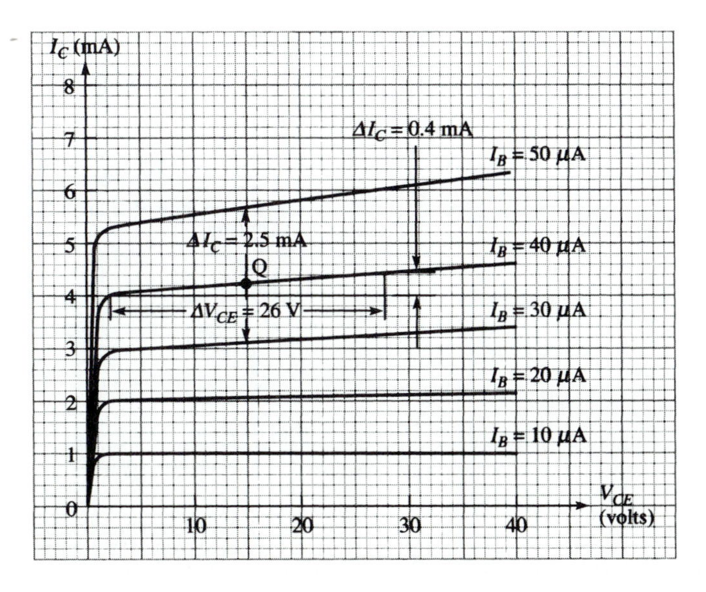

Figure 9–17
(Example 9–6)

From Table 9–2,

$$h_{fb} = \frac{-h_{fe}(1 - h_{re}) - h_{ie}h_{oe}}{(1 + h_{fe})(1 - h_{re}) + h_{ie}h_{oe}}$$

$$= \frac{-125(1 - 2 \times 10^{-4}) - 1500(15.38 \times 10^{-6})}{126(1 - 2 \times 10^{-4}) + 1500(15.38 \times 10^{-6})} = -0.992$$

$$h_{ob} = \frac{h_{oe}}{(1 + h_{fe})(1 - h_{re}) + h_{ie}h_{oe}}$$

$$= \frac{15.38 \times 10^{-6}}{126(1 - 2 \times 10^{-4}) + 1500(15.38 \times 10^{-6})} = 1.22 \times 10^{-7}\ \text{S}$$

$$h_{fc} = -(1 + h_{fe}) = -126$$

$$h_{oc} = h_{oe} = 15.38\ \mu\text{S}$$

9–5 BJT AMPLIFIER ANALYSIS USING *h* PARAMETERS

Common-Emitter Amplifiers

Figure 9–18 shows a common-emitter amplifier that is biased using the voltage-divider method. Also shown is the small-signal equivalent circuit that results when the transistor is replaced by its hybrid model. The gain and impedance equations we have already derived are applicable to this circuit, with a few minor modifications. Notice that the load impedance Z_L is now the parallel combination of R_C and R_L. The principal difference between this equivalent circuit and the general hybrid

Figure 9–18
A common-emitter amplifier and its small-signal equivalent circuit

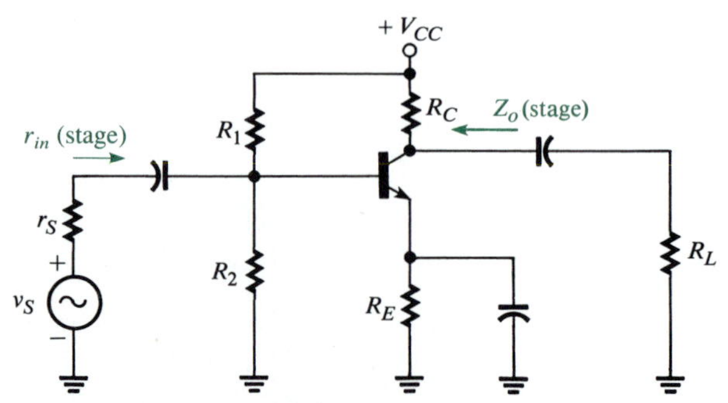

(a) A common-emitter amplifier biased using the voltage-divider method

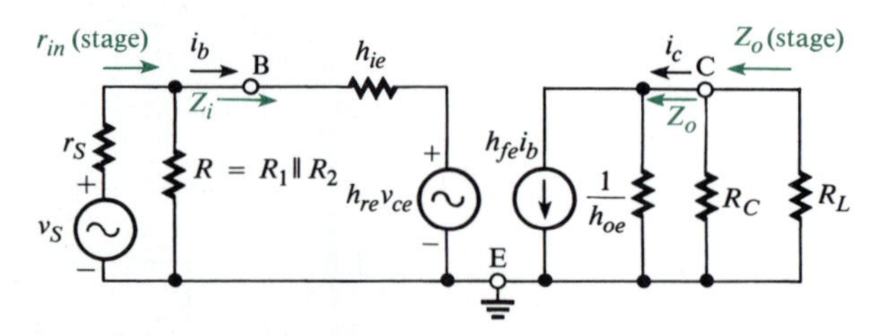

(b) The small-signal equivalent circuit of (a), with the transistor replaced by its hybrid model

amplifier we studied earlier (Figure 9–13) is the presence of $R = R_1 \| R_2$ across the input. This resistance is in parallel with the input impedance of the transistor, so the amplifier input resistance, r_{in}(stage), is smaller than it would be if R were not present. Using equation 9–21, we have

$$r_{in}(\text{stage}) = R \| Z_i = R \left\| \left(h_{ie} - \frac{h_{re}h_{fe}Z_L}{1 + h_{oe}Z_L} \right) \right. \tag{9–30}$$

The reduction in amplifier input impedance caused by R reduces the voltage gain, v_L/v_S, because of the usual voltage-divider effect at the input:

$$\begin{aligned} \frac{v_L}{v_S} &= \frac{r_{in}(\text{stage})}{r_S + r_{in}(\text{stage})} A_v \\ &= \frac{r_{in}(\text{stage})}{r_S + r_{in}(\text{stage})} \left[\frac{-h_{fe}Z_L}{h_{ie}(1 + h_{oe}Z_L) - h_{re}h_{fe}Z_L} \right] \end{aligned} \tag{9–31}$$

Note that Z_L in equations 9–30 and 9–31 is $R_C \| R_L$. See Figure 9–18(b).

Finally, the output impedance of the amplifier is affected by the fact that R_C is in parallel with Z_o and R is in parallel with r_S. Thus, using equation 9–29, we obtain

$$Z_o(\text{stage}) = R_C \left\| \left(\frac{1}{h_{oe} - \dfrac{h_{fe}h_{re}}{h_{ie} + r_S \| R}} \right) \right. \tag{9–32}$$

Example 9–7

The transistor shown in Figure 9–19 has the following *h*-parameter values: $h_{ie} = 1600\ \Omega$, $h_{fe} = 80$, $h_{re} = 2 \times 10^{-4}$, $h_{oe} = 20\ \mu S$. Find (1) Z_i, (2) r_{in}(stage), (3) $A_v = v_{ce}/v_{be}$, (4) v_L/v_S, (5) A_i, and (6) Z_o(stage).

Solution

1. $Z_L = (2.2\ k\Omega) \| (10\ k\Omega) = 1.8\ k\Omega$. From equation 9–21,

$$\begin{aligned} z_i = h_{ie} - \frac{h_{re}h_{fe}Z_L}{1 + h_{oe}Z_L} &= 1600 - \frac{(2 \times 10^{-4})(80)(1.8 \times 10^3)}{1 + (20 \times 10^{-6})(1.8 \times 10^3)} \\ &= 1600 - 27.8 = 1572\ \Omega \end{aligned}$$

2. $R = R_1 \| R_2 = (12\ k\Omega) \| (33\ k\Omega) = 8.8\ k\Omega$. From equation 9–30, $r_{in}(\text{stage}) = (8.8\ k\Omega) \| (1572\ \Omega) = 1.3\ k\Omega$.

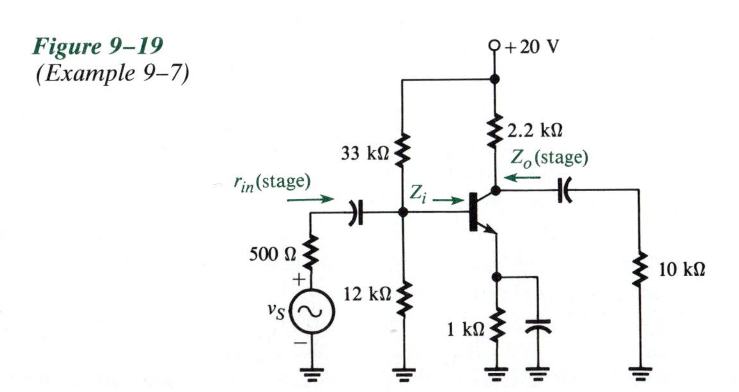

Figure 9–19
(Example 9–7)

3. From equation 9–15,

$$A_v = \frac{-h_{fe}Z_L}{h_{ie}(1 + h_{oe}Z_L) - h_{re}h_{fe}Z_L}$$
$$= \frac{-80(1.8 \times 10^3)}{1600[1 + 20 \times 10^{-6}(1.8 \times 10^3)] - (2 \times 10^{-4})(80)(1.8 \times 10^3)}$$
$$= \frac{-1.44 \times 10^5}{1600(1 + 0.036) - 28.8} = -88.56$$

The minus sign shows that the amplifier causes a phase inversion, as we expect from a CE configuration.

4. From equation 9–31,

$$\frac{v_L}{v_S} = \frac{r_{in}(\text{stage})}{r_S + r_{in}(\text{stage})}A_v = \left(\frac{1.3 \times 10^3}{500 + 1.3 \times 10^3}\right)(-88.56) = -63.96$$

5. From equation 9–18,

$$A_i = \frac{h_{fe}}{1 + h_{oe}Z_L} = \frac{80}{1 + (20 \times 10^{-6})(1.8 \times 10^3)} = \frac{80}{1.036} = 77.22$$

6. $r_S \parallel R = 500 \parallel (8.8 \times 10^3) = 473\ \Omega$. From equation 9–32,

$$Z_o(\text{stage}) = (2.2 \times 10^3) \left\| \left(\frac{1}{h_{oe} - \dfrac{h_{fe}h_{re}}{h_{ie} + r_S \| R}} \right) \right.$$
$$= (2.2 \times 10^3) \left\| \left[\frac{1}{20 \times 10^{-6} - \dfrac{80(5 \times 10^{-4})}{1600 + 473}} \right] \right.$$
$$= (2.2\ \text{k}\Omega) \parallel (1.42\ \text{m}\Omega) \approx 2.2\ \text{k}\Omega$$

Example 9–8

SPICE

Use the SPICE models for voltage-controlled voltage sources and current-controlled current sources to verify the values calculated for A_v and v_L/v_S in Example 9–7.

Solution. The SPICE circuit corresponding to the hybrid model in Figure 9–18(b) is shown in Figure 9–20(a). Note that it is necessary to insert a dummy voltage source, VIB, in series with the base (node 3) to obtain the current that controls the F1 current source. Pay particular attention to the polarities of VIB and F1. The F1 specification in the data file, F1 5 0 VIB 80, means that current flowing from node 5 to node 0 equals 80 (h_{fe}) times the current (i_b) in VIB. The specification for the voltage-controlled voltage source, E1 4 0 5 0 2E-4, means that the voltage between nodes 4 and 0 is 2×10^{-4} (h_{re}) times the voltage between nodes 5 and 0.

Since all capacitors are eliminated in the small-signal model, a .DC analysis can be performed by SPICE as well as an .AC analysis, with the same results. Note

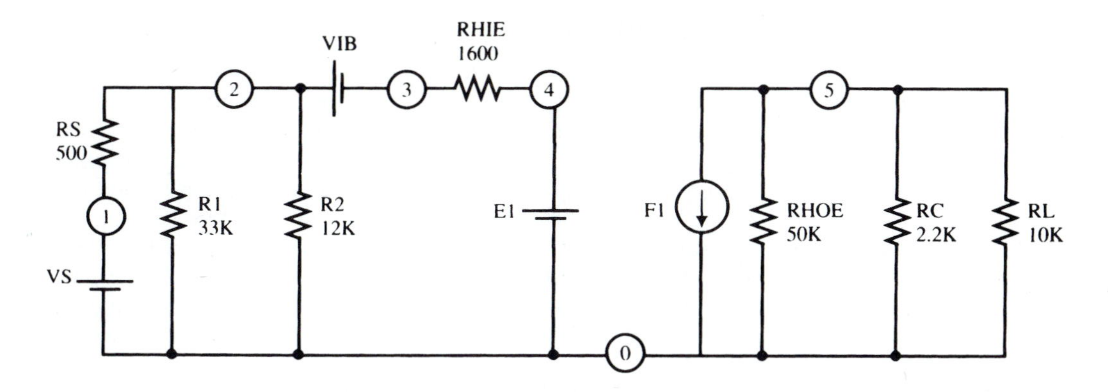


```
EXAMPLE 9.8
VS 1 0 1V
RS 1 2 500
R1 2 0 33K
R2 2 0 12K
VIB 2 3
RHIE 3 4 1600
E1 4 0 5 0 2E-4
F1 5 0 VIB 80
RHOE 5 0 50K
RC 5 0 2.2K
RL 5 0 10K
.DC VS 1 1 1
.PRINT DC V(3) V(5)
.END
```

(a)

```
EXAMPLE 9.8
****       DC TRANSFER CURVES                    TEMPERATURE =    27.000 DEG C
******************************************************************************
  VS             V(3)         V(5)
 1.000E+00      7.274E-01   -6.442E+01
```

(b)

Figure 9–20
(Example 9–8)

that VS is set equal to 1 V dc, so the voltage gain v_L/v_S is numerically equal to the dc output voltage at node 5. The gain A_v can be determined by dividing the output voltage by the input voltage at the base (node 3).

The results of the analysis are shown in Figure 9–20(b). We see that $v_L/v_S =$ V(5) $= -64.42$ and

$$A_v = \frac{-64.42}{\text{V(3)}} = \frac{-64.42}{0.7274} = -88.562$$

These results are in close agreement with the values calculated in Example 9–7.

CE h-Parameter Approximations

Certain approximations can be made to obtain simpler equations for amplifier gains and impedances in terms of h parameters. The most common approximation is $h_{re} = 0$, which is equivalent to neglecting the feedback effect of the output voltage. Under this assumption, the equations we have derived are greatly simplified. For example, when $h_{re} = 0$, the input impedance to the transistor (equation 9–21) becomes simply $Z_i = h_{ie}$. The previous example shows that this assumption gives a good approximation, since $h_{ie} = 1600\ \Omega$ and we calculated $Z_i = 1572\ \Omega$.

We will derive and present the standard approximations used in h-parameter values, and because the use of approximations is quite valid, considering the wide variation the parameters themselves are likely to have. However, it is our view that the main reason for using the hybrid model at all is that it provides an accurate representation of a transistor and is therefore useful in research applications and in the development of new devices. For routine, day-to-day analysis and design activities using commercial devices, we recommend the techniques and approximate models that were discussed in Chapters 4 and 5. We further note that the advent of computer analysis methods has greatly lessened the need for computational approximations, and has eliminated the need for the analyst to be concerned with whether or not the assumptions on which the approximations are based are valid for a particular situation.

Under the assumption $h_{re} \approx 0$, we obtain from equation 9–15

$$A_v \approx \frac{-h_{fe}Z_L}{h_{ie}(1 + h_{oe}Z_L)} \qquad \textbf{(9–33)}$$

The quantity $h_{oe}Z_L$ is generally much smaller than 1 (notice that it had value 0.036 in Example 9–7), so A_v can be further approximated by

$$A_v \approx \frac{-h_{fe}Z_L}{h_{ie}} \qquad \textbf{(9–34)}$$

Using this approximation in Example 9–7 gives $A_v = -90$, compared to the value -88.56 that we calculated using the exact equation.

Applying the assumption $h_{oe}Z_L \ll 1$ to equation 9–18, we obtain

$$A_i \approx h_{fe} \qquad \textbf{(9–35)}$$

Using this approximation in Example 9–7 gives $A_i = 80$, compared to the exact computation of 77.22.

As we have already noted, the assumption $h_{re} \approx 0$ leads to

$$Z_i \approx h_{ie} \qquad \textbf{(9–36)}$$

Some authors apply the assumption $h_{re} = 0$ to the equation for Z_o:

$$Z_o = \frac{1}{h_{oe} - \dfrac{h_{fe}h_{re}}{h_{ie} + Z_S}} \qquad \textbf{(9–37)}$$

and arrive at the approximation $Z_o \approx 1/h_{oe}$. However, this approximation is often invalid, because the quantities h_{oe} and $h_{fe}h_{re}/h_{ie}$ are usually close in value. Therefore, Z_o is generally much greater than $1/h_{oe}$, especially when Z_S is small.

Of course, the approximations given in (9–34) through (9–36) apply to the transistor alone. To determine overall amplifier gains and impedances, the effects of bias resistors must be taken into account in the usual way:

$$r_{in}(\text{stage}) \approx R \parallel h_{ie} \quad \textbf{(9–38)}$$

$$\frac{v_L}{v_S} = \frac{r_{in}(\text{stage})}{r_s + r_{in}(\text{stage})} A_v \quad \textbf{(9–39)}$$

$$Z_o(\text{stage}) = R_C \parallel Z_o \approx R_C \quad \textbf{(9–40)}$$

We should note that all of the gain and impedance equations for Figure 9–18 can be applied to the fixed-bias amplifier (Figure 5–33) simply by substituting $R = R_B$ instead of $R = R_1 \parallel R_2$. Of course, all equations are also applicable to PNP transistor amplifiers.

If the emitter resistor R_E in Figure 9–18 is not bypassed by a capacitor, then our equations must be modified to reflect the presence of R_E for ac signals. Using the methods of Section 9–3, it can be shown that

$$Z_i \approx h_{ie} + (h_{fe} + 1)R_E - \frac{h_{re}h_{fe}Z_L}{1 + h_{oe}Z_L} \quad \textbf{(9–41)}$$

Equation 9–41 shows that the input impedance is increased by an amount equal to $(\beta + 1)R_E$, as we discovered in Chapter 5. With R_E present, the term $h_{re}h_{fe}Z_L/(1 + h_{oe}Z_L)$ in equation 9–41 is even more negligible than it was when R_E was bypassed, so, to a good approximation,

$$Z_i \approx h_{ie} + (h_{fe} + 1)R_E \quad \textbf{(9–42)}$$

Similarly, the voltage gain equation with R_E unbypassed becomes

$$A_v \approx \frac{-h_{fe}Z_L}{h_{ie} + R_E(1 + h_{fe}) - \dfrac{h_{re}h_{fe}Z_L}{1 + h_{oe}Z_L}} \quad \textbf{(9–43)}$$

Assuming that $h_{re} \approx 0$ gives

$$A_v \approx \frac{-h_{fe}Z_L}{h_{ie} + R_E(1 + h_{fe})} \quad \textbf{(9–44)}$$

Since $R_E(1 + h_{fe})$ is usually much greater than h_{ie}, we obtain

$$A_v \approx \frac{-h_{fe}Z_L}{R_E(1 + h_{fe})} \approx \frac{-Z_L}{R_E} \quad \textbf{(9–45)}$$

Equation 9–45 is the same result we obtained in Chapter 5 and shows that the voltage gain is reduced to approximately the ratio of load resistance to emitter resistance.

With considerable algebraic effort, it can be shown that the output impedance when R_E is unbypassed is given by the exact equation

$$Z_o = \frac{1}{h_{oe}} + R_E - \frac{(h_{re} + h_{oe}R_E)(h_{oe}R_E - h_{fe})}{h_{oe}\,\Delta} \quad \textbf{(9–46)}$$

where $\Delta = h_{oe}(h_{ie} + r_S \parallel R) + h_{oe}R_E - h_{re}h_{fe}$

$R = R_1 \parallel R_2$ (voltage-divider bias) or R_B (fixed bias)

This is a cumbersome expression and is difficult to approximate. (It is an exercise at the end of the chapter to show that (9–46) reduces to (9–37) when $R_E = 0$.) The presence of an unbypassed emitter resistor may increase or decrease Z_o, depending on other parameter values, but in any case, it is generally true that $Z_o(\text{stage}) = Z_o \parallel R_C \approx R_C$.

Equations 9–46a provide a summary of h-parameter equations and approximations for the common-emitter amplifier.

$R_E = 0$, or R_E bypassed

$$Z_i = h_{ie} - \frac{h_{re}h_{fe}Z_L}{1 + h_{oe}Z_L} \qquad \text{where } Z_L = R_C \| R_L$$

$$\approx h_{ie}$$

$$Z_i(\text{stage}) = Z_i \| R \qquad \text{where } R = R_1 \| R_2 \text{ (voltage-divider bias) or } R = R_B \text{ (fixed bias)}$$

$$\approx h_{ie} \| R$$

$$A_v = \frac{v_{ce}}{v_{be}} = \frac{-h_{fe}Z_L}{h_{ie}(1 + h_{oe}Z_L) - h_{re}h_{fe}Z_L} \qquad \text{where } Z_L = R_C \| R_L$$

$$\approx \frac{-h_{fe}Z_L}{h_{ie}(1 + h_{oe}Z_L)}$$

$$\approx \frac{-h_{fe}Z_L}{h_{ie}}$$

$$\frac{v_L}{v_S} = \frac{r_{in}(\text{stage})}{r_S + r_{in}(\text{stage})} A_v$$

$$A_i = \frac{h_{fe}}{1 + h_{oe}Z_L}$$

$$\approx h_{fe}$$

$$Z_o = \frac{1}{h_{oe} - \dfrac{h_{fe}h_{re}}{h_{ie} + r_S \| R}} \qquad \text{where } R = R_1 \| R_2 \text{ (voltage-divider bias) or } R = R_B \text{ (fixed bias)} \qquad \textbf{(9–46a)}$$

$$Z_o(\text{stage}) = R_C \| Z_o$$

$$\approx R_C$$

R_E not bypassed

$$Z_i \approx h_{ie} + (h_{fe} + 1)R_E - \frac{h_{re}h_{fe}Z_L}{1 + h_{oe}Z_L}$$

$$\approx h_{ie} + (h_{fe} + 1)R_E$$

$$A_v \approx \frac{-h_{fe}Z_L}{h_{ie} + R_E(1 + h_{fe}) - \dfrac{h_{re}h_{fe}Z_L}{1 + h_{oe}Z_L}} \qquad \text{where } Z_L = R_C \| R_L$$

$$\approx \frac{-h_{fe}Z_L}{h_{ie} + R_E(1 + h_{fe})}$$

$$\approx \frac{-Z_L}{R_E}$$

$$Z_o = \frac{1}{h_{oe}} + R_E - \frac{(h_{re} + h_{oe}R_E)(h_{oe}R_E - h_{fe})}{h_{oe}[h_{oe}(h_{ie} + r_S \| R) + h_{oe}R_E - h_{re}h_{fe}]}$$

$$Z_o(\text{stage}) = Z_o \| R_C$$

$$\approx R_C$$

Example 9–9

The transistor in Figure 9–21 has $h_{fe} = 120$ and $h_{ie} = 2\ \text{k}\Omega$. Use approximate h-parameter equations to find (1) $r_{in}(\text{stage})$, (2) v_L/v_S, and (3) $r_o(\text{stage})$.

Figure 9–21
(Example 9–9)

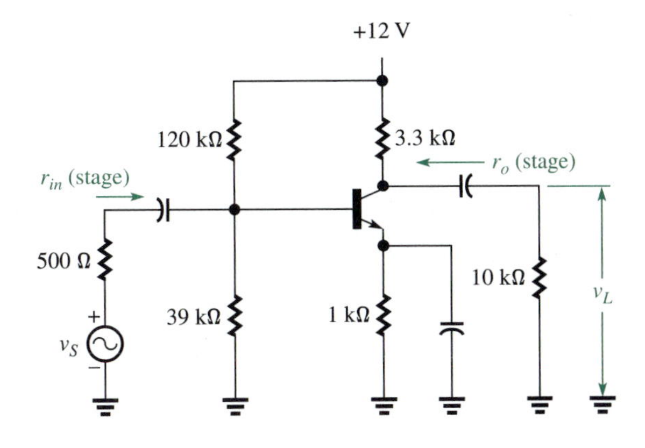

Solution

1.
$$R = (120\ \text{k}\Omega) \parallel (39\ \text{k}\Omega) = 29.43\ \text{k}\Omega$$
$$r_{in}(\text{stage}) \approx h_{ie} \parallel R = (2\ \text{k}\Omega) \parallel (29.43\ \text{k}\Omega) = 1.87\ \text{k}\Omega$$

2. $Z_L = (3.3\ \text{k}\Omega) \parallel (10\ \text{k}\Omega) = 2.48\ \text{k}\Omega$
$$A_v \approx \frac{-h_{fe}Z_L}{h_{ie}} = \frac{-(120)(2.48 \times 10^3)}{2 \times 10^3} = -148.8$$
$$\frac{v_L}{v_S} = \frac{r_{in}(\text{stage})}{r_S + r_{in}(\text{stage})}A_v = \left(\frac{1.87 \times 10^3}{500 + 1.87 \times 10^3}\right)(-148.8) = -117.4$$

3. $r_o(\text{stage}) \approx R_C = 3.3\ \text{k}\Omega$

Common-Collector and Common-Base Amplifiers

Figure 9–22 shows CC and CB amplifiers and their hybrid equivalent circuits. The general gain and impedance equations we derived in Section 9–3 are applicable to these configurations, with the usual modifications to reflect the presence of external bias resistors.

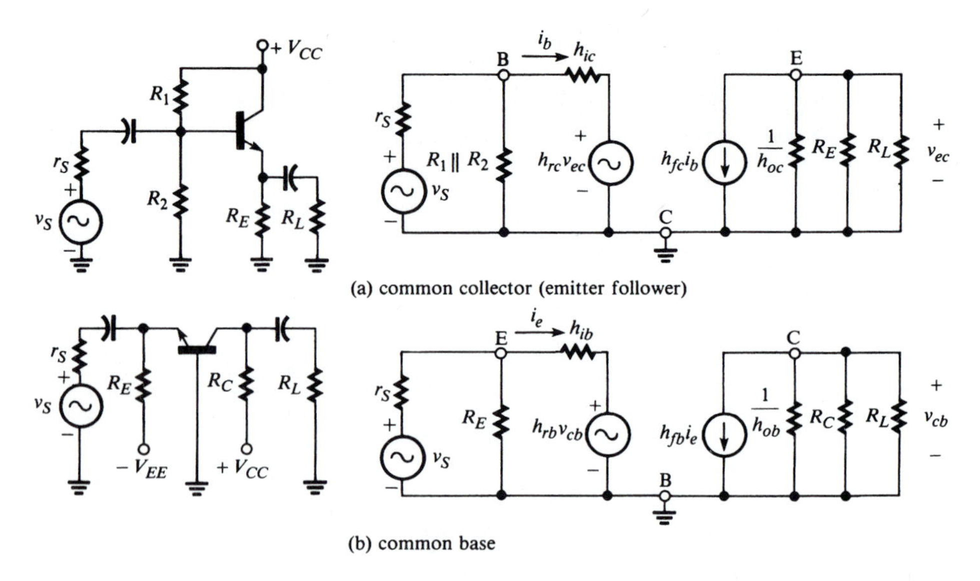

Figure 9–22
The CC and CB amplifiers and their hybrid equivalent circuits

Example 9–10

The transistor in the common-collector amplifier shown in Figure 9–23 has the following h parameters: $h_{ic} = 1400\ \Omega$, $h_{rc} = 1$, $h_{fc} = -80$, $h_{oc} = 2 \times 10^{-5}$ S. Find (1) r_{in}, (2) r_{in}(stage), (3) A_v, (4) v_L/v_S, (5) r_o, and (6) r_o(stage).

Solution

1. $Z_L = (1\text{ k}\Omega) \| (10\text{ k}\Omega) = 909\ \Omega$. From equation 9–21,

$$Z_i = r_{in} = h_{ic} - \frac{h_{rc}h_{fc}Z_L}{1 + h_{oc}Z_L} = 1400 - \frac{1(-80)(909)}{1 + (2 \times 10^{-5})(909)}$$
$$= 1400 + 71.42 \times 10^3 = 72.82\text{ k}\Omega$$

Figure 9–23
(Example 9–10)

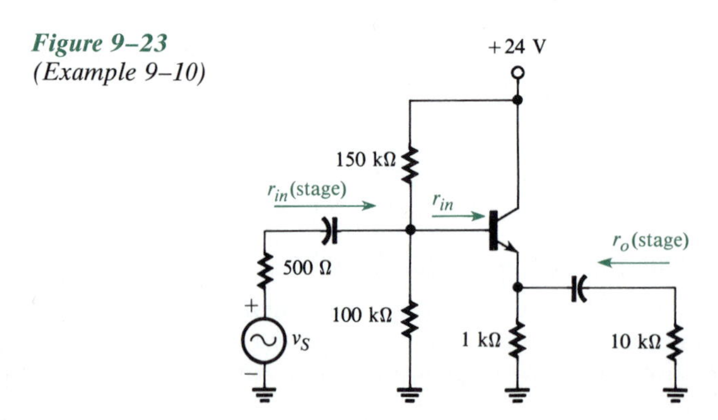

Notice the characteristically large value for the input resistance of an emitter follower.

2. $r_{in}(\text{stage}) = (100\ \text{k}\Omega) \parallel (150\ \text{k}\Omega) \parallel r_{in} = 32.9\ \text{k}\Omega$
3. From equation 9–15,

$$A_v = \frac{-h_{fc}Z_L}{h_{ic}(1 + h_{oc}Z_L) - h_{rc}h_{fc}Z_L}$$
$$= \frac{-(-80)(909)}{1400(1 + 2 \times 10^{-5} \times 909) - 1(-80)(909)} = 0.981$$

As we expect in an emitter follower, the voltage gain is positive (no phase inversion) and near 1 in value.

4.

$$\frac{v_L}{v_S} = \frac{r_{in}(\text{stage})}{r_S + r_{in}(\text{stage})}A_v = \left(\frac{32.9 \times 10^3}{500 + 32.9 \times 10^3}\right)(0.981) = 0.966$$

5. $R = R_1 \parallel R_2 = (100\ \text{k}\Omega) \parallel (150\ \text{k}\Omega) = 60\ \text{k}\Omega$. From equation 9–29,

$$Z_o = r_o = \frac{1}{h_{oc} - \dfrac{h_{fc}h_{rc}}{h_{ic} + r_S \parallel R}}$$
$$= \frac{1}{2 \times 10^{-5} - \dfrac{(-80)(1)}{1400 + 500 \parallel (60 \times 10^3)}} = 23.7\ \Omega$$

This result confirms that the output resistance of the emitter follower is quite small.

6. $r_o(\text{stage}) = (1\ \text{k}\Omega) \parallel (23.7\ \Omega) \approx 23.7\ \Omega$

It is important to include minus signs in all computations involving negative quantities. Notice, for example, that h_{fc} is a negative quantity. Failure to use the proper algebraic signs will lead to radically incorrect solutions. This precaution is equally important in the next example.

Example 9–11

The transistor shown in Figure 9–24 has the following h parameters: $h_{fe} = 100$, $h_{oe} = 25 \times 10^{-6}$ S, $h_{ib} = 12\ \Omega$, and $h_{rb} = 3 \times 10^{-5}$. Find (1) r_{in}, (2) $r_{in}(\text{stage})$, (3) A_v, (4) v_L/v_S, (5) r_o, and (6) $r_o(\text{stage})$.

Figure 9–24
(Example 9–11)

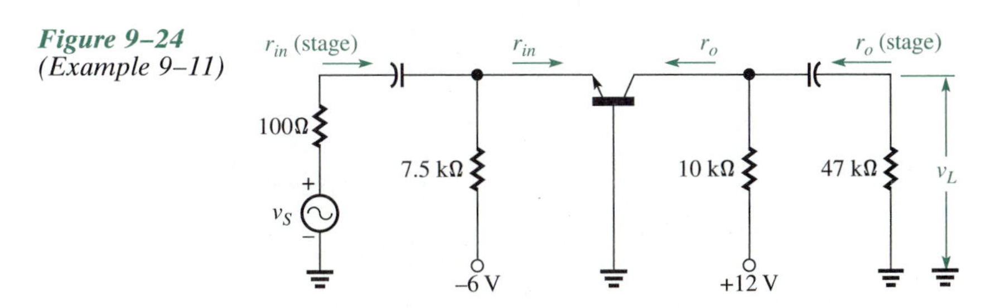

Solution. The transistor shown in the figure is in a common-base configuration, so we must use CB h parameters to analyze it. Since only two of the given parameters are CB parameters, we must use the conversion equations in Table 9–2 and the CE parameter values that are given to find h_{fb} and h_{ob}:

$$h_{fb} = \frac{-h_{fe}}{1 + h_{fe}} = \frac{-100}{101} = -0.99$$

$$h_{ob} = \frac{h_{oe}}{1 + h_{fe}} = \frac{25 \times 10^{-6}}{101} = 2.47 \times 10^{-7}\text{ S}$$

1. $Z_L = (10\text{ k}\Omega) \parallel (47\text{ k}\Omega) = 8.25\ \Omega$. From equation 9–21,

$$Z_i = r_{in} = h_{ib} - \frac{h_{rb}h_{fb}Z_L}{1 + h_{ob}Z_L} = 12 - \frac{1(3 \times 10^{-5})(-0.99)(8.25 \times 10^3)}{1 + (2.47 \times 10^{-7})(8.25 \times 10^3)}$$
$$= 12 + 0.24 = 12.24\ \Omega$$

This is the typically small value expected for the input resistance of a CB transistor.

2. $r_{in}(\text{stage}) = (7.5\text{ k}\Omega) \parallel (12.24\ \Omega) = 12.24\ \Omega$

3. From equation 9–15,

$$A_v = \frac{-h_{fb}Z_L}{h_{ib}(1 + h_{ob}Z_L) - h_{rb}h_{fb}Z_L}$$
$$= \frac{-(-0.99)(8.25 \times 10^3)}{12[1 + (2.47 \times 10^{-7})(8.25 \times 10^3)] - (3 \times 10^{-5})(-0.99)(8.25 \times 10^3)}$$
$$= \frac{8167.5}{12.02 + 0.25} = 665.64$$

As expected, the CB voltage gain is large and noninverting.

4.

$$\frac{v_L}{v_S} = \frac{r_{in}(\text{stage})}{r_S + r_{in}(\text{stage})}A_v = \left(\frac{12.24}{100 + 12.24}\right)665.64 = 72.6$$

The small input resistance of the CB amplifier can be seen to be responsible for a significant reduction in voltage gain.

5. From equation 9–29,

$$Z_o = r_o = \frac{1}{h_{ob} - \dfrac{h_{fb}h_{rb}}{h_{ib} + r_S \parallel (7.5 \times 10^3)}}$$
$$= \frac{1}{2.47 \times 10^{-7} - \dfrac{(-0.99)(3 \times 10^{-5})}{12 + 100 \parallel (7.5 \times 10^3)}} = 1.94\text{ M}\Omega$$

6. $r_o(\text{stage}) = r_o \parallel R_C = (1.94\text{ M}\Omega) \parallel (10\text{ k}\Omega) \approx 10\text{ k}\Omega$

CC and CB Approximations

The computations shown in the preceding two examples suggest some approximations that can be used in the analysis of most practical CC and CB amplifiers.

The common-collector input impedance is, to a close approximation,

$$r_{in} \approx -h_{fc}Z_L \tag{9–47}$$

Using (9–47) in Example 9–10 gives $r_{in} = 72.72$ kΩ, in comparison to the calculated value of 72.82 kΩ.

A good approximation for the common-collector voltage gain is

$$A_v \approx \frac{-h_{fc}Z_L}{h_{ic} - h_{fc}Z_L} \tag{9–48}$$

Using (9-48) in Example 9–10 gives $A_v = 0.981$, the same result (to 3 decimal places) calculated there. It is usually the case that $h_{fc}Z_L \gg h_{ic}$, in which case (9–48) reduces to the usual assumption for an emitter follower: $A_v \approx 1$.

Since $h_{rc} \approx 1$ and h_{oc} is very small, the output impedance of a CC amplifier can be approximated by

$$Z_o \approx -\frac{h_{ic} + r_S \| R}{h_{fc}}$$

Using this approximation in Example 9–10 gives 23.65 Ω, in comparison with the calculated value of 23.7 Ω.

The current gain of a CC amplifier is, from equation 9–18,

$$A_i = \frac{h_{fc}}{1 + h_{oc}Z_L} \tag{9–49}$$

To a good approximation,

$$A_i \approx h_{fc} \tag{9–50}$$

The input impedance of a CB amplifier is approximately

$$r_i \approx h_{ib} \tag{9–51}$$

Using this approximation in Example 9–11 gives $r_{in} \approx 12$ Ω, in comparison to the calculated value of 12.24 Ω.

The voltage gain of a CB amplifier is approximated by assuming that $h_{ob}Z_L$ and $h_{rb}h_{fb}Z_L$ are both negligibly small. As seen in Example 9–11, these quantities had values 0.002 and −0.25, respectively. Under that assumption,

$$A_v \approx \frac{-h_{fb}Z_L}{h_{ib}} \approx \frac{Z_L}{h_{ib}} \tag{9–52}$$

since $h_{fb} \approx 1$. Using this approximation in Example 9–11 gives $A_v \approx 687.5$, in comparison to the calculated value of 665.64.

The output impedance of a CB amplifier is always very large, so

$$r_o(\text{stage}) = r_o \parallel R_C \approx R_C \tag{9–53}$$

From equation 9–18, the current gain of a CB amplifier is

$$A_i = \frac{h_{fb}}{1 + h_{ob}Z_L} \tag{9–54}$$

Since $h_{ob}Z_L \ll 1$ and $h_{fb} \approx 1$, we have

$$A_i \approx h_{fb} \approx 1 \tag{9–55}$$

Equations 9–55a and 9–55b provide a summary of the CC and CB amplifier equations and approximations.

Common Collector

$$Z_i = h_{ic} - \frac{h_{rc}h_{fc}Z_L}{1 + h_{oc}Z_L} \qquad \text{where } Z_L = R_E \parallel R_L$$

$$\approx -h_{fc}Z_L$$

$$r_{in}(\text{stage}) = R \parallel Z_i \qquad \text{where } R = R_1 \parallel R_2 \text{ (voltage-divider bias) or } R = R_B \text{ (fixed bias)}$$

$$A_v = \frac{v_e}{v_b} = \frac{-h_{fc}Z_L}{h_{ic}(1 + h_{oc}Z_L) - h_{rc}h_{fc}Z_L}$$

$$\approx \frac{-h_{fc}Z_L}{h_{ic} - h_{fc}Z_L}$$

$$\approx 1 \quad (h_{fc}Z_L \gg h_{ic}) \tag{9–55a}$$

$$\frac{v_L}{v_S} = \left[\frac{r_{in}(\text{stage})}{r_S + r_{in}(\text{stage})}\right] A_v$$

$$A_i = \frac{h_{fc}}{1 + h_{oc}Z_L}$$

$$\approx h_{fc}$$

$$Z_o = \frac{1}{h_{oc} - \dfrac{h_{fc}h_{rc}}{h_{ic} + r_S \parallel R}} \approx -\frac{h_{ic} + r_S \parallel R}{h_{fc}}$$

$$r_o(\text{stage}) = Z_o \parallel R_E$$

$$\approx Z_o$$

Common Base

$$Z_i = h_{ib} - \frac{h_{rb}h_{fb}Z_L}{1 + h_{ob}Z_L} \qquad \text{where } Z_L = R_C \parallel R_L$$

$$\approx h_{ib}$$

$$r_{in}(\text{stage}) \approx R_E \parallel h_{ib} \approx h_{ib}$$

$$A_v = \frac{v_c}{v_b} = \frac{-h_{fb}Z_L}{h_{ib}(1 + h_{ob}Z_L) - h_{rb}h_{fb}Z_L}$$

$$\approx \frac{Z_L}{h_{ib}}$$

$$\frac{v_L}{v_S} = \left[\frac{r_{in}(\text{stage})}{r_S + r_{in}(\text{stage})}\right] A_v \qquad \textbf{(9–55b)}$$

$$A_i = \frac{h_{fb}}{1 + h_{ob}Z_L}$$

$$\approx h_{fb} \approx 1$$

$$Z_o = \frac{1}{h_{ob} - \dfrac{h_{fb}h_{rb}}{h_{ib} + r_S \parallel R_E}}$$

$$r_o(\text{stage}) = Z_o \parallel R_C$$

$$\approx R_C$$

9–6 y PARAMETERS

The y parameters of a two-port network (Figure 9–1) are derived from the following equations:

$$i_1 = y_{11}v_1 + y_{12}v_2 \qquad \textbf{(9–56)}$$

$$i_2 = y_{21}v_1 + y_{22}v_2 \qquad \textbf{(9–57)}$$

Note that each term in each equation must represent a current, since the left side of each equation is a current. Therefore, each of the parameters y_{11}, y_{12}, y_{21}, and y_{22} must be an *admittance; y* parameters are also called *admittance parameters.*

The defining equation for each y parameter is obtained from the network equations (9–56) or (9–57) following a procedure similar to that used to obtain h-parameter definitions. Setting $v_2 = 0$ in equation 9–56 gives

$$y_{11} = \left.\frac{i_1}{v_1}\right|_{v_2 = 0} \text{siemens} \qquad \textbf{(9–58)}$$

y_{11} is called the *short-circuit input admittance,* since it is the admittance at the input terminals when the output is short-circuited. With $v_2 = 0$, y_{21} can be obtained from equation 9–57:

$$y_{21} = \left.\frac{i_2}{v_1}\right|_{v_2 = 0} \text{siemens} \qquad \textbf{(9–59)}$$

y_{21} is called the *short-circuit forward transfer admittance.* Setting $v_1 = 0$ in equation 9–56 gives

$$y_{12} = \frac{i_2}{v_1}\bigg|_{v_1 = 0} \text{ siemens} \qquad \textbf{(9–60)}$$

y_{12} is called the *short-circuit reverse transfer admittance.* With $v_1 = 0$ in equation 9–57, we can solve for y_{22}:

$$y_{22} = \frac{i_2}{v_1}\bigg|_{v_1 = 0} \text{ siemens} \qquad \textbf{(9–61)}$$

y_{22} is called the *short-circuit output admittance.* We see that all of the *y*-parameter definitions are obtained by shorting either the input ($v_1 = 0$) or output ($v_2 = 0$) terminals. For that reason, *y* parameters are formally known as short-circuit admittance parameters.

Example 9–12

Find the real and imaginary parts of each of the *y* parameters of the circuit shown in Figure 9–25 when the frequency of operation is 7.958 kHz.

Solution. We must first find the capacitive reactance of the 2-pF capacitor at 7.958 kHz:

$$-jX_C = \frac{-j}{2\pi \times 7.958 \times 10^3(2 \times 10^{-12})} = -j10^7\ \Omega$$

Figure 9–26 shows the network with its output short-circuited, as necessary to compute y_{11} and y_{21}. As can be seen in the figure, the impedance z at the input terminals when the output is short-circuited is the parallel combination of the capacitive reactance and the resistance:

$$z = (-j10^7) \parallel 10^7 = \frac{(10^7\angle -90^\circ)(10^7\angle 0^\circ)}{10^7 - j10^7}$$

Figure 9–25
(Example 9–12)

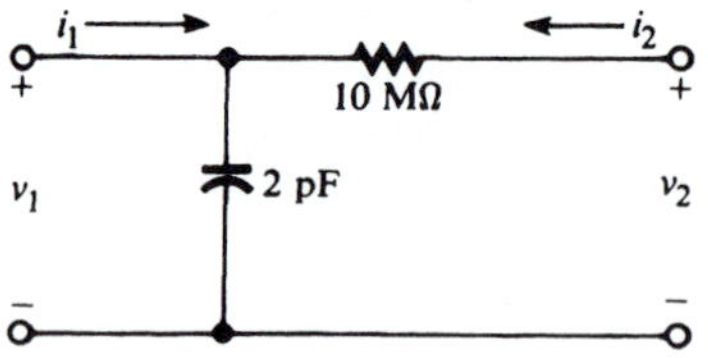

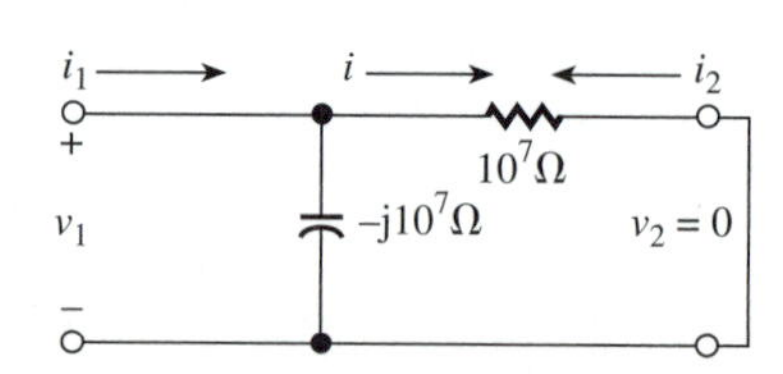

Figure 9–26
(Example 9–12) The network of Figure 9–25 with $v_2 = 0$, as necessary to compute y_{11} and y_{21}

The admittance y_{11} is the reciprocal of z:

$$y_{11} = \frac{1}{z} = \frac{10^{-7} - j10^{7}}{10^{14}\angle -90^\circ}$$
$$= \frac{1.414 \times 10^{7}\angle -45^\circ}{10^{14}\angle -90^\circ} = 1.414 \times 10^{-7}\angle 45^\circ = 10^{-7} + j10^{-7}\text{ S}$$

Thus, the real part of y_{11}, designated $\mathscr{Re}(y_{11})$, is 1×10^{-7} S, and the imaginary part, $\mathscr{Im}(y_{11})$, is 1×10^{-7} S. Recall that $\mathscr{Re}(y_{11})$ is the *conductance* component of the admittance, and $\mathscr{Im}(y_{11})$ is the *susceptance* component. (Susceptance is the reciprocal of reactance.) Also recall that the total admittance of components in parallel is the sum of the individual admittances, so the result $y_{11} = 10^{-7} + j10^{-7}$ S could have been written immediately.

The most direct way to find i_2/v_1 is to assume a value for v_1 in Figure 9–26 and then compute the current i_2. Assuming that $v_1 = 1\angle 0^\circ$, we find

$$i_1 = \frac{1\angle 0^\circ}{(-j10^{7}) \parallel 10^{7}} = v_1 y_{11} = 1.414 \times 10^{-7}\angle 45^\circ\text{ A}$$

We compute i in Figure 9–26 using the current-divider rule:

$$i = \left(\frac{-j10^{7}}{10^{7} - j10^{7}}\right) i_1 = \frac{-j10^{7}}{10^{7} - j10^{7}}(1.414 \times 10^{-7}\angle 45^\circ)$$
$$= \frac{10^{7}\angle -90^\circ(1.414 \times 10^{-7}\angle 45^\circ)}{1.414 \times 10^{-7}\angle 45^\circ} = 10^{-7}\angle 0^\circ\text{ A}$$

Now, $i_2 = -i$, so $i_2 = -10^{-7}\angle 0^\circ$ A. Then

$$y_{21} = \frac{i_2}{v_1}\bigg|_{v_2 = 0} = \frac{-10^{-7}\angle 0^\circ}{1\angle 0^\circ} = -10^{-7}\angle 0^\circ\text{ S}$$

Therefore, $\mathscr{Re}(y_{21}) = -10^{-7}$ S and $\mathscr{Im}(y_{21}) = 0$. Notice that it is possible to have negative admittance values.

To find y_{12} and y_{22}, it is necessary to short-circuit the input terminals of the network, as shown in Figure 9–27. Assume that $v_2 = 1\angle 0^\circ$ V. Then

$$i_2 = \frac{1\angle 0^\circ}{10^{7}\angle 0^\circ} = 10^{-7}\angle 0^\circ\text{ A}$$

It is clear that all of the current i_2 flows in the short-circuit around the capacitance, so $i_1 = -i_2 = -10^{-7}\angle 0^\circ$ A. Therefore,

$$y_{12} = \frac{i_1}{v_2}\bigg|_{v_1 = 0} = \frac{-10^{-7}}{1\angle 0^\circ} = -10^{-7}\text{ S}$$

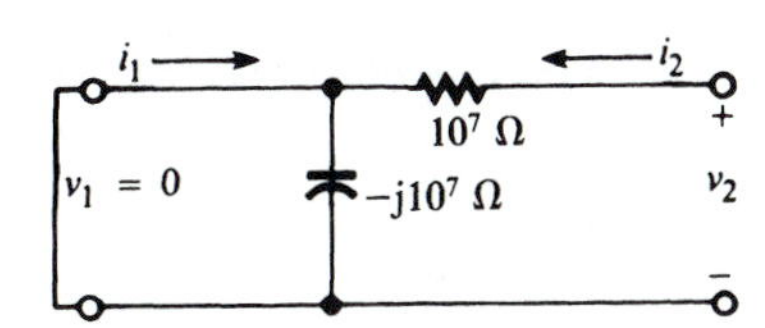

Figure 9–27
(Example 9–12) The network of Figure 9–25 with $v_1 = 0$, as necessary to compute y_{12} and y_{22}

Thus, $\mathcal{R}e(y_{12}) = -10^{-7}$ S and $\mathcal{I}m(y_{12}) = 0$. Since the capacitance is short-circuited, the output admittance is simply the reciprocal of the 10-MΩ resistance:

$$y_{22} = \frac{i_2}{v_2}\bigg|_{v_1 = 0} = \frac{1}{10^7\angle 0^\circ} = 10^{-7}\angle 0^\circ \text{ S}$$

Thus, $\mathcal{R}e(y_{22}) = 10^{-7}$ S and $\mathcal{I}m(y_{22}) = 0$.

Like h parameters, y parameters are given the letter subscripts i, r, f, and o in place of the double-number subscripts 11, 12, 21, and 22, respectively. Following is a summary of the y parameters' designations, which we will use hereafter.

$$y_{11} = \frac{i_1}{v_1}\bigg|_{v_2 = 0} = y_i \quad \text{for } input \text{ admittance}$$

$$y_{12} = \frac{i_1}{v_2}\bigg|_{v_1 = 0} = y_r \quad \text{for } reverse \text{ transfer admittance}$$

$$y_{21} = \frac{i_2}{v_1}\bigg|_{v_2 = 0} = y_f \quad \text{for } forward \text{ transfer admittance}$$

$$y_{22} = \frac{i_2}{v_2}\bigg|_{v_1 = 0} = y_o \quad \text{for } output \text{ admittance}$$

9–7 y-PARAMETER EQUIVALENT CIRCUITS

Figure 9–28 shows a y-parameter equivalent circuit that can be substituted for any network or electronic device whose y-parameter values are known. Notice that we have shown *impedance* blocks ($1/y_i$ and $1/y_o$) on the diagram, in keeping with the usual practice of representing impedance on a schematic diagram. In many books these blocks are labeled y_i and y_o, a practice that could *mistakenly* lead a reader to believe that impedances having values y_i and y_o belong in the equivalent circuit. Do not be confused by that practice.

It is easy to show that the circuit in Figure 9–28 has the same y parameters as the circuit to which it is equivalent. For example, if $v_2 = 0$, then the current source, labeled $y_r v_2$, is an *open* circuit, and the input admittance is clearly y_i. The reader should verify that the other y parameters of the circuit similarly satisfy their respective definitions.

Figure 9–29 shows the y-parameter equivalent circuit when a signal source having source impedance Z_S is connected across the input and a load impedance Z_L is connected across the output. We will obtain general expressions for the input and output impedances and for the voltage and current gains of this circuit.

Figure 9–28
The y-parameter equivalent circuit

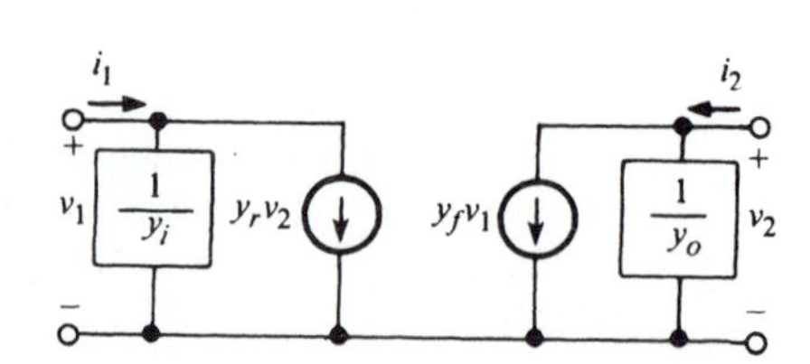

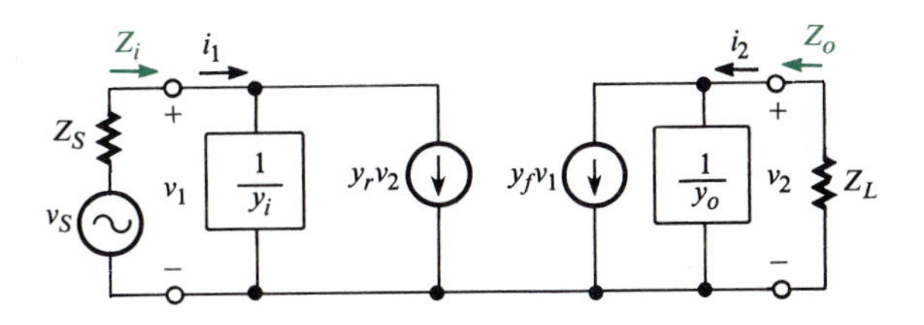

Figure 9–29
The y-parameter equivalent circuit with signal source and load connected

The parallel combination of Z_L and $1/y_o$ is

$$\frac{1}{y_o}\bigg\| Z_L = \frac{\left(\frac{1}{y_o}\right) Z_L}{\frac{1}{y_o} + Z_L} = \frac{Z_L}{1 + y_o Z_L} \tag{9–62}$$

The output voltage v_2 equals the parallel equivalent impedance of $1/y_o$ and Z_L multiplied by the current produced in the source labeled $y_{21}v_1$. Note that the source produces a voltage having polarity opposite that assumed for v_2. Thus,

$$v_2 = -y_f v_1 \left(\frac{Z_L}{1 + y_o Z_L}\right) \tag{9–63}$$

Writing Kirchhoff's current law at the input, we obtain

$$\begin{aligned} i_1 &= y_i v_1 + y_r v_2 = y_i v_1 + y_r \left[-y_f v_1 \left(\frac{Z_L}{1 + y_o Z_L}\right)\right] \\ i_1 &= v_1 \left(y_i - \frac{y_f y_r Z_L}{1 + y_o Z_L}\right) \end{aligned} \tag{9–64}$$

Solving (9–64) for v_1/i_1 gives

$$Z_i = \frac{v_1}{i_1} = \frac{1}{y_i - \dfrac{y_f y_r Z_L}{1 + y_o Z_L}} \tag{9–65}$$

Following a similar procedure, it can be shown that

$$Z_o = \frac{1}{y_o - \dfrac{y_r y_f Z_S}{1 + y_i Z_S}} \tag{9–66}$$

The current gain A_i is given by

$$A_i = \frac{-y_f/Z_L}{y_i y_o - y_f y_r + y_i/Z_L} \tag{9–67}$$

It is an exercise at the end of the chapter to show that the voltage gain is

$$A_v = \frac{-y_f}{y_o + 1/Z_L} \tag{9–68}$$

Given the y-parameter values of a circuit, the h-parameter values can be calculated, and vice versa. Table 9–3 gives the conversion relationships for the two sets of parameters.

Table 9–3
h- and y-parameter conversions

$h_i = 1/y_i$	$y_i = 1/h_i$
$h_r = -y_r/y_i$	$y_r = -h_r/h_i$
$h_f = y_f/y_i$	$y_f = h_f/h_i$
$h_o = (y_i y_o - y_r y_f)/y_i$	$y_o = (h_i h_o - h_r h_f)/h_i$

9–8 FET *y* PARAMETERS

A second letter is added to the subscript of the symbol for an FET *y* parameter to indicate the configuration for which it applies: *s* for common source, *d* for common drain, and *g* for common gate. The common-source *y* parameters are those most frequently given in specifications. These are summarized in Table 9–4.

The values of the *y* parameters for a field-effect transistor are affected by a certain small capacitance that exists between each pair of FET terminals. This capacitance is inherent in the structure of an FET, because every such device contains conducting surfaces separated from each other by dielectric material, the fundamental ingredients of capacitance. For example, the gate and channel of a JFET are separated from each other by the depletion region of a reverse-biased PN junction, and the gate and channel of a MOSFET are separated by a layer of silicon dioxide. Thus, there is a small gate-to-source capacitance, C_{gs}, in every field-effect transistor. Similarly, there are gate-to-drain capacitance, C_{gd}, and drain-to-source capacitance, C_{ds}. Figure 9–30 shows a JFET circuit in which these capacitances are identified by dashed lines, to emphasize that each is a *capacitance,* not a *capacitor.*

While the internal capacitances are small in value, on the order of a few picofarads, they are the dominant component of FET impedances at high frequencies, where their capacitive reactance is small compared to large resistive components. Therefore, the value of an FET *y* parameter is quoted at a specific frequency, often in terms of a real and an imaginary part that represent the conductance and susceptance components. The most important FET *y* parameter is y_{fs}, the forward transfer admittance, which can be expressed as

$$y_{fs} = g_{fs} + jb_{fs} \text{ siemens} \tag{9–69}$$

Table 9–4
Common-source y parameters

Symbol	Definition	Name
y_{is}	$\left.\dfrac{i_g}{v_{gs}}\right\vert_{v_{ds}=0} = \left.\dfrac{\Delta I_G}{\Delta V_{GS}}\right\vert_{V_{DS} = \text{constant}}$	Common-source short-circuit input admittance
y_{rs}	$\left.\dfrac{i_g}{v_{ds}}\right\vert_{v_{gs}=0} = \left.\dfrac{\Delta I_G}{\Delta V_{DS}}\right\vert_{V_{GS} = \text{constant}}$	Common-source short-circuit reverse transfer admittance
y_{fs}	$\left.\dfrac{i_d}{v_{gs}}\right\vert_{v_{ds}=0} = \left.\dfrac{\Delta I_D}{\Delta V_{GS}}\right\vert_{V_{DS} = \text{constant}}$	Common-source short-circuit forward transfer admittance
y_{os}	$\left.\dfrac{i_d}{v_{ds}}\right\vert_{v_{gs}=0} = \left.\dfrac{\Delta I_D}{\Delta V_{DS}}\right\vert_{V_{GS} = \text{constant}}$	Common-source short-circuit output admittance

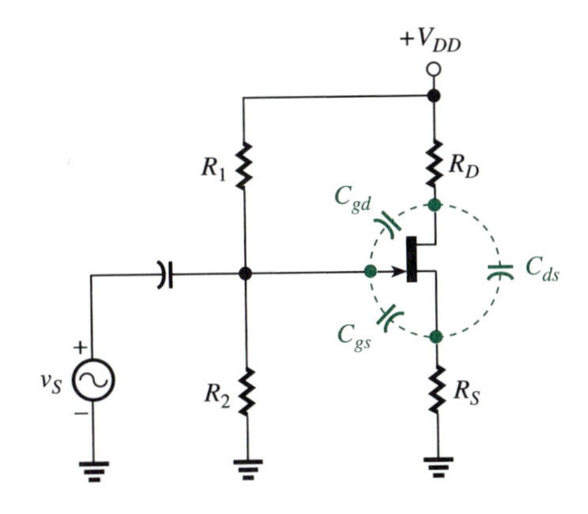

Figure 9–30
Capacitance between the terminals of an FET. Connections are shown by dashed lines to emphasize that these are inherent capacitances rather than discrete capacitors.

where g_{fs} is the real (conductance) component of y_{fs} and b_{fs} is the imaginary (susceptance) component. g_{fs} *is precisely the transconductance parameter,* g_m, that we have employed so frequently in previous discussions and examples. Thus,

$$\mathscr{Re}(y_{fs}) = g_{fs} = g_m \text{ siemens} \tag{9–70}$$

Stated another way, g_m is the dc value of y_{fs}.

The common-source short-circuit output admittance, y_{os}, can also be expressed in terms of its real and imaginary components:

$$y_{os} = g_{os} + \text{j}b_{os} \text{ siemens} \tag{9–71}$$

The real part of y_{os}, designated g_{os}, is the reciprocal of the FET parameter r_d that we have previously referred to as the ac drain resistance. Thus,

$$r_d = 1/\mathscr{Re}(y_{os}) = 1/g_{os} \text{ ohms} \tag{9–72}$$

By neglecting the very large gate-to-source resistance of an FET, as we have done in the past, it can be shown that

$$y_{is} = \text{j}\omega(C_{gs} + C_{gd}) \tag{9–73}$$
$$y_{fs} = g_m - \text{j}\omega C_{gd} \tag{9–74}$$
$$y_{rs} = -\text{j}\omega C_{gd} \tag{9–75}$$
$$y_{os} = 1/r_d + \text{j}\omega(C_{gd} + C_{ds}) \tag{9–76}$$

Example 9–13

At $f = 10$ MHz, a certain JFET has the following y parameters:

$$y_{fs} = (3 - \text{j}0.08) \times 10^{-3} \text{ S}$$
$$y_{os} = (0.02 + \text{j}0.12) \times 10^{-3} \text{ S}$$
$$y_{is} = (0 + \text{j}0.3) \times 10^{-3} \text{ S}$$

Find (1) g_m, (2) r_d, (3) C_{gd}, (4) C_{ds}, and (5) C_{gs}.

Solution

1. g_m is the real part of y_{fs}: $g_m = \mathscr{Re}(y_{fs}) = 3 \times 10^{-3}$ S.
2. r_d is the reciprocal of the real part of y_{os}: $r_d = 1/0.02 \times 10^{-3} = 50$ kΩ.

3. From equation 9–74,

$$\omega C_{gd} = \mathscr{I}m(y_{fs})$$
$$2\pi \times 10 \times 10^6 C_{gd} = 0.08 \times 10^{-3}$$
$$C_{gd} = \frac{8 \times 10^{-5}}{2\pi \times 10^7} = 1.27 \text{ pF}$$

4. From equation 9–76,

$$\omega(C_{gd} + C_{ds}) = \mathscr{I}m(y_{os})$$
$$2\pi \times 10 \times 10^6(C_{gd} + C_{ds}) = 0.12 \times 10^{-3}$$
$$C_{gd} + C_{ds} = \frac{1.2 \times 10^{-4}}{2\pi \times 10^7} = 1.91 \text{ pF}$$

Therefore, $C_{ds} = (1.91 \text{ pF}) - C_{gd} = (1.91 \text{ pF}) - (1.27 \text{ pF}) = 0.64 \text{ pF}$.

5. From equation 9–73,

$$\omega(C_{gs} + C_{gd}) = \mathscr{I}m(y_{is})$$
$$2\pi \times 10 \times 10^6(C_{gs} + C_{gd}) = 0.3 \times 10^{-3}$$
$$C_{gs} + C_{gd} = \frac{3 \times 10^{-4}}{2\pi \times 10^7} = 4.77 \text{ pF}$$

Therefore, $C_{gs} = (4.77 \text{ pF}) - C_{gd} = (4.77 \text{ pF}) - (1.27 \text{ pF}) = 3.5 \text{ pF}$.

FET specifications sometimes include values for parameters identified as C_{iss}, C_{fss}, C_{rss}, and C_{oss}. These are related to *y*-parameter values as follows:

$$y_{is} = j\omega C_{iss} \qquad \text{(9–77)}$$
$$y_{fs} = g_m - j\omega C_{fss} \qquad \text{(9–78)}$$
$$y_{rs} = -j\omega C_{rss} \qquad \text{(9–79)}$$
$$y_{os} = 1/r_d + j\omega C_{oss} \qquad \text{(9–80)}$$

Comparing equations 9–77 through 9–80 to equations 9–73 through 9–76, it is apparent that

$$C_{iss} = C_{gs} + C_{gd} \qquad \text{(9–81)}$$
$$C_{fss} = C_{gd} \qquad \text{(9–82)}$$
$$C_{rss} = C_{gd} \qquad \text{(9–83)}$$
$$C_{oss} = C_{gd} + C_{ds} \qquad \text{(9–84)}$$

Figure 9–31 shows a typical set of manufacturer's specifications for the *y* parameters of an N-channel JFET. These show the variation we would expect in the imaginary part of each *y* parameter versus frequency: As frequency increases, the capacitive susceptance increases, corresponding to the decrease in capacitive reactance. Notice that all graphs are given for the conditions $V_{DS} = 15$ V and $V_{GS} = 0$ V. Recall that the value of g_m in an N-channel JFET increases with increasing

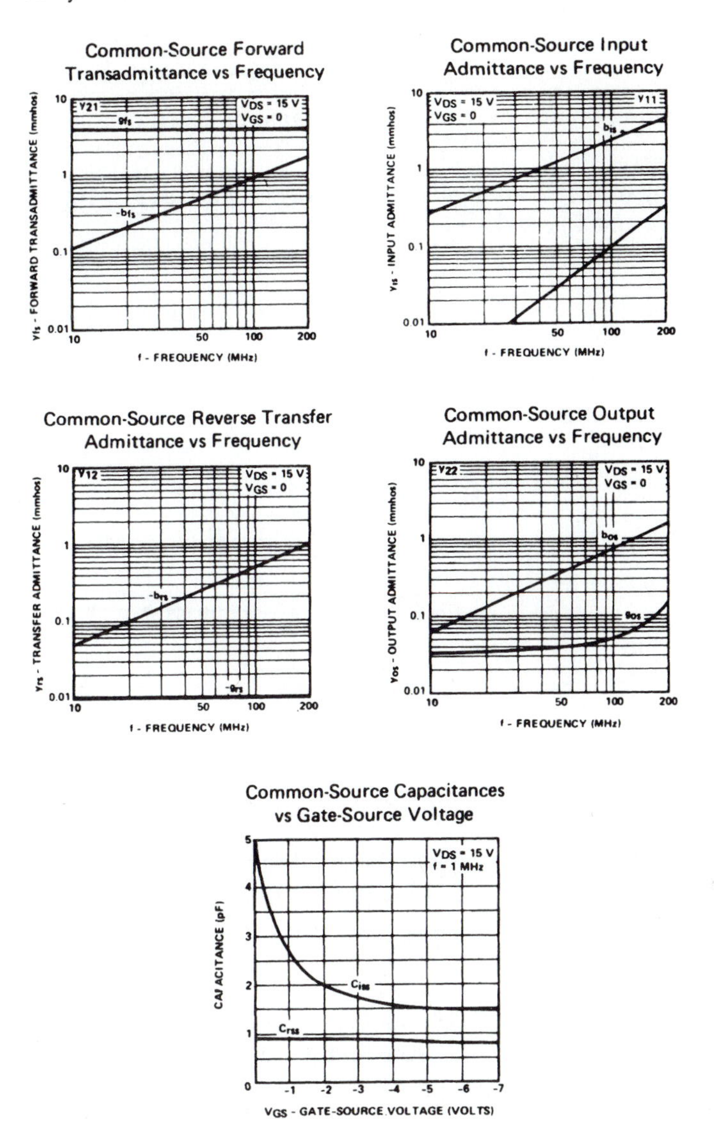

Figure 9–31
Typical JFET y-parameter and capacitance specifications (courtesy of Siliconix, Inc.)

V_{GS} and is a maximum at $V_{GS} = 0$. Therefore, the value given for $\mathcal{Re}(y_{fs})$ in the specifications is g_{mO} (see equation 8–4).

In Chapter 10, we will investigate the operation of amplifiers at high frequencies and learn how capacitance affects gain as a function of frequency. For the time being, we will be content to learn how to interpret FET parameters, obtain capacitance values, and construct y-parameter equivalent circuits. The next example shows how manufacturers' parameter specifications can be used to achieve these goals.

Example 9–14

The FET shown in Figure 9–32 has the *y*-parameter specifications shown in Figure 9–31.

1. Draw the *y*-parameter equivalent circuit of the FET at $f = 100$ MHz, assuming that $V_{GS} = 0$ and $V_{DS} = 15$ V.
2. Find the values of C_{gd}, C_{gs}, and C_{ds} under the same conditions as (1), given that $C_{OSS} = 2$ pF.
3. Draw the low-frequency *y*-parameter equivalent circuit of Figure 9–32. Assume that $V_p = -4$ V.
4. Find the low-frequency voltage gain v_L/v_S in Figure 9–32.

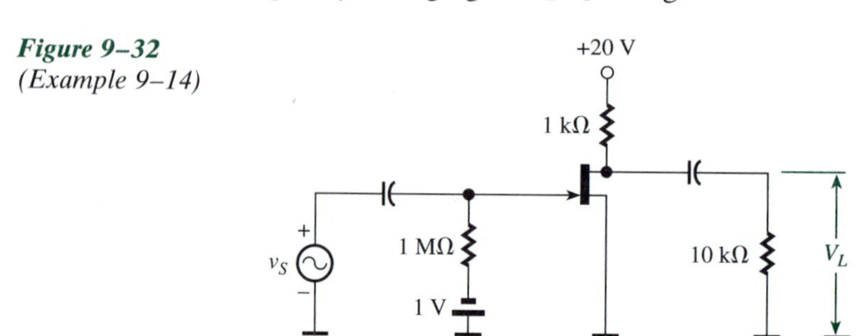

Figure 9–32
(Example 9–14)

Solution

1. From the specifications given in Figure 9–31, we obtain the following approximate values of the *y* parameters at 100 MHz:

$$y_{fs} = g_{fs} + jb_{fs} \approx (4 - j0.9) \times 10^{-3}\text{S}$$
$$y_{is} = g_{is} + jb_{is} \approx (0.1 + j2.5) \times 10^{-3}\text{S}$$
$$y_{rs} = g_{rs} + jb_{rs} \approx (-0.01 - j0.5) \times 10^{-3}\text{S}$$
$$y_{os} = g_{os} + jb_{os} \approx (0.05 + j0.8) \times 10^{-3}\text{S}$$

With reference to Figure 9–28, we can construct the *y*-parameter equivalent circuit as shown in Figure 9–33. Note that the admittances of parallel impedances are additive, so each impedance in the figure is the reciprocal of one of the conductance or susceptance values listed previously:

$$\frac{1}{g_{is}} = \frac{1}{0.1 \times 10^{-3}} = 10\text{ k}\Omega \qquad \frac{1}{b_{is}} = \frac{1}{j2.5 \times 10^{-3}} = 400\underline{/-90^\circ}\ \Omega$$
$$\frac{1}{g_{os}} = \frac{1}{0.05 \times 10^{-3}} = 20\text{ k}\Omega \qquad \frac{1}{b_{os}} = \frac{1}{j0.8 \times 10^{-3}} = 1250\underline{/-90^\circ}\ \Omega$$

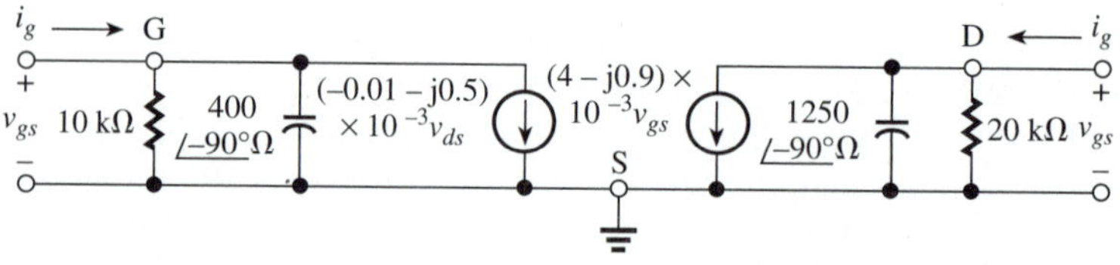

i_g → G D ← i_g
v_{gs} 10 kΩ 400 $\underline{/-90°}$Ω (–0.01 – j0.5) × $10^{-3} v_{ds}$ (4 – j0.9) × $10^{-3} v_{gs}$ S 1250 $\underline{/-90°}$Ω 20 kΩ v_{gs}

Figure 9–33
(Example 9–14) The y-parameter equivalent circuit of the JFET at $f = 100$ MHz

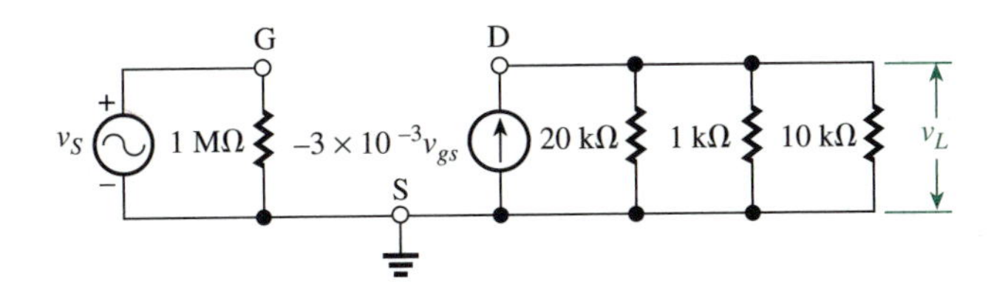

Figure 9–34
(Example 9–14) The low-frequency y-parameter equivalent circuit of Figure 9–32

2. From the specifications in Figure 9–31 at $V_{GS} = 0$ V, we find

$$C_{iss} \approx 5 \text{ pF}$$
$$C_{rss} \approx 0.9 \text{ pF}$$

From equation 9–83, $C_{gd} = C_{rss} = 0.9$ pF. From equation 9–81, $C_{gs} = C_{iss} - C_{gd} = (5 \text{ pF}) - (0.9 \text{ pF}) = 4.1$ pF. From equation 9–84, $C_{ds} = C_{oss} - C_{gd} = (2 \text{ pF}) - (0.9 \text{ pF}) = 1.1$ pF.

3. Since the specifications give y_{fs} at $V_{GS} = 0$, we have $\mathscr{Re}(y_{fs}) = g_{mO} \approx 4 \times 10^{-3}$ S. From equation 8–4,

$$g_{mO} = 4 \times 10^{-3} = \frac{2I_{DSS}}{|V_p|}$$

For the circuit of our example, $V_{GS} = -1$ V (Figure 9–32). Therefore, from equation 8–2, the transconductance at $V_{GS} = -1$ V is

$$g_m = \frac{2I_{DSS}}{|V_p|}\left(1 - \left|\frac{V_{GS}}{V_p}\right|\right) = g_{mO}\left(1 - \left|\frac{V_{GS}}{V_p}\right|\right) = 4 \times 10^{-3}\left(1 - \frac{1}{4}\right)$$
$$= 3 \times 10^{-3} \text{ S}$$

Since the imaginary parts of y_{is}, y_{rs}, and y_{os} are negligibly small at low frequency, the y-parameter equivalent circuit at low frequency is as shown in Figure 9–34. (We assume that the coupling capacitors are large enough to have negligible impedance.) Notice that the y-parameter equivalent circuit for the FET has been reduced to the form we have used in previous examples.

4. $v_L/v_S = -g_m[(20 \text{ k}\Omega) \parallel (1 \text{ k}\Omega) \parallel (10 \text{ k}\Omega)] = (-3 \times 10^{-3})(870\ \Omega) = -2.61$

EXERCISES

Section 9–2
Hybrid- (h-) Parameter Definitions

9–1. Find the values of the h parameters for each of the networks shown in Figure 9–35.

9–2. Find the values of the h parameters for each of the networks shown in Figure 9–36.

9–3. Find the values of the h parameters for the circuit shown in Figure 9–37.

Section 9–3
Hybrid Equivalent Circuits

9–4. Draw the hybrid equivalent circuit of each of the networks in Exercise 9–1. Label the values on each component of the equivalent circuits.

9–5. Draw the hybrid equivalent circuit of each of the networks in Exercise 9–2. Label the

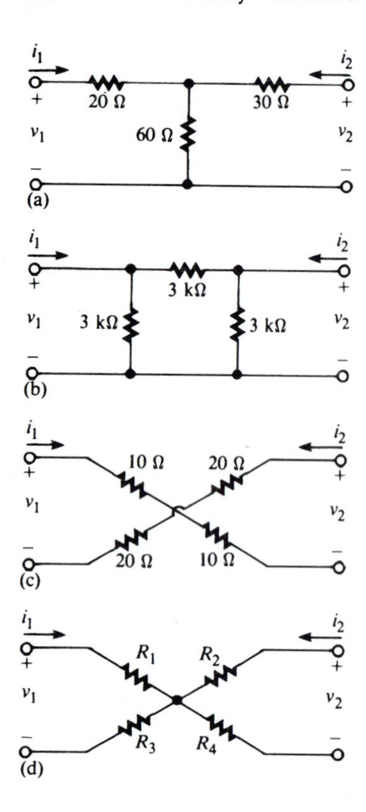

Figure 9–35
(Exercise 9–1)

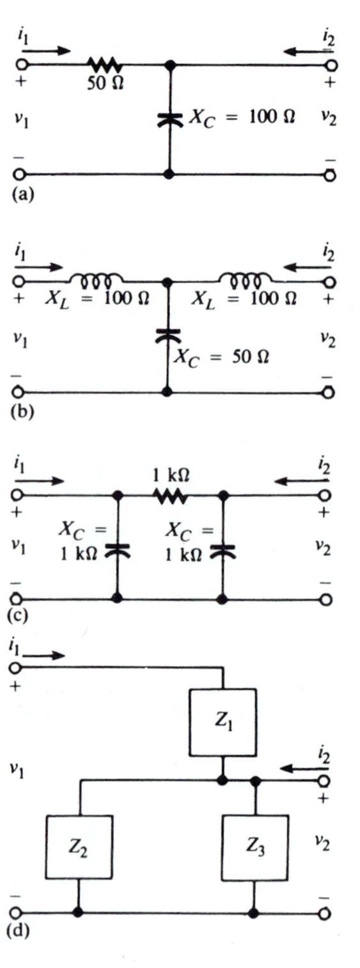

Figure 9–36
(Exercise 9–2)

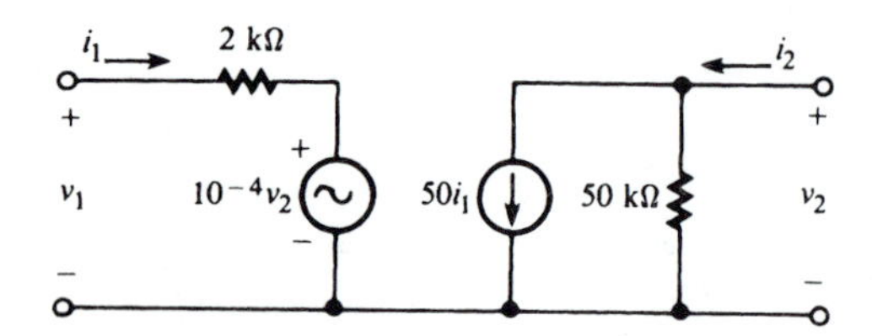

Figure 9–37
(Exercise 9–3)

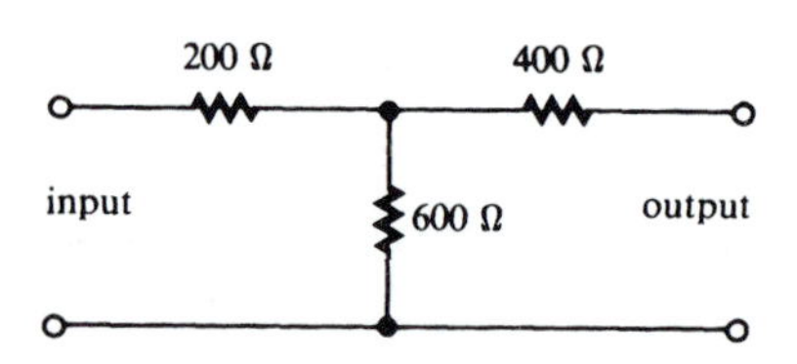

Figure 9–38
(Exercise 9–6)

values on each component of the equivalent circuits.

9–6. Find the values of the *h* parameters and draw the hybrid equivalent circuit of the network shown in Figure 9–38. Then use the hybrid circuit to find (a) the current that flows in a short connected across the output terminals when the input voltage is 22 V, (b) the voltage across a 200-Ω resistor connected across the output when the input voltage is 110 V, and (c) the current that flows in the 200-Ω resistor when the output is open and 16 V is connected across the input.

9–7. Show that the reverse voltage ratio and the

output admittance of Figure 9–9 are h_{12} and h_{22}, respectively.

9–8. Derive equation 9–24 from equation 9–23.

9–9. Find the input impedance, voltage gain v_L/v_S, current gain, and output impedance of the circuit shown in Figure 9–39.

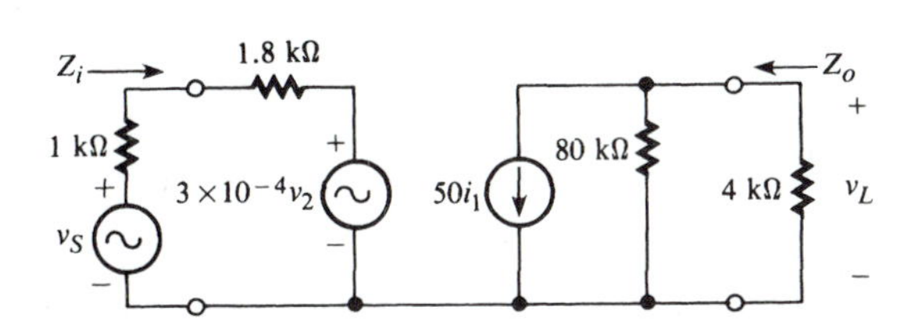

Figure 9–39
(Exercise 9–9)

9–10. A hybrid equivalent circuit has $h_r = 5 \times 10^{-5}$, $h_i = 20\ \Omega$, $h_f = -0.95$, $h_o = 10^{-6}$ S. A signal source having $r_S = 50\ \Omega$ is connected across the input and a 10-kΩ resistor is connected across its output. Find the input impedance of the circuit and its voltage gain, v_L/v_S.

Section 9–4
Transistor h *Parameters*

9–11. The transistor whose CE output characteristics are shown in Figure 9–17 has $h_{ic} = 2000\ \Omega$ and $h_{rc} = 99.99 \times 10^{-2}$ when it is biased at $V_{CE} = 30$ V and $I_B = 30\ \mu$A. Find the values of all of its CE h parameters.

9–12. Find the values of all of the CB h parameters for the transistor in Exercise 9–11.

Section 9–5
BJT Amplifier Analysis Using h *Parameters*

9–13. The transistor shown in Figure 9–40 has the following h parameters: $h_{ie} = 1500\ \Omega$, $h_{fe} = 85$, $h_{re} = 2.5 \times 10^{-4}$, $h_{oe} = 25\ \mu$S.

a. Draw the ac equivalent circuit using the hybrid model for the transistor.

b. Find the rms output voltage, v_L.

9–14. The transistor shown in Figure 9–41 has the following h parameters: $h_{ie} = 2000\ \Omega$, $h_{fe} = 90$, $h_{re} = 1.8 \times 10^{-4}$, $h_{oe} = 10\ \mu$S. Find (a) Z_i, (b) r_{in}(stage), (c) A_v, (d) v_L/v_S, (e) A_i, and (f) Z_o(stage).

9–15. Repeat Exercise 9–14 using approximate h-parameter equations.

9–16. Repeat Exercise 9–13 using approximate h-parameter equations.

9–17. In laboratory measurements made on the circuit shown in Figure 9–42, it was determined that the voltage gain v_o/v_S is -135 and that r_{in}(stage) = 800 Ω. Using approximate

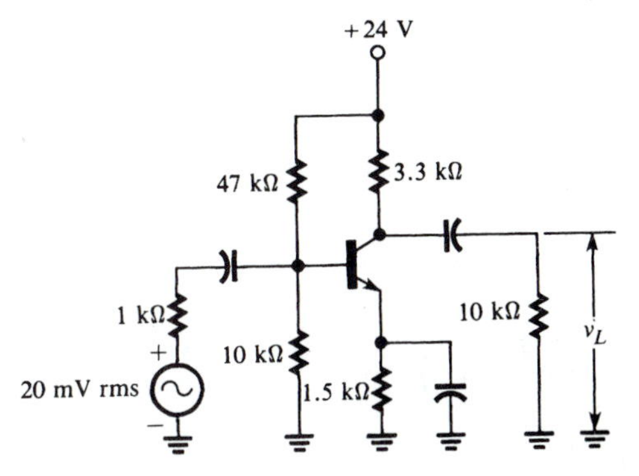

Figure 9–40
(Exercise 9–13)

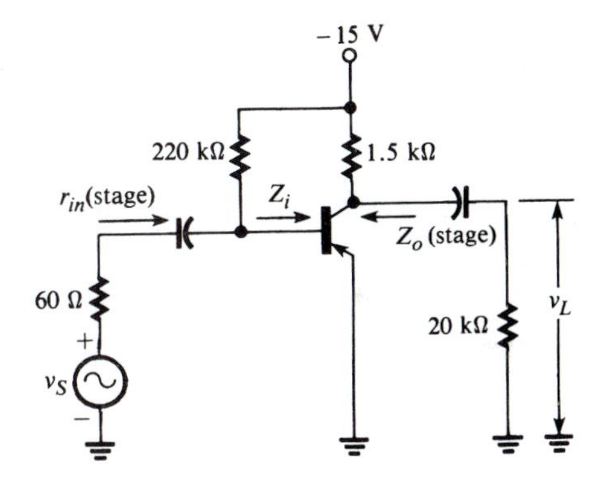

Figure 9–41
(Exercise 9–14)

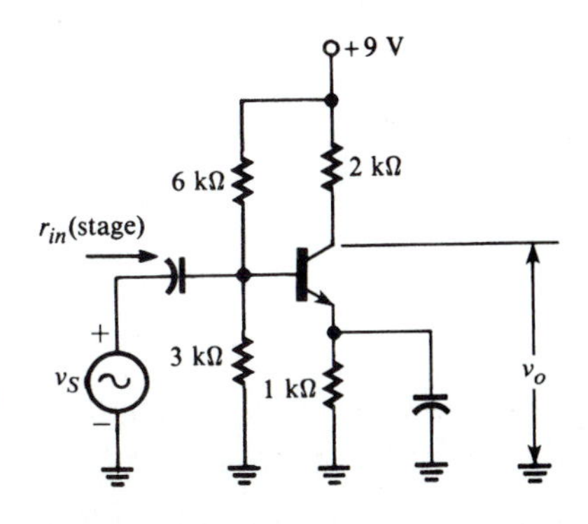

Figure 9–42
(Exercise 9–17)

h-parameter equations, find approximate values for h_{fe} and h_{ie}.

9–18. The transistor in Figure 9–43 has $h_{ie} = 1400\ \Omega$ and $h_{fe} = 120$. Find approximate values for (a) r_{in}, (b) r_{in}(stage), (c) A_v, and (d) v_L/v_S.

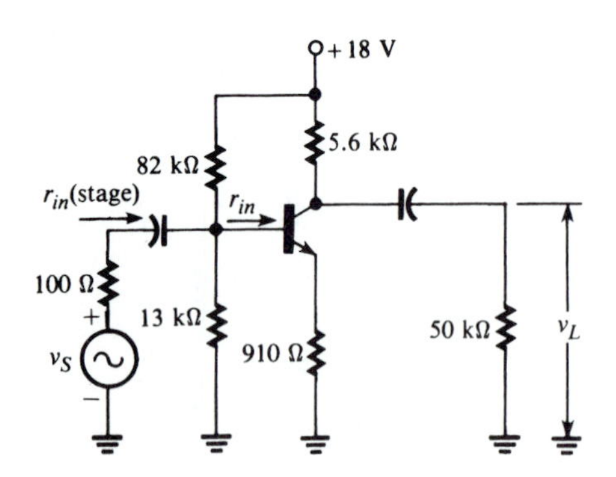

Figure 9–43
(*Exercise 9–18*)

9–19. Show that equation 9–46 reduces to equation 9–37 when the emitter resistor, R_E, is bypassed. (Assume that $r_S \parallel R = r_S$.)

9–20. The transistor in Figure 9–44 has $h_{ic} = 1600\ \Omega$, $h_{rc} = 1$, $h_{fc} = -70$, and $h_{oc} = 4 \times 10^{-5}$ S. Find (a) r_{in}, (b) r_{in}(stage), (c) v_L/v_S, (d) r_o, and (e) r_o(stage).

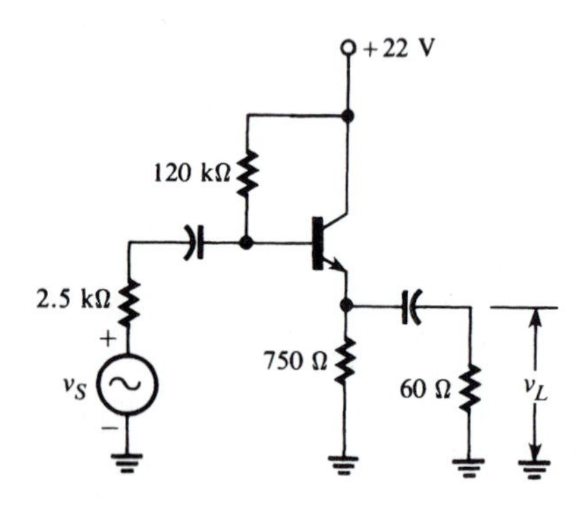

Figure 9–44
(*Exercise 9–20*)

9–21. The transistor shown in Figure 9–45 has $h_{ie} = 1500\ \Omega$, $h_{fe} = 80$, $h_{oe} = 3 \times 10^{-5}$ S and $h_{re} = 2 \times 10^{-4}$. Find (a) r_{in}, (b) r_{in}(stage), (c) v_L/v_S, (d) r_o, and (e) r_o(stage).

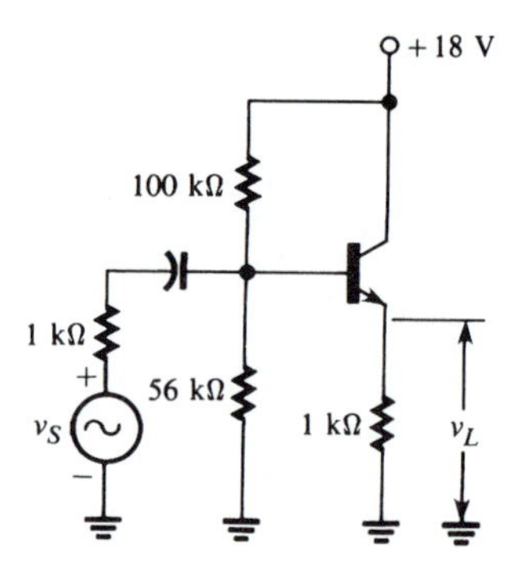

Figure 9–45
(*Exercise 9–21*)

9–22. The transistor shown in Figure 9–46 has $h_{ib} = 15\ \Omega$, $h_{rb} = 4 \times 10^{-5}$, $h_{fb} = -0.992$, and $h_{ob} = 4 \times 10^{-7}$ S. Find (a) r_{in}, (b) r_{in}(stage), (c) v_L/v_S, (d) r_o, and (e) r_o(stage).

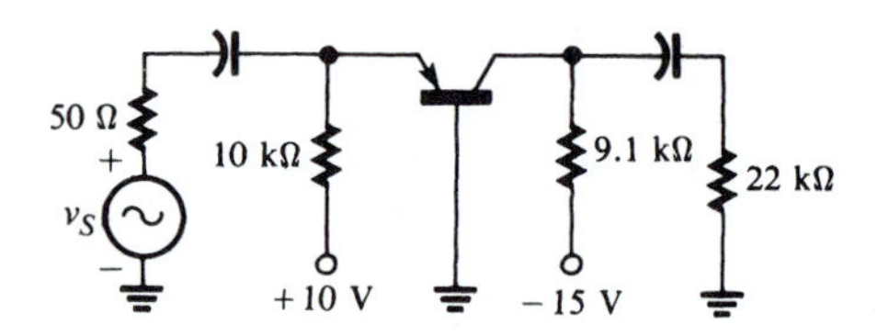

Figure 9–46
(*Exercise 9–22*)

9–23. The transistor shown in Figure 9–47 has $h_{ie} = 1400\ \Omega$, $h_{fe} = 100$, $h_{re} = 2 \times 10^{-4}$, and $h_{oe} = 20\ \mu$S. Find (a) r_{in}, (b) r_{in}(stage), (c) v_L/v_S, (d) r_o, and (e) r_o(stage).

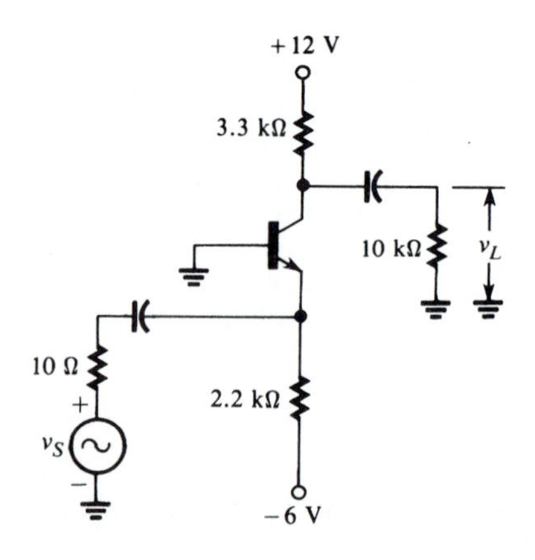

Figure 9–47
(*Exercise 9–23*)

9–24. Repeat Exercise 9–21 using h-parameter approximations.

9–25. Repeat Exercise 9–20 using h-parameter approximations.

9–26. Repeat Exercise 9–23 using h-parameter approximations.

9–27. Repeat Exercise 9–22 using h-parameter approximations.

Section 9–6
y *Parameters*

9–28. Find the values of the y parameters for each of the networks shown in Figure 9–48.

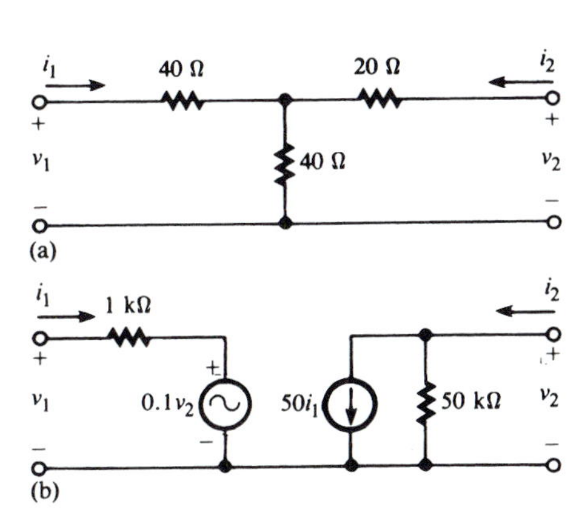

Figure 9–48
(Exercise 9–28)

9–29. Find the values of the y parameters at $f = 20$ MHz for the network shown in Figure 9–49.

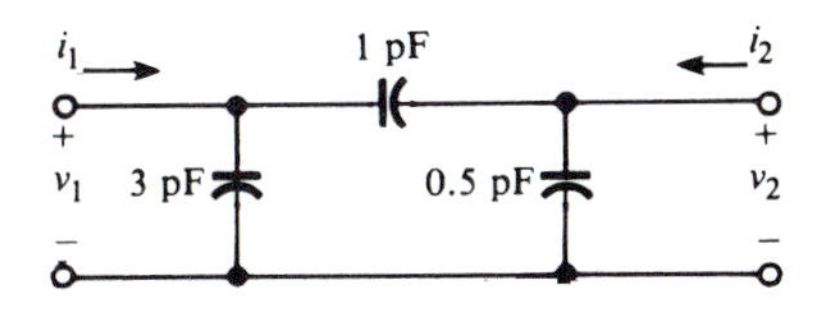

Figure 9–49
(Exercise 9–29)

Section 9–7
y*-Parameter Equivalent Circuits*

9–30. Derive equation 9–68 for the voltage gain of the y-parameter equivalent circuit shown in Figure 9–29.

9–31. A bipolar transistor has the following common-emitter h parameters: $h_{ie} = 1400\ \Omega$, $h_{re} = 2 \times 10^{-4}$, $h_{fe} = 100$, $h_{oe} = 2 \times 10^{-5}$ S. Draw the y-parameter equivalent circuit for the transistor in its common-emitter configuration. Label the values of components in the equivalent circuit.

9–32. The transistor shown in Figure 9–50 has the following h parameters: $h_{ie} = 1200\ \Omega$, $h_{re} = 2 \times 10^{-4}$, $h_{fe} = 120$, $h_{oe} = 4 \times 10^{-5}$ S. Draw the y-parameter equivalent circuit and use it to find (a) r_{in}(stage), (b) v_L/v_S, and (c) r_o(stage).

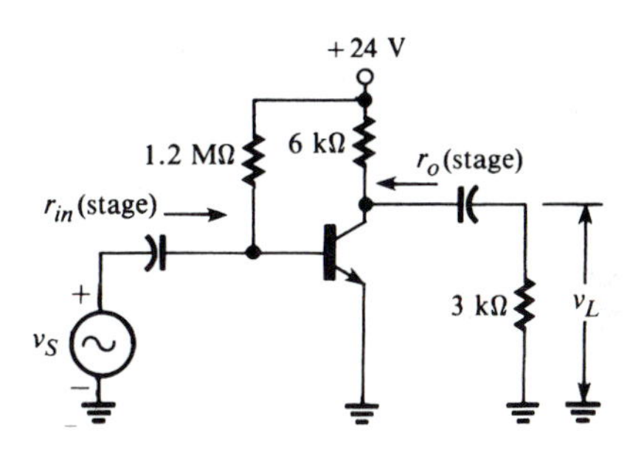

Figure 9–50
(Exercise 9–32)

Section 9–8
FET y *Parameters*

9–33. At $f = 40$ MHz, a certain JFET has the following y parameters: $y_{is} = (0 + j1.3) \times 10^{-3}$ S, $y_{fs} = (2.8 - j0.35) \times 10^{-3}$ S, $y_{os} = (0.01 + j0.5) \times 10^{-3}$ S. Find (a) g_m, (b) r_d, (c) C_{gd}, (d) C_{ds}, and (e) C_{gs}.

9–34. A certain FET has the following capacitance values: $C_{gs} = 4$ pF, $C_{gd} = 1.5$ pF, $C_{ds} = 0.8$ pF. If $g_m = 3.4 \times 10^{-3}$ S and $r_d = 80$ kΩ, find the y parameters for the FET at $f = 50$ MHz.

9–35. A certain FET has $C_{iss} = 5$ pF, $C_{rss} = 1.5$ pF, and $C_{oss} = 2$ pF. If $g_m = 3 \times 10^{-3}$ S and $r_d = 100$ kΩ, find (a) C_{gd}, C_{ds}, and C_{gs}; and (b) the y parameters at $f = 100$ MHz.

9–36. Find the values, at $f = 200$ MHz, of the y parameters of the FET whose specifications are given in Figure 9–31.

9–37. If the pinch-off voltage of the FET in Exercise 9–36 is -4.2 V, find g_m at $V_{GS} = -2$ V.

9–38. Draw the y-parameter equivalent circuit, at $f = 40$ MHz, of the FET whose specifications are given in Figure 9–31. Assume that $V_p = -4$ V and $V_{GS} = -2$ V. Label all component values in the equivalent circuit.

SPICE EXERCISES

Note: In each of the exercises that follow, use SPICE models of controlled sources.

9–39. Use SPICE to verify the values calculated in Example 9–4 (Figure 9–14) for A_v and v_L/v_S.

9–40. Use SPICE to verify the values calculated in Example 9–10 for r_{in}, r_{in}(stage), A_v, and v_L/v_S. (Note that it will be necessary to obtain appropriate voltages and currents in order to calculate values for r_{in} and r_{in}(stage).)

9–41. Use SPICE to find r_{in}, r_{in}(stage), and v_L/v_S for the circuit in Exercise 9–22.

9–42. A certain circuit has the following *y* parameters: $y_i = 0.25$ mS, $y_r = -0.01$ mS, $y_f = 4$ mS, and $y_o = 5\ \mu$S. Using SPICE to model the *y*-parameter equivalent circuit in Figure 9–29, find Z_i and v_L/v_S when $Z_S = 2\ \text{k}\Omega$ and $Z_L = 10\ \text{k}\Omega$.

10 FREQUENCY RESPONSE

10–1 DEFINITIONS AND BASIC CONCEPTS

The *frequency response* of an electronic device or system is the variation it causes, if any, in the level of its output signal when the frequency of the signal is changed. In other words, it is the manner in which the device *responds* to changes in signal frequency. Variation in the level (amplitude, or rms value) of the output signal is usually accompanied by a variation in the *phase angle* of the output relative to the input, so the term *frequency response* also refers to phase shift as a function of frequency. (Phase shift versus frequency is sometimes called *phase response.*) Figure 10–1 shows an amplifier whose frequency response causes small output amplitudes

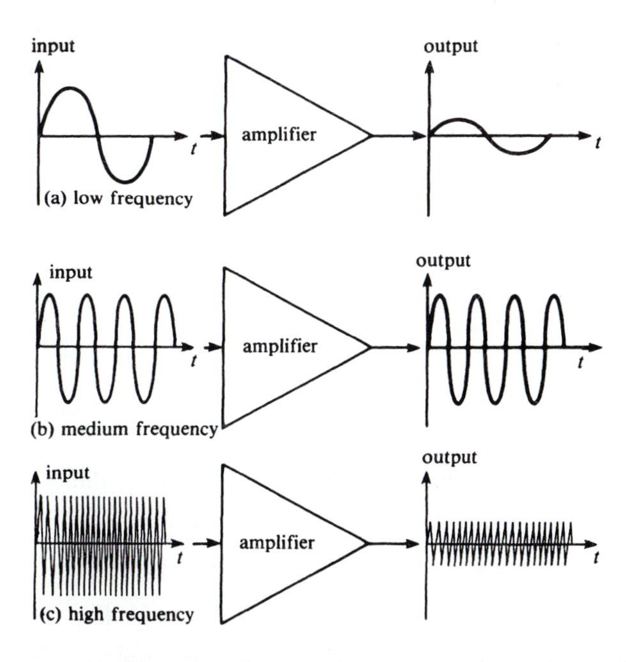

Figure 10–1
An amplifier whose frequency response is such that the output signal amplitude is small at both low and high frequencies

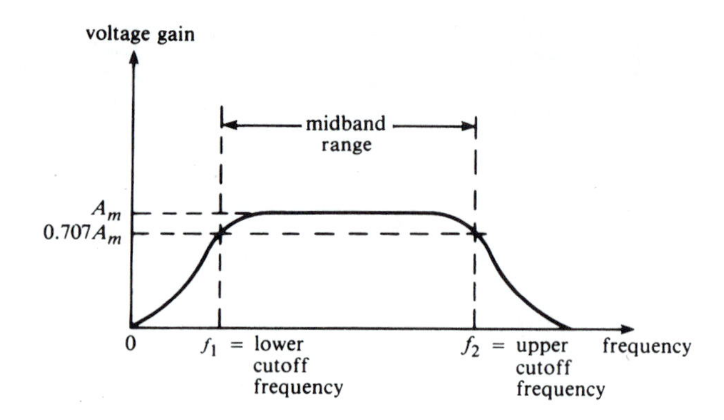

Figure 10–2
A typical amplifier's frequency response, showing gain versus frequency

at both low and high frequencies. Notice that the input signal amplitude is the same at each frequency, but the output signal amplitude changes with frequency. Thus, the *gain* of the amplifier is a function of frequency. In this example the gain is small at the low frequency and small at the high frequency.

The frequency response of an amplifier is usually presented in the form of a graph that shows output amplitude (or, more often, voltage gain) plotted versus frequency. Phase-angle variation is sometimes plotted on the same graph. Figure 10–2 shows a typical plot of the voltage gain of an ac amplifier versus frequency. Notice that the gain is 0 at dc (zero frequency), then rises as frequency increases, levels off for further increases in frequency, and then begins to drop again at high frequencies.

The frequency range over which the gain is more or less constant ("flat") is called the *midband range,* and the gain in that range is designated A_m. As shown in Figure 10–2, the low frequency at which the gain equals $(\sqrt{2}/2)\,A_m \approx 0.707A_m$ is called the *lower cutoff frequency* and is designated f_1. The high frequency at which the gain once again drops to $0.707A_m$ is called the *upper cutoff frequency* and is designated f_2. The *bandwidth* of the amplifier is defined to be the difference between the upper and lower cutoff frequencies:

$$\text{bandwidth} = \text{BW} = f_2 - f_1 \tag{10–1}$$

The points on the graph in Figure 10–2 where the gain is $0.707A_m$ are often called *half-power points,* and the cutoff frequencies are sometimes called *half-power frequencies,* because the output power of the amplifier at cutoff is one-half of its output power in the midband range. To demonstrate this fact, suppose that an rms output voltage v is developed across R ohms in the midband range. Then the output power in midband is

$$P_{(midband)} = \frac{v^2}{R}\text{ watts} \tag{10–2}$$

At each of the cutoff frequencies, the output voltage is $(\sqrt{2}/2)v$, so the power is

$$P_{(at\ cutoff)} = \frac{\left(\dfrac{\sqrt{2}}{2}v\right)^2}{R} = \frac{0.5\,v^2}{R} = 0.5\,P_{(midband)}$$

Example 10–1

An audio amplifier has a lower cutoff frequency of 20 Hz and an upper cutoff frequency of 20 kHz. (This is the frequency range of sound waves—the audio frequency range.) The amplifier delivers 20 W to a 12-Ω load at 1 kHz.

1. What is the bandwidth of the amplifier?
2. What is the rms load voltage at 20 kHz?
3. What is the rms load voltage at 2 kHz?

Solution

1. $BW = f_2 - f_1 = 20 \times 10^3 \text{ Hz} - 20 \text{ Hz} = 19{,}980 \text{ Hz}$
2. Since 1 kHz is in the midband range, the midband power is 20 W. At the 20-kHz cutoff frequency, the power is ½(20) = 10 W, so

$$\frac{v^2}{12} = 10$$

$$v^2 = 120$$

$$v = \sqrt{120} = 10.95 \text{ V rms}$$

3. Assuming that the load voltage is exactly the same throughout the midband range (not always the case in practice), the output at 2 kHz will be the same as that at 1 kHz, and we can use equation 10–2 with $P = 20$ W to solve for v. Alternatively, the midband voltage equals the voltage at cutoff *divided* by 0.707:

$$v_{(midband)} = \frac{10.95 \text{ V rms}}{0.707} = 15.49 \text{ V rms}$$

Amplitude and Phase Distortion

The signal passed through an ac amplifier is usually a complex waveform containing many different frequency components rather than a single-frequency ("pure") sine wave. For example, audio-frequency signals such as speech and music are combinations of many different sine waves occurring simultaneously with different amplitudes and different frequencies, in the range from 20 Hz to 20 kHz. As another example, any *periodic* waveform, such as a square wave or a triangular wave, can be shown to be the sum of a large number of sine waves whose amplitudes and frequencies can be determined mathematically. In previous discussions, we have analyzed ac amplifiers driven by single-frequency, sine-wave sources. This approach is justified by the fact that signals of all kinds can be regarded as sums of sine waves, as just described, so it is enough to know how an amplifier treats any single sine wave to know how it treats sums of sine waves (by the superposition principle).

In order for an output waveform to be an amplified version of the input, *an amplifier must amplify every frequency component in the signal by the same amount.* For example, if an input signal is the sum of a 0.5-V-rms, 100-Hz sine wave, a 0.2-V-rms, 1-kHz sine wave, and a 0.7-V-rms, 10-kHz sine wave, then an amplifier having gain 10 must amplify each frequency component by 10, so that the output consists of a 5-V-rms, 100-Hz sine wave, a 2-V-rms, 1-kHz sine wave, and a 7-V-rms, 10-kHz sine wave. If the frequency response of an amplifier is such that the gain

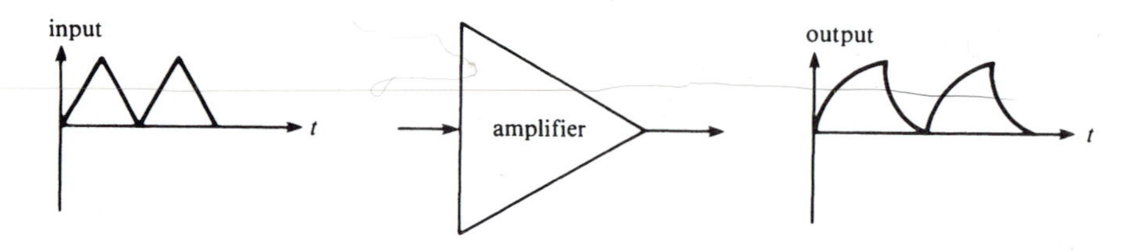

Figure 10–3
The output waveform is a distorted version of the input. The amplifier has a frequency response that is inadequate for the frequency components of the input waveform.

at one frequency is different than it is at another frequency, the output will be *distorted,* in the sense that it will not have the same shape as the input waveform. This alteration in waveshape is called *amplitude distortion.* Figure 10–3 shows the distortion that results when a triangular waveform is passed through an amplifier having an inadequate frequency response. In this example, the high-frequency components in the waveform fall beyond the upper cutoff frequency of the amplifier, so they are not amplified by the same amount as low-frequency components.

It can be seen that knowledge of the frequency response of an amplifier is important in determining whether it will distort a signal having known frequency components. The bandwidth must cover the entire range of frequency components in the signal if undistorted amplification is to be achieved. In general, "jagged" waveforms and signals having abrupt changes in amplitude, such as square waves and pulses, contain very broad ranges of frequencies and require wide-bandwidth amplifiers.

An amplifier will also distort a signal if it causes components having different frequencies to be shifted by different *times*. For example, if an amplifier shifts one component by 1 ms, it must shift every component by 1 ms. This means that the *phase* shift at each frequency must be proportional to frequency. Distortion caused by failure to shift phase in this way is called phase distortion. In most amplifiers, phase distortion occurs at the same frequencies where amplitude distortion occurs, because phase shifts are not proportional to frequency outside the midband range.

Amplitude and phase distortion should be contrasted with *nonlinear* distortion, which we discussed in an earlier chapter in connection with the nonuniform spacing of characteristic curves. Nonlinear distortion results when an amplifier's gain depends on signal *amplitude* rather than frequency. The effect of nonlinear distortion is to create frequency components in the output that were not present in the input signal. These new components are integer multiples of the frequency components in the input and are called *harmonic* frequencies. For example, if the input signal were a pure 1-kHz sine wave, the output would be said to contain third and fifth harmonics if it contained 3-kHz and 5-kHz components in addition to the 1-kHz *fundamental.* Such distortion is often called *harmonic distortion*.

10–2 DECIBELS AND LOGARITHMIC PLOTS

Decibels

Frequency-response data are often presented in *decibel* form. Recall that decibels (dB) are the units used to compare two power levels in accordance with the definition

$$\text{dB} = 10 \log_{10} \frac{P_2}{P_1} \tag{10–3}$$

The two power levels, P_1 and P_2, are often the input and output power of a system, respectively, in which case equation 10–3 defines the power gain of the system in decibels. If $P_2 > P_1$, then equation 10–3 gives a *positive* number, and if $P_2 < P_1$, the result is *negative,* signifying a reduction in power. If $P_2 = P_1$, the result is 0 dB, since $\log_{10}(1) = 0$.

Let R_1 be the resistance across which the power P_1 is developed and R_2 be the resistance across which P_2 is developed. Then, since $P = v^2/R$, we have, from equation 10–3,

$$\text{dB} = 10 \log_{10} \frac{(v_2^2/R_2)}{(v_1^2/R_1)} \tag{10–4}$$

where v_2 is the rms voltage across R_2 and v_1 is the rms voltage across R_1. If the resistance values are the *same* at the two points where the power comparison is made ($R_1 = R_2 = R$), then equation 10–4 becomes

$$\text{dB} = 10 \log_{10} \left(\frac{v_2^2/\cancel{R}}{v_1^2/\cancel{R}} \right) = 10 \log_{10} \left(\frac{v_2}{v_1} \right)^2 = 20 \log_{10} \left(\frac{v_2}{v_1} \right) \tag{10–5}$$

Equation 10–5 gives power gain (or loss) in terms of the voltage levels at two points in a circuit, but it must be remembered that the equation is valid for power comparison only if the resistances at the two points are equal. *The same equation is used to compare* voltage *levels regardless of the resistance values at the two points.* In other words, it is common practice to compute voltage gain as

$$\text{dB(voltage gain)} = 20 \log_{10} \left(\frac{v_2}{v_1} \right) \tag{10–6}$$

If the resistances R_1 and R_2 are equal, then the power gain in dB equals the voltage gain in dB.

Example 10–2

The amplifier shown in Figure 10–4 has input resistance 1500 Ω and drives a 100-Ω load. If the input current is 0.632 mA rms and the load voltage is 30 V rms, find

1. the power gain in dB; and
2. the voltage gain in dB.

Figure 10–4
(Example 10–2)

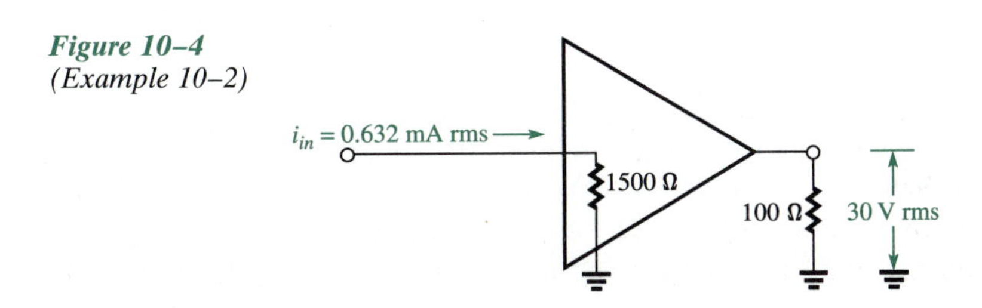

Solution

1. $P_{in} = i_{in}^2 r_{in} = (0.632 \times 10^{-3}\text{ A rms})^2(1500\ \Omega) = 0.6\text{ mW}$

$$P_L = \frac{v_L^2}{R_L} = \frac{(30\text{ V rms})^2}{100\ \Omega} = 9\text{ W}$$

From equation 10–3,

$$\text{power gain} = 10\log_{10}\left(\frac{9\text{ W}}{0.6 \times 10^{-3}\text{ W}}\right) = 10\log_{10}(15 \times 10^3) = 41.76\text{ dB}$$

2. Since $v_{in} = i_{in}r_{in} = (0.632 \times 10^{-3}\text{ A})(1500\ \Omega) = 0.948$ V rms, we have from equation 10–6,

$$\text{voltage gain} = 20\log_{10}\left(\frac{30}{0.948}\right) = 20\log_{10}(31.645) = 30\text{ dB}$$

It is helpful to remember that a two-to-one change in voltage corresponds to approximately 6 dB, and a ten-to-one change corresponds to 20 dB. The sign ($\pm$) depends on whether the change represents an increase or a decrease in voltage. Suppose, for example, that $v_1 = 8$ V rms. If this voltage is doubled ($v_2 = 16$ V rms), then

$$20\log_{10}\left(\frac{16}{8}\right) = 20\log_{10}(2) \approx 6\text{ dB}$$

If v_1 is halved ($v_2 = 4$ V rms), then

$$20\log_{10}\left(\frac{4}{8}\right) = 20\log_{10}(0.5) \approx -6\text{ dB}$$

If v_1 is increased by a factor of 10 ($v_2 = 80$ V rms), then

$$20\log_{10}\left(\frac{80}{8}\right) = 20\log_{10}(10) = 20\text{ dB}$$

If v_1 is reduced by a factor of 10 ($v_2 = 0.8$ V rms), then

$$20\log_{10}\left(\frac{0.8}{8}\right) = 20\log_{10}(0.1) = -20\text{ dB}$$

Every time a voltage is doubled, an additional 6 dB is *added* to the voltage gain, and every time it is increased by a factor of 10, an additional 20 dB is added to the voltage gain. For example, a gain of $100 = 10 \times 10$ corresponds to 40 dB and a gain of 4 corresponds to $2 \times 2 = 12$ dB. As another example, a gain of $400 = 2 \times 2 \times 10 \times 10$ corresponds to $(6 + 6 + 20 + 20)$ dB $= 52$ dB. Similarly, a reduction in voltage by a factor of $0.05 = (1/2)(1/10)$ corresponds to $-6 - 20 = -26$ dB. Common logarithms (base 10) can be computed on most scientific-type calculators, and the reader should become familar with the calculator's use for that purpose and for computing inverse logarithms. For reference and comparison purposes, Table 10–1 shows the decibel values corresponding to some frequently encountered ratios between 0.001 and 1000.

Table 10–1

(v_2/v_1)	dB = $20 \log_{10}(v_2/v_1)$	(v_2/v_1)	dB = $20 \log_{10}(v_2/v_1)$
0.001	−60	2	6
0.002	−54	4	12
0.005	−46	8	18
0.008	−42	10	20
0.01	−40	20	26
0.02	−34	40	32
0.05	−26	80	38
0.08	−22	100	40
0.1	−20	200	46
0.2	−14	400	52
0.5	−6	800	58
0.8	−2	1000	60
1.0	0		

Example 10–3

The input voltage to an amplifier is 4 mV rms. At point 1 in the amplifier, the voltage gain with respect to the input is −4.2 dB and at point 2 the voltage gain with respect to point 1 is 18.5 dB. Find

1. the voltage at point 1;
2. the voltage at point 2; and
3. the voltage gain in dB at point 2, with respect to the input.

Solution. Let v_i = the input voltage (4×10^{-3} V rms), v_1 = the rms voltage at point 1, and v_2 = the rms voltage at point 2.

1. $$20 \log_{10}\left(\frac{v_1}{4 \times 10^{-3}}\right) = -4.2$$

$$\log_{10}\left(\frac{v_1}{4 \times 10^{-3}}\right) = -0.21$$

Taking the inverse log (antilog) of both sides,

$$\frac{v_1}{4 \times 10^{-3}} = 0.617$$

$$v_1 = 2.46 \text{ mV rms}$$

(On most scientific calculators, the inverse log of −0.21 can be computed directly by entering a sequence such as *−0.21, inverse, log;* it can also be found on a calculator having the y^x function by computing $10^{-0.21}$.)

2. $$20 \log_{10}\left(\frac{v_2}{2.466 \times 10^{-3}}\right) = 18.5$$

$$\log_{10}\left(\frac{v_2}{2.466 \times 10^{-3}}\right) = 0.925$$

$$v_2 = 2.466 \times 10^{-3} \text{ antilog } (0.925) = 20.75 \text{ mV rms}$$

3. voltage gain $= 20\log_{10}\left(\dfrac{v_2}{v_i}\right) = 20\log_{10}\left(\dfrac{20.75\times10^{-3}}{4\times10^{-3}}\right) = 14.3$ dB

Notice that this result is the same as −4.2 dB + 18.5 dB; the overall gain in dB is the *sum* of the intermediate dB gains.

It must be remembered that decibels are derived from a *ratio* and therefore represent a comparison of one voltage or power level to another. It is correct to speak of voltage or power *gain* in terms of decibels, but it is meaningless to speak of output *level* in dB, unless the reference level is specified. Popular publications and the broadcast media frequently abuse the term *decibel* because no reference level is reported. Do not be confused by this practice.

It is common practice in some technical fields to use one standard reference level for all decibel computations. For example, the power level 1 mW is used extensively as a reference. When the reference is 1 mW, the decibel unit is written dBm:

$$\text{dBm} = 10\log_{10}\left(\frac{P}{10^{-3}}\right) \tag{10–7}$$

Note that 0 dBm corresponds to a power level of 1 mW. Another standard reference is 1 W:

$$\text{dBW} = 10\log_{10}\left(\frac{P}{1}\right) = 10\log_{10}P \tag{10–8}$$

When the voltage reference is 1 V, voltage gain in decibels is written dBV:

$$\text{dBV} = 20\log_{10}\left(\frac{V}{1}\right) = 20\log_{10}V \tag{10–9}$$

The *neper* is a logarithmic unit based on the natural log (ln) of a ratio:

$$\begin{aligned} A_p\,(\text{nepers}) &= \frac{1}{2}\ln\left(\frac{P_2}{P_1}\right) \\ A_v\,(\text{nepers}) &= \ln\left(\frac{v_2}{v_1}\right) \end{aligned} \tag{10–10}$$

Semilog and Log-Log Plots

It is a convenient and widely followed practice to plot the logarithm of frequency-response data rather than actual data values. If the logarithm of frequency is plotted along the horizontal axis, a wide frequency range can be displayed on a convenient size of paper without losing resolution at the low-frequency end. For example, if it were necessary to scale frequencies directly on average-sized graph paper over the range from 1 Hz to 10 kHz, each small division might represent 100 Hz. It would then be impossible to plot points in the range from 1 Hz to 10 Hz, where the lower cutoff frequency might well occur. When the horizontal scale represents logarithms of frequency values, the low-frequency end is expanded and the high-frequency end is compressed.

One way in which a logarithmic frequency scale can be obtained is to compute the logarithm of each frequency and then label conventional graph paper with those logarithmic values. An easier way is to use specially designed *log paper,* on which coordinate lines are logarithmically spaced. It is then necessary only to label each line directly with an actual frequency value. If only one axis of the graph paper has a logarithmically spaced scale and the other has conventional linear spacing, the paper is said to be *semilog* graph paper. If both the horizontal and vertical axes are logarithmic, it is called *log-log* paper.

Any ten-to-one range of values is called a *decade.* For example, each of the frequency ranges 1 Hz to 10 Hz, 10 kHz to 100 kHz, 500 Hz to 5 kHz, and 0.02 Hz to 0.2 Hz is a decade. Figure 10–5 shows a sample of log-log graph paper on which two full decades can be plotted along each axis. This sample is called *2-cycle–by–2-cycle,* or simply 2×2, log-log paper. Log-log graph paper is available with different numbers of decades along each axis, including 4×2, 5×3, 3×3,

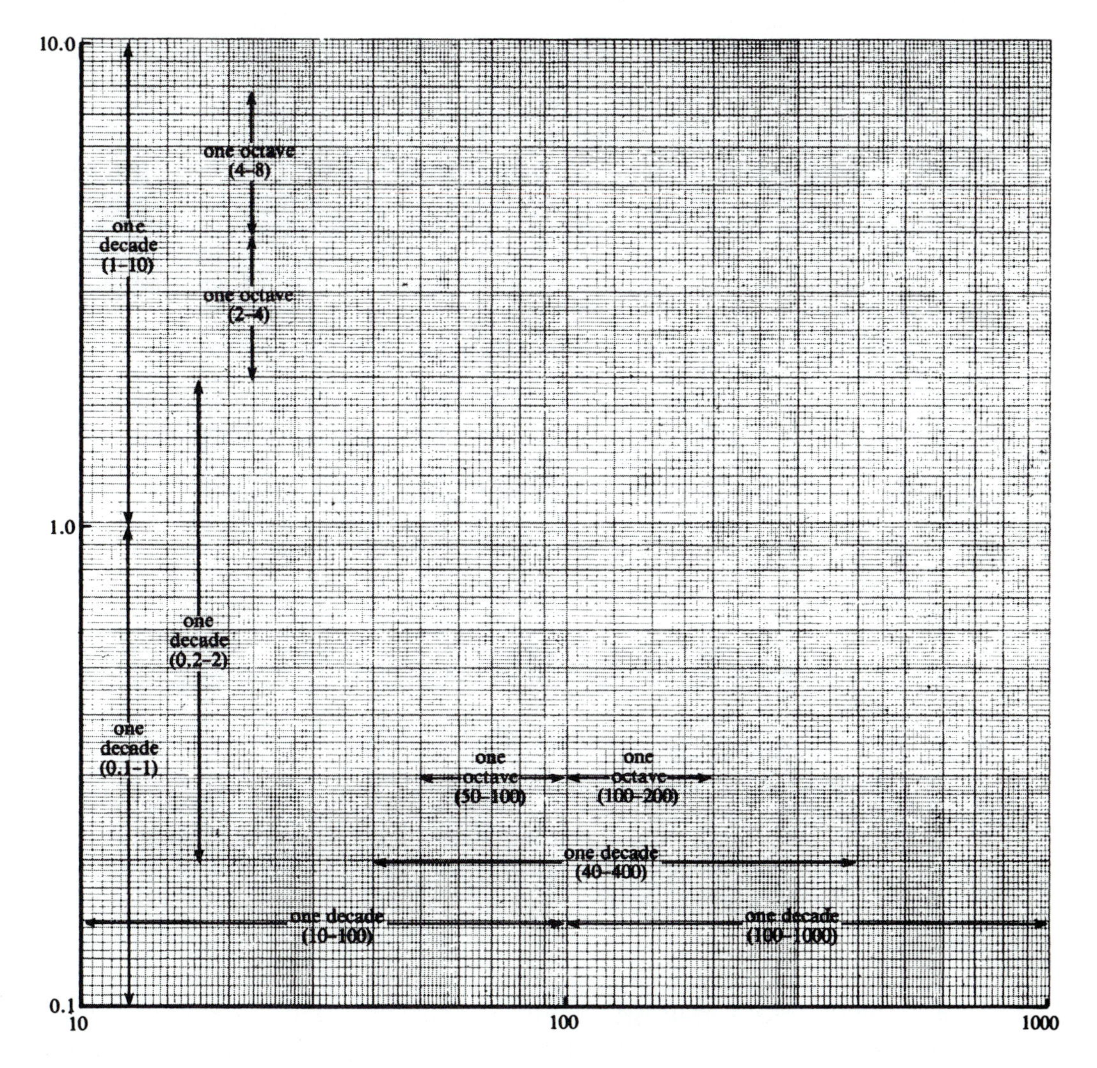

Figure 10–5
Two-cycle–by–2-cycle (2 × 2) log-log graph paper, showing some typical octaves and decades along each axis

and so forth. Notice that each decade along each axis occupies the same amount of space. Several horizontal and vertical decades are identified on the figure. Notice also that the graph paper is printed with identical scale values along each decade. The user must relabel the divisons in accordance with the actual decade values that are appropriate for the data to be plotted. Suppose, for example, that the gain of an amplifier varies from 2 to 60 over the frequency range from 150 Hz to 80 kHz. Then the frequency axis must cover the three decades 100 Hz to 1 kHz, 1 kHz to 10 kHz, and 10 kHz to 100 kHz, and the gain axis must cover the two decades 1 to 10 and 10 to 100. 3×2 graph paper would be required. In Figure 10–5, the axes are arbitrarily labeled with decades 0.1–1 and 1–10 (vertical), and 10–100 and 100–1000 (horizontal).

An *octave* is any two-to-one range of values, such as 5–10, 80–160, and 1000–2000. Notice in Figure 10–5 that every octave occupies the same length. Notice also that the value *zero* does not appear on either axis of the figure. Zero can never appear on a logarithmic scale, no matter how many decades are represented, because $\log(0) = -\infty$.

Semilog graph paper is used to plot gain in dB versus the logarithm of frequency. When the frequency axis is logarithmic and the vertical axis is linear with its divisions labeled in decibels, the graph paper is essentially the same as log-log paper. Thus, a plot of an amplifier's frequency response will have the same general shape when constructed on either type of graph paper. Semilog graph paper is also used to plot phase shift on a linear scale versus the logarithm of frequency. Graphs of frequency response plotted against the logarithm of frequency are called Bode (pronounced bō-dē) plots.

The cutoff frequency is the frequency at which the gain on a frequency-response plot is 3 dB less than the midband gain. At cutoff, the gain is said to be "3 dB down," and the cutoff frequencies are often called *3-dB frequencies.* The value is 3 dB because the output voltage is $\sqrt{2}/2$ times its value at midband, and

$$20 \log_{10}\left[\frac{(\sqrt{2}/2)v_m}{v_m}\right] = 20 \log_{10}(\sqrt{2}/2) \approx -3 \text{ dB} \tag{10–11}$$

where v_m = the midband voltage.

Example 10–4

Figure 10–6 shows the voltage gain of an amplifier plotted versus frequency on log-log paper. Find

1. the midband gain, in dB;
2. the gain in dB at the cutoff frequencies;
3. the bandwidth;
4. the gain in dB at a frequency 1 decade below the lower cutoff frequency;
5. the gain in dB at a frequency 1 octave above the upper cutoff frequency; and
6. the frequencies at which the gain is down 15 dB from its midband value.

Solution

1. As shown in Figure 10–6, the magnitude of the midband gain is 250. Therefore, $A_m = 20 \log_{10}250 = 47.96$ dB.
2. Since the gain at each cutoff frequency is 3 dB less than the gain at midband, $A_v(\text{at cutoff}) = 47.96 - 3 = 44.96$ dB.
3. The magnitude of the gain at each cutoff frequency is $(\sqrt{2}/2)250 = 176.78$. As shown in Figure 10–6, this value of gain is reached at $f_1 = 30$ Hz and at $f_2 = 10$ kHz. Therefore, $\text{BW} = f_2 - f_1 = (10 \text{ kHz}) - (30 \text{ Hz}) = 9970$ Hz.

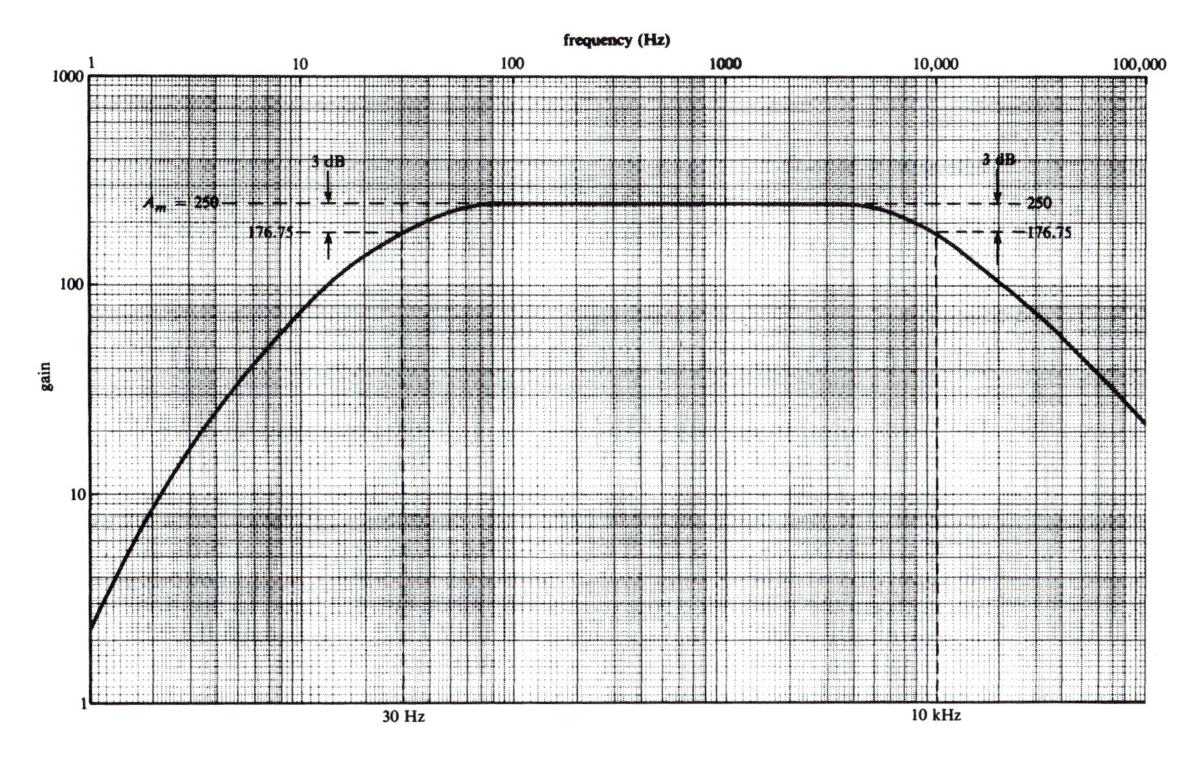

Figure 10–6
(Example 10–4)

4. The frequency 1 decade below f_1 is $(0.1)f_1 = 3$ Hz. From Figure 10–6, the magnitude of the gain at this frequency is approximately 15, so $A_v = 20 \log_{10}15 = 23.52$ dB.
5. The frequency 1 octave above f_2 is $2f_2 = 20$ kHz. From Figure 10–6, the magnitude of the gain at 20 kHz is approximately 105, so $A_v = 20 \log_{10}105 = 40.42$ dB.
6. The gain 15 dB below midband is (47.96 dB) − (15 dB) = 32.96 dB.

$$20 \log_{10}A_v = 32.96$$
$$\log_{10}A_v = 1.648$$
$$A_v = \text{antilog}(1.648) = 44.46$$

From Figure 10–6, the frequencies at which $A_v = 44.46$ are approximately 6.5 Hz and 50 kHz.

One-*n*th Decade and Octave Intervals

In many practical investigations, including computer-generated frequency-response data, it is necessary to specify logarithmic frequency intervals within one decade or within one octave. For example, we may want ten logarithmically spaced frequencies within one decade. These frequencies are said to be at *one-tenth decade* intervals. Similarly, three logarithmically spaced frequencies in one octave are said to be at *one-third octave* intervals. The frequencies in one-*n*th decade intervals beginning at frequency f_1 are

$$10^x,\ 10^{x+(1/n)},\ 10^{x+(2/n)},\ 10^{x+(3/n)},\ \ldots$$

where $x = \log_{10} f_1$.

The frequencies in one-*n*th octave intervals beginning at frequency f_1 are

$$2^x,\ 2^{x+(1/n)},\ 2^{x+(2/n)},\ 2^{x+(3/n)},\ \ldots$$

where $x = \log_2 f_1$.

10–3 SERIES CAPACITANCE AND LOW-FREQUENCY RESPONSE

The lower cutoff frequency of an amplifier is affected by capacitance connected in *series* with the signal flow path. The most important example of series-connected capacitance is the amplifier's input and output coupling capacitors. At low frequencies, the reactance of these capacitors becomes very large, so a significant portion of the ac signal is dropped across them. As frequency approaches 0 (dc), the capacitive reactance approaches infinity (open circuit), so the coupling capacitors perform their intended role of blocking all dc current flow. In previous discussions, we have assumed that the signal frequency was high enough that the capacitive reactance of all coupling capacitors was negligibly small, but we will now consider how large reactances at low frequencies affect the overall voltage gain.

Figure 10–7 shows the capacitor–resistor combination formed by the coupling capacitor and the input resistance at the input side of an amplifier. We omit consideration of any signal-source resistance for the moment. Notice that r_{in} and the capacitive reactance of C_1 form a voltage divider across the amplifier input. The amplifier input voltage, v_{in}, is found from the voltage-divider rule:

$$v_{in} = \left(\frac{r_{in}}{r_{in} - jX_{C_1}}\right) v_S \qquad \textbf{(10–12)}$$

where
$$X_{C_1} = \frac{1}{\omega C_1} \text{ ohms}$$
$$\omega = 2\pi f \text{ radians/second}$$

From equation 10–12, we can determine the *magnitude* (amplitude) of v_{in}, which we will designate $|v_{in}|$, as a function of ω:

$$|v_{in}| = \frac{r_{in}}{\sqrt{r_{in}^2 + X_{C_1}^2}}|v_S| = \frac{r_{in}}{\sqrt{r_{in}^2 + \left(\dfrac{1}{\omega C_1}\right)^2}}|v_S| = \frac{\omega r_{in} C_1}{\sqrt{1 + (\omega r_{in} C_1)^2}}|v_S| \qquad \textbf{(10–13)}$$

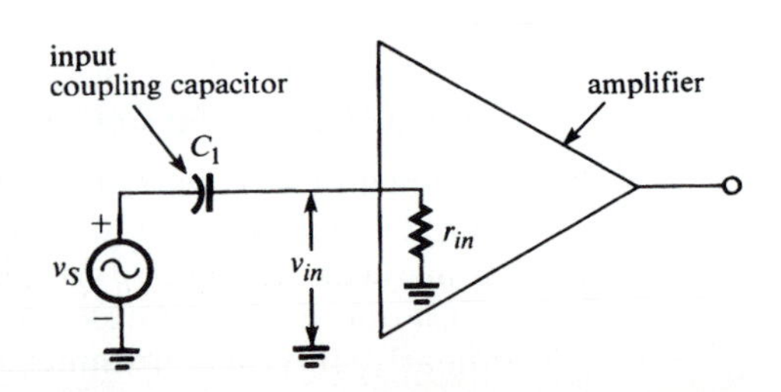

Figure 10–7
The input resistance of the amplifier and the coupling capacitor form an RC network that reduces the amplifier's input signal at low frequencies.

Equation 10–13 shows that $|v_{in}| = 0$ when $\omega = 0$ (dc) and that $|v_{in}|$ approaches $|v_S|$ in value as ω becomes very large. At the frequency $\omega = 1/(r_{in}C_1)$ rad/s, we have, from equation 10–13,

$$|v_{in}| = \frac{\left(\frac{1}{r_{in}C_1}\right)(r_{in}C_1)}{\sqrt{1 + \left[\left(\frac{1}{r_{in}C_1}\right)(r_{in}C_1)\right]^2}}|v_S| = \frac{1}{\sqrt{1+1}}|v_S| = \frac{1}{\sqrt{2}}|v_S| \approx 0.707\,|v_S|$$

This result shows that the amplifier input voltage falls to 0.707 times the source voltage when the frequency is reduced to $1/(r_{in}C_1)$ rad/s. Therefore, if there are no other frequency-sensitive components affecting the signal level, the overall gain from source to output is 0.707 times its midband value, meaning that $1/(r_{in}C_1)$ rad/s is the lower cutoff frequency. It is an exercise at the end of this chapter to show that the lower cutoff frequency is the frequency at which the capacitive reactance X_{C_1} equals the resistance r_{in}.

If source resistance r_S is present, then equation 10–13 becomes

$$|v_{in}| = \frac{\omega r_{in}C_1}{\sqrt{1 + [\omega(r_{in} + r_S)C_1]^2}}|v_S| \qquad \textbf{(10–14)}$$

and the lower cutoff frequency is

$$\omega_1 = \frac{1}{(r_{in} + r_S)C_1} \text{ rad/s} \qquad \textbf{(10–15)}$$

or

$$f_1 = \frac{1}{2\pi(r_{in} + r_S)C_1} \text{ Hz} \qquad \textbf{(10–16)}$$

The cutoff frequency defined by equations 10–15 and 10–16 is the frequency at which the ratio $|v_{in}|/|v_S|$ is 0.707 times its *midband* value, namely, 0.707 times $r_{in}/(r_S + r_{in})$. The ratio $|v_{in}|/|v_S|$ can be written in terms of the cutoff frequency f_1 and the signal frequency f as follows:

$$\frac{|v_{in}|}{|v_S|} = K\left(\frac{f}{\sqrt{f_1^2 + f^2}}\right) = K\left(\frac{1}{\sqrt{1 + (f_1/f)^2}}\right) \qquad \textbf{(10–17)}$$

where

$$K = \frac{r_{in}}{r_S + r_{in}}$$

Let us now use v_S as a phase angle reference ($\angle v_S = 0°$) and compute the phase of v_{in} as a function of frequency. From equation 10–12,

$$\angle v_{in} = -\arctan\left(\frac{-X_{C_1}}{r_{in}}\right) + \angle v_S = \arctan\left(\frac{X_{C_1}}{r_{in}}\right) = \arctan\left(\frac{1}{\omega r_{in}C_1}\right) \qquad \textbf{(10–18)}$$

Again, if source resistance r_S is present, we find

$$\angle v_{in} = \arctan\left[\frac{1}{\omega(r_{in} + r_S)C_1}\right] \qquad \textbf{(10–19)}$$

At $\omega = 1/(r_{in} + r_S)C_1$, equation 10–19 becomes

$$\angle v_{in} = \arctan(1) = 45°$$

Thus, v_{in} leads v_S by 45° at cutoff. The phase shift in terms of f_1 and the signal frequency f is

$$\underline{/v_{in}} = \arctan(f_1/f) \qquad \textbf{(10–20)}$$

Note that $\underline{/v_{in}}$ approaches 90° as f approaches 0.

Example 10–5

The amplifier shown in Figure 10–8 has midband gain $|v_L|/|v_S|$ equal to 90. Find

1. the voltage gain $|v_L|/|v_{in}|$;
2. the lower cutoff frequency; and
3. the voltage gain $|v_L|/|v_S|$, in dB, at the cutoff frequency.

Figure 10–8
(Example 10–5)

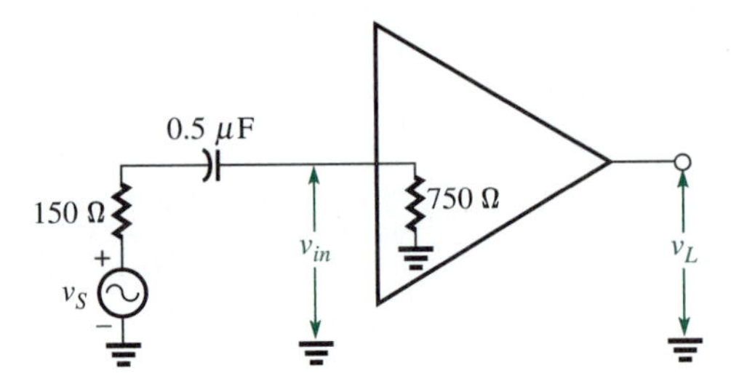

Solution

1. At midband, the reactance of the coupling capacitor is negligible, so the overall voltage gain $|v_L|/|v_S|$ is

$$\frac{|v_L|}{|v_S|} = \frac{|v_{in}|}{|v_S|}\frac{|v_L|}{|v_{in}|} = \left(\frac{r_{in}}{r_S + r_{in}}\right)\frac{|v_L|}{|v_{in}|}$$

Thus,

$$90 = \left(\frac{750\ \Omega}{150\ \Omega + 750\ \Omega}\right)\frac{|v_L|}{|v_{in}|}$$

$$\frac{|v_L|}{|v_{in}|} = \frac{90}{\left(\frac{750}{900}\right)} = 108$$

2. From equation 10–16,

$$f_1 = \frac{1}{2\pi(150 + 750)(0.5 \times 10^{-6})} = 353.67\ \text{Hz}$$

3. At midband, the gain in dB is 20 $\log_{10}(90) = 39.08$ dB. At cutoff, the gain is 3 dB less than its midband value: $39.08 - 3 = 36.08$ dB.

Figure 10–9 shows *normalized* plots of the gain of an RC network connected so that the capacitor is in series with the signal flow and the output is taken across the resistor. This is the configuration at the input of the capacitor-coupled amplifier, as shown in Figure 10–7. The gain of the network approaches 1 (0 dB) at high frequencies, which corresponds to the condition $|v_{in}|/|v_S| = 1$, when the capacitive reactance is negligibly small. Of course, the *overall* gain of an amplifier having a capacitor-coupled input may be greater than 1 at frequencies above cutoff, but the shape of the frequency response is the same as that shown in Figure 10–9. The only difference is that the gain above cutoff is A_m—the amplifier's midband gain—rather than 1 (0 dB).

Note that the plots in Figure 10–9 extend 2 decades below and 1 decade above the cutoff frequency, which is labeled f_1. Figure 10–9(b) shows that the gain is "down" approximately 6 dB at a frequency 1 octave below cutoff ($0.5f_1$) and is down 20 dB 1 decade below cutoff (at $0.1f_1$). Figure 10–9(c) shows that the phase shift is 63.4° one octave below f_1 and 87.1° one decade below f_1. The gain is down 0.04 dB and the phase shift is 5.7° at a frequency 1 decade above cutoff ($10f_1$). The gain plots also show the straight line *asymptotes* that the gain approaches at frequencies below cutoff. An asymptote is often used to approximate the gain response. Note that it "breaks" downward at f_1, where its deviation from the actual response curve is the greatest (3 dB). The frequency at which an asymptote breaks (f_1 in this example) is called a *break* frequency. *The asymptote has a slope of 6 dB/octave, or 20 dB/decade.*

When using these plots, remember that they are valid for only a *single* resistor–capacitor (RC) combination. We will study the effects of multiple RC combinations in the signal flow path in a later discussion. The plots in Figure 10–9 apply to any RC network in which the capacitor is in series with the signal path and the output is taken across the resistor. Such a network is called a *high-pass filter,* because, as the plots show, the gain is constant at high frequencies (above cutoff) and "falls off," at a constant rate, below cutoff. To find the overall gain of an RC network in which part of the resistance (r_S) is in the signal source, the gain determined from the plot must be multiplied by the factor $K = r_{in}/(r_S + r_{in})$.

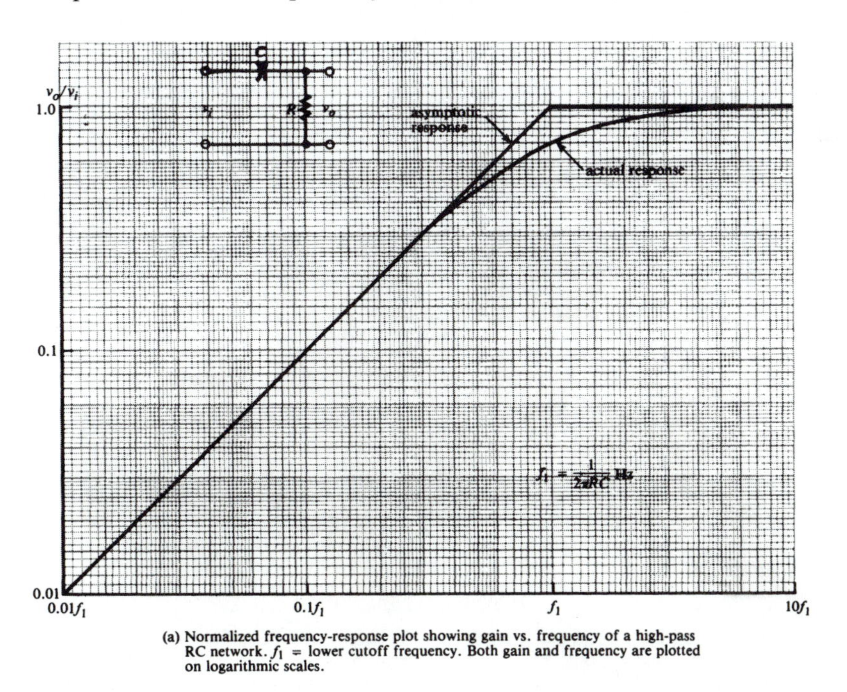

(a) Normalized frequency-response plot showing gain vs. frequency of a high-pass RC network. f_1 = lower cutoff frequency. Both gain and frequency are plotted on logarithmic scales.

Figure 10–9
Normalized gain and phase plots for a high-pass RC network

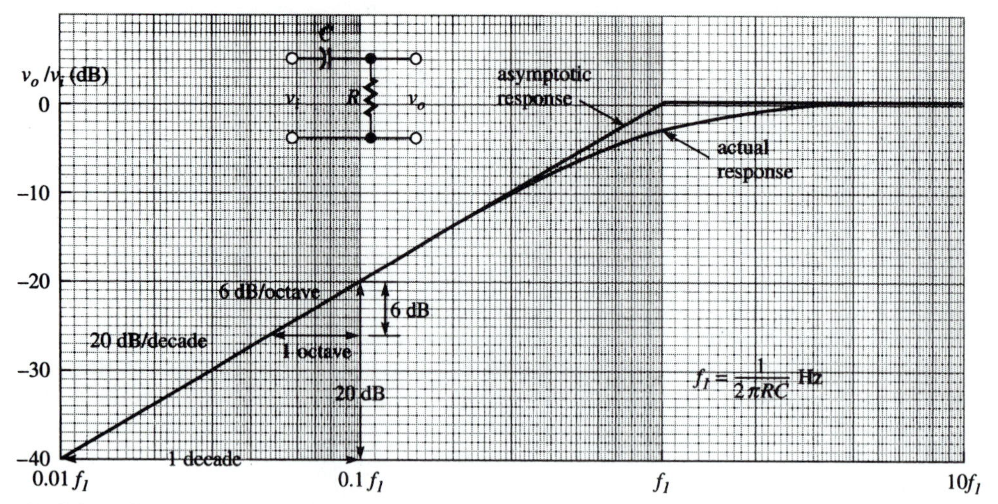

(b) Gain vs. frequency for the high-pass network, plotted on semilog graph paper. Note that the vertical axis is linear and is scaled in dB.

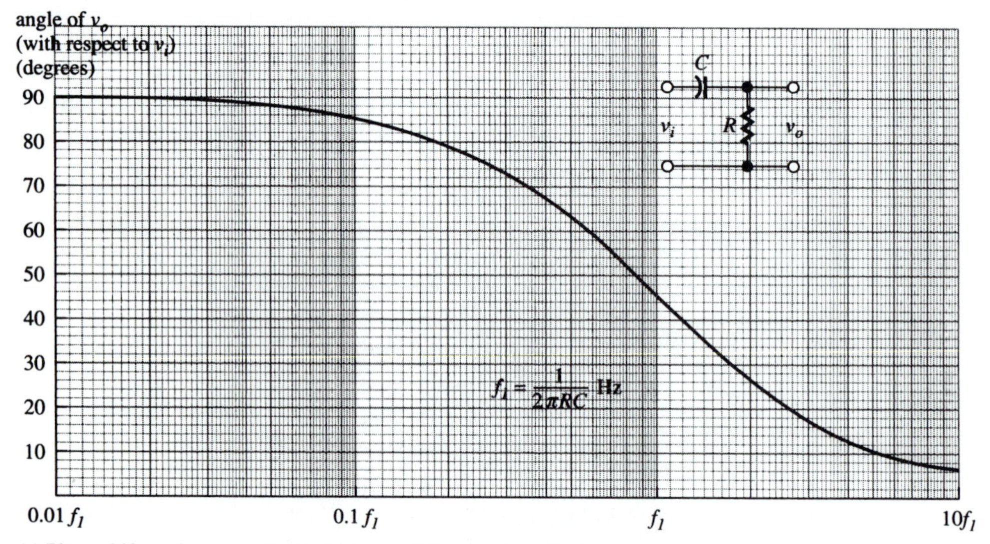

(c) Phase shift vs. frequency for the high-pass RC network. v_o leads v_i.

Figure 10–9
(Continued)

Example 10–6

The amplifier shown in Figure 10–10 has voltage gain $|v_L|/|v_{in}| = 120$. Calculate

1. the lower cutoff frequency due to the input coupling capacitor;
2. the gain $|v_L|/|v_S|$ 1 octave below cutoff; and
3. the phase shift 1 decade above cutoff.

Figure 10–10
(Example 10–6)

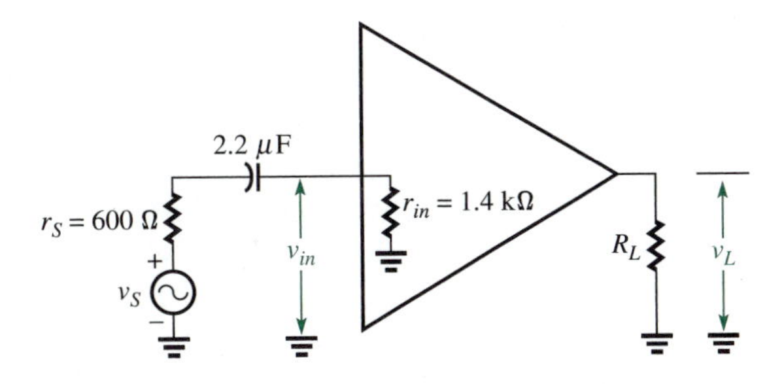

Use Figure 10–9 to find approximate values for

4. the asymptotic gain $|v_L|/|v_S|$ 1 octave below cutoff;
5. the gain $|v_L|/|v_S|$, in dB, at f = 10.85 Hz; and
6. the frequency at which the phase shift is 20°.

Solution

1. From equation 10–16,

$$f_1 = \frac{1}{2\pi(1400 + 600)(2.2 \times 10^{-6})} = 36.17 \text{ Hz}$$

2. The frequency 1 octave below cutoff is $f_1/2$ = 18.09 Hz. From equation 10–14,

$$\frac{|v_{in}|}{|v_S|} = \frac{2\pi(18.09)(1400)(2.2 \times 10^{-6})}{\sqrt{1 + [2\pi(8.09)(1400 + 600)(2.2 \times 10^{-6})]^2}} = 0.313$$

Therefore, $|v_L|/|v_S|$ at 18.09 Hz is

$$\frac{|v_L|}{|v_S|} = \frac{|v_{in}|}{|v_S|}\frac{|v_L|}{|v_{in}|} = 0.313(120) = 37.6$$

Note that this result can be computed more directly by using (10–17) and recognizing that, when f is 1 octave below f_1, the quantity f_1/f equals 2:

$$\frac{|v_{in}|}{|v_S|} = \left(\frac{r_{in}}{r_S + r_{in}}\right)\frac{1}{\sqrt{1 + (f_1/f)^2}} = \left(\frac{1400\ \Omega}{2000\ \Omega}\right)\frac{1}{\sqrt{1 + 2^2}} = 0.313$$

3. At 1 decade above cutoff, $f = 10f_1$, or f_1/f = 0.1. Assuming that the amplifier does not cause any phase shift beyond that due to the coupling capacitor, the phase shift is, from equation 10–20, arctan(0.1) = 5.71°.
4. One octave below cutoff, $f = 0.5f_1$. From Figure 10–9(a), the gain at the intersection of the asymptote and the $0.5f_1$ coordinate line is 0.5. Therefore, the overall asymptotic gain is

$$\frac{|v_L|}{|v_S|} = 0.5\left(\frac{r_{in}}{r_S + r_{in}}\right)120 = 0.5\left(\frac{1400\ \Omega}{2000\ \Omega}\right)120 = 42$$

(Compare with the actual gain of 37.6, computed in part 2.)
5. At f = 10.85 Hz, f/f_1 = 10.85/36.17 = 0.3, so $f = 0.3f_1$. From Figure 10–9(b), the gain curve intersects the 0.3 coordinate line at approximately −11 dB. The

midband gain $|v_L|/|v_S|$ in decibels is

$$20 \log_{10}\left[\left(\frac{1400\ \Omega}{600\ \Omega + 1400\ \Omega}\right)(120)\right] = 38.49\ \text{dB}$$

Therefore, the gain at 10.85 Hz is $38.49 - 11 = 27.49$ dB.

6. From Figure 10–9(c), the 20° phase coordinate intersects the curve at approximately $2.8f_1$. Therefore, the phase shift is 20° at (2.8)(36.17 Hz) = 101.28 Hz.

The smaller the desired lower cutoff frequency of an amplifier, the larger the input coupling capacitor must be. If the input resistance of the amplifier is small, the required capacitance may be impractically large. For example, if an audio amplifier having input resistance 100 Ω is to have a lower cutoff frequency of 15 Hz, then, assuming that $r_S = 0$, the coupling capacitor must have value

$$C_1 = \frac{1}{2\pi(100\ \Omega)(15\ \text{Hz})} = 106\ \mu\text{F}$$

On the other hand, if the input resistance is 10 kΩ, the required capacitance is 1.06 μF, a much more reasonable value.

Figure 10–11 shows the resistance–capacitance combination formed by the output resistance, output coupling capacitor, and load resistance of an amplifier. The effect of the output coupling capacitor on the frequency response is the same as that of the input coupling capacitor: At low frequencies, there is a significant voltage drop across the capacitive reactance, so less signal is delivered to the load. The lower cutoff frequency due to the output coupling capacitor (C_2) is found in the same way in which we found f_1 due to C_1. It is the frequency at which the capacitive reactance X_{C_2} equals the resistance $r_o + R_L$. To distinguish between the two lower frequencies, we will hereafter write $f_1(C_1)$ and $f_1(C_2)$:

$$f_1(C_1) = \frac{1}{2\pi(r_{in} + r_S)C_1} \tag{10–21}$$

$$f_1(C_2) = \frac{1}{2\pi(r_o + R_L)C_2} \tag{10–22}$$

If the lower cutoff frequency were determined by C_2 alone, then the frequency response in the vicinity of cutoff would have the same appearance as the normalized plots shown in Figure 10–9. However, when both input and output coupling capacitors are present, the low-frequency response is significantly different because both capacitors contribute *simultaneously* to a reduction in gain. Let us first consider the case where the frequencies $f_1(C_1)$ and $f_1(C_2)$ are considerably different, at least a decade apart. Then, to a good approximation, the overall frequency response will

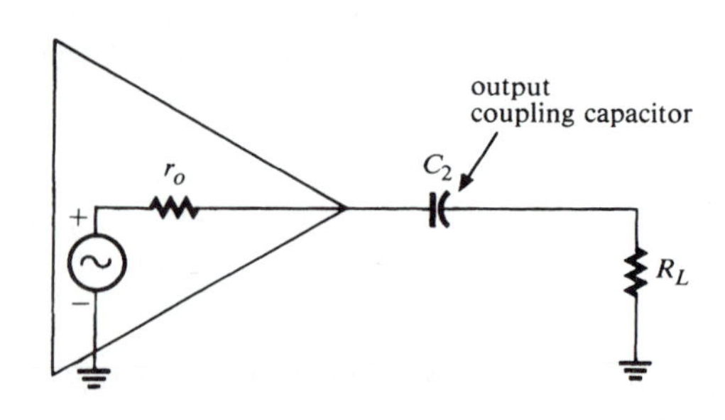

Figure 10–11
The output and load resistance of the amplifier and the output coupling capacitor form an RC network that reduces the signal reaching the load at low frequencies.

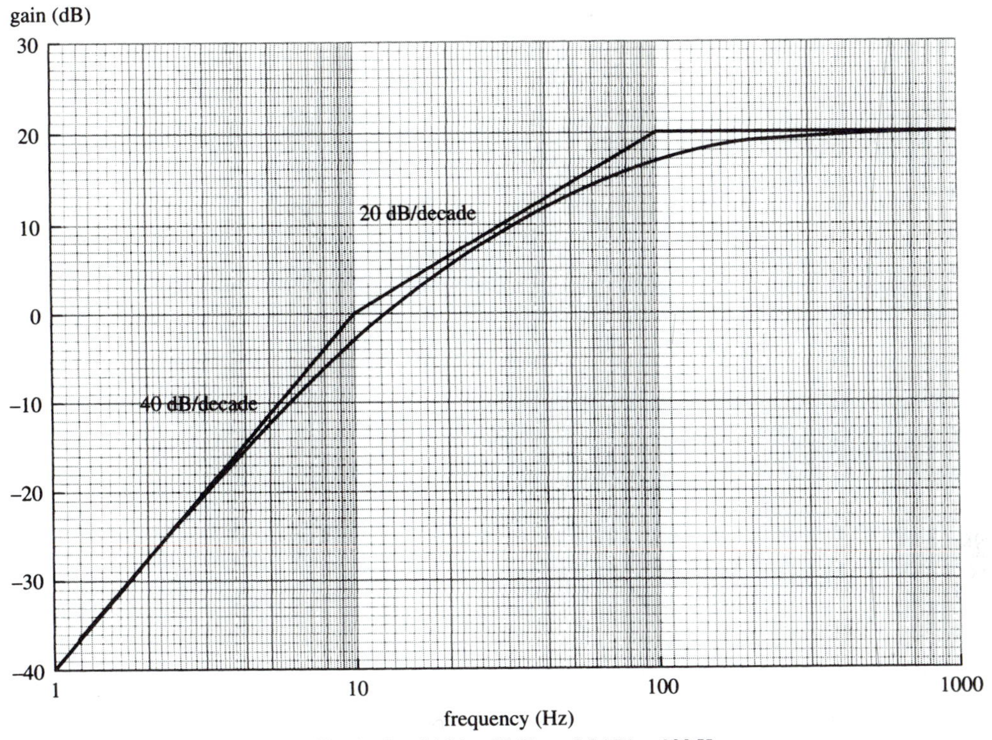

(a) Gain vs. frequency for an amplifier having $f_1(C_1) = 10$ Hz and $f_1(C_2) = 100$ Hz. The lower cutoff frequency in this case is the larger of the two, $f_1 = 100$ Hz. Note that there are two "break" frequencies. $A_m = 20$ dB.

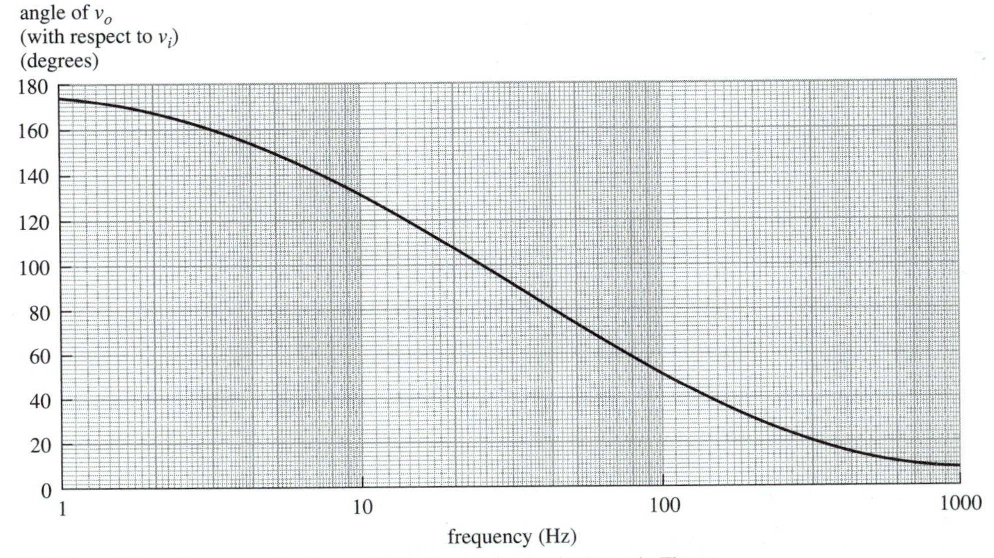

(b) Phase shift vs. frequency for the amplifier whose gain plot is shown in Figure 10.12(a). Note that the phase angle approaches 180° as frequency approaches 0.

Figure 10–12
Gain and phase versus frequency for an amplifier having two break frequencies

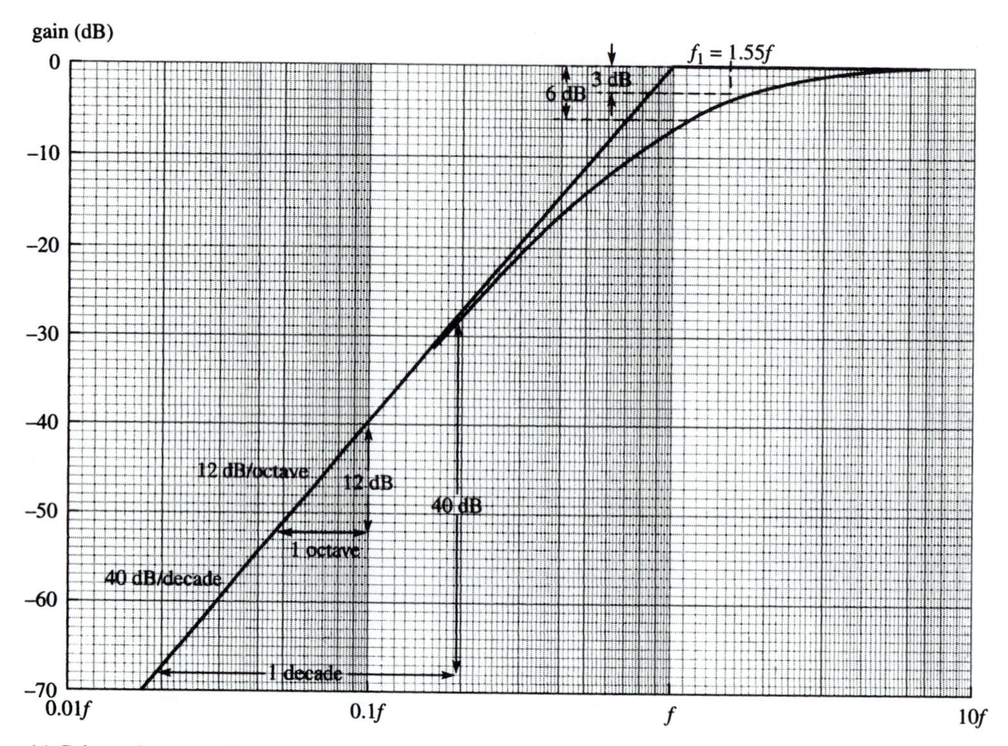

(a) Gain vs. frequency for two isolated, high-pass RC networks having the same cutoff frequency, f_1. Note that $f_1 = 1.55f$. For this plot, the midband (high-frequency) gain is 0 dB.

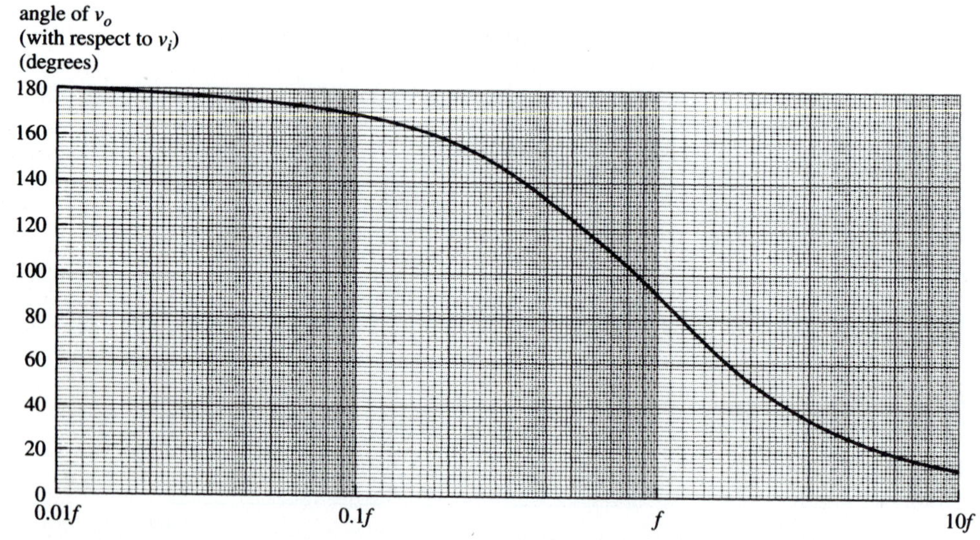

(b) Phase shift vs. frequency for the two high-pass RC networks whose gain is shown in Figure 10.13(a). Note that the output leads the input by 90° at frequency f.

Figure 10–13
Normalized gain and phase response for the case $f_1(C_1) = f_1(C_2) = f$

have a lower cutoff frequency, f_1, equal to the *larger* of $f_1(C_1)$ and $f_1(C_2)$. Figure 10–12 shows the frequency response for the case $f_1(C_1) = 10$ Hz and $f_1(C_2) = 100$ Hz. The midband gain is assumed to be 20 dB. It is clear that $f_1 = 100$ Hz, because the gain at that frequency is $20 - 3 = 17$ dB. The gain at 10 Hz is much lower (approximately -3 dB) because at frequencies below 100 Hz *both* C_1 and C_2 cause gain reduction.

Note that in Figure 10–12(a) the gain asymptote has slope 20 dB/decade (6 dB/octave) between 10 Hz and 100 Hz, but breaks downward at 10 Hz with a slope of 40 dB/decade (12 dB/octave). There are, therefore, two break frequencies in this example, i.e., two frequencies where the asymptote changes slope: 10 Hz and 100 Hz. The phase shift is approximately 51° at $f_1 = 100$ Hz and 129° at 10 Hz and approaches 180° as frequency approaches 0. The response characteristics shown in Figure 10–12 are valid for two high-pass RC networks connected in series, provided the networks are *isolated* from each other, in the sense that one does not load the other. In our illustration, the two networks are assumed to be isolated by an amplifier having gain 20 dB.

If the frequencies $f_1(C_1)$ and $f_1(C_2)$ are closer than one decade to each other, then the overall lower cutoff frequency is somewhat higher than the larger of the two. In such cases, it is usually adequate to assume that f_1 equals the larger of the two. The exact value of f_1 can be found using the following rather cumbersome equation (derived in Appendix C):

$$f_1 = \sqrt{\frac{2ab}{-(a+b) + \sqrt{(a+b)^2 + 4ab}}} \tag{10–23}$$

where

$$a = f_1^2(C_1)$$
$$b = f_1^2(C_2)$$

If $f_1(C_1) = f_1(C_2)$, equation 10–23 can be used to show the overall lower cutoff frequency is 1.55 times the value of either. For example, if $f_1(C_1) = f_1(C_2) = 100$ Hz, then $f_1 = 155$ Hz. Figure 10–13 shows the normalized gain and phase response for the special case $f_1(C_1) = f_1(C_2)$. Note that the asymptote breaks downward at $f = f_1(C_1) = f_1(C_2)$ and has a slope of 40 dB/decade, or 12 dB/octave. The actual response is 6 dB below the asymptote at that frequency. The cutoff frequency is $f_1 = 1.55f$, where the gain is down 3 dB. Note that the break frequency (f) is not the same as the cutoff frequency (f_1) in this case. The phase shift is 90° at f and approaches 180° as frequency approaches 0. Once again, the plots shown in Figure 10–13 are valid only when the two high-pass RC networks whose response they represent are isolated from each other. In our case, we assume that the isolation is provided by an amplifier (having unity gain, since the gain in Figure 10–13(a) approaches 0 dB at high frequencies).

Example 10–7

The amplifier shown in Figure 10–14 has midband gain $|v_L|/|v_S| = 140$. Find

1. the approximate lower cutoff frequency;
2. the gain $|v_o|/|v_{in}|$;
3. the lower cutoff frequency when C_2 is changed to 50 μF;
4. the approximate gain $|v_L|/|v_S|$, in dB, at 2.9 Hz with $C_2 = 50$ μF; and
5. the value that C_2 would have to be in order to obtain a lower cutoff frequency of approximately 100 Hz.

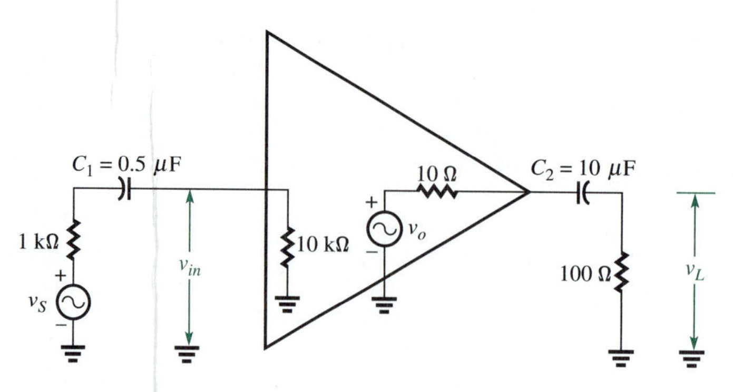

Figure 10–14
(Example 10–7)

Solution

1. From equation 10–21,

$$f_1(C_1) = \frac{1}{2\pi(1 \times 10^3 + 10 \times 10^3)(0.5 \times 10^{-6})} = 28.94 \text{ Hz}$$

From equation 10–22,

$$f_1(C_2) = \frac{1}{2\pi(10 + 100)(10 \times 10^{-6})} = 144.7 \text{ Hz}$$

Therefore, $f_1 \approx 144.7$ Hz.

2. $$\frac{|v_L|}{|v_S|} = \frac{|v_{in}|}{|v_S|}\frac{|v_o|}{|v_{in}|}\frac{|v_L|}{|v_o|}$$

$$140 = \left[\frac{10\text{ k}\Omega}{(10\text{ k}\Omega) + (1\text{ k}\Omega)}\right]\frac{|v_o|}{|v_{in}|}\left[\frac{100\ \Omega}{(10\ \Omega) + (100\ \Omega)}\right]$$

$$\frac{|v_o|}{|v_{in}|} = \frac{140}{\left(\dfrac{1 \times 10^4}{1.1 \times 10^4}\right)\left(\dfrac{1 \times 10^2}{1.1 \times 10^2}\right)} = 169.4$$

3. $$f_1(C_2) = \frac{1}{2\pi(10 + 100)(50 \times 10^{-6})} = 28.94 \text{ Hz}$$

Therefore, $f_1(C_1) = f_1(C_2) = 28.94$ Hz, so $f_1 = 1.55(28.94) = 44.86$ Hz.

4. 2.9 Hz is approximately 1 decade below the break frequency of 28.94 Hz. Since the gain falls at the rate of 40 dB/decade below the break frequency, at 2.9 Hz it will be 40 dB less than A_m:

$$A_m = 20 \log_{10} 140 = 42.9 \text{ dB}$$

$$A_v(\text{at } 2.9 \text{ Hz}) = 42.9 - 40 = 2.9 \text{ dB}$$

5. Using the approximation $f_1 \approx f_1(C_2) = 100$ Hz,

$$100 = \frac{1}{2\pi(10 + 100)C_2}$$

$$C_2 = \frac{1}{2\pi(10 + 100)(100)} = 14.5\ \mu\text{F}$$

10–4 SHUNT CAPACITANCE AND HIGH-FREQUENCY RESPONSE

Capacitance that provides an ac path between an amplifier's signal flow path and ground is said to *shunt* the signal. The most common form of shunt capacitance is that which exists between the terminals of an electronic device due to its structural characteristics. Recall, for example, that a PN junction has capacitance between its terminals because the depletion region forms a dielectric separating the two conductive regions of P and N material. Capacitance between device terminals is called *interelectrode* capacitance. Shunt capacitance is also created by wiring, terminal connections, solder joints, and any other circuit structure where conducting regions are close to each other. This type of capacitance is called *stray* capacitance.

Shunt capacitance affects the high-frequency performance of an amplifier because at high frequencies the small capacitive reactance diverts the signal to ground, or to some other point besides the load. A semiconductor device has inherent frequency limitations that may impose more severe restrictions on high-frequency operation than the effect of shunt capacitance we will study now, but these *device* considerations will be temporarily postponed. Figure 10–15 shows shunt capacitance C_A between the input side of an amplifier and ground (in parallel with r_{in}). Notice that we can now neglect series coupling capacitance, because we are considering only high-frequency operation.

In the midband frequency range, the effect of C_A in Figure 10–15 can be neglected because the frequency is not high enough to make the capacitive reactance small. In other words, the midband frequency range is that range of frequencies that are high enough to neglect coupling capacitance and low enough to neglect shunt capacitance. It is apparent in the figure that, in the midband frequency range,

$$\frac{v_{in}}{v_S} = \frac{r_{in}}{r_S + r_{in}} \tag{10–24}$$

When the frequency is high enough to consider the effect of C_A, we must replace r_{in} in (10–24) by the parallel combination of r_{in} and $-jX_{C_A}$:

$$\frac{v_{in}}{v_S} = \frac{-jX_{C_A} \| r_{in}}{r_S + (-jX_{C_A} \| r_{in})} \tag{10–25}$$

At very high frequencies, where X_{C_A} becomes very small, the parallel combination of X_{C_A} and r_{in} becomes very small, and the net effect is the same as if the input impedance of the amplifier were made small. As we know, the consequence of that result is that the overall gain is reduced due to the voltage division between r_S and the input impedance. Our goal now is to find the frequency at which that reduction in gain equals 0.707 times the midband value given by equation 10–24. From

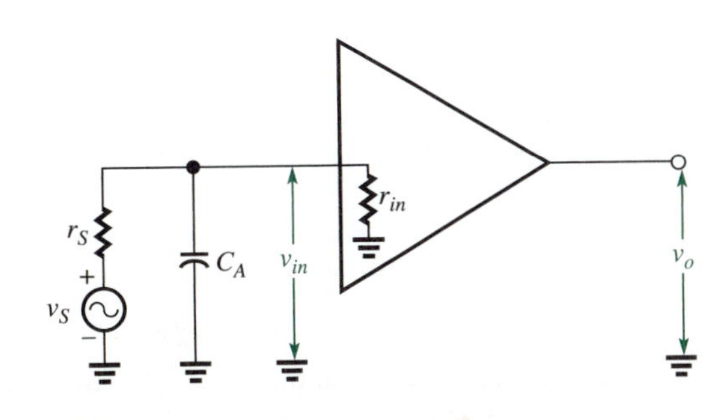

Figure 10–15
Shunt capacitance C_A in parallel with the input of an amplifier affects its high-frequency response.

equation 10–25,

$$\frac{v_{in}}{v_S} = \frac{\left(\dfrac{\dfrac{-jr_{in}}{\omega C_A}}{r_{in} - j/\omega C_A}\right)}{\left(r_S + \dfrac{\dfrac{-jr_{in}}{\omega C_A}}{r_{in} - j/\omega C_A}\right)} = \frac{\dfrac{-jr_{in}}{\omega C_A}}{r_S r_{in} - \dfrac{j(r_S + r_{in})}{\omega C_A}}$$

$$\frac{|v_{in}|}{|v_S|} = \frac{\dfrac{r_{in}}{\omega C_A}}{\sqrt{r_S^2 r_{in}^2 + \left(\dfrac{r_S + r_{in}}{\omega C_A}\right)^2}} = \frac{r_{in}}{\sqrt{(\omega C_A r_S r_{in})^2 + (r_S + r_{in})^2}}$$

$$= \frac{\dfrac{r_{in}}{r_S + r_{in}}}{\sqrt{[\omega C_A (r_S \| r_{in})]^2 + 1}} \tag{10–26}$$

The upper cutoff frequency, ω_2, is the frequency at which the expression in (10–26) equals $(\sqrt{2}/2)[r_{in}/(r_S + r_{in})]$:

$$\frac{\dfrac{r_{in}}{r_S + r_{in}}}{\sqrt{[\omega_2 C_A (r_S \| r_{in})]^2 + 1}} = \frac{\sqrt{2}}{2}\left(\frac{r_{in}}{r_S + r_{in}}\right)$$

$$\frac{1}{[\omega_2 C_A (r_S \| r_{in})]^2 + 1} = \left(\frac{\sqrt{2}}{2}\right)^2 = 0.5$$

$$\omega_2 = \frac{1}{(r_S \| r_{in}) C_A} \text{ rad/s}$$

or

$$f_2 = \frac{1}{2\pi (r_S \| r_{in}) C_A} \text{ Hz} \tag{10–27}$$

Equation 10–27 shows that the upper cutoff frequency due to C_A is inversely proportional to the parallel combination of r_S and r_{in}. Thus, to achieve a large bandwidth it is necessary to make $r_S \| r_{in}$ as small as possible, preferably by making r_S small. Theoretically, if $r_S = 0$, then $r_S \| r_{in} = 0$, and $f_2 = \infty$. In practice, it is not possible to have $r_S = 0$, but a very small value of r_S can increase the upper cutoff frequency significantly (not always a desirable practice, as we shall see in later discussions).

An RC network in which the resistor is in series with the signal flow path and the output is taken across the shunt capacitor is called a *low-pass filter.* The upper cutoff frequency f_2 is determined by $f_2 = 1/(2\pi RC)$ Hz. Comparing with equation 10–27 we see that the input of an amplifier having shunt capacitance C_A is the same as a low-pass filter having $R = r_S \| r_{in}$ and $C = C_A$. Normalized gain and phase plots for the low-pass RC network are shown in Figure 10–16. Note that the gain plot is the mirror image of that of the high-pass RC network shown in Figure 10–9. The asymptote shows that the gain falls off at the rate of 6 dB/octave, or 20 dB/decade, at frequencies above f_2. The phase shift approaches $-90°$ at frequencies above f_2. Note that negative phase means that the output voltage *lags* the input

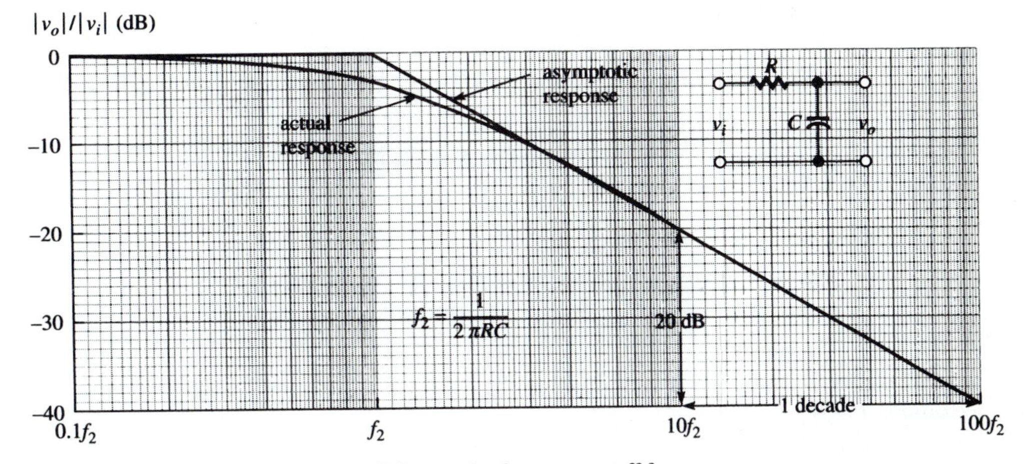

(a) Gain vs. frequency for the low-pass RC network. f_2 = upper cutoff frequency.

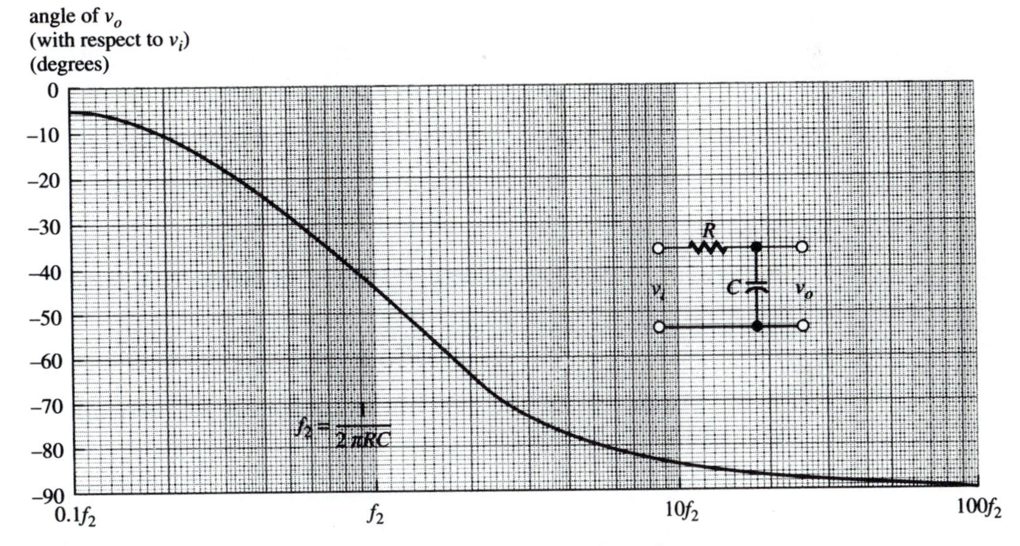

(b) Phase shift vs. frequency for the low-pass RC network. Note that v_o lags v_i.

Figure 10–16
Gain and phase versus frequency for a low-pass RC network

voltage, in contrast to the high-pass RC network, where the output leads the input. The gain and phase equations as functions of the ratio f/f_2 are

$$|A| = \frac{1}{\sqrt{1 + (f/f_2)^2}} \quad \textbf{(10–28)}$$

and

$$\underline{/A} = -\arctan(f/f_2) \quad \textbf{(10–29)}$$

where $f_2 = 1/(2\pi RC)$, $R = r_S \parallel r_{in}$, and $C = C_A$, for an amplifier having input shunt capacitance C_A.

Example 10–8

The amplifier shown in Figure 10–17 has midband gain $|v_L|/|v_S| = 40$ dB. Calculate

1. the upper cutoff frequency;
2. the gain $|v_L|/|v_S|$, in dB, at $f = 20$ MHz; and
3. the phase shift at $f = f_2$.

Use Figure 10–16 to find

4. the frequency at which $|v_L|/|v_S|$ is 35 dB; and
5. the frequency at which the output voltage lags the input voltage by 30°.

Figure 10–17
(Example 10–8)

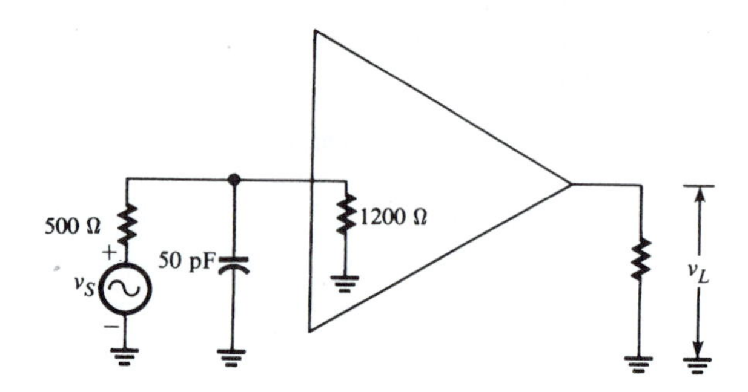

Solution

1. Since $r_S \parallel r_{in} = 500\ \Omega \parallel 1200\ \Omega = 353\ \Omega$, we have, from equation 10–27,

$$f_2 = \frac{1}{2\pi(353)(50 \times 10^{-12})} = 9.02\text{ MHz}$$

2. From equation 10–28, the gain of the equivalent RC network at the amplifier input when $f = 20$ MHz is

$$|A| = \frac{1}{\sqrt{1 + \left(\dfrac{20 \times 10^6}{9.02 \times 10^6}\right)^2}} = 0.441$$

$$20\log_{10}(0.441) = -7.11\text{ dB}$$

Therefore, the overall gain $|v_L|/|v_S|$ at $f = 20$ MHz is (40 dB) − (7.11 dB) = 32.89 dB.

3. From equation 10–29, when $f = f_2$, $\underline{/A} = -\arctan(f_2/f_2) = -\arctan(1) = -45°$. We conclude that *the output of a low-pass RC network lags the input by 45° at the cutoff frequency.* (See Figure 10–16(b).)
4. An overall gain of 35 dB corresponds to a 5-dB drop in gain from the midband value of 40 dB. From Figure 10–16(a), the gain of the equivalent RC network at the amplifier input is down 5 dB at approximately $f = 1.5f_2 = 1.5(9.02\text{ MHz}) = 13.53$ MHz.
5. From Figure 10.16(b), the frequency at which the phase shift is −30° is approximately $f = 0.57f_2 = 0.57(9.02\text{ MHz}) = 5.14$ MHz.

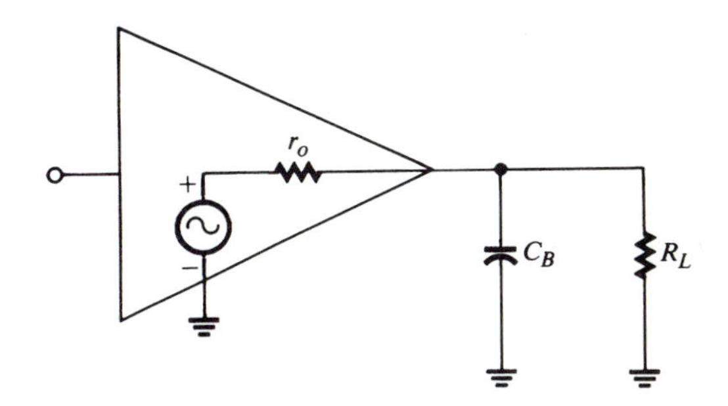

Figure 10–18
An amplifier having shunt capacitance C_B across its output

Figure 10–18 shows shunt capacitance C_B connected across the output of an amplifier. The output resistance r_o of the amplifier, the load resistance R_L, and the capacitance C_B form a low-pass RC network that has the same effect on the high-frequency response as the low-pass network at the input: the gain falls off because the impedance to ground decreases with increasing frequency. The upper cutoff frequency due to C_B is derived in the same way as that due to C_A. Hereafter, we will distinguish between the two frequencies by using the notation $f_2(C_A)$ and $f_2(C_B)$:

$$f_2(C_A) = \frac{1}{2\pi(r_S \| r_{in})C_A} \text{ hertz} \tag{10–30}$$

$$f_2(C_B) = \frac{1}{2\pi(r_o \| R_L)C_B} \text{ hertz} \tag{10–31}$$

When there is shunt capacitance at both the input and the output of an amplifier, the frequency response is different than it would be if only one were present. If $f_2(C_A)$ and $f_2(C_B)$ are not close in value, then the actual upper cutoff frequency, f_2, is approximately equal to the *smaller* of $f_2(C_A)$ and $f_2(C_B)$. The gain is asymptotic to a line that falls off at 20 dB/decade (6 dB/octave) between the smaller of $f_2(C_A)$ and $f_2(C_B)$ and is asymptotic to a line that falls off at 40 dB/decade (12 dB/octave) at frequencies above the larger of the two. See Figure 10–19(a). At the higher frequencies, the gain falls off at twice the rate it would for a single low-pass network, because both C_A and C_B contribute *simultaneously* to gain reduction. The total phase shift approaches −180° at high frequencies. The exact value of the upper cutoff frequency in terms of $f_2(C_A)$ and $f_2(C_B)$ can be found from (Appendix C):

$$f_2 = \sqrt{\frac{-(a+b) + \sqrt{(a+b)^2 + 4ab}}{2ab}} \tag{10–32}$$

where

$$a = 1/f_2^2(C_A)$$
$$b = 1/f_2^2(C_B)$$

If $f_2(C_A) = f_2(C_B) = f$, equation 10–32 can be used to show that the cutoff frequency is $f_2 = 0.645f$. In that case, there is but one break frequency, $f = f_2(C_A) = f_2(C_B)$, and the single asymptote has slope 40 dB/decade, or 12 dB/octave. See Figure 10–19(b). When the two frequencies are equal, the total phase shift is −90° at that frequency.

Thevenin Equivalent Circuits at Input and Output

We note that the same *form* of equation is used to compute both the lower and upper cutoff frequencies of an amplifier: $f = 1/2\pi RC$. In the case of f_1, C is either

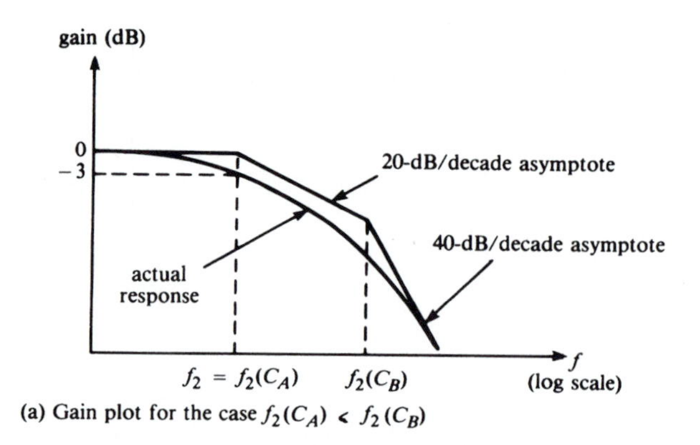

(a) Gain plot for the case $f_2(C_A) < f_2(C_B)$

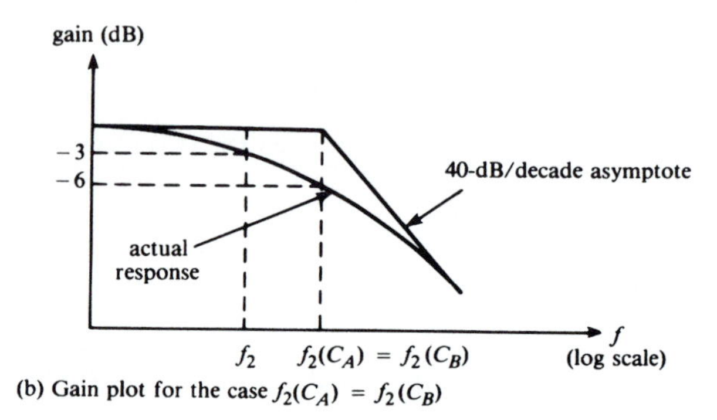

(b) Gain plot for the case $f_2(C_A) = f_2(C_B)$

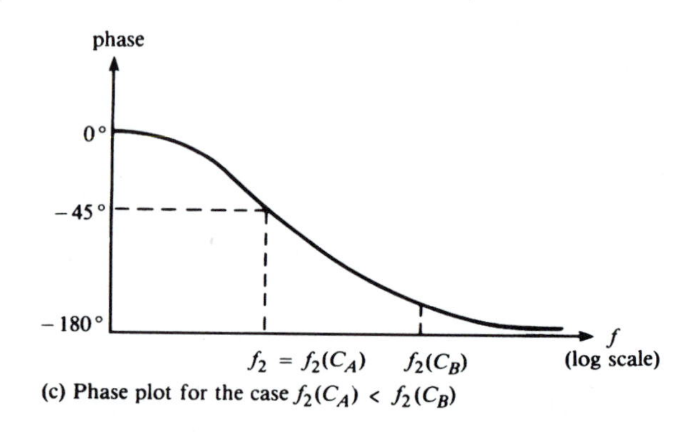

(c) Phase plot for the case $f_2(C_A) < f_2(C_B)$

Figure 10–19
High-frequency gain and phase response when shunt capacitance is present at both the input and the output of an amplifier. $f_2(C_A)$ = break frequency due to input shunt capacitance; $f_2(C_B)$ = break frequency due to output shunt capacitance.

input or output coupling capacitance and in the case of f_2, C is input or output shunt capacitance. In *both* cases, R is the Thevenin equivalent resistance seen by the capacitance. This observation provides an easy way to remember the equations for calculating f_1 and f_2:

$$f = \frac{1}{2\pi r_{TH} C} \text{ hertz} \tag{10–33}$$

where r_{TH} is the Thevenin equivalent resistance with respect to the capacitor terminals at input or output. Figure 10–20 illustrates this point. Recall that the Thevenin equivalent resistance is found by open-circuiting the capacitor terminals and com-

Figure 10–20
Examples of the use of equation 10–33 ($f = 1/2\pi r_{TH}C$) to find lower and upper cutoff frequencies

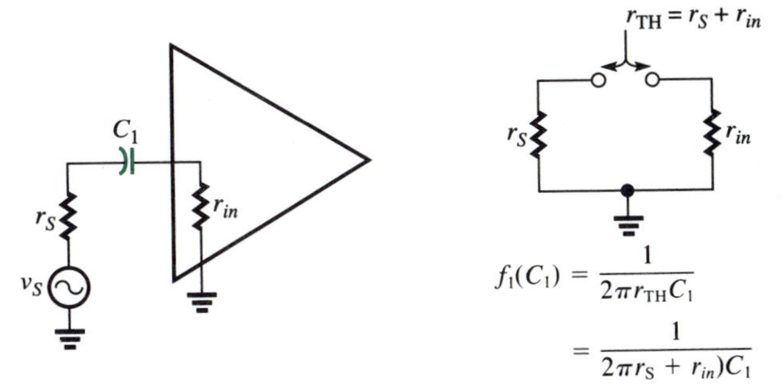

(a) Using equation 10-33 to find the lower cutoff frequency due to input coupling capacitance.

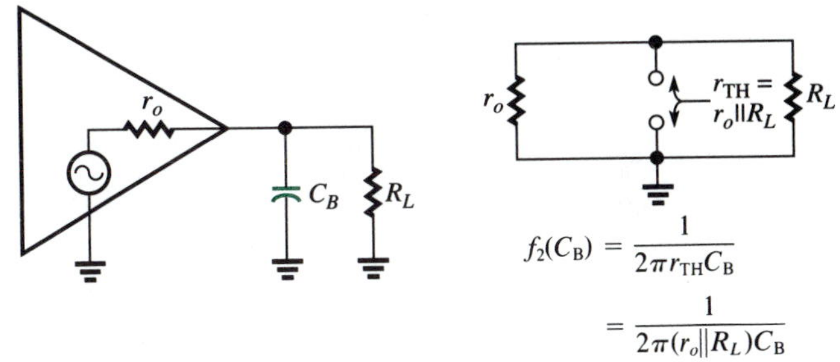

(b) Using equation 10-33 to find the upper cutoff frequency due to output shunt capacitance.

puting the total equivalent resistance looking into those terminals when all voltage sources are replaced by short circuits. Figure 10–20(a) shows how equation 10–33 is applied to calculate the lower cutoff frequency due to input coupling capacitance, C_1, and Figure 10–20(b) shows how the equation is applied to calculate the upper cutoff frequency due to output shunt capacitance, C_B. As an exercise, verify that the equations for $f_1(C_2)$ and $f_2(C_A)$ can also be found using the Thevenin equivalent resistance with respect to the capacitor terminals.

Miller-Effect Capacitance

Figure 10–21 shows an amplifier having capacitance C_C connected between its input and output terminals. The most common example of such capacitance is interelectrode capacitance, as, for example, between the base and the collector of a common-emitter amplifier or between the gate and the drain of a common-source amplifier. This capacitance forms a *feedback* path for ac signals and it can have a significant influence on the high-frequency response of an amplifier.

The conclusion in the derivation that follows is based on the method used in circuit analysis to determine the total admittance (y) of components connected in parallel, which we will review briefly now. Recall that admittance is the sum of conductance g and susceptance b: $y = g + jb$ siemens. Also recall that the total admittance of parallel-connected components is the *sum* of their individual admittances. The susceptance of capacitance C is $b = j\omega C$ S. In particular, the total

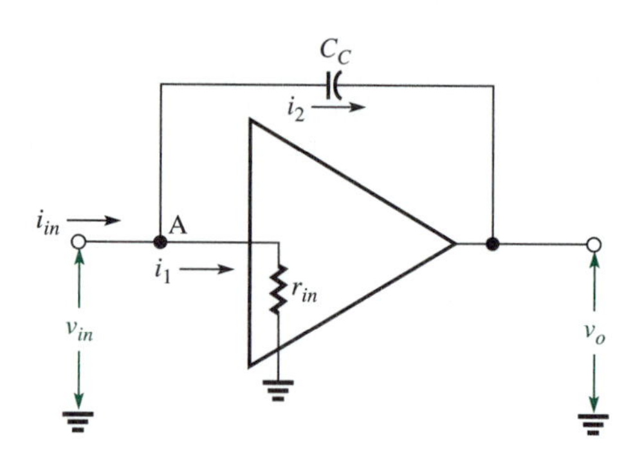

Figure 10–21
Capacitance C_C connected between the input and the output of an amplifier affects its high-frequency response due to the Miller effect.

admittance of resistance R in parallel with capacitance C at ω rad/s is

$$y = g + jb = 1/R + j\omega C \text{ siemens}$$

Thus, for example, if the *input admittance* to a circuit were 0.01 + j0.02 S at frequency ω = 1000 rad/s, we would conclude that the input circuit consisted of resistance $R = 1/g = 1/0.01 = 100\ \Omega$ in parallel with capacitance $C = b/\omega = 0.02/1000 = 20\ \mu F$.

Writing Kirchhoff's current law at the node labeled A in Figure 10–21, we have

$$i_{in} = i_1 + i_2 \qquad \textbf{(10–34)}$$

The current i_2 that flows in the capacitor is the difference in voltage across it divided by the capacitive reactance:

$$i_2 = \frac{v_{in} - v_o}{-jX_{C_C}} \qquad \textbf{(10–35)}$$

It is clear that the current i_1 flowing into the amplifier is

$$i_1 = \frac{v_{in}}{r_{in}} \qquad \textbf{(10–36)}$$

Substituting (10–35) and (10–36) into (10–34),

$$i_{in} = \frac{v_{in}}{r_{in}} + \frac{v_{in} - v_o}{-jX_{C_C}} \qquad \textbf{(10–37)}$$

Let the amplifier gain be $A_v = v_o/v_{in}$. Substituting $v_o = A_v v_{in}$ in (10–37) gives

$$i_{in} = \frac{v_{in}}{r_{in}} + \frac{v_{in} - A_v v_{in}}{-jX_{C_C}} = v_{in}\left(\frac{1}{r_{in}} + \frac{1 - A_v}{-jX_{C_C}}\right) = v_{in}\left[\frac{1}{r_{in}} + j\omega C_C(1 - A_v)\right]$$

Then

$$i_{in} = v_{in}(y_1 + y_2) \qquad \textbf{(10–38)}$$

where $y_1 = 1/r_{in}$ and $y_2 = \omega C_C(1 - A_v)$. Equation 10–38 shows that the input admittance consists of the conductance component $1/r_{in}$ and the capacitive susceptance component $\omega C_C(1 - A_v)$. Note that *this input admittance is exactly the same as it would be if a capacitance having value $C_C(1 - A_v)$ were connected between input and ground,* instead of capacitance C_C connected between input and output.

In other words, as far as the input signal is concerned, capacitance connected between input and output has the same effect as that capacitance "magnified" by the factor $(1 - A_v)$ and connected so that it shunts the input. This magnification of feedback capacitance, reflected to the input, is called the *Miller effect,* and the magnified value $C_C(1 - A_v)$ is called the *Miller capacitance,* C_M. Miller capacitance is relevant only for an *inverting* amplifier, so A_v is a negative number and the magnification factor $(1 - A_v)$ equals one *plus* the magnitude of A_v.

By a derivation similar to the foregoing, it can be shown that capacitance in the feedback path is also reflected to the output side of an amplifier. In this case, the effective shunt capacitance at the output is $(1 - 1/A_v)C_C$. Once again, A_v is negative, so the magnitude of the reflected capacitance is $(1 + 1/|A_v|)C_C$. Since the increase in capacitance is inversely proportional to gain, the effect is much less significant than that of the capacitance reflected to the input.

The computation of the value of the Miller capacitance, C_M, is complicated by the fact that the gain A_v itself depends on C_M. At high frequencies, the Miller capacitance reduces the gain, just as any other shunt capacitance does, and the gain reduction in turn reduces the Miller capacitance. As a first approximation, the midband gain can be used to compute C_M: $C_M \approx C_C(1 - A_m)$. This computation will always be conservative, in the sense that it will predict an upper cutoff frequency that is less than the actual value of f_2.

The total shunt capacitance at the input is the sum of the Miller capacitance and any other input-to-ground capacitance that may be present. Also, the total shunt capacitance at the output is the sum of the reflected capacitance and any other output-to-ground capacitance present. The effect of Miller capacitance is illustrated in the next example.

Example 10–9

The inverting amplifier shown in Figure 10–22 has midband gain $|v_L|/|v_{in}| = -200$. Find its upper cutoff frequency.

Figure 10–22
(Example 10–9)

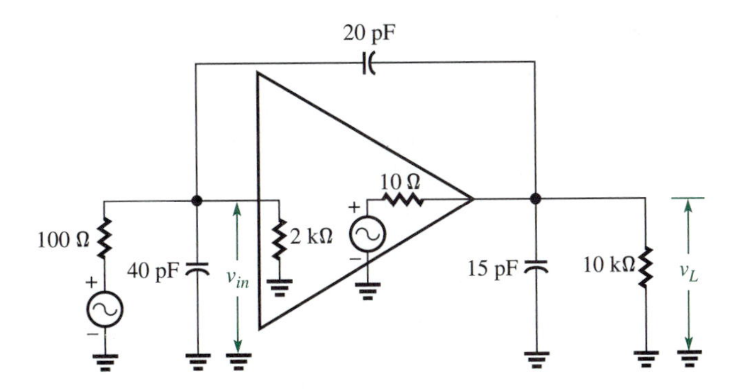

Solution. We will use the midband gain $|v_L|/|v_{in}|$ to determine the Miller capacitance. Notice that this is the gain *between the points where the 20-pF capacitance is connected.* The Miller capacitance is, therefore, $C_M \approx (20\text{ pF})[1 - (-200)] = 4020\text{ pF}$. The total capacitance C_A shunting the input is then $C_A = (40\text{ pF}) + (4020\text{ pF}) = 4060\text{ pF}$. The cutoff frequency due to C_A is

$$f_2(C_A) = \frac{1}{2\pi[(2 \times 10^3) \parallel 100]4060 \times 10^{-12}} = 411.6\text{ kHz}$$

The total capacitance shunting the output is $15\text{ pF} + (1 + 1/200)20\text{ pF} \approx 35\text{ pF}$.

The cutoff frequency due to the output capacitance is

$$f_2(C_B) = \frac{1}{2\pi[(10 \times 10^3) \parallel 10]35 \times 10^{-12}} = 455 \text{ MHz}$$

The upper cutoff frequency is the smaller of the two: $f_2 = 411.6$ kHz.

10–5 TRANSIENT RESPONSE

The *transient response* of an electronic amplifier or system is the output waveform that results when the input is a pulse or a sudden change in level. Since the transient response is a waveform, it is presented as a plot of voltage versus time, in contrast to frequency response, which is plotted versus frequency. Figure 10–23 shows a typical transient response.

The transient response of an amplifier is completely dependent on its frequency response, and vice versa. In other words, if two amplifiers have identical frequency responses, they will have identical transient responses, and vice versa. A pulse can be regarded as consisting of an infinite number of frequency components, so the transient waveform represents the amplifier's ability (or inability) to amplify all frequency components equally and to phase-shift all components equally. Theoretically, if an amplifier had infinite bandwidth, its transient response would be an exact duplicate of the input pulse. It is not possible for an amplifier to have infinite bandwidth, so the transient response is always a distorted version of the input pulse. It is necessary for an amplifier to have a wide bandwidth (to be a *wideband,* or *broadband,* amplifier) in order for it to amplify pulse or square-wave signals with a minimum of distortion. *Square-wave testing* is sometimes used to check the frequency response of an amplifier, as shown in Figure 10–24. The figure shows typical waveforms that result when a square wave is applied to an amplifier whose frequency response causes attenuation of either low- or high-frequency components. "Low" or "high" frequency in any given case means low or high in relation to the square-wave frequency, f_S.

As a concrete example of how frequency response is related to transient response, consider the low-pass RC network shown in Figure 10–25. The figure shows the output transient when the input is an abrupt change in level (called a *step* input), such as might occur when a dc voltage is switched into the network. Recall that the time constant, $\tau = RC$ seconds, is the time required for the transient output to reach 63.2% of its final value. Also recall that the upper cutoff frequency of the network is $f_2 = 1/(2\pi RC)$ Hz. Thus, $f_2 = 1/(2\pi\tau)$ Hz. Since the network passes all frequencies below f_2, down to dc, its bandwidth equals $f_2 - 0 = f_2$. Summarizing,

$$\text{BW} = f_2 = \frac{1}{2\pi\tau} \text{ Hz}$$

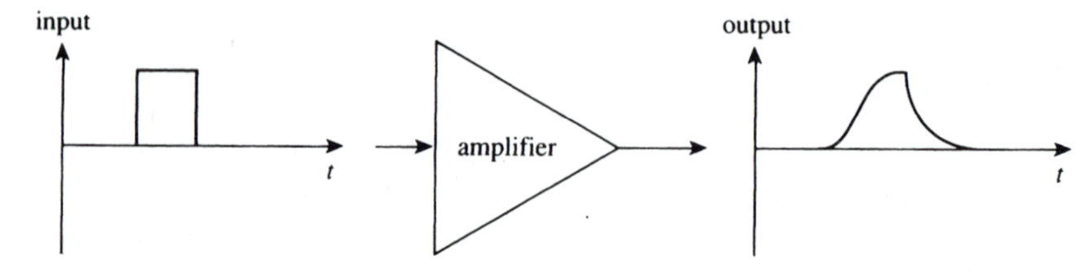

Figure 10–23
The output waveform is the transient response of the amplifier to a pulse-type input.

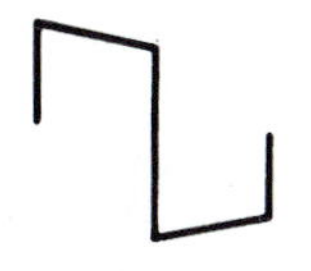
(a) Lower cutoff frequency = $0.1f_S$

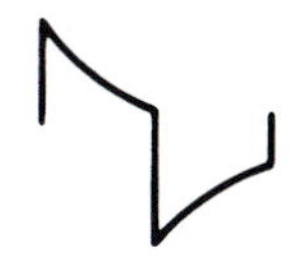
(b) Lower cutoff frequency = $0.5f_S$

(c) Upper cutoff frequency = f_S

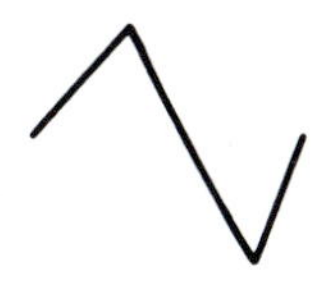
(d) Upper cutoff frequency = $0.5f_S$

Figure 10–24
Typical outputs resulting from square-wave testing of an amplifier. f_S = square-wave frequency. In (a) and (b), only low frequencies are attenuated, and in (c) and (d), only high frequencies are attenuated. In both cases, the attenuation outside cutoff is 20 dB/decade.

This equation shows that the cutoff frequency and the bandwidth, which are frequency-response characteristics, are *inversely* proportional to the time constant, which is a characteristic of the transient response. It is generally true for all electronic devices that the time required for the transient response to rise to a certain level is inversely proportional to bandwidth.

Recall that rise time, t_r, is the time required for a waveform to change from 10% of its final value to 90% of its final value. A widely used approximation that relates the rise time of the transient response of an amplifier to its bandwidth is

$$t_r \approx \frac{0.35}{\text{BW}} \text{ seconds} \tag{10–39}$$

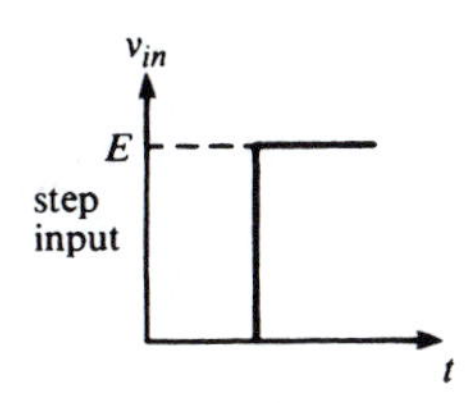

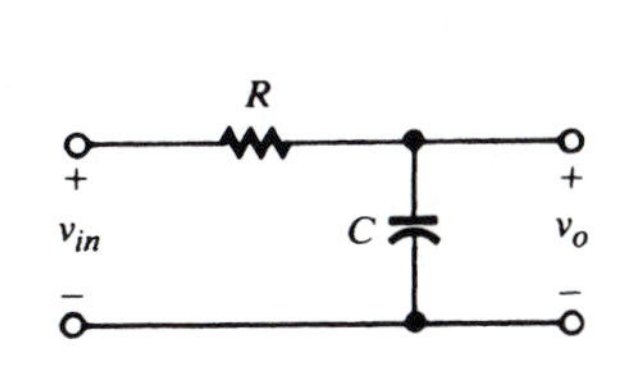

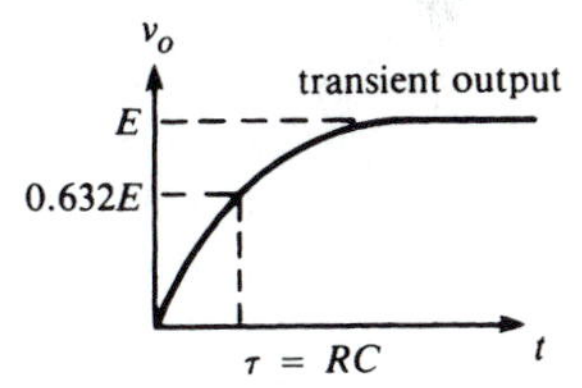

(a) Transient response to a step input

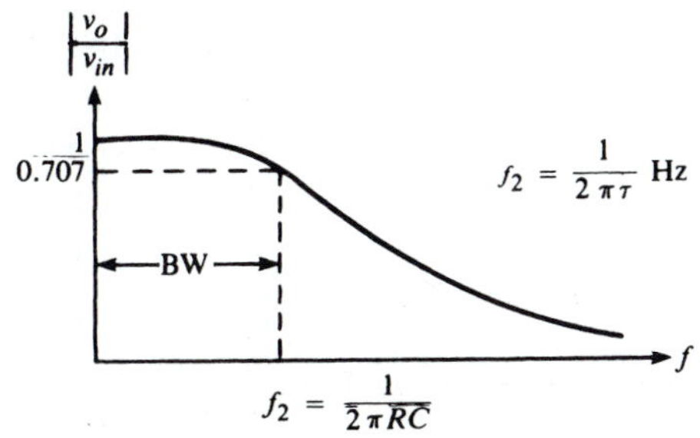

(b) Frequency response

Figure 10–25
Relationship between transient response and frequency response. Note that the bandwidth is inversely proportional to the time constant of the transient.

where BW is the bandwidth, in hertz. Equation 10–39 is used when the lower cutoff frequency is 0 (dc) or very small, so that the bandwidth is essentially the same as the upper cutoff frequency. The relationship is exact if the high-frequency response beyond cutoff is the same as that of the single RC low-pass network (Figure 10–25).

Example 10–10

The specifications for a certain oscilloscope state that the rise time of the vertical amplifier is 8.75 ns. What is the approximate bandwidth of the amplifier?

Solution. From equation 10–39,

$$\mathrm{BW} \approx \frac{0.35}{t_r} = \frac{0.35}{8.75 \times 10^{-9}} = 40 \text{ MHz}$$

10–6 FREQUENCY RESPONSE OF BJT AMPLIFIERS

Low-Frequency Response of BJT Amplifiers

We have learned that the lower cutoff frequency of an amplifier is approximately equal to the larger of $f_1(C_1)$ and $f_1(C_2)$, where

$$f_1(C_1) = \frac{1}{2\pi(r_S + r_{in})C_1} \text{ hertz} \qquad \textbf{(10–40)}$$

and

$$f_1(C_2) = \frac{1}{2\pi(r_o + R_L)C_2} \text{ hertz} \qquad \textbf{(10–41)}$$

In a BJT amplifier, the term r_{in} appearing in equation 10–40 is r_{in}(stage), the resistance seen by the source when looking into the amplifier. Its value depends on the transistor configuration, the bias resistors, and the values of the transistor parameters. Similarly, r_o in equation 10–41 is r_o(stage) and its value depends on particular amplifier characteristics. The next example illustrates the application of equations 10–40 and 10–41 to determine the lower cutoff frequency of a fixed-bias common-emitter amplifier.

Example 10–11

Find the lower cutoff frequency of the amplifier shown in Figure 10–26. Assume that the resistance looking into the base of the transistor is 1500 Ω and that the transistor output resistance at the collector, r_o, is 100 kΩ.

Solution. Since $r_{in}(\text{stage}) = (150 \text{ k}\Omega) \parallel (1.5 \text{ k}\Omega) \approx 1.5 \text{ k}\Omega$, we have, from equation 10–40,

$$f_1(C_1) = \frac{1}{2\pi(50 + 1500)(1 \times 10^{-6})} = 102.7 \text{ Hz}$$

Since $r_o(\text{stage}) = (100 \text{ k}\Omega) \parallel (1 \text{ k}\Omega) \approx 1 \text{ k}\Omega$, we have, from equation 10–41,

$$f_1(C_2) = \frac{1}{2\pi(1 \times 10^3 + 3 \times 10^3)(1 \times 10^{-6})} = 39.78 \text{ Hz}$$

Therefore, $f_1 \approx 103$ Hz.

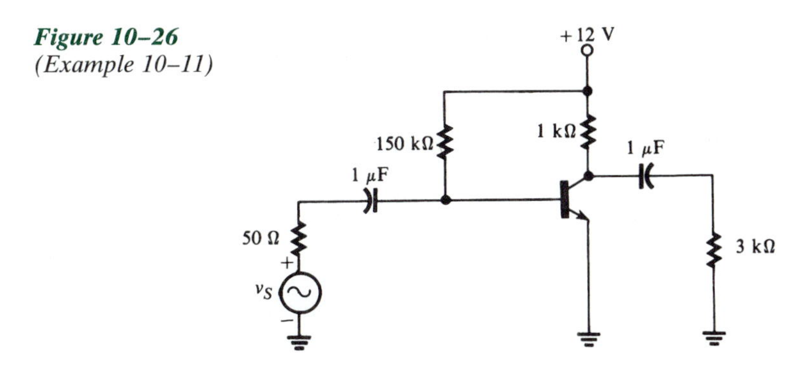

Figure 10–26
(Example 10–11)

Figure 10–27 shows a common-emitter amplifier having an emitter bypass capacitor C_E designed to eliminate signal degeneration, as discussed in Chapter 6. Recall that ac signal degeneration occurs in the absence of the bypass capacitor because of the ac voltage drop across R_E. At sufficiently high frequencies, the reactance of the capacitor is negligible, so the emitter is effectively at ac ground and there is no signal loss across R_E. At low frequencies, the reactance of C_E can become significant, to the extent that the voltage gain is reduced. Thus, we can define a lower cutoff frequency due to C_E, $f_1(C_E)$, that affects the amplifier's low-frequency response in the same way as the coupling capacitors, C_1 and C_2. If the coupling capacitors are large enough to have negligible effect, then the gain will fall to $\sqrt{2}/2$ times its midband value at the frequency where the reactance of C_E equals the resistance R_e looking into node A in Figure 10–27. R_e is the same as the output resistance of an emitter follower and was previously expressed as r_o(stage) (see equation 5–64):

$$R_e = R_E \left\| \left(\frac{r_S \| R_B}{\beta} + r_e \right) \right. \tag{10–42}$$

where $R_B = R_1 \| R_2$. Thus, $f_1(C_E)$ is the frequency at which $\frac{1}{[2\pi f_1(C_E)]C_E} = R_e$, or

$$f_1(C_E) = \frac{1}{2\pi R_e C_E} \tag{10–43}$$

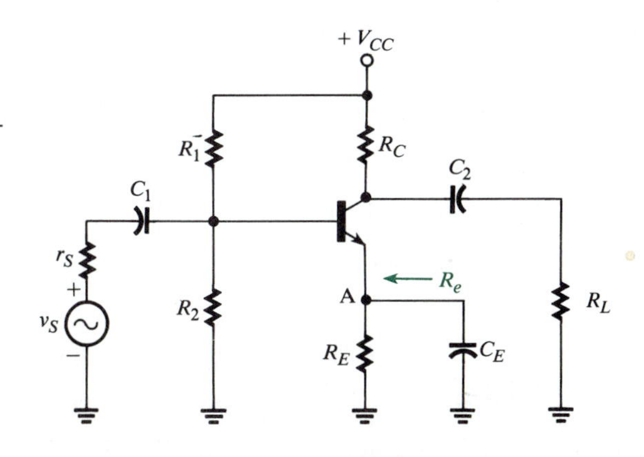

Figure 10–27
The lower cutoff frequency due to the bypass capacitor, $f_1(C_E)$, is the frequency where the reactance of C_E equals the resistance R_e.

We see that there are three possible break frequencies in the amplifier of Figure 10–27: $f_1(C_1)$, $f_1(C_2)$, and $f_1(C_E)$. If these are all distinct, and reasonably well removed from each other, the lower cutoff frequency will be the largest of the three. At frequencies below the smallest of the three, the gain will fall off at a rate of 60 dB/decade (18 dB/octave), because all three capacitors will be contributing simultaneously to gain reduction.

Example 10–12

Find the lower cutoff frequency of the amplifier shown in Figure 10–28. Assume that the transistor has $\beta = 100$, $r_e = 12\ \Omega$, and $r_o = 50\ \text{k}\Omega$.

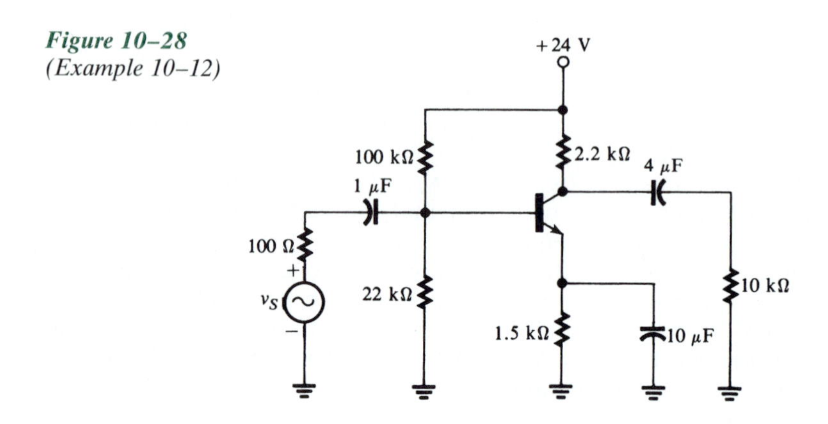

Figure 10–28
(Example 10–12)

Solution

$$r_{in}(\text{stage}) = R_1 \| R_2 \| \beta r_e = (100\ \text{k}\Omega) \| (22\ \text{k}\Omega) \| (100)(12\ \Omega) = 1.1\ \text{k}\Omega$$

$$f_1(C_1) = \frac{1}{2\pi(1.1 \times 10^3 + 100)(1 \times 10^{-6})} = 133\ \text{Hz}$$

$$r_o(\text{stage}) = (50\ \text{k}\Omega) \| (2.2\ \text{k}\Omega) = 2.1\ \text{k}\Omega$$

$$f_1(C_2) = \frac{1}{2\pi(2.1 \times 10^3 + 10 \times 10^3)(4 \times 10^{-6})} = 3.3\ \text{Hz}$$

From equation 10–42,

$$R_e = (1.5\ \text{k}\Omega) \left\| \left[\frac{(100\ \Omega) \| (100\ \text{k}\Omega) \| (22\ \text{k}\Omega)}{100} + (12\ \Omega) \right] \right.$$

$$\approx (1.5\ \text{k}\Omega) \| (13\ \Omega) \approx 13\ \Omega$$

From equation 10–43,

$$f_1(C_E) = \frac{1}{2\pi(13)(10 \times 10^{-6})} = 1.22\ \text{kHz}$$

The lower cutoff frequency is the largest of the three computed frequencies, so $f_1 = f_1(C_E) = 1.22\ \text{kHz}$.

This example demonstrates that the emitter bypass capacitor is generally the most troublesome, from the standpoint of achieving a low cutoff frequency, because of the small value of R_e. To obtain a value of $f_1(C_E)$ equal to the lowest audio frequency, 20 Hz, would require the impractically large value of $C_E = 2450\ \mu\text{F}$ in this example.

Example 10–13

SPICE

Use SPICE to obtain a plot of the frequency response of the amplifier in Example 10–12 over the frequency range from 1 Hz through 10 kHz.

Solution. Figure 10–29 shows the circuit of Figure 10–28 redrawn in the SPICE format as well as the input data file. Since the β of the transistor is 100, we can allow BETA in the .MODEL statement to have its default value (100). All other parameter values are allowed to default, since none have significant bearing on the low-frequency response. Note that the .AC statement specifies ten frequencies per decade for each of the four decades from 1 Hz through 10 kHz.

The log-log plot of the frequency response produced by SPICE is shown in Figure 10–30(a). Since we are allowing the ac source VIN to have its default value of 1 V, the output voltages are unrealistically large. Recall that SPICE does not consider practical voltage limitations and clipping when performing an ac analysis.

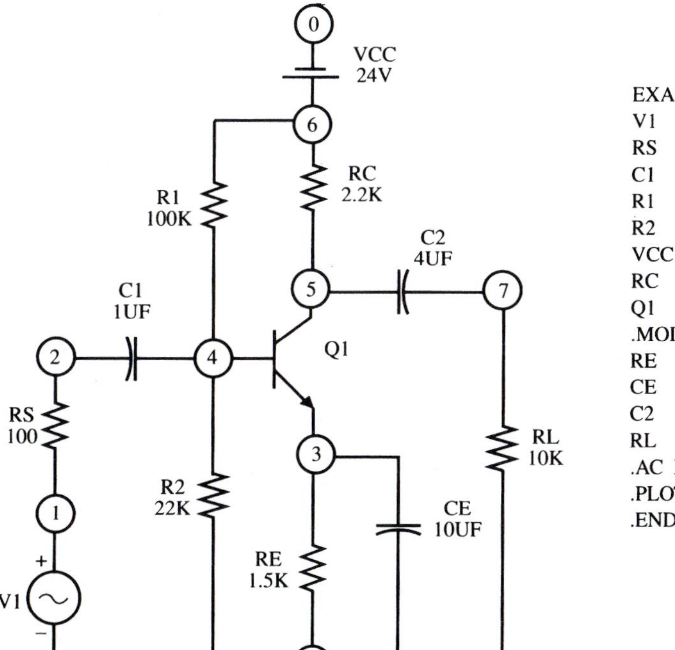


```
EXAMPLE 10.13
V1    1 0 AC
RS    1 2 100
C1    2 4 1UF
R1    6 4 100K
R2    4 0 22K
VCC 6 0 24V
RC    5 6 2.2K
Q1    5 4 3 TRAN
.MODEL TRAN NPN
RE    3 0 1.5K
CE    3 0 10UF
C2    5 7 4UF
RL    7 0 10K
.AC DEC 10 1HZ 10KHZ
.PLOT AC V(7)
.END
```

Figure 10–29
(Example 10–13)

```
EXAMPLE 10.13
****      AC ANALYSIS                              TEMPERATURE =   27.000 DEG C
*********************************************************************************
   FREQ       V(7)
                 1.000D-02      1.000D+00      1.000D+02      1.000D+04   1.000D+06
                      - - - - - - - - - - - - - - - - - - - - - - - - - - - - - -
1.000D+00  3.501D-02  .   *         .             .              .             .
1.259D+00  5.409D-02  .    *        .             .              .             .
1.585D+00  8.251D-02  .     *       .             .              .             .
1.995D+00  1.237D-01  .       *     .             .              .             .
2.512D+00  1.814D-01  .        *    .             .              .             .
3.162D+00  2.591D-01  .         *   .             .              .             .
3.981D+00  3.595D-01  .          *  .             .              .             .
5.012D+00  4.848D-01  .           * .             .              .             .
6.310D+00  6.375D-01  .            *.             .              .             .
7.943D+00  8.210D-01  .            *.             .              .             .
1.000D+01  1.042D+00  .             *             .              .             .
1.259D+01  1.310D+00  .             .*            .              .             .
1.585D+01  1.640D+00  .             . *           .              .             .
1.995D+01  2.052D+00  .             . *           .              .             .
2.512D+01  2.568D+00  .             .  *          .              .             .
3.162D+01  3.218D+00  .             .   *         .              .             .
3.981D+01  4.037D+00  .             .   *         .              .             .
5.012D+01  5.070D+00  .             .    *        .              .             .
6.310D+01  6.371D+00  .             .     *       .              .             .
7.943D+01  8.007D+00  .             .     *       .              .             .
1.000D+02  1.006D+01  .             .      *      .              .             .
1.259D+02  1.264D+01  .             .       *     .              .             .
1.585D+02  1.587D+01  .             .       *     .              .             .
1.995D+02  1.990D+01  .             .        *    .              .             .
2.512D+02  2.489D+01  .             .         *   .              .             .
3.162D+02  3.102D+01  .             .         *   .              .             .
3.981D+02  3.845D+01  .             .          *  .              .             .
5.012D+02  4.727D+01  .             .           * .              .             .
6.310D+02  5.745D+01  .             .           * .              .             .
7.943D+02  6.872D+01  .             .            *.              .             .
1.000D+03  8.053D+01  .             .            *.              .             .
1.259D+03  9.209D+01  .             .             *              .             .
1.585D+03  1.026D+02  .             .             *              .             .
1.995D+03  1.114D+02  .             .             *              .             .
2.512D+03  1.183D+02  .             .             .*             .             .
3.162D+03  1.234D+02  .             .             .*             .             .
3.981D+03  1.270D+02  .             .             .*             .             .
5.012D+03  1.294D+02  .             .             .*             .             .
6.310D+03  1.310D+02  .             .             .*             .             .
7.943D+03  1.320D+02  .             .             .*             .             .
1.000D+04  1.327D+02  .             .             .*             .             .
                      - - - - - - - - - - - - - - - - - - - - - - - - - - - - - -
```

(a)

Figure 10–30
(Example 10–13)

The results show that the midband gain is approximately 132.7 (at 10 kHz, about one decade above the lower cutoff frequency; the actual gain may be slightly greater at higher frequencies). Thus, the gain will be approximately 0.707(132.7) = 93.8 at the lower cutoff frequency. The frequency at which the output is nearest 93.8 is 1.259 kHz (92.09). A more accurate estimate of the lower cutoff frequency can be made by restricting the range and increasing the number of frequencies at which the output is calculated in the vicinity of cutoff. Figure 10–30(b) shows the results of a .PRINT statement when the .AC statement is changed to LIN 20 1KHZ 1.5KHZ, producing output at 20 linearly spaced frequencies between 1 kHz and

```
EXAMPLE 10.13
****      AC ANALYSIS                          TEMPERATURE =    27.000 DEG C
******************************************************************************
  FREQ          V(7)
 1.000E+03     8.053E+01
 1.026E+03     8.186E+01
 1.053E+03     8.315E+01
 1.079E+03     8.441E+01
 1.105E+03     8.563E+01
 1.132E+03     8.682E+01
 1.158E+03     8.797E+01
 1.184E+03     8.909E+01
 1.211E+03     9.018E+01
 1.237E+03     9.123E+01
 1.263E+03     9.226E+01
 1.289E+03     9.325E+01
 1.316E+03     9.422E+01
 1.342E+03     9.515E+01
 1.368E+03     9.606E+01
 1.395E+03     9.694E+01
 1.421E+03     9.780E+01
 1.447E+03     9.863E+01
 1.474E+03     9.944E+01
 1.500E+03     1.002E+02
```

(b)

Figure 10–30
(Continued)

1.5 kHz. We see that the lower cutoff frequency is between 1.289 kHz and 1.316 kHz. These results agree well with the value calculated in Example 10–12.

The principal consideration in the design and analysis of amplifiers in the low-frequency region is the (Thevenin equivalent) resistance in series with each capacitor used in the circuit. We have seen that a break frequency occurs whenever the frequency becomes low enough to make the capacitive reactance equal to the equivalent resistance at the point of connection. Therefore, in any BJT amplifier, the capacitor that is most critical in determining the lower cutoff frequency is the one that "sees" the smallest resistance. In the case of the common-emitter amplifier, C_E is that capacitor. In a common-base amplifier, the input resistance ($r_{in} \approx r_e$) is quite small, so the input coupling capacitor, C_1, is the most crucial. In the common-collector amplifier, the output resistance is quite small, so the output coupling capacitor, C_2, is crucial. Of course, a large source resistance in the case of a CB amplifier, or a large load resistance in the case of a CC amplifier, will mitigate these circumstances, since each is in series with the affected capacitor.

Figure 10–31 shows common-base and common-collector amplifiers and gives the equations for the break frequencies due to each coupling capacitor. These are derived in a straightforward manner by solving for the frequency at which the capacitive reactance of each capacitor equals the Thevenin equivalent resistance in series with it.

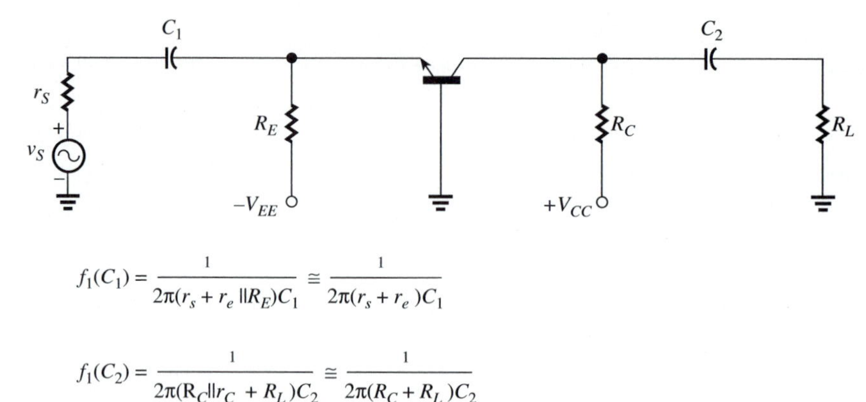

$$f_1(C_1) = \frac{1}{2\pi(r_s + r_e \| R_E)C_1} \cong \frac{1}{2\pi(r_s + r_e)C_1}$$

$$f_1(C_2) = \frac{1}{2\pi(R_C \| r_C + R_L)C_2} \cong \frac{1}{2\pi(R_C + R_L)C_2}$$

(a) Common base

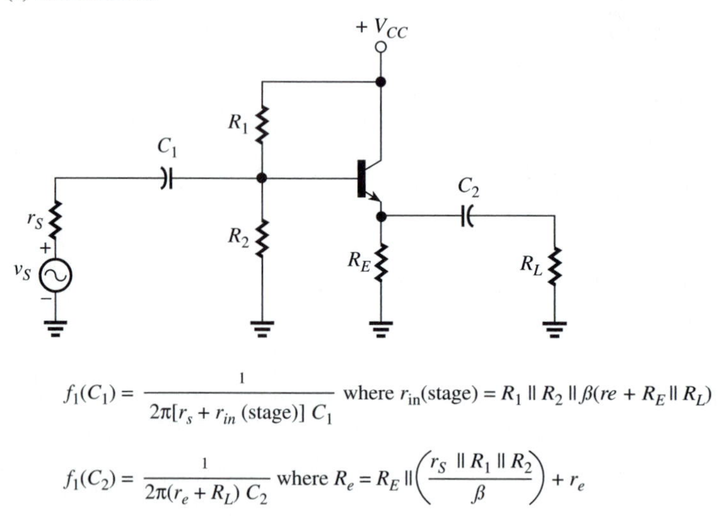

$$f_1(C_1) = \frac{1}{2\pi[r_s + r_{in}\,(\text{stage})]\,C_1} \quad \text{where } r_{\text{in}}(\text{stage}) = R_1 \| R_2 \| \beta(re + R_E \| R_L)$$

$$f_1(C_2) = \frac{1}{2\pi(r_e + R_L)\,C_2} \quad \text{where } R_e = R_E \| \left(\frac{r_S \| R_1 \| R_2}{\beta}\right) + r_e$$

(b) Common collector (emitter follower)

Figure 10–31
CB and CC amplifiers and the equations for $f_1(C_1)$ and $f_1(C_2)$ for each

Example 10–14

A certain transistor has $\beta = 100$, $r_C = 100\ \text{k}\Omega$, and $r_e = 25\ \Omega$. The transistor is used in each of the circuits shown in Figure 10–31. Find the approximate lower cutoff frequency in each circuit, given the following component values:

1. (Common-base circuit)

$r_S = 100\ \Omega \qquad R_C = 1\ \text{k}\Omega \qquad C_1 = 2.2\ \mu\text{F}$

$R_E = 10\ \text{k}\Omega \qquad R_L = 15\ \text{k}\Omega \qquad C_2 = 1\ \mu\text{F}$

2. (Common-collector circuit)

$r_S = 100\ \Omega \qquad R_E = 1\ \text{k}\Omega \qquad C_1 = 1\ \mu\text{F}$

$R_1 = 33\ \text{k}\Omega \qquad R_L = 50\ \Omega \qquad C_2 = 10\ \mu\text{F}$

$R_2 = 10\ \text{k}\Omega$

Solution

1. $$f_1(C_1) = \frac{1}{2\pi(100 + 25 \| 10 \times 10^3)(2.2 \times 10^{-6})} = 579 \text{ Hz}$$

$$f_1(C_2) = \frac{1}{2\pi(1 \times 10^3 \| 100 \times 10^3 + 15 \times 10^3)(1 \times 10^{-6})} = 9.9 \text{ Hz}$$

Therefore, $f_1 = f_1(C_1) = 579$ Hz.

2. $$r_{in}(\text{stage}) = (33 \text{ k}\Omega) \| (10 \text{ k}\Omega) \| (100[(25\ \Omega) + (1 \text{ k}\Omega) \| (50 \text{ k}\Omega)] = (7.67 \text{ k}\Omega) \| (7.26 \text{ k}\Omega) = 3.73 \text{ k}\Omega$$

$$f_1(C_1) = \frac{1}{2\pi(100 + 3.73 \times 10^3)(1 \times 10^{-6})} = 41.6 \text{ Hz}$$

$$R_e = (1 \text{ k}\Omega) \left\| \left[\frac{(100\ \Omega) \| (33 \text{ k}\Omega) \| (10 \text{ k}\Omega)}{100} + (25\ \Omega)\right] \approx 25\ \Omega\right.$$

$$f_1(C_2) = \frac{1}{2\pi(25 + 50)(10 \times 10^{-6})} = 212.2 \text{ Hz}$$

Therefore, $f_1 \approx f_1(C_2) = 212.2$ Hz.

Design Considerations

Capacitors in practical discrete circuits are generally bulky and costly, or else they are unreliable if quality is sacrificed for cost. Therefore, a principal objective in the design of discrete BJT amplifiers is selection of the smallest capacitors possible, consistent with the low-frequency response desired. Equations 10–44 through 10–46 can be used to find minimum capacitor values for CE, CB, and CC amplifiers in terms of break frequencies and circuit parameters. As noted earlier, each configuration has one capacitor whose value is the most critical: The one connected to the point where the equivalent resistance is smallest. This capacitor should be selected first to ensure that the lower cutoff frequency is at least as low as the break frequency it produces. The other capacitor(s) can then be selected to break frequencies a decade or so below cutoff. (With two identical break frequencies occurring 1 decade below the third, the gain is actually reduced by a factor of 0.7, rather than 0.707, at the third frequency.)

Capacitor values necessary to achieve specified break frequencies in a common emitter (CE) amplifier

$$C_E{}^* = \frac{1}{2\pi R_e f_1(C_E)} \tag{10–44}$$

where $R_e = R_E \left\| \left(\frac{r_S \| R_1 \| R_2}{\beta} + r_e\right)\right.$.

$$C_1 = \frac{1}{2\pi[r_{in}(\text{stage}) + r_S]f_1(C_1)}$$

where $r_{in}(\text{stage}) = R_1 \,\|\, R_2 \,\|\, \beta r_e$.

$$C_2 = \frac{1}{2\pi[r_o(\text{stage}) + R_L]\, f_1(C_2)}$$

where $r_o(\text{stage}) = r_o \,\|\, R_C$.

* Select first.

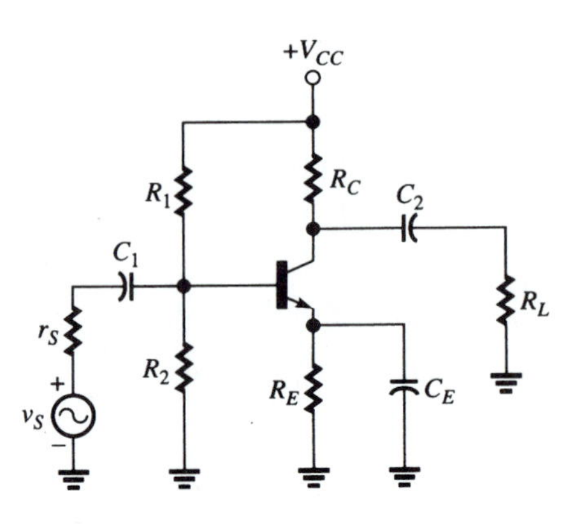

Capacitor values necessary to achieve specified break frequencies in a common base (CB) amplifier

$$C_1{*} = \frac{1}{2\pi(r_S + r_e \,\|\, R_E)\, f_1(C_1)}$$

$$C_2 = \frac{1}{2\pi(R_C \,\|\, r_C + R_L)\, f_1(C_2)}$$

* Select first. **(10–45)**

Capacitor values necessary to achieve specified break frequencies in a common collector (CC) amplifier

$$C_2{}^* = \frac{1}{2\pi(R_e + R_L)f_1(C_2)}$$

where $R_e = R_E \left\| \left(\frac{r_S \| R_1 \| R_2}{\beta} + r_e \right) \right.$.

$$C_1 = \frac{1}{2\pi[r_{in}(\text{stage}) + r_S]f_1(C_1)}$$

where $r_{in}(\text{stage}) = R_1 \| R_2 \| \beta(r_e + R_E \| R_L)$.

* Select first. **(10–46)**

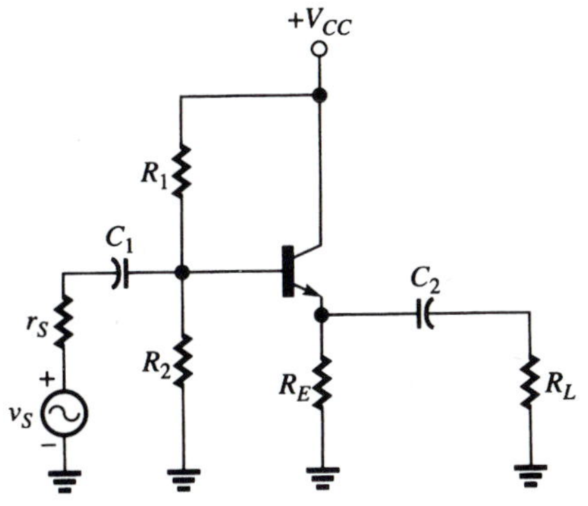

Example 10–15

DESIGN

A CE amplifier having $R_1 = 330\ \text{k}\Omega$, $R_2 = 47\ \text{k}\Omega$, $R_C = 3.3\ \text{k}\Omega$, and $R_E = 1.8\ \text{k}\Omega$ is to have a lower cutoff frequency of 50 Hz. The amplifier drives a 10-kΩ load and is driven from a signal source whose resistance is 600 Ω. The transistor has $\beta = 90$ and $r_o = 100\ \text{k}\Omega$ and is biased at $I_E = 0.5$ mA. Find coupling and bypass capacitors necessary to meet the low-frequency response requirement. Assume that capacitors must be selected from a line having standard values of 1 μF, 1.5 μF, 2.2 μF, 3.3 μF, 4.7 μF, 10 μF, 50 μF, 75 μF, and 100 μF.

Solution. To obtain a value for C_E, we must first find the value of r_e:

$$r_e \approx \frac{0.026}{I_E} = \frac{0.026}{0.5\ \text{mA}} = 52\ \Omega$$

Then, from equations 10–44,

$$R_e = 1.8\ \text{k}\Omega \left\| \left[\frac{(600\ \Omega) \| (330\ \text{k}\Omega) \| (47\ \text{k}\Omega)}{90} + 52\ \Omega \right] \right. = 58.6\ \Omega$$

and

$$C_E = \frac{1}{2\pi(58.6\ \Omega)(50\ \text{Hz})} = 54.36\ \mu\text{F}$$

To ensure that the lower cutoff frequency is no greater than 50 Hz, we choose the standard-value capacitor with the next *higher* capacitance, rather than the one closest to 54.36 μF. Thus, we choose $C_E = 75\ \mu$F.

To find C_1 and C_2, we let $f_1(C_1) = f_1(C_2) = 50$ Hz/10 $= 5$ Hz and use equations 10–44 to calculate

$$r_{in}(\text{stage}) = 330\ \text{k}\Omega \,\|\, 47\ \text{k}\Omega \,\|\, (90)(52\ \Omega) = 4.2\ \text{k}\Omega$$

$$C_1 = \frac{1}{2\pi(4.2\ \text{k}\Omega + 600\ \Omega)5\ \text{Hz}} = 6.63\ \mu\text{F}$$

$$r_o(\text{stage}) = 100\ \text{k}\Omega \,\|\, 3.3\ \text{k}\Omega = 3.19\ \text{k}\Omega$$

$$C_2 = \frac{1}{2\pi(3.19\ \text{k}\Omega + 10\ \text{k}\Omega)5\ \text{Hz}} = 2.41\ \mu\text{F}$$

Again choosing standard capacitors with the next highest values, we select $C_1 = 10\ \mu$F and $C_2 = 3.3\ \mu$F.

It is a SPICE programming exercise at the end of this chapter to find the actual reduction in gain at 50 Hz when these standard-value capacitors are used. Our conservative choices result in an actual lower cutoff frequency of about 40 Hz.

High-Frequency Response of BJT Amplifiers

Figure 10–32 shows a common-emitter amplifier having interelectrode capacitance designated C_{be}, C_{bc}, and C_{ce}. Since we are now considering high-frequency performance, the emitter bypass capacitor effectively shorts the emitter terminal to ground, so C_{be} and C_{ce} are input-to-ground and output-to-ground capacities, respectively. We can apply the general equations developed earlier to determine the upper cutoff frequency due to the interelectrode capacitances:

$$f_2(C_A) = \frac{1}{2\pi(r_S \,\|\, r_{in})C_A} \qquad \textbf{(10–47)}$$

where $C_A = C_{be} + C_M = C_{be} + C_{bc}(1 - A_v)$ and $r_{in} = r_{in}(\text{stage}) = R_1 \,\|\, R_2 \,\|\, (\beta r_e)$. Note that the Miller-effect capacitance, C_M, is due to C_{bc} because the latter is

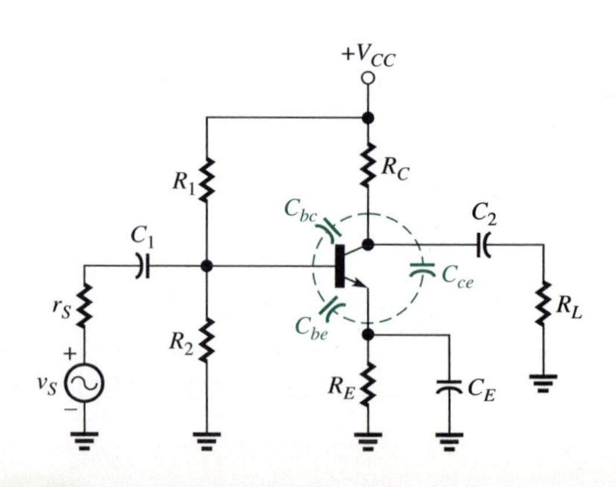

Figure 10–32
A common-emitter amplifier showing the interelectrode capacitances of the transistor

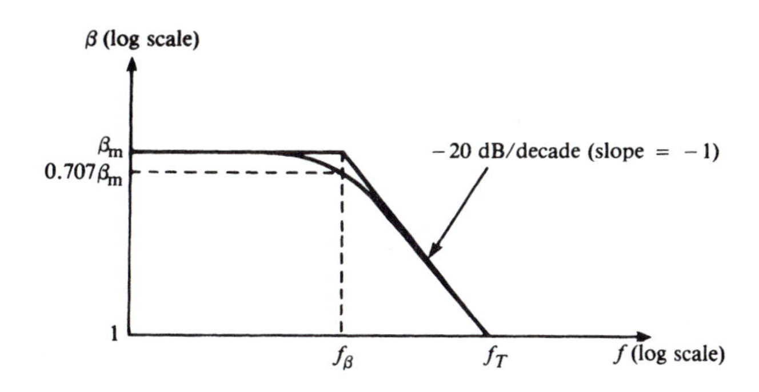

Figure 10–33
The β of a transistor becomes smaller at high frequencies. At frequencies above f_β, its value falls off at the rate of 20 dB/decade, and reaches value 1 at the frequency designated f_T.

connected between the input (base) and output (collector) of the common-emitter configuration. The cutoff frequency due to the output shunt capacitance is

$$f_2(C_B) = \frac{1}{2\pi(r_o \| R_L)C_B} \qquad \textbf{(10–48)}$$

where $C_B = C_{ce} + C_{bc}(1 - 1/A_v)$ and $r_o = r_o(\text{stage}) = r_C \| R_C$. Any wiring or stray capacitance that shunts the input or output must be added to C_A or C_B.

As mentioned in an earlier discussion, the high-frequency performance of a transistor is affected by certain internal characteristics that may or may not limit the upper cutoff to a frequency less than that predicted by equations 10–47 or 10–48. The β of a BJT is frequency dependent and its value begins to drop at the rate of 20 dB/decade at frequencies beyond a certain frequency called the *beta cutoff frequency*, f_β. By definition, f_β is the frequency at which the β of a transistor is $\sqrt{2}/2$ times its low-frequency value. Figure 10–33 shows a plot of β versus frequency as it would appear on log-log paper. The low- (or mid-) frequency value of β is designated β_m. If f_β is significantly lower than $f_2(C_A)$ and $f_2(C_B)$, then the upper cutoff frequency of an amplifier in the common-emitter configuration will be determined by f_β.

If the frequency is increased beyond f_β, β continues to decrease until it eventually reaches a value of 1. As shown in Figure 10–33, the frequency at which $\beta = 1$ is designated f_T. Note that the asymptote's slope of -20 dB/decade is the same as a slope of -1 on log-log scales, since 20 dB is a ten-to-one change along the vertical axis and a decade is a ten-to-one change along the horizontal axis. Therefore, in Figure 10–33, the difference between the logarithms of β_m and 1 must be the same as the difference between the logarithms of f_T and f_β:

$$\log \beta_m - \log(1) = \log f_T - \log f_\beta$$
$$\log(\beta_m/1) = \log(f_T/f_\beta)$$

Taking the antilog of both sides leads to the important result

$$f_\beta = \frac{f_T}{\beta_m} \qquad \textbf{(10–49)}$$

Since β decreases with frequency, we would expect α to do the same. This is, in fact, the case, and the frequency at which α falls to $\sqrt{2}/2$ times its low-frequency value is called the *alpha cutoff frequency*, f_α. Because f_α is generally much larger than f_β, the common-base amplifier is not as frequency limited by parameter cutoff as is the common-emitter amplifier. A commonly used approximation is

$$f_\alpha \approx f_T \qquad \textbf{(10–50)}$$

In specification sheets, f_T is often called the *gain-bandwidth product.* We will discuss this interpretation of a unity-gain frequency in Chapter 13. In terms of internal junction capacitances, f_T can be found from

$$f_T = \frac{1}{2\pi r_e(C_\pi + C_\mu)}$$

where C_π is the base-to-emitter junction capacitance and C_μ is the base-to-collector junction capacitance. Recall that junction capacitance depends on the width of the depletion region, which in turn depends on the bias voltage. Since r_e also depends on bias conditions, f_T is clearly a function of bias and is usually so indicated on specification sheets.

Example 10–16

The transistor in Figure 10–34 has a low-frequency β of 120, $r_e = 20\ \Omega$, $r_o = 100$ kΩ, and $f_T = 100$ MHz. The interelectrode capacitances are $C_{be} = 40$ pF, $C_{bc} = 1.5$ pF, and $C_{ce} = 5$ pF. There is wiring capacitance equal to 4 pF across the input and 8 pF across the output. Find the approximate upper cutoff frequency.

Solution. To determine the Miller-effect capacitance, we must find the midband voltage gain of the transistor. Note that the *overall* voltage gain v_L/v_S is *not* used in this computation, because the gain reduction due to voltage division at the input does not affect the Miller capacitance. (It would, if the feedback capacitance were connected all the way back to v_S.)

$$A_v \approx \frac{-r_L}{r_e} = \frac{-(4\ \text{k}\Omega) \parallel (20\ \text{k}\Omega)}{20\ \Omega} = -166.67$$

We use the midband gain as a conservative approximation in determining C_M:

$$C_M \approx C_{bc}(1 - A_v) = (1.5\text{ pF})[1 - (-166.67)] = 251.5\text{ pF}$$

The total capacitance shunting the input is, therefore,

$$C_A = C_{be} + C_M + C_{wiring} = (40\text{ pF}) + (251.5\text{ pF}) + (4\text{ pF}) = 295.5\text{ pF}$$

Then, since

$$r_{in}(\text{stage}) = R_1 \parallel R_2 \parallel \beta r_e = (100\ \text{k}\Omega) \parallel (22\ \text{k}\Omega) \parallel (2.4\ \text{k}\Omega) = 2.1\ \text{k}\Omega$$

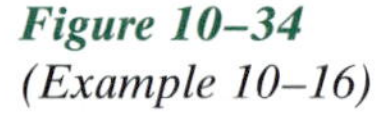
Figure 10–34
(Example 10–16)

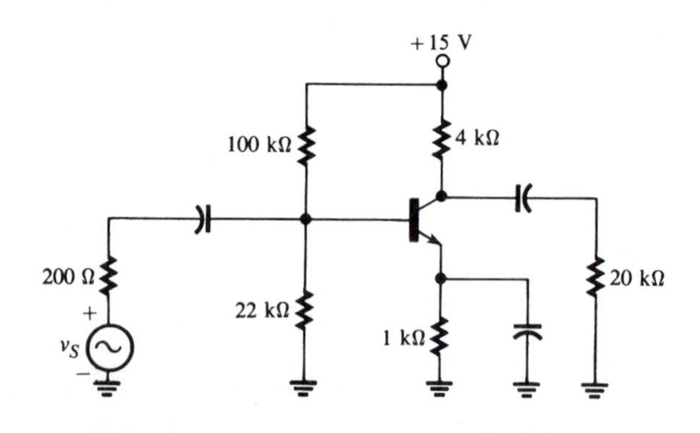

we have, from equation 10–47,

$$f_2(C_A) = \frac{1}{2\pi(200 \,\|\, 2.1 \times 10^3)295.5 \times 10^{-12}} = 2.95 \text{ MHz}$$

Since $\qquad r_o(\text{stage}) = r_o \,\|\, R_C = (100 \text{ k}\Omega) \,\|\, (4 \text{ k}\Omega) = 3.8 \text{ k}\Omega$

and $\qquad C_B = C_{ce} + C_{bc}(1 - 1/A_v) + C_{wiring}$

$$= (5 \text{ pF}) + (1.5 \text{ pF})(1 + 1/167) + (8 \text{ pF}) = 14.5 \text{ pF}$$

we have, from equation 10–48,

$$f_2(C_B) = \frac{1}{2\pi(3.8 \times 10^3 \,\|\, 20 \times 10^3)(14.5 \times 10^{-12})} = 3.44 \text{ MHz}$$

From equation 10–49,

$$f_\beta = \frac{f_T}{\beta_m} = \frac{100 \text{ MHz}}{120} = 833 \text{ kHz}$$

The upper cutoff frequency is the smallest of the three computed frequencies, in this case, f_β: $f_2 = f_\beta = 833$ kHz.

Since the common-base amplifier is noninverting, the shunt capacitance at its input does not have a Miller-effect component. Therefore, the CB amplifier is not as severely limited as its CE counterpart in terms of the upper cutoff frequency $f_2(C_A)$, and CB amplifiers are often used in high-frequency applications. We will discuss one such application, the *cascode* amplifier, in Chapter 11. We have also noted that f_α is generally much higher than f_β. The common-collector amplifier (emitter follower) is also noninverting and unaffected by Miller capacitance. It, too, is generally superior to the CE configuration for high-frequency applications.

10–7 FREQUENCY RESPONSE OF FET AMPLIFIERS

Low-Frequency Response of FET Amplifiers

Figure 10–35 shows a common-source JFET amplifier biased using the combination of self-bias and a voltage divider, as studied in Chapter 7. For our purposes now, the FET could also be a MOSFET. The capacitors that affect the low-frequency

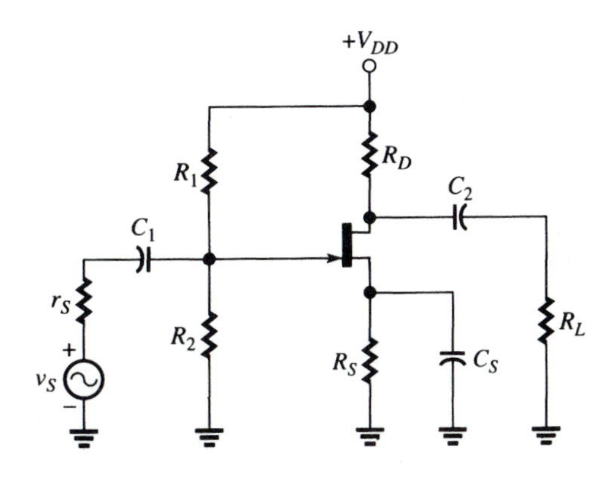

Figure 10–35
The low-frequency response of an FET amplifier is affected by capacitors C_1, C_2, and C_S.

response are the input coupling capacitor, C_1, the output coupling capacitor, C_2, and the source bypass capacitor, C_S. Applying equations 10–40 and 10–41 to this amplifier configuration, we find

$$f_1(C_1) = \frac{1}{2\pi(r_S + r_{in})C_1} \tag{10–51}$$

where $r_{in} = r_{in}(\text{stage}) \approx R_1 \parallel R_2$, and

$$f_1(C_2) = \frac{1}{2\pi(r_o + R_L)C_2} \tag{10–52}$$

where $r_o = r_o(\text{stage}) = r_d \parallel R_D$. The bypass capacitor C_S affects low-frequency response because at low frequencies its reactance is no longer small enough to eliminate degeneration. The cutoff frequency due to C_S is the frequency at which the reactance of C_S equals the resistance looking into the junction of R_S and C_S at the source terminal, namely, $R_S \parallel (1/g_m)$. Thus,

$$f_1(C_S) = \frac{1}{2\pi[R_S \parallel (1/g_m)]C_S} \tag{10–53}$$

Since the bias resistors R_1 and R_2 are usually very large in an FET amplifier, the input coupling capacitor C_1 can be smaller than its counterpart in a BJT CE amplifier, and C_1 does not generally determine the low-frequency cutoff. The small value of $R_S \parallel (1/g_m)$ makes the source bypass capacitor the most troublesome, in that a large amount of capacitance is required to achieve a small value of $f_1(C_S)$.

Example 10–17

The FET shown in Figure 10–36 has $g_m = 3.4$ mS and $r_d = 100$ kΩ. Find the approximate lower cutoff frequency.

Figure 10–36
(Example 10–17)

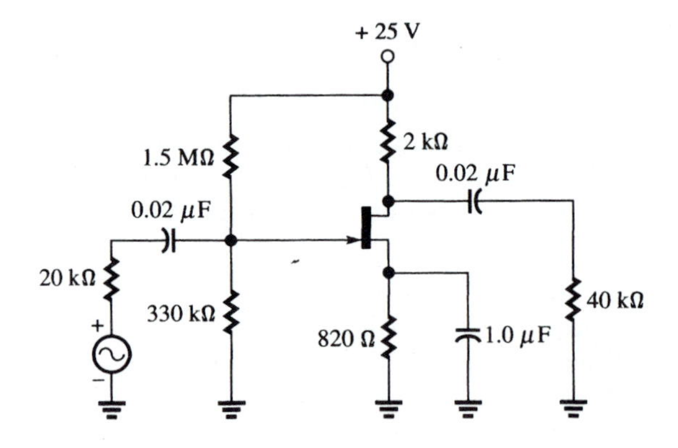

Solution

$$r_{in}(\text{stage}) \approx (1.5\ \text{M}\Omega) \parallel (330\ \text{k}\Omega) = 270\ \text{k}\Omega$$

$$f_1(C_1) = \frac{1}{2\pi(20 \times 10^3 + 270 \times 10^3)(0.02 \times 10^{-6})} = 27.4\ \text{Hz}$$

$$r_o(\text{stage}) = (100\ \text{k}\Omega) \parallel (2\ \text{k}\Omega) = 1.96\ \text{k}\Omega$$

$$f_1(C_2) = \frac{1}{2\pi(1.96 \times 10^3 + 40 \times 10^3)(0.02 \times 10^{-6})} = 189.6 \text{ Hz}$$

$$R_S \| (1/g_m) = 820 \| 294.1 = 216.5\ \Omega$$

$$f_1(C_S) = \frac{1}{2\pi(216.5)(1 \times 10^{-6})} = 735.1 \text{ Hz}$$

The actual lower cutoff frequency is approximately equal to the largest of $f_1(C_1)$, $f_1(C_2)$, and $f_1(C_S)$, namely, $f_1 \approx f_1(C_S) = 735.1$ Hz.

Figure 10–37 shows JFET and MOSFET amplifiers in the common-drain (source-follower) configuration and gives the equations for $f_1(C_1)$ and $f_1(C_2)$. These equations are obtained in a straightforward manner using the same low-frequency–cutoff theory we have discussed for general amplifiers.

For design purposes, each of the equations for $f_1(C_1)$, $f_1(C_2)$, or $f_1(C_S)$ in each of the three configurations is easily solved for C_1, C_2, or C_S, giving a set of capacitor equations similar to equations 10–44 through 10–46.

High-Frequency Response of FET Amplifiers

Like the BJT common-emitter amplifier, the FET common-source amplifier is affected by Miller capacitance that often determines the upper cutoff frequency. Figure 10–38 shows a common-source amplifier and the interelectrode capacitance that affects its high-frequency performance. Since R_S is completely bypassed at high frequencies, C_{gs} and C_{ds} are effectively shunting the input and output, respectively, to ground. The equations for $f_2(C_A)$ and $f_2(C_B)$ are obtained directly from the general high-frequency–cutoff theory we have studied:

$$f_2(C_A) = \frac{1}{2\pi(r_S \| r_{in})C_A} \quad \textbf{(10–54)}$$

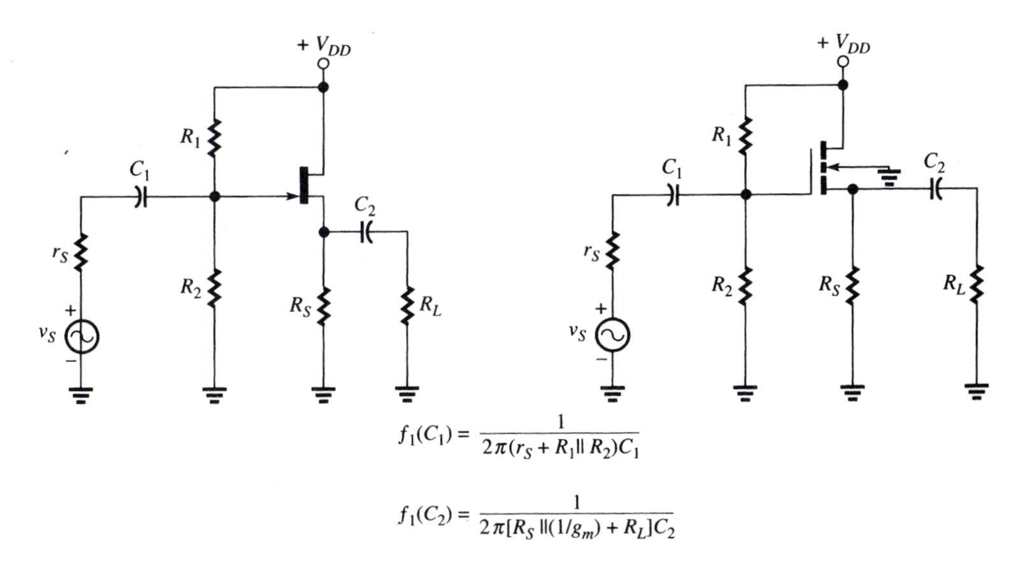

Figure 10–37
Low-frequency–cutoff equations for the common-drain amplifier

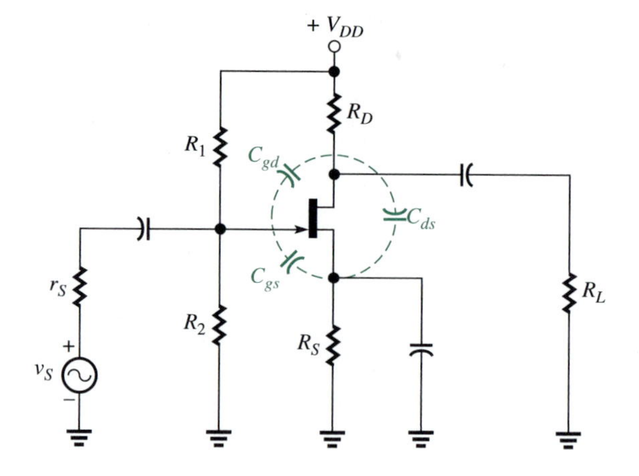

Figure 10–38
Common-source amplifier with interelectrode capacitance that affects high-frequency response

where $C_A = C_{gs} + C_M = C_{gs} + C_{gd}(1 - A_v)$ and $r_{in} = r_{in}(\text{stage}) \approx R_1 \parallel R_2$, and

$$f_2(C_B) = \frac{1}{2\pi(r_o \parallel R_L)C_B} \tag{10–55}$$

where $C_B = C_{ds} + C_{gd}(1 - 1/A_v)$ and $r_o = r_o(\text{stage}) = r_d \parallel R_D$.

Example 10–18

The JFET in Figure 10–38 has $g_m = 3.2$ mS, $r_d = 100$ kΩ, $C_{gs} = 4$ pF, $C_{ds} = 0.5$ pF, and $C_{gd} = 1.2$ pF. The wiring capacitance shunting the input is 2.5 pF and that shunting the output is 4 pF. The component values in the circuit are

$$r_S = 20\text{ k}\Omega \qquad R_D = 2\text{ k}\Omega$$
$$R_1 = 1.5\text{ M}\Omega \qquad R_S = 820\ \Omega$$
$$R_2 = 330\text{ k}\Omega \qquad R_L = 40\text{ k}\Omega$$

Find the approximate upper cutoff frequency.

Solution

$$A_v = -g_m(r_d \parallel R_D \parallel R_L) = -3.2 \times 10^{-3}[(100\text{ k}\Omega) \parallel (2\text{ k}\Omega) \parallel (40\text{ k}\Omega)] = -6$$
$$C_M = (1.2\text{ pF})[1 - (-6)] = 8.4\text{ pF}$$
$$C_A = C_{gs} + C_M + C_{(wiring)} = (4\text{ pF}) + (8.4\text{ pF}) + (2.5\text{ pF}) = 14.9\text{ pF}$$
$$r_{in}(\text{stage}) = (1.5\text{ M}\Omega) \parallel (330\text{ k}\Omega) = 270\text{ k}\Omega$$
$$f_2(C_A) = \frac{1}{2\pi(20 \times 10^3 \parallel 270 \times 10^3)(14.9 \times 10^{-12})} = 573.6\text{ kHz}$$
$$C_B = C_{ds} + C_{gd}(1 - 1/A_v) + C_{(wiring)}$$
$$= (0.5\text{ pF}) + (1.2\text{ pF})(1 + 1/6) + (4\text{ pF}) = 5.9\text{ pF}$$
$$r_o(\text{stage}) = (100\text{ k}\Omega) \parallel (2\text{ k}\Omega) = 1.96\text{ k}\Omega$$
$$f_2(C_B) = \frac{1}{2\pi(1.96 \times 10^3 \parallel 40 \times 10^3)(5.9 \times 10^{-12})} = 14.43\text{ MHz}$$

Therefore, $f_2 = f_2(C_A) = 573.6$ kHz.

Example 10–19

The MOSFET shown in Figure 10–39 has the following y parameters at 20 MHz:

$$y_{is} = 0 + j0.78 \times 10^{-3}\ \text{S}$$
$$y_{fs} = 4 \times 10^{-3} - j0.15 \times 10^{-3}\ \text{S}$$
$$y_{os} = 1.25 \times 10^{-5} + j0.25 \times 10^{-3}\ \text{S}$$

Assuming that wiring and stray capacitance is negligible, find the approximate upper cutoff frequency.

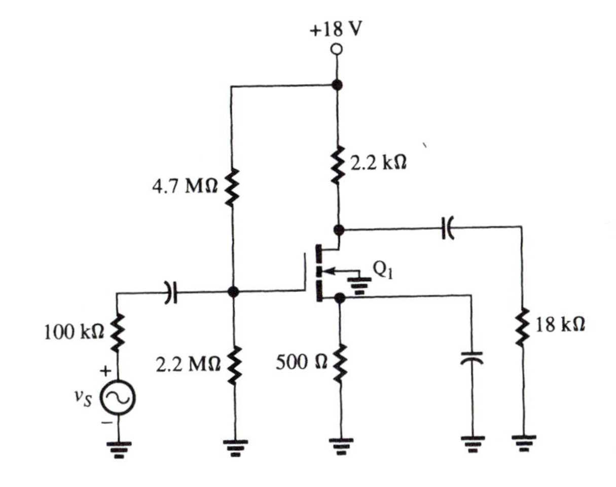

Figure 10–39
(Example 10–19)

Solution. From equation 9–74, we have

$$y_{fs} = g_m - j\omega C_{gd} = 4 \times 10^{-3} - j0.15 \times 10^{-3}\ \text{S}$$

Therefore,

$$g_m = 4 \times 10^{-3}\ \text{S}$$

and

$$(2\pi)(20 \times 10^6)C_{gd} = 0.15 \times 10^{-3}$$
$$C_{gd} = 1.2\ \text{pF}$$

From equation 9–76,

$$y_{os} = 1/r_d + j\omega(C_{gd} + C_{ds}) = 1.25 \times 10^{-5} + j0.25 \times 10^{-3}\ \text{S}$$

Therefore,

$$r_d = 1/(1.25 \times 10^{-5}) = 80\ \text{k}\Omega$$

and

$$(2\pi)(20 \times 10^6)(C_{gd} + C_{ds}) = 0.25 \times 10^{-3}$$
$$C_{gd} + C_{ds} = 2\ \text{pF}$$
$$C_{ds} = (2 - 1.2)\ \text{pF} = 0.8\ \text{pF}$$

From equation 9–73,

$$y_{is} = 0 + j\omega(C_{gs} + C_{gd}) = 0 + j0.78 \times 10^{-3}\ \text{S}$$

Therefore,

$$(2\pi)(20 \times 10^6)(C_{gs} + C_{gd}) = 0.78 \times 10^{-3}$$
$$C_{gs} + C_{gd} = 6.2\ \text{pF}$$
$$C_{gs} = (6.2 - 1.2)\ \text{pF} = 5\ \text{pF}$$

Thus,

$$A_v = -g_m(r_D \| R_D \| R_L) = -(4 \times 10^{-3}\ \text{S})[(80\ \text{k}\Omega) \| (2.2\ \text{k}\Omega) \| (18\ \text{k}\Omega)]$$
$$= -7.65$$
$$C_M = C_{gd}(1 - A_v) = (1.2\ \text{pF})[1 - (-7.65)] = 10.38\ \text{pF}$$
$$C_A = C_{gs} + C_M = (5\ \text{pF}) + (10.38\ \text{pF}) = 15.38\ \text{pF}$$
$$r_{in}(\text{stage}) = (4.7\ \text{M}\Omega) \| (2.2\ \text{M}\Omega) = 1.5\ \text{M}\Omega$$
$$f_2(C_A) = \frac{1}{2\pi(10 \times 10^3 \| 1.5 \times 10^6)(15.38 \times 10^{-12})} = 1.04\ \text{MHz}$$
$$C_B = C_{ds} + C_{gd}(1 - 1/A_v)$$
$$= 0.8\ \text{pF} + (1.2\ \text{pF})(1 + 1/7.65) = 2.16\ \text{pF}$$
$$r_o(\text{stage}) = (80\ \text{k}\Omega) \| (2.2\ \text{k}\Omega) = 2.1\ \text{k}\Omega$$
$$f_2(C_B) = \frac{1}{2\pi(2.1 \times 10^3 \| 18 \times 10^3)(2.16 \times 10^{-12})} = 38.5\ \text{MHz}$$

Therefore, $f_2 = f_2(C_A) = 1.04$ MHz.

Example 10–20

PSPICE

Use PSpice and Probe to determine the lower and upper cutoff frequencies of the MOSFET amplifier shown in Figure 10–40(a). The capacitances C_{gs}, C_{gd}, and C_{ds} shown in the figure represent stray capacitance and should be included in the input

Figure 10–40
(Example 10–20)

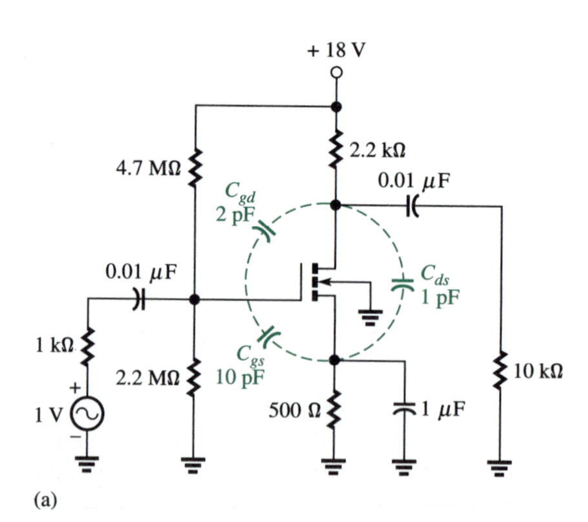

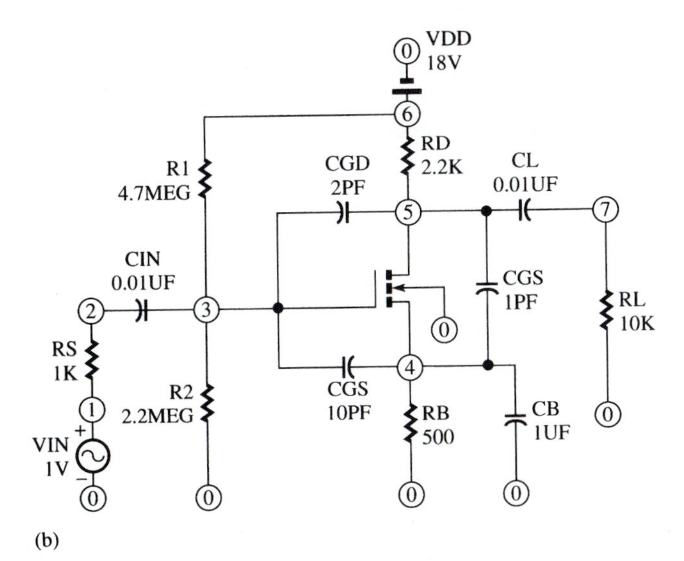


```
EXAMPLE 10-20
VIN 1 0 AC 1V
VDD 6 0 18V
RS 1 2 1K
CIN 2 3 0.01UF
R1 6 3 4.7MEG
R2 3 0 2.2MEG
RD 5 6 2.2K
RB 4 0 500
CGD 3 5 2PF
CGS 3 4 10PF
CDS 5 4 1PF
CB 4 0 1UF
CL 5 7 0.01UF
RL 7 0 10K
M1 5 3 4 0 MOSFET
.MODEL MOSFET NMOS VTO=2 KP=0.5E–3
.AC DEC 10 1K 10MEG
.END
```

(b)

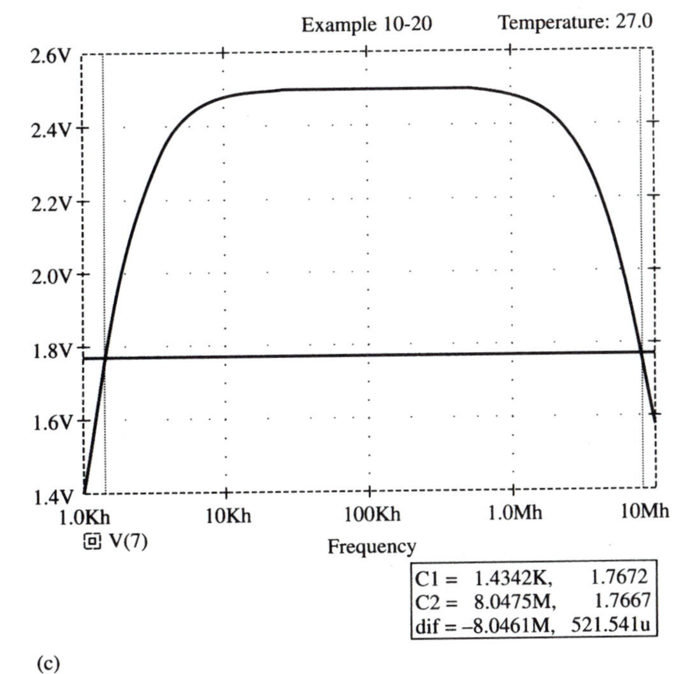

(c)

Figure 10–40
(Continued)

circuit file. The MOSFET transistor has VTO = 2 and KP = 0.5E-3. Let other parameters default.

Solution. The PSpice circuit and input circuit file are shown in Figure 10–40(b). A plot of the voltage across the load resistance, V(7), as displayed by Probe, is shown in Figure 10–40(c). We see that the maximum load voltage is 2.5 V. Therefore, the lower and upper cutoff frequencies occur where the load voltage is

0.707(2.5 V) = 1.767 V. As shown in the Probe display, cursor C1 is moved to a point on the plot where the voltage equals 1.7672 V, and we see that the lower cutoff frequency is 1.4342 kHz. C2 is moved to 1.7667 V at the high-frequency end of the plot, and we see that the upper cutoff frequency is 8.0475 MHz.

EXERCISES

Section 10–1
Definitions and Basic Concepts

10–1. The lower cutoff frequency of a certain amplifier is 120 kHz and its upper cutoff frequency is 1 MHz. The peak-to-peak output voltage in the midband frequency range is 2.4 V p–p and the output power at 120 kHz is 0.4 W. Find
a. the bandwidth,
b. the rms output voltage at 1 MHz,
c. the output power in the midband frequency range, and
d. the output power at 1 MHz.

10–2. The voltage at the input of an amplifier is 15 mV rms. The amplifier delivers 0.02 A rms to a 12-Ω load at 1 kHz. The input resistance of the amplifier is 1400 Ω. Its lower cutoff frequency is 50 Hz and its bandwidth is 9.95 kHz. Find
a. the upper cutoff frequency, f_2;
b. the output power at 50 Hz;
c. the midband power gain; and
d. the power gain at f_2.

10–3. The input signal to a certain amplifier has a 1 kHz component with amplitude 0.2 V rms and a 4-kHz component with amplitude 0.05 V rms. In the amplifier's output, the 1-kHz component has amplitude 0.5 V rms and is delayed with respect to the 1-kHz input component by 0.25 ms. If the output is to be an undistorted version of the input, what should be the amplitude and delay of the 4-kHz output component?

Section 10–2
Decibels and Logarithmic Plots

10–4. Find the power gain P_2/P_1 in dB corresponding to each of the following:
a. $P_1 = 4$ mW, $P_2 = 1$ W;
b. $P_1 = 0.2$ W, $P_2 = 80$ mW;
c. $P_1 = 5000\ \mu$W, $P_2 = 5$ mW; and
d. $P_1 = 0.4\ P_2$.

10–5. A certain amplifier has a power gain of 42 dB. The output power is 16 W and the input resistance is 1 kΩ. What is the rms input voltage?

10–6. Find the voltage gain v_2/v_1 in dB corresponding to each of the following:
a. $v_1 = 120\ \mu$V rms, $v_2 = 40$ mV rms;
b. $v_1 = 1.8$ V rms, $v_2 = 18$ V peak-to-peak;
c. $v_1 = 0.707$ V rms, $v_2 = 1$ V rms;
d. $v_1 = 5 \times 10^3 v_2$; and
e. $v_1 = v_2$.

10–7. The amplifier shown in Figure 10–41 has $v_S = 30$ mV rms and $v_L = 1$ V rms. Find
a. the power gain in decibels; and
b. the voltage gain in decibels.

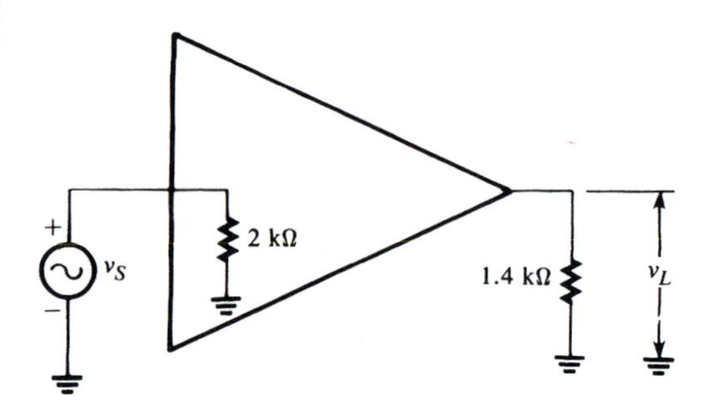

Figure 10–41
(Exercise 10–7)

10–8. Repeat Exercise 10–7 if the load resistor is 2 kΩ.

10–9. The voltage gain of an amplifier is 18 dB. If the output voltage is 6.8 V rms, and the input and load resistances both equal 800 Ω, find
a. the input voltage; and
b. the power gain in decibels.

10–10. The output voltage of an amplifier is 12.5 dB above its level at point "A" in the amplifier. The amplifier input voltage is 6.4 dB below the voltage at point "A." If the output voltage is 2.4 V rms, find

a. the voltage at point A;
b. the input voltage; and
c. the amplifier voltage gain, in decibels.

10–11. The input voltage of a certain amplifier is 0.36 V rms. Without performing any computations, estimate the output voltage that would correspond to each of the following gains: (a) 20 dB, (b) 46 dB, (c) −6 dB, and (d) 60 dB.

10–12. An amplifier has output power 28 dBm and output voltage 34 dBV. Find
a. its output power in watts; and
b. its output voltage in V rms.

10–13. In frequency response measurements made on an amplifier, the gain was found to vary from 0.06 to 75 over a frequency range from 7 Hz to 3.3 kHz. What kind of log-log graph paper would be required to plot this data (how many cycles on each axis)?

10–14. The following frequency response measurements were made on an amplifier:

Frequency	Voltage Gain	Phase Shift
2 Hz	4.0	84.3°
4 Hz	7.8	78.6°
8 Hz	14.9	68.2°
15 Hz	24.0	53.1°
30 Hz	33.3	33.7°
40 Hz	35.8	14.0°
100 Hz	39.2	6.0°
200 Hz	39.8	3.4°
500 Hz	40.0	−3.4°
1 kHz	39.2	−11.3°
2 kHz	37.1	−21.8°
4 kHz	31.2	−38.6°
8 kHz	21.2	−58.0°
10 kHz	17.9	−63.4°

Plot the gain data on log-log graph paper and the phase data on semilog graph paper. Use your plots to estimate the following:
a. the lower and upper cutoff frequencies;
b. the (positive) phase shift when the gain is 20;
c. the gain when the phase shift is −30°; and
d. the gain 2 octaves below the lower cutoff frequency.

10–15. The following frequency response measurements were made on an amplifier:

Frequency	Voltage Gain	Phase Shift
150 Hz	23.2 dB	89.6°
200 Hz	25.4 dB	68.2°
400 Hz	29.9 dB	51.3°
600 Hz	31.7 dB	39.8°
800 Hz	32.6 dB	29.0°
1 kHz	33.0 dB	22.8°
2 kHz	33.7 dB	6.4°
5 kHz	34.0 dB	−12.7°
8 kHz	32.9 dB	−28.1°
10 kHz	32.4 dB	−33.7°
12 kHz	31.9 dB	−38.7°
18 kHz	30.1 dB	−50.2°
20 kHz	29.6 dB	−53.1°
40 kHz	24.9 dB	−69.4°
100 kHz	17.4 dB	−81.5°
200 kHz	11.5 dB	−85.7°
500 kHz	3.5 dB	−88.3°
1 MHz	−2.5 dB	−89.1°

Plot the gain and phase data on semilog graph paper and use your plots to estimate the following:
a. the lower and upper cutoff frequencies;
b. the magnitude of the output voltage when the phase shift is −60° (the input voltage is 20 mV rms);
c. the (positive) phase shift when the input voltage is 15 mV rms and the output voltage is 0.336 V rms; and
d. the magnitude of the voltage gain 1 decade above the upper cutoff frequency.

Section 10–3
Series Capacitance and Low-Frequency Response

10–16. Show that the lower cutoff frequency $f_1(C_1)$ in Figure 10–7 is the frequency at which the capacitive reactance X_{C_1} equals the resistance r_{in}.

10–17. The amplifier shown in Figure 10–42 has midband voltage gain $|v_L|/|v_S|$ equal to 80. Calculate
a. the lower cutoff frequency;
b. the gain $|v_L|/|v_S|$ at 300 Hz;

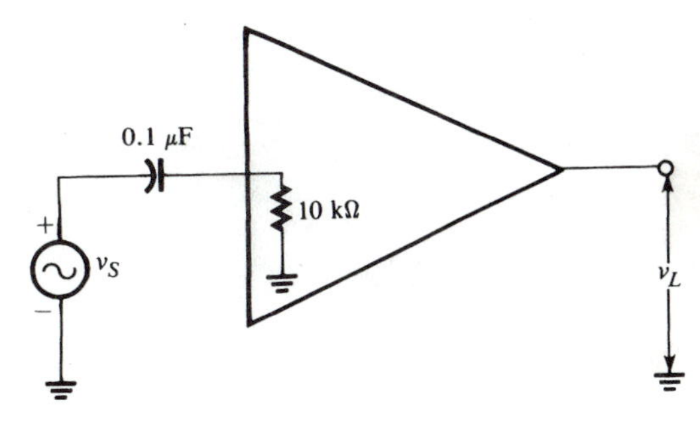

Figure 10–42
(Exercise 10–17)

c. the phase shift 1 decade below cutoff; and
d. the frequency at which the gain $|v_L|/|v_S|$ is 10 dB down from its midband value.

10–18. Repeat Exercise 10–17 using the normalized plots in Figure 10–9 for (b), (c), and (d).

10–19. The amplifier shown in Figure 10–43 has voltage gain $|v_L|/|v_{in}|$ equal to 200. Calculate
a. the lower cutoff frequency;
b. the frequency at which v_L leads v_S by 75°; and
c. the voltage gain $|v_L|/|v_S|$ 2 octaves above cutoff.

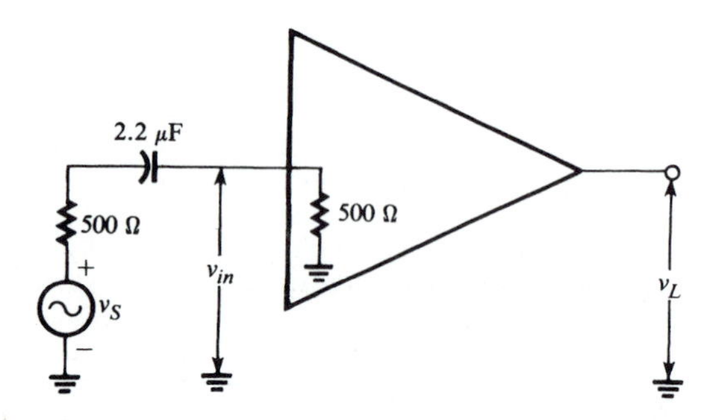

Figure 10–43
(Exercise 10–19)

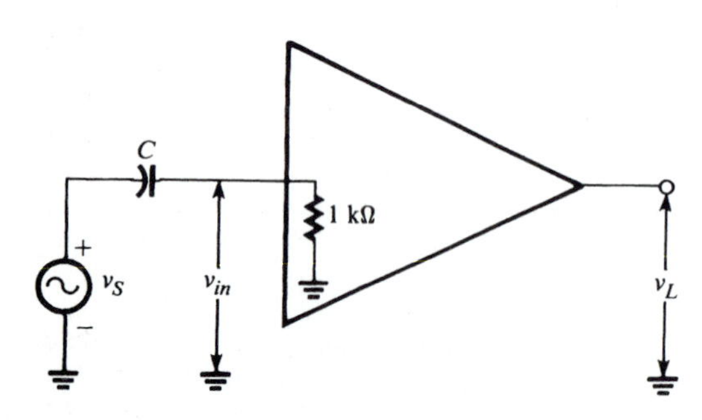

Figure 10–44
(Exercise 10–20)

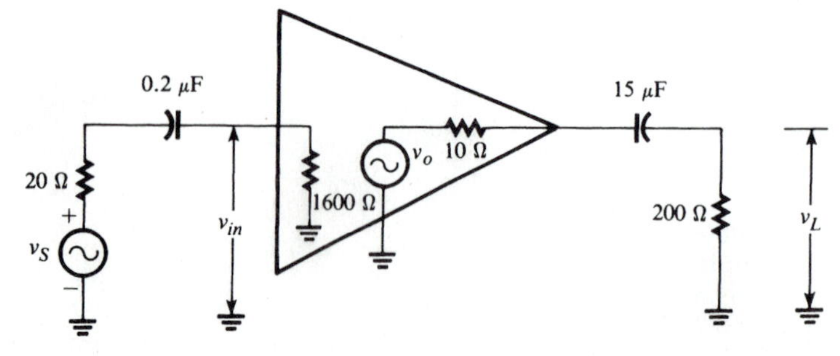

Figure 10–45
(Exercise 10–21)

10–20. The amplifier shown in Figure 10–44 has voltage gain $|v_L|/|v_{in}|$ equal to 180. What should be the value of the coupling capacitor C in order that the gain $|v_L|/|v_S|$ be no less than 100 at 20 Hz?

10–21. The amplifier shown in Figure 10–45 has midband gain $|v_L|/|v_S|$ equal to 160. Find the (approximate) values of
a. the lower cutoff frequency;
b. the gain $|v_o|/|v_{in}|$; and
c. a value for the input coupling capacitor that would cause the asymptotic gain plot to have a single break frequency.

10–22. The amplifier shown in Figure 10–46 has a very large (essentially infinite) input impedance and a very small (essentially 0) output impedance. What should be the gain $|v_o|/|v_{in}|$ if it is desired to have $|v_L| = 1$ V rms when $|v_S| = 20$ mV rms at 100 Hz?

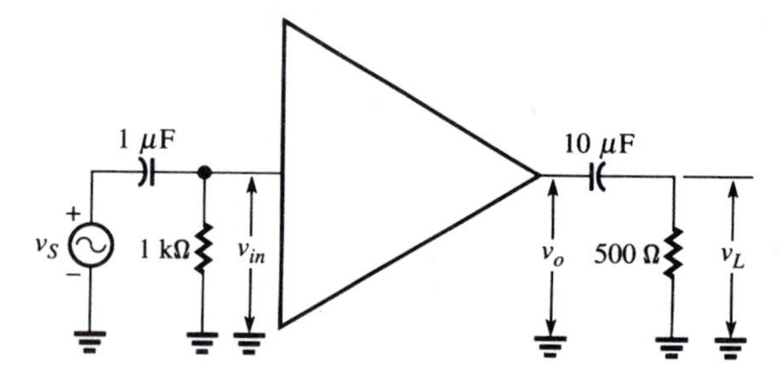

Figure 10–46
(Exercise 10–22)

10–23. The amplifier shown in Figure 10–47 has midband gain $|v_L|/|v_S| = 100$. Calculate
a. the approximate lower cutoff frequency;
b. the approximate gain $|v_L|/|v_S|$, in decibels, at $f = 120$ Hz; and
c. the approximate gain $|v_L|/|v_S|$, in decibels, at $f = 77.6$ Hz.

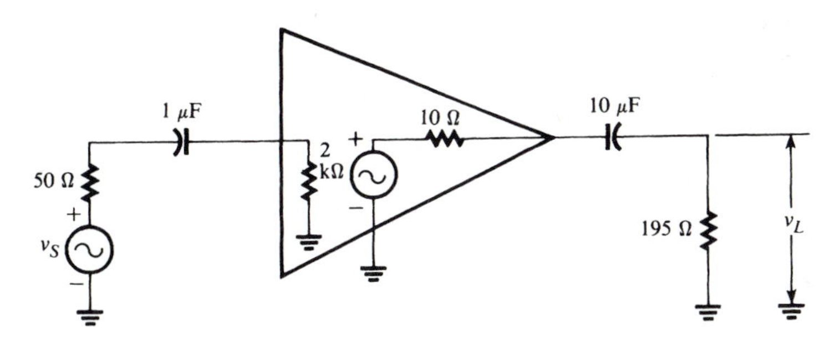

Figure 10–47
(Exercise 10–23)

Use Figure 10–13 to find

d. the phase shift 1 octave below the cutoff frequency; and

e. the frequency at which v_L leads v_S by 160°.

10–24. Find values for the input and output coupling capacitors in Exercise 10–23 that will result in a single break frequency and a lower cutoff frequency at 40 Hz.

Section 10–4
Shunt Capacitance and High-Frequency Response

10–25. The amplifier shown in Figure 10–48 has midband gain $|v_L|/|v_S| = 150$. Calculate

a. the upper cutoff frequency;

b. $|v_L|/|v_{in}|$ at midband;

c. $|v_L|/|v_S|$ at 10 MHz; and

d. the phase angle of the output with respect to v_S, at 10 MHz.

10–26. For the amplifier of Exercise 10–25, use Figure 10–16 to find approximate values of

a. $|v_L|/|v_S|$ 1 octave above cutoff;

b. the frequency at which $|v_L|/|v_S| = 75$; and

c. the frequency at which v_L lags v_S by 60°.

10–27. What is the maximum permissible value of the source resistance for the amplifier of Exercise 10–25 if the upper cutoff frequency must be at least 1 MHz?

10–28. Design a low-pass RC network, consisting of a single resistor and a single capacitor, whose output voltage is 0.5 V rms when the input is a 1-V-rms signal having frequency 120 kHz. Draw the schematic diagram, label input and output terminals, and show the component values of your design.

10–29. The amplifier shown in Figure 10–49 has midband gain $|v_L|/|v_S| = 84$.

a. Find the approximate upper cutoff frequency.

b. Find the midband value of $|v_o|/|v_{in}|$.

c. Sketch the asymptotic Bode plot of $|v_L|/|v_S|$ as it would appear if plotted on log-log graph paper. (It is not necessary to use log-log graph paper for your sketch.) Label important frequency and gain coordinates on your sketch.

10–30. **a.** What is the approximate upper cutoff frequency of the amplifier shown in Figure 10–50?

b. What value of R_L would result in the upper cutoff frequency being equal to 64.5% of that calculated in (a)?

10–31. The amplifier shown in Figure 10–51 has midband gain $|v_L|/|v_{in}|$ equal to −160. Find the approximate upper cutoff frequency.

10–32. For what value of amplifier gain $|v_L|/|v_{in}|$ in Exercise 10–31 would the frequency response have a single break frequency?

Figure 10–48
(Exercise 10–25)

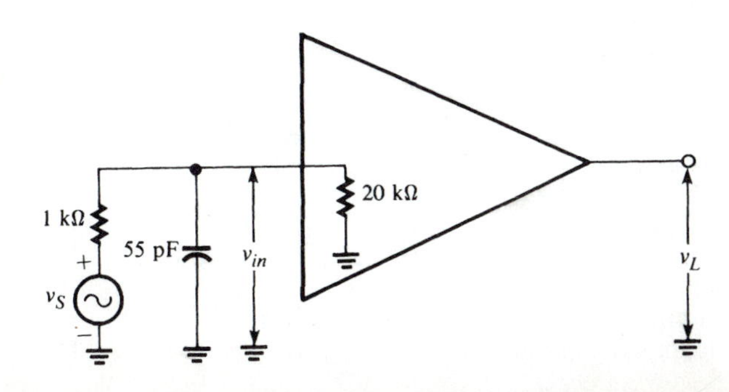

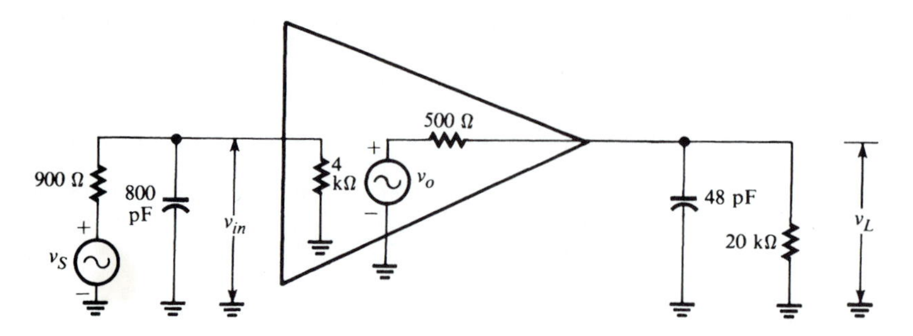

Figure 10–49
(Exercise 10–29)

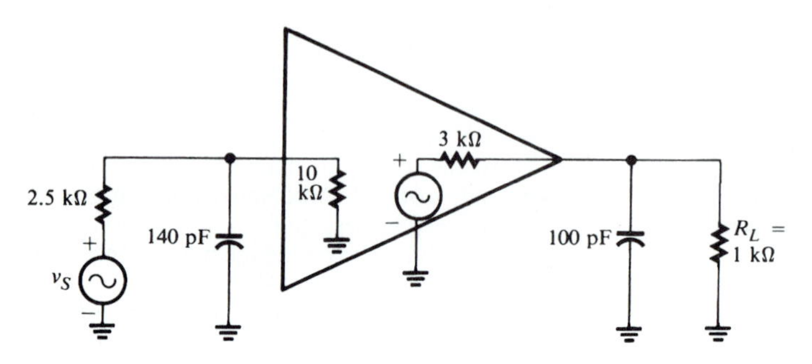

Figure 10–50
(Exercise 10–30)

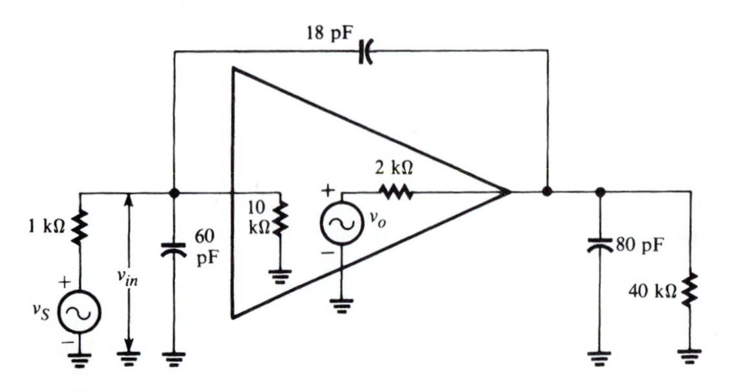

Figure 10–51
(Exercise 10–31)

Section 10–5
Transient Response

10–33. **a.** What is the bandwidth of the low-pass network shown in Figure 10–52?
b. What is the rise time at the output when the input is a step voltage?
c. How would the bandwidth be affected by an increase in capacitance? In resistance?
d. How would the rise time be affected by an increase in capacitance? In resistance?

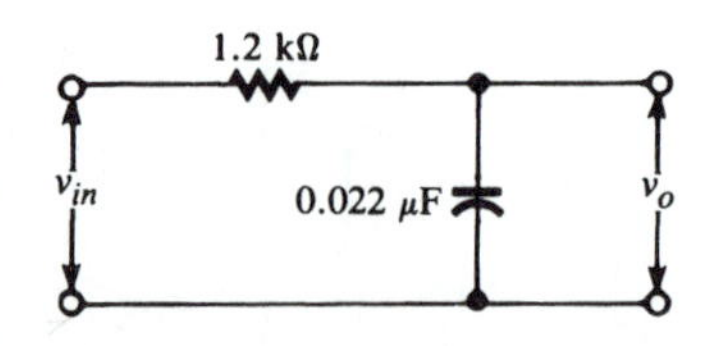

Figure 10–52
(Exercise 10–33)

10–34. The rise time of the output of the network shown in Figure 10–53 is 2.7 μs. What is the value of C?

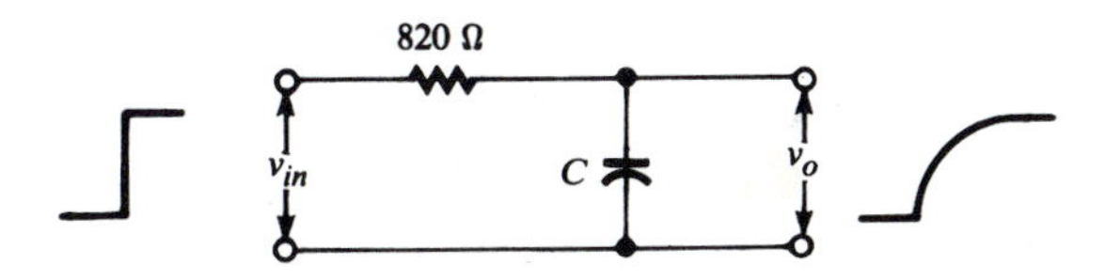

Figure 10–53
(Exercise 10–34)

Section 10–6
Frequency Response of BJT Amplifiers

10–35. The transistor shown in Figure 10–54 has $\beta = 90$, $r_e = 15\ \Omega$, and $r_o = 125\ \text{k}\Omega$. Find the approximate lower cutoff frequency.

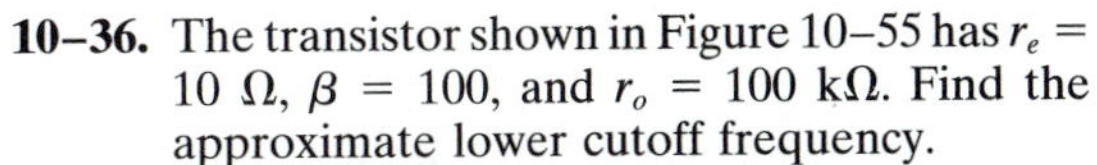

10–36. The transistor shown in Figure 10–55 has $r_e = 10\ \Omega$, $\beta = 100$, and $r_o = 100\ \text{k}\Omega$. Find the approximate lower cutoff frequency.

10–37. The transistor shown in Figure 10–56 has $r_C = 500\ \text{k}\Omega$ and $r_e = 18\ \Omega$. Find the approximate lower cutoff frequency.

10–38. The transistor shown in Figure 10–57 has $r_e = 20\ \Omega$ and $\beta = 150$. Find the approximate lower cutoff frequency.

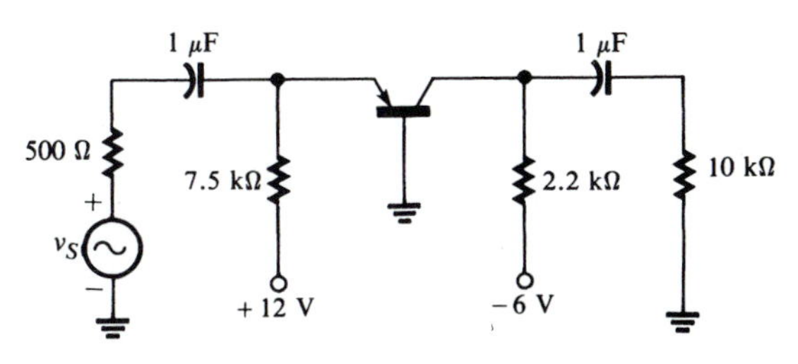

Figure 10–56
(Exercise 10–37)

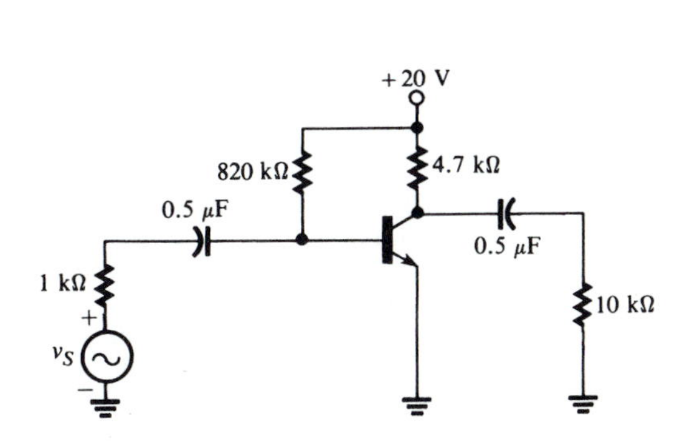

Figure 10–54
(Exercise 10–35)

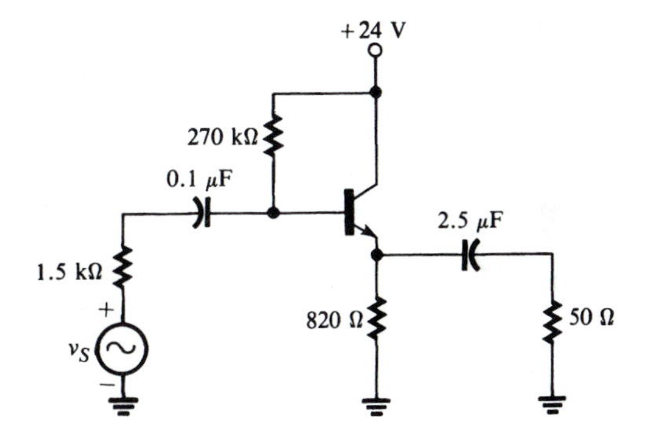

Figure 10–57
(Exercise 10–38)

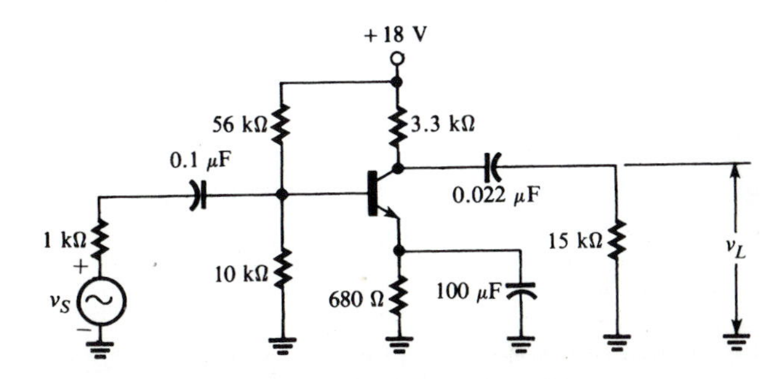

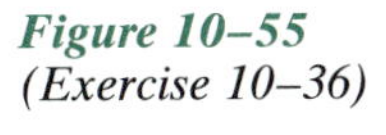

Figure 10–55
(Exercise 10–36)

DESIGN EXERCISES

10–39. Following the design procedure given in Section 10–6 (Example 10–13), redesign the amplifier in Exercise 10–36 (i.e., find new capacitor values) so that the lower cutoff frequency is approximately 250 Hz. Assume that capacitors must be selected from a line having standard values 0.1 μF, 0.2 μF, 0.4 μF, 1 μF, 2 μF, 4 μF, 10 μF, 20 μF, 40 μF, and 100 μF.

10–40. The common-base amplifier in Figure 10–31(a) has $r_S = 100\ \Omega$, $R_E = 1\ \text{k}\Omega$, $R_C = 4.7\ \text{k}\Omega$, $R_L = 10\ \text{k}\Omega$, $V_{EE} = -2$ V, and $V_{CC} = 15$ V. Select capacitor values so that the lower cutoff frequency is approximately 250 Hz. Assume that capacitors must be selected from a line having standard values 0.1 μF, 0.5 μF, 1 μF, 5 μF, 7.5 μF, 10 μF, 15 μF, 50 μF, and 100 μF.

10–41. The emitter follower in Figure 10–31(b) has $r_S = 600\ \Omega$, $R_1 = 220\ \text{k}\Omega$, $R_2 = 47\ \text{k}\Omega$, $R_E = 1\ \text{k}\Omega$, $R_L = 500\ \Omega$, and $V_{CC} = 15$ V. The β of the transistor is 100.

a. Find capacitor values so that the lower cutoff frequency is approximately 50 Hz. Assume that capacitors must be selected from a line having standard values 1 μF, 1.5 μF, 2.2 μF, 3.3 μF, 4.7 μF, 10 μF, 15 μF, 22 μF, and 33 μF.

b. Find the approximate lower cutoff frequency when the standard-value capacitors are used.

c. Find the approximate lower cutoff frequency when the standard-value capacitors are used and the β of the transistor changes to 200.

10–42. The transistor shown in Figure 10–58 has $r_e = 25\ \Omega$, $r_o = 50\ \text{k}\Omega$, $f_\beta = 0.8$ MHz, and a midband β equal to 100. Find the approximate upper cutoff frequency.

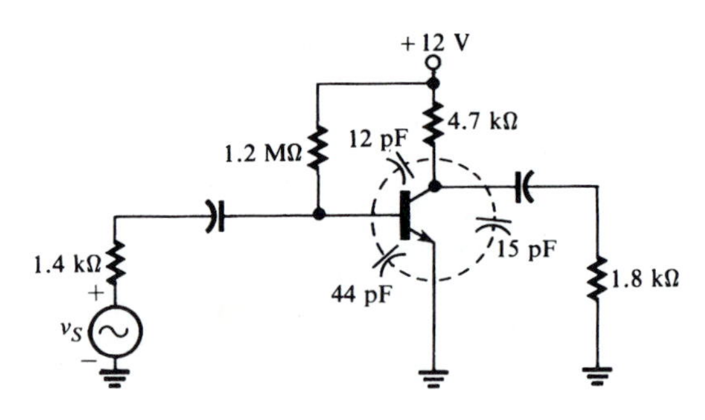

Figure 10–58
(Exercise 10–42)

10–43. Repeat Exercise 10–42 using the following parameters: $r_e = 25\ \Omega$, $r_o = 50\ \text{k}\Omega$, $f_T = 80$ MHz, $\beta_{\text{(midband)}} = 100$, $C_{bc} = 2$ pF, $C_{be} = 22$ pF, $C_{ce} = 15$ pF.

10–44. How much input-to-ground wiring capacitance could be tolerated in the amplifier of Exercise 10–42 without the upper cutoff frequency falling below 100 kHz?

Section 10–7
Frequency Response of FET Amplifiers

10–45. The JFET shown in Figure 10–59 has $g_m = 4 \times 10^{-3}$ S and $r_d = 100\ \text{k}\Omega$.

a. Find the approximate lower cutoff frequency.

b. Sketch asymptotic gain versus frequency as it would appear if plotted on log-log graph paper. Label the break frequencies on your sketch. It is not necessary to show actual gain values, but indicate the rate of gain reduction corresponding to each asymptote.

Figure 10–59
(Exercise 10–45)

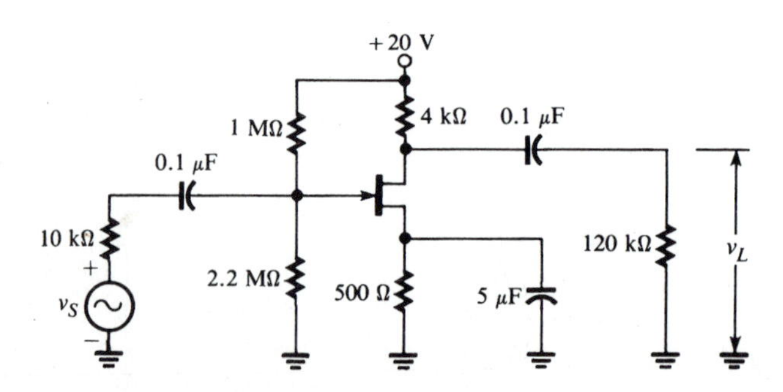

10–46. The MOSFET shown in Figure 10–60 has $g_m = 3 \times 10^{-3}$ S and $r_d = 75$ kΩ. Find the approximate lower cutoff frequency.

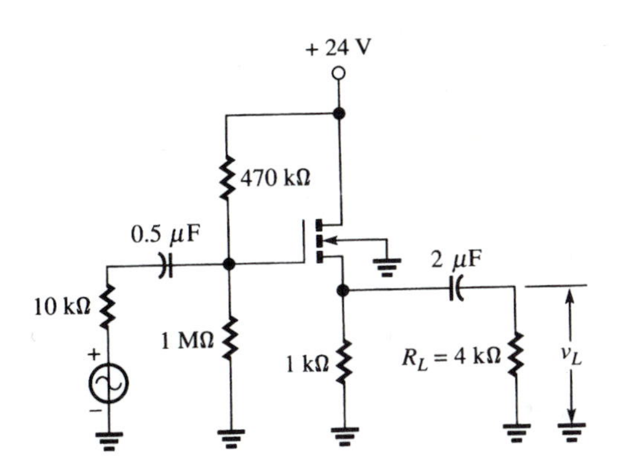

Figure 10–60
(Exercise 10–46)

10–47. What is the minimum value of R_L in Exercise 10–46 that could be used without the lower cutoff frequency being greater than 10 Hz?

10–48. The MOSFET shown in Figure 10–61 has $g_m = 4000$ μS and $r_d = 60$ kΩ. Find the approximate upper cutoff frequency.

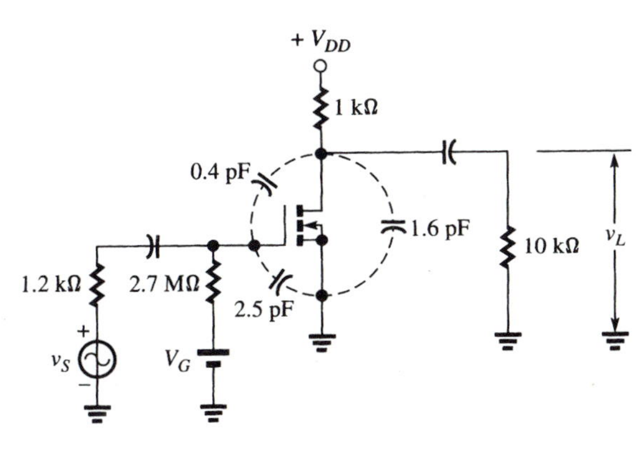

Figure 10–61
(Exercise 10–48)

10–49. What is the maximum input wiring capacitance (to ground) that could be tolerated in the amplifier of Exercise 10–48 if the upper cutoff frequency must be at least 1 MHz?

10–50. The JFET in Figure 10–62 has $C_{iss} = 6$ pF, $C_{rss} = 2$ pF, $C_{oss} = 2.8$ pF, $g_m = 3.8 \times 10^{-3}$ S, and $r_d = 100$ kΩ. The wiring and stray capacitance shunting input and output are 0.6 pF and 15 pF, respectively. Find the approximate upper cutoff frequency.

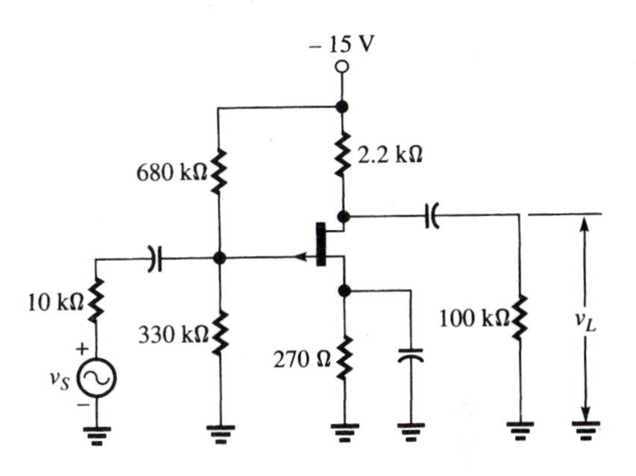

Figure 10–62
(Exercise 10–50)

SPICE EXERCISES

10–51. The amplifier in Figure 10–27 has $r_S = 1$ kΩ, $R_1 = 100$ kΩ, $R_2 = 22$ kΩ, $R_C = 3$ kΩ, $R_E = 2$ kΩ, $R_L = 20$ kΩ, $C_1 = 0.22$ μF, $C_2 = 0.22$ μF, $C_E = 15$ μF, and $V_{CC} = 24$ V. The β of the transistor is 120. Use SPICE to find the lower cutoff frequency and to obtain a plot of the low-frequency response.

10–52. Use SPICE to determine the actual factor by which the gain of the amplifier in Example 10–15 is reduced at 50 Hz. The supply voltage is $V_{CC} = 15$ V. Also find the actual lower cutoff frequency.

10–53. Use SPICE to solve Exercise 10–37.

10–54. The amplifier in Figure 10–38 has $r_S = 10$ kΩ, $R_1 = 1.5$ MΩ, $R_2 = 330$ kΩ, $R_D = 2$ kΩ, $R_S = 820$ Ω, and $R_L = 20$ kΩ. The input and output coupling capacitors are each 1 μF and the source bypass capacitor is 10 μF. The JFET has interelectrode capacitance $C_{gd} = 1.5$ pF, $C_{gs} = 6$ pF, and $C_{ds} = 5$ pF. The JFET parameters are $I_{DSS} = 10$ mA and $V_p = -2$ V. Use SPICE to obtain a plot of the high-frequency response of the amplifier and to determine its upper cutoff frequency. (Use discrete capacitors to represent interelectrode capacitance in the SPICE model.)

11 MULTISTAGE AMPLIFIERS

11–1 GAIN RELATIONS IN MULTISTAGE AMPLIFIERS

In many applications, a single amplifier cannot furnish all the gain that is required to drive a particular kind of load. For example, a speaker represents a "heavy" load in an audio amplifier system, and several amplifier *stages* may be required to "boost" a signal originating at a microphone or magnetic tape head to a level sufficient to provide a large amount of power to the speaker. We hear of *preamplifiers, power amplifiers,* and *output amplifiers,* all of which constitute stages of amplification in such a system. Actually, each of these components may itself consist of a number of individual transistor amplifier stages. Amplifiers that create voltage, current, and/or power gain through the use of two or more stages are called *multistage* amplifiers.

When the output of one amplifier stage is connected to the input of another, the amplifier stages are said to be in *cascade.* Figure 11–1 shows two stages connected in cascade. To illustrate how the overall voltage gain of the combination is computed, let us assume that the input to the first stage is 10 mV rms and that the voltage gain of each stage is $A_1 = A_2 = 20$, as shown in the figure. The output of the first stage is $A_1 v_{i1} = 20$ (10 mV rms) = 200 mV rms. Thus, the input to the second stage

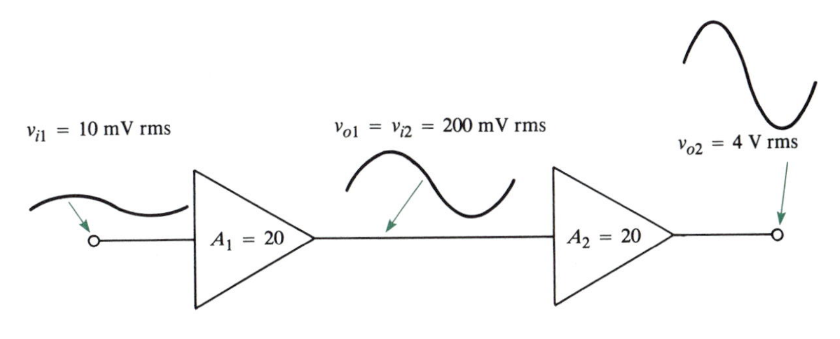

Figure 11–1
Two amplifier stages connected in cascade

is 200 mV rms. The output of the second stage is, therefore, $A_2 v_{i2} = 20(200 \text{ mV rms}) = 4 \text{ V rms}$. Therefore, the overall voltage gain is

$$A_v = \frac{v_{o2}}{v_{i1}} = \frac{4 \text{ V rms}}{10 \text{ mV rms}} = 400$$

Notice that $A_v = A_1 A_2 = (20)(20) = 400$.

Figure 11–2 shows an arbitrary number (n) of stages connected in cascade. Note that the output of each stage is the input to the succeeding one ($v_{o1} = v_{i2}$, $v_{o2} = v_{i3}$, etc.). We will derive an expression for the overall voltage gain $v_{o,n}/v_{i1}$ in terms of the individual stage gains $A_1, A_2, \ldots, A_n$. We assume that each stage gain $A_1, A_2 \ldots, A_n$ is the value of the voltage gain between input and output of a stage *with all other stages connected* (more about that important assumption later).

By definition,

$$v_{o1} = A_1 v_{i1} \tag{11–1}$$

Also,

$$v_{o2} = A_2 v_{i2} = A_2 v_{o1} \tag{11–2}$$

Substituting v_{o1} from (11–1) into (11–2) gives

$$v_{o2} = (A_2 A_1) v_{i1} \tag{11–3}$$

Similarly,

$$v_{o3} = A_3 v_{i3} = A_3 v_{o2}$$

and, from (11–3),

$$v_{o3} = (A_3 A_2 A_1) v_{i1}$$

Continuing in this manner, we eventually find

$$v_{o,n} = v_L = (A_n A_n - 1 \cdot \cdot \cdot A_2 A_1) v_{i1}$$

Therefore,

$$\frac{v_{o,n}}{v_{i1}} = A_n A_{n-1} \cdot \cdot \cdot A_2 A_1 \tag{11–4}$$

Equation 11–4 shows that the overall voltage gain of n cascaded stages is the *product* of the individual stage gains (not the sum!). In general, any one or more of the

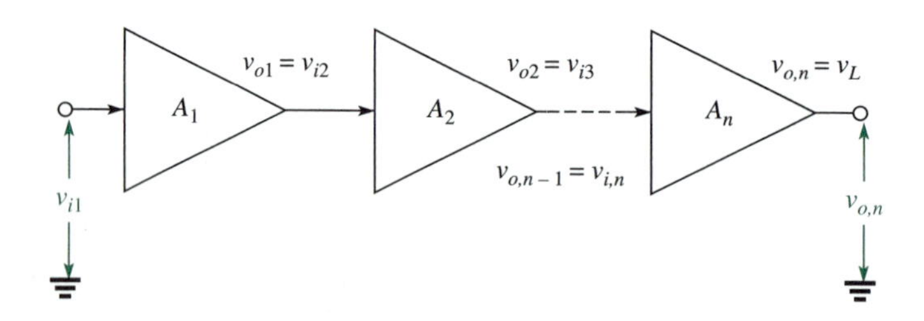

Figure 11–2
n amplifier stages connected in cascade. The output voltage of each stage is the input voltage to the next stage.

stage gains can be negative, signifying, as usual, that the stage causes a 180° phase inversion. It follows from equation 11–4 that the cascaded amplifiers will cause the output of the last stage ($v_{o,n}$) to be out of phase with the input to the first stage (v_{i1}) if there is an *odd* number of inverting stages, and will cause $v_{o,n}$ to be in phase with v_{i1} if there is an even (or zero) number of inversions.

To find the overall voltage gain of the cascaded system in decibels, we ignore the algebraic signs of each stage gain and compute

$$\begin{aligned} 20\log_{10}\left(\frac{v_{o,n}}{v_{i1}}\right) &= 20\log_{10}(A_n A_{n-1}\cdots A_2 A_1) \\ &= 20\log_{10}A_n + 20\log_{10}A_{n-1} + \cdots + 20\log_{10}A_2 + 20\log_{10}A_1 \\ &= A_n\,(\text{dB}) + A_{n-1}\,(\text{dB}) + \cdots + A_2\,(\text{dB}) + A_1\,(\text{dB}) \end{aligned} \tag{11–5}$$

Equation 11–5 shows that the overall voltage gain in dB is the *sum* of the individual stage gains expressed in decibels. Equations similar to (11–4) and (11–5) for overall current gain and overall power gain in terms of individual stage gains are easily derived.

Our derivation of equation 11–4 did not include the effect of source or load resistance on the overall voltage gain. Source resistance r_S causes the usual voltage division to take place at the input to the first stage, and load resistance r_L causes voltage division to occur between r_L and the output resistance of the last stage. Under those circumstances, the overall voltage gain between load and signal source becomes

$$\frac{v_L}{v_S} = \left(\frac{r_{i1}}{r_S + r_{i1}}\right) A_n A_{n-1}\cdots A_2 A_1 \left(\frac{r_L}{r_{o,n} + r_L}\right) \tag{11–6}$$

where r_{i1} = input resistance to first stage and $r_{o,n}$ = output resistance of last stage.

Example 11–1

Figure 11–3 shows a three-stage amplifier and the ac rms voltages at several points in the amplifier. Note that v_1 is the input voltage delivered by a signal source having zero resistance and that v_3 is the output voltage with no load connected.

1. Find the voltage gain of each stage and the overall voltage gain v_3/v_1.
2. Repeat (1) in terms of decibels.
3. Find the overall voltage gain v_L/v_S when the multistage amplifier is driven by a signal source having resistance 2000 Ω and the load is 25 Ω. Stage 1 has input resistance 1 kΩ and stage 3 has output resistance 50 Ω.

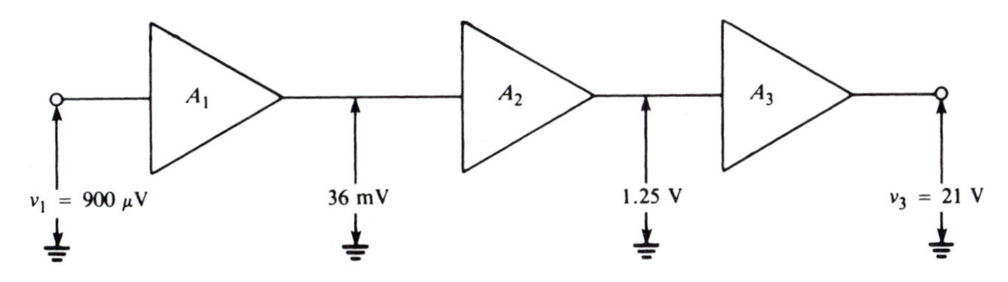

Figure 11–3
(Example 11–1)

4. What would be the gain v_L/v_S in dB for the conditions of (3) if the voltage gain of the second stage were reduced by 6 dB?
5. What is the power gain in decibels under the conditions of (3) (measured between the input to the first stage and the load)?
6. What is the overall current gain i_L/i_1 under the conditions of (3)?

Solution

1. $A_1 = (36 \text{ mV})/(900\ \mu\text{V}) = 40$

$A_2 = (1.25 \text{ V})/(36 \text{ mV}) = 34.722$

$A_3 = (21 \text{ V})/(1.25 \text{ V}) = 16.8$

$v_3/v_1 = A_1A_2A_3 = (40)(34.722)(16.8) = 23{,}333$

Note that the product of the voltage gains equals the overall voltage gain, which, in this example, can also be calculated directly: $v_3/v_1 = (21 \text{ V})/(900\ \mu\text{V}) = 23{,}333$.

2. $A_1 \text{ (dB)} = 20 \log_{10}40 = 32.04 \text{ dB}$

$A_2 \text{ (dB)} = 20 \log_{10}(34.722) = 30.81 \text{ dB}$

$A_3 \text{ (dB)} = 20 \log_{10}(16.8) = 24.51 \text{ dB}$

$v_3/v_1 \text{ (dB)} = A_1 \text{ (dB)} + A_2 \text{ (dB)} + A_3 \text{ (dB)} = 87.36 \text{ dB}$

Again, note that $20 \log_{10}(v_3/v_1) = 20 \log_{10}[21/900 \times 10^{-6})] = 20 \log_{10}(23{,}333) = 87.36$ dB.

3. From equation 11–6,

$$\frac{v_L}{v_S} = \left(\frac{1000}{2000 + 1000}\right)(23{,}333)\left(\frac{25}{50 + 25}\right) = 2592.5$$

4. The (original) voltage gain with load and source connected is $20 \log_{10}(2592.5) = 68.27$ dB. A 6-dB reduction in the gain of stage 2 therefore results in an overall gain of (68.27 dB) − (6 dB) = 62.27 dB.

5. When a signal-source resistance of 2000 Ω is inserted in series with the input, v_1 becomes

$$v_1 = \left(\frac{1000}{2000 + 1000}\right)(900\ \mu\text{V}) = 300\ \mu\text{V}$$

The input power is, therefore,

$$P_i = \frac{v_1^2}{r_{i1}} = \frac{(300 \times 10^{-6})^2}{1000} = 90 \text{ pW}$$

The voltage across the 25-Ω load is then

$$v_L = v_1(A_1A_2A_3)\left(\frac{R_L}{r_{o3} + R_L}\right)$$

$$= (300\ \mu\text{V})(40)(34.722)(16.8)\left(\frac{25}{50 + 25}\right) = 2.33 \text{ V}$$

The output power developed across the load resistance is, therefore,

$$P_o = \frac{v_L^2}{R_L} = \frac{(2.33)^2}{25} = 0.217 \text{ W}$$

Finally,

$$A_p\,(\text{dB}) = 10\log_{10}\left(\frac{P_o}{P_i}\right) = 10\log_{10}\left(\frac{0.217}{90\times 10^{-12}}\right) = 93.82\text{ dB}$$

6. Recall that $A_p = A_v A_i$. Using the results calculated in (5), the power gain between the input to the first stage and the load is

$$A_p = \frac{P_o}{P_i} = \frac{0.217\text{ W}}{90\times 10^{-12}\text{ W}} = 2.41\times 10^9$$

The voltage gain between the input to the first stage and the load is

$$A_v = \frac{v_L}{v_1} = \frac{2.33\text{ V}}{300\ \mu\text{V}} = 7766$$

Therefore,

$$A_i = \frac{A_p}{A_v} = \frac{2.41\times 10^9}{7766} = 3.1\times 10^5$$

It is important to remember that the gain equations we have derived are based on the *in-circuit* values of A_1, A_2, . . ., that is, on the stage gains that result when all other stages are connected. Thus, we have assumed that each value of stage gain takes into account the loading the stage causes on the previous stage and the loading presented to it by the next stage (except we assumed that A_1 did not include loading by r_S and A_n did not include loading by r_L). If we know the *open-circuit* (unloaded) voltage gain of each stage and its input and output resistances, we can calculate the overall gain by taking into account the loading effects of each stage on another. Theoretically, the load presented to a given stage may depend on *all* of the succeeding stages lying to its right, since the input resistance of any one stage depends on its output load resistance, which in turn is the input resistance to the next stage, and so forth. In practice, we can usually ignore this cumulative loading effect of stages beyond the one immediately connected to a given stage, or assume that the input resistance that represents the load of one stage to a preceding one is given for the condition that all succeeding stages are connected.

To illustrate the ideas we have just discussed, Figure 11–4 shows a three-stage amplifier for which the individual open-circuit voltage gains A_{o1}, A_{o2}, and A_{o3} are assumed to be known, as well as the input and output resistances of each stage. From the voltage division that occurs at each node in the system, it is apparent that the following relations hold:

$$v_1 = \left(\frac{(r_{i1}}{r_S + r_{i1}}\right) v_S$$

$$v_2 = A_{o1} v_1 \left(\frac{r_{i2}}{r_{o1} + r_{i2}}\right)$$

$$v_3 = A_{o2} v_2 \left(\frac{r_{i3}}{r_{o2} + r_{i3}}\right)$$

$$v_L = A_{o3} v_3 \left(\frac{r_L}{r_{o3} + r_L}\right)$$

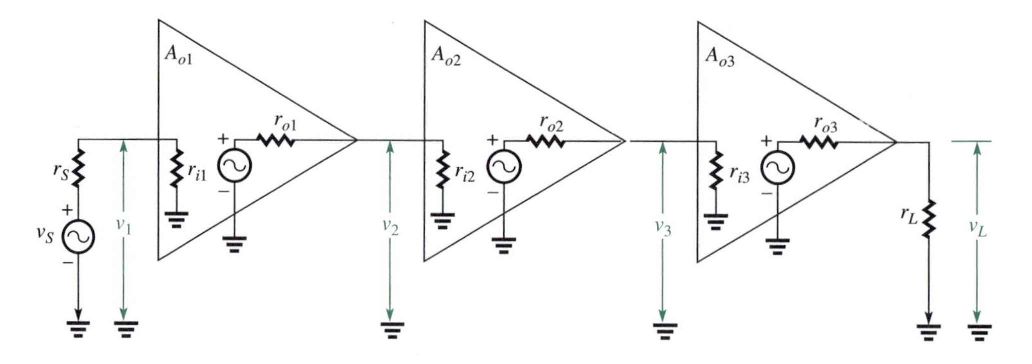

Figure 11–4
A three-stage amplifier. A_{o1}, A_{o2}, and A_{o3} are the open-circuit (unloaded) voltage gains of the respective stages.

Combining these relations leads to

$$\frac{v_L}{v_S} = \left(\frac{r_{i1}}{r_S + r_{i1}}\right) A_{o1} \left(\frac{r_{i2}}{r_{o1} + r_{i2}}\right) A_{o2} \left(\frac{r_{i3}}{r_{o2} + r_{i3}}\right) A_{o3} \left(\frac{r_L}{r_{o3} + r_L}\right) \quad \textbf{(11–7)}$$

As might be expected, equation 11–7 shows that the overall voltage gain of the multistage amplifier is the product of the open-circuit stage gains multiplied by the voltage-division ratios that account for the loading of each stage. Notice that a *single* voltage-division ratio accounts for the loading between any pair of stages. In other words, it is *not* correct to compute loading effects twice: once by regarding an input resistance as the load on a previous stage and again by regarding the output resistance of that previous stage as the source resistance for the next stage.

Example 11–2

The open-circuit (unloaded) voltage gains of three amplifier stages and their input and output resistances are shown in Table 11–1. If the three stages are cascaded and the first is driven by a 10-mV-rms signal source having resistance 12 kΩ, what is the voltage across a 12-Ω load connected at the output of the third stage?

Solution

$$20 \log_{10} A_{o1} = 24$$
$$\log_{10} A_{o1} = 1.2$$
$$A_{o1} = \text{antilog}\,(1.2) = 15.85$$

Table 11–1
(Example 11–2)

Amplifier Stage	Unloaded Voltage Gain (dB)	Input Resistance (kΩ)	Output Resistance (kΩ)
1	24	10	4.7
2	20	20	1.5
3	12	1.5	0.02

Similarly,

$$A_{o2} = \text{antilog}\,(1) = 10$$
$$A_{o3} = \text{antilog}\,(0.6) = 3.98$$

Then, from equation 11–7,

$$\frac{v_L}{v_S} = \left(\frac{10\text{ k}\Omega}{12\text{ k}\Omega + 10\text{ k}\Omega}\right)15.85\left(\frac{20\text{ k}\Omega}{4.7\text{ k}\Omega + 20\text{ k}\Omega}\right)10\left(\frac{1.5\text{ k}\Omega}{1.5\text{ k}\Omega + 1.5\text{ k}\Omega}\right)$$
$$\times\,(3.98)\left(\frac{12\ \Omega}{20\ \Omega + 12\ \Omega}\right) = 43.53$$

Therefore, $v_L = 43.53v_S = 43.53\ (10\text{ mV rms}) = 0.4353\text{ V rms}$.

Frequency Response of Cascaded Stages

In Chapter 10, we discussed the fact that the lower cutoff frequency of a single amplifier stage is influenced by as many as three different break frequencies, having values that depend on various RC components in the circuit. If the break frequencies were not close in value, we approximated the actual lower cutoff frequency of the amplifier by assuming it to equal the largest of those break frequencies. Similarly, we assumed the upper cutoff frequency to be the smallest of the break frequencies that affect the high-frequency response. The same reasoning applies to cascaded amplifier stages. If the lower cutoff frequencies of the individual stages are not close in value, the overall lower cutoff frequency is approximately equal to the largest of the stage lower cutoff frequencies. If the upper cutoff frequencies of the individual stages are not close in value, the overall upper cutoff frequency is approximately equal to the smallest of the stage upper cutoff frequencies.

In practice, a multistage amplifier may have some lower break frequencies that are equal, or close in value, and others that are not. The same is true for upper break frequencies. In these situations, computation of the actual lower and upper cutoff frequencies of a multistage amplifier is a very complex problem. From a practical standpoint, the cutoff frequencies are best determined experimentally or by use of a computer program that computes the overall frequency response.

For the special cases where all stages of a multistage amplifier have identical lower cutoff frequencies or identical upper cutoff frequencies, the overall cutoff frequencies can be calculated readily:

$$f_{1(overall)} = \frac{f_1}{\sqrt{2^{1/n} - 1}} \qquad \textbf{(11–8)}$$

$$f_{2(overall)} = f_2\sqrt{2^{1/n} - 1} \qquad \textbf{(11–9)}$$

where $f_{1(overall)}$ = overall lower cutoff frequency of the multistage amplifier
$f_{2(overall)}$ = overall upper cutoff frequency of the multistage amplifier
n = number of stages having identical lower cutoff frequencies and/or identical upper cutoff frequencies
f_1 = lower cutoff frequency of each stage
f_2 = upper cutoff frequency of each stage

Table 11–2
The lower and upper cutoff frequencies of a multistage amplifier consisting of n stages, each having lower cutoff frequency f_1 and upper cutoff frequency f_2

Number of Stages n	$f_{1(overall)}$	$f_{2(overall)}$
1	f_1	f_2
2	$1.55f_1$	$0.64f_2$
3	$1.96f_1$	$0.51f_2$
4	$2.30f_1$	$0.43f_2$
5	$2.59f_1$	$0.39f_2$

Table 11–2 shows values of $f_{1(overall)}$ and $f_{2(overall)}$ in terms of f_1 and f_2, for n ranging from 1 to 5. Note that when $n = 2$, $f_{1(overall)} = 1.55f_1$ and $f_{2(overall)} = 0.64f_2$, in agreement with our discussion in Chapter 10 for the case of two identical break frequencies in a single stage. The table confirms what should be intuitively clear: the greater the number of identical stages, the larger the lower cutoff frequency and the smaller the upper cutoff frequency. In other words, the act of cascading stages with identical frequency-response characteristics reduces the overall bandwidth of a multistage amplifier. When n stages having identical frequency response are cascaded, the overall frequency response falls off along asymptotes having slopes $20n$ dB/decade ($6n$ dB/octave) at frequencies outside the midband range. The break frequencies are in all cases equal to the cutoff frequencies of a single stage.

Example 11–3

A multistage audio amplifier is to be constructed using four identical stages. What should be the lower and upper cutoff frequencies of each stage if the overall lower and upper cutoff frequencies are to be 20 Hz and 20 kHz, respectively?

Solution. From Table 11–2, with $n = 4$, $2.3f_1 = 20$, and $0.43f_2 = 20 \times 10^3$. Therefore, $f_1 = 20/2.3 = 8.7$ Hz and $f_2 = 20 \times 10^3/0.43 = 46.5$ kHz. This example shows that each stage must have a bandwidth of approximately 46 kHz to achieve an overall bandwidth of approximately 20 kHz.

11–2 METHODS OF COUPLING

The circuitry used to connect the output of one stage of a multistage amplifier to the input of the next stage is called the *coupling* method. In previous chapters, we have discussed only one such method: capacitor coupling, also called *RC coupling* because the interstage circuitry is equivalent to a high-pass RC network. In this chapter we will consider two additional coupling methods: *direct coupling* and *transformer coupling.*

Recall that the primary reason for employing RC coupling is to block the flow of dc current. We have observed that it is often necessary to prevent the flow of dc current between the input of an amplifier and its signal source, as well as between the amplifier's output and its load. Similarly, RC coupling is used to prevent dc current from flowing between the output of one amplifier stage and the input of the next stage. The capacitor connected in the path between amplifier stages makes it possible to have a dc bias voltage at the output of one stage that is different from the dc bias voltage at the input to the next stage. This idea is illustrated in Figure

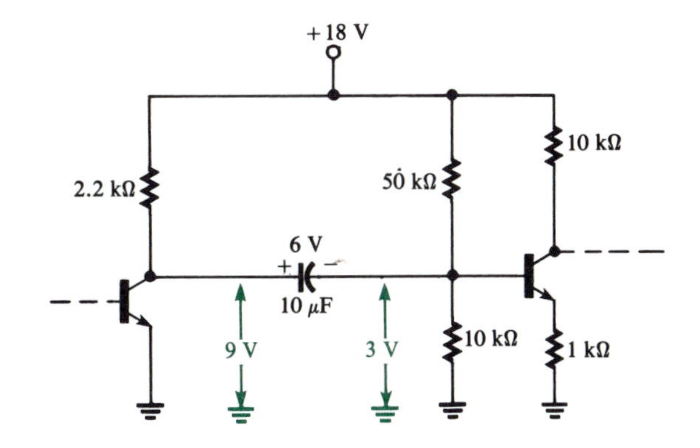

Figure 11–5
The capacitor used in the RC coupling method makes it possible to have different bias voltages on the amplifier stages. Note that the (electrolytic) capacitor has 6 V dc across it and has its positive lead connected to the more positive bias (9 V).

11–5, which shows the output of a BJT amplifier stage connected through a coupling capacitor to the input of another BJT amplifier stage. Notice that the collector of the first stage is at +9 V and that the base of the second stage is at +3 V. The dc voltage across the capacitor is therefore 9 − 3 = 6 V, so the capacitor should have a dc–working-voltage (DCWV) rating somewhat greater than 6 V. If the 10-μF coupling capacitor is of the electrolytic type, it *must* be connected with its positive terminal to the more positive bias voltage: the 9-V collector voltage in this example.

Of course, a coupling capacitor permits the flow of *ac* signal current between stages, provided the frequency is high enough to keep the capacitive reactance small. The disadvantage of the RC coupling method is that it affects the low-frequency response of the amplifier; we must sometimes choose between an impractically large capacitor value and an unreasonably large lower cutoff frequency. RC coupling is not used in integrated circuits because it is difficult and uneconomical to fabricate capacitors on a chip.

As the name implies, *direct coupling* is the coupling method in which the output of one stage is electrically connected directly to the input of the next stage. In other words, both the dc and ac voltages at the output of one stage are identical to those at the input of the next stage. The method is often referred to as *dc,* which, in the context of signal coupling, means both direct coupling and direct current. Clearly, any change in the dc voltage at the output of one stage produces an identical change in dc voltage at the input to the next stage, so a direct-coupled amplifier behaves like a direct-current amplifier. Direct coupling is used in differential and operational amplifiers, which we will study extensively in Chapter 12, and in integrated circuits. We will investigate some examples of direct-coupled discrete amplifiers later in the present chapter.

Another method of coupling an ac signal from one stage to another while maintaining dc isolation between them is through the use of a transformer. The primary winding of the transformer is in the output circuit of one stage and the secondary winding is in the input circuit of the following stage. In this way, the ac signal is passed from one stage to the next without the possibility of dc current flowing between the two. The advantages of transformer coupling include low dc power dissipation and the capability for designing a turns ratio that results in maximum power transfer between stages. We will investigate how this is accomplished in an illustrative example using bipolar transistors. The disadvantages of transformer coupling include the bulk and cost of the transformers themselves and their generally poor frequency-response characteristics. The transformer inductance and interwinding capacitance tend to reduce the usable bandwidth of these amplifiers. However, they are often used in narrowband applications, such as radio-frequency (rf) amplifiers.

11–3 RC-COUPLED BJT AMPLIFIERS

Example 11–4

Figure 11–6 shows two capacitor-coupled, common-emitter amplifier stages. Notice that the ac signal developed at the output of the first stage (the collector of Q_1) is coupled through the 0.85-μF capacitor to the input of the second stage (the base of Q_2). Assuming that the transistors are identical and have $\beta = 100$, $r_c = 1\ \text{M}\Omega$, and $r_e = 25\ \Omega$, find the small signal, midband (1) voltage gain v_L/v_S and (2) current gain i_L/i_S.

Solution. We will use the common-emitter, small-signal relations that are summarized in equations 5–44.

1. The input resistance to the first stage is

$$r_{in}(\text{stage 1}) = R_{B1} \parallel \beta r_e = (1\ \text{M}\Omega) \parallel (2.5\ \text{k}\Omega) \approx 2.5\ \text{k}\Omega$$

The output resistance of stage 1 (at the collector of Q_1) is

$$r_o(\text{stage 1}) = R_{C1} \parallel (r_c/\beta) = (3.3\ \text{k}\Omega) \parallel [(1\ \text{M}\Omega)/100] = 2.48\ \text{k}\Omega$$

The *unloaded* voltage gain of stage 1 is, therefore,

$$A_{v1} = \frac{-r_o(\text{stage 1})}{r_e} = \frac{-2.48\ \text{k}\Omega}{25\ \Omega} = -99.2$$

The input resistance to the second stage is

$$\begin{aligned} r_{in}(\text{stage 2}) &= R_1 \parallel R_2 \parallel \beta(r_e + R_{E2}) \\ &= (100\ \text{k}\Omega) \parallel (10\ \text{k}\Omega) \parallel 100\,[(25\ \Omega) + (220\ \Omega)] \\ &= (9.09\ \text{k}\Omega) \parallel (24.5\ \text{k}\Omega) = 6.63\ \text{k}\Omega \end{aligned}$$

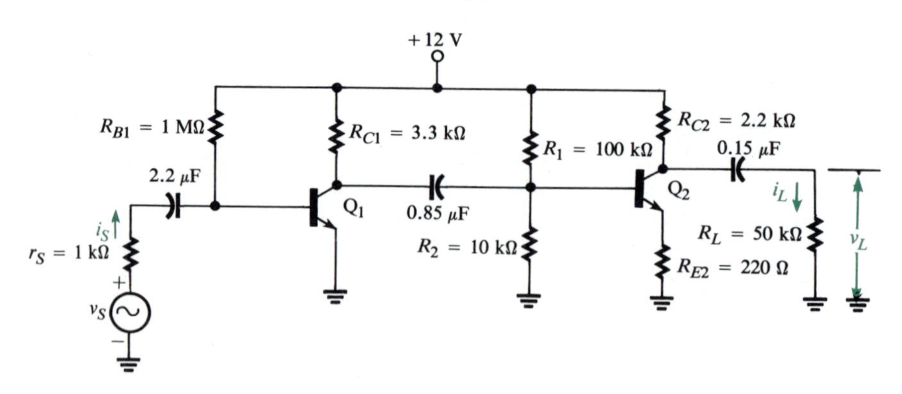

Figure 11–6
(Example 11–4)

The output resistance of stage 2 (at the collector of Q_2) is

$$r_o(\text{stage 2}) = R_{C2} \parallel (r_c/\beta) = (2.2\text{ k}\Omega) \parallel (1\text{ M}\Omega/100) = 1.8\text{ k}\Omega$$

The *unloaded* voltage gain of stage 2 is, therefore,

$$A_{v2} = \frac{-r_o(\text{stage 2})}{R_{E2} + r_e} = \frac{-1.8\text{ k}\Omega}{(220\ \Omega) + (25\ \Omega)} = -7.35$$

The two-stage amplifier can now be represented as shown in Figure 11–7. No capacitors are shown in this figure because we are assuming operation in the midband frequency range. Using (the two-stage form of) equation 11–7, we find

$$\frac{v_L}{v_S} = \left[\frac{2.5\text{ k}\Omega}{(1\text{ k}\Omega) + (2.5\text{ k}\Omega)}\right](-99.2)\left[\frac{6.63\text{ k}\Omega}{(2.48\text{ k}\Omega) + (6.63\text{ k}\Omega)}\right](-7.35)$$
$$\times\left[\frac{50\text{ k}\Omega}{(1.8\text{ k}\Omega) + (50\text{ k}\Omega)}\right] = 365.85$$

The positive result shows that v_L is in phase with v_S.

Notice that an alternative approach to finding the overall voltage gain is to find the voltage gains A_1 and A_2 *with loads connected.* Taking this approach, we compute the ac load resistance r_L of each stage:

$$r_L(\text{stage 1}) = r_o(\text{stage 1}) \parallel r_{in}(\text{stage 2}) = (2.48\text{ k}\Omega) \parallel (6.63\text{ k}\Omega) = 1.8\text{ k}\Omega$$
$$r_L(\text{stage 2}) = r_o(\text{stage 2}) \parallel R_L = (1.8\text{ k}\Omega) \parallel (50\text{ k}\Omega) = 1.74\text{ k}\Omega$$

The voltage gains with these loads connected are then

$$A_1 \approx \frac{-r_{L1}}{r_e} = \frac{-1.8\text{ k}\Omega}{25\ \Omega} = -72$$
$$A_2 \approx \frac{-r_{L2}}{r_e + R_{E2}} = \frac{-1.74\text{ k}\Omega}{(25\ \Omega) + (200\ \Omega)} = -7.1$$

The overall voltage gain is then

$$\frac{v_L}{v_S} = \left[\frac{r_{in}(\text{stage 1})}{r_{in}(\text{stage 1}) + r_S}\right]A_1A_2 = \left[\frac{2.5\text{ k}\Omega}{(2.5\text{ k}\Omega) + (1\text{ k}\Omega)}\right](-72)(-7.1) = 365.1$$

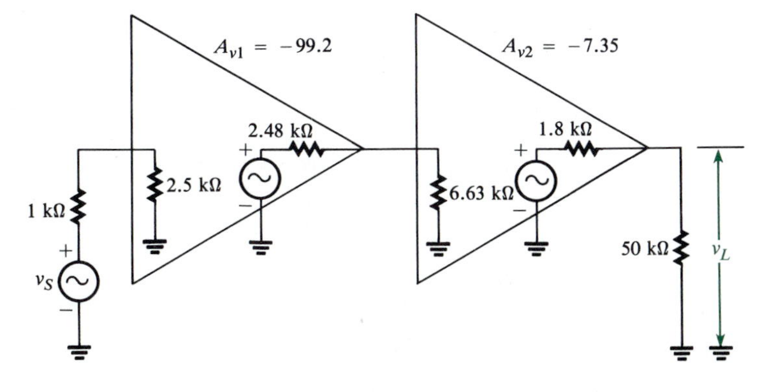

Figure 11–7
(Example 11–4) The two-stage amplifier of Figure 11–6

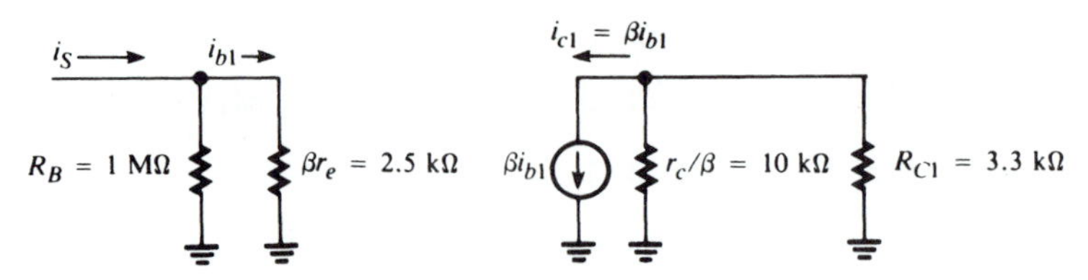

Figure 11–8
(Example 11–4) The small-signal equivalent circuit for the first stage of the amplifier in Figure 11–6. This circuit is used to compute the current i_{b1} in terms of i_S.

Except for a small rounding error, this approach produces the same result as before.

2. To determine the overall current gain, it is helpful to draw the small-signal equivalent circuit of each stage and then trace the flow of ac current through the amplifier. We will use the current-divider rule at each node to determine the portion of signal current that continues to flow toward the load. Figure 11–8 shows the small-signal equivalent of the first stage. Applying the current-divider rule to the input side of stage 1 in Figure 11–8, we obtain

$$i_{b1} = \left[\frac{1\ \text{M}\Omega}{(2.5\ \text{k}\Omega) + (1\ \text{M}\Omega)}\right] i_S = 0.9975 i_S$$

This result shows that essentially all of the source current enters the base of Q_1 ($0.0025 i_S$ is shunted to ground through R_B). At the output side of stage 1, we have

$$i_{c1} = \beta i_{b1} = 100 i_{b1} = 100(0.9975 i_S) = 99.75 i_S$$

To find the portion of βi_{b1} that flows into the base of Q_2, we must consider all the parallel paths to ground in the interstage circuitry between Q_1 and Q_2. Figure 11–9 shows the equivalent circuit of the second stage and the output side of the first stage. Note that r_o(stage 1) is the parallel combination of r_c/β and R_{C1}, and this combination is in parallel with the bias resistors R_1 and R_2 at the input to the second stage. The total equivalent resistance r_{SH} shunting the input (base) of Q_2 is, therefore,

$$r_{SH} = r_o(\text{stage 1}) \parallel R_1 \parallel R_2 = (2.48\ \text{k}\Omega) \parallel (100\ \text{k}\Omega) \parallel (10\ \text{k}\Omega) = 1.95\ \text{k}\Omega$$

The current i_{b2} into the base of Q_2 is then found by the current-divider rule:

$$i_{b2} = 99.75 i_S \left[\frac{r_{SH}}{r_{SH} + \beta(r_e + R_E)}\right]$$

$$= 99.75\, i_S \left[\frac{1.95\ \text{k}\Omega}{(1.95\ \text{k}\Omega) + (24.5\ \text{k}\Omega)}\right] = 7.35 i_S$$

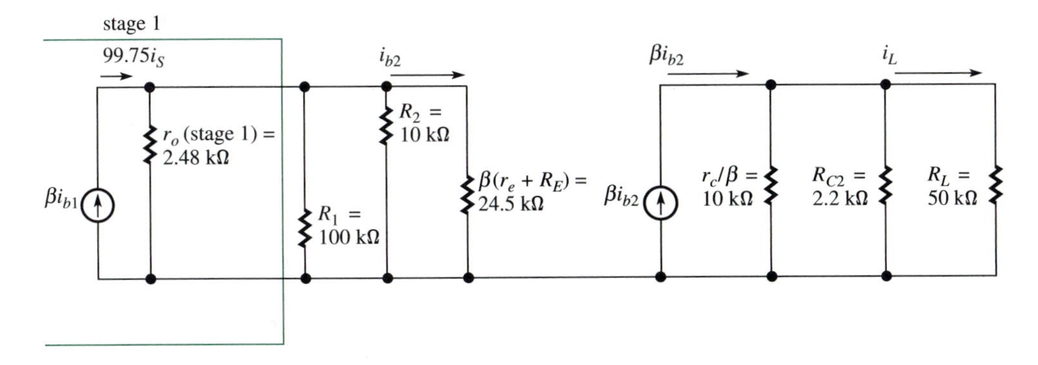

Figure 11–9
(Example 11–4) The small-signal equivalent circuit of the output side of stage 1 joined to amplifier stage 2. This circuit is used to find i_L in terms of i_S.

Notice that a significant portion of signal current is lost in the interstage circuitry due to the shunt paths to ground: Less than $\frac{1}{10}$ of i_{c1} reaches the base of Q_2. Now,

$$i_{c2} = \beta i_{b2} = 100(7.35i_S) = 735i_S$$

The parallel combination of r_c/β and R_{C2} shunts R_L, so one more application of the current-divider rule gives

$$i_L = \left[\frac{(r_c/\beta) \,\|\, R_{C2}}{R_L + (r_c/\beta) \,\|\, R_{C2}}\right]\beta i_{b2} = \left[\frac{(10\text{ k}\Omega) \,\|\, (2.2\text{ k}\Omega)}{(50\text{ k}\Omega) + (10\text{ k}\Omega) \,\|\, (2.2\text{ k}\Omega)}\right]735i_S$$

$$= \left(\frac{1.8\text{ k}\Omega}{51.8\text{ k}\Omega}\right)735i_S = 25.54i_S$$

Therefore, the current gain from source to load is

$$\frac{i_L}{i_S} = 25.54$$

This example demonstrates that the current gain of the two-stage BJT amplifier can be found using the relation

$$\frac{i_L}{i_S} = \left(\frac{R_{B1}}{r_{in1} + R_{B1}}\right)\beta_1\left(\frac{r_{SH}}{r_{SH} + r_{in2}}\right)\beta_2\left[\frac{r_o(\text{stage 2})}{R_L + r_o(\text{stage 2})}\right] \qquad \textbf{(11–10)}$$

where r_{in1}, r_{in2} = resistance looking into the input (base) terminal of Q_1 and Q_2, respectively (*not* the stage input resistance)

r_{SH} = $r_o(\text{stage 1}) \,\|\, R_{B2}$

R_{B1}, R_{B2} = total equivalent resistance shunting the input of stage 1 and stage 2, respectively

Example 11–5

Figure 11–10 shows an amplifier consisting of a common-emitter stage driving an emitter-follower stage. The transistors have the following parameter values:

$$Q_1: \quad r_{e1} = 15\ \Omega, \beta_1 = 180, r_o \approx \infty$$
$$Q_2: \quad r_{e2} = 40\ \Omega, \beta_2 = 100$$

Find (1) the midband voltage gain v_L/v_S and (2) the approximate value of the lower cutoff frequency.

Solution

1. Noting that the 1.5-kΩ emitter resistor in the first stage is bypassed by the 40-μF capacitor, the input resistance to the first stage is

$$r_{in}(\text{stage 1}) = R1 \parallel R2 \parallel \beta_1 r_{e1} = (150\text{ k}\Omega) \parallel (39\text{ k}\Omega) \parallel (180)(15\ \Omega) = 2.48\text{ k}\Omega$$

The output resistance of the first stage, with $r_{o1} \approx \infty$, is

$$r_o(\text{stage 1}) = r_{o1} \parallel R_{C1} \approx R_{C1} = 4.7\text{ k}\Omega$$

Therefore, the unloaded voltage gain of the first stage is

$$A_{v1} \approx \frac{-R_{c1}}{r_{e1}} = \frac{-4.7\text{ k}\Omega}{15\ \Omega} = -313.3$$

The ac load on the second stage is $r_L = (10\text{ k}\Omega) \parallel (50\ \Omega) \approx 50\ \Omega$, so the input resistance to the second stage is

$$\begin{aligned} r_{in}(\text{stage 2}) &= R1 \parallel R2 \parallel \beta_2(r_{e2} + r_L) \\ &= (68\text{ k}\Omega) \parallel (47\text{ k}\Omega) \parallel 100[(40\ \Omega) + (50\ \Omega)] = 6.8\text{ k}\Omega \end{aligned}$$

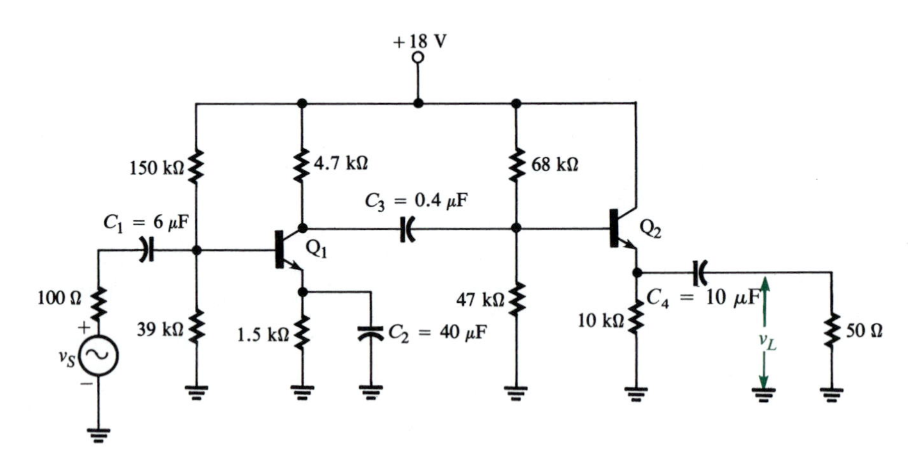

Figure 11–10
(Example 11–5)

Note this important point: The effects of the bias resistors on the gain of the amplifier will be taken into account when we calculate the voltage division between the first and second stages. Therefore, we do *not* include their effect on the voltage division that takes place at the output of the emitter follower. In other words, we do not include the term $R_B \parallel r_S/(\beta + 1)$ in the equation for the output resistance of the second stage (see equations 5–71). The only component of the output resistance that we consider at the second stage is $R_E \parallel r_{e2} = 10\ \text{k}\Omega \parallel 40\ \Omega \approx 40\ \Omega$. (Adding the term $R_B \parallel r_S/(\beta + 1)$ to 40 Ω and then calculating the voltage division caused by the load at the output of the second stage would be equivalent to considering the effects of the bias resistors twice.)

The unloaded voltage gain of stage 2 is

$$A_{v2} = \frac{R_E}{r_{e2} + R_E} = \frac{10\ \text{k}\Omega}{(40\ \text{k}\Omega) + (10\ \text{k}\Omega)} = 0.996$$

The two-stage amplifier is equivalent to that shown in Figure 11–11. It is apparent from the figure that

$$\frac{v_L}{v_S} = \left[\frac{2.48\ \text{k}\Omega}{(100\ \Omega) + (2.48\ \text{k}\Omega)}\right](-313.3)\left[\frac{6.8\ \text{k}\Omega}{(4.7\ \text{k}\Omega) + (6.8\ \text{k}\Omega)}\right] \times (0.996)\left[\frac{50\ \Omega}{(40\ \Omega) + (50\ \Omega)}\right] = -\ 98.46$$

We see that the load voltage is out of phase with the source voltage. As an exercise, repeat the computation using the loaded amplifier gains and verify that the overall gain is the same as that found above.

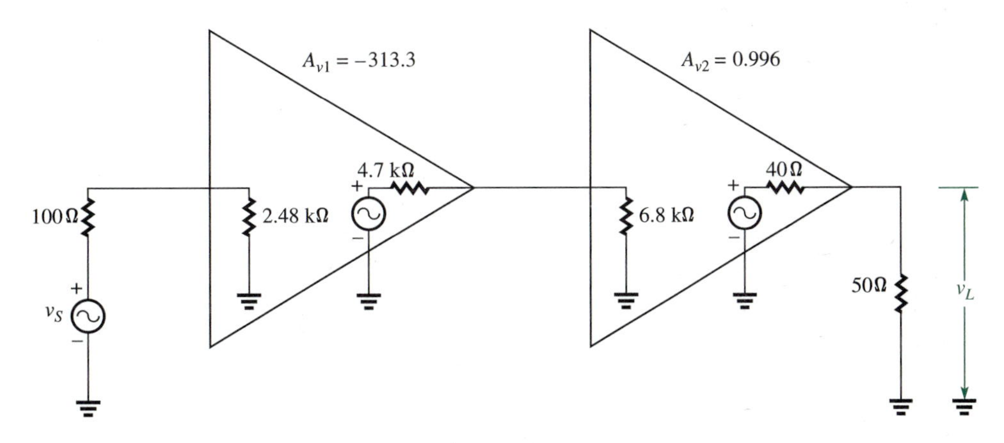

Figure 11–11
(Example 11–5) The equivalent circuit of the two-stage amplifier in Figure 11–10

2. To find the lower cutoff frequency, we must find the break frequency due to each of the capacitors C_1, C_2, C_3, and C_4 in Figure 11–10. From equation 10–16,

$$f_1(C_1) = \frac{1}{2\pi[r_{in}(\text{stage 1}) + r_S]C_1} = \frac{1}{2\pi(2.48 \times 10^3 + 100)6 \times 10^{-6}} = 10.3 \text{ Hz}$$

From equation 10–43,

$$f_1(C_2) = \frac{1}{2\pi R_e C_E}$$

where $C_E = C_2 = 40\ \mu\text{F}$ and

$$R_e = R_E \left\| \left(\frac{r_S \| R_B}{\beta_1} + r_{e1} \right) \right.$$

$$= (1.5 \times 10^3) \left\| \left[\frac{100 \| (150 \times 10^3) \| (39 \times 10^3)}{180} + 15 \right] \right. \approx 15\ \Omega$$

Thus,

$$f_1(C_2) = \frac{1}{2\pi(15)(40 \times 10^{-6})} = 265.3 \text{ Hz}$$

To find the break frequency due to C_3, we regard $r_o(\text{stage 1})$ as source resistance and $r_{in}(\text{stage 2})$ as load resistance. Therefore,

$$f_1(C_3) = \frac{1}{2\pi[r_o(\text{stage 1}) + r_{in}(\text{stage 2})]C_3}$$

$$= \frac{1}{2\pi(4.7 \times 10^3 + 6.8 \times 10^3)(0.4 \times 10^{-6})} = 34.6 \text{ Hz}$$

To find the lower cutoff frequency due to the 10-μF output coupling capacitor (C_4), we must calculate $r_o(\text{stage 2})$ and include, this time, the effects of the 68-kΩ and 47-kΩ bias resistors. Thus, we calculate

$$r_o(\text{stage 2}) = R_{e2} = R_{E2} \left\| \left(\frac{r_S \| R_{B2}}{\beta_2} + r_{e2} \right) \right.$$

where

$$R_{B2} = 68\ \text{k}\Omega \| 47\ \text{k}\Omega = 27.8\ \text{k}\Omega$$

$$r_S = r_o(\text{stage 1}) = 4.7\ \text{k}\Omega$$

$$r_o(\text{stage 2}) = 10\ \text{k}\Omega \left\| \left(\frac{4.7 \times 10^3 \| 27.8 \times 10^3}{100} + 40\ \Omega \right) \right. = 79.8\ \Omega$$

Finally,

$$f_1(C_4) = \frac{1}{2\pi[r_o(\text{stage 2}) + R_L]C_4} = \frac{1}{2\pi(79.8 + 50)(10 \times 10^{-6})} = 122.6 \text{ Hz}$$

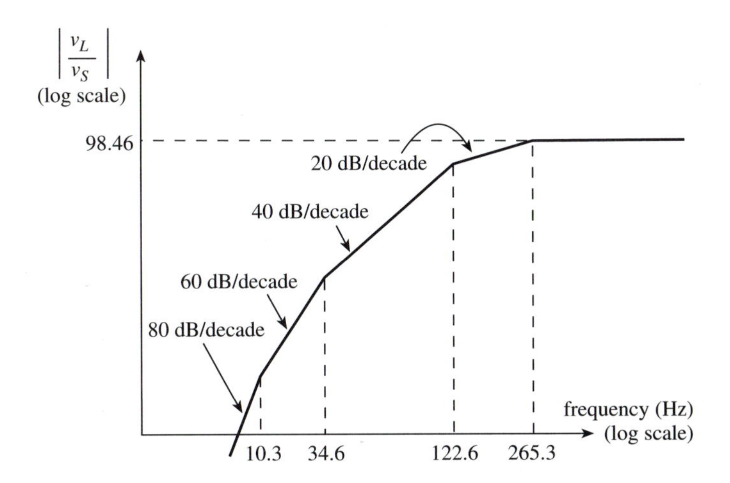

Figure 11–12
(Example 11–5) A sketch of the Bode plot for the gain of the multistage amplifier in Figure 11–10. Note that the asymptotes fall off by an additional 20 dB/decade below each break frequency.

In this example, the greatest break frequencies are $f_1(C_2) = 265.3$ Hz and $f_1(C_4) = 122.6$ Hz. While we can conclude that the lower cutoff frequency of the amplifier is no *smaller* than the larger of these two, 265.3 Hz, the two frequencies are not far enough apart to conclude that f_1 equals 265.3 Hz. Instead, we know that f_1 will be somewhat greater than 265.3 Hz. As discussed earlier, computation of the actual cutoff frequency in such cases is a complex task that is best left to a computer simulation. The next example shows how we use SPICE to find that f_1 in the present example is actually 330 Hz.

Figure 11–12 shows a sketch of the asymptotic Bode plot of the gain magnitude. Note that the gain falls off at a rate that increases by an additional 20 dB/decade below each break frequency found in the example.

Example 11–6

SPICE

Use SPICE to obtain a plot of the low-frequency response of the amplifier in Example 11–5. Find the midband voltage gain, v_L/v_S, and the lower cutoff frequency.

Solution. Figure 11–13(a) shows the SPICE format for the amplifier circuit and the input data file. Note that two .MODEL statements are required because the transistors have different values of β.

The frequency response plot in Figure 11–13(b) shows that the magnitude of the midband gain (at 10 kHz) is 99.61, in close agreement with the calculated value in Example 11–5. The lower cutoff frequency is therefore the frequency at which the gain equals 0.707(99.61) = 70.42. Figure 11–13(c) shows the results of a .PRINT

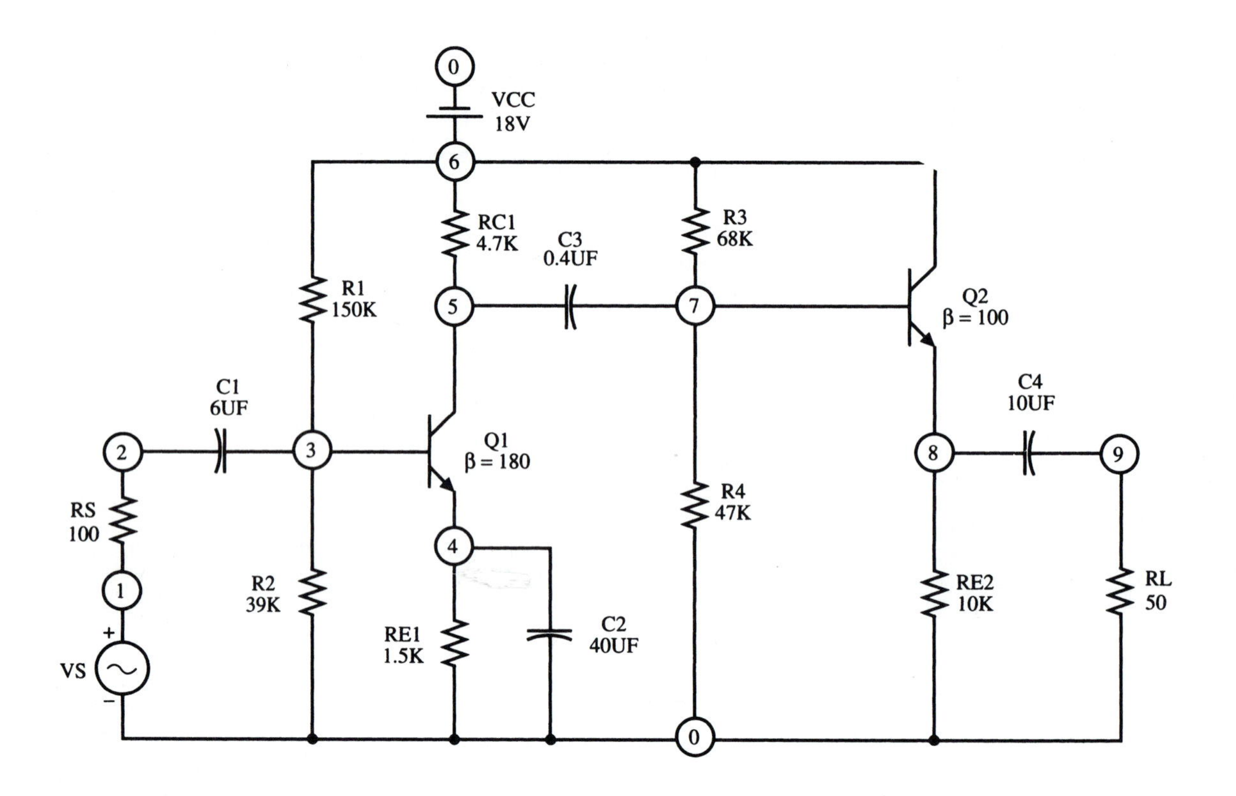


```
EXAMPLE 11.6
VS 1 0 AC
RS 1 2 100
C1 2 3 6UF
R1 6 3 150K
R2 3 0 39K
RE1 4 0 1.5K
C2 4 0 40UF
RC1 6 5 4.7K
VCC 6 0 18V
C3 5 7 0.4UF
R3 6 7 68K
R4 7 0 47K
RE2 8 0 10K
C4 8 9 10UF
RL 9 0 50
Q1 5 3 4 TRAN1
Q2 6 7 8 TRAN2
.MODEL TRAN1 NPN BF=180
.MODEL TRAN2 NPN BF=100
.AC DEC 10 10HZ 10KHZ
.PLOT AC V(9)
.END
```

(a)

Figure 11–13
(Example 11–6)

```
EXAMPLE 11.6
****      AC ANALYSIS                              TEMPERATURE =    27.000 DEG C
******************************************************************************
   FREQ       V(9)
                 1.000D-01     1.000D+00     1.000D+01     1.000D+02  1.000D+03
                 - - - - - - - - - - - - - - - - - - - - - - - - - - - - -
1.000D+01  1.769D-01 .  *          .             .             .             .
1.259D+01  3.083D-01 .      *      .             .             .             .
1.585D+01  5.253D-01 .         *   .             .             .             .
1.995D+01  8.760D-01 .            *.             .             .             .
2.512D+01  1.433D+00 .             . *           .             .             .
3.162D+01  2.308D+00 .             .    *        .             .             .
3.981D+01  3.661D+00 .             .       *     .             .             .
5.012D+01  5.719D+00 .             .          *  .             .             .
6.310D+01  8.781D+00 .             .            *.             .             .
7.943D+01  1.319D+01 .             .             . *           .             .
1.000D+02  1.927D+01 .             .             .   *         .             .
1.259D+02  2.717D+01 .             .             .     *       .             .
1.585D+02  3.674D+01 .             .             .       *     .             .
1.995D+02  4.743D+01 .             .             .        *    .             .
2.512D+02  5.834D+01 .             .             .          *  .             .
3.162D+02  6.851D+01 .             .             .           * .             .
3.981D+02  7.719D+01 .             .             .           * .             .
5.012D+02  8.405D+01 .             .             .            *.             .
6.310D+02  8.913D+01 .             .             .            *.             .
7.943D+02  9.271D+01 .             .             .             *             .
1.000D+03  9.514D+01 .             .             .             *             .
1.259D+03  9.675D+01 .             .             .             *             .
1.585D+03  9.780D+01 .             .             .             *             .
1.995D+03  9.848D+01 .             .             .             *             .
2.512D+03  9.891D+01 .             .             .             *             .
3.162D+03  9.918D+01 .             .             .             *             .
3.981D+03  9.936D+01 .             .             .             *             .
5.012D+03  9.947D+01 .             .             .             *             .
6.310D+03  9.954D+01 .             .             .             *             .
7.943D+03  9.958D+01 .             .             .             *             .
1.000D+04  9.961D+01 .             .             .             *             .
                 - - - - - - - - - - - - - - - - - - - - - - - - - - - - -
```

(b)

```
EXAMPLE 11.6
****      AC ANALYSIS                              TEMPERATURE =    27.000 DEG C
******************************************************************************
   FREQ        V(9)
 3.000E+02     6.629E+01
 3.056E+02     6.707E+01
 3.111E+02     6.783E+01
 3.167E+02     6.857E+01
 3.222E+02     6.928E+01
 3.278E+02     6.997E+01
 3.333E+02     7.065E+01
 3.389E+02     7.130E+01
 3.444E+02     7.193E+01
 3.500E+02     7.255E+01
```

(c)

Figure 11–13
(Continued)

statement over the frequency range from 300 Hz to 350 Hz, and we see that $f_1 \approx$ 330 Hz. As noted in Example 11–5, this frequency is somewhat higher than the highest break frequency calculated in the example ($f_1(C_2) = 265.3$ Hz) because of the nearness of another break frequency at 122.6 Hz.

To find the upper cutoff frequency of a multistage amplifier, we simply apply the general equations developed in Chapter 10 (equations 10–30 and 10–31) to each stage. For example, the upper cutoff frequency due to shunt capacitance in the interstage coupling circuitry between stages 1 and 2 is

$$f_2(C_I) = \frac{1}{2\pi[r_o(\text{stage 1}) \parallel r_{in}(\text{stage 2})]C_I} \tag{11–11}$$

where C_I is the sum of all the interstage shunt capacitances; $C_I = C_{out}(\text{stage 1}) + C_{in}(\text{stage 2})$. Remember to include Miller-effect capacitance, if any, in the quantity $C_{in}(\text{stage 2})$.

11–4 DIRECT-COUPLED BJT AMPLIFIERS

In Chapter 6 we discussed the current-mirroring method for eliminating coupling capacitors in integrated circuits. Figure 11–14 shows an example of direct-coupled amplifier stages using conventional bias methods. Notice that the output of the first stage (collector of Q_1) is connected directly to the input of the second stage (base of Q_2). We will first analyze the dc bias of the circuit and then consider its ac performance. Notice that the current in R_{CI} is the sum of I_{C1} and I_{B2}. To simplify the analysis, we will make the reasonable assumption that I_{B2} is negligibly small in comparison to I_{C1}.

Assuming that the first stage is well stabilized against variations in β, the base voltage V_{B1} is determined essentially by the R_1–R_2 voltage divider:

$$V_{B1} \approx \left(\frac{R_2}{R_1 + R_2}\right) V_{CC} \tag{11–12}$$

If this assumption is not valid in a particular case, then equation 6–12a can be used to determine I_{C1}. Assuming silicon transistors,

$$V_{E1} \approx V_{B1} - 0.7 \tag{11–13}$$

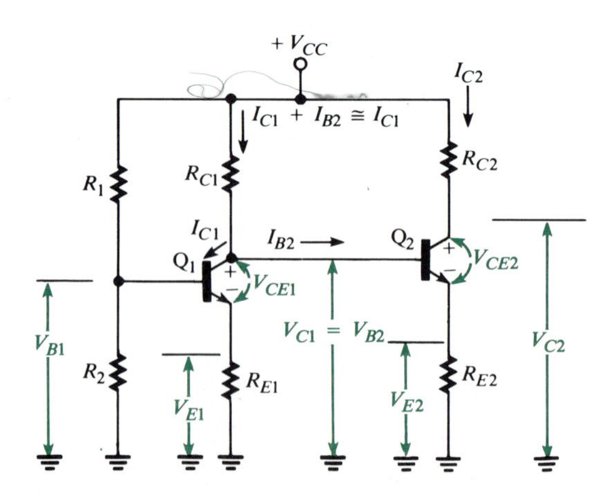

Figure 11–14
Direct-coupled CE amplifier stages using conventional bias methods

Then $I_{C1} \approx I_{E1} = V_{E1}/R_{E1}$. Under our assumption that I_{B2} is negligible in comparison to I_{C1}, the current in R_{C1} is $I_{C1} + I_{B2} \approx I_{C1}$, so the quiescent collector-to-ground voltage is

$$V_{C1} \approx V_{CC} - I_{C1}R_{C1} \tag{11–14}$$

The quiescent value of V_{CE1} is

$$V_{CE1} = V_{C1} - V_{E1} \tag{11–15}$$

Notice that $V_{C1} = V_{B2}$, so

$$V_{E2} = V_{C1} - 0.7 \tag{11–16}$$

Then $I_{C2} \approx I_{E2} = V_{E2}/R_{E2}$ and the quiescent collector voltages of stage 2 are

$$V_{C2} = V_{CC} - I_{C2}R_{C2} \tag{11–17}$$

$$V_{CE2} = V_{C2} - V_{E2} \tag{11–18}$$

We see that the dc analysis is quite straightforward, being nothing more than the application of bias principles we have already studied. The important point to note is that $V_{C1} = V_{B2}$. The ac gain analysis is similarly straightforward.

The voltage gain of the first stage is

$$A_{v1} \approx \frac{-r_o(\text{stage 1}) \parallel r_{in}(\text{stage 2})}{r_{e1} + R_{E1}} \tag{11–19}$$

where $r_o(\text{stage 1}) \approx R_{C1}$ and $r_{in}(\text{stage 2}) \approx \beta_2(r_{e2} + R_{E2})$. The voltage gain of the second stage is

$$A_{v2} \approx \frac{-r_o(\text{stage 2})}{r_{e2} + R_{E2}} \approx \frac{-R_{C2}}{r_{e2} + R_{E2}} \tag{11–20}$$

Finally, the overall gain is the product of the stage gains:

$$A_{v(overall)} = A_{v1}A_{v2} \tag{11–21}$$

If a load resistance R_L is direct-coupled between the output (collector of Q_2) and ground, then the ac load resistance on the second stage is $r_L = R_{C2} \parallel R_L$ and equation 11–20 becomes

$$A_{v2} \approx \frac{-R_{C2} \parallel R_L}{r_{e2} + R_{E2}} \tag{11–22}$$

It is important to realize that direct-coupling an output load resistance changes the dc value of V_{C2} and V_{CE2}. To demonstrate this fact, let us regard the transistor as a constant-current source, as shown in Figure 11–15(b). We can then apply the superposition principle to find the dc voltage V_L $(= V_C)$ due to each source in the circuit, as shown in Figure 11–15(c) and (d). Combining the contributions of each source leads to

$$V_L = V_C = \left(\frac{R_L}{R_L + R_C}\right)(V_{CC} - I_C R_C) \tag{11–23}$$

Equation 11–23 simply shows that the collector voltage equals its unloaded value $(V_{CC} - I_C R_C)$ divided down by the voltage divider formed by R_L and R_C. Since $V_{CE} = V_C - V_E$, it is clear that too small a value of R_L can reduce V_C to the point that V_{CE} approaches 0, severely limiting the voltage swing.

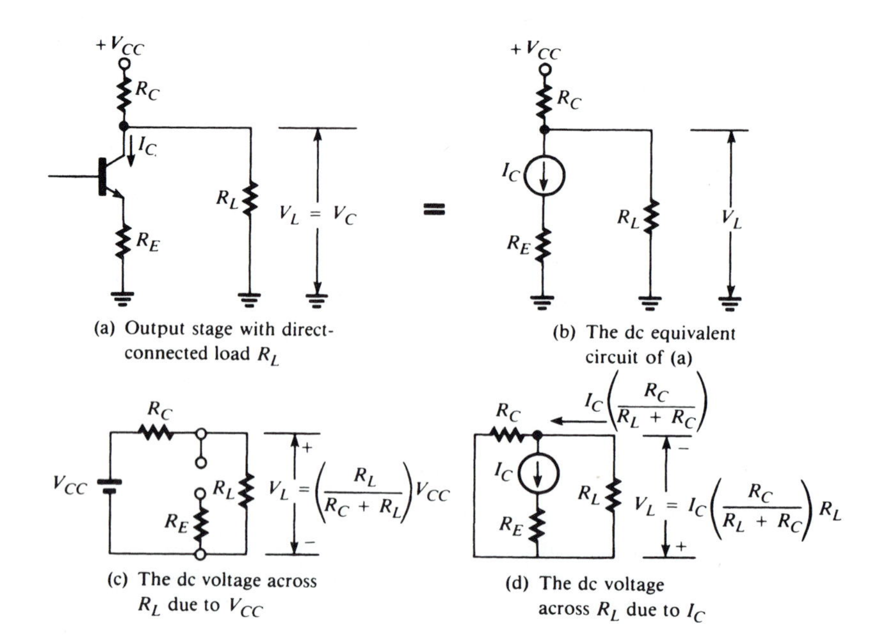

By superposition,

$$V_L = \left(\frac{R_L}{R_C + R_L}\right)V_{CC} - I_C\left(\frac{R_C}{R_L + R_C}\right)R_L = \left(\frac{R_L}{R_L + R_C}\right)(V_{CC} - I_C R_C)$$

Figure 11–15
Computing the dc voltage across a direct-connected load R_L, using the principle of superposition

Example 11–7

The silicon transistors in Figure 11–16 have the following parameters:

$$Q_1: \quad \beta_1 = 100, r_{e1} = 6\ \Omega, r_{c1} \approx \infty$$
$$Q_2: \quad \beta_2 = 60, r_{e2} = 10\ \Omega, r_{c2} \approx \infty$$

1. Find the quiescent values of V_{CE1}, I_{C2}, I_T, V_{C2}, and V_{CE2}.
2. Find the voltage gain v_o/v_{in}.
3. Repeat (1) and (2) if a 10-kΩ load is direct-coupled between the collector of Q_2 and ground.

Solution

1. The dc input resistance looking directly into the base of Q_1 is approximately

$$R_{in1} \approx \beta_1 R_{E1} = 100(75) = 7.5\ \text{k}\Omega$$

Since this resistance is not large compared to the 11 kΩ resistor in the voltage-divider network across the base, we will use equation 6–12a to find I_{C1}.

Figure 11–16
(Example 11–7)

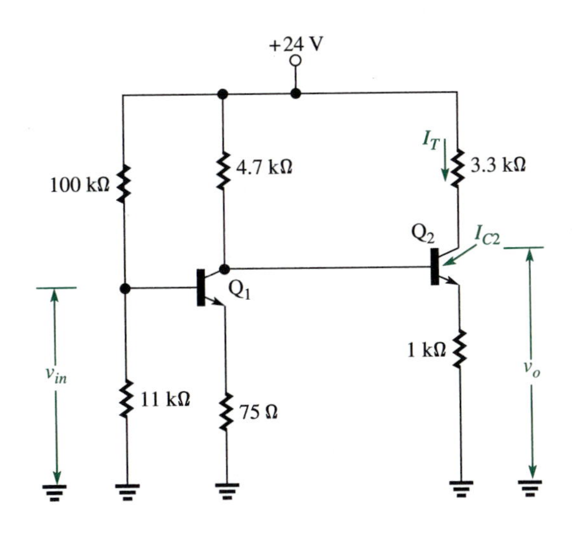

$$I_{C1} = \beta_1\left[\frac{\left(\frac{R_2}{R_1 + R_2}\right)V_{CC} - V_{BE}}{R_1 \| R_2 + (\beta_1 + 1)R_E}\right]$$

$$= 100\left[\frac{\left(\frac{11\text{ k}\Omega}{169\text{ k}\Omega + 11\text{ k}\Omega}\right)24\text{ V} - 0.7\text{ V}}{169\text{ k}\Omega \| 11\text{ k}\Omega + (101)75\ \Omega}\right] = 4.26\text{ mA}$$

Neglecting the current I_{B2}, the collector-to-ground voltage, V_{C1}, is therefore

$$V_{C1} \approx V_{CC} - I_{C1}R_{C1} = 24 - (4.26\text{ mA})\,(4.7\text{ k}\Omega) = 3.95\text{ V}$$

Thus,

$$V_{CE1} = V_{C1} - V_{E1} = 3.95 - 0.32 = 3.63\text{ V}$$

Since $V_{C1} = V_{B2}$, we have

$$V_{E2} \approx V_{C1} - 0.7 = 3.95 - 0.7 = 3.25\text{ V}$$

Therefore,

$$I_{C2} \approx I_{E2} = \frac{V_{E2}}{R_{E2}} = \frac{3.25\text{ V}}{1\text{ k}\Omega} = 3.25\text{ mA}$$

Since there is no load resistor connected to Q_2, $I_T = I_{C2} = 3.25$ mA.

Note that $I_{B2} = I_{C2}/\beta_2 = (3.25\text{ mA})/60 = 0.054$ mA, which is indeed negligibly small in comparison to $I_{C1} = 4.26$ mA.

The dc collector-to-ground voltage, V_{C2}, and the dc collector-to-emitter voltage, V_{CE2}, are

$$V_{C2} = V_{CC} - I_{C2}R_{C2} = 24 - (3.25\text{ mA})(3.3\text{ k}\Omega) = 13.3\text{ V}$$
$$V_{CE2} = V_{C2} - V_{E2} = 13.3 - 3.25 = 10.05\text{ V}$$

2.

$$A_{v1} \approx \frac{-R_{C1} \| \beta_2(R_{E2} + r_{e2})}{R_{E1} + r_{e1}} = \frac{-(4.7\text{ k}\Omega) \| 60[(1\text{ k}\Omega) + (10\ \Omega)]}{(75\ \Omega + 6\ \Omega)} = -53.8$$

$$A_{v2} \approx \frac{-R_{C2}}{R_{E2} + r_{e2}} = \frac{-3.3\text{ k}\Omega}{(1\text{ k}\Omega) + (10\ \Omega)} = -3.3$$

$$A_{v(overall)} = A_{v1}A_{v2} = (-53.8)(-3.3) = 177.5$$

3. The quiescent value of V_{CE1} is unaffected by a load resistor connected to the collector of Q_2, so $V_{CE1} = 3.63$ V, as in (1). Similarly, I_{C2} is the same as in (1): 3.25 mA. From equation 11–23,

$$V_L = V_{C2} = \left[\frac{10\text{ k}\Omega}{(10\text{ k}\Omega) + (3.3\text{ k}\Omega)}\right][24 - (3.25\text{ mA})(3.3\text{ k}\Omega)] = 10\text{ V}$$

Therefore, $V_{CE2} = V_{C2} - V_{E2} = 10 - 3.25 = 6.75$ V. Now,

$$I_L = \frac{V_L}{R_L} = \frac{10\text{ V}}{10\text{ k}\Omega} = 1\text{ mA}$$

so $I_T = I_{C2} + I_L = (3.25\text{ mA}) + (1\text{ mA}) = 4.25$ mA.

The voltage gain of stage 1 is still -53.8. The voltage gain of stage 2 is now

$$A_{v2} \approx \frac{-R_{C2} \| R_L}{R_{E2} + r_{e2}} = \frac{-(3.3\text{ k}\Omega) \| (10\text{ k}\Omega)}{(1\text{ k}\Omega) + (10\ \Omega)} = -2.48$$

Therefore, the overall voltage gain with the 10 kΩ load connected is $A_{v(overall)} = (-53.8)(-2.48) = 133.4$.

Recall that the collector voltage of an NPN transistor must always be more positive than its base voltage. When transistors are direct-connected, the collector voltage of one stage equals the base voltage of the next stage. Therefore, the collector voltage of any NPN stage must be greater than the collector voltage of a preceding NPN stage. As the number of direct-connected transistors increases, the collector voltages become progressively larger. In practice, relatively few direct-coupled stages can be cascaded before the collector voltage must be made impractically large, exceeding the power supply voltage. To overcome this problem, alternating transistor types (NPN, PNP) can be used in direct-coupled cascades. In this application, the transistors are said to be *complementary,* and Figure 11–17 shows a two-stage example. The arrangement works well because the base of a PNP transistor must be more positive than its collector—just the opposite to the condition for an NPN transistor.

Of particular interest in Figure 11–17 is the biasing of the PNP transistor. Note that the positive supply voltage is connected to the emitter side of the transistor and the collector side is grounded. Assuming a silicon transistor, the emitter-to-ground voltage of the PNP transistor is about 0.7 V more positive than its base-to-ground voltage (for forward bias of the base–emitter junction). Thus, $V_{E2} = 19.4 + 0.7 = 20.1$ V. The voltage drop across the emitter resistor is, therefore,

$$V_{CC} - V_{E2} = 24 - 20.1 = 3.9\text{ V}$$

and so

$$I_{E2} = (3.9\text{ V})/(1\text{ k}\Omega) = 3.9\text{ mA} \approx I_{C2}$$

Clearly,

$$V_{C2} = I_{C2}R_{C2} = (3.9\text{ mA})(2.7\text{ k}\Omega) = 10.5\text{ V}$$

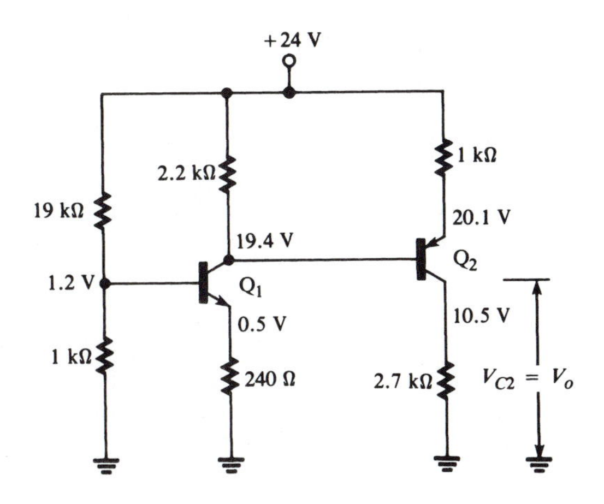

Figure 11–17
Direct-coupling using complementary transistors. Note that the PNP transistor is biased by connecting the positive supply voltage to the emitter side. (All voltages shown are dc values with respect to ground.)

The Darlington Pair

When the collectors of two BJTs are tied together and the emitter of one is direct-coupled to the base of the other, as shown in Figure 11–18, we obtain an important and highly useful configuration called a *Darlington pair.* The combination is used in amplifier circuits as if it were a single transistor having the base, collector, and emitter terminals labeled B, C, and E in the figure. We will analyze the Darlington pair to discover the effective beta (β_{DP}) of the single transistor it represents, as well as some of its small-signal characteristics.

Let β_1 and β_2 be the dc β-values of Q_1 and Q_2, respectively. Then, by definition,

$$I_{C1} = \beta_1 I_{B1} \quad \text{and} \quad I_{E1} = (\beta_1 + 1)I_{B1}$$

But $I_{E1} = I_{B2}$, so

$$I_{C2} = \beta_2 I_{B2} = \beta_2(\beta_1 + 1)I_{B1}$$

Now,

$$I_C = I_{C1} + I_{C2} = \beta_1 I_{B1} + \beta_2(\beta_1 + 1)I_{B1} = [\beta_1\beta_2 + (\beta_1 + \beta_2)]I_{B1}$$

Since $I_{B1} = I_B$, we have $I_C = [\beta_1\beta_2 + (\beta_1 + \beta_2)]I_B$, or

$$\beta_{DP} = I_C/I_B = \beta_1\beta_2 + \beta_1 + \beta_2 \tag{11–24}$$

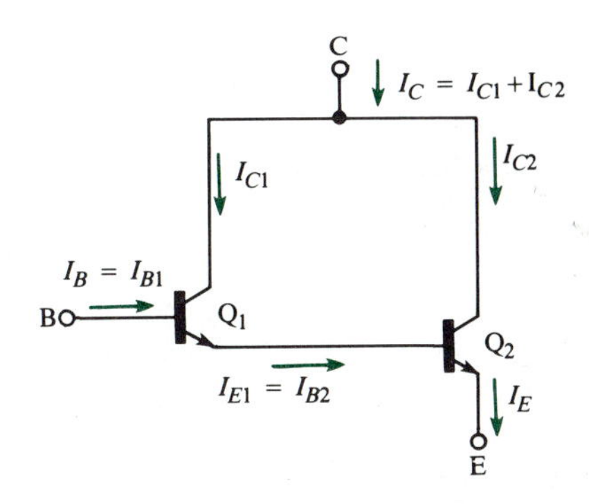

Figure 11–18
The Darlington pair is used as a single transistor having the collector, base, and emitter terminals labeled C, B, and E.

Equation 11–24 shows that the effective β of the Darlington pair is the product plus the sum of the β's of the individual transistors. It is usually true that $\beta_1\beta_2 \gg \beta_1 + \beta_2$, so

$$\beta_{DP} = \beta_1\beta_2 + \beta_1 + \beta_2 \approx \beta_1\beta_2 \tag{11–25}$$

Darlington pairs are often fabricated on a single chip to achieve matched Q_1 and Q_2 characteristics. When $\beta_1 = \beta_2 = \beta$, we have an effective β of

$$\beta_{DP} = \beta^2 + 2\beta \approx \beta^2 \tag{11–26}$$

For example, if $\beta_1 = \beta_2 = 100$, then $\beta_{DP} = 10{,}000 + 200 \approx 10{,}000$. We see that the Darlington pair can be regarded as a "super-β" transistor, and therefore enjoys all the advantages that high-β transistors have. For example, the dc resistance looking into the base (of Q_1) is the very large value $\beta_{DP}R_E$, where R_E is the resistance between the emitter (of Q_2) and ground.

While the foregoing analysis was performed for dc currents and the dc β-values, an identical small-signal analysis shows that the small-signal value of β_{DP} is the product plus the sum of the small-signal values of β_1 and β_2. Hereafter we will make the usual assumption that the dc and small-signal values of β are approximately equal and will not distinguish between the two.

We wish now to determine the effective small-signal input resistance from B to E, $r_{in(DP)}$, and the emitter resistance, $r_{e(DP)}$, of the composite transistor. Recall the general relationship (equation 5–22):

$$r_e \approx \frac{V_T}{I_E} \approx \frac{0.026}{I_E}\ \Omega \quad \text{at room temperature} \tag{11–27}$$

Since $I_{C2} \approx I_{E2}$, we have

$$r_{e2} \approx \frac{0.026}{I_{C2}}\ \Omega \tag{11–28}$$

Referring to Figure 11–18, note that $I_C = I_{C1} + I_{C2} \approx I_{C2}$ since $I_{C2} \gg I_{C1}$. Thus, (11–28) may also be written

$$r_{e2} \approx \frac{0.026}{I_C}\ \Omega \tag{11–29}$$

Recall that the ac resistance looking into the base of Q_2 is

$$r_{in(base)2} \approx \beta_2 r_{e2} \tag{11–30}$$

Now

$$r_{e1} \approx \frac{0.026}{I_{E1}}\ \Omega \tag{11–31}$$

Since $I_{E2} \approx \beta_2 I_{B2} = \beta_2 I_{E1}$, we have

$$I_{E1} \approx \frac{I_{E2}}{\beta_2} \tag{11–32}$$

Substituting (11–32) into (11–31) gives

$$r_{e1} \approx \beta_2\left(\frac{0.026}{I_{E2}}\right) = \beta_2 r_{e2} \tag{11–33}$$

The total effective resistance looking into the base of Q_1 (across the composite B–E terminals), i.e., the effective small-signal input resistance of the Darlington pair, is

$$r_{in(DP)} = \beta_1(r_{e1} + r_{in(base)2}) \approx \beta_1(r_{e1} + \beta_2 r_{e2}) \tag{11–34}$$

Substituting from equation 11–33, we have

$$r_{in(DP)} \approx \beta_1(\beta_2 r_{e2} + \beta_2 r_{e2}) = 2\beta_1\beta_2 r_{e2} \tag{11–35}$$

Since $\beta_{DP} \approx \beta_1\beta_2$, the effective emitter resistance, $r_{e(DP)}$, is

$$r_{e(DP)} \approx \frac{r_{in(DP)}}{\beta_{DP}} = \frac{2\beta_1\beta_2 r_{e2}}{\beta_1\beta_2} = 2r_{e2} \tag{11–36}$$

Example 11–8

A Darlington pair is biased so that the total collector current is 2 mA. If $\beta_1 = 110$ and $\beta_2 = 100$, find the room-temperature values of (1) β_{DP}, (2) $r_{in(DP)}$, and (3) $r_{e(DP)}$.

Solution

1. $\beta_{DP} = \beta_1\beta_2 + \beta_1 + \beta_2 = (110)(100) + 110 + 100 = 11{,}210$
2. From equation 11–29,

$$r_{e2} \approx \frac{0.026}{2 \times 10^{-3}} = 13\ \Omega$$

From equation 11–35, $r_{in(DP)} \approx 2(110)(100)13 = 286\ \text{k}\Omega$.

3. From equation 11–36, $r_{e(DP)} \approx 2(13) = 26\ \Omega$.

This example shows that the Darlington pair can be used to obtain a significant increase in base-to-emitter input resistance compared to that obtainable from a conventional BJT: 286 kΩ versus 1.3 kΩ.

The Darlington pair is most often used in an emitter-follower configuration because of the excellent buffering it provides between a high-impedance source and a low-impedance load. With an ac load resistance r_L connected to the emitter, the total input resistance to the follower is

$$r_{in} = r_{in(DP)} + \beta_{DP} r_L \tag{11–37}$$

When operated as an emitter follower, the current gain from the base of Q_1 to the emitter of Q_2 is $A_i = i_{e2}/i_{b1}$. Since $i_{e1} = i_{b2}$, we have

$$A_i = \frac{i_{e2}}{i_{b1}} = \left(\frac{i_{e2}}{i_{b2}}\right)\frac{i_{e1}}{i_{b1}} = (\beta_1 + 1)(\beta_2 + 1) \approx \beta_1\beta_2 \tag{11–38}$$

The next example illustrates an extreme case of the need for buffering, where the source resistance is 5 kΩ and the load resistance is 10 Ω.

Example 11–9

Figure 11–19 shows a common-emitter stage driving a Darlington pair connected as an emitter follower. The β-values for the silicon transistors are $\beta_1 = 200$, $\beta_2 = 100$, and $\beta_3 = 100$.

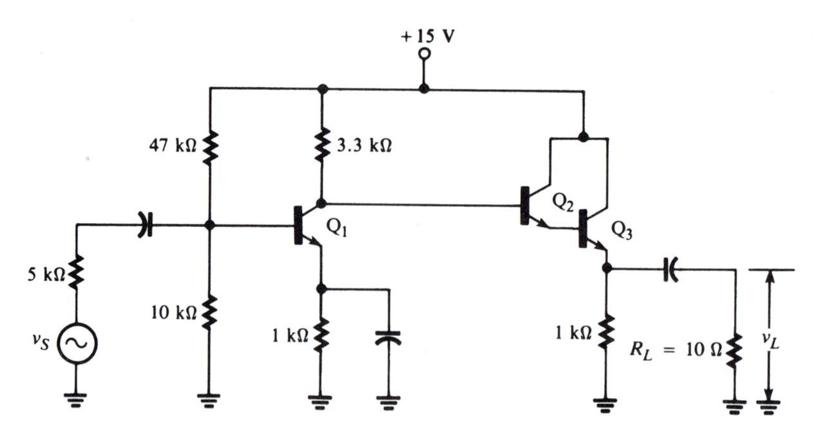

Figure 11–19
(Example 11–9)

1. Find v_L/v_S.
2. Find v_L/v_S if the Darlington pair is removed and the 10-Ω load is capacitor-coupled to the collector of Q_1.

Solution

1.

$$V_{B1} \approx \left(\frac{R_2}{R_1 + R_2}\right) V_{CC} = \left[\frac{10\ \text{k}\Omega}{(47\ \text{k}\Omega) + (10\ \text{k}\Omega)}\right](15\ \text{V}) = 2.6\ \text{V}$$

$$V_{E1} = V_{B1} - 0.7 = 2.6 - 0.7 = 1.9\ \text{V}$$

$$I_{C1} \approx I_{E1} = \frac{V_{E1}}{R_{E1}} = \frac{1.9\ \text{V}}{1\ \text{k}\Omega} = 1.9\ \text{mA}$$

$$r_{e1} \approx \frac{0.026}{I_{E1}} = \frac{0.026}{1.9\ \text{mA}} = 13.7\ \Omega$$

$$r_{in}(\text{stage 1}) = R_1 \| R_2 \| \beta_1 r_{e1} = (10\ \text{k}\Omega) \| (47\ \text{k}\Omega) \| 200(13.7\ \Omega) = 2.1\ \text{k}\Omega$$

Notice that the collector of Q_1 is direct-coupled to the base of Q_2 in the Darlington pair, so

$$V_{B2} = V_{C1} = V_{CC} - I_{C1}R_{C1} = 15\ \text{V} - (1.9\ \text{mA})(3.3\ \text{k}\Omega) = 8.7\ \text{V}$$

The dc emitter voltage of Q_3 is about 1.4 V less than the base voltage of Q_2, since there are *two* forward-biased base-emitter junctions between those two points:

$$V_{E3} = V_{B2} - 1.4 = 8.7 - 1.4 = 7.3\ \text{V}$$

$$I_{C3} \approx I_{E3} = \frac{V_{E3}}{R_{E3}} = \frac{7.3\ \text{V}}{1\ \text{k}\Omega} = 7.3\ \text{mA}$$

Noting that Q_1 and Q_2 in our previous analysis of the Darlington pair are now designated Q_2 and Q_3, we have, from equation 11–28,

$$r_{e3} \approx \frac{0.026}{I_{C3}} = \frac{0.026}{7.3\ \text{mA}} = 3.6\ \Omega$$

From equation 11–36,

$$r_{e(DP)} \approx 2(3.6\ \Omega) = 7.2\ \Omega$$

From equation 11–26,

$$\beta_{\text{DP}} = (100)^2 + 200 = 10{,}200$$

From equation 11–35,

$$r_{in(DP)} \approx 2(100)(100)(3.6) = 72\ \text{k}\Omega$$

From equation 11–37, the total resistance looking into the Darlington pair, with load connected, is

$$r_{in} = (72\ \text{k}\Omega) + (10{,}200)\ [(1\ \text{k}\Omega) \parallel (10\ \Omega)] \approx 174\ \text{k}\Omega$$

This value of input resistance is so large in comparison to the 3.3-kΩ collector resistance of Q_1 that virtually no loading occurs. Thus,

$$A_{v1} \approx \frac{-R_{c1}}{r_{e1}} \approx \frac{-3.3\ \text{k}\Omega}{13.7\ \Omega} = -241$$

The voltage gain of the emitter-follower stage is (from equation 5–59)

$$A_{v(DP)} \approx \frac{r_L}{r_{e(DP)} + r_L} = \frac{(1\ \text{k}\Omega) \parallel (10\ \Omega)}{(7.2\ \Omega) + (1\ \text{k}\Omega) \parallel (10\ \Omega)} \approx 0.58$$

Taking into account the loading caused by r_S, the overall voltage gain is

$$\frac{v_L}{v_S} = \left[\frac{r_{in}(\text{stage 1})}{r_S + r_{in}(\text{stage 1})}\right] A_{v1} A_{v(DP)}$$

$$= \left[\frac{2.1\ \text{k}\Omega}{(5\ \text{k}\Omega) + (2.1\ \text{k}\Omega)}\right](-241)(0.58) = -41.3$$

2. With the 10-Ω load connected to the collector of Q_1, the ac load on Q_1 is $r_{L1} = (3.3\ \text{k}\Omega) \parallel (10\ \Omega) \approx 10\ \Omega$. Therefore,

$$A_{v1} \approx \frac{-r_{L1}}{r_{e1}} = \frac{-10}{13.7} = -0.73$$

The overall gain is then

$$\frac{v_L}{v_S} = \left[\frac{2.1\ \text{k}\Omega}{(5\ \text{k}\Omega) + (2.1\ \text{k}\Omega)}\right](-0.73) = -0.22$$

Without the buffering provided by the Darlington pair, the output voltage turns out to be less than half the level of the input signal, i.e., we no longer have a voltage amplifier. We see that the Darlington pair increases the voltage gain by a factor of 41.3/0.22 = 187.7.

Cascode Amplifier

Another important example of direct-coupled transistors is the *cascode* amplifier (a name derived from vacuum-tube days), in which a common-emitter transistor drives a common-base transistor. A simple example of a cascode amplifier is shown

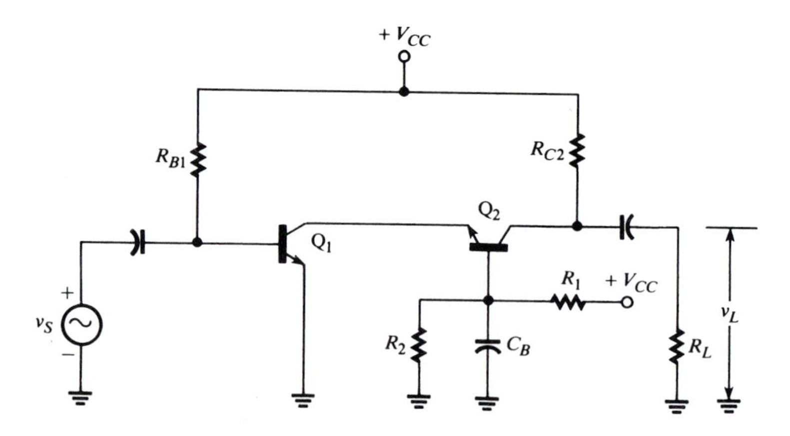

Figure 11–20
An example of a cascode amplifier. Q_1 is a common-emitter stage that is direct-coupled to the common-base stage formed by Q_2.

in Figure 11–20. Note that Q_1 is a common-emitter stage that uses R_{B1} for "fixed" bias. Since capacitor C_B grounds the base of Q_2 to ac signals, Q_2 is the common-base stage. It serves as the load on the collector of Q_1. Recall that the input to a common-base stage is at its emitter and the output is taken at its collector. Thus Q_1 is direct-coupled to the input of Q_2 and the output of the cascode amplifier is at the collector of Q_2. Resistors R_1 and R_2 form a voltage-divider bias circuit for Q_2.

The principal advantage of the cascode arrangement is that it has a small input capacitance, an important consideration in high-frequency amplifiers. The input capacitance is small because the voltage gain of Q_1 is small (near unity), which means that the Miller capacitance is minimized. Most of the voltage gain is achieved in the common-base stage. The voltage gain of Q_1 is small because the effective load resistance in its collector circuit is the small input resistance of the common-base stage.

Calculation of the bias currents and voltages is straightforward:

$$I_{B1} = \frac{V_{CC} - V_{BE}}{R_{B1}} \qquad \textbf{(11–39)}$$

$$I_{C1} = I_{E2} \approx I_{C2} = \beta_1 I_{B1} \qquad \textbf{(11–40)}$$

$$V_{B2} = \left(\frac{R_2}{R_1 + R_2}\right) V_{CC} \qquad \textbf{(11–41)}$$

$$V_{C1} = V_{E2} = V_{B2} - V_{BE} \qquad \textbf{(11–42)}$$

$$V_{C2} = V_{CC} - I_{C2}R_{C2} \qquad \textbf{(11–43)}$$

$$V_{CE2} = V_{C2} - V_{E2} \qquad \textbf{(11–44)}$$

Q_1, like any other common-emitter stage, can also be biased using the voltage-divider method and an emitter resistor.

Figure 11–21 shows a small signal equivalent circuit of the cascode amplifier. The collector, base, and emitter terminals of each transistor are labeled in the diagram (C1, B1, etc.).

Referring to the equivalent circuit, we see that

$$r_{in}(\text{stage 1}) = R_{B1} \,\|\, \beta_1 r_{e1} \qquad \textbf{(11–45)}$$

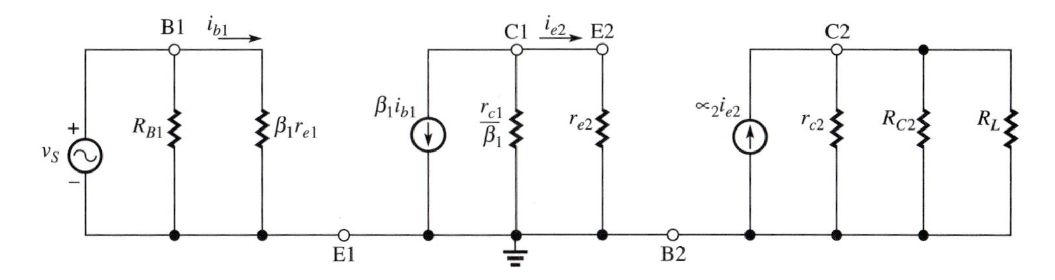

Figure 11–21
Small-signal equivalent circuit of a cascode amplifier

Note that the input resistance of Q_2 is the small emitter resistance r_{e2} of the common-base stage. The parallel combination of r_{c1}/β_1 and r_{e2} form the ac load on the collector of Q_1. Since $r_{c1}/\beta_1 >> r_{e2}$,

$$r_{L1} = \left(\frac{r_{c1}}{\beta_1}\right) \Big\| r_{e2} \approx r_{e2} \tag{11–46}$$

Thus,

$$A_{v1} \approx \frac{-r_{L1}}{r_{e1}} \approx \frac{-r_{e2}}{r_{e1}} \approx -1 \tag{11–47}$$

Equation 11–47 states that the voltage gain of Q_1 is approximately unity, which follows from the fact that $I_{E2} \approx I_{E1}$, making $r_{e2} \approx r_{e1}$. The ac load on the collector of Q_2 is seen to be

$$r_{L2} = r_{c2} \parallel R_{C2} \parallel R_L \approx R_{C2} \parallel R_L \tag{11–48}$$

Therefore, the voltage gain of the second stage is

$$A_{v2} \approx \frac{r_{L2}}{r_{e2}} \approx \frac{R_{C2} \parallel R_L}{r_{e2}} \tag{11–49}$$

The overall voltage gain of the cascode is thus

$$A_v = A_{v1}A_{v2} \approx -A_{v2} \tag{11–50}$$

Equation 11–50 shows that the common-base stage provides most of the voltage gain. The current gain of the cascode amplifier, not including current division at the input or output, is approximately

$$A_i \approx \beta_1\alpha_2 \approx \beta_1 \quad \text{since } \alpha_2 \approx 1 \tag{11–51}$$

The output resistance of the amplifier is

$$r_o(\text{stage}) = r_{c2} \parallel R_{C2} \approx R_{C2} \tag{11–52}$$

Example 11–10

The silicon transistors in Figure 11–22 have the following parameter values:

Q_1: $\beta_1 = 100$, $r_{c1} \approx \infty$, $C_{bc} = 4$ pF, $C_{be} = 10$ pF

Q_2: $\alpha_2 \approx 1$, $r_{c2} \approx \infty$

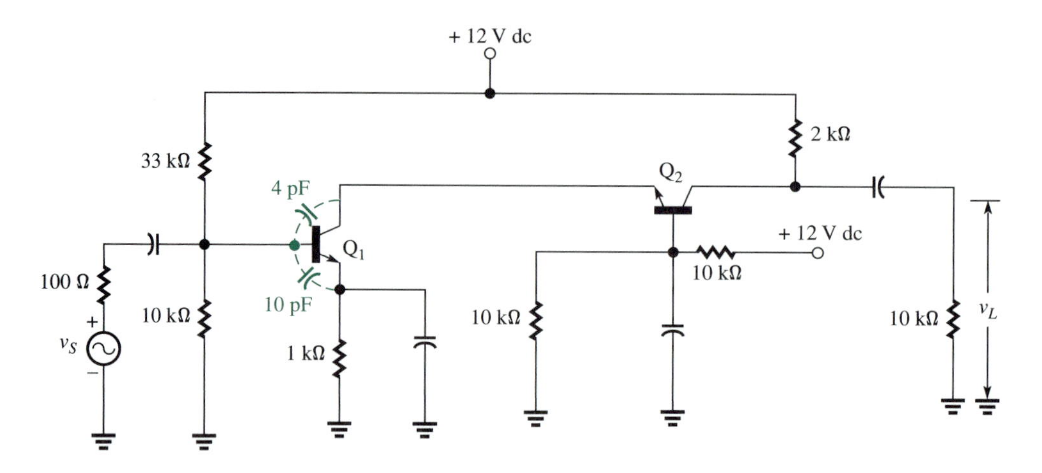

Figure 11–22
(Example 11–10)

Find approximate values for

1. the dc currents and voltages I_{C1}, I_{C2}, V_{C1}, and V_{C2};
2. the small-signal voltage gain v_L/v_S; and
3. the break frequency $f_2(C_A)$ due to shunt capacitance at the input of Q_1.

Solution

1. The base-to-ground bias voltage at the input stage (Q_1) is determined by the voltage-divider there. Since the voltage divider is loaded by $\beta R_{E1} = 100\ \text{k}\Omega$,

$$V_{B1} \approx \left[\frac{10\ \text{k}\Omega \parallel 100\ \text{k}\Omega}{(10\ \text{k}\Omega \parallel 100\ \text{k}\Omega) + (33\ \text{k}\Omega)}\right](12\ \text{V}) = 2.6\ \text{V}$$

Therefore,

$$V_{E1} = V_{B1} - 0.7 = 2.6 - 0.7 = 1.9\ \text{V}$$

$$I_{C1} \approx I_{E1} = \frac{V_{E1}}{R_{E1}} = \frac{1.9\ \text{V}}{1\ \text{k}\Omega} = 1.9\ \text{mA} = I_{E2} \approx I_{C2}$$

The base-to-ground bias voltage of Q_2 is determined by its voltage divider:

$$V_{B2} = \left[\frac{10\ \text{k}\Omega}{(10\ \text{k}\Omega) + (10\ \text{k}\Omega)}\right](12\ \text{V}) = 6\ \text{V}$$

Therefore,

$$V_{C1} = V_{E2} = V_{B2} - 0.7 = 6 - 0.7 = 5.3\ \text{V}$$

$$V_{C2} = V_{CC} - I_{C2}R_{C2} = 12 - (1.9\ \text{mA})(2\ \text{k}\Omega) = 8.2\ \text{V}$$

2. Since $I_{E1} \approx I_{E2}$,

$$r_{e2} \approx r_{e1} = \frac{0.026}{I_{E1}} = \frac{0.026}{1.9\ \text{mA}} = 13.7\ \Omega$$

$$r_{in}(\text{stage 1}) = (33\ \text{k}\Omega) \parallel (10\ \text{k}\Omega) \parallel 100(13.7\ \Omega) = 1.16\ \text{k}\Omega$$

$$A_{v1} = \frac{-r_{e2}}{r_{e1}} \approx -1$$

$$A_{v2} = \frac{r_{L2}}{r_{e2}} = \frac{(2\ \text{k}\Omega) \parallel (10\ \text{k}\Omega)}{13.7\ \Omega} = 121.6$$

$$\frac{v_L}{v_S} = \left[\frac{r_{in}(\text{stage 1})}{r_S + r_{in}(\text{stage 1})}\right] A_{v1} A_{v2}$$

$$= \left[\frac{1.16\ \text{k}\Omega}{(100\ \Omega) + (1.16\ \text{k}\Omega)}\right](-1)(121.6) = -112$$

3. The Miller capacitance at the input to Q_1 is determined by the voltage gain of stage 1 alone:

$$C_M = C_{bc}(1 - A_{v1}) = (4\ \text{pF})(2) = 8\ \text{pF}$$

The total input capacitance is, therefore,

$$C_A = C_M + C_{be} = (8\ \text{pF}) + (10\ \text{pF}) = 18\ \text{pF}$$

Then

$$f_2(C_A) = \frac{1}{2\pi[r_S \parallel r_{in}(\text{stage 1})]C_A} = \frac{1}{2\pi[100 \parallel (1.16 \times 10^3)]\,(18 \times 10^{-12})}$$
$$= 96\ \text{MHz}$$

We see that the break frequency due to input capacitance is quite high. While there may be break frequencies elsewhere in the circuit that are smaller than 96 MHz (including the f_β value of Q_1), the cascode amplifier effectively eliminates the most troublesome source of high-frequency loss—the input Miller capacitance.

11–5 TRANSISTOR ARRAYS

Multiple transistors sharing a common substrate are available in integrated-circuit packages called *transistor arrays.* Compared to medium- and large-scale integrated circuits, arrays contain relatively few devices, but they have the advantage that all or most device terminals are accessible at external pins. This accessibility allows the user to connect the components in a variety of ways and provides great flexibility in applications while retaining the integrated circuit's advantages of compactness and rugged structure. Also, like integrated-circuit components, the common fabrication processes used to create the components on one substrate make it possible to obtain devices with closely matched characteristics.

Often a transistor array will be manufactured with certain internal device connections already in place. For example, Darlington pairs are made with collector and emitter-to-base interconnections already formed (see Figure 11–18). The RCA CA3018 is an example of a transistor array containing a Darlington pair and two independent NPN transistors. A schematic diagram and manufacturer's specification sheet for the CA3018/3018A is shown in Figure 11–23. Note that the specifications show that the H_{FE} (dc β) values are matched to within $\pm 10\%$ and the dc values of V_{BE} are matched to within ± 2 mV (for the CA3018A). Also note that both emitter terminals in the Darlington pair (Q_3 and Q_4) are accessible at external pins. This

CA3018, CA3018A

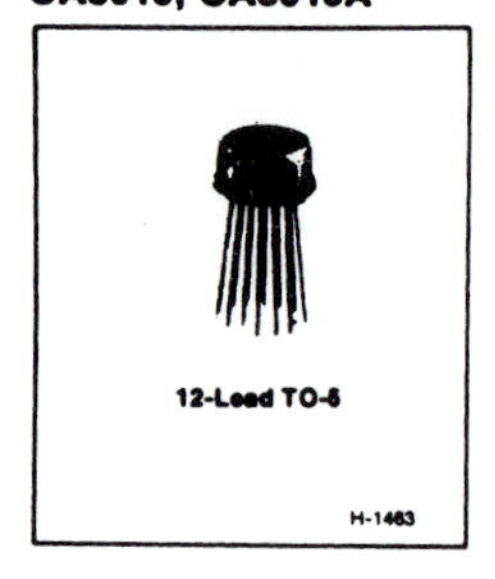

General-Purpose Transistor Arrays

Two Isolated Transistors and
a Darlington-Connected Transistor Pair

For Low-Power Applications at Frequencies
from DC Through the VHF Range

Features:

- *Matched monolithic general purpose transistors*
- *H_{FE} matched ± 10%*
- *V_{BE} matched ± 2 mV CA3018A (± 5m V CA3018)*
- *Operation from DC to 120 MHz*
- *Wide operating current range*
- *CA3018A performance characteristics controlled from 10 μA to 10 mA*
- *Low noise figure - 3.2 dB typical at 1KHz*
- *Full military temperature range capability (−55 to +125° C)*

Applications

- *General use in signal processing systems in DC through VHF range*
- *Custom designed differential amplifiers*
- *Temperature compensated amplifiers*
- *See RCA Application Note, ICAN-5296 "Application of the RCA CA3018 Integrated-Circuit Transistor Array" for suggested applications.*

The CA3018 and CA3018A consist of four general purpose silicon n-p-n transistors on a common monolithic substrate.

Two of the four transistors are connected in the Darlington configuration. The substrate is connected to a separate terminal for maximum flexibility.

The transistors of the CA3018 and the CA3018A are well suited to a wide variety of applications in low-power systems in the DC through VHF range. They may be used as discrete transistors in conventional circuits but in addition they provide the advantages of close electrical and thermal matching inherent in integrated circuit construction.

The CA3018A is similar to the CA3018 but features tighter control of current gain, leakage, and offset parameters making it suitable for more critical applications requiring premium performance.

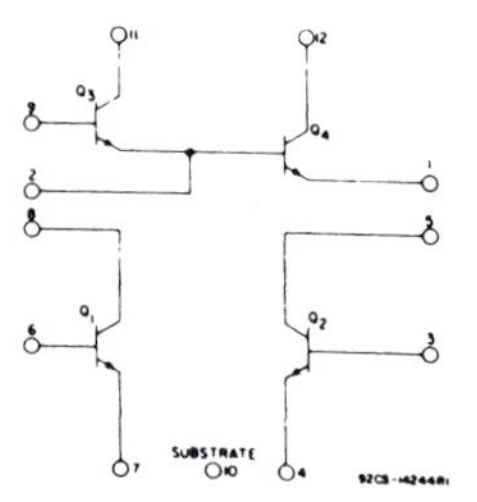

Maximum Ratings, Absolute-Maximum Values, at TA=25°C

	CA3018	CA3018A	
Power Dissipation, P:			
Any one transistor	300	300	mW
Total package	450	450	mW
Derate at 5 mW/°C for T_A > 85°C			
Temperature Range:			
Operating	-55 to + 125	-55 to + 125	°C
Storage	-65 to + 150	-65 to + 150	°C

The following ratings apply for each transistor in the device:

	CA3018	CA3018A	
Collector-to-Emitter Voltage, V_{CEO}	15	15	V
Collector-to-Base Voltage, V_{CBO}	20	30	V
Collector-to-Substrate Voltage, V_{CIO}*	20	40	V
Emitter-to-Base Voltage, V_{EBO}	5	5	V
Collector Current, I_C	50	50	mA

*The collector of each transistor of the CA3018 and CA3018A is isolated from the substrate by an integral diode. *The substrate (terminal 10) must be connected to the most negative point in the external circuit to maintain isolation between transistors and to provide for normal transistor action.*

Figure 11–23
Manufacturer's specification sheet for a transistor array (Courtesy of RCA Solid State)

arrangement makes it possible to bias each transistor separately. The individual collector terminals are also accessible. These can be tied together externally, or a resistor can be inserted in series with the collector of Q_4, as is sometimes done to reduce feedback from Q_4 to Q_3 when the Darlington pair is used in a common-emitter configuration.

RCA Application Note ICAN-5296 shows how the CA3018 can be used to construct a video (high-frequency) amplifier. A schematic diagram is shown in Figure 11–24. Note that Q_1 and Q_2 form a cascode amplifier, with Q_2 in the common-emitter configuration and Q_1 as the common-base stage. Q_3 and Q_4 are used in this example as a pair of cascaded emitter followers. The voltage gain of the amplifier is reported to be 37 dB over a midband range from 6 kHz to 11 MHz.

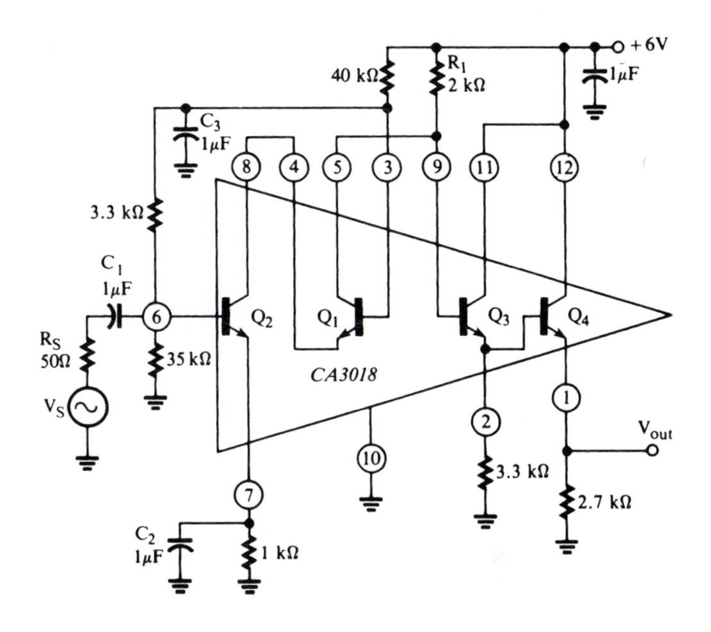

Figure 11–24
A video cascode amplifier using the CA3018 transistor array (RCA Application Note ICAN-5296; Courtesy of RCA Solid State)

11–6 MULTISTAGE FET AMPLIFIERS

The same principles we applied to the analysis of multistage BJT amplifiers can be used to analyze multistage FET amplifiers. In many ways the analysis is simpler, because the very large gate-to-source resistance of a JFET or MOSFET allows us to ignore some of the loading effects that had to be considered in BJT designs. The next example illustrates this point.

Example 11–11

The JFETs in the RC-coupled amplifier shown in Figure 11–25 have $g_{m1} = 2.7$ mS, $g_{m2} = 3.2$ mS, and $r_{d1} = r_{d2} = 100$ kΩ. Find the midband voltage gain v_L/v_S.

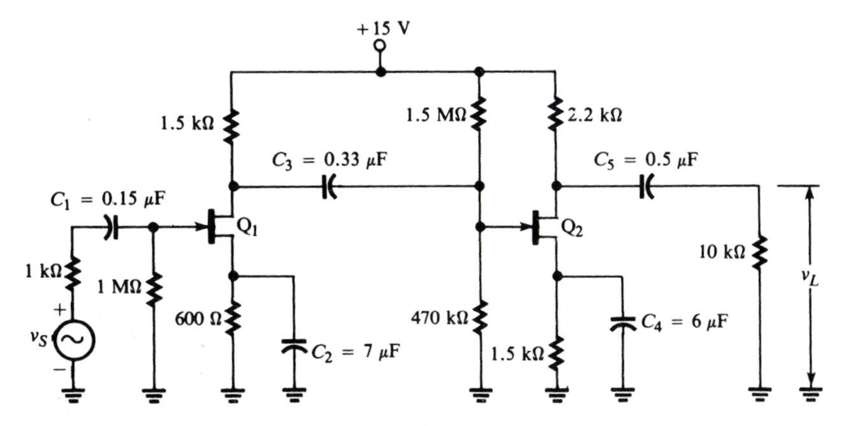

Figure 11–25
(Example 11–11)

Solution. We will compute the in-circuit (loaded) voltage gains of each stage using the general FET gain equation developed in Chapter 8:

$$A_v = -g_m r_L \tag{11–53}$$

where r_L is the ac load resistance connected between drain and ground.

Since the gate-to-source resistance of Q_2 is quite large, we neglect it in the computation of r_{L1}:

$$r_{L1} = r_{d1} \| R_{D1} \| r_{in}(\text{stage 2}) = (100\text{ k}\Omega) \| (1.5\text{ k}\Omega) \| [(1.5\text{ M}\Omega) \| (470\text{ k}\Omega)]$$
$$\approx 1.5\text{ k}\Omega$$

Therefore,

$$A_{v1} = -g_{m1} r_{L1} = -(2.7 \times 10^{-3})(1.5 \times 10^{3}) = -4.05$$

Also,

$$r_{L2} = r_{d2} \| R_{D2} \| R_L = (100\text{ k}\Omega) \| (2.2\text{ k}\Omega) \| (10\text{ k}\Omega) = 1.77\text{ k}\Omega$$

and

$$A_{v2} = -g_{m2} r_{L2} = -(3.2 \times 10^{-3})(1.77 \times 10^{3}) = -5.66$$

Taking into account the voltage division due to signal-source resistance, we have

$$A_{v(overall)} = \left(\frac{r_{in1}}{r_S + r_{in1}}\right) A_{v1} A_{v2} = \left[\frac{1\text{ M}\Omega}{(1\text{ k}\Omega) + (1\text{ M}\Omega)}\right](-4.05)(-5.66) = 22.9$$

Cutoff frequencies of multistage FET amplifiers are found using the same methods we used for multistage BJT amplifiers, as illustrated in the next example.

Example 11–12

Find the approximate lower cutoff frequency of the multistage amplifier shown in Figure 11–25.

Solution

$$f_1(C_1) = \frac{1}{2\pi[r_S + r_{in}(\text{stage 1})]C_1}$$
$$= \frac{1}{2\pi(1 \times 10^{3} + 1 \times 10^{6})(0.15 \times 10^{-6})} = 1.06\text{ Hz}$$

$$f_1(C_2) = \frac{1}{2\pi[R_{S1} \| (1/g_{m1})]C_2} = \frac{1}{2\pi\{600 \| [1/(2.7 \times 10^{-3})]\}(7 \times 10^{-6})} = 99.3\text{ Hz}$$

$$f_1(C_3) = \frac{1}{2\pi[r_o(\text{stage 1}) + r_{in}(\text{stage 2})]C_3}$$
$$= \frac{1}{2\pi[r_{d1} \| R_{D1} + (1.5 \times 10^{6}) \| (470 \times 10^{3})]C_3}$$
$$= \frac{1}{2\pi[(100 \times 10^{3}) \| (1.5 \times 10^{3}) + (357.86 \times 10^{3})](0.33 \times 10^{-6})} = 1.3\text{ Hz}$$

$$f_1(C_4) = \frac{1}{2\pi[R_{S2} \| (1/g_{m2})]C_4} = \frac{1}{2\pi\{(1.5 \times 10^3) \| [1/(3.2 \times 10^{-3})]\}(6 \times 10^{-6})}$$
$$= 102.6 \text{ Hz}$$

$$f_1(C_5) = \frac{1}{2\pi[r_o(\text{stage 2}) + R_L]C_5} = \frac{1}{2\pi(r_{d2} \| R_{D2} + R_L)C_5}$$
$$= \frac{1}{2\pi[(100 \times 10^3) \| (2.2 \times 10^3) + (10 \times 10^3)](0.5 \times 10^{-6})} = 26.2 \text{ Hz}$$

Since $f_1(C_2) \approx f_1(C_4) \approx 100$ Hz, the lower cutoff frequency is $f_1 \approx 1.55(100) = 155$ Hz.

Direct-Coupled FET Amplifiers

Direct-coupled field-effect transistors are widely used in linear integrated circuits. The next example illustrates a 3-stage, direct-coupled amplifier that uses complementary JFETs (N channel and P channel), similar in concept to the direct-coupled, complementary BJT amplifier discussed earlier. Note in particular how the P channel stage (Q_2) is biased.

Example 11–13

The JFETs in Figure 11–26 all have $I_{DSS} = 8$ mA and $|V_p| = 2$ V. Find the dc values of the drain-to-ground voltage V_D, the source-to-ground voltage V_S, and the drain-to-source voltage V_{DS}, of each transistor.

Solution. Note that Q_1 and Q_2 are both common-source stages and Q_3 is a source follower.

The gate-to-ground voltage of Q_1 is determined by the voltage divider across the input:

$$V_{G1} = \left[\frac{1 \text{ M}\Omega}{(1 \text{ M}\Omega) + (1.5 \text{ M}\Omega)}\right](20 \text{ V}) = 8 \text{ V}$$

Using equations 7–13 to find the bias point, we obtain

$$I_{D1} = 1.92 \text{ mA} \qquad \text{and} \qquad V_{DS1} = 3.1 \text{ V}$$

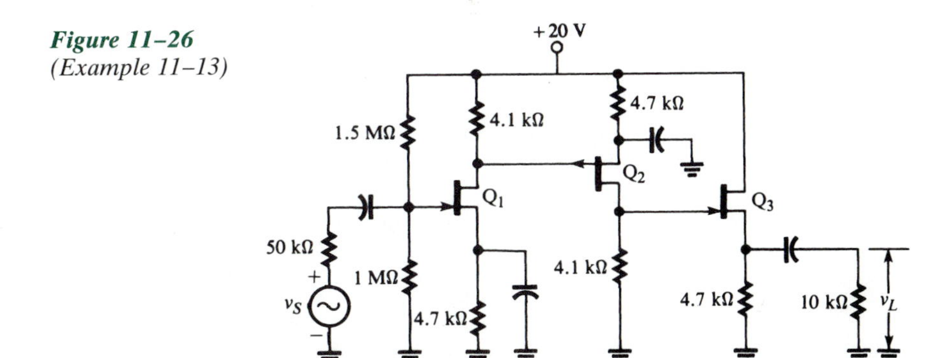

Figure 11–26
(Example 11–13)

Then

$$V_{S1} = I_{D1}R_{S1} = (1.92 \text{ mA})(4.7 \text{ k}\Omega) = 9.02 \text{ V}$$

and

$$V_{D1} = V_{DD} - I_{D1}R_{D1} = 20 - (1.92 \text{ mA})(4.1 \text{ k}\Omega) = 12.1 \text{ V}$$

Note that $|V_{DS1}| = 3.1 > |V_p| - |V_{GS}| = 0.98$ V, so Q_1 is properly biased in its pinch-off region.

Q_2 is a P-channel JFET biased with its source side connected to the positive supply voltage and its drain side grounded. Writing Kirchhoff's voltage law from V_{DD}, through R_{S2}, across the gate-to-source junction, and from the gate of Q_2 to ground, we have

$$V_{DD} - I_{D2}R_{S2} + V_{GS2} - V_{G2} = 0 \qquad \textbf{(11–54)}$$

Noting that $V_{G2} = V_{D1} = 12.1$ V, equation 11–54 becomes

$$20 - I_{D2}(4.7 \times 10^3) + V_{GS2} - 12.1 = 0$$

or

$$V_{GS2} = -7.9 + (4.7 \times 10^3)I_{D2} \qquad \textbf{(11–55)}$$

The square-law equation for the P-channel JFET is

$$I_{D2} = (8 \text{ mA})\left(1 - \frac{V_{GS2}}{2}\right)^2 \qquad \textbf{(11–56)}$$

Solving (11–55) and (11–56) simultaneously gives

$$I_{D2} = 1.9 \text{ mA} \qquad \text{and} \qquad V_{GS2} = 1 \text{ V}$$

Therefore,

$$V_{S2} = V_{DD} - (I_{D2})(4.7 \text{ k}\Omega) = 20 - (1.9 \text{ mA})(4.7 \text{ k}\Omega) = 11.1 \text{ V}$$

and

$$V_{D2} = I_{D2}R_{D2} = (1.9 \text{ mA})(4.1 \text{ k}\Omega) = 7.8 \text{ V}$$

Then

$$V_{DS2} = V_{D2} - V_{S2} = 7.8 - 11.1 = -3.3 \text{ V}$$

As an exercise, verify that Q_2 is also in its pinch-off region.

Recognizing that $V_{G3} = V_{D2} = 7.8$ V, equations 7–13 can be used to find the bias point of Q_3:

$$I_{D3} = 1.88 \text{ mA} \qquad \text{and} \qquad V_{DS3} = 11.2 \text{ V}$$

Then

$$V_{S3} = I_{D3}R_{S3} = (1.88 \text{ mA})(4.7 \text{ k}\Omega) = 8.8 \text{ V}$$

and

$$V_{D3} = V_{DD} = 20 \text{ V}$$

Example 11–14

Find the voltage gain v_L/v_S of the multistage amplifier shown in Figure 11–26.

Solution. To find the voltage gain of each stage, we must find the transconductance g_m of each JFET. Recall from equation 8–3 that

$$g_m = \frac{2I_{DSS}}{|V_p|}\sqrt{\frac{I_D}{I_{DSS}}} \tag{11–57}$$

Substituting $I_{D1} = 1.92$ mA, $I_{D2} = 1.9$ mA, and $I_{D3} = 1.88$ mA (from Example 11–13) into (11–57) gives $g_{m1} \approx g_{m2} \approx g_{m3} = 3.9$ mS. Assuming that the r_d of each JFET is very large, the ac load resistance at the drain terminal of each of Q_1 and Q_2 is simply $R_D = 4.1$ kΩ. Therefore, the in-circuit voltage gain of each stage is

$$A_{v1} = A_{v2} = -g_m R_D = -(3.9 \times 10^{-3})(4.1 \times 10^3) = -16$$

From equation 8–21, the voltage gain of the source follower (Q_3) is, neglecting r_d,

$$A_{v3} = \frac{g_{m3}(R_{S3} \| R_L)}{1 + g_{m3}(R_{S3} \| R_L)} = \frac{3.9 \times 10^{-3}[(4.7\text{ k}\Omega) \| (10\text{ k}\Omega)]}{1 + 3.9 \times 10^{-3}[(4.7\text{ k}\Omega) \| (10\text{ k}\Omega)]} = 0.93$$

Thus, the overall voltage gain is

$$\frac{v_L}{v_S} = \frac{r_{in}(\text{stage 1})}{r_S + r_{in}(\text{stage 1})}A_{v1}A_{v2}A_{v3} = \frac{(1.5\text{ M}\Omega) \| (1\text{ M}\Omega)}{(50\text{ k}\Omega) + (1.5\text{ M}\Omega) \| (1\text{ M}\Omega)} \times (-16)(-16)(0.93) = 220$$

A Bifet Amplifier

Bifet is a coined term meaning a combination of bipolar and FET, and a bifet amplifier is one that contains both types of transistors. These amplifiers are designed to exploit the most desirable characteristics of each device, such as the very large input impedance of the FET and the large voltage gain of the BJT. A good example is the cascode amplifier, the bifet version of which is shown in Figure 11–27. Note that the input stage, Q_1, is a JFET connected in a common-source configuration and the second stage is a BJT connected as a common-base amplifier.

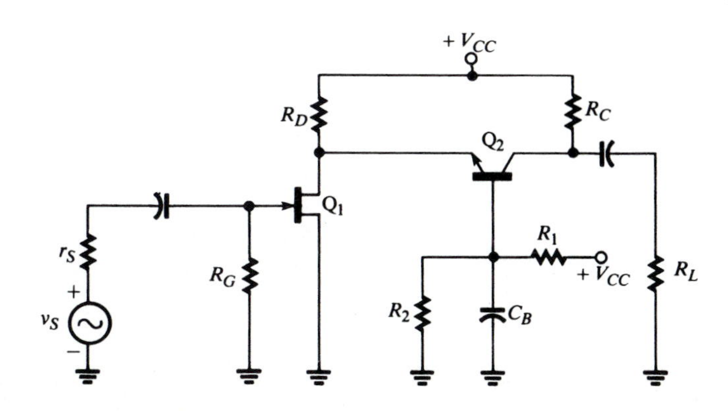

Figure 11–27
A bifet cascode amplifier

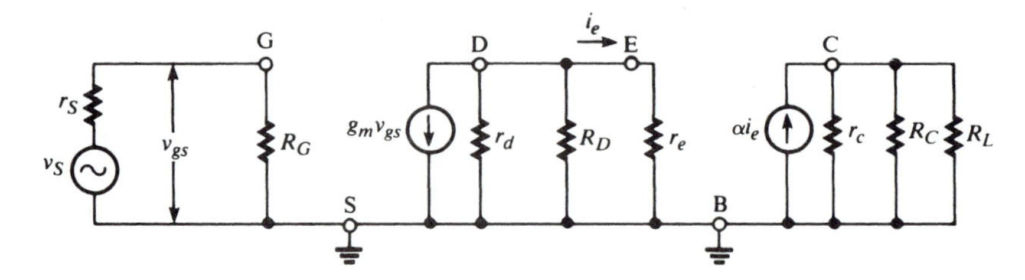

Figure 11–28
The ac equivalent circuit of the bifet cascode amplifier shown in Figure 11–27

The ac equivalent circuit of the bifet cascode is shown in Figure 11–28. The voltage gain of the first stage is

$$A_{v1} = -g_m(r_d \,\|\, R_D \,\|\, r_e) \tag{11–58}$$

Since the common-base input resistance r_e is quite small, $r_d \,\|\, R_D \,\|\, r_e \approx r_e$ and

$$A_{v1} \approx -g_m r_e \tag{11–59}$$

While A_{v1} is usually small (less than 1), the large input resistance and small Miller capacitance of the JFET stage results in better high-frequency response than that found in the BJT version. As in the BJT cascode, the voltage gain of the amplifier is achieved in the common-base stage:

$$A_{v2} = \frac{r_c \,\|\, R_C \,\|\, R_L}{r_e} \tag{11–60}$$

Since a BJT amplifier is generally capable of producing more voltage gain than its FET counterpart, we see that the bifet amplifier uses the best features of both kinds of devices.

11–7 TRANSFORMER COUPLING

Besides blocking dc, an important advantage of using a transformer for interstage coupling is that the turns ratio can be designed to achieve *impedance matching* and therefore to maximize power transfer between stages. Recall that the turns ratio of a transformer is defined by

$$\text{turns ratio} = \frac{N_p}{N_s} = \frac{e_p}{e_S} \tag{11–61}$$

where N_p, N_s = number of turns in the primary and secondary windings, respectively

e_p, e_s = primary and secondary voltages

Figure 11–29 shows a transformer whose primary is driven by a signal source v_1 having resistance r_1 and whose secondary has a load resistance r_L connected across it. Assuming an ideal transformer (zero winding resistance and no power loss), it is easy to show that the resistance r_{in} looking into the primary winding is

$$r_{in} = \left(\frac{N_p}{N_s}\right)^2 r_L \tag{11–62}$$

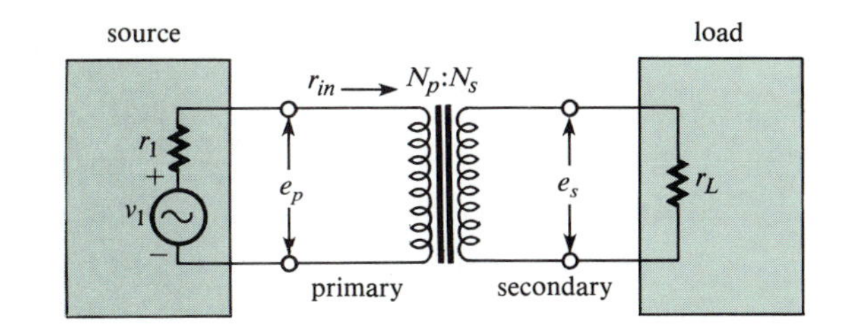

Figure 11–29
A transformer whose primary winding is driven by the signal source v_1 and whose secondary winding has load r_L connected across it

To achieve maximum power transfer from the source to the load in Figure 11–29, it is necessary (by the maximum power transfer theorem) that the resistance r_{in} seen by the signal source be equal to the source resistance r_1. Thus, from equation 11–62, we require

$$r_1 = r_{in} = \left(\frac{N_p}{N_s}\right)^2 r_L \tag{11–63}$$

Solving (11–63) for N_pN_s, we find the turns ratio necessary to achieve maximum power transfer:

$$\frac{N_p}{N_s} = \sqrt{\frac{r_1}{r_L}} \tag{11–64}$$

In multistage transistor amplifiers, the signal source driving the primary winding is often a common-emitter stage having a large output resistance, and the load resistance may be the much smaller value equal to the input resistance of the next stage. In that case, the ratio r_1/r_L is greater than 1 and therefore, by equation 11–64, the turns ratio N_pN_s is greater than 1. As a consequence, the coupling transformer is typically a *step-down* transformer, meaning the secondary voltage is less than the primary voltage.

Commercially available transformers are often specified in terms of their impedance ratios rather than their turns ratios. In these cases, the turns ratio can be determined using equation 11–64, i.e., by taking the square root of the specified impedance ratio. For example, if the specifications on a transformer state that it has primary impedance 16 kΩ and secondary impedance 16 Ω (typical values for matching a transistor output stage to a speaker), then

$$\frac{N_p}{N_s} = \sqrt{\frac{Z_P}{Z_S}} = \sqrt{\frac{16 \times 10^3}{16}} = 31.62$$

Example 11–15

Figure 11–30 shows a transformer used to match a 100-kΩ source to a 250-Ω load.

1. What should be the turns ratio of the transformer?
2. With the matching transformer in place, what is the rms voltage across the load?
3. What is the maximum power that can be delivered to the load?

Figure 11–30
(Example 11–15)

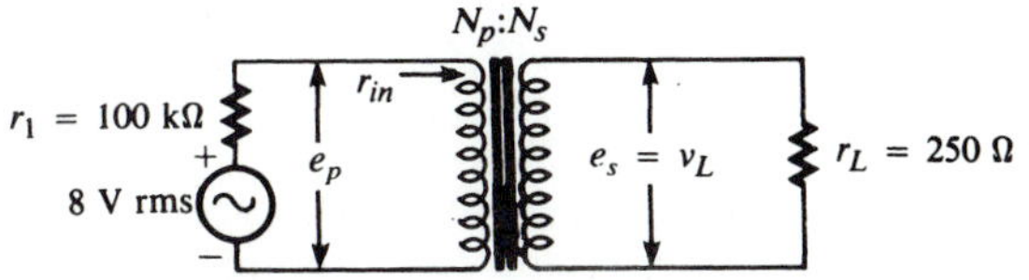

N_p:N_s r_{in} r_1 = 100 kΩ + 8 V rms − e_p $e_s = v_L$ r_L = 250 Ω

Solution

1. From equation 11–64,

$$\frac{N_p}{N_s} = \sqrt{\frac{r_1}{r_L}} = \sqrt{\frac{100 \times 10^3}{250}} = 20$$

Thus the primary winding should have 20 times as many turns as the secondary winding.

2. From equation 11–62,

$$r_{in} = (N_p/N_s)^2 r_L = (20)^2(250) = 100\ \text{k}\Omega$$

We see that the resistance looking into the primary winding is 100 kΩ, as it should be when load and source are matched by the transformer. Since the source resistance and r_{in} form a voltage divider across the 8-V-rms signal source, the voltage across the primary winding is *one-half* the signal voltage:

$$e_p = \left(\frac{r_{in}}{r_1 + r_{in}}\right) e_s = \left[\frac{100\ \text{k}\Omega}{(100\ \text{k}\Omega) + (100\ \text{k}\Omega)}\right] (8\ \text{V rms}) = 4\ \text{V rms}$$

From equation 11–61,

$$e_s = \left(\frac{1}{N_p/N_s}\right) e_p = \left(\frac{1}{20}\right) 4\ \text{V rms} = 0.2\ \text{V rms}$$

We see that the transformer steps the primary voltage down by a factor of 20. (The current is stepped up by a factor of 20.)

3. Maximum power transfer occurs when the source is matched to the load, and with the matching transformer in place, we have

$$p_{max} = P_L = \frac{e_s^2}{r_L} = \frac{v_L^2}{r_L} = \frac{(0.2\ \text{V})^2}{250} = 0.16\ \text{mW}$$

(When the source and load are matched, the same amount of power is dissipated in the source as is delivered to the load; verify that fact in this example.)

Example 11–16

SPICE

Assuming that the transformer in Figure 11–30 is ideal and that the secondary winding has inductance 50 mH, use SPICE to find the primary and secondary voltages and currents when the input voltage and the turns ratio are the same as in Example 11–15. Assume that the frequency of the input is 50 kHz.

Solution. The turns ratio is related to the primary and secondary inductance, L_p and L_s, by

$$\frac{N_p}{N_s} = \sqrt{\frac{L_p}{L_s}}, \quad \text{or} \quad L_p = \left(\frac{N_p}{N_s}\right)^2 L_s$$

Therefore, since $N_p/N_s = 20$ and $L_s = 50$ mH,

$$L_p = (20)^2(50 \times 10^{-3}\ \text{H}) = 20\ \text{H}$$

Figure 11–31(a) shows the circuit in its SPICE format. Note that we cannot isolate the secondary circuit from the primary because SPICE requires that there

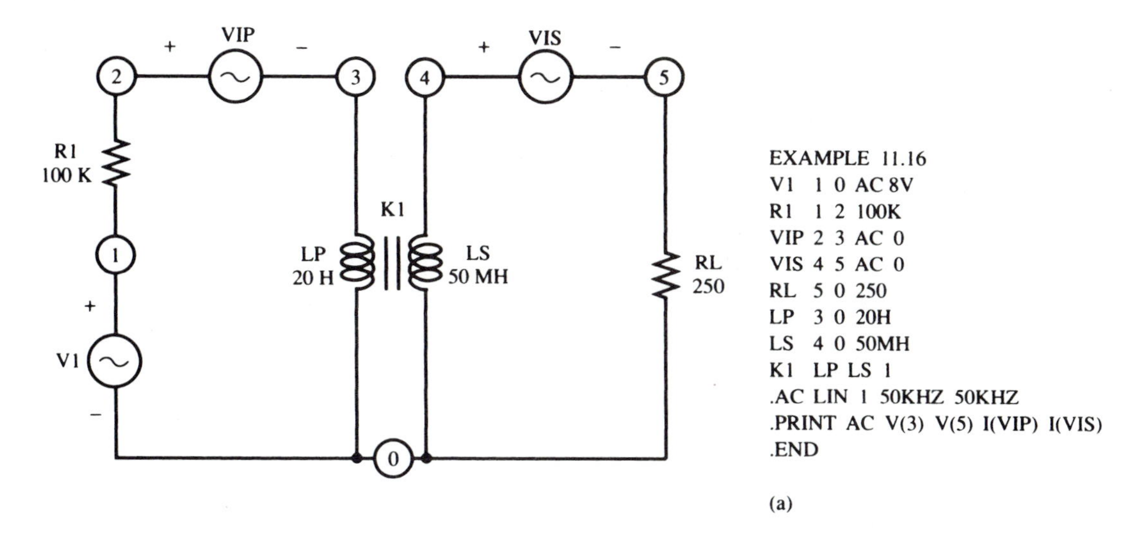
VIP
VIS
R1
100 K
K1
LP
20 H
LS
50 MH
RL
250
V1
EXAMPLE 11.16
V1 1 0 AC 8V
R1 1 2 100K
VIP 2 3 AC 0
VIS 4 5 AC 0
RL 5 0 250
LP 3 0 20H
LS 4 0 50MH
K1 LP LS 1
.AC LIN 1 50KHZ 50KHZ
.PRINT AC V(3) V(5) I(VIP) I(VIS)
.END
(a)

```
EXAMPLE 11.16
****     AC ANALYSIS                          TEMPERATURE =    27.000 DEG C
******************************************************************************
   FREQ         V(3)          V(5)          I(VIP)        I(VIS)
 5.000E+04     4.000E+00    2.000E-01     4.000E-05     8.000E-04
```

(b)

Figure 11–31
(Example 11–16)

be a dc path to ground from every node. Thus, the secondary winding must have one side connected to node 0. The statement K1 LP LS 1 makes the transformer ideal, since it specifies the coupling coefficient, k, to be 1. As is required to simulate an iron-core transformer in SPICE, the primary reactance at 50 kHz ($2\pi f L_p = 6.28$ MΩ) is much greater than the 100-kΩ source impedance, and the secondary reactance ($2\pi f L_s = 1.57$ kΩ) is much greater than the 250-Ω load impedance. Dummy voltage sources VIP and VIS have been inserted to measure the primary and secondary currents. The results of the analysis are shown in Figure 11–31(b). We see that v_p = V(3) = 4 V and v_s = V(5) = 0.2 V, in agreement with Example 11–15. The primary current is i_p = I(VIP) = 40 μA and the secondary current is i_s = I(VIS) = 0.8 mA. Note that $N_p/N_s = i_s/i_p = 20$.

Figure 11–32(a) shows a two-stage transformer-coupled amplifier. Note that there is no resistor in the collector circuit of either transistor, so the primary of each transformer is driven from a source (a collector) having relatively large resistance. The secondary winding of the interstage transformer is connected directly to the base of Q_2. The 4.7-kΩ base biasing resistor is bypassed with a capacitor, so all of the ac voltage developed in the secondary winding appears from the base of Q_2 to ground. Note that both sides of the 15-kΩ resistor are at ac ground. With the arrangement shown, the Q_2 bias resistors do not reduce the ac input resistance to the second stage, and the transformer does not short the dc bias current to ground. The 1.25-kΩ load is transformer-coupled to the output of Q_2. Note that the load is shown completely *isolated* from the rest of the circuit, in the sense that

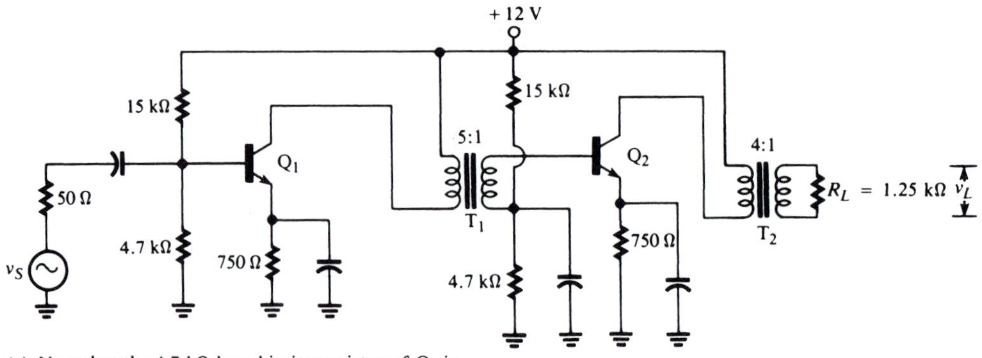

(a) Note that the 4.7-kΩ base biasing resistor of Q_2 is bypassed and that the 1.25-kΩ load is isolated from the rest of the circuit.

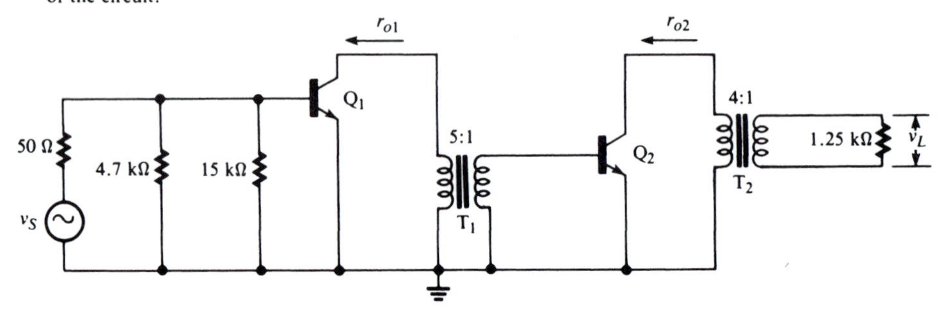

(b) The ac equivalent circuit of (a)

Figure 11–32
A two-stage, transformer-coupled amplifier

it does not share a common ground reference with Q_1 and Q_2 (although it could, if desired). This is another advantage of transformer coupling: Individual amplifier stages and loads can be isolated from each other when desired or necessary. In fact, isolation transformers having 1:1 turns ratios are sometimes used specifically for that purpose.

Figure 11–32(b) shows the ac equivalent circuit of the transformer-coupled amplifier. The next example demonstrates that the transformer turns ratios were chosen to achieve impedance matching.

Example 11–17

The transistors in Figure 11–32 have $\beta_1 = 105$, $r_{o1} = 20$ kΩ, $\beta_2 = 89$, and $r_{o2} = 20$ kΩ. Find the midband voltage gain v_L/v_S.

Solution. Q_1 and Q_2 are biased identically, so the following computations for determining r_e apply to both:

$$V_B = \left[\frac{4.7\text{ k}\Omega}{(4.7\text{ k}\Omega) + (15\text{ k}\Omega)}\right](12\text{ V}) = 2.86\text{ V}$$

$$V_E = V_B - 0.7 = 2.16\text{ V}$$

$$I_E = \frac{V_E}{R_E} = \frac{2.16\text{ V}}{750\ \Omega} = 2.88\text{ mA}$$

$$r_e \approx \frac{0.026}{I_E} = \frac{0.026}{2.88\text{ mA}} = 9\ \Omega$$

The ac input resistance at the base of Q_2 is $r_{in2} \approx \beta_2 r_{e2} = (89)(9) = 801\ \Omega$. Therefore, by equation 11–66, the resistance looking into the primary of transformer T_1 is

$$r_{in(primary\ of\ T_1)} = \left(\frac{N_p}{N_s}\right)^2 r_{in2} = 5^2(801) \approx 20\text{ k}\Omega$$

Since $r_{o1} = 20\text{ k}\Omega$, we see that the transformer matches the output of Q_1 to the input of Q_2.

Figure 11–33 shows the ac equivalent circuit of a BJT collector output with a primary winding connected across it. It is apparent that the ac load resistance seen at the collector is the parallel combination of the transistor's output resistance r_o and the effective resistance looking into the primary winding, $r_{in(primary)}$.

The ac load resistance at the collector of Q_1 is then

$$r_{L1} = r_{o1} \parallel r_{in(primary\ of\ T_1)} = (20\text{ k}\Omega) \parallel (20\text{ k}\Omega) = 10\text{ k}\Omega$$

Thus, the voltage gain from base to collector of Q_1 is

$$A_{v1(base\text{–}collector)} = \frac{-r_{L1}}{r_{e1}} = \frac{-10\text{ k}\Omega}{9\ \Omega} = -1111$$

When we compute the overall gain of the amplifier, we will take into account the fact that T_1 steps the voltage down by a factor of $N_s/N_p = 1/5$, between the collector of Q_1 and the base of Q_2.

The resistance looking into the primary of T_2 is

$$r_{in(primary\ of\ T_2)} = \left(\frac{N_p}{N_s}\right)^2 R_L = 4^2(1.25\text{ k}\Omega) = 20\text{ k}\Omega$$

We see that transformer T_2 matches the output of Q_2 to the load. The ac load resistance at the collector of Q_2 is

$$r_{L2} = r_{o2} \parallel r_{in(primary\ of\ T_2)} = (20\text{ k}\Omega) \parallel (20\text{ k}\Omega) = 10\text{ k}\Omega$$

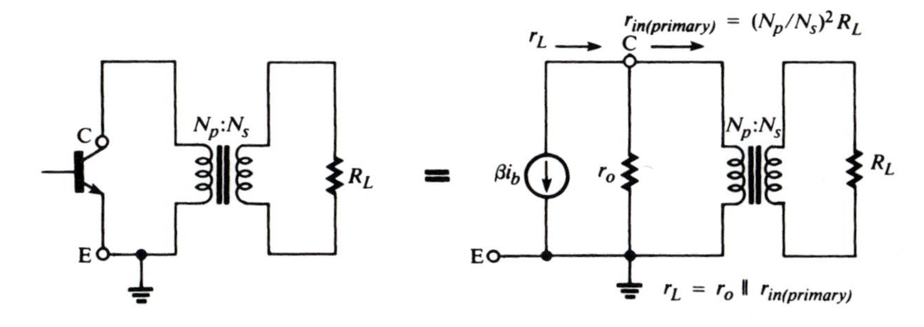

Figure 11–33
(Example 11–17) The ac resistance seen at the collector terminal is the parallel combination of r_o and $r_{in(primary)}$.

and the voltage gain of Q_2 is

$$A_{v2(base\text{-}collector)} = \frac{-r_{L2}}{r_{e2}} = \frac{-10\text{ k}\Omega}{9\ \Omega} = -1111$$

Taking into account the voltage division at the input of Q_1 and the step-down ratios of the transformers, we find the overall voltage gain to be

$$\frac{v_L}{v_s} = \left[\frac{r_{in}(\text{stage 1})}{r_S + r_{in}(\text{stage 1})}\right] A_{v1(b-c)}\left(\frac{1}{5}\right) A_{v2(b-c)}\left(\frac{1}{4}\right)$$

where $r_{in}(\text{stage 1}) = (4.7\text{ k}\Omega) \parallel (15\text{ k}\Omega) \parallel (105)(9\ \Omega) = 748\ \Omega$. Then

$$\frac{v_L}{v_S} = \left(\frac{748}{50 + 748}\right)(-1111)\left(\frac{1}{5}\right)(-1111)\left(\frac{1}{4}\right) = 57{,}849$$

The sign of the voltage gain, indicating the phase of the output with respect to the input, may be changed by the coupling transformers, depending on how each is connected in the circuit: Reversing primary or secondary connections will reverse the phase.

11–8 LIMITATIONS OF PRACTICAL TRANSFORMERS

Copper Losses

As we know, all materials have electrical resistance, including the wire wrapped around the core of a transformer to form its primary and secondary windings. The resistance in these windings is responsible for an average power loss, P_l, that can be calculated in the usual way:

$$P_l(\text{primary}) = I_p^2(rms)R_p \text{ watts} \qquad \textbf{(11–65)}$$

$$P_l(\text{secondary}) = I_s^2(rms)R_s \text{ watts} \qquad \textbf{(11–66)}$$

where R_p and R_s are the resistances of the primary and secondary windings, and I_p(rms) and I_s(rms) are the effective values of the primary and secondary currents, respectively. Since the core is usually wound with copper wire, these losses are often called *copper losses.* In most transformers, copper losses are relatively small, typically on the order of 1% of the total power transferred.

Eddy Currents

A transformer core itself has a certain inductance, so when flux in the core changes with time, as it does in normal transformer operation, electrical currents are induced in the core. If the core material is a conductor, such as the iron in an iron-core transformer, these currents may be large enough to cause noticeable power losses. The currents induced in a core this way are called *eddy currents,* and eddy current losses are losses caused by current flowing through the resistance of the core material. Eddy current losses increase as the frequency of the voltage applied to the transformer increases, because higher frequencies mean greater rates of change of flux, which in turn mean larger induced currents.

One popular construction method that is used to reduce eddy current losses is to assemble the core from *laminated* sheets. These laminations, or layers, are insulated from each other, so current flow is interrupted. Ferrite cores, which are made from a special type of ceramic that has a small electrical conductivity but large permeability, are also used for that purpose.

Hysteresis Losses

Magnetism in ferromagnetic materials is attributable to the orientation of tiny *magnetic domains.* The magnetic fields produced by these domains align themselves with an externally applied magnetic field. In an inductor or transformer core, the external field is created by current flowing through the windings. When the current is ac, as it is in a transformer, the external field is continually reversing direction, so the magnetic domains must also continually reverse their orientation. There is a type of inertia, or resistance to change, which is an inherent property of magnetic domains, and which requires energy to overcome. This energy must be supplied each time the orientation of a domain is changed, so energy is consumed in the core of a transformer as it responds to continually changing alternating current. The energy consumed in that process is called *hysteresis loss* and is responsible for still another power loss in practical transformers. Hysteresis losses increase with the frequency of the applied voltage, because higher frequencies force the domains to reverse direction more often during a given interval of time.

Transformer Efficiency

The useful output power of a transformer is that which is delivered to its load, and the input power is that which is delivered from a source to its primary side. The efficiency of a transformer is defined the same way it is for other electrical devices:

$$\eta = \frac{P_o}{P_{in}} = \frac{P_{in} - P_l}{P_{in}} = 1 - \frac{P_l}{P_{in}} \tag{11–67}$$

where P_l is the sum of all the power losses in the transformer. Eddy current and hysteresis losses are usually grouped into a category called *core losses,* because they are both associated with the core. Thus, $P_l(\text{total}) = P_{\text{core}} + P_{\text{copper}}$. Of course, the efficiency given by (11–67) is always less than 1 and is often expressed as a percentage. Practical transformer efficiencies are generally quite high in comparison to other electrical and electronic devices, on the order of 90% to 98%.

Example 11–18

The transformer shown in Figure 11–34 has a primary winding resistance of 0.5 Ω and a secondary winding resistance of 0.1 Ω. The power delivered to the load resistance is 48 W and the effective value of the primary current is 0.4 A rms. If the core losses are 0.9 W, find the efficiency of the transformer.

Solution. The magnitude of the secondary current can be found from the load power:

$$P_L = 48\text{ W} = I_s^2(\text{rms})R_L$$

$$I_s(\text{rms}) = \sqrt{\frac{48\text{ W}}{12\ \Omega}} = 2\text{ A rms}$$

Figure 11–34
(Example 11–18)

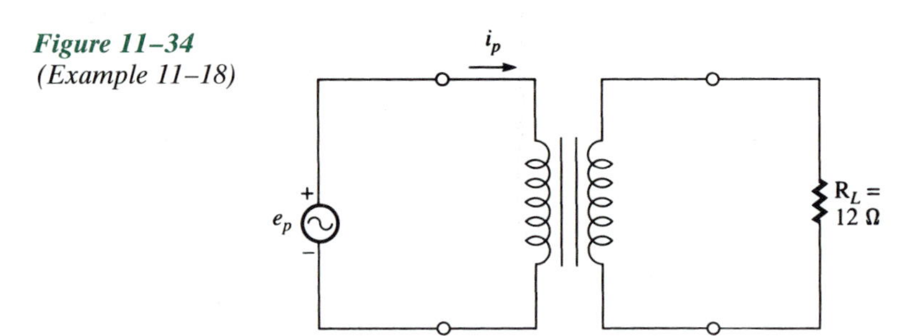

Therefore, the copper loss in the secondary winding is

$$P_l(\text{secondary}) = I_s^2(\text{rms})R_s = (2\text{ A})^2(0.1\ \Omega) = 0.4\text{ W}$$

The copper loss in the primary winding is

$$P_l(\text{primary}) = I_p^2(\text{rms})R_p = (0.4\text{ A})^2(0.5\ \Omega) = 0.08\text{ W}$$

The sum of the losses, including the 0.9-W core loss, is

$$P_l(\text{total}) = 0.9\text{ W} + 0.4\text{ W} + 0.08\text{ W} = 1.38\text{ W}$$

The input power is the sum of the output power and the losses:

$$P_{in} = P_o + P_l(\text{total}) = 48\text{ W} + 1.38\text{ W} = 49.38\text{ W}$$

Therefore, the efficiency is

$$\eta = \frac{P_o}{P_{in}} = \frac{48\text{ W}}{49.38\text{ W}} = 0.972, \quad \text{or} \quad 97.2\%$$

Leakage Flux and Coupling Coefficients

One of the assumptions made for an ideal transformer is that all of the magnetic flux generated by one winding is coupled to the other winding. In fact, some of the flux generated by the primary winding of a practical transformer "escapes," in the sense that it is not confined entirely to the core, and therefore does not reach the secondary winding. Magnetic flux is concentrated in a high-permeability material such as iron, but disperses in low-permeability air. Thus, two windings on one iron core have a good chance of sharing essentially the same flux, but are less likely to share flux in an air core. As illustrated in Figure 11–35, flux that is shared by the

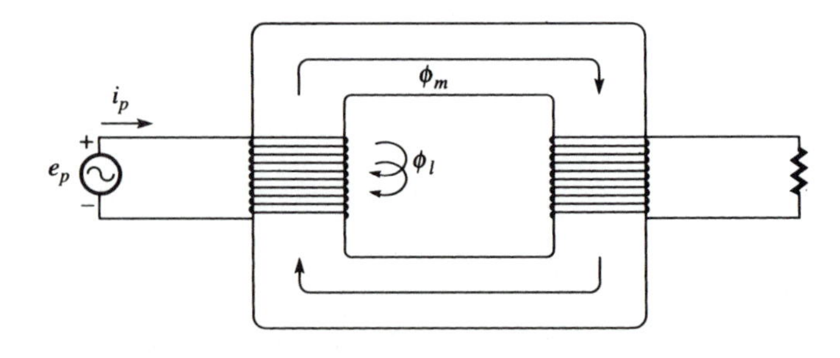

Figure 11–35
Leakage flux, ϕ_l, is flux generated by the primary winding that does not reach the secondary winding.

two windings of a transformer is called *mutual flux* ϕ_m, and that which escapes is called *leakage flux,* ϕ_l.

The *coefficient of coupling* k between the two windings of a transformer is defined to be

$$k = \frac{\phi_m}{\phi_m + \phi_l} \tag{11–68}$$

In an ideal transformer, $\phi_l = 0$ and $k = 1$. In a practical iron-core transformer, the leakage flux is very small, typically about 1% of the mutual flux. Consequently, the coupling coefficient for iron-core transformers is very nearly equal to the ideal value of 1. Such transformers are said to be *tightly coupled.* On the other hand, air-core transformers have a large leakage flux, and the coefficient of coupling may be as small as 0.01. Transformers with small values of k are said to be *loosely coupled.* The coupling coefficient can be increased by wrapping the secondary turns physically very close to the primary turns (i.e., overlapping them). However, in some high-frequency applications, loosely coupled air-core transformers are desirable.

As an aid in analyzing the effect of leakage flux on the performance of a transformer, it is convenient to think of the total inductance of the primary winding, L_p, as being composed of two components: one, the *primary magnetizing inductance,* L_{pm}, that is subjected to the mutual flux, and, two, the *primary leakage inductance,* L_{pl}, that is subjected to leakage flux only. Thus,

$$L_p = L_{pm} + L_{pl} \tag{11–69}$$

If the coefficient of coupling were equal to 1, then L_{pl} would be zero, since ϕ_l would be zero. In that case, we would have $L_p = L_{pm}$. In general,

$$L_{pm} = kL_p \tag{11–70}$$

Substituting (11–70) into (11–69) and solving for L_{pl}, we obtain

$$L_{pl} = (1 - k)L_p \tag{11–71}$$

The secondary winding is also responsible for a certain amount of leakage. For example, if it were very loosely wound on the core and had large air gaps between its turns, we would attribute the reduced coupling to it. We can therefore regard the total inductance of the secondary winding, L_s, to be similarly composed of a secondary magnetizing inductance, L_{sm}, and a secondary leakage inductance, L_{sl}. Thus, $L_s = L_{sm} + L_{sl}$, and

$$L_{sm} = kL_s \tag{11–72}$$

$$L_{sl} = (1 - k)L_s \tag{11–73}$$

Figure 11–36 shows the equivalent circuit of a practical transformer in which the winding resistances and leakage inductances are identified as separate components. In reality, these quantities are intimate, inseparable parts of the entire windings and are said to be *distributed* throughout the windings. However, we can analyze their effects on the transformer by considering them to be separate entities (called *lumped parameters*). Note that the equivalent circuit contains an *ideal* transformer that has inductances L_{pm} and L_{sm} and that has zero winding resistance. To use the transformer theory studied earlier, we now regard the primary voltage e_p as that which appears across the terminals of the ideal primary in the equivalent circuit, and the secondary voltage as that which appears across the ideal secondary.

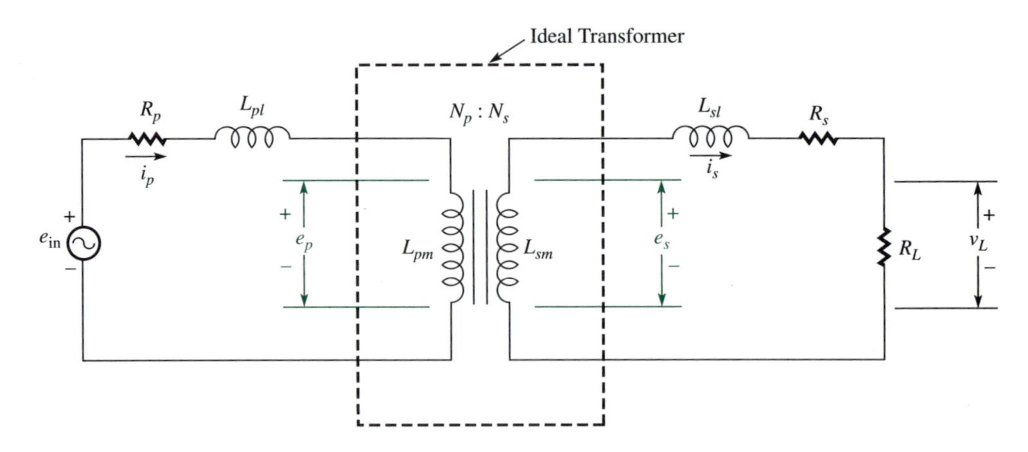

Figure 11–36
The equivalent circuit of a practical transformer is an ideal transformer that has winding resistance and leakage inductance in series with each winding.

Loading Effects

It is apparent from Figure 11–36 that the full value of e_{in} will not appear across the primary winding. The primary voltage, e_p, will be less than e_{in} because of the voltage drop across the impedance in series with the winding. Consequently, the full value of e_{in} will not be multiplied by the turns ratio, and the secondary voltage will be less than would be predicted for an ideal transformer. Similarly, the full value of e_s does not appear at the load, because of the voltage drop appearing across the impedance in series with it. We see that winding resistance and leakage inductance make the ratio e_{in}/v_L smaller than the ratio N_p/N_s. We say that R_L *loads* the circuit, in much the same way that a load on the output of a real voltage source reduces the terminal voltage.

To calculate the actual load voltage in a practical transformer, we must first consider the voltage divider formed by the impedance in series with the primary and the impedance *reflected* from the secondary. Recall that we showed how a load resistance is reflected through a transformer by the square of the turns ratio. Similarly, any *complex* impedance in series with the secondary is reflected to the primary side by the square of the turns ratio: Z (reflected to primary) = $(N_p/N_s)^2 Z_s$, where Z_s is the total impedance across the secondary winding. It is this reflected impedance that we must use in our voltage-divider computation on the primary side. The next example illustrates the computation.

Example 11–19

The transformer shown in Figure 11–37(a) has a total primary inductance of 0.3 H and a total secondary inductance of 50 mH. The winding resistances are $R_p = 2\ \Omega$ and $R_s = 1\ \Omega$. The transformer has a coupling coefficient of 0.99.

1. Draw the equivalent circuit of the transformer.
2. Find the magnitude of the load voltage.

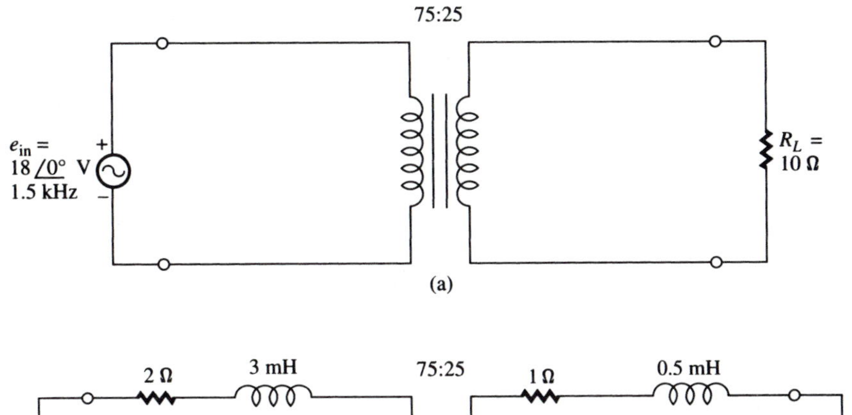

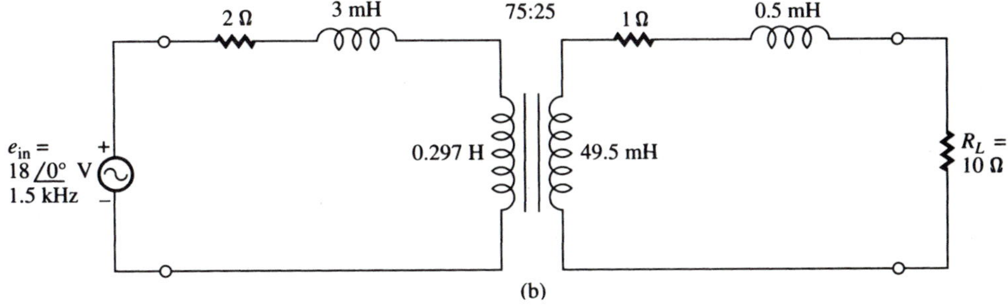

Figure 11–37
(Example 11–19)

Solution

1. We first find the mutual and leakage inductance in each winding. From equations (11–70) through (11–73),

$$L_{pm} = kL_p = 0.99(0.3\text{ H}) = 0.297\text{ H}$$
$$L_{pl} = (1 - k)L_p = (0.01)(0.3\text{ H}) = 3\text{ mH}$$
$$L_{sm} = kL_s = (0.99)(50\text{ mH}) = 49.5\text{ mH}$$
$$L_{sl} = (1 - k)L_s = (0.01)(50\text{ mH}) = 0.5\text{ mH}$$

Figure 11–37(b) shows the equivalent circuit.

2. Since the frequency of e_{in} is 1.5 kHz, the reactance corresponding to each inductive component in the circuit is $X_L = \text{j}\omega L = \text{j}(2\pi f)L = \text{j}(2\pi \times 1.5 \times 10^3)L = \text{j}9425L$ Ω. Thus,

$$X_{Lpm} = \text{j}9425(0.297\text{ H}) = \text{j}2799\ \Omega$$
$$X_{Lpl} = \text{j}9425(3 \times 10^{-3}\text{ H}) = \text{j}28.3\ \Omega$$
$$X_{Lsm} = \text{j}9425(49.5 \times 10^{-3}\text{ H}) = \text{j}467\ \Omega$$
$$X_{Lsl} = \text{j}9425(0.5 \times 10^{-3}\text{ H}) = \text{j}4.7\ \Omega$$

Figure 11–37(c) shows the equivalent circuit with the reactance values labeled. The total impedance in the secondary winding is

$$Z_s = (10 + 1 + \text{j}4.7)\ \Omega = 11 + \text{j}4.7\ \Omega$$

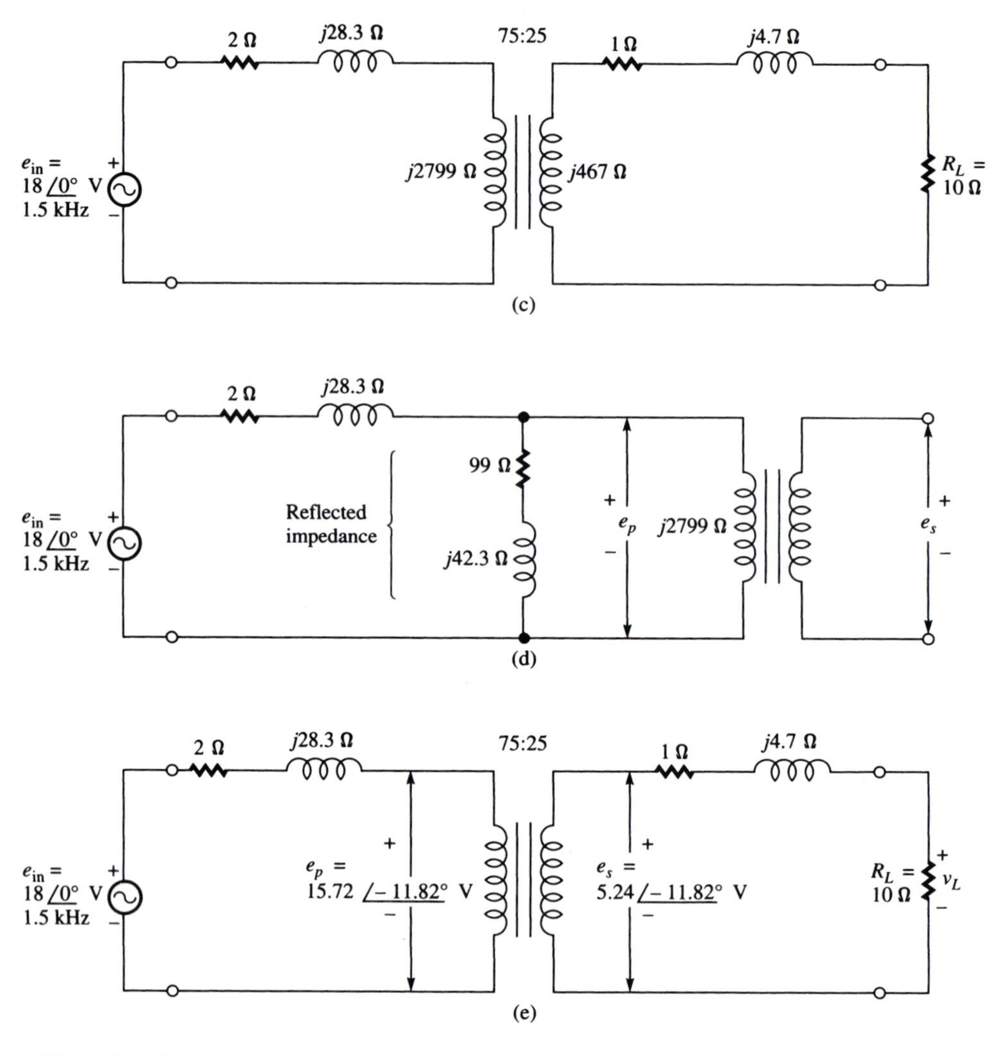

Figure 11–37
(Continued)

Therefore, the impedance reflected to the primary side is

$$Z(\text{reflected}) = \left(\frac{N_p}{N_s}\right)^2 Z_s = \left(\frac{75}{25}\right)^2 (11 + \text{j}4.7)\ \Omega$$
$$= (99 + \text{j}42.3)\ \Omega = 107.6\ \underline{/23.13^\circ}\ \Omega$$

Figure 11–37(d) shows the equivalent circuit with the reflected impedance drawn on the primary side. We can now use the voltage-divider rule to find the value of e_p. We will neglect the impedance of L_{pm} (j2799 Ω) in parallel with the reflected impedance, because it is so much greater than the reflected impedance. Thus,

$$e_p \approx \frac{Z(\text{reflected})}{(2 + j28.3) + Z(\text{reflected})} e_{in}$$
$$= \left[\frac{107.6 \angle 23.13^\circ}{(2 + j28.3) + (99 + j42.3)}\right] 18 \angle 0^\circ \text{ V} = 15.72 \angle -11.82^\circ \text{ V}$$

The secondary voltage e_s is, therefore,

$$e_s = \frac{e_p}{N_p/N_s} = \frac{15.72 \angle -11.82^\circ \text{ V}}{75/25} = 5.24 \angle -11.82^\circ \text{ V}$$

Finally, v_L can be found from the voltage divider appearing on the secondary side, as shown in Figure 11–37(e):

$$v_L = \left(\frac{10 \angle 0^\circ}{10 + 1 + j4.7}\right) e_s = \left(\frac{10 \angle 0^\circ}{11.96 \angle 23.14^\circ}\right) 5.24 \angle -11.82^\circ \text{ V}$$
$$= 4.38 \angle -34.96^\circ \text{ V}$$

If we had assumed that the entire transformer were ideal, we would have calculated a load voltage of

$$e_s = v_L = \frac{e_{in}}{N_p/N_s} = \frac{18 \angle 0^\circ \text{ V}}{75/25} = 6 \angle 0^\circ \text{ V}$$

We see that there is a discrepancy of 1.62 V, or about 37%, between the actual and ideal magnitudes of the load voltage. This rather large discrepancy is due to the small value of load resistance (10 Ω) in the example. Note that the smaller the load resistance, the greater its loading effect, due to voltage division on both sides of the transformer.

Frequency Response

The output voltage of an ideal transformer is independent of frequency. However, in a real transformer, we must consider how the inductive components affect the output when frequency changes, because the reactances of those components increase with frequency. Furthermore, every transformer has *capacitance* between the turns of its windings, and capacitive reactance is also affected by frequency. Recall that capacitance exists whenever two conductors are separated by an insulator, which is precisely the situation when the insulated conductors of a transformer winding are wrapped around the core. The capacitances of both windings are distributed throughout the windings, but it is convenient to lump their effect into equivalent capacitors in parallel with each winding. The primary and secondary capacitance, C_p and C_s, are shunt capacitances, because they divert current that would otherwise flow in the windings. Figure 11–38(a) shows the equivalent circuit of a transformer, including the lumped capacitance and the lumped inductance we discussed earlier. Also shown is the internal resistance, R_{int}, of the voltage source connected to the primary winding, because this resistance also affects frequency response.

To understand how frequency affects the response of the circuit in Figure 11–38(a), we can study the effects of low frequencies and high frequencies sepa-

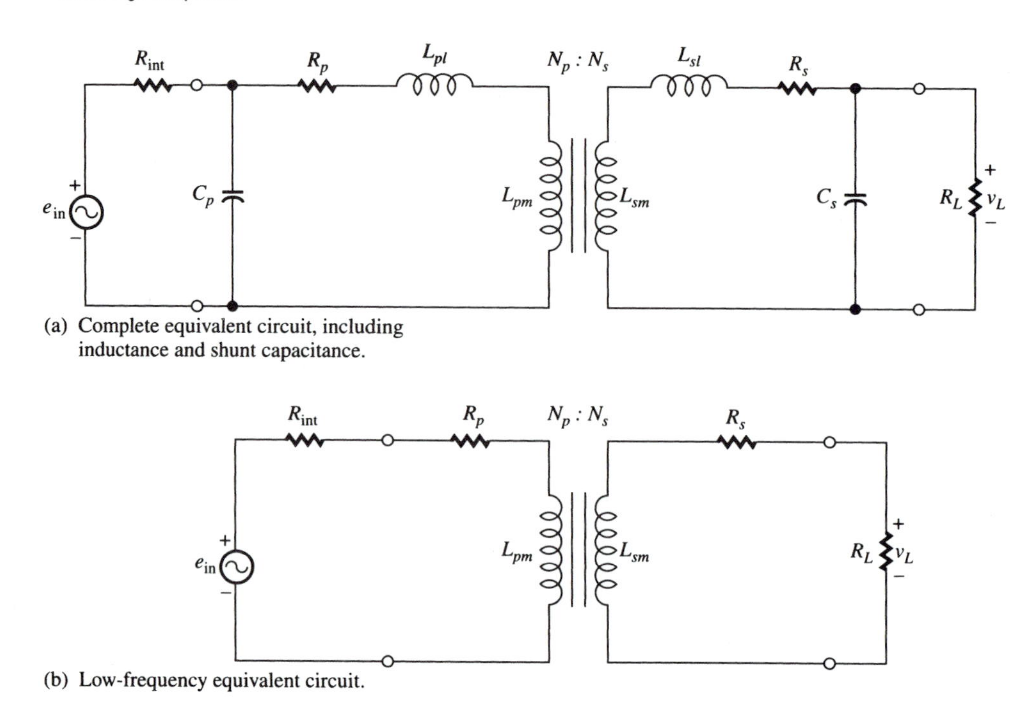

(a) Complete equivalent circuit, including inductance and shunt capacitance.

(b) Low-frequency equivalent circuit.

Figure 11–38
Equivalent circuits of a practical transformer, used to study frequency response

rately. Let us first suppose that the frequency of e_{in} is quite low. In that case, the capacitive reactances of C_s and C_p are quite large ($|X_C| = 1/\omega C$). Since the capacitances are in parallel with the input and output, their large reactances can be neglected at a sufficiently low frequency. Furthermore, the reactances of the leakage inductances, X_{Lpl} and X_{Lsl}, become very small in comparison to R_p and R_s, since inductive reactance decreases with frequency. Figure 11–38(b) shows the low-frequency equivalent circuit when the shunt capacitances and the leakage inductances are neglected (capacitances replaced by open circuits and inductances replaced by short circuits). It is clear that the inductive reactance of L_{pm} and the total series resistance ($R_{int} + R_p$) form a voltage divider across the primary winding. This voltage divider is, in fact, a high-pass filter. The lower the frequency, the smaller the voltage drop across X_{Lpm}. Consequently, as frequency decreases, the voltage appearing at the primary decreases, and the output of the transformer decreases. In the limiting case, where $f = 0$ Hz (dc), $X_{Lpm} = 0\ \Omega$, and there is zero voltage across the primary. This confirms our previous discussion of the fact that a transformer does not respond to dc.

Let us now consider the other frequency extreme, that is, the situation when the frequency of e_{in} becomes very large. In that case, the reactances of the shunt capacitances become very small and can no longer be neglected. Referring again to Figure 11–38(a), notice that R_{int} and C_p form a low-pass RC filter. As we know, the voltage across the output capacitor in a low-pass filter decreases as frequency increases. Furthermore, the reactance of L_{pl} increases with frequency, so a greater

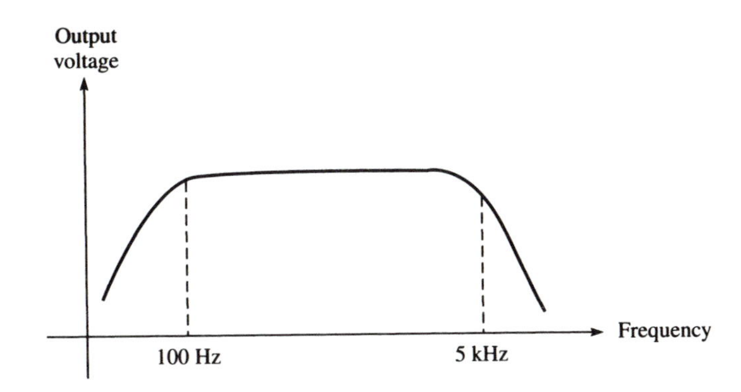

Figure 11–39
Frequency response of a typical audio transformer

portion of the input voltage is dropped across it. The combined effects of the shunt capacitance and the series inductance cause the voltage at the primary to decrease rapidly as frequency increases. Similarly, R_s and C_s form a low-pass RC filter on the secondary side of the transformer. The effect of this filter, and the series reactance of L_{sl}, further reduce the voltage v_L appearing at the load. The reduction in output voltage due to all these effects is responsible for a frequency response that falls rapidly at high frequencies.

Figure 11–39 shows a typical frequency response for a transformer used in audio circuits (an audio transformer). Note that the cutoff frequencies are about 100 Hz and 5 kHz, which means that the passband is a relatively small portion of the audio range (20 Hz to 20 kHz). In modern high-fidelity systems, audio transformers are avoided whenever possible, because of their bulk and their limited frequency response.

Mutual Inductance

Mutual inductance is a measure of the ability of one winding to induce a voltage in a second winding. Contrast this concept with that of self-inductance, which is a measure of the ability of a winding to induce a voltage in itself. Of course, in both cases, a voltage is induced only when current is changing with time. In a transformer, a change of current in the primary winding induces a voltage in the secondary winding, and vice versa. Thus, mutual inductance M relates rate of change of current in each winding to voltage induced in the other winding:

$$e_s = M\frac{\Delta i_p}{\Delta t} \tag{11–74}$$

$$e_p = M\frac{\Delta i_s}{\Delta t} \tag{11–75}$$

Like self-inductance, the units of mutual inductance are henries (H).

Mutual inductance is a parameter used most frequently in connection with loosely coupled air-core transformers, because the turns ratio in those types of transformers is of little value in predicting output voltages. As might be expected, the value of mutual inductance depends on the coefficient of coupling k between the windings, as well as upon the self-inductance of each winding:

$$M = k\sqrt{L_sL_p} \tag{11–76}$$

Example 11–20

The coupling coefficient between the primary and secondary windings of an air-core transformer is 0.05. The self-inductances of the primary and secondary windings are 8 mH and 12 mH, respectively. Find the voltage induced in the secondary winding when the current in the primary winding is changing at the rate of 600 A/s.

Solution. From equation (11–76),

$$M = k\sqrt{L_s L_p} = 0.05\sqrt{(12 \times 10^{-3})(8 \times 10^{-3})} = 0.49 \text{ mH}$$

From equation (11–74),

$$e_s = M\frac{\Delta i}{\Delta t} = (0.49 \times 10^{-3} \text{ H})(600 \text{ A/s}) = 0.294 \text{ V}$$

Example 11–21

SPICE

The loosely coupled transformer in the amplifier circuit shown in Figure 11–40(a) has primary inductance 100 μH, secondary inductance 1 mH, and a coupling coefficient of 0.1. The primary winding resistance is 0.5 Ω and the secondary winding resistance is 1 Ω. Note that the secondary winding is in a parallel-resonant (tank) circuit with tuning capacitance C_T = 2533 pF. The amplifier is operated at the resonant frequency of the tank circuit.

Assuming default parameters for the transistor, use SPICE as an aid in finding

1. the magnitude of the voltage gain of the transistor, $|v_{ce}/v_S|$;
2. the magnitude of the voltage gain of the amplifier, $|v_L/v_S|$;
3. the power gain of the amplifier; and
4. the ac power losses in the primary and secondary windings.

Solution. Figure 11–40(b) shows the SPICE circuit and input file. Notice that the winding resistances are lumped and shown as separate resistors connected in series with the windings. Since the secondary winding resistance is small, the resonant frequency is, to a close approximation, given by

$$f_r = \frac{1}{2\pi\sqrt{L_s C_T}} = \frac{1}{2\pi\sqrt{(1 \text{ mH})(2533 \text{ pF})}} = 100 \text{ kHz}$$

Execution of the program gives the following results:

$$V(3) = 0.2393 \text{ V} \qquad V(7) = 3.515 \text{ mV}$$
$$V(6) = 2.207 \text{ V} \qquad I(V1) = 36.74\ \mu\text{A}$$
$$V(5,4) = 1.834 \text{ mV}$$

Since the input voltage is assumed to be 0.1 V rms, all results are effective (rms) values.

1. The magnitude of the voltage gain of the transistor is

$$\left|\frac{v_{ce}}{v_S}\right| = \frac{V(3)}{0.1 \text{ V}} = \frac{0.2393 \text{ V}}{0.1 \text{ V}} = 2.393$$

Figure 11–40
(Example 11–21)

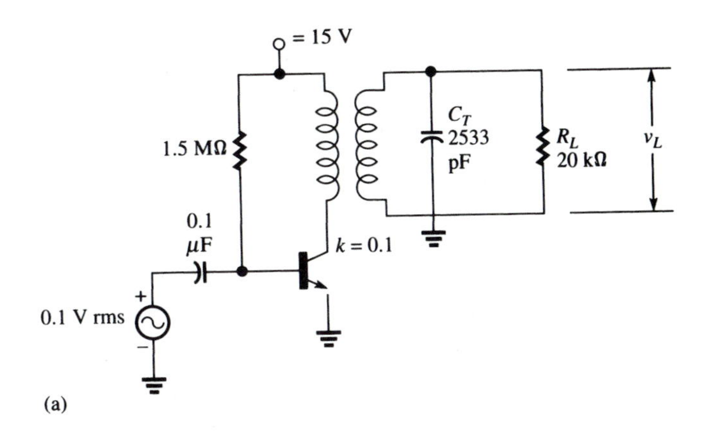

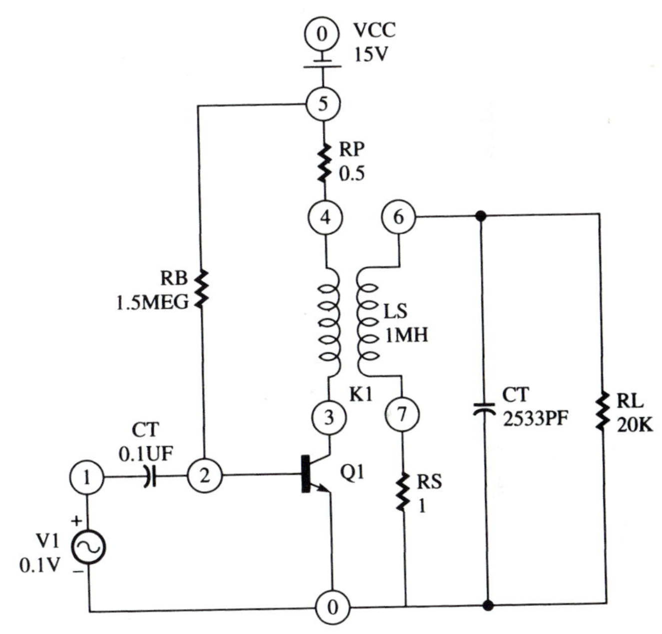


```
EXAMPLE 11.21
V1 1 0 AC 0.1V
CC 1 2 0.1UF
Q1 3 2 0 TRAN
.MODEL TRAN NPN
LP 3 4 100UH
RP 4 5 0.5
VCC 5 0 15V
RB 5 2 1.5MEG
LS 6 7 1MH
K1 LP LS 0.1
RS 7 0 1
CT 6 0 2533PF
RL 6 0 20K
.AC LIN 1 100KHZ 100KHZ
.PRINT AC V(3) V(6) V(5,4)  V(7)  I(V1)
.END
```

(b)

2. The magnitude of the voltage gain of the amplifier is

$$\left|\frac{v_L}{v_S}\right| = \frac{\text{V(6)}}{0.1\text{ V}} = \frac{2.207\text{ V}}{0.1\text{ V}} = 22.07$$

3. The phase angle between the source voltage and source current is very nearly 0°, so the power delivered by the source to the amplifier is

$$P_{in} = v_s i_s \cos 0° = (0.1\text{ V})\text{I(V1)} = (0.1\text{ V})(36.74\ \mu\text{A}) = 3.674\ \mu\text{W}$$

The power in the 20-kΩ load is

$$P_{out} = P_L = \frac{v_L^2}{R_L} = \frac{[\text{V(6)}]^2}{20\text{ k}\Omega} = \frac{(2.207\text{ V})^2}{20\text{ k}\Omega} = 243.54\ \mu\text{W}$$

Then,

$$A_p = \frac{P_{out}}{P_{in}} = \frac{243.54\ \mu\text{W}}{3.674\ \mu\text{W}} = 66.29$$

4.

$$P_l(\text{primary}) = \frac{v^2}{R_p} = \frac{[\text{V(5,4)}]^2}{0.5\ \Omega} = \frac{(1.834\text{ mV})^2}{0.5\ \Omega} = 6.73\ \mu\text{W}$$

$$P_l(\text{secondary}) = \frac{v^2}{R_s} = \frac{[\text{V(7)}]^2}{1\ \Omega} = \frac{(3.515\text{ mV})^2}{1\ \Omega} = 12.35\ \mu\text{W}$$

EXERCISES

Section 11–1
Gain Relations in Multistage Amplifiers

11–1. The in-circuit voltage gains of the stages in a multistage amplifier are shown in Figure 11–41. Find
- **a.** the overall voltage gain, v_o/v_{in}; and
- **b.** the voltage gain that would be necessary in a fifth stage, which, if added to the cascade, would make the overall voltage gain 100 dB.

11–2. The in-circuit voltage gains of the stages in a multistage amplifier are shown in Figure 11–41. (The gain of the first stage does not include loading by the signal source and that of the fourth stage does not include loading by a load resistor.) The input resistance to

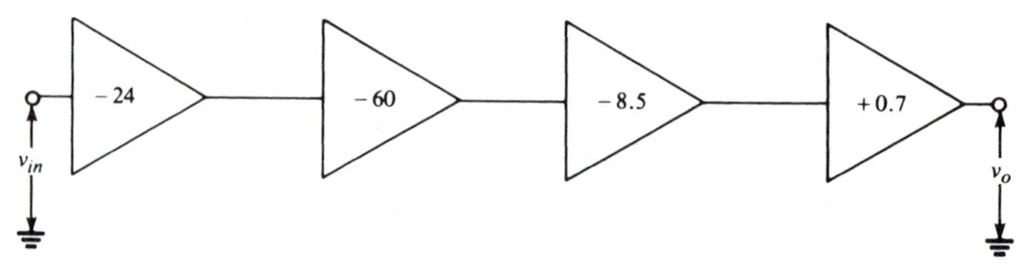

Figure 11–41
(Exercises 11–1 and 11–2)

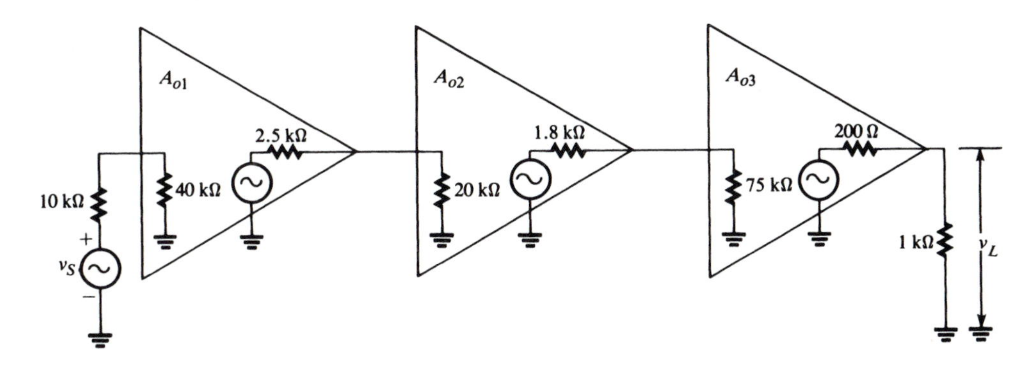

Figure 11–42
(Exercise 11–4)

the first stage is 20 kΩ and the output resistance of the fourth stage is 20 Ω. The amplifier is driven by a signal source having resistance 25 kΩ and a 12-Ω load is connected to the output of the fourth stage. If the source voltage is $v_s = 5$ mV rms, find

a. the load voltage, v_L;

b. the power gain in dB, between the input to the first stage and the load.

11–3. It is desired to construct a three-stage amplifier whose overall voltage gain is 53.98 dB. The in-circuit voltage gains of the first two stages are to be equal, and the voltage gain of the third stage is to be one-half that of each of the first two. What should be the voltage gain of each stage?

11–4. The open-circuit (unloaded) voltage gains of the stages in the multistage amplifier shown in Figure 11–42 are $A_{o1} = -42$, $A_{o2} = -26$, and $A_{o3} = 1.8$. Find the overall voltage gain v_L/v_S.

11–5. A multistage amplifier is to be constructed using four identical stages, each of which has lower cutoff frequency 15 Hz and upper cutoff frequency 30 kHz.

a. What will be the lower and upper cutoff frequencies of the multistage amplifier?

b. If the midband, in-circuit voltage gain of each stage is 8.2, what will be the asymptotic voltage gain of the multistage amplifier at 7.5 Hz? At 300 kHz?

11–6. A multistage amplifier consists of three identical stages, each of which has bandwidth 250 kHz. The bandwidth of the multistage amplifier is 40 kHz. What are the lower and upper cutoff frequencies of each stage?

11–7. A multistage amplifier is to be constructed using six identical stages, each of which has lower cutoff frequency 50 kHz and upper cutoff frequency 1 MHz. What will be the bandwidth of the multistage amplifier?

Section 11–2
Methods of Coupling

11–8. **a.** What is the minimum dc–working-voltage rating that the 1-μF capacitor shown in Figure 11–43 should have? (Assume silicon transistors and $\beta \geq 100$.)

b. If the capacitor is electrolytic, should its positive terminal be connected to the collector of Q_1 or the base of Q_2?

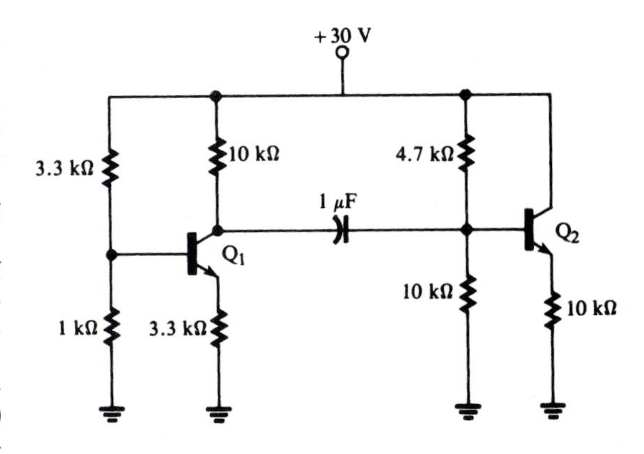

Figure 11–43
(Exercise 11–8)

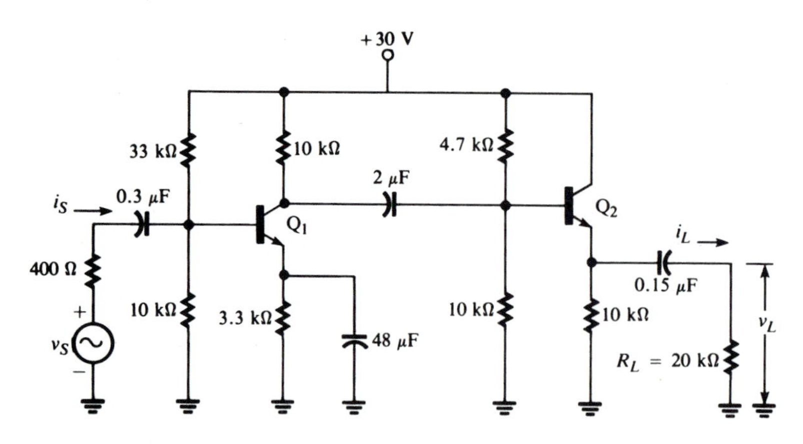

Figure 11–44
(Exercise 11–9)

Section 11–3
RC-Coupled BJT Amplifiers

11–9. The transistors in Figure 11–44 have the following parameter values:

$$Q_1:\ r_{e1} = 12\ \Omega,\ \beta_1 = 200,\ r_{c1} = 2\ \text{M}\Omega$$
$$Q_2:\ r_{e2} = 10\ \Omega,\ \beta_2 = 100$$

Find the midband values of
a. the voltage gain v_L/v_S; and
b. the current gain i_L/i_S.

11–10. Repeat Exercise 11–9 with R_L changed to 100 Ω.

11–11. a. Find the approximate lower cutoff frequency of the multistage amplifier in Exercise 11–9.
b. Sketch the asymptotic Bode plot the way it would appear if plotted on log-log paper. Label all break frequencies and asymptote slopes.

11–12. a. Find the approximate lower cutoff frequency of the multistage amplifier in Exercise 11–9 when R_L is changed to 100 Ω.
b. Sketch the asymptotic Bode plot the way it would appear if plotted on log-log graph paper. Label all break frequencies and asymptote slopes.

11–13. Find the approximate lower cutoff frequency of the multistage amplifier in Figure 11–6 (Example 11–4).

11–14. Find the approximate upper cutoff frequency of the two-stage amplifier shown in Figure 11–45. The silicon transistors have the fol-

Figure 11–45
(Exercise 11–14)

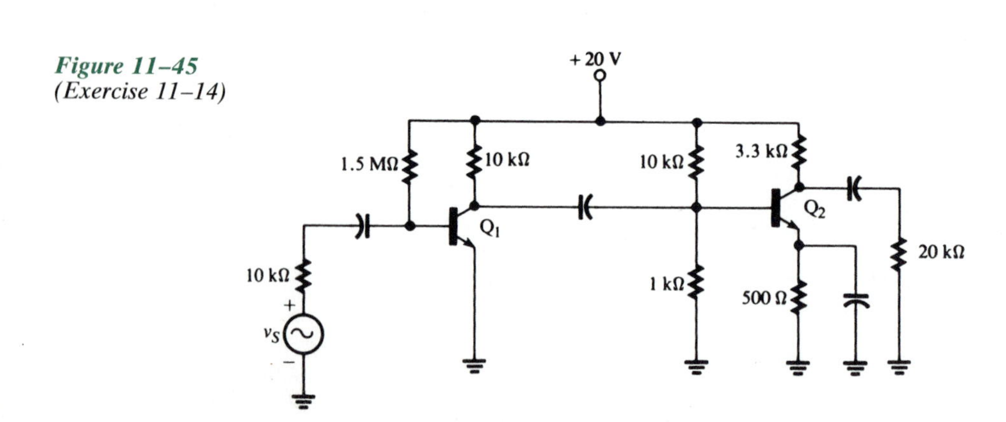

lowing parameter values:

Q_1: $\beta_1 = 100$, $f_{\beta 1} = 1$ MHz, $C_{be1} = 10$ pF, $C_{bc1} = 1.5$ pF, $C_{ce1} = 5$ pF, $r_{c1} \approx \infty$

Q_2: $\beta_2 = 100$, $f_{\beta 2} = 1$ MHz, $C_{be2} = 20$ pF, $C_{bc2} = 6.5$ pF, $C_{ce2} = 5$ pF, $r_{c2} \approx \infty$

(*Hint:* First find the bias currents I_{E1} and I_{E2} in order to determine values for r_{e1} and r_{e2}. Use these to determine stage input resistances and voltage gains. The latter are necessary to find the Miller capacitance.)

Section 11-4
Direct-Coupled BJT Amplifiers

11-15. The silicon transistors in Figure 11-46 both have $\beta = 100$. Find

a. I_{C1}, V_{C1}, V_{CE1}, I_{C2}, V_{C2}, and V_{CE2}; and
b. the small-signal gain v_o/v_i.

11-16. Repeat Exercise 11-15 when a 60-kΩ load is direct-coupled to the collector of Q_2.

11-17. Find the dc collector currents I_{C1} and I_{C2} and the dc collector-to-ground voltages V_{C1} and V_{C2} in Figure 11-47. Assume that silicon transistors are used and that $\beta_1(2\text{ k}\Omega) \gg 3.9\text{ k}\Omega$.

11-18. Repeat Exercise 11-17 for the amplifier shown in Figure 11-48.

11-19. The transistors in the Darlington pair shown in Figure 11-49 have $\beta_1 = 80$ and $\beta_2 = 100$. $I_{B1} = 1\ \mu$A.
a. Using the same approximations used in the text to analyze a Darlington pair ($I_C \approx I_E$ and $I_T \approx I_{C2}$), find I_{C1}, I_{C2}, I_{B2}, I_{E1}, I_{E2}, I_T, and V_{C2}.

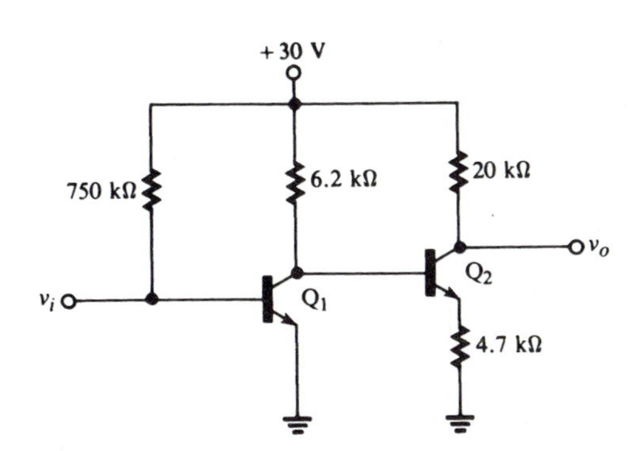

Figure 11-46
(Exercise 11-15)

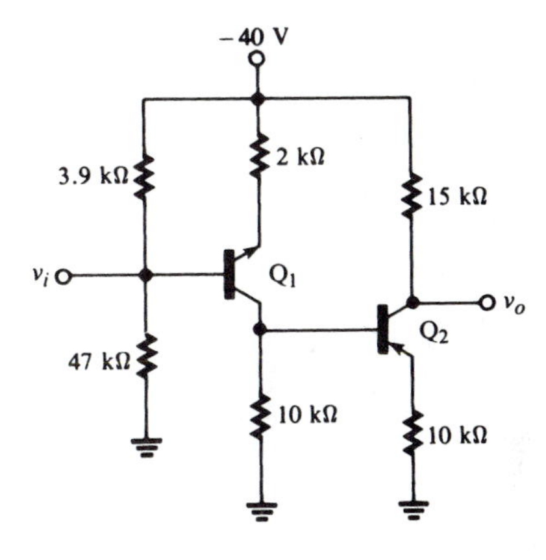

Figure 11-48
(Exercise 11-18)

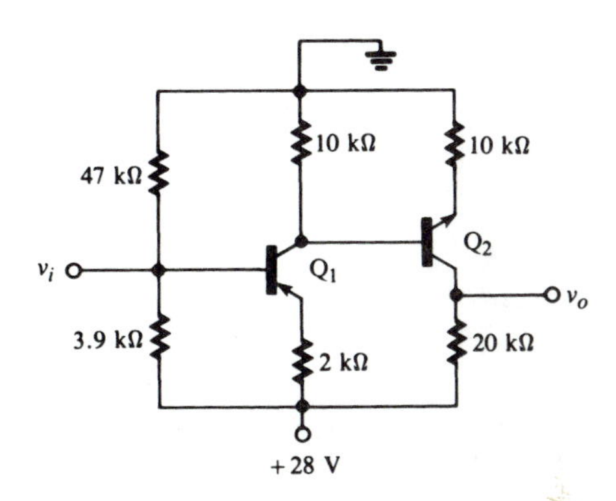

Figure 11-47
(Exercise 11-17)

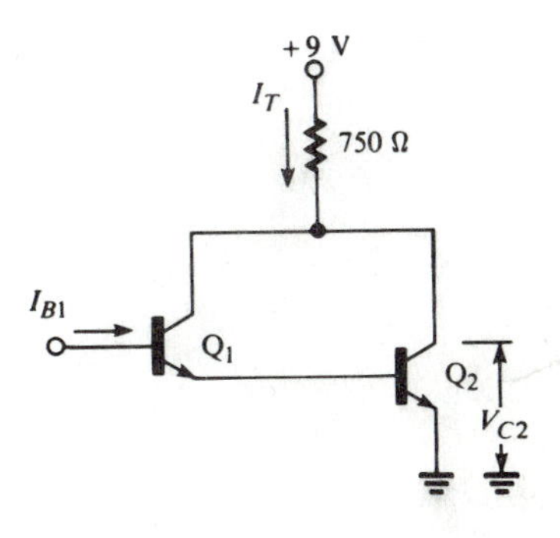

Figure 11-49
(Exercise 11-19)

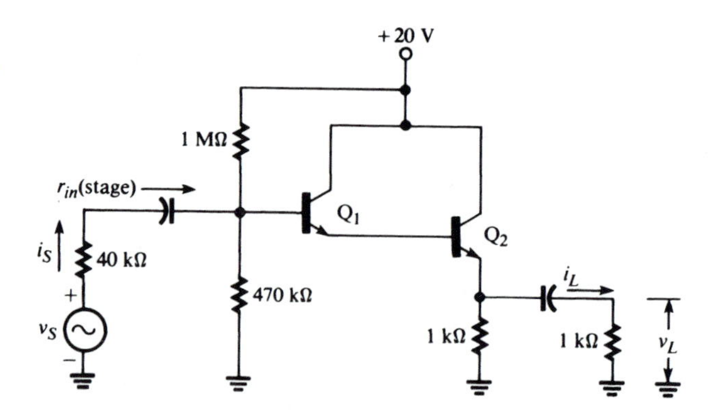

Figure 11–50
(Exercise 11–20)

b. Find r_{e1} and r_{e2} based on the values found in (a).

c. Without using the approximations, find I_{C1}, I_{C2}, I_{B2}, I_{E1}, I_{E2}, I_T, and V_{C2}. Find the percentage error in the computation of V_{C2} caused by using the approximation.

d. Find r_{e1} and r_{e2} based on the values found in (c).

11–20. The silicon transistors in Figure 11–50 have $\beta_1 = 120$ and $\beta_2 = 110$. Find (a) r_{in}(stage), (b) v_L/v_S, and (c) i_L/i_S.

11–21. The silicon transistors in Figure 11–51 have $\beta_1 = \beta_2 = 50$. The effective collector resistance (at Q_2) is 10 kΩ. Find approximate values for (a) r_{in}(stage) and (b) v_L/v_S.

11–22. The silicon transistors in Figure 11–52 have $\beta_1 = 100$ and $\alpha_2 = 0.999$. Assuming that $r_{c1} = r_{c2} \approx \infty$, find

a. the dc currents and voltages I_{C1}, I_{C2}, V_{CE1}, and V_{CE2}; and

b. the small-signal, midband voltage gain v_L/v_S.

11–23. Transistor Q_1 in Exercise 11–22 has $C_{be} = 4.2$ pF and $C_{bc} = 5.4$ pF. The wiring capacitance at the input to Q_1 is 10 pF. Find the upper break frequency due to input capacitance.

11–24. The silicon transistors in the cascode amplifier in Figure 11–53 have $\beta_1 = \beta_2 = 120$. Find the dc values I_{C1}, I_{C2}, V_{C1}, V_{C2}, V_{CE1}, and V_{CE2}. (*Hint:* The loading caused by the base of Q_1 and Q_2 on the voltage divider can be neglected; use the voltage divider to find each base voltage.)

11–25. Find the approximate midband voltage gain v_L/v_S of the cascode amplifier in Exercise 11–24.

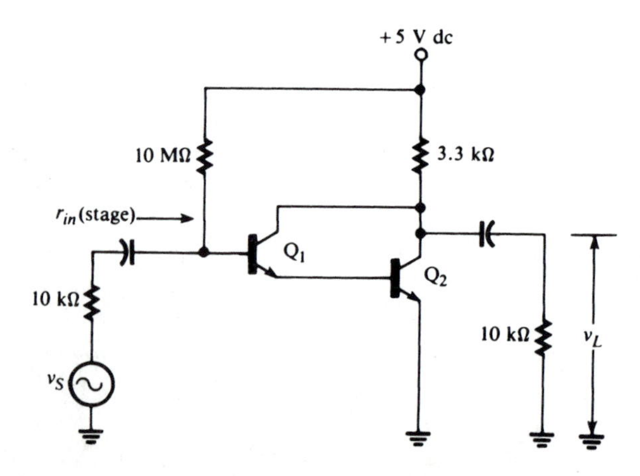

Figure 11–51
(Exercise 11–21)

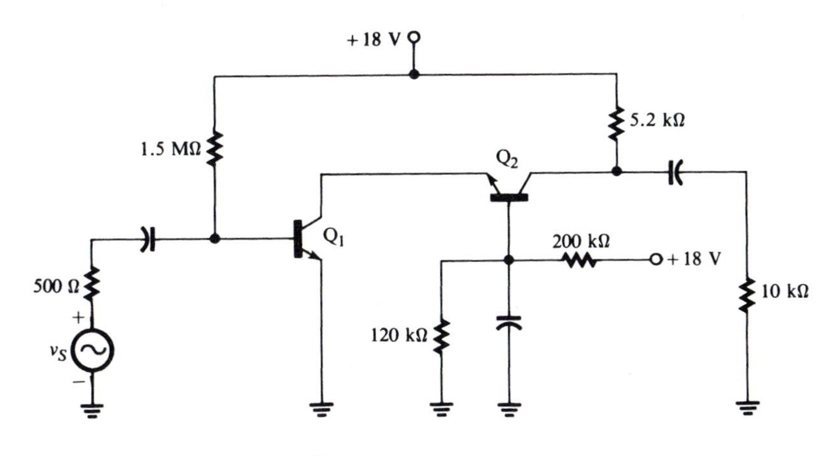

Figure 11–52
(Exercise 11–22)

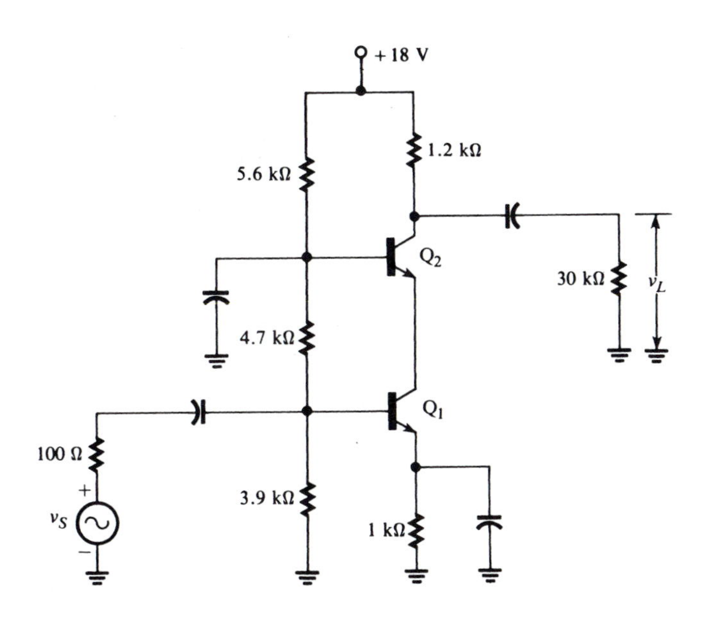

Figure 11–53
(Exercise 11–24)

Section 11–6
Multistage FET Amplifiers

11–26. The JFETs in Figure 11–54 have $g_{m1} = 3000$ μS, $r_{d1} = 100$ kΩ, $g_{m2} = 2000$ μS, and $r_{d2} =$ 60 kΩ. Find the midband voltage gain v_L/v_S.

11–27. Find the approximate lower cutoff frequency of the amplifier in Exercise 11–26.

11–28. The FETs in Figure 11–55 have $g_{m1} = 2.5$ mS, $r_{d1} = 100$ kΩ, $g_{m2} = 4$ mS, and $r_{d2} = 100$ kΩ. Find the midband voltage gain v_L/v_S.

11–29. The FETs in Figure 11–56 have $I_{DSS1} = 10$ mA, $V_{p1} = -2$ V, $I_{DSS2} = 12$ mA, and $V_{p2} =$ 2 V. Find the dc values I_{D1}, V_{GS1}, V_{DS1}, I_{D2}, V_{GS2}, and V_{DS2}.

11–30. The FETs in Figure 11–57 are the same as those in Exercise 11–29 and they are biased the same way. Assuming that $r_{d1} = r_{d2} \approx \infty$, find the midband voltage gain v_L/v_S.

11–31. The MOSFETs in Figure 11–58 are biased so that $V_{GS1} = 4.8$ V and $V_{GS2} = 6$ V. The transistors have the following parameter

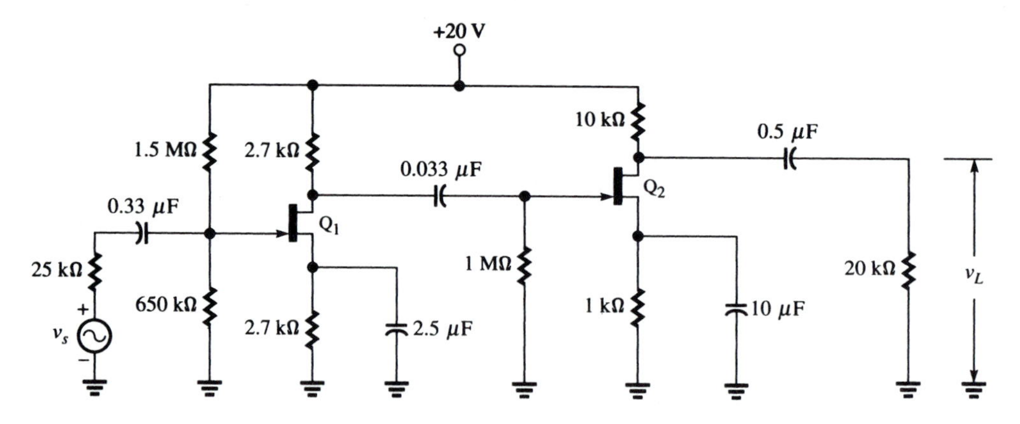

Figure 11–54
(Exercise 11–26)

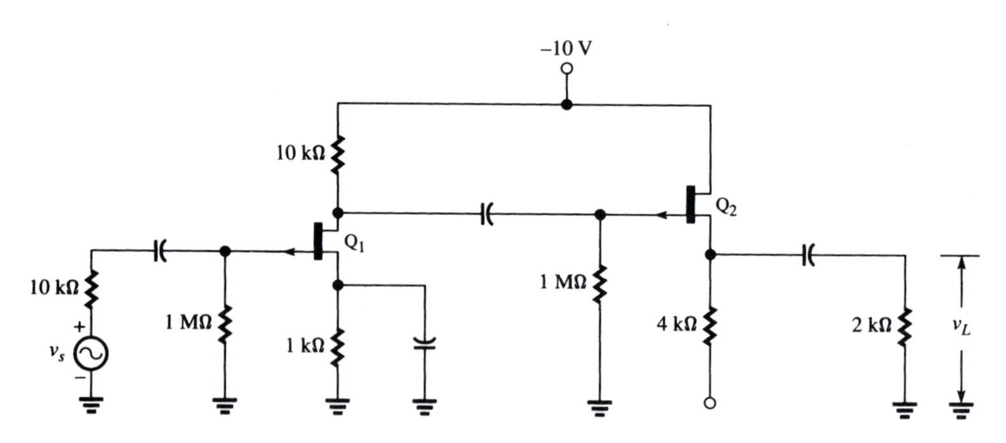

Figure 11–55
(Exercise 11–28)

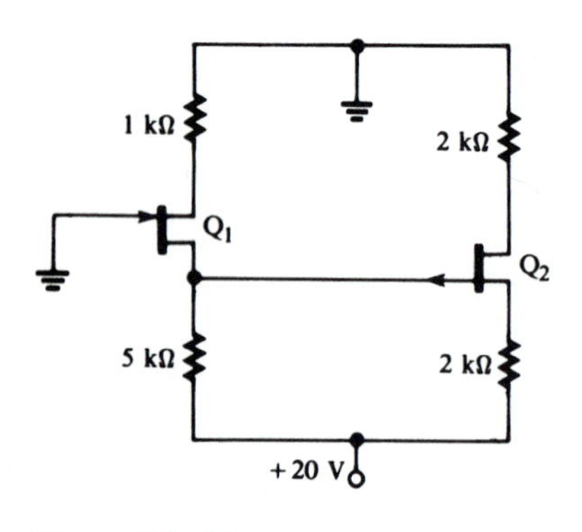

Figure 11–56
(Exercise 11–29)

values:

$$Q_1: \quad \beta_1 = 0.5 \times 10^{-3}, V_{T1} = 2\text{ V}, \quad r_{d1} = 75\text{ k}\Omega$$

$$Q_2: \quad \beta_2 = 0.3 \times 10^{-3}, V_{T2} = 2.6\text{ V}, \quad r_{d2} = 100\text{ k}\Omega$$

Find the midband voltage gain v_L/v_S.

11–32. The transistors in Figure 11–59 have the following parameter values:

$$Q_1: \quad g_{m1} = 4\text{ mS}, r_{d1} = 100\text{ k}\Omega$$

$$Q_2: \quad r_e = 30\ \Omega, r_{c2} = 2.5\text{ M}\Omega$$

Find the midband voltage gain, v_L/v_S.

11–33. The bias resistors (those bypassed with capacitors) in Figure 11–59 are adjusted to

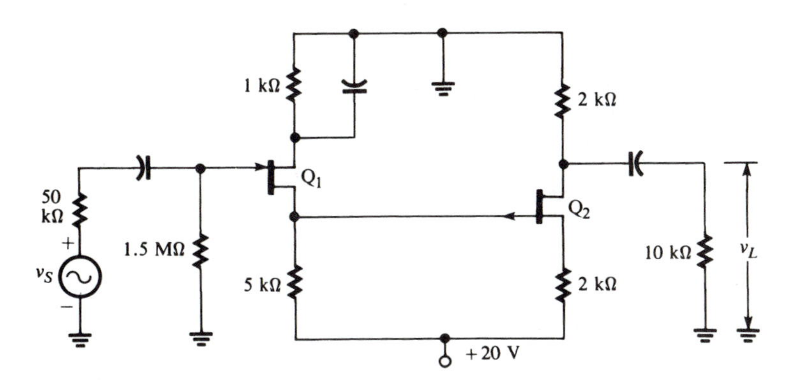

Figure 11–57
(Exercise 11–30)

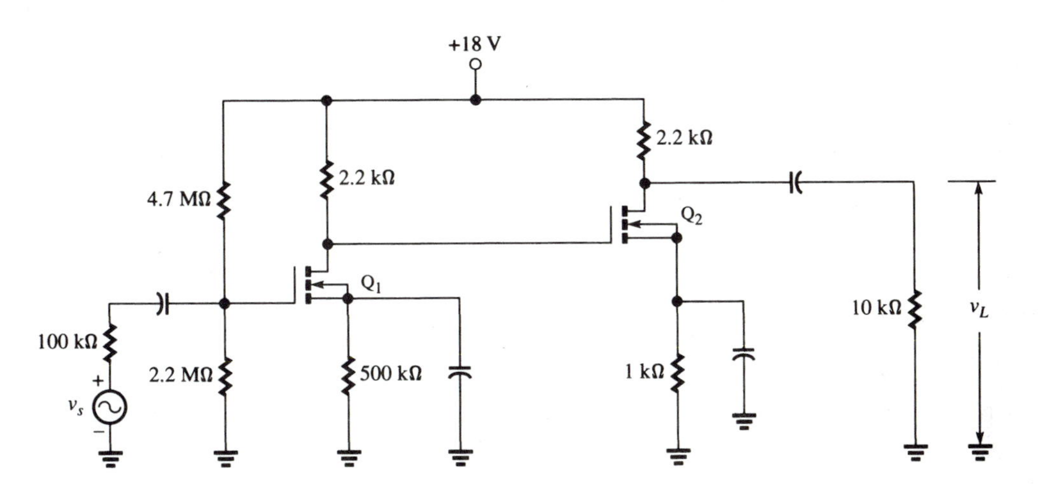

Figure 11–58
(Exercise 11–31)

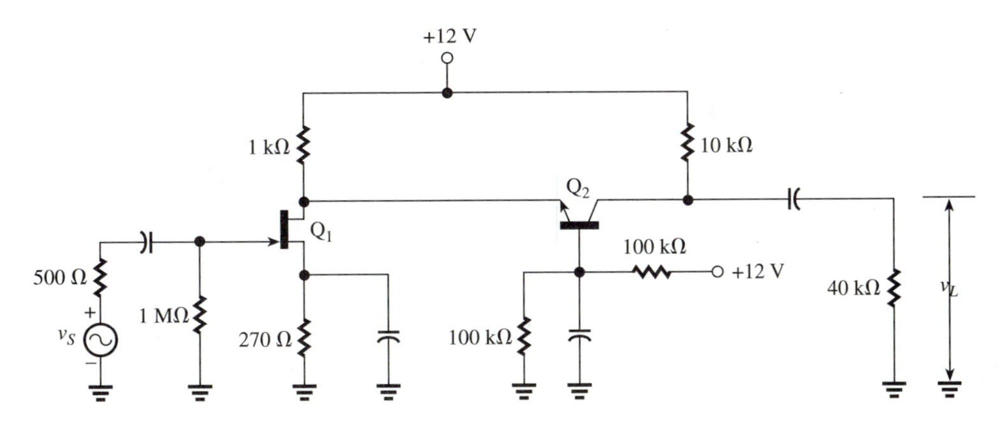

Figure 11–59
(Exercise 11–32)

make $I_{D1} = 6.3$ mA and $I_{E2} = 0.3$ mA. If $I_{DSS1} = 12$ mA and $V_{p1} = -4$ V, find the midband voltage gain v_L/v_S.

Section 11–7
Transformer Coupling

11–34. A signal source having resistance 90 kΩ drives the primary winding of a transformer whose turns ratio is 6:1. What value of load resistance across the secondary would result in maximum power delivered to the load?

11–35. A 200-Ω load resistor is connected across the secondary of a transformer whose turns ratio is 86:20. If maximum power is being transferred to the load, what is the signal-source impedance on the primary side of the transformer?

11–36. In the transformer circuit shown in Figure 11–60, find the power dissipated in r_1 and the power delivered to r_L. Are source and load matched by the transformer?

11–37. a. Find the turns ratio in Exercise 11–36 that will result in maximum power delivered to the load.

b. Using the turns ratio found in (a), find the power dissipated in r_1 and the power delivered to r_L.

11–38. a. What should be the turns ratio of the transformer in Figure 11–61 in order to deliver maximum power to r_L?

b. Using the turns ratio found in (a), find the power dissipated in r_1 and the power delivered to r_L.

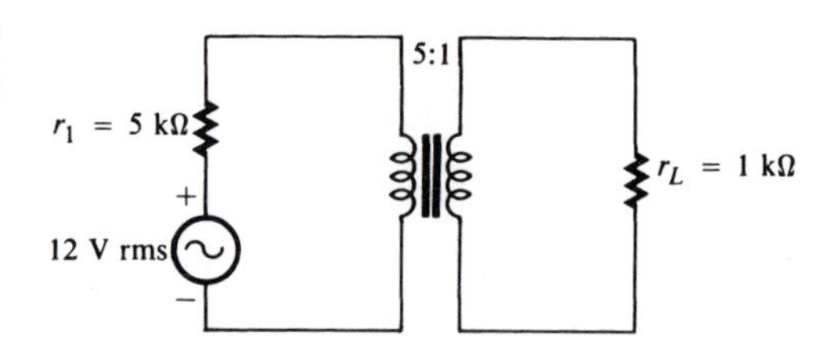

Figure 11–60
(Exercise 11–36)

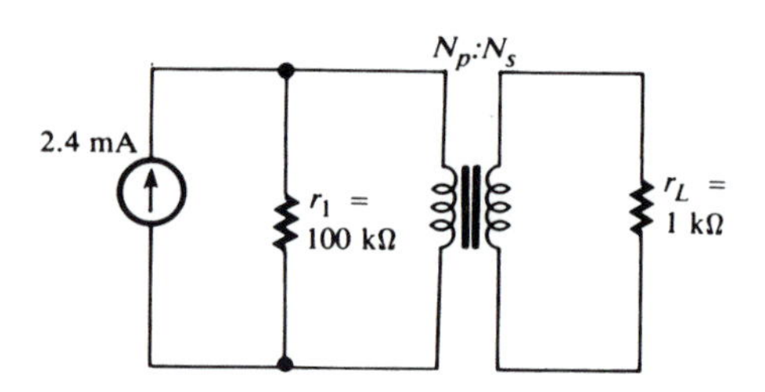

Figure 11–61
(Exercise 11–38)

11–39. The transistors in Figure 11–62 have $\beta_1 = \beta_2 = 50$ and $r_{o1} = r_{o2} = 40$ kΩ. Find the voltage gain v_L/v_S. Does the transformer match the first stage to the second?

11–40. Repeat Exercise 11–39 if the turns ratio of the transformer is changed to 4:1.

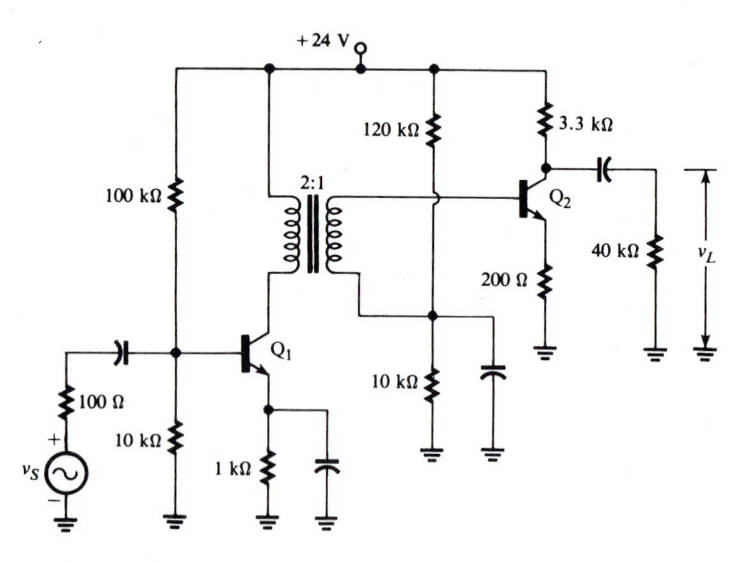

Figure 11–62
(Exercise 11–39)

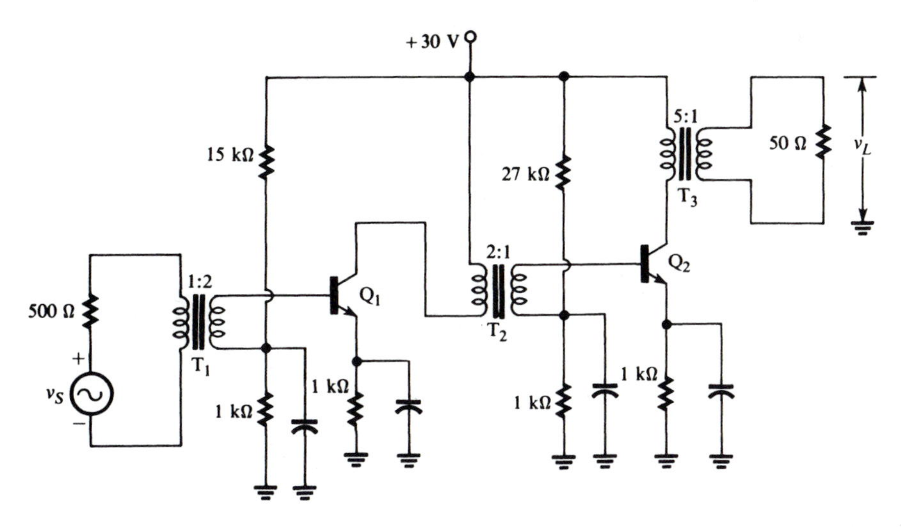

Figure 11–63
(Exercise 11–41)

11–41. The transistors in Figure 11–63 have the following parameter values:

$$Q_1:\quad \beta_1 = 100,\ r_{e1} = 20\ \Omega,\ r_{o1} = 20\ \text{k}\Omega$$
$$Q_2:\quad \beta_2 = 50,\ r_{e2} = 100\ \Omega,\ r_{o2} = 20\ \text{k}\Omega$$

a. Find the voltage gain, v_L/v_S.
b. Which transformers, if any, match the source to the load?

11–42. Repeat Exercise 11–41 when the turns ratios on all transformers are adjusted to obtain maximum power transfer.

Section 11-8
Limitations of Practical Transformers

11–43. The primary and secondary winding resistances of a transformer are 0.9 Ω and 0.5 Ω, respectively. The transformer delivers 32 W to a 64-Ω load connected across its secondary. The primary current is 0.6 A. If the core losses are 1.5% of the load power, find the efficiency of the transformer.

11–44. A transformer having an efficiency of 94.3% delivers 200 W to a load.
a. If the copper losses are 1% of the load power, what is the core loss?
b. If the magnitude of the primary current is 6 A pk, what is the magnitude of the voltage supplied by the source connected to the primary winding?

11–45. When the mutual flux in an iron-core transformer is 6×10^{-4} Wb, the leakage flux is 3.16×10^{-5} Wb. The total inductance of the primary winding is 0.22 H and the total inductance of the secondary winding is 140 mH. The resistance of the primary winding is 1.8 Ω and the resistance of the secondary winding is 0.5 Ω. Draw the equivalent circuit

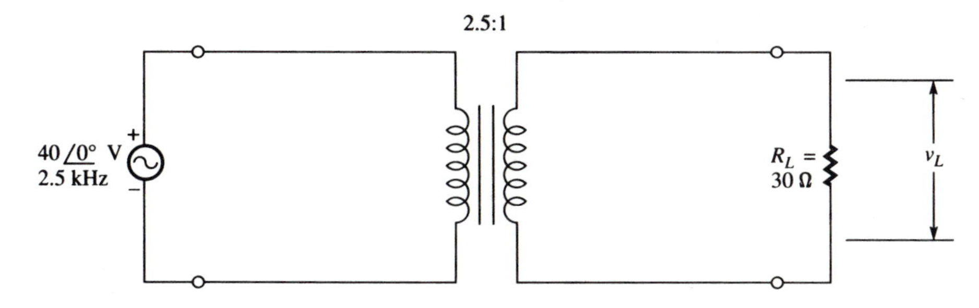

Figure 11–64
(Exercise 11–46)

of the transformer. Be certain to label the values of all components.

11–46. The transformer shown in Figure 11–64 has a coupling coefficient of 0.95. The total inductance of the primary winding is 1 H and the total inductance of the secondary winding is 80 mH. The primary and secondary winding resistances are 1.5 Ω and 0.4 Ω, respectively. Find the magnitude of the load voltage, v_L.

11–47. The coefficient of coupling in an air-core transformer is 0.08. The primary inductance is 8 mH and the secondary inductance is 18 mH. What voltage is induced in the secondary when the current in the primary is changing at the rate of 500 A/s?

11–48. The primary and secondary inductances in an air-core transformer are each equal to 14 mH. When the current in one winding is changing at the rate of 200 A/s, the voltage induced in the other winding is 0.56 V. What is the coefficient of coupling between the windings?

SPICE EXERCISES

11–49. The transistors in the direct-coupled FET amplifier in Figure 11–25 all have $I_{DSS} = 8$ mA and $|V_p| = 2$ V. The input and output coupling capacitors are 1 μF and the source bypass capacitors are 20 μF.

a. Use SPICE to find the bias point (I_D and V_{DS}) of each transistor.

b. Find the voltage gain v_L/v_S of the amplifier at 10 kHz.

11–50. The input and output coupling capacitors in the cascode amplifier shown in Figure 11–21 are each 2.5 μF. The bypass capacitors are each 20 μF. Use SPICE to find the bias points (V_{CE} and I_C) of each transistor and the voltage gain v_L/v_S at 10 kHz.

11–51. Use SPICE to find the voltage gain v_L/v_S and current gain i_L/i_S of the transformer-coupled amplifier in Figure 11–65 when the input frequency is 25 kHz. The inductance of the secondary winding of the input transformer is 10 H and the inductance of the primary winding of the output transformer is 1000 H. Both transformers can be assumed to be ideal. The transistor has $\beta = 80$.

11–52. The forward Early voltage of a transistor has a direct influence on the output resistance r_o at the collector of the transistor in a common emitter configuration. The characteristic curves slope upward (see Figure 4–21) to a greater extent as the Early voltage becomes smaller, so the output resistance $r_o = \Delta V_C/\Delta I_C$ becomes smaller as the Early voltage becomes smaller. To investigate the effect of output resistance on the output power of a

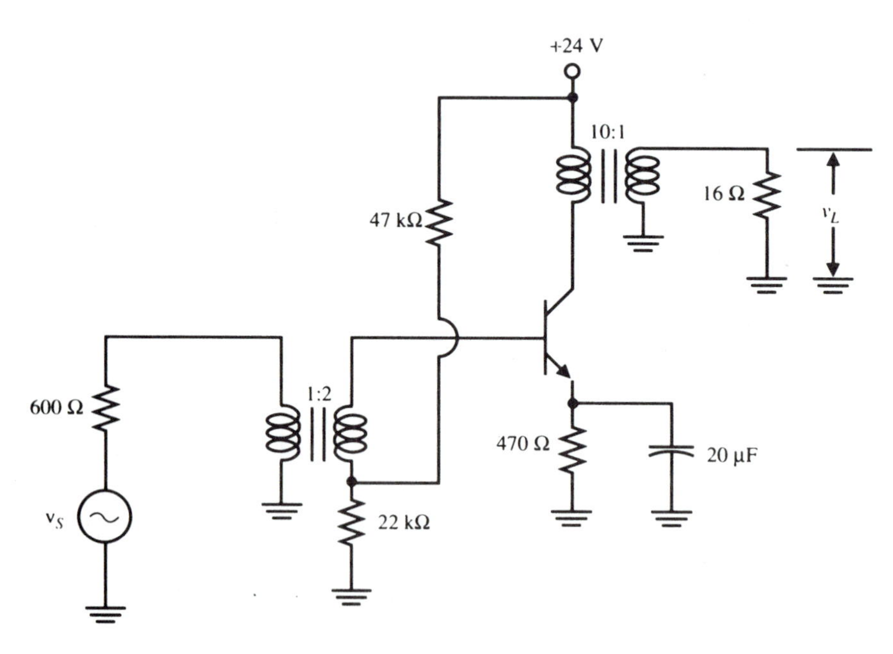

Figure 11–65
(Exercise 11–51)

transformer-coupled amplifier, use SPICE to find the power delivered to the load in the amplifier of Exercise 11–51 when the transistor's forward Early voltage is 250 V, 200 V, 150 V, 100 V, 50 V, 25 V, and 15 V. Construct a table showing load power versus output resistance. (Note that SPICE prints output resistance, RO, as part of its "operating point information.") Comment on the results of your investigation, with attention to the effective resistance seen by the transistor at its collector.

12 INTEGRATED DIFFERENTIAL AND OPERATIONAL AMPLIFIER CIRCUITS

12–1 INTRODUCTION

Differential amplifiers are widely used in linear integrated circuits. They are a fundamental component of every *operational amplifier,* which, as we shall learn, is an extremely versatile device with a broad range of practical applications. We will study the circuit theory of differential amplifiers in some detail, in preparation for a more comprehensive investigation of the capabilities (and limitations) of operational amplifiers.

Difference Voltages

A differential amplifier is also called a *difference* amplifier because it amplifies the difference between two signal voltages. Let us refine the notion of a *difference voltage* by reviewing some simple examples. We have already encountered difference voltages in our study of transistor amplifiers. Recall, for example, that the collector-to-emitter voltage of a BJT is the difference between the collector-to-ground voltage and the emitter-to-ground voltage:

$$V_{CE} = V_C - V_E \tag{12–1}$$

The basic idea here is that a difference voltage is the mathematical difference between two other voltages, each of whose values is given with respect to ground. Suppose the voltage at point A in a circuit is 12 V with respect to ground and the voltage at point B is 3 V with respect to ground. The notation V_{AB} for the difference voltage means the voltage that would be measured if the positive side of a voltmeter were connected to point A and the negative side to point B; in this case, $V_{AB} = V_A - V_B = 12 - 3 = 9$ V. If the voltmeter connections were reversed, we would measure $V_{BA} = V_B - V_A = 3 - 12 = -9$ V. Thus, $V_{BA} = -V_{AB}$.

To help get used to thinking in terms of difference voltages, consider the system shown in Figure 12–1, where two identical amplifiers are driven by two different signal voltages. Although a differential amplifier does not behave in exactly the same way as this amplifier arrangement, the concepts of input and output difference voltages are similar. The two signal input voltages, v_1 and v_2, are shown as sine waves, one greater in amplitude than the other. For illustrative purposes, their peak

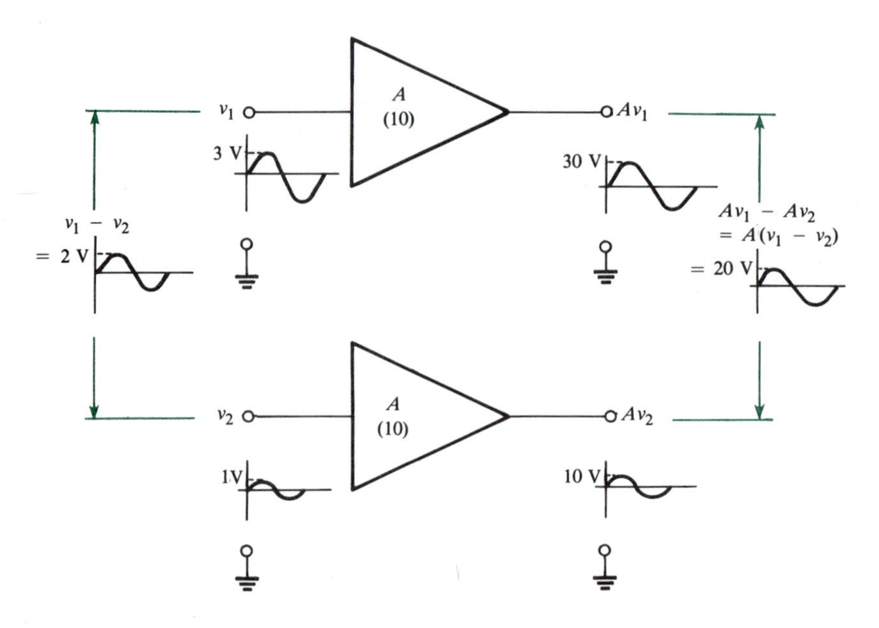

Figure 12–1
The amplification of difference voltages

values are 3 V and 1 V, respectively. If the voltage gain of each amplifier is A, then the amplifier outputs are Av_1 and Av_2. The input difference voltage, $v_{12} = v_1 - v_2$, is a sine wave and the output difference voltage, $Av_1 - Av_2 = A(v_1 - v_2)$, is seen to be an amplified version of the input difference voltage. In our illustration, the gain A is 10 and the input difference voltage is (3 V pk) − (1 V pk) = 2 V pk. The output difference voltage is $A(v_1 - v_2) = 10(3\text{ V} - 1\text{ V}) = 10(2\text{ V}) =$ 20 V pk.

12–2 THE IDEAL DIFFERENTIAL AMPLIFIER

The principal feature that distinguishes a differential amplifier from the configuration shown in Figure 12–1 is that a signal applied to one input of a differential amplifier induces a voltage with respect to ground on the amplifier's other output. This fact will become clear in our study of the voltage and current relations in the amplifier.

Figure 12–2 shows the basic BJT version of a differential amplifier. Two transistors are joined at their emitter terminals, where a constant-current source is connected to supply bias current to each. The current source is typically one of the transistor constant-current sources that we studied in Chapter 6, but for now we will represent it by an ideal current source. Note that each transistor is basically in a common-emitter configuration, with an input supplied to its base and an output taken from its collector. The two base terminals are the two signal inputs to the differential amplifier, v_{i1} and v_{i2}, and the two collectors are the two outputs, v_{o1} and v_{o2}, of the differential amplifier. Thus, the differential input voltage is $v_{i1} - v_{i2}$, and the differential output voltage is $v_{o1} - v_{o2}$.

Figure 12–3 shows the schematic symbol for the differential amplifier. Since there are two inputs and two outputs, the amplifier is said to have a *double-ended* (or double-sided) input and a double-ended output.

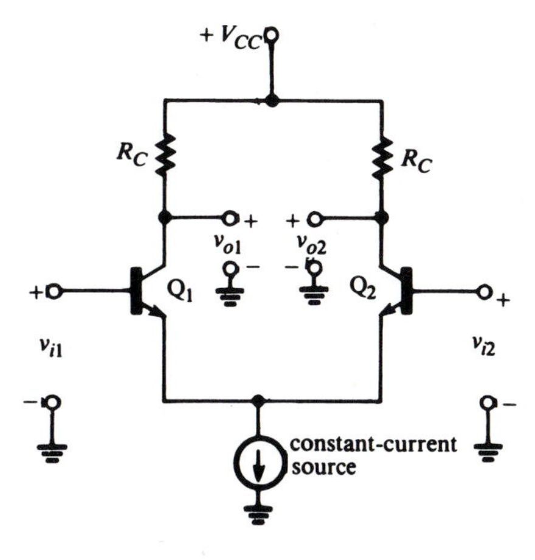

Figure 12–2
The basic BJT differential amplifier. The two transistors can be regarded as CE amplifiers having a common connection at their emitters. The base terminals are the inputs to the differential amplifier and the collectors are the outputs.

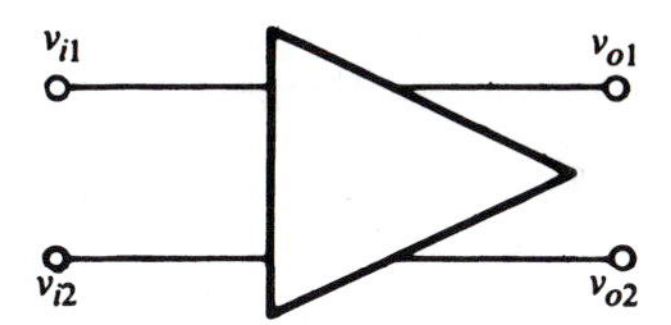

Figure 12–3
Schematic symbol for the differential amplifier

We will postpone, temporarily, our analysis of the dc bias levels in the amplifier and focus on its behavior as a small-signal amplifier. Toward that end, we will determine the output voltage at each collector due to each input voltage acting *alone,* that is, with the opposite input grounded, and then apply the superposition principle to determine the outputs due to both inputs acting simultaneously. Figure 12–4 shows the amplifier with input 2 grounded ($v_{i2} = 0$) and a small signal applied to input 1. The ideal current source presents an infinite impedance (open circuit) to an ac signal, so we need not consider its presence in our small-signal analysis. We also assume the ideal situation of perfectly matched transistors, so Q_1 and Q_2 have identical values of β, r_e, etc. Since Q_1 is essentially a common-emitter amplifier, the voltage at its collector (v_{o1}) is an amplified and *inverted* version of its input, v_{i1}.

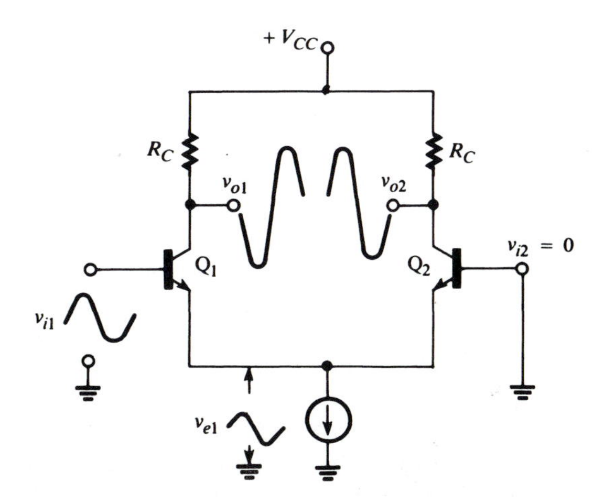

Figure 12–4
The small-signal voltages in a differential amplifier when one input is grounded. Note that v_{e1} is in phase with v_{i1}, and that v_{o1} is out of phase with v_{i1}.

Note that there is also an ac voltage v_{e1} developed at the emitter of Q_1. This voltage is in phase with v_{i1} and exists because of emitter-follower action across the base–emitter junction of Q_1.

Now, the voltage v_{e1} is developed across the emitter resistance r_e looking into the emitter of Q_2 (in parallel with the infinite resistance of the current source). Therefore, as far as the emitter-follower action of Q_1 is concerned, the load resistance seen by Q_1 is r_e. Since the emitter resistance of Q_1 is itself r_e, it follows from equation 5–58 that the emitter-follower gain is 1/2:

$$A_v = \frac{r_L}{r_L + r_e} = \frac{r_e}{r_e + r_e} = 0.5$$

Therefore, v_{e1} is in phase with, and one-half the magnitude of, v_{i1}. Now, it is clear that v_{e1} is the emitter-to-ground voltage of *both* transistors. *When v_{e1} goes positive, the base-to-emitter voltage of* Q_2 *goes negative by the same amount.* In other words, $v_{be2} = v_{b2} - v_{e1} = 0 - v_{e1}$. (Since the base of Q_2 is grounded, its base-to-emitter voltage is the same as the negative of its emitter-to-ground voltage.) We see that even though the base of Q_2 is grounded, there exists an ac base-to-emitter voltage on Q_2 that is out of phase with v_{e1} and therefore out of phase with v_{i1}. Consequently, there is an ac output voltage v_{o2} produced at the collector of Q_2 and it is out of phase with v_{o1}.

Since both transistors are identical, they have equal gain and the output v_{o2} has the same magnitude as v_{o1}. To verify this last assertion, and to help solidify all the important ideas we have presented so far, let us study the specific example illustrated in Figure 12–5. We assume that v_{i1} (which is the base-to-ground voltage of Q_1) is a 100-mV-pk sine wave, and that each transistor has voltage gain -100, where, as usual, the minus sign denotes phase inversion. By "transistor voltage gain," we mean the collector voltage divided by the *base-to-emitter* voltage.

Since the emitter-follower gain of Q_1 is 0.5, v_e is a 0.5(100 mV) = 50-mV-pk sine wave. The peak value of v_{be1} is therefore $v_{b1} - v_{e1} = (100 \text{ mV}) - (50 \text{ mV}) = 50$ mV. When v_{be1} is at this 50-mV peak, v_{o1} is $-100(50 \text{ mV}) = -5$ V, that is, an inverted 5-V-pk sine wave. At this same point in time, where v_e is at its 50-mV

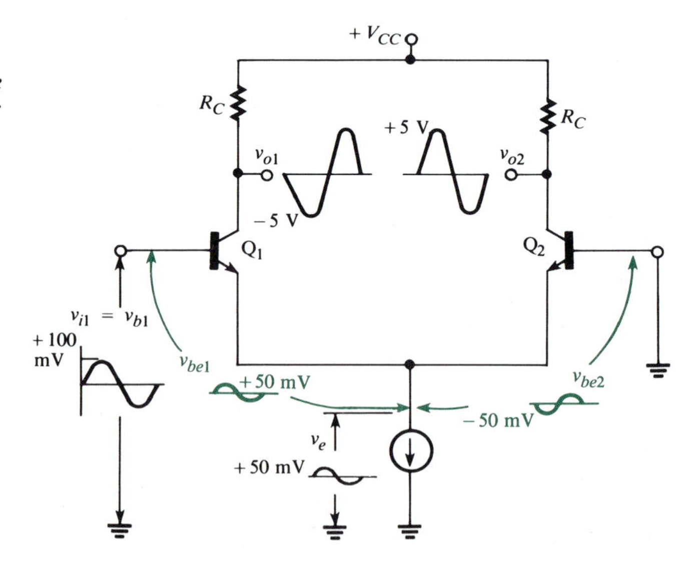

Figure 12–5
Each transistor has identical voltage gain −100, and the outputs at the collectors are −100 times their respective base-to-emitter voltages.

peak, the base-to-emitter voltage of Q_2 is $0 - (50\text{ mV}) = -50$ mV pk. Therefore, v_{o2} is $(-100)(-50\text{ mV}) = +5$ V pk; that is, v_{o2} is a 5-V-pk sine wave in phase with v_{i1} and out of phase with v_{o1}.

Note in Figure 12–5 that the input difference voltage is $v_{i1} - v_{i2} = (100\text{ mV}) - 0 = 100$ mV pk, and the output difference voltage is 10 V pk, since v_{o1} and v_{o2} are out of phase. Therefore, the magnitude of the *difference voltage gain* $(v_{o1} - v_{o2})/(v_{i1} - v_{i2})$ is 100. So, while the voltage gain v_o/v_i for each side is only 50, the difference voltage gain is the same as the gain v_c/v_{be} of each transistor. We will refine and generalize this idea later, when we finish our small-signal analysis using superposition.

In many applications, the two inputs of a differential amplifier are driven by signals that are equal in magnitude and out of phase: $v_{i2} = -v_{i1}$. Continuing our analysis of the amplifier, let us now ground input 1 ($v_{i1} = 0$) and assume that there is a signal applied to input 2 equal to and out of phase with the v_{i1} signal we previously assumed. Since the transistors are identical and the circuit is completely symmetrical, the outputs have exactly the same relationships to the inputs as they had before: v_{o2} is out of phase with v_{i2} and v_{o1} is in phase with v_{i2}. These relationships are illustrated in Figure 12–6.

When we compare Figure 12–6 with Figure 12–4, we note that the v_{o1} outputs are identical, as are the v_{o2} outputs. In other words, driving the two inputs with equal but out-of-phase signals reinforces, or duplicates, the signals at the two outputs. By superposition, each output is the sum of the voltages resulting from each input acting alone, so the outputs are exactly twice the level they would be if only one input signal were present. These ideas are summarized in Figure 12–7.

In many applications, the output of a differential amplifier is taken from just one of the transistor collectors, v_{o1}, for example. In this case the input is a difference voltage and the output is a voltage with respect to ground. This use of the amplifier is called *single-ended output* operation and the voltage gain in that mode is

$$A_{v(\text{single-ended output})} = \frac{v_{o1}}{v_{i1} - v_{i2}} \qquad \textbf{(12–2)}$$

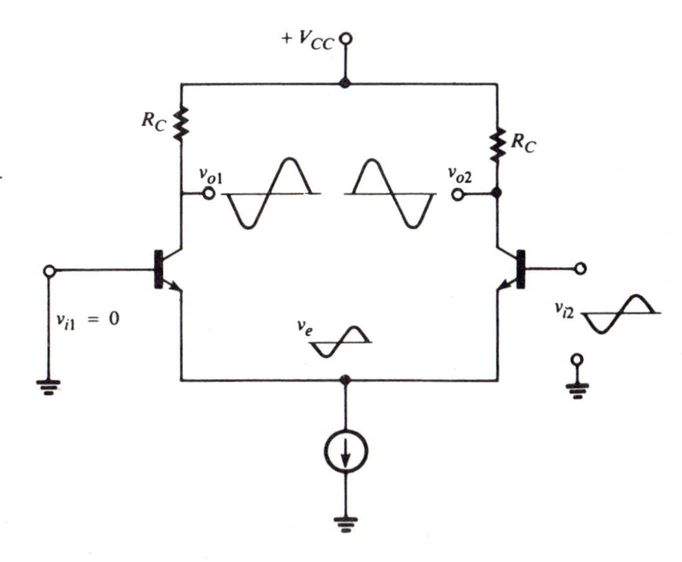

Figure 12–6
The differential amplifier with v_{i1} grounded and a signal input v_{i2}. Compare with Figure 12–4. Note that v_{i2} here is the opposite phase from v_{i1} in Figure 12–4 and that v_{o1} and v_{o2} are the same as in Figure 12–4.

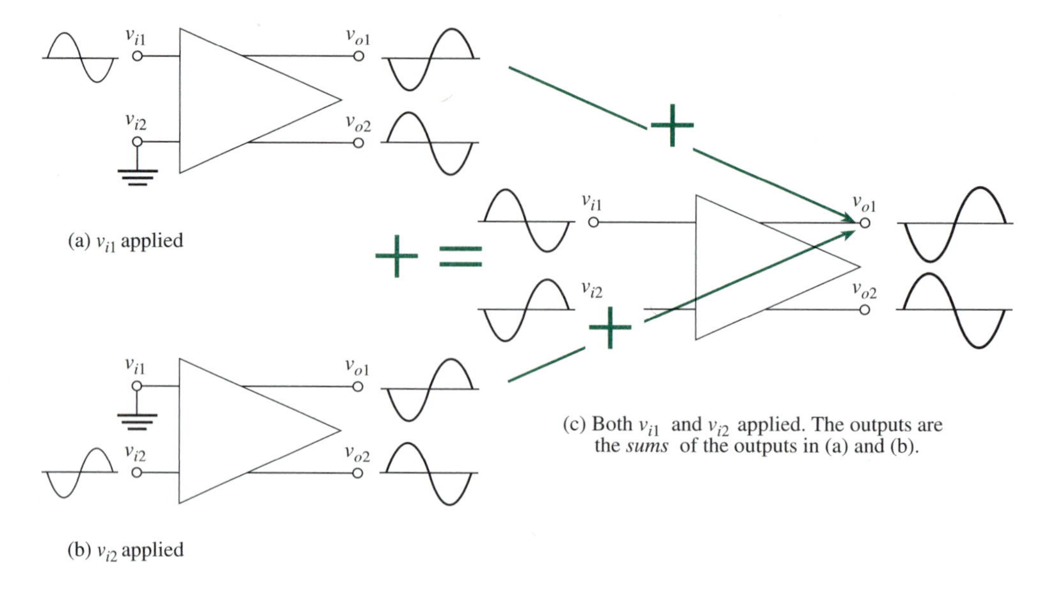

Figure 12–7
By the superposition principle, the output v_{o1} when both inputs are applied is the sum of the v_{o1} outputs due to each signal acting alone. Likewise for v_{o2}.

The next example demonstrates that the single-ended output gain is one-half the difference voltage gain. To distinguish between these terms, we will hereafter refer to $(v_{o1} - v_{o2})/(v_{i1} - v_{i2})$ as the *double-ended voltage gain.*

Example 12–1

The magnitude of the voltage gain (v_c/v_{be}) for each transistor in Figure 12–2 is 100. If v_{i1} and v_{i2} are out-of-phase, 100-mV-pk signals applied simultaneously to the inputs, find

1. the peak values of v_{o1} and v_{o2};
2. the magnitude of the double-ended voltage gain $(v_{o1} - v_{o2})/(v_{i1} - v_{i2})$; and
3. the magnitude of the single-ended output gain $v_{o1}/(v_{i1} - v_{i2})$.

Solution

1. As demonstrated in Figure 12–5, the peak value of each output is 5 V when one input is driven and the other is grounded. Since the outputs are doubled when the inputs are equal and out of phase, each output is 10 V pk.
2. Since $v_{i1} = -v_{i2}$, the input difference voltage is $v_{i1} - v_{i2} = 2v_{i1} = 200$ mV pk. Similarly, $v_{o1} = -v_{o2}$, so the output difference voltage is $v_{o1} - v_{o2} = 2v_{o1} = 20$ V pk. Therefore, the magnitude of $(v_{o1} - v_{o2})/(v_{i1} - v_{i2})$ is (20 V)/(200 mV) = 100.
3. The magnitude of the single-ended output gain is

$$|A_{v(\text{single-ended output})}| = \left|\frac{v_{o1}}{v_{i1} - v_{i2}}\right| = \frac{10\text{ V}}{200\text{ mV}} = 50$$

Since v_{o1} is out of phase with $(v_{i1} - v_{i2})$, the correct specification for the single-ended output gain is -50. If the single-ended output is taken from the *other* side (v_{o2}), which is out of phase with v_{o1}, then the gain $v_{o2}/(v_{i1} - v_{i2})$ is $+50$.

Note once again that the double-ended voltage gain is the same as the voltage gain v_c/v_{be} for each transistor. *Note also that the single-ended output gain is one-half the double-ended gain.* Since the output difference voltage $v_{o1} - v_{o2}$ is out of phase with the input difference voltage $v_{i1} - v_{i2}$, the correct specification for the double-ended voltage gain is -100.

It should now be clear that if the two inputs are driven by equal *in-phase* signals, the output at each collector will be exactly 0, and the output difference voltage will be 0. Of course, in this case, the input difference voltage is also 0. These ideas are illustrated in Figure 12–8.

We can now derive general expressions for the double-ended and single-ended output voltage gains in terms of the circuit parameters. Figure 12–9 shows one side of the differential amplifier with the other side replaced by its emitter resistance, r_e. Recall that this is the resistance in series with the emitter of Q_1 when the input to Q_2 is grounded. We are again assuming that the current source has infinite resistance.

Neglecting the output resistance r_o at the collector of Q_1, we can use the familiar approximation for the voltage gain of the transistor:

$$\frac{v_{o1}}{b_{be1}} \approx \frac{-R_C}{r_e} \tag{12–3}$$

where r_e is the emitter resistance of Q_1. It is clear from Figure 12–9 that the voltage gain v_{o1}/v_{i1} is

$$\frac{v_{o1}}{v_{i1}} \approx \frac{-R_C}{2r_e} \tag{12–4}$$

where the quantity $2r_e$ is in the denominator because we assume that the emitter resistances of Q_1 and Q_2 are equal. Equations 12–3 and 12–4 confirm our previous conclusion that the transistor voltage gain is twice the value of the gain v_{o1}/v_{i1}.

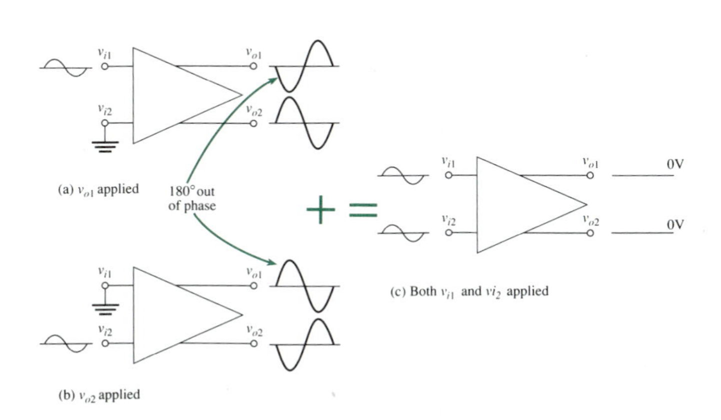

Figure 12–8
The outputs of the differential amplifier are 0 when the two inputs are equal and in phase.

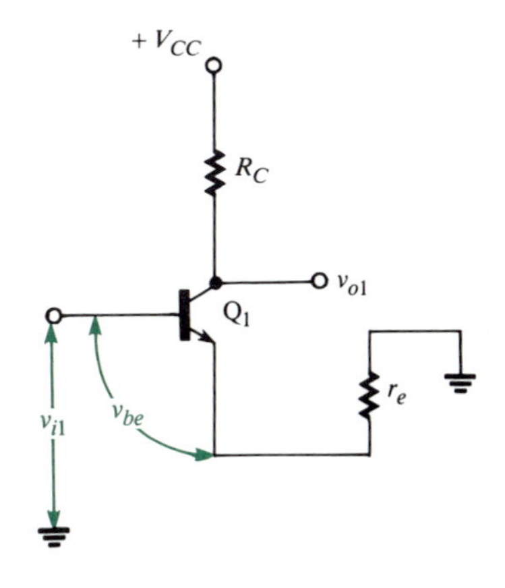

Figure 12–9
When the input to Q_2 is grounded, there is resistance r_e in series with the emitter of Q_1.

Also, we have already shown that the double-ended (difference) voltage gain equals the transistor gain and that the single-ended output gain is one-half that value. Therefore, we conclude that

$$\frac{v_{o1} - v_{o2}}{v_{i1} - v_{i2}} \approx \frac{-R_C}{r_e} \tag{12–5}$$

and

$$\frac{v_{o1}}{v_{i1} - v_{i2}} \approx \frac{-R_C}{2r_e} \tag{12–6}$$

We should note that these gain relations are valid irrespective of the magnitudes and phase relations of the two inputs v_{i1} and v_{i2}. We have considered only the two special cases where v_{i1} and v_{i2} are equal and in phase and where they are equal and out of phase, but equations 12–5 and 12–6 hold under any circumstances. Note also that v_{o1} and v_{o2} will always have the same amplitude and be out of phase with each other. Thus,

$$\frac{v_{o2}}{v_{i1} - v_{i2}} \approx \frac{R_C}{2r_e} \tag{12–7}$$

The small-signal *differential input resistance* is defined to be the input difference voltage divided by the total input current. Imagine a signal source connected *across* the input terminals, so the same current that flows out of the source into one input of the amplifier flows out of the other input and returns to the source. The signal-source voltage, which is the input difference voltage, divided by the signal-source current, is the differential input resistance. Since the total small-signal resistance in the path from one input through both emitters to the other input is $2r_e$, the differential input resistance is

$$r_{id} = 2(\beta + 1)r_e \tag{12–8}$$

Figure 12–10 shows the dc voltages and currents in the ideal differential amplifier. Since the transistors are identical, the source current I divides equally between them, and the emitter current in each is, therefore,

$$I_E = I_{E1} = I_{E2} = I/2 \tag{12–9}$$

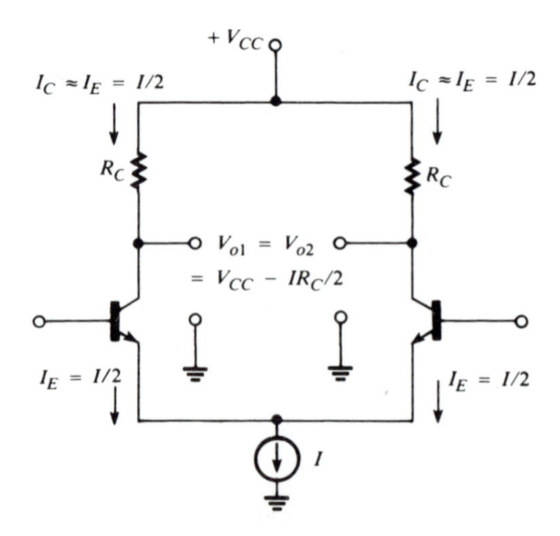

Figure 12–10
DC voltages and currents in an ideal differential amplifier

The dc output voltage at the collector of each transistor is

$$V_{o1} = V_{CC} - I_{C1}R_C$$
$$V_{o2} = V_{CC} - I_{C2}R_C$$

Since $I_C \approx I_E = I/2$ in each transistor, we have

$$V_{o1} = V_{o2} \approx V_{CC} - (I/2)R_C \tag{12–10}$$

To determine the ac emitter resistance of each transistor, we can use equation 12–9 and the familiar approximation $r_e \approx 0.026/I_E$ to obtain

$$r_{e1} = r_{e2} = r_e \approx \frac{0.026}{I_E} = \frac{0.026}{I/2} \tag{12–11}$$

Example 12–2

For the ideal differential amplifier shown in Figure 12–11, find

1. the dc output voltages V_{o1} and V_{o2};
2. the single-ended output gain $v_{o1}/(v_{i1} - v_{i2})$; and
3. the double-ended gain $(v_{o1} - v_{o2})/(v_{i1} - v_{i2})$.

Solution

1. The emitter current in each transistor is $I_E = I/2 = (2 \text{ mA})/2 = 1 \text{ mA} \approx I_C$. Therefore, $V_{o1} = V_{o2} = V_{CC} - I_C R_C = 15 - (1 \text{ mA})(6 \text{ k}\Omega) = 9 \text{ V}$.
2. The emitter resistance of each transistor is

$$r_e \approx \frac{0.026}{I_E} = \frac{0.026}{1 \text{ mA}} = 26\ \Omega$$

Therefore, from equation 12–6,

$$\frac{v_{o1}}{v_{i1} - v_{i2}} \approx \frac{-R_C}{2r_e} = \frac{-6 \text{ k}\Omega}{52\ \Omega} = -115.4$$

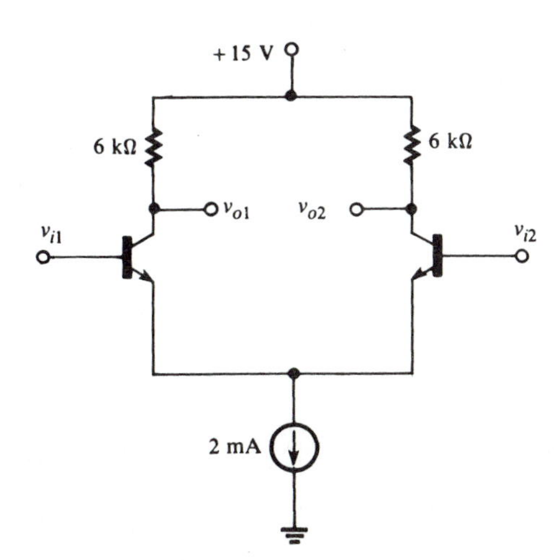

Figure 12–11
(Example 12–2)

3. From equation 12–5,

$$\frac{v_{o1} - v_{o2}}{v_{i1} - v_{i2}} = \frac{-R_C}{r_e} = \frac{-6\text{ k}\Omega}{26\ \Omega} = -230.8$$

The FET Differential Amplifier

Many differential amplifiers are constructed using field-effect transistors because of the large impedance they present to input signals. This property is exceptionally important in many applications, including operational amplifiers, instrument amplifiers, and charge amplifiers. A large voltage gain is also important in these applications, and although the FET does not produce much gain, an FET differential amplifier is often the first stage in a multistage amplifier whose overall gain is large. Because FETs are easily fabricated in integrated-circuit form, FET differential amplifiers are commonly found in linear integrated circuits.

Figure 12–12 shows a JFET differential amplifier, and it can be seen that it is basically the same configuration as its BJT counterpart. The two JFETs operate as common-source amplifiers with their source terminals joined. A constant-current source provides bias current.

The derivations of the gain equations for the JFET amplifier are completely parallel to those for the BJT version. A source-to-ground voltage is developed at the common source connection by source-follower action. With one input grounded, the output resistance and load resistance of the source follower are both equal to $1/g_m$ (assuming matched devices), so the source-follower gain is 0.5. Therefore, as shown in Figure 12–13, one-half the input voltage is developed across $1/g_m$, and the current is

$$i_d = \frac{v_i/2}{1/g_m} = \frac{v_i g_m}{2}$$

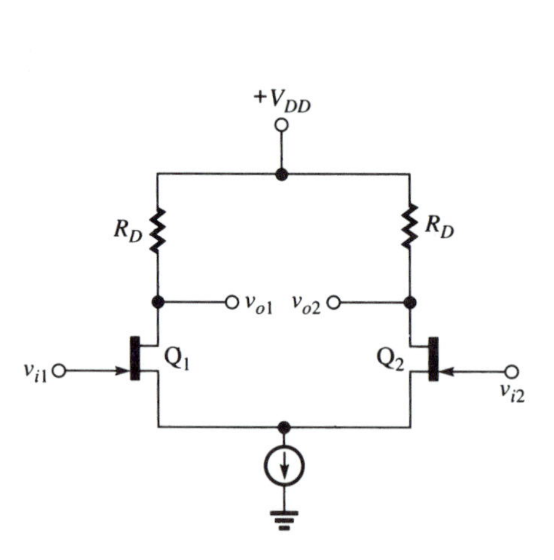

Figure 12–12
The JFET differential amplifier

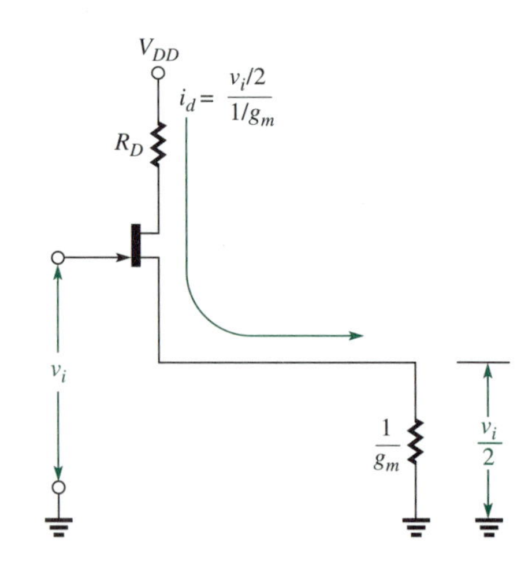

Figure 12–13
Source-follower action with gain 0.5 results in one-half the input voltage developed across ($1/g_m$) ohms

The output voltage is then

$$v_o = -i_d R_D = \frac{-v_i g_m R_D}{2} \qquad \textbf{(12–12)}$$

from which we find the voltage gain

$$\frac{v_o}{v_i} = \frac{-g_m R_D}{2} \qquad \textbf{(12–13)}$$

Applying the superposition principle in the same way we did for the BJT version, we readily find that

$$\frac{v_{o1}}{v_{i1} - v_{i2}} = \frac{-g_m R_D}{2} \qquad \textbf{(12–14)}$$

and

$$\frac{v_{o1} - v_{o2}}{v_{i1} - v_{i2}} = -g_m R_D \qquad \textbf{(12–15)}$$

Like the gain equations for the BJT differential amplifier, equations 12–14 and 12–15 show that the double-ended (difference) voltage gain is the same as the gain of one transistor, and the single-ended output gain is one-half that value.

Example 12–3

The matched transistors in Figure 12–14 have I_{DSS} = 12 mA and V_p = −2.5 V. Find

1. the dc output voltages v_{o1} and v_{o2};
2. the single-ended output gain $v_{o1}/(v_{i1} - v_{i2})$; and
3. the double-ended gain $(v_{o1} - v_{o2})/(v_{i1} - v_{i2})$.

Solution

1. The dc current in each JFET is $I_D = (1/2)(6\text{ mA}) = 3\text{ mA}$. Therefore, $V_{o1} = V_{o2} = V_{DD} - I_D R_D = 15 - (3\text{ mA})(3\text{ k}\Omega) = 6\text{ V}$.
2.

$$g_m = \frac{2I_{DSS}}{|V_p|}\sqrt{\frac{I_D}{I_{DSS}}} = \frac{2(12\text{ mA})}{2.5\text{ V}}\sqrt{\frac{3}{12}} = 4.8\text{ mS}$$

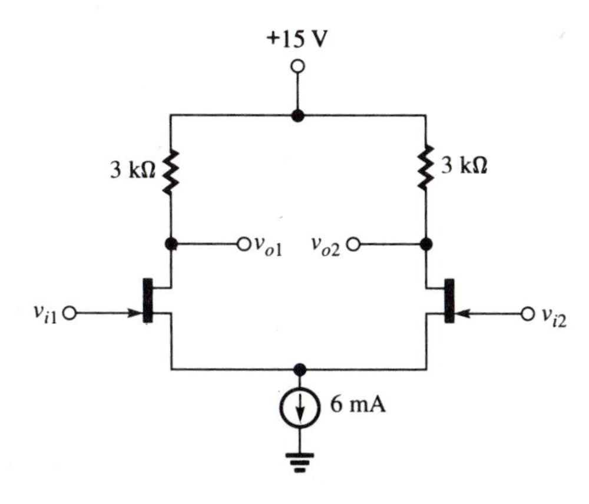

Figure 12–14
(Example 12–3)

From equation 12–14,

$$\frac{v_{o1}}{v_{i1} - v_{i2}} = \frac{-g_m R_D}{2} = \frac{-(4.8\text{ mS})(3\text{ k}\Omega)}{2} = -7.2$$

3. From equation 12–15,

$$\frac{v_{o1} - v_{o2}}{v_{i1} - v_{i2}} = -g_m R_D = -(4.8\text{ mS})(3\text{ k}\Omega) = -14.4$$

12–3 COMMON-MODE PARAMETERS

One attractive feature of a differential amplifier is its ability to reject signals that are *common* to both inputs. Since the outputs are amplified versions of the difference between the inputs, any voltage component that appears identically in both signal inputs will be "differenced out," that is, will have zero level in the outputs. (We have already seen that the outputs are exactly 0 when both inputs are identical, in-phase signals.) Any dc or ac voltage that appears simultaneously in both signal inputs is called a *common-mode* signal. The ability of an amplifier to suppress, or zero-out, common-mode signals is called *common-mode rejection.* An example of a common-mode signal whose rejection is desirable is electrical *noise* induced in both signal lines, a frequent occurrence when the lines are routed together over long paths. Another example is a dc level common to both inputs, or common dc fluctuations caused by power-supply variations.

In the ideal differential amplifier, any common-mode signal will be completely cancelled out and therefore have no effect on the output signals. In practical amplifiers, which we will discuss in the next section, mismatched components and certain other nonideal conditions result in imperfect cancellation of common-mode signals. Figure 12–15 shows a differential amplifier in which a common-mode signal v_{cm} is applied to both inputs. Ideally, the output voltages should be 0, but in fact some small component of v_{cm} may appear. The differential *common-mode gain, A_{cm}*, is defined to be the ratio of the output difference voltage caused by the common-mode signal to the common-mode signal itself:

$$A_{cm} = \frac{(v_{o1} - v_{o2})_{cm}}{v_{cm}} \tag{12–16}$$

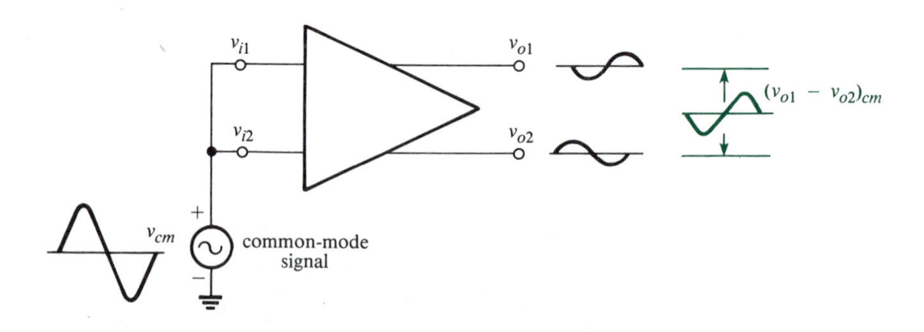

Figure 12–15
If the differential amplifier were ideal, both outputs would be 0 when the inputs have the same (common-mode) signal. In reality, there is a small common-mode output, as shown.

We can also define a single-ended common-mode gain as the ratio of $(v_{o1})_{cm}$ or $(v_{o2})_{cm}$ to v_{cm}. Obviously the ideal amplifier has common-mode gain equal to 0.

A widely used specification and figure of merit for a differential amplifier is its *common-mode rejection ratio* (CMRR), defined to be the ratio of the magnitude of its differential (difference-mode) gain A_d to the magnitude of its common-mode gain:

$$\text{CMRR} = \frac{|A_d|}{|A_{cm}|} \qquad \textbf{(12–17)}$$

The value of the CMRR is often given in decibels:

$$\text{CMRR} = 20 \log_{10}\left|\frac{A_d}{A_{cm}}\right| \qquad \textbf{(12–18)}$$

Example 12–4

When the inputs to a certain differential amplifier are $v_{i1} = 0.1 \sin \omega t$ and $v_{i2} = -0.1 \sin \omega t$, it is found that the outputs are $v_{o1} = -5 \sin \omega t$ and $v_{o2} = 5 \sin \omega t$. When both inputs are $2 \sin \omega t$, the outputs are $v_{o1} = -0.05 \sin \omega t$ and $v_{o2} = 0.05 \sin \omega t$. Find the CMRR in dB.

Solution. We will use the peak values of the various signals for our gain computations, but note carefully how the minus signs are used to preserve phase relations. The difference-mode gain is

$$A_d = \frac{v_{o1} - v_{o2}}{v_{i1} - v_{i2}} = \frac{-5 - 5}{0.1 - (-0.1)} = \frac{-10}{0.2} = -50$$

The common-mode gain is

$$A_{cm} = \frac{(v_{o1} - v_{o2})_{cm}}{v_{cm}} = \frac{-0.05 - 0.05}{2} = \frac{-0.1}{2} = -0.05$$

The common-mode rejection ratio is

$$\text{CMRR} = \frac{|A_d|}{|A_{cm}|} = \frac{50}{0.05} = 1000$$

Expressing this result in dB, we have CMRR = $20 \log_{10}(1000) = 60$ dB.

12–4 PRACTICAL DIFFERENTIAL AMPLIFIERS

Although it is possible to obtain closely matched transistors for differential amplifiers, particularly in integrated circuits, it is not reasonable to expect that the parameters of both devices will be exactly the same, or that they will track (change in the same way) under variations in temperature or bias level. Differences in parameter values cause the amplifier to be *unbalanced,* in the sense that the gain of one side is different from that of the other. For example, our gain derivations for the ideal differential amplifier were based on the assumption that both transistors had identical values of r_e. Clearly, the voltage gains of both sides will not be identical if the values of r_e are not, in which case the outputs will not truly represent amplified

Figure 12–16
Inserting a resistance R_E in series with each emitter reduces the amplifier's dependence on matched r_e values.

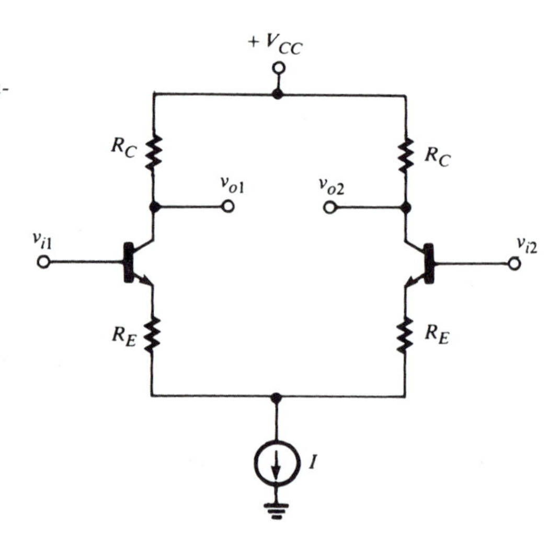

versions of the input difference voltage, and the CMRR will suffer. In discrete circuits, an adjustable resistor can be inserted between the emitter terminals to compensate for such variations, but this remedy is not practical in integrated circuits. To reduce the effect of variations in r_e, equal-valued resistors R_E can be inserted in series with the emitters, as shown in Figure 12–16. If the value of R_E is substantially larger than r_e, then variations or differences in r_e will have minimal consequence. However, the presence of additional emitter resistance decreases the voltage gain of the amplifier. Equations 12–5 and 12–6, modified for the inclusion of R_E, become

$$\frac{v_{o1} - v_{o2}}{v_{i1} - v_{i2}} \approx \frac{-R_C}{r_e + R_E} \approx \frac{-R_C}{R_E} \quad \text{if } R_E \gg r_e \tag{12–19}$$

$$\frac{v_{o1}}{v_{i1} - v_{i2}} \approx \frac{-R_C}{2(r_e + R_E)} \approx \frac{-R_C}{2R_E} \quad \text{if } R_E \gg r_e \tag{12–20}$$

A beneficial consequence of increased emitter resistance is the accompanying increase in differential input resistance. Equation 12–8 becomes

$$r_{id} = 2(\beta + 1)(r_e + R_E) \tag{12–21}$$

Another reality in practical differential amplifiers is that the current source biasing the amplifier does not have infinite resistance. Figure 12–17(a) shows the BJT amplifier biased with a current source having resistance R. We wish to investigate the effect of R on the common-mode rejection ratio, so both inputs are shown connected to the common-mode voltage v_{cm}. Since the circuit is symmetrical, we can determine v_{o1} and v_{o2} by analyzing the two equivalent half-circuits shown in Figure 12–17(b). Notice that the current-source resistance in each half-circuit must be $2R$ and the value of the current must be $I/2$ to maintain equivalence.

The voltage gain of Q_1 is

$$\frac{v_{o1}}{v_{cm}} \approx \frac{-R_C}{r_e + 2R} \tag{12–22}$$

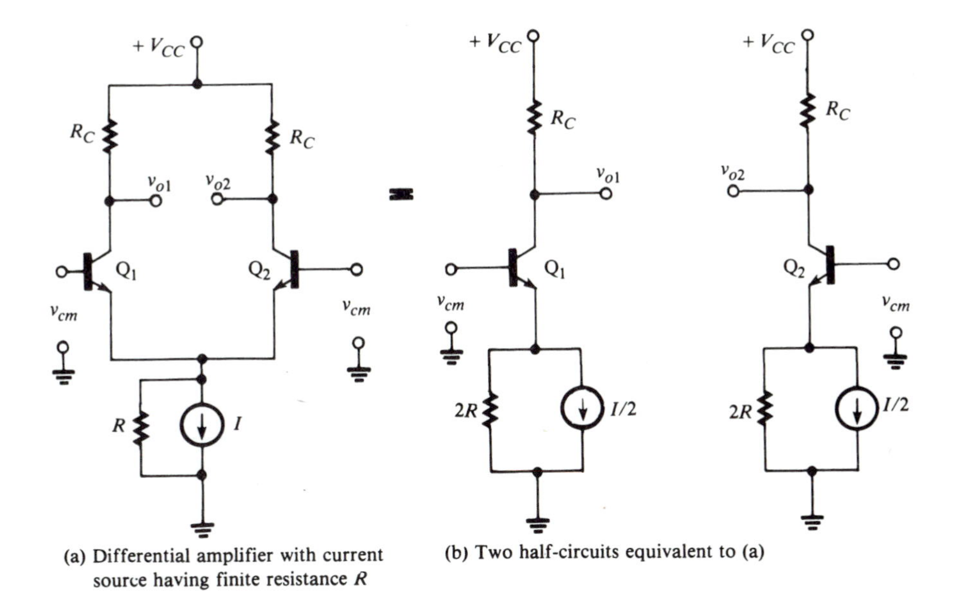

(a) Differential amplifier with current source having finite resistance R

(b) Two half-circuits equivalent to (a)

Figure 12–17
Analyzing the effect of source resistance R on common-mode behavior

and that of Q_2 is

$$\frac{v_{o2}}{v_{cm}} \approx \frac{-R_C}{r_e + 2R} \tag{12–23}$$

Therefore,

$$A_{cm} = \frac{(v_{o1} - v_{o2})_{cm}}{v_{cm}} = \frac{-R_C}{r_e + 2R} - \frac{-R_C}{r_e + 2R} = 0$$

This result shows that the *differential* common-mode gain is unaffected by source resistance R (assuming that both r_e and R_C are perfectly matched). However, the single-ended common-mode gain is given by equations 12–22 and 12–23, and from these it is clear that a large value of R is desirable.

Bias Methods in Integrated Circuits

The current source in some older designs and in some discrete amplifiers is simply a voltage source connected in series with a large resistance. The Norton equivalent circuit of this arrangement shows that the constant current is E/R and that the source resistance is R, where E is the voltage source and R is the resistance in series with it. The difficulty with this scheme is that the value of the constant current is impractically small if the resistance is made desirably large. Modern designs and integrated-circuit amplifiers use transistor constant-current sources, an example of which is illustrated in Figure 12–18. Transistor Q_3 has a large output resistance at its collector and is therefore a superior arrangement.

Assuming that $\beta_3 R_{E3} \gg R_1 \parallel R_2$, the base voltage of Q_3 is

$$V_{B3} = -V_{EE}\left(\frac{R_2}{R_1 + R_2}\right) \tag{12–24}$$

Figure 12–18
A transistor current source used to bias a differential amplifier

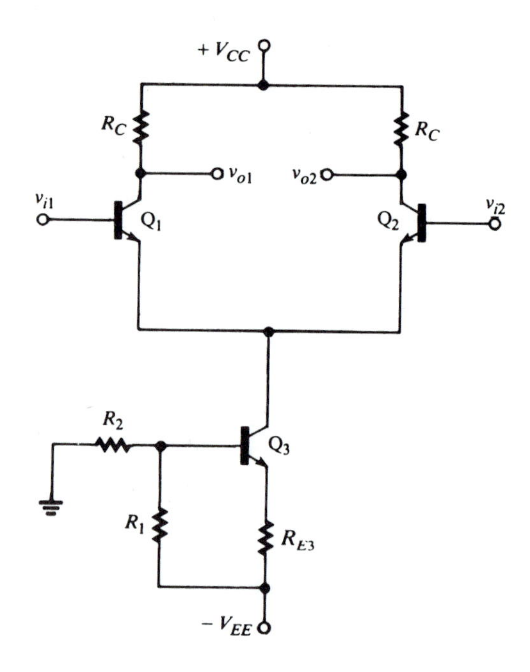

Assuming a silicon transistor, the emitter voltage of Q_3 is

$$V_{E3} = V_{B3} - 0.7 \tag{12–25}$$

The emitter current in Q_3 is then

$$I_{E3} = \frac{|V_{EE}| - |V_{E3}|}{R_{E3}} \tag{12–26}$$

Example 12–5

Transistor Q_3 in Figure 12–19 has $\beta_3 = 100$. Assuming that Q_1 and Q_2 are matched, find approximate values for

1. the emitter currents in Q_1 and Q_2; and
2. the dc output voltages V_{o1} and V_{o2}.

Solution

1. $V_{B3} = \left[\dfrac{10\text{ k}\Omega}{(10\text{ k}\Omega) + (4.7\text{ k}\Omega)}\right](-15\text{ V}) = -10.2\text{ V}$

$V_{E3} = -\ 10.2 - 0.7 = -10.9\text{ V}$

$I_{E3} = \dfrac{(15 - 10.9)\text{ V}}{4\text{ k}\Omega} = 1\text{ mA} \approx I_{C3}$

$I_{E1} = I_{E2} = I_{C3}/2 = 0.5\text{ mA}$

2. $I_{C1} = I_{C2} \approx I_{E1} = I_{E2} = 0.5\text{ mA}$

$V_{o1} = V_{o2} = 15 - (0.5\text{ mA})(10\text{ k}\Omega) = 10\text{ V}$

Figure 12–19
(Example 12–5)

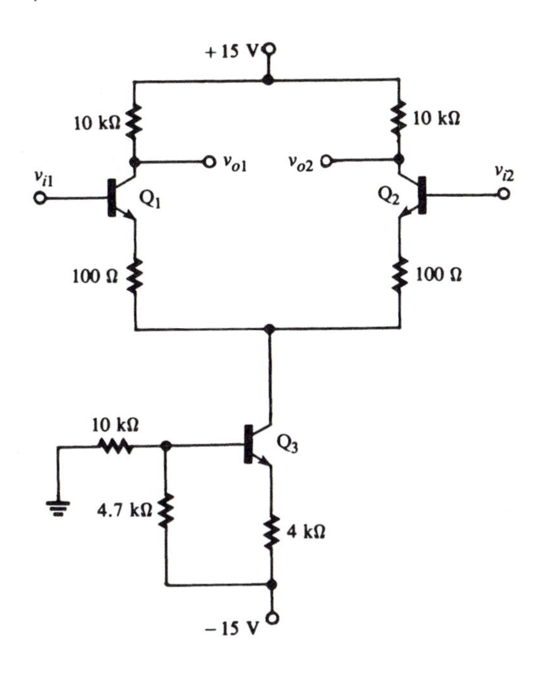

Example 12–6

SPICE

To investigate the effects of mismatched transistors and parameter variability on the balance of the differential amplifier in Figure 12–19, use SPICE to find the dc output voltages at the collectors, V_{o1} and V_{o2}, and the emitter currents, I_{E1} and I_{E2}, on each side when

1. the transistors are perfectly matched;
2. $\beta_1 = 80$ and $\beta_2 = 120$; and
3. $\beta_1 = \beta_2 = 100$; the saturation currents are $I_{s1} = 1 \times 10^{-16}$ A, $I_{s2} = 1.2 \times 10^{-16}$ A, and the temperatures are 0°C, 27°C, 50°C, and 100°C.

Assume β_3 is fixed at 250. The bases of Q_1 and Q_2 are grounded. (Thus, in a perfectly balanced amplifier, the *differential* output voltage should be 0 V.)

Solution

1. Figure 12–20(a) shows the SPICE circuit and input data file for the case where transistors Q_1 and Q_2 have identical (default) parameters. VE1 and VE2 are dummy voltage sources used to measure emitter currents in each side of the amplifier. As can be seen in Figure 12–20(b), the results of the .DC analysis show that the output voltages and emitter currents on each side are identical. Their values are very close to those calculated in Example 12–5.
2. When the .MODEL statement for Q_1 is changed so that BF = 80 and that for Q_2 is changed so that BF = 120, the results of the .DC analysis (Figure 12–20(c)) show that the amplifier is unbalanced. The difference in output voltages is seen to be 10.07 V − 10.05 V = 20 mV, and the difference in emitter currents is 0.4995 mA − 0.4988 mA = 0.7 μA.

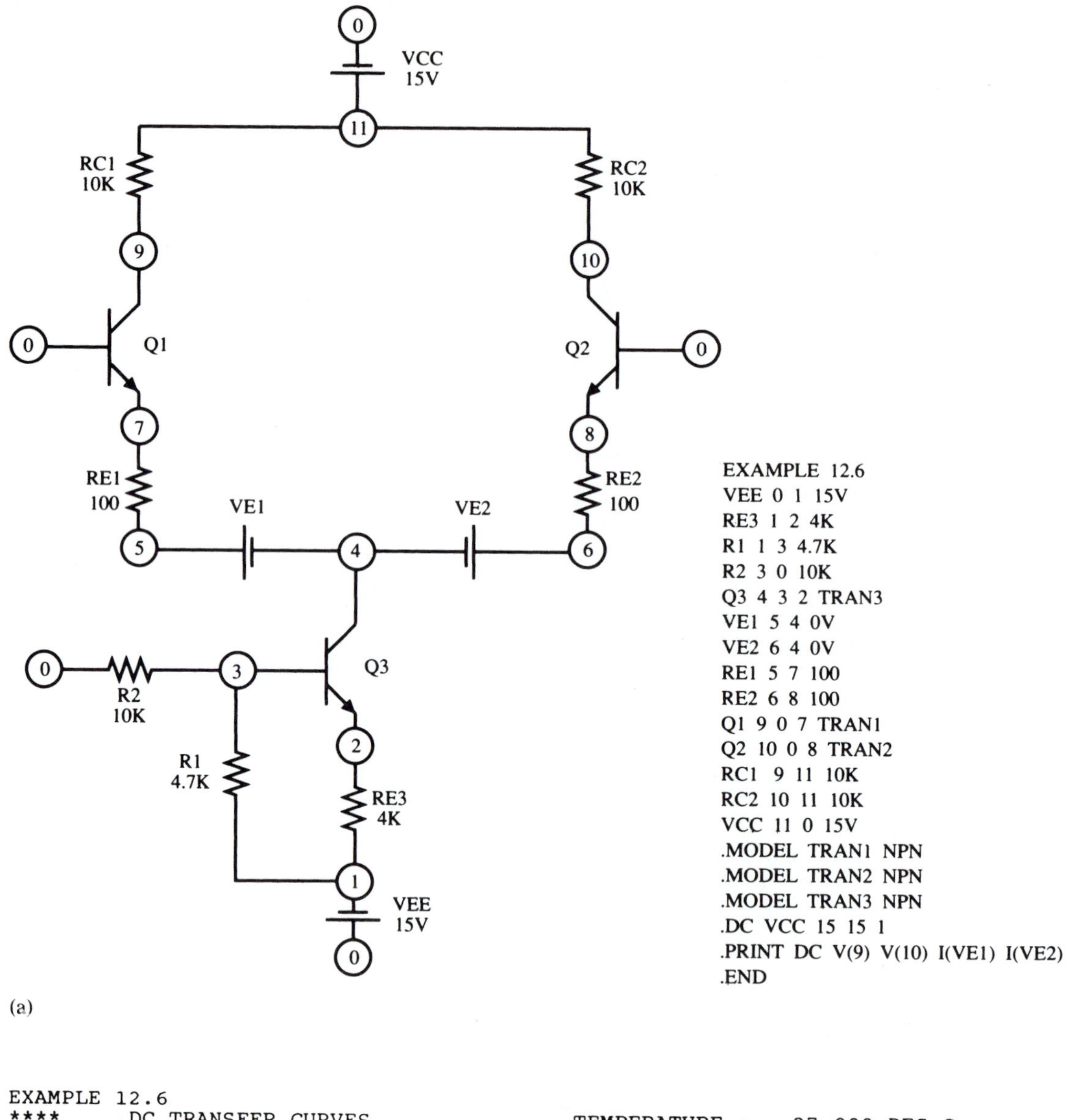
0
VCC
15V
11
RC1
10K
RC2
10K
9
10
0
Q1
Q2
0
7
8
RE1
100
RE2
100
VE1
VE2
5
4
6
0
R2
10K
3
Q3
2
R1
4.7K
RE3
4K
1
VEE
15V
0
EXAMPLE 12.6
VEE 0 1 15V
RE3 1 2 4K
R1 1 3 4.7K
R2 3 0 10K
Q3 4 3 2 TRAN3
VE1 5 4 0V
VE2 6 4 0V
RE1 5 7 100
RE2 6 8 100
Q1 9 0 7 TRAN1
Q2 10 0 8 TRAN2
RC1 9 11 10K
RC2 10 11 10K
VCC 11 0 15V
.MODEL TRAN1 NPN
.MODEL TRAN2 NPN
.MODEL TRAN3 NPN
.DC VCC 15 15 1
.PRINT DC V(9) V(10) I(VE1) I(VE2)
.END
(a)

```
EXAMPLE 12.6
****      DC TRANSFER CURVES                    TEMPERATURE =   27.000 DEG C
******************************************************************************
  VCC           V(9)         V(10)        I(VE1)       I(VE2)
 1.500E+01      1.006E+01    1.006E+01    4.991E-04    4.991E-04
```

(b)

```
****      DC TRANSFER CURVES                    TEMPERATURE =   27.000 DEG C
******************************************************************************
  VCC           V(9)         V(10)        I(VE1)       I(VE2)
 1.500E+01      1.007E+01    1.005E+01    4.995E-04    4.988E-04
```

(c)

Figure 12–20
(Example 12–6)

Temp. (°C)	V_{o1} (V)	V_{o2} (V)	$\Delta V = V_{o1} - V_{o2}$ (V)	I_{E1} (mA)	I_{E2} (mA)	$\Delta I_E = I_{E2} - I_{E1}$ (μA)
0	10.25	9.959	0.291	0.480	0.5091	29.1
27	10.21	9.904	0.306	0.4836	0.5147	31.1
50	10.18	9.857	0.323	0.4867	0.5194	32.7
100	10.11	9.755	0.355	0.4938	0.5298	36.0

(d)

Figure 12–20
(Continued)

3. To investigate the effects of temperature change on the amplifier balance, the statement .TEMP 0 27 50 100 is added to the input data file. The .MODEL statements are modified so that Q_1 and Q_2 once again have the same β, but we now specify IS = 1E-16 for Q_1 and IS = 1.2E-16 for Q_2. The table in Figure 12–20(d) summarizes the results of the SPICE analysis at each temperature. We see that both output voltages decrease and both emitter currents increase with increasing temperature. However, the most significant result from the standpoint of a differential amplifier is that the output *difference* voltage and the difference in the emitter currents both increase with temperature. Thus, the degree of unbalance increases with increasing temperature.

Example 12–7

Assuming that each of Q_1 and Q_2 in Figure 12–19 has $\beta = 100$ and that the output resistance at the collector of Q_3 is 500 kΩ, find

1. the differential input resistance;
2. the single-ended common-mode gain; and
3. the single-ended common-mode rejection ratio.

Solution

1. The small-signal emitter resistance of Q_1 and Q_2 is

$$r_e \approx \frac{0.026}{I_E} = \frac{0.026}{0.5\text{ mA}} = 52\ \Omega$$

From equation 12–21, $r_{id} = 2(\beta + 1)(r_e + R_E) = 2(101)(52 + 100) = 30.7\text{ k}\Omega$. (The resistance of the transistor current source is large enough to be neglected in this calculation.)

2. Equations 12–22 and 12–23 are for the case $R_E = 0$. When resistance R_E is included in each emitter circuit, the single-ended common-mode gain is

$$\frac{-R_C}{r_e + R_E + 2R} = \frac{-10^4}{52 + 100 + (2)(0.5 \times 10^6)} \approx -10^{-2}$$

3. From equation 12–20, the single-ended output gain is

$$\frac{-R_C}{2(r_e + R_E)} = \frac{-10^4}{2(52 + 100)} = -32.9$$

Therefore, the single-ended CMRR is

$$\frac{32.9}{10^{-2}} = 3290$$

or 70.3 dB.

Figure 12–21 shows a popular bias technique used in integrated circuits. Recall from Chapter 6 that Q_3 and Q_4 form a *current mirror,* which has the advantage that current I_X can be reflected through other transistors like Q_4 to supply bias current to other stages. Q_3 is a diode-connected transistor, and

$$I_X = \frac{V_{CC} + |V_{EE}| - 0.7}{R_B} \qquad \textbf{(12–27)}$$

Provided the β's of Q_3 and Q_4 are reasonably large, the constant current $I \approx I_X$.

Another popular means for providing bias in integrated circuits is through use of the *Widlar* current source, shown in Figure 12–22. In this circuit, the constant current I is not equal to the reference current I_X, but is instead a small fraction of I_X, as determined by resistor R. The advantage of the circuit is that small constant currents can be generated without using large resistance values. Since Q_1 is a diode-connected transistor, its base-to-emitter voltage is about 0.7 V. As can be seen in Figure 12–22, only part of this voltage is across R, so the base-to-emitter voltage of Q_2 is less than 0.7 V. Therefore, I_B is a small value near or below the knee of the base–emitter characteristic of Q_2, and I is correspondingly small. Assuming that the transistors are matched, it can be shown that I_X and I are related by the equation

$$I = I_X e^{-IR/V_T} \qquad \textbf{(12–28)}$$

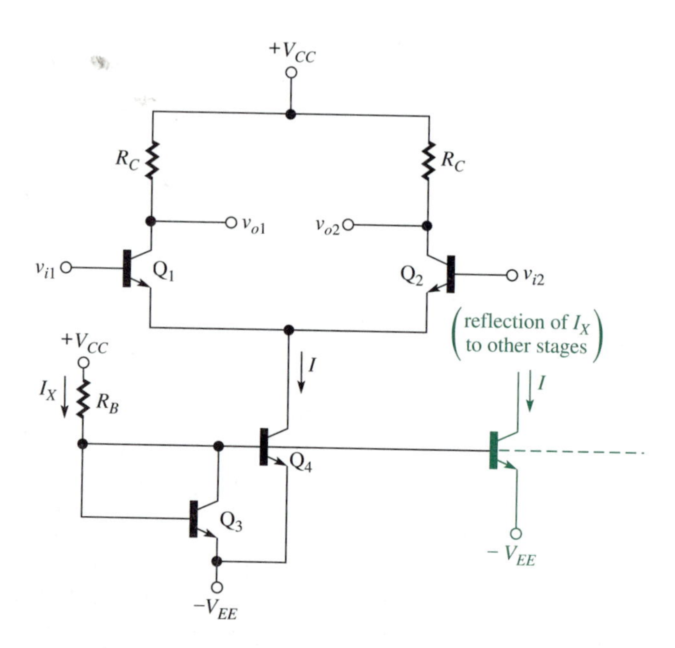

Figure 12–21
The use of a current mirror to bias a differential amplifier. $I \approx I_X$

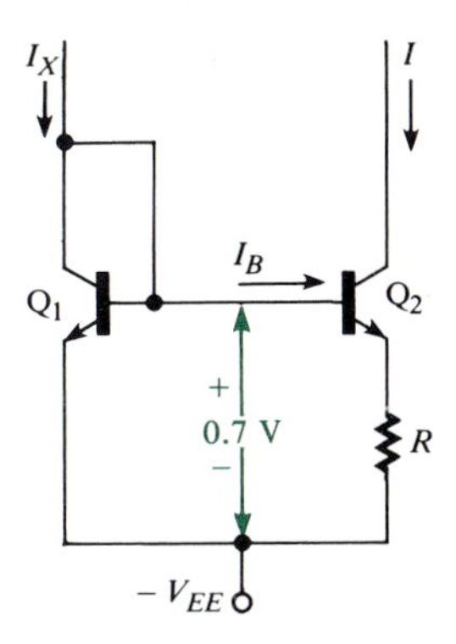

Figure 12–22
The Widlar current source, used to supply small values of bias current I. I_X is a reference current

where $V_T \approx 0.026$ at room temperature. Note that equation 12–28 cannot be solved for I using direct algebraic methods. To find the solution, a trial-and-error, or *iterative,* procedure is necessary, whereby values of I are tried until one that satisfies (12–28) is found. This kind of computational procedure is greatly facilitated by use of a computer.

Example 12–8

Resistor R in Figure 12–22 is 4.3 kΩ and the voltage drop across it is 0.18 V. If $I_B = 1.2\ \mu\text{A}$ at room temperature, what is the value of the reference current I_X?

Solution. The emitter current in Q_2 is the current through resistor R:

$$I_E = \frac{0.18\text{ V}}{4.3\text{ k}\Omega} = 41.86\ \mu\text{A}$$

The bias current I is the collector current in Q_2:

$$I = I_C = I_E - I_B = 41.86\ \mu\text{A} - 1.2\ \mu\text{A} = 40.67\ \mu\text{A}$$

From equation 12–28,

$$I = 40.67\ \mu\text{A} = I_X e^{-0.18/0.026}$$

$$I_X = \frac{40.67\ \mu\text{A}}{9.85 \times 10^{-4}} = 41.3\text{ mA}$$

Active Loads in Integrated Circuits

Instead of using resistors in series with the collectors of Q_1 and Q_2 in a differential amplifier, many integrated-circuit designs employ transistors to form *active loads* for Q_1 and Q_2. Figure 12–23 shows a widely used configuration. Here, PNP transistors Q_3 and Q_4 form a kind of current mirror, but the current in diode-connected Q_3 is determined by the input signal to Q_1. In effect, the current flowing in one side of the differential amplifier is reflected to the other. Since Q_3 acts as a forward-biased diode, the resistance in the collector circuit of Q_1 is quite small, and Q_1 has a small

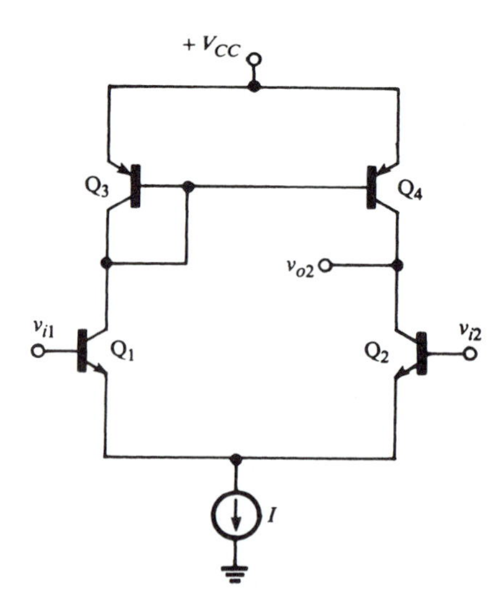

Figure 12–23
The use of PNP transistors to form active loads for Q_1 and Q_2 in a differential amplifier. The amplifier is operated with a single-ended output, v_{o2}

voltage gain. On the other hand, the resistance in the collector circuit of Q_2 is the large collector output resistance of Q_4, so the gain of Q_2 is quite large. For that reason, the amplifier is normally operated single-ended, with output v_{o2}. Note that there is no phase inversion in the single-ended gain $v_{o2}/(v_{i1} - v_{i2})$.

The common-mode rejection ratio of the amplifier in Figure 12–23 has been shown to be quite large. Assuming that the transistors are matched, the bias current I divides equally between both sides, and all devices have the same small-signal emitter resistance: $r_e \approx 0.026/(I/2)$. Since the amplifier is not balanced, the analysis to determine its gain is somewhat complex and we will not reproduce it here. It can be shown, however, that the single-ended gain is given by

$$\frac{v_{o2}}{v_{i1} - v_{i2}} = \frac{r_{o2} \parallel r_{o4}}{r_e} \qquad \textbf{(12–29)}$$

where r_{o2} and r_{o4} are the small-signal, collector output resistance of Q_2 and Q_4.

Example 12–9

Find the single-ended voltage gain $v_{o2}/(v_{i1} - v_{i2})$ of the differential amplifier shown in Figure 12–24. Assume that $r_{o2} = r_{o4} = 100\ \text{k}\Omega$ and that all transistors are matched.

Solution. From equation 12–27, the current mirror formed by Q_5 and Q_6 produces a bias current of

$$I = \frac{V_{CC} + |V_{EE}| - 0.7}{R_B} = \frac{15 + 15 - 0.7}{12 \times 10^3} = 2.44\ \text{mA}$$

Therefore, the emitter currents in Q_1 and Q_2 are

$$I_E = I/2 = (2.44\ \text{mA})/2 = 1.22\ \text{mA}$$
$$r_e \approx 0.026/(1.22\ \text{mA}) = 21.3\ \Omega$$

Figure 12–24
(Example 12–9)

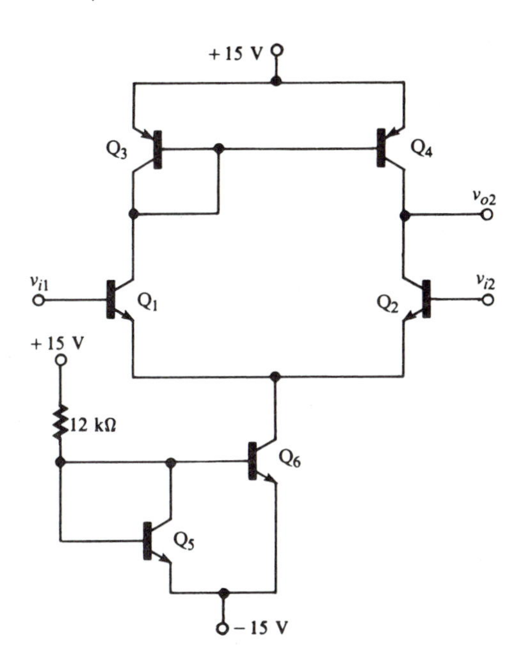

From equation 12–29,

$$\frac{v_{o2}}{v_{i1} - v_{i2}} = \frac{r_{o2} \| r_{o4}}{21.3} = \frac{(100 \times 10^3) \| (100 \times 10^3)}{21.3} = 2347$$

12–5 INTRODUCTION TO OPERATIONAL AMPLIFIERS

An operational amplifier is basically a differential amplifier modified by the addition of circuitry that improves its performance and gives it certain special features. The most important characteristics of an operational amplifier are as follows:

1. It is a dc (direct-coupled, direct-current) amplifier.
2. It should have a very large voltage gain—ideally, infinite.
3. It should have a very large input impedance—ideally, infinite.
4. It should have a very small output impedance—ideally zero.
5. The output should be exactly zero V when the inputs are zero V.
6. The output must be capable of both positive and negative voltage swings.
7. It should have a very large CMRR.
8. It is operated with a single-ended output and differential input (although one input is often grounded, as we shall presently see).
9. It should meet whatever special requirements are demanded by a particular application; these include parameters such as noise level, frequency response, and slew rate, which we will discuss in Chapter 13.

The name *operational* amplifier is derived from amplifier applications that the preceding characteristics make possible: the performance of precise mathematical operations on input signals, including voltage summation, subtraction, and integration. Characteristics 2 and 3 are particularly important for those kinds of operations, and in Chapter 13 we shall explore how these and other features contribute to many other useful applications of operational amplifiers. Our present interest is in the circuit methods used to expand a differential amplifier into an operational amplifier.

The input stage of every operational amplifier is a differential amplifier. To achieve a large input impedance, the differential stage may be constructed with field-effect transistors, or it may employ certain additional circuitry, such as emitter followers, to increase the impedance seen at each input. Components in the input stage should be very closely matched to achieve the best possible balance in the differential operation. This is important to ensure that the output of the operational amplifier is a precise representation of the input difference voltage, that the output is exactly zero when the inputs are zero, and that the CMRR is large. Ideally, component characteristics should be independent of temperature. Any changes that do occur should track one another; that is, device parameters should change in the same way and at the same rate under temperature variations. Matching and tracking of characteristics is, of course, best accomplished in integrated-circuit construction, and virtually all modern operational amplifiers are integrated circuits. Many designs include temperature-compensation circuitry to minimize the effects of temperature.

Voltage gain is achieved through the use of multistage amplifiers, at least one of which is usually another differential stage. At some point in the multistage amplification, the output becomes single-ended. We have seen that single-ended gain is one-half the double-ended value, so this conversion results in an undesirable loss of voltage gain. Some designs incorporate circuitry that eliminates this loss in a clever fashion, but we will not take time to detail the somewhat complex theory involved.

To permit the output voltage to swing through both positive and negative values, both a positive and a negative supply voltage are required. These are usually equal-valued, opposite-polarity supplies, a typical example being ± 15 V. To obtain zero output voltage when the inputs are zero, the amplifier must incorporate *level-shifting circuitry* that eliminates any nonzero bias voltage that would otherwise appear in the output. Of course, this cannot be accomplished with a coupling capacitor, because we require dc response. Level shifting is usually accomplished near or at the output stage.

12–6 CIRCUIT ANALYSIS OF AN OPERATIONAL AMPLIFIER

Figure 12–25 shows a simple operational amplifier that we can use as an example to identify and analyze the important functional components discussed in the previous section. Q_1 and Q_2 form the input differential stage. The signal inputs are shown grounded because we will presently perform a dc analysis of the entire amplifier, and we wish to verify that the output is zero V under those conditions. The 50-Ω resistors in the emitters of Q_1 and Q_2 serve to increase the amplifier's input impedance and to make the stage less sensitive to variations in r_e, as discussed in Section 12–4. Q_3 and Q_4 form an unbalanced differential stage that provides additional voltage gain. Note that its inputs are driven by the outputs of the first differential stage and that it has a single-ended output. Q_5 performs the level-shifting function, as we shall see. Q_6 is an emitter follower whose output is the output of the operational amplifier. Q_7 and Q_8 are constant-current sources that bias the two differential

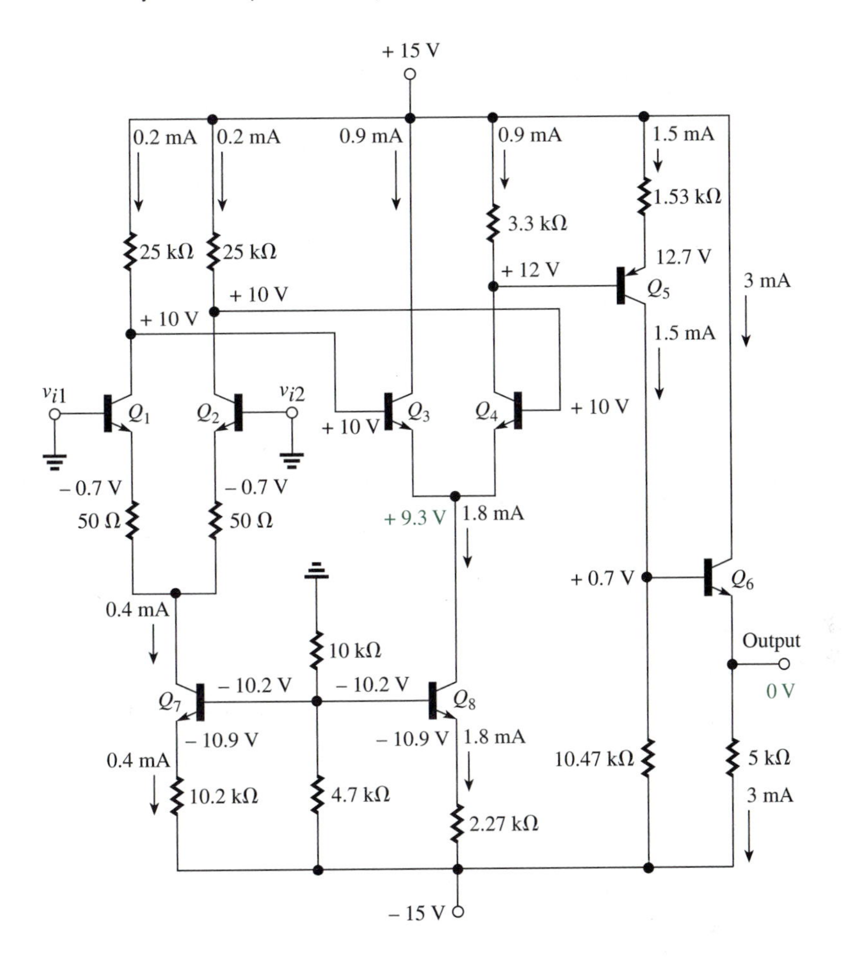

Figure 12–25
A simple operational amplifier incorporating differential, gain, and level-shifting stages. All voltages shown are dc levels with respect to ground.

stages. Note that these sources share a common voltage divider across the transistor bases.

The dc voltages (with respect to ground) and the dc currents in the amplifier are labeled in the figure. We begin our dc analysis by determining the bias current supplied by Q_7 to the input differential stage. The voltage divider across the base of Q_7 sets the base voltage to

$$V_{B7} = \left[\frac{10\text{ k}\Omega}{(10\text{ k}\Omega) + (4.7\text{ k}\Omega)}\right](-15\text{ V}) = -10.2\text{ V}$$

Therefore, the emitter voltage of Q_7 is $V_{B7} - 0.7 = -10.9$ V, and the emitter current is

$$I_{E7} = \frac{|V_{EE}| - |V_{E7}|}{R_{E7}} = \frac{(15 - 10.9)\text{V}}{10.2\text{ k}\Omega} = 0.4\text{ mA}$$

Assuming matched conditions, this current divides equally between Q_1 and Q_2, and, since $I_{C1} = I_{C2} \approx I_{E1} = I_{E2} = (0.4 \text{ mA})/2 = 0.2$ mA, the collector voltages at Q_1 and Q_2 are $V_{C1} = V_{C2} = V_{CC} - I_C R_C = 15 - (0.2 \text{ mA})(25 \text{ k}\Omega) = 10$ V. Since the bases of Q_1 and Q_2 are grounded, their emitters are at approximately $0 - 0.7 = -0.7$ V, and the small drop across each 50-Ω resistor $[(50\ \Omega) \times (0.2 \text{ mA}) = 0.01 \text{ V}]$ sets the collector of Q_7 at about the same voltage (−0.71 V).

We can now analyze the bias of the second differential stage. Since $V_{B8} = V_{B7} = -10.2$ V, the emitter of Q_8 is at $V_{E8} = V_{B8} - 0.7 = -10.9$ V. Then

$$I_{E8} = \frac{|V_{EE}| - |V_{E8}|}{R_{E8}} = \frac{(15 - 10.9)\text{V}}{2.27 \text{ k}\Omega} = 1.8 \text{ mA}$$

The 1.8 mA divides equally between Q_3 and Q_4, so the collector voltage of Q_4 is $V_{C4} = V_{CC} - I_C R_C = 15 - (0.9 \text{ mA})(3.3 \text{ k}\Omega) = 12$ V. Since the bases of Q_3 and Q_4 are direct-coupled to the collectors of Q_1 and Q_2, the base voltages are $V_{B3} = V_{B4} = 10$ V. The emitter voltages are $V_{E3} = V_{E4} = 10 - 0.7 = 9.3$ V.

The base of the level-shifting PNP transistor, Q_5, is direct-coupled to Q_4, so $V_{B5} = 12$ V. Therefore, Q_5 has emitter voltage $V_{E5} = V_{B5} + 0.7 = 12.7$ V. (Remember that the emitter of a PNP transistor is 0.7 V more *positive* than its base.) The emitter current in Q_5 is

$$I_{E5} = \frac{V_{CC} - V_{E5}}{R_{E5}} = \frac{(15 - 12.7)\text{V}}{1.53 \text{ k}\Omega} = 1.5 \text{ mA}$$

Since $I_{C5} \approx I_{E5}$, the collector of Q_5 is at $V_{C5} = (I_{C5})(10.47 \text{ k}\Omega) - V_{EE} = (1.5 \text{ mA})(10.47 \text{ k}\Omega) - 15 = +0.7$ V. Note that Q_5 accomplishes level shifting because its collector can go both positive and negative. (The collector-to-base junction of Q_5 remains reverse biased when the collector is negative and when it is up to 12 V positive.) The collector of Q_4, on the other hand, must always be more positive than its base voltage (+10 V).

Since the base of the output transistor, Q_6, is at 0.7 V, its emitter is at 0 V, and we see that the amplifier output is 0 V. The bias current in Q_6 is $(0 - V_{EE})/5 \text{ k}\Omega = (15 \text{ V})/(5 \text{ k}\Omega) = 3$ mA. This concludes our dc analysis.

In a practical operational amplifier, some of the functions we have described are accomplished in a more elaborate manner. For example, the constant-current bias sources are implemented with current mirrors and/or Widlar sources, and active loads are used instead of collector resistors. These variations were described in Section 12–4, and, in any case, the functional principles of the more elaborate designs are the same as those governing our simple example. The next example shows how to perform a small-signal analysis of the amplifier.

Example 12–10

Assume that the transistors in Figure 12–25 are matched and that all have $\beta = 100$. Neglecting the collector output resistance of each transistor (assuming it to be ∞), find

1. the voltage gain $v_o/(v_{i1} - v_{i2})$;
2. the differential input resistance of the amplifier; and
3. the output resistance of the amplifier.

Solution

1. The load driven by the input differential stage is the differential input resistance r_{id34} of the second stage. Since $I_{E3} = I_{E4} = 0.9$ mA,

$$r_{e3} = r_{e4} \approx \frac{0.026}{0.9\text{ mA}} = 28.9\ \Omega$$

Therefore, $r_{id34} = \beta(r_{e3} + r_{e4}) = 100(28.9 + 28.9) = 5.78$ kΩ.

The ac equivalent circuit of the first stage is shown in Figure 12–26. The double-ended voltage gain is given by

$$\frac{v_{o1} - v_{o2}}{v_{i1} - v_{i2}} = \frac{-(\text{resistance in collector circuit})}{\text{resistance in emitter circuit}} = \frac{-(R_{C1} + R_{C2}) \parallel r_{id34}}{2R_E + r_{e1} + r_{e2}} \tag{12–30}$$

Since $I_{E1} = I_{E2} = 0.2$ mA,

$$r_{e1} = r_{e2} \approx \frac{0.026}{0.2\text{ mA}} = 130\ \Omega$$

Thus,

$$\frac{v_{o1}}{v_{i1} - v_{i2}} = \frac{-[(25\text{ k}\Omega) + (25\text{ k}\Omega)] \parallel (5.78\text{ k}\Omega)}{2(50\ \Omega) + (130\ \Omega) + (130\ \Omega)} = -14.4$$

The ac load resistance driven by the second stage is the input resistance looking into the base of Q_5: $r_{i5} = \beta(r_{e5} + R_{E5})$. Since $I_{E5} = 1.5$ mA, $r_{e5} \approx 0.026/(1.5\text{ mA}) = 17.3\ \Omega$. Thus, $r_{i5} = 100[(17.3\ \Omega) + (1.53\text{ k}\Omega)] = 154.73$ kΩ.

The second stage is operated single-ended and its gain is

$$\frac{v_{o4}}{v_{i3} - v_{i4}} = \frac{R_{C4} \parallel r_{i5}}{r_{e3} + r_{e4}} = \frac{(3.3\text{ k}\Omega) \parallel (154.73\text{ k}\Omega)}{(28.9\ \Omega) + (28.9\ \Omega)} = 55.9$$

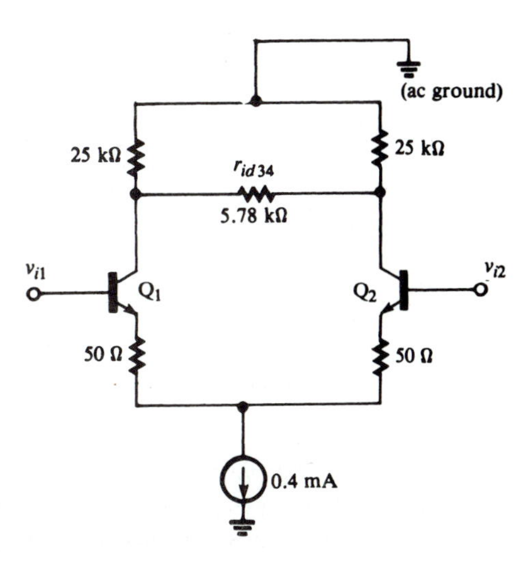

Figure 12–26
(Example 12–10) The ac equivalent circuit of the input stage, showing the differential input resistance of the second stage connected between the collectors

The resistance in the collector circuit of the level-shifting stage (Q_5) is (10.47 kΩ) $\|$ r_{i6}, where r_{i6} is the input resistance looking into the base of Q_6. Since $I_{E6} = 3$ mA, $r_{e6} \approx 0.026/(3\text{ mA}) = 8.7\ \Omega$ and $r_{i6} = \beta(r_{e6} + R_{E6}) = 100[(8.7\ \Omega) + (5\text{ k}\Omega)] \approx 500\text{ k}\Omega$. Thus, the gain of Q_5 is

$$\frac{v_{o5}}{v_{o4}} = \frac{-(10.47\text{ k}\Omega) \| r_{i6}}{R_{E5} + r_{e5}} = \frac{-(10.47\text{ k}\Omega) \| (500\text{ k}\Omega)}{(1.53\text{ k}\Omega) + (17.3\ \Omega)} = -6.6$$

Finally, the gain of the emitter-follower output stage is

$$\frac{v_{o6}}{v_{o5}} = \frac{r_{L6}}{r_{L6} + r_{e6}} = \frac{5\text{ k}\Omega}{(5\text{ k}\Omega) + (8.7\ \Omega)} \approx 1$$

The overall gain of the amplifier is the product of the gains calculated for the stages:

$$\frac{v_o}{v_{i1} - v_{i2}} = (-14.4)(55.9)(-6.6)(1) = 5312$$

(This would not be considered a very large voltage gain for modern operational amplifiers.)

2. The differential resistance looking into the first stage is $r_{id12} \approx \beta(r_{e1} + r_{e2} + 2R_E) = 100(130 + 130 + 100) = 36\text{ k}\Omega$.
3. Recall that the output resistance of an emitter-follower stage is

$$r_o(\text{stage}) = R_E \left\| \left(r_e + \frac{R_B \| r_S}{\beta + 1} \right) \right.$$

In our case, $R_B \| r_S = (10.47\text{ k}\Omega) \| r_o(Q_5) = 10.47\text{ k}\Omega$. Therefore,

$$r_o = (5\text{ k}\Omega) \left\| \left[(8.7\ \Omega) + \frac{10.47\text{ k}\Omega}{101} \right] \right. = 112\ \Omega$$

EXERCISES

Section 12–1
Introduction

12–1. For each of the following pairs of voltages, write the mathematical expression for the difference voltage $(v_1 - v_2)$ and sketch it versus time.

a. $v_1 = 4 \sin \omega t$, $v_2 = 2$
b. $v_1 = 10 \sin \omega t$, $v_2 = 6 \sin \omega t$
c. $v_1 = 0.02 \sin \omega t$, $v_2 = -0.02 \sin \omega t$
d. $v_1 = 4 + 0.01 \sin \omega t$, $v_2 = 4 - 0.01 \sin \omega t$

Section 12–2
The Ideal Differential Amplifier

12–2. The voltage gain of each transistor in the ideal differential amplifier shown in Figure 12–4 is $v_o/v_{be} = -160$. If v_{i1} is a 40-mV-peak sine wave and $v_{i2} = 0$, find

a. the peak value of v_{e1};
b. the peak value of v_{o1};
c. the peak value of v_{o2}; and
d. the voltage gain v_{o1}/v_{i1}.

12–3. A 40-mV-peak sine wave that is out of phase with v_{i1} is applied to v_{i2} in the amplifier in Exercise 12–2. Find

a. the single-ended voltage gains $v_{o1}/(v_{i1} - v_{i2})$ and $v_{o2}/(v_{i1} - v_{i2})$; and
b. the double-ended voltage gain $(v_{o1} - v_{o2})/(v_{i1} - v_{i2})$.

12–4. Repeat Exercise 12–3 if the signal applied to

v_{i2} is a 40-mV-peak sine wave that is *in phase* with v_{i1}. (Think carefully.)

12–5. The ideal BJT differential amplifier shown in Figure 12–2 is biased so that 0.75 mA flows in each emitter. If $R_C = 9.2\ \text{k}\Omega$, find

a. the single-ended voltage gains $v_{o1}/(v_{i1} - v_{i2})$ and $v_{o2}/(v_{i1} - v_{i2})$; and

b. the double-ended voltage gain $(v_{o1} - v_{o2})/(v_{i1} - v_{i2})$.

12–6. If the transistors in the differential amplifier of Exercise 12–5 each have $\beta = 120$, find the differential input resistance of the amplifier.

12–7. The β of each transistor in the ideal differential amplifier shown in Figure 12–27 is 100. Find

a. the dc output voltages V_{o1} and V_{o2};

b. the double-ended voltage gain; and

c. the differential input resistance.

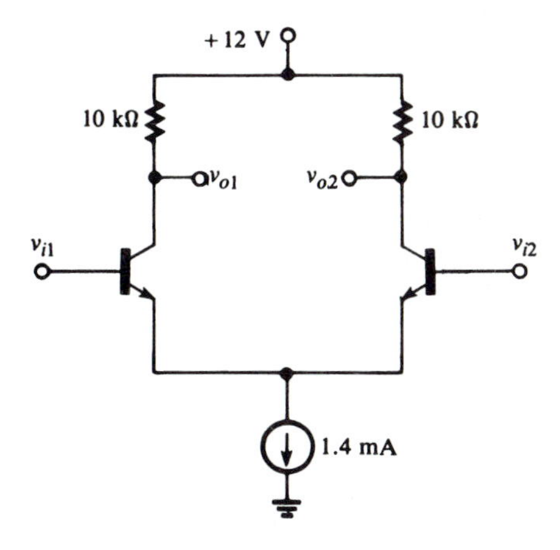

Figure 12–27
(Exercise 12–7)

12–8. The current source in Exercise 12–7 is changed to 1 mA. If v_{i1} is a 16-mV-peak sine wave and $v_{i2} = 0$, find the peak values of v_{o1} and v_{o2}.

12–9. The FETs in the ideal differential amplifier shown in Figure 12–28 have $I_{DSS} = 10$ mA and $V_p = -2$ V. Find

a. the dc output voltages V_{o1} and V_{o2};

b. the single-ended output gain $v_{o1}/(v_{i1} - v_{i2})$; and

c. the output difference voltage when the input difference voltage is 50 mV rms.

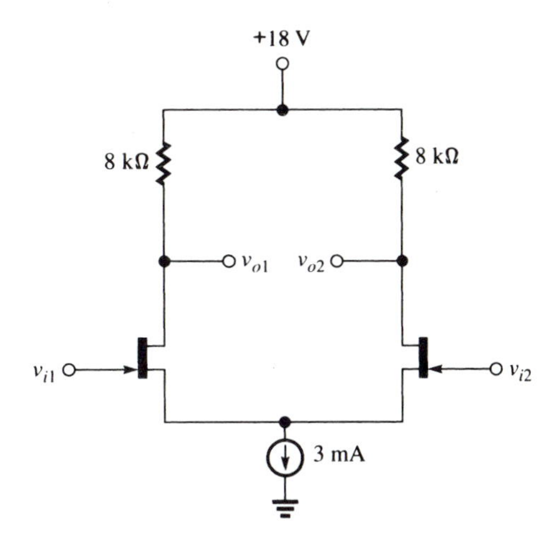

Figure 12–28
(Exercise 12–9)

12–10. The inputs to the differential amplifier in Exercise 12–9 are $v_{i1} = 65$ mV rms and $v_{i2} = 10$ mV rms. v_{i1} and v_{i2} are in phase. Find the rms values of v_{o1} and v_{o2}.

Section 12–3
Common-Mode Parameters

12–11. A differential amplifier has CMRR = 68 dB and a differential mode gain of 175. Find the rms value of the output difference voltage when the common-mode signal is 1.5 mV rms.

12–12. The noise signal common to both inputs of a differential amplifier is 2.4 mV rms. When an input difference voltage of 0.1 V rms is applied to the amplifier, each output must have 4 V rms of signal level. Assuming that the amplifier produces no noise and that the noise component in the output difference voltage must be no greater than 500 μV rms, what is the minimum CMRR, in dB, that the amplifier should have?

Section 12–4
Practical Differential Amplifiers

12–13. The transistors in Figure 12–29 are matched and have $\beta_1 = \beta_2 = 85$. Find

a. the dc values of v_{o1} and v_{o2};

b. the differential input resistance; and

c. the double-ended voltage gain $(v_{o1} - v_{o2})/(v_{i1} - v_{i2})$.

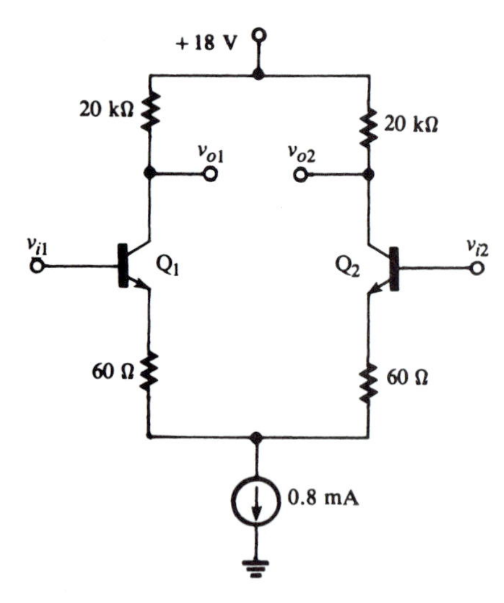

Figure 12–29
(Exercise 12–13)

12–14. If the common-mode gain v_{o1}/v_{cm} in the amplifier of Exercise 12–13 must be no greater than 0.5, what is the minimum permissible source resistance of the constant-current source?

12–15. Transistors Q_1 and Q_2 in Figure 12–30 are matched and have $\beta = 140$. The collector output resistance of Q_3 is 280 kΩ. Find

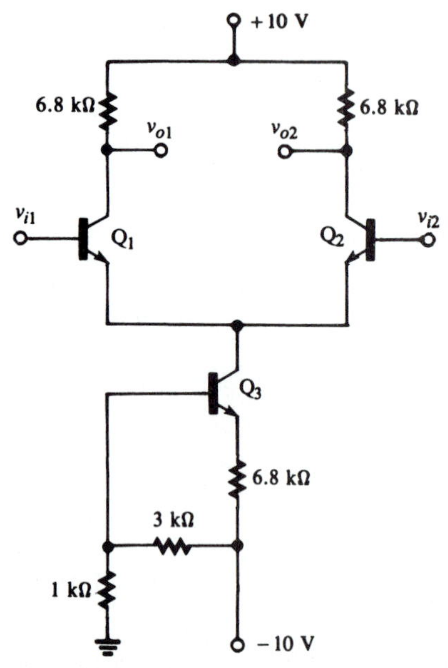

Figure 12–30
(Exercise 12–15)

a. the dc output voltages V_{o1} and V_{o2};
b. the differential input resistance;
c. the double-ended CMRR; and
d. the single-ended CMRR.

12–16. Repeat Exercise 12–15 when 40-Ω resistors are inserted in series with the emitters of Q_1 and Q_2.

12–17. The inputs to the matched transistors in Figure 12–31 are $v_{i1} = 50$ mV rms and $v_{i2} = 30$ mV rms. The inputs are in phase. Find the rms values of v_{o1} and $(v_{o1} - v_{o2})$.

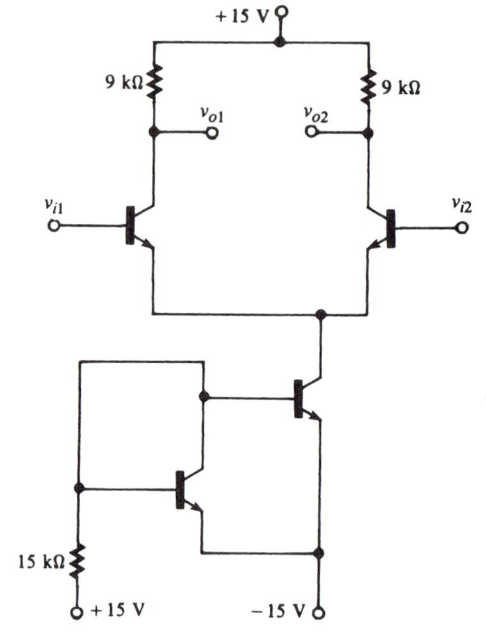

Figure 12–31
(Exercise 12–17)

12–18. In the amplifier of Exercise 12–17, how would the values of v_{o1} and $(v_{o1} - v_{o2})$ be affected by an increase in the positive supply voltage? By an increase in the negative supply voltage?

12–19. Using equation 12–28, verify that the constant current in the Widlar current source shown in Figure 12–22 is 10 μA when R = 12.22 kΩ and the reference current is 1.1 mA. Assume room-temperature conditions.

12–20. The constant current in the Widlar current source shown in Figure 12–22 is 15 μA when R = 5 kΩ. What is the reference current? Assume room temperature.

12–21. Design a Widlar current source (find the value of R) that will produce a constant current of 12 μA at room temperature when the reference current is 1 mA. (*Hint:* Solve equation 12–28 for R.)

12–22. The input signals to the amplifier shown in Figure 12–32 are v_{i1} = 0.01 V peak and v_{i2} = 0.03 V peak. The transistors are matched and Q_2 and Q_4 both have output resistance 60 kΩ. If v_{i1} and v_{i2} are out of phase, find the peak value of v_{o2}.

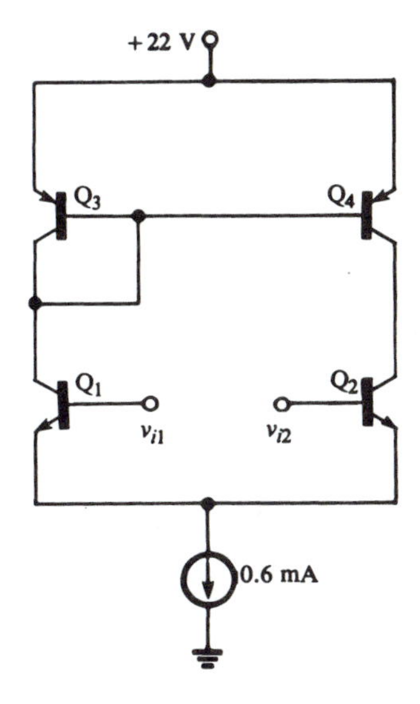

Figure 12–32
(Exercise 12–22)

Section 12–6
Circuit Analysis of an Operational Amplifier

12–23. If the 3.3-kΩ collector resistor of Q_4 in the operational amplifier shown in Figure 12–25 is changed to 4 kΩ, the output voltage of the amplifier will no longer be 0 V when the inputs are 0 V. What new resistor could be used to replace the 1.53 kΩ emitter resistor of Q_5 in order to restore the amplifier output to 0 V?

12–24. With the changes made in Exercise 12–23, what would be the new value of the amplifier voltage gain $v_o/(v_{i1} - v_{i2})$?

12–25. If the 4.7 kΩ resistor in the constant-current bias circuitry of Figure 12–25 were to increase 10%, all of the dc voltages and currents in the operational amplifier would change. Overlooking the fact that the dc output voltage of the amplifier would no longer be 0 V, what would be the new voltage gain?

12–26. If the 4.7-kΩ resistor in the constant-current bias circuitry of Figure 12–25 were to increase 10%, the output of the operational amplifier would no longer be 0 V when the inputs are 0 V.

a. What new dc value would the output be?
b. What new resistor should be used to replace the 10.47-kΩ resistor connected to the base of Q_6 in order to restore the output to 0 V?

SPICE EXERCISES

12–27. The JFET differential amplifier in Figure 12–12 has V_{DD} = 24 V, R_D = 3 kΩ, and a 6-mA constant current source supplying bias current. To investigate the effects of parameter variability on the balance of the amplifier, use SPICE to find the dc output voltages, V_{o1} and V_{o2}, and the currents in each source, I_{s1} and I_{s2}, when

a. Q_1 and Q_2 are perfectly matched, each having V_p = −2 V and I_{DSS} = 12 mA;
b. Q_1 has V_p = −1.8 V and Q_2 has V_p = −2.2 V, while I_{DSS} = 12 mA for both;
c. the conditions of (b) exist, except 100-Ω resistors are inserted in series with the source of each JFET.

For each case, use the results of the SPICE simulations to find the difference in the dc output voltages and the difference in the source currents. In particular, compare cases (b) and (c) and comment.

12–28. The differential amplifier in Figure 12–18 has R_C = 7.5 kΩ (on both sides), V_{CC} = 15 V, V_{EE} = −15 V, R_1 = 4 kΩ, R_2 = 22 kΩ and R_{E3} = 3 kΩ. The betas of the transistors are β_1 = 150, β_2 = 120, and β_3 = 180. The saturation currents are $I_{s1} = 1.2 \times 10^{-16}$ A, $I_{s2} = 0.8 \times 10^{-16}$ A, and $I_{s3} = 1 \times 10^{-16}$ A.

a. Use SPICE to find the differential-mode voltage gain and the common-mode voltage gain at 1 kHz. Use the results to calculate the CMRR in dB.
b. Repeat (a) when 100-Ω resistors are inserted in series with the emitters of Q_1 and Q_2. Comment on the effect of the resistors.

12–29. The Widlar current source in Figure 12–22 has V_{EE} = −15 V and R = 12 kΩ. The reference current I_X is supplied from a 1-mA constant current source. The collector of Q_2 is connected through a 10-kΩ load resistance to ground. Use SPICE to find the bias current I. Assume the transistors have their default parameters.

12–30. The transistors in Figure 12–23 all have $\beta = 150$ and forward Early voltage 150 V. The supply voltage is 15 V and the constant-current source is 3 mA.

a. Use SPICE to find the single-ended voltage gain $v_{o2}/(v_{i1} - v_{i2})$ at 1 kHz.

b. Using the values computed by SPICE for r_{o2}, and r_{o4}, use equation 12–29 to calculate the single-ended voltage gain. (Compute r_e from $0.026/(I/2)$, where I is the constant current from the bias source.)

13 OPERATIONAL AMPLIFIER THEORY

13–1 THE IDEAL OPERATIONAL AMPLIFIER

Recall that an operational amplifier is a direct-coupled amplifier with two (differential) inputs and a single output. In Chapter 12, we listed nine desirable characteristics of an operational amplifier; we wish now to focus on three of those. We will define an *ideal* operational amplifier to be one that has the following attributes:

1. It has infinite gain.
2. It has infinite input impedance.
3. It has zero output impedance.

Although no real amplifier* can satisfy any of these requirements, we will see that most modern amplifiers have such large gains and input impedances, and such small output impedances, that a negligibly small error results from assuming ideal characteristics. A detailed study of the ideal amplifier will therefore be beneficial in terms of understanding how practical amplifiers are used as well as in building some important theoretical concepts that have broad implications in many areas of electronics.

Figure 13–1 shows the standard symbol for an operational amplifier. Note that the two inputs are labeled "+" and "−" and the input signals are correspondingly designated v_i^+ and v_i^-. In relation to our previous discussion of differential amplifiers, these inputs correspond to v_{i1} and v_{i2}, respectively, when the single-ended output is v_{o2} (see Figure 12–2). In other words, if the inputs are out-of-phase signals, the amplifier output will be in phase with v_i^+ and out of phase with v_i^-. For this reason, the + input is called the *noninverting* input and the − input is called the *inverting input.* In many applications, one of the amplifier inputs is grounded, so v_o is in phase with the input if the signal is connected to the noninverting terminal, and v_o

* In this chapter, we will hereafter use the word *amplifier* with the understanding that operational amplifier is meant. Some authors use the term *op-amp.*

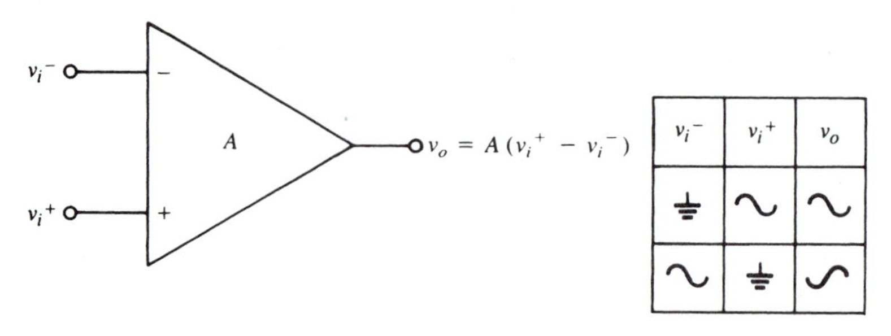

Figure 13–1
Operational amplifier symbol, showing inverting (−) and noninverting (+) inputs

is out of phase with the input if the signal is connected to the inverting input. These ideas are summarized in the table accompanying Figure 13–1.

At this point, a legitimate question that may have already occurred to the reader is this: If the gain is infinite, how can the output be anything other than a severely clipped waveform? Theoretically, if the amplifier has infinite gain, an infinitesimal input voltage must result in an infinitely large output voltage. The answer, of course, is that the gain is not truly infinite, just very large. Nevertheless, it *is* true that a very small input voltage will cause the amplifier output to be driven all the way to its extreme positive or negative voltage limit. The practical answer is that an operational amplifier is seldom used in such a way that the full gain is applied to an input. Instead, external resistors are connected to and around the amplifier in such a way that the signal undergoes vastly smaller amplification. The resistors cause gain reduction through signal *feedback,* which we will soon study in considerable detail.

The Inverting Amplifier

Consider the configuration shown in Figure 13–2. In this very useful application of an operational amplifier, the noninverting input is grounded, v_{in} is connected through R_1 to the inverting input, and feedback resistor R_f is connected between the output and v_i^-. Let A denote the voltage gain of the amplifier: $v_o = A(v_i^+ - v_i^-)$. Since $v_i^+ = 0$, we have

$$v_o = -Av_i^- \tag{13–1}$$

(Note that $v_{in} \neq v_i^-$.) We wish to investigate the relation between v_o and v_{in} when the magnitude of A is infinite.

Figure 13–2
An operational-amplifier application in which signal v_{in} is connected through R_1. Resistor R_f provides feedback. $v_o/v_i^- = -A$.

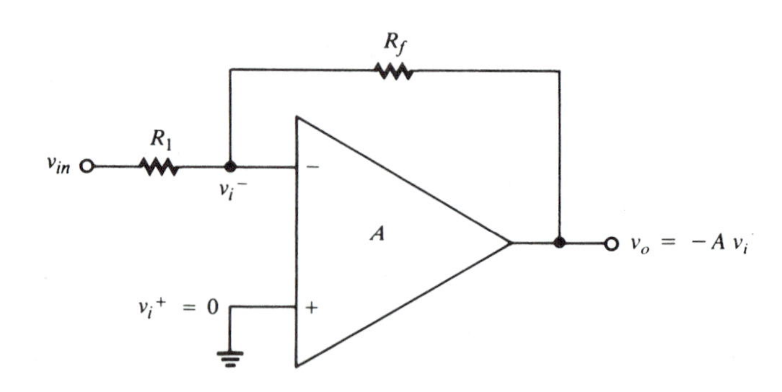

Figure 13–3 shows the voltages and currents that result when signal v_{in} is connected. From Ohm's law, the current i_1 is simply the difference in voltage across R_1, divided by R_1:

$$i_1 = (v_{in} - v_i^-)/R_1 \tag{13–2}$$

Similarly, the current i_f is the difference in voltage across R_f, divided by R_f:

$$i_f = (v_i^- - v_o)/R_f \tag{13–3}$$

Writing Kirchhoff's current law at the inverting input, we have

$$i_1 = i_f + i^- \tag{13–4}$$

where i^- is the current entering the amplifier at its inverting input. However, the ideal amplifier has infinite input impedance, which means i^- must be 0. So (13–4) is simply

$$i_1 = i_f \tag{13–5}$$

Substituting (13–2) and (13–3) into (13–5) gives

$$\frac{v_{in} - v_i^-}{R_1} = \frac{v_i^- - v_o}{R_f}$$

or

$$\frac{v_{in}}{R_1} - \frac{v_i^-}{R_1} = \frac{v_i^-}{R_f} - \frac{v_o}{R_f} \tag{13–6}$$

From equation 13–1,

$$v_i^- = -\frac{v_o}{A} \tag{13–7}$$

If we now invoke the assumption that $A = \infty$, we see that $-v_o/A = 0$ and, therefore,

$$v_i^- = 0 \qquad \text{(ideal amp, with } A = \infty) \tag{13–8}$$

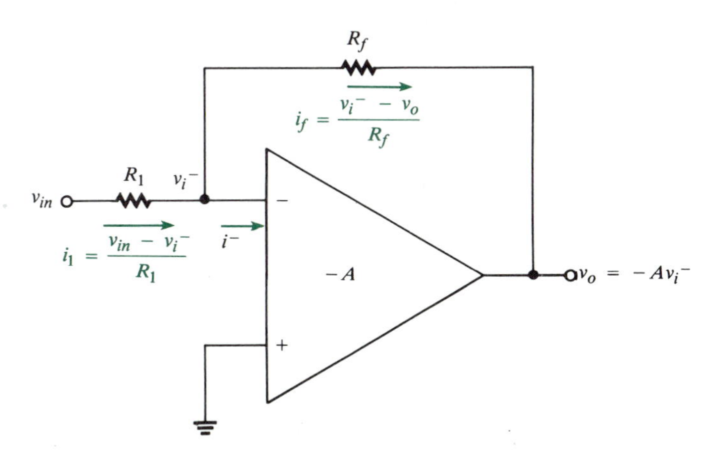

Figure 13–3
Voltages and currents resulting from the application of the signal voltage v_{in}

Substituting $v_i^- = 0$ into (13–6) gives

$$\frac{v_{in}}{R_1} = \frac{-v_o}{R_f}$$

or

$$\frac{v_o}{v_{in}} = \frac{-R_f}{R_1} \tag{13–9}$$

We see that the gain is negative, signifying that the configuration is an *inverting* amplifier. Equation 13–9 also reveals the exceptionally useful fact that the magnitude of v_o/v_{in} *depends only on the ratio of the resistor values* and not on the amplifier itself. Provided the amplifier gain and impedance remain quite large, variations in amplifier characteristics (due, for example, to temperature changes or manufacturing tolerance) do not affect v_o/v_{in}. For example, if $R_1 = 10$ kΩ and $R_f = 100$ kΩ, we can be certain that $v_o = -[(100\text{ k}\Omega)/(10\text{ k}\Omega)]v_{in} = -10\,v_{in}$, i.e., that the gain is as close to -10 as the resistor precision permits. The gain v_o/v_{in} is called the *closed-loop gain* of the amplifier, while A is called the *open-loop gain.* In this application, we see that an extremely large open-loop gain, perhaps 10^6, is responsible for giving us the very predictable, though much smaller, closed-loop gain equal to 10. This is the essence of most operational-amplifier applications: trade the very large gain that is available for less spectacular but more precise and predictable characteristics.

In our derivation, we used the infinite-gain assumption to obtain $v_i^- = 0$ (equation 13–8). In real amplifiers, having very large, but finite, values of A, v_i^- is a very small voltage, near 0. For that reason, the input terminal where the feedback resistor is connected is said to be at *virtual ground.* For *analysis* purposes, we often assume that $v_i^- = 0$, but we cannot actually ground that point. Since v_i^- is at virtual ground, the impedance seen by the signal source generating v_{in} is R_1 ohms.

Example 13–1

Assuming that the operational amplifier in Figure 13–4 is ideal, find

1. the rms value of v_o when v_{in} is 1.5 V rms;
2. the rms value of the current in the 25-kΩ resistor when v_{in} is 1.5 V rms; and
3. the output voltage when $v_{in} = -0.6$ V dc.

Figure 13–4
(Example 13–1)

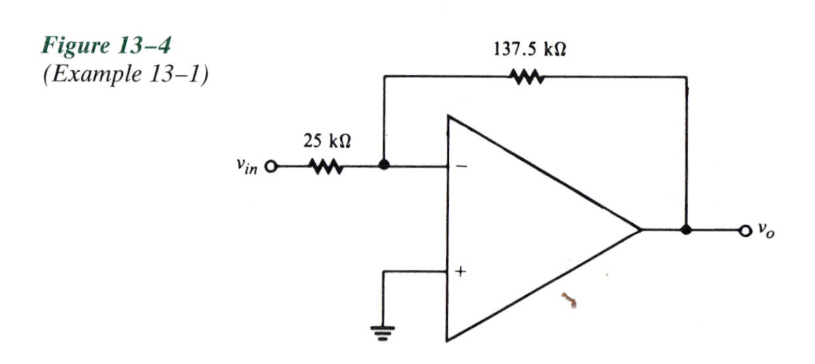

Solution

1. From equation 13–9,

$$\frac{v_o}{v_{in}} = \frac{-R_f}{R_1} = \frac{-137.5\ \text{k}\Omega}{25\ \text{k}\Omega} = -5.5$$

Thus, $|v_o| = 5.5|v_{in}| = 5.5(1.5\ \text{V rms}) = 8.25\ \text{V rms}$.

2. Since $v_i^- \approx 0$ (virtual ground), the current in the 25-kΩ resistor is

$$i = \frac{v_{in}}{R_1} = \frac{1.5\ \text{V rms}}{25\ \text{k}\Omega} = 60\ \mu\text{A rms}$$

3. $v_o = (-5.5)v_{in} = (-5.5)(-0.6\ \text{V}) = 3.3\ \text{V dc}$. Notice that the output is a positive dc voltage when the input is a negative dc voltage, and vice versa.

The Noninverting Amplifier

Figure 13–5 shows another useful application of an operational amplifier, called the *noninverting* configuration. Notice that the input signal v_{in} is connected directly to the noninverting input and that resistor R_1 is connected from the inverting input to ground. Under the ideal assumption of infinite input impedance, no current flows into the inverting input, so $i_1 = i_f$. Thus,

$$\frac{v_i^-}{R_1} = \frac{v_o - v_i^-}{R_f} \tag{13–10}$$

Now, as shown in the figure,

$$v_o = A(v_i^+ - v_i^-) \tag{13–11}$$

Solving (13–11) for v_i^- gives

$$v_i^- = v_i^+ - v_o/A \tag{13–12}$$

Letting $A = \infty$, the term v_o/A goes to 0, and we have

$$v_i^- = v_i^+ \tag{13–13}$$

Substituting v_i^+ for v_i^- in (13–10) gives

$$\frac{v_i^+}{R_1} = \frac{v_o - v_i^+}{R_f} \tag{13–14}$$

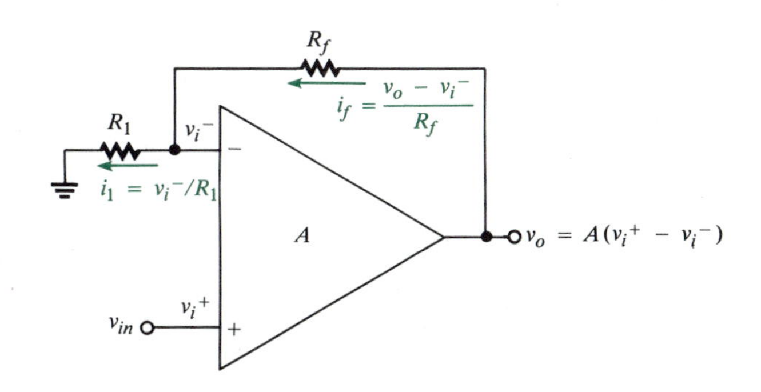

Figure 13–5
The operational amplifier in a noninverting configuration

Figure 13–6
The voltage follower

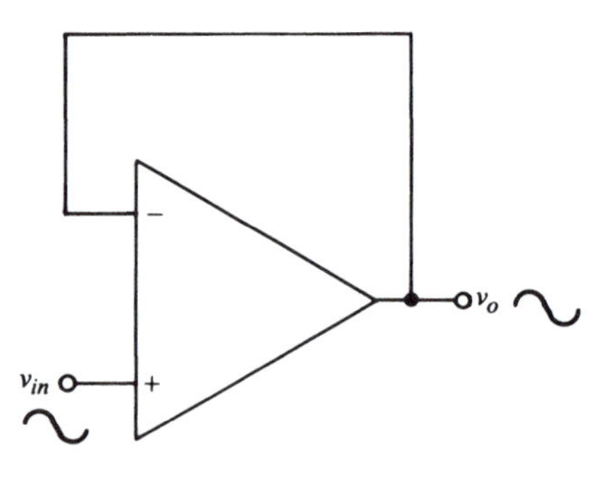

Solving for v_o/v_i^+ and recognizing that $v_i^+ = v_{in}$ lead to

$$\frac{v_o}{v_{in}} = 1 + \frac{R_f}{R_1} = \frac{R_1 + R_f}{R_1} \qquad \textbf{(13–15)}$$

We saw (equation 13–8) that when an operational amplifier is connected in an inverting configuration, with $v_i^+ = 0$, the assumption $A = \infty$ gives $v_i^- = 0$ (virtual ground), i.e., $v_i^- = v_i^+$. Also, in the noninverting configuration, the same assumption gives the same result: $v_i^- = v_i^+$ (equation 13–13). Thus, we reach the important general conclusion that feedback in conjunction with a very large voltage gain *forces* the voltages at the inverting and noninverting inputs to be approximately equal.

Equation 13–15 shows that the closed-loop gain of the noninverting amplifier, like that of the inverting amplifier, depends only on the values of external resistors. A further advantage of the noninverting amplifier is that the input impedance seen by v_{in} is infinite, or at least extremely large in a real amplifier. The inverting and noninverting amplifiers are used in voltage *scaling* applications, where it is desired to multiply a voltage precisely by a fixed constant, or scale factor. The multiplying constant in the inverting amplifier is R_f/R_1 (which may be less than 1), and it is $1 + R_f/R_1$ (which is always greater than 1) in the noninverting amplifier. A wide range of constants can be realized with convenient choices of R_f and R_1 when the gain ratio is R_f/R_1, which is not so much the case when the gain ratio is $1 + R_f/R_1$. For that reason, the inverting amplifier is more often used in precision scaling applications.

The reader may wonder why it would be desirable or necessary to use an amplifier to multiply a voltage by a number less than 1, since this can also be accomplished using a simple voltage divider. The answer is that the amplifier provides power gain to drive a load. Also, the ideal amplifier has zero output impedance, so the output voltage is not affected by changes in load impedance. Figure 13–6 shows a special case of the noninverting amplifier, used in applications where power gain and impedance isolation are of primary concern. Notice that $R_f = 0$ and $R_1 = \infty$, so, by equation 13–15, the closed-loop gain is $v_o/v_{in} = 1 + R_f/R_1 = 1$. This configuration is called a *voltage follower* because v_o has the same magnitude and phase as v_{in}. Like a BJT emitter follower, it has large input impedance and small output impedance, and is used as a buffer amplifier between a high-impedance source and a low-impedance load.

Example 13–2

DESIGN

In a certain application, a signal source having 60 kΩ of source impedance produces a 1-V-rms signal. This signal must be amplified to 2.5 V rms and drive a 1-kΩ load. Assuming that the phase of the load voltage is of no concern, design an operational-amplifier circuit for the application.

Solution. Since phase is of no concern and the required voltage gain is greater than 1, we can use either an inverting or noninverting amplifier. Suppose we decide to use the inverting configuration and arbitrarily choose $R_f = 250\text{ k}\Omega$. Then,

$$\frac{R_f}{R_1} = 2.5 \Rightarrow R_1 = \frac{R_f}{2.5} = \frac{250\text{ k}\Omega}{2.5} = 100\text{ k}\Omega$$

Note, however, that the signal source sees an impedance equal to $R_1 = 100\text{ k}\Omega$ in the inverting configuration, so the usual voltage division takes place and the input to the amplifier is actually

$$v_{in} = \left(\frac{R_1}{R_1 + r_S}\right)(1\text{ V rms}) = \left[\frac{100\text{ k}\Omega}{(100\text{ k}\Omega) + (60\text{ k}\Omega)}\right](1\text{ V rms}) = 0.625\text{ V rms}$$

Therefore, the magnitude of the amplifier output is

$$v_o = \frac{R_f}{R_1}(0.625\text{ V rms}) = \frac{250\text{ k}\Omega}{100\text{ k}\Omega}(0.625\text{ V rms}) = 1.5625\text{ V rms}$$

Clearly the large source impedance is responsible for a reduction in gain, and it is necessary to redesign the amplifier circuit to compensate for this loss. (Do this, as an exercise.)

In view of the fact that the source impedance may not be known precisely or may change if a replacement source is used, a far better solution is to design a noninverting amplifier. Since the input impedance of this design is extremely large, the choice of values for R_f and R_1 will not depend on the source impedance. Letting $R_f = 150\text{ k}\Omega$, we have

$$1 + \frac{R_f}{R_1} = 2.5$$

$$\frac{R_f}{R_1} = 1.5$$

$$R_1 = \frac{R_f}{1.5} = \frac{150\text{ k}\Omega}{1.5} = 100\text{ k}\Omega$$

The completed design is shown in Figure 13–7. Since we can assume that the amplifier has zero output impedance, we do not need to be concerned with voltage division between the amplifier output and the 1-kΩ load.

Figure 13–7
(Example 13–2)

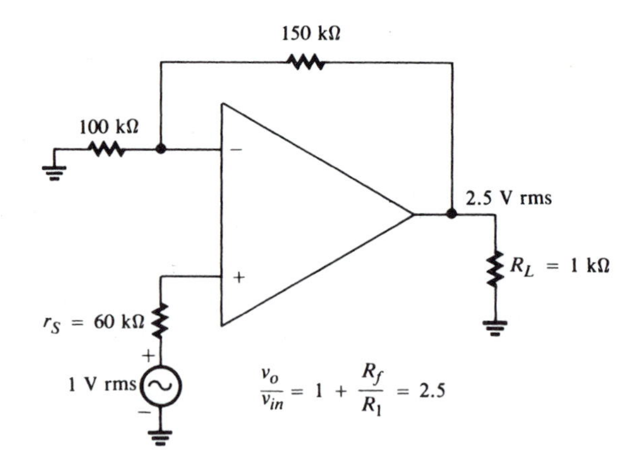

13–2 FEEDBACK THEORY

We have seen that we can control the closed-loop gain v_o/v_{in} of an operational amplifier by introducing feedback through external resistor combinations. We wish now to examine the feedback mechanism in detail and discover some other important consequences of its use. Feedback theory is widely used to study the behavior of electronic components as well as complex systems in many different technical fields, so it is important to develop an appreciation and understanding of its underlying principles.

Feedback in the Noninverting Amplifier

We begin our study of feedback principles with an analysis of the noninverting amplifier. Figure 13–8 shows that configuration along with an equivalent block diagram on which we can identify the signal and feedback paths. The block labeled A represents the amplifier and its open-loop gain, and the block labeled β is the feedback path. The quantity β is called the *feedback ratio* and represents the portion of the output voltage that is fed back to the input. For example, if $\beta = 0.5$, then a voltage equal to one-half the output level is fed back to the input. Notice the special symbol where the input and feedback paths come together. This symbol represents the differential action at the amplifier's input. It is usually called a *summing* junction, although in our case it is performing a differencing operation, as indicated by the $+$ and $-$ symbols. The output of the junction, which is the input to the amplifier, is seen to be $v_e = v_{in} - v_f$. v_e is often called the *error* voltage. Note that it corresponds to $v_{in} - v_i^-$ in the noninverting amplifier, and under ideal conditions is equal to 0 (equation 13–13). The feedback voltage $v_f = \beta v_o$ corresponds to v_i^- in the amplifier circuit. Since the feedback voltage subtracts from the input voltage, the amplifier is said to have *negative feedback.*

With reference to Figure 13–8(b), we see that

$$v_o = A(v_{in} - v_f) \tag{13–16}$$

and

$$v_f = \beta v_o \tag{13–17}$$

Figure 13–8
Block-diagram representation of the noninverting amplifier. Identify corresponding voltages in the two diagrams.

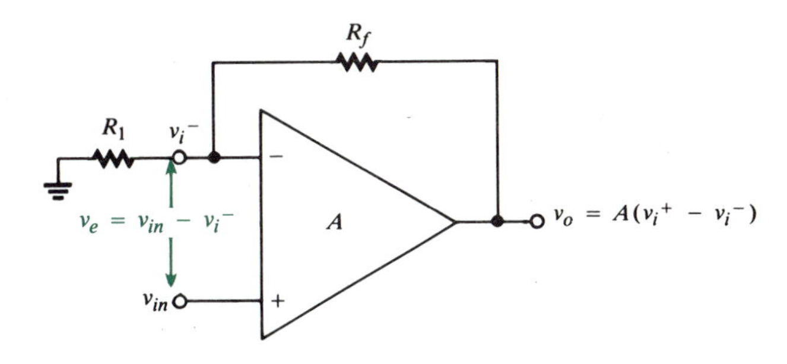

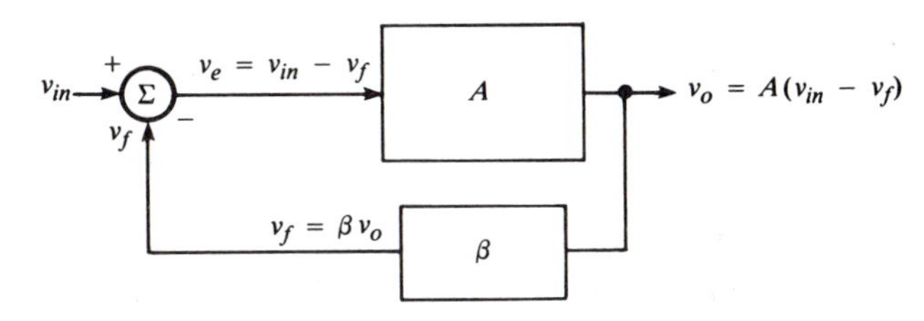

Substituting (13–17) into (13–16) gives $v_o = A(v_{in} - \beta v_o) = Av_{in} - A\beta v_o$, or $v_o(1 + A\beta) = Av_{in}$. Thus,

$$\frac{v_o}{v_{in}} = \frac{A}{1 + A\beta} = \frac{1/\beta}{1 + 1/A\beta} \qquad \textbf{(13–18)}$$

Equation 13–18 is a very important and very useful result. It expresses the closed-loop gain v_o/v_{in} as a function of the open-loop gain A and the feedback ratio β. We can now apply this result to the noninverting amplifier in Figure 13–8(a). Notice that R_f and R_1 form a voltage divider across the output of the amplifier, so

$$v_i^- = \left(\frac{R_1}{R_1 + R_f}\right) v_o \qquad \textbf{(13–19)}$$

Since v_i^- is the voltage fed back from the output, and $v_f = \beta v_o$, we conclude that

$$\beta = \frac{R_1}{R_1 + R_f} \qquad \text{(noninverting amplifier)} \qquad \textbf{(13–20)}$$

Substituting into (13–18), we find

$$\frac{v_o}{v_{in}} = \frac{(R_1 + R_f)/R_1}{1 + 1/A\beta} = \frac{(R_1 + R_f)/R_1}{1 + (R_1 + R_f)/AR_1} \qquad \textbf{(13–21)}$$

Equation 13–21 gives us the means for investigating how significant the value of open-loop gain A is in the determination of the closed-loop gain v_o/v_{in}. First, note that when $A = \infty$, (13–21) reduces to $v_o/v_{in} = (R_1 + R_f)/R_1$, which is exactly the same result we obtained in Section 13–1 for the ideal, noninverting amplifier (equation 13–15). Notice also that

$$\frac{v_o}{v_{in}} = \frac{1}{\beta} \quad \text{(ideal noninverting amplifier, } A = \infty) \qquad \textbf{(13–22)}$$

Equation 13–22 can also be obtained by letting $A = \infty$ in equation 13–18.

The next example shows how finite values of A affect the value of v_o/v_{in}.

Example 13–3

Find the closed-loop gain of the amplifier in Figure 13–9 when (1) $A = \infty$, (2) $A = 10^6$, and (3) $A = 10^3$.

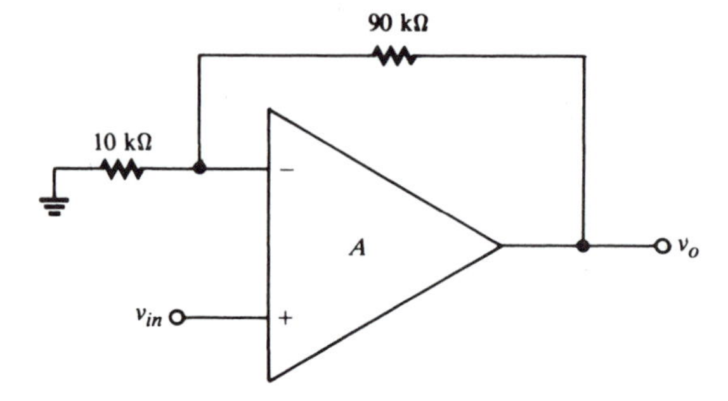

Figure 13–9
(Example 13–3)

Solution

1. The feedback ratio is

$$\beta = \frac{R_1}{R_1 + R_f} = \frac{10\text{ k}\Omega}{(10\text{ k}\Omega) + (90\text{ k}\Omega)} = 0.1$$

Therefore, the closed-loop gain when $A = \infty$ is $v_o/v_{in} = 1/\beta = 1/0.1 = 10$.

2. Using equation 13–18, the closed-loop gain when $A = 10^6$ is

$$\frac{v_o}{v_{in}} = \frac{A}{1 + A\beta} = \frac{10^6}{1 + 10^6(0.1)} = 9.99990$$

We see that v_o/v_{in} is for all practical purposes the same value when $A = 10^6$ as it is when $A = \infty$.

3. When $A = 10^3$,

$$\frac{v_o}{v_{in}} = \frac{10^3}{1 + 10^3(0.1)} = 9.90099$$

We see that reducing A to 1000 creates a discrepancy of about 1% with respect to the value of v_o/v_{in} when $A = \infty$.

Equation 13–18 shows that the closed-loop gain of a real amplifier also departs from that of an ideal amplifier when the value of β becomes very small. Small values of β correspond to large closed-loop gains.

Example 13–4

An operational amplifier has open-loop gain $A = 10{,}000$. Compare its closed-loop gain with that of an ideal amplifier when (1) $\beta = 0.1$, and (2) $\beta = 0.001$.

Solution

1. $\beta = 0.1$. For $A = \infty$, $v_o/v_{in} = 1/\beta = 10$. For $A = 10^4$,

$$\frac{v_o}{v_{in}} = \frac{A}{1 + A\beta} = \frac{10^4}{1 + 10^4(0.1)} = 9.99$$

2. $\beta = 0.001$. For $A = \infty$, $v_o/v_{in} = 1/\beta = 1000$. For $A = 10^4$,

$$\frac{v_o}{v_{in}} = \frac{A}{1 + A\beta} = \frac{10^4}{1 + 10^4(10^{-3})} = 909.09$$

We see that when $\beta = 0.001$, v_o/v_{in} departs more from the ideal case than it does when $\beta = 0.1$.

The last two examples have shown that the closed-loop gain departs from the ideal value of $1/\beta$ when A is small or when β is small. We can deduce that fact from another examination of equation 13–18:

$$\frac{v_o}{v_{in}} = \frac{1/\beta}{1 + 1/A\beta}$$

Clearly both A and β should be large if we want v_o/v_{in} to equal $1/\beta$. The product $A\beta$ is called the *loop gain* and is very useful in predicting the behavior of a feedback system. The name *loop gain* is derived from its definition as the product of the gains in the feedback model as one travels around the loop from amplifier input, through the amplifier, and through the feedback path (with the summing junction open).

Negative feedback improves the performance of an amplifier in several ways. In the case of the noninverting amplifier, it can be shown that the input resistance seen by the signal source (looking directly into the + terminal) is

$$r_{in} = (1 + A\beta)r_{id} \approx A\beta r_{id} \tag{13–23}$$

where r_{id} is the differential input resistance of the amplifier. This equation shows that the input resistance is r_{id} multiplied by the factor $1 + A\beta$, which is usually much greater than 1 and is approximately equal to the loop gain $A\beta$. For example, if $r_{id} = 20\ \text{k}\Omega$, $A = 10^5$, and $\beta = 0.01$, then $r_{in} \approx (20 \times 10^3)(10^5)(0.01) = 20\ \text{M}\Omega$, a very respectable value. In the case of the voltage follower, $\beta = 1$, and r_{id} is multiplied by the full value of A, which accounts for the extremely large input resistance it can provide in buffer applications.

The closed-loop output resistance of the noninverting amplifier is also improved by negative feedback:

$$r_o(\text{stage}) = \frac{r_o}{1 + A\beta} \approx \frac{r_o}{A\beta} \tag{13–24}$$

where r_o is the open-loop output resistance of the amplifier. Equation 13–24 shows that the output resistance is decreased by the same factor by which the input resistance is increased. A typical value for r_o is 75 Ω, so with $A = 10^5$ and $\beta = 0.01$, we have $r_o(\text{stage}) \approx 75/10^3 = 0.075\ \Omega$, which is very close to the ideal value of 0.

Finally, negative feedback *reduces distortion* caused by the amplifier itself. We will study this property later, in connection with large-signal amplifiers.

Feedback in the Inverting Amplifier

To investigate the effect of open-loop gain A and feedback ratio β on the closed-loop gain of the inverting amplifier, let us recall equations 13–6 and 13–7 from Section 13–1:

$$\frac{v_{in}}{R_1} - \frac{v_i^-}{R_1} = \frac{v_i^-}{R_f} - \frac{v_o}{R_f} \tag{13–25}$$

$$v_i^- = -v_o/A \tag{13–26}$$

Substituting (13–26) into (13–25) gives

$$\frac{v_{in}}{R_1} + \frac{v_o}{AR_1} = \frac{-v_o}{AR_f} - \frac{v_o}{R_f} \tag{13–27}$$

Exercise 13–15 at the end of this chapter is included to show that (13–27) can be solved for v_o/v_{in} with the result

$$\frac{v_o}{v_{in}} = \frac{-R_f/R_1}{1 + (R_1 + R_f)/AR_1} \tag{13–28}$$

Once again we see that the closed-loop gain reduces to the ideal amplifier value, $-R_f/R_1$, when $A = \infty$. Notice that the denominator of (13–28) is the same as that of (13–21), the equation for the closed-loop gain of the noninverting amplifier. Furthermore, the quantity $R_1/(R_1 + R_f)$ is also the feedback ratio β for the inverting amplifier. This fact is illustrated in Figure 13–10, which shows the feedback paths of both configurations when their signal inputs are grounded. Think of the amplifier output as a source that generates the feedback signal. By the superposition principle,

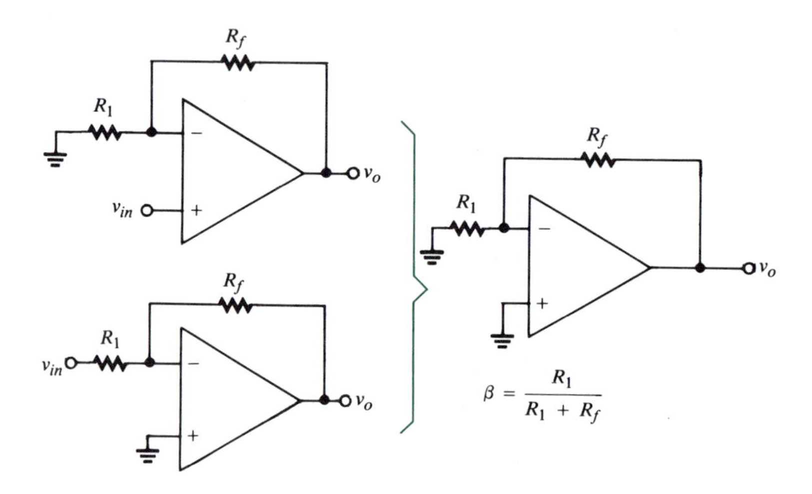

Figure 13–10
When v_{in} is grounded in both the inverting and noninverting amplifiers, it can be seen that the feedback paths are identical.

we can analyze the contribution of the feedback source by grounding all other signal sources. When this is done, as shown in Figure 13–10, we see that the feedback voltage in both configurations is developed across the R_1–R_f voltage divider, and $\beta = R_1/(R_1 + R_f)$ in both cases. In view of this fact, we can write (13–28) as

$$\frac{v_o}{v_{in}} = \frac{-R_f/R_1}{1 + 1/A\beta} \tag{13–29}$$

Toward developing a feedback model for the inverting amplifier, consider the block diagram shown in Figure 13–11. This diagram is quite similar to Figure 13–8(b) for the noninverting amplifier, except that we now denote the open loop gain by $-A$. Note also that the summing junction now *adds* its two inputs. Since v_o is inverted, so is the feedback voltage, and adding a negative voltage to v is the same as subtracting a positive one from it. In other words, we still have a negative-feedback situation. Notice that we use v to represent an arbitrary input voltage, rather than v_{in}, because we will have to make some adjustments in this model before it can truly represent the inverting amplifier.

As shown in Figure 13–11,

$$v_o = -A(v + \beta v_o) \tag{13–30}$$

Solving for v_o/v, we find

$$\frac{v_o}{v} = \frac{-A}{1 + A\beta} = \frac{-1/\beta}{1 + 1/A\beta} \tag{13–31}$$

Comparing (13–31) with the equation we have already developed for the inverting amplifier (equation 13–29), we see that they differ slightly. We must therefore adjust the model so that it produces the same result as (13–29). Equation 13–31 for the model can be written

$$\frac{v_o}{v} = \frac{\dfrac{-(R_1 + R_f)}{R_1}}{1 + 1/A\beta} \tag{13–32}$$

If the right side of (13–32) is multiplied by the factor $R_f/(R_1 + R_f)$, we obtain

$$\frac{v_o}{v} = \left[\frac{\dfrac{-(R_1 + R_f)}{R_1}}{1 + 1/A\beta}\right]\frac{R_f}{(R_1 + R_f)} = \frac{-R_f/R_1}{1 + 1/A\beta} \tag{13–33}$$

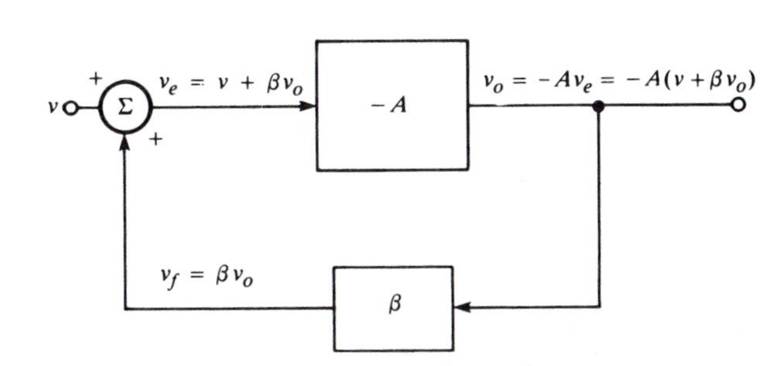

Figure 13–11
First step in the development of a feedback model for the inverting amplifier

Figure 13–12
The complete feedback model for the inverting amplifier

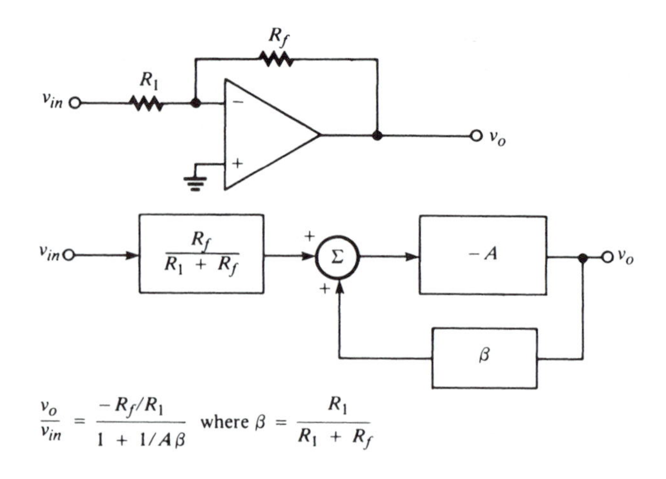

Equation 13–33 shows that multiplication of the model equation by the constant $R_f/(R_1 + R_f)$ gives us exactly the same result (equation 13–29 with $v_{in} = v$) that we obtain for the inverting amplifier. Therefore, we modify the block-diagram model in Figure 13–11 by adding a block that multiplies the input by $R_f/(R_1 + R_f)$. The complete feedback model is shown in Figure 13–12.

As can be seen in Figure 13–12, the loop gain for the inverting amplifier is $A\beta$, the same as that for the noninverting amplifier. From equation 13–29, it is apparent that the greater the loop gain, the closer the closed-loop gain is to its value in the ideal inverting amplifier, $-R_f/R_1$.

It can be shown that the input resistance seen by the signal source driving the inverting amplifier is

$$r_{in} = R_1 + \frac{R_f}{1 + A} \approx R_1 \qquad \textbf{(13–34)}$$

This equation confirms that the input resistance is R_1 for the ideal inverting amplifier, where $A = \infty$. It also shows that the input resistance decreases with increasing values of A.

As with the noninverting amplifier, the output resistance of the inverting amplifier is decreased by negative feedback. In fact, the relationship between output resistance and loop gain is the same for both:

$$r_o(\text{stage}) = \frac{r_o}{1 + A\beta} \approx \frac{r_o}{A\beta} \qquad \textbf{(13–35)}$$

Example 13–5

The amplifier shown in Figure 13–13 has open-loop gain equal to −2500 and open-loop output resistance 100 Ω. Find

1. the magnitude of the loop gain
2. the closed-loop gain;
3. the input resistance seen by v_{in}; and
4. the closed-loop output resistance.

Figure 13–13
(Example 13–5)

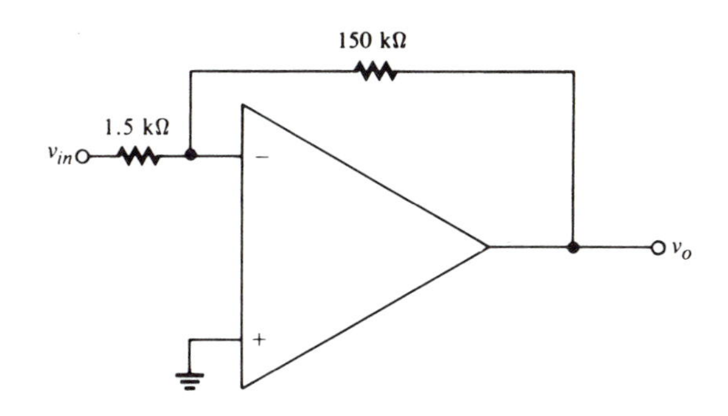

Solution

1. $\beta = R_1/(R_1 + R_f) = (1.5\text{ k}\Omega)/[(1.5\text{ k}\Omega) + (150\text{ k}\Omega)] = 9.90099 \times 10^{-3}$
loop gain $= A\beta = (2.5 \times 10^3)(9.90099 \times 10^{-3}) = 24.75$

2. From equation 13–29,

$$\frac{v_o}{v_{in}} = \frac{-R_f/R_1}{1 + 1/A\beta} = \frac{-(150\text{ k}\Omega)/(1.5\text{ k}\Omega)}{1 + 1/24.75} = -96.12$$

Note that this value is about 4% less than $-R_f/R_1 = -100$.

3. From equation 13–34,

$$r_{in} = R_1 + \frac{R_f}{1 + A} = (1.5\text{ k}\Omega) + \frac{150\ \Omega}{1 + 2500} = 1560\ \Omega$$

4. From equation 13–35,

$$r_o(\text{stage}) = \frac{r_o}{1 + A\beta} = \frac{100}{1 + 24.75} = 3.88\ \Omega$$

Example 13–6

SPICE

Use SPICE to find the closed-loop voltage gain, the input resistance seen by v_{in}, and the output resistance of the inverting amplifier in Example 13–5.

Solution. Figure 13–14 shows how we can use a voltage-controlled voltage source (EOP) to model an operational amplifier in SPICE. Notice that the inverting property of the amplifier is realized by connecting the positive terminal (N+) of EOP to ground (node 0). The voltage source is controlled by the voltage between nodes 2 and 0 (NC+ and NC−, respectively). Thus, node 2 corresponds to the inverting input of the amplifier in Figure 13–13. Notice that the simulated amplifier has infinite input impedance, since there is an open circuit between nodes 2 and 0.

Since VIN = 1 V, the output voltage at node 4, V(4), is numerically equal to the closed-loop voltage gain. The results of a program run reveal that V(4) = −96.11, in close agreement with the gain calculated in Example 13–5. SPICE computes the magnitude of I(VIN) to be 0.641 mA, so the input resistance is

$$r_{in} = \frac{v_o}{i_{in}} = \frac{\text{VIN}}{|\text{I(VIN)}|} = \frac{1\text{ V}}{0.641\text{ mA}} = 1560\ \Omega$$

Figure 13–14
(Example 13–6)

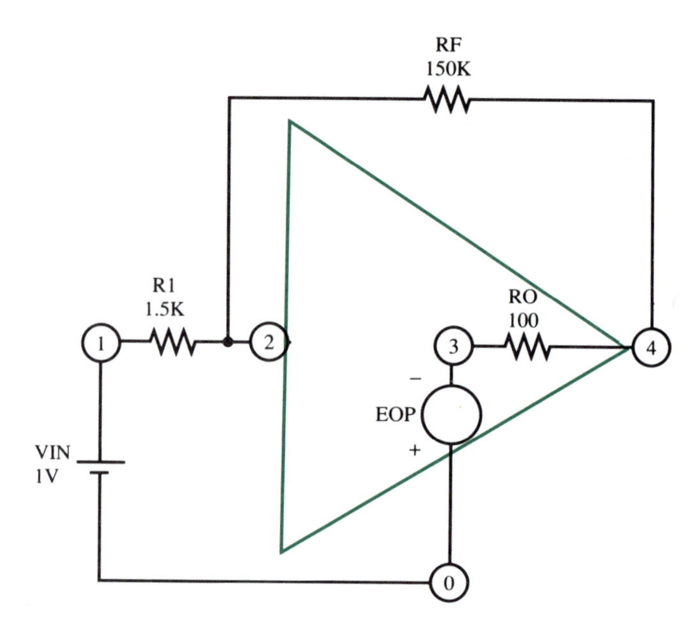


```
EXAMPLE 13.6
VIN 1 0 1V
R1 1 2 1.5K
RF 2 4 150K
EOP 0 3 2 0 2500
RO 3 4 100
.DC VIN 1 1 1
.PRINT DC V(4) I(VIN)
.END
```

To find the output resistance of the amplifier, we must compute the current that flows through a short circuit connected to the output, since $r_o = v_L$(open circuit)/i_L(short circuit). A zero-valued dummy voltage source, VDUM, connected between nodes 4 and 0 effectively shorts the output to ground. The current in VDUM is then the short-circuit output current. SPICE computes this value to be I(VDUM) = 24.7 A. Thus,

$$r_o = \frac{\text{V(4) (open circuit)}}{\text{I(VDUM) (short circuit)}} = \frac{96.11 \text{ V}}{24.7 \text{ A}} = 3.89 \ \Omega$$

This value is in close agreement with that calculated in Example 13–5.

In closing our discussion of feedback theory, we should note once again that the same relationship between actual and ideal closed-loop gain applies to inverting and noninverting amplifiers. This relationship is

$$\text{actual } \frac{v_o}{v_{in}} = \frac{\text{(ideal closed-loop gain)}}{1 + 1/A\beta} \tag{13–36}$$

where (ideal closed-loop gain) is the closed-loop gain v_o/v_{in} that would result if the amplifier were ideal. We saw this relationship in equations 13–21 and 13–29, repeated here:

$$\frac{v_o}{v_{in}} = \frac{(R_1 + R_f)/R_1}{1 + 1/A\beta} \qquad \text{(noninverting amplifier)}$$

$$\frac{v_o}{v_{in}} = \frac{-R_f/R_1}{1 + 1/A\beta} \qquad \text{(inverting amplifier)}$$

In both cases, the numerator is the closed-loop gain that would result if the amplifier were ideal. Also in both cases, the greater the value of the loop gain $A\beta$, the closer the actual closed-loop gain is to the ideal closed-loop gain.

Although we have demonstrated this relationship only for inverting and noninverting amplifiers, it is a fact that equation 13–36 applies to a wide variety of amplifier configurations, many of which we shall be examining in future discussions.

13–3 FREQUENCY RESPONSE

Stability

When the word *stability* is used in connection with a high-gain amplifier, it usually means the property of behaving like an amplifier rather than like an *oscillator.* An oscillator is a device that spontaneously generates an ac signal because of *positive* feedback. We will study oscillator theory in Chapter 14, but for now it is sufficient to know that oscillations are easily induced in high-gain, wide-bandwidth amplifiers, due to positive feedback that occurs through reactive elements. As we know, an operational amplifier has very high gain, so precautions must be taken in its design to ensure that it does not oscillate, i.e., to ensure that it remains stable. Large gains at high frequencies tend to make an amplifier unstable because those properties enable positive feedback through stray capacitance.

To ensure stable operation, most operational amplifiers have internal *compensation* circuitry that causes the open-loop gain to diminish with increasing frequency. This reduction in gain is called "rolling-off" the amplifier. Sometimes it is necessary to connect external roll-off networks to reduce high-frequency gain even more rapidly. Because the dc and low-frequency open-loop gain of an operational amplifier is so great, the gain roll-off must begin at relatively low frequencies. As a consequence, the open-loop bandwidth of an operational amplifier is generally rather small.

In most operational amplifiers, the gain over the usable frequency range rolls off at the rate of −20 dB/decade, or −6 dB/octave. Recall from Chapter 10 that this rate of gain reduction is the same as that of a single low-pass RC network. Any device whose gain falls off like that of a single RC network is said to have a *single-pole* frequency response, a name derived from advanced mathematical theory. Beyond a certain very high frequency, the frequency response of an operational amplifier exhibits further break frequencies, meaning that the gain falls off at greater rates, but for most practical analysis purposes we can treat the amplifier as if it had a single-pole response.

The Gain-Bandwidth Product

Figure 13–15 shows a typical frequency response characteristic for the open-loop gain of an operational amplifier having a single-pole frequency response, plotted on log-log scales. We use f_c to denote the cutoff frequency, which, as usual, is the

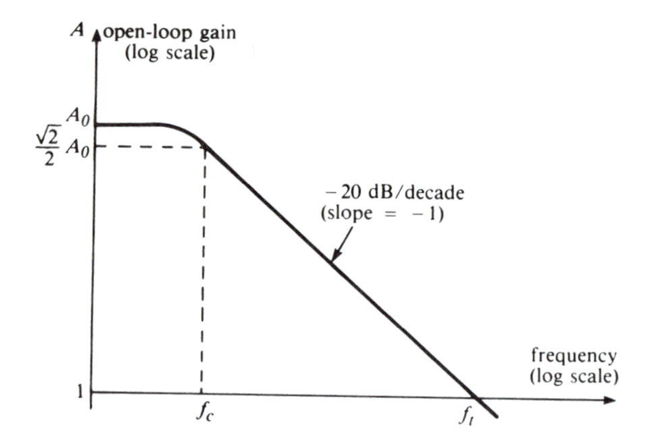

Figure 13–15
Frequency response of the open-loop gain of an operational amplifier; A_0 = dc gain, f_c = cutoff frequency, f_t = unity-gain frequency

frequency at which the gain A falls to $\sqrt{2}/2$ times its low-frequency, or dc, value (A_0). Recall that the slope of the single-pole response is -1. In Chapter 10, we studied a very similar frequency response: that of the beta of a bipolar junction transistor. Using the fact that the slope is -1, we showed that the frequency f_T at which the β falls to the value 1 (unity) is given by $f_T = \beta_m f_\beta$, where β_m is the low-frequency β and f_β is the β cutoff frequency. Using exactly the same approach, we can deduce that *the frequency at which the amplifier gain falls to value 1 equals the product of the cutoff frequency and the low-frequency gain* A_0:

$$f_t = A_0 f_c \tag{13–37}$$

where f_t = the *unity-gain frequency,* the frequency at which the gain equals 1
A_0 = the low-frequency, or dc, value of the open-loop gain
f_c = the cutoff frequency, or 3-dB frequency, of the open-loop gain

Since the amplifier is dc (lower cutoff frequency = 0), the bandwidth equals f_c. The term $A_0 f_c$ in equation 13–37 is called the *gain-bandwidth* product. In specifications, a value may be quoted either for the gain-bandwidth product, or for its equivalent, the unity-gain frequency.

The significance of the gain-bandwidth product is that it makes it possible for us to compute the bandwidth of an amplifier when it is operated in one of the more useful closed-loop configurations. Obviously, a knowledge of the upper-frequency limitation of a certain amplifier configuration is vital information when designing it for a particular application. The relationship between closed-loop bandwidth (BW_{CL}) and the gain-bandwidth product is closely approximated by

$$BW_{CL} = f_t\beta = A_0 f_c \beta \tag{13–38}$$

where β is the feedback ratio.

Example 13–7

Each of the amplifiers shown in Figure 13–16 has an open-loop, gain-bandwidth product equal to 1×10^6. Find the cutoff frequencies in the closed-loop configurations shown.

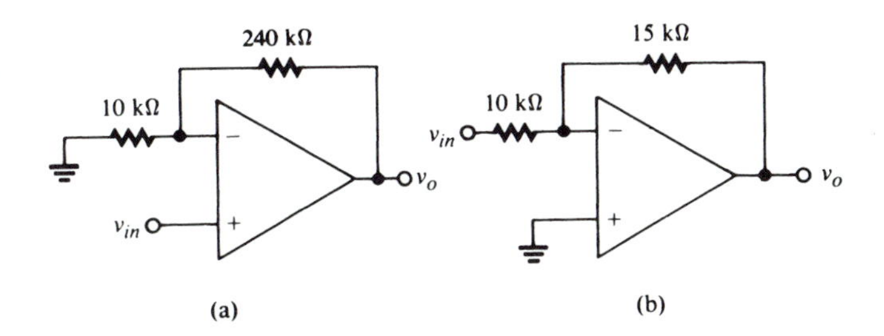

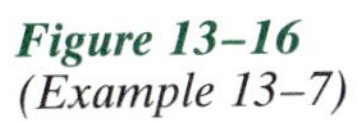

Figure 13–16
(Example 13–7)

Solution

1. In Figure 13–16(a), $\beta = R_1/(R_1 + R_f) = (10\ \text{k}\Omega)/[(10\ \text{k}\Omega) + (240\ \text{k}\Omega)] = 0.04$. From equation 13–38, $\text{BW}_{\text{CL}} = f_t\beta = (10^6)(0.04) = 40$ kHz.
   Since the amplifier is dc, the closed-loop cutoff frequency is the same as the closed-loop bandwidth, 40 kHz.
2. In Figure 13–16(b), $\beta = R_1/(R_1 + R_f) = (10\ \text{k}\Omega)/[(10\ \text{k}\Omega) + (15\ \text{k}\Omega)] = 0.4$. Then $\text{BW}_{\text{CL}} = 10^6(0.4) = 400$ kHz.

It is worthwhile noting that in the case of the *noninverting* amplifier, the fact that the ideal closed-loop gain is $1/\beta$ makes equation 13–38 equivalent to

$$\text{BW}_{\text{CL}} = f_t/(\text{ideal closed-loop gain}) \qquad \textbf{(13–39)}$$

or (ideal closed-loop gain) × (closed-loop bandwidth) = gain-bandwidth product. To illustrate the validity of this expression, refer to part 1 of Example 13–7. Here, the ideal closed-loop gain is $(R_1 + R_f)/R_1 = (250\ \text{k}\Omega)/(10\ \text{k}\Omega) = 25$, so 25 × (closed-loop bandwidth) = 10^6, which yields

$$\text{closed-loop bandwidth} = \text{BW}_{\text{CL}} = 10^6/25 = 40\ \text{kHz} \qquad \text{(correct)}$$

Equation 13–39 is *not* valid for the inverting amplifier. In part 2 of Example 13–7, we have

$$\text{ideal closed-loop gain} = R_f/R_1 = (15\ \text{k}\Omega)/(10\ \text{k}\Omega) = 1.5$$

If we now apply equation 13–39, we obtain

$$\text{BW}_{\text{CL}} = 10^6/1.5 = 666.6\ \text{kHz} \qquad \text{(incorrect)}$$

Although some authors interpret the gain-bandwidth product to be the product of closed-loop gain and closed-loop bandwidth regardless of configuration, we have seen that this interpretation yields a bandwidth for the inverting amplifier that is larger than its actual value. At large values of closed-loop gain, the bandwidths of the inverting and noninverting amplifiers are comparable, but at low gains the noninverting amplifier has a larger bandwidth. For example, when the closed-loop gain is 1, the bandwidth of the noninverting amplifier is twice that of the inverting amplifier.

Figure 13–17 shows a typical set of frequency response plots for a noninverting amplifier, as the gain ranges from its open-loop value of 10^5 to a closed-loop value of 1. This figure clearly shows how the bandwidth decreases as the closed-looped gain increases. Notice that the bandwidth at maximum (open-loop) gain is only 10 Hz.

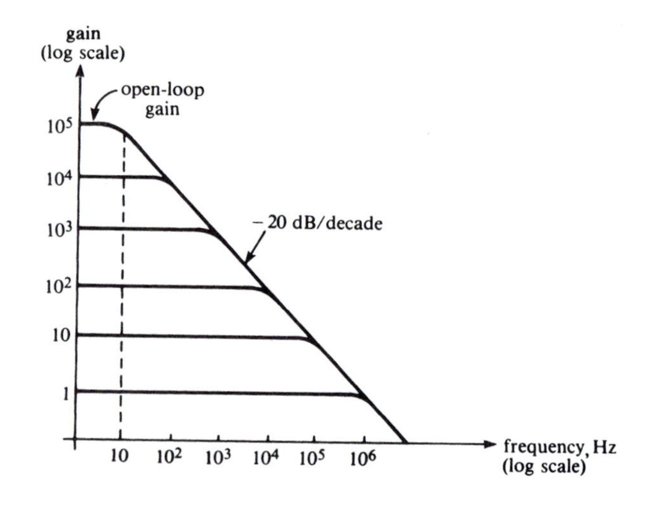

Figure 13–17
A typical set of frequency response plots for a noninverting amplifier

Example 13–8

With reference to the amplifier whose frequency response is shown in Figure 13–17, find

1. the unity-gain frequency;
2. the gain-bandwidth product;
3. the bandwidth when the feedback ratio is 0.02; and
4. the closed-loop gain at 0.4 MHz when the feedback ratio is 0.04.

Solution

1. In Figure 13–17, it is apparent that the open-loop gain equals 1 when the frequency is 1 MHz. Thus, $f_t = 1$ MHz.
2. Gain-bandwidth product $= A_0 f_c = f_t = 10^6$.
3. From equation 13–38, $BW_{CL} = f_t\beta = 10^6(0.02) = 20$ kHz.
4. $BW_{CL} = f_t\beta = 10^6(0.04) = 40$ kHz. Thus, the closed-loop cutoff frequency is 40 kHz. Since the amplifier is noninverting, the closed-loop gain is $1/\beta = 25$. Since 0.4 MHz is 1 decade above the cutoff frequency, the gain is down 20 dB from 25, or down by a factor of 1/10: $0.1(25) = 2.5$.

User-Compensated Amplifiers

As noted earlier, many commercially available amplifiers have internal compensation circuitry to make the frequency response roll off at 6 dB/octave (20 dB/decade) over the entire frequency range from f_c to f_t (Figure 13–15). Some amplifiers do not have such circuitry and must be compensated by connecting external roll-off networks. These networks, typically RC circuits, are selected by the user to ensure that the frequency response is 6 dB/octave at the closed-loop gain at which the amplifier is to be operated. Manufacturers' specifications usually include equations for determining the values of the components of external roll-off networks, based on the closed-loop gain desired.

Figure 13–18(a) shows a typical frequency response of an uncompensated amplifier. Compensation is particularly critical when the amplifier is to be operated with

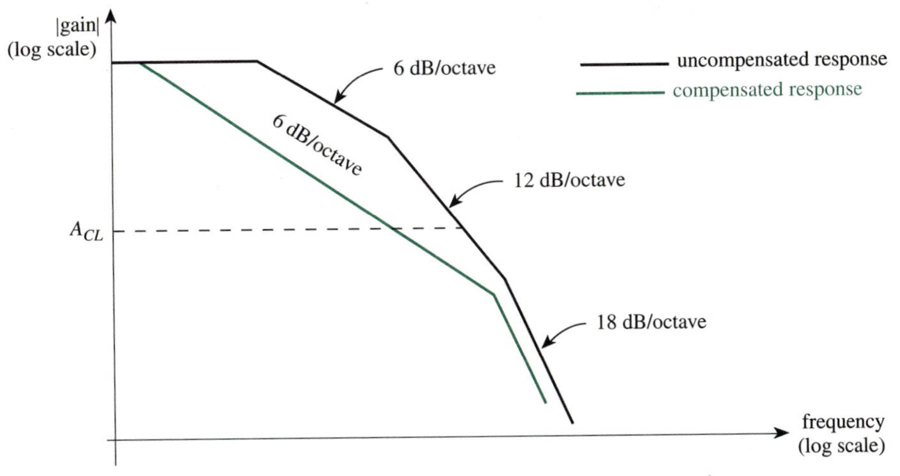

(a) Typical uncompensated and compensated frequency responses. The compensation ensures that the roll-off rate is 6 dB/octave at a desired closed-loop gain, A_{CL}.

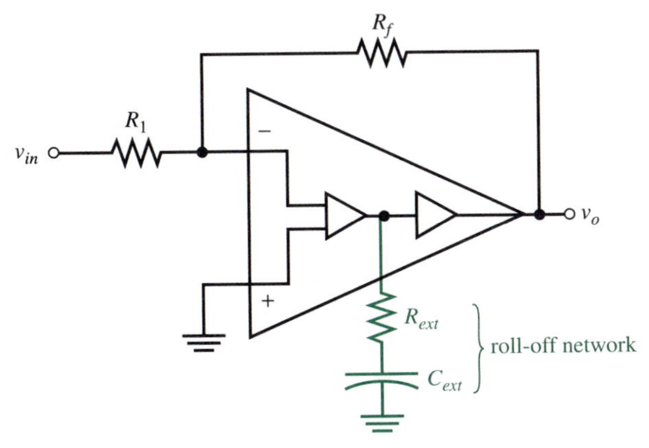

(b) Typical external roll-off network, called *lag phase compensation*, and used to create a 6 dB/octave roll-off

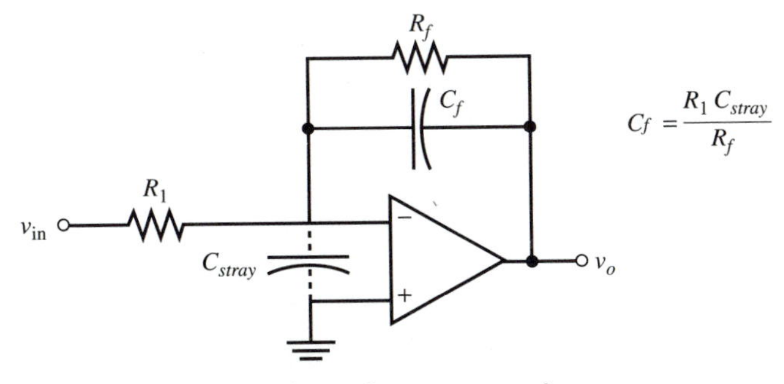

$$C_f = \frac{R_1 C_{stray}}{R_f}$$

(c) Use of feedback capacitance C_f to compensate for shunt capacitance at the input (lead compensation)

Figure 13–18
User-compensated amplifiers

a small closed-loop gain, since the bandwidth is then very large and the amplifier's roll-off rate may be 12 or 18 dB/octave, rates that jeopardize stability. Figure 13–18(a) also shows an example of a response that has been compensated to roll off at 6 dB/octave when a particular value of closed-loop gain, A_{CL}, is desired. Note that the roll-off rate would be 12 dB/octave if the uncompensated amplifier were used at that value of closed-loop gain. It is clear that compensation reduces the bandwidth of the amplifier. However, it is generally true that user compensation results in a wider bandwidth than can be achieved with an internally compensated amplifier that rolls off at 6 dB/octave over its entire range. Figure 13–18(b) shows a typical RC network used for external compensation. Note that this network is usually connected to an internal stage of the amplifier at an external terminal that may be identified as "roll-off," "phase," or "frequency compensation."

The external compensation shown in Figure 13–18(b) is called *lag phase* compensation. Figure 13–18(c) shows an example of *lead* compensation, used to offset the effects of input and stray capacitance. The feedback capacitor C_f is selected so that the break frequency due to the combination of R_f and the input shunt capacitance equals the break frequency due to R_f and C_f:

$$\frac{1}{2\pi R_1 C_{stray}} = \frac{1}{2\pi R_f C_f}$$

$$C_f = \frac{R_1 C_{stray}}{R_f} \qquad \textbf{(13–40)}$$

13–4 SLEW RATE

We have discussed the fact that internal compensation circuitry used to ensure amplifier stability also affects the frequency response and places a limit on the maximum operating frequency. The capacitor(s) in this compensation circuitry limit amplifier performance in still another way. When the amplifier is driven by a step or pulse-type signal, the capacitance must charge and discharge rapidly in order for the output to "keep up with," or track, the input. Since the voltage across a capacitor cannot be changed instantaneously, there is an inherent limit on the *rate* at which the output voltage can change. The maximum possible rate at which an amplifier's output voltage can change, in volts per second, is called its *slew rate.*

It is not possible for *any* waveform, input or output, to change from one level to another in *zero* time. An instantaneous change corresponds to an *infinite rate of change,* which is not realizable in any physical system. Therefore, in our investigation of performance limitations imposed by an amplifier's slew rate, we need concern ourselves only with inputs that undergo a total change in voltage, ΔV, over some nonzero time interval, Δt. For simplicity, we will assume that the change is linear with respect to time; that is, it is a *ramp*-type waveform, as illustrated in Figure 13–19. The rate of change of this kind of waveform is the change in voltage divided by the length of time that it takes for the change to occur:

$$\text{rate of change} = \frac{V_2 - V_1}{t_2 - t_1} = \frac{\Delta V}{\Delta t} \text{ volts/second} \qquad \textbf{(13–41)}$$

Since the value specified for the slew rate of an amplifier is the maximum rate at which its output can change, we cannot drive the amplifier with any kind of input

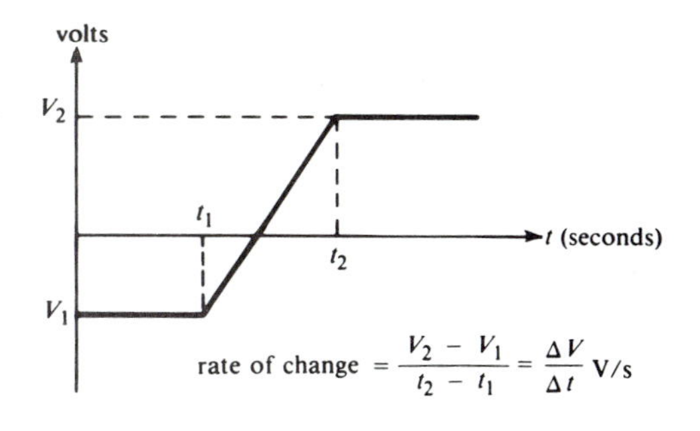

Figure 13–19
The rate of change of a linear, or ramp, signal is the change in voltage divided by the change in time.

waveform that would require the output to exceed that rate. For example, if the slew rate is 10^6 V/s (a typical value), we could not drive an amplifier having unity gain with a signal that changes from −5 V to +5 V in 0.1 μs, because that would require the output to change at the rate $\Delta V/\Delta t = (10\text{ V})/(10^{-7}\text{ s}) = 10^8$ V/s. Similarly, we could not drive an amplifier having a gain of 10 with an input that changes from 0 V to 1 V in 1 μs because that would require the output to change from 0 V to 10 V in 1 μs, giving $\Delta V/\Delta t = 10/10^{-6} = 10^7$ V/s. When we say we "could not" drive the amplifier with these inputs, we simply mean that we could not do so and still expect the output to be a faithful replica of the input.

In specifications, the slew rate is often quoted in the units volts per microsecond. Of course, 1 V/μs is the same as 10^6 V/s: $(1\text{ V})/(10^{-6}\text{ s}) = 10^6$ V/s.

Example 13–9

The operational amplifier in Figure 13–20 has a slew rate specification of 0.5 V/μs. If the input is the ramp waveform shown, what is the maximum closed-loop gain that the amplifier can have without exceeding its slew rate?

Solution. The rate of change of the input is

$$\frac{\Delta V}{\Delta t} = \frac{V_2 - V_1}{t_2 - t_1} = \frac{0.6\text{ V} - (-0.2)\text{V}}{(40 - 20) \times 10^{-6}\text{ s}} = 4 \times 10^4\text{ V/s}$$

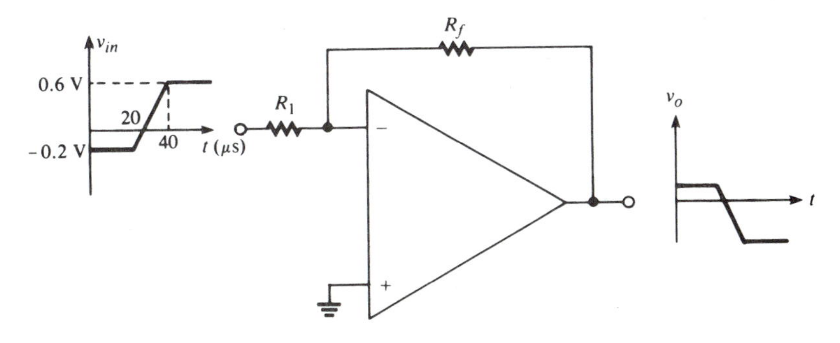

Figure 13–20
(Example 13–9)

Since the slew rate is 0.5 V/μs = 5×10^5 V/s, the maximum permissible gain is

$$\frac{5 \times 10^5 \text{ V}}{4 \times 10^4 \text{ V}} = 12.5$$

Notice that the amplifier is connected in an inverting configuration, so the output changes from positive to negative. The inversion is of no consequence as far as slew rate is concerned. With a gain of -12.5, the output will change from $(-12.5)(-0.2 \text{ V}) = +2.5$ V to $(-12.5)(0.6 \text{ V}) = -7.5$ V in 20 μs, giving

$$\frac{\Delta V}{\Delta t} = \frac{10 \text{ V}}{20 \ \mu\text{s}} = 0.5 \text{ V}/\mu\text{s}$$

the specified slew rate.

Slew rate is a performance specification used primarily in applications where the waveforms are large-signal pulses or steps that cause the output to swing through a substantial part of its total range ($\pm V_{CC}$ volts). *However,* the slew rate imposes a limitation on output rate of change regardless of the nature of the signal waveform. In particular, if the signal is sinusoidal, or a complex waveform containing many different frequencies, we must be certain that no large-amplitude, high-frequency component will require the output to exceed the slew rate. High-frequency signals change (continuously) at rapid rates, and if their amplitudes are so large that the slew rate specification is exceeded, distortion will result. It is especially important to realize that a frequency component may be within the bandwidth of the amplifier, as determined in Section 13–3, but may have such a large amplitude that it must be excluded because of slew rate limitations. The converse is also true: A high-frequency signal that does not exceed the slew rate may have to be excluded because it is outside the amplifier bandwidth. In other words, the maximum frequency at which an amplifier can be operated depends on both the bandwidth and the slew rate, the latter being a function of amplitude as well as frequency. In a later discussion, we will summarize the criteria for determining the operating frequency range of an amplifier based on both slew rate and bandwidth limitations.

When the output of an amplifier is the sine-wave voltage $v_o(t) = K \sin \omega t$, it can be shown using calculus (differentiating with respect to t) that the signal has a maximum rate of change given by

$$\text{rate of change (max)} = K\omega \text{ volts/second} \qquad \textbf{(13–42)}$$

where K is the peak amplitude of the sine wave, in volts, and ω is the angular frequency, in radians/second. (We use K to represent amplitude, rather than the conventional A, to avoid confusion with the symbol for gain.) Equation 13–42 clearly shows that the rate of change is proportional to both the amplitude and the frequency of the signal. If S is the specified slew rate of an amplifier, then we must have

$$K\omega \le S \quad \text{or} \quad K(2\pi f) \le S \qquad \textbf{(13–43)}$$

This inequality allows us to solve for the maximum frequency, f_S(max), that the slew-rate limitation permits at the output of an amplifier:

$$f_S(\text{max}) = \frac{S}{2\pi K} \text{ hertz} \quad \text{or} \quad \omega_S(\text{max}) = \frac{S}{K} \text{ radian/second} \qquad \textbf{(13–44)}$$

We emphasize again that $f_S(\text{max})$ is the frequency limit imposed by the slew rate *alone,* i.e., disregarding bandwidth limitations. Also, equation 13–44 applies to sinusoidal signals only. When dealing with complex waveforms containing many different frequency components, the slew rate should be at *least* as great as that necessary to satisfy equation 13–44 for the highest frequency component. Depending on phase relations, maximum rates of change may actually be additive.

Example 13–10

The operational amplifier in Figure 13–21 has a slew rate of 0.5 V/μs. The amplifier must be capable of amplifying the following input signals: $v_1 = 0.01 \sin(10^6 t)$, $v_2 = 0.05 \sin(350 \times 10^3 t)$, $v_3 = 0.1 \sin(200 \times 10^3 t)$, and $v_4 = 0.2 \sin(50 \times 10^3 t)$.

1. Determine whether the output will be distorted due to slew-rate limitations on any input.
2. If so, find a remedy (other than changing the input signals).

Solution

1. We must check each frequency to verify that $\omega \le \omega_S(\text{max}) = S/K$ rad/s. Note that K is the peak amplitude at the *output* of the amplifier, so each input amplitude must be multiplied by the closed-loop gain before the check is performed. Assuming that the closed-loop gain is ideal, we have $v_o/v_{in} = -R_f/R_1 = -(330\text{ k}\Omega)/(10\text{ k}\Omega) = -33$. Thus, the upper limit on the ω of each signal component will be $S/(33\,K_i)$, where K_i is the peak amplitude of the input.

$$v_1\text{:}\quad \omega_S(\text{max}) = S/(33K_i) = 0.5 \times 10^6/(33)(0.01) = 1.515 \times 10^6$$
$$\omega = 10^6 < 1.515 \times 10^6 \qquad \text{(ok)}$$
$$v_2\text{:}\quad \omega_S(\text{max}) = S/(33K_i) = 0.5 \times 10^6/(33)(0.05) = 303.03 \times 10^3$$
$$\omega = 350 \times 10^3 > 303.03 \times 10^3 \qquad \text{(not ok)}$$
$$v_3\text{:}\quad \omega_S(\text{max}) = S/(33K_i) = 0.5 \times 10^6/(33)(0.1) = 151.5 \times 10^3$$
$$\omega = 200 \times 10^3 > 151.5 \times 10^3 \qquad \text{(not ok)}$$
$$v_4\text{:}\quad \omega_S(\text{max}) = S/(33K_i) = 0.5 \times 10^6/(33)(0.2) = 75.75 \times 10^3$$
$$\omega = 50 \times 10^3 < 75.75 \times 10^3 \qquad \text{(ok)}$$

   We see that v_2 and v_3 would both cause the slew rate specification of the amplifier to be exceeded. Consequently, the output will be distorted.
2. Since we cannot change the input signal amplitudes or frequencies, there are only two remedies: (a) find an amplifier with a greater slew rate, or (b) reduce the closed-loop gain of the present amplifier. We will investigate both remedies.

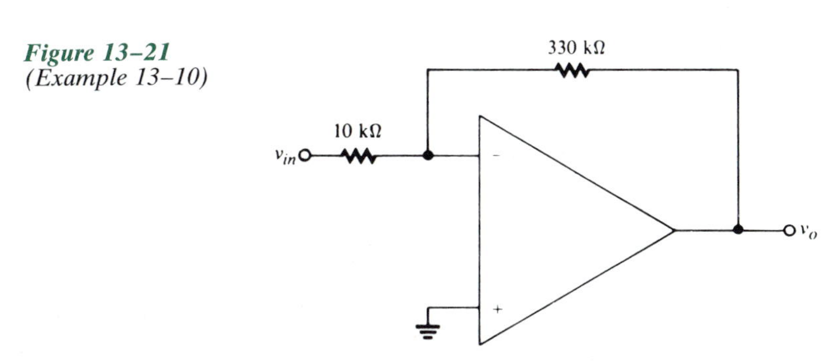

Figure 13–21
(Example 13–10)

a. The slew rate of a new amplifier must satisfy *both* $S/(33)(0.05) \geq 350 \times 10^3$ (for v_2) and $S/(33)(0.1) \geq 200 \times 10^3$ (for v_3). These inequalities are equivalent to

$$S \geq 0.5775 \times 10^6 \text{ V/s} \quad \text{and} \quad S \geq 0.66 \times 10^6 \text{ V/s}$$

Therefore, we must use an amplifier with a slew rate of at least 0.66×10^6 V/s, or 0.66 V/μs.

b. If we use the present amplifier, we must reduce its closed-loop gain G so that it satisfies *both* $0.5 \times 10^6/0.05G \geq 350 \times 10^3$ (for v_2) and $0.5 \times 10^6/0.1G \geq 200 \times 10^3$ (for v_3). These inequalities are equivalent to

$$G \leq 28.57 \quad \text{and} \quad G \leq 25$$

Therefore, the maximum closed-loop gain is 25. This limit can be achieved by changing the 330-kΩ resistor in Figure 13–21 to 250 kΩ.

To ensure that an operational-amplifier circuit will not distort a signal component having frequency f, we require that *both* of the following conditions be satisfied:

$$f \leq \text{BW}_{\text{CL}} \qquad \textbf{(13–45)}$$

$$f \leq S/2\pi K \qquad \textbf{(13–46)}$$

If the signal is a complex waveform containing multiple frequency components, the highest frequency in the signal should satisfy both conditions. Note that both conditions depend on closed-loop gain: large gains reduce BW_{CL} and increase the value of K, so the greater the closed-loop gain, the more severe the restrictions.

Example 13–11

The operational amplifier in Figure 13–22 has a unity-gain frequency of 1 MHz and a slew rate of 1 V/μs. Find the maximum frequency of a 0.1-V-peak sine-wave input that can be amplified without distortion.

Solution. The feedback ratio is

$$\beta = \frac{R_1}{R_1 + R_f} = \frac{10 \text{ k}\Omega}{(10 \text{ k}\Omega) + (490 \text{ k}\Omega)} = 0.02$$

By equation 13–38, $\text{BW}_{\text{CL}} = f_t\beta = (1 \text{ MHz})(0.02) = 20$ kHz. The closed-loop gain for the noninverting configuration is $v_o/v_{in} = 1/\beta = 1/0.02 = 50$. Therefore, the

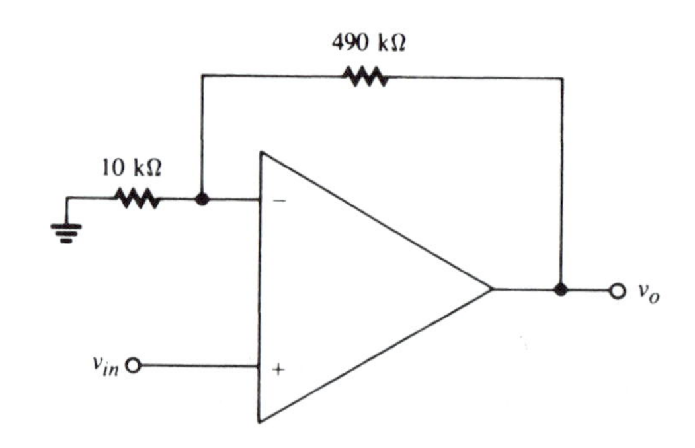

Figure 13–22
(Example 13–11)

peak value of the output is $K = 50(0.1) = 5$ V. Then $f_S(\max) = S/2\pi K = 10^6/(2\pi)(5) = 31.83$ kHz. Since we require that f satisfy both $f \leq 20$ kHz and $f \leq 31.83$ kHz, we see that the maximum permissible frequency is 20 kHz. In this case, the bandwidth sets the upper limit.

Example 13–12

DESIGN

1. Derive a design equation that imposes a limit on the value of R_f in Figure 13–23, based on bandwidth and slew-rate limitations of the amplifier. The known values that can be used in the equation are R_1, slew rate S, sinusoidal input frequency f, unity-gain frequency f_t, and peak value of the input, $V_{in}(\text{pk})$.
2. Use the design equation to find the limit on R_f when the input to the amplifier is a 0.5-V-pk sine wave with frequency 5 kHz, $R_1 = 10$ kΩ, $S = 10^6$ V/s, and $f_t = 1$ MHz.

Solution

1. Since the amplifier is in an inverting configuration, the magnitude of the closed-loop gain, G, and the feedback ratio, β, are

$$G = \frac{R_f}{R_1} \qquad \beta = \frac{R_1}{R_1 + R_f}$$

Since the input frequency, f, must be less than the closed-loop bandwidth, we have, from equation 13–38,

$$f < f_t\beta = f_t\left(\frac{R_1}{R_1 + R_f}\right)$$

Solving for R_f leads to

$$R_f < \frac{R_1(f_t - f)}{f}$$

The peak value of the output is

$$K = GV_{in}(\text{pk}) = \frac{R_f}{R_1}V_{in}(\text{pk})$$

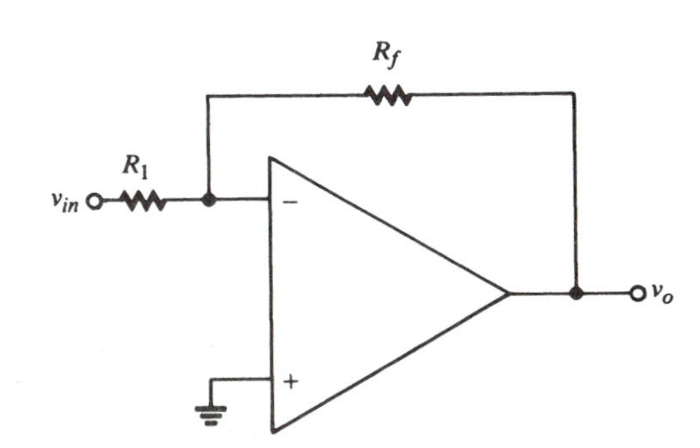

Figure 13–23
(Example 13–12)

By equation 13–44,

$$f < \frac{S}{2\pi(R_f/R_1)V_{in}(\text{pk})}$$

Solving for R_f gives

$$R_f < \frac{R_1 S}{2\pi f V_{in}(\text{pk})}$$

Since R_f must be less than both limits we have found, it must be less than the smaller of the two:

$$R_f < \min\left\{\frac{R_1(f_t - f)}{f}, \frac{R_1 S}{2\pi f V_{in}(\text{pk})}\right\}$$

2. $R_f < \min\left\{\dfrac{10^4\ \Omega(10^6\ \text{Hz} - 5 \times 10^3\ \text{Hz})}{5 \times 10^3\ \text{Hz}}, \dfrac{(10^4\ \Omega)(10^6\ \text{V/s})}{2\pi(5 \times 10^3\ \text{Hz})(0.5\ \text{V})}\right\}$

$R_f < \min\{1.99\ \text{M}\Omega, 636.6\ \text{k}\Omega\}$

$R_f < 636.6\ \text{k}\Omega$

If we choose the closest standard-value resistor that is less than 636.6 kΩ, we have $R_f = 620\ \text{k}\Omega$.

Another practical limit that is not considered in the foregoing is the maximum permissible output voltage of the amplifier. If we used $R_f = 620\ \text{k}\Omega$ in the present example, then the closed-loop gain would be $R_f/R_1 = 620\ \text{k}\Omega/10\ \text{k}\Omega = 62$, and the peak output voltage would be 62 (0.5 V pk) = 31 V pk, which is too large for many commercially available amplifiers.

We have seen that an amplifier's slew rate affects its ability to track, or follow, a rapidly changing input pulse. When the output voltage must change through ΔV volts, the minimum possible time in which that change can occur is

$$\Delta t = \frac{\Delta V}{S} \text{ seconds} \qquad \textbf{(13–47)}$$

where ΔV is the total *output* voltage change. In terms of input quantities, the minimum time allowed for an input voltage change of ΔV_{in} volts is

$$\Delta t = \frac{(A_{CL})\Delta V_{in}}{S} \text{ seconds} \qquad \textbf{(13–48)}$$

where A_{CL} is the closed-loop gain.

An amplifier's bandwidth also affects the time required for its output to change in response to a pulse input. Recall from Chapter 10 that the *rise time* t_r of a single-pole system is

$$t_r = \frac{0.35}{\text{BW}} \text{ seconds} \qquad \textbf{(13–49)}$$

We defined rise time to be the time required for a voltage to change from 10% of its final value to 90% of its final value. Therefore, if we wish the output to follow a pulse through its *entire* variation in Δt seconds, the bandwidth should be larger than that required by equation 13–49. In other words, equation 13–49 requires a bandwidth of $BW = 0.35/t_r$ to achieve a rise time of t_r, but $BW = 0.35/\Delta t$ would not be sufficient to permit a *full* voltage variation in Δt seconds.

We conclude from the foregoing remarks that both slew rate and bandwidth affect the minimum time Δt that an amplifier output can change through ΔV volts. If we now let Δt be the total time through which a given input changes value, then the amplifier must satisfy *both* of the following conditions in order to track that input without distortion:

$$\frac{\Delta V}{S} \leq \Delta t \tag{13–50}$$

$$\frac{0.35}{BW_{CL}} \ll \Delta t \tag{13–51}$$

Note that the inequality in (13–51) is "much less than."

Example 13–13

The operational amplifier shown in Figure 13–24 has a slew rate of 4 V/μs and a unity-gain frequency of 2 MHz. Determine whether the amplifier will distort the input signal shown.

Solution. The closed-loop gain of the amplifier is $v_o/v_{in} = (R_1 + R_f)/R_1 = (40\ \text{k}\Omega)/(20\ \text{k}\Omega) = 2$. Therefore, the output changes from -4 V to 10 V, giving $\Delta V = 14$ V. The voltage change occurs in $\Delta t = 5\ \mu\text{s}$, as shown in the figure. Then

$$\frac{\Delta V}{S} = \frac{14\ \text{V}}{4 \times 10^6\ \text{V/s}} = 3.5\ \mu\text{s} < 5\ \mu\text{s} = \Delta t$$

Therefore, condition 13–50 is satisfied. Then, since $\beta = R_1/(R_1 + R_f) = (20\ \text{k}\Omega)/(40\ \text{k}\Omega) = 0.5$, we have $BW_{CL} = \beta f_t = 0.5(2\ \text{MHz}) = 1$ MHz, and

$$\frac{0.35}{BW_{CL}} = \frac{0.35}{10^6} = 0.35\ \mu\text{s} \ll 5\ \mu\text{s} = \Delta t$$

Therefore, condition 13–51 is also satisfied, and we conclude that no distortion will occur.

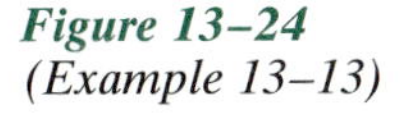
Figure 13–24
(Example 13–13)

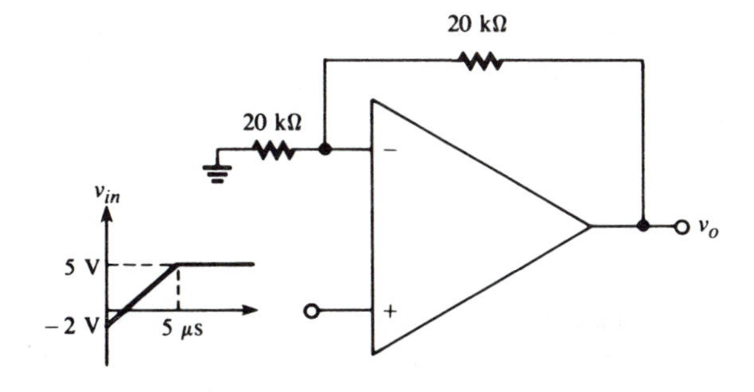

13–5 OFFSET CURRENTS AND VOLTAGES

Recall from Chapter 12 that one of the characteristics of an ideal operational amplifier is that it has zero output voltage when both inputs are 0 V (grounded). This characteristic is particularly important in applications where dc or low-frequency signals are involved. If the output is not 0 when the inputs are 0, then the output will not be at its correct dc level when the input is a dc level other than 0.

The actual value of the output voltage when the inputs are 0 is called the *output offset voltage.* Output offset is very much like a dc bias level in the output of a conventional amplifier in that it is added to whatever signal variation occurs there. If an operational amplifier is used only for ac signals, it can be capacitor-coupled if necessary or desirable to block the dc component represented by the offset. However, the capacitors may have to be impractically large if low frequencies and small impedance levels are involved. Also, a dc path must always be present between each input and ground to allow bias currents to flow. Small offsets, on the order of a few millivolts, can often be ignored if the signal variations are large by comparison. On the other hand, a frequent application of operational amplifiers is in precise, high-accuracy signal processing at low levels and low frequencies, and in these situations, very small offsets are crucial.

Manufacturers do not generally specify output offset because, as we shall see, the offset level depends on the closed-loop gain that a user designs through choice of external component values. Instead, *input* offsets are specified, and the designer can use these values to compute the output offset that results in a particular application. Output offset voltages are the result of two distinct input phenomena: input bias currents and input offset voltage. We will use the superposition principle to determine the contribution of each of these input effects to the output offset voltage.

Input Offset Current

In our discussion of differential amplifier circuits in Chapter 12, we ignored base currents because they had negligible effects on the kinds of computations that held our interest then. We know that some dc base current must flow when a transistor is properly biased, and, although small, this current flowing through the external resistors in an amplifier circuit produces a dc input voltage that in turn creates an output offset. To reduce the effect of bias currents, a *compensating resistor* R_c is connected in series with the noninverting (+) terminal of the amplifier. (R_c must provide a dc path to ground, so if a signal is capacitor-coupled to the + input, R_c must be connected between the + input and ground.) We will presently show that proper choice of the value of R_c will minimize the output offset voltage due to bias current. Figure 13–25 shows the bias currents I_B^+ and I_B^- flowing into the + and − terminals of an operational amplifier when the signal inputs are grounded. While the bias currents may actually flow into or out of the terminals, depending on the type of input circuitry, we will, for the sake of convenience, assume that the directions are as shown and that the values are always positive. These assumptions will not affect our ultimate conclusions. The figure also shows the compensating resistor R_c connected in series with the + terminal. Note that this circuit applies to both the inverting and noninverting configurations.

Figure 13–26(a) shows the equivalent circuit of Figure 13–25. Here, the bias currents are represented by current sources having resistances R_1 and R_c. Figure 13–26(b) shows the same circuit when the current sources are replaced by their Thevenin equivalent voltage sources.

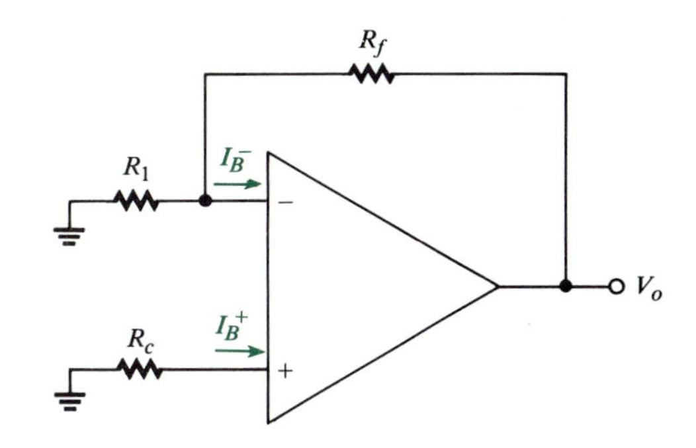

Figure 13–25
Input bias currents I_B^+ and I_B^- that flow when both signal inputs are grounded. R_c is a compensating resistor used to reduce the effect of bias current on output offset.

Using Figure 13–26(b), we can apply the superposition principle to determine the output offset voltage due to each input source acting alone. As illustrated in Figure 13–27(a), the amplifier acts as an inverter when the source connected to the + terminal is shorted to ground, so the output due to $I_B^- R_1$ is

$$V_{o1} = I_B^- R_1 \left(\frac{-R_f}{R_1} \right) = -I_B^- R_f \tag{13–52}$$

Figure 13–26
Circuits equivalent to Figure 13–25

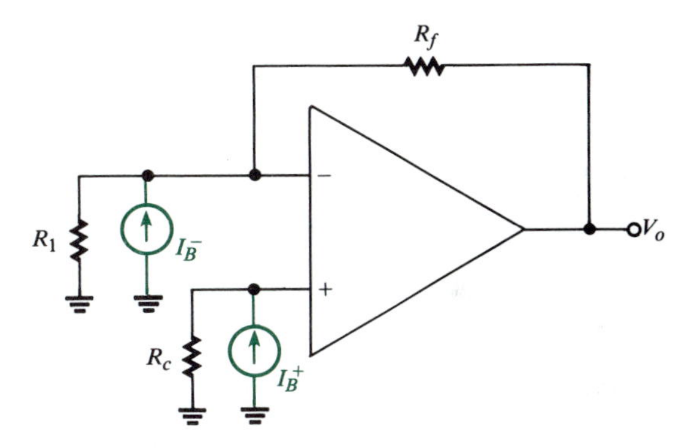

(a) The equivalent circuit of Figure 13.23

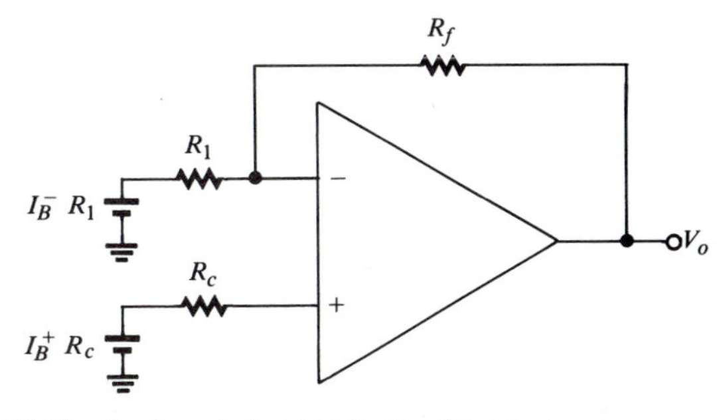

(b) The circuit equivalent to (a) when the current sources are replaced by their Thevenin equivalents

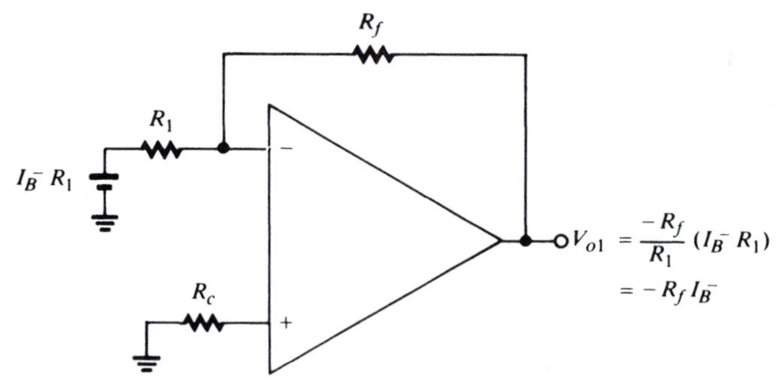

(a) When the noninverting input is grounded, the amplifier inverts and has gain $-R_f/R_1$.

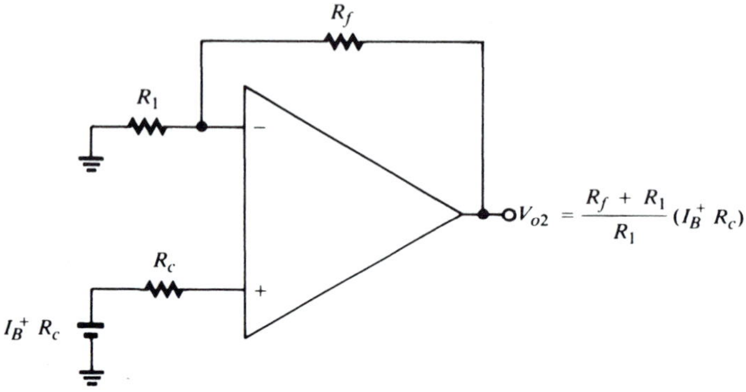

(b) When the inverting input is grounded, the noninverting amplifier has gain $(R_f + R_1)/R_1$.

Figure 13–27
Applying superposition to determine the output offset voltage due to each source in Figure 13–26(b)

When the source connected to the $-$ terminal is shorted to ground, the amplifier is in a noninverting configuration, so the output due to $I_B^+ R_c$ is

$$V_{o2} = I_B^+ R_c \left(\frac{R_f + R_1}{R_1} \right) \tag{13–53}$$

Combining (13–52) and (13–53), we obtain the total output offset voltage due to bias current, which we designate by $V_{OS}(I_B)$, as

$$V_{OS}(I_B) = I_B^+ R_c \left(\frac{R_f + R_1}{R_1} \right) - I_B^- R_f \tag{13–54}$$

Depending on which of the terms on the left side of equation 13–54 is greater, $V_{OS}(I_B)$ may be positive or negative. However, the sign of $V_{OS}(I_B)$ is of little interest, since negative offset voltage is just as undesirable as positive offset voltage. Our real interest is in finding a way to minimize the *magnitude* of $V_{OS}(I_B)$. Toward that end, let us make the reasonable assumption that the two inputs are closely matched and that, as a consequence, they have equal bias currents: $I_B^+ = I_B^- = I_{BB}$. Substituting

I_{BB} for I_B^- and I_B^+ in equation 13–54 gives

$$V_{OS}(I_B) = I_{BB}\left[R_c\left(\frac{R_f + R_1}{R_1}\right) - R_f\right] \quad \text{(13–55)}$$

If the expression enclosed by the brackets in (13–55) were equal to 0, we would have zero offset voltage. To find a value of R_c that accomplishes that goal, we set the bracketed expression equal to 0 and solve for R_c:

$$R_c\left(\frac{R_f + R_1}{R_1}\right) - R_f = 0$$

$$R_c = \frac{R_f}{\dfrac{R_f + R_1}{R_1}} = \frac{R_f R_1}{R_f + R_1} = R_f \| R_1 \quad \text{(13–56)}$$

Equation 13–56 reveals the very important result that *output offset due to input bias currents can be minimized by connecting a resistor R_c having value $R_1 \| R_f$ in series with the noninverting input.* This method of offset compensation is valid for both inverting and noninverting configurations. Notice that we say the offset can be *minimized* using this remedy, rather than being made exactly 0, because the remedy is based on the assumption that $I_B^+ = I_B^-$, which may not be entirely valid. We can compute the exact value of $V_{OS}(I_B)$ when $R_c = R_1 \| R_f$ by substituting this value of R_c back into (13–54), where the assumption is not in force:

$$\begin{aligned} V_{OS}(I_B) &= I_B^+(R_1 \| R_f)\left(\frac{R_f + R_1}{R_1}\right) - I_B^- R_f \\ &= I_B^+\left(\frac{R_1 R_f}{R_1 + R_f}\right)\left(\frac{R_f + R_1}{R_1}\right) - I_B^- R_f \\ &= (I_B^+ - I_B^-)R_f \end{aligned} \quad \text{(13–57)}$$

Equation 13–57 shows that the offset voltage is proportional to the *difference* between I_B^+ and I_B^- when $R_c = R_1 \| R_f$. Since the inputs are usually reasonably well matched, the difference between I_B^+ and I_B^- is quite small. The equation confirms the fact that V_{OS} is 0 if I_B^+ exactly equals I_B^-. The quantity $I_B^+ - I_B^-$ is called the *input offset current* and is often quoted in manufacturers' specifications. Remember that it is actually a *difference* current. Letting the input offset current $I_B^+ - I_B^-$ be designated by I_{io}, we have, from equation 13–57,

$$V_{OS}(I_B) = I_{io}R_f \qquad \text{when } R_c = R_1 \| R_f \quad \text{(13–58)}$$

$V_{OS}(I_B)$ may be either positive or negative, depending on whether $I_B^+ > I_B^-$ or vice versa. Unless actual measurements are made, we rarely know which current is larger, so a more useful form of (13–58) is

$$|V_{OS}(I_B)| = |I_{io}|R_f \qquad \text{when } R_c = R_1 \| R_f \quad \text{(13–59)}$$

Manufacturers' specifications always give a positive value for I_{io}, so it is best interpreted as an absolute value in any case.

Equation 13–59 shows that the output offset is directly proportional to the value of the feedback resistor R_f. For that reason, small resistance values should be used when offset is a critical consideration. However, to achieve large voltage gains when R_f is small may require impractically small values of R_1, to the extent that the amplifier may load the signal source driving it. In any event, large closed-loop gains are detrimental to another aspect of output offset, as we will see in a forthcoming discussion.

Another common manufacturers' specification is called simply *input bias current,* I_B. By convention, I_B is the *average* of I_B^+ and I_B^-:

$$I_B = \frac{I_B^+ + I_B^-}{2} \tag{13–60}$$

I_B is typically much larger than I_{io} because I_B is on the same order of magnitude as I_B^+ and I_B^-, while I_{io} is the difference between the two. Given values for I_B and I_{io}, we can find I_B^+ and I_B^-, provided we know which is the larger. If $I_B^+ > I_B^-$, then

$$\left.\begin{aligned} I_B^+ &= I_B + 0.5\,|I_{io}| \\ I_B^- &= I_B - 0.5\,|I_{io}| \end{aligned}\right\}(I_B^+ > I_B^-) \tag{13–61}$$

If $I_B^- > I_B^+$, the + and − signs between terms in (13–61) are interchanged. Proof of these relations is left to Exercise 13–44 at the end of the chapter.

Example 13–14

The specifications for the operational amplifier in Figure 13–28 state that the input bias current is 80 nA and that the input offset current is 20 nA.

1. Find the optimum value for R_c.
2. Find the magnitude of the output offset voltage due to bias currents when R_c equals its optimum value.
3. Assuming that $I_B^+ > I_B^-$, find the magnitude of the output offset voltage when $R_c = 0$.

Solution

1. From equation 13–56, $R_c = R_1 \parallel R_f = (10\ \text{k}\Omega) \parallel (100\ \text{k}\Omega) = 9.09\ \text{k}\Omega$.
2. From equation 13–59, $|V_{OS}(I_B)| = |I_{io}|R_f = (20 \times 10^{-9})(100 \times 10^3) = 2\ \text{mV}$.
3. When $R_c = 0$, equation 13–54 becomes $V_{OS}(I_B) = -I_B^- R_f$. From equation 13–61, $I_B^- = I_B - 0.5I_{io} = (80\ \text{nA}) - 0.5(20\ \text{nA}) = 70\ \text{nA}$. Therefore, the magnitude of the offset voltage when $R_c = 0$ is $|V_{OS}(I_B)| = (70 \times 10^{-9})(100 \times 10^3) = 7\ \text{mV}$. We see that omission of the compensating resistance more than doubles the magnitude of the offset voltage.

Figure 13–28
(Example 13–14)

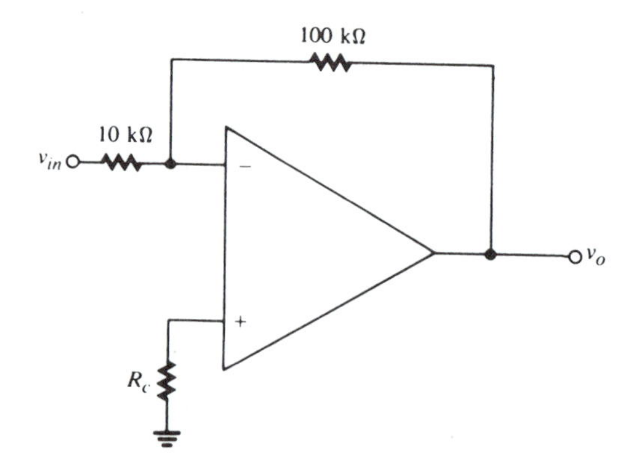

Figure 13–29
The effect of input offset voltage, V_{io}, is the same as if a dc source were connected in series with one of the inputs.

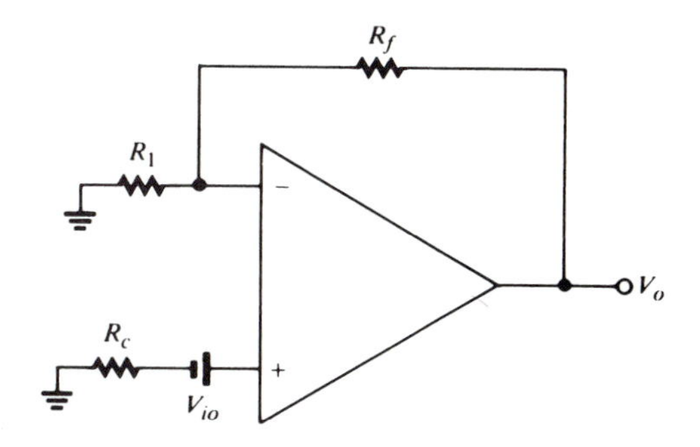

Input Offset Voltage

Another input phenomenon that contributes to output offset voltage is an internally generated potential difference that exists because of imperfect matching of the input transistors. This potential may be due, for example, to a difference between the V_{BE} drops of the transistors in the input differential stage of a BJT amplifier. Called *input offset voltage,* the net effect of this potential difference is the same as if a small dc voltage source were connected to one of the inputs. Figure 13–29 shows the equivalent circuit of an amplifier having its signal inputs grounded and its input offset voltage, V_{io}, represented as a dc source in series with the noninverting input. The effect is the same whether it is connected to the inverting or noninverting input. The polarity of the source is arbitrary, because input and output offsets may be either positive or negative. Once again, it is the magnitude of the offset that concerns us.

From Figure 13–29, it is apparent that the output voltage when the input is V_{io} is given by

$$V_{OS}(V_{io}) = V_{io}\frac{(R_1 + R_f)}{R_1} \qquad \textbf{(13–62)}$$

where $V_{OS}(V_{io})$ is the output offset voltage due to V_{io}. Note that the compensating resistor R_c is shown in Figure 13–29 for completeness' sake, but it has no effect on the output offset due to V_{io}. Equation 13–62 shows that input offset is magnified at the output by a factor equal to the closed-loop gain of the noninverting amplifier, as we would expect. If the amplifier is operated open-loop, the very large open-loop gain acting on the input offset voltage may well drive the amplifier to one of its output voltage limits. It is therefore important to have an extremely small V_{io} in any application or measurement that requires an open-loop amplifier.

Equation 13–62 is also valid for an amplifier in an inverting configuration. In fact, for a wide variety of amplifier configurations, it is true that

$$V_{OS}(V_{io}) = V_{io}/\beta \qquad \textbf{(13–63)}$$

where β is the feedback ratio.

Example 13–15

The specifications for the amplifier in Example 13–14 state that the input offset voltage is 0.8 mV. Find the output offset due to this input offset.

Solution. From equation 13–62,

$$V_{OS}(V_{io}) = V_{io}\frac{(R_1 + R_f)}{R_1} = (0.8 \times 10^{-3}\text{ V})\frac{[(10\text{ k}\Omega) + (100\text{ k}\Omega)]}{10\text{ k}\Omega} = 8.8\text{ mV}$$

The Total Output Offset Voltage

We have seen that output offset voltage is a function of two distinct input characteristics: input bias currents and input offset voltage. It may be that the polarities of the offsets caused by these two characteristics are such that they tend to cancel each other out. Of course, we cannot depend on that happy circumstance, so it is good design practice to assume a *worst-case* situation, in which the two offsets have the same polarity and reinforce each other. We can invoke the principle of superposition and conclude that the total output offset voltage is the sum of the offsets caused by the individual input phenomena, but for the worst-case situation, we assume that the total offset is the sum of the respective *magnitudes:*

$$|V_{OS}| = |V_{OS}(I_B)| + |V_{OS}(V_{io})| \qquad \text{(worst case)} \tag{13–64}$$

where V_{OS} is the total output offset voltage.

Example 13–16

The operational amplifier in Figure 13–30 has the following specifications: input bias current = 100 nA; input offset current = 20 nA; input offset voltage = 0.5 mV. Find the worst-case output offset voltage. (Consider the two possibilities $I_B^+ > I_B^-$ and vice versa.)

Solution. We first check to see if the 10-kΩ resistor in series with the noninverting input has the optimum value of a compensating resistor: $R_1 \parallel R_f = (15\text{ k}\Omega) \parallel (75\text{ k}\Omega) = 12.5\text{ k}\Omega$. $R_c = 10\text{ k}\Omega$ is not optimum, and we will have to use equation 13–54 to find $V_{OS}(I_B)$. Assuming first that $I_B^+ > I_B^-$, we have, from equation 13–61,

$$I_B^+ = I_B + 0.5I_{io} = (100\text{ nA}) + 0.5(20\text{ nA}) = 110\text{ nA}$$
$$I_B^- = I_B - 0.5I_{io} = (100\text{ nA}) - 0.5(20\text{ nA}) = 90\text{ nA}$$

Figure 13–30
(Example 13–16)

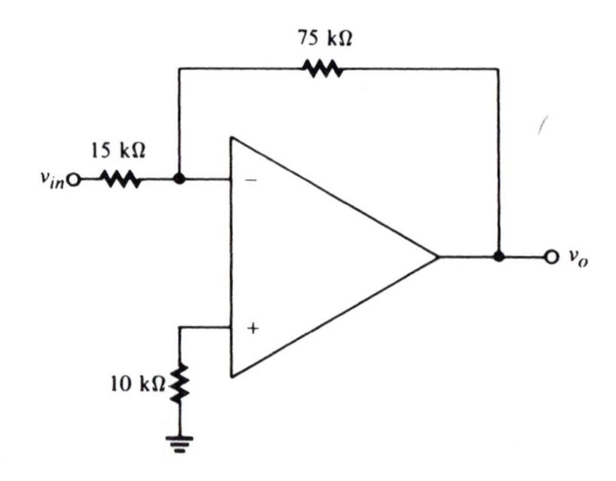

Therefore, by equation 13–54,

$$V_{OS}(I_B) = I_B^+ R_c \left(\frac{R_f + R_1}{R_1} \right) - I_B^- R_f$$

$$= (110 \times 10^{-9})(10 \times 10^3) \left(\frac{75 \times 10^3 + 15 \times 10^3}{15 \times 10^3} \right) - (90 \times 10^{-9})(75 \times 10^3)$$

$$= -0.15 \text{ mV}$$

If $I_B^- > I_B^+$, then $I_B^+ = 90$ nA and $I_B^- = 110$ nA. In that case,

$$V_{OS}(I_B) = (90 \times 10^{-9})(10 \times 10^3) \left(\frac{75 \times 10^3 + 15 \times 10^3}{15 \times 10^3} \right) - (110 \times 10^{-9})(75 \times 10^3)$$

$$= -2.85 \text{ mV}$$

We see that the worst case occurs for $I_B^- > I_B^+$, and therefore we assume that $|V_{OS}(I_B)| = 2.85$ mV.

By equation 13–62,

$$V_{OS}(V_{io}) = \frac{V_{io}(R_1 + R_f)}{R_1} = (0.5 \text{ mV}) \left[\frac{(15 \text{ k}\Omega) + (75 \text{ k}\Omega)}{15 \text{ k}\Omega} \right] = 3 \text{ mV}$$

Therefore, the worst-case offset is $V_{OS} = |V_{OS}(I_B)| + |V_{OS}(V_{io})| = (2.85 \text{ mV}) + (3 \text{ mV}) = 5.85$ mV. (Note that the "best-case" offset would be 0.15 mV.)

The values we have used for V_{io}, I_B, and I_{io} in the examples of this section are typical for general-purpose, BJT operational amplifiers. Amplifiers with much smaller input offsets are available. The bias currents in amplifiers having FET inputs are in the picoamp range.

Most operational amplifiers have two terminals across which an external potentiometer can be connected to adjust the output to 0 when the inputs are grounded. This operation is called *zeroing,* or *balancing,* the amplifier. However, operational amplifiers are subject to *drift,* wherein characteristics change with time and, particularly, with temperature. For applications in which extremely small offsets are required and in which the effects of drift must be minimized, *chopper-stabilized* amplifiers are available. Internal choppers convert the dc offset to an ac signal, amplify it, and use it to adjust amplifier characteristics so that the output is automatically restored to 0.

13–6 OPERATIONAL AMPLIFIER SPECIFICATIONS

In this section we will examine and interpret a typical set of manufacturer's specifications for an operational amplifier. The specifications shown in Figure 13–31 are those of the 741 amplifier, a popular, inexpensive, general-purpose operational amplifier that has been produced by many different manufacturers for a number of years. Let us first note that, like many integrated circuits, there are several versions of the 741 available. The different versions are identified by letter suffixes (741A, 741C, etc.), and each version has at least one performance specification or operating condition that is different from the others. The Fairchild specifications

μA741 Operational Amplifier

Linear Products

Description

The μA741 is a high performance Monolithic Operational Amplifier constructed using the Fairchild Planar epitaxial process. It is intended for a wide range of analog applications. High common mode voltage range and absence of latch-up tendencies make the μA741 ideal for use as a voltage follower. The high gain and wide range of operating voltage provides superior performance in integrator, summing amplifier, and general feedback applications.

- NO FREQUENCY COMPENSATION REQUIRED
- SHORT-CIRCUIT PROTECTION
- OFFSET VOLTAGE NULL CAPABILITY
- LARGE COMMON MODE AND DIFFERENTIAL VOLTAGE RANGES
- LOW POWER CONSUMPTION
- NO LATCH-UP

Connection Diagram
10-Pin Flatpak

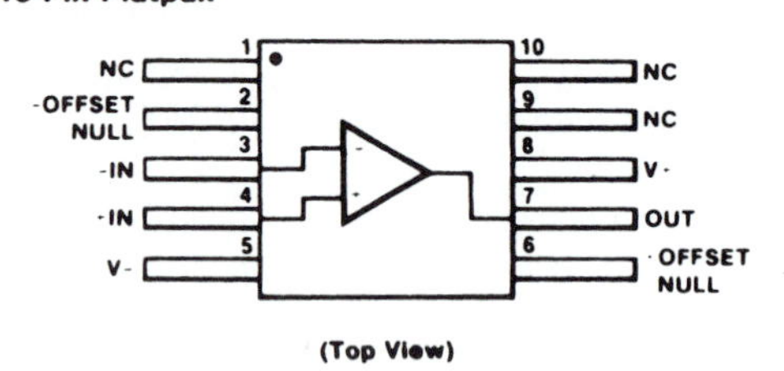

(Top View)

Order Information

Type	Package	Code	Part No.
μA741	Flatpak	3F	μA741FM
μA741A	Flatpak	3F	μA741AFM

Connection Diagram
8-Pin Metal Package

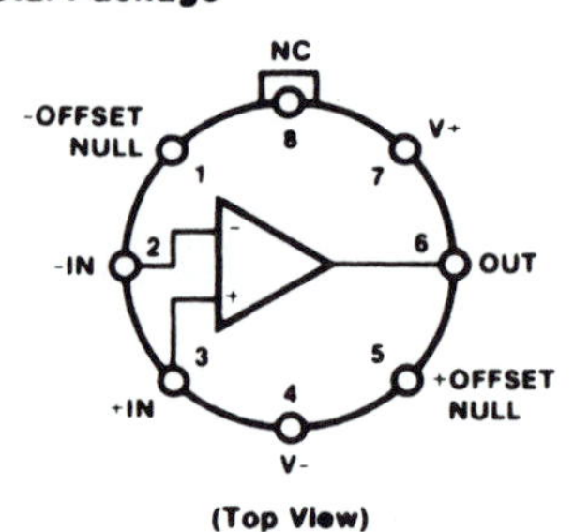

(Top View)

Pin 4 connected to case

Order Information

Type	Package	Code	Part No.
μA741	Metal	5W	μA741HM
μA741A	Metal	5W	μA741AHM
μA741C	Metal	5W	μA741HC
μA741E	Metal	5W	μA741EHC

Connection Diagram
8-Pin DIP

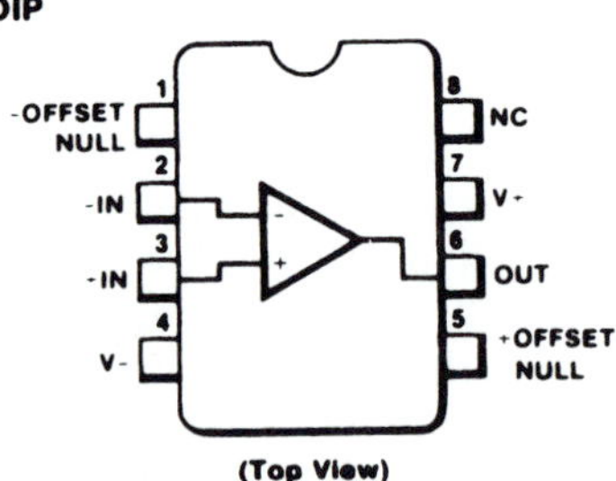

(Top View)

Order Information

Type	Package	Code	Part No.
μA741C	Molded DIP	9T	μA741TC
μA741C	Ceramic DIP	6T	μA741RC

Absolute Maximum Ratings

Supply Voltage	
μA741A, μA741, μA741E	±22 V
μA741C	±18 V
Internal Power Dissipation (Note 1)	
Metal Package	500 MW
DIP	310 mW
Flatpak	570 mW
Differential Input Voltage	±30 V
Input Voltage (Note 2)	±15 V
Storage Temperature Range	
Metal Package and Flatpak	−65°C to +150°C
DIP	−55°C to +125°C
Operating Temperature Range	
Military (μA741A, μA741)	−55°C to +125°C
Commercial (μA741E, μA741C)	0°C to +70°C
Pin Temperature (Soldering 60 s)	
Metal Package, Flatpak, and Ceramic DIP	300°C
Molded DIP (10 s)	260°C
Output Short Circuit Duration (Note 3)	Indefinite

Figure 13–31
Specifications for the μA741 operational amplifier (Courtesy of Fairchild Semiconductor)

Equivalent Circuit

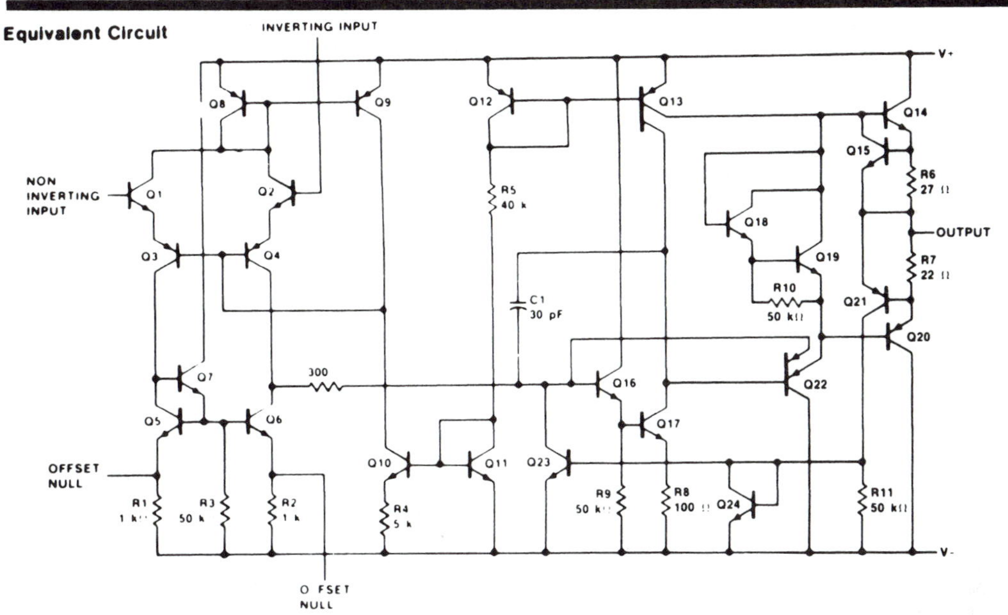

Notes

1. Rating applies to ambient temperatures up to 70°C. Above 70°C ambient derate linearly at 6.3 mW/°C for the metal package, 7.1 mW/°C for the flatpak, and 5.6 mW/°C for the DIP.
2. For supply voltages less than ± 15 V, the absolute maximum input voltage is equal to the supply voltage.
3. Short circuit may be to ground or either supply. Rating applie to +125°C case temperature or 75°C ambient temperature.

μA741 and μA741C
Electrical Characteristics V_S = ± 15 V, T_A = 25°C unless otherwise specified

Characteristic	Condition	μA741 Min	μA741 Typ	μA741 Max	μA741C Min	μA741C Typ	μA741C Max	Unit
Input Offset Voltage	$R_S \leq$ 10 kΩ		1.0	5.0		2.0	6.0	mV
Input Offset Current			20	200		20	200	nA
Input Bias Current			80	500		80	500	nA
Power Supply Rejection Ratio	V_S = +10, −20 V_S = +20, −10 V, R_S = 50 Ω		30	150		30	150	μV/V
Input Resistance		.3	2.0		.3	2.0		MΩ
Input Capacitance			1.4			1.4		pF
Offset Voltage Adjustment Range			± 15			± 15		mV
Input Voltage Range					± 12	± 13		V
Common Mode Rejection Ratio	$R_S \leq$ 10 kΩ				70	90		dB
Output Short Circuit Current			25			25		mA
Large Signal Voltage Gain	$R_L \geq$ 2 kΩ, V_{OUT} = ± 10 V	50k	200k		20k	200k		
Output Resistance			75			75		Ω
Output Voltage Swing	$R_L \geq$ 10 kΩ				± 12	± 14		V
	$R_L \geq$ 2 kΩ				± 10	± 13		V
Supply Current			1.7	2.8		1.7	2.8	mA
Power Consumption			50	85		50	85	mW

Figure 13–31
(Continued)

µA741 and µA741C
Electrical Characteristics (Cont.) V_S = ± 15 V, T_A = 25°C unless otherwise specified

Characteristic		Condition	µA741			µA741C			
			Min	Typ	Max	Min	Typ	Max	Unit
Transient Response (Unity Gain)	Rise Time	V_{IN} = 20 mV, R_L = 2 kΩ, $C_L \leq$ 100 pF		.3			.3		µs
	Overshoot			5.0			5.0		%
Bandwidth (Note 4)				1.0			1.0		MHz
Slew Rate		$R_L \geq$ 2 kΩ		.5			.5		V/µs

Notes
4. Calculated value from BW(MHz) = $\frac{0.35}{\text{Rise Time } (\mu s)}$
5. All V_{CC} = 15 V for µA741 and µA741C.
6. Maximum supply current for all devices
   25°C = 2.8 mA
   125°C = 2.5 mA
   −55°C = 3.3 mA

µA741 and µA741C
Electrical Characteristics (Cont.) The following specifications apply over the range of $-55°C \leq T_A \leq 125°C$ for µA741, $0°C \leq T_A \leq 70°C$ for µA741C

Characteristic	Condition	µA741			µA741C			
		Min	Typ	Max	Min	Typ	Max	Unit
Input Offset Voltage							7.5	mV
	$R_S \leq$ 10 kΩ		1.0	6.0				mV
Input Offset Current							300	nA
	T_A = +125°C		7.0	200				nA
	T_A = −55°C		85	500				nA
Input Bias Current							800	nA
	T_A = +125°C		.03	.5				µA
	T_A = −55°C		.3	1.5				µA
Input Voltage Range		± 12	± 13					V
Common Mode Rejection Ratio	$R_S \leq$ 10 kΩ	70	90					dB
Adjustment for Input Offset Voltage			± 15			± 15		mV
Supply Voltage Rejection Ratio	V_S = +10, −20; V_S = +20, −10 V, R_S = 50 Ω		30	150				µV/V
Output Voltage Swing	$R_L \geq$ 10 kΩ	± 12	± 14					V
	$R_L \geq$ 2 kΩ	± 10	± 13		± 10	± 13		V
Large Signal Voltage Gain	R_L = 2 kΩ, V_{OUT} = ± 10 V	25k			15k			
Supply Current	T_A = +125°C		1.5	2.5				mA
	T_A = −55°C		2.0	3.3				mA
Power Consumption	T_A = +125°C		45	75				mW
	T_A = −55°C		60	100				mW

Notes
4. Calculated value from BW(MHz) = $\frac{0.35}{\text{Rise Time } (\mu s)}$
5. All V_{CC} = 15 V for µA741 and µA741C.
6. Maximum supply current for all devices
   25°C = 2.8 mA
   125°C = 2.5 mA
   −55°C = 3.3 mA

Figure 13–31
(Continued)

μA741A and μA741E
Electrical Characteristics $V_S = \pm 15$ V, $T_A = 25°C$ unless otherwise specified.

Characteristic		Condition	μA741A/E Min	Typ	Max	Unit
Input Offset Voltage		$R_S \leq 50\ \Omega$		0.8	3.0	mV
Average Input Offset Voltage Drift					15	μV/°C
Input Offset Current				3.0	30	nA
Average Input Offset Current Drift					0.5	nA/°C
Input Bias Current				30	80	nA
Power Supply Rejection Ratio		$V_S = +10, -20$; $V_S = +20$ V, -10 V, $R_S = 50\ \Omega$		15	50	μV/V
Output Short Circuit Current			10	25	40	mA
Power Consumption		$V_S = \pm 20$ V		80	150	mW
Input Impedance		$V_S = \pm 20$ V	1.0	6.0		MΩ
Large Signal Voltage Gain		$V_S = \pm 20$ V, $R_L = 2$ kΩ, $V_{OUT} = \pm 15$ V	50	200		V/mV
Transient Response (Unity Gain)	Rise Time			0.25	0.8	μs
	Overshoot			6.0	20	%
Bandwidth (Note 4)			.437	1.5		MHz
Slew Rate (Unity Gain)		$V_{IN} = \pm 10$ V	0.3	0.7		V/μs

The following specifications apply over the range of $-55°C \leq T_A \leq 125°C$ for the 741A, and $0°C \leq T_A \leq 70°C$ for the 741E.

Characteristic	Condition			Min	Typ	Max	Unit
Input Offset Voltage						4.0	mV
Input Offset Current						70	nA
Input Bias Current						210	nA
Common Mode Rejection Ratio	$V_S = \pm 20$ V, $V_{IN} = \pm 15$ V, $R_S = 50\ \Omega$			80	95		dB
Adjustment For Input Offset Voltage	$V_S = \pm 20$ V			10			mV
Output Short Circuit Current				10		40	mA
Power Consumption	$V_S = \pm 20$ V	μA741A	−55°C			165	mW
			+125°C			135	mW
		μA741E				150	mW
Input Impedance	$V_S = \pm 20$ V			0.5			MΩ
Output Voltage Swing	$V_S = \pm 20$ V		$R_L = 10$ kΩ	±16			V
			$R_L = 2$ kΩ	±15			V
Large Signal Voltage Gain	$V_S = \pm 20$ V, $R_L = 2$ kΩ, $V_{OUT} = \pm 15$ V			32			V/mV V/mV
	$V_S = \pm 5$ V, $R_L = 2$ kΩ, $V_{OUT} = \pm 2$ V			10			V/mV

Notes

4. Calculated value from: $BW(MHz) = \dfrac{0.35}{\text{Rise Time } (\mu s)}$
5. All $V_{CC} = 15$ V for μA741 and μA741C.
6. Maximum supply current for all devices
   25°C = 2.8 mA
   125°C = 2.5 mA
   −55°C = 3.3 mA

Figure 13–31
(Continued)

Typical Performance Curves for μA741A and μA741

Open Loop Voltage Gain as a Function of Supply Voltage

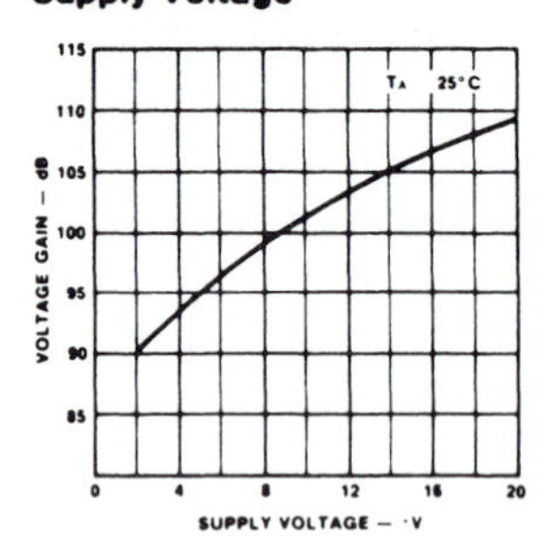

Output Voltage Swing as a Function of Supply Voltage

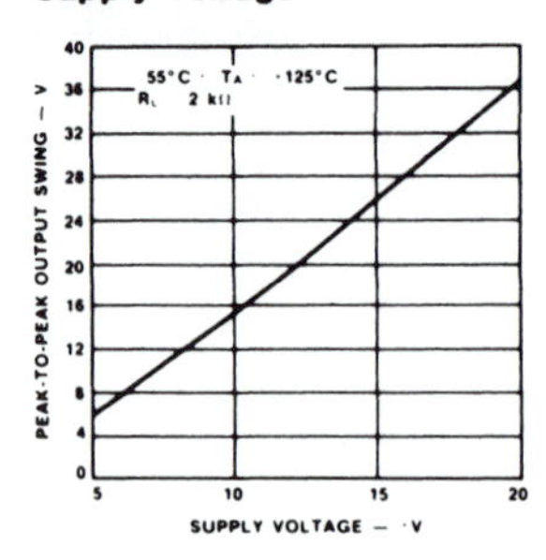

Input Common Mode Voltage as a Function of Supply Voltage

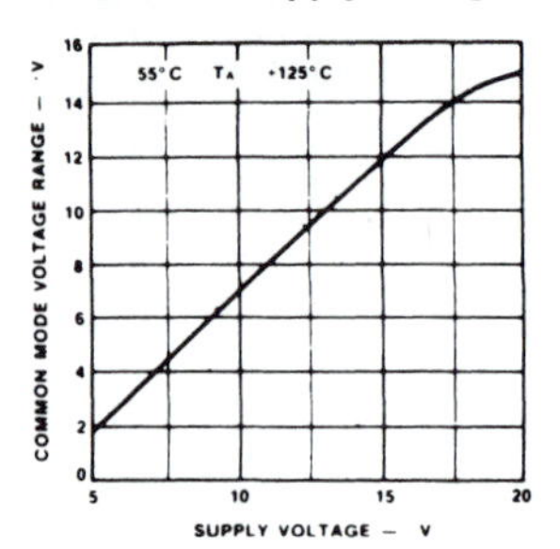

Typical Performance Curves for μA741E and μA741C

Open Loop Voltage Gain as a Function of Supply Voltage

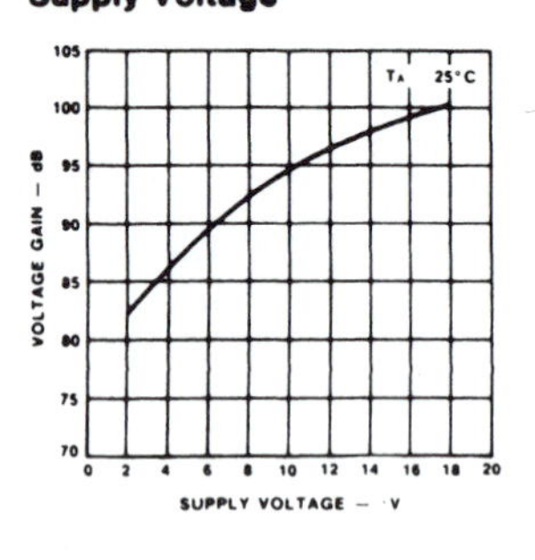

Output Voltage Swing as a Function of Supply Voltage

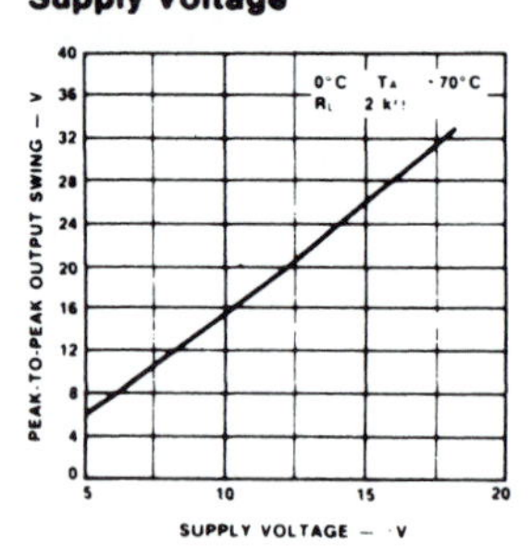

Input Common Mode Voltage Range as a Function of Supply Voltage

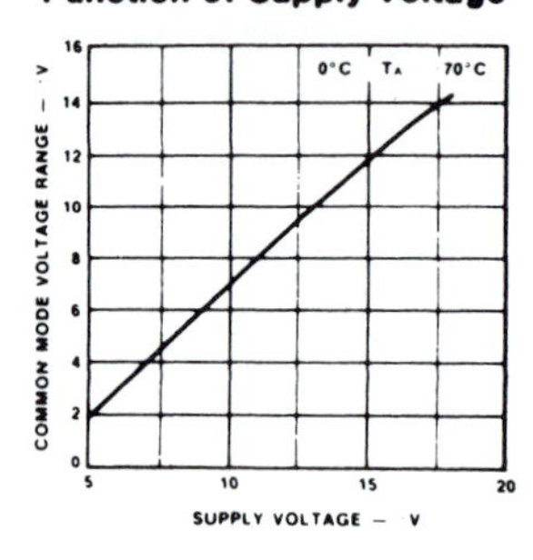

Transient Response

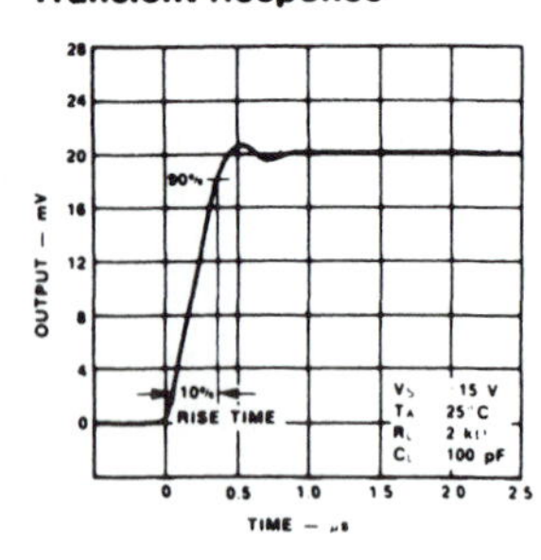

Transient Response Test Circuit

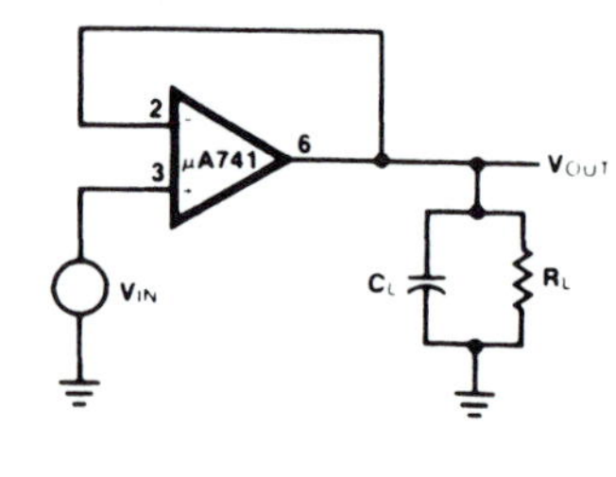

Common Mode Rejection Ratio as a Function of Frequency

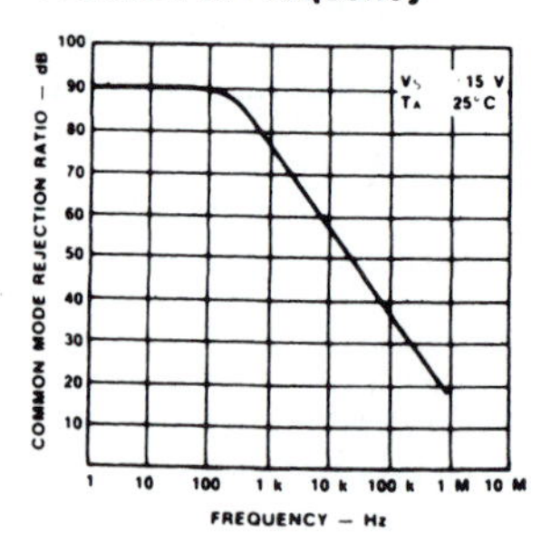

Figure 13–31
(Continued)

Typical Performance Curves for μA741E and μA741C (Cont.)

Frequency Characteristics as a Function of Supply Voltage

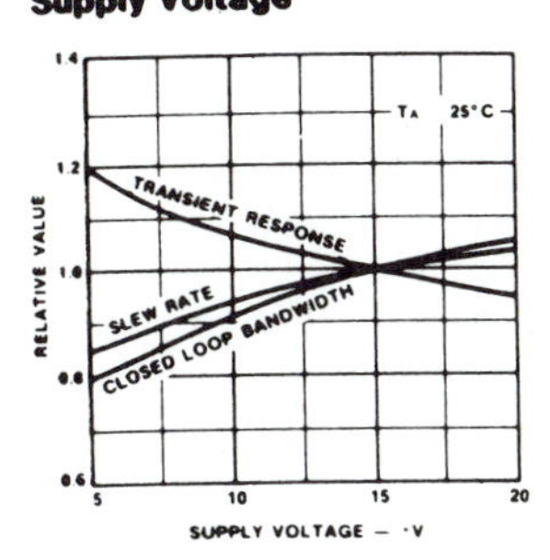

Voltage Offset Null Circuit

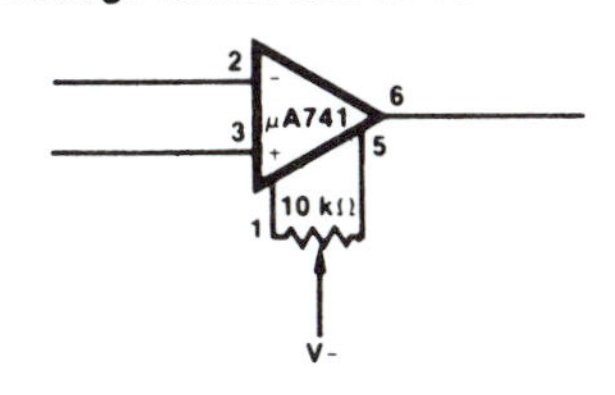

Voltage Follower Large Signal Pulse Response

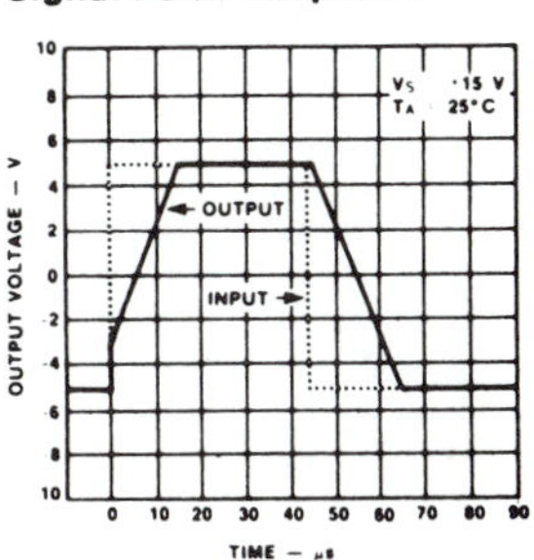

Typical Performance Curves for μA741A, μA741, μA741E and μA741C

Power Consumption as a Function of Supply Voltage

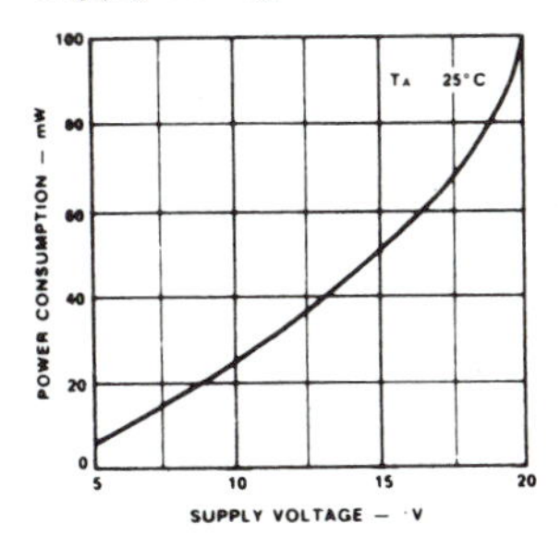

Open Loop Voltage Gain as a Function of Frequency

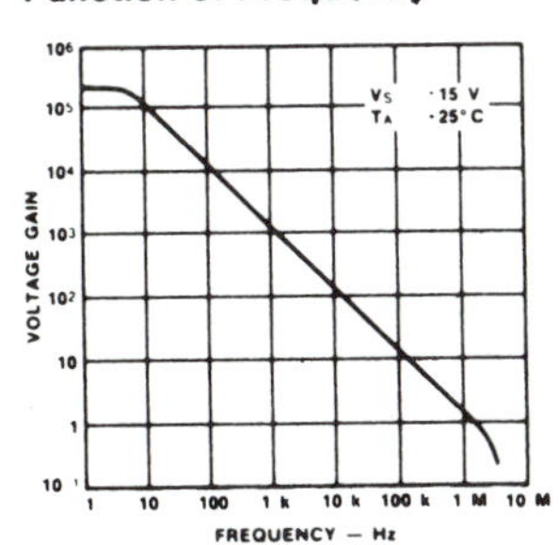

Open Loop Phase Response as a Function of Frequency

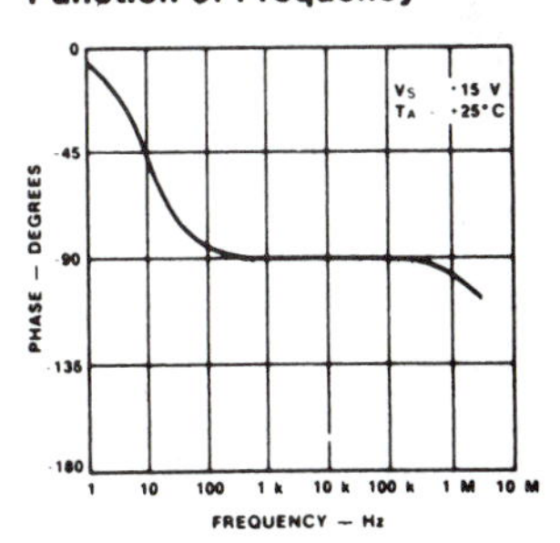

Input Offset Current as a Function of Supply Voltage

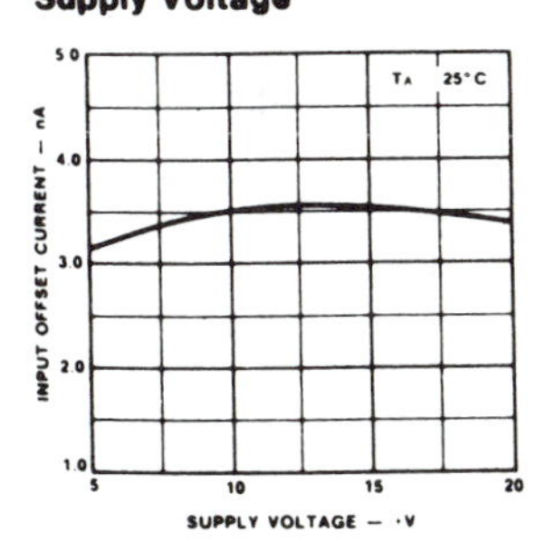

Input Resistance and Input Capacitance as a Function of Frequency

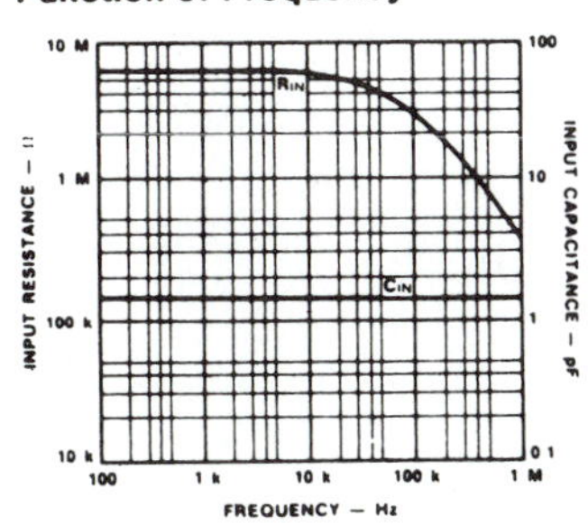

Output Resistance as a Function of Frequency

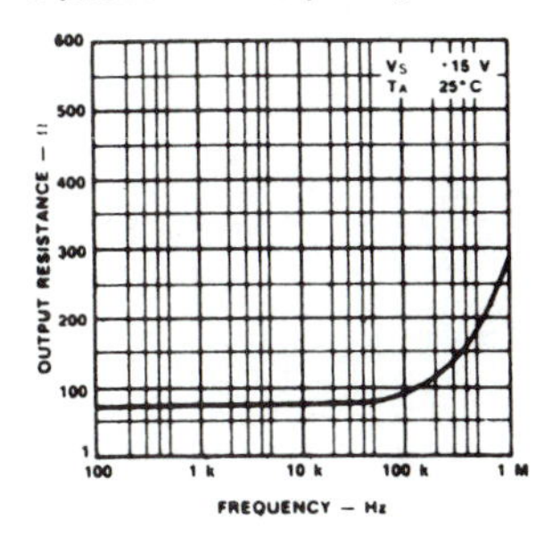

Figure 13–31
(Continued)

Typical Performance Curves for μA741A, μA741, μA741E and μA741C (Cont.)

Output Voltage Swing as a Function of Load Resistance

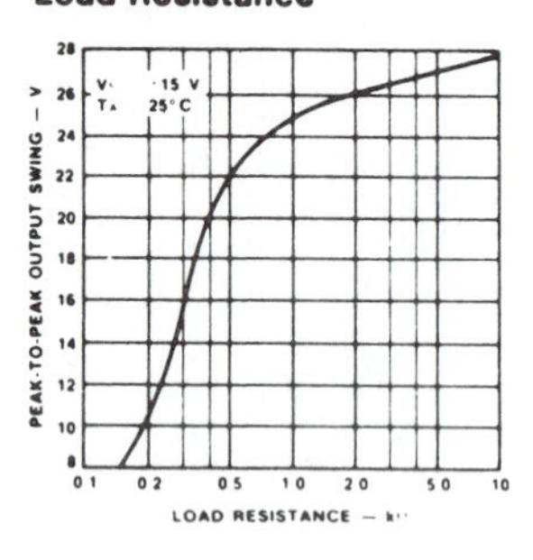

Output Voltage Swing as a Function of Frequency

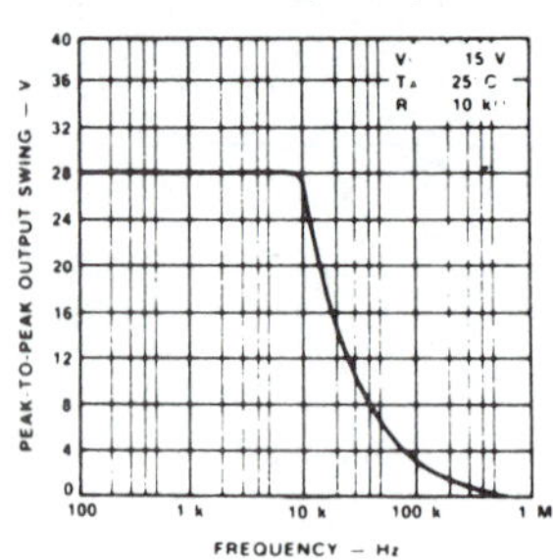

Absolute Maximum Power Dissipation as a* Function of Ambient Temperature

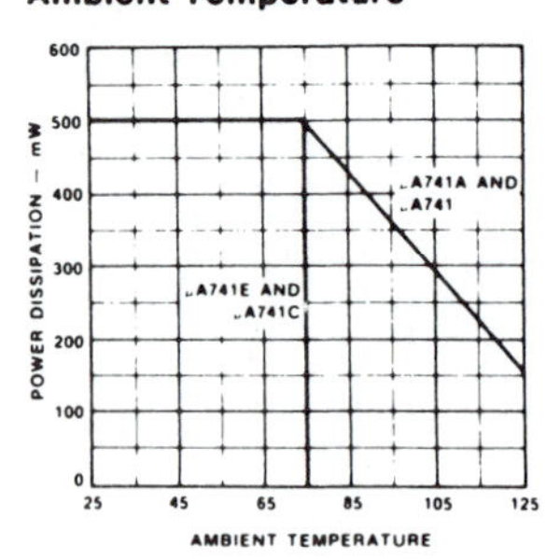

Input Noise Voltage as a Function of Frequency

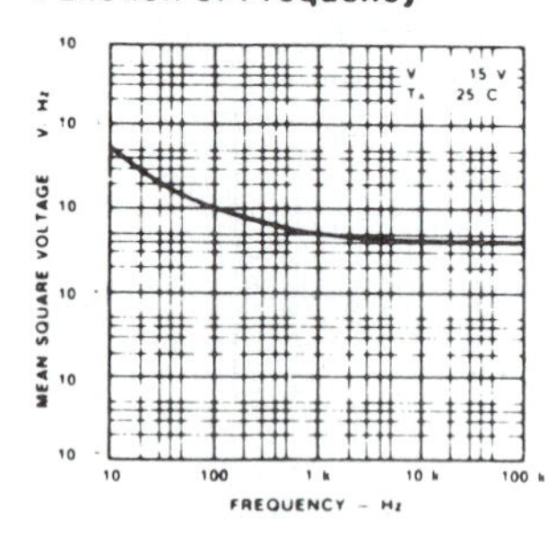

Input Noise Current as a Function of Frequency

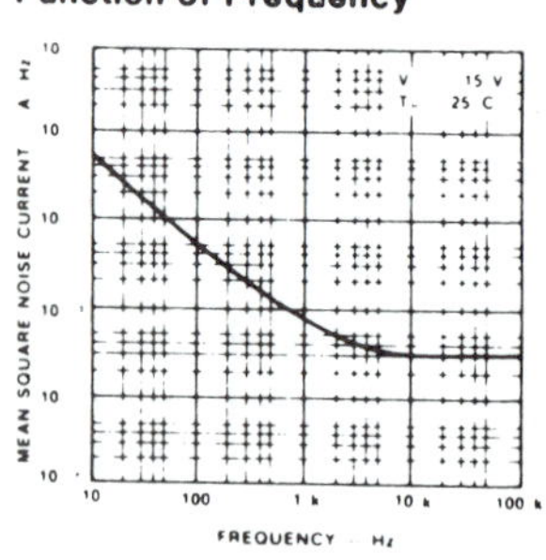

Broadband Noise for Various Bandwidths

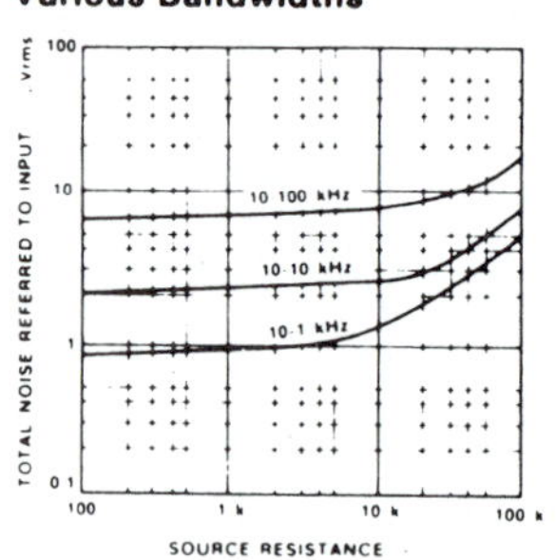

Typical Performance Curves for μA741A and μA741

Input Bias Current as a Function of Ambient Temperature

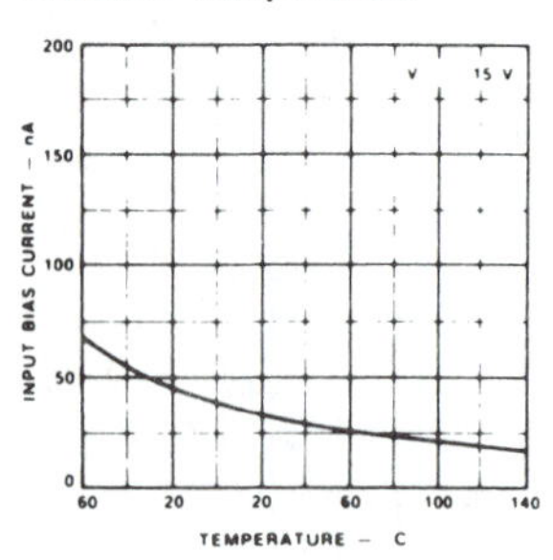

Input Resistance as a Function of Ambient Temperature

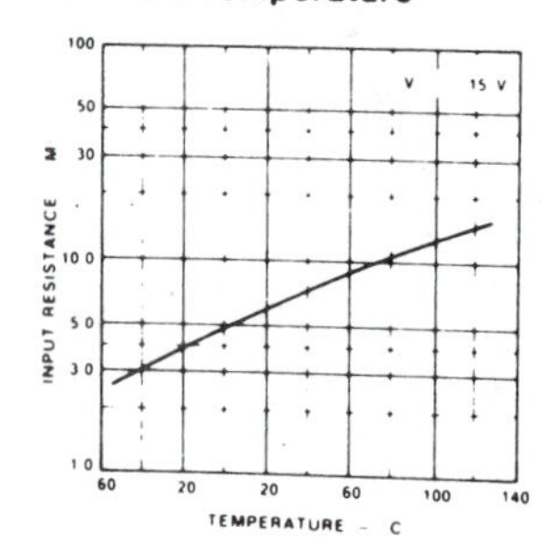

Output Short-Circuit Current as a Function of Ambient Temperature

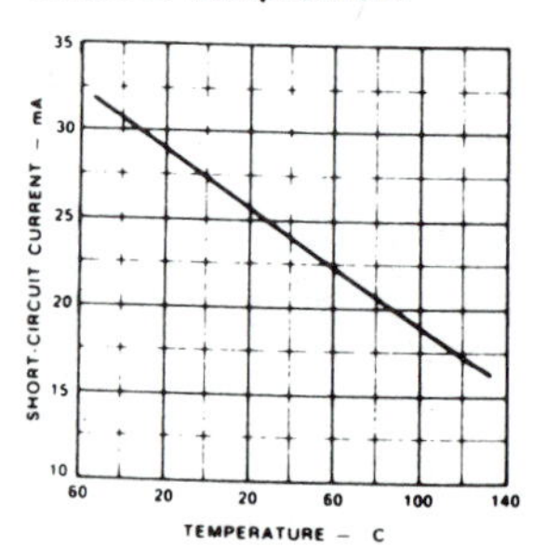

Figure 13–31
(Continued)

Typical Performance Curves for μA741A and μA741 (Cont.)

Input Offset Current as a Function of Ambient Temperature

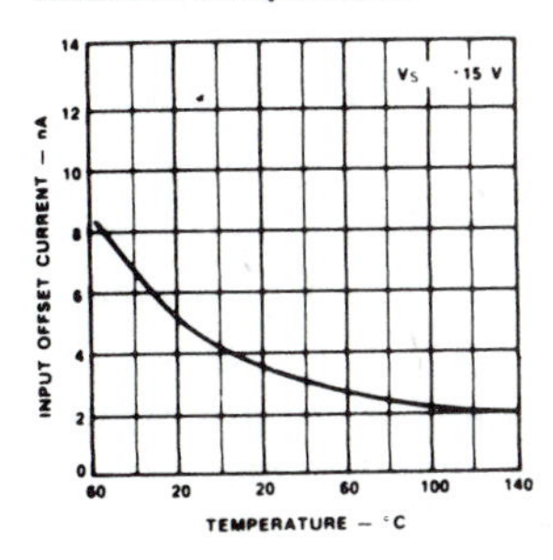

Power Consumption as a Function of Ambient Temperature

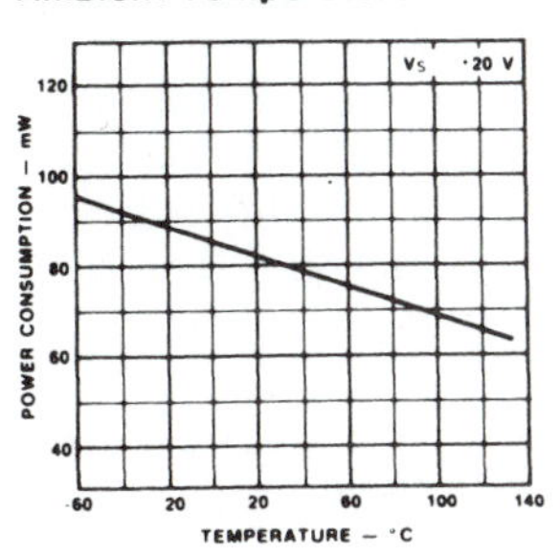

Frequency Characteristics as a Function of Ambient Temperature

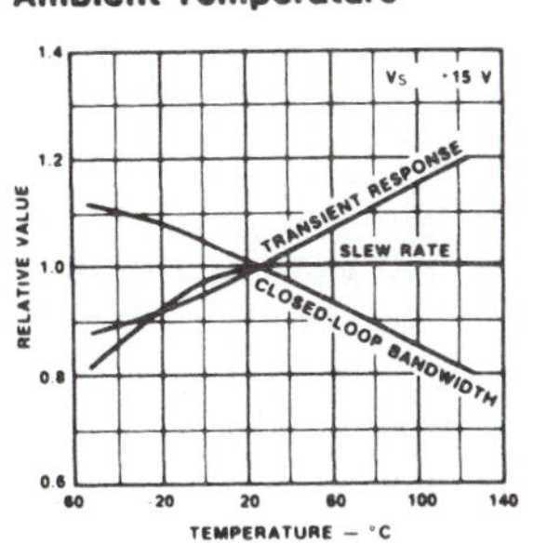

Typical Performance Curves for μA741E and μA741C

Input Bias Current as a Function of Ambient Temperature

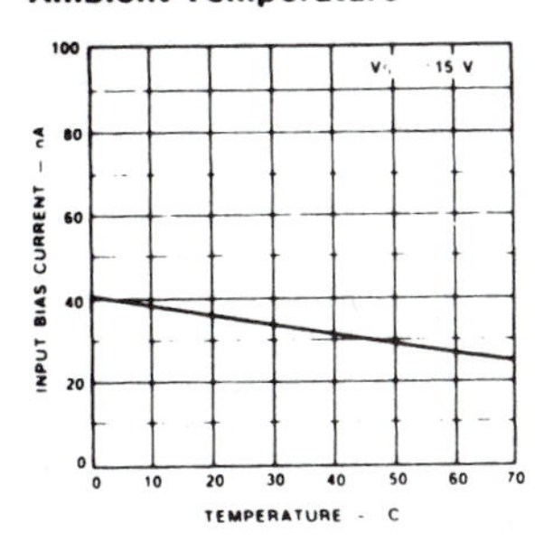

Input Resistance as a Function of Ambient Temperature

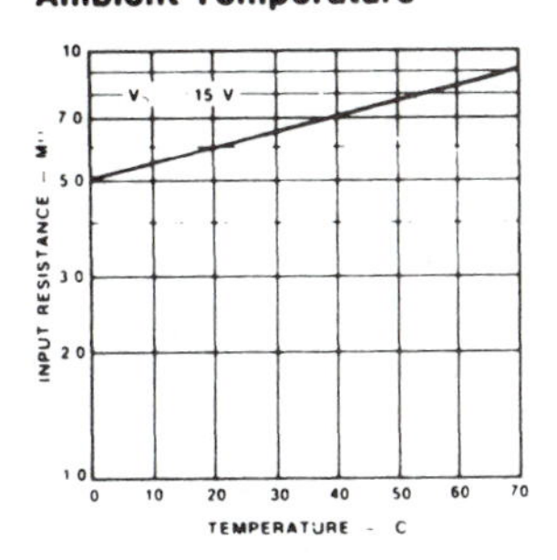

Input Offset Current as a Function of Ambient Temperature

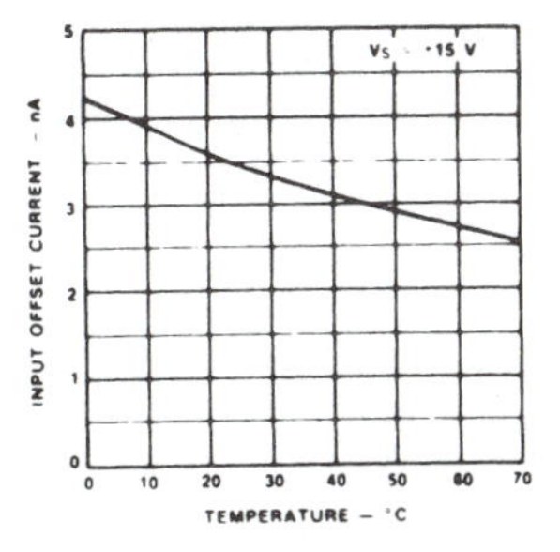

Power Consumption as a Function of Ambient Temperature

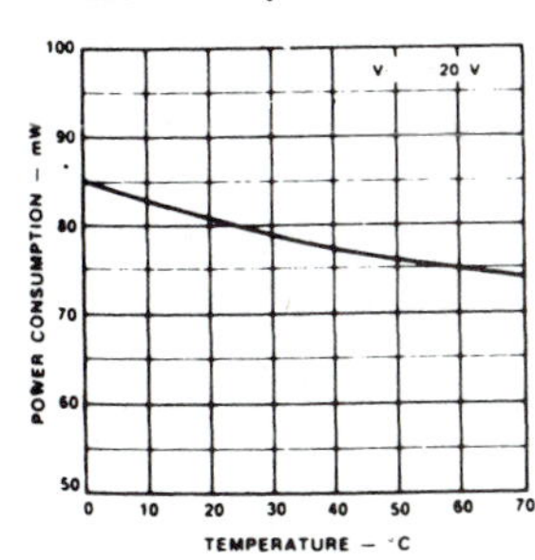

Output Short Circuit Current as a Function of Ambient Temperature

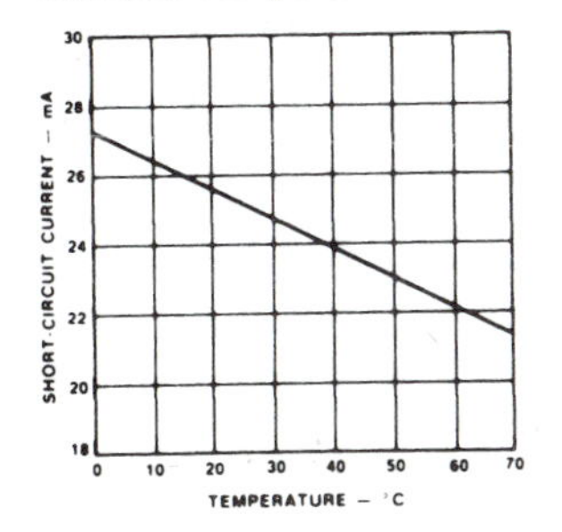

Frequency Characteristics as a Function of Ambient Temperature

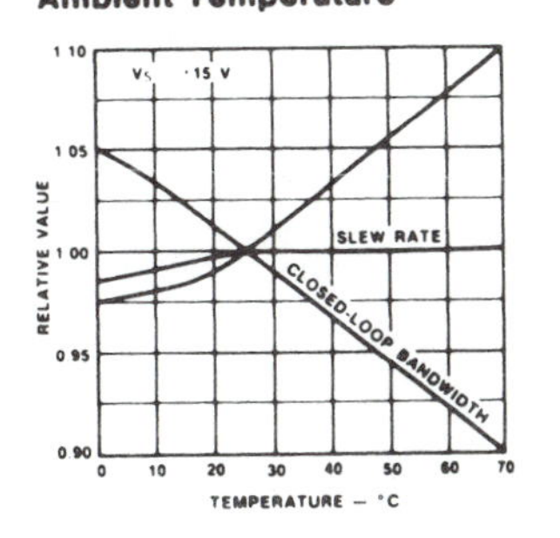

Figure 13–31
(Continued)

shown here use the μA prefix to designate a Fairchild product, and data is given for the μA741, μA741A, μA741C, and μA741E versions. Often the versions differ in respect to the intended market: military or commercial. Specifications for military versions are generally more stringent than their commercial counterparts. For example, we see that the operating temperature range for the military versions (μA741 and μA741A) is −55°C to +125°C, while that of the commercial versions (μA741C and μA741E) is 0°C to +70°C.

Reviewing the specifications, we see that parameter values are comparable to those we have used in the examples of this chapter and are representative of a general-purpose, BJT operational amplifier. Note that most entries show a typical value and a minimum or maximum value. The range of values is the manufacturer's statement of the variation that can be expected among a large number of 741 chips. Those parameters for which a large numerical value is desirable show a minimum value and those for which a small numerical value is desirable show a maximum value. For example, the input offset voltage for the μA741 at 25°C has a typical value of 1 mV but may be as great as 5 mV. Circuit designers who are using the 741, or any other operational amplifier, in the design of a product that will be manufactured in large quantities should use the worst-case specifications.

Note that the specifications include values for the common-mode rejection ratio (CMRR) in dB. The CMRR for an operational amplifier is defined in exactly the same way that we introduced it in connection with differential amplifiers in Chapter 12.

Many of the specifications vary with operating conditions such as frequency, supply voltage, and ambient temperature. Typical variations are shown in the graphs that accompany the value listings. Important examples of which the designer should be aware include the following:

1. The open-loop voltage gain for the μA741 increases from 90 dB to nearly 110 dB as the supply voltage ranges from 2 V to 20 V. Lower supply voltages mean lower values of open-loop gain.
2. Beyond about 100 Hz, the CMRR of the μA741E falls off at the rate of 20 dB/decade, typical behavior for the CMRR of an operational amplifier.
3. The closed-loop bandwidth of the μA741 decreases linearly with increasing ambient temperature, its value at 120°C being about 80% of its value at 20°C.

Example 13–17

Assuming worst-case conditions at 25°C, determine the following, in connection with the μA741 operational-amplifier circuit shown in Figure 13–32:

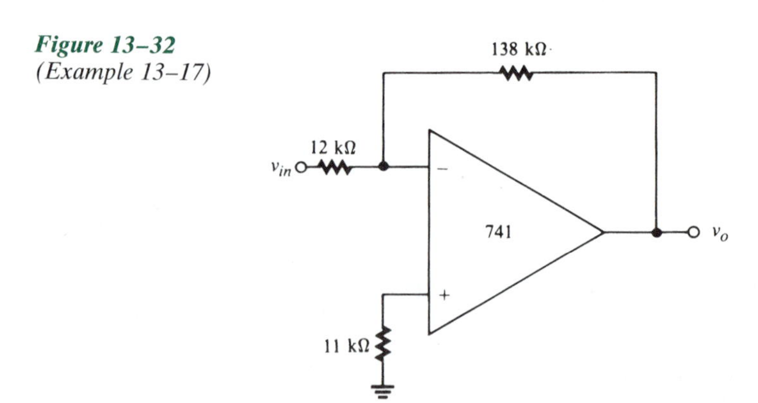

Figure 13–32
(Example 13–17)

1. the closed-loop bandwidth
2. the maximum operating frequency when the input is a 0.5 V-peak sine wave
3. the total output offset voltage $|V_{OS}|$

Solution

1. From equation 13–38, $BW_{CL} = f_t\beta$. The graph labeled "Open Loop Voltage Gain as a Function of Frequency" (the open-loop frequency response) in the 741 specifications reveals the unity-gain frequency to be approximately 1 MHz. From Figure 13–32,

$$\beta = \frac{R_1}{R_1 + R_f} = \frac{12\text{ k}\Omega}{(12\text{ k}\Omega) + (138\text{ k}\Omega)} = 0.08$$

Thus, $BW_{CL} = (1\text{ MHz})(0.08) = 80\text{ kHz}$.

2. By equation 13–44, the maximum operating frequency under the slew-rate limitation is

$$f_s(\text{max}) = \frac{S}{2\pi K}\text{ Hz}$$

The specifications show the slew rate to be 0.5 V/μs. From the figure, the closed-loop gain of the inverting configuration is

$$\frac{v_o}{v_{in}} = \frac{-R_f}{R_1} = \frac{-138\text{ k}\Omega}{12\text{ k}\Omega} = -11.5$$

Therefore, the magnitude of the peak output voltage is $K = (0.5)(11.5) = 5.75$ V. Thus,

$$f_s(\text{max}) = \frac{0.5 \times 10^6\text{ V/s}}{2\pi(5.75)\text{ V}} = 13.84\text{ kHz}$$

Since $f_s(\text{max}) < BW_{CL}$, the maximum operating frequency for a 0.5-V-peak input is 13.84 kHz.

3. $R_1 \parallel R_f = (12\text{ k}\Omega) \parallel (138\text{ k}\Omega) \approx 11\text{ k}\Omega$. Therefore, the compensating resistor has its optimum value and we can use equation 13–59 to determine the output offset due to bias currents: $|V_{OS}(I_B)| = |I_{io}|R_f$. The μA741 specifications list the maximum value of input offset current to be 200 nA. Therefore, $|V_{OS}(I_B)|_{\text{max}} = (200 \times 10^{-9})(138 \times 10^3) = 27.6$ mV. The specifications list the maximum value of input offset voltage to be 5 mV. Therefore,

$$|V_{OS}(V_{io})| = V_{io}\left(\frac{R_f + R_1}{R_1}\right) = (5\text{ mV})\left[\frac{(138\text{ k}\Omega) + (12\text{ k}\Omega)}{12\text{ k}\Omega}\right] = 62.5\text{ mV}$$

Finally, $|V_{OS}|_{\text{worst case}} = (27.6\text{ mV}) + (62.5\text{ mV}) = 90.1\text{ mV}$.

Notice that the 741 has a pair of pins across which a potentiometer can be connected for offset null (zeroing).

Example 13–18

PSPICE

Use PSpice and the PSpice library to verify the value of the closed-loop bandwidth calculated in Example 13–17 for the 741 operational amplifier. Assume a 0.1-V-ac input.

Figure 13–33
(Example 13–18)

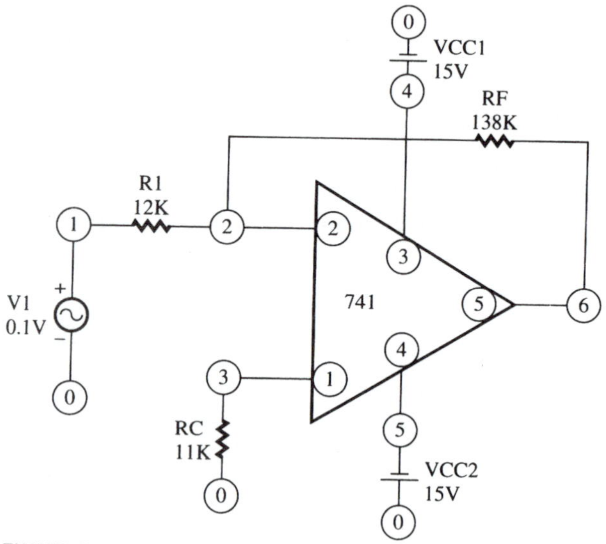

```
EXAMPLE 13.18
V1   1   0   AC   0.1V
VCC1   4   0   15V
VCC2   0   5   15V
R1   1   2   12K
RF   2   6   138K
RC   3   0   11K
X1   3   2   4   5   6   UA741
.LIB
.AC   LIN   1  80KHZ   80KHZ
.PRINT   AC   V(6)
.END
```

Solution. The PSpice circuit and input circuit file are shown in Figure 13–33. Note that subcircuit call X1 specifies the name UA741 for the subcircuit stored in the PSpice library (see Appendix Section A–17). (If the Evaluation version of PSpice is used, the .LIB statement must be written .LIB EVAL.LIB.) As shown in Example 13–17, the magnitude of the closed-loop gain is 11.5. Therefore, the midband output voltage is $|A_v|e_{in} = 11.5(0.1\text{ V}) = 1.15\text{ V}$. At cutoff, the output voltage should be $0.707(1.15\text{ V}) = 0.813\text{ V}$. To verify the calculations of Example 13–17, the frequency of the simulation is set to the theoretical bandwidth, 80 kHz, at which the output voltage should be near 0.813 V. Execution of the program reveals that the output voltage at 80 kHz is 0.822 V, which differs only 1.1% from 0.813 V.

EXERCISES

Section 13–1
The Ideal Operational Amplifier

13–1. Find the output of the ideal operational amplifier shown in Figure 13–34 for each of the following input signals:

a. $v_{in} = 120$ mV dc
b. $v_{in} = 0.5 \sin \omega t$ V
c. $v_{in} = -2.5$ V dc
d. $v_{in} = 4 - \sin \omega t$ V
e. $v_{in} = 0.8 \sin (\omega t + 75°)$ V

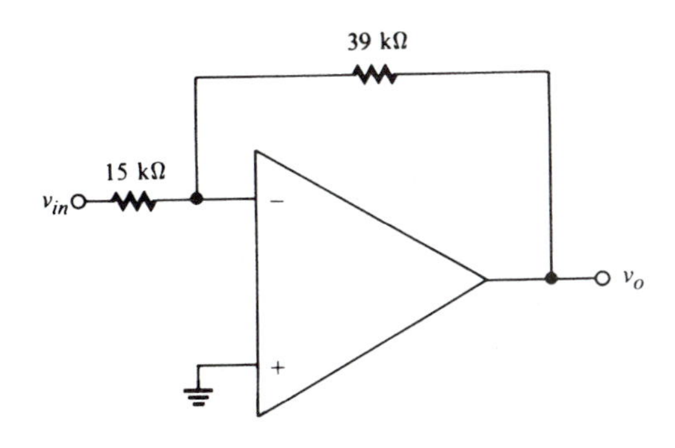

Figure 13–34
(Exercise 13–1)

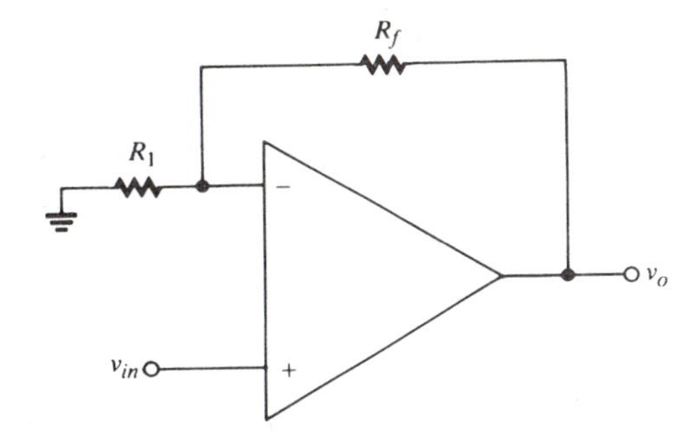

Figure 13–35
(Exercise 13–6)

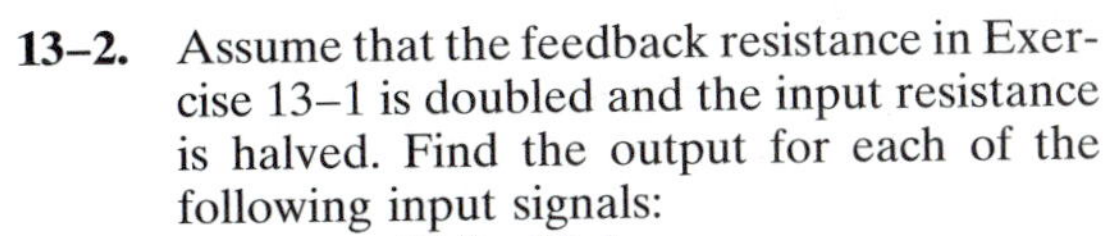

13–2. Assume that the feedback resistance in Exercise 13–1 is doubled and the input resistance is halved. Find the output for each of the following input signals:
a. $v_{in} = -60.5$ mV dc
b. $v_{in} = 500 \sin \omega t \ \mu$V
c. $v_{in} = -0.16 + \sin \omega t$ V
d. $v_{in} = -0.2 \sin (\omega t - 30°)$ V

13–3. Find the current in the feedback resistor for each part of Exercise 13–2.

13–4. The amplifier in Exercise 13–1 is driven by a signal source whose output resistance is 40 kΩ. The source voltage is 2.2 V rms. What is the rms value of the amplifier's output voltage?

13–5. Design an inverting operational-amplifier circuit that will provide an output of 10 V rms when the input is a 1-V-rms signal originating at a source having 10 kΩ source resistance.

13–6. The input to the ideal operational amplifier shown in Figure 13–35 is 0.5 V rms. Find the rms value of the output for each of the following combinations of resistor values:
a. $R_1 = R_f = 10$ kΩ
b. $R_1 = 20$ kΩ, $R_f = 100$ kΩ
c. $R_1 = 100$ kΩ, $R_f = 20$ kΩ
d. $R_f = 10R_1$

13–7. Repeat Exercise 13–6 for each of the following resistor combinations:
a. $R_1 = 125$ kΩ, $R_f = 1$ MΩ
b. $R_1 = 220$ kΩ, $R_f = 47$ kΩ
c. $R_1/R_f = 0.1$
d. $R_1/R_f = 10$

13–8. Assuming ideal operational amplifiers, find the load voltage v_L in Figure 13–36.

13–9. Assuming ideal operational amplifiers, find the load voltage v_L in each part of Figure 13–37.

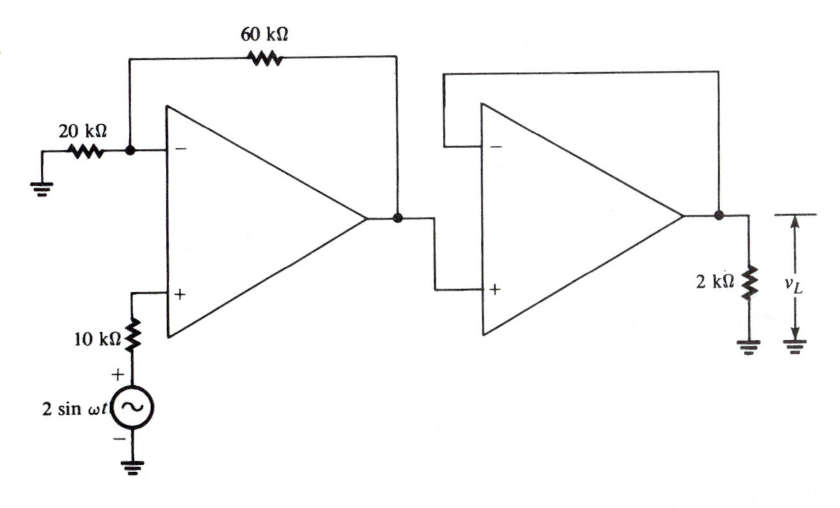

Figure 13–36
(Exercise 13–8)

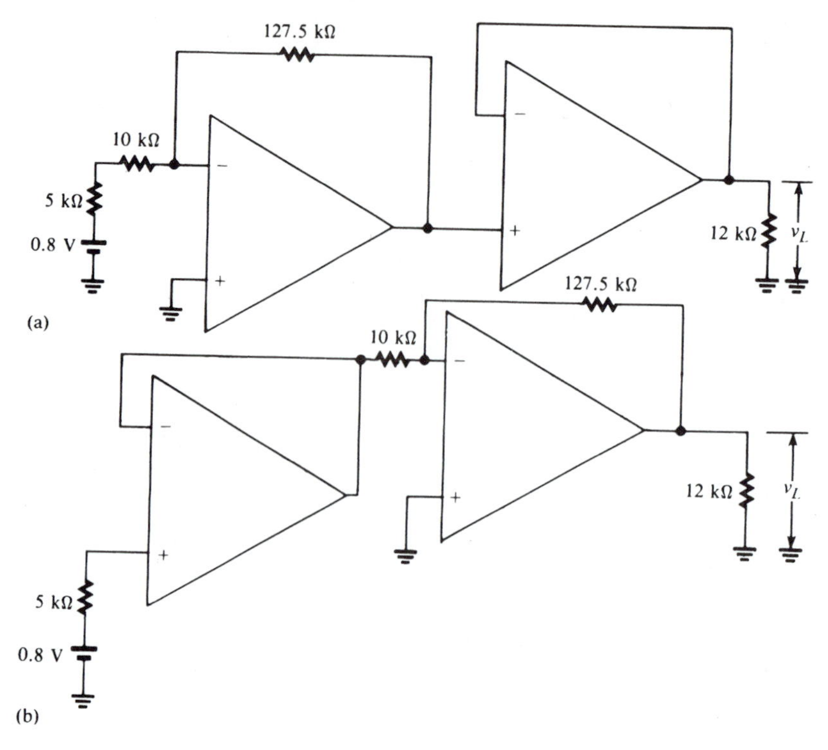

Figure 13–37
(Exercise 13–9)

Section 13–2
Feedback Theory

13–10. An operational amplifier having an open-loop gain of 5000 is used in a noninverting configuration with a feedback resistor of 1 MΩ. For each of the following values of R_1, find the closed-loop gain if the amplifier were ideal, and find the actual closed-loop gain.

a. $R_1 = 5$ kΩ
b. $R_1 = 20$ kΩ
c. $R_1 = 100$ kΩ

13–11. Repeat Exercise 13–10 when the open-loop gain of the amplifier is increased by a factor of 2.

13–12. An operational amplifier is to be used in a noninverting configuration that has an ideal closed-loop gain of 800. What minimum value of open-loop gain should the amplifier have if the actual closed-loop gain must be at least 799?

13–13. The operational amplifier in Exercise 13–10 has a differential input resistance of 40 kΩ and an output resistance of 90 Ω. Find the closed-loop input and output resistance for each of the values of R_1 listed.

13–14. An operational amplifier has open-loop gain 10^4 and output resistance 120 Ω. It is to be used in a noninverting configuration for an application in which its closed-loop output resistance must be no greater than 1 Ω. What is the maximum closed-loop gain that the amplifier can have?

13–15. Derive equation 13–28 from equation 13–27.

13–16. An operational amplifier has an open-loop gain of 5000. It is used in an inverting configuration with a feedback ratio of 0.2. What is its closed-loop gain?

13–17. The operational amplifier in Figure 13–38 has an open-loop gain of 5×10^4. Draw a block diagram of the feedback model for the configuration shown. Label each block with the correct numerical quantity.

13–18. Show that the feedback model shown in Figure 13–39 is equivalent to that shown in Figure 13–12.

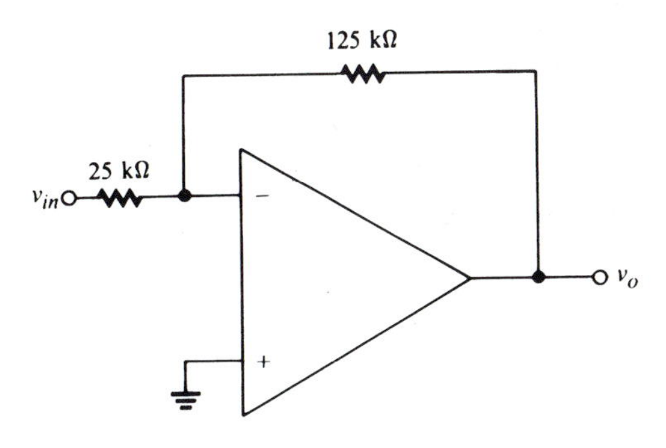

Figure 13–38
(Exercise 13–17)

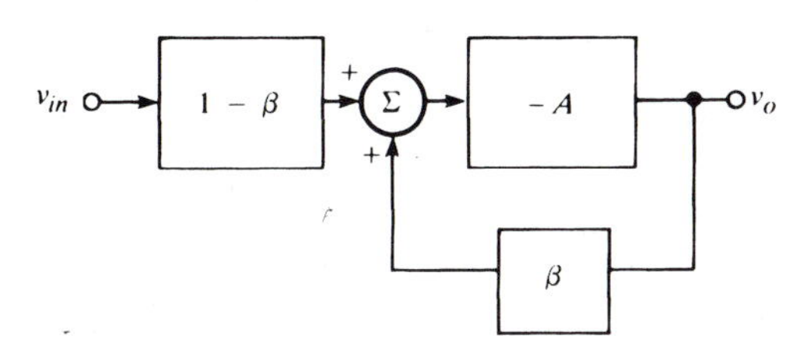

Figure 13–39
(Exercise 13–18)

13–19. An operational amplifier is to be used in an inverting configuration with feedback resistance 100 kΩ and input resistance 2 kΩ. If the closed-loop gain must be no less than −49.5,

a. what minimum value of loop gain should it have; and

b. what minimum value of open-loop gain should it have? (*Hint:* Work with gain *magnitudes.*)

13–20. The operational amplifier shown in Figure 13–40 has an open-loop gain of 8000 and an open-loop output resistance of 250 Ω. Find the closed-loop

a. input resistance; and

b. output resistance.

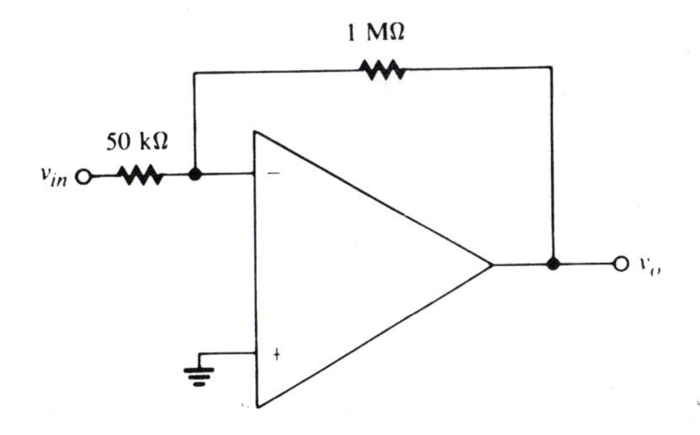

Figure 13–40
(Exercise 13–20)

Section 13–3
Frequency Response

13–21. An operational amplifier has gain-bandwidth product equal to 5×10^5 and a dc, open-loop gain of 20,000. At what frequency does the open-loop gain equal 14,142?

13–22. With reference to the operational amplifier in Exercise 13–21,

a. at what frequency is the open-loop gain equal to 0 dB, and

b. what is the open-loop gain at 2.5 kHz?

13–23. The operational amplifier in Figure 13–41 has a unity-gain frequency of 1.2 MHz.

a. What is the closed-loop bandwidth?

b. What is the closed-loop gain at 600 kHz?

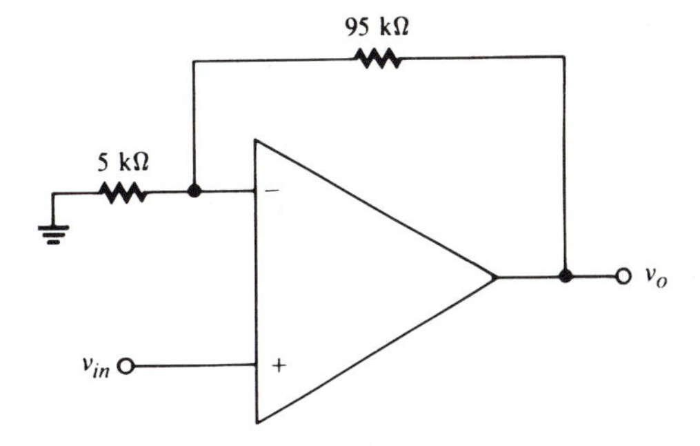

Figure 13–41
(Exercise 13–23)

13–24. An operational amplifier having a gain-bandwidth product of 8×10^5 is to be used in a noninverting configuration as an audio amplifier (20 Hz–20 kHz). What is the maximum closed-loop gain that can be obtained from the amplifier in this application?

13–25. An operational amplifier has a dc open-loop gain of 25×10^4 and an open-loop cutoff frequency of 40 Hz. It is to be used in an inverting configuration to amplify signals up to 50 kHz. What is the maximum closed-loop gain that can be obtained from the amplifier for this application?

13–26. The operational amplifier in Figure 13–42 has a unity-gain frequency of 2 MHz.

a. What is the closed-loop bandwidth?

b. What is the closed-loop gain at 2 MHz?

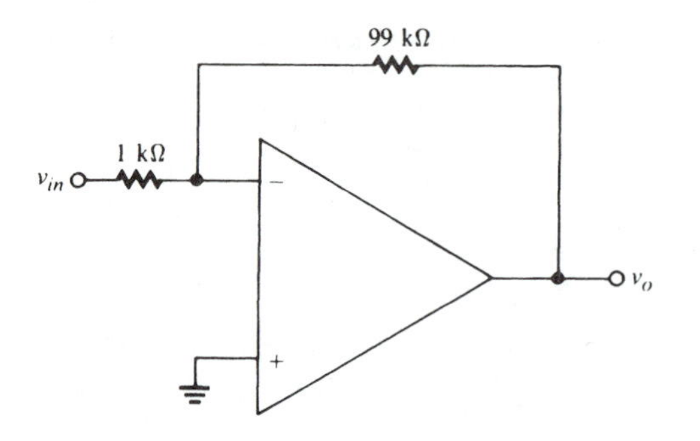

Figure 13–42
(Exercise 13–26)

13–27. Each of the operational amplifiers in Figure 13–43 has a unity-gain frequency of 750 kHz. What is the approximate upper cutoff frequency of the cascaded system?

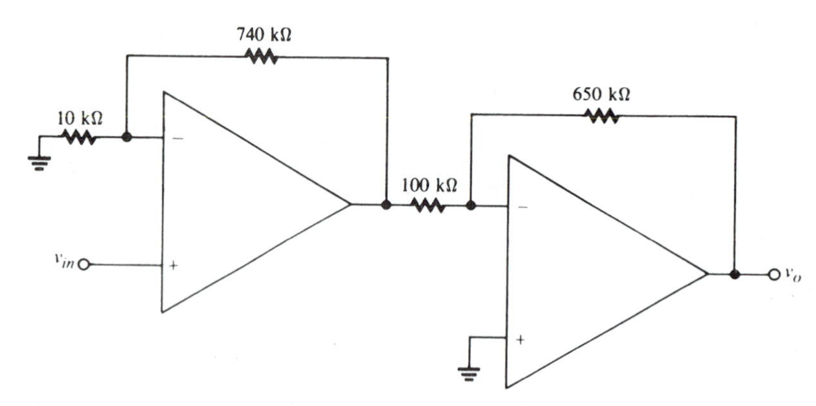

Figure 13–43
(Exercise 13–27)

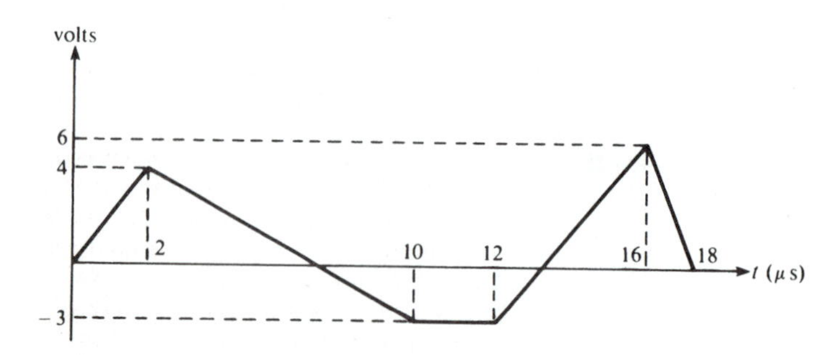

Figure 13–44
(Exercise 13–29)

Section 13–4
Slew Rate

13–28. What is the rate of change, in volts/second, of a triangular waveform that varies between 0 V and 5 V and that has frequency 10 kHz?

13–29. What minimum slew rate is necessary for a unity-gain amplifier that must pass, without distortion, the input waveform shown in Figure 13–44?

13–30. Repeat Exercise 13–29 if the amplifier is in a noninverting configuration with $R_1 = 50$ kΩ and $R_f = 100$ kΩ.

13–31. An inverting amplifier has a slew rate of 2 V/μs. What maximum closed-loop gain can it have without distorting the input waveform shown in Figure 13–44?

13–32. An operational amplifier has a slew rate of 2 V/μs. What maximum peak amplitude can a 1-MHz input signal have without exceeding the slew rate, when the amplifier is used in a voltage-follower circuit?

13–33. What minimum slew rate is required for an operational amplifier whose output must be at least 2 V rms over the audio frequency range?

13–34. The operational amplifier shown in Figure 13–45 has a slew rate of 1.2 V/μs. Determine whether the output will be distorted due to the slew-rate limitation when the input is any one of the following sinusoidal signals: v_1 = 0.7 V rms at 30 kHz; v_2 = 1.0 V rms at 15 kHz; v_3 = 0.5 V rms at 40 kHz; v_4 = 0.1 V rms at 20 kHz. If distortion occurs, determine the signals that are responsible.

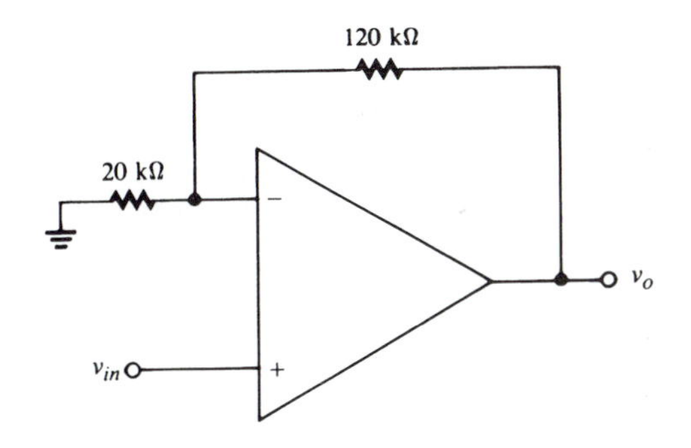

Figure 13–45
(Exercise 13–34)

13–35. If distortion occurs in Exercise 13–34, find remedies other than changing the input signals.

13–36. The operational amplifier shown in Figure 13–46 has a slew rate of 0.5 V/μs and a unity-gain frequency of 1 MHz. Find the maximum frequency of a 0.2-V-rms sine-wave input that can be amplified without distortion.

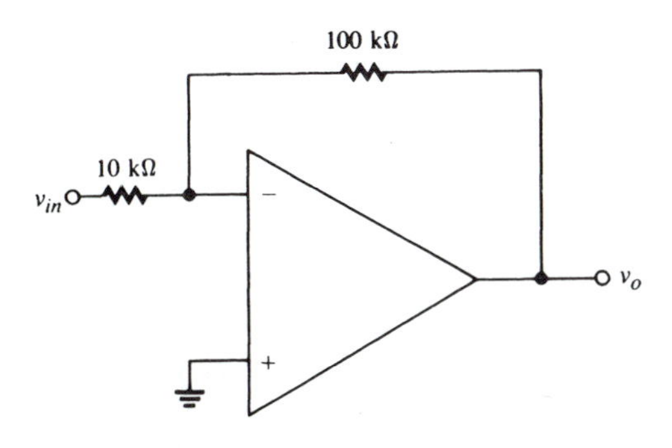

Figure 13–46
(Exercise 13–36)

13–37. To what value could the feedback resistor in Exercise 13–36 be changed if it were necessary that the maximum frequency be 50 kHz?

13–38. If the input to the amplifier in Exercise 13–36 is a triangular wave that varies between −1 V and +1 V peak and has frequency 1 kHz, determine whether distortion will occur.

13–39. If the input to the amplifier in Exercise 13–36 is a ramp that rises from −0.5 V to +1.5 V in 2.5 μs, determine whether distortion will occur.

Section 13–5
Offset Currents and Voltages

13–40. The operational amplifier shown in Figure 13–47 has I_B^+ = 100 nA and I_B^- = 80 nA. If $R_1 = R_f = R_c$ = 20 kΩ, find the output offset voltage due to the input bias currents.

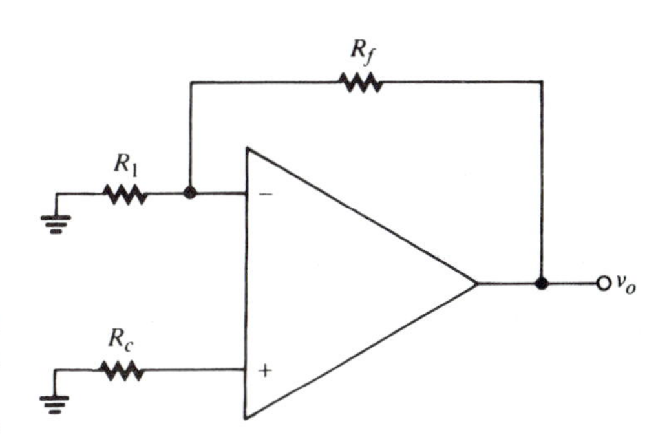

Figure 13–47
(Exercise 13–40)

13–41. Find the optimum value for R_c in Exercise 13–40 and repeat the exercise using that value.

13–42. An operational amplifier has an input offset current of 50 nA. It is to be used in an inverting-amplifier application where the output offset voltage due to bias currents cannot exceed 10 mV. If the amplifier must provide an impedance of at least 40 kΩ to the signal source driving it, what is the maximum permissible closed-loop gain of the amplifier?

13–43. An operational amplifier having an input offset current of 80 nA is to be used in a noninverting-amplifier application where the output offset voltage due to bias currents cannot exceed 5 mV. The value of R_1 is 10 kΩ and the amplifier must have the maximum possible closed-loop gain. Design the circuit and find its voltage gain.

13–44. Given $I_B = (I_B^+ + I_B^-)/2$ and $|I_{io}| = |I_B^+ - I_B^-|$, solve these equations simultaneously to show that

a. when $I_B^+ > I_B^-$, $I_B^+ = I_B + 0.5|I_{io}|$ and $I_B^- = I_B - 0.5|I_{io}|$; and

b. when $I_B^+ < I_B^-$, $I_B^+ = I_B - 0.5|I_{io}|$ and $I_B^- = I_B + 0.5|I_{io}|$. (*Hint:* When $I_B^+ > I_B^-$, $|I_B^+ - I_B^-| = I_B^+ - I_B^-$.)

13–45. The operational amplifier in Figure 13–48 has $I_B = 100$ nA and $|I_{io}| = 50$ nA. What is the maximum possible value of $|V_{OS}(I_B)|$?

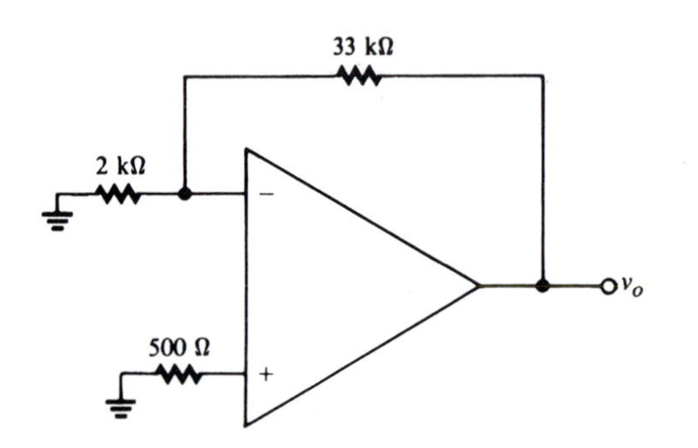

Figure 13–48
(Exercise 13–45)

13–46. An operational amplifier has an input offset voltage of 1.2 mV. It is to be used in a noninverting-amplifier application where the output offset due to V_{io} cannot exceed 6 mV. If the feedback resistor is 20 kΩ, what is the minimum permissible value of R_1?

13–47. The input offset voltage of an operational amplifier used in a voltage-follower circuit is 0.5 mV. What is the value of $|V_{OS}(V_{io})|$?

13–48. The operational amplifier shown in Figure 13–49 has $|I_{io}| = 120$ nA and $|V_{io}| = 1.5$ mV. Find the worst-case output offset voltage.

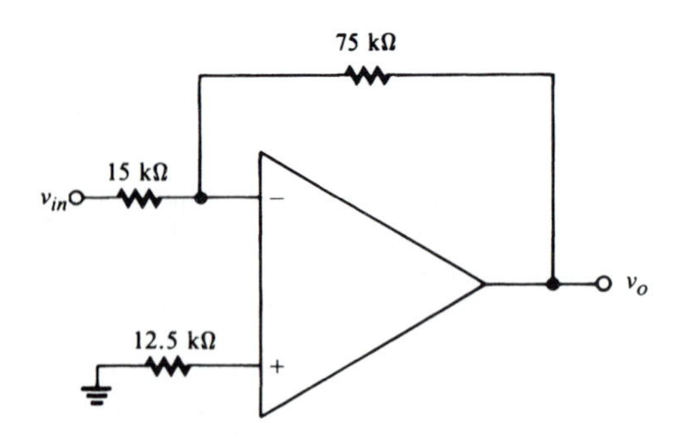

Figure 13–49
(Exercise 13–48)

13–49. The operational amplifier in Figure 13–50 has $I_B = 100$ nA, $|I_{io}| = 30$ nA, and $|v_{io}| = 2$ mV. Find the worst-case value of $|v_{OS}|$.

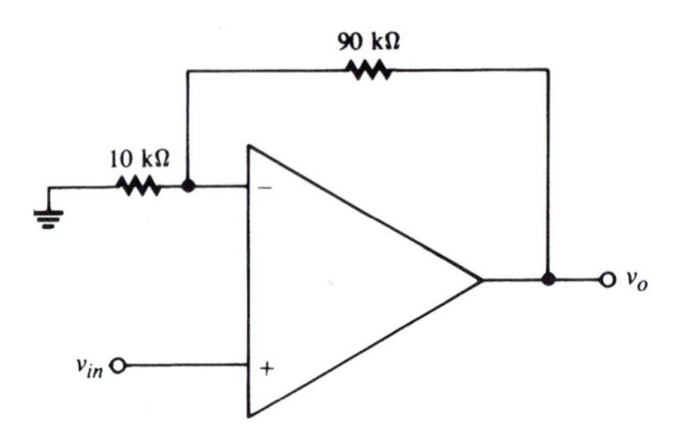

Figure 13–50
(Exercise 13–49)

Section 13–6
Operational Amplifier Specifications

13–50. Using the 741 operational amplifier specifications, determine the following:

a. the minimum slew rate of the μA741E at 25°C;

b. the output voltage swing of the μA741C at 70°C when the supply voltage is 15 V;

c. the input bias current of the μA741A at 60°C ambient temperature; and

d. the input offset current of the μA741 at 0°C ambient temperature.

13–51. Using the 741 operational amplifier specifications, determine the following:

a. the maximum input offset voltage of the μA741A at 25°C;

b. the CMRR of the μA741C at 10 kHz;

c. the input offset current of the μA741E at 25°C when the supply voltage is 10 V; and

d. the closed-loop bandwidth of the μA741 at 100°C if its closed-loop bandwidth at 20°C is 40 kHz.

13–52. Using the μA741A, design an inverting operational-amplifier circuit that will meet the following criteria at 25°C:
- **a.** maximum possible closed-loop voltage gain;
- **b.** closed-loop bandwidth of at least 20 kHz;
- **c.** worst-case output offset voltage of 50 mV; and
- **d.** minimum input resistance 10 kΩ.

13–53. Using the μA741C, design a noninverting operational-amplifier circuit that will meet the following criteria at 25°C:
- **a.** maximum possible closed-loop voltage gain;
- **b.** must not distort an input ramp voltage that changes from -1 V to $+1$ V in 150 μs;
- **c.** worst-case output offset voltage of 100 mV; and
- **d.** R_1 must be at least 1 kΩ.

SPICE EXERCISES

13–54. The ideal operational amplifier in Figure 13–2 has $R_1 = 10$ kΩ and $R_f = 25$ kΩ. The input voltage, v_{in}, is $2 \sin(2\pi \times 10^3 t)$ V. Use SPICE to obtain a plot of the output waveform versus time over one full period. Use the results to determine the closed-loop voltage gain and compare that with the theoretical value for an ideal amplifier.

13–55. An operational amplifier is connected in a noninverting configuration with $R_1 = 10$ kΩ and $R_f = 90$ kΩ. The open-loop voltage gain of the amplifier is 1×10^3 and its open-loop output resistance is 150 Ω. Assuming the input impedance is essentially infinite (10^{12} Ω), use SPICE to find the closed-loop voltage gain and output resistance. Compare your results with the theoretical values.

14 APPLICATIONS OF OPERATIONAL AMPLIFIERS

In this chapter, we will explore some useful and popular applications of operational amplifiers. In connection with several of these, we will introduce some important concepts, including oscillation theory, filtering, and waveshaping, that have broad implications beyond just their relevance to operational-amplifier circuits. Unless otherwise noted, we will assume that the operational amplifiers are ideal, or close enough to ideal that we can ignore small deviations from theory caused by finite gain, finite input impedance, and nonzero output impedance.

14–1 VOLTAGE SUMMATION, SUBTRACTION, AND SCALING

Voltage Summation

We have seen that it is possible to *scale* a signal voltage, that is, to multiply it by a fixed constant, through an appropriate choice of external resistors that determine the closed-loop gain of an amplifier circuit. This operation can be accomplished in either an inverting or noninverting configuration. It is also possible to sum several signal voltages in one operational-amplifier circuit and at the same time scale each by a different factor. For example, given inputs v_1, v_2, and v_3, we might wish to generate an output equal to $2v_1 + 0.5v_2 + 4v_3$. The latter sum is called a *linear combination* of v_1, v_2, and v_3, and the circuit that produces it is often called a *linear-combination circuit.*

Figure 14–1 shows an inverting amplifier circuit that can be used to sum and scale three input signals. Note that input signals v_1, v_2, and v_3 are applied through separate resistors R_1, R_2, and R_3 to the summing junction of the amplifier and that there is a single feedback resistor R_f. Resistor R_c is the offset compensation resistor discussed in Chapter 13.

Following the same procedure we used in Chapter 13 to derive the output of an inverting amplifier having a single input, we obtain for the three-input (ideal) amplifier

$$i_1 + i_2 + i_3 = -i_f \tag{14–1}$$

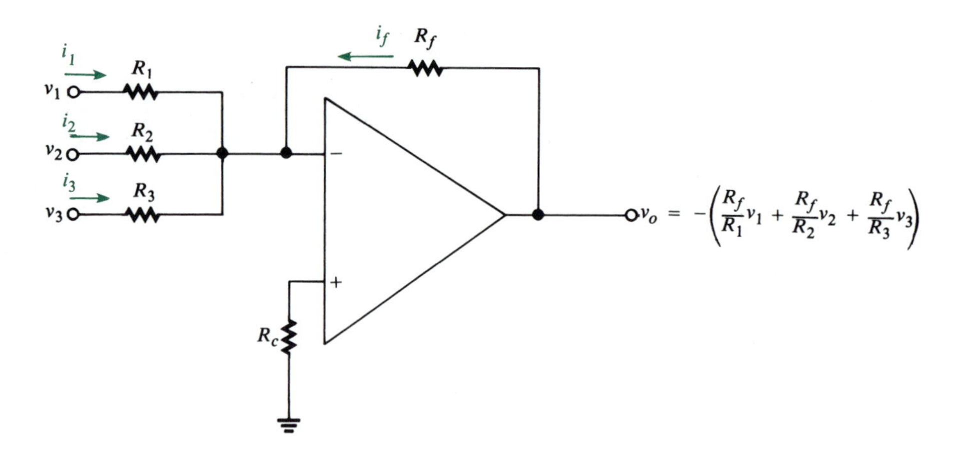

Figure 14–1
An operational-amplifier circuit that produces an output equal to the (inverted) sum of three separately scaled input signals

Or, since the voltage at the summing junction is ideally 0,

$$\frac{v_1}{R_1} + \frac{v_2}{R_2} + \frac{v_3}{R_3} = \frac{-v_o}{R_f} \qquad \textbf{(14–2)}$$

Solving (14–2) for v_o gives

$$v_o = -\left(\frac{R_f}{R_1}v_1 + \frac{R_f}{R_2}v_2 + \frac{R_f}{R_3}v_3\right) \qquad \textbf{(14–3)}$$

Equation 14–3 shows that the output is the inverted sum of the separately scaled inputs, i.e., a *weighted* sum, or linear combination of the inputs. By appropriate choice of values for R_1, R_2, and R_3, we can make the scale factors equal to whatever constants we wish, within practical limits. If we choose $R_1 = R_2 = R_3 = R$, then we obtain

$$v_o = \frac{-R_f}{R}(v_1 + v_2 + v_3) \qquad \textbf{(14–4)}$$

and, for $R_f = R$,

$$v_o = -(v_1 + v_2 + v_3) \qquad \textbf{(14–5)}$$

The theory can be extended in an obvious way to two, four, or any reasonable number of inputs. The feedback ratio for the circuit is

$$\beta = \frac{R_p}{R_p + R_f} \qquad \textbf{(14–6)}$$

where $R_p = R_1 \parallel R_2 \parallel R_3$. Using this value of β, we can apply the theory developed in Chapter 13 to determine all the performance characteristics that depend on β, including closed-loop bandwidth and output offset $V_{OS}(V_{io})$. The optimum value of the bias-current compensation resistor is

$$R_c = R_f \parallel R_p = R_f \parallel R_1 \parallel R_2 \parallel R_3 \qquad \textbf{(14–7)}$$

Example 14–1

DESIGN

1. Design an operational-amplifier circuit that will produce an output equal to $-(4v_1 + v_2 + 0.1v_3)$.
2. Write an expression for the output and sketch its waveform when $v_1 = 2 \sin \omega t$ V, $v_2 = +5$ V dc, and $v_3 = -100$ V dc.

Solution

1. We arbitrarily choose $R_f = 60$ kΩ. Then

$$\frac{R_f}{R_1} = 4 \Rightarrow R_1 = \frac{60\text{ k}\Omega}{4} = 15\text{ k}\Omega$$

$$\frac{R_f}{R_2} = 1 \Rightarrow R_2 = \frac{60\text{ k}\Omega}{1} = 60\text{ k}\Omega$$

$$\frac{R_f}{R_3} = 0.1 \Rightarrow R_3 = \frac{60\text{ k}\Omega}{0.1} = 600\text{ k}\Omega$$

By equation 14–7, the optimum value for the compensating resistor is $R_c = R_f \parallel R_1 \parallel R_2 \parallel R_3 = (60\text{ k}\Omega) \parallel (15\text{ k}\Omega) \parallel (60\text{ k}\Omega) \parallel (600\text{ k}\Omega) = 9.8\text{ k}\Omega$. The circuit is shown in Figure 14–2.

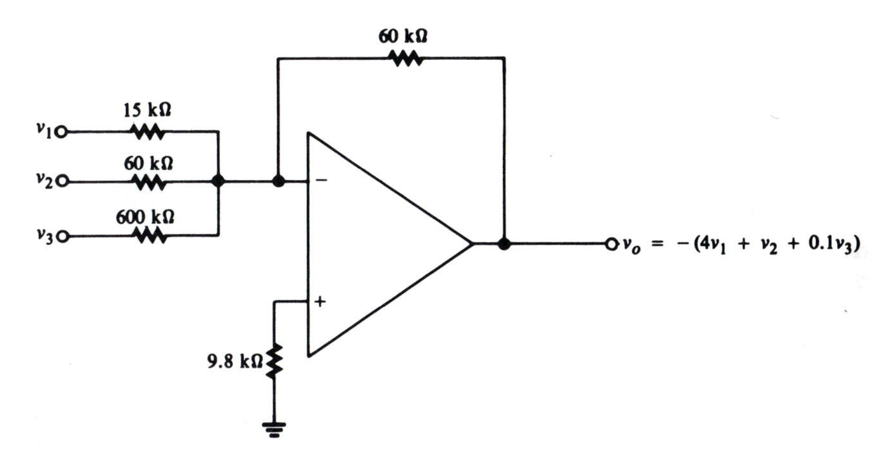

Figure 14–2
(Example 14–1)

Figure 14–3
(Example 14–1)

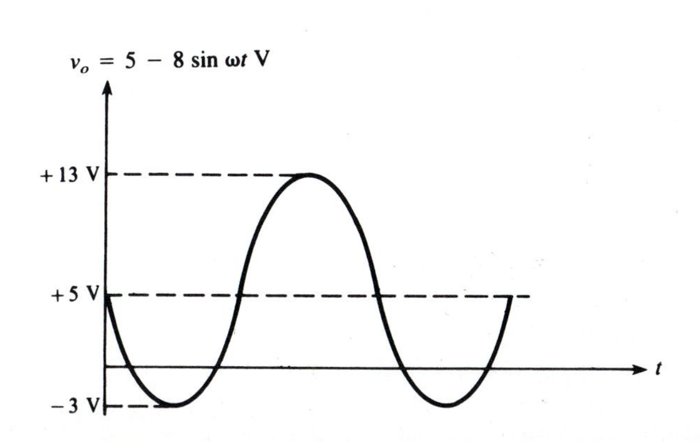

2. $v_o = -[4(2 \sin \omega t) + 1(5) + 0.1(-100)] = -8 \sin \omega t - 5 + 10 = 5 - 8 \sin \omega t$. This output is sinusoidal with a 5-V offset and varies between $5 - 8 = -3$ V and $5 + 8 = 13$ V. It is sketched in Figure 14–3.

Figure 14–4 shows a noninverting version of the linear-combination circuit. In this example, only two inputs are connected and it can be shown (Exercise 14–4) that

$$v_o = \frac{R_g + R_f}{R_g}\left(\frac{R_2}{R_1 + R_2}v_1 + \frac{R_1}{R_1 + R_2}v_2\right) \tag{14–8}$$

Although this circuit does not invert the scaled sum, it is somewhat more cumbersome than the inverting circuit in terms of selecting resistor values to provide precise scale factors. Also, it is limited to producing outputs of the form $K[av_1 + (1 - a)v_2]$ where K and a are positive constants. Phase inversion is often of no consequence, but in those applications where a noninverted sum is required, it can also be obtained using the inverting circuit of Figure 14–1, followed by a unity-gain inverter.

Voltage Subtraction

Suppose we wish to produce an output voltage that equals the mathematical difference between two input signals. This operation can be performed by using the amplifier in a *differential* mode, where the signals are connected through appropriate resistor networks to the inverting and noninverting terminals. Figure 14–5 shows the configuration. We can use the superposition principle to determine the output of this circuit. First, assume that v_2 is shorted to ground. Then

$$v^+ = \frac{R_2}{R_1 + R_2}v_1 \tag{14–9}$$

so

$$v_{o1} = \frac{R_3 + R_4}{R_3}v^+ = \left(\frac{R_3 + R_4}{R_3}\right)\left(\frac{R_2}{R_1 + R_2}\right)v_1 \tag{14–10}$$

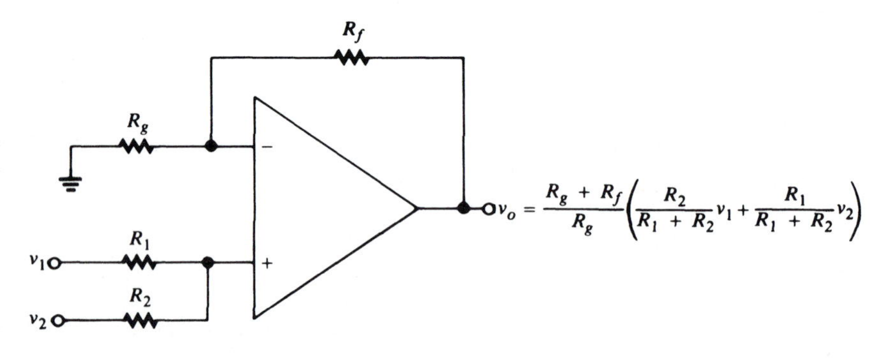

Figure 14–4
A noninverting linear combination circuit

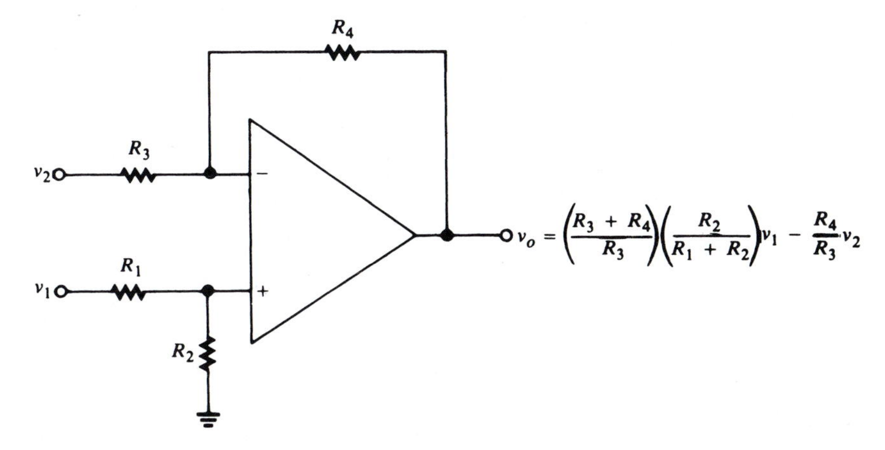

Figure 14–5
Using the amplifier in a differential mode to obtain an output proportional to the difference between two scaled inputs

Assuming now that v_1 is shorted to ground, we have

$$v_{o2} = \frac{-R_4}{R_3}v_2 \tag{14–11}$$

Therefore, with both signal inputs present, the output is

$$v_o = v_{o1} + v_{o2} = \left(\frac{R_3 + R_4}{R_3}\right)\left(\frac{R_2}{R_1 + R_2}\right)v_1 - \left(\frac{R_4}{R_3}\right)v_2 \tag{14–12}$$

Equation 14–12 shows that the output is proportional to the difference between scaled multiples of the inputs. To obtain the output

$$v_o = A(v_1 - v_2) \tag{14–13}$$

where A is a fixed constant, select the resistor values in accordance with the following:

$$R_1 = R_3 = R \quad \text{and} \quad R_2 = R_4 = AR \tag{14–14}$$

Substituting these values into (14–12) gives

$$\left(\frac{R + AR}{R}\right)\left(\frac{AR}{R + AR}\right)v_1 - \frac{AR}{R}v_2 = \frac{AR}{R}v_1 - \frac{AR}{R}v_2 = A(v_1 - v_2)$$

as required. When resistor values are chosen in accordance with (14–14), the bias compensation resistance ($R_1 \| R_2$) is automatically the correct value ($R_3 \| R_4$), namely, $R \| AR$.

Let the general form of the output of Figure 14–5 be

$$v_o = a_1v_1 - a_2v_2 \tag{14–15}$$

where a_1 and a_2 are positive constants. Then, by equation 14–12, we must have

$$a_1 = \left(1 + \frac{R_4}{R_3}\right)\left(\frac{R_2}{R_1 + R_2}\right) \tag{14–16}$$

and

$$a_2 = \frac{R_4}{R_3} \tag{14–17}$$

Substituting (14–17) into (14–16) gives

$$a_1 = (1 + a_2)\frac{R_2}{R_1 + R_2} \tag{14–18}$$

But the quantity $R_2/(R_1 + R_2)$ is always less than 1. Therefore, equation 14–18 shows that in order to use the circuit of Figure 14–5 to produce $v_o = a_1v_1 - a_2v_2$, we must have

$$(1 + a_2) > a_1 \tag{14–19}$$

This restriction limits the usefulness of the circuit.

Example 14–2

DESIGN

Design an operational-amplifier circuit that will produce the output $v_o = 0.5v_1 - 2v_2$.

Solution. Note that $a_1 = 0.5$ and $a_2 = 2$, so $(1 + a_2) > a_1$. Therefore, it is possible to construct a circuit in the configuration of Figure 14–5.

Comparing v_o with equation 14–12, we see that we must have

$$\left(1 + \frac{R_4}{R_3}\right)\left(\frac{R_2}{R_1 + R_2}\right) = 0.5$$

and

$$\frac{R_4}{R_3} = 2$$

Let us arbitrarily choose $R_4 = 100$ kΩ. Then $R_3 = R_4/2 = 50$ kΩ. Thus

$$\left(1 + \frac{R_4}{R_3}\right)\left(\frac{R_2}{R_1 + R_2}\right) = \frac{3R_2}{R_1 + R_2} = 0.5$$

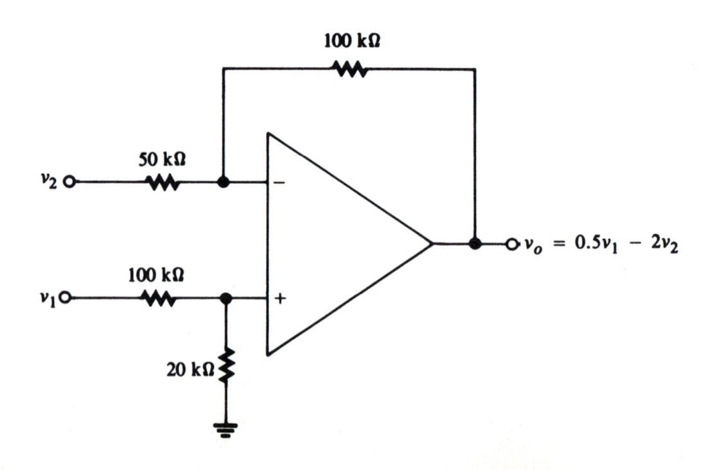

Figure 14–6
(Example 14–2)

Arbitrarily choosing $R_2 = 20\ \text{k}\Omega$, we have

$$\frac{3(20\ \text{k}\Omega)}{R_1 + (20\ \text{k}\Omega)} = 0.5$$

$$60\ \text{k}\Omega = 0.5R_1 + (10\ \text{k}\Omega)$$

$$R_1 = 100\ \text{k}\Omega$$

The completed design is shown in Figure 14–6.

In Example 14–2, we note that the compensation resistance ($R_1 \parallel R_2 = (100\ \text{k}\Omega) \parallel (20\ \text{k}\Omega) = 16.67\ \text{k}\Omega$) is not equal to its optimum value ($R_3 \parallel R_4 = (50\ \text{k}\Omega) \parallel (100\ \text{k}\Omega) = 33.33\ \text{k}\Omega$). With some algebraic complication, we can impose the additional condition $R_1 \parallel R_2 = R_3 \parallel R_4$ and thereby force the compensation resistance to have its optimum value. With $v_o = a_1v_1 - a_2v_2$, it can be shown (Exercise 14–8) that the compensation resistance ($R_1 \parallel R_2$) is optimum when the resistor values are selected in accordance with

$$R_4 = a_1R_1 = a_2R_3 = R_2(1 + a_2 - a_1) \qquad \textbf{(14–20)}$$

To apply this design criterion, choose R_4 and solve for R_1, R_2, and R_3. In Example 14–2, $a_1 = 0.5$ and $a_2 = 2$. If we choose $R_4 = 100\ \text{k}\Omega$, then $R_1 = (100\ \text{k}\Omega)/0.5 = 200\ \text{k}\Omega$, $R_2 = (100\ \text{k}\Omega)/2.5 = 40\ \text{k}\Omega$, and $R_3 = (100\ \text{k}\Omega)/2 = 50\ \text{k}\Omega$. These choices give $R_1 \parallel R_2 = 33.3\ \text{k}\Omega = R_3 \parallel R_4$, as required.

Although the circuit of Figure 14–5 is a useful and economical way to obtain a difference voltage of the form $A(v_1 - v_2)$, our analysis has shown that it has limitations and complications when we want to produce an output of the general form $v_o = a_1v_1 - a_2v_2$. An alternate way to obtain a scaled difference between two signal inputs is to use *two* inverting amplifiers, as shown in Figure 14–7. The output of the first amplifier is

$$v_{o1} = \frac{-R_2}{R_1}v_1 \qquad \textbf{(14–21)}$$

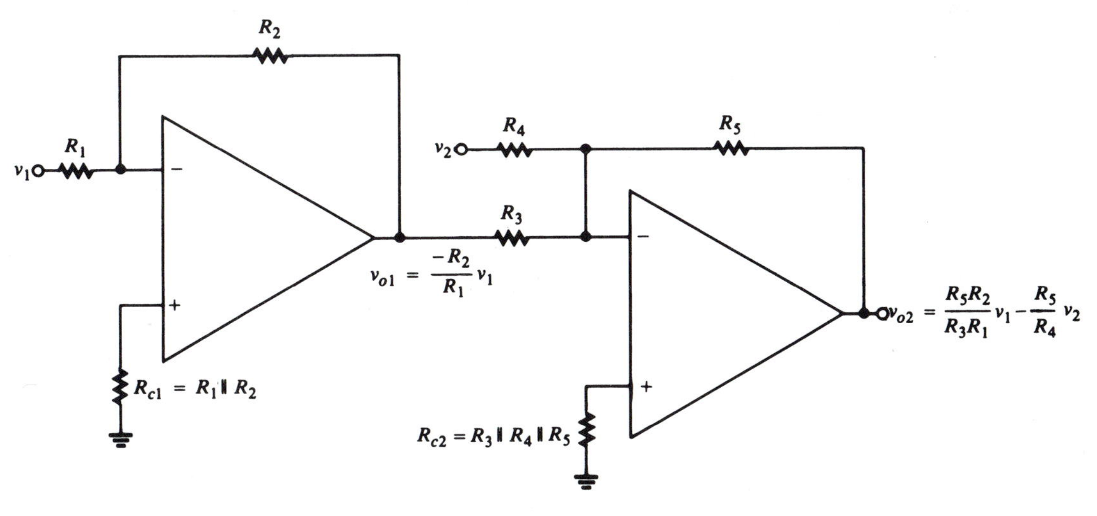

Figure 14–7
Using two inverting amplifiers to obtain the output $v_o = a_1v_1 - a_2v_2$

and the output of the second amplifier is

$$v_{o2} = -\left(\frac{R_5}{R_3}v_{o1} + \frac{R_5}{R_4}v_2\right) = \frac{R_5R_2}{R_3R_1}v_1 - \frac{R_5}{R_4}v_2 \tag{14–22}$$

This equation shows that there is a great deal of flexibility in the choice of resistor values necessary to obtain $v_o = a_1v_1 - a_2v_2$, since a large number of combinations will satisfy

$$\frac{R_5R_2}{R_3R_1} = a_1 \quad \text{and} \quad \frac{R_5}{R_4} = a_2 \tag{14–23}$$

Furthermore, there are no restrictions on the choice of values for a_1 and a_2, nor any complications in setting R_c to its optimum value.

Example 14–3

DESIGN

Design an operational-amplifier circuit using two inverting configurations to produce the output $v_o = 20v_1 - 0.2v_2$. (Note that $1 + a_2 = 1.2 < 20 = a_1$, so we cannot use the differential circuit of Figure 14–5.)

Solution. We have so many choices for resistance values that the best approach is to implement the circuit directly, without bothering to use the algebra of equation 14–20. We can, for example, begin the process by designing the first amplifier to produce $-20v_1$. Choose $R_1 = 10$ kΩ and $R_2 = 200$ kΩ. Then, the second amplifier need only invert $-20v_1$ with unity gain and scale the v_2 input by 0.2. Choose $R_5 = 20$ kΩ. Then $R_5/R_3 = 1 \Rightarrow R_3 = 20$ kΩ and $R_5/R_4 = 0.2 \Rightarrow R_4 = 100$ kΩ.

The completed design is shown in Figure 14–8(a). Figure 14–8(b) shows another solution, in which the first amplifier produces $-10v_1$ and the second multiplies that

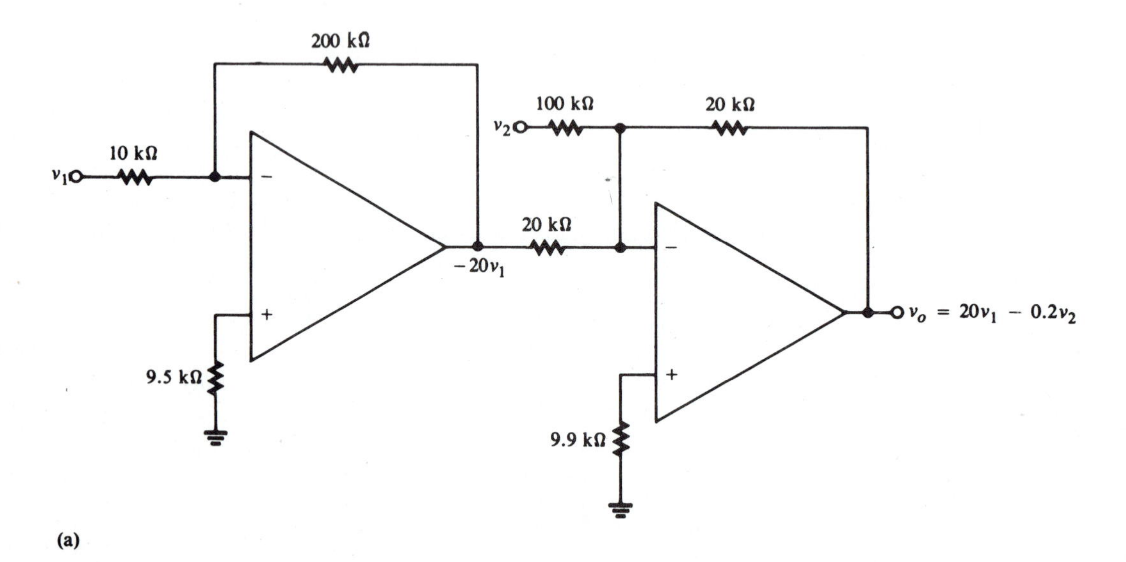

Figure 14–8
(Example 14–3) Two (of many) equivalent methods for producing $20v_1 - 0.2v_2$ using two inverting amplifiers

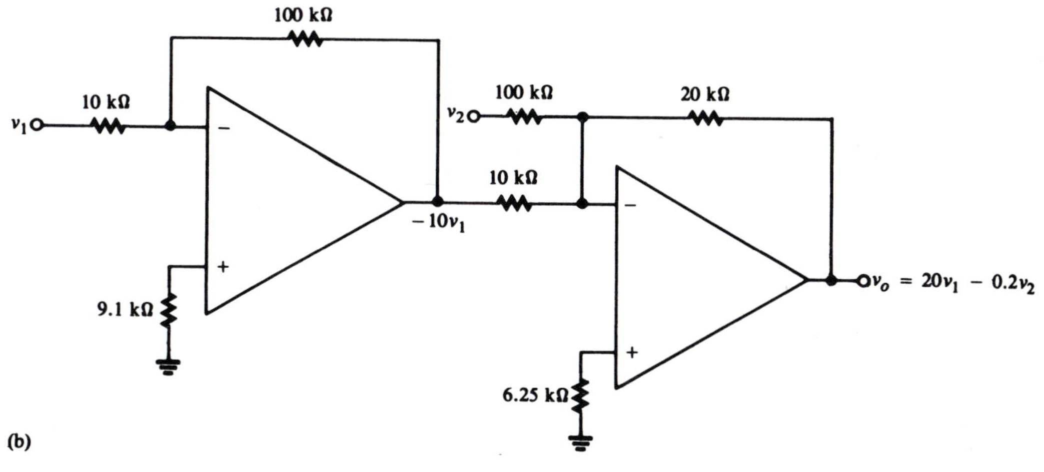

Figure 14–8
(Continued)

by the constant −2. The compensation resistors have values calculated as shown in Figure 14–7.

Although there are a large number of ways to choose resistor values to satisfy equation 14–21, there may, in practice, be constraints on some of those choices imposed by other performance requirements. For example, R_1 may have to be a certain minimum value to provide adequate input resistance to the v_1 signal source. Recall, also, that the greater the closed-loop gain of a stage, the smaller its bandwidth. Thus it may be necessary to "distribute" gain over two stages, as is done in Figure 14–8(b) to obtain $20v_1$, in order to increase the overall bandwidth. Finally, it may be necessary to minimize the gain of one stage or the other to reduce the effect of its input offset voltage. Note that the input offset voltage of the first stage is amplified by both stages.

The method used to design a subtractor circuit in Example 14–3 can be extended in an obvious way to the design of circuits that produce a linear combination of voltage sums and differences. The most general form of a linear combination is $v_o = \pm a_1v_1 \pm a_2v_2 \pm a_3v_3 \pm \ldots \pm a_nv_n$. Remember that the input signal corresponding to any term that appears in the output with a positive sign must pass through two inverting stages.

Example 14–4

1. Design an operational-amplifier circuit using two inverting configurations to produce the output $v_o = -10v_1 + 5v_2 + 0.5v_3 - 20v_4$.
2. Assuming that the unity-gain frequency of each amplifier is 1 MHz, find the approximate, overall, closed-loop bandwidth of your solution.

Solution

1. Since v_2 and v_3 appear with positive signs in the output, those two inputs must be connected to the first inverting amplifier. We can produce $-(5v_2 + 0.5v_3)$ at

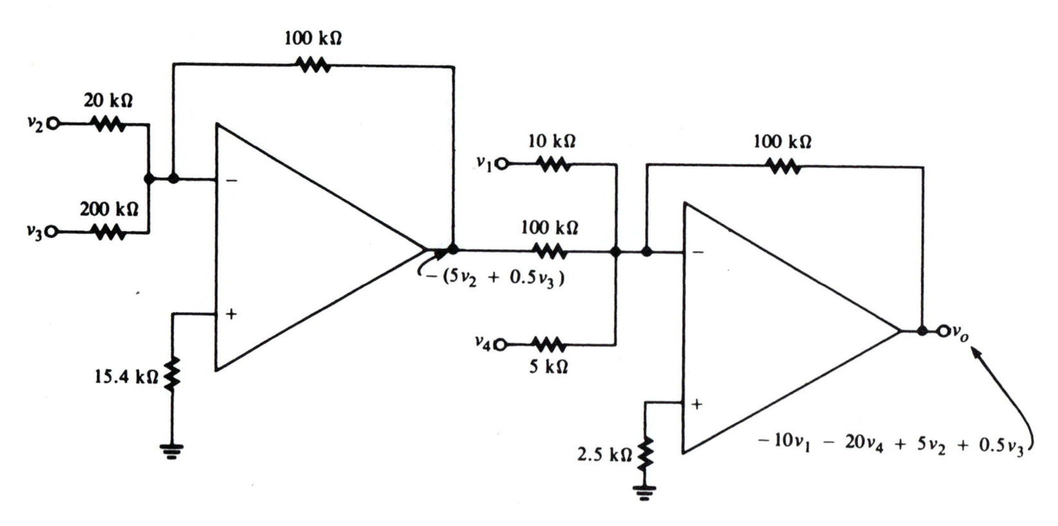

Figure 14–9
(Example 14–4)

the output of the first inverting amplifier and then invert and add it to $-(10v_1 + 20v_4)$ in the second amplifier. One possible solution is shown in Figure 14–9.

2. The feedback ratio of the first amplifier is

$$\beta_1 = \frac{(20\ \text{k}\Omega) \parallel (200\ \text{k}\Omega)}{(20\ \text{k}\Omega) \parallel (200\ \text{k}\Omega) + (100\ \text{k}\Omega)} = 0.1538$$

Therefore, the closed-loop bandwidth of the first amplifier is $\text{BW}_{\text{CL1}} = \beta_1 f_t = (0.1538)(1\ \text{MHz}) = 153.8\ \text{kHz}$. Similarly,

$$\beta_2 = \frac{(10\ \text{k}\Omega) \parallel (100\ \text{k}\Omega) \parallel (5\ \text{k}\Omega)}{(10\ \text{k}\Omega) \parallel (100\ \text{k}\Omega) \parallel (5\ \text{k}\Omega) + (100\ \text{k}\Omega)} = 0.0312$$

and $\text{BW}_{\text{CL2}} = \beta_2 f_t = (0.0312)(1\ \text{MHz}) = 31.2\ \text{kHz}$. The overall bandwidth is approximately equal to the smaller of BW_{CL1} and BW_{CL2}, or 31.2 kHz.

14–2 CONTROLLED VOLTAGE AND CURRENT SOURCES

Recall that a *controlled* source is one whose output voltage or current is determined by the magnitude of another, independent voltage or current. We have used controlled sources extensively in our study of transistor-circuit models, but those were, in a sense, fictitious devices that served mainly to simplify the circuit analysis. We wish now to explore various techniques that can be used to construct controlled voltage and current sources using operational amplifiers. As we shall see, some of these sources are realized simply by studying already-familiar circuits from a different viewpoint.

Voltage-Controlled Voltage Sources

An ideal, voltage-controlled voltage source (VCVS) is one whose output voltage V_o (1) equals a fixed constant (k) times the value of another, controlling voltage:

$V_o = kV_i$; and (2) is independent of the current drawn from it. Notice that the constant k is dimensionless. Both the inverting and noninverting configurations of an ideal operational amplifier meet the two criteria. In each case, the output voltage equals a fixed constant (the closed-loop gain, determined by external resistors) times an input voltage. Also, since the output resistance is (ideally) 0, there is no voltage division at the output and the voltage is independent of load. We have studied these configurations in detail, so we will be content for now with the observation that they do belong to the category of voltage-controlled voltage sources.

Voltage-Controlled Current Sources

An ideal, voltage-controlled current source is one that supplies a current whose magnitude (1) equals a fixed constant (k) times the value of an independent, controlling voltage: $I_o = kV_i$; and (2) is independent of the load to which the current is supplied. Notice that the constant k has the dimensions of conductance (siemens). Since it relates output current to input voltage, it is called the *transconductance*, g_m, of the source.

Figure 14–10 shows two familiar amplifier circuits: the inverting and noninverting configurations of an operational amplifier. Note, however, that we now regard the feedback resistors as *load resistors* and designate each by R_L. We will show that each circuit behaves as a voltage-controlled current source, where the load current is the current I_L in R_L.

Figure 14–10
Floating-load, voltage-controlled current sources

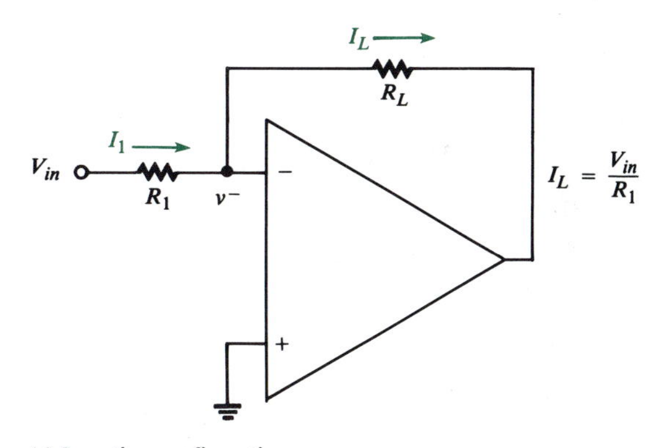

(a) Inverting configuration

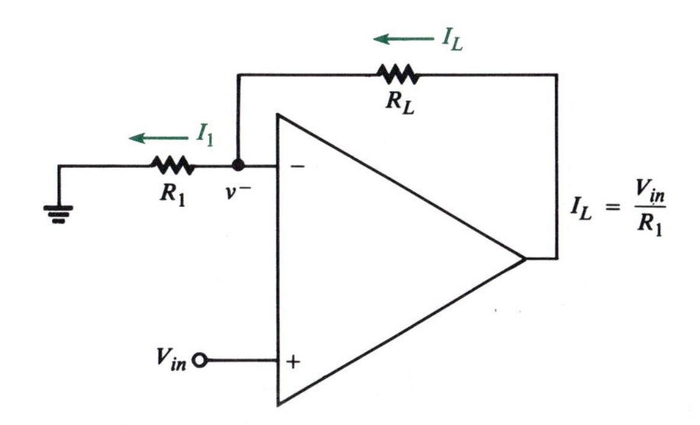

(b) Noninverting configuration

In Figure 14–10(a), v^- is virtual ground, so $I_1 = V_{in}/R_1$. Since no current flows into the inverting terminal of the ideal amplifier, $I_L = I_1$, or

$$I_L = \frac{V_{in}}{R_1} \tag{14–24}$$

Equation 14–24 shows that the load current is the constant $1/R_1$ times the controlling voltage V_{in}. Thus, the transconductance is $g_m = 1/R_1$ siemens. *Note that R_L does not appear in the equation, so the load current is independent of load resistance.* Like any constant-current source, the load voltage (voltage across R_L) will change if R_L is changed, but the current remains the same. The direction of the current through the load is controlled by the polarity of V_{in}. This version of a controlled current source is said to have a *floating load,* because neither side of R_L can be grounded. Thus, it is useful only in applications where the load is not required to have the same ground reference as the controlling voltage, V_{in}.

In Figure 14–10(b), $v^- = V_{in}$, so $I_1 = V_{in}/R_1$. Once again, no current flows into the inverting terminal, so $I_L = I_1$. Therefore,

$$I_L = \frac{V_{in}}{R_1} \tag{14–25}$$

As in the inverting configuration, the load current is independent of R_L and the transconductance is $1/R_1$ siemens. The load is also floating in this version.

Of course, there is a practical limit on the range of load resistance R_L that can be used in each circuit. If R_L is made too large, the output voltage of the amplifier will approach its maximum limit, as determined by the power supply voltages. For successful operation, the load resistance in each circuit must obey

$$R_L < \frac{R_1|V_{max}|}{V_{in}} \quad \text{(inverting circuit)} \tag{14–26}$$

$$R_L < R_1\left(\frac{|V_{max}|}{V_{in}} - 1\right) \quad \text{(noninverting circuit)} \tag{14–27}$$

where $|V_{max}|$ is the magnitude of the maximum output voltage of the amplifier.

Example 14–5

DESIGN

Design an inverting, voltage-controlled current source that will supply a constant current of 0.2 mA when the controlling voltage is 1 V. What is the maximum load resistance for this supply if the maximum amplifier output voltage is 20 V?

Figure 14–11
(Example 14–5)

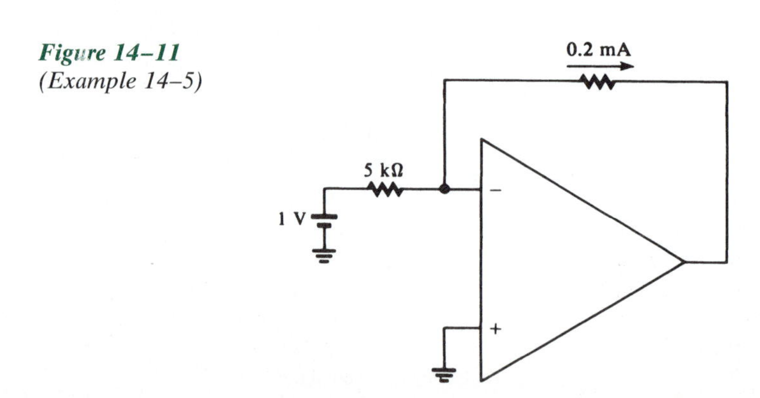

Solution. The transconductance is $g_m = (0.2 \text{ mA})/(1 \text{ V}) = 0.2 \times 10^{-3}$ S. Therefore, $R_1 = 1/g_m = 5$ kΩ. By (14–26),

$$R_L < \frac{R_1|V_{max}|}{V_{in}} = \frac{(5 \text{ k}\Omega)(20 \text{ V})}{1 \text{ V}} = 100 \text{ k}\Omega$$

The required circuit is shown in Figure 14–11.

Figure 14–12(a) shows a voltage-controlled current source that can be operated with a grounded load. To understand its behavior as a current source, refer to Figure 14–12(b), which shows the voltages and currents in the circuit. Since there is (ideally) zero current into the + input, Kirchhoff's current law at the node where R_L is connected to the + input gives

$$I_L = I_1 + I_2 \tag{14–28}$$

Figure 14–12
A voltage-controlled current source with a grounded load

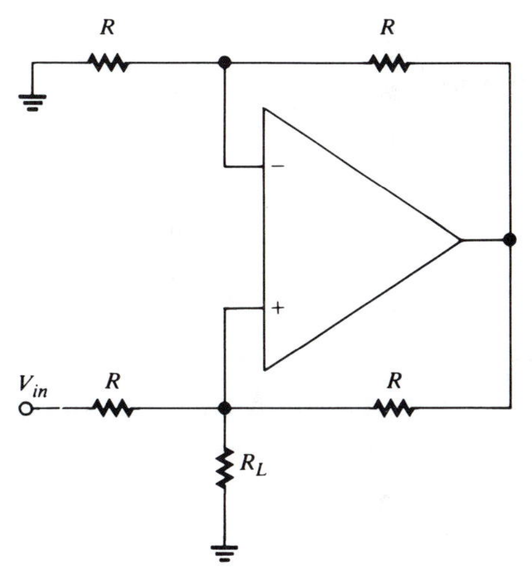

(a) The voltage-controlled current source

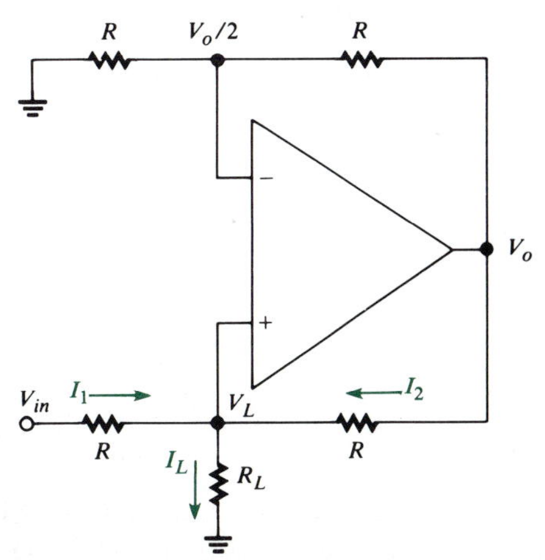

(b) Voltages and currents in the circuit of (a)

or

$$I_L = \frac{V_{in} - V_L}{R} + \frac{V_o - V_L}{R} \tag{14–29}$$

By voltage-divider action,

$$v^- = \left(\frac{R}{R+R}\right) V_o = V_o/2 \tag{14–30}$$

Since $v^- = v^+ = V_L$, we have $V_L = V_o/2$, which, upon substitution in (14–29), gives

$$I_L = \frac{V_{in}}{R} - \frac{V_o}{2R} + \frac{V_o}{R} - \frac{V_o}{2R}$$

or

$$I_L = \frac{V_{in}}{R} \tag{14–31}$$

This equation shows that the load current is controlled by V_{in} and that it is independent of R_L. Note that these results are valid to the extent that the four resistors labeled R are matched, i.e., truly equal in value. For successful operation, the load resistance must obey

$$R_L < \frac{R|V_{max}|}{2V_{in}} \tag{14–32}$$

Example 14–6

Find the current through each resistor and the voltage at each node of the voltage-controlled current source in Figure 14–13. What is the transconductance of the source?

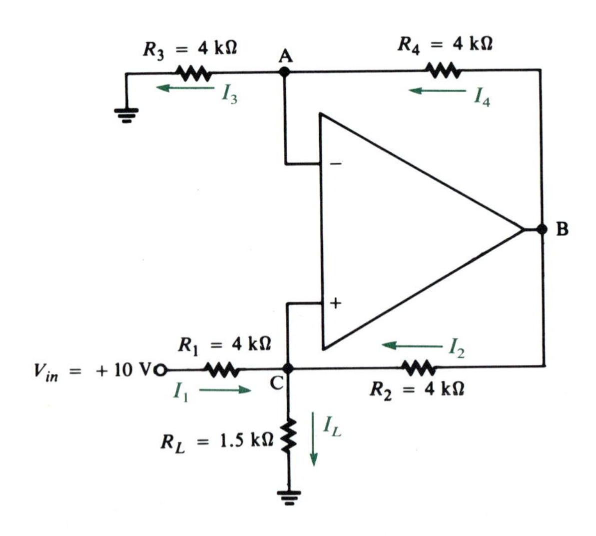

Figure 14–13
(Example 14–6)

Solution. From equation 14–31, $I_L = V_{in}/R = (10\text{ V})/(4\text{ k}\Omega) = 2.5\text{ mA}$. Therefore, the voltage at node C (V_L) is $V_C = I_L R_L = (2.5\text{ mA})(1.5\text{ k}\Omega) = 3.75\text{ V}$. We know that the voltage at node B is twice V_C ($V_o = 2V_L$): $V_B = 2V_C = 2(3.75) = 7.5\text{ V}$. The voltage at node A is one-half that at node B ($v^- = V_o/2$): $V_A = (\frac{1}{2})(V_B) = (\frac{1}{2})(7.5) = 3.75\text{ V}$. The currents I_1, I_2, I_3, and I_4 in R_1, R_2, R_3, and R_4 can then be found:

$$I_1 = (V_{in} - V_C)/R_1 = (10 - 3.75)/(4 \times 10^3) = 1.5625\text{ mA}$$
$$I_2 = (V_B - V_C)/R_2 = (7.5 - 3.75)/(4 \times 10^3) = 0.9375\text{ mA}$$
$$I_3 = V_A/R_3 = 3.75/(4 \times 10^3) = 0.9375\text{ mA}$$
$$I_4 = (V_B - V_A)/R_4 = (7.5 - 3.75)/(4 \times 10^3) = 0.9375\text{ mA}$$

The transconductance of the source is $g_m = 1/R = 1/(4\text{ k}\Omega) = 0.25\text{ mS}$.

Current-Controlled Voltage Sources

An ideal current-controlled voltage source has an output voltage that (1) is equal to a constant (k) times the magnitude of an independent current: $v_o = kI_i$, and (2) is independent of the load connected to it. Here, the constant k has the units of ohms. A current-controlled voltage source can be thought of as a *current-to-voltage converter,* since output voltage is proportional to input current. It is useful in applications where current measurements are required, because it is generally more convenient to measure voltages.

Figure 14–14 shows a very simple current-controlled voltage source. Since no current flows into the − input, the controlling current I_{in} is the same as the current in feedback resistor R. Since v^- is virtual ground,

$$V_o = -I_{in}R \tag{14–33}$$

Once again, the fact that the amplifier has zero output resistance implies that the output voltage will be independent of load.

Figure 14–15 shows a noninverting, current-controlled voltage source in which the controlling current has a return path to ground. Since $V_i^+ = I_{in}R_S$, we have

$$V_o = \left(1 + \frac{R_f}{R_1}\right) V_i^+ = \left(1 + \frac{R_f}{R_1}\right) R_S I_{in} \tag{14–34}$$

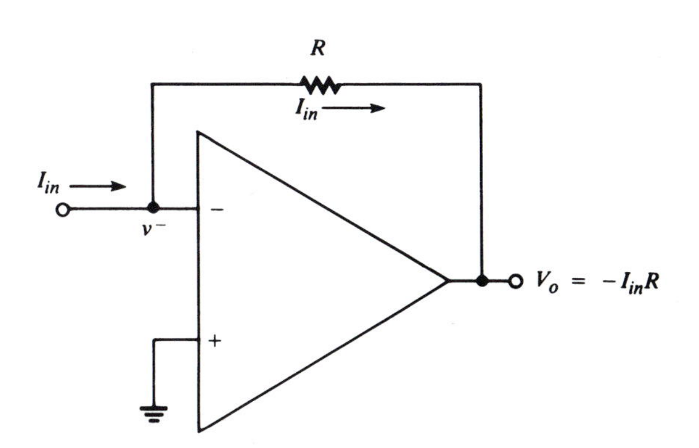

Figure 14–14
A current-controlled voltage source

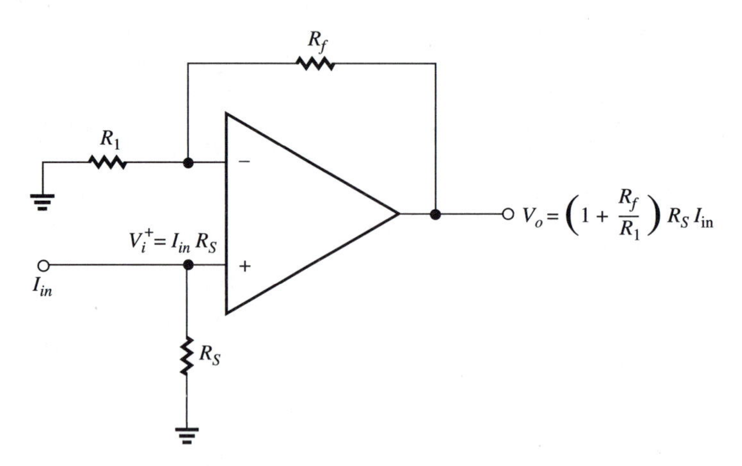

Figure 14–15
A current-controlled voltage source whose controlling current, I_{in}, has a return path to ground

Current-Controlled Current Sources

An ideal current-controlled current source is one that supplies a current whose magnitude (1) equals a fixed constant (k) times the value of an independent controlling current: $I_o = kI_i$, and (2) is independent of the load to which the current is supplied. Note that k is dimensionless, since it is the ratio of two currents.

Figure 14–16 shows a current-controlled current source with floating load R_L. Since no current flows into the − input, the current in R_2 must equal I_{in}. Since v^- is at virtual ground, the voltage V_2 is

$$V_2 = -I_{in}R_2$$

Therefore, the current I_1 in R_1 is

$$I_1 = (0 - V_2)/R_1 = I_{in}R_2/R_1 \tag{14–35}$$

Writing Kirchhoff's current law at the junction of R_1, R_2, and R_L, we have

$$I_L = I_1 + I_{in} \tag{14–36}$$

Figure 14–16
A current-controlled current source with floating load

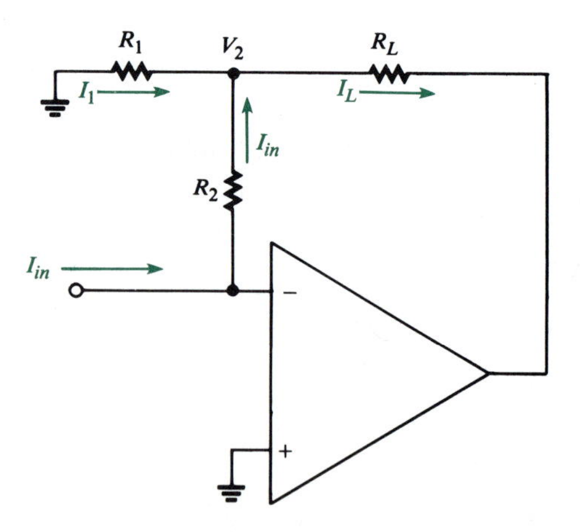

or

$$I_L = \frac{R_2}{R_1} I_{in} + I_{in} = \left(\frac{R_2}{R_1} + 1 \right) I_{in} \qquad \textbf{(14–37)}$$

This equation shows that the load current equals the constant $(1 + R_2/R_1)$ times the controlling current and that I_L is independent of R_L. For successful operation, R_L must obey

$$R_L < \left(\frac{|V_{max}|}{I_{in}} - R_2 \right) \left(\frac{R_1}{R_1 + R_2} \right) \qquad \textbf{(14–38)}$$

Note that the circuit of Figure 14–15 may be regarded as a current *amplifier,* the amplification factor being

$$k = I_L/I_{in} = 1 + R_2/R_1 \qquad \textbf{(14–39)}$$

The next example demonstrates the utility of current amplification and illustrates an application where a floating load may be used.

Example 14–7

DESIGN

It is desired to measure a dc current that ranges from 0 to 1 mA using an ammeter whose most sensitive range is 0 to 10 mA. To improve the measurement accuracy, the current to be measured should be amplified by a factor of 10.

1. Design the circuit.
2. Assuming that the meter resistance is 150 Ω and the maximum output voltage of the amplifier is 15 V, verify that the circuit will perform properly.

Solution

1. Figure 14–17 shows the required circuit. I_X is the current to be measured, and the ammeter serves as the load through which the amplified current flows. From equation 14–39, the current amplification is $I_L/I_X = 1 + R_2/R_1 = 10$. Letting $R_1 = 1$ kΩ, we find $R_2 = (10 - 1)1$ kΩ $= 9$ kΩ.

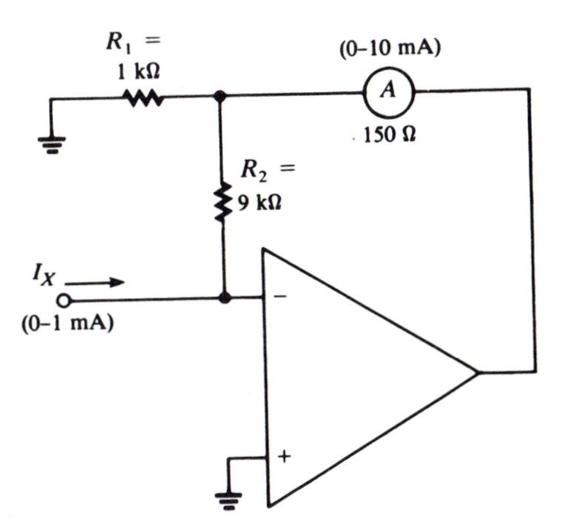

Figure 14–17
(Example 14–7) The current-controlled current source acts as a current amplifier, so a 0–1-mA current can be measured by a 0–10-mA ammeter.

2. Inequality 14–38 must be satisfied for the smallest possible value of the right-hand side, which occurs when $I_{in} = 1$ mA:

$$R_L < \left[\frac{15\text{ V}}{1\text{ mA}} - (9\text{ k}\Omega)\right]\left[\frac{1\text{ k}\Omega}{(1\text{ k}\Omega) + (9\text{ k}\Omega)}\right] = 600\ \Omega$$

Since the meter resistance is 150 Ω, the circuit operates satisfactorily.

14–3 INTEGRATION, DIFFERENTIATION, AND WAVESHAPING

Electronic Integration

An *electronic integrator* is a device that produces an output waveform whose value at any instant of time equals the total *area under the input* waveform up to that point in time. (For those familiar with mathematical integration, the process produces the time-varying function $\int_0^t v_{in}(t)\,dt$.) To illustrate this concept, suppose the input to an electronic integrator is the dc level E volts, which is first connected to the integrator at an instant of time we will call $t = 0$. Refer to Figure 14–18. The plot of the dc "waveform" versus time is simply a horizontal line at level E volts, since the dc voltage is constant. The more time that we allow to pass, the greater the area that accumulates under the dc waveform. At any time-point t, the total area under the input waveform between time 0 and time t is (height) × (width) = Et volts, as illustrated in the figure. For example, if $E = 5$ V dc, then the output will be 5 V at $t = 1$ s, 10 V at $t = 2$ s, 15 V at $t = 3$ s, and so forth. We see that the output is the *ramp* voltage $v_o(t) = Et$.

When the input to a practical integrator is a dc level, the output will rise linearly with time, as shown in Figure 14–18, and will eventually reach the maximum possible output voltage of the amplifier. Of course, the integration process ceases at that time. If the input voltage goes negative for a certain interval of time, the total area during that interval is *negative* and subtracts from whatever positive area had previously accumulated, thus reducing the output voltage. Therefore, the input must periodically go positive and negative to prevent the output of an integrator from reaching its positive or negative limit. We will explore this process in greater detail in a later discussion on waveshaping.

Figure 14–19 shows how an electronic integrator is constructed using an operational amplifier. Note that the component in the feedback path is capacitor C, and that the amplifier is operated in an inverting configuration. Besides the usual ideal-amplifier assumptions, we are assuming *zero input offset,* since any input dc level would be integrated as shown in Figure 14–18 and would eventually cause the

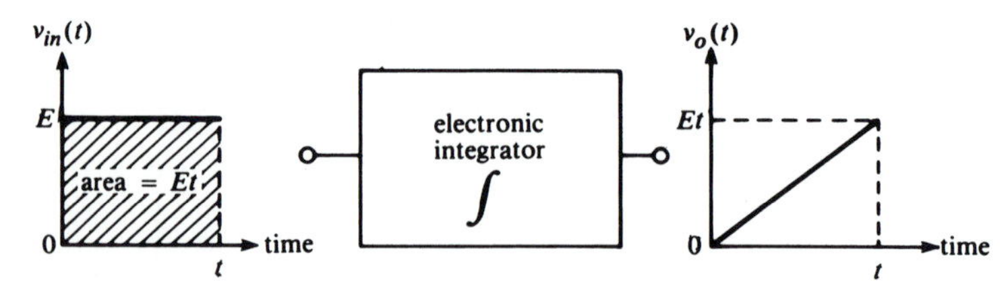

Figure 14–18
The output of the integrator at t seconds is the area, Et, under the input waveform

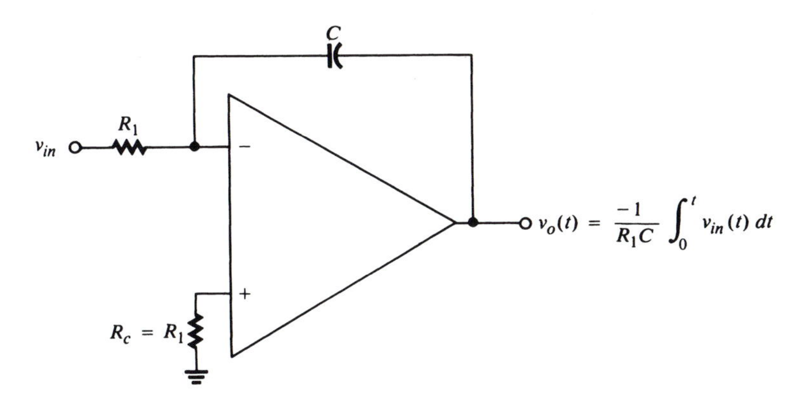

Figure 14–19
An ideal electronic integrator

amplifier to saturate. Thus, we show an *ideal-integrator* circuit. Using the standard symbol $\int_0^t v\,dt$ to represent integration of the voltage v between time 0 and time t, we can show that the output of this circuit is

$$v_o(t) = \frac{-1}{R_1C}\int_0^t v_{in}\,dt \tag{14–40}$$

This equation shows that the output is the (inverted) integral of the input, multiplied by the constant $1/R_1C$. If this circuit were used to integrate the dc waveform shown in Figure 14–18, the output would be a negative-going ramp ($v_o = -Et/R_1C$).

Readers unfamiliar with calculus can skip the present paragraph, in which we demonstrate why the circuit of Figure 14–19 performs integration. Since the current into the − input is 0, we have, from Kirchhoff's current law,

$$i_1 + i_C = 0 \tag{14–41}$$

where i_1 is the input current through R_1 and i_C is the feedback current through the capacitor. Since $v^- = 0$, the current in the capacitor is

$$i_C = C\frac{dv_o}{dt} \tag{14–42}$$

Thus

$$\frac{v_{in}}{R_1} + C\frac{dv_o}{dt} = 0 \tag{14–43}$$

or

$$\frac{dv_o}{dt} = \frac{-1}{R_1C}v_{in} \tag{14–44}$$

Integrating both sides with respect to t, we obtain

$$v_o = \frac{-1}{R_1C}\int_0^t v_{in}\,dt \tag{14–45}$$

Hereafter, we will use the abbreviated symbol $\int$ to represent integration. It can be shown, using calculus, that the mathematical integral of the sine wave $A \sin \omega t$ is

$$\int (A \sin \omega t)\, dt = \frac{-A}{\omega} \sin(\omega t + 90°) = \frac{-A}{\omega} \cos(\omega t)$$

Therefore, when the input to the inverting integrator in Figure 14–19 is $v_{in} = A \sin \omega t$, the output is

$$v_o = \frac{-1}{R_1C} \int (A \sin \omega t)\, dt = \frac{-A}{\omega R_1 C}(-\cos \omega t)$$
$$= \frac{A}{\omega R_1 C} \cos \omega t \qquad \textbf{(14–46)}$$

The most important fact revealed by equation 14–46 is that the output of an integrator with sinusoidal input is a sinusoidal waveform whose *amplitude is inversely proportional to its frequency.* This observation follows from the presence of ω ($= 2\pi f$) in the denominator of (14–46). For example, if a 100-Hz input sine wave produces an output with peak value 10 V, then, all else being equal, a 200-Hz sine wave will produce an output with peak value 5 V.

Note also that the output *leads* the input by 90°, regardless of frequency, since $\cos \omega t = \sin(\omega t + 90°)$. However, as shown by equation 14–46, this phase relation is due to both integration and the inverting action of the amplifier. In many practical applications, noninverting integrator circuits are used to introduce phase *lag* into a system:

$$\int (\sin x) dx = -\cos x = \sin(x - 90°).$$

Example 14–8

1. Find the peak value of the output of the ideal integrator shown in Figure 14–20. The input is $v_{in} = 0.5 \sin(100t)$ V.
2. Repeat, when $v_{in} = 0.5 \sin(10^3 t)$ V.

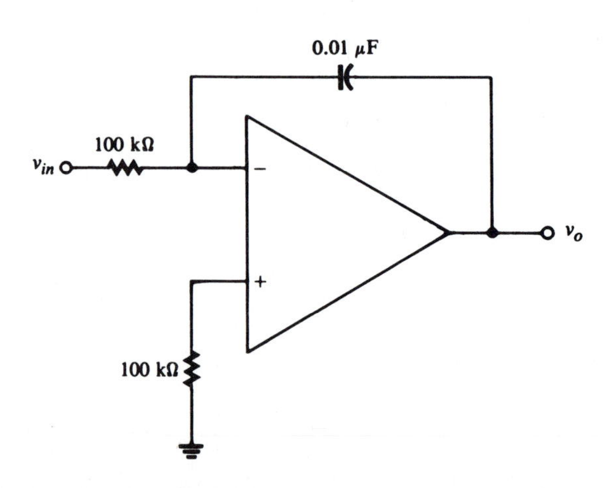

Figure 14–20
(Example 14–8)

Solution

1. From equation 14–46,

$$v_o = \frac{A}{\omega R_1 C}\cos(\omega t) = \frac{0.5}{100(10^5)(10^{-8})}\cos(100t)$$
$$= 5\cos(100t)\ \text{V} \quad \text{peak value} = 5\ \text{V}$$

2. From equation 14–46,

$$v_o = \frac{0.5}{1000(10^5)(10^{-8})}\cos(1000t)$$
$$= 0.5\cos(1000t)\ \text{V} \qquad \text{peak value} = 0.5\ \text{V}$$

Example 14–8 shows that increasing the frequency by a factor of 10 causes a decrease in output amplitude by a factor of 10. This familiar relationship implies that a Bode plot for the gain of an ideal integrator will have slope −20 dB/decade, or −6 dB/octave. Gain magnitude is the ratio of the peak value of the output to the peak value of the input:

$$\left|\frac{v_o}{v_{in}}\right| = \frac{\left(\dfrac{A}{\omega R_1 C}\right)}{A} = \frac{1}{\omega R_1 C} \tag{14–47}$$

This equation clearly shows that gain is inversely proportional to frequency. A Bode plot for the case $R_1C = 0.001$ is shown in Figure 14–21.

Because the integrator's output amplitude decreases with frequency, it is a kind of low-pass filter. It is sometimes called a *smoothing* circuit, because the amplitudes of high-frequency components in a complex waveform are reduced, thus smoothing the jagged appearance of the waveform. This feature is useful for reducing high-frequency noise in a signal. Integrators are also used in *analog computers* to obtain real-time solutions to differential (calculus) equations.

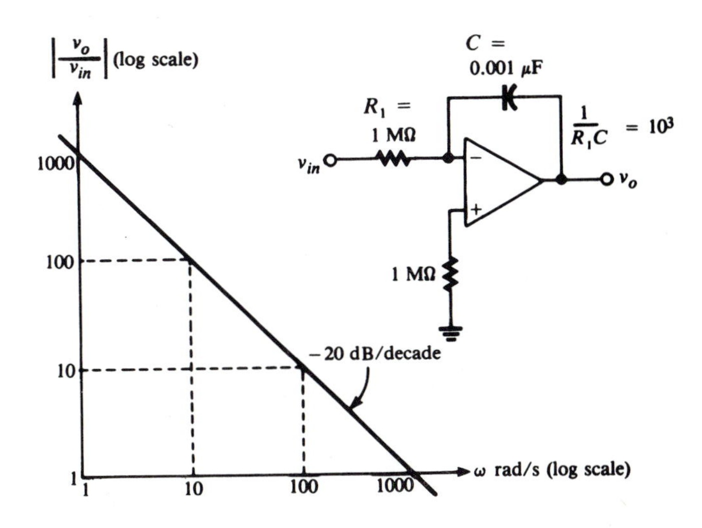

Figure 14–21
Bode plot of the gain of an ideal integrator for the case $R_1C = 0.001$

Practical Integrators

Although high-quality, precision integrators are constructed as shown in Figure 14.19 for use in low-frequency applications such as analog computers, these applications require high-quality amplifiers with extremely small offset voltages or chopper stabilization. As mentioned earlier, any input offset is integrated as if it were a dc signal input and will eventually cause the amplifier to saturate. To eliminate this problem in practical integrators using general-purpose amplifiers, a resistor is connected in parallel with the feedback capacitor, as shown in Figure 14–22(a). Since the capacitor is an open circuit as far as dc is concerned, the integrator responds to dc inputs just as if it were an inverting amplifier. In other words, the *dc closed-loop gain* of the integrator is $-R_f/R_1$. At high frequencies, the impedance of the capacitor is much smaller than R_f, so the parallel combination of C and R_f is essentially the same as C alone, and signals are integrated as usual.

While the feedback resistor in Figure 14–22(a) prevents integration of dc inputs, it also degrades the integration of low-frequency signals. At frequencies where the capacitive reactance of C is comparable in value to R_f, the net feedback impedance is not predominantly capacitive and true integration does not occur. As a rule of thumb, we can say that satisfactory integration will occur at frequencies much greater than the frequency at which $X_C = R_f$. That is, for integrator action we want

$$X_C \ll R_f$$

$$\frac{1}{2\pi fC} \ll R_f$$

or

$$f \gg \frac{1}{2\pi R_f C} \qquad \textbf{(14–48)}$$

The frequency f_c where X_c equals R_f,

$$f_c = \frac{1}{2\pi R_f C} \qquad \textbf{(14–48a)}$$

defines a *break frequency* in the Bode plot of the practical integrator. As shown in Figure 14–22(b), at frequencies well above f_c, the gain falls off at the rate of −20 dB/decade, like that of an ideal integrator, and at frequencies below f_c, the gain approaches its dc value of R_f/R_1.

Example 14–9

DESIGN

Design a practical integrator that

1. integrates signals with frequencies down to 100 Hz; and
2. produces a peak output of 0.1 V when the input is a 10-V-peak sine wave having frequency 10 kHz.

Find the dc component in the output when there is a +50-mV dc input.

Solution. In order to integrate frequencies down to 100 Hz, we require $f_c \ll 100$ Hz. Let us choose f_c one decade below 100 Hz: $f_c = 10$ Hz. Then, from equation 14–48a,

$$f_c = 10 = \frac{1}{2\pi R_f C}$$

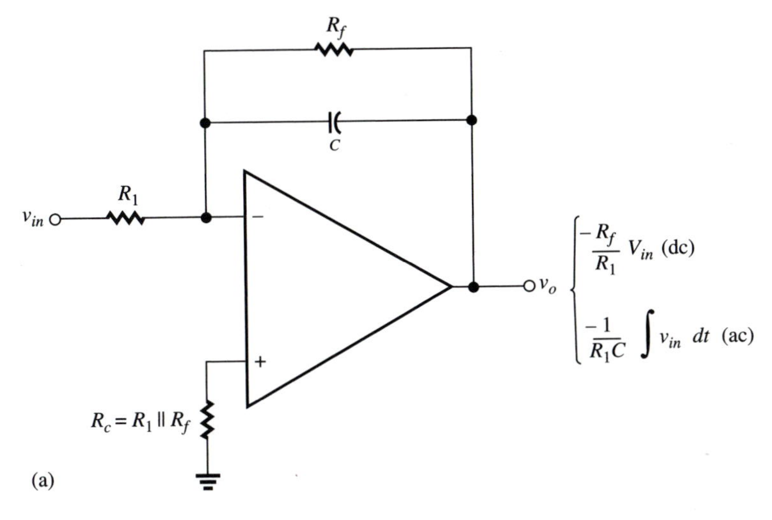

Figure 14–22(a)
A resistor R_f connected in parallel with C causes the practical integrator to behave like an inverting amplifier to dc inputs and like an integrator to high-frequency ac inputs.

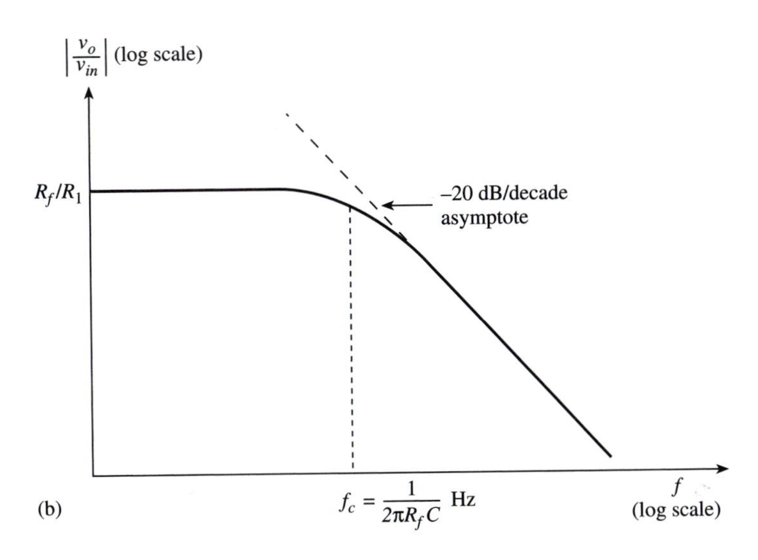

Figure 14–22(b)
Bode plot for the practical integrator, showing that integration occurs at frequencies above $1/(2\pi R_f C)$ hertz

Choose $C = 0.01\ \mu\text{F}$. Then

$$10 = \frac{1}{2\pi R_f(10^{-8})}$$

or

$$R_f = \frac{1}{2\pi(10)(10^{-8})} = 1.59\ \text{M}\Omega$$

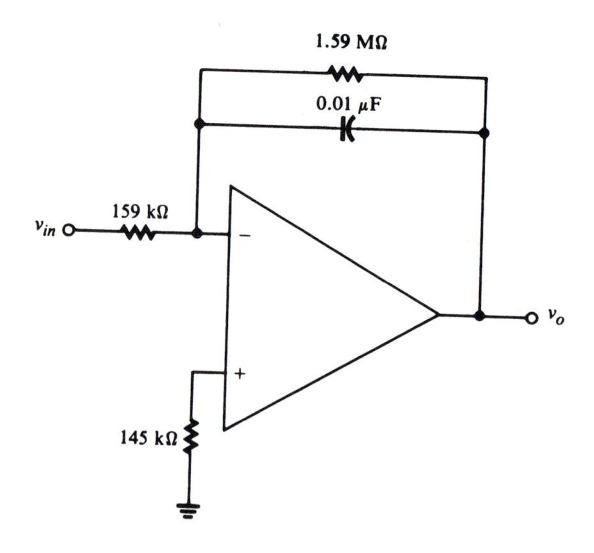

Figure 14–23
(Example 14–9)

To satisfy requirement 2, we must choose R_1 so that the gain at 10 kHz is

$$\left|\frac{v_o}{v_{in}}\right| = \frac{0.1\text{ V}}{10\text{ V}} = 0.01$$

Assuming that we can neglect R_f at this frequency (3 decades above f_c), the gain is the same as that for an ideal integrator, given by equation 14–47:

$$\left|\frac{v_o}{v_{in}}\right| = \frac{1}{\omega R_1 C} = 0.01$$

Thus

$$\frac{1}{2\pi \times 10^4 R_1(10^{-8})} = 0.01$$

or

$$R_1 = \frac{1}{2\pi \times 10^4 \times (0.01 \times 10^{-8})} = 159\text{ k}\Omega$$

The required circuit is shown in Figure 14–23. Note that R_c = (1.59 MΩ) ∥ (159 kΩ) = 145 kΩ.

When the input is 50 mV dc, the output is 50 mV times the dc closed-loop gain:

$$v_o = \frac{-R_f}{R_1}(50\text{ mV}) = \frac{-1.59\text{ M}\Omega}{159\text{ k}\Omega}(50\text{ mV}) = -0.5\text{ V}$$

In closing our discussion of integrators, we should note that it is possible to scale and integrate several input signals simultaneously, using an arrangement similar to the linear combination circuit studied earlier. Figure 14–24 shows a practical, three-input integrator that performs the following operation at frequencies above f_c:

$$v_o = -\int\left(\frac{1}{R_1C}v_1 + \frac{1}{R_2C}v_2 + \frac{1}{R_3C}v_3\right)dt \qquad \textbf{(14–49)}$$

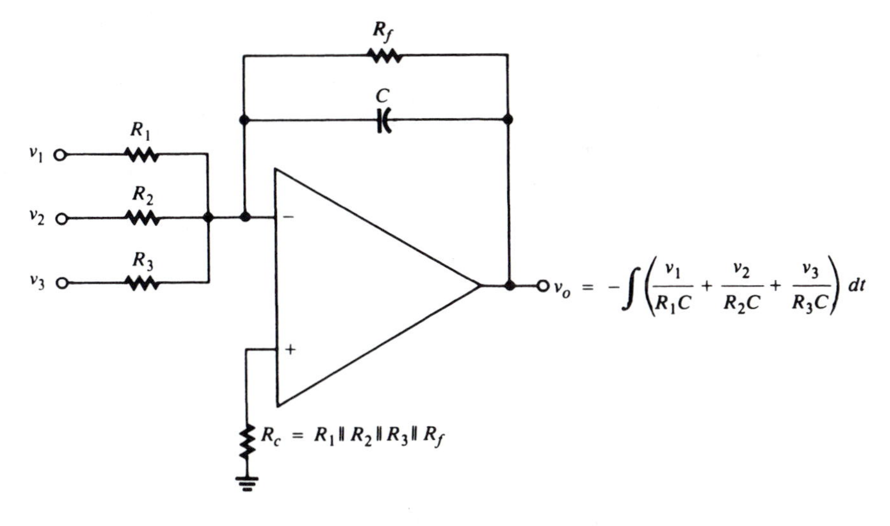

Figure 14–24
A three-input integrator

Equation 14–49 is equivalent to

$$v_o = \frac{-1}{R_1C}\int v_1\,dt - \frac{1}{R_2C}\int v_2\,dt - \frac{1}{R_3C}\int v_3\,dt \tag{14–50}$$

If $R_1 = R_2 = R_3 = R$, then

$$v_o = \frac{-1}{RC}\int (v_1 + v_2 + v_3)\,dt \tag{14–51}$$

Electronic Differentiation

An electronic differentiator produces an output waveform whose value at any instant of time is equal to the *rate of change* of the input at that point in time. In many respects, differentiation is just the opposite, or inverse, operation of integration. In fact, integration is also called "antidifferentiation." If we were to pass a signal through an ideal integrator in cascade with an ideal differentiator, the final output would be exactly the same as the original input signal.

Figure 14–25 demonstrates the operation of an ideal electronic differentiator. In this example, the input is the ramp voltage $v_{in} = Et$. The rate of change, or slope,

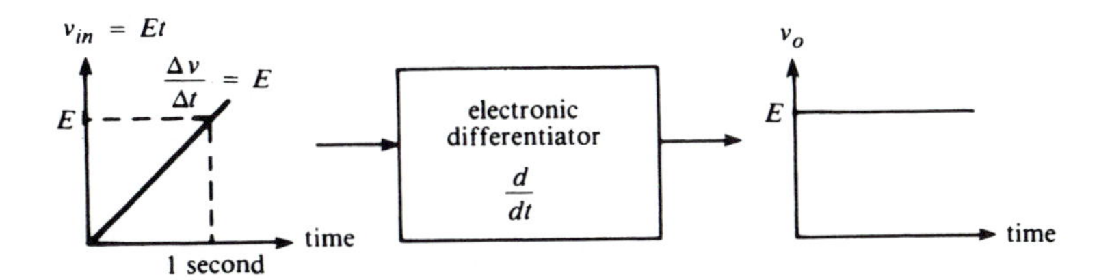

Figure 14–25
The ideal electronic differentiator produces an output equal to the rate of change of the input. Since the rate of change of a ramp is constant, the output in this example is a dc level.

of this ramp is a constant E volt/second. (For every second that passes, the signal increases by an additional E volt.) Since the rate of change of the input is constant, we see that the output of the differentiator is the constant dc level E volts.

The standard symbol used to represent differentiation of a voltage v is dv/dt. (This should not be interpreted as a fraction. The dt in the denominator simply means that we are finding the rate of change of v with respect to *time, t*.) In the example shown in Figure 14–25, we would write

$$\frac{dv_{in}}{dt} = \frac{d(Et)}{dt} = E$$

Note that the derivative of a constant (dc level) is *zero,* since a constant does not change with time and therefore has zero rate of change.

Figure 14–26 shows how an ideal differentiator is constructed using an operational amplifier. Note that we now have a capacitive input and a resistive feedback—again, just the opposite of an integrator. It can be shown that the output of this differentiator is

$$v_o = -R_f C \frac{dv_{in}}{dt} \qquad \textbf{(14–52)}$$

Thus, the output voltage is the (inverted) derivative of the input, multiplied by the constant R_fC. If the ramp voltage in Figure 14–25 were applied to the input of this differentiator, the output would be a negative dc level.

Readers not familiar with calculus can skip the present paragraph, in which we show how the circuit of Figure 14–26 performs differentiation. Since the current into the $-$ terminal is 0, we have, from Kirchhoff's current law,

$$i_C + i_f = 0 \qquad \textbf{(14–53)}$$

Since $v^- = 0$, $v_C = v_{in}$ and

$$i_C = C\frac{dv_{in}}{dt} \qquad \textbf{(14–54)}$$

Also, $i_f = v_o/R_f$, so

$$C\frac{dv_{in}}{dt} + \frac{v_o}{R_f} = 0$$

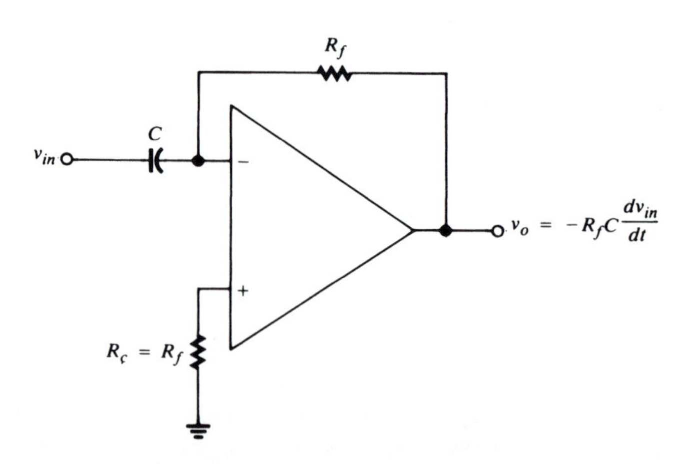

Figure 14–26
An ideal electronic differentiator

or

$$v_o = -R_f C \frac{dv_{in}}{dt} \qquad \textbf{(14–55)}$$

It can be shown, using calculus, that

$$\frac{d(A \sin \omega t)}{dt} = A\omega \cos \omega t \qquad \textbf{(14–56)}$$

Therefore, when the input to the differentiator in Figure 14–26 is $v_{in} = A \sin \omega t$, the output is

$$v_o = -R_f C \frac{d(A \sin \omega t)}{dt} = -A\omega R_f C \cos(\omega t) = A\omega R_f C \sin(\omega t - 90°) \qquad \textbf{(14–57)}$$

Equation 14–57 shows that when the input is sinusoidal, the *amplitude of the output of a differentiator is directly proportional to frequency* (once again, just the opposite of an integrator). Note also that the output lags the input by 90°, regardless of frequency. (In practice, noninverting differentiator circuits are used to introduce phase *lead* into a system.) The gain of the differentiator is

$$\left|\frac{v_o}{v_{in}}\right| = \frac{A\omega R_f C}{A} = \omega R_f C \qquad \textbf{(14–58)}$$

Practical Differentiators

From a practical standpoint, the principal difficulty with the differentiator is that it effectively amplifies an input in direct proportion to its frequency and therefore increases the level of high-frequency noise in the output. Unlike the integrator, which "smooths" a signal by reducing the amplitude of high-frequency components, the differentiator intensifies the contamination of a signal by high-frequency noise. For this reason, differentiators are rarely used in applications requiring high precision, such as analog computers.

In a practical differentiator, the amplification of signals in direct proportion to their frequencies cannot continue indefinitely as frequency increases, because the amplifier has a finite bandwidth. As we have already discussed in Chapter 13, there is some frequency at which the output amplitude must begin to fall off. Nevertheless, it is often desirable to design a practical differentiator so that it will have a break frequency even lower than that determined by the upper cutoff frequency of the amplifier, that is, to roll off its gain characteristic at some relatively low frequency. This action is accomplished in a practical differentiator by connecting a resistor in series with the input capacitor, as shown in Figure 14–27. We can understand how this modification achieves the stated goal by considering the net impedance of the R_1–C combination at low and high frequencies:

$$Z_{in} = R_1 - j/\omega C \qquad \textbf{(14–59)}$$

$$|Z_{in}| = \sqrt{R_1^2 + (1/\omega C)^2} \qquad \textbf{(14–60)}$$

At very small values of ω, Z_{in} is dominated by the capacitive reactance component, so the combination is essentially the same as C alone, and differentiator action occurs. At very high values of ω, $1/\omega C$ is negligible, so Z_{in} is essentially the resistance R_1, and the circuit behaves like an ordinary inverting amplifier (with gain R_f/R_1).

The break frequency f_b beyond which differentiation no longer occurs in Figure

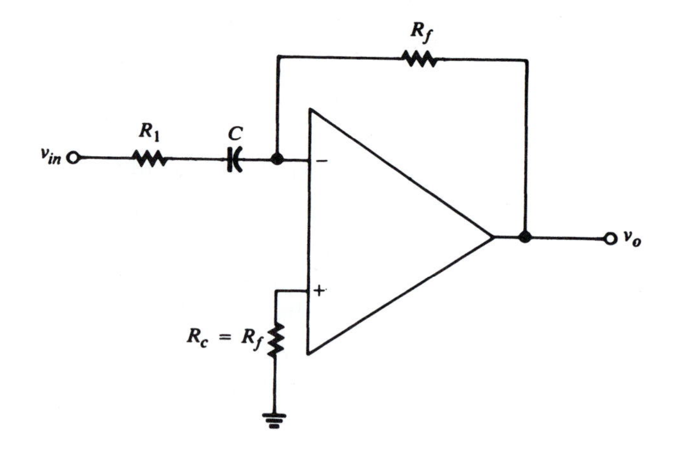

Figure 14–27
A practical differentiator. Differentiation occurs at low frequencies, but resistor R_1 prevents high-frequency differentiation.

14–27 is the frequency at which the capacitive reactance of C equals the resistance of R_1:

$$\frac{1}{2\pi f_b C} = R_1$$

$$f_b = \frac{1}{2\pi R_1 C} \text{ hertz} \tag{14–61}$$

In designing a practical differentiator, the break frequency should be set well above the highest frequency at which accurate differentiation is desired:

$$f_b \gg f_h \tag{14–62}$$

where f_h is the highest differentiation frequency. Figure 14–28 shows Bode plots for the gain of the ideal and practical differentiators. In the low-frequency region where differentiation occurs, note that the gain rises with frequency at the rate of 20 dB/decade. The plot shows that the gain levels off beyond the break frequency f_b and then falls off at −20 dB/decade beyond the amplifier's upper cutoff frequency. Recall that the closed-loop bandwidth, or upper cutoff frequency of the amplifier, is given by

$$f_2 = \beta f_t \tag{14–63}$$

where β in this case is $R_1/(R_1 + R_f)$.

In some applications, where very wide bandwidth operational amplifiers are used, it may be necessary or desirable to roll the frequency response off even faster than that shown for the practical differentiator in Figure 14–28. This can be accomplished by connecting a capacitor C_f in parallel with the feedback resistor R_f. This modification will cause the response to roll off at −20 dB/decade beginning at the break frequency

$$f_{b2} = \frac{1}{2\pi R_f C_f} \text{ hertz} \tag{14–64}$$

Obviously, f_{b2} should be set higher than f_b.

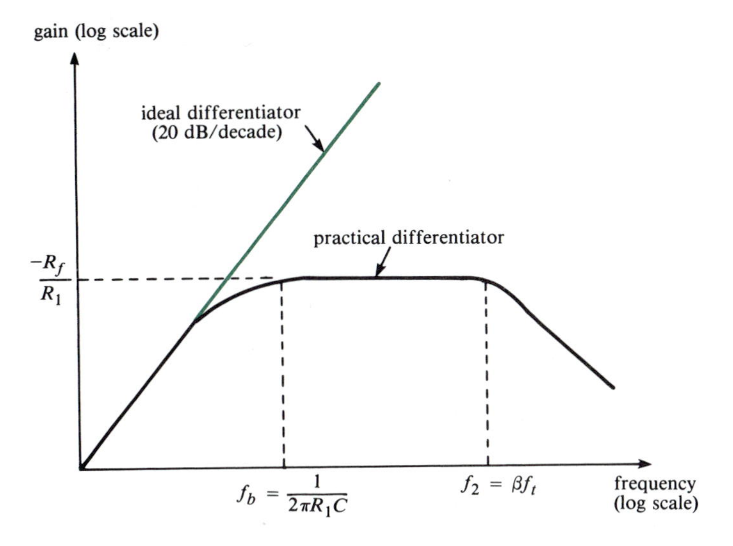

Figure 14–28
Bode plots for the ideal and practical differentiators. f_b is the break frequency due to the input R_1–C combination and f_2 is the upper cutoff frequency of the (closed-loop) amplifier.

Example 14–10

DESIGN

1. Design a practical differentiator that will differentiate signals with frequencies up to 200 Hz. The gain at 10 Hz should be 0.1.
2. If the operational amplifier used in the design has a unity-gain frequency of 1 MHz, what is the upper cutoff frequency of the differentiator?

Solution

1. We must select R_1 and C to produce a break frequency f_b that is well above f_h = 200 Hz. Let us choose f_b = 10 f_h = 2 kHz. Letting C = 0.1 μF, we have, from equation 14–61,

$$f_b = 2 \times 10^3 = \frac{1}{2\pi R_1 C}$$

$$R_1 = \frac{1}{2\pi(2 \times 10^3)(10^{-7})} = 796\ \Omega$$

In order to achieve a gain of 0.1 at 10 Hz, we have, from equation 14–58,

$$\left|\frac{v_o}{v_{in}}\right| = 0.1 = \omega R_f C = (2\pi \times 10)R_f(10^{-7})$$

$$R_f = \frac{0.1}{2\pi \times 10 \times 10^{-7}} = 15.9\ \text{k}\Omega$$

The completed design is shown in Figure 14–29.

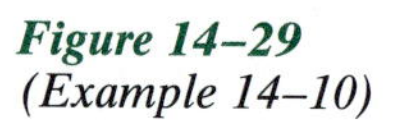

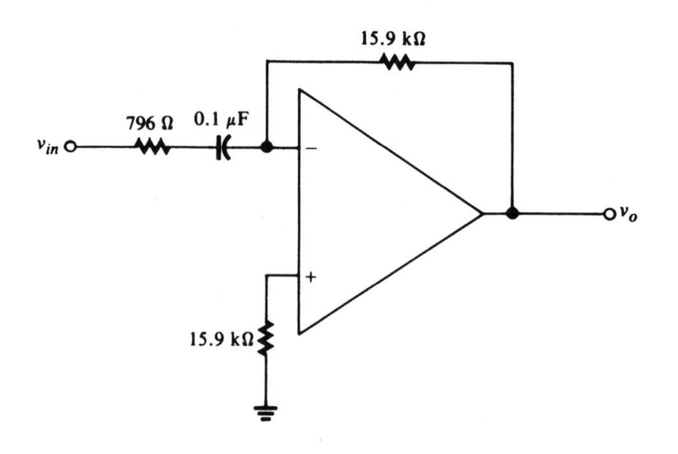

Figure 14–29
(Example 14–10)

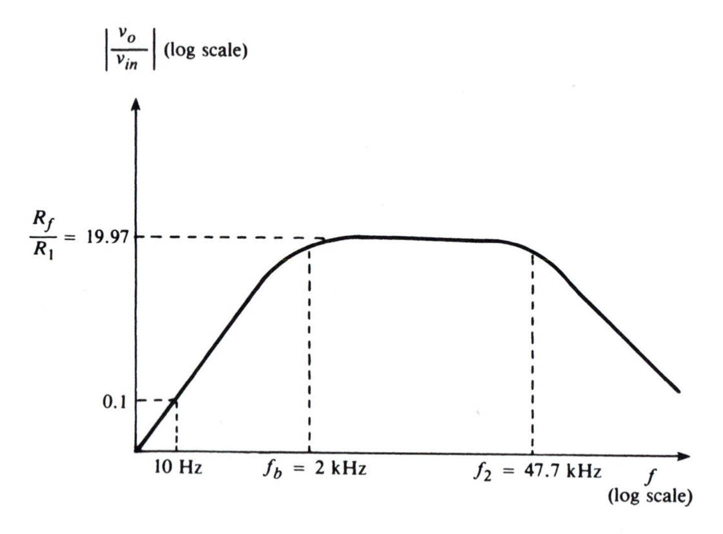

Figure 14–30
(Example 14–10)

2.

$$\beta = \frac{R_1}{R_1 + R_f} = \frac{796}{796 + 15.9 \times 10^3} = 0.0477$$

From equation 14–63, $f_2 = \beta f_t = (0.0477)(1\text{ MHz}) = 47.7\text{ kHz}$. The Bode plot is sketched in Figure 14–30.

Example 14–11

SPICE

Use SPICE to verify the design of the differentiator in Example 14–10. Assume the amplifier is ideal.

Solution. Figure 14–31(a) shows the SPICE circuit and the input data file. The inverting operational amplifier is modeled like the one in Example 13–14, except that we assume zero output resistance and a (nearly) ideal voltage gain having the very large value 1×10^9.

Figure 14–31(b) shows the frequency response computed by SPICE over the range from 1 Hz to 100 kHz. Since the input signal has default amplitude 1 V, the plot represents values of voltage gain. Notice that the gain at 10 Hz is 0.0999 ≈ 0.1, as required. The theoretical gain of the ideal differentiator at 200 Hz is

$$\omega R_f C = (2\pi \times 200)(15.9\ \mathrm{k}\Omega)(0.1\ \mu\mathrm{F}) = 1.998$$

The plot shows that the gain at a frequency close to 200 Hz (199.5 Hz) is 1.983, so we conclude that satisfactory differentiation occurs up to 200 Hz.

The plot shows that the high-frequency gain (at 100 kHz) is 19.97. At very high frequencies, the reactance of the 0.1-μF capacitor is negligible, so the gain, for all practical purposes, equals the ratio of R_f to R_1: 15.9 kΩ/796 Ω = 19.97. Since the break frequency f_b in the design was selected to be 2 kHz, the gain at 2 kHz should be approximately (0.707)(19.97) = 14.11. The plot shows that the gain at a frequency

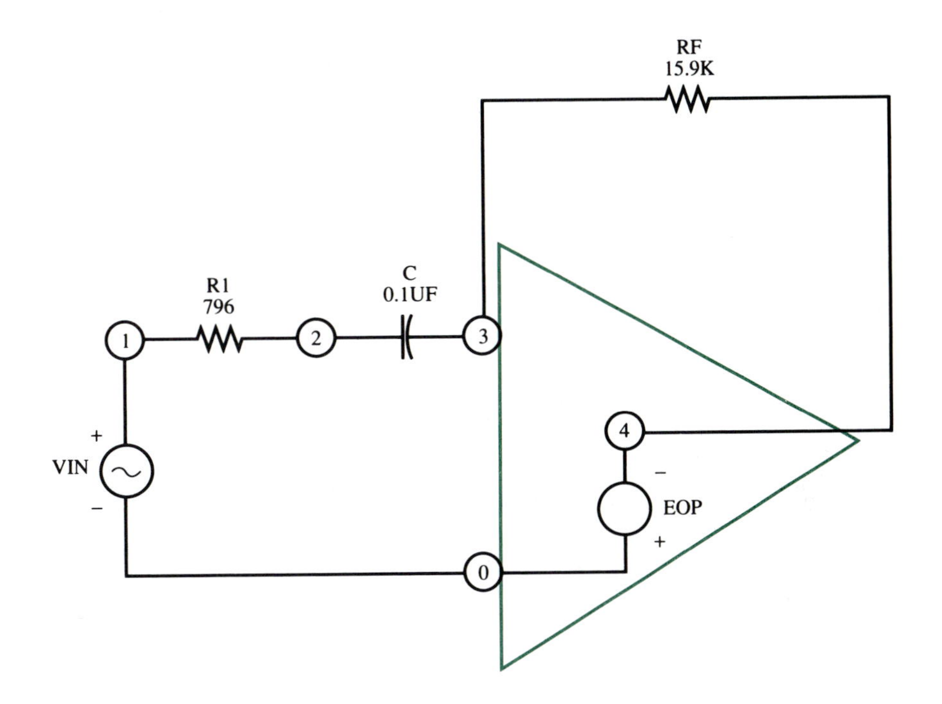


```
EXAMPLE 14.11
VIN 1 0 AC
R1 1 2 796
C 2 3 0.1UF
RF 3 4 15.9K
EOP0 4 3 0 1E9
.AC DEC 10 1 100KHZ
.PLOT AC V(4)
.END
```

(a)

Figure 14–31
(Example 14–11)

Plot scale for V(4): 1.000D-04 1.000D-02 1.000D+00 1.000D+02 1.000D+04

FREQ	V(4)
-1.000D+00	9.990D-03
1.259D+00	1.258D-02
1.585D+00	1.583D-02
1.995D+00	1.993D-02
2.512D+00	2.509D-02
3.162D+00	3.159D-02
3.981D+00	3.977D-02
5.012D+00	5.007D-02
6.310D+00	6.303D-02
7.943D+00	7.935D-02
1.000D+01	9.990D-02
1.259D+01	1.258D-01
1.585D+01	1.583D-01
1.995D+01	1.993D-01
2.512D+01	2.509D-01
3.162D+01	3.159D-01
3.981D+01	3.976D-01
5.012D+01	5.005D-01
6.310D+01	6.300D-01
7.943D+01	7.929D-01
1.000D+02	9.978D-01
1.259D+02	1.255D+00
1.585D+02	1.578D+00
1.995D+02	1.983D+00
2.512D+02	2.490D+00
3.162D+02	3.120D+00
3.981D+02	3.901D+00
5.012D+02	4.857D+00
6.310D+02	6.011D+00
7.943D+02	7.375D+00
1.000D+03	8.935D+00
1.259D+03	1.064D+01
1.585D+03	1.241D+01
1.995D+03	1.411D+01
2.512D+03	1.563D+01
3.162D+03	1.688D+01
3.981D+03	1.785D+01
5.012D+03	1.855D+01
6.310D+03	1.904D+01
7.943D+03	1.937D+01
1.000D+04	1.959D+01
1.259D+04	1.973D+01
1.585D+04	1.982D+01
1.995D+04	1.988D+01
2.512D+04	1.991D+01
3.162D+04	1.994D+01
3.981D+04	1.995D+01
5.012D+04	1.996D+01
6.310D+04	1.996D+01
7.943D+04	1.997D+01
1.000D+05	1.997D+01

(b)

Figure 14–31
(Continued)

near 2 kHz (1.995 kHz) is 14.11, and we conclude that the design is valid. Since we assumed an ideal operational amplifier (with infinite bandwidth), the cutoff frequency f_2 does not affect the computed response. (Its presence can be simulated by connecting a 210-pF capacitor between nodes 3 and 4.)

Waveshaping

Waveshaping is the process of altering the shape of a waveform in some prescribed manner to produce a new waveform having a desired shape. Examples include altering a triangular wave to produce a square wave, and vice versa, altering a square wave to produce a series of narrow pulses, altering a square or triangular wave to produce a sine wave, and altering a sine wave to produce a square wave. Waveshaping techniques are widely used in function generators, frequency synthesizers, and synchronization circuits that require different waveforms having precisely the same frequency. In this section we will discuss just two waveshaping techniques, one employing an integrator and one employing a differentiator.

Recall that the output of an ideal integrator is a ramp voltage when its input is a positive dc level. Positive area accumulates under the input waveform as time passes, and the output rises, as shown in Figure 14–32. If the dc level is suddenly made negative, and remains negative for an interval of time, then negative area accumulates, so the *net* area decreases and the output voltage begins to fall. At the point in time where the total accumulated negative area equals the previously accumulated positive area, the output reaches 0, as shown in Figure 14–32.

Figure 14–32 shows the basis for generating a triangular waveform using an integrator. When the input is a square wave that continually alternates between positive and negative levels, the output rises and falls in synchronism. Note that the *slope* of the output alternates between $+E$ and $-E$ volts/second. Figure 14–33(a) shows how a triangular wave is generated using a practical integrator. Figure 14–33(b) shows the actual waveform generated. Notice that the figure reflects the *phase inversion* caused by the amplifier in that the output decreases when the input is positive and increases when the input is negative. The *average* level (dc component) of the output is 0, assuming no input offset. Note also that the slopes of the triangular wave are $\pm E/R_1C$ volts/second, since the integrator gain is $1/R_1C$. It is important to be able to predict these slopes to ensure that they will be within the specified slew rate of the amplifier. In the figure, T represents the period of the square wave, and it can be seen that the peak value of the triangular wave is

$$|V_{peak}| = \frac{ET}{4R_1C} \text{ volts} \tag{14–65}$$

Equation 14–65 shows that the amplitude of the triangular wave decreases with increasing frequency, since the period T decreases with frequency. On the other hand, the slopes $\Delta V/\Delta t$ are (contrary to expectation) independent of frequency. This observation is explained by the fact that while an increase in frequency reduces the time Δt over which the voltage changes, it also reduces the total change in voltage, ΔV. The phenomenon is illustrated in Figure 14–34.

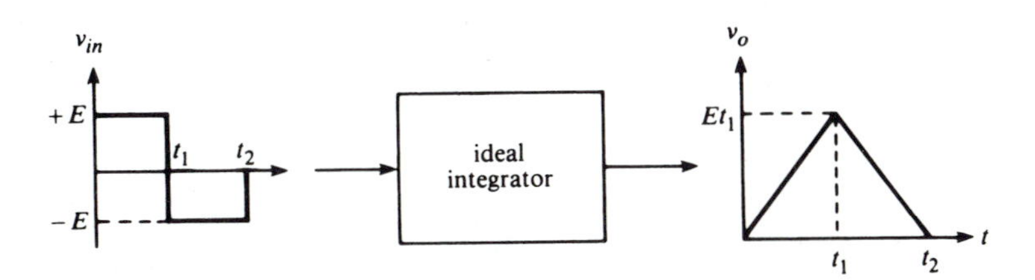

Figure 14–32
The integrator output rises to a maximum of Et_1 while the input is positive and then falls to 0 at time t_2 when the net area under the input is 0.

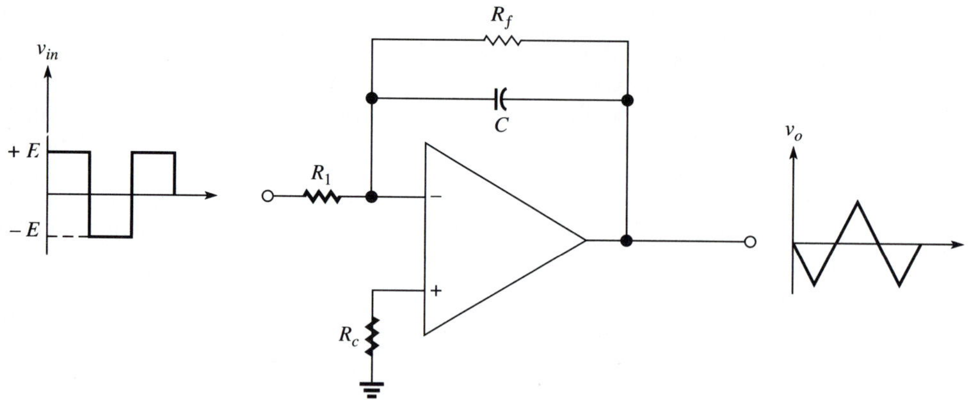

(a) Integrator used to generate a triangular waveform

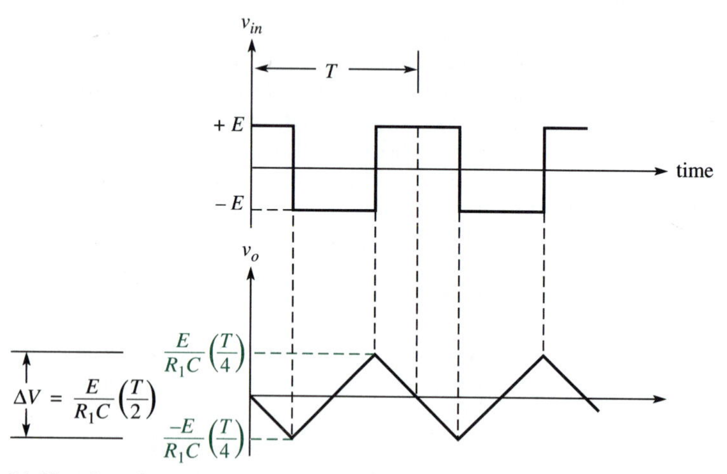

(b) The triangular wave generated by the circuit in (a), showing voltage and time relations. T is the period of the square wave.

Figure 14–33
The practical triangular-waveform generator

Figure 14–34
When the frequency of the triangular wave produced by an integrator is doubled, the amplitude is halved, so the slopes remain the same.

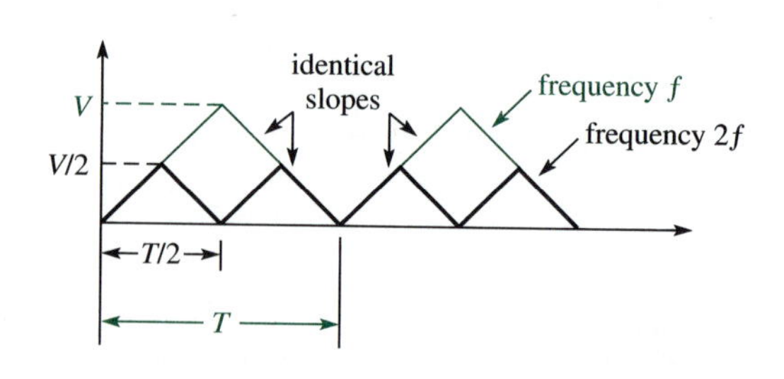

Equation 14–65 is a special case of a general relationship that can be used to find the integrator's peak output when the input is any periodic waveform symmetric about the horizontal axis:

$$V_{peak} = \frac{\text{positive area in } \frac{1}{2} \text{ cycle of input}}{2R_1C} \qquad \textbf{(14–66)}$$

For example, note that the positive area in one-half cycle of the square wave in Figure 14–33 is $E(T/2)$, so (14–66) reduces to (14–65) in that case.

Equations 14–65 and 14–66 are based on the assumption that there is *no input offset level or dc level in the input waveform.* When there is an input dc component, there will be an output dc component given by

$$V_o(\text{dc}) = \frac{-R_f}{R_1} V_{in}(\text{dc}) \qquad \textbf{(14–67)}$$

Thus, a triangular output will be shifted up or down by an amount equal to V_o(dc).

Example 14–12

The integrator in Figure 14–35 is to be used to generate a triangular waveform from a 500-Hz square wave connected to its input. Suppose the square wave alternates between ±12 V.

1. What minimum slew rate should the amplifier have?
2. What maximum output voltage should the amplifier be capable of developing?
3. Repeat (2) if the dc component in the input is −0.2 V. (The square wave alternates between +11.8 V and −12.2 V.)

Solution

1. The magnitude of the slope of the triangular waveform is

$$\frac{\Delta V}{\Delta t} = \frac{E}{R_1C} = \frac{12}{400(10^{-6})} = 3 \times 10^4 \text{ V/s}$$

Thus, we must have slew rate $S \geq 3 \times 10^4$ V/s.

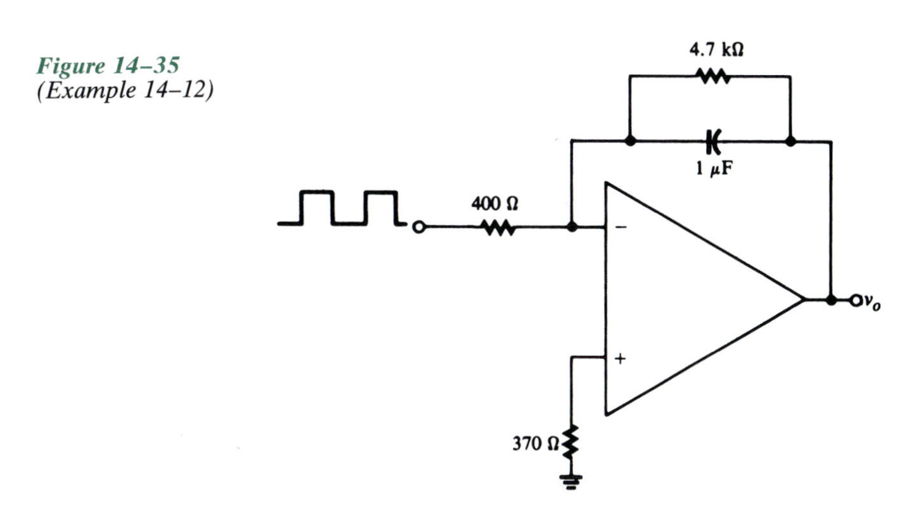

Figure 14–35
(Example 14–12)

2. The period T of the 500-Hz square wave is $T = 1/500 = 2 \times 10^{-3}$ s. By equation 14–65,

$$|V_{peak}| = \frac{ET}{4R_1C} = \frac{12(2 \times 10^{-3})}{4(400)(10^{-6})} = 15 \text{ V}$$

The triangular wave alternates between peak values of +15 V and −15 V.

3. By equation 14–67, the dc component in the output is

$$v_o(\text{dc}) = \frac{-R_f}{R_1} V_{in}(\text{dc}) = \frac{-4700}{400}(-0.2 \text{ V}) = 2.35 \text{ V}$$

Therefore, the triangular output is shifted up by 2.35 V and alternates between peak values of 15 + 2.35 = 17.35 V and −15 + 2.35 = −12.65 V. The amplifier must be capable of producing a 17.35-V output to avoid distortion.

If it were possible to generate an *ideal* square wave, it would have zero rise and fall times and would therefore change at an infinite rate every time it switched from one level to the other. If this square wave were the input to an ideal differentiator, the output would be a series of extremely narrow (actually, zero-width) pulses with infinite heights, alternately positive and negative, as illustrated in Figure 14–36. These fictional, zero-width, infinite-height "spikes" are called *impulses.* They have zero width because each change in the square wave occurs in zero time, and they have infinite height because the rate of change of the square wave is infinite at the points where changes occur. The negative impulses correspond to the negative rates of change when the square wave goes from a positive value to a negative value. Between the time points where the square wave changes, the rate of change is 0 and so is the output of the differentiator.

Of course, neither ideal square waves nor ideal differentiators exist. However, many real square waves have small rise and fall times and therefore have large rates of change at points where transitions in level occur. The output of a differentiator driven by such a square wave is therefore a series of narrow, high-amplitude pulses. These are used in *timing*-circuit applications, where it is necessary to trigger some circuit action in synchronization with another waveform. Figure 14–37 shows

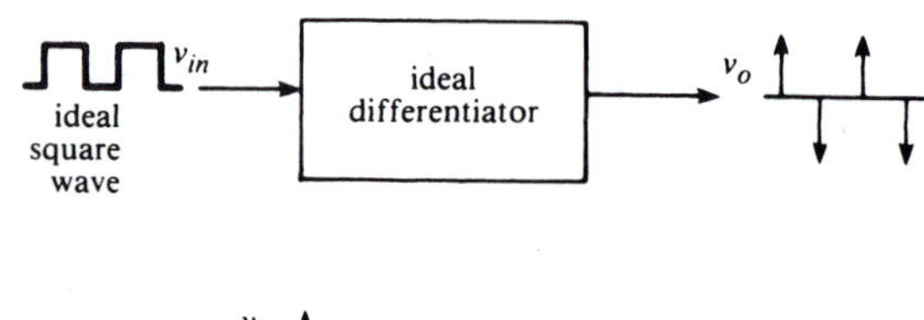

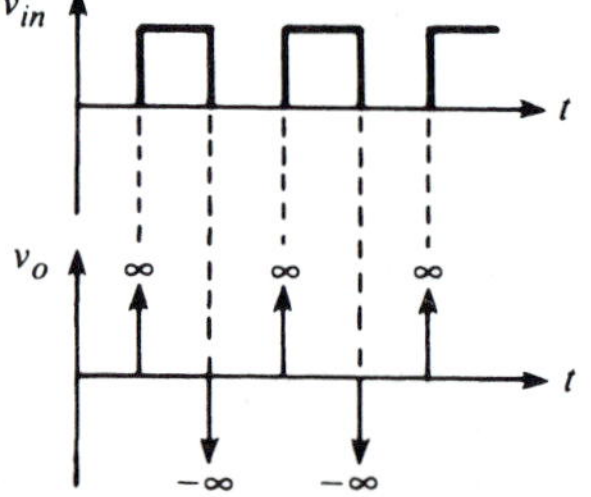

Figure 14–36
The output of an ideal differentiator driven by an ideal square wave is a series of infinite-height, zero-width impulses.

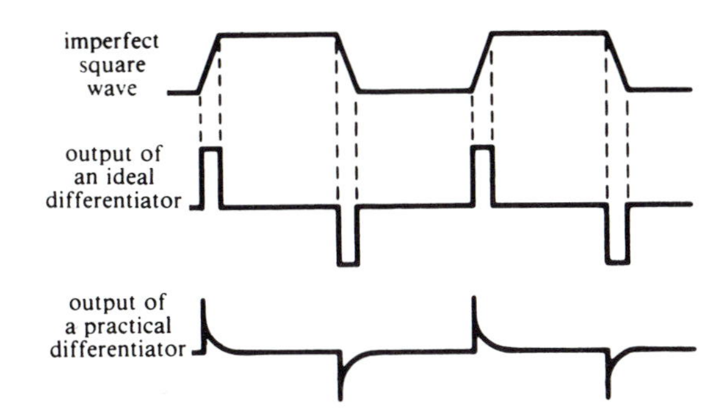

Figure 14–37
Typical outputs from ideal and practical differentiators driven by an imperfect square wave. Phase inversion is not shown.

a somewhat idealized square wave in which the nonzero rise and fall times are represented by ramps. The output that would result if this square wave were to drive an ideal differentiator is also shown. Finally, the typical output of a practical differentiator is shown. Here, the output pulse is distorted (although still usable in timing circuits), because of the restricted bandwidth of a practical differentiator (see Figure 14–28) and because of slew-rate limitations. Notice that dc level in the input to a differentiator has no effect on the output, because the rate of change of a dc component is 0. From a circuit viewpoint, the input capacitor blocks dc levels. (The figure does not show the phase inversion caused by the inverting amplifier.)

From studying the ideal output in Figure 14–37, the reader should be able to deduce that an ideal differentiator driven by a triangular wave will produce a square wave. Since the output equals the rate of change of the input, the square wave will alternate between $\pm(R_f C)|\Delta V_o/\Delta t|$ volts, where $\Delta V_o/\Delta t$ is the rate of change of the triangular input. Amplifier slew rate is a serious limitation in these waveshaping applications, because the differentiator output should theoretically have an infinite rate of change whenever the slope of the triangular input changes from positive to negative.

14–4 INSTRUMENTATION AMPLIFIERS

In Section 14–1 we saw how an amplifier can be operated in a differential mode to produce an output voltage proportional to the difference between two input signals (see Figure 14–5). Differential operation is a common requirement in instrumentation systems and other signal-processing applications where high accuracy is important. The circuit of Figure 14–5 has certain limitations in these applications, including the fact that the signal sources see different input impedances. Also, the circuit does not generally have good common-mode rejection, an important consideration in instrumentation systems, where long signal lines and high electrical noise environments are common. Figure 14–38 shows an improved configuration for producing an output proportional to the difference between two inputs. Notice that the circuit is basically the difference amplifier discussed earlier, with the addition of two input stages. Each input signal is connected directly to the noninverting terminal of an operational amplifier, so each signal source sees a very large input resistance. This circuit arrangement is so commonly used that it is called an *instrumentation amplifier* and is commercially available by that name in single-package units. These devices use closely matched, high-quality amplifiers and have very large common-mode rejection ratios.

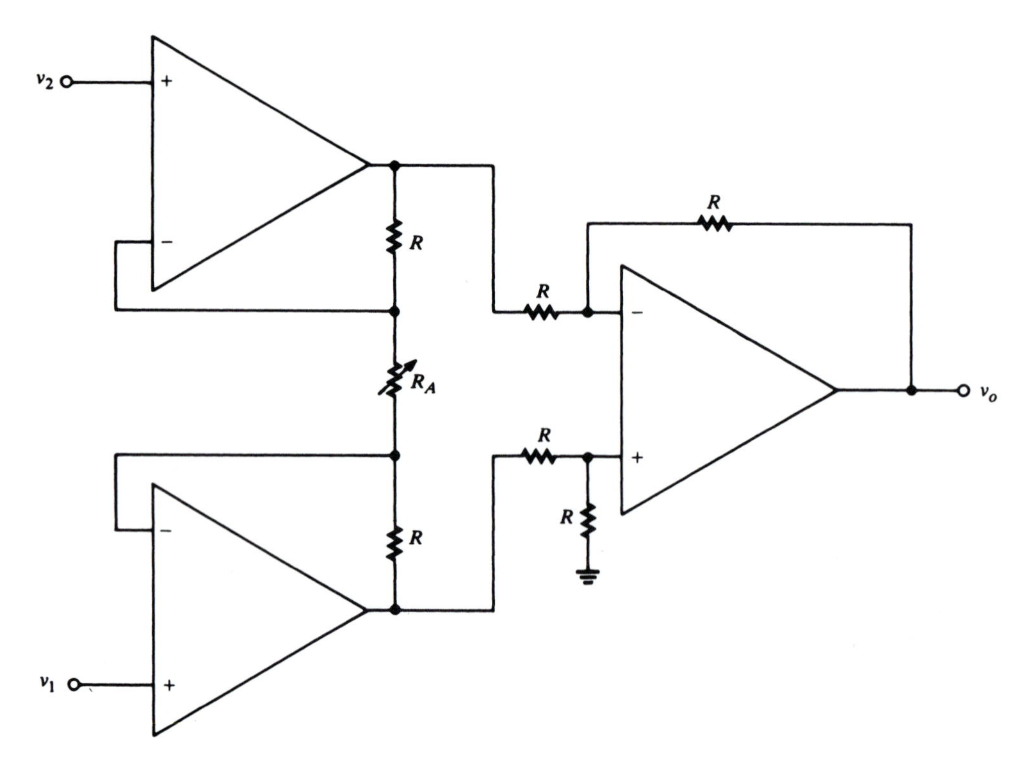

Figure 14–38
An instrumentation amplifier that produces an output proportional to $v_1 - v_2$. Adjustable resistor R_A is used to set the gain.

In our analysis of the instrumentation amplifier, we will refer to Figure 14–39, which shows current and voltage relations in the circuit. We begin by noting that the usual assumption of ideal amplifiers allows us to equate v_i^+ and v_i^- at each input amplifier ($v_i^+ - v_i^- \approx 0$), with the result that input voltages v_1 and v_2 appear across adjustable resistor R_A in Figure 14–39. For analysis purposes, let us assume that $v_1 > v_2$. Then the current i through R_A is

$$i = \frac{v_1 - v_2}{R_A} \tag{14–68}$$

Since no current flows into either amplifier input terminal, the current i must also flow in each resistor R connected on opposite sides of R_A. Therefore, the voltage drop across each of those resistors is

$$v_R = iR = \frac{(v_1 - v_2)R}{R_A} \tag{14–69}$$

The output voltages v_{o1} and v_{o2} are given by

$$v_{o1} = v_1 + v_R \tag{14–70}$$

and

$$v_{o2} = v_2 - v_R \tag{14–71}$$

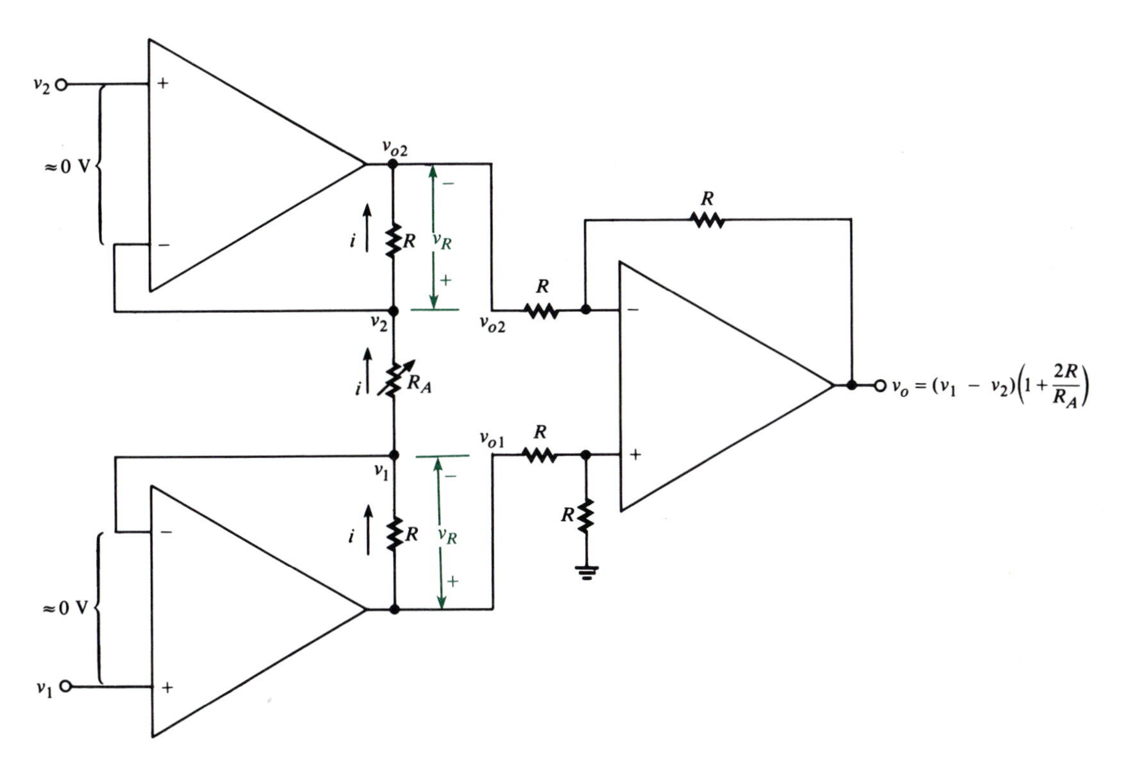

Figure 14–39
Voltage and current relations in the instrumentation amplifier. Note that the overall gain is inversely proportional to the value of adjustable resistor R_A.

Voltages v_{o1} and v_{o2} are the input voltages to the differential stage studied in Section 14–1. Since the external resistors connected to that stage are all equal to R, we recall (from equations 14–13 and 14–14, with $A = 1$) that

$$v_o = v_{o1} - v_{o2} \tag{14–72}$$

Substituting (14–70) and (14–71) into (14–72) gives

$$v_o = (v_1 + v_R) - (v_2 - v_R) = v_1 - v_2 + 2v_R \tag{14–73}$$

Substituting (14–69) into (14–73), we find

$$v_o = v_1 - v_2 + \frac{2(v_1 - v_2)R}{R_A} = (v_1 - v_2)(1 + 2R/R_A) \tag{14–74}$$

Equation 14–74 shows that the output of the instrumentation amplifier is directly proportional to the difference voltage ($v_1 - v_2$), as required. The overall closed-loop gain is clearly $(1 + 2R/R_A)$. R_A is made adjustable so that gain can be easily adjusted for calibration purposes. Note that the gain is inversely proportional to R_A.

To ensure proper operation of the instrumentation amplifier, all three of the following inequalities must be satisfied at all times:

$$\left|\left(1 + \frac{R}{R_A}\right)v_1 - \frac{R}{R_A}v_2\right| < |V_{max(1)}| \tag{14–75}$$

$$\left|\left(1+\frac{R}{R_A}\right)v_2-\frac{R}{R_A}v_1\right|<|V_{max(1)}| \tag{14–76}$$

$$\left(1+\frac{2R}{R_A}\right)|v_1-v_2|<|V_{max(2)}| \tag{14–77}$$

where $V_{max(1)}$ is the maximum output voltage of each input stage and $V_{max(2)}$ is the maximum output voltage of the differential stage.

Example 14–13

---

1. Assuming ideal amplifiers, find the minimum and maximum output voltage V_o of the instrumentation amplifier shown in Figure 14–40 when the 10-kΩ potentiometer R_p is adjusted through its entire range.
2. Find V_{o1} and V_{o2} when R_p is set in the middle of its resistance range.

Solution

1. Referring to Figure 14–38, we see that R_A in this example is the sum of R_p and the fixed 500-Ω resistor. Assuming that R_p can be adjusted through a full range

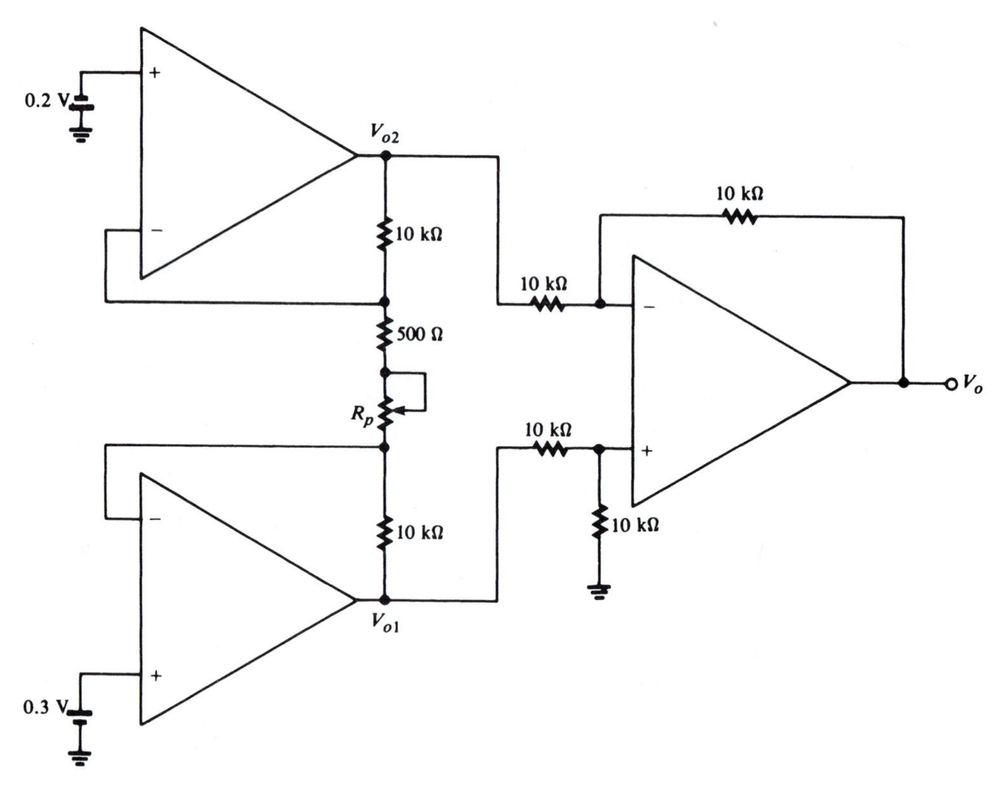

Figure 14–40
(Example 14–13)

from 0 to 10 kΩ, $R_A(\text{min}) = 500\ \Omega$ and $R_A(\text{max}) = (500\ \Omega) + (10\text{ k}\Omega) = 10.5$ kΩ. Therefore, from equation 14–74, the minimum and maximum values of V_o are

$$v_o(\text{min}) = (V_1 - V_2)\left[1 + \frac{2R}{R_A(\text{max})}\right] = [0.3\text{ V} - (-0.2\text{ V})] \times \left[1 + \frac{2(10\text{ k}\Omega)}{10.5\text{ k}\Omega}\right] = 1.45\text{ V}$$

$$v_o(\text{max}) = (V_1 - V_2)\left[1 + \frac{2R}{R_A(\text{min})}\right] = [0.3\text{ V} - (-0.2\text{ V})] \times \left[1 + \frac{2(10\text{ k}\Omega)}{500\ \Omega}\right] = 20.5\text{ V}$$

2. When $R_A = 500\ \Omega + (\frac{1}{2})(10 \times 10^3\ \Omega) = 5.5\text{ k}\Omega$, we have, from equation 14–69,

$$V_R = \frac{(V_1 - V_2)R}{R_A} = \frac{[0.3\text{ V} - (-0.2\text{ V})]10 \times 10^3\ \Omega}{5.5 \times 10^3\ \Omega} = 0.909\text{ V}$$

From equations 14–70 and 14–71,

$$V_{o1} = V_1 + V_R = 0.3\text{ V} + 0.909\text{ V} = 1.209\text{ V}$$
$$V_{o2} = V_2 - V_R = -0.2\text{ V} - 0.909\text{ V} = -1.109\text{ V}$$

Example 14–14

DESIGN

The maximum output voltages for all three operational amplifiers in an instrumentation amplifier are ±15 V. For a particular application, it is known that input signal v_1 may vary from 0 V to 0.8 V and input signal v_2 from 0 V to 1.3 V. Assuming that $R = 2\text{ k}\Omega$, design the circuit for maximum possible closed-loop gain.

Solution. Since the closed-loop gain is inversely proportional to R_A, we must find the minimum value of R_A that simultaneously satisfies inequalities 14–75 through 14–77. We must consider the worst-case condition for each inequality, that is, the combination of values for v_1 and v_2 that makes the left side of each inequality as large as possible. Thus, for inequality 14–75, we must satisfy both of the following:

$$\left|\left(1 + \frac{R}{R_A}\right) v_1(\text{max}) - \left(\frac{R}{R_A}\right) v_2(\text{min})\right| < |V_{max}|$$

and

$$\left|\left(1 + \frac{R}{R_A}\right) v_1(\text{min}) - \left(\frac{R}{R_A}\right) v_2(\text{max})\right| < |V_{max}|$$

Substituting values gives

$$\left|\left(1 + \frac{2 \times 10^3}{R_A}\right)(0.8) - \left(\frac{2 \times 10^3}{R_A}\right)0\right| < 15$$

and

$$\left|\left(1 + \frac{2 \times 10^3}{R_A}\right)0 - \left(\frac{2 \times 10^3}{R_A}\right)(1.3)\right| < 15$$

The first of these leads to

$$(0.8)\left(\frac{2 \times 10^3}{R_A}\right) < 14.2$$

or $R_A > 112.68\ \Omega$. The second inequality requires that

$$\frac{2 \times 10^3}{R_A} < 11.54$$

or $R_A > 173.3\ \Omega$.

Proceeding in a similar manner with inequality 14–76, we require that

$$\left|\left(1 + \frac{2 \times 10^3}{R_A}\right)(1.3) - \left(\frac{2 \times 10^3}{R_A}\right)0\right| < 15$$

and

$$\left|\left(1 + \frac{2 \times 10^3}{R_A}\right)(0) - \left(\frac{2 \times 10^3}{R_A}\right)(0.8)\right| < 15$$

These inequalities lead to $R_A > 189.71\ \Omega$ and $R_A > 106.67\ \Omega$.

Finally, inequality 14–77 requires that

$$\left|\left(1 + \frac{4 \times 10^3}{R_A}\right)(0 - 1.3)\right| < 15$$

and

$$\left|\left(1 + \frac{4 \times 10^3}{R_A}\right)(0.8 - 0)\right| < 15$$

which gives $R_A > 379.56\ \Omega$ and $R_A > 225.35\ \Omega$.

Summarizing, we require that *all* the following inequalities be satisfied: $R_A > 112.68\ \Omega$, $R_A > 173.3\ \Omega$, $R_A > 189.71\ \Omega$, $R_A > 106.67\ \Omega$, $R_A > 379.56\ \Omega$, and $R_A > 225.35\ \Omega$. Obviously, the only way that all inequalities can be satisfied is for R_A to be larger than the largest of the computed limits: $R_A > 379.56\ \Omega$. Choosing the closest standard 5% resistor value that is larger than 379.56 Ω, we let $R_A = 390\ \Omega$. This choice gives us the maximum permissible closed-loop gain, with a small margin for error:

$$\frac{v_o}{v_1 - v_2} = 1 + \frac{2R}{R_A} = 1 + \frac{2(2 \times 10^3\ \Omega)}{390\ \Omega} = 11.26$$

14–5 OSCILLATORS

An *oscillator* is a device that generates a periodic, ac output signal without any form of input signal required. The term is generally used in the context of a sine-wave signal generator, while a square-wave generator is usually called a *multivibrator.* A *function generator* is a laboratory instrument that a user can set to produce sine, square, or triangular waves, with amplitudes and frequencies that can be adjusted at will. Desirable features of a sine-wave oscillator include the ability to produce a low distortion ("pure") sinusoidal waveform, and, in many applications, the capability of being easily adjusted so that the user can vary the frequency over some reasonable range.

Oscillation is a form of instability caused by feedback that *regenerates,* or reinforces, a signal that would otherwise die out due to energy losses. In order for the feedback to be regenerative, it must satisfy certain amplitude and phase relations that we will discuss shortly. Oscillation often plagues designers and users of high-gain amplifiers because of unintentional feedback paths that create signal regeneration at one or more frequencies. By contrast, an oscillator is designed to have a feedback path with known characteristics, so that a predictable oscillation will occur at a predetermined frequency.

The Barkhausen Criterion

We have stated that an oscillator has no input *per se,* so the reader may wonder what we mean by "feedback"—feedback to where? In reality, it makes no difference where, because we have a closed loop with no summing junction at which any external input is added. Thus, we could start anywhere in the loop and call that point both the "input" and the "output"; in other words, we could think of the "feedback" path as the entire path through which signal flows in going completely around the loop. However, it is customary and convenient to take the output of an amplifier as a reference point and to regard the feedback path as that portion of the loop that lies between amplifier output and amplifier input. This viewpoint is illustrated in Figure 14–41, where we show an amplifier having gain A and a feedback path having gain β. β is the usual feedback ratio that specifies the portion of amplifier output voltage fed back to amplifier input. Every oscillator must have an amplifier, or equivalent device, that supplies energy (from the dc supply) to replenish resistive losses and thus sustain oscillation.

In order for the system shown in Figure 14–41 to oscillate, the loop gain $A\beta$ must satisfy the *Barkhausen criterion,* namely,

$$A\beta = 1 \quad \text{(14–78)}$$

Imagine a small variation in signal level occurring at the input to the amplifier, perhaps due to noise. The essence of the Barkhausen criterion is that this variation will be reinforced and signal regeneration will occur only if the net gain around the loop, beginning and ending at the point where the variation occurred, is unity. It is important to realize that unity gain means not only a gain magnitude of 1, but also an *in-phase* signal reinforcement. Negative feedback causes signal cancellation because the feedback voltage is out of phase. By contrast, the unity loop-gain criterion for oscillation is often called *positive feedback.*

To understand and apply the Barkhausen criterion, we must regard both the gain and the phase shift of $A\beta$ as *functions of frequency.* Reactive elements, capacitance in

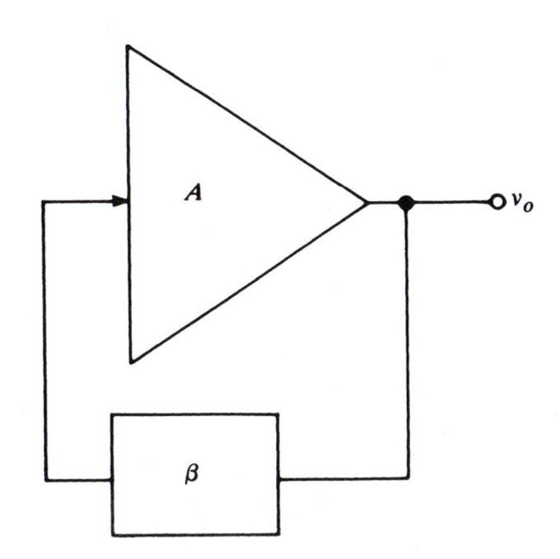

Figure 14–41
Block diagram of an oscillator

particular, contained in the amplifier and/or feedback cause the gain magnitude and phase shift to change with frequency. In general, there will be only one frequency at which the gain magnitude is unity and at which, simultaneously, the total phase shift is equivalent to 0 degrees (in phase—a multiple of 360°). *The system will oscillate at the frequency that satisfies those conditions.* Designing an oscillator amounts to selecting reactive components and incorporating them into circuitry in such a way that the conditions will be satisfied at a predetermined frequency.

To show the dependence of the loop gain $A\beta$ on frequency, we write $A\beta(j\omega)$, a complex phasor that can be expressed in both polar and rectangular form:

$$A\beta(j\omega) = |A\beta|\angle\theta = |A\beta|\cos\theta + j|A\beta|\sin\theta \tag{14–79}$$

where $|A\beta|$ is the gain magnitude, a function of frequency, and θ is the phase shift, also a function of frequency. The Barkhausen criterion requires that

$$|A\beta| = 1 \tag{14–80}$$

and

$$\theta = \pm 360° n \tag{14–81}$$

where n is any integer, including 0. In polar and rectangular forms, the Barkhausen criterion is expressed as

$$A\beta(j\omega) = 1\angle \pm 360° n = 1 + j0 \tag{14–82}$$

Example 14–15

The gain of a certain amplifier as a function of frequency is $A(j\omega) = -16 \times 10^6/j\omega$. A feedback path connected around it has $\beta(j\omega) = 10^3/(2 \times 10^3 + j\omega)^2$. Will the system oscillate? If so, at what frequency?

Solution. The loop gain is

$$A\beta = \left(\frac{-16 \times 10^6}{j\omega}\right)\left[\frac{10^3}{(2 \times 10^3 + j\omega)^2}\right] = \frac{-16 \times 10^9}{j\omega(2 \times 10^3 + j\omega)^2}$$

To determine if the system will oscillate, we will first determine the frequency, if any, at which the phase angle of $A\beta$ ($\theta = \angle A\beta$) equals 0 or a multiple of 360°. Using phasor algebra, we have

$$\theta = \angle A\beta = \angle \frac{-16 \times 10^9}{j\omega(2 \times 10^3 + j\omega)^2} = \angle -16 \times 10^9 + \angle 1/j\omega + \angle \frac{1}{(2 \times 10^3 + j\omega)^2}$$

$$= -180° - 90° - 2\arctan(\omega/2 \times 10^3)$$

This expression will equal $-360°$ if $2\arctan(\omega/2 \times 10^3) = 90°$, or

$$\arctan(\omega/2 \times 10^3) = 45°$$
$$\omega/2 \times 10^3 = 1$$
$$\omega = 2 \times 10^3 \text{ rad/s}$$

Thus, the phase shift around the loop is $-360°$ at $\omega = 2000$ rad/s. We must now check to see if the gain magnitude $|A\beta|$ equals 1 at $\omega = 2 \times 10^3$. The gain magnitude is

$$|A\beta| = \left|\frac{-16\times10^9}{j\omega(2\times10^3+j\omega)^2}\right| = \frac{|-16\times10^9|}{|j\omega|\,|(2\times10^3+j\omega)|^2}$$
$$= \frac{16\times10^9}{\omega[(2\times10^3)^2+\omega^2]}$$

Substituting $\omega = 2\times10^3$, we find

$$|A\beta| = \frac{16\times10^9}{2\times10^3(4\times10^6+4\times10^6)} = 1$$

Thus, the Barkhausen criterion is satisfied at $\omega = 2\times10^3$ rad/s and oscillation occurs at that frequency ($2\times10^3/2\pi = 318.3$ Hz).

Example 14–15 illustrated an application of the *polar* form of the Barkhausen criterion, since we solved for $\angle A\beta$ and then determined the frequency at which that angle equals $-360°$. It is instructive to demonstrate how the same result can be obtained using the *rectangular* form of the criterion: $A\beta = 1 + j0$. Toward that end, we first expand the denominator:

$$A\beta = \frac{-16\times10^9}{j\omega(2\times10^3+j\omega)^2} = \frac{-16\times10^9}{j\omega(4\times10^6+j4\times10^3\omega-\omega^2)}$$
$$= \frac{-16\times10^9}{j\omega[(4\times10^6-\omega^2)+j4\times10^3\omega]} = \frac{16\times10^9}{4\times10^3\omega^2 - j\omega(4\times10^6-\omega^2)}$$

To satisfy the Barkhausen criterion, this expression for $A\beta$ must equal 1. We therefore set it equal to 1 and simplify:

$$1 = \frac{16\times10^9}{4\times10^3\omega^2 - j\omega(4\times10^6-\omega^2)}$$
$$4\times10^3\omega^2 - j\omega(4\times10^6-\omega^2) = 16\times10^9$$
$$(4\times10^3\omega^2 - 16\times10^9) - j\omega(4\times10^6-\omega^2) = 0$$

In order for this expression to equal 0, *both the real and imaginary parts must equal 0.* Setting either part equal to 0 and solving for ω will give us the same result we obtained before:

$$4\times10^3\omega^2 - 16\times10^9 = 0 \Rightarrow \omega = 2\times10^3$$
$$4\times10^6 - \omega^2 = 0 \Rightarrow \omega = 2\times10^3$$

This result was obtained with somewhat more algebraic effort than previously. In some applications, it is easier to work with the polar form than the rectangular form, and in others, the reverse is true.

The RC Phase-Shift Oscillator

One of the simplest kinds of oscillators incorporating an operational amplifier can be constructed as shown in Figure 14–42. Here we see that the amplifier is connected in an inverting configuration and drives three cascaded (high-pass) RC sections. The arrangement is called an *RC phase-shift oscillator.* The inverting amplifier causes a 180° phase shift in the signal passing through it, and the purpose of the

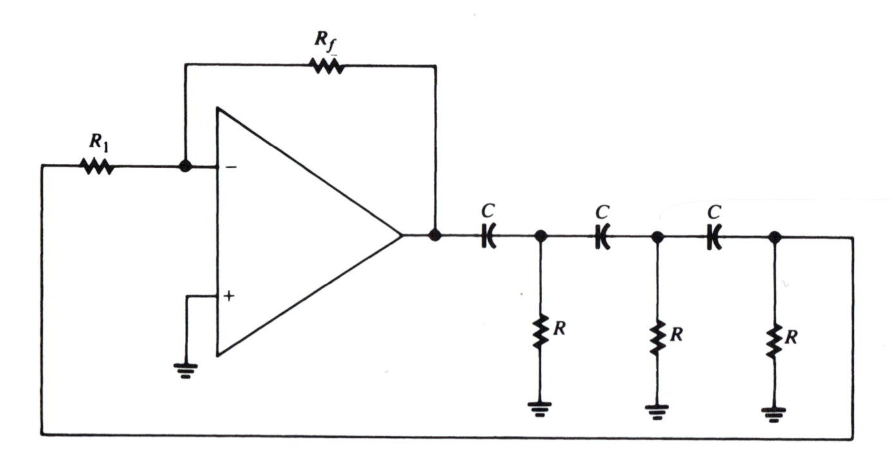

Figure 14–42
An RC phase-shift oscillator

cascaded RC sections is to introduce an additional 180° at some frequency. Recall that the output of a single, high-pass RC network leads its input by a phase angle that depends on the signal frequency. When the signal passes through all three RC sections, there will be some frequency at which the cumulative phase shift is 180°. When the signal having that frequency is fed back to the inverting amplifier, as shown in the figure, the total phase shift around the loop will equal 180° + 180° = 360° (or, equivalently, −180° + 180° = 0°) and oscillation will occur at that frequency, provided the loop gain is 1. The gain necessary to overcome the loss in the RC cascade and bring the loop gain up to 1 is supplied by the amplifier ($v_o/v_{in} = -R_f/R_1$).

With considerable algebraic effort, it can be shown (Exercise 14–40) that the feedback ratio determined by the RC cascade (with the feedback connection to R_1 opened) is

$$\beta = \frac{R^3}{(R^3 - 5RX_C^2) + j(X_C^3 - 6R^2X_C)} \tag{14–83}$$

In order for oscillation to occur, the cascade must shift the phase of the signal by 180°, which means the angle of β must be 180°. When the angle of β is 180°, β is a purely real number. In that case, the imaginary part of the denominator of equation 14–83 is 0. Therefore, we can find the oscillation frequency by finding the value of ω that makes the imaginary part equal 0. Setting it equal to 0 and solving for ω, we find

$$X_C^3 - 6R^2X_C = 0$$
$$X_C^3 = 6R^2X_C$$
$$X_C^3 = 6R^2 \tag{14–84}$$
$$\frac{1}{(\omega C)^2} = 6R^2$$
$$\omega = \frac{1}{\sqrt{6}RC} \text{ rad/s}$$

or

$$f = \frac{1}{2\pi\sqrt{6}RC}\ \text{Hz} \tag{14–85}$$

Notice that resistor R_1 in Figure 14–42 is effectively in parallel with the rightmost resistor R in the RC cascade, because the inverting input of the amplifier is at virtual ground. Therefore, when the feedback loop is closed by connecting the cascade to R_1, the frequency satisfying the phase criterion will be somewhat different than that predicted by equation 14–85. If $R_1 \gg R$, so that $R_1 \parallel R \approx R$, then equation 14–85 will closely predict the oscillation frequency.

We can find the gain that the amplifier must supply by finding the reduction in gain caused by the RC cascade. This we find by evaluating the magnitude of β at the oscillation frequency: $\omega = 1/(\sqrt{6}RC)$. At that frequency, the imaginary term in equation 14–83 is 0 and β is the real number

$$\begin{aligned} |\beta| &= \frac{R^3}{R^3 - 5RX_C^2} = \frac{R^3}{R^3 - 5R\left(\frac{\sqrt{6}RC}{C}\right)^2} \\ &= \frac{R^3}{R^3 - 30R^3} \\ &= -1/29 \end{aligned} \tag{14–86}$$

The minus sign confirms that the cascade inverts the feedback at the oscillation frequency. We see that the amplifier must supply a gain of −29 to make the loop gain $A\beta = 1$. Thus, we require

$$\frac{R_f}{R_1} = 29 \tag{14–87}$$

In practice, the feedback resistor is made adjustable to allow for small differences in component values and for the loading caused by R_1.

Example 14–16

DESIGN

Design an RC phase-shift oscillator that will oscillate at 100 Hz.

Solution. From equation 14–85,

$$f = 100\ \text{Hz} = \frac{1}{2\pi\sqrt{6}RC}$$

Let $C = 0.5\ \mu\text{F}$. Then

$$R = \frac{1}{100(2\pi)\sqrt{6}(0.5 \times 10^{-6})]} = 1300\ \Omega$$

To prevent R_1 from loading this value of R, we choose $R_1 = 20\ \text{k}\Omega$. (For greater precision, we could choose the last R in the cascade and the value R_1 so that $R \parallel R_1 = 1300\ \Omega$.) From equation 14–87, $R_f = 29R_1 = 29(20\ \text{k}\Omega) = 580\ \text{k}\Omega$. The completed circuit is shown in Figure 14–43. R_f is made adjustable so the loop gain can be set precisely to 1.

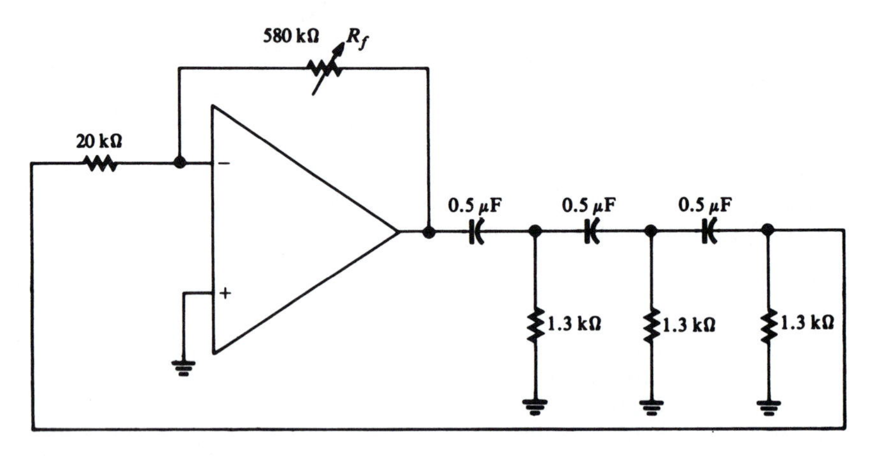

Figure 14–43
(Example 14–16)

The Wien-bridge Oscillator

Figure 14–44 shows a widely used type of oscillator called a *Wien bridge.* The operational amplifier is used in a noninverting configuration, and the impedance blocks labeled Z_1 and Z_2 form a voltage divider that determines the feedback ratio. Note that a portion of the output voltage is fed back through this impedance divider to the + input of the amplifier. Resistors R_g and R_f determine the amplifier gain and are selected to make the magnitude of the loop gain equal to 1. If the feedback impedances are chosen properly, there will be some frequency at which there is zero phase shift in the signal fed back to the amplifier input (v^+). Since the amplifier is noninverting, it also contributes zero phase shift, so the total phase shift around the loop is 0 at that frequency, as required for oscillation.

In the most common version of the Wien-bridge oscillator, Z_1 is a series RC

Figure 14–44
The Wien-bridge oscillator. Z_1 and Z_2 determine the feedback ratio to the noninverting input. R_f and R_g control the magnitude of the loop gain.

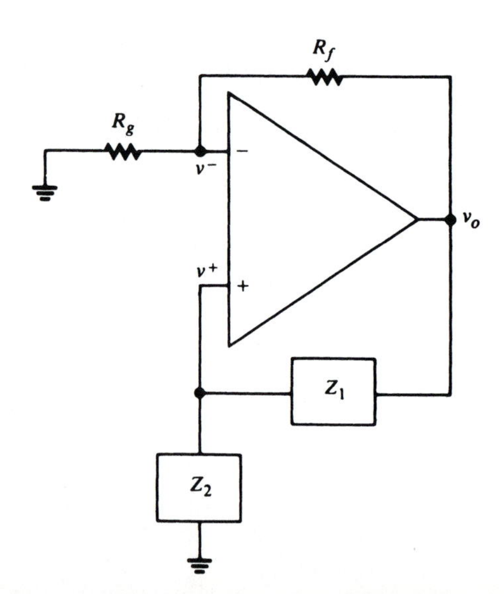

Figure 14–45
The Wien-bridge oscillator showing the RC networks that form Z_1 and Z_2

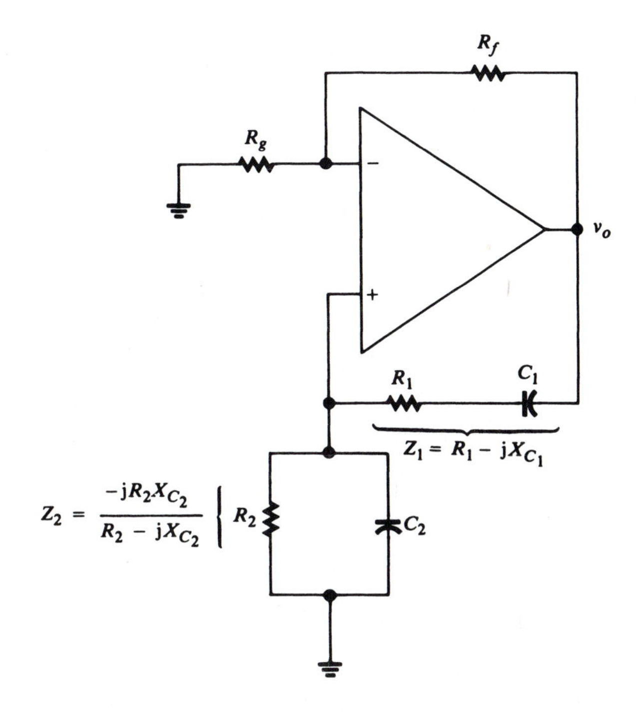

combination and Z_2 is a parallel RC combination, as shown in Figure 14–45. For this configuration,

$$Z_1 = R_1 - jX_{C_1}$$

and

$$Z_2 = R_2 \| -jX_{C_2} = \frac{-jR_2X_{C_2}}{R_2 - jX_{C_2}}$$

The feedback ratio is then

$$\beta = \frac{v^+}{v_o} = \frac{Z_2}{Z_1 + Z_2} = \frac{-jR_2X_{C_2}/(R_2 - jX_{C_2})}{R_1 - jX_{C_1} - jR_2X_{C_2}/(R_2 - jX_{C_2})} \tag{14–88}$$

which, upon simplification, becomes

$$\frac{v^+}{v_o} = \frac{R_2X_{C_2}}{(R_1X_{C_2} + R_2X_{C_1} + R_2X_{C_2}) + j(R_1R_2 - X_{C_1}X_{C_2})} \tag{14–89}$$

In order for v^+ to have the same phase as v_o, this ratio must be a purely real number. Therefore, the imaginary term in (14–89) must be 0. Setting the imaginary term equal to 0 and solving for ω gives us the oscillation frequency:

$$R_1R_2 - X_{C_1}X_{C_2} = 0$$

$$R_1R_2 = \left(\frac{1}{\omega C_1}\right)\left(\frac{1}{\omega C_2}\right)$$

$$\omega^2 = \frac{1}{R_1R_2C_1C_2} \tag{14–90}$$

$$\omega = \frac{1}{\sqrt{R_1R_2C_1C_2}} \text{ rad/s}$$

In most applications, the resistors are made equal and so are the capacitors: $R_1 = R_2 = R$ and $C_1 = C_2 = C$. In this case, the oscillation frequency becomes

$$\omega = \frac{1}{\sqrt{R^2C^2}} = \frac{1}{RC}\text{ rad/s} \tag{14–91}$$

or

$$f = \frac{1}{2\pi RC}\text{ Hz} \tag{14–92}$$

When $R_1 = R_2 = R$ and $C_1 = C_2 = C$, the capacitive reactance of each capacitor at the oscillation frequency is

$$X_{C_1} = X_{C_2} = \frac{1}{\omega C} = \frac{1}{\left(\frac{1}{RC}\right)C} = R$$

Substituting $X_{C_1} = X_{C_2} = R = R_1 = R_2$ in equation 14–89, we find that the feedback ratio at the oscillation frequency is

$$\frac{v^+}{v_o} = \frac{R^2}{3R^2 + j0} = \frac{1}{3}$$

Therefore, the amplifier must provide a gain of 3 to make the magnitude of the loop gain unity and sustain oscillation. Since the amplifier gain is $(R_g + R_f)/R_g$, we require

$$\frac{R_g + R_f}{R_g} = 3\text{, or}$$
$$1 + \frac{R_f}{R_g} = 3 \Rightarrow \frac{R_f}{R_g} = 2 \tag{14–93}$$

(Another way of reaching this same conclusion is to recognize that the operational amplifier maintains $v^+ \approx v^-$. But, from Figure 14–45, we see that $v^- = v_oR_g/(R_g + R_f)$, so $v^-/v_o = (R_g + R_f)/R_g = v^+/v_o = 1/3$.)

Example 14–17

DESIGN

Design a Wien-bridge oscillator that oscillates at 25 kHz.

Solution. Let $C_1 = C_2 = 0.001\ \mu\text{F}$. Then, from equation 14–92,

$$f = 25 \times 10^3\text{ Hz} = \frac{1}{2\pi R(10^{-9}\text{ F})}$$
$$R = \frac{1}{2\pi(25 \times 10^3\text{ Hz})(10^{-9}\text{ F})} = 6366\ \Omega$$

Let $R_g = 10\text{ k}\Omega$. Then, from equation 14–93,

$$\frac{R_f}{R_g} = 2 \Rightarrow R_f = 20\text{ k}\Omega$$

In practical Wien-bridge oscillators, R_f is not equal to exactly $2R_g$ because component tolerances prevent R_1 from being exactly equal to R_2 and C_1 from being exactly equal to C_2. Furthermore, the amplifier is not ideal, so v^- is not exactly equal to v^+. In a circuit constructed for laboratory experimentation, R_f should be made adjustable so that the loop gain can be set as necessary to sustain oscillation. Practical oscillators incorporate a nonlinear device in the R_f–R_g feedback loop to provide automatic adjustment of the loop gain, as necessary to sustain oscillation. This arrangement is a form of *automatic gain control* (AGC), whereby a reduction in signal level changes the resistance of the nonlinear device in a way that restores gain.

The Colpitts Oscillator

In the Colpitts oscillator, the impedance in the feedback circuit is a resonant LC network. See Figure 14–46(a). The frequency of oscillation is the resonant frequency of the network, which is the frequency at which the phase shift through the network is 180°. At that frequency, the impedance is a real number. Figure 14–46(b) shows the feedback network. The impedance looking into the network from the amplifier output is the parallel combination of $-\mathrm{j}X_{C_1}$ and $(\mathrm{j}X_L - \mathrm{j}X_{C_2})$:

$$Z = \frac{(-\mathrm{j}X_{C_1})(\mathrm{j}X_L - \mathrm{j}X_{C_2})}{-\mathrm{j}X_{C_1} + \mathrm{j}X_L - \mathrm{j}X_{C_2}} = \frac{X_L C_1 - X_{C_1}X_{C_2}}{\mathrm{j}(X_L - X_{C_1} - X_{C_2})}$$

In order for Z to equal a real number, the imaginary term must equal 0. Thus, at resonance,

$$\mathrm{j}(X_L - X_{C_1} - X_{C_1}) = 0 \Rightarrow X_L = X_{C_1} + \mathrm{X}_{C_2}$$

or

$$\omega L = \frac{1}{\omega C_1} + \frac{1}{\omega C_2}$$

Solving for ω gives the oscillation frequency:

$$\omega = \frac{1}{\sqrt{LC_T}} \text{ radians/second} \quad \text{or} \quad f = \frac{1}{2\pi\sqrt{LC_T}} \text{ hertz} \qquad \textbf{(14–94)}$$

where $C_T = \dfrac{C_1 C_2}{C_1 + C_2}$

Note that this computation is based on the assumption that the left-hand side of the network is open, as can be seen in Figure 14–46(b). In reality, it is loaded by input resistor R_1 (Figure 14–46(a)). Therefore, equation 14–94 is an approximation, but it is valid when $R_1 \gg X_{C_2}$.

Since v_0 appears across the voltage divider consisting of L and C_2 (see Figure 14–46(c)), the feedback factor (again neglecting R_1) is

$$\beta = \frac{-\mathrm{j}X_{C_2}}{\mathrm{j}X_L - \mathrm{j}X_{C_2}} \qquad \textbf{(14–95)}$$

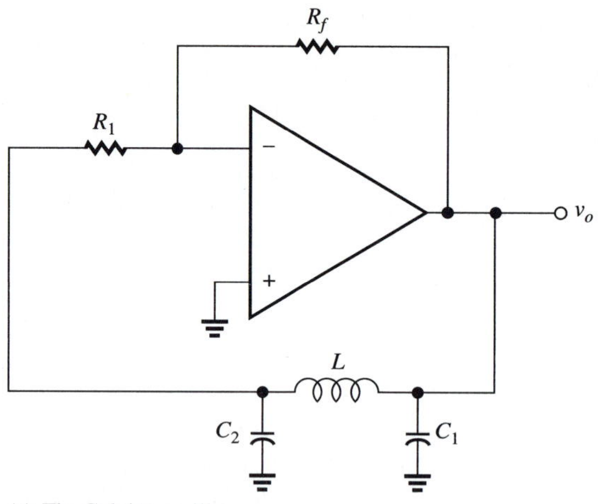

(a) The Colpitts oscillator

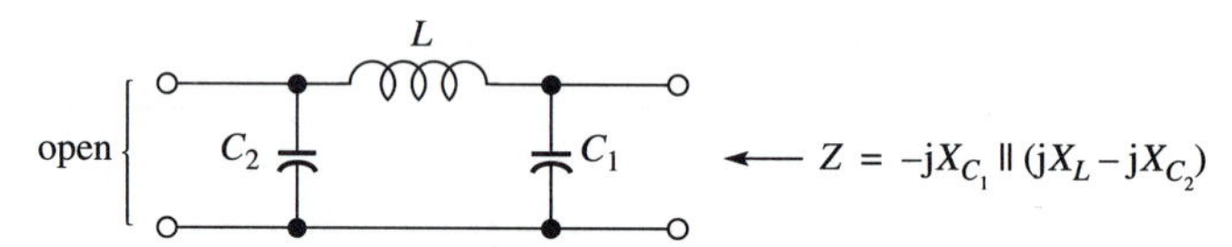

(b) The feedback network, whose impedance is derived under the assumption that the feedback path is open (valid for $R_1 >> X_{C_2}$).

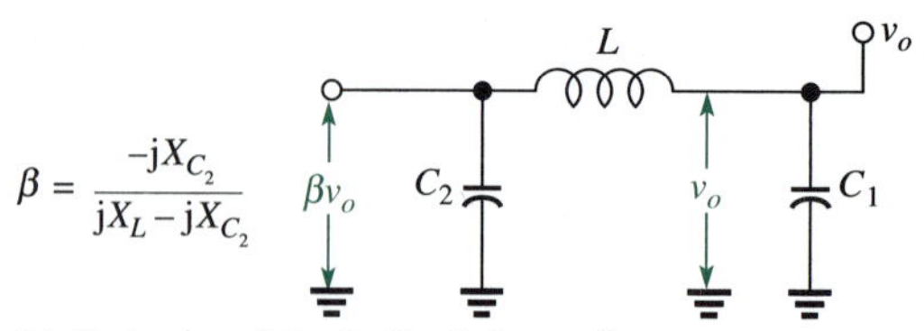

(c) Derivation of the feedback factor β.

Figure 14–46
The Colpitts oscillator

Since $X_L = X_{C_1} + X_{C_2}$ at resonance, the feedback factor at resonance is

$$\beta = \frac{-jX_{C_2}}{j(X_{C_1} + X_{C_2}) - jX_{C_2}} = \frac{-X_{C_2}}{X_{C_1}} = \frac{-C_1}{C_2} \tag{14–96}$$

Note that β is real and has angle 180°, as required. In order for the loop gain to equal 1, we require

$$|A_v\beta| = 1 \Rightarrow |A_v| = \frac{1}{|\beta|} = \frac{C_2}{C_1} \tag{14–97}$$

Thus, the closed-loop gain of the inverting amplifier must be at least C_2/C_1. In practice, $|A_v|$ is adjusted to be greater than C_2/C_1, but not so much greater that distortion results.

Example 14–18

DESIGN

Design a Colpitts oscillator that will oscillate at 100 kHz.

Solution. Let us choose $C_1 = C_2 = 0.01\ \mu\text{F}$. Then

$$C_T = \frac{C_1C_2}{C_1 + C_2} = \frac{(0.01\ \mu\text{F})(0.01\ \mu\text{F})}{0.01\ \mu\text{F} + 0.01\ \mu\text{F}} = 0.005\ \mu\text{F}$$

From equation 14–94,

$$L = \frac{1}{(2\pi)^2 f^2 C_T} = \frac{1}{(2\pi)^2(100 \times 10^3\ \text{Hz})^2(0.005 \times 10^{-6}\ \text{F})} \approx 0.5\ \text{mH}$$

We wish to make $R_1 \gg X_{C_2}$ at resonance:

$$X_{C_2} = \frac{1}{\omega C_2} = \frac{1}{2\pi(100\ \text{kHz})(0.01\ \mu\text{F})} = 159\ \Omega$$

Choose $R_1 = 10\ \text{k}\Omega$. Then, since $|A_v| = R_f/R_1$ and we require that $|A_v| \geq C_2/C_1 = 1$, we must choose $R_f \geq 10\ \text{k}\Omega$.

The Hartley Oscillator

The impedance in the feedback of a Hartley oscillator is a resonant LC network consisting of two inductors and one capacitor. See Figure 14–47. As in the Colpitts oscillator, oscillation occurs at the resonant frequency of the network, the frequency at which its impedance is real. Following the same procedure used to derive the oscillation frequency and feedback factor of the Colpitts oscillator, we can show (Exercise 14–44) that

$$f = \frac{1}{2\pi\sqrt{CL_T}}\ \text{hertz} \tag{14–98}$$

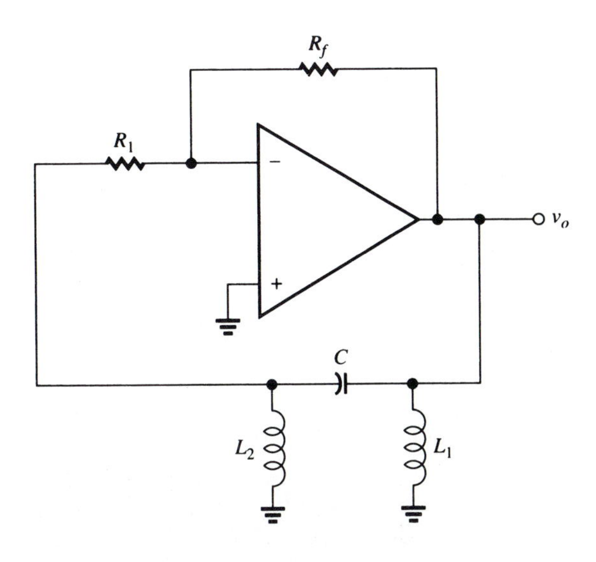

Figure 14–47
The Hartley oscillator

where $L_T = L_1 + L_2$

and

$$\beta = \frac{-L_2}{L_1} \qquad \text{(at resonance)} \tag{14–99}$$

If L_1 and L_2 are wound on the same core and therefore have mutual inductance M, then L_T in equation 14–98 is $L_T = L_1 + L_2 + 2M$. In order to satisfy $|A_v\beta| \geq 1$ in the Hartley oscillator, we require, from equation 14–99,

$$|A_v| = \frac{R_f}{R_1} \geq \frac{L_1}{L_2} \tag{14–100}$$

14–6 ACTIVE FILTERS

Basic Filter Concepts

A *filter* is a device that allows signals having frequencies in a certain range to pass through it while attenuating all other signals. An *ideal* filter has identical gain at all frequencies in its *passband* and zero gain at all frequencies outside its passband. Figure 14–48 shows the ideal frequency response for each of several commonly used filter types, along with typical responses for practical filters of the same types.

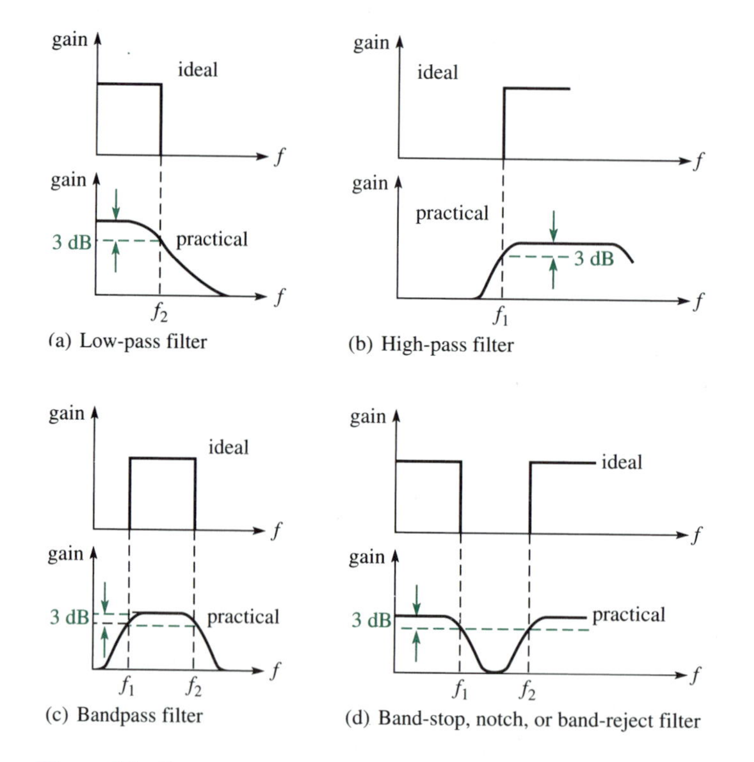

Figure 14–48
Ideal and practical frequency responses of some commonly used filter types

In each case, cutoff frequencies are shown to be those frequencies at which the response is down 3 dB from its maximum value in the passband. Filters are widely used to "extract" desired frequency components from complex signals and/or reject undesired ones, such as noise. *Passive* filters are those constructed with resistors, capacitors, and inductors, while *active* filters employ active components such as transistors and operational amplifiers, along with resistor–capacitor networks. Inductors are rarely used with active filters, because of their bulk, expense, and lack of availability in a wide range of values.

Filters are classified by their *order,* an integer number, also called the number of *poles.* For example, a second-order filter is said to have two poles. In general, the higher the order of a filter, the more closely it approximates an ideal filter and the more complex the circuitry required to construct it. The frequency response outside the passband of a filter of order n has a slope that is asymptotic to $20n$ dB/decade or $6n$ dB/octave. Recall that a simple RC network is called *first-order,* or *single-pole,* filter and its response falls off at 20 dB/decade.

The classification of filters by their order is applicable to all the filter types shown in Figure 14–48. We should note that the response of every high-pass filter must eventually fall off at some high frequency because no physically realizable filter can have infinite bandwidth. Therefore, every high-pass filter is, technically speaking, a bandpass filter. However, it is often the case that a filter behaves like a high-pass for all those frequencies encountered in a particular application, so for all practical purposes it can be treated as a true high-pass filter. The order of a high-pass filter affects its frequency response in the vicinity of f_1 (Figure 14–48(b)) and is not relevant to the high-frequency region, where it must eventually fall off.

Filters are also classified as belonging to one of several specific design types that, like order, affect their response characteristics within and outside of their pass bands. The two categories we will study are called the *Butterworth* and *Chebyshev* types. The gain magnitudes of low- and high-pass Butterworth filters obey the relationships

$$|G| = \frac{M}{\sqrt{1 + (f/f_2)^{2n}}} \qquad \text{(low-pass)} \qquad \textbf{(14–101)}$$

$$|G| = \frac{M}{\sqrt{1 + (f_1/f)^{2n}}} \qquad \text{(high-pass)} \qquad \textbf{(14–102)}$$

where M is a constant, n is the order of the filter, f_1 is the cutoff frequency of the high-pass filter, and f_2 is the cutoff frequency of the low-pass filter. M is the maximum gain approached by signals in the passband. Figure 14–49 shows low- and high-pass Butterworth responses for several different values of the order, n. As frequency f approaches 0, note that the gain of the low-pass filter approaches M. For very large values of f, the gain of the high-pass filter approaches M. The value of M is 1, or less, for passive filters. An active filter is capable of amplifying signals having frequencies in its passband and may therefore have a value of M greater than 1. M is called the "*gain*" of the filter.

The Butterworth filter is also called a *maximally flat* filter, because there is relatively little variation in the gain of signals having different frequencies within its passband. For a given order, the Chebyshev filter has greater variation in the passband than the Butterworth design, but falls off at a faster rate outside the passband. Figure 14–50 shows a typical Chebyshev frequency response. Note that the cutoff frequency for this filter is defined to be the frequency at the point of intersection of the response curve and a line drawn tangent to the lowest gain in the passband. The ripple width (RW) is the total variation in gain within the

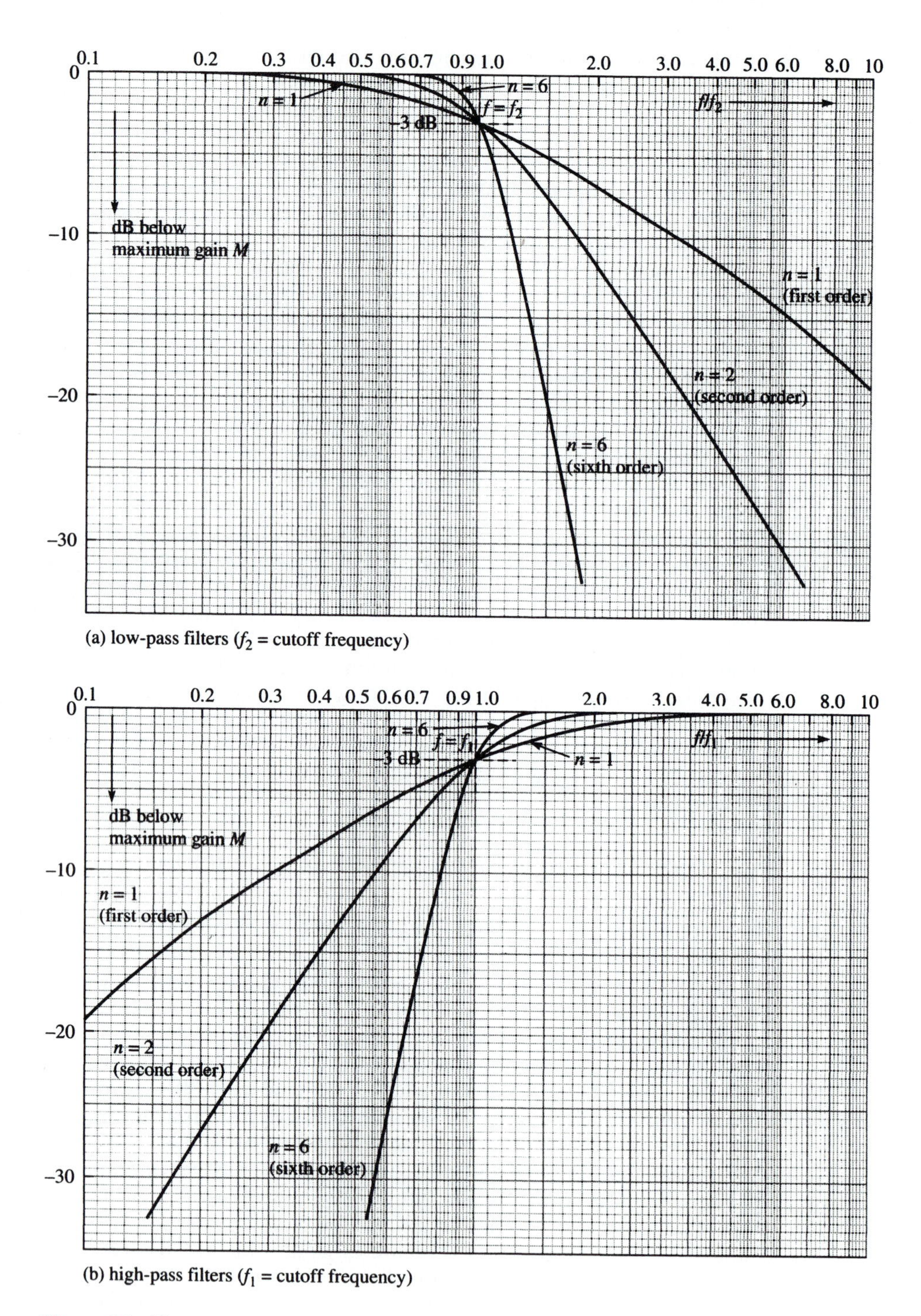

Figure 14–49
Frequency response of low-pass and high-pass Butterworth filters with different orders

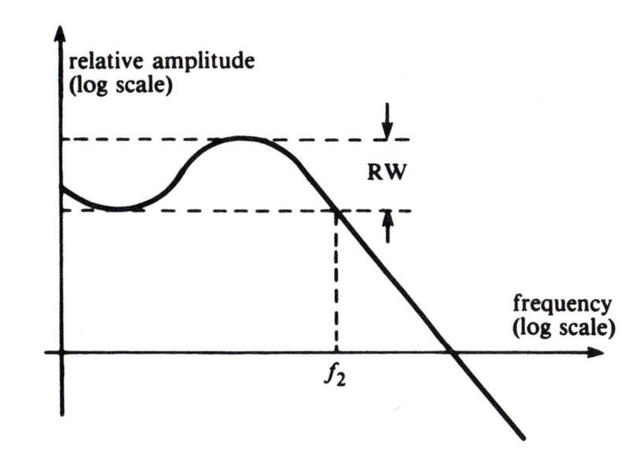

Figure 14–50
Chebyshev low-pass frequency response: f_2 = cutoff frequency; RW = ripple width

passband, usually expressed in decibels. A Chebyshev filter can be designed to have a small ripple width, but at the expense of less attenuation outside the passband.

A Chebyshev filter of the same order as a Butterworth filter has a frequency response closer to the ideal filter *outside* the passband, while the Butterworth is closer to ideal within the passband. This fact is illustrated in Figure 14–51, which compares the responses of second-order, low-pass Butterworth and Chebyshev filters having the same cutoff frequency. Note that both filters have responses asymptotic to 40 dB/decade outside the passband, but at any specific frequency beyond cutoff, the Chebyshev shows greater attenuation than the Butterworth.

A bandpass filter is characterized by a quantity called its Q (originating from "quality" factor), which is a measure of how rapidly its gain changes at frequencies outside its passband. Q is defined by

$$Q = \frac{f_o}{\text{BW}} \tag{14–103}$$

where f_o is the frequency in the passband where the gain is maximum, and BW is the bandwidth measured between the frequencies where the gain is 3 dB down from its maximum value. f_o is often called the *center* frequency because, for many high-Q filters ($Q \geq 10$), it is very nearly in the center of the passband. Figure 14–52 shows the frequency responses of two bandpass filters having the same maximum gain but different values of Q. It is clear that the high-Q filter has a narrower

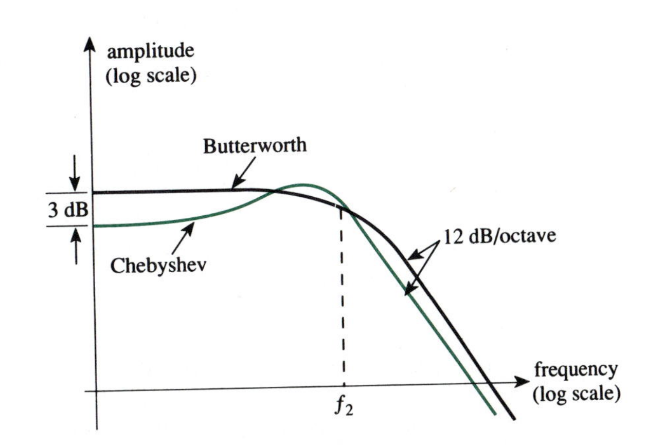

Figure 14–51
Comparison of the frequency responses of second-order, low-pass Butterworth and Chebyshev filters

Figure 14–52
Comparison of the frequency responses of low-Q and high-Q bandpass filters

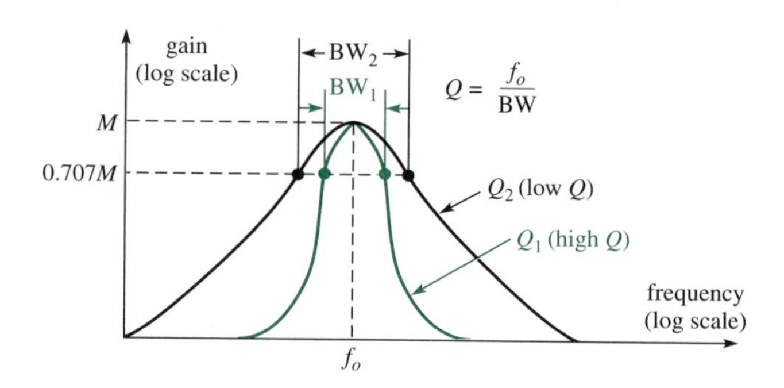

bandwidth and that its response falls off faster on either side of the passband than that of the low-Q filter.

The gain magnitude of any second-order bandpass filter is the following function of frequency f:

$$|G| = \frac{M}{\sqrt{1 + Q^2(f/f_o - f_o/f)^2}} \quad \textbf{(14–104)}$$

where M is the maximum gain in the band. The center frequency f_o is at the *geometric* center of the band, defined by

$$f_o = \sqrt{f_1 f_2} \quad \textbf{(14–105)}$$

where f_1 and f_2 are the lower and upper cutoff frequencies. As previously mentioned, f_o is very nearly midway between the cutoff frequencies (the arithmetic mean) in high-Q filters:

$$f_o \approx f_1 + \frac{\text{BW}}{2} \approx f_2 - \frac{\text{BW}}{2} \quad \textbf{(14–106)}$$

Active Filter Design

The analysis of active filters requires complex mathematical methods that are beyond the scope of this book. We will therefore concentrate on practical design methods that will allow us to construct Butterworth and Chebyshev filters of various types and orders. The discussion that follows is based on design procedures using tables that can be found in *Rapid Practical Designs of Active Filters,* by Johnson and Hilburn (Wiley, 1975). Space limitations prevent us from listing all the tables that are available in that reference, so we will simply illustrate how a few of them are used in filter design. Readers wishing a more comprehensive treatment should consult the reference.

Figure 14–53 shows a general configuration that can be used to construct second-order high-pass and low-pass filters of both the Butterworth and Chebyshev designs. The amplifier is basically operated as a noninverting, voltage-controlled voltage source (see Section 14–2), and the configuration is known as the VCVS design. It is also called a *Sallen-Key* circuit. Each impedance block represents a resistance or capacitance, depending on whether a high-pass or low-pass filter is desired. Table 14–1 shows the impedance type required for each design. We see, for example, that Z_A is a resistor, designated R_1, for a low-pass filter and is a capacitor, designated C, for a high-pass filter. The VCVS design can also be constructed as a bandpass filter using another component arrangement (not shown).

The component values in the VCVS design depend on whether a Butterworth

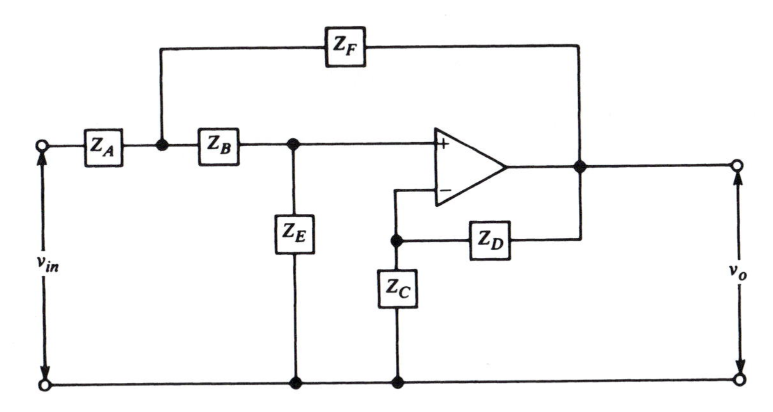

Figure 14–53
Block diagram of a second-order, VCVS low-pass or high-pass filter

or Chebyshev response is required, the gain required in the passband, and, in the case of the Chebyshev filter, on the tolerable ripple width in the passband. As we shall demonstrate shortly, different "look-up" tables are used to determine component values corresponding to these various options. The design procedure begins with selection of a value for capacitance C. Next, a constant designated K is computed as follows:

$$K = \frac{10^{-4}}{fC} \qquad \textbf{(14–107)}$$

where f is the desired cutoff frequency in Hz and C is the value of capacitance selected, in farads. We then consult an appropriate table to obtain factors by which K is multiplied to give each resistor value. Tables 14–2 through 14–4 are three such

Table 14–1
VCVS Filter Components

	Z_A	Z_B	Z_C	Z_D	Z_E	Z_F
Low-Pass Filter	R_1	R_2	R_3	R_4	C_1	C
High-Pass Filter	C	C	R_3	R_4	R_2	R_1

Table 14–2
Second-Order Low-Pass Butterworth VCVS Filter Designs

	Circuit Element Values[a]					
Gain	**1**	**2**	**4**	**6**	**8**	**10**
R_1	1.422	1.126	0.824	0.617	0.521	0.462
R_2	5.399	2.250	1.537	2.051	2.429	2.742
R_3	Open	6.752	3.148	3.203	3.372	3.560
R_4	0	6.752	9.444	16.012	23.602	32.038
C_1	0.33C	C	2C	2C	2C	2C

[a] Resistances in kilohms for a K parameter of 1.

Table 14–3
Second-Order Low-Pass Chebyshev VCVS Filter Designs (2 dB)

	Circuit Element Values[a]					
Gain	**1**	**2**	**4**	**6**	**8**	**10**
R_1	2.328	1.980	1.141	0.786	0.644	0.561
R_2	13.220	1.555	1.348	1.957	2.388	2.742
R_3	Open	7.069	3.320	3.292	3.466	3.670
R_4	0	7.069	9.959	16.460	24.261	33.031
C_1	$0.1C$	C	$2C$	$2C$	$2C$	$2C$

[a] Resistances in kilohms for a K parameter of 1.

Table 14–4
Second-Order High-Pass Chebyshev VCVS Filter Designs (2 dB)

	Circuit Element Values[a]					
Gain	**1**	**2**	**4**	**6**	**8**	**10**
R_1	0.640	1.390	2.117	2.625	3.040	3.399
R_2	3.259	1.500	0.985	0.794	0.686	0.613
R_3	Open	3.000	1.313	0.953	0.784	0.681
R_4	0	3.000	3.939	4.765	5.486	6.133

[a] Resistances in kilohms for a K parameter of 1.

Source: Reprinted from *Rapid Practical Designs of Active Filters,* D. Johnson and J. Hilburn. Copyright © 1975, John Wiley and Sons, Inc., by permission of John Wiley and Sons, Inc.

tables. These can be used to design a low-pass Butterworth, a low-pass Chebyshev with 2-dB ripple width, or a high-pass Chebyshev with 2-dB ripple width. The next example illustrates how they are used.

Example 14–19

DESIGN

Design a second-order, VCVS, low-pass Butterworth filter with cutoff frequency 2.5 kHz. The gain in the passband should be 2.

Solution. Choose $C = 0.05\ \mu\text{F}$. (If this choice results in impractical values for the other components, we can always revise the choice and try again.) From equation 14–107,

$$K = \frac{10^{-4}}{fC} = \frac{10^{-4}}{(2.5 \times 10^3)(0.05 \times 10^{-6})} = 0.8$$

From Table 14–2, in the *Gain* = *2* column, we locate the multiplying constants for each component value and obtain

$$R_1 = (1.126)(0.8)\ \text{k}\Omega = 901\ \Omega \qquad R_4 = (6.752)(0.8)\ \text{k}\Omega = 5.40\ \text{k}\Omega$$
$$R_2 = (2.250)(0.8)\ \text{k}\Omega = 1.80\ \text{k}\Omega \qquad C_1 = C = 0.05\ \mu\text{F}$$
$$R_3 = (6.752)(0.8)\ \text{k}\Omega = 5.40\ \text{k}\Omega$$

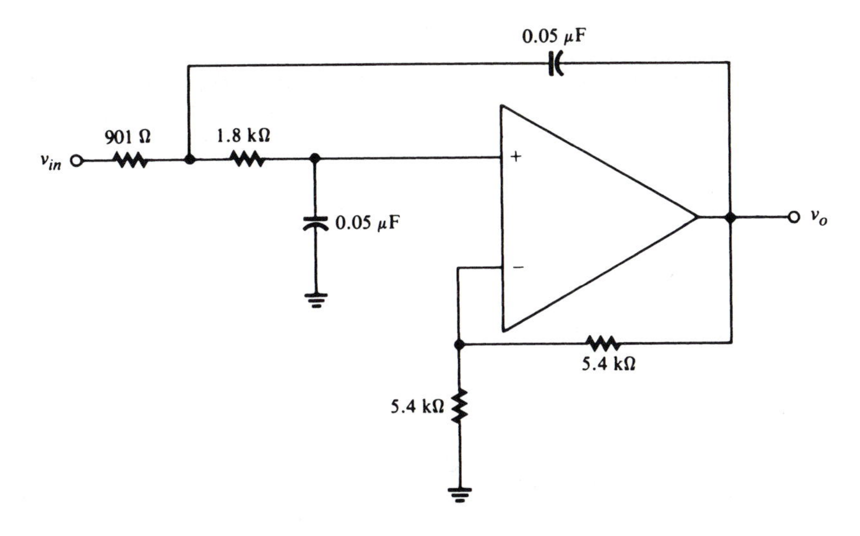

Figure 14–54
(Example 14–19)

With reference to Table 14–1 and Figure 14–53, we construct the required filter as shown in Figure 14–54. Note that precision, nonstandard component values are required if we demand exact conformity to the original design specifications. Generally speaking, satisfactory performance can be obtained (in low-order filters) by using the closest 5% standard values.

Higher-order filters of the VCVS design can be constructed by cascading VCVS stages. A fourth-order filter requires two stages, a sixth-order filter requires three stages, and so forth. Tables are available in the previously cited reference for designing each stage of VCVS filters with orders up to 8. It should be emphasized that higher-order filters are *not* constructed by cascading identical stages. This precaution applies to all the filters we will discuss in this chapter. Recall from Chapter 10 that a cascade of identical stages has an overall cutoff frequency that is different from that of the individual stages.

Another configuration that can be used to construct low-pass, high-pass, and bandpass filters is called the *infinite-gain multiple-feedback* (IGMF) design. Figure 14–55 shows the component arrangement used to obtain a second-order bandpass filter. One advantage of the IGMF design is that it requires one less component than the VCVS filter. It is also popular because it has good stability and low output impedance. The IGMF filter inverts signals in its passband.

The procedure for designing an IGMF filter is very similar to that for a VCVS filter. We first choose a value of capacitance C and then calculate

$$K = \frac{10^{-4}}{f_o C} \qquad \textbf{(14–108)}$$

where f_o is the desired center frequency of the bandpass filter. A table corresponding to the desired Q is then consulted to find the factors that multiply K to determine

Figure 14–55
The IGMF second-order bandpass filter

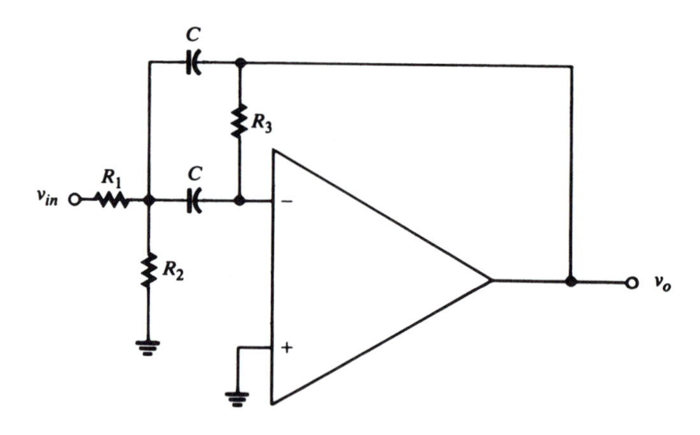

resistor values R_1, R_2, and R_3 in kΩ. Table 14–5 shows the factors used to design a second-order filter having a Q of 5. The gain headings over the columns refer to the gain of the filter at its center frequency. Tables are also available in the previously cited reference for constructing low- and high-pass IGMF filters of various orders.

Another popular active filter is the *biquad* design, which can be constructed in low-pass, high-pass, or bandpass configurations. Figure 14–56 shows the low-pass version. Although this design requires three amplifiers and a greater number of passive components than the VCVS or IGMF filters, it has the advantage that its gain and cutoff frequency are easily adjusted. Also, both an inverted and a noninverted output are available, as shown in Figure 14–56. The gain of this filter is given by

$$G = \frac{R_3}{R_1} \tag{14–109}$$

The gain is varied by adjusting R_1, while changing R_3 varies both the gain and the cutoff frequency.

The design procedure for the biquad filter begins with a choice of capacitor C and the calculation of the constant K, as before:

$$K = \frac{10^{-4}}{fC} \tag{14–110}$$

Table 14–5
Second-Order Multiple-Feedback Bankdpass Filter Designs ($Q = 5$)

	Circuit Element Values[a]					
Gain	**1**	**2**	**4**	**6**	**8**	**10**
R_1	7.958	3.979	1.989	1.326	0.995	0.796
R_2	0.162	0.166	0.173	0.181	0.189	0.199
R_3	15.915	15.915	15.915	15.915	15.915	15.915

[a] Resistances in kilohms for a K parameter of 1.

Source: Reprinted from *Rapid Practical Designs of Active Filters,* D. Johnson and J. Hilburn. Copyright © 1975, John Wiley and Sons, Inc., by permission of John Wiley and Sons, Inc.

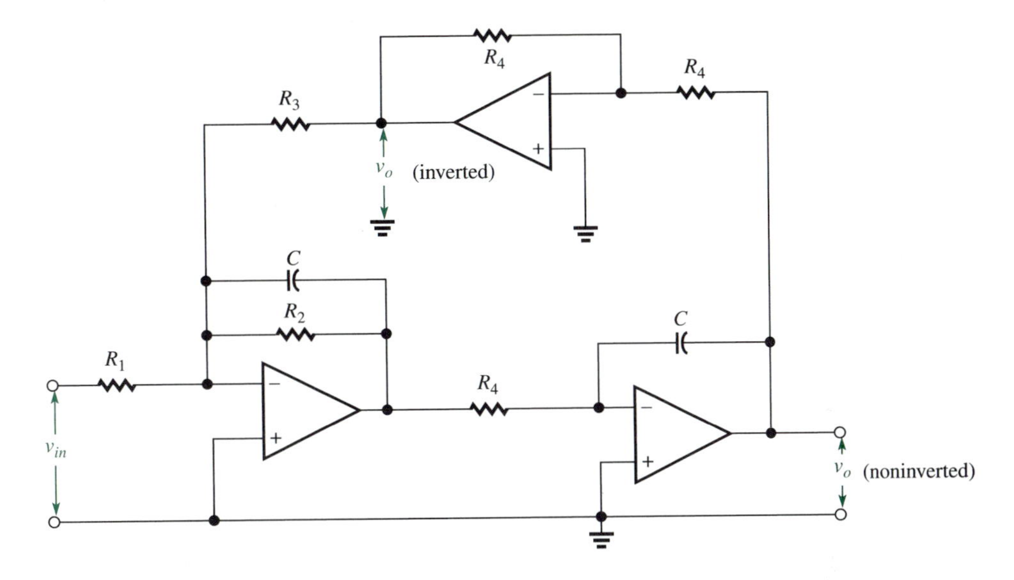

Figure 14–56
Circuit diagram of a second-order low-pass biquad filter

Table 14–6
Second-Order Low-Pass Biquad Filter Designs

	Circuit Element Values[a]					
		Chebyshev				
	Butterworth	0.1 dB	0.5 dB	1 dB	2 dB	3 dB
R_1	1.592/G	0.480/G	1.050/G	1.444/G	1.934/G	2.248/G
R_2	1.125	0.671	1.116	1.450	1.980	2.468
R_3	1.592	0.480	1.050	1.444	1.934	2.248
R_4	1.592	1.592	1.592	1.592	1.592	1.592

[a] Resistances in kilohms for a K parameter of 1, G = gain.

Source: Reprinted from *Rapid Practical Designs of Active Filters,* D. Johnson and J. Hilburn. Copyright © 1975, John Wiley and Sons, Inc., by permission of John Wiley and Sons, Inc.

A value of gain G is also chosen. The factors that multiply K to determine the other component values are then taken from a table. Table 14–6 shows the multiplying factors for second-order Butterworth responses and Chebyshev responses having several different ripple widths. Notice that the multiplying factor for R_1 is divided by the user-selected value of gain. Higher-order filters can be designed by cascading stages and using appropriate sets of tables available in the previously cited reference.

Example 14–20

SPICE

Design a low-pass biquad filter with cutoff frequency 1 kHz. The filter should have Chebyshev characteristics with a 3-dB ripple width and a gain of 2. Verify the performance using SPICE.

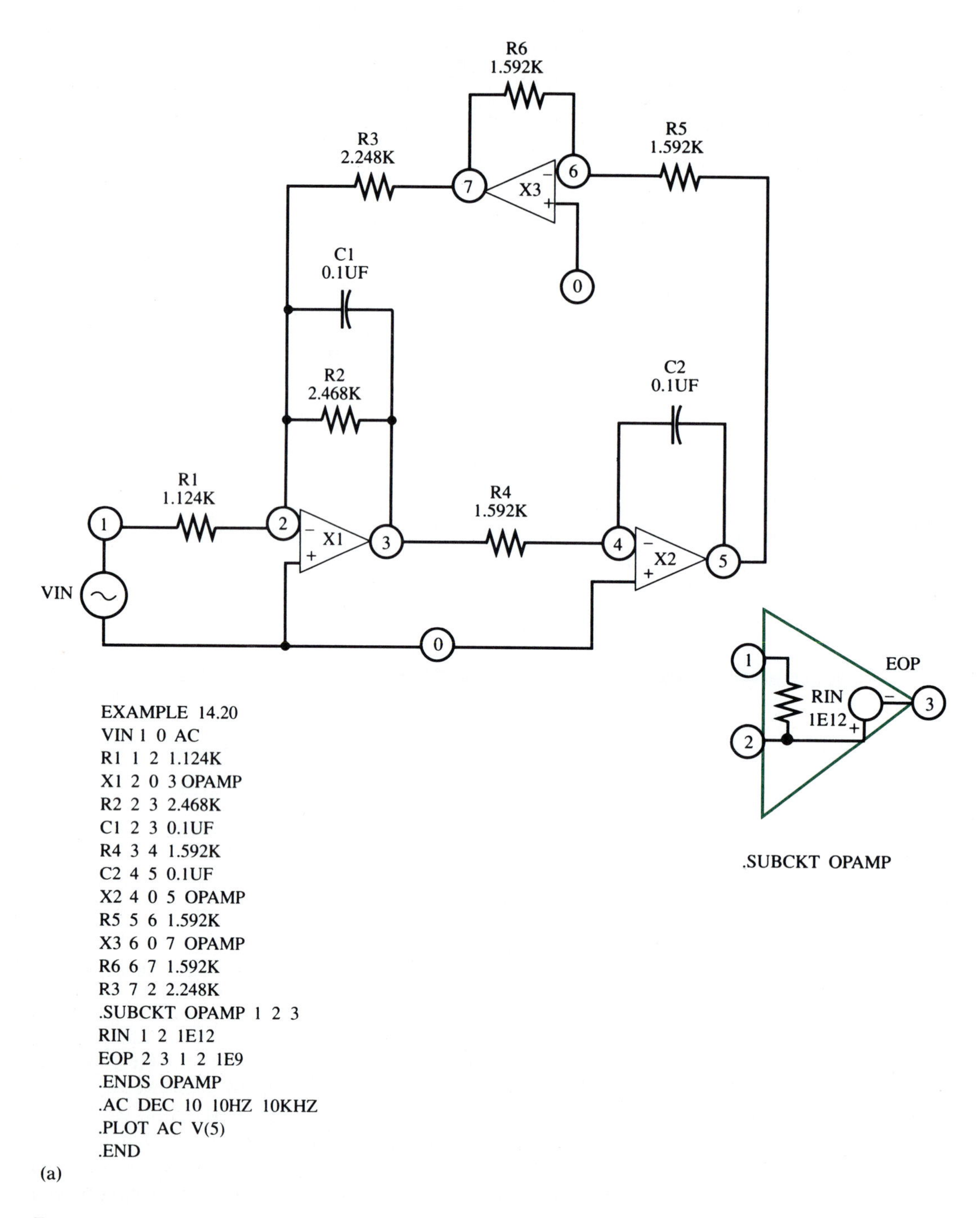

R6
1.592K
R3
2.248K
R5
1.592K
X3
C1
0.1UF
R2
2.468K
C2
0.1UF
R1
1.124K
R4
1.592K
X1
X2
VIN
EOP
RIN
1E12
EXAMPLE 14.20
VIN 1 0 AC
R1 1 2 1.124K
X1 2 0 3 OPAMP
R2 2 3 2.468K
C1 2 3 0.1UF
R4 3 4 1.592K
C2 4 5 0.1UF
X2 4 0 5 OPAMP
R5 5 6 1.592K
X3 6 0 7 OPAMP
R6 6 7 1.592K
R3 7 2 2.248K
.SUBCKT OPAMP 1 2 3
RIN 1 2 1E12
EOP 2 3 1 2 1E9
.ENDS OPAMP
.AC DEC 10 10HZ 10KHZ
.PLOT AC V(5)
.END
.SUBCKT OPAMP
(a)

Figure 14–57
(Example 14–20)

```
EXAMPLE 14.20
****     AC ANALYSIS                      TEMPERATURE  =   27.000 DEG C
********************************************************************************
   FREQ      V(5)
                 1.000D-02     1.000D-01      1.000D+00     1.000D+01  1.000D+02
                      - - - - - - - - - - - - - - - - - - - - - - - - - - - - -
1.000D+01  2.000D+00 .             .              .   *         .             .
1.259D+01  2.000D+00 .             .              .   *         .             .
1.585D+01  2.001D+00 .             .              .   *         .             .
1.995D+01  2.001D+00 .             .              .   *         .             .
2.512D+01  2.001D+00 .             .              .   *         .             .
3.162D+01  2.002D+00 .             .              .   *         .             .
3.981D+01  2.003D+00 .             .              .   *         .             .
5.012D+01  2.005D+00 .             .              .   *         .             .
6.310D+01  2.008D+00 .             .              .   *         .             .
7.943D+01  2.013D+00 .             .              .   *         .             .
1.000D+02  2.020D+00 .             .              .   *         .             .
1.259D+02  2.032D+00 .             .              .   *         .             .
1.585D+02  2.051D+00 .             .              .   *         .             .
1.995D+02  2.081D+00 .             .              .   *         .             .
2.512D+02  2.130D+00 .             .              .    *        .             .
3.162D+02  2.208D+00 .             .              .    *        .             .
3.981D+02  2.335D+00 .             .              .    *        .             .
5.012D+02  2.530D+00 .             .              .     *       .             .
6.310D+02  2.768D+00 .             .              .     *       .             .
7.943D+02  2.733D+00 .             .              .     *       .             .
1.000D+03  1.999D+00 .             .              .   *         .             .
1.259D+03  1.184D+00 .             .              .*            .             .
1.585D+03  6.827D-01 .             .            * .             .             .
1.995D+03  4.025D-01 .             .        *     .             .             .
2.512D+03  2.428D-01 .             .     *        .             .             .
3.162D+03  1.488D-01 .             .  *           .             .             .
3.981D+03  9.218D-02 .             *              .             .             .
5.012D+03  5.749D-02 .          *  .              .             .             .
6.310D+03  3.601D-02 .       *     .              .             .             .
7.943D+03  2.261D-02 .    *        .              .             .             .
1.000D+04  1.423D-02 . *           .              .             .             .
                      - - - - - - - - - - - - - - - - - - - - - - - - - - - - -
```

(b)

Figure 14–57
(Continued)

Solution. Letting $C = 0.1\ \mu\text{F}$, we have, from equation 14–110,

$$K = \frac{10^{-4}}{fC} = \frac{10^{-4}}{(1\ \text{kHz})(0.1\ \mu\text{F})} = 1$$

Then, from Table 14–6, $R_1 = (2.248/G)\ \text{k}\Omega = (2.248/2)\ \text{k}\Omega = 1.124\ \text{k}\Omega$, $R_2 = 2.468\ \text{k}\Omega$, $R_3 = 2.248\ \text{k}\Omega$, and $R_4 = 1.592\ \text{k}\Omega$. Figure 14–57(a) shows the filter circuit in its SPICE format. Note that a subcircuit (SUBCKT) is used to specify the three operational amplifiers. The subcircuit, named OPAMP, models a nearly ideal amplifier having input resistance $1 \times 10^{12}\ \Omega$ and open-loop voltage gain 1×10^9.

Figure 14–57(b) shows the results of an .AC analysis over the frequency range from 10 Hz through 10 kHz. Note the characteristic rise in the response of the Chebyshev filter. The low-frequency gain is seen to be 2, as required by the design specifications. The gain then rises and falls back to $1.999 \approx 2$ at 1 kHz, confirming that the Chebyshev cutoff frequency is 1 kHz. The gain at 5.012 kHz is 0.05749 and that at 10 kHz is 0.01423. Thus, the gain changes by

$$20 \log_{10}\left(\frac{0.01423}{0.05749}\right) = -12.12\ \text{dB}$$

over a frequency range close to one octave, confirming that the filter is of second order.

As a final example of an active filter, we will consider a three-amplifier configuration called the *state-variable* filter. This design simultaneously provides second-order low-pass, high-pass, and bandpass filtering, each of these functions being taken from a different point in the circuit. It is available in integrated-circuit form as a so-called *universal active filter.* Figure 14–58 shows the configuration. Note that it is essentially two integrators and a summing amplifier.

The following equations define the characteristics of the state-variable filter:

$$\omega_o = 2\pi f_o = \frac{1}{R_1C} \text{ radians/second} \tag{14–111}$$

$$\text{BW} = \frac{3R}{R_1C(R + R_Q)} \text{ radians/second} \tag{14–112}$$

$$Q = \frac{R + R_Q}{3R} \tag{14–113}$$

where ω_o is the center frequency, BW is the bandwidth, and Q is the Q of the bandpass output. The Q of the circuit determines whether it has Butterworth or Chebyshev characteristics. For a Q of 0.707, it is Butterworth, and for a Q of 0.885, it is Chebyshev. In the case of a Butterworth response, the cutoff frequencies for the low- and high-pass outputs are both equal to ω_o. For the Chebyshev design, the low-pass cutoff frequency is $0.812\omega_o$ rad/s and the high-pass cutoff frequency is $1.23\omega_o$ rad/s.

The resistor labeled R_Q in Figure 14–58 is used to set the Q of the filter. Solving equation 14–113 for R_Q, we find

$$R_Q = R(3Q - 1) \tag{14–114}$$

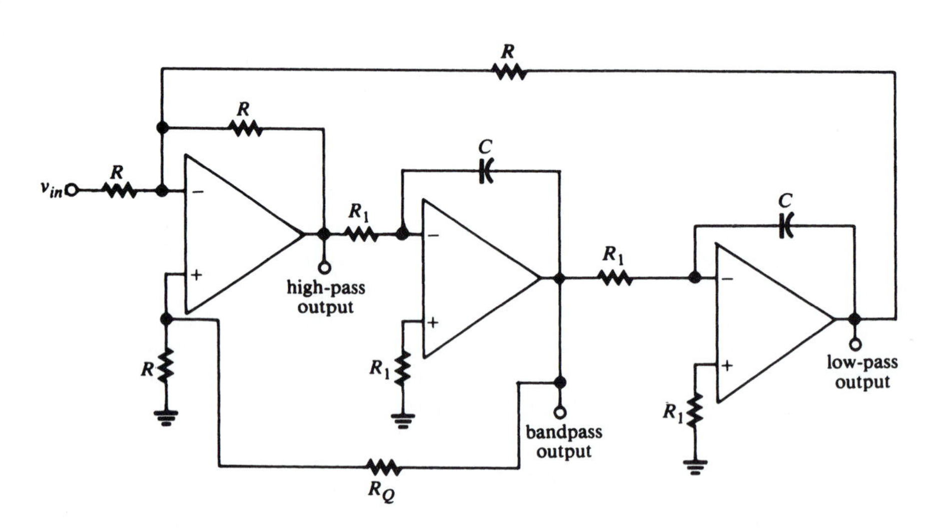

Figure 14–58
A state-variable, or universal, active filter

Resistors R_1 or capacitors C can be adjusted to change the values of ω_o. Note that *both* resistors or *both* capacitors must be adjusted simultaneously. A band-*reject* filter (see Figure 14–46(d)) can be obtained by summing the low-pass and high-pass outputs of the state-variable filter in a fourth amplifier. The summation should be performed with equal gain for each signal.

Example 14–21

DESIGN

Design a state-variable filter with Butterworth characteristics and center frequency 1.59 kHz. What are the cutoff frequencies of the low-pass and high-pass outputs? What is the bandwidth of the bandpass output?

Solution. Choose $C = 0.01\ \mu$F. Then, from equation 14–111,

$$2\pi \times (1.59 \times 10^3) = \frac{1}{R_1(10^{-8})}$$

$$R_1 = \frac{1}{2\pi(1.59 \times 10^3)(10^{-8})} = 10\text{ k}\Omega$$

For a Butterworth response, we must have $Q = 0.707$. Let $R = 10$ kΩ. Then, from equation 14–114, $R_Q = R(3Q - 1) = 10^4[3(0.707) - 1] = 11.21$ kΩ. The completed design is shown in Figure 14–59.

Since the filter has Butterworth characteristics, the cutoff frequencies of both the low- and high-pass outputs are equal to the center frequency of the bandpass output: 1.59 kHz.

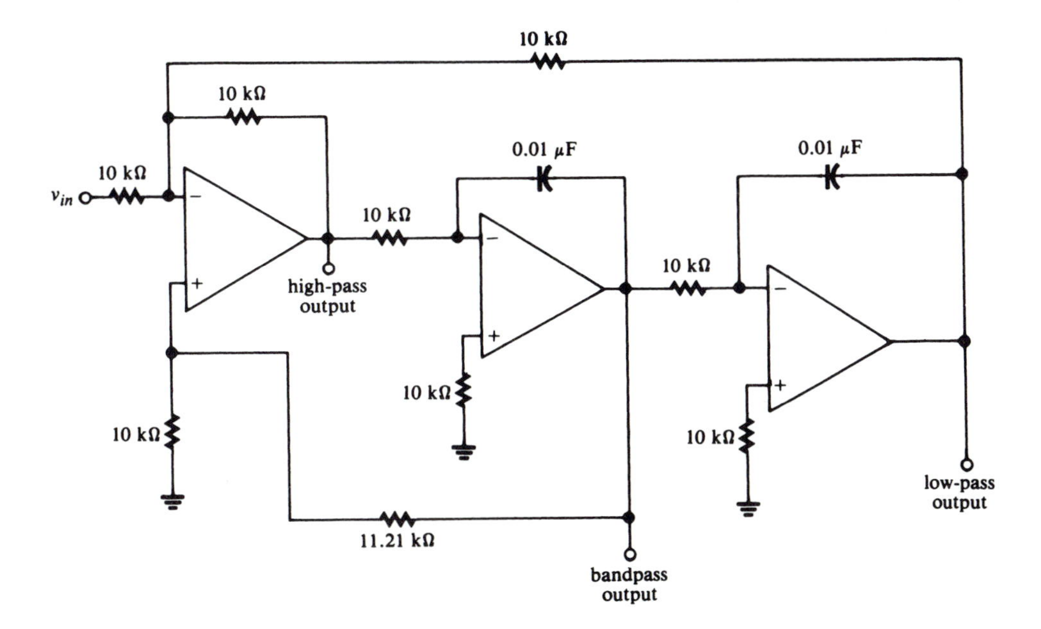

Figure 14–59
(Example 14–21)

The bandwidth of the bandpass output is

$$\text{BW} = \frac{f_o}{Q} = \frac{1.59 \text{ kHz}}{0.707} = 2.248 \text{ kHz}$$

Note that the "center" frequency in this low-Q filter is not midway between the cutoff frequencies of the bandpass output.

EXERCISES

Section 14–1
Voltage Summation, Subtraction, and Scaling

14–1. **a.** Write an expression for the output of the amplifier in Figure 14–60 in terms of v_1, v_2, v_3, and v_4.
b. What is the output when $v_1 = 5 \sin \omega t$, $v_2 = -3$ V dc, $v_3 = -\sin \omega t$, and $v_4 = 2$ V dc?
c. What value should R_c have?

14–2. **a.** Design an operational-amplifier circuit that will produce the output $v_o = -10v_1 - 50v_2 + 10$. Use only one amplifier. (*Hint:* One of the inputs is a dc source.)
b. Sketch the output waveform when $v_1 = v_2 = -0.1 \sin \omega t$ volts.

14–3. The operational amplifier in Exercise 14–1 has unity-gain frequency 1 MHz and input offset voltage 3 mV. Find
a. the closed-loop bandwidth of the configuration; and
b. the magnitude of the output offset voltage due to V_{io}.

14–4. Derive equation 14–8 for the output of the circuit shown in Figure 14–4. (*Hint:* Using superposition, write an expression for v^+. Then use the fact that $v^- = v^+$ and $v_o = v^- + v^- R_f/R_g$.)

14–5. **a.** Write an expression, in terms of v_1 and v_2, for the output of the amplifier shown in Figure 14–61.
b. Write an expression for the output in the special case in which v_1 and v_2 are equal-magnitude, out-of-phase signals.

14–6. Design a noninverting circuit using a single operational amplifier that will produce the output $v_o = 4v_1 + 6v_2$.

14–7. Using a single operational amplifier in each case, design circuits that will produce the following outputs:
a. $v_o = 0.1v_1 - 5v_2$
b. $v_o = 10(v_1 - v_2)$
Design the circuits so that the compensation resistance has an optimum value.

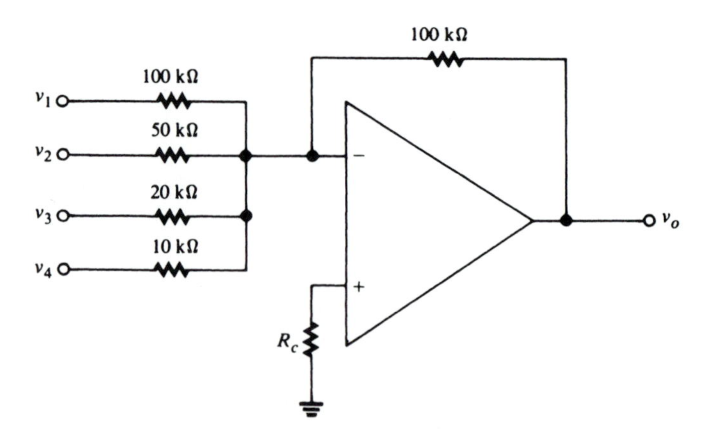

Figure 14–60
(Exercise 14–1)

Figure 14–61
(Exercise 14–5)

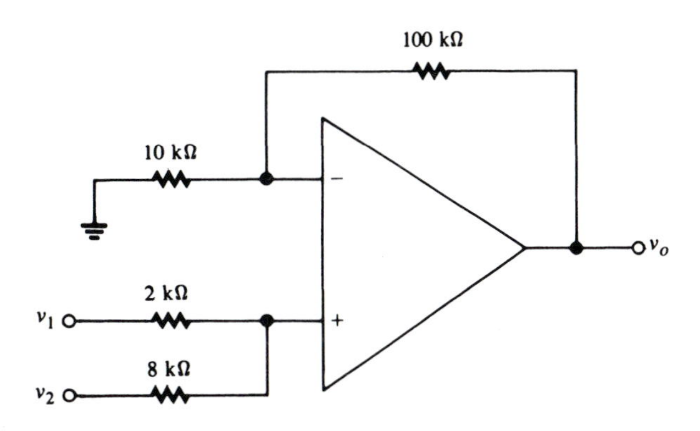

14–8. If the resistor values in Figure 14–5 are chosen in accordance with $R_4 = a_1R_1 = a_2R_3 = R_2(1 + a_2 - a_1)$, then, assuming that $1 + a_2 > a_1$, show that
a. $v_o = a_1v_1 - a_2v_2$; and
b. the compensation resistance $(R_1 \parallel R_2)$ has its optimum value $(R_3 \parallel R_4)$.

14–9. Design operational-amplifier circuits to produce each of the following outputs:
a. $v_o = 0.4v_2 - 10v_1$
b. $v_o = v_1 + v_2 - 20v_3$

14–10. Assuming that the unity-gain frequency of each amplifier used in Exercise 14–9 is 750 kHz, find the approximate, overall bandwidth of each circuit.

Section 14–2
Controlled Voltage and Current Sources

14–11. a. Design an inverting voltage-controlled current source that will supply a current of 1 mA to a floating load when the controlling voltage is 2 V.
b. If the source designed in (a) must supply its current to loads of up to 20 kΩ, what maximum output voltage should the amplifier have?

14–12. a. Design a voltage-controlled current source that will supply a current of 2 mA to a floating load when the controlling voltage is 10 V. The input resistance seen by the controlling voltage source would have to be greater than 10 kΩ.
b. If the maximum output voltage of the amplifier is 15 V, what is the maximum load resistance for which your design will operate properly?

14–13. a. Design a voltage-controlled current source that will supply a current of 0.5 mA to a grounded load when the controlling voltage is 5 V.
b. What will be the value of the amplifier output voltage if the load resistance is 12 kΩ?

14–14. The voltage-controlled current source in Figure 14–12 is to be used to supply current to a grounded 10-kΩ load when the controlling voltage is 5 V. If the maximum output voltage of the amplifier is 20 V, what is the maximum current that can be supplied to the load?

14–15. A certain temperature-measuring device generates current in direct proportion to temperature, in accordance with the relation $I = 2.5T\ \mu A$, where T is in degrees Celsius. It is desired to construct a current-to-voltage converter for use with this device so that an output of 20 mV/°C can be obtained. Design the circuit.

14–16. The circuit shown in Figure 14–14 is used with the temperature-measuring device described in Exercise 14–15. If $R = 10$ kΩ, what is the output voltage when the temperature is 75°C?

14–17. Find the currents I_1, I_2, and I_3, and the voltages V_A and V_B, in the circuit of Figure 14–62. Assume an ideal operational amplifier.

14–18. a. Design a current amplifier that will produce, in a 1-kΩ load, five times the current supplied to it.
b. If the input current supplied to the amplifier is 2 mA, what should be the magnitude of the maximum output voltage of the amplifier?

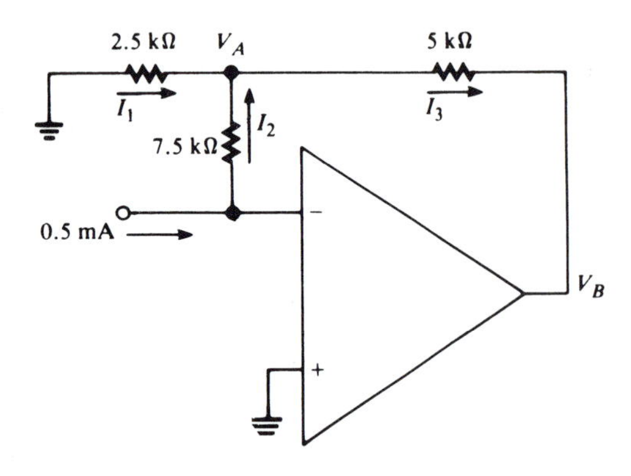

Figure 14–62
(Exercise 14–17)

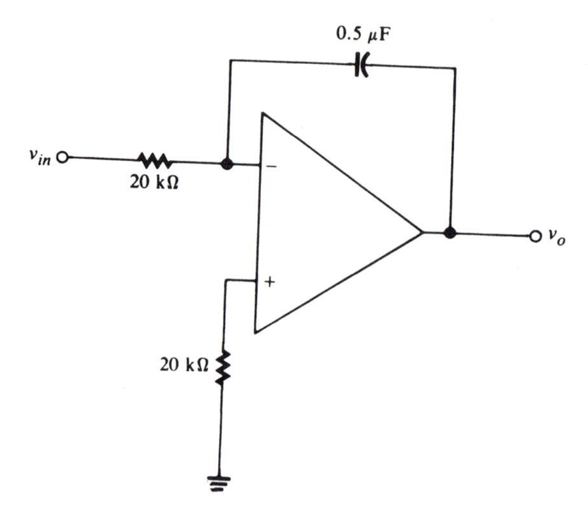

Figure 14–63
(Exercise 14–21)

Section 14–3
Integration, Differentiation, and Waveshaping

14–19. The input to an ideal electronic integrator is 0.25 V dc. Assume that the integrator inverts and multiplies by the constant 20.
- **a.** What is the output 2 s after the input is connected?
- **b.** How long would it take the output to reach −15 V?

14–20. Using an ideal operational amplifier, design an ideal integrator whose output will reach 5 V 200 ms after a −0.1 V-dc input is connected.

14–21. The input to the circuit shown in Figure 14–63 is $v_{in} = 6 \sin (500t - 30°)$ V. Write a mathematical expression for the output voltage.

14–22. The input to the integrator in Exercise 14–21 is a 100-Hz sine wave with peak value 5 V.
- **a.** What is the peak value of the output?
- **b.** What is the peak value of the output if the capacitance in the feedback is halved?
- **c.** What is the peak value of the output (with the original capacitance) if the frequency is halved?
- **d.** What is the peak value of the output (with original capacitance and original frequency) if the input resistance is halved?
- **e.** Name three ways to double the closed-loop gain of an integrator.

14–23. Using an ideal operational amplifier, design an ideal integrator that will produce the output $v_o = 0.04 \cos (2 \times 10^3 t)$ when the input is $v_{in} = 8 \sin (2 \times 10^3 t)$.

14–24. What is the closed-loop gain, in decibels, of the integrator designed in Exercise 14–23 when the angular frequency is 500 rad/s?

14–25. Design a practical integrator that will integrate signals with frequencies down to 500 Hz and that will provide unity gain to dc inputs.

14–26.
- **a.** Design a practical integrator that will integrate signals with frequencies down to 500 Hz and that will provide a closed-loop gain of 0.005 at f = 20 kHz.
- **b.** What will be the output of the integrator when the input is −10 mV dc?

14–27.
- **a.** Write an expression for the output v_o of the system shown in Figure 14–64.
- **b.** Show how the same output could be obtained using a single amplifier.

14–28. An ideal differentiator has a closed-loop gain of 2 when the input is a 50-Hz signal. What is its closed-loop gain when the signal frequency is changed to 2 kHz?

14–29. The differentiator shown in Figure 14–27 has R_1 = 2.2 kΩ, C = 0.015 μF, and R_f = 27 kΩ.
- **a.** Assuming that the circuit performs satisfactory differentiation at frequencies up to 1 decade below its first break frequency, find the frequency range of satisfactory differentiation.
- **b.** Find the magnitude of the voltage gain at 120 Hz.
- **c.** Sketch a Bode plot of the gain, as it would appear if plotted on log-log graph paper. Label all break frequencies. Assume that the operational amplifier has a gain–bandwidth product of 2×10^6.

14–30. Design a practical differentiator that will perform satisfactory differentiation of signals up to 1 kHz. The *maximum* closed-loop gain of the differentiator (at any frequency) should

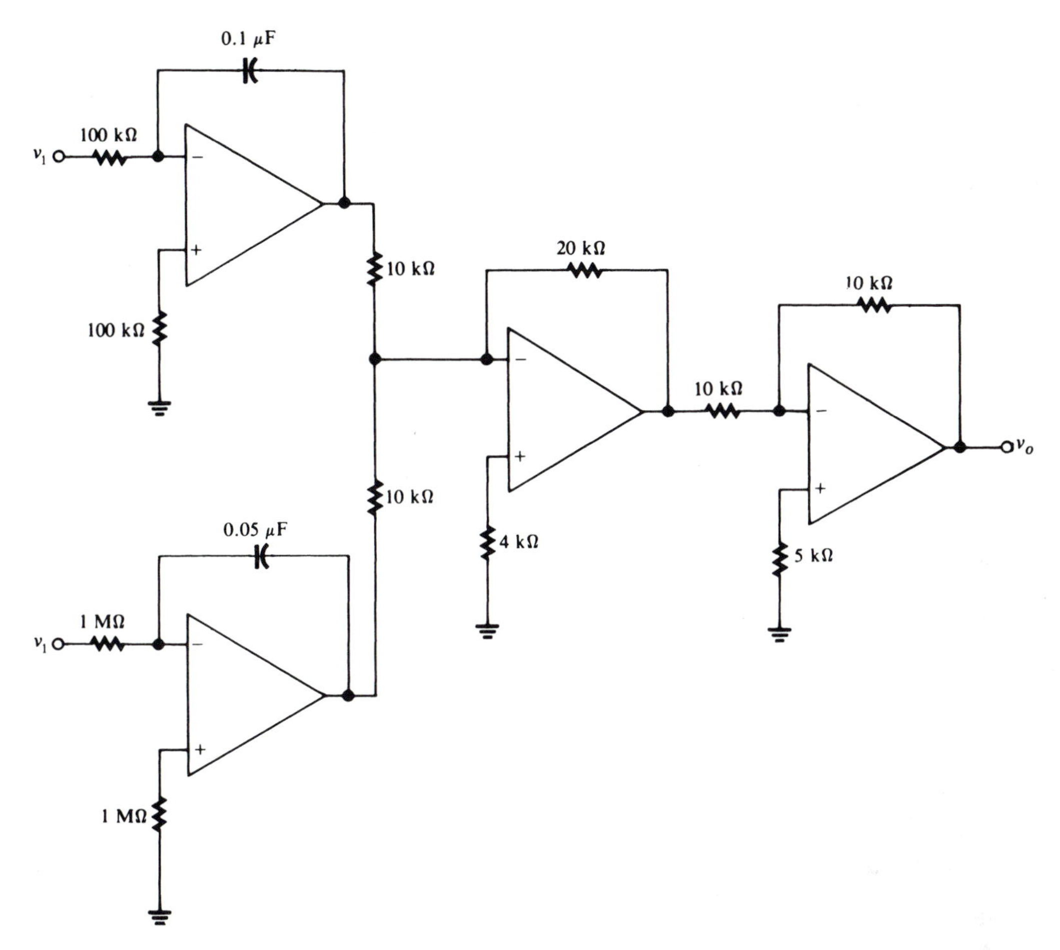

Figure 14–64
(Exercise 14–27)

be 60. (You may assume an ideal operational amplifier that has a wide bandwidth.)

14–31. Sketch the Bode plot for the gain of the circuit shown in Figure 14–65 the way it would appear if plotted on log-log graph paper. Assume that the amplifier has a unity-gain frequency of 1 MHz. On your sketch, label
a. all break frequencies, in Hz;
b. the slopes of all gain asymptotes, in dB/decade; and
c. the value of the maximum closed-loop gain.

14–32. An integrator having $R_1 = 10$ kΩ and $C = 0.02$ μF is to be used to generate a triangular wave from a square-wave input. The operational amplifier has a slew rate of 10^5 V/s. Assume that the input has zero dc component.

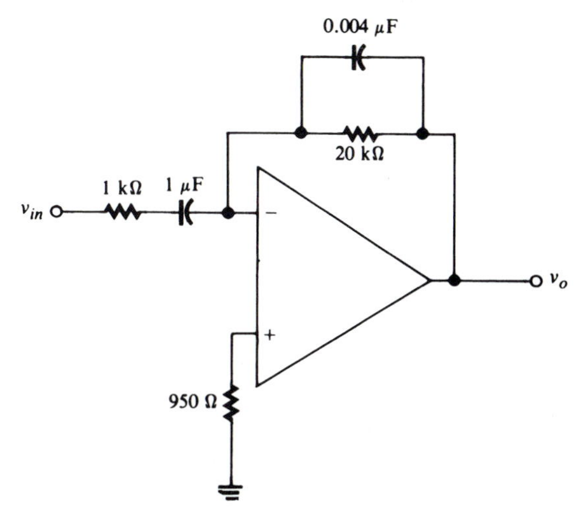

Figure 14–65
(Exercise 14–31)

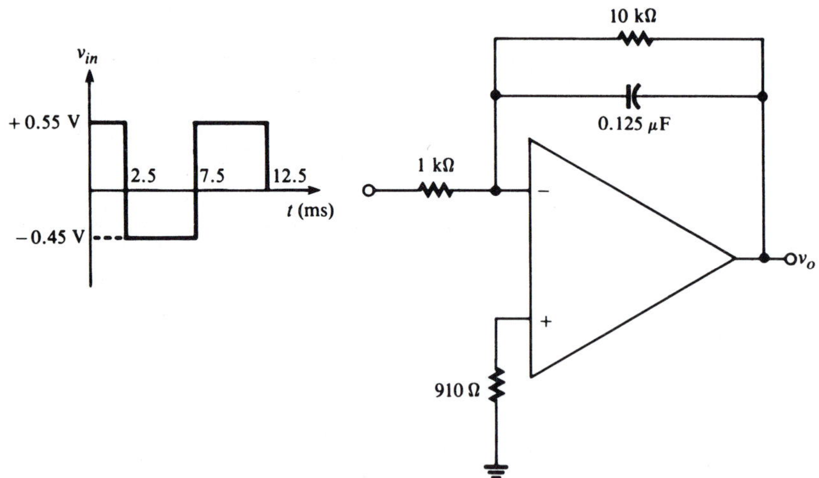

Figure 14–66
(Exercise 14–33)

a. What maximum positive level can the square wave have when its frequency is 500 Hz?

b. Repeat when the square-wave frequency is 5 kHz.

14–33. Sketch the output of the integrator shown in Figure 14–66. Label the peak positive and peak negative output voltages.

14–34. Design a triangular-waveform generator whose output alternates between +9 V peak and −9 V peak when the input is a 10-Hz square wave that alternates between +1.5 V and −1.5 V. The input resistance to the generator must be at least 15 kΩ. You may assume that there is no dc component or offset in the input.

14–35. Sketch the output of an ideal differentiator having $R_f = 40$ kΩ and $C = 0.5$ μF when the input is each of the waveforms shown in Figure 14–67. Label maximum and minimum values on your sketches.

14–36. Sketch the output of an ideal differentiator having $R_f = 60$ kΩ and $C = 0.5$ μF when the input is each of the waveforms shown in Figure 14–68. Label minimum and maximum values on your sketches.

Section 14–4
Instrumentation Amplifiers

14–37. a. Assuming that the amplifiers shown in Figure 14–69 are ideal, find V_{o1}, V_{o2}, I_1, I_2, I_3, and V_o.

b. Repeat when the inputs V_1 and V_2 are interchanged.

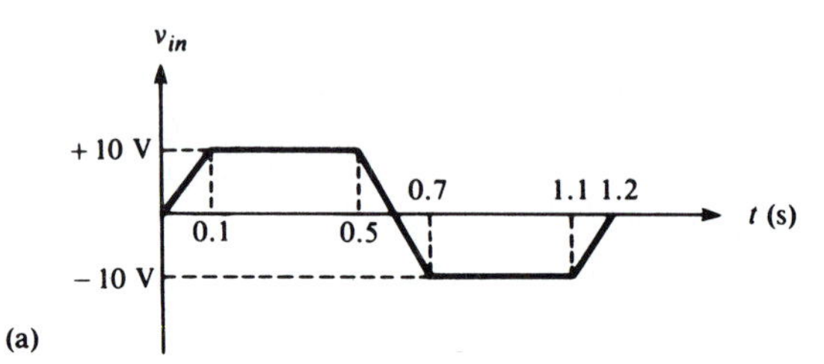

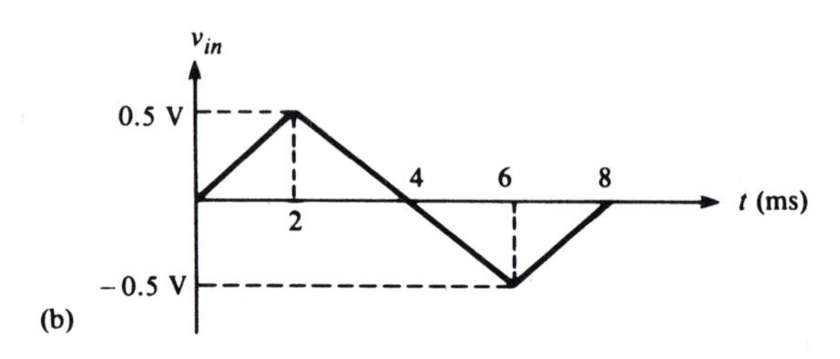

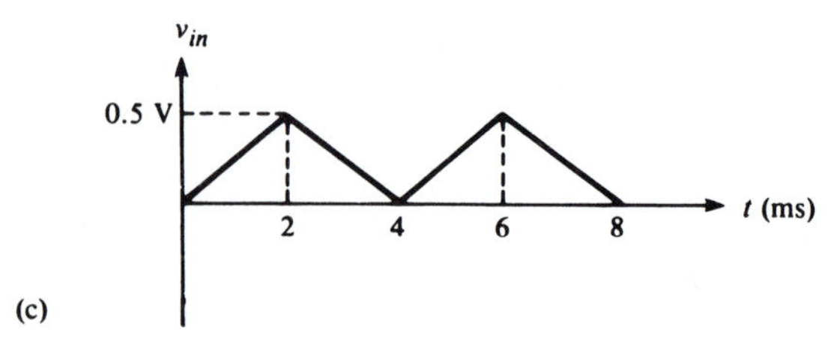

Figure 14–67
(Exercise 14–35)

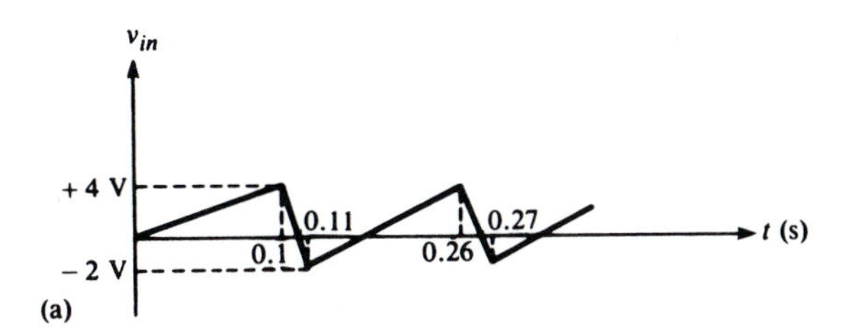

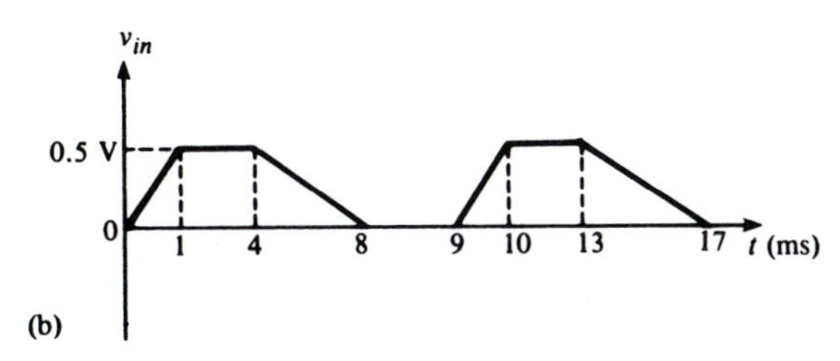

Figure 14–68
(Exercise 14–36)

14–38. a. The instrumentation amplifier in Figure 14–38 is required to have gain $|v_o/(v_1 - v_2)|$ equal to 25. Assuming that $R = 5$ kΩ, what should be the value of R_A?

b. Verify that the amplifier will perform satisfactorily if the maximum permissible output voltage of each operational amplifier is 23 V. v_1 varies from 0.3 V to 1.2 V and v_2 varies from 0.5 V to 0.8 V.

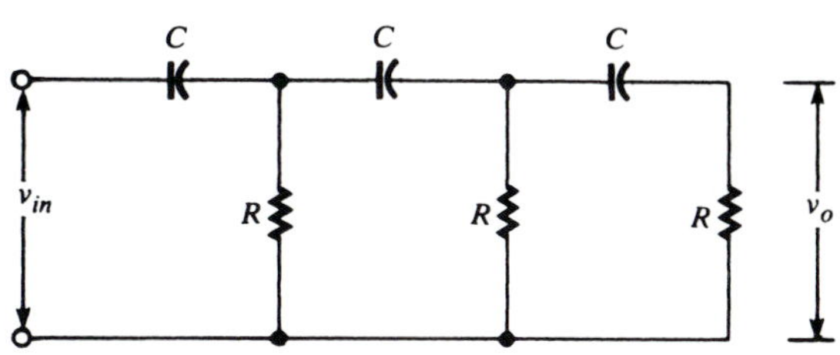

Figure 14–70
(Exercise 14–40)

Section 14–5
Oscillators

14–39. Design an RC phase-shift oscillator that will oscillate at 1.5 kHz.

14–40. Figure 14–70 shows the cascoded RC sections that form the feedback network for the RC phase-shift oscillator. Show that the feedback ratio is

$$\beta = \frac{v_o}{v_{in}} = \frac{R^3}{(R^3 - 5RX_C^2) + j(X_C^3 - 6R^2X_C)}$$

(*Hint:* Write three loop equations and solve for i_3, the current in the rightmost loop. Then $v_o = i_3R$. Solve this equation for v_o/v_{in}.)

14–41. Design a Wien-bridge oscillator that oscillates at 180 kHz.

Figure 14–69
(Exercise 14–37)

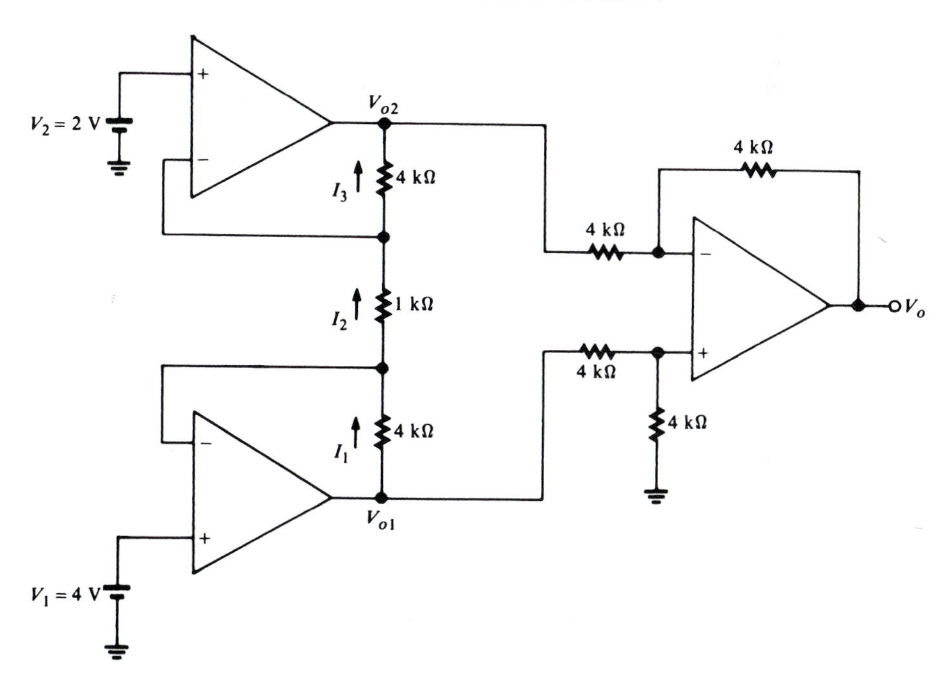

14–42. The Wien-bridge oscillator shown in Figure 14–45 has $C_1 = C_2 = 0.001\ \mu F$ and $R_1 = 10\ k\Omega$. It is desired to make the oscillation frequency variable over the range from 10 kHz to 50 kHz by making R_2 adjustable. Through what range of resistance values should it be possible to adjust R_2?

14–43. A Colpitts oscillator has $C_1 = 1000$ pF, $C_2 = 2200$ pF, and $L = 2.5$ mH.
a. At what frequency does it oscillate?
b. If $R_1 = 10\ k\Omega$, what should be the minimum value of R_f?

14–44. Derive equations 14–98 and 14–99 for the oscillation frequency and feedback factor of the Hartley oscillator.

Section 14–6
Active Filters

14–45. A low-pass Butterworth filter having a maximum low-frequency gain of 1 and a cutoff frequency of 4 kHz is to be designed so that its gain is no less than 0.92 at 3 kHz. What minimum order must the filter have? (*Hint:* Use logarithms.)

14–46. A low-pass Butterworth filter must have an attenuation of at least −20 dB at 1 octave above its cutoff frequency. What minimum order must the filter have?

14–47. A high-pass Butterworth filter having a maximum high-frequency gain of 5 and a cutoff frequency of 500 Hz must have a gain no less than 4.9 at 2 kHz and no greater than 0.1 at 100 Hz. What minimum order must the filter have?

14–48. The gain of a bandpass filter at its upper cutoff frequency is 42. The lower cutoff frequency is 10.8 kHz and the Q is 50. Assume that the center frequency is midway between the cutoff frequencies.
a. What is the gain at the center frequency?
b. What is the center frequency?
c. What is the bandwidth?

14–49. Design a second-order, high-pass, VCVS Chebyshev filter having a cutoff frequency of 8 kHz and a ripple width of 2 dB. The filter should have a gain of 4.

14–50. Design a second-order, low-pass, VCVS Butterworth filter with gain 4 and cutoff frequency 1 kHz. If the input to this filter has a 500-Hz component with amplitude 1.2 V and a 4-kHz component with amplitude 5 V, what are the amplitudes of these components in the output?

14–51. Design a second-order, IGMF bandpass filter with a center frequency of 20 kHz and a bandwidth of 4 kHz. The gain should be 2 at 20 kHz.

14–52. The input to the filter in Exercise 14–51 has the following components:
a. v_1: 6 V rms at 2 kHz
b. v_2: 2 V rms at 20 kHz
c. v_3: 10 V rms at 100 kHz
What are the rms values of these components in the filter's output?

14–53. Design a second-order, low-pass, Butterworth filter using a biquad design having a cutoff frequency 400 Hz. The gain of the filter should be 1.5.

14–54. Repeat Exercise 14–53 with the additional requirement that the input resistance to the filter must be at least 20 kΩ.

14–55. Design a state-variable filter with Chebyshev characteristics and a center frequency of 40 kHz. What is the bandwidth of the bandpass output?

14–56. If the low-pass and high-pass outputs of the filter designed in Exercise 14–55 were summed in a fourth amplifier, with equal gain for each signal, what would be the *approximate* bandwidth of the resulting band-reject filter? Why is your answer approximate? Is the actual bandwidth greater or smaller than your approximation? Why?

SPICE EXERCISES

14–57. Use SPICE to obtain a plot of the output of the amplifier designed in Example 14–1 when the inputs are $v_1 = -5 \sin(2\pi \times 1000t)$ V, $v_2 = 10 \sin(2\pi \times 1000t)$ V, and $v_3 = -50$ V dc. Compare these results with theoretically calculated values.

14–58. Use SPICE to verify the theoretical computations of the voltages and currents in the voltage-controlled current source of Example 14–6 (Figure 14–13). Assume an ideal operational amplifier.

14–59. Design an operational-amplifier integrator that integrates signals with frequencies down to 200 Hz. The magnitude of the gain at 10 kHz should be 0.02. Verify your design using SPICE. Obtain a plot of the frequency response and compare the gain at 20 Hz, 200 Hz, and 10 kHz with theoretical values.

14–60. Design an IGMF bandpass filter having center frequency 1 kHz, $Q = 5$, and gain 1. Use SPICE to verify your design. Obtain a plot of the frequency response extending from 2 decades below the center frequency to 1 decade above the center frequency. Using SPICE to determine the bandwidth of the filter, verify that its Q is 5.

15 SPECIAL-PURPOSE CIRCUITS

15–1 VOLTAGE COMPARATORS

As the name implies, a *voltage comparator* is a device used to compare two voltage levels. The output of the comparator reveals which of its two inputs is larger, so it is basically a switching device, producing a high output when one input is the larger, and switching to a low output if the other input becomes larger. An operational amplifier is used as a voltage comparator by operating it open-loop (no feedback) and by connecting the two voltages to be compared to the inverting and noninverting inputs. Because it has a very large open-loop gain, the amplifier's output is driven all the way to one of its output voltage limits when there is a very small difference between the input levels. For example, if the + input voltage is slightly greater than the − input voltage, the amplifier quickly switches to its maximum positive output, and when the − input voltage is slightly greater than the + input voltage, the amplifier switches to its maximum negative output.* This behavior is illustrated in Figure 15–1.

Note that Figure 15–1(b) is a transfer characteristic showing output voltage versus *differential* input voltage, $v^+ - v^-$. It can be seen that the output switches when $v^+ - v^-$ passes through 0.

To further clarify this behavior of the comparator, Figure 15–2 shows the output waveform when the noninverting input is a 10-V-peak sine wave and a +6-V-dc source is connected to the inverting input. The comparator output is assumed to switch between ±15 V. Notice that the output switches to +15 V each time the sine wave rises through +6 V, because $v^+ - v^- = (6\text{ V}) - (6\text{ V}) = 0\text{ V}$ at those points in time. The output remains high so long so $v^+ - v^- > 0$, i.e., $v^+ > 6\text{ V}$, and when v^+ falls below 6 V, the comparator output switches to −15 V. As an exercise, plot the output when the sine wave is connected to the inverting input and the +6-V-dc source is connected to the noninverting input.

In some applications, either the inverting or noninverting input is grounded, so the comparator is effectively a zero-crossing detector. It switches output states

* Many authors use the term *saturation voltage* to describe the minimum or maximum output voltage of the amplifier. We are avoiding this use only to prevent confusion with the saturation voltage of a transistor, which, as we have seen, is near 0.

Figure 15–1
The operational amplifier used as a voltage comparator

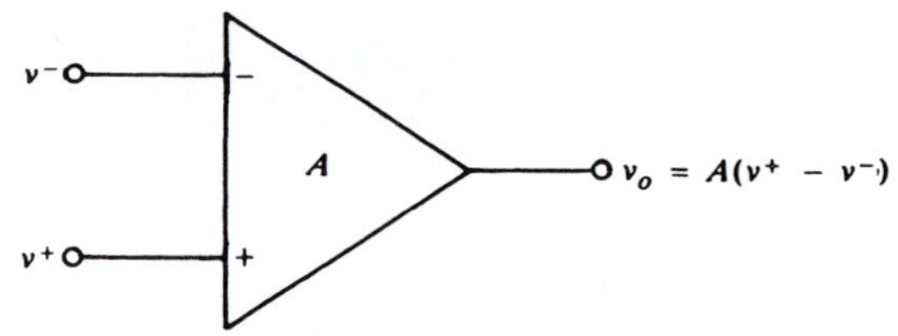

(a) Open-loop operation as a voltage comparator

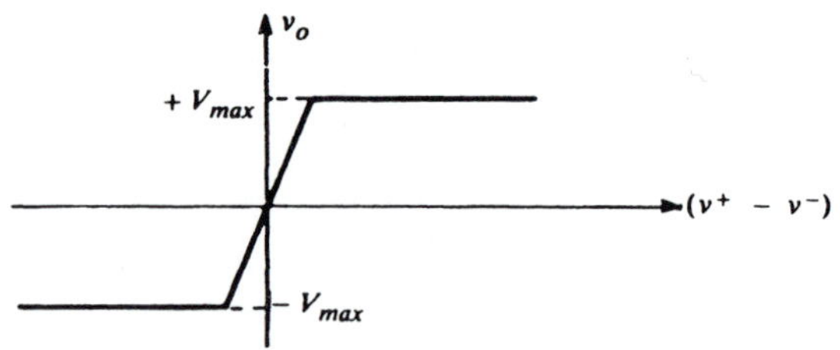

(b) Transfer characteristic of the voltage comparator. When $v^+ > v^-$, the output is at its maximum positive limit, and when $v^+ < v^-$, the output switches to its maximum negative limit.

when the ungrounded input passes through 0. For example, if the inverting input is grounded, the output switches to its maximum positive voltage when v^+ is slightly positive and to its maximum negative voltage when v^+ is slightly negative. The reverse action occurs if the noninverting input is grounded. The transfer characteristics for these two cases are shown in Figure 15–3.

In the context of a voltage comparator, the *input offset voltage* is defined to be the *minimum* differential input voltage that will cause the output to switch from

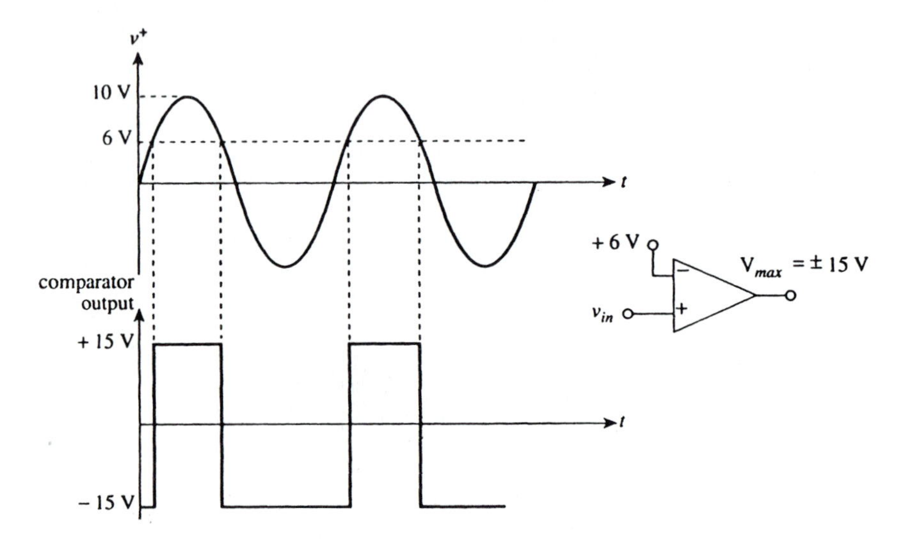

Figure 15–2
The comparator output switches to $+V_{max}$ when $v^+ - v^- > 0$ V, which corresponds to the time points where v^+ rises through +6 V. The output remains high as long as $v^+ - v^- > 0$, or $v^+ > 6$ V.

Figure 15–3
Operation of the voltage comparator as a zero-level detector

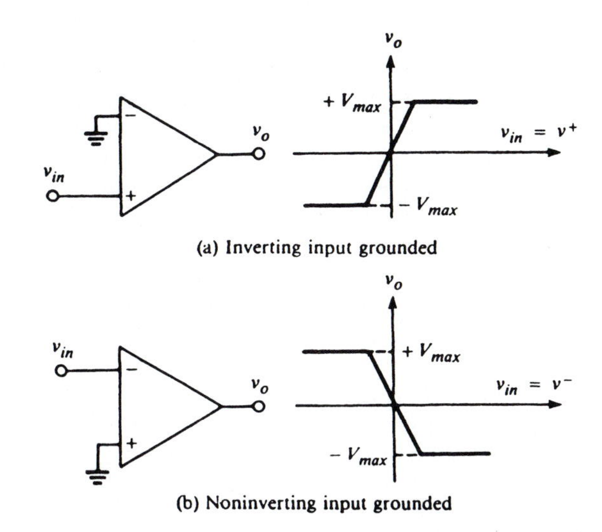

one state to the other. The smaller the input offset voltage, the more accurate the voltage comparator in terms of its ability to detect the equality of two input levels. Clearly, the greater the open-loop gain, the smaller the input offset voltage. For example, a gain of 20,000 will cause the output to switch from −10 V to +10 V when $v^+ - v^-$ is (20 V)/20,000 = 1 mV.

Two other important characteristics of a voltage comparator are its *response time* and *rise time,* illustrated in Figure 15–4. The response time is the delay between the time a step input is applied and the time the output begins to change state. It is measured from the edge of the step input to the time point where the output reaches a fixed percentage of its final value, such as 10% of $+V_{max}$. (For clarity, Figure 15–4 shows the output switching from 0 V toward $+V_{max}$; in many applications, one of the output levels actually is 0 V.) Response time is strongly dependent on the

Figure 15–4
Response time and rise time of a voltage comparator

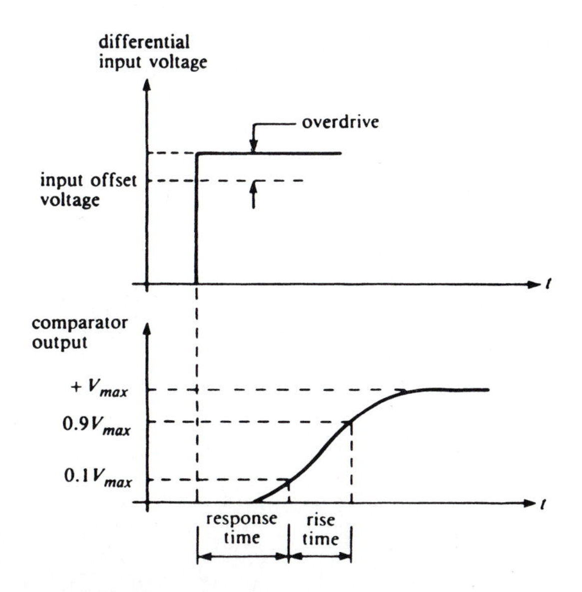

amount of *overdrive* in the input: the voltage in excess of that required to cause switching to occur. The greater the overdrive, the shorter the response time. Rise time is defined in the usual way: the time required for the output to change from 10% of its final value to 90% of its final value. Recall that rise time is inversely proportional to amplifier bandwidth: $t_r \approx 0.35/\text{BW}$. Note this important point: A large voltage gain improves the input offset voltage (reduces it), but lengthens the rise time, because large gains mean smaller bandwidths, the gain–bandwidth product being constant.

Although general-purpose operational amplifiers can be, and are, used as voltage comparators in the way we have described, there are also more elaborate, specially designed operational amplifiers manufactured and marketed specifically for voltage-comparator applications. One feature of some comparators is their ability to switch between output levels that are not necessarily related to the amplifier supply voltages. These are useful in digital systems where *level shifting* is necessary to interface logic circuitry of different types. For example, one part of a digital system may be designed to operate with logic levels of "one" (high) = +5 V and "zero" (low) = 0 V, while another part of the system uses "one" = 0 V and "zero" = −10 V. A level-shifting voltage comparator can be used to make these components compatible.

Hysteresis and Schmitt Triggers

In its most general sense, *hysteresis* is a property that means a device behaves differently when its input is increasing from the way it behaves when its input is decreasing. In the context of a voltage comparator, hysteresis means that the output will switch when the input increases to one level but will not switch back until the input falls below a *different* level. In some applications, hysteresis is a desirable characteristic because it prevents the comparator from switching back and forth in response to random noise fluctuations in the input. For example, if $v^+ - v^-$ is near 0 V, and if the input offset voltage is 1 mV, then noise voltages on the order of 1 mV will cause random switching of the comparator output. On the other hand, if the output will switch to one state only when the input rises past −1 V, and will switch to the other state only when the input falls below +1 V, then only a very large (2-V) noise voltage will cause it to switch states when the input is in the vicinity of one of these "trigger" points.

Figure 15–5(a) shows how hysteresis can be introduced into comparator operation. In this case, the input is connected to the inverting terminal and a voltage divider is connected across the noninverting terminal between v_o and a fixed reference voltage V_{REF} (which may be 0). Figure 15–5(b) shows the resulting transfer characteristic (called a *hysteresis loop*). This characteristic shows that the output switches to $+V_{max}$ when v_{in} *falls* below a lower trigger level (LTL), but will not switch to $-V_{max}$ unless v_{in} *rises* past an upper trigger level (UTL). The arrows indicate the portions of the characteristic followed when the input is increasing (upper line) and when it is decreasing (lower line). A comparator having this characteristic is called a *Schmitt trigger*.

We can derive expressions for UTL and LTL using the superposition principle. Suppose first that the comparator output is shorted to ground. Then

$$v^+ = \frac{R_2}{R_1 + R_2} V_{REF} \qquad (v_o = 0) \tag{15–1}$$

When V_{REF} is 0, we find

$$v^+ = \frac{R_1}{R_1 + R_2} v_o \qquad (V_{REF} = 0) \tag{15–2}$$

Figure 15–5
A voltage comparator with hysteresis (Schmitt trigger)

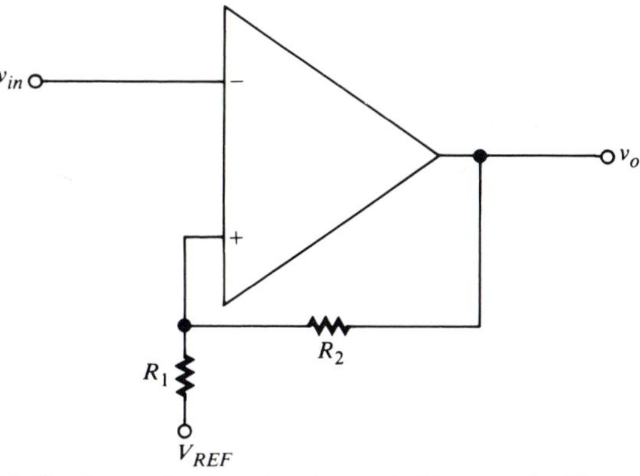

(a) Circuitry used to introduce hysteresis (V_{REF} may be 0.)

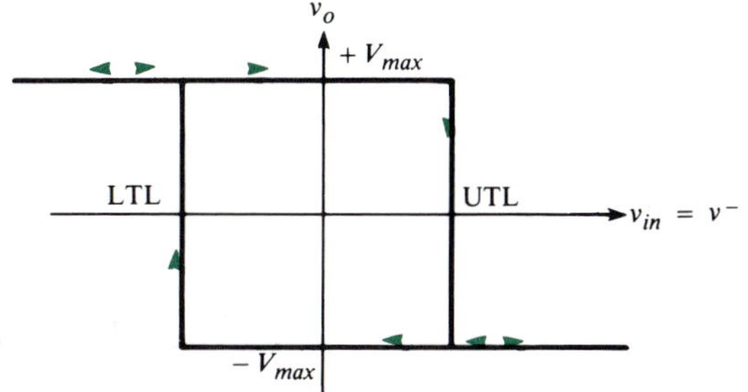

(b) Transfer characteristic of (a). The arrowheads show the portions of the characteristic followed when v_{in} is increasing (arrows pointing right) and when v_{in} is decreasing (arrows pointing left). Double-headed arrows mean that the output remains on that portion of the characteristic whether v_{in} is increasing or decreasing.

Therefore, when the output is at its negative limit ($v_o = -V_{max}$),

$$v^+ = \frac{R_2}{R_1 + R_2} V_{REF} + \frac{R_1}{R_1 + R_2}(-V_{max}) \tag{15–3}$$

As can be seen in Figure 15–5(b), v^- must *fall* to this value of v^+ before the comparator switches to $+V_{max}$. Therefore,

$$\text{LTL} = \frac{R_2}{R_1 + R_2} V_{REF} + \frac{R_1}{R_1 + R_2}(-V_{max}) \tag{15–4}$$

Similarly, when $v_o = +V_{max}$, v_{in} must *rise* to

$$\text{UTL} = \frac{R_2}{R_1 + R_2} V_{REF} + \frac{R_1}{R_1 + R_2}(+V_{max}) \tag{15–5}$$

In these equations, $+V_{max}$ is the maximum positive output voltage (a positive number) and $-V_{max}$ is the maximum negative output voltage (a negative number). The magnitudes of these quantities may be different; for example, $+V_{max} = +10$ V and $-V_{max} = -5$ V.

Quantitatively, the hysteresis of a Schmitt trigger is defined to be the difference between the input trigger levels. From equations 15–4 and 15–5,

$$\text{hysteresis} = \text{UTL} - \text{LTL}$$
$$= \left(\frac{R_1}{R_1 + R_2}\right)(+V_{max}) - \left(\frac{R_1}{R_1 + R_2}\right)(-V_{max}) \quad \textbf{(15–6)}$$

If the magnitudes of the maximum output voltages are equal, we have

$$\text{hysteresis} = \frac{2R_1V_{max}}{R_1 + R_2} \quad \textbf{(15–7)}$$

Example 15–1

1. Find the upper and lower trigger levels and the hysteresis of the Schmitt trigger shown in Figure 15–6. Sketch the hysteresis loop. The output switches between ±15 V.
2. Repeat (1) if V_{REF} = 0 V.
3. Repeat (1) if V_{REF} = 0 V and the output switches between 0 V and +15 V.

Figure 15–6
(Example 15–1)

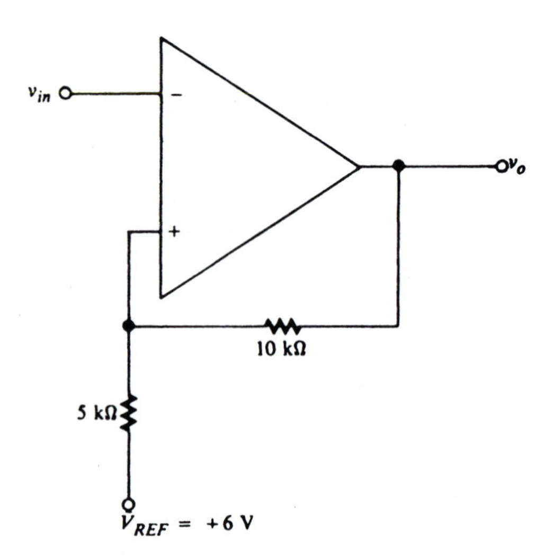

Solution

1. From equations 15–4 and 15–5.

$$\text{LTL} = \left[\frac{10\text{ k}\Omega}{(5\text{ k}\Omega) + (10\text{ k}\Omega)}\right](6\text{ V}) + \left[\frac{5\text{ k}\Omega}{(5\text{ k}\Omega) + (10\text{ k}\Omega)}\right](-15\text{ V}) = -1\text{ V}$$

$$\text{UTL} = \left[\frac{10\text{ k}\Omega}{(5\text{ k}\Omega) + (10\text{ k}\Omega)}\right](6\text{ V}) + \left[\frac{5\text{ k}\Omega}{(5\text{ k}\Omega) + (10\text{ k}\Omega)}\right](15\text{ V}) = +9\text{ V}$$

hysteresis = (9 V) − (−1 V) = 10 V

(Note also, from equation 15–7, hysteresis = 2(5 kΩ)(15 V)/(15 kΩ) = 10 V.)

2. Since $V_{REF} = 0$,

$$\text{LTL} = \left(\frac{R_1}{R_1 + R_2}\right)(-V_{max}) = \left[\frac{5\text{ k}\Omega}{(5\text{ k}\Omega) + (10\text{ k}\Omega)}\right](-15\text{ V}) = -5\text{ V}$$

$$\text{UTL} = \left(\frac{R_1}{R_1 + R_2}\right)(+V_{max}) = \left[\frac{5\text{ k}\Omega}{(5\text{ k}\Omega) + (10\text{ k}\Omega)}\right](15\text{ V}) = +5\text{ V}$$

$$\text{hysteresis} = (5\text{ V}) - (-5\text{ V}) = 10\text{ V}$$

(Again, equation 15–7 may be used.)

3. Since the output switches between 0 V and +15 V, we must use 0 in place of $-V_{max}$ in the trigger-level equations:

$$\text{LTL} = \left(\frac{R_1}{R_1 + R_2}\right)(-V_{max}) = \left[\frac{5\text{ k}\Omega}{(5\text{ k}\Omega) + (10\text{ k}\Omega)}\right]0 = 0\text{ V}$$

$$\text{UTL} = \left(\frac{R_1}{R_1 + R_2}\right)(+V_{max}) = +5\text{ V}$$

$$\text{hysteresis} = (5\text{ V}) - (0\text{ V}) = 5\text{ V}$$

(Note that equation 15–7 is not applicable in this case.)

Figure 15–7 shows the hysteresis loops for these cases, along with the output waveforms that result when v_{in} is a 10-V-peak sine wave.

The comparator we have discussed is called an *inverting* Schmitt trigger because the output is high when the input is low, and vice versa, as can be seen in Figure 15–7. Figure 15–8 shows a noninverting Schmitt trigger. For this circuit, the lower and upper trigger levels are

$$\text{LTL} = \frac{-R_1}{R_2}(+V_{max}) \quad \textbf{(15–8)}$$

$$\text{UTL} = \frac{R_1}{R_2}\left|-V_{max}\right| \quad \textbf{(15–9)}$$

Notice that these equations permit the magnitudes of $+V_{max}$ and $-V_{max}$ to be different values. For example, if $R_1 = 10\text{ k}\Omega$ and $R_2 = 20\text{ k}\Omega$, and if the output switches between +10 V and −5 V, then LTL = −(0.5)(10 V) = −5 V and UTL = 0.5 |−5 V| = +2.5 V. The derivation of equations 15–8 and 15–9 is an exercise at the end of this chapter.

An Astable Multivibrator

The word *astable* means "unstable," and, like other unstable devices, an astable multivibrator is a (square-wave) oscillator. (A *bi*stable multivibrator, also called a *flip-flop,* is a digital device with two stable stages; a *mono*stable multivibrator has one stable state, and an astable multivibrator has zero stable states.) An astable multivibrator can be constructed by using an operational amplifier as a voltage comparator in a circuit like that shown in Figure 15–9. This circuit is an example of a *relaxation* oscillator, one whose operation depends on the repetetive charging and discharging of a capacitor.

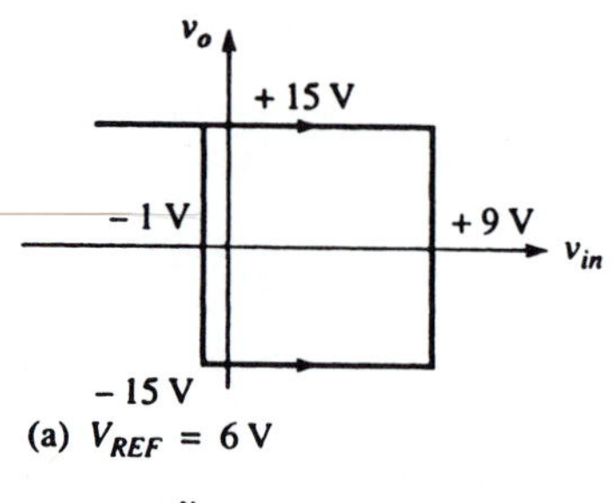

(a) V_{REF} = 6 V

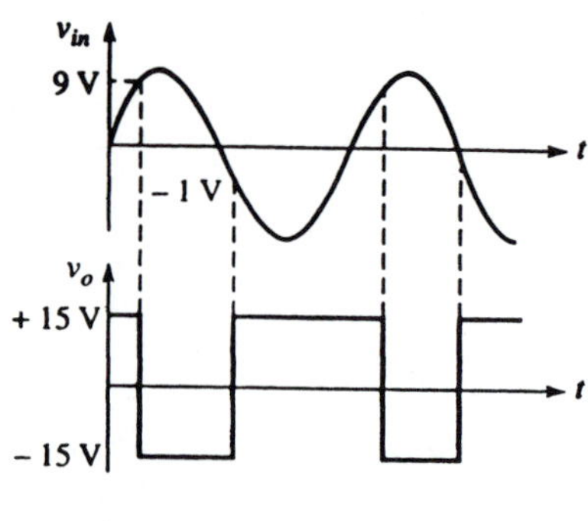

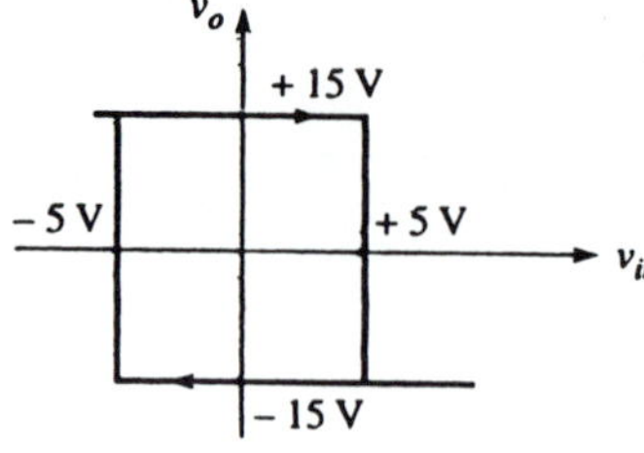

(b) V_{REF} = 0 V

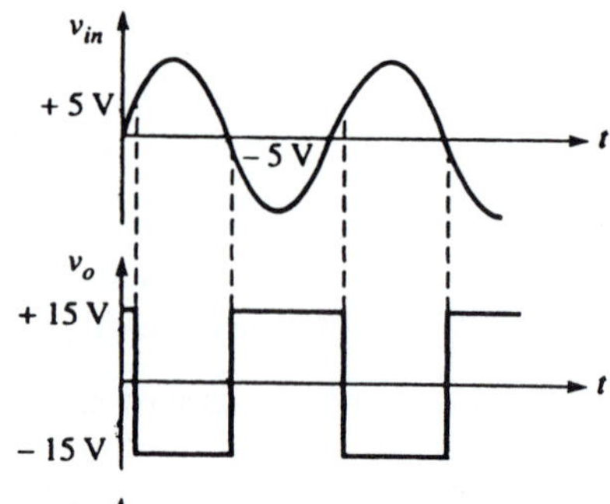

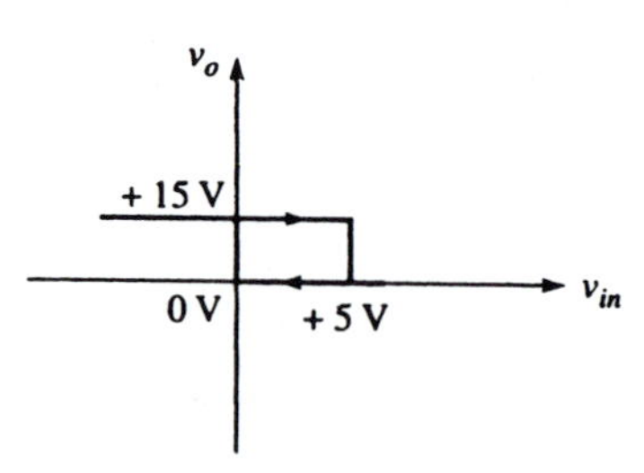

(c) V_{REF} = 0 V, $-V_{max}$ = 0 V

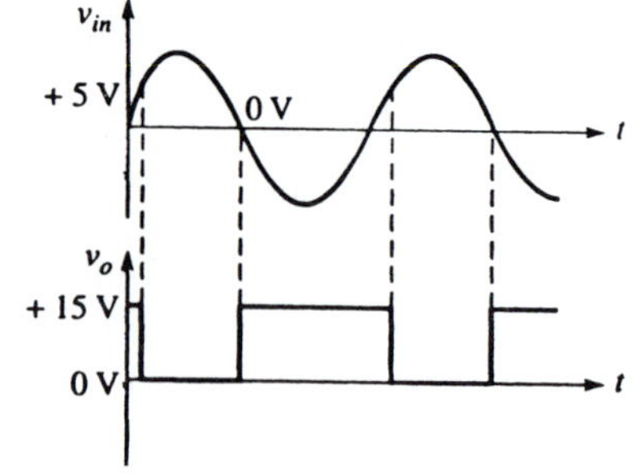

Figure 15–7
(Example 15–1)

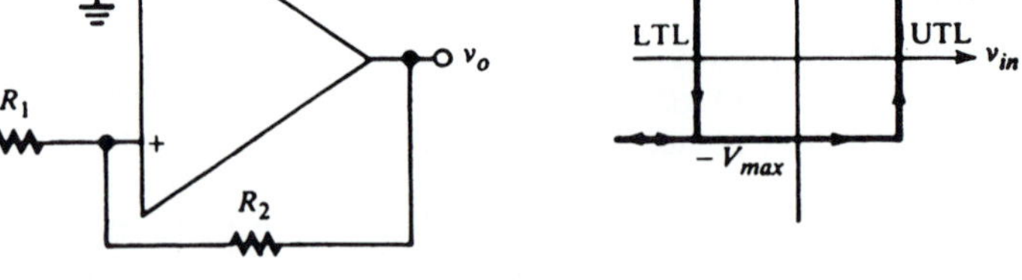
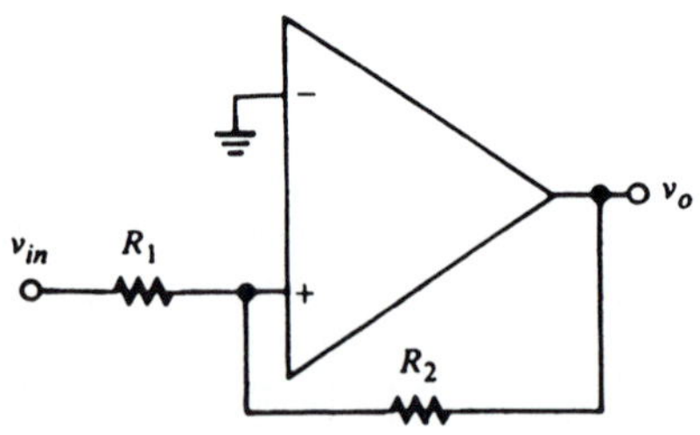

Figure 15–8
The noninverting Schmitt trigger

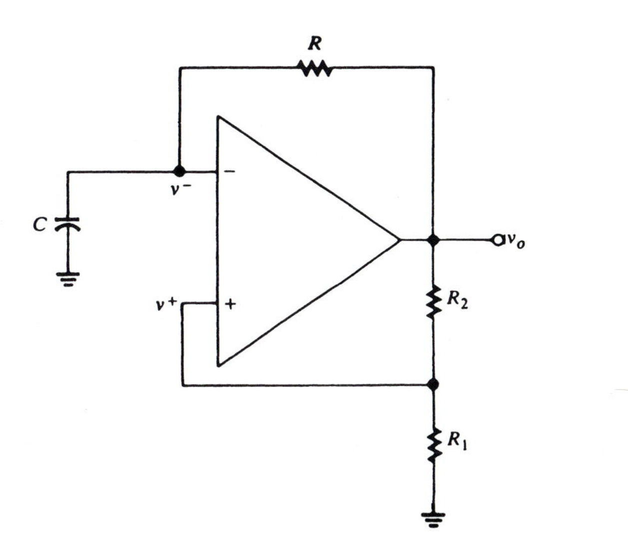

Figure 15–9
An astable multivibrator

For analysis purposes, let us assume that the output voltages of the comparator are equal in magnitude and opposite in polarity: $\pm V_{max}$. Figure 15–10 shows the voltage across capacitor C and the output waveform produced by the comparator. We begin by assuming that the output is at $+V_{max}$. Then, the voltage fed back to the noninverting input is

$$v^+ = \frac{R_1}{R_1 + R_2} V_{max} = +\beta V_{max} \tag{15–10}$$

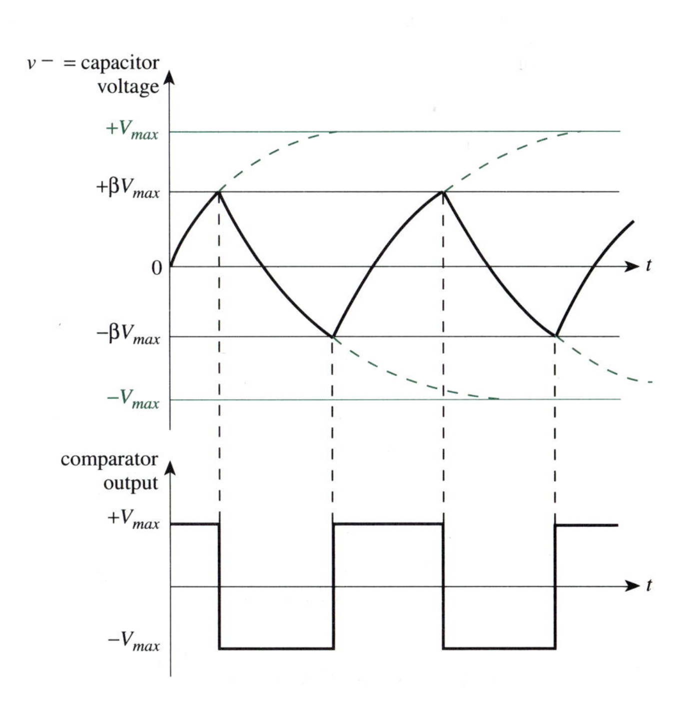

Figure 15–10
The waveforms on the capacitor and at the output of the astable multivibrator shown in Figure 15–9

Notice that v^- equals the voltage across the capacitor. The capacitor begins to charge through R *toward* a final voltage of $+V_{max}$. However, as soon as the capacitor voltage reaches a voltage equal to v^+, the comparator switches state. In other words, switching occurs at the point in time where $v^- = v^+ = +\beta V_{max}$. After the comparator switches state, its output is $-V_{max}$, and the voltage fed back to the noninverting input becomes

$$v^+ = \frac{R_1}{R_1 + R_2}(-V_{max}) = -\beta V_{max} \tag{15–11}$$

Since the comparator output is now negative, the capacitor begins to discharge through R toward $-V_{max}$. But, when that voltage falls to $-\beta V_{max}$, we once again have $v^+ = v^-$, and the comparator switches back to $+V_{max}$. This cycle repeats continuously, as shown in Figure 15–10, with the result that the output is a square wave that alternates between $\pm V_{max}$ volts.

It can be shown that the period of the multivibrator oscillation is

$$T = 2RC \ln\left(\frac{1+\beta}{1-\beta}\right) \text{ seconds} \tag{15–12}$$

15–2 CLIPPING AND RECTIFYING CIRCUITS

Clipping Circuits

In Chapter 5, we referred to *clipping* as the undesirable result of overdriving an amplifier. We have seen that any attempt to push an output voltage beyond the limits through which it can "swing" causes the tops and/or bottoms of a waveform to be "clipped" off, resulting in distortion. However, in numerous practical applications, including waveshaping and nonlinear function generation, waveforms are *intentionally* clipped.

Figure 15–11 shows how the transfer characteristic of a device is modified to reflect the fact that its output is clipped at certain levels. In each of the examples shown, note that the characteristic becomes horizontal at the output level where clipping occurs. The horizontal line means that the output remains constant regardless of the input level in that region. Outside the clipping region, the transfer characteristic is simply a line whose slope equals the gain of the device. This is the region of normal, *linear* operation. In these examples, the devices are assumed to have unity gain, so the slope of each line in the linear region is 1.

Figure 15–12 illustrates a somewhat different kind of clipping action. Instead of the positive or negative peaks being "chopped off," the output follows the input when the signal is above or below a certain level. The transfer characteristics show that linear operation occurs only when certain signal levels are reached and that the output remains constant below those levels. This form of clipping can also be thought of as a special case of that shown in Figure 15–11. Imagine, for example, that the clipping level in Figure 15–11(b) is raised to a positive value; then the result is the same as Figure 15–12(a).

Clipping can be accomplished using *biased diodes,* a technique that is more efficient than overdriving an amplifier. Clipping circuits rely on the fact that diodes have very low impedances when they are forward biased and are essentially open circuits when reverse biased. If a certain point in a circuit, such as the output of an amplifier, is connected through a very small impedance to a *constant* voltage, then the voltage at the circuit point cannot differ significantly from the constant

Figure 15–11
Waveforms and transfer characteristics of clipping circuits

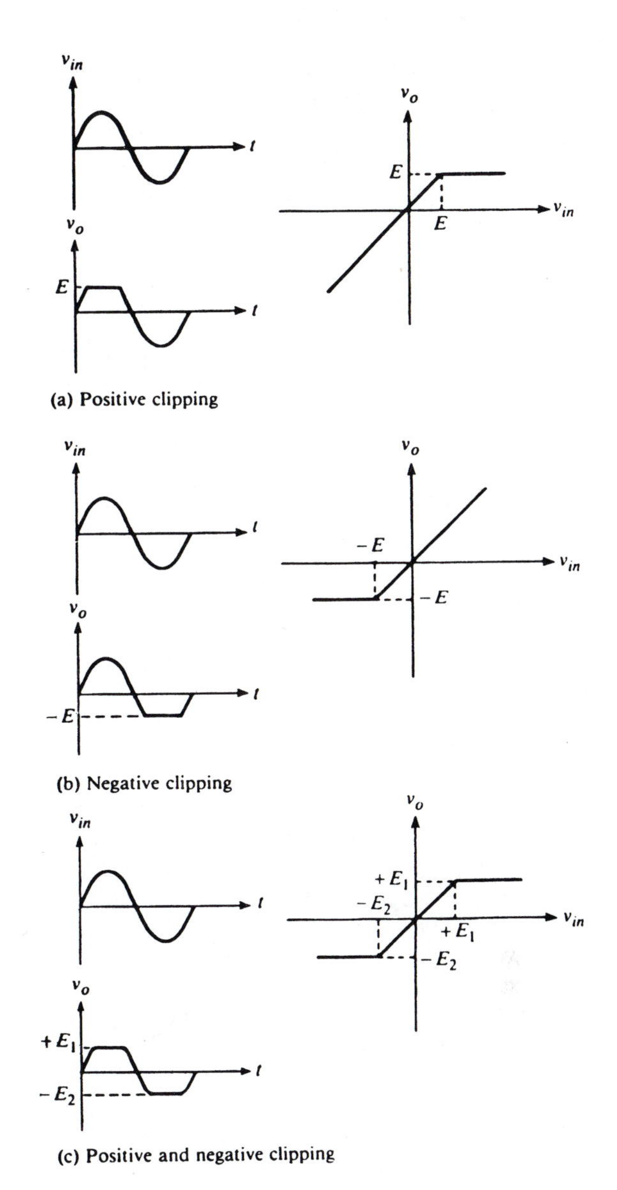

voltage. We say in this case that the point is *clamped* to the fixed voltage. (However, we will reserve the term *clamping circuit* for a special application to be discussed later.) An ideal, forward-biased diode is like a closed switch, so if it is connected between a point in a circuit and a fixed voltage source, the diode very effectively holds the point to the fixed voltage. Diodes can be connected in operational-amplifier circuits, as well as other circuits, so that they become forward biased when a signal reaches a certain voltage. When the forward-biasing level is reached, the diode serves to hold the output to a fixed voltage and thereby establishes a clipping level.

A biased diode is simply a diode connected to a fixed voltage source. The value and polarity of the voltage source determine what value of total voltage across the

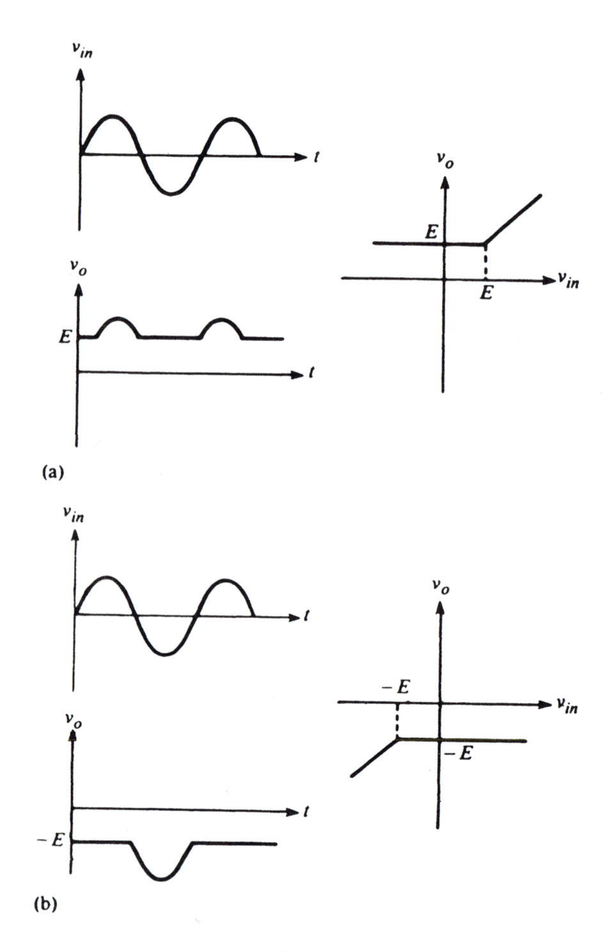

Figure 15–12
Another form of clipping. Compare with Figure 15–11

combination is necessary to forward bias the diode. Figure 15–13 shows several examples. (In practice, a series resistor would be connected in each circuit to limit current flow when the diode is forward biased.) In each part of the figure, we can write Kirchhoff's voltage law around the loop to determine the value of input voltage v_i that is necessary to forward bias the diode. Assuming that the diodes are ideal (neglecting their forward voltage drops), we determine the value of v_i necessary to forward bias each diode by determining the value of v_i necessary to make $V_D > 0$. Whenever v_i reaches the voltage necessary to make $V_D > 0$, the diode becomes forward biased and the signal source is forced to, or held at, the dc source voltage. If the forward voltage drop across the diode is not neglected, the clipping level is found by determining the value of v_i necessary to make V_D greater than that forward drop (e.g., $V_D > 0.7$ V for a silicon diode). Although these conditions can be determined through formal application of Kirchhoff's voltage law, as shown in the figure, the reader is urged to develop a mental image of circuit behavior in each case. For example, in (a), think of the diode as being reverse biased by 6 V, so the input must "overcome" that reverse bias by reaching +6 V to forward bias the diode.

Figure 15–14 shows three examples of clipping circuits using ideal biased diodes and the waveforms that result when each is driven by a sine-wave input. In each

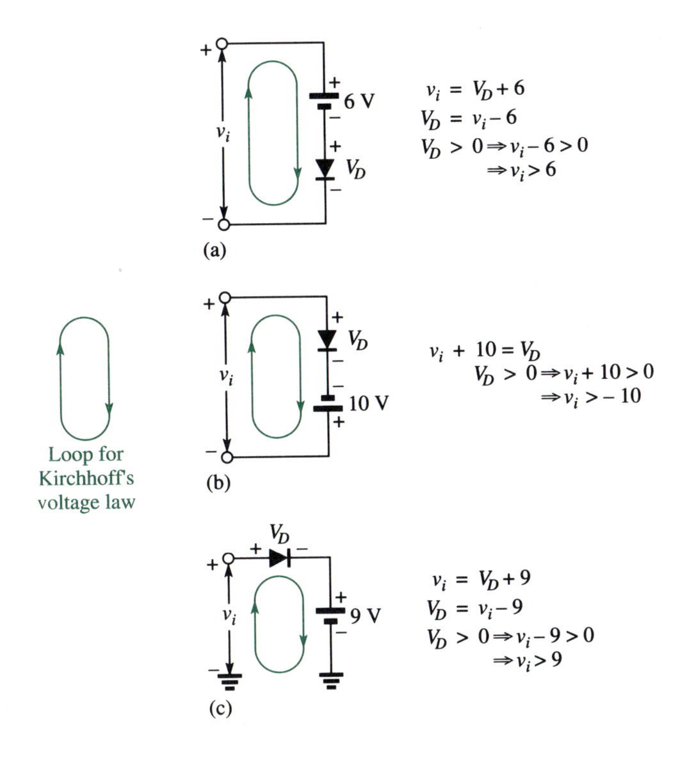

Figure 15–13
Examples of biased diodes and the signal voltages v_i required to forward bias them. (Ideal diodes are assumed.) In each case, we solve for the value of v_i that is necessary to make $V_D > 0$.

case, note that the output equals the dc source voltage when the input reaches the value necessary to forward bias the diode. Note also that the type of clipping we showed in Figure 15–11 occurs when the fixed bias voltage tends to *reverse* bias the diode, and the type shown in Figure 15–12 occurs when the fixed voltage tends to *forward* bias the diode. When the diode is reverse biased by the input signal, it is like an open circuit that disconnects the dc source, and the output follows the input. These circuits are called *parallel* clippers because the biased diode is in parallel with the output. Although the circuits behave the same way whether or not one side of the dc voltage source is connected to the common (low) side of the input and output, the connections shown in Figure 15–14(a) and (c) are preferred to that in (b) because the latter uses a floating source.

Figure 15–15(a) shows a biased diode connected in the feedback path of an inverting operational amplifier. The diode is in parallel with the feedback resistor and forms a parallel clipping circuit like that shown in Figure 15–14(a). Since v^- is at virtual ground, the voltage across R_f is the same as the output voltage v_o. Therefore, when the output voltage reaches the bias voltage E, the output is held at E volts. Figure 15–15(b) illustrates this fact for a sinusoidal input. As long as the diode is reverse biased, it acts like an open circuit and the amplifier behaves like a conventional inverting amplifier. Notice that output clipping occurs at *input* voltage $-(R_1/R_f)E$, since the amplifier inverts and has closed-loop gain magnitude

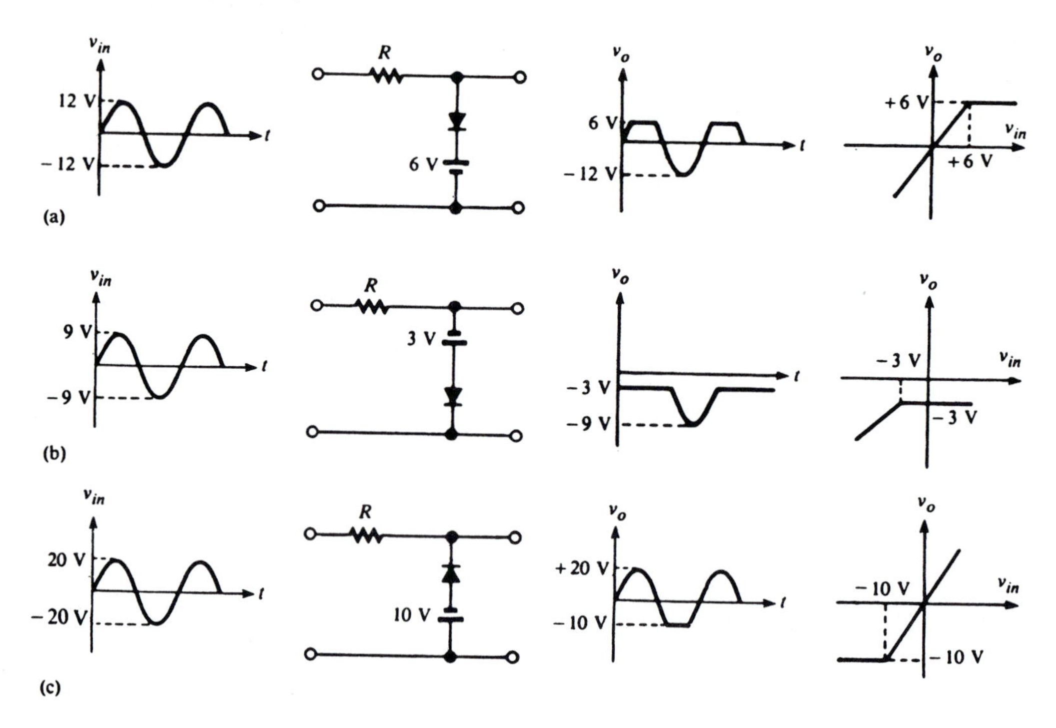

Figure 15–14
Examples of parallel clipping circuits

R_f/R_1. The resulting transfer characteristic is shown in Figure 15–15(c). This circuit is often called a *limiting* circuit because it limits the output to the dc level clamped by the diode. (In this and future circuits in this section, we are omitting the bias compensation resistor, R_c, for clarity; it should normally be included, following the guidelines of Chapter 13.)

In practice, the biased diode shown in the feedback of Figure 15–15(a) is often replaced by a *zener* diode in series with a conventional diode. This arrangement eliminates the need for a floating voltage source. We will study zener diodes in more detail in Chapter 18 (Section 18–7) and will learn that in many respects they are equivalent to biased diodes. Figure 15–16 shows two operational-amplifier clipping circuits using zener diodes. The zener diode conducts like a conventional diode when it is forward biased, so it is necessary to connect a reversed diode in series with it to prevent shorting of R_f. When the reverse voltage across the zener diode reaches V_Z, the diode breaks down and conducts heavily, while maintaining an essentially constant voltage, V_Z, across it. Under those conditions, the *total* voltage across R_f, i.e., v_o, equals V_Z plus the forward drop, V_D, across the conventional diode.

Figure 15–17 shows *double-ended* limiting circuits, in which both positive and negative peaks of the output waveform are clipped. Figure 15–17(a) shows the conventional parallel clipping circuit and (b) shows how double-ended limiting is accomplished in an operational-amplifier circuit. In each circuit, note that no more than one diode is forward biased at any given time, and that both diodes are reverse biased for $-E_1 < v_o < E_2$, the linear region.

Figure 15–18 shows a double-ended limiting circuit using *back-to-back* zener

Figure 15–15
An operational-amplifier limiting circuit

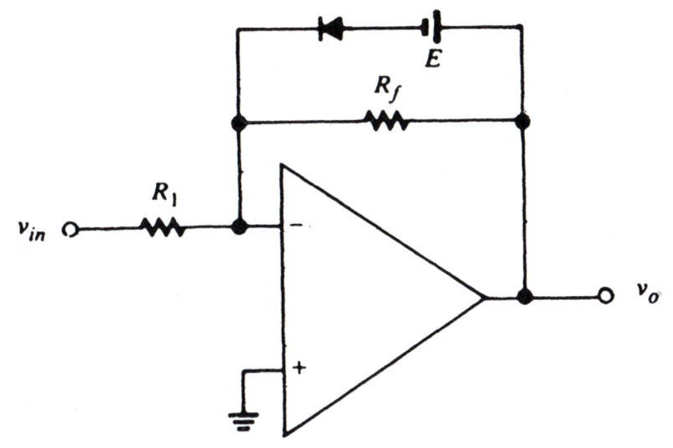

(a) The biased diode in the feedback path provides (parallel) clipping of the output at E volts.

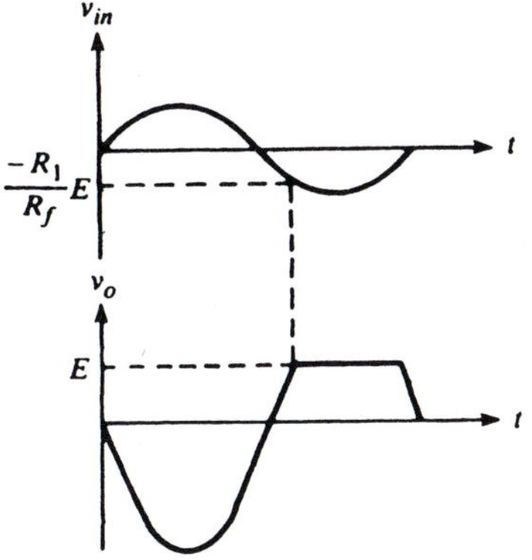

(b) The output clamps at E volts when the input reaches $\frac{-R_1}{R_f}E$ volts.

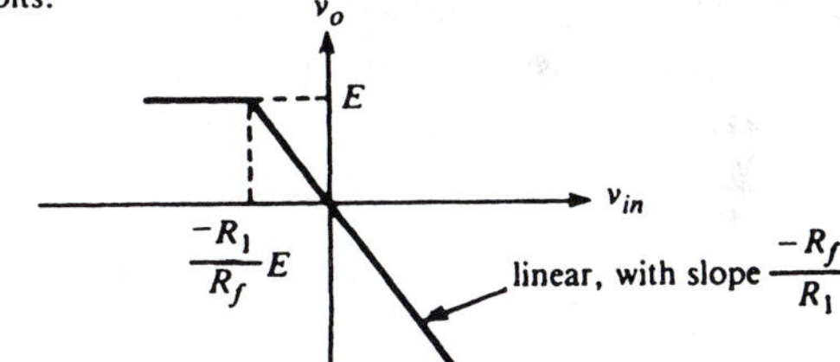

(c) Transfer characteristic

diodes. Operation is similar to that shown in Figure 15–16, but no conventional diode is required. Note that diode D_1 is conducting in a forward direction when D_2 conducts in its reverse breakdown (zener) region, while D_2 is forward biased when D_1 is conducting in its reverse breakdown region. Neither diode conducts when $-(V_{Z2} + 0.7) < v_o < (V_{Z1} + 0.7)$, which is the region of linear amplifier operation.

Precision Rectifying Circuits

Recall that a *rectifier* is a device that allows current to pass through it in one direction only (see Figure 3–17). A diode can serve as a rectifier because it permits generous current flow in only one direction—the direction of forward bias. Rectification is the same as clipping at the 0-V level: all of the waveform below (or

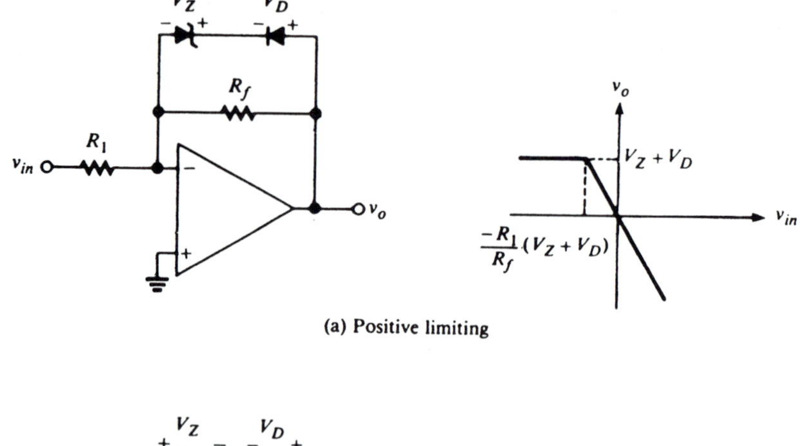

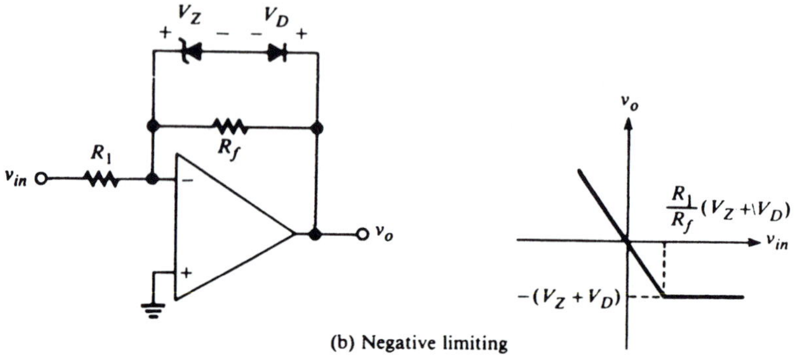

Figure 15–16
Operational-amplifier limiting circuits using zener diodes

above) the zero-axis is eliminated. Recall, however, that a diode rectifier has certain intervals of nonconduction and produces resulting "gaps" at the zero-crossing points of the output voltage, due to the fact that the input must overcome the diode drop (0.7 V for silicon) before conduction begins. See Figures 3–19 and 3–20. In power-supply applications, where input voltages are quite large, these gaps are of no concern. However, in many other applications, especially in instrumentation, the 0.7-V drop can be a significant portion of the total input voltage swing and can seriously affect circuit performance. For example, most ac instruments rectify ac inputs so they can be measured by a device that responds to dc levels. It is obvious that small ac signals could not be measured if it were always necessary for them to reach 0.7 V before rectification could begin. For these applications, *precision* rectifiers are necessary.

Figure 15–19 shows one way to obtain precision rectification using an operational amplifier and a diode. The circuit is essentially a noninverting voltage follower when the diode is forward biased. When v_{in} is positive, the output of the amplifier, v_o, is positive, the diode is forward biased, and a low-resistance path is established between v_o and v^-, as necessary for a voltage follower. The load voltage, v_L, then follows the positive variations of $v_{in} = v^+$. Note that even a very small positive value of v_{in} will cause this result because of the large differential gain of the amplifier; that is, the large gain and the action of the feedback cause the usual result that $v^+ \approx v^-$. Note also that the drop across the diode does not appear in v_L.

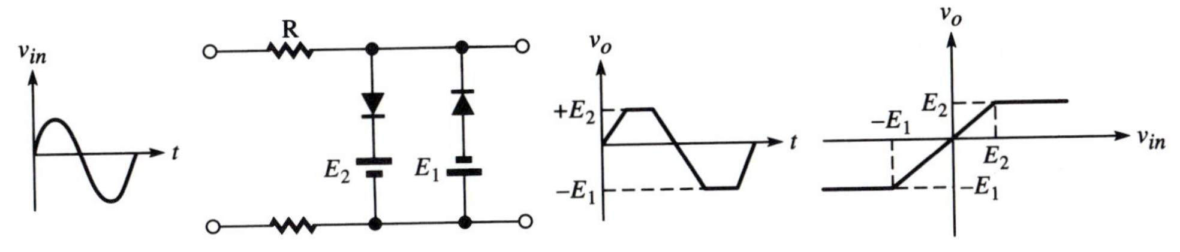

(a) Double-ended parallel clipper

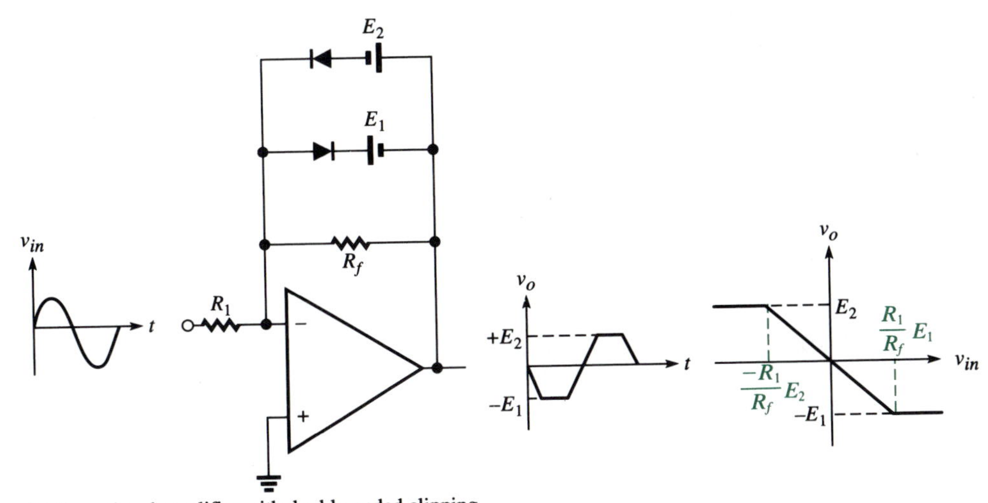

(b) Operational amplifier with double-ended clipping

Figure 15–17
Double-ended clipping, or limiting

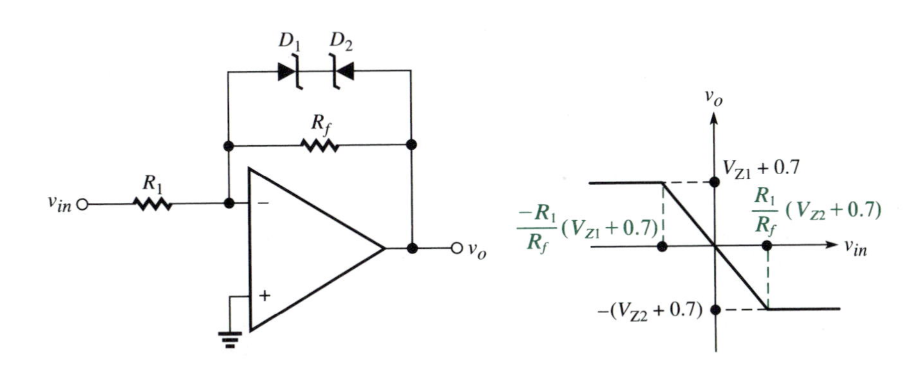

Figure 15–18
A double-ended limiting circuit using zener diodes

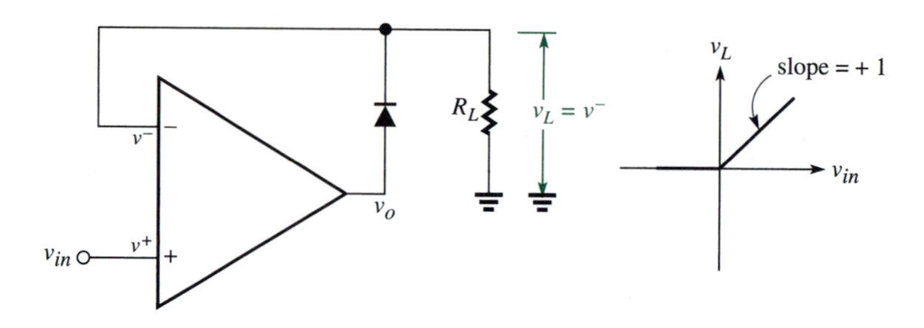

Figure 15–19
A precision rectifier. When v_{in} is positive, the diode is forward biased, and the amplifier behaves like a voltage follower, maintaining $v^+ \approx v^- = v_L$

When the input goes negative, v_o becomes negative, and the diode is reverse biased. This effectively opens the feedback loop, so v_L no longer follows v_{in}. The amplifier itself, now operating open-loop, is quickly driven to its maximum negative output, thus holding the diode well into reverse bias.

Another precision rectifier circuit is shown in Figure 15–20. In this circuit, the load voltage is an amplified and inverted version of the *negative* variations in the input signal and is 0 when the input is positive. Also in contrast with the previous circuit, the amplifier in this rectifier is not driven to one of its output extremes. When v_{in} is negative, the amplifier output, v_o, is positive, so diode D_1 is reverse biased and diode D_2 is forward biased. D_1 is open and D_2 connects the amplifier output through R_f to v^-. Thus, the circuit behaves like an ordinary inverting amplifier with gain $-R_f/R_1$. The load voltage is an amplified and inverted (positive) version of the negative variations in v_{in}. When v_{in} becomes positive, v_o is negative, D_1 is forward biased, and D_2 is reverse biased. D_1 shorts the output v_o to v^-, which is held at virtual ground, so v_L is 0. It is an exercise at the end of the chapter to analyze this circuit when the diodes are turned around.

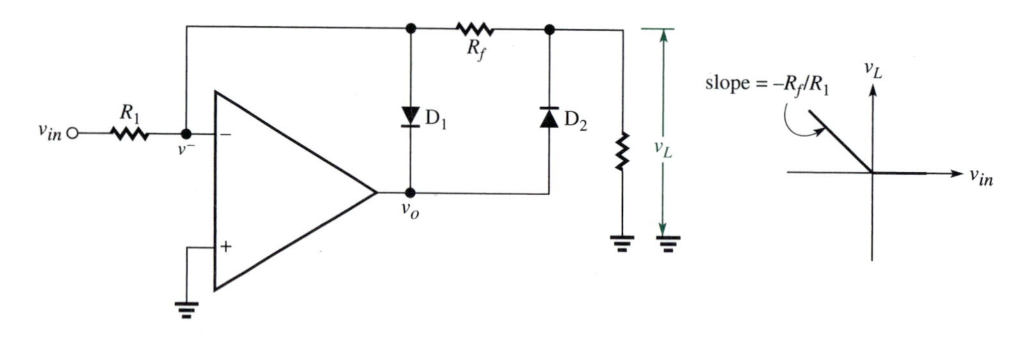

Figure 15–20
A precision rectifier circuit that amplifies and inverts the negative variations in the input voltage

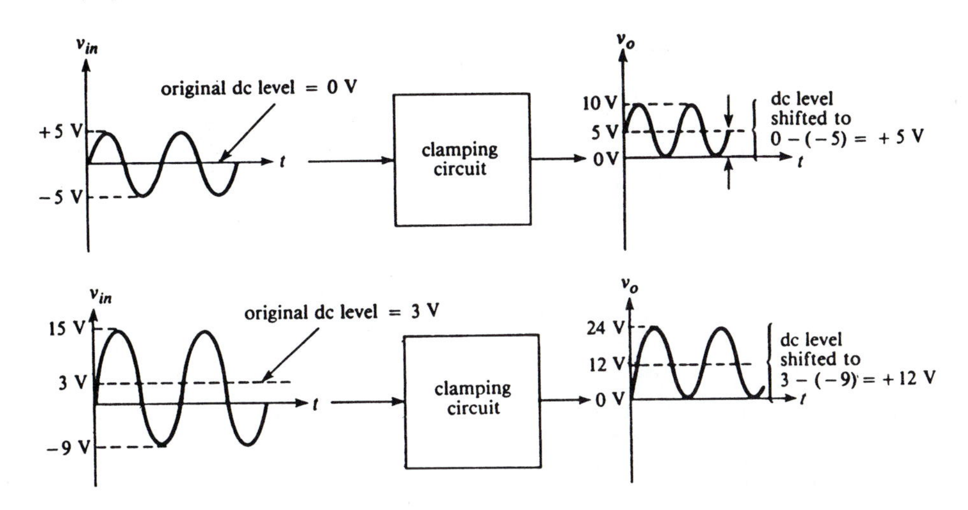

Figure 15–21
A clamping circuit that shifts a waveform up by an amount equal to the negative peak value

15–3 CLAMPING CIRCUITS

Clamping circuits are used to shift an ac waveform up or down by adding a dc level equal to the positive or negative peak value of the ac signal. In the author's opinion, "clamping" is not a particularly good term for this operation: *Level shifting* is more descriptive. Clamping circuits are also called dc *level restorers* because they are used in systems (television, for example) where the original dc level is lost in capacitor-coupled amplifier stages. It is important to recognize that the amount of dc-level shift required in these applications *varies* as the peak value of the ac signal varies over a period of time. In other words, it is not possible to simply add a fixed dc level to the ac signal using a summing amplifier. To illustrate, Figure 15–21 shows the outputs required from a clamping circuit for two different inputs. Note in both cases that the *peak-to-peak* value of the output is the same as the peak-to-peak value of the input, and that the output is shifted up (in this case) by an amount equal to the negative peak of the input.

Figure 15–22(a) shows a clamping circuit constructed from passive components. When the input first goes negative, the diode is forward biased, and the capacitor charges rapidly to the peak negative input voltage, V_1. The charging time-constant is very small because the forward resistance of the diode is small. The capacitor voltage, V_1, then has the polarity shown in the figure. Assuming that the capacitor does not discharge appreciably through R_L, the total load voltage as v_{in} begins to increase is

$$v_L = V_1 + v_{in} \tag{15–13}$$

(Verify this equation by writing Kirchhoff's voltage law around the loop.) Notice that the polarity of the capacitor voltage keeps the diode reverse biased, so it is like an open circuit during this time and does not discharge the capacitor. Equation 15–13 shows that the load voltage equals the input voltage shifted up by an amount equal to V_1, as required. When the input again reaches its negative peak, the

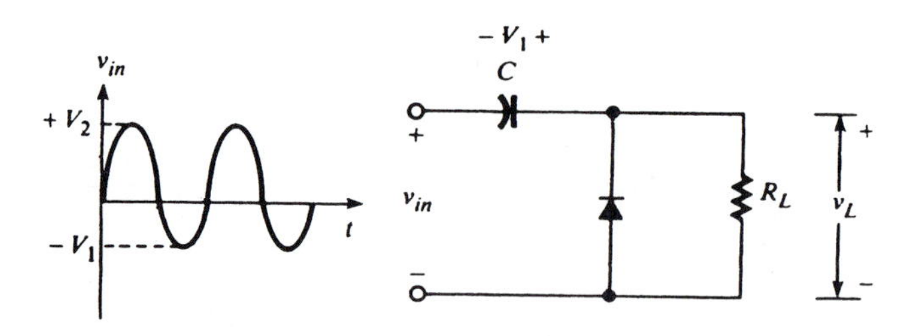

(a) The capacitor charges to V_1 volts and holds that voltage, so $v_L \cong v_{in} + V_1$.

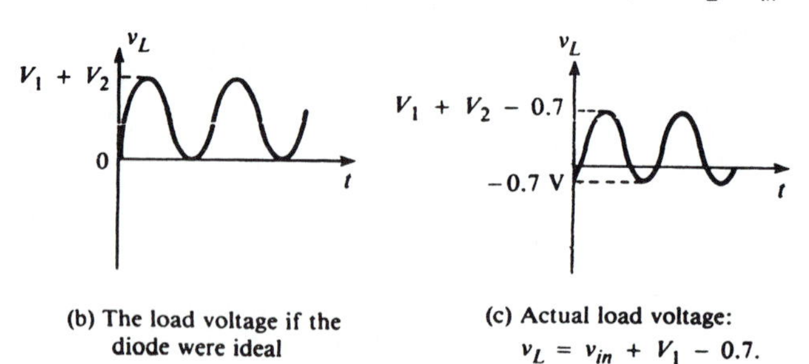

(b) The load voltage if the diode were ideal

(c) Actual load voltage: $v_L = v_{in} + V_1 - 0.7$.

Figure 15–22
A clamping circuit consisting of a diode and a capacitor

capacitor may have to recharge slightly to make up for any decay that occurred during the cycle. For proper circuit performance, the discharge time-constant, R_LC, must be much greater than the period of the input. If the diode connections are reversed, the waveform is shifted down by an amount equal to the positive peak voltage, V_2. If the diode is biased by a fixed voltage, the waveform can be shifted up or down by an amount equal to a peak value plus or minus the bias voltage. Examples are given in the exercises at the end of the chapter.

Figure 15–22(b) shows the load voltage that results if the diode is assumed to have zero voltage drop. In reality, since the capacitor charges through the diode, the voltage across the capacitor reaches only V_1 minus the diode drop. Consequently,

$$v_L = v_{in} + V_1 - 0.7 \qquad \textbf{(15–14)}$$

The waveform that results is shown in Figure 15–22(c). If the input voltage is large, this offset due to an imperfect diode can be neglected, as is the case in many practical circuits.

If precision clamping is required, the operational-amplifier circuit shown in Figure 15–23 can be used. When v_{in} in Figure 15–23 first goes negative, the amplifier output, v_o, is positive and the diode is forward biased. The capacitor quickly charges to V_1, with the polarity shown. Notice that $v_L = v_{in} + V_1$ and that the drop across the diode does not appear in v_L. With the capacitor voltage having the polarity shown, v^- becomes positive, and remains positive throughout the cycle, so the amplifier output is negative. Therefore, the diode is reverse biased and the feedback loop is opened. The amplifier is driven to its maximum negative output level and the diode remains reverse biased. During one cycle of the input, the capacitor may discharge somewhat into the load, causing its voltage to fall below V_1. If so, then when v_{in} once again reaches its maximum negative voltage, v^- will once again be

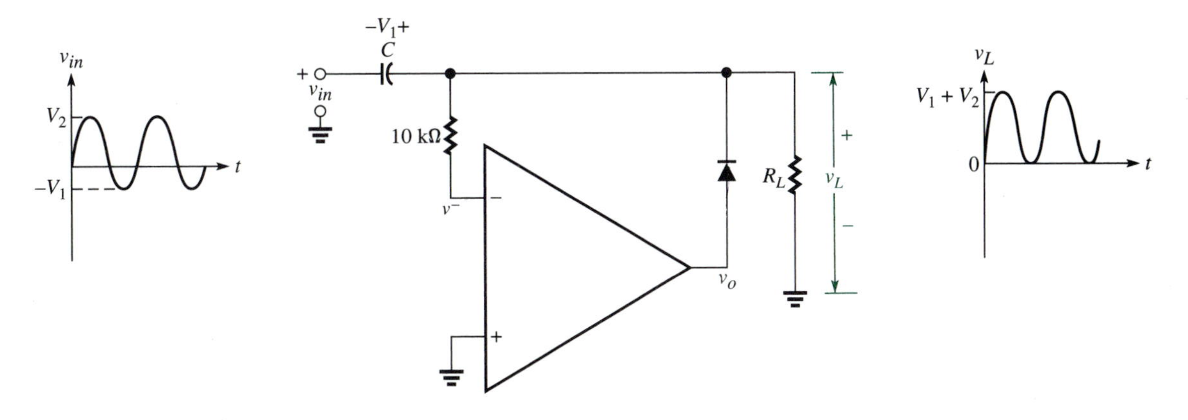

Figure 15–23
A precision clamping circuit

negative, v_o will be positive, and the capacitor will be allowed to recharge to V_1 volts, as before.

15–4 LOGARITHMIC AND ANTILOGARITHMIC AMPLIFIERS

A logarithmic (log) amplifier produces an output that is proportional to the logarithm of its input. Since the log function is nonlinear, it is clear that a log amplifier is not linear in the sense discussed in Chapter 3. A logarithmic transfer characteristic is shown in Figure 15–24. We see that the slope of the characteristic, $\Delta V_o/\Delta V_{in}$, and hence the voltage gain, is small for large values of V_{in} and large for small values of V_{in}. Since the gain decreases with increasing input signal level, the amplifier is said to *compress* signals.

One important application of log amplifiers is in the amplification of signals having a wide *dynamic range:* signals that may be very small as well as very large. Suppose, for example, that a temperature sensor generates a few millivolts at very low temperatures and a few volts at very high temperatures. To obtain good

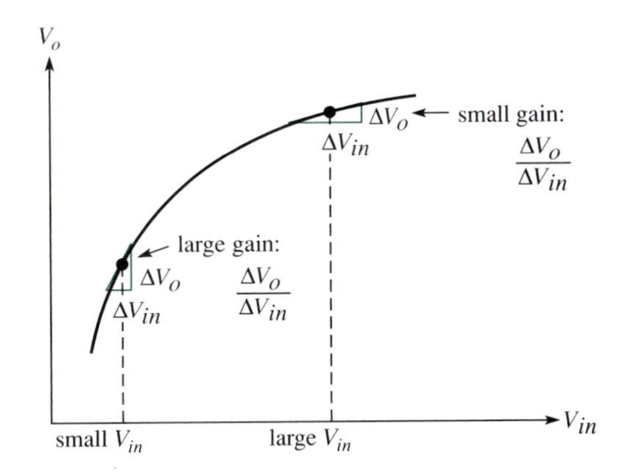

Figure 15–24
The transfer characteristic of a log amplifier shows that its voltage gain decreases with increasing values of V_{in}

resolution, we would like the small signals to undergo significant amplification. The same amplification applied to the large signals, as would occur in a linear amplifier, would overdrive the amplifier and create clipping and distortion. The log amplifier eliminates this problem.

Of course, the nonlinear characteristic of the log amplifier creates output waveforms that are distorted versions of input waveforms. If necessary for a particular application, the distortion can be removed by an *antilogarithmic* (inverse log, or exponential) amplifier, which has a transfer characteristic that is exactly the opposite of the log amplifier. On the other hand, in some applications the antilog operation is not necessary, as, for example, when it is desired to create a display of signal magnitudes on a logarithmic scale. An example is a *spectrum analyzer,* in which the frequency content of a complex signal is displayed as a plot of decibel voltage levels versus frequency. Another application of log amplifiers is in *analog computation*, where signal voltages must be multiplied or divided. For example, if we wished to generate the product voltage v_1v_2, we could sum the outputs of two log amplifiers to obtain $\log v_1 + \log v_2 = \log v_1v_2$. The output of an antilog amplifier whose input is $\log v_1v_2$ would then be a voltage proportional to v_1v_2.

The logarithmic characteristic of a log amplifier stems from the relationship between the collector current and base-to-emitter voltage of a BJT, which is similar to the diode equation (equation 2–13):

$$I_C = I_s(e^{V_{be}/V_T} - 1) \qquad \textbf{(15–15)}$$

where I_s is the reverse saturation current of the base–emitter diode and V_T is the thermal voltage, $V_T = q/kT$ (as defined in equation 2–11). When V_{be} is a few tenths of a volt, $e^{V_{be}/V_T} \gg 1$, and (15–15) becomes

$$I_C = I_s e^{V_{be}/V_T}$$

Figure 15–25 shows the basic configuration of a log amplifier, in which a BJT is connected in the feedback path of an inverting operational amplifier. Taking the logarithm of both sides of the above and solving for V_{be}, we find

$$V_{be} = V_t \ln\left(\frac{I_C}{I_s}\right) \qquad \textbf{(15–16)}$$

In Figure 15–25, note that the collector is at virtual ground (≈0 V), so

$$I_C = \frac{V_{in}}{R_1} \qquad \textbf{(15–17)}$$

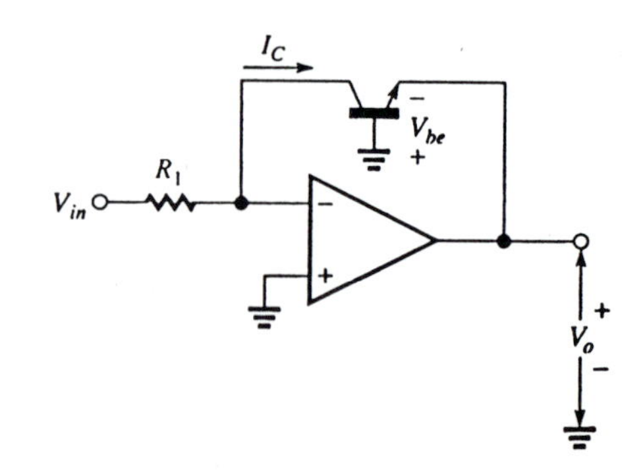

Figure 15–25
Basic configuration of a log amplifier

Also note that $V_{be} = -V_o$. Substituting these relationships in (15–16), we obtain

$$V_o = -V_T \ln\left(\frac{V_{in}}{I_s R_1}\right) \quad \textbf{(15–18)}$$

We see that V_o is a logarithmic function of V_{in}. Since the common logarithm (base 10) is related to the natural logarithm by $\ln x = 2.303\log_{10}x$, equation 15–18 can also be written in terms of $\log_{10}$ as

$$V_o = -2.303\, V_T \log_{10}\left(\frac{V_{in}}{I_s R_1}\right) \quad \textbf{(15–19)}$$

At room temperature, $V_T \approx 0.0257$ V, for which equations 15–18 and 15–19 become

$$V_o = -0.0257 \ln\left(\frac{V_{in}}{I_s R_1}\right) = -0.0592 \log_{10}\left(\frac{V_{in}}{I_s R_1}\right) \quad \textbf{(15–20)}$$

The practical difficulty of the configuration we have described is that the value of I_s cannot usually be predicted accurately and is, in any event, highly sensitive to temperaure variations. To overcome this problem, a practical log amplifier is constructed as shown in Figure 15–26. The two transistors are closely matched in an integrated circuit, so their values of I_s are essentially equal and change equally with temperature. Following the same procedure we used to obtain equation 15–20, we find

$$V_{o1} = -0.0257 \ln\left(\frac{V_{REF}}{I_s R_1}\right) \quad \textbf{(15–21)}$$

and

$$V_{o2} = -0.0257 \ln\left(\frac{V_{in}}{I_s R_1}\right) \quad \textbf{(15–22)}$$

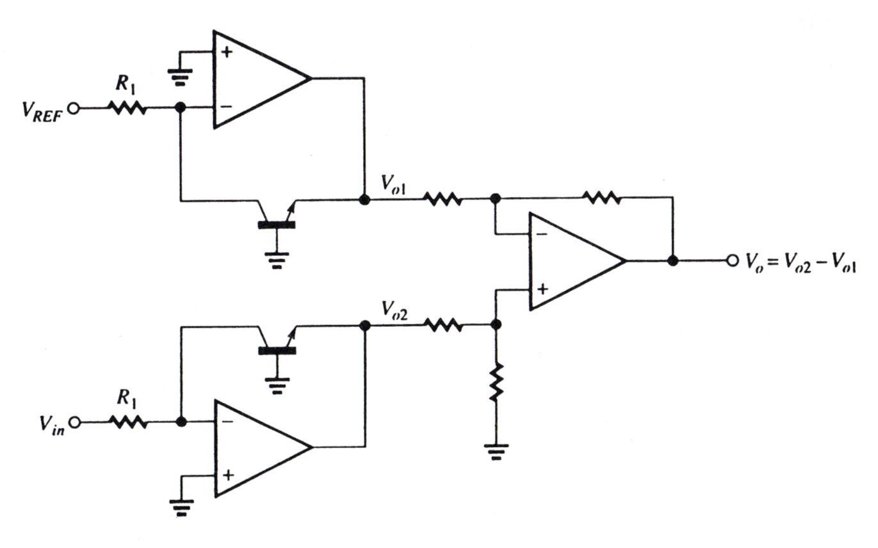

Figure 15–26
A practical log amplifier designed to compensate for the variability of I_s

As discussed in Section 14–1, the output amplifier forms the difference voltage $V_{o2} - V_{o1}$:

$$V_o = V_{o2} - V_{o1} = -0.0257\left(\ln\frac{V_{in}}{I_sR_1} - \ln\frac{V_{REF}}{I_sR_1}\right)$$

$$= -0.0257\ln\left[\frac{(V_{in}/I_sR_1)}{(V_{REF}/I_sR_1)}\right] \quad \textbf{(15–23)}$$

$$= -0.0257\ln\left(\frac{V_{in}}{V_{REF}}\right)$$

Here we see that the output is no longer dependent on the value of I_s. Also, the external voltage V_{REF} can be adjusted to control the overall sensitivity of the amplifier (as can the gain of the output difference amplifier).

Example 15–2

When V_{in} in Figure 15–26 is 1 V, it is desired that $V_o = 50$ mV. What value of V_{REF} should be used? Assume the difference amplifier has unity gain.

Solution. From equation 15–23,

$$50\text{ mV} = -0.0257\ln\left(\frac{1\ V}{V_{REF}}\right)$$

$$\frac{50\times10^{-3}}{-0.0257} = -1.9456 = \ln\left(\frac{1}{V_{REF}}\right)$$

Taking the inverse logarithm of both sides,

$$0.143 = \frac{1}{V_{REF}}$$

$$V_{REF} = 7.0\text{ V}$$

Figure 15–27 shows the basic configuration of an antilogarithmic (inverse log) amplifier. Note that the transistor in this case is connected to the input of an inverting operational amplifier. As before,

$$I_C = I_s e^{V_{be}/V_T} \quad \textbf{(15–24)}$$

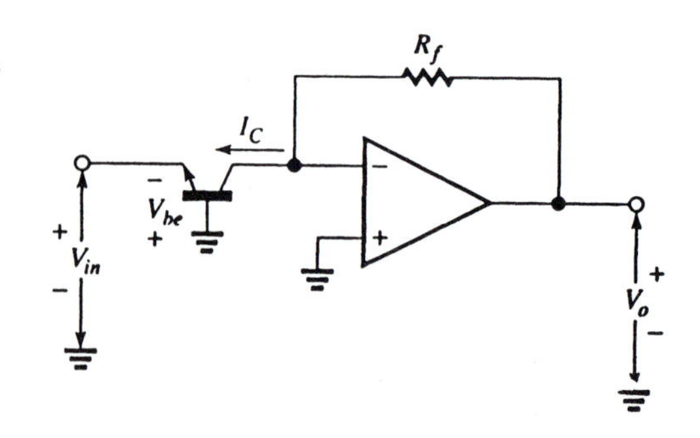

Figure 15–27
Basic configuration of an antilog amplifier

Since I_C flows through R_f and the collector is at virtual ground, we have

$$V_o = R_f I_C = R_f I_s e^{V_{be}/V_T} \qquad \textbf{(15–25)}$$

Since $V_{be} = -V_{in}$, we obtain

$$V_o = R_f I_s e^{V_{in}/V_T} \qquad \textbf{(15–26)}$$

Equation 15–26 is equivalent to

$$V_o = R_f I_s \text{ antilog } (-V_{in}/V_T) \qquad \textbf{(15–27)}$$

Note that values of V_{in} must be negative, so that $-V_{in}/V_T$ is always positive. As in practical log amplifiers, practical antilog amplifiers use two transistors and two operational amplifiers to compensate for the variability of I_s.

Examples of integrated-circuit log and antilog amplifiers are the ICL 8048 (log) and ICL 8049 (antilog) amplifiers, manufactured by Intersil. These devices can be used over a 60-dB (1000 to 1) dynamic voltage range, with output voltages up to 14 V. The 8048 does not incorporate a difference amplifier to compensate for I_s, but arranges two matched transistors and two operational amplifiers in such a way that the output of the second amplifier is proportional to the logarithm of the input difference voltage, which accomplishes the same goal.

15–5 TRANSCONDUCTANCE AMPLIFIERS

A *transconductance amplifier* is basically a voltage-controlled current source. As shown in Figure 15–28, the amplifier typically has a differential input and single-ended output. Recall that transconductance is defined to be output current divided by input voltage. The transconductance of the amplifier in Figure 15–28 is

$$g_m = \frac{i_o}{v_i^+ - v_i^-} \text{ siemens} \qquad \textbf{(15–28)}$$

Since the transconductance amplifier supplies a constant current to a load, the voltage gain of the amplifier depends directly on its load resistance. For example, if $g_m = 10$ mS and $v_i^+ - v_i^- = 50$ mV, then

$$i_o = g_m(v_i^+ - v_i^-) = (10 \times 10^{-3}\text{ S})(50\text{ mV}) = 0.5\text{ mA}$$

If $R_L = 1\text{ k}\Omega$, the voltage gain is

$$\frac{v_L}{v_i^+ - v_i^-} = \frac{i_o R_L}{v_i^+ - v_i^-} = \frac{(0.5\text{ mA})\,1\text{ k}\Omega}{50\text{ mV}} = 10$$

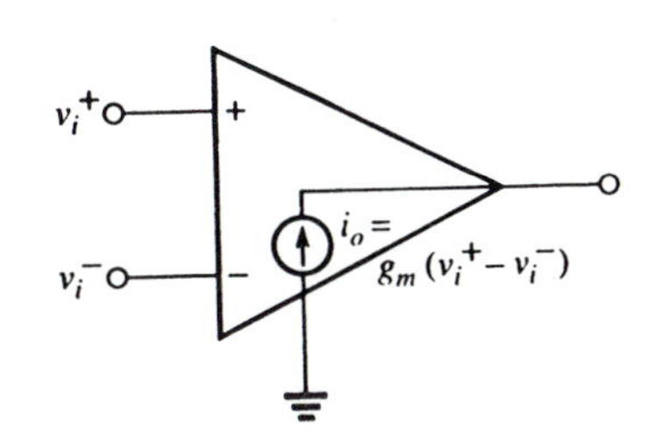

Figure 15–28
The transconductance amplifier represented as a voltage-controlled current source

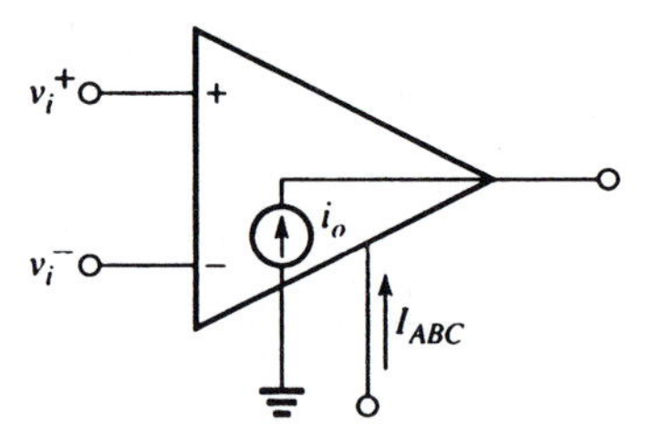

Figure 15–29
A programmable transconductance amplifier; $g_m = kI_{ABC}$

and if R_L is changed to 5 kΩ,

$$\frac{v_L}{v_i^+ - v_i^-} = \frac{(0.5\text{ mA})\, 5\text{ k}\Omega}{50\text{ mV}} = 50$$

An exercise at the end of the chapter shows that the voltage gain of a transconductance amplifier is

$$\frac{v_L}{v_i^+ - v_i^-} = g_m R_L \tag{15–29}$$

A *programmable* transconductance amplifier is one whose value of g_m is determined by the value of an externally supplied control current or voltage. In the programmable transconductance amplifier shown in Figure 15–29, the value of g_m is controlled by the value of *amplifier bias current,* I_{ABC}. A programmable transconductance amplifier can be used to construct a programmable voltage amplifier, that is, an amplifier whose voltage gain can be controlled electronically. The next example illustrates this application.

Example 15–3

The transconductance of the programmable transconductance amplifier in Figure 15–30 depends on I_{ABC} according to $g_m = 20I_{ABC}$.

1. Determine the voltage gain of the amplifier as a function of I_{ABC}.
2. For what value of I_{ABC} will the voltage gain of the amplifier equal 15?

Solution

1.

$$i_o = g_m(v_i^+ - v_i^-) = 20\, I_{ABC}(v_i^+ - v_i^-)$$

$$v_L = i_o R_L = 20\, I_{ABC} R_L (v_i^+ - v_i^-)$$

$$\frac{v_L}{v_i^+ - v_i^-} = 20\, I_{ABC} R_L = 20 \times 10\text{ k}\Omega\, I_{ABC} = 2 \times 10^5\, I_{ABC}$$

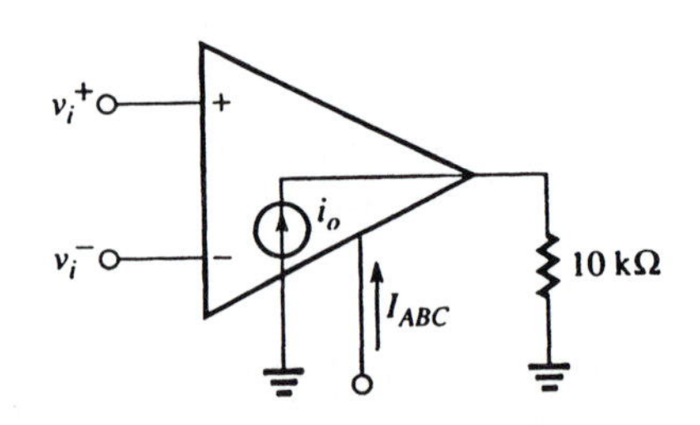

Figure 15–30
(Example 15–3)

2.

$$15 = 2 \times 10^5 \, I_{ABC}$$

$$I_{ABC} = \frac{15}{2 \times 10^5} = 75\mu\text{A}$$

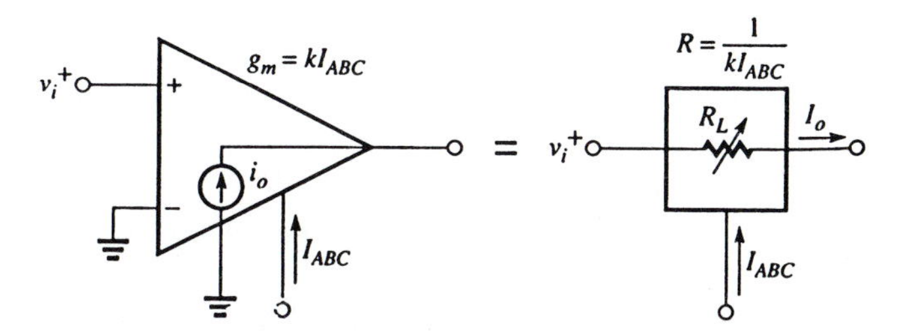

Figure 15–31
Using the programmable transconductance amplifier as a programmable resistor

Another application of a programmable transconductance amplifier is its use as a *programmable resistor:* a resistor whose resistance is determined by an external control voltage or current. Figure 15–31 illustrates this perspective. Suppose that the relationship between I_{ABC} and g_m is

$$g_m = k \, I_{ABC} \tag{15–30}$$

Then, since the inverting input is grounded,

$$i_o = g_m v_i^+ = k \, I_{ABC} \, v_i^+ \tag{15–31}$$

The effective resistance of the amplifier is

$$\frac{v_i^+}{i_o} = \frac{1}{k \, I_{ABC}} \tag{15–32}$$

We see that the programmable resistance is inversely proportional to I_{ABC}.

The 3080 Programmable Transconductance Amplifier

An example of an integrated-circuit programmable transconductance amplifier is the 3080, available from RCA as the CA3080 and National Semiconductor as the LM 3080. Figure 15–32 shows a pin diagram of the amplifier. Manufacturers' specifications show that a typical room-temperature value for g_m is 9.6 mS when I_{ABC} = 500 μA. Thus, a typical value for k (in equation 15–30) is

$$k = \frac{g_m}{I_{ABC}} = \frac{9.6 \text{ mS}}{500 \ \mu\text{A}} = 19.2$$

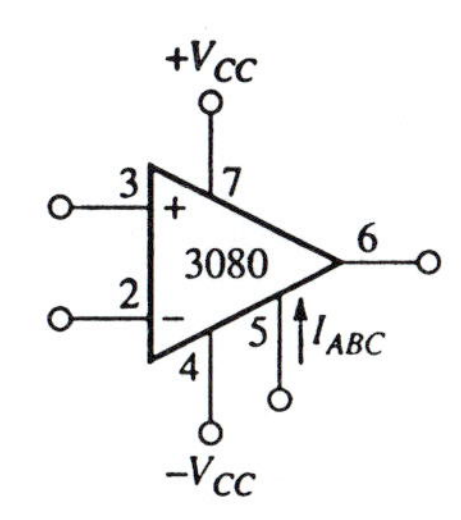

Figure 15–32
Pin diagram for the 3080 programmable transconductance amplifier

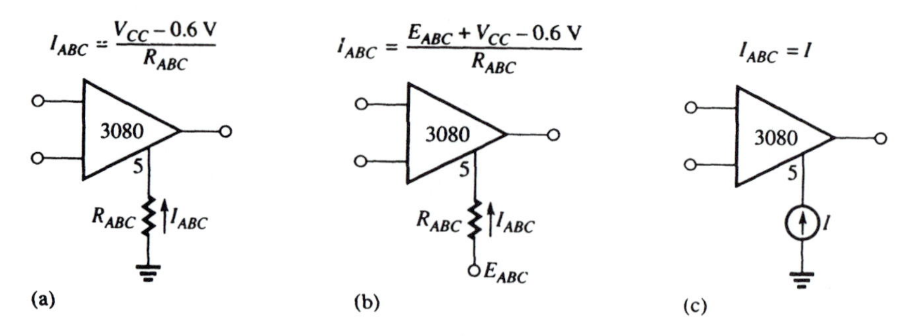

Figure 15–33
Three methods for supplying control current I_{ABC} to the 3080

However, considerable variation is possible, since g_m has a specified range of 6.7 mS to 13 mS when $I_{ABC} = 500\ \mu\text{A}$. I_{ABC} can be supplied using any one of the three methods shown in Figure 15–33. If resistor R_{ABC} is connected between pin 5 and ground, as shown in Figure 15–33(a), then

$$I_{ABC} = \frac{V_{CC} - 0.6\text{ V}}{R_{ABC}} \tag{15–33}$$

I_{ABC} can also be controlled by an external control voltage, E_{ABC}, as shown in Figure 15–33(b). In this case,

$$I_{ABC} = \frac{E_{ABC} + V_{CC} - 0.6\text{ V}}{R_{ABC}} \tag{15–34}$$

With this method of control, the previously described applications of programmable gain and programmable resistance can be implemented using control voltage E_{ABC}. Finally, I_{ABC} can be supplied directly from an external constant-current source, as shown in Figure 15–33(c).

Example 15–4

Find the range of voltage gains of the amplifier in Figure 15–34 when E_{ABC} is adjusted from 5 V to 15 V. Assume k in equation 15–30 equals 19.2.

Solution

$$v_L = i_o R_L = g_m v_i^+ R_L$$

Figure 15–34
(Example 15–4)

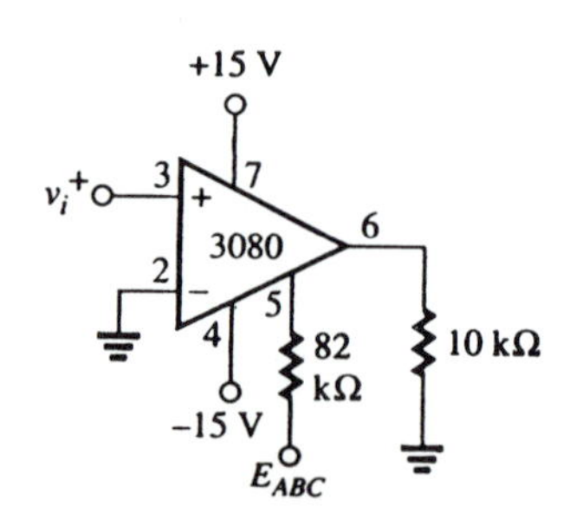

Substituting $g_m = 19.2\ I_{ABC}$,

$$v_L = 19.2\ I_{ABC}\, v_i^+ R_L$$

Substituting I_{ABC} from equation 15–34,

$$v_L = 19.2\left(\frac{E_{ABC} + V_{CC} - 0.6\text{ V}}{R_{ABC}}\right) v_i^+ R_L$$

$$\frac{v_L}{v_i^+} = 19.2\left(\frac{E_{ABC} + 15\text{ V} - 0.6\text{ V}}{82\text{ k}\Omega}\right) 10\text{ k}\Omega$$

$$= 2.34\,(E_{ABC} + 14.4)$$

When $E_{ABC} = 5$ V,

$$\frac{v_L}{v_i^+} = 2.34(5 + 14.4) = 45.4$$

When $E_{ABC} = 15$ V,

$$\frac{v_L}{v_i^-} = 2.34(15 + 14.4) = 68.8$$

15–6 PHASE-LOCKED LOOPS

A *phase-locked loop* (PLL) is a set of components that, through the use of feedback, generate a signal whose frequency tracks that of another externally connected input signal. The term *phase-locked* is derived from the fact that the (apparent) phase difference between two signals of different frequencies is used to control, or maintain, the frequency of the output, as will be discussed in more detail presently. Entire books have been written on the many applications of PLLs, particularly in signal-processing and communications equipment, but we will discuss just two of the most common: FM demodulation and frequency synthesis.

Figure 15–35 shows a block diagram of a phase-locked loop. Note the voltage-controlled oscillator (VCO), which (as will be discussed in Section 15–7) produces a signal whose frequency is proportional to input voltage. The output of the VCO is connected to a *phase comparator,* which generates an output voltage proportional to the difference in phase between the VCO signal and the externally connected input signal, v_{in}. When the frequency of the input signal fluctuates, the output of the phase comparator is a fluctuating voltage that is, in effect, an *error* voltage

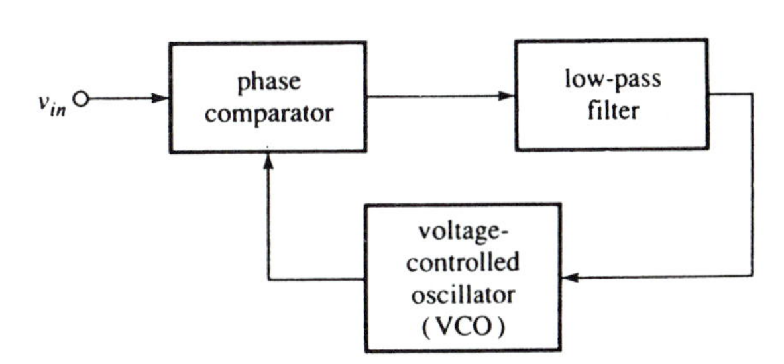

Figure 15–35
Block diagram of a phase-locked loop

Figure 15–36
Apparent phase difference between signals of different frequencies

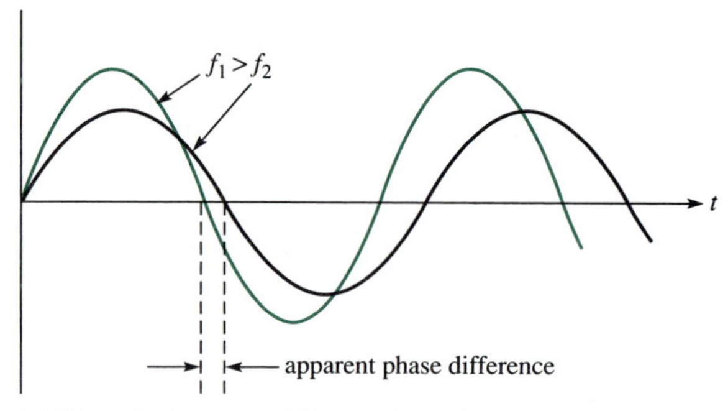

(a) When the frequency difference is small, the apparent phase difference is small.

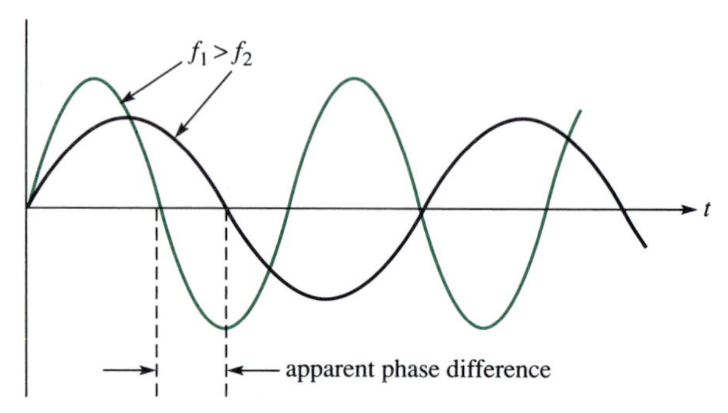

(b) When the frequency difference is larger, the apparent phase difference is larger.

proportional to the difference in *frequency* between the two inputs to the comparator. The fluctuating error voltage is smoothed by the low-pass filter and applied to the input of the VCO. The system is designed so that the error voltage causes the VCO to adjust its frequency to match the frequency of v_{in}. When the VCO frequency and the frequency of v_{in} are equal, the loop is "locked" at that frequency.

Strictly speaking, phase difference is not defined for two signals having different frequencies. However, the phase comparator "sees" an apparent phase difference when the frequency of one of its inputs changes, as illustrated in Figure 15–36. We see that a small frequency difference creates a small (apparent) phase difference and a larger frequency difference creates a larger apparent phase difference. By considering how these diagrams would appear if one frequency were very much larger than the other, we can understand why a practical PLL may not achieve lock in such a situation without some design modifications.

FM Demodulation

Recall that frequency modulation (FM) is the process in which the frequency of one signal is controlled by the magnitude of another signal. A VCO generates an FM signal, as illustrated in Figure 15–37. In this example, the input to the VCO (the *modulating* signal) is a ramp voltage whose amplitude increases with time.

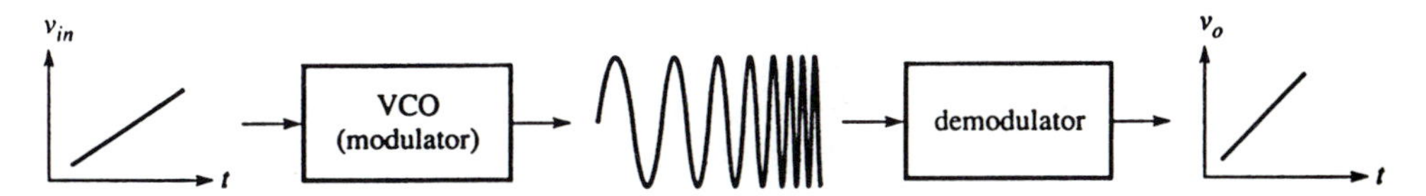

Figure 15–37
Frequency modulation and demodulation

Consequently, the output of the VCO (the *modulated* signal) is a signal whose frequency increases with time. In a communications system, the modulated signal is transmitted to a receiver, where it must be *demodulated*; that is, the original modulating signal must be recovered from the modulated signal. As shown in the example of Figure 15–37, the FM demodulator performs this function. The input to the demodulator is the FM signal and the output is the ramp voltage that created it. An FM demodulator is also called a frequency-to-voltage converter.

A phase-locked loop can serve as an FM demodulator by connecting the FM signal to the input and taking the output from the low-pass filter. As the input frequency changes, the output of the phase comparator and filter changes the same way. For example, an increasing input frequency (originally produced as the result of an increasing modulating signal) will create an increasing error voltage because the frequency of the VCO must be increased to track the input. Thus, the error voltage duplicates the original modulating signal.

Frequency Synthesizers

A frequency synthesizer is simply a signal generator whose frequency can be readily adjusted. A phase-locked loop can be used as a frequency synthesizer by inserting a frequency divider (counter) in the feedback path to the phase comparator and taking the output from the VCO. This arrangement is shown in Figure 15–38. Note that a low-pass filter is not required. The counter is a digital-type device that produces a signal having frequency f/N, where N is an integer, when the input to the counter has frequency f. As can be seen in the figure, a signal with fixed reference frequency, f_{REF}, is compared in the phase comparator to the output of the counter. Consider what happens when this arrangement is first connected: The signal fed back to the phase comparator has a frequency that is initially smaller (by the factor $1/N$) than f_{REF}. Consequently, the phase comparator produces an output that causes the VCO to increase its frequency, in the comparator's usual attempt to bring the two frequencies to equality. The VCO increases its frequency until the signal fed back to the comparator has frequency f_{REF} and lock is achieved. But, for the output

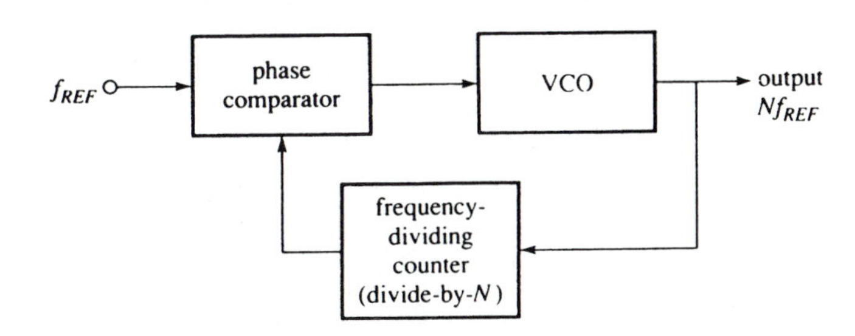

Figure 15–38
Block diagram of a PLL used as a frequency multiplier (synthesizer)

of the counter to have frequency f_{REF}, the VCO must have reached frequency Nf_{REF}, since the counter divides its input frequency by N. Thus, the output of the phase-locked loop (the output of the VCO) is a signal whose frequency is the multiple Nf_{REF} of the reference frequency.

One advantage of a frequency synthesizer using a PLL is that the signal providing the reference frequency can be very stable, as, for example, from a crystal-controlled oscillator. The higher frequency, Nf_{REF}, will then also be very stable. In some high-frequency applications, it is not possible to achieve such crystal-controlled stability at high frequencies without using such a scheme. We should note that a frequency-dividing counter can also be connected between the f_{REF} input and the phase comparator. This connection has just the opposite effect of connecting a divider in the feedback: The frequency of the VCO is *divided* by the same factor that the input divider provides.

The 565 Integrated-Circuit PLL

An example of a commercially available, integrated-circuit PLL is the LM565, manufactured by National Semiconductor. The inputs and outputs of the phase comparator and VCO are accessible at separate pins so that external components, such as filters and frequency dividers, can be inserted as required for different applications. Some important specifications of a PLL, and their values for the LM565, are the maximum VCO frequency (500 kHz), the demodulated output voltage (300 mV for a 10% change in input frequency), and the phase comparator sensitivity (0.68 V/radian).

15–7 The 8038 Integrated-Circuit Function Generator

Function generators capable of producing sinusoidal, triangular, and rectangular waveforms are commercially available in integrated-circuit form. An example is the 8038, manufactured by Intersil as the ICL8038 and also available from other manufacturers. This versatile circuit is capable of generating the aforementioned waveforms simultaneously (at three separate output terminals) over a frequency range from 0.0001 Hz to 1 MHz. Frequency is determined by externally connected resistor-capacitor combinations and can also be controlled by an external voltage. In the latter mode of operation, the generator serves as a *voltage-controlled oscillator* (VCO). (A VCO is also called a *voltage-to-frequency converter*.)

The 8038 is basically a relaxation oscillator that generates a triangular waveform. The triangular waveform is converted internally to a rectangular waveform using voltage comparators and a digital-type storage device (flip-flop). Sixteen internal transistors are used to shape the triangular wave into an approximation of a sine wave. Under ideal conditions (using external circuit connections specified in the manufacturer's literature), the total harmonic distortion of the sine-wave output can be reduced to less than 1%.

Figure 15–39(a) shows a pin diagram of the 8038 and identifies the function of each pin. Pins 1 and 12, "sine wave adjust," are used for connecting external resistors to minimize sine-wave distortion, as previously mentioned. Pins 4 and 5, "duty cycle and frequency adjust," are used for connecting external resistors that, in conjunction with an external capacitor connected to pin 10, "timing capacitor," determine the *duty cycle* and frequency of the outputs. The duty cycle of an output

Figure 15–39
The 8038 function generator

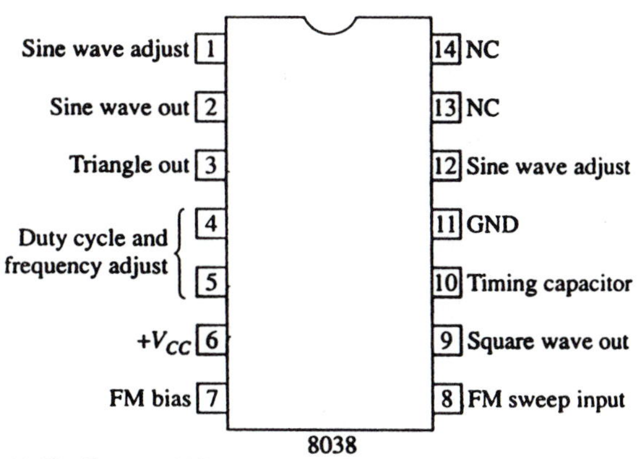

(a) Pin diagram (NC = no connection)

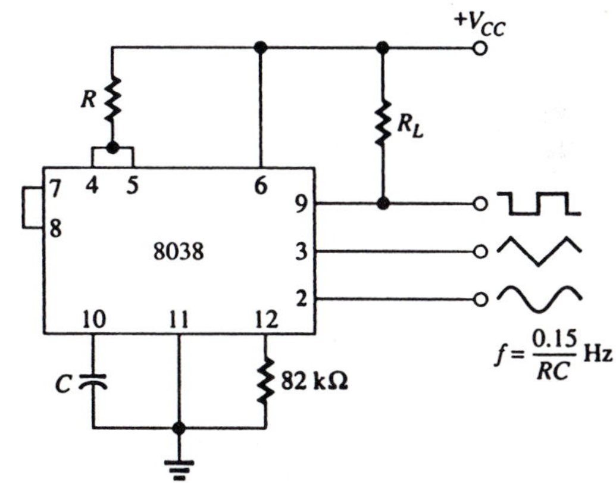

(b) Connection for operation as a 50%–duty-cycle, fixed-frequency function generator

is the time it is high, expressed as a percent of the period of oscillation. For example, a square wave, which is high for one-half of a period and low for the other half, has a 50% duty cycle. Pin 8, "FM sweep input," is the input to which an external voltage may be connected to adjust the frequency of oscillation when the 8038 is operated as a VCO. (FM refers to *frequency modulation*: As the input voltage is adjusted, the output frequency changes in direct proportion to the change in voltage.)

Figure 15–39(b) shows the simplest possible circuit connections for using the 8038 as a fixed-frequency, 50%–duty-cycle function generator. In the configuration shown, the frequency of oscillation is

$$f = \frac{0.15}{RC}\ \text{Hz} \qquad \textbf{(15–35)}$$

Manufacturer's product literature can be consulted for external circuit connections required when the 8038 is used as a VCO and for duty cycles other than 50%.

EXERCISES

Section 15-1
Voltage Comparators

15–1. The maximum output voltages of each of the voltage comparators shown in Figure 15–40 are ±15 V. Sketch the output waveforms for each when v_{in} is a 10-V-pk sine wave. (In each case, show v_{in} and v_o on your sketch, and label voltage levels where switching occurs.)

15–2. Repeat Exercise 15–1 when each of the v^+ and v^- inputs are interchanged.

15–3. The output of the comparator shown in Figure 15–41 switches between +10 V and −10 V.

a. Find the lower and upper trigger levels.
b. Find the hysteresis.
c. Sketch the hysteresis loop.
d. Sketch the output when v_{in} is a 10-V-pk sine wave.

15–4. Repeat Exercise 15–3 when V_{REF} = +4 V, under the assumption that the comparator output switches between +10 V and −5 V.

15–5. Design a Schmitt trigger circuit whose output switches to a high level when the input falls to −1 V and switches to a low level when the input rises to +1 V. Assume that the output switches between ±10 V.

15–6. Derive equations 15–8 and 15–9 for the lower and upper trigger levels of a noninverting Schmitt trigger.

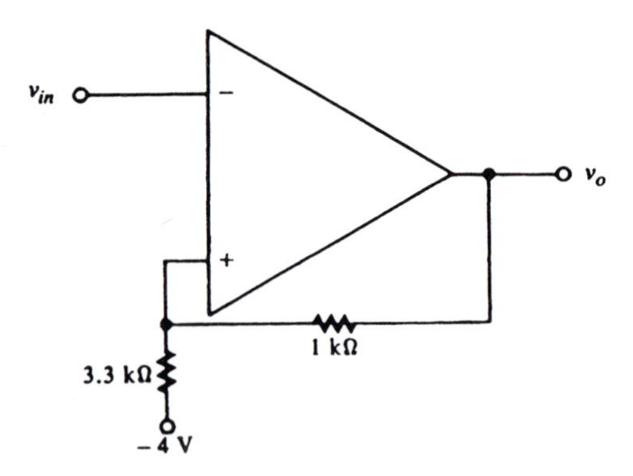

Figure 15–41
(Exercise 15–3)

15–7. Design a Schmitt trigger that switches to a high level when the input rises through +2 V and switches to a low level when the input falls through −1 V. You may choose the high and low output levels of the comparator. Sketch the output when the input is a 5-V-pk sine wave.

15–8. The output of the voltage comparator in Figure 15–42 switches between ±10 V.

a. What maximum and minimum values does the capacitor voltage reach?

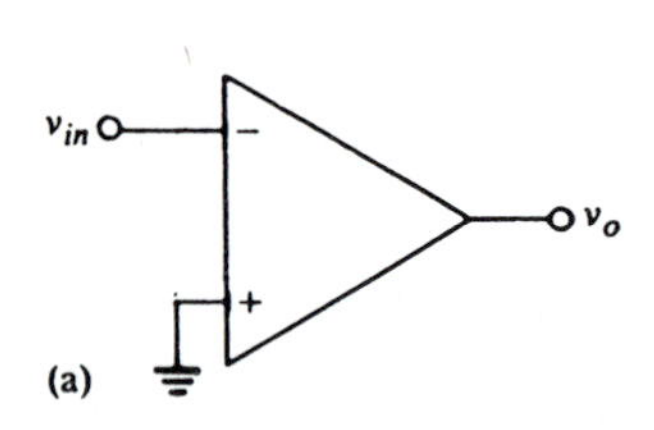

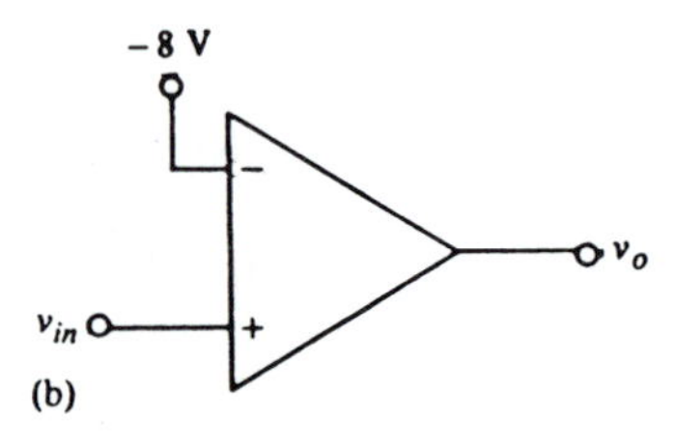

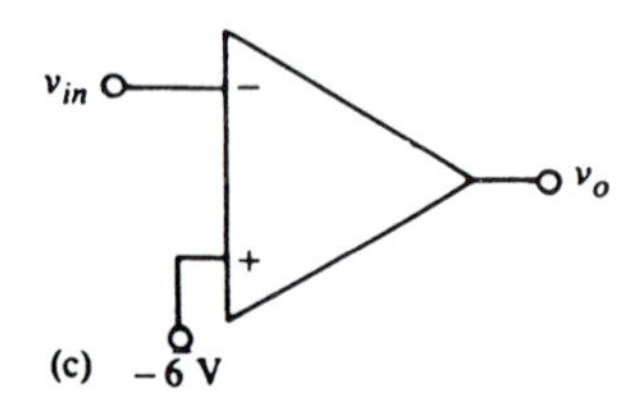

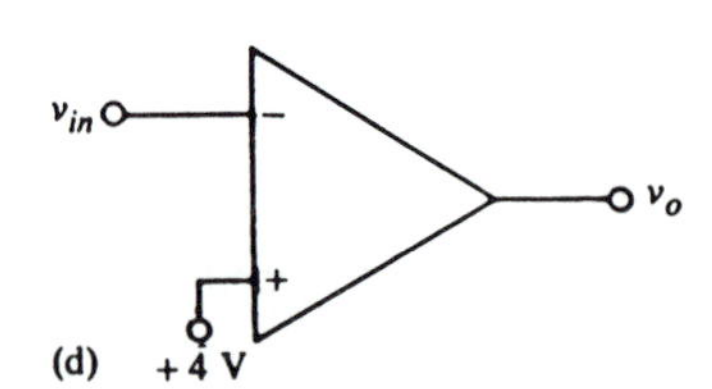

Figure 15–40
(Exercise 15–1)

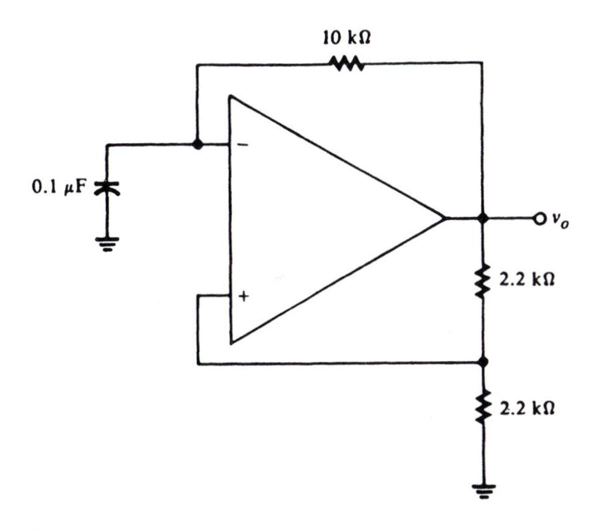

Figure 15–42
(Exercise 15–8)

b. What is the frequency of the output oscillation?

15–9. Design an astable multivibrator that produces a 1-kHz square wave.

Section 15–2
Clipping and Rectifying Circuits

15–10. Assume that the input to each device whose transfer characteristic is shown in Figure 15–43 is a 10-V-pk sine wave. Sketch input and output waveforms for each device. Label voltage values on your sketches.

15–11. Sketch the output waveform for the input shown to each circuit in Figure 15–44. Label voltage values on your sketches. Assume ideal diodes.

15–12. Design an operational-amplifier clipping circuit whose output is that shown in Figure 15–45(b) when its input is as shown in 15–45(a). Sketch the transfer characteristic.

15–13. Design an operational amplifier having the following characteristics:

a. When the input is +0.4 V, the output is −2 V.

b. When the input is a 6-V-pk sine wave, the output is clipped at +10 V and at −5 V.

c. The input resistance is at least 10 kΩ.

Sketch the transfer characteristic.

15–14. Using zener diodes, design an operational-amplifier clipping circuit that clips its output at +8.3 V and −4.8 V. (Specify the zener voltages in your design.) The gain in the linear region should be −4, and the input resistance to the circuit must be at least 25 kΩ. Sketch the transfer characteristic.

15–15. Sketch v_L in the circuit shown in Figure 15–46 when the input is a sine wave that alternates between +10 V peak and −6 V peak. Sketch the transfer characteristic of the circuit.

15–16. Repeat Exercise 15–15 for the circuit shown in Figure 15–47.

Figure 15–43
(Exercise 15–10)

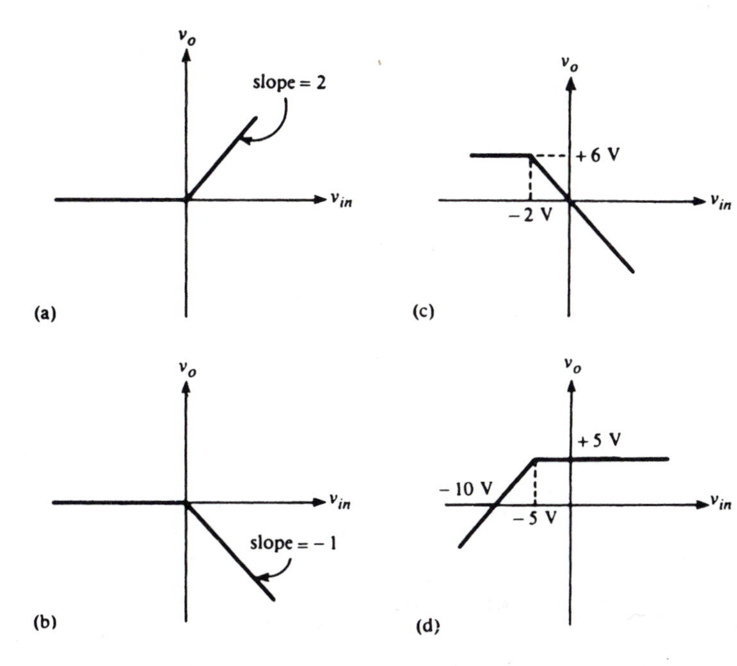

Figure 15–44
(Exercise 15–11)

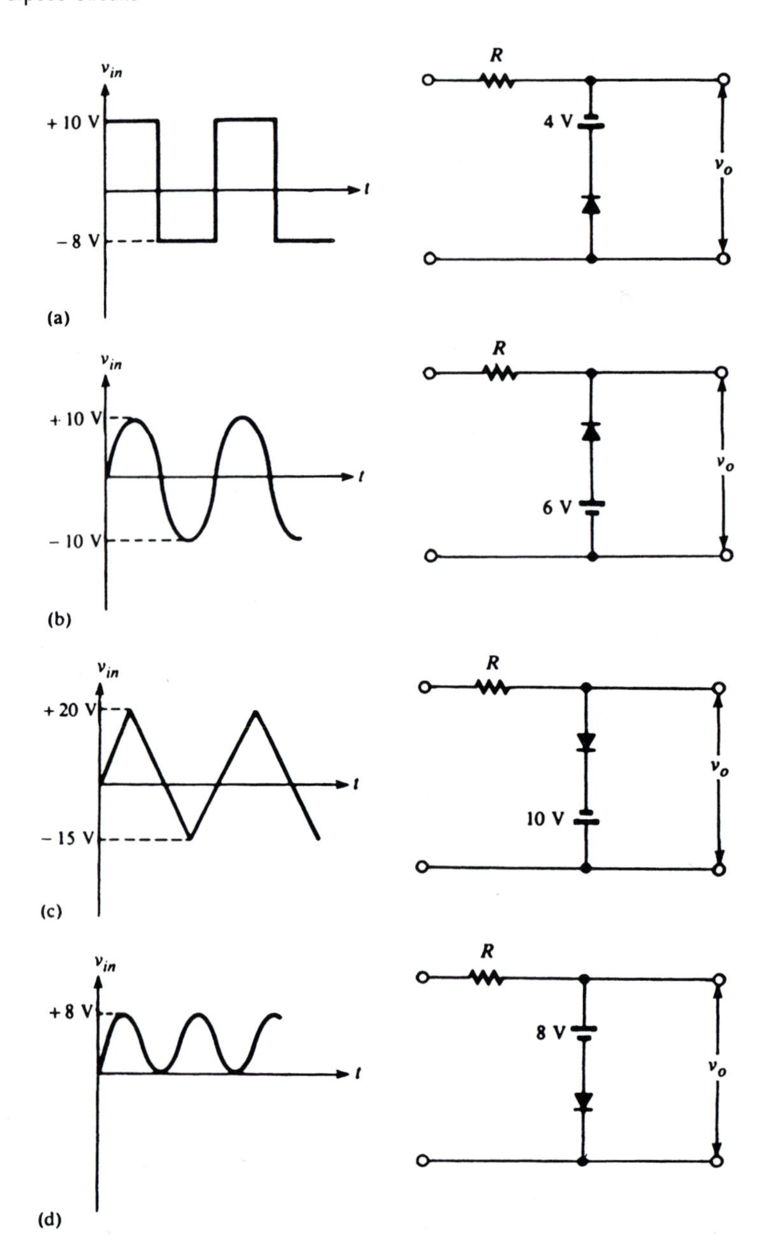

Figure 15–45
(Exercise 15–12)

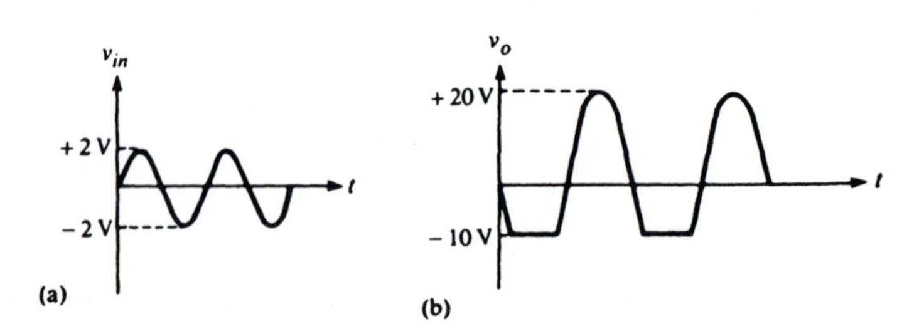

Figure 15–46
(Exercise 15–15)

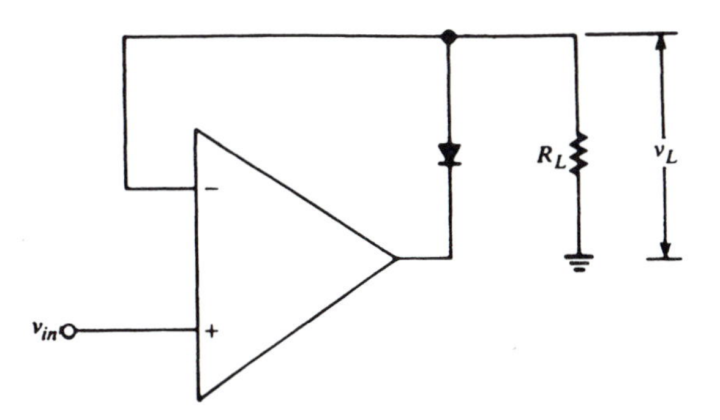

Figure 15–47
(Exercise 15–16)

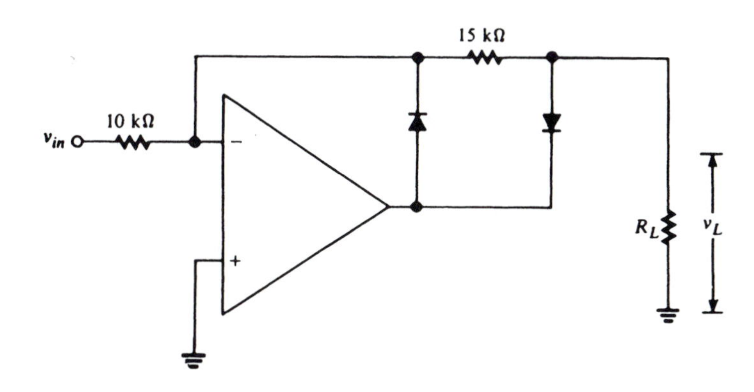

Section 15–3
Clamping Circuits

15–17. Sketch the load-voltage waveforms in each of the circuits shown in Figure 15–48. Assume ideal diodes, and label voltage values on your sketches.

15–18. Sketch the load-voltage waveforms in each of the circuits shown in Figure 15–49. Assume ideal diodes, and label voltage values on your sketches.

15–19. Sketch the load-voltage waveform in the circuit shown in Figure 15–50. Label voltage values on your sketch.

Figure 15–48
(Exercise 15–17)

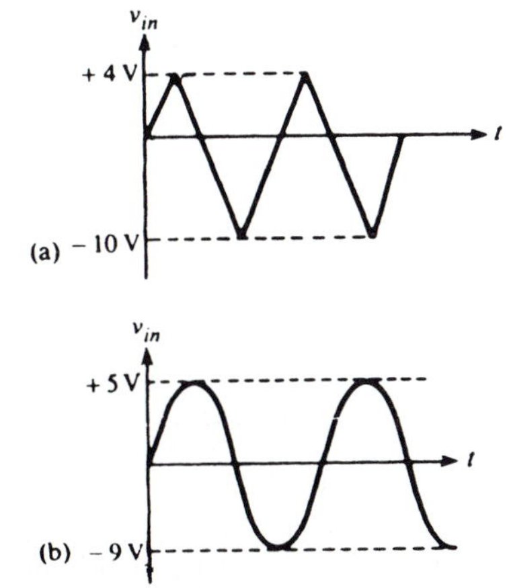

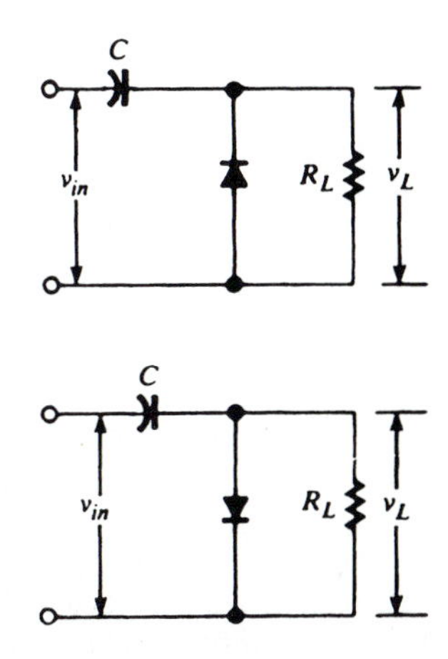

Figure 15–49
(Exercise 15–18)

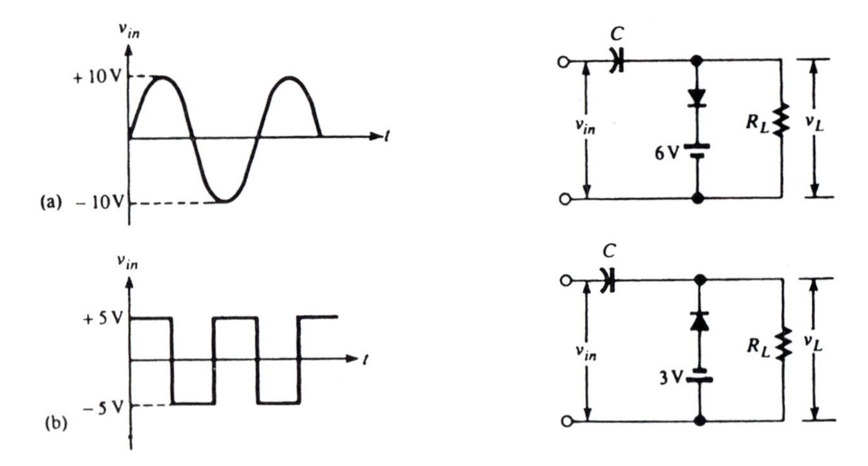

Figure 15–50
(Exercise 15–19)

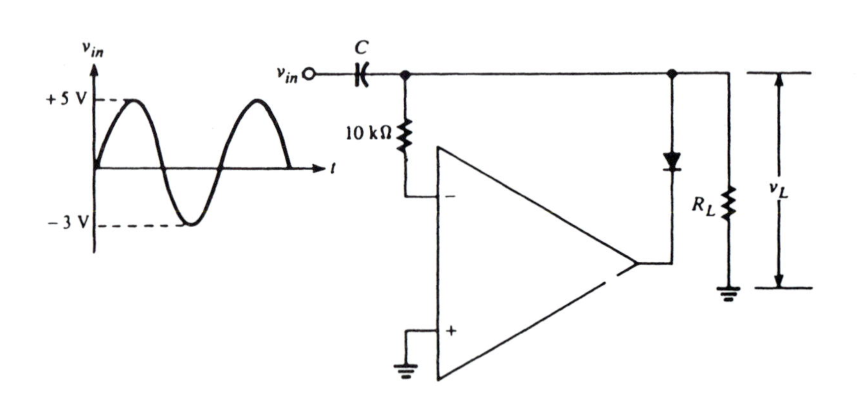

Section 15–4
Logarithmic and Antilogarithmic Amplifiers

15–20. The log amplifier in Figure 15–26 has $V_{REF} = 5$ V. It is necessary for the output voltage to be 0.25 V when the input is 0.5 V. What should be the voltage gain of the output difference amplifier?

15–21. Assuming the output difference amplifier in Figure 15–26 has unit voltage gain, what is the change in the output voltage of the log amplifier when V_{in} doubles in value?

Section 15–5
Transconductance Amplifiers

15–22. The voltage gain of a transconductance amplifier is 32 when its load resistance is 4 kΩ. What is the value of its transconductance?

15–23. Derive equation 15–29 for the voltage gain of a transconductance amplifier.

15–24. A programmable transconductance amplifier has $g_m = 16I_{ABC}$ and $R_L = 22$ kΩ. The manufacturer specifies that the permissible range of I_{ABC} is 10 μA $\leq I_{ABC} \leq$ 400 μA. What range of programmable voltage gains is possible?

15–25. A programmable transconductance amplifier having $g_m = 20I_{ABC}$ is to be used as a programmable resistor having resistance 2 kΩ. What should be the value of I_{ABC}?

15–26. A 3080 transconductance amplifier is programmed as shown in Figure 15–33(a). If $V_{CC} = 15$ V and $g_m = 19.2I_{ABC}$, what should be the value of R_{ABC} to obtain a voltage gain of 20 when $R_L = 4$ kΩ?

Section 15–6
Phase-Locked Loops

15–27. A phase-locked loop used as a frequency synthesizer is to generate an output frequency of 500 kHz. If the counter divides frequency by 10, what should be the input reference frequency to the PLL?

15–28. **a.** What phase difference (in radians) will cause the phase comparator of the 565 PLL to generate an output of 0.5 V?

b. What percent change in the frequency of the input to the 565 PLL will create a demodulated output voltage of 100 mV?

Section 15–7

The 8038 Integrated-Circuit Function Generator

15–29. A 0.1-μF timing capacitor is to be used with an 8038 function generator to provide output frequenies in the range from 200 Hz to 10 kHz. Through what range should external timing resistor R be adjustable?

16 POWER AMPLIFIERS

16–1 DEFINITIONS, APPLICATIONS, AND TYPES OF POWER AMPLIFIERS

As the name implies, a *power amplifier* is designed to deliver a large amount of power to a load. To perform this function, a power amplifier must itself be capable of *dissipating* large amounts of power; that is, it must be designed so that the heat generated when it is operated at high current and voltage levels is released into the surroundings at a rate fast enough to prevent destructive temperature build-up. Consequently, power amplifiers typically contain bulky components having large surface areas to enhance heat transfer to the environment. A *power transistor* is a discrete device with a large surface area and a metal case, characteristics that make it suitable for incorporation into a power amplifier.

A power amplifier is often the last, or *output,* stage of an amplifier system. The preceding stages may be designed to provide voltage amplification, to provide buffering to a high-impedance signal source, or to modify signal characteristics in some predictable way, functions that are collectively referred to as *signal conditioning.* The output of the signal-conditioning stages drives the power amplifier, which in turn drives the load. Some amplifiers are constructed with signal-conditioning stages and the output power stage all in one integrated circuit. Others, especially those designed to deliver very large amounts of power, have a *hybrid* structure, in the sense that the signal-conditioning stages are integrated and the power stage is discrete.

Power amplifiers are widely used in audio components—radio and television receivers, phonographs and tape players, stereo and high-fidelity systems, recording-studio equipment, and public address systems. The load in these applications is most often a loudspeaker ("speaker"), which requires considerable power to convert electrical signals to sound waves. Power amplifiers are also used in electromechanical *control systems* to drive electric motors. Examples include computer disk and tape drives, robotic manipulators, autopilots, antenna rotators, pumps and motorized valves, and manufacturing and process controllers of all kinds.

Large-Signal Operation

Because a power amplifier is required to produce large voltage and current variations in a load, it is designed so that at least one of its semiconductor components, typically a power transistor, can be operated over substantially the *entire* range of its output characteristics, from saturation to cutoff. This mode of operation is called *large-signal operation*. Recall that small-signal operation occurs when the range of current and voltage variation is small enough that there is no appreciable change in device parameters, such as β and r_e. By contrast, the parameters of a large-signal amplifier at one output voltage may be considerably different from those at another output voltage. There are two important consequences of this fact:

1. Signal distortion occurs because of the change in amplifier characteristics with signal level. *Harmonic distortion* always results from such *nonlinear* behavior of an amplifier (see Figure 5–16). Compensating techniques, such as negative feedback, must be incorporated into a power amplifier if low-distortion, high-level outputs are required.
2. Many of the equations we have developed for small-signal analysis of amplifiers are no longer valid. Those equations were based on the assumption that device parameters did not change, contrary to fact in large-signal amplifiers. Rough approximations of large-signal–amplifier behavior can be obtained by using average parameter values and applying small-signal analysis techniques. However, graphical methods are used more frequently in large-signal–amplifier design.

As a final note on terminology, we should mention that the term *large-signal operation* is also applied to devices used in digital switching circuits. In these applications, the output level switches between "high" and "low" (cutoff and saturation), but remains in those states *most of the time*. Power dissipation is therefore not a problem. Either the output voltage or the output current is near 0 when a digital device is in an ON or an OFF state, so power, which is the product of voltage and current, is near 0 except during the short time when the device switches from one state to the other. On the other hand, the variations in the output level of a power amplifier occur in the active region, *between* the two extremes of saturation and cutoff, so a substantial amount of power is dissipated.

16–2 TRANSISTOR POWER DISSIPATION

Recall that power is, by definition, the *rate* at which energy is consumed or dissipated (1 W = 1 J/s). If the rate at which heat energy is dissipated in a device is less than the rate at which it is generated, the temperature of the device must rise. In electronic devices, electrical energy is converted to heat energy at a rate given by $P = VI$ watts, and temperature rises when this heat energy is not removed at a comparable rate. Since semiconductor material is irreversibly damaged when subjected to temperatures beyond a certain limit, temperature is the parameter that ultimately limits the amount of power a semiconductor device can handle.

Transistor manufacturers specify the maximum permissible junction temperature and the maximum permissible power dissipation that a transistor can withstand. As we shall presently see, the maximum permissible power dissipation is specified as a function of *ambient temperature* (temperature of the surroundings), because the rate at which heat can be liberated from the device depends on the temperature of the region to which the heat must be transferred. Most of the conversion of electrical energy to heat energy in a bipolar junction transistor occurs at the junctions. Since power is the product of voltage and current, and since collector and emitter currents are approximately equal, the greatest power dissipation occurs at the junction where the voltage is greatest. In normal transistor operation, the collector–base junction is reverse biased and has, on average, a large voltage across it, while the base–emitter junction has a small forward-biasing voltage. Consequently, most of the heat generated in a transistor is produced at the collector–base junction. The total power dissipated at the junctions is

$$P_d = V_{CB}I_C + V_{BE}I_E \approx (V_{CB} + V_{BE})I_C = V_{CE}I_C \qquad \textbf{(16–1)}$$

Equation 16–1 gives, for all practical purposes, the total power dissipation of the transistor.

For a fixed value of P_d, the graph of equation 16–1 is a *hyperbola* when plotted on $I_C - V_{CE}$ axes. Larger values of P_d correspond to hyperbolas that move outward from the axes, as illustrated in Figure 16–1. Each hyperbola represents all possible combinations of V_{CE} and I_C that give a product equal to the same value of P_d. The figure shows sample coordinate values (I_C, V_{CE}) at points on each hyperbola. Note that the product of the coordinate values is the same for points on the same hyperbola, and that the product equals the power dissipation corresponding to the hyperbola.

Figure 16–2 shows a simple common-emitter amplifier and its dc load line plotted on $I_C - V_{CE}$ axes. Also shown are a set of hyperbolas corresponding to different values of power dissipation. Recall that the load line represents all possible combinations of I_C and V_{CE} corresponding to a particular value of R_C. As the amplifier output changes in response to an input signal, the collector current and voltage undergo variations along the load line and intersect different hyperbolas of power dissipation. It is clear that the power dissipation changes as the amplifier output changes. For safe operation, the load line must lie below and to the left of

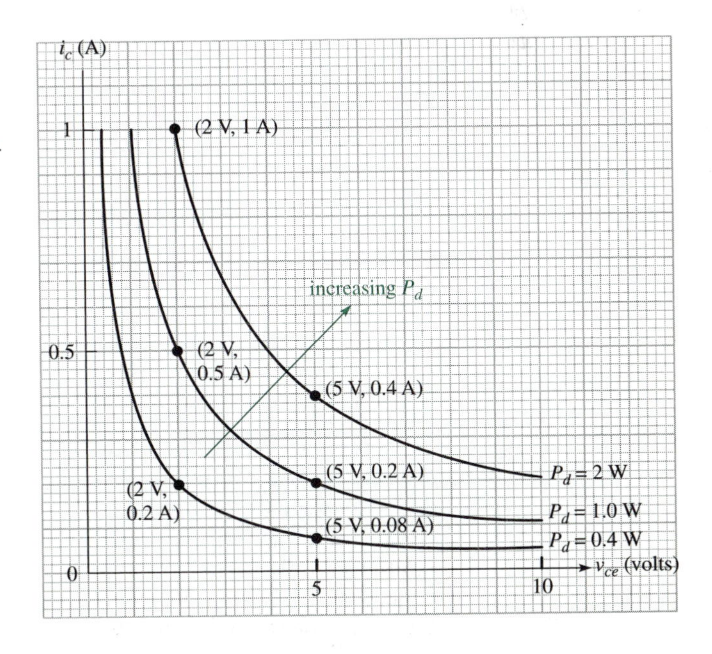

Figure 16–1
A family of graphs (hyperbolas) corresponding to $P_d = V_{CE}I_C$ for different values of P_d. Each hyperbola represents all combinations of collector voltage and collector current that result in a specific power dissipation (value of P_d).

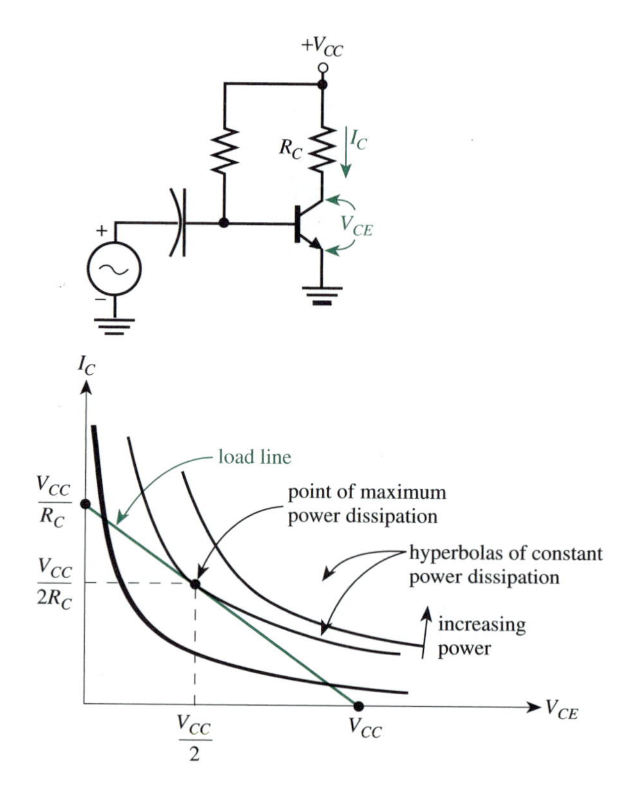

Figure 16–2
Load lines and hyperbolas of constant power dissipation

the hyperbola corresponding to the maximum permissible power dissipation. When the load line meets this requirement, there is no possible combination of collector voltage and current that results in a power dissipation exceeding the rated maximum.

It can be shown that the point of maximum power dissipation occurs at the *center* of the load line, where $V_{CE} = V_{CC}/2$ and $I_C = V_{CC}/2R_C$ (see Figure 16–2). Therefore, the maximum power dissipation is

$$P_d(\max) = \left(\frac{V_{CC}}{2}\right)\left(\frac{V_{CC}}{2R_C}\right) = \frac{V_{CC}^2}{4R_C} \quad \textbf{(16–2)}$$

To ensure that the load line lies below the hyperbola of maximum dissipation, we therefore require that

$$\frac{V_{CC}^2}{4R_C} < P_d(\max) \quad \textbf{(16–3)}$$

where $P_d(\max)$ is the manufacturer's specified maximum dissipation at a specified ambient temperature. Inequality 16–3 enables us to find the maximum permissible value of R_C for a given V_{CC} and a given $P_d(\max)$:

$$R_C > \frac{V_{CC}^2}{4P_d(\max)} \quad \textbf{(16–4)}$$

Example 16–1

The amplifier shown in Figure 16–2 is to be operated with $V_{CC} = 20$ V and $R_C = 1$ kΩ.

1. What maximum power dissipation rating should the transistor have?
2. If an increase in ambient temperature reduces the maximum rating found in (1) by a factor of 2, what new value of R_C should be used to ensure safe operation?

Solution

1. From inequality 16–3,

$$P_d(\max) > \frac{V_{CC}^2}{4R_C} = \frac{(20\ \text{V})^2}{4(10^3\ \Omega)} = 0.1\ \text{W}$$

2. When the dissipation rating is decreased to 0.05 W, the minimum permissible value of R_C is, from (16.4),

$$R_C > \frac{V_{CC}^2}{4P_d(\max)} = \frac{(20\ \text{V})^2}{4(0.05\ \text{W})} = 2\ \text{k}\Omega$$

16–3 HEAT TRANSFER IN SEMICONDUCTOR DEVICES

Conduction, Radiation, and Convection

Heat transfer takes place by one or more of three fundamental mechanisms: *conduction, radiation,* and *convection,* as illustrated in Figure 16–3. Conduction occurs when the energy that the atoms of a material acquire through heating is transferred to adjacent, less energetic atoms in a cooler region of the material. For example, conduction is the process by which heat flows from the heated end of a metal rod toward the cooler end, and the process by which heat is transferred from the heating

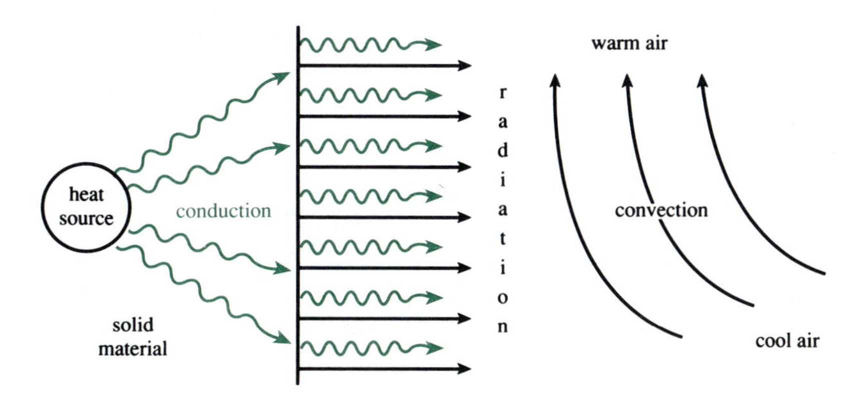

Figure 16–3
Examples illustrating heat transfer by conduction, radiation, and convection

element of a stove to a pan, and from the pan to the water it contains. Heat radiation is similar to other forms of radiation because no physical medium is required between the heat source and its destination. For example, heat is transferred by radiation through space from the sun to the earth. Convection occurs when a physical medium that has been heated by conduction or radiation *moves* away from the source of heat, i.e., is displaced by a cooler medium, so more conduction or radiation can occur. Examples include *forced* convection, where a fan is used to push air past a heated surface, and *natural* convection, where heated air rises and is replaced by cooler air from below.

Semiconductor devices are cooled by all three heat-transfer mechanisms. Heat is conducted from a junction, through the semiconductor material, through the case, and into the surrounding air. It is radiated from the surface of the case. Heating of the surrounding air creates air flow around the device, so convection cooling occurs. Various measures are taken to enhance heat transfer by each mechanism. To improve conduction, good physical contact is made between a junction and a device's metal case, or enclosure. In fact, in many power transistors the collector is in direct contact with the case, so the two are electrically the same as well as being at approximately the same temperature. To improve the conduction and radiation of heat from the case to the surrounding air, power devices are often equipped with *heat sinks.* These are attached to the device to be cooled and conduct heat outward to metal *fins* that increase the total surface area from which conduction and radiation into the air can take place. See Figure 16–4. Finally, convection cooling is enhanced by the use of fans that blow cool air past the surface of the case and/or the heat sink.

Thermal Resistance

Heat flow by conduction is very much like the conduction of electrical charge, i.e., current. Recall that current is the *rate* of flow of charge and is proportional to the

Figure 16–4
Typical heat sinks (courtesy of EG & G Wakefield Engineering)

difference in voltage across a resistance. Similarly, power is the rate of flow of heat energy and is proportional to the difference in *temperature* across the region through which heat is conducted. We can regard any impediment to heat flow as *thermal resistance,* θ. Thus, the rate of flow of heat (i.e., power) is directly proportional to temperature difference and inversely proportional to thermal resistance:

$$P = \frac{\Delta T}{\theta} = \frac{T_2 - T_1}{\theta} \text{ joules/second, or watts} \tag{16–5}$$

Insulating materials, like wool, have large thermal resistances and metals have small thermal resistances. Note the similarity of equation 16–5 to Ohm's law: $I = (V_2 - V_1)/R$. Solving for θ in (16–5), we obtain the following equivalent relation that allows us to determine the units of thermal resistance:

$$\theta = \frac{T_2 - T_1}{P} \text{ degrees/watt} \tag{16–6}$$

In a power amplifier, heat is transferred through different types of materials and across boundaries of dissimilar materials, each of which presents different values of thermal resistance. These are usually treated like series electrical circuits, so thermal resistances are added when computing the total heat flow through the system or the temperatures at various points in the system.

Example 16–2

The collector–base junction of a certain transistor dissipates 2 W. The thermal resistance from junction to case is 8°C/W, and the thermal resistance from case to air is 20°C/W. The free-air temperature (ambient temperature) is 25°C.

1. What is the junction temperature?
2. What is the case temperature?

Solution

1. The total thermal resistance between junction and ambient is the sum of that from junction to case and that from case to ambient: $\theta_T = \theta_{JC} + \theta_{CA} = 8°C/W + 20°C/W = 28°C/W$. Therefore, by equation 16–5, with $T_2 = T_J$ = junction temperature and $T_1 = T_A$ = ambient temperature,

$$T_J - T_A = P\theta_T$$
$$T_J = P\theta_T + T_A = (2\text{ W})(28°\text{C/W}) + 25°\text{C} = 81°\text{C}$$

2. The rate of heat flow, P, is constant through the series configuration, so

$$T_J - T_C = P\theta_{JC}$$
$$T_C = T_J - P\theta_{JC} = 81°\text{C} - (2\text{ W})(8°\text{C/W}) = 65°\text{C}$$

To reduce thermal resistance and improve heat conductivity, a special silicone grease is often used between contact surfaces, such as between the case and heat sink. Mica washers are used to isolate the case and heat sink *electrically* when the case is electrically common to one of the device terminals, such as a collector. The washer creates additional thermal resistance in the heat flow path between the case and heat sink and may have to be taken into account in heat-flow computations.

The values of the thermal resistance of various types of heat sinks, washers, and metals can be determined from published manufacturers' data. The data are often given in terms of thermal *resistivity* ρ (°C · in./W or °C · m/W), and thermal resistance is computed by

$$\theta = \frac{\rho t}{A} \text{ degrees Celsius/watt} \tag{16–7}$$

where

$$\rho = \text{resistivity of material}$$
$$t = \text{thickness of material}$$
$$A = \text{area of material}$$

For example, a typical value of ρ for mica is 66°C · in/W, so a 0.002-in.-thick mica washer having a surface area of 0.6 in.^{2} will have a thermal resistance of

$$\theta = \frac{(66°\text{C} \cdot \text{in./W})(2 \times 10^{-3} \text{ in.})}{0.6 \text{ in.}^2} = 0.22°\text{C/W}$$

Example 16–3

The maximum permissible junction temperature of a certain power transistor is 150°C. It is desired to operate the transistor with a power dissipation of 15 W in an ambient temperature of 40°C. The thermal resistances are as follows:

$$\theta_{\text{JC}} = 0.5°\text{C/W} \quad \text{(junction to case)}$$
$$\theta_{\text{CA}} = 10°\text{C/W} \quad \text{(case to ambient)}$$

1. Determine whether a heat sink is required for this application.
2. If a heat sink is required, determine the maximum thermal resistance it can have. Assume that a mica washer having thermal resistance $\theta_{\text{W}} = 0.5°\text{C/W}$ must be used between case and heat sink.

Solution

1.
$$\theta_{\text{T}} = \theta_{\text{JC}} + \theta_{\text{CA}} = 0.5°\text{C/W} + 10°\text{C/W} = 10.5°\text{C/W}$$
$$T_J - T_A = \theta_{\text{T}} P = (10.5°\text{C/W})(15 \text{ W}) = 157.5°\text{C}$$
$$T_J = 157.5°\text{C} + T_A = 157.5°\text{C} + 40°\text{C} = 197.5°\text{C}$$

Since the junction temperature, 197.5°C, will exceed the maximum permissible value of 150°C under the given conditions, a heat sink must be used to reduce the total thermal resistance.

2. Setting T_J equal to its maximum permissible value of 150°C, we can solve for the maximum total thermal resistance:

$$T_J - T_A = \theta_{\text{T}} P$$
$$150 - 40 = (\theta_{\text{T}})(15)$$
$$\theta_{\text{T}} = \frac{110°\text{C}}{15 \text{ W}} = 7.33°\text{C/W}$$

The thermal resistance of the heat sink, θ_H, will replace the thermal resistance θ_{CA} between case and ambient that was present in (1). Thus, taking into account the thermal resistance of the washer, we have $\theta_{\text{T}} = 7.33°\text{C/W} = \theta_{\text{JC}} + \theta_{\text{W}} + \theta_{\text{H}} = 0.5°\text{C/W} + 0.5°\text{C/W} + \theta_{\text{H}}$, or $\theta_{\text{H}} = (7.33 - 1.0)°\text{C/W} = 6.33°\text{C/W}$.

Derating

The maximum permissible power dissipation of a semiconductor device is specified by the manufacturer at a certain temperature, either case temperature or ambient temperature. For example, the maximum power dissipation of a power transistor may be given as 10 W at 25°C ambient, and that of another device as 20 W at 50°C case temperature. These ratings mean that each device can dissipate the specified power at temperatures up to and including the given temperature, but not at greater temperatures. The maximum permissible dissipation *decreases* as a function of temperature when temperature increases beyond the given value. The decrease in permissible power dissipation at elevated temperatures is called *derating,* and the manufacturer specifies a *derating factor,* in W/°C, that is used to find the *decrease* in dissipation rating beyond a certain temperature.

To illustrate derating, suppose a certain device has a maximum rated dissipation of 20 W at 25°C ambient and a derating factor of 100 mW/°C at temperatures above 25°C. Then the maximum permissible dissipation at 100°C is

$$\begin{aligned} P_d(\text{max}) &= (\text{rated dissipation at } T_1) - (T_2 - T_1)(\text{derating factor}) \qquad \textbf{(16–8)} \\ &= (20\text{ W}) - (100°\text{C} - 25°\text{C})(100\text{ mW/°C}) \\ &= (20\text{ W}) - (75°\text{C})(100\text{ mW/°C}) = (20\text{ W}) - (7.5\text{ W}) = 12.5\text{ W} \end{aligned}$$

Example 16–4

A certain semiconductor device has a maximum rated dissipation of 5 W at 50°C case temperature and must be derated above 50°C case temperature. The thermal resistance from case to ambient is 5°C/W.

1. Can the device be operated at 5 W of dissipation without auxiliary cooling (heat sink or fan) when the ambient temperature is 40°C?
2. If not, what is the maximum permissible dissipation, with no auxiliary cooling, at 40°C ambient?
3. What derating factor should be applied to this device, in watts per ambient degree?

Solution

1. We determine the case temperature when the ambient temperature is 40°C.

$$\begin{aligned} T_C - T_A &= \theta_{CA}P \\ T_C - 40°\text{C} &= (5°\text{C/W})(5\text{ W}) \\ T_C &= 40 + 25 = 65°\text{C} \end{aligned}$$

   Since the case temperature exceeds 50°C, the full 5 W of power dissipation is not permitted when the ambient temperature is 40°C.

2. The derated dissipation at $T_A = 40°\text{C}$ is

$$P_d(\text{max}) = (T_C - T_A)/\theta = \frac{(50 - 40)°\text{C}}{5°\text{C/W}} = 2\text{ W}$$

3. We must first find the maximum ambient temperature at which the device can dissipate 5 W when the case temperature is 50°C:

$$5\text{ W} = \frac{(50°\text{C} - T_A)}{5°\text{C/W}}$$

$$T_A = 50 - 25 = 25°\text{C}$$

Thus, derating will be necessary when the ambient temperature exceeds 25°C. Let D be the derating factor. Then $P_d(\text{max}) = (5\text{ W}) - (T_A - 25)D$. Substituting the results of (2), we find

$$2\text{ W} = (5\text{ W}) - (40 - 25)D$$
$$D = 0.2\text{ W/°C}$$

Power Dissipation in Integrated Circuits

All the heat transfer concepts we have described in connection with power devices apply equally to integrated circuits, including derating. Of course, a maximum junction temperature is not specified for an integrated circuit, but a maximum device temperature or case temperature is usually given. Many integrated circuits are available in a variety of case types, and the rated power dissipation may depend on the case used. For example, the specifications given in Chapter 13 for the μA741 operational amplifier show that the maximum dissipation is 500 mW for the metal package, 310 mW for the DIP, and 570 mW for the flatpak. Note that these specifications require the use of derating factors above 70°C ambient, and that the factors differ, depending on case type.

The total power dissipation of an integrated circuit can be calculated by measuring the current drawn from each supply voltage used in a particular application and forming the product of each current with the respective voltage. For example, if an operational amplifier employing ±15-V-dc power supplies draws 10 mA from the positive supply and 8 mA from the negative supply, the total dissipation is $P_d = (15\text{ V})(10\text{ mA}) + (15\text{ V})(8\text{ mA}) = 270\text{ mW}$.

16–4 AMPLIFIER CLASSES AND EFFICIENCY

Class-A Amplifiers

All the small-signal amplifiers we have studied in this book have been designed so that output voltage can vary in response to both positive and negative inputs; that is, the amplifiers are *biased* so that under normal operation the output never saturates or cuts off. An amplifier that has that property is called a *class-A* amplifier. More precisely, an amplifier is class A if its output remains in the active region during a complete cycle (one full period) of a sine-wave input signal.

Figure 16–5 shows a typical class-A amplifier and its input and output waveforms. In this case, the transistor is biased at $V_{CE} = V_{CC}/2$, which is midway between saturation and cutoff, and which permits maximum output voltage swing. Note that the output can vary through (approximately) a full V_{CC} volts, peak-to-peak. The output is in the transistor's active region during a full cycle (360°) of the input sine wave.

Efficiency

The efficiency of a power amplifier is defined to be

$$\eta = \frac{\text{average signal power delivered to load}}{\text{average power drawn from dc source(s)}} \quad (16\text{–}9)$$

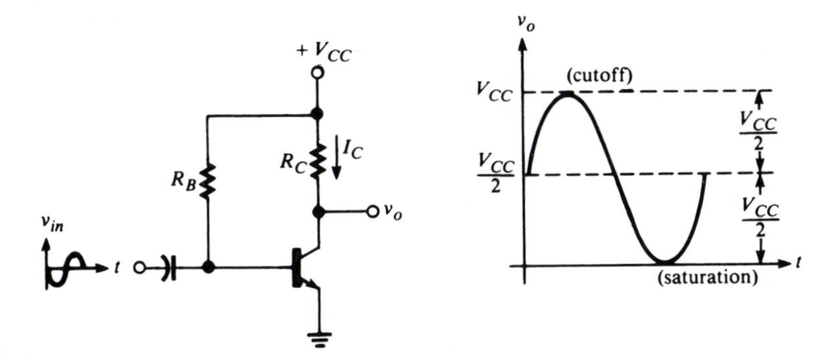

Figure 16–5
The output of a class-A amplifier remains in the active region during a full period (360°) of the input sine wave. In this example, the transistor output is biased midway between saturation and cutoff.

Note that the numerator of (16–9) is average *signal* power, that is, average *ac* power, excluding any dc or bias components in the load. Recall that when voltages and currents are sinusoidal, average ac power can be calculated using any of the following relations:

$$P = V_{rms}I_{rms} = \frac{V_P I_P}{2} = \frac{V_{PP}I_{PP}}{8} \tag{16–10}$$

$$P = I_{rms}^2 R = \frac{I_P^2 R}{2} = \frac{I_{PP}^2 R}{8} \tag{16–11}$$

$$P = \frac{V_{rms}^2}{R} = \frac{V_P^2}{2R} = \frac{V_{PP}^2}{8R} \tag{16–12}$$

where the subscripts P and PP refer to peak and peak-to-peak, respectively.

As a consequence of the definition of efficiency (equation 16–9), the efficiency of a class-A amplifier is 0 when no signal is present. (The amplifier is said to be in *standby* when no signal is applied to its input.) We will now derive a general expression for the efficiency of the class-A amplifier shown in Figure 16–5. In doing so, we will not consider the small power consumed in the base-biasing circuit, i.e., the power at the input side: $I_B^2 R_B + v_{be}i_b$. Figure 16–6 shows the voltages and currents used in our analysis. Notice that resistance R is considered to be the load. We will refer to this configuration as a *series-fed* class-A amplifier, and we will consider capacitor- and transformer-coupled loads in a later discussion.

The instantaneous power from the dc supply is

$$p_S(t) = V_{CC}i = V_{CC}(I_Q + I_P \sin \omega t) = V_{CC}I_Q + V_{CC}I_P \sin \omega t \tag{16–13}$$

Since the average value of the sine term is 0, the average power from the dc supply is

$$P_S = V_{CC}I_Q \text{ watts} \tag{16–14}$$

The average *signal* power in load resistor R is, from equation 16–11,

$$P_R = \frac{I_P^2 R}{2} \text{ watts} \tag{16–15}$$

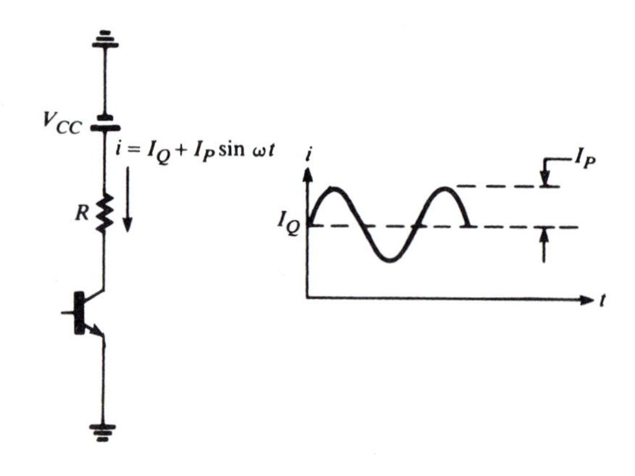

Figure 16–6
Voltages and currents used in the derivation of an expression for the efficiency of a series-fed, class-A amplifier

Therefore, by equation 16-9,

$$\eta = \frac{P_R}{P_S} = \frac{I_P^2 R}{2V_{CC}I_Q} \tag{16–16}$$

We see again that the efficiency is 0 under no-signal conditions ($I_P = 0$) and that efficiency rises as the peak signal level I_P increases. The maximum possible efficiency occurs when I_P has its maximum possible value without distortion. When the bias point is at the center of the load line, as shown in Figure 16–2, the quiescent current is one-half the saturation current, and the output current can swing through the full range from 0 to V_{CC}/R amps without distorting (clipping). Thus, the maximum undistorted peak current is also one-half the saturation current:

$$I_Q = I_P(\text{max}) = \frac{V_{CC}}{2R} \tag{16–17}$$

Substituting (16–17) into (16–16), we find the maximum possible efficiency of the series-fed, class-A amplifier:

$$\eta(\text{max}) = \frac{(V_{CC}/2R)^2 R}{2V_{CC}(V_{CC}/2R)} = 0.25 \tag{16–18}$$

This result shows that the best possible efficiency of a series-fed, class-A amplifier is undesirably small: only ¼ of the total power consumed by the circuit is delivered to the load, under optimum conditions. For that reason, this type of amplifier is not widely used in heavy power applications. The principal advantage of the class-A amplifier is that it generally produces less signal distortion than some of the other, more efficient classes that we will consider later.

Another type of efficiency used to characterize power amplifiers relates signal power to total power dissipated at the collector. Called *collector efficiency,* its practical significance stems from the fact that a major part of the cost and bulk of a power amplifier is invested in the output device itself and the means used to cool it. Therefore, it is desirable to maximize the ratio of signal power in the load to power consumed by the device. Collector efficiency η_c is defined by

$$\eta_c = \frac{\text{average signal power delivered to load}}{\text{average power dissipated at collector}} \tag{16–19}$$

The average power P_C dissipated at the collector of the class-A amplifier in Figure 16–6 is the product of the dc (quiescent) voltage and current:

$$P_C = V_Q I_Q = (V_{CC} - I_Q R) I_Q \tag{16–20}$$

Therefore,

$$\eta_c = \frac{I_P^2 R/2}{(V_{CC} - I_Q R) I_Q} \tag{16–21}$$

The maximum value of η_c occurs when I_P is maximum, $I_P = I_Q = V_{CC}/2R$, as previously discussed. Substituting these values into (16.21) gives

$$\eta_c(\max) = \frac{(V_{CC}/2R)^2 R/2}{(V_{CC} - V_{CC}/2) V_{CC}/2R} = 0.5 \tag{16–22}$$

Figure 16–7 shows the output side of a class-A amplifier with capacitor-coupled load R_L. Also shown are the dc and ac load lines that result. In this case, the average power delivered to the load is

$$P_L = I_{PL}^2 R_L/2 \tag{16–23}$$

where I_{PL} is the peak ac load current. The average power from the dc source is computed in the same way as for the series-fed amplifier: $P_S = V_{CC} I_Q$, so the efficiency is

$$\eta = \frac{I_{PL}^2 R_L}{2 V_{CC} I_Q} \tag{16–24}$$

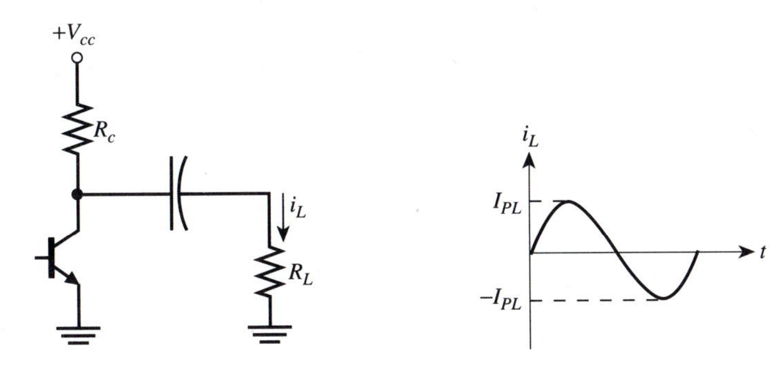

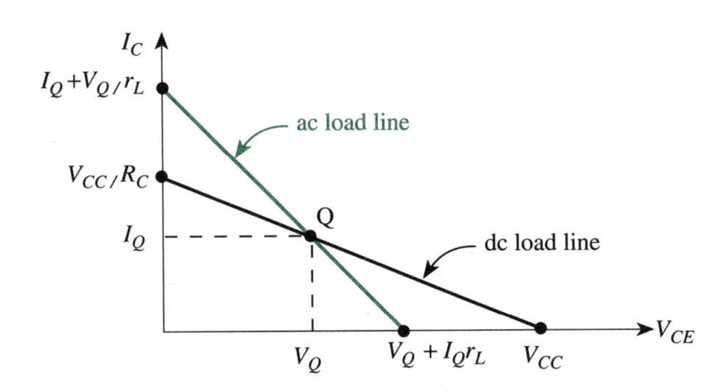

Figure 16–7
A class-A amplifier with capacitor-coupled load R_L

As in the case of the series-fed amplifier, the efficiency is 0 under no-signal (standby) conditions and increases with load current I_{PL}.

Recall that maximum output swing can be achieved by setting the Q-point in the center of the ac load line at

$$I_Q = \frac{V_{CC}}{R_C + r_L} \tag{16–25}$$

The peak collector current under those circumstances is $V_{CC}/(R_C + r_L)$. Neglecting the transistor output resistance, the portion of the collector current that flows in R_L is, by the current-divider rule,

$$I_{PL} = \left(\frac{V_{CC}}{R_C + r_L}\right)\left(\frac{R_C}{R_C + R_L}\right) \tag{16–26}$$

The average ac power in the load resistance R_L is then

$$P_L = \frac{I_{PL}^2 R_L}{2} = \left[\left(\frac{V_{CC}}{R_C + r_L}\right)\left(\frac{R_C}{R_C + R_L}\right)\right]^2 (R_L/2) \tag{16–27}$$

The average power supplied from the dc source is

$$P_S = V_{CC} I_Q = \frac{V_{CC}^2}{R_C + r_L} \tag{16–28}$$

Therefore, the efficiency under the conditions of maximum possible undistorted output is

$$\eta = \frac{P_L}{P_S} = \frac{\left[\left(\frac{V_{CC}}{R_C + r_L}\right)\left(\frac{R_C}{R_C + R_L}\right)\right]^2 R_L}{2\left(\frac{V_{CC}^2}{R_C + r_L}\right)} \tag{16–29}$$

Algebraic simplification of (16–29) leads to

$$\eta = \frac{R_C R_L}{2(R_C + 2R_L)(R_C + R_L)} = \frac{r_L}{2(R_C + 2R_L)} \tag{16–30}$$

Equation 16–30 shows that the efficiency depends on both R_C and R_L. In practice, R_L is a fixed and known value of load resistance, while the value of R_C is selected by the designer. If R_C is fixed and R_L can be selected, it can be shown using calculus (differentiating equation 16–30 with respect to R_C), that η is maximized by setting

$$R_L = \frac{R_C}{\sqrt{2}} \tag{16–31}$$

With this value of R_C, the maximum efficiency is

$$\eta(\text{max}) = 0.0858 \tag{16–32}$$

Another criterion for choosing R_L is to select its value so that maximum power is transferred to the load. Under the constraint of maintaining maximum output swing, it can be shown that maximum power transfer occurs when $R_L = R_C/2$. Under that circumstance, by substituting into (16-30), we find the maximum possible efficiency with maximum power transfer to be

$$\eta(\text{max}) = 0.0833 \quad \text{(max power transfer)} \tag{16–33}$$

It is interesting to note that the efficiency under maximum power transfer (0.0833) is somewhat less than what can be achieved (0.0858) without regard to power transfer. In either case, the maximum efficiency is substantially less than that attainable in the series-fed class-A amplifier.

Example 16–5

The class-A amplifier shown in Figure 16–8 is biased at $V_{CE} = 12$ V. The output voltage is the maximum possible without distortion. Find

1. the average power from the dc supply;
2. the average power delivered to the load;
3. the efficiency; and
4. the collector efficiency.

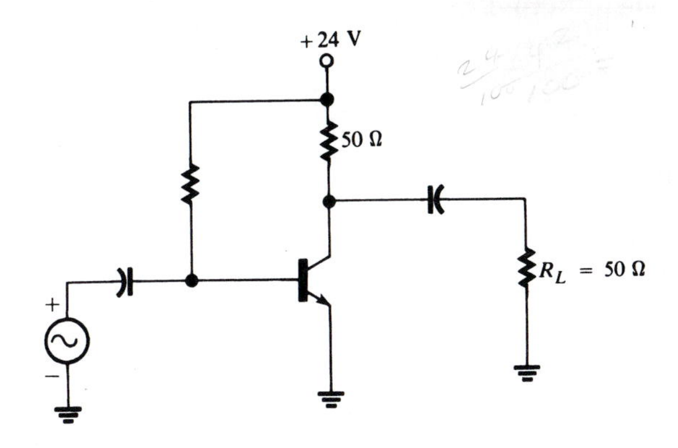

Figure 16–8
(Example 16–5)

Solution

1. $I_Q = (V_{CC} - V_Q)/R_C = (24 - 12)/50 = 0.24$ A
   $P_S = V_{CC}I_Q = (24\text{ V})(0.24\text{ A}) = 5.76$ W
2. As can be seen in Figure 16–7, the maximum value of the peak output voltage is the smaller of V_Q and $I_Q r_L$. In this case, $V_Q = 12$ V and $I_Q r_L = (0.24\text{ A})(50 \parallel 50) = 6$ V. Thus, the peak undistorted output voltage is 6 V and the ac power delivered to the 50-Ω load is

$$P_L = \frac{V_{PL}^2}{2R_L} = \frac{6^2}{100} = 0.36\text{ W}$$

$$\eta = \frac{P_L}{P_S} = \frac{0.36\text{ W}}{5.76\text{ W}} = 0.0625$$

3. The efficiency is less than the theoretical maximum because the bias point permits only a ±6-V swing. As an exercise, find the quiescent value of V_{CE} that maximizes the swing and calculate the efficiency under that condition.
4. The average power dissipated at the collector is $P_C = V_Q I_Q = (12)(0.24) = 2.88$ W. Therefore,

$$\eta_c = \frac{P_L}{P_C} = \frac{0.36}{2.88} = 0.125$$

Transformer-Coupled Class-A Amplifiers

In Chapter 11, we reviewed basic transformer theory and discussed transformer-coupled amplifiers. Transformers are also used to couple power amplifiers to their loads and, in those applications, are called *output transformers.* As in other coupling applications, the advantages of a transformer are that it provides an opportunity to achieve impedance matching for maximum power transfer and that it blocks the flow of dc current in a load.

Figure 16–9 shows a transformer used to couple the output of a transistor to load R_L. Also shown are the dc and ac load lines for the amplifier. Here we assume that the dc resistance of the primary winding is negligibly small, so the dc load line is vertical (slope $= -1/R_{dc} = -\infty$). Recall that the ac resistance r_L reflected to the primary side is

$$r_L = (N_p/N_s)^2 R_L \tag{16–34}$$

where N_p and N_s are the numbers of turns on the primary and secondary windings, respectively. As shown in the figure, the slope of the ac load line is $-1/r_L$.

Since we are assuming that there is negligible resistance in the primary winding, there is no dc voltage drop across the winding, and the quiescent collector voltage

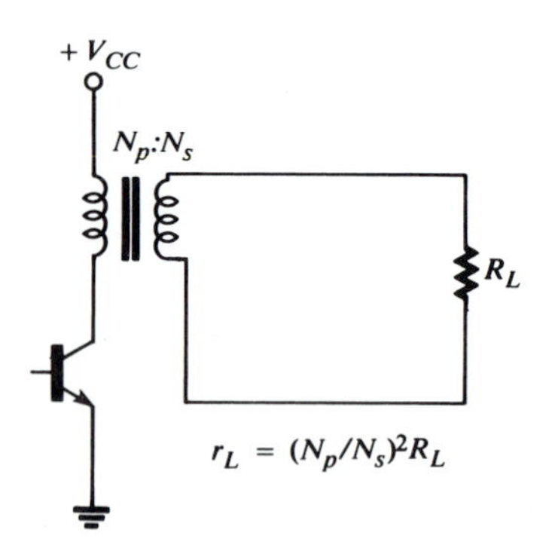

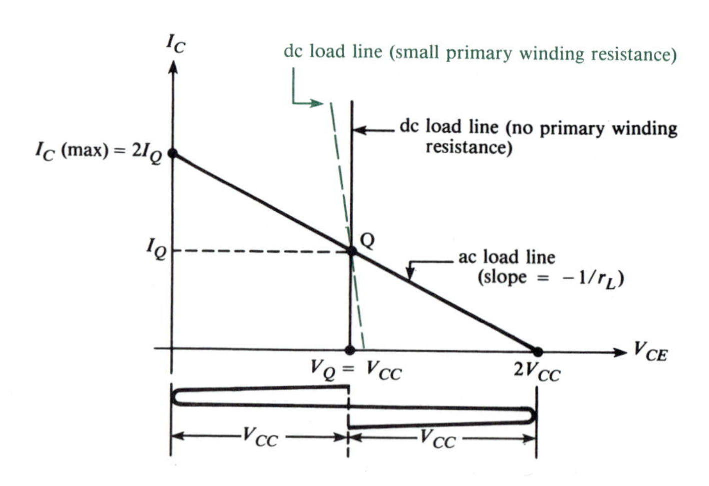

Figure 16–9
Transformer-coupled, class-A amplifier and load lines. The amplifier is biased for maximum peak-to-peak variation in V_{CE}.

is therefore V_{CC} volts, as shown in Figure 16–9. Conventional base-bias circuitry (not shown in the figure) is used to set the quiescent collector current I_Q. The Q-point is the point on the dc load line at which the collector current equals I_Q.

Since V_{CE} cannot be negative, the maximum permissible decrease in V_{CE} below its quiescent value is $V_Q = V_{CC}$ volts. Thus, the maximum possible peak value of V_{CE} is V_{CC} volts. To achieve maximum peak-to-peak output variation, the intercept of the ac load line on the V_{CE}-axis should therefore be $2V_{CC}$ volts, as shown in Figure 16–9. The quiescent current I_Q is selected so that the ac load line, a line having slope $-1/r_L$, intersects the V_{CE}-axis at $2V_{CC}$ volts.

The ac load line intersects the I_C-axis in Figure 16–9 at I_C(max). Note that there is no theoretical limit to the value that I_C may have, since there is no limiting resistance in the collector circuit. However, in practice, I_C must not exceed the maximum permissible collector current for the transistor and it must not be so great that the magnetic flux of the transformer saturates. When the transformer saturates, it can no longer induce current in the secondary winding and signal distortion results. When I_Q is set for maximum signal swing (so that $V_{CE}(\text{max}) = 2V_{CC}$), I_Q is one-half I_C(max); that is, $I_C(\text{max}) = 2I_Q$, as shown in Figure 16–9. Thus, the maximum values of the peak primary voltage and peak primary current are V_{CC} and I_Q, respectively. Since the ac output can vary through this range, below and above the quiescent point, the amplifier is of the class-A type.

Note that, unlike the case of the capacitor-coupled or series-fed amplifier, the collector voltage can exceed the supply voltage. A transistor having a collector breakdown voltage equal to at least twice the supply voltage should be used in this application.

The ac power delivered to load resistance R_L in Figure 16–9 is

$$P_L = \frac{V_s^2}{2R_L} = \frac{V_{PL}^2}{2R_L} \text{ watts} \tag{16–35}$$

where $V_s = V_{PL}$ is the peak value of the secondary, or load, voltage. The average power from the dc supply is

$$P_S = V_{CC} I_Q \text{ watts} \tag{16–36}$$

Therefore, the efficiency is

$$\eta = \frac{P_L}{P_S} = \frac{V_{PL}^2}{2R_L V_{CC} I_Q} \tag{16–37}$$

Under maximum signal conditions, the peak primary voltage is V_{CC} volts, so the peak load voltage is

$$V_{PL} = (N_s/N_p)V_{CC} \tag{16–38}$$

Also, since the slope of the ac load line is $-1/r_L$, we have

$$\frac{|\Delta I_C|}{|\Delta V_{CE}|} = \frac{1}{r_L} = \frac{I_Q}{V_Q}$$

or

$$I_Q = \frac{V_Q}{r_L} = \frac{V_{CC}}{(N_p/N_s)^2 R_L} \tag{16–39}$$

Substituting (16–38) and (16–39) into (16–37), we find the maximum possible efficiency of the class-A, transformer-coupled amplifier:

$$\eta(\text{max}) = \frac{(N_s/N_p)^2 V_{CC}^2}{(2R_L V_{CC})\dfrac{V_{CC}}{(N_p/N_s)^2 R_L}} = 0.5 \qquad \textbf{(16–40)}$$

We see that the maximum efficiency is twice that of the series-fed class-A amplifier and six times that of the capacitor-coupled class-A amplifier. This improvement in efficiency is attributable to the absence of external collector resistance that would otherwise consume dc power. Note that the collector efficiency of the transformer-coupled class-A amplifier is the same as the overall amplifier efficiency, because the average power from the dc supply is the same as the collector dissipation:

$$P_S = V_{CC} I_Q = V_Q I_Q = P_C \qquad \textbf{(16–41)}$$

In practice, a full output voltage swing of $2V_{CC}$ volts cannot be achieved in a power transistor. The device is prevented from cutting off entirely by virtue of a relatively large leakage current, and it cannot be driven all the way into saturation ($I_C = I_C(\text{max})$) without creating excessive distortion. These points are illustrated in the next example.

Example 16–6

The transistor in the power amplifier shown in Figure 16–10 has the output characteristics shown in Figure 16–11. Assume that the transformer has zero resistance.

1. Construct the (ideal) dc and ac load lines necessary to achieve maximum output voltage swing. What quiescent values of collector and base current are necessary to realize the ac load line?
2. What is the smallest value of $I_C(\text{max})$ for which the transistor should be rated?
3. What is the maximum peak-to-peak collector voltage, and what peak-to-peak base current is required to achieve it? Assume that the base current cannot go negative and that, to minimize distortion, the collector should not be driven below 2.5 V in the saturation region.
4. Find the average power delivered to the load under the maximum signal conditions of (3).

Figure 16–10
(Example 16–6)

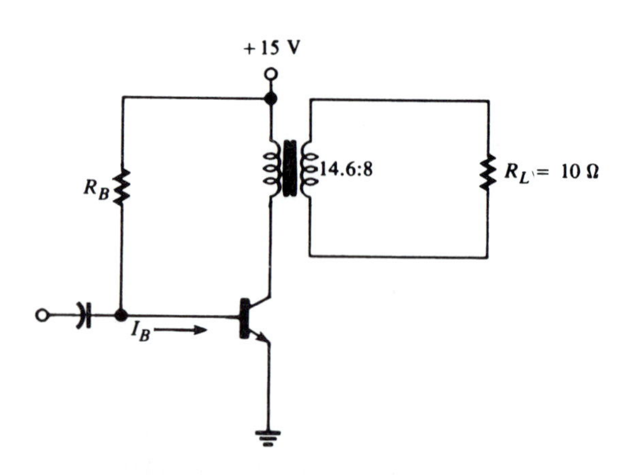

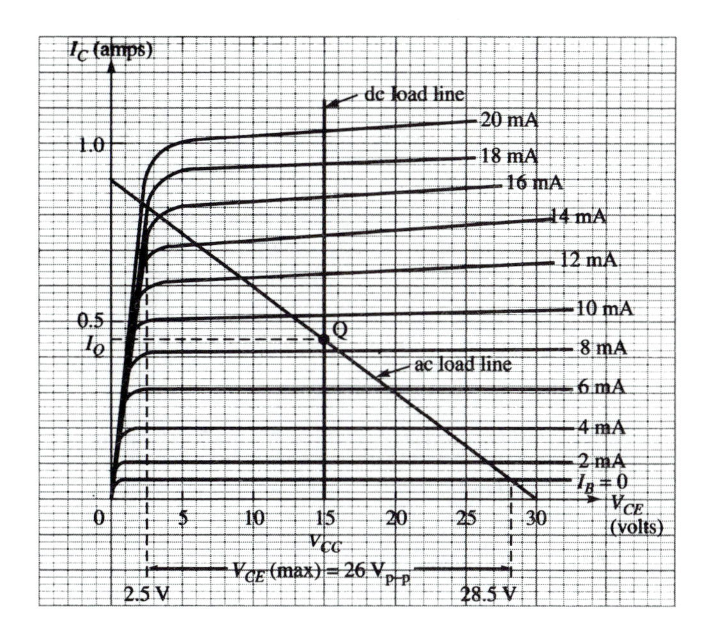

Figure 16–11
(Example 16–6)

5. Find the power dissipated in the transistor under no-signal conditions (standby).
6. Find the efficiency.

Solution

1. The vertical dc load line intersects the V_{CE}-axis at $V_{CC} = 15$ V, as shown in Figure 16–11. To find the slope of the ac load line, we must find r_L. From equation 16–34, $r_L = (N_p/N_s)^2 R_L = (14.6/8)^2(10) = 33.3\ \Omega$. Thus, the slope of the ac load line is $-1/r_L = -1/33.3 = -0.03$. To achieve the ideal maximum output swing, we want the ac load line to intercept the V_{CE}-axis at $2V_{CC} =$ 30 V. Since the slope of that line has magnitude 0.03, it will intercept the I_C-axis at $I_C = (0.03\text{ A/V})(30\text{ V}) = 0.9$ A. The ac load line is then drawn between the two intercepts (0 A, 30 V) and (0.9 A, 0 V), as shown in Figure 16–11.

   The ac load line intersects the dc load line at the Q-point. The quiescent collector current at that point is seen to be $I_Q = 0.45$ A. The corresponding base current is approximately halfway between $I_B = 8$ mA and $I_B = 10$ mA, so the quiescent base current must be 9 mA.
2. The maximum collector current is $I_C(\text{max}) = 0.9$ A, at the intercept of the ac load line on the I_C-axis. Actually, we will not operate the transistor that far into saturation, since we do not allow V_{CE} to fall below 2.5 V. However, a maximum rating of 0.9 A (or 1 A) will provide us with a margin of safety.
3. The maximum value of V_{CE} occurs on the ac load line at the point where $I_B =$ 0. As shown in Figure 16–11, this value is 28.5 V. Since the minimum permissible value of V_{CE} is 2.5 V, the maximum peak-to-peak voltage swing is $28.5 - 2.5 =$ 26 V p–p. As can be seen on the characteristic curves, the base current must vary from $I_B = 0$ to $I_B = 18$ mA, or 18 mA peak-to-peak, to achieve that voltage swing.
4. The peak primary voltage in the transformer is (1/2)(26 V) = 13 V. Therefore, the peak secondary, or load, voltage is $V_{PL} = (N_s/N_p)(13\text{ V}) = (8/14.6)13 =$

7.12 V. The average load power is then

$$P_L = \frac{V_{PL}^2}{2R_L} = \frac{(7.12)^2}{20} = 2.53 \text{ W}$$

5. The standby power dissipation is $P_d = V_Q I_Q = V_{CC} I_Q = (15 \text{ V})(0.45 \text{ A}) = 6.75 \text{ W}$.
6. The standby power dissipation found in (5) is the same as the average power supplied from the dc source, so

$$\eta = \frac{2.53 \text{ W}}{6.75 \text{ W}} = 0.375$$

(Why is this value less than the theoretical maximum of 0.5?)

Class-B Amplifiers

Transistor operation is said to be *class B* when output current varies during only one half-cycle of a sine-wave input. In other words, the transistor is in its active region, responding to signal input, only during a positive half-cycle or only during a negative half-cycle of the input. This operation is illustrated in Figure 16–12.

It is clear that class-B operation produces an output waveform that is severely clipped (half-wave rectified) and therefore highly distorted. The waveform *by itself* is not suitable for audio applications. However, in practical amplifiers, *two* transistors are operated class B: one to amplify positive signal variations and the other to amplify negative signal variations. The amplifier output is the composite waveform obtained by combining the waveforms produced by each class-B transistor. We will study these amplifiers in detail in the next section because they are more efficient and more widely used in power applications than are class-A amplifiers. An amplifier utilizing transistors that are operated class B is called a *class-B* amplifier.

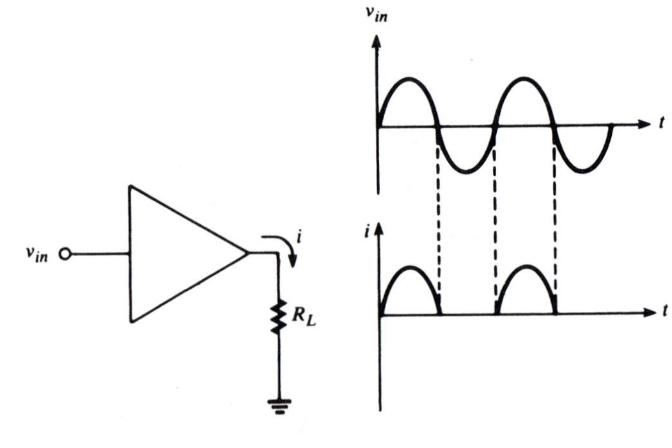

Figure 16–12
In class-B operation, output current variations occur during only each positive or each negative half-cycle of input. In the example shown, output current flows during positive half-cycles of the input, and the amplifier is cut off during negative half cycles.

16–5 PUSH-PULL–AMPLIFIER PRINCIPLES

A *push-pull* amplifier uses two output devices to drive a load. The name is derived from the fact that one device is primarily (or entirely) responsible for driving current through the load in one direction (pushing), while the other device drives current through the load in the opposite direction (pulling). The output devices are typically two transistors, each operated class B, one of which conducts only when the input is positive, and the other of which conducts only when the input is negative. This arrangement is called a *class-B, push-pull* amplifier and its principle is illustrated in Figure 16–13.

Note in Figure 16–13 that amplifying devices 1 and 2 are driven by equal-amplitude, *out-of-phase* input signals. The signals are identical except for phase. Here we assume that each device conducts only when its input is positive and is cut off when its input is negative. The net effect is that device 1 produces load current when the input is positive and device 2 produces load current, in the opposite direction, when the input is negative. An example of a device that produces output (collector) current only when its input (base-to-emitter) voltage is positive is an NPN transistor having no base-biasing circuitry, that is, one that is biased at cutoff. As we shall see, NPN transistors can be used as the output amplifying devices in push-pull amplifiers. However, the circuitry must be somewhat more elaborate than that diagrammed in Figure 16–13 since we must make provisions for load current to flow through a *complete circuit,* regardless of current direction. Obviously, when amplifying device 1 in Figure 16–13 is cut off, it cannot conduct current produced by device 2, and vice versa.

Push-Pull Amplifiers with Output Transformers

Figure 16–14 shows a push-pull arrangement that permits current to flow in both directions through a load, even though one or the other of the amplifying devices (NPN transistors) is always cut off. The *output transformer* shown in the figure is the key component. Note that the primary winding is connected between the transistor collectors and that its *center tap* is connected to the dc supply, V_{CC}. The center tap is simply an electrical connection made at the center of the winding, so there are an equal number of turns between each end of the winding and the center tap. The figure does not show the push-pull *driver* circuitry, which must produce out-of-phase signals on the bases of Q_1 and Q_2. We will discuss that circuitry later.

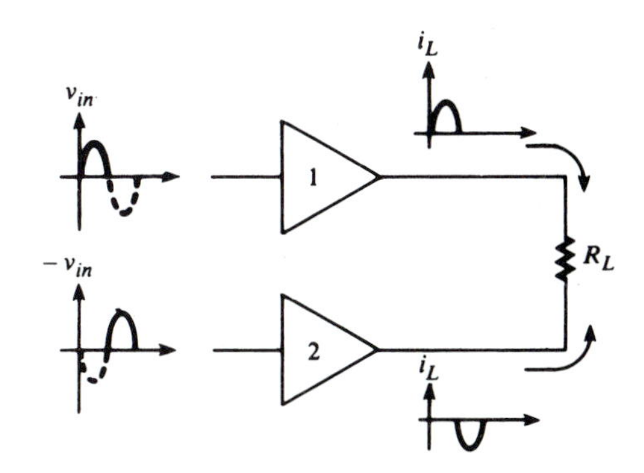

Figure 16–13
The principle of class-B, push-pull operation. Output amplifying device 1 drives current i_L through the load in one direction while 2 is cut off, and device 2 drives current in the opposite direction while 1 is cut off.

Figure 16–14
A center-tapped output transformer used in a push-pull amplifier

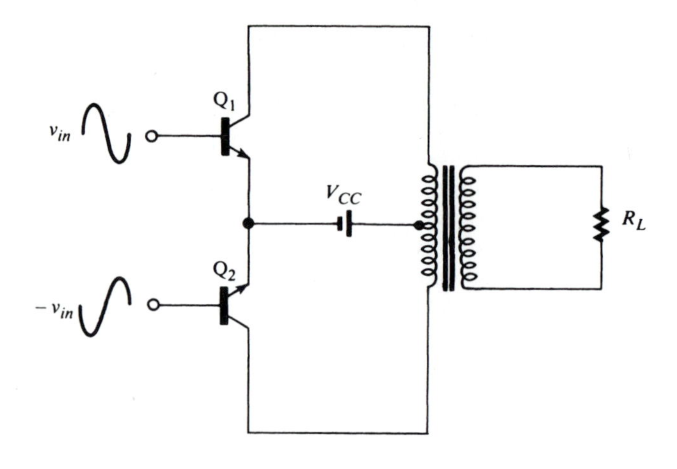

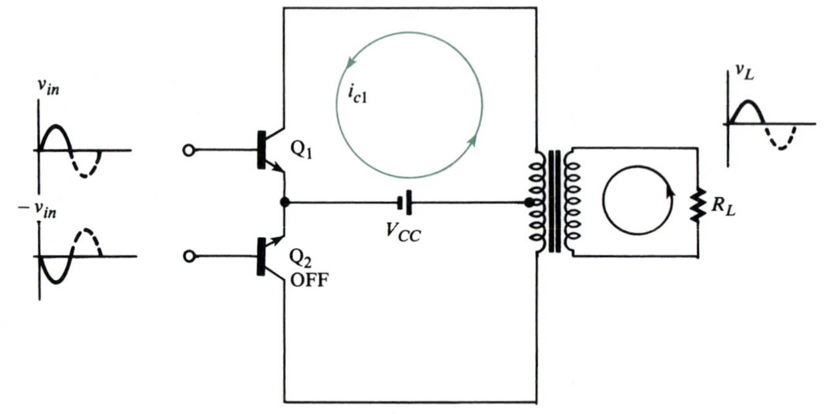

(a) When v_{in} is positive, Q_1 conducts and Q_2 is cut off. A counterclockwise current is induced in the load.

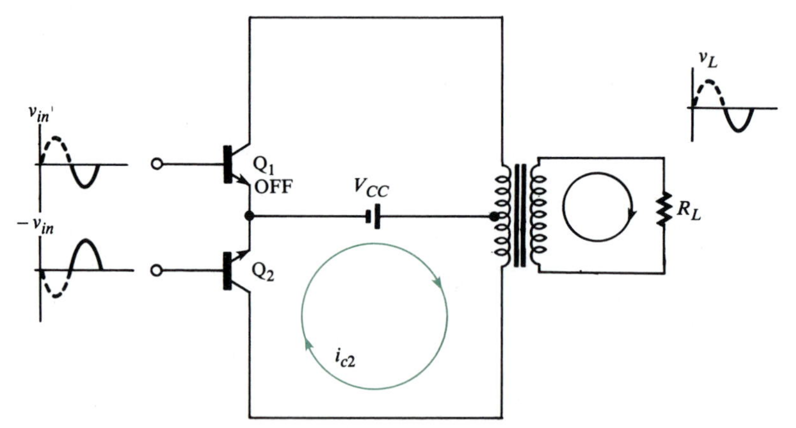

(b) When v_{in} is negative, Q_2 conducts and Q_1 is cut off. A clockwise current is induced in the load.

Figure 16–15
Current flow in a push-pull amplifier with output transformer. (Currents are those that flow during the portion of the input shown as a solid line.)

Figure 16–15 shows how current flows through the amplifier during a positive half-cycle of input and during a negative half-cycle of input. In 16–15(a), the input to Q_1 is the positive half-cycle of the signal, and since the input to Q_2 is out-of-phase with that to Q_1, Q_2 is driven by a negative half-cycle. Notice that neither class-B transistor is biased. Consequently, the positive base voltage on Q_1 causes it to turn on and conduct current in the counterclockwise path shown. The negative base voltage on Q_2 keeps that transistor cut off. Current flowing in the upper half of the transformer's primary induces current in the secondary, and current flows through the load. In 16–15(b), the input signal on the base of Q_1 has gone negative, so its inverse on the base of Q_2 is positive. Therefore, Q_2 conducts current in the clockwise path shown, and Q_1 is cut off. Current induced in the secondary winding is in the direction opposite that shown in Figure 16–15(a). The upshot is that current flows through the load in one direction when the input signal is positive and in the opposite direction when the input signal is negative, just as it should in an ac amplifier.

Figure 16–16 displays current flow in the push-pull amplifier in the form of a *timing diagram*. Here the complete current waveforms are shown over two full cycles of input. For purposes of this illustration, counterclockwise current (in Figure 16–15) is arbitrarily assumed to be positive and clockwise current is therefore negative. Note that, as far as the load is concerned, current flows during the full 360° of input signal. Figure 16–16(e) shows that the current i_S from the power supply varies from 0 to the peak value I_P every half-cycle. Because the current variation is so large, the power supply used in a push-pull amplifier must be particularly well regulated—that is, it must maintain a constant voltage, independent of current demand.

Class-B Efficiency

The principal advantage of using a class-B power amplifier is that it is possible to achieve an efficiency greater than that attainable in a class-A amplifier. The improvement in efficiency stems from the fact that no power is dissipated in a transistor during the time intervals that it is cut off. Also, like the transformer-coupled class-A amplifier, there is no external collector resistance that would otherwise consume power.

We will derive an expression for the maximum efficiency of a class-B push-pull amplifier assuming ideal conditions: perfectly matched transistors and zero resistance in the transformer windings. The current supplied by each transistor is a half-wave–rectified waveform, as shown in Figure 16–16. Let I_p represent the peak value of each. Then the peak value of the current in the secondary winding, which is the same as the peak load current, is

$$I_{PL} = (N_p/N_s)I_P \qquad \textbf{(16–42)}$$

where N_p/N_s is the turns ratio between one-half the primary winding and the secondary winding. (Note that only those primary turns between one end of the winding and its center tap are used to induce current in the secondary.) Similarly, the peak value of the load voltage is

$$V_{PL} = (N_s/N_p)V_P \qquad \textbf{(16–43)}$$

where V_P is the peak value of the primary (collector) voltage. Since the load voltage and load current are sinusoidal, the average power delivered to the load is, from

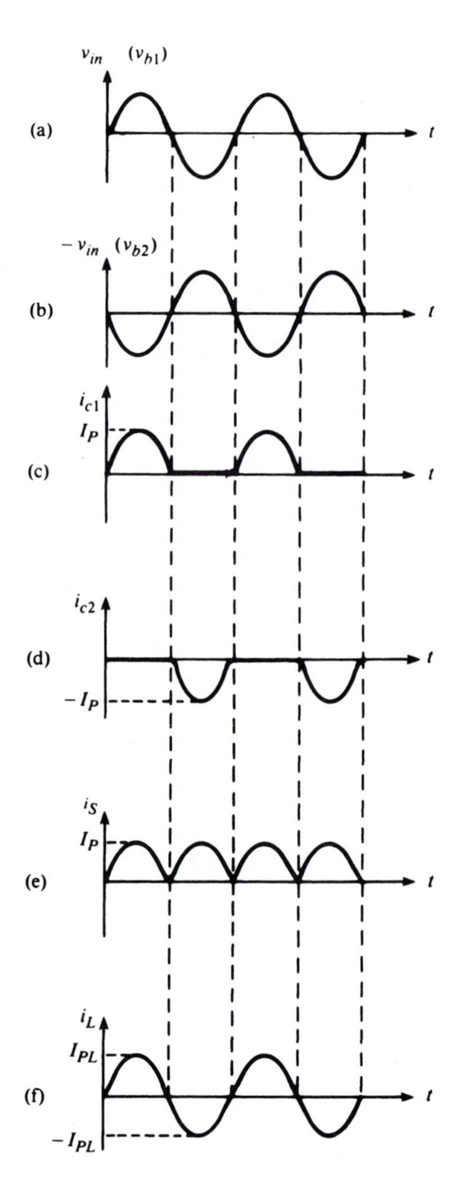

Figure 16–16
A timing diagram showing currents in the push-pull amplifier of Figure 16–15 over two full cycles of input

equation 16–10,

$$P_L = \frac{V_{PL}I_{PL}}{2} = \frac{(N_s/N_p)V_P(N_p/N_s)I_P}{2} = \frac{V_P I_P}{2} \tag{16–44}$$

As shown in Figure 16–16(e), the power-supply current is a full-wave–rectified waveform having peak value I_P. The dc, or average, value of such a waveform is known to be $2I_P/\pi$. Therefore, the average power delivered to the circuit by the dc supply is

$$P_S = \frac{2I_P V_{CC}}{\pi} \tag{16–45}$$

The efficiency is then

$$\eta = \frac{P_L}{P_S} = \frac{V_P I_P/2}{2I_P V_{CC}/\pi} = \frac{\pi V_P}{4V_{CC}} \tag{16–46}$$

Under maximum signal conditions, $V_P = V_{CC}$, and (16–46) becomes

$$\eta(\text{max}) = \frac{\pi}{4} = 0.785 \tag{16–47}$$

Equation 16–47 shows that a class-B push-pull amplifier can be operated with a much higher efficiency than the class-A amplifiers studied earlier. Furthermore, unlike the case of class-A amplifiers, the power dissipated in the transistors is 0 under standby (zero-signal) conditions because both transistors are cut off. A general expression for the total power dissipated in the transistors can be obtained by realizing that it equals the difference between the total power supplied by the dc source and the total power delivered to the load:

$$P_d = \frac{2I_P V_{CC}}{\pi} - \frac{V_P I_P}{2} \tag{16–48}$$

Using calculus, it can be shown that P_d is maximum when $V_P = 2V_{CC}/\pi = 0.636V_{CC}$. We conclude that maximum transistor dissipation does *not* occur when maximum load power is delivered ($V_P = V_{CC}$), but at the intermediate level $V_P = 0.636V_{CC}$.

Example 16–7

The push-pull amplifier in Figure 16–14 has $V_{CC} = 20$ V and $R_L = 10\ \Omega$. The *total* number of turns on the primary winding is 100 and the secondary winding has 50 turns. Assume that the transformer has zero resistance.

1. Find the maximum power that can be delivered to the load.
2. Find the power dissipated in each transistor when maximum power is delivered to the load.
3. Find the power delivered to the load and the power dissipated in each transistor when transistor power dissipation is maximum.

Solution

1. The turns ratio between each half of the primary and the secondary is $N_p/N_s = (100/2):50 = 50:50$. Therefore, the peak values of primary and secondary voltages are equal, as are the peak values of primary and secondary current.

$$V_P(\text{max}) = V_{PL}(\text{max}) = V_{CC} = 20\text{ V}$$

$$I_P(\text{max}) = I_{PL}(\text{max}) = \frac{V_{CC}}{R_L} = \frac{20\text{ V}}{10\ \Omega} = 2\text{ A}$$

Therefore,

$$P_L(\text{max}) = \frac{V_P(\text{max})I_P(\text{max})}{2} = \frac{(20\text{ V})(2\text{ A})}{2} = 20\text{ W}$$

2. From equation 16–48,

$$P_d = \frac{2I_P V_{CC}}{\pi} - \frac{V_P I_P}{2} = \frac{2(2\text{ A})(20\text{ V})}{\pi} - 20\text{ W} = 5.46\text{ W}$$

Since 5.46 W is the total power dissipated by both transistors, each dissipates one-half that amount, or 2.73 W.

3. Transistor power dissipation is maximum when $V_P = 0.636V_{CC} = (0.636)(20) = 12.72$ V. Then

$$I_P = I_{PL} = \frac{12.72\text{ V}}{10\ \Omega} = 1.272\text{ A}$$

and

$$P_L = \frac{V_P I_P}{2} = \frac{(12.72\text{ V})(1.272\text{ A})}{2} = 8.09\text{ W}$$

From equation 16–48,

$$P_d(\text{max}) = \frac{2I_P V_{CC}}{\pi} - \frac{V_P I_P}{2} = \frac{2(1.272\text{ A})(20\text{ V})}{\pi} - 8.09\text{ W} = 8.09\text{ W}$$

The maximum power dissipation in each transistor is then $8.09/2 = 4.05$ W.

The preceding example demonstrates some results that are true in general for push-pull amplifiers: (1) When transistor power dissipation is maximum, its value equals the power delivered to the load; and (2) the maximum total transistor power dissipation equals approximately 40% of the maximum power that can be delivered to the load.

16–6 PUSH-PULL DRIVERS

We have seen that the push-pull amplifier described earlier must be driven by out-of-phase input signals. Figure 16–17 shows how a transformer can be used to provide the required drive signals. Here, the *secondary* winding has a grounded center tap that effectively splits the secondary voltage into two out-of-phase signals, each having one-half the peak value of the total secondary voltage. The input signal is

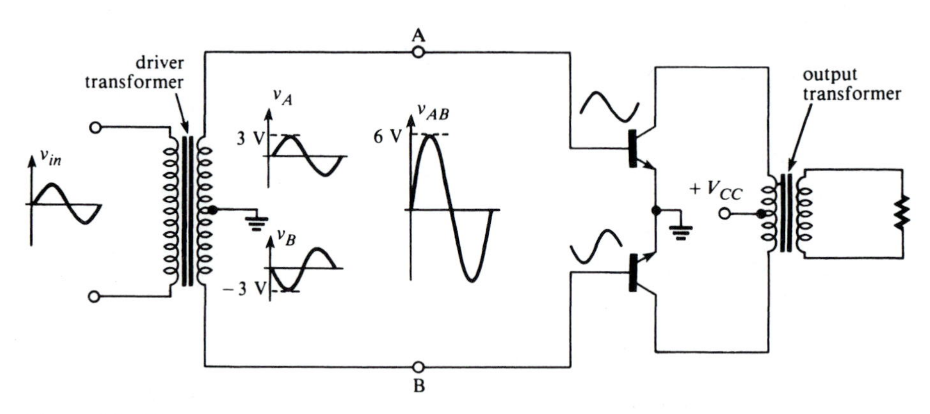

Figure 16–17
Using a driver transformer to create equal-amplitude, out-of-phase drive signals (v_A and v_B) for a push-pull amplifier

applied across the primary winding, and a voltage is developed across secondary terminals A–B in the usual transformer fashion. To understand the phase-splitting action, consider the instant at which the voltage across A–B is +6 V, as shown in the figure. Then, since the center point is at ground, the voltage from A to ground must be +3 V and that from B to ground must be −3 V: $V_{AB} = V_A - V_B = 3 - (-3) = +6$ V. The same logic applied at every instant throughout a complete cycle shows that v_B with respect to ground is always the negative of v_A with respect to ground; in other words, v_A and v_B are equal-amplitude, out-of-phase driver signals, as required.

A specially designed amplifier, called a *phase-splitter,* can be used instead of a driver transformer to produce equal amplitude, out-of-phase drive signals. Figure 16–18 shows two possible designs. Figure 16–18(a) is a conventional amplifier circuit with outputs taken at the collector and at the emitter. The collector output is out of phase with the input and the emitter output is in phase with the input, so the two outputs are out of phase with each other. With no load connected to either output, the output signals will have approximately equal amplitudes if $R_C = R_E$. However, the output impedance at the collector is significantly greater than that at the emitter, so when loads are connected, each output will be affected differently. As a consequence, the signal amplitudes will no longer be equal, and gain adjust-

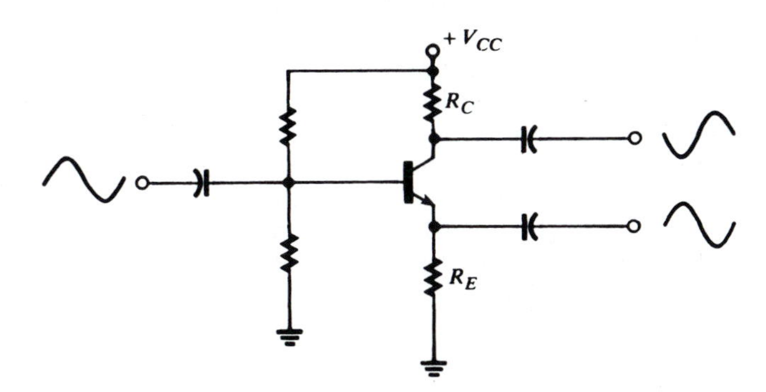

(a) A phase-splitting driver

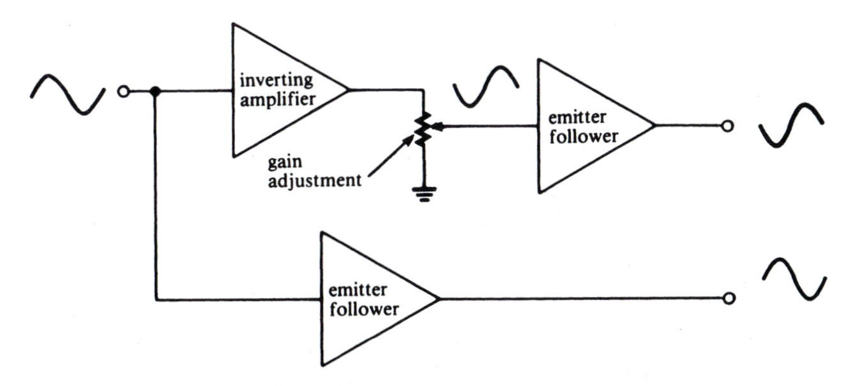

(b) An improved method for obtaining equal-amplitude, out-of-phase signals

Figure 16–18
Phase-splitting methods for obtaining push-pull drive signals

ments will be required. Furthermore, the nonlinear nature of the large-signal load (the output transistors) means that the gain of the collector output may vary appreciably with signal level. Intolerable distortion can be created as a result. A better way to drive the output transistors is from two low-impedance signal sources, as shown in Figure 16–18(b). Here, the output from an inverting amplifier is buffered by an emitter-follower stage, as is the original signal.

16–7 HARMONIC DISTORTION AND FEEDBACK

Harmonic Distortion

Recall that any periodic waveform, sinusoidal or otherwise, can be represented as the sum of an infinite number of sine waves having different amplitudes, frequencies, and phase relations. The mathematical technique called *Fourier analysis* is concerned with finding the exact amplitudes, frequencies, and phase angles of the sine-wave components that reproduce a given waveform when they are added together. Of course, if the waveform is itself a pure sine wave, then all other frequency components have zero amplitudes.

Apart from a possible dc component (the average value), the lowest frequency component in the infinite sum is a sine wave having the same frequency as the periodic waveform itself. This component is called the *fundamental* and usually has an amplitude greater than that of any other frequency component. The other frequency components are called *harmonics,* and each has a frequency that is an *integer multiple* of the fundamental frequency. For example, the second harmonic of a 3-kHz waveform has frequency 6 kHz, the third harmonic has frequency 9 kHz, and so forth. Not all harmonics need be present. For example, a square wave contains only odd harmonics (third, fifth, seventh, etc.).

As we have indicated in previous discussions, nonlinear amplifier characteristics are responsible for harmonic distortion of an output signal. This distortion is in fact the creation of harmonic frequencies that would not otherwise be present in the output when the input is a pure sine wave. The extent to which a particular harmonic component distorts a signal is specified by the ratio of its amplitude to the amplitude of the fundamental component, expressed as a percentage:

$$\% \, n\text{th harmonic distortion} = \%D_n = \frac{A_n}{A_1} \times 100\% \qquad \textbf{(16–49)}$$

where A_n is the amplitude of the nth harmonic component and A_1 is the amplitude of the fundamental. For example, if a 1-V-peak, 10-kHz sine wave is distorted by the addition of a 0.1-V-peak, 20-kHz sine wave and a 0.05-V-peak, 30-kHz sine wave, then it has

$$\%D_2 = \frac{0.1}{1} \times 100\% = 10\% \text{ second harmonic distortion}$$

and

$$\%D_3 = \frac{0.05}{1} \times 100\% = 5\% \text{ third harmonic distortion}$$

The *total harmonic distortion* (THD) is the square root of the sum of the squares of all the individual harmonic distortions:

$$\%\text{THD} = \sqrt{D_2^2 + D_3^2 + \cdots} \times 100\% \qquad \textbf{(16–50)}$$

Special instruments, called *distortion analyzers,* are available for measuring the total harmonic distortion in a waveform.

Example 16–8

The principal harmonics in a certain 15-kHz signal having a 10-Vpk fundamental are the second and fourth. All other harmonics are negligibly small. If the THD is 12% and the amplitude of the second harmonic is 0.5 V pk, what is the amplitude of the 60-kHz harmonic?

Solution

$$\text{THD} = 0.12 = \sqrt{D_2^2 + D_4^2}$$
$$0.0144 = D_2^2 + D_4^2$$
$$D_2 = 0.5/10 = 0.05 \quad (5\%)$$
$$0.0144 = (0.05)^2 + D_4^2$$
$$D_4 = \sqrt{0.0144 - 0.0025} = 0.1091 \quad (10.91\%)$$
$$A_4 = D_4A_1 = (0.1091)(10) = 1.091 \text{ V}$$

Using Negative Feedback to Reduce Distortion

We have mentioned in several previous discussions that one of the important benefits of negative feedback is that it reduces distortion caused by amplifier nonlinearities. Having defined a quantitative measure of distortion, we can now undertake a quantitative investigation of the degree to which feedback affects harmonic distortion in the output of an amplifier. Let us begin by recognizing that a nonlinear amplifier is essentially an amplifier whose gain changes with signal level. Figure 16–19(a) shows the *transfer characteristic* of an ideal, distortionless amplifier, in which the gain, i.e., the slope of the characteristic, $\Delta V_o/\Delta V_{in} = 50$, is constant. Figure 16–19(b) shows a nonlinear transfer characteristic in which the slope, and hence the gain, increases with increasing signal level. We see that the gain at $V_{in} = 0.2$ V is 50, while the gain at 0.4 V is 100. There is a 100% increase in gain over that range, so serious output distortion is to be expected.

In earlier discussions of negative feedback, we showed that the closed-loop gain A_{CL} of an amplifier having open-loop gain A and feedback ratio β can be found from

$$A_{CL} = \frac{A}{1 + A\beta} \qquad \textbf{(16–51)}$$

Let us suppose we introduce negative feedback into the amplifier with the nonlinear characteristic shown in Figure 16–19(b). Assume that $\beta = 0.05$. Then the gain at $v_{in} = 0.2$ V becomes

$$A_{CL} = \frac{50}{1 + 50(0.05)} = 14.29$$

and the gain at $v_{in} = 0.4$ V becomes

$$A_{CL} = \frac{100}{1 + 100(0.05)} = 16.67$$

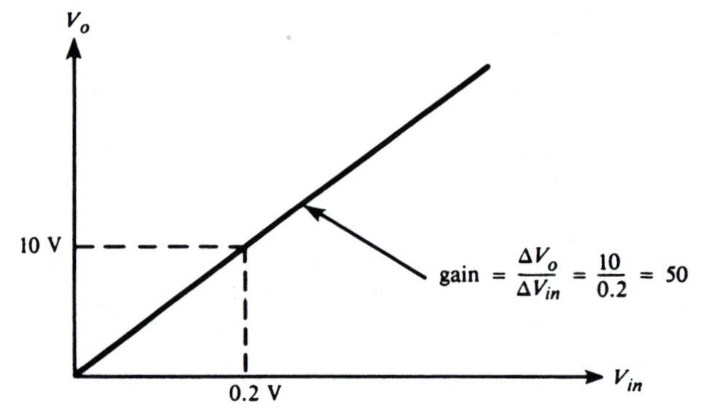

(a) Transfer characteristic of an ideal, distortionless amplifier having gain 50

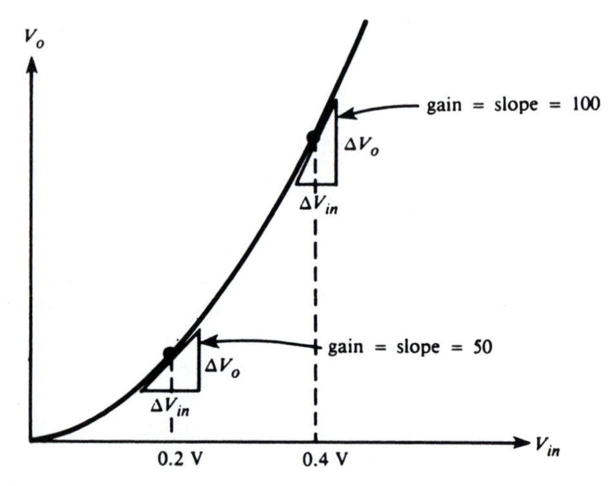

(b) Nonlinear transfer characteristic

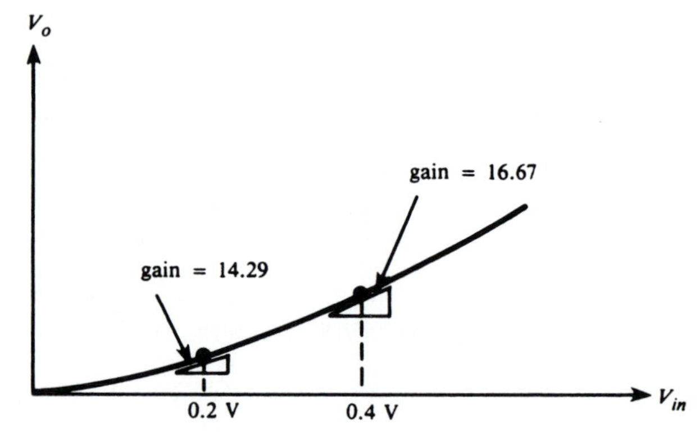

(c) Negative feedback (β = 0.05) reduces the gain but linearizes the characteristic

Figure 16–19
Transfer characteristics and the effect of negative feedback

Notice that the *change* in gain (16.7%) is now much less than it was without feedback (100%). The effect of feedback has been to "linearize" the transfer characteristics somewhat, as shown in Figure 16–19(c). We conclude that less-severe distortion will result. Of course, the penalty we pay for this improved performance is an overall reduction in gain.

To compute the reduction in harmonic distortion caused by negative feedback, consider the models shown in Figure 16–20. Figure 16–20(a) shows how amplifier distortion (without feedback) can be represented as a distortionless amplifier having a distortion component d added to its output. For example, d could represent the level of a second-harmonic sine wave added to an otherwise undistorted fundamental. The output of the amplifier is

$$v_o = Av_{in} + d \tag{16–52}$$

so the harmonic distortion is

$$D = \frac{d}{Av_{in}} \tag{16–53}$$

Figure 16–20(b) shows the amplifier when negative feedback is connected. We realize that the negative feedback will reduce the closed-loop gain, so the output level will be reduced if v_{in} remains the same as in (a). But, for comparison purposes, we want the output levels in both cases to be the same since the amount of distortion depends on that level. Accordingly, we assume that v_{in} in (b) is increased to v'_{in}, as necessary to make the distortion component d equal to its value in (a). As shown in the figure,

$$v_o = A(v'_{in} - \beta v_o) + d \tag{16–54}$$

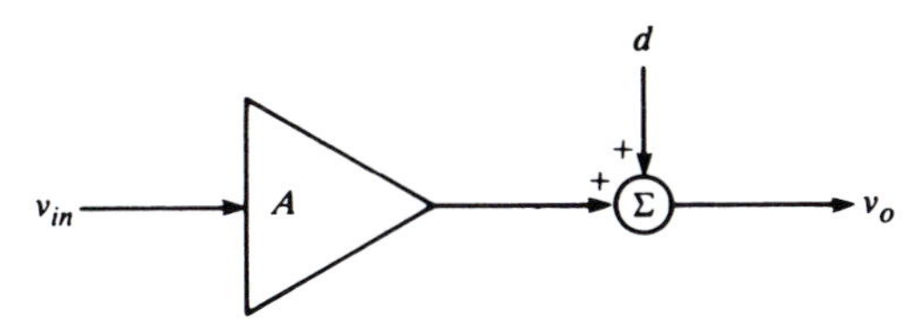

(a) Amplifier distortion represented by summing a distortion component with the amplifier output

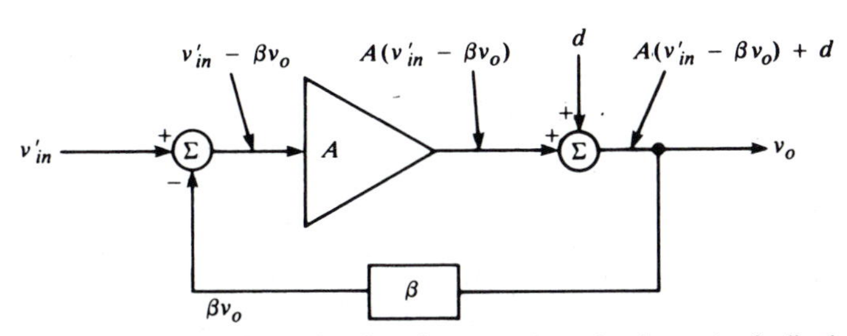

(b) Model for an amplifier with distortion component d and negative feedback ratio β

Figure 16–20
Amplifier distortion models

Solving for v_o, we find

$$v_o = \frac{Av'_{in}}{1 + A\beta} + \frac{d}{1 + A\beta} \tag{16–55}$$

When we adjusted v_{in} to v'_{in}, we did it so that the amplifier outputs in (a) and (b) of the figure would be the same:

$$Av_{in} = \frac{A}{1 + A\beta} v'_{in} \tag{16–56}$$

or

$$v'_{in} = (1 + A\beta)v_{in} \tag{16–57}$$

Substituting (16–57) into (16–55), we find

$$v_o = Av_{in} + \frac{d}{1 + A\beta} \tag{16–58}$$

The harmonic distortion is now

$$D = \frac{d/(1 + A\beta)}{Av_{in}} = \frac{d}{(1 + A\beta)Av_{in}} \tag{16–59}$$

Comparing with equation 16–53, we see that the distortion has been reduced by the factor $1/(1 + A\beta)$.

It is important to remember that it was necessary to increase the input level in our analysis to achieve the improved performance. From a practical standpoint, that means that a relatively distortion-free *preamplifier* must be used to compensate for the loss in gain caused by negative feedback around the output amplifier. Fortunately, it is generally easier to achieve low distortion in small-signal preamplifiers than in large-signal, high-power output amplifiers.

Example 16–9

An output amplifier has a voltage gain of 120 and generates 20% harmonic distortion with no feedback.

1. How much negative feedback should be used if it is desired to reduce the distortion to 2%?
2. How much preamplifier gain will have to be provided in cascade with the output amplifier to maintain an overall gain equal to that without feedback?

Solution

1. From equation 16–59,

$$D = \frac{d}{(1 + A\beta)Av_{in}}$$

where $d/Av_{in} = 0.2$ is the distortion without feedback. Then,

$$0.02 = \frac{0.2}{1 + A\beta} = \frac{0.2}{1 + 120\beta}$$

or $\beta = 0.075$. Thus, the negative feedback voltage must be 7.5% of the output voltage.

2. The closed-loop gain with 7.5% feedback is

$$A_{CL} = \frac{A}{1 + A\beta} = \frac{120}{1 + 120(0.075)} = 12$$

To maintain an overall gain of 120, the preamplifier gain, A_p, must be such that $12A_p = 120$, or $A_p = 10$.

16–8 DISTORTION IN PUSH-PULL AMPLIFIERS

Cancellation of Even Harmonics

Recall that push-pull operation effectively produces in a load a waveform proportional to the *difference* between two input signals. Under normal operation, the signals are out of phase, so their waveform is reproduced in the load. If the signals were in phase, cancellation would occur. It is instructive to view a push-pull output as the difference between two distorted (half-wave rectified) sine waves that are out of phase with each other. This viewpoint is illustrated in Figure 16–21. It can be shown that a half-wave–rectified sine wave contains only the fundamental and all *even* harmonics. Figure 16–21(a) shows the two out-of-phase, half-wave–rectified sine waves that drive the load, and 16–21(b) shows the fundamental and second-harmonic components of each. Notice that the fundamental components are out of phase. Therefore, the fundamental component is reproduced in the load, as we have already seen (Figure 16–16). However, the second-harmonic components are in phase, and therefore cancel in the load. Although not shown in this figure, the fourth and all other even harmonics are also in phase and therefore also cancel.

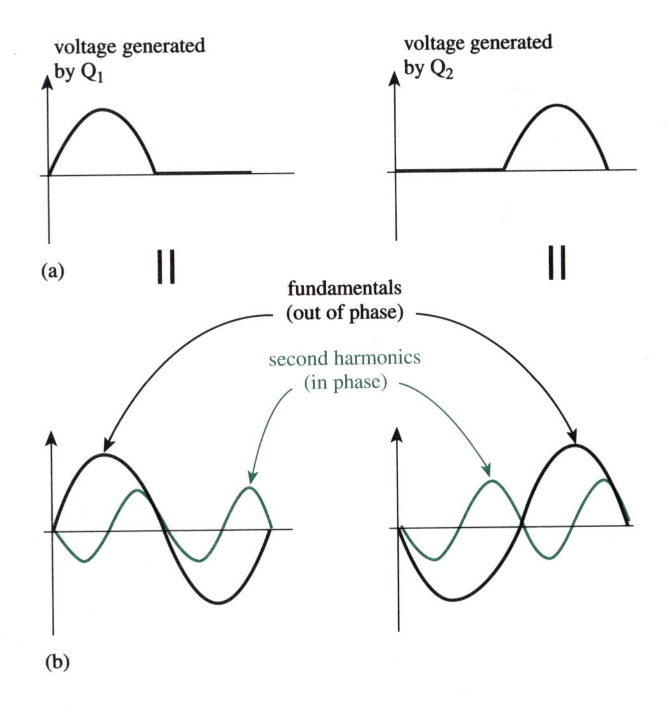

Figure 16–21
The fundamental components of the half-wave–rectified waveforms are out of phase, but the second harmonics are in phase. Therefore, second-harmonic distortion is cancelled in push-pull operation.

Our conclusion is an important property of push-pull amplifiers: *even harmonics are cancelled in push-pull operation.*

The cancellation of even harmonics is an important factor in reducing distortion in push-pull amplifiers. However, perfect cancellation would occur only if the two sides were perfectly matched and perfectly balanced: identical transistors, identical drivers, and a perfectly center-tapped transformer. Of course, this is not the case in practice, but even imperfect push-pull operation reduces even harmonic distortion. Odd harmonics are out of phase, so cancellation of those components does not occur.

Crossover Distortion

Recall that a forward-biasing voltage applied across a PN junction must be raised to a certain level (about 0.7 V for silicon) before the junction will conduct any significant current. Similarly, the voltage across the base–emitter junction of a transistor must reach that level before any appreciable base current, and hence collector current, can flow. As a consequence, the drive signal applied to a class-B transistor must reach a certain minimum level before its collector current is properly in the active region. This fact is the principal source of distortion in a class-B, push-pull amplifier, as illustrated in Figure 16–22. Figure 16–22(a) shows that the initial rise of collector current in a class-B transistor lags the initial rise of input voltage for the reason we have described. Also, collector current prematurely drops to 0 when the input voltage approaches 0. Figure 16–22(b) shows the voltage waveform that is produced in the load of a push-pull amplifier when the distortion

Figure 16–22
Crossover distortion

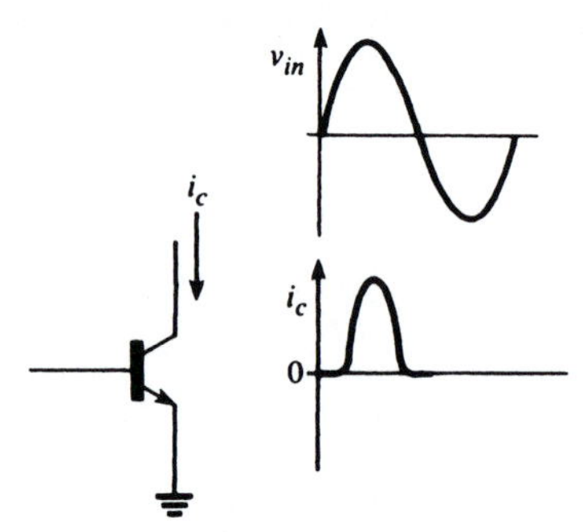

(a) Collector current in a class-B transistor does not follow input voltage in the regions near 0 (crossovers)

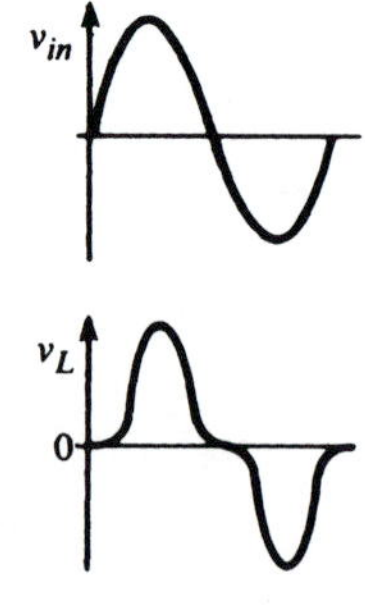

(b) The load voltage in a class-B push-pull amplifier, showing the combined effects of the distortion generated during each half-cycle of input

generated during each half-cycle by each class-B transistor is combined. This distortion is called *crossover* distortion because it occurs where the composite waveform crosses the zero-voltage axis. Clearly, the effect of crossover distortion becomes more serious as the signal level becomes smaller.

Class-AB Operation

Crossover distortion can be reduced or eliminated in a push-pull amplifier by biasing each transistor slightly into conduction. When a small forward-biasing voltage is applied across each base–emitter junction, and a small base current flows under no-signal conditions, it is not necessary for the base drive signal to overcome the built-in junction potential before active operation can occur. A simple voltage-divider bias network can be connected across each base for this purpose, as shown in Figure 16–23. Figure 16–23(a) shows how two resistors can provide bias for both transistors when a driver transformer is used. Figure 16–23(b) shows the use of two voltage dividers when the drive signals are capacitor-coupled. Typically, the

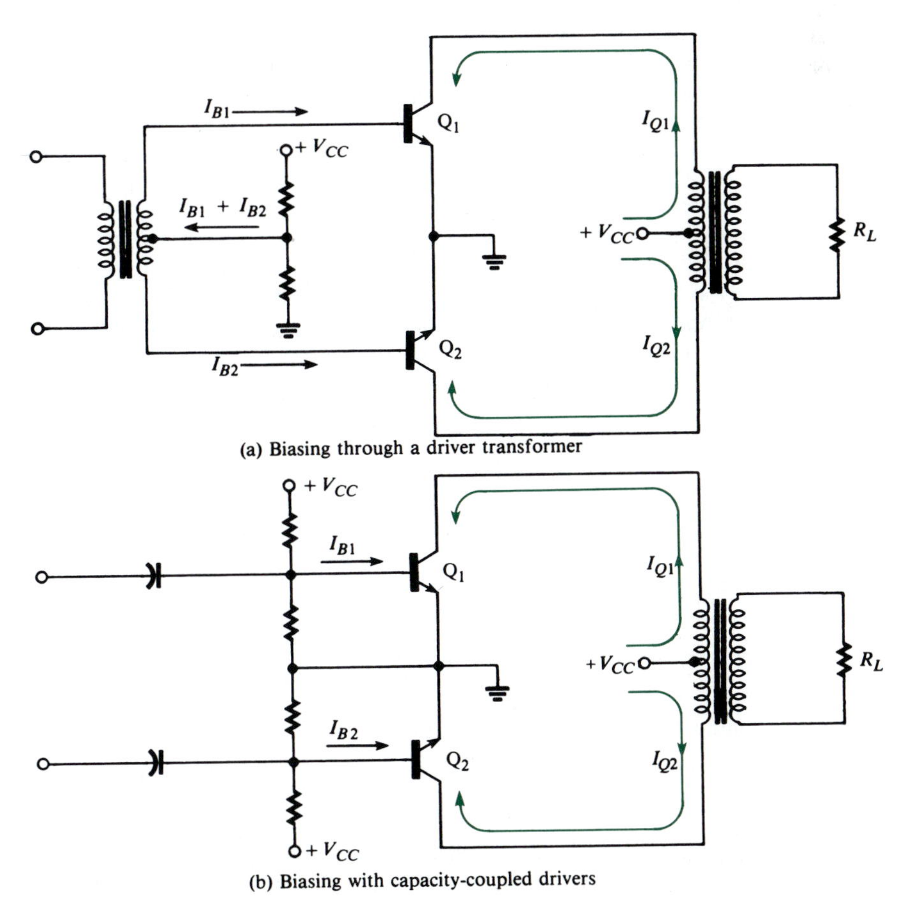

(a) Biasing through a driver transformer

(b) Biasing with capacity-coupled drivers

Figure 16–23
Methods for providing a slight forward bias for push-pull transistors to reduce crossover distortion

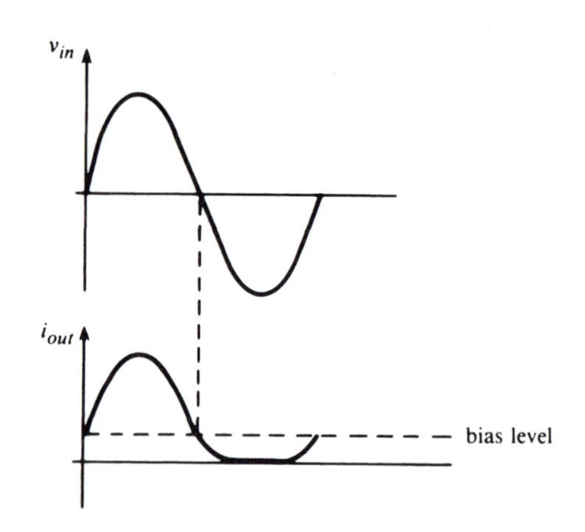

Figure 16–24
Class-AB operation. Output current i_{out} flows during more than one-half but less than a full cycle of input.

base–emitter junctions are biased to about 0.5 V for silicon transistors, or so that the collector current under no-signal conditions is about 1% of its peak signal value.

When a transistor is biased slightly into conduction, output current will flow during more than one-half cycle of a sine-wave input, as illustrated in Figure 16–24. As can be seen in the figure, conduction occurs for more than one-half but less than a full cycle of input. This operation, which is neither class A nor class B, is called *class-AB* operation.

While class-AB operation reduces crossover distortion in a push-pull amplifier, it has the disadvantage of reducing amplifier efficiency. The fact that bias current is always present means that there is continuous power dissipation in both transistors, including the time intervals during which one of the transistors would be cut off if the operation were class B. The extent to which efficiency is reduced depends directly on how heavily the transistors are biased, and the maximum achievable efficiency is somewhere between that which can be obtained in class-A operation (0.5) and that attainable in class-B operation (0.785).

In Figure 16–23, notice that the quiescent collector currents I_{Q1} and I_{Q2} flow in opposite directions through the primary of the transformer. Thus, the magnetic flux created in the transformer by one dc current opposes that created by the other, and the net flux is 0. This is an advantageous situation, in comparison with the class-A transformer-coupled amplifier, because it means that transformer current can swing positive and negative through a maximum range. If the transformer flux had a bias component, the signal swing would be limited in one direction by the onset of magnetic saturation.

16–9 TRANSFORMERLESS PUSH-PULL AMPLIFIERS

Complementary Push-Pull Amplifiers

The principal disadvantage of the push-pull amplifier circuits we have discussed so far is the cost and bulk of their output transformers. High-power amplifiers in particular are encumbered by the need for very large transformers capable of conducting large currents without saturating. Figure 16–25(a) shows a popular design using *complementary* (PNP and NPN) output transistors to eliminate the need for an output transformer in push-pull operation. This design also eliminates the need for a driver transformer or any other drive circuitry producing out-of-phase signals.

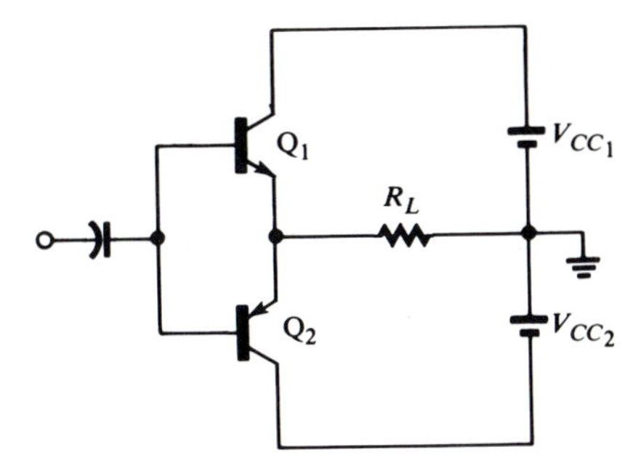

(a) Complementary push-pull amplifier

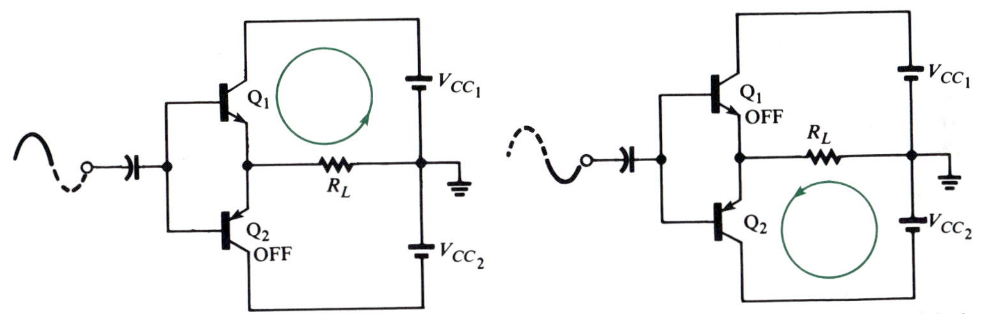

(b) When the input is positive, Q_1 conducts and Q_2 is cut off.

(c) When the input is negative, Q_2 conducts and Q_1 is cut off.

Figure 16–25
Push-pull amplification using complementary transistors. Note that the direction of current through R_L alternates each half-cycle, as required.

Figure 16–25(b) shows that current flows in a counterclockwise path through the load when the input signal on the base of NPN transistor Q_1 is positive. The positive input simultaneously appears on the base of PNP transistor Q_2 and keeps it cut off. When the input is negative, Q_1 is cut off and Q_2 conducts current through the load in the opposite direction, as shown in 16–25(c).

Note that each transistor in Figure 16–25 drives the load in an emitter-follower configuration. The advantageous consequence is that low-impedance loads can be driven from a high-impedance source. Also, the large negative feedback inherent in emitter-follower operation reduces the problem of output distortion. However, as is the case in all emitter followers, voltage gains greater than unity cannot be realized. The maximum positive voltage swing is V_{CC1} and the maximum negative swing is V_{CC2}. Normally, $|V_{CC1}| = |V_{CC2}| = V_{CC}$, so the maximum peak-to-peak swing is $2V_{CC}$ volts. Since the voltage gain is near unity, the input must also swing through $2V_{CC}$ volts to realize maximum output swing. Notice that under conditions of maximum swing, the cutoff transistor experiences a maximum reverse-biasing collector-to-base voltage of $2V_{CC}$ volts. For example, when Q_1 is off, its collector is at $+V_{CC}$ and its base voltage (the input signal) swings to $-V_{CC}$. Thus, each transistor must have a rated breakdown voltage of at least $2V_{CC}$.

Example 16–10

The amplifier in Figure 16–25 must deliver 30 W to a 15-Ω load under maximum drive.

1. What is the minimum value required for each supply voltage?
2. What minimum collector-to-base breakdown-voltage rating should each transistor have?

Solution

1. $$P_L(\text{max}) = \frac{V_P^2(\text{max})}{2R_L}$$

$$V_p(\text{max}) = \sqrt{2R_L P_L(\text{max})} = \sqrt{2(15)(30)} = 30 \text{ V}$$

Therefore, for maximum swing, $V_{CC} = V_P(\text{max}) = 30$ V; i.e., $V_{CC1} = +30$ V and $V_{CC2} = -30$ V.

2. The minimum rated breakdown is $2V_{CC} = 60$ V.

Example 16–11

SPICE

The push-pull amplifier in Figure 16–25 has $|V_{CC1}| = |V_{CC2}| = 30$ V and $R_L = 15$ Ω. The input coupling capacitor is 100 μF. Assuming the transistors have their default parameters, use SPICE to obtain a plot of one full cycle of output when the input is

1. a 1-kHz sine wave with peak value 30 V;
2. a 1-kHz sine wave with peak value 2 V.

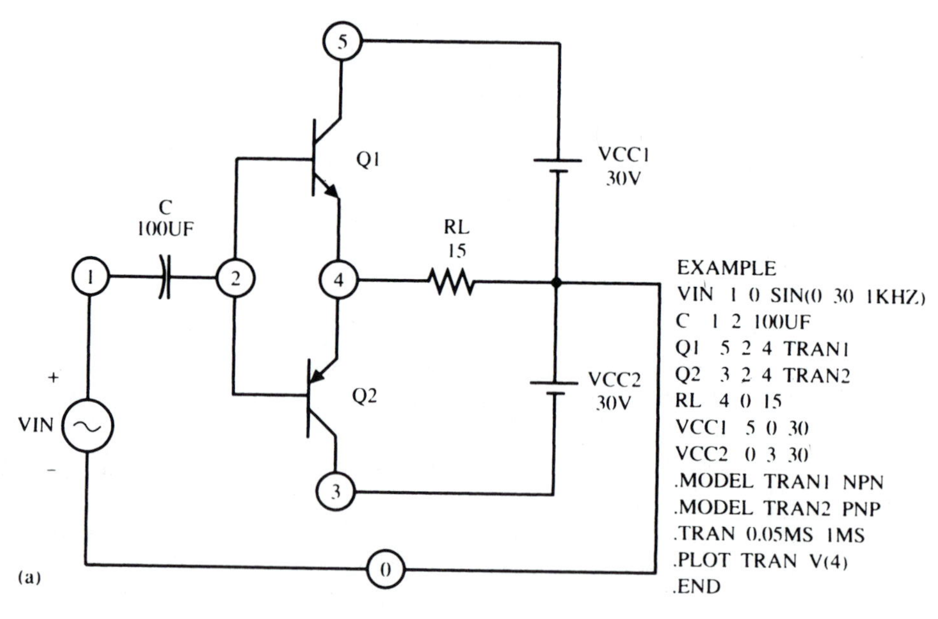
5
Q1
VCC1
30V
C
100UF
RL
15
1
2
4
+
Q2
VCC2
30V
VIN
-
3
0
(a)
EXAMPLE
VIN 1 0 SIN(0 30 1KHZ)
C 1 2 100UF
Q1 5 2 4 TRAN1
Q2 3 2 4 TRAN2
RL 4 0 15
VCC1 5 0 30
VCC2 0 3 30
.MODEL TRAN1 NPN
.MODEL TRAN2 PNP
.TRAN 0.05MS 1MS
.PLOT TRAN V(4)
.END

Figure 16–26
(Example 16–11)

```
TIME        V(4)
              -4.000D+01    -2.000D+01     0.000D+00     2.000D+01   4.000D+01
- - - - - - - - - - - - - - - - - - - - - - - - - - - - - - - - - - - - - - - -
0.000D+00 -5.841D-29 .            .            *            .            .
5.000D-05  8.326D+00 .            .            .     *      .            .
1.000D-04  1.664D+01 .            .            .          * .            .
1.500D-04  2.328D+01 .            .            .            . *          .
2.000D-04  2.749D+01 .            .            .            .    *       .
2.500D-04  2.898D+01 .            .            .            .     *      .
3.000D-04  2.747D+01 .            .            .            .    *       .
3.500D-04  2.324D+01 .            .            .            . *          .
4.000D-04  1.659D+01 .            .            .          * .            .
4.500D-04  8.270D+00 .            .            .     *      .            .
5.000D-04 -6.253D-14 .            .            *            .            .
5.500D-04 -8.374D+00 .            .      *     .            .            .
6.000D-04 -1.673D+01 .            . *          .            .            .
6.500D-04 -2.331D+01 .          * .            .            .            .
7.000D-04 -2.759D+01 .       *    .            .            .            .
7.500D-04 -2.900D+01 .      *     .            .            .            .
8.000D-04 -2.757D+01 .       *    .            .            .            .
8.500D-04 -2.327D+01 .          * .            .            .            .
9.000D-04 -1.668D+01 .            . *          .            .            .
9.500D-04 -8.317D+00 .            .      *     .            .            .
1.000D-03  3.754D-12 .            .            *            .            .
- - - - - - - - - - - - - - - - - - - - - - - - - - - - - - - - - - - - - - - -
```

(b)

```
TIME        V(4)
              -2.000D+00    -1.000D+00     0.000D+00     1.000D+00   2.000D+00
- - - - - - - - - - - - - - - - - - - - - - - - - - - - - - - - - - - - - - - -
-0.000D+00 -5.841D-29 .            .            *            .            .
5.000D-05  1.359D-03 .            .            *            .            .
1.000D-04  3.212D-01 .            .            .   *        .            .
1.500D-04  7.411D-01 .            .            .        *   .            .
2.000D-04  1.015D+00 .            .            .            *            .
2.500D-04  1.111D+00 .            .            .            . *          .
3.000D-04  1.015D+00 .            .            .            *            .
3.500D-04  7.397D-01 .            .            .        *   .            .
4.000D-04  3.200D-01 .            .            .   *        .            .
4.500D-04  4.242D-03 .            .            *            .            .
5.000D-04 -4.141D-13 .            .            *            .            .
5.500D-04 -1.922D-04 .            .            *            .            .
6.000D-04 -3.227D-01 .            .       *    .            .            .
6.500D-04 -7.428D-01 .            .   *        .            .            .
7.000D-04 -1.017D+00 .            *            .            .            .
7.500D-04 -1.113D+00 .          * .            .            .            .
8.000D-04 -1.016D+00 .            *            .            .            .
8.500D-04 -7.414D-01 .            .   *        .            .            .
9.000D-04 -3.216D-01 .            .       *    .            .            .
9.500D-04 -1.290D-04 .            .            *            .            .
1.000D-03  1.776D-18 .            .            *            .            .
- - - - - - - - - - - - - - - - - - - - - - - - - - - - - - - - - - - - - - - -
```

(c)

Figure 16–26
(Continued)

Solution. Figure 16–26(a) shows the SPICE circuit and input data file for the case where the input signal has peak value 30 V. The resulting plot is shown in Figure 16–26(b). We see that the computed output swings from −29 to +28.98 V. The output that results when the peak value of the input is changed to 2 V is shown in Figure 16–26(c). Crossover distortion is now clearly apparent in this low-amplitude case.

Figure 16–27
Variations on the basic complementary push-pull amplifier

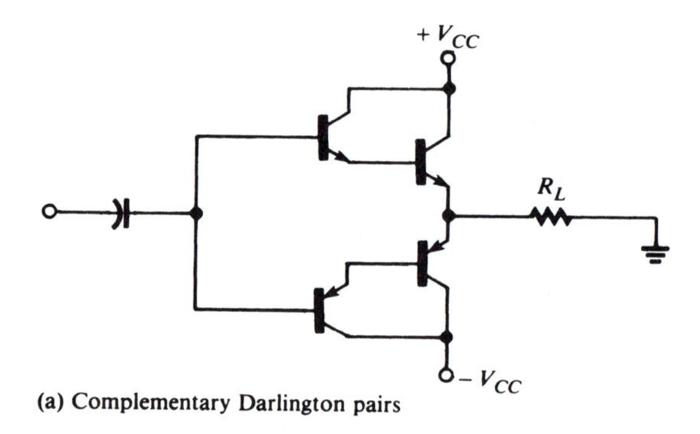

(a) Complementary Darlington pairs

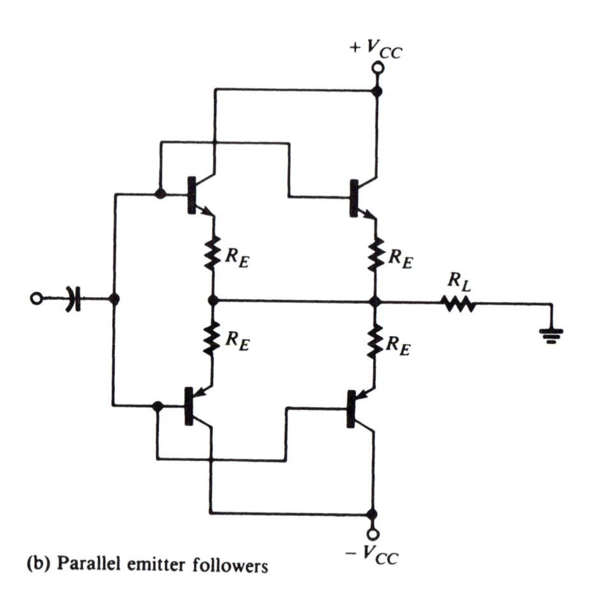

(b) Parallel emitter followers

Figure 16–27 shows two variations on the basic complementary, push-pull amplifier. In (a), the transistors are replaced by emitter-follower Darlington pairs. Since power transistors tend to have low betas, particularly at high current levels, the Darlington pair improves the drive capabilities and the current gain of the amplifier, as discussed in Chapter 11. These devices are available in matched complementary sets with current ratings up to 20 A. Figure 16–27(b) shows how emitter-follower transistors can be operated in parallel to increase the overall current-handling capability of the amplifier. In this variation, the parallel transistors must be matched closely to prevent "current hogging," wherein one device carries most of the load, thus subverting the intention of load sharing. Small emitter resistors R_E shown in the figure introduce negative feedback and help prevent current hogging, at the expense of efficiency. Amplifiers capable of dissipating several hundred watts have been constructed using this arrangement.

One disadvantage of the complementary push-pull amplifier is the need for two power supplies. Also, like the transformer-coupled push-pull amplifier, the complementary class-B amplifier produces crossover distortion in its output. Figure 16–28(a) shows another version of the complementary amplifier that eliminates these problems and that incorporates some additional features.

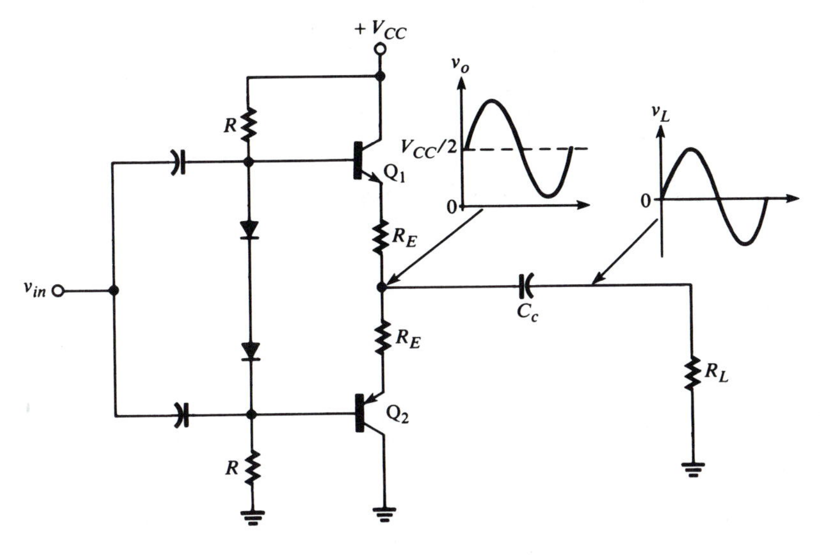

(a) The output is biased at $V_{CC}/2$ volts.

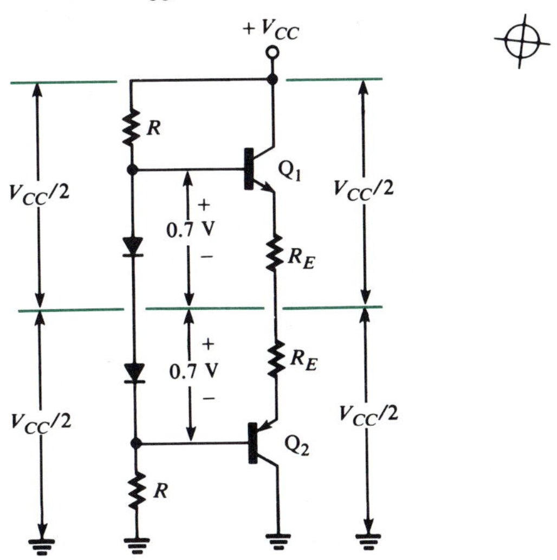

(b) DC bias voltages (All components are assumed to be perfectly matched.)

Figure 16–28
Complementary push-pull amplifier using a single power supply

The complementary amplifier in Figure 16–28 can be operated with a single power supply because the output v_o is biased at half the supply voltage and is capacitor-coupled to the load. The resistor–diode network connected across the transistor bases is used to bias each transistor near the threshold of conduction. Crossover distortion can be reduced or eliminated by inserting another resistor (not shown in the figure) in series with the diodes to bias the transistors further into AB operation. Assuming that all components are perfectly matched, the supply voltage will divide equally across each half of the amplifier, as shown in Figure 16–28(b). (In practice, one of the resistors R can be made adjustable for balance purposes.) Resistors R_E provide bias stability to prevent thermal runaway (see Section 6–1) but are made as small as possible since they adversely affect efficiency. Since each half of the amplifier has $V_{CC}/2$ volts across it, the forward-biased diode drops appear across the base–emitter junctions with the proper polarity to bias each transistor toward conduction. The diodes are selected so that their characteristics track the base–emitter junctions under temperature changes and thus ensure bias stability. The diodes are typically mounted on the same heat sinks as the transistors so that both change temperature in the same way.

In ac operation, when input v_1 is positive and Q_1 is conducting, current is drawn from the power supply and flows through Q_1 to the load. When Q_1 is cut off by a negative input, no current can flow from the supply. At those times, Q_2 is conducting and capacitor C_c discharges through that transistor. Thus, current flows from the load, through C_c, and through Q_2 to ground whenever the input is negative. The $R_L C_c$ time constant must be much greater than the period of the lowest signal frequency. The lower cutoff frequency due to C_c is given by

$$f_1(C_c) = \frac{1}{2\pi(R_L + R_E)C_c} \text{ Hz} \quad \textbf{(16–60)}$$

The peak load current is the peak input voltage V_P divided by $R_L + R_E$:

$$I_{PL} = \frac{V_P}{R_L + R_E} \quad \textbf{(16–61)}$$

Therefore, the average ac power delivered to the load is

$$P_L = \frac{I_{PL}^2 R_L}{2} = \frac{V_P^2 R_L}{2(R_L + R_E)^2} \quad \textbf{(16–62)}$$

Since current is drawn from the power supply only during positive half-cycles of input, the supply-current waveform is half-wave rectified, with peak value $V_P/(R_L + R_E)$ amperes. Therefore, the average value of the supply current is

$$I_S(\text{avg}) = \frac{V_P}{\pi(R_L + R_E)} \quad \textbf{(16–63)}$$

and the average power from the supply is

$$P_S = V_{CC} I_S(\text{avg}) = \frac{V_{CC} V_P}{\pi(R_L + R_E)} \quad \textbf{(16–64)}$$

Dividing (16–62) by (16–64), we find the efficiency to be

$$\eta = \frac{P_L}{P_S} = \frac{\pi}{2}\left(\frac{R_L}{R_L + R_E}\right)\left(\frac{V_P}{V_{CC}}\right) \quad \textbf{(16–65)}$$

The efficiency is maximum when the peak voltage V_P has its maximum possible value, $V_{CC}/2$. In that case,

$$\eta(\text{max}) = \frac{\pi}{2}\left(\frac{R_L}{R_L + R_E}\right)\left(\frac{V_{CC}/2}{V_{CC}}\right) = \frac{\pi}{4}\left(\frac{R_L}{R_L + R_E}\right) \quad \textbf{(16–66)}$$

Equations 16–65 and 16–66 show that efficiency decreases, as expected, when R_E is increased. If $R_E = 0$, then the maximum possible efficiency becomes $\eta(\text{max}) = \pi/4 = 0.785$, the theoretical maximum for a class-B amplifier.

Example 16–12

Assuming that all components in Figure 16–29 are perfectly matched, find

1. the base-to-ground voltages V_{B1} and V_{B2} of each transistor;
2. the power delivered to the load under maximum signal conditions;
3. the efficiency under maximum signal conditions; and
4. the value of capacitor C_c if the amplifier is to be used at signal frequencies down to 20 Hz.

Solution

1. Since all components are matched, the supply voltage divides equally across the resistor–diode network, as shown in Figure 16–30. Then, as can be seen from the figure, $V_{B1} = 10 + 0.7 = 10.7$ V and $V_{B2} = V_{B1} - 1.4 = 9.3$ V.

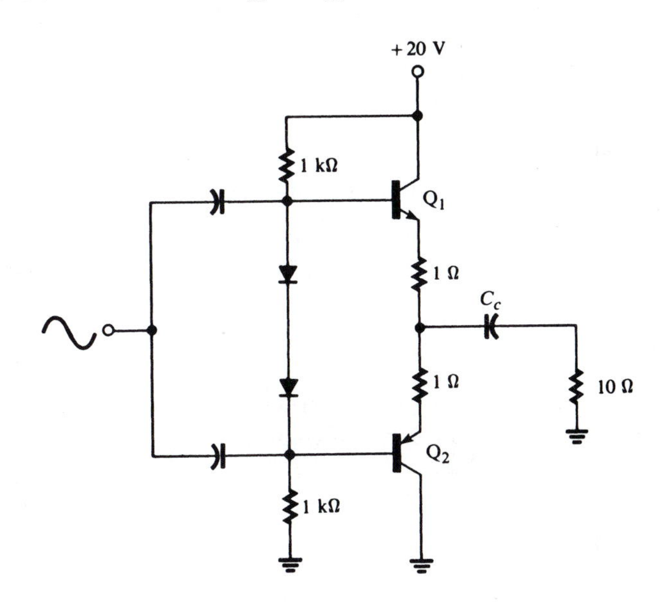

Figure 16–29
(Example 16–12)

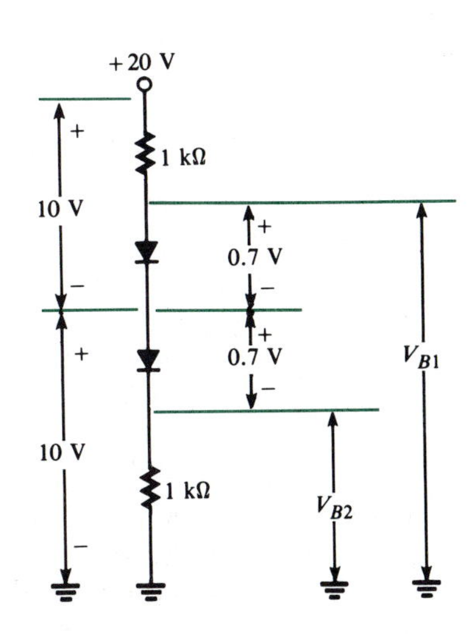

Figure 16–30
(Example 16–12)

2. From equation 16–62, with $V_P(\text{max}) = V_{CC}/2 = 10$ V,

$$P_L(\text{max}) = \frac{V_P^2(\text{max})R_L}{2(R_L + R_E)^2} = \frac{10^2(10)}{2(10 + 1)^2} = 4.132 \text{ W}$$

3. From equation 16–66,

$$\eta(\text{max}) = \frac{\pi}{4}\left(\frac{R_L}{R_L + R_E}\right) = \frac{\pi}{4}\left(\frac{10}{11}\right) = 0.714$$

4. From equation 16–60,

$$C_c = \frac{1}{2\pi(R_L + R_E)f_1(C_c)} = \frac{1}{2\pi(11)(20)} = 723 \ \mu\text{F}$$

This result demonstrates a principal disadvantage of the complementary, single-supply, push-pull amplifier: the coupling capacitor must be quite large to drive a low-impedance load at a low frequency. These operating requirements are typical for audio power amplifiers.

Quasi-Complementary Push-Pull Amplifiers

As has been discussed in previous chapters, modern semiconductor technology is such that NPN transistors are generally superior to PNP types. Figure 16–31 shows a popular push-pull amplifier design that uses NPN transistors for both output devices. This design is called *quasi-complementary* because transistors Q_3 and Q_4 together perform the same function as the PNP transistor in a complementary push-pull amplifier. When the input signal is positive, PNP transistor Q_3 is cut off, so NPN transistor Q_4 receives no base current and is also cut off. When the input is negative, Q_3 conducts and supplies base current to Q_4, which can then conduct load current. Transistors Q_1 and Q_2 form an NPN Darlington pair, so Q_1 and Q_2 provide emitter-follower action to the load when the input is positive, and Q_3 and Q_4 perform

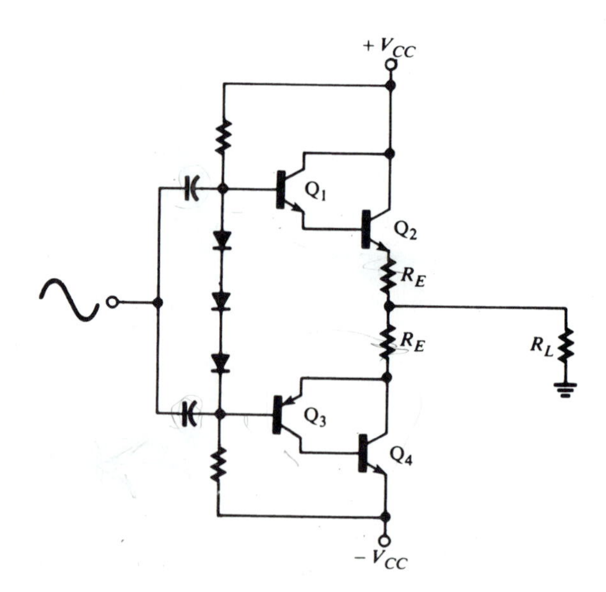

Figure 16–31
A quasi-complementary push-pull amplifier. Transistors Q_3 and Q_4 conduct during negative half-cycles of input.

that function when the input is negative. The entire configuration thus performs push-pull operation in the same manner as the complementary push-pull amplifier. The current gain of the Q_3Q_4 combination is

$$\frac{I_E}{I_B} = \beta_3\beta_4 + (1 + \beta_3) \tag{16–67}$$

This gain is very nearly that of a Darlington pair. Transistors Q_1 and Q_2 are connected as a Darlington pair to ensure that both sides of the amplifier have similar gain.

Integrated-Circuit Power Amplifiers

Low- to medium-power audio amplifiers (in the 1- to 20-W range) are available in integrated circuit form. Many have differential inputs and quasi-complementary outputs. Figure 16–32 shows an example, the LM380 2.5-W audio amplifier manufac-

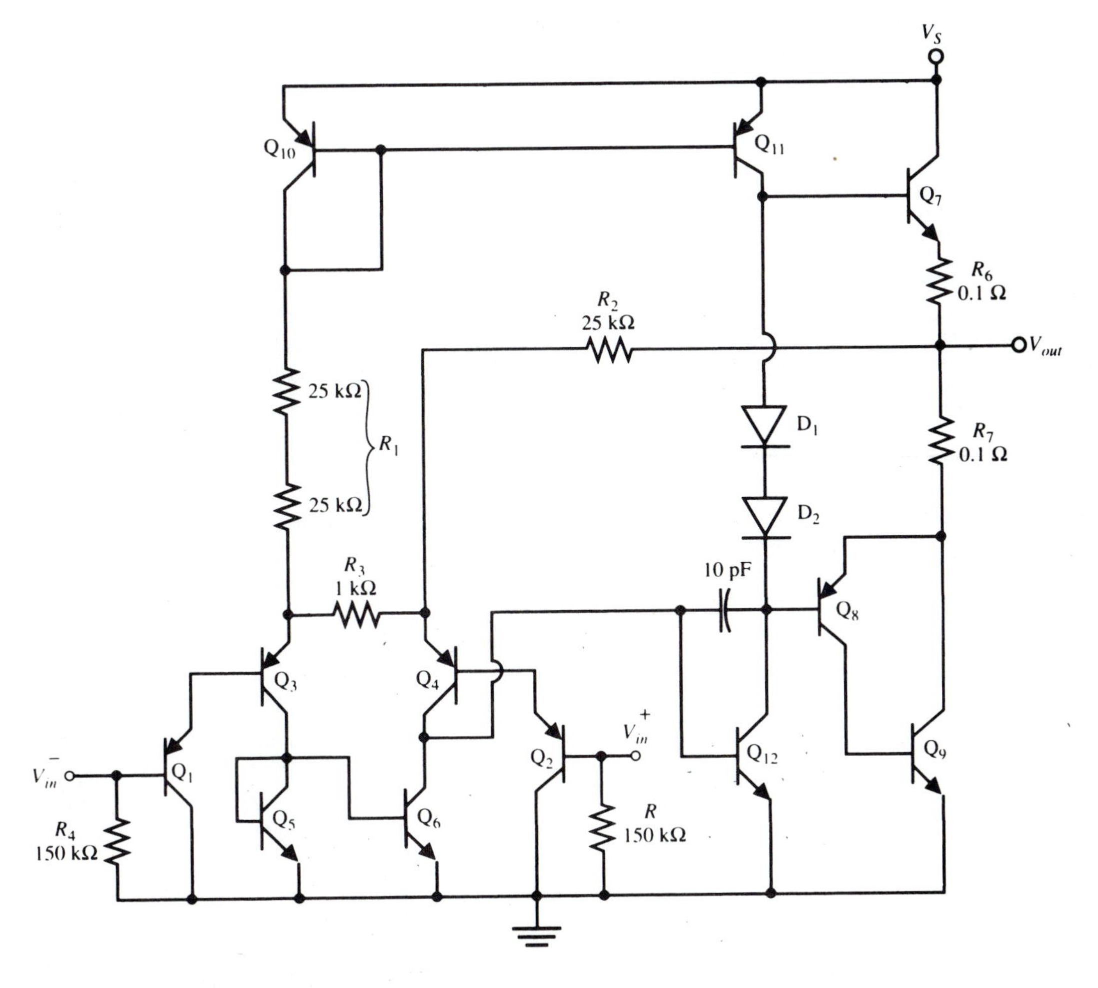

Figure 16–32
Simplified schematic diagram of the LM380 integrated-circuit amplifier

tured by National Semiconductor. Transistors Q_1 and Q_2 are input emitter followers that drive the differential pair consisting of Q_3 and Q_4. Transistors Q_5 and Q_6 are active loads in the configuration discussed in Chapter 12 (Figure 12–25). Transistors Q_{10} and Q_{11} form a current mirror supplying bias current (Figure 6–29). The output of the differential stage, at the collector of Q_4, drives a common-emitter stage (Q_{12}), which in turn drives a quasi-complementary output stage. The 10-pF capacitor provides internal frequency compensation.

One feature of the LM380 is that internal resistor networks (R_1 and R_2) are used to set the output bias level automatically at one-half the supply voltage, V_S. The output is normally capacitor-coupled to the load. The voltage gain of the amplifier is fixed at 50 (34 dB). Manufacturer's specifications state that the bandwidth is 100 kHz and that the total harmonic distortion is typically 0.2%. The supply voltage can be set from 10 to 22 V.

16–10 CLASS-C AMPLIFIERS

A class-C amplifier is one whose output conducts load current during *less* than one-half cycle of an input sine wave. Figure 16–33 shows a typical class-C current waveform, and it is apparent that the total angle during which current flows is less than 180°. This angle is called the *conduction angle, θ_c*.

Of course, the output of a class-C amplifier is a highly distorted version of its input. It could not be used in an application requiring high fidelity, such as an audio amplifier. Class-C amplifiers are used primarily in high-power, high-frequency applications, such as radio-frequency transmitters. In these applications, the high-frequency pulses handled by the amplifier are not themselves the signal but constitute what is called the *carrier* for the signal. The signal is transmitted by varying the amplitude of the carrier, using the process called *amplitude modulation* (AM). The signal is ultimately recovered in a *receiver* by filtering out the carrier frequency. The principal advantage of a class-C amplifier is that it has a very high efficiency, as we shall presently demonstrate.

Figure 16–34 shows a simple class-C amplifier with a resistive load. Note that the base of the NPN transistor is biased by a *negative* voltage, $-V_{BB}$, connected through a coil labeled *RFC*. The RFC is a *radio-frequency choke* whose inductance presents a high impedance to the high-frequency input and thereby prevents the dc source from shorting the ac input. For the transistor to begin conducting, the input must reach a level sufficient to overcome both the negative bias and the V_{BE}

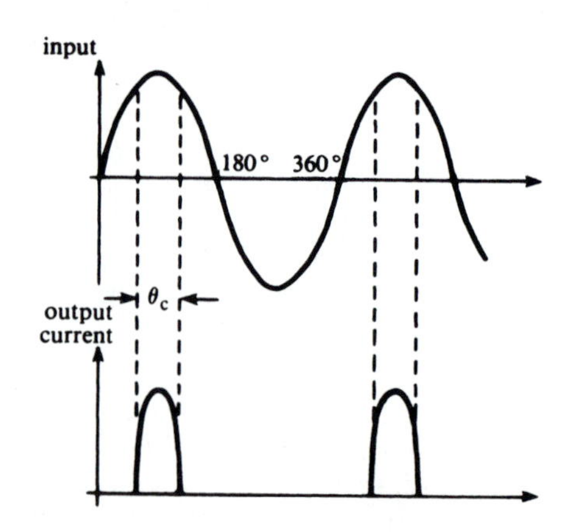

Figure 16–33
Output current in a class-C amplifier

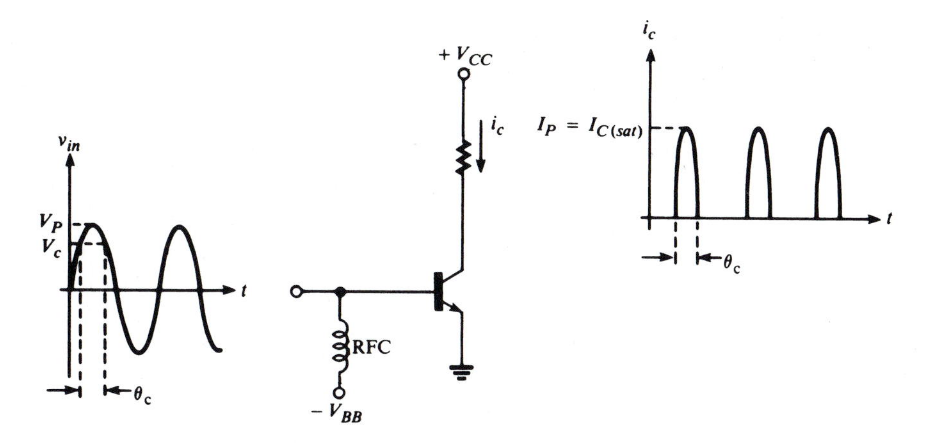

Figure 16–34
A class-C amplifier with resistive load. The transistor conducts when $v_{in} \geq |V_{BB}| + 0.7$.

drop of about 0.7 V:

$$V_c = |V_{BB}| + 0.7 \tag{16–68}$$

where V_c is the input voltage at which the transistor begins to conduct. As shown in the figure, the transistor is cut off until v_{in} reaches V_c, then it conducts, and then it cuts off again when v_{in} falls below V_c. Clearly, the more negative the value of V_{BB}, the shorter the conduction interval. In most class-C applications, the amplifier is designed so that the peak value of the input, V_P, is just sufficient to drive the transistor into saturation, as shown in the figure.

The conduction angle θ_c in Figure 16–34 can be found from

$$\theta_c = 2 \arccos\left(\frac{V_c}{V_P}\right) \tag{16–69}$$

where V_P is the peak input voltage that drives the transistor to saturation. If the peak input only just reaches V_c, then $\theta_c = 2 \arccos(1) = 0°$. At the other extreme, if $V_{BB} = 0$, then $V_c = 0.7$, $(V_c/V_P) \approx 0$, and $\theta_c = 2 \arccos(0) = 180°$, which corresponds to class-B operation.

Figure 16–35 shows the class-C amplifier as it is normally operated, with an LC *tank* network in the collector circuit. Recall that the tank is a *resonant* network whose center frequency, assuming small coil resistance, is closely approximated by

$$f_o \approx \frac{1}{2\pi\sqrt{LC}} \tag{16–70}$$

The purpose of the tank is to produce the fundamental component of the pulsed, class-C waveform, which has the same frequency as v_{in}. The configuration is called a *tuned* amplifier, and the center frequency of the tank is set equal to (tuned to) the input frequency. There are several ways to view its behavior as an aid in understanding how it recovers the fundamental frequency. We may regard the tank as a highly selective (high-Q) filter that suppresses the harmonics in the class-C waveform and passes its fundamental. We may also recall that the voltage gain of the transistor equals the impedance in the collector circuit divided by the emitter

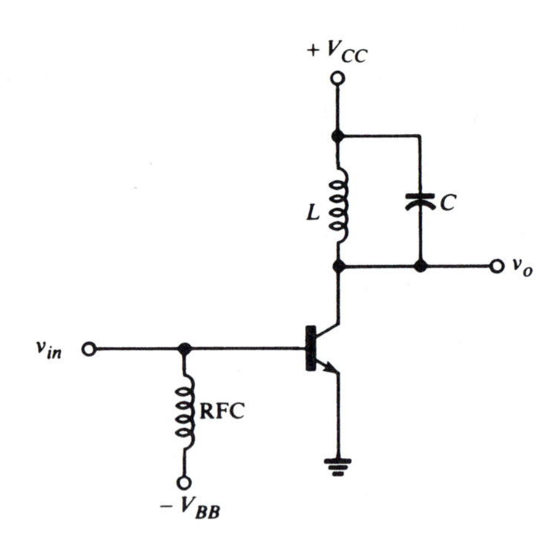

Figure 16–35
A tuned class-C amplifier with an LC tank circuit as load

resistance. Since the impedance of the tank is very large at its center frequency, the gain is correspondingly large at that frequency, while the impedance and the gain at harmonic frequencies are much smaller.

The amplitude of the fundamental component of a class-C waveform depends on the conduction angle θ_c. The greater the conduction angle, the greater the ratio of the amplitude of the fundamental component to the amplitude of the total waveform. Let r_1 be the ratio of the peak value of the fundamental component to the peak value of the class-C waveform. The value of r_1 is closely approximated by

$$r_1 \approx (-3.54 + 4.1\theta_c - 0.0072\theta_c^2) \times 10^{-3} \tag{16–71}$$

where $0° \le \theta_c \le 180°$. The values of r_1 vary from 0 to 0.5 as θ_c varies from 0° to 180°.

Let r_0 be the ratio of the dc value of the class-C waveform to its peak value. The value of r_0 can be found from

$$r_0 = \frac{\text{dc value}}{\text{peak value}} = \frac{\theta_c}{\pi(180)} \tag{16–72}$$

where $0° \le \theta_c \le 180°$. The values of r_0 vary from 0 to $1/\pi$ as θ_c varies from 0° to 180°.

The efficiency of a class-C amplifier is large because very little power is dissipated when the transistor is cut off, and it is cut off during most of every full cycle of input. The output power at the fundamental frequency under maximum drive conditions is

$$P_o = \frac{(r_1 I_P)V_{CC}}{2} \tag{16–73}$$

where I_P is the peak output (collector) current. The average power supplied by the dc source is V_{CC} times the average current drawn from the source. Since current flows only when the transistor is conducting, this current waveform is the same as the class-C collector-current waveform having peak value I_P. Therefore,

$$P_S = (r_0 I_P)V_{CC} \tag{16–74}$$

The efficiency is then

$$\eta = \frac{P_o}{P_S} = \frac{r_1 I_P V_{CC}}{2r_0 I_P V_{CC}} = \frac{r_1}{2r_0} \tag{16–75}$$

Example 16–13

A class-C amplifier has a base bias voltage of −5 V and $V_{CC} = 30$ V. It is determined that a peak input voltage of 9.8 V at 1 MHz is required to drive the transistor to its saturation current of 1.8 A.

1. Find the conduction angle.
2. Find the output power at 1 MHz.
3. Find the efficiency.
4. If an LC tank having $C = 200$ pF is connected in the collector circuit, find the inductance necessary to tune the amplifier.

Solution

1. From equation 16–68, $V_c = |V_{BB}| + 0.7 = 5 + 0.7 = 5.7$ V. From equation 16–69,

$$\theta_c = 2\arccos\left(\frac{V_c}{V_P}\right) = 2\arccos\left(\frac{5.7}{9.8}\right) = 108.9°$$

2. From equation 16–71, $r_1 \approx [-3.54 + 4.1(108.9) - 0.0072(108.9)^2] \times 10^{-3} = 0.357$. From equation 16–73,

$$P_o = \frac{(r_1 I_P)V_{CC}}{2} = \frac{(0.357)(1.8\text{ A})(30\text{ V})}{2} = 9.64\text{ W}$$

3. From equation 16–72,

$$r_0 = \frac{\theta_c}{\pi(180)} = \frac{108.9}{\pi(180)} = 0.193$$

From equation 16–75,

$$\eta = \frac{r_1}{2r_0} = \frac{0.357}{2(0.193)} = 0.925 \text{ (or 92.5\%)}$$

This result demonstrates that a very high efficiency can be achieved in a class-C amplifier.

4. From equation 16–70,

$$L = \frac{1}{(2\pi f_o)^2 C} = \frac{1}{(2\pi \times 10^6)^2(200 \times 10^{-12})} = 0.127\text{ mH}$$

Amplitude Modulation

We have mentioned that amplitude modulation is a means used to transmit signals by varying the amplitude of a high-frequency carrier. In a typical application, the signal is a low-frequency audio waveform and the amplitude of a high-frequency (radio-frequency, or rf) carrier is made to increase and decrease as the audio signal increases and decreases. Figure 16–36 shows the waveforms that are generated by

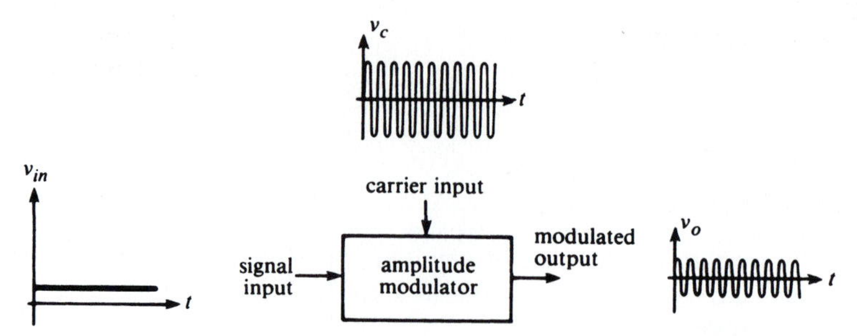

(a) The output of the modulator is a low-amplitude, high-frequency waveform when the signal input is a small dc value.

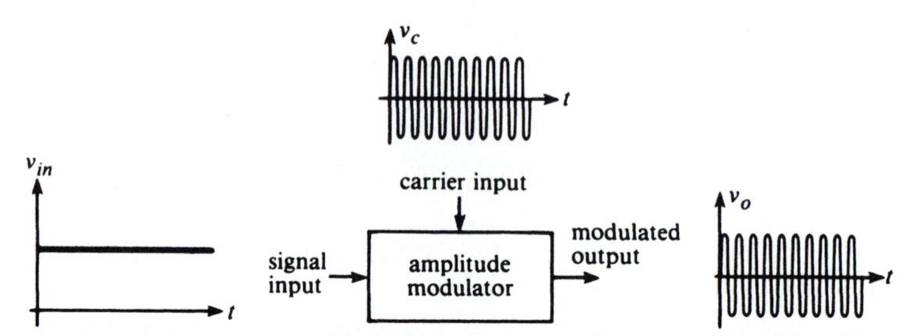

(b) The output of the modulator is a high-amplitude, high-frequency waveform when the signal input is a large dc value.

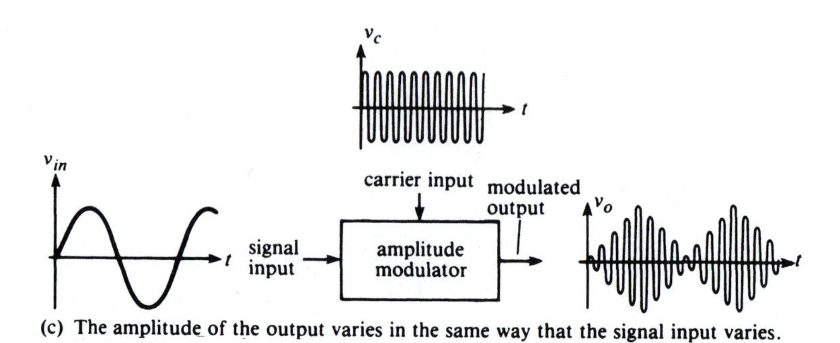

(c) The amplitude of the output varies in the same way that the signal input varies.

Figure 16–36
Amplitude-modulated (AM) waveforms

an amplitude modulator when the signal input is a small dc value, a large dc value, and a low-frequency sinusoidal wave. Notice that the carrier input to the modulator is a constant-amplitude, constant-frequency rf sine wave. It can be seen that the audio waveform is reproduced in the variations of the positive and negative peaks of the output. The audio waveform superimposed on the high-frequency peaks is called the *envelope* of the modulated wave.

It is important to note that amplitude modulation is achieved by *multiplying* two waveforms (signal $\times$ carrier), which is a *nonlinear* process. As such, the modulated output contains frequency components that are not present in either the signal or the carrier. Amplitude modulation cannot be achieved by simply adding two waveforms. Unfortunately, the term *mixing* is used in the broadcast industry to mean both the summation of signals and the multiplication of signals (some modulators are called *mixers*), but these are two very distinct processes.

We will now demonstrate that amplitude modulation creates new frequency components. Let the input signal be a pure sine wave designated

$$v_s(t) = A_s \sin \omega_s t \tag{16–76}$$

and let the carrier input be designated

$$v_c(t) = A_c \sin \omega_c t \tag{16–77}$$

where A_s and A_c are the peak values of the signal and the carrier, respectively. The signal and carrier frequencies are $\omega_s = 2\pi f_s$ rad/s and $\omega_c = 2\pi f_c$ rad/s, respectively. The envelope of the modulated output is the time-varying signal voltage added to the peak carrier voltage:

$$\text{envelope} = e(t) = v_s(t) + A_c = A_s \sin \omega_s t + A_c \tag{16–78}$$

The modulated AM waveform has frequency ω_c and has a (time-varying) peak value equal to the envelope:

$$\begin{aligned} v_o(t) &= \underbrace{e(t)}_{\text{envelope = peak value}} \sin \omega_c t \\ &= (A_s \sin \omega_s t + A_c) \sin \omega_c t \\ &= A_s (\sin \omega_c t)(\sin \omega_s t) + A_c \sin \omega_c t \end{aligned} \tag{16–79}$$

Equation 16–79 makes it apparent now that amplitude modulation involves the multiplication of two sine waves. Using a trigonometric identity, equation 16–79 may be expressed as (Exercise 16–34)

$$v_o(t) = A_c \sin \omega_c t + \tfrac{1}{2} A_s \cos(\omega_c - \omega_s)t - \tfrac{1}{2} A_s \cos(\omega_c + \omega_s)t \tag{16–80}$$

Equation 16–80 shows that the modulated waveform contains the new frequency components $\omega_c + \omega_s$ and $\omega_c - \omega_s$, called the *sum and difference frequencies,* as well as a component at the carrier frequency, ω_c. Note that there is *no* frequency component equal to ω_s. In practice, the input signal will consist of a complex waveform containing many different frequencies. Therefore, the AM output will contain many different sum and difference frequencies. The band of difference frequencies is called the *lower sideband* because its frequencies are all less than the carrier frequency, and the band of sum frequencies is called the *upper sideband,* each of its frequencies being greater than the carrier frequency. The significance of this result is that it allows us to determine the bandwidth that an amplifier must have to pass an AM waveform. The bandwidth must extend from the smallest difference frequency to the largest sum frequency.

Example 16–14

An amplitude modulator is driven by a 580-kHz carrier and has an audio signal input containing frequency components between 200 Hz and 9.5 kHz.

1. What range of frequencies must be included in the passband of an amplifier that will be used to amplify the modulated signal?
2. What frequency components are in the lower sideband of the AM output? In the upper sideband?

Solution

1. The smallest difference frequency is $2\pi(580 \times 10^3) - 2\pi(9.5 \times 10^3)$ rad/s, or (580 kHz) − (9.5 kHz) = 570.5 kHz, and the largest sum frequency is (580 kHz) + (9.5 kHz) = 589.5 kHz. Therefore, the amplifier must pass frequencies in the range from 570.5 kHz to 589.5 kHz. An amplifier having only this range would be considered a *narrowband* amplifier.
2. The lowest sideband extends from the smallest difference frequency to the largest difference frequency: [(580 kHz) − (9.5 kHz)] to [(580 kHz) − (200 Hz)], or 570.5 kHz to 579.8 kHz. The upper sideband extends from the smallest sum frequency to the largest sum frequency: [(580 kHz) + (200 Hz)] to [(580 kHz) + (9.5 kHz)], or 580.2 kHz to 589.5 kHz.

Figure 16–37 shows how a class-C amplifier can be used to produce amplitude modulation. The circuit is similar to Figure 16–35, except that the coil in the collector circuit has been replaced by the secondary winding of a transformer. The low-frequency signal input is connected to the transformer primary. The voltage at the collector is then the sum of V_{CC} and a signal proportional to v_s. As v_s increases and decreases, so does the collector voltage. In effect, we are varying the supply voltage on the collector. When the carrier signal on the base drives the transistor to saturation, the peak collector current $I_{c(sat)}$ that flows depends directly on the supply voltage: the greater the supply voltage, the greater the peak current. Since the supply voltage varies with the signal input, so does the peak collector current. As a consequence, the collector current is amplitude modulated by v_s, and the tank circuit produces an AM output voltage, as shown.

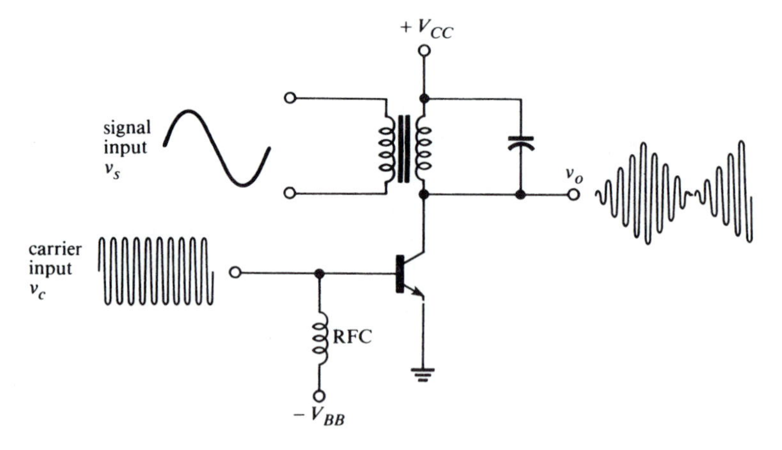

Figure 16–37
A class-C amplitude modulator

16–11 MOSFET AND CLASS-D POWER AMPLIFIERS

MOSFET Amplifiers

MOSFET devices of the VMOS design (Figure 7–47) are becoming a popular choice for power amplifiers, particularly those designed to switch large currents on and off. Examples include line drivers for digital switching circuits, switching-mode voltage regulators (discussed in Chapter 17), and class-D amplifiers, which we will discuss presently. One advantage of a MOSFET switch is the fact that turn-off time is not delayed by minority-carrier storage, as it is in a bipolar switch that has been driven deeply into saturation. Recall that current in field-effect transistors is due to the flow of majority carriers only. Also, MOSFETs are not susceptible to thermal runaway like bipolar transistors. Finally, a MOSFET has a very large input impedance, which simplifies the design of driver circuits.

Figure 16–38 shows a simple, class-A audio amplifier that uses a VN64GA VMOS transistor to drive an output transformer. Notice that the driver for the output stage is a J108 JFET transistor connected in a common-source configuration. Also notice the feedback path between the transformer secondary and the input to the JFET driver. This negative feedback is incorporated to reduce distortion, as discussed earlier. The amplifier will reportedly deliver between 3 and 4 W to an 8-Ω speaker with less than 2% distortion up to 15 kHz.

MOSFET power amplifiers are also constructed for class-B and class-AB operation. Like their bipolar counterparts, these amplifiers can be designed without the

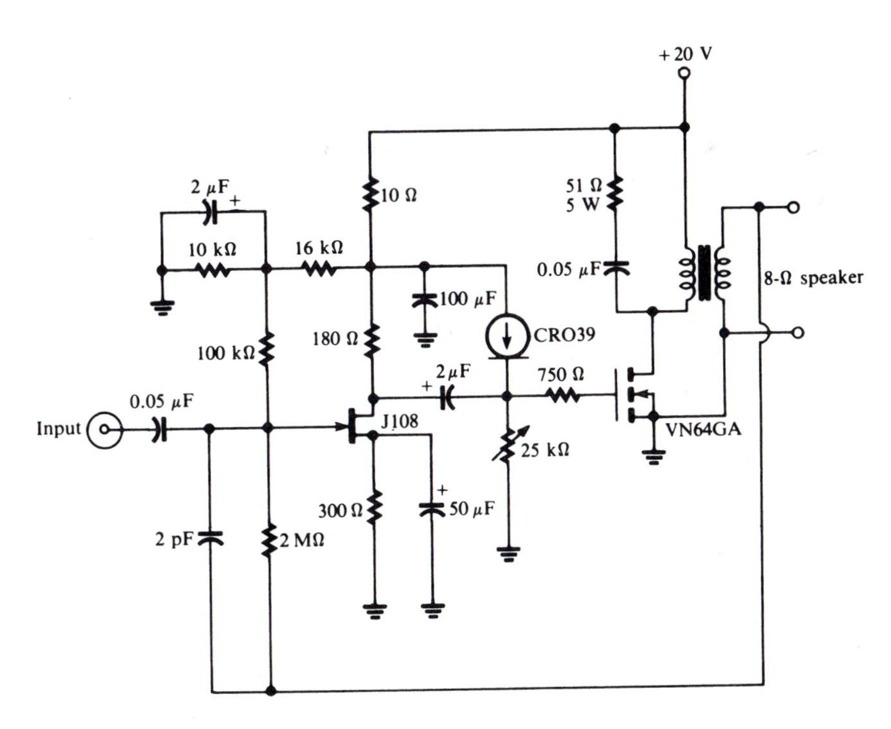

Figure 16–38
A two-stage, class-A audio amplifier with a VMOS output transistor (Courtesy of Siliconix Incorporated)

Figure 16–39
A complementary MOSFET amplifier

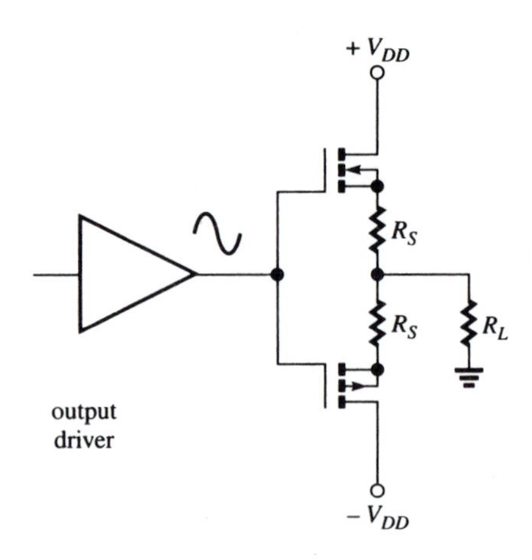

need for an output transformer, since power MOSFETs are available in complementary pairs (N-channel and P-channel). Figure 16–39 shows a typical push-pull MOSFET amplifier using complementary output devices.

Class-D Amplifiers

A class-D amplifier is one whose output is switched on and off, that is, one whose output is in its linear range for essentially *zero* time during each cycle of an input sine wave. As we progress through the letters designating the various classes of operation, A, B, C, and D, we see that linear operation occurs for shorter and shorter intervals of time, and in class D we reach the limiting case where no linear operation occurs at all. The only time that a class-D output device is in its linear region is during that short interval required to switch from saturation to cutoff, or vice versa. In other words, the output device is a digital power switch, an application ideally suited for VMOS transistors.

A fundamental component of a class-D amplifier is a *pulse-width modulator,* which produces a train of pulses having widths that are proportional to the level of the amplifier's input signal. When the signal level is small, a series of narrow pulses is generated, and when the input level is large, a series of wide pulses is generated. (See Figure 17–39, which shows some typical outputs of a pulse-width modulator used in a voltage-regulator application.) As the input signal increases and decreases, the pulse widths increase and decrease in direct proportion.

Figure 16–40 shows how a pulse-width modulator can be constructed using a *sawtooth generator* and a voltage comparator. A sawtooth waveform is one that rises linearly and then quickly switches back to its low level, to begin another linear rise, as illustrated in the figure. When the sawtooth voltage is greater than v_{in}, the output of the comparator is low, and when the sawtooth falls below v_{in}, the comparator switches to its high output. Notice that the comparator must switch high each time the sawtooth makes its vertical descent. These time points mark the beginning of each new pulse. The comparator output remains high until the sawtooth rises back to the value of v_{in}, at which time the comparator output switches low. Thus the width of the high pulse is directly proportional to the length of time it takes the sawtooth to rise to v_{in}, which is directly proportional to the level of v_{in}. As shown in the figure, the result is a series of pulses whose widths are proportional to the level of v_{in}. Notice that the peak-to-peak voltage of the sawtooth must exceed the largest peak-to-peak input voltage for successful operation. Also, the frequency

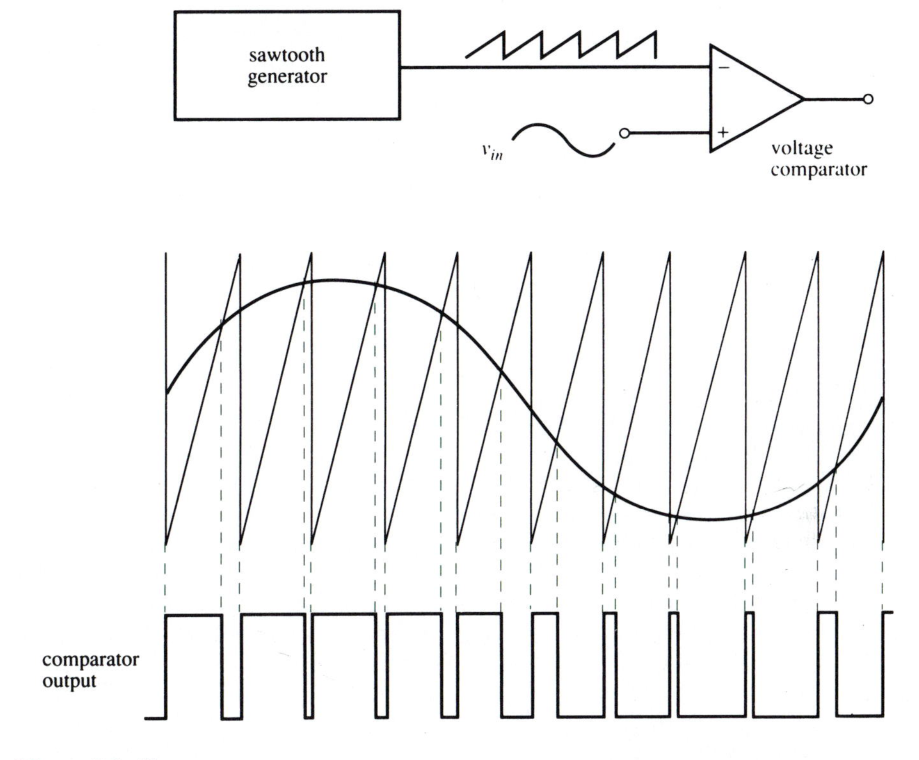

Figure 16–40
Construction of a pulse-width modulator using a sawtooth generator and a voltage comparator

of the sawtooth should be at least ten times as great as the highest frequency component of v_{in}.

The pulse-width modulator drives the output stage of the class-D amplifier, causing it to switch on and off as the pulses switch between high and low. Figure 16–41 shows a popular switching circuit, called a *totem pole,* used to drive heavy

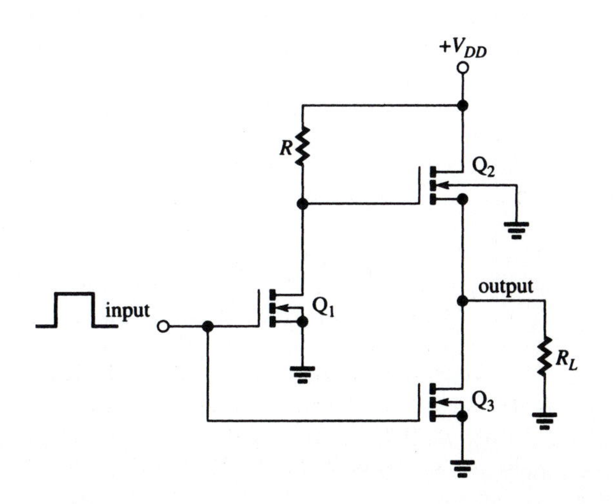

Figure 16–41
A MOSFET totem pole, used to switch heavy load currents

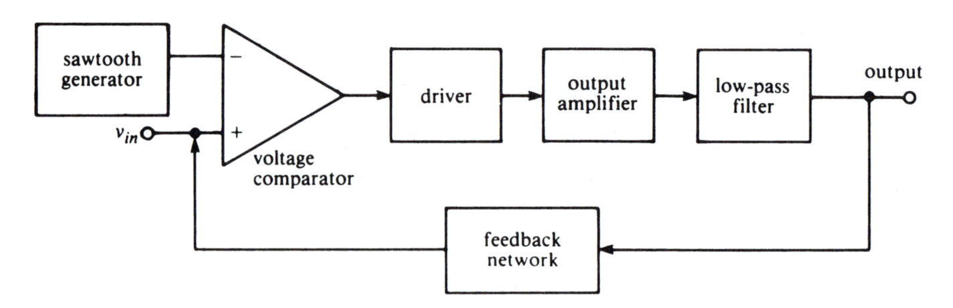

Figure 16–42
Block diagram of a class-D amplifier

loads. The MOSFET version of the totem pole is shown in the figure, because MOSFETs are generally used in class-D amplifiers. Bipolar versions of the totem pole are also widely used in digital logic circuits, particularly in the integrated-circuit family called *TTL* (transistor-transistor logic). The totem pole shown in the figure *inverts,* in the sense that the output is low when the input is high, and vice versa.

When the input in Figure 16–41 is high, Q_1 and Q_3 are on, so the output is low (R_L is effectively connected to ground through Q_3). Since Q_1 is on, the gate of Q_2 is low and Q_2 is held off. Thus, when the input is high, Q_2 is like an open switch and Q_3 is like a closed switch. When the input is low, Q_1 is off and a high voltage (V_{DD}) is applied through R to the gate of Q_2, turning it on. Thus, R_L is connected through Q_2 to the high level V_{DD}. Since Q_3 is also off, a low input makes Q_3 an open switch and Q_2 a closed switch. The advantage of this arrangement is that the

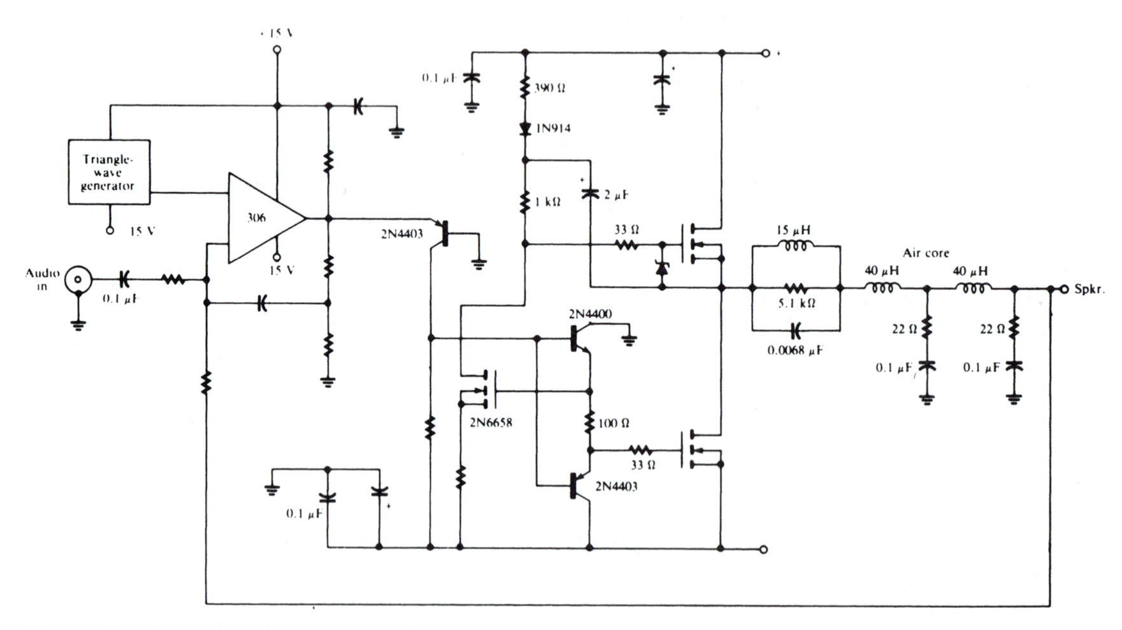

Figure 16–43
A 100-W class-D amplifier (Reprinted with permission from Power FETs and Their Applications *by Edwin S. Oxner, Prentice-Hall, 1982)*

load is driven from a low-impedance signal source, a turned-on MOSFET, both when the output is low and when the output is high.

Like the class-C amplifier, a class-D amplifier must have a *filter* to extract, or recover, the signal from the pulsed waveform. However, in this case the signal may have many frequency components, so a tank circuit, which resonates at a single frequency, cannot be used. Instead, a low-pass filter having a cutoff frequency near the highest signal frequency is used. The low-pass filter suppresses the high-frequency components of the pulse train and, in effect, recovers the *average value* of the pulse train. Since the average value of the pulses depends on the pulse widths, the output of the filter is a waveform that increases and decreases as the pulse widths increase and decrease, that is, a waveform that duplicates the input signal, v_{in}. Figure 16–42 shows a block diagram of a complete class-D amplifier, including negative feedback to reduce distortion.

The principal advantage of a class-D amplifier is that it may have a very high efficiency, approaching 100%. Like that of a class-C amplifier, the high efficiency is due to the fact that the output device spends very little time in its active region, so power dissipation is minimal. The principal disadvantages are the need for a very good low-pass filter and the fact that high-speed switching of heavy currents generates noise through electromagnetic coupling, called *electromagnetic interference,* or *EMI.*

Figure 16–43 shows a 100-W class-D amplifier using a MOSFET totem pole with a bipolar driver.

Example 16–15

A class-D audio amplifier is to be driven by a signal that varies between ±5 V. The output is a MOSFET totem pole with $V_{DD} = 30$ V.

1. What minimum frequency should the sawtooth waveform have?
2. What minimum peak-to-peak voltage should the sawtooth waveform have, assuming that the input is connected directly to the voltage comparator?
3. Assume that the amplifier is 100% efficient and is operated at the frequency found in (1). What average current is delivered to an 8-Ω load when the pulse-width modulator produces pulses that are high for 2.5 μs?

Solution

1. The nominal audio-frequency range is 20 Hz to 20 kHz, and the sawtooth waveform should have a frequency equal to at least 10 times the highest input frequency, i.e., 10(20 kHz) = 200 kHz.
2. The peak-to-peak sawtooth voltage should at least equal the maximum peak-to-peak input voltage, which is 10 V.
3. The period of the 200-kHz sawtooth is

$$T = \frac{1}{200 \times 10^3} = 5\ \mu\text{s}$$

Therefore, the pulse train is high for (2.5 μs)/(5 μs) = one-half the period. The totem-pole output therefore switches between 0 V and 30 V with pulse widths equal to one-half the period. Thus, the average voltage is

$$V_{avg} = \frac{1}{2}(30) = 15\ \text{V}$$

and the average current is

$$I_{avg} = \frac{V_{avg}}{R_L} = \frac{15 \text{ V}}{8 \, \Omega} = 1.875 \text{ A}$$

EXERCISES

Section 16–2
Transistor Power Dissipation

16–1. The transistor in Figure 16–44 is biased at $V_{CE} = 12$ V. If an increase in temperature causes the quiescent collector current to increase, will the power dissipation at the Q-point increase or decrease? Explain. What is the power dissipation at the point of maximum power dissipation?

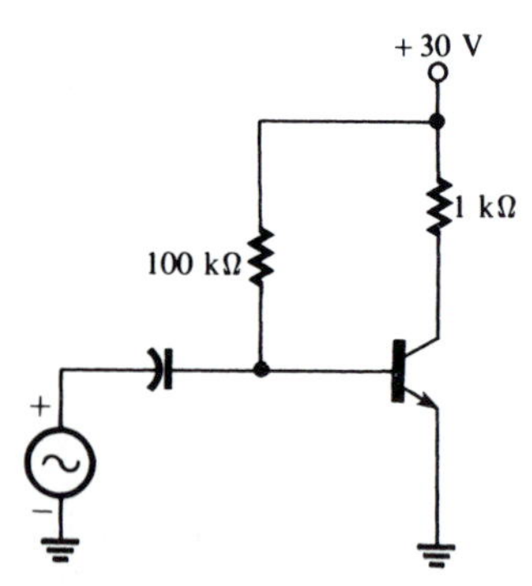

Figure 16–44
(Exercise 16–1)

16–2. Repeat Exercise 16–1 if the supply voltage is changed to 20 V and the Q-point is set at $V_{CE} = 13$ V.

16–3. The amplifier in Figure 16–45 is to be operated with a 40-V-dc supply. If the maximum permissible power dissipation in the transistor is 1 W, what is the minimum permissible value of R_C?

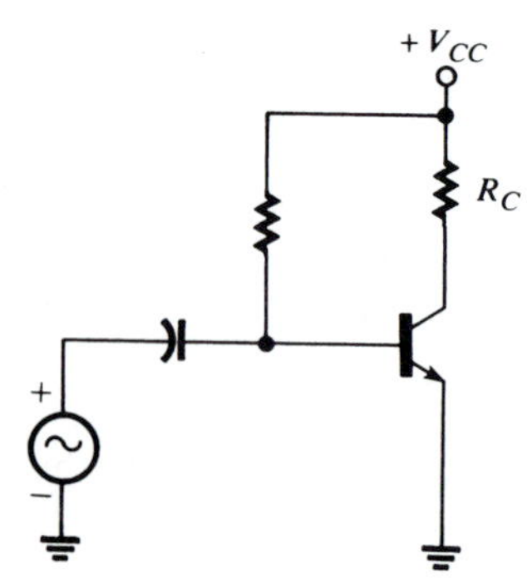

Figure 16–45
(Exercise 16–3)

16–4. If the amplifier in Exercise 16–3 has $R_C = 100 \, \Omega$, what is the maximum permissible value of the supply voltage that can be used?

Section 16–3
Heat Transfer in Semiconductor Devices

16–5. The collector–base junction of a transistor conducts 0.4 A when it is reverse biased by 21 V. At what rate, in joules/second, must heat be removed from the junction to prevent its temperature from rising?

16–6. A semiconductor junction at temperature 64°C releases heat into a 25°C ambient at the rate of 7.8 J/s. What is the thermal resistance between junction and ambient?

16–7. The thermal resistance between a semiconductor device and its case is 0.8°C/W. It is used with a heat sink whose thermal resistance to ambient is 0.5°C/W. If the device dissipates 10 W and the ambient temperature is 47°C, what is the maximum permissible thermal resistance between case and heat sink if the device temperature cannot exceed 70°C?

16–8. A semiconductor dissipates 25 W through its case and its heat sink to the surrounding air. The thermal resistances are the following: device-to-case, 1°C/W; case-to-heat sink, 1.2°C/W; and heat sink-to-air, 0.7°C/W. In what maximum air temperature can the device be operated if its temperature cannot exceed 100°C?

16–9. The mica washer shown in Figure 16–46 has a thermal resistivity of 66°C·in./W. Find its thermal resistance.

16–10. A transistor has a maximum dissipation rating of 10 W at ambient temperatures up to 40°C and is derated at 100 mW/°C at higher temperatures. At what ambient temperature

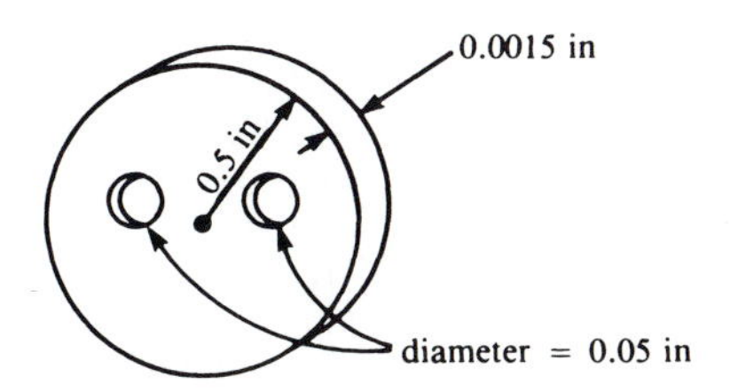

Figure 16–46
(Exercise 16–9)

is the dissipation rating one-half its low-temperature value?

16–11. The temperature of a certain semiconductor device is 100°C when the device dissipates 1.2 W. The device temperature cannot exceed 100°C. If the total thermal resistance from device to ambient is 50°C/W, above what ambient temperature should the dissipation be derated?

Section 16–4
Amplifier Classes and Efficiency

16–12. The amplifier shown in Figure 16–47 is biased at I_Q = 20 mA. Find

a. the ac power in the load resistance when the voltage swing is the maximum possible without distortion;

b. the amplifier efficiency under the conditions of (a); and

c. the collector efficiency under the conditions of (a).

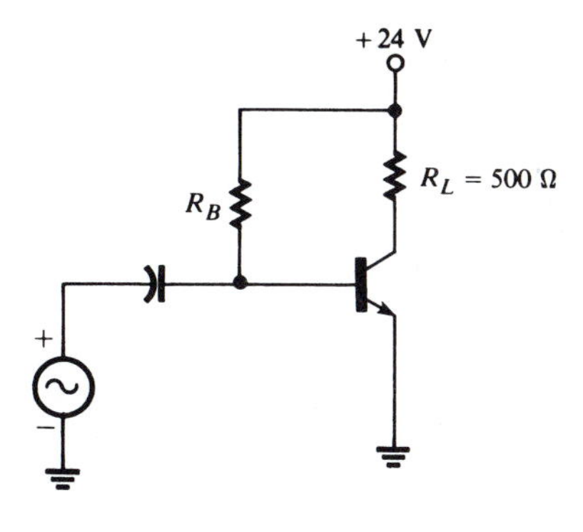

Figure 16–47
(Exercise 16–12)

16–13. a. Find the value of quiescent current in the amplifier of Exercise 16–12 that will maximize the efficiency under maximum output voltage swing.

b. Find the power delivered to the load resistance under the conditions of (a).

16–14. The amplifier shown in Figure 16–48 is biased at I_Q = 0.2 A. Find

a. the ac power in the load resistance when the voltage swing is the maximum possible without distortion;

b. the amplifier efficiency under the conditions of (a); and

c. the collector efficiency under the conditions of (a).

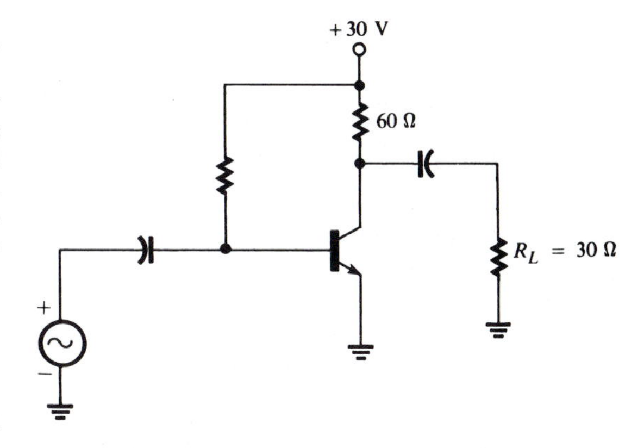

Figure 16–48
(Exercise 16–14)

16–15. a. Find the value of quiescent voltage in the amplifier of Exercise 16–14 that will maximize the efficiency under maximum output voltage swing.

b. Find the efficiency under the conditions of (a).

c. Why is the efficiency less than the theoretical maximum (0.0858)?

16–16. The class-A amplifier in Figure 16–49 is bi-

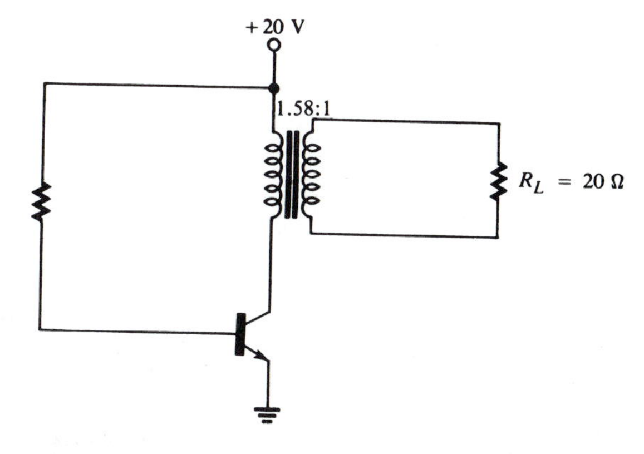

Figure 16–49
(Exercise 16–16)

ased at $I_C = 0.2$ A. The transformer resistance is negligible.

a. What is the slope of the ac load line?

b. At what value does the ac load line intersect the V_{CE}–axis?

c. What is the maximum peak-value of the collector voltage without distortion?

d. What is the maximum power delivered to the load under the conditions of (c)?

e. What is the amplifier efficiency under the conditions of (c)?

16–17. The transistor in Exercise 16–16 has a beta of 40.

a. Find the value of base current necessary to set a new Q-point that will permit maximum possible peak-to-peak output voltage.

b. What power can be delivered to the load under the conditions of (a)?

c. What are the amplifier and collector efficiencies under the conditions of (a)?

Section 16–5
Push-Pull–Amplifier Principles

16–18. Draw a schematic diagram of a class-B, push-pull, transformer-coupled amplifier using PNP transistors. Draw a loop showing current direction through the primary when the input is positive and another showing current when the input is negative.

16–19. The peak collector current and voltage in each transistor in Figure 16–50 are 4 A and 12 V, respectively. Assuming that the transformer has negligible resistance, find

a. the average power delivered to the load;

b. the average power supplied by the dc source;

c. the average power dissipated by each transistor; and

d. the efficiency.

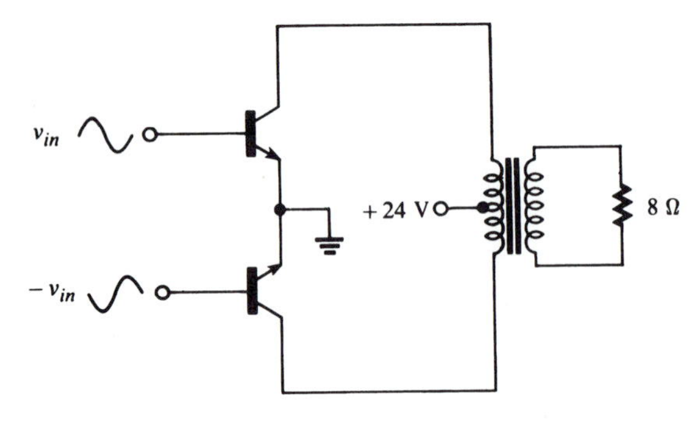

Figure 16–50
(Exercise 16–19)

16–20. Repeat Exercise 16–19 when the drive signals are increased so that the peak collector voltage in each transistor is the maximum possible. Assume that the peak current increases proportionally.

16–21. A push-pull amplifier utilizes a transformer whose primary has a total of 160 turns and whose secondary has 40 turns. It must be capable of delivering 40 W to an 8-Ω load under maximum power conditions. What is the minimum possible value of V_{CC}?

Section 16–7
Harmonic Distortion and Feedback

16–22. A square wave having peak value V_P volts and frequency ω rad/s can be expressed as the sum of an infinite number of sine waves as follows: $v(t) = 4V_P/\pi\,[\sin \omega t + (1/3) \sin 3\omega t + (1/5) \sin 5\omega t + \ldots]$. Assuming that the square wave represents a distorted sine wave whose frequency is the same as the fundamental, find its total harmonic distortion. Neglect harmonics beyond the ninth.

16–23. An amplifier has 1% distortion with negative feedback and 10% distortion without feedback. If the feedback ratio is 0.02, what is the open-loop gain of the amplifier?

16–24. An amplifier has a voltage gain of 80 when no feedback is connected. When negative feedback is connected, the amplifier gain is 20 and its harmonic distortion is 10%. What is the harmonic distortion without feedback?

Section 16–8
Distortion in Push-Pull Amplifiers

16–25. Draw schematic diagrams of class-AB amplifiers using PNP transistors and an output transformer and having

a. an input coupling transformer; and

b. capacitor-coupled inputs.

It is not necessary to show component values, but label the polarities of all power supplies used. In each diagram, draw arrows showing the directions of the quiescent base currents and the dc currents in the primary of the output transformer.

Section 16–9
Transformerless Push-Pull Amplifiers

16–26. a. What is the maximum permissible peak value of input v_{in} in the amplifier shown in Figure 16–51?

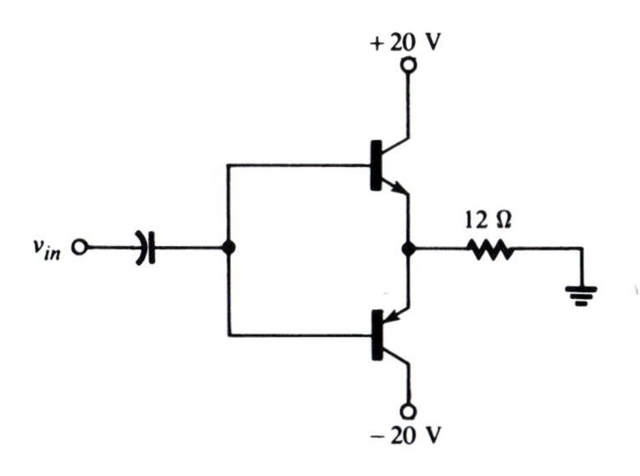

Figure 16–51
(Exercise 16–26)

b. What is the maximum power that can be delivered to the load?
c. What should be the breakdown voltage rating of each transistor?

16–27. If the amplifier in Exercise 16–26 must be redesigned so that it can deliver 36% more power to the load, by what percentage must the supply voltages be increased? What should be the minimum breakdown voltage of the transistors used in the new design?

16–28. Assuming that all the components in Figure 16–52 are perfectly matched, find
a. the current in each 680-Ω resistor;
b. the base-to-ground voltage of each transistor;
c. the efficiency when the peak voltage V_P is 10 V; and
d. the lower cutoff frequency.

Figure 16–52
(Exercise 16–28)

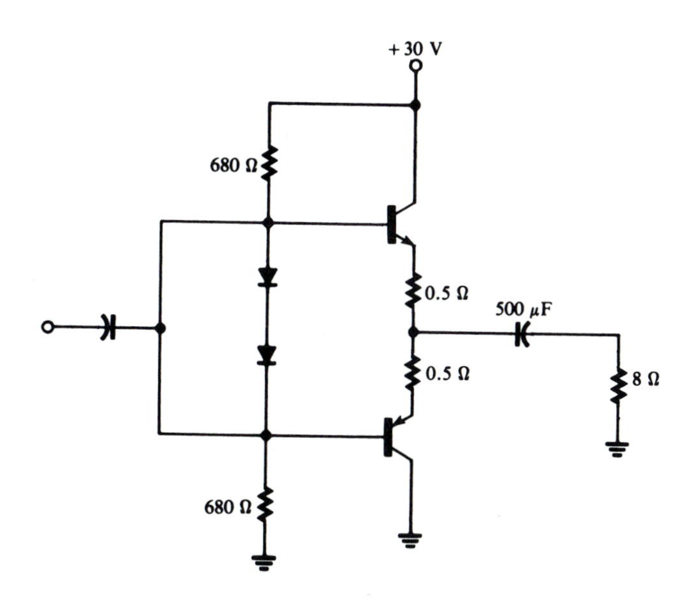

16–29. a. What new value of load resistance is required in the amplifier of Exercise 16–28 if the lower cutoff frequency must be 15 Hz?
b. What is the maximum efficiency with the new load resistance determined in (a)?

Section 16–10
Class-C Amplifiers

16–30. The peak input voltage required to saturate the transistor in Figure 16–34 is 5.5 V. What should be the value of V_{BB} in order to achieve a conduction angle of 90°?

16–31. A peak input of 8 V is required to drive the transistor in Figure 16–34 to its 1.2-A saturation current. It is desired to produce a collector current that has a dc (average) value of 0.3 A. What should be the value of V_{BB}?

16–32. A peak input of 9 V is required to drive the transistor in Figure 16–34 to its 2.0-A saturation current. It is desired to produce a collector current whose fundamental has a peak value of 0.5 A. What should be the value of V_{BB}?

16–33. The input to a class-C amplifier must be 6.8 V to drive the output to its saturation current of 1.0 A. The amplifier begins to conduct when the input voltage reaches 2.6 V.
a. What should be the value of V_{CC} if it is desired to obtain 5 W of output power at the fundamental frequency?
b. What is the efficiency of the amplifier under the conditions of (a)?

16–34. Derive equation 16–80 from equation 16–79. (*Hint:* Use the trigonometric identity for (sin *A*)(sin *B*).)

16–35. An amplitude-modulated waveform has a carrier frequency 810 kHz and an upper sideband that extends from 812.5 kHz to 829 kHz. What is the frequency range of the lower sideband?

16–36. An amplitude modulator is driven by an 8-V-rms, 10-kHz carrier and has a 2-V-rms, 10-kHz signal input. Write the mathematical expression for the modulated output.

Section 16–11
MOSFET and Class-D Power Amplifiers

16–37. The pulse-width modulator in Figure 16–40 is driven by a 100-kHz sawtooth. When v_{in} is +2 V, the modulator output is a series of pulses having widths equal to 1 μs. What are the pulse widths when $v_{in} = +3.5$ V?

16–38. The voltage comparator in Figure 16–40 switches between 0 V and +10 V. A 1-kHz sawtooth that varies between 0 V and +5 V is applied to the inverting input. A ramp voltage that rises from 0 V to 5 V in 5 ms is applied to the noninverting input. Sketch the input and output waveforms. Label the time points where the comparator output switches from low to high.

16–39. Repeat Exercise 16–38 if the inputs to the comparator are reversed. Label time points where the comparator switches from high to low instead of low to high. What would be the purpose of this input reversal, assuming that the comparator output was connected directly to a MOSFET totem-pole output stage?

16–40. Draw a schematic diagram of a MOSFET totem pole using P-channel devices. Describe an input signal that would be appropriate for driving this totem pole. Describe the operation of the totem pole, including its output levels, for each input level.

16–41. A class-D amplifier is to be designed to deliver a maximum average current of 2 A to a 12-Ω load. A 50-kHz sawtooth is to be used for the pulse-width modulator. Under maximum drive, the totem-pole output produces pulses that are 15 μs wide.

a. What should be the maximum input signal frequency?

b. Assuming 100% efficiency, what should be the value of V_{DD} used in the totem pole?

SPICE EXERCISES

16–42. Use SPICE to obtain a plot of the load voltage in the class-B amplifier shown in Figure 16–27(a). The supply voltages are ±15 V dc, the coupling capacitor is 100 μF, and the load resistance is 100 Ω. The input signal is a sine wave having frequency 1 kHz. Obtain a plot of one full cycle of output when

a. the peak input is 12.5 V. What is the (approximate) power delivered to the load, based on the SPICE results?

b. the peak input is 2.5 V. Why is the crossover distortion so much more severe than in Example 16–11?

16–43. The amplifier in Figure 16–27(b) has $R_E = 1\ \Omega$, $V_{CC} = \pm 50$ V, $R_L = 20\ \Omega$, and a 100-μF coupling capacitor. Use SPICE to obtain a plot of one full cycle of load voltage when the input is a 40-V peak sine wave having frequency 1 kHz. Also obtain plots of the collector current in each of transistors Q_3 and Q_4. Use the SPICE results to determine the power delivered to the load.

16–44. The input coupling capacitors in the amplifier in Figure 16–29 are each 100 μF and the output coupling capacitor, C_c, is 723 μF. The input signal is a 200-Hz sine wave having peak value 10 V. Use SPICE to obtain a plot of the load voltage over one full cycle and a plot of the current drawn from the power supply over one full cycle. Use the SPICE results to answer the following questions:

a. What is the dc base-to-ground voltage of each transistor? Compare with the values calculated in Example 16–12.

b. What is the power delivered to the load?

c. What is the voltage gain of the amplifier?

d. What is the peak current drawn from the power supply? Explain the appearance of the current waveform. (Why is it not sinusoidal?)

16–45. The quasi-complementary amplifier in Figure 16–31 has $R_E = 1\ \Omega$, $V_{CC} = \pm 20$ V, $R_L = 10\ \Omega$ and 100-μF-input coupling capacitors. The input signal is a 12-V-pk sine wave having frequency 1 kHz. Use SPICE to obtain a plot of the load voltage over one full cycle. Using the results, determine the peak-to-peak load current, the voltage gain of the amplifier, and the power delivered to the load.

17 POWER SUPPLIES AND VOLTAGE REGULATORS

17–1 INTRODUCTION

The term *power supply* generally refers to a source of dc power that is itself operated from a source of ac power, such as a 120-V, 60-Hz line. A power supply of this type can therefore be regarded as an *ac-to-dc converter.* A supply that produces constant-frequency, constant-amplitude, *ac* power from a dc source is called an *inverter.* Some power supplies are designed to operate from a dc source and to produce power at a different dc level; these are called *dc-to-dc converters.*

A dc power supply operated from an ac source consists of one or more of the following fundamental components:

1. A *rectifier* that converts an ac voltage to a pulsating dc voltage and permits current to flow in one direction only;
2. A low-pass *filter* that suppresses the pulsations in the rectified waveform and passes its dc (average value) component; and
3. A *voltage regulator* that maintains a substantially constant output voltage under variations in the load current drawn from the supply and under variations in line voltage.

The extent to which each of the foregoing components is required in a given supply, and the complexity of its design, depend entirely on the application for which the supply is designed. For example, a dc supply designed exclusively for service as a battery charger can consist simply of a half-wave rectifier. No filter or regulator is required, since it is necessary to supply only pulsating, unidirectional current to recharge a battery. Special-purpose, fixed-voltage supplies subjected to little or no variation in current demand can be constructed with a rectifier and a large capacitive filter, and require no regulator. On the other hand, devices such as operational amplifiers and output power amplifiers require elaborate, well-regulated supplies.

Power supplies are classified as regulated or unregulated and as adjustable or fixed. Adjustable supplies are generally well regulated and are used in laboratory applications for general-purpose experimental and developmental work. The most demanding requirement imposed on these supplies is that they maintain an extremely constant output voltage over a wide range of loads and over a continuously adjustable range of voltages. Some fixed-voltage supplies are also designed so that their output can be adjusted, but over a relatively small range, to permit periodic recalibration.

17–2 RECTIFIERS

Half-Wave Rectifiers

In Chapter 3 (Section 3–6), we showed how a single diode performs half-wave rectification. Figure 17–1 shows the circuit and the half-wave–rectified waveform developed across load R_L. The average (dc) and rms values of a half-wave–rectified *sine-wave* voltage are

$$\left.\begin{aligned} V_{avg} &= \frac{V_{PR}}{\pi} \\ V_{rms} &= \frac{V_{PR}}{2} \end{aligned}\right\} \text{(half-wave)} \qquad \textbf{(17–1)}$$

where V_{PR} is the peak value of the rectified voltage. Notice how the forward voltage drop across the diode affects the peak value of the rectified waveform shown in Figure 17–1: The peak value V_{PR} of v_L equals $V_P - 0.7$ (for a silicon diode). The 0.7-V drop can be neglected in many power-supply applications because of the large ac voltages present.

Assuming that the ac voltage is a symmetrical sine wave, the diode in Figure 17–1 must be capable of withstanding a peak reverse voltage of V_P volts, since that is the maximum reverse biasing voltage that occurs during one complete cycle. Recall that the *peak-inverse-voltage* (PIV) rating of a diode determines its maximum permissible reverse bias without breakdown.

Full-Wave Rectifiers

Figure 17–2 shows the waveform that results when a sine wave is *full-wave rectified.* In effect, the negative half-cycles of the sine wave are inverted to create a continuous

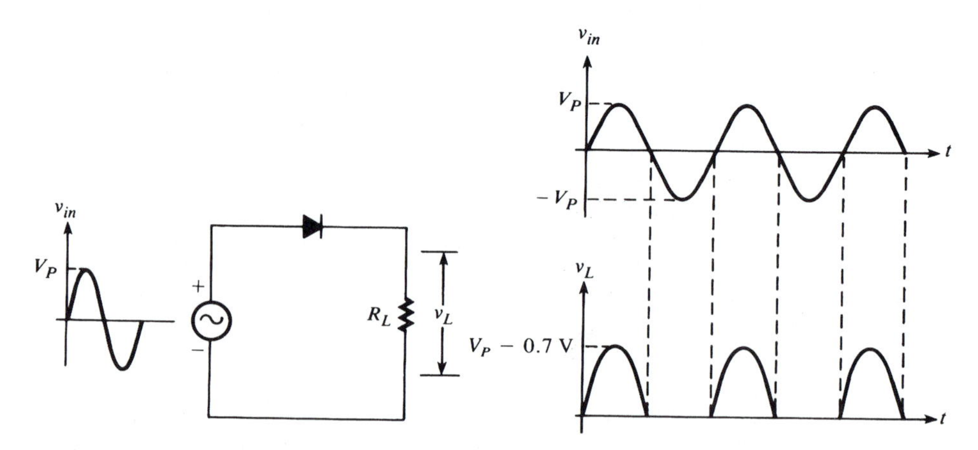

Figure 17–1
A half-wave rectifier

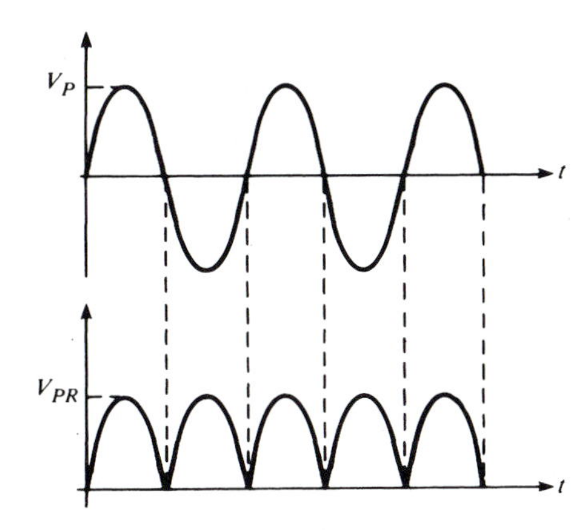

Figure 17–2
A full-wave–rectified sine wave

series of positive half-cycles. Comparing Figure 17–1 with the full-wave–rectified waveform in Figure 17–2, it is apparent that the latter has twice the average value of the former. Thus, a full-wave–rectified voltage with peak value V_{PR} has average value

$$V_{avg} = \frac{2V_{PR}}{\pi} \text{ (full-wave)} \tag{17–2}$$

The rms value of a full-wave–rectified waveform is the same as that of a sine wave: $V_{PR}/\sqrt{2}$.

Figure 17–3(a) shows how a center-tapped transformer can be connected in a circuit with two diodes to perform full-wave rectification. Assume that the transformer is wound so that terminal A on the secondary is positive with respect to terminal B at an instant of time when v_{in} is positive, as signified by the polarity symbols (dot convention) shown in the figure. Then, with the center tap as reference (ground), v_A is positive with respect to ground and v_B is negative with respect to ground. Similarly, when v_{in} is negative, v_A is negative with respect to ground and v_B is positive with respect to ground. (For a detailed discussion of the voltages developed across the secondary of a center-tapped transformer, see Section 16–6.)

Figure 17–3(b) shows that when v_{in} is positive, v_A forward biases diode D_1. As a consequence, current flows in a clockwise loop through R_L. Figure 17–3(c) shows that when v_{in} is negative, D_1 is reverse biased, D_2 is forward biased, and current flows through R_L in a counterclockwise loop. Notice that the voltage developed across R_L has the same polarity in either case. Therefore, positive voltage pulses are developed across R_L during both the positive and negative half-cycles of v_{in}, and a full-wave–rectified waveform is created.

The peak rectified voltage is the secondary voltage in the transformer, between center tap and one side, less the diode drop:

$$V_{PR} = (N_s/N_p)V_P - 0.7 \text{ V} \tag{17–3}$$

where V_P is the peak primary voltage, N_p is the total number of turns on the primary, and N_s is the number of turns between the center tap and either end of the secondary.

To determine the maximum reverse bias to which each diode is subjected, refer to the circuit in Figure 17–4. Here, we show the voltage drops in the rectifier when diode D_1 is forward biased and diode D_2 is reverse biased. Neglecting the 0.7-V

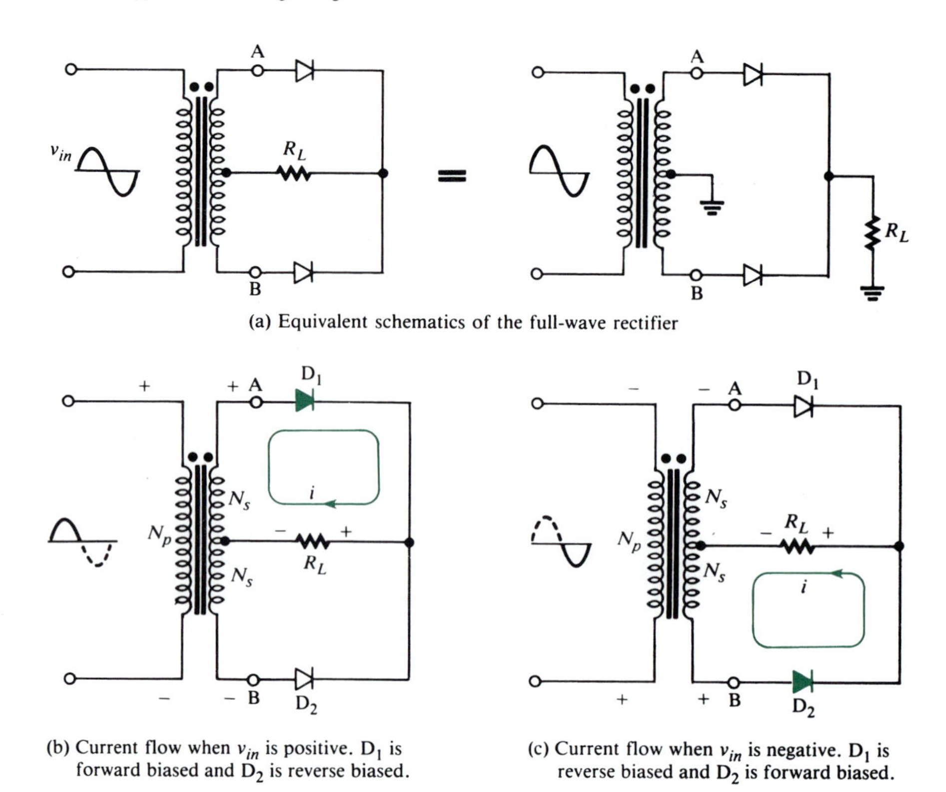

(a) Equivalent schematics of the full-wave rectifier

(b) Current flow when v_{in} is positive. D_1 is forward biased and D_2 is reverse biased.

(c) Current flow when v_{in} is negative. D_1 is reverse biased and D_2 is forward biased.

Figure 17–3
A full-wave rectifier employing a center-tapped transformer and two diodes

drop across D_1, the voltage across R_L is v_A volts. Thus the cathode-to-ground voltage of D_2 is v_A volts. Now, the anode-to-ground voltage of D_2 is $-v_B$ volts, as shown in the figure. Therefore, the total reverse bias across D_2 is $v_A + v_B$ volts, as shown. When v_A is at its positive peak, v_B is at its negative peak, so the maximum reverse bias equals *twice* the peak value of either. We conclude that the PIV rating of each

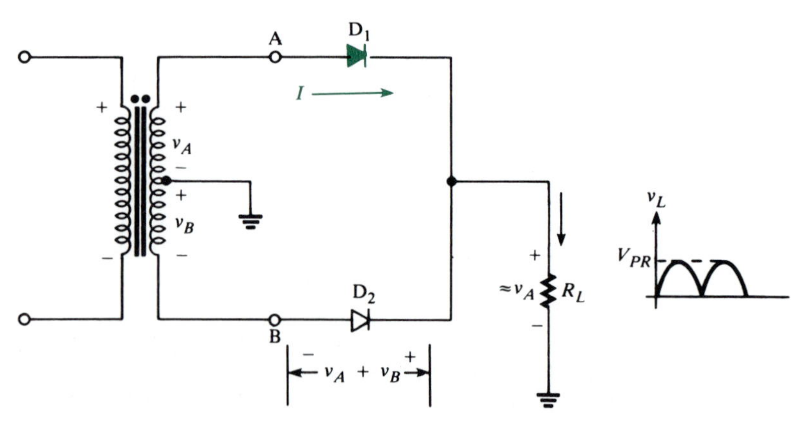

Figure 17–4
Diode D_2 is reverse biased by $v_A + v_B$ volts, which has a maximum value of $2V_{PR}$ volts

diode must be equal to at least twice the peak value of the rectified voltage:

$$\text{PIV} \geq 2V_{PR} \tag{17–4}$$

Example 17–1

The primary voltage in the circuit shown in Figure 17–5 is 120 V rms, and the transformer has $N_p:N_s = 4:1$. Find

1. the average value of the voltage across R_L;
2. the (approximate) average power dissipated by R_L; and
3. the minimum PIV rating required for each diode.

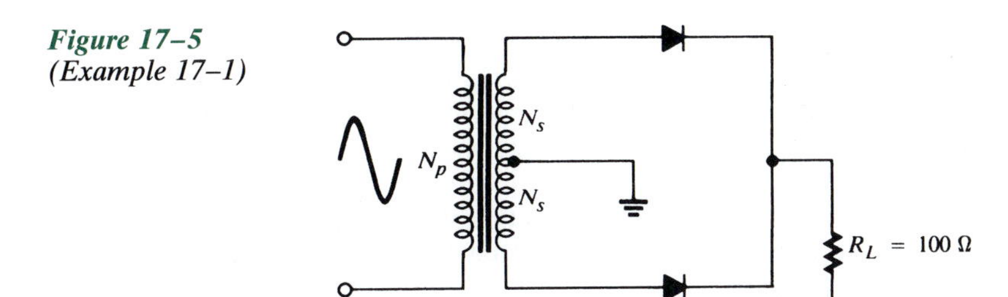

Figure 17–5
(Example 17–1)

Solution

1. $V_P = \sqrt{2}(120) = 169.7$ V

   From equation 17–3, $V_{PR} = (N_s/N_p)V_P - 0.7\text{ V} = (1/4)(169.7) - 0.7\text{ V} = 41.7$ V.

   Although equation 17–2 does not strictly apply to a rectified waveform with 0.7-V nonconducting gaps, it is a good approximation when the peak value is so much greater than 0.7 V:

$$V_{avg} \approx \frac{2V_{PR}}{\pi} = \frac{2(41.7\text{ V})}{\pi} = 26.5\text{ V}$$

2. $$V_{rms} = \frac{V_{PR}}{\sqrt{2}} = \frac{41.7\text{ V}}{\sqrt{2}} = 29.5\text{ V rms}$$

$$P_{avg} = \frac{V_{rms}^2}{R_L} = \frac{(29.5\text{ V})^2}{100\ \Omega} = 8.7\text{ W}$$

3. $\text{PIV} \geq 2V_{PR} = 2(41.7) = 83.4$ V

Full-Wave Bridge Rectifiers

Figure 17–6 shows another circuit used to perform full-wave rectification. One advantage of this circuit is that it does not require a transformer (although a transformer is often used to isolate the ac power line from the rest of the power supply). The diodes are arranged in the form of a bridge, and the circuit is called a *full-wave bridge rectifier.*

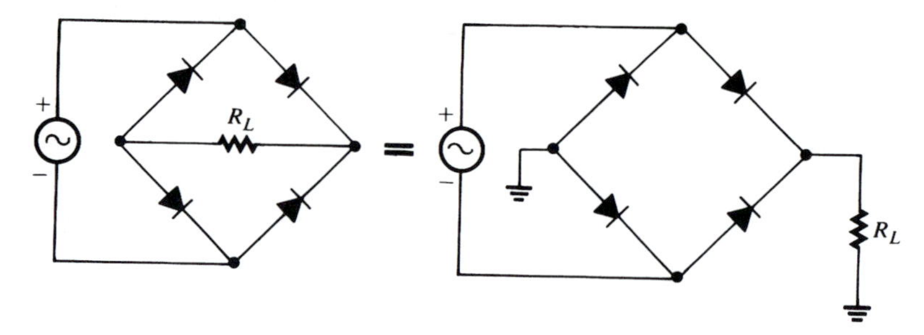

Figure 17–6
A full-wave bridge rectifier

Figure 17–7 demonstrates that current flows through R_L in the same direction when v_{in} is positive as it does when v_{in} is negative. In Figure 17–7(a), v_{in} is positive, so D_2 and D_3 are forward biased, while D_1 and D_4 are reverse biased. Therefore current flows through R_L from right to left, as shown. In (b), v_{in} is negative, so D_1 and D_4 are forward biased, and D_2 and D_3 are reverse biased. Notice that current still flows through R_L from right to left. Therefore, the polarity of the voltage across R_L is always the same, confirming that full-wave rectification occurs. Note that the common side of the ac source *must* be isolated from the common side of the load

Figure 17–7
Current flow in the full-wave bridge rectifier

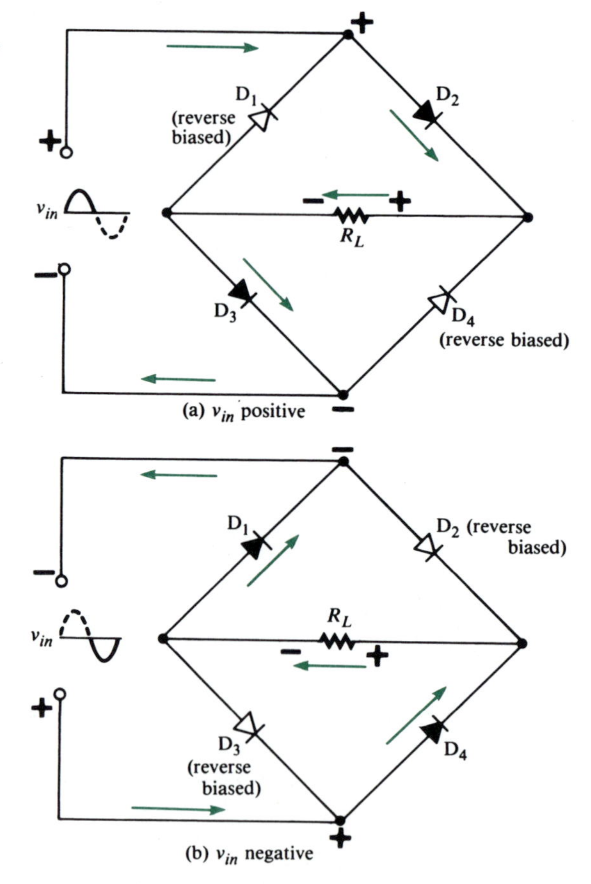

Diode Bridges

Molded diode bridges are available from 1 to 100 Amps in a variety of space-saving, easy-to-use packages.

Single ended bridges are available in three popular case styles for 1 to 6 Amps PC board-mount applications.

The 10 to 35 Amp range is covered by the JB and MB series which are isolated base, **U.L. recognized bridges** with very high surge current capability.

To make a bridge circuit into a center tap circuit, connect to terminals as shown.

SINGLE PHASE BRIDGE — CENTER TAP COMMON CATHODE — CENTER TAP COMMON ANODE

Ratings per diode are same for all three applications. NC No Connection.

SINGLE PHASE DIODE MOLDED BRIDGES — 1.0 TO 100 AMPS (I_O)

I_O (A) Output Current	1.0	1.2	1.8	1.9	2.0	3.0	6.0	10	25	35	80	100
@ T_C (°C)	25	45	50	45	50	50	50	65	65	55	85	80
I_{FSM} (A) (Surge)	30	52	52	52	60	50	125	130	350	420	800	1000
Notes	(1)	(1)	(1)	(1) (3)	(1)	(1)	(1)	(2)	(2)	(2)	(2)	(2)
Case Style	D-43	D-38	D-2	D-37	D-44	D-45	D-46	D-34	D-34	D-34	D-20-6	D-20-6
PART NUMBERS								UL	UL	UL		
V_{RRM} 50 Volts	–	–	18DB05A	–	2KBP005	KBPC1005	KBPC6005	100JB05L	250JB05L	35MB5A	800HB05U	1000HB05U
100 Volts	1DMB10	1KAB10E	18DB1A	2KBB10	–	–	–	100JB1L	250JB1L	35MB10A	800HB1U	1000HB1U
200 Volts	1DMB20	1KAB20E	18DB2A	2KBB20	2KBP02	KBPC102	KBPC602	100JB2L	250JB2L	35MB20A	800HB2U	1000HB2U
300 Volts	–	–	–	–	–	–	–	–	–	–	800HB3U	1000HB3U
400 Volts	1DMB40	1KAB40E	18DB4A	2KBB40	2KBP04	KBPC104	KBPC604	100JB4L	250JB4L	35MB40A	800HB4U	1000HB4U
600 Volts	–	1KAB60E	18DB6A	2KBB60	2KBP06	KBPC106	KBPC606	100JB6L	250JB6L	35MB60A	800HB6U	1000HB6U
800 Volts	–	1KAB80E	18DB8A	2KBB80	2KBP08	KBPC108	KBPC608	100JB8L	250JB8L	35MB80A	–	–
1000 Volts	–	1KAB100E	18DB10A	2KBB100	2KBP10	KBPC110	KBPC610	100JB10L	250JB10L	35MB100A	–	–
1200 Volts	–	–	–	–	–	–	–	100JB12L	250JB12L	35MB120A	–	–

NOTES: (1) Ambient temperature (T_A). (2) Must be used with suitable heatsink. (3) For lead configuration - ~ ~ + add 'R' to part number.

* For detailed specifications, contact your local IR Field Office or IR Distributor.

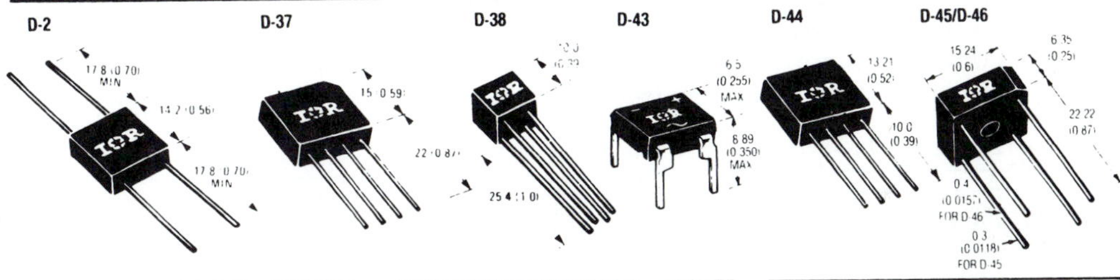

THREE-PHASE BRIDGES—MOLDED

I_O (A) Output Current	60	100
@ T_C °C	70	100
I_{FSM} (A) (Surge)	500	800
Case Style	D-20-8	D-20-8
PART NUMBERS		
V_{RRM} (V) 100 Volts	600HT1U	1000HT1U
200 Volts	600HT2U	1000HT2U
300 Volts	600HT3U	1000HT3U
400 Volts	600HT4U	1000HT4U
600 Volts	600HT6U	1000HT6U
800 Volts	600HT8U	1000HT8U
1000 Volts	600HT10U	–

SINGLE PHASE BRIDGES — FINNED

I_O (A) Output Current	24	32
@ T_C °C	50	50
I_{FSM} (A) (Surge)	250	1500
Case Style	D10	D10
PART NUMBERS		
V_{RRM} (V) 400 Volts	B12F40	B70H40
600 Volts	B12F60	B70H60
800 Volts	–	B70H80
1000 Volts	B12F100	B70H100

D-34

D-10

D-20-8

D-20-6

Figure 17–8
Diode bridge specifications (Courtesy of International Rectifier)

voltage. Thus, the negative terminal of R_L in Figure 17–7(a) cannot be the same as the negative terminal of v_{in}. As previously mentioned, a transformer is often used to provide this isolation.

Since load current flows through two forward-biased diodes during each half-cycle of v_{in}, the peak rectified voltage across R_L is the peak input voltage reduced by 2(0.7 V) = 1.4 V:

$$V_{PR} = V_P - 1.4 \text{ V} \tag{17–5}$$

Another advantage of the bridge rectifier is that the reverse bias across each diode never exceeds V_{PR}. Thus, the PIV rating of the diodes is half that required in the rectifier using a center-tapped transformer.

Diode bridges are now commonly available in single-package units. These packages have a pair of terminals to which the ac input is connected and another pair at which the full-wave–rectified output is taken. Figure 17–8 shows a typical manufacturer's specification sheet for a line of single-package bridges with current ratings from 1 to 100 A. The V_{RRM} specifications refer to the maximum repetitive reverse voltage ratings of each, i.e., the peak inverse voltage ratings when operated with repetitive inputs, such as sinusoidal voltages. These ratings range from 50 to 1200 V. Note the I_{FSM} specifications, which refer to maximum forward surge current. As we will learn in the next section, a filter is often connected to the output of a bridge, and the nature of the filter determines the amount of current that surges through the rectifier when ac power is first applied. For example, if a large capacitor is connected directly across the bridge output, then the rectifier is essentially shorted out when power is first applied. The filter must be designed so that this initial surge current does not exceed the bridge rating. The specifications show that bridges are available with I_{FSM} ratings from 30 to 1000 A.

Example 17–2

DESIGN

Figure 17–9 shows a bridge rectifier isolated from the 120-V-rms power line by a transformer.

1. What turns ratio should the transformer have to produce an average current of 1 A in R_L?
2. What is the average current in each diode under the conditions of (1)?
3. What minimum PIV rating should each diode have?
4. How much power is dissipated by each diode?

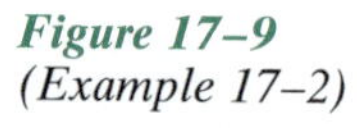
Figure 17–9
(Example 17–2)

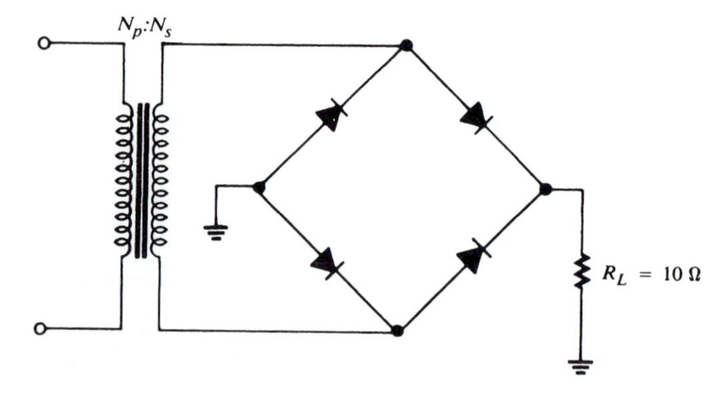

Solution

1. $I_{avg} = 2I_{PR}/\pi$, where I_{PR} is the peak value of the rectified current in R_L. Thus,

$$I_{PR} = \frac{\pi I_{avg}}{2} = \frac{\pi(1\text{ A})}{2} = 1.57\text{ A pk}$$

$$V_{PR} = (I_{PR})R_L = (1.57)(10) = 15.7\text{ V}$$

From equation 17–5, the peak value of the voltage applied across the bridge is $V_P = V_{PR} + 1.4 = 15.7 + 1.4 = 17.1$ V. Since the peak value of the 120-V-rms primary is $(\sqrt{2})(120\text{ V rms}) = 169.7$ V-pk,

$$\frac{N_P}{N_s} = \frac{169.7}{17.1} = 9.92:1$$

2. Each diode conducts during one half-cycle only, so the average current in each is the same as that of a half-wave–rectified sine wave:

$$I_{avg} = \frac{I_{PR}}{\pi} = \frac{1.57}{\pi} = 0.5\text{ A dc}$$

Note that this result is one-half the total average current in R_L, as expected.

3. PIV $\geq V_{PR} = 15.7$ V.

4. Assuming that negligible current flows through each diode when it is reverse biased, power is dissipated by a diode only when it is conducting forward current. Assuming that the forward voltage is 0.7 V during the full half-cycle that a diode conducts (a conservative, worst-case assumption), the rms diode voltage over one full cycle is *

$$V_{rms} = \frac{0.7\text{ V}}{\sqrt{2}} = 0.5\text{ V rms}$$

The rms value of one-half cycle of the sinusoidal current having peak value $I_{PR} = 1.57$ A is

$$I_{rms} = \frac{I_{PR}}{2} = \frac{1.57\text{ A}}{2} = 0.785\text{ A rms}$$

Therefore,

$$P_{avg} = V_{rms}I_{rms} = (0.5\text{ V})(0.785\text{ A}) = 0.39\text{ W}$$

17–3 CAPACITOR FILTERS

The frequency of the *fundamental* component of a half-wave–rectified waveform is the same as the frequency of its original (unrectified) ac waveform. Since the ac power source used in most dc supplies has frequency 60 Hz, a half-wave–rectified waveform contains a 60-Hz fundamental, plus *harmonic* components that are integer

* The rms value of a waveform that equals V for one half-cycle and equals 0 for the other half-cycle is $V/\sqrt{2}$.

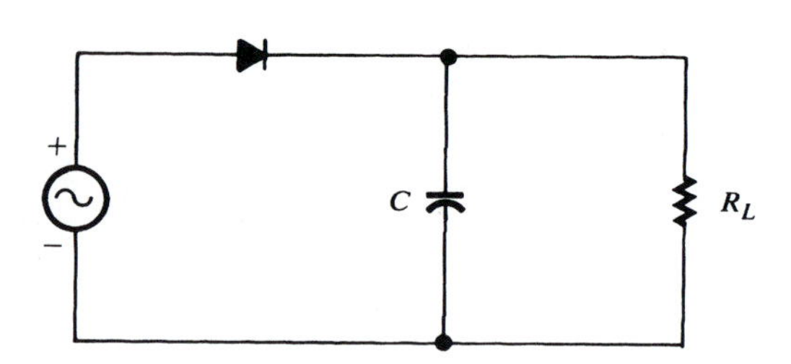

Figure 17–10
A half-wave rectifier with capacitor filter

multiples of 60 Hz, plus a dc, or average value, component. A full-wave–rectified waveform has one-half the period of a half-wave–rectified waveform, so it has twice the frequency. It therefore contains a 120-Hz fundamental, plus harmonics of 120 Hz, plus a dc component that is twice that of the half-wave–rectified waveform, as we have seen. A low-pass filter is connected across the output of a rectifier to suppress the ac components and to pass the dc component.

A rudimentary low-pass filter used in power supplies consists simply of a capacitor connected across the rectifier output, that is, in parallel with the load, as illustrated in Figure 17–10. Here, we show a simple half-wave rectifier with capacitor

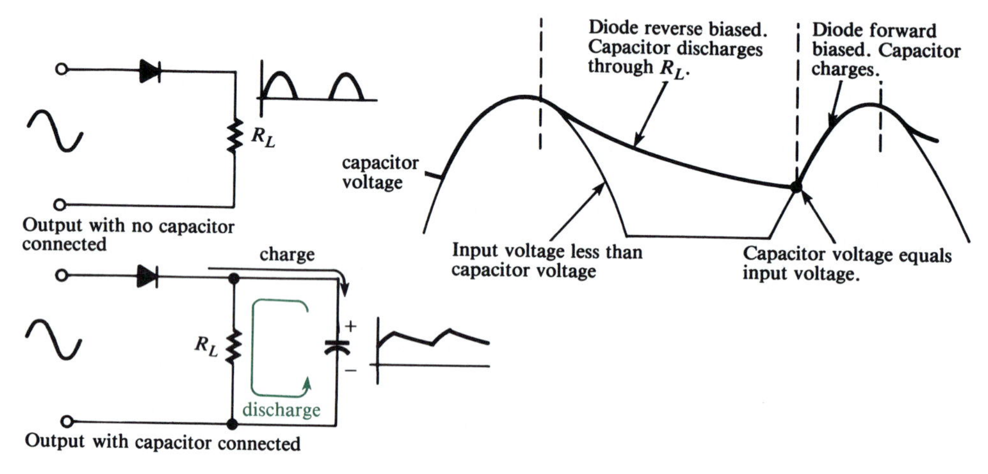

(a) The filter capacitor charges when the diode is forward biased and discharges when the diode is reverse biased.

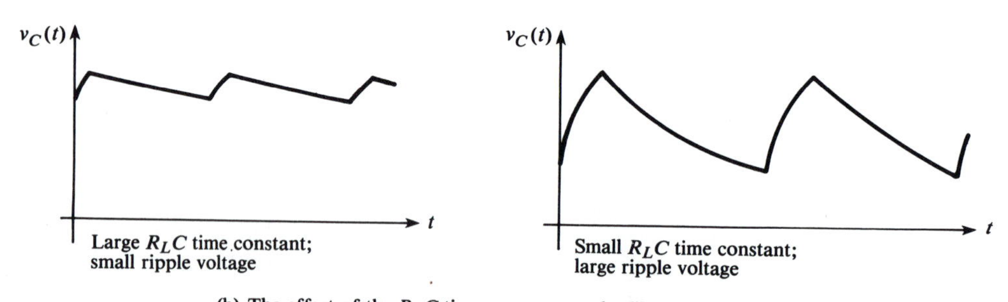

(b) The effect of the R_LC time constant on the filtered waveform

Figure 17–11
Waveforms produced by a half-wave rectifier with capacitor filter

C connected in parallel with R_L. The forward resistance of the diode is small in comparison to R_L, so during positive half-cycles, the capacitor charges to the peak value of the ac input.

It is not convenient to analyze the circuit in Figure 17–10 in terms of filter theory because the nonlinear operation of the diode effectively changes the circuit resistance as the diode is alternately forward and reverse biased. Instead, we analyze it from the standpoint of the transient voltage across the capacitor.

Figure 17–11(a) shows how the capacitor charges and discharges during a full cycle of the ac input. Notice that the capacitor charges and its voltage rises with the input voltage when the input is large enough to forward bias the diode. The capacitor discharges through R_L when the input falls to a level below which the diode is reverse biased. The smaller the R_LC time constant, the further the capacitor voltage decays before another positive pulse arrives and recharges the capacitor. Figure 17–11(b) compares the output waveforms that result when large and small time constants are used. The voltage fluctuation in the filtered waveform is called the *ripple voltage,* which in most applications should be kept as small as possible. Figure 17–11(b) shows that a heavy load (small R_L) will result in an undesirably large ripple voltage.

Figure 17–12 shows a full-wave bridge rectifier with a capacitor filter. Also shown is the filtered waveform. In this case, positive pulses are present during every half-cycle of input, so the capacitor voltage does not decay as far as it does in the half-wave circuit before another pulse is available to recharge it. As a consequence, the peak-to-peak ripple voltage is smaller, for a given R_LC time constant, than it is in the half-wave circuit.

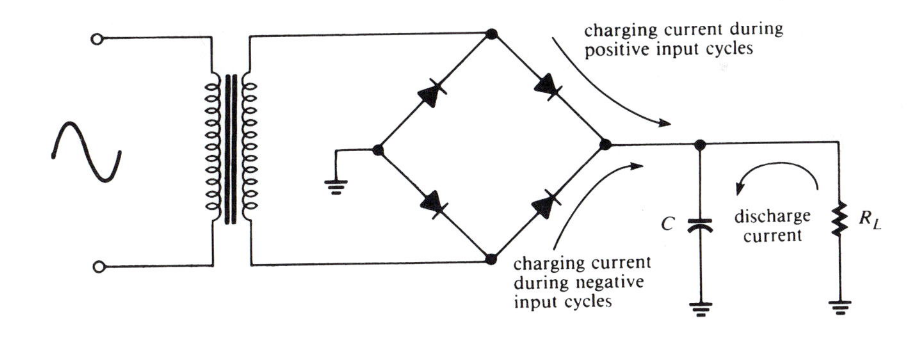

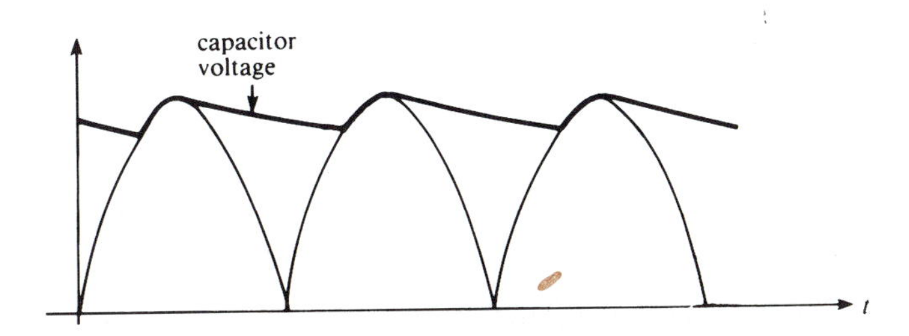

Figure 17–12
A full-wave bridge rectifier with capacitor filter

Percent Ripple

The percent ripple in a rectified waveform (also called the *ripple factor*) is defined by

$$r = \frac{\text{rms ripple voltage}}{\text{dc (average) voltage}} \times 100\% \tag{17–6}$$

Note carefully that the numerator is the rms value of just the ripple (ac components) in the rectified waveform, not the rms value of the total waveform.

Let us first compute the percent ripple in *unfiltered* half- and full-wave–rectified waveforms. Using calculus, it can be shown that the rms values of the *ac components* of half- and full-wave–rectified voltages are

$$V(\text{rms}) = 0.385V_{PR} \quad \text{(half-wave)} \tag{17–7}$$

$$V(\text{rms}) = 0.308V_{PR} \quad \text{(full-wave)} \tag{17–8}$$

where V_{PR} is the peak value of the rectified waveform in each case. (Note carefully that equations 17–7 and 17–8 give the rms values of the ac components *only, not* the rms values of the waveforms themselves.) Then

$$r(\text{half-wave, unfiltered}) = \frac{0.385V_{PR}}{V_{PR}/\pi} \times 100\% = 121\% \tag{17–9}$$

and

$$r(\text{full-wave, unfiltered}) = \frac{0.308V_{PR}}{2V_{PR}/\pi} \times 100\% = 48.4\% \tag{17–10}$$

These ripple values are quite large and demonstrate the need for a filter in most power-supply applications. If a filter is not used, or if it is improperly designed, the ac components of the ripple are superimposed on the signal lines in the device that receives power from the supply. Power-supply ripple is therefore a source of *noise* in an electronic system. It is one of the principal causes of 60-Hz (or 120-Hz) *hum* in an audio amplifier.

We will now derive an expression for the ripple in the output of a rectifier having a capacitor filter and load resistance R_L (Figures 17–11 and 17–12). Since a knowledge of ripple magnitude is important in only those applications where ripple affects the performance of a system using the power supply, we assume that the filter is well designed and that, as a consequence, the ripple voltage is small compared to the dc component. In other words, we assume that the capacitor voltage does not decay significantly from its peak value between the occurrences of the rectified pulses that recharge the capacitor. This circumstance is called *light loading* because the charge supplied to R_L by the capacitor is small compared to the total charge stored on the capacitor. To simplify the computations, we can assume that the ripple voltage in a lightly loaded filter is a sawtooth wave, as illustrated in Figure 17–13.

Figure 17–13(a) shows the waveform in a lightly loaded capacitor filter across the output of a full-wave rectifier. The derivation that follows is applicable to both half-wave and full-wave rectifiers, provided the assumption of light loading is valid for both cases. Note that the period T shown in the figure is the fundamental period of the *rectified* waveform, typically 1/60 s for half-wave rectifiers and 1/120 s for full-wave rectifiers. The light-loading assumption in the half-wave case is more restrictive than in the full-wave case because there is a longer time between pulses, during which the capacitor voltage can decay.

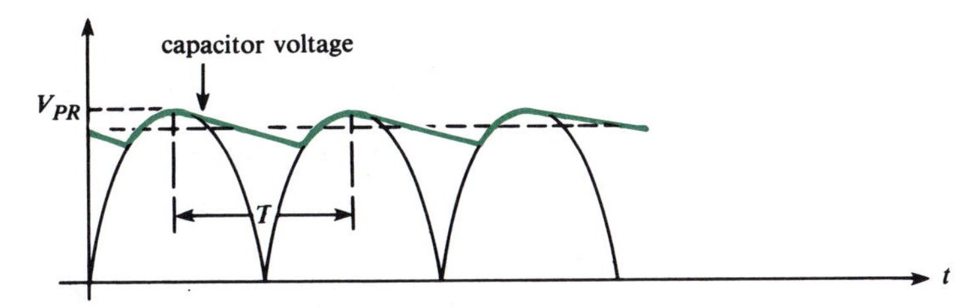

(a) In the lightly loaded filter, the capacitor voltage does not decay significantly from V_{PR} between charging pulses.

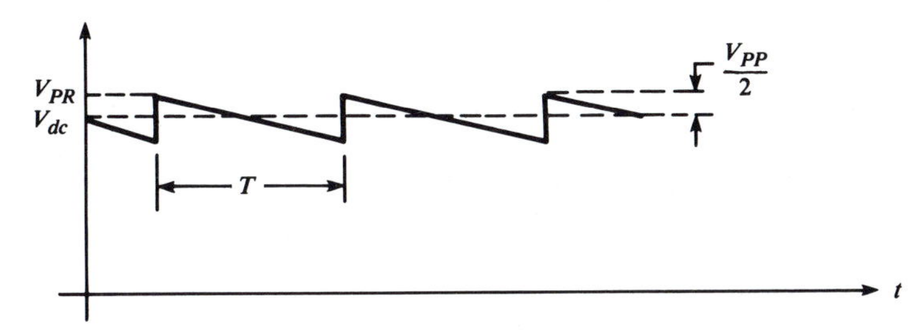

(b) The capacitor voltage in (a) approximated by a sawtooth waveform

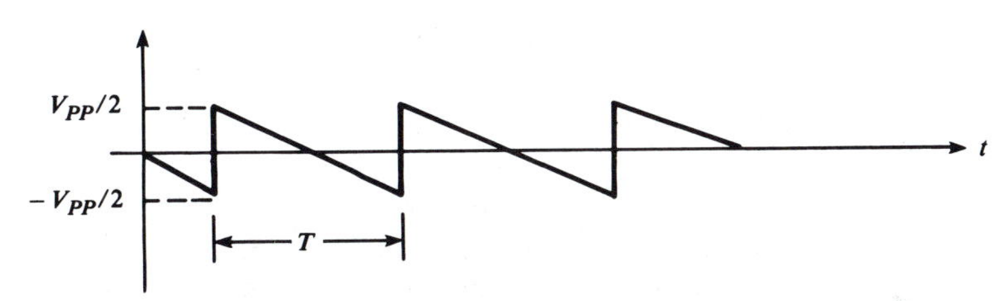

(c) The ac component (ripple voltage) in (b)

Figure 17–13
Approximating the ripple voltage in a lightly loaded capacitor filter

Figure 17–13(b) and (c) show the ripple waveform approximated by a sawtooth voltage with peak-to-peak value V_{PP}. This approximation is equivalent to assuming that the capacitor charges instantaneously and that its voltage decays linearly instead of exponentially. Assuming that the voltage decays linearly is equivalent to assuming that the discharge current is constant and equal to V_{dc}/R_L, where V_{dc} is the dc value of the filtered waveform. As shown in (c), the total change in capacitor voltage is V_{PP} volts, and this change occurs over the period of time T. Therefore, since $\Delta Q = I\Delta t$,

$$V_{PP} = \frac{\Delta Q}{C} = \frac{(V_{dc}/R_L)T}{C} \tag{17–11}$$

Since $T = 1/f_r$, where f_r is the frequency of the fundamental component of the ripple (typically 60 Hz or 120 Hz), equation 17–11 can be written

$$V_{PP} = \frac{V_{dc}}{f_r R_L C} \tag{17–12}$$

or

$$V_{dc} = V_{PP} f_r R_L C \tag{17–13}$$

From Figure 17–13(b), it is apparent that

$$V_{dc} = V_{PR} - V_{PP}/2 \tag{17–14}$$

Substituting from (17–12), we obtain

$$V_{dc} = V_{PR} - \frac{V_{dc}}{2f_r R_L C} \tag{17–15}$$

Solving for V_{dc}, we obtain an expression for V_{dc} in terms of the peak rectifier voltage:

$$V_{dc} = \frac{V_{PR}}{1 + \dfrac{1}{2f_r R_L C}} \tag{17–16}$$

It is clear that the dc value cannot exceed V_{PR} volts, and that it equals V_{PR} when $R_L = \infty$ (i.e., when the load is an open circuit).

The rms value of a sawtooth waveform having peak-to-peak value V_{PP} is known to be

$$V(\text{rms}) = \frac{V_{PP}}{2\sqrt{3}} \tag{17–17}$$

Therefore, from (17–13) and (17–17), the percent ripple is

$$\begin{aligned} r &= \frac{V(\text{rms})}{V_{dc}} \times 100\% = \frac{V_{PP}/2\sqrt{3}}{V_{PP} f_r R_L C} \times 100\% \\ &= \frac{1}{2\sqrt{3} f_r R_L C} \times 100\% \end{aligned} \tag{17–18}$$

Equation 17–18 confirms our previous analysis of the capacitor filter (Figure 17–11): A large $R_L C$ time constant results in a small ripple voltage, and vice versa. The light-load assumption on which our derivation is based is generally valid for percent ripple less than about 6.5%. From a design standpoint, the values of f_r and R_L are usually fixed, and the designer's task is to select a value of C that keeps the ripple below a prescribed value.

Example 17–3

DESIGN

A full-wave rectifier is operated from a 60-Hz line and has a filter capacitor connected across its output. What minimum value of capacitance is required if the load is 200 Ω and the ripple must be no greater than 4%?

Solution. Using the decimal form of r ($r = 0.04$), we find, from equation 17–18,

$$\frac{1}{C} = 2\sqrt{3} f_r R_L r$$

or

$$C = \frac{1}{2\sqrt{3}(120)(200)(0.04)} \approx 300\ \mu\text{F}$$

Example 17–4

The rectifier shown in Figure 17–10 is operated from a 60-Hz, 120-V-rms line. It has a 100-μF filter capacitor and a 1-kΩ load.

1. What is the percent ripple?
2. What is the average current in R_L?

Solution

1. We must first assume that the filter is lightly loaded, then perform the computation based on that assumption, and then verify from the result that the assumption is valid. From equation 17–18,

$$r = \frac{1}{2\sqrt{3}f_r R_L C} \times 100\% = \frac{100\%}{2\sqrt{3}(60)(10^3)(10^{-4})} = 4.8\%$$

We see that $r < 6.5\%$ and that our assumption of a light load is therefore appropriate.

2. Neglecting the drop across the diode,

$$V_{PR} = \sqrt{2}(120) = 169.7 \text{ V}$$

From equation 17–16,

$$V_{dc} = \frac{V_{PR}}{1 + \dfrac{1}{2f_r R_L C}} = \frac{169.7}{1 + \dfrac{1}{2(60)(10^3)(10^{-4})}} = \frac{169.7}{1.083} = 156.65 \text{ V}$$

Therefore, $I_{dc} = V_{dc}/R_L = 156.65/1000 = 156.65$ mA.

Example 17–5

SPICE

The full-wave rectifier in Figure 17–6 has $R_L = 120\ \Omega$. A 100-μF filter capacitor is connected in parallel with R_L. The input is a 60-Hz sine wave having peak value 30 V. The filter does *not* satisfy the lightly loaded criterion, so the equations for approximating the ripple voltage are not applicable. Use SPICE to determine the peak-to-peak ripple voltage.

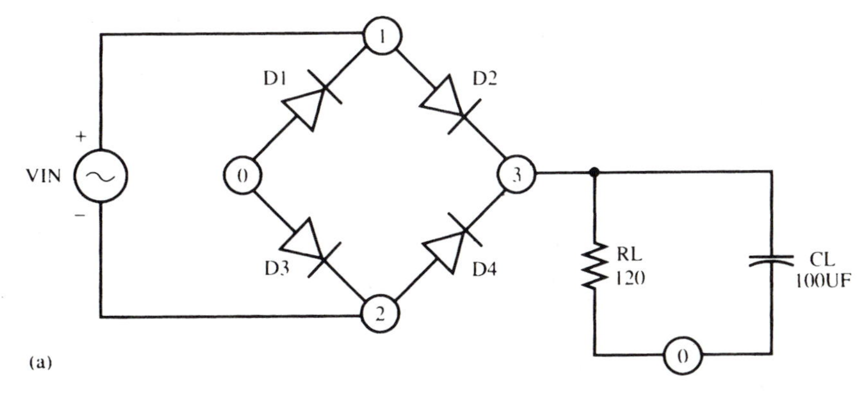


```
EXAMPLE
VIN 1 2 SIN(0 30 60HZ)
D1 0 1 DIODE
D2 1 3 DIODE
D3 0 2 DIODE
D4 2 3 DIODE
.MODEL DIODE D
RL 3 0 120
CL 3 0 100UF
.TRAN 0.5MS 17MS
.PLOT TRAN V(3)
.END
```

Figure 17–14
(Example 17–5)

```
****      TRANSIENT ANALYSIS                          TEMPERATURE =    27.000 DEG C
******************************************************************************
    TIME        V(3)
                    0.000D+00           1.000D+01           2.000D+01           3.000D+01           4.000D+01
                    - - - - - - - - - - - - - - - - - - - - - - - - - - - - - - - - - - - - - - - - -
0.000D+00  -4.899D-30 *                   .                   .                   .                   .
5.000D-04   3.946D+00 .       *           .                   .                   .                   .
1.000D-03   9.347D+00 .                  *.                   .                   .                   .
1.500D-03   1.439D+01 .                   .        *          .                   .                   .
2.000D-03   1.883D+01 .                   .                 * .                   .                   .
2.500D-03   2.259D+01 .                   .                   .    *              .                   .
3.000D-03   2.545D+01 .                   .                   .          *        .                   .
3.500D-03   2.739D+01 .                   .                   .              *    .                   .
4.000D-03   2.829D+01 .                   .                   .                *  .                   .
4.500D-03   2.817D+01 .                   .                   .               *   .                   .
5.000D-03   2.727D+01 .                   .                   .              *    .                   .
5.500D-03   2.616D+01 .                   .                   .           *       .                   .
6.000D-03   2.509D+01 .                   .                   .         *         .                   .
6.500D-03   2.407D+01 .                   .                   .       *           .                   .
7.000D-03   2.308D+01 .                   .                   .     *             .                   .
7.500D-03   2.214D+01 .                   .                   .   *               .                   .
8.000D-03   2.124D+01 .                   .                   . *                 .                   .
8.500D-03   2.037D+01 .                   .                   .*                  .                   .
9.000D-03   1.954D+01 .                   .                  *.                   .                   .
9.500D-03   1.874D+01 .                   .                *  .                   .                   .
1.000D-02   1.798D+01 .                   .               *   .                   .                   .
1.050D-02   2.018D+01 .                   .                   *                   .                   .
1.100D-02   2.364D+01 .                   .                   .      *            .                   .
1.150D-02   2.620D+01 .                   .                   .           *       .                   .
1.200D-02   2.789D+01 .                   .                   .               *   .                   .
1.250D-02   2.840D+01 .                   .                   .                *  .                   .
1.300D-02   2.791D+01 .                   .                   .               *   .                   .
1.350D-02   2.696D+01 .                   .                   .             *     .                   .
1.400D-02   2.585D+01 .                   .                   .           *       .                   .
1.450D-02   2.480D+01 .                   .                   .         *         .                   .
1.500D-02   2.379D+01 .                   .                   .       *           .                   .
1.550D-02   2.282D+01 .                   .                   .     *             .                   .
1.600D-02   2.189D+01 .                   .                   .   *               .                   .
1.650D-02   2.099D+01 .                   .                   . *                 .                   .
1.700D-02   2.013D+01 .                   .                   *                   .                   .
                    - - - - - - - - - - - - - - - - - - - - - - - - - - - - - - - - - - - - - - - - -
```

(b)

Figure 17–14
(Continued)

Solution. Figure 17–14(a) shows the SPICE circuit and input data file. Note that VIN is not connected to ground (node 0), since it must be applied across nodes 1 and 2. One-half the period of the 60-Hz input is 8.33 ms, so the .TRAN and .PLOT statements will produce a plot of the output extending over the first two half-cycles of input (0 to 17 ms). The results are shown in Figure 17–14(b). We see that the minimum and maximum values of the second pulse are 17.98 V and 28.40 V, respectively, so the peak-to-peak ripple voltage is 28.40 − 17.98 V = 10.42 V.

Repetitive Surge Currents

We have already discussed the initial surge current that flows through rectifier diodes when power is first applied and the uncharged filter capacitor behaves like a momentary short circuit. Figure 17–15 shows that current also surges through the diodes during the short time intervals that they are forward biased in normal

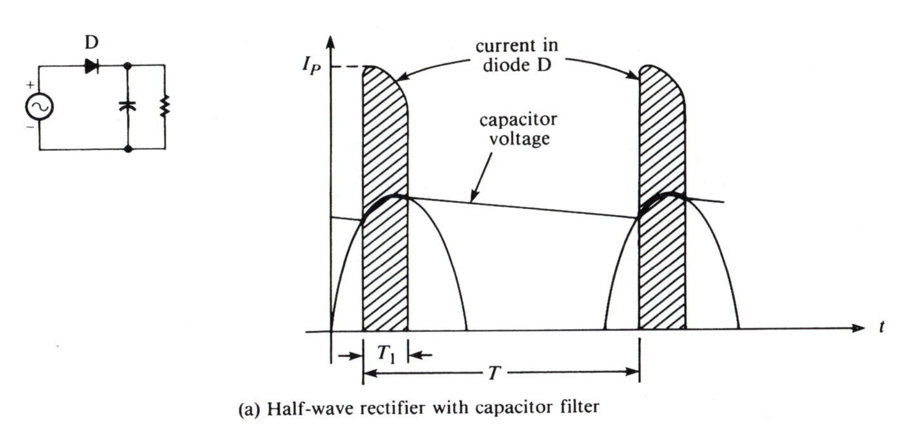

(a) Half-wave rectifier with capacitor filter

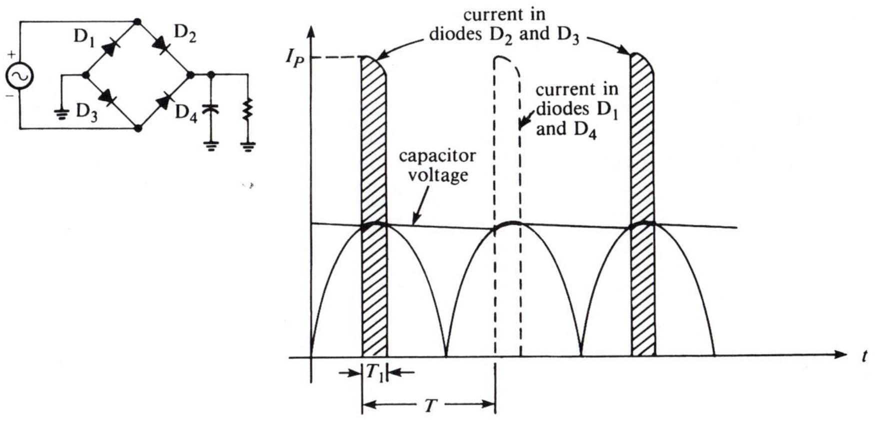

(b) Full-wave rectifier with capacitor filter

Figure 17–15
Repetitive surge currents in rectifier diodes

operation. When the capacitor voltage is below the rectifier voltage, the total amount of charge that flows through the forward-biased diode(s) to recharge the capacitor must equal the amount of charge lost from the capacitor during the time it discharged into R_L. For a lightly loaded filter, the capacitor voltage does not decay significantly, so the conduction interval T_1 shown in the figure is small compared to the period T between rectifier pulses. If we approximate the current pulse as a rectangle with height I_P and width T_1, and the discharge current as a rectangle with height I_{dc} and width T, we have

$$Q(\text{discharge}) = Q(\text{charge})$$
$$I_{dc}T = I_p T_1$$

or **(17–19)**

$$I_P = I_{dc}\frac{T}{T_1}$$

Since $T \gg T_1$, equation 17–19 shows that the peak current through the diode may be many times larger than the average current supplied to the load. The diodes must be capable of supplying this repetitive surge current.

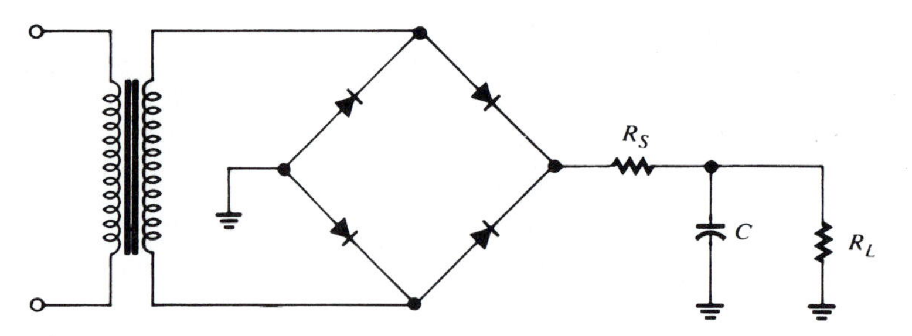

Figure 17–16
Use of series resistance R_S to limit diode surge currents

Increasing the value of filter capacitance C decreases the ripple (equation 17–18) and increases the dc voltage (equation 17–16), both of which are generally desirable outcomes. However, increased capacitance also shortens the conduction interval T_1 in Figure 17–15, because the capacitor voltage decays less when the R_LC time constant is large. Consequently, by equation 17–19, large capacitance increases the peak diode current, and the permissible repetitive surge current imposes a practical limitation on the size of the filter capacitor that can be used. To limit surge current, especially that which occurs at initial turn-on, a small resistance is sometimes connected between the rectifier and the filter capacitor, as shown in Figure 17–16. One disadvantage of this resistance is that it reduces the dc voltage level at the load, since it forms a voltage divider with R_L.

17–4 RC AND LC FILTERS

RC Filters

To obtain a greater reduction in the ripple of a capacitor-filtered waveform, a low-pass RC filter section can be connected across the capacitor, as shown in Figure 17–17. The capacitor–resistor–capacitor combination is often called an *RC* π (PI) filter. The figure shows that the low-pass RC section further attenuates the ac

Figure 17–17
Adding a low-pass RC filter section (RC_2) reduces the ripple voltage and the dc value of the filtered waveform.

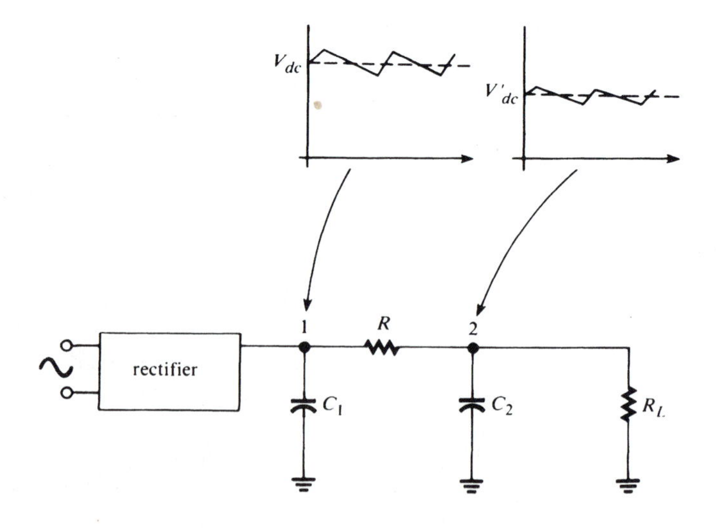

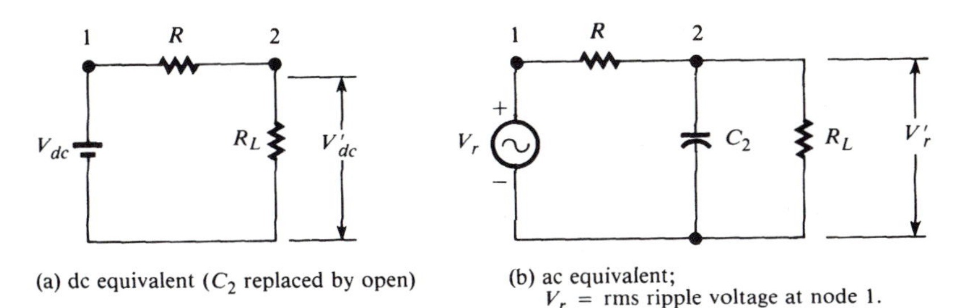

(a) dc equivalent (C_2 replaced by open)

(b) ac equivalent;
V_r = rms ripple voltage at node 1.

Figure 17–18
DC and ac equivalent circuits of Figure 17–21

components represented by the ripple and passes the dc value, although there is some reduction in dc level.

An approximate analysis of the filter shown in Figure 17–17 can be performed by assuming that the impedance looking back into node 1 is small compared to the impedance of the rest of the filter network connected across node 1. That assumption, which is valid for large C_1 and small diode on-resistance, allows us to treat node 1 as a voltage source that drives the rest of the filter. We can then apply the superposition principle to determine the dc and ac voltage components at node 2 due to the dc and ac components at node 1. Figure 17–18 shows the dc and ac equivalent circuits of the filter with node 1 replaced by dc and ac voltage sources.

It is apparent from Figure 17–18(a) that the dc load voltage is, by voltage-divider action,

$$V'_{dc} = \frac{R_L}{R + R_L} V_{dc} \qquad \textbf{(17–20)}$$

Equation 17–20 shows that the use of an RC filter causes a reduction in the dc voltage available at the load. To minimize this effect, R is usually made much smaller than R_L.

From Figure 17–18(b),

$$V'_r = \frac{Z}{R + Z} V_r \qquad \textbf{(17–21)}$$

where

$$Z = R_L \| (-jX_{C_2}) = \frac{-jR_L/\omega C_2}{R_L - jR/\omega C_2}$$

We are interested only in the magnitude of V'_r, so

$$|V'_r| = \frac{|Z|}{|R + Z|} |V_r| \qquad \textbf{(17–22)}$$

where

$$|Z| = \frac{|-jR_L/\omega C_2|}{|R_L - j/\omega C_2|} = \frac{R_L}{\sqrt{1 + (\omega R_L C_2)^2}} \qquad \textbf{(17–23)}$$

and

$$|R + Z| = \left| R - \frac{jR_L/\omega C_2}{R_L - j/\omega C_2} \right| = \sqrt{\frac{(R + R_L)^2 + (\omega R R_L C_2)^2}{1 + (\omega R_L C_2)^2}} \qquad \textbf{(17–24)}$$

Substituting (17–23) and (17–24) into (17–22), we find

$$|V'_r| = \frac{R_L}{\sqrt{(R + R_L)^2 + (\omega R R_L C_2)^2}} |V_r| \qquad \textbf{(17–25)}$$

Although the ripple voltage contains harmonics, ac calculations are usually performed using only the fundamental frequency of the ripple: $\omega = 2\pi(60)$ rad/s for a half-wave rectifier, and $\omega = 2\pi(120)$ rad/s for a full-wave rectifier.

Equation 17–25 can be written

$$|V_r'| = \frac{|V_r|}{\sqrt{\left(\frac{R}{R_L} + 1\right)^2 + (\omega RC_2)^2}} \qquad (17\text{–}26)$$

Since R is usually much smaller than R_L, as noted earlier, it follows that $R/R_L + 1 \approx 1$, and (17–26) can be approximated as

$$|V_r'| \approx \frac{|V_r|}{\sqrt{1 + (\omega RC_2)^2}} \qquad (17\text{–}27)$$

Example 17–6

1. Find the average value and the percent ripple in the load voltage in Figure 17–19(a).
2. Repeat when an RC filter is inserted between the capacitor and the load, as shown in Figure 17–19(b).

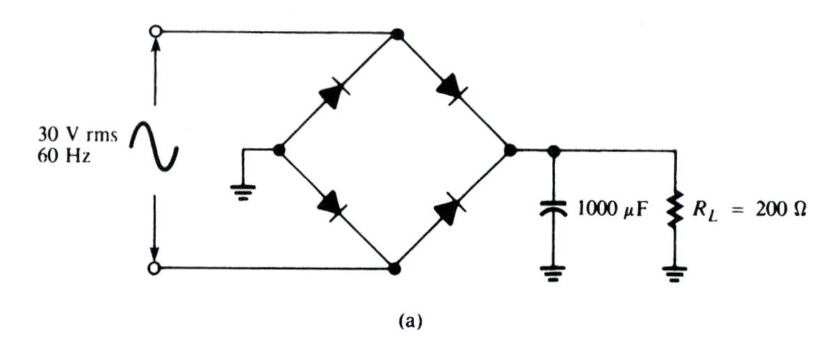

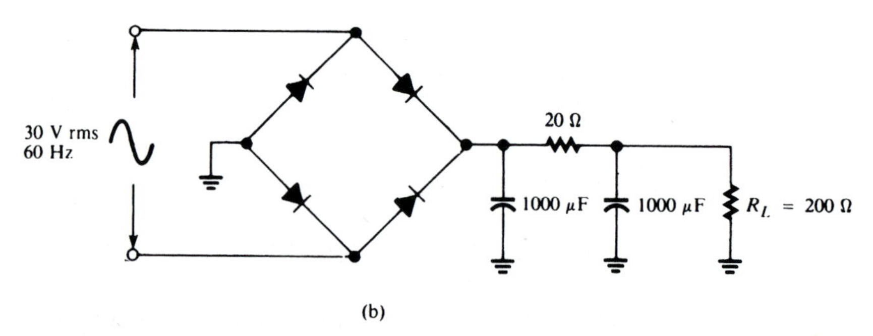

Figure 17–19
(Example 17–6)

Solution

1. Since the rectifier is full-wave, $f_r = 120$ Hz. $V_{PR} = \sqrt{2}(30\text{ V rms}) - 1.4\text{ V} = 41$ V. From equation 17–16,

$$V_{dc} = \frac{V_{PR}}{1 + \dfrac{1}{2f_rR_LC}} = \frac{41\text{ V}}{1 + \dfrac{1}{2(120)(200)(10^{-3})}} = 40.2\text{ V}$$

From equation 17–12, the peak-to-peak ripple voltage is

$$V_{PP} = \frac{V_{dc}}{f_rR_LC} = \frac{40.2\text{ V}}{(120)(200)(10^{-3})} = 1.68\text{ V}$$

From equation 17–17, the rms ripple voltage is

$$V_r = \frac{V_{PP}}{2\sqrt{3}} = \frac{1.68}{2\sqrt{3}} = 0.48\text{ V rms}$$

Therefore,

$$r = \frac{V_r}{V_{dc}} \times 100\% = \frac{0.48}{40.2} \times 100\% = 1.2\%$$

2. We will assume that V_{dc} and V_r have the same values calculated in (1). In reality, the presence of the new RC filter section affects the values of V_{dc} and V_r somewhat. In practical circuits, the effect is small and the computational error resulting from the assumption can be neglected. From equation 17–20,

$$V'_{dc} = \frac{R_L}{R + R_L}V_{dc} = \left(\frac{200}{20 + 200}\right)(40.2\text{ V}) = 36.5\text{ V}$$

From equation 17–25,

$$|V'_r| = \frac{|V_r|}{\sqrt{\left(\dfrac{R}{R_L} + 1\right)^2 + (\omega RC_2)^2}} = \frac{0.48\text{ V}}{\sqrt{\left(\dfrac{20}{200} + 1\right)^2 + [2\pi(120)(20)(10^{-3})]^2}}$$
$$= 0.032\text{ V rms}$$

Therefore,

$$r' = \frac{V'_r}{V'_{dc}} \times 100\% = \frac{0.032}{36.5} \times 100\% = 0.087\%$$

We see that the RC filter section has reduced the percent ripple from 1.2% to 0.087%, i.e., by a factor of more than 13. The penalty paid for this improvement is a 9.2% decrease in the dc voltage and a reduction in the ability of the supply to *regulate* the output (hold a constant voltage, independent of load), as we shall discuss presently.

If approximation 17–27 is used to calculate the ripple voltage, we find

$$|V'_r| \approx \frac{0.48\text{ V}}{\sqrt{1 + [2\pi(120)(20)(10^{-3})]^2}} = 0.032\text{ V rms}$$

This approximation equals the previously calculated value out to three decimal places.

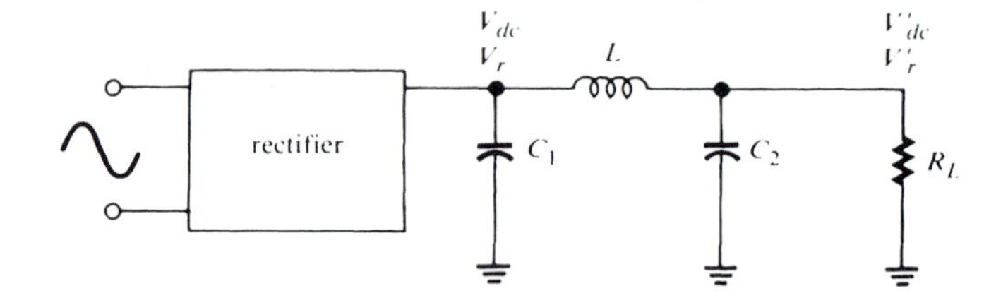

Figure 17–20
The LC π filter. The inductor presents a low resistance to the dc component and a large impedance to the ac (ripple) components.

LC π Filters

The series resistance in the RC filter causes a voltage drop that reduces load voltage, as we have seen. Furthermore, the voltage drop increases as load current increases, so the output voltage varies with load. This variation is undesirable in a power supply, which ideally should maintain a constant output voltage, independent of load current. (Series resistance always degrades the *voltage regulation* of a supply. We will discuss percent voltage regulation in detail in a subsequent discussion.) By connecting a low-resistance inductor in place of the series resistor, as shown in Figure 17–20, we can reduce the dc voltage variation with load and at the same time attenuate ripple by virtue of the inductive reactance presented to the ac components. The filter is called an LC π, and while it is an improvement over the RC π, it does require a large and usually expensive inductor.

By an analysis similar to that used for the RC filter, the following relationships can be derived:

$$V'_{dc} = \frac{R_L}{R' + R_L} V_{dc} \qquad \textbf{(17–28)}$$

where R' is the resistance of the inductor; and

$$|V'_r| \approx \frac{|V_r|}{\omega^2 L C_2} \qquad \textbf{(17–29)}$$

As in the case of the RC π, V_{dc} is the dc voltage across C_1 and V_r is the rms ripple voltage across C_1. Approximation 17–29 is valid when $R' << \omega L$, $R' << R_L$, $|X_{C_2} \| R_L| \approx |X_{C_2}|$, and $|X_L + (X_{C_2} \| R_L)| \approx |X_L|$, which is the usual situation, as illustrated in the next example.

Example 17–7

Assuming that $V_{dc} = 40.2$ V and $V_r = 0.48$ V rms in Figure 17–20 (the same values found in Example 17–6), find V'_{dc} and V'_r when $C_1 = C_2 = 1000\ \mu$F, $R_L = 200\ \Omega$, and L is a 6-H inductor with resistance 2 Ω.

Solution. By equation 17–28,

$$V'_{dc} = \frac{R_L}{R' + R_L} V_{dc} = \left(\frac{200}{2 + 200}\right) 40.2 \text{ V} = 39.8 \text{ V}$$

To verify that the conditions under which approximation 17–29 can be used are applicable in this example, we compute

$$\omega L = 2\pi(120)(6) = 4524\ \Omega$$
$$X_{C_2} = 1/(2\pi)(120)(10^{-3}) = 1.33\ \Omega$$

Thus,

$$R' = 2\,\Omega << 4524\,\Omega = \omega L$$
$$R' = 2\,\Omega << 200\,\Omega = R_L$$
$$|X_{C_2} \| R_L| = \frac{R_L}{\sqrt{1 + (\omega R_L C_2)^2}} \quad \text{(equation 17–23)}$$
$$= \frac{200}{\sqrt{1 + [2\pi(120)(200)(10^{-3})]^2}} = 1.32\,\Omega \approx |X_{C_2}|$$
$$|X_L + (X_{C_2} \| R_L)| \approx \sqrt{(4524)^2 + (1.32)^2} \approx 4524\,\Omega = |X_L|$$

The approximating conditions are therefore valid, and we have, from (17–29),

$$|V_r'| \approx \frac{|V_r|}{\omega^2 L C_2} = \frac{0.48}{(2\pi \times 120)^2(6)(10^{-3})} = 0.141 \text{ mV rms}$$

$$r' = \frac{|V_r'|}{V_{dc}'} \times 100\% = \frac{0.141 \times 10^{-3}}{39.8} \times 100\% = 3.54 \times 10^{-4}\%$$

This very substantial reduction in ripple is attributable to the series impedance presented by the inductor to the ripple voltage and to the two capacitors that shunt ripple current to ground.

17–5 VOLTAGE MULTIPLIERS

Diodes and capacitors can be connected in various configurations to produce filtered, rectified voltages that are integer multiples of the peak value of an input sine wave. The principle of operation of these circuits is similar to that of the *clamping* circuits discussed in Chapter 15. By using a transformer to change the amplitude of an ac voltage before it is applied to a voltage multiplier, a wide range of dc levels can be produced using this technique. One advantage of a voltage multiplier is that high voltages can be obtained without using a high-voltage transformer.

Half-Wave Voltage Doubler

Figure 17–21(a) shows a half-wave voltage *doubler.* When v_{in} first goes positive, diode D_1 is forward biased and diode D_2 is reverse biased. Because the forward resistance of D_1 is quite small, C_1 charges rapidly to V_P (neglecting the diode drop), as shown in (b). During the ensuing negative half-cycle of v_{in}, D_1 is reverse biased and D_2 is forward biased, as shown in (c). Consequently, C_2 charges rapidly, with polarity shown. Neglecting the drop across D_2, we can write Kirchhoff's voltage law around the loop at the instant v_{in} reaches its negative peak, and obtain

$$V_P = -V_P + V_{C_2}$$

or

$$V_{C_2} = 2V_P \qquad \textbf{(17–30)}$$

During the next positive half-cycle of v_{in}, D_2 is again reverse biased and the voltage across the output terminals remains at $V_{C_2} = 2V_P$ volts. Note carefully the polarity

Figure 17–21
A half-wave voltage doubler

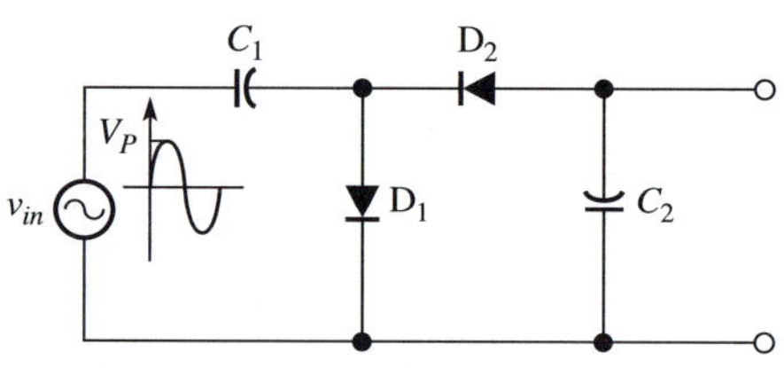

(a) Half-wave voltage-doubler circuit

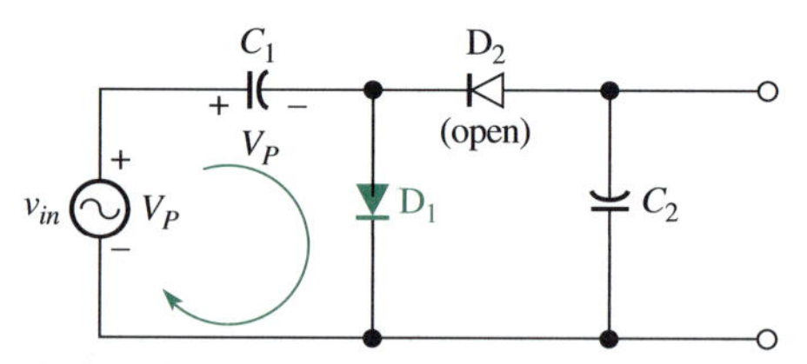

(b) C_1 charges to V_P during the positive half-cycle of v_{in}.

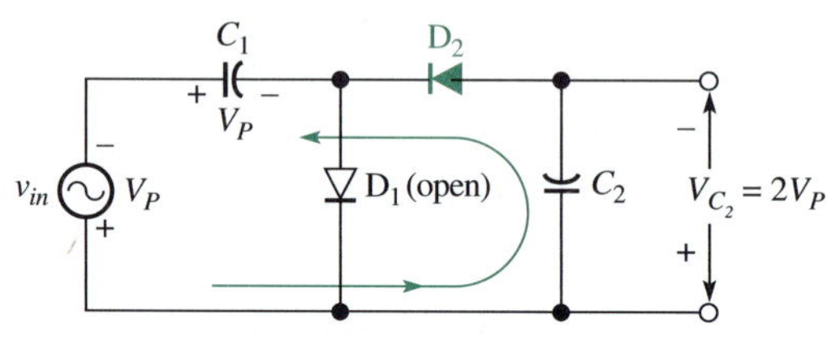

(c) C_2 charges to $2V_P$ during the negative half-cycle of v_{in}.

of the output. If a load resistor is connected across C_2, then C_2 will discharge into the load during positive half-cycles of v_{in}, and will recharge to $2V_P$ volts during negative half-cycles, creating the usual ripple waveform. The PIV rating of each diode must be at least $2V_P$ volts.

Full-Wave Voltage Doubler

Figure 17–22(a) shows a full-wave voltage doubler. This circuit is the same as the full-wave bridge rectifier shown in Figure 17–6, with two of the diodes replaced by capacitors. When v_{in} is positive, D_1 conducts and C_1 charges to V_P volts, as shown in (b). When v_{in} is negative, D_2 conducts and C_2 charges to V_P volts, with the polarity shown in (c). It is clear that the output voltage is then $V_{C_1} + V_{C_2} = 2V_P$ volts. Since one or the other of the capacitors is charging during every half-cycle, the output is the same as that of a capacitor-filtered, full-wave rectifier. Note, however, that the effective filter capacitance is that of C_1 and C_2 in series, which is less than either C_1 or C_2. The PIV rating of each diode must be at least $2V_P$ volts.

Voltage Tripler and Quadrupler

By connecting additional diode–capacitor sections across the half-wave voltage doubler, output voltages equal to three and four times the input peak voltage can be obtained. The circuit is shown in Figure 17–23. When v_{in} first goes positive, C_1 charges to V_P through forward-biased diode D_1. On the ensuing negative half-cycle, C_2 charges through D_2 and, as demonstrated earlier, the voltage across C_2 equals $2V_P$. During the next positive half-cycle, D_3 is forward biased and C_3 charges to

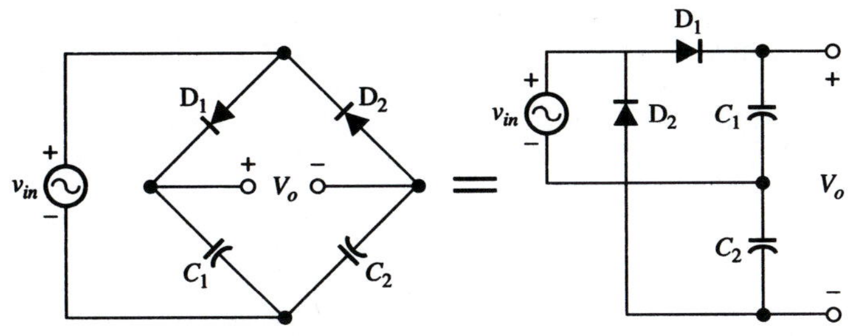

(a) The full-wave voltage-doubler circuit

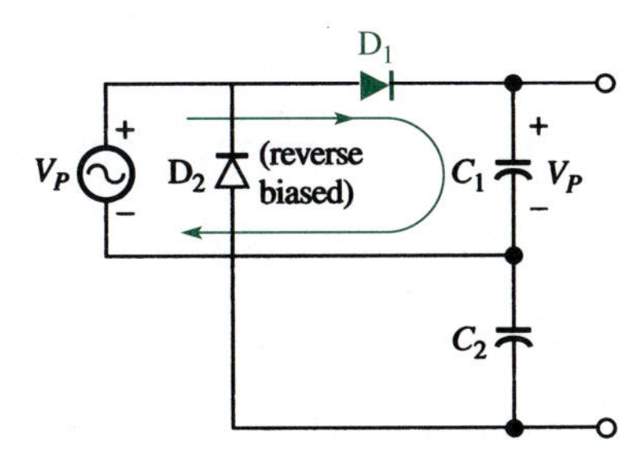

(b) C_1 charges to V_P during the positive half-cycle of V_{in}.

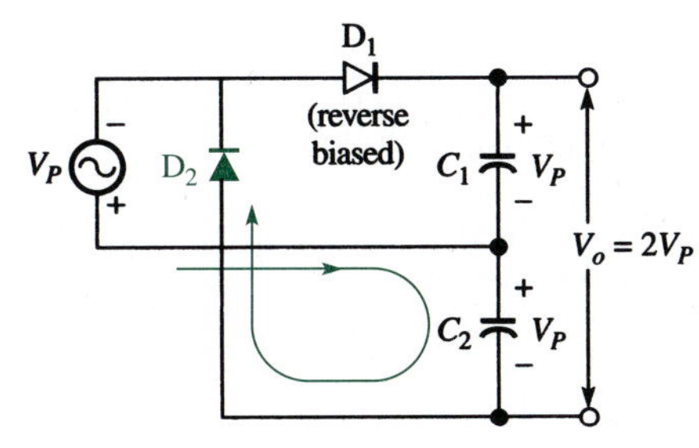

(c) C_2 charges to V_P during the negative half-cycle of V_{in}.

Figure 17–22
A full-wave voltage doubler

Figure 17–23
Voltage tripler and quadrupler

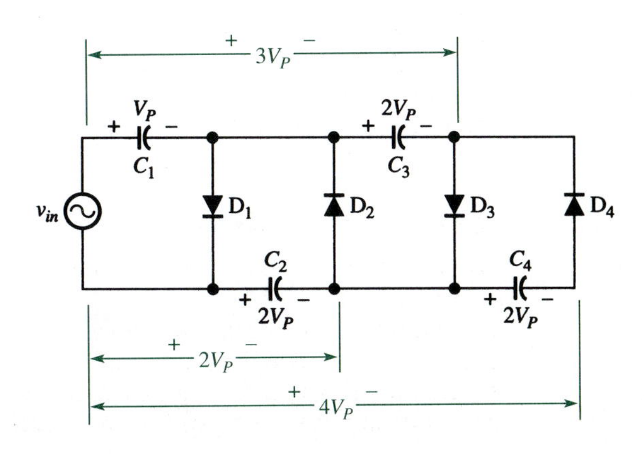

the same voltage attained by C_2: $2V_P$ volts. On the next negative half-cycle, D_2 and D_4 are forward biased and C_4 charges to $2V_P$ volts. As shown in the figure, the voltage across the combination of C_1 and C_3 is $3V_P$ volts, and that across C_2 and C_4 is $4V_P$ volts. Additional stages can be added in an obvious way to obtain even greater multiples of V_P. The PIV rating of each diode in the circuit must be at least $2V_P$ volts.

17–6 VOLTAGE REGULATION

An *ideal* power supply maintains a constant voltage at its output terminals, no matter what current is drawn from it. The output voltage of a practical power supply changes with load current, generally dropping as load current increases. Power-supply specifications include a *full-load current* (I_{FL}) rating, which is the maximum current that can be drawn from the supply. The terminal voltage when full-load current is drawn is called the *full-load voltage* (V_{FL}). The *no-load voltage* (V_{NL}) is the terminal voltage when zero current is drawn from the supply, that is, the open-circuit terminal voltage. Figure 17–24 illustrates these terms.

One measure of power-supply performance, in terms of how well the power supply is able to maintain a constant voltage between no-load and full-load conditions, is called its *percent voltage regulation:*

$$\text{VR} = \frac{V_{NL} - V_{FL}}{V_{FL}} \times 100\% \qquad (17\text{–}31)$$

More precisely, equation 17–31 defines the percent *output,* or *load,* voltage regulation, since it is based on changes that occur due to changes in load conditions, all other factors (including input voltage) remaining constant. It is clear that the numerator of equation 17–31 is the total change in output voltage between no-load and full-load, and that the ideal supply therefore has *zero* percent voltage regulation.

Figure 17–25 shows the Thevenin equivalent circuit of a power supply. The Thevenin voltage is the no-load voltage V_{NL}, and the Thevenin equivalent resistance

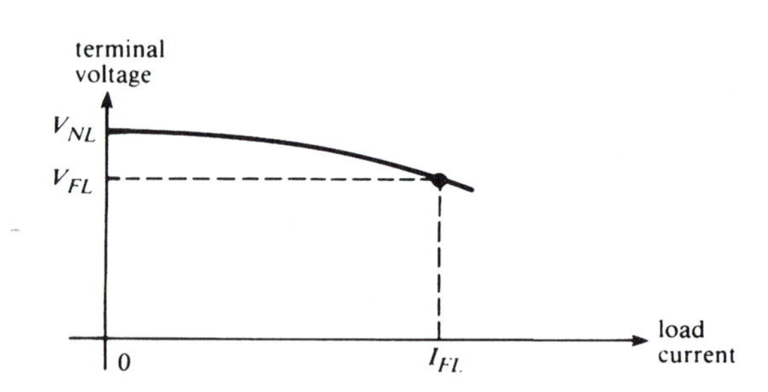

Figure 17–24
No-load and full-load conditions in a power supply

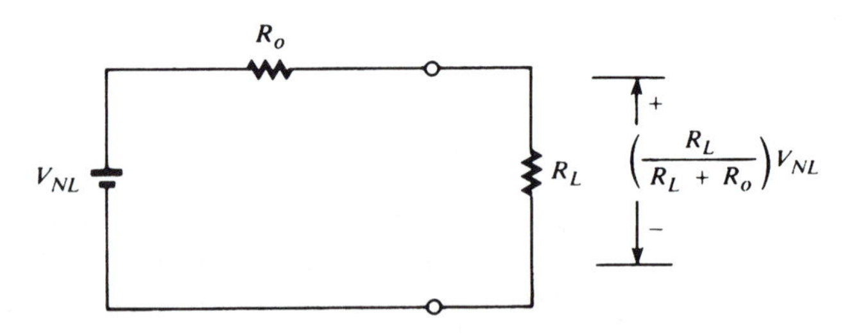

Figure 17–25
Thevenin equivalent circuit of a power supply

is called the *output resistance,* R_o, of the supply. Many power-supply manufacturers specify output resistance rather than percent voltage regulation. We will show that one can be obtained from the other, if the full-load–voltage and full-load–current ratings are known. Let the full-load resistance be designated

$$R_{FL} = \frac{V_{FL}}{I_{FL}} \tag{17–32}$$

When R_L in Figure 17–25 equals R_{FL}, we have, by voltage-divider action,

$$V_{FL} = \left(\frac{R_{FL}}{R_{FL} + R_o}\right) V_{NL} \tag{17–33}$$

Substituting (17–33) into (17–31) (and omitting the percent factor) gives

$$\mathrm{VR} = \frac{V_{NL} - \left(\frac{R_{FL}}{R_{FL} + R_o}\right) V_{NL}}{\left(\frac{R_{FL}}{R_{FL} + R_o}\right) V_{NL}} = \frac{R_o}{R_{FL}} \tag{17–34}$$

or

$$\mathrm{VR} = R_o\left(\frac{I_{FL}}{V_{FL}}\right) \tag{17–35}$$

It is clear that the ideal supply has zero output resistance, corresponding to zero percent voltage regulation. Like the output resistance we have studied in earlier chapters, R_o can also be determined as the slope of a plot of load voltage versus load current: $R_o = \Delta V_L/\Delta I_L$.

Example 17–8

A power supply having output resistance 1.5 Ω supplies a full-load current of 500 mA to a 50-Ω load.

1. What is the percent voltage regulation of the supply?
2. What is the no-load output voltage of the supply?

Solution

1. $V_{FL} = I_{FL}R_{FL} = (500\text{ mA})(50\ \Omega) = 25$ V. From equation 17–35,

$$\mathrm{VR}(\%) = R_o\left(\frac{I_{FL}}{V_{FL}}\right) \times 100\% = 1.5\left(\frac{0.5\text{ A}}{25\text{ V}}\right) \times 100\% = 3.0\%$$

(Note also that $\mathrm{VR} = R_o/R_{FL} = 1.5\ \Omega/50\ \Omega = 0.03$.)

2. From equation 17–40,

$$V_{NL} = \frac{V_{FL}(R_{FL} + R_o)}{R_{FL}} = \frac{25(50 + 1.5)}{50} = 25.75\text{ V}$$

Example 17–9

Assuming that the transformer and forward-biased–diode resistances in Figure 17–26 are negligible, find the percent voltage regulation of the power supply. The full-load current is 2 A at a full-load voltage of 15 V.

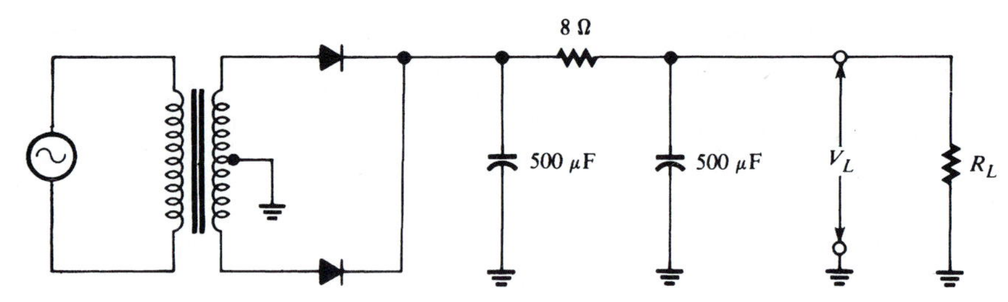

Figure 17–26
(Example 17–9)

Solution. Since the transformer and diode resistances can be neglected, the dc output resistance looking back into the load terminals is $R_o = 8\ \Omega$. Therefore, by equation 17–35,

$$\text{VR}(\%) = R_o\left(\frac{I_{FL}}{V_{FL}}\right) \times 100\% = 8\left(\frac{2\text{ A}}{15\text{ V}}\right) \times 100\% = 106.6\%$$

This example illustrates that the series resistance used in an RC π filter to reduce ripple can seriously degrade voltage regulation. Note that the no-load voltage in this example is, from equation 17–31, $V_{NL} = \text{VR}(V_{FL}) + V_{FL} = (1.066)15 + 15 = 31$ V, so there is a 16-V change in output voltage between no-load and full-load! This is an example of an *unregulated* supply, since there is no circuitry designed to correct, or compensate for, the effect of load variations. It would be unacceptable for most applications in which the load could vary over such a wide range.

Line Regulation

Percent *line regulation* is another measure of the ability of a power supply to maintain a constant output voltage. In this case, it is a measure of how sensitive the output is to changes in *input,* or line, voltage rather than to changes in load. The specification is usually expressed as the percent change in output voltage that occurs per volt change in input voltage, with the load R_L assumed constant. For example, a line regulation of 1%/V means that the output voltage will change 1% for each 1-V change in input voltage. If the input voltage to a 20-V supply having that specification were to change by 5 V, then the output could be expected to change by (5 V)(1%/V) = 5%, or 0.05(20) = 1 V.

$$\%\text{ line regulation} = \frac{(\Delta V_o/V_o) \times 100\%}{\Delta V_{in}} \qquad \textbf{(17–36)}$$

17–7 SERIES AND SHUNT VOLTAGE REGULATORS

A voltage regulator is a device, or combination of devices, designed to maintain the output voltage of a power supply as nearly constant as possible. It can be regarded as a *closed-loop control system* because it monitors output voltage and generates *feedback* that automatically increases or decreases the supply voltage, as necessary, to compensate for any tendency of the output to change. Thus, the

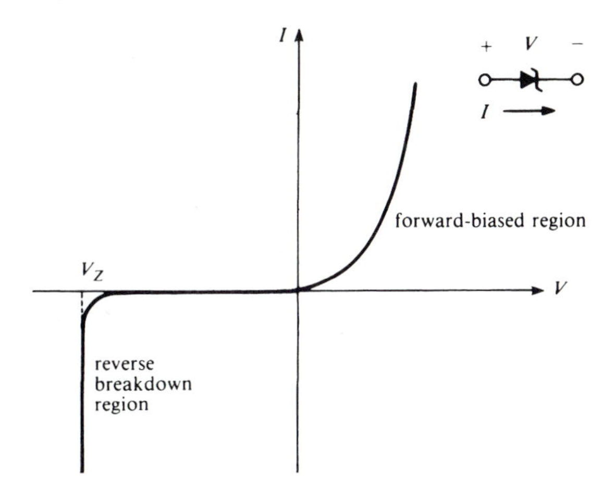

Figure 17–27
The I–V characteristic of a zener diode

purpose of a regulator is to eliminate any output voltage variation that might otherwise occur because of changes in load, changes in line (input) voltage, or changes in temperature. Regulators are constructed in numerous circuit configurations, ranging from one or two discrete components to elaborate integrated circuits.

The Zener Diode as a Voltage Reference

One of the simplest discrete regulators consists of only a resistor and a *zener diode.* We will study that circuit in detail in Chapter 18, but since zener diodes are also used in the more complex regulators that are the subject of our present discussion, we will preview the principal feature that makes them useful as *voltage references.* Figure 17–27 shows the *I–V* characteristic of a typical zener diode. Notice that the forward-biased characteristic is the same as that of a conventional diode. However, the zener is normally operated in its reverse-biased, *breakdown* region, which is the nearly vertical line drawn downward on the left side of the figure. The diode is manufactured to have a specific breakdown voltage, called the *zener voltage* V_Z, which may be as small as a few volts. The important point to notice is that the nearly vertical breakdown characteristic implies that the voltage V_Z across the diode remains substantially constant as the (reverse) current through it varies over a wide range. It therefore behaves much like an ideal voltage source that maintains a constant voltage independent of current. For this reason, the zener diode can be operated as an essentially constant voltage reference in regulator circuits, provided it remains in breakdown.

Series Regulators

Voltage regulators may be classified as being of either a *series* or a *shunt* design. Figure 17–28 is a *functional* block diagram of the series-type regulator. While the individual components shown in the diagram may not be readily identifiable in every series regulator circuit, the functional block diagram serves as a useful model for understanding the underlying principles of series regulation. V_{in} is an unregulated dc input, such as might be obtained from a rectifier with a capacitor filter, and V_o is the regulated output voltage. Notice that the *control element,* which is a device whose operating state adjusts as necessary to maintain a constant V_o, is in a series path between V_{in} and V_o. A *sampling circuit* produces a feedback voltage propor-

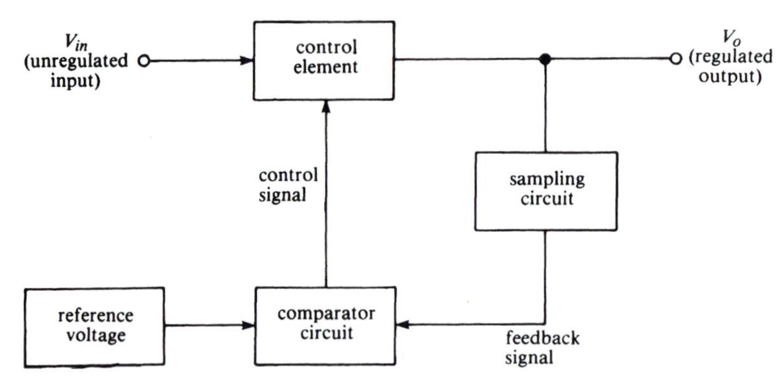

Figure 17–28
Block diagram of a series voltage regulator

tional to V_o and this voltage is compared to a reference voltage. The output of the comparator circuit is the control signal that adjusts the operating state of the control element. If V_o decreases, due, for example, to an increased load, then the comparator produces an output that causes the control element to increase V_o. In other words, V_o is automatically raised until the comparator circuit no longer detects any difference between the reference and the feedback. Similarly, any tendency of V_o to increase results in a signal that causes the control element to reduce V_o.

Figure 17–29 shows a simple transistor voltage regulator of the series type. Here, the control element is the NPN transistor, often called the *pass transistor* because it conducts, or passes, all the load current through the regulator. It is usually a power transistor, and it may be mounted on a heat sink in a heavy-duty power supply that delivers substantial current. The zener diode provides the voltage reference, and the base-to-emitter voltage of the transistor is the control voltage. In this case there is no identifiable sampling circuit, since the entire output voltage level V_o is used for feedback.

In Figure 17–29, notice that the zener diode is reverse biased and that reverse current is furnished to it through resistor R. Although V_{in} is unregulated, it must remain sufficiently large, and R must be sufficiently small, to keep the zener in its reverse breakdown region, as shown in Figure 17–27. Thus, as the unregulated input voltage varies, V_Z remains essentially constant. Writing Kirchhoff's voltage law around the output loop, we find

$$V_{BE} = V_Z - V_o \qquad (17\text{–}37)$$

Figure 17–29
A transistor series voltage regulator

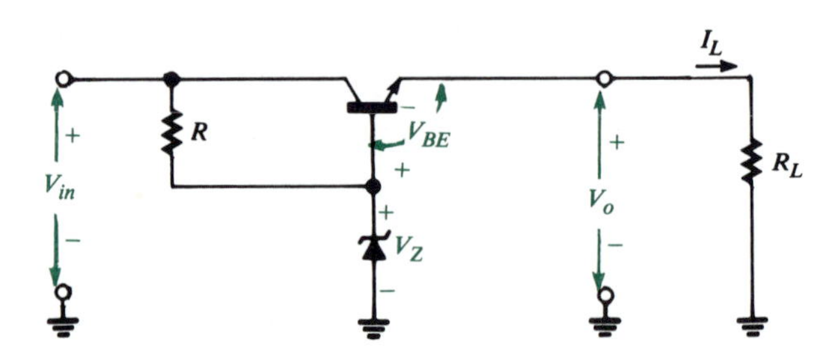

We now treat V_Z as perfectly constant. Since (17–37) is valid at all times, any change in V_o must cause a change in V_{BE}, to maintain the equality. For example, if V_o decreases, V_{BE} must increase since V_Z is constant. Similarly, if V_o increases, V_{BE} must decrease.

The behavior we have just described accounts for the ability of the circuit to provide voltage regulation. When V_o decreases, V_{BE} increases, which causes the NPN transistor to conduct more heavily and to produce more load current. The increase in load current causes an increase in V_o since $V_o = I_L R_L$. For example, suppose the regulator supplies 1 A to a 10-Ω load, so $V_o = (1\text{ A})(10\ \Omega) = 10\text{ V}$. Now suppose the load resistance is reduced to 8 Ω. The output voltage would then fall to $(1\text{ A})(8\ \Omega) = 8\text{ V}$. However, this reduction in V_o causes the transistor to conduct more heavily, and the load current increases to 1.25 A. Thus, the output voltage is restored to $V_o = (1.25\text{ A})(8\ \Omega) = 10\text{ V}$. The regulating action is similar when V_o increases: V_{BE} decreases, transistor conduction is reduced, load current decreases, and V_o is reduced.

Notice that the transistor is used essentially as an emitter follower. The load is connected to the emitter and the emitter follows the base, which is the constant zener voltage.

Example 17–10

In Figure 17–29, $V_{in} = 20$ V, $R = 200\ \Omega$, and $V_Z = 12$ V. If $V_{BE} = 0.65$ V, find

1. V_o;
2. the collector-to-emitter voltage of the pass transistor; and
3. the current in the 200-Ω resistor.

Solution

1. From equation 17–37, $V_o = V_Z - V_{BE} = 12\text{ V} - 0.65\text{ V} = 11.35\text{ V}$.
2. By writing Kirchhoff's voltage law around the loop that includes V_{in}, V_{CE}, and V_o, we find $V_{CE} = V_{in} - V_o = 20\text{ V} - 11.35\text{ V} = 8.65\text{ V}$.
3. The voltage drop across the 200-Ω resistor is $V_{in} - V_Z = 20\text{ V} - 12\text{ V} = 8\text{ V}$. Therefore, the current in the resistor is $I = (8\text{ V})/(200\ \Omega) = 0.04\text{ A}$.

For successful regulator operation, the pass transistor must remain in its active region, V_{in} must not drop to a level so small that the zener diode is no longer in its breakdown region, and the zener voltage V_Z should be highly independent of both current and temperature. Also, the feedback *loop gain,* from output back through the control element, should be very large, so that small changes in output voltage can be detected and rapidly corrected. Figure 17–30 shows an improved series regulator that incorporates an additional transistor in the feedback path to increase gain.

Note that resistors R_1 and R_2 form a voltage divider across V_o and serve as the sampling circuit that produces voltage V_2 proportional to V_o. Assuming that the divider is designed so that the zener-diode current does not load it appreciably,

$$V_2 = \frac{R_2}{R_1 + R_2} V_o \qquad (17\text{–}38)$$

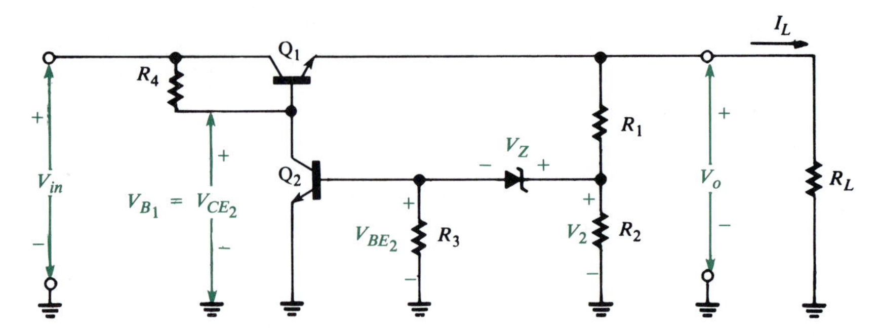

Figure 17–30
An improved series regulator that incorporates transistor Q_2 to increase sensitivity

Writing Kirchhoff's voltage law around the loop containing R_2, R_3, and the zener diode, we have

$$V_{BE_2} + V_Z = V_2 \qquad \textbf{(17–39)}$$

Any decrease in V_o will cause a decrease in V_2 and, assuming that V_Z remains constant, V_{BE_2} must therefore decrease to maintain equality in (17–39). A decrease in V_{BE_2} causes V_{CE_2} to increase, since Q_2 is an inverting common-emitter stage. Since the base voltage V_{B_1} on Q_1 experiences this same increase, Q_1 conducts more heavily, produces more load current, and increases V_o, as previously described. Another way of viewing this feedback action is to recognize that a decrease in V_{BE_2} causes Q_2 to conduct less, and therefore allows more base current to flow into Q_1. Of course, any *increase* in V_o causes actions that are just the opposite of those we have described and results in less load current.

Example 17–11

In the regulator shown in Figure 17–30, $R_1 = 50\ \text{k}\Omega$, $R_2 = 43.75\ \text{k}\Omega$, and $V_Z = 6.3$ V. If the 15-V output drops 0.1 V, find the change in V_{BE_2} that results.

Solution. From equation 17–38, when $V_o = 15$ V,

$$V_2 = \frac{R_2}{R_1 + R_2} V_o = \left[\frac{43.75\ \text{k}\Omega}{(50\ \text{k}\Omega) + (43.75\ \text{k}\Omega)}\right](15\ \text{V}) = 7\ \text{V}$$

Therefore, by (17–39), $V_{BE_2} = V_2 - V_Z = 7\ \text{V} - 6.3\ \text{V} = 0.7$ V. When $V_o = 15\ \text{V} - 0.1\ \text{V} = 14.9$ V, V_2 becomes

$$V_2' = \left[\frac{43.75\ \text{k}\Omega}{(50\ \text{k}\Omega) + (43.75\ \text{k}\Omega)}\right](14.9\ \text{V}) = 6.953\ \text{V}$$

Therefore,

$$V_{BE_2}' = V_2' - V_Z = 6.953\ \text{V} - 6.3\ \text{V} = 0.653\ \text{V}$$

$$\Delta V_{BE_2} = V_{BE_2} - V_{BE_2}' = 0.7\ \text{V} - 0.653\ \text{V} = 0.047\ \text{V}$$

Figure 17–31 shows how an operational amplifier is used in a series voltage regulator. Resistors R_1 and R_2 form a voltage divider that feeds a voltage propor-

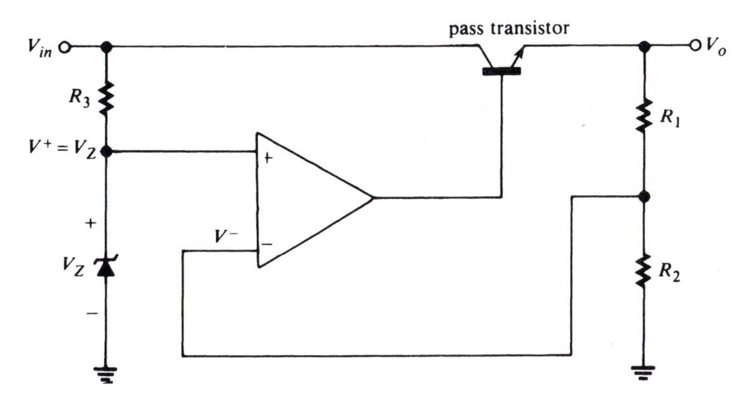

Figure 17–31
An operational amplifier used in a series voltage regulator

tional to V_o back to the inverting input: $V^- = V_oR_2/(R_1 + R_2)$. The zener voltage V_Z on the noninverting input is greater than V^-, so the amplifier output is positive and is proportional to $V_Z - V^-$. If V_o decreases, V^- decreases, and the amplifier output increases. The greater voltage applied to the base of the pass transistor causes it to conduct more heavily, and V_o increases. Similarly, an increase in V_o causes V^- to increase, $V_Z - V^-$ to decrease, and the amplifier output to decrease. The pass transistor conducts less heavily and V_o decreases.

The operational amplifier in Figure 17–31 can be regarded as a noninverting configuration with input V_Z and gain

$$A = 1 + \frac{R_1}{R_2} \tag{17–40}$$

Therefore, neglecting the base-to-emitter drop of the pass transistor, the regulated output voltage is

$$V_o = \left(1 + \frac{R_1}{R_2}\right) V_Z \tag{17–41}$$

R_3 is chosen to ensure that sufficient reverse current flows through the zener diode to keep it in breakdown.

Example 17–12

DESIGN

Design a series voltage regulator using an operational amplifier and a 6-V zener diode to maintain a regulated output of 18 V. Assume that the unregulated input varies between 20 V and 30 V and that the current through the zener diode must be at least 20 mA to keep it in its breakdown region.

Solution. From equation 17–41, 18 V = $(1 + R_1/R_2)$6 V. Thus, $(1 + R_1/R_2) = 3$. Let $R_1 = 20\ \text{k}\Omega$. Then

$$\left(1 + \frac{20\ \text{k}\Omega}{R_2}\right) = 3$$

$$R_2 = 10\ \text{k}\Omega$$

The current through R_3 is

$$I = \frac{V_{in} - V_Z}{R_3} \tag{17–42}$$

Since the current into the noninverting input is negligibly small, the current in R_3 is the same as the current in the zener diode. This current must be at least 20 mA, so in equation 17–42, I must be 20 mA for the smallest possible value of V_{in} (20 V):

$$20\text{ mA} = \frac{(20\text{ V}) - (6\text{ V})}{R_3}$$

$$R_3 = \frac{14\text{ V}}{20\text{ mA}} = 700\ \Omega$$

Current Limiting

Many general-purpose power supplies are equipped with short-circuit or overload protection. One form of protection is called *current limiting,* whereby specially designed circuitry limits the current that can be drawn from the supply to a certain specific maximum, even if the output terminals are short-circuited. Figure 17–32 shows a popular current-limiting circuit incorporated into the operational-amplifier regulator. As load current increases, the voltage drop across resistor R_{SC} increases. Notice that R_{SC} is in parallel with the base-to-emitter junction of transistor Q_2. If the load current becomes great enough to create a drop of about 0.7 V across R_{SC}, then Q_2 begins to conduct a substantial collector current. As a consequence, current that would otherwise enter the base of Q_1 is diverted through Q_2. In this way, the pass transistor is prevented from supplying additional load current. The maximum (short-circuit) current that can be drawn from the supply is the current necessary to create the 0.7-V drop across R_{SC}:

$$I_L(\text{max}) = \frac{0.7\text{ V}}{R_{SC}} \tag{17–43}$$

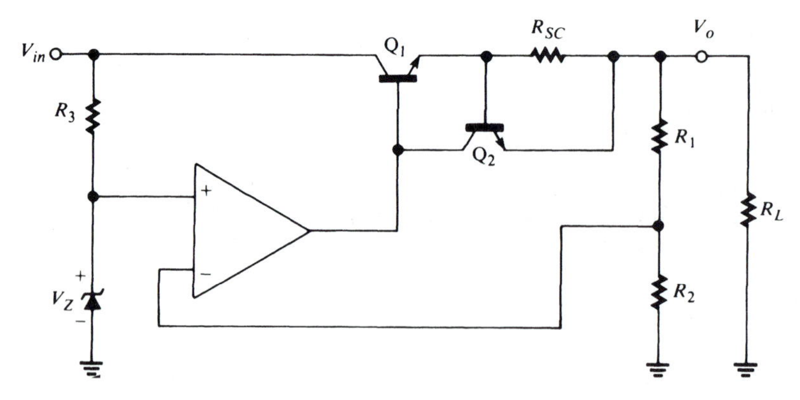

Figure 17–32
Q_2 and R_{SC} are used to limit the maximum current that can be drawn from the supply.

Example 17–13

The voltage regulator in Figure 17–32 maintains an output voltage of 25 V.

1. What value of R_{SC} should be used to limit the maximum current to 0.5 A?
2. With the value of R_{SC} found in (1), what will be the output voltage when R_L = 100 Ω? When R_L = 10 Ω?

Solution

1. From equation 17–43,

$$R_{SC} = \frac{0.7\text{ V}}{I_L(\text{max})} = \frac{0.7\text{ V}}{0.5\text{ A}} = 1.4\ \Omega$$

2. We must determine whether a 100-Ω load would attempt to draw more than the current limit of 0.5 A at the regulated output voltage of 25 V:

$$I_L = \frac{25\text{ V}}{100\ \Omega} = 0.25\text{ A}$$

Since 0.25 A $<$ I_L(max) = 0.5 A, the output voltage will remain at 25 V.
If R_L = 10 Ω, the (unlimited) current would be

$$I_L = \frac{25\text{ V}}{10\ \Omega} = 2.5\text{ A}$$

Since 2.5 A $>$ I_L(max) = 0.5 A, current limiting will occur, and the output voltage will be $V_o = I_L(\text{max})R_L$ = (0.5 A)(10 Ω) = 5 V.

Example 17–13 demonstrates that the output voltage of a current-limited regulator decreases if the load resistance is made smaller than what would draw maximum current at the regulated output voltage. Figure 17–33 shows a typical load-voltage–load-current characteristic for a current-limited regulator. The characteristic shows that load current may increase slightly beyond I_L(max) as the output approaches a short-circuit condition (V_L = 0). Note that Q_2 in Figure 17–32 supplies a small amount of additional current to the load once current limiting takes place.

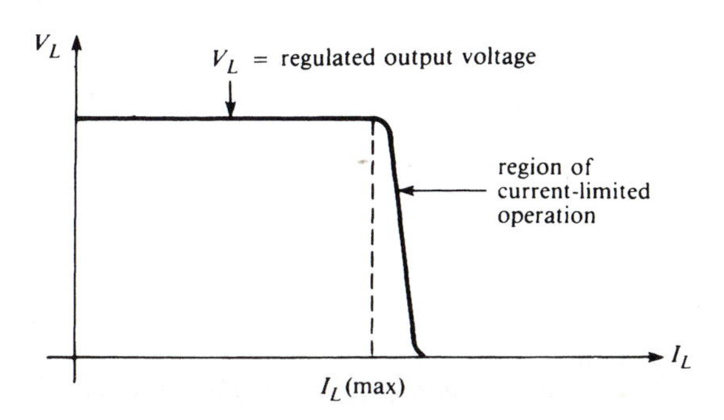

Figure 17–33
A typical load-voltage–load-current characteristic for a current-limited regulator

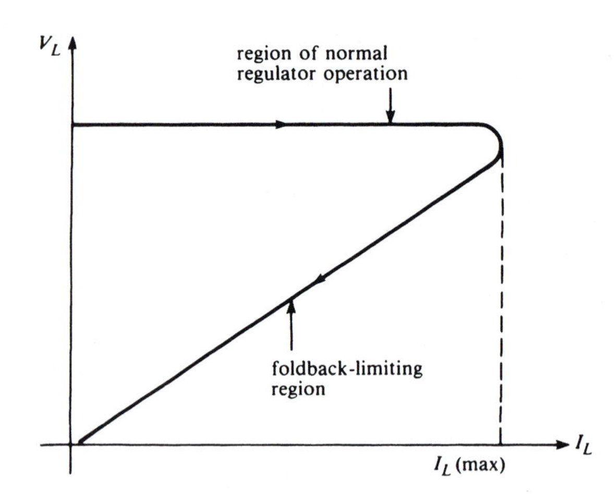

Figure 17–34
Foldback limiting (Compare with Figure 17–33.)

Foldback Limiting

Another form of overcurrent protection, called *foldback limiting,* is used to reduce both the output current and the output voltage if the load resistance is made smaller than what would draw a specified maximum current, I_L(max). We have already seen that the output voltage of a current-limited regulator decreases as the load resistance is made smaller (Example 17–13). In foldback limiting, this decrease in output voltage is sensed and is used to further decrease the amount of current that can flow to the load. Thus, as load resistance decreases beyond a certain minimum, both load voltage and load current decrease. If the load becomes a short circuit, output voltage and output current both approach 0. The foldback characteristic is shown in Figure 17–34. The principal purpose of foldback is to protect a load from overcurrent, as well as to protect the regulator itself.

Figure 17–35 shows one method used to add foldback limiting to the basic current-limited regulator. Notice the similarity of this circuit to the current-limited regulator circuit shown in Figure 17–32. The only difference is that the base of Q_2

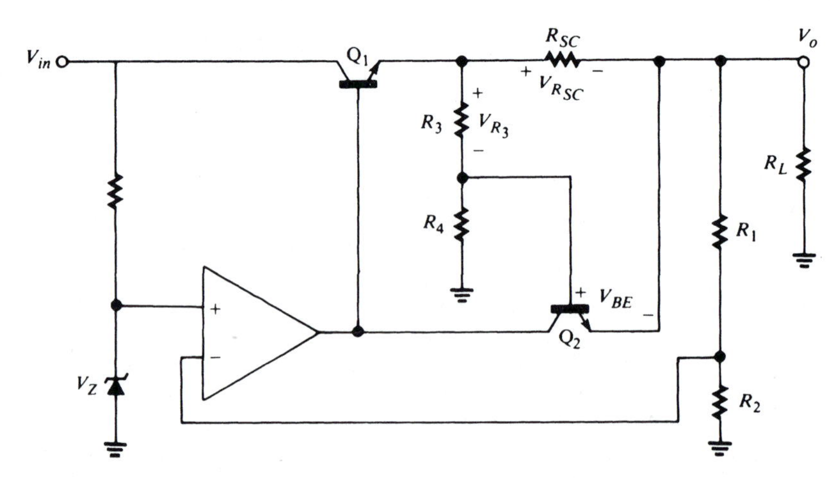

Figure 17–35
A current-limited regulator modified to provide foldback limiting

is now connected to the R_3–R_4 voltage divider. Writing Kirchhoff's voltage law around the loop, we find

$$V_{BE} = V_{R_{SC}} - V_{R_3} \tag{17–44}$$

Notice that V_{R_3} will increase or decrease if the load voltage increases or decreases. When the load current increases to its maximum permissible value, $V_{R_{SC}}$ becomes large enough to make V_{BE} approximately 0.7 V; that is, $V_{R_{SC}}$ becomes large enough to exceed the drop across V_{R_3} by about 0.7 V: $0.7 \approx V_{BE} = V_{R_{SC}} - V_{R_3}$. At this point, current limiting occurs, just as it does in the current-limited regulator. If the load resistance is now made smaller, the load voltage will drop, as we have seen previously. But when the load voltage drops, V_{R3} drops. Consequently, by equation 17–44, a smaller value of $V_{R_{SC}}$ is required to maintain $V_{BE} = 0.7$ V. Since V_{BE} remains essentially constant at 0.7 V, a smaller load current must flow to produce the smaller drop across R_3. Further decreases in load resistance produce further drops in load voltage and a further reduction in load current, so foldback limiting occurs. If the load resistance is restored to its normal operating value, the circuit resumes normal regulator action.

Example 17–14

The circuit in Figure 17–35 maintains a regulated output of 6 V. If $R_3 = 1$ kΩ and $R_4 = 9$ kΩ, what should be the value of R_{SC} to impose a maximum current limit of 1 A?

Solution. At the onset of current limiting,

$$V_{R_3} = \left(\frac{R_3}{R_3 + R_4}\right) V_o = \left[\frac{1\text{ k}\Omega}{(1\text{ k}\Omega) + (9\text{ k}\Omega)}\right] (6\text{ V}) = 0.6\text{ V}$$

From equation 17–44,

$$V_{BE} = 0.7\text{ V} = V_{R_{SC}} - 0.6\text{ V}$$

$$V_{R_{SC}} = 1.3\text{ V}$$

Thus,

$$R_{SC} = \frac{V_{R_{SC}}}{I_L(\text{max})} = \frac{1.3\text{ V}}{1\text{ A}} = 1.3\ \Omega$$

Shunt Regulators

Figure 17–36 is a functional block diagram of the shunt-type regulator. Each of the components shown in the figure performs the same function as its counterpart in the series regulator (Figure 17–28), but notice in this case that the control element is in parallel with the load. The control element maintains a constant load voltage by shunting more or less current from the load.

It is convenient to think of the control element in Figure 17–36 as a variable resistance. When the load voltage decreases, the resistance of the control element is made to increase, so less current is diverted from the load, and the load voltage rises. Conversely, when the load voltage increases, the resistance of the control element decreases, and more current is shunted away from the load. From another viewpoint, the source resistance R_S on the unregulated side of Figure 17–36 forms

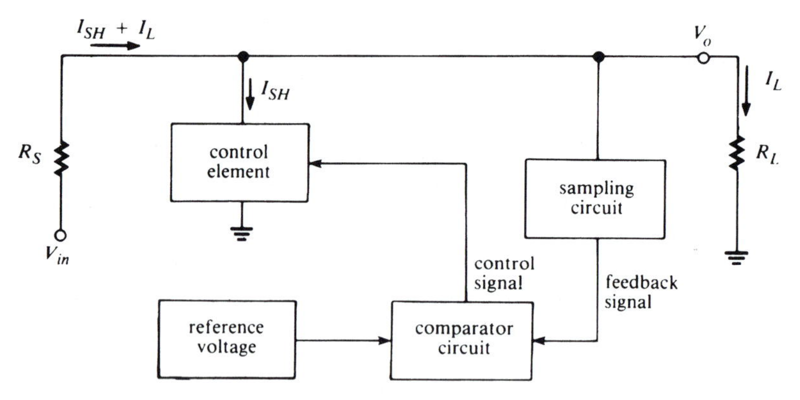

Figure 17–36
Functional block diagram of a shunt-type voltage regulator

a voltage divider with the parallel combination of the control element and R_L. Thus, when the resistance of the control element increases, the resistance of the parallel combination increases, and, by voltage-divider action, the load voltage increases.

Figure 17–37 shows a discrete shunt regulator in which transistor Q_1 serves as the shunt control element. Since V_Z is constant, any change in output voltage creates a proportional change in the voltage across R_1. Thus, if V_o decreases, the voltage across R_1 decreases, as does the base voltage of Q_2. Q_2 conducts less heavily and the current into the base of Q_1 is reduced. Q_1 then conducts less heavily and shunts less current from the load, allowing the load voltage to rise. Conversely, an increase in V_o causes both Q_1 and Q_2 to conduct more heavily, and more current is diverted from the load.

Figure 17–38 shows a shunt regulator incorporating an operational amplifier. Resistors R_1 and R_2 form a voltage divider that feeds a voltage proportional to V_o back to the noninverting input. This voltage is greater than the reference voltage V_Z applied to the inverting input, so the output of the amplifier is a positive voltage proportional to $V_o - V_Z$. If V_o decreases, the amplifier output decreases and Q_1 conducts less heavily. Thus, less current is diverted from the load and the output voltage rises.

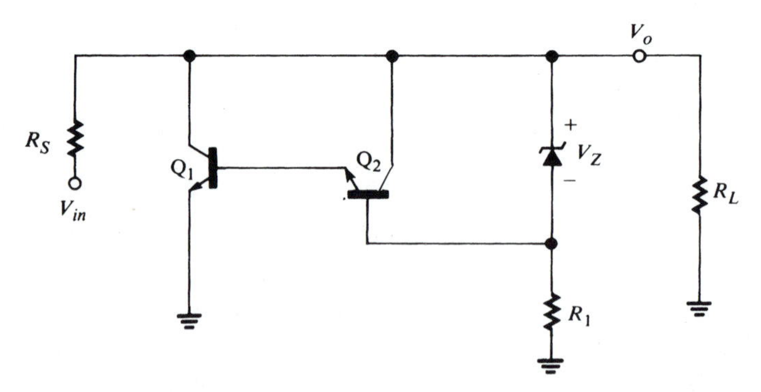

Figure 17–37
A discrete shunt regulator

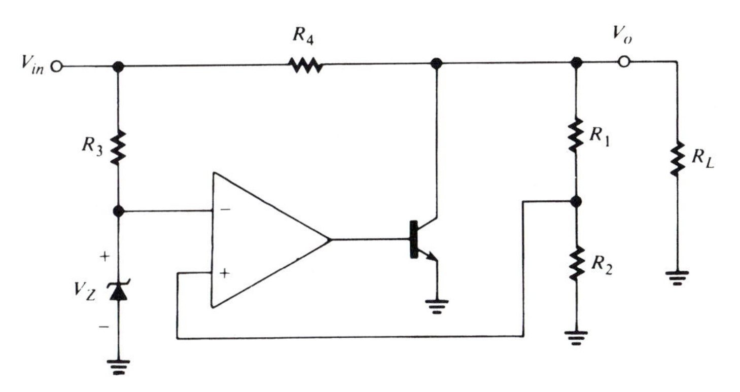

Figure 17–38
A shunt voltage regulator incorporating an operational amplifier

One advantage of the shunt-regulator circuit shown in Figure 17–38 is that it has inherent current limiting. It is clear that the load current cannot exceed V_{in}/R_4, which is the current that would flow through R_4 if the output were short-circuited. Since load current must flow through R_4, the power dissipation in the resistor may be quite large, particularly under short-circuit conditions.

17–8 SWITCHING REGULATORS

A principal disadvantage of the series and shunt regulators discussed in Section 17–7 is the fact that the control element in each type must dissipate a large amount of power. As a consequence, they have poor *efficiency;* that is, the ratio of load power to total input power is relatively small. Since the control element, such as a pass transistor, is operated in its active region, power dissipation can be quite large, particularly when there is a large voltage difference between V_{in} and V_o and when substantial load current flows. When efficiency is a major concern, the switching-type regulator is often used. In this design, the control element is switched on and off at a rapid rate and is therefore either in saturation or cutoff most of the time. As is the case with other digital switching devices, this mode of operation results in very low power consumption.

The fundamental component of a switching regulator is a *pulse-width modulator,* illustrated in Figure 17–39. This device produces a train of rectangular pulses having widths that are proportional to the device's input. The figure illustrates the pulse output corresponding to a small dc input, the output corresponding to a large dc input, and that corresponding to a ramp-type input. To produce a periodic sequence of pulses, the pulse-width modulator must employ some form of oscillator. Practical modulators used in switching regulators usually have inherent oscillation capabilities incorporated into their circuit designs. Figure 17–40 shows how a pulse-width modulator (used in power amplifiers) can be constructed using a voltage comparator driven by a sawtooth waveform.

The *duty cycle* of a pulse train is the proportion of the period during which the pulse is present, i.e., the fractional part of the period during which the output is high. For example, a square wave, which is high for one half-cycle and low for the other half-cycle, has a duty cycle of 0.5, or 50%. By definition

$$\text{duty cycle} = T_{HI}/T \tag{17–45}$$

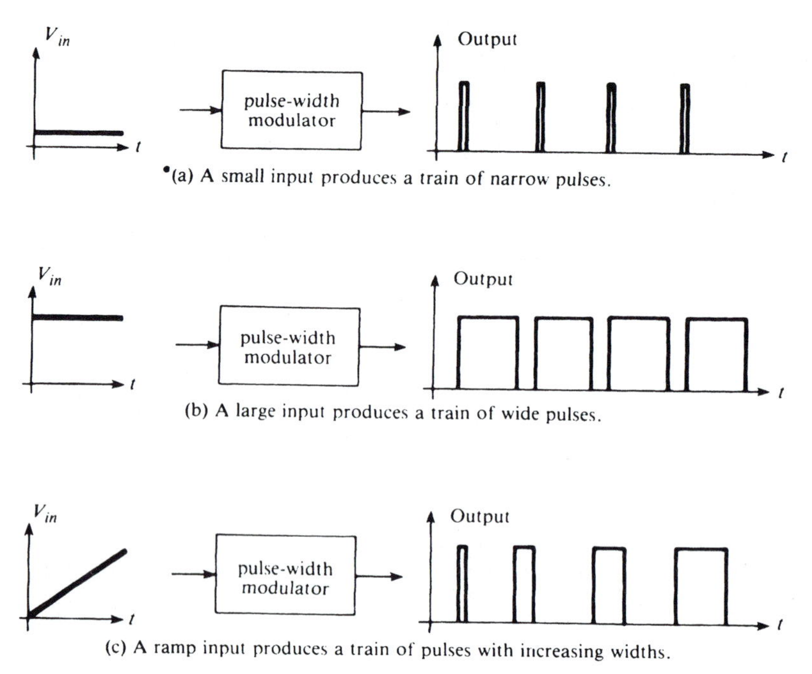
V_{in}
Output
pulse-width modulator
t
(a) A small input produces a train of narrow pulses.
(b) A large input produces a train of wide pulses.
(c) A ramp input produces a train of pulses with increasing widths.

Figure 17–39
Operation of a pulse-width modulator

where T_{HI} is the total time during period T that the waveform is high. As illustrated in Figure 17–40, the dc, or average, value of a pulse train is directly proportional to its duty cycle:

$$V_{dc} = V_{HI}\left(\frac{T_{HI}}{T}\right) \tag{17–46}$$

where V_{HI} is the high, or peak, value of the pulse. Equation 17–46 is valid for pulses that alternate between 0 volts (low) and V_{HI} volts.

A switching regulator uses a pulse-width modulator to produce a pulse train whose duty cycle is automatically adjusted as necessary to increase or decrease the dc value of the train. The basic circuit is shown in Figure 17–41. Note that the output of the pulse-width modulator drives pass transistor Q_1. When the drive pulse is high, Q_1 is saturated and maximum current flows through it. When the drive pulse is low, Q_1 is cut off and no current flows. Thus, the current through Q_1 is a series of pulses having the same duty cycle as the modulator output. As in the regulator circuits discussed earlier, the zener diode provides a reference voltage that is compared in the operational amplifier to the feedback voltage obtained from the R_1–R_2 voltage divider. If the load voltage V_o tends to fall, then the output of the operational amplifier increases and a larger voltage is applied to the pulse-width modulator. It therefore produces a pulse train having a larger duty cycle. The pulse train switches Q_1 on and off with a greater duty cycle, so the dc value of the current through Q_1 is increased. The increased current raises the load voltage to compensate for its initial decrease.

Series inductor L and shunt capacitor C in Figure 17–41 form a low-pass filter that recovers the dc value of the pulse waveform supplied to it by Q_1. The fundamen-

Figure 17–40
The dc value of a pulse train is directly proportional to its duty cycle.

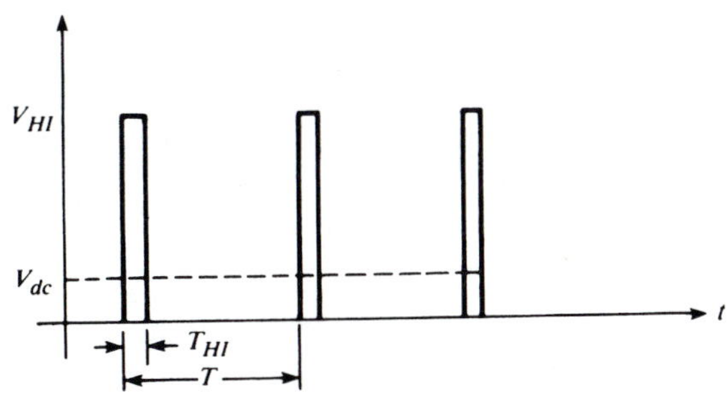

(a) A low-duty-cycle pulse train has a small dc value.

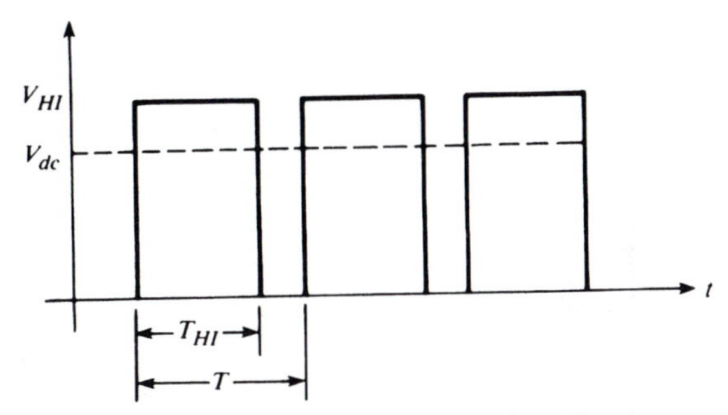

(b) A high-duty-cycle pulse train has a large dc value.

tal frequency of the pulse train and its harmonics are suppressed by the filter. Diode D_1 is used to suppress the negative voltage transient generated by the inductor when Q_1 turns off. It is called a *catcher* diode, or *free-wheeling* diode. Modern switching regulators are operated at frequencies in the kilohertz range, usually less than 100 kHz. High frequencies are desirable because the LC filter components can then be smaller, less bulky, and less expensive. However, high-frequency switching of heavy currents creates large magnetic fields that induce noise voltages in surrounding conductors. This generation of *electromagnetic interference* (EMI) is a principal limitation of switching regulators, and they require careful shielding. Also, switching

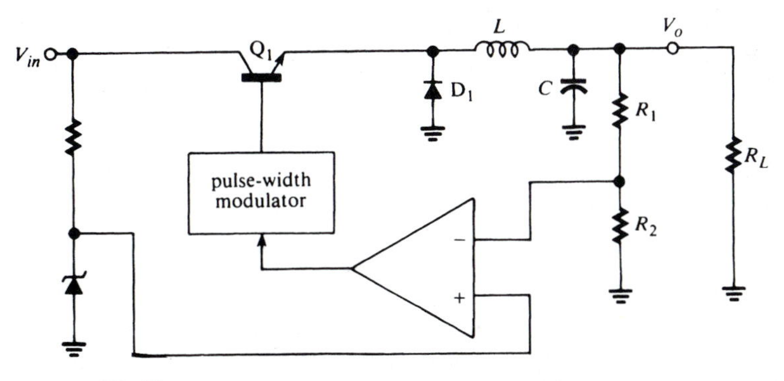

Figure 17–41
A basic switching-type regulator circuit

frequencies are limited by the inability of the power (pass) transistor to turn on and off at high speeds. Many modern regulators use power MOSFETs of the VMOS design, instead of bipolars, because of their superior ability to switch heavy loads at high frequencies (see Section 16–11 on MOSFET power amplifiers).

Example 17–15

The switching regulator in Figure 17–41 is designed to maintain a 12-V-dc output when the unregulated input voltage varies from 15 V to 24 V. When pass transistor Q_1 is conducting, its collector-to-emitter saturation voltage is 0.5 V. Assuming that the load is constant and the LC filter is ideal (so its output is the dc value of its input), find the minimum and maximum duty cycles of the pulse-width modulator.

Solution. When V_{in} has its minimum value of 15 V, the high voltage of the pulse train at the input to the filter is $15 - V_{CE(sat)} = 15 - 0.5 = 14.5$ V. The duty cycle of this pulse train must be sufficient to produce a dc value of 12 V. From equation 17–46, $T_{HI}/T = V_{dc}/V_{HI} = 12/14.5 = 0.828$. Thus, the *maximum* duty cycle, corresponding to the smallest value of V_{in}, is 0.828, or 82.8%. When V_{in} has its maximum value of 24 V, we find the minimum duty cycle to be $T_{HI}/T = 12/(24 - 0.5) = 0.51$.

17–9 THREE-TERMINAL INTEGRATED-CIRCUIT REGULATORS

A three-terminal regulator is a compact, easy-to-use, fixed-voltage regulator packaged in a single integrated circuit. To use the regulator, it is necessary only to make external connections to the three terminals: V_{in}, V_o, and ground. These devices are widely used to provide *local regulation* in electronic systems that may require several different supply voltages. For example, a 5-V regulator could be used to regulate the power supplied to all the chips mounted on one printed-circuit board, and a 12-V regulator could be used for a similar purpose on a different board. The regulators might well use the same unregulated input voltage, say, 20 V.

A popular series of three-terminal regulators is the 7800/7900 series, available from several manufacturers with a variety of output voltage ratings. Figure 17–42 shows National Semiconductor specifications for their 7800-series regulators, which carry the company's standard LM prefix and which are available with regulated outputs of +5 V, +12 V, and +15 V. The last two digits of the 7800 number designate the rated output voltage. For example, the 7805 is a +5-V regulator and the 7815 is a +15-V regulator. The 7900-series regulators provide negative output voltages. Notice that the integrated circuitry shown in the schematic diagram is considerably more complex than that of the simple discrete circuits discussed earlier. It can be seen that the circuit incorporates a zener diode as an internal voltage reference. The 7800/7900 series also has internal current-limiting circuitry.

Important points to note in the 7800-series specifications include the following:

1. The output voltage of an arbitrarily chosen device might not *exactly* equal its nominal value. For example, with a 23-V input, the 7815 output may be anywhere from 14.4 V to 15.6 V. This specification does not mean that the output voltage

Voltage Regulators

LM78XX Series Voltage Regulators

General Description

The LM78XX series of three terminal regulators is available with several fixed output voltages making them useful in a wide range of applications. One of these is local on card regulation, eliminating the distribution problems associated with single point regulation. The voltages available allow these regulators to be used in logic systems, instrumentation, HiFi, and other solid state electronic equipment. Although designed primarily as fixed voltage regulators these devices can be used with external components to obtain adjustable voltages and currents.

The LM78XX series is available in an aluminum TO-3 package which will allow over 1.0A load current if adequate heat sinking is provided. Current limiting is included to limit the peak output current to a safe value. Safe area protection for the output transistor is provided to limit internal power dissipation. If internal power dissipation becomes too high for the heat sinking provided, the thermal shutdown circuit takes over preventing the IC from overheating.

Considerable effort was expended to make the LM78XX series of regulators easy to use and minimize the number of external components. It is not necessary to bypass the output, although this does improve transient response. Input bypassing is needed only if the regulator is located far from the filter capacitor of the power supply.

For output voltage other than 5V, 12V and 15V the LM117 series provides an output voltage range from 1.2V to 57V.

Features

- Output current in excess of 1A
- Internal thermal overload protection
- No external components required
- Output transistor safe area protection
- Internal short circuit current limit
- Available in the aluminum TO-3 package

Voltage Range

LM7805C	5V
LM7812C	12V
LM7815C	15V

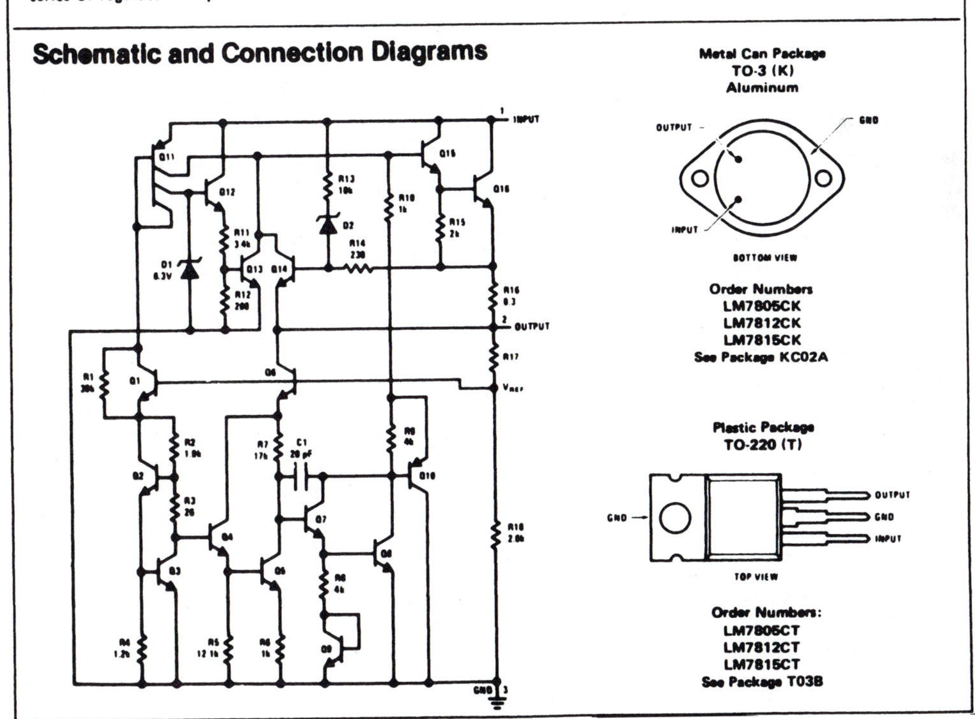

Figure 17–42
7800-series voltage-regulator specifications (Courtesy of National Semiconductor)

Absolute Maximum Ratings

Input Voltage (V_O = 5V, 12V and 15V)	35V
Internal Power Dissipation (Note 1)	Internally Limited
Operating Temperature Range (T_A)	0 °C to +70 °C
Maximum Junction Temperature	
(K Package)	150 °C
(T Package)	125 °C
Storage Temperature Range	−65 °C to +150 °C
Lead Temperature (Soldering, 10 seconds)	
TO-3 Package K	300 °C
TO-220 Package T	230 °C

Electrical Characteristics LM78XXC (Note 2) $0\,°C \le T_j \le 125\,°C$ unless otherwise noted.

OUTPUT VOLTAGE			5V			12V			15V			UNITS
INPUT VOLTAGE (unless otherwise noted)			10V			19V			23V			
PARAMETER	CONDITIONS		MIN	TYP	MAX	MIN	TYP	MAX	MIN	TYP	MAX	
V_O Output Voltage	$T_j = 25\,°C$, $5\text{ mA} \le I_O \le 1\text{A}$		4.8	5	5.2	11.5	12	12.5	14.4	15	15.6	V
	$P_D \le 15\text{W}$, $5\text{ mA} \le I_O \le 1\text{A}$		4.75		5.25	11.4		12.6	14.25		15.75	V
	$V_{MIN} \le V_{IN} \le V_{MAX}$		$(7 \le V_{IN} \le 20)$			$(14.5 \le V_{IN} \le 27)$			$(17.5 \le V_{IN} \le 30)$			V
ΔV_O Line Regulation	$I_O = 500\text{ mA}$	$T_j = 25\,°C$		3	50		4	120		4	150	mV
		ΔV_{IN}	$(7 \le V_{IN} \le 25)$			$(14.5 \le V_{IN} \le 30)$			$(17.5 \le V_{IN} \le 30)$			V
		$0\,°C \le T_j \le +125\,°C$			50			120			150	mV
		ΔV_{IN}	$(8 \le V_{IN} \le 20)$			$(15 \le V_{IN} \le 27)$			$(18.5 \le V_{IN} \le 30)$			V
	$I_O \le 1\text{A}$	$T_j = 25\,°C$			50			120			150	mV
		ΔV_{IN}	$(7.3 \le V_{IN} \le 20)$			$(14.6 \le V_{IN} \le 27)$			$(17.7 \le V_{IN} \le 30)$			V
		$0° \le T_j \le +125\,°C$			25			60			75	mV
		ΔV_{IN}	$(8 \le IN \le 12)$			$(16 \le V_{IN} \le 22)$			$(20 \le V_{IN} \le 26)$			V
ΔV_O Load Regulation	$T_j = 25\,°C$	$5\text{ mA} \le I_O \le 1.5\text{A}$		10	50		12	120		12	150	mV
		$250\text{ mA} \le I_O \le 750\text{ mA}$			25			60			75	mV
	$5\text{ mA} \le I_O \le 1\text{A}$, $0\,°C \le T_j \le +125\,°C$				50			120			150	mV
I_Q Quiescent Current	$I_O \le 1\text{A}$	$T_j = 25\,°C$			8			8			8	mA
		$0\,°C \le T_j \le +125\,°C$			8.5			8.5			8.5	mA
ΔI_Q Quiescent Current Change	$5\text{ mA} \le I_O \le 1\text{A}$				0.5			0.5			0.5	mA
	$T_j = 25\,°C$, $I_O \le 1\text{A}$				1.0			1.0			1.0	mA
	$V_{MIN} \le V_{IN} \le V_{MAX}$		$(7.5 \le V_{IN} \le 20)$			$(14.8 \le V_{IN} \le 27)$			$(17.9 \le V_{IN} \le 30)$			V
	$I_O \le 500\text{ mA}$, $0\,°C \le T_j \le +125\,°C$				1.0			1.0			1.0	mA
	$V_{MIN} \le V_{IN} \le V_{MAX}$		$(7 \le V_{IN} \le 25)$			$(14.5 \le V_{IN} \le 30)$			$(17.5 \le V_{IN} \le 30)$			V
V_N Output Noise Voltage	$T_A = 25\,°C$, $10\text{ Hz} \le f \le 100\text{ kHz}$			40			75			90		μV
$\frac{\Delta V_{IN}}{\Delta V_{OUT}}$ Ripple Rejection	$f = 120\text{ Hz}$	$I_O \le 1\text{A}$, $T_j = 25\,°C$ or	62	80		55	72		54	70		dB
		$I_O \le 500\text{ mA}$, $0\,°C \le T_j \le +125\,°C$	62			55			54			dB
	$V_{MIN} \le V_{IN} \le V_{MAX}$		$(8 \le V_{IN} \le 18)$			$(15 \le V_{IN} \le 25)$			$(18.5 \le V_{IN} \le 28.5)$			V
R_O Dropout Voltage	$T_j = 25\,°C$, $I_{OUT} = 1\text{A}$			2.0			2.0			2.0		V
Output Resistance	$f = 1\text{ kHz}$			8			18			19		mΩ
Short-Circuit Current	$T_j = 25\,°C$			2.1			1.5			1.2		A
Peak Output Current	$T_j = 25\,°C$			2.4			2.4			2.4		A
Average TC of V_{OUT}	$0\,°C \le T_j \le +125\,°C$, $I_O = 5\text{ mA}$			0.6			1.5			1.8		mV/°C
V_{IN} Input Voltage Required to Maintain Line Regulation	$T_j = 25\,°C$, $I_O \le 1\text{A}$		7.3			14.6			17.7			V

NOTE 1: Thermal resistance of the TO-3 package (K, KC) is typically 4°C/W junction to case and 35°C/W case to ambient. Thermal resistance of the TO-220 package (T) is typically 4°C/W junction to case and 50°C/W case to ambient.

NOTE 2: All characteristics are measured with capacitor across the inut of 0.22 μF, and a capacitor across the output of 0.1 μF. All characteristics except noise voltage and ripple rejection ratio are measured using pulse techniques ($t_W \le 10$ ms, duty cycle $\le 5\%$). Output voltage changes due to changes in internal temperature must be taken into account separately.

Figure 17–42
(Continued)

of a single device will vary over that range, but that one 7815 chosen at random from a large number will hold its output constant at some voltage within that range.

2. The input voltage cannot exceed 35 V and must not fall below a certain minimum value, depending on type number, if output regulation is to be maintained. The minimum specified inputs are 7.3, 14.6, and 17.7 V for the 7805, 7812, and 7815, respectively.
3. Load regulation is specified as a certain output voltage change (ΔV_o) as the load current (I_o) is changed over a certain range. For example, the output of the 7805 will change a maximum of 50 mV as load current changes from 5 mA to 1.5 A, and will change a maximum of 25 mV as load current changes from 250 mA to 750 mA.

Example 17–16

A 7815 regulator is to be used with a full-wave–rectified (120-Hz), capacitor-filtered input whose dc value may have long-term variations between 19 V and 23 V.

1. What is the maximum peak-to-peak ripple voltage that can be tolerated on the input?
2. What maximum peak-to-peak ripple voltage could appear in the output of the regulator if the input has the ripple found in (1)? (Assume that the dc value of the input is 23 V and the load current is 0.5 A.)

Solution

1. The ripple voltage must not cause the input to fall below the minimum required to maintain regulation, which is 17.7 V for the 7815. Under worst-case conditions, when the dc value of the input is 19 V, the decrease in V_{in} due to ripple cannot, therefore, exceed 19 − 17.7 = 1.3 V. Thus, the maximum tolerable peak-to-peak ripple is 2(1.3 V) = 2.6 V p-p.
2. The 7815 specifications show that ripple rejection under the given operating conditions is a minimum of 54 dB. Thus,

$$20 \log_{10}\left(\frac{\Delta V_{in}}{\Delta V_o}\right) = 54$$

$$V_o = \frac{\Delta V_{in}}{\text{antilog}(54/20)} = \frac{2.6\text{ V}}{501.19} = 5.18\text{ mV p–p}$$

17–10 ADJUSTABLE INTEGRATED-CIRCUIT REGULATORS

As the name implies, an *adjustable* voltage regulator is one that can be set to maintain any output voltage that is within some prescribed range. Unlike three-terminal regulators, adjustable IC regulators must have external components connected to them to perform voltage regulation.

The 723 integrated-circuit regulator is an example of a popular and very versatile adjustable regulator. It can be connected to produce positive or negative outputs over the range from 2 to 37 V, provide either current limiting or foldback limiting, be used with an external pass transistor to handle load currents up to 10 A, and

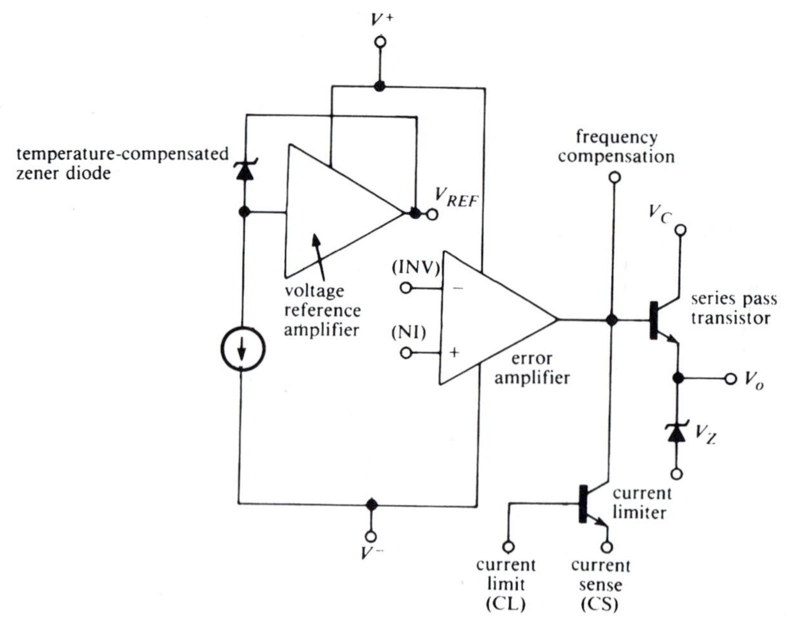

Figure 17–43
Block diagram of the 723 adjustable voltage regulator

be used as a switching regulator. Figure 17–43 is a block diagram of the regulator. Note the terminal labeled V_{REF} at the output of the voltage reference amplifier. This is an internally generated voltage of approximately 7 V that is available at an external pin. To set a desired regulator output voltage, the user connects this 7-V output, or an externally divided-down portion of it, to one of the inputs of the error amplifier. The error amplifier is a comparator that compares the externally connected reference to a voltage proportional to V_o. Depending on whether the reference is connected to the noninverting or the inverting input, the regulated output is either positive or negative. For normal positive voltage regulation, the unregulated input is connected to the terminals labeled V^+ and V_C, and V^- is connected to ground.

Note the transistor labeled "current limiter" in Figure 17–43. By making external resistor connections to the CL and CS terminals, either the current-limiting circuit of Figure 17–32 or the foldback circuit of Figure 17–35 can be implemented. The current limiter performs the function of Q_2 in each of those figures. Of course, the terminals can be left open if no limiting is desired.

Figure 17–44 shows manufacturer's specifications for the 723 regulator, along with some typical applications.

Figure 17–45 shows the 723 regulator connected to maintain its output at any voltage between +2 V and +7 V. The output is determined from

$$V_o = V_{REF}\left(\frac{R_2}{R_1 + R_2}\right) \qquad \textbf{(17–47)}$$

From the specifications in Figure 17–44, we see that V_{REF} may be between 6.95 V and 7.35 V. Therefore, the actual value produced by a given device should be measured before selecting values for R_1 and R_2, if a very accurate output voltage

Voltage Regulators

LM723/LM723C Voltage Regulator

General Description

The LM723/LM723C is a voltage regulator designed primarily for series regulator applications. By itself, it will supply output currents up to 150 mA, but external transistors can be added to provide any desired load current. The circuit features extremely low standby current drain, and provision is made for either linear or foldback current limiting. Important characteristics are:

- 150 mA output current without external pass transistor
- Output currents in excess of 10A possible by adding external transistors
- Input voltage 40V max
- Output voltage adjustable from 2V to 37V
- Can be used as either a linear or a switching regulator.

The LM723/LM723C is also useful in a wide range of other applications such as a shunt regulator, a current regulator or a temperature controller.

The LM723C is identical to the LM723 except that the LM723C has its performance guaranteed over a 0°C to 70°C temperature range, instead of −55°C to +125°C.

Schematic and Connection Diagrams*

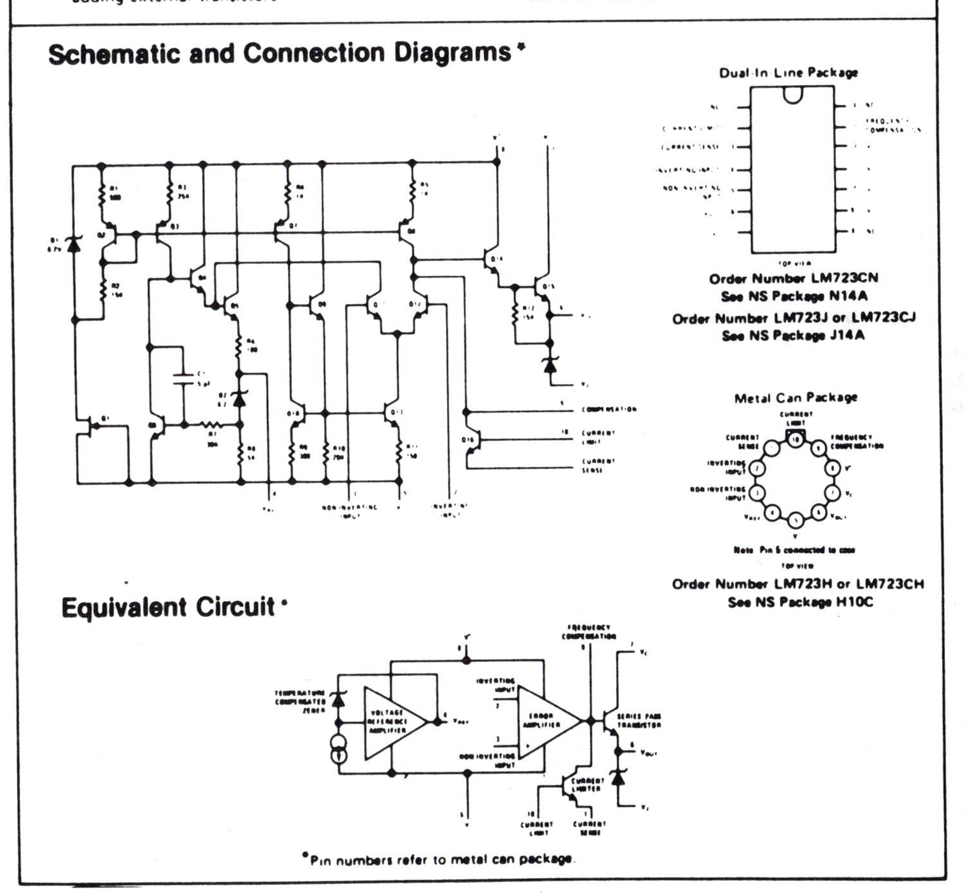

Dual-In-Line Package
Order Number LM723CN
See NS Package N14A
Order Number LM723J or LM723CJ
See NS Package J14A
Metal Can Package
Note: Pin 5 connected to case
TOP VIEW
Order Number LM723H or LM723CH
See NS Package H10C
Equivalent Circuit*
*Pin numbers refer to metal can package.

Figure 17–44
723–voltage-regulator specifications (Courtesy of National Semiconductor)

TABLE I RESISTOR VALUES (kΩ) FOR STANDARD OUTPUT VOLTAGE

POSITIVE OUTPUT VOLTAGE	APPLICABLE FIGURES	FIXED OUTPUT ±5%		OUTPUT ADJUSTABLE ±10% (Note 5)			NEGATIVE OUTPUT VOLTAGE	APPLICABLE FIGURES	FIXED OUTPUT ±5%		5% OUTPUT ADJUSTABLE ±10%		
	(Note 4)	R1	R2	R1	P1	R2			R1	R2	R1	P1	R2
+3.0	1, 5, 6, 9, 12 (4)	4.12	3.01	1.8	0.5	1.2	+100	7	3.57	102	2.2	10	91
+3.6	1, 5, 6, 9, 12 (4)	3.57	3.65	1.5	0.5	1.5	+250	7	3.57	255	2.2	10	240
+5.0	1, 5, 6, 9, 12 (4)	2.15	4.99	.75	0.5	2.2	-6 (Note 6)	3, (10)	3.57	2.43	1.2	0.5	.75
+6.0	1, 5, 6, 9, 12 (4)	1.15	6.04	0.5	0.5	2.7	-9	3, 10	3.48	5.36	1.2	0.5	2.0
+9.0	2, 4, (5, 6, 12, 9)	1.87	7.15	.75	1.0	2.7	-12	3, 10	3.57	8.45	1.2	0.5	3.3
+12	2, 4, (5, 6, 9, 12)	4.87	7.15	2.0	1.0	3.0	-15	3, 10	3.65	11.5	1.2	0.5	4.3
+15	2, 4, (5, 6, 9, 12)	7.87	7.15	3.3	1.0	3.0	-28	3, 10	3.57	24.3	1.2	0.5	10
+28	2, 4, (5, 6, 9, 12)	21.0	7.15	5.6	1.0	2.0	-45	8	3.57	41.2	2.2	10	33
+45	7	3.57	48.7	2.2	10	39	-100	8	3.57	97.6	2.2	10	91
+75	7	3.57	78.7	2.2	10	68	-250	8	3.57	249	2.2	10	240

TABLE II FORMULAE FOR INTERMEDIATE OUTPUT VOLTAGES

Outputs from +2 to +7 volts (Figures 1, 5, 6, 9, 12, (4)) $V_{OUT} = \left[V_{REF} \times \frac{R2}{R1 + R2}\right]$	**Outputs from +4 to +250 volts** (Figure 7) $V_{OUT} = \left[\frac{V_{REF}}{2} \times \frac{R2 - R1}{R1}\right]; R3 = R4$	**Current Limiting** $I_{LIMIT} = \frac{V_{SENSE}}{R_{SC}}$
Outputs from +7 to +37 volts (Figures 2, 4, (5, 6, 9, 12)) $V_{OUT} = \left[V_{REF} \times \frac{R1 + R2}{R2}\right]$	**Outputs from −6 to −250 volts** (Figures 3, 8, 10) $V_{OUT} = \left[\frac{V_{REF}}{2} \times \frac{R1 + R2}{R1}\right]; R3 = R4$	**Foldback Current Limiting** $I_{KNEE} = \left[\frac{V_{OUT}\, R3}{R_{SC}\, R4} + \frac{V_{SENSE}(R3 + R4)}{R_{SC}\, R4}\right]$ $I_{SHORT\,CKT} = \left[\frac{V_{SENSE}}{R_{SC}} \times \frac{R3 + R4}{R4}\right]$

Typical Applications

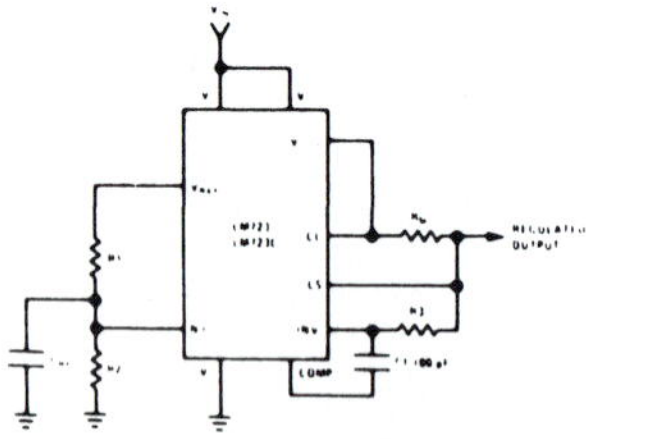

Note: $R3 = \frac{R1\,R2}{R1 + R2}$ for minimum temperature drift.

TYPICAL PERFORMANCE

Regulated Output Voltage	5V
Line Regulation (ΔV_{IN} = 3V)	0.5 mV
Load Regulation (ΔI_L = 50 mA)	1.5 mV

FIGURE 1. Basic Low Voltage Regulator (V_{OUT} = 2 to 7 Volts)

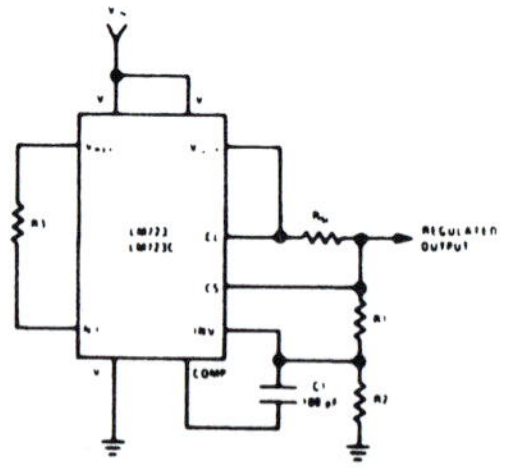

Note: $R3 = \frac{R1\,R2}{R1 + R2}$ for minimum temperature drift.

R3 may be eliminated for minimum component count

TYPICAL PERFORMANCE

Regulated Output Voltage	15V
Line Regulation (ΔV_{IN} = 3V)	1.5 mV
Load Regulation (ΔI_L = 50 mA)	4.5 mV

FIGURE 2. Basic High Voltage Regulator (V_{OUT} = 7 to 37 Volts)

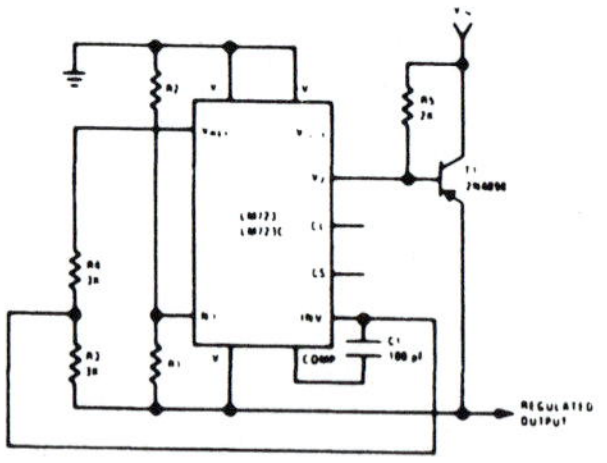

TYPICAL PERFORMANCE

Regulated Output Voltage	-15V
Line Regulation (ΔV_{IN} = 3V)	1 mV
Load Regulation (ΔI_L = 100 mA)	2 mV

FIGURE 3. Negative Voltage Regulator

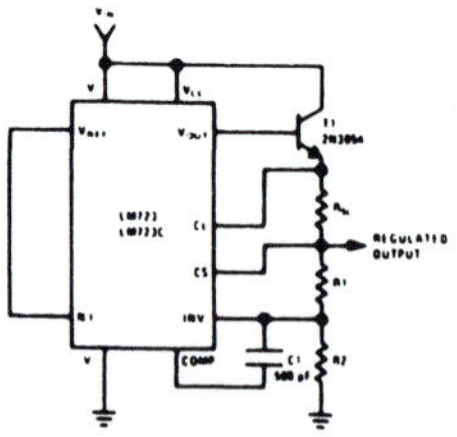

TYPICAL PERFORMANCE

Regulated Output Voltage	+15V
Line Regulation (ΔV_{IN} = 3V)	1.5 mV
Load Regulation (ΔI_L = 1A)	15 mV

FIGURE 4. Positive Voltage Regulator (External NPN Pass Transistor)

Figure 17–44
(Continued)

Absolute Maximum Ratings

Pulse Voltage from V^+ to V^- (50 ms)	50V
Continuous Voltage from V^+ to V^-	40V
Input Output Voltage Differential	40V
Maximum Amplifier Input Voltage (Either Input)	7 5V
Maximum Amplifier Input Voltage (Differential)	5V
Current from V_Z	25 mA
Current from V_{REF}	15 mA
Internal Power Dissipation Metal Can (Note 1)	800 mW
Cavity DIP (Note 1)	900 mW
Molded DIP (Note 1)	660 mW
Operating Temperature Range LM723	55 C to +125 C
LM723C	0 C to +70 C
Storage Temperature Range Metal Can	-65 C to +150 C
DIP	-55 C to +125 C
Lead Temperature (Soldering, 10 sec)	300 C

Electrical Characteristics (Note 2)

PARAMETER	CONDITIONS	LM723			LM723C			UNITS
		MIN	TYP	MAX	MIN	TYP	MAX	
Line Regulation	V_{IN} = 12V to V_{IN} = 15V		01	0 1		01	0 1	% V_{OUT}
	-55°C < T_A < +125°C			0 3				% V_{OUT}
	0°C < T_A < +70°C						0 3	% V_{OUT}
	V_{IN} = 12V to V_{IN} = 40V		02	0 2		0 1	0 5	% V_{OUT}
Load Regulation	I_L = 1 mA to I_L = 50 mA		03	0 15		03	0 2	% V_{OUT}
	-55°C < T_A < +125°C			0 6				% V_{OUT}
	0°C < T_A < +70°C						0 6	% V_{OUT}
Ripple Rejection	f = 50 Hz to 10 kHz, C_{REF} = 0		74			74		dB
	f = 50 Hz to 10 kHz, C_{REF} = 5 μF		86			86		dB
Average Temperature Coefficient of Output Voltage	-55°C < T_A < +125°C		002	015				%/°C
	0°C < T_A < +70°C					003	015	%/°C
Short Circuit Current Limit	R_{SC} = 10Ω, V_{OUT} = 0		65			65		mA
Reference Voltage		6 95	7 15	7 35	6 80	7 15	7 50	V
Output Noise Voltage	BW = 100 Hz to 10 kHz, C_{REF} = 0		20			20		μVrms
	BW = 100 Hz to 10 kHz, C_{REF} = 5 μF		2 5			2 5		μVrms
Long Term Stability			0 1			0 1		%/1000 hrs
Standby Current Drain	I_L = 0, V_{IN} = 30V		1 3	3 5		1 3	4 0	mA
Input Voltage Range		9 5		40	9 5		40	V
Output Voltage Range		2 0		37	2 0		37	V
Input Output Voltage Differential		3 0		38	3 0		38	V

Note 1: See derating curves for maximum power rating above 25°C.

Note 2: Unless otherwise specified, T_A = 25°C, V_{IN} = V^+ = V_C = 12V, V^- = 0, V_{OUT} = 5V, I_L = 1 mA, R_{SC} = 0, C_1 = 100 pF, C_{REF} = 0 and divider impedance as seen by error amplifier ≲ 10 kΩ connected as shown in Figure 1. Line and load regulation specifications are given for the condition of constant chip temperature. Temperature drifts must be taken into account separately for high dissipation conditions.

Note 3: L_1 is 40 turns of No. 20 enameled copper wire wound on Ferroxcube P36/22-3B7 pot core or equivalent with 0.009 in. air gap.

Note 4: Figures in parentheses may be used if R1/R2 divider is placed on opposite input of error amp.

Note 5: Replace R1/R2 in figures with divider shown in Figure 13.

Note 6: V^+ must be connected to a +3V or greater supply.

Note 7: For metal can applications where V_Z is required, an external 6.2 volt zener diode should be connected in series with V_{OUT}.

Figure 17–44
(Continued)

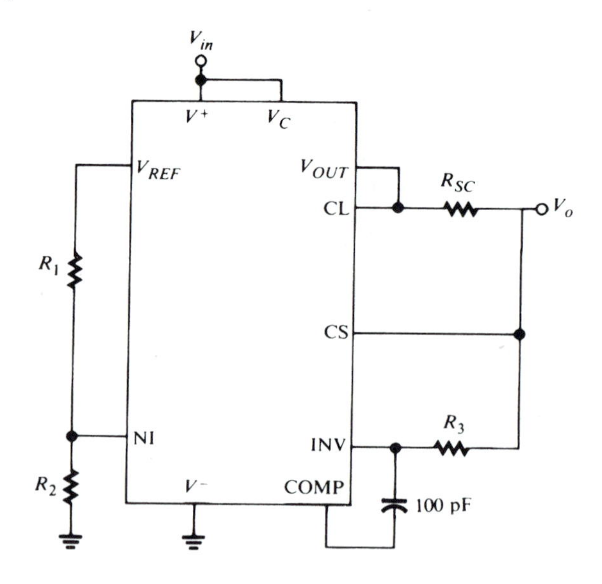

Figure 17–45
The 723 regulator connected to provide output voltages between +2 V and +7 V

is required. Notice that the full (undivided) output voltage V_o is fed back to the inverting input (INV) through R_3. For maximum thermal stability, R_3 should be set equal to $R_1 \| R_2$.

The circuit shown in Figure 17–45 is connected to provide current limiting, where

$$I_L(\text{max}) \approx \frac{0.7\text{ V}}{R_{SC}} \tag{17–48}$$

The 100-pF capacitor shown in the figure is used to ensure circuit stability. When the circuit is connected to provide foldback limiting, a voltage divider (identified as resistors R_3 and R_4 in Figure 17–35) is connected across V_{OUT} in Figure 17–45. The CL terminal on the 723 regulator is then connected to the middle of the divider, instead of to V_{OUT}.

Example 17–17

Design a 723-regulator circuit that will maintain an output voltage of +5 V and that will provide current limiting at 0.1 A. Assume that $V_{REF} = 7.0$ V.

Solution. We arbitrarily choose $R_2 = 1\text{ k}\Omega$. Then, from equation 17–47,

$$V_o = 5\text{ V} = (7\text{ V})\left[\frac{1\text{ k}\Omega}{R_1 + (1\text{ k}\Omega)}\right]$$

Solving for R_1, we find $R_1 = 400\ \Omega$. The optimum value of R_3 is

$$R_3 = R_1 \| R_2 = \frac{400(1000)}{1400} = 285.7\ \Omega$$

From equation 17–48,

$$I_L(\text{max}) = 0.1\text{ A} = \frac{0.7\text{ V}}{R_{SC}}$$

$$R_{SC} = 7\ \Omega$$

EXERCISES

Section 17–2
Rectifiers

17–1. What peak-to-peak sinusoidal voltage must be connected to a half-wave rectifier if the rectified waveform is to have a dc value of 6 V? Assume that the forward drop across the diode is 0.7 V.

17–2. What should be the rms voltage of a sinusoidal wave connected to a full-wave rectifier if the rectified waveform is to have a dc value of 50 V? Neglect diode voltage drops.

17–3. The primary voltage on the transformer shown in Figure 17–46 is 120 V rms and $R_L = 10\ \Omega$. Neglecting the forward voltage drops across the diodes, find

a. the turns ratio $N_p:N_s$, if the average current in the resistor must be 1.5 A;

b. the average power dissipated in the resistor, under the conditions of (a); and

c. the maximum PIV rating required for the diodes, under the conditions of (a).

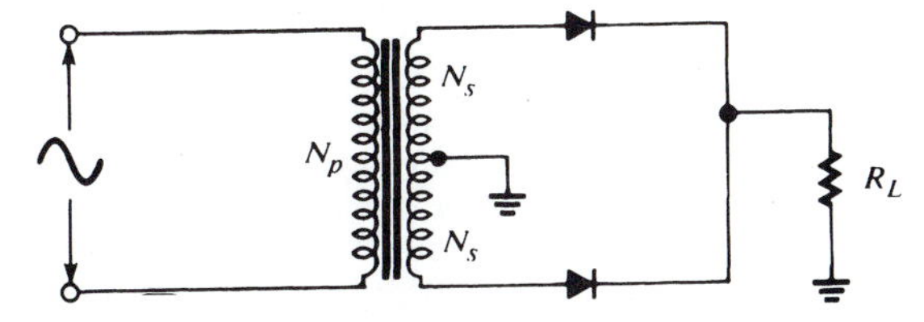

Figure 17–46
(Exercises 17–3 and 17–4)

17–4. The primary voltage on the transformer in Figure 17–46 is 120 V rms and $N_P:N_s = 15:1$. Diode voltage drops are 0.7 V.

a. What should be the value of R_L if the average current in R_L must be 0.5 A?

b. What power is dissipated in R_L under the conditions of (a)?

c. What minimum PIV rating is required for the diodes under the conditions of (a)?

17–5. The primary voltage on the transformer in Figure 17–47 is 110 V rms at 50 Hz. The diode voltage drops are 0.7 V. Sketch the waveforms of the voltage across and current through the 20-Ω resistor. Label peak values and the time points where the waveforms go to 0.

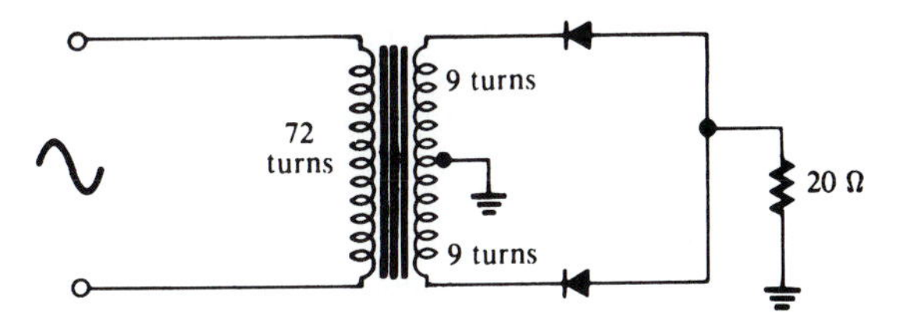

Figure 17–47
(Exercise 17–5)

17–6. Each of the diodes in Figure 17–48 has a forward voltage drop of 0.7 V. Find

a. the average voltage across R_L;

b. the average power dissipated in the 1-Ω resistor; and

c. the minimum PIV rating required for the diodes.

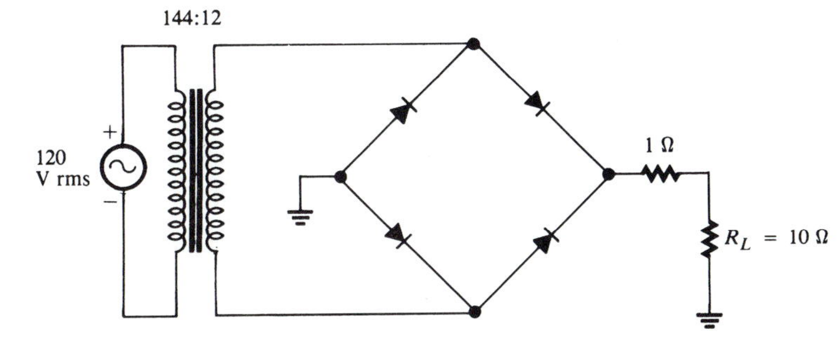

Figure 17–48
(Exercise 17–6)

17–7. Repeat Exercise 17–6 if R_L is changed to 5 Ω and the transformer turns ratio is changed to 1:1.5.

17–8. Sketch the waveform of the voltage v_L in the circuit shown in Figure 17–49. Include the ripple and show the value of its period on the sketch. Also show the value of V_{PR}. Neglect the forward drop across the diode.

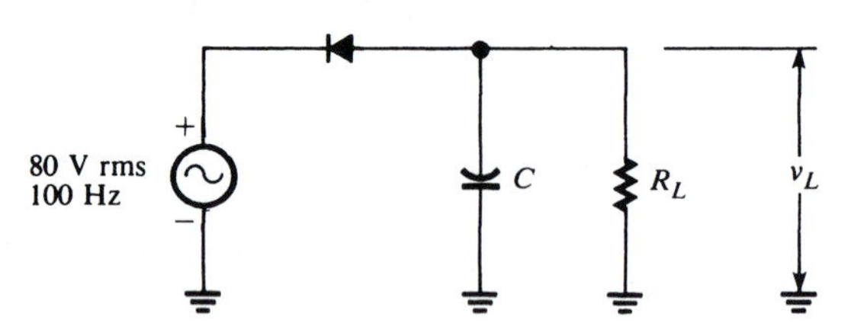

Figure 17–49
(Exercise 17–8)

17–9. What is the percent ripple of a full-wave–rectified waveform having peak value 75 V and frequency 120 Hz? What is the percent ripple if the peak value is doubled? If the frequency is halved?

17–10. A full-wave bridge is to be connected to a 240-V-rms power line. The output will be filtered and will supply an average voltage of 150 V to a 50-Ω load. The worst-case current that will flow through the bridge when power is first applied is twenty times the average load current. Using the International Rectifier specifications in Figure 17–8, select a bridge (give its part number) that can be used for this application.

Section 17–3
Capacitor Filters

17–11. A half-wave rectifier is operated from a 60-Hz line and has a 1000-μF filter capacitance connected across it. What is the minimum value of load resistance that can be connected across the capacitor if the percent ripple cannot exceed 5%?

17–12. A full-wave rectifier is operated from a 60-Hz, 50-V-rms source. It has a 500-μF filter capacitor and a 750-Ω load. Find
a. the average value of the load voltage;
b. the peak-to-peak ripple voltage; and
c. the percent ripple.

17–13. A half-wave rectifier has a 1000-μF filter capacitor and a 500-Ω load. It is operated from a 60-Hz, 120-V-rms source. It takes 1 ms for the capacitor to recharge during each input cycle. For what minimum value of repetitive surge current should the diode be rated?

17–14. Repeat Exercise 17–13 if the rectifier is full-wave and the capacitor takes 0.5 ms to recharge.

Section 17–4
RC and LC Filters

17–15. a. Find the rms ripple voltage across C_1 in Figure 17–50(a). (The filter is lightly loaded.)
b. Assuming that $R << R_L$, find the value of RC_2 in Figure 17–50(b) that will make the ripple voltage across C_2 equal to one-tenth of that found in (a).
c. Under the conditions of (b), find values for R and C_2 if it is required that the dc voltage across R_L in Figure 17–50(b) be not less than nine-tenths of the dc voltage across R_L in 17–50(a).

17–16. a. Find the percent ripple in Exercise 17–15(a) (Figure 17–50(a)).
b. Find the percent ripple in the circuit of Figure 17–50(b), using the values for R and C_2 found in Exercise 17–15(c).
c. Explain why the values of R and C_2 reduce the rms ripple voltage by a factor of 10 but do not reduce the percent ripple by the same amount.

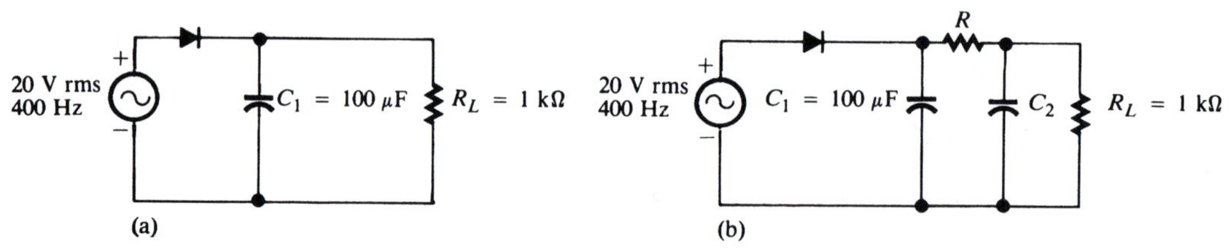

Figure 17–50
(Exercise 17–15)

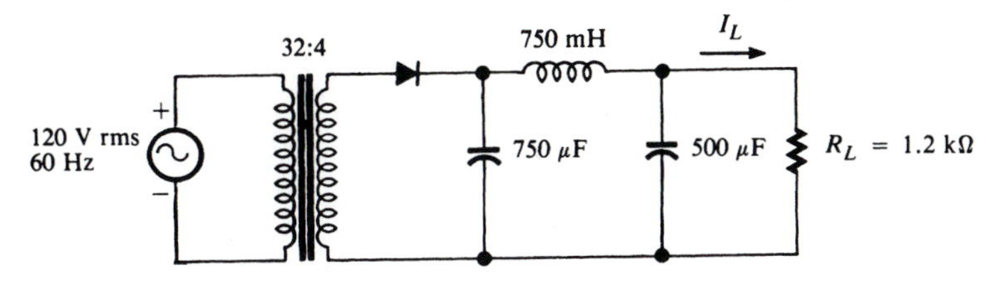

Figure 17–51
(Exercise 17–17)

17–17. The inductor in Figure 17–51 has resistance 4 Ω. Find the average load current I_L and the percent ripple in the load voltage. Be certain to confirm the validity of any approximating assumptions made in your computations.

Section 17–5
Voltage Multipliers

17–18. Assuming negligible ripple, find the average current in the 100-kΩ resistor in Figure 17–52.

17–19. The transformer shown in Figure 17–53 has a *tapped secondary,* each portion of the secondary winding having the number of turns shown. Assuming that the primary voltage is that shown in the figure, design two separate circuits that can be used with the transformer to obtain an (unloaded) dc voltage of 1200 V. It is not necessary to specify capacitor sizes. What minimum PIV ratings should the diodes in each design have?

Section 17–6
Voltage Regulation

17–20. A power supply has 4% voltage regulation and an open-circuit output voltage of 48 V dc.
a. What is the full-load voltage of the supply?
b. If a 120-Ω resistor draws full-load current from the supply, what is its output resistance?

17–21. One way to determine the output resistance of a power supply is to vary its load resistance while measuring load voltage. The output resistance of the supply equals the value of load resistance that is found to make the load voltage equal to one-half of the open-circuit voltage of the supply. Using an appropriate equation, explain why this method is valid.

17–22. A 50-V power supply has line regulation 0.2%/V. How large would the 75-V input voltage to the supply have to become for the output voltage to rise to 52 V?

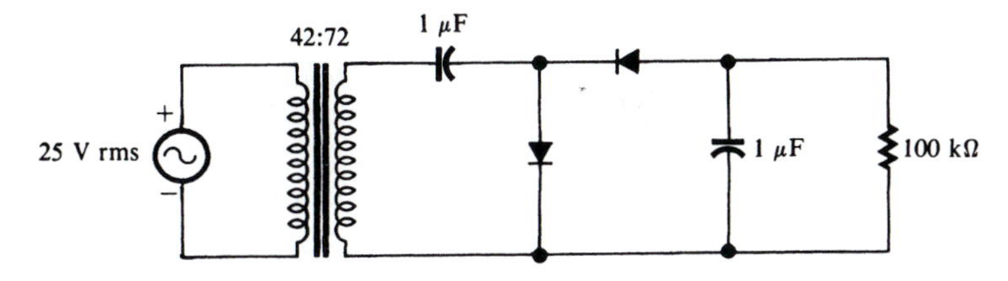

Figure 17–52
(Exercise 17–18)

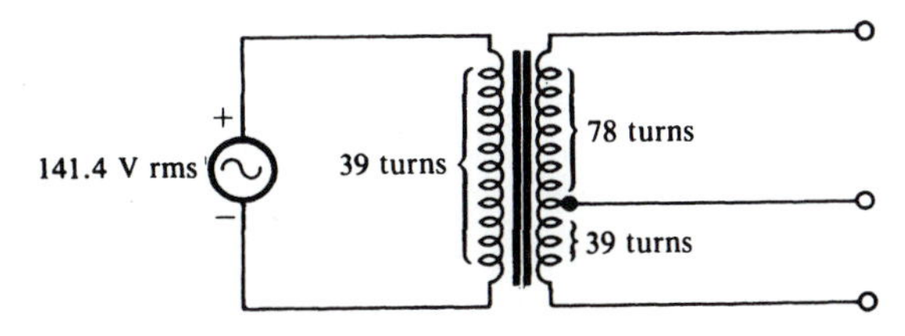

Figure 17–53
(Exercise 17–19)

17–23. The specifications for a certain 24-V power supply state that the output voltage increases 18 mV per volt increase in the input voltage. What is the percent line regulation of the supply?

Section 17–7
Series and Shunt Voltage Regulators

17–24. The base-to-emitter voltage of the transistor in Figure 17–54 is 0.7 V. V_{in} can vary from 12 V to 24 V.

a. What breakdown voltage should the zener diode have if the load voltage is to be maintained at 9 V?

b. If the zener diode must conduct 10 mA of reverse current to remain in breakdown, what maximum value should R have?

c. With the value of R found in (b), what is the maximum power dissipated in the zener diode?

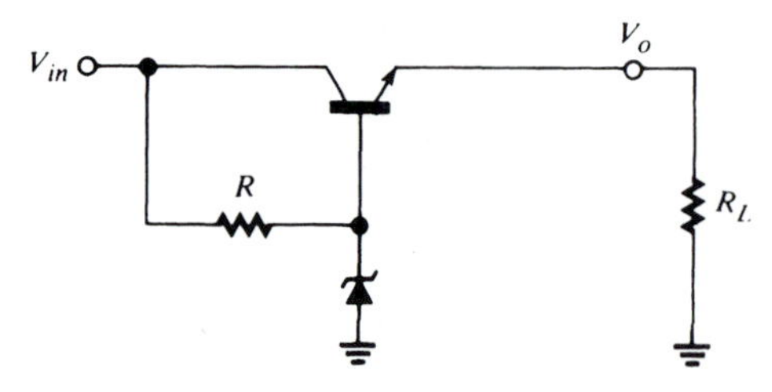

Figure 17–54
(Exercises 17–24 and 17–25)

17–25. The base-to-emitter voltage of the transistor shown in Figure 17–54 is 0.7 V. V_{in} can vary from 18 V to 30 V. If V_Z = 10 V, R = 500 Ω, and R_L = 1 kΩ, find

a. the minimum collector-to-emitter voltage of the transistor;

b. the maximum power dissipated in the transistor; and

c. the maximum current supplied by V_{in}.

17–26. Draw a schematic diagram of a series voltage regulator that employs one transistor and one zener diode and that can be operated from an unregulated negative voltage to produce a regulated negative voltage. It is not necessary to show component values.

17–27. In the regulator circuit shown in Figure 17–29, V_{BE_2} is 0.68 V when V_o is maintained at 24 V. The zener diode has V_Z = 9 V. Find values for R_1 and R_2.

17–28. In the regulator circuit shown in Figure 17–31, R_1 = 4 kΩ, R_2 = 1 kΩ, and R_3 = 2 kΩ. V_{in} can vary from 30 V to 50 V.

a. What breakdown voltage should the zener diode have so that the regulated output voltage be 25 V?

b. Using the V_Z found in (a), what is the maximum current in the zener diode?

17–29. The potentiometer in Figure 17–55 has a total resistance of 10 kΩ.

a. What range of regulated output voltages can be obtained by adjusting the potentiometer through its entire range?

b. What power is dissipated in the pass transistor when the potentiometer is set for maximum resistance?

17–30. In the circuit shown in Figure 17–32, R_1 = R_2 = 5 kΩ, V_Z = 6 V, and R_{SC} = 2 Ω. Find V_o

a. when R_L = 100 Ω; and

b. when R_L = 30 Ω.

17–31. Design a current-limited, series voltage regulator that includes an operational amplifier and a 10-V zener diode. The regulated output voltage should be 36 V and the output current

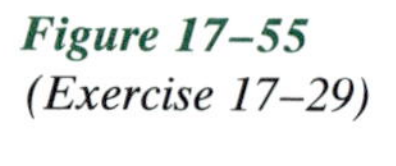

Figure 17–55
(Exercise 17–29)

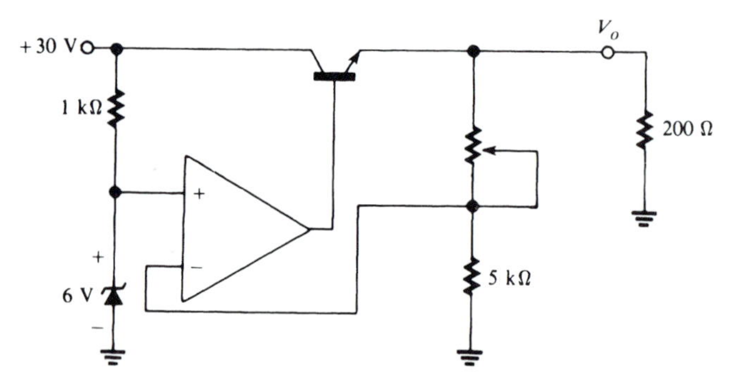

should be limited to 700 mA. The zener current should be 10 mA when $V_{in} = 40$ V.

17–32. In Figure 17–35, $R_1 = 1.5$ kΩ, $R_2 = 1$ kΩ, $R_3 = 820$ Ω, $R_4 = 8.2$ kΩ, $R_{SC} = 2$ Ω, and $V_Z = 6.3$ V.
a. Find the regulated output voltage.
b. Find the maximum load current.

17–33. When the regulated output voltage in Figure 17–37 is 18 V, V_{in} supplies a total current of 0.5 A. If $R_L = 100$ Ω, $V_Z = 6$ V, $R_1 = 200$ Ω, and the collector current in Q_2 is negligible, find the power dissipated by Q_1.

17–34. In Figure 17–38, the unregulated input voltage varies from 6 to 9 volts and $R_4 = 15$ Ω.
a. What is the maximum (short-circuit) current that can be drawn from the regulator?
b. What is the minimum power-dissipation rating that R_4 should have?

Section 17–8
Switching Regulators

17–35. What should be the duty cycle of a train of pulses that alternate between 0 V and +5 V if the dc value of the train must be 1.2 V?

17–36. **a.** In Figure 17–56(a), $T_1 = 2$ ms, $T = 5$ ms, $V_1 = 0.5$ V, and $V_2 = 6$ V. Find the dc value.
b. In Figure 17–56(b), $T_1 = 2$ ms, $T = 5$ ms, $V_1 = -0.5$ V, and $V_2 = 6$ V. Find the dc value.

17–37. Derive one general expression involving V_1, V_2, T_1, and T that can be used to determine the dc value of either of the pulse trains in Figure 17–56. (Note that V_1 would be entered in the expression as a positive number for Figure 17–56(a) and as a negative number for 17–56(b).) Use your expression to solve Exercise 17–36.

17–38. The pulse train produced by the pulse-width modulator in Figure 17–41 has a 50% duty cycle when $V_{in} = 20$ V. What is its duty cycle when $V_{in} = 30$ V? When $V_{in} = 15$ V?

Section 17–9
Three-Terminal Integrated-Circuit Regulators

17–39. A 7815 regulator is to be operated with an unregulated input whose dc value may range from 23 V to 33 V.
a. What is the maximum peak-to-peak ripple that can be tolerated on the input?
b. With the peak-to-peak ripple found in (a), what maximum percent ripple could the output have? (Assume that the ripple waveform is triangular and that the 7815 is operated at 23 V with load current 0.5 A.)

17–40. A 7812 voltage regulator is operated at 25°C with a 24-Ω load and a 19-V input.
a. What is the worst-case percent line regulation that could be expected?

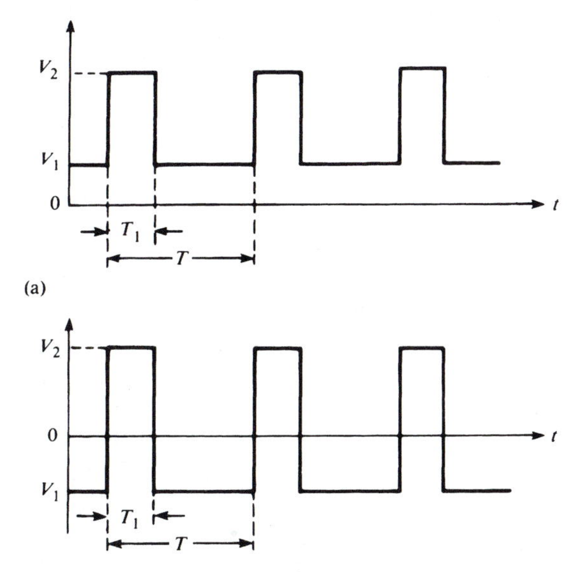

Figure 17–56
(Exercise 17–36)

b. What maximum change in output voltage could be expected if the load resistance were changed over the range from 16 Ω to 48 Ω?

Section 17–10
Adjustable Integrated-Circuit Regulators

17–41. In Figure 17–45, $R_1 = 330\ \Omega$, $R_2 = 4.7\ \text{k}\Omega$, and $R_{SC} = 3\ \Omega$. Assuming that $V_{REF} = 7$ V,
a. find the value of the regulated output voltage;
b. find the optimum value of R_3; and
c. find the maximum load current.

17–42. Design a 723-regulator circuit whose output voltage can be varied from 3 V to 6 V by adjusting a potentiometer connected between R_1 and R_2 in Figure 17–45. Assume that $V_{REF} = 7$ V.

17–43. Using the reference material given in the 723-regulator specifications (Figure 17–44), design a +20-V regulator circuit. Assume that $V_{REF} = 7$ V.

SPICE EXERCISES

17–44. The filter capacitor in Figure 17–10 is 75 μF and the input is a 50-V peak sine wave having frequency 60 Hz. By trial-and-error runs of SPICE programs, determine the smallest value that R_L can have without the peak-to-peak ripple exceeding 8.7 V ± 0.05 V.

17–45. Use SPICE to find the peak-to-peak ripple voltage of the filtered output in each of the circuits shown in Figure 17–19. Also, use the SPICE results to estimate the dc value of each filtered output. (Because of the long initial transient in each circuit, it will be necessary to examine the output after the input has been applied for about one second. Use the *TSTART* specification in a .TRAN statement to plot only the last 10 ms of a 1-s transient. The iteration limit in SPICE must be overridden using an .OPTIONS statement and a specification such as ITL5 = 20000. Considerable computer time will be required to execute this program.)

17–46. Use SPICE to determine the values of $2V_p$, $3V_p$, and $4V_p$ in Figure 17–23 when v_{in} is a 20-V peak sine wave with frequency 400 Hz. Each capacitor is 1 μF. To allow initial transients time to settle, observe only the last 10 ms of a 100-ms transient. Why are the computed values not exactly twice, three times, and four times the peak value of the input?

17–47. The voltage regulator in Figure 17–30 has $R_1 = 50\ \text{k}\Omega$, $R_2 = 43.75\ \text{k}\Omega$, $R_3 = 47\ \text{k}\Omega$, and $R_4 = 1\ \text{k}\Omega$. The zener diode has a breakdown voltage of 6.3 V and the input voltage to the regulator is 20 V dc. The regulator supplies full-load current when the load resistance is 200 Ω. Use SPICE to find the percent voltage regulation of the regulator. Model the zener diode by specifying a conventional diode having a reverse breakdown voltage of 6.3 V.

18 SPECIAL ELECTRONIC DEVICES

18–1 ZENER DIODES

In earlier chapters we discussed several zener-diode applications, perhaps the most important of which is the zener *voltage reference* used in regulator circuits. We wish now to examine the characteristics and limitations of these devices in more detail, so we can interpret and apply specifications for some of their important parameters.

Figure 18–1 shows a typical I–V characteristic for a zener diode. The forward-biased characteristic is identical to that of a forward-biased silicon diode and obeys the same diode equation that we developed in Chapter 2 (equation 2–13). The zener diode is normally operated in its reverse-biased breakdown region, where the voltage across the device remains substantially constant as the reverse current varies over a large range. Like a fixed voltage source, this ability to maintain a constant voltage across its terminals, independent of current, is what makes the device useful as a voltage reference. The fixed breakdown voltage is called the *zener voltage, V_Z*, as illustrated in the figure.

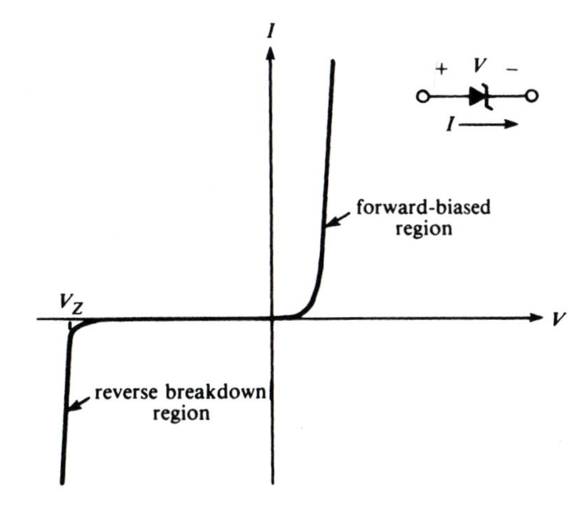

Figure 18–1
I–V characteristic of a zener diode

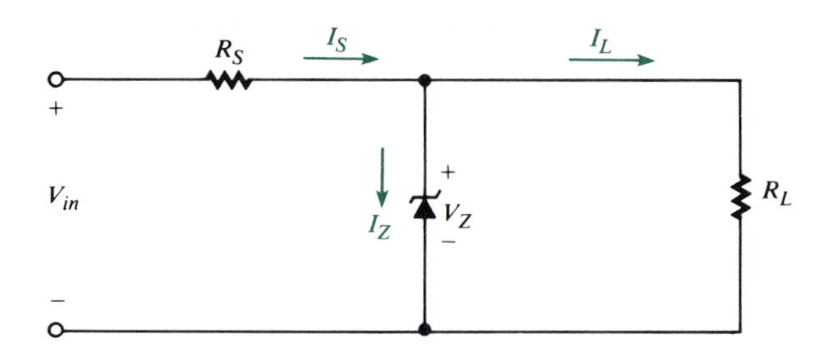

Figure 18–2
A simple voltage regulator using a zener diode

The Zener-Diode Voltage Regulator

To demonstrate how a zener diode can serve as a constant voltage reference, Figure 18–2 shows a simple but widely used configuration that maintains a constant voltage across a load resistor. The circuit is an elementary *voltage regulator* (see Section 17–7) that holds the load voltage near V_Z volts as R_L and/or V_{in} undergo changes. So the voltage across the parallel combination of the zener and R_L remain at V_Z volts, the reverse current I_Z through the diode must at all times be large enough to keep the device in its breakdown region, as shown in Figure 18–1. The value selected for R_S is critical in that respect. As we shall presently demonstrate, R_S must be small enough to permit adequate zener current, yet large enough to prevent the zener current and power dissipation from exceeding permissible limits.

It is apparent in Figure 18–2 that

$$I_S = I_Z + I_L \tag{18–1}$$

Also, I_S is the voltage difference across R_S divided by R_S:

$$I_S = \frac{V_{in} - V_Z}{R_S} \tag{18–2}$$

The power dissipated in the zener diode is

$$P_Z = V_Z I_Z \tag{18–3}$$

Example 18–1

In the circuit of Figure 18–2, $R_S = 20\ \Omega$, $V_Z = 18$ V, and $R_L = 200\ \Omega$. If V_{in} can vary from 20 V to 30 V, find

1. the minimum and maximum currents in the zener diode;
2. the minimum and maximum power dissipated in the diode; and
3. the minimum rated power dissipation that R_S should have.

Solution

1. Assuming that the zener diode remains in breakdown, then the load voltage remains constant at $V_Z = 18$ V, and the load current therefore remains constant at

$$I_L = \frac{V_Z}{R_L} = \frac{18\text{ V}}{200\ \Omega} = 90\text{ mA}$$

From equation 18–2, when $V_{in} = 20$ V,

$$I_S = \frac{(20\text{ V}) - (18\text{ V})}{20\ \Omega} = 100\text{ mA}$$

Therefore, $I_Z = I_S - I_L = (100\text{ mA}) - (90\text{ mA}) = 10$ mA. When $V_{in} = 30$ V,

$$I_S = \frac{(30\text{ V}) - (18\text{ V})}{20\ \Omega} = 600\text{ mA}$$

and $I_Z = I_S - I_L = (600\text{ mA}) - (90\text{ mA}) = 510$ mA.

2. $P_Z(\text{min}) = V_Z I_Z(\text{min}) = (18\text{ V})(10\text{ mA}) = 180$ mW
$P_Z(\text{max}) = V_Z I_Z(\text{max}) = (18\text{ V})(510\text{ mA}) = 9.18$ W

3. $P_{R_S}(\text{max}) = I_S^2(\text{max})R_S = (0.6)^2(20) = 7.2$ W

Solving equation 18–2 for R_S, we find

$$R_S = \frac{V_{in} - V_Z}{I_S} \qquad \textbf{(18–4)}$$

Substituting (18–1) into (18–4) gives

$$R_S = \frac{V_{in} - V_Z}{I_Z + I_L} \qquad \textbf{(18–5)}$$

Let $I_Z(\text{min})$ denote the minimum zener current necessary to ensure that the diode is in its breakdown region. As mentioned earlier, R_S must be small enough to ensure that $I_Z(\text{min})$ flows under worst-case conditions, namely, when V_{in} falls to its smallest possible value, $V_{in}(\text{min})$, and I_L is its largest possible value, $I_L(\text{max})$. Thus, from (18–5), we require

$$R_S \le \frac{V_{in}(\text{min}) - V_Z}{I_Z(\text{min}) + I_L(\text{max})} \qquad \textbf{(18–6)}$$

Maximum load current flows when R_L has its minimum possible value, $R_L(\text{min})$. Substituting $I_L(\text{max}) = V_Z/R_L(\text{min})$ in (18–6) gives

$$R_S \le \frac{V_{in}(\text{min}) - V_Z}{I_Z(\text{min}) + V_Z/R_L(\text{min})} \qquad \textbf{(18–7)}$$

The value of R_S must also be large enough to ensure that the zener current does not exceed the manufacturer's specified maximum or cause the zener power dissipation to exceed the rated maximum at the operating temperature. Let $I_Z(\text{max})$ denote the minimum of those two limits, i.e., the smaller of the rated current and $P_d(\text{max})/V_Z$. Then we require that R_S be large enough to ensure that I_Z does not exceed $I_Z(\text{max})$ under the worst-case conditions where $V_{in} = V_{in}(\text{max})$ and $I_L = I_L(\text{min})$, or $R_L = R_L(\text{max})$. Thus, from equation 18–5, it is necessary that

$$R_S \ge \frac{V_{in}(\text{max}) - V_Z}{I_Z(\text{max}) + I_L(\text{min})} = \frac{V_{in}(\text{max}) - V_Z}{I_Z(\text{max}) + V_Z/R_L(\text{max})} \qquad \textbf{(18–8)}$$

It is important to note that, as far as zener dissipation is concerned, the worst-case load condition in some applications may correspond to an open output; that is, $R_L = \infty$ and $I_L = 0$. In that case, all of the current through R_S flows in the zener.

Example 18–2

The reverse current in a certain 10-V, 2-W zener diode must be at least 5 mA to ensure that the diode remains in breakdown. The diode is to be used in the regulator circuit shown in Figure 18–3, where V_{in} can vary from 15 V to 20 V. Note that the load can be switched out of the regulator circuit in this application. Find a value for R_S. What power dissipation rating should R_S have?

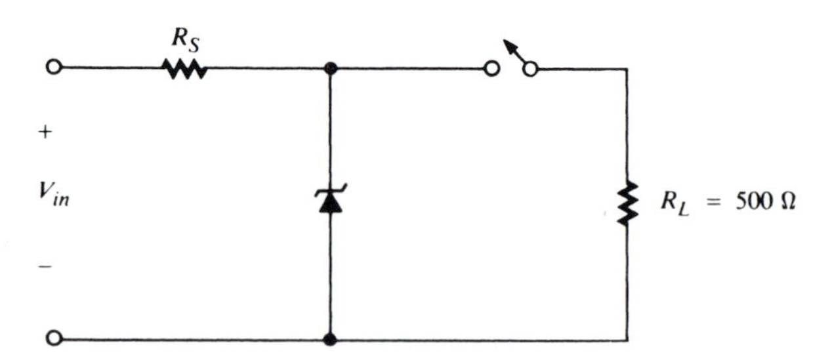

Figure 18–3
(Example 18–2)

Solution

$$V_{in}(\text{min}) = 15 \text{ V}$$
$$I_Z(\text{min}) = 5 \text{ mA}$$
$$R_L(\text{min}) = R_L = 500\ \Omega \quad \text{(when the switch is closed)}$$

Therefore, from inequality 18–7,

$$R_S \le \frac{V_{in}(\text{min}) - V_Z}{I_Z(\text{min}) + V_Z/R_L(\text{min})} = \frac{(15 - 10)\text{ V}}{(5 \text{ mA}) + (10 \text{ V})/(500\ \Omega)} = 200\ \Omega$$

$$V_{in}(\text{max}) = 20 \text{ V}$$

$$I_Z(\text{max}) = \frac{P_Z(\text{max})}{V_Z} = \frac{2 \text{ W}}{10 \text{ V}} = 200 \text{ mA}$$

$$I_L(\text{min}) = 0 \quad \text{(when the switch is open)}$$

From inequality 18–8,

$$R_S \ge \frac{V_{in}(\text{max}) - V_Z}{I_Z(\text{max}) + I_L(\text{min})} = \frac{(20 - 10)\text{ V}}{(200 \text{ mA}) + (10 \text{ A})} = 50\ \Omega$$

Thus, we require $50 \le R_S \le 200$. Choosing $R_S = 100\ \Omega$, the maximum current in R_S is

$$I_S(\text{max}) = \frac{V_{in}(\text{max}) - V_Z}{R_S} = \frac{(20 - 10)\text{ V}}{100\ \Omega} = 0.1 \text{ A}$$

Therefore, the maximum power dissipated in R_S is $P(\text{max}) = I_S^2(\text{max})R_S = (0.1)^2(100) = 1$ W.

Temperature Effects

The breakdown voltage of a zener diode is a function of the width of its depletion region, which is controlled during manufacturing by the degree of impurity doping.

Recall that heavy doping increases conductivity, which narrows the depletion region, and therefore decreases the voltage at which breakdown occurs. Zener diodes are available with breakdown voltages ranging from 2.4 V to 200 V. As noted in Chapter 2, the mechanism by which breakdown occurs depends on the breakdown voltage itself. When V_Z is less than about 5 V, the high electric field intensity across the narrow depletion region (around 3×10^7 V/m) strips carriers directly from their bonds, a phenomenon usually called *zener breakdown.* For V_Z greater than about 8 V, breakdown occurs as a result of collisions between high-energy carriers, the mechanism called *avalanching.* Between 5 V and 8 V, both the avalanching and zener mechanisms contribute to breakdown. The practical significance of these facts is that the breakdown mechanism determines how temperature variations affect the value of V_Z. Low-voltage zener diodes that break down by the zener mechanism have negative temperature coefficients (V_Z decreases with increasing temperature) and higher-voltage avalanche zeners have positive temperature coefficients. When V_Z is between about 3 V and 8 V, the temperature coefficient is also strongly influenced by the current in the diode: the coefficient may be positive or negative, depending on current, becoming more positive as current increases.

The temperature coefficient of a zener diode is defined to be its change in breakdown voltage per degree Celsius increase in temperature. For example, a temperature coefficient of +8 mV/°C means that V_Z will increase 8 mV for each degree Celsius increase in temperature. *Temperature stability* is the ratio of the temperature coefficient to the breakdown voltage. Expressed as a percent,

$$S(\%) = \frac{\text{T.C.}}{V_Z} \times 100\% \qquad \textbf{(18–9)}$$

where T.C. is the temperature coefficient. Clearly, small values of S are desirable.

In applications requiring a zener diode to serve as a highly stable voltage reference, steps must be taken to *temperature compensate* the diode. A technique that is used frequently is to connect the zener in series with one or more semiconductor devices whose voltage drops change with temperature in the opposite way that V_Z changes, i.e., devices having the opposite kind of temperature coefficient. If a temperature change causes V_Z to increase, then the voltage across the other components decreases, so the total voltage across the series combination is (ideally) unchanged. For example, the temperature coefficient of a forward-biased silicon diode is negative, so one or more of these can be connected in series with a zener diode having a positive temperature coefficient, as illustrated in Figure 18–4. The next example illustrates that several forward-biased diodes, which have relatively small temperature coefficients, may be required to compensate a single zener diode.

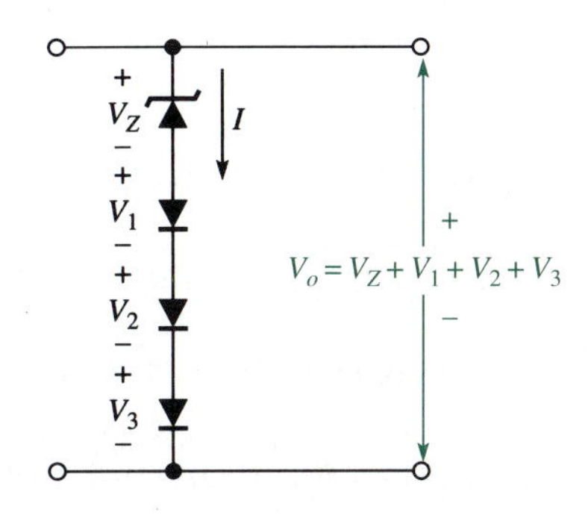

Figure 18–4
Temperature compensating a zener diode by connecting it in series with forward-biased diodes having opposite temperature coefficients

Example 18–3

A zener diode having a breakdown voltage of 10 V at 25°C has a temperature coefficient of +5.5 mV/°C. It is to be temperature compensated by connecting it in series with three forward-biased diodes, as shown in Figure 18–4. Each compensating diode has a forward drop of 0.65 V at 25°C and a temperature coefficient of −2 mV/°C.

1. What is the temperature stability of the uncompensated zener diode?
2. What is the breakdown voltage of the uncompensated zener diode at 100°C?
3. What is the voltage across the compensated network at 25°C? At 100°C?
4. What is the temperature stability of the compensated network?

Solution

1. From equation 18–9,

$$S = \frac{\text{T.C.} \times 100\%}{V_Z} = \frac{5.5 \times 10^{-3}}{10\text{ V}} \times 100\% = 0.055\%$$

2. $V_Z = (10\text{ V}) + \Delta T(\text{T.C.}) = (10\text{ V}) + (100°\text{C} - 25°\text{C})(5.5\text{ mV/°C}) = 10.4125\text{ V}$
3. As shown in Figure 18–4, $V_o = V_Z + V_1 + V_2 + V_3$. At 25°C, $V_o = 10 + 3(0.65) = 11.95$ V. At 100°C, the drop V_d across each forward-biased diode is $V_d = (0.65\text{ V}) + (100°\text{C} - 25°\text{C})(-2\text{ mV/°C}) = 0.5$ V. Therefore, at 100°C, $V_o = (10.4125\text{ V}) + 3(0.5\text{ V}) = 10.5625$ V.
4. The temperature coefficient of the compensated network is T.C. = (+5.5 mV/°C) + 3(−2 mV/°C) = (+5.5 mV/°C) − (6 mV/°C) = −0.5 mV/°C. The voltage drop across the network (at 25°C) was found to be 11.95 V, so

$$S = \frac{-0.5\text{ mV/°C}}{11.95\text{ V}} \times 100\% = -0.00418\%$$

We see that compensation has improved the stability by a factor greater than 10.

Temperature-compensated zener diodes are available from manufacturers in single-package units called *reference diodes.* These units contain specially fabricated junctions that closely track and oppose variations in V_Z with temperature. Although it is possible to obtain an extremely stable reference this way, it may be necessary to maintain the zener current at a manufacturer's specified value in order to realize the specified stability. Figure 17–43 shows a temperature-stabilized zener diode whose current is maintained by a constant-current source to ensure its stability as reference for the 723 voltage regulator.

Zener-Diode Impedance

The breakdown characteristic of an *ideal* zener diode is a perfectly vertical line, signifying zero change in voltage for any change in current. Thus, the ideal diode has zero impedance (or ac resistance) in its breakdown region. A practical zener diode has nonzero impedance, which can be computed in the usual way:

$$Z_Z = \frac{\Delta V_Z}{\Delta I_Z}\ \Omega \qquad \textbf{(18–10)}$$

Z_Z is the reciprocal of the slope of the breakdown characteristic on an I_Z–V_Z plot. The slope is not constant, so the value of Z_Z depends on the point along the characteristic where the measurement is made. The impedance decreases as I_Z increases; that is, the breakdown characteristic becomes steeper at points farther down the line, corresponding to greater reverse currents. For this reason, the diode should be operated with as much reverse current as possible, consistent with rating limitations.

Manufacturers' specifications for zener impedances are usually given for a specified ΔI_Z that covers a range from a small I_Z near the onset of breakdown to some percentage of the maximum rated I_Z. The values may range from a few ohms to several hundred ohms. There is also a variation in the impedance of zener diodes among those having different values of V_Z. Diodes with breakdown voltages near 7 V have the smallest impedances.

Example 18–4

A zener diode has impedance 40 Ω in the range from $I_Z = 1$ mA to $I_Z = 10$ mA. The voltage at $I_Z = 1$ mA is 9.1 V. Assuming that the impedance is constant over the given range, what minimum and maximum zener voltages can be expected if the diode is used in an application where the zener current changes from 2 mA to 8 mA?

Solution. From equation 18–10, the voltage change between $I_Z = 1$ mA and $I_Z = 2$ mA is $\Delta V_Z = \Delta I_Z Z_Z = [(2\text{ mA}) - (1\text{ mA})](40\ \Omega) = 0.04$ V. Therefore, the minimum voltage is $V_Z(\text{min}) = (9.1\text{ V}) + \Delta V_Z = (9.1\text{ V}) + (0.04\text{ V}) = 9.14$ V. The voltage change between $I_Z = 2$ mA and $I_Z = 8$ mA is $\Delta V_Z = [(8\text{ mA}) - (2\text{ mA})](40\ \Omega) = 0.16$ V. Therefore, the maximum voltage is $V_Z(\text{max}) = V_Z(\text{min}) + \Delta V_Z = (9.14\text{ V}) + (0.16\text{ V}) = 9.3$ V.

18–2 FOUR-LAYER DEVICES

Four-layer devices, also called *thyristors,* comprise a class of semiconductor components whose structure is characterized by alternating layers of P and N material. By altering the terminal configuration and geometry of the basic structure, it is possible to construct a variety of useful devices, including silicon controlled rectifiers, silicon controlled switches, diacs, triacs, and Shockley diodes. These are widely used in high-power switching applications where the control of hundreds of amperes and thousands of watts is not unusual. They are especially useful in the control of ac power delivered to heavy loads such as electric motors and lighting systems.

Silicon Controlled Rectifiers (SCRs)

Figure 18–5 shows the layered PNPN structure of a silicon controlled rectifier (SCR). Note that there are three PN junctions in this structure. The figure also shows the schematic symbol for an SCR and its three terminals: anode (A), gate (G), and cathode (K). In applications, the anode is made positive with respect to the cathode. As the name and symbol imply, the SCR behaves like a diode (rectifier), which is normally forward biased, but conducts from anode to cathode only if a *control* signal is applied to the gate.

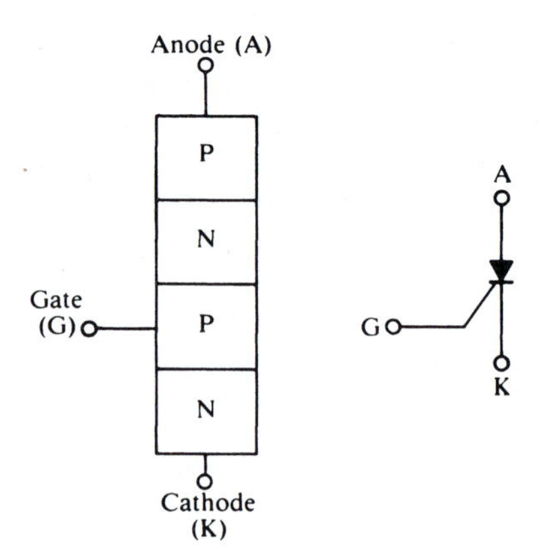

Figure 18–5
Structure and symbol of a silicon controlled rectifier (SCR)

As an aid in understanding SCR operation, Figure 18–6 shows a two-transistor equivalent of the four-layer device. Note that the N and P layers in the center of the structure are "shared" by the transistors: the N layer is the base of Q_1 and the collector of Q_2, while the P layer is the collector of Q_1 and the base of Q_2.

In our analysis of SCR operation, we will first consider the case where the gate is open and a positive voltage V_A is connected between anode and cathode, as shown in Figure 18–7. The positive voltage forward biases the emitter–base junction of Q_1. However, the base current in Q_1 is only a small leakage current because the base of Q_1 is the collector of Q_2 and the collector–base junction of Q_2 is reverse biased. The only base current in Q_2 is a small leakage current from Q_1. Thus, both the base and collector currents in Q_2 are small leakage currents. Since the emitter current in Q_2 is the same as the total current flowing through the SCR, and since $I_{E2} = I_{B2} + I_{C2}$, we see that the total SCR current is the sum of two small leakage currents, which we may regard as negligibly small. We say that the SCR itself is off (essentially nonconducting), or in a *forward blocking* state: forward biased, but blocking the flow of current.

SCR operation is based on the fundamental principle that the α and β of a transistor remain small so long as its collector current is small. As we saw in Chapter 4, collector current must be increased to a certain level before those parameters acquire the normal, essentially constant values they have in the active region. When

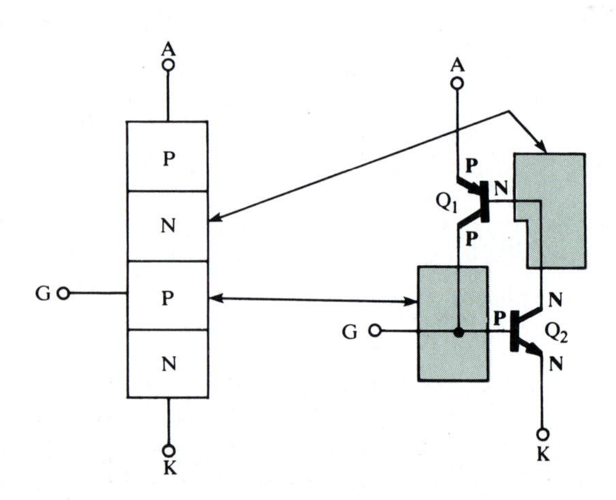

Figure 18–6
Two-transistor equivalent of an SCR

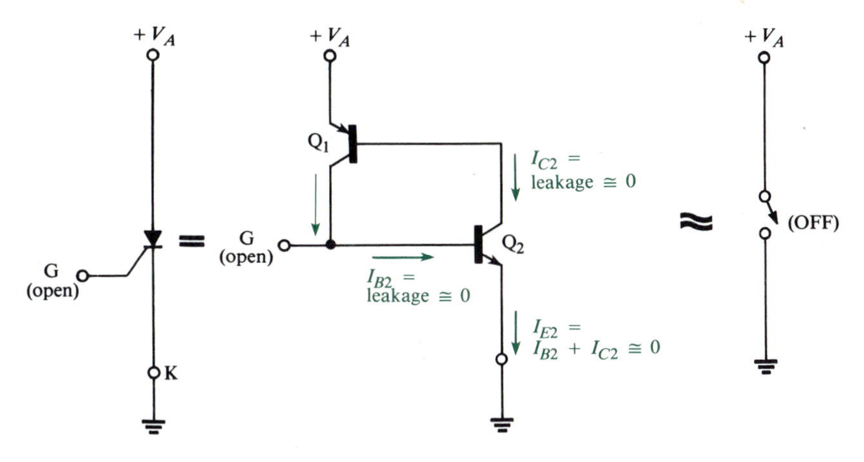

Figure 18–7
The SCR in its off, or forward blocking, state

a transistor is cut off, α and β are quite small, but they rise rapidly as the transistor comes out of cutoff.

In our analysis of Figure 18–7, we assumed that the collector currents (leakage currents) were so small that the α and β of each transistor remained small. If the value of V_A is increased, the leakage currents will increase correspondingly, and so will the values of α and β. Notice that the collector current in Q_2 has a component equal to $\beta_2 I_{B2}$, where β_2 is the beta of Q_2. Thus, as β_2 increases, so does I_{C2}. Likewise, the collector current in Q_1 increases as β_1 increases. Furthermore, as I_{C2} increases, so does base current I_{B1}, since they are identical currents. But as I_{B1} increases, so does $I_{CI} = \beta_1 I_{B1}$. This, in turn, increases I_{B2}, which further increases I_{C2}, which further increases I_{C1}, and so forth. As the currents rise, so do the β-values, which in turn increase the current values. The upshot of all this is that if V_A is made sufficiently large, the transistor α- and β-values increase enough to cause significant current flows. When that point is reached, the larger collector current in Q_2 turns Q_1 fully on, so Q_1 supplies a large base current to Q_2, which drives Q_2 fully on; in other words, both transistors saturate, and the current flowing through each ensures that the other remains in saturation. The SCR is said to be *latched.* The action we have described is an example of *regenerative* switching, where one device, as it begins to conduct, causes another device to conduct, which in turn drives the first further into conduction. Once regeneration is initiated, it quickly causes both devices to saturate. In SCRs the whole process takes a few microseconds or less.

Figure 18–8 shows the SCR after V_A has been increased sufficiently to cause regenerative switching. Since both transistors are saturated, a heavy flow of current is now possible from anode to cathode. As with a forward-biased diode, external resistance is required to limit that current. The anode-to-cathode voltage that is required to initiate regenerative switching to the on state is called the *forward breakover voltage,* $V_{BR(F)}$.

Figure 18–9(a) shows an external resistance connected to the SCR to limit current flow when breakover occurs, and (b) shows the current–voltage characteristic. When V_{AK} is small and the SCR is off, the leakage current is small, so there is only a small voltage drop across R. Most of the supply voltage V_A appears across the SCR as V_{AK}, and the SCR is in its forward blocking region. When V_{AK} reaches

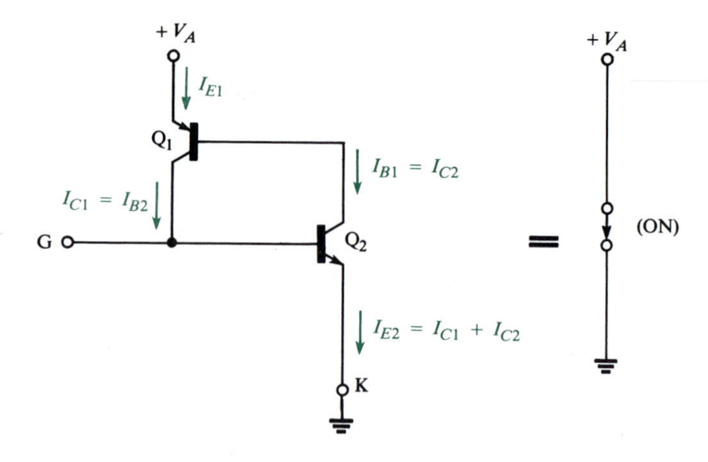

Figure 18–8
The SCR is like a closed switch when the anode-to-cathode voltage reaches the forward breakover voltage.

$V_{BR(F)}$, the SCR suddenly turns on; so, the SCR being like a closed switch, the voltage drop across it suddenly becomes small. Notice that the characteristic loops back from the point where $V_{AK} = V_{BR(F)}$ and I_A is small to the region where V_{AK} is small and I_A is large. As an aid in understanding this behavior, imagine that you are measuring the current through and voltage across an open switch that is in series with a resistor and a battery; then close the switch and observe the measured values. Once the SCR switches on, the characteristic becomes very much like that of a forward-biased silicon diode. The dc voltage drop V_{AK} across the SCR remains at around 0.9 to 1 V, depending on current. The plot also shows that if V_A is made sufficiently negative, a reverse breakdown voltage is reached, and reverse conduction occurs by the avalanching mechanism.

Once the SCR has switched on, it will remain on provided the current through it remains sufficiently large. If the current is reduced (by reducing V_A, for example) to the point that I_A falls below a certain *holding current,* then the SCR will revert to its off state. The holding current, I_H, is identified in Figure 18–9(b) as the flat portion of the characteristic corresponding to the regenerative switching region.

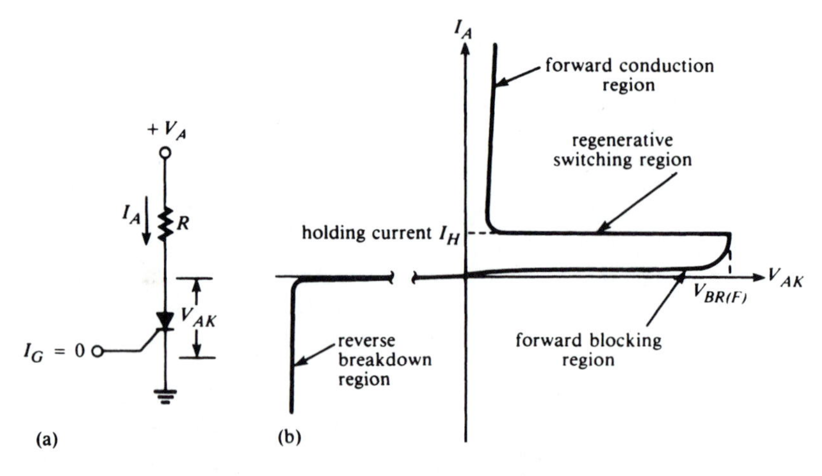

Figure 18–9
The current–voltage characteristic of an SCR with $I_G = 0$

Note that the SCR cannot remain in this region for any length of time; the horizontal line merely shows the path followed when the SCR switches on, and is often drawn as a dashed line for that reason.

Shockley Diodes

In all our analysis to this point, we have assumed that the gate terminal is open, so I_G remained 0 at all times. Certain four-layer devices are manufactured without gate terminals, so they behave exactly like an SCR with an open gate. These two-terminal devices are called *Shockley* diodes, or four-layer diodes. Figure 18–10 shows the schematic symbol for a Shockley diode and an application called a *relaxation oscillator.* Capacitor C charges through R to the breakover voltage, at which time the diode switches on and quickly discharges the capacitor. The diode then turns off and the capacitor begins charging again. The frequency of oscillation is controlled by the RC time constant.

SCR Triggering

SCRs are normally operated with an external circuit connected to the gate. The purpose of the circuit is to inject current into the base of Q_2 in Figure 18–8 and thereby "trigger" the SCR into a regenerative breakover. The additional current supplied by the gate causes I_{C2} to increase, which increases I_{C1}, and so forth, so the SCR switches to its on state, as before. Thus, the gate serves as a control input that can be used to switch the SCR from off to on when desired. It is important to realize that it is necessary to supply only a short *pulse* of current through the gate to cause the switching action; once the regeneration is initiated, latching quickly occurs and gate current is no longer required. By the same token, the SCR cannot be turned off by simply reducing gate current to 0. As described earlier, it can be turned off only by reducing the anode-to-cathode current to a value below the holding current.

Figure 18–11 is a typical family of SCR characteristics showing I_A versus V_{AK} for different values of gate current. Notice that the larger the gate current, the smaller the breakover voltage. When gate current is large, it is not necessary for

Figure 18–10
Schematic symbol and a typical application of a Shockley diode

(a) Schematic symbol for a Shockley diode

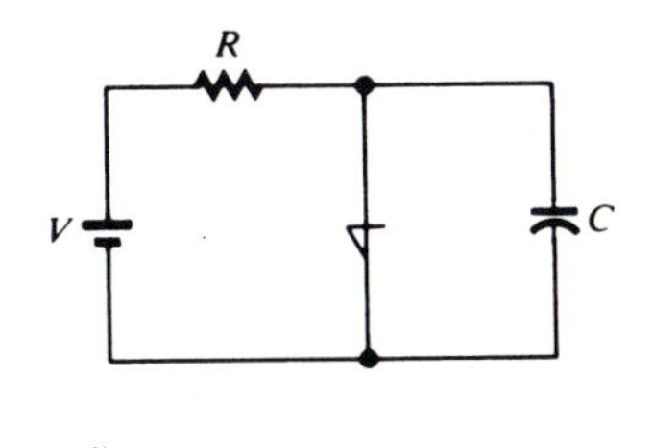

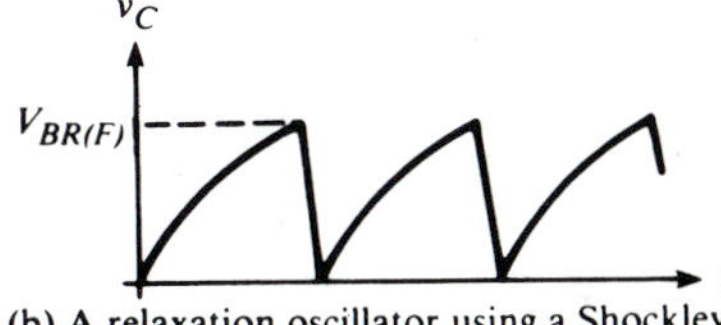

(b) A relaxation oscillator using a Shockley diode

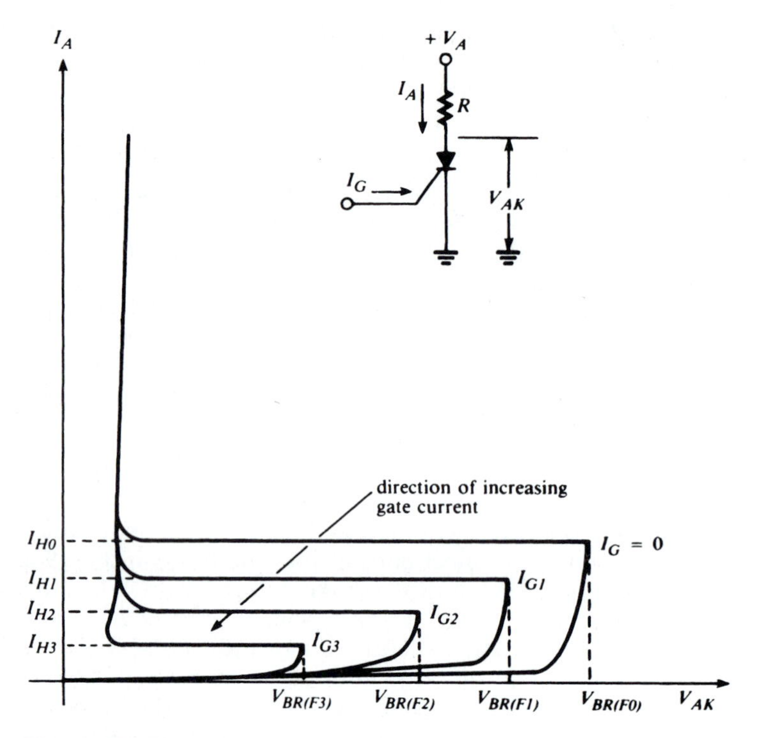

Figure 18–11
A family of SCR characteristics. As gate current increases, the breakover voltage and the holding current decrease.

V_{AK} to become large enough to induce breakover through the increase of leakage current alone. Also note that holding current decreases as gate current increases. Of course, if gate current is removed after breakover has occurred, then the holding current becomes the same as that for $I_G = 0$.

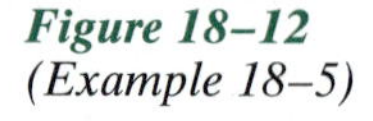

Example 18–5

The voltage drop across the SCR in Figure 18–12 is 1 V when it is conducting. It has a holding current of 2 mA when $I_G = 0$. If the SCR is triggered on by a momentary pulse of gate current, to what value must V_A be reduced to turn the SCR off?

Figure 18–12
(Example 18–5)

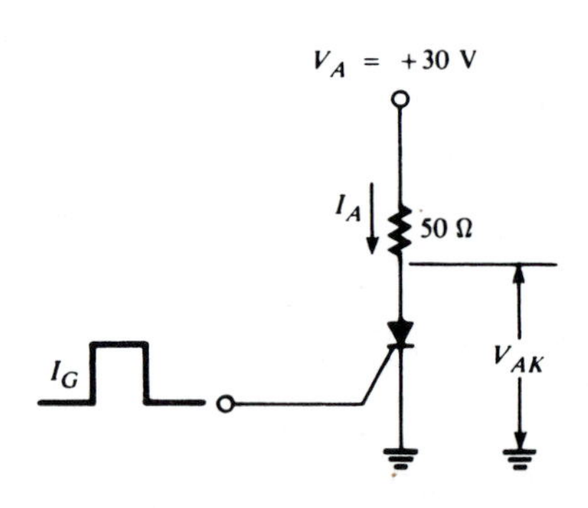

Solution. The current in the 50-Ω resistor is

$$I_A = \frac{V_A - V_{AK}}{50}$$

When the SCR is conducting, $V_{AK} = 1$ V, and

$$I_A = \frac{V_A - 1}{50}$$

Assuming that V_{AK} remains constant at 1 V as V_A and I_A are reduced (actually, V_{AK} decreases slightly), the value of V_A necessary to make $I_A = I_H = 2$ mA is found from

$$2\text{ mA} = \frac{V_A - 1}{50}$$

or $V_A = 1.1$ V.

This example shows that V_A must be reduced to a very small value to turn the SCR off. In practice, turn-off is accomplished either by switching the supply voltage off entirely or by short-circuiting the SCR.

We will now derive a general expression for the SCR current I_A in terms of gate current I_G and leakage currents I_{CBO1} and I_{CBO2}. The result will demonstrate vividly how the magnitudes of α_1 and α_2 affect the SCR current below and at the onset of breakover. In our derivation, we will use the following two relations, which are easily obtained from the standard transistor equations given in Chapter 4 (4–4 and 4–9):

$$I_C = \alpha I_E + I_{CBO} \quad \text{or} \quad I_E = \frac{I_C - I_{CBO}}{\alpha} \tag{18–11}$$

$$I_B = \frac{1-\alpha}{\alpha} I_C - \frac{I_{CBO}}{\alpha} \tag{18–12}$$

Refer to Figure 18–13. Writing Kirchhoff's current law for the external currents, as shown in Figure 18–13(a), we find

$$I_A = I_K - I_G \tag{18–13}$$

where $I_K = I_{E2}$ in Figure 18–13(b). From equation 18–11,

$$I_K = I_{E2} = \frac{I_{C2} - I_{CBO2}}{\alpha_2} \tag{18–14}$$

Note that $I_{C2} = I_{B1}$. From equation 18–12,

$$I_{C2} = I_{B1} = \frac{1-\alpha_1}{\alpha_1} I_{C1} - \frac{I_{CBO1}}{\alpha_1} \tag{18–15}$$

From (18–11),

$$\begin{aligned} I_{C1} &= \alpha_1 I_{E1} + I_{CBO1} \\ &= \alpha_1 I_A + I_{CBO1} \end{aligned} \tag{18–16}$$

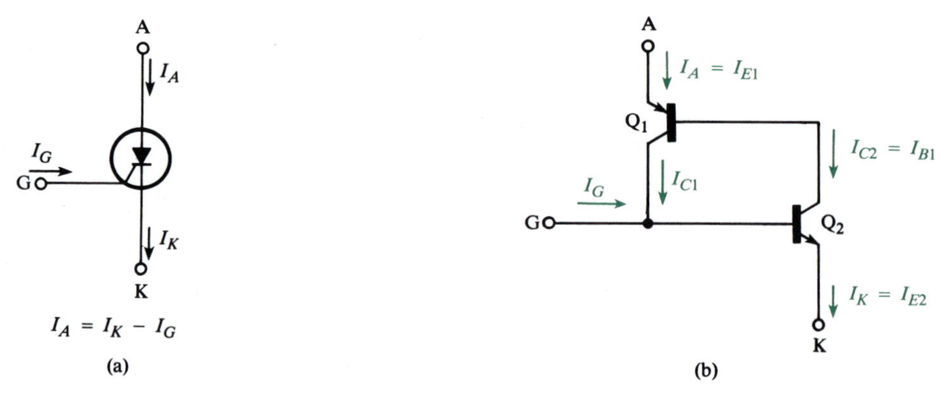

Figure 18–13
Current relations used to derive equation 18–18

Substituting (18–16) into (18–5) and the result into (18–14) gives

$$I_K = \frac{(1 - \alpha_1)I_A + \dfrac{(1 - \alpha_1)}{\alpha_1} I_{CBO1} - \dfrac{I_{CBO1}}{\alpha_1} - I_{CBO2}}{\alpha_2} \tag{18–17}$$

Substituting (18–17) into (18–13) and solving for I_A gives

$$I_A = \frac{\alpha_2 I_G + I_{CBO1} + I_{CBO2}}{1 - (\alpha_1 + \alpha_2)} \tag{18–18}$$

Equation 18–18 shows that if α_1 and α_2 are small, the current I_A is essentially the sum of two small leakage currents. However, if this current, plus the component of gate current, increases enough to raise the values of α_1 and α_2 to where $\alpha_1 + \alpha_2 = 1$, the denominator of (18–18) becomes 0. In that case I_A is theoretically infinite, limited only by external resistance.

Example 18–6

An SCR in its forward blocking region has $V_A = 30$ V, $I_G = 1\ \mu$A, and $I_{CBO1} = I_{CBO2} = 50$ nA. If $\alpha_1 = 0.5$ and $\alpha_2 = 0.3$ under those conditions, find the anode current I_A and the dc resistance of the SCR.

Solution. By equation 18–18,

$$I_A = \frac{\alpha_2 I_G + I_{CBO1} + I_{CBO2}}{1 - (\alpha_1 + \alpha_2)} = \frac{0.3(1\ \mu\text{A}) + (100\ \text{nA})}{1 - 0.8} = 2\ \mu\text{A}$$

$$R_{dc} = \frac{30\ \text{V}}{2\ \mu\text{A}} = 15\ \text{M}\Omega$$

Half-Wave Power Control Using SCRs

SCRs are commonly used to adjust, or control, the average power delivered to a load from an ac source. In these applications, the SCR is periodically switched on and off, and the average power delivered to the load is controlled by adjusting the

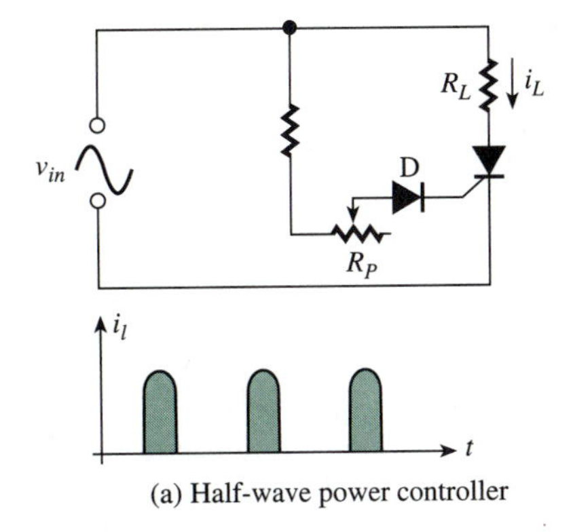

(a) Half-wave power controller

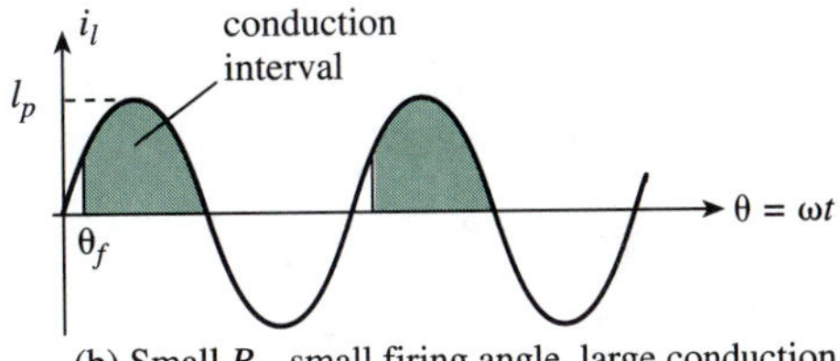

(b) Small R_p, small firing angle, large conduction interval

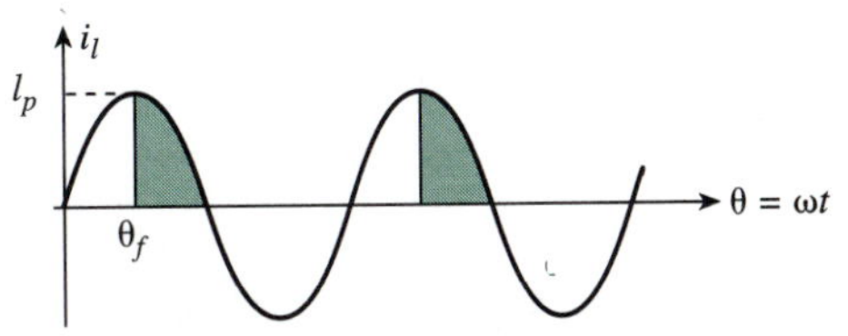

(c) Larger R_p, larger firing angle, smaller conduction interval

Figure 18–14
An SCR half-wave power controller and load-current waveforms. Current flows in the load during the shaded intervals shown on the waveforms.

total length of time the SCR conducts during each ac cycle. Figure 18–14 shows a simple half-wave power controller based on this principle.

When the ac voltage in Figure 18–14 is positive and the SCR is conducting, current flows through load resistance R_L. As shown in (b) and (c), the total length of time the SCR conducts depends on the point in each cycle at which the SCR switches on, or *fires.* After firing, the SCR continues to conduct until the ac voltage drops to near 0, causing the SCR current to fall below the holding current. It then remains off until firing again at the same point in the next positive half-cycle. The angle of the ac waveform at which the SCR switches on is called the *firing angle,* θ_f, or the *delay angle.* The firing angle is controlled by adjusting R_p. When R_p is large, the ac voltage must reach a large value to generate enough gate current to fire the SCR. Consequently, the firing angle is large, as shown in Figure 18–14(c). Conversely, when R_p is small, a relatively small voltage will cause the SCR to fire, and θ_f will be small. In the circuit shown, θ_f can be adjusted from 0° to 90°. The diode is used to protect the gate from the negative portion of the ac voltage.

When the SCR conducts, the peak load current I_P is

$$I_P = \frac{V_P - V_{AK}}{R_L} \tag{18–19}$$

where V_P is the peak ac voltage and V_{AK} is the drop across the SCR, about 1 V, when it is conducting. It can be shown that the average and rms values of the current are

$$I_{AVG} = \frac{I_P}{2\pi}(1 + \cos\theta_f)$$
$$I_{rms} = \frac{I_P}{2}\sqrt{\left(1 - \frac{\theta_f}{180°}\right) + \frac{\sin 2\theta_f}{2\pi}} \qquad \textbf{(18–20)}$$

where θ_f is in degrees. When $\theta_f = 0°$, the current is the same as that in a half-wave rectifier, and equations 18–20 show that $I_{AVG} = I_P/\pi$ and $I_{rms} = I_P/2$. When $\theta_f = 90°$, $I_{AVG} = I_P/2\pi$ and $I_{rms} = I_P/\sqrt{2}$.

One problem encountered in practical applications of SCRs, particularly high-power devices, is a tendency to "false-trigger" if there is a large rate of change of anode voltage. This tendency is attributable to the large junction capacitance associated with the large junction areas that are necessary to dissipate heavy amounts of power. Recall that capacitor current is proportional to the rate of change of capacitor voltage. The large charging current that results when the anode voltage changes at a high rate may induce regenerative breakover. To suppress a false triggering, a shunting capacitor is sometimes connected between anode and cathode.

Example 18–7

In Figure 18–14(a), v_{in} is a 120-V-rms ac voltage and $R_L = 40\ \Omega$. Assume that the drop across the SCR is 1 V when it is conducting.

1. What should be the firing angle if it is desired to deliver an average current of 1 A to the load?
2. What is the average power delivered to the load under the conditions of (1)?

Solution

1. $V_P = \sqrt{2}(120) = 169.7$ V. From equation 18–19,

$$I_P = \frac{V_P - V_{AK}}{R_L} = \frac{169.7 - 1}{40} = 4.2175 \text{ A}$$

From equations 18–20,

$$1 = \frac{4.2175}{2\pi}(1 + \cos\theta_f)$$
$$\cos\theta_f = 0.4898$$
$$\theta_f = 60.67°$$

2. From (18–20),

$$I_{rms} = \frac{4.2175}{2}\sqrt{\left(1 - \frac{60.67}{180}\right) + \frac{\sin 2(60.67)}{2\pi}} = 1.885 \text{ A rms}$$
$$P_{AVG} = I^2_{rms}R_L = (1.885)^2 40 = 142.1 \text{ W}$$

Figure 18–15 shows a typical set of SCR specifications. Notice the variety of case styles and reverse voltage ratings available. The same manufacturer supplies SCRs with forward current ratings up to 55 A rms.

On-State (RMS) Current									
8.0 AMPS		12 AMPS				12.5 AMPS	15 AMPS		
T_C = 83°C		T_C = 90°C	T_C = 85°C			T_C = 80°C	T_C = 80°C		
Case 86-01 Style 1	Case 87L-02 Style 1	Case 221A-02 TO-220AB Style 3	Case 86-01 Style 1	Case 221A-02 TO-220AB Style 3	Case 242-01 Style 1	Case 54-05 Style 2	Case 221A-02 TO-220AB Style 3		TO-220 Style 3
2N4167	2N4183		MCR67-1	MCR68-1	MCR568-1			V_{DRM} V_{RRM}	25 V
2N4168	2N4184	2N6394	MCR67-2	MCR68-2	MCR568-2				50 V
2N4169	2N4185	2N6395	MCR67-3	MCR68-3	MCR568-3	2N3668			100 V
2N4170	2N4186	2N6396				2N3669	MCR2150-4,A4		200 V
2N4171	2N4187	MCR220-5					MCR2150-5,A5		300 V
2N4172	2N4188	2N6397	MCR67-6	MCR68-6	MCR568-6	2N3670	MCR2150-6,A6		400 V
2N4173	2N4189	MCR220-7				2N4103	MCR2150-7,A7		500 V
2N4174	2N4190	2N6398					MCR2150-8,A8		600 V
		MCR220-9					MCR2150-9,A9		700 V
		2N6399					MCR2150-10,A10		800 V
100			300[1]			200	160	MAXIMUM ELECTRICAL CHARACTERISTICS	I_{TSM} (Amps) 60 Hz
30						40	50		I_{GT} (mA)
1.5						2.0	2.5		V_{GT} (V)
−40 to +100		−40 to +125				−40 to +100	−40 to +125		T_J Operating Range (°C)

Figure 18–15
SCR specifications (Courtesy of Motorola, Inc.)

Silicon Controlled Switches (SCSs)

A silicon controlled switch (SCS) is essentially an SCR equipped with a second gate terminal. Like a conventional switch, and unlike an SCR, an SCS can be switched either on or off through external gate control. Figure 18–16 shows the structure, symbol, and two-transistor equivalent of an SCS. Note that the gate terminal at the base of Q_2 is now called the *cathode gate,* and the second gate, at the base of Q_1, is called the *anode gate.*

The silicon controlled switch can be turned on in exactly the same way as an SCR: by supplying a positive pulse of current to the cathode gate, which drives Q_2 on and initiates regenerative breakover. It can also be turned on by supplying a *negative* current pulse to the anode gate. Notice that negative current turns on PNP transistor Q_1, so Q_1 initiates the regeneration in this case. The SCS can be turned off either by supplying negative current to the cathode gate, which turns off Q_2, or by supplying positive current to the anode gate, which turns off Q_1. Figure 18–17 summarizes the methods by which an SCS can be turned on and off. Silicon controlled switches can be turned on and off faster than SCRs, but they are not available with the high power, high current ratings of SCRs.

Figure 18–16
The silicon controlled switch (SCS)

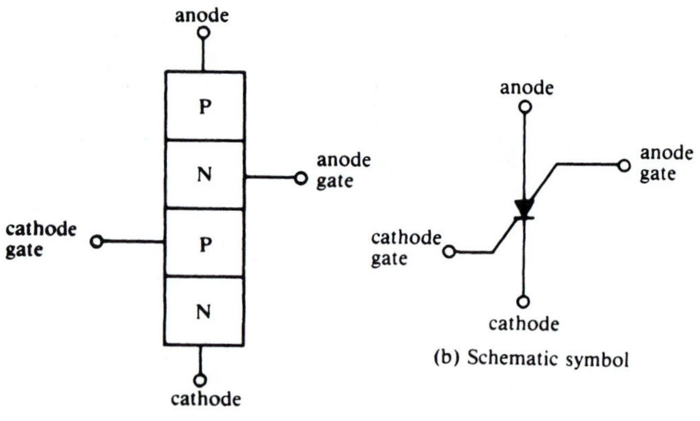

(a) Four-layer structure

(b) Schematic symbol

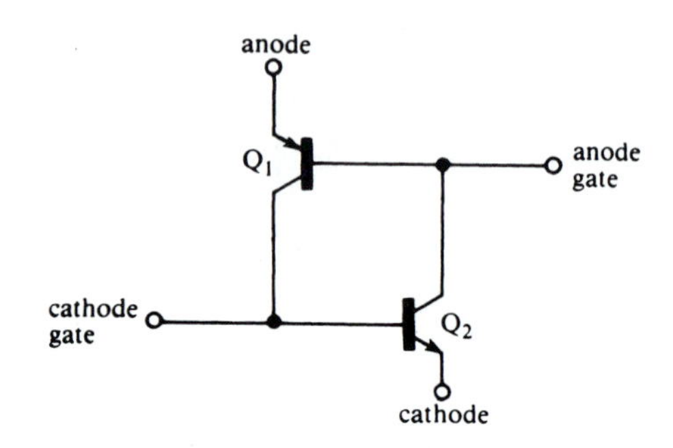

Figure 18–17
Turning an SCS on and off

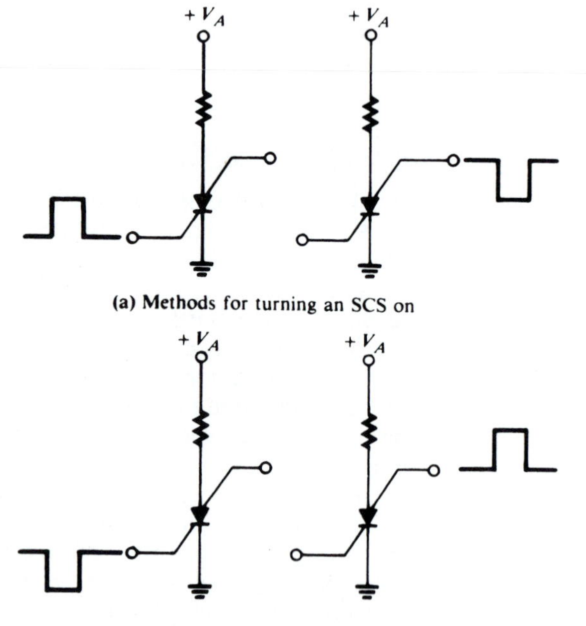

(a) Methods for turning an SCS on

(b) Methods for turning an SCS off

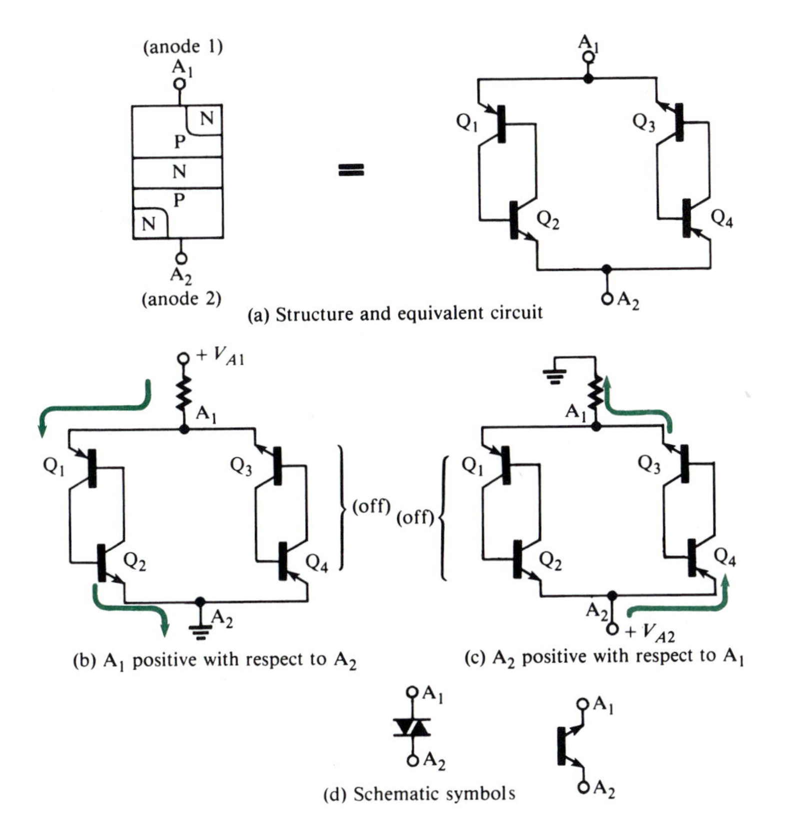

Figure 18–18
DIAC construction and operation

DIACs and TRIACs

A *DIAC* is a four-layer device whose "top" and "bottom" layers contain both N and P material, as shown in Figure 18–18. The right side of the stack can be regarded as a PNPN structure with the same characteristics as a gateless SCR, while the left side is an inverted SCR having an NPNP structure. A four-transistor equivalent circuit of a DIAC, resembling an SCR in parallel with an inverted SCR, is shown in the figure. Note that the terminals are labeled A_1 (anode 1) and A_2 (anode 2). There are no gates. The DIAC can conduct current in *either* direction: from A_1 to A_2 through Q_1 and Q_2, or from A_2 to A_1 through Q_3 and Q_4. If A_1 is made sufficiently positive with respect to A_2 to induce breakover in Q_1 and Q_2, then conduction occurs in that path, while Q_3 and Q_4 remain off. Similarly, if A_2 is sufficiently positive with respect to A_1, then Q_3 and Q_4 conduct while Q_1 and Q_2 remain off. The two cases are illustrated in Figure 18–18(b) and (c). Part (d) shows the schematic symbols.

Figure 18–19 shows the *I–V* characteristic of a DIAC. We may arbitrarily assume one direction through the DIAC as positive (from A_1 to A_2, for example), and the opposite direction as negative. The positive portion of the characteristic is the same as that of an SCR with zero gate current. The negative portion shows that breakover occurs when the reverse voltage reaches $-V_{BR(F)}$, at which time a large reverse current flows. To turn the DIAC off, the current must be reduced

Figure 18–19
The I–V characteristic of a DIAC

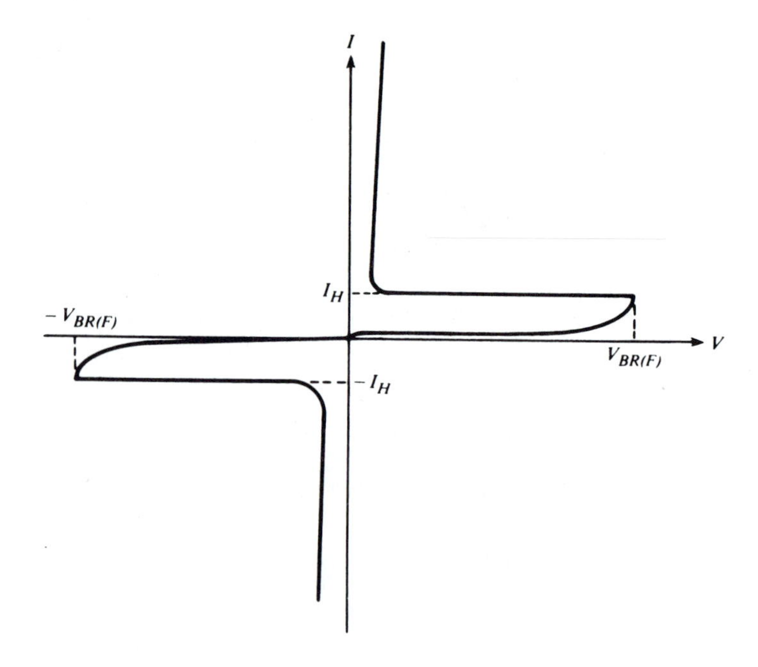

below the positive holding current I_H if conducting in the forward direction, or below the negative holding current $-I_H$ if conducting in the reverse direction.

A TRIAC is equivalent to a DIAC having a gate terminal, as shown in Figure 18–20. Note that the gate is connected to the base of Q_2 and to the base of Q_3, both of which are NPN transistors. Therefore, a pulse of current flowing into the gate will induce current flow through Q_1 and Q_2 if A_1 is positive with respect to A_2 or will induce current flow through Q_3 and Q_4 if A_2 is positive with respect to A_1.

Figure 18–20
A TRIAC equivalent to a DIAC having a gate terminal

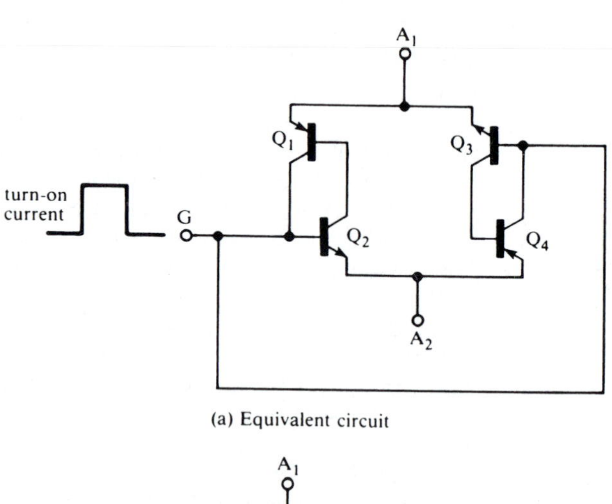

(a) Equivalent circuit

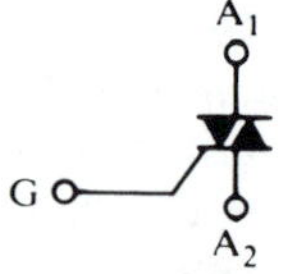

(b) Schematic symbol

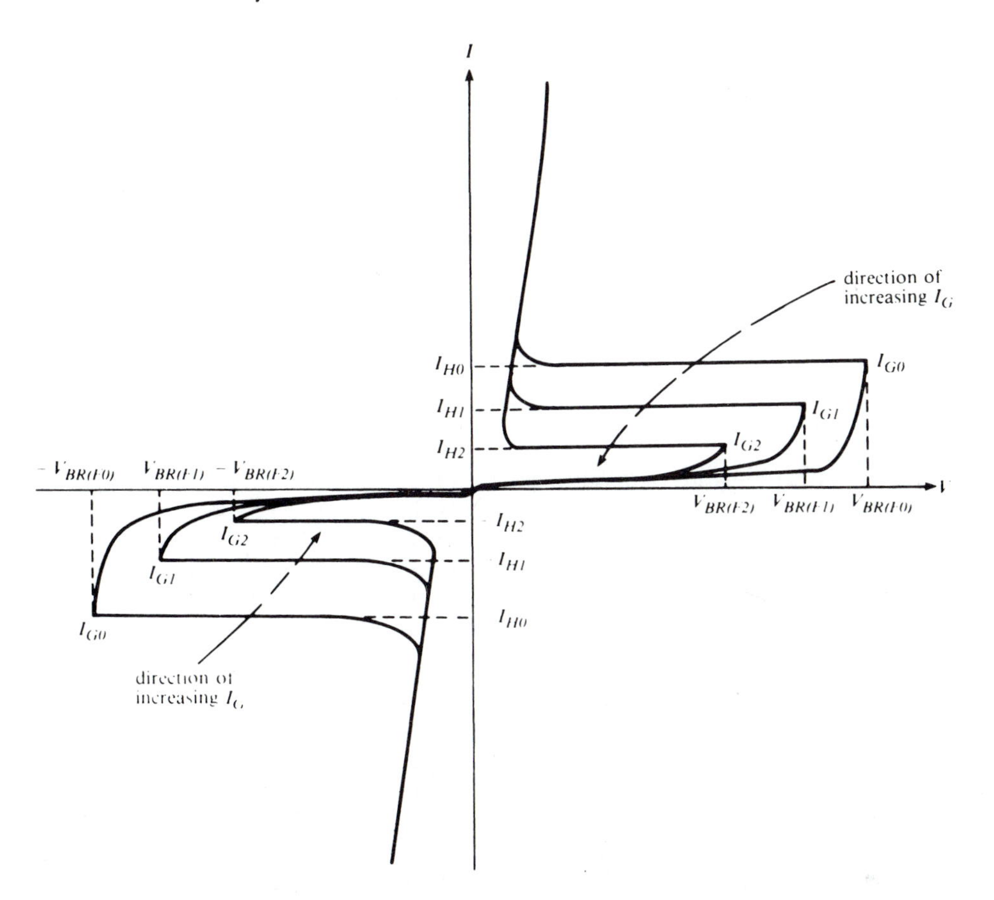

Figure 18–21
I–V characteristics of a TRIAC

The gate must be made positive with respect to A_2 to turn on Q_1 and Q_2, and must be made positive with respect to A_1 to turn on Q_3 and Q_4. Modern TRIACs have an additional N layer connected to one gate. This layer serves as the emitter of an NPN transistor that increases gate sensitivity on one side, but negative gate current is required to trigger that side on.

Figure 18–21 shows a family of characteristic curves for a TRIAC. Like the SCR, the magnitudes of the breakover voltage and holding current become smaller as the values of gate current increase. Whatever the direction of the current through the TRIAC, its value must be reduced below the holding current to turn the TRIAC off.

Figure 18–22 shows a power controller utilizing a TRIAC that is triggered by positive gate current when A_1 is positive with respect to A_2 and by negative gate current when A_1 is negative with respect to A_2. Notice that a DIAC is connected between the gate and an RC circuit. When the capacitor voltage rises far enough to overcome the sum of the breakpower voltage of the DIAC and the forward drop of the gate, the DIAC fires, the capacitor discharges rapidly into the gate, and the TRIAC is triggered into conduction. Since R_p controls the time constant at which the capacitor charges, it is used to control the firing angle, θ_f: Large values of R_p delay the charging and therefore increase the firing angle. When A_1 is negative with

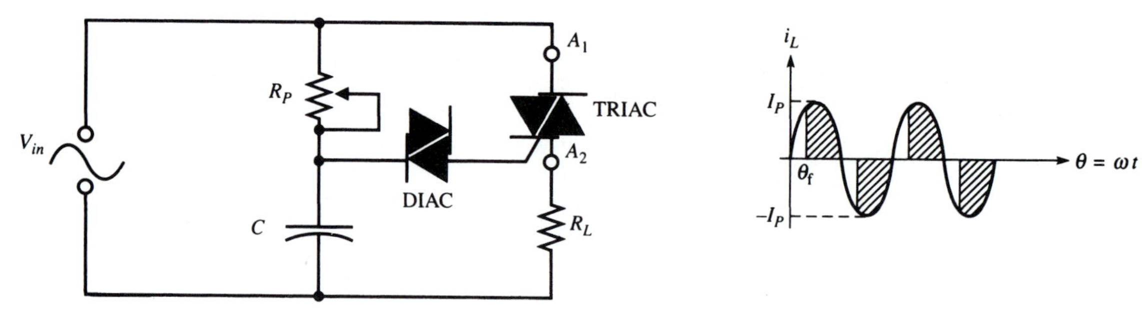

Figure 18–22
A TRIAC power controller

respect to A_2, the polarities of all voltages are reversed, and the TRIAC is triggered into conduction during a portion of the negative half-cycle. The firing angle can be adjusted in practical circuits from near 0° to near 180°, so load current can be made to flow for nearly an entire cycle of input or, at the other extreme, for a very small portion of an input cycle. Since the positive and negative areas of the current waveform are equal, the average current is 0. The rms load current is

$$I_{rms} = \frac{I_P}{\sqrt{2}}\sqrt{\left(1 - \frac{\theta_f}{180°}\right) + \left(\frac{\sin 2\theta_f}{2\pi}\right)} \tag{18–21}$$

where θ_f is in degrees.

Example 18–8

In the circuit shown in Figure 18–22, v_{in} is 60 V rms and $R_L = 15\ \Omega$. Neglecting the drop across the TRIAC when it is conducting, find

1. the maximum possible power that can be delivered to the load; and
2. the percent of maximum load power delivered when $\theta_f = 45°$.

Solution

1. Maximum power is delivered when $\theta_f = 0°$, in which case the current is a full sine wave and the load voltage is 60 V rms:

$$P_{max} = \frac{V_{rms}^2}{R_L} = \frac{60^2}{15} = 240\ \text{W}$$

2. $V_P = \sqrt{2}(60) = 84.85$ V

$$I_P = \frac{V_P}{R_L} = \frac{84.85}{15} = 5.66\ \text{A}$$

From equation 18–21,

$$I_{rms} = \frac{5.66}{\sqrt{2}}\sqrt{\left(1 - \frac{45}{180}\right) + \frac{\sin 2(45)}{2\pi}} = 3.82\ \text{A rms}$$

$$P = I_{rms}^2 R_L = (3.82)^2(15) = 218.8\ \text{W}$$

$$\frac{P}{P_{max}} \times 100\% = \frac{218.8}{240} \times 100\% = 91.2\%$$

18–3 OPTOELECTRONIC DEVICES

Optoelectronic devices are those whose characteristics are changed or controlled by light or those that create and/or modify light. In recent years, important technological advances in the design and fabrication of solid-state optoelectronic devices have been responsible for what are now widespread applications in areas such as visual displays, alarm systems, fiber-optic communications, optical readers, lasers, and solar energy converters.

Light is a form of electromagnetic radiation and, as such, is characterized by its frequency or, equivalently, by its wavelength. The fundamental relationship between frequency and wavelength is

$$f = \frac{c}{\lambda} \qquad \textbf{(18–22)}$$

where f is frequency, in hertz; λ is wavelength, in meters; and c is the speed of light, 3×10^8 m/s.

Another commonly used unit of wavelength is the angstrom (Å), where

$$1\text{Å} = 10^{-10}\ \text{m} \qquad \textbf{(18–23)}$$

Visible light occurs at frequencies in the range from about 4.3×10^{14} Hz (7000 Å) to 7.5×10^{14} Hz (4000 Å). The lowest frequency appears red and the highest appears violet. Frequencies less than that of red, down to 10^{12} Hz, are called *infrared,* and those greater than that of violet, up to 5×10^{17} Hz, are called *ultraviolet. White* light is a mixture, or composite, of all frequencies in the visible range.

A light *spectrum* is a plot of light energy versus frequency or wavelength. An *emission* spectrum is a spectrum of the light produced, or emitted, by a light source, such as an incandescent bulb or a light-emitting diode (LED). A *spectral response* shows how one characteristic of a light-sensitive device varies with the frequency or wavelength of the light to which it is exposed. Spectral response is similar in concept to the frequency response of an electronic system. For example, the spectral response of the human eye shows that it is most sensitive to yellow light, at wavelengths around 5700 Å, and least sensitive to the frequencies at opposite ends of the visible band, i.e., red and violet.

Luminous flux is the *rate* at which visible light energy is produced by a light source, or at which it is received on a surface, and is measured in *lumens* (lm), where

$$1\ \text{lm} = 1.496\ \text{mW} \qquad \textbf{(18–24)}$$

Thus, a surface receiving visible light energy at the rate of 1.496×10^{-3} J/s is said to have a luminous flux of one lumen. Luminous *intensity* is luminous flux per unit area. For example, if the luminous flux on a 0.2-m^2 surface is 1.5×10^3 lm, then the intensity is $1.5 \times 10^3\ \text{lm}/(0.2\ \text{m}^2) = 75 \times 10^3\ \text{lm/m}^2$. One lumen per square foot equals the older unit, one footcandle. Manufacturers of light-sensitive electronic devices also use the units *milliwatts per square centimeter* to specify light intensity in their product literature. Care must be exercised in the interpretation of light-intensity figures, since measured values depend on the spectral response of the instrument used to measure them.

Example 18–9

Find the horsepower equivalent of $1 \times 10^6\ \text{lm/m}^2$ (the typical intensity of visible sunlight at noon on a 10-m^2 surface).

Solution

$$P = (10^6\ \text{lm/m}^2)(10\ \text{m}^2) = 10^7\ \text{lm}$$

$$(10^7\ \text{lm})(1.496 \times 10^{-3}\ \text{W/lm}) = 1.496 \times 10^4\ \text{W}$$

$$(1.496 \times 10^4\ \text{W})\left(\frac{1\ \text{hp}}{746\ \text{W}}\right) = 20\ \text{hp}$$

Photoconductive Cells

A photoconductive cell is a passive device composed of semiconductor material whose surface is exposed to light and whose electrical resistance decreases with increasing light intensity. Recall that electrons in a semiconductor become *free* electrons if they acquire enough energy to break the bonds that hold them in a covalent structure. In conventional diodes and transistors, electrons are freed by heat energy. In a photoconductive cell, free electrons are similarly created by light energy; the greater the light intensity, the greater the number of free electrons. Since the conductivity of the material increases as the number of free electrons increases, the resistance of the material decreases with increasing light intensity. Photoconductive cells are also called *photoresistive* cells, or *photoresistors.* They are the simplest and least expensive of components belonging to the general class of *photoconductive* devices: those whose conductance changes with light intensity.

Figure 18–23 shows the typical geometry of a photoconductive cell and its schematic symbol. The light-sensitive semiconductor material is arranged in a zigzag strip whose ends are attached to external pins. A glass or transparent plastic cover is attached for protection. The light-sensitive material used in photoconductive cells is either *cadmium sulfide* (CdS) or *cadmium selenide* (CdSe). The resistance of each of these materials changes rather slowly with changes in light intensity, the CdSe cell having a response time of about 10 ms, and that of CdS being about 100 ms.

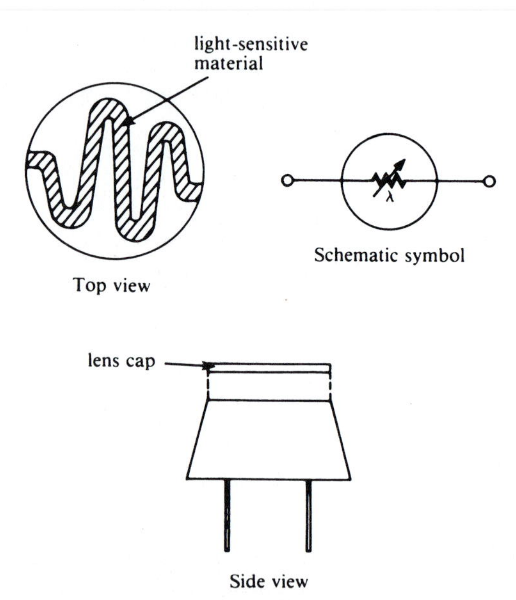

Figure 18–23
Structure and symbol for a photoconductive cell

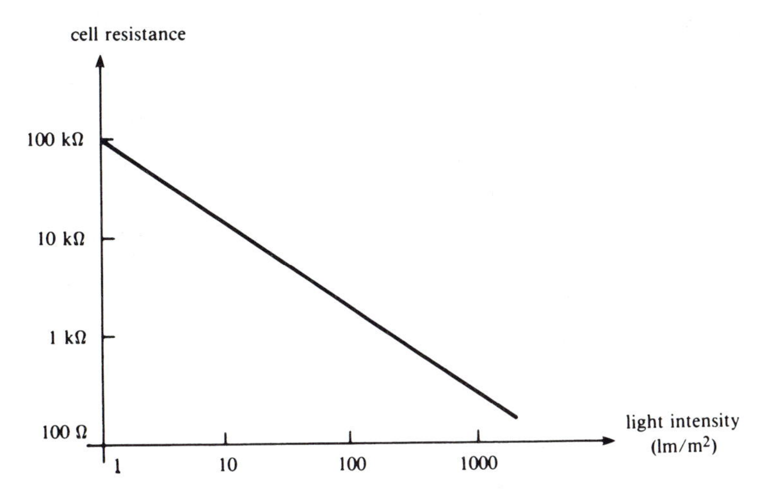

Figure 18–24
Resistance versus light intensity for a typical photoconductive cell

However, the slower CdS cell is much less sensitive to temperature changes than is the CdSe cell. The spectral response of CdS, i.e., its sensitivity as a function of the wavelength of incident light, is similar to that of the human eye, so it responds to visible light. The response of CdSe is greater at red and infrared frequencies.

Figure 18–24 shows a plot of resistance versus light intensity for a typical photoconductive cell. Note that the plot is linear on logarithmic axes. The plot demonstrates that the resistance of a typical cell can change over a very wide range, from about 100 kΩ to a few hundred ohms. The large resistance of the cell when it is not illuminated is called its *dark resistance.*

Photoconductive cells are used in light meters, lighting controls, automatic door openers, and intrusion detectors. In the latter applications, the change in resistance that occurs when incident light is interrupted causes a circuit in which the cell is connected to energize a switch or relay that in turn generates an audible or visual signal. The next example demonstrates such an application.

Example 18–10

When the photoconductive cell in Figure 18–25 is illuminated by a light beam, it has a resistance of 20 kΩ. The dark resistance is 100 kΩ. Show that the relay is de-energized when the cell is illuminated and energized when the light beam is interrupted.

Solution. When the cell is illuminated and has resistance 20 kΩ, we find V^+ using superposition:

$$V^+ = \left[\frac{50\text{ k}\Omega}{(20\text{ k}\Omega) + (50\text{ k}\Omega)}\right](-6\text{ V}) + \left[\frac{20\text{ k}\Omega}{(20\text{ k}\Omega) + (50\text{ k}\Omega)}\right](+6\text{ V}) = -2.57\text{ V}$$

Since V^+ is negative, the output of the comparator is negative and the SCR is off. No current flows in the relay coil, so it is de-energized.

Figure 18–25
(Example 18–10)

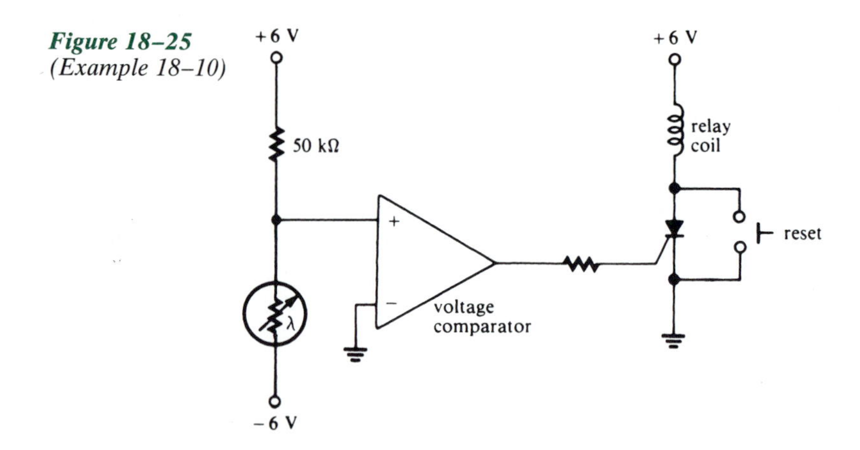

When the light beam is interrupted, the cell has dark resistance 100 kΩ, and

$$V^+ = \left[\frac{50\text{ k}\Omega}{(100\text{ k}\Omega) + (50\text{ k}\Omega)}\right](-6\text{ V}) + \left[\frac{100\text{ k}\Omega}{(100\text{ k}\Omega) + (50\text{ k}\Omega)}\right](+6\text{ V}) = +2\text{ V}$$

The output of the comparator is therefore positive, the SCR turns on, and the relay is energized. To de-energize the relay, the reset pushbutton is depressed, thus shorting the SCR and dropping its current below the holding current.

Note that a latching device such as an SCR is required in this application if the circuit is to respond to *momentary* interruptions of the light beam. Also, cell response time may be critical in detecting short-duration interruptions. The circuit can be made more sensitive by increasing the values of the positive and negative voltages applied to the comparator, but *photodiodes,* discussed in the next section, have much faster response times and are therefore more appropriate for detecting very short duration interruptions of light.

Photodiodes

A photodiode is a PN junction constructed so that it can be exposed to light. When reverse biased, it behaves as a *photoconductive* device because its resistance changes with light intensity. In this case, the change in resistance manifests itself as a change in reverse leakage current. Recall that reverse leakage current in a conventional diode is due to thermally generated minority carriers that are swept through the depletion region by the barrier voltage. In the photodiode, additional minority carriers are generated by light energy, so the greater the light intensity, the greater the reverse current and the smaller the effective resistance.

Figure 18–26 shows a family of characteristic curves for a typical photodiode. For a fixed value of reverse bias (along a vertical line), the magnitude of the reverse current increases with increasing light intensity. Along a line of constant light intensity, there is relatively little change in reverse current with increasing reverse voltage.

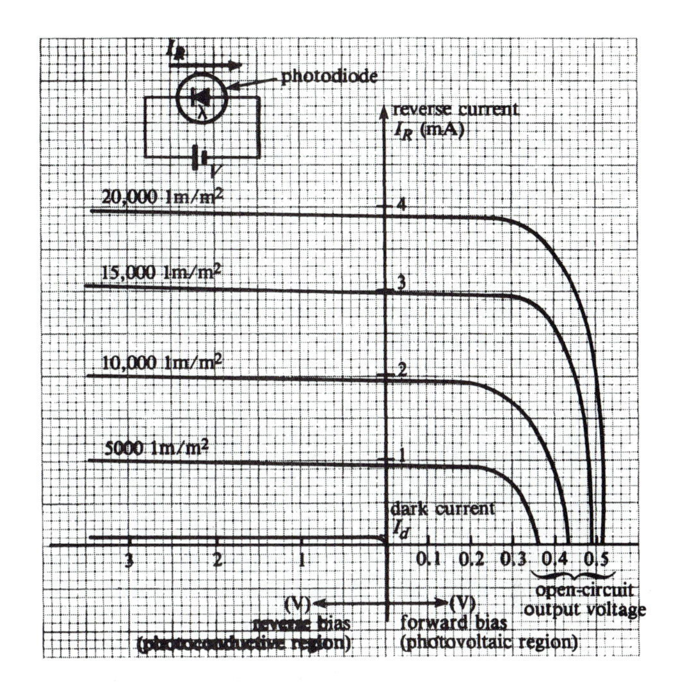

Figure 18–26
Characteristic curves of a photodiode. (Note that the horizontal scale differs along the forward and reverse axes.)

Example 18–11

The photodiode whose characteristics are shown in Figure 18–26 has a reverse bias of 2 V. Find its resistance when the light intensity is 5000 lm/m² and when it is 20,000 lm/m².

Solution. Constructing a vertical line through the 2-V reverse-bias voltage in Figure 18–26, we find that it intersects the 5000-lm/m² characteristic at approximately $I_R = 1$ mA. It intersects the 20,000-lm/m² characteristic at approximately $I_R = 3.9$ mA. Thus,

$$R\text{ (at 5000 lm/m}^2) = \frac{V_R}{I_R} = \frac{2\text{ V}}{1\text{ mA}} = 2\text{ k}\Omega$$

$$R\text{ (at 20,000 lm/m}^2) = \frac{V_R}{I_R} = \frac{2\text{ V}}{3.9\text{ mA}} = 512.8\ \Omega$$

These results confirm that the resistance of the photodiode decreases with increasing light intensity.

In the preceding example, it is worth noting that a fourfold increase in light intensity resulted in a very nearly fourfold decrease in resistance, indicating that the photodiode has good *linearity.* This property is apparent in the way the characteristic curves are nearly equally spaced, implying that the change in reverse current is proportional to the change in light intensity. But, from an applications standpoint, perhaps the most desirable feature of a photodiode is its very fast response time, on the order of nanoseconds. The ability to change resistance very quickly when there is a sudden change in light level makes the photodiode useful in high-speed digital applications, such as optical readers.

The *dark current* in the photodiode is the small reverse current that flows due to thermally generated carriers alone. Using the notation I_d for dark current, its equation is the same as that presented in Chapter 2 for the ideal diode under reverse bias*:

$$I_d = I_s(1 - e^{-V_R/\eta V_T}) \tag{18–25}$$

Letting I_p represent the reverse current produced due to carriers generated by light energy, the total reverse current is the sum of the thermal and light-generated components:

$$I_R = I_d + I_p \tag{18–26}$$

When a small forward-biasing voltage is connected across the photodiode, it is found that reverse current continues to flow. This phenomenon can be seen in the forward-bias characteristics of Figure 18–26 and is attributable to the *excess carriers* produced by the light energy. If the forward bias is made sufficiently large, the reverse current must eventually become 0. As can be seen in the forward characteristics, the greater the light intensity, the greater the value of forward bias required to reduce the current to 0. The implication of this result is that light energy creates an internal voltage across the terminals of the photodiode, positive on the anode side and negative on the cathode side, since an external voltage of equal magnitude and opposite polarity is necessary to stop the flow of current. When the terminals are left *open* and the photodiode is exposed to light, that internal voltage appears across the terminals and can be measured. The value increases with increasing light intensity and is identified as the *open-circuit output voltage* in Figure 18–26. When operated in this manner, as a voltage generator, the photodiode is said to be a *photovoltaic* device, rather than a photoconductive device.

When the photodiode terminals are open-circuited, $I_R = 0$ in equation 18–26. Also, letting V_{oc} be the open-circuit output voltage, we substitute $-V_R = V_{oc}$ in equation 18–25 and obtain

$$\begin{aligned} 0 &= I_p + I_d \\ 0 &= I_p + I_s(1 - e^{V_{oc}/\eta V_T}) \end{aligned} \tag{18–27}$$

Solving for V_{oc} gives

$$V_{oc} = \eta V_T \ln\left(1 + \frac{I_p}{I_s}\right) \tag{18–28}$$

Unless the diode is operated at very low light levels, the saturation current I_s is much smaller than I_p. Therefore, under at least moderate light levels, $(1 + I_p/I_s) \approx I_p/I_s$ and (18–28) becomes

$$V_{oc} = \eta V_T \ln(I_p/I_s) \tag{18–29}$$

Equation 18–29 shows that the open-circuit output voltage is a logarithmic function of I_p, and therefore a logarithmic function of light intensity.

---

* See equation 2–13. Since the reverse current is negative, we multiply both sides of (2–13) by -1 so we can represent I_d as a positive quantity in (18–25).

Example 18–12

A photodiode operated in the photovoltaic mode under moderate light intensity has an open-circuit output voltage of 0.4 V. Assuming that the component of reverse current produced by light energy is directly proportional to light intensity, find the output voltage when the intensity is doubled. Assume that $\eta = 1$ and $V_T = 0.026$ V.

Solution. From equation 18–29, $V_{oc} = 0.4 = (1)(0.026)\ln(I_p/I_s)$. By the linearity assumption, doubling the light intensity doubles I_p, so we must find V_{oc} from

$$V_{oc} = 0.026\ln(2I_p/I_s) = 0.026\ln 2 + 0.026\ln(I_p/I_s)$$
$$= 0.018 + 0.4 = 0.418\text{ V}$$

Phototransistors

Like a photodiode, a phototransistor has a reverse-biased PN junction that is exposed to light. In this case, the junction is the collector–base junction of a bipolar transistor. Figure 18–27(a) shows the currents that flow in an NPN transistor when the base is open. Recall from Chapter 2 that reverse leakage current I_{CBO} flows across the reverse-biased collector–base junction due to thermally generated minority carriers. Also recall that the collector current in this situation is

$$I_C = I_{CEO} = \frac{I_{CBO}}{1 - \alpha} = (\beta + 1)I_{CBO} \qquad \textbf{(18–30)}$$

When the junction is exposed to light, an additional component of reverse current, I_p, is generated due to minority carriers produced by light energy, as in the photodiode. Thus, the total collector current is

$$I_C = (\beta + 1)(I_{CBO} + I_p) \qquad \textbf{(18–30a)}$$

The thermally generated collector current $(\beta + 1)I_{CBO}$ is the dark current. In most applications, it is much smaller than the collector current created by light energy,

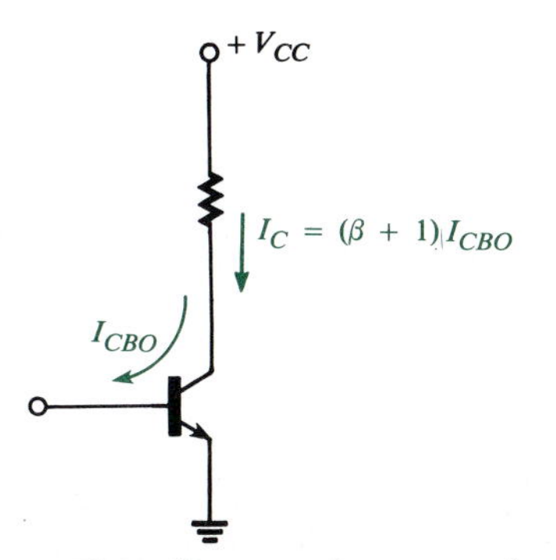

(a) Thermally generated currents under dark conditions

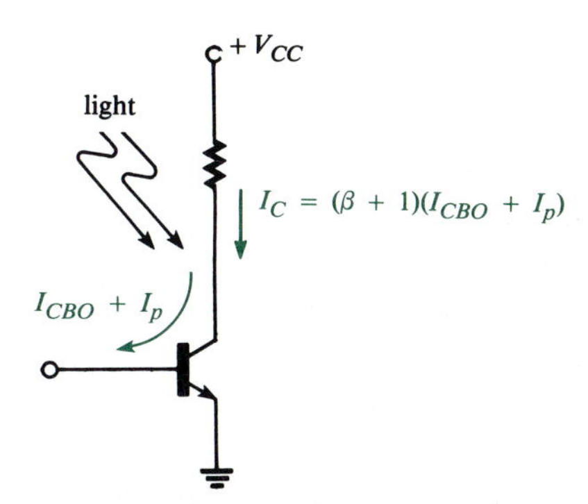

(b) Currents generated by thermal and light energy combined

Figure 18–27
Currents in a phototransistor

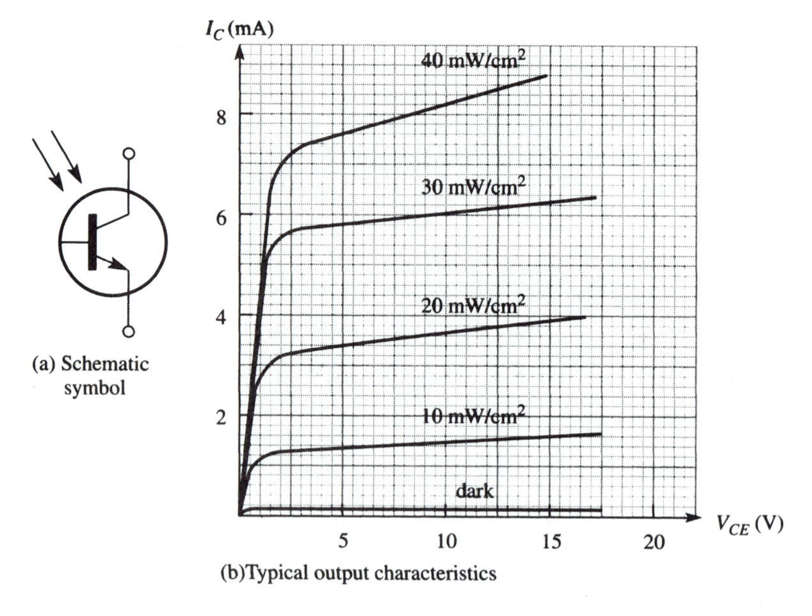

Figure 18–28
Phototransistor symbol and characteristics

so

$$I_c \approx (\beta + 1)I_p \approx \beta I_p \tag{18–30b}$$

Equation 18–30b shows that the collector current in a phototransistor is directly proportional to I_p, which, as in the photodiode, is proportional to light intensity. The advantage of the phototransistor is that it provides current gain and is therefore more *sensitive* to light than the photodiode. It is used in many of the same kinds of applications as photodiodes, but it has a slower response time, on the order of microseconds, compared to the nanosecond response of a photodiode. The phototransistor is constructed with a lens that focuses incident light on the collector–base junction. Some are constructed with no externally accessible base terminal, and others have a base connection that can be used for external bias purposes.

Figure 18–28 shows the schematic symbol and a typical set of output (collector) characteristics for a phototransistor. Notice the similarity of these characteristics to those of a conventional transistor. Light intensity in milliwatts per cubic centimeter serves as the control parameter instead of base current.

Example 18–13

The phototransistor whose characteristics are shown in Figure 18–28 is to be used in the detector circuit shown in Figure 18–29(a). When the light intensity falls below a certain level, the collector voltage rises far enough to supply the 100 μA of gate current that is necessary to turn on the SCR. Find the value of R_C that should be used if the SCR must fire when the light intensity falls to 10 mW/cm^2.

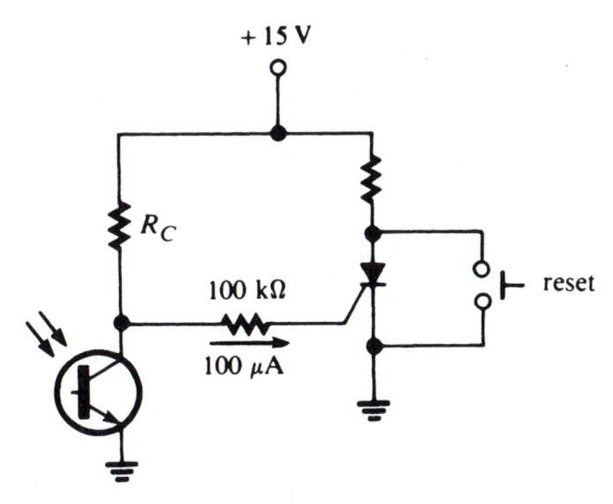

(a) When the light intensity falls, the collector voltage of the phototransistor rises and supplies the 100-μA gate current necessary to fire the SCR.

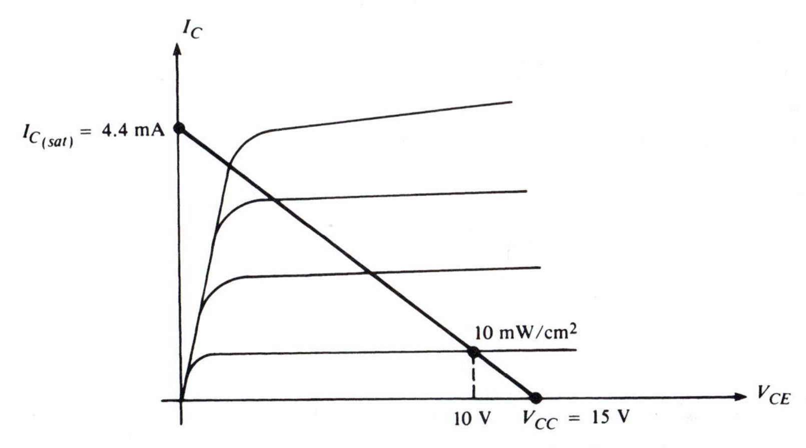

(b) The value required for R_C is found by constructing a load line on the output characteristics

Figure 18–29
(Example 18–13)

Solution. To fire the SCR, the collector voltage must rise to $V_{CE} = (100\ \mu\text{A})(100\ \text{k}\Omega) = 10$ V. Therefore, the dc load line for the phototransistor must intersect the 10-mW/cm² characteristic at $V_{CE} = 10$ V. As shown in Figure 18–29(b), the load line is drawn between that point and the point where it must intersect the V_{CE}-axis at $V_{CC} = 15$ V. (Use the output characteristic in Figure 18–28 to construct an accurate plot of the load line.) The load line is then extended to its intersection with the I_C-axis, where $I_{C(sat)} \approx 4.4$ mA. Thus,

$$R_C = \frac{V_{CC}}{I_{C(sat)}} = \frac{15\ \text{V}}{4.4\ \text{mA}} = 3410\ \Omega$$

A *photodarlington* is a phototransistor packaged with another transistor connected in a Darlington configuration, as studied in Chapter 11. Figure 18–30 shows the configuration. With its large current gain, the photodarlington can produce greater output current than either the photodiode or phototransistor and it is

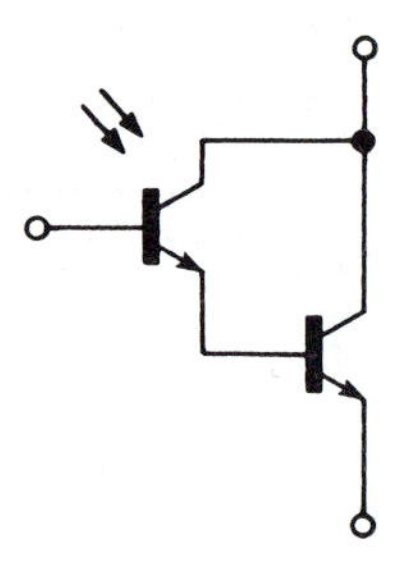

Figure 18–30
The photodarlington

therefore a more light-sensitive device. However, the photodarlington has a slower response time than either of the other two devices.

Figure 18–31 shows typical manufacturer's specifications for photodiodes, phototransistors, and photodarlingtons. Compare the response times and sensitivities for these devices. Note, for example, that the response times of the photodiodes are all 1 ns, compared to rise and fall times on the order of 2 μs for the phototransistors and 15 μs for the photodarlingtons. On the other hand, the light currents range from only 2 μA at 5 mW/cm^2 for a photodiode to 25 mA at 0.5 mW/cm^2 for a photodarlington. Note also the very small dark currents of each, in comparison to the values of light-generated current.

Solar Cells

A solar cell is simply a large photodiode operated in its photovoltaic mode. In its most common application, it is used to convert light energy to electrical energy, and it thereby serves as a source of dc power. Unlike the photodiode, the solar energy converter is usually constructed in the form of a circular wafer with a very thin layer of P material at its surface. Light penetrates the thin P layer to the PN junction where excess minority carriers are generated, as in the photodiode. Silicon is the semiconductor material most commonly used in solar-cell construction. Figure 18–32 shows the solar-cell symbol and its connection to a fixed load resistance R_L. Remember that the positive load current I_L shown here is actually the reverse current through the PN junction.

The characteristic curves of a solar cell have the same appearance as photodiode characteristics in the photovoltaic region, shown in Figure 18–26. With reference to I_L and V_L in Figure 18–32, these curves are simply plots of the photodiode equation

$$I_L = I_p + I_s(1 - e^{V_L/\eta V_T}) \qquad \textbf{(18–31)}$$

for different values of I_p, i.e., for different values of light intensity. A typical set of curves for a 3-in.-diameter cell, with axes relabeled as I_L and V_L, is shown in Figure 18–33. A nominal value for the solar intensity at noon on the earth's surface on a clear day is 100 mW/cm^2, which is often referred to as *one sun* in solar-power technology.

The intersection of each curve with the I_L-axis in Figure 18–33 is the *short-circuit current, I_{sc}*, corresponding to a given light intensity. Of course, zero power is delivered to the load when the load is a short. The intersection of each curve with the V_L-axis in Figure 18–33 is the *open-circuit voltage, V_{oc}*, corresponding to a given light intensity. Again, zero power is delivered when the load is open. On each curve of constant light intensity, there is a point where the power, $P_L = V_L I_L$, is maximum. This point is on the knee of the curve, and maximum load power is

Silicon Photodetectors

A variety of silicon photodetectors are available, varying from simple PN diodes to complex, single chip 400 volt triac drivers. They are available in packages offering choices of viewing angle and size in either economical plastic cases or rugged, hermetic metal cans. They are spectrally matched for use with Motorola infrared emitting diodes.

Photodiodes

Package	Type Number	Light Current μA Typ	@ V_R Volts	H mW/cm²	$V_{(BR)R}$ Volts Min	Dark Current nA Max	@ Volts	Response Time ns Typ
 Case 209-02 Metal Convex Lens — Actual Size	MRD500	9.0	20	5.0	100	2.0	20	1.0
 Case 210-01 Metal Flat Lens — Actual Size	MRD510	2.0	20	5.0	100	2.0	20	1.0
 Case 349-01 Plastic, Style 1 — Actual Size	MRD721	4.0	20	5.0	100	10	20	1.0

Phototransistors

Package	Type Number	Light Current mA Typ	@ V_{CC}	H mW/cm²	$V_{(BR)CEO}$ Volts Min	t_r, t_f/t_{on}, t_{off}* μs	@ Typ V_{CC}	I_L μA
 Case 173-01 Plastic, Style 1 — Actual Size	MRD150	2.2	20	5.0	40	2.5 4.0	20	1000
 Case 82-05 Metal, Style 1 — Actual Size	MRD310	3.5	20	5.0	50	2.0 2.5	20	1000
	MRD300	8.0	20	5.0	50	2.0 2.5	20	1000
	MRD3050	0.1 Min	20	5.0	[illegible]	2.0 2.5	20	1000
	MRD3051	0.2 Min	20	5.0	30	2.0 2.5	20	1000
	MRD3054	0.5 Min	20	5.0	30	2.0 2.5	20	1000
	MRD3055	1.5 Min	20	5.0	30	2.0 2.5	20	1000
	MRD3056	2.0 Min	20	5.0	30	2.0 2.5	20	1000
 Case 81A-06 Metal, Style 1* / Case 81A-07 Metal, Style 1 — Actual Size	MRD601, 611*	1.5	5.0	20	50	1.5 15	30	800
	MRD602, 612*	3.5	5.0	20	50	1.5 15	30	800
	MRD603, 613*	6.0	5.0	20	50	1.5 15	30	800
	MRD604, 614*	8.5	5.0	20	50	1.5 15	30	800
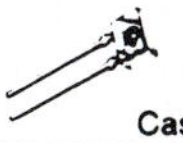 Case 349-01 Plastic, Style 2 — Actual Size	MRD701	0.5	5.0	0.5	30	10 60*	5.0*	

Photodarlingtons

Package	Type Number	Light Current mA Typ	@ V_{CC}	H mW/cm²	$V_{(BR)CEO}$ Volts Min	t_r, t_f/t_{on}, t_{off}* μs	@ Typ V_{CC}	I_L μA
 Case 82-05 Metal, Style 1 — Actual Size	MRD370	10	5.0	0.5	40	15 40	5.0	1.0
	MRD360	20	5.0	0.5	40	15 65	5.0	1.0
 Case 349-01 Plastic, Style 2 — Actual Size	MRD711(1)	25	5.0	0.5	60	125 150*	5.0*	

(1)Photodarlington with integral base-emitter resistor.

Figure 18–31
Typical specifications for light-sensitive devices (Courtesy of Motorola, Inc.)

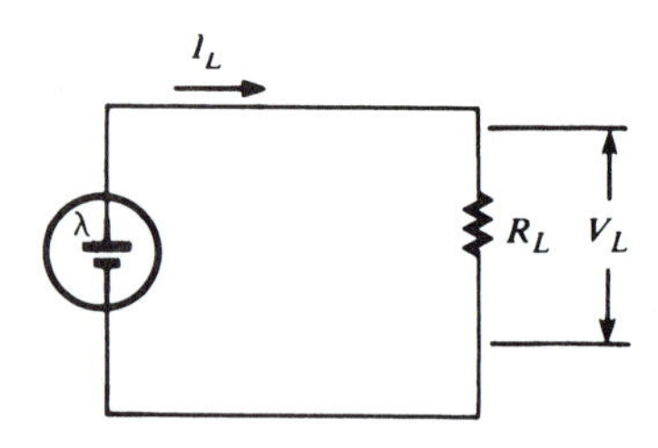

Figure 18–32
Solar cell with load connected

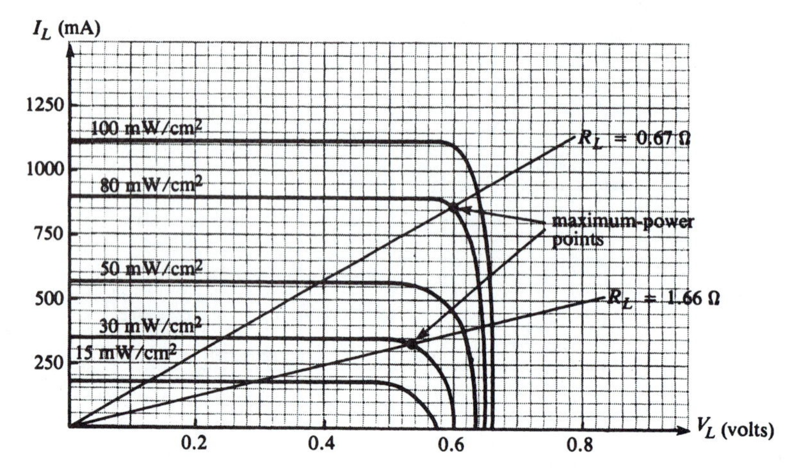

Figure 18–33
Typical characteristics of a 3-in. solar cell

delivered when the value of R_L is such that the line

$$V_L = I_L R_L \qquad \textbf{(18–32)}$$

intersects that point of maximum power. In Figure 18–33 it can be seen that a load resistance of 0.67 Ω results in maximum power when the light intensity is 80 mW/cm^2. However, a load resistance of 1.66 Ω is required to achieve maximum power at 15 mW/cm^2.

In solar-power applications, solar cells are connected in series/parallel arrays. Parallel connections increase the total current that can be supplied by the array, and series connections increase the voltage.

Example 18–14

An array of solar cells is to be constructed so it will deliver 4 A at 6 V under a certain light intensity. If each cell delivers 0.8 A at 0.6 V under the given intensity, how many cells are required and how should they be connected?

Solution. To obtain a 6-V output from cells that produce 0.6 V each, we require $6/0.6 = 10$ series-connected cells.

The current in a series string of cells is the same as the current in any one cell: 0.8 A. To obtain 4 A, we require $4/0.8 = 5$ parallel paths, each of which contributes 0.8 A.

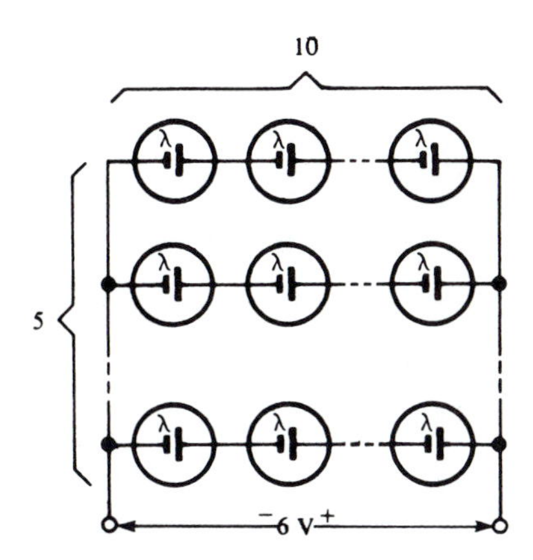

Figure 18–34
(Example 18–14)

Thus, the total number of cells required is $10 \times 5 = 50$. These are arranged in 5 parallel branches, each consisting of 10 series-connected cells, as shown in Figure 18–34.

Figure 18–35 shows specifications for a commercially available array of 36 series-connected solar cells, the Solavolt MSVM4010. Notice the dependence of the characteristics on temperature as well as solar intensity ("irradiance").

Light-Activated SCR (LASCR)

A light-activated SCR (LASCR) is constructed like a conventional SCR except that the reverse-biased NP junction in the center of the structure is exposed to light (see Figure 18–5). Light energy causes reverse current to flow across the junction in the same manner as it does in a photodiode. Recall that the SCR will regeneratively switch to its on state if the leakage current becomes sufficiently large. Thus, the LASCR can be triggered on if a sufficiently intense light falls on the exposed junction. Since the leakage current also increases with temperature, the LASCR turns on at lower light levels when temperature increases. Most LASCRs have an accessible gate terminal so they can be controlled conventionally by an external circuit.

Figure 18–36 shows the schematic symbol for an LASCR and a typical application. When switch S_1 is closed, the lamp illuminates the LASCR and triggers it on, thus allowing current to flow in the load. Momentary pushbutton switch S_2 is used to reset the LASCR by reducing its current below the holding current. The principal advantage of this arrangement is the complete electrical isolation it provides between the control circuit and the load.

Light-Emitting Diodes (LEDs)

When current flows through a forward-biased PN junction, free electrons cross from the N side and recombine with holes on the P side. Recall from Chapter 2 that free electrons are in the *conduction* band and therefore have greater energy than holes, which are in the *valence* band. When an electron in the conduction band recombines with a hole in the valence band, it releases energy as it falls into that lower energy state. The energy is released in the form of heat and light. In some materials, such as silicon, most of the energy is converted to heat, while in

ELECTRICAL SPECIFICATIONS — all tests conducted with an air mass (AM) of 1.5

		TEST CONDITIONS						
		Cell temp (°C)	Irradiance (mW/cm²)	Cell temp (°C)	Irradiance (mW/cm²)	Cell temp (°C)	Irradiance (mW/cm²)	
Specification	Symbol	25	100	48(NOCT)*	100	48(NOCT)*	80	Unit
Maximum Power Point Voltage	Vmp	16.5 ± .75		14.7 ± .75		14.7 ± .75		Vdc
Maximum Power Point Current	Imp	2.45 ± .25		2.50 ± .25		2.00 ± .20		Adc
Maximum Power (Vmp x Imp)	Pmp (typ)	40.4		36.7		29.4		W
Open Circuit Voltage	Voc (typ)	20.6		18.8		18.8		Vdc
Short Circuit Current	Isc (typ)	2.74		2.80		2.24		Adc

*Normal Operating Cell Temperature (NOCT) 48°C (118°F). This is cell temperature with 20°C air temperature, 80 mW/cm² irradiance, 1 m/s wind velocity and module tilted at 45° to the horizontal.

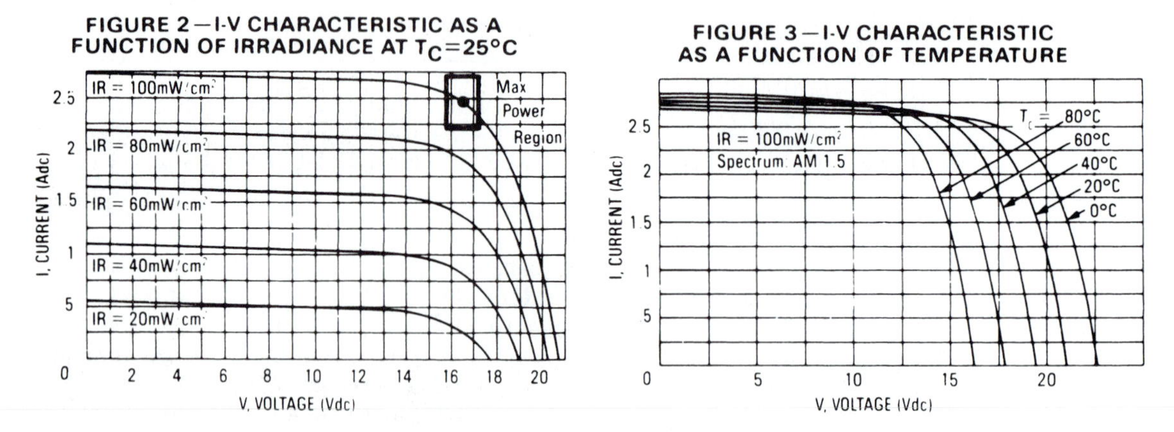

Figure 18–35
Specifications for the Solavolt model MSVM4010 solar cell module (Courtesy of Solavolt International)

others it is in the form of light. If the material is translucent and if the light energy released is visible, then a PN junction having those properties is called a *light-emitting diode* (LED). This conversion of electrical energy to light energy is an example of the phenomenon called *electroluminescence.*

Figure 18–36
The schematic symbol and an application for the light-activated SCR (LASCR)

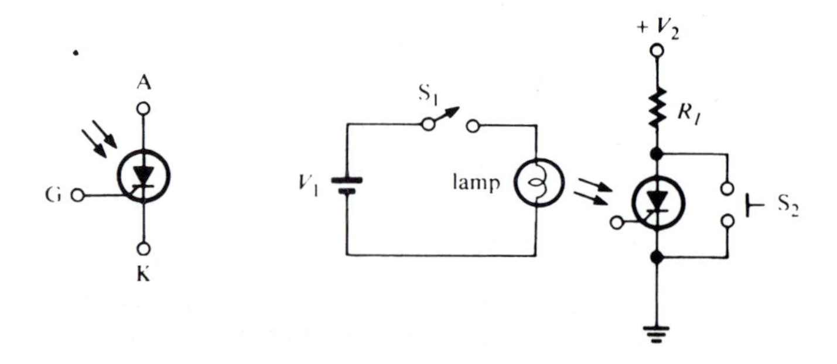

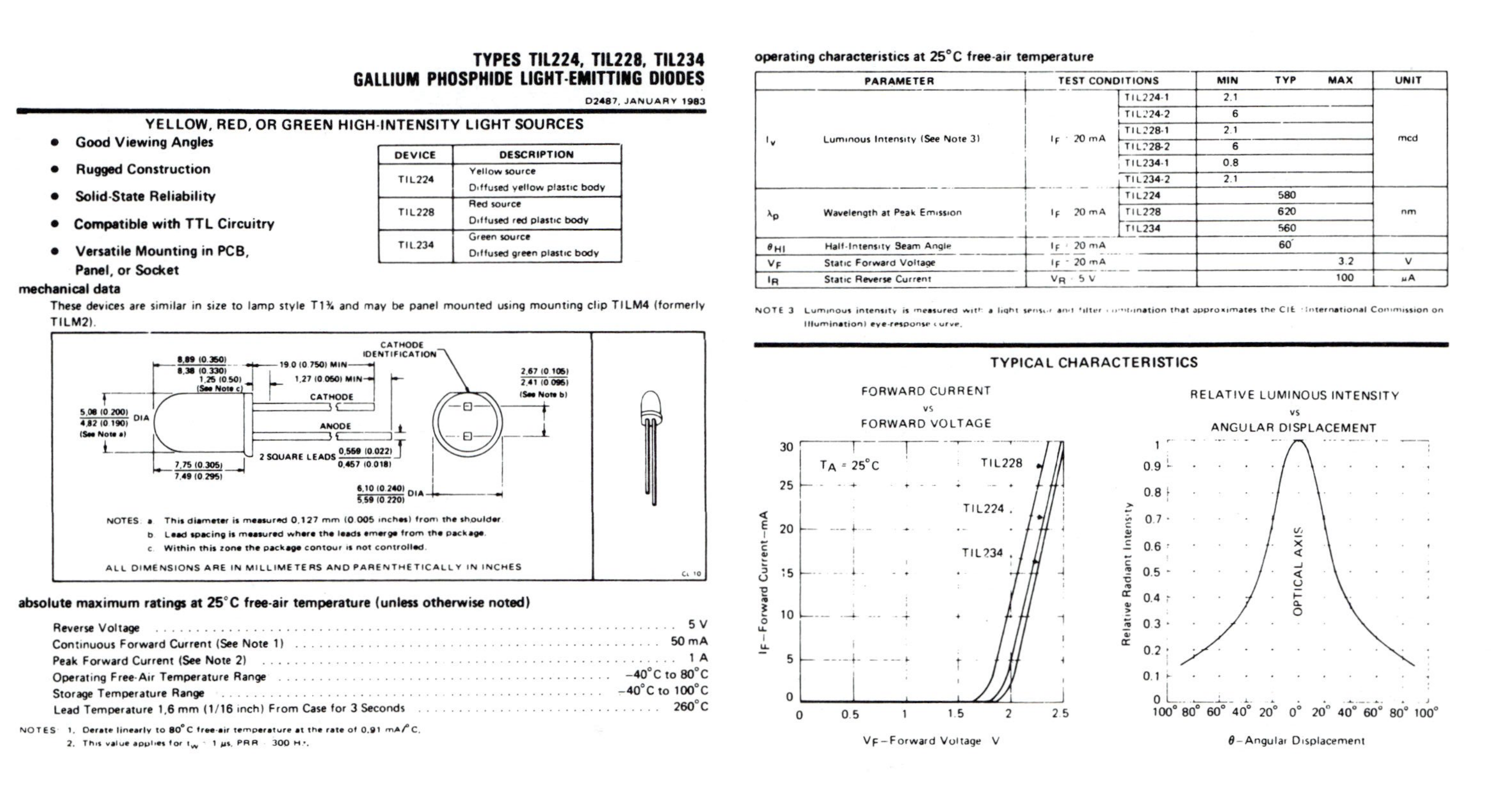

TYPES TIL224, TIL228, TIL234
GALLIUM PHOSPHIDE LIGHT-EMITTING DIODES

D2487, JANUARY 1983

YELLOW, RED, OR GREEN HIGH-INTENSITY LIGHT SOURCES

- **Good Viewing Angles**
- **Rugged Construction**
- **Solid-State Reliability**
- **Compatible with TTL Circuitry**
- **Versatile Mounting in PCB, Panel, or Socket**

DEVICE	DESCRIPTION
TIL224	Yellow source Diffused yellow plastic body
TIL228	Red source Diffused red plastic body
TIL234	Green source Diffused green plastic body

mechanical data

These devices are similar in size to lamp style T1¾ and may be panel mounted using mounting clip TILM4 (formerly TILM2).

NOTES: a. This diameter is measured 0.127 mm (0.005 inches) from the shoulder.
b. Lead spacing is measured where the leads emerge from the package.
c. Within this zone the package contour is not controlled.

ALL DIMENSIONS ARE IN MILLIMETERS AND PARENTHETICALLY IN INCHES

absolute maximum ratings at 25°C free-air temperature (unless otherwise noted)

Reverse Voltage	5 V
Continuous Forward Current (See Note 1)	50 mA
Peak Forward Current (See Note 2)	1 A
Operating Free-Air Temperature Range	−40°C to 80°C
Storage Temperature Range	−40°C to 100°C
Lead Temperature 1,6 mm (1/16 inch) From Case for 3 Seconds	260°C

NOTES: 1. Derate linearly to 80°C free-air temperature at the rate of 0.91 mA/°C.
2. This value applies for t_w = 1 μs, PRR = 300 Hz.

operating characteristics at 25°C free-air temperature

	PARAMETER	TEST CONDITIONS		MIN	TYP	MAX	UNIT
I_v	Luminous Intensity (See Note 3)	I_F = 20 mA	TIL224-1	2.1			mcd
			TIL224-2	6			
			TIL228-1	2.1			
			TIL228-2	6			
			TIL234-1	0.8			
			TIL234-2	2.1			
λ_p	Wavelength at Peak Emission	I_F = 20 mA	TIL224		580		nm
			TIL228		620		
			TIL234		560		
θ_{HI}	Half-Intensity Beam Angle	I_F = 20 mA			60°		
V_F	Static Forward Voltage	I_F = 20 mA				3.2	V
I_R	Static Reverse Current	V_R = 5 V				100	μA

NOTE 3: Luminous intensity is measured with a light sensor and filter combination that approximates the CIE (International Commission on Illumination) eye-response curve.

TYPICAL CHARACTERISTICS

Figure 18–37
Light-emitting–diode (LED) specifications (Courtesy of Texas Instruments, Inc.)

The difference in energy between the conduction band and the valence band is called the *energy gap* (see Figure 2–3), and its value, which depends on the material, governs the wavelength of the emitted light:

$$\lambda = \frac{1.24 \times 10^{-6}}{\text{E.G.}} \tag{18–33}$$

where E.G. is the energy gap in electronvolts (eV). In silicon, E.G. $\approx$ 1.1 eV, and

$$\lambda = \frac{1.24 \times 10^{-6}}{1.1} = 1.13 \times 10^{-6}\ \text{m}$$

This wavelength is in the infrared region of the light spectrum. Since infrared is not visible and since most energy is released as heat, silicon is not used in the fabrication of light-emitting diodes. For the same reasons, germanium is not used. Instead, *gallium arsenide* (GaAs), *gallium phosphide* (GaP), and *gallium arsenide phosphide* (GaAsP) are commonly used. The color of the light emitted by these materials can be controlled by the type and degree of doping. Red, green, and yellow LEDs are commonly available.

Figure 18–37 shows typical manufacturer's specifications for a line of GaP LEDs. Note that the forward voltage drop across an LED is considerably greater than that across a silicon or germanium diode. The drop is typically between 2 and 3 V, depending on forward current. Note also that the maximum permissible reverse voltage is relatively small, 5 V in this case.

Example 18–15

Figure 18–38 shows an LED *driver* circuit designed to illuminate the LED when the transistor is turned on (saturated) by a 5-V positive input at its base. Find the values of R_B and R_C that should be used if the LED is to be illuminated by 20 mA of forward current. Assume that the forward drop across the LED is 2.5 V and that the silicon transistor has $\beta = 50$.

Solution. When the transistor is saturated,

$$I_C = 20\ \text{mA} = \frac{V_{CC} - V_D}{R_C} = \frac{(5\ \text{V}) - (2.5\ \text{V})}{R_C} = \frac{2.5\ \text{V}}{R_C}$$

$$R_C = \frac{2.5\ \text{V}}{20\ \text{mA}} = 125\ \Omega$$

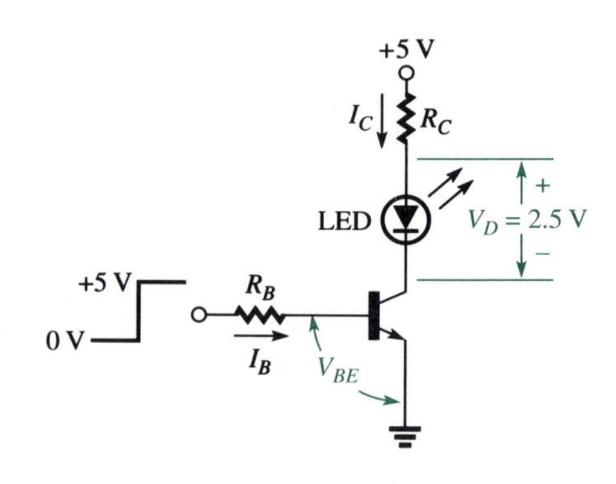

Figure 18–38
(Example 18–15)

$$I_B = I_C/\beta = 20\ \text{mA}/50 = 0.4\ \text{mA}$$

$$R_B = \frac{5 - V_{BE}}{I_B} = \frac{(5 - 0.7)\ \text{V}}{0.4\ \text{mA}} = 10.75\ \text{k}\Omega$$

LEDs are used in visual displays of all kinds. The principal disadvantage of the LED is that it draws considerable current in comparison to the types of low-power circuits with which it is typically used. For that reason, LEDs are no longer widely used in such low-power devices as calculators and watches. In some applications, power is conserved by pulsing the LEDs on and off at a rapid rate, rather than supplying them with a steady drive current. The LEDs appear to be continuously illuminated because of the eye's *persistence,* that is, its maintenance of the perception of light during the short intervals during which the LEDs are off.

A very popular use for LEDs is in the construction of *seven-segment displays,* such as those illustrated in the specification sheet shown in Figure 18–39. Notice that each of the seven segments is an LED identified by a letter from A through G. By illuminating selected segments, any numeral from 0 through 9 can be displayed. For example, the numeral 3 is displayed by illuminating only those LEDs corresponding to segments A, B, C, D, and G. The seven-segment display must be driven from logic circuitry (called a *decoder*) that supplies current to the correct combination of segments, depending on the numeral to be displayed. In some displays, the decoder circuitry is built-in. In Figure 18–39, note that both *common-anode* (TIL312) and *common-cathode* (TIL313) configurations are available. In the TIL312, a positive voltage is connected to the common anode, so selected LEDs are illuminated by making their respective cathodes low (0 volts). In the TIL313, the common cathode is held low and LEDs are illuminated by making their respective anodes high. Other LED configurations are available for *hexadecimal* displays (0 through 9 and A through F) and for *alphanumeric* displays (numerals and all alphabetic characters).

Optocouplers

An optocoupler combines a light-emitting device with a light-sensitive device in one package. One of the simplest examples is an LED packaged with a phototransistor. The LED is illuminated by an input circuit and the phototransistor, responding to the light, drives an output circuit. Thus, the input and output circuits are *coupled* by light energy alone. The principal advantage of this arrangement is the excellent electrical isolation it provides between the input and output. In fact, these devices are often called *optoisolators.* Other examples of optocouplers include LEDs packaged with photodarlingtons and LASCRs.

Figure 18–40 shows the circuit symbol and two methods used to construct an optocoupler consisting of an LED and a phototransistor. In each structure, the LED is actually a GaAs *infrared* emitter producing light with wavelengths in the vicinity of 0.9×10^{-6} m. In one device, the coupling medium is a special "infrared glass," while the other uses an air gap for better electrical isolation.

An important specification for an optocoupler is its *current-transfer ratio:* the ratio of output (phototransistor) current to input (LED) current, often expressed as a percent. The ratio varies with the value of LED current and can range from 0.1 or less to several hundred in photodarlington devices. Figure 18–41 shows a set of specification sheets for the TIL111, -114, -116, and -117 series of optocouplers. Note the high voltage isolation these devices provide: up to 2500 V. Note also that

TYPES TIL312 THRU TIL315, TIL327, TIL328, TIL333 THRU TIL335, TIL339 THRU TIL341 NUMERIC DISPLAYS

D1924, SEPTEMBER 1981 – REVISED DECEMBER 1982

SOLID-STATE DISPLAYS WITH RED, GREEN, OR YELLOW CHARACTERS

- 7.62-mm (0.300-inch) Character Height
- Continuous Uniform Segments
- Wide Viewing Angle
- High Contrast
- Yellow and Green Displays are Categorized for Uniformity of Luminous Intensity and Wavelength among Units within Each Category

	SEVEN SEGMENTS WITH RIGHT AND LEFT DECIMALS, COMMON ANODE	SEVEN SEGMENTS WITH RIGHT DECIMAL, COMMON CATHODE	PULSE/MINUS ONE WITH LEFT DECIMAL
RED	TIL312	TIL313	TIL327
GREEN	TIL314	TIL315	TIL328
RED•	TIL333	TIL334	TIL335
YELLOW	TIL339	TIL340	TIL341

Red• stands for high efficiency red

mechanical data

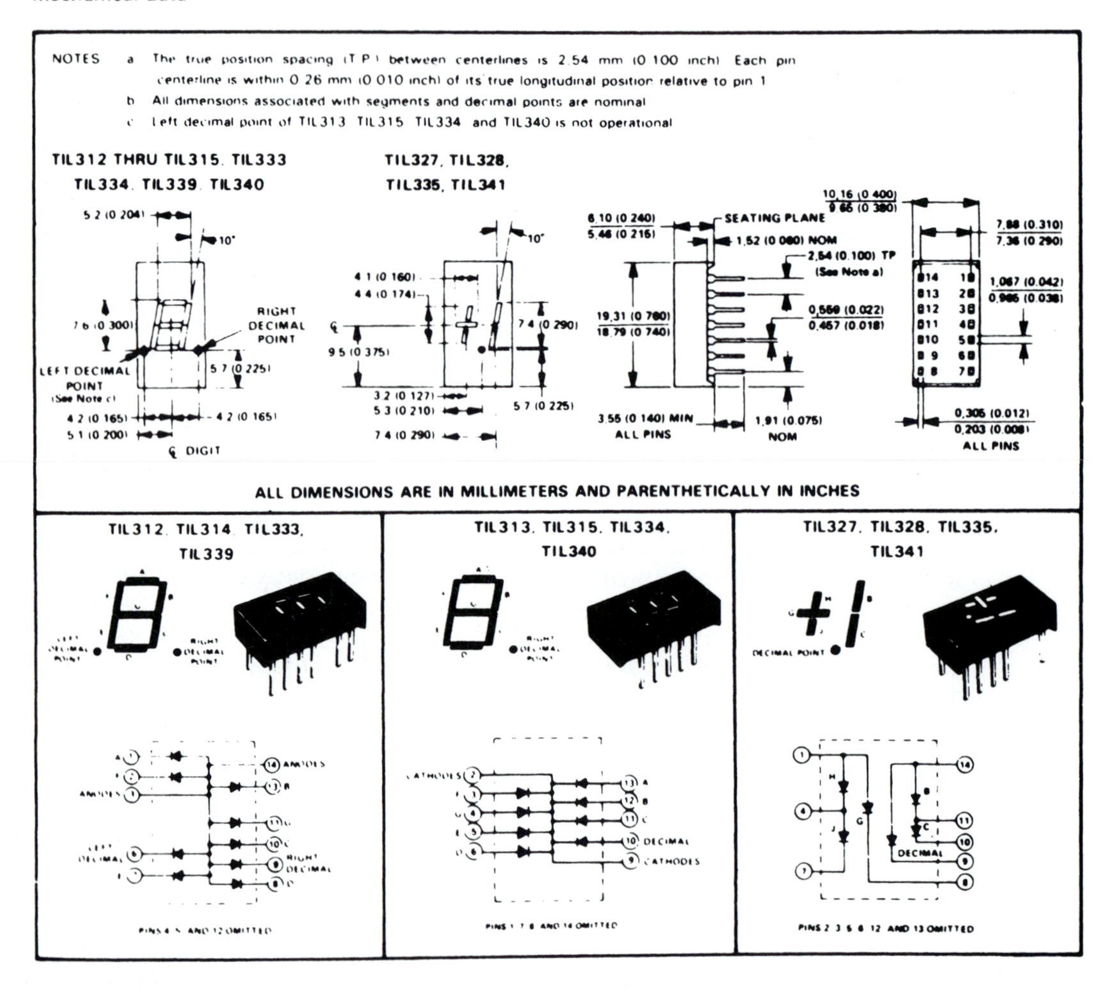

Figure 18–39

Seven-segment–display specifications (Courtesy of Texas Instruments, Inc.)

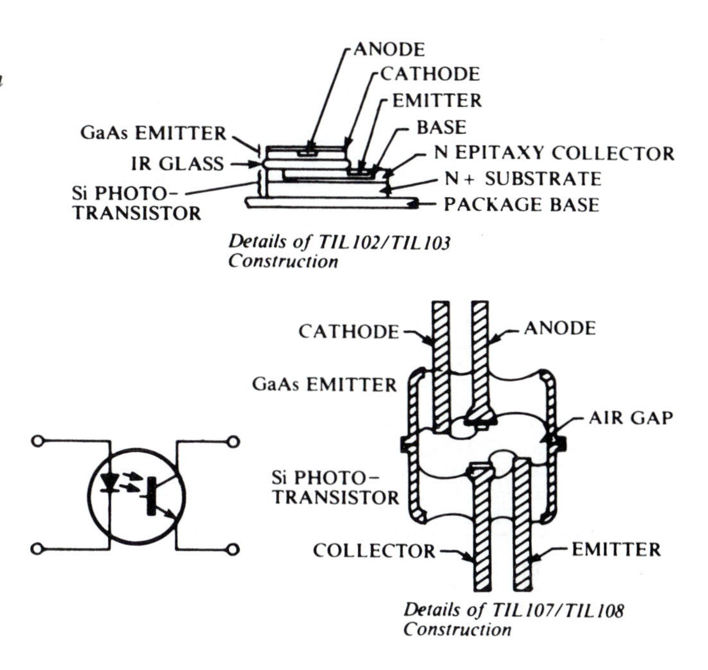

Figure 18–40
Optocoupler symbol and two construction methods (Courtesy of Texas Instruments, Inc.)

the base terminal of the phototransistor is externally accessible, so the coupler can be operated in a photodiode mode by taking the output between the base and collector terminals, as shown in Test Circuit B. The specifications for switching characteristics show that the response time is improved by a factor of 5 when in photodiode operation.

Example 18–16

Find the current-transfer ratio (CTR) of the TIL117 optocoupler when the diode current is 20 mA and the collector-to-emitter voltage of the phototransistor is 10 V. Assume that $I_B = 0$.

Solution. Referring to the log-log plot of "collector current versus input-diode forward current," shown in the "typical characteristics" section of the specifications in Figure 18–41, we see that $I_C = 20$ mA when $I_F = 20$ mA (from the TIL117 characteristic). Note that the characteristic applies to the case when $V_{CE} = 10$ V and $I_B = 0$. Thus,

$$\text{CTR}(\%) = \frac{I_C}{I_F} \times 100\% = \frac{20\text{ mA}}{20\text{ mA}} \times 100\% = 100\%$$

Liquid-Crystal Displays (LCDs)

Liquid crystals have been called "the fourth state of matter" (after solids, liquids, and gases) because they have certain crystal properties normally found in solids, yet flow like liquids. In recent years, they have found wide application in the construction of low-power visual displays, such as those found in watches and calculators. Unlike LEDs and other electroluminescent devices, LCDs do not gener-

TYPES TIL111, TIL114, TIL116, TIL117 OPTOCOUPLERS

D1607, NOVEMBER 1973—REVISED FEBRUARY 1983

COMPATIBLE WITH STANDARD TTL INTEGRATED CIRCUITS

- **Gallium Arsenide Diode Infrared Source Optically Coupled to a Silicon N-P-N Phototransistor**
- **High Direct-Current Transfer Ratio**
- **High-Voltage Electrical Isolation . . . 1.5-kV or 2.5-kV Rating**
- **Plastic Dual-In-Line Package**
- **High-Speed Switching: t_r = 5 μs, t_f = 5 μs Typical**

mechanical data

The package consists of a gallium arsenide infrared-emitting diode and an n-p-n silicon phototransistor mounted on a 6-lead frame encapsulated within an electrically nonconductive plastic compound. The case will withstand soldering temperature with no deformation and device performance characteristics remain stable when operated in high-humidity conditions. Unit weight is approximately 0.52 grams.

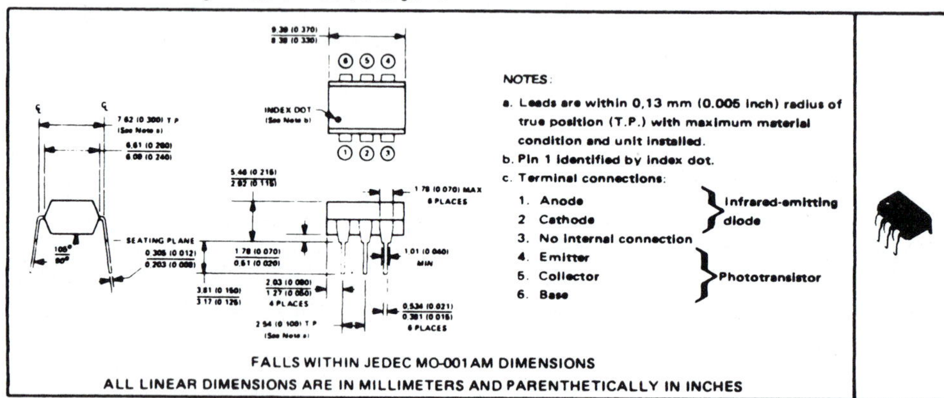

INDEX DOT
SEATING PLANE
NOTES:
a. Leads are within 0,13 mm (0.005 inch) radius of true position (T.P.) with maximum material condition and unit installed.
b. Pin 1 identified by index dot.
c. Terminal connections:
1. Anode
2. Cathode
3. No internal connection
4. Emitter
5. Collector
6. Base
Infrared-emitting diode
Phototransistor
FALLS WITHIN JEDEC MO-001AM DIMENSIONS
ALL LINEAR DIMENSIONS ARE IN MILLIMETERS AND PARENTHETICALLY IN INCHES

7

OPTOCOUPLERS

absolute maximum ratings at 25°C free-air temperature (unless otherwise noted)

Input-to-Output Voltage: TIL111	±1.5 kV
TIL114, TIL116, TIL117	±2.5 kV
Collector-Base Voltage	70 V
Collector-Emitter Voltage (See Note 1)	30 V
Emitter-Collector Voltage	7 V
Emitter-Base Voltage	7 V
Input-Diode Reverse Voltage	3 V
Input-Diode Continuous Forward Current at (or below) 25°C Free-Air Temperature (See Note 2)	100 mA
Continuous Power Dissipation at (or below) 25°C Free-Air Temperature:	
Infrared-Emitting Diode (See Note 3)	150 mW
Phototransistor (See Note 4)	150 mW
Total, Infrared-Emitting Diode plus Phototransistor (See Note 5)	250 mW
Storage Temperature Range	−55°C to 150°C
Lead Temperature 1,6 mm (1/16 Inch) from Case for 10 Seconds	260°C

NOTES
1. This value applies when the base-emitter diode is open-circuited.
2. Derate linearly to 100°C free-air temperature at the rate of 1.33 mA/°C.
3. Derate linearly to 100°C free-air temperature at the rate of 2 mW/°C.
4. Derate linearly to 100°C free-air temperature at the rate of 2 mW/°C.
5. Derate linearly to 100°C free-air temperature at the rate of 3.33 mW/°C.

TEXAS INSTRUMENTS
INCORPORATED
POST OFFICE BOX 225012 • DALLAS, TEXAS 75265

Figure 18–41
Optocoupler specifications (Courtesy of Texas Instruments, Inc.)

electrical characteristics at 25° C free-air temperature

PARAMETER			TEST CONDITIONS	TIL111 TIL114			TIL116			TIL117			UNIT
				MIN	TYP	MAX	MIN	TYP	MAX	MIN	TYP	MAX	
$V_{(BR)CBO}$	Collector-Base Breakdown Voltage		I_C = 10 µA, I_E = 0, I_F = 0	70			70			70			V
$V_{(BR)CEO}$	Collector-Emitter Breakdown Voltage		I_C = 1 mA, I_B = 0, I_F = 0	30			30			30			V
$V_{(BR)EBO}$	Emitter-Base Breakdown Voltage		I_E = 10 µA, I_C = 0, I_F = 0	7			7			7			V
I_R	Input Diode Static Reverse Current		V_R = 3 V			10			10			10	µA
$I_{C(on)}$	On-State Collector Current	Phototransistor Operation	V_{CE} = 0.4 V, I_F = 16 mA, I_B = 0	2	7								mA
			V_{CE} = 10 V, I_F = 10 mA, I_B = 0				2	5		5	9		mA
		Photodiode Operation	V_{CB} = 0.4 V, I_F = 16 mA, I_E = 0	7	20		7	20		7	20		µA
$I_{C(off)}$	Off-State Collector Current	Phototransistor Operation	V_{CE} = 10 V, I_F = 0, I_B = 0		1	50		1	50		1	50	nA
		Photodiode Operation	V_{CB} = 10 V, I_F = 0, I_E = 0		0.1	20		0.1	20		0.1	20	nA
h_{FE}	Transistor Static Forward Current Transfer Ratio		V_{CE} = 5 V, I_C = 10 mA, I_F = 0	100	300					200	550		
			V_{CE} = 5 V, I_C = 100 µA, I_F = 0				100	300					
V_F	Input Diode Static Forward Voltage		I_F = 16 mA		1.2	1.4					1.2	1.4	V
			I_F = 60 mA					1.25	1.5				V
$V_{CE(sat)}$	Collector-Emitter Saturation Voltage		I_C = 2 mA, I_F = 16 mA, I_B = 0		0.25	0.4							V
			I_C = 2.2 mA, I_F = 15 mA, I_B = 0					0.25	0.4				V
			I_C = 0.5 mA, I_F = 10 mA, I_B = 0								0.25	0.4	V
r_{IO}	Input-to-Output Internal Resistance		$V_{in\text{-}out}$ = ±1.5 kV for TIL111, ±2.5 kV for all others, See Note 6	10^{11}			10^{11}			10^{11}			Ω
C_{io}	Input-to-Output Capacitance		$V_{in\text{-}out}$ = 0, f = 1 MHz, See Note 6		1	1.3		1	1.3		1	1.3	pF

NOTE 6 These parameters are measured between both input diode leads shorted together and all the phototransistor leads shorted together

switching characteristics at 25° C free-air temperature

PARAMETER			TEST CONDITIONS	TIL111 TIL114			TIL116			TIL117			UNIT
				MIN	TYP	MAX	MIN	TYP	MAX	MIN	TYP	MAX	
t_r	Rise Time	Phototransistor Operation	V_{CC} = 10 V, $I_{C(on)}$ = 2 mA, R_L = 100 Ω, See Test Circuit A of Figure 1		5	10		5	10		5	10	µs
t_f	Fall Time				5	10		5	10		5	10	µs
t_r	Rise Time	Photodiode Operation	V_{CC} = 10 V, $I_{C(on)}$ = 20 µA, R_L = 1 kΩ, See Test Circuit B of Figure 1		1			1			1		µs
t_f	Fall Time				1			1			1		µs

Figure 18–41
(Continued)

PARAMETER MEASUREMENT INFORMATION

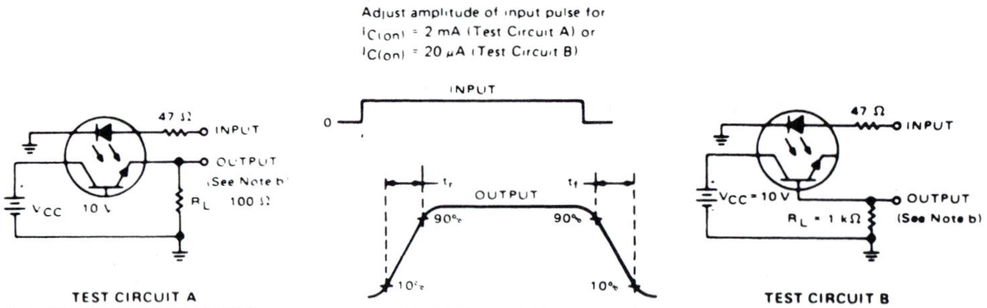

Adjust amplitude of input pulse for
$I_{C(on)}$ = 2 mA (Test Circuit A) or
$I_{C(on)}$ = 20 μA (Test Circuit B)

INPUT
0
47 Ω
INPUT
OUTPUT
(See Note b)
V_{CC} 10 V
R_L 100 Ω
t_r
t_f
OUTPUT
90%
90%
10%
10%
47 Ω
INPUT
V_{CC} = 10 V
OUTPUT
R_L = 1 kΩ
(See Note b)

TEST CIRCUIT A
PHOTOTRANSISTOR OPERATION

VOLTAGE WAVEFORMS

TEST CIRCUIT B
PHOTODIODE OPERATION

NOTES: a. The input waveform is supplied by a generator with the following characteristics: Z_{out} = 50 Ω, t_r ≤ 15 ns, duty cycle ≈ 1%, t_w = 100 μs.
b. The output waveform is monitored on an oscilloscope with the following characteristics: t_r ≤ 12 ns, R_{in} ≥ 1 MΩ, C_{in} ≤ 20 pF.

FIGURE 1–SWITCHING TIMES

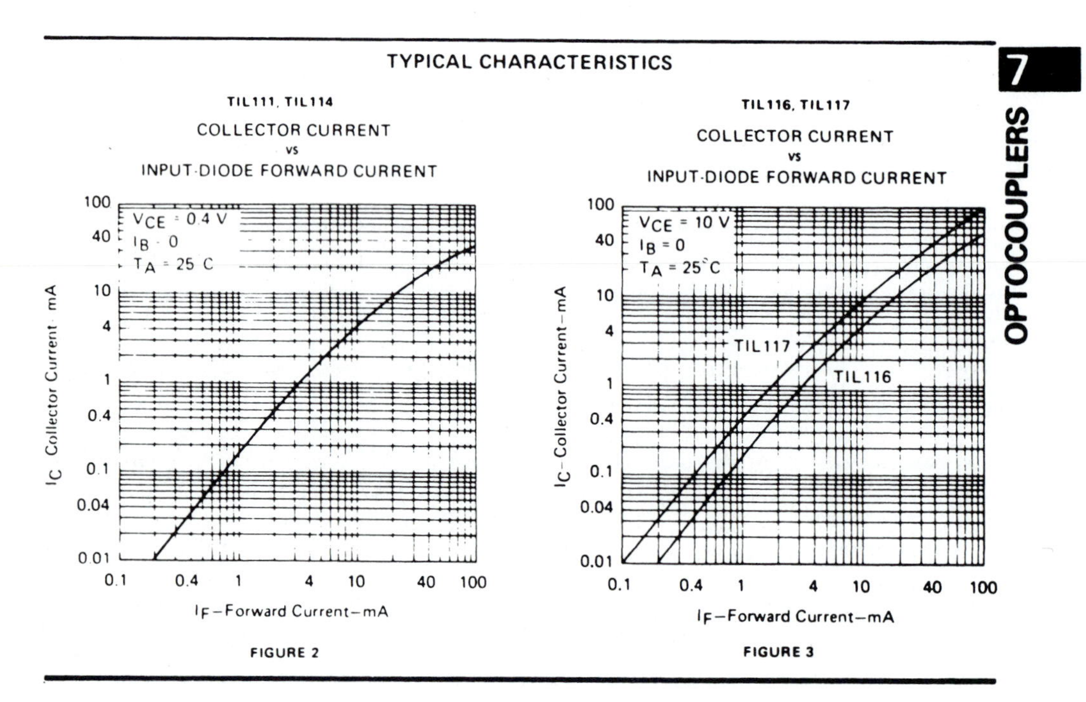

7
OPTOCOUPLERS

Figure 18–41
(Continued)

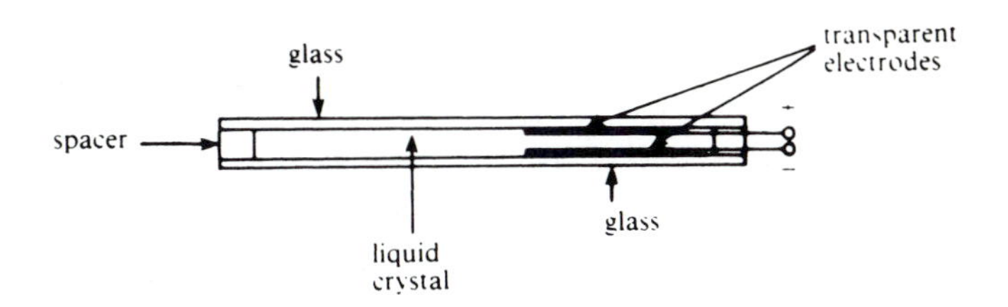

Figure 18–42
Basic construction of a liquid-crystal display

ate light energy, but simply alter or control existing light to make selected areas appear bright or dark.

There are two fundamental ways in which liquid crystals are used to control properties of light and thereby alter its appearance. In the *dynamic-scattering* method, the molecules of the liquid crystal acquire a random orientation by virtue of an externally applied electric potential. As a result, light passing through the material is reflected in many different directions and has a bright, frosty appearance as it emerges. In the *absorption* method, the molecules are oriented in such a way that they alter the *polarization* of light passing through the material. Polarizing filters are used to absorb or pass the light, depending on the polarization it has been given, so light is visible only in those regions where it can emerge from the filter.

The basic construction of a liquid-crystal display is shown in Figure 18–42. A thin layer of liquid crystal, 10–20 μm thick, is encapsulated between two glass sheets. To enable application of an electric field across the crystal, a transparent conducting material is deposited on the inside surface of each glass sheet. The conducting material forms the *electrodes* to which an external voltage is connected when it is desired to alter the molecular structure of the liquid crystal. The electrodes are etched in patterns or in individually accessible segments that can be selectively energized to create a desired display, similar to a seven-segment LED display. Materials used to form the electrodes include *stannic oxide* (SnO_2) and *indium oxide* (In_2O_3).

Some liquid-crystal displays are designed so that light passes completely through them, while others have a mirrored surface that reflects light back to the viewer. In the first instance, called the *transmissive* mode of operation, light originates from a light source on one side of the crystal and is altered in the desired pattern as it passes through to emerge on the other side. In the *reflective* mode, light enters one side, is altered within the material, and is reflected by a mirror to emerge from the same side it entered. This mode is particularly useful for small displays such as watches, where ambient light is present on one side only.

Figure 18–43 shows a dynamic-scattering LCD operated in the transmissive mode. In the region where no electric field is applied, the rod-shaped molecules have the common alignment shown, and in the region that has been activated by an externally applied field, the molecules have random orientations. The light source is displaced in such a way that the unactivated region appears dark. In the activated region, the molecules cause light to be reflected in many directions (scattered), and it escapes with a bright, diffuse appearance. Thus the pattern appears bright on a dark background. A special light-control film can be placed on the illuminated glass to permit the use of a diffuse light source.

A dynamic-scattering LCD operated in the reflective mode has essentially the same construction shown in Figure 18–43, except that a mirrored surface replaces or is added behind one of the glass sheets. However, unwanted reflections limit the readability of displays of this type, which must therefore be fitted with special *retroreflectors* in practical applications.

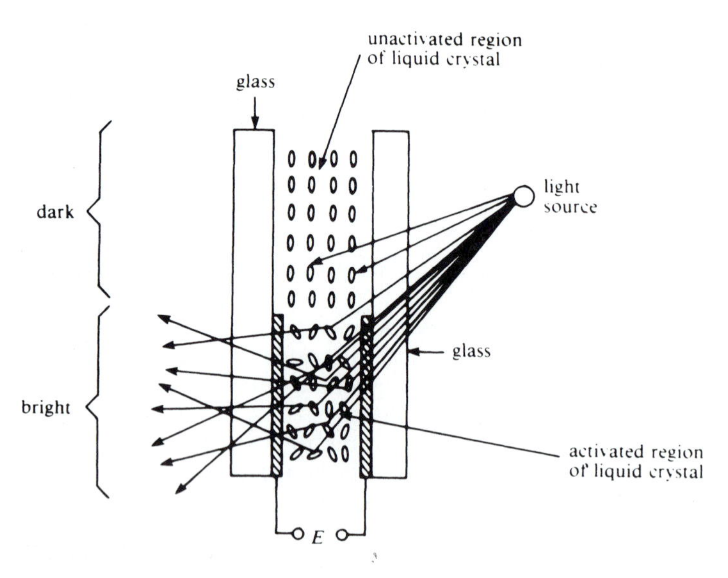

Figure 18–43
Dynamic-scattering LCD; transmissive mode

Liquid-crystal displays that depend on light absorption rather than light scattering are constructed using so-called *twisted-nematic* crystals. The electrodes in these displays are coated so that the molecules of the liquid crystal undergo a 90° rotation from one side of the cell to the other. As a result, light passing through the material undergoes a 90° change in polarization. However, if the crystal is activated by an externally applied electric field, the molecules are reoriented in such a way that the change in polarization does not occur. The absorption-type LCD operates on the principle that horizontally polarized light is not visible through a vertically polarizing filter, i.e., it is absorbed. As shown in Figure 18–44, light enters the (transmissive) LCD through a vertical polarizer and can exit through another vertical polarizer provided the liquid crystal is activated. In the region where the crystal is not activated, the vertically polarized light becomes horizontally polarized and is absorbed in the vertical filter. Thus, the unactivated region appears dark and the activated region appears bright. Since it is the electric field resulting from an externally applied voltage that reorients the molecule, the twisted-nematic display is often called a *field-effect* type.

Figure 18–45 shows a light-absorbing, field-effect LCD operated in the reflective mode. As in the transmissive LCD, incident light is vertically polarized, but in this case a horizontal polarizer is placed between one of the glass sheets and a reflector. The vertically polarized light becomes horizontally polarized when it passes through an unactivated region of the crystal. This light can pass through the horizontal polarizer and is reflected back into the crystal. The unactivated crystal then shifts the polarization back to vertical, so the light emerges through the vertical polarizer where it entered. The activated region of the crystal does not alter the vertical polarization, so light passing through it cannot pass the horizontal polarizer and is not reflected. As a consequence, the pattern created in the activated region appears dark on a light background.

Practical liquid-crystal displays are activated by ac waveforms because dc activation causes a plating effect that shortens their lifetime. The dynamic-scattering type

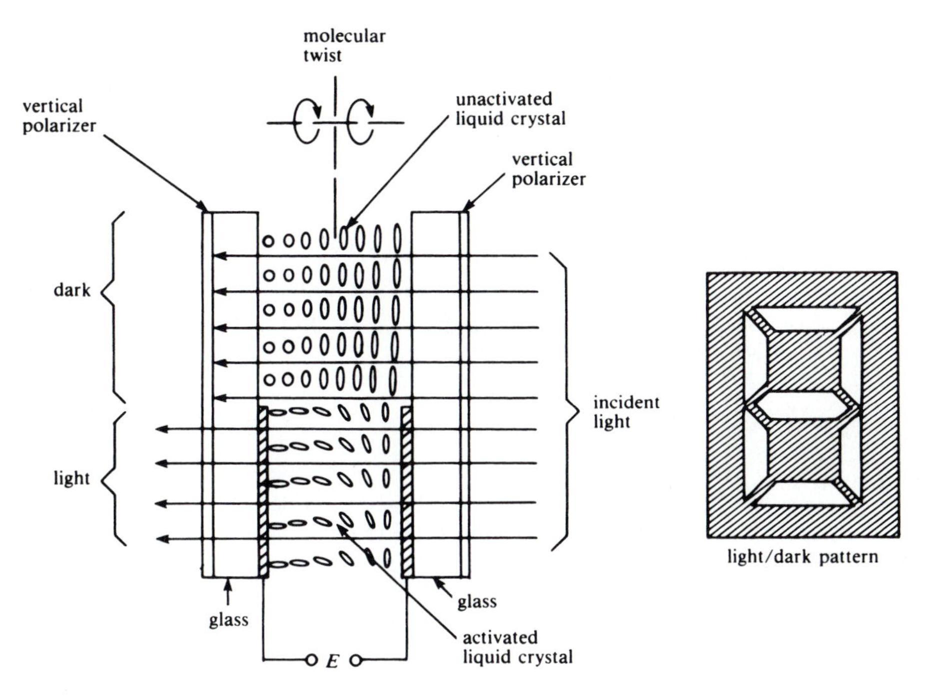

Figure 18–44
Twisted-nematic, field-effect LCD based on light absorption; transmissive-mode operation

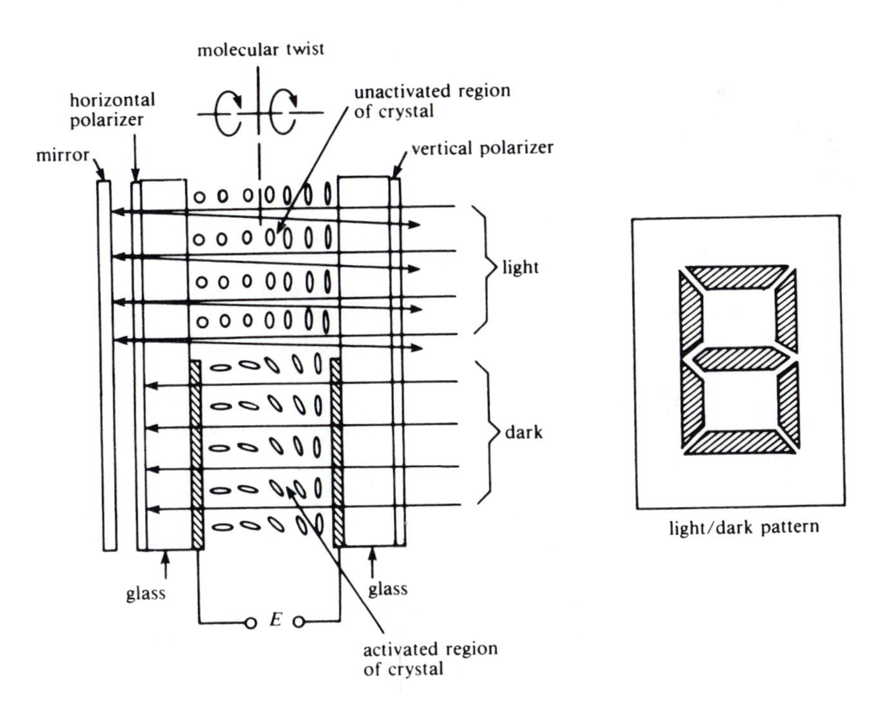

Figure 18–45
Twisted-nematic, field-effect LCD based on light absorption; reflective-mode operation

Table 18–1
Typical characteristics of liquid-crystal displays

Characteristic	Dynamic-Scattering Display (4 digits)	Twisted-Nematic Display (4½ digits)
Voltage	15–30 V	3 V
Frequency	50–60 Hz	30–1500 Hz
Operating temperature	−5°C to +65°C	−15°C to +60°C
Total current	60 μA	6 μA
Total capacitance	750 pF	2 nF
Rise time	25 ms	100 ms
Contrast ratio	20:1	50:1

requires a relatively high voltage, 25–30 V at 50–60 Hz. However, LCDs consume very little power in comparison to LED displays. One method used to activate the crystal is to drive each of the opposing electrodes with a square wave. If the square waves are in phase, then the net voltage across the crystal is zero and it remains unactivated. When the driver circuitry causes the square waves to be 180° out of phase, the net voltage across the electrodes is twice the peak-to-peak value of either wave, and the crystal is activated.

One disadvantage of a liquid-crystal display is its relatively long response time, i.e., the time required for the display to appear or disappear. LCDs are much slower in that respect than LED displays. However, in practical applications such as watches, the long response time does not impose a limitation on their usefulness. Table 18–1 shows a comparison of the principal characteristics of dynamic-scattering and twisted-nematic LCDs. Note that the twisted-nematic type requires about one-tenth the voltage and current of the dynamic-scattering type, has better contrast, but is about four times slower.

18–4 UNIJUNCTION TRANSISTORS

As shown in Figure 18–46, a unijunction transistor consists of a bar of lightly doped (high-resistivity) N material to which a heavily doped, P-type rod is attached on one side. Ohmic contacts are made at opposing ends of the N-type bar, which are called *base 1* (B_1) and *base 2* (B_2) of the transistor. The P-type rod is called the *emitter* and forms a conventional PN junction with the bar. Since this device has only one such junction, it behaves quite differently from a conventional bipolar transistor. The single junction accounts for the name *uni*junction transistor, and it is also called a *double-base diode.* Notice that the schematic symbol for a unijunction transistor (UJT) has an arrow pointing in the direction of conventional forward current through the junction. UJTs are also manufactured with N-type emitters and P-type bars.

Figure 18–47 shows external circuit connections to the UJT and its equivalent circuit. Here, the PN junction is represented by an equivalent diode. Note that base 1 is the common (ground) for both the emitter input circuit and the supply voltage, V_{BB}. The *interbase resistance* R_{BB} is defined to be the total resistance between B_1 and B_2 when the emitter circuit is open ($I_E = 0$). R_{BB} is therefore the sum of the resistances R_{B1} and R_{B2} between each base and the emitter, when $I_E = 0$:

Figure 18–46
Construction and schematic symbol of the unijunction transistor (UJT)

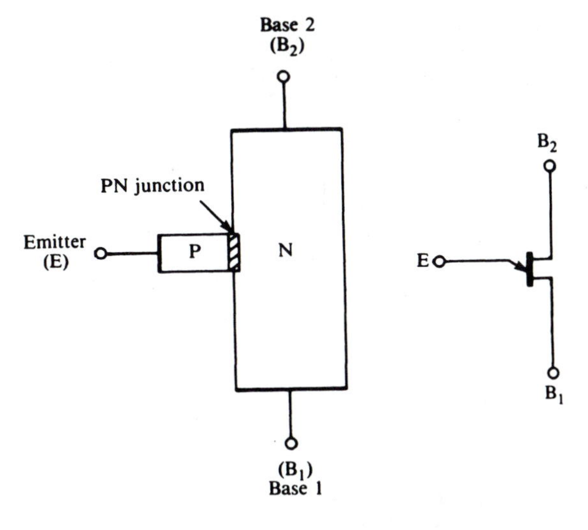

$$R_{BB} = (R_{B1} + R_{B2}) \Big|_{I_E = 0} \tag{18–34}$$

By the voltage-divider rule, the voltage V_1 across R_{B1} is (again, for $I_E = 0$)

$$V_1 = \left(\frac{R_{B1}}{R_{B1} + R_{B2}}\right) V_{BB} = \frac{R_{B1}}{R_{BB}} V_{BB} \tag{18–35}$$

The voltage-divider ratio in (18–35) is called the *intrinsic standoff ratio* η:

$$\eta = \frac{R_{B1}}{R_{BB}} \Big|_{I_E = 0} \tag{18–36}$$

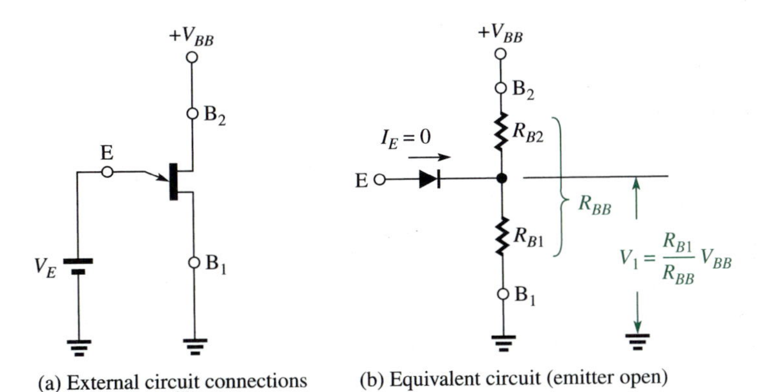

Figure 18–47
External circuit connections to a UJT

The value of η clearly depends on how close the emitter terminal is to base 2, since R_{B1} becomes larger and R_{B2} becomes smaller as the emitter is moved closer to B_2. η is typically in the range from 0.50 to 0.85, while R_{BB} may range from 4 kΩ to 12 kΩ.

Figure 18–48 shows the equivalent circuit with external emitter voltage V_E connected. Notice that R_{B1} in this case is shown as a variable resistance. If V_E is less than the positive voltage V_1 across R_{B1}, then the diode is reverse biased, and R_{B1} has essentially the same resistance as when $I_E = 0$. However, if V_E is increased to the onset of forward bias, a small amount of forward current I_E begins to flow through the emitter and into the base 1 region, causing the resistance of that region to decrease. The decrease in R_{B1} is attributable to the presence of additional charge carriers resulting from the forward current flow. If the forward bias is increased slightly, there is a sudden and dramatic reduction in R_{B1}. This phenomenon occurs because the increase in current reduces R_{B1}, which further increases the current, which further reduces R_{B1}, and so forth. In other words, a *regenerative* action occurs. The value of emitter voltage at which regeneration is initiated is called the *peak voltage*, V_P. As can be seen in Figure 18–48(a), the value that the emitter voltage must reach is V_1 plus the forward drop V_D across the diode. Since $V_1 = \eta V_{BB}$, we have

$$V_P = \eta V_{BB} + V_D \tag{18–37}$$

Figure 18–48(b) shows a characteristic curve for the unijunction transistor. Once the emitter voltage has reached V_P, emitter current increases even as V_E is made smaller. This fact is conveyed by the negative slope of the characteristic in the region to the right of the peak point (decreasing voltage accompanied by increasing current). The region is therefore appropriately referred to as the *negative-resistance* portion of the characteristic. This region is said to be *unstable* because the UJT cannot actually be operated there. The curve simply shows the combinations of V_E and I_E that the UJT undergoes during the regenerative transition. Beyond the *valley point* shown in the figure, the resistance R_{B1} has reached its minimum possible value, called the *saturation resistance.* In this saturation region, emitter

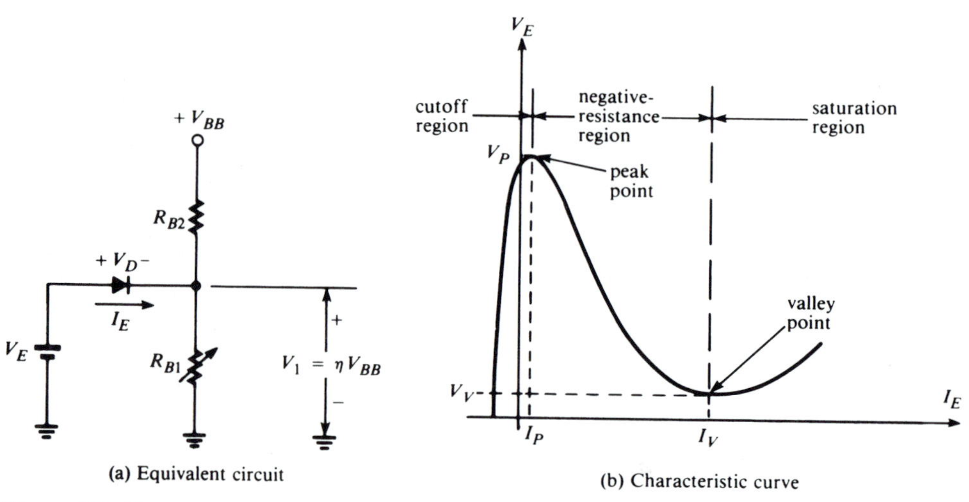

Figure 18–48
When V_E is increased to $V_P = \eta V_{BB} + V_D$, there is a sudden drop in the value of R_{B1}

Figure 18–49
A typical family of characteristic curves for a UJT

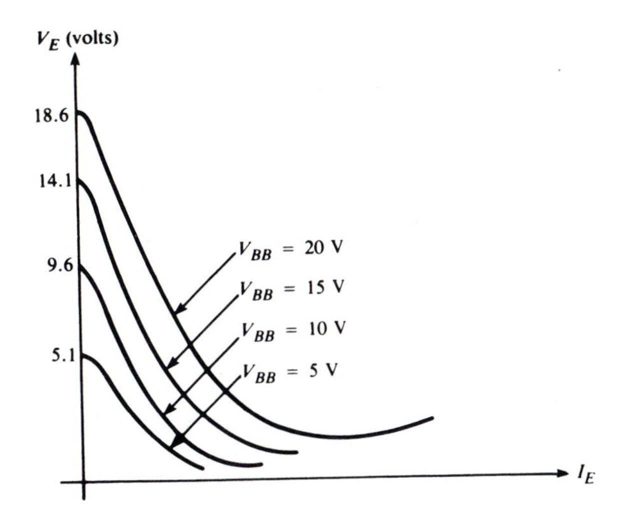

current can be increased further only by again increasing V_E. The characteristic in this region is similar to that of a conventional forward-biased diode. Note that the region to the left of the peak point is called the *cutoff* region.

In applications, the UJT is used like a voltage-controlled switch. When the input voltage is raised to V_P, the UJT "fires," or turns on, allowing a generous flow of current from V_{BB} to ground. To turn the device off, i.e., to return it to the cutoff region, the emitter current must be reduced below the valley current I_V shown in Figure 18–48(b). It must be remembered that V_P is a function of V_{BB}, as shown by equation 18–37. Thus, if V_{BB} is increased, the emitter voltage must be raised to a higher value to switch the UJT on. Figure 18–49 shows a typical family of characteristic curves, corresponding to different values of V_{BB}. (The cutoff regions are not shown.) The intersection of each characteristic curve with the V_E-axis is (approximately) the value of V_P corresponding to a particular V_{BB}, and it can be seen that V_P increases with increasing V_{BB}.

Figure 18–50(a) shows a common application of a UJT in a *relaxation* oscillator circuit. Capacitor C charges through R_E until the capacitor voltage reaches V_P. At

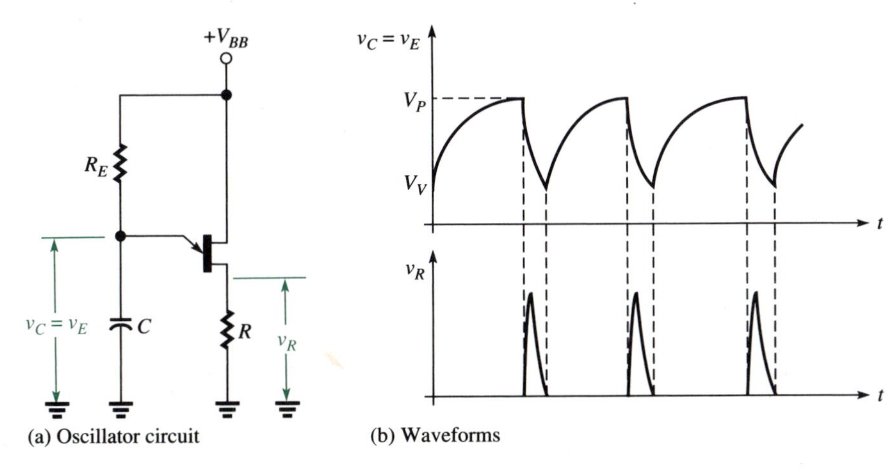

Figure 18–50
A UJT relaxation oscillator

that time, the UJT switches on, and the capacitor discharges through the emitter. The resulting surge of current through external resistor R develops a sharp voltage "spike," as shown in Figure 18–50(b). Once the capacitor has discharged, the UJT switches back to its off state, provided the emitter current drops below I_V. The capacitor then begins to recharge and the cycle is repeated. The resulting capacitor-voltage and output-voltage waveforms are shown in (b). This type of oscillator is used to generate trigger and timing pulses in a wide variety of control, synchronization, and nonsinusoidal oscillator circuits.

The period T of the oscillation in Figure 18–50 is given by

$$T = R_E C \ln\left(\frac{V_{BB} - V_V}{V_{BB} - V_P}\right) \text{ seconds} \qquad \textbf{(18–38)}$$

The choice of value for R_E in this equation is not entirely arbitrary. R_E must be small enough to permit the emitter current to reach I_P when the capacitor voltage reaches V_P, and yet be large enough to make the emitter current less than I_V when

Thyristors – Trigger Devices

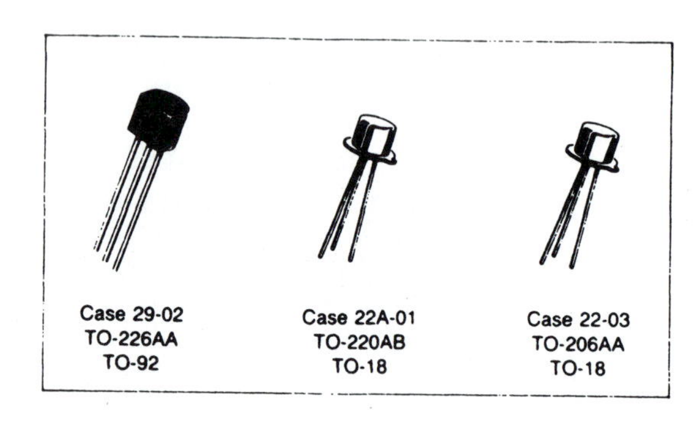

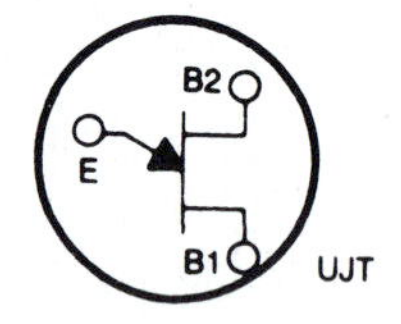

Unijunction Transistors — UJT

Highly stable devices for general-purpose trigger applications and as pulse generators (oscillators) and timing circuits. Useful at frequencies ranging (generally) from 1.0 Hz to 1.0 MHz. Available in low-cost plastic package TO-226AA (TO-92) and in hermetically sealed metal package (Case 22A).

Device Type	η Min	η Max	Ip μA Max	I_{EB20} μA Max	Iv mA Min
Plastic TO-92					
MU10	0.50	0.85	5.0	1.0	1.0
2N4870	0.56	0.75	5.0	1.0	2.0
2N4871	0.70	0.85	5.0	1.0	4.0
MU2646	0.56	0.75	5.0	12	4.0
MU4891	0.55	0.82	5.0	0.01	2.0
MU4892	0.51	0.69	2.0	0.01	2.0
MU4893	0.55	0.82	2.0	0.01	2.0
MU4894	0.74	0.86	1.0	0.01	2.0
Metal TO-18					
MU20	0.50	0.85	5.0	1.0	1.0
2N2646	0.56	0.75	5.0	12	4.0
2N2647	0.68	0.82	2.0	0.2	8.0
2N3980	0.68	0.82	2.0	0.01	1.0
2N4851	0.56	0.75	2.0	0.1	2.0
2N4852	0.70	0.85	2.0	0.1	4.0
2N4853	0.70	0.85	0.4	0.05	6.0
2N4948*	0.55	0.82	2.0	0.01	2.0
2N4949*	0.74	0.86	1.0	0.01	2.0
2N5431*	0.72	0.80	0.4	0.01	2.0

*Also available as JAN and JANTX devices

Figure 18–51
Unijunction transistor specifications (Courtesy of Motorola, Inc.)

the capacitor discharges to V_V. These constraints are necessary to ensure that the UJT switches on when V_E reaches V_P and to ensure that it switches off when V_E falls to V_V. Thus, R_E must satisfy the following two inequalities:

$$R_E < R_E(\max) = \frac{V_{BB} - V_P}{I_P} \tag{18–39}$$

$$R_E > R_E(\min) = \frac{V_{BB} - V_V}{I_V} \tag{18–40}$$

If $V_V \ll V_{BB}$, then equation 18–38 can be approximated by

$$T \approx R_E C \ln\left(\frac{V_{BB}}{V_{BB} - V_P}\right) \tag{18–41}$$

If $\eta V_{BB} >> V_D$, then $V_P \approx \eta V_{BB}$, and (18–41) can be further approximated by

$$T \approx R_E C \ln\left(\frac{V_{BB}}{V_{BB} - \eta V_{BB}}\right) \tag{18–42}$$

or

$$T \approx R_E C \ln\left(\frac{1}{1 - \eta}\right) \tag{18–43}$$

Notice that R in Figure 18–50 is in series with R_{B1} and R_{B2} in the UJT. Its value should therefore be small in comparison to R_{BB}, to prevent it from altering the value of V_P.

Figure 18–51 shows typical specifications for a number of different unijunction transistors. Notice the broad range in the value of η that a transistor of a given type may have. The specification for I_{EB2O} is the emitter reverse current.

Example 18–17

DESIGN

Design a 1-kHz relaxation oscillator using an MU4892 UJT with a 20-V-dc supply. Use the specifications given in Figure 18–51. The MU4892 has $V_D = 0.56$ V. Assume that $V_V = 2$ V.

Solution. We will use a value of η equal to the average of the minimum and maximum values given in the specifications: $\eta = (0.51 + 0.69)/2 = 0.60$. From equation 18–37, $V_P = \eta V_{BB} + V_D = (0.6)(20) + 0.56 = 12.56$ V. Since inequality 18–39 gives the maximum permissible value of R_E and I_P appears in the denominator, we should use the maximum possible value of I_P in the calculation. Similarly, since (18–40) gives the minimum permissible value of R_E, we should use the minimum specified value for I_V. From the specifications, $I_P(\max) = 2\ \mu$A and $I_V(\min) = 2$ mA. Thus, from (18–39) and (18–40),

$$R_E < \frac{V_{BB} - V_P}{I_P} = \frac{(20\text{ V}) - (12.56\text{ V})}{2\ \mu\text{A}} = 3.72\text{ M}\Omega$$

$$R_E > \frac{V_{BB} - V_P}{I_V} = \frac{(20\text{ V}) - (2\text{ V})}{2\text{ mA}} = 9\text{ k}\Omega$$

We see that we have a wide range of choices for R_E. Both inequalities are satisfied if we choose $R_E = 10$ kΩ.

Since $T = 1/f = 1/(1\text{ kHz}) = 10^{-3}$ s, we have, from equation 18–38,

$$10^{-3} = 10^4 C \ln\left(\frac{20-2}{20-12.56}\right)$$

which gives $C = 0.113\ \mu\text{F}$.

A typical value for resistance R in Figure 18–50 is 47 Ω.

Checking the accuracy of approximation 18–43, we find

$$T \approx R_E C \ln\left(\frac{1}{1-\eta}\right) = (10^4)(0.113 \times 10^{-6})\ln\left(\frac{1}{1-0.6}\right) \approx 1.04\text{ ms}$$

The approximation gives a value slightly higher than the design value of 1 ms, but, considering the variation that is possible in the parameter values, the approximation is quite good. For accurate frequency control, R_E should be made adjustable.

Unijunction transistors are often used in SCR trigger circuits, a typical example of which is shown in Figure 18–52. When the UJT switches on, a pulse of gate

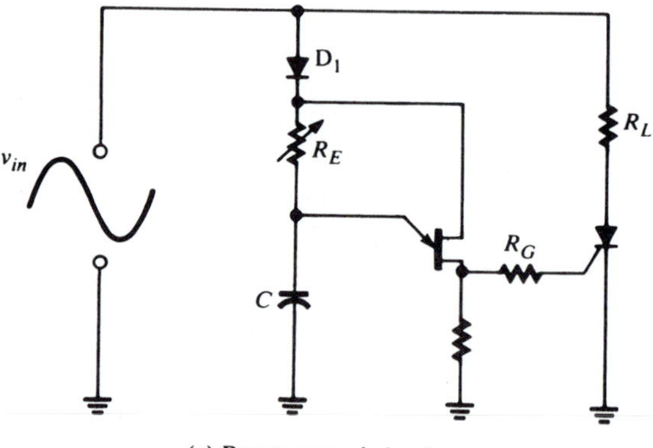

(a) Power-control circuit

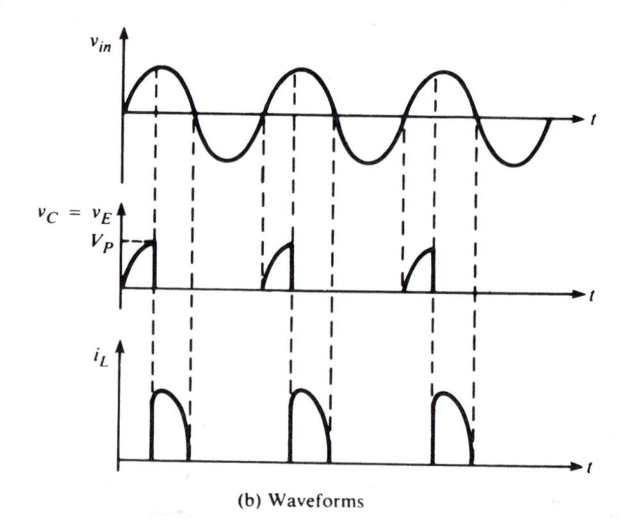

(b) Waveforms

Figure 18–52
Using a UJT to trigger an SCR in an ac power-control circuit

current is supplied to the SCR to turn it on. In this ac power-control circuit, the SCR switches off during every negative half-cycle, so the UJT must retrigger it during each positive half-cycle. Adjusting the value of R_E controls the rate at which the capacitor charges and therefore controls the point in time at which the emitter voltage reaches V_P. A small value of R_E allows the capacitor to charge quickly, which makes the UJT and SCR switch on early in the half-cycle, and allows load current to flow for a significant portion of the half-cycle. A large value of R_E causes a greater delay in switching and allows load current to flow during a smaller time interval. Diode D_1 isolates the UJT from negative half-cycles.

Programmable UJTs (PUTs)

A programmable UJT (PUT) is not a unijunction transistor at all, but a four-layer device. Its name is derived from the fact that its characteristic curve and many of its applications are similar to those of a UJT. Figure 18–53 shows the construction and schematic symbol of a PUT, which resembles an SCR more than a UJT. Like an SCR, its three terminals are designated anode (A), cathode (K), and gate (G). However, note that the gate is connected to the N region below the uppermost PN junction, like the anode gate in an SCS (Figure 18–16).

In applications, the gate of the PUT is biased positive with respect to the cathode. If the anode is then made about 0.7 V more positive than the gate, the device regeneratively switches on. Thus, the PUT is "programmed" by its gate-to-cathode bias voltage. Figure 18–54(a) shows a typical bias arrangement, where the gate voltage is obtained from a resistive voltage divider. Note that V_{AK} and V_{BB} must be separate voltage sources, since V_G, which is derived from V_{BB}, must remain constant, while V_{AK} varies. Figure 18–54(b) shows the characteristic curve. Note the resemblance of the characteristic to that of a UJT. The peak and valley points have the same significance as in a UJT. The advantage of a PUT is that it has a smaller on resistance than a UJT and can handle heavier currents.

A notation for PUTs that is consistent with that for UJTs can be developed by defining

$$\eta = \frac{R_2}{R_1 + R_2} \tag{18–44}$$

Figure 18–53
Construction and symbol for a programmable UJT (PUT)

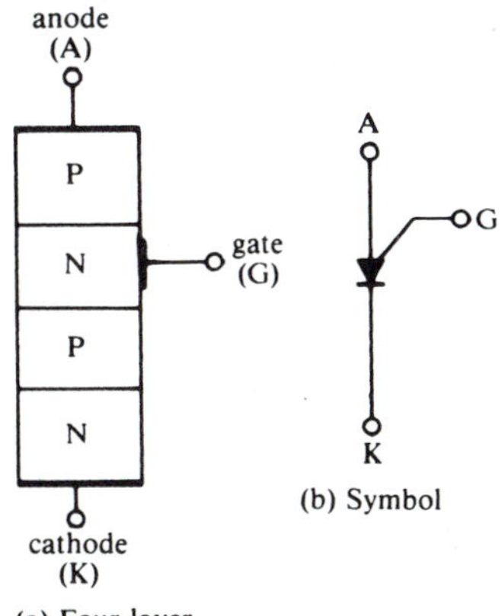

Figure 18–54
PUT biasing and characteristic curve

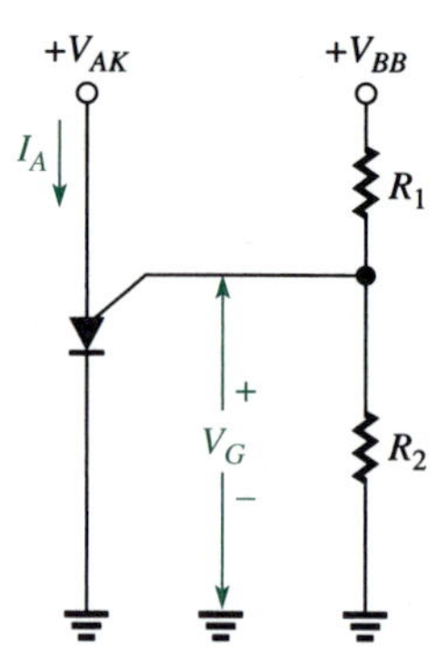

(a) Voltage-divider bias for a PUT

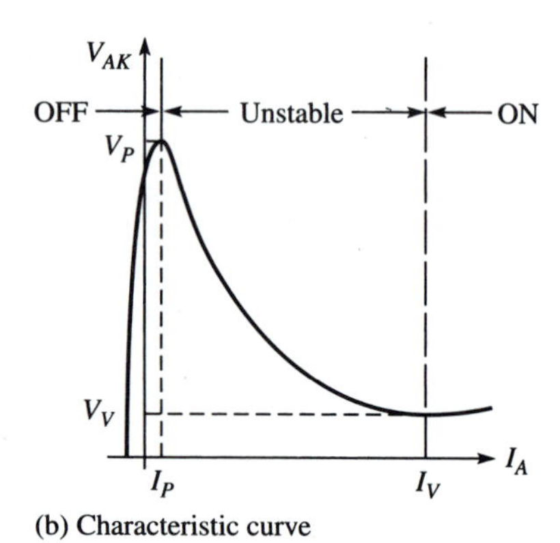

(b) Characteristic curve

where R_1 and R_2 are the resistors in the voltage divider shown in Figure 18–54(a). Then it is clear that

$$V_G = \eta V_{BB} \tag{18–45}$$

The peak-point value of V_{AK}, V_P, at which the PUT switches on, can then be written as

$$V_P = V_G + V_D = \eta V_{BB} + V_D \tag{18–46}$$

where $V_D \approx 0.7$ V. Equation 18–46 is completely consistent with equation 18–37 for the UJT.

Figure 18–55(a) shows a PUT relaxation oscillator. The capacitor charges through R_A until V_{AK} reaches V_P, at which time the PUT switches on and the capacitor discharges. Note that the anode-to-cathode voltage V_{AK} when the PUT is *off* is approximately the same as the anode-to-ground voltage V_A, which is the same as the capacitor voltage V_C, since there is very little drop across R_K. Thus, the PUT switches on when $V_C \approx V_P$. The surge of current through the PUT when it turns on creates a voltage pulse across R_K, as shown in Figure 18–55(b). When the current through the PUT falls below the holding current (I_V in Figure 18–54), the device switches back to its off state and the cycle repeats itself.

As in the UJT oscillator, R_A must be small enough to permit I_P to flow when V_C equals V_P, yet large enough to ensure that the device switches back to its off state when the capacitor discharges. The range of values of R_A that ensure oscillation

Figure 18–55
PUT relaxation oscillator

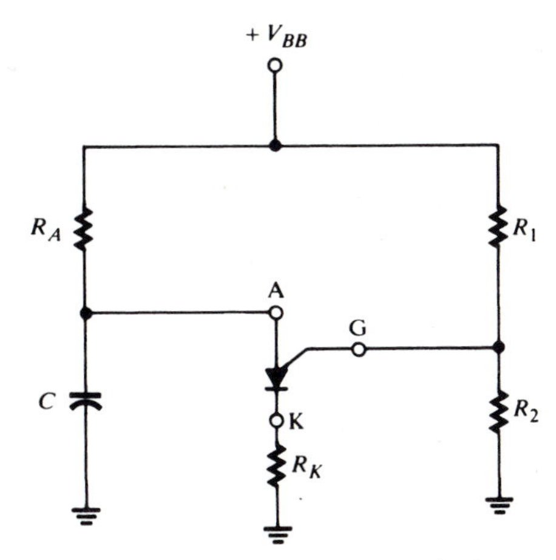

(a) PUT relaxation oscillator circuit

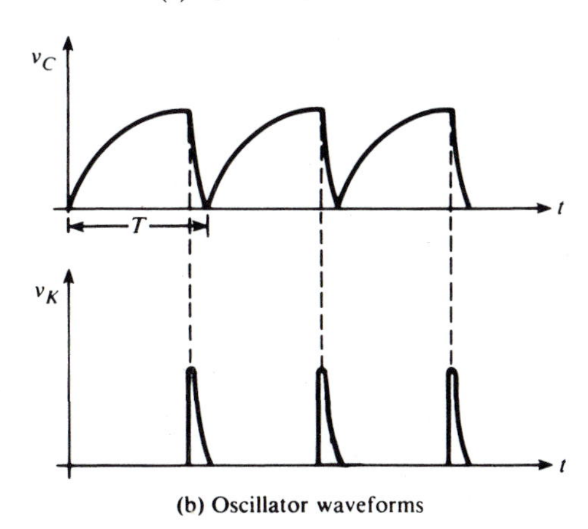

(b) Oscillator waveforms

is found in the same way as for the UJT oscillator:

$$R_A < R_A(\text{max}) = \frac{V_{BB} - V_P}{I_P} \quad \textbf{(18–47)}$$

$$R_A > R_A(\text{min}) = \frac{V_{BB} - V_V}{I_V} \quad \textbf{(18–48)}$$

Also, the period of the oscillation can be determined using the same approximation that applies to the UJT oscillator:

$$T \approx R_A C \ln\left(\frac{1}{1-\eta}\right) \quad \textbf{(18–49)}$$

Remember that η is an externally adjusted resistance ratio in this case rather than a device parameter.

18–5 TUNNEL DIODES

A tunnel diode, also called an *Esaki* diode after its inventor, is constructed with very heavily doped P and N materials, usually germanium or gallium arsenide. The doping level in these highly conductive regions may be from 100 to several thousand times that of a conventional diode. As a consequence, the depletion region at the PN junction is extremely narrow. Recall that forward conduction in a conventional diode occurs only if the forward bias is sufficient to give charge carriers the energy necessary to overcome the barrier potential opposing their passage through the depletion region. When a tunnel diode is only slightly forward biased, many carriers are able to cross through the very narrow depletion region without acquiring that energy. These carriers are said to *tunnel* through the depletion region because their energy level is lower than the level of those that overcome the barrier potential.

Because of the tunneling phenomenon, there is a substantial flow of current through a tunnel diode at the onset of forward bias. The same is true when the diode is reverse biased. This behavior is evident in the *I–V* characteristic of a tunnel diode, shown in Figure 18–56. Notice the steep slope (small resistance) of the characteristic in the region around the origin. For comparison purposes, the dashed line shows the characteristic of a conventional diode. With increasing forward bias, the tunneling phenomenon continues, until it reaches a maximum at I_P in the figure. The forward bias at this point is V_P. Because of higher energies acquired by carriers on the N side, current due to tunneling begins to diminish rapidly with further increase in forward bias. As a result, the characteristic displays a *negative-resistance region,* similar to that of a UJT, where increasing voltage is accompanied by decreasing current. When the forward bias becomes large enough to supply carriers with the energy required to surmount the barrier potential, forward current begins to increase again. As shown in the figure, the forward characteristic beyond that point coincides with the characteristic of a conventional diode.

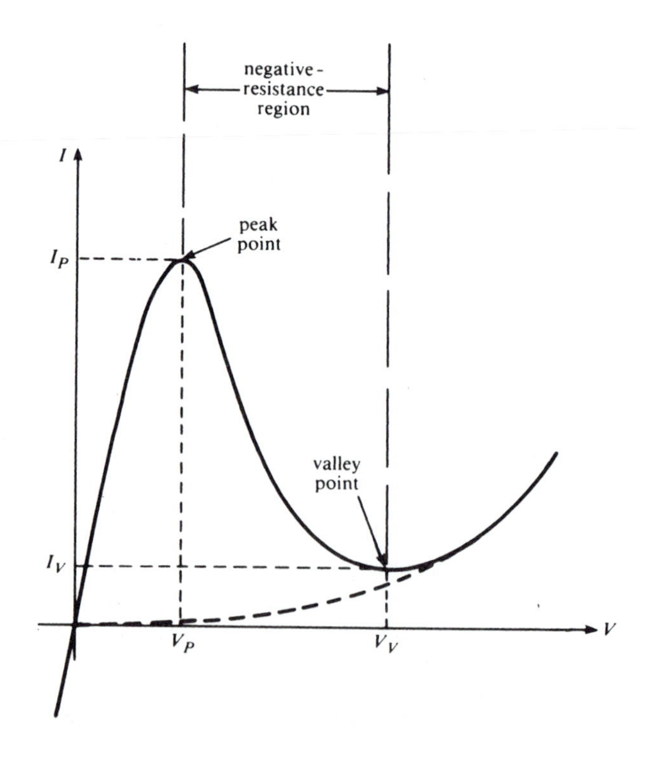

Figure 18–56
I–V characteristic of a tunnel diode

Figure 18–57
Equivalent circuit and schematic symbols for a tunnel diode

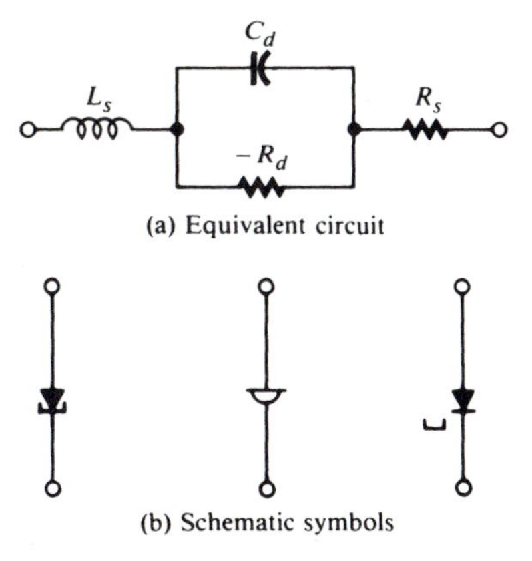

Figure 18–57 shows the equivalent circuit of a tunnel diode in its negative-resistance region and several commonly used schematic symbols. The device is most often used in high-frequency or high-speed switching circuits, so the junction capacitance C_d and the inductance of the connecting leads, L_s, are important device parameters. C_d may range from 5 to 100 pF, while L_s is on the order of a few nanohenries. Note the negative resistance labeled $-R_d$, typically $-10\ \Omega$ to $-200\ \Omega$. R_s is the resistance of the leads and semiconductor material, on the order of $1\ \Omega$.

The negative-resistance characteristic of a tunnel diode has fostered some applications that are unusual among two-terminal devices, including oscillators, amplifiers, and high-speed electronic switches. In the latter application, the tunnel diode is switched between its peak and valley points with nano- or picosecond switching times. The peak voltage is typically rather small, less than 200 mV, but the peak current may range from 1 to 100 mA.

18–6 VOLTAGE-VARIABLE CAPACITORS (VARACTOR DIODES)

A *voltage-variable capacitor* (VVC), also called a *varactor, varicap,* or *tuning diode,* is a diode constructed for use in high-frequency circuits where it is desired to control or adjust capacitance values by varying a dc voltage level. By virtue of its construction, a reverse-biased diode has all the essential ingredients of a capacitor: two conducting regions (P and N regions) separated by a dielectric (the depletion region). Recall that the capacitance of such a structure is given by

$$C = \frac{\varepsilon A}{d} \qquad \textbf{(18–50)}$$

where ε is the permittivity of the dielectric, A is the cross-sectional surface area of the conducting regions, and d is the distance separating the regions (the thickness, or width, of the dielectric). Also recall that the width of the depletion region of a reverse-biased diode increases as the reverse-biasing voltage increases. Thus, increasing the reverse bias on a diode causes d to increase and the capacitance $C = \varepsilon A/d$ to decrease. This behavior is the fundamental principle governing the operation of a varactor diode. The value of the capacitance obtained from a varactor is small, on the order of 100 pF or less, and it is used in practice only to alter the ac impedance it presents to a high-frequency signal, as, for example, in a tuned LC network. It would not be useful for applications such as a low-frequency bypass capacitor, a filter capacitor in a power supply, or a coupling capacitor in an audio amplifier.

An important characteristic of a varactor diode is the ratio of its largest to its smallest capacitance when the voltage across it is adjusted through a specified range. Sometimes called the capacitance *tuning ratio,* this value is governed by the *doping profile* of the semiconductor material used to form the P and N regions, that is, the doping density in the vicinity of the junction. In diodes having an *abrupt* junction, the P and N sides are uniformly doped and there is an abrupt transition from P to N at the junction. Abrupt-junction varactors exhibit capacitance ratios from about 2:1 to 3:1. In the *hyperabrupt* (extremely abrupt) junction, the doping level is increased as the junction is approached from either side, so the P material becomes more heavily P near the junction, and the N material becomes more heavily N. The hyperabrupt junction is very sensitive to changes in reverse voltage and this type of varactor may have a capacitance ratio up to 20:1. Figure 18–58 shows typical plots of abrupt and hyperabrupt varactor capacitance versus reverse voltage. The figure also shows the schematic symbol for a varactor diode.

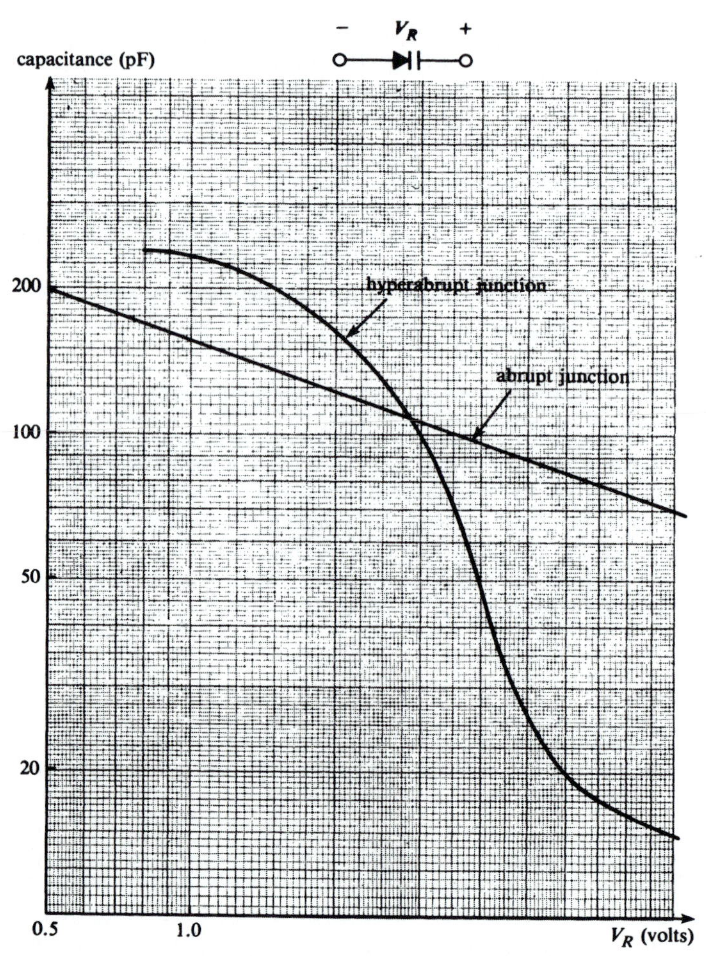

Figure 18–58
Capacitance versus reverse voltage for abrupt and hyperabrupt varactor diodes

Figure 18–59
A tuned amplifier whose resonant frequency is controlled by the reverse voltage V_R across the varactor diode

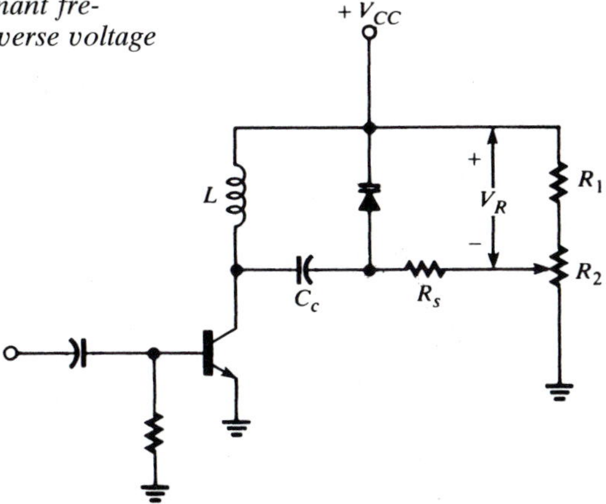

Figure 18–59 shows a *tuned amplifier* with a fixed inductor and a varactor diode that form an LC tank circuit driven by the transistor. The frequency of maximum amplification is the resonant frequency of the tank, which is closely approximated by

$$f_r = \frac{1}{2\pi\sqrt{LC_d}} \qquad \textbf{(18–51)}$$

where C_d is the varactor-diode capacitance. Capacitor C_c is a coupling capacitor that blocks the flow of dc current between the transistor and the diode circuitry. C_c is much larger than C_d and does not affect resonance. The R_1–R_2 voltage divider is used to adjust the reverse-biasing voltage across the varactor diode and thus to adjust the resonant frequency of the amplifier. Resistor R_s is a large resistance used to prevent the voltage divider from loading the tank network. Since the dc current in R_s is the very small reverse current through the varactor diode, the dc drop across R_s is negligible and the varactor voltage is, for all practical purposes, the voltage V_R shown across the divider.

Example 18–18

The varactor diode in Figure 18–59 has the abrupt-junction characteristic shown in Figure 18–58. If V_{CC} = 10 V, L = 80 μH, R_1 = 1 kΩ, and R_2 is a 10-kΩ potentiometer, find the frequency range over which the amplifier can be tuned.

Solution. The value of V_R is minimum when the wiper arm of the potentiometer is at its topmost position in Figure 18–59. In that case, by voltage-divider action,

$$V_R = \left[\frac{1\text{ k}\Omega}{(1\text{ k}\Omega) + (10\text{ k}\Omega)}\right](10\text{ V}) = 0.909\text{ V}$$

V_R has a maximum value of V_{CC} = 10 V when the wiper arm is at its bottom position.

From the abrupt-junction characteristic in Figure 18–58, the values of varactor capacitance corresponding to V_R = 0.909 V and to V_R = 10 V are found to be

$$C_d(\text{max}) \approx 160\text{ pF} \quad (\text{at } V_R = 0.909\text{ V})$$
$$C_d(\text{min}) \approx 70\text{ pF} \quad (\text{at } V_R = 10\text{ V})$$

From equation 18–51,

$$f_r(\min) = \frac{1}{2\pi\sqrt{LC_d(\max)}} = \frac{1}{2\pi\sqrt{(80 \times 10^{-6})(160 \times 10^{-12})}} = 1.41 \text{ MHz}$$

$$f_r(\max) = \frac{1}{2\pi\sqrt{LC_d(\min)}} = \frac{1}{2\pi\sqrt{(80 \times 10^{-6})(70 \times 10^{-12})}} = 2.13 \text{ MHz}$$

EXERCISES

Section 18–1
Zener Diodes

18–1. In the circuit shown in Figure 18–60, the zener diode has a reverse breakdown voltage of 10 V. If $R_S = 200\ \Omega$,

a. find V_{in} when $I_Z = 15$ mA and $I_L = 50$ mA; and

b. find I_Z when $V_{in} = 30$ V.

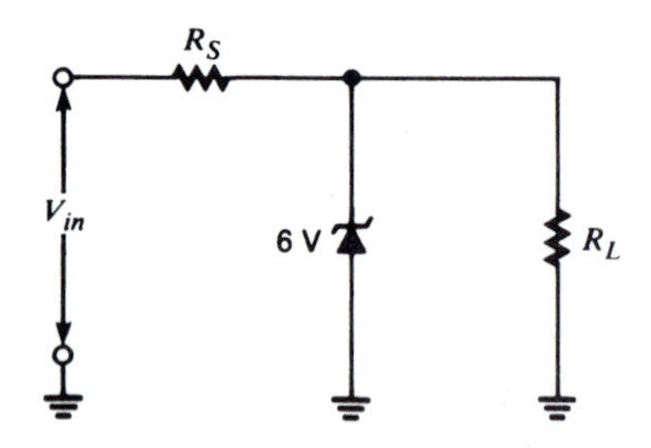

Figure 18–60
(Exercises 18–1 and 18–2)

18–2. In the circuit shown in Figure 18–60, the zener diode has a reverse breakdown voltage of 12 V. $R_S = 50\ \Omega$, $V_{in} = 20$ V, and R_L can vary from 100 Ω to 200 Ω. Assuming that the zener diode remains in breakdown, find

a. the minimum and maximum current in the zener diode;

b. the minimum and maximum power dissipated in the diode; and

c. the minimum rated power dissipation that R_S should have.

18–3. Repeat Exercise 18–2 if, in addition to the variation in R_L, V_{in} can vary from 19 V to 30 V.

18–4. The 6-V zener diode in Figure 18–61 has a maximum rated power dissipation of 0.5 W. Its reverse current must be at least 5 mA to keep it in breakdown. Find a suitable value for R_S if V_{in} can vary from 8 V to 12 V and R_L can vary from 500 Ω to 1 kΩ.

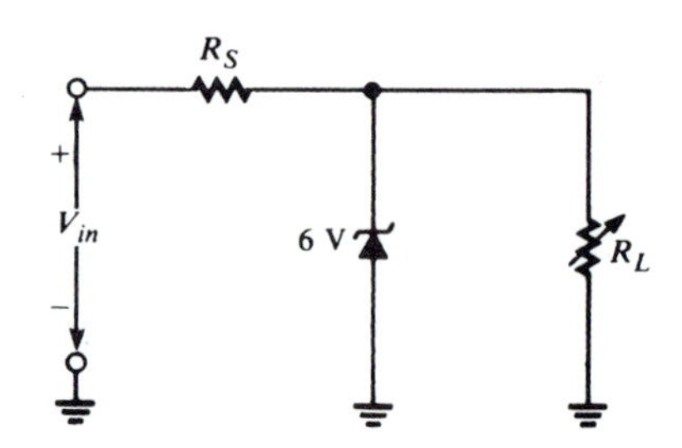

Figure 18–61
(Exercise 18–4)

18–5. **a.** If R_S in Exercise 18–4 is set to its maximum permissible value, what is the maximum permissible value of V_{in}?

b. If R_S in Exercise 18–4 is set equal to its minimum permissible value, what is the minimum permissible value of R_L?

18–6. A zener diode has a breakdown voltage of 12 V at 25°C and a temperature coefficient of +0.5 mV/°C.

a. Design a temperature-stabilizing circuit using silicon diodes that have temperature coefficients of −0.21 mV/°C. The forward drop across each diode at 25°C is 0.68 V.

b. Find the voltage across the stabilized network at 25°C and at 75°C.

c. Find the temperature stability of the stabilized network.

18–7. A zener diode has a breakdown voltage of 15.1 V at 25°C. It has a temperature coefficient of +0.78 mV/°C and is to be operated between 25°C and 100°C. It is to be temperature stabilized in such a way that the voltage across the network is never less than its value at 25°C.

a. Design a temperature-stabilizing network

using silicon diodes whose temperature coefficients are −0.2 mV/°C. The forward drop across each diode at 25°C is 0.65 V.

b. What is the maximum voltage across the stabilized network?

18–8. Following is a set of measurements that were made on the voltage across and current through a zener diode:

I_Z(mA)	V_Z(volts)
0.5	30.1
1.0	30.15
2.0	30.25
3.5	30.37
6	30.56
8	30.68
10	30.80
30	31.90
40	32.40
90	34.00

a. Find the approximate zener impedance over the range from I_Z = 3.5 mA to I_Z = 10 mA.

b. Show that the zener impedance decreases with increasing current.

18–9. The breakdown voltage of a zener diode when it is conducting 2.5 mA is 7.5 V. If the voltage must not increase more than 10% when the current increases 50%, what maximum impedance can the diode have?

Section 18–2

Four-Layer Devices

18–10. Figure 18–62 shows a Shockley diode used in a time-delay circuit to turn on an SCR.

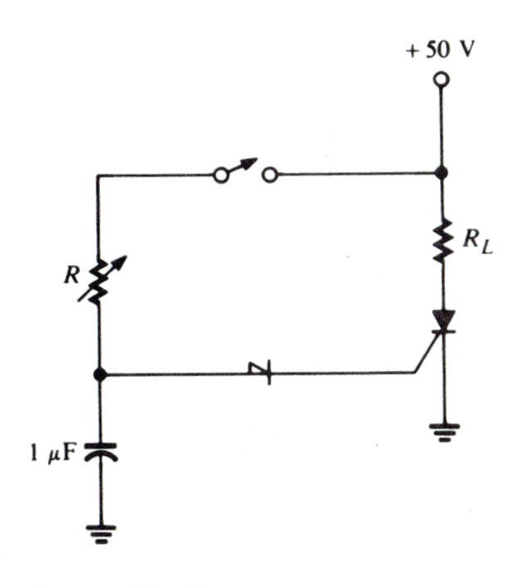

Figure 18–62
(Exercise 18–10)

After the switch is closed, the capacitor charges until its voltage reaches the forward breakover voltage of the Shockley diode. The diode then conducts gate current to the SCR and turns it on. The adjustable resistance R is used to vary the RC time constant and thus vary the time delay between switch closure and SCR turn-on. If the forward breakover voltage of the diode is 30 V, find the time delay when R = 100 kΩ.

18–11. The SCR in Figure 18–63 has $V_{BR(F)}$ = 10 V. When it is conducting, the voltage drop V_{AK} is 1 V. If the anode voltage is the sawtooth waveform shown, find the average current in R_L.

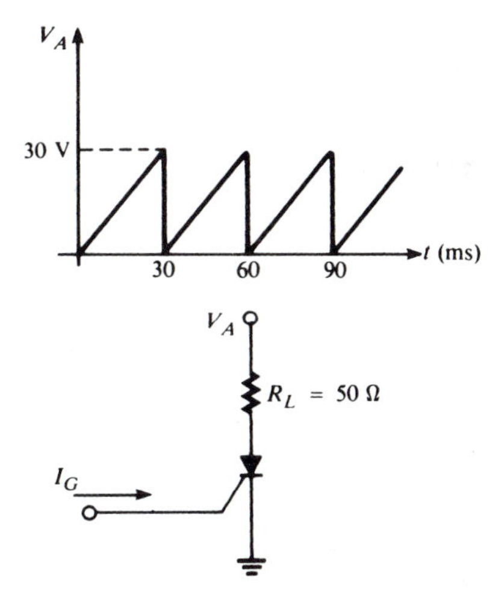

Figure 18–63
(Exercise 18–11)

18–12. Find the average voltage across the SCR in Exercise 18–11.

18–13. In the experimental circuit shown in Figure 18–64, V_A was set to +50 V and V_{GG} was adjusted until the SCR turned on. The value of V_G at that point was recorded. The value of V_A was then reduced until the SCR turned off. The value of V_A at that point was recorded. V_A was then restored to +50 V and the measurements were repeated with a new setting for V_{GG}. The procedure was repeated several times, and the following data were obtained:

V_G(volts)	V_A(volts)
0	1.15
0.2	1.075
0.4	1.00
1.0	0.95

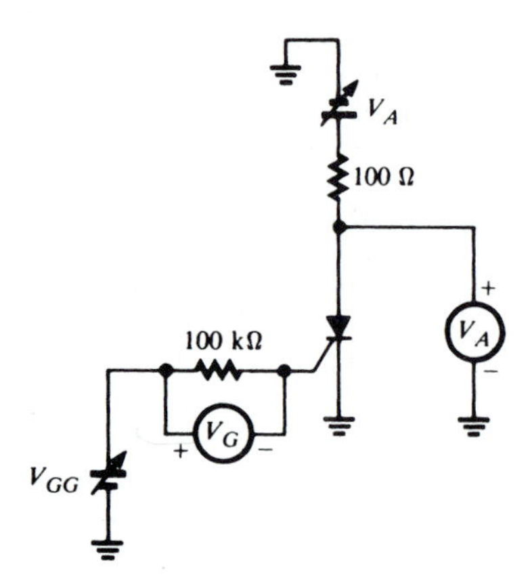

Figure 18–64
(Exercise 18–13)

Make a table showing holding current versus gate current for the SCR. Assume that the voltage drop V_{AK} across the SCR is 0.9 V when it is conducting.

18–14. Derive equation 18–17 from (18–14), (18–15), and (18–16). Then derive equation 18–18 from (18–17) and (18–13).

18–15. An SCR is connected in a half-wave power-control circuit (Figure 18–14) in which v_{in} = 100 V rms and R_L = 80 Ω. The forward voltage drop across the SCR when it is conducting is 1.0 V.

a. Find the firing angle necessary to deliver two-thirds of the theoretical maximum average current that the circuit could deliver to R_L.

b. Find the average load power under that condition.

18–16. An SCR is connected in a half-wave power-control circuit (Figure 18–14) in which v_{in} = 125 V rms and R_L = 50 Ω. Assuming that the forward voltage drop across the SCR is a constant 0.9 V when it is conducting, find the average power dissipated by the SCR when the firing angle is 30°.

18–17. The TRIAC in the power-controller circuit shown in Figure 18–22 has a voltage drop of 1 V between its anodes when it is conducting. If R_L = 30 Ω, find the rms value of v_{in} that is necessary if it is required to deliver an average power of 100 W to R_L when the firing angle is 45°.

Section 18–3
Optoelectronic Devices

18–18. Find the frequency range corresponding to radiation having wavelengths in the range from 1500 Å to 2000 Å. In what part of the light spectrum are these frequencies?

18–19. The total visible light energy received on a 1-cm² surface in 1 min is 0.042 J. What is the light intensity in 1m/m²?

18–20. The resistance of a certain photoconductive cell plotted versus light intensity is a straight line on log-log axes, as in Figure 18–24. If the cell has a resistance of 40 kΩ when the light intensity is 8 lm/m², and 30 kΩ when the intensity is 15 lm/m², what is the resistance when the intensity is 100 lm/m²?

18–21. Using the photoconductive cell whose characteristics are described in Example 18–10, redesign the circuit in Figure 18–25 so that the relay coil remains de-energized as long as the cell is dark and energizes when the cell is illuminated. Demonstrate by circuit calculations that your design works.

18–22. Assuming that the photodiode characteristics shown in Figure 18–26 are perfectly linear for light intensities between 5000 and 10,000 lm/m², calculate the light intensity required to make the effective resistance 1250 Ω when the reverse voltage is 2 V.

18–23. A photodiode has $\eta = 1$, $V_T = 0.026$ V, and $I_s = 0.5$ nA. How much light-generated reverse current is necessary to produce an open-circuit output of 0.50 V when the diode is operated as a photovoltaic device?

18–24. What light intensity will fire the SCR in Example 18–13 (Figure 18–29) if the 100-kΩ gate resistor is changed to 35 kΩ?

18–25. The phototransistor in Figure 18–65 has the characteristics shown in Figure 18–28. To what intensity must the incident light rise to cause the comparator output to switch low?

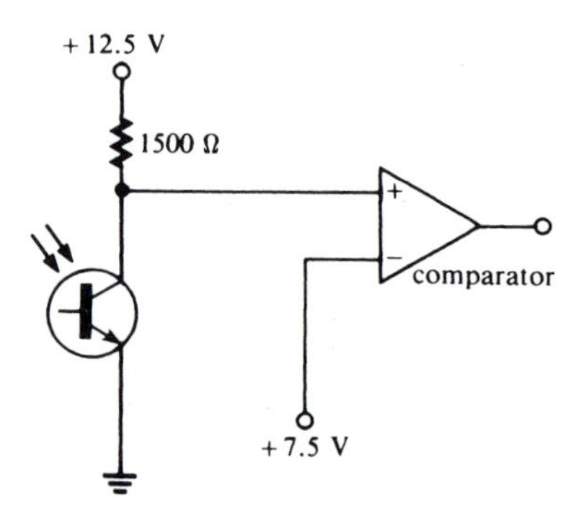

Figure 18–65
(Exercise 18–25)

18–26. What is the maximum power that can be delivered to a load by the solar cell whose characteristics are shown in Figure 18–33 when the light intensity is 80 mW/cm^2? When the light intensity is 15 mW/cm^2?

18–27. A series/parallel array of solar cells is used to provide power to a certain load. The array consists of 12 parallel branches, each of which has 8 cells in series. If each cell produces 900 mA at 0.55 V, how much power is delivered to the load?

18–28. The energy gaps for gallium arsenide and gallium phosphide are 1.37 eV and 2.25 eV, respectively. Find the wavelengths, in Å, of the light produced by LEDs fabricated from each of these materials.

18–29. If the maximum permissible forward current in the LED shown in Figure 18–38 (Example 18–15) is 40 mA, find the smallest permissible value of R_C. Assume that the transistor and LED parameters are the same as those given in the example.

18–30. The LED driver circuit shown in Figure 18–66 is designed to turn the LED off when 5 V is applied to R_B. Assuming that the current in the LED should be 15 mA when it is on and that the LED has a forward voltage drop of 2 V, find values for R_B and R_C. The silicon transistor has $\beta = 100$.

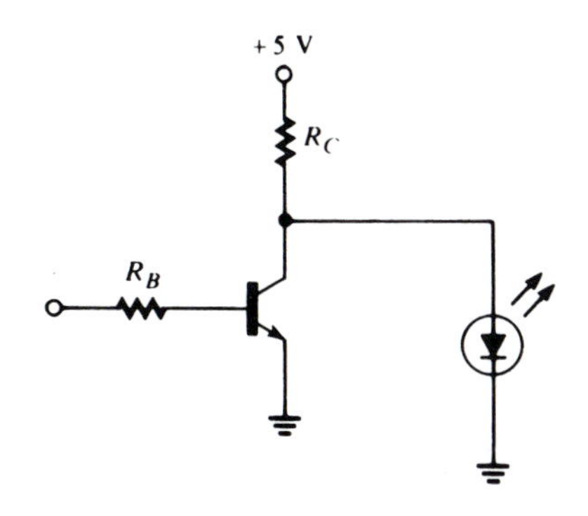

Figure 18–66
(Exercise 18–30)

18–31. Each LED in a certain seven-segment display draws 20 mA and has a 1.5-V drop when it is illuminated.

a. Which numeral (from 0 through 9) requires the most power to display? The least power?

b. How much power is required to display the numeral that requires the most power?

18–32. Figure 18–67 shows an optocoupler used as a logic inverter. When a positive pulse (high) is applied to the input, the LED illuminates and the transistor saturates, thus making the output low. To operate properly, the LED must be supplied with 15 mA, and it has a 1.2-V drop when it is illuminated. The optocoupler has an 80% current-transfer ratio. Find values for R_D and R_C. The high input level is +5 V.

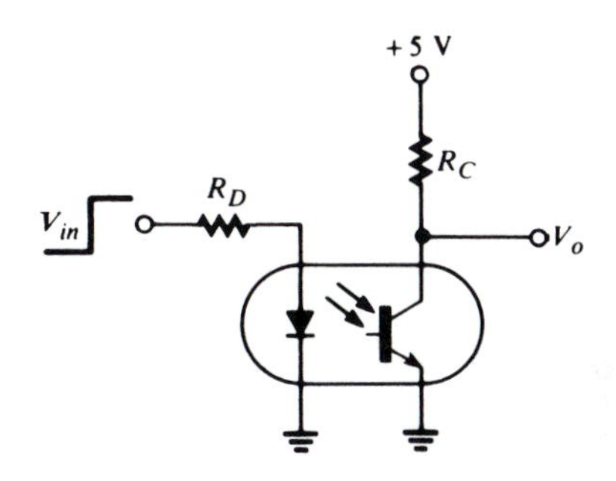

Figure 18–67
(Exercise 18–32)

18–33. Name the two methods used to alter light properties in liquid-crystal displays. In what two modes can an LCD be designed to operate using each of the light-altering methods?

18–34. What method of light alteration and what mode of operation are used in a field-effect (twisted-nematic) LCD to create a dark display on a light background?

18–35. What is the principal advantage of an LCD in comparison to an LED display? What are the principal limitations?

Section 18–4
Unijunction Transistors

18–36. In an experiment with a certain UJT, it was found that, with V_{BB} set to 8 V, it was necessary to increase the emitter voltage to 5 V to cause it to switch on. When V_{BB} was raised to 15 V, it was necessary to raise the emitter voltage to 8.85 V to cause it to switch on. What is the intrinsic standoff ratio of the UJT?

18–37. A UJT relaxation oscillator of the design shown in Figure 18–50 must oscillate somewhere in the range from 750 Hz to 1.5 kHz, no matter what UJT is used. If $R_E = 68$ kΩ and $C = 0.01$ μF, use approximation 18–43 to determine the permissible range of values of η that UJTs used in the circuit can have.

18–38. Design a bias circuit for a programmable unijunction transistor so that it will have a peak voltage of 16 V. Use a 20-V-dc supply and assume that $V_D = 0.7$ V. Draw the schematic diagram for your design.

18–39. The PUT relaxation oscillator in Figure 18–55(a) has $R_A = 10\ \text{k}\Omega$, $C = 0.1\ \mu\text{F}$, and $R_2 = 15\ \text{k}\Omega$. Through what range of values (approximately) should it be possible to adjust R_1 if it is desired to adjust the oscillation frequency from 700 Hz to 1700 Hz?

Section 18–5
Tunnel Diodes

18–40. Describe the tunneling phenomenon and explain how it accounts for the appearance of the characteristic curve of a tunnel diode in the region of small forward bias.

18–41. The negative-resistance region of a certain tunnel diode occurs for forward voltages in the range from 0.2 V to 0.4 V. Is the statement that follows correct? Why or why not? If the diode has negative resistance $-200\ \Omega$ when the voltage is 0.3 V, then the current under that amount of bias is $I = 0.3\ \text{V}/(-200\ \Omega) = -1.5$ mA.

Section 18–6
Voltage-Variable Capacitors (Varactor Diodes)

18–42. Find the capacitance ratio (tuning ratio) of each of the varactor diodes whose characteristics are given in Figure 18–58 when the reverse voltage is increased 20 dB from an initial value of 1 V.

18–43. A varactor diode having the hyperabrupt-junction characteristic shown in Figure 18–58 is to be used in the tuned amplifier circuit shown in Figure 18–59. If the amplifier is to be tuned to 1 MHz when the diode reverse voltage is 3 V, what value of inductance should be used in the circuit? With that value of inductance, what will be the tuned frequency if the diode voltage is doubled?

19 DIGITAL-TO-ANALOG AND ANALOG-TO-DIGITAL CONVERTERS

19–1 OVERVIEW

Analog and Digital Voltages

An analog voltage is one that may vary *continuously* throughout some range. For example, the output of an audio amplifier is an analog voltage that may have any value (of an infinite number of values) between its minimum and maximum voltage limits. Most of the devices and circuitry we have studied in this book are analog in nature. In contrast, a digital voltage has only two useful values: a "low" and a "high" voltage, such as 0 V and +5 V. Digital voltages are used to represent numerical values in the binary number system, which has just the two digits 0 and 1. For example, four digital voltages having the values 0 V, +5 V, 0 V, +5 V would represent the binary number 0101, which equals 5 in decimal. Digital computers perform computations using binary numbers represented by digital voltages.

Most physical variables in our environment are analog in nature. Quantities such as temperature, pressure, velocity, weight, etc., can have any of an infinite number of values. However, because of the high speed and accuracy of digital systems, such as computers, it is frequently the case that the transmission of data representing such quantities and computations performed on them are operations that are best accomplished using digital equivalents of analog values. Two modern examples of functions that have traditionally been analog in nature and that are now performed digitally include communications systems using fiber-optic cables and music reproduction using compact disks. Converting an analog value to a digital equivalent (binary number) is called *digitizing* the value. An analog-to-digital converter (ADC, or A/D converter) performs this function. Figure 19–1(a) illustrates the operation of an ADC. After a digital system has transmitted, analyzed, or otherwise processed digital data, it is often necessary to convert the results of such operations back to analog values. This function is performed by a digital-to-analog converter (DAC, or D/A converter). Figure 19–1(b) illustrates the operation of a DAC. Figure 19–1(c) shows a practical example of a system that incorporates both an ADC and a DAC. In this example, an accelerometer mounted on a vibration table (used to test components for vibration damage) produces an analog voltage proportional to the instantaneous acceleration of the table. The analog voltage is

analog input 6 V → ADC → 0 V, +5 V, +5 V, 0 V — digital (binary) output : 0110

(a) The ADC converts a 6-V analog input to an equivalent digital output. (The binary number 0110 is equivalent to decimal 6.)

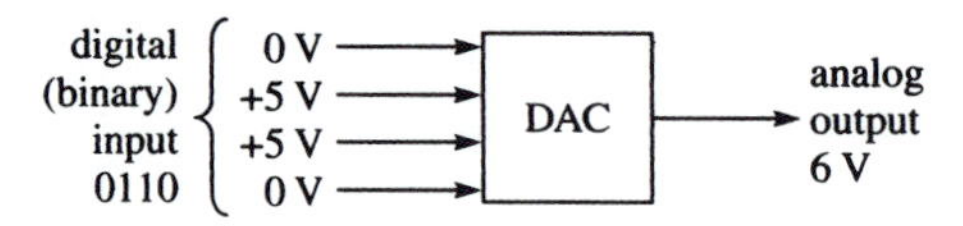

(b) The DAC converts 0110 to 6 V.

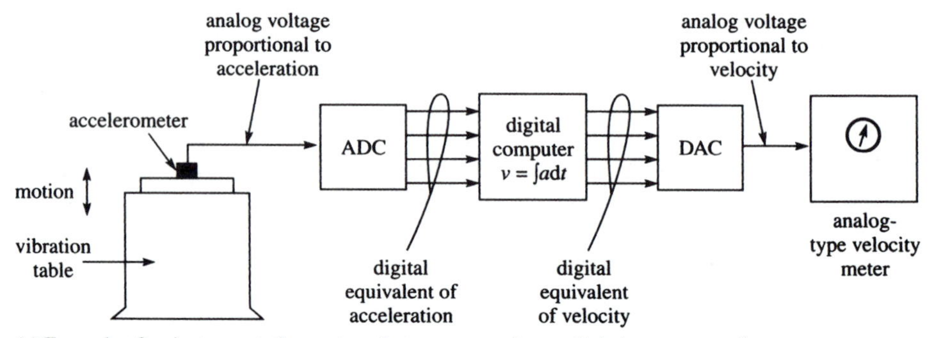

(c) Example of an instrumentation system that uses an analog-to-digital converter and a digital-to-analog converter.

Figure 19–1
The analog-to-digital converter (ADC) and digital-to-analog converter (DAC)

converted to digital form and transmitted to a computer, which calculates the instantaneous velocity of the table (by mathematical integration: $v = \int a\,dt$). The digital quantity representing the computed velocity is then converted by a DAC to an analog voltage, which becomes the input to an analog-type velocity meter.

Converting Binary Numbers to Decimal Equivalents

As an aid in understanding subsequent discussions on digital-to-analog and analog-to-digital converters, we present here for review and reference purposes a brief summary of the mathematics used to convert binary numbers to their decimal equivalents.

As in the decimal number system, the position of each digit in a binary number carries a certain *weight.* In the decimal number system, the weights are powers of 10 (the units position: $10^0 = 1$, the tens position: $10^1 = 10$, the hundreds position: $10^2 = 100$, and so forth). In the binary number system, each position carries a weight equal to a power of 2: $2^0 = 1$, $2^1 = 2$, $2^2 = 4$, $2^3 = 8$, and so forth. A binary digit is called a *bit,* so a binary number consists of a sequence of bits, each of which can be either 0 or 1. The rightmost bit in a binary number carries the least weight and is called the *least significant bit* (LSB). If the binary number is an integer (whole number), the LSB carries weight $2^0 = 1$, and each bit to the left of the LSB has a weight equal to a power of 2 that is one greater than the weight of the bit to its right. To convert a binary number to its decimal equivalent, we compute the sum of all the weights of the positions where binary 1s occur. The following is an example:

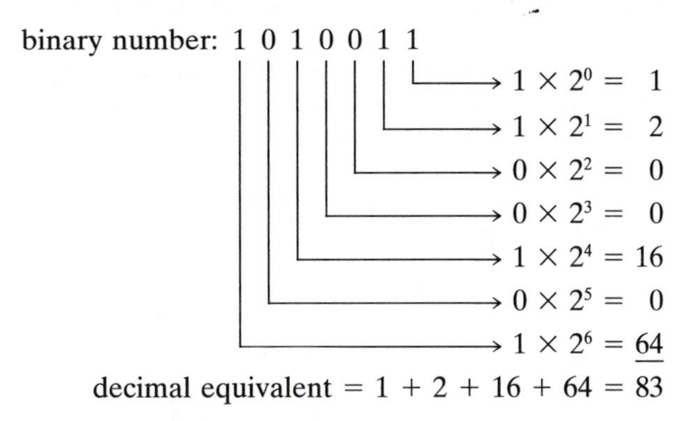

decimal equivalent $= 1 + 2 + 16 + 64 = 83$

The leftmost bit in a binary number is called the *most significant bit* (MSB). An easy way to convert a binary number to its decimal equivalent is to write the values of the powers of 2 (1, 2, 4, 8, . . .) above each bit, beginning with the LSB and proceeding through the MSB. Then we simply add those powers where 1s occur. For example:

power of 2: 16 8 4 2 1
binary number: 1 0 1 1 0

$16 + 4 + 2 = 22 =$ decimal equivalent

If a binary number has a fractional part, then a *binary point* (like a decimal point) separates the integer part from the fractional part. The powers of 2 become negative and descend as we move right past the binary point. The following is an example:

$$\begin{array}{rl} & 2^{-1}\ \ 2^{-2} \\ & \ \ \Vert \qquad \Vert \\ \text{power of 2:} & 8\ 4\ 2\ 1\ \ 0.5\ \ 0.25 \\ \text{binary number:} & 1\ 0\ 0\ 1.\ \ \ 0\ \ 1 \\ \hline & 8 + 1 + 0.25 = 9.25 \end{array}$$

Some Digital Terminology

Binary 0 and 1 are called logical 0 and logical 1 to distinguish them from the voltages used to represent them in a digital system. The voltages are called *logic levels.* For example, a common set of logic levels is 0 V (ground) for logical 0 and +5 V for logical 1.

A *binary counter* is a device whose binary output is numerically equal to the number of pulses that have occurred at its input. For example, the output of a 4-bit binary counter would be 0110 after 6 pulses had occurred at its input. The input pulses often occur at a fixed frequency from a signal called the system *clock.* The largest binary number that a 4-bit counter can contain is 1111, or decimal 15. (The counter *resets* to 0000 after the sixteenth pulse.) Thus, the total number of binary numbers that a 4-bit counter can produce is (counting 0000) 16, or 2^4. The total number of binary numbers that can be represented by n bits is 2^n. The counter we have just described is called an *up-counter* because its binary output increases by

1 (increments) each time a new clock pulse occurs. In a *down-counter*, the binary output decreases by 1 (decrements) each time a new clock pulse occurs. Thus, the sequence of outputs from a 4-bit down-counter would be 1111, 1110, 1101, . . . , 0000, 1111, ...

A *latch* is a digital device that stores the value (0 or 1) of a binary input. It is especially useful when a binary input is changing and we want to "latch onto" its value at a particular instant of time—as, for example, when performing a digital-to-analog conversion of a digital quantity that is continuously changing in value. A *register* is a set of latches used to store all the bits of a digital quantity, one latch for each bit.

Some analog-to-digital converters produce outputs that are in 8-4-2-1 *binary-coded–decimal* (BCD) form, rather than true binary. In 8-4-2-1 BCD, *each* decimal digit is represented by 4 bits. For example, if the input to an ADC having this type of output were 14 V, then the output would be

analog input: 14

8-4-2-1 output: 0001 0100

Note that this type of output is quite different from true binary. (The number 14 in binary is 1110.)

Resolution

A digital-to-analog converter having a 4-bit binary input produces only $2^4 = 16$ different analog output voltages, corresponding to the 16 different values that can be represented by the 4-bit input. The output of the converter is, therefore, not truly analog, in the sense that it cannot have an infinite number of values. If the input were a 5-bit binary number, the output could have $2^5 = 32$ different values. In short, the greater the number of input bits, the greater the number of output values and the closer the output resembles a true analog quantity. *Resolution* is a measure of this property. The greater the resolution of the DAC, the finer the increments between output voltage levels.

Similarly, an analog-to-digital converter having a 4-bit binary output produces only 16 different binary outputs, so it can convert only 16 different analog inputs to digital form. Since an analog input has an infinite number of values, the ADC does not truly convert (every) analog input to equivalent digital form. The greater the number of output bits, the greater the number of analog inputs the ADC can convert and the greater the resolution of the device. Resolution is clearly an important performance specification for both DACs and ADCs, and we will examine quantitative measures for it in later discussions.

19–2 THE *R*-2*R* LADDER DAC

The most popular method for converting a digital input to an analog output incorporates a ladder network containing series-parallel combinations of two resistor values: R and $2R$. Figure 19–2 shows an example of an R-$2R$ ladder having a 4-bit digital input.

To understand the operation of the R-$2R$ ladder, let us first determine the output voltage in Figure 19–2 when the input is 1000. We will assume that a logical-1 input is E volts and that logical 0 is 0 V (ground). Figure 19–3(a) shows the circuit with input D_3 connected to E and all other inputs connected to ground,

Figure 19–2
A 4-bit R-2R ladder network used for digital-to-analog conversion

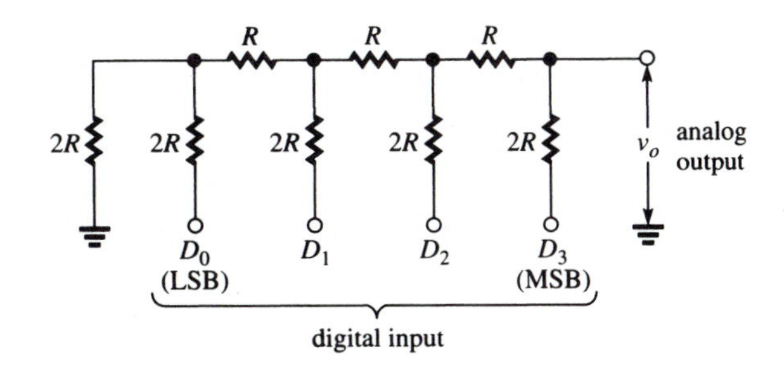

corresponding to the binary input 1000. We wish to find the total equivalent resistance, R_{eq}, looking to the left from node A. At the left end of the ladder, we see that $2R$ is in parallel with $2R$, so that combination is equivalent to R. That R is in series with another R, giving $2R$. That $2R$ is in parallel with another $2R$, which is equivalent to R once again. Continuing in this manner, we ultimately find that $R_{eq} = 2R$. In fact, we see that the equivalent resistance looking to the left from every node is $2R$. Figure 19–3(b) shows the circuit when it is redrawn with R_{eq} replacing all of the network to the left of node A. Figure 19–3(c) shows an equivalent way to draw the circuit in (b), and it is now readily apparent that $v_o = E/2$.

Figure 19–3
Calculating the output of the R-2R ladder when the input is 1000

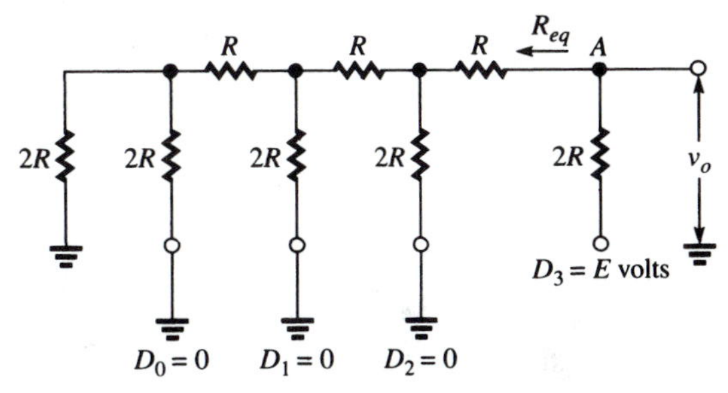

(a) When the input is 1000, D_0, D_1, and D_2 are grounded (0 V) and D_3 is E volts.

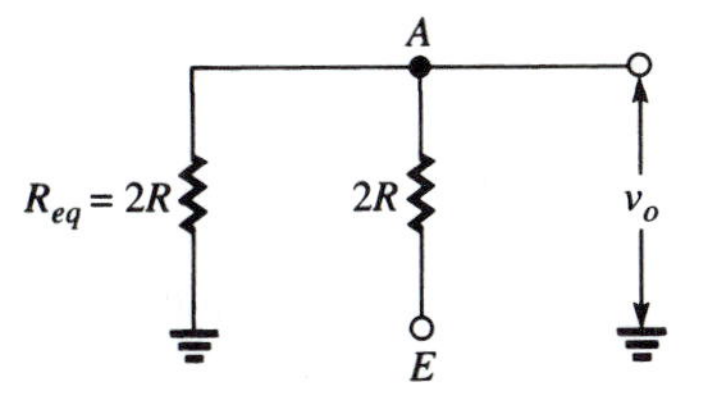

(b) The circuit equivalent to (a) when the network to the left of node A is replaced by its equivalent resistance, R_{eq}.

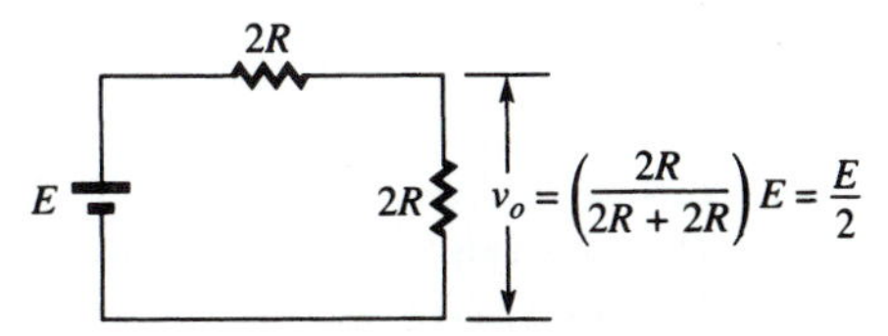

(c) Calculation of v_o using the voltage-divider rule. (Note that v_o in (b) is the voltage across $R_{eq} = 2R$.)

Figure 19–4
Calculating the output of the R-2R ladder when the input is 0100

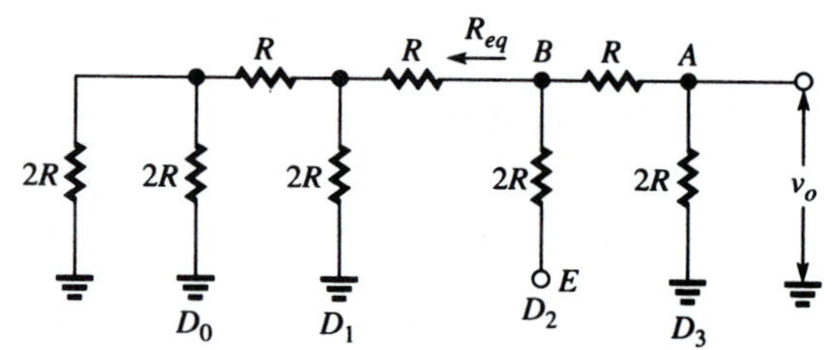

(a) When the input is 0100, D_0, D_1, and D_3 are grounded and D_2 is E volts.

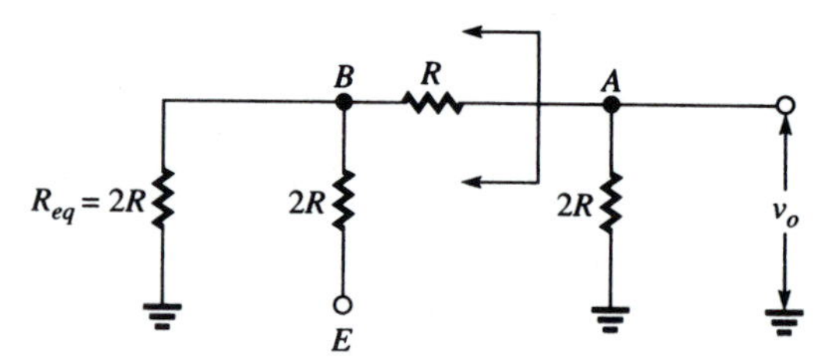

(b) The circuit equivalent to (a) when the network to the left of node B is replaced by its equivalent resistance ($2R$). The Thevenin equivalent circuit to the left of the bracketed arrows is shown in (c).

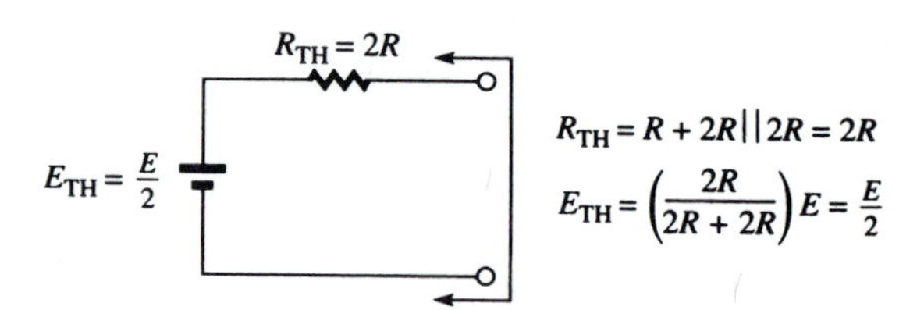

(c) The Thevenin equivalent circuit to the left of node A.

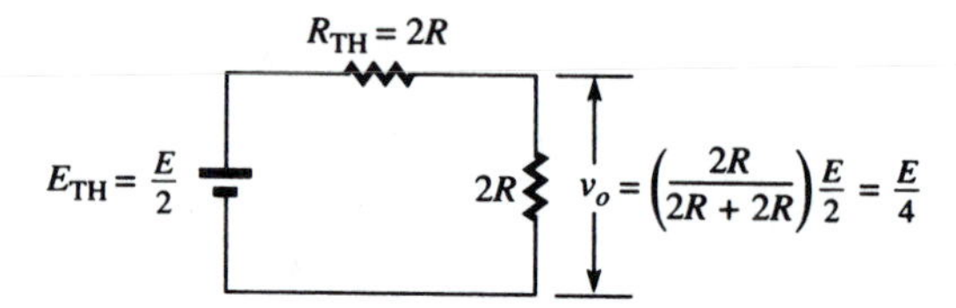

(d) Calculation of v_o using the voltage-divider rule.

Let us now find v_o when the input is 0100. Figure 19–4(a) shows the circuit, with D_2 connected to E and all other inputs grounded. As demonstrated earlier, the equivalent resistance looking to the left from node B is $2R$. The equivalent circuit with R_{eq} replacing the network to the left of B is shown in Figure 19–4(b). We proceed with the analysis by finding the (Thevenin) equivalent circuit to the left of node A, as indicated by the bracketed arrows. The Thevenin equivalent resistance (with E shorted to ground) is $R_{TH} = R + 2R \parallel 2R = R + R = 2R$. The Thevenin equivalent voltage is $[2R/(2R + 2R)]\,E = E/2$. The Thevenin equivalent circuit is shown in Figure 19–4(c). Figure 19–4(d) shows the ladder with the Thevenin equivalent circuit replacing everything to the left of node A. It is now apparent that $v_o = E/4$.

By an analysis similar to the foregoing, we find that the output of the ladder is $E/8$ when the input is 0010 and that the output is $E/16$ when the input is 0001.

In general, when the D_n input is 1 and all other inputs are 0, the output is

$$v_o = \frac{E}{2^{N-n}} \tag{19–1}$$

where N is the total number of binary inputs. For example, if $E = 5$ V, the output voltage when the input is 0010 is $(5 \text{ V})/(2^{4-1}) = (5 \text{ V})/8 = 0.625$ V.

To find the output voltage corresponding to *any* input combination, we can invoke the principle of superposition and simply add the voltages produced by the inputs where 1s are applied. For example, when the input is 1100, the output is $E/2 + E/4 = 3E/4$. The next example illustrates these computations.

Example 19–1

The logic levels used in a 4-bit R-$2R$ ladder DAC are 1 = +5 V and 0 = 0 V.

1. Find the output voltage when the input is 0001 and when it is 1010.
2. Sketch the output when the inputs are driven from a 4-bit binary up counter.

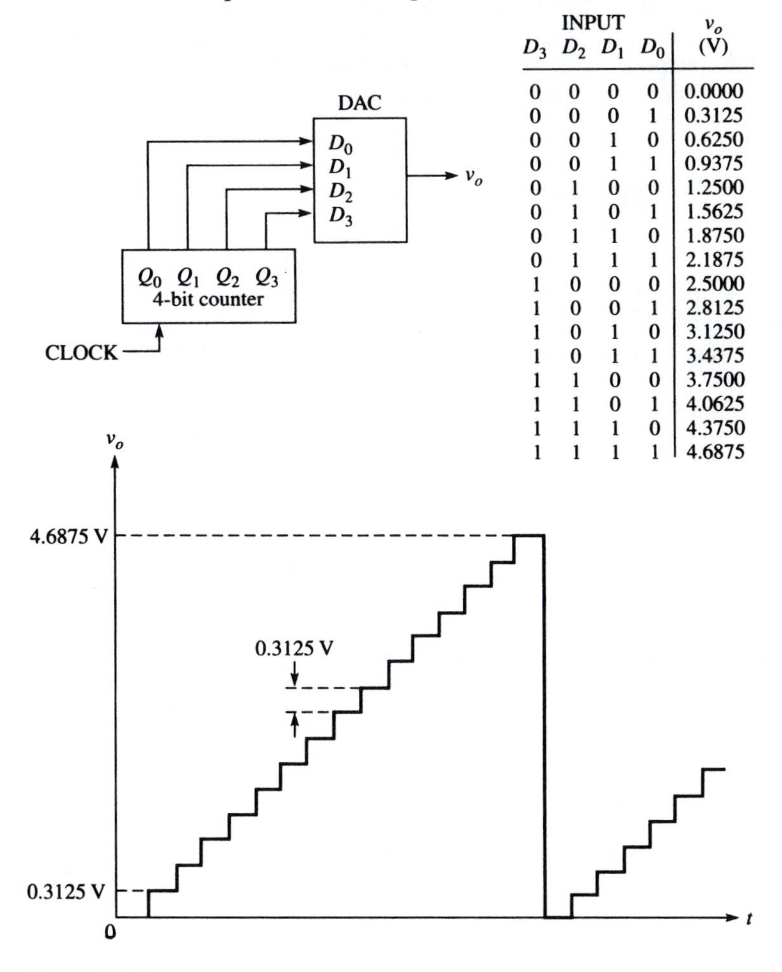

INPUT				v_o
D_3	D_2	D_1	D_0	(V)
0	0	0	0	0.0000
0	0	0	1	0.3125
0	0	1	0	0.6250
0	0	1	1	0.9375
0	1	0	0	1.2500
0	1	0	1	1.5625
0	1	1	0	1.8750
0	1	1	1	2.1875
1	0	0	0	2.5000
1	0	0	1	2.8125
1	0	1	0	3.1250
1	0	1	1	3.4375
1	1	0	0	3.7500
1	1	0	1	4.0625
1	1	1	0	4.3750
1	1	1	1	4.6875

Figure 19–5
(Example 19–1)

Solution

1. By equation 19–1, the output when the input is 0001 is

$$v_o = \frac{5\text{ V}}{2^{4-0}} = \frac{5\text{ V}}{16} = 0.3125\text{ V}$$

When the input is 1000, the output is (5 V)/2 = 2.5 V, and when the input is 0010, the output is (5 V)/8 = 0.625 V. Therefore, when the input is 1010, the output is 2.5 V + 0.625 V = 3.125 V.

2. Figure 19–5 shows a table of the output voltages corresponding to every input combination, calculated using the method illustrated in part (1) of this example. As the counter counts up, the output voltage increases by 0.3125 V at each new count. Thus, the output is the *staircase* waveform shown in the figure. The voltage steps from 0 V to 4.6875 V each time the counter counts from 0000 to 1111.

Typical values for R and $2R$ are 10 kΩ and 20 kΩ. For accurate conversion, the output voltage from the R-$2R$ ladder should be connected to a high impedance to prevent loading. Figure 19–6 shows how an operational amplifier can be used for that purpose. The output of the ladder is connected to a unity-gain voltage follower, whose input impedance is extremely large and whose output voltage is the same as its input voltage.

19–3 A WEIGHTED-RESISTOR DAC

Figure 19–7 illustrates another approach to digital-to-analog conversion. The operational amplifier is used to produce a *weighted sum* of the digital inputs, where the weights are proportional to the weights of the bit positions of the inputs. Recall from Chapter 13 that each input is amplified by a factor equal to the ratio of the feedback resistance divided by the input resistance to which it is connected. Thus, D_3, the most significant bit, is amplified by R_f/R, D_2 by $R_f/2R = 1/2(R_f/R)$, D_1 by $R_f/4R = 1/4(R_f/R)$, and D_o by $R_f/8R = 1/8(R_f/R)$. Since the amplifier sums and inverts, the output is

$$v_o = -\left(D_3 + \frac{1}{2}D_2 + \frac{1}{4}D_1 + \frac{1}{8}D_o\right)\frac{R_f}{R} \tag{19–2}$$

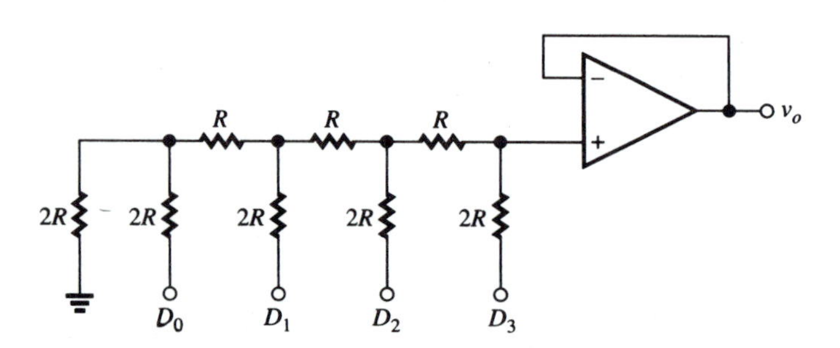

Figure 19–6
Using a unity-gain voltage follower to provide a high impedance for the R-2R ladder

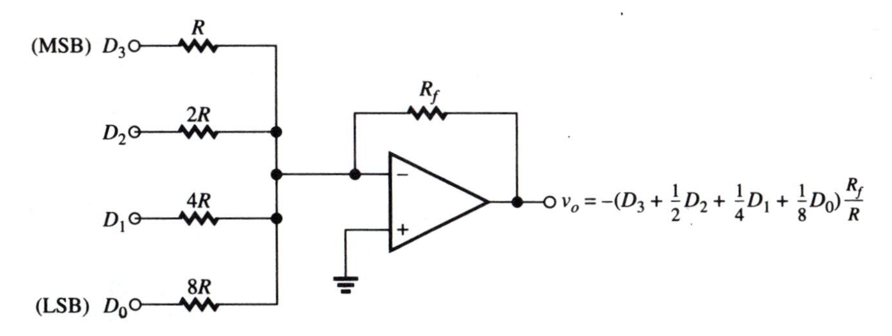

Figure 19–7
A weighted-resistor DAC using an inverting operational amplifier

The principal disadvantage of this type of converter is that a different-valued precision resistor must be used for each digital input. In contrast, the R-$2R$ ladder network uses only two values of resistance.

Example 19–2

Design a 4-bit, weighted-resistor DAC whose full-scale output voltage is −10 V. Logic levels are 1 = +5 V and 0 = 0 V. What is the output voltage when the input is 1010?

Solution. The full-scale output voltage is the output voltage when the input is maximum: 1111. In that case, from equation 19–2, we require

$$\left(5\text{ V} + \frac{5\text{ V}}{2} + \frac{5\text{ V}}{4} + \frac{5\text{ V}}{8}\right)\frac{R_f}{R} = 10\text{ V}$$

or

$$9.375\frac{R_f}{R} = 10$$

Let us choose $R_f = 10\text{ k}\Omega$. Then

$$R = \frac{9.375(10\text{ k}\Omega)}{10} = 9.375\text{ k}\Omega$$
$$2R = 18.75\text{ k}\Omega$$
$$4R = 37.50\text{ k}\Omega$$
$$8R = 75\text{ k}\Omega$$

When the input is 1010, the output voltage is

$$V_o = -\left(5\text{ V} + \frac{0\text{ V}}{2} + \frac{5\text{ V}}{4} + \frac{0\text{ V}}{8}\right)\frac{10\text{ k}\Omega}{9.375\text{ k}\Omega} = -6.667\text{ V}$$

19–4 THE SWITCHED CURRENT-SOURCE DAC

The D/A converters we have discussed so far can be regarded as switched voltage-source converters: When a binary input goes high, the high voltage is effectively switched into the circuit, where it is summed with other input voltages. Because of the technology used to construct integrated-circuit DACs, currents can be switched in and out of a circuit faster than voltages can. For that reason, most integrated-circuit DACs utilize some form of current switching, where the binary inputs are used to open and close switches that connect and disconnect internally generated currents. The currents are weighted according to the bit positions they represent and are summed in an operational amplifier. Figure 19–8 shows an example. Note that an R-$2R$ ladder is connected to a voltage source identified as E_{REF}. The current that flows in each $2R$ resistor is

$$I_n = \left(\frac{E_{REF}}{R}\right)\frac{1}{2^{N-n}} \tag{19–3}$$

when $n = 0, 1, \ldots, N - 1$ is the subscript for the current created by input D_n and N is the total number of inputs. Thus, each current is weighted according to the bit position it represents. For example, the current in the $2R$ resistor at the D_1 input in the figure ($n = 1$ and $N = 4$) is $(E_{REF}/R)(1/2)^3$. The binary inputs control switches that connect the currents either to ground or to the input of the amplifier. The amplifier sums all currents whose corresponding binary inputs are high. The amplifier also serves as a current-to-voltage converter. It is connected in an inverting configuration and its output is

$$v_o = -I_T R \tag{19–4}$$

where I_T is the sum of the currents that have been switched to its input. For example, if the input is 1001, then

$$I_T = \frac{E_{REF}}{R}\left(\frac{1}{2}\right) + \frac{E_{REF}}{R}\left(\frac{1}{16}\right) = \frac{E_{REF}}{R}\left(\frac{7}{16}\right)$$

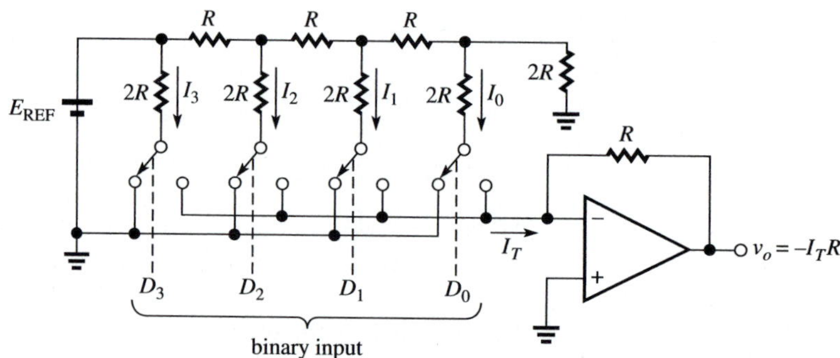

Figure 19–8
A 4-bit switched current-source DAC

and

$$v_o = -\left(\frac{E_{\text{REF}}}{R}\right)\left(\frac{7}{16}\right)R = -\frac{7}{16}E_{\text{REF}}$$

Example 19–3

The switched current-source DAC in Figure 19–8 has $R = 10\ \text{k}\Omega$ and $E_{\text{REF}} = 10\ \text{V}$. Find the total current delivered to the amplifier and the output voltage when the binary input is 1010.

Solution. From equation (19–3),

$$I_3 = \left(\frac{10\ \text{V}}{10\ \text{k}\Omega}\right)\frac{1}{2^{4-3}} = 0.5\ \text{mA}$$

and

$$I_1 = \left(\frac{10\ \text{V}}{10\ \text{k}\Omega}\right)\frac{1}{2^{4-1}} = 0.125\ \text{mA}$$

Therefore, $I_T = I_3 + I_1 = 0.5\ \text{mA} + 0.125\ \text{mA} = 0.625\ \text{mA}$. From equation (19–4),

$$v_o = -I_T R = -(0.625\ \text{mA})(10\ \text{k}\Omega) = -6.25\ \text{V}$$

The reference voltage E_{REF} in Figure 19–8 may be fixed—as, for example, when it is generated internally in an integrated circuit—or it may be externally variable. When it is externally variable, the output of the DAC is proportional to the *product* of the variable E_{REF} and the variable binary input. In that case, the circuit is called a *multiplying* D/A converter, and the output represents the product of an analog input (E_{REF}) and a digital input.

In most integrated-circuit DACs utilizing current-source switching, the output is the total current I_T produced in the R-$2R$ ladder. The user may then connect a variety of operational amplifier configurations at the output to perform magnitude scaling and/or phase inversion. If E_{REF} can be both positive and negative and if the binary input always represents a positive number, then the output of the DAC is both positive and negative. In that case, the DAC is called a *two-quadrant multiplier.* A DAC whose output can be both positive and negative is said to be *bipolar,* and one whose output is only positive or only negative is *unipolar.* A four-quadrant multiplier is one in which both inputs can be either negative or positive and in which the output (product) has the correct sign for every case. Negative binary inputs can be represented using the *offset binary code,* the 4-bit version of which is shown in Table 19–1. Note that the numbers +7 through −8 are represented by 0000 through 1111, respectively, with $0)_{10}$ represented by 1000. The table also shows the analog outputs that are produced by a multiplying D/A converter. In general, for an N-bit DAC, the offset binary code represents decimal numbers from $+2^{N-1} - 1$ through -2^{N-1} and the analog outputs range from

$$+\left(\frac{2^{N-1}-1}{2^N-1}\right)E_{\text{REF}} \quad \text{through} \quad -E_{\text{REF}}$$

in steps of size $(1/2^{N-1})E_{\text{REF}}$.

Table 19–1
The offset binary code used to represent positive and negative binary inputs to a D/A converter

Decimal Number	Offset Binary Code	Analog Output
+7	1111	$+(7/8)\,E_{REF}$
+6	1110	$+(6/8)\,E_{REF}$
+5	1101	$+(5/8)\,E_{REF}$
+4	1100	$+(4/8)\,E_{REF}$
+3	1011	$+(3/8)\,E_{REF}$
+2	1010	$+(2/8)\,E_{REF}$
+1	1001	$+(1/8)\,E_{REF}$
0	1000	0
−1	0111	$-(1/8)\,E_{REF}$
−2	0110	$-(2/8)\,E_{REF}$
−3	0101	$-(3/8)\,E_{REF}$
−4	0100	$-(4/8)\,E_{REF}$
−5	0011	$-(5/8)\,E_{REF}$
−6	0010	$-(6/8)\,E_{REF}$
−7	0001	$-(7/8)\,E_{REF}$
−8	0000	$-(8/8)\,E_{REF}$

19–5 SWITCHED-CAPACITOR DACs

The newest technology used to construct D/A converters employs weighted capacitors instead of resistors. In this method, charged capacitors form a capacitive voltage divider whose output is proportional to the sum of the binary inputs.

As an aid in understanding the method, let us review the theory of capacitive voltage dividers. Figure 19–9 shows a two-capacitor example. The total equivalent capacitance of the two series-connected capacitors is

$$C_T = \frac{C_1 C_2}{C_1 + C_2} \tag{19–5}$$

Therefore, the total charge delivered to the circuit, which is the same as the charge on both C_1 and C_2, is

$$Q_1 = Q_2 = Q_T = C_T E = \left(\frac{C_1 C_2}{C_1 + C_2}\right) E \tag{19–6}$$

The voltage across C_2 is

$$V_2 = \frac{Q_2}{C_2} = \frac{Q_T}{C_2} = \frac{\left(\dfrac{C_1 C_2}{C_1 + C_2}\right) E}{C_2} = \left(\frac{C_1}{C_1 + C_2}\right) E \tag{19–7}$$

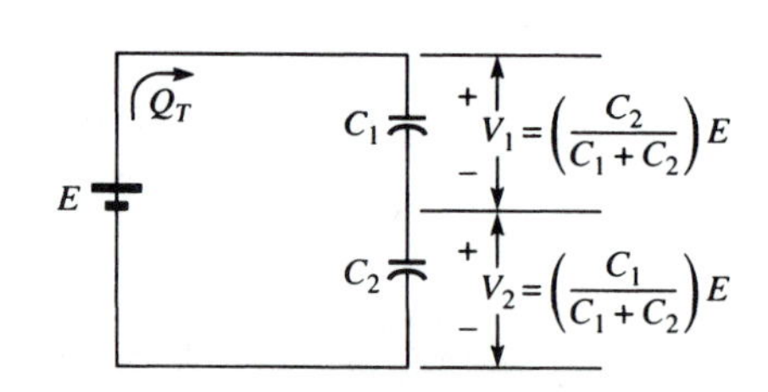

Figure 19–9
The capacitive voltage divider

Similarly,

$$V_1 = \left(\frac{C_2}{C_1 + C_2}\right) E \tag{19–8}$$

Figure 19–10(a) shows an example of a 4-bit switched-capacitor DAC. Note that the capacitance values have binary weights. A *two-phase* clock is used to control switching of the capacitors. The two-phase clock consists of clock signals ϕ_1 and ϕ_2; ϕ_1 goes high while ϕ_2 is low, and ϕ_2 goes high while ϕ_1 is low. When ϕ_1 goes high, all capacitors are switched to ground and discharged. When ϕ_2 goes high, those capacitors where the digital input is high are switched to E_{REF}, whereas those whose inputs are low remain grounded. Figure 19–10(b) shows the equivalent circuit when ϕ_2 is high and the digital input is 1010. We see that the two capacitors whose digital inputs are 1 are in parallel, as are the two capacitors whose digital inputs

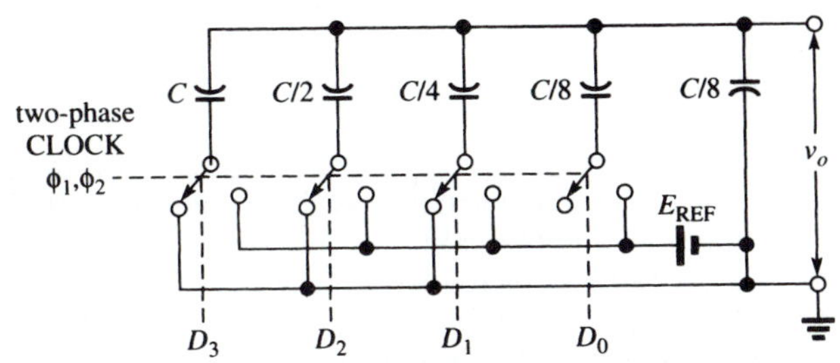

(a) All capacitors are switched to ground by ϕ_1. Those capacitors whose digital inputs are high are switched to E_{REF} by ϕ_2.

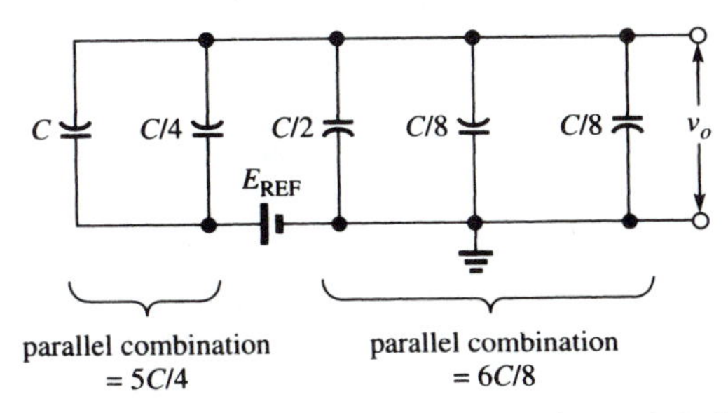

(b) Equivalent circuit when the input is 1010. The capacitors switched to E_{REF} are in parallel as are the ones connected to ground.

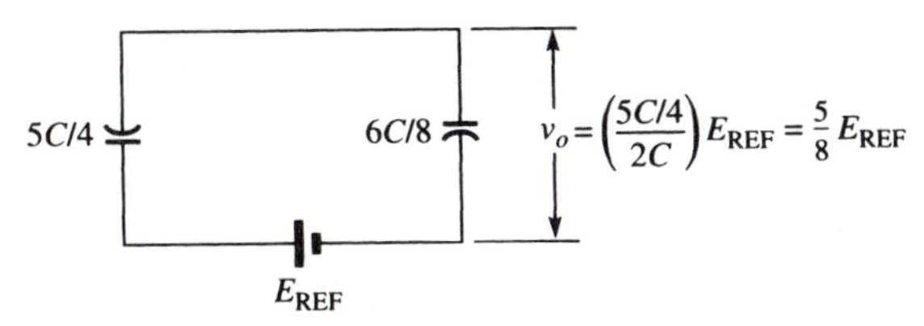

(c) Circuit equivalent to (b). The output is determined by a capacitive voltage divider.

Figure 19–10
The switched-capacitor D/A converter

are 0. The circuit is redrawn in Figure 19–10(c) with the parallel capacitors replaced by their equivalents (sums). The output of the capacitive voltage divider is

$$v_o = \left(\frac{5C/4}{5C/4 + 6C/8}\right) E_{REF} = \left(\frac{5C/4}{2C}\right) E_{REF} = \frac{5}{8} E_{REF} \quad \textbf{(19–9)}$$

The denominator in (19–9) will always be $2C$, the sum of all the capacitance values in the circuit. From the foregoing analysis, we see that the output of the circuit in the general case is

$$v_o = \left(\frac{C_{eq}}{2C}\right) E_{REF} \quad \textbf{(19–10)}$$

where C_{eq} is the equivalent (sum) of all the capacitors whose digital inputs are high. Table 19–2 shows the outputs for every possible input combination, and it is apparent that the analog output is proportional to the digital input. For the case where the input is 0000, note that the positive terminal of E_{REF} in Figure 19–10 is effectively open-circuited, so the output is 0 V.

Switched-capacitor technology evolved as a means for implementing analog functions in integrated circuits, particularly MOS circuits. It has been used to construct filters, amplifiers, and many other special devices. The principal advantage of the technology is that small capacitors, on the order of a few picofarads, can be constructed in the integrated circuits to perform the function of the much larger capacitors that are normally needed in low-frequency analog circuits. When capacitors are switched at a high enough frequency, they can be effectively "transformed" into other components, including resistors. The transformations are studied from the standpoint of sampled-data theory, which is beyond the scope of this book.

Table 19–2
Output voltages produced by a 4-bit switched-capacitor DAC

Binary Input $D_3D_2D_1D_0$	V_o
0000	0
0001	(1/16) E_{REF}
0010	(1/8) E_{REF}
0011	(3/16) E_{REF}
0100	(1/4) E_{REF}
0101	(5/16) E_{REF}
0110	(3/8) E_{REF}
0111	(7/16) E_{REF}
1000	(1/2) E_{REF}
1001	(9/16) E_{REF}
1010	(5/8) E_{REF}
1011	(11/16) E_{REF}
1100	(3/4) E_{REF}
1101	(13/16) E_{REF}
1110	(7/8) E_{REF}
1111	(15/16) E_{REF}

19–6 DAC PERFORMANCE SPECIFICATIONS

As discussed in Section 19–1, the resolution of a D/A converter is a measure of the fineness of the increments between output values. Given a fixed output voltage range, say, 0 to 10 V, a DAC that divides that range into 1024 distinct output values has greater resolution than one that divides it into 512 values. Since the output increment is directly dependent on the number of input bits, the resolution is often quoted as simply that total number of bits. The most commonly available integrated-circuit converters have resolutions of 8, 10, 12, or 16 bits. Resolution is also expressed as the reciprocal of the total number of output voltages, often in terms of a percentage. For example, the resolution of an 8-bit DAC may be specified as

$$\left(\frac{1}{2^8}\right) \times 100\% = 0.39\%$$

Some DAC specifications are quoted with reference to one or to one-half LSB (least significant bit). In this context, an LSB is simply the increment between successive output voltages. Since an n-bit converter has $2^n - 1$ such increments,

$$\text{LSB} = \frac{\text{FSR}}{2^n - 1} \tag{19–11}$$

where FSR is the full-scale range of the output voltage.

When the output of a DAC changes from one value to another, it typically overshoots the new value and may oscillate briefly around that new value before it settles to a constant voltage. The *settling time* of a D/A converter is the total time between the instant that the digital input changes and the time that the output enters a specified error band for the last time, usually ±1/2 LSB around the final value. Figure 19–11 illustrates the specification. Settling times of typical integrated-

Figure 19–11
The settling time, t_s, of a D/A converter

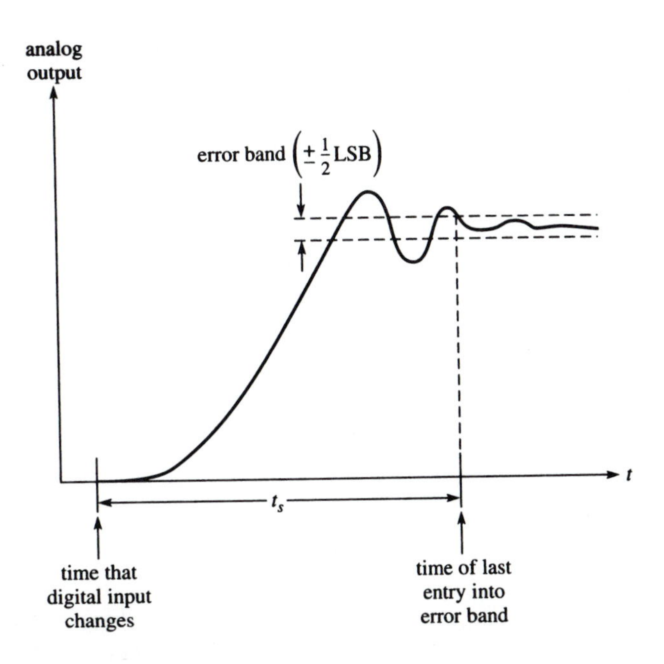

circuit converters range from 50 ns to several microseconds. Settling time may depend on the magnitude of the change at the input and is often specified for a prescribed input change.

Linearity error is the maximum deviation of the analog output from the ideal output. Since the output is ideally in direct proportion to the input, the ideal output is a straight line drawn from 0 V. Linearity error may be specified as a percentage of the full-scale range or in terms of an LSB.

Differential linearity error is the difference between the ideal output increment (1 LSB) and the actual increment. For example, each output increment of an 8-bit DAC whose full-scale range is 10 V should be $(10\text{ V})/(2^8 - 1) = 39.22$ mV. If any one increment between two successive values is, say, 30 mV, then there is a differential linearity error of 9.22 mV. This error is also specified as a percentage of the full-scale range or in terms of an LSB. If the differential linearity error is greater than 1 LSB, it is possible for the output voltage to *decrease* when there is an increase in the value of the digital input or to increase when the input decreases. Such behavior is said to be *nonmonotonic.* In a monotonic DAC, the output always increases when the input increases and decreases when the input decreases.

The input of a DAC is said to undergo a *major change* when every input bit changes, as, for example, from 01111111 to 10000000. If the switches in Figure 19–8 open faster than they close or vice versa, the output of the DAC will momentarily go to 0 or to full scale when a major change occurs, thus creating an output *glitch. Glitch area* is the total area of an output voltage glitch in volt-seconds or of an output current glitch in ampere-seconds. Commercial units are often equipped with *deglitchers* to minimize glitch area.

An Integrated-Circuit DAC

The AD7524 CMOS integrated-circuit DAC, manufactured by Analog Devices and available from other manufacturers, is an example of an 8-bit, multiplying D/A converter. Figure 19–12 shows a functional block diagram. Note that the digital input is latched under the control of $\overline{CS}$ and $\overline{WR}$. When both of these control inputs

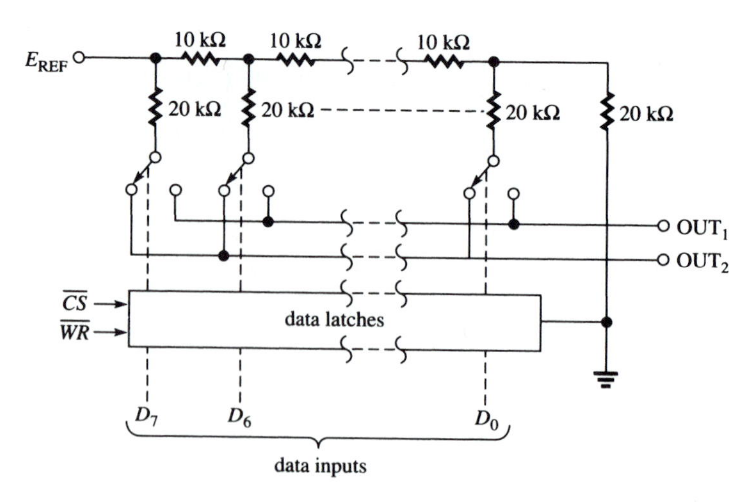

Figure 19–12
Functional block diagram of the AD7524 D/A converter

are low, the output of the DAC responds directly to the digital inputs, with no latching occurring. If either control input goes high, the digital input is latched and the analog output remains at the level corresponding to the latched data, independent of any changes in the digital input. The device is then said to be in a HOLD mode, with the data bus *locked out.* The OUT2 output is normally grounded. Maximum settling time to a ±1/2 LSB error band for the Texas Instruments version is 100 ns and maximum linearity error is ±0.2% of the full-scale range. The device can be used as a 2- or 4-quadrant multiplier and E_{REF} can vary ±25 V.

19–7 THE COUNTER-TYPE ADC

The simplest type of A/D converter employs a binary counter, a voltage comparator, and a D/A converter, as shown in Figure 19–13(a). Recall from Section 15–1 that the output of the voltage comparator is high as long as its v^+ input is greater than its v^- input. Notice that the analog input is the v^+ input to the comparator. As long as it is greater than the v^- input, the AND gate is enabled and clock pulses are passed to the counter. The digital output of the counter is converted to an analog voltage by the DAC, and that voltage is the other input to the comparator. Thus, the counter counts up until its output has a value equal to the analog input. At that time, the comparator switches low, inhibiting the clock pulses, and counting ceases. The count it reached is the digital output proportional to the analog input. Control circuitry, not shown, is used to latch the output and reset the counter. The cycle is repeated, with the counter reaching a new count proportional to whatever new value the analog input has acquired. Figure 19–13(b) illustrates the output of a 4-bit DAC in an ADC over several counting cycles when the analog input is a slowly varying voltage. The principal disadvantage of this type of converter is that

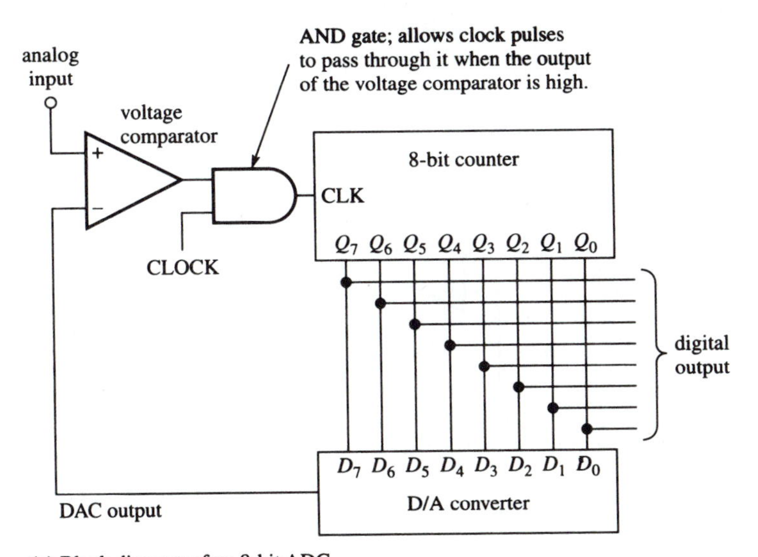

(a) Block diagram of an 8-bit ADC.

Figure 19–13
The counter-type A/D converter

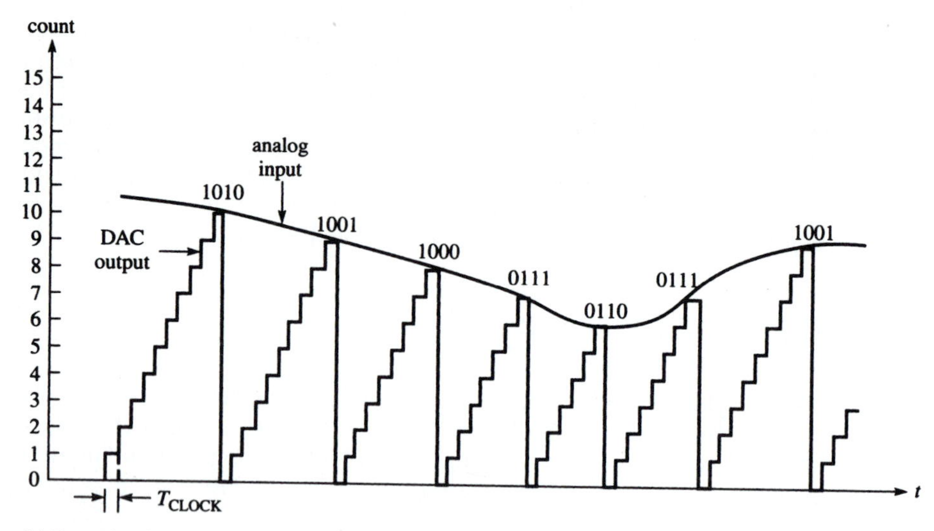

(b) Example of the output of the DAC in a 4-bit ADC.

Figure 19–13
(Continued)

the conversion time depends on the magnitude of the analog input: the larger the input, the more clock pulses must pass to reach the proper count. An 8-bit converter could require as many as 255 clock pulses to perform a conversion, so the counter-type ADC is considered quite slow in comparison to other types we will study.

Tracking A/D Converter

To reduce conversion times of the counter-type ADC, the up counter can be replaced by an up/down counter, as illustrated in Figure 19–14. In this design, the

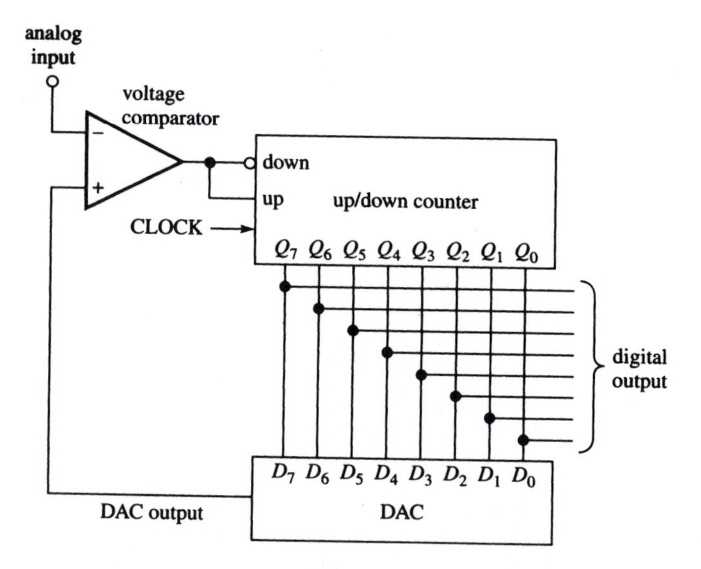

Figure 19–14
The tracking counter-type A/D converter. The counter counts up or down from its last count to reach its next count rather than resetting to 0 between counts.

counter is not reset after each conversion. Instead, it counts up or down from its last count to its new count. Thus, the total number of clock pulses required to perform a conversion is proportional to the *change* in the analog input between counts rather than to its magnitude. Since the count more or less keeps up with the changing analog input, the type of ADC is called a *tracking converter.* A disadvantage of the design is that the count may oscillate up and down from a fixed count when the analog input is constant.

19–8 FLASH A/D CONVERTERS

The fastest type of A/D converter is called the *flash* (or simultaneous, or parallel) type. As shown for the 3-bit example in Figure 19–15, a reference voltage is connected to a voltage divider that divides it into 7 ($2^n - 1$) equal-increment levels. Each level is compared to the analog input by a voltage comparator. For any given analog input, one comparator and all those below it will have a high output. All comparator outputs are connected to a *priority encoder.* A priority encoder produces a binary output corresponding to the input having the highest priority, in this case, the one representing the largest voltage level equal to or less than the analog

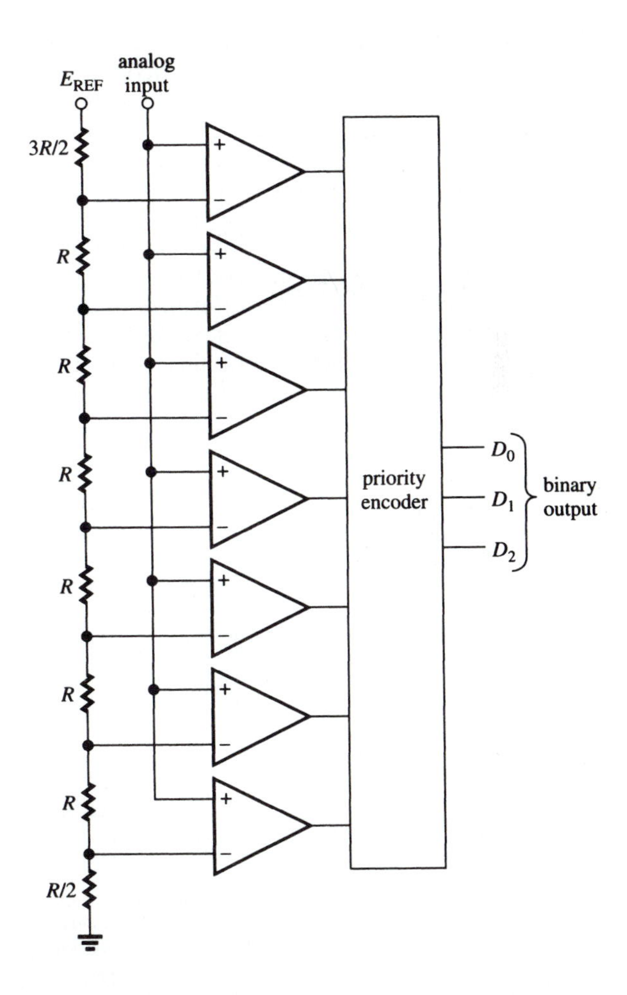

Figure 19–15
A 3-bit flash A/D converter

input. Thus, the binary output represents the voltage that is closest in value to the analog input.

The voltage applied to the v^- input of the uppermost comparator in Figure 19–15 is, by voltage-divider action,

$$\left(\frac{6R + R/2}{6R + R/2 + 3R/2}\right) E_{\text{REF}} = \frac{13R/2}{16R/2} E_{\text{REF}} = \frac{13}{16} E_{\text{REF}} \qquad \textbf{(19–12)}$$

Similarly, the voltage applied to the v^- input of the second comparator is $(11/16)E_{\text{REF}}$, that applied to the third is $(9/16)E_{\text{REF}}$, and so forth. The increment between voltages is $(2/16)E_{\text{REF}}$, or $(1/8)E_{\text{REF}}$. An n-bit flash comparator has $n - 2R$-valued resistors, and the increment between voltages is

$$\Delta V = \frac{1}{2^n} E_{\text{REF}} \qquad \textbf{(19–13)}$$

The voltage levels range from

$$\left(\frac{2^{n+1} - 3}{2^{n+1}}\right) E_{\text{REF}} \quad \text{through} \quad \left(\frac{1}{2^{n+1}}\right) E_{\text{REF}}$$

The flash converter is fast because the only delays in the conversion are in the comparators and the priority encoder. Under the control of a clock, a new conversion can be performed very soon after one conversion is complete. The principal disadvantage of the flash converter is the need for a large number of voltage comparators ($2^n - 1$). For example, an 8-bit flash ADC requires 255 comparators.

Figure 19–16 shows a block diagram of a modified flash technique that uses 30 comparators instead of 255 to perform an 8-bit A/D conversion. One 4-bit flash converter is used to produce the 4 most significant bits. Those 4 bits are converted back to an analog voltage by a D/A converter and the voltage is subtracted from the analog input. The difference between the analog input and the analog voltage corresponding to the 4 most significant bits is an analog voltage corresponding to the 4 least significant bits. Therefore, that voltage is converted to the 4 least significant bits by another 4-bit flash converter.

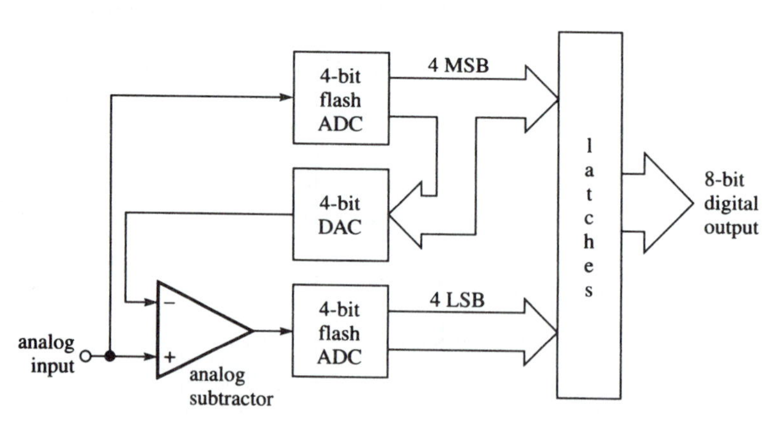

Figure 19–16
Modified flash converter that uses 30 comparators instead of 255 to produce an 8-bit output

19–9 THE DUAL-SLOPE (INTEGRATING) ADC

A dual-slope ADC uses an operational amplifier to integrate the analog input. Recall from Chapter 14 that the output of an integrator is a ramp when the input is a fixed level (Figure 14–33). The slope of the ramp is $\pm E_{in}/R_1C$, where E_{in} is the input voltage that is integrated and R_1 and C are the fixed components of the integrating operational amplifier. Since R_1 and C are fixed, the slope of the ramp is directly dependent on the value of E_{in}. If the ramp is allowed to continue for a fixed time, the voltage it reaches in that time depends on the slope of the ramp and hence on the value of E_{in}. The basic principle of the integrating ADC is that the voltage reached by the ramp controls the length of time that a binary counter is allowed to count. Thus, a binary number proportional to the value of E_{in} is obtained. In the dual-slope ADC, two integrations are performed, as described next.

Figure 19–17(a) shows a functional block diagram of the dual slope ADC. Recall that the integrating operational amplifier inverts, so a positive input generates a negative-going ramp and vice versa. A conversion begins with the switch connected to the analog input. Assume that the input is negative, so a positive-going ramp is generated by the integrator. As discussed earlier, the ramp is allowed to continue for a fixed time, and the voltage it reaches in that time is directly dependent on the analog input. The fixed time is controlled by sensing the time when the counter reaches a specific count. At that time, the counter is reset and control circuitry causes the switch to be connected to a reference voltage having a polarity *opposite* to that of the analog input—in this case, a positive voltage. As a consequence, the output of the integrator becomes a negative-going ramp, beginning from the positive value it reached during the first integration. Since the reference voltage is fixed, so is the slope of the negative-going ramp. When the negative-going ramp reaches 0 V, the voltage comparator switches, the clock pulses are inhibited, and the counter ceases to count. The count it contains at that time is proportional to the time required for the negative-going ramp to reach 0 V, which is proportional to the positive voltage reached in the first integration. Thus, the binary count is proportional to the value of the analog input. Figure 19–17(b) shows examples of the ramp waveforms generated by a small analog input and by a large analog input. Note that the slope of the positive-going ramp is variable (depending on E_{in}), and the slope of the negative-going ramp is fixed. The origin of the name *dual-slope* is now apparent.

One advantage of the dual-slope converter is that its accuracy depends neither on the values of the integrator components R_1 and C nor upon any long-term changes that may occur in them. This fact is demonstrated by examining the equations governing the times required for the two integrations. Since the slope of the positive-going ramp is E_{in}/R_1C, the maximum voltage V_M reached by the ramp in time t_1 is

$$V_M = \frac{|E_{in}|t_1}{R_1C} \tag{19–14}$$

The magnitude of the slope of the negative-going ramp is $|E_{\text{REF}}|/R_1C$, so

$$V_M = \frac{|E_{\text{REF}}|}{R_1C}(t_2 - t_1) \tag{19–15}$$

Equating (19–14) and (19–15),

$$\frac{|E_{in}|}{R_1C}t_1 = \frac{|E_{\text{REF}}|}{R_1C}(t_2 - t_1) \tag{19–16}$$

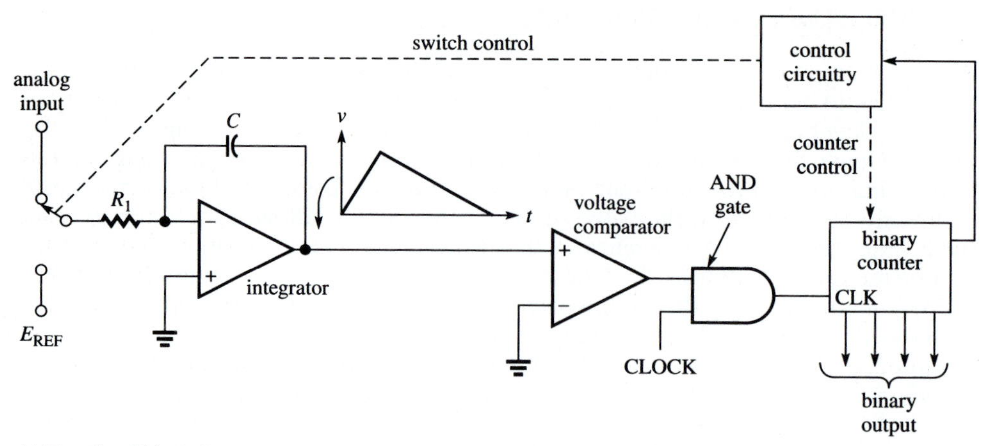

(a) Functional block diagram.

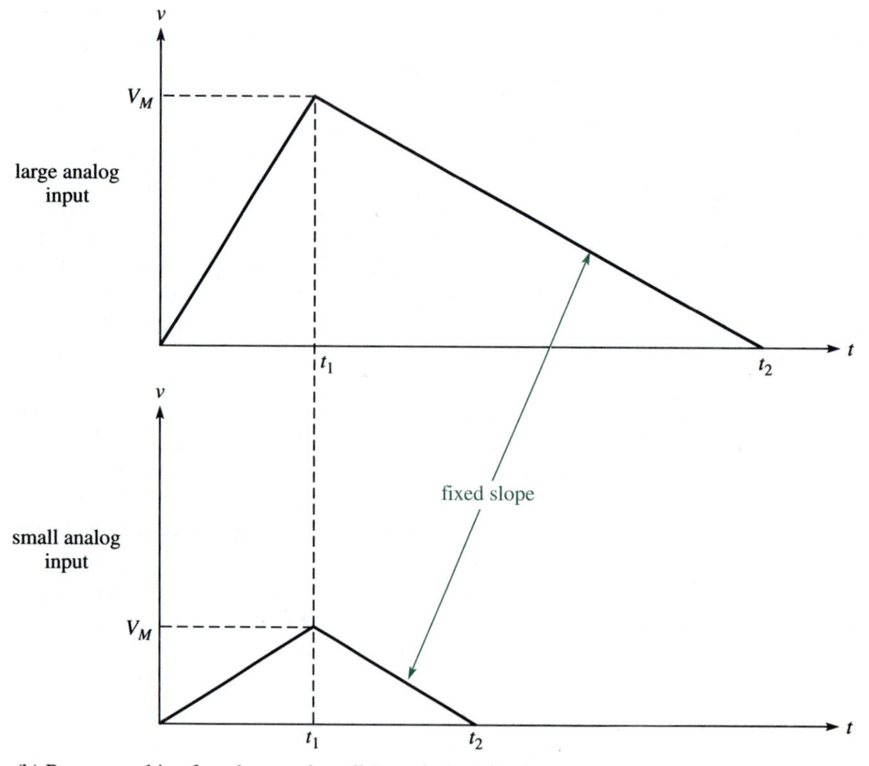

(b) Ramps resulting from large and small (negative) analog inputs.

Figure 19–17
The dual-slope integrating A/D converter

Cancelling R_1C on both sides and solving for $t_2 - t_1$ gives

$$t_2 - t_1 = \frac{|E_{in}|}{|E_{\text{REF}}|} t_1 \tag{19–17}$$

Since the counter contains a count proportional to $t_2 - t_1$ (the time required for the negative-going ramp to reach 0 V) and t_1 is fixed, equation (19–17) shows that the count is directly proportional to E_{in}, the analog input. Note that this expression does not contain R_1 or C, since those quantities cancelled out in (19–16). Thus, accuracy does not depend on their values. Furthermore, accuracy does not depend on the frequency of the clock. Equation (19–17) shows that accuracy does depend on E_{REF}, so the reference voltage should be very precise.

An important advantage of the dual-slope A/D converter is that the integrator suppresses noise. Recall from equation (14–46) that the output of an integrator has amplitude inversely proportional to frequency. Thus, high-frequency noise components in the analog input are attenuated. This property makes it useful for instrumentation systems, and it is widely used for applications such as digital voltmeters. However, the integrating ADC is not particularly fast, so its use is restricted to signals having low to medium frequencies.

19–10 THE SUCCESSIVE-APPROXIMATION ADC

The method called successive approximation is the most popular technique used to construct A/D converters, and, with the exception of flash converters, successive-approximation converters are the fastest of those we have discussed. Figure 19–18(a) shows a block diagram of a 4-bit version. The method is best explained by way of an example. For simplifying purposes, let us assume that the output of the D/A converter ranges from 0 V through 15 V as its binary input ranges from 0000 through 1111, with 0000 producing 0 V, 0001 producing 1 V, and so forth. Suppose the "unknown" analog input is 13 V. On the first clock pulse, the output register is loaded with 1000, which is converted by the DAC to 8 V. The voltage comparator determines that 8 V is less than the analog input (13 V), so on the next clock pulse, the control circuitry causes the output register to be loaded with 1100. The output of the DAC is now 12 V, which the comparator again determines is less than the analog input. Consequently, the register is loaded with 1110 on the next clock pulse.

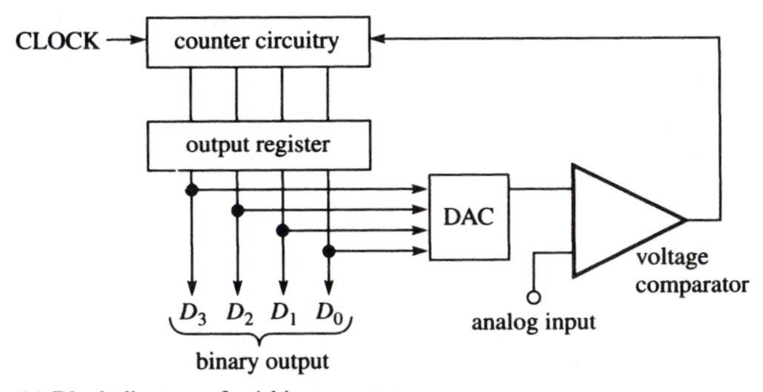

(a) Block diagram of a 4-bit converter.

Figure 19–18
The successive-approximation A/D converter

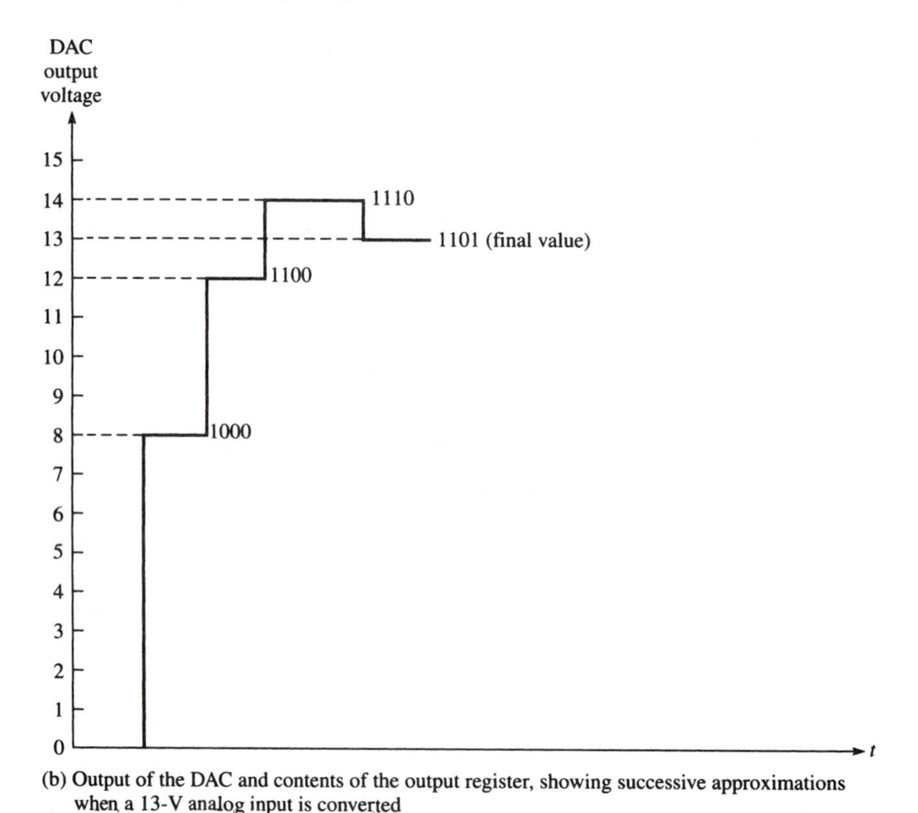

(b) Output of the DAC and contents of the output register, showing successive approximations when a 13-V analog input is converted

Figure 19–18
(Continued)

The output of the DAC is 14 V, which the comparator now determines is larger than the 13-V analog input. Therefore, the last 1 that was loaded into the register is replaced with a 0, and a 1 is loaded into the LSB. This time, the output of the DAC is 13 V, which equals the analog input, so the conversion is complete. The output register contains 1101. We see that the method of successive approximation amounts to testing a sequence of trial values, each of which is adjusted to produce a number closer in value to the input than the previous value. This "homing-in" on the correct value is illustrated in part (b) of the figure. Note that an n-bit conversion requires n clock pulses. As another example, the following is the sequence of binary numbers that would appear in the output register of an 8-bit successive-approximation converter when the analog input is a voltage that is ultimately converted to 01101001:

10000000
01000000
01100000
01110000
01101000
01101100
01101010
01101001

Some modern successive-approximation converters have been constructed using the switched-capacitor technology discussed in connection with DACs.

The primary component affecting the accuracy of a successive-approximation converter is the D/A converter. Consequently, the reference voltage connected to it and its ladder network must be very precise for accurate conversions. Also, the analog input should remain fixed during the conversion time. Some units employ a *sample-and-hold* circuit to ensure that the input voltage being compared at the voltage comparator does not vary during the conversion. A sample-and-hold is the analog counterpart of a digital data latch. It is constructed using electronic switches, an operational amplifier, and a capacitor that charges to and holds a particular voltage level.

19–11 ADC PERFORMANCE SPECIFICATIONS

The resolution of an A/D converter is the smallest change that can be distinguished in the analog input. As in DACs, resolution depends directly on the number of bits, so it is often quoted as simply the total number of output bits. The actual value depends on the full-scale range (FSR) of the analog input:

$$\text{resolution} = \frac{\text{FSR}}{2^n} \qquad \textbf{(19–18)}$$

Some A/D converters have 8-4-2-1 BCD outputs rather than straight binary. This is especially true of dual-slope types designed for use in digital instruments. The BCD outputs facilitate driving numerical displays. The resolution of a BCD converter is quoted as the number of (decimal) digits available at the output, where each digit is represented by 4 bits. In this context, the term $\frac{1}{2}$ *digit* is used to refer to a single binary output. For example, a $1\frac{1}{2}$-digit output is represented by 5 bits. The $\frac{1}{2}$ digit is used as the most significant bit, and its presence doubles the number of decimal values that can be represented by the 4 BCD bits. Some BCD converters employ multiplexers to expand the number of output digits. An example is Texas Instruments' TLC135C $4\frac{1}{2}$-digit A/D converter. Expressed as a voltage, the resolution of a BCD A/D converter is FSR/10^d, where d is the number of output digits.

The time required to convert a single analog input to a digital output is called the *conversion time* of an A/D converter. Conversion time may be quoted as including any other delays, such as access time, associated with acquiring and converting an analog input. In that case, the total number of conversions that can be performed each second is the reciprocal of the conversion time. The reason this specification is important is that it imposes a limit on the rate at which the analog input can be allowed to change. In effect, the A/D converter *samples* the changing analog input when it performs a sequence of conversions. The *Shannon sampling theorem* states that sampled data can be used to faithfully reproduce a time-varying signal provided that the sampling rate is at least *twice* the frequency of the highest-frequency component in the signal. Thus, for the sequence of digital outputs to be a valid representation of the analog input, the A/D converter must perform conversions at a rate equal to at least twice the frequency of the highest component of the input.

Example 19–4

What maximum conversion time can an A/D converter have if it is to be used to convert *audio* input signals? (The audio frequency range is considered to be 20 Hz to 20 kHz.)

Solution. Since the highest frequency in the input may be 20 kHz, conversions should be performed at a rate of at least 40×10^3 conversions/s. The maximum allowable conversion time is therefore equal to

$$\frac{1}{40 \times 10^3} = 25\ \mu\text{s}$$

One LSB for an A/D converter is defined in the same way it is for a D/A converter:

$$\text{LSB} = \frac{\text{FSR}}{2^{n-1}} \qquad \textbf{(19–19)}$$

where FSR is the full-scale range of the analog input. Other A/D converter specifications may be quoted in terms of an LSB or as a percentage of FSR.

Integrated-Circuit A/D Converters

A wide variety of A/D converters of all the types we have discussed are available in integrated circuits. Most have additional features, such as latched *three-state outputs* (0, 1, and open circuit), that make them compatible with microprocessor systems. Some have *differential* inputs, wherein the voltage converted to a digital output is the difference between two analog inputs. Differential inputs are valuable in reducing the effects of noise, because any noise signal common to both inputs is "differenced out." Recall that this property is called common-mode rejection. Differential inputs also allow the user to add or subtract a fixed voltage to the analog input, thereby offsetting the values converted. Conventional (single-ended) inputs can be accommodated simply by grounding one of the differential inputs.

The advances in MOS technology that have made high-density memory circuits possible have also made it possible to incorporate many special functions into a single-integrated circuit containing an ADC. Examples include sample-and-hold circuitry and analog multiplexers. With these functions, and versatile control circuitry, complete microprocessor-compatible *data-acquisition* systems have become available in a single-integrated circuit. The microprocessor can be programmed to control the multiplexer so that analog data from many different sources (up to 19 in some versions) can be sampled in a desired sequence. The analog data from an instrumentation system, for example, is sampled, converted, and transmitted directly to the microprocessor for storing in memory or further processing.

EXERCISES

Section 19–2
The R-2R Ladder DAC

19–1. The logic levels used in an 8-bit R-$2R$ ladder DAC are 1 = 5 V and 0 = 0 V. Find the output voltage for each input:
a. 00100000
b. 10100100

19–2. The logic levels used in a 6-bit R-$2R$ ladder DAC are 1 = +5 V and 0 = 0 V. What is the binary input when the analog output is 3.28125 V?

Section 19–3
A Weighted-Resistor DAC

19–3. Design a 5-bit weighted-resistor DAC whose full-scale output voltage is −15 V. Logic levels are 1 = +5 V and 0 = 0 V. What is the output voltage when the input is 01010?

19–4. Design an 8-bit weighted-resistor DAC using operational amplifiers and two *identical* 4-bit DACs of the design shown in Figure 19–7. The output should be +10 V when the input is 11110000. (*Hint:* Sum the outputs of two amplifiers in a third amplifier, using appropriate voltage gains in the summation.) What is the full-scale output of your design?

Section 19–4
The Switched Current-Source DAC

19–5. In the switched current-source DAC shown in Figure 19–8, R = 10 kΩ and E_{REF} = 20 V. Find the current in each 20-kΩ resistor.

19–6. An 8-bit switched current-source DAC of the design shown in Figure 19–8 has R = 10 kΩ and E_{REF} = 15 V. Find the total current I_T delivered to the amplifier and the output voltage when the input is 01101100.

19–7. An 8-bit switched current-source DAC is operated as a 2-quadrant multiplier. The binary input is positive and the reference voltage can range from −10 V to +10 V. If R = 10 kΩ, what is the total range of the output voltage?

19–8. A 4-bit switched current-source DAC is operated as a 4-quadrant multiplier. The input is offset binary code.
Find the output voltage in each case:
a. The reference voltage is −10 V and the input is 0011.
b. The reference voltage is +5 V and the input is 1010.
c. The reference voltage is +10 V and the input is 0001.

Section 19–5
Switched-Capacitor DACs

19–9. The binary input to a 4-bit switched-capacitor DAC having C = 2 pF is 0101.
a. What is the total capacitance connected to E_{REF} when the ϕ_2 clock is high?
b. What is the total capacitance connected to ground when the ϕ_2 clock is high?
c. If E_{REF} = 8 V, what is the ouptut voltage?

19–10. Draw a schematic diagram of an 8-bit switched capacitor DAC. Label all capacitor values in terms of capacitance C. What is the total capacitance when all capacitors are in parallel?

Section 19–6
DAC Performance Specifications

19–11. A 12-bit D/A converter has a full-scale range of 15 V. Its maximum differential linearity error is specified to be $\pm(\frac{1}{2})$LSB.
a. What is its percentage resolution?
b. What are the minimum and maximum possible values of the increment in its output voltage?

19–12. The LSB of a 10-bit D/A converter is 20 mV.
a. What is its percentage resolution?
b. What is its full-scale range?
c. A differential linearity error greater than what percentage of FSR could make its output nonmonotonic?

Section 19–7
The Counter-Type ADC

19–13. The A/D converter in Figure 19–13(a) is clocked at 1 MHz. What is the maximum possible time that could be required to perform a conversion?

19–14. The minimum conversion time of a tracking type A/D converter is 400 ns. At what frequency is it clocked?

Section 19–8

Flash A/D Converters

19–15. A flash-type 5-bit A/D converter has a reference voltage of 10 V.

a. How many voltage comparators does it have?

b. What is the increment between the fixed voltages applied to the comparators?

19–16. The largest fixed voltage applied to a comparator in a flash-type A/D converter is 14.824218 V when the reference voltage is 15 V. What is the number of bits in the digital output?

Section 19–9

The Dual-Slope (Integrating) ADC

19–17. In an 8-bit, dual-slope A/D converter, $R_1 = 20\ \text{k}\Omega$ and $C = 0.001\ \mu\text{F}$. An analog input of -0.25 V is integrated for $t_1 = 160\ \mu\text{s}$.

a. What is the maximum voltage reached in the integration?

b. If the input to the integrator is then switched to a reference voltage of +5 V, how long does it take the output to reach 0 V?

c. If the counter is clocked at 3.125 MHz, what is the digital output after the conversion?

19–18. An 8-bit dual-slope A/D converter integrates analog inputs for $t_1 = 50\ \mu\text{s}$. What should be the magnitude of the reference voltage if an input of 25 V is to produce a binary output of 11111111 when the clock frequency is 1 MHz?

Section 19–10

The Successive-Approximation ADC

19–19. List the sequence of binary numbers that would appear in the output register of a 4-bit successive-approximation A/D converter when the analog input has a value that is ultimately converted to 1011.

19–20. Sketch the output of the DAC in an 8-bit successive-approximation A/D converter when the analog input is a voltage that is ultimately converted to 10101011. Label each step of the DAC output with the decimal number corresponding to the binary value it represents.

Section 19–11

ADC Performance Specifications

19–21. List four types of A/D converters in descending order of speed (fastest converter type first).

19–22. List the principal advantage of each of four types of A/D converters.

19–23. The analog input of an 8-bit A/D converter can range from 0 V to 10 V. Find its resolution in volts and as a percentage of full-scale range.

19–24. The resolution of a 12-bit A/D converter is 7 mV. What is its full scale range?

19–25. The frequency components of the analog input to an A/D converter range from 50 Hz to 10 kHz. What maximum total conversion time should the converter have?

19–26. The analog input to an A/D converter consists of a 500-Hz fundamental waveform and its harmonics. If the converter has a total conversion time of 40 μs, what is the highest-order harmonic that should be in the input?

A SPICE AND PSPICE

A–1 INTRODUCTION

SPICE—Simulation Program with Integrated Circuit Emphasis—was developed at the University of California, Berkeley, as a computer aid for designing integrated circuits. However, it is readily used to analyze discrete circuits as well and can, in fact, analyze circuits containing no semiconductor devices at all. In addition to this versatility, SPICE owes its current popularity to the ease with which a circuit model can be constructed and the wide range of output options (analysis types) available to the user.

As a brief note on terminology, observe that SPICE is a computer *program,* stored in computer memory, which we do not normally inspect or alter. As users, we merely supply the program with data, in the form of an *input data file,* which describes the circuit we wish to analyze and the type of output we desire. This input data file is usually supplied to SPICE by way of a keyboard. SPICE then executes a program run, using the data we have supplied, and displays or prints the results on a video terminal or printer. The mechanisms, or *commands,* that must be used to supply the input data to SPICE and to cause it to execute a program run vary widely with the computer system used, so we cannot describe that procedure here. Furthermore, there are numerous versions of SPICE, some designed for use with microcomputers and others capable of analyzing more complex circuits at greater speed, designed for use on large mainframe computers. Depending on the version used, minor variations in *syntax* (the format for specifying the input data) may be encountered. The *User's Guide,* supplied with most versions, should be consulted if any difficulty is experienced with any of the programs in this book. All programs here have been run successfully using SPICE version 2G.6 on a Honeywell DPS 90 computer.

PSPICE

One of the most widely used versions of SPICE designed for operation on microcomputers is PSpice (a registered trademark of the MicroSim Corporation). It has numerous features and options that make it more versatile and somewhat easier to use than the original Berkeley version of SPICE. However, with a few minor exceptions, any input data file written to run on Berkeley SPICE will also run on PSpice. In the discussions that follow, features of PSpice are highlighted immediately after the corresponding capabilities or requirements of Berkeley SPICE are described.

A–2 DESCRIBING A CIRCUIT FOR A SPICE INPUT FILE

The input data file consists of successive lines, which we will hereafter refer to as *statements,* each of which serves a specific purpose, such as identifying one component in the circuit. The statements do not have to be numbered and, except for the first and last, can appear in any order.

The Title

The first statement in every input file must be a *title.* Subject only to the number of characters permitted by a particular version, the title can be anything we wish. Examples are:

AMPLIFIER
EXERCISE 2.25
A DIODE (1N54) TEST

Nodes and Component Descriptions

The first step in preparing the circuit description is to identify and number all the *nodes* in the circuit. It is good practice to draw a schematic diagram with nodes shown by circles containing the node numbers. Node numbers can be any positive integers, and one of them must be 0. (The zero node is usually, but not necessarily, the circuit common, or ground.) Node numbers can be assigned in any sequence, such as 0, 1, 2, 3, . . . or 0, 2, 4, 6, ...

Once the node numbers have been assigned, each component in the circuit is identified by a separate statement that specifies the type of component it is and the node numbers between which it is connected. The first letter appearing in the statement identifies the component type. Passive components (resistors, capacitors, and inductors) are identified by the letters *R*, *C*, and *L*. Any other characters can follow the first letter, but each component must have a unique designation. For example, the resistors in a circuit might be identified by R1, R2, RB, and REQUIV.

The node numbers between which the component is connected appear next in the statement, separated by one or more spaces. Except in some special cases, it does not matter which node number appears first. The component value, in ohms, farads, or henries, appears next. Resistors cannot have value 0. Following are some examples:

R25 6 0 100 (Resistor R25 is connected between nodes 6 and 0 and has value 100 Ω.)
R25 0 6 1E2 (Interpreted by SPICE the same as in the first case.)
CIN 3 5 22E–6 (Capacitor CIN is connected between nodes 3 and 5 and has value 22 μF.)
LSHUNT 12 20 0.01 (Inductor LSHUNT is connected between nodes 12 and 20 and has value 0.01 H.)

PSPICE

In the foregoing examples, all letters used in the input data files are capitalized, a requirement of Berkeley SPICE. In PSpice, either lowercase or capital letters (or both) may be used. For example, resistor R1 can be listed as r1 in one line and referred to again as R1 in another line, and PSpice will recognize both as representing the same component.

Node numbers in PSpice do not have to be integers; they can be identified by any set of alphanumeric characters (letters or numbers) up to 31 in length. However, one node must be node 0.

Specifying Numerical Values

Suffixes representing powers of 10 can be appended to value specifications (with no spaces in between). Following are the SPICE suffix designations:

T	(tera: 10^{12})	U	(micro: 10^{-6})
G	(giga: 10^{9})	N	(nano: 10^{-9})
MEG	(mega: 10^{6})	P	(pico: 10^{-12})
K	(kilo: 10^{3})	F	(femto: 10^{-15})
M	(milli: 10^{-3})		

Note that M represents *milli* (10^{-3}) and that MEG is used for 10^6. Following are some examples of equivalent ways of representing values, all of which are interpreted by SPICE in the same way:

$$0.002 = 2\text{M} = 2\text{E}-3 = 2000\text{U}$$
$$5000\text{E}-12 = 5000\text{P} = 5\text{N} = .005\text{U} = 0.005\text{E}-6$$
$$0.15\text{MEG} = 150\text{K} = 150\text{E}3 = 150\text{E}+3 = .15\text{E}+6$$

Any characters can follow a value specification, and unless the characters are one of the powers-of-10 suffixes just given, SPICE simply ignores them. Characters are often added to designate units. Following are some examples, all of which are interpreted by SPICE in the same way:

$$100\text{UF} = 100\text{E}-6\text{F} = 100\text{U} = 100\text{UFARADS}$$
$$56\text{N} = 56\text{NSEC} = 0.056\text{US} = 56\text{E}-9\text{SECONDS}$$
$$2.2\text{K} = 2200\text{OHMS} = 0.0022\text{MEGOHMS} = 2.2\text{E}3$$
$$0.05\text{MV} = 50\text{UVOLTS} = 50\text{E}-6\text{V} = 0.05\text{MILLIV}$$

Be careful not to use a unit that begins with one of the power-of-10 suffixes. For example, a 0.0001-F capacitor specified as 0.0001F would be interpreted by SPICE as 0.0001×10^{-15} F.

DC Voltage Sources

The first letter designating a voltage source, dc or ac, is V. As with passive components, any characters can follow the V, and each voltage source must have a unique designation. Examples are V1, VIN, and VSIGNAL. Following is the format for representing a dc source:

```
V******* N+ N- ⟨DC⟩ value
```

where ******* are arbitrary characters, N+ is the number of the node to which the positive terminal of the source is connected, and N− is the number of the node to which the negative terminal is connected. The symbol ⟨ ⟩ enclosing DC means that the specification DC is optional: If a source is not designated DC, SPICE will automatically assume that it is DC. *Value* is the source voltage, in volts. *Value* can be negative, which is equivalent to reversing the N+ and N− node numbers. The following are examples:

```
VIN 5 0 24VOLTS
VIN 0 5 DC -24
```

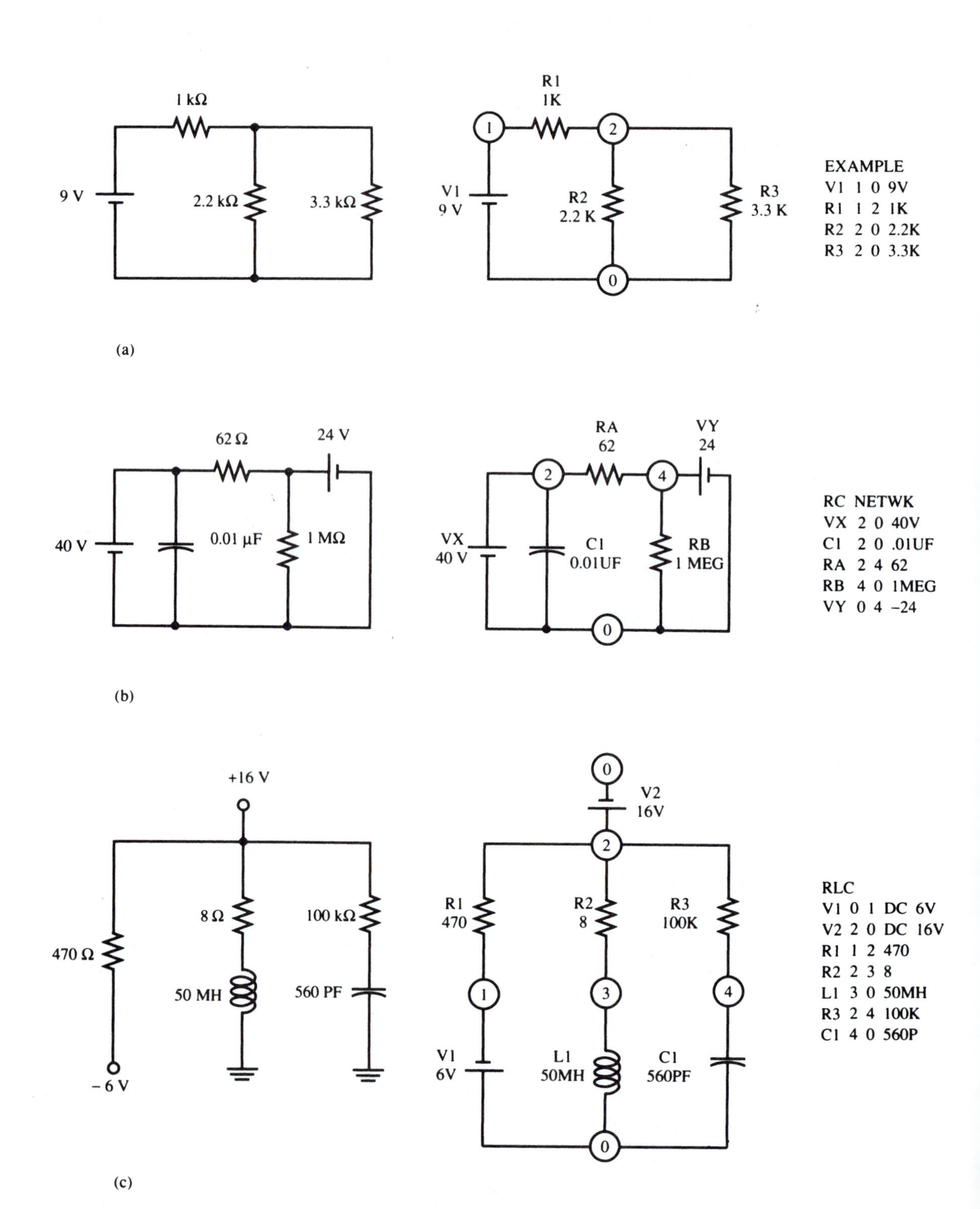

Figure A–1
Examples of circuit descriptions for a SPICE input data file

Figure A–2
Equivalent ways of specifying a dc current source. Note that the "negative" terminal is the one to which current is delivered (N– = 2).

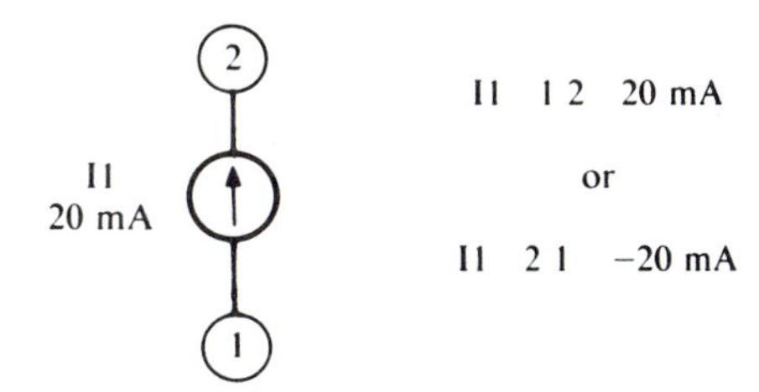

Both of these statements specify a 24-V-dc source designated VIN whose positive terminal is connected to node 5 and whose negative terminal is connected to node 0. Both statements are treated the same by SPICE.

Figure A–1 shows some examples of circuits and the statements that could be used to describe them in a SPICE input file. (These examples are not complete input files because we have not yet discussed *control* statements, used to specify the type of analysis and the output desired.)

DC Current Sources

The format for specifying a dc current source is

```
I******* N+ N- ⟨DC⟩ value
```

where *value* is the value of the source in amperes and DC is optional. Note the following *unconventional* definition of N+ and N–: *N– is the number of the node to which the source delivers current.* To illustrate, Figure A–2 shows two equivalent ways of specifying a 20-mA dc current source.

A–3 THE .DC AND .PRINT CONTROL STATEMENTS

A control statement is one that specifies the type of analysis to be performed or the type of output desired. Every control statement begins with a period, followed immediately by a group of characters that identifies the type of control statement it is.

The .DC Control Statement

The .DC control statement tells SPICE that a dc analysis is to be performed. This type of analysis is necessary when the user wishes to determine dc voltages and/or currents in a circuit. Although ac sources can be present in the circuit, there must be at least one dc (voltage or current) source present if a dc analysis is to be performed. The format for the .DC control statement is

```
.DC name1 start stop incr. ⟨name2 start2 stop 2 incr.2⟩
```

where *name1* is the name of one dc voltage or current source in the circuit and *name2* is (optionally) the name of another. The .DC control statement can be used to *step* a source through a sequence of values, a useful feature when plotting characteristic curves. For that use, *start* is the first value of voltage or current in the sequence, *stop* is the last value, and *incr.* is the value of the increment, or step, in the sequence. A second source, *name2,* can also be stepped. If analysis is desired at a *single* dc value, we set *start* and *stop* both equal to that value and arbitrarily set *incr.* equal to 1. In cases where the circuit contains several fixed-value sources, one of them (any one) *must* be specified in the .DC control statement. Following are some examples:

```
.DC  V1  24  24  1
.DC  IB  50U  50U  1
.DC  VCE  0  50  10
```

.DC V1 24 24 1	V1 is a fixed 24-V-dc voltage source.
.DC IB 50U 50U 1	IB is a fixed 50-μA-dc current source.
.DC VCE 0 50 10	VCE is a dc voltage source that is stepped from 0 to 50 V in 10-V increments.

When two sources are stepped in a .DC control statement, the first source is stepped over its entire range for each value of the second source. The following is an example:

```
.DC VCC 0 25 5 IBB 0 20U 2U
```

In this example, the dc voltage source named VCC is stepped from 0 to 25 V for *each* value of the dc current source named IBB.

Identifying Output Voltages and Currents

To tell SPICE the voltages whose values we wish to determine (the *output* voltages we desire), we must identify them in one of the following formats:

$$V(N+, N-) \quad \text{or} \quad V(N+)$$

In the first case, V(N+, N−) refers to the voltage at node N+ with respect to node N−. In the second case, V(N+) is the voltage at node N+ with respect to node 0. The following are examples:

V(5, 1)	The voltage at node 5 with respect to node 1.
V(3)	The voltage at node 3 with respect to node 0.

The only way to obtain the value of a current in Berkeley SPICE is to request the value of the current in a *voltage* source. Thus, a voltage source must be in the circuit at any point where we wish to know the value of the current. The current is identified by I(*Vname*), where *Vname* is the name of the voltage source. *We can insert zero-valued voltage sources (dummy sources) anywhere in a circuit for the purpose of obtaining a current.* These dummy voltage sources serve as ammeters for SPICE and do not in any way affect circuit behavior. Note carefully the following unconventional way that SPICE assigns polarity to the current through a voltage source: Positive current flows *into* the positive terminal (N+) of the voltage source. Figure A–3 shows an example. Here, conventional positive current flows in a

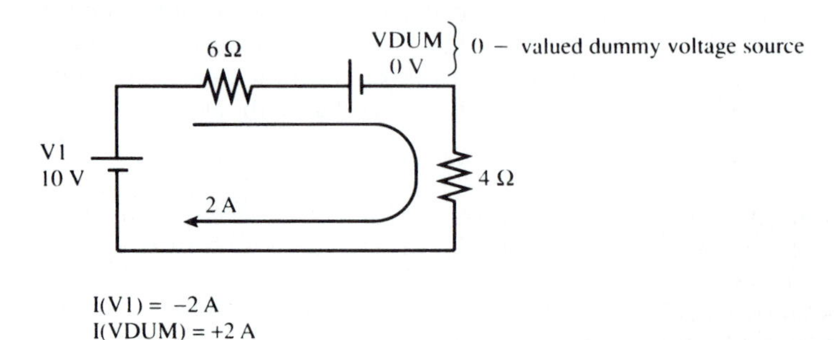

Figure A–3
SPICE treats conventional current flowing out of a positive terminal as negative current and current flowing into a positive terminal as positive current.

clockwise direction, but a SPICE output would show I(V1) equal to −2 A. On the other hand, SPICE would show the current in the zero-valued dummy source I(VDUM), to be +2 A.

The .PRINT Control Statement

The .PRINT statement tells SPICE to print the values of voltages and/or currents resulting from an analysis. The format is

```
.PRINT type out1 ⟨out2 out3 ...⟩
```

where *type* is the type of analysis and *out1, out2,* . . . , identify the output voltages and/or currents whose values we desire. So far, DC is the only analysis type we have discussed. For example, the statement

```
.PRINT DC V(1,2) V(3) I(VDUM)
```

tells SPICE to print the values of the voltages V(1, 2) and V(3) and the value of the current I(VDUM) resulting from a dc analysis. If the .DC control statement specifies a stepped source, the .PRINT statement will print all the output values resulting from all the stepped values. The number of output variables whose values can be requested by a single .PRINT statement may vary with the version of SPICE used (up to 8 can be requested in version 2G.6). Any number of .PRINT statements can appear in a SPICE input file.

PSPICE

In PSpice, the current through any component can be requested directly, without using a dummy voltage source. For example, I(R1) is the current through resistor R1. The reference polarity of the current is from the first node number of R1 to the second node number. In other words, positive current is assumed to flow from the first node given in the description of R1 to the second node. For example, if 5 A flows from node 1 to node 2 in a circuit whose input data file contains

```
R1 1 2 10
.PRINT DC I(R1)
```

then the .PRINT statement will produce 5 A. On the other hand, if the description of the same R1 were changed to R1 2 1 10, then the same .PRINT statement would produce −5 A.

Also, voltages across components can be requested directly, as, for example, V(RX), the voltage across resistor RX. The reference polarity is from the first node listed in the description of RX to the second node. For example, if the voltage across RX from node 5 to node 6 is 12 V in a circuit whose input data file contains

```
RX 5 6 1K
.PRINT DC V(RX)
```

then the .PRINT statement will produce 12 V. On the other hand, if the description of the same resistor were changed to RX 6 5 1K, the same .PRINT statement would produce −12 V.

The .END Statement

The last statement in every SPICE input file must be .END. We have now discussed enough statements to construct a complete SPICE input file, as demonstrated in the next example.

Example A–1

Use SPICE to determine the voltage drop across and the current through every resistor in Figure A–4(a).

Solution. Figure A–4(b) shows the circuit when redrawn and labeled for analysis by SPICE. Note that two dummy voltage sources are inserted to obtain currents in two branches. The polarities of these sources are such that positive

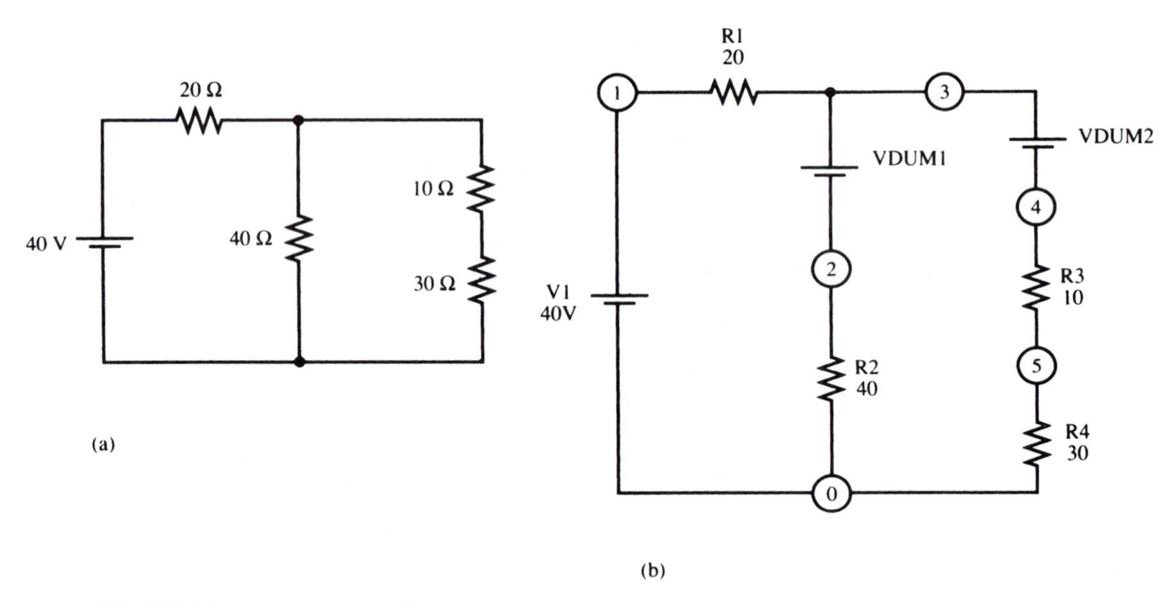


```
EXAMPLE A.1
V1 1 0
VDUM1 3 2
VDUM2 3 4
R1 1 3 20
R2 2 0 40
R3 4 5 10
R4 5 0 30
.DC V1 40 40 1
.PRINT DC I(V1) I(VDUM1) I(VDUM2) V(1,3) V(2) V(4,5) V(5)
.END
```

(c)

```
EXAMPLE A.1
****     DC TRANSFER CURVES                   TEMPERATURE =    27.000 DEG C
******************************************************************************
  V1             I(V1)        I(VDUM1)     I(VDUM2)     V(1,3)       V(2)
 4.000E+01     -1.000E+00    5.000E-01    5.000E-01    2.000E+01    2.000E+01


******************************************************************************
  V1             V(4,5)       V(5)
 4.000E+01      5.000E+00    1.500E+01
```

(d)

Figure A–4
(Example A–1)

currents will be computed. In PSpice, we can simply request I(R2) and I(R3) in the .PRINT statement. The current in R1 is the same as the current in V1, and we must simply remember that SPICE will print a negative value for that current.

Figure A–4(c) shows the SPICE input file. Note that it is not necessary to specify a voltage value in the statement defining V1, since that value is given in the .DC statement. Figure A–4(d) shows the results of a program run. The outputs appear under the heading "DC TRANSFER CURVES," which refers to the type of output obtained when the source(s) are stepped. In our case, the heading is irrelevant. Note that the analysis is performed under the (default) assumption that the circuit temperature is 27°C (80.6°F). We will see later that we can specify different temperatures. Referring to the circuit nodes in Figure A–4(b), we see that the printed results give the following voltages and currents:

	I	V
R_1	1 A	20 V
R_2	0.5 A	20 V
R_3	0.5 A	5 V
R_4	0.5 A	15 V

Circuit Restrictions

Every node in a circuit defined for SPICE must have a *dc path to ground* (node 0). A dc path to ground can be through a resistor, inductor, or voltage source but not through a capacitor or current source. Figure A–5(a) shows two examples of nodes that do not have dc paths to ground and that cannot, therefore, appear in a SPICE circuit. However, we can connect a very large resistance in parallel with a capacitor or current source to provide a dc path to ground. The resistance should be very large in comparison to other impedances in the circuit so that it will have a negligible effect on the computations. For example, if the capacitive reactances in the circuits of Figure A–5(a) are less than 1 MΩ, we can specify a 10^{12}-Ω resistor, RDUM (1 million megohms), between nodes 1 and 0:

```
RDUM 1 0 1E12
```

Figure A–5
Examples of circuits that cannot be simulated in SPICE (for the reasons cited). To force a simulation, a very large resistance can be connected in parallel with a capacitor in (a), and a very small resistance can be connected in series with either V1 or L in (b).

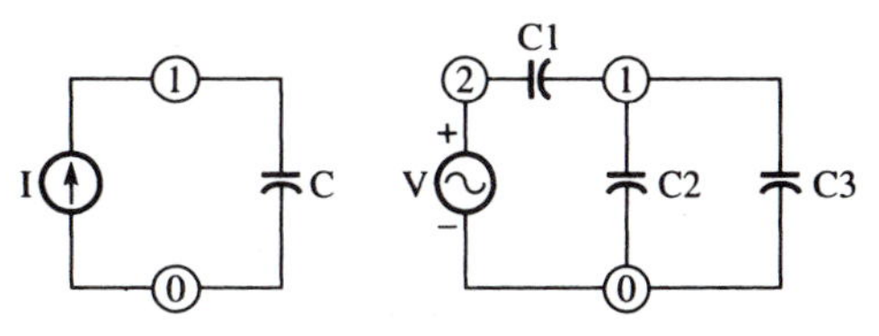

(a) Neither circuit has a dc path from node 1 to ground (node 0).

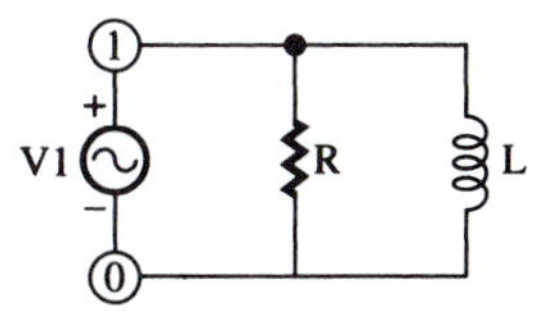

(b) Voltage source V1 and inductance L appear in a closed loop.

SPICE does not permit any loop (closed circuit path) to consist exclusively of inductance(s) and voltage source(s). Figure A–5(b) shows an example. Here, inductance L appears in a closed loop with voltage source V1. To circumvent this problem, we can insert a very small resistance in such a loop. The resistance should be much smaller than the impedances of other elements in the circuit in order to have a negligible effect on the computations. For example, if the impedances of R and L in Figure A–5(b) are greater than 1 Ω, we could insert a 1-pΩ resistor, RDUM, (10^{-12} Ω) in series with either L or V1.

A–4 THE .TRAN AND .PLOT CONTROL STATEMENTS

The .TRAN Control Statement

The .TRAN control statement (derived from "transient") is used when we want to obtain values of voltages or currents versus *time* (whether they are transients in the traditional sense or not). We must specify the total time interval over which we wish to obtain the time-varying values and the increment of time between each using the format

```
.TRAN STEP TSTOP ⟨TSTART⟩
```

where *STEP* is the time increment and *TSTOP* is the largest value of time at which values will be computed. Unless we optionally specify the start time, *TSTART,* SPICE assumes it to be 0. If we do specify *TSTART,* computations still begin at $t = 0$, but only those in the interval from t = *TSTART* through *TSTOP* are provided as output. To illustrate, the statement

```
.TRAN 5M 100M
```

will cause SPICE to produce values of the output(s) at the 21 time points 0, 5 ms, 10 ms, . . . , 100 ms. When TRAN is used as the analysis type in a .PRINT statement, 21 values of each output variable specified in the .PRINT statement will be printed. We will show an example of a .TRAN analysis (Example A–2) after discussing a few more statement types.

The .PLOT Control Statement

The .PLOT control statement can be used to obtain many different kinds of plots, depending on the analysis type specified. The format is

```
.PLOT type out1 ⟨out2 out3 ...⟩
```

where *type* is the analysis type and *out1, out2,* . . . are the outputs whose values are to be plotted. If the analysis type is TRAN, then the output variables are plotted versus time, with time increasing downward along the vertical axis. If more than one output is specified in the .PLOT statement, all will be plotted, using different symbols, on the same axes. Although it is possible to specify the scale desired, it is easier to let SPICE automatically determine the scale (using the minimum and maximum values it computes). When more than one output is plotted, SPICE automatically determines and displays all scales needed for all outputs. It also prints the time increments and the values of the points plotted. If more than one output is plotted, the values of the first output specified in the .PLOT statement are the only ones printed. Separate .PLOT statements can be used to obtain separate plots and value printouts if desired.

.DC Plots

When the analysis type is .DC, a .PLOT statement causes SPICE to plot the output(s) specified in that statement versus the values of a stepped source. The stepped source values are printed downward along the vertical axis. An example is shown in Figure 2–17. Here, the statement .PLOT DC I(VDUM) causes the computed values of I(VDUM) to be plotted versus the stepped values of V1: .DC V1 0.6 0.7 5MV. Note that the stepped values of V1 are printed down the left column, along with the computed values of I(VDUM).

When there are two stepped sources, the values of the stepped source appearing first in the .DC statement are printed downward along the vertical axis. These sets of values and the plots are repeated for each value of the stepped source appearing second in the .DC statement. Figure 4–22 is an example. Here, the combination

```
.DC VCE 0 50 5 IB 0 40U 10U
.PLOT DC I(VDUM)
```

causes SPICE to

1. plot values of I(VDUM) versus the 11 values of VCE when IB = 0;
2. plot values of I(VDUM) versus the 11 values of VCE when IB = 10 μA;

•
•
•

5. plot values of I(VDUM) versus the 11 values of VCE when IB = 40 μA.

LIMPTS and the .OPTIONS Control Statement

Normally, SPICE will not print or plot more than 201 values. We can override this limit by specifying a different limit on the number of points (LIMPTS) using the .OPTIONS control statement. The .OPTIONS statement can also be used to change many other operating characteristics and limits that are normally imposed by SPICE, most of them related to the mathematical techniques used in the computations. The majority of these will not concern us. The format for changing the limit on the number of points plotted or printed is

```
.OPTIONS LIMPTS = n
```

where *n* is the number of points.

A–5 THE SIN AND PULSE SOURCES

If we wish to obtain a printout or plot of an output versus time, as in a .TRAN analysis, at least one source in the circuit should itself be a time-varying voltage or current. In other words, we will not be able to observe an output *waveform* unless an input waveform is defined. It is not sufficient to indicate in a component definition that a particular source is AC instead of DC. A source designated AC is used by SPICE in an .AC analysis, to be discussed presently, and that analysis type does *not* cause SPICE to display time-varying outputs. For a .TRAN analysis, we use a different format to define the time-varying sources. The two sources that are most widely used for that purpose are the SIN (sinusoidal ac) and PULSE sources.

The SIN Source

The format for specifying a sinusoidal voltage source is

```
V******* N+ N- SIN(VO VP FREQ TD θ)
```

where ******* are arbitrary characters, *N+* and *N−* are the node numbers of the positive and negative terminals, *VO* is the *offset* (dc, or bias level), *VP* is the peak value in volts, and *FREQ* is the frequency in hertz. *TD* and θ are special parameters related to time delay and damping, both of which are set to 0 to obtain a conventional sine wave. A sinusoidal current source is defined by using I instead of V as the first character. Note that it is not possible to specify a phase shift (other than 180°, by reversing *N+* and *N−*). Figure A–6 shows an example of a SIN source definition. The zero values for TD and θ can be omitted in the specification, and SPICE will assume they are zero by default.

PSPICE

In PSpice, a phase angle can be specified for a SIN source. The format (for a sinusoidal voltage) is

```
V******* N+ N- SIN(VO VP FREQ TD θ PH)
```

where *PH* is phase angle in degrees and all other parameters are the same as in the SIN source description in Berkeley SPICE.

The PULSE Source

The PULSE source can be used to simulate a dc source that is switched into a circuit at a particular instant of time (a *step* input) or to generate a sequence of square, trapezoidal, or triangular pulses. Figure A–7(a) shows the parameters used to define a voltage pulse or pulse-type waveform: one having time delay (*TD*) that elapses before the voltage begins to rise linearly with time, a *rise time* (*TR*) that represents the time required for the voltage to change from *V1* volts to *V2* volts, a *pulse width* (*PW*), and a *fall time* (*TF*), during which the voltage falls from *V2* volts to *V1* volts. If the pulse is repetitive, a value for the period of the waveform (*PER*) is also specified. If *PER* is not specified, its default value is the value of *TSTOP* in a .TRAN analysis; that is, the pulse is assumed to remain at *V2* volts for the duration of the analysis, simulating a step input. The default value for the rise and fall times is the *STEP* time specified in a .TRAN analysis. The figure shows the format for identifying a pulsed voltage source. Current pulses can be obtained by using I instead of V as the first character. Figure A–7(b) shows an example of how the PULSE source is used to simulate a step input caused by

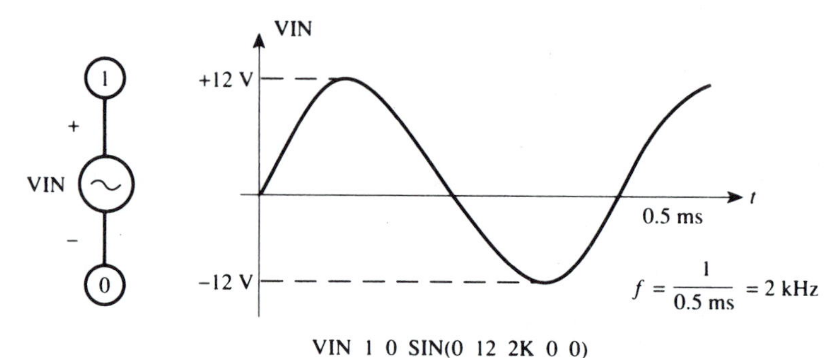

Figure A–6
An example of the specification of a sinusoidal voltage source.

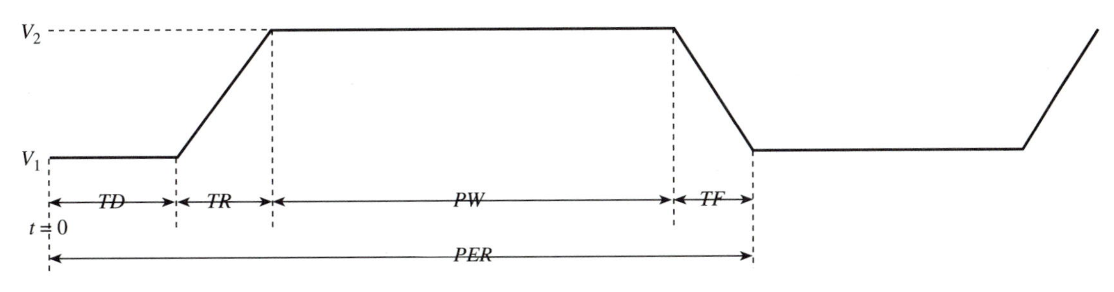

V ******* N+ N– PULSE($V_1 V_2$ < *TD TR TF PW PER* >)
Default values: *TD* = 0, *TR* = *STEP*, *TF* = *STEP*, *PW* = *TSTOP*, *PER* = *TSTOP*.

(a)

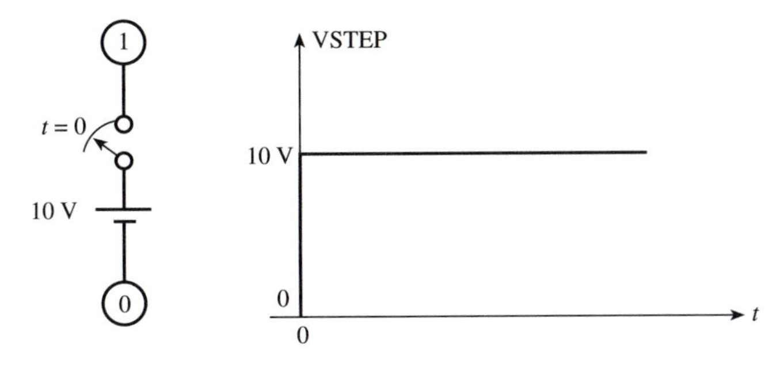

VSTEP 1 0 PULSE(0 10 0 0 0)

(b) Using the PULSE source to simulate a 10-V dc source switched into a circuit at $t = 0$. Even though *TR* and *TF* are set to 0, SPICE assigns each the default value of *STEP* specified in a .TRAN statement. *PW* and *PER* both default to the value of *TSTOP*.

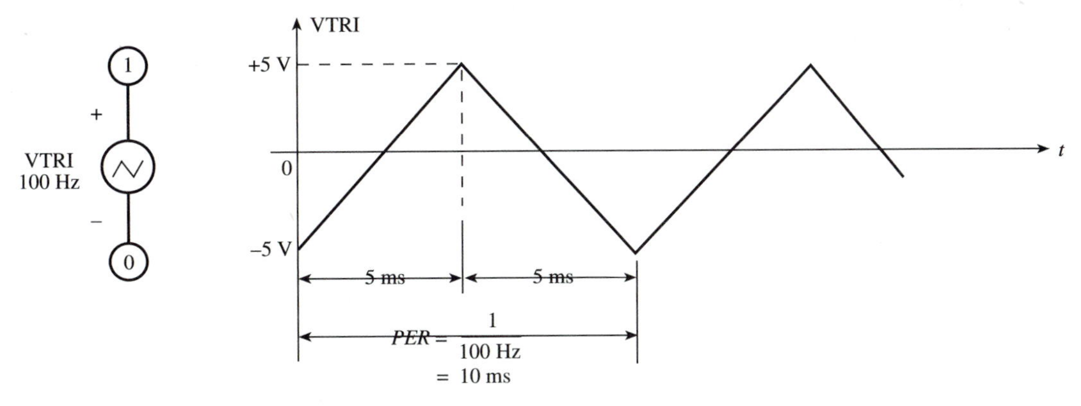

VTRI 1 0 PULSE(–5 5 0 5M 5M 1P 10M)

(c) Using the PULSE source to generate a 100-Hz triangular waveform. Note that $PW = 1$ ps $\cong 0$.

Figure A–7
The PULSE source

switching a 10-V-dc source into a circuit at $t = 0$. Figure A–7(c) shows an example of how the PULSE source is used to define a triangular waveform that alternates between ±5 V with a frequency of 100 Hz. SPICE does not accept a zero pulse width, as would be necessary to define an ideal triangular waveform, but *PW* can be made negligibly small. In this example, we set $PW = 10^{-12}$ s = 1 ps, which makes the period 10^{10} times as great as the pulse width.

A–6 THE INITIAL TRANSIENT SOLUTION

The next example demonstrates the use of the PULSE source to generate a square wave and contains some important discussion on how SPICE performs a dc analysis in conjunction with every transient analysis.

Example A–2

Use SPICE to obtain a plot of the capacitor voltage versus time in Figure A–8(a). The plot should cover two full periods of the square wave input.

Solution. Figure A–8(b) shows the circuit when redrawn for analysis by SPICE and the corresponding input data file. The period of the input is T = 1/(2.5 Hz) = 0.4 s, so *TSTOP* in the .TRAN statement is set to 0.8 s to obtain a plot covering two full periods. Note that *PW* in the definition of V1 is set to 0.2 s. The square

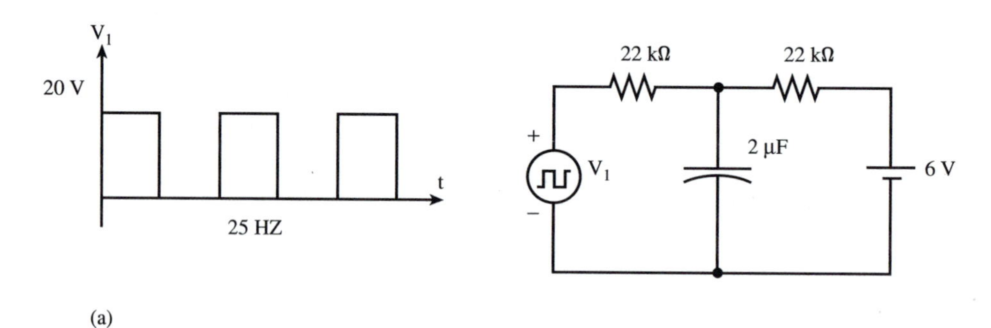

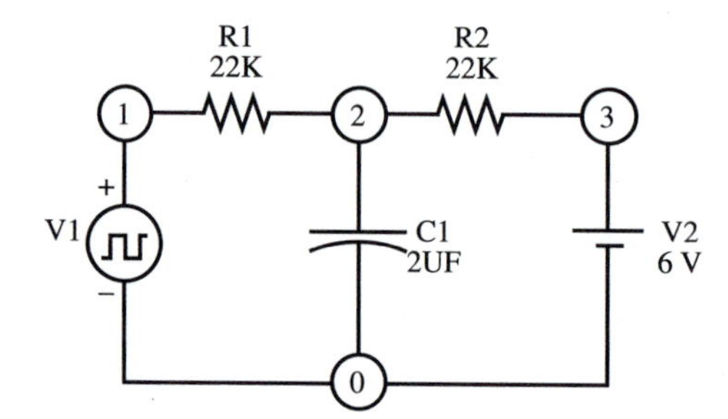


```
EXAMPLE A.2
V1 1 0 PULSE(0 20 0 0 0 0.2 0.4)
V2 3 0 6V
R1 1 2 22K
R2 2 3 22K
C1 2 0 2UF
.TRAN 0.02 0.8
.PLOT TRAN V(2)
.END
```

Figure A–8
(Example A–2)

wave is idealized by setting the rise and fall times, *TR* and *TF*, to 0, so the actual value assigned by SPICE to *TR* and *TF* is the value of *STEP*: 0.02 s.

Figure A–9 shows the results of a program run. When SPICE performs a .TRAN analysis, it first obtains an "initial transient solution." This solution is

```
EXAMPLE A.2
****     INITIAL TRANSIENT SOLUTION        TEMPERATURE =    27.000 DEG C
******************************************************************************
 NODE   VOLTAGE      NODE   VOLTAGE      NODE   VOLTAGE
(  1)     .0000     (  2)    3.0000     (  3)    6.0000
    VOLTAGE SOURCE CURRENTS
    NAME        CURRENT
    V1         1.364D-04
    V2        -1.364D-04
    TOTAL POWER DISSIPATION    8.18D-04   WATTS


EXAMPLE A.2
****     TRANSIENT ANALYSIS                  TEMPERATURE =    27.000 DEG C
******************************************************************************
   TIME        V(2)
                   0.000D+00     5.000D+00     1.000D+01     1.500D+01  2.000D+01
                - - - - - - - - - - - - - - - - - - - - - - - - - - - - - - -
0.000D+00   3.000D+00 .       *      .             .             .             .
2.000D-02   6.394D+00 .              .   *         .             .             .
4.000D-02   1.023D+01 .              .             .*            .             .
6.000D-02   1.191D+01 .              .             .    *        .             .
8.000D-02   1.259D+01 .              .             .      *      .             .
1.000D-01   1.285D+01 .              .             .       *     .             .
1.200D-01   1.294D+01 .              .             .       *     .             .
1.400D-01   1.298D+01 .              .             .       *     .             .
1.600D-01   1.299D+01 .              .             .       *     .             .
1.800D-01   1.300D+01 .              .             .       *     .             .
2.000D-01   1.300D+01 .              .             .       *     .             .
2.200D-01   1.300D+01 .              .             .       *     .             .
2.400D-01   9.591D+00 .              .            *.             .             .
2.600D-01   5.658D+00 .              . *           .             .             .
2.800D-01   4.071D+00 .           *  .             .             .             .
3.000D-01   3.420D+00 .          *   .             .             .             .
3.200D-01   3.159D+00 .        *     .             .             .             .
3.400D-01   3.059D+00 .        *     .             .             .             .
3.600D-01   3.024D+00 .       *      .             .             .             .
3.800D-01   3.009D+00 .       *      .             .             .             .
4.000D-01   3.003D+00 .       *      .             .             .             .
4.200D-01   6.410D+00 .              .   *         .             .             .
4.400D-01   1.034D+01 .              .             .*            .             .
4.600D-01   1.193D+01 .              .             .    *        .             .
4.800D-01   1.258D+01 .              .             .      *      .             .
5.000D-01   1.284D+01 .              .             .       *     .             .
5.200D-01   1.294D+01 .              .             .       *     .             .
5.400D-01   1.298D+01 .              .             .       *     .             .
5.600D-01   1.299D+01 .              .             .       *     .             .
5.800D-01   1.300D+01 .              .             .       *     .             .
6.000D-01   1.300D+01 .              .             .       *     .             .
6.200D-01   1.300D+01 .              .             .       *     .             .
6.400D-01   9.591D+00 .              .            *.             .             .
6.600D-01   5.658D+00 .              . *           .             .             .
6.800D-01   4.071D+00 .           *  .             .             .             .
7.000D-01   3.420D+00 .          *   .             .             .             .
7.200D-01   3.159D+00 .        *     .             .             .             .
7.400D-01   3.059D+00 .        *     .             .             .             .
7.600D-01   3.024D+00 .       *      .             .             .             .
7.800D-01   3.009D+00 .       *      .             .             .             .
8.000D-01   3.003D+00 .       *      .             .             .             .
                - - - - - - - - - - - - - - - - - - - - - - - - - - - - - - -
```

Figure A–9
(Example A–2)

obtained from a dc analysis with all time-varying sources set to zero. Thus, the initial solution represents the *quiescent,* or dc operating conditions in the circuit, useful information for determining the bias point(s) in circuits containing transistors. The actual time-varying outputs are computed with the initial voltages and currents as the starting points, so the outputs do not reflect initial transients associated with the charging of capacitors, such as coupling capacitors, in the circuit.

In our example, the initial transient solution gives the dc voltages and currents when V1, the square-wave generator, is set to 0. The figure shows that the dc voltages at all nodes are printed, as are the dc currents in all voltage sources and the total dc power dissipation in the circuit. (Note that dc current flows *into* V1 when it is set to 0 V.) Since the capacitor is charged to +3 VDC, the time-varying plot shows its voltage to begin at +3 V. The actual transient that would occur (beginning at $t = 0$) while the capacitor charged to 3 V does not appear in the output.

A–7 DIODE MODELS

When there is a semiconductor device in a circuit to be analyzed by SPICE, the SPICE input file must contain two new types of statements: one that identifies the device by name and gives its node numbers in the circuit and another, called a .MODEL statement, that specifies the values of the device *parameters* (saturation current and the like).

All diode names must begin with D. The format for identifying a diode in a circuit is

```
D******* NA NC Mname
```

where *NA* and *NC* are the numbers of the nodes to which the anode and cathode are connected, respectively, and *Mname* is the *model name.* The model name associates the diode with a particular .MODEL statement that specifies the parameter values of the diode:

```
.MODEL Mname D (Pval1 = n1 Pval2 = n2 ···)
```

where D, signifying diode, *must* appear as shown, and *Pval1* = *n1,* . . . specify parameter values, to be described shortly. Note that several diodes, having different names, can all be associated with the same .MODEL statement, and other diodes can be associated with a different .MODEL statement. Figure A–10 shows an example: a diode bridge in which diodes D1 and D3 are modeled by MODA and diodes D2 and D4 are modeled by MODB. MODA specifies a diode having saturation current (IS) 0.5 pA and MODB specifies a diode having saturation current 0.1 pA.

Table A–1 lists the diode parameters whose values can be specified in a .MODEL statement and the default values of each. The default values are typical, or average, values and are acceptable for most electronic circuit analysis at our level of study. In some examples in the book, we specify parameter values different from the default values to illustrate certain points, but average diode behavior is adequate for most of our purposes and can be realized without knowledge of specific values. In practice, some of these diode parameters are very difficult to obtain or measure and are necessary only when a highly accurate model is essential. For example, if we were designing a new diode to have specific low-noise

```
D1 5 6 MODA
D2 6 7 MODB
D3 4 7 MODA
D4 5 4 MODB
.MODEL MODA D IS = 0.5P
.MODEL MODB D IS = 0.1P
```

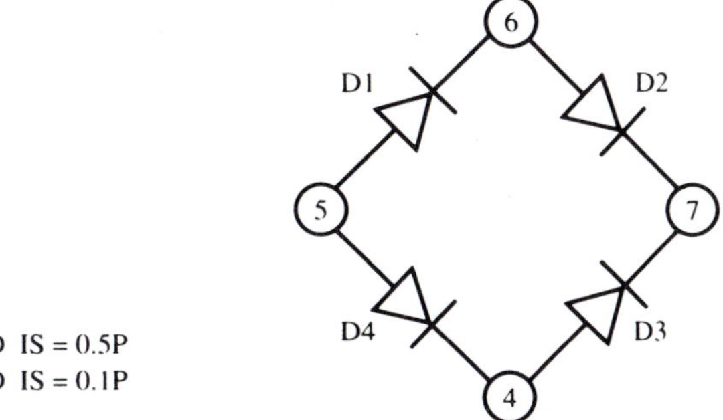

Figure A–10
The parameter values of diodes D1 and D3 are given in the model whose name is MODA, and the parameter values of D2 and D4 are given in the model whose name is MODB.

characteristics, we would want to know the noise parameters, KF and AF, very accurately. Examples 2–8 and 3–5 illustrate the .DC and .TRAN analysis of circuits containing a diode.

PSPICE

In PSpice, the current through and/or the voltage across a diode can be printed or plotted by specifying output variables I(D*******) and V(D*******), where D******* is the diode name. For example, the following statements request the value of the dc current through diode D1 and a plot of the ac voltage across diode D2:

```
.PRINT DC I(D1)
.PLOT AC V(D2)
```

The reference polarity in each case is from the anode node to the cathode node.

Table A–1
Diode parameters

Parameter	Identification	Units	Default Value
Saturation current	IS	A	1×10^{-14}
Ohmic resistance	RS	Ω	0
Emission coefficient	N	—	1
Transit time	TT	s	0
Zero-bias junction capacitance	CJO	F	0
Junction potential	VJ	V	1
Grading coefficient	M	—	0.5
Activation energy	EG	eV	1.11
Saturation current temperature exponent	XTI	—	3
Flicker noise coefficient	KF	—	0
Flicker noise exponent	AF	—	0
Coefficient for forward-bias depletion capacitance equation	FC	—	0.5
Reverse breakdown voltage	BV	V	Infinite
Current at breakdown voltage	IBV	A	1×10^{-3}

A–8 BJT MODELS

The format for identifying a bipolar junction transistor in a circuit is

```
Q******* NC NB NE Mname
```

where *NC, NB,* and *NE* are the numbers of the nodes to which the collector, base, and emitter are connected and *Mname* is the name of the model that specifies the transistor parameters. The format of the .MODEL statement for a BJT is

```
.MODEL Mname type (Pval1 = n1 Pval2 = n2 ···)
```

where *type* is either NPN or PNP. Figure A–11 shows an example of how an NPN transistor and a PNP transistor in a circuit are identified and modeled. The NPN model specifies that the "ideal maximum forward beta" (BF) of the transistor is 150, and the PNP model allows that parameter to have its default value of 100.

As shown in Table A–2, we can specify up to 40 different parameter values for a BJT. The mathematical model used by SPICE to simulate a BJT is very complex but very accurate, provided that all parameter values are accurately known. However, to an even greater extent than in the diode model, many of the parameters are very difficult to measure, estimate, or deduce from other characteristics. Once again, for routine circuit analysis, it is generally sufficient to allow most of the parameters to have their default values. This approach is further justified by the fact that there is usually a wide variation in parameter values among transistors of the same type. Therefore, it is unrealistic and unnecessary in many practical applications to seek a highly accurate analysis of a circuit containing transistors. On the other hand, there are situations, such as the design and development of a new integrated circuit required to have certain properties, where an accurate determination of parameter values is warranted.

PSPICE

PSpice has a *library* option that allows access to model statements for many commonly used diodes and transistors. These model statements specify values for the many device parameters that we could not ordinarily determine from a manufacturer's specifications. The PSpice library is discussed in Section A–17.

In PSpice, the voltage across any pair of BJT terminals and/or the current into any BJT terminal can be printed or plotted using the letters B, E, and C to represent the base, emitter, and collector terminals. Following are some examples of output

Figure A–11
An example of how an NPN transistor and a PNP transistor are identified and modeled

```
Q1 4 3 2 TYPE1
Q2 0 1 2 TYPE2
.MODEL TYPE1 NPN BF=150
.MODEL TYPE2 PNP
```

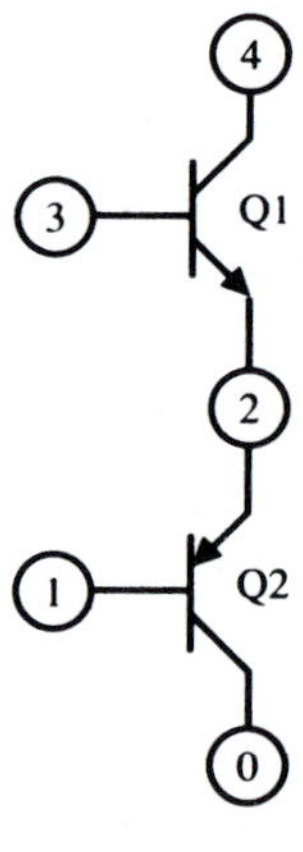

Table A–2
BJT parameters

Parameter	Identification	Units	Default Value
Transport saturation current	IS	A	10^{-16}
Ideal maximum forward beta	BF	—	100
Forward current emission coefficient	NF	—	1
Forward Early voltage	VAF	V	Infinity
Corner for forward beta high-current roll-off	IKF	A	Infinity
Base-emitter leakage saturation current	ISE	A	0
Base-emitter leakage emission coefficient	NE	—	1.5
Ideal maximum reverse beta	BR	—	1
Reverse current emission coefficient	NR	—	1
Reverse Early voltage	VAR	V	Infinity
Corner for reverse beta high-current roll-off	IKR	A	Infinity
Base-collector–leakage saturation current	ISC	A	0
Base-collector–leakage emission coefficient	NC	—	2
Zero-bias base resistance	RB	Ω	0
Current where base resistance falls to half of its minimum value	IRB	A	Infinity
Minimum base resistance at high currents	RBM	Ω	RB
Emitter resistance	RE	Ω	0
Collector resistance	RC	Ω	0
Base-emitter zero-bias depletion capacitance	CJE	F	0
Base-emitter built-in potential	VJE	V	0.75
Base-emitter–junction exponential factor	MJE	—	0.33
Ideal forward transit time	TF	s	0
Coefficient for bias dependence of TF	XTF	—	0
Voltage describing V_{BC}-dependence of TF	VTF	V	Infinity
High-current parameter for effect on TF	ITF	A	0
Excess phase at $f = 1/(2\pi \text{ TF})$ Hz	PTF	Degrees	0
Base-collector zero-bias depletion capacitance	CJC	F	0
Base-collector built-in potential	VJC	V	0.75
Base-collector–junction exponential factor	MJC	—	0.33
Fraction of base-collector depletion capacitance to internal base node	XCJC	—	1
Ideal reverse transit time	TR	s	0
Zero-bias collector-substrate capacitance	CJS	F	0
Substrate-junction built-in potential	VJS	V	0.75
Substrate-junction exponential factor	MJS	—	0
Forward and reverse beta temperature exponent	XTB	—	0
Energy gap for temperature effect on IS	EG	eV	1.11
Temperature exponent for effect on IS	XTI	—	3
Flicker-noise coefficient	KF	—	0
Flicker-noise exponent	AF	—	1
Coefficient for forward-bias depletion capacitance formula	FC	—	0.5

variables that could be specified with a .PRINT or .PLOT statement:

```
VBE(Q1)   The base-to-emitter voltage of transistor Q1
IC(Q2)    The collector current of transistor Q2
VC(QA)    The collector-to-ground voltage of transistor QA
```

A-9 THE .TEMP STATEMENT

As noted earlier, all SPICE computations are performed under the assumption that the temperature of the circuit is 27°C, unless a different temperature is specified. The .TEMP statement is used to request an analysis at one or more different temperatures:

```
.TEMP T1 ⟨T2 T3 ...⟩
```

where *T1* is the temperature in degrees Celsius at which an analysis is desired. *T2*, *T3*, . . . are optional additional temperatures at which SPICE will repeat the analysis, once for each temperature. Temperature is a particularly important parameter in circuits containing semiconductor devices, since their characteristics are temperature-sensitive. However, SPICE will not adjust all device characteristics for temperature unless the parameters in the .MODEL statement that relate to temperature sensitivity are given specific values. A case in point is β in a BJT. The parameter BF is called the "ideal maximum forward beta." The actual value of β used in the computations depends on other factors, including the forward Early voltage (VAF) and the temperature. However, no temperature variation in the value of BF will occur unless the parameter XTB (forward and reverse beta temperature exponent) is specified to have a value other than 0 (its default value). When SPICE performs an analysis at a temperature other than 27°C, it prints a list of temperature-adjusted values of device parameters that are temperature-sensitive.

The values of resistors in a circuit are not adjusted for temperature unless either first- or second-order temperature coefficients of resistance, tc_1 and tc_2 (or both) are given values in the statements defining resistors. The format is

```
R******** N1 N2 value TC = tc1, tc2
```

The temperature-adjusted value is then computed by

$$R_T = R_{27}[1 + tc_1(T - 27°) + 1 + tc_2(T - 27°)]$$

where R_T is the resistance at temperature T and R_{27} is the resistance at temperature 27°C. The default values of tc_1 and tc_2 are 0.

A-10 AC SOURCES AND THE .AC CONTROL STATEMENT

The format for identifying an ac voltage source in a circuit is

```
V******* N+ N- AC ⟨mag⟩ ⟨phase⟩
```

where ******* is an arbitrary sequence of characters, *N+* and *N−* are the numbers of the nodes to which the positive and negative terminals are connected, *mag* is the magnitude of the voltage in volts, and *phase* is its phase angle in degrees. All ac sources are assumed to be sinusoidal. If *mag* is omitted, its default value is 1 V, and if *phase* is omitted, its default value is 0°. An ac current source is identified by making the first character I instead of V. Note that *mag* may be regarded as either a peak or an rms value, since SPICE output from an ac analysis does not consist of instantaneous (time-varying) values.

The .AC Control Statement

The .AC control statement is used to compute ac voltages and currents in a circuit *versus frequency.* A single frequency or a range of frequencies can be specified. The circuit must contain at least one source designated AC, and all AC sources are assumed to be sinusoidal and to have identical frequencies or to undergo the same frequency variation, if any. The format is

```
.AC vartype N fstart fstop
```

where *vartype* specifies the way frequency is to be varied in the range from *fstart* through *fstop.* N is a number related to the number of frequencies at which computations are to be performed, as will be discussed shortly. The *vartype* is one of DEC, OCT, or LIN (decade, octave, or linear). If analysis at a single frequency is desired, we can use any *vartype,* set *fstart* equal to *fstop,* and let $N = 1$.

When *vartype* is LIN, the frequencies at which analysis is performed vary linearly from *fstart* through *fstop.* In that case, N is the total number of frequencies at which the analysis is performed (counting *fstart* and *fstop*). Thus, the interval between frequencies is

$$\Delta f = \frac{fstop - fstart}{N - 1}$$

For example, the statement

```
.AC LIN 21 100 1K
```

will tell SPICE to perform an ac analysis at 21 frequencies from 100 Hz through 1 kHz. The frequency interval will be $(1000 - 100)/20 = 45$ Hz, so computations will be performed at 100 Hz, 145 Hz, 190 Hz, . . . , 1 kHz.

When the *vartype* is DEC or OCT, the analysis is performed at logarithmically spaced intervals and N is the total number of frequencies *per decade or per octave.* For example, the statement

```
.AC DEC 10 100 10K
```

will cause SPICE to analyze the circuit at ten frequencies in each of the decades 100 Hz to 1 kHz and 1 kHz through 10 kHz. The frequencies will be at one-tenth–decade intervals, so each interval will be different. The frequencies at one-tenth–decade intervals are 10^x, $10^{x+0.1}$, $10^{x+0.2}$, . . . , where $x = \log_{10}(fstart)$. In general, the frequencies at which SPICE performs an ac analysis using the DEC *vartype* are 10^x, $10^{x+1/N}$, $10^{x+2/N}$, . . . , where $x = \log_{10}(fstart)$. The last frequency in this sequence is not necessarily *fstop,* but SPICE will compute at frequencies in the sequence up through the first frequency that is equal to or greater than *fstop.* The frequencies at which SPICE performs an ac analysis when the *vartype* is OCT are 2^x, $2^{x+1/N}$, $2^{x+2/N}$, . . . , where $x = \log_2(fstart)$.

AC Outputs

AC voltages and currents whose values are desired from an .AC analysis are specified the same way as dc voltages and currents, using $V(N1, N2)$, $V(N1)$, or I(*Vname*) in a .PRINT or .PLOT statement. The values of the magnitudes of these quantities are printed or plotted. In addition, we can request certain other values, as indicated in the following list of voltage characteristics:

VR	real part
VI	imaginary part
VM	magnitude, $\lvert V \rvert$
VP	phase, degrees
VDB	$20 \log_{10} \lvert V \rvert$

The same values for ac currents can be obtained by substituting I for V. To illustrate, the statement

```
.PRINT AC V(1) VLP(2,3) II(VDUM) IR(VX)
```

causes SPICE to print the magnitude of the ac voltage between nodes 1 and 0, the phase angle of the ac voltage VL between nodes 2 and 3, the imaginary part of the current in VDUM, and the real part of the current in VX. Note that ac must be listed as the analysis type.

AC Plots

When ac voltages or currents are specified in a .PLOT statement, we obtain a linear, semilog, or log-log plot, depending on the type of voltage or current output requested and the *vartype*. Table A–3 summarizes the types of plots produced for each combination. "Log" in the table means that the scale supplied by SPICE has logarithmically spaced values.

Small-Signal Analysis and Distortion

When a circuit contains active devices such as transistors, an .AC analysis by SPICE is assumed to be a small-signal analysis. That is, ac variations are assumed to be small enough that the values of device parameters do not change. As in a .TRAN analysis, SPICE performs an initial dc analysis to determine quiescent voltages and currents so that the values of those device parameters affected by dc levels can be computed. In the ac analysis, the device parameters are assumed to retain those initial values, regardless of the actual magnitudes of the ac variations. *Thus, in an .AC analysis, SPICE does not take into account any distortion, even clipping, that would actually occur if we were to severely overdrive a transistor by specifying very large ac inputs.* For example, if the output swing of an actual transistor circuit were

Table A–3

Vartype	Output Requested	Frequency (Vertical) Axis	Output (Horizontal) Axis
LIN	Magnitude	Linear	Log
	Phase	Linear	Linear
	Imaginary part	Linear	Linear
	Real part	Linear	Linear
	Decibels	Linear	Linear
DEC or OCT	Magnitude	Log	Log
	Phase	Log	Linear
	Imaginary part	Log	Linear
	Real part	Log	Linear
	Decibels	Log	Linear

limited to 10 V, this fact would not be "known" to SPICE, and by overdriving the computer-simulated circuit, we could obtain outputs of several hundred volts from an .AC analysis.

To observe the effects of distortion, such as clipping, it is necessary to perform a .TRAN analysis and obtain a plot of the output waveform versus time. There is also a .DISTO (distortion) control statement that can be used to obtain limited information on harmonic distortion, but we will not have occasion to use that statement.

Example 5–7 illustrates an .AC analysis at a single frequency. The next example illustrates an .AC analysis over a range of frequencies.

Example A–3

Use SPICE to perform an ac analysis of the transistor amplifier in Figure A–12(a) over the frequency range from 100 Hz through 100 kHz. Obtain a plot of the

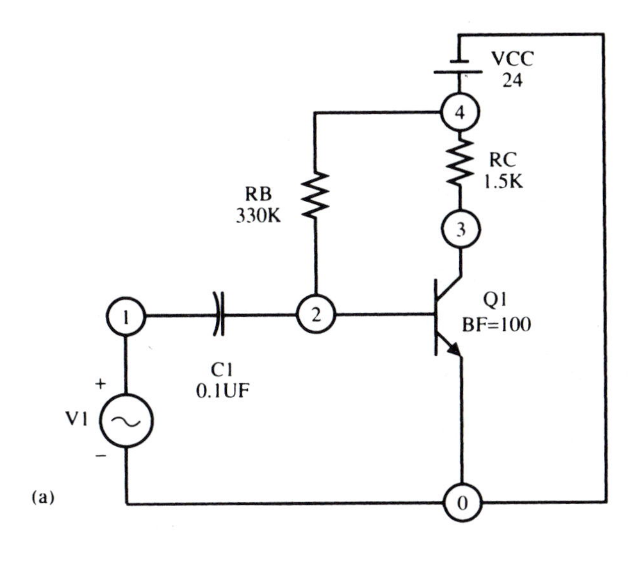


```
EXAMPLE A.3
VCC 4 0 24
V1 1 0 AC
RC 4 3 1.5K
RB 4 2 330K
C1 1 2 0.1UF
Q1 3 2 0 TRANS
.MODEL TRANS NPN BF=100
.AC DEC 10 100 100K
.PLOT AC V(3) VP(3)
.END
```

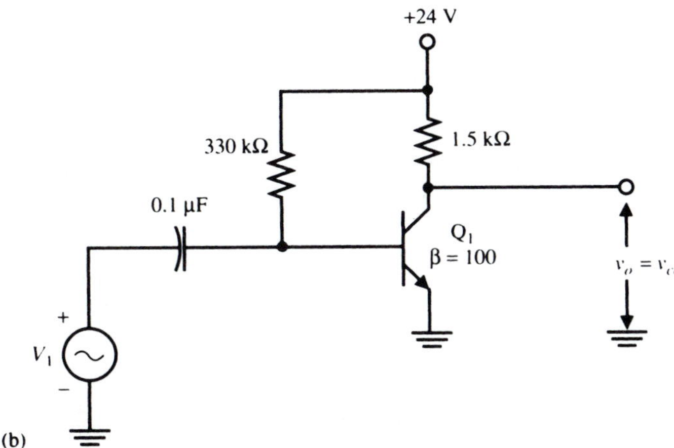

Figure A–12
(Example A–3)

magnitude and phase angle of the collector-to-emitter voltage over the frequency range, with 10 frequencies per decade.

Solution. Figure A–12(b) shows the circuit redrawn for analysis by SPICE and the corresponding input data file. Note that we allow the magnitude and phase of the ac source (V1) to have the default values 1 V and 0°, respectively.

Figure A–13 shows the results of a program run. Note the following points in connection with an .AC analysis of a circuit containing a transistor:

1. SPICE prints a list of the values of the transistor model parameters, which in this case are default values. (Since the default value of BF is 100, we could have omitted that specification in the .MODEL statement.)
2. SPICE obtains a "small-signal bias solution" to determine the dc voltages and currents in the circuit with the ac source set to 0. This is similar to the "initial transient solution" obtained in a .TRAN analysis.
3. Using the dc values obtained from the small-signal bias solution, SPICE computes "operating point information." This information is provided in the form of a list of important bias-dependent parameter values.
4. The two outputs are plotted using * to represent V(3), the magnitude of V_{CE}, on a log scale and + to represent VP(3), the phase angle of V_{CE} on a linear scale. The values of V(3) are printed along the vertical scale, along with the logarithmically spaced frequencies. Thus, the plot is a log-log plot of voltage magnitude and a semilog plot of phase angle.
5. The plots intersect at $f = 1$ kHz, and SPICE prints an X where that occurs.
6. The maximum output voltage (at 100 kHz) is 406.9 V, which is clearly impossible in the actual circuit. As noted earlier, SPICE does not take output voltage limits (clipping) into account during an .AC analysis. Since the input signal magnitude is 1 V, the output magnitude also represents the voltage gain at each frequency.
7. At high frequencies, the phase angle approaches −180°, confirming the phase inversion caused by a common-emitter amplifier.

```
EXAMPLE A.3
****     BJT MODEL PARAMETERS                  TEMPERATURE =    27.000 DEG C
******************************************************************************
              TRANS
TYPE          NPN
IS          1.00D-16
BF           100.000
NF             1.000
BR             1.000
NR             1.000
*******02/22/89 ********  SPICE 2G.6     3/15/83 ********13:15:17*****
EXAMPLE A.3
****      SMALL SIGNAL BIAS SOLUTION           TEMPERATURE =    27.000 DEG C
******************************************************************************
 NODE    VOLTAGE      NODE    VOLTAGE      NODE    VOLTAGE      NODE    VOLTAGE
(  1)      .0000     (  2)      .8246     (  3)    13.4657     (  4)    24.0000
   VOLTAGE SOURCE CURRENTS
   NAME         CURRENT
   VCC       -7.093D-03
   V1         0.000D+00
   TOTAL POWER DISSIPATION   1.70D-01   WATTS
*
```

Figure A–13
(Example A-3)

```
EXAMPLE A.3
****      OPERATING POINT INFORMATION        TEMPERATURE =    27.000 DEG C
******************************************************************************
**** BIPOLAR JUNCTION TRANSISTORS
              Q1
MODEL        TRANS
IB           7.02E-05
IC           7.02E-03
VBE              .825
VBC           -12.641
VCE            13.466
BETADC        100.000
GM           2.72E-01
RPI          3.68E+02
RX           0.00E+00
RO           1.00E+12
CPI          0.00E+00
CMU          0.00E+00
CBX          0.00E+00
CCS          0.00E+00
BETAAC        100.000
FT           4.32E+18
*

EXAMPLE A.3
****      AC ANALYSIS                        TEMPERATURE =    27.000 DEG C
******************************************************************************
LEGEND:
*: V(3)
+: VP(3)
   FREQ        V(3)
*)-------------  1.000D+00     1.000D+01     1.000D+02     1.000D+03  1.000D+04
                - - - - - - - - - - - - - - - - - - - - - - - - - - - - - - - -
+)------------- -2.000D+02    -1.500D+02    -1.000D+02    -5.000D+01  0.000D+00
                - - - - - - - - - - - - - - - - - - - - - - - - - - - - - - - -
1.000D+02  9.412D+00 .             *             . +           .            .
1.259D+02  1.185D+01 .             .*            . +           .            .
1.585D+02  1.491D+01 .             . *           . +           .            .
1.995D+02  1.876D+01 .             .   *         . +           .            .
2.512D+02  2.361D+01 .             .    *        . +           .            .
3.162D+02  2.969D+01 .             .      *      . +           .            .
3.981D+02  3.732D+01 .             .       *     .+            .            .
5.012D+02  4.687D+01 .             .        *    .+            .            .
6.310D+02  5.878D+01 .             .          *  +             .            .
7.943D+02  7.355D+01 .             .           * +             .            .
1.000D+03  9.172D+01 .             .            X.             .            .
1.259D+03  1.138D+02 .             .           + .*            .            .
1.585D+03  1.401D+02 .             .          +  .  *          .            .
1.995D+03  1.706D+02 .             .         +   .   *         .            .
2.512D+03  2.045D+02 .             .       +     .    *        .            .
3.162D+03  2.403D+02 .             .      +      .     *       .            .
3.981D+03  2.758D+02 .             .    +        .      *      .            .
5.012D+03  3.083D+02 .             .  +          .       *     .            .
6.310D+03  3.359D+02 .             .+            .       *     .            .
7.943D+03  3.577D+02 .             +             .        *    .            .
1.000D+04  3.738D+02 .           + .             .        *    .            .
1.259D+04  3.852D+02 .          +  .             .        *    .            .
1.585D+04  3.929D+02 .         +   .             .        *    .            .
1.995D+04  3.980D+02 .        +    .             .        *    .            .
2.512D+04  4.014D+02 .       +     .             .        *    .            .
3.162D+04  4.035D+02 .       +     .             .        *    .            .
3.981D+04  4.049D+02 .      +      .             .         *   .            .
5.012D+04  4.058D+02 .      +      .             .         *   .            .
6.310D+04  4.063D+02 .      +      .             .         *   .            .
7.943D+04  4.067D+02 .     +       .             .         *   .            .
1.000D+05  4.069D+02 .     +       .             .         *   .            .
                - - - - - - - - - - - - - - - - - - - - - - - - - - - - - - - -
```

Figure A–13
(Continued)

A–11 JUNCTION FIELD-EFFECT TRANSISTORS (JFETs)

An N-channel or P-channel JFET in a circuit to be analyzed by SPICE must be given a name that begins with J, such as JX or JFET1. The format for specifying both P-channel and N-channel devices is

```
J******* ND NG NS Mname
```

where *ND* is the node number of the drain, *NG* the node number of the gate, *NS* the node number of the source, and *Mname* is the model name. The model name appears in the MODEL statement for a JFET using the format

```
.MODEL Mname type ⟨Pval1 = n1 Pval2 = n2 ···⟩
```

where *type* is NJF for an N-channel JFET and PJF for a P-channel JFET. Table A–4 shows the JFET parameters that can be specified in a .MODEL statement and their default values.

The two principal parameters related to the dc characteristics of a JFET are VTO (the pinch-off voltage, V_p) and BETA (the transconductance parameter, β). In the pinch-off region, these are related by

$$\beta = \frac{I_D}{(V_{GS} - V_p)^2} \quad \text{A/V}^2$$

Since $I_D = I_{DSS}$ when $V_{GS} = 0$, we have

$$\beta = \frac{I_{DSS}}{V_p^2}$$

When modeling a JFET for which values of I_{DSS} and V_p are known, we must calculate β using this equation, so that the parameter BETA can be specified in the .MODEL statement. *Note this important point: The value entered for the parameter VTO is always negative, whether the JFET is an N-channel or a P-channel device.*

Table A–4
JFET parameters

Parameter	Identification	Units	Default Value
Threshold voltage	VTO	V	−2
Transconductance parameter	BETA	A/V^2	10^{-4}
Channel-length-modulation parameter	LAMBDA	1/V	0
Drain ohmic resistance	RD	Ω	0
Source ohmic resistance	RS	Ω	0
Zero-bias gate-to-source–junction capacitance	CGS	F	0
Zero-bias gate-to-drain–junction capacitance	CGD	F	0
Gate-junction saturation current	IS	A	10^{-14}
Gate-junction potential	PB	V	1
Flicker-noise coefficient	KF	—	0
Flicker-noise exponent	AF	—	1
Coefficient for forward-bias depletion capacitance formula	FC	—	0.5

The parameter LAMBDA (λ, the channel-length-modulation factor) controls the extent to which the characteristic curves rise for increasing values of V_{DS} in the pinch-off region. If λ is 0 (the default value), the characteristic curves are perfectly flat. A typical value for λ is 10^{-4}. Examples 7–2 and 7–9 show sample programs used to analyze circuits containing N-channel and P-channel JFETs.

A–12 MOS FIELD-EFFECT TRANSISTORS (MOSFETs)

An N-channel or P-channel MOSFET must be identified in a SPICE data file by a name beginning with the letter M, using the format:

```
M******* ND NG NS NSS Mname
```

where *ND, NG, NS,* and *NSS* are the node numbers of the drain, gate, source, and substrate, respectively, and *Mname* is the model name. The values of certain geometric parameters, such as the length and width of the channel, can be optionally specified with the MOSFET identification, but these will not concern us and we can allow all of them to default.

The MOSFET model is very complex. Like other semiconductor models in SPICE, it involves many parameters whose values are difficult to determine and many that are beyond the scope of our treatment. There are actually three built-in models, referred to as *levels* 1, 2, and 3. LEVEL is one of the parameters that can be specified in a MOSFET .MODEL statement, and its value prescribes the particular model to be used. The default level is 1, which we can assume for all purposes in this book. The format of the MOSFET .MODEL statement is

```
.MODEL Mname type ⟨Pval1 = n1 Pval2 = n2 ···⟩
```

where *type* is NMOS for an N-channel MOSFET and PMOS for a P-channel MOSFET. A MOSFET can be of either the depletion-mode type or of the enhancement-mode type, as is discussed shortly.

Since the MOSFET parameters in the three-level SPICE model are so numerous and complex, we will not present a table showing their identifications, units, and default values. (Such information should be available in a user's guide furnished with the SPICE software used.) In any case, there is a significant variation among the several versions of SPICE in the number and type of MOSFET parameters that can be specified. However, there are two fundamental parameters whose values should probably be specified in every MOSFET simulation: β and V_T. In SPICE, β is called the *intrinsic transconductance parameter* and is identified in a .MODEL statement by KP. β = KP provided the channel length and width parameters, L and W, are allowed to default; otherwise, KP = β(L/W). Its value is always positive. V_T is called the zero-bias threshold voltage and is identified by VTO. Following is an example of a .MODEL statement for an N-channel MOSFET having $\beta = 0.5 \times 10^{-3}$ A/V^2 and $V_T = 2$ V:

```
.MODEL M1 NMOS KP = 0.5E-3 VTO = 2
```

The threshold voltage, VTO, is positive or negative, according to the following table:

Mode	Channel	Sign of VTO
Depletion	N	−
Depletion	P	+
Enhancement	N	+
Enhancement	P	−

For example, the value specified for the VTO of an N-channel, enhancement-mode MOSFET should be positive.

Example 7–13 illustrates a SPICE simulation of a circuit containing a MOSFET.

PSPICE

In PSpice, the voltage across any pair of JFET or MOSFET terminals and/or the current into any JFET or MOSFET terminal can be printed or plotted using the letters D, G, and S to represent the drain, gate, and source terminals. The following are some examples of output variables that could be specified with a .PRINT or .PLOT statement:

VGS(J1)	gate-to-source voltage of the JFET named J1
ID(MXY)	drain current of the MOSFET named MXY
VD(M25)	drain-to-ground voltage of the MOSFET named M25

A–13 CONTROLLED (DEPENDENT) SOURCES

A controlled voltage source is one whose output voltage is controlled by (depends on) the value of a voltage or current elsewhere in the circuit. The simplest and most familiar example is a voltage amplifier: It is a voltage-controlled voltage source because its output voltage *depends* on its input voltage. The output voltage equals the input voltage multiplied by the gain. A current-controlled voltage source obeys the relation $v_o = ki$, where i is the controlling current and k is a constant having

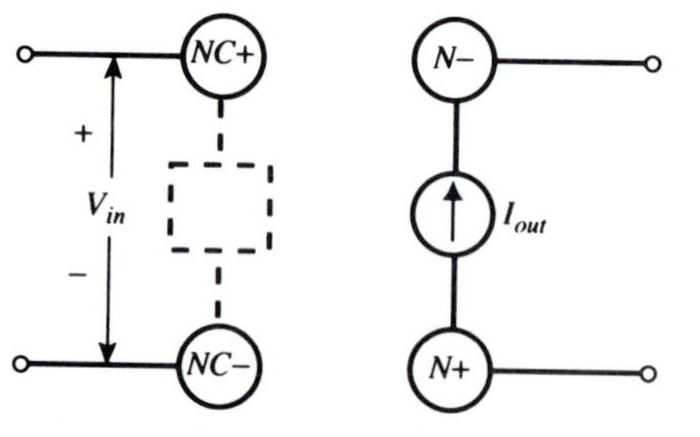

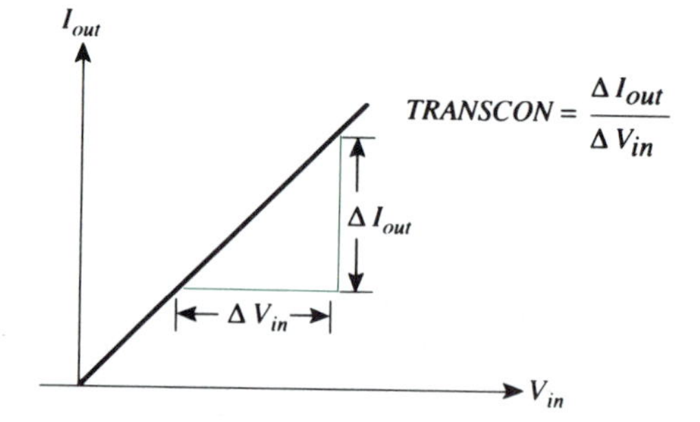

G******* *N+ N– NC+ NC– TRANSCON*

(a) Voltage-controlled current source

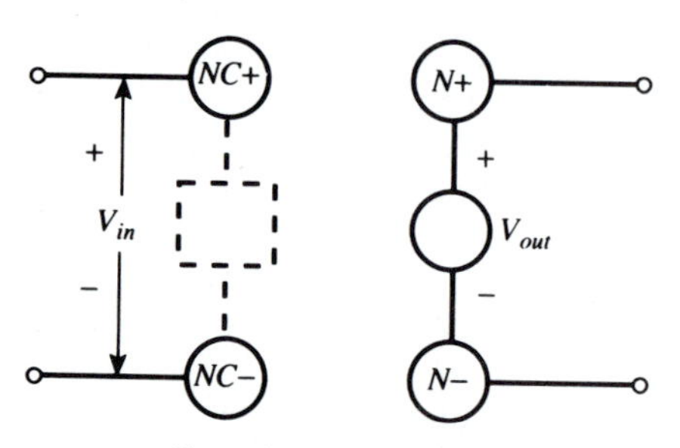

V_{out}

$VGAIN = \frac{\Delta V_{out}}{\Delta V_{in}}$

ΔV_{out}

ΔV_{in}

V_{in}

E******* *N+ N– NC+ NC– VGAIN*

(b) Voltage-controlled voltage source

Figure A–14
Specification of controlled sources in SPICE

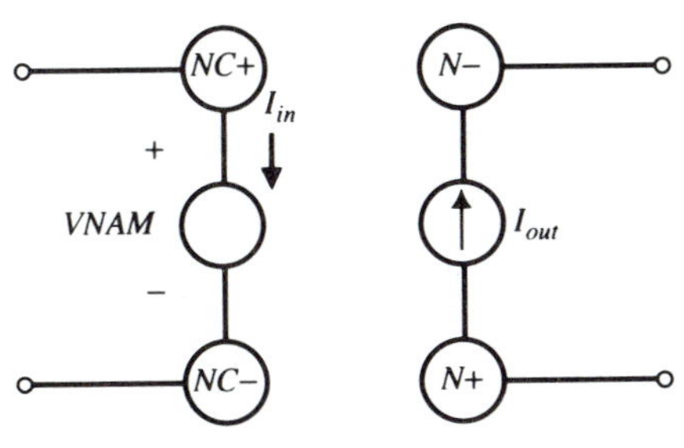

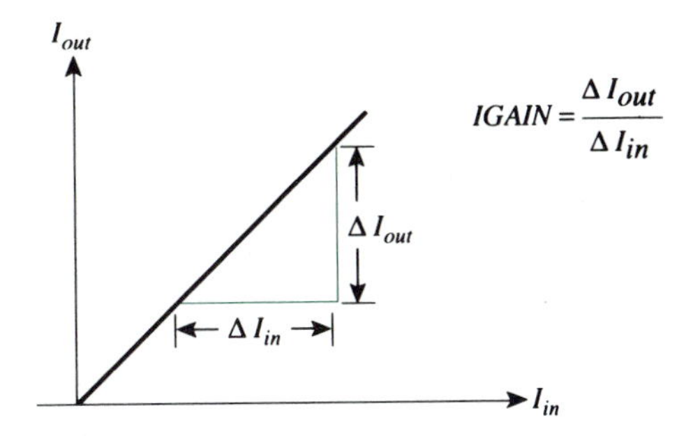

F******* *N+ N− VNAM IGAIN*

(c) Current-controlled current source

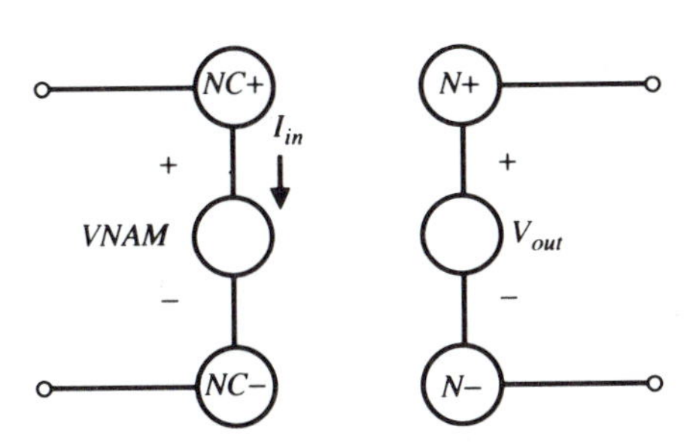

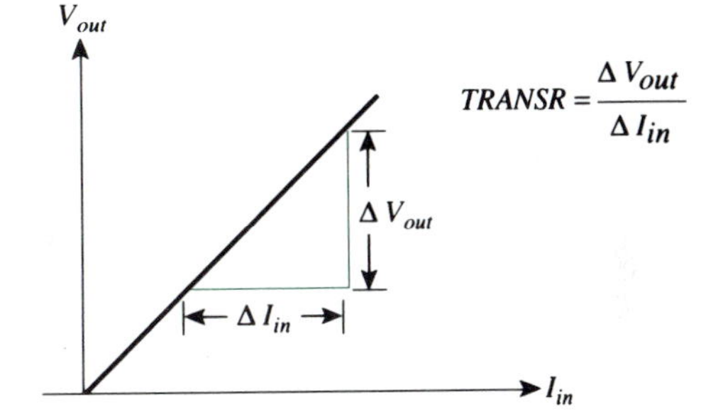

H******* *N+ N− VNAM TRANSR*

(d) Current-controlled voltage source

Figure A–14
(Continued)

the units of resistance: $k = v_o/i$ volts per ampere, or ohms. In the context of a current-controlled voltage source, k is called a *transresistance.*

Similarly, a controlled current source produces a current whose value depends on a voltage or a current elsewhere in the circuit. A voltage-controlled current source obeys the relation $i_o = kv$, where v is the controlling voltage and k has the units of conductance: $k = i_o/v$ amperes per volt, or siemens. In the context of a controlled source, k is called a *transconductance.*

The four types of controlled sources, voltage-controlled voltage sources, current-controlled voltage sources, voltage-controlled current sources, and current-controlled current sources, can be modeled in SPICE. Figure A–14 shows the format used to model each type. Note that the controlling voltage in voltage-controlled sources (the voltage between NC+ and NC− in (a) and (b) of the figure) can be the voltage between any two nodes; it is not necessary that a component be connected between those nodes. Also note that the controlling current in controlled current sources (parts (c) and (d) of the figure) is always the current in a voltage source. Thus, it may be necessary to insert a dummy voltage source in a circuit in order to specify a controlling current at a desired point in the circuit. Observe how the "plus" and "minus" nodes are defined in the figure, in connection with the polarity assumptions made by SPICE for the currents and current sources. Example 9–8 illustrates the use of a voltage-controlled voltage source (E1) and a current-controlled current source (F1) in a SPICE program.

A–14 TRANSFORMERS

A transformer can be modeled in SPICE using three statements: one to specify the nodes and inductance of the primary winding, one to specify the nodes and inductance of the secondary winding, and one to specify the *coefficient of coupling* between the windings. The primary and secondary windings are specified exactly as ordinary inductors are, using an L prefix, node numbers, and the inductance (in henries). The coefficient of coupling is specified in a statement that must begin with K:

```
K******* LNAM1 LNAM2 k
```

where *LNAM1* and *LNAM2* are the inductors comprising the primary and secondary windings and k is the value of the coefficient of coupling ($0 < k < 1$). For an ideal transformer, in which all the magnetic flux in the primary is coupled to the secondary, the coefficient of coupling equals 1. (In PSpice, k cannot be set equal to 1, but can be made arbitrarily close to it, for example, 0.999.) To simulate an ideal iron-core transformer in SPICE, it is necessary to specify inductance values so that the reactance of the primary winding is much greater than the impedance of the signal source driving it and so that the reactance of the secondary winding is much greater than the load impedance. (The primary and secondary windings of an ideal transformer have infinite inductance.) Thus, it may be necessary to specify unrealistically large values for the inductances of the primary and secondary windings. (Such will be the case in all examples found in this book.) The following is an example showing the specification of a transformer (KXFRMR) having primary and secondary windings named LPRIM and LSEC:

```
LPRIM 5 0 20H
LSEC 6 0 0.2H
KXFRMR LPRIM LSEC 1
```

If operation of this transformer is to be simulated at 1 kHz, the source impedance should be much smaller than $2\pi f(\text{LPRIM}) = 2\pi(10^3 \text{ Hz})(20 \text{ H}) = 125.7 \text{ k}\Omega$, and the load impedance should be much smaller than $2\pi f(\text{LSEC}) = 2\pi(10^3 \text{ Hz})(0.2 \text{ H}) = 1.257 \text{ k}\Omega$. In SPICE, an inductor cannot appear in a loop isolated from ground, so it is *not* possible to isolate the primary and secondary windings from each other, as is done in a real transformer.

The turns ratio of an ideal transformer is determined by the inductance values of the primary and secondary windings, according to

$$\frac{N_p}{N_s} = \sqrt{\frac{L_p}{L_s}}$$

where N_p/N_s is the turns ratio and L_p and L_s are the primary and secondary inductances, respectively. For example, the turns ratio of the transformer in the foregoing example is

$$\frac{N_p}{N_s} = \sqrt{\frac{20 \text{ H}}{0.2 \text{ H}}} = \sqrt{100} = 10$$

Thus, the transformer is a step-down type whose secondary voltage is one-tenth of its primary voltage. Example 11–16 demonstrates the use of a transformer in a SPICE program.

A–15 SUBCIRCUITS

A complex electronic circuit will often contain several components, or subsections of circuitry, that are identical to each other. Examples include active filters containing several identical operational amplifiers, multistage amplifiers consisting of identical amplifier stages, and digital logic systems containing numerous identical logic gates. When modeling such circuits in SPICE, it is a tedious and time-consuming task to write numerous sets of identical statements describing identical circuitry. Furthermore, the input data file for such a system may become so long and cumbersome that it is difficult to interpret or modify. To alleviate those kinds of problems, SPICE allows a user to create a *subcircuit:* circuitry that can be defined just once and then, effectively, inserted into a larger system (which we will call the *main* circuit) at as many places as desired. The concept is similar to that of a *subroutine* in conventional computer programming. In SPICE, it is possible to define several different subcircuits in one data file, and, in fact, one subcircuit can contain other subcircuits.

The first statement of a subcircuit is

```
.SUBCKT NAME N1 ⟨N2 N3 ...⟩
```

where *NAME* is any name chosen to identify the subcircuit and *N1, N2,* . . . are the numbers of the nodes in the subcircuit that will be joined to other nodes in the main circuit. None of these can be node 0. Components in a subcircuit are defined in exactly the same way they are in any SPICE data file, using successive statements following the .SUBCKT statement. A subcircuit can contain .MODEL statements, but it cannot contain any control statements, such as .DC, .TRAN, .PRINT, or .PLOT. The node numbers in a subcircuit do not have to be different from those in the main program. SPICE treats a subcircuit as a completely separate (isolated) entity, and a node having the same number in a subcircuit as another node in the main circuit will still be treated as a different node. The exception is node 0: If node 0 appears in a subcircuit, it is treated as the same node 0 as in the main circuit. (In the language of computer science, subcircuit nodes are said to be *local,* except node 0, which is *global.*) The last statement in a subcircuit must be

```
.ENDS⟨NAME⟩
```

If a subcircuit itself contains subcircuits, the *NAME* must be given in the .ENDS statement to specify which subcircuit definition has been ended.

In order to "insert" a subcircuit into the main circuit, we must write a subcircuit *call* statement in the main program. A different call statement is required for each location where the subcircuit is to be inserted. The format of a call statement is

```
X******* N1⟨N2 N3 ...⟩ NAME
```

where ******* are arbitrary characters that must be different for each call; *N1, N2,* . . . , are the node numbers in the main circuit that are to be joined to the subcircuit nodes specified in the .SUBCKT statement; and *NAME* is the name of the subcircuit to be inserted. The node numbers in the call statement will be joined to the nodes in the .SUBCKT statement in exactly the same order as they both appear. That is, *N1* in the subcircuit will be joined to *N1* in the main circuit, and so forth. The following is an example:

```
.SUBCKT OPAMP 1 2 3
-]
-} Statements describing components in the subcircuit
-]
.ENDS
X1 8 4 12 OPAMP
X2 1 4 3  OPAMP
 -
 -
```

In this example, the subcircuit named OPAMP is inserted at two locations in the main program. The first call statement (X1) connects subcircuit nodes 1, 2, and 3 to main-circuit nodes 8, 4, and 12, respectively. The X2 call statement connects subcircuit nodes 1, 2, and 3 to main-circuit nodes 1, 4, and 3, respectively. (In this case some of the subcircuit and main-circuit node numbers are the same.)

The next example illustrates the use of a subcircuit to model an RC filter containing three identical stages. Although this circuit could not be considered complex enough to warrant the use of a subcircuit, it does serve to demonstrate the syntax we have described. Another example can be found in Example 14–20.

Example A–4

Using a SPICE subcircuit, determine the magnitude and phase angle of the output of the RC filter in Figure A–15(a) when the input is a 1-kHz sine wave with peak value 10 V.

Solution. Figure A–15(b) shows the RC subcircuit, named STAGE, and the way that it is inserted into the main circuit in three locations. Its locations are identified by rectangles labeled X1, X2, and X3. The node numbers inside the rectangles are the subcircuit node numbers and those outside are the main-circuit node numbers.

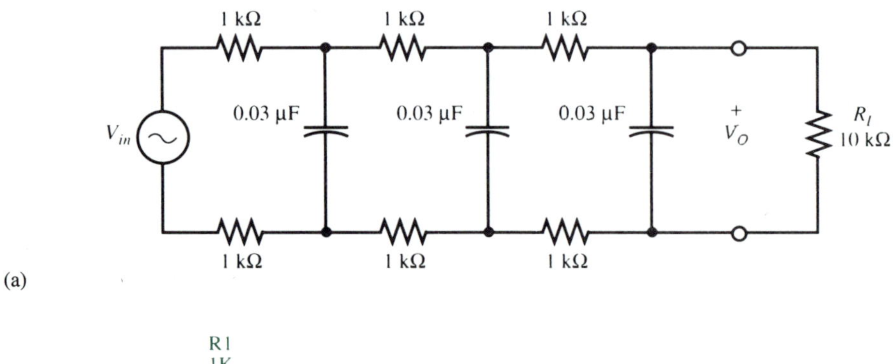

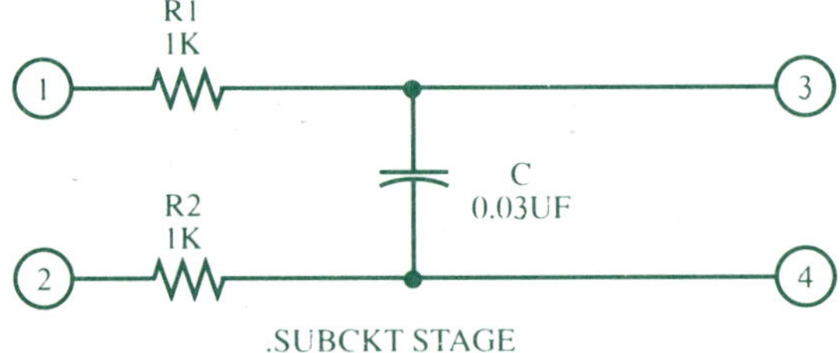

Figure A–15
(Example A–4)

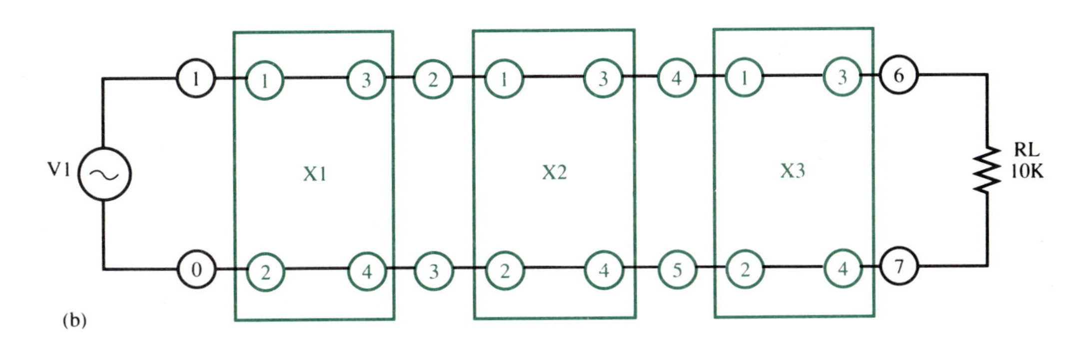

(b)

```
EXAMPLE A.4
V1 1 0 AC 10V
.SUBCKT STAGE 1 2 3 4
R1 1 3 1K
R2 2 4 1K
C 3 4 0.03UF
.ENDS
X1 1 0 2 3 STAGE
X2 2 3 4 5 STAGE
X3 4 5 6 7 STAGE
RL 6 7 10K
.AC LIN 1 1K 1K
.PRINT AC VM(6,7) VP(6,7)
.END
```

(c)

```
EXAMPLE A.4
****      AC ANALYSIS                              TEMPERATURE =    27.000 DEG C
******************************************************************************
   FREQ        VM(6,7)      VP(6,7)
 1.000E+03     3.769E+00   -7.107E+01
```

(d)

Figure A–15
(Continued)

In the input data file, shown in part (c) of the figure, note that the node numbers in the three call statements are the main-circuit nodes that are connected to subcircuit nodes 1, 2, 3, and 4, in that order. The results of a program run are shown in Figure A–15(d). We see that the output voltage has peak value 3.769 V and lags the input by 71.07°.

A–16 PROBE AND CONTROL SHELL

PSPICE

Probe

Probe is a PSpice option that makes it possible to obtain output plots having greater resolution than those produced by .PLOT statements. Instead of plotting a sequence of asterisks, Probe produces a virtually solid line on a high-resolution monitor. Also, Probe has other features that make it very useful for analyzing output data.

It behaves very much like a high-quality oscilloscope that allows the user to position a cursor on a trace and to obtain a direct readout of the value of the variable displayed at the position of the cursor.

If the statement

```
.PROBE
```

appears anywhere in an input data file that requests a .DC, .AC, or .TRAN analysis, then Probe automatically generates plotting data for the voltage (with respect to ground) at every node in the circuit and for the current entering every device in the circuit. Probe stores the data in an *output data file* named PROBE.DAT. To initiate a Probe run, the user enters the command PROBE directly from the keyboard. If more than one analysis type appears in the input data file (called the *circuit file* in PSpice), a "start-up menu" appears, and the user selects a single analysis type for the Probe run. If only one analysis type appeared in the circuit file, then a set of axes and a menu are displayed immediately after Probe is entered. Selecting the option ADD TRACE from this menu prompts the user to enter the variable to be plotted, using the same format used to specify outputs in PSpice (such as V(2), I(R1), etc.). The plot is scaled automatically and displayed on the monitor. Additional variables can be plotted simultaneously by repeatedly selecting the ADD TRACE option. Also, plots can be deleted by selecting the DELETE TRACE option.

Selecting the CURSOR option from the Probe menu creates two sets of cross hairs, identified as cursor 1 (C1) and cursor 2 (C2), on the display. These cross hairs can be moved along the plot using direction keys (← and →) on the keyboard. Using the direction keys alone moves C1 and using the direction keys with SHIFT depressed moves C2. The numerical values of the variable at the positions of the cross hairs on the plot are displayed in a window, along with the difference in the values. Figure A–16 shows an example. Example 5–8 shows an actual Probe display.

Figure A–16
Example of a PSpice Probe display

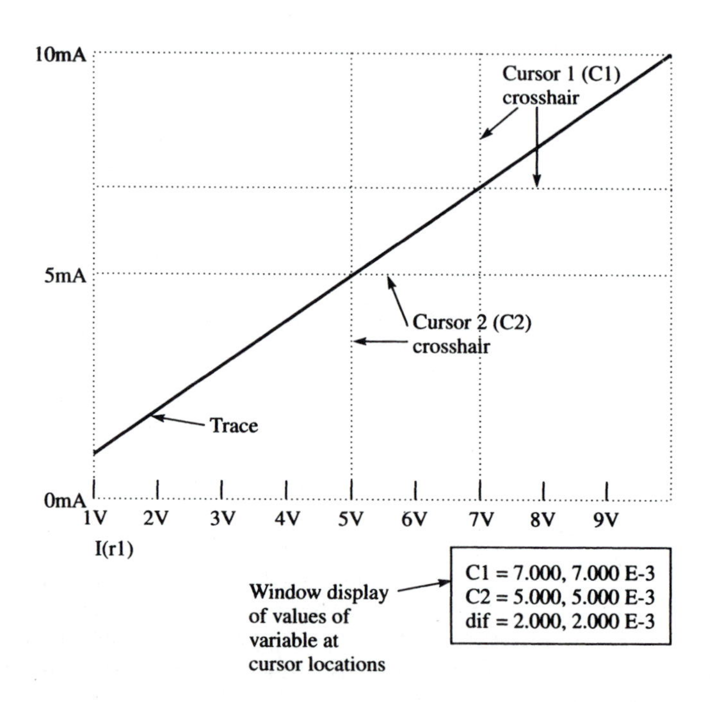

If desired, Probe plots can be limited to specific variables by giving their names in the .PROBE statement in the circuit file. For example, the statement

```
.PROBE V(2)
```

causes Probe to create an output data file containing only plotting data for the voltage at node 2. This option is useful when memory capacity is too small to store all voltages and currents.

Another feature of Probe is that it can create plots of mathematical functions of the variables. For example, after selecting ADD TRACE, entering the expression

```
I(R1)*I(R1)*1K
```

creates a plot of the power (I^2R) dissipated by 1-kΩ resistor R1. Numerous mathematical functions, including trigonometric functions, logarithmic functions, average values, square roots, and absolute values, of the variables can also be plotted. One difference between Probe and PSpice is that suffixes in expressions written for Probe must use lowercase m to represent milli and capital M to represent meg (as opposed to M for milli and MEG for meg).

It should be noted that the high-resolution plots created by Probe are *not* the result of Probe increasing the number of output values computed in a circuit simulation. Rather, Probe *interpolates* values between the data points that a circuit file specifies are to be computed. Thus, if the output variable has a sudden change in value (such as a sharp peak) and the circuit file does not specify a fine-enough increment between data points to detect that change, Probe cannot be expected to detect it either.

Control Shell

Control Shell is a PSpice option that coordinates and simplifies writing, editing, and running input circuit files and using the various options and features available in PSpice. For example, Control Shell can be used to make Probe run automatically after every circuit simulation. Options are selected by the user from various menus displayed by Control Shell.

In a system equipped with Control Shell, the first (main) menu displayed contains, among others, selections entitled Files, Analysis, and Probe. These are the principal choices that will be used in creating and running most circuit simulations. When the user selects Files, another menu is displayed with options that include Edit, Browse Output, and Current File. Edit and Current File allow the user to edit an existing circuit file or to create and name a new circuit file. Selecting Browse Output allows the user to scroll through the entire output created after a circuit simulation has been run.

To perform a program run of an input circuit file that has already been created, the option Analysis is selected from the main menu. This selection creates another menu that allows the user to specify the analysis type (AC, DC, or transient). Since at least one analysis type is (presumably) already specified in the input circuit file, an arrow in this menu points to one analysis type and the user can simply press the ENTER key to initiate the run. Alternatively, another analysis type specified in the circuit file can be selected, and the user is prompted to enter (or change) analysis parameters (such as start and stop voltages and step size in a .DC analysis). It should be noted that if there is an error in the input circuit file, Control Shell does not allow the user to select the Analysis option.

Selecting Probe from the main menu creates another menu that allows the user to request automatic running of Probe after a PSpice simulation run. In this option, it is not necessary to include a .PROBE statement in the input circuit file.

A-17 THE PSPICE LIBRARY

PSPICE

The MicroSim Corporation has created models and subcircuits for over 3500 standard analog devices, including transistors, diodes, and operational amplifiers. These can be stored in a computer system as a *library* that can be accessed by users who wish to specify any one or more of the devices in an input circuit file. It is not necessary to write the entire model or subcircuit defining a particular device in the input file; it is necessary only to refer to it by name. PSpice automatically retrieves the description from the library and uses it in the circuit simulation.

Transistors and diodes are stored in the library as model statements. If we wish to incorporate a particular transistor in an input circuit file, we specify the device name in place of the model name in the statement defining the transistor in our circuit. The statement .LIB is used to inform PSpice that we are accessing the library. The following is an example:

```
Q1 3 5 8 Q2N2222A
.LIB
```

These statements cause PSpice to use the model statement in the library for the 2N2222A transistor as the model statement for Q1. That model statement is as follows:

```
.model Q2N2222A NPN (Is=14.34f Xti=3 Eg=1.11 Vaf=74.03 Bf=255.9 Ne=1.307
          Ise=14.34f Ikf=.2847 Xtb=1.5 Br=6.092 Nc=2 Isc=0 Ikr=0 Rc=1
          Cjc=7.306p Mjc=.3416 Vjc=.75 Fc=.5 Cje=22.01p Mje=.377 Vje=.75
          Tr=46.91n Tf=411.1p Itf=.6 Vtf=1.7 Xtf=3 Rb=10)
```

We should note that the PSpice library actually consists of a number of library files, each having a name and containing devices of a certain type. For example, DIODE.LIB is the name of a library file that contains model statements for diodes, and LINEAR.LIB contains subcircuits for operational amplifiers. PSpice can be directed to a particular library file by including the name of the file in the .LIB statement. For example,

```
.LIB LINEAR.LIB
```

directs PSpice to the linear library file. In particular, in the *evaluation version* (student version) of PSpice, the library file is called EVAL.LIB, and this must be included in the .LIB statement. Linear devices in EVAL.LIB include the following:

Q2N2222A	NPN transistor
Q2N2907A	PNP transistor
Q2N3904	NPN transistor
Q2N3906	PNP transistor
D1N750	zener diode
MV2201	voltage variable-capacitance diode
D1N4148	switching diode
J2N3819	N-channel JFET
J2N4398	N-channel JFET
LM324	operational amplifier

UA741	operational amplifier
LM111	voltage comparator

Example 5–8 illustrates use of the PSpice library to access the model statement for the 2N2222A transistor.

Since operational amplifiers are stored in the library as subcircuits, they must be accessed in the input circuit file by subcircuit *calls* (statements beginning with X; see Section A–15). Figure A–17 shows an example of an input circuit file containing a call for the 741 operational amplifier. Note that the subcircuit nodes are

1	noninverting input
2	inverting input
3	positive supply voltage
4	negative supply voltage
5	output

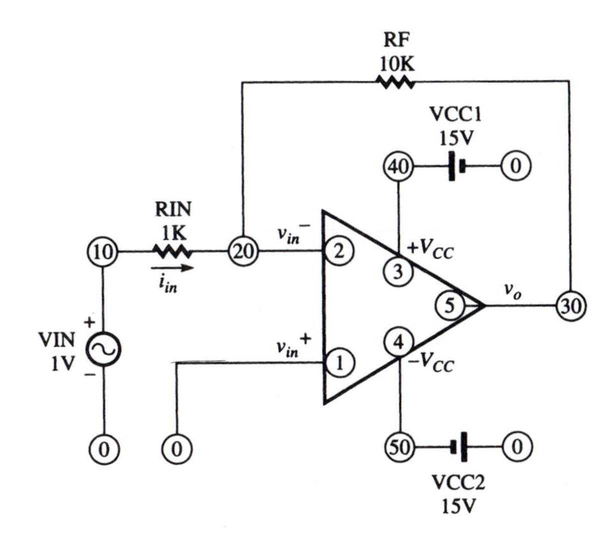


```
OPAMP EXAMPLE
VIN 10 0 AC 1V
RIN 10 20 1K
RF 20 30 10K
VCC1 40 0 15V
VCC2 0 50 15V
X1 0 20 40 50 30 UA741
.LIB EVAL.LIB
.AC DEC 1 1KHZ 1KHZ
.PRINT AC V(30) VP(30) I(RIN)
.END

(EVAL.LIB for Evaluation version of PSpice only)
```

Figure A–17
Example of a call for an operational amplifier subcircuit from the PSpice library

In this example, the amplifier is connected in an inverting configuration with voltage gain $-R_f/R_1 = -10\ \text{k}\Omega/1\ \text{k}\Omega = -10$. The input is a 1-V, 1-kHz sine wave. Execution of the program reveals that v_o = V(30) = 9.999 V, $\angle v_o$ = VP(30) = 179.4°, and i_{in} = I(RIN) = 0.9999 mA.

The subcircuits for operational amplifiers are the *functional equivalents* of the devices they represent. That is, the circuits do not contain the transistors, diodes, resistors, etc., that are in the actual amplifier circuits. These functionally equivalent circuits are designed to exhibit nominal amplifier characteristics (such as bandwidth), not the worst-case values given in manufacturers' specifications. These characteristics do *not* change with temperature, as they do in actual amplifiers. Finally, we should note that component names used in library subcircuits are all *local.* In other words, PSpice will recognize that R1, for example, used in a library subcircuit is different from another resistor named R1 in the input circuit file.

B STANDARD VALUES OF 5% AND 10% RESISTORS

Resistors with 5% tolerance are available in all values shown. Resistors with 10% tolerance are available only in the boldfaced values.

Ohms (Ω)					Kilohms ($k\Omega$)		Megohms ($M\Omega$)	
0.10	**1.0**	**10**	**100**	**1000**	**10**	**100**	**1.0**	**10.0**
0.11	1.1	11	110	1100	11	110	1.1	11.0
0.12	**1.2**	**12**	**120**	**1200**	**12**	**120**	**1.2**	**12.0**
0.13	1.3	13	130	1300	13	130	1.3	13.0
0.15	**1.5**	**15**	**150**	**1500**	**15**	**150**	**1.5**	**15.0**
0.16	1.6	16	160	1600	16	160	1.6	16.0
0.18	**1.8**	**18**	**180**	**1800**	**18**	**180**	**1.8**	**18.0**
0.20	2.0	20	200	2000	20	200	2.0	20.0
0.22	**2.2**	**22**	**220**	**2200**	**22**	**220**	**2.2**	**22.0**
0.24	2.4	24	240	2400	24	240	2.4	
0.27	**2.7**	**27**	**270**	**2700**	**27**	**270**	**2.7**	
0.30	3.0	30	300	3000	30	300	3.0	
0.33	**3.3**	**33**	**330**	**3300**	**33**	**330**	**3.3**	
0.36	3.6	36	360	3600	36	360	3.6	
0.39	**3.9**	**39**	**390**	**3900**	**39**	**390**	**3.9**	
0.43	4.3	43	430	4300	43	430	4.3	
0.47	**4.7**	**47**	**470**	**4700**	**47**	**470**	**4.7**	
0.51	5.1	51	510	5100	51	510	5.1	
0.56	**5.6**	**56**	**560**	**5600**	**56**	**560**	**5.6**	
0.62	6.2	62	620	6200	62	620	6.2	
0.68	**6.8**	**68**	**680**	**6800**	**68**	**680**	**6.8**	
0.75	7.5	75	750	7500	75	750	7.5	
0.82	**8.2**	**82**	**820**	**8200**	**82**	**820**	**8.2**	
0.91	9.1	91	910	9100	91	910	9.1	

C DERIVATION OF FREQUENCY-RESPONSE EQUATIONS

C–1 CUTOFF FREQUENCY OF TWO ISOLATED HIGH-PASS RC FILTERS IN CASCADE

Let

$$f_A = \text{cutoff frequency of first filter}$$

$$f_B = \text{cutoff frequency of second filter}$$

Since the filters are assumed to be isolated (by an ideal, unity-gain amplifier), the overall voltage gain of the cascade at frequency f is

$$\frac{|v_o|}{|v_{in}|} = \frac{1}{\sqrt{1 + (f_A/f)^2}}\,\frac{1}{\sqrt{1 + (f_B/f)^2}}$$

At cutoff ($f = f_1$),

$$\frac{1}{\sqrt{1 + (f_A/f_1)^2}}\,\frac{1}{\sqrt{1 + (f_B/f_1)^2}} = \frac{1}{\sqrt{2}}$$

Let

$$x = \frac{1}{f_1^2},\ a = f_A^2,\ b = f_B^2$$

Then

$$\frac{1}{\sqrt{1 + ax}\,\sqrt{1 + bx}} = \frac{1}{\sqrt{2}}$$

$$\frac{1}{(1 + ax)(1 + bx)} = \frac{1}{2}$$

$$(1 + ax)(1 + bx) = 2$$

$$abx^2 + (a + b)x - 1 = 0$$

$$x = \frac{-(a + b) \pm \sqrt{(a + b)^2 + 4ab}}{2ab}$$

$$f_1 = \frac{1}{\sqrt{x}} = \sqrt{\frac{2ab}{-(a + b) + \sqrt{(a + b)^2 + 4ab}}}$$

C–2 CUTOFF FREQUENCY OF TWO ISOLATED LOW-PASS RC FILTERS IN CASCADE

Let

$$f_A = \text{cutoff frequency of first filter}$$
$$f_B = \text{cutoff frequency of second filter}$$

Since the filters are assumed to be isolated (by an ideal, unity-gain amplifier), the overall voltage gain of the cascade at frequency f is

$$\frac{|v_o|}{|v_{in}|} = \frac{1}{\sqrt{1 + (f/f_A)^2}} \frac{1}{\sqrt{1 + (f/f_B)^2}}$$

At cutoff ($f = f_2$)

$$\frac{1}{\sqrt{1 + (f_2/f_A)^2}} \frac{1}{\sqrt{1 + (f_2/f_B)^2}} = \frac{1}{\sqrt{2}}$$

Let

$$x = f_2^2, a = 1/f_A^2, b = 1/f_B^2$$

Then

$$\frac{1}{\sqrt{1 + ax}\sqrt{1 + bx}} = \frac{1}{\sqrt{2}}$$

$$\frac{1}{(1 + ax)(1 + bx)} = \frac{1}{2}$$

$$(1 + ax)(1 + bx) = 2$$

$$abx^2 + (a + b)x - 1 = 0$$

$$x = \frac{-(a + b) \pm \sqrt{(a + b)^2 + 4ab}}{2ab}$$

$$f_2 = \sqrt{x} = \sqrt{\frac{-(a + b) + \sqrt{(a + b)^2 + 4ab}}{2ab}}$$

ANSWERS TO ODD-NUMBERED EXERCISES

NOTE: Solutions to SPICE programming exercises may be found in the *Solutions Manual* accompanying the text.

CHAPTER 2

2–1. N; 32 electrons

2–3.

K	L		M		
s	s	p	s	p	d
2	2	6	2	3	0

2–5. Si: 1.7622×10^{-19} J; Ge: 1.0733×10^{-19} J
2–7. 0.365 $m^2/(V \cdot s)$
2–9. (a) 48 mA/m^2; (b) 180 mA/m^2; (c) 228 mA/m^2
2–11. 1.026 mm
2–13. (carriers/m^3)[$m^2/(V \cdot s)$](C/carrier) = $C/(m \cdot V \cdot s) = A/(m \cdot V) = S/m$
2–15. 620 Ω
2–17. 2.2×10^{17} electrons
2–19. 288 A/m^2
2–21. 12 mm
2–23. $V_T = 26.996$ mV; $V_0 = 0.649$ V
2–25. 0.2118 mA
2–27. 0.6432 V
2–29. A cathode; B anode
2–31. −0.06 pA
2–33. 3.79
2–35. 0.02 μW
2–37. (a) 0.546 V; (b) 0.721 V
2–39. High-speed switching circuits in which the diode must change rapidly between its low- and high-resistance states; only majority carriers (electrons in N-type silicon) are involved in the process.
2–41. silicon and germanium; the ability to form crystals having special electrical properties
2–43. positively; free electron falling into a hole
2–45. doping; donors and acceptors
2–47. barrier voltage; from N to P
2–49. (a) −1.97 pA; (b) −2.4 pA; (c) 3.404 mA; (d) 337.9 mA
2–51. 12 pA

CHAPTER 3

3–1. slope = 10^{-4} S
3–3. 320 Ω (approx.); 16 Ω (approx.); silicon
3–5. 5.4 kΩ; 183 Ω
3–7. 260 Ω; 7.43 Ω
3–9. 18.18 MΩ
3–11. (a) 16.78 Ω, 11.12 Ω; (b) 0.464 Ω
3–13. 0.6414 V; 12.42 Ω
3–15. (a) 7.6 mA; (b) 8 mA
3–17. 11.11%
3–19. (a) 1.08 mA; (b) 24.07 Ω; (c) $i(t) = (1.08 + 0.1962 \sin \omega t)$ mA, $v_D(t) = (0.65 + 0.00472 \sin \omega t)$ V; (d) $i_{max} = 1.276$ mA; $i_{min} = 0.8838$ mA
3–21. (a) $I = -8 \times 10^{-4} V + 1.6 \times 10^{-3}$; (b) $I_D \approx 1.12$ mA, $V_D \approx 0.62$ V; (c) 554 Ω; (d) $I_{max} \approx 1.3$ mA, $I_{min} \approx 0.82$ mA; (e) 31.25 Ω
3–23. $I_{PK} = -15.3$ mA, $V_{R_{PK}} = -15.3$ V
3–25. (a) 167.4 V; (b) 1.86 V pk–pk; (c) doubles V_{pp}: 3.72 V pk–pk
3–27. (a), (c), (d) forward; (b) reverse
3–29. (a) 1.76 V; (b) 3.158 V; (c) 3.158 V; (d) 10.588 V
3–31. (a) −0.7 V; (b) −0.7 V; (c) −0.7 V; (d) −5.7 V
3–33. $3.00

3–35. 1N482A or B
3–37. 1N4004
3–39. 1860 Ω
3–41. 8.03 V

CHAPTER 4

4–1. 0.12 mA
4–3. (a) 0.995; (b) 22.1055 mA; (c) 0.9952
4–5. $I_B = I_E - I_C \approx I_E - \alpha I_E = (1 - \alpha)I_E$
4–7. 7.6 mA (approx.)
4–9. 0.99375
4–11. $\alpha \approx 1$ [so $\alpha/(1 - \alpha) \approx 1/(1 - \alpha)$]
4–13. (a) 0.95 mA (approx.); (b) 95 (approx.)
4–15. (a) 156; (b) 168; (c) 70 V
4–17. $I_E = (\beta + 1)I_B = [\alpha/(1 - \alpha) + 1]I_B = [(\alpha + 1 - \alpha)/(1 - \alpha)]I_B = I_B/(1 - \alpha)$
4–19. (a) 1.515 mA; (b) 11.7 V
4–21. (a) 19.5 mA, 4.2 V (approx.); (b) 20 mA, 4 V
4–23. (a) 16.95 V; (b) 2.55 mA; (c) 24 V
4–25. (a) 42.5 mA, 3.8 V (approx.); (b) 1 mA (approx.); (c) 42 mA, 3.8 V
4–27. 209.86 kΩ
4–29. 6.35 V, 5.65 mA
4–31. (a) R_E = 4.7 kΩ, R_C = 5.6 kΩ; (b) $I_{E(min)}$ = 1.88 mA, $I_{E(max)}$ = 2.08 mA, $V_{CB(min)}$ = 7.77 V, $V_{CB(max)}$ = 10 V
4–33. (a) R_E = 1.8 kΩ, R_B = 110 kΩ; (b) 9.89 V
4–35. $I_C = I_{C(sat)} = V_{CC}/R_C = \beta I_B$ = 4.545 mA
4–37. 1.97 kΩ
4–39. 4.8 V
4–41. (a) 20; (b) 80
4–43. 0.01 μA
4–45. (a) 50 (approx.); (b) 58.3 (approx.)

CHAPTER 5

5–1. (a) 8.3; (b) 3 mA rms; (c) 0.6 V rms; (d) 33 V rms; (e) 1.325 kΩ; (f) 0.8215 W
5–3. $r_{in} \geq$ 3 kΩ
5–5. 11.25 mW
5–7. (a) 8 V, 3 mA (approx.); (b) −137.5 (approx.)
5–9. 10 μA to 25 μA
5–11. (b) $A_v \approx -58.3$
5–13. (a) 23.06 Ω; (b) 10 kΩ; (c) 433.65; (d) 0.99
5–15. 195.27
5–17. (a) 0.59 V rms; (b) 1.91 V rms; (c) 2.46 V rms
5–19. Substituting eq. 5.47 into (5.48) and solving for I_Q gives $I_Q = V_{CC}/(r_L + R_C)$. Then substitute for I_Q in (5.48).
5–21. (a) 601.7 kΩ; (b) 58.08 mV p–p
5–23. 0.605
5–25. (a) 54.2 mS; (b) 162.6 mS; (c) 81.29 mS
5–27. See Figure 5.43. βr_e = 1.08 kΩ, r_o = 45.59 kΩ, $g_m = 92.47 \times 10^{-3}$. Add: r_S = 1 kΩ (in series with v_S); R_B = 470 kΩ (in parallel with input); R_C = 1 kΩ, R_L = 1 kΩ (in parallel with output).
5–29. h_{fe}: 18.75%; h_{re}: −77.3%; h_{ie}: −69.4%; h_{oe}: 56.25%
5–31. β: 30–150; g_m: 8.57 mS–150 mS

CHAPTER 6

6–1. -9.33×10^{-4}
6–3. $S(V_{BE}) = -1.09 \times 10^{-3}$, $S(I_{CBO}) = 19.5$, $S(\beta) = 6.86 \times 10^{-6}$
6–5. (a) 2.01 mA, 9.975 V; (b) 1.98 mA, 10.05 V; (1.5% error)
6–7. 2 mA, 1.01%
6–9. (a) −183.8; (b) 9.8%
6–11. (a) 1.12 V rms; (b) 18.25 mV rms
6–13. (One solution): R_1 = 20 kΩ, R_2 = 2.2 kΩ, R_C = 3.3 kΩ, R_E = 500 Ω
6–15. (a) 4.67 mA, 5.38 V; (b) 411; (c) 169 Ω; (d) 103.8
6–17. (One solution): R_E = 1.5 kΩ; R_C = 8.2 kΩ, R_1 = 100 kΩ, R_2 = 15 kΩ; V_Q = 9.22 V, $A_v = -5.47$
6–29. (a) 266.67 Ω/square; (b) 1600 Ω
6–31. $l = 2.5 \times 10^{-4}$ m, $a = 10$
6–37. 8.84 mA
6–39. 4.224 mA
6–41. I_B = 11.94 μA, I_C = 2.388 mA, I = 2.41188 nA
6–43. (a) V_{CE} = 13.03 V; (b) I_C = 2.2623 mA, V_{CE} = 13.1771 V

CHAPTER 7

7–1. (a) 156.3 Ω (approx.); (b) 250 Ω (approx.); (c) 500 Ω (approx.). Resistance varies with voltage.
7–3. 5 V
7–5. (a) 2.6 mA (approx.); (b) 4 mA (approx.). Nonlinear: The same ΔV_{GS} produces different values of ΔI_D.
7–7. (a) 4.508 mA; (b) 0.828 mA
7–9. (a) $I_D \approx$ 2.5 mA, $V_{DS} \approx$ 5.3 V; (b) $I_D \approx$ 4.1 mA, $V_{DS} \approx$ 1 V; (a) is in pinchoff, (b) is not.
7–11. (a) I_D = 1.098 mA, V_{DS} = 13.02 V; (b) I_D = 2.24 mA, V_{DS} = 10.97 V; (c) I_D = 6.09 mA, V_{DS} = 4.04 V (not valid)
7–13. $I_D \approx$ 2.3 mA, $V_{GS} \approx$ −1.15 V; V_{DS} = 4.59 V (valid)
7–15. (a) $I_D \approx$ 6.6 mA, $V_{GS} \approx$ 4 V; V_{DS} = −11.52 V (valid); (b) I_D = 6.52 mA, V_{GS} = 3.91 V, V_{DS} = −11.74 V
7–17. $I_D \approx$ 9 mA, $V_{DS} \approx$ −6.12 V
7–19. (a) $I_D \approx$ 6.1 mA (valid); (b) I_D = 6.09 mA, V_{DS} = 6.73 V
7–21. R_S = 146 Ω, R_D = 854 Ω

7–23. $R_S = 684\ \Omega$, $R_D = 2.07\ k\Omega$, let $R_2 = 470\ k\Omega$, $R_1 = 3.06\ M\Omega$
7–25. (a) 15 mA; (b) −6 V; (c) −30 V; (d) 2N4220, 2N4220A
7–27. (a) $I_D \approx 6$ mA, $V_{DS} \approx 9.2$ V; (b) $I_D \approx 17$ mA, $V_{DS} \approx 3.4$ V; (c) (a) is in pinchoff; (d) (b) is in enhancement mode.
7–29. 5.35 V
7–31. (a) 3.3 mA; (b) 0
7–33. (a) 7 V (approx.); (b) 7.06 V
7–35. (a) 9.68 V; (b) 271.3 kΩ
7–37. $I_D = 2.05$ mA, $V_{GS} = 4.86$ V, $V_{DS} = 14.1$ V
7–39. $I_D = 2.14$ mA, $V_{GS} = -5.03$ V, $V_{DS} = -14.21$ V
7–41. $V_{DS} = 9.6$ V, $V_{GS} = 9.6$ V, $I_D = 14.4$ mA; yes
7–43. Complementary metal-oxide semiconductor. It contains both NMOS and PMOS transistors.
7–45. close control of channel length; conservation of surface space; high current-handling capability

CHAPTER 8

8–1. (a) 0; (b) 0
8–3. (a) 4.38×10^{-3} S (approx.); (b) 4.375×10^{-3} S
8–5. −3.67 V
8–7. (a) 3.16 mA, 7.57 V; (b) 3.16×10^{-3} S;

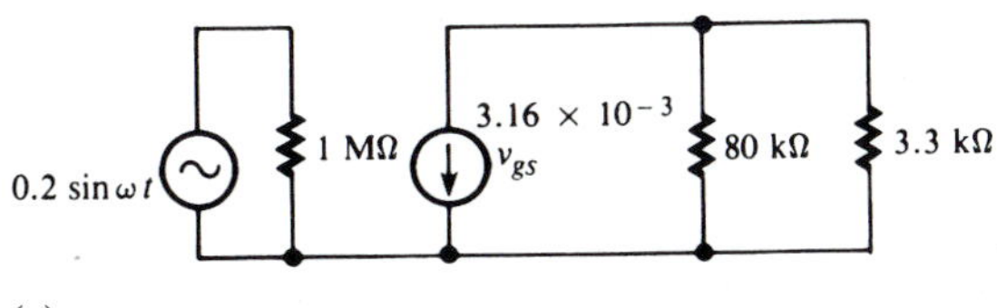

(c)

(d) 2 V
8–9. (a) 414 kΩ; (b) −18.8
8–11. −7.32
8–13. −5.98
8–15. −1.7
8–17. (a) 0.77; (b) 181 Ω
8–19. −3.74
8–21. $500\ \Omega < R_L < 1.65\ k\Omega$
8–23. (a) 0.1584 V; (b) 11.85 mV
8–25.

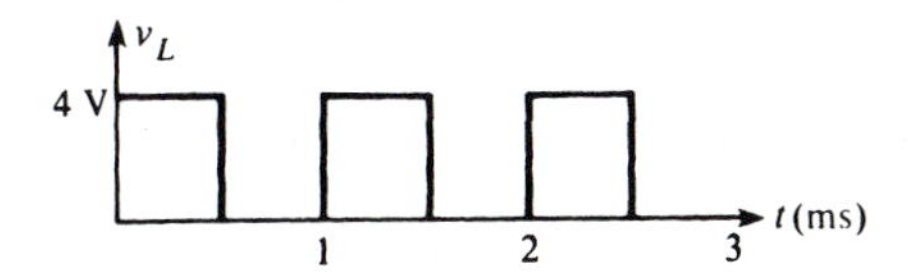

8–27. 2.6×10^{-3} S (approx.)
8–29. (a) See Figure 8.31. $r_S = 20\ k\Omega$, $R_1 \parallel R_2 = 1.24\ M\Omega$, $-g_m v_{gs} = -3 \times 10^{-3} v_{gs}$, $r_d = 100\ k\Omega$, $R_D \parallel R_L = 1998\ \Omega$ (b) −5.79
8–31. 13 A
8–33. (a) 1.8 mA; (b) 1250 Ω; (c) 3125 Ω
8–35. Solving for V in eq. 8.34 gives $V = (2I/\beta)^{1/2} + V_T$. Dividing through by I gives $V/I = R = (2/\beta I)^{1/2} + V_T/I$.
8–37. (a) 0.3 V; (b) 0.351 mA
8–39. (a) 7.8 V; (b) 0.498 V
8–41. 9 ns
8–43. $t_r \approx 359.6$ ns, $t_f \approx 3.92$ ns
8–45. (a) 0 V; (b) 10 V; (c) 10 V; (d) 0 V

CHAPTER 9

9–1. (a) $h_{11} = 40\ \Omega$, $h_{21} = -2/3$, $h_{12} = 2/3$, $h_{22} = 0.0111$ S; (b) $h_{11} = 1.5\ k\Omega$, $h_{21} = -0.5$, $h_{12} = 0.5$, $h_{22} = 0.5 \times 10^{-3}$ S; (c) $h_{11} = 60\ \Omega$, $h_{21} = 1$, $h_{12} = -1$, $h_{22} = 0$; (d) $h_{11} = R_1 + R_3$, $h_{21} = 0$, $h_{12} = 0$, $h_{22} = 1/(R_2 + R_4)$
9–3. $h_{11} = 2\ k\Omega$, $h_{21} = 50$, $h_{12} = 10^{-4}$, $h_{22} = 2 \times 10^{-5}$ S
9–5. See Figure 9.9 and use h-parameter values calculated in Exercise 9.2.
9–7. (a) $i_1 = 0 \Rightarrow 0$ drop across $h_{11} \Rightarrow v_1 = h_{12}v_2 \Rightarrow (v_1/v_2)\,|_{i_1=0} = h_{12}$ (b) $i_1 = 0 \Rightarrow h_{11}i_1 = 0 \Rightarrow$ current source is open $\Rightarrow (i_2/v_2)\,|_{i_1=0} = h_{22}$
9–9. $Z_i = 1.74\ k\Omega$, $A_v = -69.4$, $A_i = 47.62$, $Z_o = 1.4 \times 10^5\ \Omega$
9–11. $h_{oe} \approx 8 \times 10^{-6}$ S, $h_{fe} \approx 130$, $h_{ie} = 2000\ \Omega$, $h_{re} = 10^{-4}$
9–13. (a) See Figure 9.18. $v_S = 20$ mV rms, $r_S = 1\ k\Omega$, $R = 8.25\ k\Omega$, $h_{ie} = 1.5\ k\Omega$, $h_{re}v_{ce} = 2.5 \times 10^{-4} v_{ce}$, $h_{fe}i_b = 85 i_b$, $1/h_{oe} = 40\ k\Omega$, $R_C = 3.3\ k\Omega$, $R_L = 10\ k\Omega$; (b) 1.509 V rms
9–15. (a) 2 kΩ; (b) 2 kΩ; (c) −63; (d) −63; (e) 90; (f) 1.5 kΩ
9–17. $h_{ie} \approx 1333\ \Omega$, $h_{fe} \approx 89.98$
9–19. Substituting $R_E = 0$ and $r_S \parallel R = r_S$ into eq. 9.46 and simplifying gives eq. 9.37.
9–21. (a) 80.125 kΩ; (b) 24.8 kΩ; (c) 0.942; (d) 30.51 Ω; (e) 29.6 Ω
9–23. (a) 14.05 Ω; (b) 14 Ω; (c) 101.89; (d) 294 kΩ; (e) 3.3 kΩ
9–25. (a) 3.94 kΩ; (b) 3.8 kΩ; (c) 0.425; (d) 57.72 Ω; (e) 53.6 Ω
9–27. (a) 15 Ω; (b) 15 Ω; (c) 99.1; (d) 989.6 kΩ; (e) 9.1 kΩ
9–29. $y_{11} = 0 + j0.503$ mS; $y_{21} = 0 - j0.1256$ mS; $Y_{12} = 0 - j0.1256$ mS; $y_{22} = 0 + j0.188$ mS
9–31. $y_i = 7.14 \times 10^{-4}$ S, $y_r = -1.43 \times 10^{-7}$ S, $y_f = 71.4 \times 10^{-3}$ S, $y_o = 5.71 \times 10^{-6}$ S
9–33. (a) 2.8×10^{-3} S; (b) 100 kΩ; (c) 1.39 pF; (d) 0.6 pF; (e) 3.78 pF

9–35. (a) $C_{gd} = 1.5$ pF, $C_{ds} = 0.5$ pF, $C_{gs} = 3.5$ pF; (b) $y_{is} = 0 + j3.14 \times 10^{-3}$ S, $y_{fs} = 3 \times 10^{-3} - j0.942 \times 10^{-3}$ S, $y_{rs} = 0 - j0.942 \times 10^{-3}$ S, $y_{os} = 10^{-5} + j1.257 \times 10^{-3}$ S
9–37. 2.46×10^{-3} S

CHAPTER 10

10–1. (a) 0.88 MHz; (b) 0.6 V rms; (c) 0.8 W; (d) 0.4 W
10–3. 0.125 V rms, 0.25 ms
10–5. 1 V rms
10–7. (a) 32 dB; (b) 30.46 dB
10–9. (a) 0.856 V rms; (b) 18 dB
10–11. (a) 3.6 V rms; (b) 72 V rms; (c) 0.18 V rms; (d) 360 V rms
10–13. 4×4
10–15. (a) $f_1 \approx 500$ Hz, $f_2 \approx 15$ kHz; (b) 0.5 V rms; (c) 63°; (d) 5
10–17. (a) 159.15 Hz; (b) 70.67; (c) 84.29°; (d) 52.63 Hz
10–19. (a) 72.34 Hz; (b) 19.38 Hz; (c) 97.01
10–21. (a) 491.2 Hz; (b) 170.1; (c) 1.94 μF
10–23. (a) 120.3 Hz; (b) 37 dB; (c) 34 dB; (d) 126°; (e) 13.2 Hz
10–25. (a) 3.04 MHz; (b) 157.5; (c) 43.62; (d) −73.1°
10–27. 3383 Ω
10–29. (a) 271 kHz; (b) 105.47;

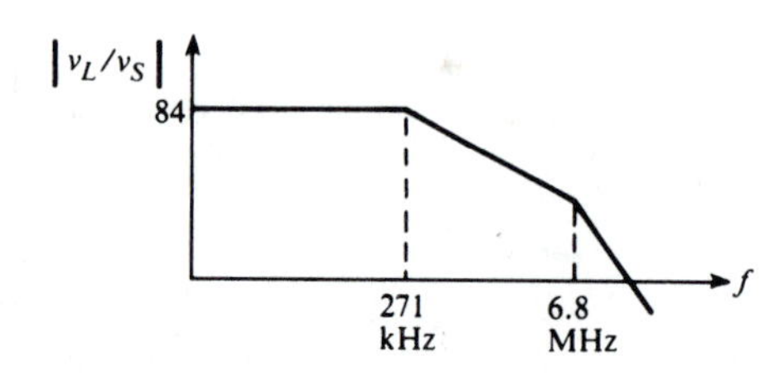

(c)

10–31. 59.19 kHz
10–33. (a) 6029 Hz; (b) 58 μs; (c) decreases with increases in either C or R; (d) increases with increases in either C or R
10–35. 135.45 Hz
10–37. 307 Hz
10–39. $C_E = 40\ \mu$F, $C_1 = 4\ \mu$F, $C_2 = 0.4\ \mu$F
10–41. (a) $C_1 = 1.5\ \mu$F, $C_2 = 10\ \mu$F; (b) 30.2 Hz; (c) 30.6 Hz
10–43. 0.8 MHz
10–45. (a) 191 Hz

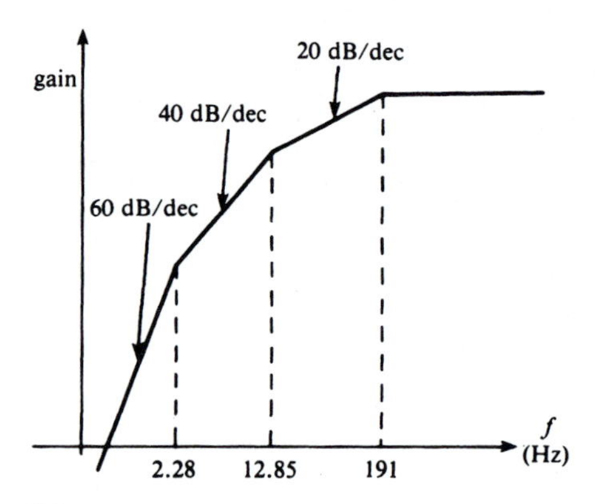

(b)

10–47. 7708 Ω
10–49. 128.3 pF

CHAPTER 11

11–1. (a) −8568; (b) 11.6
11–3. $A_1 = 10$, $A_2 = 10$, $A_3 = 5$
11–5. (a) $f_1 = 34.5$ Hz, $f_2 = 12.9$ kHz; (b) 285.25 at 7.5 Hz, 0.45 at 300 kHz
11–7. 207.1 kHz
11–9. (a) −133; (b) 14.8
11–11. (a) 367 Hz

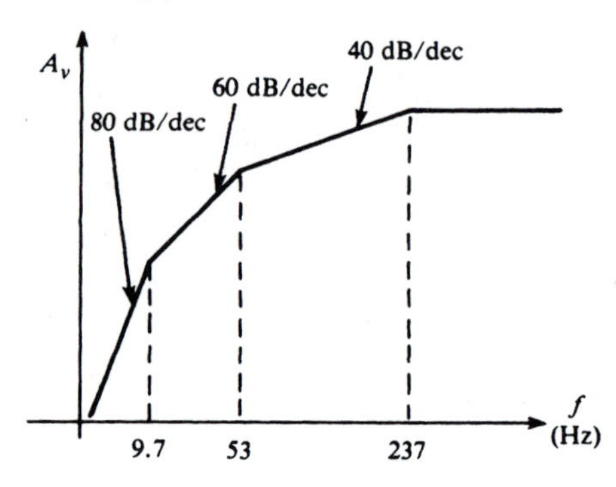

(b)

11–13. 40.4 Hz
11–15. (a) $I_{C1} = 3.91$ mA, $V_{C1} = V_{CE1} = 5.8$ V, $I_{C2} = 1.09$ mA, $V_{C2} = 8.2$ V, $V_{CE2} = 3.1$ V; (b) 3950
11–17. $I_{C1} = 0.725$ mA, $V_{C1} = 7.25$ V, $I_{C2} = 0.655$ mA, $V_{C2} = 14.9$ V
11–19. (a) $I_{C1} = 80\ \mu A \approx I_{E1} = I_{B2}$, $I_{E2} \approx I_{C2} = 8$ mA $\approx I_T$, $V_{C2} = 3$ V; (b) $r_{e1} = 325\ \Omega$, $r_{e2} = 3.25\ \Omega$; (c) $I_{C1} = 80\ \mu$A, $I_{E1} = I_{B2} = 81\ \mu$A, $I_{C2} = 8.1$ mA, $I_{E2} = 8.181$ mA, $I_T = 8.18$ mA, $V_{C2} = 2.865$ V, 4.7%; (d) $r_{e1} = 321\ \Omega$, $r_{e2} = 3.18\ \Omega$

11–21. (a) 144 kΩ; (b) −32.1
11–23. 15.55 MHz
11–25. −157
11–27. 215 Hz
11–29. I_{D1} = 1.28 mA, V_{GS1} = −1.28 V, V_{DS1} = 12.32 V, I_{D2} = 3.65 mA, V_{GS2} = 0.9 V, V_{DS2} = −5.4 V
11–31. 5.1
11–33. −32.3
11–35. 3698 Ω
11–37. (a) 2.236; (b) $P_1 = P_L$ = 7.2 mW
11–39. 5732; yes
11–41. (a) 590; (b) T_1 and T_2
11–43. 96.8%
11–45.

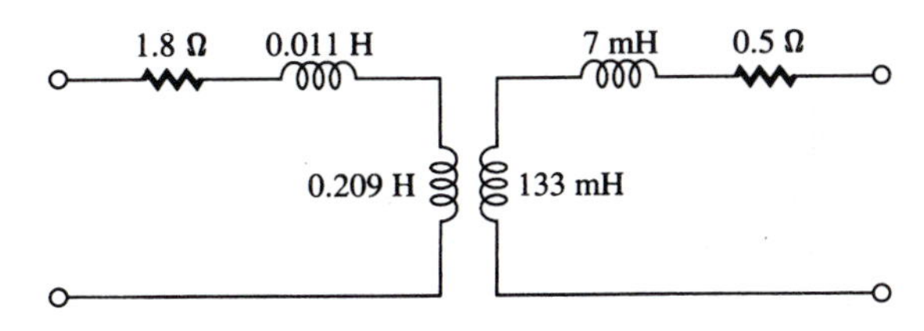

11–47. 0.48 V

CHAPTER 12

12–1. (a) (4 sin ωt) − 2; (b) 4 sin ωt; (c) 0.04 sin ωt; (d) 0.02 sin ωt
12–3. (a) −80, +80; (b) −160
12–5. (a) −132.7, +132.7; (b) −265.4
12–7. (a) $V_{o1} = V_{o2}$ = 5 V; (b) −269.5; (c) 7.49 kΩ
12–9. (a) $V_{o1} = V_{o2}$ = 6 V; (b) −15.48; (c) 1.55 V rms
12–11. 0.104 mV rms
12–13. (a) $V_{o1} = V_{o2}$ = 10 V; (b) 21.5 kΩ; (c) −160
12–15. (a) $V_{o1} = V_{o2}$ = 6.6 V; (b) 14.66 kΩ; (c) ∞; (d) 74.6 dB
12–17. v_{o1} = 3.37 V rms, $v_{o1} - v_{o2}$ = 6.74 V rms
12–19. $(1.1\text{ mA})e^{-(10\mu A)(12.22\text{ k}\Omega)/0.026} = 10\ \mu A$
12–21. 9.58 kΩ
12–23. 1.93 kΩ
12–25. 5880

CHAPTER 13

13–1. (a) −0.312 V dc; (b) −1.3 sin ωt V; (c) 6.5 V dc; (d) −10.4 + 2.6 sin ωt V; (e) −2.08 sin (ωt + 75°) V
13–3. (a) 8.06 μA dc; (b) −0.067 sin ωt μA; (c) (21.3 − 133.3 sin ωt) μA; (d) 26.6 sin(ωt − 30°) μA
13–5. See Figure 13.2. (One solution): R_1 = 10 kΩ, R_f = 200 kΩ
13–7. (a) 4.5 V rms; (b) 0.6068 V rms; (c) 5.5 V rms; (d) 0.55 V rms
13–9. (a) −6.8 V; (b) −10.2 V
13–11. (a) ideal: 201; actual: 197.04; (b) ideal: 51; actual: 50.741; (c) ideal: 11; actual: 10.988
13–13. (a) r_{in} = 1.035 MΩ, r_o = 3.48 Ω; (b) r_{in} = 3.96 MΩ, r_o = 0.908 Ω; (c) r_{in} = 18.22 MΩ, r_o = 0.197 Ω
13–15. From eq. 13.37, $v_o(1/AR_1 + 1/AR_f + 1/R_f) = -v_{in}/R_1$. Solving for v_o/v_{in} and simplifying gives eq. 13.38.
13–17.

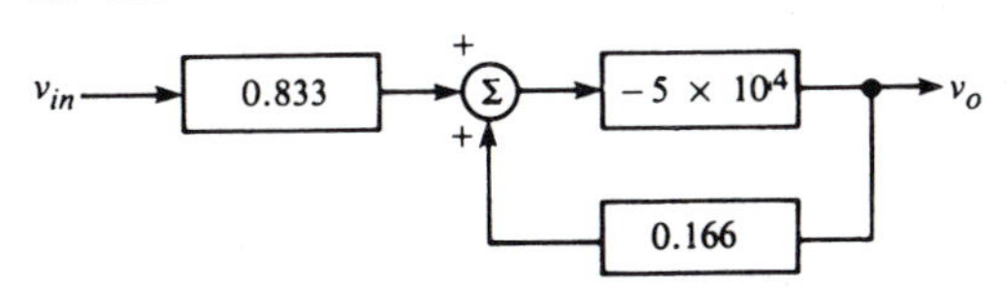

13–19. (a) 99; (b) 5049
13–21. 25 Hz
13–23. (a) 60 kHz; (b) 2
13–25. 199
13–27. 10 kHz
13–29. 3 V/μs
13–31. 0.667
13–33. 355,376 V/s
13–35. new slew rate of 1.31 × 10^6 V/s or new gain of 6.4
13–37. 56.2 kΩ
13–39. Distortion occurs.
13–41. 0.4 mV
13–43. See Figure 13.23. R_f = 62.5 kΩ, R_1 = 10 kΩ, R_c = 8.62 kΩ, v_o/v_{in} = 7.25
13–45. 3.469 mV
13–47. 0.5 mV
13–49. 30.35 mV
13–51. (a) 3 mV; (b) 58 dB (approx.); (c) 3.5 nA; (d) 34 kHz (approx.)
13–53. R_f = 18.27 kΩ, R_c = 948 Ω

CHAPTER 14

14–1. (a) $-(v_1 + 2v_2 + 5v_3 + 10v_4)$; (b) −14 V dc; (c) 5.26 kΩ
14–3. (a) 53 kHz; (b) 57 mV
14–5. (a) $8.8v_1 + 2.2v_2$; (b) $6.6v_1$ (or) $-6.6v_2$
14–7.

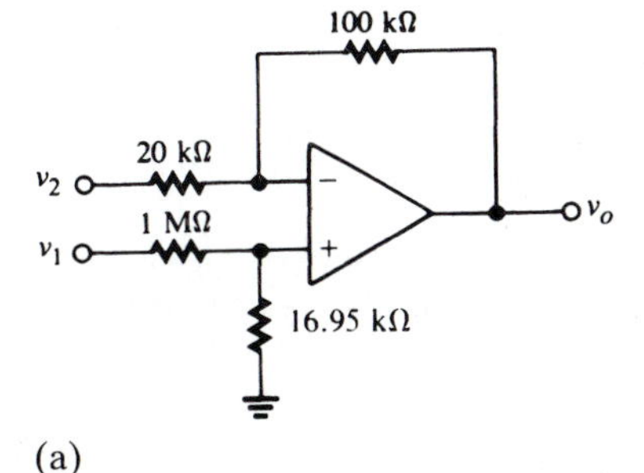

(a)

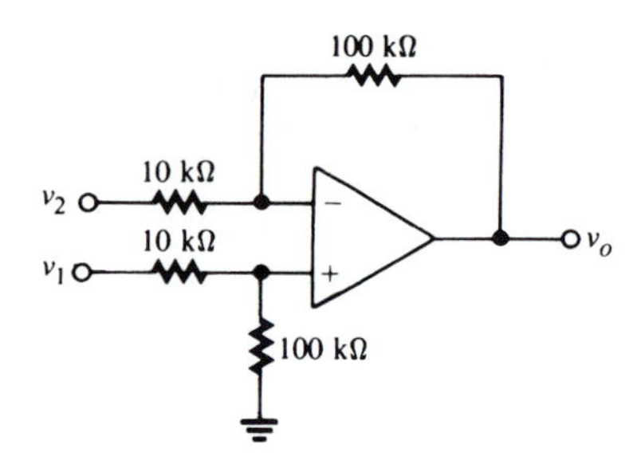

(b)

14–9. (a) (One solution):

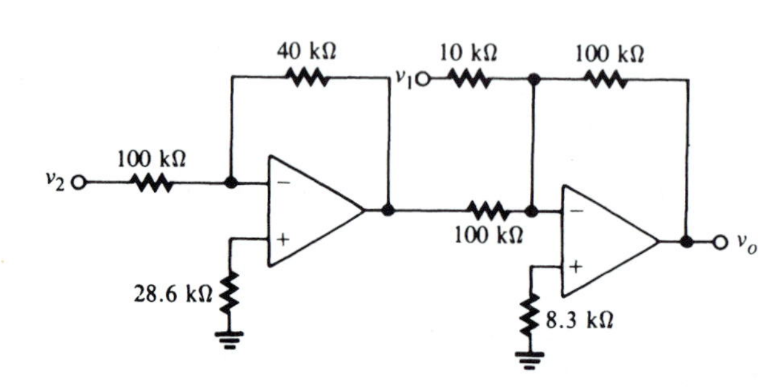

(b) (One solution):

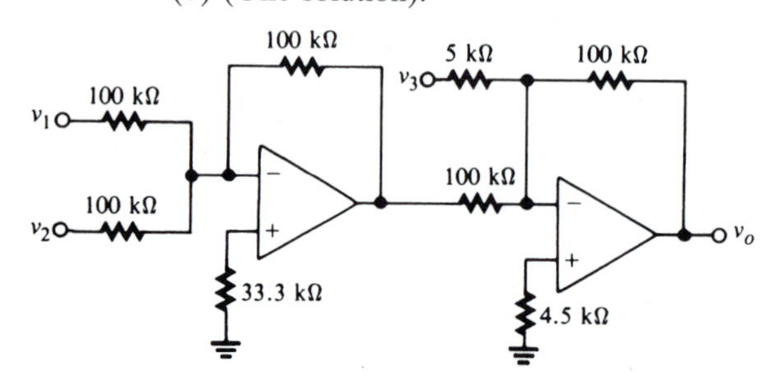

14–11.

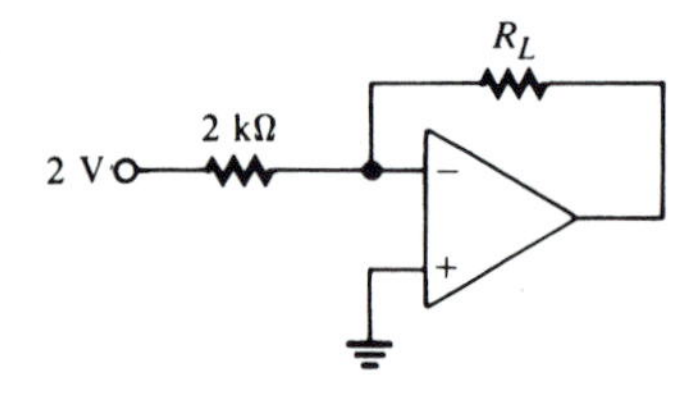

(a)

(b) 20 V

14–13.

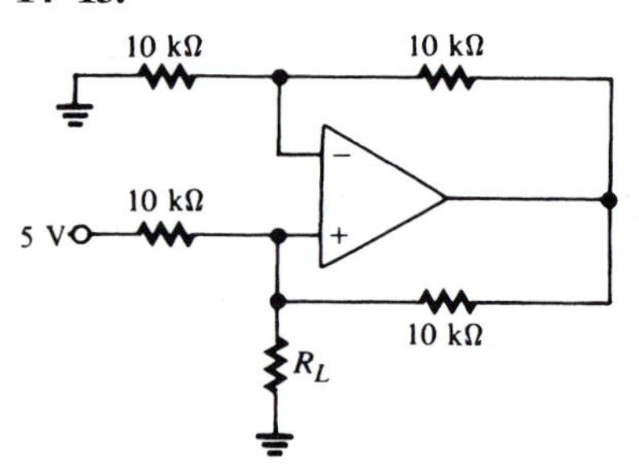

(a)

(b) 12 V

14–15.

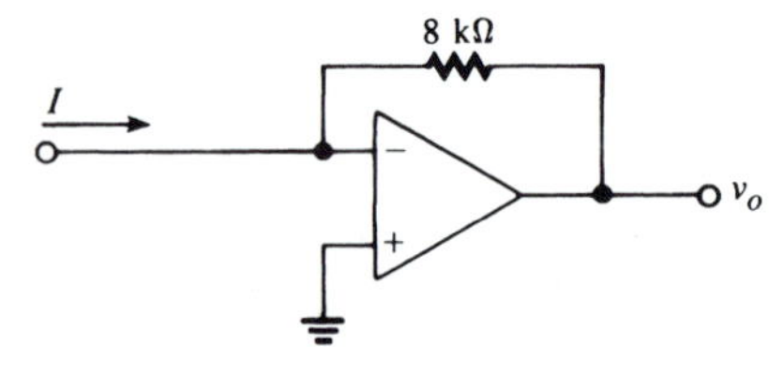

14–17. $I_1 = 1.5$ mA, $I_2 = 0.5$ mA, $I_3 = 2$ mA, $V_A = -3.75$ V, $V_B = -13.75$ V

14–19. (a) −10 V; (b) 3 s

14–21. $1.2\cos(500t - 30°) = 1.2\sin(500t + 60°)$ V

14–23. Use $R_1C = 0.1$.

14–25. See Figure 14.21. One solution is $R_1 = R_f = 318$ kΩ, $C = 0.01$ μF, $R_C = 159$ kΩ.

14–27. (a) $v_o = -\int(200v_1 + 40v_2)\,dt$

(b) (One solution):

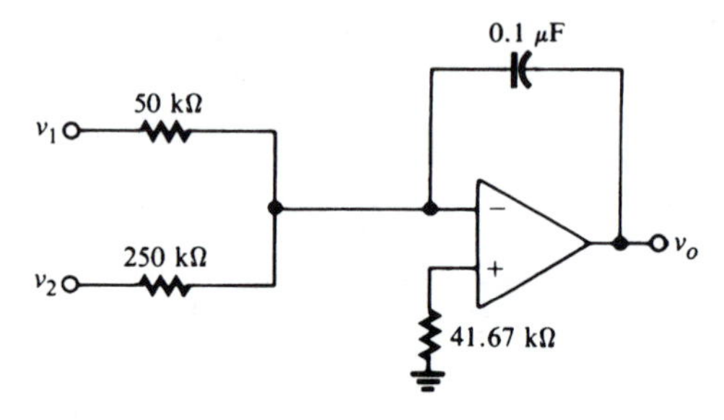

14–29. (a) $f_h = 482$ Hz $= 0.1f_b$; (b) 0.305; (c) See Figure 14.28; $f_b = 4.82$ kHz, $f_2 = 150.7$ kHz, $R_f/R_1 = 12.27$.

14–31.

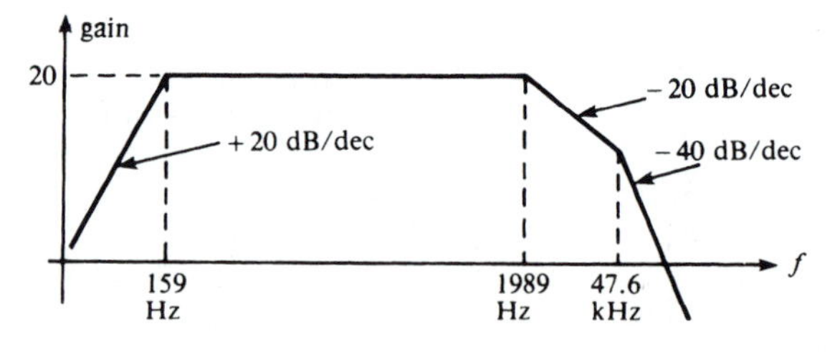

14–33.

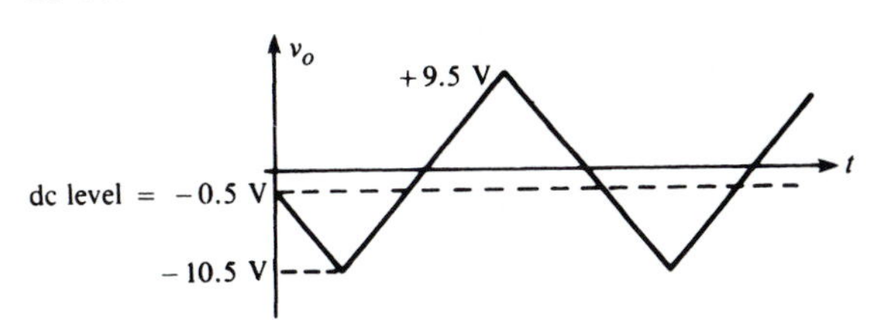

14–35.

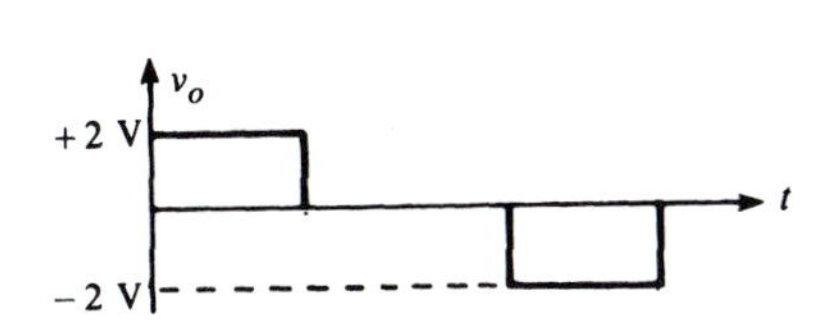

(a)

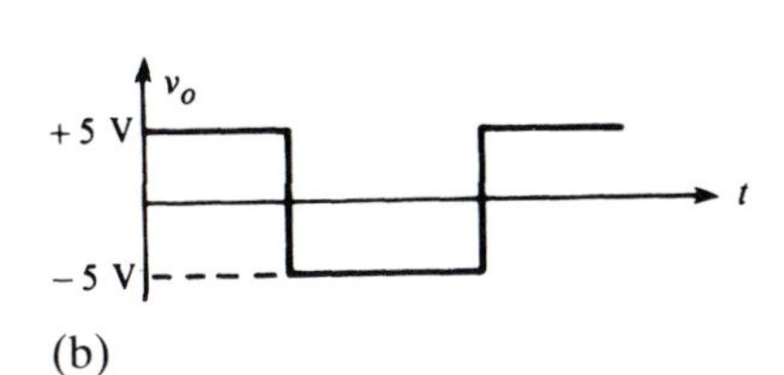

(b)

(c) Same as (b)

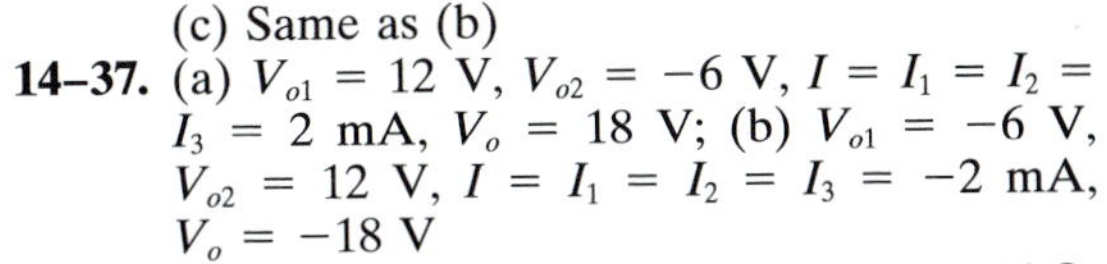

14–37. (a) $V_{o1} = 12$ V, $V_{o2} = -6$ V, $I = I_1 = I_2 = I_3 = 2$ mA, $V_o = 18$ V; (b) $V_{o1} = -6$ V, $V_{o2} = 12$ V, $I = I_1 = I_2 = I_3 = -2$ mA, $V_o = -18$ V

14–39. See Figure 14.41. (One solution): $R_1 = 5$ kΩ, $R_f = 145$ kΩ, $C = 0.1$ μF, $R = 433$ Ω

14–41. See Figure 14.44. (One solution): $R_g = 10$ kΩ, $R_f = 20$ kΩ, $R_1 = R_2 = 8842$ Ω, $C_1 = C_2 = 100$ pF

14–43. (a) 121.4 kHz; (b) 22 kΩ

14–45. 3

14–47. 3

14–49. See Figure 14.52. (One design): $C = 0.01$ μF, $R_1 = 2.65$ kΩ, $R_2 = 1.23$ kΩ, $R_3 = 1.64$ kΩ, $R_4 = 4.92$ kΩ

14–51. See Figure 14.54. (One design): $C = 0.01$ μF, $R_1 = 1.99$ kΩ, $R_2 = 83$ Ω, $R_3 = 7.96$ kΩ

14–53. See Figure 14.55. (One design): $C = 0.1$ μF, $R_1 = 2.65$ kΩ, $R_2 = 2.81$ kΩ, $R_3 = 3.98$ kΩ, $R_4 = 3.98$ kΩ

14–55. See Figure 14.56. (One design): $C = 0.001$ μF, $R_1 = 3.98$ kΩ, $R_Q = 16.55$ kΩ, BW = 45.198 kHz

14–57.

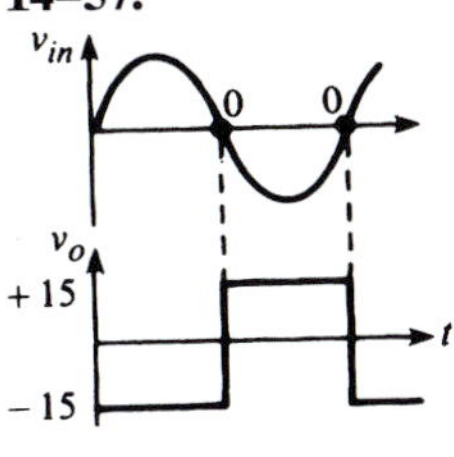

(a)

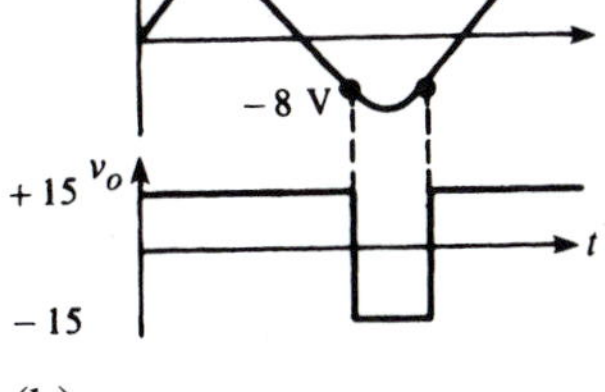

(b)

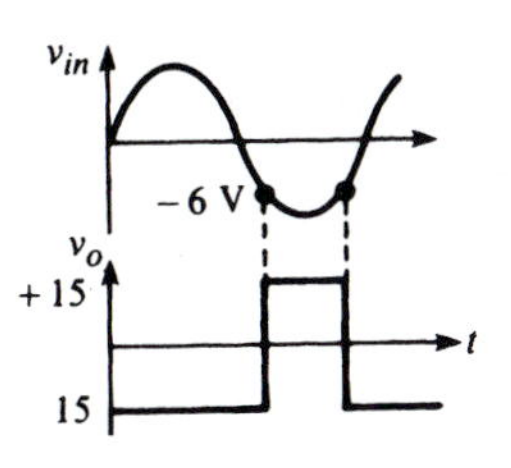

(c)

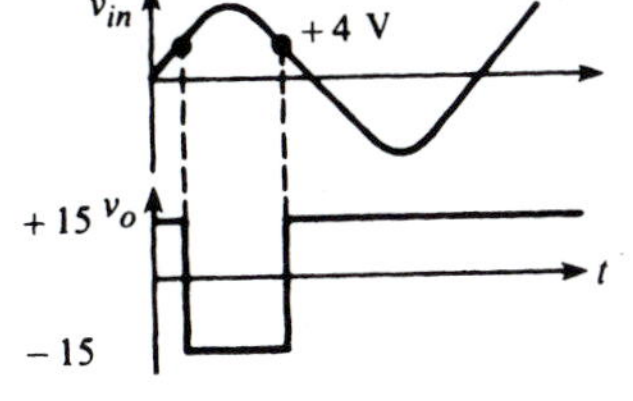

(d)

14–59. (a) LTL = −8.6 V, UTL = +6.74 V; (b) 15.34 V;

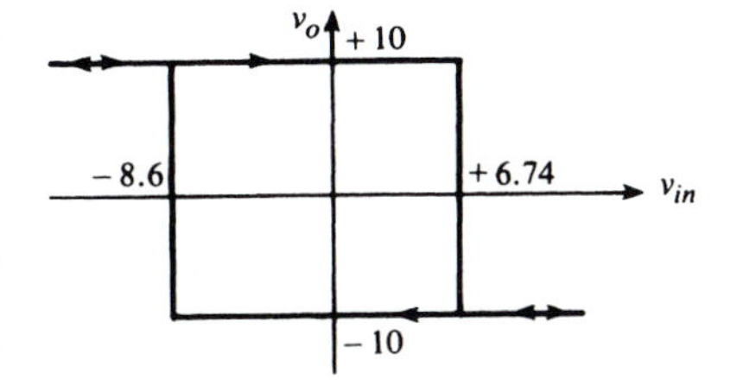

(c)

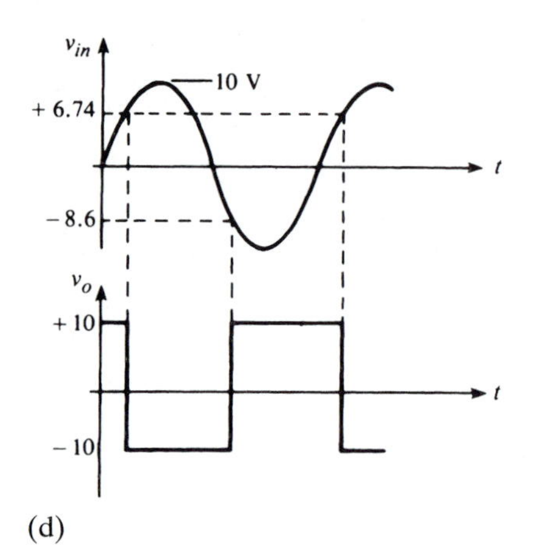

(d)

14–61. (One design):

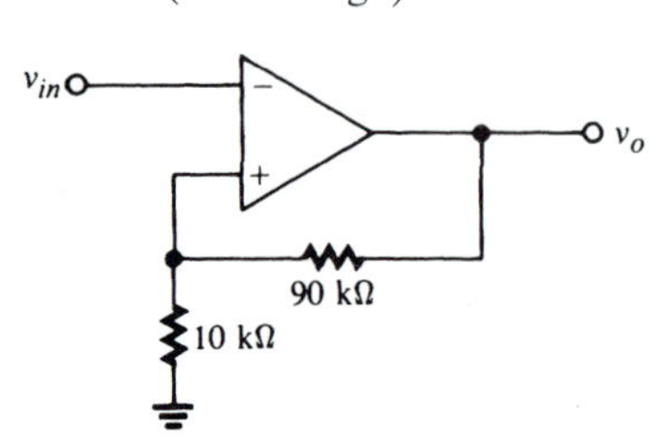

14–63. (One design):

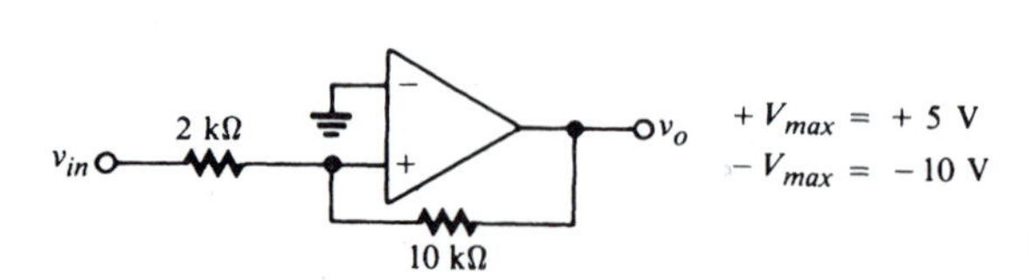

v_{in}

2 V

−1 V

t

v_o

+5

−10

t

14–65. See Figure 14.66. (One design): $C = 0.1\ \mu F$, $R = 4.55\ k\Omega$, $R_1 = R_2 = 10\ k\Omega$

14–67.

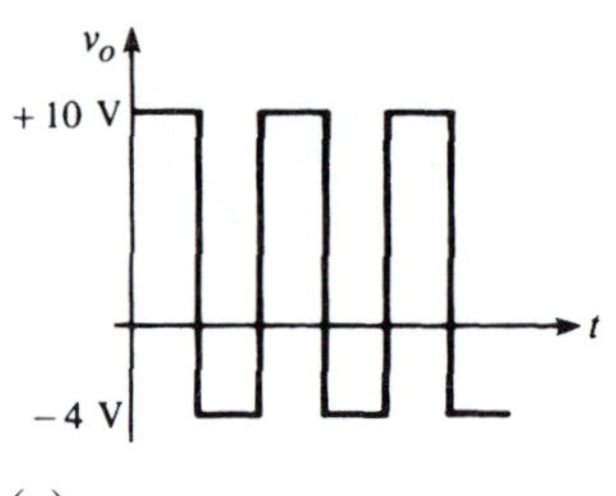

(a)

v_o

+ 10 V

+ 6 V

t

(b)

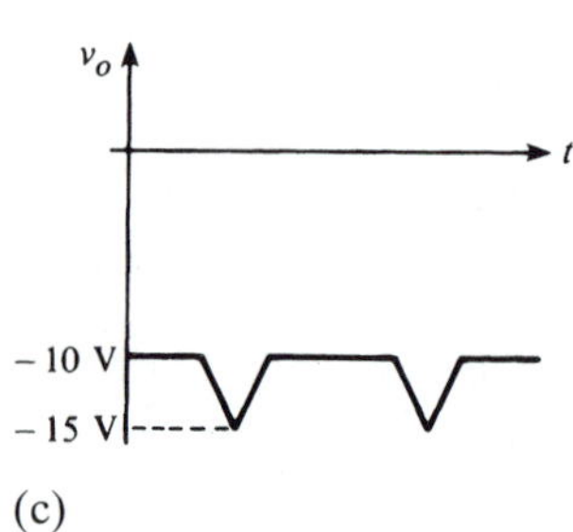

(c)

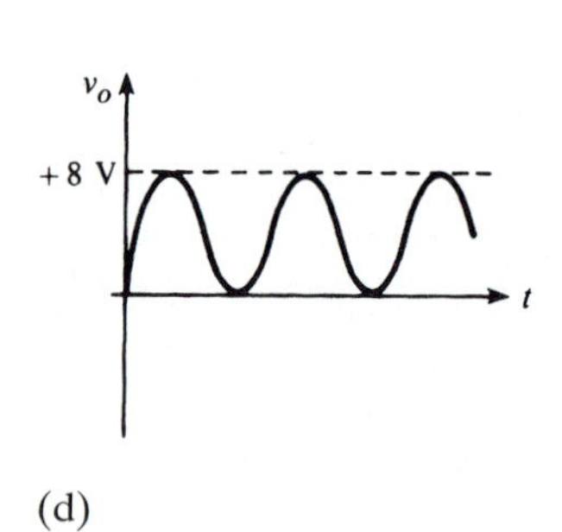

(d)

14–69.

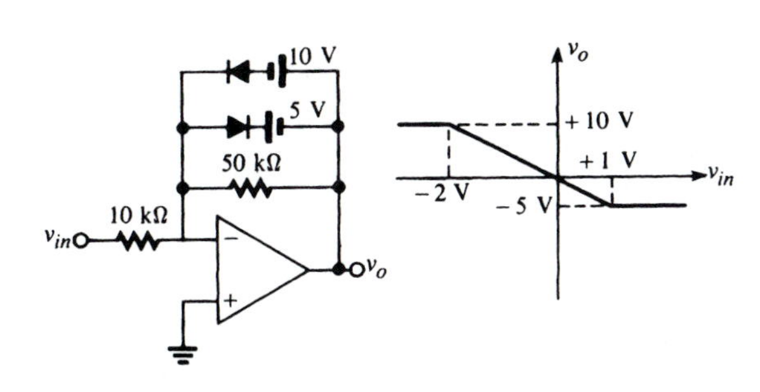

14–71.

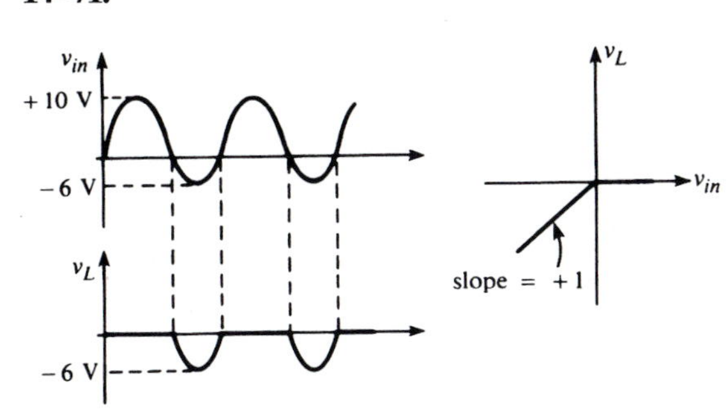

14–73.

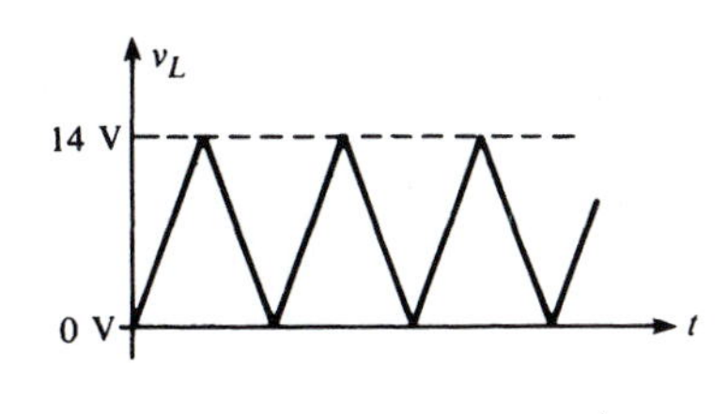

(a)

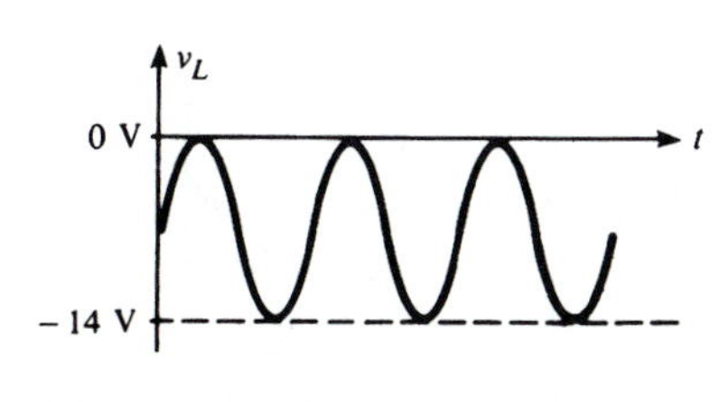

(b)

14–75.

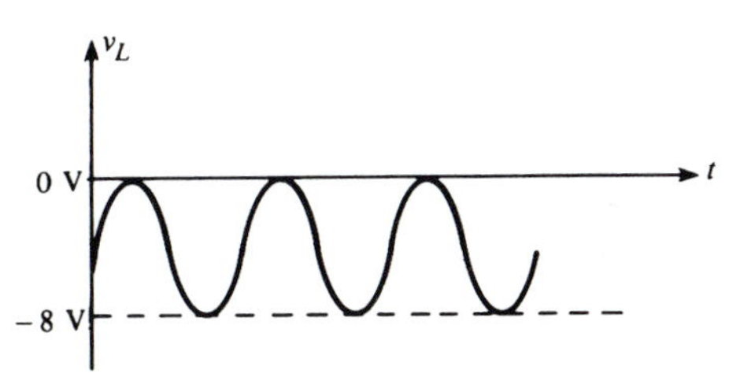

CHAPTER 15

15–1.

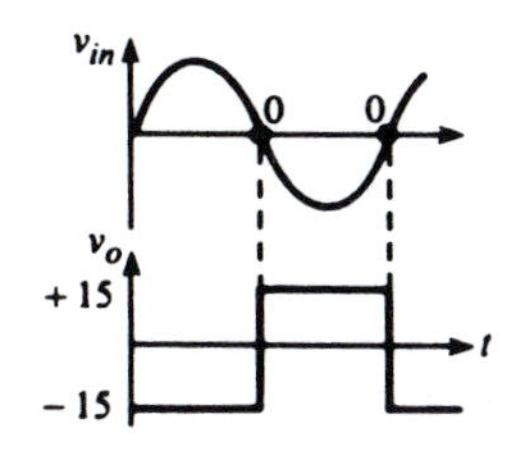

(a)

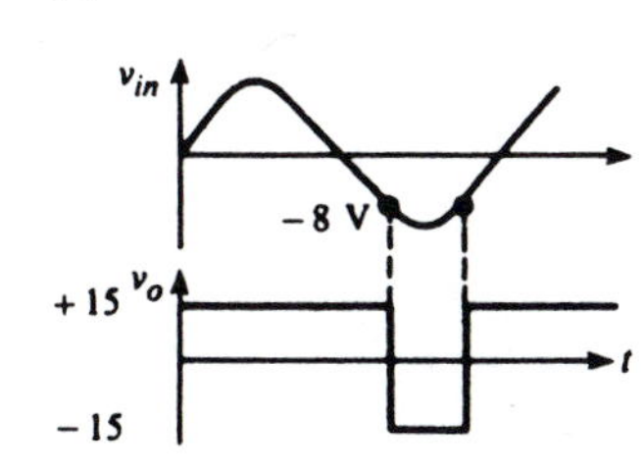

(b)

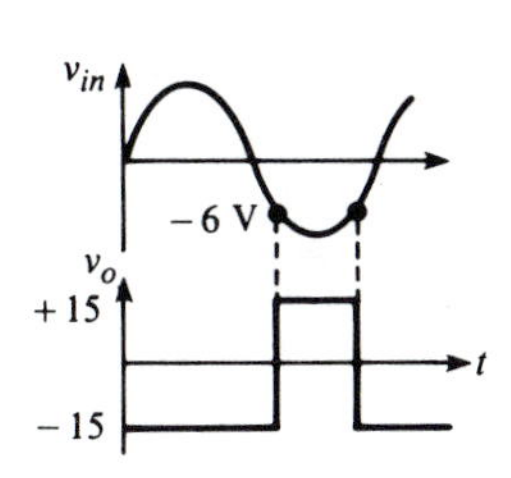

(c)

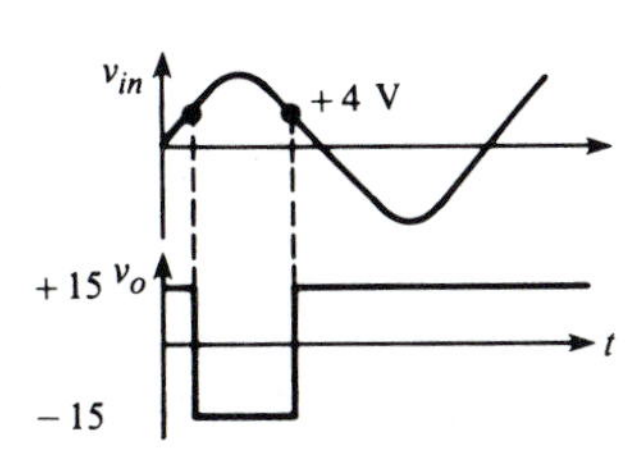

(d)

15–3. (a) LTL = −8.6 V, UTL = +6.74 V; (b) 15.34 V;

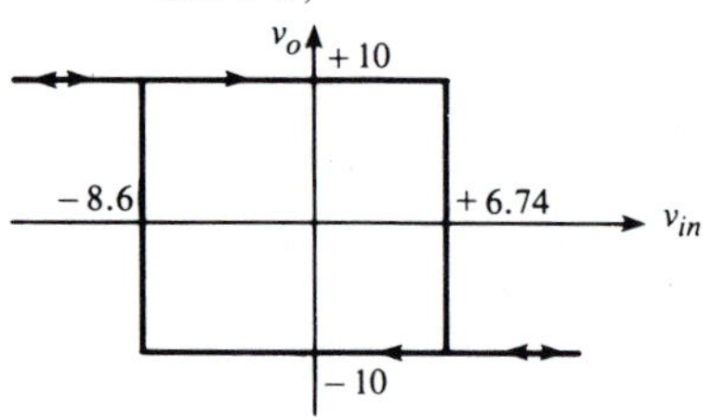

(c)

15–5. (One design):

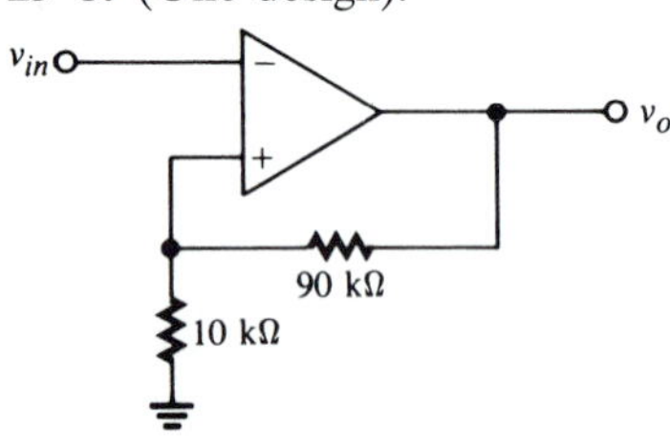

15–7. (One design);

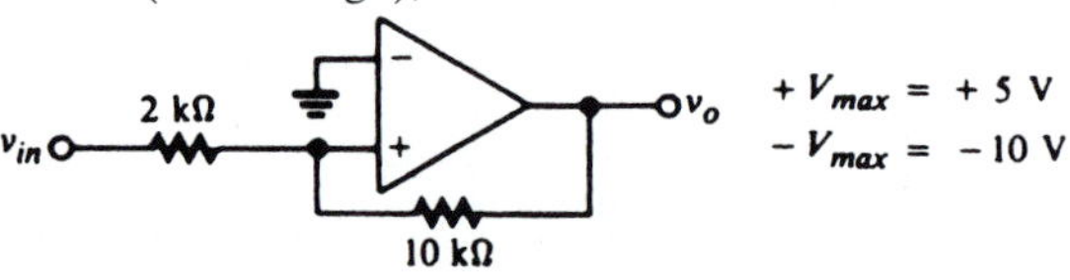

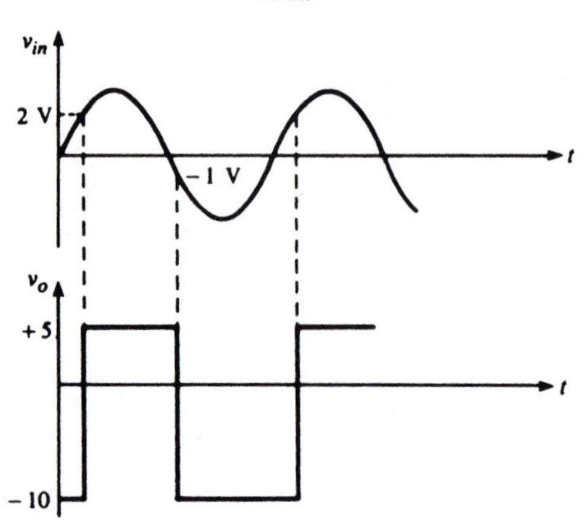

15–9. See Figure 15–5. (One design): $C = 0.1\ \mu F$, $R = 4.55\ k\Omega$, $R_1 = R_2 = 10\ k\Omega$

15–11.

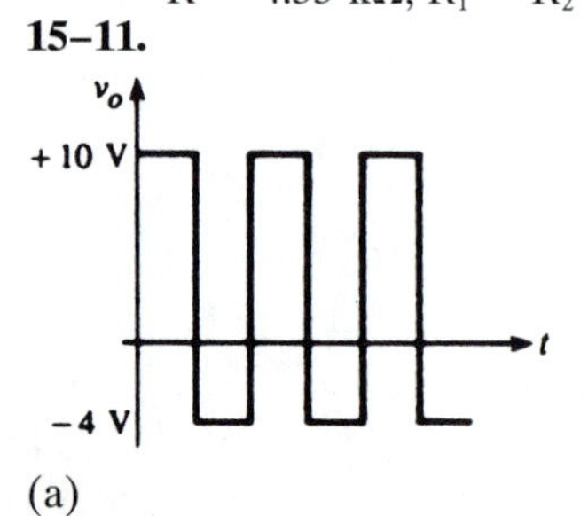

(a)

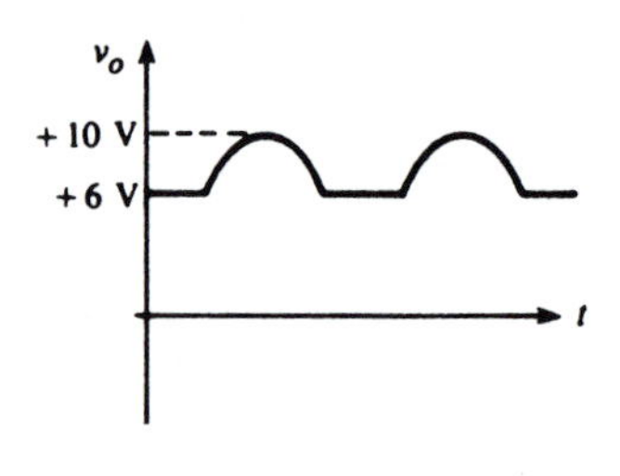

(b)

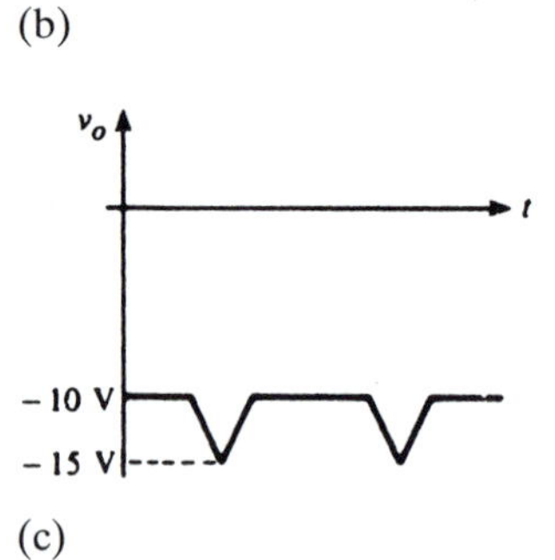

(c)

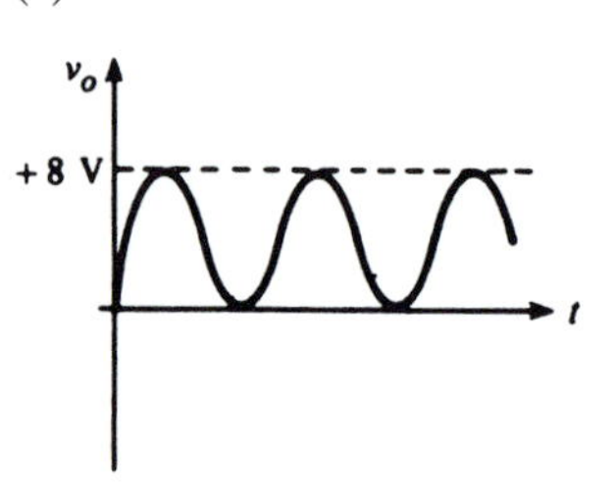

(d)

15–13.

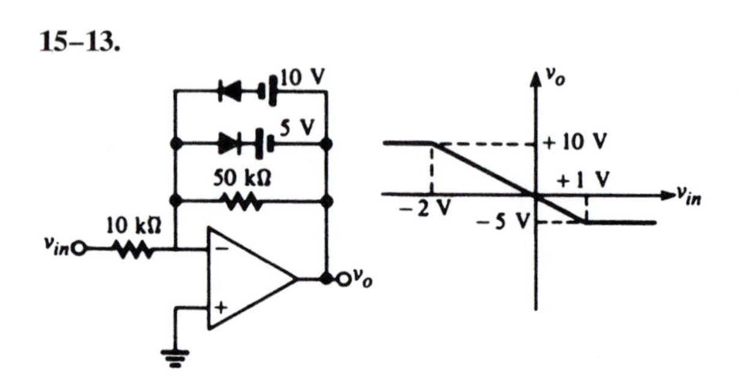

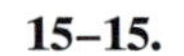

15–15.

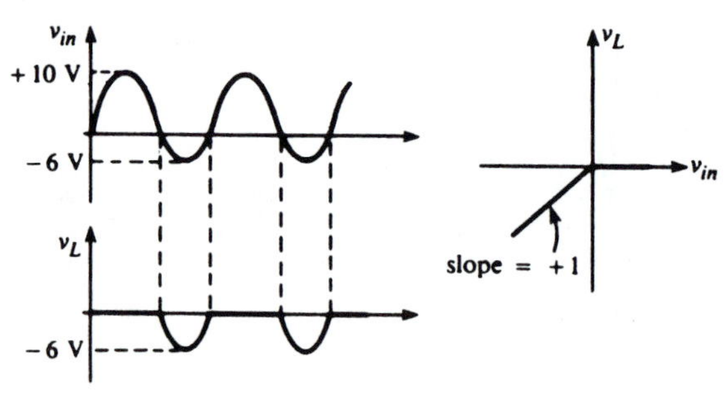

15–17.

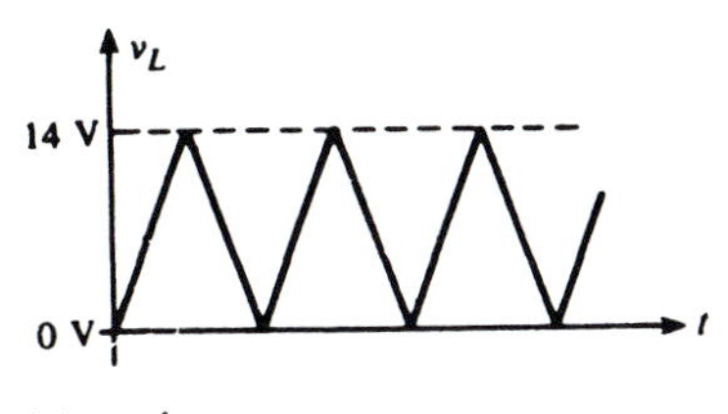

(a)

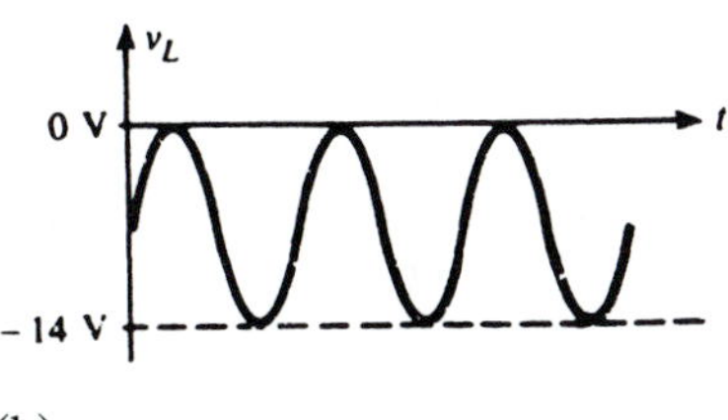

(b)

15–19.

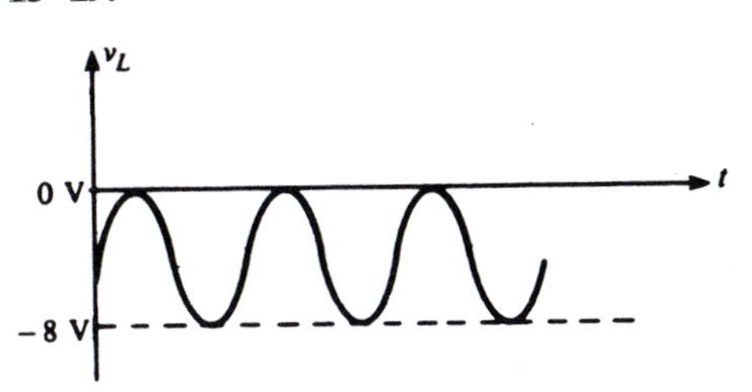

15–21. 17.8 mV

15–23. $v_L = i_o R_L = g_m(v_i^+ - v_i^-)R_L \Rightarrow \dfrac{v_L}{v_i^+ - v_i^-} = g_m R_L$

15–25. 25 μA
15–27. 50 kHz
15–29. 150 Ω – 7.5 kΩ

CHAPTER 16

16–1. decrease; 225 mW
16–3. 400 Ω
16–5. 8.4 J/s
16–7. 1°C/W
16–9. 0.127°C/W
16–11. 40°C
16–13. (a) 24 mA; (b) 0.144 W
16–15. (a) 7.5 V; (b) 0.08333; (c) $R_C \neq \sqrt{2}\ R_L$
16–17. (a) 0.01 A; (b) 4 W; (c) $\eta = \eta_C = 0.5$
16–19. (a) 24 W; (b) 61.12 W; (c) 18.56 W; (d) 0.3927
16–21. 50.6 V
16–23. 450

16–25.

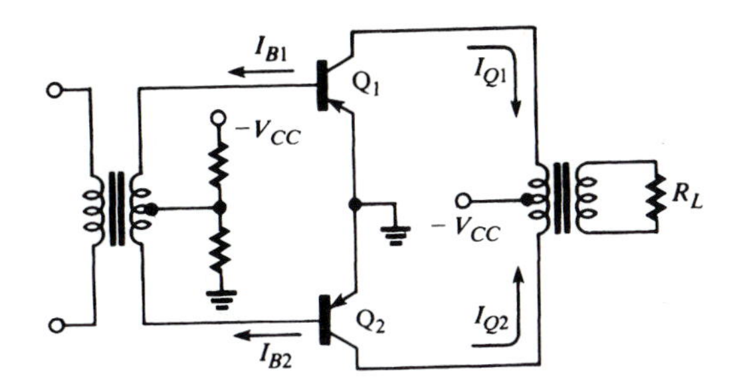

(a)

16–27. 16.62%, 46.66 V
16–29. (a) 20.72 Ω; (b) 0.767
16–31. 1.95 V
16–33. (a) 23.9 V; (b) 0.877
16–35. 791 kHz–807.5 kHz
16–37. 1.75 μs
16–39.

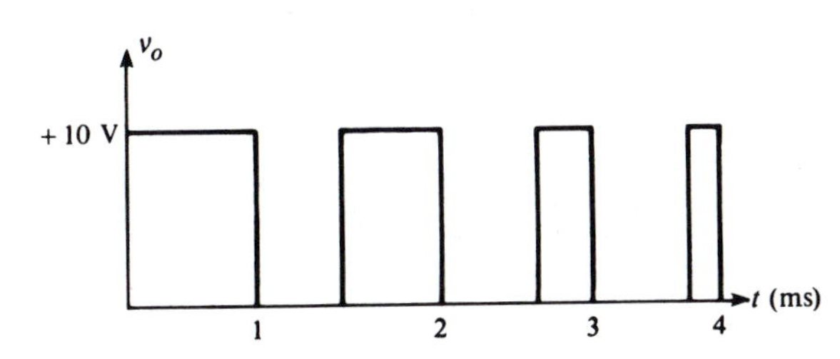

16–41. (a) 5 kHz; (b) 32 V

CHAPTER 17

17–1. 39.1 V p–p
17–3. (a) 120/16.66 = 7.2 : 1; (b) 27.76 W; (c) 47.12 V
17–5.

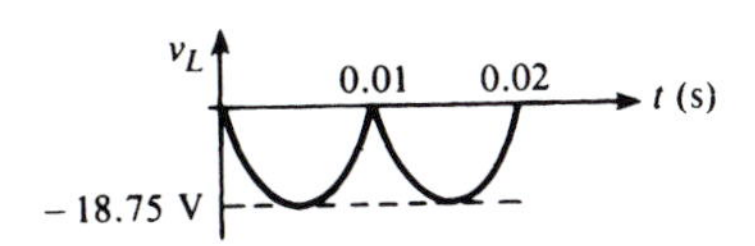

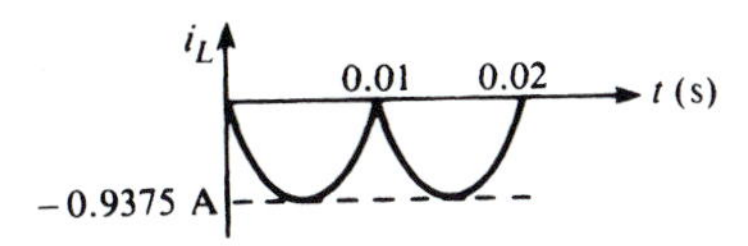

17–7. (a) 134.3 V; (b) 890 W; (c) 253.15 V
17–9. 48.4% in every case
17–11. 96.23 Ω
17–13. 5.54 A
17–15. (a) 0.202 V rms; (b) 3.96×10^{-3}; (c) 111.11 Ω, 35.6 μF

17–17. 17.5 mA, 0.01%
17–19.

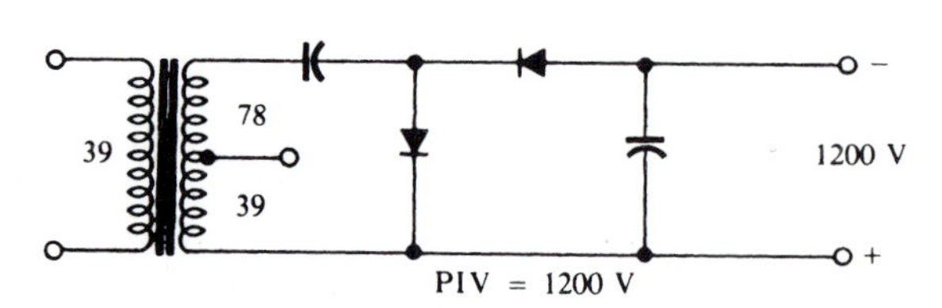

(a)

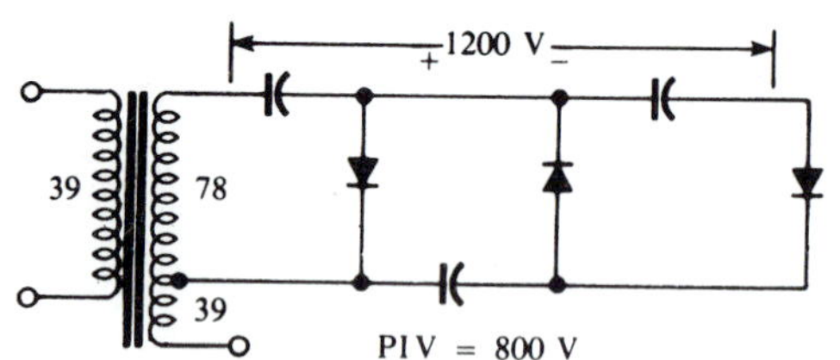

(b)

17–21. $V_L = [R_L/(R_L + R_o)]V_{NL}$; $R_L = R_o \Rightarrow V_L = 0.5V_{NL}$
17–23. 0.075%
17–25. (a) 8.7 V, (b) 0.192 W; (c) 49.3 mA
17–27. (a) Let $R_2 = 1\text{k}\Omega \Rightarrow R_1 = 1.48\text{ k}\Omega$
17–29. (a) 6 V–18 V; (b) 1.08 W
17–31. See Figure 17.36. (One solution): $R_1 = 26$ kΩ, $R_2 = 10$ kΩ, $R_3 = 3$ kΩ, $R_{SC} = 1$ Ω
17–33. 4.68 W
17–35. 0.24
17–37. $V_{dc} = [(V_2 - V_1)T_1 + V_1T_1]/T$
17–39. (a) 4 V p–p; (b) 0.0154%
17–41. (a) 6.54 V; (b) 308 Ω; (c) 0.233 A
17–43. See Figure 2 in *Typical Applications* section of specifications. (One solution): $R_1 = 1.86$ kΩ, $R_2 = 1$ kΩ, $R_3 = 650$ Ω

CHAPTER 18

18–1. (a) 23 V; (b) 50 mA
18–3. (a) $I_{Z(min)} = 20$ mA, $I_{Z(max)} = 300$ mA; (b) $P_{Z(min)} = 0.24$ W, $P_{Z(max)} = 3.6$ W; (c) 6.48 W
18–5. (a) 16.51 V; (b) 242.18 Ω
18–7.

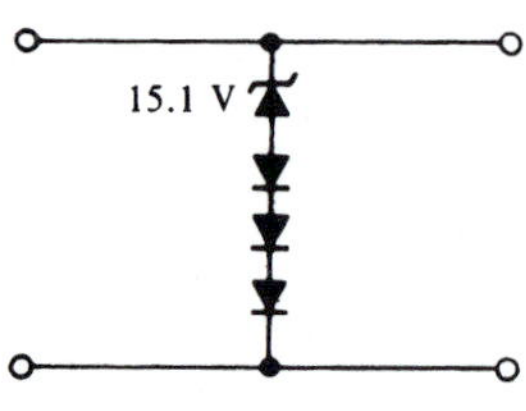

(a)

(b) 17.0635 V

18–9. 600 Ω
18–11. 0.2533 A
18–13.

V_G (V)	V_A (V)	I_G (μA)	I_H (mA)
0	1.15	0	2.5
0.2	1.075	2	1.75
0.4	1.00	4	1
1.0	0.95	10	0.5

18–15. (a) 70.53°; (b) 43.62 W
18–17. 58.16 V rms
18–19. 4679 lm/m²
18–21. Interchange the V^+ and V^- inputs in Figure 18.25
18–23. 112.4 mA
18–25. 20 mW/cm²
18–27. 47.52 W
18–29. 62.5 Ω
18–31. (a) most: 8; least: 1; (b) 210 mW
18–33. dynamic scattering and absorption; reflective and transmissive
18–35. small power consumption; require higher voltages and have slower response times
18–37. $0.625 \le \eta \le 0.859$
18–39. 4737 Ω–18.783 kΩ
18–41. incorrect
18–43. $L = 253$ μH; $f_r = 2.24$ MHz

CHAPTER 19

19–1. (a) 0.625 V; (b) 3.203125 V
19–3. One solution (see Figure 19–7): $R_f = 10$ kΩ, $R = 6.4583$ kΩ, $2R = 12.917$ kΩ, $4R = 25.833$ kΩ, $8R = 51.667$ kΩ, $16R = 103.33$ kΩ; output = −4.8387 V
19–5. $I_o = 0.125$ mA; $I_1 = 0.250$ mA; $I_2 = 0.500$ mA; $I_3 = 1.00$ mA
19–7. −9.9609375 V to +9.9609375 V
19–9. (a) 1.25 pF; (b) 2.75 pF; (c) 2.5 V
19–11. (a) 0.0244%; (b) minimum = 1.8315 mV; maximum = 5.4945 mV
19–13. 255 μs
19–15. (a) 31; (b) 0.3125 V
19–17. (a) 2 V; (b) 8 μs; (c) 00011001
19–19. 1000, 1100, 1010, 1011
19–21. (1) flash; (2) successive approximation; (3) dual-slope; (4) tracking counter; (5) counter
19–23. 39.06 mV; 0.3906%
19–25. 50 μs

INDEX

Abrupt junction, 904
AC control statement, in SPICE, 959
A_{cm}, 536
Ac resistance
  emitter, 163
  of a diode, 48
  of a MOSFET resistor, 328
Acceptor atom, 17, 33, 36
Active filters, 666–680
Active loads, 545
Active region, 97, 104
ADC, 927–936
  counter type, 927
  dual slope, 931
  flash, 929
  integrated circuit, 936
  modified flash, 930
  performance specifications, 935
  successive approximation, 933
  tracking, 928
A/D converter. *See* ADC
Admittance parameters, 377–387. *See also* y *parameters*
AGC, 663
Alpha (α), 92
  -cutoff frequency, f_α, 437
Amplitude distortion, 396
Amplitude modulation (AM), 772
Analog switch
  CMOS, 338
  JFET, 319
Analog-to-digital converter. *See* ADC
Analog voltage, 911
Angstrom, 867
Anode
  gate, of an SCS, 861
  of a diode, 24, 34, 36
  of a PUT, 899
  of an SCR, 851
Antilogarithmic amplifiers, 710
Arrays, transistor, 487
Aspect ratio, 227
Astable multivibrator, 693
Asymptotes, 407
Atom
  acceptor, 17, 33, 36
  donor, 16, 33, 37
  structure of, 5
Automatic gain control, 663
Avalanching, 26, 251, 849

Bandpass filter
  frequency response of, 669
  ideal, 666
  state-variable, 678
Band-reject filter, 666, 679
Bandwidth, 394
  closed-loop, 574
  relation to gain, 438, 573
  relation to rise time, 424
Barkhausen criterion, 655
Barrier potential, 20, 34, 36
Barrier region, 20
Base, BJT, 88
Base characteristics, 104
Batch production, 220
BCD, 914
Beta (β)
  -cutoff frequency, f_β, 437
  Darlington pair, 479
  feedback ratio, 564
  of a BJT, 102
  of a MOSFET, 282
Bias
  amplifier, purpose of, 149
  BJT, 89, 100, 109, 111, 204
  circuit design, BJT, 199–218
  common-base, 112–116
  common-collector, 121–122
  common-emitter, 116–121
  differential amplifier, 526, 538
  effect on small-signal parameters, 187
  fixed, 199
  integrated circuit, 231–237
  JFET, 254–270

Bias (*cont.*)
MOSFET, 282–288
of the stabilized BJT circuit, 204
PN junction, 21, 34
voltage divider, BJT, 200
JFET, 261
Biased diodes, 696
Bifet amplifier, 493
Bilateral transmission gate, 338
Binary
coded decimal (BCD), 914
counter, 913
numbers, 912
offset code, 921
Bipolar junction transistors, 87–142
Biquad filter, 674
Bit, 912
BJT. *See* Bipolar junction transistors
Blocking capacitors. *See* Coupling capacitors
Bode plot, 402
of a multistage amplifier, 471
Boltzmann's constant, 21, 34, 36
Break frequency, 407
of a practical differentiator, 640
of a practical integrator, 634
Breakdown region
of CB output characteristics, 99
of CE output characteristics, 105
of JFET output characteristics, 251
Breakdown voltage, of a diode, 25, 36
Bridge, diode, 66, 793
Bulk resistance, 49
Buried layer, 224
Butterworth filter, 667
BV_{CBO}, 99, 133
BV_{CEO}, 105, 133
BW_{CL}, 574

Capacitor
coupling, 150
emitter bypass, 208
filters, power supply, 64, 797
integrated-circuit, 228
source bypass, 305
Carriers, 12
Cascade, 455
Cascode amplifier, 483
bifet, 493
Cathode
gate, of an SCS, 861
of a diode, 24, 34, 36
of a PUT, 899
of an SCR, 851
CB. *See* Common-base
CC. *See* Common-collector
CE. *See* Common-emitter
Center frequency, 669
Center tap, 747
C_{fss}, 384
Channel
induced, 279
JFET, 244
MOSFET, 273
Chebyshev filter, 667
Chemical vapor deposition, 224
Chopper, JFET, 322
Chopper-stabilized amplifier, 593
C_{iss}, 384
Clamping circuits, 705
Class-A amplifiers, 736
Class-AB operation, 761
Class-B amplifiers, 746
Class-C amplifiers, 772
Class-D amplifiers, 780
Clipping
circuits, 696
in ac amplifiers, 150, 156–157
Closed-loop
bandwidth, 574
gain, 560
CMOS, 289, 336
CMRR. *See* Common mode, rejection ratio
Collector
efficiency, 738
feedback, 214
region of a BJT, 88
resistance, 164
Collector characteristics, 104
Colpitts oscillator, 663
Common-base (CB)
bias, 112–116
characteristics, 93–100
small-signal amplifier, 164–170, 371
Common-collector (CC)
bias, 121–122, 125
characteristics, 109–111
small-signal amplifier, 179–186, 371
Common-drain amplifier, 313
Common-emitter (CE)
bias, 116–121
characteristics, 100–109
h parameters, 360
small-signal amplifier, 151–162, 170–179, 364
Common-gate amplifier, 316
Common mode
gain, 536
parameters, 536–537
rejection, 536
rejection ratio, 537
Common-source amplifier
JFET, 302
MOSFET, 324
y parameters for, 382
Compensating resistor, 586, 589, 614
Compensation circuitry, 573, 576
Complementary
MOS integrated circuits, 289, 336
MOSFET power amplifiers, 780
multistage BJT amplifiers, 478
push-pull amplifiers, 762
Component designations, in SPICE, 940
Conduction, heat, 731
Conduction band, 10
Conductivity, 14
Controlled sources, in SPICE, 966
Convection, heat, 731
C_{oss}, 384
Counter-type ADC, 927
Coupling capacitors
effect on frequency response, 404, 410
use of, in ac amplifiers, 150

Coupling methods, in multistage amplifiers, 462
Covalent bonding, 8, 32, 37
C_π, 438
Crossover distortion, 748
C_{rss}, 384
Crystal, 8
  growth of, 219
Current, in semiconductors, 9
  diffusion, 15
  drift, 12
Current density, 13
Current feedback, 201
Current gain, 144, 147
  α, 163
  β, 163
  common-base, 166, 377
  common-collector, 182, 376
  common-emitter, 164
  computation of, using *h* parameters, 356
  computation of, using *y* parameters, 381
  Darlington pair, 479
  graphical analysis, CE, 154
Current limiting, in power supplies, 822
Current mirrors, 234, 544
Current sources
  current-controlled, 627, 967
  integrated circuit, 231, 539
  JFET, 317
  voltage-controlled, 623, 967
  Widlar, 544
Current-transfer ratio, 883
Curve tracers, 133–135
Cutoff frequency, 394, 407, 667
Cutoff region
  common-base, 98
  common-emitter, 106
CVD. *See* Chemical vapor deposition
Czochralski method, 219

DAC, 914–927
  AD7524 integrated circuit, 926
  multiplying, 921
  performance specifications, 925
  switched capacitor, 922
  switched current source, 920
  weighted resistor, 916
D/A converter. *See* DAC
Dark current, 872
Darlington pair, 479
  in power amplifiers, 766
dB. *See* Decibel
dBm, 400
dBV, 400
dBW, 400
Dc
  amplifiers, 231
  control statement, in SPICE, 943
  coupling, 462
  current sources, in SPICE, 943
  resistance, 48
  voltage sources, in SPICE, 941
Decade, 401
  one-*n*th interval, 403
Decibel
  power gain, 397
  voltage gain, 397
Degeneration, 208, 305
Demodulation, FM, 716
Density, hole and electron, 12, 17, 37, 38
Dependent sources, in SPICE, 966
Depletion region, 20
  of a JFET, 246
Derating, 132, 735
Design
  BJT bias, 123–126, 215–218
  BJT inverter, 128
  circuit, 3
  JFET bias, 267–270
  of coupling and bypass capacitors, 433–436
DIAC, 863
Difference
  amplifier, 525
  voltage gain, 529
Differential amplifiers, 525–547
  BJT, 526
  FET, 534
Differential input resistance, 532
Differential linearity error, 926
Differentiator, 637
  use in waveshaping, 648
Diffusion, 15, 220
Digital analog switch (DAS), 319
Digital-to-analog converter. *See* DAC
Digital voltage, 911
Diode, 24, 34
  as a circuit element, 45–85
  biased, in clipping circuits, 696
  bridge, 66, 793
  equation, ideal, 23, 34
  models, in SPICE, 954
  Schottky, 30
  switching circuits, 67–74
  types, ratings, and specifications, 74–79
  varicap, 228, 903
  zener. *See* Zener diode
DIP, 230
Direct coupling, 462
  of BJT amplifiers, 474
  of FET amplifiers, 491
Discrete
  circuits, 218
  diode, 45
Distortion, 156
  amplitude and phase, 395
  harmonic, 396, 728, 754
  in a push-pull amplifier, 759
  model, 757
DMOS
  amplifier, 326
  transistor, 291
D_n, 742
Donor atom, 16, 33, 37
Dopant, 16
Doping, 16, 33, 37
Double-base diode, 892
Double-diffused resistor, 227
Double-ended
  amplifier operation, 526
  limiting, 700
Doubler, voltage
  full-wave, 812
  half-wave, 811

Drain
- characteristics, 249, 275, 280
- JFET, 245
- MOSFET, 273, 278
- resistance, 301

Drift, 12
Dual in-line package, 230
Dual-slope ADC, 931
Dummy voltage sources, in SPICE, 944
Duty cycle, 827
Dynamic resistance, 48
Dynamic scattering, 889

Early voltage, 106
Efficiency
- class-A, 738
- class-B, 749, 768
- collector, 738
- power amplifier, 736
- voltage regulator, 827

Electroluminescence, 880
Electromagnetic interference, 783, 829
Electrons, 6
- free, 8, 32, 38

Electron-volt (eV), 10, 37
EMI, 783, 829
Emitter
- bypass capacitor, 208
- follower, 181, 766
- of a UJT, 892
- region of a BJT, 88
- resistance, 163
- stabilization, 201

Emitter characteristics, 95
END statement, in SPICE, 945
Energy band, 10
Energy gap, 882
Enhancement-type MOSFET, 278
Envelope, of an AM waveform, 776
Epitaxial layer, 224
Equivalent circuit
- of a diode switching circuit, 70–74
- of a forward-biased diode, 52
- of a tunnel diode, 903

Error voltage, 564
Excess carriers, 872
Extrinsic, 16

Fall time, 332
Feedback
- block diagrams, 565, 569
- current, 201
- in the inverting amplifier, 568
- in the noninverting amplifier, 564
- in voltage regulators, 816
- negative, 564
- positive, 573, 655
- ratio, 564
- theory, 564–573
- use of, to reduce distortion, 755
- voltage, 214

FET, 87, 245. *See also* JFET and MOSFET
- circuits and applications, 299–346

f_h, 640
Field-effect transistor. *See* FET
Firing angle, 859
Fixed bias circuit
- BJT, 199
- FET, 254, 302

Flash ADC, 929
Foldback limiting, 824
Forbidden band, 10
Forward-biased junction, 23
Forward blocking state, 852
Forward breakover voltage, 853
Four-layer devices, 851–866
Free electrons, 8, 32, 38
Frequency modulation (FM), 716
Frequency response, 393–453
- of BJT amplifiers, 426
- of cascaded stages, 461
- of FET amplifiers, 439
- of operational amplifiers, 573–578
- single-pole, 573, 667

Frequency synthesizers, 717
FSR, 925, 935
f_T, 437, 574
Full load, 814
Full-wave
- bridge, 66, 793, 799
- rectifiers, 66, 790
- voltage doubler, 812

Function generator (8038), 718
Fundamental frequency component, 754

Gain, 87. *See also* Current gain, Power gain, Voltage gain
- amplifier, 143
- computation of, using h parameters, 355
- of a multistage amplifier, 455

Gain-bandwidth product, 437, 574
Gate
- anode and cathode, of an SCS, 861
- bilateral transmission, 338
- insulated, 273
- JFET, 245
- SCR, 851

Geometric center, 670
Germanium, 7, 9, 31
Glitch area, 926
g_m. *See* Transconductance
g_{mo}, 300
Graphical analysis
- of a JFET bias circuit, 255, 257–259
- of a MOSFET bias circuit, 284
- of a small-signal CE amplifier, 151
- to determine g_m, 299, 323

h parameters, 190–193, 347–392
- amplifier analysis using, 364
- approximations, 368
- conversions, between BJT parameters, 362
- conversions, to y parameters, 382
- transistor, 359

Half-power frequency, 394

Half-wave
  power control, using SCRs, 858
  rectifiers, 64, 790
  voltage doubler, 811
Harmonic distortion, 396, 728, 754
Hartley oscillator, 665
Heat sinks, 732
Heat transfer, in semiconductor devices, 731–736
High-pass filter
  ideal, 666
  RC, 407
Holding current, 854, 864
Holes, 11
Hybrid
  equivalent circuits, 353
  integrated circuit, 219
  models, 190
  parameters, 190–193, 347–377
Hyperabrupt junction, 904
Hyperbola of constant power dissipation, 729
Hysteresis, 690

I_{ABC}, 712
I_{BB}, 588
I_B^-, 586
I_B^+, 586
I_{CBO}, 92, 98
I_{CEO}, 101
$I_{C(INJ)}$, 92
$I_{C(sat)}$, 126
Ideal
  distortionless amplifier, 755
  operational amplifier, 557
  power supply, 814
  square wave, 648
$I_{DS(reverse)}$, 306
I_{DSS}, 247
I_{FSM}, 77, 796
IGFET, 273
IGMF filter, 673
I_{GSS}, 271
I_{io}, 589
Impedance, zener-diode, 850
Impulse, 648
Impurity diffusion, 220
Induced-channel MOSFET, 279
Infinite-gain multiple-feedback (IGMF) filter, 673
Ingot, 219
Initial transient solution, in SPICE, 952
Input bias current, 586
Input characteristics
  CB, 94
  CC, 110
  CE, 104
Input data file, in SPICE, 939
Input offset
  current, 586–590
  voltage, 591–593
Input resistance
  amplifier, 144
  common-base, 164
  common-collector, 181
  common-emitter, 170
  computation of, using h parameters, 356
  Darlington pair, 480
  JFET, common-drain, 314
  JFET, common-source, 303
  MOSFET, common-source, 324
  of an inverting operational amplifier, 560, 570
  of a noninverting operational amplifier, 561, 567
  (stage), 167
Instrumentation amplifiers, 649–654
Integrated circuits, 218–231
  MOSFET, 289
  power dissipation in, 736
Integrator, 630
  multiple-input, 636
  use in waveshaping, 645
Interbase resistance, 892
Interconnection, of IC components, 229
Interelectrode capacitance, 415, 436
Intrinsic, 12, 38
Intrinsic standoff ratio, 893
Inverted mode, of BJT operation, 100
Inverter
  BJT, 126
  CMOS, 336
  dc to ac, 789
  MOSFET, 327
Inverting input, of an operational amplifier, 557
Ion implantation, 223

JAN, 131
JEDEC, 77
JFET, 245–273

Knee, 48

Large-signal
  amplifiers, 728
  diode circuits, 61–74
LASCR, 879
Latch, 914
LC filters, 810
LCDs, 885–892
Leakage current
  in a BJT, 93, 101
  in a PN junction, 25
LEDs, 879–883
Level shifting, 548, 690, 705
Light-activated SCR, 879
Light-emitting diodes, 819–883
Limiting circuits, 700
LIMPTS control statement, in SPICE, 949
Line regulation, 816
Linear combination circuit, 613
Linearity, 46
Linearity error, of a DAC, 926
Liquid-crystal displays, 885–892
Load-line
  ac, 158
  common-base, 112
  common-emitter, 117, 158
  dc, 158
  diode, 56–60
  nonlinear, MOSFET, 329
Load resistance, 148, 158
  MOSFET, 327–331
Logarithmic amplifiers, 707
Logic levels, 913
Log-log plots, 401
Loop gain, 567, 819

Lower sideband, 777
Lower trigger level, 690
Low-pass filter
  ideal, 666
  in a power supply, 798, 806
  integrator as a, 633
  LC, 810
  RC, 416, 426, 573, 806
LSB, 912
LTL, 690
Lumen, 867
Luminous
  flux, 867
  intensity, 867
LV_{CEO}, 105

Major change, in a DAC, 926
Majority carriers, 17, 33, 38
Mask, 221
Maximally flat filter, 667
Maximum forward current, 76
Maximum reverse voltage. *See* Peak inverse voltage
Metallization, 221
Micron, 226
Midband range, 394
Miller-effect capacitance, 421, 436
Minority carriers, 17, 33, 38
  suppression of, 18
Mixers, 777
Mobility, 12, 38
Models
  BJT, in SPICE, 956
  diode, in SPICE, 954
  JFET, in SPICE, 964
  MOSFET, in SPICE, 965
  small-signal, BJT, 163–186
  transconductance, BJT, 188
  transformer, in SPICE, 968
  two-port parameter, 347–348
Modulator
  amplitude, 775
  pulse-amplitude, 322
  pulse-width, 780, 827
MOSFET, 273–288
  depletion type, 273–278
  enhancement type, 278–288
  power amplifiers, 779
MS junction, 30
MSB, 913
Multiplying DAC, 921
Multistage amplifiers, 455–523

N material, 16, 33, 38
N^+ material, 33, 38, 88
Negative feedback, 214. *See also* Feedback
Negative resistance region
  of a tunnel diode, 902
  of a UJT, 894
Neper, 400
Neutrons, 6, 38
NMOS, 281
Node numbering, in SPICE, 940
Noise, 338, 800
Noninverting input, of an operational amplifier, 557
Nonlinear, 47, 156
  load line, MOSFET, 329
  modulation, 776
NPN, 88
Nucleus, 6, 38

Octave, 401
  one-*n*th interval, 403
Offset binary code, 921
Offset current, 586–590
Offset voltage, 591–593
Ohmic contacts, 31
Ohmic region, 249
One-*n*th decade and octave intervals, 403
Open-circuit output admittance, 350
Open-circuit reverse voltage ratio, 349
Open-loop gain, 560
Operating point. *See* Quiescent point
Operational amplifiers
  applications of, 613–686
  circuit analysis of, 548–552
  introduction to, 547
  theory of, 557–611
OPTIONS control statement, in SPICE, 949
Optocouplers, 883–885
Optoelectronic devices, 867–892
Optoisolators, 883–885
Orbit, 6, 38
Order, of a filter, 667
Oscillators, 654–666
  relaxation, 855, 895, 900
Output characteristics
  CB, 96
  CC, 110
  CE, 104
Output offset voltage, 592
Output resistance
  amplifier, 145
  common base, 164, 165, 168
  common-collector, 182, 183, 186
  common-emitter, 170, 173
  of an inverting operational amplifier, 570
  of a noninverting operational amplifier, 567
  power supply, 815
  (stage), 167
Output transformer, 742, 747
Overdriving
  amplifiers, 150
  voltage comparators, 690

P material, 16, 33, 39
P^+ material, 31, 33, 39
Packaging, of integrated circuits, 230
Pads, 229
Parallel clipper, 699
Parameters
  common mode, 536
  h, 190–193, 348–377
  small-signal BJT, 163
  small-signal JFET, 291
  y, 377–387
Passband, 666
Pass transistor, 818
Peak point
  of a PUT, 899
  of a tunnel diode, 902
  of a UJT, 894
Peak inverse voltage (PIV)
  rating of a diode, 76, 790
  ratings for rectifiers, 790
Phase distortion, 396
Phase inversion
  in a CE amplifier, 154

in a multistage amplifier, 457
in an operational amplifier, 557
Phase-locked loops, 715
the 565 integrated circuit, 718
Phase response, 393
Phase-splitter, 753
Photoconductive cells, 868–870
Photodarlington, 875
Photodiodes, 870–873
Photolithographic process, 220–224
Photoresist (PR), 221
Photoresistors, 868
Phototransistors, 873–876
Photovoltaic device, 872–876
PI (π) filter
LC, 810
RC, 806
Pinch-off, 247
Pins, integrated circuit, 230
PIV. *See* Peak inverse voltage
Planar transistor, 222
PLOT control statement, in SPICE, 948
PMOS, 281
PN
device fabrication, 220
junction, 18, 33, 39
PNP, 88
PNPN, 851
Polarizing, of light, 889
Polycrystalline, 219
Port, 347
Power
amplifiers, 727–788
control, using SCRs, 858
control, using TRIACs, 865
dissipation, 199, 728, 736
supplies, 64, 789–844
Power gain, 144
common-collector, 182
in decibels, 397
PR. *See* Photoresist
Preamplifier, 455, 758
PRINT control statement, in SPICE, 945
Priority encoder, 929
Probe, in PSpice, 971
Programmable resistor, 713
Programmable transconductance amplifier, 712
Programmable UJTs, 899–901
PSpice, 939, 940, 945, 950, 955, 966, 971–976
PULSE source, in SPICE, 950
Pulse-width modulator, 780, 827
Punch through, 99
Push-pull
amplifiers, 747
distortion, 759
drivers, 752
quasi-complementary amplifiers, 770
transformerless amplifiers, 762
PUTs, 899–901

Q, of a bandpass filter, 669, 678
Quadrupler, voltage, 812
Quasi-complementary push-pull amplifiers, 770
Quiescent (*Q*) point
BJT, 115, 118, 152–158
diode, 58
effect of temperature on, 199

Radiation, heat, 731
Radio-frequency choke, 772
Ramp waveform, 578
R_{BB}, 892
r_c, 164
R_c, 586, 589, 614
RC
coupling of BJT amplifiers, 464
high-pass filter, 407
low-pass filter, 416, 425, 573, 806
phase-shift oscillator, 657
power-supply filters, 806
r_d, r_{ds}, 301
$R_{D(ON)}$, 321
r_e, 164
Recombination, 12, 32, 39
Rectifiers, 62, 790
full-wave, 66, 790
full-wave bridge, 793
half-wave, 64
precision, 702
$r_{e(DP)}$, 483
Reflective mode, 889
Regulation
line, 816
voltage, 814
Regulators. *See* Voltage regulators
Relaxation oscillator, 855, 895, 900
Resistivity, 14
Resistor, integrated-circuit, 224
Resolution, 914
ADC, 935
DAC, 925
Reverse-biased junction, 23, 34, 39
Reverse current
in a BJT, 91, 102
in a PN junction, 24–26
RFC, 772
r_{id}, 532
r_{in}, 144. *See also* Input resistance
$r_{in(DP)}$, 481
$r_{in(stage)}$, 167. *See also* Input resistance
Ripple, in a power supply, 65, 799
Ripple width, of a Chebyshev filter, 667
Rise time, 332, 425, 584, 689
r_o, 145. *See also* Output resistance
(stage), 167
Roll-off, 573
r_π, 170
r_S, 146, 306
R-2*R* ladder DAC, 914
RW, 667

Saturation
current of a JFET, 247
region of CB output characteristics, 97
region of CE output characteristics, 106
Saturation current, 24

Scaling of voltages, by an operational amplifier, 562, 613
Schmitt triggers, 690
Schottky diode, 30
SCRs, 851–855
  light-activated, 879
Self-bias, JFET, 257
Semiconductor, 1, 8, 31, 39
Semilog plots, 401
Series-fed amplifier, 739
Series regulators, 817
Settling time, 925
Seven-segment display, 883
Shannon sampling theorem, 935
Sheet resistance, 226
Shell, 5, 39
Shockley diodes, 855
Short-circuit forward current ratio, 350
Short-circuit forward transfer admittance, 378
Short-circuit input admittance, 377
Short-circuit input resistance, 349
Short-circuit output admittance, 378
Short-circuit reverse transfer admittance, 378
Shunt capacitance, 414
Shunt regulators, 825
Sidebands, 777
Silicon, 5
Silicon controlled rectifiers, 851–860
  light-activated, 879
Silicon controlled switches, 861
SIN source, in SPICE, 950
Single-ended, 529
Single-pole frequency response, 573, 667
Slew rate, 578–585
Small-signal
  BJT amplifers, 143–198
  diode circuits, 54–61
  equivalent circuit of a cascode amplifier, 485
  equivalent circuit of a multistage amplifier, 467
  JFET amplifiers, 299–317
  models, 163, 302
  parameters, 163, 190, 291
  performance of a stabilized BJT, 207–214
Solar cells, 876–879
Source
  bypass capacitor, 305
  follower, 314
  JFET, 245
  MOSFET, 273, 278
Source resistance (r_S), 146, 306
Space charge, 19
Spectral response, 867
SPICE, 939–976
Square-law characteristic
  JFET, 252
  MOSFET, 277
Square-wave testing, 424
Stability, of an operational amplifier, 573
Stability factors, 201–204
Stabilization, of BJT bias circuits, 200–207
State-variable filter, 678
Static resistance, 49
Stray capacitance, 415
Subcircuits, in SPICE, 969
Subshell, 7
Substrate, 224, 273
Subtraction, of voltages, 616
Successive-approximation ADC, 933
Sum and difference frequencies, 777
Summation, voltage, 613
Supply voltage, 112
Suppression, minority carrier, 18
Surge current
  in rectifiers, 792
  rating, for a diode, 77, 796
Swing, 156
Switch
  analog, 319
  BJT, 126–130
Switched-capacitor DAC, 922
Switched-current source DAC, 920
Switching circuits
  BJT, 126–130
  capacitive loading of, 332
  CMOS, 336–339
  diode, 67–74
  MOSFET, 327–331
Switching regulators, 827

Tank network, 773
TEMP control statement, in SPICE, 958
Temperature
  ambient, 728
  coefficient of JFET characteristics, 273
  coefficient of resistance for semiconductors, 11
  effects on β, 132
  effects on BJTs, 132
  effects on diodes, 27, 36
  effects on Q-point, 199
  effects on V_{BE}, 200
  effects on zener voltage, 848
  stability of a zener diode, 849
t_f, 332
THD, 754
Thermal resistance, 733
Thermal runaway, 199
Thermal voltage, 21, 39
Three-terminal regulators, 830
Threshold voltage, 279
Thyristors, 851–866
Timing diagram, 749
Title statement, in SPICE, 940
Total harmonic distortion, 754
Totem pole, 781
t_r, 332
Tracking ADC, 928
TRAN control statement, in SPICE, 948
Transconductance
  amplifiers, 711
  as a y parameter, 383
  BJT, 187
  definition, 187, 383
  JFET, 299

model, for a BJT, 188–190
MOSFET, 323
Transfer characteristic
of a depletion MOSFET, 277
of a distortionless amplifier, 755
of an enhancement MOSFET, 282
of a JFET, 251
of a voltage comparator, 687
Transformer
copper losses, 500
coupled class-A amplifiers, 742
coupling, of multistage amplifiers, 463, 494
coupling coefficients, 503
eddy currents, 500
efficiency, 501
frequency response, 507
hysteresis losses, 501
leakage flux, 502
loading effects, 504
model, in SPICE, 968
mutual inductance, 509
Transient response, 423
Transistor
BJT, 87–142
curve tracer, 133–135
FET, 245–297
planar, 222
types, ratings, and specifications, 130–133
Transmissive mode, 889
Transresistance, 967
TRIAC, 864
Tripler, voltage, 812
Tuned amplifier, 773, 905
Tuning ratio, 904
Turns ratio, 494
Twisted-nematic crystal, 890

UJTs, 892–899
Unijunction transistors, 892–899
Universal active filter, 678
Upper sideband, 777
Upper trigger level, 690
User-compensated amplifiers, 576
UTL, 690

Valence shell, 8, 39
Valley point
of a PUT, 899
of a tunnel diode, 902
of a UJT, 894
Varactor, 228, 903
Varicap, 228, 903
$V_{BR(F)}$, 853
$V_{CE(sat)}$, 126, 129
VCVS filter, 670
$V_{DS(sat)}$, 249, 281
$V_{GS(cutoff)}$, 250
Virtual ground, 560
VLSI, 289
VMOS
amplifiers, 326, 779
structure, 290
Voltage comparators, 687–696
Voltage-controlled resistance, 249, 281, 320, 713
Voltage feedback, 214
Voltage follower, 562
Voltage gain, 143
at cutoff, 402, 405
bifet, 494
cascode, 485
common-base, 165, 375
common-collector, 181, 375
common-drain, 314
common-emitter, 172, 208, 215, 365, 368
common-mode, 536
common-source, 303, 306
computation of, using *h* parameters, 356
computation of, using *y* parameters, 381
differential amplifier, 529
graphical analysis of, CE, 154
in decibels, 397
MOSFET, 324
of direct-coupled amplifiers, 475
of filters, 667
of a multistage amplifier, 456, 460
of RC networks, 405, 416
Voltage multipliers, 811–814
Voltage regulation, 814
Voltage regulators, 816–840
adjustable, 833
series, 817
shunt, 825
switching, 827
three-terminal, 830
zener-diode, 846
Voltage sources
current-controlled, 627, 966
voltage-controlled, 622, 966
Voltage tripler, 812
Voltage-variable capacitor, 903
V_{OS}, 588, 591–592
V_p, 247
V_{RRM}, 796
V_T, 21, 39, 279

Wafer, 219
Wavelength, 867, 882
Waveshaping, 645–649
Weighted-resistor DAC, 918
Widlar current source, 544
Wien-bridge oscillator, 660
Windows, 221

y parameters, 377–387
conversions to *h* parameters, 382
equivalent circuits for, 380
FET, 382

Zener diode, 26, 845–851
impedance of, 850
temperature effects on, 848
use as a voltage reference, 817, 845
use in limiting circuits, 700
voltage regulator, 846
Zero-crossing detector, 687